VAN NOSTRAND REINHOLD
ENCYCLOPEDIA OF
CHEMISTRY

VAN NOSTRAND REINHOLD ENCYCLOPEDIA OF CHEMISTRY

Fourth Edition

DOUGLAS M. CONSIDINE, P.E.
Editor-in-Chief
(Fellow, American Association for the Advancement of Science;
Instrument Society of America; Senior Member, American
Institute of Chemical Engineers)

GLENN D. CONSIDINE
Managing Editor
(Member, American Society for Metals; Institute of Food Technologists)

VNR VAN NOSTRAND REINHOLD COMPANY
NEW YORK   CINCINNATI   TORONTO   LONDON   MELBOURNE

Manufactured in the United States of America

Published by Van Nostrand Reinhold Company Inc.
135 West 50th Street
New York, New York 10020

Van Nostrand Reinhold Company Limited
Molly Millars Lane
Wokingham, Berkshire RG11 2PY, England

Van Nostrand Reinhold
480 Latrobe Street
Melbourne, Victoria 3000, Australia

Macmillan of Canada
Division of Gage Publishing Limited
164 Commander Boulevard
Agincourt, Ontario M1S 3C7, Canada

15 14 13 12 11 10 9 8 7 6 5 4 3 2 1

Library of Congress Cataloging in Publication Data

Main entry under title:

Van Nostrand Reinhold encyclopedia of chemistry.

Rev. ed. of: The Encyclopedia of chemistry. 3rd. ed.
1973.

1. Chemistry—Dictionaries. I. Considine, Douglas
Maxwell. II. Considine, Glenn D. III. Encyclopedia of
Chemistry.
QD5.V37 1984 540'.3'21 83-23336
ISBN 0-442-22572-5

CONTRIBUTORS

Many specialists have made this Fourth Edition of *The Encyclopedia of Chemistry* a reality. Authorities from the various subdivisions of chemistry and related scientific disciplines assisted in the compilation of this work, which ranged from editorial counsel to the preparation of comprehensive articles on complex subjects. It is impractical to acknowledge individually many of the persons who helped in some way to produce this volume. Principal contributors are listed. Although not mentioned here, grateful acknowledgment is also expressed for the earlier work of people who contributed to the prior editions of this book.

Otto J. Adlhart, Englehard Minerals and Chemicals Corp., Iselin, New Jersey.

Alberta Oil Sands Technology and Research Authority, Calgary, Alberta, Canada.

Penrose S. Albright, Wichita, Kansas.

Edward S. Amis, University of Arkansas, Fayetteville, Arkansas.

Ralph S. Armstrong, The Sherwin-Williams Co., Cleveland, Ohio.

Robert Q. Barr, Molybdenum Company, Greenwich, Connecticut.

Wayne T. Barrett, Foote Mineral Co., Exton, Pennsylvania.

A. A. Bondi, Shell Development Co., Houston, Texas.

H. P. Burchfield, Gulf South Research Institute, New Iberia, Louisiana.

L. H. Busker, Beloit Corp., Beloit, Wisconsin.

Richard D. Cadle, National Center for Atmospheric Research, Boulder, Colorado.

S. C. Carapella, Jr., ASARCO Inc., South Plainfield, New Jersey.

C. Sharp Cook, The University of Texas, El Paso, Texas.

M. A. Cook, Salt Lake City, Utah.

P. H. Cook, Dow Chemical U.S.A., Freeport, Texas.

Donald A. Corrigan, Handy & Harman, Fairfield, Connecticut.

Robert A. Daene, Beloit Corp., Beloit, Wisconsin.

Ralph Daniels, University of Illinois, Chicago, Illinois.

D. F. Decraene, Chemetals Corp., Baltimore, Maryland.

Suman C. Desai, Davy McKee Iron & Steel Div., Ashmore House, Stockton-on-Tees, England.

Marcia L. Dilling, The Dow Chemical Co., Midland, Michigan.

Wendell L. Dilling, The Dow Chemical Co., Midland, Michigan.

M. B. Edwards, Longview, Texas.

Stanley B. Elliott, Bedford, Ohio.

T. E. Ferrington, Clarksville, Maryland.

L. F. Fieser, Harvard University, Cambridge, Massachusetts.

M. Fieser, Harvard University, Cambridge, Massachusetts.

Frederic C. Flint, Monsanto Textile Co., Decatur, Alabama.

John A. Garman, Great Lakes Chemical Corp., West Lafayette, Indiana.

E. D. Goddard, Lever Bros. Co., Edgewater, New Jersey.

K. A. Gschneidner, Jr., Iowa State University, Ames, Iowa.

Reigh C. Gunderson, The Dow Chemical Co., Midland, Michigan.

Louis H. Goodson, Midwest Research Institute, Kansas City, Missouri.

H. J. Hagemeyer, Jr., Longview, Texas.

Wilbur S. Hall, Amchem Products, Inc., Ambler, Pennsylvania.

W. B. Hardy, American Cyanamid Company, Bound Brook, New Jersey.

W. D. Hatfield, Decatur, Illinois.

R. H. Hausler, UOP Inc., Des Plaines, Illinois.

B. W. Heinemeyer, The Dow Chemical Co., Freeport, Texas.

Joel H. Hildebrand, University of California, Berkeley, California.

Stephen E. Hluchan, Calcium Metal Products, Pfizer Inc., Wallingford, Connecticut.

Herbert S. Hopkins, Olin Corp., New Haven, Connecticut.

E. W. Horvick, Zinc Institute Inc., New York, New York.

Charles D. Hurd, Northwestern University, Evanston, Illinois.

M. L. Jackson, University of Wisconsin, Madison, Wisconsin.

Edwin C. Jahn, State College of Forestry, Syracuse, New York.

V. A. Johnson, Purdue University, West Lafayette, Indiana.

Mark M. Jones, Vanderbilt University, Nashville, Tennessee.

Samuel Kaufman, Naval Research Laboratory, Washington, D.C.

Robert W. Keyes, IBM Corporation, Yorktown Heights, New York.

Charles J. Knuth, Pfizer Inc., New York, New York.

M. I. Kohan, E. I. DuPont de Nemours & Co., Inc., Wilmington, Delaware.

Kenneth D. Kopple, Illinois Institute of Technology, Chicago, Illinois.

Ihor A. Kunasz, Foote Mineral Co., Exton, Pennsylvania.

Robert T. LaLonde, State University of New York, Syracuse, New York.

John Lamartire, Monsanto Textile Co., Decatur, Alabama.

Joseph B. Lambert, Northwestern University, Evanston, Illinois.

Walter C. Lapple, Alliance, Ohio.

G. G. Lauer, Pittsburgh, Pennsylvania.

Walter William Lawrence, Jr., Ethyl Corp., Baton Rouge, Louisiana.

J. M. Lee, M. W. Kellogg Co., Houston, Texas.

Linton Libby, Simmons Refining Co., Chicago, Illinois.

M. H. Lietzke, Oak Ridge National Laboratory, Oak Ridge, Tennessee.

William F. Little, University of North Carolina, Chapel Hill, North Carolina.

John H. Lupinski, General Electric Co., Schnectady, New York.

Fred T. Mackenzie, Northwestern University, Evanston, Illinois.

Henry E. Mahncke, King of Prussia, Pennsylvania.

Henry Margenau, Yale University, New Haven, Connecticut.

Hugh J. McDonald, Loyola University, Maywood, Illinois.

W. F. McIlhenny, The Dow Chemical Co., Freeport, Texas.

George L. McNew, Boyce Thompson Laboratories, Yonkers, New York.

L. Dow Moore, PPG Industries, Pittsburgh, Pennsylvania.

Marguerite K. Moran, M & T Chemicals Inc., Rahway, New Jersey.

William T. Nearn, Wyerhaeuser Co., Seattle, Washington.

Eskel Nordell, Edison, New Jersey.

Hauromi Oeda, Ajinomoto Co., Inc., Kawasaki, Japan.

E. A. Ogryzle, University of British Columbia, Vancouver, B.C. (Canada).

Lloyd Osipow, New York, New York.

W. L. Peticolas, University of Oregon, Eugene, Oregon.

A. W. Petrocelli, Westerley, Rhode Island.

L. V. Pfaender, Owens-Illinois, Toledo, Ohio.

Hermann Pohland, E. I. DuPont de Nemours & Co., Inc., Wilmington, Delaware.

Howard W. Post, Williamsville, New York.

Duane B. Priddy, The Dow Chemical Co., Midland, Michigan.

Howard Reiss, University of California at Los Angeles, Los Angeles, California.

R. P. Rich, Eastman Chemical Products, Inc., Kingsport, Tennessee.

H. S. Richardson, Proctor & Gamble Co., Cincinnati, Ohio.

John A. Riddick, Baton Rouge, Louisiana.

George R. Romovacek, Koppers Company, Inc., Monroeville, Pennsylvania.

Alex T. Rowland, Gettysburg College, Gettysburg, Pennsylvania.

Elmer B. Rowley, Union College, Schenectady, New York.

R. T. Sanderson, Arizona State University, Tempe, Arizona.

S. John Sansonetti, Reynolds Metals Co., Richmond, Virginia.

Joseph W. Schappel, Avtex Fibers, Inc., Front Royal, Virginia.

C. E. Schildknecht, Gettysburg College, Gettysburg, Pennsylvania.

M. Schussler, Fansteel, North Chicago, Illinois.

Raymond B. Seymour, University of Houston, Houston, Texas.

W. G. Shequin, Bausch & Lomb, Sunland, California.

E. C. Shuman, State College, Pennsylvania.

D. G. Sleeman, Davy McKee Ltd., London.

M. J. Sterba, formerly with UOP Inc., Des Plaines, Illinois.

R. W. Stoughton, Oak Ridge National Laboratory, Oak Ridge, Tennessee.

Ulrich P. Strauss, Rutgers University, New Brunswick, New Jersey.

L. F. Urry, Battery Products, Union Carbide Corp., Parma, Ohio.

Ann C. Vickery, University of South Florida, College of Medicine, Tampa, Florida.

R. C. Vickery, Hudson Laboratories, Hudson, Florida.

D. Warschauer, Naval Weapons Center, China Lake, California.

Phillip A. Waitkus, Plastics Engineering Co., Sheboygan, Wisconsin.

Byron H. Webb, U.S. Dept. of Agriculture, Washington, D.C.

Kenneth A. Walsh, Brush Wellman Inc., Elmore, Ohio.

J. Y. Welsh, Chemetals Corp., Baltimore, Maryland.

F. Williams, University of Delaware, Newark, Delaware.

W. Williams, Cobalt Information Centre, London.

PREFACE

The first edition of *The Encyclopedia of Chemistry* was introduced in 1956. Subsequent editions (1966 and 1973) have reflected progress and change during their respective periods. This Fourth Edition is fully updated and contains over 85% new text. The expansion of chemistry into new fields and the impact of related sciences on chemistry during the past decade have required the reporting of many new topics and a considerable reorganization of format to accommodate the rapidly growing interdisciplinary character of chemistry. The science of chemistry has continued in its trend toward greater compartmentalization and specialization. Concurrently, countless thousands of scientists, engineers, and technologists have been drawn to a knowledge of chemistry because of the influences of chemistry on so many of the modern sciences and technologies—electronics and communications, energy sources and conservation, waste handling and pollution abatement, biotechnology, molecular biology, and the development of pharmaceuticals, biologicals, and chemotherapeutic methodologies. This list of the contributions of chemistry is long and continuing to grow with the passage of each year. Thus, it is evident that while there is a growing interest in chemistry, there is also the need for the traditional chemist to relate to fields that one day were considerably beyond the boundaries of chemistry.

In designing this Fourth Edition, the editors have given very serious consideration to the foregoing observations. For example, to accommodate the much wider areas of interest in chemistry and to provide substantive coverage of the progress made during the past decade, the number of specific entries in the volume has increased from just over 600 to approximately 1300 topics. To avoid an unwieldy long volume, this has required tighter wording and condensation of the text and greater use of tabular and graphic presentations. The alphabetical index has been expanded and restructured and many more cross references have been added within the text to make the volume easy to use. In addition to including topics which the user of an encyclopedia of chemistry would normally expect to find, the editors have given particular emphasis to the following:

1. *Advanced processes*—catalytic conversion, cryogenics, dialysis, exomosis, freeze-concentrating, drying, and preserving, molecular distillation, photonuclear reactions, reverse osmosis, semipermeable membranes, molecular sieves, solvolysis, supercooling, superfluidity, thermoelectric cooling, and ultrafiltration.
2. *Strategic raw materials*—the addition of several hundred economic minerals and important raw materials used in chemical processing.
3. *Chemistry of metals*—greater stress on metallurgical phenomena and processes.
4. *Energy sources and conversion*—biomass; batteries; fuel cells; hydrogen as a fuel; liquid and gaseous fuels from coal, oil shale, tar sands; nuclear fission and fusion; lithium for thermonuclear reactors; and insulating materials.
5. *Wastes and pollution*—carcinogens, cytotoxic chemicals, dioxin, biphenyls, air pollution, water treatment and pollution, radioactive waste handling.
6. *Analytical instrumentation*—new tools for determining ppm and ppb trace materials, as by chromatography.
7. *Growing use of food chemicals*—descriptions of all major food additives (anticaking agents, antimicrobial agents, bodying and bulking agents, coating materials, flavorings and flavor enhancers and potentiometers, humectants, intermediate-moisture food technology, polymeric food additives, among others).
8. *Structure of matter*—once in the province of chemistry and later annexed by physics, research in molecular biology and biotechnology has once again narrowed the gap between chemistry and physics. Considerable emphasis is given in this volume to structure, including comprehensive coverage of subatomic particles, at one end of the spectrum, to macromolecules at the other end.
9. *New and improved materials*—metal alloys, glass fibers for fiber optic communications, graphite structures for aircraft, xanthan gum for the food industry, YIGs and YAGs from the rare-earth metals for the electronics industry, cavitands, metallobiomolecules, metalloids, metalloproteins, electroconductive polymers, and superconductors, among others.
10. *Plant chemistry*—allopathic substances, anthocyanins, betalaines, gibberellic acid and gibberrelin plant growth hormones, maleic hydrazide growth inhibitors, herbicides, insecticides, and other agricultural control chemicals, including new concepts in insect control.
11. *Biochemistry and biotechnology*—enzymes, coenzymes, fermentation, recombinant DNA, drugs, hormones, contractile proteins, enkephalins and endorphins, antimetabolites, immunochemistry, dietary minerals, vitamins, amino acids, proteins are among topical coverage in this important topical area. Notable are detailed discussions of the effects of the chemical elements in biological systems, including calcium, chloride ion, cobalt, copper, fluorine, iodine, iron, magnesium, molybdenum, phosphorus, sodium, potassium, sulfur, and zinc.

Douglas M. Considine, P.E.
Glenn D. Considine

VAN NOSTRAND REINHOLD
ENCYCLOPEDIA OF
CHEMISTRY

A

ABSOLUTE ZERO. Conceptually that temperature where there is no molecular motion, no heat. On the Celsius scale, absolute zero is −273.15°C; on the Fahrenheit scale, −459.67°F; and zero degrees Kelvin (0K). The concept of absolute zero stems from thermodynamic postulations.

Heat and temperature were poorly understood prior to Carnot's analysis of heat engines in 1824. The Carnot cycle became the conceptual foundation for the definition of temperature. This led to the somewhat later work of Lord Kelvin, who proposed the Kelvin scale based upon a consideration of the second law of thermodynamics. This leads to a temperature at which all the thermal motion of the atoms stops. By using this as the zero point or absolute zero and another reference point to determine the size of the degrees, a scale can be defined. The Comité Consultatif of the International Committee of Weights and Measures selected 273.16K as the value for the triple point for water. This set the ice-point at 273.15K.

From the standpoint of thermodynamics, the thermal efficiency E of an engine is equal to the work W derived from the engine divided by the heat supplied to the engine, $Q2$. If $Q1$ is the heat exhausted from the engine,

$$E = (W/Q2) = (Q2 - Q1)/Q2 = 1 - (Q1/Q2)$$

where W, $Q1$, and $Q2$ are all in the same units. A Carnot engine is a theoretical one in which all the heat is supplied at a single high temperature and the heat output is rejected at a single temperature. The cycle consists of two adiabatics and two isothermals. Here the ratio $Q1/Q2$ must depend only on the two temperatures and on nothing else. The Kelvin temperatures are then defined by the relation

$$\frac{Q1}{Q2} = \frac{T1}{T2}$$

where $Q1/Q2$ is the ratio of the heats rejected and absorbed, and $T1/T2$ is the ratio of the Kelvin temperatures of the reservoir and the source. If one starts with a given size for the degree, then the equation completely defines a thermodynamic temperature scale.

A series of Carnot engines can be postulated so that the first engine absorbs heat Q from a source, does work W, and rejected a smaller amount of heat at a lower temperature. The second engine absorbs all the heat rejected by the first one, does work and rejects a still smaller amount of heat which is absorbed by a third engine, and so on. The temperature at which each successive engine rejects its heat becomes smaller and smaller, and in the limit this becomes zero so that an engine is reached which rejects no heat at a temperature which is absolute zero. A reservoir at absolute zero cannot have heat rejected to it by a Carnot engine operating between a higher temperature reservoir and the one at absolute zero. This can be used as the definition of absolute zero. Absolute zero is then such a temperature that a reservoir at that temperature cannot have heat rejected to it by a Carnot engine which uses a heat source at some higher temperature.

ABSORPTION. This is a process commonly used in the chemical process industries for separating materials, notably a specific gas from a mixture of gases; and in the production of solutions of gases, such as found in the manufacture of hydrochloric and sulfuric acids. Absorption is a key operation in many air pollution abatement systems, where it is required to remove noxious gases, such as sulfur dioxide and hydrogen sulfide, from an effluent gas prior to release to the atmosphere.

The absorption medium is most frequently a liquid in which (1) the gas to be removed, i.e., absorbed, is soluble in the liquid, or (2) a chemical reaction occurs between the gas and the absorbing liquid. Sometimes a chemical reagent is added to the absorbing liquid to increase the ability of the solvent to absorb.

It is desirable to select an absorbing liquid that can be regenerated and thus recycled and used over and over. An example of absorption with chemical reaction is the absorption of carbon dioxide from a flue gas with aqueous sodium hydroxide. In this reaction, sodium carbonate is formed. This reaction is irreversible, but continued absorption of the CO_2 with the sodium carbonate solution results in the formation of sodium acid carbonate. The latter compound can be decomposed upon heating to form CO_2, water, and sodium carbonate, and thus the sodium carbonate can be recycled.

There are several types of equipment frequently used in industrial absorption systems: (1) A packed tower filled with packing material, with the absorbent liquid flowing down through the packing, or with the gas flowing upward in a countercurrent fashion. In either arrangement, the equipment will be designed to provide a maximum contact surface between gas and solution. (2) A spray tower in which the absorbing liquid is sprayed into an essentially empty tower through which the gas flows upward. (3) A tray tower which contains bubble caps, sieve trays, valve trays, or other means for insuring maximum solution-gas contact. (4) A falling-film absorber or wetted-wall column, wherein the solution-gas contact is made with a film of liquid continuously clinging to the sides of the vessel, thus promoting rapid achievement of gas saturation in the absorbing liquid. (5) A stirred vessel in which the mixing action brings about the desired absorption.

A typical packed tower is shown in Fig. 1. In addition to obtaining good absorption efficiency, the designer also must minimize the pressure drop through the tower. The principal points of pressure loss are shown to the right of the diagram. Numerous packings, often ceramic, are available in a wide range of operating parameters. Packing is selected with the same objectives—providing maximum absorption and minimum pressure drop. In noncorrosive applications, metal packing is sometimes used. Various types of packing are shown in Fig. 2.

In the purification of natural gas, the gas is fed into the bottom of an absorption tower where the gas is contacted countercurrently by a lean absorption oil. Hydrochloric acid is produced by absorbing gaseous hydrogen chloride in water, usually in a spray-type tower. Unreacted ammonia in the manufacture of hydrogen cyanide is absorbed in dilute sulfuric acid. In the production of nitric acid, ammonia is catalytically oxidized and

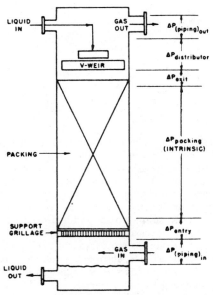

Fig. 1.   Sectional view of packed absorption tower, with major points of pressure drop indicated at the right of diagram.

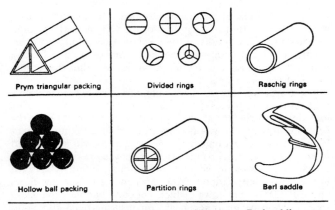

Fig. 2.   Types of packing used in absorption towers. Berl saddles range in size from 1/4-in. (6 mm) up to 2-in. (51 mm); raschig rings, 1/4-in. (6 mm) up to 4-in. (102 mm); lessing rings, 1-in. (25 mm) up to 6-in. (152 mm). It is interesting to note that over 100,000 of the 1/4-in. (6 mm)-size packing shapes can be contained in each cubic foot of tower space where the densest packing configuration may be desired.

the gaseous products are absorbed in water. The ethanolamines are widely used in scrubbing gases for removal of acid compounds. Hydrocarbon gases containing hydrogen sulfide can be scrubbed with monoethanolamine, which combines with it by salt formation and effectively removes it from the gas stream. In plants synthesizing ammonia, hydrogen and carbon dioxide are formed. The hydrogen can be obtained by countercurrently scrubbing the gas mixture in a packed or tray column with monoethanolamine which absorbs the carbon dioxide. The latter can be recovered by heating the monoethanolamine. In a non-liquid system, sulfur dioxide can be absorbed by dry cupric oxide on activated alumina, thus avoiding the disadvantages of a wet process. Sulfuric acid is produced by absorbing sulfur trioxide in weak acid or water.

ABSORPTION COEFFICIENT.   For the absorption of one substance or phase in another, as in the absorption of a gas in a liquid, the absorption coefficient is the volume of gas dissolved by a specified volume of solvent; thus a widely used coefficient is the quantity α in the expression $\alpha = V_0/V_p$, where V_0 is the volume of gas reduced to standard conditions, V is the volume of liquid, and p is the partial pressure of the gas.

ABS RESINS (Acrylonitrile-Butadiene-Styrene).   **Resins, Acrylonitrile-Butadiene-Styrene.**

ACARICIDE.   A substance, natural or synthetic, used to destroy or control infestations of members of species (*Arachnida, Acarina*), principally mites and ticks, several forms of which are very injurious to both plants, livestock, pets, and poultry, and often annoying to humans. Some substances are effective both as acaricides and insecticides; other compounds of a more narrow spectrum are strictly acaricides. See also **Insecticides.**

ACCELERATOR (Chemical).   **Rubber (Natural).**

ACETALDEHYDE. CH_3CHO, formula wt 44.05, colorless, odorous liquid, mp $-123.5°C$, bp $20.2°C$, sp gr 0.783. Also known as *ethanal*, acetaldehyde is miscible with H_2O, alcohol, or ether in all proportions. Because of its versatile chemical reactivity, acetaldehyde is widely used as a commencing material in organic syntheses, including the production of resins, dye-stuffs, and explosives. The compound is also used as a reducing agent, preservative, and as a medium for silvering mirrors. In resin manufacture, paraldehyde $(CH_3CHO)_3$ is sometimes preferred because of its higher boiling and flash points.

In large-scale production, acetaldehyde may be manufactured by (1) direct oxidation of ethylene, requiring a catalytic solution of copper chloride plus small quantities of palladium chloride; (2) oxidation of ethyl alcohol with sodium dichromate; and (3) dry distillation of calcium acetate with calcium formate.

Chemistry.   Acetaldehyde reacts with many chemicals in a marked manner: (1) with ammonio-silver nitrate (Tollen's solution) to form metallic silver, either as a black precipitate or as an adherent mirror film on glass; (2) with alkaline cupric solution (Fehling's solution) to form cuprous oxide (red to yellow) precipitate; (3) with rosaniline (fuchsine, magenta) which has been decolorized by sulfurous acid (Schiff's solution), the pink color of rosaniline is restored; (4) with NaOH, upon warming, a yellow-to-brown resin of unpleasant odor separates (this reaction is given by aldehydes immediately following acetaldehyde in the series, but not by formaldehyde, furfuraldehyde, or benzaldehyde); (5) with anhydrous ammonia, to form aldehyde-ammonia $CH_3 \cdot CHOH \cdot NH_2$, white solid, mp 97°C, bp 111°C, with decomposition; (6) with concentrated H_2SO_4, heat is evolved and, with rise of temperature, paraldehyde $(C_2H_4O)_3$ or

$$CH_3 \cdot CH \overset{\text{\Large<}}{} \begin{matrix} OCH(CH_3) \\ OCH(CH_3) \end{matrix} \overset{\text{\Large>}}{} O$$

colorless liquid bp 124°C, slightly soluble in H_2O, is formed; (7) with acids below 0°C, forms metaldehyde $(C_2H_4O)_x$, white solid, sublimes at about 115°C without melting, but with partial conversion to acetaldehyde; (8) with dilute HCl or dilute NaOH, aldol $CH_3 \cdot CHOH \cdot CH_2 \cdot CHO$ slowly forms; (9) with phosphorus pentachloride, forms ethylidene chloride $CH_3 \cdot CHCl_2$, colorless liquid, bp 58°C; (10) with ethyl alcohol and dry hydrogen chloride, forms acetal, 1,1-diethyl-oxyethane $CH_3 \cdot CH(OC_2H_5)_2$, colorless liquid, bp 104°C; (11) with hydrocyanic acid, forms acetaldehyde cyanhydrin $CH_3 \cdot CHOH \cdot CN$, readily converted into alphahydroxypropionic acid $CH_3 \cdot CHOH \cdot COOH$; (12) with sodium hydrogen sulfite, forms acetaldehyde sodium bisulfite $CH_3 \cdot CHOH \cdot SO_3 \cdot Na$, white solid, from which acetaldehyde is readily recoverable by treatment with sodium carbonate solution; (13) with hydroxylamine hydrochloride, forms acetaldoxime $CH_3 \cdot CH:NOH$, white solid, mp 47°C; (14) with phenylhydrazine, forms acetaldehyde phenylhydrazone $CH_3 \cdot CH:N \cdot NH \cdot C_6H_5$, white solid, mp 98°C; (15) with magnesium methyl iodide in anhydrous ether (Grignard's solution), yields, after reaction with water, isopropyl alcohol $(CH_3)_2CHOH$, a secondary alcohol; (16) with semicarbazide, forms acetaldehyde semicarbazone $CH_3 \cdot CH:N \cdot NH \cdot CO \cdot NH_2$, white solid, mp 162°C; (17) with chlorine, forms trichloroacetal dehyde $CH_3 \cdot CHS$ or $(CH_3 \cdot CHS)_3$.

Acetaldehyde stands chemically between ethyl alcohol—to which it can be reduced—and acetic acid—to which it can be oxidized. These reactions of acetaldehyde, coupled with its ready formation from acetylene by mercuric sulfate solution as a catalyzer, open up numerous possibilities in organic synthesis with acetaldehyde as raw material. Acetaldehyde can be hydrogenated to ethyl alcohol; oxygenated to acetic acid—then to acetone, acetic anhydride, vinyl acetate, and vinyl alcohol. Acetaldehyde is also formed by the regulated oxidation of ethyl alcohol by a reagent such as sodium dichromate in H_2SO_4 (chromic sulfate also produced). Reactions (1), (3), (14), and (16) given above are most commonly used in the detection of acetaldehyde.

See also **Aldehydes.**

ACETAL GROUP. An organic compound of the general formula $RCH(OR')(OR'')$ is termed an *acetal* and is formed by the reaction of an aldehyde with an alcohol, usually in the presence of small amounts of acids or appropriate inorganic salts. Acetals are stable toward alkali, are volatile, insoluble in H_2O, and generally are similar structurally to ethers. Unlike ethers, acetals are hydrolyzed by acids into their respective aldehydes. $H(R)CO + (HO \cdot C_2H_5)_2 \rightarrow H(R)COC_2H_5)_2 + H_2O$. Representative acetals include: $CH_2(OCH_3)_2$, methylene dimethyl ether, bp 42°C; $CH_3CH(OCH_3)_2$, ethylidene dimethyl ether, bp 64°C; and $CH_3CH(OC_2H_5)_2$, ethylidene diethyl ether, bp 104°C.

ACETAL RESINS. (Resins, Acetal).

ACETATE FIBERS. Fibers (Acetate).

ACETATES. Acetic Acid; Fibers (Acetate).

ACETIC ACID. CH_3COOH, formula wt 60.05, colorless, acrid liquid, mp 16.7°C, bp 118.1°C. sp gr 1.049. Also known as ethanoic acid or vinegar acid, this compound is miscible with H_2O, alcohol, and ether in all proportions. Acetic acid is available commercially in several concentrations. The CH_3COOH content of glacial acetic acid is approximately 99.7%, with H_2O as the major impurity. Reagent acetic acid generally contains 36% CH_3COOH by weight. Standard commercial aqueous solutions are 28, 56, 70, 80, 85, and 90% CH_3COOH. Acetic acid is the active ingredient in vinegar, in which the content ranges from 4 to 5% CH_3COOH. Acetic acid is classified as a weak, monobasic acid. The three hydrogen atoms linked to one of the two carbon atoms are not replaceable by metals.

In addition to large quantities of vinegar produced, acetic acid in its more concentrated forms is a high-tonnage industrial chemical, both as a reactive raw and intermediate material for numerous organic syntheses. Acetic acid is also an excellent industrial solvent. Acetic acid is required in the manufacture of several synthetic resins and fibers, pharmaceuticals, photographic chemicals, flavorants, and bleaching and etching compounds.

Chemistry. Acetic acid solutions react with alkalis to form *acetates*, e.g., sodium acetate, calcium acetate; similarly, with some oxides, e.g., lead acetate; with carbonates, e.g., sodium acetate, calcium acetate, magnesium acetate; with some sulfides, e.g., zinc acetate, manganese acetate. Ferric acetate solution, upon boiling, yields a red precipitate of basic ferric acetate. Acetic acid attacks several metals, liberating hydrogen and forming the metal acetate, e.g., magnesium, zinc, and iron acetate.

Acetic acid is an important organic substance: with alcohols, forming esters (acetates); with phosphorus trichloride, forming acetyl chloride $CH_3 \cdot CO \cdot Cl$, which is an important reagent for transfer of the acetyl ($CH_3CO—$) group.

Acetic anhydride is also an acetyl reagent, forming acetone and calcium carbonate when passed over lime and a catalyzer (barium carbonate) or when calcium acetate is heated; forming methane (and sodium carbonate) when sodium acetate is heated with NaOH; forming mono-, di-, trichloroacetic (or bromoacetic) acids by reaction with chlorine (or bromine), from which hydroxy- and amino-, aldehydic, and dibasic acids, respectively, may be made; forming acetamide when ammonium acetate is distilled.

Acetic acid dissolves sulfur and phosphorus and is an important solvent for many organic substances, causing painful wounds, for example, when it contacts the skin. Normal acetates are soluble, basic acetates, insoluble. The latter are important in their compounds with lead, copper (*verdigris*).

Organic Acetates. Some of the commercially available acetate esters of simple alcohols are those of the following: ethyl, propyl, isopropyl, butyl, isobutyl, amyl, and benzyl. The mono-, di-, and triesters of glycerol are also commercially available. These esters are all liquids, most of them have low boiling points, low toxicity, and are relatively inexpensive. Hence they find extensive application as organic solvents in the chemical processing industry and in some consumer products, e.g., lacquers, paints, degreasing solvents, etc. The lower molecular weight products have a characteristic pleasant, "fruity" odor which leads to their use in flavors, essences, and perfumes.

These esters are generally produced by direct esterification of the alcohol with acetic acid, using a strong acid, such as sulfuric, as a catalyst. Since the esterification reaction is an equilibrium, the reaction is forced by using excess alcohol, and removing the ester and the by-product water. In the case of ethyl acetate this is done by azeotropic distillation, the overhead of the esterification column being the ternary azeotrope: 82.2% ester, 8.4% alcohol and 9% water. The alcohol is then washed out of the ester with excess water, and the water removed via a rectification column, the overhead of which is a binary azeotrope.

Cellulose acetate is described in entry on **Fibers (Acetate).**

Another acetate of considerable commercial importance is vinyl acetate, which serves as a raw and intermediate material for many processes, such as polyvinyl acetate, polyvinyl alcohol, and copolymers of vinyl acetate with other vinyl monomers.

Acetates may be detected by formation of foul-smelling cacodyl (poisonous) on heating with dry arsenic trioxide. Other tests for acetate are the lanthanum nitrate test, in which a blue or bluish-brown ring forms when a drop of 2.5% $La(NO_3)_3$ solution, a drop of 0.01 N iodine solution, and a drop of 0.1% NH_4OH solution are added to a drop of a neutral acetate solution; the ferric chloride test in which a reddish color is produced by the addition of 1 N ferric chloride solution to a neutral solution of acetate; and the ethyl acetate test in which ethyl alcohol and H_2SO_4 are added to the acetate solution and warmed to form the odorous compound.

Production. Early commercial sources of acetic acid included: (1) the combined action of *Bacterium aceti* and air on ethyl alcohol in an oxidation-fermentation process ($C_2H_5OH + O_2 \rightarrow CH_3COOH + H_2O$), the same reaction which occurs when weakly alcoholic beverages, such as beer or wine, are exposed to air for a prolonged period and turn sour because of the formation of acetic acid; (2) the destructive distillation of wood. A number of natural vinegars are still made by fermentation and marketed as the natural product, but diluted commercially and synthetically produced acetic acid is much more economical. The wood distillation route was phased out a number of years ago because of shortages of raw materials and the much more attractive economy of synthetic processes.

The most important synthetic processes are (1) oxidation of acetaldehyde; and (2) direct synthesis from methyl alcohol and carbon monoxide. The latter reaction must proceed under very high pressures (about 650 atm) and at about 250°C. The reaction takes place in the liquid phase and dissolved cobaltous iodide serves as catalyst: $CH_3OH + CO \rightarrow CH_3COOH$ and $CH_3OCH_3 + H_2O + 2CO \rightarrow 2CH_3COOH$. The crude acid produced first is separated from the catalyst and then dehydrated and purified in an azerotropic distillation column. The final product is approximately 99.8% pure CH_3COOH.

ACETOACETIC ESTER CONDENSATION.

A class of reactions occasioned by the dehydrating power of metallic sodium or sodium ethoxide on the ethyl esters of monobasic aliphatic acids and a few other esters. It is best known in the formation of acetoacetic ester:

$$2CH_3 \cdot COOC_2H_5 + 2CH_3 \cdot COOC_2H_5 + 2Na \rightarrow$$

$$2CH_3 \cdot C(ONa){:}CH \cdot COOC_2H_5 + 2C_2H_5OH + H_2$$

The actual course of the reaction is complex. By the action of acids the sodium may be eliminated from the first product of the reaction and the free ester obtained. This may exist in the tautomeric enol and keto forms ($CH_3 \cdot COH{:}CH \cdot COOC_2H_5$ and $CH_3 \cdot CO \cdot CH_2 \cdot COOC_2H_5$).

On boiling ester with acids or alkalies it will split in two ways, the circumstances determining the nature of the main product. Thus, if moderately strong acid or weak alkali is employed, acetone is formed with very little acetic acid (ketone splitting). In the presence of strong alkalies however, very little acetone and much acetic acid result (acid splitting). Derivatives of acetoacetic ester may be decomposed in the same fashion, and this fact is responsible for the great utility of this condensation in organic synthesis. This is also due to the reactivity of the $\cdot CH_2 \cdot$ group, which reacts readily with various groups, notably halogen compounds. Usually the sodium salt of the ester is used, and the condensation is followed by decarboxylation with dilute alkali, or deacylation with concentrated alkali.

$$CH_3 \cdot CO \cdot CHNa \cdot COOC_2H_5 + RI \rightarrow$$

$$CH_3 \cdot CO \cdot CHR \cdot COOC_2H_5 + NaI$$

$$CH_3 \cdot CO \cdot CHR \cdot COOC_2H_5 \xrightarrow[\text{Dilute alkali}]{H_2O}$$

$$CH_3 \cdot CO \cdot CH_2R + C_2H_5OH + CO_2$$

$$CH_3 \cdot CO \cdot CHR \cdot COOC_2H_5 \xrightarrow[\text{Concentrated alkali}]{2H_2O}$$

$$HOOC \cdot CH_2 \cdot R + C_2H_5OH + CH_3COOH$$

ACETONE. $CH_3 \cdot CO \cdot CH_3$, formula wt 58.08, colorless, odorous liquid ketone, mp −94.6°C, bp 56.5°C, sp gr 0.792. Also known as dimethyl ketone or propanone, this compound is miscible in all proportions with H_2O, alcohol, or ether. Acetone is an important solvent and widely used in the manufacture of plastics and lacquers. For storage purposes, acetylene may be dissolved in acetone. A high-volume chemical, acetone is the starting ingredient or intermediate for numerous organic syntheses. A closely related, industrially important compound, diacetone alcohol (DAA), $CH_3 \cdot CO \cdot CH_2 \cdot COH(CH_3)_2$, is used as a solvent for cellulose acetate and nitrocellulose, as well as for various resins and gums, and as a thinner for lacquers and inks. Sometimes DAA is mixed with castor oil for use as a hydraulic brake fluid, for which its physical properties are well suited (mp −54°C, bp 166°C, sp gr 0.938). A product known as synthetic methyl acetone is prepared by mixing acetone (50%), methyl acetate (30%), and methyl alcohol (20%) and is used widely for coagulating latex and in paint removers and lacquers.

Chemistry. Acetone reacts with many chemicals in a marked manner: (1) with phosphorus pentachloride, yields acetone chloride $(CH_3)_2CCl_2$; (2) with hydrogen chloride (dry), yields both mesityl oxide $CH_3COCH{:}C(CH_3)_2$, liquid, bp 132°C and phorone $(CH_3)_2C{:}CHCOCH{:}C(CH_3)_2$, yellow solid, mp 28°C; (3) with concentrated H_2SO_4, yields mesitylene $C_6H_3(CH_3)_3$ (1, 3,5); (4) with NH_3, yields acetone amines, e.g., diacetoneamine $C_6H_{12}ONH$; (5) with HCN, yields acetone cyanhydrin $(CH_3)_2CHOH \cdot CN$, readily converted into the alpha-hydroxy acid $(CH_3)_2CHOH \cdot COOH$; (6) with sodium hydrogen sulfite, forms acetonesodiumbisulfite $(CH_3)_2COH \cdot SO_3Na$, white solid, from which acetone is readily recoverable by treatment with sodium carbonate solution; (7) with hydroxylamine hydrochloride, forms acetoxime $(CH_3)C{:}NOH$, solid, mp 60°C; (8) with phenylhydrazine, yields acetonephenylhydrazone $(CH_3)_2C{:}NNHC_6H_5 \cdot H_2O$, solid, mp 16°C, anhydrous compound, mp 42°C; (9) with semicarbazide, forms acetonesemicarbazone $(CH_3)C{:}NNHCONH_2$, solid, mp 189°C; (10) with magnesium methyl iodide in anhydrous ether (Grignard's solution), yields, after reaction with H_2O, trimethylcarbinol $(CH_3)_3COH$, a tertiary alcohol; (11) with ethyl thioalcohol and hydrogen chloride (dry), yields mercaptol $(CH_3)_2C(SC_2H_5)_2$; (12) with hypochlorite, hypobromite, or hypoiodite solution, yields chloroform $CHCl_3$, bromoform $CHBr_3$, or iodoform CHI_3, respectively; (13) with most reducing agents, forms isopropyl alcohol $(CH_3)_2CHOH$, a secondary alcohol, but with sodium amalgam forms pinacone $(CH_3)_2COH \cdot COH(CH_3)_2$; (14) with sodium dichromate and H_2SO_4, forms acetic acid CH_3COOH plus CO_2. When acetone vapor is passed through a tube at a dull red heat, ketene $CH_2{:}CO$ and methane CH_4 are formed.

Production. In older processes, acetone was prepared (1) by passing the vapors of acetic acid over heated lime. In a first step, calcium acetate is produced, followed by a breakdown of the acetate into acetone and calcium carbonate:

$$CH_3 \cdot CO \cdot O \cdot Ca \cdot OOC \cdot CH_3 \rightarrow CH_3 \cdot CO \cdot CH_3 + CaCO_3;$$

and (2) by fermentation of starches, such as maize, which produce acetone along with butyl alcohol. Modern industrial processes include (3) the use of cumene as a chargestock, in which cumene first is oxidized to cumene hydroperoxide (CHP), this followed by the decomposition of CHP into acetone and phenol; and (4) by the direct oxidation of propylene, using air and catalysts. The catalyst solution consists of copper chloride and small amounts of palladium chloride. The reaction: $CH_3CH{=}CH_2 + 1/2O_2 \rightarrow CH_3COCH_3$. During the reaction, the palladium chloride is reduced to elemental palladium and HCl. Reoxidation is effected by cupric chloride. The cuprous chloride resulting is reoxidized during the catalyst regeneration cycle. The process is carried out under moderate pressure at about 100°C.

ACETYL CHLORIDE. Chlorinated Organics.

ACETYLENE. CH ⫶ CH, formula wt 26.04, mp −81.5°C, bp −84°C, sp gr 0.905 (air = 1.00). Sometimes referred to as *ethyne*, *ethine*, or *gaseous carbon* (92.3% of compound is carbon), acetylene is moderately soluble in H_2O or alcohol, and exceptionally soluble in acetone (300 volumes of acetylene in 1 volume of acetone at 12 atm). The gas burns when ignited in air with a luminous sooty flame, requiring a specially devised burner for illumination purposes. An explosive mixture is formed with air over a wide range (about 3 to 80% acetylene), but safe handling is improved when the gas is dissolved in acetone. The heating value is 1455 Btu/ft³ (8.9 Cal/m³).

Although acetylene still is used in a number of organic syntheses on an industrial scale, its use on a high-tonnage basis has diminished because of the lower cost of other starting materials, such as ethylene and propylene. Acetylene has been widely used in the production of halogen derivatives, acrylonitrile, acetaldehyde, and vinyl chloride. Within recent years, producers of acrylonitrile switched to propylene as a starting material.

Commercially, acetylene is produced from the pyrolysis of naphtha in a two-stage cracking process. Both acetylene and ethylene are end-products. The ratio of the two products can be changed by varying the naphtha feed rate. Acetylene also has been produced by a submerged-flame process from crude oil. In essence, gasification of the crude oil occurs by means of the flame which is supported by oxygen beneath the surface of the oil. Combustion and cracking of the oil take place at the boundaries of the flame. The composition of the cracked gas includes about 6.3% acetylene and 6.7% ethylene. Thus, further separation and purification are required. Several years ago when procedures were developed for the safe handling of acetylene on a large scale, J. W. Reppe worked out a series of reactions that later became known as "Reppe chemistry." These reactions were particularly important to the manufacture of many high polymers and other synthetic products. Reppe and his associates were able to effect synthesis of chemicals that had been commercially unavailable. An example is the synthesis of cyclooctatetraene by heating a solution of acetylene under pressure in tetrahydrofuran in the presence of a nickel cyanide catalyst: $C_2H_2 \rightarrow C_8H_8$. In another reaction, acrylic acid was produced from CO and H_2O in the presence of a nickel catalyst: $C_2H_2 + CO + H_2O \rightarrow CH_2 \text{⫶} CH \cdot COOH$. These two reactions are representative of a much larger number of reactions, both those that are straight-chain only, and those involving ring closure.

Acetylene reacts (1) with chlorine, to form acetylene tetrachloride $C_2H_2Cl_4$ or $CHCl_2 \cdot CHCl_2$ or acetylene dichloride $C_2H_2Cl_2$ or $CHCl \text{⫶} CHCl$, (2) with bromine, to form acetylene tetrabromide $C_2H_2Br_4$ or $CHBr_2 \cdot CHBr_2$ or acetylene dibromide $C_2H_2Br_2$ or $CHBr \text{⫶} CHBr$, (3) with hydrogen chloride (bromide, iodide), to form ethylene monochloride $CH_2 \text{⫶} CHCl$ (monobromide, monoiodide), and 1,1-dichloroethane, ethylidene chloride $CH_3 \cdot CHCl_2$ (dibromide, diiodide), (4) with H_2O in the presence of a catalyzer, e.g., mercuric sulfate, to form acetaldehyde $CH_3 \cdot CHO$, (5) with hydrogen, in the presence of a catalyzer, e.g., finely divided nickel heated, to form ethylene C_2H_4 or ethane C_2H_6, (6) with metals, such as copper or nickel, when moist, also lead or zinc, when moist and unpurified. Tin is not attacked. Sodium yields, upon heating, the compounds C_2HNa and C_2Na_2. (7) with ammonio-cuprous (or silver) salt solution, to form cuprous (or silver) acetylide C_2Cu_2, dark red precipitate, explosive when dry, and yielding acetylene upon treatment with acid, (8) with mercuric chloride solution, to form trichloromercuric acetaldehyde $C(HgCl)_3 \cdot CHO$, precipitate, which yields with HCl acetaldehyde plus mercuric chloride.

ACETYLSALICYLIC ACID. $C_6H_4(COOH)CO_2CH_3$, formula wt, 180.06, mp 133.5°C, colorless, crystalline, slightly soluble in water, soluble in alcohol and ether, commonly known as aspirin, also called orthoacetoxybenzoic acid. The substance is commonly used as a relief for mild forms of pain, including headache and joint and muscle pain. The drug tends to reduce fever. Aspirin and other forms of salicylates have been used in large doses in acute rheumatic fever, but must be administered with extreme care in such cases by a physician. Commercially available aspirin is sometimes mixed with other pain relievers as well as buffering agents.

ACHLORHYDRIA. The cessation of acid production by the stomach. The condition is relatively common among people of about 50 years of age and older, affecting 15–20% of the population in this age group. A well-balanced diet of easily digestible foods minimizes the discomforting effects of complete absence of hydrochloric acid in the stomach. The condition does not preclude full digestion of fats and proteins, the latter being attacked by intestinal and pancreatic enzymes. In rare cases, where diarrhea may result from achlorhydria, dilute HCl may be administered by mouth.

Commonly, achlorhydria may not be accompanied by other diseases, but in some cases, there is a connection. For example, achlorhydria is an abnormality which sometimes occurs with severe iron deficiency. Histalog-fast achlorhydria, resulting from intrinsic factor deficiency in gastric juice, may be an indication of pernicious anemia. Hyperplastic polyps are sometimes found in association with achlorhydria.

ACIDIC SOLVENT. A solvent which is strongly protogenic, i.e., which has a strong capacity to donate protons and little tendency to accept them. Liquid hydrogen chloride and hydrogen fluoride are acidic solvents, and in them even such normally strong acids as nitric acid do not exhibit acidic properties, since there are no molecules which can accept protons, but, on the contrary, behave to some extent as bases by accepting protons yielded by the dissociation of the HCl or the H_2F_2. See **Acids and Bases.**

ACIDIMETRY. An analytical method for determining the quantity of acid in a given sample—often by titration against a standard solution of a base.

ACIDITY. The amount of acid present, expressed for a solution either as the molecular concentration of acid, in terms of normality, molality, etc., or the ionic concentration (hydrogen ions or protons) in terms of pH (the logarithm of the reciprocal of the hydrogen ion concentration). The acidity of a base is the number of molecules of monoatomic acid which one molecule of the base can neutralize. See **Acids and Bases;** and **pH (Hydrogen Ion Concentration).**

ACID NUMBER. A term used in the analysis of fats or waxes to designate the number of milligrams of potassium hydroxide (KOH) required to neutralize the free fatty acids in 1 gram of substance. The determination is performed by titrating an alcoholic solution of the wax or fat with 0.1 or 0.5N alkali, using phenolphthalein as indicator.

ACIDOSIS. A condition of excess acidity (or depletion of alkali) in the body, in which acids are absorbed or formed in excess of their elimination, thus increasing the pH of the blood, exceeding the normal value of 7.4. The acidity-alkalinity ratio in body tissue normally is delicately controlled by several mechanisms, notably the regulation of carbon dioxide-oxygen transfer in the lungs, the presence of buffer compounds in the blood, and the numerous sensing areas that are part of the central nervous system. See **Blood.** Normally, acidic materials are produced in excess in the body, this excess being neutralized by the presence of free alkaline elements, such as sodium, occurring in plasma. The combination of sodium with excess acids produces carbon dioxide, which is exhaled. Acidosis may result from severe exercise, sleep, especially under narcosis, where elimination of CO_2 is depressed, heart failure, diabetes and starvation, kidney failure, and severe diarrhea accompanied by loss of alkaline substances.

ACIDS AND BASES. The conventional definition of an acid is that it is an electrolyte that furnishes protons, i.e., hydrogen ions, H^+. An acid is sour to the taste and usually quite corrosive. A base is an electrolyte that furnishes hydroxyl ions, OH^-. A base is bitter to the taste and also usually quite corrosive. These definitions are formulated in terms of water solutions and consequently do not embrace situations where some ionizing medium other than water may be involved. In the definition of Lowry and Brønsted, an acid is a proton donor and a base is a proton acceptor.

Acidification is the operation of creating an excess of hydrogen ions, normally by the addition of an acid to a neutral or alkaline solution until a pH below 7 is reached, thus indicating an excess of hydrogen ions. In *neutralization*, a balance between hydrogen and hydroxyl ions exists. An acid solution may be neutralized by the addition of a base; and vice versa. The products of neutralization are a salt and water.

Some of the inorganic acids, such as hydrochloric acid (HCl), nitric acid (HNO_3), and sulfuric acid (H_2SO_4), are high-production chemicals and are considered a basic raw material for many industries. The most common inorganic bases (or alkalis) include sodium hydroxide (NaOH) and potassium hydroxide (KOH), both of which are high-production materials, notably NaOH.

Several classes of organic substances are classified as acids, particularly the carboxylic acids, the amino acids, and the nucleic acids. These acids are described in separate entries under those names in this Encyclopedia.

In terms of the definition that an acid is a proton donor and a base is a proton acceptor, hydrochloric acid, water, and ammonia (NH_3) are acids in the reactions:

$$HCl \rightleftharpoons H^+ + Cl^-$$

$$H_2O \rightleftharpoons H^+ + OH^-$$

$$NH_3 \rightleftharpoons H^+ + NH_2^-$$

Note that this definition is different in at least two major respects from the conventional definition of an acid as a substance dissociating to give H^+ in water. The Lowry-Brønsted definition states that for every acid there be a "conjugate" base, and vice versa. Thus, in the examples cited above, Cl^-, OH^-, and NH_2^- are the conjugate bases of HCl, H_2O, and NH_3. Furthermore, since the equations given above should more properly be written:

$$HCl + H_2O \rightleftharpoons H_3O^+ + Cl^-$$

$$H_2O + H_2O \rightleftharpoons H_3O^+ + OH^-$$

$$NH_3 + H_2O \rightleftharpoons H_3O^+ + NH_2^-$$

it can be seen that every acid-base reaction involving transfer of a proton will involve two conjugate acid-base pairs, e.g., in

the last equation NH_3 and H_3O^+ are the acids and NH_2^- and H_2O the respective conjugate bases. On the other hand, in the reaction:

$$NH_3 + H_2O \rightleftharpoons NH_4^+ + OH^-$$

H_2O and NH_4^+ are the acids and NH_3 and OH^- the bases. In other reactions, e.g.,

$Base_1$	$Acid_2$	$Acid_1$	$Base_2$
$C_2H_3O_2^-$	$+ H_2O$	$\rightleftharpoons HC_2H_3O_2$	$+ OH^-$
HCO_3^-	$+ HCO_3^-$	$\rightleftharpoons H_2CO_3$	$+ CO_3^{-2}$
$N_2H_5^+$	$+ N_2H_5^+$	$\rightleftharpoons N_2H_6^{+2}$	$+ N_2H_4$
H_2O	$+ Cr(H_2)_6^{+3}$	$\rightleftharpoons N_3O^+$	$+ Cr(H_2O)_5OH^{2+}$

the conjugate acids and bases are as indicated. The theory is not limited to the aqueous solution; for example, the following reactions can be considered in exactly the same light:

$Base_1$	$Acid_2$	$Acid_1$	$Base_2$
NH_3	$+ HCl$	$\rightleftharpoons NH_4^+$	$+ Cl^-$
CH_3CO_2H	$+ HF$	$\rightleftharpoons CH_3CO_2H_2^+$	$+ F^-$
HF	$+ HClO_4$	$\rightleftharpoons H_2F^+$	$+ ClO_4^-$
$(CH_3)_2O$	$+ HI$	$\rightleftharpoons (CH_3)_2OH^+$	$+ I^-$
C_6H_6	$+ HSO_3F$	$\rightleftharpoons C_6H_7^+$	$+ SO_3F^-$

Acids may be classified according to their charge or lack of it. Thus, in the reactions cited above, there are "molecular" acids and bases, such as HCl, H_2CO_3, $HClO_4$, etc., and N_2H_4, $(CH_3)_2O$, C_6H_6, etc., and also cationic acids and bases, such as H_3O^+, $N_2H_5^+$, $N_2H_6^{+2}$, NH_4^+, $(CH_3)_2OH^+$, etc., as well as anionic acids and bases, such as HCO_3^-, Cl^-, NH_2^-, CO_3^{-2}, etc. In a more general definition, Lewis calls a base any substance with a free pair of electrons that it is capable of sharing with an electron pair acceptor, which is called an acid. For example, in the reaction:

$$(C_2H_5)_2O: + BF_3 \rightarrow (C_2H_5)_2O:BF_3$$

the ethyl ether molecule is called a base, the boron trifluoride, an acid. The complex is called a *Lewis salt*, or *addition compound.*

Acids are classified as monobasic, dibasic, tribasic, polybasic, etc., according to the number (one, two, three, several, etc.) of hydrogen atoms, replaceable by bases, contained in a molecule. They are further classified as (1) organic when the molecule contains carbon; (1a) carboxylic, when the proton is from a —COOH group; (2) normal, if they are derived from phosphorus or arsenic, and contain three hydroxyl groups; (3) ortho, meta, or para, according to the location of the carboxyl group in relation to another substituent in a cyclic compound; or (4) ortho, meta, or pyro, according to their composition.

Superacids. Although mentioned in the literature as early as 1927, *superacids* were not investigated aggressively until the 1970s. Prior to the concept of superacids, scientists generally regarded the familiar mineral acids (HF, HNO_3, H_2SO_4, etc.) as the strongest acids attainable. Relatively recently, acidities up to 10^{12} times that of H_2SO_4 have been produced.

In very highly concentrated acid solutions, the commonly used measurement of pH is not applicable. See also **pH (Hydrogen Ion Concentration).** Rather, the acidity must be related to the degree of transformation of a base with its conjugate acid. The *Hammett acidity function*, developed by Hammett and Deyrup is:

$$H_0 = pK_{BH^+} - \log \frac{BH^+}{B}$$

where pK_{BH^+} is the dissociation constant of the conjugate acid (BH^+), and BH^+/B is the ionization ratio, measurable by spectroscopic means (UV or NMR). In the Hammet acid function, acidity is a logarithmic scale wherein H_2SO_4 (100%) has an H_0 of -11.9; and HF, an H_0 of -11.0.

As pointed out by Olah et al. (1979), "The acidity of a sulfuric acid solution can be increased by the addition of solutes that behave as acids in the system: $HA + H_2SO_4 \rightleftharpoons H_3SO_4^+ + A^-$. These solutes increase the concentration of the highly acidic $H_3SO_4^+$ cation just as the addition of an acid to water increases the concentration of the oxonium ion, H_3O^+. Fuming sulfuric acid (oleum) contains a series of such acids, the polysulfuric acids, the simplest of which is disulfuric acid, $H_2S_2O_7$, which ionizes as a moderately strong acid in sulfuric acid: $H_2S_2O_7 + H_2SO_4 \rightleftharpoons H_3SO_4^+ + HS_2O_7^-$. Higher polysulfuric acids, such as $H_2S_3O_{10}$ and $H_2S_4O_{13}$, also behave as acids and appear somewhat stronger than $H_2S_2O_7$."

Hull and Conant in 1927 showed that weak organic bases (ketones and aldehydes) will form salts with perchloric acid in nonaqueous solvents. This results from the ability of perchloric acid in nonaqueous systems to protonate these weak bases. These early investigators called such a system a superacid. Some authorities believe that any protic acid that is stronger than sulfuric acid (100%) should be typed as a superacid. Based upon this criterion, fluorosulfuric acid and trifluoromethanesulfonic acid, among others, are so classified. Acidic oxides (silica and silica-alumina) have been used as solid acid catalysts for many years. Within the last few years, solid acid systems of considerably greater strength have been developed and can be classified as *solid superacids*.

Superacids have found a number of practical uses. Fluoroantimonic acid, sometimes called *Magic Acid*, is particularly effective in preparing stable, long-lived carbocations. Such substances are too reactive to exist as stable species in less acidic solvents. These acids permit the protonation of very weak bases. For example, superacids such as Magic Acid can protonate saturated hydrocarbons (alkanes) and thus can play an important role in the chemical transformation of hydrocarbons, including the processes of isomerization and alkylation. See also **Alkylation; and Isomerization.** Superacids also can play key roles in polymerization and in various organic syntheses involving dienonephenol rearrangement, reduction, carbonylation, and oxidation, among others. Superacids also play a role in inorganic chemistry, notably in the case of halogen cations and the cations of nonmetallic elements, such as sulfur, selenium, and tellurium.

An excellent summary of superacids, their nature and utility, is given in "Superacids" by G. A. Olah, G. K. Surya Prakash, and J. Sommer in *Science*, **206**, 13–20 (1979).

ACID-BASE REGULATION (Blood). Blood.

ACIDULANTS AND ALKALIZERS (Foods).

Well over 50 chemical additives are commonly used in food processing or as ingredients of final food products, essentially to control the pH (hydrogen ion concentration) of the process and/or product. An excess of hydrogen ions, as contributed by acid substances, produces a sour taste, whereas an excess of hydroxyl ions, as contributed by alkaline substances, creates a bitter taste. Soft drinks and instant fruit drinks, for example, owe their tart flavor to acidic substances, such as citric acid. Certain candies, chewing gum, jellies and jams, salad dressing are among the many other products where a certain degree of tartness contributes to the overall taste and appeal.

Taste is only one of several qualities of a process or product which is affected by an excess of either of these ions. Some raw materials are naturally too acidic, others too alkaline—so that neutralizers must be added to adjust the pH within an acceptable range. In the dairy industry, for example, the acid in sour cream must be adjusted by the addition of alkaline compounds in order that satisfactory butter can be churned. Quite often, the pH may be difficult to adjust or to maintain after adjustment. Stability of pH can be accomplished by the addition of buffering agents which, within limits, effectively maintain the desired pH even when additional acid or alkali is added. For example, orange-flavored instant breakfast drink has just enough "bite" from the addition of potassium citrate (a buffering agent) to regulate the tart flavor imparted by another ingredient, citric acid. In some instances, the presence of acids or alkalis assists mechanical processing operations in food preparation. Acids, for example, make it easier to peel fruits and tubers. Alkaline solutions are widely used in dehairing animal carcasses.

The pH values of various food substances cover a wide range: plant tissues and fluids (about 5.2); animal tissues and fluids (about 7.0 to 7.5); lemon juice (2.0 to 2.2); acid fruits (3.0 to 4.5); fruit jellies (3.0 to 3.5).

Acidulants commonly used in food processing include: acetic acid (glacial), citric acid, fumaric acid, glucono-delta-lactone, hydrochloric acid, lactic acid, malic acid, phosphoric acid, potassium acid tartrate, sodium bisulfate, sulfuric acid, and tartaric acid. Alakalis commonly used include: ammonium bicarbonate, ammonium hydroxide, calcium carbonate, calcium oxide, magnesium carbonate, magnesium hydroxide, magnesium oxide, potassium bicarbonate, potassium carbonate, potassium hydroxide, sodium bicarbonate, sodium carbonate, sodium hydroxide, and sodium sesquicarbonate. Among the buffers and neutralizing agents favored are: adipic acid, aluminum ammonium sulfate, ammonium phosphate (di- or monobasic), calcium citrate, calcium gluconate, sodium acid pyrophosphate, sodium phosphate (di-, mono-, and tribasic), sodium pyrophosphate, and succinic acid.

See also **Buffer (Chemical);** and **pH (Hydrogen Ion Concentration).**

ACMITE-AEGERINE.

Acmite is a comparatively rare rock-making mineral, usually found in nephelite syenites or other nephelite or leucite bearing rocks, as phonolites. Chemically it is a soda-iron silicate, and its name refers to its sharply pointed monoclinic crystals. Bluntly terminated crystals form the variety aegerine, named for Aegir, the Icelandic sea god.

Acmite has a hardness of 6 to 6.5, specific gravity 3.5, vitreous, color brown to greenish-black (aegerine); red-brown to dark green and black (acmite). Acmite is synonymous with aegerine, but usually restricted to the long slender-crystalled variety of brown color.

The original acmite locality is in Greenland, Norway; U.S.S.R., Kenya, India, and Mt. St. Hilaire, Quebec, Canada furnish fine specimens. United States localities are Magnet Cove, Arkansas, and Libby, Montana, where a variety carrying vanadium occurs.

ACRYLAMIDE POLYMERS.

Using acid or base catalysts, acrylamide is derived from acrylonitrile by a hydration reaction. Although catalytic processes for the direct hydration of acrylonitrile over solid surface of metals, metallic salts, and oxides have been reported, commercial processes for its production use sulfuric acid. The acrylamide-sulfate salt initially formed may be isolated. When aqueous solutions of this product are neutralized with bases, such as ammonia, sodium hydroxide, or lime, acrylamide is yielded. Acrylamide is a white crystalline solid, mp 84.5°C.

Several techniques have been used to polymerize acrylamide

under controlled conditions. Radiation, photopolymerization, and ultrasonic methods are among these. Standard free-radical initiation of acrylamide in aqueous solution is most frequently used. The catalysts used include azo catalysts and a number of inorganic redox couples. Ionizing radiation in the solid state may also effect polymerization of acrylamide. Polyacrylamide may be "grown" to very high molecular size, with molecular weights above 10 million frequently obtained.

Acrylamide copolymerizes with several polar vinyl monomers. The vinyl polymer of acrylamide is a white solid with a high glass transition temperature (165°C) and softening temperature in excess of 200°C. The material is soluble in water, but not in most organic solvents. Polyacrylamides undergo the typical reactions of simple aliphatic amides.

Industrially, polyacrylamides are usually used in aqueous solution. Among the main uses for aqueous polyacrylamide solutions are in flocculation, in which suspended matter is removed or concentrated in aqueous systems.

ACRYLATES AND METHACRYLATES.

A wide range of plastic materials that date back to the pioneering work of Redtenbacher before 1850, who prepared acrylic acid by oxidizing acrolein

$$CH_2\text{=}CHCHO \xrightarrow{O} CH_2\text{=}CHCOOH.$$

At a considerably later date, Frankland prepared ethyl methacrylate and methacrylic acid from ethyl-α-hydroxyisobutyrate and phosphorus trichloride. Tollen prepared acrylate esters from 2,3-dibromopropionate esters and zinc. Otto Rohm, in 1901, described the structures of the liquid condensation products (including dimers and trimers) obtained from the action of sodium alkoxides on methyl and ethyl acrylate. Shortly after World War I, Rohm introduced a new acrylate synthesis, noting that an acrylate is formed in good yield from heating ethylene cyanohydrin and sulfuric acid and alcohol. A major incentive for the development of a clear, tough plastic acrylate was in connection with the manufacture of safety glass.

Ethyl methacrylate went into commercial production as early as 1933. The synthesis proceeded in the following steps:

(1) Acetone and hydrogen cyanide, generated from sodium cyanide and acid, gave acetone cyanohydrin

$$HCN + CH_3COCH_3 \rightarrow (CH_3)_2C(OH)CN$$

(2) The acetone cyanohydrin was converted to ethyl α-hydroxyisobutyrate by reaction with ethyl alcohol and dilute sulfuric acid

$$(CH_3)_2C(OH)CN + C_2H_5OH \xrightarrow{H_2SO_4} (CH_3)_2C(OH)COOC_2H_5$$

(3) The hydroxy ester was dehydrated with phosphorus pentoxide to produce ethyl methacrylate

$$(CH_3)_2C(OH)COOC_2H_5 \xrightarrow{P_2O_5} CH_2\text{=}C(CH_3)COOC_2H_5$$

In 1936, the methyl ester of methacrylic acid was introduced and used to produce an "organic glass" by cast polymerization. Methyl methacrylate was made initially through methyl α-hydroxyisobutyrate by the same process previously indicated for the ethyl ester. Over the years, numerous process changes have occurred and costs lowered, making these plastics available on a high production basis for many hundreds of uses. For example, the hydrogen cyanide required is now produced catalytically from natural gas, ammonia, and air.

As with most synthetic plastic materials, they commence with the monomers. Any of the common processes, including bulk, solution, emulsion, or suspension systems may be used in the free-radical polymerization or copolymerization of acrylic monomers. The molecular weight and physical properties of the products may be varied over a wide range by proper selection of acrylic monomer and monomer mixes, type of process, and process conditions.

In bulk polymerization, no solvents are employed and the monomer acts as the solvent and continuous phase in which the process is carried out. Commercial bulk processes for acrylic polymers are used mainly in the production of sheets, rods and tubes. Bulk processes are also used on a much smaller scale in the preparation of dentures and novelty items and in the preservation of biological specimens. Acrylic castings are produced by pouring monomers or partially polymerized sirups into suitably designed molds and completing the polymerization. Acrylic bulk polymers consist essentially of poly(methyl methacrylate) or copolymers with methyl methacrylate as the major component. Free radical initiators soluble in the monomer, such as benzoyl peroxide, are the catalysts for the polymerization. Aromatic tertiary amines, such as dimethylaniline, may be used as accelerators in conjunction with the peroxide to permit curing at room temperature. However, colorless products cannot be obtained with amine accelerators because of the formation of red or yellow colors. As the polymerization proceeds, a considerable reduction in volume occurs which must be taken into consideration in the design of molds. At 25°C, the shrinkage of methyl methacrylate in the formation of the homopolymer is 21%.

Solutions of acrylic polymers and copolymers find wide use as thermoplastic coatings and impregnating fluids, adhesives, laminating materials, and cements. Solutions of interpolymers convertible to thermosetting compositions can also be prepared by inclusion of monomers bearing reactive functional groups which are capable of further reaction with appropriate crosslinking agents to give three-dimensional polymer networks. These polymer systems may be used in automotive coatings and appliance enamels, and as binders for paper, textiles, and glass or nonwoven fabrics. Despite the relatively low molecular weight of the polymers obtained in solution, such products are often the most appropriate for the foregoing uses. Solution polymerization of acrylic esters is usually carried out in large stainless steel, nickel, or glass-lined cylindrical kettles, designed to withstand at least 50 psig. The usual reaction mixture is a 40–60% solution of the monomers in solvent. Acrylic polymers are soluble in aromatic hydrocarbons and chlorohydrocarbons.

Acrylic emulsion polymers and copolymers have found wide acceptance in many fields, including sizes, finishes and binders for textiles, coatings and impregnants for paper and leather, thermoplastic and thermosetting protective coatings, floor finishing materials, adhesives, high-impact plastics, elastomers for gaskets, and impregnants for asphalt and concrete.

Advantages of emulsion polymerization are rapidity and production of high-molecular-weight polymers in a system of relatively low viscosity. Difficulties in agitation, heat transfer, and transfer of materials are minimized. The handling of hazardous solvents is eliminated. The two principal variations in technique used for emulsion polymerization are the redox and the reflux methods.

Suspension polymerization also is used. When acrylic monomers or their mixtures with other monomers are polymerized while suspended (usually in aqueous system), the polymeric product is obtained in the form of small beads, sometimes called pearls or granules. Bead polymers are the basis of the production of molding powders and denture materials. Polymers derived from acrylic or methacrylic acid furnish exchange resins of the carboxylic acid type. Solutions in organic solvents furnish lacquers, coatings and cements, while water-soluble hydrolysates are used as thickeners, adhesives, and sizes.

The basic difference between suspension and emulsion processes lies in the site of the polymerization, since initiators insoluble in water are used in the suspension process. Suspensions are produced by vigorous and continuous agitation of the monomer and solvent phases. The size of the drop will be determined by the rate of agitation, the interfacial tension, and the presence of impurities and minor constituents of the recipe. If agitation is stopped, the droplets coalesce into a monomer layer. The water serves as a dispersion medium and heat-transfer agent to remove the heat of polymerization. The process and resulting product can be influenced by the addition of colloidal suspending agents, thickeners, and salts.

Product Groupings. The principal acrylic plastics are cast sheet, molding powder, and high-impact molding powder. The cast acrylic sheet is formable, transparent, stable, and strong. Representative uses include architectural panels, aircraft glazing, skylights, lighted outdoor signs, models, product prototypes, and novelties. Molding powders are used in the mass production of numerous intricate shapes, such as automotive lights, lighting fixture lenses, and instrument dials and control panels for autos, aircraft, and appliances. The high-impact acrylic molding powder yields a somewhat less transparent product, but possesses unusual toughness for such applications as toys, business machine components, blow-molded bottles, and outboard motor shrouds. The various acrylic resins find numerous uses, with varied and wide use in coatings. Acrylic latexes are composed mainly of monomers of the acrylic family, such as methyl methacrylate, butyl methacrylate, methyl acrylate, and 2-ethylhexylacrylate. Additional monomers, such as styrene or acrylonitrile, can be polymerized with acrylic monomers. Acrylic latexes vary considerably in their properties, mainly affected by the monomers used, the particle size, and the surfactant system of the latex. Generally, acrylic latexes are cured by loss of water only, do not yellow, possess a good exterior durability, are tough, and usually have good abrasion resistance. The acrylic polymers are reasonably costly and some latexes do not have very good color compatibility. Acrylic latex paints can be used for concrete floors, interior flat and semigloss finishes, and exterior surfaces.

ACRYLIC ACID. $CH_2:CH \cdot COOH$, formula wt 72.06, colorless liquid monocarboxylic acid, mp 12°C, bp 141°C, sp gr 1.062. Also called prepenoic acid, this compound is miscible in all proportions with H_2O or alcohol. The acid forms esters and metallic salts and forms addition products. The compound is of particular interest because of the large number of synthetic plastics and resins which are made as the result of polymerizing various acrylic derivatives, notably the esters of acrylic acid. The anhydrous monomer, glacial acrylic acid, contains less than 2% H_2O. It yields esters when reacted with alcohols, including ethyl acrylate and methyl acrylate. See also **Acrylic and Modacrylic Fibers; Acrylonitrile;** and **Resins (Acrylonitrile-Butadiene-Styrene).**

ACRYLIC AND MODACRYLIC FIBERS. The U.S. Federal Trade Commission defines an acrylic fiber as one "in which the fiber-forming substance is any long-chain synthetic polymer composed of at least 85% by weight of acrylontrile units (—CH_2—CH—)." Further, when a fiber is "composed of less
  |
  CN

than 85%, but at least 35% by weight of acrylonitrile units, it is properly known as a modacrylic fiber."

Acrylonitrile polymers were reported in the German patent literature as early as the 1920s, but because of the instability of the fibers at the melting point and the lack of appropriate solvents, conversion into synthetic fibers by either melt spinning or solution spinning was not practical at that time. H. Rein of I. G. Farbenindustrie, in 1938, described fibers obtained from polymer dissolved in aqueous solutions of quaternary ammonium compounds, such as benzylpyridinium chloride, or of metal salts, such as lithium bromide, sodium thiocyanate, or aluminum perchlorate. In the early 1940s, Dupont, after studying the suitability of many solvents, chose dimethyl formamide and commenced development of "Fiber A," which became *Orlon*®, the first acrylic fiber to be commercially manufactured (1949) primarily for industrial and apparel use. Monsanto (1952) chose a different solvent, and *Acrilan*® acrylic fiber was introduced for broad use in textiles and home furnishings.

Subsequently, other U.S. firms entered into acrylic fiber production, including Dow (*Zefran*®, 1958) and American Cyanamid (*Creslan*®, 1959). *Zefran*® acrylic fiber is now produced by Badische Corp. Modacrylic fibers are produced in the U.S. by Tennessee Eastman (*Verel*®), and, Monsanto (*SEF*®), and Badische (*Zefran*® type M-281).

Acrylic fiber production also commenced in various other locations in the world as well. As of 1983, about 60 plants were producing acrylic and modacrylic fibers worldwide. Principal uses of acrylic and modacrylic fibers are given in the accompanying table.

ACRYLIC AND MODACRYLIC FIBERS—
MAJOR USES

USES	PERCENT OF TOTAL
Sweaters	13.1
Craft yarns	16.0
Half hose	13.8
Pile	12.1
Single knit	15.6
Double knit	2.4
Broadwovens	1.1
Blankets	8.8
Draperies	2.2
Upholstered furniture	1.5
Other home furnishings	2.0
Carpet face yarn	9.2
Industrial end-uses	2.2

The basic raw material for acrylic fibers is acrylonitrile, a colorless liquid that, through the process of bulk, emulsion, solution, or suspension polymerization, is converted into large linear molecules ("polyacrylonitrile") of the type:

$$n(CH_2 = CHCN) \rightarrow \left(\begin{array}{c} -CH_2CH- \\ | \\ CN \end{array} \right)_n .$$

The value of n is normally in the range of 600–2000 for commercial fibers. This is equivalent to weight-average molecular weights of 100,000 to 150,000 or number-average molecular weights of 35,000 to 50,000.

Because of difficulties with dyeing and other limiting characteristics, fibers are seldom made from 100% polyacrylonitrile, but are usually copolymerized with one or more different monomers.

Commercially important comonomers include methylacrylate, methylmethacrylate, vinylacetate, and vinylbenzene. Terpolymers and tetrapolymers based on these and other comonomers, plus other compounds, particularly those containing dye-receptive groups, also are commercially important.

There are, however, some 100% polyacrylonitrile fibers available, such as Badische's *Zefran* A-405, *Zefran* A-507, and Bayer's *Dralon* T-100. The A-405 is a producer-colored product for use in apparel knit goods, while the others are uncolored fibers for industrial uses, such as filter cloths. Homopolymer fibers are more mechanically and thermally stable and more chemical-resistant than the copolymer types.

Sites for attaching dyestuffs may be introduced either by polymerizable additives or by the generation of residual end groups through the selection of suitable radical-generating catalyst systems.

Fiber properties can be strongly influenced during the manufacturing step as a result of changing: (1) the inherent fiber structure; (2) the kind and amount of modifier used; (3) the spinning method used; and (4) the degree of stretching during fiber formation. A schematic representation of an acrylic (or modacrylic) fiber production process is shown in Fig. 1.

Acrylic fibers are produced by wet or dry spinning methods. Melt spinning is not commercially feasible because of polymer degradation at temperatures near or at the melting point. Both wet and dry spinning require that the polymer be dissolved in a suitable solvent to form a viscous solution that is forced through a spinnerette. The spinnerette is a metal plate or disk perforated with holes typically ranging from 0.002 to 0.010 in. (0.05–0.25 mm) in diameter. Spinnerettes for dry spinning have 300–900 holes, while those for wet spinning have 10,000 to 60,000 (or more) holes. About 60% of acrylic fiber production in the United States is by the wet spinning process.

Upon extrusion of the fiber from the spinnerette hole, solvent is removed from the plastic mass, thereby regenerating the acrylic polymer in filamentary form. If solvent is removed by hot gases, the system is called dry spinning. If the solvent is leached out by another liquid, the process is called wet spinning. In either case, the extruded fibers must be drawn (extended to impart satisfactory strength and elongation). Then, the fibrous strand is crimped to make processing on textile-staple equipment possible and to give products made from acrylic staple or tow better aesthetic and performance properties.

Acrylic fibers usually are manufactured in the form of staple and tow. Staple length varies to suit the type of yarn spinning system used to convert the fiber into spun yarn. In the United States, lengths range from $1\frac{1}{8}$ in. (29 mm) for the cotton system to 6 in. (153 mm) for the worsted system. Variable cuts are also produced for special uses, such as blankets, pile, and fleece. For certain applications, even shorter fiber lengths, e.g., $\frac{3}{4}$ in. (19 mm) have been provided. Acrylic fibers are also marketed as tow for processing on stretch-breaking equipment (*Seydel, Turbo, Tematex*) and straight-cutting units (*Pacific Converter*). A small amount of continuous-filament yarn is produced in Japan.

Fiber fineness, reported by denier (d = weight in grams of

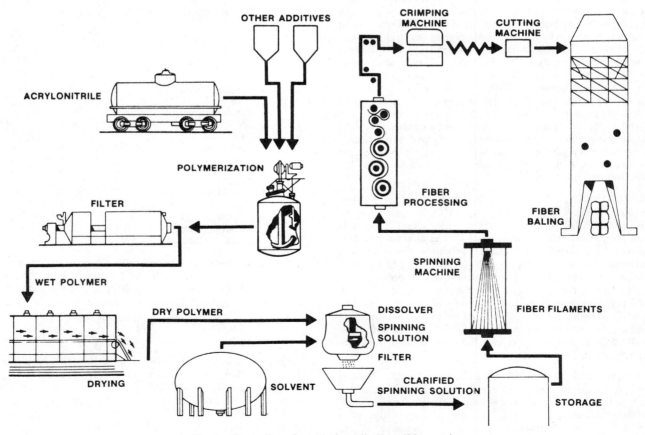

Fig. 1.   Production of a typical acrylic fiber. (*Monsanto*)

9,000 meters of fiber) or tex (weight of 1000 meters of fiber), varies commercially from the apparel range, 1–6 d, up to the carpet range, 8–20 d. Microfibers (less than 0.75 d per filament) also have been produced.

Modacrylic fiber processes are similar to the acrylic manufacturing methods.

In general, all acrylics are resistant to ordinary chemicals, but are degraded readily by hot, concentrated alkalis. The acrylics, as a class, are very resistant to light, insects, and to microbiological attack. Moth and carpet beetle larvae, for example, have virtually no effect on acrylic fibers. The fibers are not weakened by mold or mildew. Modacrylics have lower acrylonitrile content (<85%) and more comonomer. As a result certain physical properties are altered. In particular, behavior toward heat and solvents differs from that of their acrylic counterparts. Lower softening temperatures and solubility in a few more solvents are often noted in modacrylic fibers although these effects are found to varying degrees in fibers from different producers. For example, see Fig. 2.

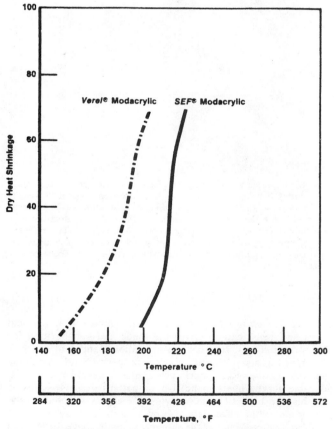

Fig. 2. Shrinkage comparison under dry heat. Note that the conventional modacrylic fiber undergoes severe shrinkage from 140°C upward, whereas Type S-06 SEF^R modacrylic does not show appreciable shrinkage at temperatures below 190°C. (*Monsanto*)

Cross sections of acrylic and modacrylic fibers can vary, but in general, dry-spun fibers are dog-bone [Fig. 3(b)], ribbon [Fig. 3(d)], or peanut-shaped [Fig. 3(e)]. Modified cross sections can, however, be obtained from either spinning process. Different cross sections can influence the luster, sparkle, and other proper-

ties of a fabric. Fibers may be produced in bright form or delustered with titanium dioxide (TiO_2).

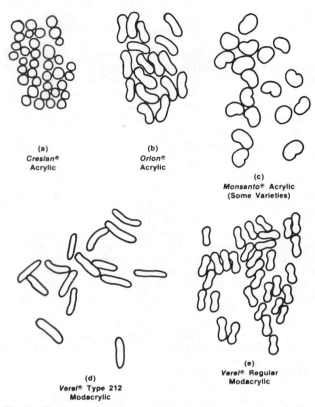

Fig. 3. Representative cross sections of acrylic and modacrylic fibers.

Fabrics made from acrylic fibers have a warm, soft touch, launder readily, dry rapidly and are dimensionally stable. A low specific gravity coupled with a high bulk procedure results in thick bulky fabrics of low weight. The inherent characteristics of acrylic fibers, especially resilience, coloration potential, hand, and blending potential with other fibers, make them the fiber of choice for certain end uses, such as sweaters, craft yarns, pile fabrics, and blankets. Significant use is also evident in circular and flat knit goods, hosiery, carpets, and broadwovens. Use of acrylic fibers in fabrics for upholstered furniture and draperies is also increasing.

Of all the synthetic fibers, acrylics and modacrylics have aesthetics closest to wool in many applications. In contrast to wool, most acrylics are supplied with a good white that does not generally need bleaching. Where light hues require a better white base, a sodium chlorite bleach and addition of the proper optical brightener will allow virtually any clear bright color to be obtained on bright acrylics. Semi-dull acrylics readily accept the muted shades characteristic of wool. The different lusters are frequently blended to achieve the desired effect.

Like wool, acrylics are colored on a broad variety of conventional textile dyeing equipment to optimize the excellent aethetic qualities of the fibers. For regular-dyeing acrylics, disperse or cationic (basic) dyes are the dyestuffs of choice. Dyes may be applied to staple (stock dyeing) or the staple may be spun into yarn for skein or package dyeing. The dyeing of fabrics or semi-finished garments (piece dyeing) is also practiced to get the desired bulk level. The dyeing of acrylics can be done under pressure or at atmospheric conditions, depending on the equipment available. Acrylic fabrics and garments are often printed on standard textile equipment. These and other coloration techniques allow textile designers to utilize acrylics in a wide range of styles and applications in apparel, home furnishings, or industrial areas.

To enhance performance or styling, acrylics are often blended with polyester, nylon, cotton, rayon, or wool. Dyeing techniques are available to color even three-way and four-way blends.

Fiber Modifications. Certain properties can be built into or excluded from basic forms of the fiber. Some properties that are manipulated include crimp, residual shrinkage, elongation to break, strength, hand, and dyeability. Of special significance are the variants listed below that have become so important to particular end uses.

High-Bulk Fibers. It is well known that for every high polymer there is a narrow temperature region in which the polymer characteristically changes from a more or less highly resilient state to a viscoelastic state. If acrylic fiber is heated to 140°F (60°C) or above, depending upon the individual fiber, stretched 10–20%, and then cooled under tension, the fiber is put into a metastable state. In this form, the fiber producer calls it a "hi-bulk" fiber. It is stable at ordinary temperatures and humidities. This modified fiber may be processed either alone or, preferably, in blends with regular acrylic or other fibers. When the high-bulk fiber in the blended yarn is relaxed in a high-temperature dyebath or by steaming, it shrinks to its original length and thus shortens the entire yarn. This causes the nonshrinking fibers to pucker. The unusual softness, lightness, pleasing appearance, and high cover thus obtained are important to certain styles of garments.

The fiber producer may use this principle to produce regular, low-bulk, and high-bulk variants. The fiber processor may also use it to produce bulky yarns and fabrics by stretch-breaking or other processes, as mentioned earlier.

Bicomponent Acrylic Fibers. Introduced in 1959 (by Dupont), these fibers are increasing in commercial importance. In one common type of bicomponent acrylic, a homopolymer and a copolymer are brought together as separate entities, the two sides being joined together along the entire length of the fiber. When such a fiber shrinks, it develops a helical twist because the two sides shrink by a different amount. Skin-core, biconstituent, and other types of two-element fibers also are known. When the two elements belong to different fiber classes, the product must be designated a biconstituent fiber, according to a Federal Trade Commission regulation. Bicomponent or biconstituent fibers have improved bulk, cover, and resilience, generally because of built-in, permanent crimp (Fig. 4). The newer bicomponent acrylic fibers, produced by either wet or dry spinning, have cross sections ranging from bean to popcorn to worm shapes, or round. These fiber types are particularly suitable for hand craft yarns and sweaters.

One very special and important type of bicomponent acrylic fiber uses selected polymer pairs that respond differently to a combination of heat and moisture. This approach leads to bulkier garments with greatly improved shape restorability after wearing in comparison with apparel made from other acrylic fiber types.

Producer-Colored Fibers. Several producers of acrylic fibers also offer colored fibers. The coloration is imparted to the fiber during the spinning step and is of value for several practical reasons. By using producer-colored fiber, fabrics can be produced that require no dyeing. If the producer-colored fiber is not delustered, then fabrics with enhanced luster can be obtained. Excellent color continuity is also assured by this technique.

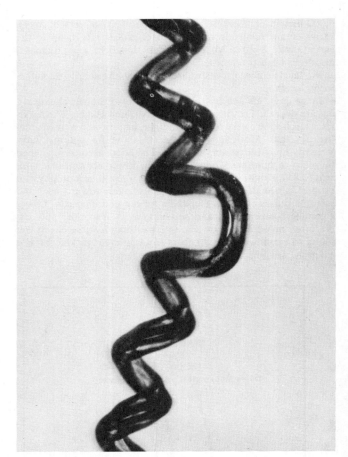

Fig. 4. Representation of *Monsanto*[R] bicomponent acrylic carpet fiber after crimp development. Note the helix direction reversal. Reversals of this type add greatly to softness, bulkiness, and resilience. The mechanism of crimp development is similar to that causing curvature in a bimetallic thermostat. Due to large recovery forces, helical bicomponent crimp is much more energetic in resisting and recovering from deformation than the planar crimp mechanically induced in monocomponent fiber. (*Monsanto*)

Acid-Dyeability Fibers. Acrylic fibers which will accept acid dyes and resist cationic dyes are made and can be used for cross-dyed effects. In cross dyeing, cationic and acid dyes are combined in a single bath set at the proper pH and containing suitable auxiliaries. A fabric prepared from regular and acid-dyeable acrylic fibers in heather or pattern designs is treated in this dye bath to achieve two color effects simultaneously. Color and white effects are also possible in this way.

Modacrylics. Many of the statements made above for acrylics apply equally well to modacrylics, but there are some special differences that should be noted. Modacrylic fibers have lower heat stability and greater solubility in some solvents, as noted earlier and shown in Fig. 2. Although the definition of modacrylic does not specify the other raw materials to be used besides acrylonitrile, in actual fact the acrylonitrile is usually combined with one or more halogenated monomers to produce so-called inherently flame-resistant fibers. When these products are converted into properly constructed fabrics, the fabrics are then able to comply with specific flammability test criteria. As a result, modacrylics have found their main usage in those areas where flammability is of particular concern, for example, children's sleepwear, work uniforms, draperies in places of public assembly, pile fabrics, and some other specialty items. Because these flame-resistant characteristics are the result of polymer structure, they are unaffected by age, repeated laundering or dry cleaning.

Some of the newer modacrylics, e.g. SEF® modacrylic fiber (Monsanto), can be bleached, dyed, or printed in the same way as acrylic fibers. More conventional modacrylics may require special techniques for satisfactory dyeing, such as use of carriers to obtain dark shades. In the wet processing of modacrylics, the recommendations of individual fiber producers should be obtained and followed, since they may vary considerably. A good range of shades can be obtained on modacrylic fibers with cationic (basic) or disperse dyestuffs, depending on the particular requirements for fastness to light, washing, perspiration, etc. of a particular end use. Blends of modacrylics with other fibers, particularly polyester, are quite common and techniques are available to color such blends, usually to anion shades.

See also **Fibers**.

References

Corbman, B. P.: "Textiles: Fiber to Fabric," 8th Edition, McGraw-Hill, New York (1983).
Staff: For current United States and world statistical information, check publications generated by U.S. Census Bureau, Textile Economics Bureau, and Man-Made Fiber Association.
Wingate, I. B.: "Textile Fabrics and Their Selection," Prentice-Hall, Hightstown, New Jersey (1976).

—John Lomartire and Frederic C. Flindt, Monsanto Textiles Company.

ACRYLIC PAINT.   Paint and Finishes.

ACRYLONITRILE.
$CH_2:CHCN$, formula wt 29.04, liquid, bp 78°C. Also called vinyl cyanide or propene nitrile, this compound is a high-production chemical used as an intermediate in the production of plastics, nitrile rubbers, acrylic fibers, insecticides, and numerous other synthetic materials. Several large-scale process for making acrylonitrile have appeared since the early 1960s. The majority of these use propylene, ammonia, and air as raw materials in what may be termed an ammonoxidation or oxyamination reaction:

$$CH_3CH:CH_2 + NH_3 + 1\tfrac{1}{2}O_2 \rightarrow CH_2:CHCN + 3H_2O.$$

In one process, the starting ingredients are mixed with steam, preheated, and fed to the reactor. There are two main byproducts, acetonitrile ($CH_3 \cdot CN$) and HCN, with accompanying formation of small quantities of acrolein, acetone, and acetaldehyde. The acrylonitrile is separated from the other materials in a series of fractionation and absorption operations. A number of catalysts have been used, including phosphorus, molybdenum, bismuth, antimony, tin, and cobalt.

ACRYLONITRILE-BUTADIENE RUBBER (ABR).   Elastometers.

ACTH.
The adrenocorticotrophic hormone of the anterior lobe of the pituitary gland, which specifically stimulates the adrenal cortex to secrete cortisone, and hence has effects identical with those of cortisone. ACTH differs in its chemistry, absorption, and metabolism from the other adrenal steroids. Chemically, it is a water-soluble polypeptide having a molecular weight of about 3000. Its complete amino acid sequence has been determined. ACTH produces its peripheral physiological effects by

causing discharge of the adrenocortical steroids into the circulation. ACTH has been extracted from pituitary glands. In purified form, ACTH is useful in treating some forms of arthritis, lupus erythematosus, and severe skin disorders. The action of ACTH injections parallels the result of large quantities of naturally formed cortisone if they were released naturally. See also **Hormones**; and **Steroids**.

ACTIN.
One of the two proteins that make up the myofibrils of striated muscles. The other protein is myosin. The combination of these two proteins is sometimes spoken of as actinomyosin. The banded nature of the myofibrils is due to the fact that both proteins are present where the bands are dark and only one or the other is present in the light bands. Since these bands lie side by side in the different myofibrils that go to make up a muscle fiber, the entire muscle fiber shows a banded or striated appearance. See also **Contractile Proteins**.

ACTINIDE CONTRACTION.
An effect analogous to the lanthanide contraction, which has been found in certain elements of the actinide series. Those elements from thorium (at. no. 90) to curium (at. no. 96) exhibit a decreasing molecular volume in certain compounds, such as those which the actinide tetrafluorides form with alkali metal fluorides, plotted in the accompanying graph. The effect here is due to the decreasing crystal radius of the tetrapositive actinide ions as the atomic number increases. Note that in the actinides the tetravalent ions are compared instead of the trivalent ones as in the case of the lanthanides, in which the trivalent state is by far the most common. The behavior is attributed to the entrance of added electrons into an (inner) shell ($4f$ for the lanthanides; $5f$ for the actinides), so that the increment they produce in atomic volume is less than the reduction due to the greater nuclear charge.

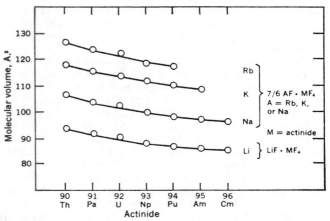

Plot of molecular volume versus atomic number of the tetravalent actinides.

ACTINIDES.
The chemical elements with atomic numbers 90 to 103, inclusively, commencing with 90 (thorium) and through 103 (lawrencium) frequently are termed, collectively, the *actinides*. The term derives from actinium (at. no. 89), which is considered the anchor element of the series, also appearing in group 3a of the periodic table. Members of the series are listed in the accompanying table. Some authorities place actinium in the series per se. This series is somewhat analogous to the lanthanide series, which is described in a separate alphabetical entry in this volume.

ELECTRONIC CONFIGURATIONS FOR NEUTRAL ATOMS OF THE ACTINIDE ELEMENTS

Element	Atomic Number (Z)	Electronic Configuration
Actinium	89	$6d7s^2$
Thorium	90	$6d^27s^2$
Protactinium	91	$5f^26d7s^2$
Uranium	92	$5f^36d7s^2$
Neptunium	93	$5f^46d7s^2$
Plutonium	94	$5f^67s^2$
Americium	95	$5f^77s^2$
Curium	96	$5f^76d7s^2$
Berkelium	97	$(5f^86d7s^2$ or $5f^97s^2)$
Californium	98	$(5f^{10}7s^2)$
Einsteinium	99	$(5f^{11}7s^2)$
Fermium	100	$(5f^{12}7s^2)$
Mendelevium	101	$(5f^{13}7s^2)$
Nobelium	102	$(5f^{14}7s^2)$
Lawrencium	103	$(5f^{14}6d7s^2)$

Justification for the grouping is found in the higher elements of (III) oxidation states similar to actinium, and (IV) oxidation states similar to thorium. Certain similarities also exist between the atomic spectra and magnetic properties in the actinide and lanthanide series. Note that actinium $(Z = 89)$ and thorium $(Z = 90)$ differ in electronic configuration from their immediate predecessor in atomic number (radium, $Z = 88$) in having, respectively, 1 and 2 electrons in their $6d$ subshells. The next element, protactinium $(Z = 91)$ is the first to have an electron of the $5f$ subshell. Note also that the configurations of the last seven elements are enclosed in parentheses in the table; this is to indicate that they are predicted, rather than determined, configurations. Two major methods have been used in making these determinations for the first eight elements: emission spectroscopy for actinium, thorium, uranium, and americium; and atomic-beam experiments for protactinium, neptunium, plutonium, and curium.

While in many respects the electronic configurations and chemical properties of the actinide elements are similar to those of the lanthanide series, the $4f$, $5d$ and $6s$ subshells of the latter corresponding to the $5f$, $6d$ and $7s$ of the latter, there are, however, significant differences. Cerium in the lanthanide series, unlike its analog thorium in the actinide series, has an electron in its $4f$ subshell. Moreover, for the first few members of each series, the $5f$ and $6d$ electrons are less energetically bound to the atomic nucleus than the $4f$ and $5d$ ones, so that the first few actinide elements (except actinium) in general have higher oxidation states (lose electrons more readily) than the corresponding lanthanides. Thus uranium, neptunium, plutonium, and americium have all four of the oxidation states 3, 4, 5 and 6. Later in the series, the actinides correspond more closely to the lanthanides in this respect.

In their electronic configurations the actinide elements all have their innermost 86 electrons arranged in the configuration of radon, and their additional electrons as shown in the accompanying table.

ACTINIUM. Chemical element, symbol Ac, at. no. 89, at. wt. 227 (mass number of the most stable isotope), periodic table group 3b, classed in the periodic system as a higher homologue of lanthanum. The electronic configuration for actinium is $1s^22s^22p^63s^23p^63d^{10}4s^24p^64d^{10}4f^{14}5s^25p^65d^{10}6s^26p^66d^17s^2$. The ionic radius (Ac^{3+}) is 1.11Å.

Presently, 24 isotopes of actinium, with mass numbers ranging from 207 to 230, have been identified. All are radioactive. One year after the discovery of polonium and radium by the Curies, A. Debierne found an unidentified radioactive substance in the residue after treatment of pitchblende. Debierne named the new material *actinium* after the Greek word for ray. F. Giesel, independently in 1902, also found a radioactive material in the rare-earth extracts of pitchblende. He named this material *emanium*. In 1904, Debierne and Giesel compared the results of their experimentation and established the identical behavior of the two substances. Until formulation of the law of radioactive displacement by Fajans and Soddy about ten years later, however, actinium definitely could not be classed in the periodic system as a higher homologue of lanthanum.

The isotope discovered by Debierne and also noted by Giesel was ^{227}Ac, which has a half-life of 21.7 years. The isotope results from the decay of ^{235}U (AcU—*actinouranium*) and is present in natural uranium to the extent of approximately 0.715%. The proportion of Ac to U in uranium ores is estimated to be approximately 2×10^{-10} at radioactive equilibrium. O. Hahn established the existence of a second isotope of actinium in nature, ^{228}Ac, in 1908. This isotope is a product of thorium decay and logically also is referred to as *meso*-thorium, with a half-life of 6.13 hours. The proportion of mesothorium $(MsTh_2/Th)$ in thorium ores is about 5×10^{-14}. The other isotopes of actinium were found experimentally as the result of bombarding thorium targets. The half-life of 10 days for ^{225}Ac is the longest of the artificially produced isotopes. Although occurring in nature as a member of the neptunium family, ^{225}Ac is present in extremely small quantities and thus is very difficult to detect.

^{227}Ac can be extracted from uranium ores where present to the extent of 0.2 mg/ton of uranium and it is the only isotope that is obtainable on a macroscopic scale and that is reasonably stable. Because of the difficulties of separating ^{227}Ac from uranium ores, in which it accompanies the rare earths and with which it is very similar chemically, fractional crystallization or precipitation of relevant compounds no longer is practiced. Easier separations of actinium from lanthanum may be effected through the use of ion-exchange methods. A cationic resin and elution, mainly with a solution of ammonium citrate or ammonium-α-hydroxyisobutyrate, are used. To avoid the problems attendant with the treatment of ores, ^{227}Ac now is generally obtained on a gram-scale by the transmutation of radium by neutron irradiation in the core of a nuclear reactor. Formation of actinium occurs by the following process:

$$^{226}Ra(n, \gamma)^{227}Ra \xrightarrow{\beta^-} {}^{227}Ac$$

In connection with this method, the cross section for the capture of thermal neutrons by radium is 23 barns $(23 \times 10^{-24}$ cm²). Thus, prolonged radiation must be avoided because the accumulation of actinium is limited by the reaction $(\sigma = 500$ barns):

$$^{227}Ac(n, \gamma)^{228}Ac(MsTh_2) \rightarrow {}^{228}Th(RdTh)$$

In 1947, F. Hagemann produced 1 mg actinium by this process and, for the first time, isolated a pure compound of the element. It has been found that when 25 g of $RaCO_3$ (radium carbonate) are irradiated at a flux of 2.6×10^{14} ncm^{-2}s^{-1} for a period of 13 days, approximately 108 mg of ^{227}Ac (8 Ci) and 13 mg of ^{228}Th (11 Ci) will be yielded. In an intensive research program by the Centre d'Etude de l'Energie Nucléaire Belge, Union Minière, carried out in 1970–1971, more than 10 g of actinium were produced. The process is difficult for at least two reasons: (1) the irradiated products are highly radioactive, and (2) radon gas, resulting from the disintegration of radium, is evolved.

The methods followed in Belgium for the separation of ^{226}Ra, ^{227}Ac, and ^{228}Th involved the precipitation of $Ra(NO_3)_2$ (radium nitrate) from concentrated HNO_3, after which followed the elimination of thorium by adsorption on a mineral ion exchanger (zirconium phosphate) which withstands high levels of radiation without decomposition.

Metallic actinium cannot be obtained by electrolytic means because it is too electropositive. It has been prepared on a milligram-scale through the reduction of actinium fluoride in a vacuum with lithium vapor at about 350°C. The metal is silvery white, faintly emits a blue-tinted light which is visible in darkness because of its radioactivity. The metal takes the form of a face-centered cubic lattice and has a melting point of $1050 \pm 50°C$. By extrapolation, it is estimated that the metal boils at about 3300°C. An amalgam of metallic actinium may be prepared by electrolysis on a mercury cathode, or by the action of a lithium amalgam on an actinium citrate solution (pH = 1.7 to 6.8).

In chemical behavior, actinium acts even more basic than lanthanum (the most basic element of the lanthanide series). The mineral salts of actinium are extracted with difficulty from their aqueous solutions by means of an organic solvent. Thus, they generally are extracted as chelates with trifluoroacetone or diethylhexylphosphoric acid. The water-insoluble salts of actinium follow those of lanthanum, namely, the carbonate, fluoride, fluosilicate, oxalate, phosphate, double sulfate of potassium. With exception of the black sulfide, all actinium compounds are white and form colorless solutions. The crystalline compounds are isomorphic.

In addition to its close resemblance to lanthanum, actinium also is analogous to curium ($Z = 96$) and lawrencium ($Z = 103$), both of the group of trivalent transuranium elements. This analogy led G. T. Seaborg to postulate the actinide theory, wherein actinium begins a new series of rare earths which are characterized by the filling of the $5f$ inner electron shell, just as the filling of the $4f$ electron shell characterizes the Lanthanide series of elements. However, the first elements of the Actinide series differ markedly from those of actinium. Notably, there is a multiplicity of valences for which there is no equivalent among the lanthanides. See **Chemical Elements** for other properties of actinium.

ACTINOLITE. The term for a calcium-iron-magnesium amphibole, the formula of which is $Ca_2(Mg,Fe)_5Si_8O_{22}(OH)_2$. The amount of iron varies considerably. Actinolite occurs as bladed crystals or in fibrous or granular masses. Its hardness is 5–6, sp gr 3–3.2, color green to grayish green, transparent to opaque, luster vitreous to silky or waxy. Iron in the ferrous state is believed to be the source of its green color. Actinolite derives its name from the frequent radiated groups of crystals. Essentially, it is an iron-rich tremolite; the division between the two is quite arbitrary, with color the macroscopic definitive factor: white for tremolite, green for actinolite.

Actinolite is found in schists, often with serpentine, and in igneous rocks, probably as the result of the alteration of pyroxene. The schists of the Swiss Alps carry actinolite. It is also found in Austria, Saxony, Norway, Japan, and Canada in the provinces of Quebec and Ontario. In the United States, actinolite occurs in Massachusetts, Pennsylvania, Maryland, and as a zinc-manganese bearing variety in New Jersey.

ACTINON. The name of the isotope of radon (emanation), which exists in the naturally occurring actinium series, being produced by alpha-decay of actinium X, which is itself a radium isotope. Actinon has an atomic number of 86, a mass number of 219, and a half-life of 3.92 seconds, emitting an alpha particle to form polonium-215 (actinium A).

ACTIVATED SLUDGE. This is the biologically active sediment produced by the repeated aeration and settling of sewage and/or organic wastes. The dissolved organic matter acts as food for the growth of an aerobic flora. This flora produces a biologically active sludge which is usually brown in color and which destroys the polluting organic matter in the sewage and waste. The process is known as the activated sludge process.

The activated sludge process, with minor variations, consists of aeration through submerged porous diffusers or by mechanical surface agitation, of either raw or settled sewage for a period of 2–6 hours, followed by settling of the solids for a period of 1–2 hours. These solids, which are made up of the solids in the sewage and the biological growths which develop, are returned to the sewage flowing into the aeration tanks. As this cycle is repeated, the aerobic organisms in the sludge develop until there is 1000–3000 ppm of suspended sludge in the aeration liquor. After a while more of the active sludge is developed than is needed to purify the incoming sewage, and this excess is withdrawn from the process and either dried for fertilizer or digested anaerobically with raw sewage sludge. This anaerobic digestion produces a gas consisting of approximately 65% methane and 35% CO_2, and changes the water-binding properties so that the sludge is easier to filter or dry.

The activated sludge is made up of a mixture of zoogleal bacteria, filamentous bacteria, protozoa, rotifera, and miscellaneous higher forms of life. The types and numbers of the various organisms will vary with the types of food present and with the length of the aeration period. The settled sludge withdrawn from the process contains from 0.6 to 1.5% dry solids, although by further settling it may be concentrated to 3–6% solids. Analysis of the dried sludge for the usual fertilizer constituents show that it contains 5–6% of slowly available N and 2–3% of P. The fertilizing value appears to be greater than the analysis would indicate, thus suggesting that it contains beneficial trace elements and growth-promoting compounds. Recent developments indicate that the sludge is a source of vitamin B_{12}, and has been added to mixed foods for cattle and poultry.

The quality of excess activated sludge produced will vary with the food and the extent of oxidation to which the process is carried. In general, about 1 part sludge is produced for each part organic matter destroyed. Prolonged or over-aeration will cause the sludge to partially disperse and digest itself. The amount of air or more precisely oxygen that is necessary to keep the sludge in an active and aerobic condition depends on the oxygen demand of the sludge organisms, the quantity of active sludge, and the amount of food to be utilized. Given sufficient food and sufficient organisms to eat the food, the process seems to be limited only by the rate at which oxygen or air can be dissolved into the mixed liquor. This rate depends on the oxygen deficit, turbulence, bubble size, and temperature and at present is restricted by the physical methods of forcing the air through the diffuser tubers and/or mechanical agitation.

In practice, the excess activated sludge is conditioned with 3–6% $FeCl_3$ and filtered on vacuum filters. This reduces the moisture to about 80% and produces a filter cake which is dried in rotary or spray driers to a moisture content of less than 5%. It is bagged and sold direct as a fertilizer, or to fertilizer manufacturers who use it in mixed fertilizer.

The mechanism of purification of sewage by the activated sludge is two-fold i.e., (1) absorption of colloidal and soluble organic matter on the floc with subsequent oxidation by the organisms, and (2) chemical splitting and oxidation of the soluble carbohydrates and proteins to CO_2, H_2O, NH_3, NO_2, NO_3, SO_4, PO_4 and humus. The process of digestion proceeds by hydrolysis, decarboxylation, deaminization and splitting of S and P from the organic molecules before oxidation.

The process is applicable to the treatment of almost any type of organic waste waters which can serve as food for biological growth. It has been applied to cannery wastes, milk products wastes, corn products wastes, and even phenolic wastes. In the treatment of phenolic wastes a special flora is developed which thrives on phenol as food.

—W. D. Hatfield, Decatur, Illinois.

ACTIVATION (Molecular). When a molecule which is a Lewis base forms a coordinate bond with a metal ion or with a molecular Lewis acid (such as $AlCl_3$ or BF_3), its electronic density pattern is altered, and with this, the ease with which it undergoes certain reactions. In some instances the polarization that ensues is sufficient to lead to the formation of ions by a process of the type

$$A\!:\!B\!: + M = A^+ + :\!B\!:\!M^-$$

More commonly the molecule is polarized by coordination in such a manner that A bears a partial positive charge and B a partial negative charge in a complex $A\!:\!B\!:\!M$. Where M is a reducible or oxidizable metal ion, an electron-transfer process may result in which a free radical is generated from $A\!:\!B\!:$ or its fragments and the metal assumes a different oxidation state. In any case the resulting species is often in a state in which it undergoes one or more types of chemical reaction much more readily.

Theoretical Basis. The theoretical basis underlying these activation processes is in the description of the bonding which occurs between the ligand (Lewis base) and the coordination center. This consists of variable contributions from two types of bond. The first is from the sigma bond in which both the electrons in the bonding orbital come from the ligand. This kind of bonding occurs with ligands with available lone pairs (as NH_3, H_2O and their derivatives) and leads to a depletion of electronic charge from the substrate. The second is from the pi bond; here the electrons may come from the metal (if it is a transition metal with suitably occupied d orbitals) or from the ligand (where it has filled p orbitals or molecular orbitals of suitable symmetry). In this case the electronic shifts may partially compensate those arising from the sigma bond if the metal "back-donates" electrons to the ligand. Such shifts may also accentuate those of sigma bonding where ligand electrons are used for both bonds. It is generally found that the net drift of the electronic density is *away* from the ligand.

Activation of Electrophiles. The activation of electrophiles by coordination is a direct result of the weakening of the bond between the donor atom and the rest of the ligand molecule after the donor atom has become bonded to the coordination center. There is considerable evidence to support the claim that this bond need not be broken heterolytically prior to reaction as an electrophile. When this does not occur the literal electrophile is that portion of the ligand which bears a partial positive charge. In such a case the activation process can be more accurately represented by:

$$A\!:\!B\!: + M = \overset{\delta+ \ \ \delta-}{A\!:\!B\!:\!M}$$

Examples of this type of process are:

A:B	+	M	→	$\overset{\delta+ \ \delta-}{A\!:\!B\!:\!M}$
Cl:Cl:		$FeCl_3$		Cl:Cl:$FeCl_3$
NC:Cl:		$AlCl_3$		NC:Cl:$AlCl_3$

RC:Cl: ‖ O	$AlCl_3$	RC:Cl:$AlCl_3$ ‖ O
O_2N:Cl: O	$AlCl_3$	O_2N:Cl:$AlCl_3$ O
RS:Cl: ‖ O	$AlCl_3$	RS:Cl:$AlCl_3$ ‖ O
R:Cl:	BF_3	R:Cl:BF_3

The resultant electrophiles are effective attacking species and can be used to replace an aromatic hydrogen by the group A. When coordination is used to activate an electrophile which has additional donor groups not involved in the principal reaction, and if these are sufficiently effective as donors, they will react with the coordination centers initially added and stop the activation process. In these cases a larger amount of the Lewis acid must be added so that there is more than enough to complex with all the uninvolved donor groups. The extra reagent then provides the Lewis acid needed for the activation process. This is encountered in the Fries reaction and in Friedel-Crafts reactions where the substrate has additional coordination sites. This particular procedure is also used in the "swamping catalyst" procedure for the catalytic halogenation of aromatic compounds.

The usual activation of carbon monoxide by coordination appears to involve complexes in which the carbon atom bonded to the metal is rendered slightly positive, and thus more readily attacked by electron rich species such as ethylenic or acetylenic linkages. An example is seen in the reaction of nickel carbonyl and aqueous acetylene, which results in the production of acrylic acid.

Activation of Free Radicals. The formation of free radicals results from a very similar process when the species M can be oxidized or reduced in a one-electron step with the resultant heterolytic splitting of the $A\!:\!B$ bond. The basic reaction in an oxidation reaction of this sort is

$$A\!:\!B\!: + M^{+x} = A\cdot + :\!B\!:\!M^{+x+1}$$

where A and B may be the same or different. The most thoroughly characterized of these reactions is the one found with Fenton's reagent:

$$Fe^{2+} + HOOH = HO\cdot + Fe(OH)^{2+}.$$

The resultant hydroxyl radicals are effective in initiating many chain reactions. The number of metal ions and complexes which are capable of activating hydrogen peroxide in this manner is quite large and is determined in part by the redox potentials of the activator. Related systems in which free radicals are generated by the intervention of suitable metallic catalysts include many in which oxygen is consumed in autoxidations. Cobalt(II) compounds which act as oxygen carriers can often activate radicals in such systems by reactions of the type:

$$Co(II)L + O_2 = Co(III)L + O_2^-, \text{ etc.}$$

Processes of this sort have been used for catalytic oxidations and in cases where a complex with O_2 is formed, the reversibility of the reaction has been studied as a potential process for separating oxygen from the atmosphere.

Radical generating systems of this sort may be used for the initiation of many addition polymerization reactions including those of acrylonitrile and unsaturated hydrocarbons. The information on systems other than those derived from hydrogen peroxide is very meager.

The activation of O_2 by low oxidation states of platinum has also been demonstrated in reactions such as

$$2P(C_6H_5)_3 + O_2 \xrightarrow{Pt(P(C_6H_5)_3)_3} 2(C_6H_5)_3PO.$$

Ligands such as ethylene are Lewis bases by virtue of the availability of the electrons of their pi bonds to external reagents and the coordination of unsaturated organic compounds to species such as Cu(I), Ag(I), Pd(II), and Pt(II) is a well established phenomenon. The coordination process with such ligands usually involves a considerable element of back bonding from the filled d orbitals of the metal ion. The coordination process activates olefins towards cis-trans isomerizations and attack by reagents such as hydrogen halides. Coordination to palladium(II) facilitates attack of olefins by water via a redox process as seen in the Smid reaction:

$$PdCl_2 + H_2C{=}CH_2 + H_2O = Pd + CH_3CHO + 2HCl$$

$$Pd + 2CuCl_2 = 2CuCl + PdCl_2$$

$$2HCl + 2CuCl + \tfrac{1}{2}O_2 = 2CuCl_2 + H_2O$$

A similar reaction also occurs for carbon monoxide:

$$Pd^{2+} + CO + H_2O = Pd + CO_2 + 2H^+.$$

Activation of nucleophiles by coordination, best exemplified by various complexes used as catalysts for hydrogenation, and coordination assistance to photochemical activation, may be similarly demonstrated.

— Mark M. Jones, Vanderbilt University, Nashville, Tennessee.

ACTIVATOR. A substance that renders a material or system reactive; commonly, a catalyst. A special use of this term occurs in the flotation process, where an activator assists the action of the collector. An activator also may be an impurity atom, present in a solid, that makes possible the effects of luminescence, or markedly increases their efficiency. Examples are copper in zinc sulfide; and thallium in potassium chloride.

ACTIVITY COEFFICIENT. A fractional number which when multiplied by the molar concentration of a substance in solution yields the chemical activity. This term provides an approximation of how much interaction exists between molecules at higher concentrations. Activity coefficients and activities are most commonly obtained from measurements of vapor-pressure lowering, freezing-point depression, boiling-point elevation, solubility, and electromotive force. In certain cases, activity coefficients can be estimated theoretically. As commonly used, activity is a relative quantity having unit value in some chosen standard state. Thus, the standard state of unit activity for water, a_w, in aqueous solutions of potassium chloride is pure liquid water at one atmosphere pressure and the given temperature. The standard state for the activity of a solute like potassium chloride is often so defined as to make the ratio of the activity to the concentration of solute approach unity as the concentration decreases to zero.

In general, the activity coefficient of a substance may be defined as the ratio of the effective contribution of the substance to a phenomenon to the actual contribution of the substance to the phenomenon. In the case of gases the effective pressure of a gas is represented by the fugacity f and the actual pressure of the gas by P. The activity coefficient, γ, of the gas is given by $\gamma = f/P$. (1)

One method of calculating fugacity and hence γ is based on the measured deviation of the volume of a real gas from that of an ideal gas. Consider the case of a pure gas. The free energy F and chemical potential μ changes with pressure according to the equation

$$dF = d\mu = V\,dP. \tag{2}$$

but by definition

$$d\mu = V\,dP = RT\,d\ln f \tag{3}$$

If the gas is ideal, the molal volume V_i is given by

$$V_i = \frac{RT}{P} \tag{4}$$

but for a nonideal gas this is not true. Let the molal volume of the nonideal gas be V_n and define the quantity α by the equation

$$\alpha = V_i - V_n = \frac{RT}{P} - V_n \tag{5}$$

Then V of Eq. (2) is V_n of Eq. (5) and hence from Eq. (5)

$$V = \frac{RT}{P} - \alpha \tag{6}$$

Therefore from Eqs. (2), (3), and (6)

$$RT\,d\ln f = dF = d\mu = RT\,D\ln P - \alpha\,dP \tag{7}$$

and

$$RT\ln f = RT\ln P - \int_0^P \alpha\,dP \tag{8}$$

Thus knowing PVT data for a gas it is possible to calculate f. The integral in Eq. (8) can be evaluated graphically by plotting α, the deviation of gas volume from ideality, versus P and finding the area under the curve out to the desired pressure. Also it may be found by mathematically relating α to P by an equation of state, or by using the method of least squares or other acceptable procedure the integral may be evaluated analytically for any value of P. The value of f at the desired value of P may thus be found and consequently the activity coefficient calculated. Other methods are available for the calculation of f and hence of γ, the simplest perhaps being the relationship

$$f = \frac{P^2}{P_i} \tag{9}$$

where P_i is the ideal and P the actual pressure of the gas.

In the case of nonideal solutions, we can relate the activity α_A of any component A of the solution to the chemical potential μ_A of that component by the equation

$$\mu_A = \mu_A^{0\prime} + RT\ln a_A \tag{10}$$

$$= \mu_A^{0\prime} + RT\ln \gamma_A X_A \tag{11}$$

where $\mu_A^{0\prime}$ is the chemical potential in the reference state where a_i is unity and is a function of temperature and pressure only, whereas γ_A is a function of temperature pressure and concentration. It is necessary to find the conditions under which γ_A is unity in order to complete its definition. This can be done using two approaches—one using Raoult's law which for solutions composed of two liquid components is approached as $X_A \to 1$; and two using Henry's law which applies to solutions, one component of which may be a gas or a solid and which is approached at $X_A \to 0$. Here X_A represents the mole fraction of component A.

For liquid components using Raoult's law

$$\gamma_A \rightarrow 1 \text{ as } X_A \rightarrow 1 \qquad (12)$$

Since the logarithmic term is zero in Eq. (11) under this limiting condition, $\mu_A^{o\,\prime}$ is the chemical potential of pure component A at the temperature and pressure under consideration. For ideal solutions the activity coefficients of both components will be unity over the whole range of composition.

The convention using Henry's law is convenient to apply when it is impossible to vary the mole fraction of both components up to unity. Solvent and solute require different conventions for such solutions. As before, the activity of the solvent, usually taken as the component present in the higher concentration, is given by

$$\gamma_A \rightarrow 1 \text{ as } X_A \rightarrow 0 \qquad (14)$$

Thus, $\mu_A^{o\,\prime}$ for the solute in Eq. (11) is the chemical potential of the solute in a hypothetical standard state in which the solute at unit concentration has the properties which it has at infinite dilution.

γ_A is the activity coefficient of component A in the solution and is given by the expression

$$\gamma_A = \frac{a_A}{X_A} \qquad (15)$$

In Eq. (15) a_A is the activity or in a sense the effective mole fraction of component A in the solution.

The activity a_A of a component A in solution may be found by considering component A as the solvent. Then its activity at any mole fraction is the ratio of the partial pressure of the vapor of A in the solution to the vapor pressure of pure A. If B is the solute, its standard reference state is taken as a hypothetical B with properties which it possesses at infinite dilution.

The equilibrium constant for the process

$$B \text{ (gas)} \leftrightarrow B \text{ (solution)} \qquad (16)$$

is

$$K = \frac{a_{\text{solution}}}{a_{\text{gas}}} \qquad (17)$$

Since the gas is sufficiently ideal its activity a_{gas} is equivalent to its pressure P_2. Since the solution is far from ideal, the activity a_{solute} of the liquid B is not equal to its mole fraction N_2 in the solution. However,

$$K' = N_2/P_2 \qquad (18)$$

and extrapolating a plot of this value versus N_2 to $N_2 = 0$ one obtains the ratio where the solution is ideal. This extrapolated value of K' is the true equilibrium constant K when the activity is equal to the mole fraction

$$K = a_2/P_2 \qquad (19)$$

Thus a_2 can be found. The methods involved in Eqs. (16) through (19) arrive at the activities directly and thus obviate the determination of the activity coefficient. However, from the determined activities and known mole fractions γ can be found as indicated in Eq. (15).

In the case of ions the activities, a_+ and a_- of the positive and negative ions, respectively, are related to the activity, a, of the solute as a whole by the equation

$$a_+^p \times a_-^q \qquad (20)$$

and the activity coefficients γ_+ and γ_- of the two charge types of ions are related to the molality, m, of the electrolyte and ion activities a_+ and a_- by the equations

$$\gamma_+ = \frac{a_+}{pm}; \quad \gamma_- = \frac{a_-}{qm} \qquad (21)$$

Also the activity coefficient of the electrolyte is given by the equation

$$\gamma = (\gamma_+^p \times \gamma_-^q)^{(1/p+q)} \qquad (22)$$

In Eqs. (20), (21) and (22) p and q are numbers of positive and negative ions, respectively, in the molecule of electrolyte. In dilute solutions it is considered that ionic activities are equal for uniunivalent electrolytes, i.e., $\gamma_+ = \gamma_-$.

Consider the case of $BaCl_2$.

$$\gamma = (\gamma_+ \times \gamma_-^2)^{(1/1+2)} = (\gamma_+ \times \gamma_-^2)^{1/3} \qquad (23)$$

or

$$\gamma^3 = \gamma_+ \gamma_-^2 \qquad (24)$$

also

$$a = a_+ \times a_-^2 = (m\gamma_+)(2m\gamma_-)^2 \qquad (25)$$

$$= 4m^3\gamma_+\gamma_-^2 = 4m^3\gamma^3 \qquad (26)$$

Activity coefficients of ions are determined using electromotive force, freezing point, and solubility measurements or are calculated using the theoretical equation of Debye and Hückel.

The solubility, s, of AgCl can be determined at a given temperature and the activity coefficient γ determined at that temperature from the solubility and the solubility product constant K. Thus

$$K = a_+ a_- = \gamma_+ c_+ \gamma_- c_- \qquad (27)$$

where c_+ and c_- are the molar concentrations of the positive silver and negative chloride ions, respectively. The solubility s of the silver chloride is simply $s = c_+ = c_-$. The expression for K is then

$$K = \gamma^2 s^2 \qquad (28)$$

and

$$\gamma = \frac{K^{1/2}}{s} \qquad (29)$$

By measuring the solubility, s, of the silver chloride in different concentration of added salt and extrapolating the solubilities to zero salt concentration, or better, to zero ionic strength, one obtains the solubility when $\gamma = 1$, and from Eq. (29) K can be found. Then γ can be calculated using this value of K and any measured solubility. Actually, this method is only applicable to sparingly soluble salts. Activity coefficients of ions and of electrolytes can be calculated from the Debye-Hückel equations. For a uni-univalent electrolyte, in water at 25°C, the equation for the activity coefficient of an electrolyte is

$$\log \gamma = -0.509 z_+ z_- \sqrt{\mu} \qquad (30)$$

where z_+ and z_- are the valences of the ion and μ is the ionic strength of the solution, i.e.,

$$\mu = \tfrac{1}{2} \Sigma\, c_i z_i^2 \qquad (31)$$

where c_i is the concentration and z_i the valence of the ith type of ion.

To illustrate a use of activity coefficients, consider the cell without liquid junction

$$Pt, H_2(g); \ HCl(m); \ AgCl, Ag \qquad (32)$$

for which the chemical reaction is

$$\tfrac{1}{2} H_2(g) + AgCl \text{ (solid)} = HCl \text{ (molality, m)}$$
$$+ Ag \text{ (solid.)} \qquad (33)$$

The electromotive force, E, of this cell is given by the equation

$$E = E° - \frac{2.303RT}{n\mathbf{F}} \log \frac{a_{HCl}}{P_{H_2}}$$

$$= E° - 0.05915 \log m^2\gamma^2 \qquad (34)$$

where $E°$ is the standard potential of the cell, n is the number of electrons per ion involved in the electrode reaction (here $n = 1$), $\mathbf{F}$ is the coulombs per faraday, a (equal to $m^2\gamma^2$) is the activity of the electrolyte HCl, P_{H_2} is the pressure (1 atm) and is equal to the activity of the hydrogen gas, and AgCl (solid) and Ag (solid) have unit activities. Transferring the exponents in front of the logarithmic term in Eq. (34), the equation can be written,

$$E = E° - 0.1183 \log m - 0.1183 \log \gamma \qquad (35)$$

which by transposing the log m term to the left of the equation becomes

$$E + 0.1183 \log m = E° - 0.1183 \log \gamma \qquad (36)$$

For extrapolation purposes, the extended form of the Debye-Hückel equation involving the molality of a dilute univalent electrolyte in water at 25°C is used:

$$\log \gamma = -0.509\sqrt{m} + bm \qquad (37)$$

where b is an empirical constant.
Substitution of log γ from Eq. (37) into Eq. (36) gives

$$E + 0.1183 \log m - 0.0602 m^{1/2}$$

$$= E° - 0.1183bm \qquad (38)$$

A plot of the left hand side of Eq. (38) versus m yields a practically straight line, the extrapolation of which to $m = 0$ gives $E°$ the standard potential of the cell. This value of $E°$ together with measured values of E at specified m values can be used to calculate γ for HCl in dilute aqueous solutions at 25° for different m − values. Similar treatment can be applied to other solvents and other solutes at selected temperatures.

Activity coefficients are used in calculation of equilibrium constants, rates of reactions, electrochemical phenomena, and almost all quantities involving solutes or solvents in solution.

—Edward S. Amis, University of Arkansas, Fayetteville, Arkansas.

ACTIVITY SERIES. Also referred to as the *electromotive series* or the *displacement series*, this is an arrangement of the metals (other elements can be included) in the order of their tendency to react with water and acids, so that each metal displaces from solution those below it in the series and is displaced by those above it. See accompanying table. Since the electrode potential of a metal in equilibrium with a solution of its ions cannot be measured directly, the values in the activity series are, in each case, the difference between the electrode potential of the given metal (or element) in equilibrium with a solution of its ions, and that of hydrogen in equilibrium with a solution of its ions. Thus, in the table, it will be noted that hydrogen has a value of 0.000. In experimental procedure, the hydrogen electrode is used as the standard with which the electrode potentials of other substances are compared. The theory of displacement plays a major role in electrochemistry and corrosion engineering. See also **Corrosion;** and **Electrochemistry.**

ACYL. An organic radical of the general formula, RCO—. These radicals are also called acid radicals because they are often produced from organic acids by loss of a hydroxyl group. Typical acyl radicals include acetyl (CH_3CO—) and benzyl (C_6H_5CO—).

STANDARD ELECTRODE POTENTIALS (25°C)

REACTION		VOLTS
$Li^+ + e^-$	$\rightarrow Li$	−3.045
$K^+ + e^-$	$\rightarrow K$	−2.924
$Ba^{2+} + 2e^-$	$\rightarrow Ba$	−2.90
$Ca^{2+} + 2e^-$	$\rightarrow Ca$	−2.76
$Na^+ + e^-$	$\rightarrow Na$	−2.711
$Mg^{2+} + 2e^-$	$\rightarrow Mg$	−2.375
$Al^{3+} + 3e^-$	$\rightarrow Al$	−1.706
$2H_2O + 2e^-$	$\rightarrow H_2 + 2OH^-$	−0.828
$Zn^{2+} + 2e^-$	$\rightarrow Zn$	−0.763
$Cr^{3+} + 3e^-$	$\rightarrow Cr$	−0.744
$Fe^{2+} + 2e^-$	$\rightarrow Fe$	−0.41
$Cd^{2+} + 2e^-$	$\rightarrow Cd$	−0.403
$Ni^{2+} + 2e^-$	$\rightarrow Ni$	−0.23
$Sn^{2+} + 2e^-$	$\rightarrow Sn$	−0.136
$Pb^{2+} + 2e^-$	$\rightarrow Pb$	−0.127
$2H^+ + 2e^-$	$\rightarrow H^2$	0.000
$Cu^{2+} + 2e^-$	$\rightarrow Cu$	+0.34
$I_2 + 2e^-$	$\rightarrow 2I^-$	+0.535
$Fe^{3+} + e^-$	$\rightarrow Fe^{2+}$	+0.77
$Ag^+ + e^-$	$\rightarrow Ag$	+0.799
$Hg^{2+} + 2e^-$	$\rightarrow Hg$	+0.851
$Br_2 + 2e^-$	$\rightarrow 2Br$	+1.065
$O_2 + 4H^+ + 4e^-$	$\rightarrow 2H_2O$	+1.229
$Cr_2O_7^{2+} + 14H^+ + 6e^-$	$\rightarrow 2Cr^{3+} + 7H_2O$	+1.33
$Cl_2(gas) + 2e^-$	$\rightarrow 2Cl$	+1.358
$Au^{3+} + 3e^-$	$\rightarrow Au$	+1.42
$MnO_4^- + 8H^+ + 5e^-$	$\rightarrow Mn^{2+} + 4H_2O$	+1.491
$F_2 + 2e^-$	$\rightarrow 2F^-$	+2.85

ACYLATION. A reaction or process whereby an acyl radical, such as acetyl, benzoyl, etc., is introduced into an organic compound. Reagents often used for acylation are the acid anhydride, acid chloride, or the acid of the particular acyl radical to be introduced into the compound.

ADAMANTINE COMPOUND. A compound having in its crystal structure an arrangement of atoms essentially that of diamond, in which every atom is linked to its four neighbors mainly by covalent bonds. An example is zinc sulfide, but it is to be noted that the eight electrons involved in forming the four bonds are not provided equally by the zinc and sulfur atoms, the sulfur yielding its six valence electrons, and the zinc, two. This is the structure typical of semiconductors, e.g., silicon and germanium.

ADDITIVES (Foods). The use of food additives is centuries old. One of the earliest and still most widely practiced example is the addition of common salt to foods, either by a food preparer or processor, or by the consumer at the table. Normally, the food producing professionals and the lay public do not view salt as a food additive, probably because salt is regarded as a substance widely found in nature, and particularly in seawater. Nevertheless, within the last 10–20 years there has been a growing awareness that salt is truly an additive and to certain controlled diets can have an adverse role as well as the numerous positive roles that it plays in food processing and preservation.

The term *food additive* is most frequently associated with the more complex substances, the presence or function of which the average consumer seldom notes unless the package labels are read. Particularly where additives contribute to the pleasure of the consumer, they are commonly passed over unnoticed. By far, the majority of consumers prefer breads that do not stale readily, salad dressings that do not separate into layers after a period of shelf life, dairy products that do not become rancid after a few days, meats that retain their color and wholesome appearance and taste after relatively long periods of refrigeration. These food qualities, of course, are made possible because of the many advances over the years in food additive technology.

Additives also contribute to the ease of processing and marketing foods and, even though indirectly, have played a positive role in terms of food transportation, distribution, and inventorying costs. Conversely, some additives in the past have been found to be damaging to health. Some of the colorants formerly used, for example, have been banned because of their carcinogenic qualities when tested in laboratory animals. Considering the vast quantities of additives used, the majority used to date have not been shown to be harmful.

There are tens of categories of food chemicals, most of which are described in various entries throughout this book. Consult the following entries:

Acidulants and Alkalizers (Foods)
Anticaking Agents
Antimicrobial Agents (Foods)
Antioxidant
Bleaching Agents
Bodying and Bulking Agents (Foods)
Buffer (Chemical)
Carrier (Food Additive)
Clarifying Agents
Coating Agents
Colorants (Foods)
   Annatto Food Colors
   Anthocyanins
   Betalaines
   Carotenoids
Enzyme Preparations
Firming Agents (Foods)
Flavorings
   Flavor Enhancers and Potentiators
Humectants and Moisture-Retaining Agents
Leavening Agents
Masticatory Substances
Oxidation and Oxidizing Agents
Pectins
Polymeric Food Additives
Sweeteners

Dispersing, emulsifying, gelling, thickening, and stabilizing agents are covered under **Colloid Systems.** Dough conditioners are described under **Oxidation and Oxidizing Agents.** Nonnutritive sweeteners are included in entry on **Sweeteners.**

References

Cole, M. S., and J. F. Wintermantel, Co-Chairpersons: "Food Additives and Ingredients," *Institute of Food Technologists Symposium,* Saint Louis, Missouri (1979).

Davis, R. B.: "Regulatory Aspects of Migration of Indirect Additives to Food," *Food Technology,* 33, 4, 55–60 (1979).

FAO–UN: "Additives Evaluated for Their Safety-in-Use in Foods," Codex Alimentarius Commission, World Health Organization (WHO) and Food and Agriculture Organization (FAO), United Nations, Rome (Revised periodically).

Fazio, T.: "Extraction Testing Methods for Evaluation of Food Packing Materials," *Food Technology,* 33, 4, 61–62 (1979).

Gilbert, S. G.: "Modeling the Migration of Indirect Additives to Food," *Food Technology,* 33, 4, 63–65 (1979).

Grice, H. C.: "Food Additive Evaluation and Regulation: The Canadian Approach," *Food Product Development,* 11, 4, 28–32 (1977).

Kermode, G. L.: "Food Additives," *Sci. American,* 226, 5, 15–21 (1972).

Kirschman, J. C.: "Evaluating the Safety of Food Additives," *Food Technology,* 37, 3, 75–79 (1983).

Kohl, P.: "Food and Additives," General Foods Corporation, White Plains, New York (Revised periodically).

Koros, W. J., and H. B. Hopfenberg: "Scientific Aspects of Migration of Indirect Additives from Plastics to Food," *Food Technology,* 33, 4, 55–60 (1979).

Lowenstein, M.: "What Does 'Natural' Mean?" *Dairy Record* (May 1977).

Staff: "Food Chemicals Codex," National Academy of Sciences, Washington, D.C. (2nd Edition, 1972, but with updated supplements issued periodically).

ADENINE. A prominent member of the family of naturally occurring purines with the accompanying formula. Adenine occurs not only in ribonucleic acids (RNA), and deoxyribonucleic acids (DNA), but in nucleosides, such as adenosine, and nucleotides, such as adenylic acid, which may be linked with enzymatic functions quite apart from nucleic acids. Adenine, in the form of its ribonucleotide, is produced in mammals and fowls endogenously from smaller molecules and no nutritional essentiality is ascribed to it. In the nucleosides, nucleotides, and nucleic acids, the attachment or the sugar moiety is at position 9.

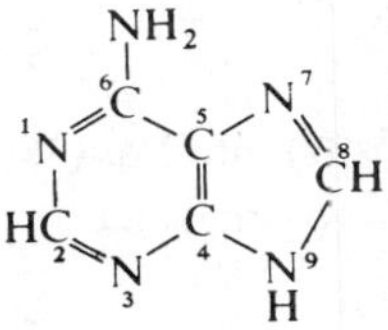

Structure of adenine.

The purines and pyrimidines absorb ultraviolet light readily, with absorption peaks at characteristic frequencies. This has aided in their identification and quantitative determination.

ADENOSINE. An important nucleoside. The upper portion of the accompanying formula represents the adenine moiety, and the lower portion of the pentose, D-ribose.

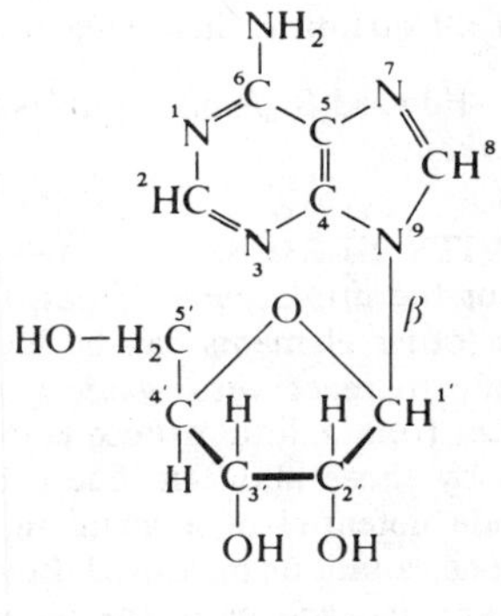

Structure of adenosine.

ADENOSINE DI- AND TRIPHOSPHATE. Carbohydrates; Phosphorylation (Oxidative); Phosphorylation (Photosynthetic).

ADENOSINE PHOSPHATES. The adenosine phosphates include *adenylic acid* (adenosine monophosphate, AMP) in which adenosine is esterified with phosphoric acid at the 5'-position; *adenosine diphosphate* (ADP) in which esterification at the same position is with pyrophosphoric acid,

$$\underset{\substack{\\ (HO)_2}}{}{-}\overset{\overset{\displaystyle O}{\displaystyle \|}}{P}{-}O{-}\overset{\overset{\displaystyle O}{\displaystyle \|}}{P}{-}(OH)_2$$

and *adenosine triphosphate* (ATP) in which three phosphate residues

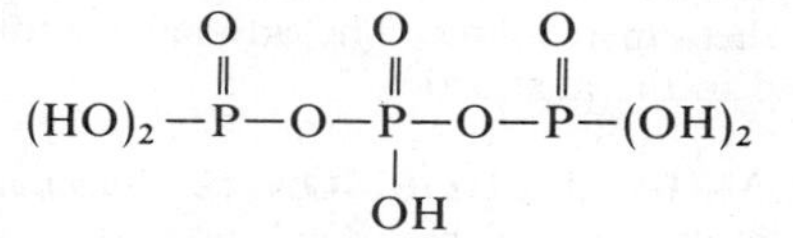

$$(HO)_2-\overset{\overset{O}{\|}}{P}-O-\overset{\overset{O}{\|}}{P}-O-\overset{\overset{O}{\|}}{P}-(OH)_2$$
$$\underset{OH}{|}$$

are attached at the 5'-position. Adenosine-3'-phosphate is an isomer of adenylic acid, and adenosine-2',3'-phosphate is esterified in two positions with the same molecules of phosphoric acid and contains the radical

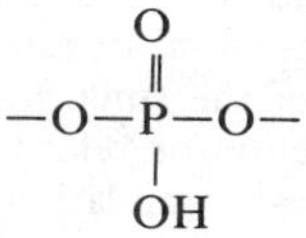

$$-O-\overset{\overset{O}{\|}}{P}-O-$$
$$\underset{OH}{|}$$

ADHESIVES. Materials which enable two surfaces to become united are known as adhesives. Years ago the terms *adhesive* and *glue* were almost synonymous in usage, but the general term glue now implies a sticky substance, whereas many adhesives are not sticky. Adhesives may be broadly classified into two main groups: organic and inorganic adhesives, with the organic materials being subdivided into those of animal, vegetable, and synthetic origin. A more useful classification is based upon the chemical nature of adhesives: (1) protein or protein derivatives; (2) starch, cellulose, or gums and their derivatives; (3) thermoplastic synthetic resins; (4) thermosetting synthetic resins; (5) natural resins and bitumens; (6) natural and synthetic rubbers; and (7) inorganic adhesives.

Adhesives also may be classified according to the purpose for which they are use: (1) bonding rigid surfaces, such as wood, glass, porcelain, rigid plastics, and metals; and (2) bonding flexible surfaces, such as paper, textiles, leather, flexible plastics, thin metallic sheets, among others. In the latter category, it is necessary that the adhesive be at least as flexible as the surface to be bonded.

The protein and protein derivative adhesives include those made from casein, zein, and soybean proteins, among others, as well as the important groups of glues made from hides, bones, etc., fish glues from fish offal, and those made from blood albumen.

The group of adhesives made from materials such as starch and the vegetable gums are also known as vegetable adhesives. They comprise adhesives made from starch and processed starch, the dextrins, and the water-soluble gums, such as gum arabic, ghatti, tragacanth, Indian gum, and the like. These are water soluble or water-suspendable materials. The starch and dextrin products are used for bonding paper, wood, and textiles; the gum products are used as adhesives for paper, postage stamps and other stamps.

There are a number of different types of rubber adhesives. Some are simply solutions of rubber latex, rubber, or synthetic and modified rubbers in a solvent, and others are compositions consisting of one of the aforementioned and casein or a synthetic resin. Such compositions are very widely used from the bonding of flexible materials like paper, textiles, leather and rubber to rigid materials like metals and plastics.

The inorganic adhesives are a very important group. They comprise: sodium silicate, principally used for corrugated paper and other paper products, laminating metal foils, plywood bonding, and in the production of building and insulating boards; plaster of Paris for ceramic and similar products; magnesium oxychloride for ceramics; litharge-glycerin for joints in plum-

bing fixtures; and Portland cement for bonding mineral aggregates. See also **Silicates (Soluble).**

While it is convenient to classify adhesives on the basis of materials used, in actual practice, many adhesives are compounded from several substances. The term *cement* is sometimes used as a synonym for adhesive. A cement is a particular kind of adhesive which consists of any material that can be prepared in a plastic form which hardens to bond together various surfaces.

Cellulose adhesives are principally cellulose derivatives such as methylcellulose, ethylcellulose, cellulose acetate and nitrate, and sodium carboxymethyl cellulose. These products are used as components in adhesive compositions such as those used for the bonding of leather, cloth, paper, and many other materials.

Thermoplastic synthetic resin adhesives comprise a variety of polymerized materials such as polyvinyl acetate, polyvinyl butyral, polyvinyl alcohol, and other polyvinyl resins; polystyrene resins; acrylic and methacrylic acid ester resins; cyanoacrylates; and various other synthetic resins such as polyisobutylene, polyamides, coumarone-idene products, and silicones. Such thermoplastic resins usually have permanent solubility and fusibility so that they creep under stress and soften when heated. They are used for the manufacture of tapes, safety glass, and shoe cements and for the bonding of foils, metals, woods, rubber, paper, and many other materials.

Thermosetting synthetic resin adhesives comprise a variety of phenol-aldehyde, urea-aldehyde, melamine-aldehyde, and other condensation-polymerization materials like the furane and polyurethane resins. Thermosetting adhesives are characterized by being converted to insoluble and infusible materials by means of either heat or catalytic action. Adhesive compositions containing phenol-, resorcinol-, urea-, melamine-formaldehyde, phenolfurfuraldehyde, and the like are used for the bonding of wood, textiles, paper, plastics, rubbers, and many other materials.

The adhesives of the natural resin and bitumen group consist of those made from asphalts, shellac, rosin and its esters, and similar materials. They are used for the bonding of various materials including minerals, linoleum, and the like.

ADRENAL CORTICAL HORMONES. The hormones elaborated by the adrenal cortex are steroidal derivatives of cyclopentanoperhydrophenanthrene related to the sex hormones. The structural formulas of the important members of this group are shown in accompanying figure. With the exception of *aldosterone*, the compounds may be considered derivatives of *corticosterone*, the first of the series to be identified and named. The C_{21} steroids derived from the adrenal cortex and their metabolites are designated collectively as *corticosteroids*. They belong to two principal groups: (1) those possessing an O or OH substituent at C_{11} (*corticosterone*) and an OH group at C_{17} (*cortisone* and *cortisol*) exert their chief action on organic metabolism and are designated as *glucocorticoids*. (2) Those lacking the oxygenated group at C_{17} (*desoxycorticosterone* and *aldosterone*) act primarily on electrolyte and water metabolism and are designated as *mineralocorticoids*. In humans, the chief glucocorticoid is *cortisol*. The chief mineralocorticoid is *aldosterone*.

Glucocorticoids. These hormones are concerned in organic metabolism and in the organism's response to stress. They accelerate the rate of catabolism (destructive metabolism) and inhibit the rate of anabolism (constructive metabolism) of protein. They also reduce the utilization of carbohydrate and increase the rate of gluconeogenesis (formation of glucose) from protein. They also exert a lipogenic as well as lipolytic action, potentiat-

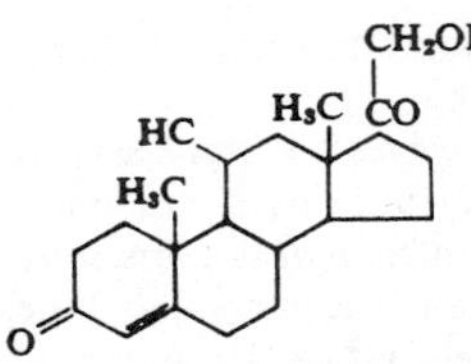

Corticosterone

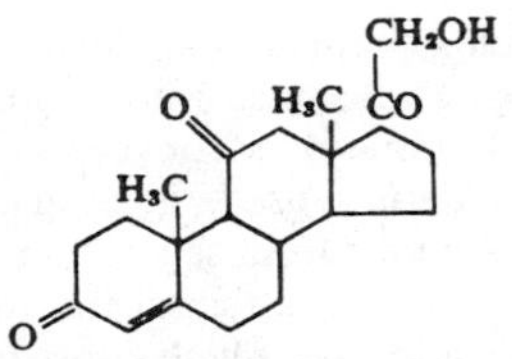

11-Dehydrocorticosterone

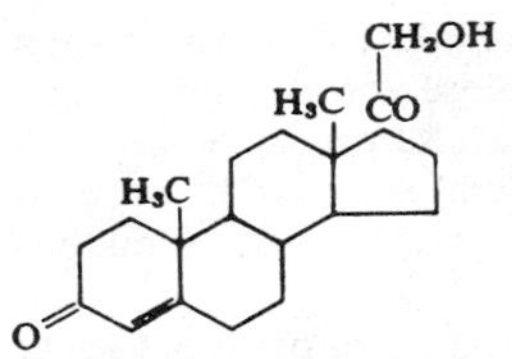

Desoxycorticosterone

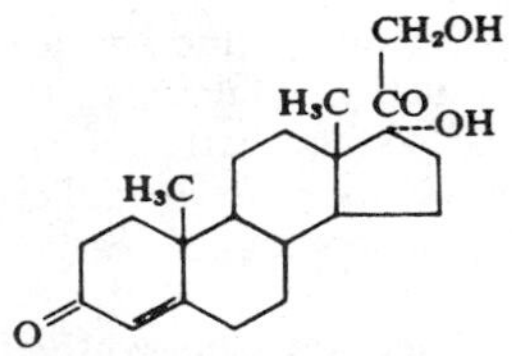

17-Hydroxy-desoxycorticosterone

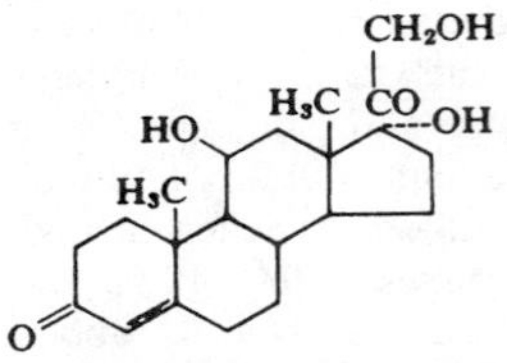

17-Hydroxycorticosterone
(Cortisol)

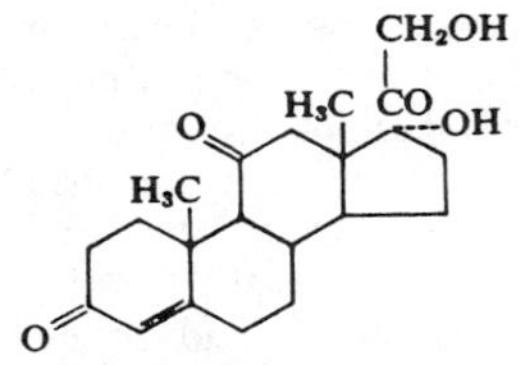

17-Hydroxy-dehydrocorticosterone
(Cortisone)

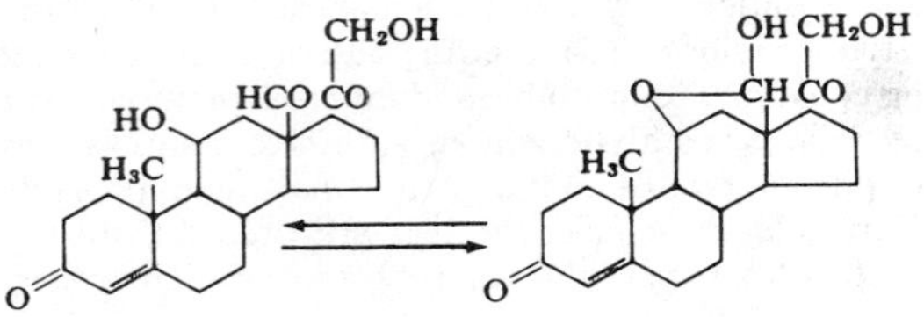

Aldosterone

Adrenal cortical hormones.

ing the release of fatty acids from adipose tissue. In addition to these effects on the organic metabolism of the basic foodstuffs, the glucocorticoids affect the body's allergic, immune, inflammatory, antibody, and anamnestic responses, and the general responses of the organism to environmental disturbances. It is these reactions which are the basis for the wide use of the corticosteroids therapeutically.

Aldosterone. This hormone exerts its main action in controlling water and electrolyte metabolism. Its presence is essential for the reabsorption of sodium by the renal tube, and it is the loss of salt and water which is responsible for the acute manifestations of adrenocortical insufficiency. The action of aldosterone is not limited to the kidney, but is manifested on the cells generally, this hormone affecting the distribution of sodium, potassium, water, and hydrogen ions between the cellular and extracellular fluids independently of its action on the kidney.

Differentiation of Glucocorticoids and Mineralocorticoids. The differentiation in the action of these two classes of hormones is not an absolute one. Aldosterone is about 500 times as effective as cortisol in its salt and water retaining activity, but is one-third as effective in its capacity to restore liver glycogen in the adrenalectomized animal. Cortisol in large doses, conversely, exerts a water and salt retaining action. Cortisterone is less active than cortisol as a glucocorticoid, but exerts a more pronounced mineralocorticoid action than does the latter. See also **Steroids.**

In addition to the aforementioned corticosteroidal hormones, the adrenal glands produce several oxysteroids and small amounts of testosterone and other androgens, estrogens, progesterone, and their metabolites. The adrenal medulla hormones are described in the next article.

ADRENAL MEDULLA HORMONES. *Adrenaline (epinephrine)* and its immediate biological precursor *noradrenaline (norepinephrine,* levarternol) are the principal hormones of the adult adrenal medulla. See the accompanying figure. Some of the physiological effects produced by adrenaline are contraction of the dilator muscle of the pupil of the eye (mydriasis); relaxation of the smooth muscle of the bronchi; constriction of most small blood vessels; dilation of some blood vessels, notably those in skeletal muscle; increase in heart rate and force of ventricular contraction; relaxation of the smooth muscle of the intestinal tract; and either contraction or relaxation, or both, of uterine smooth muscle. Electrical stimulation of appropriate sympathetic (adrenergic) nerves can produce all of the aforementioned effects with exception of vasodilation in skeletal muscle.

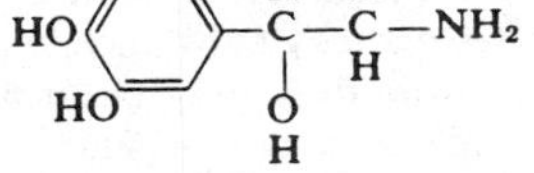

Noradrenaline

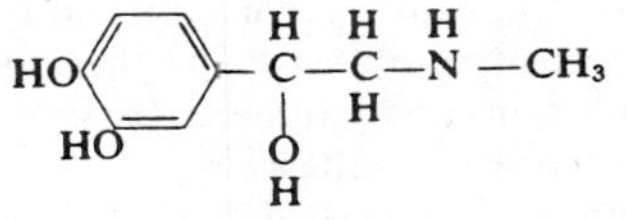

Adrenaline

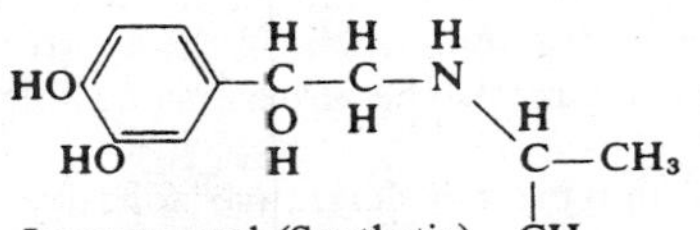

Isoproterenol (Synthetic)

Adrenal medulla hormones.

Noradrenaline, when administered, produces the same general effects as adrenaline, but is less potent. Isoproternol, a synthetic analogue of noradrenaline, is more potent than adrenaline in relaxing some smooth muscle, producing vasodilation and increasing the rate and force of cardiac contraction.

ADSORPTION. Adsorption is a kind of adhesion which takes place at the surface of a solid or a liquid in contact with another medium, resulting in an accumulation or increased concentration of molecules from that medium in the immediate vicinity of the surface. For example, if freshly heated charcoal is placed in an enclosure with ordinary air, a condensation of certain gases occurs upon it, resulting in a reduction of pressure; or if it is placed in a solution of unrefined sugar, some of the impurities are likewise adsorbed, and thus removed from the solution. Charcoal, when activated (i.e., freed from adsorbed matter by heating), is especially effective in adsorption, probably because of the great surface area presented by its porous structure. Its use in gas masks is dependent upon this fact. Penicillin is recovered in one state of the process by adsorption on activated carbon. Scanning electron micrographs of charcoal magnified several times are shown in the accompanying figure.

When colloidal hydroxides, notably aluminum hydroxide, are

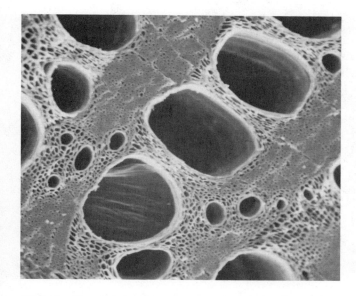

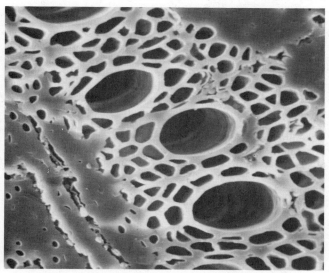

Scanning electron micrographs of charcoal magnified 100× (top) and 500× (bottom). (*Polaroid Type 105 Land Film*)

precipitated in a solution of acid dyes, i.e., those containing the groups —OH or —COOH, the dye adheres to the precipitate, yielding what is termed a *lake*. The "adsorption" of dirt on one's hands results from the unequal distribution of the dirt between the skin of the hands and the air or solid with which the skin comes in contact. Water is frequently ineffective in removing the dirt. The efficacy of soap in accomplishing its removal is due to the unequal distribution of dirt between skin and soap "solution," this time favoring the soap and leaving the hands clean.

At a given fixed temperature, there is a definite relation between the number of molecules adsorbed on a surface and the pressure (if a gas) or the concentration (if a solution), which may be represented by an equation, or graphically by a curve called the *adsorption isotherm*.

The Freundlich or classical adsorption isotherm is of the form:

$$\frac{x}{m} = kp^{1/n}$$

in which x is the mass of gas adsorbed, m is the mass of adsorbent, p is the gas pressure, and k and n are constants for the temperature and system. In certain systems, it is necessary to express this relationship as:

$$\frac{x}{m} = k(h\gamma)^{1/n}$$

where h is the relationship of the partial pressure of the vapor to its saturation value, and γ is the surface tension. Numerous isotherm equations have been proposed in the chemical literature in the last fifty years. The Langmuir adsorption isotherm is of the form:

$$\frac{x}{m} = \frac{k_1 k_2 p}{1 + k_1 p}$$

The Brunauer, Emmett and Teller equation is more general than those of Freundlich or Langmuir; for among other limitations, those two equations apply only to the adsorption of gases. Even in those cases, the degree of adsorption depends upon five factors: (1) the composition of the adsorbing material, (2) the condition of the surface of the adsorbing material, (3) the material to be adsorbed, (4) the temperature, and (5) the pressure (if a gas). A notable case in point is carbon. Of the finely divided varieties of carbon there are important sugar charcoal, bone black or animal black, blood charcoal, wood charcoal, coconut-shell charcoal, activated carbon. The temperature of preparation of adsorbent charcoal is an important factor, high temperatures being deleterious, and the removal (or non-removal) of gases by passing steam over the heated carbon, which operation increases the adsorptive power. Bone black is used for removing the coloring matter from raw sugar solutions. Fusel oil is removed from whiskey and poison gases from air by adsorption with the proper form of carbon. By cooling carbon in a vacuum to the temperature of liquid air, the concentration of residual gas is greatly decreased. Dewar (1906) found that 5 grams of charcoal (presumably coconut-shell charcoal) at the temperature of liquid air reduced the pressure of air in a 1-liter container from 1.7 to 0.00005 millimeters. This is now a standard method for the production of a high vacuum, as in the neon lighting tube.

Besides carbon, other important adsorbents in use are infusorial or diatomaceous earth, fuller's earth, clay, activated silica (silica gel), and activated alumina. All surfaces that behave indifferently towards non-electrolytes have the ability to adsorb electrolytes.

In comparing the adsorption properties of silica gel and activated carbon it may be noted that the latter is nonpolar and since it has no affinity for water will adsorb organic compounds in preference to water. Silica gel, on the other hand, is polar. It retains water and may thus reject an organic compound. It can also discriminate more selectively than activated carbon and consequently can be used for the fractionation of organic solvents.

Adsorption may be classified into types in various ways. On the basis of the adsorbate (or the substance which is adsorbed), adsorption may be *polar*, when the material adsorbed consists of positive or negative ions, so that the adsorbed film has an overall electrical charge. The term polar adsorption is also applied to adsorption chiefly attributable to attraction between polar groups of adsorbate and adsorbent. *Specific* adsorption is the preferential adsorption of one substance over another, or the quantity of adsorbate held per unit area of adsorbent.

On the basis of the process involved, adsorption may be classified as *chemical adsorption* (or chemisorption) where forces of chemical or valence nature between adsorbate and adsorbent are involved; and *van der Waals adsorption*, involving chiefly van der Waals forces. The difference is usually indicated experimentally by the greater heat of adsorption and more specific nature of the chemical process.

Other types of adsorption are *oriented adsorption*, in which the adsorbed molecules or other entities are directionally arranged on the surface of the adsorbent; and *negative adsorption*, a phenomenon exhibited by certain solutions, in which the concentration of solute is less on the surface than in the body of the solution.

Adsorption plays an important role in the process of dyeing, and in contact catalytic processes such as the conversion of sulfur dioxide to trioxide, and of nitrogen plus hydrogen to ammonia. In the case of insoluble organic acids (containing —COOH group) and substances containing hydroxyl (—OH) groups on the surface of water, the film is oriented so that the —COOH or —OH groups are attracted into the surface of the water, while their hydrocarbon ends project away from the surface of the water showing no tendency to dissolve (Langmuir).

Industrial Adsorption Operations. A common requirement in the chemical and process industries is that of separating materials. In addition to filtering, evaporating, distilling, and screening, adsorption is frequently used.

In a gas-solids system, the simplest form of contactor is a static bed in which particulate solids are held in fixed positions, one particle resting upon another, with no relative motion among particles. Gas is made to flow over, impinge upon, or flow through the voids among the particles. Contacting is confined to the interface between the gas and solid phases. Fixed beds usually require cyclic operation because it is difficult to add or remove solids to replenish the adsorption effectiveness while a unit is operating. Fixed beds tend to channel, resulting in a portion of the gas flowing through the bed without contacting the adsorbent solid.

Solid adsorbents used industrially include active alumina, alumina impregnated with calcium chloride, activated bauxite, fuller's earth, silica gel, shell-base carbon, wood-base carbon, coal-base carbon, petroleum-base carbon, and anhydrous calcium sulfate. See accompanying table.

To avoid some of the problems of fixed adsorption beds, fluid beds are frequently used. In a moving-bed (fluid bed) system, adsorbent solid particles move countercurrently by gravity through a rising pressurized-gas stream. As the particles adsorb the desired materials from the gas stream, they become heavier and ultimately gravitate to the bottom of the vessel where they are removed and regenerated externally from the fluid bed per se. The operation is continuous. Moving-bed systems also are used for heat transfer, the mixing of solids with gases, drying solids and gases, and implementing chemical reactions as in a fluid catalytic cracker.

Adsorption operations also are used to remove impurities from liquids. In purifying the products of organic chemical syntheses, for example, decolorizing carbons and clays often are used as contact adsorbents. They are stirred directly into a liquid-phase mixture or solution and subsequently removed by filtration. In some cases, as in contact filtration, the adsorbents are incorporated in the filtering medium. Contact filtration is widely used for removing colored and carbon-forming materials from lubricating oils. In another configuration, a liquid may be allowed to percolate by gravity or under pressure through a bed of solid adsorbents. Packed columns or towers filled with adsorbents also are used, particularly for drying gases.

AERATION. A process of contacting a liquid with air, often for the purpose of releasing other dissolved gases, or for increasing the quantity of oxygen dissolved in the liquid. Aeration is commonly used to remove obnoxious odors or disagreeable tastes from raw water. The principle of aeration is also used in the treatment of sewage by a method known as the activated sludge process. See **Activated Sludge.** Aeration is also important in industrial fermentation processes. In the manufacture of baker's yeast, penicillin, and many other antibiotics, an adequate air supply is required for optimum yields in certain submerged fermentation operations.

Aeration can be accomplished by allowing the liquid to fall in a thin film or by spraying it in the form of droplets in air at atmospheric pressure; or the air, under pressure, may be bubbled into the liquid, for example, by means of a sparger or other device that creates thousands of small bubbles, thus providing maximum contact area between the air and the liquid.

AEROSOL. A colloidal system in which a gas, frequently air, is the continuous medium and particles of solids or liquid are dispersed in it. Aerosol thus is a common term used in

PHYSICAL PROPERTIES OF REPRESENTATIVE ADSORBENTS

Adsorbent Substance	Internal Porosity (%)	External Void Fraction (%)	Average Pore Diameter (10^{-10} meter)	Surface Area (square meters/gram)	Adsorptive Capacity (gram/gram of dry solid)
Activated alumina	25	49	34	250	0.14[a]
Activated bauxite	35	40	~50	—	0.04–0.2[b]
Fuller's earth	~54	40	—	130–250	—
Silica gel	~70	30–40	25–30	~320	1.0[c]
Shell-base carbon	~50	~37	20	800–1100	45
Wood-base carbon	55–75	~40	20–40	625–14,000	6–9[d]
Coal-base carbon	65–75	45–70	20–38	500–1200	~0.4[e]
Petroleum-base carbon	70–85	26–34	18–22	800–1100	0.6–0.7[f]
Anhydrous calcium sulfate	38	~45	—	—	0.1[g]

[a] Water at 60% relative humidity.
[b] Water; test condition not specified.
[c] Water at 100% relative humidity.
[d] Phenol value.
[e] Benzene at 20°C; 7.5 millimeters partial pressure.
[f] Test conditions not specified.
[g] Water; test conditions not specified.

connection with air pollution control. Studies of the particle size distribution of atmospheric aerosols have shown a multimodal character, usually with a bimodal mass, volume, or surface area distribution, and frequently trimodal surface area distribution near sources of fresh combustion aerosols. The *coarse mode* (2 micrometers and greater) is formed by relatively large particles generated mechanically or by evaporation of liquid from droplets containing dissolved substances. The *nuclei mode* (0.03 micrometer and smaller) is formed by condensation of vapors from high-temperature processes, or by gaseous reaction products. The *intermediate* or *accumulation mode* (from 0.1 to 1.0 micrometer) is formed by coagulation of nuclei. Study of the behavior of the particles in each mode has led to the opinion that the particles tend to form a stable aerosol having a size distribution ranging from about 0.1 to 1 micrometer. The larger, settleable particles (in excess of 1 micrometer) fall out, whereas the very fine particles (smaller than 0.1 micrometer) tend to agglomerate to form larger particles which remain suspended.

Both settleable and suspended atmospheric particulates have deleterious effects upon the environment. The settleable particles can affect health if assimilated and also can cause adverse effects on materials, crops, and vegetation. Further, such particles settle out in streams and upon land where soluble substances, sometimes including hazardous materials, are dissolved out of the particles and thus become pollutants of soils and surface and ground waters. Suspended atmospheric particulate matter has undesirable effects on visibility and, if continuous and of sufficient concentration, possible modifying effects on the climate. Importantly, it is particles within a size range from 2 to 5 micrometers and smaller that are considered most harmful to health because particles of this size tend to penetrate the body's defense mechanisms and reach most deeply into the lungs.

Aerosol Packaging. The word aerosol is also applied to a form of packaging in which a gas under pressure, or a liquefied gas which has a pressure greater than atmospheric pressure at ordinary temperatures, is used to spray a liquid. The result of the spraying process is to produce a mist of small liquid droplets in air, although not necessarily a stable colloidal system. Numerous products, such as paints, clear plastic solutions, fire-extinguishing compounds, insecticides, waxes, and cleaners are packaged in this fashion for convenience. Food products, such as topping and whipped cream, also are packaged in aerosol cans.

For a number of years, chlorofluorocarbons were the most popular source of pressure for these cans. Because of concern in recent years over the reactions of chlorofluorocarbons in the upper atmosphere of the earth that appear to be leading to a deterioration of the ozone layer, some countries have banned their use in aerosol cans. Manufacturers have turned to other gases or to conveniently operated hand pumps. See also **Ozone.**

Particle Dynamics.[1] A spherical particle suspended in a gas quickly attains a constant velocity. If the gas can be assumed to be continuous, and viscous drag is the only restraining force on the particle, the settling velocity (μ) can be calculated from the Stokes equation:

$$\mu = \frac{2gr^2(\rho - \rho^1)}{g\eta}$$

where g is the acceleration of gravity, η is the viscosity of the gas, r is the particle radius, and ρ and ρ^1 are the densities of the particle and gas, respectively. These assumptions cannot

[1] The remainder of this article prepared by Richard D. Cadle, National Center for Atmospheric Research, Boulder, Colorado.

be made when the particles are so large that a turbulent wake forms, or so small that the diameters are approximately equal to or less than the mean free path of the gas. In the latter case, the particles "slip" between the atoms or molecules of the gas and settle at a higher velocity than predicted by the Stokes equation. The correction introduced by Cunningham is satisfactory for most purposes:

$$\mu = \frac{2gr^2(\rho - \rho^1)}{g\eta}\left(1 + \frac{Al}{r}\right)$$

where A is a constant close to unity (0.9 is often used) and l is the mean free path of the fluid. Turbulence develops behind a particle when the Reynolds number ($2\mu r/\nu$) exceeds about 1.2 (ν is the kinematic viscosity of the gas), in which case the Stokes equation no longer applies.

The aerosols in which particles are settling are usually somewhat turbulent, and a random motion is superposed on the downward movement of the particles. Such turbulence does not affect the average rate of fall but tends to maintain a uniform concentration throughout the aerosol which decreases with time.

Coagulation of aerosol particles can occur by several processes. When the particles are not all the same size (a "polydisperse" aerosol), the different settling rates can lead to collisions. This does not occur to an appreciable extent in aerosols as defined above, but is an important process in natural clouds and fogs. A much more important mechanism when the particles are small is coagulation by Brownian motion (diffusion). The Smoluchowski equation describing this coagulation for spherical particles in a monodisperse aerosol, and corrected for slippage, is:

$$\frac{1}{n} - \frac{1}{n_o} = \frac{4RT}{3\eta N}\left(1 + \frac{Al}{r}\right)t$$

where n_o and n are the particle number concentrations before and after elapsed time t, R is the gas constant, T is the absolute temperature, and N is Avogadro's number. Coagulation is faster for polydisperse aerosols and if the particles are not spheres.

Electrically charging aerosol particles may markedly affect the rate of coagulation, but even the direction of the effect is often difficult to predict. For example, the rate of coagulation in a unipolar electrically charged aerosol is not necessarily zero even if all the particles are charged, since the close approach of the particles may induce charges of opposite sign. When an aerosol is charged in a bipolar symmetrical manner, the charging may have little effect on the coagulation rate, but when the charging is unsymmetrical, there may be a large increase in the rate.

When an aerosol is accelerated (or decelerated), the inertia of the particles results in their moving relative to the suspending air. If the flow of an aerosol is diverted by an object immersed in the aerosol, particles will tend to impact on the object. If an aerosol flows around a bend in a duct, particles will tend to collect on the walls of the duct. This behavior plays a major role in the laws governing many methods for the collection of particles from aerosols, such as filtration and the impingement of aerosol jets on microscope slides. The collection efficiency for a jet of aerosol impinging on a surface or for an object such as a rod immersed in a flowing aerosol is usually expressed in terms of the ratio of the particle "stopping distance" to the width of the jet or rod. The stopping distance is defined as the distance a particle will continue to travel when the moving aerosol is stopped, and equals $2r^2p\mu/9\eta$. A convenient "rule of thumb" is that the collection efficiency usually is low when this ratio is much less than unity.

When a thermal gradient exists in an aerosol, a force (F) is exerted on the particles such that they move in the direction of decreasing temperature (thermophoresis). The theory of this behavior varies with r/l. When the value is large, the rather complicated equations developed by J. R. Brock of the University of Texas seem to be reasonably satisfactory. When this value is much less than unity, the Waldmann equation is valid:

$$F = -\frac{32}{15}\, r^2\, \lambda_{\text{trans}}\, \Delta T / \bar{c}$$

where λ_{trans} is the translational part of the thermal conductivity of the gas, ΔT is the temperature gradient, and $\bar{c}$ is the mean thermal velocity of the gas molecules. No really satisfactory quantitative theory has been developed for the intermediate region where the ratio is close to unity.

A closely related phenomenon is photophoresis, the particle movement that may occur when an aerosol is irradiated with light. The motion is believed to result from the nonuniform heating of the particles when they absorb the light.

Diffusiophoresis can be defined as the motion imparted to particles suspended in a mixture of gases in which there is a concentration gradient. Several situations are possible. An example is the motion of a small nonvolatile particle in the vicinity of a drop of evaporating liquid.

Industrial Production of Aerosols. Methods for producing aerosols usually follow one of two approaches—dispersion and condensation. However, a great variety of techniques exists, and only a few important examples can be mentioned.

One of the oldest methods of controlled aerosol production is by means of aspirators. These utilize the pressure drop produced by an airstream passing by an orifice to draw a liquid to be dispersed into the airstream where it breaks up into droplets. The droplet size distribution is very wide, but by forcing the resulting aerosol to pass around a series of baffles or through a bed of beads, the larger droplets are removed by impaction. Aerosols containing droplets all of which are smaller than one micron diameter, can be prepared in this way. If the liquid is a solution of a nonvolatile solute in a volatile solvent, an aerosol of the solute is obtained with median particle size largely controlled by the concentration of the solution. Individual particles of a powder can also be dispersed in this manner using a suspension of the powder in a volatile liquid. This suspension must be so dilute that there is only a small probability that a droplet will contain more than one particle; the suspension must also be free of dissolved solids such as dispersing agents.

A method for producing nearly monodisperse aerosols from high-boiling liquids was devised by Sinclair and LaMer and has been used in many modifications. Filtered air or other gas is passed through a condensation nuclei generator, over the heated liquid, into a "super heater" at a somewhat higher temperature than the liquid, and finally through a long, double walled condenser where the vapor condenses largely on the nuclei to form the aerosol. It is then diluted to slow coagulation. Additional filtered air may be passed over or bubbled through the heated liquid to dilute that coming from the nuclei generator. The latter may be simply a coil of electrically heated resistance wire coated with sodium chloride. Ions produced from an arc or spark can also be used. The particle size can be varied within certain limits by varying the concentration of nuclei and of vapor. Low concentrations of nuclei and high ones of vapor produce relatively large particles. Modifications of this approach have been used extensively in aerosol research.

Disk atomizers also produce nearly monodisperse aerosols. They consist of horizontal disks rotating about the vertical axis and may be compressed-air driven at very high speeds. The liquid to be dispersed flows onto the center of the disk and moves as a film to the edges where droplets break away. Very small satellite droplets are also produced and must be removed. The retained droplets are tens or hundreds of microns in diameter, much larger than those usually produced by the Sinclair-LaMer generators.

Many generators have used a vibrating capillary. One consists of a glass capillary 20 cm long and 0.6 mm diameter. With part of a hypodermic needle attached to the capillary and an electromagnet, the capillary is forced to vibrate, throwing out droplets. As with the use of the spinning disk, the droplets obtained are quite large and most of the dispersions obtained should probably not be considered to be aerosols.

Aerosols can be produced by a number of gas-phase chemical reactions, although they are usually polydisperse. A classic example is the reaction of hydrogen chloride with ammonia to form clouds of ammonium chloride. Another example is the hydrolysis of titanium tetrachloride vapor in air to produce copious white clouds. This reaction has been used to form smoke screens. Photochemical smog is a chemically-produced aerosol, an important environmental problem in some areas. A closely related method is to produce aerosols as "smoke" from combustion, as by burning magnesium ribbon.

Various methods have been developed for the direct dispersion of powders. They usually consist of some method for feeding the powder into a jet of air. The greatest difficulties are achieving a uniform rate of feeding and achieving the dispersion of ultimate particles instead of aggregates.

Finally, it should be emphasized that nature is a great producer of aerosols and our atmosphere is a huge aerosol system. The importance of this can hardly be overemphasized. For example, all rainfall depends on the existence of fine particles serving as condensation or freezing nuclei.

AGAR. Sometimes called *agar-agar*, this is a gelatine-like substance which is prepared from various species of red algae growing in Asiatic waters. The prepared product appears in the form of cakes, coarse granules, long shreds, or thin sheets, and is used extensively alone, or in combination with various other nutritive substances, as a medium for culturing bacteria and various fungi. See also **Gums and Mucilages.**

AGATE. Agate is a variety of chalcedony, whose variegated colors are distributed in regular bands or zones, in clouds, or in dendritic forms, as in moss agate.

The banding is often very delicate, with parallel lines of different colors, sometimes straight, sometimes undulating or concentric. The parallel bands represent the edges of successive layers of deposition from solution in cavities in rocks, which generally conform to the shape of the enclosing cavity.

As agate is an impure variety of quartz, it has the same physical properties as that mineral. It is named from the river Achates in Sicily where it has been known from the time of Theophrastus.

Agate is found in many localities—India, Brazil, Uruguay, and Germany are notable for fine specimens.

Onyx is a variety of agate in which the parallel bands are perfectly straight and can be used for the cutting of cameos. Sardonyx has layers of dark reddish-brown carnelian alternating with light and dark colored layers of onyx.

AGGLOMERATION. This word connotes a gathering together of smaller pieces or particles into larger-size units. It is a frequently used operation in the chemical and process industries. Specific advantages of agglomeration include increasing the bulk density of a material, reducing storage-space needs, improving handling qualities of bulk materials, improving heat-transfer properties, improving control over solubility, reducing material loss, and lessening of pollution particularly of dust, converting waste materials into a more useful form, and reducing labor costs because of improved handling efficiency.

The principal methods used for agglomerating materials include (1) compaction, (2) extrusion, (3) agitation, and (4) fusion.

Tableting is an excellent example of compaction. In this operation, loose material, such as a powder, is compressed between two opposing surfaces, or compacted in a die or cavity. Some tableting machines use the action of two opposing plungers which operate within a cavity. Resulting tablets may range from $\frac{1}{8}$ to 4 inches in diameter. Uniformity and dimensional precision are outstanding. Numerous pharmaceutical products are formed in this manner, as well as some metallic powders and industrial catalysts.

Pellet mills exemplify the use of extrusion. In some designs the charge material is forced out of cylindrical or other shaped holes located on the periphery of a cylinder within which rollers and spreaders force the bulk materials through the openings. A knife cuts the extruded pellets to length as they are forced through the dies.

The *rolling drum* is the simplest form of aggregation using agitation. Aggregates are formed by the collision and adherence of the bulk particles in the presence of a liquid binder or wetting agent to produce what essentially is a "snow ball" effect. As the operation continues, the spheroids become larger. The strength and hardness of the enlarged particles is determined by the binder and wetting agent used. The operation is followed by screening, with recycling of the fines.

The *sintering process* utilizes fusion as a means of size-enlargement. This process, used mainly for ores and minerals and some powdered metals, employs heated air which is passed through a loose bed of finely ground material. The particles partially fuse together without the assistance of a binder. Sintering frequently is accompanied by the volatilization of impurities and the removal of undesired moisture.

The *spray-type* agglomerator utilizes several principles. Loosely bound clusters or aggregates are formed by the collision and coherence of the fine particles and a liquid binder in a turbulent stream. The mixing vessel consists of a vertical tank, around the lower periphery of which are mounted spray nozzles for introduction of the liquid. A suction fan draws air through the bottom of the tank and creates an updraft within the mixing vessel. Materials spiral downward through the mixing chamber, where they meet the updraft and are held in suspension near the portion of the vessel where the liquids are injected. The liquids are introduced in a fine mist. Individual droplets gather the solid particles until the resulting agglomerate overcomes the force of the updraft and falls to the bottom of the vessel as finished product.

AGGREGATION. This overall operation may be considered to include the more specific designations of *agglutination*, *coagulation*, and *flocculation*. These terms imply some change in the state of dispersion of sols or of macromolecules in solution.

Agglutination generally refers to the aggregation of particulate matter mediated by an interaction with a specific protein. More specifically, the term refers to antigen-antibody reactions characterized by a clumping together of visible cells, such as bacteria or erythrocytes. The distinguishing feature of these reactions appears to be the presence of special areas where the orientation of active groups permits specific interaction of antigen with antibody. There is evidence that the forces involved may include hydrogen bonding, electrostatic attraction, and London-van der Waals forces. The clumping of the fat globules in milk has been described as an agglutination by the proteins of the eu-globulin fraction. When the fat globules are dispersed in a dilute salt solution, the addition of this protein fraction induces normal creaming. Although this action is nonspecific, the same protein fraction causes agglutination of certain bacteria when they are added to milk.

Coagulation of a hydrophobic sol may be brought about by the addition of small amounts of electrolytes. Coagulation may be rapid, occurring in seconds, or slow, requiring months for completion. The resultant coagula contain relatively small proportions of the dispersion medium, in contrast with jellies formed from hydrophilic systems. Sometimes it is found that what at first appeared to be a homogeneous liquid becomes turbid and distinctly nonhomogeneous. Systems which are intermediate between true hydrophobic sols and hydrophilic sols are encountered frequently, so that in common usage the word *coagulation* is applied to such diverse phenomena as the clotting of blood by thrombin or the clotting of milk by rennin.

Flocculation is generally considered synonymous with coagulation, but is widely used in connection with certain kinds of applications. If one considers only hydrophilic systems, it is apparent that an important factor in flocculation is the solvation of the particles, despite the common presence of an electric charge. Since stability appears to depend upon solute-solvent interactions and solubility properties, flocculation can frequently be brought about by either of two pathways. The addition of salts may compress the double layer, leaving the macromolecules stabilized by a diffuse solvation shell. The addition of alcohol or acetone will dehydrate the particles, leading to instability and flocculation. Alternatively, the alcohol or acetone may be added first, which will convert the particle to one of hydrophobic character stabilized largely by the electric double layer. Such a sol can be coagulated by the addition of small amounts of electrolytes.

See also **Colloid Systems.**

AIR. In addition to being the principal substance of the earth's atmosphere, air is a major industrial medium and chemical raw material. The average composition of dry air at sea level, disregarding unusual concentrations of certain pollutants, is given in Table 1. The amount of water vapor in the air varies seasonally and geographically and is a factor of large importance where air in stoichiometric quantities is required for reactions processes, or where water vapor must be removed in air-conditioning and compressed-air systems. The water content of air for varying conditions of temperature and pressure is shown

TABLE 1.   COMPOSITION OF AIR

CONSTITUENT	PERCENT BY WEIGHT	PERCENT BY VOLUME
Oxygen (O_2)	23.15	20.95
Ozone (O_3)	1.7×10^{-6}	0.00005
Nitrogen (N_2)	75.54	78.08
Carbon dioxide (CO_2)	0.05	0.03
Argon (Ar)	1.26	0.93
Neon (Ne)	0.0012	0.0018
Krypton (Kr)	0.0003	0.0001
Helium (He)	0.00007	0.0005
Xenon (Xe)	5.6×10^{-5}	0.000008
Hydrogen (H_2)	0.000004	0.00005
Methane (CH_4)	trace	trace
Nitrous oxide (N_2O)	trace	trace

TABLE 2.   WATER CONTENT OF SATURATED AIR

TEMPERATURE		WATER CONTENT (Pounds in 1 Pound of Air, or Kilograms in 1 Kilogram of Air)
(°F)	(°C)	
40	4.44	0.00520
45	7.22	0.00632
50	10	0.00765
55	12.8	0.00920
60	15.6	0.01105
65	18.3	0.01322
70	21.1	0.01578
75	23.9	0.01877
80	26.7	0.02226
85	29.4	0.02634
90	32.2	0.03108
95	35.0	0.03662
100	37.8	0.04305
105	40.6	0.05052

in Table 2. The water content of saturated air at various temperatures is shown in the accompanying figure. See also **Oxygen;**

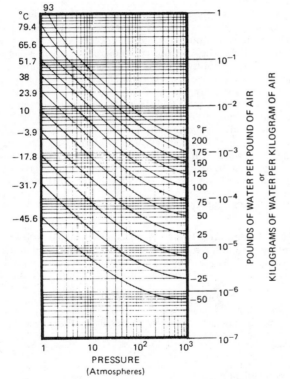

Water content of saturated air at various temperatures and pressures.

Nitrogen; and **Pollution (Air).**

ALABANDITE. Manganese sulfide, MnS. Associated with pyrite, sphalerite and galena in metallic sulfide vein deposits.

ALABASTER. A fine-grained variety of the mineral gypsum, formerly much used for vases and statuary. It is usually white in color or may be of other light, pleasing tints.

The word *alabaster* is derived from the Greek name for this substance. See also **Gypsum.**

ALANINE.   Amino Acids.

ALBUMIN. An albumin is a member of a class of proteins which is widely distributed in animal and vegetable tissues. Albumins are soluble in water and in dilute salt solutions, and are coagulable by heat.

Albumin is of great importance in animal physiology; in man it constitutes about 50% of the plasma proteins (blood) and is responsible to a great extent for the maintenance of osmotic equilibrium in the blood. The high molecular weight (68,000) of the albumin molecule prevents its excretion in the urine; the appearance of albumin in the urine may be an indication of kidney damage.

ALCOHOL. A type of organic compound. Also, a word sometimes used to designate ethyl alcohol. See **Alcohols;** and **Ethyl Alcohol.**

ALCOHOLATE. Replacement of one hydrogen of an alcohol by a metal, particularly a metal that forms a strong base, results in formation of an alcoholate. An example is sodium ethylate, C_2H_5ONa.

ALCOHOLS. The alcohols may be regarded as hydrocarbon derivatives in which the hydroxyl group (OH) replaces hydrogen on a saturated hydrocarbon molecule. Alcohols are classified as *primary*, *secondary*, or *tertiary*, according to the number of hydrogen atoms that are bonded to the carbon atom with the hydroxyl substituent. Alcohols also may be regarded as alkyl derivatives of water. Thus, alcohols with a small hydrocarbon group tend to be more like water in their properties than like hydrocarbons of the same number. Alcohols with a large hydrocarbon group are found to have physical properties similar to a hydrocarbon of the same structure. Some comparisons are evident from the accompanying table. Structures are summarized by:

$$
\begin{array}{ccc}
\text{H} & \text{H} & \text{R} \\
| & | & | \\
\text{R}'\text{--C--OH} & \text{R--C--OH} & \text{R--C--OH} \\
| & | & | \\
\text{H} & \text{R} & \text{R} \\
\text{Primary} & \text{Secondary} & \text{Tertiary}
\end{array}
$$

where $R' = H$, alkyl or aryl; R = alkyl or aryl.

In addition to the basic classification as primary, secondary, or tertiary, alcohols may be further grouped according to other structural features. Aromatic or aryl alcohols contain an aryl group attached to the carbon having the hydroxyl function; aliphatic or alkyl alcohols contain only aliphatic groups. The prefix *iso-* usually indicates branching of the carbon chain.

Alcohols containing two hydroxyl groups are called *dihydric alcohols* or *glycols*. Ethylene glycol, $HOCH_2CH_2OH$, trimethylene glycol, $HOCH_2CH_2CH_2OH$, and 1,4-butanediol are examples of industrially important glycols. Glycerol, $HOCH_2CHOHCH_2OH$, has three hydroxyl groups per molecule and is a *trihydric alcohol*. Physical properties of alcohols containing more than one hydroxyl group can be estimated by considering the number of carbons for each hydroxyl group, as in the case of the simple alcohols.

Reactions of Alcohols. Alcohols undergo a large number of reactions. However, these reactions may be grouped into a few general types. Reactions of alcohols may involve the O—H or C—O bonds. Ester formation and salt formation are examples of the former class, while conversion to halides is an example of the latter type.

Many industrially important substitution reactions of alcohols are conducted in the vapor phase over a catalyst. Only primary

COMPARISON OF PHYSICAL PROPERTIES OF ALCOHOLS AND HYDROCARBONS

ALCOHOL	HYDROCARBON	FORMULA	PROPERTIES
Methanol		CH_3OH	Liquid, water soluble, bp, 65°C
	Methane	CH_3—H	Gas, water insoluble
Ethanol		CH_3CH_2OH	Liquid, water soluble, bp, 78.5°C
	Ethane	CH_3CH_2—H	Gas, water insoluble
Tetradecanol		$CH_3(CH_2)_{12}CH_2OH$	Liquid, water insoluble, bp, 263.2°C
	Tetradecane	$CH_3(CH_2)_{12}CH_2$—H	Liquid, water insoluble, bp, 253.5°C

—O—H Bond Cleavage

$$ROH + CH_3COOH \underset{}{\overset{H_+}{\rightleftharpoons}} CH_3COOR + H_2O$$
$$ROH + K \rightarrow RO^-K^+ + \tfrac{1}{2}H_2$$

C—O Bond Cleavage

$$RCH_2OH + HCl \xrightarrow[170°C]{ZnCl_2} RCH_2Cl + H_2O$$

alcohols give satisfactory yields of product under these conditions:

$$CH_3OH + H_2S \xrightarrow{K_2WO_4} CH_3SH + H_2O$$
$$RCH_2OH + (CH_3)_2NH \xrightarrow{Al_2O_3} RCH_2N(CH_3)_2 + H_2O$$

Production of Alcohols. Lower alcohols (amyl and below) are prepared by: (1) hydrogenation of carbon monoxide (yields methanol); (2) olefin hydration (yields ethanol, isopropanol, secondary and tertiary butanol); (3) hydrolysis of alkyl chlorides; (4) direct oxidation; and (5) the OXO process.

$$C{=}O + 2H_2 \rightarrow CH_3OH$$
$$CH_3CH{=}CH_2 + H_2O \xrightarrow{H^+} CH_3CHOHCH_3$$
$$C_5H_{11}Cl + H_2O \rightarrow \quad C_5H_{11}OH$$
$$\text{(mixture of isomers)}$$

Most higher alcohols (hexanol and higher) and primary alcohols of three carbons or more are synthesized by one of four general processes, or derived from a structurally related natural product. See **Organic Chemistry.**

1. *OXO Process.* An olefin may be hydroformylated to a mixture of aldehydes. The aldehydes are readily converted to alcohols by hydrogenation. Many olefins from ethylene to dodecenes are used in the OXO reaction. OXO alcohols are typically a mixture of linear and methyl branched primary alcohols.

$$RCH{=}CH_2 + CO + H_2$$
$$\rightarrow [RCH_2CH_2CHO + RCH(CHO)CH_3]$$

2. *Aldol Condensation.* Aldehydes may also be dimerized by an aldol condensation reaction to give a branched unsaturated aldehyde. This may be converted to a branched alcohol by hydrogenation.

$$2CH_3CH_2CH_2CHO \rightarrow$$
$$CH_3CH_2CH_2CH{=}\underset{\underset{C_2H_5}{|}}{C}CHO \xrightarrow{H_2} CH_3(CH_2)_3\underset{\underset{C_2H_5}{|}}{C}HCH_2OH$$

Alcohols from an aldol reaction may be linear if acetaldehyde is a reactant, but usually aldol alcohols are branched primary alcohols. An aldol condensation sometimes is done with an OXO reaction. The combined process is called the ALDOX process.

3. *Oxidation of Hydrocarbons.* Using air, the oxidation of hydrocarbons generally results in a mixture of oxygenated compounds and is not a useful synthesis of alcohols except under special circumstances. Cyclohexanol may be prepared by air oxidation of cyclohexane inasmuch as only one isomer can result.

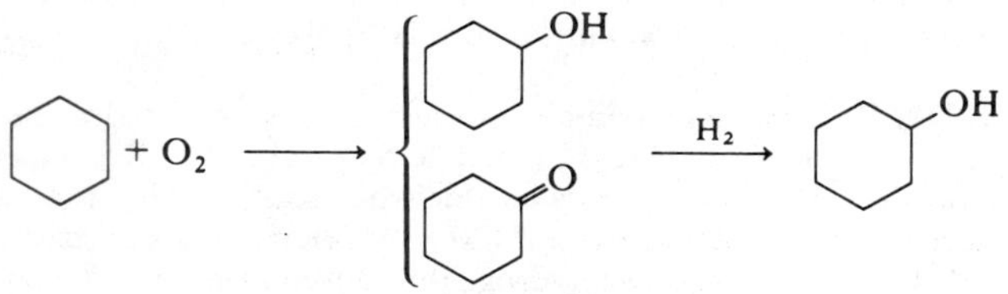

The yield of alcohol from normal paraffin oxidation may be improved to a commercially useful level by oxidizing in the presence of boric acid.

$$3RH + \tfrac{3}{2}O_2 + H_3BO_3 \rightarrow (RO)_3B + 3H_2O$$
$$(RO)_3B + 3H_2O \rightarrow H_3BO_3 + 3ROH$$

A borate ester is formed which is more stable to further oxidation than the free alcohol. This is easily hydrolyzed to recover the alcohol. These alcohols which are predominately secondary are used in surfactant manufacture.

4. *Synthesis from Alkylaluminums.* Fundamental work on organoaluminum chemistry by Prof. Karl Ziegler and coworkers at the Max Planck Institute provided the basis for a commercial synthesis of even-carbon-numbered straight chain primary alcohols. These alcohols are identical with products derived from naturally occurring fats. In this process, ethylene is reacted with aluminum triethyl to form a higher alkylaluminum which then is oxidized and hydrolyzed to give the corresponding alcohols.

$$(C_2H_5)_3Al \xrightarrow{CH_2{=}CH_2}$$
$$[CH_3(CH_2CH_2)_nCH_2]_3Al \xrightarrow[(2)\ H_2O]{(1)\ O_2}$$
$$3CH_3(CH_2CH_2)_nCH_2OH + Al(OH)_3$$

Commercialization of this route to higher alcohols is the most significant development in this area in recent years.

5. *Synthesis from Natural Products.* Many alcohols are prepared by reduction of the corresponding methyl esters which are derived from animal or vegetable fats. These alcohols are straight chain even-carbon-numbered compounds. Tallow and coconut oil are two major raw materials for higher alcohol manufacture.

Because of concern with the long-range outlook for petroleum products, a rekindled awareness of the possible commercial use of the simpler alcohols (ethanol and to a lesser extent methanol) as vehicle fuels (so-called *gasahol*) developed during the 1970s

$$(RCOO)_3C_3H_5 + 3CH_3OH \rightarrow 3RCOOCH_3$$

Triglyceride

$$+ CH_2OHCHOHCH_2OH$$

$$RCOOCH_3 + 2H_2 \xrightarrow{\text{Catalyst}} RCH_2OH + CH_3OH$$

and continues well into the 1980s. The technology, of course, is available, but making the product available is largely a matter of economics and world politics. A few countries, notably Brazil, have made large strides in gasahol production (sugar cane is used as a raw material in Brazil). See also **Biomass and Wastes as Energy Sources.**

Very significant quantities of ethanol, of course, are produced in the form of alcoholic beverages, including wine, whiskey, and beer.

It is the opinion of most authorities that so long as natural gas supplies remain reliable and relatively abundant, gas will continue as a major starting material for the synthesis of alcohol.

A number of the industrially important alcohols are described in separate alphabetical entries in this Encyclopedia.

ALDEHYDES. The homologous series of aldehydes (like ketones) has the formula $C_nH_{2n}O$. The removal of two hydrogen atoms from an alcohol yields an aldehyde. Thus, two hydrogens taken away from ethyl alcohol $CH_3 \cdot C(H_2)OH$ yields acetaldehyde $CH_3 \cdot CHO$; and two hydrogens removed from propyl alcohol $C_2H_5 \cdot C(H_2)OH$ yields propaldehyde $C_2H_5 \cdot CHO$. The trivial names of aldehydes derive from the fatty acid which an aldehyde will yield upon oxidation. Thus, formaldehyde is named from formic acid, the latter being the oxidation product of formaldehyde. Similarly, acetaldehyde is oxidized to acetic acid. Or the aldehyde may be named after the alcohol from which it may be derived. Thus formaldehyde, which may be derived from methyl alcohol, may be named methaldehyde; or acetaldehyde may be named ethaldehyde, since it may be derived from ethyl alcohol. In still another system, the aldehyde may take its name from the parent hydrocarbon from which it theoretically may be derived. Thus propan*al* (not to be confused with propan*ol*) may signify propaldehyde (as a derivative of propane).

Essentially aldehydes exhibit the following properties: (1) with exception of the gaseous formaldehyde, all aldehydes up to C_{11} are neutral, mobile, volatile liquids. Aldehydes above C_{11} are solids under usual ambient conditions; (2) formaldehyde and the liquid aldehydes have an unpleasant, pungent, irritating odor, (3) although the low-carbon aldehydes are soluble in H_2O, the solubility decreases with formula weight, and (4) the high-carbon aldehydes are essentially insoluble in H_2O, but are soluble in alcohol or ether.

The presence of the double bond (carbonyl group $C:O$) markedly determines the chemical behavior of the aldehydes. The hydrogen atom connected directly to the carbonyl group is not easily displaced. The chemical properties of the aldehydes may be summarized by: (1) they react with alcohols, with elimination of H_2O, to form *acetals*; (2) they combine readily with HCN to form *cyanohydrins*, (3) they react with hydroxylamine to yield *aldoximes*; (4) they react with hydrazine to form *hydrazones*; (5) they can be oxidized into *fatty acids* which contain the same number of carbons as in the initial aldehyde; (6) they can be reduced readily to form *primary alcohols*. When benzaldehyde is reduced with sodium amalgam and H_2O, benzyl alcohol $C_6H_5 \cdot CH_2 \cdot OH$ is obtained. The latter compound also may be obtained by treating benzaldehyde with a solution of cold KOH in which benzyl alcohol and potassium benzoate are produced. The latter reaction is known as Cannizzaro's reaction.

In the industrial production of higher alcohols (above butyls), aldehydes play the role of an intermediate in a complete process that involves aldol condensation and hydrogenation. In the OXO process, olefins are catalytically converted into aldehydes that contain one more carbon than the olefin in the feedstock. Aldehydes also serve as starting materials in the synthesis of several amino acids. See also **Acetaldehyde; Aldol Condensation; Benzaldehyde;** and **Furfuraldehyde.**

ALDOL CONDENSATION. Alcohols.

ALDOSES. Carbohydrates.

ALDOSTERONE. Steroids.

ALDOXIMES. Hydroxylamine.

ALEXANDRITE. A variety of chrysoberyl, originally found in the schists of the Ural Mountains. It absorbs yellow and blue light rays to such an extent that it appears emerald green by daylight, but columbine-red by artificial light. It is used as a gem and was named in honor of Czar Alexander II of Russia.

ALGICIDE. A substance, natural or synthetic, used for destroying or controlling algae. The term is also sometimes used to describe chemicals used for controlling aquatic vegetation, although these materials are more properly classified as aquatic herbicides. See **Herbicide.**

ALGIN. A hydrophilic colloidal polysaccharide obtained from several species of brown algae. The term is used both in reference to the pure substance, alginic acid, extracted from the algae, and also to the salts of this acid, such as sodium or ammonium alginate, in which forms it is used commercially. The alginates currently find a number of applications in the paint, rubber, pharmaceutical, and food industries. See also **Gums and Mucilages.**

ALIPHATIC COMPOUND. An organic compound that can be regarded as a derivative of methane, CH_4. Most aliphatic compounds are open carbon chains, straight or branched; saturated, or unsaturated. Originally, the term was used to denote the higher (fatty) acids of the $C_nH_{2n}O_2$ series. The word is derived from the Greek term for oil (fatty). See also **Compound (Chemical);** and **Organic Chemistry.**

ALKALI. A term that was originally applied to the hydroxides and carbonates of sodium and potassium, but since has been extended to include the hydroxides and carbonates of other alkali metals and ammonium. Alkali hydroxides are characterized by ability to form soluble soaps with fatty acids, to restore color to litmus which has been reddened by acids, and to unite with carbon dioxide to form soluble compounds. See also **Acids and Bases.**

ALKALI METALS. The elements of group 1a of the periodic classification. In order of increasing atomic number, they are hydrogen, lithium, sodium, potassium, rubidium, cesium, and francium. With exception of hydrogen, which is a gas and which frequently imparts a quality of acidity to its compounds, the other members of the group display rather striking similarities of chemical behavior, all reactive with H_2O to form strongly alkaline solutions. The elements in the group, including hydrogen, are characterized by a valence of one, having one electron in an outer shell available for reaction. Because of their chemical similarities, these elements, along with *ammonium* and sometimes magnesium, are considered the sixth group in classical qualitative chemical analysis separations.

ALKALINE EARTHS. The elements of group 2a of the periodic classification. In order of increasing atomic number, they are beryllium, magnesium, calcium, strontium, barium, and radium. The members of the group display rather striking similarities of chemical behavior, including stable oxides and carbonates, with hydroxides that are less alkaline than those of group 1a. The elements of this group are characterized by a valence of two, having two electrons in an outer shell available for reaction. Because of their chemical similarities, these elements are considered the fifth group in classical qualitative chemical analysis separations.

ALKALOIDS. The beginnings of alkaloid chemistry date back to the very early 1800s. In 1805, Sertürner announced the isolation of morphine. He prepared several salts of morphine and demonstrated that it is the principal agent responsible for the physiological effects of opium. In 1810, Gomes treated an alcoholic extract of cinchona bark with alkali and obtained a crystalline precipitate which he named *cinchonino*. Subsequent studies by Pelletier and Caventou (Faculty of Pharmacy, Paris) showed

rated into a heterocyclic ring. It apparently qualifies as an alkaloid because of its particular pharmacological activity and limited distribution in the plant world.

Over 2,000 alkaloids are known and it is estimated that they are present in only 10–15% of all vascular plants. They are rarely found in cryptogamia (exception, ergot alkaloids), gymnosperms, or monocotyledons. They occur abundantly in certain dicotyledons and particularly in the following families: *Apocynaceaae* (dogbane, quebracho, pereiro bark); *Papaveraceae* (poppies, chelidonium); *Papilionaceae* (lupins, butterfly-shaped flowers); *Ranunculaceae* (aconitum, delphinium); *Rubiaceae* (cinchona bark, ipecacuanha); *Rutaceae* (citrus, fagara); and *Solanaceae* (tobacco, deadly nightshade, tomato, potato, thorn apple). Well-characterized alkaloids have been isolated from the roots, seeds, leaves or bark of some 40 plant families. *Papaveraceae* is an unusual family, in that all of its species contain alkaloids.

Brief descriptions, in alphabetical order, of alkaloids of commercial or medical importance or of societal concern (alkaloid narcotics) are given later in this article. See also **Amphetamine; Morphine;** and **Pyridine and Derivatives.**

The nomenclature of alkaloids has not been systemized, both because of the complexity of the compounds and for historical

GENERAL CLASSIFICATION OF ALKALOIDS

GENERAL CLASS	EXAMPLES
Derivatives of aryl-substituted amines	**Adrenaline, amphetamine, ephedrine, phenylephrine tyramine**
Derivatives of pyrrole	Carpaine, hygrine, nicotine
Derivatives of imidazole	Pilocarpine
Derivatives of pyridine and piperidine	Anabasine, coniine, ricinine
Containing fusion of two piperidine rings	Isopelletierine, pseudopelletierine
Pyrrole rings fused with other rings	Gelsemine, physostigmine, vasicine, yohimbine
Aporphone alkaloids	Apomorphine corydine, isothebaine
Berberine alkaloids	Berberine, emetine
Bis-benzylisoquinoline alkaloids	Bebeering, trilobine
Cinchona alkaloids	Cinchonine, quinidine, quinine
Cryptopine alkaloids	Cryptopine, protopine
Isoquinoline alkaloids	Anhalidine, pellotine, sarsoline
Lupine alkaloids	Lupanine, sparteine
Morphine and related alkaloids	Codeine, morphine, thebaine
Papaverine alkaloids	Codamine, homolaudanosine, papeverine
Phthalide isoquinoline alkaloids (also known as narcotine alkaloids)	Hydrastine, narceine, narcotine
Quinoline alkaloids	Dictamine, galipoline, lycorine
Tropine alkaloids	Atropine, cocaine, ecgonine, scopolamine, tropine
Other alkaloids	Brucine, solanidine, strychnine

that cinchonino was a mixture, which they separated into two new alkaloids called *quinine* and *cinchonine*. The term *alkaloid* was first proposed by the pharmacist, W. Meissner, in 1819 to indicate "alkali-like" (basic) nitrogen-containing compounds of plant origin. Two further qualifications usually are added to the definition: (1) the compounds have complex molecular structures, and (2) the compounds manifest significant pharmacological activity. Such compounds occur only in certain plant genera and families, rarely being universally distributed in larger groups of plants.

Many widely distributed bases of plant origin, such as methyltrimethyl- and other open-chain simple alkylamines, the cholines, and the phenylalkylamines, are not classed as alkaloids. Alkaloids usually have a rather complex structure with the nitrogen atom involved in a heterocyclic ring. However, thiamine, a heterocyclic nitrogenous base, is not regarded as an alkaloid mainly because of its almost universal distribution in living matter. Colchicine, on the other hand, is classed as an alkaloid even though it is not basic and its nitrogen atom is not incorpo-

reasons. The two commonly used systems classify alkaloids either according to the plant genera in which they occur, or on the basis of similarity of molecular structure. Important classes of alkaloids containing generically related members are the aconitum, cinchona, ephedra, lupin, opium, rauwolfia, senecio, solanum, and strychnos alkaloids. Chemically derived alkaloid names are based upon the skeletal feature which members of a group possess in common. Thus, indole alkaloids (e.g., psilocybin, the active principle of Mexican hallucinogenic mushrooms) contain an indole or modified indole nucleus, and pyrrolidine alkaloids (e.g., hygrine) contain the pyrrolidine ring system. Other examples of this type of classification include the pyridine, quinoline, isoquinoline, imidazole, pyridine-pyrrolidine, and piperidine-pyrrolidine type alkaloids. Several alkaloids are summarized along these general terms in the accompanying table.

Sometimes, there is confusion between alkaloids and narcotics. It should be stressed that all alkaloids are not narcotics; and all narcotics are not alkaloids. A narcotic has the general definition of a drug which produces sleep or stupor, and also

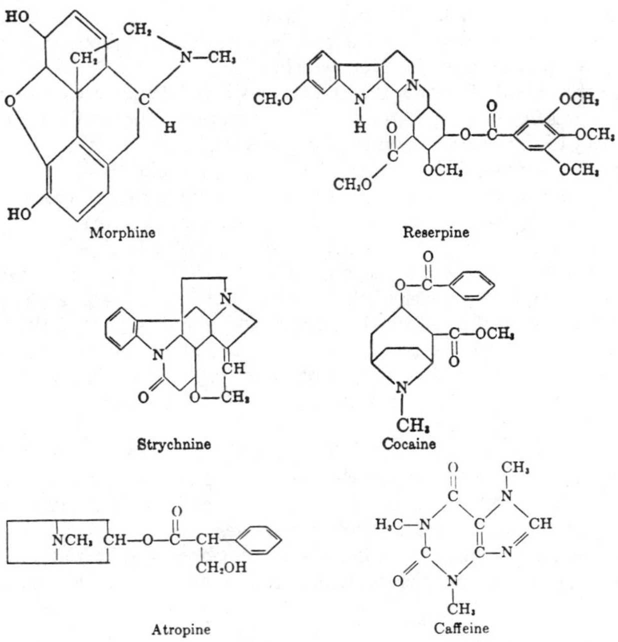

Morphine

Reserpine

Strychnine

Cocaine

Atropine

Caffeine

Structures of representative alkaloids. The carbon atoms in the rings and the hydrogen atoms attached to them are not designated by letter symbols. However, there is understood to be a carbon atom at each corner (except for the cross-over in the structure of morphine) and each carbon atom has four bonds, so that any bonds not shown or represented by attached groups are joined to hydrogen atoms.

relieves pain. Many alkaloids do not meet these specifications.

The molecular complexity of the alkaloids is demonstrated by the accompanying figure. Alkaloids react as bases to form salts. The salts used especially for crystallization purposes are the hydrochlorides, sulfates, and oxalates which are generally soluble in water or alcohol, insoluble in ether, chloroform, carbon tetrachloride, or amyl alcohol. Alkaloid salts unite with mercury, gold, and platinum chlorides. Free alkaloids lack characteristic color reactions but react with certain reagents, as follows, with (1) iodine in potassium iodide solution, forming chocolate brown precipitate; (2) mercuric iodide in potassium iodide solution (potassium mercuriiodide), forming precipitate; (3) potassium iodobismuthate, forming orange-red precipitate; (4) bromine-saturated concentrated hydrobromic acid forming yellow precipitate; (5) tannic acid, forming precipitate; (6) molybdophosphoric acid, forming precipitate; (7) tungstophosphoric acid, forming precipitate; (8) gold(III) chloride, forming crystalline precipitate of characteristic melting point; (9) platinum(IV) chloride, forming crystalline precipitate of characteristic melting point; (10) picric acid, forming precipitate; (11) perchloric acid, forming precipitate. Many alkaloids form more or less characteristic colors with acids, solutions of acidic salts, etc.

The function of alkaloids in the source plant has not been fully explained. Some authorities simply regard them as byproducts of the plant metabolism. Others conceive of alkaloids as reservoirs for protein synthesis; as protective materials discouraging animal or insect attacks; as plant stimulants or regulators in such activities as growth, metabolism, and reproduction; as detoxicating agents, which render harmless (by processes such as methylation, condensation, and ring closure) substances whose accumulation might otherwise cause damage to the plant. While these theories are of interest, it is also of interest to observe that from 85–90% of all plants manage well without the presence of alkaloids in their structures.

Adrenaline®. See *Epinephrine*, later in this article.

Atropine. Also known as *daturine*, $C_{17}H_{23}NO_3$ (see structural formula in accompanying diagram), white, crystalline substance, optically inactive, but usually contains levorotatory hyoscyamine. Compound is soluble in alcohol, ether, chloroform, and glycerol; slightly soluble in water; mp 114–116°C. Atropine is prepared by extraction from *Datura stramonium*, or synthesized. The compound is toxic and allergenic. Atropine is used in medicine and is an antidote for cholinesterase-inhibiting compounds, such as organophosphorus insecticides and certain nerve gases. Atropine is commonly offered as the sulfate. Atropine is used in connection with the treatment of disturbances of cardiac rhythm and conductance, notably in the therapy of sinus bradycardia and sick sinus syndrome. Atropine is also used in some cases of heart block. In particularly high doses, atropine may induce ventricular tachycardia in an ischemic myocardium. Atropine is frequently one of several components in brand-name prescription drugs.

Caffeine. Also known as theine, or methyltheobromine, 1,2,7-trimethyl xanthine (see structural formula in accompanying diagram), white, fleecy or long, flexible crystals. Caffeine effloresces in air and commences losing water at 80°C. Soluble in chloroform, slightly soluble in water and alcohol, very slightly soluble in ether, mp 236.8°C, odorless, bitter taste. Solutions are neutral to litmus paper.

Caffeine is derived by extraction of coffee beans, tea leaves, and kola nuts. It is also prepared synthetically. Much of the caffeine of commerce is a byproduct of decaffeinized coffee manufacture. The compound is purified by a series of recrystallizations. Caffeine finds use in medicine and in soft drinks. Caffeine is also available as the hydrobromide and as sodium benzoate, which is a mixture of caffeine and sodium benzoate, containing 47–50% anhydrous caffeine and 50–53% sodium benzoate. This mixture is more soluble in water than pure caffeine. A number of nonprescription (pain relief) drugs contain caffeine as one of several ingredients. Caffeine is a known cardiac stimulant and in some persons who consume significant amounts, caffeine can produce ventricular premature beats.

Cocaine. Also known as *methylbenzoylecgonine*, $C_{17}H_{21}NO_4$, is a colorless-to-white crystalline substance, usually reduced to powder. Cocaine is soluble in alcohol, chloroform, and ether; slightly soluble in water, giving a solution slightly alkaline to litmus. The hydrochloride is levorotatory, mp 98°C. Cocaine is derived by extraction of the leaves of coca (*Erythroxylon*) with sodium carbonate solution, followed by treatment with dilute acid and extraction with ether. The solvent is evaporated, after which the substance is redissolved and subsequently crystallized. Cocaine also is prepared synthetically from the alkaloid ecgonine. Cocaine is *highly toxic* and *habit forming*. While there are some medical uses of cocaine, usage must always be under the direction of a physician. It is classified as a narcotic in most countries. Society's major concern with cocaine is its use (increasing in recent years) as a narcotic.

Cocaine has been known as a very dangerous material since the early 1900s. When use of it as a narcotic increased during the early 1970s, serious misconceptions concerning its "safety" as compared with many other narcotics led and continue to lead to many deaths from its use. For example, health authorities in Dade County, Florida reported in 1977 that over 50% of drug-related overdose deaths were attributable to cocaine (Wetli and Wright, 1979).

Addicts use cocaine intravenously or by snorting the powder. After intravenous injection, coma and respiratory depression can occur rapidly. It has been reported that fatalities associated with snorting usually occur shortly after the abrupt onset of major motor seizures, which may develop within minutes to an hour after several nasal ingestions. Similar results occur if the substance is taken by mouth. Treatment is directed toward ventilatory support and control of seizures—although in many instances a victim may not be discovered in time to prevent death. It is interesting to note that cocaine smugglers, who have placed cocaine-filled condoms in their rectum or alimentary tract, have died (Suarez et al, 1977). The structural formula of cocaine is given in the accompanying diagram.

Codeine. Also known as *methylmorphine*, $C_{18}H_{21}NO_3 \cdot H_2O$, codeine is a colorless white crystalline substance, mp 154.9°C, slightly soluble in water, soluble in alcohol and chloroform, effloresces slowly in dry air. Codeine is derived from opium by extraction or by the methylation of morphine. For medical use, codeine is usually offered as the dichloride, phosphate, or sulfate. Codeine is *habit forming*. Codeine is known to exacerbate *urticaria* (familiarly known as *hives*). Since codeine is incorporated in numerous prescription medicines for headache, heartburn, fatigue, coughing, and relief of aches and pains, persons with a history of urticaria should make this fact known to their physician. Codeine is sometimes used in cases of acute *pericarditis* to relief severe chest pains in early phases of disease. Codeine is sometimes used in drug therapy of renal (kidney) diseases.

Colchicine. An alkaloid plant hormone, $C_{22}H_{25}NO_6$, colchicine is yellow, crystalline or powdered, nearly odorless, mp 135–150°C, soluble in water, alcohol, and chloroform, moderately soluble in ether. Solutions are levorotatory and deteriorate under light. The substance is highly toxic (0.02 gram may be fatal if ingested). Colchicine is extracted from the plant *Colchicum autumnale*, after which it is crystallized. The compound also has been synthesized. Biologists have used colchicine to induce chromosome doubling in plants. Colchicine finds a number of uses in medicine.

Although colchicine has been known for many years, interest in the drug has been revitalized in recent years as the result of the discovery that it interferes with cell division by destroying the spindle mechanism. The two chromatids which represent one chromosome at the metaphase stage fail to separate and do not migrate to the poles (ends) of the cell. Each chromatid becomes a chromosome in situ. The entire group of new chromosomes now form a resting nucleus and the next cell division reveals twice as many chromosomes as before. The cell has changed from the diploid to the tetraploid condition. Applied to germinating seeds or growing stem tips in concentrations of about 1 gram in 10,000 cubic centimeters of water for 4 or 5 days, colchicine may thus double the chromosome number of many or all of the cells, producing a tetraploid plant or shoot. Offspring from such plants may be wholly tetraploid and breed true. Tetraploid plants are larger than diploid plants and often more valuable. The alkaloid has also been used to double the chromosome number of sterile hybrids produced by crossing widely separated species of plants. Such plants, after colchicine treatment, contain in each cell two complete diploid sets of chromosomes, one from each of the parent species, and become fertile, pure-breeding hybrid species.

In medicine, colchicine is probably best known for its use in connection with the treatment of gout. Acute attacks of gout are characteristically and specifically aborted by colchicine. The response noted after administration of the drug also can be useful in diagnosing gout cases where synovial fluid cannot be aspirated and examined for the presence of typical urate crystals. However, colchicine does not affect the course of acute synovitis in rheumatoid arthritis. Kaplan (1960) observed that colchicine

may produce objective improvement in the periarthritis associated with *sarcoidosis* (presence of noncaseating granulomas in tissue). Colchicine is sometimes used in the treatment of *scleroderma* (deposition of fibrous connective tissues in skin or other organs); it may assist in preventing attacks of Mediterranean fever; and it is sometimes used as part of drug therapy for some renal (kidney) diseases.

Colchicine can cause diarrhea as the result of mucosal damage and it has been established that colchicine interferes with the absorption of vitamin B_{12}.

Emetine. An alkaloid from ipecac, $C_{29}H_{40}O_4N_2$, emetine is a white powder, mp 74°C, with a very bitter taste. The substance is soluble in alcohol and ether; slightly soluble in water. Emetine darkens upon exposure to light. The compound is derived by extraction from the root of *Cephalis ipecacuanha* (ipecac). It is also made synthetically. Medically, ipecac is useful as an emetic (induces vomiting) for emergency use in the treatment of drug overdosage and in certain cases of poisoning. Ipecac should not be administered to persons in an unconscious state. It should be noted that emesis is not the proper treatment in all cases of potential poisoning. It should not be induced when such substances as petroleum distillates, strong alkali, acids, or strychnine are ingested.

Ephedrine. Also known as 1-phenyl-2-methylaminopropanol, $C_6H_5CH(OH)CH(NHCH_3)CH_3$, ephedrine is a white-to-colorless granular substance, unctuous (greasy) to the touch, and hygroscopic. The compound gradually decomposes upon exposure to light. Soluble in water, alcohol, ether, chloroform, and oils; mp 33–40°C, bp 255°C and decomposes above this temperature. Ephedrine is isolated from stems or leaves of *Ephedra* spp., especially Ma huang (found in China and India). Medically, it is usually offered as the hydrochloride. In the treatment of bronchial asthma, ephedrine is known as a *beta agonist*. Compounds of this type reduce obstruction by activating the enzyme adenylate cyclase. This increases intracellular concentrations of cAMP (cyclic 3'5'-adenosine monophosphate) in bronchial smooth muscle and mast cells. Ephedrine is most useful for the treatment of mild asthma. In severe asthma, ephedrine rarely maintains completely normal airway dynamics over long periods. Ephedrine also has been used in the treatment of cerebral transient ischemic attacks, particularly with patients with vertebro-basilar artery insufficiency who have symptoms associated with relatively low blood pressure, or with postural changes in blood pressure. Ephedrine sulfate also has been used in drug therapy in connection with urticaria (hives).

Epinephrine. A hormone having a benzenoid structure, $C_9H_{13}O_3N$, also called *adrenaline*. It can be obtained by extraction from the adrenal glands of cattle and also prepared synthetically. Its effect on body metabolism is pronounced, causing an increase in blood pressure and rate of heart beat. Under normal conditions, its rate of release into the system is constant, but emotional stresses such as fear or anger rapidly increase the output and result in temporarily heightened metabolic activity. Epinephrine is used for the symptomatic treatment of bronchial asthma and reversible bronchospasm associated with chronic bronchitis and emphysema. The drug acts on both alpha and beta receptor sites. Beta stimulation provides bronchodilator action by relaxing bronchial muscle. Alpha stimulation increases vital capacity by reducing congestion of the bronchial mucosa and by constricting pulmonary vessels.

Epinephrine is also used in the management of anaesthetic procedures in connection with noncardiac surgery of patients with active ischemic heart disease. The drug is useful in the treatment of severe urticarial (hives) attacks, especially those accompanied by angioedema.

Epinephrine has numerous effects on intermediary metabolism. Among these are promotion of hepatic glycogenolysis, inhibition of hepatic gluconeogenesis, and inhibition of insulin release. The drug also promotes the release of free fatty acids from triglyceride stores in adipose tissue. Epinephrine produces numerous cardiovascular effects. Epinephrine is particularly useful in treating conditions of immediate hypersensitivity—interactions between antigen and antibody. These mechanisms cause attacks of anaphylaxis, hay fever, hives, and allergic asthma. Anaphylaxis can occur after bee and wasp stings, venoms, etc. Although the mechanism is not fully understood, epinephrine can play a life-saving role in the treatment of acute systemic anaphylaxis.

In some instances, epinephrine can be a cause of a blood condition involving the leukocytes and known as neutrophilia. In very rare cases, an intramuscular injection of epinephrine can be a cause of clostridial myonecrosis (gas gangrene).

Heroin. Also known as diacetylmorphine, $C_{17}H_{17}NO$ $(C_2H_3O_2)_2$, heroin is a white, essentially odorless, crystalline powder with bitter taste. Soluble in alcohol; mp 173°C. Heroin is derived by the acetylation of morphine. The substance is highly toxic and is a habit-forming narcotic. One-sixth grain (0.0108 gram) can be fatal. Although emergency facility personnel in some areas during recent years have come to regard heroin overdosage as approaching epidemic statistics, it is nevertheless estimated that the majority of persons with heroin overdose die before reaching a hospital. The initial crisis of an overdose is a severe respiratory depression and sometimes *apnea* (cessation of breathing). In emergency situations, the victim may be ventilated with a self-inflating resuscitative bag with delivery of 100% oxygen. Then, an endotracheal tube attached to a mechanical ventilator may be inserted. Naloxone (*Narcan*®), a narcotic antagonist, then may be administered intravenously, often with repeated dosages over short intervals, until an improvement is noted in the respiratory rate or sensorial level of the victim. If a victim does not respond, this is usually indication that the situation is not opiate related, or that other drugs also have been taken. Inasmuch as the antagonizing action of naloxone persists for only a few hours, a heroin overdose patient should be observed in the hospital for an indeterminate period. In heroin overdose cases, pulmonary edema (as the result of altered capillary permeability) may occur. This is directly associated with the overdose and not with subsequent treatment. Aside from the severe overdose situation, use of the drug causes or contributes to a number of ailments. These include chronic renal (kidney) failure and nephrotic syndrome. Septic arthritis caused by *Pseudomonas* and *Serratia* infections, is sometimes found as the result of intraveneous heroin abuse. Drug-induced immune platelet destruction also may occur.

Morphine. See separate entry on **Morphine**.

Neo-Synephrine®. See *Phenylephrine hydrochloride* later in this entry.

Nicotine. Also known as beta-pyridyl-alpha-N-methylpyrrolidine, $C_5H_4NC_4H_7NCH_3$, nicotine is a thick, water-white, levorotatory oil that turns brown upon exposure to air. The compound is hygroscopic; soluble in alcohol, chloroform, ether, kerosene, water, and oils; bp 247°C, at which point it decomposes. Specific gravity is 1.00924. Nicotine is combustible, with an autoignition temperature of 243°C. Nicotine is derived by distilling tobacco with milk of lime and extracting with ether. Nicotine is used in medicine, as an insecticide, and as a tanning agent. Nicotine is commercially available as the dihydrochloride, salicylate, sulfate, and bitartrate. Nicotinic acid (pyridine-3-carboxylic acid) is a vitamin in the B complex. See also **Vitamins**.

Phenylephrine Hydrochloride. L-1-(meta-hydroxyphenyl-2-methylaminoethanol) hydrochloride, $HOC_6H_4CH(OH)CH_2$-$NHCH_3 \cdot HCl$, white or nearly white crystalline substance, odorless, bitter taste. Solutions are acid to litmus paper, freely soluble in water and in alcohol; mp 140–145°C. Levorotatory in solution. Phenylephrine hydrochloride is used medically as a vasoconstrictor and pressor drug. It is chemically related to epinephrine and ephedrine. Actions are usually longer-lasting than the latter two drugs. The action of phenylephrine hydrochloride contrasts sharply with epinephrine and ephedrine, in that its action on the heart is to slow the rate and to increase the stroke output, inducing no disturbance in the rhythm of the pulse. In therapeutic doses, it produces little if any stimulation of either the spinal cord or cerebrum. The drug is intended for the maintenance of an adequate level of blood pressure during spinal and inhalation anesthesia and for the treatment of vascular failure in shock, shocklike states, and drug-induced hypotension, or hypersensitivity. It is also used to overcome paroxysmal supraventricular tachycardia, to prolong spinal anesthesia, and as a vasoconstrictor in regional analgesia. Caution is required in the administration of phenylephrine hydrochloride to elderly persons, or in patients with hyperthyroidism, bradycardia, partial heart block, myocardial disease, or severe arteriosclerosis. The brand name *Neo-Synephrine*® is also used to designate another product (nose drops) which does not contain phenylephrine hydrochloride. The nose drops contain xylometazoline hydrochloride.

Quinine. $C_{20}H_{24}N_2O_2 \cdot 3H_2O$, a bulky, white, amorphous powder or crystalline substance, with very bitter taste. It is odorless and levorotatory. Soluble in alcohol, ether, chloroform, carbon disulfide, oils, glycerol and acids; very slightly soluble in water. Quinine is derived from finely ground cinchona bark mixed with lime. This mixture is extracted with hot, high-boiling paraffin oil. The solution is filtered, shaken with dilute sulfuric acid and then neutralized while hot with sodium carbonate. Upon cooling, quinine sulfate crystallizes out. Pure quinine is obtained by treating the sulfate with ammonia. In addition to medical uses, quinine and its salts are used in soft drinks and other beverages.

Quinine derivatives are used in therapy for mytonic dystrophy (usually weakness and wasting of facial muscles) and in the treatment of certain renal (kidney) diseases. Quinine and derivatives are best known for their use in connection with malaria. Acute attacks of malaria are usually treated with oral chloroquine phosphate. The drug is given intramuscularly to patients who cannot tolerate oral medication. Combined therapy is indicated for treating *Plasmodium falciparum* infections, using quinine sulfate and pyrimethamine. A weekly oral dose of chloroquinone phosphate is frequently prescribed for persons who travel in malarious regions. The drug is taken one week prior to travel into such areas and continued for 6 weeks after leaving the region. Chloroquine phosphate has not proved fully satisfactory in the treatment of babesiosis, a malarialike illness caused by a parasite.

Strychnine. $C_{21}H_{24}ON_2$, hard, white crystals or powder of a bitter taste. Soluble in chloroform; slightly soluble in alcohol and benzene; slightly soluble in water and ether; mp 268–290°C; bp 270°C (5 millimeters pressure). Strychnine is obtained by extraction of the seeds of *Nux vomica* with acetic acid, followed by filtration, precipitation by an alkali, followed by final filtration. The compound is highly toxic by ingestion and inhalation. The phosphate finds limited medical use. Strychnine is also used in rodent poisons. Strychnine acts as a powerful stimulant to the central nervous system. At one time, strychnine was used in a very carefully controlled way in the treatment of some cardiac disorders. Acute strychnine poisoning resembles fully developed generalized tetanus.

References

Kaplan, H.: "Sarcoid Arthritis with a Response to Colchicine: Report of Two Case," *N. Engl. J. Med.*, **263**, 778 (1960).

Pelletier, S. W.: "The Chemistry of Certain Imines Related to Diterpene Alkaloids," *Experimentia*, **20**, 1–10 (1964).

Raffauf, R. F.: "A Handbook of Alkaloids and Alkalid-Containing Plants," Wiley, New York (1970).

Rao, T. K. S., et al.: "Natural History of Heroin-Associated Nephropathy," *N. Eng. J. Med*, **290**, 19 (1974).

Suarez, C. A., et al.: "Cocaine-Condom Ingestion: Surgical Treatment," *J. Amer. Med. Assn.*, **238**, 1391 (1977).

Swan, G. A.: "An Introduction to Alkaloids," Halsted Press, London (1967).

Wetli, C. V., and R. K. Wright: "Death Caused by Recreational Cocaine Use," *J. Amer. Med. Assn.*, **241**, 2519 (1979).

ALKALOSIS. A condition of excess alkalinity (or depletion of acid) in the body, in which the acid-base balance is upset. The hydrogen ion concentration of the blood drops below the normal level, increasing the pH above the normal 7.4 value. Alkalosis can result from the ingestion or formation in the body of an excess of alkali or loss of acid. Common causes include overbreathing (hyperventilation), ingestion of excessive alkali (for example, an overdosage of sodium bicarbonate), and excessive vomiting, which leads to loss of chloride and retention of sodium ions.

ALKANE. One of the group of hydrocarbons of the paraffin series, e.g., methane, ethane, and propane.

ALKENE. One of a group of hydrocarbons having one double bond and the formula $C_n H_{2n}$, e.g., ethylene and propylene.

ALKYD. Paint.

ALKYD RESINS. The esterification of a polybasic acid with a polyhydric alcohol yields a thermosetting hydroxycarboxylic resin, commonly referred to as an alkyd resin. Some common uses of alkyds are military switchgear, electrical terminal strips, electrical relay housings and bases, and television tuner segments. The resin frequently is combined with organic and inorganic fillers. These impart desired electrical and physical properties and advantageously influence the molding characteristics. Among the advantages of alkyds are rapid curing, with no volatiles emitted during the cure cycle; low molding pressures; and high production rates on compression or transfer presses and in injection molding machines. The resins have very good dimensional stability and electrical properties.

The mineral-filled grades, sometimes modified with cellulose for reducing specific gravity and cost, are used in small switch housings, automotive ignition parts, and electronic component bases.

Alkyd resins are furnished in three major forms: (1) *fibrous*, in which the resins are compounded with long glass fibers (about 0.5 in.; 12 mm) and have medium strength; (2) *rope*, which is a medium-impact material and conveniently handled and processed; and (3) *granular*, in which the resins are compounded with other fibers, such as glass and cellulose (length about $\frac{1}{16}$ in.; 2 mm). A commonly used member of the alkyd resin family is made from phthalic anhydride and glycerol. These resins are hard and possess very good stability. Where maleic acid is used as a starting ingredient, the resin has a higher melting point. Use of azelaic acid produces a softer and less brittle resin. Very tough and stable alkyds result from the use of adipic and other long-chain dibasic acids. Pentaerythritol may be substitute for glycerol as a starting ingredient.

Alkyd resins are extensively used in paints and coatings. Some advantages include good gloss retention and fast drying characteristics. However, most unmodified alkyds have low chemical and alkali resistance. Modification with esterified rosin and phenolic resins improves hardness and chemical resistance. Styrene and vinyl toluene improve hardness and toughness. For high-temperature coatings (up to about 450°F (232°C)), copolymers of silicones and alkyds are used. Such coatings include stove and heating equipment finishes. To obtain a good initial gloss, improved adhesion, and exterior durability, acrylic monomers can be copolymerized with oils to modify alkyd resins. Aromatic acids, such as benzoic or butylbenzoic, also have been used with alkyd resins in coatings.

ALKYL. A generic name for any organic group or radical formed from a hydrocarbon by elimination of one atom of hydrogen and so producing a univalent unit. The term is usually restricted to those radicals derived from the aliphatic hydrocarbons, those owing their origin to the aromatic compounds being termed "aryl." Usually, saturated radicals, such as methyl, ethyl, propyl, etc., containing, respectively, one hydrogen atom less than the corresponding saturated hydrocarbons—methane, ethane, propane, etc.—are understood.

ALKYLATION. A process involving the addition of an alkyl group. Alkylation reactions are important throughout synthetic organic chemistry. For many years, alkylation has been used in the production of gasolines with high antiknock ratings for automobiles and aircraft.

The nature of the products of these reactions, as well as the yields, depend upon the catalysts and physical conditions. The reactions below have been written to show two combination reactions of two isobutene molecules, one yielding diisobutene, which reduces to isooctane, and the other yielding a trimethylpentane by a direct reduction reaction.

Specifically, the term is applied to various methods, including both thermal and catalytic processes, for bringing about the union of paraffin hydrocarbons with olefins. The process is especially effective in yielding gasolines of high octane number and low boiling range (aviation fuels).

$$2CH_3-\underset{\underset{Isobutene}{}}{\overset{\overset{CH_3}{|}}{C}}=CH_2 + CH_3-\underset{\underset{Isobutene}{}}{\overset{\overset{CH_3}{|}}{C}}=CH_2 \longrightarrow CH_3-\underset{\underset{\underset{Diisobutene}{}}{CH_3}}{\overset{\overset{CH_3}{|}}{\underset{|}{C}}}-CH_2-\overset{\overset{CH_3}{|}}{C}=CH_2$$

$$\overset{H_2}{\longrightarrow} CH_3-\underset{\underset{\underset{Isooctane}{}}{CH_3}}{\overset{\overset{CH_3}{|}}{\underset{|}{C}}}-CH_2-\overset{\overset{CH_3}{|}}{CH}-CH_3$$

$$CH_3-\underset{\underset{\underset{Isobutene}{}}{CH_2}}{\overset{\overset{CH_3\quad CH_3}{|\quad\quad|}}{\underset{\|}{C}}} + CH=CH-CH_3 \overset{H_2}{\longrightarrow} CH_3-\underset{\underset{\underset{2-2-3-Trimethylpentane}{}}{CH_3}}{\overset{\overset{CH_3\quad CH_3}{|\quad\quad|}}{\underset{|}{C}}}-CH-CH_2-CH_3$$

In the petroleum industry, catalytic cracking units provide the major source of olefinic fuels for alkylation. A feedstock from a catalytic cracking units it typified by a C_3/C_4 charge with an approximate composition of: propane, 12.7%; propylene, 23.6%; isobutane, 25.0%; n-butane, 6.9%; isobutylene, 8.8%; 1-butylene, 6.9%; and 2-butylene, 16.1%. The butylenes will produce alkylates with octane numbers approximately three units higher than those from propylene.

One possible arrangement for a hydrofluoric acid alkylation unit is shown schematically in the accompanying figure. Feedstocks are pretreated, mainly to remove sulfur compounds. The hydrocarbons and acid are intimately contacted in the reactor to form an emulsion, within which the reaction occurs. The reaction is exothermic and temperature must be controlled by cooling water. After reaction, the emulsion is allowed to separate in a settler, the hydrocarbon phase rising to the top. The acid phase is recycled. Hydrocarbons from the settler pass to a fractionator which produces an overhead stream rich in isobutane. The isobutane is recycled to the reactor. The alkylate is the bottom product of the fractionater (isostripper). If the olefin feed contains propylene and propane, some of the isostripper overhead goes to a depropanizer where propane is separated as an overhead product. A hydrofluoric acid (HF) stripper is required to recover the acid so that it may be recycled to the reactor. HF alkylation is conducted at temperature in the range of 24–38°C (75–100°F).

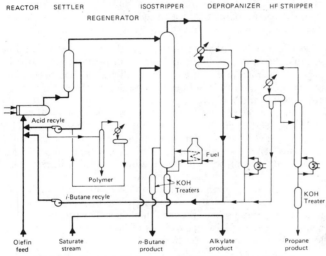

Hydrofluoric acid alkylation unit. (*UOP Process Division*)

Sulfuric acid alkylation also is used. In addition to the type of acid catalyst used, the processes differ in the way of producing the emulsion, increasing the interfacial surface for the reaction. There also are important differences in the manner in which the heat of reaction is removed. Often, a refrigerated cascade reactor is used. In other designs, a portion of the reactor effluent is vaporized by pressure reduction to provide cooling for the reactor.

ALKYNES. A series of unsaturated hydrocarbons having the general formula $C_n H_{2n-2}$, and containing a triple bond between two carbon atoms. The simplest compound of this series is acetylene HC : CH. Formerly, the series was named after this compound, i.e., the *acetylene series*. The latter term remains in popular usage. Particularly, the older names of specific compounds, such as acetylene, allylene CH_2C : CH, and crotonylene CH_3C : CCH_3, persist. These compounds are sometimes called *acetylenic hydrocarbons*. In the alkyne system of naming, the -yl termination of the alcohol radical corresponding to the carbon content of the alkyne is changed to -yne. Thus, C_2H_2 (acety-

lene by the earlier system) becomes *ethyne* [the eth- from ethyl (C_2)]; and C_4H_6 (crotonylene by the earlier system) becomes *butyne* [the but- from butyl (C_4)].

ALLANITE. This is a rather rare monoclinic mineral of somewhat variable, but quite complex chemical composition, perhaps represented satisfactorily by the formula $(Ce,Ca,Y)_2(Al,Fe)_3Si_3O_{12}(OH)$. The color of the fresh mineral is black, but it usually is brown or yellow with a coating of some alteration product; often the altered crystals have the appearance of small rusty nails. Allanite occurs characteristically in plutonic rocks like granite, syenite, or diorite and is found in large masses in pegmatites. Localities in the United States are Essex and Orange Counties, New York; Franklin, New Jersey; Amherst County, Virginia, and Llano County, Texas. The slender prismatic crystals are sometimes called *orthite*. Allanite was named for its discoverer, T. Allan. Orthite was so named from the Greek word meaning straight, in reference to the straight prisms, a common habit of this mineral.

ALLELOPATHIC SUBSTANCE. A material contained within a plant that tends to suppress the growth of other plant species. The alkaloids present in several seed-bearing plants are believed to play an allelopathic role. Other suspected allelopathic substances contained in some plants include phenolic acids, flavonoids, terpenoid substances, steroids, and organic cyanides.

ALLOBAR. A form of an element differing in atomic weight from the naturally occurring form, hence different in isotopic composition.

ALLOCHROMATIC. With reference to a mineral that, in its purest state is colorless, but that may have color due to submicroscopic inclusions, or to the presence of a closely related element that has become part of the chemical structure of the mineral. With reference to a crystal that may have photoelectric properties due to microscopic particles occurring in the crystal, either present naturally, or as the result of radiation.

ALLOMERISM. A property of substances that differ in chemical composition, but have the same crystalline form.

ALLOMORPHISM. A property of substances that differ in crystalline form, but have the same chemical composition.

ALLOTROPES. Chemical Elements.

ALLOYS. Although the term has grown into wider usage in recent years, fundamentally an alloy is an intentional mixture of two or more chemical elements which have metallic properties. Most metals are soluble in one another in the liquid state and alloying procedures usually involve melting; however, alloying by treatment in the solid state without melting is accomplished by the methods of powder metallurgy. When molten alloys solidify, they may remain soluble in one another or they may separate into intimate mechanical mixtures of the pure constituent metals. More often, there is partial solubility in the solid state and the structure consists of a mixture of the saturated solid solutions. Another important type of solid phase is the intermetallic compound which is characterized by hardness and brittleness and usually has only limited solid solubility with the other phases present. Many possibilities exist and these can be portrayed by diagrams showing the relationships between the solid and liquid phases at various temperatures under equilibrium conditions.

It is well known that many important alloy combinations have properties which are not easy to predict on the basis of the properties of the constituent metals. For example, copper and nickel, both having good electrical conductivity, form solid-solution alloys having very low conductivity, or high resistivity, making them useful as electrical resistance wires. In some cases, very small amounts of an alloying element produce remarkable changes in properties, as in steel, containing less than 1% carbon with the balance principally iron. Steels and the age-hardening alloys depend upon heat treatment to develop special properties, such as great strength and hardness. Other properties which can be developed to a much higher degree in alloys than in pure metals include corrosion resistance, oxidation resistance at elevated temperatures, abrasion or wear resistance, good bearing characteristics, creep strength at elevated temperatures, and impact toughness. Solid-state physics has been successful in explaining many of these properties of metals and alloys.

There are various types of alloys. Thus, the atoms of one metal may be able to replace the atoms of the other on its lattice sites, forming a substitutional alloy, or solid solution. If the sizes of the atoms, and their preferred structures, are similar, such a system may form a continuous series of solutions; otherwise, the miscibility may be limited. Solid solutions, at certain definite atomic proportions, are capable of undergoing an order-disorder transition into a state where the atoms of one metal are not distributed at random through the lattice sites of the other, but form a superlattice. Again, in certain alloys systems, intermetallic compounds may occur, with certain highly complicated lattice structures, forming distinct crystal phases. It is also possible for light, small atoms to fit into the interstitial positions in a lattice of a heavy metal, forming an interstitial compound.

In this Encyclopedia, alloys of chemical elements of alloying importance are discussed under that particular element, or in an entry immediately following.

Commercially, the major categories of alloys include:

CAST FERROUS METALS

Gray, Ductile, and High-Alloy Irons

In gray iron, most of the contained carbon is in the form of graphite flakes, dispersed throughout the iron. In ductile iron, the major form of contained carbon is graphite spheres which are visible as dots on a ground surface. In white iron, practically all contained carbon is combined with iron as iron carbide (cementite), a very hard material. In malleable iron, the carbon is present as graphite nodules. High-alloy irons usually contain an alloy content in excess of 3%.

Malleable Iron

The two main varieties of malleable iron are ferritic and pearlitic, the former more machinable and more ductile; the latter stronger and harder. Carbon in malleable iron ranges between 2.30 and 2.65%. Ranges of other constituents are: manganese, 0.30 to 0.40%; silicon, 1.00 to 1.50%; sulfur, 0.07 to 0.15%; and phosphorus, 0.05 to 0.12%.

Carbon and Low-Alloy Steels

Low-carbon cast steels have a carbon content less than 0.20%; medium-carbon steels, 0.20 to 0.50%; and high-carbon steels have in excess of 0.50% carbon. Ranges of other constituents are: manganese, 0.50 to 1.00%; silicon, 0.25 to 0.80%; sulfur, 0.060% maximum; and phosphorus, 0.050% maximum.

Low-alloy steels have a carbon content generally less than 0.40% and contain small amounts of other elements, depending upon the desired end-properties. Elements added include aluminum, boron, chromium, cobalt, copper, manganese, molybdenum, nickel, silicon, titanium, tungsten, and vanadium.

High-Alloy Steels

When "high-alloy" is used to describe steel castings, it generally means that the castings contain a minimum of 8% nickel and/or chromium. Commonly thought of as stainless steels, nevertheless *cast grades* should be specified by ACI (Alloy Casting Institute) designations and not by the designations that apply to similar *wrought alloys*.

WROUGHT FERROUS METALS

Carbon Steels

These steels account for over 90% of all steel production. There are numerous varieties, depending upon carbon content and method of production. In one classification, there are *killed* steels, *semikilled* steels, *rimmed* steels, and *capped* steels. These are described in considerable detail under **Iron Metals, Alloys, and Steels.**

High-Strength Low-Alloy Steels

There are several varieties, with high-yield strength depending mainly on the precipitation of martensitic structures from an austenitic field during quenching. Small additions of alloy elements, such as manganese and copper, are dissolved in a ferritic structure to obtain high strength and corrosion resistance.

Low and Medium-Alloy Steels

The two basic types are (1) *through* hardenable, and (2) *surface* hardenable. Subcategories of surface hardenable alloys include carburizing alloys, flame and induction-hardening alloys, and nitriding alloys.

Stainless Steels

A stainless steel is defined as iron-chromium alloy that contains at least 11.5% chromium. There are three major categories: (1) austenitic, (2) ferritic, and (3) martensitic, depending upon the metallurgical structure. There are scores of varieties. Type 302 is the base alloy for austenitic stainless steels. Representative stainless steels in this category provide some insight as to why so many varieties are made and of how rather small changes in composition and production can bring about significant differences in the final properties of the various stainless steels. A slightly lower carbon content improves weldability and inhibits carbide formation. An increase in nickel content lowers the work hardening. By increasing both chromium and nickel, better corrosion and scaling resistance is achieved. The addition of sulfur or selenium increases machinability. The addition of silicon increases scaling resistance at high temperature. Small amounts of molybdenum improve resistance to pitting corrosion and temperature strength.

High-Temperature, High-Strength, Iron-Base Alloys

There are two general objectives in making these alloys: (1) they can be strengthened by a martensitic type of transformation, and (2) they will remain austenitic regardless of heat treatment and derive their strength from cold working or precipitation hardening. Again, there are numerous types. Considering the main types, the carbon content may range from 0.05% to 1.10%; manganese, 0.20 to 1.75%; silicon, 0.20 to 0.90%; chromium, 1.00 to 20.75%; nickel, 0 to 44.30%; cobalt, 0 to 19.50%; molybdenum, 0 to 6.00%; vanadium, 0 to 1.9%; tungsten, 0 to 6.35%; copper, 0 to 3.30%; columbium (niobium), 0 to 1.15%; tantalum, 0 to < 1%; aluminum, 0 to 1.17%; and titanium, 0 to 3%.

Ultrahigh-Strength Steels

Normally a steel is considered in this category if it has a yield strength of 160,000 psi or more. The first of these steels to be produced was a chromium-molybdenum alloy steel, shortly followed by a stronger chromium-nickel-molybdenum grade.

Free-Machining Steels

Normally, the carbon content is kept under 0.10%, but as

much as 0.25% carbon has little deleterious effect on machinability. Aluminum and silicon are held to a minimum (aluminum not used as a deoxidizer where machinability is extremely important). Lead, sulfur, bismuth, selenium, and tellurium (0.04%) improve machinability when in the proper combination. Sulfur improves machinability by combining with any manganese and oxygen present to form oxysulfides.

NONFERROUS METALS

Aluminum Alloys

These alloys are available as wrought or cast alloys. The principal metals alloyed with aluminum include copper, manganese, silicon, magnesium, and zinc. These alloys are discussed in considerable detail under **Aluminum Alloys.**

Copper Alloys

These alloys are available as wrought or cast alloys. The principal wrought copper alloys are the brasses, leaded brasses, phosphor bronzes, aluminum bronzes, silicon bronzes, beryllium coppers, cupro nickels, and nickel silvers. The major cast copper alloys include the red and yellow brasses, manganese, tin, aluminum, and silicon bronzes, beryllium coppers, and nickel silvers. The chemical compositions range widely. For example, a leaded brass will contain 60% copper, 36 to 40% zinc, and lead up to 4%; a beryllium copper is nearly all copper, containing 2.1% beryllium, 0.5% cobalt, or nickel, or in another formulation, 0.65% beryllium, and 2.5% cobalt.

Nickel Alloys

Although nickel is present in varying amounts in stainless steel *commercially* a high-nickel stainless steel is not categorized as a nickel *alloy*, but rather as a stainless steel. Most nickel alloys are proprietary formulations and hence designated by trade names, such as *Duranickel, Monel* (several), *Hastelloy* (several), *Waspaloy, Rene 41, Inco, Inconel* (several), and *Illium* G. The nickel content will range from about 30% to nearly 95%.

Magnesium Alloys

It is the combination of low density and good mechanical strength which provides magnesium alloys with a high strength-to-weight ratio. Again, these generally are proprietary formulations. Aluminum, manganese, thorium, zinc, zirconium, and some of the rare-earth metals are alloyed with magnesium.

Zinc Alloys

Zinc alloys are available as die-casting alloys or wrought alloys. The principal alloys used for die casting contain low percentages of magnesium, from 3.5 to 4.3% aluminum, and carefully controlled amounts of iron, lead, cadmium, and tin.

Titanium Alloys

The titanium-base alloys are considerably stronger than aluminum alloys and superior to most alloy steels in several respects. Several types are available. Alloying metals include aluminum, vanadium, tin, copper, molybdenum, and chromium.

References

Baker, H., and D. Benjamin (editors): "Glossary of Metallurgical Terms and Engineering Tables," American Society for Metals, Metals Park, Ohio, 1979.

Bardes, B. P. (editor): "Metals Handbook," 9th Edition, Vol. 1 (Properties and Selection: Irons and Steels); and Vol. 2 (Properties and Selection: Nonferrous Alloys and Pure Metals), American Society for Metals, Metals Park, Ohio, 1979.

Chin, G. Y.: "New Magnetic Alloys," *Science*, **208**, 888–894 (1980).

Coyne, J. E.: "Technology is Key to Production of Superalloy Disc Forgings," *Metal Progress*, **118**, 6, 34–41 (1980).

Eiselstein, H. S.: "Monel Alloys Mark 75th Anniversary," *Metal Progress*, **118**, 8, 11–16 (1980).

Furrer, A. Editor: "Crystal Field Effects in Metals and Alloys," Plenum, New York, 1977.

Ketcham, S. J., and E. J. Jankowsky: "How Aluminum Alloys Fare in Shipboard Exposure Tests," *Metal Progress*, **119**, 4, 38–44 (1981).

Peckner, D., and I. M. Bernstein: "Handbook of Stainless Steels," McGraw-Hill, New York, 1977.

Pickering, F. B.: "The Metallurgical Evolution of Stainless Steels," American Society for Metals, Metals Park, Ohio, 1979.

Rashid, M. S.: "High-Strength, Low-Alloy Steels," *Science*, **208**, 862–869 (1980).

Schetky, L. M.: "Shape-Memory Alloys," *Sci. Amer.*, **241**, 5, 74–82 (1979).

Schillmoller, C. M., and H. P. Klein: "Selecting and Using Some High Technology Stainless Steels," *Metal Progress*, **119**, 2, 22–29 (1981).

Smith, D. F., and E. F. Clatworthy: "The Development of High Strength, Low Expansion Alloys," *Metal Progress*, **119**, 4, 32–35 (1981).

Smith, R. D.: "Temper Designations for Copper and Copper Alloys," *Metal Progress*, **119**, 2, 47–49 (1981).

Staff: "Unified Numbering System for Metals and Alloys," 2nd Edition, Society of Automotive Engineers, Warrendale, Pennsylvania, 1977.

Staff: "Source Book on Selection and Fabrication of Aluminum Alloys," American Society for Metals, Metals Park, Ohio, 1983.

Staff: "Source Book on Copper and Copper Alloys," American Society for Metals, Metals Park, Ohio, 1979.

Staff: "Worldwide Guide to Equivalent Irons and Steels," and "Worldwide Guide to Equivalent Nonferrous Metals and Alloys," American Society for Metals, Metals Park, Ohio, 1983.

Staff: "Trends in Superalloy Technology," *Metal Progress*, **119**, 1, 44–46 (1981).

Tien, J. K., et al. (editors): "Superalloys 1980," American Society for Metals, Metals Park, Ohio, 1980.

ALLYL ALCOHOL. CH_2=$CHCH_2OH$, formula wt 58.08, colorless liquid with pungent mustardlike odor, mp −129°C, bp 96.6°C, sp gr 0.854. Also known as propenyl alcohol or 2-propen-1-ol, this alcohol is miscible with water, alcohol, chloroform, ether, and petroleum ether. It is prepared by the hydrolysis of allyl chloride (from propylene) with dilute caustic; or by isomerization of propylene oxide over lithium phosphate catalyst at 230–270°C. It may also be prepared by dehydration of propylene glycol. A high-production chemical, allyl alcohol esters are used in resins and plasticizers. Allyl alcohol is also used as an intermediate for organic syntheses, for the manufacture of glycerol and acrolein, and herbicides.

ALLYL CHLORIDE. Chlorinated Organics.

ALLYLIC RESINS. Among the thermosetting allylic resins are diallyl phthalate, diallyl isophthalate, diallyl maleate, and diallyl chlorendate. The allylic monomers serve as nonvolatile cross-linking agents in polyester compounds. Allylic resins find numerous applications because of their good temperature range, dimensional stability, and fine electrical properties. Allylic prepolymers are used in compounds for either compression or transfer molding. Generally, these prepolymers may be classified as medium-to-soft-flow molding materials. The addition of fibrous fillers improves strength, but at the sacrifice of electrical properties. The use of *Orlon* fibers enhances electrical properties, even under adverse humidities. *Dacron* provides impact resistance and strength in thin sections. The allylic prepolymers are especially suited for electronic gear that is subjected to adverse environmental conditions. Because of their chemical inertness, the prepolymers also are used for molding pump impellers and other chemical equipment.

Usually allylic resin parts are compression molded when long fiber glass fillers are used. Allylic molding compounds flow readily around inserts, fill complex cavities, knit well, and have low mold shrinkage. Parts can have small draft angles without causing mold-release problems. In addition to good chemical resistance, fungus resistance, thermal stability, high heat distortion temperature, and moisture resistance are among the advantages of the allylic resins. By comparison with polyesters, allylic systems have no styrene odor, low toxicity, low evaporation losses during evacuation cycles, and no subsequent bleed out.

ALUM. A series (*alums*) of usually isomorphous crystalline (most commonly octahedral) compounds in which sulfate is usually the negative ion (although it may often be replaced by selenate or, in some compounds, partly by such complexes as $BeF_4^=$ or $ZnCl_4^=$), and two positive ions are present. Alums have the general formula $M^IM^{III}(AX_4) \cdot 12H_2O$. The first is potassium or a higher alkali metal (sodium alums are rare) or thallium (I) or ammonium (or a substituted ammonium ion) and the second is a tripositive ion of relatively small ionic radius (0.5–0.7 Å). The tripositive ions of larger ionic radius such as the lanthanides form double sulfates, but not alums. The tripositive ions in alums, in roughly the order of known compounds, are aluminum, chromium, iron, manganese, vanadium, titanium, cobalt, gallium, rhodium, iridium, and indium.

ALUMINUM. Symbol Al, at. no. 13, at. wt. 26.98154, periodic table group 3b, mp 660 ± 1°C, bp 2452 ± 15°C, sp gr 2.699. In Canada and several other English speaking countries, the spelling of the element is *aluminum*. The element is a silver-white metal, with bluish tinge, capable of taking a high polish. Commercial aluminum has a purity of 99% (minimum), is ductile and malleable, possesses good weldability, and has excellent corrosion resistance to many chemicals and most common substances, including foods. The electrical conductivity of aluminum (on a volume basis) is exceeded only by silver, copper, and gold. First ionization potential, 5.984 eV; second, 18.823 eV; third, 28.44 eV. Oxidation potentials, $Al \rightarrow Al^{3+} + 3e^-$, 1.67 V; $Al + 4OH^- \rightarrow AlO_2^- + 2H_2O + 3e^-$, 2.35 V. See also **Chemical Elements** for other physical properties.

Aluminum occurs abundantly in all ordinary rocks, except limestone and sandstone; it is third in abundance of the elements in the earth's crust (8.1% of the solid crust), exceeded only by oxygen and silicon, with which two elements aluminum is generally found combined in nature. Aluminum is present in igneous rocks and clays as aluminosilicates; in the mineral cryolite in Greenland as sodium aluminum fluoride, Na_3AlF_6; in the minerals corundum and emery and the gems ruby and sapphire as aluminum oxide Al_2O_3; in the mineral bauxite, which contains 30–75% Al_2O_3, 9–31% H_2O, 3–25% Fe_2O_3, 2–9% SiO_2, and 1–3% TiO_2. Bauxite occurs in southern France, Hungary, Yugoslavia, Greece, the Guianas of South America, Arkansas, Georgia, and Alabama of the United States, Italy, the U.S.S.R., Australia, and New Zealand. Aluminum occurs as hydrated oxide, $Al_2O(OH)_4$, in Indonesia; in the mineral alunite or alum stone in Utah as aluminum potassium sulfate, $Al_2(SO_4)_3 \cdot K_2SO_4 \cdot 4Al(OH)_3$. See also **Bauxite**; and **Cryolite**.

Uses. Because of its corrosion resistance and relatively low cost, aluminum is used widely for food-processing equipment, food containers, food-packaging foils, and numerous vessels for the processing of chemicals. Because of the presence of aluminum in soils and rocks, there are natural traces of the element in nearly all foods. Whereas the processing of foods in copper vessels will cause destruction of vitamins, aluminum does not accelerate the degradation of vitamins. Aluminum foil has been used to cover severe burns as a means to enhance healing. Because of its comparatively good electrical conductivity compared with other metals except silver and copper, aluminum is used as an electrical conductor, particularly for high-voltage transmission lines. For the same conductance, the weight of aluminum required is about one-half that of annealed copper. The greater diameter of aluminum conductors also reduces corona loss. But, because aluminum has about 1.4 times the linear temperature coefficient of expansion as compared with annealed copper, the changes in sag of the cable are greater with temperature changes. For long spans requiring high strength, the center strand may be a steel cable, or supporting steel cables may be used. Aluminum finds use for bus bars because of its large heat-dissipating surface available for a given conductance.

Aluminum alloys readily with copper, manganese, magnesium, silicon, and zinc. Many aluminum alloys are commercially available. See also **Aluminum Alloys.**

Production. Although aluminum was predicted by Lavoisier (France) as early as 1782 when he was investigating the properties of aluminum oxide (alumina), the metal was not isolated until 1825 by H. C. Oersted (Denmark). Oersted obtained an impure aluminum metal by heating potassium amalgam with anhydrous aluminum chloride, followed by distilling off the mercury. Using similar methods, Woehler (Germany) produced an aluminum powder in 1827. In 1854, Deville (France) and Bunsen (Germany) separately but concurrently found that aluminum could be isolated by using sodium instead of potassium in the amalgam of Oersted and Woehler. Deville exhibited his product at the Paris Exposition of 1855, after which Napoleon III commissioned Deville to improve the process and lower the cost. Because of improved processing techniques, the price declined from $115 To $17 per pound ($254 to $38 per kilogram) by 1859, with numerous plants throughout France. The price was further reduced to $8 per pound ($17.60 per kilogram) by 1885. The first breakthrough for production of aluminum on a large scale and at a much lower cost occurred as the result of experimentation by Hall (Oberlin, Ohio) who found that metallic aluminum could be produced by dissolving Al_2O_3 (alumina) in molten $3NaF \cdot AlF_3$ (cryolite) at a temperature of about 960°C and then passing an electric current through the bath. Heroult (France) independently discovered the same process. Most modern aluminum production plants utilize this electrolytic process.

The major differences found in aluminum production plants relate to the design of the electrolytic cells. Basically, the cells are large carbon-lined steel boxes. The carbon lining serves as the cathode. Separate anodes are immersed in the bath. The electric circuit is completed by passage of current through the bath. Thus, the alumina is decomposed into aluminum and oxygen by action of the electric current. The molten aluminum collects at the bottom of the cell from which it is siphoned off periodically. The released oxygen combines with carbon at the electrodes to form CO_2. The bath is replenished with alumina intermittently. The carbon anodes must be replaced periodically. Additional cryolite also is required to make up for volatilization losses. The several reactions which occur during electrolysis still are not fully understood.

Energy requirements for aluminum production in modern plants approximate 6 to 8 kilowatthours per pound (13.2 to 17.6 kilowatthours per kilogram). Per weight unit of aluminum produced, 0.4 to 0.6 weight units of carbon, 1.9 weight units of alumina, and 0.1 weight unit of cryolite are required. One form of electrolytic cell is shown in Fig. 1.

As of the early 1980s, there are 32 aluminum smelters within the United States and 2 smelters in the planning stage. Smelters originally were located mainly for considerations of low-cost

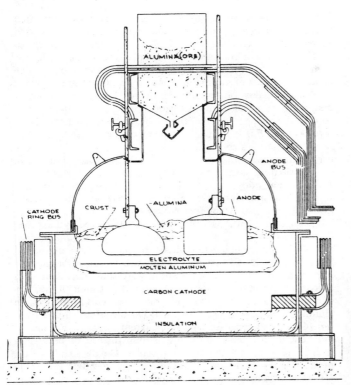

Fig. 1. Aluminum electrolytic reduction cell of the prebake type. The anodes are constructed of separate blocks of carbon which have been prebaked. There is a lead to the main bus bar from each block. (*Reynolds Metals Company*)

energy and transportation—and thus are found in Washington, Oregon, western Montana, Texas, New York, and along the Mississippi River. The capacity in the United States is 5.2 million short tons (about 4.9 million metric tons) per year. The United States and Canada account for 30.6% of the world production, which is 19.6 million short tons (17.6 million metric tons). This figure is expected to increase to 22.2 million short tons (about 20 million metric tons) by the mid-1980s or earlier. The developing areas for new production capacity are Venezuela, New Zealand, Australia, Brazil, and possibly the Arabian Peninsula.

By 1995, Australia will probably be one of the world's three largest aluminum producers. Australia has long been known for its vast deposits of bauxite and its significant potential capacity to produce aluminum. Australia presently supplies one-third of the world's bauxite. New smelter construction is expected to boost the world production by 2 million tons (1.8 million metric tons). The rapid development in Australia can be traced to a number of circumstances, including increasing demand for the metal as compared with existing production capacity. Electrical energy costs have increased markedly, especially in regions dependent upon oil and gas. For example, in the United States, political and environmental restraints placed on new power-generating facilities have stymied growth. Australia has coal reserves estimated for 1500 years.

It is also projected that Venezuela, because of its abundance of low-cost hydropower, will become a major aluminum producer. During the past decade, Brazil has attracted considerable interest because of its vast reserves of good-quality bauxite and its potential to develop attractive hydropower.

The People's Republic of China is planning to modernize its existing capacity of about 300,000 short tons (270,000 metric tons) to about 400,000 short tons (360,000 metric tons) during the next decade.

Thus, most of the increasing capacity for aluminum production during the next several years will be outside the continental United States, principally in the developing countries. As of 1984, developing countries account for 14% of total free world aluminum capacity, but this is expected to increase to 22% by 1990.

Secondary Recovery. In the early 1980s, the secondary recovery of aluminum through recycling programs in the United States was 1.75 million short tons (1.6 million metric tons), amounting to the equivalent of one-third of the primary production. Secondary recovery of scrap will continue to play a major role in the aluminum industry. Larger quantities of aluminum are being used in the beverage, automotive, and several other consumer industries, making it easy to accomplish effective recycling. See Fig. 2.

Fig. 2. Recycling of aluminum beverage cans and other scrap saves 95% of the energy required when starting from ore. Approximately 20% of the metal processed in 1980 was recycled from scrap. (*Reynolds Aluminum Recycling Company*)

The lowest price for aluminum occurred in the early 1940s as the result of improved production methodologies and increased volume. Since then, the price has risen from the low of $0.15 per pound ($0.33 per kilogram) because of the increasing costs of raw materials and electrical energy. Because of the very large capital investment represented by an aluminum reduction plant, accompanied by very heavy electric power consumption, there has been a continuing search for lower-cost processes.

New Processes. A relatively new process may reduce the electrical energy requirements by as much as 30% and involves the elimination of fluoride in the cells. The cells are operated at a low temperature and are completely enclosed. The process combines alumina (refined from bauxite) with chlorine to form aluminum chloride, after which the chloride is electrolyzed in a closed cell, releasing molten aluminum and chlorine. The chlorine gas is recycled to the chemical reactor. A 15,000-ton

(13,500-metric ton) plant scaling up this process is located in Anderson County, Texas. Reports from Alcoa indicate that the 30% less electricity required (as compared with the most efficient Hall process) previously expected has been substantiated. Aluminum production at the facility will be cut back to 7500 tons (6750 metric tons) per year, with redesign, construction, and testing of a new segment of the plant to achieve further efficiency. This program is scheduled for completion during the mid- 1980s.

Reynolds has announced a process based upon the electrolysis of aluminum from a cell using a fused bath of 50% sodium chloride (NcCl), 45% lithium chloride (LiCl), and 5% aluminum chloride ($AlCl_3$) at 700°C. The process requires 5 kwh per pound (11 kwh per kilogram) or 25% less energy than present reduction cells require. The process also requires less space for production.

Thermal Reduction Processes. Thermal processes, including the use of electric arc furnaces, make it possible to achieve a greater concentration of electric energy in the refining process. Commercial electric furnace processes are fed with bauxite and clays (kaolin or kyanite or partially refined bauxite mixed with carbon or charcoal in briquetted form). An electric dc or ac arc is used. To date, this process has been practical only for the production of impure aluminum alloys, ranging in content from 65–70% aluminum, 25–30% silicon, and 1% or more of iron. Other impurities may include titanium and the carbides and nitrides of aluminum and iron. This impure product may be refined to yield pure aluminum by means of a subsequent distillation process in which the electric-furnace product is dissolved at high temperatures in zinc, magnesium, or mercury baths and then distilled. Aluminum with a purity of 99.99+% may be produced in this manner.

A gas-reaction purification process also has attracted attention over the years. This is based on reacting aluminum trichloride gas with molten aluminum at about 1000°C to produce aluminum monochloride gas. Aluminum fluoride gas may be substituted for the aluminum monochloride gas in the process. Raw materials for the process may be scrap aluminum, aluminum from thermal-reduction processes, aluminum carbide, or aluminum nitride.

For the production of superpurity aluminum on a large scale, the Hoopes cell is used. This cell involves three layers of material. Impure (99.35 to 99.9% aluminum) metal from conventional electrolytic cells is alloyed with 33% copper (eutectic composition) which serves as the anode of the cell. A middle, fused-salt layer consists of 60% barium chloride and 40% $AlF_3$1.5NaF (chiolite), mp 720°C. This layer floats above the aluminum-copper alloy. The top layer consists of superpurity aluminum (99.995%). The final product usually is cast in graphite equipment because iron and other container metals readily dissolve in aluminum. For extreme-purity aluminum, zone refining is used. This process is similar to that used for the production of semiconductor chemicals and yields a product that is 99.9996% aluminum.

Aluminum Chemistry and Chemicals. Since aluminum has only three electrons in its valence shell, it tends to be an electron acceptor. Its strong tendency to form an octet is shown by the tetrahedral aluminum compounds involving *sp³* hybridization. Aluminum halides include the trifluoride, an ionic crystalline solid; the trichloride and tribromide, both of which are dimeric in the vapor and in nonpolar solvents, with the halogen atoms arranged in tetrahedra about each aluminum atom, giving a bridge structure; and the triiodide. They combine readily with many other molecules, especially organic molecules having donor groups, and aluminum chloride is an active catalyst. Certain

of the hydrates have a saltlike nature; thus $AlCl_3 \cdot 6H_2O$ acts as a salt, even though Al_2Cl_6 is covalent.

Common water-soluble salts include the sulfate, selenate, nitrate and perchlorate, all of which are hydrolyzed in aqueous solution. Double salts are formed readily by the sulfate and by the chloride.

There are many fluorocomplexes of aluminum. The general formula for the fluoroaluminates is $M_3^1[Al_yF_{x+3y}]$, based upon AlF_6 octohedra, which may share corners to give other ratios of Al:F than 1:6. Chloroaluminates of the type $M^1[AlCl_4]$ are obtainable from fused melts. Aluminum ions form chloro-, bromo- and iodo-complexes containing tetrahedral $[AlX_4]^-$ ions. However, in sodium aluminum fluoride $NaAlF_4$, the aluminum atoms are in the centers of octohedra of fluorine atoms in which the fluorine atoms are shared with neighboring aluminum atoms.

Aluminum hydroxide is about equally basic and acidic, the pK_b being about 12 and the pK_a being 12.6. Sodium aluminate seems to ionize as a uni-univalent electrolyte: $NaAl(OH)_4(H_2O)_2 \leftrightharpoons Na^+ + [Al(OH)_4(H_2O)_2]^-$. The high viscosity of sodium aluminate solutions is explained by hydrogen bonding between these hydrated ions, and between them and water molecules.

By reaction of aluminum and its chloride or bromide at high temperature, there is evidence of the existence of monovalent aluminum. Here the aluminum atom is apparently in the *sp* state, with an electron-pair on the side away from the chlorine atom, whereby the single pairs on the two chlorine atoms are shared to form two weak π-bonds.

Aluminum forms a polymeric solid hydride, $(AlH_3)x$, unstable above 100°C, extremely reactive, e.g., giving hydrogen explosively with H_2O and igniting explosively in air. From it are derived the tetrahydroaluminates, containing the ion AlH_4^-, which are important, powerful but selective reducing agents, e.g., reducing chlorophosphines R_2PCl to phosphines R_2PH; reactive alkyl halides and sulfonates to hydrocarbons, epoxides, and carboxylic acids; aldehydes and ketones to alcohols; nitrites, nitro compounds and N,N-dialkyl amides to amines. The salt most commonly used is lithium (tetra)hydroaluminate ("lithium aluminum hydride").

Aluminum compounds are generally made starting with bauxite, which is reactive with acids and with bases. With acids, e.g., H_2SO_4, any iron contained in the bauxite is dissolved along with the aluminum and silicon is left in the residue, whereas with bases, e.g., NaOH any silicon is dissolved and iron left in the residue.

Acetate. Aluminum acetate $Al(C_2H_3O_2)_3$, white crystals, soluble, by reaction of aluminum hydroxide and acetic acid and then crystallizing. Used (1) as a mordant in dyeing and printing textiles, (2) in the manufacture of lakes, (3) for fireproofing fabrics, (4) for waterproofing cloth.

Aluminates. Sodium aluminate $NaAlO_2$, white solid, soluble, (1) by reaction of aluminum hydroxide and NaOH solution, (2) by fusion of aluminum oxide and sodium carbonate; the solution of sodium aluminate is reactive with CO_2 to form aluminum hydroxide. Used as a mordant in the textile industry, in the manufacture of artificial zeolites, and in the hardening of building stones. See silicates below and calcium aluminates.

Alundum. See oxide (below).

Carbide. Aluminum carbide Al_4C_3, yellowish-green solid, by reaction of aluminum oxide and carbon in the electric furnace, reacts with H_2O to yield methane gas and aluminum hydroxide.

Chlorides. Aluminum chloride $AlCl_3 \cdot 6H_2O$, white crystals, soluble, by reaction of aluminum hydroxide and HCl, and then crystallizing; anhydrous aluminum chloride $AlCl_3$, white powder, fumes in air, formed by reaction of dry aluminum oxide plus carbon heated with chlorine in a furnace, used as a reagent in petroleum refining and other organic reactions.

Fluoride. Aluminum fluoride AlF_3, white solid, soluble, by reaction of aluminum hydroxide plus hydrofluoric acid and then crystallizing $2AlF_3 \cdot 7H_2O$, used in glass and porcelain ware.

Hydroxide. Aluminum hydroxide $Al(OH)_3$, white gelatinous precipitate, by reaction of soluble aluminum salt solution and an alkali hydroxide, carbonate or sulfide (sodium aluminate is formed with excess NaOH but no reaction with excess NH_4OH), upon heating aluminum hydroxide the residue formed is aluminum oxide. Used as intermediate substance in transforming bauxite into pure aluminum oxide.

Nitrate. Aluminum nitrate $Al(NO_3)_3$, white crystals, soluble, by reaction of aluminum hydroxide and HNO_3, and then crystallizing.

Oleate. Aluminum oleate $Al(C_{18}H_{33}O_2)_3$, yellowish-white powder, by reaction of aluminum hydroxide, suspended in hot H_2O, shaken with oleic acid, and then drying, the product is used (1) as a thickener for lubricating oils, (2) as a drier for paints and varnishes, (3) in waterproofing textiles, paper, leather.

Oxide. Aluminum oxide, alumina Al_2O_3, white solid, insoluble, melting point 2020°C, formed by heating aluminum hydroxide to decomposition; when bauxite is fused in the electric furnace and then cooled there results a very hard glass ("alundum"), used as an abrasive (hardness 9 Mohs scale) and heat refractory material. Aluminum oxide is the only oxide which reacts both in H_2O medium and at fusion temperature, to form salts with both acids and alkalis.

Palmitate. Aluminum palmitate $Al(C_{16}H_{31}O_2)_3$, yellowish-white powder, by reaction of aluminum hydroxide, suspended in hot H_2O, shaken with palmitic acid, and then drying, the product is used (1) as a thickener for lubricating oils, (2) as a drier for paints and varnishes, (3) in waterproofing textiles, paper, leather, (4) as a gloss for paper.

Silicates. Many complex aluminosilicates or silicoaluminates are found in nature. Of these, clay in more or less pure form, pure clay, kaolinite, kaolin, china clay $H_4Si_2Al_2O_9$ or $Al_2O_3 \cdot 2SiO_2 \cdot 2H_2O$ is of great importance. Clay is formed by the weathering of igneous rocks, and is used in the manufacture of bricks, pottery, porcelain and Portland cement.

Stearate. Aluminum stearate $Al(C_{18}H_{35}O_2)_3$, yellowish-white powder, by reaction of aluminum hydroxide suspended in hot H_2O, shaken with stearic acid, and then drying the product, used (1) as a thickener for lubricating oils, (2) as a drier for paints and varnishes, (3) in waterproofing textiles, paper, leather, (4) as a gloss for paper.

Sulfate. Aluminum sulfate $Al_2(SO_4)_3$, white solid, soluble, by reaction of aluminum hydroxide and H_2SO_4, and then crystallizing, used (1) as a clarifying agent in water purification, (2) in baking powders, (3) as a mordant in dyeing, (4) in sizing paper, (5) as a precipitating agent in sewage disposal; aluminum potassium sulfate, see **Alums** and **Alunite.**

Sulfide. Aluminum sulfide Al_2S_3, white to grayish-black solid, reactive with water to form aluminum hydroxide and H_2S, formed by heating aluminum powder and sulfur to a high temperature.

Aluminum in solution of its salts is detected by the reaction (1) with ammonium salt of aurin tricarboxylic acid ("aluminon"), which yields a red precipitate persisting in NH_4OH solution, (2) with alizarin red S, which yields a bright red precipitate persisting in acetic acid solution.

Organoaluminum Compounds. By the action of alpha olefins on AlH_3, the trialkyls of aluminum can be prepared:

$$AlH_3 + 3CH_2{=}CHR \xrightarrow{120°C} (CH_2CH_2R)_3Al$$

In the presence of ethers, magnesium-aluminum alloys react with alkyl halides to yield trialkyls, R_3AlOEt_2. The trimethyl aluminum compound is a dimer, mp 15°C. Triethyl aluminum is a liquid. The alkyls react readily with H_2O:

$$Et_3Al + 3H_2O \rightarrow Al(OH)_3 + 3EtH$$

When aluminum is reacted with diphenyl mercury $(C_6H_5)_2Hg$, triphenylaluminum $(C_6H_5)_3Al$ is yielded. The aluminum organometallic alkyls and aryls act as Lewis acids to form compounds with electron-donating substances. The resulting compounds are the organometallic basis for producing polymers of the Al-N type.

References

Aluminum Association: "Aluminum Standards and Data" and "Aluminum Statistical Review" (Issued periodically).

Anderson, W. A., and H. C. Stumpf: "Forming Characteristics of Aluminum Alloys for Rigid Containers," *Metal Progress*, **119**, (6), 60 (1981).

ASM: "Metals Handbook," 9th Edition, Vol. 2 (Properties and Selection: Nonferrous Alloys and Pure Metals), American Society for Metals, Metals Park, Ohio (1979).

Charles, J., Berghezan, A., and A. Lutts: "New Cryogenic Materials: Fe-Mn-Al Alloys," *Metals Progress*, **119**(6), 71 (1981).

Erich, D. L.: "Benefits of Mechanically Alloyed Aluminum," *Metal Progress*, **121**(2) 22 (1982).

Reyss, J. L., Yokoyama, Y., and S. Tanaka: "Aluminum-26 in Deep-Sea Sediment," *Science*, **193**, 1119–1121 (1976).

Staff: "Filler Metal Selection for Welding Aluminum Alloys," *Metal Progress*, **119**(5), 52 (1981).

Staff: "Materials & Processing Databook," *Metal Progress* (Issued annually).

Stoffyn, M.: "Biological Control of Dissolved Aluminum in Seawater: Experimental Evidence," *Science*, **203**, 651–653 (1979).

Van Horn, K. R., Editor: "Aluminum," Volumes 1–3, American Society for Metals, Metals Park, Ohio (1967).

—S. John Sansonetti, Reynolds Metals Company, Richmond, Virginia.

ALUMINUM ALLOYS. The intrinsic properties of aluminum—light weight, strength, good electrical conductivity, nonmagnetic properties, high reflectivity, transparency to neutrons, good thermal conductivity, weather resistance, nontoxicity, decorative appeal, weldability, machinability, corrosion resistance, fracture toughness, strength at cryogenic temperatures, and nonsparking characteristics, among others, make it one of the most versatile engineering and construction materials. The aluminum industry offers more than 300 combinations of alloys and tempers, from which engineers and designers may select the most suitable alloy for their needs.

Aluminum is unique among the metals because it responds to nearly all of the known finishing processes. It can be finished in the softest, most delicate textures, as exemplified by tableware and jewelry. Aluminum can be anodized and dyed to appear like gold or in numerous colors. It can be made as specular as a silver mirror and jet black. The metal also can be anodized to an extremely hard, wear- and abrasion-resistant surface that approaches the hardness of a diamond. Aluminum is available in many convenient shapes and forms—sheet, plate, ingot, wire, rod, and bar as well as foil, castings, forgings, powdered metals, and extrusions.

In the production of these products, new high-speed automated equipment has been developed over the past decade. Rolling speeds as high as 8000 feet (2438 meters) per minute are possible on some products. New continuous casting processes have been developed where molten metal can be converted directly into coiled sheet, using roll casting, belt casting, and

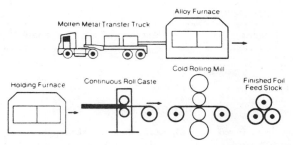

Fig. 1. Schematic portrayal of continuous aluminum roll casting process. (*Reynolds Metals Company*)

articulated block casting. See Figures 1 and 2. Many of the common alloys respond well to these new processes. Electromagnetic casting of the conventional giant ingots used for the conventional rolling mills has been developed in the fully automated mode. This casting process eliminates the need for scalping ingots prior to processing into final products, such as can stock, foil, auto body sheet, siding, and many other products. In this process, an electrical field rather than a solid mold wall is used to shape the ingot. The process is based upon the fact that liquids with good electrical conductivity, such as aluminum, can be constrained where an alternating magnetic field is applied. The liquid metal remains suspended by the electrical forces while it is being solidified.

Fig. 2. The continuous roll casting process is the most recent, economical, and energy efficient method for producing aluminum foil feed stock and other products. Liquid metal at 1350°F (732°C) is fed directly between water-cooled rolls to emerge as 0.4 × 78-inch (10.2 millimeter × 198.1-centimeter) strip to be coiled into 35,000-pound (15,876-kilogram) coils. (*Reynolds Metals Company*)

The casting alloy products comprise sand castings, permanent mold castings, and die castings. Aluminum is the basic raw material for more than 20,000 businesses in the United States. Aluminum is an indispensable metal for aircraft, for example. See Fig. 3. A guide to the selection of aluminum alloys for a broad classification of uses is given in Table 1.

The free world's demand trend for aluminum over a 28-year period to 1980 shows an average growth of 7.9% per year. Projections through 1985 are estimated at a growth rate of 5% per year. Aluminum is quite well established as an energy saver and a cost-effective material in numerous applications. Particularly strong growth is occurring in the automotive and transportation fields, housing, and food packaging. For example, a number of breakthroughs in foil pouch packaging have been achieved for replacing conventional cans for shelf-stable foods. Steady growth has continued in other areas, including aircraft and space applications, electrical wire and cable, marine, architectural, and consumer durable products. See Fig. 4.

Pure aluminum is soft and ductile. In its annealed form, the tensile strength is approximately 13,000 pounds per square inch (90 megapascals). This can be work hardened by rolling, drawing into wire, or by other cold-working techniques to increase its strength to approximately 24,000 pounds per square inch (165 megapascals). Pure aluminum is particularly useful in the food and chemical industries where its resistance to corrosion and high thermal conductivity are desirable characteristics, and in the electrical industry, where its electrical conductivity of about 62% that of copper and its lightweight make it desirable for wire, cable and bus bars. Large quantities are rolled to thin foils for the packaging of food products, for collapsible tubes, and for paste and powders used for inks and paints.

Most commercial uses require greater strength than pure aluminum affords. Higher strength is achieved by alloying other metal elements with aluminum and through processing and thermal treatments. The alloys can be classified into two categories, nonheat treatable and heat treatable.

Nonheat-Treatable Alloys. The initial strength of alloys in this group depend upon the hardening effect of elements such as manganese, silicon, iron, and magnesium, singly or in various combinations (see first three alloys in Table 2). The nonheat-treatable alloys are usually designated therefore in the 1xxx, 3xxx, 4xxx, and 5xxx series (Table 3). Since these alloys are

Fig. 3. Final operations on machined aluminum plates for wings of large jet aircraft. (*Reynolds Metals Company*)

TABLE 1. ALUMINUM ALLOY SELECTION GUIDE

PRODUCT	ALLOY	FORM	PRODUCT	ALLOY	FORM
Architectural and			*Electrical*		
Building Products			Bus Bar	6063, 6201	Extrusions or Rolled Rod
Awnings	3003	Sheet	Cable	E.C., 5005, 6201	Wire
Fence Wire	6061	Wire	Conduit	6063	Tubing
Fittings	A514.0	Castings	Motors	319.0, 355.0, 360.0, 380.0	Castings
Gutters	Alc. 3004	Sheet			
Nails	5056, 6061	Wire	Transmission Towers and Substations	6061, 6063, 7005	Extrusions
Panels	3003, 6063	Sheet and Extrusions			
Roofing	Alc. 3004	Sheet	*Consumer Durables*		
Screens	Alc. 5056	Wire	Appliances	3003, 4343, 5052, 5252, 5357, 5457, 6063, 6463	Sheet and Extrusions
Siding	3003, Alc. 3004	Sheet			
Transportation				356.0, 390.0	Castings
Aircraft	2014, 2024, 2048, 2124, 2219, 7075, 7079, 7175, 7178, 7179, 7475.	Sheet, Plate, Extrusions, and Forgings	Cooking Utensils	1100, 3003, Alc. 3003, 3004, Alc. 3004, 5052, 5454	Sheet
	B295.0, 355.0, 356.0, 518.0, 520.0	Castings		B443.0, A514.0	Castings
			Furniture	3003, 3004, 5005, 5457, 6061, 6063, 6463	Sheet and Tubing
Automotive					
Auto and Truck Bodies	2036, 2037, 5052, 5182, 5252, 5657, 6009, 6010, 6061, 6063, 6463, 7016, 7021, 7029, 7046, 7129	Sheet and Extrusions	Refrigerators	3003, Alc. 3003, 3004, Alc. 3004, 4343, 5005, 5050, 5052, 5252, 5457, 6061, 6463	Sheet and Extrusions
Buses	2036, 5083, 5086, 5182, 5252, 5457, 5657, 6061, 6063, 6463, 7016, 7046	Sheet and Extrusions	Water Heaters	Alc. 6061, 3003	Sheet
			Machinery and Equipment		
			Chemical Processing	1060, 1100, 3003, Alc. 3003, 3004, Alc. 3004, 5083, 5086, 5154, 5454, 6061, 6063	Sheet, Plate, Extrusions, and Tubes
	242.0, B295.0, 355.0, 356.0, 360.0, 380.0, 390.0, B443.0	Castings			
				356.0, 360.0, B514.0	Castings
Marine			Heat exchangers and Solar Panels	3003, Alc. 3003, 3004, Alc. 3004, 4343, 5005, 5052, 6061, 6951	Sheet and Tubing
Barges, Small Craft, Ships, Tankers	5052, 5083, 5086, 5454, 5456, 6061, 6063	Sheet, Plate, Extrusions, and Tubing			
	360.0, 413.0, B443.0	Castings	Sheet for Vacuum Brazing Heat Exchangers	4003, 4004, 4005, 4044, 4104	
Railroad	Alc. 2024, 5052, 5083, 5086, 5454, 5456, 6061, 6063, 7005	Sheet, Plate, and Extrusions	Irrigation	3003, Alc. 3003, 3004, Alc. 3004, 5052, 6061, 6063	Tubing
	B295.0, 356.0, 520.0	Castings	Sewage Plants	3003, Alc. 3003, 3004, Alc. 3004, 5052, 5083, 5086, 5454, 5456, 6061, 6063	Sheet, Extrusions, and Tubes
Containers and Packaging					
Foils	1100, 1235, 3003, 5005, 8079, 8111	Foils		B514.0	Castings
			Screw Machine Parts	2011, 2024, 6262	Rod and Bar
Cans	3003, 3004, 5182	Sheet	Textile Machinery	2014, 2024, 6061, 6063	Extrusions, Sheet and Plate

work hardenable, further strengthening is made possible by various degrees of cold work denoted by the H series tempers.

Heat-Treatable Alloys. This group of alloys includes the alloying elements, copper, magnesium, zinc, and silicon (-T tempered alloys in Table 2). Since these elements singly or in combination show increasing solid solubility in aluminum with increasing temperature, it is possible to subject them to thermal treatments which will impart pronounced strengthening. The first step called heat treatment or solution heat treatment, is an elevated temperature process designed to put the soluble element or elements in solid solution. This is followed by a rapid quenching which momentarily "freezes" the structure and for a short time renders the alloy very workable. At room or elevated temperatures, the alloys are not stable after quenching, and precipitation of the constituent from the supersaturated solid solution begins. After several days at room temperature, termed aging or room temperature precipitation, the alloy is considerably stronger. Many alloys approach a stable condition at room temperature but some alloys, especially those containing magnesium and

silicon or magnesium and zinc continue to age harden for long periods of time at room temperature. By heating for a controlled time at slightly elevated temperatures, even further strengthening is possible and the properties are stabilized.

Clad Alloys. The heat-treatable alloys, in which copper or zinc are major alloying constituents, are less resistant to corrosive attack than a majority of the nonheat-treatable alloys. To increase the corrosion resistance of these alloys in sheet and plate form, they are often clad with a high-purity aluminum, a low-magnesium-silicon alloy, or an aluminum alloy containing 1% zinc.

The cladding, usually from $2\frac{1}{2}$ to 5% of the total thickness on each side, not only protects the composite due to its own inherent excellent corrosion resistance, but also exerts a galvanic effect which further protects the core alloy.

Special composites may be obtained such as clad nonheat-treatable alloys for extra corrosion protection, for brazing purposes, and for special surface finishes. See Fig. 5.

Composites. The most commonly used metal elements alloyed

Fig. 4. Four aluminum sections of the type shown here make up the fuel tank used for propelling the NASA space shuttle into orbit. Each section is produced from computer-machined plate and weighs 4750 pounds (2155 kilograms). (*Reynolds Metals Company*)

with aluminum are magnesium, manganese, silicon, copper, zinc, iron, nickel, chromium, titanium, and zirconium. The strength of aluminum can be tailored to the specific end applications ranging from the very soft ductile foils to the high-strength aircraft and space alloys equal to steels in the 90,000 pounds per square inch (621 megapascals) tensile strength range. In addition to the homogeneous aluminum alloys, dispersion hardened and advanced filament-aluminum composites provide even higher strengths. Boron filaments in aluminum can provide strengths in the range of 150,000–300,000 pounds per square inch (1034–2068 megapascals) tensile strength.

Superplasticity. Eutectoid and near eutectoid alloy chemistry research has resulted in alloys exhibiting superplasticity with unusual elongation (> 1000%) and formability and with 60,000 pounds per square inch (414 megapascals) tensile strength. Three examples of these alloys are: 94.5% Al-5% Cu-0.5% Zr; and 22% Al-78% Zn; and 90% Al, 5% Zn, 5% Ca.

Casting Alloys. During the last two decades, the quality of castings has been improved substantially by the development of new alloys and better liquid-metal treatment and also by improved casting techniques. Casting techniques include sand casting, permanent mold casting, pressure die casting, and others. Today sand castings can be produced in high-strength alloys and are weldable. Die casting permits large production outputs per hour on intricate pieces that can be cast to close dimensional tolerance and have excellent surface finishes; hence, require minimum machining. Since aluminum is so simple to melt and cast, a large number of foundry shops have been established to supply the many end products made by this method of fabrication. See Table 3.

Casting Semi-solid Metal. A new casting technology is based on vigorously agitating the molten metal during solidification. A very different metal structure results when this metal is cast. The vigorously agitated liquid-solid mixture behaves as a slurry still sufficiently fluid (thixotropic) to be shaped by casting. The shaping of these metal slurries is termed "Rheocasting."

The slurry nature of "Rheocast" metal permits addition and retention of particulate nonmetal (e.g., Al_2O_3, SiC, T, C, glass beads) materials for cast composites.

The importance of this new technology has yet to be commercialized.

Alloy and Temper Designation Systems for Aluminum. The aluminum industry has standardized the designation systems for wrought aluminum alloys, casting alloys and the temper designations applicable. A system of four-digit numerical designations is used to identify wrought aluminum alloys. The first digit indicates the alloy group as shown in Table 4. The 1xxx series is for minimum aluminum purities of 99.00% and greater; the last two of the four digits indicate the minimum aluminum percentage; i.e., 1045 represents 99.45% minimum aluminum, 1100 represents 99.00% minimum aluminum. The 2xxx through 8xxx series group aluminum alloys by major allowing elements. In these series the first digit represents the major alloying element, the second digit indicates alloy modification, while the third and fourth serve only to identify the different alloys in the group. Experimental alloys are prefixed with an X. The prefix is dropped when the alloy is no longer considered experimental.

Cast Aluminum Alloy Designation System. A four-digit number system is used for identifying aluminum alloys used for castings and foundry ingot (see Table 5). In the 1xx.x group for aluminum purities of 99.00% or greater, the second and third digit indicate the minimum aluminum percentage. The last digit to the right of the decimal point indicates the product form: 1xx.0 indicates castings and 1xx.1 indicates ingot. Special control of one or more individual elements other than aluminum is indicated by a serial letter before the numerical designation. The serial letters are assigned in alphabetical sequence starting with A but omitting I, O, Q, and X, the X being reserved for experimental alloys.

In the 2xx.x through 9xx.x alloy groups, the second two of the four digits in the designation have no special significance but serve only to identify the different aluminum alloys in the group. The last digit to the right of the decimal point indicates the product form: .0 indicates casting and .1 indicates ingot. Examples: Alloy 213.0 represents a casting of an aluminum alloy whose major alloying element is copper. Alloy C355.1 represents the third modification of the chemistry of an aluminum alloy ingot whose major alloying elements are silicon, copper, and magnesium.

Temper Designation System. A temper designation is used for all forms of wrought and cast aluminum alloys. The temper designation follows the alloy designation, the two letters being separated by a hyphen. Basic designations consist of letters followed by one or more digits. These designate specific sequences of basic treatments but only operations recognized as significantly influencing the characteristics of the product. Basic tempers are -F (as fabricated), -O annealed (wrought products only), -H strain-hardened (degree of hardness is normally quarter hard, half hard, three-quarters hard, and hard designated by the symbols H12, H14, H16, and H18, respectively), -W solution heat treated and -T thermally treated to produce stable tempers. Examples: 1100-H14 represents commercially pure aluminum cold rolled to half-hard properties. 2024-T6 represents an aluminum alloy whose principal major element is copper which has

TABLE 2. NOMINAL CHEMICAL COMPOSITION[1] AND TYPICAL PROPERTIES OF SOME COMMON ALUMINUM WROUGHT ALLOYS

		1100	3003	5052	2014T6[10]	2017T4[9]	2024T4[9]	6061T6[10]	7075T6[10]	6101T6[10]
Nominal chemical composition[1]		99% min. Alum.	1.2% Mn	2.5% Mg 0.25% Cr	4.4% Cu 0.8% Si 0.8% Mn 0.4% Mg	4.0% Cu 0.5% Mn 0.5% Mg	4.5% Cu 1.5% Mg 0.6% Mn	1.0% Mg 0.6% Si 0.25% Cu 0.25% Cr	5.5% Zn 2.5% Mg 1.5% Cu 0.3% Cr	0.5% Mg 0.5% Si
Tensile strength, psi	A	13,000[7]	16,000	28,000	—	—	—	—	—	—
	H	24,000[7]	29,000	42,000	70,000	62,000	68,000	45,000	83,000	32,000
Tensile strength, MPa	A	90	110	193	—	—	—	—	—	—
	H	165	200	290	483	427	469	310	572	221
Yield strength, psi[2]	A	5000	6000	13,000	—	—	—	—	—	—
	H	22,000	27,000	37,000	60,000	40,000	47,000	40,000	73,000	28,000
Yield strength, MPa	A	34	41	90	—	—	—	—	—	—
	H	152	186	255	414	276	324	276	503	193
Elongation percent in 2 in. (5.1 cm)[11]	A 45 H 15		A 40 H 10	A 30 H 8	13	22	19	17	11	15
Modulus of elasticity[3]		10	10	10.2	10.6	10.5	10.6	10	10.4	10
Brinnell hardness[8]		23–44	28–55	45–85	135	105	120	95	150	71
Melting range, °C		643–657	643–654	593–649	510–638	513–640	502–638	582–652	477–638	616–651
Melting range, °F		1190–1215	1190–1210	1100–1200	950–1180	955–1185	935–1180	1080–1250	890–1180	1140–1205
Specific gravity		2.71	2.73	2.68	2.80	2.79	2.77	2.70	2.80	2.70
Electrical resistivity[4]		2.9	3.4	4.93	4.31	5.75	5.75	4.31	5.74	3.1
Thermal conductivity[5]		0.53	A 0.46	A 0.33	0.37	0.29	0.29	0.37	0.29	0.52
SI units		221.9	A 192.6	A 138.2	154.9	121.4	121.4	154.9	121.4	217.7
Coefficient of expansion[6]		23.6	23.2	23.8	22.5	23.6	22.8	23.4	23.2	23

[1] Aluminum plus normal inpurities is the remainder.
[2] 0.2% permanent set.
[3] Multiply by 10[6].
[4] Microhms per cm (room temperature).
[5] C.g.s. units (at 100°C).
[6] Per °C (20–100°C); multiply by 10[-6].
[7] A = annealed; H = hard.
[8] 500 kg load, 10 mm ball.

[9] Solution heat-treated and naturally aged.
[10] Solution heat-treated and artifically aged.
[11] Round specimens, $\frac{1}{2}$-m diameter.
Conversion factors used: 1 psi = 6.894757×10^{-3} megapascals (MPa)
C.g.s. = (cal)(cm[2])/(sec)(cm)(°C)
SI unit = Watts/meter·°K
1 c.g.s. unit = 418.68 SI units

been solution heat treated and then artifically aged to develop stable full-strength properties of the alloy.

References

Ballance, J. B.: "Aluminum Recycling Comes of Age," *Journal of Metals*, **32**, 2, 55–62 (1980).

Brondyke, K. J.: "Energy Savings in Aluminum Production," *Industrial Heating*, **XLVI**, 7, 29–32 (1979).

Canby, T. Y.: "Aluminum the Magic Metal," *National Geographic*, **154**, 2, 186–211 (1978).

EPA: "Environmental Considerations of Selected Energy Conserving Manufacturing Process Options," VIII (Alumina/Aluminum Industry Report). EPA-600/7-76-034h, Environmental Protection Agency, Office of Research and Development, Cincinnati, Ohio (1976).

Froberg, H.: "Status and Outlook for the Aluminum Industry," *Journal of Metals*, **32**, 4, 58–59 (1980).

Mazel, J. L.: "Expansion-Minded Aluminum Producers are Eyeing Australia," *33 Metal Producing*, **18**, 1, 56–61 (1980).

Regan, R. J.: "Windfall Profits Still Elude Caribbean Bauxite," *Iron Age*, Reprint 3532, 31–33 (1980)

TABLE 3. NOMINAL CHEMICAL COMPOSITION[1] AND TYPICAL PROPERTIES OF SOME ALUMINUM CASTING ALLOYS
Properties for alloys 195, B195, 220, 355, and 356 are for the commonly used heat treatment.

	413.0[2]	B443.0[2]	208.0[3]	308.0[4]	295.0[3]	B295.0[4]	514.0[3]	518.0[2]	520.0[3]	355.0[3]	356.0[3]	380.0[2]
Nominal chemical composition	12% Si	5% Si	4% Cu 3% Si	5.5% Si 4.5% Cu	4.5% Cu 0.8% Si	4.5% Cu 2.5% Si	3.8% Mg	8% Mg	10% Mg	5% Si 1.3% Cu 0.5% Mg	7% Si 0.3% Mg	8.5% Si 3.5% Cu
Tensile strength, psi[5]	37,000	19,000	21,000	28,000	36,000	45,000	25,000	42,000	46,000	35,000	33,000	45,000
Tensile strength, MPa[5]	255	131	145	193	248	310	172	290	317	241	228	310
Yield strength, psi[5]	18,000	9000	14,000	16,000	24,000	33,000	12,000	23,000	25,000	25,000	24,000	25,000
Yield strength, MPa[5]	124	62	97	110	165	228	83	159	174	174	165	174
Elongation, percent[5]	1.8	6	2.5	2	5	5	9	7	14	2.5	4	2
Brinnell hardness[6]	—	40	55	70	75	90	50	—	75	80	70	—
Melting range, °C	574–585	577–630	521–632	—	549–646	527–627	580–640	540–621	449–621	580–627	580–610	521–588
Melting range, °F	1065–1085	1070–1165	970–1170	—	1020–1195	980–1160	1075–1185	1005–1150	840–1150	1075–1160	1075–1130	970–1090
Specific gravity	2.66	2.69	2.79	2.79	2.81	2.78	2.65	2.53	2.58	2.70	2.68	2.76
Electrical resistivity	4.40	4.66	5.56	4.66	4.66	3.45	4.93	7.10	8.22	4.79	4.42	6.50
Thermal conductivity[7]	0.37	0.35	0.29	0.34	0.35	0.45	0.33	0.24	0.21	0.34	0.36	0.26
SI units	154.9	146.5	121.4	142.4	146.5	188.4	138.2	100.5	87.9	142.4	150.7	108.9
Coefficient of expansion[8]	20.0	22.8	22.8	22.7	23.9	22.8	24.8	24.0	25.4	22.8	22.8	20.0

[1] Remainder is aluminum plus minor impurities.
[2] Die cast.
[3] Sand cast.
[4] Permanent mold cast.
[5] For separately cast test bars.
[6] 500 kg/load, 10 mm. ball.

[7] C.g.s. units.
[8] Multiply by 10[-6]. Per °C, for temperature range 20 to 200°C.
Conversion factors used: 1 psi = 6.894757×10^{-3} megapascals (MPa)
SI unit = Watts/meter·°K
1 c.g.s. unit = 418.68 SI units

Staff: "Superplastic Aluminum Alloy Contains Zinc and Calcium," *Metal Progress*, **116**, 6, 73 (1979).

Staff: "Alcoa Pushes Ahead on Aluminum Process," *Chemical & Engineering News*, **57**, 23, 8 (1979).

Staff: "Successful Ocean Energy Test Stirs Interest," *Chemical & Engineering News*, **58**, 18, 24 (1980).

—S. John Sansonetti, Reynolds Metals Company, Richmond, Virginia.

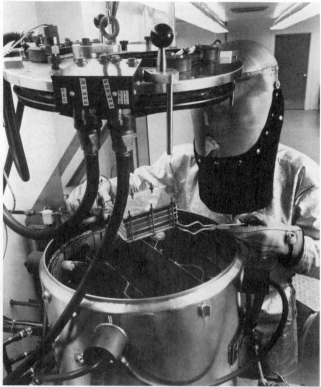

Fig. 5. Operator placing experimental assembled heat exchanger, fabricated of special aluminum alloys which respond to brazing, into vacuum tank. After pressure is reduced to 10^{-5} millimeters, radiant heaters are turned on to obtain a temperature of 1000°F (593°C), whereupon unit is instantly brazed. (*Reynolds Metals Company, Research and Development Laboratories*)

ALUNITE.   The mineral alunite, $KAl_3(SO_4)_2(OH)_6$, is a basic hydrous sulfate of aluminum and potassium; a variety called natroalunite is rich in soda. Alunite crystallizes in the hexagonal system and forms rhombohedrons with small angles, hence resembling cubes. It may be in fibrous or tabular forms, or massive. Hardness, 3.5–4; sp gr, 2.58–2.75; luster, vitreous to pearly; streak white; transparent to opaque; brittle; color, white to grayish or reddish.

Alunite is commonly associated with acid lavas due to sulfuric vapors often present; it may occur around fumaroles or be associated with sulfide ore bodies. It has been used as a source of potash. Alunite is found in Czechoslovakia, Italy, France, and Mexico; in the United States, in Colorado, Nevada, and Utah.

Alunite is also known as *alumstone*.

AMALGAM.   1. An alloy containing mercury. Amalgams are formed by dissolving other metals in mercury, when combination takes place often with considerable evolution of heat. Amalgams are regarded as compounds of mercury with other metals, or as solutions of such compounds in mercury. It has been demonstrated that products which contain mercury and another metal in atomic proportions may be separated from amalgams. The most commonly encountered amalgams are those of gold and silver. See also **Gold; Mercury; Silver.**

2. A naturally occurring alloy of silver with mercury, also referred to as mercurian silver, silver amalgam, and argental mercury. The natural amalgam crystallizes in the isometric system; hardness, 3–3.5; sp gr, 13.75–14.1; luster, metallic; color, silver-white; streak, silver-white; opaque. Amalgam is found in Bavaria, British Columbia, Chile, Czechoslovakia, France, Nor-

TABLE 4.   DESIGNATIONS FOR WROUGHT ALUMINUM ALLOY GROUPS

		Alloy No.
Aluminum—99.00% minimum and greater		1xxx
Major Alloying Element		
Aluminum	Copper	2xxx
Alloys	Manganese	3xxx
grouped	Silicon	4xxx
by Major	Magnesium	5xxx
Alloying	Magnesium and Silicon	6xxx
Elements	Zinc	7xxx
	Other Element	8xxx
Unused Series		9xxx

(1) For codification purposes an alloying element is any element which is intentionally added for any purpose other than grain refinement and for which minimum and maximum limits are specified.

(2) Standard limits for alloying elements and impurities are expressed to the following places:

Less than 1/1000%	0.000X
1/1000 up to 1/100%	0.00X
1/100 up to 1/10%	
Unalloyed aluminum made by a refining process	0.0XX
Alloys and unalloyed aluminum not made by a refining process	0.0X
1/10 through 1/2%	0.XX
Over 1/2%	0.X, X.X, etc.

TABLE 5.   DESIGNATIONS FOR CAST ALUMINUM ALLOY GROUPS

		Alloy No.
Aluminum	99.00% minimum and greater	1xx.x
	Major Alloy Element	
	Copper	2xx.x
Aluminum	Silicon, with added Copper and/or Magnesium	3xx.x
Alloys	Silicon	4xx.x
grouped	Magnesium	5xx.x
by Major	Zinc	7xx.x
Alloying	Tin	8xx.x
Elements	Other Element	9xx.x
Unused Series		6xx.x

(1) For codification purposes an alloying element is any element which is intentionally added for any purpose other than grain refinement and for which minimum and maximum limits are specified.

(2) Standard limits for alloying elements and impurities are expressed to the following places:

Less than 1/1000%	0.000X
1/1000 up to 1/100%	0.00X
1/100 up to 1/10%	
Unalloyed aluminum made by a refining process	0.0XX
Alloys and unalloyed aluminum not made by a refining process	0.0X
1/10 through 1/2%	0.XX
Over 1/2%	0.X, X.X, etc.

way, and Spain. In some areas, it is found in the oxidation zone of silver deposits and as scattered grains in cinnabar ores.

AMBER. Amber is a fossil resin which has been known since early times because of its property of acquiring an electric charge when rubbed. In modern times it has been used largely in the making of beads, cigarette holders, and trinkets. Its amorphous non-brittle nature permits it to be carved easily and to acquire a very smooth and attractive surface. Amber is soluble in various organic solvents, such as ethyl alcohol and ethyl ether.

It occurs in irregular masses showing a conchoidal fracture. Hardness, 2.25; sp gr, 1.09; luster, resinous; color, yellow to reddish or brownish; it may be cloudy. Some varieties will exhibit fluorescence. Amber is transparent to translucent, melts between 250 and 300°C.

Amber has been obtained for over 2,000 years from the lignite-bearing Tertiary sandstones on the coast of the Baltic Sea from Danzig to Memel; also from Denmark, Sweden and the other Baltic countries. Sicily furnishes a brownish-red amber that is fluorescent.

The association of amber with lignite or other fossil woods, as well as the beautifully preserved insects that are occasionally in it, is ample proof of its organic origin.

AMBERGRIS. A fragrant waxy substance formed in the intestine of the sperm whale and sometimes found floating in the sea. It has been used in the manufacture of perfumes to increase the persistence of the scent.

AMBLYGONITE. A rather rare compound of fluorine, lithium, aluminum, and phosphorus $(Li,Na)AlPO_4(F,OH)$. It crystallizes in the triclinic system; hardness, 5–5.6; sp gr 3.08; luster, vitreous to greasy or pearly; color, white to greenish, bluish, a yellowish or grayish; streak, white; translucent to sub-transparent.

Amblygonite occurs in pegmatite dikes and veins associated with other lithium minerals. It is used as a source of lithium salts. The name is derived from two Greek words meaning blunt and angle, in reference to its cleavage angle of 75°30'.

Amblygonite is found in Saxony; France; Australia; Brazil; Varutrask, Sweden; Karibibe, W.W. Africa; and in the United States in Maine, Connecticut, South Dakota (Black Hills), and California (Pala).

AMERICIUM. Chemical element symbol Am, at no. 95, at. wt. 243 (mass number of most stable isotope), radioactive metal of the actinide series, also one of the transuranium elements. All isotopes of americium are radioactive; all must be produced synthetically. The element was discovered by G. T. Seaborg and associates at the Metallurgical Laboratory of the University of Chicago in 1945. At that time, the element was obtained by bombarding uranium-238 with helium ions to produce ^{241}Am which has a half-life of 475 years. Subsequently, ^{241}Am has been produced by bombardment of plutonium-241 with neutrons in a nuclear reactor by the series of nuclear reactions:

$$^{239}Pu + n \longrightarrow {}^{240}Pu + \gamma$$
$$^{240}Pu + n \longrightarrow {}^{241}Pu + \gamma$$
$$^{241}Pu + n \longrightarrow {}^{241}Am + e^-$$

Of the isotopes, ^{243}Am is an alpha emitter with a half-life of 7950 years; others have shorter half-lives. Known isotopes include ^{237}Am, ^{238}Am, ^{240}Am, ^{241}Am, ^{242}Am, ^{244}Am, ^{245}Am, and ^{246}Am. Electronic configuration is $1s^2 2s^2 2p^6 3s^2 3p^6 3d^{10} 4s^2 4p^6 4d^{10} 4f^{14} 5s^2 5p^6 5d^{10} 5f^7 6s^2 6p^6 7s^2$. Ionic radii are: Am^{4+}, 0.85 Å; Am^{3+}, 1.00 Å.

This element exists in acidic aqueous solution in the (III), (IV), (V), and (VI) oxidation states with the ionic species probably corresponding to Am^{3+}, Am^{4+}, AmO_2^+, and AmO_2^{2+}. The oxidation potentials in acidic aqueous solution are summarized in the following diagram in which the americium(IV)-americium(V) and americium(III)-americium(V) couples are calculated from the others.

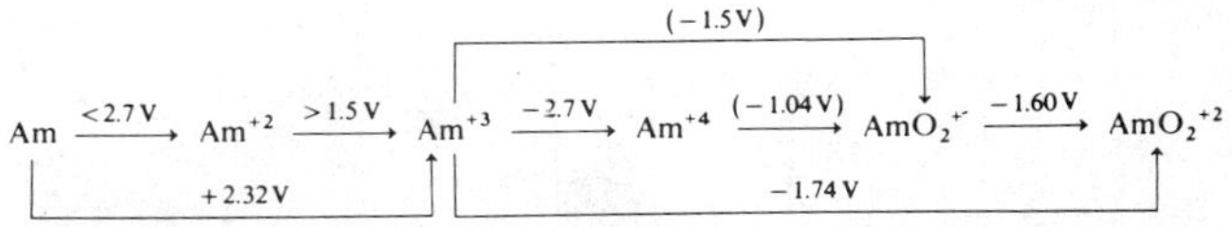

The colors of the ions are: Am^{3+}, pink; Am^{4+}, rose; AmO_2^+, yellow; and AmO_2^{2+}, rum-colored.

It can be seen that the (III) state is highly stable with respect to disproportionation in aqueous solution and is extremely difficult to oxidize or reduce. There is evidence for the existence of the (II) state since tracer amounts of americium have been reduced by sodium amalgam and precipitated with barium chloride or europium sulfate as carrier. The (IV) state is very unstable in solution: the potential for americium(III)-americium(IV) was determined by thermal measurements involving solid AmO_2. Americium can be oxidized to the (V) or (VI) state with strong oxidizing agents, and the potential for the americium(V)-americium(VI) couple was determined potentiometrically. The value of the potential for the americium(III)-americium(VI) couple has been measured as -1.9 V, but the value of -1.74 V/V is used in the above potential diagram in order to make the known disproportionation of americium(V) consistent with the other more accurately measured potentials. The value for the americium-americium(III) couple is calculated from heat of solution measurements using estimations for the entropy change.

In its precipitation reactions americium(III) is very similar to the other tripositive actinide elements and to the rare earth elements. Thus the fluoride and the oxalate are insoluble and the phosphate and iodate are only moderately soluble in acid solution, whereas the nitrates, halides, sulfates, sulfides, and perchlorates are all soluble. Americium(VI) can be precipitated with sodium acetate giving crystals isostructural with sodium uranyl acetate,

$$NaUO_2(C_2H_3O_2)_3 \cdot xH_2O$$

and the corresponding neptunium and plutonium compounds.

Of the hydrides of americium, both AmH_2 and Am_4H_{15} are black and cubic.

When americium is precipitated as the insoluble hydroxide from aqueous solution and heated in air, a black oxide is formed which corresponds almost exactly to the formula AmO_2. This may be reduced to Am_2O_3 through the action of hydrogen at elevated temperatures. The AmO_2 has the cubic fluorite type structure, isostructural with UO_2, NpO_2, and PuO_2. The sesquioxide, Am_2O_3 is allotropic, existing in a reddish brown and a tan form, both hexagonal. As in the case of the preceding actinide elements, oxides of variable composition between $AmO_{1.5}$ and AmO_2 are formed depending upon the conditions.

All four of the trihalides of americium have been prepared and identified. These are prepared by methods which are similar to those used in the preparation of the trihalides of other actinide elements. AmF_3 is pink and hexagonal, as in $AmCl_3$; $AmBr_3$ is white and orthorhombic; while a tetrafluoride, AmF_4 is tan and monoclinic.

In research at the Institute of Radiochemistry, Karlsruhe, West Germany during the early 1970s, investigators prepared alloys of americium with platinum, palladium, and iridium. These alloys were prepared by hydrogen reduction of the americium oxide in the presence of finely divided noble metals according to:

$$Am_2O_3 + 10Pt \xrightarrow[1100°C]{H_2} 2AmPt_5 + H_2O$$

The reaction is called a *coupled reaction* because the reduction of the metal oxide can be done only in the presence of noble metals. The hydrogen must be extremely pure, with an oxygen content of less than 10^{-25} torr.

References

Asprey, L. B., Stephanou, W. E., and R. A. Penneman: "Hexavalent Americium," *Amer. Chem. Soc. J.*, **73**, 5715–5717 (1951).
Asprey, L. B.: "New Compounds of Quadrivalent Americium, AmF$_4$, KAmF$_5$," *Amer. Chem. Soc. J.*, **76**, 2019–2020 (1954).
Cunningham, B. B.: "Chemistry of Element 96," Metallurgical Laboratory Report CS-3312, Univ. of Chicago, Chicago, 1945.
Ghiorso, A., James, R. A., Morgan, L. O., and G. T. Seaborg: "Preparation of Transplutonium Isotopes by Neutron Irradiation," *Phys. Rev.*, **78**, 4, 472 (1950).
Graf, P., et al.: "Crystal Structure and Magnetic Susceptibility of American Metal," *Amer. Chem. Soc. J.*, **78**, 2340 (1956).
Keller, C., and B. Erdmann: "Preparation and Properties of Transuranium Element—Noble Metal Alloy Phases," *Proc. of the Moscow Symposium on Chemistry of Transuranium Elements*, Pergamon, Elmsford, New York, 1976.
Roof, P. B., et al.: "High-Pressure Phase in Americium Metal," *Science*, **207**, 1353–1354 (1980).
Seaborg, G. T.: "The Chemical and Radioactive Properties of the Heavy Elements," *Chemical & Engineering News*, **23**, 2190–2193 (1945).
Seaborg, G. T., Editor: "Transuranium Elements," Dowden, Hutchinson & Ross, Stroudsburg, Pennsylvania, 1978.
Smith, J. L., and R. G. Haire: "Superconductivity of Americium," *Science*, **200**, 535–537 (1978).

AMETHYST. A purple- or violet-colored quartz having the same physical characteristics as quartz. The source of color is not definite but thought to be caused by ferric iron contamination. Oriental amethysts are purple corundum.

Amethysts are found in the Ural Mountains, India, Sri Lanka, the Malagasy Republic, Uruguay, Brazil, the Thunder Bay district of Lake Superior in Ontario, and Nova Scotia; in the United States, in Michigan, Virginia, North Carolina, Montana, and Maine.

The name amethyst is generally supposed to have been derived from the Greek word meaning not drunken. Pliny suggested that the term was applied because the amethyst approaches but is not quite the equivalent of a wine color.

See also **Quartz.**

AMIDES. An amide may be defined as a compound that contains the $CO \cdot NH_2$ radical, or an acid radical(s) substituted for one or more of the hydrogen atoms of an ammonia molecule (NH_3). Amides may be classified as follows: (1) *Primary amides*, contain one acyl radical, such as $-CO \cdot CH_3$ (acetyl) or $-CO \cdot C_6H_5$ (benzoyl), linked to the *amido* group ($-NH_2$). Thus, acetamide NH_2COCH_3 is a combination of the acetyl and amido groups. (2) *Secondary amides* contain two acyl radicals and the *imido* group ($=NH_2$). Diacetamide $HN(COCH_3)_2$ is an example. (3) *Tertiary amides* contain three acyl radicals attached to the N atom. Triacetamide $N(COCH_3)_3$ is an example.

A further structural analysis will show that amides may be regarded as derivatives of corresponding acids in which the amido group substitutes for the hydroxyl radical (OH) of the carboxylic group COOH. Thus, in the instance of formic acid HCOOH, the amide is $HCONH_2$ (formamide); or in the case of acetic acid CH_3COOH, the amide is CH_3CONH_2 (acetamide). Similarly, urea may be regarded as the amide of carbonic acid (theoretical) $O:C$, that is, NH_2CONH_2 (urea). The latter represents a dibasic acid in which two H atoms of the hydroxyl groups have been replaced by amido groups. A similar instance, malamide

$$NH_2CO \cdot CH_2CH(OH) \cdot CONH_2$$

is derived from the dibasic acid, malic acid,

$$OHCO \cdot CH_2CH(OH) \cdot COOH.$$

Aromatic amides, sometimes referred to as *arylamides*, exhibit the same relationship. Note the relationship of benzoic acid C_6H_5COOH with benzamide $C_6H_5CONH_2$. *Thiamides* are derived from amides in which there is substitution of the O atom by a sulfur atom. Thus, acetamide $NH_2 \cdot CO \cdot CH_3$ becomes thiacetamide $NH_2 \cdot CS \cdot CH_3$; or acetanilide $C_6H_5 \cdot NH \cdot CO \cdot CH_3$ becomes thiacetanilide $C_6H_5 \cdot NH \cdot CS \cdot CH_3$. Sulfonamides are derived from the sulfonic acids. Thus, benzenesulfonic acid $C_6H_5 \cdot SO_2 \cdot OH$ becomes benzene-sulfonamide $C_6H_5 \cdot SO_2 \cdot NH_3$. See also **Sulfonamide Drugs.**

Amides may be made in a number of ways. Prominent among them is the acylation of amines. The agents commonly used are, in order of reactivity, the acid halides, acid anhydrides, and esters. Such reactions are:

$$R'COCl + HNR_2 \rightarrow R'C(=O)NR_2 + HCl$$

$$R'C(=O)OC(=O)R' + HNR_2 \rightarrow R'C(=O)NR_2 + R'COOH$$

$$R'C(=O)OR'' + HNR_2 \rightarrow R'C(=O)NR_2 + R''OH$$

The hydrolysis of nitriles also yields amides:

$$RCN + H_2O \xrightarrow{OH} RCONH_2$$

Amides are resonance compounds, having an ionic structure for one form:

$$R-C(=O)NR_2 \qquad R-C(-O^-):N^+R_2$$

Evidence for the ionic form is provided by the fact that the carbon-nitrogen bond (1.38 Å) is shorter than a normal C—N bond (1.47 Å) and the carbon-oxygen bond (1.28 Å) is longer than a typical carbonyl bond (1.21 Å). That is, the carbon-nitrogen bond is neither a real C—N single bond nor a C=N double bond.

The amides are sharp-melting crystalline compounds and make good derivatives for any of the acyl classes of compounds, i.e., esters, acids, acid halides, anhydrides, and lactones.

Amides undergo hydrolysis upon refluxing in H_2O. The reaction is catalyzed by acid or alkali.

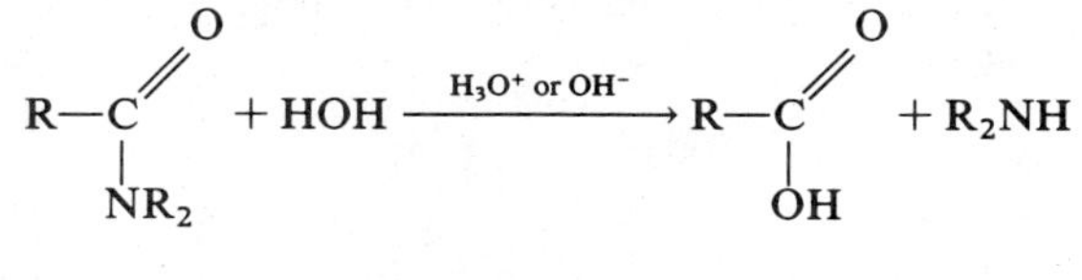

Primary amides may be dehydrated to yield nitriles.

$$R-CONH_2 + C_6H_5SO_2Cl \xrightarrow[70°]{pyridine} R-CN$$
$$+ C_6H_5SO_3H + HCl$$

The reaction is run in pyridine solutions.

Primary and secondary amides of the type $RCONH_2$ and

RCONHR react with nitrous acid in the same way as do the corresponding primary and secondary amines.

$$RCONH_2 + HONO \rightarrow RCOOH + N_2 + HOH$$

$$RCONHR + HONO \rightarrow RCON(NO)R + HOH$$

When diamides having their amide groups not far apart are heated, they lose ammonia to yield imides. See also **Imides.**

AMINATION. The process of introducing the amino group (—NH$_2$) into an organic compound is termed *amination.* An example is the reduction of aniline $C_6H_5 \cdot NH_2$ from nitrobenzene $C_6H_5 \cdot NO_2$. The reduction may be accomplished with iron and HCl. Only about 2% of the calculated amount of acid (to produce H$_2$ by reaction with iron) is required because of the fact that H$_2$O plus iron in the presence of ferrous chloride solution (ferrous and chloride ions) functions as the primary reducing agent. Such groups as nitroso (—NO), hydroxylamine (—NH·NH—), and azo (—N:N—) also yield amines by reduction. Amination also may be effected by the use of NH$_3$, a process sometimes referred to as *ammonolysis.* An example is the production of aniline from chlorobenzene:

$$C_6H_5Cl + NH_3 \rightarrow C_6H_5 \cdot NH_2 + HCl.$$

The reaction proceeds only under high pressure. In the ammonolysis of benzenoid sulfonic acid derivatives, an oxidizing agent is added to prevent the formation of soluble reduction products, such as NaNH$_4$SO$_3$, which commonly form. Oxygen-function compounds also may be subjected to ammonolysis: (1) methanol plus aluminum phosphate catalyst yields mono-, di-, and trimethylamines; (2) β-naphthol plus sodium ammonium sulfite catalyst (Bucherer reaction) yields β-naphthylamine; (3) ethylene oxide yields mono-, di-, and triethanolamines; (4) glucose plus nickel catalyst yields glucamine; and (5) cyclohexanone plus nickel catalyst yields cyclohexylamine.

AMINES. An amine is a derivative of NH$_3$ in which there is a replacement of one or more of the H atoms of NH$_3$ by an alkyl group, such as —CH$_3$ (methyl) or —C$_2$H$_5$ (ethyl); or by an aryl group, such as —C$_6$H$_5$ (phenyl) or —C$_{10}$H$_7$ (naphthyl). Mixed amines contain at least one alkyl and one aryl group, as exemplified by methylphenylamine CH$_3 \cdot$N(H)$\cdot$C$_6$H$_5$. When one, two, and three H atoms are thus replaced, the resulting amines are known as *primary, secondary,* and *tertiary,* respectively. Thus methylamine CH$_3$NH$_2$ is a primary amine; dimethylamine (CH$_3$)$_2$NH is a secondary amine; and trimethylamine (CH$_3$)$_3$N is a tertiary amine. Secondary amines sometimes are called *imines.* Tertiary amines are sometimes called *nitriles.*

Quaternary amines consist of four alkyl or aryl groups attached to an N atom and, therefore, may be considered substituted ammonium bases. Commonly, they are referred to as *quaternary ammonium compounds.* An example is tetramethyl ammonium iodide:

$$\begin{array}{c} H_3C \qquad\qquad CH_3 \\ \diagdown \qquad\qquad \diagup \\ N - CH_3 \\ \diagup \qquad\qquad \diagdown \\ I \qquad\qquad CH_3 \end{array}$$

The amines and quaternary ammonium compounds, exhibiting such great versatility for forming substitution products, are important starting and intermediate materials for organic syntheses. Important industrially are the ethanolamines which are excellent absorbents for certain materials. Hexamethylene tetramine is a high-production material used in plastics production. See **Hexamine.** Phenylamine (aniline), although not as important industrially as it was some years ago, still is produced in quantity. Melamine is produced on a large scale and is the base for a series of important resins. See **Melamine.** There are numerous amines and quaternary ammonium compounds that are not well known because of their importance as intermediates rather than as final products. Examples of these include acetonitrile and acrylonitrile.

Primary amines react (1) with nitrous acid, yielding (a) with alkylamine, nitrogen gas plus alcohol, (b) with warm arylamine, nitrogen gas plus phenol (the amino-group of primary amines is displaced by the hydroxyl group to form alcohol or phenol) (c) with cold arylamine, diazonium compounds, (2) with acetyl chloride or benzoyl chloride, yielding substituted amides, thus, ethylamine plus acetyl chloride forms N-ethylacetamide $C_2H_5NHOCCH_3$, (3) with benzene-sulfonyl chloride $C_6H_5SO_2Cl$, yielding substituted benzene sulfonamides, thus, ethylamine forms N-ethylbenzenesulfonamide $C_6H_5SO_2$—NHC$_2$H$_5$, soluble in sodium hydroxide, (4) with chloroform CHCl$_3$ with a base, yielding isocyanides (5) with HNO$_3$ (concentrated), yielding nitramines, thus, ethylamine reacts to form ethylnitramine C_2H_5—NHNO$_2$.

Secondary amines react (1) with nitrous acid, yielding nitrosamines, yellow oily liquids, volatile in stream, soluble in ether. The secondary amine may be recovered by heating the nitrosamine with concentrated HCl, or hydrazines may be formed by reduction of the nitrosamines, e.g., methylaniline from methylphenylnitrosamine $CH_3(C_6H_5)NNO$, reduction yielding unsymmetrical methylphenylhydrazine, $CH_3(C_6H_5)NHNH_2$, (2) with acetyl or benzoyl chloride, yielding substituted amides, thus, diethylamine plus acetyl chloride to form N, N-diethylacetamide $(C_2H_5)_2$—NOCCH$_3$, (3) with benzene sulfonyl chloride, yielding substituted benzene sulfonamides, thus, diethylamine reacts to form N,N-diethylbenzenesulfonamide $C_6H_5SO_2N$-$(C_2H_5)_2$, insoluble in NaOH.

Tertiary amines do not react with nitrous acid, acetyl chloride, benzoyl chloride, benzenesulfonyl chloride, but react with alkyl halides to form quaternary ammonium halides, which are converted by silver hydroxide to quaternary ammonium hydroxides. Quaternary ammonium hydroxides upon heating yield (1) tertiary amine plus alcohol (or, for higher members, olefin plus water). Tertiary amines may also be formed (2) by alkylation of secondary amines, e.g., by dimethyl sulfate, (3) from amino acids by living organisms, e.g., decomposition of fish in the case of trimethylamine.

AMINO ACIDS. The scores of proteins which make up about one-half of the dry weight of the human body and are so vital to life functions are made up of a number of amino acids in various combinations and configurations. The manner in which the complex protein structures are assembled from amino acids is described in the entry on **Protein** in this Encyclopedia. For some users of this book, it may be helpful to scan that portion of the protein entry which deals with the chemical nature of proteins prior to considering the details of this immediate entry on amino acids.

Although the proteins resulting from amino acid assembly are ultimately among the most important chemicals in the animal body (as well as plants), the infrastructure, so-called, of the proteins is dependent upon the amino acid building blocks. Although there are many hundreds of amino acids, only about 20 of these are considered very important to living processes, and of these some 6–10 are classified as essential. Another 3 or 4 may be classified as quasi-essential, and 10–12 may be categorized as nonessential. As more is learned about the fundamentals of protein chemistry, the scientific importance attached to specific amino acids varies. Usually, as the learning process continues, the findings tend to increase the importance of specific amino acids. Actually, the words *essential* and *nonessential* are not very good choices for naming categories of amino acids. Generally, those amino acids which the human body cannot synthesize at all or at a rate commensurate with its needs are called essential amino acids (EAA). In other words, for the growth and maintenance of a normal healthy body, it is essential that these amino acids be ingested as part of the diet and in the necessary quantities. To illustrate some of the indefinite character of amino acid nomenclature, some authorities classify histidine as an essential amino acid; others do not. The fact is that histidine is essential for the normal growth of the human infant, but to date it is not regarded as essential for adults. By extension of the preceding explanation, the term nonessential is taken to mean those amino acids that are really synthesized in the body and hence need not be present in food intake. This classification of amino acids, although amenable to change as the results of new findings, has been quite convenient in planning the dietary needs of people as well as of farm animals and pets, and also in terms of those plants that are of economic importance. The classification has been particularly helpful in planning the specific nutritional content of food substances involved in various aid and related programs for the people in needy and underdeveloped areas of the world.

Food Fortification with Amino Acids. In a report of the World Health Organization, the following observation has been made: "To determine the quality of a protein, two factors have to be distinguished, namely, the proportion of essential to nonessential amino acids and, secondly, the relative amounts of the essential amino acids. . . . The best pattern of essential amino acids for meeting human requirements was that found in whole egg protein or human milk, and comparisons of protein quality should be made by reference to the essential amino acid patterns of either of these two proteins." The ratio of each essential amino acid to the total sum is given for hen's egg and human and cow's milk in Table 1.

TABLE 1.   REPRESENTATIVE ESSENTIAL AMINO ACID PATTERNS

A/E* RATIO (Milligrams per gram of total essential amino acids)

	Hen's Egg (Whole)	Human Milk	Cow's Milk
Total "aromatic" amino acids	195	226	197
Phenylalanine	(114)	(114)	(97)
Tyrosine	(81)	(112)	(100)
Leucine	172	184	196
Valine	141	147	137
Isoleucine	129	132	127
Lysine	125	128	155
Total "S"	107	87	65
Cystine	(46)	(43)	(17)
Methionine	(61)	(44)	(48)
Threonine	99	99	91
Tryptophan	31	34	28

SOURCE: World Health Organization; FAO Nutrition Meeting Report Series, No. 37, Geneva, 1965.

* A/E Ratio equals ten times percentage of single essential amino acid to the total essential amino acids contained.

In the human body, tyrosine and cysteine can be formed from phenylalanine and methionine, respectively. The reverse transformations do not occur. Human infants have an ability to synthesize arginine and histidine in their bodies, but the speed of the process is slow compared with requirements.

Several essential amino acids have been shown to be the limiting factor of nutrition in plant proteins. In advanced countries, the ratio of vegetable proteins to animal proteins in foods is 1.4:1. In underdeveloped nations, the ratio is 3.5:1, which means that people in underdeveloped areas depend upon vegetable proteins. Among vegetable staple foods, wheat easily can be fortified. It is used as flour all over the world. L-Lysine hydrochloride (0.2%) is added to the flour. Wheat bread fortified with lysine is used in several areas of the world; in Japan it is supplied as a school ration.

The situation of fortification in rice is somewhat more complex. Before cooking, rice must be washed (polished) with water. In some countries, the cooking water is allowed to boil over or is discarded. This significant loss of fortified amino acids must be considered. L-Lysine hydrochloride (.02%) and L-threonine (0.1%) are shaped like rice grain with other nutrients and enveloped in a film. The added materials must hold the initial shape and not dissolve out during boiling, but be easily freed of their coating in the digestive organs.

The amino acids are arranged in accordance with essentiality in Table 2. Each of the four amino acids at the start of the table are all limiting factors of various vegetable proteins. Chick feed usually is supplemented with fish meals, but where the latter is in limited supply, soybean meals are substituted. The demand for DL-methionine, limiting amino acid in soybean meals, is now increasing. When seed meals, such as corn and sorghum, are used as feeds for chickens or pigs, L-lysine hydrochloride must be added for fortification. Lysine production is increasing upward to the level of methionine.

Early Research and Isolation of Amino Acids. Because of such rapid strides made within the past few decades in biochemistry and nutrition, these sciences still have a challenging aura about them. But, it is interesting to note that the first two natural amino acids were isolated by Braconnet in 1820. As shown in Table 3, these two compounds were glycine and leucine. Bopp isolated tyrosine from casein in 1849. Additional amino acids were isolated during the 1800s, but the real thrust into research in this field commenced in the very late 1800s and early 1900s with the work of Emden, Fischer, Mörner, and Hopkins and Cole. It is interesting to observe that Emil Fischer (1852–1919), German chemist and pioneer in the fields of purines and polypeptides, isolated three of these important compounds, namely, proline from gelatin in 1901, valine from casein in 1901, and hydroxyproline from gelatin in 1902. As an understanding of the role of amino acids in protein formation and of the function of proteins in nutrition progressed, the pathway was prepared for further isolation of amino acids. For example, in 1907, a combined committee representing the American Society of Biological Chemists and the American Physiological Society, proposed a formal classification of proteins into three major categories: (1) simple proteins, (2) conjugated proteins, and (3) derived proteins. The last classification embraces all denatured proteins and hydrolytic products of protein breakdown and no longer is considered as a general class.

Very approximate annual worldwide production of amino acids, their currant method of preparation (not exclusive), and general characteristics are given in Table 2.

The *isoelectric point* is very important in the preparation and separation of amino acids and proteins. Protein solubility varies markedly with pH and is at a minimum at the isoelectric point.

TABLE 2.   IMPORTANT NATURAL AMINO ACIDS AND PRODUCTION

Amino Acid	World Annual Production, tons	Present Mode of Manufacture	Characteristics
ESSENTIAL AMINO ACIDS			
DL-Methionine	10^4	Synthesis from acrolein and mercaptan	First limiting amino acid for soybean
L-Lysine·HCl	10^3	Fermentation (AM)*	First limiting amino acid for all cereals
L-Threonine	10	Fermentation (AM)	Second limiting amino acid for rice
L-Tryptophan	10	Synthesis from acrylonitrile and resolution	Second limiting amino acid for corn
L-Phenylalanine	10	Synthesis from phenyl-acetaldehyde and resolution	
L-Valine	10	Fermentation (AM)	Rich in plant protein
L-Leucine	10	Extraction from protein	
L-Isoleucine	10	Fermentation (WS)**	Deficient in some cases
QUASI-ESSENTIAL AMINO ACIDS			
L-Arginine·HCl	10^2	Synthesis from L-ornithine Fermentation (AM)	Essential to human infants
L-Histidine·HCl	10	Extraction from protein	
L-Tyrosine	10	Enzymation of phenol and serine	Limited substitute for phenylalanine
L-Cysteine L-Cystine	10	Extraction from human hair	Limited substitute for methionine
NONESSENTIAL AMINO ACIDS			
L-Glutamic acid	10^5	Fermentation (WS) Synthesis from acrylonitrile and resolution	MSG, taste enhancer
Glycine	10^3	Synthesis from formaldehyde	Sweetener
DL-Alanine	10^2	Synthesis from acetaldehyde	
L-Aspartic acid	10^2	Enzymation of fumaric acid	Hygienic drug
L-Glutamine	10^2	Fermentation (WS)	Anti-gastroduodenal ulcer drug
L-Serine	< 10	Synthesis from glycolonitrile and resolution	Rich in raw silk
L-Proline	< 10	Fermentation (AM)	Rich in gelatin
L-Hydroxyproline	< 10	Extraction from gelatin	
L-Asparagine	< 10	Synthesis from L-aspartic acid	Neurotropic metabolic regulator
L-Alanine	< 10	Enzymation of L-aspartic acid	Rich in degummed white silk
L-Dihydroxy-phenylalanine	10^2	Synthesis from piperonal, vanillin, or acrylonitrile and resolution	Specific drug for Parkinson's disease
L-Citrulline	< 10	Fermentation (AM)	Ammonia detoxicant
L-Ornithine	< 10	Fermentation (AM)	

* AM, artificial mutant; ** WS, wild strain. MSG, monosodium glutamate.

By raising the salt concentration and adjusting pH to the isoelectric point, it is often possible to obtain a precipitate considerably enriched in the desired protein and to crystallize it from a heterogenous mixture.

Chemical Nature of Amino Acids. In a very general way, an amino acid is any organic acid which incorporates one or more amino groups. This definition includes a multitude of substances of most diverse structure. There are seemingly limitless related compounds of differing molecular size and constitution which incorporate varying kinds and numbers of functional groups. Most extensive study has centered around the relatively small group of alpha-amino acids which are combined in amide linkage to form proteins. With few exceptions, these compounds possess the general structure NH_2CHRCO_2H, where the amino group occupies a position on the carbon atom *alpha* to that of the carboxyl group, and where the side chain R may be of diverse composition and structure.

Few products of natural origin are as versatile in their behav-

TABLE 3. FIRST ISOLATION OF AMINO ACIDS

Abbreviation	Name and Formula	First Isolation and (Source)	Isoelectric Point
	Neutral Amino Acids—Aliphatic Type		
Ala	Alanine $CH_3-CH-COOH$ $\quad\quad\vert$ $\quad\quad NH_2$	1879 by Schutzenberger 1888 by Weyl (silk fibroin)	6.0
Gly	Glycine NH_2-CH_2-COOH	1820 by Braconnot (gelatin)	6.0
Ile	Isoleucine $\quad\quad\quad H$ $\quad\quad\quad\vert$ $C_2H_5-C-CH-COOH$ $\quad\quad\vert\quad\vert$ $\quad\quad H_3C\ NH_2$	1904 by Ehrlich (fibrin)	6.0
Leu	Leucine $(CH_3)_2CH-CH_2-CH-COOH$ $\quad\quad\quad\quad\quad\quad\vert$ $\quad\quad\quad\quad\quad\quad NH_2$	1820 by Braconnot (muscle fiber; wool)	6.0
Val	Valine $(CH_3)_2CH-CH-COOH$ $\quad\quad\quad\quad\vert$ $\quad\quad\quad\quad NH_2$	1901 by Fischer (casein)	6.0
	Neutral Amino Acids—Hydroxy Type		
Ser	Serine $HO-CH_2-CH-COOH$ $\quad\quad\quad\quad\vert$ $\quad\quad\quad\quad NH_2$	1865 by Cramer (sericine)	5.7
Thr	Threonine $CH_3-CH-CH-COOH$ $\quad\quad\vert\quad\vert$ $\quad\quad OH\ NH_2$	1925 by Gortner and Hoffman 1925 by Schryver and Buston (oat protein)	6.2
	Neutral Amino Acids—Sulfur-Containing Type		
Cys	Cysteine $HS-CH_2-CH-COOH$ $\quad\quad\quad\quad\vert$ $\quad\quad\quad\quad NH_2$	----------------------------	5.1
C͡ys Cys	Cystine $(-SCH_2-CH-COOH)_2$ $\quad\quad\quad\quad\vert$ $\quad\quad\quad\quad NH_2$	1899 by Mörner (horn) 1899 by Emden	4.6
Met	Methionine $CH_3-S-CH_2-CH_2-CH-COOH$ $\quad\quad\quad\quad\quad\quad\quad\quad\vert$ $\quad\quad\quad\quad\quad\quad\quad\quad NH_2$	1922 by Mueller (casein)	5.7

(continued)

TABLE 3. FIRST ISOLATION OF AMINO ACIDS (continued)

ABBREVIATION	NAME AND FORMULA	FIRST ISOLATION AND (SOURCE)	ISOELECTRIC POINT
	Neutral Amino Acids—Amide Type		
Asn	Asparagine $H_2NOC-CH_2-CH-COOH$ $\mid$ NH_2	1932 by Damodaran (edestin)	5.4
Gln	Glutamine $H_2NOC-CH_2-CH_2-CH-COOH$ $\mid$ NH_2	1932 by Damodaran, Jaaback and Chibnall (gliadin)	5.7
	Neutral Amino Acids– Aromatic Type		
Phe	Phenylalanine 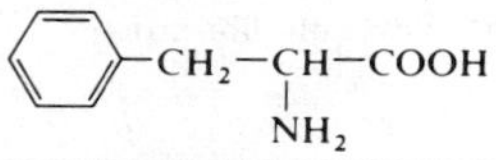	1881 by Schulze and Barbieri (lupine seedlings)	5.5
Trp	Tryptophan 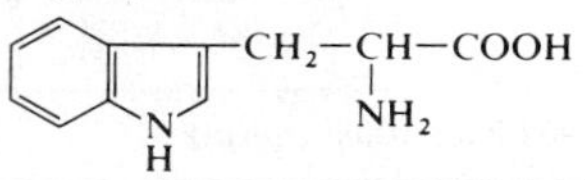	1902 by Hopkins and Cole (casein)	5.9
Tyr	Tyrosine 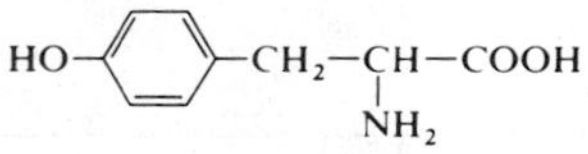	1849 by Bopp (casein)	5.7
	Acidic Amino Acids		
Asp	Aspartic Acid $HOOC-CH_2-CH-COOH$ $\mid$ NH_2	1868 by Ritthausen (conglutin; legumin)	2.8
Glu	Glutamic acid $HOOC-CH_2-CH_2-CH-COOH$ $\mid$ NH_2	1866 by Ritthausen (gluten-fibrin)	3.2
	Basic Amino Acids		
Arg	Arginine $H_2N-C-NH(CH_2)_3-CH-COOH$ $\;\;\; \| \;\;\;\;\;\;\;\;\;\;\;\; \|$ $\;\;\; NH \;\;\;\;\;\;\;\;\;\;\;\;\; NH_2$	1895 by Hedin (horn)	11.2
His	Histidine 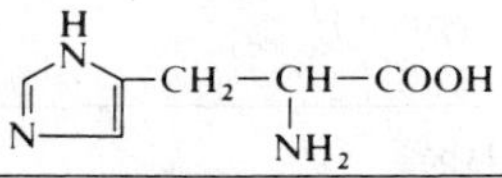	1896 by Kossel (sturine) 1896 by Hedin (various protein hydrolysates)	7.6
Lys	Lysine $H_2N-(CH_2)_4CH-COOH$ $\mid$ NH_2	1889 by Dreschel (casein)	9.7
	Imino Acids		
Hyp	Hydroxyproline 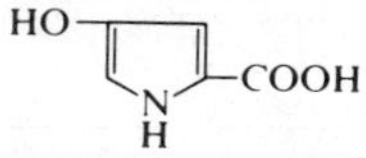	1902 by Fischer (gelatin)	5.8
Pro	Proline 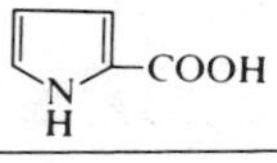	1901 by Fischer (casein)	6.3

ior and properties as are the amino acids, and few have such a variety of biological duties to perform. Among their general characteristics would be included:

(a) Water-soluble and amphoteric electrolytes, with the ability to form acid salts and basic salts and thus act as buffers over at least two ranges of pH (hydrogen ion concentration).

(b) Dipolar ions of high electric moment with a considerable capacity to increase the dielectric constant of the medium in which they are dissolved.

(c) Compounds with reactive groups capable of a wide range of chemical alterations leading readily to a large variety of degradation, synthetic, and transformation products, such as esters, amides, amines, anhydrides, polymers, polypeptides, diketopiperazines, hydroxy acids, halogenated acids, keto acids, acylated acids, mercaptans, shorter- or longer-chained acids, and pyrrolidine and piperidine ring forms.

(d) Indispensable components of the diet of all animals including humans.

(e) Participants in crucial metabolic reactions on which life depends, and substrates for a variety of specific enzymes *in vitro*.

(f) Binders of metals of many kinds.

(g) Absorbers of ultraviolet and infrared radiation within specific ranges of wavelength.

(h) Possessors with one exception of optical rotatory power related to the configuration of asymmetric centers. The exception is glycine.

(i) Essential constituents of protein molecules whose biological and chemical specificities are determined in part by the number, distribution, and spatial interrelations of the amino acids of which they are composed.

They reveal at once uniformity and diversity. Uniformity, because with rare exceptions they are α-amino acids with all the physical consequences which flow from this fact and because, for those that are constituents of proteins and hence of living tissues, the same optical configuration at the α-carbon atom is common to all. Diversity, because each possesses a different side chain which confers upon it unique properties distinguishing it physically, chemically, and biologically from the others. In this duality, the array of the amino acids is a partial reflection of the larger biological world which is "always the same and yet always different."

Optical Properties. With the exception of glycine ($NH_2CH_2CO_2H$) and amino-malonic acid [$NH_2CH(CO_2H)_2$], all α-amino acids which are classifiable according to the general formula previously given exist in at least two different optically isometric forms. The optical isomers of a given amino acid possess identical empirical and structural formulas, and are indistinguishable from each other on the basis of their chemical and physical properties, with the singular exception of their effect on plane polarized light. This may be illustrated with the two optically active forms as shown by

$$
\begin{array}{ccccc}
\text{COOH} & & & \text{HOOC} \\
| & & & | \\
\text{H}_2\text{N---C---H} & & & \text{H---C---NH}_2 \\
| & & & | \\
\text{R} & & & \text{R} \\
\text{L-form} & & & \text{D-form}
\end{array}
$$

mirror
image

One form (L-form) exhibits the ability to rotate the plane of polarization of plane polarized light to the left (levorotatory),

whereas the other form (D-form) rotates the plane to the right (dextrorotatory). Although the direction of optical rotation exhibited by these optically active forms is different, the magnitude of their respective rotations is the same. If equal amounts of *dextro* and *levo* forms are admixed, the optical effect of each isomer is neutralized by the other, and an optically inactive product known as a *racemic modification* or *racemate* is secured.

The ability of the alanine molecule, for example, to exist in two stereoisomeric forms can be attributed to the fact that the α-carbon atom of this compound is attached to four different groups which may vary in their three-dimensional spatial arrangement. Compounds of this type do not possess complete symmetry when viewed from a purely geometrical standpoint and hence are generally referred to as *asymmetric*. As a consequence of this molecular asymmetry, the four covalent bonds of an asymmetric carbon atom can be aligned in a manner such that a regular tetrahedron is formed by the straight lines connecting their ends. Hence, two different tetrahedral arrangements of the groups about the asymmetric carbon atom can be devised so that these structures relate to one another as an object relates to its mirror image, or as the right hand relates to the left hand. Molecules of this type are endowed with the property of optical activity and together with their nonsuperimposable mirror images, are generally referred to as *enantiomorphs*, *enantiomers*, *antimers*, or *optical antipodes*.

Classification. In accordance with the structure of the R-group, the amino acids of primary importance can be classified into eight groups. Additional amino acids composing protein are not included in this classification, because they occur infrequently. See Table 4.

Normally, amino acids exist as dipolar ions. $RCH(NH_3^+)COO^-$, in a neutral state, where both amino and carboxyl groups are ionized. The dipolar form, $RCH(NH_2)COOH$ may be considered, but the dipolar form predominates for the usual

TABLE 4. STRUCTURAL CLASSIFICATION OF AMINO ACIDS

NEUTRAL AMINO ACIDS

Aliphatic-type	*Hydroxy-type*	*Sulfur-containing*
Glycine	Serine	Cysteine
Alanine	Threonine	Cystine
Valine		Methionine
Leucine		
Isoleucine		
Amide-type	*Aromatic-type*	
Asparagine	Phenylalanine	
Glutamine	Tryptophan	
	Tyrosine	

ACIDIC AMINO ACIDS

Aspartic acid
Glutamic acid

BASIC AMINO ACIDS

Histidine
Lysine
Arginine

IMINO ACIDS

Proline
Hydroxyproline

monoamino monocarboxylic acid and it is estimated that these forms occur 10^5 to 10^6 times more frequently than the nonpolar forms. Amino acids decompose thermally at what might be considered a relatively high temperature (200–300°C). The compounds are practically insoluble in organic solvents, have low vapor pressure, and do not exhibit a precisely defined melting point.

The ionic states of a simple α-amino acid are given by

$$RCH(NH_3^+)COOH \underset{+H^+}{\overset{-H+(K_1)}{\rightleftharpoons}}$$
(Cationic form; acidic)

$$RCH(NH_3^+)COO^- \underset{+H^+}{\overset{-H+(K_2)}{\rightleftharpoons}} RCH(NH_2)COO^-$$
(Dipolar form; neutral)   (Anionic form; basic)

In accordance with the change of the ionic state, dissociation constants are

$$K_1(COOH) = \frac{[H^+][RCH(NH_3^+)COO^-]}{[RCH(NH_3^+)COOH]}$$

$$K_2(NH_3^+) = \frac{[H^+][RCH(NH_2)COO^-]}{[RCH(NH_3^+)COO^-]}$$

Inasmuch as pK = −log K, the values for glycine are $pK_1 = 2.34$ and $pK_2 = 9.60$ (in aqueous solution at 25°C). The homologous amino acids indicate similar values. The pH at which acidic ionization balances basic ionization is termed the *isoelectric point* (pH_I), corresponding to

$$[RCH(NH_3^+)COOH] = [RCH(NH_2)COO^-]$$

Thus, from these formulas, the pH_I is

$$pH_I = 1/2(pK_1 + pK_2)$$

Formation of Salts. Amino acids have certain characteristics of both organic bases and organic acids because they are amphoteric. As amines, the amino acids form stable salts, such as hydrochlorides or aromatic sulfonic acid salts. These are used as selective precipitants of certain amino acids. As organic acids, the amino acids form complex salts with heavy metals, the less soluble salt being used for amino acid separation.

Esters. When heated with the equivalent amount of a strong acid, usually hydrochloric acid in absolute alcohol, amino acids form esters. These are obtained as hydrochlorides.

Acylation. In alkaline solution, amino acids react with acid chlorides or acid anhydrides to form acyl compounds of the type

$$\underset{\underset{NHCOR'}{|}}{RCH-COONa}$$

Van Slyke Reaction (Deamination). With excess nitrous acid, -amino acids react to form -hydroxyl acids on a quantitative basis. Nitrogen gas is generated.

$$\underset{\underset{NH_2}{|}}{RCH-COOH} + HNO_2 \rightarrow \underset{\underset{OH}{|}}{RCH-COOH} + N_2 + H_2O$$

The reaction is completed within five minutes at room temperature. Thus, measurement of the volume of nitrogen generated can be used in amino acid determinations.

Decarboxylation. When heated with inert solvents, such as kerosene, amino acids form amines

$$\underset{\underset{NH_2}{|}}{RCH-COOH} \rightarrow \underset{\underset{NH_2}{|}}{RCH_2} + CO_2$$

Decarboxylative enzymes may react specifically with amino acids having free polar groups at the ω position. Cadaverine can be produced from lysine; histamine from histidine, and tyramine from tyrosine.

Formation of Amides. When condensed with ammonia or amines, amino acid esters form acid amides:

$$\underset{\underset{NH_2}{|}}{RCH-CONH_2}$$

Oxidation. Oxidizing agents easily decompose α-amino acids, forming the corresponding fatty acid with one less carbon number:

$$\underset{\underset{NH_2}{|}}{RCH-COOH} \overset{O}{\rightarrow} RCHO \overset{O}{\rightarrow} RCOOH + NH_3 + CO_2$$

Ninhydrin Reaction. A neutral solution of an amino acid will react with ninhydrin (triketohydrindene hydrate) by heating to cause oxidative decarboxylation. The central carbonyl of the triketone is reduced to an alcohol. This alcohol further reacts with ammonia formed from the amino acid and causes a red-purplish color. Since the reaction is quantitative, measurement of the optical density of the color produced is an indication of amino acid concentration. Imino acids, such as hydroxyproline and proline, develop a yellow color in the same type of reaction.

Maillard Reaction. In amino acids, the amino group tends to form condensation products with aldehydes. This reaction is regarded as the cause of the browning reaction when an amino acid and a sugar coexist. A characteristic flavor, useful in food preparations, is evolved along with the color in this reaction.

Ion-exchange Separations. Because amino acids are amphoteric, they behave as acids or bases, depending upon the pH of the solution. This makes it possible to adsorb amino acids dissolved in water on either a strong-acid cation exchange resin; or a strong-base anion exchange resin. The affinity varies with the amino acid and the solution pH. Ion-exchange resins are widely used in amino acid separations.

Production of Amino Acids. There are three means available for making (or separating) amino acids in large quantity lots: (1) *extraction* from natural proteins; (2) *fermentation*; and (3) chemical *synthesis*. During the early investigations of amino acids, the first method was widely used and still applies to four amino acids. See Table 3.

L-Leucine is easily extracted in quantity from almost any type of vegetable protein hydrolyzates. Cystine is extracted from the human-hair hydrolyzate. L-Histidine is obtainable from the blood of animals, but future yields may stem from fermentation inasmuch as some artificial mutants of bacteria have been discovered. Gelatin is the prime source of L-hydroxyproline.

Natural amino acids, normally not contained in proteins, but which are effective in medicine, include citrulline, ornithine, and dihydroxyphenylalanine. These are not listed in Tables 2 and 3. Citrulline (Cit) with an isoelectric point of 5.9 was isolated by Koga in 1914; by Odake in 1914; and by Wada in 1930. It has the formula

$$\underset{\underset{O}{\|}}{H_2NC-NH}-(CH_2)_3\underset{\underset{NH_2}{|}}{CH-COOH}$$

Dihydroxyphenylalanine (Dopa) with an isoelectric point of 5.5 was isolated by Torquati in 1913; and by Guggenheim in 1913. It has the formula

$$HO-\text{benzene ring}-CH_2-CH-COOH$$
$$\underset{NH_2}{|}$$

with OH at top of ring.

Ornithine (Orn) with an isoelectric point of 9.7 was isolated by Riesser in 1906 from arginine. It has the formula

$$CH_2-CH_2-CH_2-CH-COOH$$
$$\underset{NH_2}{|}\qquad\qquad\underset{NH_2}{|}$$

Fermentation Methods. Numerous microorganisms can synthesize the amino acids required to support their life from a simple carbon source and an inorganic nitrogen source, such as ammonium or nitrate salts, or nitrogen gas.

Japanese microbiologists, in 1956, first succeeded in developing industrial production of L-glutamic acid by a microbiological process. As of the present, nearly all common amino acids can be produced on a low cost industrial scale by fermentation. From microbiological studies, it has been ascertained that some microbial stains isolated from natural sources serve to excrete and accumulate a large amount of a particular amino acid in the cultural broth under carefully controlled conditions. The production of glutamic acid is produced by adding a selected bacterial strain and culturing aerobically for one to two days in a chemically defined medium which contains carbon sources, such as sugar or acetate, and nitrogen sources, such as ammonium salts. About 50% (wt) of the carbon sources can be converted to glutamate.

Genetic techniques have been used to improve the ability of microorganisms to accumulate amino acids. Several amino acids are manufactured from their direct precursors by the use of microbially produced enzymes. For example, bacterial L-aspartate β-carboxylase is used for the production of L-alanine from L-aspartic acid.

In isolating the amino acids from the fermentation broth, chromatographic separations using ion-exchange resins are the most important commercial method. Precipitation with compounds which yield insoluble salts with amino acids also are used. Purification is possible by crystallization through careful adjustment of the isoelectric point, at which point the amino acid is least soluble.

There are several laboratory-size methods for synthesizing amino acids, but few of these have been scaled up for industrial production. Glycine and DL-alanine are made by the Strecker synthesis, commencing with formaldehyde and acetaldehyde, respectively. In the Strecker synthesis, aldehydes react with hydrogen cyanide and excess ammonia to give amino nitriles which, in turn, are converted into α-amino acids upon hydrolysis.

$$RCH \xrightarrow{HCN} RCH-CN \xrightarrow{NH_3} RCH-CN \xrightarrow{NaOH} RCH-COONa$$
$$\underset{O}{\|}\qquad\underset{OH}{|}\qquad\qquad\underset{NH_2}{|}\qquad\qquad\underset{NH_2}{|}$$

The Hydantoin Process. Hydantoins are produced by reacting aldehydes with sodium cyanide and ammonium carbonate. Upon hydrolysis, α-amino acids will be yielded

$$RCHO \xrightarrow{NaCN,\ (NH_4)_2CO_3} RCH-CO \xrightarrow{NaOH}$$
$$\qquad\qquad\qquad HN\qquad NH$$
$$\qquad\qquad\qquad\quad CO$$

$$RCH-COONa + (NH_4)HCO_3$$
$$\underset{NH_2}{|}$$

The production of α-amino acids by chemical synthesis yields a mixture of DL forms. The D-form of glutamic acid has no flavor-enhancing properties and thus requires transformation into the optically active form insofar as monosodium glutamate is concerned. The three methods for separating the optical isomers are: (1) preferential inoculation method; (2) the diasteroisomer method; and (3) the acylase method.

References

Adams, C. P.: "Nutritive Value of American Foods," *Agriculture Handbook 456*, U.S. Department of Agriculture, Washington, D.C. (1975).
Kolata, G. B.: "Protein Structure: Systematic Alteration of Amino Acid Sequences," *Science*, **191**, 373 (1976).
Orr, M. L., and B. K. Watt: "Amino Acid Content of Foods," *Home Economics Research Report 4*, U.S. Department of Agriculture, Washington, D.C. (Revised periodically).
Paul, P. C., and H. H. Palomer, Editors: "Food Theory and Applications," Wiley, New York (1972).
Staff: "Amino-Acid Content of Foods and Biological Data on Proteins," *Nutritional Studies No. 24*, Food and Agriculture Organization (United Nations), Rome (1976).
Staff: "Catalog of Food and Nutrition Information and Educational Materials Center," U.S. Department of Agriculture, National Agricultural Library, Beltsville, Maryland (Revised periodically).
Tabor, H., and C. W. Tabor, Editors: "Metabolism of Amino Acids and Amines," Academic, New York (1971).
Villee, C. A.: "Biology," 7th Edition, Saunders, Philadelphia (1977).
Weissbach, H., and S. Pestka, Editors: "Molecular Mechanisms of Protein Biosynthesis," Academic, New York (1977).

—Hauromi Oeda, Ajinomoto Co., Inc., Kawasaki, Japan.

AMINO RESINS. A family of resins resulting from an addition reaction between formaldehyde and compounds such as aniline, ethylene urea, dicyandiamide, melamine, sulfonamide, and urea. The resins are thermosetting and have been used for many years in such products as textile-treating agents, laminating coatings, wet-strength paper coatings, and wood adhesives. The urea and melamine compounds are most widely used. Both of these basic resins are water white (transparent). However, the resins readily accept pigments and opacifying agents. The addition of cellulose filler can be used to reduce light transmission. Where color is unimportant, various materials are added to the melamine resin compounds, including macerated fabric, glass fiber, and wood flour. Wood flour frequently is added to the urea resins to yield a low-cost industrial material.

Advantages claimed for amino resins include: (1) good electrical insulation characteristics, (2) no transfer of tastes and odors to foods, (3) self-extinguishing burning characteristics, (4) resistance to attack by oils, greases, weak alkalis and acids, and organic solvents, (5) abrasion resistance, (6) good rigidity, (7) easy fabrication by economical molding procedures, (8) excellent resistance to deformation under load, (9) good subzero characteristics with no tendency to become brittle, and (10) marked hardness.

Amino resins are fabricated principally by transfer and compression molding. Injection molding and extrusion are used on a limited scale. Urea resins are not recommended for outdoor exposure. The resins show rather high mold shrinkage and some shrinkage with age. The melamines are superior to the ureas insofar as resistance to heat and boiling water, acids, and alkalis is concerned.

Some of the hundreds of applications for amino resins include: Closures for glass, metal, and plastic containers; electrical wiring devices; appliance knobs, dials, handles, and push buttons; lamp shades and lighting diffusers; organ and piano keys; dinnerware; food service trays; food-mixer housings; switch parts; decorative buttons; meter blocks; aircraft ignition parts; heavy duty switch gear; connectors; and terminal strips. Not all of the urea or melamine amino resins are suited to all of the foregoing uses. Because of the large number of fillers and additives available, the overall range of use of this family of resins is large.

AMMINES. Dry ammonia gas reacts with dehydrated salts of some of the metals to form solid ammines (not to be confused with amines). Ammines, upon warming, evolve ammonia, sometimes with final decomposition of the salt itself, in a manner analogous to the decomposition of certain hydrates. The ammines of chromium(III) (Cr^{3+}), cobalt(III), platinum(IV), and other metals have been studied in detail. Two series of ammines are shown in the accompanying diagram, the first being one in which the neutral ammonia group is replaced step by step by the negative nitro group (NO_2^-) and the second, one in which the neutral ammonia group is replaced step by step by the neutral H_2O group.

The neutral group of the complex may be replaced step by step by the following negative groups: Cl^-, Br^-, I^-, F^-, OH^-, NO_2^-, NO_3^-, CN^-, CNS^-, SO_4^{2-}, CO_3^{2-}, $C_2O_4^{2-}$; or by the following neutral groups: H_2O, NO, NO_2, SO_2, S, N_2H_4, H_2NOH, CO, C_2H_5OH, C_6H_6. All neutral groups are of substances capable of independent existence.

In the ammines, trivalent metals, such as cobalt(III) and chromium(III) and iron(III), possess a coordination number of 6, this number being the sum of the unit replacements on the metal in the complex ion. Since a regular octahedron has six corners equidistant from the center, it is assumed that the metal occupies the center and each of the six replacing groups occupies a corner of a a regular octahedron. Support for this assumption is offered by the x-ray examination of these ammines. When there is only one of the six groups replaced by a second group, as in $[Co(NH_3)_5(NO_2)]Cl_2$, and in $[Cr(NH_3)_5(H_2O)]X_3$ the octahedral placement of groups supplies only one form, but when two of the six groups are replaced by a second group, as in $[Co(NH_3)_4(NO_2)_2]Cl$, and in $[Cr(NH_3)_4(H_2O)_2]X_3$, two

$[Co(NH_3)_6]Cl_3$
410

$[Co(NH_3)_5(NO_2)]Cl_2$
240

$[Co(NH_3)_4(NO_2)_2]Cl$
95

$[Co(NH_3)_3(NO_2)_3]$
1.5

$K[Co(NH_3)_2(NO_2)_4]$
95

$K_2[Co(NH_3)(NO_2)_5]$
240

$K_3[Co(NO_2)_6]$
420

(a)

$[Cr(NH_3)_6]X_3$
$[Cr(NH_3)_5(H_2O)]X_3$
$[Cr(NH_3)_4(H_2O)_2]X_3$
$[Cr(NH_3)_3(H_2O)_3]X_3$
$[Cr(NH_3)_2(H_2O)_4]X_3$

X = unit anion

(b)

Two series of ammines: (a) Square bracket contains the ion. The equivalent electrical conductivity is shown below each compound. The number of neutral groups, e.g., (NH_3), on metal, e.g., Co, is varied from 6 to 0. (b) The number of neutral groups is constant, but the groups are varied.

different octahedral corner arrangements are possible depending upon whether the two replacing groups are adjacent (cisform) or opposite (transform). Two substances differing in physical properties and corresponding to these two forms are know. Fur-

ther, when three divalent groups, e.g., $3C_2O_4^2\text{Å}$ are present in the complex, two arrangements—not identical but mirror-images of each other—are possible. Two optically active sustances are known in such cases corresponding to these two stereoisomeric forms.

Six is the ordinary coordination number for metallic ammines and similar complexes. Additional examples are $K_2[Pt(NH_3)_2(CN)_4]$, $[Ni(NH_3)_6]Cl_2$, $K_4[Fe(CN)_6]$, $K_3[Fe(CN)_6]$, $K_2[Fe(CN)_5(NO)]$, $K_2[SiF_6]$, $[Ca(NH_3)_6)]Cl_2$. But, for the elements boron, carbon, and nitrogen four is the coordination number, e.g., $[BH_4]Cl$, $[CH_4]$, $[NH_4]Cl$, and in these substances the groups are assumed to occupy the corners of a regular tetrahedron; in $K_4[Mo(CN)_8]$ and $[Ba(NH_3)_8]Cl_2$ the coordination number is eight, and the groups are assumed to occupy the corners of a cube.

AMMONIA. Known since ancient times, ammonia, NH_3, has been commercially important for well over 100 years and has become the second largest chemical in terms of tonnage and the first chemical in value of production. The first practical plant of any magnitude was built in 1913. Worldwide production of NH_3 as of the early 1980s is estimated at 100 million metric tons per year or more, with the United States accounting for about 15% of the total production. A little over three-fourths of ammonia production in the United States is used for fertilizer, of which nearly one-third is for direct application. An estimated 5.5% of ammonia production is used in the manufacture of fibers and plastics intermediates.

Properties. At standard temperature and pressure, NH_3 is a colorless gas with a penetrating, pungent-sharp odor in small concentrations which, in heavy concentrations, produces a smothering sensation when inhaled. Formula weight is 17.03, mp $-77.7°C$, bp $-33.35°C$, and sp. gr. 0.817 (at $-79°C$) and 0.617 (at $15°C$). Ammonia is very soluble in water, a saturated solution containing approximately 45% NH_3 (weight) at the freezing temperature of the solution and about 30% (weight) at standard conditions. Ammonia dissolved in water forms a strongly alkaline solution of ammonium hydroxide, NH_4OH. The univalent radical NH_4^+ behaves in many respects like K^+ and Na^+ in vigorously reacting with acids to form salts. Ammonia is an excellent nonaqueous electrolytic solvent, its ionizing power approaching that of water. Ammonia burns with a greenish-yellow flame.

Ammonia derives its name from sal ammoniac, NH_4Cl, the latter material having been produced at the Temple of Jupiter Ammon (Libya) by distilling camel dung. During the Middle Ages, NH_3 was referred to as the spirits of hartshorn because it was produced by heating the hoofs and horns of oxen. The composition of ammonia was first established by Claude Louis Berthelot (France, ca. 1777). The first significant commercial source of NH_3 (during the 1880s) was its production as a by-product in the making of manufactured gas through the destructive distillation of coal. See also **Coal Tar and Derivatives.**

Nitrogen fixation is a term assigned to the process of converting nitrogen in the air to nitrogen compounds. Although some bacteria in soil are capable of this process, N_2 as an ingredient of fertilizer is required for soils that are depleted by crop production. The production of synthetic NH_3 is the most important industrial nitrogen-fixation process. See also **Fertilizers.**

Synthesis of Ammonia. The first breakthrough in the large-scale synthesis of ammonia resulted from the work of Fritz Haber (Germany, 1913), who found that ammonia could be produced by the direct combination of two elements, nitrogen and hydrogen, $(N_2 + 3H_2 \rightleftharpoons 2NH_3)$ in the presence of a catalyst (iron oxide with small quantities of cerium and chromium) at a relatively high temperature (550°C) and under a pressure of about 200 atmospheres, representing difficult processing conditions for that era. Largely because of the urgent requirements for ammonia in the manufacture of explosives during World War I, the process was adapted for industrial-quality production by Karl Bosch, who received one-half of the 1931 Nobel Prize for chemistry in recognition of these achievements. Thereafter, many improved ammonia-synthesis systems, based on the Haber-Bosch process, were commercialized, using various operating conditions and synthesis-loop designs.

The principal features of an NH_3 synthesis process system are the converter designs, operating conditions, method of product recovery, and type of recirculation equipment. Most current systems operate at or above the pressure used in the original Haber-Bosch process. Converter designs have either a single continuous catalyst bed, which may or may not have heat-exchange cooling for controlling reaction heat, or several catalyst beds with provision for temperature control between the beds.

Claude Process. The original Claude process was one of the first systems to use a high operating pressure (1000 atmospheres), achieving 40% conversion without recycling. This system used multiple converters in a series-parallel arrangement. The present Claude process[1] operates at 340–650 atmospheres, using a single converter with continuous catalyst-charged tubes externally cooled to remove the heat of reaction. Approximate hydrogen conversion is 30–34 mole percent per pass. The pressure is increased gradually to compensate for catalyst aging and loss in activity. Product recovery is by simple condensation in a water-cooled condenser. Unreacted gas is recycled by compressor.

Casale Process. This is another high-pressure conversion system, using synthesis pressures of 450–600 atmospheres, which also permits hydrogen conversions in the 30 mole percent range. As in the Claude process, the high pressure allows NH_3 to be recovered from the converter effluent by water cooling. The Casale converter uses a single catalyst bed with internal heat-exchange surfaces. Reaction rate and temperature rise across the catalyst is controlled by the internal exchanger and retaining 2–3 mole percent NH_3 in the converter feed. This eliminates the need for a mechanical recycle compressor, but requires high feed-gas pressures to supply the energy required for the ejector.

Low-Pressure Processes. Several systems use low synthesis pressures with hydrogen conversion below 30 mole percent and product recovery by water and refrigeration.

Synthesis-Gas-Production Processes. These processes were improved and developed as a result of changes in feedstock availability and economics. Before World War II, most NH_3 plants obtained H_2 by reacting coal or coke with steam in the water-gas process. A small number of plants used water electrolysis or coke-oven byproduct hydrogen. The subsequent low-cost availability of natural gas brought about steam-hydrocarbon reforming as the major source of H_2 for the NH_3 synthesis gas.

Partial oxidation processes to produce H_2 from natural gas and liquid hydrocarbons were also developed after World War II and accounted for 15% of the synthetic NH_3 capacity by 1962. The steam-hydrocarbon reforming process[2] was developed in 1930. In this process, methane was mixed with an excess of steam at atmospheric pressure, and the mixture reformed inside nickel-catalyst-filled alloy furnace tubes. The heat of reaction was supplied by externally heating the catalyst-filled tubes to about 871°C. Since the late 1950s, improvements in the tubu-lar-reforming technology and metallurgy have brought about the utilization of high-pressure (> 24 atmospheres) reforming, which cut synthesis-gas-compression costs and increased heat recovery. The first pressure reformer[3] was built in 1953. In addition, the higher pressures allowed improvements in the efficiency of synthesis-gas-purification systems. High-pressure steam-reforming technology also has been extended to cover heavier hydrocarbon gases, including propane, butane, reformer gases, and streams containing a high amount of olefins. In 1962, a process[4] for reforming straight-run liquid distillates (naphthas) was commercialized. This process is based on the use of an alkali oxide-promoted nickel catalyst[5] which permits reforming of desulfurized naphthas at low (~ 3.5:1) steam-to-carbon ratios, without significant carbon deposition problems.

Noncatalytic partial oxidation processes designed to produce H_2 from a wide range of hydrocarbon liquids, including heavy fuel oils, crudes, naphthas, coal tar, and pulverized bituminous coal, were commercialized in 1954[6] and 1956.[7] In both these processes, the hydrocarbon feed is oxidized and reformed in a refractory-lined pressure vessel. The required oxygen usually is supplied by an air separation plant from which nitrogen also is used as feed for the synthesis gas. The main differences between the two processes are in the reactor design, feeding method, burner design, and carbon and heat recovery. The partial oxidation processes and the steam-naphtha reforming process are favored in areas with short supplies of natural gas.

The source of nitrogen for the synthesis gas has always been air, either supplied directly from a liquid-air separation plant or by burning a small amount of the hydrogen with air in the H_2 gas. The need for air separation plants has been eliminated in modern ammonia plants by use of secondary reforming, where residual methane from the primary reformer is adiabatically reformed with sufficient air to produce a 3:1 mole ratio hydrogen-nitrogen synthesis gas.

Most ammonia plants built since the early 1960s are in the 600–1500 short tons/day (540–1350 metric tons/day) range and are based on new integrated designs that have cut the cost of ammonia manufacture in half. The plants of the early 1980s, in fact, have reached the best combination in terms of plant overall efficiency and cost by combining all the separate units (e.g., synthesis-gas preparation, purification, and ammonia synthesis) in one single train. High-pressure reforming has reduced the synthesis-gas compression load and front-end plant equipment size. This compactness in design has also led to increased plant size at reduced investment and operating costs.

Use of Multistage Centrifugal Compressors. One of the major factors contributing to the improved economics of ammonia plants is the application of multistage centrifugal compressors, which have replaced the reciprocating compressors traditionally used in the synthesis feed and recycle service. A single centrifugal compressor can do the job of several banks of reciprocating compressors, thus reducing equipment cost, floor space, supporting foundations, and maintenance.

The use of multistage centrifugal compressors was made possible by redesigning the synthesis loop to operate at low pressures (150–240 atmospheres) and by increasing plant capacity to above the compressor's minimum-flow restriction in order to obtain a reasonable compressor efficiency. (Most synthesis loops using reciprocating compressors had been operating at intermediate

---

[1] Developed by Grande Pariosse and L'Air Liquide.
[2] Originally developed by Standard Oil Company of New Jersey.

[3] Built by M. W. Kellogg Co. for Shell Chemical Corp. (Ventura, California).
[4] M. W. Kellogg Co. and Imperial Chemical Industries.
[5] Developed by M. W. Kellogg Co.
[6] Texaco partial oxidation process.
[7] Shell gasification process.

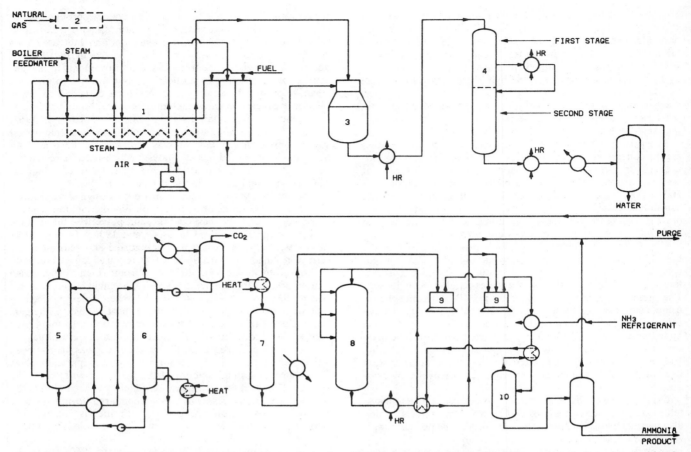

Ammonia production process: (1) Primary reformer, (2) desulfurization, (3) secondary reformer, (4) CO shift converter (in two stages), (5) CO₂ absorber, (6) CO₂ stripper, (7) methanator, (8) NH₃ converter, (9) compressor, (10) separator. HR = heat recovery. (*M. W. Kellogg Co.*)

pressures of 300–350 atmospheres.) Centrifugal compressors capable of developing pressures up to 340 atmospheres already are being offered and used in some large-capacity (1000 short tons/day; 900 metric tons/day) plants, where the increasing compressor horsepower is partially offset by reduction of the refrigeration horsepower requirement.

An operating ammonia plant using the aforementioned improvements is shown schematically in the accompanying figure. This plant[8] has a capacity of 1000 short tons/day (900 metric tons/day) and uses natural gas as feedstock. The plant can be divided into the following integrated-process sections: (a) synthesis-gas preparation; (b) synthesis-gas purification; and (c) compression and ammonia synthesis.

Synthesis Gas Preparation. The desulfurized natural gas mixed with steam is fed to the primary reformer, where it is reacted with steam in nickel-catalyst-filled tubes to produce a major percentage of the hydrogen required: The principal reactions taking place are[9]

$$CH_4 + H_2O \rightleftharpoons CO + 3H_2 \tag{1}$$

$$\Delta H_{298} = 49.3 \text{ kcal/mole}$$

$$CO + H_2O \rightleftharpoons CO_2 + H_2 \tag{2}$$

$$\Delta H_{298} = -9.8 \text{ kcal/mole.}$$

[8] Designed by The M. W. Kellogg Company, Houston, Texas, for which Kellogg received the 1967 Kirkpatrick Chemical Engineering Achievement Award.

[9] Heats of reaction at 198 K (25°C), 1 atmosphere pressure, gaseous substances in ideal state.

Reaction (1) is the principal reforming reaction, and reaction (2) is the water-gas shift reaction. The net reactions are highly endothermic. The partially reformed gas leaves the primary reformer containing approximately 10% methane, on a mole dry-gas basis, at 27–34 atmospheres and up to 816°C. The required heat of reaction is supplied by natural-gas-fired arch burners, which are designed to also burn purge and flash gases from the synthesis resection. Waste heat from the primary flue gas is recovered by generating high-pressure superheated steam, which along with waste-heat process boilers and an appended auxiliary boiler assure a steam system that is always in balance, while providing high-pressure steam to compressor turbine drivers and low-pressure steam to pump drivers. Further waste heat is recovered by preheating the natural-gas-steam feed mixture, steam-air for secondary reforming, and fuel.

The primary reforming step is followed by conversion of the residual methane to hydrogen and carbon oxides over a bed of high-temperature chrome and nickel catalysts in the secondary reformer. The secondary reforming step not only achieves a great degree of overall reforming economically possible, but also reduces fuel-gas input and overall reforming costs by shifting part of the required hydrocarbon conversion from the high-cost primary reformer to the lower-cost secondary reformer. It also permits an increase in the residual methane level at the primary effluent, which results in lower operating temperatures, reduced steam requirements, and milder tube-metal conditions.

Process waste-heat boilers then cool the reformed gas to about

371°C while generating high-pressure steam. The cooled gas-stream mixture enters a two-stage shift converter. The purpose of shift conversion is to convert CO to CO_2 and produce an equivalent amount of H_2 by the reaction: $CO + H_2O \rightleftharpoons CO_2 + H_2$. Since the reaction rate in the shift converter is favored by high temperatures, but equilibrium is favored by low temperatures, two conversion stages, each with a different catalyst provide the optimum conditions for maximum CO shift. Gas from the shift converter is the raw synthesis gas, which, after purification, becomes the feed to the NH_3 synthesis section.

Purification of Synthesis Gas. This involves the removal of carbon oxides to prevent poisoning of the NH_3 catalyst. An absorption process is used to remove the bulk of the CO_2, followed by methanation of the residual carbon oxides in the methanator. Modern ammonia plants use a variety of CO_2-removal processes with effective absorbent solutions. The principal absorbent solutions currently in use are hot carbonates and ethanolamines. Other solutions used include methanol, acetone, liquid nitrogen, glycols, and other organic solvents.

The partially purified synthesis gas leaves the CO_2 absorber containing approximately 0.1% CO_2 and 0.5% CO. This gas is preheated at the methanator inlet by heat exchange with the synthesis-gas compressor interstage cooler and the primary-shift converter effluent and reacted over a nickel oxide catalyst bed in the methanator. The methanation reactions are highly exothermic and are equilibrium favored by low temperatures and high pressures:

$$CO + 3H_2 \rightleftharpoons CH_4 + H_2O$$

$$CO_2 + 4H_2 \rightleftharpoons CH_4 + 2H_2O.$$

The methanator effluent is cooled by heat exchange with boiler feedwater and cooling water. The synthesis gas leaves the methanator containing less than 10 parts per million (ppm) of carbon oxides.

Compression and Synthesis. The purified synthesis gas, containing H_2 and N_2 in a 3:1 mole ratio and with an inert gas (methane and argon) content of about 1.3 mole percent, is delivered to the suction of the synthesis-gas compressor. Anhydrous ammonia is catalytically synthesized in the converter. The effluent from the converter, after taking off a small purge stream, is recycled for eventual conversion to ammonia. Reaction takes place at approximately 427–482°C. Ammonia liquid, separated from the loop in the separator and from the purge, contains dissolved synthesis gas, which is released when the combined stream is flashed into the letdown drum. The flashed gas is then separated in the letdown drum and combined with the vapors from the purge separator to form a stream of purge fuel gases. Liquid ammonia in the letdown drum still contains some dissolved gases which must be disengaged. This is effected in the ammonia-refrigeration cycle.

Future Considerations in Ammonia Production. The development of a modern and more energy efficient ammonia process has behind it a long history. Great progress has been made in the past 20 years. Presently, the demand for a more energy efficient ammonia plant is greater than before, due to the economics and short supply of light hydrocarbon feedstock. As hydrocarbon supplies diminish, costs will rise even higher. Unquestionably, the development of new ammonia processes aimed at high energy efficiency and alternate feedstocks will continue to be the primary area of ammonia technology.

In the past, coal or heavy hydrocarbon feedstock ammonia plants were not economically competitive with plants where the feedstocks were light hydrocarbons (natural gas to naphtha). Because of changing economics, however, plants that can handle heavy hydrocarbon feedstock are now attracting increasing attention. In addition, the continuous development and improvement of partial oxidation processes at higher pressure have allowed reductions in equipment size and cost. Therefore, the alternate feedstock ammonia plants based on a partial oxidation process may become economically competitive in the near future.

The ultimate goal of any ammonia process will be the direct fixation of nitrogen by reaction of water with air, $1.5H_2O + 0.5N_2 \rightarrow NH_3 + 0.75O_2$. The theoretical energy requirement of the reaction is about 18 MMBTU(LHV)/ST of ammonia. This feed energy requirement is the same as a natural-gas-feed ammonia plant. The major difference is that there is no short supply of water and air. However, the technology required for this route is not expected to be available any time soon. It is believed that, in the near future, ammonia plant designs will be based essentially on the present-day process with modifications to reduce energy consumption.

References

Axelrod, L. C., and T. E. O'Hare: "Production of Synthetic Ammonia," Chap. 5 in "Fertilizer Nitrogen: Its Chemistry and Technology," (V. Sauchelli, editor), Van Nostrand, New York (1964).
Hargette, H. L., and J. T. Berry: "Fertilizer Summary Data," Tennessee Valley Authority, Knoxville, Tennessee (Published annually).
"Kirk-Othmer Encyclopedia of Chemical Technology," Chap. 470, Vol. 2, 3rd Edition, Wiley, New York (1978).
Livingstone, J. G., and A. Pinto: "New Ammonia Process Reduces Costs," *Chem. Eng. Progress*, **79**, 5, 62–66 (1983).
MacLean, D. L., Prince, C. E., and Y. C. Chae: "Energy-Saving Modifications in Ammonia Plants," *Chem. Eng. Progress*, **76**, 3, 98–104 (1980).
Quartulli, O. J., et al.: "Best Pressure for Ammonia Plants," *Hydrocarbon Processing*, **47**, 11 (November 1968).
Stokes, K. J.: "Compression Systems for Ammonia Plants," *Chem. Eng. Progress*, **75**, 7, 88–91 (1979).
Yost, C. C., Curtis, C. R., and C. J. Ryskamp: "Advanced Control at Sycon's Ammonia Plant," *Chem. Eng. Progress*, **76**, 4, 31–36 (1980).

—J. M. Lee, M. W. Kellogg Co., Houston, Texas.

AMMONIUM CHLORIDE. NH_4Cl, formula wt 53.50, white crystalline solid, decomposes at 350°C, sublimes at 520°C under controlled conditions, sp gr 1.52. Also known as *sal ammoniac*, the compound is soluble in H_2O and in aqueous solutions of NH_3; slightly soluble in methyl alcohol. Ammonium chloride is a high-production chemical, finding uses as an ingredient of dry cell batteries, as a soldering flux, as a processing ingredient in textile printing and hide tanning, and as a starting material for the manufacture of other ammonium chemicals.

The compound can be produced by neutralizing HCl with NH_3 gas, or with liquid NH_4OH, evaporating the excess H_2O, followed by drying, crystallizing, and screening operations. The product also can be formed in the gaseous phase by reacting hydrogen chloride gas with NH_3. Ammonium chloride generally is not attractive as a source of nitrogen for fertilizers because of the buildup and damaging effects of chloride residuals. See also **Nitrogen**.

AMMONIUM COMPOUNDS. Numerous ammonium compounds are important industrially. Several of these are described in separate entries in this Encyclopedia. In particular, see **Ammonium Chloride; Ammonium Nitrate; Ammonium Phosphates; and Ammonium Sulfate**.

Principal factors concerning several other ammonium compounds, several of which are important commercially, are given in the following summary.

Acetate. Ammonium acetate $NH_4C_2H_3O_2$, white solid, soluble, formed by reaction of ammonia or NH_4OH and acetic acid, reacts upon heating to yield acetamide.

Alum. Ammonium alums are those alums, such as aluminum ammonium sulfate $Al_2(NH_4)_2(SO_4)_4 \cdot 24H_2O$, ferric ammonium sulfate $Fe_2(NH_4)_2(SO_4)_4 \cdot 24H_2O$, chromium ammonium sulfate $Cr_2(NH_4)_2(SO_4)_4 \cdot 24H_2O$ where ammonium sulfate is crystallized with the heavier metal sulfate.

Benzoate. Ammonium benzoate $NH_4C_7H_5O_2$, white solid, soluble, formed by reaction of NH_4OH and benzoic acid. Used (1) as a food preservative, (2) in medicine.

Borate. Ammonium borate, ammonium tetraborate

$$(NH_4)_2B_4O_7 \cdot 4H_2O,$$

white solid, soluble, formed by reaction of NH_4OH and boric acid. Used (1) in fireproofing fabrics, (2) in medicine.

Bromide. Ammonium bromide NH_4Br, white solid, soluble, sublimes at $542°C$, formed by reaction of NH_4OH and hydrobromic acid. Used in photography.

Carbonates. Ammonium caronate, volatile $(NH_4)_2CO_3$, white solid, soluble, formed by reaction of NH_4OH and CO_2 by crystallization from dilute alcohol, loses NH_3, CO_2, and H_2O at ordinary temperatures, rapidly at $58°C$; ammonium hydrogen carbonate, ammonium bicarbonate, ammonium acid carbonate NH_4HCO_3, white solid, soluble, formed by reaction of NH_4OH and excess CO_2. This salt is the important reactant in the ammonia soda process for converting sodium chloride in solution into sodium hydrogen carbonate solid.

Chloroplatinate. Ammonium chloroplatinate $(NH_4)_2PtCl_6$, yellow solid, insoluble, formed by reaction of soluble ammonium salt solutions and chloroplatinic acid. Used in the quantitative determination of ammonium.

Cobaltinitrite. Diammonium sodium cobaltinitrite

$$(NH_4)_2NaCo(NO_2)_6 \cdot H_2O$$

golden yellow precipitate, formed by reaction of sodium cobaltinitrite solution in acetic acid with soluble ammonium salt solution. Used in the detection of ammonium.

Cyanate. Ammonium cyanate NH_4CNO, white solid, soluble, formed by fractional crystallization of potassium cyanate and ammonium sulfate (ammonium cyanate is soluble in alcohol), when heated changes into urea.

Dichromate. Ammonium dichromate $(NH_4)_2Cr_2O_7$, red solid, soluble, upon heating evolves nitrogen gas and leaves a green insoluble residue of chromic oxide.

Fluoride. Ammonium fluoride NH_4F, white solid, soluble, formed by reaction of NH_4OH and hydrofluoric acid, and then evaporating. Used (1) as an antiseptic in brewing, (2) in etching glass; ammonium hydrogen fluoride, ammonium bifluoride, ammonium acid fluoride NH_4F_2, white solid, soluble.

Iodide. Ammonium iodide NH_4I, white solid, soluble, formed by reaction of NH_4OH and hydriodic acid, and then evaporating. Used (1) in photography, (2) in medicine.

Linoleate. Ammonium linoleate $NH_4C_{18}H_{31}O_2$. Used (1) as an emulsifying agent, (2) as a detergent.

Nitrite. Ammonium nitrite NH_4NO_2. When ammonium sulfate or chloride and sodium or potassium nitrite are heated, the mixture behaves like ammonium nitrite in yielding nitrogen gas.

Oxalate. Ammonium oxalate $(NH_4)_2C_2O_4$, white solid, soluble, formed by reaction of NH_4OH and oxalic acid, and then evaporating. Used as a source of oxalate; ammonium binoxalate $NH_4HC_2O_4 \cdot H_2O$, white solid, soluble.

Perchlorate. Ammonium perchlorate NH_4ClO_4, white solid, soluble, formed by reaction of NH_4OH and perchlorate acid, and then evaporating. Used in explosives and pyrotechnics.

Periodate. Ammonium periodate NH_4IO_4, white solid, moderately soluble.

Persulfate. Ammonium persulfate $(NH_4)_2S_2O_8$, white solid, soluble, formed by electrolysis of ammonium sulfate under proper conditions. Used (1) as a bleaching and oxidizing agent, (2) in electroplating, (3) in photography.

Phosphomolybdate. Ammonium phosphomolybdate

$$(NH_4)_3PO_4 \cdot 12MoO_3$$

(or similar composition), yellow precipitate, soluble in alkalis, formed by excess ammonium molybdate and HNO_3 with soluble phosphate solution. Used as an important test for phosphate (similar product and reaction when arsenate replaces phosphate).

Salicylate. Ammonium salicylate $NH_4C_7H_5O_3$, white solid, soluble, formed by reaction of NH_4OH and salicylic acid, and then evaporating. Used in medicine.

Sulfide. Ammonium sulfide $(NH_4)_2S$, colorless to yellowish solution, formed by saturation with hydrogen sulfide of one-half of a solution of NH_4OH, and then mixing with the other half of the NH_4OH. Dissolves sulfur to form ammonium polysulfide, yellow solution. Used as a reagent in analytical chemistry; ammonium hydrogen sulfide, ammonium bisulfide, ammonium acid sulfide NH_4HS, colorless to yellowish solution, formed by saturation with H_2S of a solution of NH_4OH.

Tartrate. Ammonium tartrate $(NH_4)_2C_4H_4O_6$, white solid, moderately soluble, formed by reaction of NH_4OH and tartaric acid, and then evaporating. Used in the textile industry; ammonium hydrogen tartrate, ammonium bitartrate, ammonium acid tartrate $NH_4HC_4H_4O_6$, white solid, slightly soluble, formation sometimes used in detection of ammonium or tartrate.

Thiocyanate. Ammonium thiocyanate, ammonium sulfocyanide, ammonium rhodanate NH_4CNS, white solid, soluble, absorbs much heat on dissolving with consequent marked lowering of temperature, mp $150°C$, formed by boiling ammonium cyanate solution with sulfur, and then evaporating. Used (1) as a reagent for ferric, (2) in making cooling solutions, (3) to make thiourea.

Ammonium compounds liberate NH_3 gas when warmed with $NaOH$ solution.

AMMONIUM HYDROXIDE. NH_4OH, formula wt 35.05, exists only in the form of an aqueous solution. The compound is prepared by dissolving NH_3 in H_2O and usually is referred to in industrial trade as aqua ammonia. For industrial procurements, the concentration of NH_3 in solution normally is specified in terms of the sp gr (degrees Baumé, °Be). Common concentrations are 20 and 26°Be. The former is equivalent to a sp gr of 0.933, or a concentration of about 17.8% NH_3 in solution; the later is equivalent to a sp gr of 0.897, or a concentration of about 19.4% NH_3. These figures apply at a temperature of 60°F (15.6°C). Reagent grade NH_4OH usually contains approximately 58% NH_4OH (from 28 to 30% NH_3 in solution).

Ammonium hydroxide is one of the most useful forms in which to react NH_3 (becoming the NH_4^+ radical in solution) with other materials for the creation of ammonium salts and other ammonium and nitrogen-bearing chemicals. Ammonium hydroxide is a direct ingredient of many products, including saponifiers for oils and fats, deodorants, etching compounds, and cleaning and bleaching compounds. Because aqua ammonia is reasonably inexpensive and a strongly alkaline substance, it finds wide application as a neutralizing agent. See also **Nitrogen.**

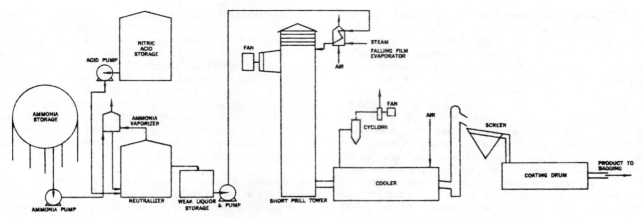

Process for making ammonium nitrate on a large scale. (*C & I/Girdler Incorporated*)

AMMONIUM NITRATE. NH_4NO_3, formula wt 80.05, colorless crystalline solid, occurs in two forms:

α-NH_4NO_3, tetragonal crystals, stable between -16 and $32°C$, sp gr 1.66.

β-NH_4NO_3, rhombic or monoclinic crystals, stable between 32 and $84°C$, sp gr 1.725.

The melting point generally ascribed to the alpha form is $169.6°C$, with decomposition occurring above $210°C$. Upon heating, ammonium nitrate yields nitrous oxide (N_2O) gas and can be used as an industrial source of that gas. Ammonium nitrate is soluble in H_2O, slightly soluble in ethyl alcohol, moderately soluble in methyl alcohol, and soluble in acetic acid solutions containing NH_3.

As shown in the accompanying figure, in making ammonium nitrate on a large scale, NH_3, vaporized by waste steam from neutralizer, is sparged along with HNO_3 into the neutralizer. A ratio controller automatically maintains the proper proportions of NH_3 and acid. The heat of neutralization evaporates a part of the H_2O and gives a solution of 83% NH_4NO_3. Final evaporation to above 99% for agricultural prills, or to approximately 96% for industrial prills is accomplished in a falling-film evaporator located at the top of the prilling tower. The resultant melt flows through spray nozzles and downward through the tower. Air is drawn upward by fans at the top of the tower. The melt is cooled sufficiently to solidify, forming round pellets or prills of the desired range of sizes. The prills are removed from the bottom of the tower and fed to a rotary cooler. Where industrial-type prills are produced, a pre-drier and drier precede the cooler. Fines from the rotary drums are collected in wet cyclones. This solution eventually is returned to the neutralizer. After cooling, the prills are screened to size and the over- and undersize particles are sent to a sump and returned to the neutralizer. Intermediate or product-size prills are dusted with a coating material, usually diatomaceous earth, in a rotary coating drum and sent to the bagging operation. The process can be adapted to other types of materials and mixtures of ammonium nitrate and other fertilizer materials. Mixtures include the incorporation of limestone and ammonium phosphates.

Ammonium nitrate is a high-production industrial chemical, finding major applications in explosives and fertilizers, and additional uses in pyrotechnics, freezing mixtures (for obtaining low temperatures), as a slow-burning propellant, as an ingredient of rust inhibitors, particularly for vapor-phase corrosion, and as a component of some insecticides.

Amatol, an explosive developed by the British, is a mixture of ammonium nitrate and TNT. A special explosive for tree-trunk blasting consists of ammonium nitrate coated with TNT. In strip mining, an explosive consisting of ammonium nitrate and carbon black is used. The explosive ANFO is a mixture of ammonium nitrate and fuel oil. The typical reaction is: $3NH_4NO_3 + CH_2 \rightarrow 3N_2 + 7H_2O + CO_2 + 82$ kcal/mole. ANFO accounts for about 50% of the commercial explosives used in the United States. Slurry explosives consist of oxidizers (NH_4NO_3 and $NaNO_3$), fuels (coals, oils, aluminum, other carbonaceous materials), sensitizers (TNT, nitrostarch, and smokeless powder), and water mixed with a gelling agent to form a thick, viscous explosive with excellent water-resistant properties. Slurry explosives may be manufactured as cartridged units, or mixed on-site. Although Nobel introduced NH_4NO_3 into his dynamic formulations as early as 1875, the tremendous explosive power of the compound was not realized until the tragic Texas City, Texas disaster of 1947 when a shipload of NH_4NO_3 blew up while in harbor. See also **Explosive.**

As a fertilizer, NH_4NO_3 contains 35% nitrogen. Because of the explosive nature of the compound, precautions in handling are required. This danger can be minimized by introducing calcium carbonate into the mixture. Depending upon amount used, this will reduce the effective nitrogen content per weight unit. Clay coatings and moistureproof bags are required to preclude damage of the product during shipment and storage because of its highly hygroscopic nature.

AMMONIUM PHOSPHATES. There are two ammonium phosphates, both produced on a very high-tonnage scale:

Monoammonium phosphate, $NH_4H_2PO_4$, white crystals, sp gr 1.803
Formula weight 115.04, N = 12.17%, P_2O_5 = 61.70%
Diammonium phosphate, $(NH_4)_2HPO_4$, white crystals, sp gr 1.619
Formula weight 132.07, N = 21.22%, P_2O_5 = 53.74%

Both compounds are soluble in H_2O; insoluble in alcohol or ether. A third compound, triammonium phosphate $(NH_4)_3PO_4$ does not exist under normal conditions because, upon formation, it immediately decomposes, losing NH_3 and reverting to one of the less alkaline forms.

Large quantities of the ammonium phosphates are used as fertilizers and in fertilizer formulations. The compounds furnish both nitrogen and phosphorus essential to plant growth. The compounds also are used as fire retardants in wood building materials, paper and fabric products, and in matches to prevent afterglow. Solutions of the ammonium phosphates sometimes are air dropped to retard forest fires, serving the double purpose of fire fighting and fertilizing the soil to accelerate new plant growth. The compounds are used in baking powder formulations, as nutrients in the production of yeast, as nutritional supplements in animal feeds, for controlling the acidity of dye baths, and as a source of phosphorus in certain kinds of ceramics.

Ammonium phosphates usually are manufactured by neutralizing phosphoric acid with NH_3. Control of the pH (acidity/alkalinity) determines which of the ammonium phosphates will be produced. Pure grades can be easily made by crystallization of solutions obtained from furnace-grade phosphoric acid. Fertilizer grades, made from wet-process phosphoric acid, do not crystallize well and usually are prepared by a granulation technique. First, a highly concentrated solution or slurry is obtained by neutralization. Then the slurry is mixed with from $6 \times$ to $10 \times$ its weight of previously dried material, after which the mixture is dried in a rotary drier. The dry material is then screened to separate the desired product size. Oversize particles are crushed and mixed with fines from the screen operation and then returned to the granulation step where they act as nuclei for the production of further particles. Other ingredients often are added during the granulation of fertilizer grades. The ratio of nitrogen to phosphorus can be altered by the inclusion of ammonium nitrate, ammonium sulfate, or urea. Potassium salts sometimes are added to provide a 3-component fertilizer (N, P, K). A typical fertilizer grade diammonium phosphate will contain 18% N and 46% P_2O_5 (weight). See also **Fertilizer; Nitrogen.**

AMMONIUM SULFATE. $(NH_4)_2SO_4$, formula wt 132.14, colorless crystalline solid, decomposes above 513°C, sp gr 1.769. The compound is soluble in H_2O and insoluble in alcohol. Ammonium sulfate is a high-production industrial chemical, but can be considered as a byproduct as well as an intended end-product of production. Large quantities of ammonium sulfate result from a variety of industrial neutralization operations required for alleviation of stream pollution by free H_2SO_4. The ammonium sulfate so produced is not always recovered and marketed, but a significant commercial source of ammonium sulfate is its creation as a byproduct in the manufacture of caprolactam which yields several tons of the compound per ton of caprolactam made. See also **Caprolactam.** Ammonium sulfate is also a byproduct of coke oven operations where the excess NH_3 formed is neutralized with H_2SO_4 to form $(NH_4)_2SO_4$. However, as a major fertilizer and ingredient of fertilizer formulations, additional production is required, largely depending upon the proximity of consumers to byproduct $(NH_4)_2SO_4$ sources.

In the Meresburg reaction, natural or byproduct gypsum is reacted with ammonium carbonate to yield ammonium sulfate and calcium carbonate:

$$CaSO_4 \cdot 2H_2O + (NH_4)_2CO_3 \rightarrow CaCO_3 + (NH_4)_2SO_4 + H_2O.$$

The product is stable, free-flowing crystals. As a fertilizer, $(NH_4)_2SO_4$ has the advantage of adding sulfur to the soil as well as nitrogen. By weight, the compound contains 21% N and 24% S. Ammonium sulfate also is used in electric dry cell batteries, as a soldering liquid, as a fire retardant for fabrics and other products, and as a source of certain ammonium chemicals. See also **Fertilizer; Nitrogen.**

AMORPHOUS. As opposed to a crystalline substance which exhibits an orderly structure, the behavior of an amorphous substance is similar to a very viscous, inelastic liquid. Examples of amorphous substances include amber, glass, and pitch. An amorphous material may be regarded as a liquid of great viscosity and high rigidity, with physical properties the same in all directions (may be different for crystalline materials in different directions). Usually upon heating, an amorphous solid gradually softens and acquires the characteristics of a liquid, but without a definite point of transition from solid to liquid state. In geology, an amorphous mineral lacks a crystalline structure, or has an internal arrangement so irregular that there is no characteristic external form. This does not preclude, however, the existence of any degree of order. The word *amorphous* is used in connection with amorphous graphite and amorphous peat, among other naturally occurring substances.

AMOSITE. Amosite is a long-fiber gray or greenish asbestiform mineral related to the cummingtonite-grunerite series, and is of economic importance. It occurs within both regional and contact metamorphic rocks in the Republic of South Africa. The name amosite is a product of the initial letters of its occurrence at the Asbestos Mines of South Africa.

See also **Asbestos.**

AMPHETAMINE. Also called methylphenethylamine; l-phenyl-2-aminopropane; *Benzedrine*; formula $C_6H_5CH_2CH(NH_2)CH_3$, amphetamine is a colorless, volatile liquid with a characteristic strong odor and slightly burning taste. Boils and commences decomposition at 200–203°C. Low flash point, 26.7°C. Soluble in alcohol and ether; slightly soluble in water. Amphetamine is the basis of a group of hallucinogenic, habit-forming drugs which affect the central nervous system. The drug also finds medical application, notably in appetite suppressants. It should be emphasized that administration of amphetamines for prolonged periods in connection with weight-reduction programs may lead to drug dependence. Particular attention must be paid by professionals to the possibility of persons obtaining amphetamines for nontherapeutic use or distribution to others.

Amphetamines for obesity control may be in the form of amphetamine adipate, the sulfate, the saccharate, or phosphate. Amphetamines are sympathomimetic amines with central nervous system stimulant activity. Peripheral actions include elevation of systolic and diastolic blood pressures and weak bronchodilator and respiratory stimulant action. When the drugs are used for obesity control, they are sometimes called *anoretics* or *anorexigenics*. The effectiveness of amphetamines, in numerous clinical trials, as appetite depressants is not fully convincing. It has been found, for example, that the magnitude of increased weight loss of drug-treated patients over placebo-treated patients is only a fraction of a pound per week. The rate of weight loss is greatest during the first few weeks of treatment. The natural history of obesity is measured in years, whereas studies have been restricted to comparatively few weeks duration. Thus, the total impact of drug-induced weight loss over that of diet alone must be considered clinically limited. One should stress that there are varying opinions among the authorities in this area.

In connection with the treatment of narcolepsy (a sleep disorder), the use of amphetamines is not suggested because of the risks of habituation, abuse, and the development of tolerance.

AMPHIBOLE. This is the name given to a closely related group of minerals all showing in common a prismatic cleavage of 54–56% as well as similar optical characteristics and chemical composition.

The amphiboles may be said to represent chemically a series of metasilicates corresponding to the general formula $RSiO_3$

where R may be calcium, magnesium, iron, aluminum, titanium, sodium or potassium. The crystals of the amphibole family group fall within both the monoclinic and orthorhombic systems.

There is a clear parallelism between the amphiboles and the pyroxenes. There are two basic differences between the minerals of these two family groups: amphiboles have cleavage angles of 56° and 124°, with essential (OH) groups in their structure while pyroxenes have cleavage angles of 87° and 93° and are anhydrous, with no (OH) content. Amphibole crystals are usually long and slender and tend to be simple while pyroxene crystals tend to be complex, short and stout prisms.

Amphibole is common in both lavas and deep-seated rocks, though less so in the basic lavas than pyroxene. Many of the amphiboles may be developed as metamorphic minerals. The following members of the amphibole group are described under their own headings: actinolite, anthophyllite, cummingtonite, glaucophane, grünerite, hornblende, riebeckite and tremolite. Amphibole was so named by Haüy from the Greek word, meaning doubtful, because of the many varieties of this mineral.

See also **Pyroxene**; and terms listed under **Mineralogy**.

AMPHIBOLITE. The amphibolites form a large group of rather important rocks of metamorphic character. As the name implies they are made up very largely of minerals of the amphibole group. There may be also a variety of other minerals present, such as quartz, feldspar, biotite, muscovite, garnet, or chlorite in greater or less amounts.

Depending upon the particular amphibole present these rocks may be light to dark green or black, the amphibole usually being in long slender prisms or laths, often quite coarse, sometimes in acicular or fibrous forms.

Because the mineral constituents are arranged parallel to the schistosity, amphibolites may have a strongly developed cleavage.

The occurrence of amphibolites accompanying gneisses, schists, and other metamorphic rocks of probable sedimentary origin strongly suggests a similar derivation. Yet some amphibolites cut other metamorphic rocks in the manner of dikes or sills. It is very likely that they have been derived from both original igneous and sedimentary rocks. Large masses of amphibolite suggest gabbroic stocks. Well-known areas in which amphibolites are found are New England, New York State, Canada, Scotland, and the Alps.

AMPHIPROTIC. Capable of acting either as an acid or as a base, i.e., as a proton donor or acceptor, according to the nature of the environment. Thus, aluminum hydroxide dissolves in acids to form salts of aluminum, and it also dissolves in strong bases to form aluminates. Solvents like water which can act to give protons or accept them, are amphiprotic solvents.

AMPOULE. Sometimes spelled *ampule*, a small sealed glass container for drugs that are to be given by injection. As they are completely sealed, the contents are kept in their original sterile condition until used.

AMYGDALOID. A vesicular rock, commonly a lava, whose cavities have become filled with a secondary deposit of mineral material such as quartz, calcite, and zeolites. The term is derived from the Greek word meaning almond, in reference to the frequent almondlike appearance of the filled vesicles which are called amygdales or amygdules.

AMYLOSE. Starch.

ANABOLIC AGENTS. Steroids.

ANAEROBE. An organism that can grow in the absence of free oxygen is referred to as an anaerobic organism. Subdivided into *facultative* anaerobes, which can grow and utilize oxygen when it is present; and *obligatory* anaerobes, which cannot tolerate even a trace of oxygen in their surroundings. Yeast is an example of a facultative anaerobe. Yeast can grow and utilize oxygen in its metabolism, in which case it utilizes all the energy in a carbohydrate and yields water and carbon dioxide. In the absence of oxygen, the yeast cells turn to fermentation and anaerobic metabolism wherein the carbohydrate is converted into alcohol and carbon dioxide. The bacterium which produces botulism in "preserved" foods is an obligatory anaerobe (*Clostridium botulinum*).

In addition to botulism, other species of *Clostridium* are the etiological agents of tetanus, gas gangrene, and a variety of animal diseases including struck, lamb dysentery, pulpy kidney, and enterotoxemia. The pathogenicity of the clostridia depends largely on the production of potent toxins. Generally, these are heat labile exotoxins and often several pharmacologically and immunologically different toxins may be produced by the same species. In recent years, certain of the toxins have been purified to the point of crystalline purity and revealed to be high-molecular-weight proteins. The availability of the pure toxins has permitted more precise studies both of the chemical nature of the toxin and also of their properties as enzymes. From such toxins, it is possible to produce efficient toxoids for use in prophylactic immunization measures for tetanus and botulism. Some clostridia, in contrast, are beneficial. In addition to the nitrogen-fixing ability of certain species, C. *sporogenes* is active in the decomposition of proteins and contributes to biodegradability of substances, particularly of plant remains. C. *acetobutyleium* is active in production of acetone and butyl alcohol by fermentation.

Glucose can be broken down through glycolysis into pyruvic acid, but in the absence of oxygen as a final acceptor for the hydrogen, the pyruvic acid cannot enter the tricarboxylic acid cycle for further breakdown. Instead the pyruvic acid itself serves as an acceptor for the hydrogen split off in glycolysis. Hence, much less energy is obtained from food in anaerobic metabolism. Lactic acid, alcohol, and some extremely poisonous substances are products of anaerobic metabolism.

ANALCIME. A common zeolite mineral, $NaAlSi_2O_6 \cdot H_2O$, a hydrous soda-aluminum silicate. It crystalizes in the isometric system, hardness, 5-5.5; specific gravity, 2.2; vitreous luster; colorless to white; but may be grayish, greenish, yellowish or reddish. Its trapezohedral crystals resemble garnet crystals but are softer, and are distinguished from leucite only by chemical tests.

There are many excellent European localities. Magnificent crystals occur at Mt. St. Hilaire, Quebec, Canada. In the United States at Bergen Hill and West Paterson, N.J.; Keweenaw County, Mich.; and Jefferson County, Colorado. Nova Scotia furnishes beautiful specimens.

Analcime is a relatively common mineral and occurs with other zeolites in cavities and fissures in basic igneous rocks, occasionally in granites or gneisses. It seems to occur as a replacement and perhaps in some cases as a primary mineral crystallizing from a magma rich in soda and water vapor under pressure. The name analcime is derived from the Greek word meaning weak, in reference to the weak electric charge developed when heated or subjected to friction.

ANALGESICS. Drugs which diminish sensitivity to pain without impairing consciousness. These drugs act on the central nervous systems and include opiates, coal tar analgesics, aminopyrine, salicyliates, and phenylbutazone. Some of the analgesics also act in other ways pharmacologically. In the case of phenylbutazone, the excretion of uric acid is promoted. Salicylates reduce fever. Of the various alkaloids, only morphine and codeine are analgesics. Although still a valuable drug for some situations, particularly for relieving intense pain, morphine has several objectionable characteristics, including its toxicity and addicting nature. Codeine is a much less powerful drug and with much less objectionable aftereffects. See also **Alkaloids; and Morphine.**

Colchicine, an alkaloid, is used for the abatement of swelling and pain in acute attacks of gout. Normally, colchicine is not considered an analgesic, but in this one circumstance, it does play this role as well as serving as an antipyretic and antiphlogistic (counteraction of inflammation and fever). See also **Alkaloids.**

For the relief of ordinary aches and pains, the coal-tar analgesics are commonly used. These include acetanilid, acetophenetidin (phenacetin), and N-acetyl-p-aminophenol. They frequently are mixed with caffeine and aspirin, or a barbiturate. Acetanilid is somewhat more toxic than acetophenetidin and thus is used less frequently. A side effect of N-acetyl-p-aminophenol may be minor gastrointestinal distress.

Aminopyrine, an analgesic and antipyretic, is used less frequently because it can cause agranulocytosis.

Of the salicylate drugs (derivatives of salicylic acid), sodium salicylate and acetylsalicylic acid (aspirin) are the most widely used. The latter is poorly soluble in water and hydrolyzes into salicylic and acetic acids. Aspirin is combined with such compounds as phenacetin (acetophenetidin) and caffeine in a number of proprietary preparations. Aspirin is the most widely used medicine in the United States. Generally, aspirin is much more effective for pain originating in joints and muscles than in the internal organs. Aspirin also performs well in the treatment of acute rheumatic fever and is a uricosuric drug (stimulates excretion of uric acid). The latter causes symptoms of gout. Large quantities of aspirin also are consumed by arthritics. Aspirin is not totally harmless even though used widely. It contributes to accidental poisoning of children and can cause allergic reactions among some individuals among other side effects.

ANALYSIS (Chemical). Analytical chemistry is that branch of chemistry which is concerned with the detection and identification of the atoms, ions, or radicals (groups of atoms which react as a unit) of which a substance is composed, the compounds which they form, and the proportions of these compounds which are present in a given substance. The work of the analyst begins with sampling, since analyses are performed upon small quantities of material. The validity of the result depends upon the procurement of a sample that is representative of the bulk of material in question (which may be as large as a carload or tankload). See accompanying photo.

For clarity and convenience of description, analytical chemistry may be divided into:

1. *Qualitative chemical analysis,* in which one is concerned simply with the identification of the constituents of a compound or components of a mixture, sometimes accompanied by observations (rough estimates) of whether certain ingredients may be present in major or trace proportions.

2. *Quantitative chemical analysis,* in which one is concerned with the amounts (to varying degrees of precision) of all or frequently of only of some specific ingredients of a mixture or compound. Classically, quantitative chemical analysis is divided into (a) *gravimetric analysis* wherein weight of sample, precipitates, etc., is the underlying basis of calculation, and (b) *volumetric analysis* (titrimetric analysis) wherein solutions of known concentration are reacted in some fashion with the sample to determine the concentration of the unknown. Obviously, the figures from either gravimetric or volumetric determinations are convertible and the two methodologies frequently are combined in a multistep analytical procedure.

Analytical methods are affected not only by the type of material under examination, but also by the objectives of the analysis. *Ultimate analysis,* for example, is a term used to describe the determination of the proportions in which the elements are present in an organic substance, such as the amounts of and the ratio of carbon to hydrogen. This is normally a relatively simple procedure and often will provide answers sufficient to the need—as contrasted with the usually much more complex procedures required to identify specific compounds in a mixture of organic compounds. The term *functional group analysis,* where the analyst is concerned with the determination of specific organic groups and radicals, such as the carboxyl or carbonyl groups, also is used. In the case of solid fuels, such as coal, the term *proximate analysis* is used to describe the determination of moisture, volatile matter, fixed carbon, and ash. Although similar objectives (ultimate analysis and functional group analysis) also arise in connection with the analysis of inorganic materials, no special term applies.

Analytical chemistry also may be classified in terms of kinds of procedures in addition to those already mentioned:

3. *Classical laboratory, manual methods* conducted on a macroscale where sample quantities are in the range of grams and several milliliters. These are the techniques that developed from the earliest investigations of chemistry and which remain effective for teaching the fundamentals of analysis. However, these methods continue to be widely used in industry and research, particularly where there is a large variety of analytical work to be performed. The equipment, essentially comprised of analytical balances and laboratory glassware, tends to be of a universal nature and particularly where budgets for apparatus are limited, the relative modest cost of such equipment is attractive.

4. *Microchemical methods* which essentially extend macro-scale techniques so that they may be applied for determinations involving very small (milligram) quantities of samples. These methods have required fully new approaches or extensive modifications of macro-scale equipment. Consequently, the apparatus usually is sophisticated, relatively costly, and requires greater manipulative skills. Nevertheless, microchemical methods opened up entirely new areas of research, making possible the determination of composition where the availability of samples, as in many areas of biochemistry, was confined to very small quantities.

5. *Semi-automated apparatus* which introduces an interim step between (a) macro-scale and microchemical analysis techniques on the one hand and (b) fully instrumented and automated analytical methods on the other hand. Significant design changes in chemical balances that greatly increase the speed of weighing samples and reagents and automatic and self-refilling burettes are examples of ways in which an analytical procedure can be "tooled" to conserve manpower, reduce drudgery, and often contribute to more reliable and precise results.

6. *Analytical instrumentation* which first appeared in a major way in the 1930s and which has been expanding at a fast pace ever since, has revolutionized analytical chemistry. In-

strumental procedures range from essentially the replication on a miniature scale of a highly automated laboratory analytical procedure involving chemical reagents and reactions (represented by automatic chromatographs) to numerous techniques which essentially infer the presence and quantities of substances from the manner in which the unknowns react with various forms of energy. The latter procedures are extremely *unlike* former conventional laboratory techniques. Of course, over the years, there were harbingers of later-generation analytical instruments in the form of early polariscopes and spectroscopes. Modern analytical instruments range from small, inexpensive table-top black boxes, such as simple electrometric pH and electric-conductivity devices, up to very large, costly mass spectrometers and other forms of electronic, photometric, and spectrophotometric apparatus. Analytical instrumentation also ranges widely from the standpoint of manual assistance and intervention required in its application. Usually as the equipment becomes more sophisticated, less (if any) direct attention is required in the performance of the analyses, but greater skill and time are required in the initial planning of an analytical procedure.

7. *Process analyzers* represent the ultimate extension of analytical instrumentation, moving the chemical control laboratory from a central location, dependent upon grab samples, out to the process where the chemical composition changes to be measured are occurring within process vessels and pipelines. Many of the laboratory-type analytical instruments, at least in principle, are applicable to on-line situations, but new problems arise, such as effective sampling systems, the maintenance of delicate equipment in severe environments, and arranging for back-up control when there are equipment failures. On-line analytical equipment consequently is usually costly, both in terms of initial investment and of maintenance. Nevertheless, the improved control wherein long time lags are virtually eliminated has justified the installation of tens of thousands of such systems.

Analytical Instrumentation. Chemical-composition variables are measured by observing the interactions between matter and energy. See accompanying table. That such measurements are possible stems from the fundamental that all known matter is comprised of complex, but systematic arrangements of particles which have mass and electric charge. Thus, there are neutrons which have mass but no charge; protons which have essentially the same mass as neutrons with a unit positive charge; and electrons which have a negligible mass with a unit negative charge. The neutrons and protons comprise the nuclei of atoms. Each nucleus ordinarily is provided with sufficient orbital electrons, in what is often visualized as a progressive shell-like arrangement of different energy levels, to neutralize the net positive charge on the nucleus. The total number of protons plus neutrons determines the atomic weight. The number of protons which, in turn, fixes the number of electrons, determines the chemical properties and the physical properties, except mass, of the resulting atom.

The chemical combinations of atoms into molecules involve only the electrons and their energy states. Chemical reactions involving both structure and composition generally occur by loss, gain, or sharing of electrons among the atoms. Thus, every configuration of atoms in a molecule, crystal, solid, liquid, or gas may be represented by a specific system of electron energy states. Also, the particular physical state of the molecules, as resulting from their mutual arrangement, also is reflected upon these energy states. Fortunately, these energy states, characteristic of the composition of any particular substance, can be in-

ferred by observing the consequences of interaction between the substance and an external source of energy.

External energy sources used in analytical instrumentation include:

1. electromagnetic radiation
2. electric or magnetic fields
3. chemical affinity or reactivity
4. thermal energy
5. mechanical energy

The interaction of electromagnetic radiation with matter yields fundamental information as the result of the fact that photons of electromagnetic radiation are emitted or absorbed whenever changes take place in the quantized energy states occupied by the electrons associated with atoms and molecules. X-rays (photons or electromagnetic wave packets with relatively high energy) penetrate deeply into electron orbits of an atom and provide, upon absorption, the large quantity of energy required to excite one of the innermost electrons. Thus, the pattern of x-ray excitation or absorption is relative to the identity of those atoms whose orbital electrons are excited, ideally suiting x-ray techniques for determining atoms and elements in dense samples. But, because of the penetrating powers of x-rays, they are not suited to the excitation of low-energy states which correspond to outer-shell or valence electrons; or of the interatomic bonds which involve vibration or rotation.

In contrast, the relatively longer wavelengths of infrared radiation (photons having relatively low energy) correspond to the energy transformations involved in the vibration of atoms in a molecule as resulting from stretching or twisting of the interatomic bonds. Thus, because the penetrating power of electromagnetic radiation varies over the total spectrum, an instrumental irradiation technique can be developed for almost any analytical instrumentation requirement.

The interaction of matter with electric or magnetic fields is widely applied for determining chemical composition. The mass spectrometer, for example, which uses a combination of electric and magnetic fields to sort out constituent ions in a sample, takes full advantage of this interaction. A simple electric-conductivity apparatus determines ions in solution as the result of applying an electric potential difference across an electrolyte.

Application of Chemical-Composition Measurements

There are few phases of industrial operations where chemical composition variables are not important. Following are some of the applications of major significance:

Raw Materials:
1. Composition analysis to check purchase specifications.
2. Detection of contamination by trace impurities.
3. Analysis check on materials priced on an active-ingredient basis.
4. Continuous analysis of materials delivered by pipeline; water analysis.

Process Control:
1. Speed up and improve control by automatization of, or replacement of, control laboratory tests on "grab" samples.
2. Improve control by replacing or augmenting inferential measurements, such as temperature or pressure, with more significant composition data.
3. Permit use of continuous processes that could not be controlled except by continuous analysis instrumentation.

Process Trouble Shooting:
1. Temporary use of analysis instruments for process studies aimed at overcoming occasional upsets.

ENERGY-MATTER INTERACTIONS UTILIZED IN ANALYTICAL INSTRUMENTATION

Analytical Instrumentation Techniques *See separate editorial entries with the approximate titles shown below.	Interaction with Electromagnetic Radiation			Reaction to Electric Magnetic Fields		Interaction with Other Chemicals			Interaction with Thermal and Mechanical Energy	
	Emitted Radiation		Transmission and Reflection Measurements	Electrical Properties	Magnetic Properties	Sample or Reactant Consumed	Thermal Energy Liberated	Equilibrium Solution Potential	Thermal Energy	Mechanical Energy
	Thermally Excited	Electromagnetically Excited								
Amperometer*				X						
Aulyzers, Automatic distillation type									X	
Analyzers, reaction-product*						X	X			
Analyzers, reagent-tape*						X				
Atomic absorption Spectrometry*	X									
Beta-ray chemical analyzers*			X							
Beta-ray spectrometry			X							
Bioluminescence and chemiluminescence meters*						X				
Boiling-point chemical analyzers*									X	
Chromatography*									X	
Colorimetry*			X							
Dielectric-constant chemical analyzers*				X						
Ebulliometer*									X	
Electric-conductivity measurements*				X						
Electrolysis-type chemical analyzers*				X						
Electrophoresis*				X						
Flame photometry and spectrometry*	X									
Fluorometers*		X								
Gamma-ray spectroscopy*			X							
Gas analyzers, combustion-type*							X			
Gas analyzers, thermal-conductivity type*									X	
Indicator, chemical*						X				
Infrared analytical techniques*			X							
Magnetic resonance spectroscopy					X					
Mass spectrometry*				X	X					
Nephelometry*			X							
Optical-emission spectrochemical analysis	X									
Orsat analyzers						X				
Oscillometers*				X						
Oxygen analyzers				X	X	X		X	X	X
pH (hydrogen ion concentration)*								X		
Polarimetry*			X							
Polarographic analyzers*								X		
Radioactivation analysis (See Radioactivity)		X								
Radiometric analysis			X							
Radioactive tracer techniques (See Radioactivity)		X								
Raman spectrometry*		X								
Redox (oxidation-reduction potential)								X		
Refractometers*			X							
Saccharimeter*			X							
Specific gravity*										X
Spectro instruments*			X							
Spectrochemical analysis, visible*			X							
Spectrography*			X							
Spectroscope*			X							
Thermoanalyzers									X	
Titrators, automatic*						X				
Titration, thermometric*						X				
Turbidimetry*			X							
Ultraviolet analytical techniques*			X							
X-ray analytical instruments*		X	X							

Note: Some kinds of analytical instruments, such as oxygen analyzers, utilize more than one of the energy-matter interactions. This table is not exhaustive. A number of techniques of lesser importance are not indicated.

Yield Improvement:
1. Continuous analysis of process streams to measure effects of variables influencing yields.
2. Analysis of overflow or purge streams, recirculated material, sumps, and the like to determine product losses and detect buildup of undesirable by-products that affect yield.

Inventory Measurements:
1. Analytical monitoring of material flowing between process steps and plant areas to establish consumption and in-process inventory on the basis of active or essential ingredients.

Product Quality:
1. Determination of product composition.
2. Assess structurally dependent attributes, such as color, melting or boiling points, and refractive index.
3. Assist in adjustment of product to meet specifications.

Safety:
1. Detection of leaks in equipment.
2. Survey operating areas for escape of toxic materials from leaks or spills, especially materials not readily detected by human senses.
3. Detection of flammable or explosive mixtures in atmosphere or process lines.

Waste Disposal:
1. Monitoring plant stacks for accidental discharge of toxic or nuisance gases, vapors, or smokes.
2. Analysis of waste streams for toxic or other objectionable materials.
3. Control of waste treatment or product recovery facilities.

Research and Development:
1. Continuous analysis to speed up research and optimize results.
2. Provide structural and compositional information not otherwise obtainable.
3. Produce results in a more directly usable form.

In considering the application of instrumental methods for the determination of chemical composition, it is important to bear in mind that measurement is the first step toward control and that the closer the information can be brought to the process, the better will be the control. This is true whether the information is merely presented continuously to the operator or actually used to control the process itself automatically. Complete automatic control becomes more desirable as process throughout is increased and holdup is decreased in modern high-speed processing equipment. This trend also places a premium on high speeds of response in analytical instruments and their associated control equipment.

Practical Considerations in the Measurement of Chemical Composition

Any practical appraisal of the merits of chemical-composition variables for process-control purposes must recognize certain inherent physical limitations in their measurement. Generally speaking, these limitations include the following:
1. *Sample must be representative.* Although the requirement may appear obvious, it is a factor that is frequently overlooked. In the first place, the sample must be gathered or drawn off in such a fashion that it will consist of the same composition as the body of the processed material. Moreover, there must be assurance that any changes in conditions, such as temperature or pressure, between the sampling and measuring points cannot influence sample composition. In addition, in nearly all cases, the probable composition of the sample must be known ahead of time through some independent method before an analysis

technique can be selected for on-line use. These conditions differ markedly from those of the laboratory.
2. *Physical state of sample.* The technique must provide for interaction between the applied energy and the entire sample, as well as for the observation of the total result. This seldom can be accomplished. It is for this reason that a large majority of the techniques are applicable to gases, where the molecules are widely spaced and free to react in a characteristic manner, and that fewer techniques are applicable to liquids, and still fewer can be applied to solids.
3. *Uniqueness or specificity of method.* The selection of the method must be customized to the sample composition and to the information requirements. Some methods or techniques involving atomic and molecular structure are rather universal in that they permit exact identification and measurement of every elemental or molecular constituent present in the sample. These methods are usually the most complex and costly. They are sometimes considerably less sensitive than simpler methods whose only drawbacks are an inability to distinguish between related substances having similar gross interactions with energy. Where the related substances are known not to be present in the sample, these simpler, less specific methods should always be considered.

Trends in Analytical Instrumentation

Much progress has been made in chemical analysis instrumentation in the past and further is expected in the future. Several factors contribute to the need for continuing developments and refinements. These include:

(1) Conservation of energy has increased efforts to use on-line analytical instruments. Fuel costs have risen over the past decade to a point where on-line instrumentation has become more cost effective. One immediate area affected by this is the application of the oxygen and combustibles analyzer for optimum fuel-to-air ratios in the combustion process and the resulting heat generation. Likewise, calorimetric analysis of fuel quality is becoming even more important, as are density and specific gravity.

(2) Requirements for monitoring pollution generated additional needs for high quality analytical devices that have the capability to provide good records of the various pollutants and particulate emissions. Stack gas monitors are widely used for pollution control and monitoring with design features for long-term unattended operation within stacks or through sampling systems.

(3) Pressing demands for more accurate, thorough, and rapid means for testing materials and products—from the raw materials receiving and inspection throughout manufacturing to the completion of production, warehousing, and distribution.

Just as the availability of better analysis instruments since the 1950s has moved the formerly isolated chemical control laboratory *on-stream* in the form of continuous instrumental analyzers, the advent of improved nondestructive inspection techniques and fully automated testing procedures will complete a revolution already underway in the quality control of production, particularly in the discrete-piece manufacturing industries.

In the past 5–10 years, the replacement of mechanical timers and logic sequencers with microprocessors for automated sequencing and data reduction has not only improved reliability, accuracies, and ease of operation and maintenance, but also has allowed further development in enhanced production strategies through interfacing with hierarchial computer systems, such as *data acquisition and management information systems*. Automated calibration, self-diagnostics of electronics and sequencing operation have further improved reliability and have elevated the importance of on-stream analyzers as primary measuring devices for process control. Microprocessors allow easy programming of all functions required for each individual application and can, of course, be reprogrammed for a much wider field of applications as compared with a single analyzer with conventional analog control function, timer, cams, sequencers, and the like.

Chemists of the mid-1980s who are concerned with analysis in the laboratory or on-line in processing have routinely mastered the subscience of chemical analysis instrumentation. Unfortunately, the scope of this volume does not permit more detailed descriptions of this important topic. For those persons who are just entering the field, reference to the following publications as a bridge to further knowledge is suggested.

References

Considine, D. M. (editor): "Process Instruments & Controls Handbook," 3rd Edition, McGraw-Hill, New York, 1984. (Several hundred pages of this publication are devoted to chemical analytical techniques, Handbook Section 6).
Dean, J. A.: "Chemical Separation Methods," Van Nostrand Reinhold, New York, 1969.
Dilts, R. V.: "Analytical Chemistry," Van Nostrand Reinhold, New York, 1974.
Elving, P. J., and I. M. Kolthoff (editors): "Chemical Analysis," series of monographs, Wiley, New York, various dates (1959–1972).
Gouw, T. H. (editor): "Guide to Modern Methods of Instrumental Analysis," Wiley, New York, 1972.
Kodama, K.: "Methods of Quantitative Inorganic Analysis: An Encyclopedia of Gravimetric, Titrimetric, and Colorimetric Methods," Wiley, New York, 1963.
Meites, L.: "Handbook of Analytical Chemistry," McGraw-Hill, New York, 1963.
Snell, Foster Dee: "Encyclopedia of Industrial Chemical Analysis," seventeen volumes, Wiley, New York, various dates (1966–1972).

ANA-POSITION. The position of two substituent groups on atoms diagonally opposite, in alpha-positions on symmetrical rings, as the 1,5 or the 4,8 positions (which are identical) of the naphthalene ring.

ANATASE. The mineral anatase, TiO_2 crystallizing in the tetragonal system is relatively uncommon. It occurs as a trimorphous form of TiO_2 with rutile and brookite: rutile and anatase with tetragonal crystallization, brookite with orthorhombic. Originally named *octahedrite* from its pseudo-octahedral, acute pyramidal crystal habit. Hardness, 5.5–6; specific gravity, 3.82–3.97; brittle with subconchoidal fracture; color, shades of brown, into deep blue to black; also colorless, grayish and greenish. Transparent to opaque with adamantine luster.

Occurs as an accessory mineral in both igneous and metamorphic rocks; gneisses and schists. Fine crystals have been found in Arkansas in the United States, and in Switzerland.

ANDALUSITE. An aluminum silicate corresponding to the formula Al_2SiO_5, and is one of a three-member polymorphous group consisting of andalusite, sillimanite, and kyanite. Andalusite occurs in contact-metamorphic shales, and in rocks of regional metamorphic origin in association with sillimanite and kyanite. Andalusite crystallizes in the orthorhombic system, developing coarse prisms of approximately square cross-section, but may be massive or granular. It shows a distinct cleavage parallel to the prism; hardness 6.5–7.5; specific gravity, 3.13–3.16; vitreous luster; colorless to white, gray, brown, greenish or reddish; streak, white; transparent to opaque.

This mineral is named for its original locality, Andalusia, Spain. A variety of andalusite, chiastolite, has carbonaceous impurities so oriented that they produce a cross or a tesselated figure at right angles to the prism. Chiastolite comes from the Greek word meaning a cross. Localities are the Urals, the Alps, the Tyrol, the Pyrenees, Australia, and Brazil; in the United States, at Standish, Maine; Sterling and Lancaster, Massachusetts; Delaware County, Pennsylvania; and Madera County, California.

When clear it is used as a gem, and it has also been used to manufacture porcelain for spark plugs.

ANDESITE. A term originally applied to a porphyritic lava from the Andes Mountains by Leopold Van Buch. In modern terminology andesite is an extrusive igneous rock, the surface equivalent of diorite. In other words, it is composed chiefly of plagioclase, corresponding in chemical composition to oligoclase or andesine together with biotite, hornblende, or pyroxene in varying quantities.

Andesites are of rather widespread occurrence, being found in the Rocky Mountains, California, Alaska, South America, and in many other localities.

ANDROGENS. Male sex hormones. The androgenic hormones are steroids and are synthesized in the body by the testis, the cortex of the adrenal gland, and to a lesser degree, by the ovary. See also **Steroids.**

The adrenal cortex produces hydroisoandrosterone which is found in blood and urine largely conjugated as the sulfate ester. The amounts of androgen secreted by the normal adrenal cortex are insufficient to maintain reproductive function in the male. The normal human ovary and placenta also produce small amounts of androgenic steroids that serve as precursors for the estrogens in these tissues. In the human, little testosterone is excreted into the urine and virtually none into the feces. The principal metabolic transformation products are androsterone and 5β-androsterone, with small amounts of other reduced compounds. These substances are excreted in the urine in the form of esters with sulfuric acid or glycosides with glucuronic acid. See Fig. 1.

Like other classes of steroid hormones, the androgens are synthesized from acetyl coenzyme A *via* mevalonic acid, isopentenyl pyrophosphate, farnesyl pyrophosphate, squalene, lanosterol, and cholesterol. Enzyme systems in the testis then catalyze the cleavage of the sidechain of cholesterol to pregnenolone which can give rise to testosterone by the two pathways shown in Fig. 2.

Testosterone is formed by the interstitial or Leydig cells of the testes which develop under the influence of gonadotrophic hormones discharged into the bloodstream by the anterior pitu-

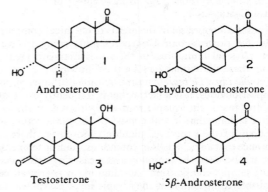

Androsterone   Dehydroisoandrosterone

Testosterone   5β-Androsterone

Fig. 1.   Androsterone and related hormones.

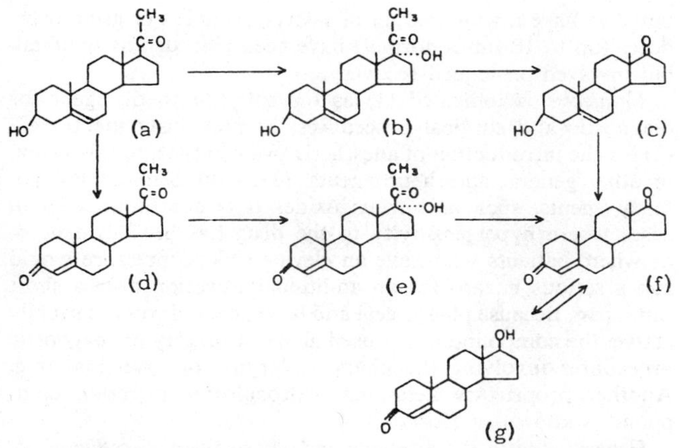

Fig. 2. Biosynthesis of testosterone: (a) Prognenolone; (b) 17-hydroxy pregnenoline; (c) dehydroisoandrosterone; (d) progesterone; (e) 17-hydroxprogesterone; (f) androstenedione; (g) testosterone.

itary gland. In pituitary insufficiency, this hormonal stimulus is lacking and, as a consequence, the Leydig cells do not secrete testosterone. In such instances, the male secondary sex characteristics fail to develop. However, interstitial cell tumors may occur, leading to excessive androgen production and precocious puberty. In women, tumors or excessive function of the adrenal cortex and, rarely, of the ovary, result in the production of large amounts of androgens with associated virilization.

ANDROSTERONE. Steroids.

ANESTHETICS. Anesthetics are agents which, when suitably applied, cause a general or localized loss of feeling or sensation. Traditionally, anesthetics have been classified as follows: (1) *General anesthetics* exert their action on the higher nerve centers and produce involuntary loss of consciousness. General anesthetics may be further classified according to their physical properties as volatile (gases or low-boiling liquids) and nonvolatile (high-boiling liquids and solids). (2) *Local, block*, or *regional anesthetics* produce a loss of sensation within a restricted or readily predictable large segment of the body. These anesthetics may be further classified according to the manner of application. Thus, *topical anesthetics* are applied to peripheral nerve endings; *intra-neural anesthetics* infiltrate into the nerve fiber; *para-neural anesthetics* act around the nerve sheath; *intra-spinal anesthetics* are injected into the spinal canal. Depending upon usage, some of these anesthetics are further categorized as caudal or rectal anesthetics. Anesthetics also may be categorized by their chemical structure, their immediate and residual functions or persistence, their side-effects, and their safety in usage, among other factors. It is not uncommon to use a combination of anesthetics in various procedures—for example, taking advantage of the characteristics of one compound for its rapid action coupled with excellent control, while using another compound to maintain a state of anesthesia over comparatively long periods where complex procedures may be involved. The recovery aspects of various anesthetics range rather widely. Anesthetics may also be classified in accordance with their mode of administration—inhalation, injection, surface application, etc. Some compounds possess sedative, hypnotic, and analgesic properties in addition to producing anesthesia. In some cases, minor structural alteration of some compounds may accentuate one or more of these properties while diminishing others.

General Anesthetics

Although general anesthetics had been sought by surgeons for centuries, until the 19th century pain was only inadequately overcome by wine, whiskey, and opium. Although ether (diethyl ether) had been discovered by Valerius Cordus in 1540 as a sleep-producing substance, the possible use of it as an anesthetic for use in surgery did not occur until 1842 when Crawford Long, an American surgeon, first used it during an operative procedure for the removal of tumors on the neck of a patient. Long did not publish these results, however, and thus W. T. G. Morton, an American dentist, who publicly demonstrated the efficacy of ether during a tooth extraction, was generally accredited as the father of ether anesthesia. Nitrous oxide, originally prepared by Priestley in 1772 and known as "laughing gas," was first used for its anesthetic properties by Humphrey Davy in 1800. J. Y. Simpson of Edinburgh first introduced chloroform for relieving the pain in childbirth during the early 1800s. Inhalation-type general anesthetics introduced and used over various spans of time have included hydrocarbons, such as cyclopropane and ethylene; halogenated hydrocarbons, such as chloroform, ethyl chloride, and trichloroethylene; ethers, such as diethyl ether and vinyl ether; and a miscellaneous category, including tribromoethanol (*Avertin*), nitrous oxide, and barbiturates.

Over the years a number of these compounds has been eliminated, mainly for reasons of unsafe usage. Some, such as ether, cyclopropane, ethylene, trichlorethylene, are flammable and capable of producing explosive mixtures unless very carefully monitored. Others, such as chloroform, have been found carcinogenic.

As of the early 1980s and for some period in the past, *halothane* has been widely used. In this category, the proprietary compound *Fluothane®* is 2-bromo-2-chloro-1,1,1-trifluroroethane, with a boiling point of about 51°C (760 millimeters mercury pressure). The compound is nonflammable and its vapors mixed with oxygen in proportions from 0.5 to 50% (volume) are not explosive. Stability of the compound is maintained by the addition of 0.01% thymol (weight) and by storage in amber-colored bottles. *Fluothane®* is an inhalation anesthetic. Induction and recovery are rapid and depth of anesthesia can be rapidly altered. The compound progressively depresses respiration. There may be tachypnea with reduced tidal volume and alveolar ventilation. The compound is not an irritant to the respiratory tract, and no increase in salivary or bronchial secretions ordinarily occurs. Pharyngeal and laryngeal reflexes are rapidly obtunded. It causes bronchodilation. Hypoxia, acidosis, or apnea may develop during deep anesthesia. *Fluothane®* is not recommended for obstetrical anesthesia except when uterine relaxation is required.

Another general, inhalation-type anesthetic in prominent usage today is *methoxyflurane*, which is 2,2-dichloro-1,1-difluoroethyl methyl ether. The proprietary compound, *Penthrane®*, has a boiling point of about 105°C. The flash point in air is 62.8°C; in oxygen (closed system), 32.8°C; and in nitrous oxide 50% with 50% oxygen, 28.2°C. The compound has a mildly pungent odor. An antioxidant, butylated hydroxytoluene (0.01% weight) is added to insure stability on standing. After surgical anesthesia with *Penthrane®*, analgesia and drowsiness may persist after consciousness has returned and this may obviate or reduce the need for narcotics in the immediate postoperative period. When used alone in safe concentration, the compound will not produce appreciable skeletal muscle relaxation. Thus a muscle relaxing agent, such as succinylcholine chloride, may be used as an adjunct.

Penthrane® is indicated, usually in combination with oxygen and nitrous oxide, to provide anesthesia for surgical procedures in which the total duration of *Penthrane®* is anticipated to be 4 hours or less, and in which the compound is not used in concentrations that will provide skeletal muscle relaxation. *Penthrane®* may be used alone with hand-held inhalers or in combination with oxygen and nitrous oxide for analgesia in obstetrics and in minor surgical procedures. To date, the safe use of the compound other than for obstetrics has not been established with respect to adverse effects upon fetal development. Therefore, the compound is not used in women of childbearing potential and particularly during early pregnancy unless, in the judgment of the physician, the potential benefits outweigh the possible hazards.

Numerous side-effects are attributable to the aforementioned compounds and, in fact, to all kinds of anesthetics, but such effects occur in a minority of cases. The expertese of the physician and anestheologist are one's best protection against these effects. These effects are described in some detail in the *Physicians' Desk Reference*, listed with other references at the end of this entry.

One of the most serious of side-effects of methoxyflurane is acute and chronic renal failure (Halpren, et al., 1973).

Noted during the last decade, *malignant hyperthermia* is described as a rare and dramatic complication of anesthesia. This syndrome can develop during the induction with practically any anesthetic agent, but notably halothane or succinylcholine. Development of muscular rigidity, rapidly rising body temperature, acidosis, cardiac arrhythmias, circulatory collapse, coma, and death characterize the syndrome. A body temperature of 43.5°C (110.3°F) may develop. Present in about 80% of the cases, the muscular rigidity may disallow the use of intubation and assisted ventilation. Causation is poorly understood, but some authorities believe that this may be due to uncoupled mitochondrial respiration. There may be a genetically determined susceptibility and some authorities suggest that screening for an elevated serum CPK level may be useful in assessing risk. Mortality is high and appears to be correlated with the degree of hyperthermia. Upon indication of malignant hyperthermia, anesthesia is ceased immediately, efforts are made to cool the body, oxygen is administered, along with infusion with intravenous bicarbonate. Intravenous procaine also has been advocated.

For diagnostic and surgical procedures that do not require skeletal muscle relaxation, ketamine hydrochloride injection is frequently used. One of the proprietary products, *Ketaject®*, is a nonbarbiturate anesthetic chemically designated DL-2-(*o*-chlorophenyl)-2-(methylamino) cyclohexanone hydrochloride. It is formulated as a slightly acid (pH = 3.5–5.5) solution for intravenous or intramuscular injection. *Ketaject®* is a rapid-acting general anesthetic producing an anesthetic state characterized by profound analgesia, normal pharyngeal-laryngeal reflexes, normal or slightly enhanced skeletal muscle tone, cardiovascular and respiratory stimulation, and occasionally a transient and minimal respiratory depression. A patent airway is maintained partly by virtue of unimpaired pharyngeal and laryngeal reflexes. The anesthetic state produced by this compound has been termed by some authorities as "dissociative anesthesia," in that it appears to selectively interrupt association pathways of the brain before producing somesthetic sensory blockade. It may selectively depress the thalamoneocortical system before significantly obtunding the more ancient cerebral centers and pathways (recticular-activating and limbic systems). Elevation of blood pressure begins shortly after injection, reaches a maximum within a few minutes, and usually returns to preanesthetic values within 15 minutes after injection. The compound is attri-

buted to have a wide margin of safety. Instances of gross overdoses (up to 10 times normal) have been recorded as nonfatal, but involved prolonged recovery.

Ketaject® is indicated (1) as the sole anesthetic agent for diagnostic and surgical procedures, as previously mentioned; (2) for the introduction of anesthesia prior to the administration of other general anesthetic agents; (3) to supplement low-potency agents, such as nitrous oxide. It is contraindicated in cases where hypersensitivity to the drug has been evidenced, or where patients who have an elevated blood pressure would run a serious hazard for an additional elevation over a short time span. Because pharyngeal and laryngeal reflexes are usually active, the compound is not used alone in surgery or diagnostic procedures involving the pharynx, larynx, or bronchial tree. Another proprietary ketamine hydrochloride injection compound is known as *Ketalar®*.

General Anesthesia Adjuncts and Short-Term Anesthetics. A number of compounds have been developed that are used to enhance and assist in the performance of general anesthetics. The expertise of the anesthesiologist is required in selecting the numerous combinations of compounds available because all such compounds are not without conflict and certain combinations may put the patient in jeopardy.

Certain *injectable barbiturates* act as an ultrashort-acting depressant of the central nervous system which induce hypnosis (within 30 to 40 seconds of intravenous injection) and anesthesia, but not analgesia. One of these compounds is thiopental sodium (*Pentothal®*), which is the sulfur analogue of sodium pentobarbital. Chemically, the compound is sodium 5-ethyl-5-(1-methylbutyl)-2-thiobarbiturate. Recovery after a small dose is rapid, with some somnolence and retrograde amnesia. Repeated intravenous doses lead to prolonged anesthesia because fatty tissues act as a reservoir; they accumulate the drug in concentrations 6–12 times greater than the plasma concentrate, and then release the drug slowly to cause prolonged anesthesia. Thiopental sodium is indicated: (1) As the sole anesthetic agent for brief (15-minute) procedures; (2) for induction of anesthesia prior to administration of other anesthetic agents; (3) to supplement regional anesthesia; (4) to provide hypnosis during balanced anesthesia with other agents for analgesia or muscle relaxation; (5) for the control of convulsive states during or following inhalation anesthesia, local anesthesia, or other causes; (6) in neurosurgical patients with increased intracranial pressure, if adequate ventilation is provided; and (7) for narcoanalysis and narcosynthesis in psychiatric disorders. Contraindications include: (1) Absence of suitable veins for intravenous administration; (2) hypersensitivity to barbiturates; (3) status asthmaticus; and (4) latent or manifest porphyria.

Another injectable barbiturate, thiamyial sodium (*Surital®*), has the formula sodium-5-allyl-5-(1-methylbutyl)-2-thiobarbiturate. This compound is indicated for induction of anesthesia, for supplementing other anesthetic agents, as an intravenous anesthesia for short surgical procedures with minimal painful stimuli, and as an agent for inducing a hypnotic state. Contraindications are similar to those previously described for *Pentothal®*.

Injectable *succinylcholine chloride solution* is a diquaternary base consisting of the dichloride salt of the dicholine ester of succinic acid. Sterile solutions of the drug lose potency unless refrigerated. The solutions have a pH of about 4.0. Succinylcholine is indicated as an adjunct to anesthesia to induce skeletal muscle relaxation. It may be used to reduce the intensity of muscle contractions of pharmacologically or electrically induced convulsions. The safe use of succinylcholine has not been established with respect to the possible adverse effects upon fetal development. Therefore, it is not used in women of childbearing potential and particularly during early pregnancy unless in the judgment of the physician the potential benefits outweigh the potential hazards. Extra precautions are taken in considering the drug in patients with severe burns, who are digitalized, or those who are recovering from severe trauma because serious cardiac arrhythmias or cardiac arrest may result. Caution is also observed in patients with preexisting hyperkalemia and those who are paraplegic, who have suffered spinal cord injury, or have degenerative or dystrophic neuromuscular disease, as such patients tend to become severely hyperkalemic when succinylcholine is given.

Succinylcholine is a depolarizing skeletal muscle relaxant. Like acetylcholine, it combines with the cholinergic receptors of the motor endplate to produce depolarization followed by an initial muscle contraction often visible as fasciculations. Neuromuscular transmission is then inhibited and remains so as long as an adequate concentration of the compound remains at the receptor site. The neuromuscular block caused by succinylcholine produces a flaccid paralysis. It is hydrolyzed by plasma pseudocholinesterase at such a rate that the effect of a single paralyzing dose normally disappears within 8–10 minutes. Depolarization of the motor endplate gradually fades away as the concentration of succinylcholine decreases at the receptor sites. Then the endplate assumes its state of normal polarity and can again generate an electrical potential of sufficient intensity to initiate muscular contraction.

Following intravenous injection of an effective dose, complete muscular relaxation occurs within one minute, persists for about two minutes, and returns to normal within six minutes. Following intramuscular injection, onset of action may be delayed for up to three minutes.

Proprietary preparations of succinylcholine include *Anectine®*, *Quelicin®*, and *Sucostrin®*.

Regional Anesthetics

Several compounds are available for production of local anesthesia by caudal, epidural, or peripheral nerve block. *Bupivacaine hydrochloride*, 1-butyl-2',6'-pipecoloxylidide hydrochloride, has an amide linkage between the aromatic nucleus and the amino or piperidine group. Anesthetics with this type of linkage are sometimes referred as amide-type local anesthetics. An injection-type anesthetic, bupivacaine hydrochloride stabilizes the neuronal membrane and prevents the initiation and transmission of nerve impulses, thereby effecting local anesthesia. The compound is not used for spinal anesthesia. Safe use of the drug in pregnant women, other than those in labor, has not been established. Onset is rapid and action is prolonged. Anesthesia may persist for several hours. The duration of action is sometimes extended by the addition of a dilute concentration of epinephrine or other appropriate vasoconstriction drugs. Analgesia often persists after sensation has returned, thereby reducing the need for additional potent analgesics. Peak blood levels of bupivacaine hydrochloride are reached within 30–45 minutes following injection for caudal, epidural, or peripheral nerve block, and then decline substantially during the next three to six hours.

Adverse reactions may occur from use of the drug. These are characteristic of those associated with amide-type local anesthetics and are reported in some detail by Baker (1980).

Mepivacaine hydrochloride has a similar chemical structure and is an amide-type local anesthetic, 1-methyl-2',6'-pipercoloxylidide monohydrochloride. The compound is used much as bupivacaine hydrochloride. Safe use of the compound has not been established with respect to adverse effects on fetal development. Thus, careful consideration should be given to this before administration during pregnancy. Obstetricians are particularly careful because severe persistent hypertension, and even a rupture of a cerebral blood vessel, may occur after administration of certain oxytocic drugs when vasopressors have already been used during labor (e.g., in the local anesthetic solution or to correct hypotension).

Prilocaine is a local anesthetic, chemically designated as 2-propylamino-*o*-propionotoluidide. This drug stabilizes the neuronal membrane and prevents the initiation and transmission of nerve impulses, thereby effecting local anesthetic action. When used for infiltration anesthesia in obstetrical patients, the time of onset averages 1–2 minutes with an approximate duration of 60 minutes or longer. For major nerve blocks (e.g., epidural block), the onset of analgesia is approximately two minutes longer than for lidocaine (described later), whereas the duration of action is at least 30–60 minutes longer than for lidocaine. One proprietary preparation (*Citanest®*) contains no epinephrine, making it particularly useful for patients with hypertension, diabetes, thyrotoxicosis, or other cardiovascular diseases. Contraindications include use in patients with a known history of hypersensitivity to amide-type local anesthetics. The drug is not used in patients with congenital or idiopathic methemoglobinemia. Also, the drug is not used in patients with severe shock or heart block.

Etidocaine hydrochloride is another amide-type local anesthetic with the chemical structure (±)-2-(N-ethylpropylamino)-2',6'butroxylidide. Its mode of operation is essentially similar to the drugs just described, stabilizing the neuronal membrane. The initial onset of sensory analgesia and motor blockade is rapid (3–5 minutes), similar to that produced by lidocaine (described next). Duration of sensory analgesia is 1.5 to 2 times longer than that of lidocaine by peridural route. Duration of analgesia in excess of nine hours is not infrequent when the compound is used for peripheral nerve blocks, such as brachial plexus blockade. The drug produces a profound degree of motor blockade and abdominal muscle relaxation when used for peridural analgesia. The compound is indicated for percutaneous infiltration anesthesia, peripheral nerve blocks, and central neural blocks, i.e., caudal or epidural blocks. Contraindications are similar to other amide-type local anesthetics.

Lidocaine hydrochloride is a local anesthetic chemically designated as diethylaminoacet-2,6-xylidide. Lidocaine is best known for its use in the control of ventricular arrhythmias in acute myocardial infarction. Lidocaine hydrochloride is indicated for production of local or regional anesthesia by infiltration techniques, including percutaneous injection and intravenous regional anesthesia, by peripheral nerve block techniques, such as brachial plexus and intercostal blocks, and by central neural techniques, including epidural and caudal blocks. The safe use of lidocaine has not been established with respect to adverse effects upon fetal development and thus it is not given to women of childbearing potential, particularly during early pregnancy. The drug is used, however, for obstetrical analgesia. Contraindications are similar to other amide-type local anesthetics. Proprietary formulations of lidocaine include *LTA®*, *Xylocaine®*, and *Duranest®*. Lidocaine is also used as one of several ingredients in anesthetics for surface applications.

Chloroprocaine hydrochloride, with the chemical designation β-diethylaminoethyl-2-chloro-4-aminobenzoate hydrochloride, is used in caudal and epidural anesthesia. The parenteral administration of the drug stabilizes the neuronal membrane and prevents the initiation and transmission of nerve impulses. The decision whether or not to use chloroprocaine hydrochloride in certain instances depends upon the judgment of advantages and risks by the physician. Such situations include: (1) paracervical block when fetal distress is anticipated or when predisposing factors causative of fetal distress are present (i.e., toxemia, prematurity, diabetes, etc.); (2) injection of solutions containing epinephrine in areas where the blood supply is limited (i.e., ears, nose, digits, etc.) or when peripheral vascular disease is present; (3) when there is potential for serious cardiac arrhythmias, which may occur if preparations containing a vasopressor are employed in patients during or following the administration of chloroform, halothane, cyclopropane, trichloroethylene, or other related agents.

Procaine hydrochloride (Novocain®) is a local anesthetic used for major and minor surgery. This compound is indicated for the production of local anesthesia by infiltration injection, nerve block, caudal, or other epidural blocks. The anesthetic can be used safely in nearly all cases if suitable precautions are observed. However, as with any local anesthetic, uncommon idiosyncrasy manifested by the rapid appearance of symptoms, including nausea, vomiting, rapid pulse, talkativeness, syncope, respiratory difficulty, and convulsions may be encountered. Extreme caution is imperative when any local anesthetic, even in relatively small amounts, is injected into the traumatized urethra or under conditions in which trauma is likely to occur. Procaine hydrochloride does not irritate tissues and has a low relative toxicity.

Topical (Local) Anesthetics

It has long been known that lowering the temperature of a portion of the body will make the perception of pain originating in the area of lowered temperature less apparent. Likewise, it is known that pressure applied to a nerve fiber or trunk will interfere with the transmission of pain impulses originating peripherally to the site of the pressure. The first principle is still occasionally used to produce local anesthesia. Advantage is taken of the low temperatures produced by the vaporization of such compounds as ethyl chloride, ether, and solid carbon dioxide to produce local anesthesia in restricted segments of the body.

The beginning of the modern period of local anesthesia by physiologically active compounds can be traced to the isolation and characterization of the alkaloid cocaine. See also **Alkaloids.** Its local anesthetic effect on the tongue was noted by Wöhler in the mid-1800s. Its use in surgery dates from the work of Koller who, in 1884, described the use of cocaine in surgery of the eye, nose, and throat. Koller also used cocaine by infiltration techniques.

A systematic investigation of the chemical nature of cocaine led to its synthesis by Willstätter in 1902 and to the development of a large number of synthetic compounds, including the procaine group of which *Novocain®* is a member. As a result of much work relating to the chemical constitution of cocaine analogs and local anesthetic activity, a number of general relationships were traced. It was found that alkyl esters of aromatic acids, such as benzoic and naphthoic acids, had potential local anesthetic activity. This was found to be enhanced by the presence of substituent groups in the aromatic ring, such as alkyl, amino, hydroxy, alkoxy, and alkylthio. Isosteric compounds, such as substituted amides and amidines, as well as thiocarboxylic esters and urethanes, retain the anesthesiophore.

Many of the simple alkyl esters of aromatic acids were only useful as topical anesthetics, since their water solubility was too low to permit their use in parenteral solutions. Of these, benzocaine (ethyl-4-aminobenzoate) is a prototype, with the *n*-propyl and *n*-butyl esters subsequently developed. It was found that the inclusion of a tertiary or secondary aliphatic

or cycloalkylamine in their structure made it possible to prepare water-soluble salts with acids which could be administered by parenteral pathways. This was also possible in the case of amides, thicarboxylic esters, and urethanes. Amidines which are of moderately high molecular weight are sufficiently basic to form salts which can be dissolved and used. A large number of individual compounds, many of which are proprietary, have been synthesized and a considerable number have been marketed. Many experimental compounds are rejected because of their locally irritant effects or systemic toxicity or both. Other compounds which, structurally, are potentially local anesthetics have found their greatest clinical use for other effects, such as antispasmodics.

A number of mildly anesthetic preparations are available for temporarily relieving pain in connection with itches, rashes, bites, sunburn, plant poisons, hives, hemorrhoids, and minor burns. These are marketed in a number of forms—ointments, creams, pastes, balms, suppositories, and sprays, among others. Typical compounds popular as of the early 1980s contain one or more of the following anesthetic substances: alkyl and aryl phenols, benzalkonium chloride, benzocaine, benzethonium chloride, benzyl alcohol, cetyl dimethyl ethyl, cetylpyridinium chloride, chloroprocaine hydrochloride, chlorothymol, cyclomethycaine sulfate, dibucaine hydrochloride, dimethisoquin hydrochloride, lidocaine hydrochloride, methylbenzethonium chloride, methyl salicylate, piperocaine hydrochloride, pramoxine hydrochloride, proparacaine hydrochloride (ophthalmic use), and sodium phenolate, among others.

References

Adriana, J.: "The Chemistry and Physics of Anesthesia," 2nd Edition, Charles C. Thomas, Springfield, Illinois (1979).

Baker, C. E., Jr.: "Physicians' Desk Reference," 34th Edition, Medical Economics Company, Oradell, New Jersey (1980).

Bennett, E. J.: Fluids for Anesthesia and Surgery in the Newborn and the Infant," Charles C. Thomas, Springfield, Illinois (1975).

Brechner, V. L.: "Pathological and Pharmacological Considerations in Anesthesiology," Charles C. Thomas, Springfield, Illinois (1973).

Cohen, D., and J. B. Dillon: "Anesthesia for Outpatient Surgery," Charles C. Thomas, Springfield, Illinois (1979).

deJong, R. H.: "Local Anesthetics," 2nd Edition, Charles C. Thomas, Springfield, Illinois (1977).

Eng, G. D., et al.: "Malignant Hyperthermia and Central Core Disease in a Child with Congenital Dislocating Hips," *Arch. Neurol.*, **35**, 189 (1978).

Gravenstein, J. S. et al.: "Monitoring Surgical Patients in the Operating Room," Charles C. Thomas, Springfield, Illinois (1980).

Halpren, B. A., et al.: "Interstitial Fibrosis and Chronic Renal Failure Following Methoxyflurane Anesthesia," *Jrnl. of Amer. Med., Assn.*, **223**, 1239 (1973).

Harrison, G. G.: "Anesthetic-Induced Malignant Hyperpyrexia," *Br. Med. Jrnl.*, **3**, 454 (1971).

Issacs, H., and M. B. Barlow: "Malignant Hyperpyrexia," *Jrnl. Neurol., Neurosurg., Psychiatry*, **36**, 228 (1973).

Lorhan, P. H.: "Anesthesia for the Aged," Charles C. Thomas, Springfield, Illinois (1971).

Moore, D.C.: "Regional Block: A Handbook for Use in the Clinical Practice of Medicine and Surgery," 4th Edition, Charles C. Thomas, Springfield, Illinois (1979).

Snow, J. C.: "Anesthesia: In Otolaryngology and Ophthalmology," Charles C. Thomas, Springfield. Illinois (1972).

ANGLESITE. Naturally occurring lead sulfate ($PbSO_4$), which crystallizes in the orthorhombic system and may be found mixed with galena, from which it is usually formed by oxidation. Hardness, 3; sp gr 6.12–6.39; luster, adamantine to vitreous or resinous; transparent to opaque; streak, white; colorless to white or green, but may be rarely yellow or blue. This mineral is a source of lead.

Anglesite, whose name derives from Anglesey, England, is found in many European localities. In the United States, it has been found in large crystals in the Wheatley Mine, Phoenixville, Pennsylvania; also in Missouri, Utah, Arizona, and Idaho.

ANHYDRIDE. A chemical compound derived from an acid by elimination of a molecule of water. Thus sulfur trioxide, SO_3, is the anhydride of sulfuric acid; CO_2 is the anhydride of carbonic acid; phthalic acid, $C_6H_4(CO_2H)_2$ minus water gives phthalic anhydride. The term should not be confused with *anhydrous*.

ANHYDRITE. The mineral anhydrous calcium sulfate, $CaSO_4$, occurs in granular, scaly, or fibrous masses, is rarely crystallized in orthorhombic tabular or prismatic forms. Hardness, 3–3.5; specific gravity, 2.9–2.98; translucent to opaque; streak white; color, white, gray, bluish or reddish. Anhydrite has three cleavages at right angles to one another. It is similar to gypsum and occurs under the same conditions, often with the latter mineral. It is usually found in sedimentary rocks associated with limestones, salt, and gypsum, into which it changes slowly by the absorption of water. See also **Gypsum.**

Anhydrite is found in Poland, Saxony, Bavaria, Württemberg, Switzerland, and France; in the United States in So. Dakota, New Mexico, Texas, New Jersey, and Massachusetts; in Canada in Nova Scotia, New Brunswick and exceptional specimens from the Faraday Uranium Mine near Bancroft, Ontario.

ANHYDROUS. Descriptive of an inorganic compound that does not contain water either adsorbed on its surface or combined as water of crystallization.

ANILINE. Also known as phenylamine or aminobenzene, $C_6H_5NH_2$, aniline is a colorless, odorous liquid, an amine, mp $-6°C$, bp $184°C$, slightly soluble in H_2O, miscible in all proportions with alcohol or ether, poisonous, turns yellow to brown in air, is a weak base, forming salts with acids, e.g., aniline hydrochloride, from which aniline is reformed by the addition of sodium hydroxide solution. Aniline reacts (1) with hypochlorite solution to form a transient violet coloration; (2) with nitrous acid (a) warm, to form nitrogen gas plus phenol, or (b) cold, to form diazonium salt (benzene diazonium chloride, $C_6H_5N \cdot Cl$); (3) with acetyl chloride, acetic anhydride, or acetic acid (glacial), to form N-phenylacetamide

acetanilide, "antifebrin," $C_5H_6N \diagdown_{OCCH_3}^{H}$

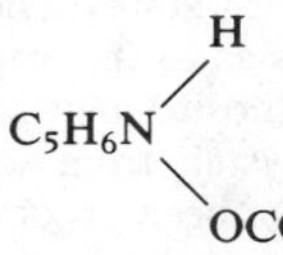

(4) with benzoyl chloride, to form N-phenylbenzamide

benzanilide, $C_5H_5N \diagdown_{OCC_6H_5}^{H}$

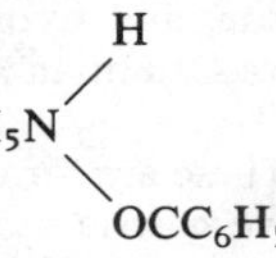

(5) with benzenesulfonyl chloride, to form N-phenylbenzene sulfonamide $C_6H_5SO_2NHC_6H_5$, soluble in sodium hydroxide; (6) with chloroform $CHCl_3$ plus alcohol plus sodium hydroxide, to form phenyl isocyanide C_6H_5NC, very poisonous; (7) with H_2SO_4 at 180° to 200°C, to form para-aminobenzene sulfonic acid (sulfanilic acid, $H_2N \cdot C_6H_4 \cdot SO_2H(1,4)$); (8) with HNO_3, when the amine group is protected, e.g., using acetanilide, to

form mainly paranitroacetanilide $CH_3CONH \cdot C_6H_4 \cdot NO_2(1.4)$, from which paranitroaniline $H_2N \cdot C_6H_4 \cdot NO_2(1.4)$ is obtained by boiling with concentrated hydrochloric acid; (9) with chlorine in an anhydrous solvent, such as chloroform or acetic acid glacial, to form 2,4,6-trichloroaniline $(1)H_2N \cdot C_6H_2Cl_3(2,4,6)$; (10) with bromine water, to form white solid 2,4,6-tribromoaniline $(1)H_2H \cdot C_6H_2Br_3(2,4,6)$; (11) with potassium dichromate in sulfuric acid, to form aniline black dye, and, by further oxidation, benzoquinone $O:C_6H_4:O(1,4)$; (12) with potassium permanganate in sodium hydroxide, to form azobenzene $C_6H_5N:NC_6H_5$ along with some azoxybenzene $C_6H_5NO:NC_6H_5$; (13) with reducing agents, to form aminohexahydrobenzene (cyclohexylamine, $H_2N \cdot C_6H_{11}$); (14) with alkyl halides or alcohols heated, to form alkyl anilines, e.g., methylaniline $C_6H_5NHCH_3$, dimethylaniline $C_6H_5N(CH_3)_2$.

Aniline may be made (1) by the reduction, with iron or tin in HCl, of nitrobenzene, and (2) by the amination of chlorobenzene by heating with ammonia to a high temperature corresponding to a pressure of over 200 atmospheres in the presence of a catalyst (a mixture of cuprous chloride and oxide). Aniline is the end-point of reduction of most mononitrogen substituted benzene nuclei, as nitrosobenzene, beta-phenylhydroxylamine, azoxybenzene, azobenzene, hydrazobenzene. Aniline is detected by the violet coloration produced by a small amount of sodium hypochlorite.

Aniline is used (1) as a solvent, (2) in the preparation of compounds as illustrated above, (3) in the manufacture of dyes and their intermediates, (4) in the manufacture of medicinal chemicals. See also **Amines.**

ANION. A negatively charged atom or radical. In electrolysis, an anion is the ion which deposits on the anode; that portion of an electrolyte which carries the negative charge and travels against the conventional direction of the electric current in a cell. Within the category of anions are included the nonmetallic ions and the acid radicals, as well as the hydroxyl ion (OH^-). In electrochemical reactions, they are designated by the minus sign placed above and behind the symbol, such as Cl^- and SO_4^{2-}, the number of the minus sign indicating the magnitude, in electrons, of the electrical charge carried by the anion. In a battery, it is the deposition of negative anions that makes the anode negative. See also **Ion.**

ANISODESMIC STRUCTURE. A type of ionic crystal in which some of the ions tend to form tightly bound groups, e.g., nitrate and chlorate.

ANISOTROPIC MEDIUM. An anisotropic medium has different optical or other physical properties in different directions. Wood and calcite crystals are anisotropic, while fully annealed glass and, in general, fluids at rest are isotropic.

ANNABERGITE. The mineral annabergite is a rather rare nickel arsenate with the formula $Ni_3(AsO_4)_2 \cdot 8H_2O$., crystallizing in the monoclinic system. It is of secondary origin, resulting from the alteration of preexisting nickel minerals, commonly found as surface alteration crust on nickeline. Annabergite has been found in Saxony (Germany); in France, including Annaberg, from which its name is derived; and as exceptional crystals at Laurium, Greece, and in Cobalt, Province of Ontario, Canada.

ANNATTO FOOD COLORS. These colors are natural carotenoid colorants derived from the seed of the tropical annatto tree (*Bixa orellana*). The surface of the seeds contains a highly colored resin, consisting primarily of the carotenoid *bixin*. The bixin is extracted from the seed by a special process to produce a pure, soluble colorant. Bixin, one of the relatively few naturally occurring *cis* compounds, has a chemical structure similar to the nucleus of carotene with a free and esterified carboxyl group as end groups. Its formula is $C_{25}H_{30}O_4$.

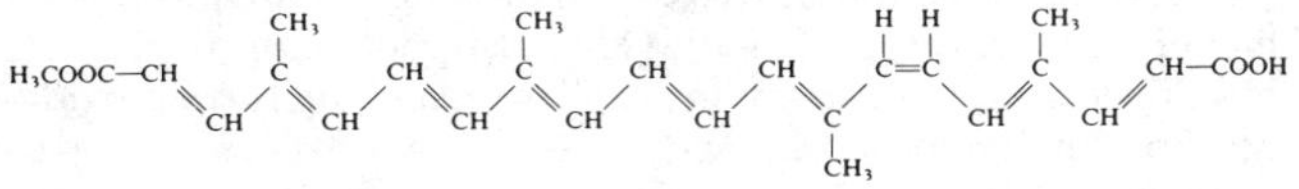

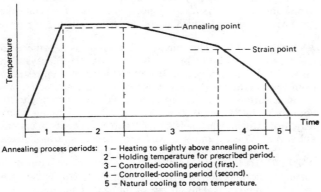

Structure of bixin.

Bixin is an oil-soluble, highly stable coloring ingredient. The saponification of the methyl ester group to form the dicarboxylic acid yields the water-soluble form of bixin, sometimes called *norbixin*. Annatto colorants date back into antiquity. The colorant has been used for centuries in connection with various textiles, medicinals, cosmetics, and foods. Annatto colors have also been used to color cheese, butter, and other dairy products for over a century. See also **Carotenoids**.

Processors make annatto colors available as a refined powder, soluble in water at pH values above 4.0 (solubility about 10 grams in 100 milliliters of distilled water at 25°C), in an acid-soluble form, in an oil-soluble form, in a water- and oil-soluble form, and in a variety of hues ranging from delicate yellows to hearty orange. Annatto extract is frequently mixed with turmeric extract to obtain various hues.

ANNEALING. The process of holding a solid material at an elevated temperature for a specified length of time in order that any metastable condition, such as frozen-in strains, dislocations, and vacancies may go into thermodynamic equilibrium. This may result in recrystallization and polygonization of cold-worked materials.

Annealing generally falls into the technology of heat treatment and varies with materials and the intended end uses of the materials, as well as the prior processing of them. In the case of nonferrous alloys, annealing is primarily a heat treatment for the purpose of removing the hardening due to cold work. Annealing also may be used with nonferrous precipitation hardening alloys to cause softening through agglomeration of the hardening constituents into fewer and larger particles.

As with metals, glass is fabricated at high temperatures and is annealed to relieve stresses which would develop if the glass were permitted to cool in an uncontrolled fashion. If not annealed, products made from high-expansion glasses can break spontaneously as they cool freely in air. In annealing glasses, they are raised to an annealing-point temperature and then cooled gradually to a temperature that is somewhat below the

strain point. Usually, the rate of cooling within this range determines the magnitude of residual stresses after the glass arrives at room temperature. Once below the strain point, the cooling rate is limited only by any transient stresses that may develop. A typical time-temperature glass-annealing curve is shown in the accompanying diagram.

ANODE. An anode is the electrode via which current enters a device. The anode is the positively charged electrode of an electrolytic cell.

ANODIC OXIDATION. Since oxidation is defined not only as reaction with oxygen, but as any chemical reaction attended by removal of electrons, then when current is applied to a pair of electrodes so as to make them anode and cathode, the former can act as a continuous remover of electrons and hence bring about oxidation (while the latter will favor reductions since it supplies electrons). This anodic oxidation is utilized in industry for various purposes. One of the earliest to be discovered (H. Kolbe, 1849) was the production of hydrocarbons from aliphatic acids, or more commonly, from their alkali salts. Many other substances may be produced, on a laboratory scale or even, in some cases, on an economically sound production scale, by anodic oxidation. The process is also widely used to impart corrosion resistant or decorative films to metal surfaces. For example, in the anodization or Eloxal process, the protection afforded by the oxide film ordinarily present on the surface of aluminum articles is considerably increased by building up this film through anodic oxidation. Also, one process for coloring the surface of aluminum and retaining a metallic luster is by adding substances to the metal and subsequently oxidizing the surface anodically.

ANODIZED COATINGS. Conversion Coatings.

ANORTHOSITE. The name *anorthosite* was given by T. Sterry Hunt to rocks of gabbroid nature which were essentially free from pyroxene, hence almost wholly plagioclase *usually* labradorite. The term is derived from the French word for plagioclase, anorthose. Small quantities of pyroxene may be present as well as magnetite or ilmenite. The rock is commonly white to gray, bluish, greenish, or perhaps nearly black. A variety from the Province of Quebec is purplish-brown due to the inclusion of ilmenite dust within the feldspars. Although not a common rock in the ordinary sense of the word, occurrences of great areal extent are known in Canada, Norway, and Russia and in the United States in northern New York State and Minnesota. Opinions as to the origin of this rock differ. The development of anorthosite may have been due to the settling out of labradorite crystals from a gabbro magma as many believe, or there may have been an original anorthosite magma.

A study of anorthosite occurrences brings out two very curious circumstances, first, that there is no extrusive (lava) equivalent of anorthosite, and second, that most anorthosite masses seem to be of pre-Cambrian age.

ANTACIDS. These are formulations widely used in the treatment of excessive gastric secretions and peptic ulcer. Several factors determine efficacy of antacids: (1) the ability and capacity of the stomach to secrete acid; (2) the duration of time the antacid is retained in the stomach; and (3) the nature of the gastric response upon eating.

Five principal active ingredients are used in antacid preparations: (1) *Sodium bicarbonate* is a rapid and effective neutralizer. The compound yields large amounts of absorbable sodium, un-

Annealing process periods: 1 — Heating to slightly above annealing point.
2 — Holding temperature for prescribed period.
3 — Controlled-cooling period (first).
4 — Controlled-cooling period (second).
5 — Natural cooling to room temperature.

Typical glass annealing curve.

desirable in some persons (heart disease; hypertension). The compound also may induce milk-alkali syndrome. (2) *Calcium carbonate* is a strong, effective neutralizer, but can cause constipation, hypercalcemia, acid rebound, and milk-alkali syndrome. (3) *Aluminum hydroxide* provides slow and not potent action. The compound causes constipation, and absorbs phosphates, as well as certain drugs, such as tetracyclines. (4) *Magnesium hydroxide* provides a slow and prolonged action with no major side reactions. (5) *Magnesium trisilicate* acts like magnesium hydroxide, but is poorly absorbed and acts as an osmotic laxative. In cases of renal insufficiency, the serum magnesium should be monitored.

The foregoing compounds are frequently used in combination and, in some, simethicone is added to relieve flatulence. There are striking differences of commercial antacids in terms of their neutralizing capacity. Among compounds having a high neutralizing capacity are Gelusil-II®, Maalox® Therapeutic Concentrate, and Mylanta II. The relative potency of antacid tablets is less than that of liquid formulations.

The physician is concerned with at least three factors when prescribing antacids: (1) Acid rebound (associated with calcium carbonate); (2) milk-alkali syndrome (caused by ingestion of large quantities of alkali); and (3) phosphorus depletion (by aluminum salts). The mechanism of acid rebound, especially in the long-term use of calcium carbonate, is poorly understood. It has been established that there is an excessive reacidification of the antrum (pyloric gland area) a number of hours after ingestion of calcium carbonate.

The ingestion of a quart of milk or more while taking large amounts of alkali, as from antacids, sets up conditions favorable to milk-alkali syndrome. Generally, with withdrawal of the milk or the antacid, the condition is self-correcting. Symptoms of milk-alkali syndrome include nausea, vomiting, anorexia, weakness, polydipsia, and polyuria. Abnormal calcifications also may occur in the chronic stage and other symptoms include mental changes, asthenia, aching muscles, band keratopathy, and nephocalcinosis. Symptoms of milk-alkali syndrome sometimes tend to mimic hyperparathyroidism and vitamin D intoxication.

Cimetidine, a histamine H_2 receptor antagonist and frequently used in the treatment of duodenal ulcer, among other actions, inhibits acid secretion rather than neutralizing acid once secreted.

ANTHOCYANINS. A group of water-soluble pigments which account for many of the red, pink, purple, and blue colors found in higher plants. Most plants contain more than one of these pigments and they occur most prevalently as glycosides. Several hundred different anthocyanins are known. Anthocyanins have been isolated and some have been found to be acylated with substituted cinnamic acids. The site of attachment of these acids to the anthocyanins has not been fully definitized. The natural role of the anthocyanins in plants to date has not been related to any factor of plant metabolism and many authorities believe that the pigments play more of an ecological role in regard to pollination and seed dispersal through their ability to act as an insect and bird attractant.

The anthocyanins are part of the larger group of aromatic, oxygen-containing, heterocyclic compounds known as *flavonoids*, most of which have a 2-phenylbenzopyran skeleton as their basic ring system. Although widely distributed among higher plants, including ferns and mosses, they are not found in algae, fungi, bacteria, or lichens.

There has been considerable interest and research activity in connection with anthocyanins during the past decade or so, stemming principally from the tighter restrictions, including banning, of several synthetic colorants. See also **Colorants.** Representative of the food processing industry's desire to find colorants that are beyond suspicion as health deterrents, scientists have been investigating various sources of anthocyanins. They have found that pigments from roselle plants (*Hibiscus sabdariffa*) native to the West Indies can be used for coloring apple and pectin jellies. A cranberry pomace extract has been found useful in coloring cherry pie filling. The potential of blueberry as a source of anthocyanin pigments also has been investigated. The berry is rich in nonacylated anthocyanins, but presently appears to be too costly as a coloring substitute.

Grape anthocyanins have been intensely investigated and have been found reasonably satisfactory, for example, in carbonated beverages. Although to date the grape anthocyanins are not as stable as Red No. 2, research will continue, encouraged by the large amounts of grape wastes produced in the production of wine and grape juice. Red cabbage also has been seriously considered as a source of anthocyanin pigments. The anthocyanins are most stable at a pH range of 1.0–4.0; this acidity dicates the products in which they can be used.

Much more detail on this topic can be found in references listed.

References

Ballinger, W. E., Maness, E. P., and L. J. Kushman: "Anthocyanins in Ripe Fruit of the Highbush Blueberry (*Vaccinium corymbosum* L.)," *Jrnl. of Amer. Society Horticultural Science*, **95**, 283 (1970).

Clydesdale, F. M., et al.: "Concord Grape Pigments as Colorants for Beverages and Gelatin Desserts," *Jrnl. of Food Science*, **43**, 6, 1687–1692 (1978).

Considine, D. M., Editor: "Foods and Food Production Encyclopedia," Van Nostrand Reinhold, New York (1981).

Shewfelt, R. L., and E. M. Ahmed: "Anthocyanin Extracted from Red Cabbage Shows Promise as Coloring for Dry Beverage Mixes," *Food Product Development*, **11**, 4, 52–58 (1977).

Volpe, T.: "Cranberry Juice Concentrate as a Red Food Coloring," *Food Product Development*, **10**, 9, 13 (1976).

ANTHOPHYLLITE. The mineral anthophyllite is an orthorhombic amphibole essentially $(Mg, Fe)_7Si_8O_{22}(OH)_2$ with aluminum sometimes present. This mineral corresponds to enstatite and hypersthene in the pyroxene group. It has a prismatic cleavage; hardness, 5.5–6; specific gravity, 2.8–3.57; luster, vitreous; color, gray, yellow, brown, green or brownish-green; transparent to translucent; probably always a metamorphic mineral in magnesium-rich rocks, often associated with talc; very common in schists. Found in Norway, Austria, Greenland, Pennsylvania, Georgia and elsewhere. The name is derived from the Latin *anthophyllum*, clove, because of its usual brownish shades. See also **Amphibole.**

ANTHRACENE. A colorless solid, mp 218°C, blue fluorescence when pure; insoluble in H_2O, slightly soluble in alcohol or ether, soluble in hot benzene, slightly soluble in cold benzene; transformed by sunlight into para-anthracene $(C_{14}H_{10})_2$.

$$C_{14}H_{10} \text{ or }$$

Anthracene reacts (1) with oxidizing agents, e.g., sodium dichromate plus sulfuric acid to form anthraquinone $C_6H_4(CO)_2C_6H_4$; (2) with chlorine in water or in dilute acetic acid below 250°C to form anthraquinol and anthraquinone, and at higher temperatures 9,10-dichloroanthracene. The reaction varies with the temperature and with the solvent used. The reaction has been studied, using as solvent benzene, chloroform, alcohol, carbon bisulfide, ether, glacial acetic acid; and also without solvent by heating. Bromine reacts similarly to chlorine. Anthracene also reacts (3) with concentrated H_2SO_4 to form various anthracene sulfonic acids; (4) with nitric acid, to form nitroanthracenes and anthraquinone; (5) with picric acid $(1)HO \cdot C_6H_2(NO_2)_3(2,4,6)$ to form red crystalline anthracene picrate, mp 138°C.

Anthracene is obtained from coal tar in the fraction distilling between 300 and 400°C. This fraction contains 5–10% anthracene; by fractional crystallization followed by crystallization from solvents, such as oleic acid, and washing with such solvents as pyridine, relatively pure anthracene is obtained. It may be detected by the formation of a blue-violet coloration on fusion with mellitic acid. Anthracene derivatives, especially anthraquinone, have been important in dye chemistry for many years.

ANTHRAQUINONE. A yellow solid, mp 286°C; can be sublimed; forms monoxime, mp 224°C, by heating under pressure at 180°C with hydroxylamine chloride; with strong oxidizing agents, reacts with difficulty to yield phthalic acid; with reducing agents, e.g., sodium hyposulfite, zinc in sodium hydroxide solution, tin or stannous chloride in hydrochloric acid (but not sulfurous acid), is reduced to anthraquinol, anthrone, dianthrol, and dianthrone, depending upon conditions.

Anthraquinone (9,10)

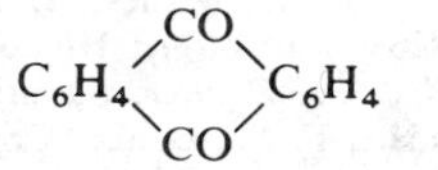

anthraquinol

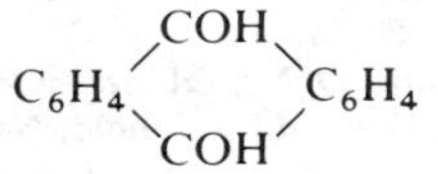

anthrone

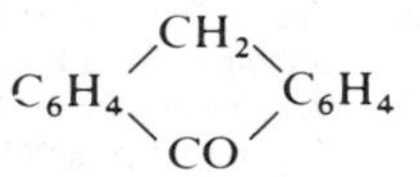

dianthrol

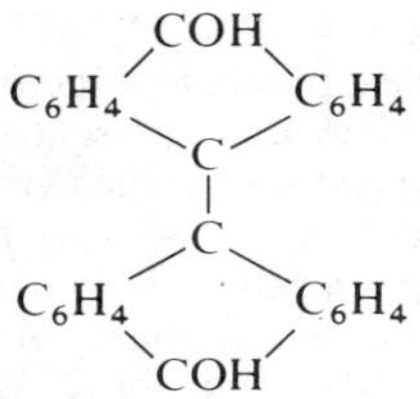

dianthrone

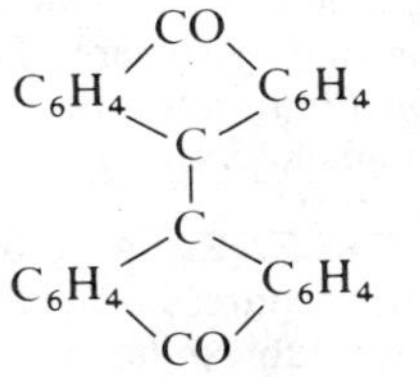

Anthraquinone is obtained by oxidation of anthracene using sodium dichromate plus sulfuric acid, and is purified by dissolving in concentrated sulfuric acid at 130°C and pouring into boiling water, whereupon anthraquinone separates as pure solid, and is recovered by filtration. Further purification may be accomplished by sublimation or crystallization from nitrobenzene, aniline or tetrachloroethane. Anthraquinone is used as the material from which many dyes are made, notably alizarin $C_6H_4(CO)_2C_6H_2(OH)_2$ and related substances. These are vat dyes, that is, insoluble colored substances which are readily reduced to a substance having marked affinity for the fiber to be dyed and which upon exposure to the air are readily reoxi-

dized to the original dye. Anthraquinone may be detected by the appearance of a red color on treatment with alkali, zinc powder and water.

See also **Coal Tar and Derivatives.**

ANTHRAXOLITE. A coal-like metamorphosed bitumen, often closely associated with igneous rocks. Commonly associated with "Herkimer Diamond" type quartz crystals in dolomitic limestones in Herkimer and Montgomery Counties in New York State.

ANTIBIOTIC. A chemical substance derived from one or more types of microorganisms which has the ability of inhibiting the growth of, or of killing, other microorganisms. In *bacteriostatic* drugs, very little killing, if any, of the target microorganisms occurs at the minimal inhibitory concentration (MIC). In *bactericidal* drugs, the objective is that of killing bacteria (or other target microorganisms), i.e., the minimal bactericidal concentration (MBC) is equal to or very close to the minimal inhibitory concentration (MIC). In bacteriostatic drugs, the objective is to cause bacteriostasis, a condition that is reversed if the chemotherapeutic agent is withdrawn. These rather fine dividing lines are much more clearly evident and measurable *in vitro* than *in vivo*. The principal antibiotic drugs in use today, classified by these two categories, are listed in Table 1. Sulfonamides are described in a separate entry on **Sulfonamide Drugs.**

TABLE 1. CLASSIFICATION OF DRUGS USED IN THE CHEMOTHERAPY OF MICROBIAL DISEASES

BACTERIOSTATIC AGENTS	BACTERICIDAL AGENTS
Sulfonamides	Penicillins (penicillin G, penicillin V, methicillin, oxacillin, cloxacillin, nafcillin, ampicillin, amoxicillin, carbenicillin)
Trimethoprim	
Tetracyclines	
Chloramphenicol	
Erythromycin	Cephalosporins
Lincomycin or clindamycin	Aminoglycosides (streptomycin, neomycin, kanamycin, gentamicin, tobramycin, amikacin)
	Vancomycin
	Polymyxins (polymyxin B, colistin)
	Bacitracin

Background. Although well established for about 50 years, by comparison with many other types of drugs used in medicine, the use of antibiotics as chemotherapeutic agents is relatively recent. However, the existence of antibiotic-type drugs dates back over 100 years.

The first scientific demonstration of microbial antagonism was made by Pasteur and Joubert in 1877, when they observed that certain common bacteria inhibited the growth of anthrax bacilli. This basic phenomenon by which one microorganism destroys another to preserve its own life was, at that time, called *antibiosis* by Vuillemin (1889). In the decades that followed, the therapeutic efficacy of antibiotics to control infectious disease was eventually demonstrated, after which the pursuit of microbial antagonists rapidly became an organized applied science.

Pyocyanase was the first microbially derived antibiotic product to be used in treating bacterial infections in humans. Although it had only limited clinical use, it is interesting historically because it demonstrated as early as 1906 the principle of selective toxicity, i.e., specificity of action against the invading pathogen and a correlative lack of toxic action in the host. Following the decline in use of pyocyanase, it was almost a quarter of a century before interest in anti-infective agents from microbial sources was renewed.

In 1929, the British bacteriologist Alexander Fleming published his observations on the inhibition of a staphylococcus culture by growing colonies of *Penicillium notatum*. This report went largely unpursued for a decade, after which Florey and Chain reinvestigated Fleming's work and, in 1941, demonstrated the clinical usefulness of penicillin. In 1939, Dubos, by careful, well-planned studies, obtained the antibiotic tyrothricin from the soil organism *Bacillus brevis*. Although tyrothricin found only limited use, the work of Dubos on the chemical, biological, and physical properties of this antibiotic contributed immensely toward forcing a realization of the potentialities of antibiotic substances. Similarly, Waksman undertook a systematic search for antimicrobial substances in a group of soil-inhabiting microbes known as *Streptomyces* and announced the discovery of streptomycin in 1944.

The foregoing discoveries stimulated worldwide interest, and the discovery of useful new antibiotics during the period 1939–1959 was prolific. During this period, the major classes of antibacterial antibiotics were recognized. Many specific drugs which presently occupy places in therapeutic practice were discovered directly in microbial fermentations. The later work in the field had consisted principally of chemical modifications of antibiotic substances previously known.

Classification The antibiotics comprise an unusually diverse group of substances, differing not only in chemical structure, but also in their mode of action, antibacterial spectra, origin, and other features. In spite of the variety of chemical structures involved, most antibiotics appear to arise from a limited number of biogenetic themes, and may be divided into a few main groups in accordance with their potential derivation from amino acids, sugars, and acetate or propionate units. See Table 2.

TABLE 2. CLASSES OF ANTIBIOTICS ON BASIS OF BIOGENETIC ORIGIN AND CHEMICAL STRUCTURE

Amino Acid Units	Acetate/Propionate	Sugar Units
Amino acid cogeners	Fused-ring systems	Aminoglycosides
D-Cycloserine	Tetracyclines	Streptomycin
Chloramphenicol	Oxytetracycline	Kanamycins
Beta-Lactams	Chlortetracycline	Gentamicins
Penicillins	Steroidal antibiotics	Neomycins
Cephalosporins	Fusidic acid	Tobramycins
Polypeptides	Griseofulvin	Amikacins
Bacitracins	Antibacterial macrolides	
Polymixins	Erythromycin	
Viomycin	Oleandomycin	
Capreomycin	Leucomycins	
Vancomycin	Spiramycins	
	Polyene macrolides	
	Nystatin	
	Amphotericins	
	Ansa-macrolides	
	Rifamycins	

Note: Some of the foregoing are mainly of historical or research interest.

In general, antibiotics exert their toxic action on bacteria by impairing one of the following processes: (1) synthesis of the bacterial cell wall; (2) synthesis of intracellular protein; (3) synthesis of deoxyribonucleic acid; and (4) function of the cytoplasmic membrane. Grouping antibiotics by these criteria bears little resemblance to their classification according to biogenetic origin and chemical structure. Thus, while cycloserine and chloramphenicol are biogenetically derivable from a single amino acid, cycloserine inhibits cell-wall synthesis, while chloramphenicol acts on protein synthesis. In contrast, the mechanisms of action of the macrolide antibiotics and lincomycin are very similar, but their chemical structures are markedly different.

Much information concerning the use of antibiotics was known before their detailed mechanisms were explained in part. Research has shown that the cell wall of bacteria is very different from the membrane surrounding mammalian cells, and the probable reason for the selectively toxic effect on bacteria rather than mammalian cells rests on this difference. The bacterial cell wall protects the microbial cell from the osmotic difference that exists between the inside and outside of the bacterial cell. Thus, when the cell wall is sufficiently damaged, the cell will disrupt and die unless the osmotic strength of the medium is greatly increased to minimize the osmotic difference. Under these special conditions, cells free of cell wall (protoplasts) are formed which are no longer sensitive to penicillin. Further evidence for the mechanism of action of penicillin was obtained by Park and Strominger, who observed the accumulation of a conjugate of uridine diphosphate and N-acetylmuramyl peptide in the medium of penicillin-inhibited cells. The condensation into the cell walls of this latter compound, which is a constituent of bacterial cell walls but not of mammalian cell membranes, was prevented by the drug.

Antibiotics are used to combat a large number of pathogens that range in complexity from viruses to protozoa. Bacteria constitute the largest single class of microorganisms which are susceptible to antibiotics. There are also antifungal, antiprotozoal, anthelmintic, antiviral, antineoplastic, and animal-growth promoting antibiotics. The microbial origin of antibiotics serves as yet another means of classification. For example, there are over 2000 antibiotics that have been discovered and studied since 1940, of which about 68% have been isolated from the family Streptomycetaceae, 12% from Bacillaceae, and about 20% from various fungi. These, in turn, may be further subclassified into various genera, species, and varieties of microbes. Thus, antibiotics may be classified on the basis of many different criteria, each of which represents an integral part of antibiotics technology.

Among the most important of the antibiotics used in current medical practice are: (1) the beta-lactams, which include the penicillins and cephalosporins; (2) the tetracyclines; (3) the macrolides; and (4) the aminoglycosides.

Penicillins

The penicillins are chemically characterized by a four-membered lactam ring fused to a thiazolidine ring and are differentiated by the side-chain (R) attached to the bicyclic nucleus. See Table 3. Penicillins are sometimes named by attaching the chemical name of the R-substituent as a prefix to the word *penicillin*. Thus, in the case where R is $C_6H_5CH_2—$, the compound may be called *benzylpenicillin*. However, this compound in commerce is more commonly referred to as *Penicillin G*. Similarly, in the case where R is $C_6H_5OCH_2—$, the compound may be called *phenoxymethylpenicillin*, although commercially it is more commonly called *Penicillin V*. Most frequently, the commercial names bear no resemblance to structure.

The naturally occurring penicillins have been discovered in the fermentation broths of *Penicillium* and *Cephalosporium* cultures.

The earliest of the penicillins, simply called penicillin, was benzylpenicillin (Penicillin G). This early product, still used, was shown to have several limitations, including acid instability, allergenicity, and susceptibility to enzymatic inactivation by penicillinases.

In 1947, it was discovered that addition of phenylacetic acid to penicillin fermentation media increased the yield of benzylpenicillin at the expense of other less desirable natural penicillins. Following this observation, a new generation of *biosynthetic* penicillins was prepared by addition of monosubstituted acetic

TABLE 3. STRUCTURES OF SOME OF THE
PRINCIPAL PENICILLINS

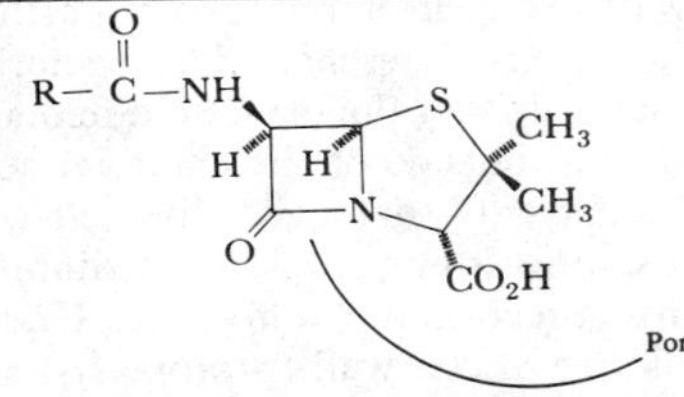

Portion of β-lactam ring cleaved by penicillinase
(β-lactamase)

GENERIC OR CHEMICAL NAME	SUBSTITUENT SIDE CHAIN (R)
Penicillin G (Benzylpenicillin)	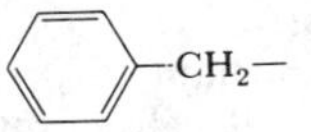
Penicillin V (Phenoxymethylpenicillin)	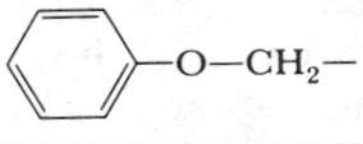
Methicillin	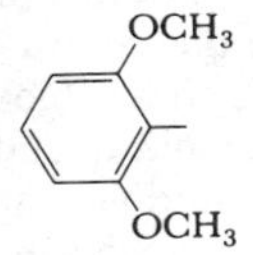
Oxacillin	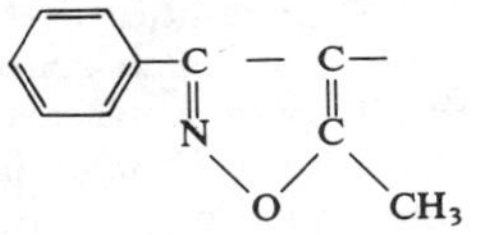
Cloxacillin	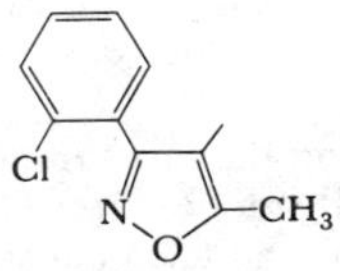
Nafcillin	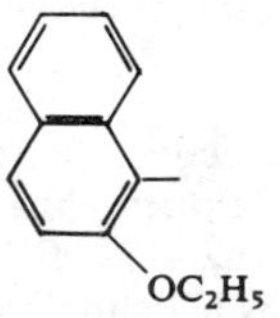
Ampicillin	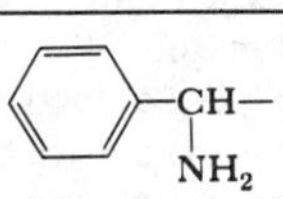
Amoxicillin	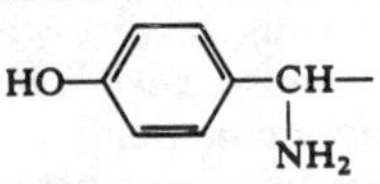
Carbenicillin	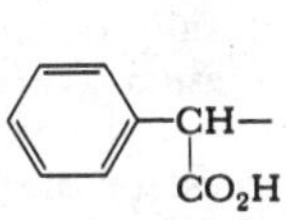

acid derivatives to penicillin fermentations. The most important of these biosynthetic derivatives is penicillin V, obtained by adding phenoxyacetic acid to penicillin growth media. This widely used antibiotic is relatively stable in dilute acid, is not destroyed by the acidic contents of the stomach, and consequently can be effective by oral administration.

Although the biosynthetic approach created several new penicillins, it was limited in the type of side chain (R) that could be introduced. Only derivatives with an unsubstituted methylene adjacent to the amide carbonyl (X as shown below) could be generated.

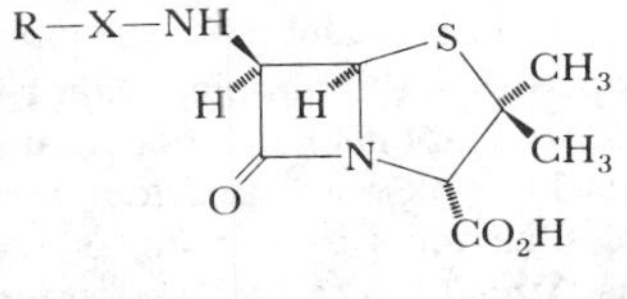

The next major breakthrough in penicillin research came in 1959 with the isolation of the penicillin nucleus 6-aminopenicillanic acid (6-APA) from fermentation mixtures to which no side-chain precursor had been added. Although chemical synthesis of 6-APA and its utility for the preparation of new penicillins by acylation were announced by Sheehan in 1958, the fermentation method provided the first practical means of obtaining large quantities of 6-APA. Chemical acylation of 6-APA allowed introduction of almost unlimited varieties of side chains and gave rise to a third generation of penicillins called *semisynthetic penicillins*.

The nature of the acyl side chain has been found to have a profound effect on the properties of the penicillins, influencing such therapeutically important properties as acid stability, oral absorption, serum protein binding, penicillinase resistance, and gram-negative activity.

One of the major developments resulting from the availability of 6-APA was the creation of semisynthetic penicillins that resist destruction by the penicillinases. The empirical finding that triphenylmethylpenicillin was resistant to penicillinase led to the screening of other penicillins with sterically hindered side chains, partly because the presence of a bulky group near the beta-lactam ring resulted in reduced affinity of these substances for the enzyme. Methicillin and cloxacillin are compounds with such side chains which have proved clinically useful.

Administration of Penicillins. Blood levels of penicillin V are more predictable and are usually from 2 to 5 times higher than as achieved by an equal dosage of penicillin G. Although all forms of penicillin introduced since penicillin G are prescribed by weight, penicillin G is still commonly prescribed by unitage. One milligram of penicillin G = ~1600 units. Penicillin G procaine is almost painless when administered intramuscularly and is slowly absorbed, usually requiring only one injection per each 12 to 24 hours for treatment of highly susceptible infections. In the treatment of meningitis and of endocarditis due to highly susceptible organisms, high doses of penicillin G are usually administered intravenously at the shorter intervals of 2 to 4 hours because of the special problems of penetration across the blood-brain barrier and into vegetations. Where renal dysfunction is present, neurotoxicity may be produced and enhance hyperkalemia because of high potassium content. Where penicillin G may present problems, the sodium salt of penicillin G or a sodium salt of a similar penicillin, such as ampicillin or carbenicillin, may be substituted.

Ampicillin has a broader range of activity than that of penicillin G. The spectrum of ampicillin encompasses not only pneumococci, meningococci, gonococci, and a number of streptococci, but also several gram-negative bacilli, including *Haemophilus influenzae, Salmonella* species, *Shigella* species, *Escherichia coli*, and *Proteus mirabilis*. Because ampicillin is readily cleaved by β-lactamase, the drug is of no value in the treatment of infections caused by *Staphylococcus aureus* and other organisms that elaborate the enzyme. Within recent years, plasmids conferring ampicillin resistance have appeared in *Salmonella typhi, H. influenzae*, and *Neisseria gonorrhoeae*. These organisms previously were uniformly susceptible to ampicillin. The increasing use of ampicillin also has been evidenced by increasing resistance among strains of *E. coli* and nontyphoidal *Salmonella* strains.

Amoxicillin, as evident from Table 3, is structurally similar to ampicillin with exception of an OH instead of an H in one of the positions of the side chain. Although this difference does not alter the spectrums of the two drugs, amoxicillin is better absorbed from the gastrointestinal tract, resulting in longer effective concentrations of the drug present in the circulation. The prolonged blood level and fewer problems with diarrhea sometimes permit more effective oral therapy than can be achieved with ampicillin.

Carbenicillin, in which there is a carboxyl rather than an amino substituent, has greater activity against gram-negative bacilli. These include *Pseudomonas aeruginosa*, species of *Proteus* (other than *mirabilis*), as well as some strains of *Enterobacter*, plus the strains that are susceptible to ampicillin. It has been found that over 80% of *Pseudomonas* isolates are susceptible. Carbenicillin is not absorbed from the gastrointestinal tract, although the indanyl carbenicillin ester is acid stable and thus suitable for oral administration. It is used in the treatment of urinary tract infections due to susceptible gram-negative bacilli, including *Pseudomonas*. Blood levels are too low for use of the drug in infections located elsewhere. Another semisynthetic penicillin, *ticarcillin*, has essentially the same spectrum as carbenicillin and, in the case of *Ps. aeruginosa*, is about twice as active as carbenicillin.

The penicillinase-resistant (not affected by cleavage problem previously mentioned) penicillins include *methicillin, oxacillin, nafcillin, cloxacillin*, and *dicloxacillin*. Methicillin was the first of these drugs to be introduced (early 1960s). When this drug was introduced, infections due to penicillin-resistant *Staph. aureus* had reached epidemic proportions. Methicillin is equally effective against penicillinase-producing and non-penicillinase-producing *Staph. aureus*. It is, however, less effective against pneumococci and streptococci than is penicillin G. Thus, its major use has been against penicillinase-producing *Staph. aureus*.

Methicillin is administered intravenously or intramuscularly. Because the latter route causes pain and discomfort, it is not used where large doses are required. In recent years, the semisynthetic, penicillnase-resistant *oxacillin* and *nafcillin* have markedly supplanted methicillin for many situations. These more recent antibiotics can be administered orally or parenterally. They are sometimes used in the treatment of mixed infections involving *Staph. aureus* and other gram-positive cocci or of infections of unknown etiology, but where the aforementioned organisms are suspected.

Cloxacillin and *dicloxacillin* are available as oral preparations. These compounds are sometimes preferred over oxacillin for treatment of mild staphylococcal infections. However, oxacillin and its two derivatives are usually effective in the primary treatment of soft tissue infections. They also may be used as an extension to prior intravenous therapy.

It should be mentioned that the side effects of penicillin are generally shared by the various derivatives. Allergy to one penicillin compound almost always indicates allergy to all other penicillin compounds.

Cephalosporins

These substances constitute another major class of β-lactam antibiotics and are chemically characterized by a β-lactam fused to a dihydrothiazine ring. In contrast to the penicillins, where the side chain of the antibiotic varies, depending upon precursors present in the fermentation mixture, fermentation-derived cephalosporins contain the same side chain. *Cephalosporin C*, the parent antibiotic of this class, is not useful clinically. For example, it is about 0.1% as active as benzylpenicillin against staphylococci. However, in early research, cephalosporin C exhibited certain interesting properties which provoked further study. Cephalosporin C was more stable toward acid than penicillin; it was unaffected by penicillinases; it exhibited appreciable activity against some gram-negative bacteria, and it appeared to have no cross-allergenicity with the penicillins. Consequently, in the late 1950s, many laboratories investigated both chemical and microbiological methods for removing the aminoadipoyl side chain of cephalosporin C to obtain the cephalosporin nucleus, 7-aminocephalosporanic acid (7-ACA).

A practical chemical process for accomplishing this transformation was announced in 1962. Like 6-APA, previously described, 7-ACA can be readily acylated, and a large number of semisynthetic cephalosporins thus have been possible. Cephalothin was the first clinically useful broad-spectrum cephalosporin to emerge from synthetic studies. Cephaloridine soon followed cephalothin and was found to be 2–8 times more active than the latter against gram-positive organisms. In 1970, cephaloglycin became commercially available as the first orally effective broadspectrum cephalosporin. Cephalexin is metabolically more stable than cephaloglycin. More recently available cephalosporins have included cephapirin, cephradine, and cefazolin.

The basic structure of the cephalosporins and structures of several of the antibiotics in this family are shown in Table 4.

TABLE 4.   STRUCTURES OF REPRESENTATIVE CEPHALOSPORINS

Generic or Chemical Name	R	X	Y
Cephalosporin C (parent of class—essentially inactive)	H_2N—(CH₂)₃—, CO_2H	C=O	CH_3CO_2—
7-Aminocephalosporanic acid	—	H	CH_3CO_2—
Cephalothin	(thiophene)—CH₂—	C=O	CH_3CO_2—
Cephaloridine	(thiophene)—CH₂—	C=O	(pyridinium)
Cephaloglycin	(phenyl)—CH—NH₂	C=O	CH_3CO_2—
Cephalexin	(phenyl)—CH—NH₂	C=O	H

Note: Important cephalosporins not included in table include Cefamandole, cefazolin, Cefoxitin, cephapirin, cephradine.

Administration of Cephalosporins. The antibacterial spectrum of the current cephalosporins is fundamentally broad and includes most gram-positive bacteria (*Staph. aureus, Staph. epidermidis*, pneumococci, streptococci (except enterococci), *Actinomyces*, most gram-positive anaerobes, as well as numerous gram-negative organisms, including *E. coli, Prot. mirabilis*, and *Klebsiella*. Although the cephalosporins are resistant to the penicillinase of *Staph. aureus*, they are not effective against many gram-negative organisms because these organisms produce chromosomally determined, membrane-bound cephalosporinases.

Cephalothin, when it first became available, was used as an alternative to penicillin in treating infections arising from *Staph. aureus* and some other gram-positive infections. Later, use of the drug was extended to infections arising from *K. pneumoniae*, *E. coli*, and *Prot. mirabilis*. Currently, in the treatment of major infections, cephalothin is frequently used in combination with a second antimicrobial drug. The drug usually is administered intravenously because of pain associated with intramuscular injections. Where renal dysfunction is present, the dosage must be reduced. Cephalothin is not the drug of choice in treatment of bacterial meningitis because this specific drug and the cephalosporins in general do not penetrate the cerebrospinal fluid very well. It should be noted, however, that the cephalosporins do cross the placenta and penetrate effectively into the pericardium and the joints.

Cephapirin and *cephradine* are later introductions, but cephalothin still is generally preferred because of the experience factor.

Cefazolin is also a later introduction to cephalosporin therapy. The spectrum is quite similar to that of cephalosporin, but two advantages are relative lack of pain on intramuscular administration and ability to use at higher blood levels.

Cefamandole and *cefoxitin* are even more recent members of this family of drugs. These drugs extend somewhat the spectrum of the cephalosporins against bacteria. They are less effective against some strains and more effective against others. For example, cafamandole is less active against *Staph. aureus* and other gram-positive cocci than cephalothin, but is more active against strains of *Proteus* species (not including *mirabilis*) and *Enterobacter*. Cefamandole is also effective against *H. influenzae*. Cefoxitin is effective against strains of *E. coli, Klebsiella, Salmonella*, and *Shigella* and, additionally, some strains of *Serratia* and *Proteus*, and *Bact. fragilis*. Cefoxitin is effective against enteric gram-negative bacilli and *Bacteroides*, but less active against gram-positive organisms.

The three principal oral cephalosporins are cephaloglycin, cephalexin, and cephradine. These drugs are used in the treatment of urinary tract infections that may arise from susceptible gram-negative bacilli, but frequently the drug of first choice may be sulfonamides, ampicillin, or tetracyclines.

The principal side reactions of the cephalosporins include local pain, renal impairment, and allergic reactions.

Aminoglycosides

These substances comprise a class of potent broad-spectrum antibiotics which are chemically characterized by basic carbohydrate moieties glycosidically bound to a cyclitol unit. In general, the aminoglycosides are effective against most gram-positive and gram-negative bacteria, as well as *Mycobacterium tuberculosis*. Because of their highly ionic nature, the aminoglycosides are not absorbed from the gastrointestinal tract and must be administered parenterally. In a small percentage of patients, prolonged use of this class of antibiotics can adversely affect the eighth cranial nerve, causing some impairment of hearing and balance.

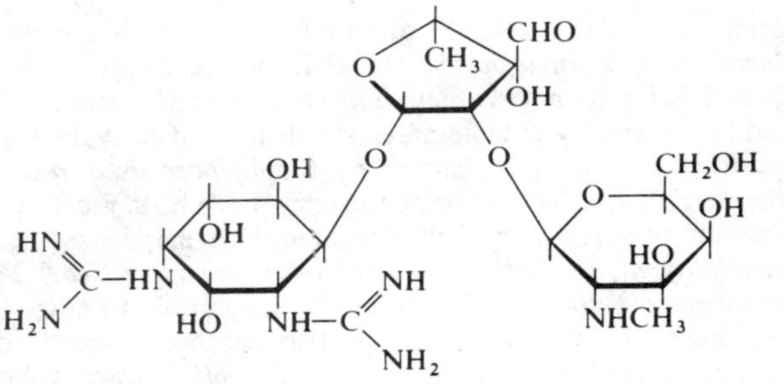

Streptomycin. Discovery in 1944 of streptomycin (structure shown below) drew immediate interest because it was the least toxic of the broad-spectrum antibiotics known at that time. Indeed, streptomycin was used to treat many gram-negative microbial infections, but because of the ease with which organisms developed resistance to it during treatment, many of these applications were abandoned when the tetracyclines, discussed later, became available. Streptomycin was the first parenterally administered antibiotic active against many microorganisms, but during the last several years, its use is limited essentially to three situations: (1) the initial treatment of serious tuberculous infections when the principal drugs of choice (isoniazid, rifampin) cannot be used because of their adverse effects on a particular patient; (2) treatment of enterococcal and other infections in which synergism between a penicillin and an aminoglycoside is desired; and (3) treatment of certain uncommon infections (plague and tularemia).

Kanamycin. Considerably broader in spectrum than streptomycin, kanamycin is more effective against gram-negative bacilli (other than *Pseudomonas*) and also is effective to a degree against *Staph. aureus*. However, it is ineffective against streptococci and pneumococci. The availability of penicillinase-resistant penicillins and cephalosporins essentially obsoleted kanamycin as the primary drug in the treatment of staphylococcal infections. Kanamycin has been essentially replaced by gentamicin and other aminoglycosides which are less ototoxic (adverse to hearing), and which also have a wider range of antibacterial activity.

Gentamicin. One of the successors of kanamycin, gentamicin possesses essentially the same spectrum as kanamycin, but is also active against *Pseudomonas aeruginosa*. An advantage of gentamicin is its penetration into pleural, ascitic, and synovial fluids where there is inflammation. Although not necessarily the drug of choice, gentamicin has been used in the treatment of acute cholecystitis, acute septic arthritis, anaerobic infections, *Bacillus* infections, gram-negative bacteremia, infective endocarditis, meningitis, osteomyelitis, peritonitis, staphylococcal infections, and tularemia, among others. In some situations, gentamicin acts synergistically with penicillin.

Tobramycin. Pharmacologically, tobramycin is quite similar to gentamicin. The drug is somewhat more active against *Ps. aeruginosa* than gentamicin. Tobramycin also acts synergistically with penicillin, but to a lesser degree than gentamicin.

Amikacin. This drug is a semisynthetic derivative of kanamycin. It is much less sensitive to the enzymes that inactivate aminoglycoside antibiotics. The spectrum is similar to that of gentamicin. Amikacin principally finds use in the treatment of infections arising from bacteria that are resistant to gentamicin and/or tobramycin.

Tetracyclines

These substances comprise a family of broad-spectrum antibiotics possessing a common perhydronaphthacene skeleton. They have a wider range of antimicrobial activity than other classes of clinically useful antibiotics. See Table 5. The tetracyclines are active against many species of gram-positive and gram-negative bacteria, spirochetes, rickettsiae, and some of the larger viruses.

TABLE 5. STRUCTURES OF REPRESENTATIVE TETRACYCLINES

Generic or Chemical Name	R_1	R_2	R_3	R_4	R_5
Tetracycline	H	OH	CH_3	H	H
Chlortetracycline	Cl	OH	CH_3	H	H
Oxytetracycline	H	OH	CH_3	OH	H
Demethylchlortetracycline	Cl	OH	H	H	H
Methacycline	H	$=CH_2$	$=CH_2$	OH	H
Doxycycline	H	H	CH_3	OH	H
Rolitetracycline	H	OH	CH_3	H	$N\!-\!CH_2\!-$

Note: Minocycline is not included in table.

Chlortetracycline. The first member of this class to be isolated, chlortetracycline, was discovered in 1948 among the metabolites of *Streptomyces aureofaciens.* Oxytetracycline was isolated two years later from a *S. rimosus* fermentation. Both antibiotics quickly found wide medical use, not only because they were effective orally, but because they were useful against a much wider spectrum of bacteria than penicillin G.

Chemical studies on chlortetracycline and oxytetracycline, which provided a basis for structure assignment, in general led to products with diminished or no antibacterial activity. In 1953, the first scientific reports appeared describing an active tetracycline prepared by chemical modification of a fermentation product. This was *tetracycline,* the parent member of this family of antibiotics, prepared by catalytic hydrogenolysis of chlortetracycline. It was more stable and better tolerated than its fermentation-produced progenitor, and almost completely displaced chlortetracycline from medical practice. Interestingly, tetracycline was later found in fermentation broths of a mutant strain of *S. aureofaciens* and also may be manufactured by this method.

Following the discovery of tetracycline, useful new drugs from chemical modification of tetracycline antibiotics were slow in coming, for the complexity and chemical lability of the tetracyclines did not render them amenable to facile systematic studies of the relationships between chemical structure and biological properties. Unlike the β-lactam antibiotics, where structural modifications were being sought mainly to improve their antibacterial spectra and potency, superior semisynthetic tetracyclines were obtained, in general, with improved pharmacokinetic properties, i.e., such factors as rate of oral absorption, degree of serum protein binding, rate of urinary excretion, and biological half-life.

An effective approach to the discovery of superior tetracycline antibiotics stemmed from studies yielding tetracyclines modified at the C-6 position. Thus, demethylchlortetracycline and methacycline were found to be somewhat superior to tetracycline in terms of a longer serum half-life, and later compounds such as doxycycline were shown to exhibit near-ideal pharmacokinetics. Doxycycline, among a number of other tetracyclines, remains in wide use today.

Among the many diseases treated with tetracyclines are urinary tract infections, gonorrhea, nongonococcal urethritis, Rocky Mountain spotted fever, other rickettsioses, mycoplasmal pneumonia, chlamydial diseases (psittacosis, trachoma, lymphogranuloma venereum), brucellosis, plague, cholera, granuloma inguinale, syphilis, and gonococcal pelvic inflammatory diseases, particularly in a number of instances where a patient may be allergic to penicillin. Tetracyclines also have been used in the treatment of cystic acne.

Doxycycline has been reported as effective in the prophylaxis of traveler's diarrhea in Kenya. The tetracycline, minocycline, is sometimes used instead of the sulfonamides against nocardial infections. It is not used in connection with meningococcal infections.

Macrolides

The macrolides comprise a family of antibiotics chemically characterized by a macrocyclic lactone to which one or more sugars are attached. The compounds are often divided into various subgroupings, but the group with antibacterial properties are known as *antibacterial macrolides.* They are distinguished chemically by having, in addition to the large lactone, various ketonic and hydroxyl functions and glycosidically bound deoxy sugars. A second grouping of commercial importance is known as the *polyene macrolides,* chemically characterized by extended conjugated double-bond systems. The polyenes are devoid of antibacterial activity, but are potent antifungal agents.

A number of the antibacterial macrolides have been found to be clinically useful chemotherapeutic substances, falling generally under the title of *medium-spectrum antibiotics.* This term is taken to mean that these substances are effective against most gram-positive bacteria and have a degree of activity against certain gram-negative organisms, such as *Haemophilus, Brucella,* and *Neisseria* species. The antibacterial macrolides also appear to inhibit certain pleuro-pneumonialike organisms.

Erythromycin. This is the principal drug in this category. Although available as the parent entity, semisynthetic derivaties have proved to be clinically superior to the natural cogener. Like the tetracyclines, synthetic transformations in the macrolide series have not significantly altered their antibacterial spectra, but have improved the pharmacodynamic properties. For example, the propionate ester of erhthromycin lauryl sulfate (erythromycin estolate) has shown greater acid stability than the unesterified parent substance. Although the estolate appears in the blood somewhat more slowly, the peak serum levels reached are higher and persist longer than other forms of the drug. However, cholestatic hepatitis may occasionally follow administration of the estolate and, for that reason, the stearate is often preferred.

Erythromycin is effective against Group A and other non-enterococcal streptococci, *Corynebacterium diphtheriae, Legionella pneumophila, Chlamydia trachomatis, Mycoplasma pneumoniae,* and *Flavobacterium.* Because of the extensive use of erythromycin in hospitals, a number of *Staph. aureus* strains have become highly resistant to the drug. For this reason, erythromycin has been used in combination with chloramphenicol. This combination is also used in the treatment of severe sepsis when etiology is unknown and patient is allergic to penicillin.

A structural representation of erythromycin is shown on next page.

Other Antibiotics

Chloramphenicol. This compound is derived from *Streptomyces venezuelae* or by organic synthesis. It was the first substance of natural origin shown to contain an aromatic nitro group. Although the drug is a valuable broad-spectrum antibiotic, its use has been somewhat limited because of the occasional development of aplastic anemia in the patient. Thus, its use has been largely confined to its administration as the drug of choice in patients allergic to penicillin in connection with typhoid fever, nontyphoidal salmonelloses (due to ampicillin-resistant strains), *H. influenzae* meningitis, meningitis arising from *N. meningitidis*, and *Str. pneumoniae*. Because effective and safer drugs are not available, it is used for infections arising from *Bact. fragilis*. The drug is administered orally or intravenously.

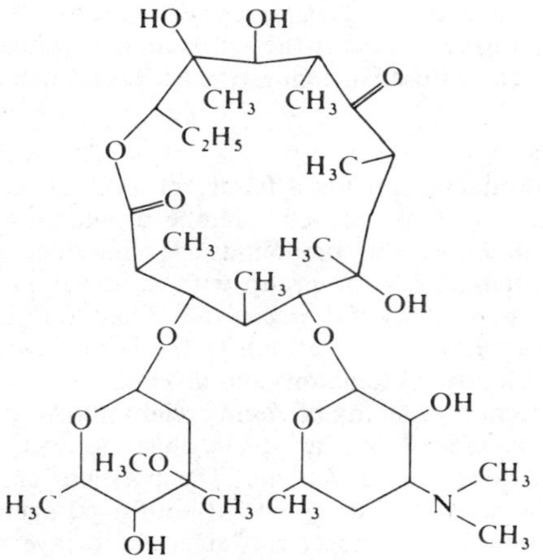

Erythromycin

Vancomycin. This is a narrow-spectrum antibiotic and produced by *Streptomyces orientalis* or synthetically. Its effectiveness is essentially confined to the treatment of streptococci (including enterococci), pneumococci, staphylococci, and a few other gram-positive bacteria. Serious side effects include possible hearing loss and renal insufficiency, particularly when the drug is administered with an aminoglycoside.

Polymyxins. This is a generic term for a series of antibiotic substances produced by strains of *Bacillus polymyxa*. Various polymyxins are differentiated by letters A, B, C, D, and E. All are active against certain gram-negative bacteria. Polymyxin B and E (colistin) have been the most important in the past, but currently are only rarely used—because they have been replaced by more effective aminoglycosides. The B and E drugs are effective against most of the common aerobic gram-negative bacilli, but not *Proteus*, *Providencia*, and *Serratia*. Prior to their replacement by aminoglycosides, the polymyxins were used mainly in connection with infections arising from *Ps. aeruginosa*.

Spectinomycin. This drug finds principal application in the treatment of gonorrhea. It should be noted that the antibiotic resistance among *N. gonorrhoeae* has caused a number of therapeutic problems. It has been found that only by escalating the antibiotic doses and using probenecid to retard the excretion of penicillin and ampicillin (the drugs of choice) has the continued effective use of penicillin, ampicillin, and tetracycline been possible. Even with modifications in the therapy, 3–8% of cases fail to respond to the usual regimens for uncomplicated gonorrhea. Thus, the treatment of uncomplicated gonorrhea that fails to respond to the usual regimen is spectinomycin therapy.

Chemoprophylaxis with Antibiotics

In addition to their use in treating infections arising from bacteria and other microorganisms, antibiotics are also used in chemoprophylaxis, i.e., treating a patient before or shortly after the entry of pathogenic organisms. There are three common situations: (1) preventing infection following exposure to known pathogens; (2) preventing specific infections in highly susceptible individuals; and (3) preventing postoperative infectious complications.

Antibiotics in Feedstuffs

For several years, antibiotics have been used in feedstuffs, not only to lower the incidence of certain diseases in livestock, but also because antibiotics play a function in the rate of growth of animals. It is estimated that, as of the late 1970s, nearly 40% of the antibacterials produced in the United States are used in animal feeds or for other nonhuman purposes. In 1977, the U.S. Food and Drug Administration commenced an investigation because it is observed that, at low levels of use of antibiotics in animal husbandry provides an "almost ideal" environment for breeding antibiotic-resistant strains of bacteria that could eventually infect humans. A strain resistant to both penicillin and tetracycline, for example, could overcome two of the most effective, commonly used drugs in the antibiotic therapy of humans. The problem for the authorities is to prove convincingly that such suspicions are a forecast of real truths.

Manufacture of Antibiotics

Generally, most antibiotics and the starting materials for semisynthetic antibiotics are manufactured by fermentation, with accompanying extraction, purification, crystallization, and packaging operations. Commercial fermenting vessels are stainless- or carbon-steel enclosed tanks with capacities up to several tens of thousands of gallons (many hundreds of hectoliters). Such factors as aeration, agitation, temperature, and pH must be monitored and controlled carefully.

The antibiotic-producing microorganism is grown in submerged culture in a fermentation medium which contains various carbon, nitrogen, and trace-metal sources, required by the organism for its nutrition. The organism is grown under conditions of pure culture, that is, other microorganisms are excluded from the fermentation inasmuch as the latter will compete for nutrients and may contribute undesirable contamination and reduce yield of the desired product. When the fermentation has reached peak potency (which varies with each product), the antibiotic may be recovered by an extraction technique, such as distribution into a water-immiscible solvent, ion-exchange chromatography, or precipitation. Following extraction, purification and crystallization are carried out by procedures compatible with the physicochemical properties of the particular antibiotic being produced. A representative flowsheet is shown on next page.

References

Abramowicz, M., Editor: "Adverse Interactions of Drugs," *The Medical Letter on Drugs and Therapeutics*, **19**, 5 (1977).

Abramowicz, M., Editor: "Antimicrobial Prophylaxis: Prevention of Wound Infection and Sepsis after Surgery," *The Medical Letter on Drugs and Therapeutics*, **19**, 37 (1977).

Abramowicz, M., Editor: "The Choice of Antimicrobial Drugs," *The Medical Letter on Drugs and Therapeutics*, **20**, 1 (1978).

Braude, A. I.: "Antimicrobial Drug Therapy," Vol. 8 "Major Problems in Internal Medicine," Saunders, Philadelphia (1976).

Cook, F. V., and W. E. Varrar: "Vancomycin Revisited," *Ann. Intern. Med.*, **88**, 813 (1978).

Krogstad, D. J., et al.: "Plasmid-mediated Resistance to Antibiotic Synergism in Enterococci," *Jrnl. Clin. Invest.*, **61**, 1645 (1978).

Kunin, C. M., et al.: "Veterans Administration Ad Hoc Interdisciplinary Advisory Committee on Antimicrobial Drug Usage: Prophylaxis in Surgery," *Jrnl. Amer. Med. Assn.*, **237**, 1003 (1977).

Moellering, R. C., Jr., and M. N. Swartz: "Drug Therapy: The Newer Cephalosporins," *N. Engl. Jrnl. Med.*, **294**, 24 (1976).

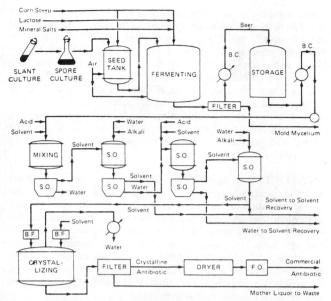

Materials flow in a representative commercial antibiotic manufacturing process. B.C. = brine cooler; S.O. = separating operation; B.F. = bacteriological filter.

Rahal, J. J.: "Antibiotic Combinations: The Clinical Relevance of Synergy and Antagonism," *Medicine*, **57**, 179 (1978).

Siegel, M. S., et al.: "Penicillinase-Producing *Neisseria gonorrhoeae*: Results of Surveillance in the United States," *Jrnl. Infect. Dis.*, **137**, 170 (1978).

Sivonen, A., et al.: "The Effect of Chemoprophylactic Use of Rifampin and Minocycline on Rates of Carriage of *Neisseria meningitidis* in Army Recruits in Finland," *Jrnl. Infect. Dis.*, **137**, 238 (1978).

Staff: "Missed Chance" in "Science and the Citizen," *Sci. Amer.*, **239**, 5, 90–91 (1978).

ANTIBODIES. In immunity to infectious agents, some time after a foreign macromolecule has entered the body, induced mechanisms come into play which result in the synthesis of specially adapted molecules (*antibodies*) capable of combining with the foreign substances which have elicited them. Most macromolecules (proteins, carbohydrates, nucleic acids) can function as antigens (the provoking agents), provided that they are different in structure from autologous macromolecules, i.e., from the macromolecules of the responding organisms.

Antibodies are proteins with a molecular weight of 150,000–1,000,000 and with electrophoretic mobility predominantly of gamma globulins. The combination between antigen and antibody results in inhibition of the biological activity of the antigen and leads to increase rate of ingestion (*opsonization*) of the antigen by phagocytic cells. In addition, combination of antigen and antibody results in the activation of a complex chain of interacting constitutive molecules—the *complement system*—leading to lysis of the cell membranes to which antibody (directed against cellular antigens) is attached.

Antigens are excluded from the body by skin and mucous membranes. If these barriers are penetrated, the foreign organism may be ingested by phagocytic cells (monocytes, polymorphs, macrophages) and subsequently destroyed by cytoplasmic enzymes. Sometime after a foreign macromolecule has entered the body, induced mechanisms come into play. There are two basic biological manifestations of the immune reaction: (1) Immunity to infectious agents, as previously mentioned; and (2) specific hypersensitivity. Hypersensitivity, or the heightened response to an agent, can be divided into anaphylactic, allergic, and bacterial. Anaphylaxis, which can be produced by either active or passive sensitization, is a laboratory tool for studying the fundamental nature of hypersensitivity. The amounts of antigen and antibody involved, as well as the nature and source of the antibody, govern the extent of the hypersensitive reaction.

ANTICAKING AGENTS. Some products, particularly foods that contain one or more hygroscopic substances, require the addition of an *anticaking agent* to inhibit formation of aggregates and lumps and thus retain the free-flowing characteristic of the products. Calcium phosphate, for example, is commonly used in instant breakfast drinks and lemonade and other soft drink mixes.

The general function of an anticaking agent can be described by using silica gel as an example. Generally anticaking agents are available as very small particles (ranging from 2 to 9 micrometers in diameter). A typical application for silica gel is admixture with orange-juice crystals to assure a free-flowing product, avoiding formation of crystal cakes and hard lumps. The very high adsorption properties of the anticaking substance removes moisture that can cause fusion. The billions of extremely fine, inert particles coat and separate each grain of powder (product) to keep it free-flowing. Many anticaking agents, including silica gel, also act as dispersants for powdered products. Many food products, when stirred into water, tend to form lumps which are difficult to disperse or dissolve. The agent not only improves flow properties, but also increases speed of dispersion by keeping the food particles separated and permitting the water to wet them individually instead of forming lumps. As is true with so many food additive chemicals, anticaking agents serve multiple functions. In addition to acting as an anticaking and dispersing agent, silica gel also can be used as a moisture scavenger and carrier. Some additives, when they are capable of serving several functions, may be called *conditioning agents*.

Anticaking agents commonly used include: calcium carbonate, phosphate, silicate, and stearate; cellulose (microcrystalline); kaolin; magnesium carbonate, hydroxide, oxide, silicate, and stearate; myristates; palmitates; phosphates; silica (silicon dioxide); sodium ferrocyanide; sodium silicoaluminate; and starches.

ANTICOAGULANTS. These are substances which prevent coagulation of the blood. For blood investigations made outside the body, sodium or potassium citrates, oxalates, and fluorides are sometimes used. For blood which is to be used for transfusions, sodium citrate is used.

Organic anticoagulants are used in vivo in the treatment of numerous conditions where blood coagulation can be dangerous, as in cerebral thrombosis, coronary heart disease, among others, which will be described later. The main anticoagulants used are heparin and coumarin compounds, such as warfarin.

Heparin. A complex organic acid (mucopolysaccharide) present in mammalian tissues and a strong inhibitor of blood coagulation. Although the precise formula and structure of heparin are uncertain, it has been suggested that the formula for sodium heparinate, generally the form of the drug used in anticoagulant therapy, is $(C_{12}H_{16}NS_2Na_3)_{20}$ with a molecular weight of about 12,000. The commercial drug is derived from animal livers or lungs.

Heparin is considered a hazardous drug. Heparin may be the leading cause of drug-related deaths in hospitalized patients who are relatively well (Porter and Jick, 1978). It has been reported (Bell et al., 1976) that some patients who receive continuously infused intravenous heparin develop *thrombocytopenia* (condition where the platelet count is less than 100,000/cubic millimeter). Some authorities believe that the risk of thrombocytopenia associated with porcine heparin may be less than the risk associated with heparin of bovine origin. (Babcock et al., 1976; Powers et al., 1979; Hrushesky, 1978).

Heparin, in addition to inhibiting reactions which lead to blood clotting, also inhibits the formation of fibrin clots, both in vitro and in vivo. Heparin acts at multiple sites in the normal coagulation system. Small amounts of heparin in combination with antithrombin III (heparin cofactor) can prevent the development of a hypercoagulable state by inactivating activated factor X, preventing the conversion of prothrombin to thrombin. Once a hypercoagulable state exists, larger doses of heparin, in combination with antithrombin II, can inhibit the coagulation process by inactivating thrombin and earlier clotting intermediates, thus preventing the conversion of fibrinogen to fibrin. Heparin also prevents the formation of a stable fibrin clot by inhibiting the activation of the fibrin stabilizing factor. The half-life of intravenously administered heparin is about 90 minutes.

Coumarin. Oral anticoagulants can be prepared from compounds with coumarin as a base. Coumarin has been known for well over a century and, in addition to its use pharmaceutically, it is also an excellent odor-enhancing agent. However, because of its toxicity, it is not permitted in food products in the United States (Food & Drug Administration). The commercial drug *Sintrom®* is a 3-(alpha-acetonyl-4-nitrobenzyl)-4-hydroxycoumarin. This drug reduces the concentration of prothrombin in the blood and increases the prothrombin time by inhibiting the formation of prothrombin in the liver. The drug also interferes with the production of factors VII, IX, and X, so that their concentration in the blood is lowered during therapy. The inhibition of prothrombin involves interference with the action of vitamin K, and it has been postulated that the drug competes with vitamin K for an enzyme essential for prothrombin synthesis.

The commercial drug *Dicumarol®* is a bis-hydroxycoumarin, $C_{19}H_{12}O_6$. The actions of this drug are similar to those just described.

Warfarin. This compound is also of the coumarin family. The formula is 3-(alpha-acetonylbenzyl)-4-hydroxycoumarin. In addition to use in anticoagulant therapy in medicine, the compound also has been used as a major ingredient in rodenticides, where the objective is to induce bleeding and, when used in heavy doses, is thus lethal. The compound can be prepared by the condensation of benzylideneacetone and 4-hydroxycoumarin.

The anticoagulant action of warfarin is through interference with the gamma-carboxylation of glutamic acid residues in the polypeptide chains of several of the vitamin K-dependent factors. The carboxylation reaction is required for the calcium-binding activity of the K-dependent factors. Because of the reserve of procoagulant proteins in the liver, usually several days are required to effect anticoagulation with warfarin.

Warfarin antagonists include vitamin K, barbiturates, gluthethimide, rifampin, and cholestyramine. Warfarin potentiators include phenylbutazone, oxyphenbutazone, anabolic steroids, clofibrate, aspirin, hepatotoxins, disulfiram, nalidixic acid, and metronidazole. In patients undergoing anticoagulation therapy with warfarin, it has been found that cimetidine (used in therapy of duodenal ulcer) may increase anticoagulant blood levels and consequently prolong the prothrombin time (Serlin et al., 1979).

Anticoagulation Therapy. Prior to administration of anticoagulant drugs, patients must be carefully evaluated. Anticoagulant drugs are to be avoided if any of the following conditions prevail: a history of abnormal bleeding, recent corticosteroid therapy, recent intraocular or intracranial bleeding, recent pericarditis, and recent peptic ulcer or esophageal bleeding. A history of the individual's use of antiplatelet agents, such as aspirin, dipyridamole, phenylbutazone, and indomethacin, should be obtained and evaluated. Anticoagulant drugs should be administered with particular care during pregnancy. Heparin does not anticoagulate the fetus because it does not cross the placenta. Warfarin, on the other hand, anticoagulates both the mother and the fetus. Problems which may arise from the administration of warfarin during pregnancy, particularly during the first trimester, are described by Hirsch and Gallus (1972) and Ravio et al. (1977).

ANTIFOULING AGENTS. Various chemical substances added to paints and coatings to combat mildew and crustaceous formations, such as barnacles on the hull of a ship. In the past, large quantities of mercury compounds have been used in this manner. With growing environmental concern over possible mercury pollution, manufacturers have been turning to other, sometimes less efficacious compounds. Research continues to find compounds of a less toxic, but equally effective power of the mercury compounds. Bis(tributyltin) fluoride has been used on ship bottoms. See also **Mercury.**

ANTIFREEZE AGENT. A substance that lowers the freezing point (essentially of water). At one time, sodium chloride and magnesium chloride were widely used, but their extremely corrosive properties made them a liability in automotive and industrial cooling systems. Methyl alcohol, which requires about 27% by volume to protect against freezing to 0°F (−17.8°C), has a tendency to evaporate rapidly at operating temperatures and, coupled with its flammability and low boiling point of 147°F (63.9°C) limits its practical use in cooling systems. Many years ago, alcohols were replaced by glycol derivatives, which are relatively noncorrosive, nonflammable, have very low evaporation rates, and are effective heat-exchange media. A concentration of about 35% ethylene or propylene glycol antifreeze provides protection against freezing to 0°F (−17.8°C). Because of their overall properties, it is now common practice to retain the antifreeze concentrations in cooling systems the year around.

Antifreeze agents are also used in fuels where severe environmental conditions are encountered. For example, a mixture of methyl alcohol, isopropyl alcohol, and sometimes proprietary substances are used to inhibit the formation of ice from water vapor in hydrocarbon fuels. These additives are toxic and flammable.

ANTIGEN. A substance, usually a protein, a polysaccharide, or a lipid which, when introduced into the body, stimulates the production of antibodies. Bacteria, their toxins, erythrocytes, tissue extracts, pollens, dusts, and many other substances may act as antigens.

ANTIHISTAMINE. A synthetic substance essentially structurally analogous to histamine, the presence of which in minute amounts prevents or counteracts the action of excess histamine formed in body tissues. See also **Histamine.** Antihistamines are usually complex amines of various types. They find a number of medical uses.

In immediate hypersensitivity situations (reaction between antigen and antibody as encountered in hay fever, hives (urticaria), allergic (extrinsic) asthma, bites, and drug injections, among others), antihistamines can be part of the effective therapy. Although widely used, antihistamines and steroids are not always the drugs of choice. In atopic dermatitis (chronic skin disorder), antihistamines may assist in breaking the itch-scratch cycle, particularly in persons whose sleep may be interrupted by pruritus. Antihistamine compounds for urticaria are also effective for atopic dermatitis therapy. Frequently, shifting from one antihistamine to another is effective and helps to reduce side effects of the drugs. Antihistamines are sometimes effective in the treatment of autoerythroyce purpura, a rare disease. Antihistamines are also used in connection with mild penicillin reactions; also in cases of penicillin desensitization procedures. Certain antihistamines find application to control mild parkinsonism.

Some antihistamines are particularly effective in alleviating the onset of motion sickness. Some antihistamines have been found helpful in relieving persistent, unproductive coughs that frequently accompany bronchitis or coughs associated with allergy. They are used in connection with perennial and seasonal allergic rhinitis.

Most antihistamines have anticholinergic (drying) and sedative side effects, sometimes producing marked drowsiness and reduction of mental alertness, and thus should not be used by persons who operate machinery, drive vehicles, or otherwise must react quickly. Because of their similar structure, antihistamines appear to compete with histamine for cell receptor sites. Although conventional antihistaminic drugs, such as mepyramine, block the allergic and smooth muscle effects caused by histamine, the structure of these drugs is not sufficiently similar to histamine to inhibit histamine-stimulated gastric acid secretion. However, during the last few years, so-called histamine-blocking drugs have been developed which appear to be effective. It has been found that such compounds must contain the imidazole ring of histamine, with their potency enhanced by extension of the side chain. Among these new drugs are metiamide and cimetidine.

Some drugs in the antihistamine series play markedly different roles. Hydroxyzine hydrochloride and hydroxyzine pamoate have been used in the total management of anxiety, tension, and psychomotor agitation in conditions of emotional stress, usually requiring a combined approach of psychotherapy and chemotherapy. Hydroxyzine has been found to be particularly useful for making the disturbed patient more amenable to psychotherapy in long-term treatment of the psychoneurotic and the psychotic. The drug is not used as the only treatment of psychosis or of clearly demonstrated cases of depression. Hydroxyzine has also been found useful in alleviating the manifestations of anxiety and tension in acute emotional problems, and in such situations as preparation for dental procedures. Hydroxyzine therapy has been used in treatment of chronic alcoholism where anxiety withdrawal symptoms or delirium tremens may be present. Hydroxyzine may potentiate narcotics and barbiturates.

Most conventional antihistamines are available for both oral and intravenous or intramuscular administration. In serious cases of urticaria (hives), for example, the injection rather than oral route is most effective. The major excretion route for most antihistamines is hepatic (liver), occurring within 4 to 15 hours.

In addition to the side effects previously mentioned, some antihistamines may cause neutropenia (neutrophil count in the blood is less than 1800/cubic millimeter). Some antihistamines also may cause a modification of normal platelets in the blood.

Some of the more commonly used antihistamine compounds are listed below:

Ethanolamines:
Diphenhydramine hydrochloride (*Benadryl®*)
Dimenhydrinate (*Dramamine®*)

Ethylenediamines:
Tripelennamine hydrochloride (*Pyribenzamine®*)
Alkylamines:
Chlorpheniramine maleate (*Chlor-Trimeton®*)
Piperazines:
Cyclizine hydrochloride (*Marezine®*)
Phenothiazines:
Promethazine hydrochloride (*Phenergan®*)
Others:
Cyproheptadine hydrochloride (*Periactin®*)
Hydroxyzine hydrochloride (*Atarax®*)
Hydroxyzine pamoate (*Vistaril®*)

References

Forno, L. S., and E. C. Alvord, Jr.: "The Pathology of Parkinsonism," in "Recent Advances in Parkinson's Disease," (F. H. McDowell and C. H. Markham, Editors), F. A. Davis, Philadelphia (1971).

Hanifin, J. M., and W. C. Lobitz: "Newer Concepts of Atopic Dermatitis," *Arch. Dermatol.*, **113**, 663 (1977).

Mathews, K. P.: "A Current View of Urticaria," in "Symposium on Allergy in Adults: Review and Outlook," (M. Samter, Editor), *Med. Clin. North. Amer.* (1974).

Middleton, E. Jr., Reed, C. E., and E. F. Ellis: "Allergy Principles and Practice," C. V. Mosby, Saint Louis, Missouri (1978).

Monroe, E. W., and H. E. Jones: "Urticaria: An Updated Review," *Arch. Dermatol.*, **113**, 80 (1977).

Rajka, G.: "Atopic Dermatitis," in "Major Problems in Dermatology," (A. Rook, Editor), Vol. 3, Saunders, Philadelphia (1975).

ANTIMETABOLITES. These substances fall into the general class of cytotoxic chemicals, i.e., agents that damage cells to which they are applied. Antimetabolites are so similar to normal enzymatic substrate molecules or metabolites as to gain entry into the cellular machinery of intermediary metabolism, but once there they differ enough to cause enzymatic inhibition. If incorporated into protein, nucleic acids, or coenzymes, for example, they will diminish the biological worth of those substances. Spectacular agents of this sort include the antifolic acids, such as aminopterin and amethopterin, various other vitamin analogs, and analogs of the naturally occurring purines, pyrimidines, nucleosides, and amino acids. Effective action against the integrity of the cell appears to be exerted at a number of points of intermediary metabolism in these multifarious antimetabolites. Of particular interest with many of them is an interference in normal nucleic acid metabolism. 5-Fluoro-2'-deoxyuridine, for example, acts to inhibit the synthesis of thymidylate, a necessary precursor of DNA, and the related 5-bromo-2'-deoxyuridine is actually incorporated into new DNA in the place of thymidine. Both of these agents increase the frequency of chromosomal disturbances. Various other base analogs, if incorporated into DNA, can lead to gene mutation by alteration of the normal sequence of nucleotides during replication through incorrect base pairing. 2-Aminopurine is an example of such a mutagen. 8-Azaguanine can be incorporated into ribonucleic acids, which are thus rendered defective. Among the actions of 6-mercaptopurine is an interference in the biochemical activity of coenzyme A, with resultant mitochondrial damage. Such amino acid analogs as β-fluorophenylalanine can effectively halt cellular activities by being incorporated into new proteins, which thereupon fail to attain their proper enzymatic or other functions.

Advantage is taken of the properties of antimetabolites in chemotherapy. In cancer chemotherapy, several antimetabolites are used. These include methotrexate, 6-mercaptopurine, 6-thioguanine, 5-fluorouracil, and cystine arabinoside. In the chemotherapy of metastatic breast cancer, 5-fluorouracil and methotrexate, in combination with cyclophosphamide, have been used. Antimetabolites, sometimes along with corticosteroids, are used in the therapy of various autoimmune diseases, such as thrombocytemic purpura, thyroiditis, and Goodpasture's syndrome, among others.

Metabolites are implicated as agents that produce marrow aplasia as found in leukemia.

References

Carbone, P. P., et al.: "Chemotherapy of Disseminated Breast Cancer," *Cancer*, **39** (supplement), 2916 (1977).

Clarkson, B., Dowling, M. D., and T. S. Gee: "Treatment of Acute Leukemia in Adults," *Cancer*, **36**, 775 (1975).

Crowther, D.: "Blood and Neoplastic Diseases: A Rational Approach to the Chemotherapy of Human Malignant Disease," *I. Br. Med. Jrnl.*, **4**, 156 (1974).

Haskell, C. M., et al.: "Systemic Therapy for Metastatic Breast Cancer," *Ann. Intern. Med.*, **86**, 68 (1977).

Keiser, L. W., and R. L. Capizzi: "Principles of Combination Chemotherapy," in "Cancer, A Comprehensive Treatise," (J. F. Becker, Editor), Vol. 5, Plenum, New York (1977).

Lipton, A.: "Chronic Idiopathic Neutropenia: Treatment with Corticosteroids and Mercaptopurine," *Arch. Intern. Med.*, **123**, 694 (1969).

Talal, N., Editor: "Autoimmunity: Genetic Immunologic, Virologic and Clinical Aspects," Academic, New York (1978).

ANTIMICROBIAL AGENTS (Foods). These are substances added to food products, often in minute quantities, to destroy or inhibit the activity of microorganisms. These organisms are responsible for a high percentage of food spoilage and frequently severely limit the shelf-life of food substances.

The fundamental mechanisms used by antimicrobial agents in acting against microorganisms are: (1) adverse influence on the cellular membranes; (2) interference with genetic mechanisms; and (3) interference with cellular enzymes.

Among the principal antimicrobial agents used in foods are: benzoic acid and sodium benzoate; the parabens; sorbic acid and sorbates; propionic acid and propionates; sulfur dioxide and sulfites; acetic acid and acetates; nitrates and nitrites; antibiotics; diethyl pyrocarbonate; epoxides; hydrogen peroxide; and phosphates. In addition to incorporation of antimicrobial agents directly into processed foods, the agents also are used in various processing stages, but are not always transferred to the food products per se.

Benzoic Acid and Sodium Benzoate. These compounds are most active against yeasts and are less effective against molds. They are best suited for foods with a natural or adjusted pH below 4.5. The average dosage in foods ranges between 0.05 and 0.1% (weight), depending upon product. Benzoic acid occurs naturally in cinnamon, ripe cloves, cranberries, greengage plums, and prunes. See also **Benzoic Acid.** For a number of years the sodium salt has been preferred over the acid by food processors.

Common applications for these compounds include carbonated and noncarbonated beverages, but excluding beers and wines because of their action against yeasts. They are also used in salted margarines, jams, jellies, and preserves, pie fillings, salads and salad dressing, pickles, relishes, and other condiments, as well as olives and sauerkraut. In terms of human metabolism of these substances, some authorities have suggested that benzoate is conjugated with glycine to produce hippuric acid, which is excreted, possibly accounting for 65–95% of benzoate ingested. It has been postulated that the remainder is detoxified by conjugation with glycuronic acid.

Parabens. These compounds include the methyl, ethyl, propyl, and butyl esters of para-hydroxybenzoic acid. In the United States and a number of other countries, the methyl and propyl esters are preferred, while European food processors favor the ethyl and butyl esters. The parabens were first described in 1924 as having antimicrobial activity and initially were used in cosmetic and pharmaceutical products.

The parabens are most effective against molds and yeasts, but less active against bacteria, particularly gram-negative bacteria. The antimicrobial activity of the parabens is directly related to the molecular chain length (methyl is weakest; butyl is strongest). However, the solubility of these compounds is in inverse relationship with chain length. These characteristics give rise to the use of two or more esters in combination and sometimes in combination with entirely different antimicrobial agents, such as sodium benzoate. Below a pH of 7, the parabens are only weakly effective.

The parabens are used in carbonated beverages and other soft drinks, including cider. In lieu of pasteurizing or using Millipore filtration, some brewers use the parabens for controlling secondary yeast formation. Because of their activity against yeasts, they are not used in bread and rolls, but they find use in other bakery products, such as pie crusts, certain pastries, icings, toppings, fillings, and cakes. The parabens are particularly effective in preserving fruit cakes. Usage also includes creams and pastes, fruit products, flavor extracts, pickles and olives, and artificially sweetened jams, jellies, and preserves. Average dosage ranges from 0.03 to 0.6% (weight).

Sorbic Acid and Sorbates. Sorbic acid and its potassium and sodium salts are effective against molds, but less effective against bacteria. These compounds may be incorporated directly into the food product, but they are frequently applied by spraying, dipping, or coating. The compounds are effective up to a pH of about 6.5. This is higher than propionates and sodium benzoate, but not so high as the parabens. Metabolism in humans parallels that of other fatty acids.

Because sorbates affect yeasts, the compounds are not directly useful in yeast-raised goods. Sorbates are particularly favored for use in chocolate syrups. They can be used in wine production in conjunction with sulfur dioxide against bacteria and are effective in inhibiting development of unwanted yeasts. They are also used in artificially sweetened jellies, jams, and preserves; in pickles and related products; in nonsalted margarines; in dried and smoked fish products; in semi-moist pet foods; in dry sausage casings; in fruit-filled toaster pastries; and in cheese and cheese products. In the latter products, the agents usually are applied by dipping or spraying. Wrappers also may be impregnated with sorbates.

Propionic Acid and Propionates. The antimicrobial properties of propionic acid and its calcium and sodium salts were first noted in 1913. Today, the calcium and sodium salts are most commonly used. These compounds are more active against molds than sodium benzoate, but have little if any activity against yeasts. The propionates are well known for their effectiveness against *Bacillus mesentericus*, a "rope"-forming microorganism. These compounds are effective up to a pH of 5 or slightly higher. Metabolism in the human body parallels that of other fatty acids.

An early application for the propionates was that of dipping cheddar cheese in an 8% propionic acid solution. This increased mold-free life by 4 to 5 times that when no preservative was added. For pasteurized process cheese and cheese products, propionates can be added before or with emulsifying salts. Research has indicated that propionate-treated parchment wrappers provide protection for butter.

Use of propionates in breads can extend mold-free life by 8 days or more. Propionates are favored by bakers because of their effectiveness against ropy mold in breads up to pH levels of 6. For cakes and unleavened bakery goods, the sodium salt is usually preferred; for bread, the calcium salt is favored. This additive also contributes to the mineral enrichment of the product.

Acetic Acid and Acetates. Acetic acid (pure and as vinegar) and calcium, potassium, and sodium acetates, as well as sodium diacetate, serve as antimicrobial agents. In the United States, vinegar can contain no less than 4 grams of acetic acid per 100 milliliters of product. Acetic acid and calcium acetate are most effective against yeasts and bacteria, and to a lesser extent, molds. The diacetate is effective against both rope and mold in bread. It is interesting to note that the antimicrobial effectiveness of acetic acid and its salts is increased as the pH is lowered.

Optimal pH range varies with products and target microorganisms, but generally falls between 3.5 and 5.5. These agents are particularly effective against *Salmonella aertrycke*, *Staphylococcus aureus*, *Phytomonas phaseoli*, *Bacillus cereus*, *B. mesentericus*, *Saccharomyces cervisiae*, and *Aspergillus niger*.

Unfortunately, to be effective against microorganisms in bakery products, acetic acid concentrations must be so high that an overly sour taste is imparted to the products. Sodium diacetate, however, can be used in small concentrations in bread and rolls to control rope and molds. Traditional concentrations of the acetate are 0.4 part to 100 parts of flour. During recent years, the propionates have largely displaced sodium diacetate for this use.

Vinegar or acetic acid is used in a number of products as much for its sour taste as for its antimicrobial properties. Such products include catsup, mayonnaise, pickles, salad dressing, and various condiment sauces. These agents also have been used to a lesser extent in malt syrups and concentrates, cheeses, and in the treatment of parchment wrappers for products, such as butter, to inhibit mold.

Nitrates and Nitrites. For many decades, sodium nitrate and nitrite, and potassium nitrate and nitrite have been used to cure, preserve, and provide a characteristic flavor to such meats as bacon, corned beef, frankfurters, ham, and various sausages. This tradition continues into the 1980s, but was seriously threatened in the mid- and late 1970s. Some researchers reported that N-nitrosopyrolidine (NPyr) formed in bacon upon application of heat during preparation for consumption. It was observed that there was a greater concentration of the NPyr in adipose tissue than in the lean portion. A connection was proposed that involved serious implications of the ultimate carcinogenic risk involved in meat treated with the nitrates and nitrites. Numerous tests proceeded. One of the main factors learned during this period was how little knowledge food scientists had concerning the fate of the nitrates and nitrites. Although the precursors of the nitrosamines formed under certain conditions of cooking were known, the mechanism of formation was unknown. As of the early 1980s, there is a period of quietude, with regulatory officials in the United States not fully convinced of taking serious measures that would result in serious breaks in tradition on the parts of meatpackers and consumers alike. Perhaps with further investigations in a less tense atmosphere, solid scientific information will be developed to prove or disprove a number of points that remain unanswered.

Sulfur Dioxide and Sulfites. The use of sulfur dioxide gas and with it the production of sulfites differs somewhat from the other antimicrobial agents thus far described. Historical records show that burning sulfur to produce sulfur dioxide (SO_2 gas) dates back to the ancient Egyptians and Romans who used it in connection with wine making. See **Sulfur.** Action by sulfur dioxide is accomplished in the gaseous phase. The effect of SO_2 is markedly determined by concentration and pH conditions of the target product. Research has demonstrated that most bacteria are inhibited by HSO_3^- at concentrations of 200 parts per million (ppm) or less. With few exceptions, yeasts are also similarly inhibited. There are, however, some strains of molds that are considerably more resistant. The sulfite salts tend to be unstable and oxidize during long periods of storage, thus decreasing the availability of SO_2. This process is aggravated by the presence of moisture.

The most effective range for optimal microbial inhibition with sulfites is a pH of 2.5–3. It has been found that from 2 to 4 times greater concentrations of SO_2 are needed to inhibit the growth of microorganisms at a pH of 3.5 than at a pH of 2.5. It also has been demonstrated that, at a pH of 7, SO_2 has little if any effect on yeasts and molds, even at concentrations up to 1000 ppm. The inhibitory effects against bacteria are also considerably less when pH rises above 3.5. Some researchers believe that at higher pH levels, penetration of cell walls is much more difficult.

Residual levels of sulfites in excess of 500 ppm impart a noticeable taste to food substances; this fact, regardless of any regulations toward limiting concentration, requires SO_2 levels to be controlled.

The use of SO_2 for preserving fruit juices, syrups, concentrates, and purees is particularly attractive in regions with warm climates and where products must be stored in bulk prior to processing. In these situations, the SO_2 concentration will range between 350 and 600 ppm. High sugar concentrations require higher levels of SO_2. For optimal effectiveness, the pH of some products has to be reduced.

It is a common practice to expose many fruits to SO_2 prior to dehydration. The SO_2 also extends storage life of raw fruit prior to dehydration. The optimal temperature for exposure to the gas is from 43 to 49°C. Unlike fruit, vegetables are usually dipped in solutions of neutral sulfites and bisulfites. Suggested levels of SO_2 in some dried and dehydrated fruits and vegetables, in parts per million, are:

Apricots, peaches, and nectarines	2000
Raisins	800–1500
Pears	1000
Apples	800
Cabbage	750–1000
Carrots and potatoes	200– 250

It is important to note, however, that bulk-treated fruits intended for canning should not have a residual level in excess of 20 ppm SO_2 because of possible sulfide (black precipitate) forming in the can as the result of hydrogen sulfide generation.

Many countries do not allow use of SO_2 or sulfite salts for use on meats, fish, or processed meat and fish products. Where permitted, sulfite is helpful in eliminating "black spot" formation in shrimp.

A major use of sulfites is in wine making. It is used for sanitizing equipment and, prior to fermenting, the grape musts have to be treated with sulfites to inhibit the growth of any natural microbial flora present. This is done prior to the addition of pure cultures of the appropriate wine-making yeasts. During fermentation, SO_2 also can function as an antioxidant, clarifier, and dissolving agent. Sulfur dioxide is often used after fermentation to prevent undesirable postfermentation alterations by various microorganisms. Levels of SO_2 during fermentation range from 50 to 100 ppm, depending upon condition of the grapes, temperature, pH, and sugar concentration. The wine industry uses sulfur dioxide dissolved in water, vaporized SO_2 and sulfite salts. An SO_2 level of 50–75 ppm assists the prevention of bacterial spoilage during the bulk storage of wine after fermentation.

Antibiotics. Much attention has been given to the use of antibiotics in food-associated applications since the introduction of penicillin in the 1940s. Their use has been limited. The use of antibiotics in animal feedstuffs continues to remain a topic of controversy in many countries; in other countries they have been banned. Antibiotics carry into the meat produced and further into human diets, thus possibly reducing their effectiveness in the treatment of human diseases.

Diethyl Pyrocarbonate. The preservative qualities of this compound was not recognized until the late 1930s and research on the compound continues to date. Also called pyrocarbonic acid diethyl ester, the compound is extremely effective against yeasts. It is also active against bacteria, such as *Lactobacillus pastorianus*, and various molds. The substance is generally used in still wines, fermented malt beverages, and noncarbonated soft drinks, as well as fruit-based beverages. Regulations on its use vary from one country to the next. Effective inhibition by diethyl pyrocarbonate is largely confined to acid products of low microorganism count. Some researchers point out that the pH should be less than 4 and that the microorganism count should not exceed 500 per milliliter. Some authorities observe that because of its rapid hydrolysis, no toxicity or residue problems should occur in products where the compound is permitted.

Epoxides. Two compounds are included in the category of antimicrobials. One is the gas, *ethylene oxide*; the other, *propylene oxide*, is a colorless liquid with a boiling point of 35°C. Ethylene is highly reactive and must be used carefully and only with proper equipment. Somewhat less hazardous from an explosion standpoint, propylene oxide also as an explosive range of 2–22%. Consequently, these materials are usually mixed with inert substances, such as carbon dioxide or organic diluents.

Ethylene oxide is a universal antimicrobial, in that it is lethal to all microorganisms. However, it is not universal from the standpoint of application. Propylene oxide is considered a broad-range microbiocide. In practically all aspects, propylene oxide is a considerably less effective agent, requiring longer exposures and greater concentrations because of its less penetrating power. However, propylene oxide is less toxic to humans.

The use of these gases (propylene oxide is volatilized) has been called *cold sterilization*, and is frequently useful in sterilizing a number of low-moisture ingredients which end up in high-moisture foods. This prior sterilization lessens the total load on later thermal processing. The ability to kill microorganisms in low-moisture foods is an outstanding advantage of the gases. At the same time, macroorganisms also are killed. The gases find application in connection with spices, starches, nut meats, dried prunes, and glacé fruit. They are not used on peanuts (groundnuts).

Hydrogen Peroxide. Although usually regarded as a bleaching and oxidizing agent, this compound, H_2O_2, can be an effective antimicrobial and can be particularly useful in sterilizing processing equipment and packaging materials, notably prior to the aseptic packaging process. Regulations regarding the use of hydrogen peroxide vary from one country to the next.

Phosphates. The various phosphates are effective multipurpose food additive chemicals and functions other than their antimicrobial properties are usually given the greatest stress. The antimicrobial properties of the phosphates have been investigated over the years and are reasonably well documented.

References

Crocco, S. C.: "Sorbate Plant," *Food Engineering* (April 1977).
Gilliland, S. E., and M. L. Speck: "Inhibition of Psychrotrophic Bacteria by Lactobacilli and Pediococci in Nonfermented Refrigerated Foods," *Journal of Food Science*, 40, 903 (1975).
Kraft, A. A., and C. R. Rey: "Psychotrophic Bacteria in Foods," *Food Technology*, 33, 1, 66–71 (1979).
Miller, M. W.: "Yeasts in Food Spoilage," *Food Technology*, 33, 1, 76–80 (1979).
Segner, W. P.: "Mesophilic Aerobic Sporeforming Bacteria in the Spoilage of Low-Acid Canned Foods," *Food Technology*, 33, 1, 55–59, 80 (1979).
Staff: "Shelf Life Increased by Sorbates," *Baking Industry* (May 1977).
Troller, J. A.: "Food Spoilage by Microorganisms Tolerating Low-a_w Environments," *Food Technology*, 33, 1, 72–75 (1979).

ANTIMONY. Chemical element symbol Sb, at. no. 51, at. wt. 121.75, periodic table group 5a, mp 630.5°C, bp 1380°C, sp gr 6.62 (vacuum-distilled solid at 20°C) and 6.73 (single crystal). Naturally occurring isotopes are ^{121}Sb and ^{123}Sb. Antimony metal is a lustrous, silvery, blue-white solid, extremely brittle and exhibiting a scalelike or flaky crystalline texture. The metal is easy to pulverize. The pure metal has a hardness of 3.0–3.3 on the Mohs scale and 55 on the Brinell scale. Of the more common metals, antimony is the poorest conductor (4.5 on a scale of 100 for copper). From careful studies, it has been observed that Sb contracts upon solidification rather than expanding. The element was first described by Thölden (Valentine) in 1450.

There are two natural isotopes, ^{121}Sb and ^{123}Sb; and ten radioactive isotopes, ^{116}Sb through ^{120}Sb, ^{122}Sb, and ^{124}Sb through ^{127}Sb. ^{124}Sb is used as a radiation source in industrial instruments for the measurement of flow of slurries and interface measurements in pipelines. See also **Radioactivity.**

First ionization potential 8.64 eV; second 16.5 eV; third 25.3 eV; fourth 44.1 eV; fifth 56 eV. Oxidation potentials $Sb + H_2O \rightarrow SbO^+ + 2H^+ + 3e^-$, −0.212 V; $2SbO^+ + 3H_2O \rightarrow Sb_2O_5 + 6H^+ + 4e^-$, −0.581 V; $Sb + 4OH^- \rightarrow SbO_2^- + 2H_2O + 3e^-$, 0.66 V. Other important physical properties of antimony are given under **Chemical Elements.**

Antimony exists in a number of allotropic forms. Gray or metallic antimony, density 6.79 g/per cm³, is the stable form, forming rhombohedral crystals. Its vapor is that of Sb_4 up to 800°C, where dissociation to Sb_2 commences. Yellow antimony, Sb_4, density 5.3 g/cm³ is less stable than yellow arsenic. It is produced by oxidation of stibine (see below) at very low temperatures, above which it is unstable. It changes even in the dark to black antimony at −90°C (in the light at −180°C). Black antimony, produced most readily by cooling antimony vapor or oxidizing stibine at 40°C, density 5.3, is metastable with respect to the gray form. It is also more reactive, igniting in air at room temperatures or above. Explosive antimony is produced by rapid electrodeposition of antimony from its halides. When heated or scratched it undergoes an exothermic transformation to gray antimony. Its structure is amorphous, and differs somewhat from that of gray antimony.

Antimony is used in alloys, with lead for storage battery plates, with lead and tin in type metals and body solders, with tin and copper in bearing or anti-friction metals. Antimony occurs chiefly as the sulfide (stibnite, Sb_2S_3) which is produced mainly in China, only small amounts in Mexico and Bolivia. Stibnite is (1) melted and reduced to antimony by iron metal and separated from fused ferrous sulfide; See also **Stibnite** (2) is roasted in air and sublimed antimonous oxide collected and reduced by heating to fusion with carbon and sodium carbonate.

Antimony is also leached from tetrahedryte ore and recovered by electrowinning.

Antimony is scarcely tarnished in dry air but oxidized slowly in moist air; burns at a red heat in air or oxygen with incandescence forming antimonous oxide; insoluble in HCl; converted by HNO$_3$ into antimonous oxide or antimonic oxide, depending upon the concentration of acid; by chlorine into trichloride or pentachloride, by NaOH solution into antimonite.

Stibine. SbH$_3$ is formed by hydrolysis of some metal antimonides or reduction (with hydrogen produced by addition of zinc and HCl) of antimony compounds, as in the Gutzeit test. It is decomposed by aqueous bases, in contrast with arsine. It reacts with metals at higher temperatures to give the antimonides. The antimonides of elements of group 1a, 2a, and 3a usually are stoichiometric, with antimony trivalent. With other metals, the binary compounds are essentially intermetallic, with such exceptions as the nickel series, Ni$_2$Sb$_3$, NiSb, Ni$_5$Sb$_2$ and Ni$_4$Sb.

Trihalides. SbF$_3$, SbCl$_3$, SbBr$_3$, and SbI$_3$ are solids, and have pyramidal structures. Except for the fluoride, which is not hydrolyzed, they undergo partial hydrolysis only (in contrast with the phosphorus trihalides) on contact with water to yield insoluble oxyhalides, either of composition SbOX or varying somewhat from this composition to give such compounds as Sb$_4$O$_5$Cl$_2$. The antimony pentahalides, SbF$_5$ and SbCl$_5$ can be prepared, but the pentabromide exists only in double compounds, known as bromoantimonates, those for monovalent metals being of the type MSbBr$_6$, plus water of hydration, and yielding SbBr$_6^-$ ions. SbCl$_6^-$ and SbF$_6^-$ ions are also known. Mixture of antimony(III) chloride, SbCl$_3$, in HCl solution with antimony(V) chloride, SbCl$_5$, in equimolar proportions yields a dark colored solution. While antimony(IV) chloride cannot be isolated from it, compounds such as cesium antimony(IV) chloride, Cs$_2$SbCl$_6$ are formed by addition of cesium chloride, CsCl, and they are isomorphous with similar compounds of lead, tin and other metals. However, tetravalent antimony should be paramagnetic because of the unpaired electron, whereas compounds of SbCl$_6^{2-}$ are diamagnetic. Therefore it may be that these compounds contain equimolar mixtures of SbCl$_6^-$ and SbCl$_6^{-3}$. The existence of these higher halide complexes with tin (and bismuth) but not with phosphorus or arsenic, may be due to steric considerations.

Antimony (III) oxide. Sb$_2$O$_3$ or Sb$_4$O$_6$, formed by melting antimony in air, or from the hydroxide Sb (OH)$_3$. The Sb$_2$O$_3$ of commerce is produced from the oxidation of stibnite ore. Antimony is below arsenic in the periodic table, and Sb(OH)$_3$ is more definitely amphiprotic than As(OH)$_3$, forming not only antimony(III) salts and antimonites (containing the ion SbO$_2^-$ or Sb(OH)$_4^-$), but also basic salts, especially the antimonyl salts, containing the ion SbO$^+$. Antimony(V) oxide, formed by oxidation of the metal with HNO$_3$, is less soluble in H$_2$O than As$_2$O$_5$. Antimonic acid cannot be obtained by hydration, and the product resulting upon hydrolysis of pentahalides has a variable H$_2$O content. The salts of the acid, the antimonates, are of the type MISb(OH)$_6$, as Pauling showed to be necessary to conform to accepted ionic radius ratios. Although the strength of antimonic acid has not been accurately determined, it appears to be comparable to acetic acid.

Antimony(IV) Oxide. Obtained by heating in air the trioxide or the hydrated pentoxide.

There is a marked structural difference between the phosphates and the antimonates. Thus sodium pyroantimonate, Na$_2$H$_2$Sb$_2$O$_7$ · 5H$_2$O contains the ion Sb(OH)$_6^-$ rather than Sb$_2$O$_7^{4-}$, and the magnesium compound (hydrated) which has a 12:1 ratio of oxygen to antimony, and would thus be a hexahydroxyantimonate, has the (X-ray determined) structure [Mg(H$_2$O)$_6$][Sb(OH)$_6$]$_2$.

Sulfides. Sb$_2$S$_3$ and Sb$_2$S$_5$ may be obtained from the elements or by precipitation, respectively, of Sb(III) and Sb(V) solutions

with H$_2$S. The Sb$_2$S$_3$ dissolves in alkaline solutions to form thioantimonites, containing the ion SbS$_3^{3-}$, or Sb(SH)$_6^-$, while Sb$_2$S$_5$ forms the thioantimonates, containing SbS$_4^{3-}$. The latter is probably present as [SbS$_2$(SH)$_2$-(OH)$_2$]$^{3-}$ or [SbS$_4$(H$_2$O)$_2$]$^{3-}$.

In alloys, antimony is easily detected by its formation of a white solid upon treatment with concentrated HNO$_3$ and subsequent separation from tin, which is the only other metal thus forming a white solid.

Both trivalent and pentavalent antimony form several organoantimony compounds. Some of these include methylstibine CH$_3$SbH$_2$ and the substitution product, methyldichlorostibine CH$_3$SbCl$_2$; phenylstibine C$_6$H$_5$SbH$_2$ and the substitution product, phenyldichlorostibine C$_6$H$_5$SbCl$_2$; methylantimony tetrachloride CH$_3$SbCl$_4$; phenylantimony tetrachloride C$_6$H$_5$SbCl$_4$; sodium methylantimonate Na[CH$_3$Sb(OH)$_5$]; sodium trifluoromethyl antimonate Na[(CF$_3$)$_3$Sb(OH)$_3$]; triethylstibine sulfide (C$_2$H$_5$)$_3$SbS; tetraphenylstibonium tetraphenylborate [(C$_6$H$_5$)$_4$Sb][B(C$_6$H$_5$)$_4$]; stibiobenzene C$_6$H$_5$Sb=SbC$_6$H$_5$; and lithium hexaphenylantimonate LiSb(C$_6$H$_5$)$_6$.

Uses. Representative alloys containing antimony are described in the accompanying table.

ANTIMONY CONTENT OF REPRESENTATIVE ANTIMONY-CONTAINING ALLOYS

Hard lead	Up to 12% Sb
Antimony reduces mp of Pb and hardens resulting alloy. Alloy has better abrasion resistance than chemical Pb at temperatures below 140°C. Alloy is age-hardenable.	
Tin-lead solders	Up to 1% Sb
Type metals	3–19% Sb
These Pb-base alloys also contain from 3–9% Sn.	
Lead-base diecasting alloys:	
^{a}ASTM No. 4	14–16% Sb
ASTM No. 5	9.25–10.75% Sb
Bearing alloy	15% Sb
CT metal	12.5% Sb
Tin-free alloy	10% Sb
Babbitt (bearing) metals:	
^{b} SAE 10	4–5% Sb
SAE 11	6–7.5% Sb
SAE 12	7–8.5% Sb
SAE 13	9.25–10.25% Sb
SAE 14	14–16% Sb
SAE 15	14.5–16% Sb
Britannia metal	5% Sb
This alloy also contains 93% Sn and 2% Cu. Very useful for spinning utensils.	
Pewter	Up to 7% Sb
Pewter also contains up to 20% Pb and 4% Cu with the remainder made up by Sn.	

^{a} American Society for Testing and Materials
^{b} Society of Automative Engineers

Metallic antimony is an effective pearlitizing agent for producing pearlitic cast iron. The principal use of antimony, however, is in the form of the oxide. Its major application is as a flame retardant for plastics and textiles. Other applications of importance are in glass, pigments, and catalysts.

Toxicity. The threshold limit value of antimony and its compounds is 0.5 milligram/cubic meter (as Sb). Antimony and its compounds used under conditions giving rise to dust, fume, and vapor should be carried out under proper ventilation. In handling antimony and its compounds, appropriate hygienic practices and good housekeeping should be observed. Stibine, SbH$_3$, requires extreme caution in handling because it is very toxic. When using antimony and its compounds, reducing conditions, which may give rise to the undesired formation of stibine, must be avoided.

References

Carapella, S. C., Jr.: "Antimony" in "Kirk-Othmer Encyclopedia of Chemical Technology," Vol. 3, 3rd Edition, 105–128, Wiley, New York (1978).
Sneed, M. C., and R. C. Brasted: "Comprehensive Inorganic Chemistry," Vol. 5, Van Nostrand Reinhold, New York (1956).
Wang, C. Y.: "Antimony," Chas. Griffin, London (1952).

—S. C. Carapella, Jr., ASARCO Incorporated, South Plainfield, New Jersey.

ANTIOXIDANT. Usually an organic compound added to various types of materials, such as rubber, natural fats and oils, food products, gasoline and lubricating oils, for the purposes of retarding oxidation and associated deterioration, rancidity, gum formation, reduction in shelf-life, etc.

Rubber antioxidants are commonly of an aromatic amine type, such as di-beta-naphthyl-para-phenylenediamine and phenyl-beta-naphthylamine. Usually, only a small fraction of a percent affords adequate protection. Some antioxidants are substitute phenolic compounds (butylated hydroxyanisole, di-*tert*-butyl-para-cresol, and propyl gallate). See **Rubber (Natural)**.

When used in foods, antioxidants are highly regulated to extremely small percentages in most countries—down to the low fractions of one percent. Composition of the substrate, processing conditions, impurities, and desired shelf-life are among the most important factors in selecting the best antioxidant system for a given food product. The desirable features of antioxidants may be summarized as: (1) effectiveness at low concentrations; (2) compatibility with the substrate; (3) nontoxicity to consumers; (4) stability in terms of conditions encountered in processing and storage, including temperature, radiation, pH, etc.; (5) nonvolatility and nonextractability under the conditions of use; (6) ease and safety in handling; (7) freedom from off-flavors, off-odors, and off-colors that might be imparted to the food products; and (8) cost effectiveness.

Mechanism of Oxidative Degradation. It could appear that inasmuch as oxidative degradation occurs in a variety of organic materials that are dissimilar in appearance and have entirely different applications and different properties, with degradation producing different effects, the oxidation mechanism itself might be different. Current knowledge indicates, however, that the mechanism of oxidative degradation is the same for all organic substances. They appear to degrade by the same free-radical mechanism.

Common examples of food oxidative degradation include products that contain oils and fats. For example, some antioxidants have made it possible to store groundnuts (peanuts) and other nuts, maize (corn) products, and bakery and cereal products on the shelf for periods well in excess of the four months that was considered the traditional limiting period prior to the appearance of such additives. Other examples of food products that tend to become rancid by way of oxidation include various meat-flavor stuffing mixes, cake mixes, unbaked cheesecake mix, and essentially all foods that incorporate lipids. The stability of natural fats and oils present in raw materials varies over a wide range and hence the amount of antioxidant required must be tailored to each product situation. Enzymatic "browning" is another example of oxidative degradation. The enzymes in fruits and vegetables cause apples, apricots, potatoes, among others, to darken when they are exposed to air after being cut, bruised, or allowed to overmature. Some antioxidants can prevent or delay enzymatic browning much in the same manner as dipping freshly cut fruits in lemon, orange, or pineapple juice. Limonene and ascorbic acid naturally present in these juices serve as antioxidants. Oxidative changes may affect carbohydrate, protein, and fat substances, the primary building blocks of foodstuffs, but generally the oxidative rancidity problem results mainly from the *autoxidative* degradation of fatty (glyceridic) components.

Some authorities describe oxidation as a free-radical, chain-type reaction. At usual processing temperatures and more slowly at room temperature, organic free radicals ($R\cdot$) are formed. These react with oxygen to form peroxy radicals ($ROO\cdot$), which can abstract a hydrogen atom from the affected substance to form a hydroperoxide ($ROOH$) and another organic free radical. The cycle repeats itself with the addition of oxygen to the new free radical. The unstable hydroperoxides left along with the substance are the major source of degradation. Under the influence of heat, light, and any metals if present, the hydroperoxides decompose to form carbonyl groups. When this happens, the organic molecule breaks and splits off another organic free radical. Ultimately, this type of degradation can lead to rancidity and color deterioration in oils and fats.

An antioxidant ties up the peroxy radicals so that they are incapable of propagating the reaction chain or to decompose the hydroperoxides in such a manner that carbonyl groups and additional free radicals are not formed. The former, which are called *chain-breaking antioxidants*, *free-radical scavengers*, or *inhibitors*, are usually hindered phenols or amines. The latter, called *peroxide decomposers*, are generally sulfur compounds or organophosphites. A number of antioxidants useful in rubber and plastics, for example, are not suited to food products because of their toxicity.

A mixture of two antioxidants often will display synergism. Probably the most generally effective mixtures of antioxidants are those in which one compound functions as a decomposer of peroxides (sulfides, thiodipropionate) and the other as an inhibitor of free radicals (hindered phenols, amines). Although the latter retards the formation of reaction chains, some hydroperoxide is nevertheless formed. If this hydroperoxide then reacts with a decomposer of peroxides, instead of decomposing into free radicals, the two antioxidants act together to complement each other. Moreover, the peroxide decomposer may itself be subject to oxidation by peroxy radicals, and its efficiency will therefore be increased in the presence of an inhibitor of free radicals. In the case of phenol-sulfide mixtures, the sulfide (peroxide decomposer) also continuously regenerates the phenol (radical scavenger) to accentuate the synergistic nature of the mixture. Metal chelators or deactivators, such as citric and phosphoric acids, of prooxidant metals (iron, copper, nickel, tin), ultraviolet-light absorbers (carbon black, substitute benzophenones, benzotriazoles, and salicylates), and antiozonants (substituted phenylenediamines) also develop synergistic effects with antioxidants.

Applications of Antioxidants. The use of antioxidants in foods, pharmaceuticals, and animal feeds (direct feed additives), as well as their use in food-contact surfaces (indirect additives) is closely regulated by the governments of several countries. Antioxidants are approved only after extensive extraction, toxicological, and feeding studies. The list is relatively limited. Although antioxidants have been used for several decades and some occur naturally in food substances, intensive research is continuing, partly accelerated by the growing use of unsaturated oils in numerous food products.

Butylated hydroxyanisole (BHA) was first used in food products in 1940. This continues as one of the commonly used antioxidants, sometimes in combination with butylated hydroxytoluene (BHT), propyl gallate, or citric or phosphoric acids, to obtain a synergistic effect. In food-contact surfaces, BHT has been used by itself or in combination with thiodipropionates and/or phosphoric acids, to obtain a synergistic effect. Well over $50 million of antioxidants are produced per year commercially in the United States alone.

The value of antioxidant protection by way of natural food sources has been pointed out in the literature with considerable frequency. Among the components of soy flour known to have some antioxidant properties are isoflavones and phospholipids. Amino acids and peptides in soybean flour also possess some antioxidant activity. There also may be some antioxidant impact from aromatic amines and sulfhydryl compounds.

Rosemary and sage have been shown to have effective antioxidant properties. The extracts in the past have been of strong odor and bitter taste and thus unsuited for use in most food products. However, solvent extraction procedures have been developed to produced purified antioxidants from rosemary and sage.

For many years, in connection with certain food products, a barrier to freeze-drying has been the problems associated with the storage stability of foods that are susceptible to lipid oxidation. In order for such foods to have a reasonable shelf life and acceptable flavor characteristics, protective additives which retard oxidation are often added before dehydration. Such antioxidants must carry through the process and not be lost due to volatilization. For these applications, BHA and BHT and *tert*-butylhydroquinone (TBHQ) have been found quite effective. See also **Polymeric Food Additives.**

References

Chang, S. S., et al: "Natural Antioxidants from Rosemary and Sage," *Jrnl. of Food Science*, 42, 4, 1102–1106 (1977).
Hammerschmidt, P. A., and D. E. Pratt: "Phenolic Antioxidants of Dried Soybeans," *Jrnl. of Food Science*, 32, 2, 556–559 (1978).
Regnarsson, J. O., et al.: "Accelerated Temperature Study of Antioxidants," *Jrnl. of Food Science*, 42, 6, 1536–1539 (1977).
Regnarrson, J. O., et al.: "Accelerated Shelf Life Testing for Oxidative Rancidity in Foods," *Jrn. of Food Chemistry* (1978).
Waletzko, P. T., and T. P. Labuza: "Accelerated Shelf Life Testing of an Intermediate Moisture Food in Air and in an Oxygen-free Atmosphere," *Jrnl. of Food Science*, 41, 1338 (1976).

ANTIPROTON. An elementary particle having a mass equal to that of the proton, differing from the proton only in the sign of its charge, which is negative, and a magnetic moment oppositely directed with respect to its spin. Positive identification of the antiproton was first made at the University of California. In 1959 Segre and Chamberlain received the Nobel Prize in physics for this discovery. Protons which had been accelerated to an energy of 6200 MeV in the bevatron, the proton synchrotron of the University of California Radiation Laboratory at Berkeley, were allowed to collide with a copper target. Negatively charged particles coming out in a forward direction from this collision were selected and separated in momentum by a focusing and analyzing magnet system to provide a beam of negative particles of known momentum. After a time of flight of about one-tenth of a microsecond, this beam may be expected to consist mainly of negative pions and muons, with some negative kaons (mass about 965 electron masses) and possibly negative protons. These particles were then distinguished both by measurement of their time of flight from the target (since particles of different mass have different velocities for given momentum) and by means of a device measuring the velocity of each particle passing through by the angle of its Cerenkov radiation. In this way the presence of negative particles with protonic mass (within about 10%) and distinct from the known kaons and hyperons was established. Their rate of production for the momentum and direction of this experiment was about one negative proton for every 50,000 negative pions with the same momentum and direction. See also **Particles (Subatomic);** and **Proton.**

ANTIPYRETIC. Any physical agent or drug that lowers the temperature of the body. Among antipyretics used are aspirin, antipyrine, acetanalid, and phenacetin. See also **Analgesics.**

ANTITOXIN. (1) A substance made and elaborated in the body to neutralize a specific bacterial, plant, or animal toxin; (2) one of the class of specific antibodies.

The history of the development of antitoxins in combating bacterial infection dates back to the early beginnings of organized bacteriology. Behring was the first to show that animals that were immune to diphtheria contained, in their serum, factors which were capable of neutralizing the poisonous effects of the toxins derived from the diphtheria bacillus. While this work was carried out in 1890, prior to many of the great discoveries of mass immunization, and much later the antibiotics, it is interesting to note that there remains a place for some antitoxins in medical treatment or prophylaxis for some diseases, such as tetanus and botulism.

The more important approach to immunity to infectious disease now is the development of *active immunity* by injection of a vaccine. The vaccine may be either an attenuated live infectious agent, or an inactivated or killed product. In either case, protective substances are generated in the bloodstream called *antibodies* which help to neutralize the infectious agent when it is introduced. The principle of *passive immunization*, on the other hand, involves the development of the antibodies in another host and most frequently a different species as well. The antiserum or antitoxin (from the other host) is used in preventing the onset of the disease, or in actual treatment of the active infection in subjects who have not had the advantage of becoming actively immunized due either to neglect or unavailability of an effective vaccine. The use of antitoxins prepared in another species (for example, horse) is not without some element of risk.

Antitoxins are prepared by injecting the donor animals with frequent and increasing doses of toxin while maintaining a level at each injection that the animal can tolerate. The initial doses are critical since these toxins may be among the most poisonous agents known. In one technique, the toxin is diluted so that the first injection contains less than the minimum lethal dose. Other programs use toxins that are inactivated with formaldehyde so that they are no longer poisonous, but may still elicit an immune response and result in antitoxin that will neutralize the unaltered toxin. This method of inactivation also was developed prior to the twentieth century for preparing many of the important vaccines against diseases, such as influenza, tetanus, and diphtheria.

A means for enhancing the potency of the antitoxin in horses is to use an agent called an *adjuvant*. Several adjuvants have been used, including tapioca, mineral oil, and aluminum hydroxide. The mechanism by which these agents increase the intensity of the immune response is not fully understood, but local inflammatory reaction and the resulting slower release of the injected material from the original site appear to play a role. Adjuvants also are sometimes used with vaccines for human use. It has been recorded that one horse, during an eleven-year period, gave 657 gallons (~25 hectoliters) of blood from which tetanus antitoxin and, at different times, pneumococcus antiserum, was prepared. The volume of serum removed from the horse can be increased by the return of the red cells after removal of the plasma or liquid component. It has been shown that if the red cell level can be maintained in human donors of special serums, they can safely give as much as one liter of serum per week.

ANTIVIRAL DRUGS. Compounds for use in the treatment of viral infections are quite limited. A problem in the development of antiviral drugs centers around the relationship between the replication function of the virus and functions of host cells. Obviously, a drug must target the virus and virus-infected cells with no destruction of healthy cell functions. Among antiviral drugs currently used are amantadine, idoxuridine, and adenine arabinoside. Research activity in the antiviral drug field is vigorous because of the obvious great need for them.

Amantadine Hydrochloride. Chemically, this drug is 1-adamantanamine hydrochloride. A commercial preparation is known as *Symmetrel*®. This drug is indicated in the preventing (prophylaxis) and symptomatic management of respiratory tract illness caused by influenza A virus strains. The drug is particularly considered in connection with high-risk patients, close household or hospital ward contacts of index cases, and patients with severe influenza A virus illness. In the prophylaxis of influenza due to A virus strains, early immunization as periodically recommended by public health authorities is the method of choice. When early immunization is not feasible, or when the vaccine is contraindicated or not available, amantadine hydrochloride is sometimes used chemoprophylactically with inactivated influenza A virus vaccine until protective antibody responses develop. Principal contraindications include hypersensitivity to the drug and patients with a history of epilepsy, congestive heart failure or peripheral edema.

To date, amantadine hydrochloride has not been used extensively clinically, although excellent controlled trials indicating that it has a prophylactic effect have been reported. Infected persons have been reported to suffer less cough, sore throat, and fever than other persons not taking the drug. Reports also indicate the drug accelerates recovery from peripheral airway abnormalities in normal persons who have uncomplicated influenza.

In the treatment of mild parkinsonism, amantadine has been used to advantage, particularly in conjunction with L-dopa. It is believed that amantadine stimulates the release of dopamine from nerve terminals.

Side-effects from the continuous use of amantadine include dizziness, drowsiness, difficulty in thinking, hallucinations, convulsions, and, rarely, psychosis. The effects are more commonly experienced by middle-age groups and the elderly.

Idoxuridine. Chemically, this drug is 5-iodo-2'-deoxyuridine. A commercial preparation is known as *Stoxil*®. This drug is indicated for the treatment of herpes simplex keratitis. The drug inhibits replication of herpes simplex virus by irreversibly inhibiting the incorporation of thymidine into the viral DNA. Although tests with rabbits of known genetic ancestry have been completed with no malformations resulting from idoxuridine, the drug is still administered with caution in pregnancy or in women of childbearing potential. The commercial compound is available in the form of a solution or as an ophthalmic ointment, with petrolatum used as an inactive ingredient. While the compound will frequently control infection, it apparently has no effect on the accumulated scarring, vascularization, or on the resultant progressive loss of vision.

Adenine Arabinoside. Also known as vidarabine, or *Vira-A*®, has the empirical formula, $C_{10}H_{13}N_5O_4 \cdot H_2O$. The chemical name is 9-β-D-arabinofuranosyladenine monohydrate. This compound is indicated in the treatment of herpes simplex virus encephalitis. Controlled studies have indicated that therapy with

this drug may reduce the mortality caused by herpes simplex virus encephalitis from 70 to 28%. The therapy does not appear to alter morbidity and resulting serious neurological sequelae in the comatose patient and thus early diagnosis and treatment are essential. Herpes simplex virus encephalitis should be suspected in patients with a history of an acute febrile encephalopathy associated with disordered mentation, altered level of consciousness, and focal cerebral signs. Licensing for this use (United States) was made in December 1978. The drug was licensed in 1977 for treatment of acute herpes simplex keratoconjunctivitis and recurrent epithelial keratitis caused by herpes simplex types 1 and 2. Experimentally, under controlled conditions, the drug has been used for treatment of herpes simplex type 1 encephalitis, herpes simplex type 2 infections, herpes zoster, smallpox, progressive multifocal encephalopathy, and chronic hepatitis B virus and cytomegalovirus infections. Large doses of adenine arabinoside administered parenterally have been noted to suppress bone marrow function, particularly if administered over long periods.

Still considered experimental among antiviral drugs are methisazone, human leukocyte interferon, acyclovir (*Zovirax*®), and cytosine arabinoside.

References

Baker, C. E.: "Physicians' Desk Reference," 34th Edition, Medical Economics Co., Oradell, New Jersey (1980).
Bauer, D. J., et al.: "Prophylaxis of Smallpox with Methisazone," *Amer. Jrnl. Epidemiol.*, **90**, 130 (1969).
Greenberg, H. B., et al.: "Effect of Human Leukocyte Interferon on Hepatitis B Virus Infection in Patients with Chronic Active Hepatitis," *New Engl. Jrnl. Med.*, **295**, 517 (1976).
Jones, B. R., et al.: "Efficacy of Acyloguanosine Against Herpes Simplex Corneal Ulcers," *Lancet*, **1**, 243 (1979).
Merigan, T. C., Editor: "Antivirals with Clinical Potential," Univ. of Chicago Press, Chicago, Illinois (1976).
Merigan, T. C., et al.: "Human Leukocyte Interferon Therapy of Herpes Zoster in Patients with Cancer," *New Engl. Jrnl. Med.*, **298**, 981 (1978).
Pavan-Langston, D., et al., Editors: "Adenine Arabinoside: An Antiviral Agent," Raven Press, New York (1975).
Whitley, R. J., et al.: "Adenine Arabinoside Therapy of Biopsy-Proven Herpes Simplex Encephalitis," *New Engl. Jrnl. Med.*, **297**, 289 (1977).

ANTLERITE. This is a relatively uncommon mineral found within the oxidized zones of copper deposits in arid regions. It is a basic sulfate of copper $Cu_3(SO_4)(OH)_4$, crystallizing in the orthorhombic system. Hardness of 3.5, sp gr 3.88, with vitreous luster and emerald-green to black-green color. Originally found in Arizona, it is the principal copper ore mineral at Chuquicamata, Chile.

APATITE. The mineral apatite is a phosphate of calcium with either fluorine or chlorine or sometimes both, hence the distinction between fluor-apatite and chlor-apatite. Sometimes both fluorine and chlorine are present. Most apatite is, however, fluorapatite, $Ca_5(PO_4)_3F$.

Apatite crystallizes in the hexagonal system in prismatic and tabular forms. Hardness, 4.5–5; specific gravity, 3.17–3.23; luster, vitreous to resinous; transparent to opaque; streak, white; cleavage, imperfect basal and prismatic; color, white, green, yellow, red, brown and purple; sub-conchoidal fracture. The variety called asparagus stone is yellow-green and manganapatite which is a dark bluish-green may contain as much as 10% manganese dioxide replacing the calcium. Werner devised the name apatite from the Greek word meaning to deceive, as it was frequently mistaken for beryl and other species. Apatite has been found widely distributed both geographically and petrologically as it occurs in many sorts of rocks, metamorphic limestones, gneisses, schists, granites and syenites, pegmatite veins and even with iron ores. It has been prepared artifically. It has been mined for the manufacture of fertilizers and to a slight extent for jewelry.

Apatite occurs extensively in Europe and America, especially in New England, New Jersey, New York, North Carolina, California, and in the provinces of Ontario and Quebec in Canada.

API Gravity. Petroleum; Specific Gravity.

APOPHYLLITE. The mineral is a hydrous silicate of potassium, calcium, and fluorine, corresponding to the formula $KCa_4Si_8O_{20}(F,OH) \cdot 8H_2O$. The true crystallographic symmetry is evident on crystals by the luster difference between the basal pinacoid facial planes and other crystal faces; also prism faces show vertical striations; basal planes do not. It crystallizes in the tetragonal system in square prisms resembling cubes terminated by based pinacoids or pyramids, often with both. Prism faces show vertical striations; basal pinacoid either dull or rough. Cleavage is perfect, parallel to the base; hardness, 4.5–5; specific gravity, 2.3–2.4; luster, vitreous to pearly; transparent to translucent or nearly opaque; color may be white, grayish, greenish, yellowish, or reddish. This mineral was named by Haüy from the Greek words meaning *from a leaf*, referring to its exfoliation when heated with the blow pipe.

Apophyllite is a secondary mineral found with the zeolites and has been classed with them by some writers, but it contains no aluminum, which element is understood to be an essential in a zeolite. It occurs in cavities in basalts and less often filling openings in granites or other crystalline rocks; it also is a gangue mineral in certain veins.

There are many localities for apophyllite: Bohemia, Trentino, Italy, the Hartz Mountains, and Iceland. Fine specimens have been obtained from the Ghats Mountains in India. The Triassic trap rocks of New Jersey, Connecticut and Nova Scotia have also furnished many specimens.

APPARENT MOLAR QUANTITY. For a solution containing n_1 moles of solvent and n_2 moles of solute, an apparent molar quantity is defined as

$$\frac{X - n_1 x_1}{n_2}$$

where X is the value of the quantity for the whole solution and x_1 the molar quantity for the pure solvent; e.g., the apparent molar volume of the solute is

$$\frac{V - n_1 v_1}{n_2}$$

V being the total volume and v_1 the volume per mole of the solvent.

See also **Molal Concentration; Molar Concentration; Mole (Stochiometry); Mole Fraction; Mole Volume.**

AQUAMARINE. A form of gem beryl. See **Beryl.**

AQUA REGIA. Also known as nitrohydrochloric acid, aqua regia is made up of three parts hydrochloric acid and one part nitric acid, each of the usual concentrated laboratory form. Aqua regia will dissolve all metals except silver. The latter is converted to silver chloride. The reaction of metals with nitrohydrochloric acid typically involves oxidation of the metal to a metallic ion and the reduction of the nitric acid to nitric oxide. Aqua regia also dissolves the common oxides and hydroxides of metals with the exception of silver, the ignited oxides of tin, aluminum, chromium, and iron, and the higher oxides of lead, cobalt, nickel, and manganese, the latter dissolving effectively in hydrochloric acid alone.

ARACHIDIC ACID. Also known as eicosanoic acid, formula $CH_3(2)_{18}COOH$. A widely distributed, but minor component of the fats of certain edible vegetable oils. Shiny, white crystalline leaflets; soluble in ether; slightly soluble in water. Specific gravity 0.8240 (100/4°C); mp 75.4°C; bp 205°C (1 millimeter pressure). Decomposes at 328°C. Commercial product derived from groundnut (peanut) oil. Used in organic synthesis; lubricating greases; waxes and plastics. Source of arachidyl alcohol. See also **Vegetable Oils (Edible).**

ARAGONITE. The mineral aragonite is calcium carbonate, $CaCO_3$, chemically identical with calcite but crystallizing in the orthorhombic system, with acicular crystals. By repeated twinning, pseudo-hexagonal forms result. Aragonite may be columnar or fibrous, occasionally in branching stalactitic forms called flosferri (flowers of iron) from their association with the ores at the Carinthian iron mines. Its hardness is 3.5–4; specific gravity, 2.93–2.95; luster, vitreous to resinous; colors, white, gray, green-yellow or purple; transparent to translucent. Aragonite forms at temperatures of 80–100°C and is relatively unstable at ordinary temperatures and pressures. It alters to calcite, although very slowly. There are many localities for aragonite in Europe, Bolivia, Pennsylvania, Iowa, Missouri, South Dakota, New Mexico, Arizona and Colorado. Its name is derived from Aragon in Spain. See also **Calcite.**

ARGENTITE. The mineral argentite, sometimes called silver glance, is naturally occurring silver sulfide, corresponding to the formula Ag_2S. It crystallizes in the isometric system in cubes, octahedrons and dodecahedrons, or may be massive. Hardness, 2–2.5; specific gravity, 7.2–7.34; luster, metallic; streak, gray; color, black, blackish-gray or gray; opaque and sectile to such an extent that it cuts like wax with a knife. Heated upon charcoal it yields a malleable mass of silver. The name is derived from the Latin word for silver, *argentum*.

Localities for fine crystals are Sonora, Mexico, and Freiberg, Saxony; in the United States, at Butte, Montana; Tonopah, Nevada; and Aspen, Colorado.

Argentite is probably the most important primary silver mineral. However, it maintains its cubic (isometric) characteristic only above 179°C (354°F). On cooling the inward structure inverts to a non-isometric form, usually orthorhombic, yet retaining its original outward form. It is, therefore, a paramorph after argentite, known as acanthite.

ARGININE. Amino Acids.

ARGON. Chemical element symbol Ar, at. no. 18, at. wt. 39.948, periodic table group 0 (inert or noble gases), mp −189.2°C, bp −185.7°C, density 1.78 (solid at −233°C). Solid argon has a face-centered cubic crystal structure. At standard conditions, argon is a colorless, odorless gas; it does not form stable compounds with any other element under *normal* conditions. Due to its low valence forces, argon is unable to form diatomic molecules, except in discharge tubes. It does form compounds under highly favorable conditions, as excitation in discharge tubes, or pressure in the presence of a powerful dipole. As an example of the first, argon forms amorphous compounds of the type FeA in a discharge tube having iron electrodes. An example of the second is furnished by the hydrates which argon forms with H_2O at 150 atmospheres and 0°C. Argon forms compounds, possibly clathrates, with a number of organic substances, such as a compound with hydroquinone containing 9% argon, in which the amount of argon may vary from this proportion. The compounds are made by crystallization of the aqueous solution of the hydroquinone under argon gas pressure on the order of 40 atmospheres.

Argon occurs in the atmosphere to the extent of approximately 0.935%. In terms of abundance, argon does not appear on lists of elements in the earth's crust because it does not exist in stable compounds. However, argon is 2.5 × more soluble in H_2O than nitrogen and thus is found in seawater to the extent of approximately 2,800 tons per cubic mile (605 metric tons per cubic kilometer). Commercial argon is derived from air by liquefaction and fractional distillation. There are three natural isotopes, ^{36}Ar, ^{38}Ar, and ^{40}Ar, and four radioactive isotopes, ^{35}Ar, ^{37}Ar, ^{39}Ar, and ^{41}Ar. The lengths of half-lives of the isotopes vary widely, the shortest ^{35}Ar with a half-life of about 2s; the longest ^{39}Ar with a half-life of about 260 years. The first ionization potential of Ar is 15.755 eV; second, 27.76 eV; third, 40.75 eV. Other important physical characteristics of argon are given under **Chemical Elements**.

The presence of argon in air was suspected by Cavendish as early as 1785, but was not positively identified until 1894 by Lord Rayleigh and Sir William Ramsay. Argon exhibits a characteristic series of lines in the red end of the spectrum. Commercially, argon gas is used in incandescent lamps and fluorescent lamps as an inert gas to minimize vaporization of the filaments and, for this, is preferable to nitrogen. The gas also is used for shielding electrodes in arc welding. A gas of about 99.995% purity is required for lamps. Argon also has found effective use in certain lasers.

Regarding argon in meteorites, see **Krypton**. See also references listed at end of entry on **Chemical Elements**.

AROMATIC COMPOUND. An organic compound that incorporates a closed-chain or (ring) nucleus in its structure. This is in contrast with the aliphatic compound which is comprised of an open-chain structure. The classical example of an aromatic compound is benzene. Aromatic compounds are also called *benzenoids*. A few ring-type compounds generally are not classified as aromatic. These include the cycloparaffins and cycloolefins, which are considered to be derivatives of methane. Bonding characteristics of aromatic compounds are described under **Compound (Chemical); and Organic Chemistry.**

ARSENIC. Chemical element symbol As, at. no. 33, at. wt. 74.9216, periodic table group 5a, mp 817°C (28 atmospheres), sublimes at 618°C, density 5.72 g/cm³. One naturally occurring stable isotope, ^{75}As. Various studies indicate that arsenic exists in several allotropic forms. The metallic form has a steel-gray color in the crystalline form and is brittle. Although the red form of arsenic sulfide As_2S_2 was observed by Aristotle as early as 400 B.C., the first attempt to isolate the metal was not made until 1250 by Albertus Magnus. Later documentation on the preparation of the element was given by J. Schroder and N. Lemery in the 1600s. First ionization potential, 9.8 eV; second, 18.63 eV; third, 28.34 eV; fourth, 50.1 eV; fifth, 62.5 eV. Oxidation potentials $AsH_3 \rightarrow As + 3H^+ + 3e^-$, 0.54 V; $As + 2H_2O \rightarrow HAsO_2 + 3H^+ + 3e^-$, −0.2475 V; $HAsO_2 + 2H_2O \rightarrow H_3AsO_4 + 2H^+ + 2e^-$, −0.559 V; $AsO_2^- + 4OH^- \rightarrow AsO_4^{3-} + 2H_2O + 2e^-$, 0.71 V; $As + 4OH^- \rightarrow AsO_2^- + 2H_2O + 3e^-$, 0.68 V. Other important physical properties of arsenic are given under **Chemical Elements**.

Gray or metallic arsenic, density 5.73 g/cm³, which sublimes

on heating, and has the vapor composition As_4, becoming As_2 at higher temperatures, is the ordinary variety. On rapid cooling, the vapor condenses to yellow arsenic, density 1.97 g/cm³, which reverts to the gray variety on warming. An intermediate form in the transition is black (amorphous β) arsenic, density 4.6–5.2, also obtained by the thermal decomposition of arsine. Brown arsenic, density 3.7–4.2, obtained by reduction of acid solutions of trivalent arsenic, is probably a finely-divided form of black arsenic.

Arsenic sublimes on heating; is unchanged in dry air but a film of oxide is formed in moist air; heated in air at 180°C forms arsenic trioxide of the odor of garlic, poisonous; insoluble in HCl but soluble in concentrated HNO_3 or concentrated H_2SO_4 to form arsenic acid; soluble in hot NaOH solution; heated with chlorine forms arsenic trichloride; heated with metals forms metallic arsenides. When arsenic is heated in a tube and the vapor cooled (1) slowly (that is, in the hot part of the tube) black arsenic is formed, and this form is converted into the gray at 360°C, (2) rapidly (that is, in the cold part of the tube) yellow arsenic is formed, and this form is quickly converted into the gray by the action of light. Yellow arsenic is soluble in CS_2.

Arsenic occurs in nature as the arsenide of iron, cobalt, nickel, and as the mineral sulfides, *realgar* (arsenic monosulfide, AsS), red colored; *orpiment* (arsenic trisulfide, As_2S_3), yellow colored—these two minerals when powdered once were used as paint pigments—*arsenopyrite, mispickel* (iron arsenosulfide, FeAsA); *enargite*, Cu_3AsS_4; and *tennantite*, $Cu_8As_2S_7$.

The primary arsenic-containing material is arsenious oxide obtained by separation from roaster or smelter flue gases, and is produced in Tacoma, Washington. Metallic arsenic is obtained as sublimate by heating the oxide with carbon.

Arsine. AsH_3 is formed by hydrolysis of arsenides, or reduction (by zinc and HCl or aluminum and NaOH) of arsenic compounds, as in the Gutzeit test. It reacts with metals at higher temperatures or in solution to give the arsenides. Diarsine, As_2H_4, is produced by reduction of arsenic trichloride, $AsCl_3$, by lithium aluminum hydride, $LiAlH_4$ in ether at −190°C. It melts below −50°C, but begins to decompose into AsH_3 and brown polymeric $(AsH)_x$ about −100°C. It is more stable in the gas phase than in the solid or liquid phases.

Arsenides. These are prepared by fusion from the elements. Their properties vary across the periodic table, those of the alkalies and alkaline earths being readily hydrolyzed by H_2O or acids and are stoichiometric, while the arsenides of the other metals show an increasingly intermetallic character and resist hydrolysis.

Trihalides. The trifluoride and trichloride, AsF_3 and $AsCl_3$ are liquidsat room temperature and the tribromide and triiodide, $AsBr_3$ and AsI_3 are solids, although the former melts at 31°C. Like the analogous phosphorus compounds, they have pyramidal structures. Their hydrolysis in aqueous solution is not quite complete, consistent with their greater ionic character (than the phosphorus halides), as is the fact that As^{3+} is precipitated from their solutions as the sulfide. The only stable binary pentahalogen compound of arsenic is the pentafluoride, AsF_5, a colorless gas, which like the trihalides, is less readily hydrolyzed than the corresponding phosphorus compound. A very unstable pentachloride, $AsCl_5$, has been reported. The mixed halide AsF_3Cl_2 can be made by passing chlorine into ice-cold arsenic trifluoride.

Arsenic(III) Oxide. As_4O_6 exists as tetraarsenic hexoxide in the solid state and in the vapor to above 800°C, where dissociation to As_2O_3 commences. It is somewhat soluble in H_2O (about 20 g/l at 25°C), and its solutions have some acidic properties, although the acid has not been isolated and its formula is probably not As(OH)$_3$, the form used for convenience in writing reactions. It is an amphiprotic substance, since, as stated above, As^{3+} is precipitated by H$_2$S from acid solutions as the sulfide, while the salts, the arsenites (containing the ion AsO$_3^{3-}$), are readily formed. Their solubility in H$_2$O varies across the periodic table, those of the alkali metals being very soluble, those of the alkaline earth metals less so, and those of the heavy metals essentially insoluble. Arsenite ion probably exists as As(OH)$_4^-$ in solution.

Arsenic(V) Oxide. As$_4$O$_{10}$ is a white solid, decomposes at 315°C, isomorphous with phosphorus pentoxide, P$_2$O$_5$, but not produced by simple oxidation of As$_2$O$_3$. It is made by dehydration of arsenic acid or As$_4$O$_{10}\cdot$4H$_2$O. It hydrates to give arsenic acid, H$_3$AsO$_4\cdot\frac{1}{2}$H$_2$O. This acid is only slightly weaker than phosphoric acid, which it resembles in forming a wide variety of polyacids. It also forms primary, secondary, and tertiary (ortho) arsenates. Raman spectral studies of concentrated arsenic acid solutions in H$_2$O have a strong band assigned to the —OH group, whence it is inferred that the acid is present in different forms in concentrated and dilute solutions. Many arsenates are converted by ignition into pyroarsenates, e.g., calcium pyroarsenate, Ca$_2$As$_2$O$_7$, and metaarsenates are also known.

Direct fusion of the elements yields a number of arsenic sulfides, including As$_4$S$_3$, As$_4$S$_4$, As$_2$S$_3$, and As$_2$S$_5$, the last two being obtained also by precipitation from arsenic(III) and arsenic(V) solutions, respectively. The trisulfide dissolves in alkali sulfide solutions to form thioarsenites:

$$As_2S_3 + 3S_2^{2-} \rightarrow 2AsS_3^{3-} + S$$

while with polysulfides it forms thioarsenates:

$$As_2S_3 + 2S_2^{2-} + S \rightarrow 2AsS_4^{2-}$$

Organoarsenic Compounds. The largest group of organic arsenic-containing compounds is the arsenic acids RAsO(OH)$_2$, where the R may be alkyl, aryl, or heterocyclic groups and their salts. In addition to specific compounds mentioned under the uses of arsenic, some organoarsenic compounds include methylarsine CH$_3$sH$_2$; methylarsenic tetrachloride CH$_3$AsCl$_4$; diphenylarsenic peroxide (C$_6$H$_5$)$_2$AsOOAs(C$_6$H$_5$)$_2$; triphenylarsenic dihydroxide

$$(C_6H_5)_3As(OH)_2;$$

dimethylarsine borane (CH$_3$)$_2$AsHBH$_3$; and ethoxydichloroarsine C$_2$H$_5$OAsCl$_2$.

Uses. In the mid-1970s, the worldwide production of As$_2$O$_3$ was estimated to be about 50,000 metric tons. Future production of As$_2$O$_3$ will be influenced by the ability to handle ores in the manner required to comply with environmental restrictions. As$_2$O$_3$ is available in two grades: (1) crude, 95% As$_2$O$_3$; and (2) refined arsenic, 99% As$_2$O$_3$. Domestic supplies of the United States are supplemented by imports from Sweden, France, and Mexico. Commercial arsenic metal is produced chiefly by the United States and Sweden.

Arsenic trioxide finds major use in the preparation of other compounds, notably those used in agricultural applications. The compounds monosodium methylarsonate, disodium methylarsonate, and methane arsenic acid (cacodylic acid) are used for weed control, while arsenic acid, H$_3$AsO$_4$, is used as a desiccant for the defoliation of cotton crops. Other compounds once widely used in agriculture are calcium arsenate for control of boll weevils, lead arsenate as a pesticide for fruit crops, and sodium arsenite as a herbicide and for cattle and sheep dip. In some areas, arsenilic acid has been used as a feed additive

for swine and poultry. Restrictions on these compounds vary from one country and region to the next.

Refined arsenic trioxide is used both as a fining and decolorizing agent in glass. As$_2$O$_5$ and arsenic acid are in the manufacture of chromated copper arsenate, which is used extensively as a wood preservative.

Indium arsenide, gallium arsenide, and gallium arsenide phosphide find use as semiconductors. For these materials, the starting arsenic source must be extremely pure. Arsenic trichloride and arsenic hydride (very high purity) find application in the production of epitaxial gallium arsenide. Also, in various combinations with iodine, germanium, selenium, sulfur, tellurium, and thallium, arsenic will form a group of glasses with very low melting points.

The applications of arsenic as a metal are quite limited. Metallurgically, it is used mainly as an additive. The addition of from $\frac{1}{2}$ to 2% of arsenic improves the sphericity of lead shot. Arsenic in small quantities improves the properties of lead-base bearing alloys for high-temperature operation. Improvements in hardness of lead-base battery grid metal and cable-sheathing alloys can be obtained by slight additions of arsenic. Very small additions (0.02–0.05%) of arsenic to brass reduce dezincification.

Toxicity. Although metallic arsenic and arsenic trisulfide may be handled, as in the case of most arsenical compounds, skin contact should be avoided. Arsine requires extreme caution in handling because of its very high toxicity. In handling arsenic and its compounds, reducing conditions should be avoided because these may give rise to the undesired formation of arsine. Epidemiological studies indicate an association between high and lengthy exposures to inorganic arsenic compounds and cancer. Wherever arsenic and its compounds are present as dusts or vapors, proper ventilation and respirators are mandatory. Good housekeeping and appropriate hygienic practices should also be observed.

Arsenic is commonly found in small amounts in the tissues of plants and animals. A human body may contain as much as 20 mg (As$_2$O$_3$). No role in natural biological phenomena has been found for As. Although the element may be present in seawater to the extent of 0.006–0.03 ppm, it may be ten times as high in estuaries. Shellfish tend to accumulate the arsenic from the large amount of seawater with which they come in contact. Oysters may contain 3–10 ppm. However, shellfish of the same species grown in different localities show wide variations in arsenic content, suggesting that it is an accidental constituent which the organisms learn to tolerate. Its lack of function in the human body is suggested by the fact that it tends to accumulate in the hair and nails which are essentially nonliving.

References

Carapella, S. C., Jr.: "Arsenic," in "Kirk-Othmer Encyclopedia of Chemical Technology," Vol. 3, 3rd Edition, 251–266, Wiley, New York (1978).
Liddell, D. M., Editor: "Handbook of Nonferrous Metallurgy," Vol. 2, 94–103, McGraw-Hill, New York (1945).

—S. C. Carapella, Jr., ASARCO Incorporated, South Plainfield, New Jersey.

ARSENOPYRITE. The mineral arsenopyrite is a sulfarsenide of iron corresponding to the formula FeAsS. A variety in which some of the iron is replaced by cobalt is known as danaite. It crystallizes in the monoclinic system but twinning produces pseudo-orthorhombic crystals. Its hardness is 5.5–6; specific gravity, 6.07; luster, metallic color, silvery-white to steel-gray, but usually with a yellow to gray tarnish; streak, black. Arsenopyrite is a common mineral with tin and lead ores and in pegmatites, probably having been deposited by action of both vapors and hydrothermal solutions. It is a widespread mineral, well-known deposits occurring in Austria, Saxony, Switzerland, Sweden, Norway; Cornwall and Devonshire, England; Bolivia; in the United States at Roxbury, Connecticut; Franklin, New Jersey; Paris, Maine; Emery, Montana; and Leadville, Colorado. Danaite was first found in Franconia, New Hampshire, by J. D. Dana, for whom it was later named. Arsenopyrite also is known as *mispickel*, an old German term whose exact derivation is unknown.

ASBESTOS. A group of impure magnesium silicate minerals which occur in fibrous form. Colors may be white, gray, green, or brown, sp gr 2.5, noncombustible.

Serpentine asbestos is the mineral chrysotile, a magnesium silicate. The fibers are strong and flexible. Spinning is possible with the longer fibers.

Amphibole asbestos includes various silicates of magnesium, iron, calcium, and sodium. The fibers are generally brittle and cannot be spun, but are more resistant to chemicals and to heat than serpentine asbestos.

Because asbestos has long been indicated in asbestosis (similar to silicosis) and, in more recent years, considered a carcinogen, several countries have issued regulations that restrict its use. For many years, it has been used in fireproof fabrics, brake lining, gaskets, roofing, insulation, paint fillers, reinforcing agents in rubber and plastics, and in electrolytic diaphragm cells.

ASCORBIC ACID (Vitamin C). Infrequently referred to as the antiscorbutic vitamin and earlier called cevitamic acid or hexuronic acid; the present terms, *ascorbic acid* and *vitamin C*, are synonymous. Ascorbic acid was one of the first, if not the first nutrient to be associated with a major disease. Lind first described *scurvy* in 1757. However, this vitamin C-deficiency disease had been recognized by Hippocrates in about 400 B.C. and was a curse during the time of the Crusaders. In time of war, the disease killed untold numbers in armies and navies and besieged towns. During the early days of the sailing ships, often requiring months between port calls accom-

in 1907. About 6 months were required to produce scurvy experimentally, as individual susceptibility and the quantity of vitamin C previously stored in the body affects the onset of scurvy. The earliest sign of scurvy is usually a sallow or muddy complexion, a feeling of listlessness, general weakness, and mental depression. Soon the bones are affected and increasing pain and tenderness develop. Teeth easily decay and become loose and often fall out, while the gums bleed easily and are sore. Changes in the blood vessels occur, producing hemorrhages in different parts of the body. In infants, irritability, loss of appetite, fever, and anemia also occur. An infant between 6 and 12 months of age, who has not had sufficient intake of vitamin C (as from fruit juices, supplements, etc.) may show abnormal irritability and tenderness and pain in the legs, often accompanied by pain and swelling of joints (elbows and knees). Immediate administration of vitamin C is indicated in such cases.

In 1928, Zilva first described antiscorbutic agents in lemon juice, although the importance of fresh fruit or vegetables for preventing scurvy had been established a century or more earlier. Also in 1928, Szent-Györgyi isolated hexuronic acid (vitamin C) from lemon juice. In 1932, Waugh and King identified hexuronic acid as an antiscorbutic agent. Haworth, in 1933, established the configuration of hexuronic acid and, in that same year, Reichstein first synthesized hexuronic acid. Later in that year, Haworth and Szent-Györgyi changed the name of hexuronic acid to ascorbic acid.

In 1950, King et al., by the use of glucose labeled with radiocarbon in known positions, traced glucose through intermediate steps in the formation of ascorbic acid in plant and animal tissues, and then by using ascorbic acid with radiocarbon-labeled positions, it was possible to determine with considerable accuracy the metabolic distribution, storage, and chemical changes characteristic of the vitamin molecule. That experimentation made it clear that the carbon atoms in glucose or galactose all retain their original positions along the carbon chain in the vitamin when it is formed biologically. No rupture or replacement in the chain during conversion was noted. It was also found that the synthesis can be considerably enhanced by feeding livestock small amounts of *chloretone* or any of a score or more of organic compounds. Reactions just described are indicated below:

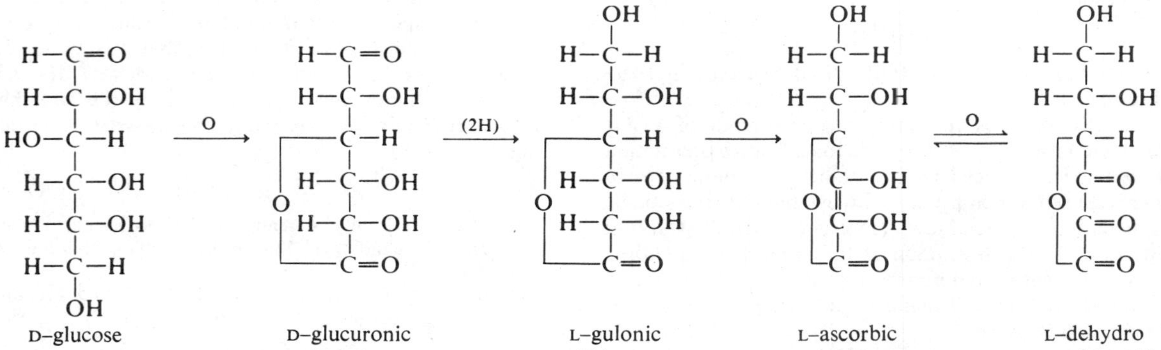

D–glucose   D–glucuronic acid lactone   L–gulonic acid lactone   L–ascorbic acid   L–dehydro ascorbic acid

panied by a lack of fresh food for long periods, the disease affected the crew as a plague. Scurvy was of some importance as recently as World War II. Currently, the disease is of prime concern in pediatrics. It is rarely seen in breast-fed children, but pasteurization of cow's milk degrades the vitamin and an addition to the diet of ascorbic acid must be provided for infants under 1 year of age.

Experimental scurvy was first produced by Holst and Frolich

Biological Role of Ascorbic Acid. Apparently all forms of life, both plant and animal, with the possible exception of simple forms such as bacteria that have not been studied thoroughly, either synthesize the vitamin from other nutrients or require it as a nutrient. Dormant seeds contain no measurable quantity of the vitamin, but after a few hours of soaking in water, the vitamin is formed.

Ascorbic acid is easily oxidized to dehydroascorbic acid. The latter is less stable than ascorbic acid and tends to yield products such as oxalate, threonic acid, and carbon dioxide. When administered to animals or consumed in foods, dehydroascorbic acid has nearly the same antiscorbutic activity as ascorbic acid, and it can be quantitatively reduced to ascorbic acid.

In its biochemical functions, ascorbic acid acts a regulator in tissue respiration and tends to serve as an antioxidant in vitro by reducing oxidizing chemicals. The effectiveness of ascorbic acid as an antioxidant when added to various processed food products, such as meats, is described in entry on **Antioxidants.** In plant tissues, the related glutathione system of oxidation and reduction is fairly widely distributed and there is evidence that electron transfer reactions involving ascorbic acid are characteristic of animal systems. Peroxidase systems also may involve reactions with ascorbic acid. In plants, either of two copper-protein enzymes are commonly involved in the oxidation of ascorbic acid.

In animal tissues, it is easily demonstrated that, as the vitamin content of tissues is depleted, many enzyme systems in the body are decreased in activity. Full explanation of this decreased activity still requires further research. In the total animal and in isolated tissues from animals with scurvy, there is an accelerated rate of oxygen consumption even though the animal becomes very weak in mechanical strength and many physiologic functions are disorganized. With the onset of scurvy, the most conspicuous tissue change is the failure to maintain normal collagen. Sugar tolerance is decreased and lipid metabolism is altered. There is also marked structural disorganization in the odontoblast cells in the teeth and in bone-forming cells in skeletal structures. In parallel with the foregoing changes, there is a decrease in many hydroxylation reactions. The hydroxylation of organic compounds is one of the most characteristic features disturbed by a vitamin C deficiency. These reactions relate to the vitamin's regulation of respiration, hormone formations, and control of collagen structure.

A partial list of physiological functions that have been determined to be affected by vitamin C deficiencies include: (1) Absorption of iron; (2) cold tolerance, maintenance of adrenal cortex; (3) antioxidant; (4) metabolism of tryptophan, phenylalanine, and tyrosine; (5) body growth; (6) wound healing; (7) synthesis of polysaccharides and collagen; (8) formation of cartilage, dentine, bone, teeth; and (9) maintenance of capillaries.

Requirements. Species known to require exogenous sources of ascorbic acid include the primates, guinea pig, Indian fruit bat, red vented bulbul, trypanosomes, and yeast. Species capable of endogenous sources include the remainder of vertebrates, invertebrates, plants, and some molds and bacteria. Estimates of requirements of vitamin C by humans has been approached in several ways: (1) by direct observation in human studies; (2) by analogy to experimentation with guinea pigs; (3) by analogy to experimental studies in monkeys and other primates; and (4) by analogy to animals, such as the albino rat, that normally synthesize the vitamin in accordance with physiological need. It is relatively easy to maintain intakes at recommended levels by use of mixed practical dietaries that include nominal quantities of fresh, canned, or frozen vegetables or fruits. Generally, ascorbic acid is considered as nontoxic to humans. Possible exceptions include kidney stones (in gouty individuals); inhibitory in excess doses on cellular level (mitosis inhibitor); possible damage to beta-cells of pancreas and decreased insulin production by dehydroxyascorbic acid.

For the species where the ascorbic acid is synthesized endogenously, the precursors include D-mannose, D-fructose, glycerol, sucrose, D-glucose, and D-galactose. Intermediates include uridine diphosphate glucose, D-glucuronic acid, gulonic acid,

L-gulonolactone, (Mn^{2+} cofactor). Production sites in animals are the kidney and liver in most instances. In rat, it is the intestinal bacterial supply. In plants, the production sites are found in green leaves, and fruit skins. Cell sites include microsomes, mitochondria, and golgi.

Supplements. Commercially available ascorbic acid still includes isolation from natural sources, such as rose hips, but large-scale production will involve the microbiological approach, i.e., *Acetobacter suboxidans* oxidative fermentation of calcium D-gluconate; or the chemical approach, i.e., the oxidation of L-sorbose.

Bioavailability of Ascorbic Acid. The general causes of reduced availability of vitamin C include damage to adrenal cortex, presence of antagonists, and food preparation practices (oxidation, storage, leaching, cooking). Excepting the use of supplements, the almost universal requirement for fresh foods as a source of vitamin C is readily explained by the sensitivity of the vitamin to destruction by reaction with oxygen. This is accelerated by the presence of minute quantities of enzymes that occur in most living tissues, in which copper or iron is combined with a protein to form a catalyst for the oxidation reaction. Other chemicals, such as quinones or high-valence salts of manganese, chromium, and iodine can also oxidize the vitamin readily in aqueous solutions. Most of these reactions increase rapidly in proportion to exposure to air and rising temperature. In the dry crystalline state, however, and in many dried plant tissues, particularly if acidic in reaction, the vitamin is quite stable at room temperature over a period of several months.

Freshly-cut oranges or their juices may be exposed in an open glass for several hours without appreciable loss of the vitamin because of the protective effect of the acids present and the practical absence of enzymes that catalyze its destruction. In potatoes, when baked or boiled, there is a slight loss of the vitamin, but if they are whipped up with air while hot, as in the production of mashed potatoes, a large fraction of the initial vitamin content usually will be lost. In freezing foods, it is common practice to dip them in boiling water or to treat them briefly with steam to inactivate enzymes, after which they are frozen and stored at very low temperatures. In this state, the vitamin is reasonably stable.

Factors which increase the bioavailability of ascorbic acid include the presence of antioxidants and synergists in the diet.

References

King, C. G.: "Ascorbic Acid (Vitamin C) and Scurvy," in "The Encyclopedia of Biochemistry" (R. J. Williams and E. M. Lansford, Jr., Editors), Van Nostrand Reinhold, New York (1967).

Kutsky, R. J.: "Handbook of Vitamins and Hormones," Van Nostrand Reinhold, New York (1973).

Lee, S. H., and T. P. Labuza: "Destruction of Ascorbic Acid as a Function of Water Activity," *Journal of Food Science*, **40**, 370 (1975).

Linkswiler, H.: "The Effect of the Ingestion of Ascorbic Acid and Dehydroascorbic Acid Upon the Blood Levels of These Two Components in Human Subjects," *Journal of Nutrition*, **64**, 43 (1958).

Staff: "Food Chemicals Codex," National Academy of Sciences, Washington, D.C. (1979).

Watt, B. K., and A. L. Merrill: "Composition of Foods," *Agriculture Handbook 8*, U.S. Department of Agriculture, Washington, D.C. (1975).

ASPARAGINE. Amino Acids.

ASPARTIC ACID. Amino Acids.

ASPHALT. Also known as asphaltum, a semisolid mixture of several hydrocarbons, probably formed because of the evaporation of the lighter and more volatile constituents. Asphalt is amorphous, of low specific gravity, with a black or brownish-black color and pitchy luster. Notable localities for asphaltum are the Island of Trinidad and the Dead Sea region, where Lake Asphaltites were known to the ancients. See also **Coal Tar and Derivatives.**

ASSOCIATION (Chemical).

At one time, association was conceived as a reversible reaction between like molecules that distinguished it from polymerization, which is not reversible. Association is characterized by reversibility or ease of disassociation, low energy of formation (usually about 5 and never more than 10 kcal per mole), and by the coordinate covalent bond which Lewis called the acid-base bond. Association takes place between like and unlike species. The most common type of this phenomenon is hydrogen bonding. Association of like species is demonstrable by one or more of the several molecular weight methods. Association between unlike species is demonstrable by deviation of the system from Raoult's law.

The strength of the coordinate covalent bond is a function of polarity of the associating molecules. Hence, associated molecules vary in stability from very unstable to very stable. The argon-boron trifluoride complex is quite unstable, whereas calcium sulfate dihydrate (gypsum) is very stable. The bond strength associated with stability has been measured for a number of combinations. The strengths of some hydrogen bond types decrease in the order FHF, OHO, OHN, NHN, CHO but they are dependent upon the geometry of the combination and upon the acid-base characteristics of the group. Steric effects can have a marked effect on the strength of the coordinate covalent bond. This was demonstrated by a study of the strength of the series NH_3, $C_2H_5NH_2$, $(C_2H_5)_2NH$, $(C_2H_5)_3N$ as bases toward an acid in solution and in the gaseous state and the comparison of the base strength of triethylamine and quinuclidine. The latter is, in effect, triethylamine in which the 2-carbon atoms of each ethyl group are tied together by another carbon. The geometry of the ethyls around the nitrogen is drastically changed, and the cyclic is a stronger base than the triethyl compound. The factors affecting the strength of the hydrogen bond also influence the degree of association.

Association within the same species accounts for the high boiling points of, for example, water, ammonia, hydrogen fluoride, alcohols, amines and amides. Ethyl ether and butanol contain the same number of atoms of each element but butanol has a boiling point 83° above that of ethyl ether as a result of more extensive hydrogen bonding. Some substances associate completely to two or more formula weights per molecule. Carboxylic acids, by a hydrogen-to-oxygen association, form dimers with a six-membered ring. N-Unsubstituted amides dimerize in the same manner, whereas N-substituted amides dimerize in a chain form in a *trans* configuration.

Hydrogen bonding is so common that coordinate bonds between other elements are sometimes overlooked. Antimony (III) halides form very few complexes with other halides, whereas aluminum halides readily form complexes. The octet of electrons is complete in all atoms of the antimony halides but is incomplete in the aluminum atom of aluminum halides:

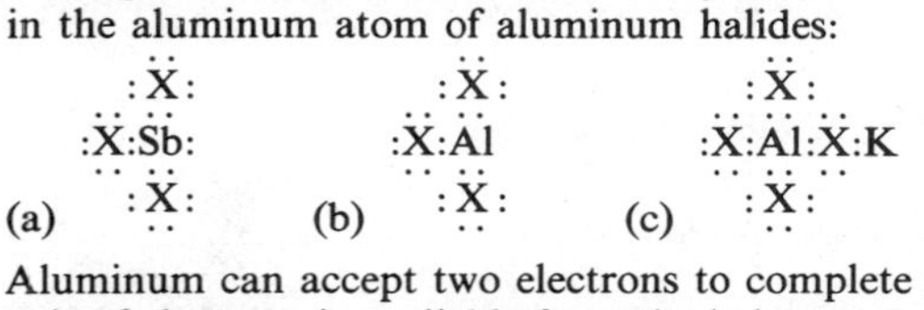

Aluminum can accept two electrons to complete its octet. The pair of electrons is available from the halogen. An alkali halide can supply the electrons and form a complex (c, above); or

the electron pair may come from the halogen of another aluminum chloride. Association with other aluminum halides accounts for the higher melting point of aluminum halides over antimony (III) halides which have a formula weight of 95 or more. The association of aluminum sulfate, alkali metal sulfate and water to form the stable alums is one of the more complex examples.

The formation of solvates is association between unlike species. Solvation is more frequent between substances of high polarity than those of low polarity. This is illustrated by the decrease in the tendency to form solvates with decrease in dipole moment and dielectric constant (in parentheses) for N-methylacetamide (3.59, 172), to water (1.84, 78.4), to ethanol (1.70, 24.6), to ammonia (1.48, 17.8), to methylcyclohexane (0, 2.02) for which few associations are known.

—John A. Riddick, Baton Rouge, Louisiana.

ASTATINE.

Chemical element symbol At, at. no. 85, at. wt. 210 (mass number of the most stable isotope), periodic table group 7a, classed in the periodic system as a halogen, mp 302°C, bp 337°C. All isotopes are radioactive. This element occurs in nature only in minute amounts as a result of minor branching in the naturally occurring alpha decay series: $^{218}At(t_{1/2} = $ ca. 2 sec.) is produced to the extent of 0.03% by the beta decay of ^{218}Po(radium A), 99,97% going by alpha decay to ^{214}Pb(RaB); $^{216}At(t_{1/2} = 3 \times 10^{-4}$ sec.) 0.013% by beta decay from ^{216}Po(thorium A); $^{215}At(t_{1/2} = 0.018$ sec.) 0.0005% by beta decay from ^{215}Po(actinium A). Astatine-217 $(t_{1/2} = 0.020$ second) is a principal member of the neptunium $(4n + 1)$ series, all members of which occur only to that extent to which the parent ^{237}Np is produced by naturally occurring slow neutrons from uranium.

The first isotope to be discovered was ^{211}At made by Carson, Mackenzie and Segrè by bombardment of a bismuth target with α-particles from the 60-inch cyclotron at Berkeley in 1940. The reaction is $^{209}Bi(\alpha, 2n)$ ^{211}At. The half-life of ^{211}At is 7.2 hr. It decays in two modes, 60% by K-electron capture and 40% by α-particle emission. The longest-lived isotope is $^{210}At(t_{1/2} = 8.3$ hr.); other isotopes having half-lives longer than 1 hour are 206, 207, 208 and 209. Various of the collateral radioactive series, involving bombardment reactions contain other astatine isotopes, such as ^{214}At, and ^{216}At. All these isotopes have half-lives that are only fractions of a second. The total number of isotopes is at least nineteen, including spallation reaction products as well as bombardment ones. They also include two short-lived isotopes, ^{215}At and ^{218}At, occurring in very small amount in the branched β-disintegration of ^{215}Po (actinium A) and ^{218}Po(radium A), respectively, as noted above.

The chemistry of astatine determined by tracer techniques, is in keeping with the regular transition of properties of the halogens. The acid properties of astatine are less marked than those of iodine, while its electropositive character is more marked than that of iodine. After reduction by SO_2 or metallic zinc, the astatine activity is carried by silver iodide or thallium iodide, so it evidently forms insoluble silver and thallium salts. This represents astatine in the univalent negative state characteristic of the halogens. However, astatine is very readily oxidized by bromine and ferric ions, giving indications of two higher oxidation states. Although there is no evidence from migration experiments of the presence of positive ions in the solution, astatine deposits on the cathode, as well as on the anode, in the electrolysis of oxidized solutions. Elemental astatine can be volatilized, although not so readily as iodine, and it has a specific affinity for metallic silver. The similarity to iodine is also shown by the observation that astatine concentrates in the thyroid glands of animals.

See list of references at end of entry on **Chemical Elements.**

ASYMMETRIC TOP. A model of a molecule which has no threefold or higher-fold axis of symmetry, so that during rotation all three principal moments of inertia are in general, different. Examples are the water molecule and the ethylene molecule.

ASYMMETRY (Chemical). Asymmetry involves the presence of four different atoms, or substituent groups bonded to an atom. Its existence was discovered in 1815 by the French physicist, J. B. Biot (1774–1867). Biot found that oil of turpentine and solutions of sugar, camphor, and tartaric acid all rotate the plane of plane-polarized light when placed between two Nicol prisms. This phenomenon is called optical rotation and is indicated in symbols, such as: $[\alpha]_D^{20°} = +53.4$ aq., this signifying that the substance gives a rotation of 53.4° to the right (clockwise, or plus) in water solution at 20°C using sodium D line as the light source. Substances in solution that rotate light to the right are designated D and are called *dextrorotatory*; substances rotating light to the left are designated L and are called *levorotatory*. See also **Isomerism.**

ATACAMITE. This mineral is a basic chloride of copper corresponding to formula $Cu_2Cl(OH)_3$. Crystallizes in thin, orthorhombic prisms, may occur massive. Hardness, 3–3.5; specific gravity, 3.76–3.78; luster, adamantine to vitreous; color, green; streak, green; transparent to translucent.

It is a secondary mineral found associated with malachite and cuprite; originally found at Atacama, Chile, whence its name. Other localities are Bohemia, South Australia, and in the United States in Arizona, Utah and Wyoming.

See also **Cuprite; Malachite.**

ATOM. An atom is a basic structural unit of matter, considered today to be the smallest particle of an element that can enter into chemical combination. Each atomic species has characteristic physical and chemical properties that are determined by the number of constituent particles (protons, neutrons, and electrons) of which it is composed; especially important is the number Z of protons in the nucleus of each atom. To be electrically neutral the number of electrons in an atom must also be Z. The arrangement of these electrons in the internal structure of an atom determines its chemical properties. All atoms having the same atomic number Z have the same chemical properties, but differ in greater or lesser degree from atoms having any other value of Z. Thus, for example, all atoms of sodium ($Z = 11$) exhibit the same characteristic properties and undergo those reactions which chemists have found for the element sodium. While these reactions are similar in some degree to those reactions characteristic of certain other elements, such as closely related potassium and lithium, they are not precisely the same and thus can be distinguished chemically. See **Chemical Elements.** Individual atoms can usually combine with other atoms of either the same or another species to form molecules.

The word *atom* has a long history, going back as far as the Greek philosopher Democritus. The concept of the atomic nature of matter was revived near the beginning of the nineteenth century. It was used to explain and correlate the advancing knowledge of chemistry and to establish many of the basic principles of chemistry—this despite the fact that conclusive experimental verification for the existence of atoms was not forthcoming until late in the nineteenth century. It was on this early basis that Mendeleev first prepared a periodic table. See **Periodic Table of Elements.**

Several qualifying terms are used commonly to refer to specific types of atoms. Examples of some of the terms are given in the following paragraphs.

An *excited atom* is an atom which possesses more energy than a normal atom of that species. The additional energy commonly affects the electrons surrounding the atomic nucleus, raising them to higher energy levels.

An *ionized atom* is an ion, which is an atom that has acquired an electric charge by gain or loss of electrons surrounding its nucleus.

A *labeled atom* is a tracer which can be detected easily, and which is introduced into a system to study a process or structure. The use of those labeled atoms is discussed at length in the entry on isotopes.

A *neutral atom* is an atom which has no overall, or resultant, electric charge.

A *normal atom* is an atom which has no overall electric charge, and in which all the electrons surrounding the nucleus are at their lowest energy levels.

A *radiating atom* is an atom which is emitting radiation during the transition of one or more of its electrons from higher to lower energy states.

A *recoil atom* is an atom which undergoes a sudden change or reversal of its direction of motion as the result of the emission by it of a particle or radiation in a nuclear reaction.

A *stripped atom* is an atomic nucleus without surrounding electrons; also called a nuclear atom. It has, of course, a positive electric charge equal to the charge on its nucleus.

During recent years, much research has been directed toward a better understanding of subatomic particles and, from this knowledge, further insights into the behavior of atoms at the "atomic" level are being gained. See also **Particles (Subatomic).**

ATOMIC ENERGY LEVELS. (1) The values of the energy corresponding to the stationary states of an isolated atom. (2) The set of stationary states in which an atom of a particular species may be found, including the ground state, or normal state, and the excited states.

ATOMIC HEAT. The product of the gram-atomic weight of an element and its specific heat. The result is the atomic heat capacity per gram-atom. For many solid elements, the atomic heat capacity is very nearly the same, especially at higher temperatures and is approximately equal to $3R$, where R is the gas constant (Law of Dulong and Petit).

ATOMIC HEAT OF FORMATION. Of a substance, the difference between the enthalpy of one mole of that substance and the sum of the enthalpies of its constituent atoms at the same temperature; the reference state for the atoms is chosen as the gaseous state. The atomic heat of formation at 0 K is equal to the sum of all the bond energies of the molecule, or to the sum of all the dissociation energies involved in any scheme of step-by-step complete dissociation of the molecule.

ATOMIC ORBITALS. Orbitals.

ATOMIC PERCENT. The percent by atom fraction of a given element in a mixture of two or more elements.

ATOMIC SPECTRA. An atomic spectrum is the spectrum of radiation emitted by an excited atom, due to changes within the atom; in contrast to radiation arising from changes in the condition of a molecule. Such spectra are characterized by more or less sharply defined "lines," corresponding to pronounced maxima at certain frequencies or wavelengths, and representing radiation quanta of definite energy.

The lines are not spaced at random. In the spectrum of hydrogen, for example, there is a prominent red line (H_α) and, far from it, another (H_β) in the greenish-blue, then after a shorter wavelength interval a blue-violet line (H_γ), and after a still shorter interval another violet line (H_δ), etc. One has only to plot the frequencies of these lines as a function of their ordinal number in the sequence, to get a smooth curve which shows that they are spaced in accordance with some law. In 1885, Balmer studied these lines, now called the Balmer series, and arrived at an empirical formula which in modern notation reads

$$\nu = Rc \left(\frac{1}{n_1^2} - \frac{1}{n_2^2} \right)$$

It gives the frequency of successive lines in the Balmer series if R is the Rydberg constant, c the velocity of light, $n_1 = 2$, $n_2 \sim 3, 4, 5, \ldots$ As n_2 becomes large, the lines become closer together and eventually reach the series limit of $\nu = Rc/4$. Ritz, as well as Rydberg suggested that other series might occur where n_1 has other integral values. These, with their discoverers and the spectral region in which they occur are as follows:

Lyman series, far ultraviolet, $n_2 = 2, 3, 4, \cdots, n_1 = 1$
Paschen series, far infrared, $n_2 = 4, 5, 6, \cdots, n_1 = 3$
Brackett series, far infrared, $n_2 = 5, 6, 7, \cdots, n_1 = 4$
Pfund series, far infrared, $n_2 = 6, 7, 8, \cdots, n_1 = 5$.

ATTRITION. A process used to reduce the size of various substances and frequently used in the chemical and process industries. Attrition connotes a rubbing action, although this action usually is combined with other forces, including shear and impact. Attrition mills also are referred to as disk mills and normally comprise two vertical disks mounted on horizontal shafts, with adjustable clearance between the vertical disks.

AUGITE. This mineral is a common monoclinic variety of pyroxene whose name is derived from the Greek word meaning luster, in reference to its shining cleavage faces. Chemically it is a complex metasilicate of calcium, magnesium, iron and aluminum. Color, dark green to black, may be brown or even white; hardness, 5–6; specific gravity, 3.23–3.52. Augite is important as a primary mineral in the igneous rocks and also as secondary mineral. The white augite is called leucaugite from the Greek word meaning white. Chemical analysis reveals this variety as containing little or no iron. Augite is of widespread occurrence.

See also **Pyroxene**.

AUTOCATALYSIS. A word used to describe the experimentally observable phenomenon of a homogeneous chemical reaction that shows a marked increase in rate with time, reaching a peak at about 50% conversion and then dropping off. The temperature has to remain constant and all ingredients mixed at the start for proper observation.

This definition excludes those exothermic reactions which shown an increase in rate with time (like explosions) caused by the rapidly rising temperature.

AUTODEPOSITION. A generic term coined to describe a fairly recent (late 1970s) development in which conversion coatings and organic coatings are applied to metal substrates in a single stage. The process is analogous to the electrodeposition of organic coatings, but in autodeposition, the coatings are deposited by means of chemical action rather than under the influence of an electric current and there is no need for a separate conversion coating stage. See also entry on **Conversion Coatings**.

Autodeposition baths are comprised of colloidal dispersions of film-forming coating materials, such as latexes or polymer emulsions, acids and oxidizers. Cleaned metal surfaces are immersed in the coating composition where the substrate metal is lightly attacked by the activating system of acid and oxidizer. This results in the formation of metal ions which overcome the stabilizing charges on the polymer particles and cause them to deposit on the metal surface. The thickness and type of conversion coating which forms simultaneously is determined by the kind and degree of activation.

Autodeposition is characterized by the growth of the coating with time of immersion and the ability to withstand water rinsing immediately upon removal from the coating bath and before fusion or curing of the coating by heat without loss of coating.

The elimination of the conversion coating sequence used in conventional industrial coating processes means that autodeposited coatings can be applied in fewer steps and in smaller finishing areas. There are no solvents in the commercial baths. This reduces air pollution to virtually zero and completely eliminates fire hazard. The coatings exhibit excellent adhesion, impact resistance, flexibility, and chemical resistance. There is no "throwing power" limit, such as that in electrodeposition. The coatings can be applied to any partially enclosed surface that can be contacted by the bath and are not attacked by "solvent wash." As of the early 1980s, the largest application of autodeposition systems is in the automotive industry.

—Wilbur S. Hall, Amchem Products, Inc., Ambler, Pennsylvania.

AUTOLYSIS. The energy derived from biological oxidations in living cells serves to promote anabolic processes, i.e., to produce relatively complex, highly ordered molecules and structures, and thus normally keeps living cells in a steady state remote from equilibrium. In organisms that lack cellular nutrients or oxygen (or in dead organisms or cells that have been disrupted so as to destroy much subcellular organization), the opposing catabolic tendency toward equilibrium, including the tendency toward degradation of macromolecules to simpler monomeric subunits, is not counterbalanced. These degradative processes, many of them enzymatically catalyzed, are collectively termed autolysis. Autolytic processes may include, for example, hydrolysis of proteins catalyzed by proteolytic enzymes or hydrolysis of nucleic acids catalyzed by nucleases. Autolysis of tissues (e.g., liver homogenate) has sometimes been used as a method for releasing bound molecules (e.g., vitamins or coenzymes) into free soluble form.

AUTOXIDATION. A word used to describe those spontaneous oxidations which take place with molecular oxygen or air at moderate temperatures (usually below 150°C) without visible combustion. Autoxidation may proceed through an ionic mechanism, although in most cases the reaction follows a free radical-induced chain mechanism. The reaction is usually autocatalytic and may be initiated thermally, photochemically, or by addition of either free radical generators or metallic catalysts. Being a chain reaction, the rate of autoxidation may be greatly increased or decreased by traces of foreign material.

Many organic and a variety of inorganic compounds are susceptible to autoxidation.

AUTUNITE. This mineral is a hydrous phosphate of calcium and uranium, crystallizing in the tetragonal system, usually in thin tabular crystals. Good basal cleavage; hardness, 2–2.5; specific gravity, 3.1; luster, subadamantine to pearly on the base; color, lemon yellow; streak, yellow; transparent to translucent; strongly fluorescent.

Originally from near Autun in France, whence the name, it is a secondary mineral associated commonly with uraninite. In the United States, it occurs sparsely in the pegmatites of Connecticut, New Hampshire and North Carolina. Autunite also is known as *calco-uranite*.

See also **Uraninite**.

AVOGADRO CONSTANT. The number of molecules contained in one mole or gram-molecular weight of a substance. The most recent value is 6.022045×10^{23} mol^{-1} (uncertainty = 5.1 ppm). In measurements made by scientists at the U.S. National Bureau of Standards and announced in the mid-1970s, the uncertainty, as compared with prior determinations, was reduced by a factor of 30.

AVOGADRO LAW. The well-recognized principle known by this name was originally a hypothesis suggested by the Italian physicist Avogadro, in 1811, to explain the puzzling rule of proportional volumes observed in chemical reactions of gases and vapors. It states simply that equal volumes of all gases and vapors at the same temperature and pressure contain the same number of molecules. Though this assumption accords with the facts and aids the kinetic theory of gases, just why it should be true is by no means self-evident, unless one starts with the much more recent Maxwell-Boltzmann law of equipartition of energy, which also requires proof. That Avogadro's law is true cannot be said to have been positively established until the experiments of J. J. Thomson, Millikan, Rutherford, and others determined the value of the electron as an electric charge and thereby made it possible to count the number of atoms of different elements in a gram. The actual number of molecules contained in one mole (gram-molecular weight) of a substance is the Avogadro Constant.

At any fixed temperature and pressure, the density of carbon dioxide gas, for example, is approximately 22 times greater than the density of hydrogen gas. Thus, the mass of 1 liter of carbon dioxide is 22 times the mass of 1 liter of hydrogen gas. According to Avogadro's principle, the number of molecules in 1 liter of carbon dioxide is the same as the number of molecules in 1 liter of hydrogen. Thus, it follows that a carbon dioxide molecule must have a mass that is 22 times larger than the mass of a hydrogen molecule. Since the molecular weight of hydrogen (H_2) was set equal to 2, carbon dioxide was assigned a molecular weight of 22×2, or 44. Cannizzaro was the first to use gas densities to assign atomic and molecular weights. Avogadro's principle also may be used to assign molecular weights in a slightly different way. At standard temperature and pressure, the volume of a mole of any gas is 22.4 liters. The molecular weight of a gas, therefore, is the mass (in grams) of 22.4 liters of the gas under standard conditions. For most gases, the deviation from this ideal value is less than 1%. See also **Avogadro Constant**.

AXINITE. This mineral is an aluminum-boron-calcium silicate with iron and manganese, $(Ca, Mn, Fe)_3Al_2BSi_4O_{15}(OH)$. It crystallizes in the triclinic system, yielding broad sharp-edged forms, which has led to its name, derived from the Greek word meaning axe. It breaks with a conchoidal fracture; hardness, 6.5–7; specific gravity, 3.22–3.31; luster, vitreous; colors, brown, blue, yellow and gray; transparent to translucent.

Axinite occurs in granites or more basic rocks along contacts and in cavities in Saxony, Switzerland, France, England, Tasmania, and Japan; in the United States, in New Jersey, Pennsylvania, and California.

AZEOTROPIC SYSTEM. A system of two or more components which has a constant boiling point at a particular composition. If the constant boiling point is a minimum, the system is said to exhibit *negative azeotropy*; if it is a maximum, *positive azeotrypy*.

Consider a mixture of water and alcohol in the presence of the vapor. This system of two phases and two components is divariant (see **Phase Rule**). Now choose some fixed pressure and study the composition of the system at equilibrium as a function of temperature. The experimental results are shown schematically in the accompanying figure.

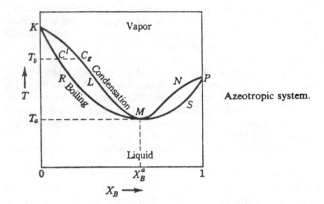

Azeotropic system.

The vapor curve KLMNP gives the composition of the vapor as a function of the temperature T, and the liquid curve KRMSP gives the composition of the liquid as a function of the temperature. These two curves have a common point M. The state represented by M is that in which the two states, vapor and liquid, have the same composition x_B^a on the mole fraction scale. Because of the special properties associated with systems in this state, the Point M is called an azeotropic point and the system is said to form an azeotrope. In an azeotropic system, one phase may be transformed to the other at constant temperature, pressure and composition without affecting the equilibrium state. This property justifies the name azeotropy, which means a system which boils unchanged.

AZIDES. The salts of hydrazoic acid are termed *azides*. Metallic azides can be prepared from barium azide and the metal sulfate, or from potassium azide and the metal perchlorate.

Soluble azides react with iron(III) salt solutions to produce a red color, similar to that of iron(III) thiocyanate. Sodium azide is not explosive, even on percussion, and nitrogen may be evolved upon heating. With iodine dissolved in cold ether silver azide forms iodine azide (IN_3), yellow explosive solid.

Sodium azide is a slow oxidizing agent. It has a selective action in inhibiting the growth of gram-negative organisms. It has been used as a component in selective media such as azide glucose broth or azide blood agar base for the isolation of mastitis and fecal *Streptococci*.

A number of alkyl and aryl azides are known, such as CH_3N_3, $C_2H_5N_3$ and $C_6H_5N_3$. The nonmetallic inorganic azides include ClN_3, an explosive gas, BrN_3, an orange liquid, mp $-45°C$, and IN_3, a yellow solid, decomposing above $-10°C$. The gas FN_3 is more stable than ClN_3, decomposing only slowly at room temperature.

Lead and silver azides are widely used as initiating, or primary explosives because they can be readily detonated by heat, impact, or friction. As such, these materials, particularly lead azide, are used in blasting caps, percussion caps, and delay initiating devices. The function of the azides is similar to that of mercury fulminate or silver fulminate.

AZO AND DIAZO COMPOUNDS. Characteristically, these are compounds containing the group —N:N— (azo) or >N $\vdots$ N (diazo). They are closely related to the substituted hydrazines. The N_2 group may be covalently attached to other groups at both ends, as in the azo compounds, or at only one end, as in the diazo compounds or diazonium salts. Although organic chemistry furnishes the most numerous examples, many inorganic azo compounds also exist.

Compounds related to aniline, either directly or by oxidation, and to nitrobenzene by reduction, are numerous and important. When nitrobenzene is reduced in the presence of hydrochloric acid by tin or iron, the product is aniline (colorless liquid); in the presence of water by zinc, the product is phenylhydroxylamine (white solid); in the presence of methyl alcohol by sodium alcoholate or by magnesium plus ammonium chloride solution, the product is azoxybenzene (pale yellow solid); by sodium stannite, or by water plus sodium amalgam, the product is azobenzene (red solid); in the presence of sodium hydroxide solution by zinc, the product is hydrazobenzene (pale yellow solid). The behavior of other nitro-compounds is similar to that of nitrobenzene.

Hydrazobenzene is converted by oxygen of the air or by ferric chloride solution into azobenzene, and by strong acids into benzidine hydrochloride $(4')H_2N \cdot C_6H_4 \cdot C_6H_4 \cdot NH_2(4) \cdot HCl)$. Benzidine is prepared by reducing nitrobenzene to hydrazobenzene as above, and then treating the product with acid. Benzidine and its toluene relative, orthotolidine, are important intermediates for dyes. The counterpart of aniline is toluidine in its three forms, *ortho*, *meta*, *para*.

Azoxybenzene is converted by distillation with iron into azobenzene (ferrous oxide also formed), and by concentrated sulfuric acid warm into *para*-hydroxyazobenzene $(4)HO \cdot C_6H_4N:NC_6H_5$, which is a dye.

Diazonium salts are usually colorless crystalline solids, soluble in water, moderately soluble in alcohol, and when dry are violently explosive by percussion or upon heating. These salts are generally used in cold (near 0°C) acid solution, without separation of the salt, and are prepared by reaction of the desired benzenoid primary amine with nitrous acid (from sodium nitrite plus hydrochloric acid). Alkyl amines with nitrous acid yield the corresponding alcohol.

(A) By treatment of benzene diazonium chloride

$$\left(\begin{matrix} C_6H_5N{-}Cl \\ \vdots \\ N \end{matrix} \right)$$

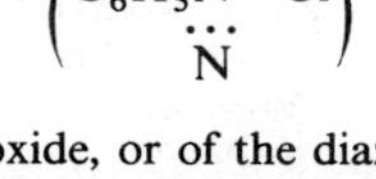

solution with silver oxide, or of the diazonium sulfate solution with barium hydroxide, the hydroxide (benzene diazo hydroxide, $C_6H_5N:N{-}OH$) is obtained, which is intermediate in basicity between ammonium hydroxide and sodium hydroxide. Most diazo hydroxides are unstable, and are spontaneously transformed into nitrosoamines (group —NH·NO), yellow neutral compounds. With sodium hydroxide, diazonium salt solutions yield sodium benzenediazoate, more active chemically when first formed than upon standing, due to change from syn-diazoate

$$\left(\begin{matrix} C_6H_5N \\ \vdots \\ NaON \end{matrix} \right)$$

which evolves nitrogen readily, to anti-diazoate

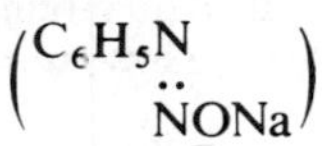

$$\left(\begin{matrix} C_6H_5N \\ \vdots \\ NONa \end{matrix} \right)$$

which is more stable. Sodium benzene-syn-diazoate reacts with phenols in alkaline solution to give azo-dyes, e.g., *para*-hydroxyazobenzene, wherein hydrogen *para* (or *ortho* but not *meta*) to the hydroxyl group is reactive. Other reactions of diazonium salts are:

(B) Replacement of the diazo-group with loss of nitrogen, (1) by hydroxyl-group, forming phenols by warming with water, (2) by alkoxy-group, forming ethers, by warming with an alcohol, (3) by acyl-group, forming esters, with an acid, (4) by hydrogen, forming hydrocarbons or substituted hydrocarbons, e.g., tribromobenzenediazonium chloride forms tribromobenzene, with alcohol or sodium stannite solution, (5) by chlorine, forming, for example, chlorobenzene, by warming with cuprous chloride (Sandmeyer's reaction), (6) by bromine, forming, for example, bromobenzene, by warming with cuprous bromide (Sandmeyer's reaction), (7) by iodine, forming, for example, iodobenzene, by warming a solution of the diazonium iodide, (8) by cyanide, forming, for example, cyanobenzene, by warming with cuprous cyanide (Sandmeyer's reaction), (9) by fluorine, forming, for example, fluorobenzene, by warming a solution of the diazonium borofluoride. Gatterman's modificaton of Sandmeyer's reactions above is to use copper powder plus the corresponding sodium or potassium salt instead of the cuprous salt. Nitro-compounds, since they are readily reduced to the corresponding amine, form an important group of compounds for the preparation of various derivatives by means of the diazoreaction.

(C) Reduction, to form the corresponding hydrazine, for example, benzene diazonium chloride to form betaphenylhydroxylamine hydrochloride.

(D) Bromination followed by treatment with ammonia to form azides, for example, benzene diazonium bromide plus bromine forms benzene diazonium perbromide $C_6H_5Br_3$, which upon treatment with ammonia forms phenylazide

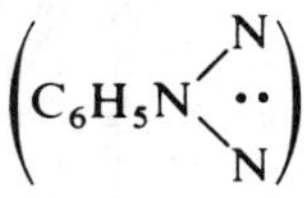

$$\left(\begin{matrix} & & N \\ & & \diagup \\ C_6H_5N & \cdots & \\ & & \diagdown \\ & & N \end{matrix} \right)$$

(E) Amines, (1) primary or (2) secondary amine to form diazoamino-compounds, e.g., benzenediazonium chloride (a) plus aniline forms diazoaminobenzene, benzene diazoaniline $C_6H_5N:N{-}NHC_6H_5$, (b) plus methylaniline forms benzenediazomethylaniline

$$\left(\begin{matrix} & & C_6H_5 \\ & & \diagup \\ C_6H_5N:N{-}N & & \\ & & \diagdown \\ & & CH_3 \end{matrix} \right)$$

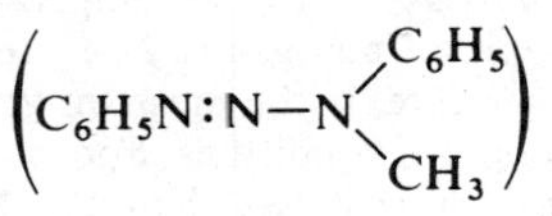

Diazoamino-compounds readily change into aminoazo-compounds upon standing in alcohol solution or in the presence of amine hydrochloride, thus, benzenediazoaniline changes into *para*-aminoazobenzene, benzeneazoaniline-4 $C_6H_5N:NC_6H_4NH_2(4)$, benzenediazomethylaniline changes into methyl-*para*-aminobenzene, benzeneazomethylaniline-4

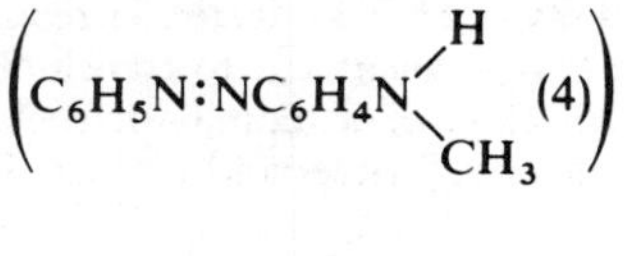

$$\left(\begin{matrix} & & H \\ & & \diagup \\ C_6H_5N:NC_6H_4N & & (4) \\ & & \diagdown \\ & & CH_3 \end{matrix} \right)$$

(3) tertiary amine to form aminoazo-compounds directly, e.g., benzenediazonium chloride plus dimethylaniline forms dimethyl-*para*-aminoazobenzene, benzeneazodimethylaniline-4

$$\left(C_6H_5N:NC_6H_4N \begin{matrix} CH_3 \\ \\ CH_3 \end{matrix} (4) \right)$$

In this manner are prepared the azo-dyes, which contain the chromophore azo-group —N : N— plus an auxochrome amino-group

$$\left(-NH_2, -N \begin{matrix} H \\ \\ CH_3 \end{matrix}, -N \begin{matrix} CH_3 \\ \\ CH_3 \end{matrix} \right)$$

The simplest azo-dyes are yellow, but by increasing the number of auxochrome groups or by increasing the percentage of carbon, the color darkens to red, violet, blue, and in some cases brown. Naphthalene residues darken to red, violet, blue and finally black. These amino-azo-dyes, together with the hydroxyazo-dyes (containing auxochrome hydroxyl-group —OH), are generally only slightly soluble in water. In order that the dye may be soluble it is desirable that it contain one or more sulfonic acid groups —SO_2OH. This group may be introduced either by treating the dye with concentrated sulfuric acid, or by using sulfonic acid derivatives in preparing the dye, e.g., methyl orange, sodium dimethyl-*para*-aminoazobenzene-*para*-sulfonate

$$(4)(CH_3)_2NC_6H_4N:NC_6H_4SO_2ONa(4)$$

from dimethylaniline and diazotized sulfanilic acid (*para*-amino-benzene sulfonic acid, $(1)H_2N \cdot C_6H_4 \cdot SO_2OH(4)$, and then the sodium salt made from the product. Other azo-dyes are

chrysoidine $\left(C_6H_5N:NC_6H_3 \begin{matrix} NH_2(2) \\ \\ NH_2(4) \end{matrix} \right)$

Bismarck brown $\left((3)H_2N \cdot C_6H_4N:NC_6H_3 \begin{matrix} NH_2(2) \cdot HCl \\ \\ NH_2(4) \end{matrix} \right)$

Congo red $\left(\begin{matrix} (4)HOO_2S \\ (1) \quad H_2N \end{matrix} C_{10}H_5:NC_6H_4 \right)$

$$C_6H_4N \cdot NC_{10}H_5 \begin{matrix} SO_2OH(4) \\ \\ NH_2(1) \end{matrix}$$

See also **Aniline**; **Dyes (Textile)**; **Hydrazine**; **Nitro- and Nitroso-Compounds**.

AZURITE. This mineral is a basic carbonate of copper, crystallizing in the monoclinic system, with the formula $Cu_3(CO_3)_2(OH)_2$, so called from its beautiful azure-blue color. It is a brittle mineral with a conchoidal fracture; hardness, 3.5–4; specific gravity, 3.773; luster, vitreous; color and streak, blue; transparent to translucent. Azurite like malachite is a secondary mineral, but far less common than that mineral. It has been formed by the action of carbonated waters on compounds of copper or solutions of copper compounds, probably most abundantly by rich solutions reacting with limestones.

Azurite almost always occurs associated with malachite. It is found in Siberia, Greece, Rumania, at Chessy, France (whence the name Chessylite), in Southwest Africa, Australia and elsewhere; in the United States, at Bisbee, Arizona, and Kelly, New Mexico. Azurite also is known as *Chessylite*.

B

BABINGTONITE. This mineral is a relatively rare calcium-iron-manganese silicate, occurring in small black triclinic crystals, found in Italy, Norway and in the United States at Somerville and Athol, Massachusetts, and in Passaic County, New Jersey. It was named for Dr. William Babington.

BAGASSE. In the manufacture of sugar from sugar cane, the crushed fibers from which the sap has been expressed are called *bagasse*. The principal use of byproduct is as a fuel to power the mills which crush the cane. Bagasse is mixed with petroleum oil for this purpose. Bagasse is also used as a fertilizer and to some extent in the manufacture of heavy insulation board and coarse paper.

BALL, PEBBLE, AND ROD MILLS. Basically, these mills are used for the size reduction of materials, frequently prior to processing. They are made up of a rotating drum which operates on a horizontal axis and is filled partially with a free-moving grinding medium which is harder and tougher than the material to be ground. The tumbling action of the grinding medium (balls, pebbles, rods) crushes and grinds the material by combination of attrition and impact. A conical ball mill is shown in Fig. 1. In the case of a pebble mills, a nonmetallic

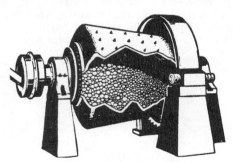

Fig. 1.   Conical ball mill. (*Hardinge.*)

medium, such as flint, pebbles, or even large pieces of the material being ground, will be used. In a rod mill (Fig. 2), the grinding

Fig. 2.   Rod mill (*Hardinge.*)

medium is a series of metallic rods essentially as long as the mill cylinder. Grinding generally requires several hours to assure the necessary fineness within particle-size limits.

BARBITURATES. Sedative drugs derived from barbituric acid. These drugs depress the central nervous system and act especially on the sleep center of the brain, thus their sedative and sometimes hypnotic effects. Barbital and phenobarbital are relatively long-lasting in effects. Other drugs, such as *Amytal* and *Nembutal*, are more powerfully hypnotic and have a shorter action. A third group, including thiopentone and hexobarbital, are very powerful and short-lived in action and sometimes are used for the production of intravenous anesthesia. In the treatment of epilepsy, phenobarbital is sometimes used. It is an effective anticonvulsant, but produces drowsiness when given in large amounts. *Dilantin* suppresses grand mal convulsions and can be taken in larger doses than phenobarbital without causing lethargy. Phenobarbital and *Dilantin* make petit mal worse, so *Tridione* is sometimes used in this type of epilepsy.

Barbituric acid, also known as malonylurea, pyrimidinetrione, and 2,4,6-trioxohexahydropyrimidine, has the formula

$$OCNHCOCH_2CONH \cdot 2H_2O.$$

It is a white, crystallize, odorless substance, mp 245°C with some decomposition. It is slightly soluble in water and alcohol; soluble in ether. The acid forms salts with metals. It is used in the preparation of barbiturates and as a polymerization catalyst.

Barbital, $C_8H_{12}N_2O_3$ is diethylmalonylurea or diethylbarbituric acid. The compound is derived by the interaction of the diethyl ester of diethylmalonic acid and urea. Phenobarbital is phenylethylmalonylurea, $C_{12}H_{12}N_2O_3$. It is derived by condensation of phenlethylmalonic acid derivatives and urea. It is available as the sodium salt, which has good water solubility.

The Expert Committee on Drug Addition of the World Health Organization has advised that barbiturates "must be considered drugs liable to produce addiction."

BARITE. The mineral barite is barium sulfate, $BaSO_4$, crystallizing in the orthorhombic system. It may occur as tabular crystals, in groups, or lamellar, fibrous and massive. Barite has two perfect cleavages, basal and prismatic; hardness, 3–3.5; specific gravity, 4.5, which has led to the term heavy spar, occasionally used for this mineral. Its luster is vitreous; streak, white; color, white to gray, yellowish, blue, red and brown; transparent to opaque. It sometimes yields a fetid odor when broken or when pieces are rubbed together, due probably to the inclusion of carbonaceous matter. It is used as a source of barium compounds.

Barite is a frequently occurring gangue mineral and is found also in large masses in sedimentary rocks. It occurs in many places in Europe, including Czechoslovakia, Germany, France, Spain and England; in the United States, New York, Connecticut, Pennsylvania, Virginia, Michigan, Missouri, New Mexico, Oklahoma, Utah, Colorado, South Dakota, Georgia and Tennessee. In Canada it occurs in Ontario and in Nova Scotia.

The name of this mineral derives from the Greek word meaning heavy.

BARIUM. Chemical element symbol Ba, at. no. 56, at. wt. 137.34, periodic table group 2a (alkaline earths), mp 725°C, bp 1640°C, density, 3.5 g/cm³ (²⁰°C). Body-centered cubic crystal form. Naturally occurring isotopes are ^{130}Ba, ^{132}Ba, ^{134}Ba, ^{135}Ba, ^{136}Ba, ^{137}Ba, and ^{138}Ba. Barium metal is comparatively soft and ductile and capable of mechanical working. Barium metal and all barium compounds are highly toxic to humans, although barium sulfate (because of its insolubility in H_2O and body fluids) can be ingested without harm and is widely used as an opaque medium in X-ray diagnostic studies of the body. First ionization potential, 5.21 eV, second, 9.95 eV. Oxidation potentials $Ba \rightarrow Ba^{2+} + 2e^-$, 2.90 V, $Ba + 2OH^- + 8H_2O \rightarrow Ba(OH)_2 \cdot 8H_2O + 2e^-$, 2.97 V. Other important physical properties of barium are given under **Chemical Elements.**

Barium occurs chiefly as sulfate (barite, barytes, heavy spar, $BaSO_4$), and, of less importance, carbonate (witherite, $BaCO_3$). Georgia and Tennessee are the principal producing states. The sulfate is transformed into chloride, and the electrolysis of the fused chloride yields barium metal. See also **Barite; Witherite.** Barium ores are mined chiefly as a source of barium compounds because very little metallic barium is consumed commercially. The metal is obtained by thermal reduction of the oxide, using aluminum metal at a high temperature and under vacuum in a closed retort: $4BaO + 2Al \rightarrow BaOAl_2O_3 + 3Ba$. The gaseous barium produced is recovered by condensation.

As is to be expected from its high electrode potential (2.90 V) barium, like strontium and calcium, reacts readily with the halogens, oxygen and sulfur to form halides, oxide and sulfide, as well as with nitrogen and hydrogen at higher temperatures to form the nitride and hydride. In all its stable compounds it is divalent. It reacts vigorously with water, displacing hydrogen to form the hydroxide. Barium peroxide is formed on treatment of the hydroxide with hydrogen peroxide in the cold and also by direct combination of oxygen and barium oxide or metal. The peroxide prepared in the latter way is frequently paramagnetic because of the presence of some superoxide, $Ba(O_2)_2$. Barium exhibits little tendency to form complexes, the ammines formed with NH_3 being unstable and the β-diketones and alcoholates are not well characterized. Barium metal solutions in liquid NH_3 solution yield $Ba(NH_3)_6$ upon evaporation. Common compounds of barium are:

Barium acetate, $Ba(C_2H_3O_2)_2$, white crystals, solubility 76.4 g/100 ml H_2O at 26°C, formed by reaction of barium carbonate or hydroxide and acetic acid.

Barium carbide (acetylide), BaC_2, black solid, by reaction of barium oxide and carbon at electric furnace temperatures, reacts with H_2O; yielding acetylene gas and barium hydroxide.

Barium carbonate, $BaCO_3$, white solid, insoluble ($K_{sp} = 5.13 \times 10^{-9}$), formed (1) by reaction of barium salt solution and sodium carbonate or bicarbonate solution; (2) by reaction of barium hydroxide solution and CO_2. With excess CO_2 barium hydrogen carbonate, $Ba(HCO_3)_2$, solution is formed. Barium carbonate decomposes at 1,450°C.

Barium chloride, $BaCl_2 \cdot 2H_2O$, white crystals, solubility 31 g/100 ml H_2O at 0°C, formed by reaction of barium carbonate or hydroxide and HCl.

Barium chromate, $BaCrO_4$, yellow precipitate, $K_{sp} = 1.17 \times 10^{-10}$, formed by reaction of barium salt solution and potassium chromate solution.

Barium cyanamide, $BaCN_2$, formed in a mixture with barium cyanide, $Ba(CN)_2$ by heating barium carbide at 800°C with nitrogen gas. Fusion of the cyanamide-cyanide mixture with sodium carbonate converts it entirely to cyanide.

Barium nitrate, $Ba(NO_3)_2$, white crystals, solubility 8.7 g/100 ml H_2O at 20°C, formed by reaction of barium carbonate or hydroxide and HNO_3.

Barium oxide, BaO, white solid, mp about 1,900°C, reactive with H_2O to form barium hydroxide. Barium peroxide,

$BaO_2 \cdot 8H_2O$, white precipitate, formed by reaction of barium salt solution and hydrogen or sodium peroxide, yields anhydrous barium peroxide upon heating at 100°C in a current of dry air. Anhydrous barium peroxide is also formed by heating barium oxide in air or oxygen under pressure (at somewhat over one atmosphere pressure) and temperature of 400°C.

Barium oxalate, BaC_2O_4, white precipitate, $K_{sp} = 1.1 \times 10^{-7}$, formed by reaction of barium salt solution and ammonium oxalate solution.

Barium sulfate, $BaSO_4$, white precipitate, $K_{sp} = 8.7 \times 10^{-11}$, formed by reaction of barium salt solution and H_2SO_4 or sodium sulfate solution, insoluble in acids, by heating with carbon yields barium sulfide.

Barium sulfide, BaS, grayish-white solid, formed by heating barium sulfate and carbon, reactive with H_2O to form barium hydrosulfide, $Ba(SH)_2$, solution. The latter is also made by saturation of barium hydroxide solution with H_2S. Barium polysulfides are formed by boiling barium hydrosulfide with sulfur.

Uses of Barium. The major use of barium metal for a number of years has been as a getter for oxygen in electronic vacuum tubes. A layer of the metal is deposited inside the glass envelope of the tube. Minute quantities of gases which leak into the tube react with the barium layer to form compounds. If the gases remained free, they would alter the conductance of the tube and cause deterioration of its performance. The addition of barium prolongs the useful life of vacuum tubes.

References

Carter, G. C.: "Barium" in "Metals Handbook," 9th Edition, Vol. 2 (D. Benjamin, Senior Editor), American Society for Metals, Metals Park, Ohio, 1979.

Hampel, C. A.: "Encyclopedia of the Chemical Elements," Van Nostrand Reinhold, New York, 1968.

Mantell, C. L., Editor: "Engineering Materials Handbook," McGraw-Hill, New York, 1970.

Pearson, W. B.: "Handbook of Lattice Spacings and Structures of Metals," Pergamon Press, New York, 1967.

Roberts, B. W.: "Superconductive Materials and Some of Their Properties," NBS Technical Note 724, National Bureau of Standards, Washington, D.C., 2970.

Weat, R. C.: "Handbook of Chemistry and Physics," CRC Press, Boca Raton, Florida (Published annually).

—Stephen E. Hluchan, Calcium Metal Products, Minerals, Pigments & Metals Division, Pfizer Inc., Wallingford, Connecticut.

BARYONS. A class of subatomic particles including the proton, the neutron, and several heavier particles, such as the lambda, the sigma (plus, minus, and neutral), and the omega (minus) particles. Baryons are particles that interact with the strong nuclear force. Each baryon is given a baryon number 1, each corresponding antibaryon is given a baryon number −1, while the light particles (photons, electrons, neutrinos, muons, and mesons) are given baryon number 0. The total baryon number in a given reaction is found by algebraically adding up the baryon numbers of the particles entering into the reaction. During any reaction among particles, the baryon number cannot change. This rule ensures that a proton cannot change into an electron, even though a neutron can change into a proton. Similarly, to create an antiproton in a reaction, one must simultaneously create a proton or other baryon. Baryon conservation ensures the stability of the proton against decaying into a particle of smaller mass. See also **Particles (Subatomic); Neutron;** and **Proton.**

BARYTOCALCITE. This mineral is a carbonate of barium and calcium which crystallizes in the monoclinic system but occurs massive as well. It has a perfect cleavage parallel to the prism and one, less perfect, parallel to the base; fracture, subconchoidal; brittle; hardness, 4; specific gravity 3.66–3.71; luster, vitreous; color, white or gray or may be greenish or yellowish; transparent to translucent. Barytocalcite is found in Cumberland, England, associated with barite and fluorite.

BASAL METABOLISM. The metabolism of a living cell or organism refers to the total turnover of chemical material and energy. It consists of *anabolism*, or assimilation, mostly of substances, of high potential energy (primarily protein, fat, and carbohydrate), and *catabolism* or dissimilation. Metabolic rate refers to the metabolism in a given period of time. The "basal" rate refers to the fundamental energy requirement for maintenance and continued functioning of the organism (aside from external muscular work and work of digestion), such as respiration, contraction of the heart, function of the kidney, the liver, and of the cells in general. Basal metabolic rate (BMR) in humans refers to the determination of metabolic rate under certain standardized conditions, including complete physical rest (but not sleep), a fasting state, and an ambient temperature that does not require energy expenditure for physiological temperature regulation. The BMR is below normal in sleep, starvation, anesthesia, and certain endocrine disturbances (hypothyroidism), and is elevated in fever, athletic training, under the influence of certain drugs (such as caffeine) and endocrines (adrenaline and thyroid hormones).

In studies of animals, it becomes technically difficult to make observations under standard conditions which include rest. With exception of a few animals that are well adapted to the laboratory, the BMR of other animals has been estimated from long observations of their eating, exercise, and resting habits.

Methods of Determination. In principle, the metabolic rate is determined in three different ways: (a) determination of the energy value of all food less the energy value of excreta (mainly feces and urine) should give the energy turnover of the organism. However, the result must be corrected for any change in the composition of the body, mainly deposition or utilization of body fats. The method is cumbersome and is accurate only if the period of observation is sufficiently long. (b) Measurement of total heat production of the organism. This is fundamentally the most accurate method. The value obtained must be corrected for any external work performed, including such items as heating of the foodstuffs taken in, vaporization of water, etc. The determinations are made with the organism in a calorimeter, technically a rather difficult procedure, but it yields very accurate results. (c) The amount of oxygen used in oxidation processes can be used to determine the metabolic rate. (In theory, the carbon dioxide production could also be used, but it is less accurate, mainly because there is a large pool of carbon dioxide in the organism that undergoes changes relatively easily.) The reason that oxygen can be used is that similar amounts of heat are produced for each liter of oxygen, irrespective of whether fat, carbohydrate or protein is oxidized. The figures are: fat 4.7 kcal; carbohydrate, 5.0 kcal; and protein, 4.5 kcal, per liter oxygen. It is customary to use an average value, 4.8 kcal/liter oxygen consumed. The use of oxygen consumption for the determination of metabolic rate is so common that the two concepts have become practically synonymous. Obviously, the oxygen consumption cannot be used for determinations of metabolic rate in, for example, anaerobic organisms.

Metabolic Rate in Relation to Body Size. If a uniform group of animals, such as mammals, is used for a comparison of metabolic rates, an interesting relationship is revealed. The smaller the animal, the higher is the metabolic rate per gram of body weight. If, on logarithmic coordinates, the metabolic rate is plotted against body size, we obtain a straight line (see accompanying figure) which corresponds to the equation: log metabolic rate = k + 0.74 log body weight (k being a constant whose

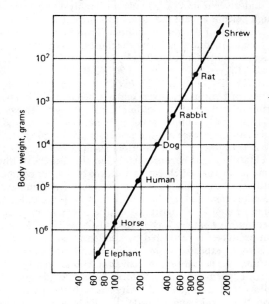

Body weight versus metabolic rate plotted on logarithmic coordinates.

numerical value depends on the units used). It has been suggested that this relationship expresses the need for a higher heat production in the smaller animal, which, because of its larger relative surface, must produce heat at a higher rate than a large animal in order to maintain its body temperature. However, similar relationships between metabolic rate and body size have been found in numerous groups of cold-blooded animals as well as plants, where the need for heat regulation cannot be the fundamental explanation of this interesting relationship.

BASE (Chemistry). Acids and Bases.

BASIC SALT. A compound belonging to the categories of both salts and bases, because it contains OH (hydroxyl) or O (oxide) as well as the usual positive and negative radicals of normal salts. Among the best examples are bismuth subnitrate, often written $BiONO_3$; and basic copper carbonate, $Cu_2(OH)_2CO_3$. Most basic salts are insoluble in water and many are of variable composition.

BASIC SOLVENT. A solvent that accepts protons (hydrogen ions) from the solute.

BASTNASITE. A wax-yellow to reddish-brown, greasy mineral of the composition, $(Ce,La)(CO_3)F$, usually found in contact zones or associated with zinclodes. Sometimes spelled *bastnaesite.* See also **Rare-Earth Elements and Metals.**

BATTERY. One or more galvanic cells in a finished package constitute a battery. The cells may be electrically connected in various series, parallel, and series-parallel combinations, including discrete groupings with separate terminals. In each individual cell of the battery, two electrodes of dissimilar materials are normally joined by an ion-carrying path, but are separated by an electron insulator. The electron-transfer-carrying path is external, via some conductor, such as a metal wire, through a using device where the useful work is done. The ion path is internal, via an electrolyte which is an ionizable salt dissolved in a solvent. The electrolyte may be solid or liquid, although the former is usually limited to low discharge rates.

The active material of the solid electrodes may be solid, gaseous, or liquid and, in some special cases, such as with certain types of liquid cathode cells, the cathode serves a dual role as electrolyte solvent and depolarizer.

Control of the discharge is usually by a switch in the electron-carrying circuit, but in some special cases, it is done by making or breaking the ion-carrying path.

The cells may be rechargeable and such are often called *secondary* or *storage* batteries. Secondary batteries are reversed by passing a current through the cells in the opposite direction of current flow when the cells were working. The chemical conditions of the undischarged battery are restored and the cells are ready to be discharged again. Primary cells are meant to be discharged for exhaustion only once and then discarded. Some of the electrochemical systems used in "primary" cells are reversible, but the cells do not have the physical features needed to render them *safely* rechargeable. Various schemes or devices have been proposed over the years to recharge "primary" batteries, but generally, they are impractical and sometimes potentially *dangerous*.

Both primary and secondary types are produced in wet and dry cell versions. *Wet cells* are liquid electrolyte cells sensitive to orientation, whereas *dry cells* are not. Dry cells can also be made with liquid electrolytes, but in such cases, the electrolyte is immobilized by some mechanism, such as gelling or holding it in absorbent mediums, such as paper. Solid electrolyte cells are, by definition, also called dry cells.

As a further variation, both primary and secondary cells can be made as *reverse cells*. Activation is held off until just prior to use of the battery. This may involve holding off addition of liquid electrolyte or the solvent, such as with dry charged car storage batteries; internal storage of liquid electrolyte or solvent apart from the active materials; or, as with some thermally activated batteries, heating of the batteries to change a solid inactive material into a molten active electrolyte. An example of a novel approach is with a seawater-activated battery, where the salt water of the ocean serves as the electrolyte. Activation is achieved by immersing rigid electrode assemblies in seawater.

Both primary and secondary batteries are marketed in a wide variety of sizes and shapes, from the little button cells used in watches and hearing aids to the huge rectangular cells for missile batteries and load-leveling power storage units. Some applications also involve stringent conditions, such as very low (−50°C) or very high (above 200°C) operating temperatures, underwater usage, outer-space usage, continuous or intermittent discharge, or where discharge may be spread out over several years, or over a very short period, such as a few seconds. With the larger sizes, energy per unit weight is usually the important criterion. For example, in rockets or in electric vehicles. With the smaller sizes, output per unit volume is usually more significant.

There are also cells which convert light and heat into electricity (solar cells and thermal cells), which are described under **Solar Energy.**

The number of electrochemical systems, the variations in cell sizes and internal configurations and the specialized applications for commercially available batteries all have increased dramatically during the past 40 years and change is continuing at a

rapid pace in the early 1980s. A good example of change is the practical nonaqueous batteries now coming on the market. All of those offered thus far are with lithium anodes, but they are available in a wide range of sizes and in solid cathode, liquid cathode, and solid electrolyte versions. The rapidly changing price of raw materials is a significant factor. For example, a rise in the price of silver resulted in a switch from aqueous, silver oxide/zinc cells to nonaqueous types for many applications, such as some types of wrist watches. The development and design of galvanic cells and batteries has therefore become an increasingly complex science, far removed from the near-art it was in pre-World War II years.

Cell Electrochemical Systems and Designs

The simplest cell would consist of two dissimilar metals in an ionically conductive medium, provided that the combination yielded a voltage above that needed to continuously breakdown the electrolyte at the positive electrode. Metal would go into solution at the negative electrode, releasing electrons to travel via the external circuit to the positive electrode, doing work during the transit. Such a cell, however, is impractical for most uses, due to shortcomings, such as gas formation, loss of electrolyte solvent, poor storageability, among other factors. Thus, in practice, material is included in the positive electrode which will go through a valency drop on electrochemical discharge and, in so doing, will accept the electrons coming from the negative electrode. This material is known as the *depolarizer*. Some of the depolarizers now in use in practical cells are liquids, some are solids, and some are gases.

The depolarizer or cathode is positioned in the positive electrode in combination with some electron-carrying matrix, and usually is porous to allow access of the ions carrying electrolyte to a large area of the depolarizer. The anode is the electrode that goes into solution in the electrolyte as ions, releasing electrons to do the work. It also can be in the form of solid, liquid, or gas, although in the majority of the presently marketed systems, it is in the form of a solid and usually is a metal. The negative electrode contains the anode material and, like the cathode, it is assembled with an electron-carrying matrix and usually of a design to provide large area contact with the electrolyte. In some cases, where the anode material is a metal, an additional conductive grid is not required. The electrolyte forms the ion bridge between the anode and the cathode materials. It is usually an ion-forming salt dissolved in a solvent. Some cell types have the electrolyte as a solid, but the majority are as liquids—at least when the cell is being discharged, since the solid electrolytes will accommodate only a very low rate of discharge.

A relatively few years ago, all the liquid electrolyte systems employed water as the electrolyte solvent, but in present practice, there are many nonaqueous systems in use. Thus cell systems are divided into two main categories—aqueous and nonaqueous. Electrolytes can be further subdivided into acidic, mildly acidic, and alkaline.

Aqueous Acidic Systems. The best known example of this type is the lead-sulfuric acid, rechargeable system in wide use for starting automobiles. In the car storage battery, the configuration is usually several flat-plate electrodes per cell with negative and positive electrodes interspaced, separated by porous plastic, wood, or other suitable separators to give a high electrode interface area. In the car storage battery, the electrolyte is not immobilized—so this is a wet cell rather than a dry cell. For present-day automobiles which operate on 12 volts nominal, six cells are connected in series per battery. A typical lead-sulfuric acid storage battery is shown in Fig. 1. This system

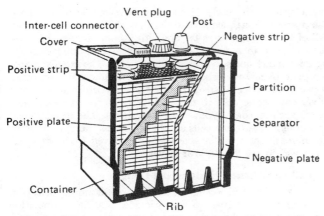

Fig. 1.   Cutaway view of automotive-type storage battery cell.

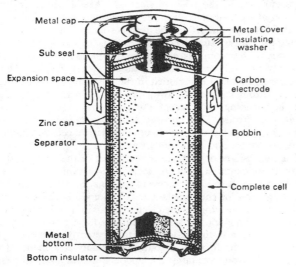

Fig. 2.   Round carbon-zinc cell.

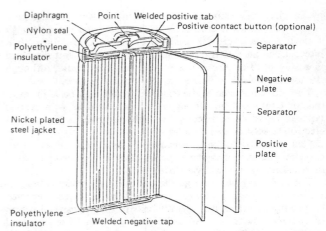

Fig. 3.   Cutaway view of sealed, coiled-type, sintered nickel-cadmium cell.

is also marketed as a rechargeable, cylindrical dry cell battery. The latter configuration is achieved by winding two electrodes, one positive and one negative, separated by an absorbent material, into a helix or "jelly-roll" inside a cylindrical can.

Aqueous Mildly Acidic Cells. Probably the best known cell of this type is the commonly called carbon-zinc or Leclanche cell, widely employed because of its low cost. A flashlight battery version is shown in Fig. 2. The positive electrode in this case is referred to as a *bobbin*, and is a mechanical mixture of manganese dioxide powder, acetylene black particles, and an aqueous solution of zinc chloride and ammonium chloride. This electrode is molded about a carbon rod. The manganese dioxide is the depolarizer; the acetylene black is the electron-distributing conductive matrix; the salt solution is the ion-distributing electrolyte. The carbon rod is the conductive path to an outside terminal (the *cap*), since economically acceptable metals will be chemically attacked by the manganese dioxide. The anode is zinc and, in this case, it serves a dual role, both as the anode and as the container for the actual cell. A layer of electrolyte, gelled with material such as starch, separates the positive and negative electrodes electronically, but connects them with an ion path. The actual cell is enclosed with a package imprinted with descriptive data and decorations, thus converting the single cell into a battery. The nominal voltage of a single cell is 1.5 V and it discharges with a sloping voltage-versus-output discharge curve.

Carbon-zinc cells are marketed in many different sizes and shapes. In some cylindrical cell versions, the positive electrode is on the outside. These have sometimes been called "inside-out" cells. High-voltage stacks of cells have been marketed which use flat cells, wherein the zinc is coated on one side with an electrolyte-impervious carbon coating, and this serves as the positive collector for the cathode mix of the next cell. A series connection of the cells in the stack is thus achieved. Encapsulating materials are kept to a minimum in order to achieve a high energy density. This design approach accounted for a significant proportion of the Leclanche system cells, particularly in years when there was little rural electrification, but it represents only a small portion of the market today, mainly in applications requiring small, high-voltage batteries, such as personal portable radios.

Aqueous Alkaline Systems. The majority of these systems use potassium hydroxide as the electrolyte solute, with the remainder using sodium hydroxide. Potassium hydroxide imparts higher rate capabilities, but sodium hydroxide is cheaper and easier to contain and sometimes imparts better storageability. The aqueous alkaline systems, as a rule, have higher rate capabilities, better low-temperature performance, and higher energy densities than the aqueous, mildly acidic cell designs. However, the aqueous alkaline systems are usually more complex and costly.

Aqueous alkaline systems are used in wet and dry cells and in secondary as well as primary batteries. An example of a secondary dry cell is the jelly-roll cylindrical nickel-cadmium rechargeable battery shown in Fig. 3.

As the size of batteries becomes relatively small, such as for use with electric wrist watches, the cost of the materials is a small portion of the total cost. Assembly cost for a given size remains relatively constant regardless of the aqueous system used, and since cost per hour of service usually controls in these applications, systems that yield a high output per unit volume of cells, even if made with expensive electrochemical materials, become preferable. Silver oxide/zinc is a good example. The miniature cell designs for the several aqueous alkaline systems in use, i.e., MnO_2/Zn, Ag_2O/Zn, AgO/Zn, HgO/Zn, CuO/Zn, and Air/Zn are quite similar. A typical miniature (sometimes called *button*) silver oxide cell is shown in Fig. 4. The larger the cell, the more important the cost of the materials. Thus, in flashlight battery sizes, the MnO_2/Zn system is the most commonly used. More costly systems are usually limited

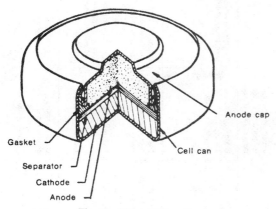

Fig. 4.   Silver oxide cell.

to special applications, such as space vehicles. In retail stores, cells labeled simply as alkaline cells are usually MnO_2/Zn cells.

Gas Electrodes. Good examples of the wide variety of systems of this type being used or being considered for future use are the air-zinc and chlorine-zinc cells. Both of these cells use a gas as the positive electrode-active material. The first is an alkaline aqueous system which uses oxygen from the air, activated on a special material, to act as the oxidant or depolarizer. Button air-cells for special applications, recently introduced to the market, provide significantly higher volumetric energy densities than some of the systems marketed earlier. The second is a mildly acidic aqueous system being considered for rechargeable batteries for electric cars and load leveling in electric power generation. Chlorine is ionized at an activating electrode on discharge and formed as a gas to be transported to and stored in another location as a solid hydrate, on charge.

Nonaqueous Batteries. These cover almost as wide a range as do the aqueous systems. They employ active metals that are incompatible with water, and thus must be water free—hence the term *nonaqueous.* A relatively few years ago, nonaqueous batteries were unknown in the marketplace, but rapid strides have been made in recent years, due to the incentives brought about by the need for better storageability, higher energy densities, better performance at very low temperatures and, in some cases, such as silver oxide, alternatives to increasing costs of materials.

The higher a metal on the electromotive (activity) series, the higher the voltage versus a given positive electrode and, therefore, the higher the Wh for the same Ah. With water-free systems, metals such as lithium, sodium, calcium, and magnesium, which are more negative than those compatible only in aqueous systems, can be used. To date, the nonaqueous systems reaching the market use lithium as the anode.[1]

---

[1] *Editor's Note*: In mid-1980, researchers at the General Motors Research Laboratories announced new findings pertaining to lithium's use in batteries. High electropositivity and low equivalent weight make lithium an ideal battery reactant, capable of supplying the specific energy needed to operate an electric vehicle. The source of the abundant energy available in lithium, however, is exactly what makes it almost impossible to manage. The challenge is to prepare alloys and find materials stable enough to contain the aggressiveness of lithium without greatly suppressing its activity. Work at GM has been related to the search for an advanced molten salt battery cell.

Specific energies greater than 180 Wh/kg, about 5 times that of the lead-acid battery, have been demonstrated by electrochemical cells utilizing LiCl-KCl electrolyte and electrodes of metal sulfide and lithium alloy. But operating temperatures of 723 K and the aggressive nature of the chemical reactants pose serious new challenges to cell construction

Positive electrode materials under consideration cover a wide range. Where the battery must be substitutable for a currently available single-cell aqueous type, a low-voltage match must be used. Examples are PbO, CuO, FeS, FeS_2 versus Li. These are all solid depolarizers. For existing or new high-voltage applications, MnO_2 and fluorinated carbon are being used as solid depolarizers, and SO_2, SO_2Cl_2 and $SOCl_2$ are being used as liquid cathodes. The latter two have the advantage of being good solvents for the needed salt and of reducing to another liquid which will also dissolve the salt. Therefore, they serve a dual role as cathode and electrolyte solvent. They are active only on a special activating electrode surface—so they do not combine directly with the lithium anode. The solid depolarizers all use organic solvents for the electrolyte. Several nonaqueous systems are being investigated for high-energy-density rechargeable batteries. Those showing promise are all thermally activated types, i.e., salt is heated to the melting point to form a liquid electrolyte. This usually means that the lithium or sodium, the usual choices for the anode, are also molten. These involve high operating temperatures, such as 300°C, and thus unique design and operating problems must be solved.

The high-temperature batteries as just described use a liquid salt for the ion transporter; no solvent is used. There is a family of cells that use salt as the sole ion transporter, but the salt is solid at operating temperatures. These are known as *solid-electrolyte systems.* Because of very poor ion mobility, they are used only for extremely low current discharge, such as constant current drain over periods of years. One application, for example, is in heart pacemakers. An example is the iodine lithium system, in which a layer of lithium iodide forming on the surface of the lithium serves as the separator and ion carrier. Water is a contaminant for nonaqueous systems and, therefore, special environments for cell assembly are required and these add significantly to cost. Also, the organic solvent electrolytes are poorer in ion transportability than the aqueous electrolytes and thus they are poorer on high current discharge.

ble batteries are described in the accompanying table. The table is abridged and included mainly to illustrate the present diversity in battery technology.

Standard Cells

Primary cells are also used as standards of electromotive force (emf) and are suitable for precise measurements, such as those made with a potentiometer which requires a constant, known emf. Over the years two principal standard cells have been used, the Clarke cell and the Weston cell. The Clarke cell has a mercury cathode and an amalgamated zinc anode. The cathode is submerged in a mercurous sulfate paste, and the electrolyte is a solution of zinc sulfate saturated at 0°C. The system is sealed in a glass tube and develops an emf of 1.440 volts at 15°C with a decrease of 0.00056 volt per °C rise in temperature. The Weston standard cell, a diagram of which is given in Fig. 5, is made in two types, the normal cell containing a saturated

materials. Of particular concern is the lithium attack upon separators and seal components. Most inorganic insulators, including the refractory oxides and nitrides, are destroyed or rendered conductive by this attack. Boron nitride, one of the more resistant materials, has been the subject of recent research. Earlier research had also indicated that silicon reduces the activity of lithium without substantially increasing its weight and produces a manageable solid at 723 K. Studies of the stability of boron nitride with Li-Si alloys of differing composition are continuing apace.

CHARACTERISTICS OF SOME COMMERICALLY AVAILABLE BATTERIES

Battery System	Basic Type	Overall Chemical Reaction	Negative Electrode	Positive Electrode	Electrolyte	Nominal Voltage per Cell	Energy Density Watt-Hours		Capacity	Input for Recharging	Advantages	Disadvantages
							Per Pound	Per Cubic Inch				
Zinc-manganese dioxide (usually called Leclanché or carbon-zinc)	Primary	$2MnO_2 + 2NH_4Cl + Zn \rightarrow ZnCl_2 \cdot 2NH_3 + H_2O + Mn_2O_3$	Zinc	Manganese dioxide (Often natural ore)	Aqueous solution of ammonium chloride, zinc chloride	1.5	5-40	1-3	Several hundred mAh to 30 Ah*	—	Low cost; variety of shapes and sizes; excellent shelf life	Efficiency decreases at high current drains; poor low-temperature performance
Zinc–zinc chloride manganese dioxide	Primary	$4Zn + 8MnO_2 + ZnCl_2 + 8H_2O + xH_2O \rightarrow 8MnOOH + ZnCl_2 \cdot 4Zn(OH)_2 \cdot xH_2O$	Zinc	Manganese dioxide (usually synthetic)	Aqueous solution of zinc chloride	1.5	15-30	1-3	Several hundred mAh to several Ah	—	Good service at high current drain, leak resistant, good low-temperature performance	Relatively expensive for low drains
Zinc-alkaline-manganese dioxide	Primary	$2Zn + 3MnO_2 \rightarrow 2ZnO + Mn_3O_4$	Zinc	Manganese dioxide (usually synthetic)	Aqueous solution of potassium hydroxide	1.5	20-40	2-3	Several hundred mAh to 23 Ah	—	High efficiency under moderate continuous drain conditions; good low-temperature performance; low impedance; long shelf life	Expensive for low drains
	Rechargeable	$Zn + 2MnO_2 \rightleftharpoons ZnO + Mn_2O_3$					10	1.0-1.2		Approximately 100% of energy withdrawn		Rechargeable-limited cycle life; voltage-limited taper current charging
Zinc-mercuric oxide	Primary	$Zn + HgO \rightarrow ZnO + Hg$	Zinc	Mercuric oxide	Aqueous solution of potassium hydroxide	1.35	10-50	4-8	16 mAh-14 Ah	—	High service capacity/volume ratio; flat voltage discharge characteristic; good high-temperature performance; good storage life	Poor low-temperature performance on some types
Zinc-silver oxide	Primary	$Zn + Ag_2O \rightarrow ZnO + 2Ag$	Zinc	Monovalent silver oxide	Aqueous solution of potassium hydroxide or sodium hydroxide	1.5	30-60	4-8	38-190 mAh	—	Good high current pulsing; flat voltage discharge	Silver is expensive; poor storageability
	Rechargeable	$Zn + Ag \rightleftharpoons ZnO + Ag$		Divalent silver oxide	Aqueous solution of potassium hydroxide	1.8/1.5 (two-step)	40-70	2-8	Vented, 5 Ah to several thousand Ah	Minimum of 110% of energy withdrawn	High-energy density; high rate capability	Expensive; two-step discharge; poor charge maintenance; limited cycle life
Lead-lead dioxide (usually called lead-acid)	Rechargeable	$2Pb + 2PbO_2 + 2H_2SO_4 + H_2O \rightleftharpoons PbSO_4 + 2PbO + 3H_2O$	Lead	Lead dioxide	Aqueous solution of sulfuric acid	2	Sealed, 10-15 Vented, 7-12	0.8-1.1 0.5-2	Vented, 1-10,000 Ah	Minimum of 110% of energy withdrawn	Spill-resistant Low cost	Limited low-temperature performance; vented cells require servicing
Lithium-iron sulfide	Primary solid cathode	$FeS_2 + 4Li \rightarrow Fe + 2Li_2S$	Lithium	Iron sulfide	Nonaqueous solution of lithium salts in ethers	1.0	—	—	38 mAh to 120 mAh	—	Good storageability; lower cost substitute for silver cells.	Limited high rate capabilities

Type	Type	Reaction	Negative electrode	Positive electrode	Electrolyte							Advantages	Disadvantages
Sulfur dioxide–lithium	Primary liquid cathode	$2SO_2 + 2Li \rightarrow Li_2S_2O_4$	Lithium	Sulfur dioxide	Nonaqueous solution of lithium bromide in mixture of SO_2 and acetonitrile	3.0	—	—	—	—		High energy and power densities; good storageability; good at low temperatures	Complex manufacturing facilities required
Nickel-cadmium	Rechargeable	$Cd + 2NiOOH + KOH + 2H_2O \rightleftarrows$ $Cd(OH)_2 + 2Ni(OH)_2 + KOH$	Cadmium	Nickelic hydroxide	Aqueous solution of potassium hydroxide	1.25	Sealed, 12–17	Sealed, 1–1.5	Sealed, 20 mAh–100 Ah	Sealed, minimum of 140% of energy withdrawn		Excellent cycle life; flat voltage discharge characteristic; good high- and low-temperature performance; high resistance to shock and vibration; can be stored indefinitely in any charge state	High initial cost; only fair charge retention
							Vented, 12–20	Vented, 1–1.5	Vented—few Ah to over 500 Ah	Vented, minimum of 125–150% of energy withdrawn			
Silver-cadmium	Rechargeable	$Cd + AgO + KOH \cdot H_2O \rightleftarrows$ $Cd(OH)_2 + Ag + KOH$	Cadmium	Divalent silver oxide	Aqueous solution of potassium hydroxide	1.4	22–34	1.8–2.5	Sealed, up to 300 Ah	Minimum of 110% of energy withdrawn		Good energy/weight ratio; good charge retention; long wet-stand life	Expensive; poor low-temperature performance; two step discharge curve
Zinc-air (oxygen)	Primary	$2Zn + O_2 + 4KOH + 2H_2O \rightarrow 2K_2Zn(OH)_4$ (large cells) $2Zn + O_2 \rightarrow 2ZnO$ (small cells)	Zinc	Oxygen	Aqueous solution of potassium hydroxide	1.25	80–100	3.2	Vented, ½–2,000 Ah	—		Flat voltage discharge characteristic; high input per unit volume in small cells	Drying out; carbonation

• Ah = ampere-hours; mAh = milliampere-hours   Source: Union Carbide Corporation.

The two electrodes are termed negative and positive here to avoid confusion which can result from use of the terms anode and cathode, the latter often being used loosely in the battery industry. The negative electrode of the primary (or charged secondary) cell is metallic and is oxidized (increased in valence) during discharge, giving up electrons to the external circuit. The positive electrode in an aqueous system initially is an oxygen donor, usually a metal oxide, which is reduced as it receives electrons from the external circuit. Charge transfer from one electrode to the other within the cell is via the ions of the electrolyte salt. In certain cells, the electrolyte enters further into reactions and actually changes in composition during use.

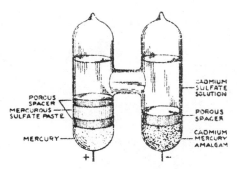

POROUS SPACER
MERCUROUS SULFATE PASTE
MERCURY

+

CADMIUM SULFATE SOLUTION
POROUS SPACER
CADMIUM MERCURY AMALGAM

−

Fig. 5.   Weston cadmium standard cell.

cadmium sulfate solution and another type which is used as a working standard in which the solution is less than saturated above 4°C. The saturated cell is the basic standard for maintaining the value of the volt, and is used in this manner in national standards laboratories. Its rather high temperature coefficient must always be taken into account. The legal voltage of the standard-cell group in the United States is 1.018636 volts absolute at 20°C and this value is accepted by international agreement.

The emf of the unsaturated Weston cell is within 1.0188 to 1.0198 volts absolute, with the exact voltage of each cell being established by comparison with the normal or saturated cell. The unsaturated cell is a useful working standard because of its negligible temperature coefficient.

References

Fleischer, A., and J. J. Lander, Editors: "Zinc-Silver Oxide Batteries," Wiley, New York (1971).
Heise, G. H., and N. C. Cahoon, Editors: "The Primary Battery," Vol. 1, Wiley, New York (1971).
Kordesch, K. V., Editor: "Batteries," Vol. 1 (Manganese Dioxide), Dekker, New York (1974).
Murphy, D. W., and P. A. Christian: "Solid State Electrodes for High Energy Batteries," *Science*, **205**, 651–656 (1979).
Robinson, A. L.: "Advanced Storage Batteries," *Science*, **192**, 541–543 (1976).
Weinstein, J. N., and F. B. Leitz: "Electric Power from Differences in Salinity; The Dialytic Battery," *Science*, **191**, 557–559 (1976).
Falk, S. O., and A. J. Salkind: "Alkaline Storage Batteries," Wiley, New York (1969).

—L. F. Urry, Battery Products Division, Union Carbide Corporation, Parma, Ohio.

BAUMÉ SCALE.   Specific Gravity.

BAUXITE.   There are two main types of bauxite ores used as the primary sources for aluminum metal and aluminum chemicals: $Al(OH)_3$ (gibbsite) and $AlO(OH)$ (boehmite). Thus, bauxite is a term for a family of ores rather than a substance of one definite composition. The first bauxite ore was found near Les Baux in the south of France by P. Berthier (1821). Deposits are found worldwide except in Antarctica. Secondary sources of aluminum include a large array of clays that are rich in alumina (30–40%) found in large abundance throughout the world. Bauxite ores range in color from white to dark red or brown, largely depending upon the iron content. An average composition of the ores used by industry today would be: Al_2O_3, 35–60%; SiO_2, 1–15%; Fe_2O_3, 5–40%; TiO_2, 1–4%, H_2O; 10–35%; other substances, 0–2%.

Although developed as early as 1888 by the Austrian chemist, Karl Josef Bayer, the Bayer process still is used almost exclusively for the extraction of alumina from ores. The bauxite first

is reacted under pressure with hot caustic which dissolves the $Al_2O_3 \cdot xH_2O$ to form sodium aluminate. The solution is filtered hot, then cooled and agitated with the addition of a small quantity of aluminum hydrate to enhance the precipitation of the crystalline hydrate. After filtration, the cake is kiln-dried at 1100°C to remove H_2O and yield Al_2O_3.

The purity of aluminum produced by the electrolytic process (see **Aluminum**) is determined mainly by the purity of the Al_2O_3 used. Thus, commercial grades of Al_2O_3 are 99–99.5% pure with traces of H_2O, SiO_2, Fe_2O_3, TiO_2, ZnO, and very minute quantities of other metal oxides.

As of the mid-1980s, there are nine alumina plants in the United States and territories. Six plants are located on the Gulf Coast; two plants are in Arkansas; and one plant is in the Virgin Islands. Total capacity is estimated at 7.7 million short tons (approximately 6.9 million metric tons). The aluminum industry in the United States has depended on imports from the Caribean, South America, and Australia. Carribean exporting countries have high levies on bauxite exports and, politically, have moved to nationalize and expropriate the bauxite mines developed by U.S. industry. It is doubtful that new Bayer alumina plants will be built in the United States. The technology in the United States is focusing on extracting alumina from abundant alumina-containing clays. New Bayer plants are being built in Australia and Venezuela.

Uses and Grades of Alumina. Although aluminum production is a major consumer of alumina, the compound is used widely elsewhere. The properties of alumina and hydrated aluminas may be varied, ranging from a talc-like softness to the hardness of a ruby or sapphire. Some of the uses for alumina include water purification, glassmaking, production of steel alloys, waterproofing of textiles, coatings for ceramics, abrasives and refractory materials, cosmetics, and electronics. *Hydrated aluminas* may be represented: α-$Al_2O_3 \cdot 3H_2O$ or α-$Al(OH)_3$. The compounds are dry, snow-white, free-flowing crystalline powders and may be obtained in a wide range of particle sizes. The compounds are widely used in the production of aluminum salts because of their reactions with strong acids and alkalies. Some of the salts prepared in this manner include aluminum chloride, aluminum phosphate, and aluminum sulfate. *Activated aluminas* are very porous aluminum oxide. γ-Al_2O_3 is made by heating the hydrate to drive off nearly all of the combined water. The final products are granules or fine powders with a large surface area to provide absorptive capacity per unit volume. Applications of the aluminas are enhanced by virtue of their inertness chemically and nontoxic qualities. They are extensively used for drying gases and for dehydrating liquids, such as alcohol, benzol, carbon tetrachloride, ethyl acetate, gasoline, toluol, and vegetable and animal oils. They also are used as filter aids in the manufacture of lubricating and other oil products. Their large surface area qualifies the aluminas as catalysts for numerous reactions. The compounds also find extensive application in ceramics, particularly in abrasive and cutting wheels, polishing compounds, additives to glass, tank linings, spark plugs, electrical substrates, and linings for high-temperature furnaces. *Corundum* is an aluminum oxide that possesses a hexagonal crystal structure. The compound is extremely hard (2000 on the Knoop scale), sp gr 3.95, and is widely used in abrasives and refractories. Corundum is manufactured by fusing alumina or bauxite in an electric arc furnace operated at about 2200°C.

See also **Aluminum; Corundum;** and terms listed under **Mineralogy.**

—S. J. Sansonetti, Consultant, Reynolds Metals Company, Richmond, Virginia.

BECKE TEST. A microscope of moderate or high magnification is used to compare the indices of refraction of two contiguous minerals (or of a mineral and a mounting medium or immersion liquid), in a thin section or other mount. When the two substances differ substantially in refractive index, they are separated by a bright line called the *Becke line*. The line moves toward the least refractive of the two materials when the tube of the microscope is lowered.

BECKMANN METHOD. A procedure for measuring elevation of the boiling point or depression of the freezing point of a solution. It may be used to measure concentration if the nature of the solute is known, or the molecular weight of the solute if the volume concentration is known.

BÉNARD CONVECTION CELLS. When a layer of liquid is heated from below, the onset of convection is marked by the appearance of a regular array of hexagonal cells, the liquid rising in the center and falling near the wall of each cell. The criterion for the appearance of the cells is that the Rayleigh number should exceed 1700 (for rigid boundaries).

BENEDICT SOLUTION. In its original, classical form, this was an alkaline solution of copper hydroxide and sodium citrate in sodium carbonate used either as a mild oxidizing agent or as a test for easily oxidizable groups such as aldehyde groups. The formation of cuprous oxide is a positive test, its color red, but often yellow at first. Many other forms of this solution have been developed. Glucose reacts with Benedict solution to form cuprous oxide.

BENTONITE. The term applied to altered fine-grained volcanic ashes which have been blown considerable distance from their origin and deposited in marine waters. The resulting material is usually a white, but sometimes a colored, clay-like sediment which may contain bits of volcanic glass but is composed mainly of colloidal silica which will absorb large quantities of water. Since bentonites are wind-blown deposits they are useful as definite datum planes in stratigraphy, especially in helping to determine the contemporaneity of the different facies of marine sediments.

BENZALDEHYDE. C_6H_5CHO, formula wt 106.12, colorless liquid, mp $-26°C$, bp $179°C$, sp gr 1.046. Sometimes referred to as artificial almond oil or *oil of bitter almonds*, benzaldehyde has a characteristic nutlike odor. The compound is slightly soluble in H_2O, but is miscible in all proportions with alcohol or ether. Upon standing in air, benzaldehyde oxidizes readily to benzoic acid. Commercially, benzaldehyde may be produced by (1) heating benzal chloride $C_6H_5CHCl_2$ with calcium hydroxide; (2) heating calcium benzoate and calcium formate:

$$(C_6H_5COO)_2Ca + (HCOO)_2Ca \rightarrow 2C_6H_5CHO + 2CaCO_3;$$

or (3) boiling glucoside amygdalin of bitter almonds with a dilute acid.

Benzaldehyde reacts with many chemicals in a marked manner, (1) with ammonio-silver nitrate ("Tollen's solution") to form metallic silver, either as a black precipitate or as an adherent mirror film on glass, but does not reduce alkaline cupric solution ("Fehling's solution"), (2) with rosaniline (fuchsine, magenta), which has been decolorized by sulfurous acid ("Schiff's solution"), the pink color of rosaniline is restored, (3) with NaOH solution, yields benzyl alcohol and sodium benzoate, (4) with NH_4OH, yields tribenzaldeamine (hydrobenzam-

ide, $(C_6H_5CH)_3N_2)$, white solid, mp $101°C$, (5) with aniline, yields benzylideneaniline ("Schiff's base" $C_6H_5CH:NC_6H_5)$, (6) with sodium cyanide in alcohol, yields benzoin $C_6H_5 \cdot CHOHCOC_6H_5$, white solid, mp $133°C$, (7) with hydroxylamine hydrochloride, yields benzaldoximes $C_6H_5CH:NOH$, white solids, antioxime, mp $35°C$, syn-oxime, mp $130°C$, (8) with phenylhydrazine, yields benzaldehyde phenylhydrazone $C_6H_5CH:NNHC_6H_5$, pink solid, mp $156°C$, (9) with concentrated HNO_3, yields metanitrobenzaldehyde $NO_2 \cdot C_6H_4CHO$, white solid, mp $58°C$, (10) with concentrated H_2SO_4 yields metabenzaldehyde sulfonic acid $C_6H_4CHO(SO_3H)(2)$, (11) with anhydrous sodium acetate and acetic anhydride at $180°C$, yields sodium cinnamate C_6H_5COONa, (12) with sodium hydrogen sulfite, forms benzaldehyde sodium bisulfite $C_6H_5CHOHSO_3Na$, white solid, from which benzaldehyde is readily recoverable by treatment with sodium carbonate solution, (13) with acetaldehyde made slightly alkaline with NaOH, yields cinnamic aldehyde $C_6H_5CH:CHCHO$, (14) with phosphorus petachloride, yields benzylidine chloride $C_6H_5CHCl_2$.

Benzaldehyde may be detected by the appearance of a blue color on treating with acenaphthene and H_2SO_4, followed by heating. Benzaldehyde is used (1) as a flavoring material, (2) in the production of cinnamic acid, (3) in the manufacture of malachite green dye.

BENZENE. C_6H_6, formula wt 78.11, colorless, highly flammable liquid that burns with a smoky flame, mp. $5.5°C$, bp $80.1°C$, sp gr 0.879 (20° referred to H_2O at 4°C). Sometimes called *benzol, phenyl hydride*, or *cyclohexatriene*, benzene is practically insoluble in H_2O (0.07 part in 100 parts at 22°C); and fully miscible with alcohol, ether, and numerous organic liquids. Benzene is of much importance industrially, mainly as a starting ingredient for many reactions and as a solvent. The compound also is of much theoretical interest, being the simplest hydrocarbon of the aromatic group. Forming an explosive mixture with air, benzene can be used as a fuel or fuel component for internal combustion engines. Benzene is the first member of a homologous series of compounds, C_nH_{2n-6}. Methylbenzene or toluene, $C_6H_5 \cdot CH_3$, is the only homolog with the formula C_7H_8. Next in order of the homologous series, C_8H_{10}, exists in four isomeric forms, namely ethylbenzene $C_6H_5 \cdot C_2H_5$ and ortho-, meta-, and paradimethylbenzene, $C_6H_4(CH_3)_2$. Of the formula C_9H_{12}, eight isomerides are possible. Possible isomerides increase rapidly as the carbon count increases.

Many of the C_nH_{2n-6} series of hydrocarbons occur in coal gas and coal tar, from which they may be extracted. These were the early sources for benzene and related compounds. Much of the benzene currently is produced from petroleum sources. The separation of benzene from coal tar is difficult because of the presence of scores of isomerides with close boiling points. In one process for making high-purity benzene (99.94% or higher) from coke-oven light oil (the cut boiling between 60–150°C), the light oil and a stream of hydrogen are heated to reaction temperature and passed through fixed-bed reactors that contain a proprietary catalyst. In this reaction, the nonaromatics present are converted to light hydrocarbon gases. Sulfur compounds present are converted to H_2S. Some dealkylation of the higher aromatics present also produces benzene in addition to that contained in the feedstock. Vapors from the reactor are cooled and passed to a stabilizer tower where dissolved H_2S and light hydrocarbons (with boiling points lower than benzene) are removed. The bottoms from the stabilizer containing benzene, toluene, and xylene, are clay-treated. Then follows a series of fractionations to produce benzene, toluene, and xylene, in addition to higher-boiling hydrocarbons. If a portion of the hydrocarbons in the product fuel gas is reformed, no external hydrogen is required.

The synthetic production of benzene generally involves the dealkylation of toluene. In one noncatalytic process, a hydrogen-rich gas is mixed with liquid toluene feed and preheated prior to charging to the reactor. Toluene reacts with the hydrogen to form benzene and methane. The reaction is exothermic. Operating conditions approximate 34–68 atm and 595–760°C. The process provides about 98% yield of benzene. The toluene is recycled.

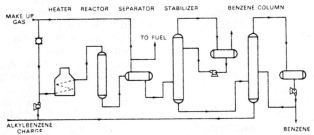

Process for converting toluene or C_8 aromatics (catalytic hydrodealkylation) to high-purity benzene. (*UOP Process Division.*)

In a catalytic dealkylation process, shown in the accompanying figure, toluene or C_8 aromatics (alkylbenzenes) are fed to a reactor, together with a hydrogen-containing gas. The hydrogen source is not critical and may be manufactured hydrogen or off-gas from a reforming or other refining unit. Effluent from the reactor, after cooling, is charged to a separator, from which hydrogen is removed and recycled to the reactor. Liquid phase from the separator is stripped of hydrocarbons (boiling lower than benzene) in a stabilizing column. One further fractionating step yields product benzene overhead. The bottoms from the tower are recycled to the reactor for dealkylation. Yields of 98% of theoretical are claimed.

In another process, mixtures of aromatics and nonaromatic hydrocarbons comprise the charge. In a first step, aromatics are continuously extracted from the feed by using an aqueous solution of *N*-methylpyrollidone. A multistage countercurrent extraction tower is used. The operation is carried out at modest temperatures and pressures. The rich aromatic extract phase then proceeds to a stripper, where pentane and a part of the benzene are removed overhead and recycled to the extractor. The bottoms from the stripper are free of nonaromatics and enter a second stripper for further separation. The distillate from the second stripper contains aromaticfree solvent, which is returned to the extractor. One or more further fractionations yield benzene, toluene, and xylenes of desired specification. A typical feedstock may contain an aromatics mixture of about the following ranges: benzene, 26–60%; toluene, 14–22%, xylenes plus ethylbenzene, 15–50%. A similar process uses dimethyl sulfoxide as a solvent.

Styrene is a major consumer of benzene, followed by the production of cyclohexane. At one time, phenol production was the second largest consumer of benzene. Benzene and cyclohexene are closely related economically because cyclohexane can be produced by reacting benzene with hydrogen. Although the foregoing represent the major tonnage uses of benzene, the compound is critically important to the production of hundreds of other compounds. The halogen derivatives of benzene are particularly important. See also **Chlorinated Organics.** These

include chlorobenzene, C_6H_5Cl; bromobenzene, C_6H_5Br; benzal chloride, $C_6H_5 \cdot CHCl_2$; benzyl chloride, $C_6H_5 \cdot CH_2Cl$; and benzotrichloride, $C_6H_5 \cdot CCl_3$. Important nitro derivatives of benzene include nitrobenzene, $C_6H_5 \cdot NO_2$; and metadinitrobenzene, $C_6H_4(NO_2)_2$. Amino compounds of large importance derived from benzene include aminobenzene, $C_6H_5 \cdot NH_2$ (aniline); diaminobenzene, $C_6H_4(NH_2)_2$; and triaminobenzene, $C_6H_3(NH_2)_3$.

Phenol, C_6H_5OH, is hydroxybenzene. Resorcinol, catechol, and quinol, $C_6H_4(OH)_2$, may be considered to be dihydroxybenzenes. Pyrogallol and phloroglucinol, $C_6H_3(OH)_3$, may be considered as trihydroxybenzenes. The benzene-related alcohols, aldehydes, and ketones include benzyl alcohol, $C_6H_5 \cdot CH_2 \cdot OH$; benzaldehyde, $C_6H_5 \cdot CHO$; benzoin, $C_6H_5 \cdot CO \cdot C_6H_5$; salicylaldehyde, $C_6H_4(OH) \cdot CHO$; anisaldehyde, $C_6H_4(OCH_3) \cdot CHO$; acetophenone, $C_6H_5 \cdot CO \cdot CH_3$; benzophenone, $C_6H_5 \cdot CO \cdot C_6H_5$; and quinone, $C_6H_4O_2$. The benzene-related acids and salts include benzoic acid, $C_6H_5 \cdot COOH$; ethyl benzoate, $C_6H_5 \cdot COOC_2H_5$; benzoyl chloride, $C_6H_5 \cdot COCl$; benzoic anhydride, $(C_6H_4 \cdot CO)_2O$; benzamide, $C_6H_5 \cdot CO \cdot NH_2$; benzonitrile, $C_6H_5 \cdot CN$; anthranilic acid, $C_6H_4(NH_2) \cdot COOH$; phthalic acid, $C_6H_4(COOH)_2$; phthalic anhydride, $C_6H_4(CO)_2O$; phthalimide, $C_6H_4(CO)_2NH$; isophthalic acid, $C_6H_4(COOH)_2$; terephthalic acid, $C_6H_4(COOH)_2$; benzenehexacarboxylic acid, $C_6H_4(COOH)_6$; and phenylacetic acid, $C_6H_5 \cdot CH_2 \cdot COOH$.

Benzene has long been known to consist of six CH groups connected by valence bonds between the carbon atoms into the form of a hexagon. The precise nature of this bonding has been studied for generations. The numbering system used for benzene compounds is given in the article on **Organic Chemistry.**

Benzene reacts (1) with chlorine, to form (a) substitution products (one-half of the chlorine forms hydrogen chloride) such as chlorobenzene, C_6H_5Cl; dichlorobenzene, $C_6H_4Cl_2(1,4)$; and (1,2), trichlorobenzene, $C_6H_3Cl_3(1,2,4)$; tetrachlorobenzene (1,2,3,5); and (b) addition products, such as benzene dichloride $C_6H_6Cl_2$; benzene tetrachloride, $C_6H_6Cl_4$; and benzene hexachloride, $C_6H_6Cl_6$. The formation of substitution products of the benzene nucleus, whether in benzene or its homologues is favored by the presence of a catalyzer, e.g., iodine, phosphorus, iron; (2) with concentrated HNO_3, to form nitrobenzene, $C_6H_5NO_2$; 1,3-dinitrobenzene, $C_6H_4(NO_2)_2(1,3)$, 1,3,5-trinitrobenzene, $C_6H_3(NO_2)_3(1,3,5)$; (3) with concentrated H_2SO_4, to form benzene sulfonic acid, $C_6H_5SO_3H$, benzene disulfonic acid, $C_6H_4(SO_3H)_2(1,3)$, benzene trisulfonic acid, $C_6H_3(SO_3H)_3(1,3-5)$; (4) with methyl chloride plus anhydrous aluminum chloride (Friedel-Crafts reaction) to form toluene, monomethyl benzene, $C_6H_5CH_3$; dimethyl benzene $C_6H_4(CH_3)_2$; trimethyl benzene $C_6H_3(CH_3)_3$; (5) with acetyl chloride plus anhydrous aluminum chloride (Friedel-Crafts reaction) to form acetophenone (methylphenyl ketone), $C_6H_5COCH_3$; (6) with hydrogen in the presence of a catalyzer, e.g., finely divided nickel, heated, to form dihydrobenzene (cyclohexadiene), (1,3) C_6H_8; a cyclic diolefin hydrocarbon, tetrahydrobenzene, cyclohexene, C_6H_{10}; a cyclic mono-olefin, hexhydrobenzene (cyclohexane), C_6H_{12}, a cycloparaffin; (7) with ozone, to form benzene triozonide, $C_6H_6(O)_3)_3$. See also **Organic Chemistry.**

Although additional research is indicated, benzene has been implicated as a causative factor in *marrow aplasia* and some forms of *leukemia* (Forni and Vigliani, "Chemical Leukemogenesis in Man," *Series Haematologica*, 7, 2, 214, 1974). Benzene also has been identified as a cause of *neutropenia.* See also **Carcinogens.**

BENZENE HEXACHLORIDE (BHC). A commercial mixture of isomers of 1,2,3,4,5,6-hexachlorocyclohexane. Once used extensively as an insecticide and fumigant, but now subject to regulations in a number of countries. The mixture is effective in the control of chiggers, ticks, fleas, cockroaches, lice, and *Acarus scabiei*. The mixture is sometimes called *gammexane*; more commonly called *lindane*. Quite toxic.

BENZIDINE REARRANGEMENT. Rearrangement (Organic Chemistry).

BENZIL REARRANGEMENT. Rearrangement (Organic Chemistry).

BENZINE. A product of petroleum boiling between 120° and 150°F (49 and 66°C) and composed of aliphatic hydrocarbons. Not to be confused with benzene, previously described.

BENZOIC ACID. $C_6H_5 \cdot COOH$, formula wt 122.12, white crystalline solid, mp 121.7°C, bp 249.2°C, sublimes readily at 100°C and is volatile in steam, sp gr 1.266. Sometimes referred to as phenylformic acid, the compound is insoluble in cold H_2O, but readily soluble in hot H_2O, or in alcohol or ether. Commercially, benzoic acid finds major use as a starting or intermediate material in various organic syntheses, notably in the preparation of terephthalic acid, a high-production chemical. See **Terephthalic Acid.** Benzoic acid forms benzoates, e.g., sodium benzoate or calcium benzoate which, when heated with calcium oxide, yield benzene and calcium. With phosphorus trichloride, benzoic acid forms benzoyl chloride C_6H_5COCl, an important agent for the transfer of the benzoyl group ($C_6H_5CO—$). Benzoic acid reacts with chlorine to form *m*-chlorobenzoic acid and reacts with HNO_3 to form *m*-nitrobenzoic acid. Benzoic acid forms a number of industrially useful esters, including methyl benzoate, ethyl benzoate, glycol dibenzoate, and glyceryl tribenzoate.

Although benzoic acid occurs naturally in some substances, such as gum benzoin, dragon's blood resin, Peru and Tolu balsams, cranberries, and the urine of the ox and horse, the product is made on a large scale by synthesis from other materials. Benzoic acid can be prepared from toluene and air in a process that takes place in the liquid-phase in a continuous oxidation reactor operated at moderate pressure and temperature: $C_6H_5 \cdot CH_3 + 1\frac{1}{2}O_2 \rightarrow C_6H_5 \cdot COOH + 2H_2O$. The acid also can be obtained as a by-product of the manufacture of benzaldehyde from benzal chloride or benzyl chloride.

See also **Antimicrobial Agents (Foods).**

BENZOYL CHLORIDE. Chlorinated Organics.

BENZYL ALCOHOL. $C_6H_5CH_2OH$, formula wt 108.13, water-white liquid with slight odor, sharp burning taste, mp −15.3°C, bp 204.7°C, sp gr 1.043. Also known as alpha-hydroxytoluene, phenylmethanol, phenylcarbinol, benzyl alcohol is somewhat soluble in water; miscible with alcohol, ether, and chloroform. Benzyl alcohol is used in perfumes and flavors, dyes, textiles, and sheet plastics manufacture. It is generally useful as a solvent.

BENZYL BENZOATE. A water-white liquid with formula $C_6H_5CH_2OOCC_6H_5$. Sharp, burning taste with faint aromatic odor. Supercools easily. Insoluble in water and glycerin; soluble in alcohol, chloroform, and ether. Specific gravity 1.116–1.120 (25/25°C); bp 325°C; mp 18.8°C; flash point ~150°C. This compound is produced by the Cannizzaro reaction from benzaldehyde; or by esterifying benzyl alcohol with benzoic acid; or by treating sodium benzoate with benzyl chloride. It is purified by distillation and crystallization. Benzyl benzoate is used as a fixative and solvent for musk in perfumes and flavors, as a plasticizer, miticide, and in some external medications. The compound has been found effective in the treatment of *scabies* and *pediculosis capitis* (head lice, *Pediculus humanus* var. *capitis*).

BENZYL CHLORIDE. Chlorinated Organics.

BENZYNE. The concept of benzyne intermediates has largely stemmed from work by Dr. Georg Wittig of the University of Heidelberg, Germany, and Dr. J. D. Roberts of the California Institute of Technology. Dr. Roberts postulated that benzynes form when a substituted benzene (such as bromobenzene) reacts with a nucleophilic reagent, such as potassium amide in liquid ammonia. He and other workers have shown that strong nucleophiles add readily to arynes. If the nucleophile is attached to a side chain on the aryne, a new ring fused to the original aromatic nucleus forms by intramolecular addition. This was shown by Dr. Bunnett and Dr. B. F. Hrutfiord and also by Dr. R. Huisgen and co-workers of the University of Munich, Germany. In this work, heterocyclic compounds were usually obtained. However, in later work Dr. J. F. Bunnett and Dr. J. A. Skorez, then at Brown University, used the synthesis to obtain homocyclic ring closures, producing derivatives of such compounds as benzocyclobutene, indane, tetralin and benzocycloheptane. Their type reaction may be written as

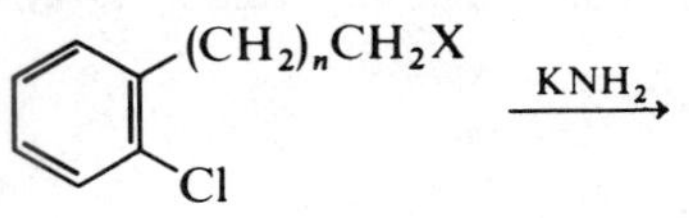

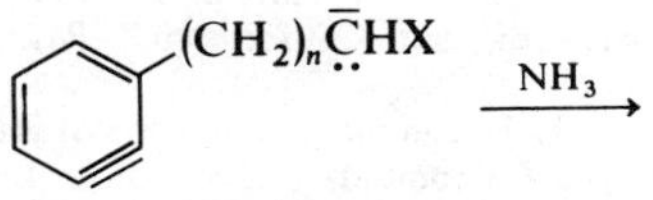

Intermediate benzyne compound

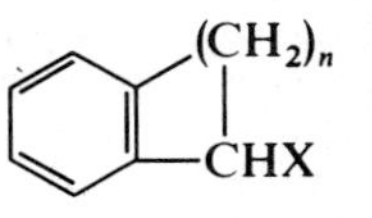

where $n = 1, 2, 3,$ or 4 $X =$ cyano, acyl, carbethoxy or sulfonyl.

Dr. Lester Friedman and Francis M. Logullo prepared substituted benzynes by diazotizing substituted anthranilic acid. This is a mild, room-temperature reaction which permits simultaneous reactions of the benzynes with suitable acceptors to prepare halogen, —NO$_2$, —CH$_3$, and —OCH$_3$ derivatives. They have also prepared new heterocyclic arynes, such as 3-pyridyne from 3-amino-isonicotinic acid.

BERKELIUM. Chemical element symbol Bk, at. no. 97, at. wt. 247 (mass number of the most stable isotope), radioactive metal of the actinide series, also one of the transuranium elements. All isotopes are radioactive; all must be produced synthetically. The element was discovered by G. T. Seaborg and associates at the Metallurgical Laboratory of the University of Chicago in 1949. At that time, the element was produced by bombarding ^{241}Am with helium ions. ^{247}Bk is an alpha emitter and may be obtained by alpha bombardment of ^{244}Cm, ^{245}Cm, or ^{246}Cm. Other nuclides include those of mass numbers 243–246 and 248–250. The probable electronic configuration is

$$1s^2 2s^2 2p^6 3s^2 3p^6 3d^{10} 4s^2 4p^6 4d^{10} 4f^{14} 5s^2 5p^6 5d^{10} 5f^9 6s^2$$
$$6p^6 7s^2.$$

Ionic radius Bk^{3+} 0.99 Å. Longest-lived isotope, ^{247}Bk($t_{1/2}$ = 7000 years).

Berkelium is known to exist in aqueous solution in two oxidation states, the (III) and the (IV) states, and the ionic species presumably correspond to Bk^{3+} and Bk^{4+}. The oxidation potential for the berkelium(III)-berkelium(IV) couple is about −1.6V on the hydrogen scale (hydrogen-hydrogen ion couple taken as zero).

The solubility properties of berkelium in its two oxidation states are entirely analogous to those of the actinide and lanthanide elements in the corresponding oxidation states. Thus in the tripositive state such compounds as the fluoride and the oxalate are insoluble in acid solution, and the tetrapositive state has such insoluble compounds as the iodate and phosphate in acid solution. The nitrate, sulfate, halides, perchlorate, and sulfide of both oxidation states are soluble.

The first compound of berkelium of proven molecular structure was isolated in 1962 by Cunningham and Wallman. A small quantity (0.004 microgram) of berkelium (as berkelium-249) dioxide was used to determine structure by x-ray diffraction.

References

Ghiorso, A., Thompson, S. G., Higgins, G. H., and G. T. Seaborg: "Transplutonium Elements in Thermonuclear Test Debris," *Phys. Rev.*, 102, 1, 180–182 (1956).

Hulet, E. K., Thompson, S. G., Ghiorso, A., and K. Street, Jr.: "New Isotopes of Berkelium and Californium," *Phys. Rev.*, 84, 2, 366–367 (1951).

Peterson, J. R., and B. B. Cunningham: "Crystal Structures and Lattice Parameters of the Compounds of Berkelium: I. Berkelium Dioxide and Cubic Berkelium Sesquioxide," *Inorg. Nucl. Chem. Lett.*, 3, 9, 327–336 (1967).

Samhoun, K., and F. David: "Radiopolarography of Am., Cm, Bk, Cf, Es and Fm," Proceedings of the *4th International Transplutonium Element Symposium*, Baden Baden, W. Germany (1975).

Seaborg, G. T., Editor: "Transuranium Elements," Dowden, Hutchinson & Ross, Stroudsburg, Pennsylvania (1978).

Thompson, S. G., Ghiorso, A., and G. T. Seaborg: "Element 97," *Phys. Rev.*, 77, 6, 838–839 (1950).

Thompson, S. G., and M. L. Muga: "Methods of Production and Research on Transcurium Elements," *Proceedings of the Second United Nations International Conference on Peaceful Uses of Atomic Energy*, Geneva, Switzerland (1958).

BERTHELOT EQUATION. A form of the equation of state, relating the pressure, volume, and temperature of a gas, and the gas constant R. The Berthelot equation is derived from the Clausius equation and is of the form

$$PV = RT\left(1 + \frac{9PT_c}{128P_c T}\left[1 - 6\frac{T_c^2}{T^2}\right]\right)$$

in which P is the pressure, V is the volume, T is the absolute temperature, R is the gas constant, T_c is the critical temperature, and P_c the critical pressure.

BERTHOLLIDE COMPOUNDS. Chemical Composition; Compound (Chemical).

BERYL. The mineral beryl is a silicate of beryllium (glucinium) and aluminum corresponding to the formula Be$_3$Al$_2$Si$_6$O$_{18}$. Crystalizing in the hexagonal system the 6-sided prisms of beryl may be very small or range up to several feet in length and a yard or so in diameter. Terminated crystals are relatively rare. Its fracture is conchoidal; hardness, 7.5–8; specific gravity, 2.6–2.9; colors, emerald green, green, blue-green, blue, yellow, red, white and colorless; luster, vitreous; transparent to translucent.

Beryl has long been used as a gem, the emeralds being a rich green variety, colored probably by minute amounts of some chromium compound. A beautiful bluish sort is called aquamarine; morganite is pink, and the golden beryl is a clear bright yellow. Other shades like honey yellow and yellowish-green are common. Metallic beryllium is obtained from beryl. Its lightness and strength make it very valuable for industrial purposes.

Beryl is found in granite rocks and especially in pegmatites, but it occurs also in mica schists in the Urals. In addition to the many European localities, including Austria, Germany, and Ireland, beryls of gem quality are found in Africa, the Malagasy Republic (especially for morganite), and Brazil. The most famous place in the world for emeralds is at Muso, Colombia, South America, where they form a unique occurrence in limestones. Emeralds are also obtained in the Transvaal and near Mursinsk, in Siberia. In the United States, New England has furnished much beryl from its pegmatites, and for a long time the huge crystals from Acworth and Grafton, New Hampshire, were the largest known. Later, however, giant crystals even larger than those from New Hampshire were discovered in Albany, Maine, the largest of which was 18 feet by 4 feet, and weighed about 18 tons. Other localities are Paris and, elsewhere, in Oxford County, Maine; Royalston, Massachusetts; North Carolina; Colorado; South Dakota; and California. See also terms listed under **Mineralogy**.

—Elmer B. Rowley, Union College, Schenectady, N.Y.

BERYLLIUM. Chemical element symbol Be, at. no. 4, at. wt. 9.0122, periodic table group 2a, mp 1287–1292 ± 3°C, bp 2970 ± 5°C. The vapor pressure at the melting point calculates to be 55 N/m^2 from the equations for the vapor pressure:

Solid log $P_{(bar)}$ = 6.266 + 1.473 × 10^{-4}T − 16,950T^{-1} (1)
Liquid log $P_{(bar)}$ = 6.578 − 11,860T^{-1} (2)

Considerable variation exists in the reported specific heat data. The following appear to be representative:

TEMPERATURE °C	SPECIFIC HEAT kJ/(kg · K)
−13	1630
+25	1970
100	2130
200	2340
800	1970

Similar scatter exists in the reported thermal conductivity data. An average value lies around 125 kW/(m·K).

Density of beryllium is 1.847 g/cm³ (based upon average values of lattice parameters at 25°C (a = 22.856 nm and c = 35.832 nm). Beryllium products generally have a density around 1.850 g/cm³ or higher because of impurities, such as aluminum and other metals, and beryllium oxide. The crystal structure is close-packed hexagonal. The alpha-form of beryllium transforms to a body-centered cubic structure at a temperature very close to the melting point.

First ionization potential 9.32 eV; second, 18.4 eV. Oxidation potentials $Be \rightarrow Be^{2+} + 2e^-$, 1.70 V; $2 Be + 6 OH^- \rightarrow Be_2O_3^{2-} + 3 H_2O + 4e^-$, 2.28 V.

All naturally occurring beryllium compounds are made up of the ^9Be isotope. Artificially produced isotopes occur during some nuclear reactor operations and include ^6Be, ^7Be, ^8Be, and ^{10}Be.

The thermal neutron absorption cross section is 0.0090 barns/atom.

The electrical conductivity of beryllium is dependent upon both temperature and metal purity. It varies at room temperature between 38 and 42% International Annealed Copper Standard. Electrical resistivity of 4.266×10^{-8} ohm·m at 25°C has been reported.

Background. In 1797, Vauquelin discovered beryllium to be a constituent of the minerals beryl and emerald. Soluble compounds of the new element tasted sweet, so it was first known as glucinium from the corresponding Greek term. Quarrels over the name of the element were perpetuated by the simultaneous and independent isolations of metallic beryllium in 1828 by Wohler and Bussy. Both reduced beryllium chloride with metallic potassium in a platinum crucible. The name beryllium and symbol Be were officially recognized by the IUPAC in 1957.

Hope for the emergence of beryllium beyond the laboratory curiosity status resulted from publication of the work of the French scientist Lebeau in 1899. His paper described the electrolysis of fused sodium fluoberyllate to produce small hexagonal crystals of beryllium. Lebeau also reported the direct reduction of a beryllium oxide-copper oxide mixture with carbon to yield a beryllium-copper alloy. In 1926, Lebeau's alloy was rediscovered and found to have remarkable age-hardenable mechanical properties. Copper-beryllium alloy was first marketed in 1931, and this market remains important today.

Commercial development of beryllium in the United States was begun in 1916 by Hugh S. Cooper with the production of the first significant metallic beryllium ingot. This was followed by formation of the Brush Laboratories Company, which started its development work under the direction of Dr. C. B. Sawyer in 1921. In Germany, the Siemens-Halske Konzern began commercial development work in 1923.

Occurrence. Occurrences of beryllium in the earth's crust are widely distributed and estimates of the amount fall in the 4–6 ppm range. Forty-five beryllium-containing minerals have been identified. Only two are commercially important—beryl, $3BeO \cdot Al_2O_3 \cdot 6SiO_2$, for its high beryllium content, and bertrandite, $Be_4Si_2O_7(OH)_2$, for its large quantities located in the United States.

In 1959, beryllium was found in the rhyolitic tuffs of Spor Mountain, Utah, containing from 0.1 to 1.0% beryllium oxide. The practical processing limit requires an average beryllium oxide content of the ore of 0.6% to compete with beryl ore processing of material with more than 10% beryllium oxide. Deposits of this processable grade are adequate for the industrial requirements of the United States for several decades at present levels of consumption. Although the ore grade is much lower

than that of beryl ore, the beryllium values in the rhyolitic tuffs are acid soluble and recoverable by established processing technology.

In pure form, beryl mineral contains nearly 14% beryllium oxide, as found in its precious forms, emerald and aquamarine. Industrial grades of the mineral contain 10–12% beryllium oxide. Beryl occurs as a minor constituent of pegmatic dikes and is mined primarily as a byproduct of feldspar, spodumene, and mica operations. Only the relatively large crystals are recovered by hand-picking or cobbing to supply industrial requirements of 3000 tons (2700 metric tons) per year. In 1979, the main producers were the U.S.S.R., China, Brazil, and Argentina. The opening of mines and a mill to process bertrandite-bearing ores in Utah has decreased beryl requirements in the United States to 1000–2000 tons (900–1800 metric tons) per year.

Extractive and Process Metallurgy. The production of metallic beryllium, its alloys, or its ceramic products centers around the recovery of an intermediate, partially purified concentrate from ore processing. The usual intermediate is beryllium basic carbonate or hydroxide. The mill in Utah is the only one in the Western world which extracts beryllium from its ores. The processes used to extract the beryllium are based on sulfuric acid. The sulfate solutions from beryl or bertrandite sources are partially purified by solvent extraction before yielding beryllium hydroxide as the end product. The hydroxide is converted to beryllium fluoride by reaction with ammonium bifluoride. Thermal reduction with magnesium metal forms beryllium pebbles. Final purification is accomplished by vacuum melting the beryllium pebbles to remove fluorides and magnesium impurities and casting into graphite molds. Standard powder metallurgy processes are generally used to convert the cast billets to solid shapes. The prevalent final consolidation step is hot pressing.

Important Commercial Properties. Beryllium has several unique properties which have given it a position of commercial significance. Its low atomic mass, low absorption cross section, and high scattering cross section are neutronic properties of importance. These properties spurred the expansion of beryllium production beyond the pilot scale immediately after the formation of the United States Atomic Energy Commission and the initiation of nuclear reactor development programs. About 1960, structural applications using beryllium began to utilize its modulus of elasticity of 2.93×10^5 MPa, its low density, and its relatively high melting point. Beryllium has good thermal conductivity and excellent thermal capacity properties, which gave rise to its use as a thermal barrier and heat sink for reentry vehicles and other aerospace applications. The latter properties have been coupled with favorable ductility properties at elevated temperatures for the development of aircraft brakes.

Chemical Properties. Many chemical properties of beryllium resemble aluminum, and to a lesser extent, magnesium. Notable exceptions include solubility of alkali metal fluoride-beryllium fluoride complexes and the thermal stability of solutions of alkali metal beryllates.

All of the common mineral acids attack beryllium metal readily, with the exception of nitric acid. It is also attacked by sodium hydroxide and potassium hydroxide, but not by ammonium hydroxide.

Beryllium interacts with most gases. Polished beryllium surfaces retain their brilliance for years on exposure to air at ambient temperatures. The oxidation rate in air increases parabolically at temperatures above 850°C with the formation of a loosely adherent, white oxide.

Compounds of Beryllium. Ammonium beryllium carbonate solutions are prepared by dissolving the hydroxide or the basic carbonate in warm (50°C) aqueous mixtures of NH_4HCO_3 and

$(NH_4)_2CO_3$. After filtering to remove insoluble impurity hydroxides and adding a chelating agent, heating above 88°C evolves NH_3 and CO_2 and precipitates a high-purity, basic beryllium carbonate. If the aqueous system has the stoichiometry of $(NH_4)_4Be(CO_3)_3$, analogous to the ammonium uranyl carbonate system, the basic beryllium carbonate product of hydrolysis is $2BeCO_3 \cdot Be(OH)_2$. This compound is readily dissolved in all mineral acids, making it a valuable starting material for laboratory synthesis of beryllium salts of high purity.

Beryllium hydroxide, $Be(OH)_2$, is precipitated as an amorphous, gelatinous material by addition of ammonia or alkali to a solution of a beryllium salt at slightly basic pH values. A pure hydroxide can be prepared by pressure hydrolysis of a slurry of beryllium basic carbonate in water at 165 C. All forms of beryllium hydroxide begin to decompose in air or water to beryllium oxide at 190°C.

Beryllium sulfate, $BeSO_4 \cdot 4H_2O$, is an important salt of beryllium used as an intermediate of high purity for calcination to beryllium oxide powder for ceramic applications. A saturated aqueous solution of beryllium sulfate contains 30.5% $BeSO_4$ by weight at 30°C and 65.2% at 111°C.

Beryllium fluoride, BeF_2, is readily soluble in water, dissolving in its own water of hydration as $BeF_2 \cdot 2H_2O$. The compound cannot be crystallized from solution and is prepared by thermal decomposition of ammonium fluoberyllate, $(NH_4)_2BeF_4$.

Beryllium chloride, $BeCl_2$, with a melting point of 440°C, is used as a component of molten salt baths for electrowinning or electrorefining of the metal. The compound hydrolyzes readily with atmospheric moisture, evolving HCl, so protective atmospheres are required during processing.

Basic beryllium acetate, $Be_4O(C_2H_3O_2)_6$, is the best known of the beryllium salts of organic acids which can be divided into normal beryllium carboxylates, $Be(RCOO)_2$, and beryllium oxide carboxylates, $Be_4O(RCOO)_6$. The basic acetate is soluble in glacial acetic acid and can readily be crystallized therefrom in very pure form. It is also soluble in chloroform and other organic solvents. It has been used as a source of pure beryllium salts.

Applications. In the foregoing sections on properties, the early applications dependent on the nuclear characteristics of beryllium were described. Structural uses of beryllium in the aerospace industry have been realized because no other known material exceeds its modulus-to-weight ratio while supplying significant ductility.

Examples of applications which fully exploit the nuclear, rigidity, and high-temperature properties include heat shields, guidance system parts such as gimbals, gyroscopes, stable platforms and accelerometers, housings, mirrors, aircraft brakes, and formable grades of beryllium used in drawing and related methods of fabrication. Recent applications have been in structures which are loaded in compression. Wrought products are being developed with yield strengths approaching 690 MPa (100000 psi) and 20% elongation at room temperature.

Additions of beryllium to commercial copper- and nickel-base alloys enable these materials to be precipitation-hardened to strengths approaching those of heat-treated steels. Yet, copper-beryllium alloys retain the corrosion resistance, electrical and thermal conductivities, and spark-resistant properties of copper-base alloys.

Beryllium oxide particles take advantage of the high thermal conductivity and heat capacity, high melting point, very low electrical conductivity, and high transparency to microwaves in microelectronic substrate applications.

Biology and Toxicology. Beryllium can be handled safely with reasonable controls, but it can cause serious illness if these controls are not observed. Skin and respiratory reactions can be experienced. There is, however, no ingestion problem.

The hazards are generally classified as (a) acute respiratory disease, (b) chronic pulmonary disease, and (c) dermatitis.

Dermatitis is produced by skin contact with soluble salts of beryllium, especially the fluoride. It is controlled by a program of good personal hygiene and frequent washing of the exposed parts of the body, as well as by a clothing program where clothing is laundered on the plant site.

Acute pulmonary disease is due exclusively to inhalation of soluble beryllium salts and is not caused by exposure to the oxide, the metal, or its alloys. The exact forms of beryllium causing the chronic pulmonary disease and the degree of exposure necessary to induce it are not precisely known. It is known that under the completely uncontrolled conditions existing in beryllium extraction plants before the establishment of air-count standards in 1949, when beryllium air-counts were in milligrams per cubic meter of air rather than micrograms, only about 1% of the exposed workers became ill. This would indicate a sensitivity of a limited number of individuals to beryllium.

Investigations by medical, toxicological, and engineering personnel led to the promulgation of safe limits of exposure by the Atomic Energy Commission in 1949. It should be noted that no disabling cases of chronic berylliosis from exposure after 1949 have been documented. The disease is believed to be avoidable when air-counts are held within average limits of 2 micrograms per cubic meter of air for an 8-hour exposure, with a maximum at any time of 25 micrograms per cubic meter of air.

The Occupational Safety and Health Administration issued a proposed new occupational standard for beryllium air-counts. This proposal is highly controversial and remains under review.

Local exhaust ventilation is the major engineering control used to limit concentrations of airborne beryllium. Modern air cleaners allow control within recommended out-plant levels of 0.01 microgram beryllium per cubic meter of air, averaged over one month periods.

References

Ballance, J., Stonehouse, A. J., Sweeney, R., and K. Walsh: "Beryllium and Beryllium Alloys," in "Encyclopedia of Chemical Technology," 3rd Edition (Kirk-Othmer, Editors), Vol. 3, pages 803–823, Wiley, New York (1978).

Busch, L. S.: "Beryllium," in "Encyclopedia of the Chemical Elements," (C. A. Hampel, Editor), pages 49–56, Van Nostrand Reinhold, New York (1968).

Pinto, N. P., and J. Greenspan: "Beryllium," in "Modern Materials—Advances in Development and Applications," (B. W. Gonser, Editor), Vol. 6, Academic, New York (1968).

Staff: "Beryllium in Aerospace Structures," Brush Wellman Inc., Cleveland, Ohio (1963).

Stokinger, H. E.: "Beryllium, Its Industrial Hygiene Aspects," Academic, New York (1966).

Walsh, K., and G. Rees: "Beryllium Compounds," in Encyclopedia of Chemical Technology," 3rd Edition (Kirk-Othmer, Editors), Vol. 3, page 824–829, Wiley, New York (1978).

Webster, D., et al., Editors: "Beryllium Science and Technology," 2 volumes, Plenum, New York (1979).

—Kenneth A. Walsh, Ph.D., Brush Wellman Inc., Elmore, Ohio.

BETA DECAY. The process that occurs when beta particles are emitted by radioactive nuclei. The name *beta particle* or beta radiation was applied in the early years of radioactivity investigations, before it was fully understood what beta particles are. It is known now, of course, that beta particles are electrons.

When a radioactive nuclide undergoes beta decay its atomic number Z changes by $+1$ or -1, but its mass number A is unchanged. When the atomic number is increased by 1, negative beta particle (negatron) emission occurs; and when the atomic number is decreased by 1, there is positive beta particle (positron) emission or orbital electron capture.

Because atomic nuclei contain only protons and neutrons, beta particles must be created at the moment of emission, just as photons are created at the time of emission of electromagnetic radiation. Because of this creation process, the amount of energy equal to the rest energy, $m_e c^2$ of an electron, must be consumed when beta decay occurs. Any remaining energy can be given to the beta particle as kinetic energy. The nuclear transitions producing beta decay are between discrete energy states differing by a definite amount of energy W_0, so we expect the total energy of a beta-decay transition to be W_0. However, emitted beta particles are experimentally found to have a continuous range of total (rest plus kinetic) energies W of such magnitude that $m_e c^2 < W < W_0$, rather than all having a single energy W_0. This distribution as a function of energy (or momentum) forms what is known as a beta-ray spectrum. The shape of the spectrum depends on the sign of the charge on the beta particle (positive or negative), the energy W_0, and the degree of forbiddenness of the transition (explained below). Unless energy and momentum are not conserved in the process, the energy not carried away by the beta particle must be given to some other particle. Furthermore, since the beta particle has a spin quantum number $\frac{1}{2}$, angular momentum cannot be conserved unless another $\frac{1}{2}$ unit of angular momentum can be disposed of. Both of these possible discrepancies in the conservation laws have been taken care of in the Fermi theory of beta decay through postulation of a massless particle, a neutrino or an antineutrino, which has a spin quantum number $\frac{1}{2}$ and also carries away the remaining energy and momentum. Neutrinos were difficult to find experimentally but, even before they were experimentally detected, so much evidence had been developed to show their existence that the Fermi theory of beta decay was generally accepted.

Beta-decay processes are classified as allowed or forbidden but, as in many other physical processes, the term forbidden does not mean nonoccurrence, just a significant retardation relative to the rate for allowed transitions. The degree of forbiddenness is determined by the magnitude of the difference in angular momentum between the initial and final nuclear states as well as the parity of these states. If more than one unit of angular momentum must be carried away by the decay products ($\frac{1}{2}$ unit by the beta particle and $\frac{1}{2}$ unit by the neutrino or antineutrino), the transition must be forbidden. Allowed transitions give straight-line Kurie plots, as do some forbidden transitions but some forbidden transitions have distinct shapes other than straight lines for their Kurie plots.

A negatron emitted during beta decay has its spin aligned away from the direction of its emission (its angular momentum vector is antiparallel to its momentum vector) and hence has negative helicity, but an emitted positron has positive helicity. It is because of the absence of beta particles with both positive and negative helicity in both types of beta-emission processes that parity is not conserved in beta decay.

See also **Particles (Subatomic)**; and **Radioactivity**.

BETA-LACTAM RING. Antibiotic.

BETALAINES. Sometimes referred to as *beetroot pigments*, the *betalaines*[1] are made up of two main groups: (1) *betacyanins*,

[1] Sometimes spelled without the last *e*.

the principal component of which is betanin. It is estimated that this pigment contributes from 75–95% of the total red color. (2) *betaxanthins*, the principal component of which is vulgaxanthin-I, contribute about 95% of the yellow color. Another yellow pigment, betalamic acid, derives directly from cleavage of betanin and probably the key intermediate in the biogensis of all betalaines.

There has been considerable interest and research activity in connection with the betalaines during the past decade or so, stemming principally from tighter restrictions, including banning, of several synthetic colorants for foods. See also **Colorants**.

The red and golden cultivars of beetroot (*Beta vulgaris* L.) appear to be excellent sources of both red and yellow, water-soluble colorants. A factor of concern in connection with the betalaines as possible coloring agents is their earthy flavor, directly reminiscent of beet taste. The principal contributor of this flavor is a substance known as *geosmin*, a complex organic alcohol. Some problems encountered to date in preparing suitable red and yellow colorants from beet raw materials, in addition to the flavor, are rather poor yields and lack of stability of the extracted substances.

References

Acree, T. E., et al.: "Geosmin, the Earthy Component of Table Beet Odor," *Jrnl. of Ag. Food Chem.*, **24**, 430 (1976).
Considine, D. M., Editor: "Foods and Food Production Encyclopedia," Van Nostrand Reinhold, New York (1982).
Driver, M. G., and F. J. Francis: "Stability of Phytolaccanin, Betanin, and FD&C Red #2 in Dessert Gels," *Jrnl. of Food Science*, **44**, 2, 518–520 (1979).
Pasch, J. H., and J. H. von Elbe: "Sensory Evaluation of Betanine and Concentrated Beet Juice," *Jrnl. of Food Science*, **43**, 5, 1624–1625 (1978).
Williams, M., and G. Hrazdina: "Anthocyanins as Food Colorants," *Jrnl. of Food Science*, **44**, 1, 66–68 (1979).

BILE. A bitter alkaline fluid secreted by the liver into the duodenum, which aids in the digestion of food. The chief components of bile are bile salts and bile pigments. Because of its strong alkalinity, bile neutralizes the acid coming into the duodenum from the stomach. The bile not only performs important functions in the process of digestion, but also serves as a vehicle for the excretion of waste products from the body.

Bile salts help in the breakdown of fat in the intestines and in fat absorption through the intestinal wall. The bile salts are injected into the digestive canal at the duodenum. They are not excreted, but are almost totally absorbed through the walls of the intestine, to be used over and over again. Bile pigments are derived from the hemoglobin of broken-down red blood cells and are excreted with the feces. When the pigments appear in excessive amounts in the blood, the mucous membranes and conjunctiva of the eye become stained a pale yellow, and the patient is said to be jaundiced.

Bile is continually secreted by the liver and stored in the gallbladder. Here the bile is concentrated by the absorption of water through the walls of the gallbladder. Bile is released from the gallbladder into the intestine when food passes through the pyloric valve from the stomach into the small intestine. Gallstones are formed of constituents of the bile which have settled out of solution. The stones vary in size, color, and structure, according to the materials composing them.

An inadequate supply of bile contributes to vitamin A deficiency because of disturbances of the intestinal tract which prevent the effective absorption of the vitamin. In an average adult, from one-half to one liter of bile is secreted every 24 hours, the quantity depending upon the amount and kind of food eaten.

Part of the cholesterol newly synthesized in the liver is excreted into bile in a free non-esterified state (in constant amount).

Cholesterol in bile is normally complexed with bile salts to form soluble choleic acids. Free cholesterol is not readily soluble and with bile stasis or decreased bile salt concentration may precipitate as gall-stones. Most common gallstones are built of alternating layers of cholesterol and calcium bilirubin and consist mainly (80–90%) of cholesterol. Normally, 80% of hepatic cholesterol arising from blood or lymph is metabolized to cholic acids and is eventually excreted into the bile in the form of bile salts.

The C_{24} bile acids arise from cholesterol in the liver after saturation of the steroid nucleus and reduction in length of the side chain to a 5-carbon acid; they may differ in the number of hydroxyl groups on the sterol nucleus. The four acids isolated from human bile include *cholic acid* (3,7,12-trihydroxy), as shown in Fig. 1; *deoxycholic acid* (2,12-dihydroxy); *chenodeoxy-*

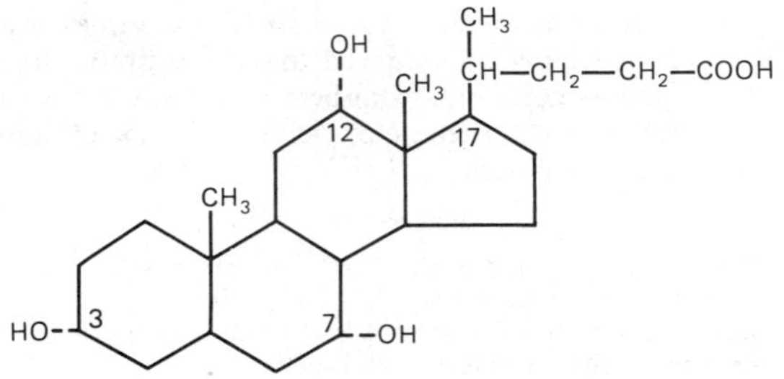

Fig. 1. Cholic acid.

cholic acid (3,7-dihydroxy); and *lithocholic acid* (3-hydroxy). The bile acids are not excreted into the bile as such, but are conjugated through the C_{24} carboxylic acid with glycine or taurine, $NH_2-CH_2-CH_2-SO_3H$. This esterification of the bile acids to soluble conjugates occurs in the microsomes and requires coenzyme A, magnesium ion, and ATP (adenosine triphosphate). Although taurocholic acid predominates at birth, the most abundant of the bile acids in the adult is glycocholic acid. In alkaline bile, the conjugated bile acids exist in their ionized form as the bile salts, glycocholate or taurocholate. Bile salts can function as effective product feedback inhibitors of hepatic cholesterol synthesis. Because of their detergent action, bile salts play an important role in the absorption of cholesterol, fats, and fat-soluble vitamins. The bile salts are believed to facilitate absorption of these compounds by the formation of micelles or aggregates of low osmotic pressure. The bile salts themselves are not absorbed during this process. Their absorption from the intestine occurs at a different site and at an entirely different rate from that of the lipids. Approximately 95% of the bile salts are reabsorbed, enter the enterohepatic circulation, and are ultimately re-excreted into the bile for further utilization in lipid absorption.

Most of the hormones that are normally conjugated in the liver to form glucuonides or sulfates, such as the steroids, thyroxine, epinephrine, and norepinephrine, are secreted into the bile, but to a varying degree, may be re-absorbed in the intestine and eventually excreted in the urine. The 17-hydroxysteroids, including cortisol, are secreted into the bile primarily as reduced glycuronide conjugates. More than 70% of these conjugates enter the enterohepatic circulation and are eventually excreted into the urine; less than 30% are found in the feces. Progesterone, after its conversion to pregnanediol, is also excreted into the bile primarily as the glucuronide, some 75% of which is eventually excreted in the feces. Most androgens are excreted as sulfates in the urine, part of which are of non-hepatic or non-biliary origin. Significant amounts of estrogens are excreted into bile

as estriol glucuronide or estrone sulfates. Many derivatives of epinephrine and norepinephrine are eventually conjugated with either glucuronide or sulfate at the 4-hydroxy position and excreted into the bile. Thyroxine is predominantly conjugated with glucuronic acid and is excreted as such into the bile. The bile, however, is not a significant route for the net disposal of thyroxine, since this hormone is rapidly re-absorbed, enters the enterohepatic circulation and is eventually excreted as urinary metabolites.

The major components of bile, the bile pigments, can account for 15–20% of the total solids. *Bilirubin* comes primarily from the degradation of heme in the reticuloendothelial system in the spleen, bone marrow, and to a lesser extent, the liver. The initial step in the metabolism of heme is the cleavage of the porphyrin ring and elimination of the alpha methylene carbon to produce an open tetrapyrrole. This may exist as a complex with iron and globin called choleglobin. After removal of the iron and globin, the resulting tetrapyrrole, biliverdin, is rapidly reduced to bilirubin, the major pigment in human bile. Not all bilirubin results from the breakdown of hemoglobin from mature red cells. The early appearance of labeled bilirubin after injection of precursor glycine-^{14}C indicates that some bilirubin (approximately 10%) may arise from: (1) the rapid breakdown of immature red cells in the bone marrow; (2) from heme that had not entered hemoglobin; or (3) from the destruction of newly formed red cells in the peripheral circulation. This "shunt" pathway for bilirubin formation may predominate in pernicious anemia and some porphyrias. A small amount of bilirubin may also arise from other heme pigments, such as myoglobin or the cytochromes. The bilirubin that enters the blood is rapidly and solely bound to albumin. Normal circulating levels of bilirubin are less than 1 mg/100 ml. Free bilirubin, which readily crosses the blood-brain barrier in the newborn, and to a lesser extent in the adult, is an effective uncoupler of oxidative phosphorylation in the brain and is highly toxic.

The hepatic transport of bilirubin from plasma to bile involves three independent, but related mechanism, i.e., uptake, conjugation, and secretion. Plasma bilirubin is dissociated from plasma albumin in the liver and is rapidly concentrated in the cytoplasm of the hepatic cells by an unknown mechanism which precedes and is relatively independent of any subsequent hepatic conjugation. After concentration in the liver, bilirubin is conjugated with 2 moles of glucuronic acid to form bilirubin diglucuronide, the glucuronic acid moieties being attached in ester linkage to the carboxyl groups on the propionic acid side chains. See Figs. 2 and 3.

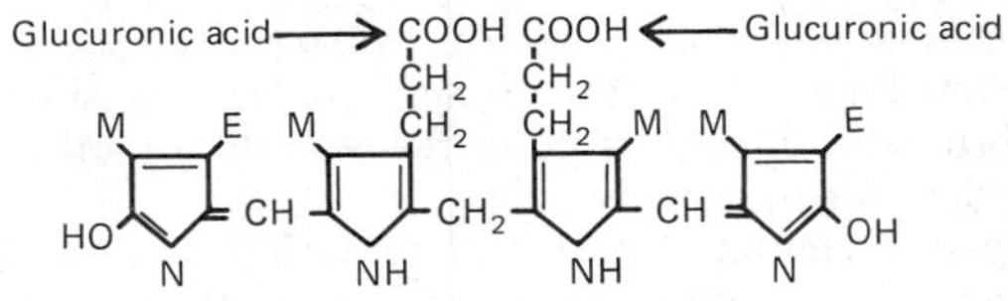

Fig. 2. Structure of "direct" bilirubin.

Glucuronyl transferase, the enzyme catalyzing the final step, is located in the smooth endoplasmic reticulum of liver and to a lesser extent in kidney and gastric mucosa, where a small amount of extrahepatic conjugation may occur. This enzyme has not been purified and it is unclear whether it nonspecifically catalyzes glucuronide conjugation of many non-bilirubinoid substrates, or is bilirubin specific and a member of a large group of closely related glucuronyl transferases. Its activity can be induced by a variety of drugs and can be inhibited with steroids or steroid glucuronides found in plasma of pregnant women.

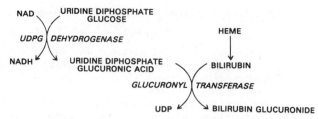

Fig. 3.   Conjugation of bilirubin.

Crigler-Najjar's disease in humans is characterized by increased levels of unconjugated bilirubin in the serum. A genetic impairment of glucuronyl transferase, the enzyme responsible for the transfer of glucuronic acid from uridine diphosphate glucuronic acid, exists not only in the liver, but in the kidney as well. Gilbert's disease is characterized by a mild increase of unconjugated bilirubin in the plasma, which may result from a partial impairment of glucuronyl transferase, from a defect in bilirubin transport in the blood, or a defect in hepatic uptake. In subjects with Dubin-Sprinz or Dubin-Johnson disease, the serum contains high levels of both unconjugated and conjugated bilirubin, and an unidentified brown pigment is present in the liver. A defect in the secretion of the bilirubin conjugates from the hepatic cell is a probable causative factor. Rotor's disease is also characterized by increased serum levels of both unconjugated and conjugated bilirubin, but it differs from Dubin-Sprinz disease in that the hepatic brown pigment is not found.

The mild nonhemolytic jaundice often present in the newborn (physiological jaundice) or the more severe jaundice and kernicterus in premature infants may result in part from an inability of the immature liver to conjugate bilirubin; low hepatic levels of both glucuronyl transferase and uridine diphosphate glucuronic acid dehydrogenase (the enzyme that catalyzes the synthesis of uridine diphosphate glucose glucuronic acid from uridine diphosphate glucose) are found in fetus and newborn. Hepatic secretion of conjugated bilirubin may also be impaired.

References

Elias, E.: Cholangiography in the Jaundiced Patient," *Gut*, **17**, 801 (1976).

Metzger, A. L., et al.: "Diurnal Variation in Biliary Lipid Composition: Possible Role in Cholesterol Gallstone Formation," *New Eng. Jrnl. Med.*, **288**, 333 (1973).

Small, D. M.: "The Formation and Treatment of Gallstones," in "Diseases of the Liver," (L. Schiff, Editor), 4th Edition, Lippincott, Philadelphia (1975).

Thistle, J. L., et al.: "Chemotherapy for Gallstone Dissolution," *Jrnl. Amer. Med. Assn.*, 239, 1041 (1978).

Warren, W. K., and E. G. C. Tan: "Diseases of the Gallbladder and Bile Ducts," in "Diseases of the Liver," (L. Schiff, Editor), 4th Edition, Lippincott, Philadelphia (1975).

Wenckert, A., and B. Robertson: "The Natural Course of Gallstone Disease," *Gastroenterology*, **50**, 376 (1966).

BINDING ENERGY. This term is used in atomic physics with two closely related meanings: the binding energy of a particle (or other entity) is the energy required to remove the particle from the system to which it belongs; the binding energy of a system is the energy required to disperse the system into its constituent entities. Explicit definitions are obviously necessary.

Some explicit definitions for the binding energies of particles are the following:

1) The *electron binding energy* is the energy necessary to remove an electron from an atom. It is identical with the ionization potential.

1a) The *total electron binding energy* is the energy necessary to remove all the electrons from an atom to infinite distances, so that only the nucleus remains. It is equal to the sum of the successive ionization potentials of that atom.

2) The *proton binding energy* is the energy necessary to remove a single proton from a nucleus. Most known proton binding energies are in the range 5–12 MeV, although that for ^2H is 2.23 MeV, that for ^4He is 19.81 MeV, and those for ^5Li and ^9Be are negligible.

3) The *neutron binding energy* is the energy required to remove a single neutron form a nucleus. Most known neutron binding energies are in the ranges 5–8 MeV, though that for ^2H is 2.23 MeV, that for ^9Be is 1.67 MeV, and that for ^{12}C is 18.7 MeV.

4) The *alpha-particle binding energy* is the energy required to remove an alpha-particle from a nucleus. For most light nuclides the alpha-particle binding energy is positive and is equal to several MeV. For nuclides of mass number about 125, it is approximately zero. For nuclides of mass number about 150 to 200, it is negative by about 1 to 3 MeV, but the magnitude of the Coulomb potential barrier at the nucleus is sufficiently large that penetration by an alpha particle is so improbable that lifetimes for alpha disintegration are generally too long for detection of alpha activity. For most nuclides of mass number exceeding 200, the alpha-particle binding energy is negative by about 4 to 8 MeV, which is a negative binding energy of sufficiently large magnitude to give a measurable probability of penetration of the potential barrier by an alpha particle, hence an observable alpha activity.

Some explicit definitions for the binding energies of systems are:

(1) The *nuclear binding energy* is the energy that would be necessary to separate an atom of atomic number Z and mass number A into Z hydrogen atoms and $A—Z$ neutrons. This energy is the energy equivalent of the difference between the sum of the masses of the product hydrogen atoms and neutrons, and the mass of the atom; it includes the effect of electronic binding. (See *total electron binding energy* above.)

(2) The binding energy of a solid is the energy required to disperse a solid into its constituent atoms, against the forces of cohesion. In the case of ionic crystals, it is given by the Born-Mayer equation. See **Crystal**.

BIOCHEMICAL INDIVIDUALITY. The possession of biochemical distinctiveness by individual members of a species, whether plant, animal, or human. The primary interest in such distinctiveness has centered in the human family, and in the distinctiveness within animal species as it might illuminate some of the questions on human biochemistry.

While it has been known for centuries that bloodhounds, for example, can tell individuals apart even by the attenuated odors from their bodies left on a trail, the first scientific work which hinted at the existence of substantial biochemical distinctiveness in human specimens was the discovery of blood groups by Landsteiner about 1900.

A few years later Garrod noted what he called "inborn errors of metabolism"—rare instances where individuals gave evidence of being abnormal biochemically in that they were albinos (lack of ability to produce pigment in skin, hair and eyes), or excreted some unusual substance in the urine or feces. To Garrod these observations suggested the possibility that the biochemistry of all individuals might be distinctive.

About 50 years later serious attention to the phenomenon of biochemical individuality resulted in the publication of several articles and a book on this subject. (Williams, R. J.: "Biochemical Individuality," Wiley, New York, 1956.) These reported evidence indicating that every human being, including all those designated as "normal," possesses a distinctive metabolic pattern which encompasses everything chemical that takes place in his or her body. That these patterns, like the abnormalities discussed by Garrod, have genetic roots is indicated by the pioneer explorations of Beadle and Tatum in the field of biochemical genetics in which they established the fact that the potentiality for producing enzymes resides in the genes.

Biochemical individuality, which is genetically determined, is accompanied by, and in a sense based upon, anatomical individuality, which must also have a genetic origin. Substantial differences, often of large magnitude, exist between the digestive tracts, the muscular systems, the circulatory systems, the skeletal systems, the nervous systems, and the endocrine systems of so-called normal people. Similar distinctiveness is observed at the microscopic level, for example in the size, shape and distribution of neurons in the brain and in the morphological "blood pictures," i.e., the numbers of the different types of cells in the blood.

Individuality in the biochemical realm is exhibited with respect to (1) the composition of blood, tissues, urine, digestive juices, cerebrospinal fluid, etc.; (2) the enzyme levels in tissues and in body fluids, particularly the blood; (3) the pharmacological responses to numerous specific drugs; (4) the quantitative needs for specific nutrients—minerals, amino acids, vitamins—and in miscellaneous other ways including reactions of taste and smell and the effects of heat, cold, electricity, etc. Each individual must possess a highly distinctive pattern, since the differences between individuals with respect to the measurable items in a potentially long list are by no means trifling. Often a specific value derived from one "normal" individual of a group will be several times as large as that derived from another.

BIODEGRADABILITY. Detergents.

BIOLOGICAL ENERGY TRANSFER. When ionization is produced in a substance such as a protein, the net charge produced in the protein probably migrates throughout a large region of the molecule, with various probabilities favoring its occurrence in one part of the molecule or another. Eventually, after approximately 10^{-14} second, the excess (or deficiency) of charge probably settles in an *s-s* bond or in the hydrogen atom attached to the carbon of the peptide bond which is opposite to one or other of the amino acid residues. Thus, regardless of the site of the original ionization in the molecule, there is considerable transfer of energy throughout a large portion of the molecule. However, the phrase *energy transfer* is generally meant to include those cases where it might occur in addition to this; for example, intermolecularly, either between adjacent protein molecules or between protein and solvent molecules. It can also apply to excitation.

BIOLUMINESCENCE. Many living organisms exhibit the unique property of producing visible light, a phenomenon referred to as bioluminescence. Known light-emitting organisms have either oxidative or peroxidative enzymes that couple the chemical energy released from the enzyme reaction to give electronic excitation of a luminescent compound. The compound that is oxidized with subsequent light emission is usually referred to as *luciferin* and the enzyme which catalyzes the reaction as *luciferase*. Most luciferins and luciferases that have been isolated from unrelated species are different in molecular structure. With one known exception, combinations of luciferin and luciferase from different species do not exhibit bioluminescence.

The light-producing reaction in a number of organisms can be represented simply by: Luciferin $+ O_2 \xrightarrow{Luciferase}$ Light. Some luminous organisms catalyzing this reaction are: (1) *Cypridina* (a crustacean); (2) *Apogon* (a fish), and (3) *Gonyaulax* (a protozoan). The latter organism is mainly responsible for the phosphorescence (so-called) of the sea.

In other instances, some luciferins must first undergo a luciferase-catalyzed activation reaction prior to their being catalytically oxidized by the enzyme to produce light. There are two well-known cases:

(1) The firefly:

Luciferin + Adenosine Triphosphate (ATP)

$$\xrightarrow{Luciferase;\ Mg^{2+}} \text{Activated Luciferin}$$

Activated Luciferin $+ O_2 \xrightarrow{Luciferase}$ Light

(2) The sea pansy (*Renilla*):
Luciferin $+ 3',5'$-Diphosphoadenosine (DPA)

$$\xrightarrow{Luciferase;\ Ca^{2+}} \text{Activated Luciferin}$$

Activated Luciferin $+ O_2 \xrightarrow{Luciferase}$ Light

Both of these activation reactions are linked to adenine-containing nucleotides of great biological importance. Since the measurement of light can be made an extremely sensitive and rapid technique, the most sensitive and rapid assays known have been developed for ATP and DPA, using the foregoing luminescent systems. Nucleotide concentrations of less than 1×10^{-9} M are easily detectable using electronic instrumentation. Firefly luciferase-luciferin preparations for ATP assays are commercially available.

The structure of firefly luciferin has been confirmed by total synthesis. The firefly emits a yellow-green luminescence, and luciferin in this case is a benzthiazole derivative. Activation of the firefly luciferin involves the elimination of pyrophosphate from ATP with the formation of an acid anhydride linkage between the craboxyl group of luciferin and the phosphate group of adenylic acid forming luciferyladenylate.

All other systems that have been extensively studied emit light in the blue-green region of the spectrum. In these cases, the luciferins appear to be indole derivatives.

Some animals, such as the marine acorn worms (Balanoglossus), produce light via a peroxidation reaction and appear not to require molecular oxygen for luminescence. The luciferase in this case is a peroxidase of the classical type and catalyzes

the reaction: Luciferin $+ H_2O_2 \xrightarrow{Luciferase}$ Light.

Commercially available horseradish peroxidase (crystalline) will substitute for luciferase in the foregoing reaction. In addition, a compound of known structure, 5-amino-2,3-dihydro-1,4-phthalazinedione (also known as *luminol*), will substitute for luciferase. The mechanisms appear to be the same regardless of the way in which the crosses are made. Thus, a model bioluminescent system is available and can be used as a sensitivity assay for H_2O_2 at neutral pH. The identification of luciferase as a peroxidase is of interest since this represents the only demonstration of a bioluminescent system in which the catalytic nature of a luciferase molecule has been defined.

Most of the luminescent systems mentioned appear to be under some nerve control. Normally, a luminous flash is observed after mechanical or electrical stimulation of most of the aforementioned species. A number of these also exhibit a diurnal rhythm of luminescence.

Among the lower forms of life, there are two well-known examples of luminescence which are not under nerve control, giving a continuous glow of visible light. These are the luminous bacteria, frequently found growing on dead fish, and luminous fungi which grow abundantly on rotting wood. These cells apparently depend upon the oxidation of an organic molecule and hydrogen which is transferred through diphosphopyridine nucleotide (DPN; also termed NAD, nicotinamide adenine dinucleotide) and the enzyme system to drive the luminescent reaction. Known details of these luminescent reactions are represented as follows. For bacteria:

$$DPNH + H^+ + Flavin\ Mononucleotide\ (FMN)$$
$$\xrightleftharpoons{Oxidase} FMNH_2 + DPN$$

$$FMNH_2 + Long\text{-}chain\ Aliphatic\ Aldehyde$$
$$+ O_2 \xrightarrow{Luciferase} Light$$

and for fungi:

$$DPNH + H^+ + Unknown\ Compound\ (X) \xrightleftharpoons{Oxidase} XH_2 + DPN$$

$$XH_2 + O_2 \xrightarrow{Luciferase} Light$$

Both of these systems are apparently closely linked to respiratory processes and in this sense are analogous to one another. Luciferase from a luminous bacterium, *Photobacterium fischeri*, has been crystalled in high yield.

See also **Luminescence**.

References

Cormier, M. J. and J. R. Totter: *Ann. Rev. Biochem.* **33**, 431–458 (1964).
Dure, L. S. and M. J. Cormier: *J. Biol. Chem.*, **239**, 2351–2359 (1964).
Firth, F. E.: "The Encyclopedia of Marine Resources," Van Nostrand Reinhold, New York (1969).
Herring, P. J.: "Bioluminescence in Action," Academic, New York, 1979.
Levandowsky, M., and S. H. Hutner, Editors; "Biochemistry and Physiology of Protozoa," 3 volumes, Academic, New York (1979–1980).

BIOMASS AND WASTES AS ENERGY SOURCES.

Initially, *biomass* was defined as the amount of living organisms in a particular area, stated in terms of the weight or volume of organisms per unit area or of the volume of the environment. This definition still applies very well to ecological and geophysical assessments of land areas or regions and depths of the seas and lakes. Within the past decade or two, *biomass* as a word has come to describe the exploitation of plants (terrestrial mainly, but also marine) that may be grown and harvested as crops, all or parts of which may be combusted directly for their heat energy or processed into fuels.[1] When such materials serve other primary purposes (as foods, fibers and construction members), waste is created—straw, sawdust, sewage sludge, etc.—that possesses value as an energy source. The majority

---

[1] The generation of biomass from carbon dioxide is called "primary production" because it is the first, fundamental step in turning inorganic material into organic compounds and cell constituents. This reduction of carbon dioxide uses light as the source of energy (solar). See entry on **Photosynthesis**.

of agricultural, commercial, industrial, and urban or municipal wastes are of a biological rather than mineral nature and thus fall under the umbrella of biomass. The simple burning of wood for heat illustrates one of the simplest ways to convert biomass to energy. All biomass represents an indirect form of solar energy. Biomass, as a source of energy, differs from coal, natural gas, and petroleum in one major way—biomass is renewable. Some potential biomass energy crops can be renewed as frequently as two or three times per year, depending upon location, while other materials such as trees have a renewal cycle of several years. Anthropogenic wastes are renewed on a daily basis. Interest in biomass over the last several years has stemmed from the overall concern with ultimate exhaustion of nonrenewable energy sources as well as the desire for a degree of political independence on the part of many nations that either do not have any fossil fuel resources, or that have insufficient supplies to maintain a strong economic and industrial position.

Much of the technology required to exploit biomass is available. As of the early 1980s, the principal constraints are economic and political. Aside from biomass wastes that are created anyway (and thus available for exploitation), biomass energy technology is largely a fundamental matter of agriculture and forestry (silviculture). Since modern agriculture is by itself a large consumer of fossil fuels, it is necessary to find crops that provide ample return of energy (output) for the energy invested. When analyzed critically, some previous proposals have failed to meet this criterion. Thus, the search continues for crops and regions in which to grow them that will provide a good energy return. Biomass fuels, especially when combusted directly, also create their share of environmental problems. Biomass energy sources have an advantage, on the one hand, that fuels produced from them can be reasonably close to the consuming markets and thus minimize transportation costs, but, on the other hand, highly populated urban areas are often not adjacent to good agricultural land, thus requiring long transportation of fuels—as is traditionally the case with fossil fuels. Proposals have been made that would utilize crops, such as sugarcane (quite successful to date in Brazil), cassava leaves, and pineapple, that are best grown in tropical regions, but the fuels from which would have to be transported long distances to consuming centers.

When considering the potential of biomass as an energy source, there is also a confrontation with the production of food. Agriculturists, particularly in some regions of the world, have been predicting for a number of years that the available arable land is shrinking while the population is increasing. Extensive use of good land for energy-yielding plants would, of course, aggravate this situation. Particularly in some regions of the world there is a potential crisis in terms of available irrigation water for crops of any purpose. However, even these kinds of serious problems may be amenable to solution if a total systems approach to a biomass energy plan for a given region, i.e., considering not only the contribution of such a plan to energy needs, but also all of its positive as well as negative interactions with food production, water needs, animal feedstuff needs, economic situations (role as a buffer in terms of traditional surpluses and shortages of crops), political implications (diplomatic and military advantages of energy independence), and other factors that may or may not be cost-related.

In a study of corn (maize) biomass as a source of chemicals and fuels, Lipinsky (1978) compared the traditional usage of corn grain and corn stover with an alternative routing of materials that would essentially perform the same functions as traditional routing, but furnish chemicals and fuels as well. This is evidence of how application of a *systems approach* may overcome what initially appears to be large negative concerns over using biomass as an energy source. See Fig. 1.

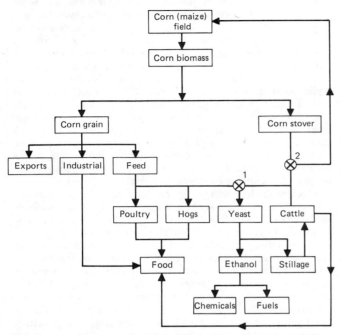

Fig. 1. Traditional procedures and alternative scheme for handling corn (maize) crops. In the conventional method, corn grain goes into (1) export, (2) industrial, and (3) animal feed channels. The feed is used for poultry, hogs, and cattle. Corn stover (stalks, leaves, etc.) is left in the field and turned under (except in no-till operations, where it is left in place) for the next crop. Thus, for the traditional method, valve 1 in the flowsheet is closed to the yeast pathway and valve 2 is closed to the return of stover to the field. In an alternative scheme (Lipinsky, 1978), some or all of the feed would be routed to the yeast pathway, with valve 1 closed to the feeding of cattle. In this scheme, stover would be used as cattle feed (preferably after treatment to improve digestibility) and part of the grain would be used in the manufacture of chemicals and fuels. Within the last few years, researchers at Purdue University have made a major breakthrough in the treatment of corn stover to obtain simple sugars for fermentation to ethanol. An organic acid treatment facilitates the attack of fungal cellulases on lignocellulose. The probability that this process would also increase the digestibility of corn stover by cattle is high because the low digestibility of raw stover arises from crystalline cellulose or cellulose protected by lignin. It is envisioned that the digestibility of corn stover could rise from 45 to 60% or more. More details are given in Lipinsky (1978).

Some successes have been realized in the biomass energy field. (1) most automobiles in Brazil burn *gasohol* (in Brazil, there is a 20% content of alcohol made from sugarcane). By the mid-1980s, Brazil plans to convert about 80,000 automobiles for using gasohol and some 250,000 new cars by the mid-1980s will be built to use gasohol. (2) It is reported that as of 1981, over 7.5 million biogas generators (methane) have been built, with obvious emphasis upon decentralized activities. Particularly in the southern area of China, the direct burning of wood is being replaced by gas generated from human and animal wastes. (3) Countries, such as Sweden, which have no coal, oil, or natural gas, are seriously planning the expansion of biomass energy systems. Sweden, a neutral nation during World War II, was cut off from fossil fuel supplies and developed a high biomass technology for that time. Sweden, which fortunately is highly forested, is taking steps toward increasing use

of wood as a basic fuel. Research has been directed toward fast-rotation trees with harvesting on a 3- to 5-year schedule. Emphasis is being placed on willow and birch trees.

During the past few years in the United States, there have been several hundred research and development projects aimed toward biomass energy systems. For example, a machine has been developed that can reduce a substantial tree, including branches, to dollar-size chips in less than one minute.

Principal Biomass Materials for Energy

Biomass-to-energy systems fall into two principal categories: (1) materials for direct combustion that will generate heat for processing, for warming living and working spaces, for steam and hence also for generating electricity; and (2) materials from which both fuels and chemicals can be obtained through biochemical or thermochemical conversion processes. Resulting fuels must have an ample caloric content per unit of weight and thus be rich in carbon and hydrogen and poor in atoms such as oxygen and nitrogen which do not contribute to the caloric value of the fuel. Fuels of all physical states are envisioned—solids, liquids, and gases. In searching for new biomass raw materials, scientists have found it helpful to study the various biosynthetic pathways followed by plants from seed to maturation.

In studying the potential of various biomass materials, Weisz and Marshall (1979) have compared the characteristics of several materials that have been considered for energy production. See Table 1.

Sugarcane, Pineapple, and Cassava. The optimal zone for growing sugarcane extends over a range of latitude that lies 30°N and 30°S of the equator. The growing season varies considerably—10 months or less in Louisiana and up to 2 years or longer in Hawaii, and parts of South America. Sugarcane requires considerable water—either from rain or irrigation. In several areas where rainfall is less than 1000–4000 millimeters (39–157 inches) per year, irrigation is required. Sugarcane requires a mean annual temperature of 25°C (77°F) or higher—growth essentially ceases when soil temperature drops to 16°C (61°F). Soil requirements are not rigid so long as the pH lies between 4.5 and 5.5. Yields of sugarcane can be increased by the application of fertilizer. The greatest experience gained from a noncereal crop as a source of biomass for fuels to date has been from sugarcane.

In recent years, cassava leaves and, even more recently, pineapple have been considered seriously as biomass energy crops. For pineapple, the growing period extends to about 14 months, but can be as long as 3 years at the northern and southern limits of its growing area. These roughly parallel the characteristics of sugarcane. A second (ratoon) crop of pineapple can be raised in about half the time required to grow the mother plant. Cassava, on the other hand, is raised as an annual crop and is essentially limited to a region 15°N and 15°S of the equator. Cassava roots keep well if left in the soil and during a period of about 21 months continue to accumulate starch. Pineapple and cassava require about the same temperatures as sugarcane. Cassava is similar to sugarcane in terms of water requirements, whereas pineapple can tolerate drier conditions, although at sacrifice of yield. Cassava and pineapple require about the same soil types as sugarcane and react favorably to fertilizers.

As indicated by Table 2, a number of comparative studies have been made of these three biomass plants. As reported by Marzola and Bartholomew (1979), cassava and sugarcane are good candidates for alcohol production because of their poten-

TABLE 1. COMPARISON OF BIOMASS MATERIALS

CROP	AGRICULTURAL ENERGY INPUT[1] kcal per Pound of Crop	Kilogram of Crop	RATIO OF CROP TO TOTAL BIOMASS (WEIGHT)	AGRICULTURAL ENERGY INPUT kcal per Pound of Total Biomass	Kilogram of Total Biomass	RATIO OF A TO G[3]
Alfalfa	425	937	~0.9	382	842	0.21
Sorghum (dryland)	443	977	~0.9	399	880	0.22
Sorghum (irrigated)	506	1116	~0.9	455	1003	0.25
Wheat (dryland)	393	867	~0.5	200	441	0.11
Wheat (irrigated)	790	1742	~0.5	395	871	0.22
Corn (maize)	622	1372	~0.5[4]	331	730	0.18
Sugar beets[2]	320–700	706–1544	~0.6	195–420	430–926	0.15–0.23

[1] From E. S. Lipinksy, et al.: "Systems Study of Fuels from Sugarcane, Sweet Sorghum and Sugar Beets," National Technical Information Service, Springfield, Virginia, 1976.

[2] Variation over eight states of the United States.

[3] Biomass energy content assumed to be 1800 kcal/pound (3969 kcal/kilogram).

[4] Value from V. J. Kavlick, paper presented at the 42nd American Petroleum Institute Refining Department Mid-Year Meeting, Chicago, 1977.

NOTE: Format from Weisz and Marshall (1979). See references listed at end of entry. Metric equivalents added. A = Agricultural energy input; G = biomass energy content.

tially high productivity and because they accumulate large amounts of starch or sucrose. Sucrose can be fermented without pretreatment, but a preliminary degradation step is needed for starch. Sucrose levels in sugarcane stalks are low during rapid vegetative growth, but ripening of the plant with chemicals (Alexander, 1974) or by withholding nitrogen and water (Clements, 1953) results in sucrose accumulation to concentrations of 15–20%. When stored for several months, cassava roots eventually reach a starch concentration of about 33% (fresh weight basis). In contrast, sucrose and reducing sugars accumulate in pineapple fruit to a concentration of about 16%, although the plant and ratoon crop stems also contain 30–40% starch (dry weight basis) a short while after the ratoon crop fruit has been harvested. Thus, both sugar and starch substrates are available from pineapple. Pineapple is well adapted to the subhumid or semiarid tropics and thus is particularly well suited for exploit-

ing large areas not under cultivation with other crops of commercial value.

In commenting on the potential of sugarcane as a biomass energy source, Lipinsky (1978) observes that conventional sugarcane bagasse is a source of fiber for papermaking, especially in countries that have considerable sugarcane and few pulpwood-type trees. To make paper of relatively high quality, it is necessary to depith the bagasse. Depithing is a relatively costly process, with considerable loss of fibers. A relatively new process (Canadian Separator Equipment Process) generates rind fiber that already has been depithed without the injury to the fibers that results from traditional crushing and milling operations. It is estimated that the value of rind fiber will be about double that of conventional bagasse. Thus, the new process may make it possible to generate a coproduct of significant value along with fuels from the sugarcane. Excess rind fibers could be used

TABLE 2. COMPARISON OF THREE BIOMASS ENERGY PLANTS

TIME AFTER PLANTING (months)	CASSAVA[1] LEAVES (% of Plant Dry Weight)	SUGARCANE[2] Total Plant Fresh Weight (g)	Percentage Green Top	PINEAPPLE[3] Total Plant Dry Weight (g)	Percentage Leaves
3	16			64	83
6	22			96	88
9	8	249	32	296	90
12	2	434	21	551	88
14		433	18		
15	7			812	80
16½		401	16		
18	7	390	15	1176	65
20		386	13	1278	47
21	1				

[1] From C. N. Williams: Exp. Agric. 7, 49 (1972).

[2] From J. L. Monteith: Exp. Agric. 14, 1 (1978).

[3] From Pineapple Research Institute of Hawaii, Honolulu.

NOTE: Format from Marzola and Bartholomew (1979). See references listed at end of entry.

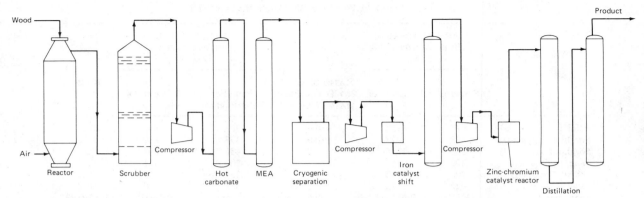

Fig. 2. Schematic flowsheet of process for producing methanol from wood wastes. MEA = monoethanolamine. (*After Hokanson and Katzen, 1978.*)

as a source of heat for distillation required during the manufacture of alcohol. Other byproducts, such as the leftover pith (used as an absorbant in feedstuffs made from stillage), are envisioned. As Lipinsky observes, the fact that the millable sugarcane stalk might yield so much salable coproduct that there would be no fiber left to make steam and electricity to run the process is not a serious disadvantage, because very little power is required by the new process as compared with the traditional grinding and milling process.

Direct Wood Burning. Wood was the major fuel of the United States until about 1886, when the consumption of coal equaled that of wood. Oil did not appear on the chart until about 1900 and gas in 1910–1920. The use of wood tapered off while other fuels climbed at amazing rates, but wood never ceased to be a factor, even if small. It is interesting to note that about a million modern woodburning stoves have been purchased since the late 1970s and that about 40% of the wood products industry is furnished by combusting bark and mill wastes. This amounts to about 1 quad (10^{15} Btu). Wood burning has been sufficiently extensive during the past few years to cause environmental concerns in some regions. Although wood has a low sulfur content and produces minimal amounts of nitrogen oxides even by the hottest fires (1370°C; 2500°F), nevertheless it contains air pollutants in the form of particulates, gases, and tars. Environmentalists in the New England region have estimated that the 300,000–400,000 tons of wood burned per year (New Hampshire only), if the fuel is very dry red oak, will add 1000 tons of particulates to the air; if dry white pine is burned, the total may be over

5000 tons. Since a mixture of woods usually is used, the figure lies somewhere between the two aforementioned quantities.

Fuels from Wood Wastes. Where economics may be favorable, factories that produce wood wastes may consider converting the wastes into fuels rather than to use them for firing boilers. A number of processes have been suggested for producing methanol and ethanol from such wastes. Two such proprietary processes are shown schematically in Figs. 2 and 3.

Biomass from the Oceans. As pointed out by Ryther (1980), the oceans are the largest uncultivated and underutilized pastures on earth. Seaweeds which have limited uses (hydrocolloids, such as agar, alginic acid, and carrageenan) could become a major source of biomass for energy. Although quite long-range, studies are being made of the phenomenon of bacterial chemosynthesis—a topic that has been spurred by the deep-ocean findings of the Galapagos Rift Thermal Springs. In *green plant photosynthesis*, water serves as the source of electrons and light as the source of energy. In *bacterial photosynthesis*, reduced sulfur compounds serve as the source of electrons and light as the source of energy. In *bacterial chemosynthesis*, reduced sulfur compounds serve as the source of electrons and chemical oxidation with free oxygen serves as the source of energy. Jannasch (1980) reports on very early research studies of bacterial chemosynthesis which ultimately could lead to an advantageous way of creating biomass.

Euphorbia. Considerable interest has been shown by some scientists in the genus *Euphorbia*, which yields a significant quantity of a milklike emulsion of hydrocarbons in water. Na-

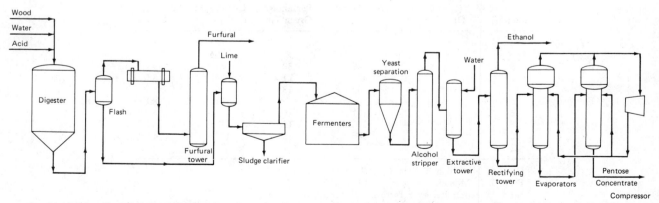

Fig. 3. Schematic flowsheet of process for producing ethanol from wood waste. Furfural and pentose concentrate are coproducts. (*After Hokanson and Katzen, 1978.*)

tives in the forests of Brazil have been familiar with the species *Cobaifera langsdorfii* for a number of years. Trees are tapped about twice per year, with the collection of 15–20 liters of hydrocarbon. Tests have shown that the liquid can be placed directly in the fuel tank of a diesel-powered vehicle. It is estimated that a plot of 100 acres (40.4 hectares) will yield about 25 barrels of fuel per year. The Brazilian government has established experimental plantations. Very few regions in the United States, possibly including southern Florida, would support this tropical tree. In addition to work at the University of California at Berkeley and an experimental plantation operated by the Japanese in Okinawa, most attention has been concentrated in Brazil. More detail is given by Maugh (1979).

Utilization of Urban and Industrial Wastes as Energy Sources

At a minimum, the solid waste production in the United States alone is estimated at about 566×10^9 kilograms (624×10^6 tons), representing an energy content of about 2.37×10^{18} Calories (9.42×10^{15} Btu). Although impractical, if all solid wastes could be converted to useful energy, this would represent about 10% of the energy requirements of the country. The types of dry combustible solids discarded are summarized in Table 3. The quantities shown estimate only the combustible portions of the solid waste—inert materials are not included. In terms of converting waste to energy, unfortunately, solid waste materials are not available in concentrations sufficient to fuel a solid waste-energy-producing plant except in certain

locations. Agricultural wastes, in particular, are not concentrated. An estimate of the quantity of concentrated waste is given in Table 4. The figures do not include wastes that presently are being recycled for use as energy sources.

Urban waste includes household, sewage, commercial, institutional, manufacturing plant, and demolition waste. The availability of this waste is directly related to the population living in urban areas of adequate size to support a given size system.

Manufacturing and processing wastes include all residuals generated from material inputs that leave the plant as product output. Office and packaging wastes associated with this sector are included in the urban waste sector. The majority of these wastes are from pulp and paper manufacturing, primary and secondary wood manufacturing, and the construction industry. Agricultural wastes include animal manures, crop wastes and forest and logging residues. The only available wastes in this sector are those animal wastes generated on large feedlots and dairies, and a portion of those crop wastes (bagasse and fruit tree prunings) not readily recycled back to the soil.

The energy recovery system selected dictates the extent that solid waste must be prepared. Some systems require nothing more than the removal of massive noncombustibles, such as kitchen appliances from the refuse, while other processes require extensive shredding, air classification, reshredding, and drying. In conjunction with fuel preparation, it is usually worthwhile to reclaim metals and glass for recycling.

One-stage shredding is often used to reduce waste to a nominal size as small as one inch (2.5 centimeters). When finer-sized fuel is required, a second shredding step is usually used after air classification has removed many of the noncombustibles. Both vertical and horizontal air classifiers depend on the heavy noncombustibles settling out by gravity in a moving air stream, while the lighter combustibles are pneumatically transferred through the air classifier. Denser combustibles, such as rubber and leather, may be removed with the heavy fraction, while some of the fine glass and metal foils are carried with the combustibles. Thus, desired separation may not always be achieved on one pass through an air classifier.

Some energy recovery systems require drying to remove excess moisture in the waste. This is required almost without exception when animal manures or sewage sludge is used as a fuel. Usually,

TABLE 3. SUMMARY OF COMBUSTIBLE SOLID WASTE DISCARDED IN THE UNITED STATES

Source of Waste	Quantity Discarded			Energy Value	
	% of Total Weight	10^9 Kilograms	10^6 Tons	10^{15} Cal	10^{15} Btu
Refuse and sewage/urban	16.3	92	101	413	1.64
Manufacturing and processing	5.8	33	36	144	0.57
Agricultural	77.9	441	487	1,817	7.21
Total	100.0	566	624	2,374	9.42

TABLE 4.   SUMMARY OF SOLID WASTE AVAILABILITY

Source of Waste	Number of Plants	Waste per Year		Energy Value per Year	
		10^9 Kilograms	10^6 Tons	10^{15} Cal	10^{12} Btu
To Solid Waste Plants capable of processing 90,700 kg (100 tons)/day					
Urban	1,598	52.9	58.3	23.8	943
Manufacturing	579	19.1	21.1	80	319
Agricultural	1,026	34.0	37.5	134	532
Sub-Total	3,203	106.0	116.9	237.8	1,794
To Solid Waste Plants capable of processing 453,000 kg (500 tons)/day					
Urban	252	41.7	46.0	187	744
Manufacturing	9	1.6	1.8	7	27
Agricultural	5	0.9	1.0	4	15
Sub-Total	266	44.2	48.8	198	786
Grand Total	3,469	150.2	165.7	435.8	2,580

NOTE: Plant size in terms of processing combustibles.

waste heat from the total process can be used for the drying system.

Pyrolysis. One process uses low-temperature flash pyrolysis to produce char and a highly viscous, highly oxygenated fuel oil, having a heating value of about 10,500 Btu/pound (5838 Calories/kilogram). The system uses two stages of shredding, air classification, and drying to produce a minus-24 mesh fuel for the pyrolysis reactor. The heat required for pyrolysis is derived from the combustion of the pyrolysis off-gas and from a portion of the char produced; it is transferred by means of a heat exchanger. From the pyrolysis unit, the gases are exhausted through a cyclone to remove the char and then scrubbed to remove the oil, water, and other solids and liquids. Major properties of the pyrolytic oil include: By percent (weight): Carbon, 57.5%; oxygen, 33.4%; hydrogen, 7.6%; nitrogen, 0.9%; chlorine, 0.3%; sulfur, 0.1 to 0.3%; ash, 0.2 to 0.4%. The oil has a heating value of 10,500 Btu/pound (5838 Calories/kilogram); specific gravity, 1.30; pour point, 32°C; flash point, 56°C; and a viscosity of 3,150 (SSU at 88°C).

In another system, municipal refuse is charged at the top of a shaft furnace and is pyrolyzed as it passes downward through the furnace. Oxygen enters the furnace through tuyeres near the furnace bottom and passes upward through a (1,425 to 1,650°C) combustion zone. The products of combustion then pass through a pyrolysis zone and exit at about 93°C. The offgas then passes through an electrostatic precipitator to remove flyash and oil formed during pyrolysis. The latter are recycled to the furnace combustion zone. The gas then passes through an acid absorber and a condenser. The clean fuel gas has a heating value of about 300 Btu/cubic foot (2670 Calories/cubic meter) and a flame temperature equivalent to that of natural gas. The solid waste that remains is a slag at the furnace bottom.

Biological Methane Production. This process involves the anaerobic digestion of a solid waste and water or sewage sludge slurry at 60°C for five days to produce a methane-rich gas. Solid waste is prepared by shredding and air classification, followed by blending with water to produce a mixture of 10 to 20% solids concentration. The slurry is heated and placed in a mixed digester at 60°C for five days detention. The digester gas is drawn off and separated into carbon dioxide and methane. The spent slurry from the digester is pumped through a heat exchanger to partially heat the incoming slurry prior to filtration. The filtrate is returned to the blender and the sludge is used for landfill. Heat addition to the refuse slurry is required to maintain the required digester temperature. The process is well suited for use on sewage sludge, animal manures, and other high-moisture-content solid wastes. It is estimated that the process can reduce the volume of volatile solids by 75%, while producing about 3,000 cubic feet (85 cubic meters) of methane per ton of incoming solid waste. The major residue is used for land fill or incinerated. About 10% of the methane is required to heat the digester feed.

Direct Steam Process. One process uses a rotary kiln pyrolizer followed by an afterburner and boiler to produce steam from shredded waste. The pyrolysis process in a kiln is operated counter-currently. Solid waste enters at one end and pyrolyzed residue is discharged at the other. External fuel and air are introduced at the residue discharge area and combustion products and pyrolysis gases leave the kiln at the feed opening. This arrangement causes the solid waste to be exposed to progressively higher temperatures as it passes through the kiln. The kiln off-gases pass through a refractory-lined afterburner into which air is introduced to allow complete combustion prior to passing through the waste heat boiler. A wet scrubber is used for air pollution control, while an induced draft fan is used to draw the gases through the system. One ton of solid waste, augmented by 1.25 million Btu (0.3 million Calories) from auxiliary fuel and 55 kilowatt-hours of electricity will produce about 4,800 pounds (2177 kilograms) of steam at 330 psig (22.4 atmospheres) along with 200 pounds (91 kilograms) of char.

Waterwall Incinerators. These devices generate steam by burning unprepared solid waste on a grate and passing the hot products of combustion through a boiler. Numerous waterwall incinerators have been built in Europe and the United States. Unprepared refuse is taken from storage pits and charged directly into the incinerator feed hopper. From there, the refuse drops onto a feed chute and then is fed automatically onto

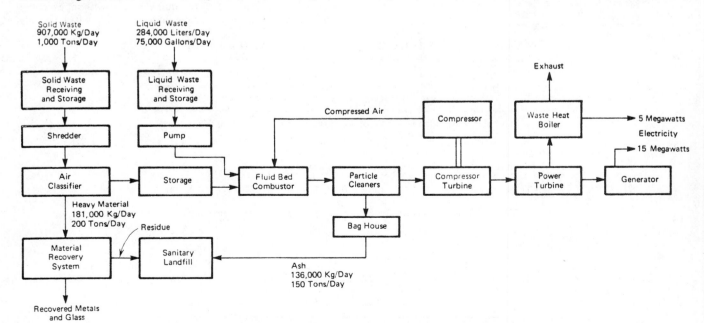

Fig. 4.   Process for converting solid and liquid wastes to electrical power. (*Combustion Power Company, Inc.*)

the stoker by means of a hydraulic feed ram. Temperatures in the 870°C range effectively burn the solid waste. Before the flue gas enters the boiler, secondary air is added to produce a temperature in the 1,090°C range. The boiler is constructed of membrane waterwalled tubes with extruded fins. After passing through the boiler, the gases travel through an economizer section and then into an electrostatic precipitator for particle removal. A typical 1,000 tons/day (900 metric tons) waterwall incinerator produces about 300,000 pounds (136,080 kilograms) of steam per hour.

Fluid Bed/Turbine Systems. Several variations of a system involving combustion of solid wastes in a fluid bed combustor and using the gases produced for expansion through a gas turbine are available. A system of this type is shown in block diagram format in Fig. 4. Shredded and air classified waste, together with liquid waste, are fired in a fluidized bed combustor to produce hot products of combustion for expansion through the gas turbine. The turbine compressor provides pressurized air for combustion and for transporting the shredded waste from beneath the rotary air lock feeder valves into the combustor. Three stages of cyclones are used to remove fly ash from the hot gas stream prior to expansion through the gas turbine. A waste heat boiler is used to produce steam for the generation of additional electricity. The primary control loop regulates the fuel feed rate in response to turbine temperatures.

Liquid waste firing is not required, but is available as a desired option because the addition of water to the system allows more solid waste to be consumed and greater amounts of electricity to be produced at a lower-than-normal inlet temperature in the same size of equipment. System efficiency is reduced somewhat by the addition of water, but with a negative value fuel that is not undesirable. The energy used to evaporate the water is partially recovered by the expansion of the resulting steam through the gas turbine. The system is flexible in terms of the type of fuel that can be burned (up to 60% water) and the mix of steam and electricity produced.

References

Alexander, A. G.: "Sugarcane Physiology," Elsevier, New York, 1974.
Anderson, R. E.: "Biological Paths to Self-Reliance—Biological Solar Energy Conversion," Van Nostrand Reinhold, New York (1979).
Bolton, J. R.: "Solar Fuels," *Science*, **202**, 705–711 (1978).
Clements, H. F.: "International Society of Sugar Cane Technologists; Proceedings of the 8th Congress," Elsevier, New York, 1953.
Hammond, A. L.: "Photosynthetic Solar Energy—Biomass," *Science*, **197**, 745–746 (1977).
Hayes, D.: "Biological Sources of Commercial Energy," *Bioscience*, **27**, 540–546 (1977).
Hokanson, A. E., and R. Katzen: "Chemicals from Wood Waste," *Chem. Eng. Prog.*, **74**, 67–71 (1978).
Jannasch, H. W., and C. O. Wirsen: "Chemosynthetic Primary Production at East Pacific Seafloor Spreading Centers," *Bioscience*, **29**, 591–598 (1979).
Jannasch, H. W.: "Chemosynthetic Production of Biomass," *Oceanus*, **22**, 4, 59–62 (1980).
Kos, P., Meier, P. M., and J. M. Joyce: "Economic Analysis of the Processing and Disposal of Refuse Sludges," Curran Associates, Inc., Washington, D.C., 1974.
Lipinsky, E. S.: "Fuels from Biomass: Integration with Food and Materials Systems," *Science*, **199**, 644–651 (1978).
Lowe, R. A.: "Energy Recovery from Waste, Solid Waste as Supplementary Fuel in Power Plant Boilers," Pubn. SW-36d, U.S. Environmental Protection Agency, Washington, D.C., 1973.
Marzola, D. L., and D. P. Bartholomew: "Photosynthetic Pathway to Biomass Energy Production," *Science*, **205**, 555–559 (1979).
Mattill, J. I.: "Particulates from Wood Burning in New Hampshire," *Technology Rev. (MIT)*, **81**, 2, 18 (1978).
Maugh, T. H., II: "Diesel Fuel Grows on Trees," *Science*, **206**, 436 (1979).
Nystrom, J. M., and S. M. Barnett: "Biochemical Engineering: Renewable Sources of Energy and Chemical Feedstocks," Amer. Inst. of Chem. Engrs., New York, 1978.
Rau, G. H., and J. I. Hedges: "Carbon-13 Depletion in a Hydrothermal Vent Mussel: Suggestion of a Chemosynthetic Food Source," *Science*, **203**, 648–649 (1979).
Rhyther, J. H., DeBoer, J. A., and B. E. Lapointe: "Cultivation of Seaweeds for Hydrocolloid, Waste Treatment and Biomass for Energy Conversion," *Proc. of 9th Intl. Seaweed Symposium*, Santa Barbara, California (A. Jensen and J. R. Stein, editors), Science Press, Princeton, New Jersey, 1979.
Sarkanen, K. V., and D. A. Tillman (editors): "Progress in Biomass Conversion," Vol. 1, Academic, New York, 1979.
Shuster, W. W., Editor: "Proceedings of 2nd Annual Fuels and Biomass Symposium, Rensselaer Polytechnic Institute, Troy, New York, June 1978.
Tillman, D. A.: "Wood as an Energy Resource," Academic, New York, 1978.
Weisz, P. B., and J. F. Marshall: "High-Grade Fuels from Biomass Farming: Potentials and Constraints," *Science*, **206**, 24–29 (1979).

BIOTECHNOLOGY. This is a relatively recent word to enter the scientific lexicon. Variously defined and, as of the mid-1980s, still not fully stabilized in terms of exact meaning, biotechnology is generally considered an "umbrella" term to embrace the many technologies from numerous disciplines, including chemical, that deal essentially with the commercialization of products from *living matter*, as contrasted with products made from mineral and other lifeless substances. The *biosciences*, so-called, fundamentally stem from biology. Over the past few decades, biology has been split into numerous special disciplines—biochemistry, biomechanics, bioelectronics, molecular biology, industrial microbiology, genetic engineering, among others.

Aside from a few of these special disciplines, biotechnology tends to be synonymous with *industrial microbiology*. The latter field, of course, extends back into time well beyond the early days of Louis Pasteur (France), John Tyndall (Britain), Charles Cagniard de la Tour (France), Theodore Schwann (Germany), and Friedrich Kützing (Germany), among many other early investigators. Although crudely, biotechnology has been practiced for many centuries as exemplified by the production of wine, cheese, yogurt, vinegar, and other food products that result from probably the most important of all biotechnical operations, *fermentation*. See **Fermentation**.

During the last few years, biotechnology has been somewhat disproportionately associated with genetic engineering. See **Recombinant DNA**. Outstanding progress in biotechnology has been made during the past decade, notably in the development and production of antibiotics, immunological products, and interferons, among others.

BIOTIN. Infrequently referred to as Bios IIB, protective factor X, vitamin H, egg-white injury factor, and CoR, biotin is required by most vertebrates, invertebrates, higher plants, most fungi and bacteria and falls into the general classification of vitamins. In certain species, a deficiency of this substance is a cause of desquamation of the skin; lassitude, somnolence, and muscle pain; hyperesthesia; seborrheic dermatitis; alopecia, spastic gait and kangaroolike posture (rats and mice); dermatitis and perosis (chicks and turkeys); progressive paralysis and K^+ deficiency (dogs); alopecia, spasticity of hind legs (pigs); and thinning and depigmentation of hair (monkeys).

Biotin reacts with an oxidized carbon fragment (denoted as CO_2) and an energy-rich compound, adenosine triphosphate (ATP), to form carboxy biotin, which is "activated carbon dioxide." Biotin is firmly bound to its enzyme protein by a peptide linkage. Structurally, biotin and carboxy biotin are:

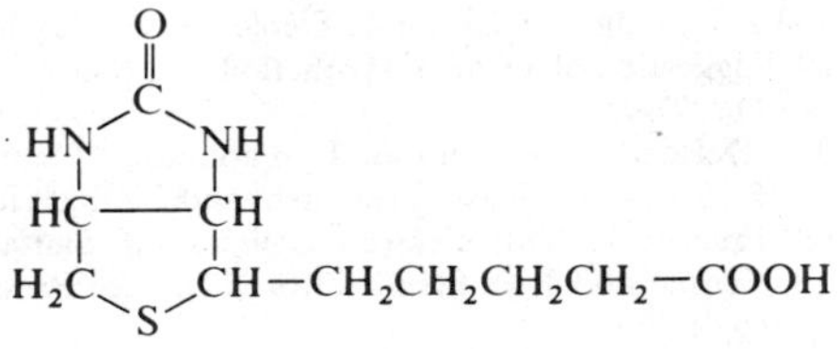

Biotin

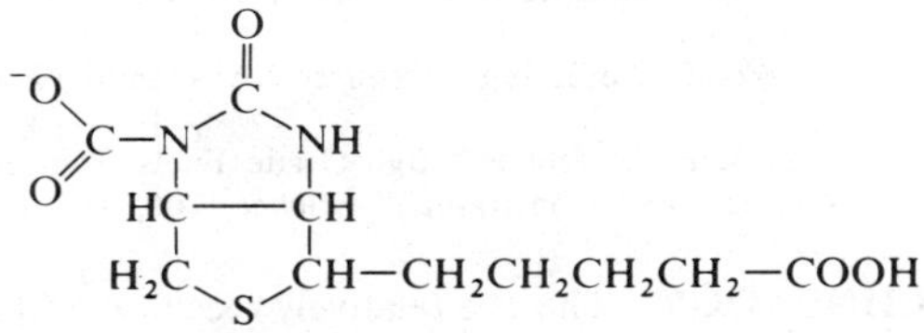

"Activated" carboxy-biotin

Biotin enzymes are believed to function primarily in reversible carboxylation-decarboxylation reactions. For example, a biotin enzyme mediates the carboxylation of propionic acid to methylmalonic acid which is subsequently converted to succinic acid, a citric acid cycle intermediate. A vitamin B_{12} coenzyme and coenzyme A are also essential to this overall reaction, again pointing out the interdependence of the B vitamin coenzymes. Another biotin enzyme-mediated reaction is the formation of malonyl-CoA by carboxylation of acetyl-CoA ("active acetate"). Malonyl-CoA is believed to be a key intermediate in fatty acid synthesis.

Bios (in Greek means *life*) was a word coined to describe a growth-promoting substance for yeast and discovered by Wildiers in 1901. When added in small amounts to sugar and salts medium, it permitted rapid growth of yeast even from a small seeding. Subsequent investigations proved that there was not merely a single substance involved, but that depending upon the strain of yeast and the circumstances of testing, a number of different substances could act, often synergistically, to promote the rapid growth of yeast. Pantothenic acid, biotin, inositol, thiamine, and pyridoxine all have "bios" properties when appropriately tested. Even an amino acid may be a limiting factor for yeast growth when other needs are supplied. The term *bios* has fallen from use in the literature.

Biotin required for growth and normal function by animals, yeast, and many bacteria is seldom found in deficiency in humans because the intestinal bacteria synthesize it in sufficient quantity to meet requirements. Biotin deficiency does occur, however, in animals fed raw whites of eggs. The egg white contains the protein *avidin* which combines with biotin, and this complex is not broken down by enzymes of the gastrointestinal tract. Hence, a deficiency develops.

Biotin was first isolated in pure form in 1936 by two Dutch chemists, Koegel and Tonnis, who obtained 1.1 milligrams from 250 kilograms of dried egg yolk. They showed that the compound was necessary for the growth of yeast and gave it the name *biotin*. Five years later, in America, György and coworkers found that the same compound prevented the toxicity of raw egg white in animals and, in 1942, du Vigneaud and collaborators determined the structure of the compound.

Bioavailability of Biotin. Factors which cause a decrease in bioavailability include: (1) presence of avidin in food; (2) cooking losses; (3) presence of antibiotics; (4) presence of sulfa drugs; and (5) binding in foods (such as yeast). Availability can be increased by stimulating synthesis by intestinal bacteria.

Antagonists of biotin include desthiobiotin in some forms, urelene phenyl, homobiotin, urelenecyclohexyl butyric and valeric acid, norbiotin, avidin, lysolecithin, and biotin sulfone. Synergists include vitamins B_2, B_6, B_{12}, folic acid, pantothenic acid, somatotrophin (growth hormone), and testosterone.

Precursors for the biosynthesis of biotin include pimelic acid, cysteine, and carbamyl phosphate. Desthiobiotin acts as an intermediate. In plants, the production sites are seedlings and leaves. In most animals, production is in the intestine. Storage site is the liver.

Some of the unusual features of biotin noted by investigators include: (1) binding and inactivation by avidin protein found in egg white; (2) fetal tissues and cancer tissues are higher in biotin than normal adult tissues; (3) biotin deficiency increases severity and duration of some diseases, notably some of the protozoan infections; and (4) oleic acid and related compounds act to replace biotin as unspecific stimulatory compounds in bacteria.

References

Du Vigneaud, V., et al: *J. Biol. Chem.*, **146**, 475 (1942).
Greenberg, D. M. (Editor): "Metabolic Pathways," 3rd Edition, Volumes I–IV, Academic, New York, 1967.
Haresign, W., and D. Lewis: "Recent Advances in Animal Nutrition," Butterworth Group, Woburn, Massachusetts, 1977.
Koegel, F., and B. Tonnis: *Z. Physiol. Chem.*, **242**, 43 (1963).
Kutsky, R. J.: "Handbook of Vitamins and Hormones," Van Nostrand Reinhold, New York, 1973.
Preston, R. L.: "Typical Composition of Feeds for Cattle and Sheep—1977–1978," *Feedstuffs*, A-2-A (October 3, 1977); and 3A-10A (August 18, 1978).
Staff: "Food Chemicals Codex," National Academy of Sciences, Washington, D.C., 1972.
Staff: "Nutritional Requirements of Domestic Animals," separate publications on beef cattle, dairy cattle, poultry, swine, and sheep, periodically updated, National Academy of Sciences, Washington, D.C.
White, C.: "Biotin Requirements of Broilers on Litter," World's Poultry Congress, Rio de Janeiro, Brazil, 1978.
Wood, H. G.: "Biotin," in "The Encyclopedia of Biochemistry," (R. J. Williams and E. M. Lansford, Jr., editors), Van Nostrand Reinhold, New York, 1967.

BIOTITE. A common silicate mineral containing potassium, magnesium, iron and aluminum. Biotite is found in granitic rocks, gneisses, and schists. Although actually monoclinic, it often assumes a pseudohexagonal form. Like others of the mica group, it shows a highly perfect basal cleavage. Hardness, 2.5–3; specific gravity, 2.7–3.4; luster, pearly to vitreous or sometimes submetallic when very black in color; cleavage sheets are elastic; color, greenish to brown or black; transparent to opaque. A general formula is $K(Mg,Fe)_3(Al,Fe)Si_3O_{10}(OH,F)_2$.

Biotite is occasionally found in large sheets, especially in pegmatite veins. It also occurs as a contact metamorphic mineral or the product of the alteration of hornblende, augite, wernerite and similar minerals.

Biotite is found in the lavas of Vesuvius, at Monzoni, and in many other European localities; in the United States, especially in the pegmatites of New England, Virginia, and North

Carolina, and the granite of Pikes Peak, Colorado. The mineral was named in honor of the French physicist, J. B. Biot. Biotite is also known as *iron mica*.

BIPHENYLS. A lay term usually used for polychorinated biphenyls (PCBs), which have been implicated in several instances of food contamination. In 1973, food and feedstuffs were contaminated with polybrominated biphenyls in Michigan. Similarly, contamination of the James River was found in 1975. In a more recent instance, in Montana a power transformer stored in a shed at a packing company leaked PCB coolant into slaughterhouse sewage. These wastes were collected, rendered, and added to animal feed for distribution throughout nine states. A complete listing of the various PCBs can be found in "Dangerous Properties of Industrial Materials," 5th Edition, N. I. Sax, Editor, Van Nostrand Reinhold, New York (1979).

BISMUTH. Chemical element symbol Bi, at. no. 83, at. wt. 208.981, periodic table group 5a, mp 271.3°C, bp 1562 ± 3°C, density 9.78 g/cm³ (20°C). Elemental bismuth has a rhombohedral crystal structure. The metal is of a silvery-white color with limited ductility. Like gallium, bismuth is one of the few metals that increases its volume (3.32%) upon solidifying from the molten state. It is the most diamagnetic of all the metals. All isotopes of the element (^{205}Bi through ^{215}Bi) are radioactive. See **Radioactivity.** However, the naturally occurring isotope ^{209}Bi generally is not regarded in this category because of its extremely long half-life (2×10^{17} years). Although described by Basil Valentine in the fifteenth century, the element was not defined as a new element until its characteristics were published in 1773 by C. Geoffroy and T. Bergman.

First ionization potential, 7.287 eV; second, 16.6 eV; third, 25.56 eV; fourth, 45.1 eV; fifth, 55.7 eV. Oxidation potentials Bi + $H_2O \rightarrow BiO^+ + 2H^+ + 3e^-$, −0.32 V; Bi + $3OH^- \rightarrow BiOOH + H_2O + 3e^-$, 0.46 V. Other important physical characteristics of bismuth are given under **Chemical Elements.**

Bismuth occurs as native bismuth in Bolivia and Saxony and frequently is associated with lead, copper, and tin ores—the sulfide (bismuthinite, bismuth glance, Bi_2S_3) is also found in nature. Separation of bismuth from lead takes place during the electrolytic refining of the latter, with bismuth remaining in the anode mud, or by prometallurgical methods by which it is removed from the lead as a calcium-magnesium compound. See also **Bismuthnite.**

Alloys. Metallurgically, bismuth is used in the production of low melting point fusible alloys and as an additive to steel, cast iron, and aluminum. The fusible alloys contain about 50% bismuth in combination with lead, tin, cadmium, and indium and are used in a variety of ways, including fire-protective devices, joining and sealing hardware, and in short-life dies. Because of the special volume-increase property with solidification, bismuth is used to manufacture alloys with a zero liquid-to-solid volume change. Alloy compositions are given in the accompanying table. The addition of about 0.2% bismuth, along with a similar quantity of lead, improves the machinability of aluminum. Very small quantities (0.02%) of bismuth are used in the production of malleable cast iron for stabilization of carbides upon solidification, particularly desirable for castings with heavy cross sections. Combinations of bismuth and tin and bismuth and cadmium have found use as counterelectrode alloys in the manufacture of selenium rectifiers. Bismuth telluride Bi_2Te_3 and bismuth selenide Bi_2Se_3 display thermoelectric properties. With modification, these compounds are used for certain commercial and military solid-state devices, including small units for portable power generation and refrigeration.

SOME REPRESENTATIVE LOW-MELTING-POINT ALLOYS CONTAINING BISMUTH

Fusible alloy, melting at 96°C	53% Bi	32% Pb	15% Sn
Fusible alloy, melting at 91.5°C	52% Bi	40% Pb	8% Cd
Fusible alloy, melting at 100°C	50% Bi	30% Sn	20% Pb
Fusible alloy, melting at 70°C. (Wood's metal)	50% Bi	25% Pb	12.5% Sn 12.5% Cd
Fusible alloy, melting at 70°C. (Lipowitz' alloy)	50% Bi	27% Pb	13% Sn 10% Cd
Rose metal	50% Bi	27% Pb	23% Sn
Bismuth solder, melting at 111°C.	40% Bi	40% Pb	20% Sn

Chemistry and Compounds. Generally, the chemical behavior of bismuth parallels that of arsenic and antimony, but bismuth is the most metallic of the group. Bismuth is not soluble in cold H_2SO_4 or cold HCl, but is attacked by these acids when hot and also by cold aqua regia. Elemental bismuth is not attacked by cold alkalies. The metal is soluble in HNO_3 and forms nitrates. When heated with chlorine, bismuth yields a chloride.

Some of the salts of bismuth are used in medicines for the relief of digestive disorders because of the smooth, protective coating the compounds impart to irritated mucous membranes. Like barium, bismuth also is used as an aid in x-ray diagnostic procedures because of its opacity to x-rays. At one time, certain bismuth compounds were used in the treatment of syphilis. Bismuth oxychloride, which is pearlescent, finds wide use in cosmetics, imparting a frosty appearance to nail polish, eye shadow, and lipstick. Bismuth phosphomolybdate has been used as a catalyst in the production of acrylonitrile for use in synthetic fibers and paints. Bismuth oxide or subcarbonate are used as fire retardants for plastics.

Bismuth trihalides exhibit an increased tendency toward hydrolysis, usually forming bismuthyl compounds, also called bismuth oxyhalides, which are often assumed to contain the ion BiO^+. This is not a discrete ion, however, and the crystal lattices of the "bismuthyl" compounds actually are comprised of Bi(III), O(−II) and X(−I) units. For example, BiOCl has the same crystal structure as PbFCl. The trihalides also form halobismuthates, with halogen ions, such as the chlorobismuthates, which contain the ions $BiCl_4^-$ and $BiCl_5^{2-}$. The BiI_4^- ion is precipitated analytically as the cinchonine salt.

Bismuth(III) oxide, Bi_2O_3, is the compound produced by heating the metal, or its carbonate, in air. It is definitely a basic oxide, dissolving readily in acid solutions, and unlike the arsenic or antimony compounds, not amphiprotic in solution, although it forms stoichiometric addition compounds on heating with oxides of a number of other metals. It exists in three modifications, white rhombohedral, yellow rhombohedral and gray-black cubical. Bismuth(II) oxide, BiO, has been produced by heating the basic oxalate.

Bismuth(III) hydroxide also is not significantly amphiprotic in solution, dissolving only in acids. Its formula is given as $Bi(OH)_3$ but it is difficult to isolate, due to adsorption of acid anions and to its dehydration to BiO(OH). The action of strong oxidants in concentrated alkalies on the hydroxide yields alkali bismuthates, such as $NaBiO_3$, sodium metabismuthate, which

$NaBi(OH)_6$ is initially produced. Other metal bismuthates may be made from them or directly from the oxides and Bi_2O_3, and bismuth(V) oxide is obtained by the action of HNO_3 on the alkali bismuthates; however, some oxygen is lost and the product is a mixture of Bi_2O_5 and BiO_2.

Bismuth(III) sulfide, Bi_2S_3, is precipitated by H_2S from bismuth solutions. Complex sulfide ions form only slowly, so bismuth sulfide may be separated from the arsenic and antimony sulfides by this difference in properties. Like the oxide, bismuth sulfide forms double compounds with the sulfides of the other metals.

Bismuth forms a number of complex compounds, including the sulfatobismuthates, e.g., $NaBi(SO_4)_2$ and $Na_3Bi(SO_4)_3$; and the thiocyanatobismuthates, e.g., $Na_3Bi(SCN)_6$, by the interaction of sodium thiocyanate and $Bi(SCN)_3$. The salts of bismuth tend to lose part of their acid readily, especially on heating, to form basic salts.

True pentavalent compounds of bismuth are rare, but include bismuth pentafluoride, BiF_5 (subl. 550°C), and $KBiF_6$; pentaphenylbismuth, $(C_6H_5)_5Bi$; various compounds $(C_6H_5)_3BiX_2$, where X = F, Cl, Br, N_3, NCO, CH_3CO_2, $\frac{1}{2}CO_3$; and tetraphenylbismuthonium salts, $[(C_6H_5)_4Bi]X$, where X = Cl, — $[B(C_6H_5)_4]$, etc.

Organobismuth Compounds. Numerous bismuth organic compounds have been prepared. Some of these include methylbismuthine CH_3BiH_2; phenyldibromobismuthine $C_6H_5BiBr_2$; potassium diphenylbismuthide $K[Bi(C_6H_5)_2]$; triphenylbismuthdihydroxide $(C_6H_5)_2Bi(OH)_2$; tetraphenylbismuthonium tetraphenylborate $[(C_6H_5)_4Bi][B(C_6H_5)_4]$; and pentaphenylbismuth $(C_6H_5)_5Bi$.

References

Carapella, S. C., Jr.: "Bismuth" in "Kirk-Othmer Encyclopedia of Chemical Technology," Vol. 3, 3rd Edition, 921–937, Wiley, New York (1978).

Staff: "The Bulletin of The Bismuth Institute," Bismuth Institute, Brussels, Belgium (Issued periodically).

—S. C. Carapella, Jr., ASARCO Incorporated, South Plainfield, New Jersey.

BISMUTHINITE. A mineral containing a sulfide of bismuth, Bi_2S_3, and sometimes copper and iron; a variety from Mexico contains about 8% antimony. Bismuthinite is orthorhombic, although its thin, needlelike crystals are rare, as it usually occurs in foliated or fibrous masses. It has one good cleavage parallel to the prism; hardness, 2; specific gravity, 6.78; metallic luster; streak, lead gray; color, similar but often with iridescent tarnish; opaque.

Bismuthinite is a rather rare mineral, although somewhat widely distributed. European localities are in Norway, Sweden, Saxony, Rumania, and England. It is found also in Bolivia, Australia, and in the United States in Utah. It is used as an ore of bismuth. Bismuthinite also is known as *bismuth glance*.

BISULFITE PROCESS. Pulp (Wood) Production and Processing.

BITUMEN. Tar Sands.

BIXIN. Annatto Food Colors.

BLAGDEN LAW. The depression of the freezing point of a solution is, for small concentrations, proportional to the concentration of the dissolved substance.

BLEACHING AGENTS. The end-result of bleaching action is decolorizing although decolorizing can be accomplished by means other than bleaching. Several types of compounds are used for bleaching to satisfy a wide range of requirements. Bleaching is used in a positive way to remove color imperfections (grayness, offwhiteness, etc.) from raw materials, such as cotton, wool, and other natural fibers; to produce a white, pleasing laundered product; to bleach flour and other foodstuffs; and to bleach wood pulp prior to the preparation of paper. Bleaching also occurs in a negative fashion through the action of the rays of the sun which cause fading of large numbers of fabrics, paints, and other coatings; or the action of washing and chemicals on surfaces and fabrics. The specific use of bleaching agents in connection with detergents is described under **Detergents**.

Bleaching agents fall roughly into two categories: (1) the hypochlorite or chlorine-type bleach; and (2) the peroxy compounds.

Hypochlorites. These compounds frequently are used to provide the sanitizing and bleaching property of chlorine without requiring the handling of liquid or gaseous chlorine. The term "liquid chlorine" is used in the swimming pool trade to describe sodium hypochlorite solutions, and the term "dry chlorine" is part of the registered trademark of a proprietary calcium hypochlorite product containing 70% available chlorine. Although nonscientific terms, their usage is well established in practice.

Sodium Hypochlorite: This product generally is available in one of two strengths. The household liquid bleach contains about 5.25% (wt) NaClO. The commercial product (sometimes called 15% bleach) contains 150 grams per liter of available chlorine. This is equivalent to about 13% (wt) sodium hypochlorite.

Calcium Hypochlorites: The forerunner of bleaching agents was patented as early as 1799 and termed *bleaching powder*. It is produced by passing chlorine gas over slaked lime and the resulting powder usually contains about 30% available chlorine. Although the original compound was quite unstable, it became of immense value to the bleaching of textiles and later for sanitizing.

In the United States, bleaching powder has largely been supplanted by an improved calcium hypochlorite product containing about 70% available chlorine. The compound, available under several brand names, is essentially calcium hypochlorite dihydrate.

In another form, calcium hypochlorite, $Ca(ClO)_2$, containing from 20 to 40 grams per liter of available chlorine is commonly produced by pulp mills for pulp bleaching.

Sodium perborate is the least expensive and most commonly used of the peroxy type bleaches and is incorporated in some household and commercial detergent formulations.

Hydrogen peroxide. A significant portion of the production of hydrogen peroxide, H_2O_2, goes into the bleaching of cotton, wool, and groundwood pulp, as well as use in hair bleaching preparations.

Organic peroxides also find use as bleaching agents, notably dibenzoyl peroxide, $[C_6H_5C(O)\cdot]_2$, which is still the preferred bleaching agent for flour. Peroxyacetic acid, CH_3CO_3H, also finds use in specialized bleaching situations.

Sodium bromide, in combination with hypochlorites, is sometimes used in bleaching systems, particularly for cellulosics.

See also **Oxidation and Oxidizing Agents**.

BLENDING. Mixing and Blending.

BLOOD. A major tissue of the human body, blood is a characteristically red, mobile fluid with an average sp gr of about 1.058. Slightly sticky and somewhat viscous, blood has a viscosity between 4.5 and 5.5 times greater than that of water at the same temperature. Thus, blood flows somewhat more sluggishly than water. The odor of blood is characteristic; the taste is slightly saline. The pH of blood ranges between 7.35 and 7.45. The complex acid-base regulatory system is described a bit later in this entry. Under normal conditions, the blood circulates through the body at a temperature of 100.4°F (38.0°C). This is slightly higher than the body temperature as determined by mouth, 98.6°F (37.0°C). An adult human of average age and size has just over 6 quarts (5.7 liters) of blood.

In very general terms, blood serves as a chemical transport and communications system for the body (i.e., it carries chemical messengers as well as nutrients, wastes, etc.). Circulated by the heart through arteries, veins, and capillaries, blood carries oxygen and a variety of chemicals to all cells, acting as a delivery agent to serve the needs of the cells. Blood also takes away waste products, including carbon dioxide, from the various tissues to organs, such as the kidneys and lungs, which ultimately dispose these wastes to the environment. Thus, the blood serves as a collecting agent.

Unlike a simple liquid, such as water, or a simple solution, such as salt water, blood is a complex fluid made up of several components, each of which is, in turn, extremely complex and even after decades of research it is still not fully understood. Many of these substances are solids in suspension. Unlike most simple liquids that are not easily changed when exposed to air or to slight alterations in their environment, the physical and biochemical properties of blood undergo marked changes (*hemostatic responses*) when blood is removed from the body's circulatory system and, for example, placed in a test tube. Separation of blood from its unusual environment immediately initiates a biochemical process which alters its properties and causes it to release its components. When so removed, blood shortly becomes viscid and forms a soft, jellylike substance, then soon separates into a firm solid mass (*clot*) and liquid (*serum*). This extremely important property of *clotting* is unique to blood among known inorganic and organic fluids and solutions. Were it not for this property, a person would bleed (*hemorrhage*) to death if a blood vessel were opened by accident or as the consequence of disease. Thus, blood may be described as a living fluid and most accurately as a living tissue, like the other tissues of the body.

Illustrative of the complex constitution of blood is the partial list of constituents given in Table 1.

Principal Components of Human Blood. Not considering the numerous substances, other than oxygen, carried by the blood where the main function of the blood is one of transport, the main functional components of the blood are indicated by Table 2.

Erythrocytes. It is estimated that an adult man will have about 5 million erythrocytes per cubic millimeter of blood. This is equivalent to about 82 billion erythrocytes per cubic inch of blood. In an adult woman, there are about 4.5 million erythrocytes per cubic millimeter. Erythrocytes are homogeneous circular disks with no nucleus. These red cells are about 0.0077 millimeter in diameter. When viewed singly by transmitted light, the erythrocyte has a yellowish red tinge, but when viewed in great numbers, the erythrocytes have the distinctly blood red coloration. Erythrocytes possess a certain degree of elasticity, so that they can pass through tiny apertures and passages on their way to reach tissue supplied by the capillaries.

The prime function of the erythrocytes is to deliver oxygen to peripheral tissues. This oxygen is furnished to these cells by an exchange-diffusion system brought about in the lungs. The color of the erythrocytes is derived from a red iron-containing pigment called *hemoglobin*. This is a conjugated protein

TABLE 1. REPRESENTATIVE CONSTITUENTS OF HUMAN BLOOD
(Values are per 100 milliliters)

Constituent	Plasma or Serum	Whole Blood	Constituent	Plasma or Serum	Whole Blood
Adenosine	1.09 mg		Mucopolysaccharides	175–225 mg	
Adenosine triphosphate (total)		31–57 mg	Mucoproteins	86.5–96 mg	
Amino acids (total)		38–53 mg	Nicotinic acid	0.02–0.15 mg	0.5–0.8 mg
Ammonia N	0.1–1.1 mg	0.1–0.2 mg	Nitrogen (total)		3.0–3.7 g
Ascorbic acid	0.7–1.5 mg	0.1–1.3 mg	Non-protein nitrogen	18–30 mg	25–50 mg
Base (total)	145–160 meq/liter		Nucleotide (total)		31–52 mg
Bicarbonate	25–30 meq/liter	19–23 meq/liter	Nucleotide phosphorus		2–3 mg
Bile acids		0.2–3.0 mg	Oxygen (arterial)		17–22 vol %
Biotin		0.7–1.7 μg	Oxygen (venous)		11–16 vol %
Blood volume		2990–6980 ml	pH	7.38–7.42	7.36–7.40
—adult men	33.7–43.7 ml/kg	66.2–97.7 ml/kg	Pantothenic acid	6–35 μg	15–45 μg
—adult women	32.0–42.0 ml/kg	46.3–85.5 ml/kg	Polysaccharides (total)	73–131 mg	
—infants	36.3–46.3 ml/kg	79.7–89.7 ml/kg	Protein (total)	6.0–8.0 g	19–21 g
Carbon dioxide			Protein (albumin)	4.0–4.8 g	
—Arterial blood (total)		45–55 vol %	Protein (globulin)	1.5–3.0 g	
—Venous blood (total)		50–60 vol %	Purines (total)		9.5–11.5 mg
Cholesterol (total)	120–250 mg	115–225 mg	Pyruvic acid	0.7–1.2 mg	0.5–1.0 mg
Cholesterol esters	75–150 mg	48–115 mg	Riboflavin	2.6–3.7 μg	15–60 μg
Cholesterol (free)	30–60 mg	82–113 mg	Ribonucleic acid	4–6 mg	50–80 mg
Choline (total)	26–35 mg	11–31 mg	Sphingomyelin	10–47 mg	150–185 mg
Fat (neutral)	25–260 mg	85–235 mg	Thiamine	1–9 μg	3–10.7 μg
Fatty acids	190–450 mg	250–390 mg	Urea	28–40 mg	20–40 mg
Fibrinogen	200–400 mg	120–160 mg	Urea N	8–28 mg	5–28 mg
Fructose	7–8 mg	0–5 mg	Uric acid (male)	2.5–7.2 mg	0.6–4.9 mg
Glucose (adult)	65–105 mg	80–120 mg	Vitamin A (carotenol)	15–60 μg	9–17 μg
Hemoglobin	trace	14.8–15.8 g	Vitamin A (carotene)	40–540 μg	20–300 μg
Histamine		6.7–8.6 μg	Vitamin B$_{12}$ (cyanocobalamin)	0.01–0.07 μg	0.06–0.14 μg
Ketone bodies (total)	0.15–1.36 mg	0.23–1.00 mg	Vitamin D$_2$ (as calciferol)	1.7–4.1 μg	
Lactic acid	30–40 mg	5–40 mg	Vitamin E	0.9–1.9 mg	
Lecithin	100–225 mg	110–120 mg	Water	93–95 g	81–86 g
Lipids (total)	400–700 mg	445–610 mg			

NOTE: *Plasma* is the liquid portion of whole blood. *Serum* is the liquid portion of blood after clotting, the fibrinogen having been removed.

TABLE 2.   BASIC COMPONENTS OF BLOOD

OXYGEN TRANSPORT*

Erythrocytes	Sometimes called *red cells*.

IMMUNE SYSTEM*

Lymphocytes		
Monocytes (macrophages)		Sometimes collectively called *white cells*. Also, sometimes collectively called *leukocytes*.
Granulocytes (granular leukocytes)	Leukocytes	
Neutrophils (polymorphonuclear leukocytes)		
Eosinophils		
Basophils		

BLOOD CLOTTING*

Platelets	

MULTIFUNCTIONAL—PROVIDES VOLUME AND FLUIDITY TO BLOOD

Plasma	

* Predominant, but not exclusive role.

that consists of a globin (a protein) and *hematin* (a nonprotein pigment), the latter containing iron. Hemoglobin contains 0.33% iron. When hemoglobin combines with oxygen, *oxyhemoglobin* is formed. When oxygen is given up to the tissues, it is then reduced back to hemoglobin. The erythrocytes also carry some carbon dioxide from the tissues and function to maintain a normal acid-base balance (pH) of the blood. When the hemoglobin has its full complement of oxygen, it is a bright red. This scarlet blood is found in the arteries which carry the blood to organ tissues throughout the body. As the oxyhemoglobin gives up oxygen, it takes on a darker crimson hue, and this is found in the veins which return the blood to the lungs for reoxygenation. See also **Hemoglobin.**

Megaloblasts, cells that are the precursors of erythrocytes, are noted in the blood islands of the yolk sac of the human embryo. By the end of the embryo's second month of life, manifesting the second step in the erythrocytic series, *erythroblasts* are found in the liver and spleen. These cells are somewhat smaller and possess a smaller nucleus than the megaloblasts. At about the fifth month, centers of blood formation appear in the middle regions of the bones, with an accompanying progressive expansion of the marrow cavities. At this stage the marrow assumes nearly exclusively the function of producing the erythrocytes (red cells) required by the body—a process which continues throughout the life of the individual. At the time of birth, essentially all bone marrow is engaged in blood formation (not exclusively red cells). As the individual progresses toward maturity, much of the marrow of the long bones is converted into a fatty tissue in which blood cell formation (*hematopoiesis*) is no longer apparent. In adults, bone marrow active in the formation of blood cells is found in the ribs, vertebrae, skull, and the proximal ends of the humerus (upper arm) and femur (upper thigh).

Once erythrocytes enter the blood, it is estimated that they have an average lifetime of about 120 days. In an average person, this indicates that about 1/120th or 0.83% of the red cells are destroyed each day. At least three important mechanisms are involved in the death of erythrocytes: (1) *Phagocytosis*, defined as the ingestion of solid particles by living cells—in this case, by cells of the reticuloendothelial system. (2) *Hemolysis* by spe-

cific agents in the blood plasma. The erythrocytes are protected by a membrane. If this membrane is broken, the hemoglobin goes into solution in the plasma. Numerous substances (*hemolytic agents*) may cause this action and these include hypotonic solutions, foreign blood serums, snake venom, various bacterial metabolites, chloroform, bile salts, ammonia and other alkalis, among others. In this condition, the erythrocytes no longer can serve as oxygen carriers. (3) *Mechanical damage* and destruction, brought about by simple wear and tear as the reasonably fragile red cells circulate and recirculate through the body.

The stimulus for production of new red cells is provided by *erythropoietin*, a hormone that is apparently produced by the kidneys. The actual production is accomplished almost entirely by the red portions of the bone marrow, but certain substances necessary for their manufacture must be furnished by the liver. Surplus red cells, needed to meet an emergency, are stored in the body, mainly in the spleen. The spleen also breaks down old and worn red cells, conserving the iron during the process.

When a sudden loss of a large amount of blood occurs, the spleen releases large numbers of red cells to make up for the loss, and the bone marrow is stimulated to increase its rate of manufacture of blood cells. When a donor gives a pint of blood, it usually requires about seven weeks for the body reserve of red corpuscles to be replaced, although the circulating red cells may be back almost to normal within a few hours. Repeated losses of blood within a short time, however, may easily deplete the red cell reserves.

In addition to hemoglobin, it has been found that there are least two other alternative oxygen carriers—*hemerythrin* and *hemocyanin*. In overall terms, as presently understood, these carriers are minor. Unlike hemoglobin, these two blood proteins do not incorporate an iron-porphyrin ring. The three blood proteins are strikingly colored in their oxygenated states—the familiar red of hemoglobin; the unusual reddish-tinted violet of hemerythrin; and the cupric-bluish color of hemocyanin. Klotz and colleagues (Northwestern University and other locations) have made a detailed study of the alternative oxygen carriers and suggest that an understanding of the three-dimensional structure of hemerythrin and of the electronic state of the active site is approaching, in refinement, that which is currently known about hemoglobin.

White Cells. There are several types of white cells, which are sometimes collectively called *leukocytes*, although some authorities reserve that term to identify only the granulocytes. White cells are irregular in shape and size, but generally are larger than the red cells. They differ from the red cells in that each white cell contains a nucleus. Adult humans have from 5000 and 9000 leucocytes per cubic millimeter of blood. In infants, the number is essentially doubled. There is roughly a ratio of 1 white to every 700 red cells. When white cells increase in number, the condition is called *leukocytosis*, a situation that is presented in pneumonia, appendicitis, and abscesses, among other conditions. A decrease in the number of leukocytes below normal is called *leukopenia*. In *leukemia*, there is an uncontrolled increase in the number of leucocytes.

In general terms, the white cells, each type with a specific function, accomplish the following actions: (1) Protection of the body from pathogenic organisms; and (2) participation in tissue repair and regeneration. Over the years, an increasingly detailed understanding of the white cells has occurred. Generally, whenever bacteria or other foreign substances enter the tissues, large numbers of white cells immediately travel through the walls of the blood vessels and to the site of disturbances. They take the bacteria and any other foreign materials into their own bodies, where they are digested. White cells are able to break up and carry away even as large an object as a splinter or thorn in the skin. They also help in carrying away dead tissue and blood clots which remain after a wound. *Pus* is largely composed of white cells which have been drawn to the infected area, as well as the dead and disintegrating tissue and bacteria. During severe infections, the white cells may be increased in the blood five- or tenfold. Because of this, a white cell count is made on the blood in order to confirm diagnosis in many infections.

Lymphocytes generally comprise between 25–30% of the white cells in human blood. These immunologically active cells are comprised of several classes, each of which has specific properties and functions. Lymphocytes are derived from stem cells located in the yolk sac and fetal liver. Later, some stem cells originate from the bone marrow. These cells then differentiate into lymphocytes in the primary lymphoid organs, principally the thymus and lymph nodes.

Monocytes (macrophages) are part of the mononuclear phagocytic system. They are large, mononuclear cells and comprise 3–8% of the leukocytes found in the peripheral blood. Monocytes originate in the bone marrow. When the mature cells enter the peripheral blood, they are called monocytes; when they leave the blood and infiltrate tissues, they are called macrophages. These cells play an important role in induction of the immune response. They present antigen to the lymphocytes that bear specific receptors for the antigen and also act as effector cells, attacking certain microorganisms and neoplastic cells.

Granulocytes contain specifically identifiable granules, including the neutrophils, eosinophils, and basophils. The *neutrophils* comprise 60–70% of all leukocytes in the blood. Neutrophils arise from precursors in the bone marrow and have a half-life of 4 to 8 hours in the blood, with about a day of life in the tissues. Neutrophils hasten to inflammatory sites by a number of different and poorly understood chemotactic (response to chemical stimulation) factors. Neutrophils have a marked capacity to phagocytize and destroy microorganisms. These cells also contain a number of degradative enzymes and small proteins. The cells are endowed with receptors for IgC and for a complement component (C3b). The *eosinophils* are named by virtue of the fact that the granules of cytoplasm are stainable with acid dyes, such as eosin. These cells are present in small numbers (2–4% of the blood), but under certain pathological conditions they show a marked increase. The exact function of eosinophils has been a mystery for many years. Some studies commenced in the mid-1970s have indicated a number of different functions. Many eosinophils have been found in tissues at sites of immune reactions that have been triggered by IgE antibodies (as found in nasal polyps or in the bronchial wall of some patients with asthma). Eosinophils have been found to contain several enzymes that can degrade mediators of immediate hypersensitivity, such as histamine, suggesting that they may control or diminish some hypersensitivity reactions. These cells have been found associated with infections caused by helminths (worms). The *basophils* are formed in the bone marrow and have a polymorphic nucleus. They occur only to the extent of about 1% of the leukocytes. The function of these cells is poorly understood. They are known to play a role in immediate hypersensitivity reactions and in some cell-mediated delayed reactions, such as contact hypersensitivity in humans and skin graft or tumor rejections and hypersensitivity to certain microorganisms in animals.

Platelets are the smallest of the formed elements of the blood. Every cubic millimeter of blood contains about 250 million platelets, as compared with only a few thousand white cells. There are about a trillion platelets in the blood of an average human adult. Platelets are not cells, but are fragments of the giant bone-marrow cells called *megakaryocytes*. When a megakaryocyte matures, its cytoplasm (substance outside cell nucleus) breaks up, forming several thousand platelets. Platelets are roughly disk-shaped objects between one-half and one-third the diameter of a red cell, but containing only about one-thirteenth the volume of the red cell. Platelets lack DNA and have little ability to synthesize proteins. When released into the blood, they circulate and die in about ten days. However, they do possess an active metabolism to supply their energy needs.

Because platelets contain a generous amount of contractile protein (*actomyosin*), they are prone to contract much as muscles do. This phenomenon explains the shrinkage of a fresh blood clot after it stands for only a few minutes. The shrinkage plays a role in forming a hemostatic plug when a blood vessel is cut. The primary function of platelets is that of forming blood clots. When a wound occurs, numbers of platelets are attracted to the site where they activate a substance (*thrombin*) which starts the clotting process. *Prothrombin* is the precursor of thrombin. Thrombin, in addition to converting fibrinogin into fibrin, also makes the platelets sticky. Thus, when exposed to collagen and thrombin, the platelets aggregate to form a plug in the hole of an injured blood vessel. Persons with a low platelet count (*thrombocytopenia*) have a long bleeding time. Platelet counts may be low because of insufficient production in the bone marrow (from leukemia or congenital causes, or from chemotherapy used in connection with cancer), among other causes. Also, individuals may manufacture antibodies to their own platelets to the point where they are destroyed at about the same rate they are produced. A major symptom of this disorder is purpura. Aspirin may aggreavate this condition. The bleeding of hemophilia results from a different cause. Transfusion of blood is a major therapy used in treating platelet disorders.

Platelets not only tend to stick to one another, but to the walls of blood vessels as well. Obviously because they promote clotting, they have a key role in forming thrombi. The dangerous consequences of thrombi are present in cardiovascular and cerebrovascular disorders. Many attempts have been made to explain the process of atherogenesis, that is, the creation of plaque which narrows arteries and, of particular concern, the coronary arteries. Recently, there has been increasing interest in the possible role of platelets in atherosclerosis. Evidence from experimentation with laboratory animals has provided some evidence of a role for platelets in this process. This is covered in some detail by Zucker (1980).

As reported by Turitto and Weiss (1980), red blood cells may have a physical and chemical effect on the interaction between platelets and blood vessel surfaces. Under flow conditions in which primarily physical effects prevail, it has been found that platelet adhesion increases fivefold as *hematocrit*[1] values increase from 10 to 40%, but undergoes no further increase from 40 to 70%, implying a saturation of the transport-enhancing capabilities of red cells. For flow conditions in which platelet surface reactivity is more dominant, platelet adhesion and thrombus formation increase monotonically as hematocrit values increase from 10 to 70%. Thus, the investigators suggest that red cells may have a significant influence on hemostasis and thrombosis; the nature of the effect is apparently related to the flow conditions.

Plasma. Normal blood plasma is a clear, slightly yellowish fluid which is approximately 55% of the total volume of the

---

[1] A hematocrit is a tube calibrated to facilitate determination of the volume of erythrocytes (red cells) in centrifuged, oxalated blood, expressed as corpuscular volume percent.

TABLE 3.   INORGANIC CONSTITUENTS
OF HUMAN PLASMA OR SERUM.

Constituent	Value/100 ml
Aluminum	45 μg
Bicarbonate	24–31 meq/liter
Bromine	0.7–1.0 μg
Calcium	9.8 (8.4–11.2) mg
Chloride	369 (337–400) mg
Cobalt	10 (3.7–16.6) μg
Copper	8–16 μg
Fluorine	109 (75–145) μg
Iodine, total	7.1 (4.8–8.6) μg
Protein bound I	6.0 (3.5–8.4) μg
Thyroxine I	4–8 μg
Iron	105 (39–170) μg
Lead	2.9 μg
Magnesium	2.1 (1.6–2.6) mg
Phosphorus, total	11.4 (10.7–12.1) mg
Inorganic P	3.5 (2.7–4.3) mg
Organic P	8.2 (7–9) mg
ATP P	0.16 (0–6.4) mg
Lipid P	9.2 (6–12) mg
Nucleic acid P	0.54 (0.44–0.65) mg
Potassium	16.0 (13–19) mg
Rubidium	0.11 mg
Silicon	0.79 mg
Sodium	325 (312–338) mg
Sulfur	
Ethereal S	0.1 (0–0.19) mg
Inorganic S	0.9 (0.8–1.1) mg
Non-protein S	2.8 (2.4–3.6) mg
Organic S	1.7 (1.4–2.6) mg
Sulfate S	1.1 (0.9–1.3) mg
Tin	4 μg
Zinc	300 (0–613) μg

blood. The plasma is a water solution in which are transported the digested food materials from the walls of the small intestine to the body tissues, as well as the waste materials from the tissues to the kidneys. Consequently, this solution contains several hundred different substances. In addition, the plasma carries antibodies, which are responsible for immunity to disease, and hormones. The plasma transports most of the waste carbon dioxide from the tissues back to the lungs. Plasma consists of about 91% water, 7% protein material, and 0.9% various mineral salts. The remainder consists of substances already mentioned. The salts and proteins are important in keeping the proper balance between the water in the tissues and in the blood. Disturbances in this ratio may result in excessive water in the tissues (swelling or edema). The mineral salts in the plasma all serve other vital functions in the body and must be supplied through diet. See Table 3.

Some of the blood plasma, as well as some of the white cells, filters through the walls of the blood vessels and out into the tissues. This filtered plasma (lymph) is a clear and colorless fluid which returns to the blood through a series of canals referred to as the *lymphatic system*. This system contains filters (*lymph nodes*) which remove bacteria and other debris from the lymph. These nodes, especially those located in the neck, armpit, and groin, may become swollen when an infection occurs in a nearby site. Blood clots do not occur normally while the blood is in the vessels. But in an injury, one of the plasma proteins (*fibrin*) forms a mesh in which the blood cells are trapped, and this mesh is the clot. Blood serum is the yellowish fluid left after the cells and fibrin have been removed from the blood.

Acid-Base Regulation.   The hydrogen ion concentration of the blood is maintained at a constant level (pH = 7.35–7.45) by a complex system of physicochemical processes involving, among others, neutralization, buffering, and excretion by the lungs and kidneys.

Most physiological activities, and especially muscular exercise, are accompanied by the production of acid, to neutralize which, a substantial alkali reserve, mainly in the form of bicarbonate, is maintained in the plasma, and so long as the ratio of carbon dioxide to bicarbonate remains constant, the hydrogen ion concentration of the blood does not alter. Any non-volatile acid, such as lactic or phosphoric, entering the blood reacts with the bicarbonate of the alkali reserve to form carbon dioxide, which is volatile, and which combines with hemoglobin by which it is transported to the lungs and eliminated by the processes of respiration. It will also be evident from this that no acid stronger than carbon dioxide can exist in the blood. The foregoing neutralizing and buffering effects of bicarbonate and hemoglobin are short-term effects; to insure final elimination of excess acid or alkali, certain vital reactions come into play. The rate and depth of respiration are governed by the level of carbon dioxide in the blood, through the action of the respiratory center in the brain; by this means the pulmonary ventilation rate is continually adjusted to secure adequate elimination of carbon dioxide. In the kidneys two mechanisms operate; ammonia is formed, whereby acidic substances in process of excretion are neutralized, setting free basic ions such as sodium to return to the blood to help maintain the alkali reserve. Where there is a tendency to the development of increased acidity in the blood, the kidneys are able selectively to re-absorb sodium bicarbonate from the urine being excreted, and to release into it acid sodium phosphate; where there is a tendency to alkalemia, alkaline sodium phosphate is excreted, the hydrogen ions thus liberated being re-absorbed to restore the diminishing hydrogen ion concentration. See also **Potassium** and **Sodium (In Biological Systems).**

Blood Osmotic Pressure.   The presence of solute molecules and ions in relatively high concentrations in blood establishes an osmotic pressure which tends to transport water from the exterior, through the semipermeable membranes of the blood vessel walls, into the bloodstream. This osmotic transport of water inward is opposed by the effect of hydrostatic pressure within the blood vessels, tending to force water (and soluble substances) out through the capillary walls. The loss through leakage of some of these solutes is indirectly restored through the action of the lympatic system. Among the blood constituents important in maintaining blood osmotic pressure (and thus helping to regulate the volume of fluid in the blood) are the blood proteins. Among these, the protein fraction termed *albumins*, being relatively low in molecular weight, makes the greatest contribution to the total osmotic effect.

Blood Processing and Transfusion Therapy.   Blood transfusion practice has changed markedly in recent years. At one time, units of *whole blood* were administered to patients with a variety of requirements stemming from different conditions. These conditions ranged from acute blood loss (hemorrhage, bleeding from injuries, etc.) to aplastic anemia, among other blood-related problems.

The outer portion of the erythrocyte (red cell) is a very complex material composed of proteins, polysaccharides, and lipids, many of which are *antigens*, sometimes referred to as blood group substances. The presence of most, if not all, of these antigens is genetically determined, and their number is such that there may be few, if any, individuals in the world with an identical set of antigens on the red cells—monozygotic twins excepted. These differences in whole blood were learned early in the development of transfusion technology. Fundamental to the refinement of the technology was the discovery by Karl Landsteiner (Nobel Prize winner in 1930) for his observations of the four hereditary blood groups. Landsteiner developed the ABO blood-typing system which serves as a principal guideline in determining the suitability of donors and recipients. This system consists of three allelic genes, dividing all humans into four groups, A, AB, B, and O. In a few rare individuals, the presence of a suppressor gene may prevent the expression of the A, B, O group character. The products of these genes are the A, B, and O antigens or substances. These antigens not only are located on red cells, but are widely distributed in the body, occurring in the endothelium of capillaries, veins, and arteries, and in numerous cells throughout the body. In addition to a cell-associated form, these antigens occur in soluble form in many body fluids, such as the saliva, gastric juice, urine, amniotic fluid, and in very high concentrations is pseudomucinous ovarian cyst fluid. All individuals possess cell-associated A, B, and O antigens. The presence of the soluble form, however, is governed by a recessive gene called the secretor gene which exists as two alleles, Se and se. Individuals who possess at least one Se gene secrete the antigens, while those with two se genes do not. The A, B, and O antigens are not uniquely human, but are quite widely distributed in nature. They are found on primate erythrocytes and in the stomach lining of pigs and horses. Intensive investigation has produced considerable information concerning the chemical composition of these antigens. They are extremely stable substances, which is attested by the fact that they can be extracted from Egyptian mummies, thus making it possible to obtain the blood groups of this ancient people. Specific antigenic activity is associated with the carbohydrate moiety, and since the A, B, and O substances possess the same four sugars, the difference between them lies in their arrangement. Analysis of purified A, B, and O substances reveals that about three-fourths of the weight is accounted for by four sugars: L-fructose, D-galactose, N-acetyl-D-glucosamine, and N-acetyl-D-galactosamine. The remainder consists of amino acids. See **Immune System and Immunochemistry.**

Cross-Matching. Upon receipt of a tube of clotted blood at a blood bank for typing and cross-matching, procedures are undertaken to determine which antigens are on patient's red cells and which antibodies against red cell antigens are present in the patient's serum. Typing is routine for red blood cell antigens in the ABO system and for a single specificity in the Rh system, namely, the D phenotype.

The Rh group, so denoted because the antigen was first found in the red cells of Rhesus monkeys, is very complex, consisting of perhaps 20 antigens. The Rh_0D antigen is the most important of these antigens because of its possible involvement in the induction of *hemolytic disease of the newborn.* Today, Rh_0D immune globulin (RhoGam® and Gamulin-Rh®, among others) is available to alleviate this danger. This danger is brought about when an Rh-negative woman and an Rh-positive man have an Rh-positive child. There is the grave risk that the woman will become sensitized to the Rh factor in her infant's blood and begin to produce anti-Rh antibodies. The first child is not usually affected, but with subsequent pregnancies, the mother may send sufficient damaging antibodies into the child's blood to threaten its life. When this occurs, in the absence of using the Rh_0D immune globulin, an exchange blood transfusion with almost complete replacement of the infant's blood by Rh-negative blood of the proper ABO group is necessary.

Component Therapy. Frequently, patients do not require all of the blood components and, in fact, their presence can cause many problems. From experience with whole blood therapy over a number of years, *component therapy* emerged. In component therapy, which has many advantages, the patient is given specifically what is needed by way of blood components. Further, separate blood fractions can be stored under those special conditions best suited to assure their biological activity at the time of transfusion. Component therapy also avoids the introduction of foreign antigens and antibodies. It is seldom that fresh whole blood is the treatment of choice providing that specific components are readily available within the time needed.

Processing of Donor Blood. When donor blood is received at a processing center, it is first tested for syphilis and hepatitis B antigen. One unit of whole blood is 500 milliliters. It is then separated into: (1) A unit of *packed erythrocytes* (volume of 300 milliliters and hematocrit value of 70 to 90). This substance is storable in citrate-phosphate-dextrose at 4°C (39.2°F) for up to 3 weeks. (2) A unit of *platelets* (packet) with a volume of 50 milliliters containing about 80 billion platelets). This substance is storable (while being gently mixed) at room temperature for 2 or 3 days. (3) A unit of *cryoprecipitate* (volume of about 10 milliliters containing from 80 to 120 units of Factor VIII and from 300 to 400 milligrams of fibrinogen). This substance is stable in a frozen condition for about one year. (4) One unit of *plasma* (a volume of about 200 milliliters, from which about half of the fibrinogen has been removed). This substance contains platelets, Factor VIII, as well as all remaining procoagulants, albumin, salt, and antibodies which were a part of the original plasma. This substance may be (a) stored in the frozen state, (b) refrigerated, or (c) further processed for individual globulin classes and albumin.

Thus, somewhat analogous to obtaining specific drugs for different conditions, the physician can order up specifically those blood components required for a given need. Platelets, which survive poorly in whole blood stored in acid-citrate-dextrose solution under refrigeration, can be obtained in platelet packets as previously mentioned, or as washed platelets. Factor VIII, for the management of hemophilia, can be obtained in a lyophilized or other purified form. In whole blood, by contrast, procoagulants stored in whole blood decay so that, after a few days, the availability of Factor VIII and other allied components is extremely low. As the result of additional processing, blood centers can furnish prothrombin complex concentrate (*Proplex*), each batch bearing a specific analysis. Preparations of peripheral white blood cells are available from centers with specific blood processing equipment. Additionally, substances for expanding plasma volume, such as fresh frozen plasma, albumin solutions, and Dextran, among others, are available.

Blood Substitutes. Researchers in Japan (Fukushima Medical Center) and in other institutions in Europe and North America have been investigating substances that, in major characteristics, may serve as a substitute for blood, particularly in emergency situations where rare blood types are not immediately available to severely ill patients who require transfusions. For example, in early 1979, a Japanese patient with a rare O-negative blood was given an infusion of one liter of a new, oxygenated perfluorocarbon emulsion. This compound carried oxygen through the patient's circulatory system until the rare blood could be obtained. Later that year, eight additional patients survived infusions with artificial blood. As early as 1966, investigators at the University of Cincinnati demonstrated that life could be sustained when rodents were immersed in perfluorochemicals for long periods. This class of chemicals can dissolve as much as 60% oxygen by volume, as contrasted with whole blood

(20%), or salt water or blood plasma (3%). Initially, a major problem existed because pure perfluorochemicals are not miscible with blood. In the late-1960s, researchers (University of Pennsylvania; Harvard School of Public Health) demonstrated that perfluorochemicals could be emulsified. Research is continuing along these lines. Also, new chemicals of this class are being sought. Initially, the research was done with perfluorobutyltetrahydrofuran and perfluorotripopylamine, both superior carriers of oxygen, but prone to concentrate in some organs of the body, notably the liver and spleen. In 1973, perfluorodecalin was found to be completely eliminated from the body. The approach in Japan has differed somewhat, in that research has been directed to add other chemicals which will increase the half-life of the chemicals in the body. As summarized by Maugh (1979), it appears that the perfluorochemical emulsions are beginning to fulfill some of the promise they first showed and that it may only be a matter of time and research effort before they can be used to sustain life in emergency situations.

Blood Recycling. In a process known as *autotransfusion*, introduced in the early 1980s, blood lost during operative procedures, particularly in heart surgery, is recycled back to the patient. Some reports indicate the need for donor blood can be reduced by as much as 60%. Instead of discarding blood lost during surgery, as has been the traditional practice, the blood is collected in a plastic bag with a special filter to cleanse impurities before the blood is returned to the patient. The procedure has many advantages, including costs of transfusion and elimination of risks from hepatitis, errors in mismatching blood types, and other complications that may arise with donor blood. Although results appear to be positive thus far, a few additional years may be required before the procedure is fully accepted as standard practice.

Blood as an Indicator of Disorders and Diseases. Since the blood performs many services for all parts of the body, it will reflect disturbances that occur as the result of many widely divergent diseases. This had led to the development of a variety of blood tests, either to confirm a diagnosis or to follow the effectiveness of treatment in the patient. *Immunological* or *serological* tests are performed to confirm the diagnosis of selected types of infectious diseases, and are based upon the principle that in certain diseases there appear in the blood specific substances (antibodies) which are produced by the body in resisting invasion by specific disease-producing media. One of the more widely used tests is the Kolmer test for syphilis. Blood typing tests are also serological in nature. A second group of blood tests are known as *hematological*. These tests determine the number of each type of circulating blood cell (*blood count*), the total volume of red cells in a blood sample (*hematocrit*), and the hemoglobin content of the blood. A *differential* blood count is one in which selected dyes are used to distinguish better the different kinds of white blood cells. These tests are important in diagnosing and treating illnesses, such as infections, the anemias, and the leukemias.

Another group of blood tests involves *bacteriological* techniques. Blood and bone marrow samples are obtained under aseptic precautions and introduced into a variety of artificial culture media, with subsequent isolation and identification of the specific microorganism responsible for the illness. Relative susceptibility of the specific strain of bacteria to the available chemotherapeutic and antibiotic agents may then be determined and the effectiveness of such agents in sterilizing the bloodstream can be determined by further blood cultures.

Many *chemical* tests are performed on blood samples to determine the quantitative relationships between circulating globulins, abulmin, sugar, nonprotein nitrogen, minerals, and other normal and abnormal constituents. Such chemical tests are important in diabetes, kidney diseases, the failing heart, and in pancreatic and liver diseases. In all of these disorders, pronounced changes in the relative amounts of the various chemical constituents of the blood occur. Chemical tests also may be performed on urine, spinal fluid, and saliva for some special purpose, and since most of these fluids are derived from the blood plasma, their chemical analysis frequently reflects changes in the blood itself. During prolonged therapy with certain drugs, it may be desirable to measure chemically the concentration of the drug in the blood plasma.

Sophisticated instrumentation and procedures are used in research involving blood and its functions. Phase contrast microscopy has the advantage that living cells can be studied for long periods of time; chromatin, mitochondria, centrosomes and specific granules can be seen and photographed at magnifications of 2,500×. The method is excellent for the study of granules of the matrix of cells which is unseen in traditionally fixed and stained cells. It is an excellent aid for those who wish to use the electron microscope, because areas demonstrated by light can be compared with those visualized by the electron beam. The study of blood by motion pictures (*microcinematography*) has been used for many years. With the invention of the phase microscope, this approach to the study of blood cells has been an important tool. Studies of the movements of the lymphocytes in rats showed a softening of the membrane at the forward moving end, and pseudopod formation; contractions of the cell force the inner plasma forward, while the external plasmagel remains fixed except at the posterior end, then it becomes softer and passes through the stiffer ring of plasmagel to become more gelated at the anterior end. This is an example of the type of detailed investigation that can be made with microcinematography. Using speed photography at 3,200 frames per second, the red blood cell has been observed to have an interior velocity of 30 × that of water at 38°C. In the dog's mesentery, red blood cells passing into capillaries from larger arterioles take the form of an inverted cap or parachute; when blood flow is stopped they become biconcave disks. The cub shape is suggested as bringing more surface close to the capillary endothelium.

Other blood research techniques include the use of physical and chemicals agents, ultracentrifugation, cytochemical methods, microincineration, and autoradiography.

References

Albert, S. N.: "Blood Volume and Extra-cellular Fluid Volume," 2nd edition, Charles C. Thomas, Springfield, Illinois, 1971.

Bank, A., Mears, J. G., and F. Ramirez: "Disorders of Human Hemoglobin," *Science*, **207**, 486–493 (1980).

Boggs, D. R.: "Neutrophils in the Blood Bank," *N. Engl. J. Med.*, **296**, 748 (1977).

Cash, J. D.: "Blood Transfusion and Blood Products," *Clin. Haematol.* (May 1976).

Grady, G. F., et al.: "Transfusions and Hepatitis," *N. Engl. J. Med.*, **298**, 1413 (1978).

Hoffman, R., et al.: "Diamond-Blackfan Syndrome," *Science*, **193**, 899–900 (1976).

Holmsen, H., Salganicoff, L., and M. H. Fukami: "Platelet Behavior and Biochemistry," in "Haemostasis: Biochemistry, Physiology, and Pathology," (D. Ogston and B. Bennett, editors), Wiley, New York, 1977.

Klotz, I. M., et al.: "Hemerythrin: Alternative Oxygen Carrier," *Science*, **192**, 335–344 (1976).

Maines, M. D., and A. Kappas: "Metals as Regulators of Heme Metabolism," *Science*, **198**, 1215–1221 (1977).

Maugh, T. H., II: "Blood Substitute Passes Its First Test," *Science*, **206**, 205 (1979).

Moffat, K., Deatherage, J. F., and D. W. Seybert: "A Structural Model for the Kinetic Behavior of Hemoglobin," *Science*, **206**, 1035–1042 (1979).

Mustard, J. F., et al.: "Platelets, Thrombosis and Atherosclerosis," *Prog. Biochem. Pharmacol.*, **13**, 312–325 (1977).

Perutz, M. F.: "Hemoglobin Structure and Respiratory Transport," *Sci. Amer.*, **239**, 6, 92–125 (1978).

Simmons, A.: "Basic Hematology," Charles C. Thomas, Springfield, Illinois, 1973.

Staff: "Symposium on Blood Groups: Their Genetics, Function, and Relation to Disease," *Mayo Clin. Proc.*, **52**, 135 (1977).

Sussman, L. N.: "Paternity Testing by Blood Grouping," 2nd edition, Charles C. Thomas, Springfield, Illinois, 1976.

Tullis, J. L.: "Clot," Charles C. Thomas, Springfield, Illinois, 1976.

Turitto, V. T., and H. J. Weiss: "Red Blood Cells: Their Dual Role in Thrombus Formation," *Science*, **207**, 541–543 (1980).

Weiss, H. J.: "Platelet Physiology and Abnormalities of Platelet Functions," in two issues of *N. Engl. J. Med.*, **293**, 531–541; and 580–588 (1975).

Zucker, M. B.: "The Functioning of Blood Platelets," *Sci. Amer.*, **242**, 6, 86–103 (1980).

BLOODSTONE. A massive variety of quartz of greenish color with small spots of red jasper somewhat resembling blood drops. It is used as a semi-precious stone. When placed in water in full sunlight bloodstone will frequently give a general reddish reflection, hence the name *heliotrope*, derived from the Greek words meaning sun and to turn, which is also given to bloodstone. See also **Chalcedony**; and **Quartz**.

BODYING AND BULKING AGENTS (Foods). These terms tend to be self-defining. Additives in these classifications are frequently described together because many substances will serve one or both purposes.

Bodying Agents. The *body* of a food substance is generally associated with the textural qualities of the substance, notably with mouthfeel or chewiness. Some food products, particularly those of a fabricated nature, may possess a full complement of desirable consumer appeals (taste, odor, color, nutritive value, etc.) and yet lack the desirable textural quality of body. Thus, soups, gravies, sauces, cheese foods and spreads, dressings, snack dips and spreads, margarines, among others, can be improved through the addition of bodying agents. For example, formulations for frozen desserts can be improved in this respect by the addition of low levels of a material such as microcrystalline cellulose (about 0.25% weight), in combination with soluble hydrocolloids, such as guar, locust bean gum, alginates, or carrageenans. See also **Gums and Mucilages.** Microcrystalline cellulose achieves about the same degree of body and substance in frozen desserts that is normally achieved only in well-emulsified products with a 2–4% higher fat content. This is the result of the ability of microcrystalline cellulose to stabilize the serum solids. Microcrystalline cellulose imparts body and smoothness to ice cream and ice milk, and tends to make them less "cold tasting." There are no off-flavors associated with the substance, and frozen desserts melt to smooth, creamy consistencies. Bodying agents play an effective, if not exclusive role, in improving freeze-thaw properties of numerous products. In another example, when whey or sugar solids are used to reduce or replace portions of the milk solids nonfat (MSNF), there is a definite loss of functionality of the mix, resulting in reduced body and texture. Problems, such as stickiness, gumminess, and weak body can be corrected by the addition of a bodying agent such as microcrystalline cellulose in very small amounts (0.25–0.4% weight).

Bulking Agents. These substances are added to semiliquid and solid food products to add bulk to the end product over and beyond the bulk resulting from the use of conventional ingredients only. For example, when added as an ingredient of baked foods, microcrystalline cellulose accomplishes two functions—weight is added, thus reducing the effective caloric content of a given portion; and water is tied up, so that considerably more liquid can be incorporated into the formulation. However, it should be stressed that microcrystalline cellulose can only partially substitute for fat, which is needed for air entrapment, or flour, which provides the elastic gluten structure. As an additional advantage, the cellulose also increases the fiber content of the product.

In the currently very important field of manufacturing low-calorie foods, a bulking agent essentially can be considered as a diluent even though it may play other important roles. Thus, the diet-conscious consumer can eat cookies, doughnuts, or portions of cake of traditional size and yet consume considerably fewer calories. The important factor in selecting a bulking agent for low-calorie foods is that of finding a substance that combines noncaloric qualities with other functional capabilities so that lower amounts of relatively high-calorie ingredients can be reduced or replaced without detracting drastically from the consumer appeals of the finished product.

In addition to microcrystalline cellulose, there are several other bodying and bulking compounds used. These include glycerin, methylcellulose, polyvinylpyrrolidone (PVP), sodium carboxymethylcellulose, whey solids, and xanthan gum.

References

Shama, F., and P. Sherman: "The Texture of Ice Cream: Rheological Properties of Frozen Ice Cream," *Jrnl. of Food Science*, **31**, 699 (1966).

Shama, F., and P. Sherman: "Identification of Stimuli Controlling the Sensory Evaluation of Viscosity: Oral Methods," *Jrnl. of Texture Studies*, **4**, 111 (1973).

Staff: "Food Chemicals Codex," National Academy of Sciences, Washington, D.C. (Revised periodically).

Staff: "Microcrystalline Cellulose in Low Calorie Foods," FMC Corporation, Philadelphia, Pennsylvania (Revised periodically).

Voisey, P. W.: "Rheology and Texture in Food Quality," AVI, Westport, Connecticut (1976).

BOILING CURVE AND CONDENSATION CURVE. Consider the phase diagram of a binary system forming a liquid and a vapor phase at constant pressure. Curve I is the boiling curve, which gives the coexistence temperature as a function of liquid composition; and curve II is the condensation curve,

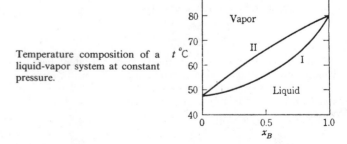

Temperature composition of a liquid-vapor system at constant pressure.

which gives the coexistent temperature as a function of the composition of the vapor phase. If the temperature is increased, vaporization begins when the boiling curve is crossed. Inversely, condensation begins when the temperature is decreased below the condensation curve.

BOILING POINT. The normal boiling point of a liquid is the temperature at which its maximum or "saturated" vapor pressure is equal to the normal atmospheric pressure, 760 mm of mercury. If the pressure on the liquid varies, the actual boiling point varies in accordance with the relation between the vapor pressure and the temperature for the liquid in question. (See **Vapor.**) Water, for example, with a normal boiling point of 100°C or 212°F, boils at ordinary room temperature when the pressure is reduced to about 17 mm; and inhabitants of elevated regions often find difficulty in cooking food by boiling, because of the low boiling point. On the other hand, the boiling water and steam in a "pressure cooker" are so hot that such foods as meat and rice are cooked tender in a very short time. If a solid is dissolved in the liquid, or if another, less volatile liquid is mixed with it, the boiling point is raised to a degree expressed by the boiling point laws of Van't Hoff, Raoult, and others.

A liquid does not necessarily begin boiling when the temperature reaches the boiling point. If kept perfectly quiet, and especially if covered with a film of oil, water may be raised several degrees above its normal boiling point, before it suddenly boils with explosive violence; it then returns to its true boiling point.

To prevent this superheating, it is customary to add to laboratory distillation flasks small pieces of inert material having sharp corners; the latter favor the formation of "bubbles" of vapor when the liquid reaches the boiling point.

The *maximum boiling point* is that temperature corresponding to a definite composition of a two-component or multi-component system at which the boiling point of the system is a maximum. At this temperature the liquid and vapor have the same composition and the solution distills completely without change in temperature. Binary liquid systems which show negative deviations from Raoult's law have maximum boiling points. See **Raoult's Law; Van't Hoff Law.**

The *minimum boiling point* is that temperature corresponding to a definite composition of a two-component or multi-component liquid system at which the boiling temperature is the lowest for that particular system. At the minimum boiling point the liquid and vapor have the same composition.

BOILING POINT CONSTANT. Consider a dilute solution in which all solute species may be regarded as nonvolatile. The vapor in equilibrium with the solution is then formed from the solvent only. Call T^0 the boiling point of the pure solvent at the pressure concerned, and T the boiling point of the solution. For a dilute solution, the difference

$$\theta = T - T^0$$

will be small compared with T^0.

If the solution is also ideal, one has

$$\theta = \frac{R(T^0)^2}{\Delta_e h^0} \frac{M_1}{1000} \sum_s m_s = \theta_e \sum_s m_s$$

M_1 is the molar mass of the solvent, $\Delta_e h^0$ its latent heat of vaporization in kcal per mole at temperature T^0, m_s the molality of solute s; θ_e is called the *boiling point constant*, or *ebullioscopic constant*. It depends only on the properties of the solvent. For water,

$$\theta_e = 0.51°C$$

BOILING POINT ELEVATION. The boiling point of a solution is, in general, higher than that of pure solvent, and the elevation is proportional to the active mass of the solute for dilute (ideal) solutions.

$$\Delta T = Km$$

where ΔT is the elevation of the boiling point, K is the *boiling point constant* or the *ebullioscopic constant* and m, the molality of the solution.

There are several methods for measuring elevation: In the Beckmann method a Beckmann-type thermometer is immersed in a weighed amount of solvent and the boiling point determined by gentle heating until a steady temperature is reached. A weighed amount of solute is then added and the boiling point redetermined. The difference gives the elevation of the boiling point. The glass vessel containing the liquid is provided with a platinum wire sealed through the bottom to promote steady boiling and to prevent overheating, and reflux condensers are used to minimize loss of liquid.

In the Landsberger method, vapor from boiling solvent is passed through the solvent contained in another vessel and by giving up its latent heat will eventually raise the liquid to the boiling point. At this stage a weighed amount of solute is added to the second vessel and the boiling point is again determined.

In the Cottrell method, the thermometer is placed in the vapor phase above the surface of the liquid and the apparatus so designed that boiling liquid is pumped continuously over the bulb of the thermometer.

BOILING WATER REACTOR. Nuclear Reactor.

BOMB CALORIMETER. Calorimetry.

BOND (Chemical). Chemical Elements.

BONDING ELECTRON. Electron.

BORACITE. This mineral is a magnesium borate containing some chlorine, $Mg_3B_7O_{13}Cl$. It appears to be isometric but probably becomes so only at 265°C, below which temperature it is believed to be orthorhombic. Its hardness is 7; specific gravity, 2.9; luster, vitreous; color, white to gray, sometimes yellow or green; translucent to subtransparent. It occurs in beds with gypsum and salt in Germany, particularly at Strassfurt in Saxony.

BORAX. This hydrated sodium borate mineral, $Na_2B_4O_7 \cdot 10H_2O$, is a product of evaporation from shallow lakes and plays. Borax crystallizes in the monoclinic system, usually in short prismatic crystals. Its color grades from colorless through gray, blue to greenish. Vitreous to resinous luster of translucent to opaque character. Hardness of 2–2.5, and specific gravity of 1.715.

Borax from the salt lakes of Kashmir and Tibet has been known since early history of man. India, the U.S.S.R., and Persia possess small deposits. Extensive deposits are known in the United States, notably in Lake, San Bernardino, Inyo, and Kern Counties in California, and Esmerelda and Dona Ana Counties in New Mexico.

Its use include antiseptics, medicines, a flux in smelting, soldering and welding operations, a deoxidizer in nonferrous metals, a neutron absorber for atomic energy shields, in rocket fuels, and as extremely hard abrasive boron carbide (harder than corundum). See also **Boron.**

BORIC ACID. Boron.

BORNITE. Named for the German mineralogist of the eighteenth century, Ignatius von Born, this mineral is a sulfide of copper and iron corresponding to the formula Cu_5FeS_4. It is isometric with a cubic habit, although crystals are rare, usually occurring as granular or compact masses. Its fracture is conchoidal to uneven; brittle; hardness, 3; specific gravity, 5.079; color, copper-red to reddish-brown (hence the name horseflesh ore) when freshly fractured; it soon assumes an iridescent tarnish (hence the name peacock ore); luster, metallic; streak, grayish-black; opaque.

Bornite as a primary mineral has been observed in pegmatite veins and in igneous rocks and is also a common secondary mineral.

Bornite crystals have been obtained in Austria and England. As an ore it is important in Tasmania, Chile, Peru and in Montana. In the United States, bornite also has been found in Connecticut, and in Canada, in the Province of Quebec.

Bornite also is known as *peacock ore* and *horseflesh ore*.

BORON. Chemical element symbol B, at. no. 5, at. wt. 10.81, periodic table group 3a, mp 2300°C, sublimes at approximately 2550°C, density 2.35 g/cm³ (amorphous form). There are four principal crystal modifications of boron: (1) α-rhombohedral, (2) β-rhomobohedral, (3) I-tetragonal, and (4) II-tetragonal. There are two natural isotopes, ^{10}B and ^{11}B. In 1807, Davy first produced elemental boron in amorphous form by electrolyzing boric acid. A year later, Gay-Lussac and Thénard produced elemental boron by reducing boric acid with potassium. However, it was not until 1892 that boron with a purity of over 90% was produced by Moissan, who reduced the element from B_2O_3. Moissan observed that the earlier claims of producing elemental boron were in effect compounds of boron. First ionization potential, 8.296 eV; second, 23.98 eV; third, 37.75 eV. Oxidation potential $B + 3H_2O \rightarrow H_3BO_3 + 3H^+ + 3e^-$, 0.73 V; $B + 4OH^- \rightarrow H_2BO_3^- + H_2O + 3_e^-$, 2.5 V. Other important physical properties of boron are given under **Chemical Elements.**

Boron is (1) a yellowish-brown crystalline solid, (2) an amorphous greenish-brown powder. Both forms are unaffected by air at ordinary temperatures but when heated to high temperatures in air form oxide and nitride. Crystalline boron is unattacked by HCl or HNO_3, or by NaOH solution, but with fused NaOH forms sodium borate and hydrogen; reacts with magnesium but not with sodium.

Boron occurs as rasorite or kernite (sodium tetraborate tetrahydrate, $Na_2B_4O_7 \cdot 4H_2O$) and colemanite (calcium borate, $Ca_2B_6O_{11} \cdot 5H_2O$) in California, as sassolite (boric acid, H_3BO_3) in Tuscany, Italy, and also locally in Chile, Turkey, and Tibet. See also **Colemanite; Kernite; Ulexite.**

Production. Commercial boron is produced in several ways: (1) by reduction with metals from the abundant B_2O_3, using lithium, sodium, potassium, magnesium, beryllium, calcium, or aluminum. The reaction is exothermic. Magnesium is the most effective reductant. With magnesium, a brown powder of approximately 90–95% purity is produced. (2) By reduction with compounds, such as calcium carbide or tungsten carbide, or with hydrogen in an electric arc furnace. The starting boron source may be B_2O_3 or BCl_3; (3) reduction of gaseous compounds with hydrogen. In an atmosphere of a boron halide, metallic filaments or bars at a surface temperature of about 1,200°C will receive depositions of boron upon admission of hydrogen to the process atmosphere. Although the deposition rate is low, boron of high purity can be obtained because careful control over the purity of the starting ingredients is possible. (4) Thermal decomposition of boron compounds, such as the boranes (very poisonous). Boranes in combination with oxygen or H_2O are very reactive. In this process, boron halides, boron sulfide, some borides, boron phosphide, sodium bornate and potassium bornate also can be decomposed thermally; and (5) electrochemical reduction of boron compounds where the smelt-

ings of metallic fluoroborates or metallic borates are electrolytically decomposed. Boron oxide alkali metal oxide-alkali chloride compounds also can be decomposed in this manner.

Both chemical methods and float zoning are used to purify the boron product from the foregoing processes. In the latter method, a boron of 99.99% purity can be obtained.

Although the chemistry of boron is extremely interesting, there is no substantial market for elemental boron. Some boron compounds are high-tonnage products. Elemental boron has found limited use to date in semiconductor applications, although it does possess current-voltage characteristics that make it suitable for use as an electrical switching device. In a limited way, boron also is used as a dopant (*p*-type) for *p-n* junctions in silicon. The principal problem deterring the larger use of boron as a semiconductor is the high lattice-defect concentration in the crystals currently available.

Uses. Boron formed on hot tungsten wire finds use for reinforcement of metals and plastics and for weight reduction of materials and products. Deoxidizers of alloys are prepared from borides. ^{10}B and its compounds are used for neutron absorption. Various boron compounds are used as rocket fuels, diamond substitutes, and additives to aluminum alloys to improve electrical and thermal conductivity, as well as for grain refining of aluminum alloys. Boron hydrides are very sensitive to shock and can detonate easily. Boron halides are corrosive and toxic.

Biological Functions. Although boron is required by plants, there is little solid evidence to date that it is required for the nutrition of livestock or humans. Boron deficiency may alter the levels of vitamins or sugars in plants owing to the effect of boron upon the synthesis and translocation of these compounds within the plant. The addition of boron to some boron-deficient soils has increased the carotene or provitamin A concentration in carrots and alfalfa.

As with several of the other trace elements, while concentrations of very low levels are desirable, high levels of boron are toxic to plants. Different plant species vary widely in their requirement for this element and in their tolerance for high levels. Application of boron-containing fertilizer must be carefully adjusted for different crops. An application of boron-containing fertilizer to improve the yields of alfalfa or beets may be toxic to such boron-sensitive crops as tomatoes and grapes. In the southwestern United States, serious boron toxicity to plants has resulted from using irrigation waters that are high in boron.

Boric Acid. Boric oxide B_2O_3 is acidic. It exists in two forms, a glassy form obtained by high-temperature dehydration of boric acid, and crystalline form obtained by slow heating of metaboric acid.

The oxyacids of boron are of two types: (A) the boric acids, based upon boric oxide, and (B) the lower oxyacids based upon boron-to-boron structural linkages.

The really acidic boric acids consist essentially of metaboric acid (HBO_2), a polymer, and boric or orthoboric acid H_3BO_3, ($pK_a = 9.24$). There is no compound corresponding to the formula for tetraboric acid $H_2B_4O_7$, although there are a number of salts which may be based upon this composition. Sometimes called *boracic acid*, H_3BO_3 is a high-tonnage material, the main uses being in the medical and pharmaceutical fields. A saturated solution of H_3BO_3 contains about 2% of the compound at 0°C, increasing to about 39% at 100°C. The compound also is soluble in alcohol. In preparations, solutions of boric acid are nonirritating and slightly astringent with antiseptic properties. Although no longer used as a preservative for meats, boric acid finds extensive use in mouthwashes, nasal sprays, and eye-hygiene formulations. Boric acid (sometimes with borax) is used as a fire-retardant. A commercial preparation of this type (*Minalith*) consists of diammonium phosphate, ammonium sulfate, sodium tetraborate, and boric acid. The tanning industry uses boric acid in the deliming of skins where calcium borates, soluble in H_2O, are formed. As sold commercially, boric acid is $B_2O_3 \cdot 3H_2O$, prepared by adding HCl or H_2SO_4 to a solution of borax.

Borates. Sodium tetraborate $Na_2B_4O_7 \cdot 10H_2O$ is a very high-tonnage material. Natural borax has a hardness of 2–2.5, mp 75°C, sp gr 1.75. An aqueous solution of borax is mildly alkaline and antiseptic. The compound finds many uses, including: (1) cleaning compounds of numerous types; (2) important ingredient of glass and ceramics, notably for heat-resistant glass where as much as 40 pounds of borax may be required per 100 pounds of finished glass; (3) source of elemental boron and other boron compounds; (4) flux for soldering and welding; (5) constituent of fertilizers; (6) filler in paper and paints; and (7) corrosion inhibitor in antifreeze formulations. Borax also is used in fire retardants.

Chemistry of Boron and Other Boron Compounds. In 1901, the German chemist, Alfred Stock, stated, "It was evident that boron, the close neighbor of carbon in the periodic system, might be expected to form a much greater variety of interesting compounds than merely boric acid and the borates, which were almost the only ones known." In 30 years of research which followed that statement, Stock synthesized almost all of the important *boranes* (hydrogen and boron). Some of these compounds now find use in glass, ceramics, synthetic lubricants, and as ingredients of high-energy rocket fuels and jet-engine and automotive fuels. Further pioneering of borohydride chemistry was carried on by Schlesinger and Burg of the University of Chicago in the late 1940s. Boron carbide B_4C is used as a neutron-absorbing material in nuclear reactors. Sodium borohydride $NaBH_4$ is applied as a reducing agent in the manufacture of certain synthetics. Although not ultimately selected, because of the greater volatility of uranium hexafluoride UF_6, both uranium borohydride $U(BH_4)_4$ and its methyl derivative $U(CH_3BH_3)_4$ were considered for use in separating the isotopes of uranium during the Manhattan Project. ^{10}B is used in brain tumor research. When injected intravenously, borax concentrates in the areas of tumors and its presence can be detected by radiation techniques. With further research, the tendency of boron to link with itself may comprise the foundation of future inorganic polymeric materials. Although they have poor mechanical strength, boron-phosphorus polymers, prepared by reacting diborane with phosphone derivatives, do exhibit excellent heat-resistance.

X-ray diffraction studies show five general types of structures in solid borates:

1. Discrete anions containing individual BO_3^{3-} groups, or a limited number of other groups combined by sharing oxygen atoms. (The simplest is $B_2O_5^{4-}$, which is called pyroborate.)

2. Extended anions in which individual BO_3 groups are linked into rings or chains, such as $B_3O_6^{3-}$ or $B_2O_4^{2-}$ (metaborate).

3. Sheet structures in which all the oxygen atoms are shared between borate groups, as in $B_5O_{10}^{5-}$ (pentaborate).

4. Structures containing the tetrahedral $B(OH)_4^-$ ion, which is the principal ion found in alkaline aqueous solutions.

5. Extended anions containing tetrahedral BO_4 units, usually linked with triangular BO_3 groups.

The lower oxyacids of boron may be considered to be derived from the various boron hydrides, whence their boron-boron linkages result. These compounds include the hypoborates, which may be produced by reactions of tetraborate with strong alkali, and which may be formulated from the structure $H_2[H_6B_2O_2]$; the subborates, derived from $H_4[B_2O_4]$, which

is called subboric acid; and the borohydrates, which are derived from acids of various compositions, such as $H_2[B_4O_2]$, $H_2[B_2O_2]$ and $H_2[H_4B_2O_2]$. The last of these compounds contains a double bonded boron-boron linkage, and exhibits *cis-trans* isomerism.

The borides are binary compounds of boron with metals or electropositive elements in general. Except in isolated cases their compositions depart from the stoichiometry of trivalent boron compounds and are determined more by the requirements of metal and boron lattices than by valencies. On the basis of composition, they may be classified into types based respectively upon zig-zag chains (MB) represented by CoB; isolated boron atoms (M_2B) represented by Co_2B; double chains (M_3B_4) represented by Mo_3B_4; hexagonal layers (MB_2) represented by CoB_2; three dimensional frameworks (MB_6 or MB_{12}) represented by SiB_6 or UB_{12}. It is apparent that these borides are interstitial compounds existing primarily with the metals of main groups, II, III, IV, V and VI.

There are at least six definitely characterized boron hydrides, as follows: diborane(6), B_2H_6; tetraborane(10) B_4H_{10}; pentaborane(9) (stable) B_5H_9; pentaborane(11) (unstable) B_5H_{11}; hexaborane(10) B_6H_{10} and decaborane(14) $B_{10}H_{14}$. In these names, note that the prefix denotes the number of boron atoms, while the figure in parentheses denotes the number of hydrogen atoms. In addition to these compounds, which are all gases or volatile liquids except decaborane(14), decomposition of the lower boron hydrides yields colorless or yellow solid boron hydrides, ranging in composition from $(BH_{1.5})_x$ to $(BH)_x$. This readiness to polymerize is evidence of the reactivity of these borane compounds, which readily form additional products with ammonia, with the amalgams of the active metals, and with many organic compounds, as well as with CO.

In addition to BH_4^- there exist a number of hydroborate anions which may be considered to be derived from real or hypothetical boron hydrides by addition of hydride ion. These include $B_2H_7^-$, formed by the reaction of B_2H_6 and BH_4^- in organic solvents, and the extremely stable ions $B_{10}H_{10}^{2-}$ and $B_{12}H_{12}^{2-}$, unaffected by either acidic or alkaline aqueous solutions or by atmospheric oxygen. Free halogens merely cause substitution of halogen for hydrogen. The structure of $B_{10}H_{10}^{2-}$ is based on the square antiprism, while that of $B_{12}H_{12}^{2-}$ is a regular icosahedron.

In 1976, the Nobel Prize for Chemistry was awarded to William Nunn Lipscomb, Jr., of Harvard University, for original research on the structure and bonding of boron hydrides and their derivatives. As pointed out by Grimes (1976), the insight into electron-deficient borane structures originally provided by Lipscomb carries over not only to the carboranes, but also to their organic cousins, the so-called "nonclassical" carbonium ions. The three-center bond descriptions given by Lipscomb to B_5H_9 and B_6H_{10} can as easily be applied to their hydrocarbon analogs, the pyramidal ions $C_5H_5^+$ and $C_6H_6^{2+}$, both presently known as alkyl derivatives. Also, molecules usually not so considered, such as metallocenes, organometallics, such as $(C_4H_4)Fe(CO)_3$ or $[(CO)_3Fe]_5C$, metal clusters, and others, can be considered from the perspective of borane analogs. The boranes, once considered peculiar, over the years have provided insight to many cluster-type molecules, for which classical Lewis bond descriptions do not fit. Lipscomb's lecture given in Stockholm on December 11, 1976 provides an excellent overview of the boranes and their relatives. See Lipscomb (1977) reference.

In 1979, the Nobel Prize for Chemistry was received by Herbert C. Brown of Purdue University (shared with Georg Wittig for research in another field) for the discovery of the *hydroboration reaction*. This reaction, depicted below, has made the organoboranes readily available as chemical intermediates. The boron atom adds to the less substituted carbon atom. As pointed out by Brewster and Negishi (1980), depending upon steric factors, mono-, di-, or trialkylboranes may be formed. These products comprise synthetically useful reactions whereby the boron atom is replaced, but the mono- and dialkylboranes are also useful as reducing or hydroborating agents.

$$RCH{=}CH_2 + B_2H_6 \longrightarrow (RCH_2CH_2-)_3B \quad (1)$$

$$CH_3-\underset{\underset{CH_3}{|}}{C}{=}CHCH_3 + B_2H_6 \longrightarrow (CH_3-\underset{\underset{CH_3}{|}}{CH}-\underset{\underset{CH_3}{|}}{CH}-)_2BH \quad (2)$$

$$CH_3-\underset{\underset{CH_3}{|}}{C}{=}\underset{\underset{CH_3}{|}}{C}-CH_3 + B_2H_6 \longrightarrow CH_3-\underset{\underset{CH_3}{|}}{CH}-\underset{\underset{CH_3}{|}}{\overset{\overset{CH_3}{|}}{C}}-BH_2 \quad (3)$$

Among the other inorganic compounds of boron are the following:

Borides. Carbon boride CB_6 and silicon borides SiB_3 and SiB_6 are hard, crystalline solids, produced in the electric furnace; magnesium boride Mg_3B_2, brown solid, by reaction of boron oxide and magnesium powder ignited, forms boron hydrides with HCl; calcium boride Ca_3B_2, forms boron hydrides and hydrogen gas with HCl.

Nitride. Boron nitride BN, white solid, insoluble, reacts with steam to form NH_3 and boric acid, formed by heating anhydrous sodium borate with ammonium chloride, or by burning boron in air.

Sulfide. Boron sulfide B_2S_3, white solid, unpleasant odor, irritating to the eyes, reactive with water to form boric acid and hydrogen sulfide, formed by reaction of boron oxide plus carbon heated in a current of CS_2 at red heat.

The great number of compounds of boron are due to the readiness with which boron atoms form, to some extent, chain structures with other boron atoms, and to a far greater extent, cyclic compounds both with other boron atoms, and with atoms of carbon, oxygen, nitrogen, phosphorus, arsenic, the halogens, and many other elements. Examples of them are shown below, beginning with the two pentaboranes B_5H_9 and B_5H_{11}:

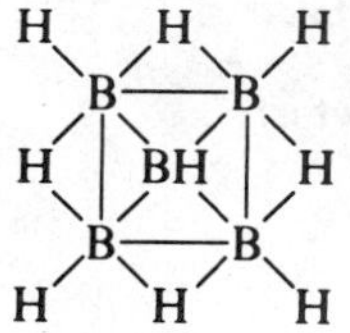

B_5H_9 Pentaborane(9)

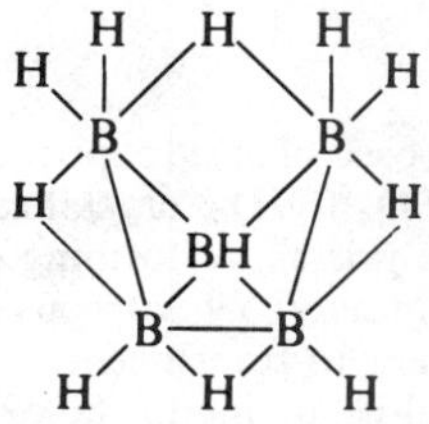

B_5H_{11} Pentaborane(11)

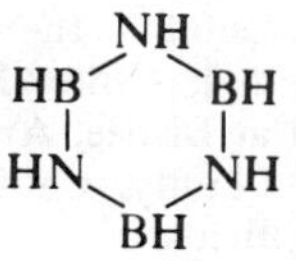

hexahydro-*s*-triazatriborine (also called *borazine*)

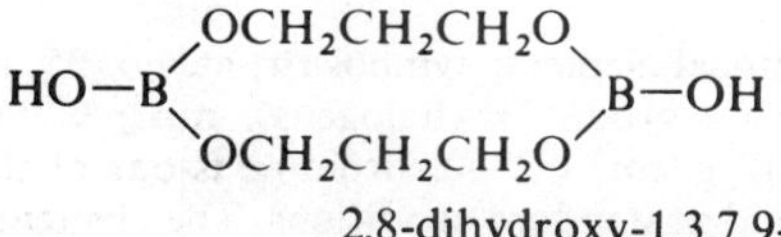

2,8-dihydroxy-1,3,7,9-tetroxa-2,8-diboracyclododecane

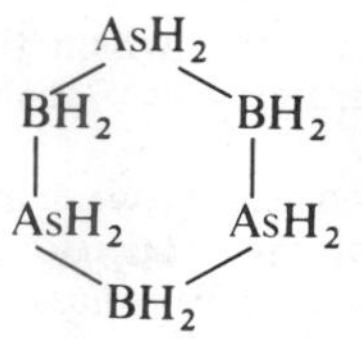

dodecahydro-*s*-triarsatriborine (also called *s*-triphosphatriborane or *borarsane*)

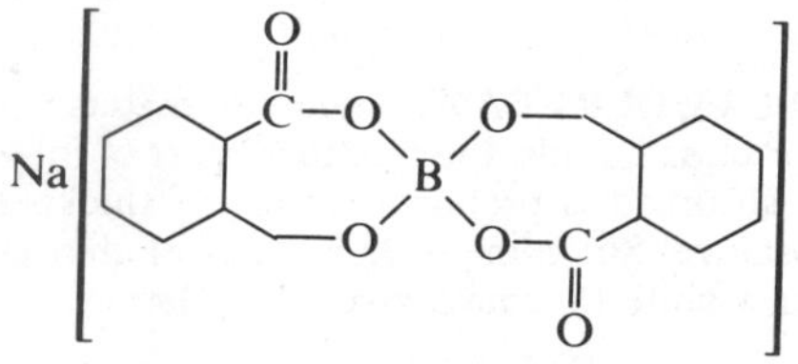

sodium bis(salicylato-*O,O'*)borate (1⁻)

Halides. Since simple boron compounds have only three electron-pairs in the valence shell of boron, they tend to be electron acceptors. Its simple molecules are formed by sp^2 hybrid sigma bonds lying in a plane. Its strong tendency to form an octet is shown by the tetrahedral boron compounds involving sp^3 hybridization. Boron halides include the trifluoride, BF_3, the trichloride, BCl_3, the tribromide, BBr_3 and the triiodide, BI_3, which range in mp from -127 to $+43°C$. Typical methods of forming the boron halides are treatment of boron oxide with hot concentrated H_2SO_4 in a reaction mixture with calcium fluoride to produce BF_3, and by heating boron, or boron oxide plus carbon with chlorine to produce the chloride.

In addition to the simple halides, boron forms fluorine complexes containing the fluoroborate ion (BF_4^-). Subhalides of boron are known (B_2X_4) of the structure:

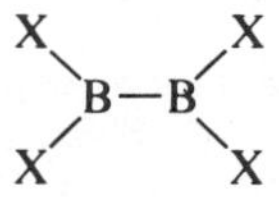

and B_4X_4 of the structure

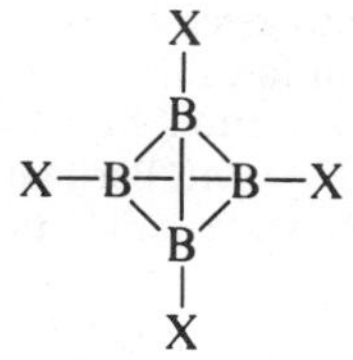

References

Armington, A. F., Bufford, J. T., and R. J. Starks: "Boron," Plenum, New York (1965).

Brewster, J. H., and E. Negishi: "The 1979 Nobel Prize for Chemistry," *Science*, **207**, 44–46 (1980).

Brown, H. C.: "From Little Acorns to Tall Oaks: From Boranes through Organoboranes," *Science*, **210**, 485–492 (1980).

Grimes, R. N.: "The 1976 Nobel Prize for Chemistry," *Science*, **194**, 709–710 (1976).

Lipscomb, W. N.: "The Boranes and Their Relatives," *Science*, **196**, 1047–1055 (1977).

Pulich, W. M., Jr.: "Photocontrol of Boron Metabolism in Sea Grasses," *Science*, **200**, 319–320 (1978).

BOSONS. Those elementary particles for which there is symmetry under intra-pair production. They obey Bose-Einstein statistics. Included are photons, pi mesons, and nuclei with an even number of particles. (Those particles for which there is antisymmetry are *fermions*.) See **Mesons; Particles (Subatomic);** and **Photon.**

Recent progress toward a complete theory of the weak interactions has led to sharper predictions for the properties of the hypothetical weak-force particles known as *intermediate bosons*. This is described by Hung and Quigg (*Science*, **210**, 1205–1211, 1980).

BOUGUER AND LAMBERT LAW. In homogeneous materials, such as glass or clear liquids, the fractional part of intensity or radiant energy absorbed is proportional to the thickness of the absorbing substance. Summing over a series of thin layers or integration over a finite thickness gives the relation

$$\log I_0/I = k_1 b$$

where I_0 is the intensity or radiant power incident on a sample b centimeters thick and I is the intensity of the transmitted beam. The constant k_1 depends on the wavelength of the incident radiation, the nature of the absorbing material and other experimental conditions. Verification of the law fails unless appropriate corrections are made for reflection, convergence of the light beam and spectral slit width, as well as possible scattering, fluorescence, chemical reaction, inhomogeneity, and anisotropy of the sample. Formerly, the constant k_1 was called the absorption coefficient. It is now preferable to avoid this term and to call the ratio I/I_0 the transmittance. The law was first expressed by Bouguer in 1729 but it is often attributed to Lambert, who restated it in 1768.

BOULANGERITE. A mineral compound of lead-antimony sulfide, $Pb_5Sb_4S_{11}$. Crystallizes in the monoclinic system; hardness, 2.5–3; specific gravity, 6.23; color, lead gray.

BOURNONITE. An antimony-copper-lead sulfide corresponding to the formula $PbCuSbS_3$. It is orthorhombic, and repeated twinning often produces crosses or wheel-shaped crystals. It is brittle; fracture, subconchoidal; hardness, 2.5–3; specific gravity, 5.83; luster metallic; color and streak, dark gray to black; opaque.

Bournonite is found with galena, chalcopyrite, and sphalerite. There are many European localities; it was first found in Cornwall, England, by Count Bournon, for whom it was later named. Bournonite occurs in Bolivia and Peru and in the United States in Arizona, Montana, Nevada and Utah. Bournonite is also known as *wheel ore*.

BOYLE-CHARLES LAW. This law states that the product of the pressure and volume of a gas is a constant which depends only upon the temperature. This law may be stated mathematically as

$$p_2 v_2 = p_1 v_1 [1 + \alpha(t_2 - t_1)]$$

where p_1 and v_1 are the pressure and volume of a body of gas at temperature t_1, p_2 and v_2 are the pressure and volume of the same body of gas at another temperature t_2, and a is the volume coefficient of expansion of the gas. If the temperature is expressed in degrees absolute, this expression becomes

$$\frac{P_2 v_2}{T_2} = \frac{p_1 v_1}{T_1}$$

which is the ideal gas law, so-called because all real gases depart from it to a greater or lesser extent. See also **Characteristic Equation.**

BOYLE'S LAW. This law, attributed to Robert Boyle (1662) but also known as Mariotte's law, expresses the isothermal pressure-volume relation for a body of ideal gas. That is, if the gas is kept at constant temperature, the pressure and volume are in inverse proportion, or have a constant product. The law is only approximately true, even for such gases as hydrogen and helium; nevertheless it is very useful. Graphically, it is represented by an equilateral hyperbola. If the temperature is not constant, the behavior of the ideal gas must be expressed by the Boyle-Charles law.

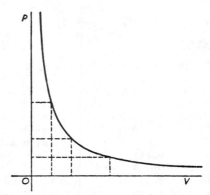

Equilateral hyperbola representing Boyle's law. The rectangular areas (PV) are all equal.

The Boyle temperature is that temperature, for a given gas, at which Boyle's law is most closely obeyed in the lower pressure range. At this temperature, the minimum point (of inflection) in the $pV - T$ curve falls on the pV axis. See **Compression (Gas);** and **Ideal Gas Law.**

BRAIN CHEMICALS. Enkephalins and Endorphins.

BRASS. Copper.

BRAVAIS-MILLER INDICES. Crystal.

BREEDER REACTOR. Nuclear Reactor.

BRIGHTENERS. Detergents.

BRILLOUIN ZONE. Fermi Surface.

BRIX SCALE. Specific Gravity.

BROCHANTITE. A mineral composed of basic copper sulfate corresponding to the formula $Cu_4(SO_4)(OH)_6$, crystallizing in the monoclinic system in needle-like prisms, or forming druses or masses. Hardness, 3.5–4; specific gravity, 3.9; vitreous luster; color, green; streak, green; transparent to translucent.

Brochantite is a secondary mineral occurring in the oxidized zones with other copper minerals, and is found in the Urals, in Rumania, in Sardinia; Cornwall, England; Chile. In the United States this mineral has been found at Bisbee, Arizona, Utah, in the Tintic District and in Inyo County, California. Brochantite was named for Brochant de Villiers.

BROMINE. Chemical element, symbol Br, at. no. 35, at. wt. 79.904, periodic table group 7a (halogens), mp $-7.2°C$, bp $58.8°C$, density, 3.12 g/cm^3 ($20°C$). Bromine is one of the few elements that is liquid at standard conditions. The element volatilizes readily at room temperature to form a red vapor which is very irritating to the eyes and throat. Liquid bromine causes painful lesions upon contact with the flesh. Bromine has two stable isotopes, ^{79}Br and ^{81}Br. Elemental bromine finds limited application as a chemical intermediate and as a sanitizing, disinfecting, and bleaching agent. Both the inorganic and organic compounds of the element find extensive commercial usage.

Bromine was discovered in 1826 by Antoine-Jérôme Balard, who identified the element as a component of seawater bitterns. Electronic configuration is $1s^22s^22p^63s^23\ p^63d^{10}4s^24p^5$. Ionic radius Br^- 1.97 Å, Br^{7+} 0.39 Å. Covalent radius 1.193_5. First ionization potential 11.84 eV; second, 19.1 eV; third 25.7 eV. Oxidation potentials $2Br^- \rightarrow Br_2(l) + 2e^-$, -1.065 V; $2Br^- \rightarrow Br_2(aq) + 2e^-$, -1.087 V; $Br^- + H_2O \rightarrow HBrO + H^+ + 2e^-$, -1.33 V; $Br^- + 3H_2O \rightarrow BrO_3^- + 6H^+ + 6e^-$, -1.44 V; $\frac{1}{2}Br_2 + 3H_2O \rightarrow BrO_3^- + 6H^+ + 5e^-$, -1.52 V; $\frac{1}{2}Br_2 + H_2O \rightarrow HBrO + H^+ + e^-$, -1.59 V; $Br^- + 6OH^- \rightarrow BrO_3^- + 3H_2O + 6e^-$, -0.61 V; $Br^- + 2OH^- \rightarrow BrO^- + H_2O + 2e^-$, -0.70 V.

Other important physical properties of bromine are described under **Chemical Elements.**

Bromine is only moderately soluble in H_2O (3.20 g/100 ml) but markedly so in nonpolar solvents, e.g., carbon tetrachloride, as is consistent with the covalent character of the Br—Br bond. It dissolves more readily in alkali bromide solutions due to formation of the tribromide ion (Br_3^-), and in certain associated solvents, such as concentrated H_2SO_4 and ethyl alcohol. Its aqueous solution is more stable than that of chlorine, since the tendency of Br_2 to hydrolyze to unstable hypobromous acid and hydrogen bromide is less than the corresponding reaction for chlorine. Bromine exhibits in common with the other halogens a marked readiness to form singly charged negative ions, as would be expected from the fact that these atoms need only one electron to acquire an inert gas configuration. Its electron affinity (3.53 eV) is between that of chlorine and iodine. The bromides range in character from ionic to covalent compounds, many of them having bonds of intermediate nature. In addition to its negative univalence, bromine forms essentially covalent linkages with negative elements, in which it has positive valences 1, 3, and 5.

Bromine occurs as bromide in seawater (0.188% Br), in the mother liquor from salt wells of Michigan, Ohio, West Virginia, Arkansas, and in the potassium deposits of Germany and France.

Production. In the United States, nearly all bromine is derived from natural brines. The Arkansas brines which contain a minimum of 4,000 ppm bromide account for over half of this production. Recovery is effected by a *steaming-out* process. After heating fresh brine, the solution is fed to the top of a tower. Chlorine and steam are injected at the bottom of the tower. The chlorine oxidizes the bromide and displaces one resultant bromine from solution. For brines of lower concentration, air instead of steam is used to sweep out the bromine vapors after chlorination.

Hydrogen Bromide and Hydrobromic Acid. HBr, is formed directly from the elements, effectively when catalyzed by sunlight, by heated charcoal or platinum, or more conveniently by hydrolysis of phosphorus tribromide. Treatment of bromides with H_2SO_4 yields mixtures of HBr and bromine. The H—Br bond is considered to be partly covalent. Hydrobromic acid is a strong acid in aqueous solution. Its salts are the bromides, all of which are water-soluble except those of copper(I), silver, gold(I), mercury(I), thallium(I) and lead(II), the divalent ions of the elements of the second and third transition series, and the salts of the heavy alkali ions with many bromo-complex anions, e.g., Cs_2PtBr_6, $RbAuBr_4$, etc. The main uses of HBr

and hydrobromic acid are in the production of alkyl bromides (by replacement of alcoholic hydroxyl groups or by addition to olefins) and inorganic bromides.

Sodium Bromide. This is a high-tonnage chemical and one of the most important of the bromide salts commercially. High-purity grades are required in the formulation of silver bromide emulsions for photography. The compound, usually in combination with hypochlorites, is used as a bleach, notably for cellulosics. The production of sodium bromide simply involves the neutralization of HBr with NaOH or with sodium carbonate or bicarbonate.

Calcium Bromide. Because of its ready solubility calcium bromide forms solutions of high density, which when properly formulated are finding increasing use as functional fluids in oil well completion and packing applications.

Lithium Bromide. LiBr finds use as a desiccant in industrial air-conditioning systems.

Zinc Bromide. $ZnBr_2$ is used as a rayon-finishing agent, as a catalyst, as a gamma-radiation shield in nuclear reactor viewing windows, and as an absorbent in humidity control. It too finds use in high density formulated functional fluids in oil well applications. Zinc bromide is prepared either by the direct reduction of bromine with zinc, or by reacting HBr with zinc oxide or carbonate.

Other Bromides. Aluminum bromide $AlBr_3$ is used as a catalyst and parallels $AlCl_3$ in this role. Strontium and magnesium bromides are used to a limited extent in pharmaceutical applications. Ammonium bromide is used as a flame retardant in some paper and textile applications; potassium bromide is used in photography. Phosphorus tribromide PBr_3 and silicon tetrabromide $SiBr_4$ are used as intermediates and catalysts, notably in the production of phosphite esters.

Hypobromous Acid and Hypobromites. Hypobromous acid HOBr results from the hydrolysis of bromine with H_2O and exists only in aqueous solution. The compound finds limited use as a germicide and in water treatment; also it can be used as an oxidizing or brominating agent in the production of certain organic compound. Although hypobromous acid is low in bromine content, concentrated hypobromite solutions can be formed by adding bromine to cooled solutions of alkalis.

Bromic Acid and Bromates. Bromic acid, $HBrO_3$, can exist only in aqueous solution. Bromic acid and bromates are powerful oxidizing agents. Bromic acid decomposes into bromine, oxygen and water. Many oxidizing agents, e.g., hydrogen peroxide, hypochlorous acid, and chlorine convert Br_2 or Br^- solutions to bromates. The decomposition reactions of bromates vary considerably. Lead(II) bromate and copper(II) bromate give the metal oxides and Br^-; silver, mercury(II) and potassium bromates give the metal ion, Br^- and oxygen, while zinc, magnesium and aluminum bromates give the metal oxide, Br_2 and oxygen.

Halogen Compounds. Bromine forms a number of compounds with the other halogens. Its binary iodine compounds are discussed under iodine; other interhalogen compounds of bromine include bromine monochloride, bromine monofluoride, bromine trifluoride, and bromine pentafluoride. The non-existence of higher chlorides of bromine, differing from iodine, can readily be explained in terms of the oxidation potential of Br(III) and Br(V). The monochloride, bromine chloride, BrCl, exists in pure state only at very low temperatures in the solid form. Dissociation in the gas phase is approximately 40% at 25°C and increases slowly with increasing temperature; less than 20% occurs in the liquid phase. With many substrates, bromine chloride reacts much more rapidly than does bromine itself to introduce bromine substituents.

Bromine monofluoride, BrF, is also somewhat unstable, decomposing spontaneously at 50°C to Br_2, BrF_3 and BrF_5. It has never been prepared pure since it is always in equilibrium with Br_2 and BrF_5. It is a gas at room temperature, reacting readily with water, phosphorus, and the heavy metals. Bromine trifluoride, BrF_3, is much more stable than the monofluoride. It is obtained directly from the elements at 10°C, or by fluorination of univalent heavy metal bromides. It is a liquid, bp 127.6°C, mp 8.8°C. There is evidence (high Trouton constant) that it undergoes self-ionization to form BrF_2^+ and BrF_4^-. The former is found in the acidic addition products it forms with gold(III), antimony(V) and tin(IV) fluorides, BrF_2AuF_4, BrF_2SbF_6 and $(BrF_2)_2SnF_6$. The latter occurs in the tetrafluorobromates, such as $KBrF_4$ and $Ba(BrF_4)_2$. The solvent properties of BrF_3 are consistent with its indicated dissociation, i.e., reactions involving the two classes of compounds mentioned take place as if BrF_3 acts as a fluoride ion donor or acceptor as H_2O is a proton donor or acceptor in the H_2O system. For example, potassium dihydrogen phosphate, KH_2PO_4 gives KPF_6, a mixture of HNO_3 and B_2O_3 gives NO_2BF_4, etc. Bromine trifluoride fluorinates many of the metal halides and oxides. Bromine pentafluoride, BrF_5, is prepared from BrF_3 and fluorine. It is thermally stable. It is a very active fluorinating agent, converting to fluorides most metals, their oxides and other halides, and being hydrolyzed by H_2O probably to hydrofluoric and bromic acids.

The polyhalide complexes of bromine include $(PBr_4)(IBr_2)$ formed by reaction of phosphorus pentabromide and iodine monobromide, and dissociating in certain organic polar solvents to the ions PBr_4^+ and IBr_2^-. Other polyhalides include NH_4IBr_2, $[(CH_3)_4N][IBr_2]$, $Cs[IFBr]$, $Rb[IClBr]$, and $Cs[IClBr]$. Most of these compounds hydrolyze readily, ionize in polar nonreacting solvents to the corresponding polyhalide ions, and decompose on heating to give the metal halide of greatest lattice energy.

Oxides. In binary combination with oxygen, bromine forms at least three compounds. Bromine(I) oxide, Br_2O, is a dark brown solid that is stable only in the dark below −40°C. It is prepared by passing dry gaseous bromine through dry mercury-(II) oxide and sand. Bromine(I) oxide, in carbon tetrachloride at low temperatures, reacts with alkali hydroxide to give hypobromites. Bromine(IV) oxide, BrO_2, is obtained by reaction of the elements in a cooled electric discharge tube. It is yellow, and is stable only at low temperatures. The compound appearing in the older literature as tribromine octoxide, Br_3O_8, is actually bromine trioxide, BrO_3, or dibromine hexoxide, Br_2O_6 (cf. chlorine). It is obtained from the low temperature, low pressure reaction of ozone and bromine; it is stable only at low temperatures and is soluble in H_2O with decomposition. Bromine(VII) oxide may be present among the decomposition products of BrO_2, or BrO_3, but no other evidence for its existence has been found.

Organic Bromine Compounds. Commercially important organic bromine compounds include the following: (1) Methyl bromide CH_3Br is formed by reacting methanol with HBr or hydrobromic acid. The compound is a highly toxic gas at standard conditions. Because of its toxicity, it is used as a soil and space fumigant. In many organic syntheses, the compound is used as a methylating agent. (2) Ethylene dibromide (1,2-dibromoethane) is used in combination with lead alkyls as an antiknock agent for gasoline. The compound also is used as a fumigant. (3) Methylene chlorobromide (bromochloromethane) is a low-boiling liquid of low toxicity and is useful as a fire-extinguishing agent in portable equipment and aircraft. (4) Bromotrifluoromethane is increasingly employed as a fire extinguishant in permanently installed systems protecting high-cost installa-

tions, such as computer rooms, where its low toxicity and especially its freedom from corrosivity are important considerations. (5) Acetylene tetrabromide (1,1,2,2-tetrabromoethane) is made by adding bromine to acetylene. The compound is comparatively dense and finds use as a gage fluid and in specific gravity separations of solids. It is also used as part of the catalyst system for the oxidation of p-xylene to terephthalic acid. (6) Tris(2,3-dibromopropyl)phosphate may be prepared by the reaction of phosphorus oxychloride with 2,3-dibromopropanol, or by the addition of bromine to triallyl phosphate. This viscous fluid was used as a flame retardant in a number of polymer systems, but has been displaced from most, or all, of these uses because it is a mutagen and suspect carcinogen. (7) Tetrabromobisphenol A is produced by the direct bromination of bisphenol A. The compound is used extensively as a flame retardant and usually is incorporated into the polymer backbone structure of epoxy resins, unsaturated polyesters, and polycarbonates. (8) Tetrabromophthalic anhydride, made by the catalytic bromination of phthalic anhydride in fuming H_2SO_4, also finds use as a reactive flame retardant in the formulation of polyol systems (for polyurethane foams) and in unsaturated polyesters. (9) Decabromodiphenyl ether, and others of the lower brominated diphenyl ethers, are finding increasing use as flame retardants in a variety of thermoplastic polymer systems. (10) Vinyl bromide has also found major use in flame-retarding modacrylic textile fibers when introduced as a comonomer in the synthesis of the polymer itself.

Alkanes and arenes, e.g., ethane and benzene, respectively, react with bromine by *substitution* of bromine for hydrogen (hydrogen bromide also formed)—ethane to yield ethyl bromide C_2H_5Br plus further substitution products; benzene, in the presence of a catalyst, e.g., iodine, phosphorus, iron, to yield bromobenzene C_6H_5Br plus further substitution products; toluene, under like conditions to benzene, to yield orthobromotoluene and parabromotoluene $CH_3C_6H_4Br$ plus further substitution products, but at the boiling temperature, in sunlight, dry, and in the absence of a catalyst, to yield alkyl side-chain substitution products, benzyl bromide $C_6H_5CH_2Br$, benzal bromide $C_6H_5CHBr_2$, and benzotribromide $C_6H_5CBr_3$.

Alkenes, alkynes, and arenes, e.g., ethylene, acetylene and benzene, respectively, react (1) with bromine by *addition*, e.g., ethylene dibromide $C_2H_4Br_{2(1,2)}$, acetylene tetrabromide $C_2H_2Br_{41,1,2,2}$, hexabromocyclohexane $C_6H_6Br_6$; also carbon monoxide yields carbonyl bromide $COBr_2$; (2) with hypobromous acid by *addition*, e.g., olefins form, for example, ethylene bromohydrin $CH_2Br \cdot CH_2OH$; (3) with hydrogen bromide by *addition*, to form, for example, ethyl bromide $CH_3 \cdot CH_2Br$ from ethylene. When the two olefin carbons have unequal numbers of hydrogens, the carbon to which one bromide or one hydroxyl attaches can be controlled by the reaction conditions.

Oxygen-function compounds, e.g., ethyl alcohol, acetaldehyde, acetone, acetic acid, react (1) with bromine, to form *bromosubstituted* corresponding or related *compounds*, e.g., ethyl alcohol or acetaldehyde to yield bromal $CBr_3 \cdot CHO$, acetone to yield bromoacetone $CH_2Br \cdot CO \cdot CH_3$; acetic acid to yield, at the boiling temperature, dry, and in the absence of a catalyst, monobromoacetic acid $CH_2Br \cdot COOH$, dibromoacetic acid $CHBr_2 \cdot COOH$, tribromoacetic acid $CBr_3 \cdot COOH$, the substitution taking place on the alpha-carbon (the carbon next to the carboxyl-group —COOH), (2) with phosphorus bromides, to form corresponding *bromides*, e.g., ethyl bromide C_2H_5Br, ethylidene dibromide CH_3CHBr_2, acetone bromide $(CH_3)_2CBr_2$, acetyl bromide CH_3COBr, (3) with hydrobromic acid, concentrated, alcohol forms the corresponding bromide.

Bromoform is made by reaction of acetone or ethyl alcohol

with sodium hypobromite; carbon tetrabromide by reaction of CS_2 plus bromine Br_2 in the presence of iron, heated; or by one reaction of bromoform with aqueous hypobromide solutions. Use is made of the diazo-reaction to introduce bromine into aryl compounds.

Many of the bromo-compounds are used as reagents or as intermediate compounds in organic chemistry. When alkyl bromocompounds are treated (1) with NaOH dissolved in alcohol, hydrogen bromide is removed, e.g., ethyl bromide $CH_3 \cdot CH_2Br$ yields ethylene $CH_2 : CH_2$, ethyl dibromide $CH_2Br \cdot CH_2Br$ yields acetylene $CH : CH$; (2) with magnesium or zinc and alcohol, bromine is removed, e.g., ethylene dibromide $CH_2Br \cdot CH_2Br$ yields ethylene $CH_2 : CH_2$, acetylene tetrabromide $CHBr_2 \cdot CHBr_2$ yields acetylene $CH : CH$.

References

Faith, W. L., Keyes, D. B., and R. L. Clark: "Industrial Chemicals," 3rd Edition, Wiley, New York (1965).
Jolles, Z. E., Editor: "Bromine and Its Compounds," Academic, London (1966).
Robinson, A. L.: "Infrared Photochemistry—Laser-Catalyzed Reactions," *Science*, **193**, 1230–1231 (1976).
Theiler, R., et al.: "Hydrocarbon Synthesis by Bromoperoxidase," *Science*, **202**, 1094–1096 (1978).
Theiler, R., et al.: "Drugs from the Sea," University of Oklahoma Press, Norman, Oklahoma (1979).

—John A. Garman, Director, Research and Development, Great Lakes Chemical Corporation, West Lafayette, Indiana.

BROOKITE. Brookite, composed of titanium dioxide, TiO_2, is an orthorhombic mineral of the same chemical composition as rutile and octahedrite. It was named for the English mineralogist H. J. Brooke. See also **Rutile; Titanium Dioxide.**

BRUCITE. The mineral brucite is magnesium hydroxide corresponding to the formula $Mg(OH)_2$; iron and manganese may occasionally be present. The crystals are usually tabular rhombohedrons of the hexagonal system; it may also occur fibrous or foliated. Brucite has one perfect cleavage parallel to the prism base; hardness, 2.5; specific gravity, 2.39; luster, pearly to vitreous; commonly white but may be gray, bluish or greenish; transparent to translucent. Brucite is a secondary mineral found with serpentine and metamorphic dolomites. It has been found in Italy, Sweden, and the Shetland Islands; and in the United States in New York, Pennsylvania, Nevada and California. Brucite was named in honor of Archibald Bruce, and American physician.

BUBBLE CAP. Distillation.

BUFFER (Chemical). When acid is added to an aqueous solution, the pH (hydrogen ion concentration) falls. When alkali is added, it rises. If the original solution contains only typical salts without acidic or basic properties, this rise or fall may be very large. There are, however, many other solutions which can receive such additions without a significant change in pH. The solutes responsible for this resistance to change in pH, or the solutions themselves, are known as *buffers*. A weak acid becomes a buffer when alkali is added, and a weak base becomes a buffer upon the addition of acid. A simple buffer may be defined, in Brönsted's terminology, as a solution containing both a weak acid and its conjugate weak base.

Buffer action is explained by the mobile equilibrium of a reversible reaction:

$$A + H_2O \rightleftharpoons B + H_3O^+$$

in which the base B is formed by the loss of a proton from the corresponding acid A. The acid may be a *cation*, such as NH_4^+, a *neutral molecule*, such as CH_3COOH, or an *anion*, such as $H_2PO_4^-$. When alkali is added, hydrogen ions are removed to form water, but so long as the added alkali is not in excess of the buffer acid, many of the hydrogen ions are replaced by further ionization of A to maintain the equilibrium. When acid is added, this reaction is reversed as hydrogen ions combine with B to form A.

The pH of a buffer solution may be calculated by the mass law equation

$$pH = pK' + \log\frac{C_B}{C_A}$$

in which pK' is the negative logarithm of the apparent ionization constant of the buffer acid and the concentrations are those of the buffer base and its conjugate acid.

A striking illustration of effective buffer action may be found in a comparison of an unbuffered solution, such as 0.1 M NaCl with a neutral phosphate buffer. In the former case, 0.01 mole of HCl will change the pH of 1 liter from 7.0 to 2.0, while 0.01 mole of NaOH will change it from 7.0 to 12.0. In the latter case, if 1 liter contains 0.06 mole of Na_2HPO_4 and 0.04 mole of NaH_2PO_4, the initial pH is given by the equation:

$$pH = 6.80 + \log\frac{0.06}{0.04} = 6.80 + 0.18 = 6.98$$

After the addition of 0.01 mole of HCl, the equation becomes:

$$pH = 6.80 + \log\frac{0.05}{0.05} = 6.80$$

while, after the addition of 0.01 mole of NaOH, it is

$$pH = 6.80 + \log\frac{0.07}{0.03} = 6.80 + 0.37 = 7.17.$$

The buffer has reduced the change in pH from ±5.0 to less than ±0.2.

The accompanying diagram shows how the pH of a buffer varies with the fraction of the buffer in its more basic form. The buffer value is greatest where the slope of the curve is least. This is true at the midpoint, where $C_A = C_B$ and pH $= pK'$. The slope is practically the same within a range of 0.5 pH unit above and below this point, but the buffer value is slight at pH values more than 1 unit greater or less than pK'. The curve has nearly the same shape as the titration curve of a buffer acid with NaOH or the titration curve of a buffer base

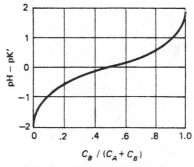

The pH of a simple buffer solution. Abscissas represent the fraction of the buffer in its more basic form. Ordinates are the difference between pH and pK'.

with HCl. Sometimes buffers are prepared by such partial titrations instead of by mixing a weak acid or base with one of its salts. Certain "universal" buffers consisting of mixed acids partly neutralized by NaOH, have titration curves which are straight over a much wider pH interval. This is also true of the titration curves of some polybasic acids, such as citric acid, with several pK' values not much more than 1 or 2 units apart. Other polybasic acids, such as phosphoric acid, with pK' values farther apart, yield curves having several sections, each somewhat similar to the accompanying curve. At any pH, the buffer value is proportional to the concentration of the effective buffer substances or groups. See also **pH (Hydrogen Ion Concentration).**

The accompanying table gives approximate pK' values, obtained from the literature, for several buffer systems.

REPRESENTATIVE BUFFER
SOLUTIONS

Constituents	pK'
H_3PO_4; KH_2PO_4	2.1
HCOOH; HCOONa	3.6
CH_3COOH; CH_3COONa	4.6
KH_2PO_4; Na_2HPO_4	6.8
HCl; $(CH_2OH_3)CNH_2$	8.1
$Na_2B_4O_7$; HCl or NaOH	9.2
NH_4Cl; NH_3	9.2
$NaHCO_3$; Na_2CO_3	10.0
Na_2HPO_4; NaOH	11.6

Buffer substances which occur in nature include phosphates, carbonates, and ammonium salts in the earth, proteins of plant and animal tissues, and the carbonic-acid-bicarbonate system in blood. See also **Blood.**

Buffer action is especially important in biochemistry and analytical chemistry, as well as in many large-scale processes of applied chemistry. Examples of the latter include the manufacture of photographic materials, electroplating, sewage disposal, agricultural chemicals, and leather products.

BULKING AGENTS (Foods). Bodying and Bulking Agents (Foods).

BUTADIENE. $CH_3CH:C:CH_3$, 1,3-butadiene (methylallene), formula wt 54.09, bp $-4.41°C$, sp gr 0.6272, insoluble in H_2O, soluble in alcohol and ether in all proportions. Butadiene is a very reactive compound, arising from its conjugated double-bond structure. Most butadiene production goes into the manufacture of polymers, notably SBR (styrene-butadiene rubber) and ABS (acrylonitrile-butadiene-styrene) plastics. Several organic syntheses, such as the Diels-Alder reaction, commence with the double-bond system provided by butadiene.

Butadiene came into prominence as an important industrial chemical during World War II as the result of the natural rubber shortage. Originally, butadiene was made by the dehydrogenation of butylenes. Later, the naphtha cracking for ethylene and propylene, with a byproduct C_4 stream, created another source of butadiene. The basis for one butadiene recovery process is the change in relative volatility of C_4 hydrocarbons in the presence of acetonitrile solvent. The latter makes the separation easier. The C_4 mixed charge goes to an extractive distillation column, where it is separated in a solvent environment into a solvent-butadiene stream and a byproduct butane-butylenes stream (overhead). The acetonitrile is recovered from the butane-butylenes. Butadiene is stripped from the fat solvent, after which it goes to a postfractionator for recovery as a 99.5% pure product.

Other solvents used in the extractive distillation include *n*-methyl pyrrolidone, dimethyl formamide, furfural, and dimethyl acetamide.

BUTYLATED HYDROXYANISOLE (BHA). Antioxidant.

BUTYLATED HYDROXYTOLUENE (BHT). Antioxidant.

BUTYL RUBBER. Elastomers.

BUTYRATE PLASTICS. Cellulse Ester Plastics (Organic).

C

CADMIUM. Chemical element, symbol Cd, at. no. 48, at. wt. 112.41, periodic table group 2b, mp 320.9°C, bp 766°C, density, 8.65 g/cm³ (20°C). Elemental cadmium has a hexagonal crystal structure. Cadmium is a silvery-white metal, malleable and ductile, but at 80°C it becomes brittle. Cadmium remains lustrous in dry air and is only slightly tarnished by air or H_2O at standard conditions. The element may be sublimed in a vacuum at a temperature of about 300°C, and when heated in air burns to form the oxide. Cadmium dissolves slowly in hot dilute HCl or H_2SO_4 and more readily in HNO_3. The element first was identified by M. Stromeyer in 1817. Naturally occurring isotopes 106, 108, 110–114, and 116. ^{113}Cd is unstable with respect to beta decay (0.3 MeV) into ^{113}In ($t_{1/2} = 10^{13}$ years). Electronic configuration is $1s^2 2s^2 2p^6 3s^2 3p^6 3d^{10} 4s^2 4p^6 4d^{10} 5s^2$. Ionic radius Cd^{2+} 0.99 Å. Metallic radius 1.489 Å. First ionization potential 8.99 eV; second, 16.84 eV; third, 38.0 eV. Oxidation potentials $Cd \rightarrow Cd^{2+} + 2e^-$, 0.402 V; $Cd + 2OH^- \rightarrow Cd(OH)_2 + 2e^-$, 0.915 V; $Cd + 4CN^- \rightarrow Cd(CN)_4 + 2e^-$, 0.90 V. Other important physical properties of cadmium are given under **Chemical Elements.**

Although ranking 57th in abundance in the earth's crust (0.15 ppm), cadmium is not encountered alone, but is always associated with zinc. The only known cadmium minerals are greenockite (sulfide) and otavite (carbonate), both minor constituents of sphalerite (zinc oxide) and smithsonite (zinc carbonate), respectively. See also **Greenockite; Smithsonite; Sphalerite Blende.**

Production. Two major processes are used for producing cadmium: (1) pyrohydrometallurgical and (2) electrolytic. Zinc blende is roasted to eliminate sulfur and to produce a zinc oxide calcine. The latter is the starting material for both processes. In the pyrohydrometallurgical process, the zinc oxide calcine is mixed with coal, pelletized, and sintered. This procedure removes volatile elements such as lead, arsenic, and the desired cadmium. From 92–94% of the cadmium is removed in this manner, the vapors being condensed and collected in an electrostatic precipitator. The fumes are leached in H_2SO_4 to which iron sulfate is added to control the arsenic content. The slurry then is oxidized, normally with sodium chlorate, after which it is neutralized with zinc oxide and filtered. The cake goes to a lead smelter, while the filtrate is charged with high-purity zinc dust to form zinc sulfate or zinc carbonate and cadmium sponge. The latter is briquetted to remove excess H_2O and melted under caustic to remove any zinc. The molten metal then is treated with zinc ammonium chloride to remove thallium, after which it is cast into various cadmium metal shapes. The process just described is known as the *melting under caustic process*. In a *distillation process*, regular rather than high-purity zinc is used to make the sponge. Then, after washing and centrifuging to remove excess H_2O, the sponge is charged to a retort. The heating and distillation process is under a reducing atmosphere. Lead and zinc present in the vapors contaminate about the last 15% of the distillate. Thus, a redistillation is required.

The cadmium vapors produced are collected and handled as previously described.

Reactions which occur in the foregoing processes are: (Leaching): $CdO + H_2SO_4 \rightarrow CdSO_4 + H_2O$; (Oxidation): $3As_2O_3 + 2NaClO_3 \rightarrow 3As_2O_5 + 2NaCl$; and $6FeSO_4 + NaClO_3 + 3H_2SO_4 \rightarrow 3Fe_2(SO_4)_3 + NaCl + 3H_2O$; (Neutralization): $Fe_2(SO_4)_3 + As_2O_5 + 3ZnO + 8H_2O \rightarrow 2FeAs(OH)_8 + 3ZnSO_4$; (Cadmium Precipitation): $CdSO_4 + Zn \rightarrow Cd + ZnSO_4$; (Melting Under Caustic): $Zn + 2NaOH + \frac{1}{2}O_2 \rightarrow Na_2ZnO_2 + H_2O$.

In the electrolytic process, the calcine first is leached with H_2SO_4. Charging the resultant solution with zinc dust removes the cadmium and other metals which are more electronegative than zinc. The sponge which results is digested in H_2SO_4 and purified of all contaminants except zinc. Nearly-pure cadmium sponge is precipitated by the addition of high-purity, lead-free zinc dust. The cadmium sponge then is redigested in spent cadmium electrolyte, after which the cadmium is deposited by electrolysis onto aluminum cathodes. The metal is then stripped from the electrodes, melted, and cast into various shapes. Reactions which occur during the electrolytic process are: (Roasting): $ZnS + 1\frac{1}{2}O \rightarrow ZnO + SO_2$; (Leaching): $ZnO + H_2SO_4 \rightarrow ZnSO_4 + H_2O$; (Neutralization): $Fe_2(SO_4)_3 + 3ZnO + 3H_2O \rightarrow 2Fe(OH)_3 + 3ZnSO_4$; (Cadmium Precipitation): $CdSO_4 + Zn \rightarrow Cd + ZnSO_4$; (Electrolysis): $CdSO_4 + H_2O \rightarrow Cd + H_2SO_4 + O_2$.

Industrial specifications normally require that impurities in cadmium metal not exceed the following: zinc, 0.035%; copper, 0.015%; lead, 0.025%; tin, 0.01%; silver, 0.01%; antimoy, 0.001%; arsenic, 0.003%; and tellurium, 0.003%. The metal is available in numerous forms. Electroplaters generally prefer balls 2 in. (5 cm) in diameter.

Uses. A major use of cadmium is for electroplating steel to improve its corrosion resistance. It is also used in low-melting-point alloys, brazing alloys, bearing alloys, nickel-cadmium batteries, and nuclear control rods, and as an alloying ingredient to copper to improve hardness. Cadmium, unfortunately, is limited in its usefulness because fumes and dusts containing cadmium are quite toxic. Melting and handling conditions that create dust or fumes must be equipped with exhaust ventilation systems.

Biological Aspects. Traditionally, cadmium has been regarded by animal feedstuff scientists as a toxic minor substance. Normally, the element is not reported in feedstuff ingredient tables because of its infrequent occurrence. When it does occur in feeds, it is believed to be derived from the fat of cereals, nuts, and vegetables. Feed researchers have found that cadmium tends to reduce iron absorption. Little information is available to indicate possible nutritional benefits of cadmium in extremely minute quantities. Cadmium-plated equipment is not used for working parts in food processing equipment.

Chemistry and Compounds. In virtually all of its compounds, cadmium exhibits the +2 oxidation state, although compounds of cadmium (I) containing the ion Cd_2^{2+}, have occasionally been reported. Cadmium hydroxide is more basic than zinc hydroxide, and only slightly amphiprotic, requiring very strong alkali to dissolve it, and forming $Cd(OH)_3^-$, or $Cd(OH)_4^{2-}$, depending upon the pH.

Cadmium Oxide. CdO, formed by burning the metal in air or heating the hydroxide or carbonate, is soluble in acids, ammonia, or ammonium sulfate solution, and is more readily reduced on heating with carbon, carbon monoxide or hydrogen than zinc oxide. Cadmium suboxide, Cd_2O, formed by thermal reduction of cadmium oxalate with carbon monoxide, is believed to be a mixture of CdO and finely divided cadmium. CdO_2 and Cd_4O have been reported. Sodium hydroxide solution precipitates cadmium hydroxide, $Cd(OH)_2$, from solutions of the sulfate or nitrate, but with the chloride the $Cd(OH)_2$ precipitate is mixed with CdOHCl and other hydroxychlorides. $Cd(OH)_2$ exists in two forms, an "active" and an "inactive" one, which have different solubility products. Cadmium(I) hydroxide $Cd_2(OH)_2$, prepared by hydrolysis of Cd_2Cl_2, is, like Cd_2O, believed to be a mixture of the metal and the divalent compound. The Cd_2^{2+} ion is definitely established, however, in such compounds as $Cd_2(AlCl_4)_2$.

Cadmium Halides. These compounds can be prepared by the action of the corresponding hydrohalic acids upon the carbonate; or by direct union of the elements. If bromine water is used, some hydrobromic acid must be added to prevent hydrolysis of the bromide to the oxybromide, CdOHBr.

In general, the cadmium halides show in their crystal structure the relation between polarizing effect and size of anion. The fluoride has the smallest and least polarizable anion of the four and forms a cubic structure, while the more polarizable heavy halides have hexagonal layer structures, increasingly covalent and at increasing distances apart in order down the periodic table. In solution the halides exhibit anomalous thermal and transport properties, due primarily to the presence of complex ions, such as CdI_4^{2-} and $CdBr_4^{2-}$, especially in concentrated solutions or those containing excess halide ions.

Cadmium Sulfide. CdS is the most extensively used of cadmium compounds and generally is prepared by precipitation from cadmium salts. The wide range of colors, varying from lemon yellow through the oranges and deep red, coupled with the stability and intensity of these colors, qualify CdS as a most desirable pigment for paints, plastics, and other products. The range of colors of CdS precipitates results from differing conditions in their formulation, including the temperature and acidity of the salt solutions from which they are precipitated. The particular salt, such as nitrate, chloride, sulfate, etc., also affects the resulting color. The rate of addition of hydrogen sulfide to the liquor affects particle size and color of the precipitates. Cadmium sulfide is insoluble in H_2O, is dimorphous, and sublimes at 1,350°C. Several crystalline forms exist. When precipitated from normal H_2SO_4 and HNO_3 solutions, the crystals are cubic. From other media, stable alpha hexagonal and unstable beta cubic forms may be formed, these ranging in specific gravity from 3.9 to 4.5, respectively. Pigment colors are not due to crystal form, but rather derive from the particle size and dispersion of the precipitates. Of total cadmium production, pigments account for 20–25% of the total.

Other cadmium compounds used as pigments in ceramics, glass, and paints include cadmium nitrate, selenide, sulfoselenide, and tungstate. Cadmiopone ($BaSO_4$ + CdS) ranges from yellow to crimson and has been used for coloring plastics and rubber goods. Cadmium stearate, when combined with barium stearate, is used as a stabilizer in thermosetting plastics and

accounts for well over 20% of the total cadmium produced. The use of cadmium compounds has been under increasing regulation because of possible carcinogenic effects of some of them. See Sax reference listed.

Cadmium Carbonate. $CdCO_3$, $pK_{sp} = 11.3$, is formed by the hydroxide upon absorption of CO_2, or upon precipitation of a cadmium salt with ammonium carbonate. With alkali carbonates, the oxycarbonates are produced.

Cadmium nitrate tetrahydrate, solubility 215 g/100 ml H_2O at 0°C, is obtained by action of HNO_3 upon the carbonate. It is ionized completely only in solutions weaker than about tenth molar. However, it does not form hydroxy compounds as readily as the zinc salt, requiring the action of NaOH, which in moderate concentration gives $Cd(NO_3)_2 \cdot 3Cd(OH)_2$ and $Cd(NO_3)_2 \cdot Cd(OH)_2$; excess sodium hydroxide precipitates the hydroxide.

Cadmium forms a wide variety of other salts, many by reaction of the metal, oxide, or carbonate with the acids, although some can be obtained only by fusion of the oxides or hydroxides. They include the antimonates (pyro and meta), the arsenates (ortho, meta, and pyro, including acid salts as well as normal), the arsenites, the borates $Cd(BO_2)_2$, $Cd_2B_6O_{11}$, $Cd_3(BO_3)_3$ and $Cd_3B_2O_6$ have been identified), the bromate, the bicarbonate, the chlorate, the chlorite, chromates and dichromates, the cyanide, the ferrate, the iodate, the molybdate, $CdMoO_4$, the nitrate, the perchlorate, various periodates, the permanganate, various phosphates (ortho, meta, and para, including acid salts as well as normal), the selenates and selenites, various silicates, the stannate, the sulfate (which reacts with limited amounts of NaOH or NH_3 solution to give various hydroxy sulfates), the thiosulfate, the titanate, the tungstate, and the uranate.

Cadmium arsenide, nitride, selenide, and telluride are known, the first and third obtainable from the elements, while the nitride is obtained by heating the amide (obtained by reaction of cadmium thiocyanate and potassium amide in liquid NH_3), and the telluride is obtainable by reduction of the tellurate with hydrogen. Cadmium arsenide is used as a semiconductor.

One of the features of the chemistry of cadmium is that it forms a relatively large number of complexes. A number of solid double halides of compositions $MCdX_3$, M_2CdX_4, M_3CdX_5 and M_4CdX_6 where M is an alkali metal and X a halogen are known, the last two probably existing only in the solid state. Conductance studies of solutions indicate the presence of such ions as CdX^+, CdX_3^- and CdX_4^{2-}. The donor ability of oxygen is less toward cadmium than zinc, fewer oxygen complexes and organic oxygen-linked complexes being known. Sulfur is a better donor than oxygen; additives of the type $(R_2S)_2 \cdot CdX_2$ are formed from dialkyl sulfides and cadmium halides. The ready reactions with NH_3, as with amines, give large numbers of complexes; those with ammonia include tetrammines and hexammines, containing $[Cd(NH_3)_4]^{2+}$ and $[Cd(NH_3)_6]^{2+}$, respectively. Ethylenediamine forms 6-coordinate compounds containing $[Cd(en)_3]^{2+}$. Prominent among the carbon donor complexes are the cyanides, principally compounds of $Cd(CN)_4^{2-}$, although $Cd(CN)_3^-$ is also known. Other carbon donor compounds are the organometallic compounds CdR_2, where R may be methyl, ethyl, propyl, butyl, isobutyl, isoamyl, amylthio, phenyl, octylthio, decylthio, and higher organic radicals.

References

Campbell, J. K., and C. F. Mills: "Effects of Dietary Cadmium and Zinc on Rats Maintained on Diets Low in Copper," *Proceedings Nutr. Society*, 33, 1, 15A (Abstract) (1974).

Carapella, S. C., Jr.: "Cadmium," Metals Handbook, 9th Edition, Volume 2, American Society for Metals, Metals Park, Ohio 44073 (1979).

Chizhikov, D. M.: "Cadmium," Pergamon, London (1966).

Chowdhury, P., and D. B. Louria: "Influence of Cadmium and Other Trace Metals on Human Alpha l-Antitrypsin: An in vitro Study," Science, 191, 480–481 (1976).

Farnsworth, M.: "Cadmium Chemicals," International Lead Zinc Research Organization, Inc., New York (1980).

Kirchgessner, M., Editor: "Trace Element Metabolism in Man and Animals," Institut für Ernährungsphysiologie, Technische Universität Munchen, Freising-Weihenstephan, Germany (1978).

Pleban, P.: "Cadmium Concentration in Blood," Science, 208, 520 (1980).

Sax, N. I.: "Dangerous Properties of Industrial Materials," 5th Edition, Van Nostrand Reinhold, New York (1979).

Underwood, E. J.: "Trace Elements in Human and Animal Nutrition," 4th Edition, Academic, New York (1977).

CAFFEINE.   Alkaloids.

CAIRNGORM STONE.   The name given to the smoky brown variety of quartz, particularly when transparent, from Cairngorm, Scotland, a well-known locality. See also **Quartz.**

CALAVERITE.   A gold telluride, $AuTe_2$, associated with quartz in low-temperature veins. A valuable gold ore from Kalgoorlie, Western Australia and the Cripple Creek region of Colorado. The ore occurs in bladed to lathlike monoclinic crystals with striations parallel to the long axis of the crystals. The ore has a metallic luster of brass-yellow to silver-white color, a hardness of 2.5–3, a specific gravity of 9.24–9.31, and a yellowish to greenish-gray streak.

CALCINATION.   Subjection of a substance to a high temperature below its normal fusion point, often to make the substance friable. Calcination frequently is carried out in long, rotating, cylindrical vessels, known as *kilns*. Material so treated may (1) lose moisture, e.g., the heating of silicic acid or ferric hydroxide, resulting in formation of silicon oxide or ferric oxide, respectively; (2) lose a volatile constituent, e.g., the heating of limestone (calcium carbonate), resulting in the formation of carbon dioxide gas and calcium oxide residue—destructive distillation of many organic substances is of this type; (3) be oxidized or reduced, e.g., the heating of a pyrite (iron disulfide) in air, resulting in formation of sulfur dioxide gas and ferric oxide residue. When calcination involves oxidation, as in the preceding cases, the operation is sometimes called *roasting*. When heating involves a reduction of metals from their ores, with separation from the gangue of the liquid metal, and slags, the process may be called *smelting*.

CALCITE.   The mineral calcite, carbonate of calcium corresponding to the formula $CaCO_3$, is one of the most widely distributed minerals. Its crystals are hexagonal-rhombohedral although actual calcite rhombohedrons are rare as natural crystals. However, they show a remarkable variety of habit including acute to obtuse rhombohedrons, tabular forms, prisms, or various scalenohedrons. It may be fibrous, granular, lamellar or compact. The cleavage in three directions parallel to rhombohedron is highly perfect; fracture, conchoidal but difficult to obtain; hardness, 3; specific gravity, 2.7; luster, vitreous in crystallized varieties; color, white or colorless through shades of gray, red, yellow, green, blue, violet, brown, or even black when charged with impurities; streak, white; transparent to opaque; it may occasionally show phosphorescence or fluorescence.

Calcite is perhaps best known because of its power to produce strong double refraction of light such that objects viewed through a clear piece of calcite appear doubled in all of their parts. A beautifully transparent variety used for optical purposes comes from Iceland, for that reason is called Iceland spar.

Acute scalenohedral crystals are sometimes referred to as dogtooth spar. Calcite represents the stable form of calcium carbonate; aragonite will go over to calcite at 470°C (878°F). Calcite is a common constituent of sedimentary rocks, as a vein mineral, and as deposits from hot springs and in caverns as stalactites and stalagmites.

Localities which produce fine specimens in the United States include the Tri-State area of Missouri, Oklahoma and Kansas, Wisconsin, Tennessee, and Michigan with inclusions of native copper; several areas in Mexico, notably Charcas and San Luis Potosi; Iceland; Cumberland and Durham regions in England; and at various regions in S.W. Africa, notably Tsumeb. The exceptionally fine sand-calcite crystals from South Dakota and Fontainebleau in France are well known. See also list of terms under **Mineralogy.**

CALCIUM.   Chemical element symbol Ca, at. no. 20, at. wt. 40.08, periodic table group 2a (alkaline earths), mp 837–841°C, bp 1484°C, density, 1.54 g/cm³ (single crystal). Elemental calcium has a face-centered cubic crystal structure when at room temperature, transforming to a body-centered cubic structure at 448°C.

Calcium is a silvery-white metal, somewhat malleable and ductile; stable in dry air, but in moist air or with water reacts to form calcium hydroxide and hydrogen gas; when heated, calcium burns in air to form calcium oxide, emitting a brilliant light. Discovered by Davy in 1808.

There are six stable isotopes, ^{40}Ca, ^{42}Ca, ^{43}Ca, ^{44}Ca, ^{46}Ca, and ^{48}Ca, with a predomination of ^{40}Ca. In terms of abundance, calcium ranks fifth among the elements occurring in the earth's crust, with an average of 3.64% calcium in igneous rocks. In terms of content in seawater, the element ranks seventh, with an estimated 1,900,000 tons of calcium per cubic mile (400,000 metric tons per cubic kilometer) of seawater. Electronic configuration $1s^2 2s^2 2p^6 3s^2 3p^6 4s^2$. Ionic radius Ca^{2+} 1.06 Å. Metallic radius 1.874 Å. First ionization potential 6.11 eV; second, 11.82 eV; third, 50.96 eV. Oxidational potentials $Ca \rightarrow Ca^{2+} + 2e^-$, 2.87 V; $Ca + 2OH^- \rightarrow Ca(OH)_2 + 2e^-$, 3.02 V.

Other important physical properties of calcium are given under **Chemical Elements.**

Calcium occurs generally in rocks, especially limestone (average 42.5% CaO) and igneous rocks; as the important minerals limestone (calcium carbonate, $CaCO_3$), gypsum (calcium sulfate dihydrate, $CaSO_4 \cdot 2H_2O$), phosphorite, phosphate rock (calcium phosphate, $Ca_3(PO_4)_2$), apatite (calcium phosphate-flouride, $Ca_3(PO_4)_2$ plus CaF_2), fluorite, fluorspar (calcium fluoride, CaF_2); in bones and bone ash as calcium phosphate, and in egg shells and oyster shells as calcium carbonate. See also **Apatite; Calcite; Fluorite; Gypsum.**

In the United States and Canada, calcium metal is produced by the thermal reduction of lime with aluminum. Before World War II, most elemental calcium was made by electrolysis of fused calcium chloride. In the thermal reduction process, lime and aluminum powder are briquetted and charged into high-temperature alloy retorts which are maintained at a vacuum of 100 μm or less. Upon heating the charge to 1,200°C, the reaction takes place slowly, releasing Ca vapor. The latter is removed continuously by condensation, thus permitting the reaction to proceed to completion. High-purity lime is required as a starting ingredient if resulting calcium metal of high purity is desired. Aluminum contamination of the resulting calcium is removed by an additional vacuum-distillation step. Other impurities also are reduced by this distillation step.

Uses of Elemental Calcium. The very active chemical nature of calcium accounts for its major uses. Calcium is used in tonnage quantities to improve the physical properties of steel and iron. Tonnage quantities also are used in the production of automotive and industrial batteries. Other major uses include refining of lead, aluminum, thorium, uranium, samarium, and other reactive metals.

Calcium treatment of steel results in improved yields, cleanliness, and mechanical properties. Because it is a very strong deoxidizer and sulfide former, calcium will improve the deoxidation and desulfurization of steel; in addition, it alters the morphology and size of inclusions, reduces internal and surface defects and reduces macrosegregation. Hydrogen-induced cracking of line pipe steels by high-sulfur fuels is reduced with calcium treatment. Several grades of calcium-treated steel are used in automotive, industrial, and aircraft applications; oil line pipe, heavy plate, and deep drawing sheet were first treated in Japan, and additional uses have been developed in the U.S. and Europe.

The high vapor pressure and reactivity of calcium limited its use in steel and iron making prior to the development of injection systems and mold nodularization processes. There are two types of injection systems. One consists of the use of a holding furnace, a sealed vessel, a carrier gas, and a lance through which calcium or calcium compounds are blown into the molten metal. This system is effective for massive desulfurization of large quantities of steel; it is a ladle process. A second injection process is wire feeding. A steel-jacketed calcium-core wire is fed through a delivery system which drives the composite wire below the surface of the liquid metal bath. The steel jacket protects the solid metallic calcium from reacting at the surface and allows it to penetrate deep in the bath; because the reaction occurs below the surface, high and reproducible calcium recoveries are possible. This process is used in both ladle additions and in tundish additions for continuous casting. It provides inclusion shape control, deoxidation, final desulfurization, and reduction of macrosegregation.

Ladle and mold processes using calcium ferroalloys are important in the production of nodular iron castings. The principal calcium alloy used is magnesium ferrosilicon. Calcium reduces the reactivity of the alloy; with the molten iron it enhances nucleation and improves morphology. The calcium content of the alloy is proportional to the magnesium content, typically in the range of 15% to 50%. In ladle or sandwich treatment techniques, pieces of the ferroalloy are placed in a pocket cut in the refractory lining of the ladle and the molten iron is then poured into the ladle. The treated, nodularized iron is then cast from the ladle into molds.

In the mold addition process, a granular form of the alloy is placed in a small reaction chamber in the molds. The nodularization treatment occurs in the mold when the iron is cast, rather than in the ladle. The reaction is contained in the mold and high recoveries result. The production of nodular iron castings is over 3 million tons/year.

A calcium-lead alloy is used in maintenance-free automotive and industrial batteries. The use of calcium reduces gassing and improves the life of the battery; 0.1–0.5% calcium is alloyed with the lead prior to the fabrication of the battery plates either by casting or through the production of coiled sheet. With calcium present these lead-acid batteries can be sealed and do not require the service of conventional batteries. They have a higher energy-to-weight ratio. The U.S. market is over 50 million

batteries/year, of which 40% are maintenance free.

Calcium is used in refining battery-grade lead for removing bismuth. Calcium also is used as an electrode material in high-energy thermal batteries.

The production of samarium cobalt magnets requires the use of calcium. The reaction is:

$$3\ Sm_2O_3 + 10\ Co_3O_4 + 49Ca\ (vapor) \xrightarrow[\Delta]{850-1150°C} 6SmCo_5 + 49CaO.$$

$$0.73\ lb\ Ca \longrightarrow 1\ lb\ SmCo_5$$

The production of 1 kilogram of $SmCo_5$ requires 0.73 kilogram of calcium. The samarium cobalt magnets have 3–6 times greater magnetic energy than alnico.

Calcium serves as a reductant for such reactive metals as zirconium, thorium, vanadium, and uranium. In zirconium reduction, zirconium fluoride is reacted with calcium metal, the high heat of reaction melting the zirconium. The zirconium ingot resulting is remelted under vacuum for purification. Thorium and uranium oxides are reduced with an excess of calcium in reactors or trays under an atmosphere of argon. The resulting metals are leached with acetic acid to remove the lime.

Calcium also is used in aluminum alloys and as an addition in a magnesium alloy used for etching. An alloy of 80% Ca–20% Mg is used to deoxidize magnesium castings. The metal also is used in the production of calcium pantothenate, a B-complex vitamin.

Chemistry and Compounds. Calcium exhibits a valence state of +2 and is slightly less active than barium and strontium in the same series. Calcium reacts readily with all halogens, oxygen, sulfur, nitrogen, phosphorus, arsenic, antimony, and hydrogen to form the halides, oxide, sulfide, nitride, phosphide, arsenide, antimonide, and hydride. It reacts vigorously with water to form the hydroxide, displacing hydrogen. Calcium oxide (quicklime) adds water readily and with the evolution of much heat (slaked lime) to form the hydroxide. Calcium hydroxide forms a peroxide on treatment with hydrogen peroxide in the cold. Calcium exhibits little tendency to form complexes, the ammines formed with ammonia are unstable, although a solid of composition $Ca(NH_3)_6$ can be isolated from solutions of the metal in liquid ammonia.

Calcium Acetate. $Ca(C_2H_3O_2)_2 \cdot H_2O$, white solid, solubility: at 0°C, 27.2 g; at 40°C, 24.9 g, at 80°C, 25.1 g of anhydrous salt per 100 g saturated solution, formed by reaction of calcium carbonate or hydroxide and acetic acid.

Calcium Aluminates. Four in number, have been prepared by high temperature methods and identified, $3CaO \cdot Al_2O_3$, at 1,535°C, decomposes with partial fusion; $5CaO \cdot 3Al_2O_3$, mp 1,455°C, $CaO \cdot Al_2O_3$, mp 1,590°C, $3CaO \cdot 5Al_2O_3$, mp 1,720°C.

Calcium Aluminosilicates. Two in number, have been prepared by high temperature methods and identified $2CaO \cdot Al_2O_3 \cdot SiO_2$, gehlinite; $CaO \cdot Al_2O_3 \cdot 2SiO_2$, anorthite.

Calcium Arsenate. $Ca_3(AsO_4)_2$, white precipitate, formed by reaction of soluble calcium salt solution and sodium arsenate solution. $pK_{sp} = 18.17$.

Calcium Arsenite. $Ca_3(AsO_3)_2$, white precipitate, formed by reaction of soluble calcium salt solution and sodium arsenite solution.

Calcium Borates. Found in nature as the minerals colemanite, $Ca_2B_6O_{11} \cdot 5H_2O$, borocalcite, $CaB_4O_7 \cdot 4H_2O$, and pandermite $Ca_2B_6O_{11} \cdot 3H_2O$. See also **Colemanite.**

Calcium Bromide. $CaBr_2 \cdot 6H_2O$, white solid, solubility,

1,360 g/100 ml H_2O at 25°C, formed by reaction of calcium carbonate or hydroxide and hydrobromic acid.

Calcium Carbide. CaC_2, grayish-black solid, reacts with water yielding acetylene gas and calcium hydroxide, formed at electric furnace temperature from calcium oxide and carbon.

Calcium Carbonate. $CaCO_3$, found in nature as calcite, Iceland spar, marble, limestone, coral, chalk, shells of mollusks, aragonite. $pK_{sp} = 8.32$. It is (1) readily dissolved by acids forming the corresponding calcium salts, (2) converted to calcium oxide upon heating. Aragonite is an unstable form at room temperature, although no change is observable until heated, when at 470°C, it is quickly converted into calcite; calcium hydrogen carbonate, calcium bicarbonate, $Ca(HCO_3)$, known only in solution, formed by reaction of calcium carbonate and carbonic acid. See also **Aragonite; Calcite.**

Calcium Chloride. $CaCl_2 \cdot 6H_2O$, white solid, solubility 536 g/100 g H_2O at 20°C, absorbs water from moist air, formed by reaction (1) of calcium carbonate or hydroxide and HCl, (2) of calcium hydroxide and ammonium chloride.

Calcium Chromate. $CaCrO_4$, yellow solid, formed by the reaction of chrome ores and calcium oxide heated to a high temperature in a current of air. $pK_{sp} = 3.15$.

Calcium Citrate. $Ca_3(C_6H_5O_7)_2 \cdot 4H_2O$, white solid, solubility: at 18°C 0.085 g/100 g H_2O, formed by reaction of calcium carbonate or hydroxide and citric acid solution.

Calcium Cyanamide. $CaCN_2$, white solid, formed (1) by heating cyanamide or urea with calcium oxide, sublimes. at 1,050°C, (2) by heating calcium carbide at 1,100–1,200°C in a current of nitrogen. Decomposes in water with evolution of NH_3.

Calcium Fluoride. CaF_2, white precipitate, formed by reaction of soluble calcium salt solution and sodium fluoride solution. $pK_{sp} = 10.40$. See also **Fluorite.**

Calcium Formate. $Ca(CHO_2)_2$, white solid, solubility at 0°C 13.90 g, at 40°C 14.56 g, at 80°C 15.22 g of anhydrous salt per 100 g saturated solution, formed by reaction of calcium carbonate or hydroxide and formic acid. Calcium formate, when heated with a calcium salt of a carboxylic acid higher in the series, yields an aldehyde.

Calcium Furoate. $Ca(C_4H_3O \cdot COO)_2$, formed by reaction of calcium carbonate or hydroxide and furoic acid.

Calcium Hydride. CaH_2, white solid, reacts with water yielding hydrogen gas and calcium hydroxide; when electrolyzed in fused potassium lithium chloride, hydrogen is liberated at the anode.

Calcium Hypochlorite. $CaOCl_2$ or $Ca(ClO)_2 \cdot 4H_2O$, white solid, contains 60%–65% "available chlorine" and sufficient calcium hydroxide to stabilize, formed by reaction of calcium hydroxide and chlorine. Very soluble in water.

Calcium Hypophosphite. $Ca(H_2PO_2)_2$, white solid, solubility 15.4 g/100 g H_2O at 25°C, formed (1) by boiling calcium hydroxide suspension in water and yellow phosphorus, (2) by reaction of calcium carbonate or hydroxide and hypophosphorous acid.

Calcium Iodide. CaI_2, yellowish-white solid, solubility 66 g/100 g H_2O at 10°C, formed by reaction of calcium carbonate or hydroxide and hydriodic acid. The hexahydrate, $CaI_2 \cdot 6H_2O$, is soluble to the extent of 1.680 g/100 g H_2O at 30°C.

Calcium Lactate. $Ca(C_3H_5O_3)_2 \cdot 5H_2O$, white solid, solubility; at 0°C 3.1 g, at 30°C 7.9 g of anhydrous salt per 100 g H_2O, formed by reaction of calcium carbonate or hydroxide and lactic acid.

Calcium Malate. $CaC_4H_4O_5 \cdot 2H_2O$, white solid, solubility; at 0°C 0.670 g, at 37.5°C 1.011 g of anhydrous salt per 100 g saturated solution. Formed (1) by reaction of calcium carbon-

ate or hydroxide and malic acid, (2) by precipitation of soluble calcium salt solution and sodium malate solution.

Calcium Nitrate. $Ca(NO_3)_2 \cdot 4H_2O$, white solid, solubility 660 g/100 g H_2O at 30°C, formed by reaction of calcium carbonate or hydroxide and HNO_3.

Calcium Oxalate. CaC_2O_4, white precipitate, insoluble in weak acids, but soluble in strong acids, formed by reaction of soluble, calcium salt solution and ammonium oxalate solution. Solubility at 18°C 0.0056 g anhydrous salt per liter of saturated solution.

Calcium Oxide. CaO, (quicklime), white solid, mp 2,570°C, reacts with H_2O to form calcium hydroxide with the evolution of much heat; reacts with H_2O vapor and CO_2 of the atmosphere to form calcium hydroxide and carbonate mixture (slaked lime); formed by heating limestone at high temperature (800°C) and removal of CO_2. This process is conducted industrially in a lime kiln.

Tricalcium Phosphate. $Ca_3(PO_4)_3$, white solid, insoluble in water; reactive with silicon oxide and carbon at electric furnace temperature yielding phosphorus vapor; reactive with H_2SO_4 to form, according to the proportions used, phosphoric acid, or dicalcium hydrogen phosphate, $CaHPO_4$, white solid, insoluble; or calcium dihydrogen phosphate, $Ca(H_2PO_4)_2 \cdot H_2O$, white solid, soluble. $pK_{sp} = 28.70$. See also **Apatite.**

Calcium Silicates. Four in number, have been prepared by high temperature methods and identified, $3CaO \cdot SiO_2$, prepared by heating the constituents to a temperature below the mp (mp is 1,700°C but substance unstable); $2CaO \cdot SiO_2$, mp 2,080°C, but upon slow cooling changes to forms of different volume; $3CaO \cdot 2SiO_2$, mp 1,475°C; $CaO \cdot SiO_2$, wollastinite, mp approximately 1,400°C. See also **Clinozoisite; Datolite; Diopside; Feldspar; Lawsonite; Tremolite; Wernerite; Wollastoiite.**

Calcium Sulfate. Gypsum $CaSO_4 \cdot 2H_2O$ plaster of Paris $CaSO_4 \cdot \frac{1}{2}H_2O$, anhydrite $CaSO_4$, white solid, slightly soluble (about 0.2 g per 100 ml of H_2O), formed by reaction of soluble calcium salt solution with a sulfate solution. pK_{sp} of $CaSO_4 = 4.6_{25}$. See also **Anhydrite; Gypsum.**

Calcium Sulfide. CaS, grayish-white solid, reactive with H_2O, formed by reaction of calcium sulfate and carbon at high temperatures. Calcium hydrogen sulfide $Ca(HS)_2$, formed in solution by saturating calcium hydroxide suspension with H_2S. pK_{sp} of $CaS = 7.24$.

Calcium Sulfite. $CaSO_3 \cdot 2H_2O$ white precipitate, $pK_{sp} = 7.9$, formed by reaction of soluble calcium salt solution and sodium sulfite solution, or by boiling calcium hydrogen sulfite solution; calcium hydrogen sulfite, $Ca(HSO_3)_2$, formed in solution by saturating calcium hydroxide or carbonate suspension with sulfurous acid.

Calcium Tartrate. $CaC_4H_4O_6 \cdot H_2O$, white solid, solubility: at 0°C 0.0875, at 80°C 0.180 g anhydrous salt in 100 ml saturated solution, formed by reaction of calcium carbonate or hydroxide and tartaric acid, or by precipitation of Ca^{2+} with a tartrate solution.

For the role of calcium in biological systems, see **Calcium (In Biological Systems).**

References

Bansal, H. N., and Z. S. Naigamwalla: "Calcium Wire Treatment of Tube Steel at Algoma," McMaster Symposium, Toronto, Canada, *Paper 11,* 1–17 (May 1979).

Bienvenu, Y., et al.: "Desoxydation et Desulfuration, Calcium et Baryum," *C.I.T.* (French), **6,** 1183 (1978).

Dunks, C. M., Hobman, G., and G. Mannion: *AFS Trans.,* **82,** 391 (1974).

Emi, T., et al.: "Mechanism of Sulfides Precipitation during Solidification of Calcium and Rare Earth-treated Ingots," (Japan), *3rd Interna-*

tional Iron and Steel Congress, Chicago, Illinois (April 1978).

Faulring, G. M., Farrell, J. W., and D. C. Hilty: "Steel Flow through Nozzles—Influence of Calcium," 37th Electric Furnace Conference, Detroit, Michigan (December 1979).

Forster, E., et al.: "Desoxidation und Entschwefelung durch Einblasen von Calciumverbindungen in Stahlschmelzen," Stahl und Eisen, 94, 11, 474–485 (May 1974).

Haida, O.: "Sulphides Shape Control in Continuous Casting by Addition of Ca, R.E. (rare earths), Ca + R.E.," 95th I.S.I.J. Meeting, Paper No. 94, (1978).

Ikeshima, T.: Transactions of the Iron and Steel Institute of Japan, 19, 589–594 (1979).

Kataura, Y., and D. Oeschlagel: "Die Behandlung von Stahlschmelzen mit Calcium," Stahl und Eisen, 100, 1, 20–29 (1980).

Riboud, P. V., et al.: "Steel Desulphurization in the Ladle and Calcium Treatment," McMaster Symposium, Toronto, Canada, pp. 10-1–20-32 (May 1979).

Scott, W., and R. A. Swift: "Advantages of Ladle Injection of Calcium and Magnesium Reagents for Steel Desulfurization," I.S.S.-A.I.M.E. Meeting, Paper 36, 128–142 (1978).

Smith, J. F.: "Calcium," in "Metals Handbook," 9th Edition, Vol. 2, American Society for Metals, Metals Park, Ohio (1979).

Staff: Foundry Management and Technology (January 1980).

Staff: "Proceedings of Scaninject II," 2nd International Conf. on Injection Metallurgy, Lulea, Sweden (June 12–13, 1980).

—Stephen E. Hluchan, Business Manager, Calcium Metal Products, Minerals, Pigments & Metals Division, Pfizer, Inc., Wallingford, Connecticut.

CALCIUM HYPOCHLORITE. Bleaching Agents.

CALCIUM (In Biological Systems). The biological role and, consequently, the importance of calcium in foods for humans and feedstuffs for livestock is well established. Although about 99% of the calcium in the bodies of animals is found in bones and teeth, the element is an essential constituent of all living cells.

Various calcium salts and organic compounds fall into this category of dietary supplements and are frequently used in feeds and foods. Some of the more important additives include calcium carbonate, calcium glycerophosphate, calcium phosphate (di- and monobasic), calcium pyrophosphate, calcium sulfate, and calcium pantothenate.

Limestone is frequently used to augment animal feedstuffs. When used, it must be low in fluorine. Calcite limestone is preferred. Calcium is also supplied in the form of crushed oyster shells, marl, gypsum (calcium sulfate), bone meal, and basic slag. In compounding feedstuffs, the specific selection of calcium source is dependent upon the species to be fed. The requirements differ, for example, between cattle, swine, and poultry. The quantity required also varies with the life stage of the animal. For example, laying hens require a much higher percentage of calcium in their diet than starting poultry.

In the mammalian body, calcium is required to insure the integrity and permeability of cell membranes, to regulate nerve and muscle excitability, to help maintain normal muscular contraction, and to assure cardiac rhythmicity. Calcium plays an essential role in several of the enzymatic steps involved in blood coagulation and also activates certain other enzyme-catalyzed reactions not involved in any of the foregoing processes. Calcium is the most important element of bone salt. Together with phosphate and carbonate, calcium confers on bone most of its mechanical and structural properties.

Calcium Metabolism

The aggregate of the various processes by which calcium enters and leaves the body and its various subsystems can be summarized by the term *calcium metabolism*. The principal pathways of calcium metabolism are intake, digestion and absorption, transport within the body to various sites, deposition in and removal from bone, teeth, and other calcified structures, and excretion in urine and stool.

Pathways. The principal pathways involve three subsystems of the body: (1) the oral cavity where ingestion occurs and the gastrointestinal tract where digestion and absorption take place and from which the feces is excreted; (2) the body fluids, including blood, which transport calcium; the soft tissues and body organs to which calcium is transported and where many of its physiological functions are carried out. Some of the organs, like the kidney, the liver, and sweat glands, are also responsible for calcium excretion; (3) the skeleton, including the teeth, where calcium is deposited in the form of bone salt and from where it is removed (resorbed) after destruction of the bone salt.

Calcium Intake. This varies in different populations and is related to the food supply and to the cultural and dietary patterns of a given population. The intake of a substantial fraction of the world population falls between 400 and 1,100 mg/day, but a range encompassing 95% of all people would undoubtedly be even wider. Most populations derive half or more of their calcium intake from milk and dairy products. Calcium intakes of domestic and laboratory animals are higher than those of humans. For example, rats typically ingest 250 mg Ca/kg body weight, and cattle 100 mg/kg, whereas humans ingest only 10 mg/kg. Ingestion falls with age in all species. The average percentage concentration of minerals in the lean body mass of vertebrates ranges from 1.1 to 2.2%.

Calcium Absorption. In most animals, including the human body, this occurs mainly in the upper portion of the small intestine. The amount and, therefore, the fraction of calcium absorbed from the gut are a function of intake, age, nutritional status, and health. Generally, the fraction absorbed decreases with age and intake and as the nutritional status improves. The absolute amount absorbed increases with intake and may or may not decrease with age. The mechanisms by which calcium is absorbed are not well understood. Active transport of the ion against an electrochemical gradient seems to be involved, but not all of the calcium appears to be absorbed by way of this process, because calcium absorption continues under conditions when active transport is severely depressed, as in vitamin D deficiency. Calcium absorption can be enhanced by the administration of large doses of vitamin D and is depressed in vitamin D deficiency. There is uncertainty regarding the effect on calcium absorption of the parathyroid hormone, the major endocrine control of the blood calcium level. Patients with hyperparathyroidism have been shown to have higher than normal absorption and patients with hypoparathyroidism to have lower than normal absorption. Similar effects have been observed in acute animal experiments, but in most of these instances a possible indirect effect has not been excluded.

Interrelationship with Phosphorus and Vitamin D. The interdependence of calcium, phosphorus, and vitamin D is exemplary of how synergistic effects can occur from combinations of feed and food components, either with a positive or negative result in the animal body. The relative concentrations (proportions) of each component in such a combination can be quite critical. Much research has gone into these particular interrelationships; much further research is required. The relationship between phosphorus and calcium nutrition has been known since the early 1840s, when Chossat in France first discovered that pigeons develop a poor bone structure when fed diets low in calcium. A few years later, the fundamental relationship of calcium and phosphorus in animal diets was developed by French and German researchers. It was not until 1922, however, with the discovery of vitamin D, that a triangular relationship was observed. See also **Bone; Phosphorus;** and **Vitamin D.**

Calcium in Blood Plasma. The concentration of calcium in the blood plasma of most mammals and many vertebrates is quite constant at about 2.5 mM (10 milligrams per 100 milliliters plasma). In the plasma, calcium exists in three forms: (1) as the free ion, (2) bound to proteins, and (3) complexed with organic (e.g., citrate) or inorganic (e.g., phosphate) acids. The free ion accounts for about 47.5% of the plasma calcium; 46% is bound to proteins; and 6.5% is in complexed form. Of the latter, phosphate and citrate account for half.

The mechanism involved in the regulation of the plasma calcium level is not fully understood. The parathyroid glands regulate both level and constancy; when these glands are removed, the plasma level drops and tends to stabilize at about 1.5 mM, but variations in calcium intake may induce fairly wide fluctuations in the plasma level. In the intact organism, wide variations in intake produce essentially no variations in the plasma calcium value which is stabilized at about 2.5 mM. The equilibrium between bone and plasma is believed to determine the level of the plasma calcium in parathyroidectomized animals, but this reasonable hypothesis requires further experimental support.

The problem of whether parathyroid regulation is due to a single hormone with hypercalcemic properties or to two hormones, one hypocalcemic, termed calcitonin, the other hypercalcemic, termed parathyroid hormone, continues under investigation.

When the calcium ion concentration is lowered in the fluids bathing nerve axons—fluids which are in very rapid equilibrium with the blood plasma—the electrical resistance of the axon membrane is lowered, there is increased movement of sodium ions to the inside, and the ability of the nerve to return to its normal state following a discharge is slowed. Thus, on the one hand, there is hyperexcitability. But, the ability for synaptic transmission is inhibited because the rate of acetylcholine liberation is a function of the calcium ion concentration. The neuromuscular junction is affected in a similar fashion; hence, the end plate potential is lowered below the muscle membrane potential and the muscle membrane is in a hyperexcitable state. These events are reversed when the calcium ion concentration is raised above the normal in the blood plasma and in the fluids bathing muscle and nerve. It is for these reasons that hypocalcemia is associated with hyperexcitability and ultimately tetany and hypercalcemia with sluggishness and bradycardia.

Muscular Contraction and Relaxation. The role of calcium in this function is not fully understood. Some researchers have proposed that calcium is the link between the electrical and mechanical events in contraction. It has been shown *in vitro* that when calcium ions are applied locally, muscle fibers can be triggered to contract. It has further been postulated that relaxation of muscle fibers is brought about by an intracellular mechanism for reducing the concentration of calcium ions available to the muscle filaments. Others postulate that contraction occurs because calcium inactivates a relaxing substance which is released from the sarcoplasmic reticulum in the presence of ATP (adenosine triphosphate).

Bone. This is the most important reservoir of calcium in the animal body. Accounting for the largest portion of the body's calcium, bone calcium also constitutes about 25% (weight) of fat-free, dried bones. Calcium occurs in bone mostly in the form of a complex, apatitic salt, so named for its structural resemblance to a family of calcium phosphates of which hydroxyapatite $[Ca_{10}(PO_4)_6(OH)_2]$ is the best-known mineralogical example. Since calcium occurs also as the carbonate, there is dis-

cussion as to whether bone salt contains the carbonate as a separate phase, whether some of the surface phosphate in apatite has been substituted for by carbonate, or whether bone mineral is a carbonato-apatite, such as dahlite. It is important to recognize that the crystal lattice of the bone mineral, when first laid down, does not and probably cannot have all possible calcium positions occupied. Whether stability is derived from hydrogen and/or organic bonds to which the mineral may be attached is not fully determined. It has been proposed that bone salt is a lamellar mixture of octocalcium phosphate and hydroxyapatite. This hypothesis has to account for the amount of pyrophosphate formed when bone salt is heated and also for its evolution with age, i.e., the increase with age in the calcification of bone and the corresponding drop in its induced pyrophosphate content, observations for which the apatitic structure can account. The proponents of the octocalcium phosphate hypothesis explain this by showing that octocalcium phosphate breaks down to apatite and anhydrous dicalcium phosphate which upon further heating give rise to pyrophosphate. Finally, it is postulated that octocalcium phosphate may be present in young and presumably newly formed bone, whereas in older bone an apatitic phosphate admittedly dominates the equilibrium.

Calcium enters and remains in bone as a result of calcification processes which involve two steps: (1) deposition of bone salt of a minimum calcium content and specific gravity. Deposition occurs by way of nucleation, probably an epitactic process on the collagen fibers, with the ground substance (mostly mucopolysaccharides) between the fibers exerting either a positive or an inhibitory effect on the nucleation process; and (2) subsequent further mineralization of the bone mineral, leading to an increase in its calcium content and its specific gravity.

Calcium removal, in contrast, involves destruction of the calcified structure *in toto*. There is no evidence that only particular structures are resorbed, e.g., those with a given degree of mineralization.

The amount of calcium deposited in bone at any moment may be determined from experiments with radioactive calcium. In growing individuals, it exceeds the amount removed by bone destruction. In adults, it is about the same as the amount removed. Such individuals are considered to be in "zero" calcium balance. In older persons, the amount deposited is less than the amount removed.

Because of the high incidence of osteoporosis in elderly people, a number of investigators have been looking into the effects of phosphorus consumption on calcium metabolism (Lutwak, 1969, 1974; Albanese et al., 1975; Jowsey, 1977). This particular concern comes from the fact that the American diet, for example, may have a calcium:phosphorus ratio of about 1:4. This ratio is much less in certain regions. The calcium:phosphorus ratio is particularly important when one of the elements is low in the diet (Wasserman, 1960). See also **Phosphorus.**

Excretion of Calcium. The principal routes of excretion are stool and urine. Calcium in the stool may be considered as made up of unabsorbed food calcium and nonreabsorbed digestive juice calcium. The latter is termed the fecal endogenous calcium. The proportion of fecal endogenous calcium to urinary calcium varies in different species. It is approximately 1:1 in humans and 10:1 in the rat and in cattle. The calcium in the urine may have a dual origin—calcium that was filtered at the glomerulus and failed to get reabsorbed along the length of the nephron, and calcium that may have originated from transtubular movement in certain regions of the nephron. The amount of calcium that may be lost in sweat can be large, but there is no convincing evidence that sweat is a habitual route of significant loss.

Calcium Occurrence. The soils of humid regions are commonly low in calcium and ground limestone is usually applied to add the element, reduce the toxicity of aluminum and manganese, and to correct soil acidity. The soils of dry areas are frequently rich in calcium. There is little evidence to indicate a strong relationship between human nutrition and calcium excesses or deficiencies in the soil. Even with farm livestock, most calcium deficiencies are not related to levels of available calcium in the soil. The reason for this anomaly is evident when one examines some of the controls over the movement of calcium in the food chain.

At the step in the food chain when calcium moves from the soil to the plant, controls based upon the genetic nature of the plant are very important. Because of these controls, certain plant species always accumulate fairly high concentrations of calcium, while other plants accumulate rather low concentrations. Among the forage crops, red clover grown, for example, on the low-calcium soils of the northeastern United States, contains more calcium than grasses grown on the high-calcium soils of the western United States. Among the food crops, snap beans and peas normally contain about 3–5 times as much calcium as corn (maize) and tomatoes. Thus, the level of calcium in the diets of people or of animals depends more on what kinds of plants are included in the diet than it does on the supply of available calcium in the soil where these plants are grown.

Adding limestone to soils to correct soil acidity and to supplement available calcium will, of course, indirectly affect human and calcium nutrition, but this is a difficult quantity to measure.

References

Albanese, A. A., et al.: "Problems of Bone Health in the Elderly," *New York State Jrnl. of Medicine,* **75,** 326 (1975).

Allaway, W. H.: "The Effect of Soils and Fertilizers on Human and Animal Nutrition," *Agriculture Information Bulletin 378,* Cornell University Agricultural Experiment Station and U.S. Department of Agriculture, Washington, D.C. (1975).

Jowsey, J.: "Osteoporosis: Dealing with a Crippling Bone Disease of the Elderly," *Geriatrics,* **32,** 41 (1977).

Kirkgessner, M., Editor: "Trace Element Metabolism in Man and Animals," Institut for Ernährungsphysiologie, Technische Universität München, Freising-Weihenstephan, Germany (1978).

Lutwak, L.: "Current Concepts of Bone Metabolism," *Ann. Internal Medicine,* 80, 630 (1974).

Mahoney, A. W., and D. G. Hendricks: "Some Effects of Different Phosphate Compounds on Iron and Calcium Absorption," *Jrnl. of Food Science,* **43,** 5, 1473–1475 (1978).

Staff: "Dietary Levels of Households in the United States," U.S. Department of Agriculture, Washington, D.C. (1968).

Stewart, A. K., and A. C. Magee: "Effect of Zinc Toxicity on Calcium, Phosphorus, and Magnesium Metabolism of Young Rats," *Jrnl. of Nutrition,* **82,** 287 (1964).

Underwood, E. J.: "Trace Elements in Human and Animal Nutrition," 4th Edition, Academic, New York (1977).

Wallace, G. W., et al.: "Calcium Binding and Its Effects on Properties of Food Protein Sources," *Jrnl. of Food Science,* **42,** 2, 473–474 (1977).

Wasserman, R. H.: "Calcium and Phosphorus Interactions in Nutrition and Physiology," *Fed. Proc.,* **19,** 636 (1960).

CALIFORNIUM. Chemical element symbol Cf, at. no. 98, at. wt. 251 (mass number of the most stable isotope), radioactive metal of the actinide series, also one of the transuranium elements. All isotopes of californium are radioactive; all must be produced synthetically. See **Radioactivity.** The isotope ^{245}Cf was first produced by S. G. Thompson, K. Street, Jr., A. Ghiorso, and G. T. Seaborg at the University of California at Berkeley in 1950 by bombarding microgram quantities of ^{242}Cm with helium ions. The reaction: ^{242}Cm (α, n) ^{245}Cf. The isotope has a half-life of 44 minutes. A number of other isotopes of Cf have been made, one of which, ^{254}Cf, half-life 55 days, is of interest because it decays predominantly by spontaneous fission. The longest-lived isotope is ^{251}Cf ($t_{1/2}$ = about 700 yrs), the next is ^{249}Cf ($t_{1/2}$ = 470 yrs). Except for ^{250}Cf ($t_{1/2}$ = 10 yrs), and ^{252}Cf ($t_{1/2}$ = 2.2 yrs), all other isotopes have half-lives less than one year. Several other isotopes (246, 248, 249, 250, 252) also decay by spontaneous fission, but with fission half-lives much longer than the half-lives for alpha-decay. Californium is considered to occur in its compounds only in the tripositive state.

Studied through the use of tracer quantities, the chemical properties of californium indicate that its chemical properties are analogous to the tripositive actinides and lanthanides, showing the fluoride and the oxalate to be insoluble in acid solution, and the halides, perchlorate, nitrate, sulfate, and sulfide to be soluble.

The probable electronic configuration is $1s^2 2s^2 2p^6 3s^2 3p^6 3d^{10} 4s^2 4p^6 4d^{10} 4f^{14} 5s^2 5p^6 5d^{10} 5f^{10} 6s^2 6p^6 7s^2$. Ionic raidus: Cf^{3+}, 0.98 Å.

In 1960, Cunningham and Wallmann isolated 0.3 microgram of californium (as californium-249) oxychloride. The best isotope for the study of californium is ^{249}Cf, which can be isolated in pure form through its beta-particle-emitting parent, ^{249}Bk.

Californium-252 is an intense neutron source. One gram emits 2.4×10^{12} neutrons per second. This isotope shows promise for applications in neutron activation analysis, neutron radiography, and as a portable source for field use in mineral prospecting and oil well logging. The isotope also is being investigated for medical research applications. It may find use as a neutron source for irradiation of certain tumors for which gamma-ray treatment is inadequate.

References

Choppin, G. R., Thompson, S. G., Ghiorso, A., and B. G. Harvey: "Nuclear Properties of Some Isotopes of Californium, Elements 99 and 100," *Phys. Rev.,* **94,** 4, 1080–1081 (1954).

Conway, J. G., et al.: "The Solution Absorption Spectrum of Cf^{3+}," *J. Inorg. Nucl. Chem.,* **28,** 3064–3066 (1966).

Cunningham, B. B., and T. C. Parsons: "Preparation and Determination of the Crystal Structure of Californium and Einsteinium Metals," *Lawrence Berkeley Laboratory Nuclear Chemistry Annual Report,* UCRL-20426, University of California, Berkeley, California (1970).

Fields, P. R., et al.: "Transplutonium Elements in Thermonuclear Test Debris," *Phys. Rev.,* **102,** 1, 180–182 (1956).

Ghiorso, A., Thompson, S. G., Choppin, G. R., and B. G. Harvey: "New Isotopes of Americium, Berkelium and Californium," *Phys. Rev.,* **94,** 4, 1081 (1954).

Ghiorso, A., et al.: "New Elements Einsteinium and Fermium, Atomic Numbers 99 and 100," *Phys. Rev.,* **99,** 3, 1048–1049 (1955).

Green, J. L., and B. B. Cunningham: "Crystallography of the Compounds of Californium: I. Crystal Structure and Lattice Parameters of Californium Sesquioxide and Californium Trichloride," *Inorg. Nucl. Chem. Lett.,* **3,** 9, 343–349 (1967).

Hulet, E. K., Thompson, S. G., Ghiorso, A., and K. Street, Jr.: "New Isotopes of Berkelium and Californium," *Phys. Rev.,* **84,** 2, 366–367 (1951).

Peterson, J. R., and R. D. Baybarz: "The Stabilization of Divalent Californium in the Solid State: Californium Dibromide," *Inorg. Nucl. Chem. Lett.,* **8,** 4, 423–431 (1972).

Samhoun, K., and F. David: "Radiopolarography of Am, Cm, Bk, Cf, Es, and Fm," *Proceedings of the 4th International Transplutonium Element Symposium,* Baden-Baden, W. Germany (1975).

Seaborg, G. T., Editor: "Transuranium Elements," Dowden, Hutchinson & Ross, Stroudsburg, Pennsylvania (1978).

Thompson, S. G., Street, K., Jr., Ghiorso, A., and G. T. Seaborg: "Element 98," *Phys. Rev.,* **78,** 3, 298–299 (1950).

Thompson, S. G., and M. L. Muga: "Methods of Production and Research on Transcurium Elements," *Proceedings of the Second United Nations Conference on Peaceful Uses of Atomic Energy*, Geneva, Switzerland (1958).

CALORESCENCE. A term designating the production of visible light by means of energy derived from invisible radiation of frequencies below the visible range. Tyndall found it possible to raise a piece of blackened platinum foil to a red heat by focusing upon it infrared radiation from an arc or from the sun, the visible wavelengths having been filtered out. It is to be noted that the transformation is indirect, the light being produced by heat and not by any direct stepping up of the infrared frequency. A somewhat analogous phenomenon is the production of visible sparks or the glowing of a fine platinum wire in a resonant circuit energized by long-wave Hertzian radiation.

CALORIMETRY. The study of heat as contrasted with temperature. The oxygen bomb calorimeter, which is used to determine the heat of combustion of fuels, is only one of many types of calorimeters. Steam calorimeters, for example, are used to measure heat capacities, heat of reaction, or energy changes in biological processes. Instruments for differential thermal analysis are sometimes referred to as differential scanning calorimeters. The bomb calorimeter is a batch-type instrument which requires a discrete sample and, therefore, is used only for solid and liquid materials. Gaseous fuels (nondiscrete) are analyzed in flow-type calorimeters.

One of the most important characteristics of any combustible fuel is the quantity of energy or heat that it releases as it is burned. This value is referred to as either the *heat of combustion*, or the *calorific value* of the fuel and is usually expressed in *British thermal units (Btu)* per pound or ton, or in *calories per gram*. The heat of combustion of solid and liquid fuels is routinely determined in order to establish the price of the fuel, as well as to serve as a basis for calculating the overall efficiency of a power generating facility or engine.

To determine the heat of combustion of a fuel, a representative sample is burned in a high-pressure oxygen atmosphere within a metal bomb or pressure vessel. The energy released by this combustion is adsorbed within the calorimeter and measured in terms of temperature change within the calorimeter. The heat of combustion of the sample is obtained by multiplying the temperature rise of the calorimeter by a previously determined energy equivalent or heat capacity for the instrument. Corrections are applied to adjust these values for any heat transfer occurring in the calorimeter as well as for any side reactions which are unique to the bomb combustion process.

The reliability of results obtained with bomb calorimetry depends upon a truly representative sample as well as a reliable calorimeter and proper operating techniques. A typical load of coal will include large lumps, fine powders, and particles varying in size between the two extremes. During loading and transit, the fines and smaller particles will work their way to the bottom of the shipment. A sample taken from the bottom would not be representative of the entire shipment inasmuch as it would be rich in fines and deficient in the larger particles. A sample from the top of the shipment obviously would be biased in favor of the larger particles. Similar problems can occur with liquid fuels.

Any oxygen bomb calorimeter consists of four essential parts: (1) A bomb or vessel in which the combustible charge is burned; (2) a bucket or container which holds the bomb as well as a measured quantity of water to absorb the heat released from the bomb and a stirring device to assure thermal equilibrium; (3) a jacket for protecting the bucket from transient thermal stresses; and (4) a calorimeter thermometer for measuring temperature changes within the bucket. The cross section of such a calorimeter is shown in Fig. 1. A photo of the actual bomb is given in Fig. 2.

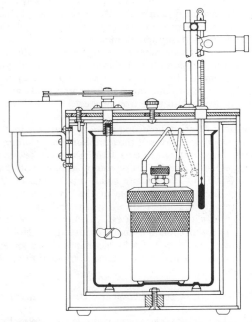

Fig. 1. Cross section of plain jacket oxygen bomb calorimeter. (*Parr Instrument Co.*)

Fig. 2. Bomb portion of oxygen bomb calorimeter. (*Parr Instrument Co.*)

The bomb consists of a strong, thick-walled, metal vessel which can be opened for inserting the sample, for cleaning, and for recovering the products of combustion. Valves must be provided for filling the bomb with oxygen under pressure and for releasing residual gases after the combustion is complete. Electrodes to carry the ignition current to the fuse wire also are required. Since an internal pressure up to 1,500 psig (102 atmospheres) can be developed during combustion, most bombs are constructed to withstand pressures of at least 3,000 psig (204 atmospheres).

CALORIZING. Production of a protective coating of iron-aluminum alloy on iron or steel. The articles are ordinarily coated by heating to a high temperature in a closed container packed with powdered aluminum. Other processes include impregnation at high temperature with an aluminum chloride vapor and spraying with molten aluminum from a spray gun and then heating to a high temperature. When the aluminum coating is held at high temperatures, an iron-aluminum alloy forms which is resistant to oxidation and corrosion by hot combustion gases, especially those containing sulfur compounds which are particularly corrosive to bare iron or steel.

CAMPHOR. A crystalline compound occurring in various parts of the wood and leaves of the camphor tree, (*Cinnamonum camphora*), a large evergreen tree with light-green leaves growing in many warm regions of southeastern Asia, notably Taiwan. Camphor, $C_{10}H_{16}O_7$, is a white solid, mp 179°C, bp 209°C, of a characteristic odor, insoluble in H_2O, soluble in alcohol or ether. Camphor may be produced synthetically by converting pinene into bornyl chloride with HCl, thence to isobornyl acetate, thence to isoborneol, and finally oxidizing borneol to camphor. Camphor has found use in medicines, insecticides, and moth preventives. Earlier uses included manufacture of plastics and lacquers.

CANCER (Drugs). Chemotherapeutic Drugs (Cancer).

CANCRINITE. The mineral cancrinite is a complex hydrous silicate (see **Silicon**) corresponding approximately to the formula $(Na, K, Ca)_{6-8}(Al, Si)_{12}O_{24}(SO_4, CO_3, Cl)_{1-2} \cdot nH_2O$. It is hexagonal, with prismatic cleavage; hardness, 5–6; specific gravity, 2.42–2.50; color, white to gray or may be greenish, bluish, yellow, or flesh red; colorless streak; luster, subvitreous to greasy; transparent to translucent. Cancrinite is found only in the nephelite-syenites and related rock types and is commonly associated with sodalite. It is believed to be in part primary, having crystallized direct from the magma, and in part secondary as a result of alteration of nephelite by solutions of calcium carbonate. It is found in the Ilmen Mountains of the U.S.S.R., in Rumania, in Norway, in Canada in Hastings County, Ontario, and in the United States in Kennebec County, Maine. This mineral was named for Count Georg Cancrin, a Russian statesman who died in 1845.

CAPRIC ACID. Also called decanoic, decoic, and decyclic acid, formula $CH_3(CH_2)_8COOH$. The acid occurs as a glyceride in natural oils. Usual form is white crystals having an unpleasant odor. Soluble in most organic solvents and dilute nitric acid; insoluble in water. Specific gravity 0.8858 (40°C); mp 32.5°C; bp 270°C. Combustible. A component of some edible vegetable oils. See also **Vegetable Oils (Edible)**. Capric acid is derived from the fractional distillation of coconut oil fatty acids. The acid is used in esters for perfumes; fruit flavors; a base for wetting agents; as an intermediate in organic synthesis; plasticizer; resins; and used in food-grade additives.

CAPROIC ACID. Also called hexanoic, hexylic, or hexoic acid, formula $CH_3(CH_2)_4COOH$. Present in milk fats to extent of about 2%. Also a constituent of some edible vegetable oils. See **Vegetable Oils (Edible)**. The acid is oily, colorless or slightly yellow, and liquid at room temperature. Odor is that of lim-

burger cheese. Soluble in alcohol and ether; slightly soluble in water. Specific gravity 0.9276 (20/4°C; mp −4.0°C; bp 205°C. Combustible. Caproic acid is derived from the crude fermentation of butyric acid; or by fractional distillation of natural fatty acids. Used in various flavorings; manufacture of rubber chemicals; varnish driers; resins; pharmaceuticals.

CAPROLACTAM. $NH(CH_2)_5CO$, formula weight 112.15, liquid ingredient used in the manufacture of type 5 nylon. See also **Fibers**. Several hundred million pounds of the compound are produced annually. There are a number of proprietary processes for caprolactam production. In one process, the charge-stock is nitration-grade toluene, air, hydrogen, anhydrous NH_3, and H_2SO_4. The toluene is oxidized to yield a 30% solution of benzoic acid, plus intermediates and by-products. Pure benzoic acid, after fractionation, is hydrogenated with a palladium catalyst in stirred reactors operated at about 170°C under a pressure of 10 atmospheres. The resultant product, cyclohexane-carboxylic acid is mixed with H_2SO_4 and then reacted with nitrosylsulfuric acid to yield caprolactam. The nitrosylsulfuric acid is produced by absorbing mixed nitrogen oxides N_2O_3 in $H_2SO_4 : N_2O_3 + H_2SO_4 \rightarrow SO_3 + 2NOHSO_4$. The resulting acid solution is neutralized with NH_3 to yield $(NH_4)_2SO_4$ and a layer of crude caprolactam which is further purified. The overall process reaction is:

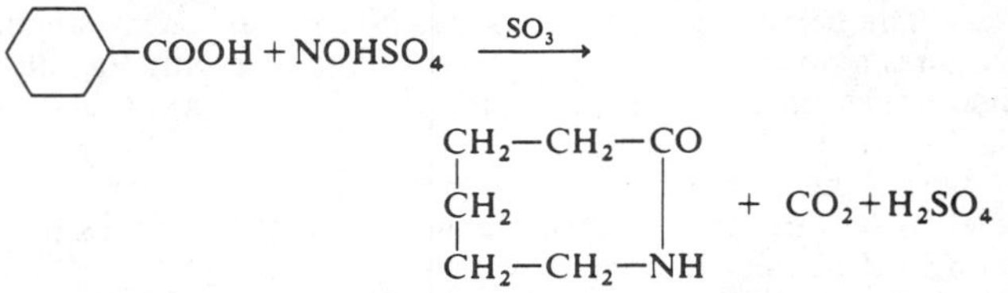

$$\text{\Large\bigcirc}\!\!-COOH + NOHSO_4 \xrightarrow{SO_3}$$

$$\begin{array}{c} CH_2\!-\!CH_2\!-\!CO \\ CH_2 \qquad\qquad | \\ CH_2\!-\!CH_2\!-\!NH \end{array} + CO_2 + H_2SO_4$$

A comparatively recent process utilizes a photochemical reaction in which cyclohexane is converted into cyclohexanone oxime hydrochloride: $C_6H_{12} = NOCl \xrightarrow[\text{light}]{HCl} C_6H_{10}NOH \cdot 2HCl$. The yield of cyclohexanone is estimated at about 86% by weight. Then, in a Beckmann rearrangement, the cyclohexanone oxime hydrochloride is converted to ε-caprolactam:

$$C_6H_{10}NOH \cdot 2HCl \xrightarrow{H_2SO_4} \begin{array}{c} CH_2\!-\!(CH_2)_4\!-\!C\!=\!O \\ |\!\!-\!\!-\!\!-\!\!NH\!-\!\!-\!\!-\!\!| \end{array} + 2HCl$$

To obtain the nitrosyl chloride for producing the oxime, the following reactions are required: (1) NH_3 is burned in air to produce NO_x: $2NH_3 + 3O_2 \rightarrow N_2O_3 + 3H_2O$; (2) nitrosylsulfuric acid is made by reacting nitrogen trioxide with H_2SO_4: $2H_2SO_4 + N_2O_3 \rightarrow 2HNOSO_4 + H_2O$; (3) nitrosyl chloride then is made by adding HCl to the nitrosylsulfuric acid: $HNOSO_4 + HCl \rightarrow NOCl + H_2SO_4$. High-pressure mercury lamps are used to effect the photochemical reaction. Any radiation of a wavelength shorter than 3,650 Å must be filtered out to avoid formation of tarry products. Critical factors that determine the yield of the process include temperature, the distance between the lamps and reactor, and the volume of the reactor. The crude caprolactam solution is neutralized with NH_3. The resulting mixture separates into an upper layer of crude caprolactam; the lower layer contains aqueous $(NH_4)_2SO_4$. An advantage claimed for this process is that only about one-half as much by-product $(NH_4)_2SO_4$ is formed as compared to other processes.

CARAT. Diamond.

CARBAMATES. Derivatives of the hypothetical carbamic acid, H_2NCOOH, which is not known to exist. The ethyl derivative urethane is prepared by heating urea in alcohol under pressure by the reaction

$$H_2NC(=O)NH_2 + C_2H_5OH \rightarrow H_2NCOOC_2H_5 + NH_3.$$

Other derivatives include methyl-, propyl-, phenyl-, and benzyl-carbamate.

CARBAMIC ACID. Herbicide; Insecticide.

CARBANION. An ion of the general formula

$$B{-}\overset{\overset{\displaystyle A}{|}}{\underset{\underset{\displaystyle D}{|}}{C}}{:}^{-},$$

where A, B and D are substituent groups. Their importance in elucidating the mechanism of organic reactions is because a considerable proportion of all organic reactions involve carbanions, as others do carbonium ions and carbon free radicals (including carbene radicals). Many carbanion reactions involve removal of a proton from a carboxylic acid to form a carbanion. Many electrophilic substitution reactions involve carbanions. Carbanions are strong bases or nucleophiles. Many electrophilic substitution reactions that have carbanion intermediates are base-catalyzed since the basic reagent produces the basic carbanion. Because of the negative charge on carbanions, their structures are affected by cations, by attached substituents and particularly by the solvent.

CARBENE. The name quite generally used for the methylene radical, $:CH_2$. It is formed during a number of reactions. Thus the flash photochemical decomposition of ketene ($CH_2=C=O$) has been shown to proceed in two stages. The first yields carbon monoxide and $:CH_2$, the latter then reacting with more ketene to form ethylene and carbon monoxide. Carbene reacts by insertion into a C—H bond to form a C—CH_3 bond. Thus carbene generated from ketene reacts with propane to form n-butane and isobutane. Carbene generated by pyrolysis of diazomethane reacts with diethyl ether to form ethylpropyl ether and ethylisopropyl ether.

Substituted carbenes are also known; chloroform reacts with potassium t-butoxide to form dichlorocarbene $:CCl_2$, which adds to double or triple carbon-carbon bonds to form cyclopropane derivatives.

CARBIDES. Carbon; Iron Metals, Alloys, and Steels.

CARBOCYCLIC COMPOUNDS. Organic Chemistry.

CARBOHYDRATES. These are compounds of carbon, hydrogen, and oxygen that contain the saccharose grouping (below), or its first reaction product, and in which the ratio of hydrogen to oxygen is the same as in water.

$$H{-}\overset{\overset{\displaystyle |}{|}}{\underset{\underset{\displaystyle OH}{|}}{C}}{-}\overset{\overset{\displaystyle |}{|}}{\underset{\underset{\displaystyle O}{||}}{C}}{-}$$

Carbohydrates are the most abundant class of organic compounds, representing about three-fourths of the dry weight of all vegetation. Carbohydrates are also widely distributed in animals and lower life forms. These compounds comprise one of the three major components (others are protein and fat) of the human diet, and indeed that of most other animals. In a nutrition-conscious era, advocates for both more and fewer carbohydrate calories in the human diet can be found.

Classification of Carbohydrates. Because carbohydrates as components of foods and feedstuffs are not limited to just a few specific classes or types, but essentially run the gamut of the carbohydrate spectrum, it is in order here to review briefly the organization of carbohydrate chemistry, with some examples from the various classes. See also entry on **Organic Chemistry.**

Elementary Terminology. A term synonymous with carbohydrate is *saccharide* (sometimes *saccharose*). When referring to saccharides, the basic molecular formula is considered to be $C_6H_{12}O_6$. Compounds with this general formula, such as glucose, mannose, and galactose, are known as *monosaccharides* because they contain one $C_6H_{12}O_6$. A *disaccharide*, as typified by sucrose, lactose, and maltose, has the general molecular formula, $C_{12}H_{22}O_{11}$ and may be considered as containing two $C_6H_{12}O_6$ groupings that have been joined by one atom of oxygen, with the elimination of one molecule of water. Similarly, the *trisaccharides*, such as raffinose, have the molecular formula, $C_{18}H_{32}O_{16}$. Any larger molecules of the $C_x(H_2O)_y$ configuration are termed *polysaccharides*, and include the starches, celluloses, dextrin, and glycogen. See also **Starch.** An *oligosacchharide* is a carbohydrate containing from 2 up to 10 simple sugars linked together (e.g., sucrose, composed of dextrose and fructose). Beyond ten, the term polysaccharide is used. Gums and mucilages are complex carbohydrates. See **Gums and Mucilages.**

Both the terms carbohydrate and saccharide are significant only by way of classifying these compounds, because neither term appears in whole or in part in any of the widely used names of these compounds. About the only point of nomenclature enjoyed in common by several of the saccharides is the termination -ose, as found, for example, in cellulose, dextrose, sucrose, and glucose. Any saccharides having the structure of an aldehyde is termed an *aldose*; any saccharide with the structure of a ketone is termed a *ketose*. For those saccharides that contain 4–6 carbons, the number of carbons forms a nomenclature base, as a *tetrose*, $C_4H_8O_4$, a *pentose*, $C_5H_{10}O_5$, and a *hexose*, $C_6H_{12}O_6$.

To be consistent with the relationship between a mono- and a disaccharide, some authorities do not term a tetrose or a pentose a monosaccharide. By combining the *ald-* and *ket-* prefixes, certain compounds then may be called aldohexoses, such as glucose and galactose, or ketohexoses, such as fructose and sorbose.

The mono-, di-, and trisaccharides are also commonly termed *sugars*. A sugar generally is considered to possess the properties of a crystalline solid with a relatively low melting point (below 150°C), of being soluble in water, and of possessing a sweet taste. Thus, the common names of several saccharides incorporate the term sugar, preceded by the common raw source of the substance, as glucose (grape sugar), sucrose (cane or beet sugar), maltose (malt sugar), and lactose (milk sugar). The crosscurrents of the nomenclature employed for the carbohydrates will be evident from the accompanying table.

Important Carbohydrates in Foods and Biological Systems

The properties of several carbohydrates that are of particular importance in foods and biological systems are described in the following paragraphs.

Glucose. This may be considered the key carbohydrate. It is the leading member of the aldohexose group, and is formed as one of the products or the only product when the following carbohydrates are hydrolyzed, sucrose, lactose, maltose, cellulose, glycogen. In many of its properties and its structural forms, it is representative of the sugars, and it is therefore discussed in detail here. Glucose is a colorless solid ($C_6H_{12}O_6$), less sweet

CLASSES OF CARBOHYDRATES (With examples)

Monosaccharides (sugars):
 crystalline solids, soluble in water, sweet taste; those that occur in nature are hydrolyzed by certain enzymes.
 Tetrose, $C_4H_8O_4$
  1. Erythrose
 Pentoses, $C_5H_{10}O_5$
  2. Arabinose
   By boiling gum arabic, cherry gum, corn pith, elder pith with dilute sulfuric acid.
  3. Xylose
   By boiling substances mentioned under arabinose above.
  4. Ribose
  5. Lyxose
 Hexoses, $C_6H_{12}O_6$
  Aldohexoses
  6. Glucose, dextrose ("grape sugar"), melting point 146°C (anhydrous). With the enzyme zymase (of yeast) yields ethyl alcohol plus carbon dioxide. Specific rotatory power—see glucose below.
  7. Galactose
   Specific rotatory power +83.9°.
  8. Mannose
   Specific rotatory power +14.1°.
  9. Gulose
  10. Idose
  11. Talose
  12. Altrose
  13. Allose
  Ketohexoses
  14. Fructose, levulose ("fruit sugar"), melting point 95°C. Specific rotatory power −88.5°.
  15. Sorbose
  16. Tagatose
Disaccharides (sugars), $C_{12}H_{22}O_{11}$:
 crystalline solids, soluble in water, sweet taste.
  17. Sucrose ("cane sugar," "beet sugar"), melting point 170–186°C (de-composes). With the enzyme invertase, yields glucose plus fructose. Specific rotatory power +66.4°.
  18. Lactose ("milk sugar"), melting point 202°C (anhydrous). With the enzyme lactase yields glucose plus galactose. Specific rotatory power +52.4°.
  19. Maltose ("malt sugar"), melting point of $C_{12}H_{22}O_{11}$. H_2O: 100°C. With the enzyme maltase yields glucose plus glucose. Specific rotatory power +138.5°.
  20. Melibiose
   With enzymes or dilute acid yields glucose plus galactose.
  21. Cellobiose
   With the enzymes maltase, or cellase, yields glucose plus glucose.
  22. Trehalose
Trisaccharide, $C_{18}H_{32}O_{16}$:
 crystalline solid, soluble in water, tasteless.
  23. Raffinose, melitose, melting point 118°C (anhydrous). With the enzyme invertase, yields fructose plus melibiose. With the enzyme emulsin, yields sucrose plus galactose.
Polysaccharides (non-sugars), $(C_6H_{10}O_5)_n$:
 noncrystalline solids, insoluble in water, tasteless.
  24. Starches
   With the enzyme diastase yield maltose.
  25. Celluloses
   With hydrochloric acid, heated, yield glucose.
   With acetic anhydride plus concentrated sulfuric acid, yield cellobiose.
  26. Dextrin
   With the enzyme diastase yields maltose.
   With the enzyme maltase or with acids yields glucose.
  27. Inulin, melting point 178°C (decom.) $(C_6H_{10}O_5)_n$.
   With the enzyme inulase (but not with diastase) yields fructose.
  28. Glycogen, melting point 240°C.
   With the enzyme diastase (or ptyalin), yields glucose plus maltose.
  29. Pentosans

than sucrose, soluble in water from which it may be crystallized $C_6H_{12}O_6 \cdot H_2O$. Glucose reacts (1) with alkaline cupric salt solution (Fehling's solution or Benedict's solution) to form cuprous oxide, (2) with ammonio-silver salt solution (Tollens' solution) to form finely divided or mirror film of silver, (3) with phenylhydrazine in acetic acid, to form glucose phenylhydrazone $CH_2OH(CHOH)_4CH\!:\!NNHC_6H_5$, white solid, melting point alpha 159–160°C, beta 140–141°C, with excess phenylhydrazine to form glucosazone

$$CH_2OH(CHOH)_3C\!:\!(NNHC_6H_5) \cdot CH\!:\!NNHC_6H_5$$

yellow solid, melting point 205°C decom., (4) with acetic anhydride, to form glucose pentacetate $C_5H_6(OOCCH_3)_5CHO$, melting point alpha 112 to 113°C, beta 131 to 134°C, (5) with sodium amalgam, to form sorbitol $CH_2OH(CHOH)_4CH_2OH$, (6) with hydriolic acid, to form 2-iodo-normal-hexane $CH_3(CH_2)_3CHICH_3$, (7) with sodium hydroxide solution, to form yellowish-brown solutions upon warming, (8) with calcium hydroxide solution, to form calcium glucosate $CH_2OH(CHOH)_4COCa(OH)$, slightly soluble solid from which glucose is recoverable by action of carbon dioxide (calcium carbonate formed simultaneously). Strontium hydroxide and barium hydroxide react similarly. Any of these three reactions may be utilized to recover glucose, with the limitation that barium soluble compounds are poisonous, (9) with hydroxylamine hydrochloride, to form glucoseoxime $CH_2OH(CHOH)_4CH\!:\!NOH$, melting point 138°C, (10) with hydrocyanic acid, to form glucosecyanhydrin

$$CH_2OH(CHOH)_4CHOHCN,$$

(11) by oxidation, to yield with bromine gluconic acid $CH_2OH(CHOH)_4COOH$, and with nitric acid saccharic acid $COOH(CHOH)_4COOH$, (12) with alpha-naphthol dissolved in chloroform and then forming a layer of concentrated sulfuric acid beneath the mixture, to form a red coloration at the junction of the two liquid layers (Molisch's test for carbohydrates). Upon standing, the color changes to purple. (13) With methyl alcohol in the presence of hydrogen chloride, to form methyl glucoside (methyl ether of glucose). See also **Glycosides.**

If a sample of glucose is recrystallized from water, it is found that a freshly prepared aqueous solution of this sample has a specific rotation of +113°, and upon standing, the value steadily changes to +52° and remains there. On the other hand, if a sample of the same glucose is recrystallized from pyridine, a freshly prepared aqueous solution has a specific rotation of +19°, which steadily increases upon standing and levels off at a constant value of +52°. This changing of optical rotation with time is referred to as mutarotation. The fact that the two portions of glucose when recrystallized from different solvents mutarotate and stop at the same position suggests the formation of some equilibrium mixture.

To explain this situation, it must be recognized that glucose contains an aldehyde (—CHO) group and four alcohol groups (—OH). These two kinds of groups can react to form a hemiacetal just as if they were present in different molecules (Fig. 1).

Glucose and fructose are present in sweet fruits, such as grapes and figs, and in honey. These two are the only hexoses found in nature in the free state. Glucose is normally present in human urine to the extent of about 0.1%, but in the case of those suffering from diabetes glucose is excreted in large amount. Glucose is formed, as previously mentioned, by the reaction of polysaccharides and water, the reaction with starch in the presence of very dilute hydrochloric acid serving as the industrial source (the hydrochloric acid acts as a catalyzer, and the

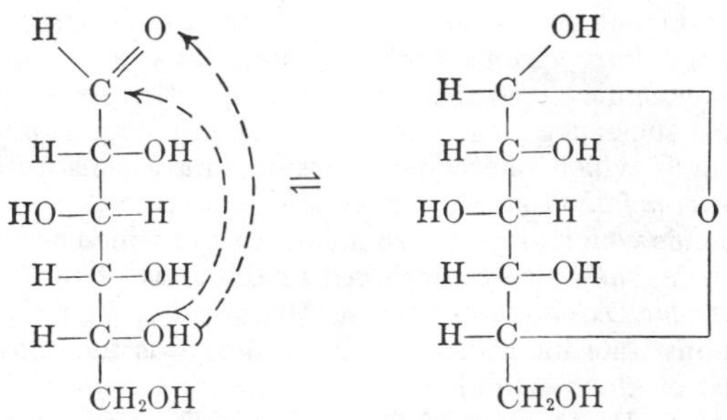

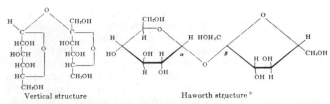

Fig. 1.   Mutarotational aspects of glucose.

Fig. 2.   Sucrose. * When the oxygen atom is drawn at the top of the furanose ring, OH groups are drawn downward to correspond to those on the left side of the vertical structure.

small percentage present is later neutralized to form sodium chloride). The solution is evaporated to a syrup or to crystallization, and is used in the manufacture of sweets, and (usually) alcohol, and in foods. The reaction of glucosides with water, by enzymes or acids, produces glucose as one of the products. With sodium hydroxide, under carefully defined conditions, glucose forms lactic acid. Glucose is used as food and for the production of alcohol (wines) from fruit juices. Glucose may be detected by formation of glucosazone, and determination of its melting point.

Industrial processes for converting starch into dextrose (glucose) are described under **Starch.**

Fructose. This sugar is present with glucose in sweet fruits and honey, and may be obtained free by reaction of inulin of dahlia tubers or artichokes with water, and with glucose by reaction of sucrose with water, the product being known as invert sugar. Fructose differs from glucose in structure in being a pentahydroxy-2-ketone,

$$CH_2OH(CHOH)_3COCH_2OH$$

instead of aldehyde. The specific rotary power of fructose is −88.5°. Fructose forms the same identical osazone as glucose, and sorbitol plus mannitol by reduction. Fructose may be used as sugar by diabetic patients to advantage instead of glucose or sucrose. Fructose is detected by the violet color its alkaline solution gives with meta-dinitrobenzene.

Sucrose. This is a colorless solid, when heated melts to 170–186°C, and upon cooling forms barley sugar, which gradually crystallizes. Upon heating above the melting point, it forms caramel, a brown liquid, with decomposition. Caramel is used in confectionery, and in coloring beverages and foods. At higher temperatures decomposition into gaseous and tarry substances occurs, finally leaving a residue of carbon ("sugar charcoal"). Other sugars behave similarly. Sugars are also carbonized by concentrated sulfuric acid. Sucrose is very soluble in water, and is obtained from solution by crystallization, usually by vacuum evaporation. The solution has a specific rotatory power of +66.4°, does not exhibit mutarotation, but is converted by acids or invertase into invert sugar (glucose plus fructose), specific rotatory power −19.7°. Sucrose forms with calcium hydroxide calcium sucrosate, a 1% solution of sugar dissolves about 18 times as much calcium hydroxide as does pure water. This behavior is utilized to recover sugar from solutions, as in the case of glucose, and also to determine free calcium oxide in burnt lime, due to the reactivity of calcium hydroxide and nonreactivity of calcium carbonate. Sucrose is nonreactive with dilute sodium hydroxide, with phenylhydrazine, with ammonio-silver salt solution, but, when inverted to glucose plus fructose, these reactions may be obtained. Sucrose forms with acetic anhydride sucrose octaacetate. The suggested structural formula is as shown in Fig. 2. Sucrose is an important food preservative,

food flavor, and a raw material for confectionery and for industrial alcohol.

Sucrose is extensively distributed in the seeds and leaves of plants, and is the most abundant of the sugars. The commercial sources of sucrose are the stems of sugar-cane (11 to 16% sucrose, average 13%), the root of the sugar-beet (average 16% sucrose, selection having raised the sucrose content from 5% to a maximum of 20%), the sap of the sugar maple, and the stems of sorghum-cane. Sucrose is pressed from the stems of sugar-cane or sorghum-cane, and extracted with the water from the sliced roots of sugar-beets. The solutions are purified, evaporated and crystallized to such a degree that commercial sucrose is practically chemically pure (about 99% sucrose). The purity of sugar and the concentration or strength of sugar solutions is determined by the rotatory power of the solution, the special polariscope usually used being called a saccharimeter. Sucrose is reduced with Fehling's solution only after inversion.

The sugar content of some common fruits have been reported by Kulisch:

	SUCROSE	HEXOSES
Apple	1.0–5.4	7.0–13.0
Apricot	6.0	2.7
Banana, ripe	5.0	10.0
Pineapple	11.3	2.0
Strawberry	6.3	5.0

Lactose. This sugar is obtained from the residual water solution (whey) of milk after removal of fat and casein for making butter and cheese. Milk contains about 4.5% of lactose. Lactose forms hard gritty crystals ("sand sugar") $C_{12}H_{22}O_{11} \cdot H_2O$, loses water at 140°C, melting point 202°C (anhydrous) with decomposition; is less sweet than sucrose, reduces ammoniocupric salt solution, ammoniosilver salt solution, forms osazone, melting point 200°C, turns yellow when warmed with sodium hydroxide solution. Lactose is the source of galactose, and undergoes, with the proper enzymes, fermentation into lactic acid and butyric acid.

Maltose. This sugar is found in soybean, and is produced by the action of the enzyme diastase of germinated barley (malt) on starch at 50°C, and is thus an intermediate product in the transformation of starch into alcohol. Maltose $C_{12}H_{22}O_{11} \cdot H_2O$, melting point 100°C, when rapidly heated, may be crystallized from the concentrated malt syrup after removal of proteins and insoluble material. Maltose reduces ammonio-cupric salt solution, and forms osazone.

Starch. This is a white powder, odorless and tasteless, insoluble in cold water, forming an emulsion ("starch paste") or gel with hot water, the consistency of which depends upon the ratio of starch to water used. When boiled starch emulsion is cooled and treated with a solution of iodine in alcohol or potassium iodide, a blue coloration is produced, which is a sensitive and characteristic test. The blue color is associated with the adsorption of iodine on the surface of the starch, and disappears in the presence of alkalis. When boiled with dilute acid, starch is first changed into a soluble gummy mixture known as dextrin, and finally into glucose. When starch, either alone or in the presence of a slight amount of nitric acid, is heated to 120° to 200°C, dextrin is formed; at higher temperatures starch behaves similarly to sucrose. With concentrated nitric acid, starch forms esters, similar to cellulose nitrates. By the action of the enzyme diastase, starch is converted into maltose, which with the enzyme maltase yields glucose. Starch is nonreactive with ammonio-cupric salt solution, and with phenylhydrazine. See also **Starch**.

Dextrin. This is a white-to-yellow solid, forming an adhesive with water, nonreactive with ammonio-cupric salt solution, reactive with iodine in alcohol or potassium iodide, usually forming red, brown, or blue color. Formed when starch is heated to 120° to 200°C either alone or in the presence of a slight amount of nitric acid. Dextrin is formed when bread is toasted and is present in well-baked bread crust, and on the surface of starched goods that have been ironed hot. Dextrin is used in adhesives.

Inulin. This is a white solid, soluble in warm water, specific rotatory power −40°, with iodine in alcohol or potassium iodide gives yellow color. Inulin is present in tubers of dahlia to the extent of about 10%. Inulin reacts with water in the presence of the enzyme inulase or of acids to form fructose. The enzyme diastase does not produce this change.

Glycogen. Also known as *animal starch*, this is a white solid, soluble in water, specific rotatory power +197°, with iodine in alcohol or potassium iodide solution, forming brown color. Glycogen is found as reserve carbohydrates in the animal body, more particularly in the liver. Horseflesh, oysters and beef are sources of glycogen.

Pentosans. These compounds are polysaccharides which may be considered as anhydrides of pentose sugars, after the manner of the hexosans, sucrose, starch, from glucose, fructose. When pentosans or pentoses are heated with hydrochloric or sulfuric acid, furfural $C_4H_3O \cdot CHO$ is formed, and addition of aniline produces a red color. Pentosans are present in gummy carbohydrates, in bran of wheat seed, and in woods.

By means of the cyanhydrin reaction, higher sugars of the heptose, octose, and nonose types have been prepared. A monosaccharide such as an aldohexose may be converted into the next lower monosaccharide, such as an aldopentose, by oxidation to the acid, which corresponds to the aldohexose, then treating the calcium salt solution of this acid with a solution of ferrous acetate plus hydrogen peroxide. Carbon dioxide is evolved and aldopentose formed.

For a description of cellulose, see **Cellulose**.

Carbohydrate Metabolism

Carbohydrates are utilized by the cells as a source of energy and as precursors for the manufacture of many of their structural and metabolic components. In the mammal, for example, D-glucose is the carbohydrate primarily used for this purpose. Certain microorganisms, in contrast, can grow on a medium containing some other hexose or a pentose as the principal source of carbon. Green plants obtain their carbohydrates by photosynthesis, while animals receive most of their carbohydrates by ingestion and digestion. See also **Photosynthesis**.

The complete oxidation of glucose to carbon dioxide and water yields 689 kcal of heat per mole of glucose. When this oxidation occurs in a cell, the energy is not all dissipated as heat. Some of the evolved energy is conserved in biochemically utilizable form of "high-energy" phosphates, such as adenosine triphosphate (ATP) and guanosine triphosphate (GTP). In addition to enzymes concerned with energy metabolism, there are enzymes in biological systems which catalyze the transformation of glucose into various carbohydrates, fatty acids, steroids, amino acids, nucleic acid components, and other necessary biochemical substances. The entire network of reactions involving compounds which interconvert carbohydrates constitutes *carbohydrate metabolism*. By convention, some reactions involving compounds which are not carbohydrates, but which are derived from them, may also be included in this area of metabolism.

Anaerobic Oxidation of Glucose. Historically, the first system of carbohydrate metabolism to be studied was the conversion by yeast of glucose to alcohol (fermentation) according to the equation: $C_6H_{12}O_6 \rightarrow 2CH_3CH_2OH + 2CO_2$. The biochemical process is complex, involving the successive catalytic actions of 12 enzymes and known as the *Embden-Meyerhof pathway*. The series of reactions is summarized in the entry on **Glycolysis**.

In order for the cell to carry out a "controlled" oxidation of D-glucose and conserve some of the energy derived from the process, it is first necessary to add phosphate to the hexose with the expenditure of energy. The necessary energy and the phosphate per se is supplied by ATP in two separate reactions of the system. Since each molecule of glucose can yield two molecules of triose phosphate for oxidation, the conversion of glucose to pyruvic acid nets two molecules of ATP per molecule of hexose utilized.

Approximately 30% of the evolved energy is conserved as ATP, but only about 8% of the total energy in glucose is made available in this anaerobic oxidation of glucose to pyruvic acid. Since nicotinamide adenine dinucleotide (NAD⁺), also called diphosphoryidine nucleotide, DPN⁺, which is involved in the oxidation of glyceraldehyde-3-phosphate, is present in the cell in small quantities only, the coenzyme must constantly be regenerated for the oxidative process to continue. This regeneration is accomplished by the reduction of *acetaldehyde* to *ethanol*. Since oxygen plays no role in this process, the system can obviously proceed anaerobically. In fact, the presence of oxygen decreases the net disappearance of glucose (*Pasteur effect*).

Fermentation occurs in many microorganisms, but not all organisms reoxidize the reduced nicotinamide adenine dinucleotide (NADH) through the formation of ethanol. In certain organisms, for example, *pyruvic acid* is converted to *acetoin* which is then reduced with NADH to 2,3-butylene glycol. In other organisms and in animal tissues, NADH is oxidized in the reduction of *pyruvic acid* to *lactic acid*. In insects, and possibly in some animal tissues, the reduction of *dihydroxyacetone phosphate* to *alpha-glycerol phosphate* may serve to regenerate NAD⁺. The conversion of glucose to lactic acid in animal tissues is termed *glycolysis*. This term arose from the initial understanding that this process was markedly different from the microbial fermentation process. Fermentation and glycolysis are now known to differ primarily in the further anaerobic utilization of pyruvic acid.

Aerobic Oxidation of Pyruvic Acid. Pyruvic acid can be oxidized completely to carbon dioxide and water in a cyclic enzymatic system known as the *Krebs citric acid cycle*, or the *tricarboxylic acid cycle* (*TCA cycle*). In this system, a 2-carbon unit in the form of a acetyl coenzyme A, derived from the NAD⁺-mediated oxidative decarboxylation of pyruvic acid in the presence of coenzyme A, is condensed with oxalacetic acid to form citric acid. This tricarboxylic acid is then converted back to oxalacetic acid in a stepwise manner, with the formation of $2CO_2$ and $2H_2O$. In addition to this formation of CO_2, one reduced nicotinamide adenine dinucleotide phosphate (NADPH), two NADH, one reduced flavin, and one GTP arise per 2-carbon unit oxidized in the cycle. Since in the aerobic oxidation of the reduced flavin and the reduced nicotinamide adenine nucleotides, ATP is formed, the oxidation of a molecule of "acetate" results in the conservation of energy in the form of 12 molecules of triphosphate. In the complete oxidation of glucose through glycolysis and the citric acid cycle, about 40% of the energy originally present in the glucose can be retained as triphosphate. The ubiquitous distribution of this cycle in nature suggests that the citric acid cycle is a major energy-yielding pathway in biological systems.

Certain microorganisms have a modification of this cycle in which isocitric acid is cleaved to succinic acid and glyoxylic acid. The latter acid is condensed with acetyl-CoA to form malic acid. In this modification (the *glyoxylic acid cycle*), oxal-succinic acid and alpha-ketoglutaric acid are not involved. This is sometimes referred to as the *glyoxylate shunt* pathway.

Since in the citric acid cycle there is no net production of its intermediates, mechanisms must be available for their continual production. In the absence of a supply of oxalacetic acid, "acetate" cannot enter the cycle. Intermediates for the cycle can arise from the carboxylation of pyruvic acid with CO_2 (e.g., to form malic acid), the addition of CO_2 to phosphenolpyruvic acid to yield oxalacetic acid, the formation of succinic acid from propionic acid plus CO_2, and the conversion of glutamic acid and aspartic acid to alpha-ketoglutaric acid and oxalacetic acid, respectively. See Fig. 3.

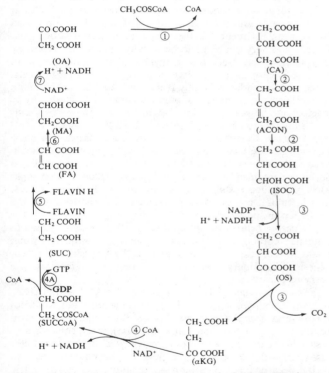

Fig. 3. Krebs citric acid cycle. *Enzymes involved*: (1) Condensing enzyme; (2) aconitase; (3) isocitric acid; (4) α-ketoglutaric acid dehydrogenase; (4A) succinic acid thiokinase; (5) succinic acid dehydrogenase; (6) fumarase; (7) malaic acid dehydrogenase. *Abbreviations*: CA = citric acid; ACON = *cis*-aconitic acid; KG = α-ketoglutaric acid; SUC = succinic acid; FA = fumaric acid; MA = malic acid; OA = oxalacetic acid.

The utilization of carbohydrate intermediates for the biosynthesis of amino acids, fatty acids, steroids, et al. occurs at various stages of the cycle and its related reactions. See Fig. 4. See also **Coenzymes**.

Other Carbohydrate Interconversions. Two systems, as shown in Fig. 5, are available for the synthesis of ribose-5-phosphate,

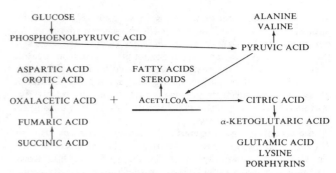

Fig. 4. Representative conversions of carbohydrates to other substances.

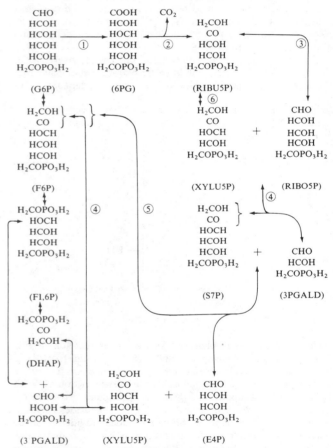

Fig. 5. Pentose phosphate cycle. *Enzymes involved*: (1) Glucose-6-phosphate dehydrogenase; (2) 6-phosphogluconic acid dehydrogenase; (3) pentose phosphate isomerase; (4) transketolase; (5) transaldolase; (6) pentose phosphate epimerase. *Abbreviations*: G6P = glucose-6-phosphate; 6PG = 6-phosphogluconic acid; RIBU5P = ribulose-5-phosphate; 3PGALD = glyceraldehyde-3-phosphate; E4P = erythrose-4-phosphate; F1,6P = fructose-1-6-diphosphate; DHAP = dihydroxyacetone phosphate; F6P = fructose-6-phosphate. Enzymes not named are those of glycolysis. $NADP^+$ is reduced in reactions (1) and (2).

a precursor of the pentose moiety of ribonucleic acid, ATP, and other substances. The formation of ribose-5-phosphate from glucose-6-phosphate by formation and decarboxylation of 6-phosphogluconic acid and isomerization of the resulting ribulose-5-phosphate is termed the *hexose monophosphate oxidative pathway*. The scheme, together with the system involving the enzymes *transketolase* and *transaldolase* (which also can synthesize pentose) that act to form hexose phosphate from pentose phosphate, is called the *pentose phosphate cycle*. This cycle represents an alternative pathway to glycolysis for the formation of triose phosphate from glucose-6-phosphate. The relative importance of the two pathways seems to be different among the various organisms and tissues.

In a certain group of bacteria, still another pathway (*Entner-Doudoroff pathway*) for the utilization of glucose has been studied. Here glucose-6-phosphate is oxidized to 6-phosphogluconic acid which is dehydrated to 2-keto-3-deoxy-6-phosphogluconic acid. This substance is then split to pyruvic acid and glyceraldehyde-3-phosphate (which also can be converted to pyruvic acid).

The formation of deoxyribose, the pentose moiety of deoxyribonucleic acid, can occur directly from ribose while the latter is in the form of a nucleotide diphosphate. Deoxyribose-5-phosphate can also be formed by condensation of acetaldehyde and glyceraldehyde-3-phosphate.

Transglycosylation. An enzymatic process, transglycosylation, plays an important role in carbohydrate metabolism. Figure 6 represents the formation of the disaccharide, sucrose, as

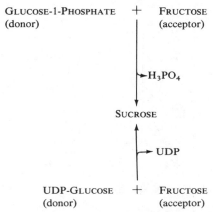

Fig. 6. Examples of transglycosylation.

an example of this mechanism. In the upper reaction of Fig. 6, glucose-1-phosphate is the glycosyl donor and fructose is the acceptor. In the lower reaction, the sugar nucleotide, uridine diphosphoglucose (UDP-glucose), is the glycosyl donor. With UDP-glucose as donor and glucose-6-phosphate as acceptor, trehalose-6-phosphate may be formed. Polysaccharides may also be formed by this process. The donor residues provided by sugar nucleotides are added to preexisting polysaccharide chains (known as *primers*) acting as glycosyl acceptors. In the formation of glycogen, for example, UDP-glucose donates the glucose moiety which is added to the end of a previously synthesized chain by a 1,4-linkage, thereby lengthening the chain by one glucose unit.

Digestion, Absorption and Storage of Carbohydrates

In the mammal, complex polysaccharides which are susceptible to such treatment, are hydrolyzed by successive exposure to the amylase of the saliva, the acid of the stomach; and the disaccharases (e.g., maltase, invertase, amylase, etc.) by exposure to juices of the small intestine. The last mechanism is very important. Absorption of the resulting monosaccharides occurs primarily in the upper part of the small intestine from which the sugars are carried to the liver by the portal system. The

absorption across the intestinal mucosa occurs by a combination of active transport and diffusion. For glucose, the active transport mechanism appears to involve phosphorylation. The details are not yet fully understood. Agents which inhibit respiration (e.g., azide, fluoracetic acid, etc.) and phosphorylation (e.g., phlorizin), and those which uncouple oxidation from phosphorylation (e.g., dinitrophenol) interfere with the absorption of glucose. See also **Phosphorylation (Oxidative).** Once the various monosaccharides pass through the mucosa, interconversion of the other sugars to glucose can begin, although the liver is probably the chief site for such conversions. Even though many organs and tissues store carbohydrates as glycogen for their own use, the liver provides the main source of glucose for all tissues through conversion of its glycogen (and other substances) to glucose-6-phosphate, hydrolysis of this ester by the specific liver glucose-6-phosphatase, and transport of the free glucose in the bloodstream throughout the body.

A common cause of *osmotic diarrhea* is the ingestion of carbohydrates that a person cannot digest. Retention of a disaccharide, such as lactose or sucrose, within the intestinal lumen occurs because of the absence of the appropriate disaccharidase at the intestinal surface membrane. Unless they are converted to monosaccharides, these sugars cannot be transported. Their retention in the lumen can cause a significant diarrheal water loss per day. As an added complication, bacteria in the lower small intestine and colon may catabolize the 12-carbon sugars to 3-carbon fragments, further aggravating the osmotic effect. Infants and very young children usually have sufficient lactase and sucrase, but there is a tendency among some people to become lactase deficient between the age of 3 and 14 years. Once such a condition is fully recognized, the ingestion of milk and other dairy products should be eliminated. It should be pointed out, however, that the so-called *irritable bowel syndrome* is attributable to lactase deficiency in a relatively small percentage of cases. Carbohydrates that can cause diarrhea in some persons include lactose and sucrose, already mentioned; stachyose and raffinose contained in many legumes; mannitol and sorbitol, contained in artificial sweeteners (which contain sugar alcohols); glucose and galactose present in all dietary sugars; and lactulose, contained in nondietary disaccharides as parts of certain medications.

Endocrine Influences. A number of hormones are known to influence carbohydrate metabolism in the mammal. Insulin seems to increase oxidation of glucose, lipogenesis, and glycogenesis. Its primary mode of action may be to facilitate the entry of glucose into the cell.

Vitamin Influences. The involvement of NAD+ and NADP+ in many carbohydrate reactions explains the importance of nicotinamide in carbohydrate metabolism. Thiamine, in the form of thiamine pyrophosphate (cocarboxylase), is the cofactor necessary in the decarboxylation of pyruvic acid, in the trans-ketolase-catalyzed reactions of the pentose phosphate cycle, and in the decarboxylation of alpha-ketoglutaric acid in the citric acid cycle, among other reactions. Biotin is a bound cofactor in the fixation of carbon dioxide to form oxalacetic acid from pyruvic acid. Pantothenic acid is a part of the CoA molecule. There are separate alphabetical entries in this volume on the various specific vitamins as well as a review entry on **Vitamin.**

Photosynthesis. The formation of carbohydrates in green plants by the process of photosynthesis is described in the entry on **Photosynthesis.** The synthetic mechanism involves the addition of carbon dioxide to ribulose-1,5,-diphosphate and the subsequent formation of two molecules of 3-phosphoglyceric acid which are reduced to glyceraldehyde-3-phosphate. The triose phosphates are utilized to again form ribulose-5-phosphates by enzymes of the pentose phosphate cycle. Phosphorylation of ribulose-5-phosphate with ATP regenerates ribulose-1,5-diphosphate to accept another molecule of carbon dioxide. See also **Phosphorylation (Photosynthetic).**

Carbohydrates in Foods

Sugar is also discussed under **Fiber; and Gums and Mucilages.** Statistics on the carbohydrate content of diets of various peoples throughout the world have not been very reliable because of the scores of variables involved, the great difficulties in establishing reliable sampling procedures, and lack of past records, among other factors. One summary, for example, that breaks down food energy from protein, fat, and carbohydrates shows a downward trend for carbohydrates in the American diet—from 56% in 1911 to 46% in the mid-1970s. These figures were based upon U.S. Department of Agriculture statistics of food disappearance at the retail level, but they do not take into consideration food spoilage, cooking waste, plate waste, and other factors which affect actual consumption. Since protein remained quite constant at 11–12% throughout this time span, the drop in carbohydrates was made up by an increase in fats—from 32% in 1911 to 42% in the mid-1970s. In another study, of the 46% carbohydrate energy intake as of 1977, 24% is attributed to sugar and 22% to complex carbohydrates. In a controversial U.S. Government study, which attempts to set new dietary goals for the nation, it was suggested that the traditional 12% protein be retained, but that fat be reduced from 42% and carbohydrates upped to 58%, but with a major difference, namely, cutting the sugar portion of carbohydrates from 24% to 15%. Thus, the dietary goal would require 40–45% complex carbohydrates in the diet. It has been suggested that to achieve the projected carbohydrate goals, there would have to be a 66% increase in the consumption of grain products; a 25% increase of vegetables and fruit; and a 50% reduction in sugar and sweets.

Even though much visibility has been given by the various news media to the dietary role of sugar, it is obvious that, as of the early 1980s, a great deal of fundamental research remains to prove or disprove many conclusions, often conflicting and confusing, in order to establish reliable dietary guidance in this area.

References

Celender, I. M., et al.: "Dietary Trends and Nutritional Status in the United States," *Food Technology*, **32**, 9, 39–41 (1978).
Cornblath, M., and R. Schwartz: "Disorders of Carbohydrate Metabolism in Infancy" in "Major Problems in Clinical Pediatrics," Vol. 3 (A. J. Schaffer and M. Markowitz, editors), Saunders, Philadelphia, 1976.
Gray, G. M.: "Intestinal Digestion and Maldigestion of Dietary Carbohydrates," *Ann. Rev. Med.*, **22**, 391 (1971).
Greenberg, D. M.: "Metabolic Pathways," 3rd edition, Vol. 1, Academic, New York, 1967.
Hollingsworth, D. F.: "Translating Nutrition into Diet," (The British Nutritional Foundation), *Food Technology*, **31**, 2, 38–44 (1977).
Hood, L. F., Wardrip, E. K., and G. N. Bollenback: "Carbohydrates and Health," AVI, Westport, Connecticut, 1977.
Mottram, R. F.: "Human Nutrition," 3rd edition, Food and Nutrition Press, Westport, Connecticut, 1979.
Ralph, C. L.: "Carbohydrate Metabolism," in "The Encyclopedia of the Biological Sciences," (P. Gray, editor), Van Nostrand Reinhold, New York, 1970.
Scala, J.: "Responsibilities of the Food Industry to Ensure an Optimum Diet," *Food Technology*, **32**, 9, 77–79 (1978).
Segal, S., and L. F. Hood, Co-Chairpersons: "Carbohydrates," *Institute of Food Technologists Symp.*, Saint Louis, Missouri, 1979.
Sharon, N.: "Carbohydrates," *Sci. Amer.*, **243**, 5, 90–117 (1980).
Siperstein, M. D., Foster, D. W., et al.: "Control of Glucose and Diabetic Vascular Disease" (editorial), *N. Engl. J. Med.*, **296**, 1060 (1977).
Staff: "Food and Nutrient Intake of Individuals in the United States," Report 11, U.S. Department of Agriculture, Washington, D.C., 1972.
Staff: "A Guide for Professionals: The Effective Application of Exchange Lists for Meal Planning," Amer. Diabetes Assn. and Amer. Dietetic Assn., New York, 1977.
Sussman, K. E.: "Juvenile-Type Diabetes and Its Complications. Theoretical and Practical Considerations," Charles C. Thomas, Springfield, Illinois, 1971.
Sussman, K. E., and R. J. S. Metz (editors): "Diabetes Mellitus," 4th edition, Amer. Diabetes Assn., New York, 1975.
Sutherland, H. W., and J. M. Stowers: "Carbohydrate Metabolism in Pregnancy and the Newborn," Churchill Livingstone, New York, 1975.
Watt, B. K., and A. L. Merrill: "Composition of Foods," Agriculture Handbook 8, U.S. Department of Agriculture, Washington, D.C., 1975.
Weser, E., et al.: "Lactose Deficiency in Patients with 'Irritable-Colon' Syndrome," *N. Engl. J. Med.*, **273**, 1070 (1965).

CARBON. Chemical element symbol C, at. no. 6, at. wt. 12.011, periodic table group 4a, mp 3550°C (approximate), bp 4287°C (approximate), density 3.52 g/cm³ (diamond at 20°C), 2.25 g/cm³ (graphite at 20°C). The specific gravity of amorphous carbon at 20°C ranges from 1.8 to 2.1. There are two stable isotopes of the element, ^{12}C and ^{13}C and four known radioactive isotopes, ^{10}C, ^{11}C, ^{14}C, and ^{15}C. Isotope ^{14}C occurs in nature as the result of interaction of cosmic rays with ^{14}N. Inasmuch as the half-life of ^{14}C has been established (about 5760 years), the occurrence of this isotope in ancient documents, artifacts, and materials makes it a useful diagnostic tool in archeological and paleoscientific investigations. See **Radioactivity.** The first ionization potential of carbon is 11.264 eV; second, 24.28 eV; third, 47.7 eV; fourth, 64.19 eV. Other important physical characteristics of carbon are given under **Chemical Elements.**

Principal Elemental Forms. There are two allotropic forms (1) diamond, and (2) graphite. Diamond, the hardest of natural substances, consists of a lattice of carbon atoms arranged in a tetrahedral structure at equal distance apart (1.544 Å) and bonded by electron pairs in localized molecular orbitals formed by overlapping of the sp^3 hybrids. In graphite, one of the softest substances, the carbon atoms are arranged in laminar sheets, 3.40 Å apart and composed of carbon atoms in hexagonal arrangement 1.42 Å apart, with each atom bonded to three others in its sheet by electron pairs in localized molecular orbitals formed by overlapping of the sp^2 hybrids. The remaining p-electrons form a mobile system of nonlocalized π-bonds that permits of electrical conductivity within the lamina. The various carbon blacks formed by such methods as combustion of carbon-containing materials with insufficient oxygen are found to have x-ray diffraction patterns suggestive of graphite, but with more diffuse rings, indicating a much lower degree of crystallinity. When carbon black is heated its diffraction pattern develops new rings indicative of a structure more like graphite. At the same time, its properties as an absorbent deteriorate.

The uses of carbon are dependent upon the form and variety: diamonds for jewels and as abrasives, graphite in lubricants and as an electrical conductor, cocoanut charcoal for adsorbing gases at low temperature in an enclosed space to produce a high vacuum, activated carbon to absorb color from solutions and to remove odor from water, coke and wood charcoal as fuels. See also **Carbon Black; Coal; Diamond; Graphite.**

Revised Views of High-Temperature Behavior of Carbon. The probable importance of the —C≡C— bond to the high-temperature behavior of carbon (most familiarly encountered in acetylene, which itself is stable at high temperatures) was not proposed until the late 1960s (Gorsey and Donnay, 1968; Sladkov and Koudrayatsev, 1969). It was proposed that high-temperature carbon forms are made up of chains of —C≡C—

units, called *carbynes* by the Soviet scientists. A proposed mechanism for the transformation of a graphite basal plane sheet of atoms in —C≡C— units is shown in Fig. 1. It will be noted

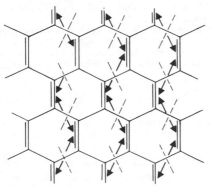

Fig. 1. Possible mechanism for transformation of a graphite basal plane sheet of atoms into carbyne chains. (*After Whittaker, 1978.*)

that at high temperatures, a single bond in the structure may break. This shifts an electron into each of the adjacent double bonds, forming a triple bond. Completion of the process transforms the sheet of atoms into a chain of carbynes.

The chains, of course, can be variously stacked. Kasatochkin et al. (1973) reported at least five such forms. As explained at a conference on carbon held in Irvine, California in 1977, the transformation from the carbyne form to graphite involves a reaction between acetylenelike molecules (acting rapidly and exothermically), whereas the reverse reaction (breaking of single bonds) can be expected to be a much slower process. Thus, it is observed that the conventional carbon phase diagram may be lacking because it does not consider carbyne forms. Whittaker (1978) pointed out that for years it has been difficult to reconcile high-pressure results with the low-pressure data on the vapor pressure of carbon. As shown by Fig. 2, there is inclusion of

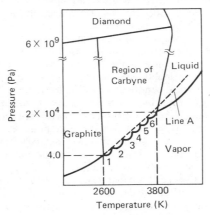

Fig. 2. Proposed carbon phase diagram to accommodate region of carbyne. Dashed line A is vapor pressure graphite would have if it were the stable form above 2600 K. (*After Whittaker, 1978.*)

a region for carbynes which accommodates experimental findings. Whitakker further notes: (1) Graphite is not stable above 2600 K at any pressure; (2) the solid-liquid-vapor triple point occurs at 3800 K and 2×10^4 Pa; and (3) carbyne forms are stable between 2600 and 3800 K, and their stability region extends to the diamond transition line.

Compounds of Carbon. With exception of hydrogen, carbon forms the largest number of known compounds (hundreds of thousands) among the chemical elements. Traditionally, carbon compounds fall into two fundamental classes (1) *inorganic* compounds and (2) *organic* compounds, although the line of demarcation is not always precise, differing from one authority to the next.

The main subclasses of inorganic carbon compounds include:
1. The *carbon oxides*, notably CO (carbon monoxide) and CO_2 (carbon dioxide). (In attempts to resolve spectral discrepancies from observations of the planet Mars, it appears that the red coloration of the planet may be due at least partially to the presence of carbon suboxide C_3O_2 in the Martian atmosphere. It is believed that the linear molecules of C_3O_2 polymerize, forming several heavier molecules having color gradations ranging from pale yellow to orange, reddish brown, violet, and nearly black. *Science*, **166**, 1141–1142 (November 28, 1969).
2. The *carbonates* —CO_3, which occur widely in nature—minerals, rocks, ores, and mineral waters—and include such compounds as Na_2CO_3, $CaCO_3$ (limestone), $MgCO_3$ (magnesite), $MgCa(CO_3)_2$ (dolomite), etc.
3. Some carbon–sulfur compounds, such as CS_2 (carbon disulfide), the thiocyanides, and thiocyanates —CNS, such as HCNS (thiocyanic acid), $Pb(CNS)_2$ (lead thiocyanate), etc.
4. The *carbides*, such as Na_2C_2, Cu_2C_2, WC, ZrC, etc.
5. The *carbonyls* —CO, such as $Cr(CO)_6$, $Fe(CO)_5$, $Ni(CO)_4$, etc.
6. The *halides*, such as CCl_4 (carbon tetrachloride), CBr_4, etc.

The subclasses of organic compounds comprise the realm of organic chemistry and are described under **Organic Chemistry.** Compounds made up of carbon and hydrogen are described there under *Hydrocarbons*. Several subclasses of organic compounds include oxygen along with hydrogen and carbon in their structure—e.g., acid anhydrides, alcohols, aldehydes, carbohydrates, carboxylic acids, esters, ethers, fatty acids, furans, ketones, lactides, lactones, phenols, quinones, and terpenes. Some of the main subclasses of nitrogen-bearing organic compounds include the amides, amines, amino acids, anilides, azo and diazo compounds, carbamates, cyanamides, hydrazines, polypeptides, proteins, purines, pyridines, pyrroles, quaternary ammonium compounds, semicarbazones, ureas, and ureides. The addition of the halogens to the structure yields chlorine organics, brominated compounds, fluorocarbons, etc. Most of the metals combine with carbon compounds to form organometallics. Sulfur-bearing organics include the sulfonic acids, sulfonyls, sulfones, thioalcohols, thioaldehydes, sulfoxides, etc. Silicones are silicon-bearing carbon compounds. See also **Chlorinated Organics.**

Carbides. As might be expected from its position in the periodic table, carbon forms binary compounds with the metals in which it exhibits a negative valence, and binary compounds with the nonmetals in which it exhibits a positive valence. A convenient classification of the binary compounds of carbon is into ionic or salt-like carbides, intermediate carbides, interstitial carbides, and covalent binary carbon compounds.

The *ionic* or *salt-like carbides* are formed directly from the elements, or from metallic oxides and carbon, carbon monoxide, or hydrocarbons. This last reaction is reversible, and this group of carbides may be further subdivided into acetylides, e.g., Li_2C_2, Na_2C_2, K_2C_2, Rb_2C_2, Cs_2C_2, Cu_2C_2, Ag_2C_2, Au_2C_2, BeC_2, Mg_2C_2, CaC_2, SrC_2, BaC_2, ZnC_2, CdC_2, Al_2C_6, Ce_2C_6, and ThC_4; methanides, e.g., Be_2C and Al_4C_3; and the allylides, primarily magnesium allylide, Mg_2C_3, according to the hydrocarbon or the principal hydrocarbon formed upon hydrolysis. By the term *intermediate carbides* is meant compounds intermediate in character between the ionic carbides and the interstitial carbides. The intermediate carbides, such as Cr_3C_2, Mn_3C, Fe_3C, Co_3C, and Ni_3C are similar to the ionic carbides in that they react with water or dilute acids to give hydrocarbons, and they resemble the interstitial carbides in their electrical conductivity, opacity, and metallic luster. The *interstitial carbides* have these properties, and are uniformly chemically inert. They include those having cubic close-packed structures, such as TiC, ZrC, HfC, VC, NbC, TaC, MoC, and WC, and those having hexagonal close-packed structures such as V_2C, Mo_2C, and W_2C. In both, the carbon atoms occupy interstitial positions in the crystal lattices of the metals, giving hardness, high melting points, and chemical inertness, as well as electrical conductivity with a positive temperature coefficient and metallic luster. The *covalent binary compounds of carbon* range in character from hard, chemically inert solids, such as silicon carbide, SiC, to volatile liquids, such as carbon disulfide and carbon tetrachloride, CS_2 and CCl_4, and even to gases such as carbon tetrafluoride, carbon dioxide, and methane, CF_4, CO_2, and CH_4, varying in thermal stability. With several of these elements carbon forms a series of compounds, or as with hydrogen, a number of series of hydrocarbons, consisting of both compounds based upon chains and branched chains of carbon atoms, variously saturated (i.e., joined by single, double, or triple bonds), and also of ring-connected carbon atoms, with or without side chains, with varying degrees of saturation, and capable of replacement of the hydrogen atoms with other atoms or radicals.

Carbonates. Carbonic acid H_2CO_3 is present to the extent of 0.27% of the total CO_2 present in the solution that is formed by dissolving CO_2 in H_2O at room temperature. The CO_2 may be expelled fully upon boiling. The solution reacts with alkalis to form carbonates, e.g., sodium carbonate, sodium hydrogen carbonate, calcium carbonate, calcium hydrogen carbonate. The acid ionization constant usually cited for carbonic acid (4.2×10^{-7}) is actually for the equilibrium $CO_2(ap) + H_2O \rightleftharpoons H^+ + HCO_3^-$. The true ionization constant, i.e., for the equilibrium $H_2CO_3 \rightleftharpoons H^+ + HCO_3^-$ is about 1.5×10^{-4}. The carbonate ion is a resonance hybrid of the three structures shown *a*, *b*, and *c* as well as structures of the type *d* which give a partial ionic character to bonds. This resonance is somewhat inhibited in the acid and its esters, but is complete, or much more nearly complete, in many other derivatives and in the carbonate ion. Esters of both metacarbonic, $(RO)_2CO$, and orthocarbonic acid, $(RO)_4C$, are known. The esters also exhibit resonance.

Metallic carbonates are (1) soluble in H_2O, e.g., sodium carbonate, potassium carbonate, ammonium carbonate (2) insoluble in H_2O and excess alkali carbonate, e.g., calcium carbonate, strontium carbonate, barium carbonate, magnesium carbonate, ferrous carbonate (3) insoluble in H_2O but soluble in excess alkali carbonate forming carbonato complexes, e.g., compounds of uranium and ytterbium $U(CO_3)_2$, UO_2CO_3, $Yb_2(CO_3)_3$. Metallic bicarbonates are known in solution and on warming are converted into ordinary or normal carbonates, e.g., bicarbonates of sodium, potassium, calcium, barium. These are preferably named as "hydrogen carbonates," e.g., $NaHCO_3$ = sodium hydrogen carbonate. Basic carbonates are important in such cases as lead ("white lead"), zinc, magnesium, and copper. Carbonates of very weak bases, such as aluminum, iron(III), and chromium (III), are now known.

The carbonates are found in nature as the carbonates, calcite, iceland spar, limestone and various forms of impure calcium carbonate $CaCO_3$, as magnesite (magnesium carbonate, $MgCO_3$), as dolomite (various compositions of calcium and magnesium carbonates), as witherite $SrCO_3$ as strontianite $SrCO_3$, as azurite and malachite (various compositions of cupric hydroxycarbonates), in various natural waters as carbonic acid, calcium and magnesium hydrogen carbonates, in blood, as sodium hydrogen carbonate.

Many esters of carbonic acid are known, e.g., diethyl carbonate, ethyl ester of metacarbonic acid, $(C_2H_5)_2CO$, made by reaction of ethyl alcohol and carbonyl chloride; dimethyl carbonate, $(CH_3O)_2CO$; methyl ethyl carbonate, $(CH_3O)CO(OC_2H_5)$; dipropyl carbonate, $(C_3H_7O)_2CO$; tetraethyl carbonate, ethyl ester of orthocarbonic acid, $(C_2H_5)_4C$, bp 158°C.

Peroxycarbonic acid exists only in its compounds. Alkali peroxycarbonates are obtained by electrolysis of concentrated solutions of the carbonates, the anodic reaction being written as

$$2CO_3^{2-} \rightarrow C_2O_6^{2-} + 2e^-$$

The peroxycarbonates are relatively stable only in concentrated alkaline solutions. On dilution they decompose to give the bicarbonate and hydrogen peroxide

$$Na_2C_2O_6 + 2H_2O \rightarrow 2NaHCO_3 + H_2O_2$$

when acidified, the peroxycarbonate ion gives, correspondingly, CO_2 and hydrogen peroxide

$$C_2O_6^{2-} + 2H^+ \rightarrow 2CO_2 + H_2O_2$$

Carbonyls. The metal carbonyls are strongly covalent in character, as shown by their volatility, their solubility in many nonpolar solvents, and their insolubility in polar solvents. They also behave in many reactions like mixtures of carbon monoxide, CO, and the metal. Those of group 6b elements, $Cr(CO)_6$, $MO(CO)_6$, and $W(CO)_6$ are more stable and less reactive than the others, especially those of group 8 elements. Group 7b carbonyls are $Mn_2(CO)_{10}$, $Tc_2(CO)_{10}$, and $Re_2(CO)_{10}$, while group 8 elements form $Fe(CO)_5$, $Fe_2(CO)_9$, $Fe_3(CO)_{12}$, $Co_2(CO)_8$, $Co_4(CO)_{12}$, $Ni(CO)_4$, $Ru(CO)_5$, $Ru_2(CO)_9$, $Ru_2(CO)_{12}$, $Rh_2(CO)_8$, $Rh_3(CO)_9$ (and multiples), $Rh_4(CO)_{11}$ (and multiples), $Os(CO)_5$, $Os_2(CO)_9$, $Ir_2(CO)_8$, and $Ir_3(CO)_9$ (and multiples). The carbonyls form a wide variety of addition compounds; they also dissolve in alcoholic potassium hydroxide or other strong alkalies to form hydrides which are acids, and can be used to form a wide variety of more complex compounds. Although $H_2Fe(CO)_4$ is a moderately weak acid, $pK_1 = 4.44$, $pK_2 = 14.0$, $HCo(CO)_4$ appears to be comparable with HCl in acidity. The carbonyl compounds have zero charge number on the metal. The mononuclear carbonyls are spin-paired complexes, and are formed only by metals having even atomic numbers. However, metals having odd atomic numbers can form carbonyl compounds with other atoms or radicals, as exemplified by the nitrosyl compound of cobalt carbonyl, $Co(CO)_2NO$, where the —NO radical contributes the electron necessary to complete the $3d$ level of the cobalt atom. More than one NO group may occur in a metal carbonyl, as, for example, in $Fe(CO)_2(NO)_2$. This is isostructural with $Co(CO)_3NO$ and $Ni(CO)_4$.

Halides. The four tetrahalides of carbon are symmetrical, planar compounds, with the general property of marked stability to chemical reactions, although the tetraiodide undergoes slow hydrolysis in contact with water to form iodoform and iodine. It also decomposes under the action of light and heat. The stability of these four compounds decreases in order of descending periodic table position. Their properties are given below.

NAME	FORMULA	MP	BP
Carbon tetrafluoride	CF_4	$-184°C$	$-128°C$
Carbon tetrachloride	CCl_4	$-23.0°$	$76.8°$
Carbon tetrabromide	CBr_4	$\begin{cases} \alpha 48.4° \\ \beta 90.1° \end{cases}$	$189.5°$
Carbon tetraiodide	CI_4	$171°$ dec.	

The same relation of reactivity and stability to periodic position is exhibited by such other carbon halides as hexachloroethane $CCl_3 \cdot CCl_3$ and hexabromoethane, $CBr_3 \cdot CBr_3$, as well as by hexachloroethylene, $CCl_2 = CCl_2$ and hexabromoethylene, $CBr_2 = CBr_2$. Carbon also forms halides containing more than one halogen. See also **Carbon Tetrachloride.**

It is well established that hydrogen forms more than one covalent binary compound with carbon. Fluorine behaves similarly. Thus, fluorine forms CF_4, C_2F_4, C_2F_6, C_3F_8 and many higher homologs, as well as the definitely interstitial compound $(CF)_n$. The other halogens form some similar compounds, although to more limited extent, and various polyhalogen compounds have been prepared. They exhibit the maximum covalency of four and are therefore inert to hydrolysis and most other low-temperature chemical reactions.

Carbon Oxides: See **Carbon Dioxide; Carbon Monoxide; Carbon Suboxide.**

Revised Theories on Structure of Carbon Compounds. For over a century, a principle stated by H. van't Hoff and Joseph A. LeBel (1874) has permeated organic chemistry, namely, that a carbon atom with substituents bonded to it prefers a tetrahedral geometry.[1] Experimental work, until the mid-1970s, came up with no fundamental exceptions to the principle. Organic compounds in which all four substituents lie in a plane had not been found. But, as the result of work by Schleyer (University of Erlangen-Nürnberg) and Pople (Carnegie-Mellon University), aided by advanced computer techniques in making molecular orbital calculations, a number of familiar organic compounds have been proposed as having preferred confirmations that differ markedly from accepted norms. As pointed out by Maugh (1976), the technique used by Schleyer and Pople essentially comprises the selection of two or more likely geometries for a molecule, followed by calculation of their relative energies. If all appropriate structures have been included in the calculations, then the geometry with the lowest calculated energy should be the most stable. The approach has been validated for numerous molecules whose structures are well known. This is exemplified for calculations of the methane molecule. A tetrahedral configuration is approximately 150 kcal per mole lower than the energy calculated for a structure in which all five atoms lie in the same plane. On the other hand, the energy is less for a planar configuration of CF_2Li_2 than for the tetrahedral configuration. Although practical application of this new information remains to be developed, it is believed that this new structural perspective will contribute to a better understanding of many organic reactions.

References

El Gorsey, A., and G. Donnay: *Science*, **161**, 363 (1968).

[1] Double-bonded molecules, such as ethylene, and triple-bonded molecules, such as acetylene conventionally have been considered as generally linear.

Field, J. E., Editor: "The Properties of Diamond," Academic, New York (1979).

Kasatochkin, V. I., et al.: *Carbon*, **11**, 70 (1973).

Maugh, T. H., II: "Unusual Structures Predicted for Carbon Compounds," *Science*, **194**, 413 (1976).

Sladkov, A. M., and Y. P. Koudrayatsev, *Priroda*, **5**, 37 (1969).

Whittaker, A. G.: "Carbon: A New View of Its High-Temperature Behavior," *Science*, **200**, 763–764 (1978).

CARBONADO. The mineral carbonado is an opaque massive black variety of diamond, often crystalline to granular or compact and without cleavage. In thin splinters it appears greenish-black by transmitted light. It is found chiefly in Bahia, Brazil. Carbonado is used for rock-drilling apparatus. Carbonado also is known as *black diamond*.

CARBON BLACK. Finely divided carbonaceous pigments of a wide variety are termed carbon blacks. Over 90% of the carbon black manufactured is consumed as reinforcing and compounding agents for rubber, mainly for motor vehicle and aircraft tires. Most users of tires do not realize that the effective use of these agents extends the life of a tire in normal usage by eight to ten times. The addition of as little as 1 to 2% carbon black to plastics greatly minimizes the effects of sunlight in degrading the materials. Most carbon blacks are derived from the pyrolysis of hydrocarbon gases and oils. The permanent and penetratingly deep black coloration obtainable with carbon blacks also makes the materials attractive for paints, inks, protective coatings, and as colorants for paper and plastics.

Two properties of carbon blacks are most significant for commercial applications: (1) particle size and (2) surface area. The particle sizes range from 100 to 5,000 micrometers. Surface areas will range from 6 to 1,100 m^2/g of material. Under electron microscopic examination, the carbon particles appear as rough spheres, usually as clusters of spheres rather than as individual spheres. The clustering characteristics stem from both chemical and physical bonding forces. Classically, the arrangement of the carbon particles may be likened to hexagonal nets of carbon atoms which are *paracrystalline* in nature. The particle size and surface area characteristics essentially are at the microscopic level—hence control over carbon black production is exacting. In terms of coloration, for example, the human eye can resolve 260 shades of blackness. The blackest of commercially produced carbon particles will have a diameter of about 100 micrometers. The grayest particle will have a diameter of about 5,000 micrometers. The blackness characteristic sometimes is referred to as *masstone* (mass-tone). The particles with the smaller diameters and hence greater surface area exhibit the highest masstone.

Lampblacks have been made for many centuries. Early methods involved the burning of petroleum-like substances or coal-tar residues with a minimum of air, thus producing large amounts of unoxidized carbon particles. The earlier settling chambers in which the particles collected have been replaced by cyclones, bag filters, or electrical precipitators. Modern installations use oil furnaces to create the particles.

Channel or *impingement carbons* are produced from burning natural gas (sometimes containing oil vapors) in many hundreds of small burners. The flames from the burners impinge upon flat surfaces called channels. The carbon deposits are periodically removed by scraping into a collector. The burning equipment is contained within a large burner house which has means for carefully regulating both bottom and top drafting of air.

Thermal blacks are also derived from natural gas, but by thermal decomposition completely in the absence of air. Large furnaces first are preheated to a temperature ranging from

1,100–1,650°C. When the checkerwork is at the proper temperature, natural gas is bled into the furnace, whereupon the gas decomposes into carbon and hydrogen. This is a batch process, requiring pairs of furnaces, one furnace preheating, while the other furnace is decomposing the gas feed. Frequently, the hydrogen by-product is recycled as fuel to heat the furnaces. Where very fine thermal blacks are produced, the byproduct hydrogen is used as a diluent for the gas feed.

Furnace carbons also are derived from natural gas, but in a process in which a slight excess of air is introduced to support combustion. The hydrocarbon feedstock or liquid oil is injected into the furnace at a location where the so-called blast-flame gases are circulating at their greatest velocity. Injection of the feed at this point causes an instant high rise in temperature which results in practically instantaneous decomposition of the feed into carbon black. For coarse particles, the oil/air ratio is greater, furnace gas velocities are lower, and residence time in the furnace is longer. There is a wide range of furnace carbon particle sizes. The very fine particles go into tire treads, whereas the coarser particles are used in tire carcasses.

Acetylene black is derived from feeding acetylene into high-temperature retorts whereupon the acetylene dissociates into carbon and hydrogen. This reaction is exothermic (other carbon black processes are endothermic). Temperature control of the furnace is effected by throttling the acetylene feed.

CARBON COMPOUNDS. Organic Chemistry.

CARBON DIOXIDE. CO_2, formula weight 44.01, colorless, odorless, nontoxic gas at standard conditions. High concentrations of the gas do cause stupefaction and suffocation because of the displacement of ample oxygen for breathing. Density 1.9769 g/1 (0°C, 760 torr), sp gr 1.53 (air = 1.00), mp −56.6°C (5.2 atmospheres), solid CO_2 sublimes at −79°C (760 torr), critical pressure 73 atmospheres, critical temperature 31°C. Carbon dioxide is soluble in H_2O (approximately 1 volume CO_2 in 1 volume H_2O at 15°C, 760 torr), soluble in alcohol, and is rapidly absorbed by most alkaline solutions. The solubility of CO_2 in H_2O for various pressures and temperatures is given in the accompanying table.

SOLUBILITY OF CARBON DIOXIDE IN WATER

Pressure (atmospheres)	Parts (weight) CO_2 Soluble in 100 Parts Water				
	18°C	35°C	50°C	75°C	100°C
25	3.7	2.6	1.9	1.4	1.1
50	6.3	4.4	4.0	2.5	2.0
75	6.7	5.5	4.5	3.4	2.8
100	6.8	5.8	5.1	4.1	3.5
200	—	6.3	5.8	5.3	5.1
300	7.4	—	6.2	5.8	5.7
400	7.8	7.1	6.6	6.3	6.4
700	—	—	7.6	7.4	7.6

Carbon dioxide plays several roles: (1) as a *raw material* for several processes, as in the Solvay process for the manufacture of sodium bicarbonate and sodium carbonate, (2) as a *by-product* from many processes, notably as a product of combustion of fossil fuels, (3) as an *ingredient* of products, for example, carbonated beverages, (4) as a *product* for direct consumption, for example, CO_2 fire extinguishers and dry ice refrigerants, and (5) as a *pollutant* of the atmosphere. Although not toxic, the presence of CO_2 in the atmosphere disturbs the environmental

energy balance. The latter aspects of CO_2 are discussed under **Pollution (Air)**. Normally, CO_2 is present in the air at sea level to the extent of about 0.05% by weight.

Solid carbon dioxide (dry ice) is an effective refrigerant for transportation uses. Refrigeration of moving vehicles may be derived from (1) mechanical systems which, of course, require a continuous input of energy, (2) water ice and ice–salt mixes which require water (often briny) removal, and are corrosive and subject to algae formations, and (3) dry ice, the end-product of which is simply gaseous CO_2, which is easily removed. To maintain a cool temperature in a railroad refrigerator car for a trip between California and New York, about 1,000 pounds of dry ice would be required. To maintain the same conditions with water ice and salt would require 10,000 pounds of ice. The fact that CO_2 is heavier than air makes it particularly effective for fighting fires in low places, such as pipe trenches and hard-to-reach low corners and basements, where the CO_2 tends to roll under the air required to maintain combustion. Both manually- and automatically-controlled CO_2 fire-fighting systems are available. These can be actuated by heat-sensitive systems—just as a conventional water-sprinkling system. CO_2 is effective for fires involving electrical and electronic gear because, if a fire is not fully out-of-hand, the CO_2 often can quickly quench the fire source without leaving any residual damage, as often is the disastrous consequences of using water or sand.

Although carbon dioxide must be generated on site for some processes, there is a trend toward CO_2 recovery where it is a major reaction byproduct and, in the past, vented to the atmosphere. For example, very large quantities of CO_2 are generated by various fermentation processes and in cement production. If the CO_2 must be removed from stack gases because of pollution control regulations, it is only one more step to purify the gas and sell it, usually in compressed liquid form. There are, of course, several economic tradeoffs which must be considered. Where the gas is recovered, it usually is first absorbed in sodium or potassium carbonate solutions, followed by steam-heating the solutions to free a reasonably pure CO_2. The last step is compression of the gas into steel cylinders. The ethanolamines also are excellent absorbents of CO_2.

In some dry ice manufacturing plants, CO_2 is produced by burning fuel oil. The heat generated is used to strip the resulting CO_2 from the monoethanolamine absorbing solution. The gas then is scrubbed with potassium permanganate solution to remove odorous impurities prior to compressing.

Structure. Carbon dioxide offers one of the classical examples of the role of resonance in interpreting the structure and properties of substances. The heat of formation from the atoms of the linkage of a $C=O$ bond is 150 kcal per mole, but the heat of formation of CO_2 from one C and two O atoms is not $2 \times 150 = 300$ kcal, but 336 kcal per mole. Moreover, the observed $C=O$ bond length in carbon dioxide is not the usual 1.22 Å, but 1.15 Å. The greater than expected heat of formation, and the shorter than expected bond length are explained by assigning to CO_2 not only the structure $O=C=O$, but two others also, $^-O-C\equiv O^+$ and $^+O\equiv C-O^-$, the structural picture being a hybrid (not an interchange) of the three, and more stable than any one alone. The contributions of the last two are equal in amount, so the molecule has a dipole moment of zero. The molecular orbital explanation is in terms of two fractional nonlocalized π-bonds.

Direct measurement gave a value of -4×10^{-26} electrostatic unit for the molecular quadrupole moment of CO_2. This corresponds to a charge on each oxygen atom of 0.32 electronic charge.

Carbon Dioxide in Biological Systems. Carbon dioxide, which is a byproduct of the metabolic activity of all cells, is one of the most important chemical regulators in the human body. It can be said that human life without carbon dioxide would be impossible. In less specialized forms of life, carbon dioxide is essentially a waste product. In the more highly developed animals, such as humans, the gas is used to regulate the activity of the heart, the blood vessels, and the respiratory system.

As mentioned, CO_2 is normally present in air at sea level at about 0.05% (weight). A poorly ventilated room may contain as much as 1% (volume). Concentrations of the gas from about 0.1 to 1% (volume) induce languor and headaches; concentrations of 8–10% (volume) bring about death by asphyxiation. High concentrations of the gas are toxic. See also **Basal Metabolism.**

As a general rule, the respiration of individual cells decreases as the concentration of carbon dioxide in the medium increases. Fish show a lessened capacity to extract oxygen from their environment with increasing amounts of carbon dioxide present. On the other hand, many invertebrates show marked increases in respiratory rate (or ventilation) with increased amounts of the gas in their surroundings.

Photosynthetic and autotrophic bacteria reduce carbon dioxide which is assimilated into complex molecules for use in synthesizing various cellular constituents. The gas is apparently assimilated, at least to a small extent, by the heterotrophic bacteria. Certainly it is required for any growth in these forms. Many pathogenic bacteria require increased carbon dioxide tension for growth immediately after they are isolated from the body. The production of hemolysins and like substances is greatly enhanced by adding 10–20% of CO_2 in the air which comes in contact with the cultures.

The oxygen dissociation curve for blood is shifted to the right when the partial pressure of carbon dioxide is increased. This is referred to as the *Bohr Effect*. It means that for a given partial pressure of oxygen, hemoglobin holds less oxygen at high concentration of carbon dioxide than at a lower concentration. It is evident, then, that the production of carbon dioxide by actively metabolizing tissues favors the release of oxygen from the blood to the cells where it is urgently needed. Moreover, at the alveolar surfaces in the lungs, the blood is losing carbon dioxide rapidly, which loss favors the combination of oxygen with hemoglobin. In males, the average amount of CO_2 in the alveolar air is about 5.5% (volume); during the breathing cycle, this concentration varies only slightly. In females and children, somewhat lower mean values obtain.

In every 100 milliliters of arterial blood, there is a total of 48 milliliters of free and combined CO_2. In venous blood of resting humans, there is about 5 milliliters more than this. Only about one-twentieth of the carbon dioxide is uncombined, a fact which indicates that there is a specialized mechanism, aside from simple solution, for the transport of CO_2 in the blood.

About 20% of the CO_2 in the blood is carried in combination with hemoglobin as *carbaminohemoglobin*. The balance of the combined carbon dioxide is carried as bicarbonate. A CO_2 dissociation curve for blood can be prepared just as for oxygen, but the shape is not the same as for the latter. As the partial pressure of CO_2 in the air increases, the amount in the blood increases; the increase is practically linear in the higher ranges. Oxygen exerts a negative effect on the amount of CO_2 which can be taken up by the blood.

In working muscles large amounts of CO_2 are produced. This causes local vasodilation. The diffusion of some of the CO_2 into the bloodstream slightly raises the concentration there. It circulates through the body and the capillaries of the vasoconstrictor center, where it excites the cells of the center, resulting in an increase of constrictor discharges. Regarding the stimulating effect of CO_2 on cardiac output, it is evident that a most effective mechanism exists for increasing circulation through active muscles. More blood is pumped by the heart per minute and the arterial pressure is increased by the general vasoconstriction; blood is forced from the inactive regions, under increased pressure, through the widely dilated vessels of the active muscles.

The partial pressure of CO_2 is important in connection with a number of physiological problems. For example, respiratory acidosis is the result of an abnormally high p_aCO_2. The value of arterial pCO_2 varies directly with changes in the metabolic production of CO_2 and indirectly with the amount of alveolar ventilation. The problem is more commonly the result of decreased alveolar ventilation caused by abnormally low CO_2 excretion by the lungs (alveolar *hypo*ventilation).

On the other hand, primary respiratory alkalosis occurs as a result of alveolar *hyper*ventilation. This condition is associated with a number of pulmonary diseases, but also may appear during pregnancy, liver disease, and salicylate intoxication, among others. The sequence of events proceeds along these lines: (1) Ventilation removes CO_2 faster than the gas is produced by metabolism, causing a decrease in pCO_2 in the blood and body fluids, including a reduction of venous pCO_2. This reduces the gradient for excretion of CO_2 by the lungs. (2) Pulmonary excretion and metabolic production ultimately balance out at a lower pCO_2 level for all body fluids. (3) The lower pCO_2 level causes a lower carbonic acid concentration and consequently an increase in pH. The latter is relative to the reduced level of pCO_2, but the pH change also alters bicarbonate concentration. The steplike process is quite complex.

Narcosis due to CO_2 is characterized by mental disturbances which may range from confusion, mania, or drowsiness to deep coma; headache; sweating; muscle twitching; increased intracranial pressure; bounding pulse; low blood pressure; hypothermia; and sometimes papilloedema. The basic mechanisms by which carbon dioxide induces narcosis is probably through interference with the intracellular enzyme systems, which are all sensitive to pH changes.

See also **Photosynthesis.**

CARBON GROUP (The). The elements of group 4a of the periodic classification sometimes are referred to as the Carbon Group. In order of increasing atomic number, they are carbon, silicon, germanium, tin, and lead. The elements of this group are characterized by the presence of four electrons in an outer shell. The similarities of chemical behavior among the elements of this group are less striking than that for some of the other groups, e.g., the close parallels of the alkali metals or alkaline earths. However, as more knowledge is gained of silicon, including the element's ability to form "carbon-like" chains with alternating silicon and oxygen atoms, to polymerize, and to form silicones, silanes, etc., the similarity of silicon and carbon emerges more sharply. The seimiconductor properties of silicon and germanium in this group are striking, but of course such properties are not limited to elements in this group. Although some of the elements of the group have valences in addition to +4, all do have the +4 valence in common. Unlike the alkali metals or alkaline earths, for example, the elements of the carbon group are not so similar chemically that they comprise a separate group in classical qualitative chemical analysis separations.

CARBONITRIDING. A surface hardening process for steels involving the introduction of carbon and nitrogen into steels by heating in a suitable atmosphere containing various combinations of hydrocarbons, ammonia, and carbon monoxide followed by a quenching to harden the case.

CARBONIZATION (Coal). Coal.

CARBONIUM ION. An ion of the general formula $B-\overset{\overset{A}{|}}{\underset{\underset{D}{|}}{C^+}}$,

where A, B and D are substituent groups. Its importance in elucidating the mechanism of organic reactions is because a considerable proportion of all organic reactions involve carbonium ions, as others do carbanions and carbon free radicals (including carbene radicals). Nucleophilic substitution at saturated carbon atoms includes most of carbonium ion chemistry. Carbonium ions are usually powerful acids or electrophiles, and thus many nucleophilic substitution reactions that involve carbonium ions are acid-catalyzed. For example, the tertiary-butyl carbonium ion offers a clear understanding of the probable course of the conversion of isobutylene to its dimers and trimers.

$$(CH_3)_2C{=}CH_2 + H^+ \leftrightarrows (CH_3)_3C^+$$

$$(CH_3)_3C^+ + (CH_3)_2C{=}CH_2 \leftrightarrows (CH_3)_2\overset{+}{C}{-}CH_2C(CH_3)_3$$

The larger carbonium ion thus formed cannot continue to exist, but may depolymerize, unite with the catalyst, or stabilize itself by the attraction of an electron pair from a carbon atom adjacent to the electronically deficient carbon (C^+) with its proton. This establishes a double bond involving the formerly deficient atom. Thus a proton is expelled to the catalyst or attracted to the catalyst. If this takes place with one of the methyl groups, the product is

$$CH_2{=}\overset{\overset{}{|}}{\underset{\underset{CH_3}{|}}{C}}{-}CH_2C(CH_3)_3 .$$

If the methylene group is involved, the product is $(CH_3)_2C{=}CHC(CH_3)_3$.

CARBON MONOXIDE. CO, formula weight 28.01, colorless, odorless, very toxic gas at standard conditions, density 1.2504 g/l (0°C, 760 torr), sp gr 0.968 (air = 1.000), mp −207°C, bp −192°C, critical temperature −139°C, critical pressure 35 atmospheres. Carbon monoxide is virtually insoluble in H_2O (0.0044 part CO in 100 parts H_2O at 50°C). The gas is soluble in alcohol or solutions of cupric chloride. Because carbon monoxide has an affinity for blood hemoglobin that is 300 times that of oxygen, exposure to the gas greatly reduces or fully hinders the ability of hemoglobin to carry oxygen throughout the body, causing death in excessive concentrations. Engines and stoves in poorly ventilated areas are especially hazardous.

Carbon monoxide plays several roles: (1) as a *raw material* for chemical processes (a) particularly as an effective reducing agent in various metal smelting operations, (b) in the manufacture of formates: $CO + NaOH \rightarrow HCOONa$, (c) in the production of carbonyls, such as $Ni(CO)_4$ and $Fe(CO)_5$, which are useful intermediate compounds in the separation of certain metals, (d) in combination with chlorine to form $COCl_2$ (phosgene), (e) as an ingredient of several synthesis gases, as for the produc-

tion of methanol and ammonia; (2) as a *fuel* where CO is a major ingredient of such artificial fuels as coal gas, producer gas, blast-furnace gas, and water gas; (3) as a *by-product* of numerous chemical reactions, notably combustion processes where there is insufficient oxygen for complete combustion—the fumes from internal-combustion engines may contain in excess of 7% CO, and (4) as a dangerous *air pollutant*, particularly in industrial areas and where there are high concentrations of automotive vehicles and aircraft. The latter aspects of CO are discussed under **Pollution (Air).**

Summary of Chemical Reactivity. Chemically, carbon monoxide (1) reacts with oxygen to form CO_2 accompanied by a transparent blue flame and the evolution of heat, but the fuel value is low (320 Btu per ft³); (2) reacts with chlorine, forming carbonyl chloride $COCl_2$ in the presence of light and a catalyzer; (3) reacts with sulfur vapor at a red heat, forming carbonyl sulfide COS; (4) reacts with hydrogen, forming methyl alcohol, CH_3OH or methane CH_4 in the presence of a catalyzer; (5) reacts with nickel (also iron, cobalt, molybdenum, ruthenium, rhodium, osmium, and iridium) to form nickel carbonyl, $Ni(CO)_4$ (and carbonyls of the other metals named); (6) reacts with fused NaOH, forming sodium formate, HCOONa; (7) reacts with cuprous salt dissolved in either ammonia solution or concentrated HCl, which solutions are utilized in the estimation of carbon monoxide in mixtures of gases, e.g., flue gases of combustion, coal gas, exhaust gases of internal combustion engines; (8) reacts with iodine pentoxide at 150°C. For the reaction of carbon monoxide with oxygen to form CO_2 finely divided iron or palladium wire is used as a catalyst; for the reaction of carbon monoxide with H_2O vapor to form CO_2 plus hydrogen ("water gas reaction") important studies have been made of the conditions; and for the reaction of CO_2 plus carbon (hot) similar important studies have been made (at 675°C, 50% CO_2 plus 50% CO; at 900°C, 5% CO_2 plus 95% CO). The reaction of carbon plus oxygen at such a temperature as produces carbon monoxide (say 900°C, 95% CO plus 5% CO_2) and *evolves heat*; while the reaction of carbon plus CO_2, producing carbon monoxide at the same temperature *absorbs heat*. Accordingly, it is possible to arrange the oxygen (free or as air) and CO_2 supply ratio in such a way that the desired temperature may be continuously maintained. The reduction of CO_2 by iron forms carbon monoxide plus ferrous oxide.

In valence bond terms, carbon monoxide is considered as a resonance compound with the structures

$$:\overset{+}{C}:\overset{..}{O}:^- \quad :C::\overset{..}{O}: \quad :\overset{-}{C}::\overset{..}{O}: \quad :\overset{+}{C}:::O:$$

In molecular orbital terms the CO molecule is described as $CO(KK(z\sigma)^2 (y\sigma)^2(x\sigma)^2(w\pi)^4)$, one $(z\sigma)^2$ pair being formed from the oxygen $2s$ electrons, and one $(y\sigma)$ pair held by the carbon sp hybrid. This $(y\sigma)^2$ pair offsets the dipole moment of the π electrons, and also accounts for the readiness with which the CO molecule coordinates with metals to form the carbonyls.

CARBON SUBOXIDE. C_3O_2, formula weight 68.03, colorless, toxic, gas at room temperature, very unpleasant odor, sp gr 2.10 (air = 1.00), 1.24 (liquid at −87°C), mp −107°C, bp 7°C (760 torr), burns with a blue smoky flame, producing CO_2. When condensed to liquid, the oxide slowly changes at ordinary temperature to a dark red solid, soluble in water to a red solution. Reacts with water to form malonic acid, with hydrogen chloride to form malonyl chloride, with ammonia to form malonamide. Made by heating malonic acid or its ester at 300°C under diminished pressure, and separation from simultaneously formed carbon dioxide and ethylene by condensation and fractional distillation.

Carbon suboxide has a linear structure, probably a resonance of four structures, of which the last two below probably make a smaller contribution to the normal state of the molecule than the first two.

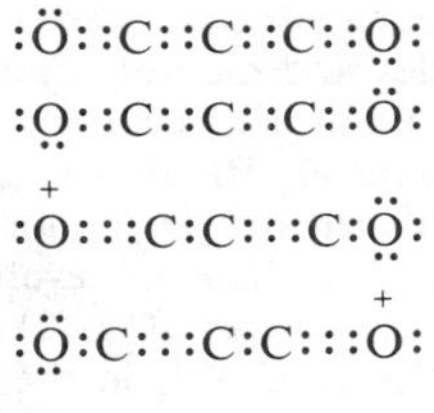

$$:\overset{..}{O}::C::C::C::\overset{..}{O}:$$

$$:\overset{..}{O}::C::C::C::\overset{..}{O}:$$

$$\overset{+}{:O}:::C:C:::C:\overset{..}{\underset{..}{O}}:\overset{-}{}$$

$$\overset{-}{:}\overset{..}{\underset{..}{O}}:C:::C:C:::\overset{+}{O}:$$

CARBON TETRACHLORIDE. CCl_4, formula weight 82.82, heavy, colorless, nonflammable, noncombustible liquid, mp $-23°C$, bp $76.75°C$, sp gr 1.588 ($25°C/25°C$), vapor density 5.32 (air = 1.00), critical temperature $283.2°C$, critical pressure 661 atmospheres, solubility 0.08 g in 100 g H_2O, odor threshold 80 ppm. Dry carbon tetrachloride is noncorrosive to common metals except aluminum. When wet, CCl_4 hydrolyzes and is corrosive to iron, copper, nickel, and alloys containing those elements. About 90% of all CCl_4 manufactured goes into the production of chlorofluorocarbons:

$$2CCl_4 + 3HF \xrightarrow{catalyst} CCl_2F_2 + CCl_3F + 3HCl.$$

Carbon tetrachloride was first made by chlorinating chloroform (1839). Later, CCl_4 was made by chlorinating carbon disulfide CS_2 in the first commercial process, developed by Müller and Dubois (1893). Large-scale production commenced in the early 1900s at which time carbon tetrachloride became a popular metal-degreasing solvent, dry-cleaning fluid, fabric-spotting fluid, grain fumigant, and fire extinguishing fluid. Many of these uses now have been displaced by other less toxic chlorinated hydrocarbons. The carbon disulfide process consists of: (1) $3C + 6S \rightarrow 3CS_2$; (2) $2CS_2 + 6Cl_2 \rightarrow 2CCl_4 + 2S_2Cl_2$; (3) $CS_2 + 2S_2Cl_2 \rightarrow CCl_4 + 6S$. The reaction must be carried out in a lead-lined reactor in a solution of CCl_4 at $30°C$ in the presence of iron filings as catalyst. The chlorination of methane is now the principal production route to CCl_4: $CH_4 + Cl_2 \rightarrow CH_3Cl + CH_2Cl_2 + CHCl_3 + CCl_4 + HCl + excess CH_4$. The reaction is carried out in the liquid phase at about $35°C$. Ultraviolet light is used as a catalyst. The same reaction can be carried out at $475°C$ without catalyst. The unreacted methane and partially-chlorinated products are recycled to control the yield of CCl_4.

Toxicity. The experimental exposure of laboratory animals to the vapors of CCl_4 has shown it to be very toxic by inhalation at concentrations which are easily obtainable at ambient temperatures. An overexposure to carbon tetrachloride has been known to cause acute but temporary loss of renal function.

CARBONYLS. Carbon.

CARBONYLS (Chlorinated). Chlorinated Organics.

CARBORUNDUM. Silicon.

CARBOXYLIC ACIDS. The general formula for a carboxylic

acid is 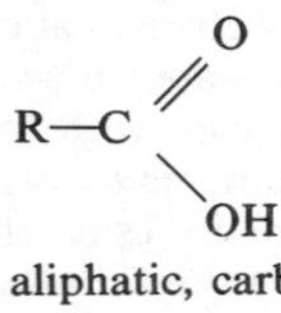
$$R-C \overset{O}{\underset{OH}{\big\langle}}$$
In terms of structure, a carboxylic acid

may be aliphatic, carbocyclic, or heterocyclic:

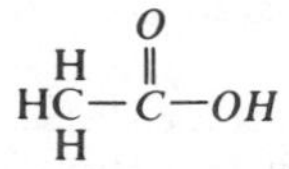

Aliphatic acetic acid 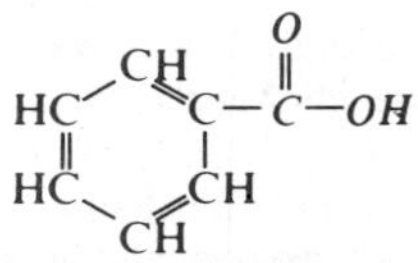 · Carbocyclic or aromatic benzoic acid 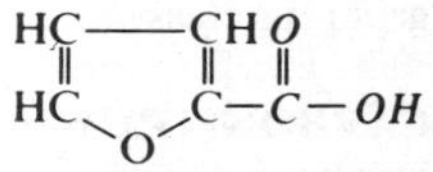 · Heterocyclic pyromucic or furoic acid

Or, a carboxylic acid may be classified in terms of the number of carboxyl (—COOH) groups which it contains. If one carboxyl group, it is designated as *mono*carboxylic; if two groups, as *di*carboxylic; if three groups, as *tri*carboxylic; and if four groups, as *tetra*carboxylic:

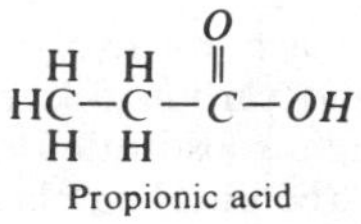

Propionic acid (*mono*)

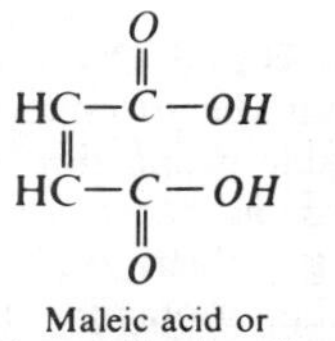

Maleic acid or *cis*-ethylene dicarboxylic acid (*di*)

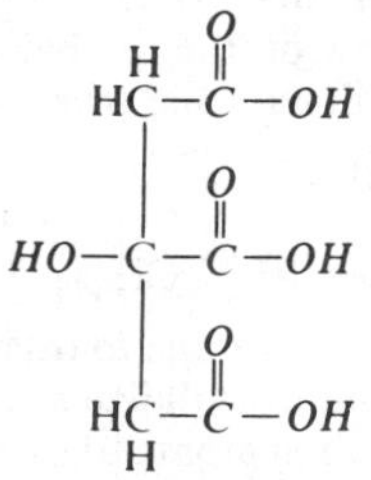

Citric acid (*tri*)

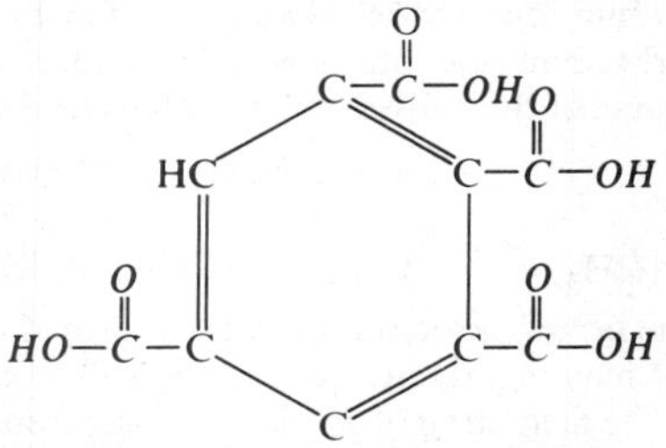

1,2,3,5-Benzenetetracarboxylic acid or mellophanic acid (*tetra*)

When a carboxylic acid contains a hydroxyl group in addition to that of the principal —COOH grouping, the term *hydroxy* is sometimes used. If there is only one additional hydroxyl group, the acid may be designated simply as a *hydroxycarboxylic* acid; if two groups, a *di*hydroxycarboxylic acid; if three groups, a *tri*hydroxycarboxylic acid.

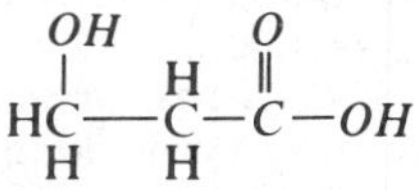

Hydracrylic acid or β-hydroxypropionic acid A hydroxymonocarboxylic acid

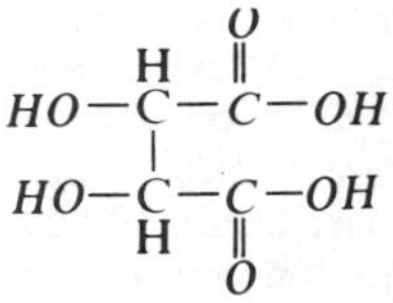

Tartaric acid A dihydroxycarboxylic acid

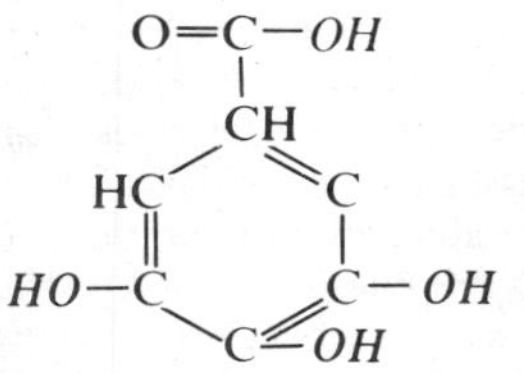

Gallic or pyrogallol carboxylic acid or 3,4,5-trihydroxybenzoic acid A trihydroxymonocarboxylic acid

A carboxylic acid may be classified in accordance with the number of available hydrogens for salt formation. If only one hydrogen is available, the acid is *monobasic*; if two hydrogens are available, the acid is *dibasic*; if three or more hydrogens are available, the acid is *polybasic*.

A carboxylic acid also may be classified from the standpoint of other groups which it contains. An *aldehydic* carboxylic acid contains the CHO group. An example is glyoxalic acid, $CHO \cdot COOH$. An *amino* carboxylic acid contains the NH_2 group. An example is carbamic or amino-formic acid, NH_2COOH. A *ketonic* carboxylic acid contains the CO group. An example is benzoylacetic acid, $C_6H_5 \cdot CO \cdot CH_2 \cdot COOH$. In the case of a *phenolic* carboxylic acid, the acid is structurally derived from benzoic acid, with uniting of the OH group with a carbon of the nucleus.

There are several homologous series of carboxylic acids, including:

$C_nH_{2n}O_2$	Saturated monobasic fatty acids
$C_nH_{2n-2}O_2$	Unsaturated monobasic fatty acids
$C_nH_{2n-4}O_2$	Propioloic acid series
$C_nH_{2n}(COOH)_2$	Dicarboxylic acids, where $n = 0$ for oxalic acid
$C_nH_{2n}(OH)(COOH)$	Hydroxymonocarboxylic acids, where $n = 0$ for carbonic acid

Fatty Acids. The simplest or lowest member of the fatty acid series is formic acid, HCOOH, followed by acetic acid, CH_3COOH, propionic acid with three carbons, butyric acid with four carbons, valeric acid with five carbons, and upward to palmitic acid with sixteen carbons, stearic acid with eighteen carbons; and melissic acid with thirty carbons. Fatty acids are considered to be the oxidation product of saturated primary alcohols. These acids are stable, being very difficult (with the exception of formic acid) to convert to simpler compounds; they easily undergo double decomposition because of the carboxyl group; they combine with alcohols to form esters and water; they yield halogen-substitution products; they convert to acid chlorides when reacted with phosphorus pentachloride; and their acidic qualities decrease as their formula weight increases.

Monohydroxy Fatty Acids. Structurally, these acids may be considered as the monohydroxy derivatives of the fatty acids. Included among these acids are hydroxyacetic acid (glycollic acid) and β-hydroxypropionic acid (β-lactic acid). These acids generally are syrupy liquids that tend to give up water readily and form crystalline anhydrides, they decompose when volatilized, and they are soluble in water and usually in alcohol and ether.

Polyhydric Monobasic Acids. Structurally, these acids are considered to be the oxidation products of polyhydric alcohols. However, a number of them can be formed from the oxidation of sugars. The careful oxidation of glycerol will yield a syrupy liquid, glyceric acid, an example of a dihydroxymonobasic carboxylic acid.

Aromatic Carboxylic Acids. In many ways, these acids are similar to the fatty acids. Generally, they are crystalline solids which are only slightly soluble in water, but most often they dissolve easily in alcohol or ether. The simpler aromatic acids may be distilled (or sublimed) without decomposition. The more complex acids, such as the phenolic and polycarboxylic aromatic acids, break down when heated, yielding carbon dioxide and a simpler compound. As an example, salicylic acid degrades

to carbon dioxide and phenol. In nature, the aromatic acids are found in balsams, animal organisms, and resins.

The monobasic saturated aromatic acids include benzoic, hippuric, toluic acids (three structures), phenylacetic, phenylchloracetic, and dimethylbenzoic acid. Among the monobasic unsaturated acids are cinnamic, atropic, and phenylpropionic acids. The saturated phenolic acids include gallic and salicylic acids. The alcohol acids include amygdalic, tropic, and mandelic acids. One example of an unsaturated monobasic phenolic acid is coumaric acid.

Formation of Carboxylic Acids. Commercially, these acids are produced in several ways: (1) oxidation of relevant alcohol—e.g., acetic acid from ethyl alcohol; (2) oxidation of relevant aldehyde—e.g., acetic acid from acetaldehyde; (3) bacterial fermentation of dilute alcohols; (4) reaching a methyl ketone with sodium hypochlorite (haloform reaction); (5) carbonation of Grignard reagents; (6) hydrolysis of nitriles; (7) malonic ester synthesis route; (8) oxidation of relevant alkylaromatic—e.g., benzoic acid from toluene; (9) reaction of an alkali metal phenolic with carbon dioxide; and (10) hydrocarboxylation of olefins—e.g., butyric acid from propylene.

See also **Organic Chemistry.**

—Duane B. Priddy, The Dow Chemical Company, Midland, Michigan.

CARBURETION. The fuel for an internal combustion engine must be well mixed with the air required for combustion. This is particularly true of the Otto cycle engine, inasmuch as thorough distribution of particles of fuel in the air is essential to rapid and complete explosive combustion of the fuel in that cycle. One of the most effective means of mixing the particles of a liquid fuel with air is by vaporization. The vaporizing and mixing of a liquid fuel with air in the correct proportions is called *carburetion*; the device used is called a *carburetor*.

By using multiple jets, adjustable orifices, and other intricacies, commercial carburetors attain mixture control approximating the desirable performance for a gasoline engine as shown by accompanying diagram. Increasing air pollution standards

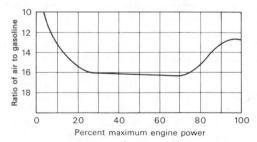

Optimal carburetor performance for gasoline engine.

have exacted greater demands on carburetor performance and most of these demands have been met by utilizing solid-state controls over carburetor optimization.

CARBURIZING. Machine parts requiring high strength, hardness, and toughness can often be made by one of two methods: (1) Use of a medium-carbon steel (0.30–0.50% C) heat treated to the required properties; (2) use of a low-carbon steel (0.08–0.25% C) subjected to *carburizing*. The latter process yields a high-carbon surface layer after heat treating. In carburizing, parts are heated in an atmosphere rich in CO or hydrocarbon gases at a temperature in the range of 1650–1800°F (899–982°C). Reactions at the surface liberate atomic carbon which is readily dissolved by the steel and diffuses inward from the surface. In a typical carburized case a depth of penetration of 0.05 in. (0.13 cm) will be obtained in about 4 hours at 1700°F (927°C).

CARCINOGENS. Cancer-causing substances are called *carcinogens*. During the past decade, there has been a growing awareness of the presence of carcinogenic materials in the environment, both air and water. The causes of carcinogenesis are still poorly understood. Because the numbers of substances with which a person comes in contact are in the tens of thousands, and because science has not achieved that position of understanding where carcinogens can be easily grouped, such as by chemical structure or other properties, the testing of so many substances becomes a formidable task. Understood even less are the long-term effects of these substances in their possible propensity to cause genetic errors that ultimately lead to carcinogenesis. Some progress, albeit a beginning, in classification has been made and reference to this is cited here. By and large, however, it would be an exaggeration to state that a great deal of progress has been made as of the early 1980s. Further complicating the social, economic, and scientific interfaces are compounds which are known to be carcinogenic above certain levels, but which are of great value, if not indispensable, to many aspects of society. Some chemical pesticides, for example, obviously fall into that category where risks and benefits must be weighed.

Until the mid-1970s, most work in chemical carcinogenesis was dominated by experimentalists, i.e., the testing of suspect chemicals on the skins of or in the diets for animals and waiting to see if the chemicals induced tumor formations. Such work is valuable in identifying certain materials that should be removed from the environment and otherwise avoided by humans, but such activities have not led in a major way to the understanding of the mechanism or chemistry of chemical carcinogenesis. Starting in a rather small way in the early 1970s, new emphasis now is being placed on the molecular biology of carcinogenesis. Some of the first steps in the interaction between carcinogen and cell have been demonstrated, including observations that most carcinogens must be activated by the host's metabolism. Cell culture methodologies have been developed which enable the transformation of healthy cells into malignant cells by chemicals. These methods have several advantages over work *in vivo*. A major step, of course, remains—the transformation of human cells by chemicals in culture.

There is some evidence that the form of the chemical carcinogen that ultimately reacts with cellular macromolecules must contain a reactive electrophilic center, that is, an electron-deficient atom which can attack the numerous electron-rich centers in polynucleotides and proteins. As examples, significant electrophilic centers include free radicals, carbonium ions, epoxides, the nitrogen in esters of hydroxylamines and hydroxamic acids, and some metal cations. It is believed that carcinogens which in themselves are not electrophiles are metabolized to electrophilic derivatives that then become the "ultimate" carcinogens.

The high incidence of skin cancer in coal tar workers was recognized as early as 1880. The carcinogenic activity of coal tar was demonstrated in 1915, when Yamagiwa and Ichikawa obtained epitheliomas (malignant tumor originating from epithelial cells) by its prolonged application to the ears of rabbits. Identification of the active material (in 1933) as the polycyclic aromatic hydrocarbon 3,4-benzopyrene (III, Fig. 1) is due to Cook, Kennaway, Hieger and their co-workers. This discovery was followed up by the synthesis and testing of a considerable variety of polycyclic aromatic hydrocarbons. All compounds of this class may be regarded as composed of condensed benzene rings. The arrangement of the hexagonal rings in various patterns results in a variety of compounds having different physical, chemical, and biological properties. However, not all polycyclic aromatic hydrocarbons possess carcinogenic activity; certain requirements of molecular geometry must be met.

For maximum activity, the molecule must have (Fig. 1): (a) an optimum size; (b) a coplanar molecular configuration, meaning that all hexagonal rings must lie flatly in one plane; in fact, hydrogenation of many of the active hydrocarbons results in buckled molecular conformation and this is concomitant with partial or total loss of activity; (c) at least one meso-phenanthrenic double bond, also called the K-region (as indicated by arrows in Fig. 1) of high π-electron density (i.e., of high chemical reactivity). In addition to III, 1,2,5,6-dibenzanthracene (IV), and 20-methylcholanthrene (V) are commonly used to study the experimental induction of tumors. The activity of most hydrocarbon carcinogens was tested on the skin of mice and the subcutaneous connective tissue of mice and rats. There is a vast body of evidence indicating that 3,4-benzopyrene and other carcinogenic hydrocarbons are formed during pyrogenation or incomplete burning of almost any kind of organic material. For example, carcinogenic hydrocarbons have been identified in overheated fats, broiled and smoked meats, coffee, burnt sugar, rubber, commercial paraffin oils and solids, soot, the tar contained in the exhaust fumes of internal combustion engines, cigarette smoke, etc.

Attention to the carcinogenic aromatic amines was drawn by the high incidence of urinary bladder tumors in dye works exposed to 2-naphthylamine (VII, Fig. 2), and benzidine (IX). The carcinogenic activity of VII, IX, and 4-aminobiphenyl (X) toward the bladder of the dog and the mouse has been demonstrated. In the rat, however, there is a change in target specificity, and tumors are induced by IX and X in the liver, mammary gland, ear duct, and small intestine. Carcinogenic activity is considerably heightened in 2-acetylaminofluorene (XI) without change of target specificity. Increased activity is due to the fact that XI is more coplanar than X, because of the internuclear methylene —CH$_2$— bridge in the former. 2-Acetylaminofluorene was proposed as an insecticide before its carcinogenic activity was accidentally discovered; it is a ubiquitous, potent carcinogen in a variety of species. Changing the internuclear bridge of XI to a —CH=CH—, as in 2-aminophenanthrene (XII), causes a shift in target specificity; thus, in the rat, XII is inactive toward the liver, but in addition to inducing tumors in the mammary gland, ear duct, and small intestine, it produces leukemia. Compound XII represents a structural link between the aromatic amine and polycyclic hydrocarbon carcinogens (compare XII with I); it is also interesting in this respect that 2-aminoanthracene (VIII), which is a higher homologue of VII, is inactive toward the bladder, but is able to induce skin tumors in rats.

4-Dimethylaminoazobenzene (XIII) is the parent compound of the aminoazo dye carcinogens; it is also known in the earlier literature as Butter Yellow, because it was used to color butter and vegetable oils before its carcinogenic activity was discovered. Many derivatives of XIII have been prepared and tested for carcinogenic activity. In the rat, the aminoazo dye carcinogens, administered in the diet, specifically induce hepatomas. Tumor induction by most of the aminoazo dyes is delayed or inhibited by high dietary levels of riboflavin (vitamin B$_2$) or protein. Replacement of the —N=N—azo linkage by —CH=CH—, as in 4-dimethylaminostilbene (XIV), results in widening the target tissue spectrum; XIV induces tumors in the liver, mammary gland, and ear duct. Mice are much more resistant than rats

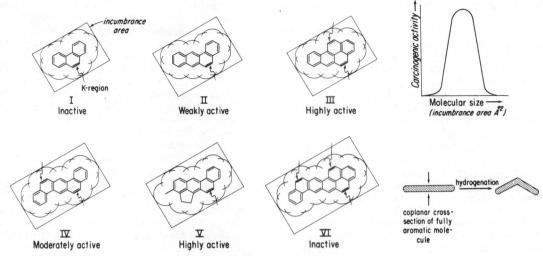

Fig. 1.   Polycyclic aromatic hydrocarbons.

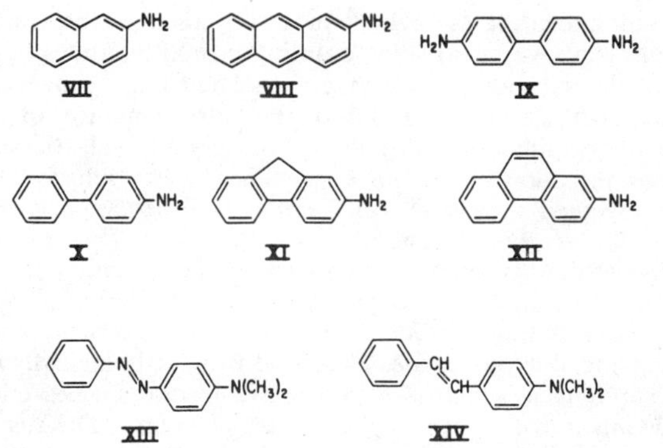

Fig. 2.   Carcinogenic aromatic amines.

to the carcinogenic activity of both aminoazo dyes and amino-stilbenes.

Figure 3 illustrates some aliphatic carcinogens. N-Methyl-bis-β-chloroethylamine (XV), a nitrogen mustard, produces local sarcomas, lung, mammary, and hepatic tumors upon injection in mice; because of its tumor-inhibitory properties, XV has also been used in the therapeutic treatment of certain types of human cancers. Bisepoxybutane (XVI), β-propiolactone (XVII) and N-lauroylethyleneimine (XVIII) produce local sarcomas in rats upon injection. Ethylcarbamate (XIX), the parent compound of several hypnotic drugs used in humans, produces

malignant lung adenomas in rats, mice, and chickens. Dimethyl-nitrosamine (XX) is a potent carcinogen toward the liver, lung, and kidney, and ethionine (XXI) toward the liver of the rat; the former is an intermediate in the manufacture of the rocket-fuel component, dimethylhydrazine $(CH_3)_2N—NH_2$, while the latter is the S-ethyl analogue of the natural amino acid, methionine.

Chemically identified carcinogens may be grouped in many ways, including a division into inorganic ions and organic compounds. The inorganic carcinogens are the elements beryllium, cadmium, iron, cobalt, nickel, silver, lead, zinc, and possibly arsenic; these can form coordination compounds and/or react with sulfhydryl groups. Also asbestos powder is a powerful carcinogen toward the lung upon inhalation (asbestos cancer in miners). In recent years, the characteristics of asbestos and related substances have caused much controversy in connection with the pollution of certain waters, notably Lake Superior by taconite processing waste products which contain fibers that have been compared with asbestos fibers. Most likely this situation will require some years for full scientific and legal resolution.

The organic carcinogens may be subdivided in several ways, including: (a) condensed polycyclic aromatic hydrocarbons and heteroaromatic polycyclic compounds; (b) aromatic amines and N-aryl hydroxylamines; (c) aminoazo dyes and diaryl-azo compounds; (d) aminostilbenes and stilbene analogues of sex hormones. Further breaking down the aliphatic carcinogens, these include: (1) alkylating agents (such as sulfur and nitrogen mustards, derivatives of ethyleneimine, lactones, epoxides, alkane-

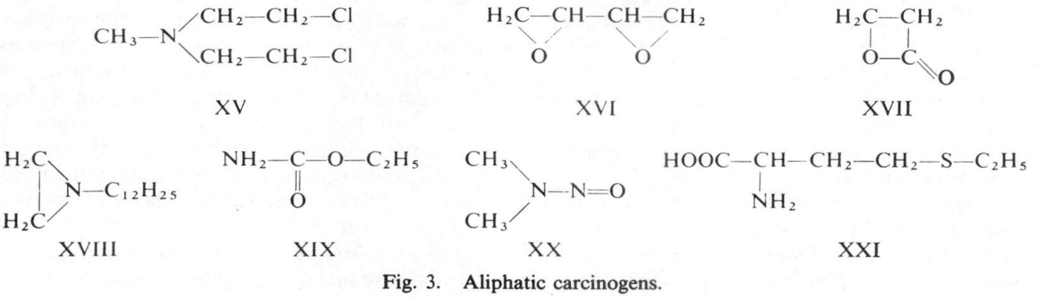

Fig. 3.   Aliphatic carcinogens.

α-ω-bis-methanesulfonates, certain dialkylnitrosamines, and ethionine; (b) lipophilic agents and hydrogen-bond reactors; this class comprises a wide variety of agents, such as chlorinated hydrocarbons (chloroform, carbon tetrachloride, and compounds used as pesticides under the names aldrin and dieldrin), bile acids, certain water-soluble high polymers, certain phenols, urethane and some of its derivatives, thiocarbonyls, and cycloalkynitrosamines; (c) naturally occurring carcinogens.

Until the early 1950s, the concept prevailed that the activity of carcinogenic chemicals was somehow related to the fact that they were synthetic "unnatural" substances which, since they were not present in the natural environment, were not factors of selection during developing life processes and hence contemporary living organisms were not equipped for effective metabolic "detoxification" of these compounds. However, during the last 25 years, a number of carcinogenic compounds of plant and fungal origin have been identified, including safrole in sassafras, capsaicine in chili peppers, various tannins, cycasin in the cycad groundnut, parasorbic acid in mountain ash berry, pyrollizidine alkaloids in *Senecio* shrubbery, and patulin, griseofulvin, penicillin G, aflatoxin and actinomycin produced by various molds, and many others too numerous to detail. The number and variety of identified naturally occurring carcinogens continues to increase at a rapid rate.

Carcinogen Mechanisms. The biochemical pathways in the cell are closely interconnected and are in a state of dynamic equilibrium (homeostasis). This equilibrium is maintained by feedback relationships existing between a great number of pathways. Chemical "communication" between subcellular organelles, such as the nucleus (within which the chromosomes contain the genetic "blueprints" for cell reproduction and the synthetic processes of cell life), the mitochondria (the "powerhouse" of the cell, which assures the synthesis of the universal cellular fuel, ATP, through the metabolism of carbohydrates and fatty acids), and the endoplasmic reticulum (synthesizing the proteins of the cell and assuring the metabolic breakdown—"detoxification"—of a multitude of endogenous and foreign compounds), depends on the constant interchange of a large variety of metabolic products and inorganic ions between them. There are probably a very great number of loci (receptor sites) upon which these regulatory chemical "stimuli" act. The receptor sites are of an enzymic and nucleic acid nature. Other control points of protein character regulate the morphology of the intracellular lipoprotein membranes which serve as "floor space" to the organized arrangements of multienzyme systems. The specificity of compounds of chemical control toward given receptor sites is due to a three-dimensional geometric "fit" following the lock and key analogy. Such is the general scheme of functional interrelationships in monocellular organisms which, hence, in a favorable medium multiply unchecked to the limit of the availability of nutrients.

In multicellular organisms, the subordination of the individual cells to the whole is assured by the existence of additional receptor sites which enable the cells to be response to chemical "stimuli" emitted by neighboring cells in the tissue and to hormonal regulation by the endocrine system in higher organisms. Hence, depending on the requirements of the moment, cells may remain stationary or may undergo cell division because of the need for repair of tissue injury, they may secrete different products, or they may perform some other specialized function depending on the nature of the particular tissue.

Carcinogenic substances are nonspecific cell poisons which cause the alterations and hence functional deletion of a large number of metabolic control sites. Present evidence suggests that these alterations are produced by the accumulation of the carcinogen in subcellular organelles, by covalent binding of the carcinogen to cellular macromolecules (proteins and nucleic acids) through metabolism, and by denaturation (i.e., destruction of the three-dimensional geometry) of the control sites through secondary valence interactions (hydrogen bonds, hydrophobic bonding, etc.) with the carcinogen. Early stages of tumor induction generally coincide with extensive cell death (necrosis) in the target tissues because a number of the biochemical lesions cause the irreversible blocking of metabolic pathways essential for cell life. However, because of the random distribution of the biochemical lesions in the cell population, in a small number of cells vital pathways are only slightly damaged and the lesions involve those sites and pathways which are not essential for cell life proper, but are necessary for organismic control. Thus, due to the action of the carcinogen, these cells escape physiological control and revert to a simpler, less specialized cell type (i.e., dedifferentiate). Such cells respond to continuous nutrition with continuous growth, which is an essential characteristic of malignant tumor cells.

Testing. Because of their short life span (average 3 years) small rodents (mice, rats, and hamsters) are frequently used for the testing of chemicals for carcinogenic activity; occasionally testing is done with rabbits, dogs, fowls, monkeys, etc. While a great variety of ways of administration have been used, a common method is to introduce substances to be tested in the following ways: (a) skin "painting"; small volumes of solution of the substance in an inactive solvent (e.g., benzene) are applied to the shaved surface of the skin (generally of mice, in the intersapular region) daily or at longer intervals; (b) subcutaneous injection of the pure substance or its solution (once or at repeated intervals); (c) feeding; the substance is mixed in the diet at given levels, or dissolved in the drinking water. Testing of new substances for possible carcinogenic activity is conducted for a minimum of 1 year to be meaningful. At the end of the testing period, all animals are autopsied, and all tumors and dubious tissues examined histopathologically.

A carcinogen which is highly active in one species may be totally inactive in another species, and vice versa. The susceptibility of a species to a given carcinogen also depends on the genetic strain, sex, and dietary conditions. Moreover, carcinogenic substances generally show a rather selective specificity toward certain target tissues; e.g., certain compounds produce exclusively hepatomas in the susceptible species. For these reasons, no chemical compound may be stated safely to be devoid of carcinogenic activity toward humans unless it has been found inactive when tested in a variety of mammalian species and by a variety of routes of administration for a length of time corresponding to half of the life span of each species.

But, the foregoing observation emphasizes one of the main problems in cancer research, namely, *extrapolation* of research findings. To be fully safe, literally tens of thousands of commonly encountered natural materials and synthetic materials covering the complete spectrum of products with which people are in contact over their lifespan would have to be tested and regarded suspect until thoroughly tested on the species of most importance to humans, namely, the testing of reactions among people themselves. But, this alone would not suffice because, as stressed throughout biochemical studies of human systems, there is individuality (see also **Biochemical Individuality**). The problem of developing improved (vastly improved) testing systems and attention to the problem of extrapolation of findings rivals in terms of difficulty the basic problem of identifying the nature of cancer itself.

The interesting observation has been made that, because of the vast amounts of money going into cancer research, the field has become a large business for numerous suppliers. Animal cells now can be procured by the kilogram from suppliers. Because of this availability (mostly two kinds, 3T3 and W138), much information has been accumulated concerning the biology of these cells. But, is much of what has been learned in this regard, meaningful? Researchers have found that viruses and chemicals can transform 3T3 cells to a neoplastic state and that these cells can produce tumors when inoculated in suitable hosts. It should be noted, however, that the tumors so produced are sarcomas (derived from fibroblasts), which are very rare in human beings. Ninety percent of human tumors are carcinomas. With the exception that epithelial cells and fibroblasts are both animal cells, they have little in common. They stem from two embryonic sources with different functions. The tumors they produce are different.

Meaningful research in genetic techniques as applied to cancer research is just getting underway as of the mid-1980s. See also **Wastes and Water Pollution.**

References

Ames, B. N., McCann, J., and E. Yamasaki: "Methods for Detecting Carcinogens and Mutagens with the Salmonella/Mammalian-Microsome Mutagenecity Test," *Mutation Res.*, **31**, 6, 347–364 (1975).

Ames, B. N.: "Identifying Environmental Chemicals Causing Mutations and Cancer," *Science*, **204**, 587–593 (1979).

Devoret, R.: "Bacterial Tests for Potential Carcinogens," *Sci. Amer.*, **241**, 2, 40–49 (1979).

Fajen, J. M., et al.: "*N*-Nitrosamines in the Rubber and Tire Industry," *Science*, **205**, 1262–1264 (1979).

Gelboin, H. V. (editor): "Polycyclic Hydrocarbons and Cancer," Academic, New York, 1978.

Gori, G. B.: "The Regulation of Carcinogenic Hazards," *Science*, **208**, 256–261 (1980).

Hartline, B. K.: "Cancer and Environment," *Science*, **205**, 1363–1364 (1979).

Kolata, G. B.: "Testing for Cancer Risk," *Science*, **207**, 967–969 (1980).

Land, C. E.: "Estimating Cancer Risks from Low Doses of Ionizing Radiation," *Science*, **209**, 1197–1203 (1980).

Sax, N. I.: "Dangerous Properties of Industrial Materials," 5th edition, Van Nostrand Reinhold, New York, 1979.

Stein, M. W., and E. B. Sansone: "Degradation of Chemical Carcinogens: An Annotated Bibliography," Van Nostrand Reinhold, New York, 1979.

Witkin, E. M.: "Ultraviolet Mutagenesis and Inductible DNA Repair in *Escherichia coli*," *Bacteriol. Rev.*, **40**, 4, 869–907 (1976).

CARNALLITE. This mineral is a product of evaporation of saline deposits rich in potash content, as a hydrated chloride of potassium and magnesium, $KMgCl_3 \cdot 6H_2O$. Hardness, 2.5; specific gravity, 1.602. It crystallizes in the orthorhombic system usually as massive, granular aggregates. Luster greasy, with indistinct cleavage and conchoidal fracture. Color grades from colorless to white, into reddish from included hematite scales. Transparent to translucent with bitter taste, deliquesces readily in moist environment.

Found associated with sylvite, halite and polyhalite at Stassfurt, Germany; Abyssinia; the U.S.S.R.; and in southeastern New Mexico and adjacent areas in Texas. It is an important source of potash for use in fertilizers. See also **Potassium.**

CARNELIAN. The mineral carnelian is a red or reddish-brown chalcedony; the word is derived from the Latin word meaning flesh, in reference to the flesh color sometime exhibited. See also **Chalcedony.**

CARNOTITE. This mineral is a vanadate of potassium and uranium with small amounts of radium. Its formula may be written $K_2(UO_2)_2(VO_4) \cdot 3H_2O$. The amount of water, however, seems to be variable. It occurs as a lemon-yellow earthy powder disseminated through cross-bedded sandstones with rich concentrations around petrified and carbonized trees. Hardness, soft; specific gravity, 4.7. It was mined in Colorado and Utah as a source of radium. Other localities are in Arizona, Pennsylvania, and Zaire.

CAROTENOID PIGMENTS. Pigmentation (Plants).

CAROTENOIDS. Lipid-soluble, yellow-to-orange-red pigments universally present in the photosynthetic tissues of higher plants, algae, and the photosynthetic bacteria. They are spasmodically distributed in flowers, fruit, and roots of higher plants, in fungi, and in bacteria. They are synthesized *de novo* in plants. Carotenoids are also widely, but spasmodically distributed in animals, especially marine invertebrates, where they tend to accumulate in gonads, skin, and feathers. All carotenoids found in animals are ultimately derived from plants or protistan carotenoids, although because of metabolic alteration of the ingested pigments, some carotenoids found in animals are not found in plants or protista.

Carotenoids are tetraterpenoids, consisting of eight isoprenoid residues, and can be regarded as being synthesized by the tail-

to-tail dimerization of two 20-carbon units, themselves each produced by the head-to-tail condensation of four isoprenoid units. Hydrocarbon carotenoids are termed *carotenes* and oxygenated carotenoids are known as *xanthophylls*. The structure of the best-known carotene, β-carotene, is

α-Carotene is widely distributed in trace amounts, together with β-carotene in leaves; γ-carotene is found in many fungi, and lycopene is the main pigment of many fruits, such as the tomato.

See also **Annatto Food Colors;** and **Pigmentation (Plants).**

CARRIER (Food Additive). A substance well named because its primary function is that of conveying and distributing other substances throughout a food substance. The role parallels that of a carrier in paint, wherein a vehicle (carrier) holds and distributes pigment throughout the entire paint product. Silica gel and magnesium carbonate serve as carriers in food substances. For example, the high porosity of silica gel enables it to adsorb internally up to three times its own weight of many liquids. This property is used to convert various liquid ingredients, such as flavors, vinegar, oils, vitamins, and other nutritional additives, into easy-to-handle powders. These powders, in turn, can be measured easily and blended effectively with other constituents to provide a uniform food substance. Advantage is taken of the properties of carriers in the convenience food field, where flavors remain entrapped inside silica particles until the food product is mixed with water, at which time the flavors are released just prior to consumption, giving the product an aura of richness and freshness.

CASCADE COOLING.   Cryogenics; Natural Gas.

CASE HARDENING.   Hardening of the surface layer or case of a ferrous alloy while leaving the core or center in a softer, tougher condition. There are two basic methods: (1) Gaseous elements, such as carbon or nitrogen, are introduced into the surface layer, thereby forming a hardening or hardenable alloy at the surface. Specific processes include carburizing, nitriding, and carbonitriding. (2) The surface may be given a hardening heat treatment that does not affect the core. This may be accomplished by flame hardening or induction heating, whereby the surface is rapidly heated into the austenite range and the specimen quenched before the center has obtained a temperature high enough to allow it to be hardened.

CASEIN.   Casein is the phosphoprotein of fresh milk; the rennin-coagulated product is sometimes called paracasein. British nomenclature terms the casein of fresh milk caseinogen and the coagulated product casein. As it exists in milk it is probably a salt of calcium.

Casein is not coagulated by heat. It is precipitated by acids and by rennin, a proteolytic enzyme obtained from the stomach of calves. Casein is a conjugated protein belonging to the group of phosphoproteins. The enzyme trypsin can hydrolyze off a phosphorus-containing peptone.

The commercial product which is also known as casein is used in adhesives, binders, protective coatings, and other products.

The purified material is a water-insoluble white powder. While it is also insoluble in neutral salt solutions, it is readily dispersible in dilute alkalis and in salt solutions such as those of sodium oxalate and sodium acetate.

CASSITERITE.   The mineral cassiterite, chemically tin dioxide, SnO_2, is almost the sole ore of tin. It is a noticeably heavy mineral crystallizing in the tetragonal system, as low pyramids, prisms, often very slender, and as twinned forms. It is a brittle mineral, hardness, 6.0–7.0; specific gravity, 6.99; luster, adamantine; color, generally brown to black, but may be red, gray to white, or yellow; streak whitish, grayish, or brownish; may be almost transparent to opaque. A fibrous variety somewhat resembling wood is called wood tin. Cassiterite occurs in widely scattered areas, but deposits of a size to be commercially important are few. It is associated with granites and rhyolites.

Cassiterite is heavily concentrated in bands and layers of varying thickness, forming economically valuable deposits, such as those found in the Malay States of southeastern Asia. Bolivia, Nigeria, and the Belgian Congo are also major producers of tin ore. Cassiterite is also known as *tin stone*.

CAST IRON.   Iron Metals, Alloys, and Steels.

CATALYSIS.   The process of changing the velocity of a chemical reaction by the presence of a substance that remains apparently chemically unaffected throughout the reaction. Berzelius (1836) applied this term to those reactions that do not progress unless a catalyst is present in the mixture. "Contact actions" (Mitscherlich) and "cyclic actions" (Brodie) have been suggested as names for the phenomenon.

In general, the following rules hold true for all catalytic processes:
1. The catalyst has the same composition at the beginning as at the end of the reaction.
2. A small quantity of the catalyst is capable of effecting the transformation of an indefinitely large quantity of the reacting substance.
3. No catalytic agent has power to start a chemical reaction; it may merely modify the velocity of the reaction.
4. The catalyst has no effect upon the final state of equilibrium of the forward reaction with any opposing reactions.
5. The velocity of two inverse reactions is affected in the same degree by a catalyst.

There are two general types of catalytic processes: (1) homogeneous, and (2) heterogeneous. In the *homogeneous* type, the chemical reaction is said to take place in a single phase, usually a liquid environment. In the *heterogeneous* type, the process takes place in a multiple phase, usually in a gaseous environment, and usually in the presence of a solid catalyst phase, which is either present as an undiluted material or on the surface of an inert substance, such as charcoal, the latter called a *support*.

A catalyst must be active and selective and have chemical and physical stability. Initial consideration must be given to selectivity, which is a measure of the degree the reaction is made to go in the direction that is desired. It is also desirable to convert the reactant(s) as rapidly as possible. The catalyst must be active, for an active catalyst will have a desirably high production rate. The concept of productivity is measured by the product of activity or conversion and selectivity per unit time and may be written Yt. The product of conversion (C) times selectivity (S) equals yield (Y).

In a heterogeneous catalytic reaction (by a solid surface), the following sequence of steps occurs: (1) diffusion of reactants from the bulk or gas phase to the catalytic surface, (2) adsorption of reactants, (3) diffusion of adsorbed species to active sites, (4) electron transfer processes at active sites, (5) chemical interaction of neutral and charged species, (6) desorption of reaction products from active sites, and (7) diffusion of products into the bulk phase.

Inasmuch as catalyst activity is proportional to the ability of the catalyst to chemisorb reacting species, activity will normally increase as surface area increases. For very fast reactions (high-activity catalysts), only the external surface is usually involved, and the overall rate of reaction may become controlled by the rate of transfer of reagents to the catalyst surface. In contrast, some reactions with porous catalysts may involve lesser reaction rates than rate of external diffusion. Hence overall reaction rate will be noticeably affected by diffusion within catalyst pores.

Solid catalysts which contain networks of pores and create a large surface area will allow for diffusion of reacting molecules and for high catalytic activity. Selectivity, on the other hand, may or may not be affected adversely, depending on the nature of the reaction. As a rule, oxidation catalysts usually perform best with large-diameter pores and low surface areas. Hydrogenation catalysts perform better with high surface areas.

The concept of protonic acid (hydrogen ion) and base (hydroxyl ion) catalysis has been found applicable in discussions on heterogeneous catalysis, wherein the mechanism is seen as the transfer of a proton from the catalyst to the reactant (acid catalysis), or from the reactant to the catalyst (base catalysis). Intermediates may be of two types: (Type I) in which the reversible reaction that forms the intermediate is fast compared with further change into final products, so that the intermediate is always present in its equilibrium concentration; and (Type II) in which the intermediate is never present in appreciable concentration, and the velocity of the reaction is determined by the speed of formation of the complex. A number of apparently different catalysts behave similarly because of their acidic nature, i.e., mineral acids, Friedel-Crafts catalysts, silica-alumina materials, and zeolite-cracking catalysts.

In recent years, the electronic concepts of solids have been introduced as a means toward understanding catalytic action. It is possible to divide heterogeneous catalysts into groups based upon their electronic properties. For example, metals (conductors) can be used in hydrogenation, dehydrogenation, and hydrogenolysis reactions; metal oxides or their sulfides (semiconductors) can be used in oxidation, reduction, dehydrogenation, or cyclization reactions; and salts or acid-site (insulators) catalysts can be used in cracking, dehydration, isomerization, polymerization, alkylation, dehalogenation, halogenation, and hydrogen transfer reactions.

The effects of different catalysts on the same reagent is shown by the reaction of ethyl alcohol. When reacted over copper (a conductor), either acetaldehyde or ethyl acetate plus hydrogen is obtained. When reacted over alumina (an insulator), ethylene plus water or diethyl ether plus water is recovered.

In general, a catalyst problem is either one of development of a new catalyst for a novel process, or of a catalyst modification for an existing process. A catalyst must be considered as part of an overall process, for when studying a chemical process, not only is the precise nature of a catalyst studied, but also the chemical-reaction mechanism, practical processing details, and production techniques. Advances in physical chemistry, instrumental analysis, radioactive techniques, solid-state physics, and computer applications have assisted catalytic work to a large degree. However, published discovery of new processes and continued application of old catalysts to new reactions indicates that catalytic chemistry remains an experimental art and is dependent to a large extent on the intuition, ingenuity, and perseverance of both chemist and chemical engineer.

Chemical and petroleum processing require large production of catalysts. In cracking operations, where hydrocarbons are changed from such substances as tar and asphalt to fuels, such as light oils and gasoline, the consumption of silica, aluminas, processed clays, and zeolite catalysts is very high. Isomerization operations, where straight-chain paraffinic molecules are rearranged to branched-chain molecules in order to improve octane number of gasoline, consume large quantities of aluminum chloride catalyst. In alkylation reactions, where an alkyl radical is introduced into a molecule by addition or substitution into an organic compound in order to make such substances as gasoline, rubber antioxidants, dyes, flavors, and other compounds, sulfuric acid, as a catalyst, is used extensively.

As shown by the table (next page), scores of reaction types and catalyst materials are involved in industrial processes.

Trends. In the mid-1960s, scientists (*Exxon*) made the fundamental discovery that supported bimetallic catalysts are superior to supported single-metallic catalysts. These catalysts have made it possible to produce high-octane gasoline without tetraethyl lead. Further research has led to the realization that all catalyst support materials modify to some extent the structure, electronic properties, and chemical behavior of catalysts.

Breakthroughs also have occurred in instrumentation for the measurement of catalytic reactions. An X-ray photoelectron spectrometer coupled to a high-pressure catalytic reactor is capable of measuring the presence and strength of a metal support interaction and also tests catalytic activity. Using a dedicated computer to analyze large amounts of information generated in catalytic reaction studies, much more is being learned concerning the structure and reactivity of catalytically active sites. Electron microscopes are also being used to determine the rela-

tionship between the structures of catalytic surfaces and the chemistry taking place at such surfaces.

Schrock (M.I.T.) reported in 1983 that several well-characterized transition metal catalysts contain a metal-carbon double bond or a metal-carbon triple bond. It has been postulated that in homogeneous or heterogeneous catalyst systems in which the metal is likely to be in a relatively high oxidation state, such as molybdenum (V) or tungsten (VI), metal-carbon multiple bonds may play an important role. Recent research has suggested that even supposedly well understood reactions, such as ethylene polymerization, may actually involve catalysts that behave as if they contained a metal-carbon double bond instead of a metal-carbon single bond. It has been further observed that the chemistry of metal-carbon double and triple bonds may eventually complement and perhaps overlap the known chemistry of complexes containing metal-oxygen double bonds or metal-nitrogen triple bonds, respectively. Thus, unique catalytic reactions involving carbon, nitrogen, and oxygen ligands multiply bonded to transition metals may be possible. (*Science*, **219**, 13, 1983.) See also **Carbon.**

References

Baker, R. T. K.: "Catalysts in Action," *Chem. Eng. Progress*, **73**, 4, 97–99 (1977).
Kay, E., and P. S. Bagus, Editors: "Topics in Surface Chemistry," Plenum, New York (1978).
Lundberg, W. C.: "Extending Catalyst Life," *Chem. Eng. Progress*, **75**, 6, 81–86 (1979).
Parshall, G. W.: "Organometallic Chemistry in Homogeneous Catalysis," *Science*, **208**, 1221–1224 (1980).
Robinson, A. L.: "Homogeneous Catalysis: Transition Metal Clusters," *Science*, **194**, 1150–1152 (1976).
Sinfelt, J. H.: "Heterogeneous Catalysis," *Science*, **195**, 641–646 (1977).
Sleight, A. W.: "Heterogeneous Catalysts," *Science*, **208**, 895–900 (1980).
Voorhoeve, R. J. H., et al.: "Perovskite Oxides: Materials Science in Catalysis," *Science*, **195**, 827–833 (1977).

CATALYTIC CONVERTER. These devices have been used on automotive vehicles (gasoline-fueled) in a number of countries since the mid-1970s in response to clean air regulations. Converters in the United States use noble metals on a ceramic substrate (e.g., platinum dispersed on alumina). The converter is located in the exhaust system in two possible locations—underfloor and close-coupled near the manifold. Operating temperature range for noble metal catalysts is from 600 to 1200°F (316–649°C), which is similar to the exhaust-pipe skin temperatures normally encountered on standard automotive engines. Catalytic materials can be physically mounted on either pelleted or monolithic substrates. In the case of the pelleted catalyst, the support is an activated alumina. Pelleted catalyst is confined by screens (Fig. 1). In the case of a typical monolithic catalyst,

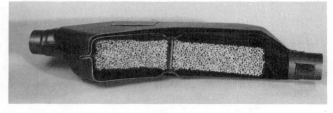

Fig. 1.   UOP-designed converter to use pelleted catalyst.

this is composed of a channeled ceramic (cordierite) support having 300–400 square channels per square inch on which an activated alumina layer is applied. The active agents (platinum,

EXAMPLES OF CATALYTIC PROCESSES[a]

PROCESS AND PRODUCT	CATALYST	REACTANTS	YIELD
Amination			
Amines	$Al_2O_3(Co)$	Alcohols + Ammonia	90+
Ammoxidation			
Acrylonitrile	Bi-Mo-P CuO, SbSn	Propylene + O_2 + NH_3	60–80
Benzonitrile	V_2O_5-Sb	Toluene + O_2 + NH_3	90+
Phthalonitrile	V_2O_5	o-Xylene + O_2 + NH_3	90+
Chlorination			
Chlorobenzene	Fe	Benzene + Cl_2	70–75
Chloroacetic acid	red P	Acetic Acid + Cl_2	90
Benzoyl chloride	UV Light	Toluene + Cl_2	95+
Hydration			
Acetaldehyde	Hg_2SO_4	Acetylene + H2O	95
Ethanol (alcohols)	H_3PO_4, WO_3	Ethylene + H2O	95
Dehydration			
Styrene	TiO_2	Methylethyl carbinol	80+
Ethylene (olefins)	Al_2O_3, ThO_2	Ethanol (alcohols)	90+
Acrylonitrile	Al_2O_3	Ethylene cyanohydrin	90+
Hydrogenation			
Aniline	Fe-HCl, Cu·SiO_2	Nitrobenzene	90–95
Butanol	Co, Ni-SiO_2	Butyraldehyde	98+
Cyclohexane	Ni-Al_2O_3, PtO_2	Benzene	96+
Ethylene	Fe, Ni, Cu, Pd-$BaSO_4$	Acetylene	99
Methanol	ZnO-CrO_3, Ni-Co	Carbon monoxide	60
Dehydrogenation			
Acetaldehyde	Cu, Ag, $FeMoO_4$	Ethanol + H_2	85–95
Benzene	Nu/Al_2O_3, Pt-Al_2O_3	Cyclohexane	95+
Butadiene	Fe, Cr, K, $CaNiPO_4$	Butenes	75–85+
Butene	Cr_2O_3-Al_2O_3	Butane	
Methyl ethyl ketone	ZnO-ZnCu	Sec-Butanol	85–90
Styrene	$ZnCrO_2$-FeMgO $CaNiPO_4$	Ethyl benzene	86–92
Styrene	TiO_2	Phenyl methyl carbinol	80+
Oxidation			
Acetaldehyde	$PdCl_2$-MgO-Cu	Ethylene	95+
Acetic acid	Mn^{2+}	Acetaldehyde	88–95
Acetic acid	Co, Bi	Butane	20–40
Acetic anhydride	Cu, Co	Acetaldehyde	70–75
Acetone	Cu, Ag, Zno	Isopropanol	85–90
Adipic acid	Cu-Mn, V-Cu	Cyclohexanone	70–90
Benzoic acid	Co^{2+}	Toluene	90
Benzoic acid	Cu	Phenol	90
Benzaldehyde	UO_2-MoO_3-Cu	Toluene	30–50
Ethylene oxide	Ag, AgO	Ethylene	70
Propylene oxide	Mo, W, Ti, V	Propylene	90
Phthalic anhydride	V_2O_5-K_2SO_4	Napthalene o-Xylene	70–80
Maleic anhydride	Mo-V-P-Na	Benzene	85
Maleic anhydride	V-P	Butene	60
Terephthalic acid	Mn-Co	p-Xylene	90+
Reductive Dehydration			
Butane	Ni-Al_2O_3	Butanol + H_2	90+
Reforming			
Aromatic	Mo-Al_2O_3 Pt-Al_2O_3-halides	Naphthenes + H_2	
Desulfurization			
Butane	Co-Mo-Al_2O_3	Thiophene + H_2S + H_2	

SOURCE: Catalyst Development Corporation.
[a] Discussed in detail in Baker reference listed.

palladium, rhodium, etc.) are then highly dispersed on the alumina. See Fig. 2.

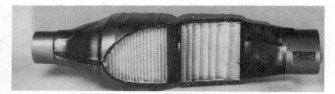

Fig. 2.   UOP-designed converter to use monolithic catalyst.

Noble metal containing catalysts require use of essentially lead-free gasoline to avoid catalyst poisoning.

CATENATION COMPOUND.   Compound (Chemical).

CATHODE.   The electrode at which positive current leaves a device which employs electrical conduction other than that through solids. The negative terminal of an electroplating cell, i.e., the electrode from which electrons enter the cell, and thus at which positively charged ions (cations) are discharged. The positive terminal of a battery.

CATHODIC PROTECTION.   Corrosion.

CATION.   A positively charged ion. Cations are those ions that are deposited, or which tend to be deposited, on the cathode of an apparatus. They travel in the nominal direction of the current. In electrochemical reactions, they are designated by a dot or a plus sign placed above and behind the atomic or radical symbol, as $H^{\cdot}$ or H^+, the number of dots or plus signs indicating the valence of the ion. See also **Cathode.**

CAT'S-EYE.   This name is applied to varieties of several mineral and gemstone species that enclose fine fibers or cellular structures in parallel arrangement, causing, particularly when cut and polished *en cabochon*, a band of reflected light to play on the surface of it. Because of fancied resemblance to the eyes of cats, such stones are called cat's-eyes, and the effect is referred to as chatoyancy. The stone is said to be chatoyant. True cat's-eye is a variety of chrysoberyl, but tourmaline and quartz are also found which show this same effect. Ordinary quartz cat's-eyes are a pale yellowish or greenish, but a beautiful golden-yellow sort is known from South Africa called tiger's-eye which probably represents a replacement of crocidolite by quartz.

When the term *tiger's eye* is used, this applies only to chrysoberyl. Other gemstones that exhibit this phenomenon include sillimanite, scapolite, cordierite, orthoclase, albite, and beryl. See also **Chrysoberyl; Crocidolite.**

CAUSTIC.   A corrosive substance, almost always of an alkaline nature, such as sodium hydroxide (NaOH); potassium hydroxide (KOH), or calcium oxide (CaO). Such substances attack many metals, plastics, and other materials, including human tissue, and generally fall in the category of *corrosives*.

CAUSTIC POTASH.   Potassium.

CAUSTIC SODA.   Sodium.

CAVITAND.   As defined by Cram (*Science*, **219**, 1177, 1983), a cavitand is a synthetic organic compound that contains an enforced cavity of dimensions at least equal to those of the smaller ions, atoms, or molecules. As Cram points out, if organic compounds that contain enforced (rigid) cavities are to be designed and prepared, they must be composed of units that are concave on parts of their surfaces. Very few organic compounds have concave surfaces of any size. Among the most studied of the naturally occurring compounds that contain rigid cavities are the cyclodextrins. In these cyclic oligomers of the 1,4-glucopyranoside unit, from 6 to 8 monosaccharide units are contained in a torus-shaped cavity. Organic hosts are now being designed and synthesized which contain enforced cavities sufficiently large to complex and even surround simple inorganic or organic guest compounds. This new field of investigation is of interest in enzyme and catalytic systems.

CELESTITE.   The mineral celestite (also known as celestine) is composed of strontium sulfate $SrSO_4$, occasionally with calcium and barium. It crystallizes in the orthorhombic system in tabular or prismatic crystals. More rarely it may be pyramidal or simply fibrous or granular. Two essentially perfect cleavages may be observed, one parallel to the base, the other parallel to the prism. Its fracture is uneven; hardness, 3–3.5; specific gravity 3.97; luster, vitreous; color, white, but may be slightly reddish or bluish; transparent to translucent.

Celestite may occur with gypsum and salt associated with beds of limestone, or by itself in large, commercially important veins. It sometimes occurs with sulfur in volcanic localities and is often a gangue mineral in veins of galena, sphalerite and similar metallic minerals. In Europe there are many localities for fine crystals, especially in England. In the United States celestite is found in New York, Pennsylvania, West Virginia, Tennessee, Kansas, Colorado, and California. The first celestite described was the delicate blue material from Blair County, Pennsylvania. Its "celestial" tints suggested the name. Celestite resembles barite.

CELLULOSE.   The formula for cellulose is sometimes given as $(C_6H_{10}O_5)_n$. This is an oversimplification inasmuch as the cellulose present in natural substances, such as wood and cotton fibers, usually is combined with other constituents, such as fats and gums. Cellulose is found almost exclusively in plants and accounts for about 30% of all vegetable matter. Cellulose is the principal substance of which the walls of vegetable cells are constructed. The term *cellulose* is derived from the Latin *cellula*, meaning little cell. Relatively pure cellulose can be obtained from cotton fibers (90% cellulose) and flax fibers. Very small amounts of cellulose are found in insects and none in animal tissues. Digestive juices and enzymes present in animal systems do not appear to attack cellulose and thus ingestion by humans is relatively limited. By means of other biological processes, such as amoeboid protozoa present in the digestive tract, herbivora and insects digest and absorb some cellulose.

Cellulose is a polysaccharide of glucose. See Fig. 1. Cellulose

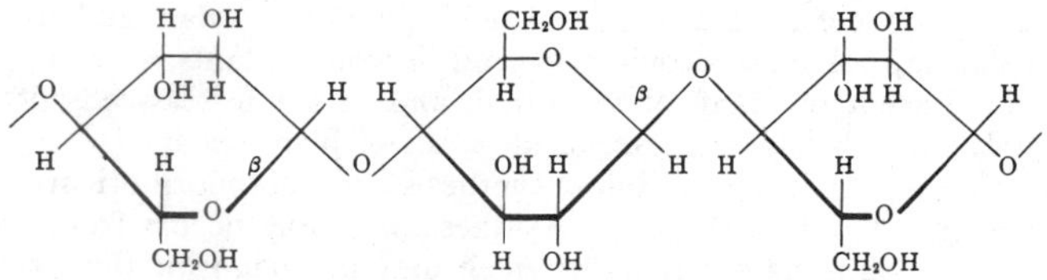

Fig. 1.   A segment of the cellulose molecule.

is a white solid, odorless and tasteless, insoluble in cold or hot water, and chemically nonreactive except when treated with strongly corrosive materials. If heated with water at 260°C and under rather high pressure, however, cellulose dissolves, but with decomposition. Concentrated sulfuric acid dissolves cellulose, the solution upon dilution and boiling yielding glucose. When treated with sodium hydroxide (15 to 25% NaOH) cellulose fibers swell up and upon washing and drying possess a lustrous appearance. This is the mechanism of *mercerization*. With iodine in potassium iodide solution plus zinc chloride (Schulze's solution), cellulose produces a dark blue color. When treated with an 80% sulfuric acid solution and rapidly washed and dried, cellulose yields a parchment-like surface.

In biochemical terms, cellulose is the name given both to a specific polysaccharide, consisting of β-D-glucose residues joined end-to-end by linkage through —O— of C_1 of one residue

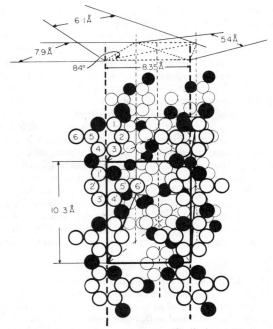

Fig. 2. Diagrammatic representation of the unit cell of cellulose. The monoclinic cell, dimensions 10.3 Å × 8.35 Å = 84°, is delineated by solid lines with one cellulose chain at each vertical edge and one (antiparallel) in the center. Open circles = carbon atoms; solid circles = oxygen atoms. For clearness of the diagram, the hydroxyl groups on carbons 2, 3, and 6 are omitted; and hydrogen atoms are omitted. Two spacing at 6.1 Å and 5.4 Å are included, inasmuch as these are strongly represented in the x-ray diffraction diagram. ("*The Encyclopedia of Biochemistry,*" *Van Nostrand Reinhold Co.*)

to C_4 of the next (see Fig. 2); and to a resistant family of polysaccharides (containing cellulose in the strict sense) isolated from plants by specific chemical treatment. Cellulose appears to be always associated in plants with other polysaccharides and polysaccharide derivatives, such as mannan, xylan, araban, galactan, polygalacturonic acid, and in woody plants, with lignin. Except in many of the fungi, and in a few seaweeds, it forms the skeletal polysaccharide walls of plant cells.

The mechanism of cellulose synthesis and microfibril orientation is not fully understood. Synthesis probably occurs from a glucose/phosphate precursor which may be guanosine diphosphate glucose. The enzyme system involved can be extracted from the plant (e.g., from *Acetobacter xylinum*) and synthesis by cell-free extracts has been achieved. The synthetic mechanism in plants higher than bacteria is thought to be located on the

cell surface, and both granular aggregates and microtubules seen in the electron microscope are considered as possible sites.

The "brown rots" (e.g., *Coniphora casebella, Poria monticola*) and the "white rots" (e.g., *Polystictus versicola*) are both basidiomycetes. Both attack cellulose, but only the latter takes lignin to any large extent. "Soft rot fungi," recognized relatively recently as important in this regard are members of the *Ascomycetes* and *Fungi imperfecti* (e.g., *Chaetomium globosum*). They all attack the cellulose of wood, producing characteristic angular cavities. The evidence is that all of these fungi attack the paracrystalline component of cellulose more rapidly than the crystalline component.

Compound celluloses are widely distributed in plants, the two principal types being:

(a) Lignocelluloses, of woods, cereal straws, jute. These cellulose materials yield lignin by treatment (1) with 43% hydrochloric acid, cold for 12 hours, (2) with 8 to 12% sodium hydroxide at 140 to 160°C for 6 to 10 hours, (3) with 72% sulfuric acid at ordinary temperature for 18 hours (the common method). Wood yields about 25% of lignin by the last treatment.

(b) Pectocelluloses, of flax, hemp, ramie. These cellulose materials yield pectic substances by treatment with oxalic acid or ammonium oxalate at 85°C for 24 hours, followed by carefully defined treatment with alcohol, acetic acid and calcium chloride. Pectic substances are most abundant in leaves, e.g., ivy, sycamore, and in apples or oranges, especially the white peel of the latter.

Cellulose dissolves in Schweitzer's reagent, an ammoniacal solution of cupric oxide. After treatment with an alkali, the addition of carbon disulfide causes formation of sodium xanthate, a process used in the production of rayon. See also **Fibers (Textile)**. The action of acetic anhydride in the presence of sulfuric acid produces cellulose acetates, the basis for a line of synthetic materials. See **Fibers (Acetate)**; and **Cellulose Ester Plastics (Organic)**.

On a high-production basis, cellulose is important as a raw material in the production of wood pulp for paper manufacture. See **Papermaking and Finishing**; and **Pulp Production and Processing**.

CELLULOSE ACETATE. Acetic Acid; Fibers (Acetate).

CELLULOSE ESTER PLASTICS (Organic). The cellulosics are unique among the plastics in that the basic materials used in their manufacture are not synthetic polymers. Rather, they are derivatives of a natural polymer, *cellulose*. See also **Cellulose**. The preparation of an organic cellulose ester plastic involves the formation of a suitable cellulose derivative, followed by processing steps that convert the cellulose derivative into a plastic.

Cellulose, with its many hydroxyl groups, can react with organic reagents such as acids, anhydrides, and acid chlorides to form organic esters. The first reported organic ester of cellulose was cellulose acetate, prepared by Schützenberger in 1865 by heating cotton and acetic anhydride to about 180°C in a sealed tube until the cotton dissolved. Franchimont, in 1879, accomplished this reaction at a lower temperature with the aid of sulfuric acid as a catalyst. The product in both cases was very nearly the triester. Miles, in 1903, first described partially hydrolyzed (generally called "secondary") cellulose acetate and distinguished it from the triacetate by its acetone solubility. The solubility of secondary cellulose acetate in such inexpensive and relatively nontoxic solvents as acetone contributed greatly to the development and commercialization of this material.

Cellulose esters of the 2-, 3-, and 4-carbon acids are readily prepared by the cellulose–anhydride reaction; the acetate ester and the mixed acetate butyrate and acetate propionate esters are manufactured and used in large amounts. Esters of higher acids require different synthesis techniques and tend to be prohibitively expensive except as specialty products. Some are in commercial production, however. Cellulose acetate phthalate, for example, is manufactured for use as an enteric coating on pills.

Most commercial preparations of cellulose esters still follow, basically, the methods described by Franchimont and Miles—esterification with sulfuric acid catalyst followed by hydrolysis. The principal steps in this process are shown in the accompanying figure.

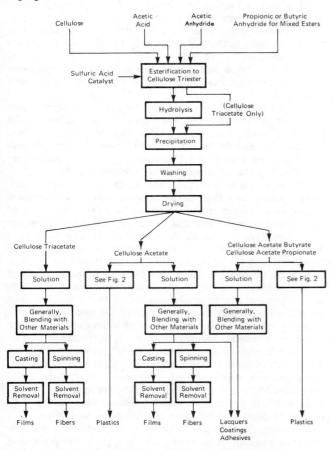

Production and end-uses of organic cellulose esters.

Esterification. The nature of cellulose is such that its esterification does not occur randomly. Even when the DS (degree of substitution—the average number of hydroxyl groups replaced per anhydroglucose unit in the cellulose chain) is approaching 3 (complete reaction), many anhydroglucose units have a DS of zero. If the ester is recovered before the reaction is complete, the product will not be homogeneous and it will be hazy. Regardless of the desired DS of the final product, therefore, the reaction must be allowed to proceed to virtual completion if a homogeneous material is to be produced. In most processes, the ester dissolves in the reaction mixture as the reaction approaches completion.

The cellulose used to manufacture cellulose esters is highly purified cotton linters or wood pulp. It is generally treated to reduce its crystallinity and make it more reactive, then agitated at somewhat elevated temperatures with the appropriate acids,

anhydrides, and catalyst until it dissolves. Some of the polymeric chains of the cellulose are broken during the reaction, and consequently the molecular weight decreases; thus, the catalyst concentration, reaction temperature, and reaction time must be controlled very carefully to give a product of the desired molecular weight.

Some acetate ester is recovered at the completion of reaction and marketed as commercial cellulose triacetate; it has a DS very close to 3. Other triesters have found no commercial applications.

Hydrolysis. The compatibility of a cellulose ester with other materials is influenced by the acid or acids used in its preparation, but it is controlled primarily by the hydroxyl content of the ester, which varies reciprocally with the degree of substitution. Cellulose triacetate, for example, does not dissolve in acetone; its solution requires very strong solvents such as methylene chloride. Plasticizers can be added to the triacetate in solution and will remain in the ester if the solvent is removed. Plasticizer added in this manner will affect some of the properties of the triester, but they will not appreciably reduce its softening temperature, which is higher than its decomposition temperature. The hydroxyl groups in secondary cellulose acetate, however, have for other polar materials an affinity that is lacking in the triacetate. The ester will dissolve in acetone, and plasticizers will both affect its mechanical properties and reduce its softening temperature sufficiently for the mixture to be processed as a thermoplastic. It is necessary, then, that cellulose esters to be used for plastics contain a significant number of hydroxyl groups, and these are produced by partial hydrolysis of the triester formed in the reaction step. Since the triester is in solution, hydrolysis is random and produces a homogeneous product.

Hydrolysis is initiated by the addition of water and stopped at the desired point by neutralization of the catalyst.

Precipitation, Washing, and Drying. The cellulose ester in solution in the reaction mixture is precipitated by the addition of water. Some precipitation processes produce a flake precipitate; some produce a powder. The precipitate is removed from the slurry, washed with water until it is free of acids, and dried.

Cellulose Esters

The cellulose esters that result from the process described are chemical raw materials and are used by many branches of the chemical industry. They are generally characterized by acyl content in weight percent and viscosity in seconds, the viscosity being obtained by timing the fall of a steel ball through a solution of the cellulose ester in accordance with ASTM Method D 1343. Low-viscosity esters are generally used in solution processes; high-viscosity esters are used in the production of plastics.

Cellulose Triacetate. It has already been implied that cellulose triacetate will not produce a thermoplastic, as its softening point cannot be reduced appreciably by plasticizers. It is used in solution processes, however, to produce films and fibers. Triacetate films absorb less water than films of secondary cellulose acetate, and they are therefore more dimensionally stable in environments where the humidity is not controlled. Triacetate fibers, with a similar resistance to water, impart to fabrics wrinkle resistance, dimensional stability, and the ability to dry rapidly. Under United States federal regulations, a fiber must be made from a cellulose acetate having at least 92% of its hydroxyl groups acetylated if it is to be called "triacetate."

Cellulose Acetate. Like the triacetate, secondary cellulose acetate (CA) is used in solution processes to produce fibers and films. CA fibers were originally called "rayon," the name that was already in use for regenerated cellulose fibers. In 1951, however, the regulatory authorities formally acknowledged the chemical distinction between CA and cellulose, and the term "rayon" was reserved for fibers of regenerated cellulose. CA fibers are officially called "acetate," and they are used in a wide variety of fabrics. They also are used for cigarette filters. However, the majority of CA produced is used for manufacture of plastics.

Cellulose Acetate Butyrate and Cellulose Acetate Propionate. These two cellulose esters are somewhat similar in properties and applications. Cellulose acetate butyrate is commonly referred to in the chemical industry as CAB, while cellulose acetate propionate is simply termed "cellulose propionate" and referred to as CAP or as CP.

The major usage of CAB and CP is in plastics. Additionally, CAB and CP are mixed with a variety of synthetic polymers to produce lacquers for wood, metal, and plastics. CAB finds relatively small usage in hot-metal coatings and in optical-grade cast film used in the manufacture of sunglass lenses.

Organic Cellulose Ester Plastics

Although the first cellulosic plastic (cellulose nitrate plastic—based on an *inorganic* ester of cellulose) was developed in 1865, the first *organic* cellulose ester plastic was not offered commercially until 1927. In that year, cellulose acetate plastic became available as sheets, rods, and tubes. Two years later, in 1929, it was offered in the form of granules for molding. It was the first thermoplastic sufficiently stable to be melted without excessive decomposition, and it was the first thermoplastic to be injection molded. Cellulose acetate butyrate plastic became a commercial product in 1938 and cellulose propionate plastic followed in 1945. The latter material was withdrawn after a short time because of manufacturing difficulties, but it reappeared and became firmly established in 1955.

Since the cellulose esters CA, CAB, and CP are chemical raw materials, the word "plastic" was used in the preceding paragraph to differentiate the product from the raw material. The cellulose ester plastics are commonly called simply "acetate," "butyrate," and "propionate," and these names will be used in the text that follows.

Commercial Production. In the manufacture of cellulose ester plastic, the appropriate ester is blended with plasticizer and other additives, such as stabilizers, ultraviolet inhibitors, dyes, and pigments, commonly in a large sigma-blade mixer. The mixture thus obtained is heated to its softening temperature and kneaded until it is homogeneous. This is done on hot milling rolls, in a compounding extruder, or in a Banbury mixer. The molten mass of plastic that results is formed into small rods or strips that are then cut into cylindrical or cubical pellets, which ordinarily have dimensions of about $\frac{1}{8}$ inch.

Properties. The concentration of plasticizer in a cellulose ester plastic determines its "flow temperature," which in turn determines its "flow designation," as defined by ASTM Method D 569. Flow designations range from various degrees of hardness (H4, H3, H2, H) through medium-hard (MH), medium (M), and medium-soft (MS) to various degrees of softness (S, S2, S3, S4, etc.). At any given flow designation, the characteristics of the plastic will vary somewhat with the identity of the plasticizer used. Some plasticizers, for example, give very hard materials, some give very low water absorption, and some permit unusual ease of processing. The flow temperature is used for quality control and the corresponding flow designation is a part of the purchase specification when material is ordered.

Representative properties of general-purpose formulations of acetate, butyrate, and propionate are shown in the accompanying table.

Depending upon plasticizer content, cellulose ester plastics range from soft, extremely tough materials to hard, strong, stiff compositions that still retain a considerable degree of toughness over a wide range of temperatures. They are basically transparent and virtually colorless, which makes it possible for them to be manufactured in almost any desired transparent, translucent, or opaque color. They are resistant to water and aqueous salt solutions, but they are attacked by aqueous solutions that are strongly acidic or basic. They resist several types of organic solvents, such as ethers and aliphatic hydrocarbons, but they are dissolved or swollen by strongly polar liquid organic compounds such as aromatic hydrocarbons, chlorinated hydrocarbons, ketones, and esters. The susceptibility of the plastics to attack decreases as the molecular weight of the attacking compound increases.

All three cellulose ester plastics are available in formulations that meet the regulatory requirements for use in contact with food.

Butyrate and propionate are available in special formulations for continuous use outdoors, where they generally remain useful for several years. Formulations other than the special outdoor-type materials should not be used in this manner. Acetate formulations are not suggested for outdoor use, as CA does not respond well to the addition of protective compounds.

Although acetate, butyrate, and propionate resemble each other in many ways, there are a number of significant differences among them. Butyrate and propionate are generally easier to process than acetate, and this factor, also, subtracts from the price advantage of acetate. Acetate is available in very hard flows, so it can be obtained with higher stiffness, hardness, and tensile strength than can the other two cellulose ester plastics. Butyrate is available in the softest flows, so it can be obtained in the toughest and easiest-processing formulations. Butyrate and propionate are generally considered to be tougher than acetate, even though in some instances the measured impact strengths may be similar. Butyrate retains its toughness better than does propionate at low temperatures. Butyrate and propionate use higher-boiling, less-water-soluble plasticizers than does acetate, which leads to better retention of plasticizer by butyrate and propionate when exposed to elevated temperatures or to the leaching action of water. Better plasticizer retention, in turn, leads to better permanence characteristics in the plastic—i.e., smaller changes in dimensions and properties with time.

Processing. The organic cellulose ester plastics are versatile materials and can be processed by almost any hot-processing technique used for thermoplastics. The principal techniques for all three plastics are injection molding and extrusion. Blow molding is also possible. Butyrate and propionate powder are used in fluidized-bed and electrostatic coating processes, as well as in the rotational molding process.

The toughness of cellulosics nearly always enters into the selection of one of these plastics for a particular application, but if the potential toughness of cellulosics is to be realized, the materials must be processed correctly. Correct processing involves heating the plastic sufficiently for it to flow freely (*not* forcing half-melted plastic through a die or into a mold) and cooling it slowly. Fast cooling causes the outside of the finished product to harden while the inside is still molten. When the inside cools and contracts, powerful stresses form within the plastic, and these frozen-in stresses can detract very significantly from the toughness of the plastic. Sprues, runners, trim, and other scrap from molding and extrusion operations, if kept clean,

REPRESENTATIVE PROPERTIES OF CELLULOSE ESTER PLASTICS

	PROPERTY, UNIT [special conditions of test][a]	ASTM TEST METHOD	ACETATE		BUTYRATE			PROPIONATE	
			H2 Flow	M Flow	H2 Flow	M Flow	S2 Flow	H4 Flow	H Flow
Injection-Molded Specimens ⅛-inch thick (0.32 centimeter)	Flow Temperature, °F (°C)	D 569	320(160)	293(145)	320(160)	293(145)	266(130)	338(170)	311(155)
	Specific Gravity	D 792	1.28	1.27	1.21	1.19	1.17	1.22	1.20
	Hardness, Rockwell, R Scale	D 785	85	55	96	63	—	99	51
	Tensile Strength at Yield, psi	D 638	4,800	3,400	5,400	3,800	2,250	5,200	3,600
	Tensile Strength at Fracture, psi at 73°F (23°C) at 158°F (70°)	D 638	6,300 3,200	4,700 1,950	6,300 3,900	5,050 2,300	3,600 1,350	6,100 4,000	4,500 2,400
	Flexural Strength at Yield, psi	D 790	7,500	4,800	7,750	5,400	3,200	6,800	4,400
	Stiffness in Flexure, 10^5 psi	D 474	1.90	1.45	1.60	1.20	0.80	1.70	1.20
	Impact Strength, Izod, ft-lb/in. of notch at 73°F (23°C) at −40°F (−40°C)	D 256 D 758	3.5 1.1	5.1 1.2	4.2 1.2	6.5 1.6	9.2 2.5	2.7 1.1	9.5 1.4
	Deformation Under Load, % [24 hours at 122°F (50°)] at 1000 psi at 2000 psi	D 621	1 3	2 23	1 2	3 14	25 45	1.0 1.0	2.0 21
	Water Absorption, % [24-hour immersion] Soluble Matter Lost, %	D 570	2.3 0.2	2.3 0.6	1.8 0.1	1.5 0.1	1.3 0.1	2.0 0.1	1.9 0.1
	Weight Loss on Heating, % [72 hours at 180°F (82°C)]	See Note	1.2	3.6	0.3	0.9	2.0	0.3	1.2
Compression-Molded Specimens ½-inch thick (1.27 centimeters)	Hardness, Rockwell, R Scale	D 785	101	75	109	89	—	107	83
	Impact Strength, Izod, ft-lb/in. of notch at 73°F (23°C) at −40°F (−40°C)	D 256 D 758	1.6 0.5	2.5 0.6	1.4 0.6	2.5 0.9	4 5 1.3	1 7 0.6	4.0 0.8
	Compressive Strength at Yield, psi	D 695	7,200	4,900	7,800	5,500	3,400	6,600	4,400
	Deflection Temperature, °F (°C) at 264-psi fiber stress at 66-psi fiber stress	D 648	151(66) 171(77)	128(53) 144(62)	173(78) 199(93)	143(62) 165(74)	123(51) 140(60)	172(78) 200(93)	136(58) 167(75)

[a] All tests run at 73°F (23°C) and 50 percent relative humidity unless otherwise specified.
NOTE: D 706 for Acetate; D 707 for Butyrate; D 1562 for Propionate.
Degrees Celsius are approximate.

can be reground, mixed with new feed stock, and reused. Butyrate and propionate are sometimes compatible with each other, but acetate is not compatible with either of the others and must not be allowed to contaminate them. Synthetic polymers must be rigorously excluded from all cellulosics.

Injection Molding. Flows of acetate and butyrate most commonly used for injection molding are MS, M, and MH; flows of propionate, H and H_2. Melt temperatures are generally in the 350 to 500°F (177 to 260°C) range; pressures, between 10,000 and 30,000 psi (680 and 2040 atmospheres) on the plastic. Molding shrinkage varies from about 0.003 to 0.008 inch per inch (~0.08 to 0.2 millimeter per millimeter). Molding cycles can be as low as about 15 seconds for thin sections, increasing with the thickness of the molded item.

Extrusion. Extruders can process higher-melting materials than injection molding machines, and the average flow of cellulosic plastic for extrusion is 2 or 3 flows harder than the average flow used for most injection molding. Extrusion temperatures are ordinarily between about 335 and 450°F (~168 and 232°C). Extruded products are generally sized by building the die slightly oversize and drafting the extrudate, but heavy sections should

not be drafted more than a few percent. Excessive draft will cause stresses to form in the plastic, which can cause the product to become brittle.

Blow Molding. Processing of cellulosics by the normal extruded-parison method can be done, but it requires special techniques and is not widely practiced. Considerable success has been achieved, however, with cutting extruded butyrate pipe into short lengths, reheating, and blow molding.

Thermoforming. Thin sheeting of all three plastics is widely thermoformed for blister packaging. Heavy sheeting of special outdoor-type formulations, principally of butyrate, is formed into outdoor signs. Excellent detail can be obtained. Drawdown ratios as high as 3:1 are practical with adequate draft in the mold. Molds can be relatively inexpensive.

Cementing. Any of the three organic cellulose ester plastics can be cemented to itself, and propionate and butyrate generally can be cemented to each other using solvents. A cement produced by dissolving (in an appropriate solvent) a small amount of the plastic to be cemented will ordinarily produce a better bond. Solvents most commonly used are mixtures of an active solvent, such as acetone or methylethyl ketone, a latent solvent, such as alcohol, and a diluent, such as benzene. Most solvents are flammable and toxic and must be handled accordingly.

R.P. Rich, Plastics Division, Eastman Chemical Products, Inc., Kingsport, Tennessee.

CEMENT. Cement is a finely powdered substance which possesses strong adhesive powers when combined with water. Gypsum plaster (see **Calcium**), common lime, hydraulic limes, Puzzolan, natural and Portland cements are a few of the materials which are used for cementing purposes.

Portland cement, which is the most important of these materials since it is a basic ingredient of concrete, was first manufactured in England in the early part of the nineteenth century. It derived its name from the fact that this newly discovered cement resembled a building stone that was quarried near Portland, England.

There are three fundamental stages in the process of manufacture of Portland cement, namely, (1) preparation of the raw mixture, (2) production of the clinker, (3) preparation of the cement. Whether the process used is wet or dry, the raw materials are selected, analyzed, and mixed so that, after treatment, the product, or clinker, has a desired, narrowly specified composition. A factory analysis of slurry, where the wet process is in use, is as follows: calcium oxide 44%, aluminum oxide 3.5%, silicon oxide 14.5%, ferric oxide 3%, magnesium oxide 1.6%, loss on ignition about 33% (largely carbon dioxide), showing that the composition of the resulting burned clinker is essentially a calcium aluminosilicate. The system calcium oxide-aluminum oxide-silicon oxide has been determined by Rankin and co-workers. In some places the composition of the rock is practically of the desired composition, and in other places clay and limestone are mixed in the desired proportions.

The raw mixture is heated in a continuously operated, long, almost horizontal, slowly rotated furnace or kiln at a high temperature. The temperature is regulated so that the product consists of sintered but not fused lumps. This is clinker. Too low a temperature causes insufficient sintering, and too high a temperature results in a molten mass or glass, the product in either of these cases being valueless for cement purposes. Clinker is unaffected by water, and may be stored indefinitely without detriment. In 1824, Joseph Aspin, an English bricklayer, took out a patent for the manufacture of an improved cement, which he called Portland cement. The cement thus produced was not what is known now as Portland cement as the temperature of burning was not sufficiently high. The value of burning at a temperature sufficiently high to cause incipient fusion was soon afterwards discovered.

In order to obtain the desired setting qualities in the finished cement, there is added to the clinker about 2% of gypsum (calcium sulfate, $CaSO_4 \cdot 2H_2O$), and the mixture is pulverized very finely. For every ton of Portland cement shipped, over two and one-half tons of raw materials *and* cement clinker must be ground very finely. See Table 1.

The wet process for making cement is shown in the accompanying flowsheet. Selective quarry of raw materials for specific cement requirements is based upon chemical data. A television receiver, mounted on the quarry control panel, oversees these operations. Rocks sent to the crushing plant are reduced to pieces smaller than 2 inches. These are dropped onto a reversible shuttle conveyor that stockpiles the materials according to chemical composition in the raw materials storage located above a reclaiming tunnel. From the stockpiles, vibrating feeders then withdraw specified amounts of raw materials and discharge them over a 1,230-foot (375-meter) belt conveyor that passes through

TABLE 1. ANALYSIS OF MATERIALS USED FOR MANUFACTURE OF LIME AND CEMENT

MATERIAL	SiO_2	Fe_2O_3 Al_2O_3	CaO	MgO	CO_2	SO_3	USED FOR
Limestone	0.36	0.45	54.45	0.54	43.24		Portland cement
Limestone	3.30	1.30	52.15	1.58	40.98		,, ,,
Limestone	0.74	0.13	52.94	1.87	43.68		,, ,,
Marl	1.78	1.21	49.55	1.30	40.35		,, ,,
Cement rock	13.44	6.60	41.84	1.94	32.94		,, ,,
Cement rock	11.11	6.31	42.51	2.89	36.57		,, ,,
Clay	61.09	26.97	2.51	0.65	—	1.42	,, ,,
Clay	58.44	26.50	1.70	1.88	—		,, ,,
Cement rock	15.37	11.38	25.50	12.35	34.20		Natural cement
Limestone	0.89	0.47	54.68	0.32	43.44		Lime
Limestone	0.78	0.48	31.15	20.78	45.76		,,
Oyster shells	3.30	0.25	52.14	0.25	41.61		,,

the reclaiming tunnel. A television camera at the tunnel entrance observes the materials on their way to the next crushing stage. A second television camera supervises the screening operation, and a third camera observes the unloading of crushed material into the storage silos.

Raw materials, withdrawn from the silos, are conveyed to the raw-grinding mill, and water is added. A sonic system which "listens" to the sound of steel balls impacting the inside of the mill indicates the amount and fineness of the material being ground. Feedback from this system controls the feed rate to the mill. Raw-mill discharge (a slurry of about two-thirds solids content) is screened over a 50-mesh screen cloth and pumped to two slurry basins where it is homogenized. Each basin (44 feet high × 85 feet diameter) (13.2 meters high × 25.5 meters diameter) holds about a 3-day supply for the kiln.

The rotary kiln is 510 feet (153 meters) long and is fired by oil. Maximum daily output of the kiln is 1,760 tons (1,584 metric tons) of clinker, equivalent to 9,700 barrels of cement. Operations are automatically controlled, and two television cameras, one at each end of the kiln, observe all material flow.

In its downward path through the kiln, the slurry passes first into a 91-foot long drying zone (maintained at 2,000°F) (1,093°C) and then into a hotter (above 2,800°F) (1,538°C) calcining and burning section. The drying zone is fitted with steel chains to improve fuel efficiency through better heat transfer. Feed rate of the oil is controlled by the gas temperature in the calcining zone. Thermocouples placed at intervals inside the kiln measure the temperature of solid particles and kiln gases. The critical parameters are relayed to the plant central control room.

The initial quarrying, primary, and secondary crushing and screening of raw materials are not shown in the flowsheet. See also **Gypsum.**

When Portland cement is mixed with water, the product sets in a few hours and hardens over a period of weeks. The initial setting is caused by the interaction of water and tricalcium aluminate $3CaO \cdot Al_2O_3$, present in the cement, accompanied by the separation of gelatinous hydrated product. The later hardening and the development of cohesive strength are due to the interaction of water and tricalcium silicate $3CaO \cdot SiO_2$, also present in the cement, accompanied by the separation of gelatinous hydrated product. In each case the gelatinous material surrounds and cements together the individual grains. The hydration of dicalcium silicate $2CaO \cdot SiO_2$, also present in the cement, proceeds still more slowly than that of the above compounds. The ultimate cement agent is probably gelatinous hydrated silica SiO_2. See also **Concrete.** The analyses of some typical Portland cements are given in Table 2.

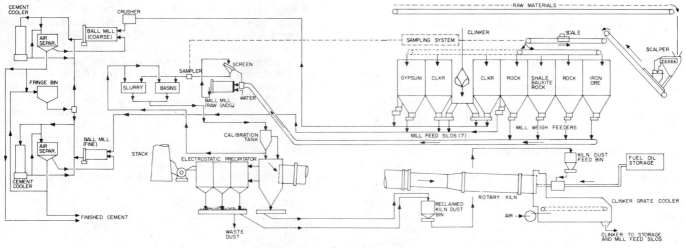

Wet process for making cement.

TABLE 2. ANALYSIS OF PORTLAND CEMENTS[a]

Where Made	Made from	SiO_2	Fe_2O_3	Al_2O_3	CaO	MgO	SO_3	Loss
New Jersey	Cement rock and	21.82	2.51	8.03	62.19	2.71	1.02	1.05
Pennsylvania	limestone	21.94	2.37	6.87	60.25	2.78	1.38	3.55
Michigan	Marl and clay	22.71	3.54	6.71	62.18	1.12	1.21	1.58
Ohio		21.86	2.45	5.91	63.09	1.16	1.59	2.98
Virginia		21.31	2.81	6.54	63.01	2.71	1.42	2.01
Missouri	Limestone and clay	23.12	2.49	6.18	63.47	0.88	1.34	1.81
Pennsylvania[b]		23.56	0.30	5.68	64.12	1.54	1.50	2.92
Illinois		22.41	2.51	8.12	62.01	1.68	1.40	1.02
Germany	Blast furnace slag	20.48	3.88	7.28	64.03	1.76	2.46	
Belgium	limestone	23.87	2.27	6.91	64.49	1.04	0.88	
France		22.30	3.50	8.50	62.80	0.45	0.70	
England		19.75	5.01	7.48	61.39	1.28	0.96	
Germany[c]	Iron ore and limestone	20.5	11.0	1.5	63.5	1.5	1.0	

[a] From Meade's "Portland cement."
[b] White Portland cement.
[c] Seawater cement.

Deductions regarding the mechanism of setting and hardening, and the identity of the substances concerned are the results of extensive studies, involving the use of the microscope in the examination of thin sections, on the individual compounds, the clinker, and the resulting concrete. Elaborate researches have also been conducted to determine the best way of incorporating the ingredients of concrete, the nature of the aggregate (sand, gravel, crushed rock) to be used, and the proportions of cement, water, and aggregate, in order that the resulting concrete, really an artificial rock, shall possess the greatest possible strength.

Special applications of cements and lutes are frequently demanded, as for example in floor covering, in tank lining, and in the closure of joints. The difference between a cement and a lute is that the former sets to a rigid solid mass whereas the latter retains some plasticity so that some movement of the lute is possible without cracking. A lute must have support in order that it be retained in position.

A somewhat crude though convenient classification can be made on the basis of the principal ingredients, thus, (1) Portland cement, (2) high alumina cement, (3) sodium silicate, (4) magnesium oxychloride plus copper powder, (5) litharge or red lead plus glycerol, (6) rubber latex, and (7) synthetic resins. Supplementary materials to be considered are asbestos, white lead, plaster of Paris, sulfur, graphite, sand, pitch, tar, rosin, and boiled linseed oil.

The choice to be made depends upon the kind of material to which the cement or lute must adhere; what it must withstand in the way of acid, base, sulfate, or organic liquid; also what temperature is involved; and finally the matter of resistance to vibration and shock.

Portland and high alumina cements do not withstand acids but are resistant to bases. High alumina cement attains its maximum strength more quickly than Portland, and has the extra advantage that it withstands solutions of sulfates.

Sodium silicate cement does not withstand bases, but is resistant to acids except hydrofluoric. This cement sets to a very rigid solid, so that when subjected to mechanical shock or to temperature change it is liable to crack.

A cement containing 90% magnesium oxychloride and 10% copper powder is strong, resistant to abrasion, and can be bonded to Portland cement.

Litharge or red lead plus glycerol has been used, especially for pipe joints.

Rubber latex cement withstands silute acids and dilute bases, and adheres well to ceramic materials such as stoneware. This cement remains somewhat pliable, thus resisting mechanical shock and temperature change. Organic liquids in general attack this cement.

Synthetic resin cements withstand hydrochloric acid, dilute nitric acid, dilute sulfuric acid, and dilute bases, and are frequently more resistant to organic liquids than is rubber latex cement. The adherence to ceramic materials is good, and the liability to cracking less than for sodium silicate cement.

As for the miscellaneous ingredients mentioned, some are used as fillers or extenders as in the case of sand, and some are used in their own right as when pitch, tar, rosin, molten sulfur, or packed asbestos can be used. See also **Adhesives.**

CENTRIFUGING. A separation technique based upon the application of centrifugal force to a mixture or suspension of materials of closely similar densities. The smaller the difference in density, the greater is the force required. The equipment used (centrifuge) is a chamber revolving at high speed to impart a force up to 17,000 times that of gravity (much higher in the *ultracentrifuge*). The materials of higher density are thrown toward the outer portion of the chamber, while those of lower density are concentrated at or near the inner portion. A common cream separator is a type of centrifuge in which the flow is continuous. The centrifugal force throws the heavier milk into a different chamber from the lighter cream. Many hundreds of products, particularly in the chemical and food processing industries, require centrifugation at some stage in their manufacture. For example, important applications are found in the beet sugar and sugar cane industries, in the processing of corn (maize) products, in soybean processing, and in milk and dairy products manufacture.

Basic Principle of Centrifuging. The force acting on a particle within a centrifugal field is defined by Newton's fundamental force equation, $F = ma$. Acceleration acting upon the particle, directed toward the center of rotation is $a = r\omega^2$. Therefore the centrifugal force acting on the particle is $F = mr\omega^2$, or, expressed as multiples of gravity,

$$F = 14.2 \times 10^{-6} DN^2 (D \text{ in inches})$$
$$= 5.59 \times 10^{-6} DN^2 (D \text{ in centimeters})$$

where m = mass of particle, g
a = acceleration, cm/s^2
r = radial distance of a particle in a centrifugal field from axis of rotation, cm
ω = angular velocity, rad/s
D = inner diameter of centrifugal bowl
N = bowl speed, rpm

A particle and a mixture introduced, confined, and rotated within a circular enclosure accelerates as it moves from a neutral center toward the maximum diameter (inner periphery) of the enclosure. Thus, if a mixture is introduced to the center of a 24-inch (61-centimeter) diameter solid-bowl centrifuge rotating at 1500 rpm, the particle will be caused to move at a speed of 32.2 feet (9.8 meters) per second at the center. At the maximum diameter, the particle will have a terminal velocity of 766.8 × 32.2 feet per second, or 24,690 feet (7525 meters) per second. Essentially, the separation occurs at 766.8 × g.

Industrial centrifuges are available in a variety of designs.

Bowl sizes range from less than 0.5 to over 4 feet (15.2 to 122 centimeters) in diameter. Forces applied generally range from about 500 to 14,200 times the force of gravity. Operating temperatures range up to 260°C or higher; pressures up to about 8.5 atmospheres. Liquid flow rates range from 5 to 500 gallons (19 to 1893 liters) per minute; solid rates from 0.2 to 68 metric tons per hour. Depending upon design, the range of size of particles which can be handled is from 1 micrometer to as large as 0.6 centimeter. Centrifuges often are equipped with filters to provide a highly clarified effluent. The majority of industrial centrifuges are designed for continuous operation, although batch designs are obtainable.

One of several types of industrial designs is shown in the accompanying figure. This is a continuous horizontal, pusher

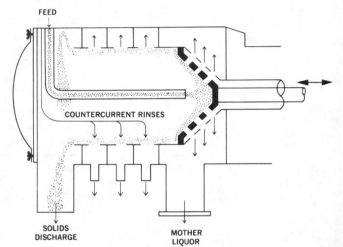

Continuous horizontal, pusher (reciprocating) filter centrifuge. Double arrows at right indicate motion of ram.

(reciprocating) filter centrifuge. The feed is continuously introduced through a stationary pipe. The solids settle to the screen surface and form a filter bed through which the liquid passes. The liquid continues through the basket and is discharged from the machine, while the solids, having formed a uniform filter-bed thickness in the cylinder basket, are assisted to discharge by a pusher ram. Washing of cake solids, following the initial dewatering, is a feature of this design.

Among other types are the gas centrifuges which have been used for the separation of isotopes. In the concurrent type, one or more streams of gas enters at one end of the centrifuge and the partially-separated isotopes are removed in two or more streams at the other end. In the countercurrent type, countercurrent circulation is established in a centrifuge either thermally or mechanically. By the circulation of the gas the radial concentration gradient is converted into an axial gradient. The evaporation type operates on volatile liquids, which evaporate within the apparatus. Two streams of vapor are removed from a point near the axis of the centrifuge, having been separated by diffusion through the centrifugal field.

Ultracentrifuge. Centrifuges which operate at very high speeds and which find applications in colloid chemistry and biochemical research are sometimes called *ultracentrifuges*. In research laboratories where proteins, polymers, and other substances with high molecular weights are studied, the ultracentrifuge is effectively used. The sedimentation of large molecules in a strong centrifugal field enables the determination of both average molecular weights and the distribution of molecular weights in various systems. When a solution containing polymer or other large molecules is centrifuged at forces up to 250,000 times gravity, the molecules begin to settle, leaving pure solvent above a boundary which progressively moves toward the bottom of the cell. This boundary is a rather sharp gradient of concentrations for molecules of uniform size, such as globular proteins. For polydisperse systems, the boundary is diffuse, the lowest molecular weights lagging behind the larger molecules. An optical system can be provided for viewing this boundary, and a study as a function of time of centrifuging yields the rate of sedimentation for the single component, or for each of many components of a polydisperse system. These sedimentation rates may then be related to the corresponding molecular weights of the species present after the diffusion coefficients for each species are determined by independent experiments. Both the sedimentation and diffusion rates are affected by interactions between molecules, so that each must be studied as a function of concentration and extrapolated to infinite dilution, as is done for the colligative properties. The result of this detailed work is the distribution of molecular weights in the sample, which is available by few other methods. Extrapolation of diffusion coefficients to infinite dilution is difficult for high-molecular-weight linear polymers, and so alternate means are used to relate sedimentation constants to molecular weights in these important applications.

For research applications, ultracentrifuges may be equipped with microprocessors for programming, as well as automated handling of samples.

References

Charm, S. E.: "Fundamentals of Food Engineering," 3rd edition, AVI, Westport, Connecticut, 1978.
Hall, C. W., and D. C. Davis: "Processing Equipment for Agricultural Products," AVI, Westport, Connecticut, 1979.
Crosby, H. L.: "Suppressing Centrifugal Chatter," *Chem. Eng. Prog.,* 75 (10), 92–94, 1979.
Staff: "Water Removal Processes: Drying and Concentration of Foods and Other Materials," Amer. Inst. of Chem. Eng., New York, 1977.
Staff: "Food, Pharmaceutical and Bioengineering," *Pubn. S–172,* Amer. Inst. of Chem. Eng., New York, 1978.

CERAMICS. Derived from the Greek word *keramos* ("burnt stuff"), ceramics comprise a wide variety of materials which constitute a major industry. The principal facets of the ceramic industry, in order of increasing value of annual production, are: (1) abrasives; (2) porcelain enamel coatings; (3) refractories; (4) whitewares; (5) structural clay products; (6) electronic and technical ceramic products; and (7) glass. Glass accounts for about 45% of all ceramics produced. See also **Glass.**

Porcelain enamels are used to protect and decorate steel and aluminum metals. Actually, they are glasses especially designed to have high thermal expansions to match the base metal and to mature (to become glassy) at temperatures low enough to prevent distortion of the underlying metal sheet. Glazes perform a similar function on ceramic substrates, again, as special glasses matching the thermal expansion of the base and maturing at the desired temperature. Although not relatively large in terms of production value, the preparation of ceramic composites is important and growing. This area covers a variety of combinations, such as sapphire, Al_2O_3, whiskers in metals, metal-bonded carbides used in the machine tool industry, and directionally solidified two-phase ceramic systems. In a system of this type, an oriented fibrous second phase is grown in a primary matrix phase to maximize in a selected direction various important characteristics, such as minimum long-term, high-temperature creep. An example of use is in gas turbine rotor blades.

Conventional Ceramics (Structurals). The essential raw material is clay. Clay is essentially a hydrated compound of aluminum and silicon $H_2Al_2Si_2O_9$, containing more or less foreign matter, such as (1) ferric oxide Fe_2O_3, which contributes the reddish color frequently associated with clay; (2) silica SiO_2 as sand; (3) calcium carbonate $CaCO_3$ as limestone. Since clay is formed by the decomposition of igneous rocks, followed by transportation of the fine particles by running water and later deposition of these particles by sedimentation when the flow of water diminishes in speed, the quality of clays shows a wide range. When clay is wet, it is plastic and can be shaped according to the desire and skill of the operator. The shape is retained on drying, and subsequent heating produces a coherent, hard mass, which suffers in the process more or less shrinkage and deformation depending upon the composition of the raw materials, and the method and temperature of treatment. Common bricks are made of crude materials without careful regulation of the conditions of treatment.

Bricks and plain clay products possess an earthy surface and fracture, are porous, and the strength depends upon the materials and treatment. Porcelain, on the other hand, possesses a glasslike or vitreous surface and fracture, and is not porous. Porcelain is made by mixing with the clay some powdered feldspar mineral, potassium aluminosilicate ($KAlSi_3O_8$ approximately). At the temperature of firing, feldspar undergoes a gradual change from the crystalline to the glassy state, the rate depending upon the time of heating and the temperature to which it is subjected. The fusion point of feldspar is of the order of 1300°C, whereas that of kaolin (pure clay) is of the order of 1700°C. Subjection of the porcelain raw material to the latter temperature would result in the formation of a glass. But when the temperature used is below the melting point of the clay portion and about the melting point of the feldspar, the latter produces a glass cement which binds together the particles of the former. When ground quartz SiO_2 is added to the original clay mixture the shrinkage of the material in the processes of drying and firing is reduced, the resistance to deformation during firing is increased, and the temperature coefficient of expansion of the product is affected.

The range of clay, feldspar, and quartz, as to the ratios in the mixture and as to individual composition of each (see accompanying table), as well as the available range of temperature of firing makes possible the production of products of a wide variety of physical structure. There has been proposed an arbitrary line of demarcation, namely that the unglazed product, such as has been described, which absorbs not more than 1% of its weight upon and after immersion in water, shall be termed porcelain, otherwise it shall be called earthenware. Such as nonporous material as porcelain, which includes chinaware, is also distinctly translucent in thicknesses of a few millimeters, whereas earthenware is nontranslucent and somewhat porous.

Materials that are to be glazed are dipped in a slip (the mixture of raw materials and water), dried, and refired. The glaze mixture is made up so that its fusion temperature is lower than that of the body of the ware, and the firing temperature is such that a surface of glass is formed over the body of the ware.

Designs and colors may be placed, as is commonly done, on the glaze and refired, or, as less commonly and more recently with fine effect, directly on the body under the glaze, in which case the glaze when produced covers and protects both the body of the ware and the decoration.

The properties of ceramics depend mainly on how the atoms are arranged and the interatomic bonds they form. The most important bonding force in the crystalline phases in most ceramics is ionic bonding, the metallic atoms losing an outer electron

ANALYSES OF CERTAIN CERAMIC RAW MATERIALS

	SILICA	ALUMINA	LIME	MAGNESIA	IRON OXIDES	SODA	POTASH
	%	%	%	%	%	%	%
Kaolin	58	29	0.2	0.3	1	1	1
Fire clay	61	26	0.3	0.4	1	1	1
Common brick clay	58	14	7	1.5	4	3	3
Feldspar	71	16	0.3	0.0	0.5	4	7
Quartz	100	—	—	—	—	—	—

to become positive ions; and the nonmetallic atoms gaining an outer electron to form a negative ion. Ionic crystals are brittle and hard, melt at high temperatures, and have low electrical conductivity at room temperature. Compounds of metals with oxygen ions that are largely ionic are MgO, Al_2O_3, and ZrO_2. Covalent bonding is also found in ceramic crystalline materials. In this case, a pair of electrons is shared by two atoms. Covalent crystals, such as diamond and silicon carbide, have high hardness, high melting points, and low electrical conductivities at low temperatures.

The basic building block for the silicate crystal structures is the silicon–oxygen tetrahedron with a silicon atom at the center and four oxygen atoms at the corners. The silicates are classed by the types of bonding existing between the tetrahedra in their crystal structures. In neosilicates, the tetrahedra are independent of each other. These structures make good refractories because of their high melting points. This group includes the olivine minerals, garnets, zircon, kyanite, and mullite. When the tetrahedra are joined at only one corner (oxygen atom), they form sorosilicates, which are rare. In cyclosilicates and inosilicates, the tetrahedra share two corners to form a variety of ring or chain structures. Minerals of this type include the pyroxines, such as spodumene, and the amphiboles, such as asbestos. Sharing three corners, the tetrahedra form phyllosilicates, which exist as sheets or planes, forming such minerals as mica. In the various forms of silica, such as quartz and crystobalite, all four tetrahedron corners are shared.

Most ceramic shapes do not consist of one single crystal, but are composed of numerous crystals joined together to form polycrystalline structures. The characteristics of the grain boundaries between crystals can influence the strength, chemical stability, and electrical properties as much as do the crystalline structures within the individual grains.

Glass is a very important part of ceramics. The glass industry is the largest single element of the entire ceramic industry, and the glassy portions of many ceramic bodies are the bond that hold many ceramics together. Probably the majority of the ceramics produced are a mixture of crystalline grains and a glassy phase. The glass frequently acts as the bond. This is the basis of the vitrified-grinding wheel industry and much of the structural and whiteware branches of the ceramics industry.

The rare-earth elements have found application in the ceramics field. In one example, a mixture of about 90% yttrium oxide powder and 10% thorium oxide powder is pressed into the desired shape and then sintered at about 2200°C. This heat treatment removes the microscopically small pores from between the powder particles. The result is a single-phase, polycrystalline material with a grain size normally between 10 and 50 micrometers in diameter. Yttrium oxide has a cubic crystal structure, thus light is not scattered at grain boundaries. This property, combined with the absence of a second phase and pores, imparts exceptional transparency (with polishing) to visible and infrared light. The transmission cutoff in the ultraviolet range occurs at 0.24 micrometers and in the infrared range at about 9 mi-

crometers. Although the index of refraction of the ceramic is high, about 1.91 at the sodium D line wavelength, the optical dispersion of the material is very low. See also **Rare-Earth Elements and Metals.**

CERIUM. Chemical element symbol Ce, at. no. 58, at. wt. 140.12, first in the Lanthanide Series in the periodic table, mp. 798°C, bp 3433°C, density 6.770 g/cm^3 (20°C). Elemental cerium has a face-centered cubic crystal structure at 25°C. Cerium is the most abundant element of the rare-earth group and is 28th in ranking of the naturally occurring elements in the earth's crust. The element is a silver-gray metal which oxidizes readily at room temperature, particularly in moist air, to form the oxide CeO_2, which is of a pale yellowish-green color. Above 300°C, the element may ignite and burn with a bright red glow. Of the nineteen isotopes of cerium, only four occur in nature, ^{136}Ce, ^{138}Ce, ^{140}Ce, and ^{142}Ce. The thermal neutron-absorption cross section of the element is low. The element has a low toxicity rating. Electronic configuration is $1s^22s^2 2p^63s^23p^63d^{10}4s^24p^64d^{10}4f^15s^25p^65d^16s^2$. Ionic radius Ce^{3+} 1.034 Å, Ce^{4+} 0.92 Å. Metallic radius 1.825 Å. First ionization potential 5.47 eV; second 10.85 eV.

Other important physical properties of cerium are given under **Rare-Earth Elements and Metals.**

Cerium was first identified by M. H. Klaproth in 1803 and, independently, in the same year by J. J. Berzelius and W. Hisinger. The element occurs in four source minerals, allanite, bastnasite, serite, and monazite. Bastnasite, which is a rare-earth fluorocarbonate, is found in southern California. Monazite, a phosphate that contains thorium and the light lanthanides, is distributed widely throughout the world. See also **Bastnasite; Monazite.** Cerium is recovered from the minerals through an extractive process using H_2SO_4, followed by precipitation with oxalic acid which separates the light lanthanides from thorium, yttrium, and the heavy lanthanides. Cerium metal is produced from its salts, such as CeF_3 or $CeCl_3$, by thermal reduction in a tantalum or molybdenum crucible. Alternative processes include the electrolysis of $CeCl_3$ or CeO_2. The latter compounds are soluble in a complex molten halide flux. The Ce^{3+} is reduced to metal at a molybdenum electrode. The process is carried out at from 800 to 1,000°C.

A major use for CeO_2 is in decolorizing soda-lime container glass. The compound also is used for polishing gemstones and glass, notably precision optical glasses. Cerium is particularly useful in glass that is subject to α-, γ-, and x-radiation, and the impingement of light and electrons because the cerium prevents discoloration that may arise from the presence of Fe(II) by oxidizing the Fe(II) as it is formed to Fe(III). This is an important factor in color television tubes. Cerium dioxide also is used in cathodes, capacitors, phosphors, ceramic coatings, refractory oxides, semiconductors, and photochromic glasses. The compound also is used as a catalyst and as an opacifying agent in porcelain enamels. Because of its low nuclear cross section, CeO_2 may be applied as a diluent in oxide nuclear fuels.

Cerium metal finds wide application in *mischmetal*, which is a rare-earth metal comprised of 50% Ce, 25% La, 18% Nd, 5% Pr, and 2% other rare earths. This alloy is used in shell linings for military projectiles, as an alloying agent for improving the malleability of ductile iron, and in lighter "flints" where the alloy is compounded with a 30% iron alloy. The pyrophoric and incendiary nature of cerium are evident when cerium-base alloys are machined. Mischmetal also improves the creep resistance of magnesium alloys, the resistance to oxidation of nickel alloys, the hardness of copper alloys, and the strength of aluminum alloys. Both cerium metal and mischmetal are used as *getters* to remove traces of oxygen in vacuum tubes and equipment. When alloyed with cobalt, cerium is gaining importance as a magnet material. $CeCo_5$, as a permanent magnet material, has properties which exceed those of the alnicos and ferrites. Mixed rare-earth oxides and fluorides containing up to 50% cerium are used as cores for carbon arcs which, for illuminating purposes, have much greater intensity and color balance. The mixed oxides with cerium also are used as catalysts (petroleum cracking and chemical oxidation reactions) and in a variety of waterproofing agents, fungicides, and polishing materials.

See references listed at ends of entries on **Chemical Elements**; and **Rare-Earth Elements and Metals**.

NOTE: This 4th Edition entry was revised and updated by K. A. Gschneidner, Jr., Director, and B. Evans, Assistant Chemist, Rare-Earth Information Center, Energy and Mineral Resources Research Institute, Iowa State University, Ames, Iowa.

CERMET.   Chromium; Nuclear Reactor.

CERUSSITE. The mineral cerussite, lead carbonate, $PbCO_3$, is orthorhombic with tabular, prismatic and pyramidal crystals, with twinned forms very common. If not in crystal aggregates it may occur in granular or compact masses. Cerussite is very brittle with a conchoidal fracture; hardness 3–3.5; specific gravity, 6.55 (a heavy mineral); luster, adamantine but may be vitreous to resinous, pearly, or even submetallic. Its color is variable, white to gray, grayish-black or blue or green, transparent to translucent.

Cerussite is of secondary origin being found associated with other lead minerals, and is widely distributed. There are many European and American localities. Fine crystals have been obtained from Phoenixville, Pennsylvania; Joplin, Missouri; Leadville, Colorado; Pima County, Arizona, and Dona Ana County, New Mexico. It is an ore of lead, and frequently carries values of silver. Derived from the Latin *cerussa*, white lead.

CESIUM. Chemical element symbol Cs, at. no. 55, at. wt. 132.905, periodic table group 1a, mp 28.40°C, bp 678°C, density, 1.88 g/cm³ (20°C). Elemental cesium has a body-centered cubic crystal structure. Cesium is a silvery-white, very soft metal, one of the softest of all metals. The element tarnishes instantly upon exposure to air, soon igniting spontaneously with flame to form the oxide. Generally, the element is preserved under kerosene. Cesium reacts vigorously with H_2O, forming cesium hydroxide and hydrogen gas. In these respects, the element behaves much like sodium. The element was first identified by Bunsen and Kirchhoff in 1860 through spectroscopic observations. Cesium occurs in nature as the ^{133}Cs isotope. There are 15 radioactive isotopes, ^{125}Cs through ^{132}Cs and ^{134}Cs through ^{139}Cs. The half-life of ^{137}Cs is 33 years. This isotope is used as a source of gamma radiation, particularly in radiography

and therapy, but also in as number of industrial applications, such as level and continuous weight measurement. See also **Radioactivity.** First ionization potential, 3.89 eV; second, 23.4 eV. Oxidation potential $Cs \rightarrow Cs^+ + e^-$, 3.02 V. Other important physical characteristics of cesium are given under **Chemical Elements.**

The main source of cesium is carnallite $KCl \cdot MgCl_2 \cdot 6H_2O$ which contains a small percentage of cesium compounds. See also **Carnallite.** Cesium also occurs in pollucite (cesium aluminosilicate, 35% Cs_2O) and lepidolite (lithium aluminosilicate). See also **Lepidolite; Pollucite.** In early processes, cesium metal was obtained by the reduction of cesium salts, such as the hydroxide or chloride. In current practice, the metal is produced by electrolyzing the cyanide. The latter compound usually is fused cesium barium cyanide mixture.

The uses for cesium and its compounds are limited. Cesium is used in photoelectric devices because of its high sensitivity to light, finding applications in television, motion picture, radar, and instrumentation equipment. Cesium also has been used in luminescent tubes and screens. Certain processes for the manufacture of synthetic resins, such as chloroprene, use cesium as a catalyst. Some interest has been indicated in cesium as a fuel for ion-propulsion engines of low thrust for spacecraft. Like sodium, cesium also has been considered as a heat-transfer medium for special applications. The function of cesium in time measurement is important. As officially defined in 1967 by the International Bureau of Weights and Measures, the atomic second is equivalent to 9,192,631,770 oscillations of the atom of ^{133}Cs. This value expresses the ephemeris time (ET) second as closely as practical in terms of an atomic standard. To derive this value, scientists at Great Britain's National Physical Laboratory and the United States Naval Observatory used a dual-rate moon-position camera and a cesium-beam clock.

Cesium forms several solid solutions with rubidium. These alloys are used as *getters* for eliminating residual gases from vacuum tubes and systems. Because of their extreme reactivity in air, the alloys are difficult to apply. For easier handling, cesium can be alloyed with calcium, barium, or strontium. The ternary alloys of cesium, aluminum, and barium or strontium are employed in photoelectric cells. Cesium alloyed with antimony, silver, bismuth, and gold also displays photoelectric properties.

Chemistry and Compounds. Cesium is more electropositive than rubidium (or the lower alkali metals), as is consistent with its position in group 1a.

Because of the ease of removal of its single $6s$ electron (3.89 eV) and the difficulty of removing a second electron (23.4 eV) cesium is monovalent in its compounds, which are ionic.

In its solutions in liquid NH_3, cesium is like the other alkali metals, a powerful reducing agent, so that in such solutions, titrations of cesium polysulfide with cesium are made by electrometric methods. The solubility of cesium salts in liquid NH_3 increases markedly with the radius of an anion (the chloride, CsCl, 0.0227 moles per kg, the bromide, CsBr, 0.215 moles per kg, and the iodide, CsI, 5.84 moles per kg), though the values are less than for the corresponding rubidium compounds.

As in the case of the other alkali metals, cesium forms compounds generally with the inorganic and organic anions. For a general discussion of these compounds (see also **Sodium**) because the sodium compounds differ principally in their greater extent of hydration and greater number of hydrates. However, cesium coordinates with large organic molecules, such as salicylaldehyde, even though it does not with H_2O.

One respect in which cesium (and rubidium) are outstanding among the alkali metals is the readiness with which it forms alums. Cesium alums are known for all of the trivalent cations that form alums, Al^{3+}, Cr^{3+}, Fe^{3+}, Mn^{3+}, V^{3+}, Ti^{3+}, Co^{3+}, Ga^{3+}, Rh^{3+}, Ir^{3+}, and In^{3+}.

As in the case of potassium and rubidium, cesium forms a superoxide on reaction of the metal with oxygen. The compound is orange in color and paramagnetic because it contains the O_2^- ion with an odd electron in an antibonding orbital, and has the formula CsO_2. On heating, this compound loses oxygen to form black Cs_2O_3, which contains both CsO_2 and Cs_2O_2 (peroxide), which is the product of further heating. A series of suboxides of cesium is known, Cs_7O, Cs_4O (uncertain), Cs_7O_2, Cs_3O, and Cs_2O. Moreover the normal oxide, Cs_2O, can be prepared by heating cesium nitrite with metallic cesium. It reacts explosively with oxygen to form CsO_2.

Cesium hydroxide, $CsOH$, is the strongest of the five alkali metal hydroxides, as would be expected from its position in the periodic table (francium hydroxide, when prepared, would be expected to be stronger). For the same reason, it has the lowest lattice energy of the five (135.6 kcal per mole).

The most numerous organic compounds of cesium are the oxygen-connected ones, such as the salts of organic acids, and the alkoxy and aryloxy compounds (alcoholates, phenates, etc.). Among the carbon-connected compounds, an ethyl cesium, CsC_2H_5, and a phenyl cesium, CsC_6H_5, have been reported.

Rogowski and Tamura (1970) studied the environmental chemistry of ^{137}Cs. Later studies by other investigators (Alberts et al., 1979) have shown that ^{137}Cs introduced into a watershed is attached to soil particles, which are removed by erosion and runoff. Some of the eroded soil particles comprise the sediments of the catchment basins in the watersheds and act as "sinks" for ^{137}Cs. Other investigators have reported an almost irreversible fixation of this element in clay interlattice sites in freshwater environments and that it is unlikely that this nuclide will be removed from these sediments under normal environmental conditions other than by exposure to solutions of high ionic strength, such as may occur in estuarine environments. Studies of ^{137}Cs have been important because the element can be introduced into a water system from a leak in a nuclear fuel element. These findings are reported in some detail by Alberts et al. in *Science*, **203**, 649–651 (1979).

See list of references at end of entry on **Chemical Elements**.

CETANE NUMBER.   Petroleum.

CHABAZITE.   The mineral chabazite is a member of that group of hydrous silicates, the zeolites, and corresponds to the formula $CaAl_2Si_4O_{12} \cdot 6H_2O$ with sodium sometime replacing a part of the calcium. Potassium, barium, and strontium may be present in very small amounts. Chabazite is hexagonal, usually in rhombohedrons that tend to resemble cubes. It has a rhombohedral clevage; is brittle; hardness 4–5; specific gravity 2.05–2.10; luster vitreous; color white to flesh-red; streak white; translucent to transparent. Chabazite is found in the amygdaloidal cavities of basalts often associated with other zeolites. It is occasionally found in such crystalline rocks as syenites, gneisses and schists. Chabazite is a rather common zeolite, being found in many localities in Europe. In the United States it occurs in the Triassic traps of New Jersey and Maryland. The Triassic lavas of Nova Scotia have yielded fine specimens. The name chabazite is derived from the Greek word meaning a precious stone.

CHAIN REACTION (Free Radical).   Free Radical.

CHALCANTHITE.   This mineral, of triclinic crystallization, is found only as a rare secondary mineral in the oxidized zones of sulfide copper ores within arid regions. It is a hydrous copper sulfate, $CuSO_4 \cdot 5H_2O$, and is a most unstable mineral in moist atmospheric environments, altering readily to a powder-blue dust. The mineral possesses a vitreous luster of deep azure-blue color, ranging from transparent to translucent. Chalcanthite is found in abundance only in Chuquicamata and other arid regions of Chile, where it is an important copper ore.

CHALCEDONY.   One of the cryptocrystalline varieties of the mineral quartz, having a waxy luster. It may be semitransparent or translucent and is usually white to gray or grayish-blue or some shade of brown, sometimes nearly black. Light colored clear red chalcedony is known as carnelian; deep reddish brown as sardonyx; a green variety colored by nickel oxide is called chrysoprase. Prase is a dull green. Plasma is a bright to emerald-green chalcedony which sometimes is found with small spots of jasper resembling blood drops; it is then referred to as bloodstone or heliotrope.

Chalcedony and agate are essentially porous, which permits their being dyed various colors by artificial means. Red color is produced by iron nitrate solution; nickel nitrate produces vivid green color; ammonium bichromate produces blue-green; and ferrocyanide salts a vivid blue. The black onyx used extensively in rings is a product of soaking chalcedony in sugar solutions and later in sulfuric acid.

The term chalcedony is derived from the Greek word meaning Chalkedon, a town in Asia Minor.

CHALCOCITE.   This mineral is cuprous sulfide, Cu_2S, crystallizing in the orthorhombic system, often in pseudo-hexagonal forms. Above a temperature of 91°C, chalcocite changes into an isometric form. It has conchoidal fracture; hardness, 2.5–3; specific gravity, 5.5–5.8; metallic luster; color dark gray to blackish-gray, frequently with bluish-green tarnish. Chalcocite is of widespread occurrence and a valuable copper ore. It seems in some cases to be definitely secondary in origin, in other cases primary. It may have been formed from bornite by the action of alkaline solutions. It sometimes carries valuable amounts of silver.

Among the many European localities might be mentioned Cornwall, England, the Ural Mountains, and Rumania. It occurs also in the Congo, South West Africa, Peru, Mexico, and Alaska. In the United States it is found at Bristol, Connecticut, in fine crystals, Montana, Tennessee, Arizona, Nevada, and California.

The word chalcocite is derived from the Greek word meaning copper. Chalcocite also is known as *copper glance*.

CHALCOPYRITE.   The mineral chalcopyrite (also known as copper pyrites) is a sulfide of copper and iron corresponding to the formula $CuFeS_2$. Its tetragonal crystals are often complex with repeated twinning; massive chalcopyrite is common. It has an uneven fracture; is brittle; hardness, 3.5–4; specific gravity, 4.1–4.3; luster, metallic; color, brass-yellow, may be iridescent from tarnish; streak, greenish-black; opaque. Chalcopyrite is the most common copper-bearing mineral known and it is the most important ore of copper. It is a primary mineral in many igneous rocks and from it a host of secondary copper minerals have been derived.

Among the many localities where fine specimens of this mineral have been obtained might be mentioned: Freiburg, Saxony; Alsace: Rio Tinto, Spain; Cornwall, England; Australia; Chile, Peru, and Bolivia, South America; and in the United States, Ellenville, New York; Chester County, Pennsylvania; Joplin, Missouri; Gilpin County, Colorado; Arizona, Montana, Utah,

Nevada, California, New Mexico and Tennessee. In Canada there are notable deposits of chalcopyrite in the Provinces of British Columbia, Ontario, and Quebec. The name chalcopyrite is derived from the Greek word meaning copper, and the word pyrites.

CHALK. Chalk is a soft, porous limestone of white, grayish-white or buff color made up of the minute shells of foraminifera and fragments of cocospheres. It occurs extensively in England and France and less so in the United States.

Chalk consists almost entirely of calcite which has formed principally by shallow-water accumulation of (1) calcareous tests of floating microorganisms and (2) comminuted remains of calcareous algae. The most widely distributed chalks are of Cretaceous age, as exemplified by the cliffs on both sides of the English Channel. Although an unaltered deposit, chalk masses may contain noddules of chert and pyrite.

CHARACTERISTIC EQUATION. A class of equations connecting those variables, such as temperature, pressure, and volume, which define the physical condition of a given substance (variables of state).

The ideal gas law and the Boyle-Charles law represent approximately the behavior of all gases, but if accuracy is required, some modification of these must be sought which will take into account the differences between individual gases. The best known characteristic equation for gases is that of van der Waals. Using the same notation as for the ideal gas law, this may be written

$$\left(p + \frac{n^2 a}{v^2}\right)(v - nb) = nRT$$

where n is the number of moles of gas, and a and b are constants characteristic of the gas in question. They are very small; if they were zero we should have the ideal gas law. Following are their approximate values for certain gases, where a is expressed in atmosphere (liter/gram-mole)^2 and b in liter/gram-mole:

Gas	a	b
Ammonia	4.170	0.03707
Helium	0.034	0.03412
Hydrogen	0.244	0.02661
Nitrogen	1.390	0.03913
Oxygen	1.360	0.03183

CHARLES LAW. Although the coefficients of expansion of different solids or of different liquids are notably different, the coefficients of expansion of all gases are nearly the same, namely, about $\frac{1}{273}$ of the volume at 0°C per centigrade degree. The law, stated by Charles in 1787 and independently by Gay-Lussac in 1802 (hence sometimes called Gay-Lussac's law) is not strictly true. Regnault obtained the following values of the volume coefficient for various gases:

Air	0.0036706
Hydrogen	0.0036613
Carbon dioxide	0.0037099
Sulfur dioxide	0.0039028
Carbon monoxide	0.0036688
Nitrous oxide	0.0037195
Cyanogen	0.0038767

None of these is far from $\frac{1}{273}$ = 0.003663, which is therefore

commonly taken as the expansion coefficient for gases; especially as the value for hydrogen, commonly used in the standard gas thermometer, is very near it. If the pressure as well as the volume is allowed to vary, the behavior of the ideal gas must be expressed by the Boyle-Charles law or the ideal gas law; and the behavior of a real gas by one of the other equations of state. See **Ideal Gas Law.**

CHARNOCKITE. Charnockite is a granular variety of hypersthene granite which was first described from the gravestone of Job Charnock, who founded the city of Calcutta, India whence the derivation of the name charnockite.

CHELATION COMPOUNDS. These are coordination compounds in which a single ligand occupies more than one coordination position. Such ligands are called *chelating agents* (the word derived from the Greek, meaning *crab's claw*). Thus, ethylenediamine $H_2N—CH_2—CH_2—NH_2$, abbreviated as *en*, forms a $Cr(en)_3^{3+}$ ion having three molecules of ethylenediamine, each occupying two coordination positions, as shown below.

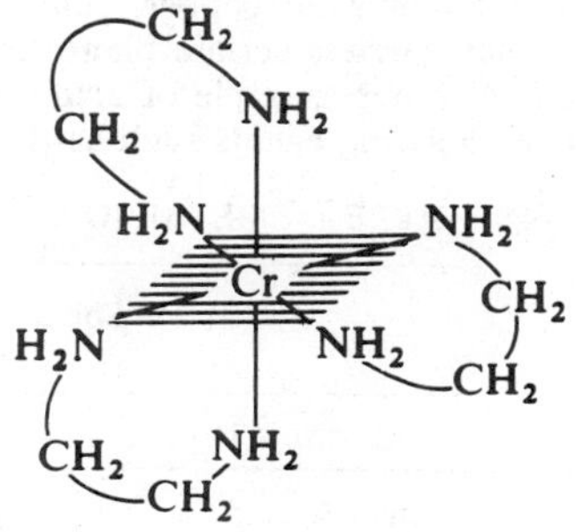

Ethylenediamine is therefore called a bidentate group, as are many other ligands, such as the β-diketones, which form chelation compounds of the type where M is a metal ion:

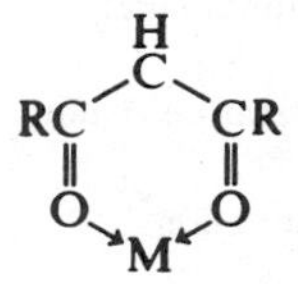

Although bidentate ligands are more common, there are polydentate ligands which occupy more than two coordination positions; ethylenediaminetetraacetic acid is such a polydentate ligand.

While ethylenediamine, and many other chelating agents, form only covalent bonds, there are others which attach by both covalent and ionic bonds. Thus glycine forms with cupric ions (Cu^{2+}) the compound copper bisaminoacetate:

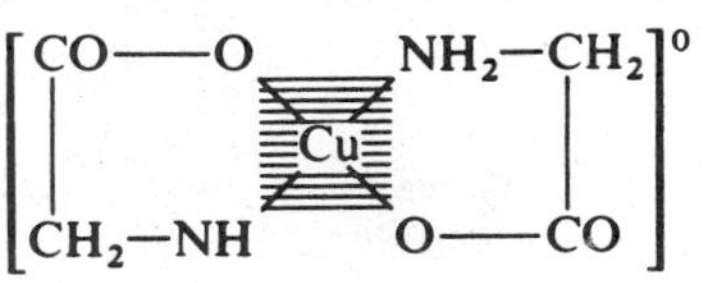

A number of synthetic chelating agents have been developed. They are substances like ethylenediaminetetraacetic acid (EDTA) and N-hydroxyethylethylenediaminetriacetic acid (HEDTA) and their salts, usually sodium salts. Many of these compounds and mixtures of these compounds are sold under trademarks.

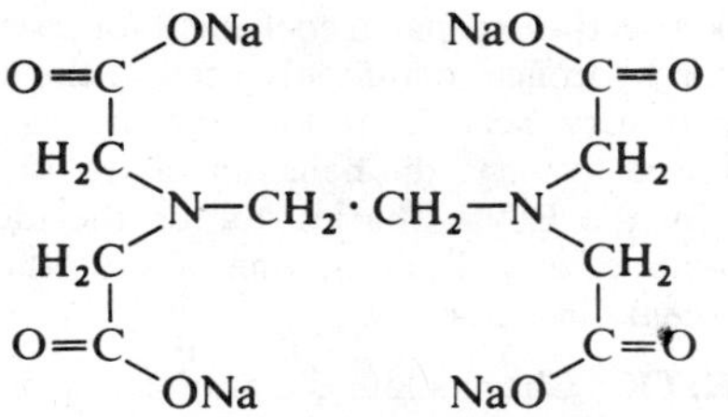

Tetrasodium Ethylenediaminetetraacetate (Tetrasodium EDTA)

Chelating agents are being used in increasing amounts for a number of important purposes. These uses may be put into two important categories: first, artificial trace metal carriers and, second, sequestering agents.

As artificial carriers for trace metals, chelating agents can be used as aids in agriculture by supplying the metals for soils which are deficient in their trace metal content. Both EDTA and HEDTA are adequate iron carriers and EDTA can be used as the carrier for bivalent copper, zinc, manganese, and cobalt. By use of such carriers certain plant deficiency diseases can be controlled. Another example of artificial carrier use is the employment of chelating agents such as the EDTA deriva-

tives or mixtures of such derivatives with pyrophosphates or a mixture of EDTA and the sodium salt of N,N-di(2-hydroxy-ethyl)glycine in controlling polymerization reactions in synthetic rubber manufacture by the controlled release of trace metal catalysts.

As sequestering agents, chelating compounds have a wide variety of uses, for instance, for water softening in both soaps and synthetic detergents; in textile processing as in kier boiling operations where iron, copper, zinc, etc., ions are inactivated so that discoloration of cloth is prevented; in the stabilization of hydrogen peroxide; in boiler and heat exchanger cleaning.

Chelation in Biological Systems. Chemical reactions in biological systems are usually mediated by selective catalysts called *enzymes.* The high efficiencies and stereospecificities achieved require that enzymes have definite and characteristic geometries, whereby specific functional groups coordinated to the metal ion are held in definite spatial positions relative to each other and relative to the substances on which they exert their catalytic effects. The incorporation of metal ions into enzyme structures can assist in the maintaining of a definite geometrical relationship between ionic and polar groups, through the geometric requirements of the coordinate bonds of the metal ion. Certain metal ions may also participate in the catalytic properties of enzymes through ionic and coordinate bonding between the

TABLE 1. METAL ION AND METAL CHELATE CATALYSIS OF CHEMICAL REACTIONS

Solvolysis and Other Reactions Involving Acid Catalysis
by the Metal Ion

REACTION TYPE	SUBSTRATE	CATALYST
Solvolysis	Amino acid esters, peptides, and amides	Cu^{2+}, Co^{2+}, Mn^{2+}
	Phosphate esters	La^{3+}, Cu^{2+}, VO^{2+}
	Fluorophosphates	Cu^{2+}, UO_2^{2+} diamine-Cu(II) complexes
	Polyphosphates	Ca^{2+}, Mg^{2+}
	Schiff bases	Cu^{2+}, Ni^{2+}
Transamination	Schiff bases of pyridoxal and α-amino acids	Fe^{3+}, Cu^{2+}, Al^{3+}, Zn^{2+}, Ni^{2+}, Co^{2+}
Decarboxylation	α-Keto polycarboxylic acids (e.g., oxalacetic and oxalsuccinic acids)	Cu^{2+}, Zn^{2+}, Ni^{2+}, Co^{2+} Mn^{2+}, Fe^{2+}
Acylation	Acetylacetone	Co(III), Rh(III) or Cr(III) chelates of acetylacetone

Catalysis of Oxidation Reactions by Electron Exchange
with Metal Ions or Metal Complexes

REACTION	SUBSTRATE	METAL ION OR COMPLEX
Oxidation by molecular O_2	Ascorbic acid, catechols, quinoline, salacylic acid	Fe(III), Fe(III)-EDTA, Cu(II), Cu(II)-EDTA, V(IV)
Oxidation by H_2O_2	Phenol, anisole	Fe(II) (Fenton's reagent), Fe(II)-hydroquinone Fe(II)-EDTA-ascorbic acid
Formation of oxygen	Hydrogen peroxide	Fe^{3+}, Fe(III)-phthalocyanine chelate
Formation of disulfides from mercaptides	Thioglycolic acid	Fe^{3+}, Cu^{2+}

metal ion and electron donating groups of the enzyme and substrate, and through the ability of the metal ion to initiate oxidation-reduction reactions. Because of these chemical and steric effects, coordinated metal ions in the complex compounds that

catalyze biological reactions frequently are found. See also **Metalloproteins.**

Most of the metal ions that have biological functions have a coordination number of six, with the donor groups arranged in an octahedral fashion. There are a few metals, such as Mg^{2+} and Zn^{2+}, that frequently coordinate only four donor groups tetrahedrally, and Cu^{2+}, which has four coordinations directed to the corners of a square plane with the metal ion at the center of the plane.

Many simple acid-base reactions are catalyzed by both metal

ing enzyme. This high activity of the enzyme is ascribed to the special environment of the substrate around the active site of the enzyme, through which additional binding of the substrate by adjacent organic groups of the enzyme takes place.

The enzyme aconitase, which contains the Fe^{2+} ion at the reactive center, catalyzes the interconversion of citric, isocitric,

TABLE 2. BIOLOGICALLY ACTIVE METAL CHELATES

METAL	METALLOENZYME	OTHER BIOLOGICAL FUNCTIONS
Mg	Polynucleotide phosphorylase, ATPase, choline acylase, deoxyribonuclease, acetate kinase, adenosine phosphokinase, fructokinase, glyceric kinase, hexokinase	Chlorophyll
Ca	α-Amylase, aldehyde dehydrogenase, lipase	
V		Green algae, blood of marine worm (ascidian)
Cr		Glucose tolerance factor
Mn	Arginase, carnosinase, prolinase, enolase, isocitricdehydrogenase, 3-phosphoglycerate kinase, glucose-1-P kinase	
Fe	Aconitase, formic hydrogenylase, phenylalanine hydroxylase, peroxidase, catalase, cytochromes	Hemoglobin, ferritin, hemosiderin, siderophilin
Co	Aspartase, acetylornithinase	Vitamin B_{12}
Cu	Lactase, phenolase, tyrosinase, uricase	Ceruloplasmin, cytochrome
Zn	Carbonic anhydrase, carboxypeptidase, alcohol dehydrogenase, glutamic dehydrogenase, acylase	
Mo	Nitrate reductase, xanthine oxidase	

ions and hydrogen ions. Because of small size, the electronic interaction of the hydrogen ion with a substrate is much greater than that of a metal ion. The latter, however, has properties not possessed by hydrogen ions, which are useful in catalysis, i.e., the ability to coordinate a large number of electron donor groups simultaneously, the specific geometric orientation of the coordinate bonds of certain metal ions, and the ability of metal ions to undergo oxidation-reduction reactions. Many of these reactions are models of the more complex catalytic effects that occur in biological systems. Since these reactions of simple coordination compounds aid in the understanding of biological reactions, a few of the more common examples are given in Table 1.

The function of the metal ions in the reactions listed is to attract electrons from the substrate. When this effect takes the form of simple polarization of the functional groups of the substrate, charge variations and electron shifts in these groups facilitate the chemical reactions listed under solvolysis and acid catalysts. When the metal ion removes completely one or more electrons from the substrate, the first step in an oxidation reaction occurs. This type of catalysis can be accomplished only by metals capable of existing in more than one valence state.

There is a saturation effect in the coordination of a metal ion by donor groups of both the enzyme and the substrate. Therefore, one would expect that the interaction of a free metal ion with the substrate would be greater than that of the metalloenzyme (in which the metal is already partially coordinated). If this were true, the metal ion would have a greater catalytic effect than the metalloenzyme. The reverse is always the case; thus far, no metal ions, or metal complex enzyme models, have been found to approach the catalytic activities of the correspond-

and aconitic acids. The reaction has been shown to occur through the formation of a single intermediate carbonium ion structure in which the Fe^{2+} ion is always bound to the same donor atoms, while the interconversion of the substrate occurs through the migration of only protons and electrons.

Some of the more important biological reactions that are catalyzed by metal ions are summarized in Table 2.

CHEMICAL AFFINITY. The entropy production due to a chemical reaction has the form

$$\frac{d_i S}{dt} = \frac{1}{T} A v \geq 0 \qquad (1)$$

where A is the chemical affinity and v, the reaction rate. A is related to the characteristic functions U, H, A, G, and to the chemical potentials μ by the relations:

$$A = -\left(\frac{\partial U}{\partial \xi}\right)_{S,V} = -\left(\frac{\partial H}{\partial \xi}\right)_{S,p}$$

$$= -\left(\frac{\partial A}{\partial \xi}\right)_{T,V} = -\left(\frac{\partial G}{\partial \xi}\right)_{T,p} \qquad (2)$$

$$= -\sum_i v_i \mu_i$$

when ξ is the extent of reaction and v_i the stoichiometric coefficient.

The basic properties of the affinity A are that it is always of the same sign as the reaction rate, and that if the affinity is zero the reaction rate is also zero, i.e., the system is in equilibrium.

This definition of affinity is essentially due to De Donder and is called De Donder's fundamental inequality. In the notation used by G. N. Lewis and his school, it is supposed that ξ increases by unity, therefore the relations of (2) are written in the form:

$$\mathbf{A} = -(\Delta U)_{S,V} = -(\Delta H)_{S,p} = -(\Delta A)_{T,V} = -(\Delta G)_{T,p}. \quad (3)$$

Note that in this entry, $\mathbf{A}$ is the affinity and A, the Helmholtz function (work function).

See also **Chemical Reaction Rate.**

CHEMICAL COMPOSITION. Matter is composed of the chemical elements, which may be in the free or elementary state, or in combination. In the former case, as exemplified by iron, tin, lead, sulfur, iodine, and the rare gases, matter commonly exhibits the properties of the atoms of the particular element, including the chemical properties whereby they combine to form molecules. Molecules may (1) be monoatomic; (2) they may consist of atoms of one element only, such as nitrogen or hydrogen molecules (N_2 or H_2), (3) they may be composed of atoms of more than one element, called compounds, which usually have distinctive properties.

The molecular formulas of gaseous compounds are obtained from a study of the composition by elements and the density,

by a method introduced by the Italian chemist, Cannizzaro, in 1858. Later, in 1872, in the course of his Faraday Lecture before the Chemical Society (London) on the subject "Some Points in the Teaching of Chemistry" Cannizzaro stated that "Symbols and formulas, in my opinion, constitute the introduction, preparation, and base of the study of the transformations of matter, which is the true object of our science." The simplest way to understand the method is to arrange in tabular form (1) the individual gases, (2) the weight in grams of 1 liter (at 0°C, 760 millimeters of mercury pressure) of each gas, (3) the weight in grams of *each element* present in the above volume (1 standard liter) found by exact analysis (percentage composition by chemical elements using the methods of analytical chemistry). See Table 1.

Careful examination of the figures in the last six columns reveals the experimental fact that (1) in each separate vertical column the figures represent a minimum weight or a small multiple (approximately) of this weight, (2) the smallest of the six minimum weights is that for hydrogen, namely, 0.045 gram in 1 standard liter of hydrogen gas.

The next step involves changing 0.045 gram of hydrogen to exactly 1.000 gram and finding arithmetically the volume of hydrogen chloride containing this weight (1.000 gram hydrogen). The volume is found to be 22.2 standard liters.

Therefore, 1.000 gram minimum weight of hydrogen is contained in 22.2 standard liters of hydrogen chloride.

Using this standard volume of 22.2 liters, the next step is

TABLE 1. CANNIZZARO METHOD OF COMPOUND COMPUTATION

Gas	Grams per Standard Liter	Percentage Composition by Chemical Elements	Grams per Standard Liter by Chemical Elements					
			Hydrogen	Oxygen	Carbon	Nitrogen	Sulfur	Chlorine
1. Hydrogen chloride	1.639	Hydrogen 2.76% / Chlorine 97.24	0.045					1.594
2. Ammonia	0.771	Hydrogen 17.75 / Nitrogen 82.25	0.137			0.634		
3. Carbon dioxide	1.977	Oxygen 72.73 / Carbon 27.27		1.438	0.539			
4. Carbon monoxide	1.250	Oxygen 57.14 / Carbon 42.86		0.714	9.536			
5. Methane	0.717	Hydrogen 25.14 / Carbon 74.86	0.180		0.537			
6. Ethylene	1.260	Hydrogen 14.38 / Carbon 85.62	0.181		1.079			
7. Acetylene	1.173	Hydrogen 7.75 / Carbon 92.25	0.091		1.082			
8. Oxygen	1.429	Oxygen 100.00		1.429				
9. Hydrogen	0.090	Hydrogen 100.00	0.090					
10. Nitrogen	1.251	Nitrogen 100.00				1.251		
11. Chlorine	3.214	Chlorine 100.00						3.214
12. Sulfur dioxide	2.927	Oxygen 49.95 / Sulfur 50.05		1.462			1.465	
13. Hydrogen sulfide	1.539	Hydrogen 5.91 / Sulfur 94.09	0.091				1.448	
14. Nitrous oxide	1.978	Oxygen 36.35 / Nitrogen 63.65		0.719		1.259		
15. Nitric oxide	1.340	Oxygen 53.32 / Nitrogen 46.68		0.715		0.625		
Minimum weight (approximate)			0.045	0.715	0.538	0.626	1.45	1.60

NOTE: Data are displayed in this table to illustrate the Cannizzaro method of arriving at the symbol and symbol weight of chemical elements; and the formula and formula weight of chemical compounds.

to ascertain the minimum weight of the other elements in this volume.

Chemical Element	Approximate Minimum Weight in Grams of Each of the Six Chemical Elements in the Standard Volume, 22.2 Liters
Hydrogen	1
Oxygen	16
Carbon	12
Nitrogen	14
Sulfur	32
Chlorine	35.5

Then, the abbreviation is introduced by the representation:

SYMBOL WEIGHTS OF EACH ELEMENT BY THE SYMBOLS

1 gram of hydrogen by the symbol H
16 grams of oxygen by the symbol O
12 grams of carbon by the symbol C
14 grams of nitrogen by the symbol N
32 grams of sulfur by the symbol S
35.5 grams of chlorine by the symbol Cl

By setting up again the second half of the table for the 15 gases, this time for 22.2 standard liters instead of 1 standard liter, the results obtained may be observed in Table 2.

Thus, it is seen, the chemical formulas and formula weights (last column) of 15 gaseous chemical compounds have been arrived at, using the Cannizzaro method, by purely experimental and rational means, involving no theoretical considerations. Extension of the method serves to ascertain the chemical formula of all gases and vaporizable substances. For compounds which are neither gases nor vaporizable, other methods are available. Of these the most used are those of Raoult depending upon the depression of the freezing point or the elevation of the boiling point of a compound dissolved in a given solvent.

It remains to be noted that, when there is no method available for ascertaining the formula weight of a compound, the *simplest* formula, based on chemical analysis and the use of symbol weights of the contained elements, is used, e.g., ferric oxide, Fe_2O_3, ferroferric oxide, Fe_3O_4, ferrous oxide, FeO, cupric oxide (black copper oxide), CuO, suprous oxide (red copper oxide), Cu_2O. The customary formula of water is H_2O, which is correct at temperatures above 100°C—actually, liquid water is mainly dihydrol $(H_2O)_2$.

It should be understood from the above discussion that a chemical formula is no chance throwing together of chemical symbols, but represents the results of careful analysis, and the scrutiny and deduction of the most skillful workers in the field. On this score alone, chemical formulas demand the greatest respect in understanding and use.

Symbol weights and atomic weights are used synonymously, as are formula weights and molecular weights. Unless otherwise stated, symbol weights and formula weights are expressed in grams, and the numbers used are those taken from the accepted list of atomic weights. See **Chemical Elements.**

One formula volume of a gas is 22.242 liters. It is necessary to state that actual gases under ordinary conditions show some variation from this value, so that for accurate work the records should be consulted in each case.

Summarizing, the formula "HCl" states that "36.5 grams of hydrogen chloride gas occupies a standard volume of 22.2 liters and is composed of 1 gram of hydrogen element chemically united with 35.5 grams of chlorine element." The reason for the formulas of the simple gases, oxygen, O_2, hydrogen, H_2, nitrogen, N_2, chlorine, Cl_2, is apparent from the general method of deduction. The formula O_2 represents 22.2 liters or 32 grams of oxygen *gas*, whereas O represents 16 grams of oxygen *element* in any substance, or more precisely, 15.9994 grams.

It has become customary in chemical literature to use the formula of a substance as an accepted abbreviation for the name of the substance, especially in cases of frequent repetition.

Up to this point, the discussion in this entry has related to substances which are either elements, or single compounds of

TABLE 2. DERIVATION OF FORMULAS AND FORMULA WEIGHTS OF GASES

Gas Symbol Weight Symbol	In 22.2 Liters						Formula of Gas	Grams of Same Gas in 22.2 Liters
	1 g. H	16 g. O	12 g. C	14 g. N	32 g. S	35.5 g. Cl		
1. Hydrogen chloride	1					1	HCl	36.5
2. Ammonia	3			1			NH_3	17
3. Carbon dioxide		2	1				CO_2	44
4. Carbon monoxide		1	1				CO	28
5. Methane	4		1				CH_4	16
6. Ethylene	4		2				C_2H_4	28
7. Acetylene	2		2				C_2H_2	26
8. Oxygen		2					O_2	32
9. Hydrogen	2						H_2	2
10. Nitrogen				2			N_2	28
11. Chlorine						2	Cl_2	71
12. Sulfur dioxide		2			1		SO_2	64
13. Hydrogen sulfide	2				1		H_2S	34
14. Nitrous oxide		1		2			N_2O	44
15. Nitric oxide		1		1			NO	30

NOTE: Derivation assumes data available on the percentage composition by chemical elements of each gas and the symbols and symbol weights of the elements contained.

elements combined in proportions that can be represented by the ratio of small whole numbers. Such compounds are called *stoichiometric compounds* or *Daltonide compounds* (after the British chemist Dalton). There exist, however, some compounds in which the ratios of the amounts of elements present are not integral. Such compounds are called *nonstoichiometric compounds* or *Berthollide compounds* (after the French chemist Berthollet), and are exemplified by some oxides of the transition elements, by many intermetallic compounds, by the copper sulfide $Cu_{1.7}S$, the copper selenide $Cu_{1.6}Se$ and the cerium hydride $CeH_{2.7}$. Some such compounds vary over a range of composition, depending upon their method of preparation.

In spite of these departures of some compounds from whole number formulas, the fact remains that the great majority of compounds with which the chemist is concerned do contain their constituent elements in integral multiples of their atomic weights. In fact, there is even a further uniformity in the behavior of many of the elements. Thus the great majority of the compounds of the alkali elements (Group 1A in the periodic table) contain equal atomic weight proportions of hydrogen or its equivalent in other elements. Thus the hydrides of this group have compositions corresponding to the formulas LiH, NaH, KH, etc.; the halogen compounds of the group have the compositions, LiF, NaF, KF, LiCl, NaCl, KCl, etc.; while their simple sulfur compounds (since in many of its compounds sulfur combines with two hydrogen equivalents as represented by the formual H_2S) have the compositions Li_2S, Na_2S, K_2S, etc. However, there also exist more complex binary sulfur compounds of these elements which contain higher proportions of sulfur, so that they combine with sulfur in more than one atomic proportion. Thus this relative combining power, which is called valence, has more than one value for many elements, but is still useful in organizing the data of chemistry. It is discussed at length in the entry on valence, and is explained in structural terms in the entry on molecule.

Radicals. In many chemical compounds there are groups of two or more elements that frequently have the properties of or enter into chemical reaction as a unit. Of those which are of outstanding importance the following are cited:

1. Ammonium NH_4— behaves as a unit in ammonium compounds and in some of these compounds is very similar to potassium K— in potassium compounds.

2. Hydroxyl —OH which behaves as a unit in bases (e.g., sodium hydroxide, NaOH), alcohols (e.g., methyl alcohol, CH_3OH), and phenols (e.g., phenol, C_6H_5OH).

3. Anion-groups of acids, their salts and their esters: Sulfate $>SO_4$, sulfite $>SO_3$, nitrate —NO_3, nitrite —NO_2, phosphate →PO_4, perchlorate —ClO_4, chlorate —ClO_3, chlorite —ClO_2, hypochlorite —OCl, carbonate $>CO_3$, formate —CHO_2, acetate —$C_2H_3O_2$, palmitate —$C_{16}H_{31}O_2$, stearate —$C_{18}H_{35}O_2$, oleate —$C_{18}H_{33}O_2$, oxalate $>C_2O_4$, lactate —$C_3H_5O_3$, malate $>C_4H_4O_5$, tartrate $>C_4H_4O_6$, citrate →$C_6H_5O_7$, benzoate —$C_7H_5O_2$, cinnamate —$C_9H_7O_2$, phthalate $>C_8H_4O_4$, salicylate —$C_7H_5O_3$.

4. Alkyl- and aryl-groups of alcohols, phenols, their esters and their alcoholates and phenolates: (a) Alkyl (non-benzenoid)-methyl CH_3—, ethyl C_2H_5—, propyl C_3H_7—, butyl C_4H_8— and similar radicals of alcohols; (b) Aryl (benzenoid)-phenyl C_6H_5—, tolyl C_7H_7—, xylyl C_8H_9—, naphthyl $C_{10}H_7$— and similar radicals of phenols.

5. Acyl-groups of organic acids: acetyl CH_3CO—, benzoyl C_6H_5CO—.

6. Miscellaneous radicals, for example, cacodyl $(CH_3)_2As$—, celebrated on account of the investigations by Bunsen (1838).

All of the above radicals are associated with a corresponding radical or element in a compound. While a radical frequently and rather generally enters into chemical reaction as a unit, it is not implied that this is always so, the stability in each case is characteristic of each radical and each reaction in which it is involved. Thus, ammonium hydroxide NH_4OH yields ammonia gas NH_3 and water H_2O at room temperature; ammonium nitrate NH_4NO_3 is decomposed, upon heating, with the accompanying disruption of both the ammonium and nitrate radicals to yield nitrous oxide N_2O gas and water H_2O.

Radicals enter widely into reactions involving electrolytic dissociation of salts, acids, bases in water solution.

Radicals exist most commonly in combination with atoms or other radicals. However, they can be produced "free," and can so exist for a finite period. Even when it is very short, the radical itself is often of great interest in elucidating reaction mechanisms. The first free radical discovered was triphenylmethyl.

Gomberg, by treating triphenylmethyl chloride in carbon dioxide, with zinc, silver, or mercury, obtained the free radical, triphenylmethyl. On dissolving the colorless solid in organic solvents a yellow solution is obtained, and the reactivity (due to unsaturation) of the yellow solution is marked towards oxygen, dissolved iodine, ether. Triphenylmethyl is present in solution in two forms, (1) monomolecular $(C_6H_5)_3C$ yellow, in equilibrium with (2) dimolecular $((C_6H_5)_3C)_2$ colorless. But tribiphenylmethyl $(C_6H_5—C_6H_4)_3C$ occurs only in the monomolecular form, purple. The action of alkali metals on ketones in some cases produces metallic ketyl (Schlenk, 1913) thus:

$$\begin{matrix} R' \\ \diagdown \\ C{—}ONa, \\ \diagup \\ R'' \end{matrix}$$

which is a free radical, or contains trivalent carbon as does monomolecular triphenylmethyl. Many other free radicals are known. See **Free Radical.**

This entry has dealt with two types of chemical composition—elements and compounds. Many materials, including the great majority of those found in nature, are mixtures of compounds and often elements. Practically all biochemical materials and rocks are complex mixtures. Obviously the first step in the determination of the composition of such substances is their separation into the individual compounds, and elements if any which they contain. See also **Organic Chemistry.**

CHEMICAL ELEMENTS. A chemical element may be defined as a collection of atoms of one type which cannot be decomposed into any simpler units by any chemical transformation, but which may spontaneously change into other types by radioactive processes. A chemical element is a substance that is made up of but one kind of atom. Of the over-100 chemical elements known, only 90 are found in nature. The remaining elements have been produced in nuclear reactors and particle accelerators. Theoretical physicists do not all agree, but some consider that fission-stable nuclei should exist at atomic number 114. Claims thus far have been made for the discovery, isolation, or "creation" of elements up to 106. The element with the highest atomic number officially named and entered into the formal table of atomic weights is lawrencium (Lr) with an atomic number of 103.

Each of the chemical elements is described in a separate alphabetical entry in this encyclopedia.

Some of the principal characteristics of the elements are given in Table 1. The lanthanide series elements are described in fur-

TABLE 1.   PRINCIPAL CHARACTERISTICS OF CHEMICAL ELEMENTS

Name	Symbol	Atomic Number	Atomic Weight	Periodic Group	Valency	Density[b] g/cm³	Melting Point, °C	Boiling Point, °C	Discovery (year)
Actinium	Ac	89	227[a]	3b	3	—	1050	3500	1899
Aluminum	Al	13	26.98	3b	3	2.699	660	2467	1827
Americium	Am	95	243[a]	Actinides	3	—	990–998	2600–2608	1945
Antimony	Sb	51	121.75	5a	3,5	6.68	630.7	1587	Early
Argon	Ar	18	39.948	0	0	1.78	−189.2	−185.7	1894
Arsenic	As	33	74.9216	5a	5, ±3	5.73	613	—	Early
Astatine	At	85	210[a]	7a	—	—	302	337	1940
Barium	Ba	56	137.34	2a	2	3.5	725	1640	1808
Berkelium	Bk	97	247[a]	Actinides	3, 4	—	—	—	1950
Beryllium	Be	4	9.012	2a	2	1.848	1287–1292 ± 5	2970 ± 5	1798
Bismuth	Bi	83	208.981	5a	3, 5	9.8	271.3	1562 ± 3	1753
Boron	B	5	10.81	3a	3	2.35	2300	2550[i]	1808
Bromine	Br	35	79.904	7a	±1, 5	3.12	−7.2	58.8	1826
Cadmium	Cd	48	112.41	2b	2	8.65	321	765	1817
Calcium	Ca	20	40.08	2a	2	1.54	837–841	1484	1808
Californium	Cf	98	251[a]	Actinides	3	—	—	—	1950
Carbon	C	6	12.011	4a	±4, 2	(c)	(c)	4827	Early
Cerium	Ce	58	140.12	Lanthanides	3, 4	6.770	798	3433	1803
Cesium	Cs	55	132.905	1a	1	1.88	28.4	678	1860
Chlorine	Cl	17	35.453	7a	±1, 5, 7	3.214[d]	−101	−34.6	1774
Chromium	Cr	24	51.996	6b	2, 3, 6	7.2	1837–1873	2671–2673	1797
Cobalt	Co	27	58.9332	8	2, 3	8.832	1495	2869–2871	1735
Copper	Cu	29	63.546	1b	1, 2	8.92	1083	2566 ± 0.5	Early
Curium	Cm	96	247[a]	Actinides	3	—	1300–1380	—	1944
Dysprosium	Dy	66	162.50	Lanthanides	3	8.551	1412	2567	1886
Einsteinium	Es	99	254[a]	Actinides	—	—	—	—	1955
Erbium	Er	68	167.26	Lanthanides	3	9.066	1529	2868	1843
Europium	Eu	63	151.96	Lanthanides	2,3	5.244	822	1529	1896
Fermium	Fm	100	257[a]	Actinides	—	—	—	—	1955
Fluorine	F	9	18.9984	7a	−1	1.696[d]	−219.62	−188.1	1771
Francium	Fr	87	223[a]	1a	1	2.4	26.28	676–678	1939
Gadolinium	Gd	64	157.25	Lanthanides	3	7.901	1313	3273	1880
Gallium	Ga	31	69.72	3a	3	5.9	29.78	2403 ± 0.5	1875
Germanium	Ge	32	72.59	4a	4	5.36	937	2830	1886
Gold	Au	79	196.967	1b	1, 3	19.32	1064.43	2805–2809	Early
Hafnium	Hf	72	178.49	4b	4	13.3	2207–2247	4601–4603	1923
Helium	He	2	4.0026	0	0	0.15–0.18	−272.2	−268.93	1895
Holmium	Ho	67	164.93	Lanthanides	3	8.795	1472	2567	1879
Hydrogen	H	1	1.0080	1a	1	0.0899	−259.14	−252.87	1766
Indium	In	49	114.82	3a	3	7.31	156.6	2078–2082	1863
Iodine	I	53	126.9045	7a	−1, 5, 7	4.94	113.5	184.35	1811
Iridium	Ir	77	192.20	8	3, 4, 6	22.42	2410	4130	1803
Iron	Fe	26	55.847	8	2, 3	7.874	1536	2745–2755	Early
Krypton	Kr	36	83.80	0	0	3.4[e]	−156.6	−152.3	1898
Lanthanum	La	57	138.91	3b	3	6.146	918	3464	1839
Lawrencium	Lr	103	257[a]	Actinides	—	—	—	—	1961
Lead	Pb	82	207.19	4a	2,4	11.35	327.5	1740	Early
Lithium	Li	3	6.939	1a	1	0.534	180.54	1315–1319	1817
Lutetium	Lu	71	174.98	Lanthanides	3	9.841	1663	3402	1907
Magnesium	Mg	12	24.312	2a	2	1.74	649	1106–1108	1755
Manganese	Mn	25	54.9380	7b	2, 3, 4, 6, 7	7.3	1241–1247	1962	1774
Mendelevium	Md	101	256[a]	Actinides	—	—	—	—	1955
Mercury	Hg	80	200.59	2b	1, 2	13.546	−38.87	356.58	Early
Molybdenum	Mo	42	95.94	6b	3, 5, 6	9.0–10.2	2610	5560	1778
Neodymium	Nd	60	144.24	Lanthanides	3	7.004	1021	3074	1885
Neon	Ne	10	20.183	0	0	1.204[f]	−248.68	−245.9	1898
Neptunium	Np	93	237.0482	Actinides	3, 4, 5, 6	18.0–20.5	629–631	3900	1940
Nickel	Ni	28	58.71	8	2, 3	8.9	1454–1456	2725–2735	1751
Niobium (Columbium)	Nb	41	92.906	5b	3, 5	8.6	2458–2478	4740–4744	1801
Nitrogen	N	7	14.0067	5a	−3, 2, 5	1.25[d]	−209.86	−195.8	1772
Nobelium	No	102	259[a]	Actinides	—	—	—	—	1957
Osmium	Os	76	190.2	8	4, 6, 8	22.5	2700 ± 5.0	>5300	1803
Oxygen	O	8	15.9994	6a	−2	1.429[d]	−218.4	−182.96	1774
Palladium	Pd	46	106.4	8	2, 4	12.16	1550–1552	3139–3141	1803
Phosphorus	P	15	30.9738	5a	±3, 5	1.82	44.1	280	1669
Platinum	Pt	78	195.09	8	2, 4	21.4	1772	3725–3925	1735
Plutonium	Pu	94	244[a]	Actinides	3, 4, 5, 6	—	—	—	1940
Polonium	Po	84	210[a]	6a	2, 4	9.4	254	.962	1898
Potassium	K	19	39.098	1a	1	0.87	63.7	774	1807
Praseodymium	Pr	59	140.91	Lanthanides	3	6.773	931	3520	1879
Promethium	Pm	61	145[a]	Lanthanides	3	7.264	1042	3000	1947
Protactinium	Pa	91	231.036	Actinides	5	—	—	—	1917
Radium	Ra	88	226.026	2a	2	5	700	1140	1898
Radon	Rn	86	222[a]	0	0	9.72[d]	−71	−61.8	1900

Continued

TABLE 1—*Continued*

NAME	SYMBOL	ATOMIC NUMBER	ATOMIC WEIGHT	PERIODIC GROUP	VALENCY	DENSITY[b] g/cm³	MELTING POINT, °C	BOILING POINT, °C	DISCOVERY (year)
Rhenium	Re	75	186.2	7b	−1,4,7	20.5–21.0	3178–3182	5600–5900	1925
Rhodium	Rh	45	102.905	8	3,4	12.44	1963–1969	3625–3825	1803
Rubidium	Rb	37	85.466	1a	1	1.53	38.9	689	1861
Ruthenium	Ru	44	101.07	8	3, 4, 6, 8	12.1–12.3	2310	3900–4000	1844
Samarium	Sm	62	150.35	Lanthanides	3	7.520	1074	1794	1879
Scandium	Sc	21	44.956	3b	3	2.985	1541	2831	1879
Selenium	Se	34	78.96	6a	−2, 4, 6	4.82	217	684–686	1817
Silicon	Si	14	28.086	4a	4	2.3	1408–1412	2355	1823
Silver	Ag	47	107.868	1b	1	10.49	961.93	2212	Early
Sodium	Na	11	22.9898	1a	1	0.9721	97.82	882.9	1807
Strontium	Sr	38	87.62	2a	2	2.6	770 ± 1.0	1384	1790
Sulfur	S	16	32.064[j]	6a	−2, 4, 6	2.07[k]	112.8	444.7	Early
Tantalum	Ta	73	180.948	5b	5	16.63	2996	5325–5525	1802
Technetium	Tc	43	98.906	7b	7	11.5	2171–2172	4875–4877	1937
Tellurium	Te	52	127.60	6a	−2, 4, 6	6.25	450–452	1387–1393	1782
Terbium	Tb	65	158.92	Lanthanides	3	8.230	1365	3230	1843
Thallium	Tl	81	204.37	3a	1, 3	11.85	303.3	1460 ± 7.0	1861
Thorium	Th	90	232.038	Actinides	4	11.5–11.9	1740–1760	4780–4800	1828
Thulium	Tm	69	168.93	Lanthanides	3	9.321	1545	1950	1879
Tin	Sn	50	118.69	4a	2, 4	7.29[g]	231.97	2270	Early
Titanium	Ti	22	47.90	4b	3, 4	4.507	1650–1670	3290	1791
Tungsten	W	74	183.85	6b	6	19.3	3410	5660	1781
Uranium	U	92	238.03	Actinides	3, 4, 5, 6	19.05	1131–1133	3818	1789
Vanadium	V	23	50.941	5b	2, 4, 5	6.0–6.10	1880–2000	3390–3400	1830
Xenon	Xe	54	131.30	0	0	3.5[h]	−112	−107.1 ± 2.5	1898
Ytterbium	Yb	70	173.04	Lanthanides	2, 3	6.966	819	1196	1878
Yttrium	Y	39	88.9058	3b	3	4.469	1522	3338	1794
Zinc	Zn	30	65.38	2b	2	7.1	419.57	907	1746
Zirconium	Zr	40	91.22	4b	4	6.5	1853 ± 1.0	4376 ± 1.0	1789

[a] Denotes mass number of isotope of longest known half-life (or a better known one for Bk, Cf, Po, Pm and Tc).
[b] Densities for solids at 20°C unless otherwise specified.
[c] For diamond at 20°C. Graphite is 2.25 g/cm³ at 20°C.
[d] Denotes grams/liter at 0°C.
[e] Denotes solid at −273°C.
[f] Denotes density of liquid.
[g] For white tin at 15°C; 5.77 for gray tin at 13°C; 6.97 for liquid tin at melting point.
[h] For liquid at −109°C.
[i] Denotes sublimes.
[j] Atomic weight varies slightly because of naturally occurring isotopes 32, 33, 34, and 36, the total possible variation amounting to ±0.003.
[k] For rhombic sulfur; 1.96 for monoclinic sulfur; 2.046 for amorphous sulfur.
GENERAL NOTE: Atomic weights are believed to have the following uncertainty in specific instances: Br, ±0.002; Cl, ±0.001; Cr, ±0.001; Fe, ±0.003; Ag, ±0.003. For other elements, the last digit given for atomic weight is believed correct and reliable to ±0.5. Values are current as of early 1981. Values for all rare-earth elements furnished by Rare-Earth Information Center, Energy and Mineral Resources Research Institute, Iowa State University, Ames, Iowa.

ther detail under **Rare-Earth Elements and Metals;** the refractory metals are delineated under **Niobium;** the platinum group of metals are summarized in **Platinum and Platinum Group;** the exceptional diversity of carbon for forming compounds is illustrated in the entry on **Organic Chemistry.**

Periodicity. The chemical elements display a periodicity or repeating pattern of physical and chemical properties when they are arranged in order of increasing atomic number. This discovery, generally attributed to Dimitri Mendeleev (1869), although some of the relationships were previously known, led to the development of the Periodic Law and the resulting matrix arrangement known as the Periodic Table. See **Periodic Table of the Elements.**

The first listing of the elements is generally attributed to Lavoisier in 1789. Of the twenty elements then listed, the discovery of five had been the result of research conducted by Scheele of Gothenberg. With the development of nuclear physics and the application of these principles to astronomy and cosmology, in recent years the chemical elements have been viewed from new vantage points, with concentrated attention to the physical and nuclear characteristics as well as the chemical properties. An excellent summary of progress made in the study of the origin of the elements was given by Penzias in a lecture delivered in Stockholm when he received the Nobel Prize in Physics in 1978. See the reference list at end of this entry.

Abundance of the Chemical Elements. Considering the large number of chemical elements, it is interesting to note that insofar as the earth's crust, the oceans, and human knowledge of the cosmos to date are taken into consideration, there is far from a uniform distribution of the elements. The distribution, in fact, is exceedingly unbalanced, with, for example, only nine elements making up 99.25% of the composition of the earth's crust. In terms of abundance, many materials which are often considered quite common make up in reality quite a small percentage of the total materials in the earth's crust. In order of descending occurrence in the crust, all but the scarcest elements are listed in Table 2. The occurrence of elements in seawater is also given where such data are available.

In terms of cosmic abundance of the elements, the situation is quite different, but again the abundance heavily favors a comparatively few of the total number of elements. In 1952, Harold C. Urey made an estimate of the elements in the cosmos, using earlier values of V. M. Goldschmidt and Harrison S. Brown in the calculations. A base figure of 10,000 was established for silicon; the abundance of the other elements was related to that figure. The study showed that hydrogen (3.5×10^8) is the super-

TABLE 2.   ABUNDANCE OF THE CHEMICAL ELEMENTS
(In Grams/Metric Ton)*

Element	Terrestrial Abundance	Occurrence in Seawater	Element	Terrestrial Abundance	Occurrence in Seawater
Oxygen	466,000	850,000	Beryllium	6	—
Silicon	277,200	2.98	Praseodymium	5.53	—
Aluminum	81,300	0.01	Arsenic	5	0.003
Iron	50,000	0.01	Scandium	5	4×10^{-5}
Calcium	36,300	404	Hafnium	4.5	—
Sodium	28,300	10,550	Dysprosium	4.47	—
Potassium	25,900	380	Uranium	4	—
Magnesium	20,900	1,290	Boron	3	4.9
Titanium	4,400	0.001	Thallium	3	—
Hydrogen	1,300	108,200	Ytterbium	2.66	—
Phosphorus	1,180	0.07	Erbium	2.47	—
Manganese	1,000	0.002	Tantalum	2.1	—
Fluorine	900	1.27	Bromine	1.62	66
Sulfur	520	894	Holmium	1.15	—
Carbon	320	27.6	Europium	1.06	—
Chlorine	314	19,050	Antimony	1	0.0005
Rubidium	310	0.121	Terbium	0.91	—
Strontium	300	8.1	Lutetium	0.75	—
Barium	250	0.006	Mercury	0.50	3×10^{-5}
Zirconium	220	—	Iodine	0.30	0.05
Chromium	200	5×10^{-5}	Thulium	0.20	—
Vanadium	150	0.002	Bismuth	0.20	0.0002
Zinc	132	0.01	Cadmium	0.15	6×10^{-5}
Nickel	80	0.0005	Silver	0.10	0.0003
Copper	70	0.003	Indium	0.10	0.02
Tungsten	69	0.0001	Selenium	0.09	4×10^{-5}
Lithium	65	0.2	Argon	0.04	0.595
Nitrogen	46.3	0.51	Palladium	0.01	—
Cerium	46.1	0.00038	Tellurium	0.002	0.00001
Tin	40	0.003	Gold	0.005	4×10^{-6}
Yttrium	28.1	0.0003	Osmium	0.005	—
Niobium (Columbium)	24	5×10^{-6}	Platinum	0.005	—
Neodymium	23.9	—	Ruthenium	0.004	—
Cobalt	23	0.0005	Rhodium	0.001	—
Lanthanum	18.3	0.0003	Iridium	0.001	—
Lead	16	0.003	Neon	7×10^{-5}	0.0003
Gallium	15	3×10^{-5}	Radium	13×10^{-6}	996×10^{-13}
Molybdenum	15	0.01	Krypton	9.8×10^{-6}	0.0003
Thorium	11.5	0.0007	Xenon	1.2×10^{-6}	0.0001
Cesium	7	0.0005	Protactinium	8×10^{-7}	0.003
Germanium	7	6×10^{-5}	Actinium	3×10^{-10}	—
Samarium	6.47	—	Polonium	3×10^{-10}	—
Gadolinium	6.36	—			

* Presented in order of diminishing terrestrial abundance.

abundant element in the cosmos, closely followed by helium (3.5×10^7). Other elements near the top of the list are oxygen (220,000), nitrogen (160,000), carbon (80,000), and neon (between 9000 and 240,000). Magnesium, silicon, and iron are also relatively high.

Researchers at the California Institute of Technology have been studying the occurrence of various isotopes in the earth's crust as a possible lead to understanding the mechanics of continental crust formation. A means for determining the time of formation of new crustal segments is a key to understanding how the continental crust evolved. In this work, they have studied samarium-neodymium and rubidium-strontium isotopic systematics.

Critical Importance of Certain Elements.   The uneven distribution of the chemical elements in the earth's crust causes critical imbalances in their availability for use in manufacturing. It is not unusual for the major consuming nations to be long

distances from the principal suppliers. Further, there are political differences between supplier and consumers that, at times, can upset the normal exchange of trade. See Fig. 1.

There are many "hidden" uses for the elements and their compounds, as exemplified by Table 3, which lists the elements used in a telephone handset.

Nuclides, Isotopes, and Isobars.   A nuclide may be defined as a species of atom with specified atomic number and mass number. The term *nuclide* is used—*not* isotope. Different nuclides having the same atomic number are *isotopes*; different nuclides having the same mass number are *isobars*.

A comprehensive listing of the nonradioactive nuclides of the chemical elements is given in Table 4. The isotopic abundance and mass number are given. In all but a few instances, the isotopes listed are stable. The elements also have a number of radioisotopes each, the number varying considerably from one to the next. Extensive tables listing the radioactive isotopes,

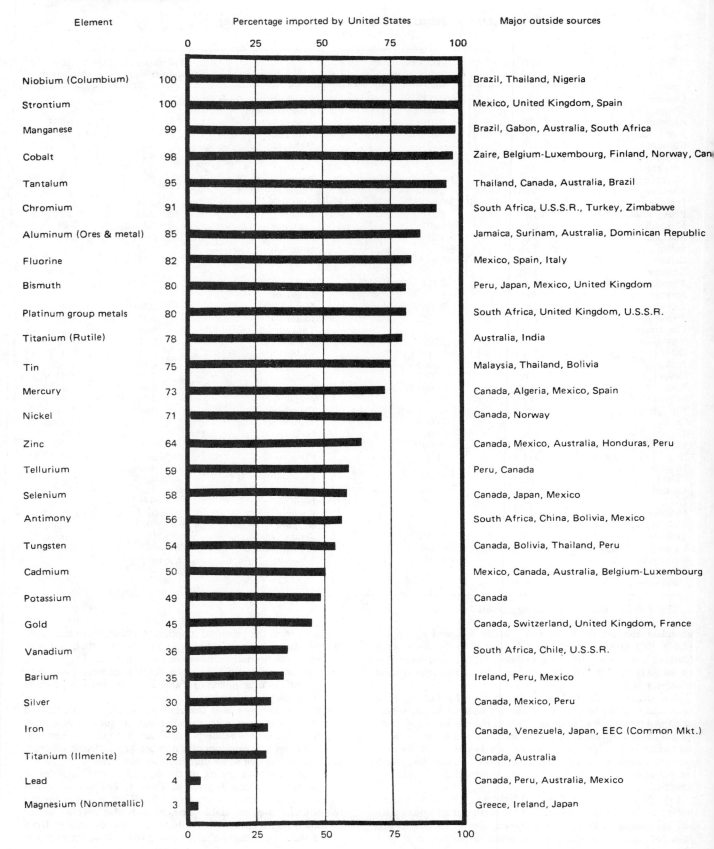

Element	Percentage imported by United States	Major outside sources
Niobium (Columbium)	100	Brazil, Thailand, Nigeria
Strontium	100	Mexico, United Kingdom, Spain
Manganese	99	Brazil, Gabon, Australia, South Africa
Cobalt	98	Zaire, Belgium-Luxembourg, Finland, Norway, Can
Tantalum	95	Thailand, Canada, Australia, Brazil
Chromium	91	South Africa, U.S.S.R., Turkey, Zimbabwe
Aluminum (Ores & metal)	85	Jamaica, Surinam, Australia, Dominican Republic
Fluorine	82	Mexico, Spain, Italy
Bismuth	80	Peru, Japan, Mexico, United Kingdom
Platinum group metals	80	South Africa, United Kingdom, U.S.S.R.
Titanium (Rutile)	78	Australia, India
Tin	75	Malaysia, Thailand, Bolivia
Mercury	73	Canada, Algeria, Mexico, Spain
Nickel	71	Canada, Norway
Zinc	64	Canada, Mexico, Australia, Honduras, Peru
Tellurium	59	Peru, Canada
Selenium	58	Canada, Japan, Mexico
Antimony	56	South Africa, China, Bolivia, Mexico
Tungsten	54	Canada, Bolivia, Thailand, Peru
Cadmium	50	Mexico, Canada, Australia, Belgium-Luxembourg
Potassium	49	Canada
Gold	45	Canada, Switzerland, United Kingdom, France
Vanadium	36	South Africa, Chile, U.S.S.R.
Barium	35	Ireland, Peru, Mexico
Silver	30	Canada, Mexico, Peru
Iron	29	Canada, Venezuela, Japan, EEC (Common Mkt.)
Titanium (Ilmenite)	28	Canada, Australia
Lead	4	Canada, Peru, Australia, Mexico
Magnesium (Nonmetallic)	3	Greece, Ireland, Japan

Fig. 1. Imports of strategic elements (ores) by the United States. (*U.S. Department of the Interior, Washington, D.C.*)

TABLE 3. ELEMENTS IN THE TELEPHONE HANDSET

ELEMENT	APPLICATION
Aluminum	Metal alloy in dial mechanism, transmitter, and receiver
Antimony	Alloy in dial mechanism
Arsenic	Alloy in dial mechanism
Beryllium	Alloy in dial mechanism
Bismuth	Alloy in dial mechanism
Boron	Touch-Tone® dial mechanism
Cadmium	Color in yellow plastic housing
Calcium	In lubricant for moving parts
Carbon	Plastic housing, transmitter steel parts
Chlorine	Wire insulation
Chromium	Color in green plastic housing, metal plating, stainless steel parts
Cobalt	Magnetic material in receiver
Copper	Wires, plating, brass parts
Fluorine	Plastic parts
Germanium	Transistors in some dial mechanisms
Gold	Electrical contacts
Hydrogen	Plastic housing, wire insulation
Indium	Touch-Tone® dial mechanism
Iron	Steel and magnetic materials
Krypton	Ringer in Touch-Tone® set
Lead	Solder in connections
Lithium	In lubricant for moving parts
Magnesium	Die castings in transmitter, ringer
Manganese	Steel in various parts
Mercury	Color in red plastic housing
Molybdenum	Magnet in receiver
Nickel	Magnet in receiver, stainless steel parts
Nitrogen	Hardened heat-treated steel parts
Oxygen	Plastic housing, wire insulation
Palladium	Electrical contacts
Phosphorus	Steel in various parts
Platinum	Electrical contacts
Silicon	Touch-Tone® dial mechanism
Silver	Plating
Sodium	In lubricant for moving parts
Sulfur	Steel in various parts
Tantalum	Integrated circuit in some sets (Trimline®)
Tin	Solder in connections, plating
Titanium	Color in white plastic housing
Tungsten	Lights in Princess® and key sets
Vanadium	Receiver
Zinc	Brass, die casting in transmitter, ringer

Source: Committee on Survey of Materials Science and Engineering, Appendix to the COMSAT Report (national Academy of Sciences, Washington, D.C., 1975), Volume 2.

giving lifetime, modes of decay, decay energy, particle energies, particle intensities, and thermal neutron capture cross section, among other factors, can be found in the literature. For example, see the Heath and Weast references listed at end of this entry.

An important fact pertaining to naturally-occurring elements is that many of them consist of several isotopes and that these isotopes are present in nearly all cases in the same proportion by weight. These constant isotopic compositions have enabled scientists over the years to analyze materials by weight, making reference to a table of atomic weights. In fact, for many years the atomic weight of an element was regarded as its most distinguishing characteristic, even though some discrepancies were noticed in the periodic table. It was not until the work of Moseley on characteristic x-ray spectra and the development of positive-ray analysis that the nuclear charge was recognized as the fundamental chemical characteristic of an element. The use of mass spectrography, with other methods of determining the masses of the atoms in a given element and the proportions in which they are present, has permitted the determination of the isotopic composition of the elements.

Sulfur is one of the few exceptions to the constancy of isotopic proportions, in that there is sufficient variation, dependent upon the source of the sulfur, to cause a variation in its atomic mass by approximately $\pm 0.01\%$. For normal stochiometric calculations, however, this small variation is unimportant.

Naturally-occurring elements which do not display any isotopic behavior include aluminum, arsenic, beryllium, bismuth, cobalt, fluorine, gold, helium, holmium, iodine, manganese, niobium (columbium), phosphorus, praseodymium, rhodium, scandium, sodium, terbium, thulium, and yttrium. Elements which have one predominating isotope (in excess of 98%) include argon, carbon, lanthanum, lutetium, nitrogen, oxygen, tantalum, and vanadium. Elements which have several isotopes and in which no one isotope is in excess of 80% of the total include antimony, barium, bromine, chlorine, copper, dysprosmium, erbium, gadolinium, gallium, germanium, hafnium, iridium, krypton, lead, magnesium, mercury, molybdenum, nickel, osmium, palladium, platinum, rhenium, rubidium, ruthenium, selenium, silver, tellurium, tin, titanium, tungsten, xenon, ytterbium, zinc, and zirconium. Tin leads with a total of ten isotopes; xenon has nine isotopes; cadmium and tellurium each have eight isotopes.

Radioactive Elements. There are (1) naturally-occurring radioactive and (2) artificially-produced radioactive elements. There are three series of naturally-occurring radioactive elements:

The actinium series. This series commences with ^{225}U and ends with the stable isotope, ^{207}Pb. The decay scheme is represented by: ^{235}U $\xrightarrow{\alpha}$ ^{231}Th $\xrightarrow{\beta}$ ^{231}Pa $\xrightarrow{\alpha}$ ^{227}Ac $\xrightarrow{\beta \text{ and } \alpha}$ ^{227}Th $\xrightarrow{\alpha}$ ^{223}Fr $\xrightarrow{\beta}$ ^{223}Ra $\xrightarrow{\alpha}$ ^{219}Rn $\xrightarrow{\alpha}$ ^{215}Po $\xrightarrow{\alpha \text{ and } \beta}$ ^{211}Pb $\xrightarrow{\beta}$ ^{215}At $\xrightarrow{\alpha}$ ^{211}Bi $\xrightarrow{\beta \text{ and } \alpha}$ ^{211}Po $\xrightarrow{\alpha}$ ^{207}Tl $\xrightarrow{\beta}$ ^{207}Pb (stable).

The thorium series. This series commences with ^{232}Th and ends with the stable isotope, ^{208}Pb. The decay scheme is represented by: ^{232}Th $\xrightarrow{\alpha}$ ^{228}Ra $\xrightarrow{\beta}$ ^{228}Ac $\xrightarrow{\beta}$ ^{228}Th $\xrightarrow{\alpha}$ ^{224}Ra $\xrightarrow{\alpha}$ ^{220}Rn $\xrightarrow{\alpha}$ ^{216}Po $\xrightarrow{\alpha}$ ^{212}Pb $\xrightarrow{\beta \text{ and } \alpha}$ ^{216}At $\xrightarrow{\alpha}$ ^{212}Bi $\xrightarrow{\beta \text{ and } \alpha}$ ^{212}Po $\xrightarrow{\alpha}$ ^{208}Tl $\xrightarrow{\beta}$ ^{208}Pb (stable).

The uranium series. This series commences with ^{238}U and ends with the stable isotope, ^{206}Pb. The decay scheme is represented by: ^{238}U $\xrightarrow{\alpha}$ ^{234}Th $\xrightarrow{\beta}$ ^{234}Pa $\xrightarrow{\beta}$ ^{234}U $\xrightarrow{\alpha}$ ^{230}Th $\xrightarrow{\alpha}$ ^{226}Ra $\xrightarrow{\alpha}$ ^{222}Rn $\xrightarrow{\alpha}$ ^{218}Po $\xrightarrow{\alpha \text{ and } \beta}$ ^{214}Pb $\xrightarrow{\beta}$ ^{218}At $\xrightarrow{\alpha}$ ^{214}Bi $\xrightarrow{\beta \text{ and } \alpha}$ ^{214}Po $\xrightarrow{\alpha}$ ^{210}Tl $\xrightarrow{\beta}$ ^{210}Pb $\xrightarrow{\beta}$ ^{210}Bi $\xrightarrow{\beta \text{ and } \alpha}$ ^{210}Po $\xrightarrow{\alpha}$ ^{206}Tl $\xrightarrow{\beta}$ ^{206}Pb (stable).

In the foregoing series, the type of radiation given off during the decay process is indicated above the arrows.

The production of artificially-produced radioactive elements dates back to the early work of Rutherford in 1919 when it was found that alpha particles reacted with nitrogen atoms to yield protons and oxygen atoms. Curie and Joliot found (1933) that when boron, magnesium, or aluminum were bombarded with alpha particles from polonium, the elements would emit neutrons, protons, and positrons. They also found that upon cessation of bombardment the emission of protons and neutrons stopped, but that the emission of positrons continued. The targets remained radioactive. They also found that the radiation emitted dropped off exponentially as would be expected from a naturally-occurring radioactive element. Further investigation

TABLE 4. THE NUCLIDES (ISOTOPES AND ISOBARS)

ELE-MENT	MASS No. A	ISOTOPIC ABUNDANCE %	ELE-MENT	MASS No. A	ISOTOPIC ABUNDANCE %	ELE-MENT	MASS No. A	ISOTOPIC ABUNDANCE %	ELE-MENT	MASS No. A	ISOTOPIC ABUNDANCE %
H	1	99.985	Zn	66	27.8	Sn	112	0.96	Dy	162	25.5
	2	0.015	(cont.)	67	4.1		114	0.66	(cont.)	163	<25.0
He	3	1.3×10^{-4}		68	18.6		115	0.35		164	28.2
	4	~100		70	0.63		116	14.30	Ho	165	100
Li	6	7.5	Ga	69	60.1		117	7.61	Er	162	0.136
	7	92.5		71	39.9		118	24.03		164	1.56
Be	9	100	Ge	70	20.52		119	8.58		166	>33.4
B	10	18.7		72	27.43		120	32.85		167	22.9
	11	81.3		73	7.76		122	4.72		168	27.1
C	12	98.9		74	36.54		124	5.94		170	14.9
	13	1.1		76	<7.76	Sb	121	57.25	Tm	169	100
N	14	99.62	As	75	100		123	42.75	Yb	168	0.14
	15	0.38	Se	74	0.87	Te	120	0.088		170	3.03
O	16	99.76		76	9.02		122	2.83		171	14.3
	17	0.04		77	7.58		123	0.85		172	>21.8
	18	0.20		78	23.52		124	4.59		173	16.2
F	19	100		80	49.82		125	6.93		174	>31.8
Ne	20	90.8		82	9.19		126	18.71		176	12.7
	21	<0.3	Br	79	50.54		128	<31.86	Lu	175	97.5
	22	8.9		81	49.46		130	<34.52		176	2.5
Na	23	100	Kr	78	0.342	I	127	100	Hf	174	0.18
Mg	24	77.4		80	2.23	Xe	124	0.094		176	5.2
	25	11.5		82	11.50		126	0.088		177	>18.4
	26	11.1		83	11.48		128	1.92		178	27.1
Al	27	100		84	57.02		129	>26.23		179	13.8
Si	28	92.21		86	<17.43		130	4.05		180	35.3
	29	4.70	Rb	85	72.2		131	>21.14	Ta	180	0.012
	30	3.09		87	27.8		132	26.93		181	99.988
P	31	100	Sr	84	0.56		134	10.52	W	180	0.122
S	32	95.0		86	9.86		136	8.93		182	26.20
	33	<0.8		87	7.02	Cs	133	100		183	14.26
	34	4.2		88	82.56	Ba	130	0.101		184	<30.74
	36	<0.02	Y	89	100		132	0.097		186	<28.82
Cl	35	75.4	Zr	90	51.5		134	2.42	Re	185	37.1
	37	24.6		91	11.2		135	6.59		187	62.9
Ar	36	0.337		92	17.1		136	7.81	Os	184	0.018
	38	0.061		94	17.4		137	>11.32		186	1.59
	40	99.602		96	2.8		138	>71.66		187	1.64
K	39	93.1	Nb	93	100	La	138	0.089		188	13.3
	40	<0.012	Mo	92	15.84		139	99.911		189	16.1
	41	<6.9		94	9.04	Ce	136	0.19		190	26.4
Ca	40	96.96		95	15.72		138	0.26		192	<41.0
	42	0.64		96	16.53		140	88.47	Ir	191	38.5
	43	0.15		97	9.46		142	11.08		193	61.5
	44	2.06		98	23.78	Pr	141	100	Pt	190	0.012
	46	0.0033		100	9.63	Nd	142	27.11		192	0.78
	48	<0.019	Tc	(all radioactive)			143	12.17		194	32.8
Sc	45	100	Ru	96	5.51		144	23.85		195	>33.7
Ti	46	7.93		98	1.87		145	8.30		196	25.4
	47	7.28		99	12.72		146	17.22		198	7.2
	48	73.94		100	12.62		148	5.73	Au	197	100
	49	5.51		101	17.07		150	5.62	Hg	196	0.15
	50	5.34		102	31.61	Pm	(all radioactive)			198	10.1
V	50	0.25		104	18.58	Sm	147	15.0		199	17.0
	51	99.75	Rh	103	100		148	11.2		200	23.3
Cr	50	4.31	Pd	102	1.0		149	13.8		201	13.2
	52	83.76		104	11.0		150	7.4		202	<29.6
	53	9.55		105	22.2		152	26.8		204	6.7
	54	2.38		106	27.2		154	22.7	Tl	203	29.5
Mn	55	100		108	26.8	Eu	151	47.8		205	70.5
Fe	54	5.82		110	11.8		153	52.2	Pb	204	1.37
	56	91.66	Ag	107	51.4	Gd	152	0.2		206	26.26
	57	2.19		109	48.6		154	2.15		207	20.8
	58	0.33	Cd	106	1.23		155	<14.7		208	>51.55
Co	59	100		108	0.88		156	20.5	Bi	209	100
Ni	58	67.88		110	12.32		157	15.7			
	60	26.22		111	12.67		158	<24.9			
	61	1.18		112	24.15		160	21.9			
	62	3.66		113	12.21	Tb	159	100			
	64	<1.08		114	28.93	Dy	156	0.052			
Cu	63	69.09		116	7.61		158	0.090			
	65	30.91	In	113	4.2		160	2.29			
Zn	64	<48.9		115	95.8		161	>18.9			

Following elements in increasing mass number are all radioactive: Po, At, Rn, Fr, Ra, Ac, Th, Pa, U, Np, Pu, Am, Cm, Bk, Cf, Es, Fm, Md, No, and Element 104 and heavier.

indicated that nuclear reactions lead to the formation of radioactive isotopes. This and subsequent work by several investigators led to the formulation of another series of radioactive elements, namely, the Neptunium Series, which commences with ^{245}Cm (curium) and ends with the stable isotope, ^{209}Bi. The decay scheme is represented by: ^{245}Cm $\xrightarrow{\alpha}$ ^{241}Pu $\xrightarrow{\beta}$ ^{241}Am $\xrightarrow{\alpha}$ ^{237}Np $\xrightarrow{\alpha}$ ^{233}Pa $\xrightarrow{\beta}$ ^{233}U $\xrightarrow{\alpha}$ ^{229}Th $\xrightarrow{\alpha}$ ^{225}Ra $\xrightarrow{\beta}$ ^{225}Ac $\xrightarrow{\alpha}$ ^{221}Fr $\xrightarrow{\alpha}$ ^{217}At $\xrightarrow{\alpha}$ ^{213}Bi $\xrightarrow{\beta\text{ and }\alpha}$ ^{213}Po $\xrightarrow{\alpha}$ ^{209}Tl $\xrightarrow{\beta}$ ^{209}Pb $\xrightarrow{\beta}$ ^{209}Bi (stable).

A *radioactive element* is an element that disintegrates spontaneously with the emission of various rays and particles. Most commonly, the term denotes radioactive elements such as radium, radon (emanation), thorium, promethium, uranium, which occupy a definite place in the periodic table because of their atomic number. The term radioactive element is also applied to the various other nuclear species (which are produced by the disintegration of radium, uranium, etc.) including the members of uranium, actinium, thorium, and neptunium families of radioactive elements, which differ markedly in their stability, and are isotopes of elements from thallium (atomic number 81) to uranium (atomic number 92), as well as the partly artificial actinide group, which extends from actinium (atomic number 89) to lawrencium (atomic number 103), and includes the following transuranic elements: neptunium (atomic number 93), plutonium (atomic number 94), americium (atomic number 95), curium (atomic number 96), berkelium (atomic number 97), californium (atomic number 98), einsteinium (atomic number 99), fermium (atomic number 100), mendelevium (atomic number 101), nobelium (atomic number 102). The radioactive nuclides produced from nonradioactive ones are discussed under **Radioactivity.**

A radioactive element may be designated as being in a *collateral series*. In addition to the three main natural and one artificial disintegration series of radioactive elements, each has been found to have at least one parallel or collateral series. The main series and the collateral series have different parents, but become identical in the course of disintegration, when they have a member in common.

Superheavy and Transactinide Elements. Those elements with an atomic number above 103 are sometimes referred to as the *transactinide elements*; and those with atomic (proton) numbers $Z \geq 110$ are sometimes called *superheavy elements*) (SHEs). Considerable research has been concentrated on synthesizing elements heavier than 103. It will be recalled that the last of the elements to be synthesized with positive proof of identity, timing of research, and place and persons associated with discovery—and thus relatively little controversy pertaining to the naming of the element—was lawrencium (103), discovered by Ghiorso, Sikkeland, Larsh, and Latimer in March 1961 at Berkeley. Research in this direction had been underway in the early 1960s at the Joint Nuclear Research Institute at Dubna (U.S.S.R.). The half-lives of the Lr isotopes range from 8 to 35 seconds, enabling researchers to use solvent extraction techniques for determining the chemical characteristics and atomic number of the element. However, with the possible exception of a predicted *island of stability*, as one goes up the scale of heavier elements, the half-lives appear to become progressively shorter, making chemical separation, needed for identifying the atomic number of any laboratory-produced superheavy nucleus, increasingly difficult. Much research has been directed toward improving these techniques and, as of the early 1980s, methodology is available to cope with nuclei having a half-life of 1 second or greater. Separation techniques include the ion exchange behavior of the bromide complexes of the elements; and the ease with which the elements coprecipitate with cupric sulfide (Seaborg-Loveland-Morrissey, 1979).

Applying modern theories of nuclear structure, scientists working in the late 1960s and early 1970s made calculations that showed for superheavy elements in the vicinity of $Z = 114$ (proton number) and $N = 184$ (neutron number), ground states of nuclei were stabilized against fission. As pointed out in the aforementioned reference, "This stabilization was due to the complete filling of major proton and neutron shells in this region and is analogous to the stabilization of chemical elements, such as the noble gases by the filling of their electronic shells." Some of the calculations indicated that the half-lives of some of these superheavy nuclei might be on the order of the age of the universe. This observation, of course, was stimulating to further research. Calculations indicated that there should be an island of relative stability which would extend above the Z and N figures previously given. The calculations also showed that between the presently known elements and these stable superheavy elements, there would be an intervening region of instability.

Although beyond the scope of this volume, in the early 1930s, a new mechanism for the interaction of heavy ions was discovered. (See Lefort-Ngo and the Schröder-Huizenga references listed.) The method, known as *deep inelastic scattering*, involves a massive transfer of energy and nucleons between the projectile and the target. More detail is also given in the Seaborg-Loveland-Morrissey reference.

A team of researchers from Oak Ridge National Laboratory, the University of California (Davis, California), and Florida State University reported an x-ray spectra in June 1976 that appeared to confirm the existence of superheavy elements with atomic numbers near 126. For a short period, this finding created much interest in the scientific community—and mainly because the atomic numbers reported were much greater than might be expected at that stage of research and, possibly of even greater interest, the findings were based upon the elements being part of monazite crystals believed to be about 1 billion years old, thus giving rise to the previous mention of their life on the order of the age of the universe. Further confirmatory proof was lacking, capped by the finding of a researcher at Florida State University who showed that a gamma ray with the same energy as the x-ray peak for "element 126" is emitted when an excited praseodymium nucleus relaxes after being created from cerium during bombardment by protons. It is noteworthy that cerium is a major constituent of monazite. More details on subsequent study of the initial findings are given in the Robinson reference listed.

Element 104. As of the mid-1980s, claims for discovery and thus the procedure for officially naming element 104 remain unresolved. Researchers at Dubna (U.S.S.R.), in 1964, bombarded plutonium with accelerated 113–115 MeV neon ions. During this process, an isotope that decayed by spontaneous fission was observed. It was reported that the isotope had a half-life of 0.3 ± 0.1 second and it was reasoned that the isotope was 104^{260}, resulting from: $_{94}Pu^{242} + {}_{10}Ne^{22} \rightarrow 104^{260} + 4n$. Although subsequent work toward chemically separating the new element from all others has not been conclusive, considerable evidence for evaluation has been obtained. Estimates of the half-life of the element have been reduced from the prior 0.3 second to 0.15 second. Ghiorso, Nurmia, Harris, K. Eskola, and P. Eskola (University of California, Berkeley) reported, in 1969, the positive identification of two and possibly three isotopes of the element. The Berkeley discovery resulted from bombarding a target of Cf^{249} with C^{12} nuclei of 71 MeV, and

C^{13} nuclei of 69 MeV. The first combination resulted in the instant emission of four neutrons to produce 104^{257}. The isotope was reported to have a half-life of 4–5 seconds. Decay was by emission of an alpha particle into No^{253}, with a half-life of 105 seconds. In further research, several thousand atoms of 104^{257} and 104^{259} were produced. Thus far, the Dubna workers have proposed the name *kurchatovium* (Ku); and the Berkeley group has suggested *rutherfordium* (Rf).

Element 105. In 1967, workers at Dubna (U.S.S.R.) reported producing a few atoms of element 105^{260} and 105^{261}, as the result of bombarding Am^{243} with Ne^{22}. Appropriate confirmations of identification, however, were lacking, the evidence being based upon time-coincidence measurements of alpha energies. In 1970, it was reported that the Dubna researchers had investigated all the types of decay of the new element and had determined its chemical properties. As of that time, the Dubna group had not proposed a name for element 105. Ghiorso, Nurmia, Harris, K. Eskola, and P. Eskola (University of California, Berkeley) reported a positive identification of element 105. This resulted from bombarding a target of Cf^{249} with 84 MeV nitrogen nuclei. Upon absorption of N^{15} nuclei by a Cf^{249} nucleus, four neutrons are emitted, forming element 105^{260}, with a half-life of 1.6 seconds. The Berkeley group has proposed *hahnium* (Ha) as a name for the element.

Element 106. In late 1974, two groups announced the synthesis of the fourteenth trans-urium element, namely, element number 106 (eka-tungsten). A Soviet group at the Joint Institute for Nuclear Research in Dubna bombarded a target of lead atoms with ions of various weights. Argon ions were used to form a short-lived isotope of fermium which decayed by spontaneous fission. After this trial, the Soviet group used ions of titanium to produce a similarly short-lived isotope of element 104 (tentatively named kurchatovium or rutherfordium—the official name to be selected after settlement of Soviet and American claims on the priority of the synthesis). Then, the Soviet scientists finally used ions of chromium to produce what they believe to be element 106. This presumed element is described as having 151 or 152 neutrons, which decays by spontaneous fission with a half-life of about 4 to 10 milliseconds. The Soviet scientists claim that the chromium and lead will combine to form element 106.

An American group at the University of California's Lawrence Berkeley Laboratory, using the modified super-HILAC accelerator (heavy ion linear accelerator) to bombard a target of californium-249 with ions of oxygen-18, caused the oxygen ions to combine with molecules in the target, releasing four neutrons per collision and become eka-tungsten-263. It is reported that this isotope of element 106 has a half-life of 0.9 second. Contrary to earlier predictions, it does not decay by spontaneous fission, but emits an alpha particle with an energy of 9.06 MeV to become an isotope of element 104. The previously observed daughter isotope emits an alpha particle with an energy of 8.8 MeV, becoming nobelium-255. The latter, in turn, emits an alpha particle with an energy of 8.11 MeV. The American scientists believe that observation of this complete sequence of transmutations is conclusive proof of the formation of eka-tungsten. The American team first observed element 106 in 1970, but lead impurities in the target presented difficulties in providing conclusive evidence of the existence of the new element. The HILAC accelerator was shut down shortly after the experiment and two years were required for the improvements that resulted in the super-HILAC. Then other experiments were given a higher priority. Thus there was an approximately four-year delay in the American experiments.

The isotope of element 106, synthesized at Berkeley, contains 157 neutrons. As of the mid-1980s, no name had been accepted for element 106.

Element 107. Investigators at Dubna announced the synthesis of element 107 in 1976, as the result of bombarding Bi^{204} with heavy nuclei of Cr^{54}. Prior experiments had suggested the probable very brief (0.002 second) observation of the element. Research reports on the element remain sketchy and it is considered premature to fully presume that 107 exists.

Allotropes. Some of the elements exist in two or more modifications distinct in physical properties, and usually in some chemical properties. Allotropy in solid elements is attributed to differences in the bonding of the atoms in the solid. Various types of allotropy are known. In *enantiomorphic allotropy*, the transition from one form to another is reversible and takes place at a definite temperature, above or below which only one form is stable, e.g., the alpha and beta forms of sulfur. In *dynamic allotropy*, the transition from one form to another is reversible, but with no definite transition temperature. The proportions of the allotropes depend upon the temperature. In *monotropic allotropy*, the transition is irreversible. One allotrope is metastable at all temperatures, e.g., explosive antimony.

Examples of allotropes include:
Arsenic with four forms, metallic, yellow, gray, and brown.
Boron with two forms, crystalline and amorphous.
Carbon with three forms, amorphous, diamond, and graphite.
Phosphorus with four forms, two white forms, a violet, and a black form. Red phosphorus is a mixture of the white and violet forms.
Selenium with four forms, amorphous, two crystalline monoclinic forms (red), and the stable, crystalline gray metallic form.
Sulfur with two forms, alpha-rhombic sulfur with a density of 2.07 and a mp of 112.8°C, and beta-monoclinic sulfur with a density of 1.96 and a mp of 119°C. The beta form changes to the alpha form below 96°C.

In the case of some of the less common elements, impure forms have been mistaken in the past as allotropic forms. A number of elements that once were considered to exist in both crystalline and amorphous forms have been found to exist in only one form when perfectly pure.

Gaseous Elements. Several gaseous elements (at standard conditions of temperature and pressure) form molecules of two atoms each. These are known as *diatomic gases*. Included in this category are hydrogen, H_2, nitrogen, N_2, oxygen, O_2, and chlorine, Cl_2. The inert gases, helium, neon, argon, krypton, and xenon are *monatomic gases* and their symbols do not carry a subscript.

Atomic Structure of the Elements

The internal structure of an atom consists of electrons moving within a region having a diameter slightly greater than 10^{-8} cm and of protons and neutrons that are confined to a nucleus at the center of the electron distribution in a region having a diameter of about 10^{-12} cm. The electrons of a neutral atom are sufficient in number so that their total negative charge is equal to the positive charge on the nucleus. For example, atoms of the element hydrogen have in their neutral state a single electron moving about a nucleus which has a positive charge equal to the negative charge of an electron. Helium, which has two electrons moving about its nucleus, has a positive charge on its nucleus equal to twice one electronic charge and lithium, which has three electrons moving about its nucleus, has a positive charge on its nucleus of three electronic charges. One basis for the classification of atoms is by those numbers which correspond to the number by which the charge on the hydrogen nucleus must be multiplied to equal the nuclear charge of the atom in question. These numbers, ranging from one, for hydrogen, to 103, for lawrencium, are called atomic numbers.

The atomic number may be defined as the number of protons in an atomic nucleus, or the positive charge of the nucleus, expressed in terms of the electronic charge. Atomic number usually is denoted by the symbol Z. In the symbolic designation of individual nuclides, the atomic number sometimes is written as a subscript to the left of the chemical symbol of the atomic species, such as $^{16}_{8}O$ for the oxygen isotope of mass number 16. This usage is redundant, in that the chemical symbol per se specifies the atomic number of the nuclide.

Besides nuclear charge, atoms also differ in their masses. The mass is determined by the number of protons Z and the number of neutrons N in the atomic nucleus. The total number of nuclear particles, $Z + N$, in an atomic species is known as its mass number A.

The nuclear model for an atom is of relatively recent origin for, at the beginning of the 20th century, J. J. Thomson's "plum-pudding" model of an atom was the more generally accepted version. In this model Thomson supposed that the positive charge forms a plasma that is distributed throughout the atomic volume and that the electrons are mixed into this plasma with a relatively uniform distribution. E. Rutherford proposed the nuclear model on the basis of experimental work by H. Geiger and E. Marsden in which they observed that a small number of alpha particles from a naturally occurring radioactive source are scattered through angles greater than 90° by thin foils of gold and silver. Although some small angle scattering is predicted by the Thomson model, such large angle scattering is not at all expected. Large angle scattering is however completely consistent with the idea that the alpha particles interact with point positively charged objects of large mass at the center of each gold and silver atom.

Subsequently (1913), N. Bohr found that he could use the Rutherford nuclear model to explain in almost complete detail the observed spectrum of hydrogen (see **Atomic Spectra**). Bohr proposed that, in a neutral hydrogen atom, a single electron revolves in a stationary orbit around a point nucleus that has a charge $+e$. This electron is held in its orbit by the electrostatic force between the positive charge on the nucleus and its own negative charge. The stationary-orbit assumption was classically unsatisfactory but necessary to account for the behavior of the electron. If the laws of classical electrodynamics, according to which accelerated charges must be sources of electromagnetic radiation, were strictly obeyed, the electron would gradually lose energy; hence it would revolve in orbits of smaller and smaller radii and eventually fall into the nucleus. Furthermore, Bohr postulated that the magnitude of the orbital angular momentum of each stationary orbit is $L = nh/2\pi = n\hbar$, where n is an integer, h is Planck's constant, and $\hbar$ is simply an abbreviated form for $h/2\pi$.

The angular momentum of an electron moving in an orbit of the type described by Bohr is an axial vector $\mathbf{L} = \mathbf{r} \times \mathbf{p}$, formed from the radial distance $\mathbf{r}$ between electron and nucleus and the linear momentum $\mathbf{p}$ of the electron relative to a fixed nucleus. Figure 2 shows the customary method used to illustrate the axial vector $\mathbf{L}$ in terms of the orbital motion of any object, of which the electron of the Bohr atom is only one example. Although Bohr's planetary model needed only circular orbits to explain the spectral lines observed in the spectrum of a hydrogen atom, subsequent development of similar models for other atoms containing more than a single electron needed elliptically shaped orbits to explain the observed spectra. For such orbits $\mathbf{r}$ and $\mathbf{p}$ are usually not perpendicular to each other so that the magnitude $|\mathbf{L}| = |\mathbf{r} \times \mathbf{p}|$ is $rp \sin \theta$.

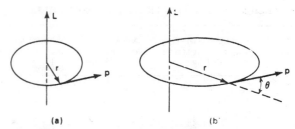

Fig. 2. Direction of angular momentum vector is perpendicular to plane formed by radial vector r and momentum vector p.

A significant change in the theoretical treatment of atomic structure occurred in 1924 when Louis deBroglie proposed that an electron and other atomic particles simultaneously possess both wave and particle characteristics and that an atomic particle, such as an electron, has a wavelength $\lambda = h/p = h/mv$. Shortly thereafter, C. J. Davisson and L. H. Germer showed experimentally the validity of this postulate. DeBroglie's assumption that wave characteristics are inherent in every atomic particle was quickly followed by the development of quantum mechanics. In its most simple form, quantum mechanics introduces the physical laws associated with the wave properties of electromagnetic radiation into the physical description of a system of atomic particles. By means of quantum mechanics a much more satisfactory explanation of atomic structure can be developed.

The quantity n introduced by Bohr in his description of the hydrogen atom is what is called a quantum number. These numbers enter quite naturally from quantum-mechanical descriptions of atomic energy states, including nuclear states. In quantum-mechanical descriptions of atoms, the number n characterizes a limited number of electron states that have very nearly the same energy. A group of electrons in an atom with a common value of n are usually said to be in a single shell of the atom. In the simple two-body problem of the hydrogen atom, all states having the same value of n have the same energy, but in multielectron atoms there are interactions between individual pairs of electrons as well as between electrons and the atomic nucleus. The result is a limited spread in energy between the most-tightly bound and the least-tightly bound electron with the same value of n. Except for $n = 1$, more than one orbital angular momentum state is possible for each shell. Each such state is described by a quantum number l, which can have only whole-number values between zero and $n - 1$. The number n is usually called the principal quantum number and l the orbital angular momentum, or sometimes azimuthal, quantum number. In addition to its characteristic angular momentum, each electron spins on an axis, such that it also has a spin angular momentum, described by a quantum number s. Because all electrons are identical, s has only one value, $\frac{1}{2}$.

In the development of the concepts of atomic structure much of the experimental evidence came from optical and x-ray spectroscopy. From this work certain notations have arisen that are now an accepted part of the language. For example, the $n = 1$ shell is sometimes known as the K-shell, the $n = 2$ shell as the L-shell, the $n = 3$ shell as the M-shell, etc., with consecutively following letters of the alphabet being used to designate those shells with successively higher principal quantum numbers. A Roman numeral subscript further subdivides the shells in accordance with the n, l and j quantum numbers of the electrons, as shown in Table 5.

A letter notation has been given to the different orbital angular momentum states by optical spectroscopists. In this notation an $l = 0$ state is called an s state (not to be confused with the s used as a quantum number for spin, and which can usually

TABLE 5. IDENTIFICATION OF ATOMIC SHELLS

X-RAY NOTATION	CORRESPONDING QUANTUM NUMBERS			X-RAY NOTATION	CORRESPONDING QUANTUM NUMBERS		
	n	l	j		n	l	j
K	1	0	$\frac{1}{2}$	N_{I}	4	0	$\frac{1}{2}$
L_{I}	2	0	$\frac{1}{2}$	N_{II}	4	1	$\frac{1}{2}$
L_{II}	2	1	$\frac{1}{2}$	N_{III}	4	1	$\frac{3}{2}$
L_{III}	2	1	$\frac{3}{2}$	N_{IV}	4	2	$\frac{3}{2}$
M_{I}	3	0	$\frac{1}{2}$	N_{V}	4	2	$\frac{5}{2}$
M_{II}	3	1	$\frac{1}{2}$	N_{VI}	4	3	$\frac{5}{2}$
M_{III}	3	1	$\frac{3}{2}$	N_{VII}	4	3	$\frac{7}{2}$
M_{IV}	3	2	$\frac{3}{2}$				
M_{V}	3	2	$\frac{5}{2}$				

be distinguished from the way it is used) and $l = 1$ state a p state, an $l = 2$ a d state, an $l = 3$ state an f state, with consecutively higher letters above f in the alphabet designating each succeeding value of l. A number often accompanies the state designation to indicate the appropriate value of n. For example, a $3p$ level is an $n = 3$, $l = 1$ state.

Electrically neutral atoms with nuclear charge $Z > 1$ are not hydrogen-like, but have more than a single orbital electron. With more than one electron in an atom, it is necessary to determine the relationship of one electron to its neighbors. A clue to this relationship is found from the observation that the energy required to remove a second electron from an atom is greater than that required to remove the first electron and the energy so required increases with each succeeding electron that is removed. In other words, the potential energy that holds some electrons in an atom is much greater than the energy that holds other electrons in the same atom. From the results of the detailed observations, Pauli formulated a principle, the Pauli Exclusion Principle, which states that no two electrons within the same atom may have exactly the same wave function and, on this basis, that no two electrons in the same atom may have the same set of quantum numbers. Thus, if a particular configuration state in an atom is occupied by an electron, that state is forbidden to all other electrons. If each electron falls into the lowest possible energy state, the next electron to enter must remain in some higher energy state. Acceptance of this principle allows us to classify observed spectral characteristics of the radiation emitted by atoms in such a way that we can determine within reasonable limits the number of electrons in each shell or subshell of the atom producing the radiation. The electron configurations in the known atomic species are shown in Table 6, in which the atoms are ordered in accordance with their atomic numbers Z. Generally the innermost (lower quantum number) shells are filled with all the electrons allowed. For example, neon has two $1s$ electrons, two $2s$ electrons, and six $2p$ electrons, such that its $K(n = 1)$ and $L(n = 2)$ shells are both filled. An atom with Z having a magnitude one greater than neon, which is sodium, also has both the K and L shells filled; thus there is no space for another electron to go into either of these shells, and its additional electron must go into the $M(n = 3)$ shell. Shells with higher values of the orbital angular momentum quantum number l are sometimes partially shielded electrically from the nucleus by other electrons in such a way that their binding energies are not as great as the lower orbital angular momentum subshells associated with the next higher principal quantum number. Thus, as can be seen in Table 6, the $5s$ subshell is filled in atomic species of lower Z than any of those that have any $4f$ electrons.

The periodic table of elements can be described in terms of the similarities of the various atomic species. For example, simi-

lar but not identical chemical properties are observed for a number of elements that have all but one of their electrons in filled shells, with the extra electron being a single s electron. In the periodic table, lithium, sodium, potassium, rubidium, and cesium all have this characteristic and thus are placed in a single column. Other groups of atomic species that have similar electron configurations in their outermost shell are arranged in the periodic table in the same chemical grouping. See **Periodic Table of the Elements.**

Motions of charged particles are expected to establish magnetic fields that interact with each other. As a result, coupling between spin and orbital angular momenta of a single electron or coupling between angular momenta of different electrons in a single atom is expected. The spin-orbit coupling of a single electron results in a total angular momentum state, described by a total quantum number j, such that either $j = l + \frac{1}{2}$ or $j = l - \frac{1}{2}$. Quantum-mechanical solutions show that the magnitude of the orbital angular momentum is $l^*\hbar = [l(l + 1)]^{1/2}\hbar$, not $l\hbar$, as would be expected if we had followed the simpler assumption of the Bohr model. Similarly, the magnitude of the spin and total angular momenta are $s^*\hbar = [s(s + 1)]^{1/2}\hbar$ and $j^*\hbar = [j(j + 1)]^{1/2}\hbar$, respectively. Coupled spin and orbital angular momenta thus do not align themselves in a linear pattern but in such a way that the spin and orbital angular momenta precess around the direction of the total angular momentum, as shown in Fig. 3. A similar precession around the direction of the resultant angular momentum state is found, as we shall find later, in the coupling of any two or more angular momentum states into a single common system.

Without some external influence, angular momentum states have no preferred orientation in space because the only interacting magnetic fields are entirely within the individual atom. An external magnetic field, however, may couple to the magnetic field of an atomic angular momentum state of the type described in the preceding paragraphs. In coupling to the total angular momentum, for example, only a limited number of orientations of $j^*\hbar$ are possible, these being such that their projections in the direction of the magnetic field have magnitudes $m_j\hbar$ in which m_j, a magnetic quantum number, can have only those whole-number values for which $m_j \leq |j|$. Thus $2j + 1$ orientations, as illustrated in Fig. 4 for $j = \frac{3}{2}$, are possible. Note that the vector representing the total angular momentum is not parallel to the direction of the applied magnetic field. As a result, the total angular momentum vector precesses around the direction of the magnetic field such that the component of the total angular momentum that is perpendicular to the direction of the magnetic field has a time-averaged value of zero and the only component that can be observed with an external detecting device is that component parallel to the direction of the external magnetic field. A magnetic field of this type splits a set of levels characterized by a quantum number j, all of which initially have the same energy, into $2j + 1$ components. The nature of this level splitting is deduced from observations of the splitting of characteristic line spectra, in which case an originally monoenergetic radiation is split by the magnetic field into several components that are usually relatively closely spaced in wave length. The observed effect is known as the Zeeman effect. See also **Zeeman Effect.**

The use of axial vectors to describe states of an atom in terms of a coupling between the angular momentum inherent in the electrons of the atom forms what is commonly called the *vector model* of the atom. The type of coupling just described forms the basis for the j-j coupling model. Such coupling is found between electrons that produce the optical radiation in the higher Z atomic systems. In these systems the total angular momentum $j^*\hbar$ formed by the coupling of the spin and orbital

TABLE 6. ELECTRON STRUCTURE OF ATOMS (NORMAL STATE)

Element	Atomic Number	Chemical Symbol	K	L		M			N				O				
			1s	2s	2p	3s	3p	3d	4s	4p	4d	4f	5s	5p	5d	5f	5g
Hydrogen	1	H	1														
Helium	2	He	2														
Lithium	3	Li	2	1													
Beryllium	4	Be	2	2													
Boron	5	B	2	2	1												
Carbon	6	C	2	2	2												
Nitrogen	7	N	2	2	3												
Oxygen	8	O	2	2	4												
Fluorine	9	F	2	2	5												
Neon	10	Ne	2	2	6												
Sodium	11	Na	2	2	6	1											
Magnesium	12	Mg	2	2	6	2											
Aluminum	13	Al	2	2	6	2	1										
Silicon	14	Si	2	2	6	2	2										
Phosphorus	15	P	2	2	6	2	3										
Sulfur	16	S	2	2	6	2	4										
Chlorine	17	Cl	2	2	6	2	5										
Argon	18	Ar	2	2	6	2	6										
Potassium	19	K	2	2	6	2	6		1								
Calcium	20	Ca	2	2	6	2	6		2								
Scandium	21	Sc	2	2	6	2	6	1	2								
Titanium	22	Ti	2	2	6	2	6	2	2								
Vanadium	23	V	2	2	6	2	6	3	2								
Chromium	24	Cr	2	2	6	2	6	5	1								
Manganese	25	Mn	2	2	6	2	6	5	2								
Iron	26	Fe	2	2	6	2	6	6	2								
Cobalt	27	Co	2	2	6	2	6	7	2								
Nickel	28	Ni	2	2	6	2	6	8	2								
Copper	29	Cu	2	2	6	2	6	10	1								
Zinc	30	Zn	2	2	6	2	6	10	2								
Gallium	31	Ga	2	2	6	2	6	10	2	1							
Germanium	32	Ge	2	2	6	2	6	10	2	2							
Arsenic	33	As	2	2	6	2	6	10	2	3							
Selenium	34	Se	2	2	6	2	6	10	2	4							
Bromine	35	Br	2	2	6	2	6	10	2	5							
Krypton	36	Kr	2	2	6	2	6	10	2	6							
Rubidium	37	Rb	2	2	6	2	6	10	2	6			1				
Strontium	38	Sr	2	2	6	2	6	10	2	6			2				
Yttrium	39	Y	2	2	6	2	6	10	2	6	1		2				
Zirconium	40	Zr	2	2	6	2	6	10	2	6	2		2				
Niobium	41	Nb	2	2	6	2	6	10	2	6	4		1				
Molybdenum	42	Mo	2	2	6	2	6	10	2	6	5		1				
Technetium	43	Tc	2	2	6	2	6	10	2	6	(5)		(2)				
Ruthenium	44	Ru	2	2	6	2	6	10	2	5	7		1				
Rhodium	45	Rh	2	2	6	2	6	10	2	6	8		1				
Palladium	46	Pd	2	2	6	2	6	10	2	6	10						
Silver	47	Ag	2	2	6	2	6	10	2	6	10		1				
Cadmium	48	Cd	2	2	6	2	6	10	2	6	10		2				
Indium	49	In	2	2	6	2	6	10	2	6	10		2	1			
Tin	50	Sn	2	2	6	2	6	10	2	6	10		2	2			
Antimony	51	Sb	2	2	6	2	6	10	2	6	10		2	3			
Tellurium	52	Te	2	2	6	2	6	10	2	6	10		2	4			
Iodine	53	I	2	2	6	2	6	10	2	6	10		2	5			
Xenon	54	Xe	2	2	6	2	6	10	2	6	10		2	6			
Cesium	55	Cs	2	8	18	2	6	10		2	6			1			
Barium	56	Ba	2	8	18	2	6	10		2	6			2			
Lanthanium	57	La	2	8	18	2	6	10		2	6	1		2			
Cerium	58	Ce	2	8	18	2	6	10	2	2	6			2			
Praseodymium	59	Pr	2	8	18	2	6	10	3	2	6			2			
Neodymium	60	Nd	2	8	18	2	6	10	4	2	6			2			
Promethium	61	Pm	2	8	18	2	6	10	5	2	6			2			
Samarium	62	Sm	2	8	18	2	6	10	6	2	6			2			
Europium	63	Eu	2	8	18	2	6	10	7	2	6			2			
Gadolinium	64	Gd	2	8	18	2	6	10	7	2	6	1		2			
Terbium	65	Tb	2	8	18	2	6	10	8 or	2	6			1 or	2		

(Continued)

TABLE 6 (Continued)

ELEMENT	ATOMIC NUMBER	CHEMICAL SYMBOL	K	L	M	N				O					P						Q
						4s	4p	4d	4f	5s	5p	5d	5f	5g	6s	6p	6d	6f	6g	6h	7s
Dysprosium	66	Dy	2	8	18	2	6	10	10	2	6				2						
Holmium	67	Ho	2	8	18	2	6	10	11	2	6				2						
Erbium	68	Er	2	8	18	2	6	10	12	2	6				2						
Thulium	69	Tm	2	8	18	2	6	10	13	2	6				2						
Ytterbium	70	Yb	2	8	18	2	6	10	14	2	6				2						
Lutetium	71	Lu	2	8	18	2	6	10	14	2	6	1			2						
Hafnium	72	Hf	2	8	18	2	6	10	14	2	6	2			2						
Tantalum	73	Ta	2	8	18	2	6	10	14	2	6	3			2						
Tungsten	74	W	2	8	18	2	6	10	14	2	6	4			2						
Rhenium	75	Re	2	8	18	2	6	10	14	2	6	5			2						
Osmium	76	Os	2	8	18	2	6	10	14	2	6	6			2						
Iridium	77	Ir	2	8	18	2	6	10	14	2	6	7			2						
Platinum	78	Pt	2	8	18	2	6	10	14	2	6	9			1						
Gold	79	Au	2	8	18	2	6	10	14	2	6	10			1						
Mercury	80	Hg	2	8	18	2	6	10	14	2	6	10			2						
Thallium	81	Tl	2	8	18	2	6	10	14	2	6	10			2	1					
Lead	82	Pb	2	8	18	2	6	10	14	2	6	10			2	2					
Bismuth	83	Bi	2	8	18	2	6	10	14	2	6	10			2	3					
Polonium	84	Po	2	8	18	2	6	10	14	2	6	10			2	4					
Astatine	85	At	2	8	18	2	6	10	14	2	6	10			2	5					
Radon	86	Rn	2	8	18	2	6	10	14	2	6	10			2	6					
Francium	87	Fr	2	8	18	2	6	10	14	2	6	10			2	6					(1)
Radium	88	Ra	2	8	18	2	6	10	14	2	6	10			2	6					(2)
Actinium	89	Ac	2	8	18	2	6	10	14	2	6	10			2	6	(1)				(2)
Thorium	90	Th	2	8	18	2	6	10	14	2	6	10			2	6	(2)				(2)
Protactinium	91	Pa	2	8	18	2	6	10	14	2	6	10	(2)		2	6	(1)				(2)
Uranium	92	U	2	8	18	2	6	10	14	2	6	10	(3)		2	6	(1)				(2)
Neptunium	93	Np	2	8	18	2	6	10	14	2	6	10	(5)		2	6					(2)
Plutonium	94	Pu	2	8	18	2	6	10	14	2	6	10	(6)		2	6					(2)
Americium	95	Am	2	8	18	2	6	10	14	2	6	10	(7)		2	6					(2)
Curium	96	Cm	2	8	18	2	6	10	14	2	6	10	(7)		2	6	(1)				(2)
Berkelium	97	Bk	2	8	18	2	6	10	14	2	6	10	(9)		2	6					(2)
Californium	98	Cf	2	8	18	2	6	10	14	2	6	10	(10)		2	6					(2)
Einsteinium	99	Es	2	8	18	2	6	10	14	2	6	10	(11)		2	6					(2)
Fermium	100	Fm	2	8	18	2	6	10	14	2	6	10	(12)		2	6					(2)
Mendelevium	101	Mv	2	8	18	2	6	10	14	2	6	10	(13)		2	6					(2)
Nobelium	102	No	2	8	18	2	6	10	14	2	6	10	(14)		2	6					(2)
Lawrencium	103	Lw	2	8	18	2	6	10	14	2	6	10	(14)		2	6	(1)				(2)

angular momentum states of individual electrons are further coupled within any single shell of an atom to form a total angular momentum $J^*\hbar$, characterized by a quantum number J, which then is the angular momentum for the system of several electrons. According to customary usage, a lower case letter as a

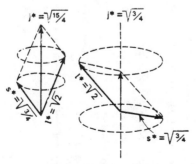

Fig. 3. Magnitudes and directions of the angular momentum vectors for an $l = 1$, $s = \frac{1}{2}$ electron in an atomic energy state.

designator for a quantum number indicates that the quantum number is that of a single electron, while a capital letter indicates that the quantum number represents a state formed by several electrons.

For the lower Z part of the periodic table of elements, the appropriate coupling system for angular momentum states in an atom is the L-S, or Russell-Saunders, coupling. In this description the orbital angular momenta of individual electrons in any single shell of an atom are coupled to form a resultant orbital angular momentum described by the quantum number L, and the spin angular momenta of the same individual electrons are coupled to form a resultant spin angular momentum, described by the quantum number S. Coupling of the individual orbital angular momentum states is not shown but possible coupling schemes for the spins of 2, 3, and 4 electrons is shown in Fig. 5. Note that as many possible couplings exist as there are electrons. Only one of these possible schemes will be the lowest energy state, the others being at higher energies. The resultant spin and orbital angular momentum vectors for a group of electrons in a single shell of an atom can then be used to

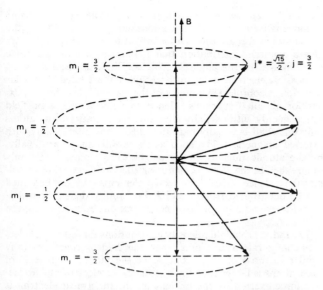

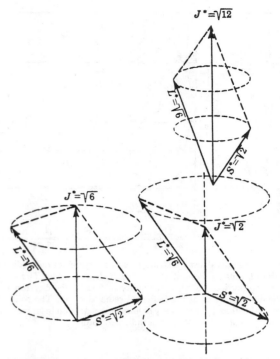

Fig. 4. Possible orientations of the total angular momentum vector *i* relative to the direction of an externally applied magnetic field **B** and the magnitudes of the associated magnetic quantum state vectors m_j.

Fig. 6. Vector model coupling of spin and orbital angular momenta for which $L = 2$, $S = 1$.

describe the coupling that gives the total angular momentum J^*h for these electrons. In Fig. 6 is shown the coupling according to the vector model between the orbital and spin angular momenta for which $L = 2$ and $S = 1$. There are $2J + 1$ possible states. All $2J = 1$ states have the same energy unless under the influence of an external magnetic field, in which case they are broken into components, each of which is designated by a magnetic quantum number M_J. In principle this splitting is the same as for the splitting of the total angular momentum states for a single electron, as shown in Fig. 4, except for the use of capital letters M_J and J.

The Pauli Exclusion Principle states that no two electrons of any single atom may simultaneously occupy a state described by only a single set of quantum numbers. Five such numbers are needed to describe fully the quantum-mechanical conditions of an electron. For j-j coupling this set is generally n, l, s, j, m_j, and for L-S coupling it is n, l, s, m_l, m_s. From the coupling of the angular momentum associated with the latter sets a full description of the multielectron state, described by n, L, S, J, M_j, is determined.

Part of the outgrowth of the determination that atomic particles have wave properties and of the subsequent development of quantum mechanics is the Heisenberg Uncertainty Principle, which states that an electron cannot be located exactly in terms of both its space and momentum coordinates, or in terms of both its time and energy coordinates. If the energy of a particular atomic state is precisely defined, the time at which an electron is found in that energy state cannot be so defined. Likewise, if the momentum of an electron is precisely defined, its position in space cannot be so defined. On the other hand, the probability for finding an electron in a particular location relative to the atomic nucleus can be determined. After substitution of the appropriate potential energy terms needed to describe the atomic system, these probability density distributions may be found from solutions of the Schroedinger wave equation, which is a quantum-mechanical description of the system. This distribution, which is directly related to the wave function that describes the atomic state in a quantum-mechanical manner, must be distributed through a region of space and is hence a function of all three spatial variables, which are r, θ and ϕ in the spherical coordinate system. Because only two variables are available on the plane of a page to describe these functions, they are usually described by a series of graphs, which must then be assembled in the imagination of the reader to picture the complete three-dimensional distribution. The radial part of the distribution is dependent only on the quantum numbers n and l and is usually represented by the terminology $|R_{nl}|^2$. The radial distribution for selected states of a hydrogen atom are shown in Fig. 7. The dashed lines are proportional to the probability of finding the electron in the appropriate nl state in an incremental volume dv of constant magnitude at the indicated radial distance from the atomic nucleus. The solid lines are proportional to the probability of finding the electron in an incremental shell between the radial distances r and $r + dr$, with a volume $4\pi r^2 \, dr$ at

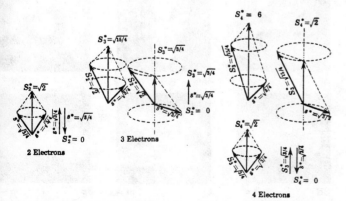

Fig. 5. Vector model coupling of the spin angular momenta of two, three, and four electrons.

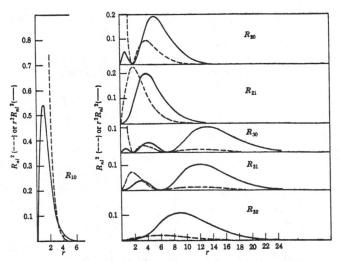

Fig. 7. Probability density distributions as a function of radial distance from the nucleus for several states of a hydrogen atom. The dashed lines are proportional to the probability of finding the electron in an incremental volume dv at the indicated radial distance. The solid lines are proportional to the probability of finding the electron in an incremental shell of volume $4\pi r^2\,dr$ at the indicated radius.

the indicated radius. Radial distances in Fig. 7 are given in units of Bohr radii, the distance from the nucleus of the $n = 1$ orbit in the Bohr planetary model of the hydrogen atom.

In the quantum-mechanical description of a hydrogen atom, the radial portion of the probability density distribution is the same in all directions from the nucleus, but only for the case $l = 0$ is the magnitude of the distribution the same in all radial directions. For all other values of l, the magnitude of the distribution is a function of the angular direction, defined by the coordinates θ and ϕ. However, as in the case of the discussion of the vector model of an atom, we cannot define an angular direction unless an axis exists to provide a reference direction. To provide this axis some force or torque external to the electron configuration must be found. The only external force strong enough to interact with the electron configuration is that provided by certain magnetic fields. The interaction with the magnetic moment of the electron configuration can then be described in terms of a magnetic quantum number m_l, which is the magnetic quantum number associated with orbital angular momentum quantum number. When $l = 1$, m_l may have any of three values, $+1$, 0, or -1. Three possible angular distributions are then possible for the electrons described by a quantum number $l = 1$, but the distributions for $m_l = +1$ and $m_l = -1$ are identical, thereby providing some simplification. If we define $\theta = 0$ in the direction of the applied magnetic field, the probability density distribution in the $r\theta$-plane for an electron of a hydrogen atom described by $l = 1$, $m_l = 0$ is given in the $2p$, $m = 0$ part of Fig. 8. This distribution is symmetric in ϕ; it may thus be rotated around an axis perpendicularly directed through the center of the figure (the $\theta = 0$ axis) to give the full three-dimensional distribution. For either $m_l = +1$ or $m_l = -1$ the distribution is given by that part of Fig. 8 labeled $2p$, $m = 1$. When rotated about the $\theta = 0$ axis the $m_l = 0$ distribution gives a dumbbell like distribution and the $|m_l| = 1$ distributions give ring-like distributions, all of which fade away to negligible magnitude at large distances from the center of the distribution. For the $2p$, $m_l = 0$ configuration, the angular distribution in the $r\theta$-plane is given by the equation $\frac{3}{2}\cos^2\theta$,

and for the $2p$, $m_l = \pm1$ configuration by the equation $\frac{3}{4}\sin^2\theta$. Since two $|m_l| = 1$ configurations exist for each $l = 1$, the distributions for all three m_l configurations, when summed, give a resultant distribution that is independent on the angle θ, as well as of the angle ϕ. This condition is reached when all possible electron states in the 2 subshell are filled. A similar situation, a complete symmetry of the electron distribution, exists for all filled subshells. Hence an external magnetic field interacts only with partially filled shells. In multielectron atoms only one shell is usually partially filled; thus interactions with external magnetic fields, such as those introduced artificially, by the atomic nucleus, or by neighboring atoms, occur only in one shell of the atom, sometimes called in chemical terminology the valence shell. For specific elements, the unfilled shell can be determined from Table 3. Probability density distributions for several other electron configurations besides $2p$ is given in Fig. 8.

The radial probability density distributions for individual electron configurations in atoms other than hydrogen are generally similar to those of Fig. 7 but, since the nuclear charge Z of these atoms is larger than for hydrogen, the electrostatic attractive force exerted by the nucleus on the innermost electron is stronger than for hydrogen and hence pulls its distribution closer to the nucleus. Outer electrons, however, are partially shielded electrostatically from the nucleus by the inner electrons and hence are influenced by a weaker force than that which would be provided by a bare nucleus. The least tightly bound electron

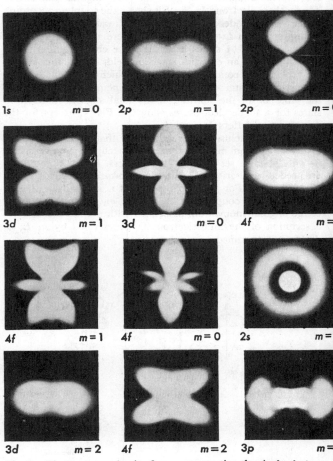

Fig. 8. Electron wave density figures representing the single electron states of the hydrogen atom.

is held by a force that has an effective Z of about 1 but for the more tightly bound electrons the effective Z is much higher. The probability density distribution for the least tightly bound electron then extends to a distance comparable to the distribution for hydrogen, but the main part of the distribution is much closer to the nucleus. A typical distribution for all the electrons of a multielectron atom, in this case rubidium, is shown in Fig. 9.

As stated earlier in this entry, there are three p orbitals, with magnetic quantum numbers of 0, 1 and −1. The orbital for which $m = 0$ corresponds to the direction of the applied magnetic field, so that its effect is zero. This direction is conventionally taken as that of the z-axis, so that p_z, which is symmetrical about that axis, coincides with p_0. The other p orbitals, p_x and p_y, are perpendicular to p_z and to each other and correspond to standing waves built up by mixing of p_{+1} and p_{-1}, which are oppositely directed waves.

An analogous situation for the d orbitals, for which $l = 2$, and hence $m_l = 2, 1, 0, -1$ and -2, gives rise to five d orbitals, designated as $d_{xz}, d_{xy}, d_{yz}, d_{x_2{}^2-y_2{}^2}$ and $d_{9z_2{}^2}$.

It follows from the earlier discussion that each orbital may be occupied, in accordance with the Pauli Exclusion Principle, by two electrons of opposed spin.

Atomic Radius

The radius of an atom may be defined as the distance of closest approach to another atom and is the distance at which the mutual repulsion of the electron clouds and the mutual attraction of the nuclear charge of each for the electrons of the other are in equilibrium under specified circumstances. If the two atoms are the same, the radius of each is one half the internuclear distance; if they are unlike, the internuclear distance is the sum of the individual radii. Atomic radii fall roughly into four categories, which, however, merge into one another. These are Van der Waals and ionic radii, which are radii of equilibrium approach for mutually nonbonded atoms (neutral and charged, respectively), and covalent and metallic radii, which are for mutually bonded atoms (where the bonding electrons are, respectively, largely localized between the bonded atoms, or highly delocalized). Radii of all types vary in essentially the same manner from one part of the periodic table to another. In general, the radii decrease from the beginning to the end of any period, the rate of decrease being less the higher the number of the period. There are discontinuities in the decrease at points at which the quantum levels become half or completely filled. For example, Mn^{2+} ($r = 0.83$ Å) is larger

than either Cr^{2+} (0.80 Å) or Fe^{2+} (0.80 Å), and Zn^{2+} (0.75 Å) is larger than Cu^{2+} (0.72 Å). In this description, angstroms (Å) are used for convenience of notation. (1 angstrom = 10^{-10} meter). Again, in general, the radii increase going down any column in the periodic table except in the region of hafnium, which results in a far greater decrease between these two elements (average metallic radii: La 1.87 Å, Hf 1.58 Å) than between yttrium and zirconium (Y 1.80 Å, Zr 1.60 Å), where no such series intervenes. Because of the slower rate of decrease of the radius in the sixth period as compared to the fifth, however, the radii in the sixth period soon become larger again than those in the fifth (cf. Nb 1.429 Å, Ta 1.43 Å; Mo 1.36 Å, W 1.37 Å).

Van der Waals Radius. This is the radius of closest approach of one atom to another with which it does not form a chemical bond. It represents the distance at which the mutual repulsion of the nonbonding electrons of each exactly balances the attraction of the nucleus of each for the electrons of the other. See also **Van der Waals Forces.** Experimentally determined and calculated values (indicated by asterisk) of Van der Waals radii in angstrom units of a number of elements are: H 1.2, He 1.78$_5$, B 2.17*, C 1.85, N 1.54, O 1.40, F 1.35, Ne 1.60, Al 2.53*, Si 2.24*, P 1.9(1.96*), S 1.85, Cl 1.80, Ar 1.92, Ge 2.02, As 2.0, Se 2.00, Br 1.95, Kr 2.01, Sb 2.2, Te 2.20, I 2.15, Xe 2.20, Hg 1.50, Rn ~ 2.3*. The often cited value of 1.70 Å for the Van der Waals radius of carbon is derived from the interlayer distance in crystalline graphite (3.40 Å) and is too short because the weak interaction of the pi-electron systems draws the layers together.

Van der Waals radii of some groups are: CH_3 2.0, C_6H_5 (interplanar) 1.85, BH_4^- 2.08, SiH_3^- 2.25.

Ionic Radius. This is the radius of closest approach of a charged atom or group (i.e., an ion) to another atom or group with which it does not form a covalent bond. It represents the distance at which the mutual repulsion of the nonbonding electrons of each exactly balances the mutual attraction of oppositely charged ions, or the attraction of a positive ion for the electrons of a neutral atom or group or the attraction of a negative ion for the nucleus or nuclei of a neutral atom or group. Ionic radii for the cations of the least electronegative metals and of the most electronegative nonmetals are clearly defined, but for the elements of intermediate electronegativity (including the transition elements) they are much more doubtful because of the varying degrees of covalency in the compounds of these elements. Published ionic radii for these latter elements can therefore not be considered to be accurate measures of the true sizes of the ions, but are useful for comparison of relative sizes. In particular, there are certainly no simple cations of charge greater than 4+, and the simple tetrapositive cations are limited to Th^{4+}, probably the other tetrapositive actinide cations, and possibly Ce^{4+}. "Ionic radii" cited for such "cations" as Si^{4+}, P^{5+}, S^{6+}, Cl^{7+}, etc., are fictions arrived at by subtracting from an observed interatomic distance an arbitrary anionic radius for the second atom or else are extrapolated or theoretically calculated radii. However, inasmuch as all compounds of such elements have a high degree of covalency, such radii have no real meaning in normal compounds. Simple anions of absolute charge greater than 3 undoubtedly do not exist and the existence of monatomic trinegative ions is open to question. However, in their binary compounds with the lanthanide elements, phosphorus, arsenic, antimony and possibly nitrogen and bismuth apparently have radii very close to their Van der Waals radii and may therefore be considered essentially ionic. (See Table 7.)

The monatomic ionic radii of a given element become smaller

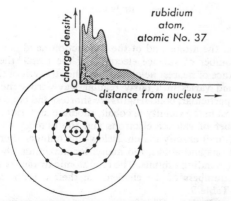

Fig. 9.   Radial probability density distribution, derived from quantum-mechanical predictions, for rubidium, along with Bohr planetary model for the same atom.

TABLE 7. IONIC CRYSTAL RADII (IN ANGSTROM UNITS)*

Element	Ion	Radius	Element	Ion	Radius	Element	Ion	Radius	Element	Ion	Radius
Actinium	Ac^{3+}	1.11		Cu^{2+}	0.72	Molybdenum	Mo^{4+}	0.68	Selenium	Se^{2-}	1.960
Aluminum	Al^{3+}	0.57	Curium	Cm^{3+}	1.00		(Mo^{6+})	0.65		(Se^{6+})	0.42
Americium	Am^{3+}	1.00	Dysprosium	Dy^{3+}	0.908	Neodymium	Nd^{3+}	0.995	Silicon	(Si^{4+})	0.40
	Am^{4+}	0.85	Einsteinium	Es^{3+}	0.97	Neptunium	Np^{3+}	1.02	Silver	Ag^+	0.97
Antimony	Sb^{3-}	2.170	Erbium	Er^{3+}	0.881		Np^{4+}	0.88	Sodium	Na^+	1.00
	Sb^{3+}	0.90	Europium	Eu^{2+}	1.137	Nickel	Ni^{2+}	0.74	Strontium	Sr^{2+}	1.18
	(Sb^{5+})	0.62		Eu^{3+}	0.950	Niobium	Nb^{4+}	0.67	Sulfur	S^{2-}	1.855
Arsenic	As^{3-}	1.991	Fermium	Fm^{3+}	0.97		(Nb^{5+})	0.70?		(S^{6+})	0.29
	(As^{3+})	0.69	Fluorine	F^-	1.36	Nitrogen	N^{3-}	1.56	Tantalum	(Ta^{5+})	0.73
	(As^{5+})	0.47		(F^{7+})	0.07		(N^{5+})	0.11	Technetium	Tc^{4+}	0.5
Astatine	At^-	2.2	Francium	Fr^+	1.9	Nobelium			Tellurium	Te^{2-}	2.21
Barium	Ba^{2+}	1.38	Gadolinium	Gd^{3+}	0.938	Osmium	Os^{4+}	0.65		(Te^{4+})	0.84
Berkelium	Bk^{3+}	0.99	Gallium	Ga^{3+}	0.65	Oxygen	O^{2-}	1.40		(Te^{6+})	0.56
Beryllium	Be^{2+}	0.31	Germanium	Ge^{2+}	0.65		(O^{6+})	0.09	Terbium	Tb^{3+}	0.923
Bismuth	Bi^{3-}	2.217		(Ge^{4+})	0.55	Palladium	Pd^{2+}	0.50	Thallium	Tl^+	1.50
	Bi^{3+}	1.20	Gold	Au^+	1.37	Phosphorus	P^{3-}	1.920		Tl^{3+}	0.95
	(Bi^{5+})	0.74	Hafnium	Hf^{4+}	0.86		(P^{5+})	0.34	Thorium	Th^{4+}	0.95
Boron	(B^{3+})	0.20	Holmium	Ho^{3+}	0.894	Platinum	Pt^{2+}	0.52	Thulium	Tm^{3+}	0.869
Bromine	Br^-	1.97	Hydrogen	H^-	2.08		(Pt^{4+})	0.55	Tin	Sn^{2+}	1.02
	(Br^{7+})	0.39	Indium	In^{3+}	0.95	Plutonium	Pu^{3+}	1.01		(Sn^{4+})	0.65
Cadmium	Cd^{2+}	0.99	Iodine	I^-	2.16		Pu^{4+}	0.86	Titanium	Ti^{2+}	0.76
Calcium	Ca^{2+}	1.06		(I^{7+})	0.50	Polonium	(Po^{4+})	0.9		(Ti^{4+})	0.60
Californium	Cf^{3+}	0.98	Iridium	Ir^{4+}	0.66	Potassium	K^+	1.33	Tungsten	W^{4+}	0.68
Carbon	(C^{4-})	2.60	Iron	Fe^{2+}	0.80	Praseodymium	Pr^{3+}	1.013		(W^{6+})	0.65
	(C^{4+})	0.15		Fe^{3+}	0.67		Pr^{4+}	0.87	Uranium	U^{3+}	1.04
Cerium	Ce^{3+}	1.034	Lanthanum	La^{3+}	1.071	Promethium	Pm^{3+}	0.98		U^{4+}	0.89
	Ce^{4+}	0.941	Lead	Pb^{2+}	1.18	Protoactinium	Pa^{3+}	1.06	Vanadium	V^{2+}	0.82
Cesium	Cs^+	1.70		(Pb^{4+})	0.70	Radium	Ra^{2+}	1.42		V^{3+}	0.75
Chlorine	Cl^-	1.81	Lithium	Li^+	0.70	Rhenium	(Re^{6+})	0.52		(V^{5+})	0.59
	(Cl^{7+})	0.26	Lutetium	Lu^{3+}	0.83	Rhodium	Rh^{3+}	0.75	Ytterbium	Yb^{2+}	1.02
Chromium	Cr^{2+}	0.80	Magnesium	Mg^{2+}	0.75		Rh^{4+}	0.65		Yb^{3+}	0.858
	Cr^{3+}	0.70	Manganese	Mn^{2+}	0.83	Rubidium	Rb^+	1.52	Yttrium	Y^{3+}	0.910
	(Cr^{6+})	0.52		Mn^{3+}	0.52	Ruthenium	Ru^{4+}	0.60	Zinc	Zn^{2+}	0.75
Cobalt	Co^{2+}	0.78		(Mn^{7+})	0.46	Samarium	Sm^{2+}	1.143	Zirconium	Zr^{4+}	0.80
	Co^{3+}	0.65	Mendelevium	Mv^{3+}	0.96		Sm^{3+}	0.964			
Copper	Cu^+	0.96	Mercury	Hg^{2+}	1.12	Scandium	Sc^{3+}	0.83			

* 1 angstrom = 10^{-10} meter.

as the oxidation state increases, provided this implies actual removal of electrons. For example, the radius of Fe^{2+} is 0.80 Å, whereas that of Fe^{3+} is 0.67 Å; $Cu^+ = 0.96$, $Cu^{2+} = 0.72$.

The ionic radii discussed above are properly "crystal radii," i.e., the radii exhibited by the ions in ionic crystals. Although these radii are probably reasonable representations of the radii of contact of the ions in solution with the nearest atoms of solvate molecules, especially for ions of low charge density, nevertheless, most ions in solution have far larger effective radii because they carry with them a sheath of solvent molecules, the tenacity and thickness of which is a function of the charge density and electronic structure of the ion. In consequence, ions of small crystal radius (e.g., Li^+) may act larger (e.g., have lower mobility) in solution than ions of large crystal radius (e.g., Cs^+).

The effective radius of an ion changes with coordination number. An ion of coordination number 4 (tetrahedral) has a radius 0.93–0.95 times as large as the same ion with coordination number 6, while an ion of coordination number 8 is about 1.03 times as large as the same ion with coordination number 6.

Effective spherical crystal radii for some polyatomic ions are given in Table 8.

Metallic Radius. Metals may be considered to be composed of cations bonded together by a cement of mobile electrons which are located in the conduction bands. Since the number of available energy levels in the conduction bands is a function of the number of available orbitals of the atoms making up the metal, and since the electron population of the nonbonding

TABLE 8. CRYSTAL RADII OF POLY-ATOMIC IONS (Å)*

NH_4^+	1.48		OH^-	1.53
OH_3^+	1.35		SH^-	2.00
PH_4^+	1.61		SeH^-	2.15
BH_4^-	2.08		SiH_3^-	2.25
CN^-	1.92		TeH^-	(2.35)
NO_3^-	2.3			

* 1 angstrom = 10^{-10} meter.

orbitals of the atoms and of the conduction band is a function of the number of valence electrons of the metal, the number and distance of nearest neighbors and the strength of the metal-metal bond varies in a fairly regular way across the periodic table. In particular, bond lengths are shortest and bond strengths are greatest in the vicinity of cobalt, rhodium and iridium, where the number of valence electrons and the number of available orbitals (nine) exactly match. Below this number there are too few electrons and above, too many, for maximum sharing and use of the bonding orbitals. The interatomic distances and coordination numbers of the elements in their metallic states are given in Table 9.

Covalent Radius. This is the radius of closest approach for atoms bonded together by electrons which are localized in the region between the atoms. It represents the distance at which

TABLE 9.   INTERATOMIC DISTANCES IN METALS (IN ANGSTROM UNITS)*

Element	Interatomic Distances	Coordination Number
Actinium (room temp.)	3.756	12
Aluminum (25°C)	2.8635	12 (f.c.c.)
Americium		
Antimony (25 C)	2.90, 3.36	3,3 (rhombohedral)
Arsenic	2.49$_5$, 3.33	3,3 (rhombohedral)
Barium (room temp.)	4.347	8 (b.c.c.)
Berkelium		
Beryllium (α-form, 20°C)	2.2260, 2.2856	6,6 (c.p. hex.)
Bismuth (25°C)	3.09$_5$, 3.47	3,3 (rhombohedral)
Boron	1.75–1.80	
Cadmium (21°C)	2.9788, 3.2933	6,6 (c.p. hex.)
Calcium (α-form, 18°C)	3.947	12 (f.c.c.)
(γ-form, 500°C)	3.877	8 (b.c.c.)
Californium		
Cerium (room temp.)	3.650	12 (f.c.c.)
	3.620, 3.652	6,6 (c.p. hex?)
(15,000 atm.)	3.42	(f.c.c.)
Cesium (−100°C)	5.264	8 (b.c.c.)
(−10°C)	5.309	8
Chromium (α-form, 20°C)	2.4980	8 (b.c.c.)
(β-form, > 1850°C)	2.61	12 (f.c.c.)
Cobalt (18°C)	2.5061	12 (f.c.c.)
(room temp.)	2.505–2.498, 2.505–2.507	6,6 (c.p. hex.)
Copper (20°C)	2.5560	12 (f.c.c.)
Curium		
Dysprosium (room temp.)	3.503, 3.590	6,6 (c.p. hex.)
Einsteinium		
Erbium (room temp.)	3.468, 3.559	6,6 (c.p. hex.)
Europium (room temp.)	3.989	8 (b.c.c.)
Fermium		
Francium		
Gadolinium (20°C)	3.573, 3.636	6,6 (c.p. hex.)
Gallium (20°C)	2.44$_2$, 2.71$_2$, 2.74$_2$, 2.80	1,2,2,2 (orthorhombic)
Germanium (20°C)	2.4498	4 (diamond)
Gold (25°C)	2.8841	12 (f.c.c.)
Hafnium (α-form, 24°C)	3.1273, 3.1947	6,6 (c.p. hex.)
Holmium (room temp.)	3.486, 3.577	6,6 (c.p. hex.)
Indium (20°C)	3.2511, 3.3730	4,8 (f.c.t.)
Iridium (room temp.)	2.714	12 (f.c.c.)
Iron (α-form, 20°C)	2.4823	8 (b.c.c.)
(γ-form, 916°C)	2.578	12 (f.c.c.)
(δ-form, 1394°C)	2.539	8 (b.c.c.)
Lanthanum (α-form, room temp.)	3.739, 3.770	6,6 (c.p. hex.)
(β-form, room temp.)	3.745	12 (f.c.c.)
Lead (25°C)	3.5003	12 (f.c.c.)
Lithium (20°C)	3.0390	8 (b.c.c.)
(78°K)	3.111, 3.116	6,6 (c.p. hex.)
Lutetium (room temp.)	3.435, 3.503	6,6 (c.p. hex.)
Magnesium (25°C)	3.1971, 3.2094	6,6 (c.p. hex.)
Manganese (γ-form, 1095°C)	2.7311	12 (f.c.c.)
(δ-form, 1134°C)	2.6679	8 (b.c.c.)
Mendelevium		
Mercury (−46°C)	3.005	6 (rhombohedral)
Molybdenum (20°C)	2.7251	8 (b.c.c.)
Neodymium (room temp.)	3.628, 3.658	6,6(?) (modified c.p. hex.)
Neptunium (α-form, 20°C)	2.60–2.64	4 (orthorhombic)
(β-form, 313°C)	2.76	4 (tetragonal)
(δ-form, 600°C)	3.05	8 (b.c.c.)
Nickel (18°C)	2.4916	12 (f.c.c.)
Niobium (20°C)	2.8584	8 (b.c.c.)
Nobelium		
Osmium (20°C)	2.6754, 2.7354	6,6 (c.p. hex.)
Palladium (25°C)	2.7511	12 (f.c.c.)
Phosphorus (black)	2.18, —	3,3 (orthorhombic)
Platinum (20°C)	2.7746	12 (f.c.c.)
Plutonium (γ-form, 235°C)	3.026, 3.159, 3.287	4,2,4 (f.c.c.)

Continued

TABLE 9—*Continued*

ELEMENT	INTERATOMIC DISTANCES	COORDINATION NUMBER
Polonium (α-form, 10°C)	3.345	6 (cubic)
(β-form, 75°C)	3.359	6 (rhombohedral)
Potassium (78°K)	4.544	8 (b.c.c.)
Praseodymium (α-form, room temp.)	3.640, 3.673	6 (hexagonal)
(β-form, room temp.)	3.649	12 (f.c.c.)
Promethium		
Protactinium (room temp.)	3.212, 3.238	8,2 (b.c.t.)
Radium		
Rhenium (room temp.)	2.741, 2.760	6,6 (c.p. hex.)
Rhodium (20°C)	2.6901	12 (f.c.c.)
Rubidium (20°C)	4.95	8 (b.c.c.)
(-196°C)	4.860	
Ruthenium (25°C)	2.6502, 2.7058	6,6 (c.p. hex.)
Samarium		
Scandium (room temp.)	3.256, 3.309	6,6 (c.p. hex.)
(room temp.)	3.212	12 (f.c.c.?)
Selenium (20°C)	2.321, 3.464	2,4 (hexagonal)
Silicon (20°C)	2.3517	4 (diamond)
Silver (25°C)	2.8894	12 (f.c.c.)
Sodium (20°C)	3.7157	8 (b.c.c.)
Strontium (α-form, 25°C)	4.3026	12 (f.c.c.)
(β-form, 248°C)	4.32, 4.324	6,6 (c.p. hex.)
(γ-form, 614°C)	4.20	8 (b.c.c.)
Tantalum (20°C)	2.86	8 (b.c.c.)
Technetium (room temp.)	2.703, 2.735	6,6 (c.p. hex.)
Tellurium (25°C)	2.864, 3.468	2,4 (hexagonal)
Terbium (room temp.)	3.525, 3.601	6,6 (c.p. hex.)
Thallium (α-form, 18°C)	3.4076, 3.4566	6,6 (c.p. hex.)
(β-form, 262°C)	3.362	8 (b.c.c.)
Thorium (α-form, 25°C)	3.595	12 (f.c.c.)
(β-form, 1450°C)	3.56	8 (b.c.c.)
Thulium (room temp.)	3.447, 3.538	6,6 (c.p. hex.)
Tin (α-form, 20°C)	2.8099	4 (diamond)
(β-form, 25°C)	3.022, 3.181	4,2 (tetragonal)
Titanium (α-form, 25°C)	2.8956, 2.9505	6,6 (c.p. hex.)
(β-form, 900°C)	2.8636	8 (b.c.c.)
Tungsten (25°C)	2.7409	8 (b.c.c.)
Uranium (α-form, room temp.)	2.77, 2.86, 3.28, 3.37	2,2,4,4
(γ-form, 805°C)	3.058	8 (b.c.c.)
Vanadium (30°C)	2.6224	8 (b.c.c.)
Ytterbium (room temp.)	3.880	12 (f.c.c.)
Yttrium (room temp.)	3.551, 3.647	6,6 (c.p. hex.)
Zinc (25°C)	2.6649, 2.9129	6,6 (c.p. hex.)
Zirconium (α-form, 25°C)	3.1790, 3.2313	6,6 (c.p. hex.)
(β-form, 862°C)	3.1254	8 (b.c.c.)

* 1 angstrom $= 10^{-10}$ meter.

the attraction of each nucleus for the bonding electrons is in equilibrium with the mutual repulsion of the two nuclei and the repulsion of the inner electrons of each atom for the inner electrons of the other.

The length of the covalent radius is a function of several factors, among which are (1) bond order (multiplicity), (2) electronegativity, (3) hybridization, (4) orbital overlap, (5) steric factors, (6) special electronic effects.

(1) The effect of bond order is exemplified by the familiar shortening of the carbon-carbon bond in ethane (1.543 Å), graphite (1.4210 Å), benzene (1.397 Å), ethylene (1.353 Å), and acetylene (1.207 Å) as the bond order goes from 1 to $1\frac{1}{3}$ to $1\frac{1}{2}$ to 2 to 3. Bond orders affect the covalent radii of other elements similarly.

The determination of the standard single bond radius is not always a simple matter. In those cases where two like atoms are joined by an unquestionable single bond (as the carbon atoms in ethane) the standard single bond radius is one-half the internuclear distance. Similarly the single bond radii for nitrogen, oxygen, and fluorine are half the interatomic distances in N_2H_4, H_2O_2 and F_2. On the other hand, acceptance of half the interatomic distances in P_4, H_2S_2 and Cl_2 for the corresponding single bond radii is open to question, since in these and similar cases the possibility exists of multiple bond formation by overlap of the electron-filled p-orbitals of each atom with the empty d-orbitals of the other. This will result in an increase in strength and a decrease in length for these bonds compared with what they would have if they were exactly single bonds. In support of this idea, it can be seen from Table 10 that although the homoatomic bond energy for silicon (where no electrons are available for multiple bonding) is less than that for carbon, the bond energies for phosphorus, sulfur, and chlorine are signif-

TABLE 10. SINGLE BOND ENERGIES
(kcal/mole)

C—C	78.9	Si—Si	53
N—N	39	P—P (in P_4)	48
O—O	35	S—S	58.1
F—F	38	Cl—Cl	57.87

icantly greater than for nitrogen, oxygen, and fluorine, respectively. Representative bond strengths are given in Table 11.

In consequence of this effect, the single bond radii for such elements are better derived from the alkyl derivatives (with an electronegativity correction) in which only single bonding is possible. The single bond radii in Table 12, covalent radii of the elements, were derived insofar as possible from such

TABLE 11. REPRESENTATIVE SINGLE-BOND ENERGIES (In kcal/mole)

H—H	104.18	O—Sb	71		C—Si	72	Cl—As	70
H—B	ca 93	O—I	ca 48		C—P	63	Cl—Se	58
H—C	98.7	F—F	38		C—S	65.6	Cl—Br	52.7
H—N	93.4	F—Si	135		C—Cl	78.2	Cl—Sn	76
H—O	110.6	F—P	117		C—Zn	40	Cl—Sb	74
H—F	135	F—S	68		C—Ge	ca 44	Cl—I	51
H—Si	76	F—Cl	ca 61		C—As	48	Cl—Hg	54
H—P	ca 77	F—As	111		C—Se	58	Cl—Bi	67
H—S	83	F—Se	68		C—Br	68	K—K	12.6
H—Cl	103.1	F—Br	61		C—Cd	32	Ge—Ge	45
H—As	ca 59	F—Te	80		C—Sn	54	Ge—Br	66
H—Se	ca 66	F—I	63		C—Sb	47	Ge—I	51
H—Br	87.4	Na—Na	18.4		C—I	51	As—As	35
H—Te	ca 57	Si—Si	53		C—Hg	23	As—Br	58
H—I	71.4	Si—S	60.9		C—Pb	31	As—I	43
Li—Li	27.2	Si—Cl	91		C—Bi	31	Se—Se	41
B—C	89	Si—Br	74		N—N	39	Br—Br	46.08
B—N	106.5	Si—I	56		N—O	48	Br—Sn	65
B—O	128	P—P	48		N—F	65	Br—I	43
B—F	154	P—Cl	78		N—Cl	46	Br—Hg	44
B—Cl	109	P—Br	63		O—O	35	Rb—Rb	11.5
B—Br	90	P—I	44		O—F	45.3	Sn—Sn	39
C—C	78.9	S—S	58.1		O—Si	108	Sn—I	65
C—N	72.8	S—Cl	61		O—P	ca 80	Sb—Sb	ca 29
C—O	85.5	S—Br	ca 52		O—Cl	ca 49	I—I	36.06
C—F	116	Cl—Cl	57.87		O—As	72	I—Hg	35
C—Al	61	Cl—Ge	81		O—Br	ca 48	Cs—Cs	10.4

TABLE 12. COVALENT RADII OF THE ELEMENTS (IN ANGSTROM UNITS)*

Element	Valence	Bond Order	Hybridization	Radius		Element	Valence	Bond Order	Hybridization	Radius
H	1	1	s	0.3754_5		O	2	1	p	0.745
Li	1	1	s	1.336				2	p	0.654
Be	2	1	s	0.86			(6)	2	sp	0.58
		1	sp^3	1.07				3	p	0.599
B	1	1	p	0.79		F	1	1	p	0.709
	3	1	sp^2	0.84		Na	1	1	s	1.539
		1	sp^3	0.92		Mg	2	1	s	1.20
C	4	1	sp	0.691		Al	1	1	p	1.22
		1	sp^2	0.74			3	1	sp^2	1.24
		1	sp^3	0.772				1	sp^3	1.26
		2	sp	0.643				1	sp^3d^2	1.44
		2	sp^2	0.666		Si	4	1	sp^3	1.176
		3	p	0.60				1	sp^3d^2	1.31
		3	sp	0.602				2	sp^3	1.00
N	3	1	p	0.73_7		P	3	1	p	1.113_5
	5	1	sp	0.700			5	1	sp^2	1.11
		1	sp^2	0.727				1	sp^3	1.12_5
		1	sp^3	0.74_9				1	pd	1.23
	3	2	p	0.61			5	1	sp^3d^2	1.20
	5	2	sp	0.617				2	sp^3	0.872
	3	3	p	0.60		S	2,4	1	p	1.06_8
	5	3	sp	0.638			6	1	sp^3	1.03

Continued

TABLE 12—*Continued*

ELE-MENT	VALENCE	BOND ORDER	HYBRIDI-ZATION	RADIUS	ELE-MENT	VALENCE	BOND ORDER	HYBRIDI-ZATION	RADIUS
S		1	sp^3d^2	1.10	Ru	4	1	d^2sp^3	1.38
(*cont.*)	2	2	$p(p\pi)$	0.914		7	2	d^3s	1.23
	4	2	$p(pd\pi)$	0.868		8	2	d^3s	1.14
	6	2	sp^3	0.757	Rh	3	1	d^2sp^3	1.48
Cl	1	1	p	1.050	Pd	2	1	dsp^2	1.30
	3	1	pd	1.135	Ag	1	1	s	1.32
	7	2	sp^3	0.681			1	sp	1.42
K	1	1	s	1.962			1	sp^3	1.48
Ca	2	1	s	1.39	Cd	2	1	sp	1.47
Sc							1	sp^3d^2	1.62
Ti	4	1	d^3s	1.25	In	1	1	p	1.47
		1	d^5s	1.43		3	1	sp^2	1.47
V	4	1	d^3s	1.10			1	sp^3	1.43
	5	1	d^3s	1.17			1	sp^3d^2	1.66
		2	d^3s	1.05	Sn	2	1	p	1.45
Cr	3	1	d^2sp^3	1.45		4	1	sp^3	1.405
	6	1	d^3s	1.16			1	sp^3d^2	1.47
		2	d^3s	1.03	Sb	3	1	p	1.376
Mn	2	1	d^2sp^3	1.6		5	1	sp^2	1.43
	4	1	d^2sp^3	1.20			1	sp^3d^2	1.42(?)
	7	1	d^3s	1.13			1	pd	1.48
		2	d^3s	1.02	Te	2	1	p	1.39
Fe	3	1	d^2sp^3	1.39		4	1	p^3d	1.34(?)
Co	2	1	sp^3	1.55			1	p^3d^3	1.55
	3	1	d^2sp^3	1.35		6	1	sp^3d^2	1.36(?)
Ni	2	1	dsp^2	1.28		2	2	p	1.279
		1	d^2sp^3	1.54	I	1	1	p	1.360
Cu	1	1	s	1.22		5	1	pd	1.42
		1	sp^3	1.38	Cs	1	1	s	2.18
	2	1	dsp^2	1.29	Ba	2	1	s	1.52
		1	d^2sp^3	1.43	La				
		1	sp	1.34	Hf				
	3	1	dsp^2	1.39	Ta	5	1	d^4s	1.37(av.)
Zn	2	1	sp	1.15(?)			1	d^5sp	1.47(av.)
		1	sp^3	1.34			1	d^5sp^2	1.48(av.)
		1	sp^3d^2	1.46	W	6	1	d^5s	1.32
Ga	1	1	p	1.29	Re	4	1	d^2sp^3	1.44
	3	1	sp^2	1.23		6	1	d^2sp^3	1.37
		1	sp^3	1.27		7	1	d^3s	1.25
Ge	4	1	sp^3	1.225			2	d^3s	1.22
		1	sp^3d^2	1.3	Os	4	1	d^2sp^3	1.40
As	3	1	p	1.218		8	2	d^3s	1.171
	5	1	sp^3	1.19	Ir	4	1	d^2sp^3	1.50(?)
		1	sp^3d^2	1.33(?)	Pt	2	1	dsp^2	1.35
Se	2	1	p	1.21_5		4	1	d^2sp^3	1.34
	4	1	pd	1.38	Au	1	1	s	1.24
	6	1	sp^3d^2	1.22(?)		4	1	sp^3d^2	1.51
	2	2	p	1.075	Hg	2	1	$s(Hg_2^{2+})$	1.27
Br	1	1	p	1.193_5			1	sp	1.33
	5	1	pd	1.28			1	sp^3	1.54
Rb	1	1	s	2.06	Tl	1	1	p	1.54
Sr	2	1	s	1.49		3	1	sp^3d^2	1.58
Y	3				Pb	2	1	p	1.50
Zr	4	1	d^3s	1.42		4	1	sp^3	1.44
		1	d^5s	1.53	Bi	3	1	p	1.53
Nb	5	1	d^4s	1.37(av.)			1	p^3d^3	1.58
		1	d^5s	1.5	Po	4	1	p^3d^3	1.58
Mo	5	1	d^4s	1.30(av.)	Th	4	1	d^3s	1.69
	6	1	d^3s	1.31	U	6	1	d^2sp^3	1.50
		1	d^5s	1.27			2	dp	1.41
		2	d^3s	1.23	Pu	4	1	d^2sp^3	1.72
Tc					Am	5	2	dp	1.42

* 1 angstrom $= 10^{-10}$ meter.

compounds. The multiple bond radii were derived similarly. For example, the oxygen double bond radius may be obtained from acetone by using the double bond radius for carbon taken from ethylene and applying the electronegativity correction discussed below.

(2) When two atoms of different electronegativity are connected by a covalent bond, the bond length is always shorter than the sum of the individual homatomic covalent radii for the atoms in question. The shortening is proportional to the difference in electronegativity of the two elements and is expressed by the relationship due to Stevenson and Schomaker:

$$R = (r_A + r_B) - 0.09|x_A - x_B|$$

where R = observed bond length

r_A, r_B = standard covalent radii of atoms A and B

x_A, x_B = electronegativities of A and B

(3) It has been pointed out by a number of workers that for bonds of the same multiplicity, the lower the average value of the l quantum number in a hybrid bonding orbital, the shorter the bond should be. For example, the carbon-carbon single bond in $HC\equiv C-C\equiv CH$ which involves orbitals of sp hybridization is 1.37 Å long compared with the carbon-carbon bond in ethane (sp^3 hybridization) which is 1.543 Å long. Similarly the B—C bond in $B(C_6H_5)_3$ (sp^2) is shorter than in $B(C_6H_5)_4^-$ (sp^3). Thus it is found that a bond of given multiplicity has a length which is characteristic not only of the bond order, but also of the individual states of hybridization of the atoms. For example, the C—C bond in CH_3CN ($sp^3 + sp$) (1.46 Å) is almost exactly the average of the bonds in CH_3CH_3 (sp^3) and in $N\equiv C-C\equiv N$ (sp) (1.37 Å). Though the data for other elements are less extensive than for carbon, and the interpretation frequently is much more complicated, the same general principles seem to apply to other elements as well.

(4) The effectiveness of overlap of bonding orbitals of the same symmetry appears to decrease as the principal quantum number increases and as the difference between the principal quantum numbers increases. This is reflected in the bond strengths shown in Table 11. The covalent radius of hydrogen is especially subject to effects of this kind, and has the values 0.3707, 0.362, 0.306, 0.284 and 0.293 Å respectively in H_2, HF, HCl, HBr and HI. The apparent anomaly of the P—P, S—S, and Cl—Cl bonds being stronger than the N—N, O—O, and F—F bonds has been considered in paragraph (1).

(5) In the case of very large atoms or groups bonded to small atoms, the spatial requirements of the large groups may result in a bond lengthening. For example, it is probable that the C—I bond in CI_4 is longer and weaker than in CH_3I because of the steric repulsions of the large iodine atoms. Again, the N—N bond in $[(CH_3)_3NN(CH_3)_3]^{2+}$ may be longer than in $H_3NNH_3^{2+}$.

(6) Special effects of electronic or orbital structure may result in either lengthening or shortening a bond. The first situation is exemplified by such compounds as O_2N-NO_2, O_2N-X, $(C_6H_5)_3C-C(C_6H_5)_3$ and the like, where the long bond is indicated in the formula. Cases of this sort involve molecules having a pielectron system capable of accepting additional electrons (frequently in low-lying antibonding orbitals) so that the electrons required for the bond in question are partially drained away from it, leaving the bond weak and long. Thus the N—N bond in N_2O_4 has a length of 1.75 Å and a dissociation energy of 12.9 kcal compared with 1.47 Å and 60 kcal for N_2H_4.

Bond shortening, on the other hand, may occur in compounds of the most electronegative elements, notably fluorine. Typical examples are the fluoromethanes, in which the C—F bond lengths are CH_3F 1.391 Å, CH_2F_2 1.358 Å, CHF_3 1.332 Å, and CF_4 1.323 Å. This has been explained in terms of electronegativity; i.e., that the polar C—F bond requires a high degree of p-character, thus releasing the s-orbital for the bonds to the less electronegative atoms, making them shorter and stronger. As more fluorine atoms are added, the s-orbital is more equally divided among the bonds, resulting in a regular shortening of the C—F bonds. This theory has been used to explain the supposed shortening of the C—C bond from 1.543 Å in C_2H_6 to 1.52 Å in C_2F_6. However, the experimental error attached to the latter value does not allow it to be considered really different from the former (a more recent value is 1.56 Å), and furthermore the effect is not observed in other halosubstituted ethanes for which more accurate data are available: e.g., the C—C bond length is CF_3CN (1.464 Å) is if anything slightly longer than that in CH_3CN (1.458 Å), and microwave determinations on C_2H_5Br, C_2H_5Cl and C_2H_5 give 1.5508, 1.5508 and 1.540, respectively, for the C—C bonds. In addition, the C—H bond lengths in the fluoromethanes appear to be essentially constant: CH_4 1.092, CH_3F 1.109, CH_2F_2 1.092 and CHF_3 1.093. It should also be noted that the C—C stretching force constants in the two cyanides (4.50 × 10^{-5} and 4.55 × 10^{-5} dyne cm^{-1}, respectively) do not differ appreciably.

A theory which more satisfactorily explains all the known facts has been suggested by J. F. A. Williams (*Trans. Faraday Soc.*, **57**, 2089 (1961)). This proposes that the highly electronegative fluorine atom drains electron density away from the carbon atom in a C—F group sufficiently to make the lobe of the σ-antibonding (σ*) orbital which is concentrated beyond the carbon atom available for π-bonding. In CH_3F where the hydrogen atoms have no nonbonding electrons to interact with the σ* orbital, there is little or no effect. However, in CH_2F_2, where each fluorine atom has nonbonding electrons in p-orbitals of favorable disposition, the p-electrons of each interact with the σ* orbital associated with the other to give a p_π-σ_π^* bond, which results in a strengthening and shortening of both bonds.

This effect is observed in the shortening of such bonds as C—N in CCl_3NO_2 and $(CF_3)_3N$, C—P in $(CF_3)_3P$, C—S in $(CF_3)_2S$ and so forth. Such a mechanism, on the other hand, could not result in a shortening of the C—C bond in C_2F_6.

Table 12 gives standard covalent radii for most of the elements of the periodic table. Most of these have been calculated from the best data available in the literature using the considerations of paragraphs 1–3 above. Thus when radii of the appropriate multiplicity and hybridization are added and corrected for difference in electronegativity by the Stevenson and Schomaker relationship, the observed bond length will be obtained. For example, the O—F bond in OF_2 would have the value $0.745 + 0.709 - 0.9(0.5) = 1.409$ Å. The experimental value is 1.41 Å.

The sp^3 double bond radii for Si, P, S and Cl are taken from the paper of D. W. J. Cruickshank, *J. Chem. Soc.*, 5486 (1961). A calculation of the Cl=O bond length gives

$$0.681 + 0.654 - 0.9(0.5) = 1.290 \text{ Å}$$

while for the Cl—O bond we calculate

$$1.050 + 0.745 - 0.9(0.5) = 1.750 \text{ Å}$$

The observed Cl—O bond length in ClO_4^- (1.48 Å) indicates that it has a bond order somewhat greater than 1.5.

The hybridization designations given in the table are not intended to be exact. In particular it must be recognized that the nitrogen bonding orbitals in the ammonia molecule, for example, have considerable s character. Nevertheless, such orbitals, for simplicity's sake, have been designated merely as p.

This practice has been used uniformly when a nonbonding pair of electrons is found in the valence level. In the case of "sp^3d" hybridization, the radii are given as "sp^3d" if only average lengths were available, but are separated into sp^2 and pd if the data differentiated the two types of bonds. The values given for radii of the transition elements and for the less common hybridization states of the other elements, especially where the bond order is not accurately known, must be considered to be only approximate.

Valence

The capacity of an atom to combine with other atoms to form a molecule. Valence is specified as the number of hydrogen atoms or twice the number of oxygen atoms with which one atom of the element under question will combine. Thus, nitrogen has the valence 3,2,4,5 in the compounds NH_3, NO, NO_2, N_2O_5. A further distinction is made by considering positive and negative valences. If the hydrogen is assigned the valence of plus one, and oxygen that of minus two, and if the valences in a compound are made to total up to zero, we have a formal scheme of positive and negative valences. In ammonia, NH_3, the three hydrogen atoms each with a valence of plus one exactly balance the one nitrogen atom with the valence of negative three. Many atoms possess more than one valence but the principal valence is correlated with the periodic table and the atomic structure of the atom. The principal positive valence is the number of the group in which the element falls in the periodic table. Thus hydrogen is one, lithium also one, boron three, etc. The negative valence is eight minus the number of the group in the periodic table. Negative valences greater than four do not occur. For example, oxygen has the valence of eight minus six, that is two negative in H_2O (water); and nitrogen has the valence of eight minus five, that is three negative in NH_3 (ammonia).

On the basis of modern electronic theory of atomic structure we can classify the different types of valence. The guiding principle is that the atoms tend to assume an inert gas electronic structure of eight electrons in the outer shell (in the case of hydrogen it is two). To do this, the atom either loses to, gains from, or shares with other atoms, electrons. This process leads to molecule formation. The following are the principal types of valences and their electronic interpretation.

Electrovalence or polar valence is associated with a transfer of an electron from one element to the other in order to complete by such a transfer the octet of each element. Thus in sodium chloride the sodium atom has one valence electron outside a closed octet of eight. By loss of this electron the sodium atom becomes positively charged sodium ion because the nuclear positive charge exceeds that of the electrons by one. On the other hand, the chlorine atom has a grouping of seven electrons in the outer shell. It picks up another electron to complete its outer shell to an octet, but in so doing obtains a total charge of one minus, becoming a chloride ion. The result is that in sodium chloride we are not dealing with sodium atoms and chlorine atoms but with sodium and chloride ions. This is experimentally substantiated. The forces holding the ions together are the electrostatic forces, which are equal to the product of the electronic charges on the ions divided by the product of the separation squared times the dielectric constant of the medium. Thus when the sodium chloride crystal is placed in solvent of high dielectric constant such as water, the forces between the ions are weakened and the ions float away from each other. In other words, electrolytic dissociation takes place. It must be noted that polar valences have no specific directional effects in space. The electrostatic attraction is best satisfied by a close packing of the ions. Inasmuch as there are large stray electric fields present in polar compounds, they possess a high melting point and considerable hardness.

Homopolar or covalent bonds are formed by a different mechanism. Here again we have as the basis the tendency of each atom to complete its outer shell of electrons to eight, or in the case of hydrogen to a doublet. In contrast to polar valence, in covalence we have no direct transfer of electrons, but merely a sharing. In the case of molecular hydrogen each hydrogen atom with its one electron shares this electron with the other hydrogen. The result is that each atom in the molecule has at least part of the time a complete shell of two electrons. The electrons can be visualized as traveling in orbits encompassing the two hydrogen nuclei. It is a property of the covalent bond that it is not weakened by electrolytic solvents and that it has a definite direction in space. These directional effects of covalent bonds are expressed in stereochemistry. Thus, for example, the four valence bonds of the carbon atoms are arranged to extend from the center of a tetrahedron to the four corners. Furthermore, since there is a one-to-one saturation of the electron forces, the stray electric fields are negligible, the melting points are low, and the crystals are soft.

Intermediate in properties between the electrovalent and covalent bonds discussed above is the *semi-covalent bond* (also called *dative* or *polarized ionic bond*). It is formed when both electrons that constitute the bonding pair are supplied by one of the atoms. An example is the formation of amine oxides between tertiary amines and oxygen, in which both electrons are donated by the nitrogen atom. Such bonds naturally exhibit electrical polarity. They are members of the large class of heteropolar bonds which are characterized by an unequal distribution of charge due to a displacement of the electron-pair so that the effect of the bond is to make the atoms differ in polarity. In fact, atomic bonds are best described, not qualitatively, but in terms of bond angles and distances. In water, for example, the bond angle is 109.5°, indicating that the lines joining the two hydrogen atoms to the oxygen atom meet at this angle. However, there are two special types of bonds which deserve individual mention.

The *hydrogen bond* is actually two bonds, whereby two electronegative atoms are joined through a hydrogen atom. Since a stable hydrogen atom cannot be associated with more than two electrons, the hydrogen bond may be regarded as a resonance phenomenon, whereby the hydrogen atom is periodically attached to each of the two other atoms in turn, so that its behavior is a composite of the two structures.

Another type of bond which occurs in solids is the metallic bond. It can be considered as an extreme case of sharing of electrons in that an electron gas (present in the crystal lattice) is shared not by two ions but by all the ions in the lattice. This electron gas is responsible for the metallic properties of certain solids, especially for thermal and electrical conductivity.

As a consequence of the fact that many valence bonds leave residual electrical fields, many molecules in which the "primary" valences are satisfied can combine further with other molecules or with atoms. These higher combinations enter into many important areas of chemical science. They are the basis of the formation of coordination compounds, discussed under that heading. They cause molecular association. They are responsible for the formation of hydrates. They are in many cases the binding forces in nonionic solids, and are of great importance in explaining the structure of larger material aggregates.

The foregoing discussion of valence is, of course, a simplified one. From the development of the quantum theory and its application to the structure of the atom, there has ensued a quantum theory of valence and of the structure of the molecule, discussed in this book under **Molecule**. Topics that are basically important to modern views of molecular structure include, in addition to those already indicated, the Schrodinger wave equation, the molecular orbital method (introduced in the article on **Molecule**) as well as directed valence bonds, bond energies, hybrid orbitals, the effect of Van der Waals forces, and electron-deficient molecules. Some of these subjects are clearly beyond the space available in this book and its scope of treatment. Even more so is their use in interpretation of molecular structure. See **Ligand**.

There are a number of terms used in describing the individual valence bonds. The *bond angle* is the angle between two bonds in a molecule, e.g., in water the bond angle is 109.5°, indicating that the lines joining the two hydrogen atoms to the oxygen atom meet at this angle.

The term *bond direction* arises from the fact that certain covalent bonds prefer to lie in particular directions with respect to the bonded atoms. For example, the bonds from carbon point from the center to the vertices of a regular tetrahedron.

Atoms sharing two pairs of electrons are said to be connected by a *double bond*. The bond energy of the C—C bond is 80 kcal and that of the C=C bond is 145 kcal. The second bond is formed by *p*-electrons and, while its energy effect is considerable, it does not produce a double bond having twice the energy of the single bond. Moreover, its electrons, being less firmly held between the carbon atoms, are available for addition reactions. These bonds between *p*-electrons, or π-bonds, tend to delocalize in many cases, i.e., the electronic charges "spread" over other atoms than those furnishing them.

The diagram below, which shows the bonding of the carbon atoms in the ethylene molecule, shows the carbon atoms connected by one of the sp^2 hybrid bonds (solid line) the other two being used for the hydrogen atoms. The dotted lines show the π-bond formed between the two *p*-electrons. Since they occupy *p*-orbitals which are perpendicular to the plane of the sp^2 bonds, they cannot form a bond without considerable overlapping. From the figures of 80 and 145 kcal for single and double bonds, the π-bond accounts for 44% of the energy of the double bond, indicating extensive overlapping.

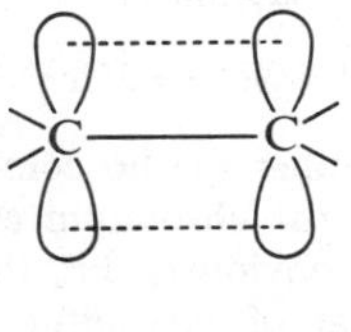

Conjugated double bonds are two double bonds in positions connecting alternate pairs of carbon atoms. For example, the compound

$$CH_2 = CH - CH = CH_2$$

has conjugated double bonds. In addition reactions, the conjugated double bond system commonly changes to a single double bond between the second and third carbon atoms, accompanied by the addition of atoms or groups to the first and fourth carbon atoms.

Triple Bond. A single C—C bond involves *sp-sp* overlapping of orbitals, while a C—H bond is the result of *sp-s* overlapping. The other two valences on carbon atoms are represented by two remaining π-electrons, occupying mutually perpendicular *p*-orbitals. In the case of *triple bonding*, —C≡C—, overlapping of these four *p*-orbitals gives two π molecular orbitals. Thus the carbon-carbon *triple bond* is conveniently represented as —C:C—, in which the π-electrons are shown occupying posi-

tions on the periphery of the carbon-carbon single bond. Their mutual repulsion reduces their bonding effectiveness, so that the bond energies for single, double and triple carbon-carbon bonds are 80, 145, and 198 calories, respectively.

Bond Energies. It has been suggested that ΔH^0_{298}, the heat of formation of a molecule from its constituent atoms (see **Atomic Heat of Formation**), could be computed from a table of average bond energies, and the assumption of additivity:

$$\Delta H^0_{298} = \sum_{\substack{all\ types \\ of\ bonds}} n_{Xi-Xj} \cdot E_{Xi-Xj}.$$

n_{Xi-Xj} is the number of bonds between the two atomic species, X_i and X_j, in the molecule. E_{Xi-Xj} is the average bond energy associated with each of these bonds. Fairly accurate predictions of the heats of formations of organic molecules can be made in this way, particularly for the *larger* hydrocarbons, alcohols and other aliphatic derivatives.

The differences between the observed heats of formation of *cistrans* isomers and of branched and unbranched hydrocarbon chains show that the additivity rule is not strictly rigorous. Various improvements have been suggested.

Single bond energies are given directly by the heat of dissociation of the corresponding molecules into neutral atoms. In cases where the molecule has no independent existence, other data may often be used. For example, the complete dissociation energy of a binary compound containing more than two atoms may be divided by the number of bonds broken in the dissociation, that is, the energy of the A—B bond may be taken as $\frac{1}{2}$ the dissociation energy of A—B—A into 2A and B, or as $\frac{1}{3}$

$$A$$
$$/$$
the dissociation energy of A—B\ \ \ \ into 3A and B. This multiple
$$\backslash$$
$$A$$

bond calculation yields, of course, an average value for the energy of the bonds involved. In the simple case of the H_2O molecule, the dissociation energies of the two successive steps $H_2O \rightarrow H + OH \rightarrow H + H + O$, differ by about 10%.

Representative single bond values were given in Table 11. Summation of such energies to obtain *average* values for molecules applies only when the constituent atoms exhibit their normal covalences and is subject to the exceptions already stated.

Electrides. In late 1979, researchers at Michigan State University reported the assembly of a curious salt wherein an electron, instead of chloride for example, furnishes the negative charge. These unusual compounds have been called *electrides* and apparently can be built around alkali metal ions, such as lithium, potassium, and cesium cations. In the synthesis, positively charged metal ions are trapped. Large organic molecules called *cryptate complexes* are used as the traps. Using various solvents, such as liquid ammonia, the investigators have solvated the electrons on the outside. Upon removal of the solvent, the remaining electron appears to serve as an anion to the alkali metal-cryptate complex. Most research to date has concentrated on lithium. Thus far, the electride has not been crystallized. However, several properties have been determined. For example, lithium electride, found in two forms, transmits intense blue light. The researchers observed that although electrons, as yet, have no place on the periodic table, they appear to have a demonstrated chemistry.

References

Christiansen, B., and T. H. Clack, Jr.: "A Western Perspective on Energy: A Plea for Rational Energy Planning," *Science*, **194**, 578–584 (1976).

Chynoweth, A. G.: "Electronic Materials: Functional Substitutions," *Science*, **191**, 725–732 (1976).

Harvey, B. G., et al.: "Criteria for the Discovery of Chemical Elements," *Science*, **193**, 1271–1272 (1976).

Hayes, E. T.: "Energy Implications of Materials Processing," *Science*, **191**, 661–665 (1976).

Heath, R. L.: "Table of the Isotopes," in "Handbook of Chemistry and Physics," 61st Edition, CRC Press, Boca Raton, Florida (1980).

Heath, R. L.: "Gamma Energies and Intensities of Radionuclides," in "Handbook of Chemistry and Physics," 61st Edition, CRC Press, Boca Raton, Florida (1980).

Holden, C.: "Critical Minerals," *Science*, **212**, 305 (1981).

Lefort, M., Ngo, C., Peter, J., and B. Tamain: *Nucl. Phys.*, **A 216**, 166 (1973).

Lefort, M., and C. Ngo: *Ann. Phys.* (Paris), **3**, 5, (1978).

McCulloch, M. T., and G. J. Wasserburg: "Sm-Nd and Rb-Sr Chronology of Continental Crust Formation," *Science*, **200**, 1003–1011 (1978).

Metz, W.: "The Essential Trace Elements," *Science*, **213**, 1332–1338 (1981).

Meyerhof, W. E.: "X-rays from Coalescing Atoms," *Science*, **193**, 839–848 (1976).

Morgan, J. D., Jr.: "The Mineral Position of the United States," *Chemical Engineering Progress*, **73**, 2, 51–56 (1977).

Naldrett, A. J., and J. M. Duke: "Platinum Metals in Magmatic Sulfide Ores," *Science*, **208**, 1417–1424 (1980).

Nozette, W., and W. V. Boynton: "Superheavy Elements: An Early Solar System Upper Limit for Elements 107 to 110," *Science*, **214**, 331–332 (1981).

Penzias, A. A.: "The Origin of the Elements," *Les Prix Nobel en 1978*, Nobel Foundation, Stockholm; also "Nobel Lectures," (in English), Elsevier, Amsterdam and New York (1979); also reprinted in *Science*, **205**, 549–554 (1979).

Rampacek, C.: "Impact of Research and Development on Utilization of Low-Grade Resources," *Chemical Engineering Progress*, **73**, 2, 57–68 (1977).

Robinson, A. L.: "Superheavy Elements: Confirmation Fails to Materialize," *Science*, **195**, 473–474 (1977).

Schröder, W. U., and J. R. Huizenga: *Annu. Rev. Nucl. Sci.*, **27**, 465 (1977).

Seaborg, G. T., Editor: "Transuranium Elements: Products of Modern Alchemy," Academic, New York (1979).

Seaborg, G. T., Loveland, W., and D. J. Morrissey: "Superheavy Elements: A Crossroads," *Science*, **203**, 711–717 (1979).

Staff: "Electrons as Chemical Elements," *Science News*, **116**, 427 (1979).

Weast, R. C., Editor: "Handbook of Chemistry and Physics," CRC Press, Boca Raton, Florida (Published annually).

CHEMICAL EQUATION. By means of chemical formulas, the changes occurring during a chemical reaction can be expressed as an equation. Thus the reaction of 1 mole of sulfur with 1 mole of oxygen to produce 1 mole of sulfur dioxide is written as

$$S + O_2 \rightarrow SO_2$$

The arrow is preferred to the equality sign, which does not emphasize the direction of the reaction. In addition to the identity of the reactants and products, the equation shows the number of atoms entering into the reaction, either in the atomic state or as constituents of molecules. It also shows the number of moles of each reactant and product, so that by use of the table of atomic weights, the relative masses can be computed.

Since the principle of conservation of masses applies to chemical reactions, coefficients must often be used in writing chemical reactions so that the number of atoms of products is equal to the number of atoms of reactants. An example is the reaction of *two* moles of hydrogen with *one* mole of oxygen to form *two* moles of water

$$2H_2 + O_2 \rightarrow 2H_2O$$

In writing such equations, a convenient procedure is to write first an expression containing only the formulas, and then to add the smallest coefficients that will give the same number of atoms of products as of reactants. This operation is called balancing the equation.

Equilibrium reactions, such as that of acetic acid and ethyl alcohol to form ethyl acetate and water, which is cited in the entry on **Chemical Reaction Rate**, are indicated by use of the double arrow

$$CH_3COOH + C_2H_5OH \leftrightharpoons HOH + CH_3COOC_2H_5$$

Reactions which result in the precipitation of a solid or the evolution of a gas are sometimes denoted by vertical arrows

$$AgNO_3 + NaCl \rightarrow AgCl \downarrow + NaNO_3$$

$$Na_2CO_3 + 2HCl \rightarrow CO_2 \uparrow + H_2O + 2NaCl$$

This information about the state of the products may also be denoted by writing after their formulas the expressions (s), (l), or (g).

In some cases, as in reactions in electrochemical cells or other reactions involving oxidation-reduction, the half reactions of the ions are useful. Consider the Daniell cell, which consists of a zinc electrode in a zinc sulfate solution, and a copper electrode in a copper solution, the two solutions being separated by a porous partition. The half reactions are

$$Zn \rightarrow Zn^{2+} + 2e^-$$

$$Cu^{2+} + 2e^- \rightarrow Cu$$

so that the overall reaction is

$$CuSO_4 + Zn(s) \rightarrow ZnSO_4 + Cu(s)$$

The more difficult oxidation reduction equations can often be written more easily by use of the Stock system of oxidation numbers, which are positive or negative valences or charges. Consider the reaction of potassium dichromate $K_2Cr_2O_7$ with potassium sulfite K_2SO_3 in acid solution to form chromium(III) sulfate $Cr_2[SO_4]_3$ and potassium sulfate K_2SO_4. The unbalanced expression for the ionic reaction is

$$Cr_2O_7^{2-} + SO_3^{2-} \rightarrow 2Cr^{3+} + SO_4^{2-}$$

Since the oxidation number of the combined oxygen atom is 2− throughout, that of the chromium atom in $Cr_2O_7^{2-}$ is 6+, that of the Cr^{3+} ion is obviously 3+, that of the sulfur atom is SO_3^{2-} is 4+, and that of the sulfur atom in SO_4^{2-} is 6+. The total loss in oxidation number by the two chromium atoms is therefore $(2 \times 6) - (2 \times 3) = 6+$. Since this loss must be offset by a gain made by the sulfur atoms, and since one sulfur atom gains 2+, the reaction must require 3 sulfur atoms. Therefore, the next partially balanced equation is written as

$$Cr_2O_7^{2-} + 3SO_3^{2-} \rightarrow 2Cr^{3+} + 3SO_4^{2-}$$

Counting the charges in this expression shows that there are 8 negative charges on the left-hand side and a net total of 0 charges on the right-hand side. Therefore, since the reaction occurs in acid solution, requiring that hydrogen ions are needed and must be present, $8H^+$ are added to the right-hand to balance the expression electronically, giving

$$Cr_2O_7^{2-} + 3SO_3^{2-} + 8H^+ \rightarrow 2Cr^{3+} + 3SO_4^{2-}$$

Now it is balanced in number of atoms by counting the hydrogen ions (8 on the left-hand side), and the oxygen atoms (an excess of 4 on the left-hand side). Therefore 4 H_2O is added to the right-hand side

$$Cr_2O_7^{2-} + 3SO_3^{2-} + 8H^+ \rightarrow 2Cr^{3+} + 3SO_4^{2-} + 4H_2O$$

If the molecular equation is wanted, it can be written by grouping the ions, and adding those that did not enter into the ionic equations, i.e., the potassium ions of the salts and the anions of the acid

$$K_2Cr_2O_7 + 3K_2SO_3 + 4H_2SO_4 \rightarrow$$
$$Cr_2(SO_4)_3 + 4H_2O + 4K_2SO_4$$

CHEMICAL EQUILIBRIUM. The fundamental law of chemical equilibrium is that enunciated by Le Chatelier (1884), and may be stated as follows: If any stress or force is brought to bear upon a system in equilibrium, the equilibrium is displaced in a direction which tends to diminish the intensity of the stress or force. This is equivalent to the principle of least action. Its great value to the chemist is that it enables him to predict the effect upon systems in equilibrium of changes in temperature, pressure, and concentration.

The chemical system, hydrogen-nitrogen-ammonia, furnishes a notable example of the application of the principle:

nitrogen + hydrogen ⇆ ammonia +   heat
1 vol.   3 vol.   2 vol.   12,000 calories
                               per mole ammonia
       4 vol.

At the temperature 700°C and pressure 1 atmosphere, the equilibrium percentage of ammonia is 0.03 in the above system, and at 100 atmospheres 2.5. Increase of pressure shifts the equilibrium towards the side of the smaller total volume, at a constant temperature. Decrease of pressure shifts the equilibrium towards the side of the larger total volume, at a constant temperature. Systems of the same initial and final volumes are unaffected, as to equilibrium amounts of materials, by change of pressure.

At the pressure 100 atmospheres, and temperature 700°C, the equilibrium percentage of ammonia is 2.5 in the above system, at 600°C it is 5, at 500°C it is 10. Increase of temperature shifts the equilibrium in the direction which absorbs heat, at a constant pressure. Decrease of temperature shifts the equilibrium in the direction which evolves heat (van't Hoff's principle, 1884).

At constant pressure and temperature, the equilibrium is shifted away from the side subjected to an increase in concentration of any constituent, or towards the side subjected to a decrease in concentration of any constituent. (See **Chemical Reaction Rate.**) For the qualitative effect of temperature change, one may visualize the heat of an equilibrium reaction as material, and an increase of temperature (heat intensity) as operating to increase the concentration of "heat material" thus shifting the equilibrium away from the side of its increased concentration, and conversely. It is possible, knowing the heat of reaction, Q, on the assumption that the heat of reaction is constant between two given (absolute) temperatures, T_1 and T_2, to calculate the equilibrium constant K_2 (at T_2) when the equilibrium constant K_1 (at T_1) and the gas constant, R (equals 2 calories per mole), are known, by the application of van't Hoff's equation:

$$\log_{10} K_2 - \log_{10} K_1 = \frac{Q}{2.3 \times R}\left(\frac{1}{T_2} - \frac{1}{T_1}\right)$$

In this way the quantitative effect of temperature change on the state of equilibrium may be calculated.

In reactions of the ammonia synthesis type, to which sulfur trioxide from sulfur dioxide plus oxygen also belongs, the rate of reaction decreases with lowering of the temperature as the conversion is increased. There is, in such types of reactions, a limit to the practicable lowering of the temperature. The finding of a positive catalyzer for a given reaction of this sort permits the operation to gain the advantage of equilibrium conversion at the lower temperature as well as the increased rate of reaction at that temperature due to the presence of the catalyzer. (See **Chemical Reaction Rate.**) The time yield of product is, therefore, very important, and, with a catalyzer, the space-time yield.

Systems in equilibrium are divided into two great divisions, according to whether they are (A) homogeneous, that is, chemically and physically uniform throughout, or (B) heterogeneous, that is, not uniform throughout but consisting of two or more phases. Each phase is a homogeneous, physically distinct, and mechanically separable portion of a system. For example, ice, water, water vapor are three different phases (solid, liquid, gas) of the substance water. There can be only one gas phase of a system, and only one liquid phase where a *single* homogeneous solution is present. But the number of liquid and of solid phases in general is limited by the number of components (not constituents) of a system. The number of components is the least number of constituents, *independently* variable, and requisite to compose each and every phase. For example, the system consisting of saturated solution in water H_2O of sodium sulfate Na_2SO_4 plus solid sodium sulfate decahydrate $Na_2SO_4 \cdot 10H_2O$ plus water vapor consists of three phases, (a) gas, (b) solution, (c) solid sodium decahydrate. The *least* number of constituents, independently variable in amount *and* requisite to compose each and every phase is two, namely, Na_2SO_4 and H_2O. These, therefore, are the two components of this system. Since zero and negative as well as positive amounts of compounds are permitted in expressing the composition of each phase of any system, the three phases of this system are composed of the following components:

gas phase, zero	Na_2SO_4 plus H_2O	
liquid phase,	Na_2SO_4 plus H_2O	
solid phase,	Na_2SO_4 plus H_2O	

The number of components in the ice-water-water vapor system is one, namely, H_2O.

To systems in which equilibrium depends solely upon the following variables, namely, (1) composition of each and every phase, (2) temperature, and (3) pressure, the phase rule (Willard Gibbs, 1874) applies: The number of variables, that is (1) the number of components, C, plus (2) temperature plus (3) pressure, above, equals the number of phases, P, plus the number of degrees of freedom, F. The number of degrees of freedom of a system is the least number of the above variables which must be arbitrarily fixed in order to define the condition of the system:

$$C + 2 = P + F$$

The phase rule applies to true equilibrium systems, where the equilibrium can be reached from either side, and, furthermore, takes no account of the time involved to attain equilibrium. The phase rule is a qualitative statement, whereas the law of mass action (concentration effect) is quantitatively applicable to those equilibrium systems where the reaction which occurs may be considered to take place in a homogeneous system, e.g., gas phase, or solution phase. (See **Chemical Reaction Rate.**)

In a one-component system, $P + F = 3$, and physical changes only occur. When only one phase is present, for example, liquid water (no vapor, no solid) the system is bivariant, that is, two variables—temperature and pressure—may be independently changed over a range. When a second phase, either vapor or solid appears through a sufficient change of temperature or pressure or both, or when two phases are originally present, the system is univariant, that is, one variable—either temperature or pressure—may be independently changed over a range. When the third phase appears or when three phases are originally present, the system is invariant, that is, a change of either temperature or pressure destroys the equilibrium, and the disappearance of one of the phases occurs. A system of one component in three phases is invariant and the conditions are represented by a point known as the triple point. The triple point for water is 0.007°C, 4.6 millimeters mercury pressure. When the total pressure is one atmosphere (760 millimeters) the equilibrium temperature of water-ice is 0.000°C, and when the water vapor pressure is one atmosphere the equilibrium temperature of water-water vapor is 100.000°C.

If, in dealing with any system, the gas phase or pressure may be neglected, on account of constancy or slightness of effect, the phase rule is simplified for practical purposes to $C + 1 = P + F$, and, if both may be neglected, to $C = P + F$.

Many two- and three-component systems have been studied and recorded in detail. The iron-carbon system is one that has attracted much attention and been of great value in iron metallurgy.

In 1977, Professor Ilya Prigogine of the Free University of Brussels, Belgium was awarded the Nobel Prize in chemistry for his central role in the advances made in irreversible thermodynamics over the last three decades. Prigogine and his associates investigated the properties of systems far from equilibrium where a variety of phenomena exist that are not possible near or at equilibrium. These include chemical systems with multiple stationary states, chemical hysteresis, nucleation processes which give rise to transitions between multiple stationary states, oscillatory systems, the formation of stable and oscillatory macroscopic spatial structures, chemical waves, and the critical behavior of fluctuations. As pointed out by I. Procaccia and J. Ross (*Science*, **198**, 716–717, 1977), the central question concerns the conditions of instability of the thermodynamic branch. The theory of stability of ordinary differential equations is well established. The problem that confronted Prigogine and his collaborators was to develop a thermodynamic theory of stability that spans the whole range of equilibrium and nonequilibrium phenomena.

CHEMICAL FORMULA. The formulas of chemistry constitute a shorthand notation used to represent the composition by weight, the molecular properties, the characteristic chemical reactions or at times even the ordering of the atoms in space of the elements which go to make up the chemical compound. Chemical formulas are classified into empirical, molecular, structural, or configurational, the order given being that of increasing content of information. The meaning of empirical and the molecular formulas is explained in the entry on **Chemical Composition,** which also describes methods for determining the formulas for some simple compounds. Their determination for compounds in general, especially if they are present in mixtures, requires considerable experimental work. The first step consists of the isolation of a pure chemical compound. Chemical purification can be obtained by methods such as crystallization, distillation adsorption, and sublimation. Some of the criteria of purity which a substance must satisfy are constancy and sharpness of melting point and boiling point on repeated purification. As an example, let us assume that we have succeeded in purifying

a solid compound which we shall call tartaric acid and whose formula we wish to determine.

The second step consists in a qualitative and quantitative analysis of the compound. In the case of tartaric acid, qualitative analysis tells us that the compound contains carbon, oxygen and hydrogen, while quantitative analysis shows that the porportions are 48 parts by weight of carbon, 96 of oxygen, and 6 of hydrogen. To obtain the empirical formula, one divides each proportion by the atomic weight of the particular element, obtaining in this way a set of numbers which can be represented by a ratio of small integers. The simplest ratio of integers is commonly used to indicate as subscripts on the right of the chemical symbol of the element to represent the empirical formula. In the case of tartaric acid, the atomic weights are approximately 12 for carbon, 16 for oxygen, and 1 for hydrogen. Dividing the percentages as determined by analysis by the atomic weights, we get:

$$\text{carbon} \quad \tfrac{48}{12} = 4.00$$
$$\text{oxygen} \quad \tfrac{96}{16} = 6.00$$
$$\text{hydrogen} \quad \tfrac{6}{1} = 6.00$$

The set of numbers is 4,6,6 and can be presented, in this case, by the ratio of integers $2:3:3$. The empirical formula is therefore $C_2O_3H_3$. Empirical formula is thus only a convenient method for representing the percentage composition by weight of the different elements in the compound. The third step is the determination of the molecular weight of the compound in question. This allows us to assign to the compound a molecular formula. The molecular weight can be determined in a variety of methods, such as by the determination of the weight of 22.242 liters of the vapor of the substance at 1 atmosphere pressure and 0°C, temperature. Other methods are based on the differences in the boiling point or freezing point of solutions of known concentration and those of the pure solvent. To determine the molecular formula from the knowledge of the empirical formula and the molecular weight, the following procedure must be followed: Multiply the atomic weight of each element by its subscript, as indicated in the empirical formula, and add the result. On comparison of such a sum with the molecular weight it will be found that the molecular weight is equal to the sum times an integer. To obtain the molecular formula multiply each subscript in the empirical formula by this integer and obtain a new set of subscripts. We found the empirical formula of tartaric acid was $C_2O_3H_3$. The sum mentioned above is

$$12 \times 2 + 16 \times 3 + 1 \times 3 = 75$$

The molecular weight determined experimentally is 150. The integer multiple is 2, and the molecular formula becomes $C_4O_6H_6$.

The molecular weight of the compound can be obtained from the molecular formula by summing the products obtained by multiplication of the atomic weights of the elements times their subscripts in the molecular formula. The latter contains all the information that the empirical formula contains but in addition specifies the number of atoms in the molecule and also the molecular weight of the substance.

Important as the molecular formula is, it does not describe fully the properties, or even in some cases the identity, of chemical compounds. For example, there are two compounds we have the molecular formula C_2H_6O. They are different in all their properties, both chemical and physical. This difference is due to a difference in the manner in which the atoms are connected in the molecules of the two substances. These differences can be shown only by the use of structural formulas, such as those

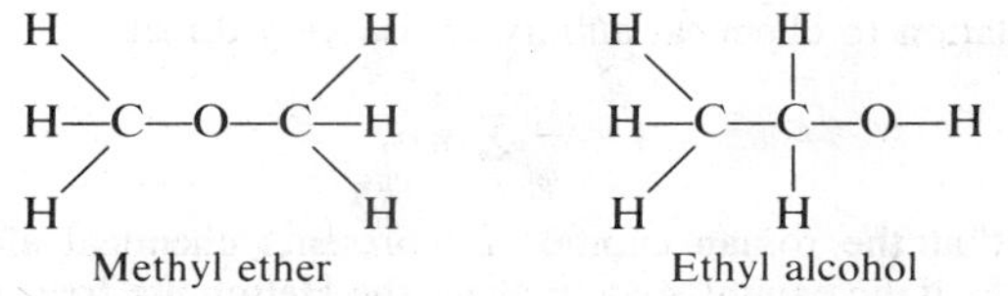

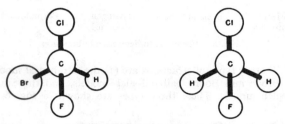

Fig. 1.   Examples of structural formulas.

shown in Fig. 1, in which the valence bonds between the atom are shown. These structural formulas are determined circumstantially, that is, by the chemical reactions into which the compounds enter. (However, their arrangments have been confirmed in many cases by a direct instrumental means such as spectrometric methods, x-ray studies, etc.) These reactions differ markedly for ethyl alcohol and methyl ether. Such compounds which have the same molecular formula but differ due to the arrangements or positions of their atoms are called isomers, and the type just cited, in which the difference is in the grouping of the atoms, are called *functional isomers*. These, and many other types of isomers, are treated in the entry on **Isomerism.**

The structural formula is also a shorthand notation for the important chemical reactions of the compound. It can be considered as being built up of a group of organic radicals, i.e., groups of atoms which retain their individuality in the course of certain reactions. Each radical has reactions which are characteristic of its presence in the molecule.

For instance, the carboxyl radical $-\overset{\overset{O}{\|}}{C}-OH$ will react with

alkali such as sodium hydroxide to form salts $-\overset{\overset{O}{\|}}{C}-ONa,$ with phosphorous pentachloride to form acid chlorides

$-\overset{\overset{O}{\|}}{C}-Cl$; with alcohols to form esters; with reducing agents under certain conditions to form successively the aldehyde radi

cal $-\overset{\overset{O}{\|}}{C}-H$ and the alcohol radical. Any compound which undergoes such reactions is said to contain a carboxyl group. The number of such carboxyl groups in a molecule can be determined by studying the above reactions quantitatively. On the other hand if the compound will react with sodium to give off hydrogen; with phosphorus trichloride to give a halogen substitution product which can be reduced to hydrocarbon; with an oxidizing agent to give an aldehyde or ketone, with organic acids to form esters; with alcohols to form esthers; then the molecule is said to contain a hydroxyl group $-OH$. Analogously there are similar characteristic reactions for a variety of radicals. It often happens that the presence of one type of a radical near another type mutually influences their reactivity, but one can consider to the first approximation that the radicals act independently of each other. The structural formula is considered completely established if one can synthesize the compound by simple clear-cut reactions involving no rearrangements on the basis of the proposed formula.

Just as it was stated above that two compounds may have the same molecular formula and yet have quite different structural formulas and properties, so there are also many instances in chemistry of compounds which have the same planar structural formula and yet differ in properties. In such cases their differences in structure can be shown only by three-dimensional formulas or their projections which portray differences in the

arrangement in space of the atoms or radicals that make up the molecules of the two compounds.

Thus, in cases where four different atoms or groups are attached to the same atom, it is possible to have two arrangements in space which cannot be made to coincide geometrically. This situation can be demonstrated by use of a special type of formula, shown in Fig. 2 for the two forms of the compound fluorochlorobromomethane.

Fig. 2.   Formulas to indicate spatial geometry of compounds.

This existence of two forms due to a difference in orientation in space is called *stereoisomerism*, and is discussed in the entry on isomerism. It also follows that for compounds containing more than one atom bonded to four unlike groups, the number of different forms increases rapidly, as is shown by the three possible forms of tartaric acid, HOOC—CHOH—CHOH—COOH, as portrayed by the three formulas shown in Fig. 3.

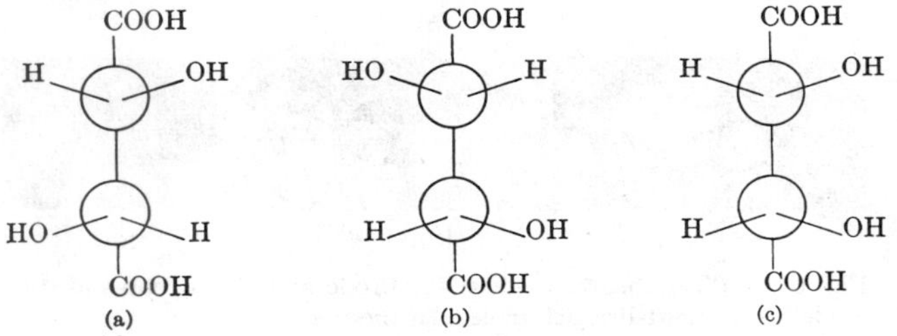

Fig. 3.   Formulas to demonstrate stereoisomerism.

Still other types of formulas showing the spatial positions of atoms and groups are perspective and projection formulas, as shown in Fig. 4 for the compound 1,2-dichloroethane. These

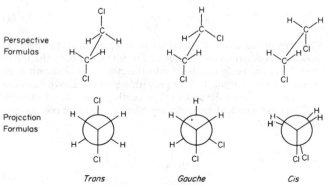

Fig. 4.   Perspective- and projection-type formulas.

differences are due to differences in conformation, that is, to the various configurations of a molecule which differ in space by the rotation of two atoms about a single bond. See also **Isomerism.**

Another type of formula which is often written for compounds is the electronic formula, showing the distribution of the valence electrons among the atoms of the molecule, as shown in Fig. 5 and explained under valence and molecule.

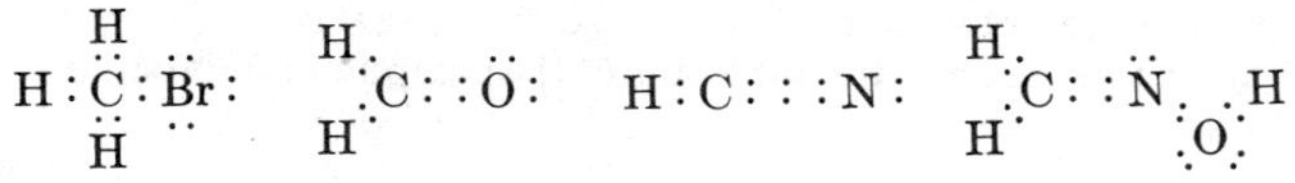

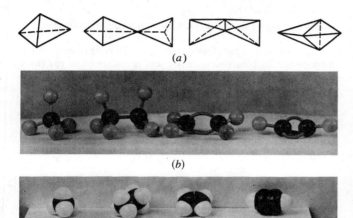

| Methyl bromide | Formaldehyde | Hydrogen cyanide | Formaldoxime |

Fig. 5.   Formulas to demonstrate bonding.

Still other types of structures are (1) the tetrahedral models, the Brode and Boord ball-and-stick models, and the Stuart-Brieglib models. These three types are shown in Fig. 6 for

Fig. 6.   (a) Tetrahedral models; (b) Brode and Boord ball-and-stick models; (c) Stuart-Brieglib molecular models.

the compounds methane CH_4, ethane H_3C-CH_3, ethylene $H_2C=CH_2$ and acetylene $HC\equiv CH$.

CHEMICAL POTENTIALS.   Chemical potentials are defined in terms of the entropy by the relationship

$$\mu_i = -T\left(\frac{\partial S}{\partial n_i}\right)_{U,V} \tag{1}$$

Apart from the factor T (the absolute temperature) the chemical potential is equal to the change of the entropy due to the introduction of the mole number **i** into the system, at constant total energy U and volume V. The parentheses in the above equation contain the partial derivative representing this rate of change.

Other independent variables are often much more convenient. One has also

$$\mu_i = \left(\frac{\partial U}{\partial n_i}\right)_{S,\,v} = \left(\frac{\partial H}{\partial n_i}\right)_{S,\,p}$$
$$= \left(\frac{\partial A}{\partial n_i}\right)_{T,\,V} = \left(\frac{\partial G}{\partial n_i}\right)_{T,\,p} \tag{2}$$

The last member of Equation (2) shows that μ_i is the partial molar quantity associated with the Gibbs free energy, G. Euler's theorem gives then

$$G = \sum_i n_i \mu_i \tag{3}$$

The relation to chemical affinity is also very direct

$$A = -\sum_i \nu_i \mu_i \tag{4}$$

(Note that the roman capital A represents chemical affinity, while the italic capital A symbolizes the Helmholtz free energy (also called work function.) It follows that the condition for chemical equilibrium is

$$\sum_i \nu_i \mu_i = 0 \tag{5}$$

where the ν_i are stoichiometric coefficients. This formula expresses the law of mass action. Similarly the condition for two phases α and β to be in equilibrium with respect to species i is

$$\mu_i^\alpha - \mu_i^\beta \tag{6}$$

The chemical potential has then the same value in the two phases.

CHEMICAL REACTION RATE.   The chemical composition of a substance is subject to various changes under various conditions, depending upon (1) the nature of the specific substance, (2) the nature of other substances present, and (3) the environment in which it exists (the physical and chemical ambient conditions). Similarly, chemical reaction rates are affected.

The majority of reactions take place between two substances—occasionally one substance only, and sometimes three or more different substances. There are many cases when simple contact of the substances is sufficient to bring about the chemical change, e.g., the rusting of iron in oxygen. In many other cases, the change is not spontaneous, but must be induced, frequently by raising the temperature, as in the burning of fuels. The conditions that are considered important and fundamental are (1) temperature, (2) pressure, (3) medium, if any, (4) catalyzer, if any (5) electric direct current, (6) light. In a given reaction, the change in composition of the substance or substances involved is inherently connected with a change in energy. Thermal, electrical or light energy of a certain potential or intensity and in definite amounts, is requisite to initiate and carry on the reaction, and thermal, electrical or light energy of definite amount is liberated or consumed in the reaction. Every reaction, properly speaking, has both a matter and an energy aspect. While the energy aspect is frequently neglected directly, the conditions must always be in accord with the energy demand, even if apparently not considered. See **Electrochemistry; Photochemistry and Photolysis.** Chemical changes require consideration of three topics, namely, (A) natural rate of chemical reactions, (B) acceleration of the natural rate in the presence of a catalyzer, and (C) the end-point of chemical reactions.

A.   *Natural Rate of Chemical Reactions.* Various factors operate to affect the rate of chemical reactions. By natural rate is understood the rate of a reaction in the absence of a catalyzer. Excluding electrochemical and photochemical reactions, and giving attention to thermochemical reactions only, there are four factors or conditions to be considered, namely, (1) concentration of constituents, (2) temperature, and (3) pressure—important where a gas is involved, (4) nature of the medium, if any.

The general mathematical definition of the rate of a chemical reaction v is

$$v = \frac{d\xi}{dt} \tag{1}$$

where ξ (Greek letter xi) is the extent of reaction, t is time, and the derivative thus represents the rate of change of the extent of reaction.

1. Relation between concentration of reactants and rate of reaction. The rate of a given reaction, at constant temperature and pressure under stated conditions of concentration of the reacting substances, is quantitatively expressible by a velocity constant, which is the fraction of the substances transformed in a unit of time. Many reactions occur instantaneously—true for most reactions in solution in inorganic chemistry—and many others are complicated in subsidiary reactions, so that the velocity constant is measurable in comparatively few cases. The principle, however, holds as stated, whether or not the desired value can be ascertained experimentally.

A simple reaction that was studied by Wilhelmy, and since then by various investigators, is the transformation (hydrolysis) of sucrose $C_{12}H_{22}O_{11}$ in water solution into glucose ($C_6H_{12}O_6$, a polyhydroxy aldehyde) plus fructose ($C_6H_{12}O_6$, a polyhydroxy ketone), which proceeds at a measurable, steady rate in the presence of acid (hydrogen ion). The rate of reaction at any instant is found to be proportional to the amount of sucrose present at that instant.

When a dilute water solution of an ester, such as methyl acetate, is similarly hydrolyzed in the presence of hydrogen ion, the reaction is of the same type. And this statement also applies to the decay of radioactive elements. One of the important radioactive constants is the half-life, that is, the time required for the decay of one-half of the element present at a given instant. See **Nuclear Reactor.**

The preceding cases are instances of *first order reactions*, that is, reactions in which the rate depends only upon the concentration of a single molecular or atomic species. They are also *monomolecular reactions*, that is, reactions in which the initial reactant is only of one species, that is, sucrose, or an ester, or a radioactive element. Note that first order reactions are not necessarily monomolecular; thus $H_2 + D_2 \leftrightarrows 2HD$, is a first order reaction, even though it is called bimolecular because a hydrogen-1 molecule reacts with a hydrogen-2 (deuterium) molecule to form hydrogen deuteride molecules.

We can now introduce a general treatment of the concept of the order of a chemical reaction. A chemical reaction is said to be of the nth order if its rate is directly proportional to the product of n concentrations. Therefore the decomposition of A, if described by the equation

$$\frac{dC_A}{dt} = -kC_A \qquad (2)$$

is a *first order* reaction. Similarly if it is described by the equation

$$\frac{dC_A}{dt} = -kC_A C_B \text{ or } \frac{dC_A}{dt} = -kC_A^2 \qquad (3)$$

it is a *second order reaction.*

The coefficient k which appears in (2) or (3) is called the *rate constant.* Generally the temperature variation of a rate constant may be expressed by

$$k = Pe^{-Q/kT} \qquad (4)$$

where k is the rate constant and k is the Boltzmann constant. This equation is called the *Arrhenius equation.*

We now introduce a general treatment of rate of reaction, which is often called the *absolute reaction rate theory,* because its purpose is to calculate the rate in terms of molecular quantities only.

Consider the reaction

$$A + BC \rightarrow AB + C \qquad (5)$$

To simplify the discussion, assume that A, B and C always remain in a straight line. The course of the reaction may then be followed by noting the values of the two interatomic distances r_{AB} and r_{BC}. At the beginning of the reaction r_{AB} is large and r_{BC} is small while at the end of the reaction r_{AB} is small and r_{BC} is large.

Let us introduce the potential energy surface. The representative point of the system moves on this surface along the so-called *reaction coordinate.* The potential energy along the reaction coordinate is represented schematically in the figure. The maximum of the curve corresponds to a situation where three atoms are very close to one another. Moreover this point is a *maximum* along the reaction coordinate but a *minimum* for the direction normal to the reaction coordinate. Indeed the most probable path is the path involving the minimum potential energy in going from the initial to the final state.

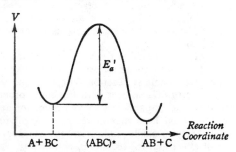

Potential energy along the reaction coordinate.

Therefore the point considered corresponds to a *saddle point* of the energy surface. It is called the *activated complex.*

One may now assume that the reaction rate is the product of the following three factors: (1) the average number of activated complexes; (2) the characteristic frequency of the activated complex (that is, the inverse of its lifetime); (3) the *transmission coefficient,* K, which is the probability that a chemical reaction takes place after the system has reached the activated state.

Moreover the number of activated complexes is calculated by the equilibrium assumption.

Using this description of the reaction process one derives the following expression for the reaction constant

$$k = K \frac{\phi_t(T)}{\phi_A(T)\phi_{BC}(T)} \frac{kT}{h} \exp\left(-\frac{E^x}{kT}\right) \qquad (6)$$

Here the ϕ terms are the partition function $f(T, V)$, the volume factor being removed

$$f = V\phi \qquad (7)$$

ϕ_t corresponds to the activated complex, the degree of freedom associated with the reaction coordinate being removed; k is Boltzmann's constant, h is Planck's constant, E^x is the energy associated with activated complex, or *activation energy* of the reaction.

This expression may also be written in the thermodynamic form

$$k = K \frac{kT}{h} \exp\left(-\frac{\Delta G\ddagger}{kT}\right)$$
$$= K \frac{kT}{h} \exp[-(\Delta H\ddagger - T\Delta S\ddagger)/kT] \qquad (8)$$

where $\Delta G\ddagger$ is a suitable free energy of activation and $\Delta H\ddagger$, $\Delta S\ddagger$ the corresponding enthalpy and entropy of activation.

When a reaction involves two different phases, that is, when the system is not homogeneous but heterogeneous, as in reactions between a solid phase, such as zinc or calcium carbonate, and a liquid phase, such as hydrochloric acid solution, the rate of reaction involves consideration of (1) the area of the surface of contact of the solid with the solution, and (2) the rate of diffusion from the surface of the solid, as well as (3) the concentration of hydrogen ion of the acid solution.

When the rate of a chemical process is dependent upon (1) two or more *consecutive* reactions, the observed rate is limited by the rate of the slowest reaction in the series, (2) two or more *concurrent* reactions, the products are in the same ratio at any instant only when the reactions themselves are of the same rate.

2. Relation between temperature of reactants and rate of reaction. The rate of chemical reaction is increased by an increase in temperature, as is evident from the fact that temperature occurs in the numerators in the foregoing equations.

3. Relation between pressure of reactants, if gaseous, and rate of reaction. Since pressure changes amount to concentration changes in such systems, the behavior is as described above under concentration.

4. Relation between nature of the medium and rate of reaction. Very slight changes in the nature of the medium greatly affect the rate of a chemical reaction, but attempts to relate any physical property of a solvent with the effect observed on the rate of a given reaction appear to have proved unsuccessful.

One should mention here that in reactions involving ions, the effects of electrolytes can be put into two principal categories: (a) primary salt effect and (b) secondary salt effects.

Primary salt effects refer to the effects of electrolyte concentration on the activity coefficients. Secondary salt effects are those concerned with the actual changes in concentration of the reacting species resulting from the addition of electrolytes.

Brønsted has shown that the variation in specific rate with ionic strength depends on the magnitude and sign of the ionic charges. Thus three main types of primary salt effect in reactions of two species can be distinguished. If the products of the signs of the charges are positive, the velocity of the reaction increases with increasing ionic strength:

$$Co[(NH_3)_5Br]^{2+} + Hg^{2+} \qquad (+2 \times +2 = +4)$$

or

$$S_2O_8^{2-} + I^- \qquad (-2 \times -1 = +2)$$

or

$$BrCH_2COO^- + S_2O_3^{2-} \qquad (-1 \times -2 = +2)$$

If the products of the charges are negative the velocity of the reaction decreases with increasing ionic strength:

$$Co[(NH_3)_5Br]^{2+} + OH^- \qquad (+2 \times -1 = -2)$$

or

$$H_2O_2 + H^+ + Br^- \qquad (+1 \times -1 = -1)$$

In the third category where the product of the ionic charges is zero, the ionic concentration has no or very little effect on the velocity of the reaction, particularly in dilute solution, as in the case of

$$[Cr(NH_2CONH_2)_6]^{3+} + 6H_2O \rightarrow [Cr(H_2O)_6]^{3+}$$
$$+ 6NH_2CONH_2 \qquad (+ 3 \times 0 = 0)$$

Brønsted showed the inadequacy of both the classical and the activity theories of rate of certain reactions. He proposed a new theory postulating that when ions or molecules react, they first form an unstable critical complex which then decomposes to give the reaction products. The reaction which determines the velocity of a chemical change consists in the formation of that unstable critical complex.

Summarizing, the rates of chemical reactions are subject to highly specific influences in each case, as has been abundantly demonstrated by experimental investigations, and recognized in numerous legal battles in chemical patent suits.

B. *Acceleration of the Natural Rate of Chemical Reactions* in the presence of a positive or negative catalyzer. When, in the presence of a given substance, the natural rate of a chemical reaction is changed, either increased or decreased, the given substance is called a catalyzer. Examples are numerous. (1) When a gas-lighter of the type known as platinum black, the active part of which consists of very finely divided platinum, is held in a stream of hydrogen or city gas, the gas is ignited in air. Platinum is a catalyzer for this reaction, and causes ignition to take place at a temperature much lower than by subjecting to fire. (2) The changing of sulfur dioxide into sulfur trioxide is accomplished by passing a mixture of sulfur dioxide and air (one-fifth oxygen) over asbestos coated with finely divided platinum. The temperature required is much lower by the use of platinum catalyzer than without its use. (3) Solutions of sulfites are subject to oxidation to sulfates by oxygen upon allowing to stand in air. The addition of sugar or glycerol retards the speed of this reaction. These substances act in this case as negative catalyzers. (4) The combination of nitrogen and hydrogen gases under high pressure to form ammonia gas is accomplished at a lower temperature in the presence of a catalyzer than in its absence, thus increasing the yield of ammonia (see **Equilibrium**). One of the catalyzers is composed of iron, intimately mixed with 1% aluminum oxide and 1% potassium oxide. Iron is a catalyzer for this reaction, but is more active as such in the presence of aluminum oxide and potassium oxide, which are spoken of as promoters, a sort of catalyzer of a catalyzer. (5) The hydrogenation of liquid fatty oils and of oleic acid is conducted in the presence of finely divided nickel as a catalyzer. (6) Enzymes are very specific catalyzers, "the most selective and deicate of all known catalysts (Hilditch)," at ordinary temperatures, say 25 to 30°C. Dextroglucose is converted into ethyl alcohol in the presence of the enzyme (zymase) of yeast, and ethyl alcohol into acetic acid (vinegar) in the presence of the enzyme of *Mycoderma aceti*. (7) Nitric acid reacts slowly with copper metal, but the rate of reaction is accelerated more and more as nitrogen tetroxide (catalyzer) is formed in the solution. This is an example of autocatalysis, wherein the reaction brings about the formation of its own catalyzer. (8) Arsenic-containing substances are extreme negative catalyzers, called inhibitors or poisons, of platinum catalyzer.

When the catalyzer is a solid substance, the greatest difficulty in use is to maintain a clean surface. The presence of a positive catalyzer enables a reaction to proceed more rapidly at a lower temperature than corresponds to the natural rate of the reaction. This increases the amount of substances converted in a given time, decreases the demands as to temperature resistance of materials of construction of the apparatus, and frequently makes possible a state of equilibrium more favorable to the yield of desired material.

C. *The End-point of Chemical Reactions.* If a chemically reactive system is isolated from the rest of the universe at a constant temperature and pressure, a definite end-point is often attained short of the complete transmutation of reactants into resultants. In order to be certain that this end-point (short of complete transmutation) is what is known as the equilibrium point, the equilibrium must be approached from both directions, e.g., A + B → C + D and C + D → A + B. If the equilibrium constant (see treatment below) is the same when approached from both directions, then the reaction is one of true chemical equilibrium. Such equilibrium reactions are also referred to as balanced or reversible reactions. In such reactions the extent of the chemical change is proportional to the concentrations of all the reactants—reactants and resultants being interchangeable, depending upon the direction of the reaction. (Generalization of Guldberg and Waage, 1864, called Law of Mass Action, or more correctly Law of Concentration Effect. Reaction studied by Guldberg and Waage (1867): Barium sulfate plus potassium carbonate plus barium carbonate plus potassium sulfate.)

A classical case, frequently cited, is that investigated by Berthelot in 1963. When 1 mole (60 grams) of acetic acid CH_3COOH and 1 mole (46 grams) of ethyl alcohol C_2H_5OH, both of which substances are soluble in water, are mixed, a reaction takes place which results in the formation of water and ethyl acetate ester, which is likewise in the ratio of 1 mole (18 grams) of water, and 1 mole (88 grams) of ethyl acetate $CH_3COOC_2H_5$. On the other hand, when 1 mole of water and 1 mole of ethyl acetate ester are mixed, a reaction takes place which results in the formation of acetic acid and ethyl alcohol in the ratio of 1 mole of acetic acid and 1 mole of ethyl alcohol. Three important observations have resulted from the detailed study of this reaction, namely, (1) the reaction between acetic acid and ethyl alcohol as reactants proceeds at such a rate that the fraction 0.005/5 of the amount present at any instant reacts, at 6 to 9°C, in 1 day to form equivalent amounts of water and ethyl acetate ester, (2) the reaction between water and ethyl ester as reactants proceeds at such a rate that the fraction 0.00144 of the amount present at any instant reacts, at 6 to 9°C, in 1 day to form equivalent amounts of acetic acid and ethyl alcohol, and (3) the end-point of each reaction is the same, that is, the reaction is one of true chemical equilibrium, and the resulting equilibrium mixture contains, in each case, 0.33 mole acetic acid plus 0.33 mole ethyl alcohol plus 0.67 mole water plus 0.67 mole ethyl acetate ester. This system attains practical equilibrium, at 6 to 9°C in about 1 year, at 100°C in about 8 days, and at 200°C in about 24 hours.

The equilibrium constant is calculated numerically as follows:

Equation:	CH_3COOH	+ C_2H_5OH	$\leftrightarrows$ HOH	+ $CH_3COOC_2H_5$
Reaction weights:	60	46	18	88
Molar ratio at equilibrium:	0.33	0.33	0.67	0.67
Weights at equilibrium:	0.33×60	0.33×46	0.67×18	0.67×88

$$\frac{\text{Equilibrium}}{\text{constant at 9°C}} = \frac{\text{conc. HOH} \times \text{conc. } CH_3COOC_2H_5}{\text{conc. } CH_3COOH \times \text{conc. } C_2H_5OH}$$

$$= \frac{0.67 \times 0.67}{0.33 \times 0.33} = 4$$

Knowing the equilibrium constant at any stated temperature enables one to calculate the equilibrium end-point at that temperature for any ratio of reactants. Thus, when 1 mole (60) grams of acetice acid and 10 moles (460) grams of ethyl alcohol at 9°C are taken:

$$\text{Equilibrium constant at 9°C} = 4 = \frac{X \times X}{(1 - X) \times (10 - X)}$$

where X is the number of moles of water and also the number of moles of ethyl acetate ester formed (1 mole of each is formed by reaction of 1 mole acetic acid plus 1 mole ethyl alcohol). Solution of this equation shows $X = 0.97$. Therefore, by taking the above ratio of acetic acid (1 mole) to ethyl alcohol (10 moles) 0.97 (or 97%) of the acetic acid, the excess reactant being ethyl alcohol (9 moles), is converted at equilibrium into

water plus ethyl acetate ester. In practice, the reaction is conducted by the use of a catalyzer, e.g., sulfuric acid concentrated, zinc chloride.

In cases where one of two resultants can be separated from the reactants and the other resultant, by precipitation as a solid, by condensation as a liquid, or by volatilization as a gas or vapor, the yield of the desired substance from a given amount of reactants can sometimes be materially increased. In the case of heterogeneous systems, whenever a solid participant is present, the *concentration* of said solid is considered constant. The precipitation and solution of solids are in this category, as well as the reactions between a gas and a solid, e.g., the system ferroferric oxide plus hydrogen gas plus iron plus water vapor.

The effect of change of temperature on a system in chemical equilibrium is that the equilibrium point is shifted (1) towards the side *away* from that which evolves heat when the temperature is *raised*, and (2) towards the side which evolves heat when the temperature is lowered. It is *as if* the amount of heat were a *material* reactant and its concentration (temperature or intensity of heat) increased, in respect to the *direction* of the shift of the equilibrium point. The amount of the shift at constant pressure can be calculated in cases where one possesses the proper data.

The effect of change of pressure on a system in chemical equilibrium is that the equilibrium point is shifted (1) towards the side possessing the smaller aggregate volume when the pressure is increased, and (2) towards the side possessing the larger aggregate volume when the pressure is decreased. The amount of the shift at constant temperature can be calculated by means of the equilibrium constant (above) recalling that increase of pressure is equivalent to increase of concentration of gases (temperature constant). When the volume of resultants equals the volume of reactants, no effect is produced on the equilibrium point by change of pressure. See **Chemical Equilibrium**.

In some chemical reactions, the use of concentrations does not give calculated results that agree with those observed, because of the departure from ideality of real gases and solutions. In such reactions, concentrations are replaced by apparent effective concentrations, or activities, as explained in that entry.

CHEMICALS (Number of). With over 100 chemical elements from which chemical compounds can be built, the large numbers of atoms which may be present in various compounds, and the many ways in which the atoms can be linked (straight chains, branching chains, rings, half-rings, etc.), the number of "possible" chemical compounds is a very high number indeed and, of course, is much larger than the millions of "known" compounds, which can be described fully in terms of constituents and structure. Keeping track of so many compounds—in the interest of fundamental research and, in more recent years, the identification of chemical compounds in terms of both beneficial and adverse effects on biological systems, health, and the environment—commenced many years ago. Thousands of compounds are described in numerous handbooks, various societies have compiled lists and tabulations, as for example the Chemical Abstracts Service (CAS) of the American Chemical Society, and special encyclopedias which describe inorganic and organic compounds. Possibly the most outstanding example is the "Handbuch der Organishe Chemie," first undertaken in Germany in 1918. The first 27 volumes of this book were published between 1928 and 1938. Supplements have been released periodically since that time. Part II, consisting of 29 volumes was published between 1941 and 1957; Part III, 14 volumes, was published between 1958 and 1973. Part IV was commenced in 1972 and continues. The publisher is Springer-Verlag, Berlin.

In 1978, the CAS, in connection with the Toxic Substances Control Act, was engaged in a project for the Environmental Protection Agency (United States). It was estimated that some 33,000 chemicals are in common use. The problem of identification is complicated by the fact that, because of various common and trade names for the same substance, there are approximately 183,000 designations for these 33,000 chemicals. Of all chemicals cataloged to date, by CAS and similar organizations throughout the world, it is estimated that about 3.5 to 4 million have been identified. Statistics indicate that of these about 96% are organic, i.e., they contain carbon. This translates into about 3.4 to 3.8 million organic compounds and from 100,000 to 200,000 inorganic compounds. Although essentially meaningless, a statistical summary of cataloged chemicals indicate that the "average chemical," a highly fictious entity, contains some 43 atoms, 22 of which are hydrogen. Of the organic chemicals, it is estimated that nearly 300,000 are coordination compound and about 60,000 compounds are substances whose structures have not been completely defined. The CAS registry also lists some 72,000 alloys, 120,000 polymers, and 10,000 mixtures with definite names.

CHEMISORPTION. The formation of bonds between the surface molecules of a metal (or other material of high surface energy) and another substance (gas or liquid) in contact with it. The bonds so formed are comparable in strength to ordinary chemical bonds, and much stronger than the van der Waals type characterizing physical adsorption. Chemisorbed molecules are almost always altered, e.g., hydrogen is chemisorbed on metal surfaces as hydrogen atoms; chemisorption of hydrocarbons may result in the formation of chemisorbed hydrogen atoms and hydrocarbon fragments. Even when dissociation does not occur, the properties of the molecules are changed by the surface in important ways. This mechanism is the activating force of catalysis. Solids with high surface energies are necessary for chemisorption to occur, e.g., nickel, silver, platinum, and iron.

CHEMOTHERAPEUTIC DRUGS (Cancer). Literally thousands of chemical compounds have been tested for their antitumor activity, both *in vitro* and *in vivo*.

For most cancer drugs, the level of dangerous or lethal toxicity and the level of dosage required for efficient therapeutic action are disturbingly close. Thus cancer themotherapy requires the utmost care in monitoring and controlling the administration of drugs to human patients. This largely explains the sometimes extremely long periods required to develop successful regimens. To make matters more complex, a new drug with superior characteristics may appear at just about the time the regimen for an older drug has been fully refined.

In developing cancer drugs, an attempt is always made to match the performance with what is currently known about the sequence of biochemical and genetic events that occur during the development of both normal and malignant cells. Certain drugs are effective only during certain specific phases of the cell development cycle, whereas other drugs are relatively independent of the cycle. Thus, some authorities classify cancer drugs into two broad categories: (1) drugs that are *phase-specific*; and (2) drugs that are *phase-nonspecific*.

Although probably the majority of details of normal cell and malignant cell development remain to be learned, or at least may be considered to be poorly understood, for guidance at the present juncture of the technology some authorities have developed a profile of the cell cycle. It is useful to match cur-

rently used as well as experimental cancer drugs against the stages of this cycle. The cycle is divided into five phases or periods. These phases are of unequal time spans.

M phase (mitosii)—This is the very beginning of the cell cycle. This period lasts for only 30 to 60 minutes. No DNA synthesis is assumed during this phase. Synthetic processes are comparatively inactive. Drugs found to be effective during this phase include vincristine and vinblastine.

G₁ phase—This is essentially an interval of a few to many hours and is the first time gap observed in the cell cycle. Not by confirmed observation, but by definition, it is assumed that DNA synthesis does not take place during this phase. Drugs found to be effective during this phase include actinomycin D, mitomycin, 6-mercaptopurine, and 6-thioguanine.

G₀ period—Regarded as an extension of the G_1 phase and as a "resting period." The cells are assumed not to be actively dividing during the period. It is assumed that a cell in the G_0 period may be stimulated to reenter the G_1 phase. It is believed that none of the chemotherapeutic agents are effective during this period.

S phase (DNA synthesis)—Regarded as the period of active DNA synthesis. A doubling of the DNA content of the cell is assumed during this period, estimated to span between 6 and 12 hours. It is assumed that once the DNA synthesis is initiated, the cell will divide. Drugs found to be effective during this phase include 6-mercaptopurine, 6-thioguanine, methotrexate, 5-fluorouracil, doxorubicin, daunorubicin, mitomycin, cyclophosphamide, and cytosine arabinoside.

G₂ phase—Estimated to span about 2 hours, this is stipulated as the second gap between DNA synthesis and cell division. Drugs found to be effective during this period include bleomycin and cyclophosphamide.

Combination chemotherapy has made major improvements in the treatment of cancer during the past several years. This is based upon the use of several drugs at one time, sometimes up to as many as six different agents. The mechanisms of action and side effects are different. By carefully determining quantities and timing of administration, the therapeutic advantages of each drug can be maximized while toxicity is minimized. This procedure, of course, requires a great deal of experience based upon empirical data gained through trials and case experimentation. Often there is only a rudimentary understanding of the mechanisms of each drug. From an overall standpoint, the multiple administration of drugs also may lessen the development of drug resistance. The prolonged administration of a single drug in ever increasing concentrations, which is retained in the environment, is believed by some authorities to be precisely that form of administration most likely to result in amplification of genes in a stable state, thereby imparting stable resistance.

There are some precautions applicable, however, to the use of combination chemotherapy. One group of researchers reported that the most frequently used antitumor agent, methotrexate, interferes with the activity of another antitumor drug, 5-fluorodeoxyuridine when given at the same time in tests on tumor cells grown in culture. Their evidence suggested that it may also interfere with the activity of the more frequently used antitumor drug, 5-fluorouracil. Methotrexate and 5-fluorouracil have been used together in therapy of many forms of cancer—breast, head, neck, liver, colon, and rectum.

Some of the most widely used of the cancer drugs are listed in the accompanying table; the structures of some of these are given in the accompanying diagram.

The first advance in cancer chemotherapy came in 1941, when researchers found that the female sex hormone, estrogen, was useful in the treatment of prostatic cancer in men. The discovery

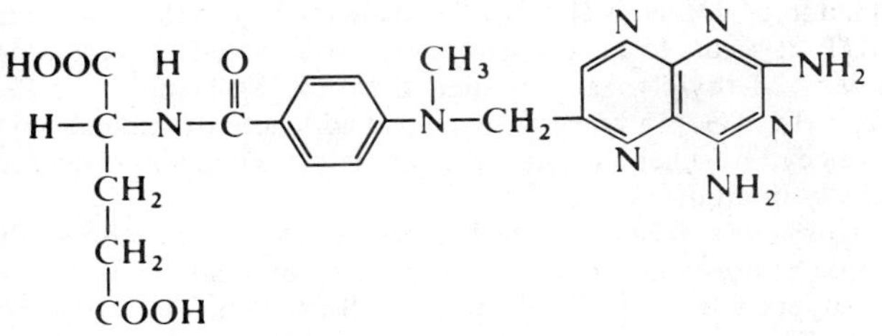

Methotrexate

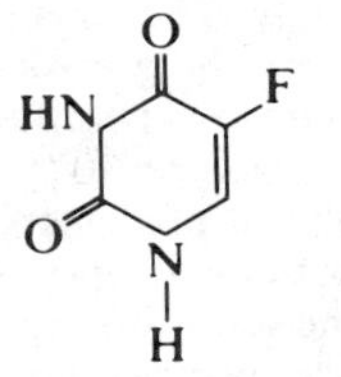

5-Fluorouracil

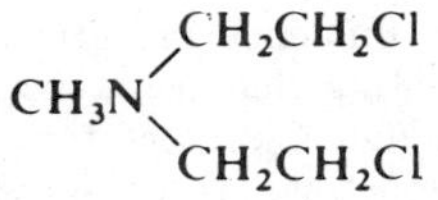

Mechlorethamine

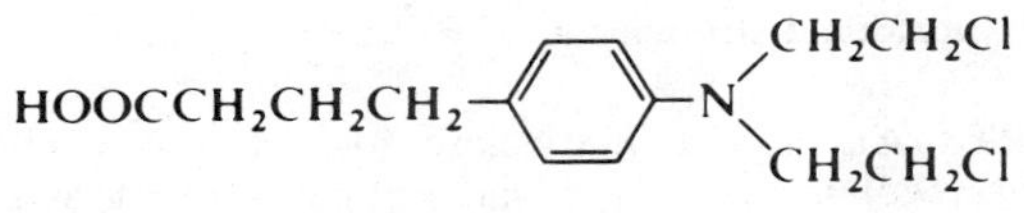

Chlorambucil

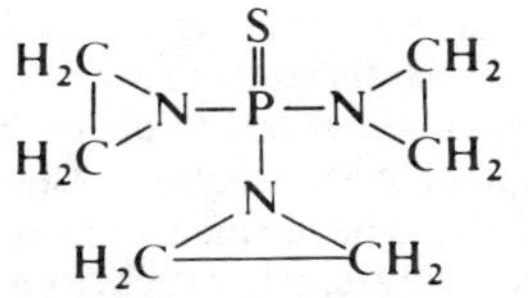

Triethylenethiophosphoramide

Chemical structures of representative drugs used in cancer chemotherapy.

of nitrogen mustard's effectiveness as a cancer drug was the product of investigating chemical warfare agents. In 1948, the first of the antimetabolites, an antivitamin called aminopterin, was reported to be useful in the treatment of leukemia. In the following year, the effectiveness of a related drug, methotrexate, was reported. The availability of cancer drugs and their use is now widespread. Methotrexate has proven strikingly useful, particularly against a rare uterine cancer known as choriocarcinoma. Before the effectiveness of this drug was discovered, five of every six women with this or a related kind of cancer, even when treated early with surgery, died within a year of diagnosis. Similarly, the effectiveness of several other chemotherapeutic agents for different kinds of cancers has been demonstrated.

The present consensus among scientists is that these drugs interfere with cell division at the core of the cell, within the nucleus, in the chemicals that contain the cell's genetic machinery, the nucleic acids DNA and RNA. Most cancer drugs are limited in their usefulness primarily by two factors: (1) toxicity—

damage to normal cells and tissues as well as cancer cells and tissues; and (2) decreased effectiveness—gradual loss of drug effectiveness due to host resistance. When the drug administered is a hormone, toxicity is sometimes manifested as changes in secondary sex characteristics, such as voice and facial hair. Other drugs may temporarily produce nausea, loss of appetite, loss of hair, hypertension, or diabetes. These side effects are usually reversed when drug treatment is halted.

Chemotherapists working in the clinic can usually anticipate a drug's toxicity from data gathered in animal tests. Rats, dogs, and monkeys are ordinarily reliable indicators of the human response to potential new drugs. Initial clinical trials of a drug begin at one-tenth the dose level tolerated by the larger laboratory animals. The final dosage formula is measured in milligrams of drug per kilogram of body weight, or even more efficiently, by milligrams of drug per square meter of body surface.

As an additional measure of regulating dose, means have been devised to circulate the drugs in a closed circuit through the bloodstream of the cancer-affected region, while a tourniquet prevents the drug from reaching and damaging sensitive organs beyond the cancerous area. The drug is injected through an artery to the cancerous area and is withdrawn from a vein by special tubes, and then recirculated through the artery and vein by means of a pump oxygenator. This technique, known as regional perfusion, is especially adapted to treating certain cancers of the arms and legs.

Infusion methods, by which a drug is dripped slowly into a patient's bloodstream and travels throughout the entire circulatory system, have also been modified to focus drug effects on cancerous areas, for example, cancers of the head and neck. In another system, developed for treating cancer of the liver, a plastic tube carries a continuous supply of drug directly to the cancer at a uniform rate regulated by an infusion pump. The tiny pump and a 7-day supply of the prescribed drug constitute a small package that can be strapped to the chest of a nonhospitalized patient for round-the-clock treatment.

It is generally believed that the cancerous change is induced in the cell's nucleic acid by a chemical, by radiation, by a virus, or by some combination of these factors. Medical scientists are attempting to employ the same factors to block growth of cancer cells. As pointed out in the following paragraphs, each class of drugs functions in a somewhat different way. See accompanying table.

Alkylating Agents. This group of quick-acting, highly reactive compounds includes nitrogen mustard and other close relatives of the wartime poison gases. Often referred to as "cell poisons," these agents are rich in electrons when they are in solution. For this reason they combine rapidly with many of the constituents of a cell. Many scientists believe that the aklylating agents exert their anticancer effects by a direct chemical interaction with the DNA of the cell.

In addition to nitrogen mustard, among the best known alkylating agents are cyclophosphamide (*Endoxan, Cytoxan*), chlorambucil (*Leukeran*), triethylene thiophosphoramide (thio-TEPA), and dimethanesulfonoxybutane (*Myleran*). They act primarily on tissues that are being quickly replaced, such as bone marrow and cells lining the intestine. They are used primarily in the treatment of Hodgkin's disease, lymphosarcoma, the chronic leukemias, and in some cancers of the lung, ovary, and breast.

Antimetabolites. These drugs structurally resemble metabolites, the nutrients a cell needs for growth. Antimetabolites mimic normal nutrients so closely that they are taken up by the cell through mistaken identity. Once inside the cell they interfere competitively with the production of nuclei acids and thereby prevent cell growth.

DRUGS USED IN CHEMOTHERAPY OF CANCER

GROUP AND GENERIC NAME	TRADE NAMES AND ABBREVIATIONS
ALKYLATING AGENTS	
Mechlorethamine hydrochloride	Mustargen (nitrogen mustard; HN2)
Cyclophosphamide	Cytoxan, Endoxan
Chlorambucil	Leukeran
Melphalan	Alkeran (phenylalanine mustard; L-sarcolysin; L-PAM)
Busulfan	Myleran
Triethylenethiophosphoramide	Thiotepa
ANTIMETABOLITES	
Methotrexate	Methotrexate (amethopterin; MTX)
6-Mercaptopurine	Purinethol (6-MP)
6-Thioguanine	Thioguanine (6-TG)
5-Fluorouracil	Fluorouracil (5-FU)
Cytosine arabinoside	Cytosar (Ara-C)
PLANT ALKALOIDS	
Vincristine	Oncovin
Vinblastine	Velban, Velbe
ANTIBIOTICS	
Actinomycin D	Cosmegen
Doxorubicin	Adriamycin
Daunorubicin	Daunomycin
Bleomycin	Blenoxane
Mithramycin	Mithracin
Mitomycin C	Mutamycin
OTHER AGENTS	
Hydroxyurea	Hydrea
Carmustine	Bi CNU; BCNU
Lomustine	CeeNU; CCNU
Procarbazine	Matulane
Decarbazine	DITC-Dome
Cisplatin	Platinol
L-Asparaginase	Elspar
Streptozotocin	—

NOTE: See also entry on **Hormones.**

Among the antimetabolites are antagonists of purines and pyrimidines, essential components of the cell's nucleic acids. The drugs 6-mercaptopurine and 5-fluorouracil are examples, respectively, of the antipurines and antipyrimidines. 6-Mercaptopurine was developed by Elion, Burgi, and Hitchings of the Wellcome Research Laboratories; 5-fluorouracil was synthesized in the laboratory by Heidelberger and his associates at the University of Wisconsin.

One of the most widely used antimetabolites is amethopterin (methotrexate), synthesized in 1948 by Seeger and associates. By inhibiting the enzyme folic acid reductase, methotrexate acts as an antagonist to a needed B vitamin, folic acid. This, in turn, interferes with both purine and pyrimidine synthesis.

The antimetabolites are useful in leukemia and in several types of solid tumors. See also **Antimetabolites.**

Plant Akaloids. Several compounds, derived from extracts of the common periwinkle plant, seem to act through interference with a phase of cell division. Best known are vinblastine sulfate and vincristine sulfate, the latter of which is useful in treating acute lymphocytic leukemia. Both are effective in certain lymphomas.

Antibiotics. Notable in this group is actinomycin D, a drug believed to achieve its effectiveness by locking itself onto a base of the DNA molecule, thereby blocking cell growth.

Other Agents. A drug used in treating cancer of the adrenal gland, *o,p'* -DDD, is closely related chemically to the insecticide DDT. It seems to have a selective destructive effect on adrenal cells. Methylglyoxal-*bis*-(guanylhydrazone), often called Methyl-GAG, is a synthetic chemical with activity against acute myelocytic leukemia, the type of acute leukemia occurring mostly in adults.

Hormones. Also described in the entry on **Hormones,** the action of hormones in cancer chemotherapy tends to accelerate or suppress the growth of specific cells, tissues, and target organs. They are thought to derive their effectiveness by altering or reversing a hormonal imbalance in the body that encourages the cancer cells to multiply. The female hormone estrogen, for example, helps to suppress the growth of disseminated cancer of the prostate. Conversely, male hormones or androgens cause temporary regression of disease in 20% of breast cancer patients, and are especially helpful in premenopausal women.

Among the other hormonal types, corticosteroids such as cortisone and prednisone seem to suppress the growth of white blood cells known as lymphocytes. For this reason these drugs are frequently useful in acute lymphocytic leukemia, which is characterized by an abundance of abnormal lymphocytes.

Scientists are uncertain as to the exact chemical mechanism by which hormones influence the growth of cells. However, evidence is accumulating that here, too, nucleic acid may be implicated. Karlson and Clever at the Max Planck Institute of Munich have demonstrated through experiments with insects that a hormone controls certain phases of insect growth by controlling its production of the nucleic acid RNA. Studies of hydrocortisone in rat cells by Tomkins and associates at the National Institute of Arthritis and Metabolic Diseases also suggest hormonal stimulation of RNA production.

CHERT. An impure, flinty hard rock composed chiefly of cryptocrystalline silica. Chert varies in color from gray through brown to black according to the kind and amount of coloring matter. It occurs principally as concretions, nodules or bands in limestones and dolomites, and unlike flint its fracture tends to be splintery instead of conchoidal. A great deal has been written on the occurrence and origin of chert and there is no doubt but that it may be formed in several different ways. Many of the nodular and concretionary cherts have grown around siliceous sponge spicules or radiolaria. Chert may be either sygenetic or epigenetic. The former type is supposed by some authors to be chemically precipitated from river waters on the bottom of the sea as a colloid contemporaneously with the limestones or dolomites. On the other hand certain cherts are obviously secondary although they may have been formed previous

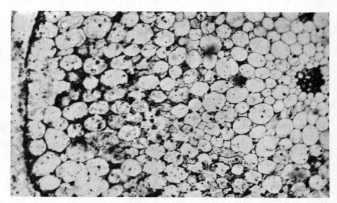

Transverse section from chert bed of Rhynie, UK. (*Photomicrograph by Brian J. Ford; copyright.*)

to the final lithification of the formations in which they occur. Cherts which contain relatively large amounts of iron are called Jasper.

CHINA CLAY. A commercial term, more or less identical with kaolin, as applied to the relatively pure clay concentrated by washing from a thoroughly kaolinized granite. England is the chief exporter of china clay. France has unique clays from which are made the famous Sèvres and Limoges potteries.

China-clay rock is a kaolinized granite made up chiefly of quartz and kaolin, with sometimes the presence of muscovite and tourmaline. The rock crumbles easily in the fingers. *China stone* is (1) a partially kaolinized granite, which contains quartz, kaolin, and sometimes mica and fluorite, is harder than china-clay rock and is used as a glaze in the production of china; or (2) a fine-grained, compact mudstone or limestone found in England and Wales.

CHIRALITY.   Isomerism.

CHLORAL.   Chlorinated Organics.

CHLORARGYRITE. Also known as *horn silver*, chlorargyrite is silver chloride, AgCl. The mineral crystallizes in the isometric system but is usually massive, appearing like wax or horn, hence the name. It has no cleavage, is highly sectile, yielding bright surfaces; hardness 2.5; specific gravity, 5.55; luster, resinous to adamantine; color gray, white to colorless. May be blue, violet-brown after exposure to light; transparent to translucent. Chlorargyrite is largely a secondary mineral, usually associated with other silver minerals as well as with compounds of lead, zinc, and copper. Saxony and the Harz Mountains are European localities. The Broken Hill district of New South Wales is a well-known occurrence, but probably the most important deposits are found in Atacama, Chile. The mineral also is found in Bolivia and Mexico. In the United States, chlorargyrite comes from Colorado, Idaho, Utah, Nevada, Arizona, and New Mexico.

CHLORIDE (Biological Aspects). Sodium chloride, potassium chloride, and other chloride salts, when ingested by animals from feedstuffs and humans from various food substances, reduce to a consideration of the cation involved (Na^+, K^+, etc.) and the Cl^- (chloride) ion. Generally, in terms of animal and human nutrition, more research has been conducted and more is known about the role of cations in metabolism than that of the chloride ion. Some physiologists and nutritionists in the past have described chloride as playing a "passive role" in maintaining the body's ionic and fluid balance. With exception of the "chloride shift" in venous blood, the movements of chloride have usually been considered secondary to those of the cations.

Much is known, of course, concerning the effects of excessive sodium chloride and of deficient sodium chloride in human and animal diets, but the physiological and nutritional roles of chloride have not been thoroughly studied and fully explained. C. E. Coppock, Department of Animal Science, Texas A & M University, College Station Texas; and M. J. Fettman, Cornell University College of Veterinary Medicine, undertook a study of chloride (targeted to chloride as a required nutrient for lactating dairy cows) and also carefully reviewed the prior work in this area of other researchers. (Interested readers are referred to the bibliography at the end of the article, "Chloride as a Required Nutrient for Lactating Dairy Cows," *Feedstuffs* (February 10, 1978).)

Despite important physiological functions and its presence in milk at about 0.11%, chloride is a neglected element in large animal nutrition. The practice of adding sodium chloride to concentrate mixtures and free-choice feeding seem to have precluded the possibility of a practical deficiency problem. When salt was omitted from the diet, researchers found that under the conditions used in their study, sodium was the first limiting element. This was true because sodium is present in most natural ingredients at much lower levels, relative to the cow's requirements, than is chloride.

Those who formulate diets usually ignore sodium levels in the natural ingredients (which are usually low in forages, but may be appreciable in certain concentrate ingredients) because of the traditional value of salt as a condiment. In addition, other sodium salts are often included in concentrate mixtures: sodium sulfate as a sulfur replacement, the sodium phosphates as phosphorus supplements, and sodium bicarbonate as a buffer. Even when these supplements are used, salt is still often included because of tradition. Under many conditions, gross overfeed of both sodium and chloride occurs. High dietary levels of sodium and chloride do not increase the levels of these elements in milk. Excess will be excreted in urine and manure. Excess salt intake may result in greater water consumption, waste transport, bedding requirements, and transfer of sodium and chloride to the soil.

Of the seven macro mineral elements required by dairy cattle, five can be considered fertilizer elements (potassium, calcium, phosphorus, magnesium, and sulfur), but sodium and chloride are both toxic to plants at high concentrations and present practical problems in areas with saline soils. High salt intakes have also been shown to increase udder edema in heifers. Because of the importance of chloride in nutrition and metabolism, research is needed to define the chloride requirements of lactating cows and clarify mineral relationships, especially between chloride and potassium plus sodium.

Chloride and Plants. Chloride is one of the most recent elements to be shown essential for plant growth. In 1954, Broyer et al. presented evidence that tomato plants grown in a low-chloride solution developed wilting of leaflet blade tips, which progressed to chlorosis, bronzing, and necrosis. Growth was proportional to chloride concentration (up to a point) in the culture medium. Chloride additions to the medium prevented the deficiency symptoms and caused their disappearance in deficient plants. Later, the same team showed that often other species, including barley and alfalfa, displayed severe deficiency symptoms when grown in a low-chloride medium. Although buckwheat, corn (maize), and beans did not exhibit these obvious symptoms, yield effects were apparent.

Despite the presence of 250 parts per million (ppm) in the leaves of chloride-deficient tomato plants compared to a 0.1 ppm for molybdenum, chloride was classified as a micronutrient by plant physiologists. The reasons why the essentiality of chloride escaped detection for so long was its wide distribution in nature, the high solubility of most chloride salts, and difficulties of purification. Because chloride is a principal ion of sea water, it is picked up by winds from sea spray and carried far inland. For example, Geneva, New York has been estimated to receive 18 kilograms/hectare of chloride annually; Mount Vernon, Iowa, 73 kilograms/hectare annually. Deposits of more than 45 kilograms/hectare have been reported near coastlines. It was also suggested that leaf structures were capable of capturing airborne chloride. The popularity of potassium chloride as a potassium fertilizer and manure from cows fed excessive levels of chloride relative to their requirement are additional sources of soil chloride.

According to Stout and Jackson, chloride differs from other

nutrient elements present in native rocks because it is not fixed by colloids; it is repelled by negatively charged clay surfaces, and all chloride compounds formed in soils are highly soluble. In addition to leaching, chloride is lost from soils through crop removal.

Several factors affect chloride uptake by plants: (1) age or advancing maturity, shown in avocado, apricot, and grape leaves; (2) chloride concentration in the soil; (3) soil oxygen levels; (4) plant species; and (5) competition from other anions. Muraka and others have shown a specific antagonism between nitrate nitrogen and chloride; increasing chloride levels in the growth medium reduces nitrogen uptake, but this reduction is primarily in the nitrate-nitrogen fraction, with little if any effect on the protein-nitrogen fraction. The reciprocal effect was also observed between chloride and sulfate accumulation. Chloride is essential in the plant for photosynthetic reactions in chloroplasts which produce oxygen.

Chloride is also treated as a toxic element as well as a part of saline toxicity. Over the toxic range, the reduction in growth is approximately linear. For example, at about 3,500 ppm chloride in the cultural medium, alfalfa growth will be depressed to about 60% that of normal. Obviously, in areas with saline soils, there is concern about excessive levels of salt returned to the soil via manure.

Gastrointestinal Absorption of Chloride. Since 1952, when in goats, sheep, and dairy cattle, the observation was made that chloride could be absorbed from ruminal fluid into the blood against a ten-fold concentration difference (normal rumen fluid chloride concentrations may range from 10–30 mEq/l, while those of the plasma may be 100–110 mEq/l), numerous researchers have attempted to describe accurately and explain the processes by which chloride might move across the reticulorumen epithelium against its apparent chemical gradients. For a number of years, it was assumed by some workers that the observed electrical potential difference across the forestomach epithelium, making the plasma approximately 30 mV positive to the contents of the reticulorumen, could adequately account for the otherwise anomalous movements. If chloride's movement into the blood was truly attributable solely to the combined electrochemical gradient acting upon it, then its distribution across the gastric epithelium should have been describable by the Nernst equation.

In rumen-fistulated experimental animals, it has been observed that for certain distribution ratios of chloride in the ruminal fluids and blood plasma, the calculated equilibrium potential for chloride is relatively the same as that measured directly with KCl-agar bridges and calomel electrodes. However, in many circumstances, the calculated and measured values have been found to be significantly different, an observation that could only be accounted for by the presence of an active transport mechanism responsible for the movement of chloride out of its equilibrium distribution.

The active transport of chloride has also been demonstrated across the wall of the frog stomach, rat ileum, dog ileum, and the human ileum. Turnberg et al. produced a double exchange model by which bicarbonate secretion and chloride absorption are linked by an isoelectric mechanism to hydrogen ion secretion and sodium absorption across the human ileum. Chien and Stevens, in 1972, proposed a similar model of coupled transport across the reticulorumen epithelium, and further concluded that active anion and cation transport in the rumen cannot function efficiently unless both components are intact, i.e., the net transport of the body's major cations from the rumen to the blood, appeared to be dependent in part on the activity of chloride in the system.

Chloride in Cerobrospinal Fluid. A similar story has unfolded concerning the distribution and movement of chloride between the blood and cerebrospinal fluid (CSF). The first indication for the possible existence of an active mechanism responsible for the maintenance of chloride levels in the CSF came almost 40 years ago when in dogs, Hiatt demonstrated the persistence of CSF chloride concentrations at 44% of normal, despite the reduction of chloride levels to 30% of normal in all body fluids during a nitrate-induced diuresis. Over the next 20 years, researchers recorded chloride ion concentration and electrical potential differences across CSF-ECF (extracellular fluid) and CSF-blood barriers ranging from 15 to 20% and from 5 to 30 mV, respectively, in such varied subjects as dogfish, rats, cats, dogs, monkeys, and humans. In all cases, the CSF was both higher in chloride ion concentration and negative in potential with respect to the reference body fluid.

In 1970, Bourke et al. studied the distribution and kinetics of chloride in cats following the isoosmotic replacement of body chloride with isethionate via extracorporeal hemodialysis. They found that when the plasma chloride concentration was reduced by approximately 93%, the cerebral cortex, corpus callosum, and CSF chloride concentrations were reduced by approximately only 26.5, 35, and 21%, respectively. Other body tissues and fluids showed reductions in chloride concentration closer to those of the plasma (skeletal muscle and liver, 73%). The influx of chloride into the CSF at various plasma chloride concentrations was then plotted as a Lineweaver-Burk plot, and was shown to behave as a carrier-mediated process, as described by Michaelis-Menten kinetics. This information, combined with the observations made by Abbot et al. that reduction of the plasma chloride concentration by isethionate replacement did not produce a change in electrical potential "commensurate with or even in the same direction" as that expected by Nernst equation predictions, led workers in the field to ascribe the bulk of chloride movement from the blood into the CSF to an active transport process. It is possible that control of the rare of chloride transport is a factor in the regulation of the secretion of CSF, the medium that bathes, protects, and nourishes the central nervous system.

Chloride in the Humoral Regulation of Sodium and Potassium. Conventional presentations of the regulatory mechanisms involved in body fluid and electrolyte homeostasis usually have considered maintenance of the sodium/potassium ratio in the ECF both the prime means and end toward a functional electrolyte balance. Certainly, the sodium/potassium ratio provides the axis about which the body's humoral mechanism of electrolyte homeostasis revolves, represented mainly by the renin-angiotensin-aldosterone system. Aldosterone increases the activity of sodium retaining processes in the body. These include the active uptake of sodium from the gastrointestinal tract and the reabsorption of sodium from the renal tubes in exchange for potassium or hydrogen ion.

Given chloride's role in ruminal absorption and CSF secretion, perhaps its participation in the aldosterone mechanism of regulating the sodium/potassium "axis" should come as no surprise.

Upon detection of decreased blood pressure, volume, and/or sodium, the juxtaglomerular apparatus in the kidney secretes renin, an enzyme which then cleaves a decapeptide, angiotensin I, from a plasma a_2-globulin, angiotensinogen. A converting enzyme present in the plasma, and most abundant in the pulmonic circulation, cleaves a dipeptide from angiotensin I, thus forming angiotensin II. The effect of angiotensin II, potentiated by ACTH and high plasma potassium, is to induce the secretion of aldosterone by cells in the zona glomerulosa (arcuata) of the adrenal gland cortex. Aldosterone then exerts its effects on the sweat glands, salivary glands, intestinal mucosa, and distal convoluted tubules of the kidneys, promoting sodium absorption and retention. Because of its potent vasoconstrictive properties (40 times that of norepinephrine) angiotensin II can effectively reduce both renal blood flow and glomerular filtration rate, leading to an immediate decrease in excretion of water and electrolytes, before aldosterone can affect tubular sodium reabsorption.

Research has demonstrated that chloride is not only responsible in part for angiotensin II formation, but also for its deactivation or catabolism by the major angiotensinase of the body. Furthermore, chloride ion's relations to angiotensin II may not be its only route to affecting aldosterone secretion and sodium/potassium balance. Chloride's role in the metabolism of ACTH has led to support for the hypothesis that fluid and electrolyte homeostatic mechanisms may revolve not just around the sodium/potassium ratio, but also around the levels of chloride in the body.

See also **Blood; Sodium;** and **Sodium Chloride.**

CHLORINATED ORGANICS.

Organic compounds containing chlorine are valued as reagents and intermediates in chemical synthesis and for their commercial and industrial importance. Several are produced in high tonnages. The large volume market for these compounds is in plastics, including vinyl chloride for polyvinyl chloride (PVC), or as a copolymer with vinyl acetate; vinylidene chloride for *Saran;* and chloroprene for neoprene. Other important uses include agricultural chemicals, solvents, plasticizers, and medicines. Uses as intermediates to produce other chemicals also are important and varied. The largest volume chlorine organic is ethylene dichloride (EDC). About 11 billion pounds (5 billion kilograms) per year of EDC are produced, but over half of this is consumed by the producers to make vinyl chloride monomer. Nearly all of the ethyl chloride produced is used in making tetraalkyl leads. Methyl chloride is the intermediate for many chemicals, as are benzyl chloride, phosgene, and chloroform.

The chlorine on certain compounds is used as a facile leaving group for the introduction of another functional group. The displacement of a halogen atom by a cyano group to form a nitrile is one of the most useful reactions of halogen compounds. This opens a route to carboxylic acids having one carbon atom more than the original halide, aside from the importance of the nitriles themselves. Adiponitrile, used in the manufacture of nylon, can be made by treatment of 1,4-dichlorobutane with cyanide:

$$\begin{array}{ccc} CH_2CH_2Cl & & CH_2CH_2CN \\ | & \xrightarrow{NaCN} & | \\ CH_2CH_2Cl & & CH_2CH_2CN \end{array}$$

Long-chain alkyl chlorides can be used for the synthesis of various amines, while benzyl chloride is used for production of quaternary ammonium compounds. Alkyl chlorides are used for the formation of organometallics, including the Grignard reagents as well as for alkylation of aromatics. One of the important reactions of phosgene is with diamines for production of diisocyanates (polyurethanes).

Synthesis of Chlorinated Organics. Chlorine derivatives or organic compounds are obtained by substitution, addition, or displacement. Substitution reactions of Cl on hydrocarbons involve radical attack to remove a hydrogen, forming the hydrocarbon radical as an intermediate: $R—H + Cl\cdot \rightarrow R\cdot + HCl$.

Since a tertiary carbon radical $—\overset{|}{\underset{|}{C}}\cdot$ is most stable, it chlorinates more readily than a secondary carbon $—\overset{|}{C}H_2$ and that

more readily than a methyl group. Due to inductive effects, the presence of chlorine in a molecule reduces the activity of the hydrogens on adjacent carbons more than on the chlorinated carbon. Thus, a second radical $(Cl\cdot)$ will preferentially attack a hydrogen on the same carbon. For example, chlorination of ethyl chloride will produce nearly twice as much 1,1-dichloroethane as 1,2-dichloroethane. Specificity of this sort is decreased at higher temperatures, leading to more random substitution. Hydrogens further away on a longer-chained molecule are essentially unaffected by the first chlorine, therefore little selectivity occurs for subsequent substitutions.

Chlorine can be substituted onto an aromatic ring in the presence of a catalyst, such as ferric chloride, $FeCl_3$, or aluminum chloride, $AlCl_3$. The simplest case would be chlorination of benzene. Substitution of a second Cl onto the ring preferentially goes to the para position, but the ortho and meta isomers can be formed with the latter least favored. If the chlorination is carried out in the presence of a radical source, such as ultraviolet light, addition occurs instead. See *Chlorinated Aromatics* described later in this entry. When a functional group is present on the aromatic ring, Cl attack will depend upon the type of group present. Phenol and benzoic acid will chlorinate on the ring to give chlorophenol and p-chlorobenzoic acid. Alkyl benzenes will chlorinate on the alkyl group if a radical source is present. In the presence of an iron catalyst, the product is a mixture of the ortho- and para-chloroalkylbenzenes:

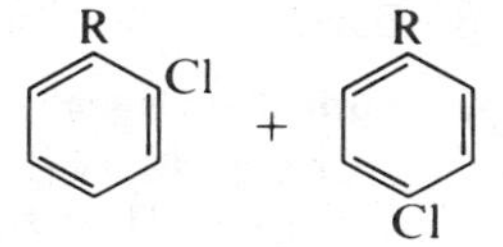

With higher aromatics, such as naphthalene, chlorine successively substitutes all of the hydrogens. The first product is α-chloronaphthalene and the final compound is perchloronaphthalene.

Chlorine addition occurs on unsaturated hydrocarbons having double or trible bonds. Addition can occur by use of Cl_2, HCl, or HOCl:

where X = Cl, H, or OH.

With ethylene, the products would be 1,2-dichloroethane, ethyl chloride, or ethylene chlorohydrin. When the unsaturated molecule has three or more carbons, HCl will add, preferentially, with the Cl on the carbon having the fewest hydrogens. For addition of HOCl, the opposite is favored.

Displacement occurs when a functional group is replaced. Chlorine can displace groups, such as hydroxyl, OH, in an acid-catalyzed reaction. For example, methyl chloride can be prepared from the reaction of HCl on methanol. Other alkyl chlorides can be made from their corresponding alcohols. Another type of displacement would be the exchange of one halogen for another, e.g., Cl can be substituted for bromine or iodine in a molecule. An acyl chloride can be formed by the reaction of a strong dehydrating Cl carrier, such as PCl_3, PCl_5, $POCl$, or $SOCl_2$, with an organic acid or its salt.

Industrially, chlorinations are carried out in five ways: (1) radical substitution of hydrogens; (2) molecular Cl addition across unsaturated (double or triple) bonds; (3) HCl addition across an unsaturated bond; (4) HCl reaction with an alcohol; and (5) oxychlorination. The latter reaction is similar to producing molecular chlorine, *in situ*, from HCl and air in the presence of a catalyst. An example of this is the production of ethylene dichloride, 1,2-dichloroethane, from ethylene, HCl, and oxygen.

Characteristics of Chlorinated Organics

The presence of Cl in an organic molecule increases the density, viscosity, and chemical reactivity, while decreasing the specific heat, solubility in water, and flammability. Chlorine is normally an excellent leaving group, particularly in base-catalyzed reactions, which makes it important for syntheses. Toxicity is the principal hazard. Threshold Limit Values, established by the American Conference of Governmental Industrial Hygienists, for tetra- and pentachloroethane are 5 ppm (vol.) in the atmosphere. Corresponding values for CCl_4, $CHCl_3$, and perchloroethylene are 10, 50, and 100 ppm, respectively. Chloroacetylenes are highly explosive, especially in contact with caustic, e.g., NaOH.

Safety and Handling. Chlorinated organics are absorbed through the skin and lungs and can seriously damage vital organs, especially the liver. Therefore, they should be handled with rubber gloves and in well-ventilated areas. When these materials are subject to burning, they have the potential of forming hydrochloric acid and phosgene, $COCl_2$, besides carbon monoxide. In highly chlorinated compounds, such as carbon tetrachloride and perchloroethylene, there is some danger of forming phosgene in a fire, or from high heat. Compounds that have sufficient hydrogen to combine with any Cl released, such as methyl chloride, vinyl chloride, and ethyl chloride, will form large amounts of hydrochloric acid. Although some small amounts of phosgene may be produced, the hydrogen chloride will naturally drive personnel away from such a fire.

Types or Families of Chlorinated Organics

Chlorinated Paraffins. Cl will displace one, two, three, or more hydrogens from the paraffins. These substitution products are referred to as *mono* ($C_nH_{2n+1}Cl$), *di* ($C_nH_{2n}Cl_2$), *tri* ($C_nH_{2n-1}Cl_3$), and so on. *Monochloro* derivatives include methyl chloride, CH_3Cl, ethyl chloride, C_2H_5Cl, and propyl chloride, C_3H_7Cl. These are also called *alkyl chlorides*. Examples of *dichloro* compounds include methylene dichloride, CH_2Cl_2, and ethylene dichloride, $C_2H_4Cl_2$. Chloroform, $CHCl_3$, and 1,1,1-trichloroethane, $C_2H_3Cl_3$, are *trichloro* derivatives, while carbon tetrachloride, CCl_4, is a *tetrachloro* molecule. When all the hydrogens are substituted by Cl, the term *perchloro* is sometimes used.

Chlorinated Carbonyls. Chlorination of an aldehyde or ketone occurs most readily on a carbon next to the carbonyl function. This is due to proton interaction with the carbonyl and is acid catalyzed. Reaction of Cl with acetone yields chloroacetone. Substitution of a second Cl on chloroacetone occurs with no preference for sites. Thus, equal amounts of 1,1-dichloro and 1,3-dichloroacetone are produced. (The opposite is true when brominating, since it is possible to form nine parts of 1,3-dibromo to one of 1,1-dibromacetone.) Chloral is produced from acetaldehyde. It is also produced by hydrolysis of trichlorodiethyl ether. Acrolein reacts with dry HCl at low temperatures to give β-chloropropionaldehyde.

The chlorinated acetones are strong lachrymators. Tear gas contains chloroacetophenone, which is also a component of the nonlethal disabling spray chemical, *Mace.*

Chlorination of diketene yields α-chloroacetoacetyl chloride:

$$H_2C=C-O \atop H_2C-C=O \quad + \; Cl_2 \; \longrightarrow \; Cl-CH_2-\overset{\overset{\displaystyle O}{\|}}{C}-CH_2\overset{\overset{\displaystyle O}{\|}}{C}Cl$$

This product is both a vesicant (blistering agent) and a lachrymator.

Chlorinated Fatty Acids. Chlorination of carboxylic acids is much more difficult because the contribution of the carbonyl group toward proton removal is offset by the electron donation effect from the hydroxyl group. This hindrance is obviated by reaction with the acid chloride or anhydride. Chlorination is normally accomplished by use of a catalyst, such as phosphorus trichloride. Monochloroacetic acid is an important industrial chemical. Dichloro- and trichloroacetic acids can be produced by further chlorination, although the latter can be produced conveniently by nitric acid oxidation of chloral. Higher chlorinated fatty acids can be produced by treatment of the hydroxy carboxylic acid or ester with HCl or PCl_5:

$$CH_3CHOHCH_2COOR + PCl_5 \rightarrow CH_3CHClCH_2COOR$$

Amino fatty acids can be treated with a mixture of nitric oxide and chlorine to produce the corresponding chloroacid. Mono- and dichlorosuccinic acids are examples of chlorinated dicarboxylic acids.

Chlorinated Ethers. Ethylene chlorohydrin reacts with sulfuric acid to form β,β'-dichloroethyl ether. It is a by-product of ethylene glycol production. The chlorines on this ether are inert making it a good solvent. Further chlorination at 20–30°C gives α,β,β'-trichloro diethyl ether which hydrolyzes to chloroacetaldehyde and ethylene chlorohydrin. Ethylene and sulfur monochloride react to give β,β'-dichlorodiethyl sulfide (mustard gas), which is a thioether.

Chlorinated Aromatics. Chlorination of benzene in the presence of a catalyst ($FeCl_3$ or $AlCl_3$) yields chlorobenzene as the first product. Substitution with a second Cl yields ortho, para, or meta dichlorobenzene. Eventually all the hydrogens can be substituted to give hexachlorobenzene, C_6Cl_6. In the presence of ultraviolet light, the chlorination of benzene yields benzene hexachloride, $C_6H_6Cl_6$, a derivative of cyclohexane. Under the same conditions toluene chlorinates on the methyl group to give one, two or three substitutions (benzyl chloride, benzal chloride or benzotrichloride), while in the presence of an iron catalyst, one obtains ortho- and parachlorotoluene.

Chlorinated Heterocyclics. Substitution in pyridine is more difficult than in benzene but Cl will enter the β position slowly. Chlorine will not add to furan to give stable addition products but substitution occurs to give 2-chloro- or 3-chlorofuran, 2,5-dichlorofuran, and 2,3,5-trichlorofuran.

Important Specific Chlorinated Organic Compounds

Several thousand chlorine-containing compounds are known and have been synthesized. A select group is included for description here to provide a cross section of the most important of these compounds. The number in brackets following each heading, where appropriate, is the *Chemical Abstracts Service Registration* number.

Acetyl chloride ($CH_3\overset{\overset{\displaystyle O}{\|}}{C}Cl$) [75–36–5]. Acetyl chloride can be prepared by treatment of acetic acid with various reagents, such as PCl_3, $SOCl_2$ or $COCl_2$. It can be prepared by chlorination of acetic anhydride in several different ways, by reaction of methyl chloride with carbon monoxide in the presence of catalysts, by reaction of ketene ($H_2C=C=O$) with HCl, or by partial hydrolysis of 1,1,1-trichloroethane. Acetyl chloride hydrolyzes in the presence of water to give acetic acid. It reacts with ammonia and amines to give acetamides:

$$CH_3\overset{\overset{\displaystyle O}{\|}}{C}Cl + RNH_2 \rightarrow CH_3\overset{\overset{\displaystyle O}{\|}}{C}NHR.$$

Reaction with alcohols gives the corresponding acetate esters. Acetyl chloride will add across unsaturated bonds in the presence of suitable catalysts to give halogenated ketones:

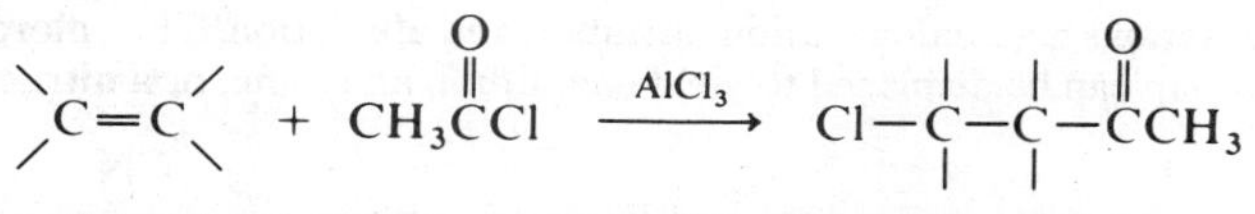

$$\text{C=C} + CH_3\overset{O}{\overset{\|}{C}}Cl \xrightarrow{AlCl_3} Cl-\overset{|}{\underset{|}{C}}-\overset{|}{\underset{|}{C}}-\overset{O}{\overset{\|}{C}}CH_3$$

Allyl Chloride (3-chloropropene-1) [107–05–1]. Allyl chloride can be synthesized by reaction of allyl alcohol with HCl or by treatment of allyl formate with HCl in the presence of a catalyst (ZnCl$_2$). Commercial production is by chlorination of propylene at high temperatures, about 500°C, using a large excess of propylene. It is used in the synthesis of glycerol, allyl alcohol and epichlorohydrin. Since the chlorine is situated alpha to a double bond, it is particularly reactive. Thus, hydrolysis to allyl alcohol occurs rapidly in dilute caustic at about 150°C. Addition of HOCl followed by treatment with an alkali yields epichlorohydrin:

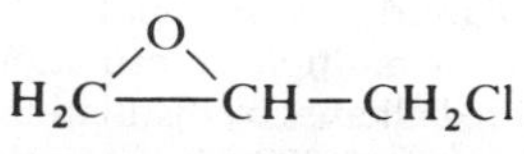

$$H_2C\overset{O}{\diagdown\diagup}CH-CH_2Cl$$

Addition of HBr in the presence of an oxidizing agent yields 1-chloro-3-bromopropane, which is used to prepare cyclopropane. Allyl chloride is one of the most toxic of the chlorinated organics.

Benzoyl Chloride [98–88–4].

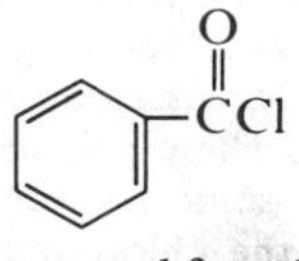

[structure: benzene ring with $\overset{O}{\overset{\|}{C}}Cl$ substituent]

Benzoyl chloride can be prepared from benzoic acid by reaction with PCl$_5$ or SOCl$_2$, from benzaldehyde by treatment with POCl$_3$ or SO$_2$Cl$_2$, from benzotrichloride by partial hydrolysis in the presence of H$_2$SO$_4$ or FeCl$_3$, from benzal chloride by treatment with oxygen in a radical source, and from several other miscellaneous reactions. Benzoyl chloride can be reduced to benzaldehyde, oxidized to benzoyl peroxide, chlorinated to chlorobenzoyl chloride and sulfonated to *m*-sulfobenzoic acid. It will undergo various reactions with organic reagents. For example, it will add across an unsaturated (alkene or alkyne) bond in the presence of a catalyst to give the phenylchloroketone:

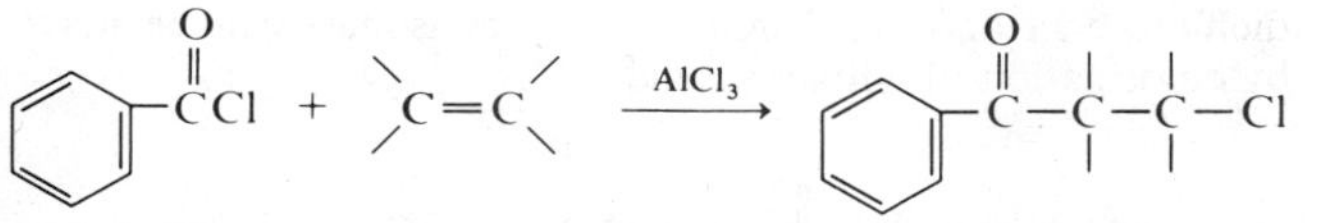

[structure: benzene ring–$\overset{O}{\overset{\|}{C}}Cl$ + C=C $\xrightarrow{AlCl_3}$ benzene ring–$\overset{O}{\overset{\|}{C}}$–$\overset{|}{\underset{|}{C}}$–$\overset{|}{\underset{|}{C}}$–Cl]

Reaction with benzene yields benzophenone while toluene gives phenyl-*p*-tolyl ketone. Reaction of benzoyl chloride with monohydric alcohols gives the corresponding alkyl ester, but with phenols the product can either be the phenylbenzoate or a phenolic ketone:

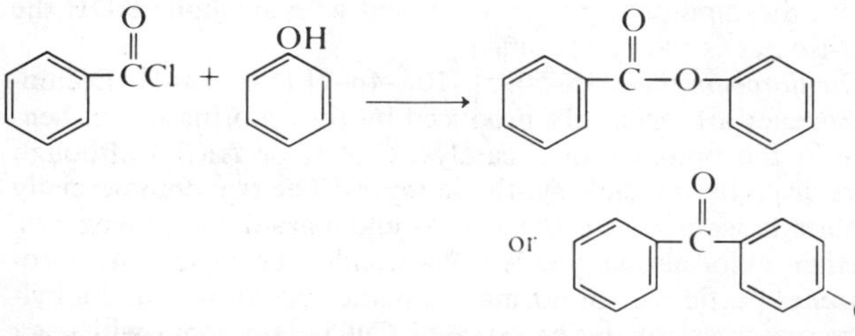

[structure: benzene ring–$\overset{O}{\overset{\|}{C}}Cl$ + phenol–OH → benzene ring–$\overset{O}{\overset{\|}{C}}$–O–phenol or benzene ring–$\overset{O}{\overset{\|}{C}}$–phenol–OH]

With ammonia and various primary and secondary amines the corresponding amide is formed.

Benzyl Chloride (α-chlorotoluene) [100–44–7]. Benzyl chloride can be synthesized by chloromethylation of benzene in the presence of a catalyst (ZnCl$_2$) or by treatment of benzyl alcohol with SO$_2$Cl$_2$. Commercially it is produced by chlorina-

tion of boiling toluene in the presence of light. Benzyl chloride can be oxidized to benzoic acid or benzaldehyde, or substituted to give the halogenated, sulfonated or nitrated product:

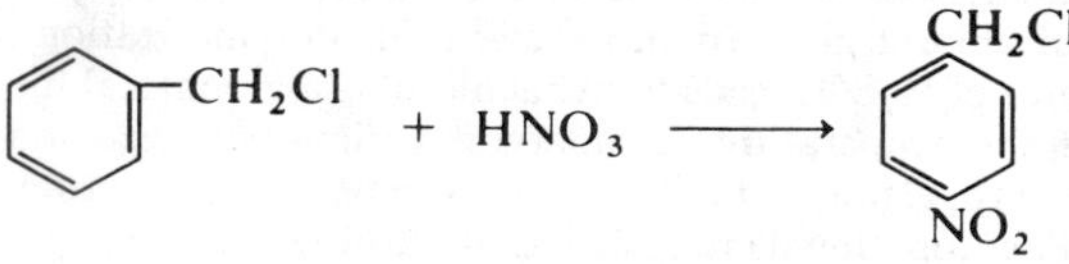

[structure: benzene ring–CH$_2$Cl + HNO$_3$ → benzene ring with CH$_2$Cl and NO$_2$ substituents]

With NH$_3$ it yields mono-, di- or tribenzyl amine. With alcohols in base the benzylalkyl ether is formed

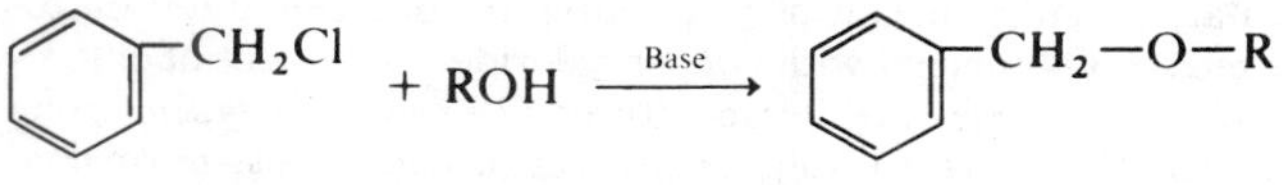

[structure: benzene ring–CH$_2$Cl + ROH $\xrightarrow{Base}$ benzene ring–CH$_2$–O–R]

With phenols either the phenolic or nuclear hydrogens can react to give benzylaryl ether or benzylated phenols. Reaction with NaCN gives benzyl cyanide (phenylacetonitrile); with aliphatic primary amines the product is the N-alkylbenzylamine, and with aromatic primary amines N-benzylaniline is formed. Total capacity (United States) exceeds 125 million pound per year. Benzyl chloride is converted to butyl benzyl phthalate plasticizer and other chemicals.

Carbon Tetrachloride (tetrachloromethane) [56–23–5]. Carbon tetrachloride can be synthesized by the chlorination of CS$_2$, acetylene and other higher hydrocarbons but the primary source is the exhaustive chlorination of methane. It can be pyrolyzed to yield hexachloroethane, oxidized to phosgene and carbonylated with CO in the presence of AlCl$_3$ to give trichloroacetylchloride:

$$CCl_4 + CO \xrightarrow{AlCl_3} Cl_3C-\overset{O}{\overset{\|}{C}}Cl.$$

Some of the more important commercial uses involve fluorine displacement to yield chlorofluoromethane refrigerants, such as trichlorofluoromethane (R-11) and dichlorodifluoromethane (R-12).

Chloroacetic Acid (ClCH$_2$COOH) [79–11–8]. Chloroacetic acid can be synthesized by the radical chlorination of acetic acid, treatment of trichloroethylene with concentrated H$_2$SO$_4$, oxidation of 1,2-dichloroethane or chloroacetaldehyde, amine displacement from glycine, or chlorination of ketene. It behaves as a very strong monobasic acid and is used as a strong acid catalyst for diverse reactions. The Cl function can be displaced in base-catalyzed reactions. For example, it condenses with alkoxides to yield alkoxyacetic acids: ClCH$_2$COOH + KOR → ROCH$_2$COOH. Oxidation of chloroacetic acid leads to formation of methylene chloride. Treatment with ammonia yields

glycine (ClCH$_2$COOH $\xrightarrow{NH_3}$ H$_2$NCH$_2$COOH) while use of

amines leads to formation of substituted glycines. Commercially, chloroacetic acid is an intermediate in the production of herbicides (2,4-D, 2,4,5-T and others) and cellulose ethers.

Chloroacetylene (HC≡CCl) [593–63–51]. This compound is a gas with a very unpleasant odor. It ignites spontaneously in air and may detonate during handling. It can be synthesized by dehydrochlorination of dichloroethylenes with a strong base. It will react with silver or mercury to give explosive salts. Addition occurs across the unsaturated bond—for example, bromination yields 1-chloro-1,1,2,2-tetrabromoethane.

Chloral (trichloroacetaldehyde) [75–87–6]. Chloral can be prepared by action of Cl$_2$ on ethanol, chlorination of acetaldehyde, oxidation of 1,1,2-trichloroethylene in the presence of a catalyst ($FeCl_3$, $AlCl_3$, $TiCl_4$ or $SbCl_3$), and by reaction of CCl_4 with formaldehyde. Chloral can be reduced either at the —CCl_3 group or the —CHO group. In the first case the product is acetaldehyde while the second gives Cl_3C—CH_2OH. Oxidation of chloral gives trichloracetic acid. Polymerization in the presence of H_2SO_4 leads to metachloral or parachloral, depending on the temperature. It undergoes various condensation reactions with alcohols to yield hemiacetals. With organic acids and other functional groups a multitude of reactions are possible. It is used in medicine as a hypnotic.

Chlorobiphenyls. These compounds can be synthesized by direct chlorination of biphenyl in the presence of iron or other catalysts. Other means of preparation include reaction if diazotized aminobiphenyl with copper chloride. Treatment of chlorobiphenyls at elevated temperatures (300–400°C) with strong caustic yields hydroxybiphenyls. Various reactions, normal to aromatic systems, will occur—usually on the unsubstituted ring.

Chloroform (trichloromethane) [67–66–3]. Although chloroform can be prepared by various means it is almost exclusively produced by the chlorination of methane. It can be oxidized to phosgene, substituted with various halogens, nitrated to chloropicrin (Cl_3CNO_2), hydrolyzed to formic acid ($\overset{\overset{\displaystyle O}{\|}}{H}COH$) and carbonylated to dichloroacetic acid. It will react with unsaturated halohydrocarbons in the presence of $AlCl_3$:

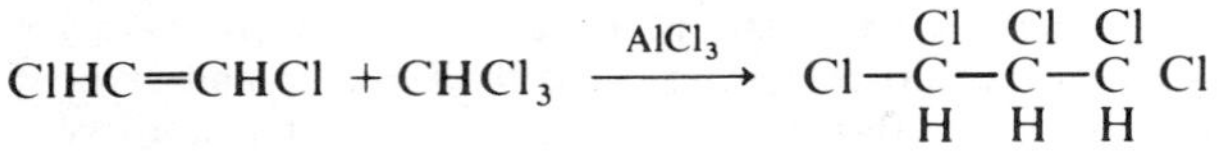

$$ClHC=CHCl + CHCl_3 \xrightarrow{AlCl_3} Cl-\overset{\overset{\displaystyle Cl}{|}}{\underset{\underset{\displaystyle H}{|}}{C}}-\overset{\overset{\displaystyle Cl}{|}}{\underset{\underset{\displaystyle H}{|}}{C}}-\overset{\overset{\displaystyle Cl}{|}}{\underset{\underset{\displaystyle H}{|}}{C}} Cl$$

With aromatic aldehyde or ketones base catalyzed additions occur, while it condenses with primary amines to yield isocyanides: $C_2H_5NH_2 + CHCl_3 \xrightarrow{NaOH} C_2H_5N\equiv C$. A special type of addition can occur when chloroform is reacted with an unsaturated molecule in the presence of potassium alkoxide or sodium hydroxide in a polymer medium. The strong base removes HCl and produces dichlorocarbene ($\cdot CCl_2$) which adds across the double bond:

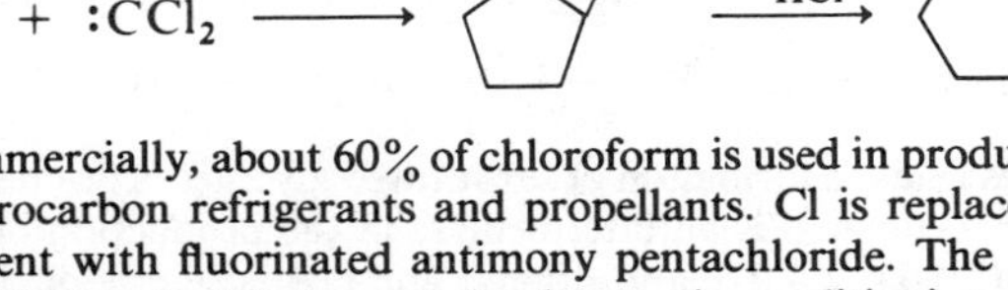

This type of synthesis is particularly useful for ring expansions with specific stereochemistry:

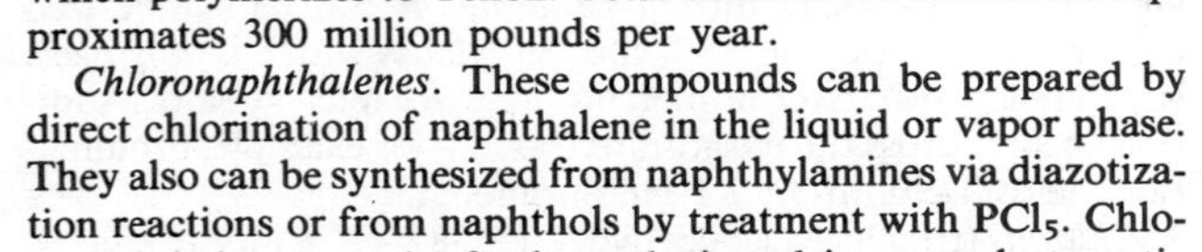

Commercially, about 60% of chloroform is used in production of fluorocarbon refrigerants and propellants. Cl is replaced by treatment with fluorinated antimony pentachloride. The product, $CHClF_2$ (R-22) is used for home air-conditioning units. It is also used as a feed for production of tetrafluoroethylene, which polymerizes to Teflon. Total demand for chloroform approximates 300 million pounds per year.

Chloronaphthalenes. These compounds can be prepared by direct chlorination of naphthalene in the liquid or vapor phase. They also can be synthesized from naphthylamines via diazotization reactions or from naphthols by treatment with PCl_5. Chloronaphthalenes can be further substituted in normal aromatic

reactions e.g., halogenation, nitration and alkylation. The chloro group can be displaced to yield anaphthol, an amine, or a nitrile:

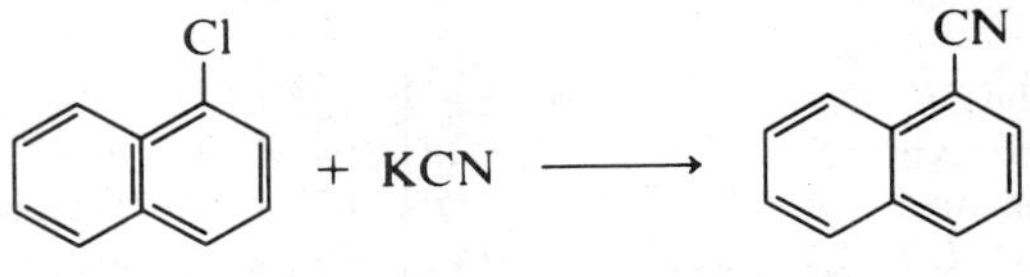

Chlorparaffins. These are produced by the random chlorination of various mixed long chain paraffins. They are used as secondary plasticizers for polyvinyl chloride, lubricating oil additives, resinous materials for coatings, and in flame-retardants.

A particularly valuable use for chlorparaffins is in preparation of linear, primarily internal olefins as feedstock for long-chain synthetic oxo-alcohols. Typically, *n*-paraffins (C_{11}–C_{14}) are chlorinated in a fluidized bed at about 300°C. Conversion is maintained low to limit multiple chlorination. After separation of the monochlorinated alkanes by distillation, dehydrochlorination over nickel acetate at 300°C yields the desired internal olefins. Unreacted paraffins are recycled.

Chloroprene (2-chlorobutadiene-1,3) [126–99–8]. Chloroprene can be synthesized by addition of HCl to vinyl acetylene ($H_2C=CH—C\equiv CH + HCl \rightarrow H_2C=CH—CCl=CH_2$) and by dehydrochlorination of dichlorobutenes or 2,2,3-trichlorobutane. It undergoes the normal addition reactions across the double bond and readily polymerizes or copolymerizes with other unsaturated compounds. These polymers resemble natural rubber but are superior in some respects, such as oil resistance (neoprene). Almost all the chloroprene produced is used for the manufacture of these polychloroprene rubbers. Chloroprene is a volatile, toxic, flammable liquid and is especially susceptible to oxidation and polymerization.

Chlorostyrene (chlorovinylbenzene) [1331–28–8]. The alpha isomer can be prepared by PCl_5 reaction on acetophenone

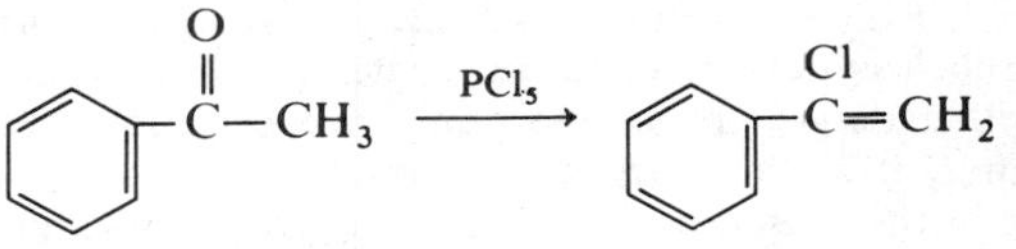

heating of acetophenone dichloride, or hydrolysis of styrene dichloride in aqueous NaOH. The beta isomer can be made by chlorination of cinnamic acid:

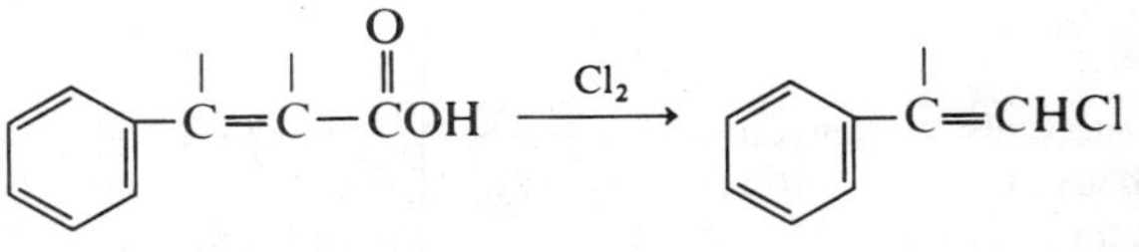

or dehydration of styrene chlorohydrin. Chlorostyrene will add Cl_2 to give α,β,β-trichloroethylbenzene or Br_2 to give the chloro-dibromo product. On treatment with alcoholic KOH the beta isomer is partly resinified.

Dichlorobenzenes [95–50–1] [106–46–7] [541–73–1]. Dichlorobenzenes are primarily produced by the chlorination of benzene in the presence of a catalyst ($FeCl_3$ or $AlCl_3$) although there are other possible synthetic routes. The two commercially important isomers are the ortho- and para-dichlorobenzenes. Further chlorination yields 1,2,4-trichlorobenzene. Dichlorobenzenes participate in normal aromatic substitution and alkylation reactions. In the presence of $CuCl_2$, ammonia will react with the dichlorobenzenes to yield chloroanilines. The ortho-dichlorobenzene is used for pesticides, moth control, as a solvent and for dyestuff manufacture. About half of the para is used as a space odorant.

Epichlorohydrin (γ-chloropropylene oxide) [106–89–8]. This compound can be prepared from 1,3-dichloropropanol-2,2,3-dichloropropanol-1, or allyl chloride. Commercially it is prepared as an intermediate in glycerol synthesis via alkaline hydrolysis of glycerol dichlorohydrin. Both come from allyl chloride. Epichlorohydrin reacts with monohydric alcohols to give ethers by opening the oxide ring. It will react with ethers, aldehydes, ketones, organic acids and amines to give a wide variety of useful syntheses.

Commercially the most important use is production of glycerine. Large volumes are consumed in nonglycerine areas, which largely consist of the various epoxy resins. It has use as a solvent and in the production of epichlorohydrin rubber.

Ethyl Chloride (chloroethane) [75–00–3]. This compound can be synthesized by treatment of ethyl alcohol with HCl, cleavage of diethyl-ether with HCl in the presence of a catalyst (ZnCl$_2$), chlorination of ethane or hydrochlorination of ethylene. The latter is the choice of industry. The reaction is carried out at 125°F and 125 psi in the presence of AlCl$_3$, which is dissolved in ethyl chloride. It will undergo all the reactions of a typical alkyl chloride—halogenation, hydrolysis, amination, alkylation, and will form the magnesium Grignard reagent. The compound is used in production of tetraethyllead (TEL) by reaction with sodium-lead alloy:

$$4PbNa + 4C_2H_5Cl \rightarrow Pb(C_2H_5)_4 + 3Pb + 4NaCl.$$

Ethyl cellulose is produced by treating alkali cellulose (cotton linter digested in dilute caustic) with ethyl chloride. Up to three ethyl ether stages can be made, giving various grades. These are used as synthetic gums and thickeners in the lacquer and plastics industries. Ethyl chloride is also used in the Friedel-Crafts alkylation of benzene and other aromatics. Additional uses include solvent, refrigerant, heat-transfer medium, aerosol propellant and anesthetic. Much is used captively by the producers; total demand is estimated at about 700 million pounds per year.

Ethylene Dichloride (1,2-dichloroethane) [107–06–2]. Ethylene dichloride (EDC) is produced by reacting ethylene and chlorine in the presence of ferric chloride, using the liquid product as solvent. It is also produced by oxychlorination—ethylene, hydrogen chloride, and air are reacted at about 250°C with a copper chloride catalyst. This latter is the reaction of choice only when cheap by-product HCl is available. EDC reacts with Cl$_2$ to give derivatives of ethylene or ethane, depending on conditions and catalysts. It will dehydrochlorinate to give vinyl chloride, which is its principal commercial use. EDC hydrolyzes to ethylene glycol and reacts with aromatic hydrocarbons in the presence of AlCl$_3$ to give polyarylethylene plastics. The largest use is for vinyl chloride; next is its use as a solvent intermediate; third is the use as a lead scavenger in lead antiknock fluids—these fluids normally contain EDC at about 30% of the weight of TEL, along with some ethylene dibromide (EDB). Other uses include the manufacture of ethylenediamine and succinic acid, by way of the nitrile. Reaction of EDC with sodium tetrasulfide is used to produce thiokol rubbers.

Methyl Chloride [74–87–3]. This compound is produced by direct chlorination of methane. Since methyl chloride adds chlorine faster than methane, the yield of methyl chloride is increased by using a large excess of methane in the feed, i.e., about ten volumes of methane to one volume of chlorine. The reaction is carried out at about 450°C with very short contact times. Methyl chloride is also commercially produced by reaction of HCl on methanol in the presence of zinc chloride. Methyl chloride is mainly used in the production of silicone resins and rubbers. Silicon metal is reacted with an excess of methyl chloride at 300°C in the presence of a copper catalyst. The product includes mono-, di-, and trichloromethyl silanes. Hydrolysis of the chloro groups converts them into the corresponding hydroxymethylsilanes. These are then polymerized to silicones. Nearly equal amounts of methyl chloride are used in making these rubbers and the other principal user, production of tetramethyllead. Production of methyl chloride approximates 440 million pounds per year.

Methylene Chloride (dichloromethane) [75–09–2]. As with the other members of the methyl series of chlorinated hydrocarbons, methylene chloride can be produced by direct chlorination of methane. The usual procedure involves a modification of the simple methane process. The product from the first chlorination passes through aqueous zinc chloride, contacting methanol at about 100°C. Thus, HCl from chlorination is used to displace the alcohol group, producing additional methyl chloride. This is further chlorinated to methylene chloride. Methylene chloride reacts violently in the presence of alkali or alkaline earth metals and will hydrolyze to formaldehyde in the presence of an aqueous base. Alkylation reactions occur at both functions, thus di-substitutions result. For example, reaction with benzene plus AlCl$_3$ yields diphenyl methane:

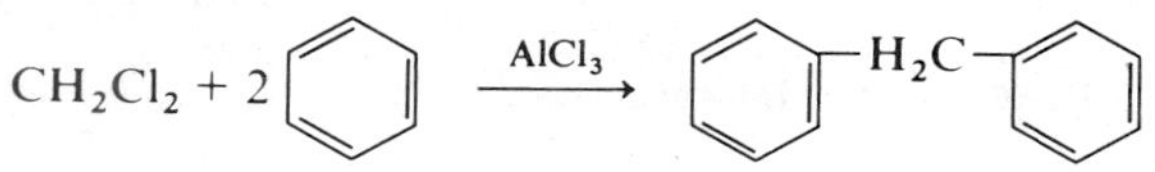

Catalyzed carbonylation (CO) reactions lead to formation of either chloroacetyl chloride or malonyl dichloride.

Methylene chloride is used in refrigeration, aerosol propellants, paint stripping, urethane foam-blowing agents, adhesive, and food extractants. It has low toxicity compared with other chlorinated hydrocarbons.

Monochlorobenzene (phenyl chloride) [108–90–7]. Benzene is chlorinated at 80°C in the presence of FeCl$_3$ catalyst. By using low conversions very little dichlorobenzene is produced. The chlorine on this compound is quite inactive, but hydrolysis can be effected by use of a strong caustic at high temperature and pressure, especially in the presence of a catalyst. This has been an important commercial route to phenol. With concentrated aqueous ammonia heated at high temperatures in the presence of a copper catalyst, aniline or diphenylaniline can be synthesized. Reaction with nitric acid yields chloronitrobenzenes—the para isomer predominates. Treatment with hot sulfuric acid leads to formation of *p*-chlorobenzensulfonic acid:

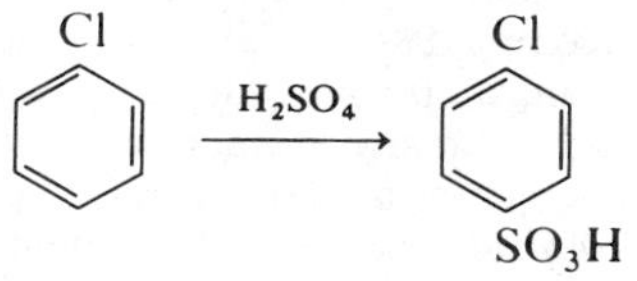

Monochlorobenzene is used commercially as a solvent and to produce phenol and nitrochlorobenzenes.

p-Nitrochlorobenzene [100–00–5].

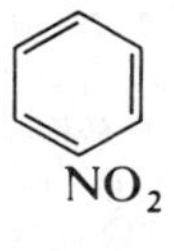

This compound is made by the nitration of chlorobenzene and is largely used to produce *p*-nitrophenol with smaller production of *p*-nitroaniline:

Various agricultural pesticides, rubber chemicals, phenacetin, and *p*-aminophenol consume about 30% of the total. Most of the production is used captively as an intermediate in the production of other chemicals.

Pentachlorophenol [87–86–5]. This compound can be produced by the chlorination of phenol in the presence of AlCl₃, or by hydrolysis of hexachlorobenzene with NaOH in methanol. Pentachlorophenol is used as a wood preservative for poles, crossarms, and pilings, and thus competes with creosote.

Vinyl Chloride [75–01–4]. This compound is produced by alkaline dehydrochlorination of ethylene dichloride, or by thermal cracking of EDC, or 1,1-dichloroethane. Vinyl chloride is polymerized in various ways to polyvinyl chloride (PVC). It is also copolymerized with various other monomers to make a variety of useful resins. The copolymers with about 3 to 20% vinyl acetate are the most important. Demand for vinyl chloride is high, approximating 7 billion pounds (3.2 billion kilograms) per year.

See also **Chlorofluorocarbons.**

References

ACS: "Chemical Abstracts Service," American Chemical Society, Washington, D.C. (Continuing).
CRC: "Handbook of Chemistry and Physics," CRC Press, Inc., Boca Raton, Florida, 1980.
Huntress, E. M.: "Organic Chlorine Compounds," Wiley, New York, 1948.
ITII: "Toxic and Hazardous Industrial Chemicals Safety Manual," The International Technical Information Institute, Tokyo, 1976.
Kirk, R. E., and D. F. Othmer: "Encyclopedia of Chemical Technology," 3rd edition, Wiley, New York, 1979.

—Walter Wm. Lawrence, Jr., Ethyl Corporation, Baton Rouge, Louisiana.

CHLORINATION (Process). Chlorinated Organics.

CHLORINATION (Water). A principal means for disinfecting municipal water supplies as well as public swimming pools, and some municipal and industrial wastes is by liquid- or gas-phase chlorination. Liquid chlorine is packaged in several types of containers to accommodate a wide range of uses, which may vary from a few hundred pounds during a season (in the case of a swimming pool) to many thousands of tons per year for water supplies. Liquid chlorine is obtainable in pressurized 100- and 150-pound (~45- and 68-kilogram) cylinders, 1-ton (0.9-metric-ton) containers, and for large users, is shipped by railroad tank cars, tank barges, and tank trailers. Large users, of course, must provide local storage means.

Chlorine cylinders are equipped with a single valve. Gas is delivered when the tank is in an upright position; liquid when the cylinder is in an inverted position. However, liquid withdrawal from cylinders is not usually practiced. In the case of ton containers, two valves are provided, permitting easy withdrawal of either gaseous or liquid chlorine. Bulk shipments almost always are unloaded in the liquid phase.

In the case of gaseous withdrawal, the vaporization of the liquid chlorine lowers the temperature surrounding the valve and hence withdrawal rates are limited, ranging up to a maximum of about 1.75 pounds (0.8 kilogram) per hour for a 150-

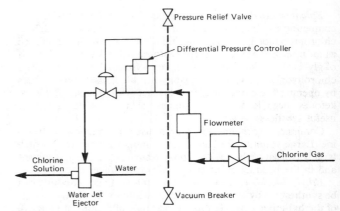

Fig. 1. Chlorinator arrangement commonly used by municipal water and waste water treatment plants.

pound (~68-kilogram) cylinder; 15 pounds (6.8 kilograms) per hour for a ton container. Sometimes, cylinders are manifolded to increase the capacity of the system. In the case of liquid withdrawal, the rate ranges up to 400 pounds (181 kilograms) per hour for ton containers; up to 7,000 pounds (3175 kilograms) per hour for a tank car where discharge is from one valve. Usually the liquid is forced out of the container or tank by its own vapor pressure. However, air pressure up to 200 psi (13.6 atmospheres) may be superimposed to increase withdrawal rates.

In municipal water and waste water treatment installations, the chlorine usually is introduced into the main water system by way of a concentrated water solution of chlorine. The schematic of such a system is shown in Fig. 1. Chlorine is metered under a vacuum created by the ejector. The chlorine is dissolved in water in the ejector, and then discharged into the water system as a high-strength solution. Where the chlorine feeding is done automatically in proportion to the flow of water being chlorinated, a system of the type shown in Fig. 2 can be used.

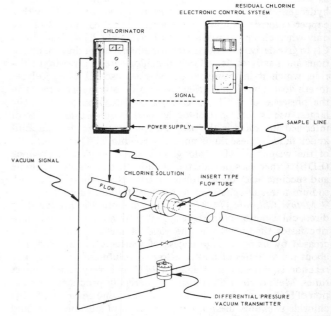

Fig. 2. Closed-loop chlorination control system. (*Fischer & Porter Co.*)

In the control of chlorine disinfectant systems, the effective use of the chlorine for its intended purpose is assumed if the treated water considerably downstream from the chlorinator contains a residual of chlorine. Depending upon use, full-contact time may be assumed after ten minutes, or the interval may be extended to several hours. The systems also are usually carefully monitored by bacteriological testing. Normally a dose of 1 to 2 miligrams of chlorine per liter is adequate to destroy all bacteria and leave an effective residual. Residuals of 0.1 to 0.2 milligrams per liter are usually maintained in the effluent streams from water-treatment plants as a factor of safety for consumers.

Surface waters require in most instances more extensive treatment, including chlorination, than do groundwaters. By the time some river water reaches some consuming communities it will have received large inputs of organics. Because of its great oxidizing power, chlorine is highly reactive and can combine in a variety of ways with both inorganic and organic pollutants. Thus, there is concern over the possible formation of carcinogens or otherwise harmful compounds in waters that are heavily chlorinated, particularly waters that have been recycled a number of times along a waterway. Major halogenated compounds found in water supplies suspected of posing health hazards to humans include: (1) chloro-esters, such as *bis*-(2-chloroethyl)ether and *bis*-(2-chloroisopropyl)ether; (2) halobenzenes, such as chlorobenzenes, bromobenzenes, and chlorobromo benzenes; and (3) haloforms, such as chloroform, bromodichloromethane, dibromochloromethane, and bromoform. More information on this topic can be found in "Chlorine in the Marine Environment," by J. C. Goldman. **Oceanus, 22,** 2, 36–43 (1979).

The fundamentals of chlorination for potable water, waste water, cooling water, industrial process water, and swimming pools are well covered in the "Handbook of Chlorination," by G. C. White (Van Nostrand Reinhold, New York, 1972).

CHLORINE. Chemical element symbol Cl, at. no. 17, at. wt. 35.435, periodic table group 7a (halogens), mp $-101°C$, bp $-34.1°C$, density (chlorine gas) 3.209 grams/liter (0°C and 1 atm. pressure). Chlorine gas is approximately 2.5 times heavier than air at standard conditions. Chlorine in the gaseous phase is diatomic (mol. wt. 70.906), pale greenish yellow of marked odor, irritating to the eyes and throat, poisonous. At 10°C and one atmosphere pressure 9.8 grams of Cl_2 will dissolve in one liter of water; at 30°C and one atmosphere pressure 5.6 grams will dissolve. Critical pressure is 1118.4 psia (7.7 MPa), critical temperature 144°C. CAS Registry No. 7782-50-5. Other important physical characteristics of chlorine are given under **Chemical Elements.**

Chlorine was discovered by Scheele in 1774 and confirmed as an element by Davy in 1810. It is a high-tonnage industrial chemical with many uses.

Naturally occurring isotopes* 35, 37. Electronic configuration $1s^2 2s^2 2p^6 3s^2 3p^5$. Ionic radius Cl^{7+} 0.26 Å, Cl^- 1.81 Å. Covalent radius 1.050 Å. First ionization potential 13.01 eV; second, 23.70 eV; third, 39.69 eV; fourth, 53.16 eV; fifth, 67.4 eV. Oxidation potential $ClO_3^- + H_2O \rightarrow ClO_4^- + 2H^+ + 2e^-$, -1.00 V; $HClO_2 + H_2O \rightarrow ClO_3^- + 3H^+ + 2e^-$, -1.23 V; $\frac{1}{2}Cl_2 + 4H_2O \rightarrow ClO_4^- + 8H^+ + 7e^-$, -1.34 V; $Cl^- \rightarrow \frac{1}{2}Cl_2 + e^-$, -1.3583 V; $Cl^- + 3H_2O \rightarrow ClO_3^- + 6H^+ + 6e^-$, -1.45 V; $\frac{1}{2}Cl_2 + 3H_2O \rightarrow ClO_3^- + 6H^+ + 5e^-$, -1.47 V; $Cl^- + H_2O \rightarrow HClO + H^+ + 2e^-$, -1.49 V; $Cl^- + 2H_2O \rightarrow HClO_2 + 3H^+ + 4e^-$, -1.56 V; $\frac{1}{2}Cl_2 + H_2O \rightarrow HClO + H^+ + e^-$, -1.63 V; $\frac{1}{2}Cl_2 + 2H_2O \rightarrow HClO_2 + 3H^+ + 3e^-$, -1.67 V; $ClO_3^- + 2OH^- \rightarrow ClO_4^- + H_2O + 2e^-$, -0.17 V; $ClO_2^- + 2(H^- \rightarrow ClO_3^- + H_2O + 2e^-$, -0.35 V; $ClO^- + 2OH^- \rightarrow ClO_2^- + H_2O + 2e^-$, -0.59 V; $Cl^- + 6OH^- \rightarrow ClO_3^- + 3H_2O + 6e^-$, -0.62 V; $Cl^- + 4OH \rightarrow ClO_2^- + 2H_2O + 4e^-$, -0.76 V; $Cl^- + 2OH^- \rightarrow ClO^- + H_2O + 2e^-$, -0.94 V; $ClO_2^- \rightarrow ClO_2 + e^-$, -1.15 V.

Production. Most of the chlorine produced in the world is manufactured by electrolysis of sodium chloride brine. Two processes are in common use: the mercury cell process and the diaphragm cell process. Since 1969 when international concern about the effect of mercury in the envionment became widespread, some mercury cell plants have been shut down. Most expansion of chlorine production has been in diaphragm cell plants. Indeed, all mercury cell plants in Japan are now required to be converted to diaphragm cell plants. However, it should be noted that existing mercury cell plants operate well within strict standards as to mercury discharge both into the air and into waterways. They produce chlorine (and caustic soda) of the most exacting quality suitable for all uses including food preparation. There is no reason to believe that mercury cell technology is obsolescent in this country or, generally speaking, worldwide.

Technology. In production of chlorine by the diaphragm cell process, salt is dissolved in water and stored as a saturated solution. Chemicals are added to adjust the pH and to precipitate impurities from both the water and the salt. Recycled salt solution is added. The precipitated impurities are removed by settling and by filtration. The purified, saturated brine is then fed to the cell, which typically is a rectangular box. It uses vertical anodes (ruthenium dioxide with perhaps other rare metal oxides deposited on an expanded titanium support). The cathode is perforated metal which supports the asbestos diaphragm. This is vacuum deposited in a separate operation. The diaphragm serves to separate the anolyte (the feed brine) from the catholyte (brine containing caustic soda). Chlorine is evolved at the anode. It is collected under vacuum, washed with water to cool it, dried with concentrated sulfuric acid, and further scrubbed, if necessary. It is then compressed and sent to process as a gas or liquefied and sent to storage for transfer to shipping containers and, ultimately, shipment to consumers.

A cell of this type is called a monopolar cell. In a cell bank, several cells have their negative electrodes and their positive electrodes connected by means of external bus bars. Some companies use a bipolar cell in which the electrodes are internally connected. This results in a configuration like a plate and frame filter press.

The latest development in cell technology is the so-called membrane cell. This uses a cation exchange membrane in place of an asbestos diaphragm. It permits the passage of sodium ions into the catholyte but effectively excludes chloride ions. Thus the concept permits the production of high-purity, high-concentration sodium hydroxide directly. The chlorine side of the cell is identical to existing technology. Research in membrane cells is preceeding at a rapid rate. Some companies are known to be operating this process commercially, but as of 1983 not all problems have been solved.

In the mercury cell process chlorine is liberated from a brine solution at the anodes which are, today, typically metal anodes (Dimensionally Stable Anodes or DSA). Collection and processing of the chlorine is similar to the techniques employed when diaphragm cells are used. However the cathode is a flowing bed of mercury. When sodium is released by electrolysis it is immediately amalgamated with the mercury. The mercury amalgam is then decomposed in a separate cell to form sodium hydroxide and the mercury is returned for reuse.

---

* Information in this paragraph furnished by encyclopedia staff.

Uses. The principal use of chlorine is in the production of organic compounds. In particular, the production of PVC (polyvinylchloride) has been the largest single consumer, although chlorinated solvents as a class account for larger tonnage. See **Chlorinated Organics.**

In many cases chlorine is used as a route to a final product which contains no chlorine. For instance propylene oxide has traditionally been manufactured by the chlorohydrin process. Modern technology permits abandoning this route in favor of direct oxidation, thus eliminating a need for chlorine.

Large quantities of chlorine are used in bleaching. Pulp bleaching for paper manufacture consumes about 13% of all chlorine produced in the United States. Since none of the chlorine used for this purpose winds up in the finished product, it must all be discharged as chlorides or chlorinated organics or be reprocessed. At the present time there is no proven, wholly satisfactory technique for removing chlorine compounds from pulp mill bleach plant wastes. It is doubtful that existing mills will be converted to a bleaching technique which does not require chlorine but future mills may be designed to minimize the use of chlorine.

Substantial quantities of chlorine go into household bleaches. It is used also in laundry and other commercial bleaches. It is the active element in most swimming pool sanitizers.

Large quantities of chlorine are used for treating municipal and industrial water supplies and this use will probably continue. However some concern has been felt that traces of organic compounds in all water supplies react with the chlorine to form chlorinated organics which are suspected of being carcinogenic. Further the usefulness of chlorination of municipal wastes has been questioned in some quarters in the light of the fact that such treatment adds chlorinated organics to the waterways.

Safety and Handling. Although chlorine is a hazardous substance, it can be handled safely. All persons who handle chlorine should be thoroughly trained in its properties; in correct use of safety equipment; and in the operation of all other equipment including containers. The Chlorine Institute, 342 Madison Ave., New York, NY 10017 publishes the *Chlorine Manual* (available from the Institute at nominal cost) which provides useful information on these matters. In addition the Chlorine Institute has designed emergency kits capable of capping off certain types of leaks which can occur in chlorine containers.

There have been several recent studies of the physiologic effects of chlorine. These have considered chlorine both as an occupational exposure and as an environmental pollutant (see references). The National Institute of Occupational Safety and Health study recommended a 0.5 ppm concentration of chlorine in air for any 15-minute sampling period as the maximum permissible ceiling value. This contrasts with the generally accepted value of 1 ppm TLV (time weighted average for an eight hour exposure).

Chlorine is primarily a respiratory irritant. When the concentration in the air is sufficient, chlorine irritates the mucous membranes, the respiratory system and the skin. It causes irritation of the eyes, coughing, and labored breathing. It may cause vomiting. In extreme cases, the difficulty of breathing may increase to the point where death can occur from suffocation. Liquid chlorine in contact with the eyes or skin will cause local irritation or severe burns.

Persons who have been overcome by chlorine should be removed to an uncontaminated area, contaminated clothing should be removed, and the victim should be kept warm. Medical help should be provided. If breathing appears to have ceased, artificial respiration should begin immediately. If breathing is labored, the administration of oxygen may be helpful.

Chlorine Chemistry. Chlorine exhibits in common with the other halogen elements a marked readiness to form singly charged negative ions, as would be expected from the fact that these atoms need only one electron to acquire an inert gas configuration. Thus, chlorine behaves in its normal chemical reactions as an electron acceptor. While there are many compounds in which chlorine has a positive valence, there are no simple compounds of positively charged chlorine (contrast the I^+ of iodine). The positively charged chlorine forms part of a radical, as in combination with oxygen. The electron affinity of chlorine (4.02 eV) is the greatest of all the halogens, and is greater than that of oxygen.

Chlorine reacts readily with hydrogen to form hydrogen chloride, with metals and many non-metals to give chlorides, with metal oxides to give chlorides or oxychlorides, and with many salts of metals to give chlorides. These include the iodides and bromides, whose halogen is displaced by chlorine.

Four isolatable oxides of chlorine are known, Cl_2O, ClO_2, $Cl_2O_6(\rightleftharpoons 2ClO_3)$, and Cl_2O_7. Chlorine(I) oxide, Cl_2O, obtained by passing Cl_2 over mercury(II) oxide and sand, is a gas, bp 2°C, somewhat soluble in H_2O to form hypochlorous acid. The Cl—O—Cl bond angle is 111° and the Cl—O distance 1.71 Å. Cl_2O is an active oxidizing agent. Chlorine(II) oxide, ClO, is produced by reaction of Cl_2O with atomic chlorine, or as an intermediate product in the decomposition of the Cl_2O. The ClO then decomposes into chlorine and oxygen. In view of this instability the properties of the compound are not established. Chlorine(IV) oxide, ClO_2, is produced by treatment of sodium chlorate, $NaClO_3$, with mixed HCl, oxalic acid or other mild reducing agent, and H_2SO_4 (and H_2O). Cl(IV) oxide is a greenish yellow gas, having an odd electron in its molecule and is consequently paramagnetic. Electron diffraction studies indicate its structure to be

$$:\overset{..}{\underset{}{Cl}}::\overset{..}{\underset{}{O}}$$
$$\overset{..}{\underset{}{:O:}}$$

with the odd electron in an antibonding orbital. The O—Cl—O bond angle is 116.5° and Cl—O distance 1.49 Å. It is readily hydrolyzed, but is stable when dry. The mechanism of its hydrolysis is complex, yielding all four of the oxychloric acids, and it is widely used as a heavy-duty oxidizing agent. It reacts with metal hydroxides to give the mixture of chlorate and chlorite, with metal peroxides to give chlorite and oxygen and with metals to give chlorite alone. ClO_2 is photosensitive, decomposing when illuminated at about 8°C, to give some Cl_2O_6, bp 3.5°C. Chlorine hexoxide, Cl_2O_6, has a molecular weight corresponding to the formula ClO_3—ClO_3. Its vapor pressure in the liquid state is 0.31 mm at 0°C against values of 23.7 for Cl_2O_7, 490 for ClO_2 and 699 mm for Cl_2O. This is consistent with a bitrigonal-pyramidal structure in which the two pyramids have three oxygen atoms at their base corners, and are joined by the two chlorine atoms at the apices. In contrast, the additional oxygen atom of Cl_2O_7 would separate the two chlorine atoms, preventing close packed structure. Chlorine heptoxide, Cl_2O_7, the anhydride of perchloric acid, is obtained by heating the latter with phosphorus pentoxide, and consists of two chlorine atoms, each bonded to three oxygen atoms, and jointly bonded to the seventh. All of the oxides of chlorine are thermodynamically unstable with respect to decomposition into the elements.

Hypochlorous acid, HClO, is formed by hydrolysis of chlorine (I) oxide. It is present in aqueous solutions of chlorine because of the equilibrium

$$Cl_2 + 2H_2O \rightleftharpoons HClO + H_3O^+ + Cl^-$$

and can be freed by the addition of any substance that combines with the Cl^-, such as mercury(II) oxide, or with the H^+, such as calcium carbonate or other weak bases which do not react with HClO, HClO is a weak acid (K = 3×10^{-8} at 25°C). It reacts with hydrochloric acid to give chlorine and H_2O. On warming or irradiation it undergoes this reaction as well as two other decompositions, i.e., to oxygen, H^+ and Cl^-, and to ClO_3^- and H^+. The presence of oxygen favors the last reaction. HClO and its salts are strong oxidizing agents, oxidizing iodine and bromine to iodates and bromates. Covalent hypochlorites are known, such as the alkyl esters, ROCl. In common with other esters of oxidizing acids, these are unstable if R is a primary or secondary alkyl group. However, t-butyl hypochlorite is quite stable. Reduction of chlorine dioxide, ClO_2, with hydrogen peroxide, yields oxygen and chlorous acid, $HClO_2$, which exists only in solution. It is stronger than hypochlorous acid (K = 1.01×10^{-2} at 23°C). It is also a strong oxidizing agent and its sodium salt is widely used for this purpose, generally as a source of ClO_2.

Chloric acid, $HClO_3$, is readily prepared by passing chlorine into hot caustic solutions, since these conditions favor the formation of chlorate. Chloric acid is a more active oxidizing agent than HClO, reacting explosively with organic matter. Its alkali salts undergo on heating two modes of decomposition, one (catalyzed, e.g., by manganese dioxide) into the chloride and oxygen, and the other (uncatalyzed) into the chloride and perchlorate.

Perchloric acid, $HClO_4$, is obtained in anhydrous form from perchlorates by H_2SO_4 distillation, or from ammonium perchlorate by aqua regia distillation. Perchlorates are also obtained by electrolysis of chlorides or chlorates. Perchloric acid is explosive unless properly handled; it is of course a powerful oxidizing agent, but has a higher activation energy than the lower acids.

The chlorides range in character from ionic to covalent compounds, many of them having bonds of intermediate character. There are also a number of interhalogen compounds containing chlorine. Those of iodine and bromine are discussed under those entries. With fluorine, chlorine forms ClF, chlorine monofluoride, which is also obtained (along with ClF_3), when mixtures of the elements are subjected to spark discharge. It may also be obtained by heating a mixture of chlorine and chlorine trifluoride. It is a reactive gas, bp $-100.8°C$, and with its bond having 20–30% ionic character. Chlorine trifluoride, ClF_3, is also obtained from the elements or from chlorine monofluoride and fluorine, by varying the conditions. It is a gas, bp of liquid 11.3°C, and is a more powerful fluorinating agent than the monofluoride. Present views on its structure suggest a trigonal bi-pyramid having chlorine in the center, a fluorine atom at each apex and the third fluorine and two non-bonding pairs of electrons in the three equatorial positions, giving a T-shaped molecule. ClF_3 reacts with all elements except the noble gases, nitrogen, chromium, and certain noble metals, although some metals (e.g., copper) require elevated temperature. It does not react with oxides or salts as readily as fluorine, but nevertheless ignites such materials as asbestos.

See also **Chlorinated Organics; Halides; Hypochlorites;** and **Sodium Chloride.**

References

Sconce, J. S.: "Chlorine: Its Manufacture, Properties and Uses," ACS Monograph 154, Van Nostrand Reinhold, New York (1962).
Somers, H. A.: "The Chlor-Alkali Industry," *Chem. Eng. Progress,* **61,** 3 (March 1965) (Covers mercury cells only).
Staff: "Chlorine Manual," The Chlorine Institute, New York (1969) (Updated periodically).
Staff: "Exceeding All Expectations: A Short History of Chlorine," The Chlorine Institute, New York (1968).
Staff: "Criteria for a Recommended Standard: Occupational Exposure to Chlorine," National Institute for Occupational Safety and Health, *HEW Publication No.* (*NIOSH*) *76-170,* Washington, D.C. (1976).
Staff: "Medical and Biologic Effects of Environmental Pollutants: Chlorine and Hydrogen Chloride," National Academy of Sciences, Washington, D.C. (1976).
Staff: "Diaphragm Cells for Chlorine Production," *Proceedings of Symposium* held at City University, London, England (June 16–17, 1977).
Weast, R. C.: "Handbook of Chemistry and Physics," 61st Edition, CRC Press, Boca Raton, Florida (1981).

—Herbert S. Hopkins, Research Center, Olin Corporation, New Haven, Connecticut.

CHLORINITY. A measure of the chloride content, by mass, of seawater (grams per kilogram of seawater, or per cubic mile). Originally, chlorinity was defined as the weight of chlorine in grams per kilogram of seawater after the bromides and iodides had been replaced by chlorides. To make the definition independent of atomic weights, chlorinity is now defined as 0.3285233 times the weight of silver equivalent to all the halides present.

CHLORITE. Chlorite is an ubiquitous mineral usually a product of secondary origin from the alteration of silicates containing aluminum, ferrous iron, and magnesium. Pyroxenes, amphiboles, biotite garnet, and idocrase within rocks which have undergone metamorphism are common source minerals for chlorite. Distinct crystals are extremely rare; more often found as foliated masses or fine scaly aggregates. Color includes various shades of green. Hardness of 2–2.5, and specific gravity of 2.6–2.9, with vitreous to pearly luster. Individual folia characterized by flexible, not elastic property. A general formula is $(Mg,Fe^{2+}, Fe^{3+}, Mn)_6 \ AlSi_3O_{10} \ (OH)_8$.

CHLORITOID. A mineral which occurs as tabular crystals, probably triclinic, foliated masses or scattered scales and plates of a greenish-tray to greenish-black color. It is characteristic of the less intensely altered metamorphic rocks such as phyllites and quartzites. Chemically it is a hydrous iron-aluminum silicate, $Fe_2Al_4Si_2O_{10}(OH)_4$. Ottrelite contains some manganese as well. Chloritoid was originally noted as from the Ural Mountains and named for its greenish color from the Greek word meaning *green.* Ottrelite was named from Ottrez in Luxemburg.

CHLOROACETIC ACID. Chlorinated Organics.

CHLOROACETYLENE. Chlorinated Organics.

CHLOROBIPHENYLS. Chlorinated Organics.

CHLOROFORM. Anesthesia; Chlorinated Organics.

CHLOROFLUOROCARBONS. Methanes, ethanes, and ethylenes which contain at least one fluorine atom per molecule. Because of the large number of compounds in these series and to avoid the complexities of organic chemistry nomenclature insofar as the commercial user of the products is concerned, an abbreviated designation system is used. The system comprises a four-element designation, preceded by a generic term that describes the main application of the product. Thus, *Refrigerant ABCD.* Actually, this system is also complex for users—hence, a strong reliance upon strictly trade names.

In the *ABCD* system, *A* equals the number of double bonds in the molecule; *B* equals the number of carbon atoms minus 1; *C* equals 1 (one) plus the number of hydrogen atoms in the molecule; and *D* equals the number of fluorine atoms in the molecule. If A and B are zero, the digits simply are omitted. Thus, *Refrigerant 12* is dichlorodifluoromethane.[1]

Most uses of saturated chlorofluorocarbons capitalize on the volatility, stability, and safety of this class of compounds. Refrigerant 11, Refrigerant 12, and Refrigerant 22 are applied to a variety of basic jobs. Refrigerant 11 is used in large centrifugal air-conditioning units in office buildings and industrial plants; Refrigerant 12 is usually selected for household refrigerators and freezers, as well as for automobile air conditioners; Refrigerant 22 is used extensively in residential air conditioning, where high capacity and small unit size are important. A number of binary azeotropes are used in special situations. Components of some of the common binary azeotropes, with their numerical designations, are:

Refrigerant 500	R-12/difluoroethane
Refrigerant 502	R-22/R-115
Refrigerant 503	R-13/R-23

Additionally, chlorofluorocarbons have solvent-cleaning properties that are particularly attractive in the aerospace, electronics, optical, and miniature, precision mechanism manufacturing fields. Their selective solvent properties are advantageous. Trichlorotrifluoroethane is especially suitable because of its lack of attack on paint, gaskets, and wire insulation. Binary azeotropes containing methylene chloride, ethanol, methanol, or acetone provide variations of the solvent properties obtainable with trichlorotrifluoroethane alone.

Substantial quantities of Fluorocarbon 11 and Fluorocarbon 12 are used in plastic foams. Flexible polyurethane foams are commonly expanded with Fluorocarbon 11, while rigid foams frequently are prepared from polystyrene and Fluorocarbon 12.

Dichlorodifluoromethane has been adapted to food freezing, in which food particles are frozen upon contact with boiling chlorofluorocarbon. The resulting rapid heat transfer reduces the freezing time to seconds for most foods.

Fluoroolefins, such as chlorotrifluoroetyylene, tetrafluoroethylene, vinylidene fluoride, and vinyl fluoride are used extensively in the synthesis of high-performance lubricants, plastics, and elastomers.

Carbon tetrachloride or chloroform and anhydrous hydrogen fluoride are the usual starting ingredients in the manufacture of chlorofluorocarbons. A catalyst, such as AlF_3 or $SbCl_5$, is used. Examples include:

$$CCl_4 + HF \rightarrow CCl_xF_y \qquad x + y = 4$$

$$CHCl_3 + HF \rightarrow CHCl_xF_y \qquad x + y = 3$$

Perchloroethylene, chlorine, and anhydrous hydrogen fluoride are used in the preparation of ethane derivatives:

$$CCl_2{=}CCl_2 + HF + Cl_2 \rightarrow CClF_2{-}CCl_2F + [CClF_2 + CClF_2]$$

$$\text{Fluorocarbon} \qquad \text{Fluorocarbon}$$
$$113 \qquad\qquad 114$$

By further fluorination, Fluorocarbon 115 and Fluorocarbon 116 can be prepared from Fluorocarbon 114:

$$[CClF_2 + CClF_2] \xrightarrow{HF} CClF_2{-}CF_3 + CF_3{-}CF_3$$

$$\text{Fluorocarbon} \qquad \text{Fluorocarbon} \quad \text{Fluorocarbon}$$
$$114 \qquad\qquad 115 \qquad\qquad 116$$

Such syntheses usually are accomplished in the liquid or vapor phase at moderate temperature and pressure.

[1] These materials also are referred to by commonly known trade names, such as *Freon* (E. I. DuPont DeNemours & Co., Inc.); *Genetron* (Allied Chemical Corp.); *Ucon* (Union Carbide Corp.); *Isotron* (Pennwalt Corp.); and others.

Fluoroolefins result from a number of steplike reactions, typified by:

$$CClF_2{-}CCl_2F + Zn \rightarrow CClF{=}CF_2 + ZnCl_2, \text{ or}$$

$$CH{\equiv}CH + HF \rightarrow CH_2{=}CHF + CH_3{-}CHF_2, \text{ or}$$

$$CH_3{-}CClF_2 \rightarrow CH_2{=}CF_2 + HCl$$

Starting materials and byproducts are separated by fractional distillation. The products are further purified by washing, followed by drying over suitable desiccants. Because of extensive purification, chlorofluorocarbons rank among the highest-purity organic materials commercially marketed.

CHLOROPHYLLS. A group of closely related green pigments in leaves, bacteria and organisms capable of photosynthesis. The major chlorophylls in land plants are designated a and b. Chlorophyll c occurs in certain marine organisms. Because of the over-whelming percentage of the total photosynthesis which is performed by marine organisms, It is possible that chlorophyll c is equivalent in importance to chlorophyll b. Chlorophyll a is several times as abundant as chlorophyll b. Bacteriochlorophyll contains two more hydrogens than the plant chlorophylls and has the vinyl group altered to an acetyl.

The canonical form for chlorophyll a is $R = CH_3$. For chlorophyll b, $R = CHO$. These structures have been established by a long series of degradation studies mainly by R. Willstätter, Hans Fischer and their collaborators, and by synthetic studies in the laboratories of Fischer.

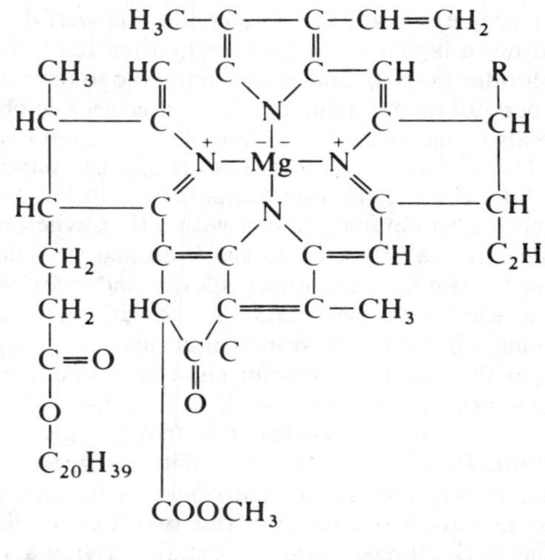

The laboratories of Woodward and of Strell announced the complete synthesis of chlorophyll a almost simultaneously.

The biological significance of the chlorophylls stems from their role in photosynthesis, the process by which plants fix the sun's energy in the form of organic matter. This process corresponds to the reversal of the combustion of hydrogen. The oxygen liberated is set free in the air. Under special conditions, some organisms are also capable of liberating the hydrogen, but usually this is used for chemical reductions in the plant. Atmospheric carbon dioxide is fixed enzymatically and is thus used as the source of the carbon in the synthetic process but is not reduced directly. The path of the carbon from carbon dioxide in photosynthesis has been elucidated largely by the studies of Calvin and his collaborators. While it is known that most of the energy fixed in photosynthesis is absorbed originally by the chlorophylls, the exact reactions which they undergo to initiate the process of reduction are not yet understood. It is known, however, that the photosynthetic sequence requires a high degree of organization within the plant cells where it occurs, and that destruction of the organization of the chloroplasts by processes like grinding are sufficient to bring photosynthesis to a stop, even when the chlorophyll and the soluble enzymes participating in the process are still presumably intact.

Chlorophyll derivatives with the phytyl group intact are oil-soluble and form a series of green dyes which have found wide commercial application in the coloring of oils and waxes. The chlorophyll soaps, resulting from combined saponification and cleavage of the isocyclic ring, form valuable water-soluble dyes, useful in the coloring of soaps and similar products.

Both the medical and the cosmetic literature are replete with claims of therapeutic or physiological activity of "chlorophyll" The substances utilized in this work range from partially purified chloroplasts to mixtures of materials which have undergone deep-seated chemical alteration. Some of the types of activity claimed can be shown to be due to incidental impurities. The field for investigation of the action of pure chemical individuals produced by the action of various reagents upon chlorophyll or its derivatives is unexplored. It is known, however, that neither chlorophyll nor hemoglobin in the diet is utilized by the body in the formation of the physiologically active pyrrole pigments. These are derived, instead, from such simple building blocks as glycine and acetate ion. Only the iron in dietary blood pigment can be utilized by the body.

The work of Granick has shown that, in the physiological processes of plants, chlorophyll is formed from protoporphyrin, which can be obtained in the laboratory by the removal of iron from hemin. The pathways to heme and to chlorophyll diverge at protoporphyrin. To form heme, an organism introduces iron into protoporphyrin. To form chlorophyll from protoporphyrin, an oxidation, a reduction, a ring closure and esterifications are performed and the magnesium is introduced. The end product of the enzymatic synthetic chain is presumably protochlorophyll, the magnesium derivative of the porphyrin corresponding in structure to chlorophyll. The addition of the two hydrogens necessary to convert the red compound to the green is accomplished under the influence of light.

CHOLEIC ACIDS. Bile.

CHOLESTEROL. Also known as cholesterin or 5-cholesten-3-beta-ol), cholesterol is the most common animal sterol; a monohydric secondary alcohol of the cyclopentenophenanthrene (4-ring fused) system, containing one double bond. It occurs in part as the free sterol and in part esterified with higher fatty acids as a lipid in the human blood system. The primary precursor in biosynthesis appears to be acetic acid or sodium acetate. Cholesterol itself in the animal system is the precursor of bile acids, steroid hormones, and provitamin D3. Cholesterol is a white, or faintly yellow, almost odorless substance and may take the form of pearly granules or crystals. The substance is affected by light; mp 148.5°C; bp 360°C, but tends to decompose at lower temperatures. Specific gravity 1.067 (20/4°C); insoluble in water; slightly soluble in alcohol; soluble in fat solvents, vegetable oils, and in aqueous solutions of bile salts. Cholesterol occurs in egg yolk, liver, kidneys, saturated fats, and oils. In addition to the importance of cholesterol in medicine and biology in general, cholesterol is used as an emulsifying agent in cosmetic and pharmaceutical products. It is the source of estradiol.

Research over the years has shown that cholesterol is carried in the bloodstream in complexes with other lipids and proteins. Based upon their density, there are four classes of lipoproteins: the chylomicrons; the VLDLs (very low density; the LDLs (low density); and the HDLs (high density). It is estimated that about 80% of the total blood cholesterol is carried by the LDLs. Most of the remainder is carried by the HDLs. Large quantities of triglycerides are carried by the chylomicrons and VLDLs, but very little cholesterol.

As early as 1951, Barr (Cornell University Medical College) observed that in males with coronary heart disease, the HDL concentrations were low. There were several confirmations of this during the 1950s and 1960s. However, for many years, the principal criterion in connection with heart attack and stroke with relation to cholesterol was considered to be total cholesterol and LDL levels, with little attention given to the HDLs. It was not until 1975 that Miller (Royal Infirmary, Edinburgh, Scotland) reported an inverse correlation between blood concentration of HDLs and total body cholesterol. Miller hypothesized that HDLs may lessen body cholesterol by facilitating its excretion. Epidemiological studies were made shortly thereafter on Japanese-Hawaiian males, Israeli males, and black sharecroppers in Georgia. These studies showed that risk of heart attack increases as blood HDL level decreases. The average value for HDL levels in human males is 45 milligrams per deciliter (55 milligrams for females). The generally higher level of HDLs in females may partially account for their lower heart attack rates. For both sexes, it has been estimated that a 5 milligram drop in the aforementioned HDL levels may increase the risk of heart attack by about 25%.

In other research efforts, Glueck et al. (University of Cincinnati College of Medicine) have identified two groups of people who are genetically endowed either with high HDL or low LDL concentrations and who have lifespans some 5 to 10 years greater than the average. The usual dietary and smoking restrictions for heart attack and stroke prevention do not appear to apply to these groups to any degree approaching that of the average persons. However, it also has been noted that the general dietary and nonsmoking recommendations given for many years also tend to increase HDL concentration.

In recent years, it has been found that there is little correlation between total blood cholesterol and heart attacks in people over age 50. But, the HDL level is more meaningful with this group.

The status of research in this field as of late 1979 is well summarized in *Science*, **205**, 677–679 (1979).

Cholesterol enters into several discussions in this book. In particular, see **Bile**; and **Steroids**. The structural formula for cholesterol, $C_{27}H_{15}OH$, is given in the latter entry.

CHOLESTERIC LIQUID CRYSTALS. Liquid Crystals.

CHOLIC ACID. Bile.

CHOLINE AND CHOLINESTERASE. An enzyme (acetylcholinesterase) is specific for the hydrolysis of acetylcholine to acetic acid and choline in the animal body. It is found in the brain, nerve cells and red blood cells and is important in the mechanism of nerve action. Acetylcholine was first synthesized in 1867. It consists of a combination of choline and acetic acid in an ester linkage. The components parts of the acetylcholine molecule are both normal constituents of the animal body. Acetylcholine has the structure:

Acetylcholine assumed no importance to biologists until 1899 when Hunt identified the presence of choline in extracts of the adrenal glands and suggested that some derivative of choline was capable of causing a fall in blood pressure. This stimulated interest in studying the physiological effects of various choline derivatives and, in 1906, Hunt and Taveau found that acetylcholine was 100,000 times more effective than choline in causing a fall in blood pressure. Shortly thereafter, acetylcholine was identified in extracts of ergot, a fungus that grows on rye and other cereal grains. The first real proof of the role of acetylcholine in transmitting the effects of nerve stimulation did not come until 1921. The acetylcholine-cholinesterase system has served as the basis for the development of a number of drugs needed to alter the activity of the autonomic nervous system in certain disease states. Inhibitors of cholinesterase are used in insecticides. Parathion and malathion are examples of organic phosphate cholinesterase inhibitors that are effective for this purpose. Cholinesterase inhibitors are capable of producing poisoning and death in humans and domestic animals by the same mechanism.

Choline is an essential metabolic substance for building and maintaining cell structure. Choline is usually described along with B complex vitamins, although it is essentially a structural component of tissue rather than a metabolic catalyst. Choline is a part of the structure of phospholipids and acetylcholine.

Choline participates in normal fat metabolism and interrelates with methionine in a biochemical manipulation referred to as transmethylation. Choline, when in adequate quantity, can replace the essential amino acid, methionine, when the latter is in limited quantity; or the reverse may occur wherein methionine can be dismantled to replace choline. Choline deficiencies result in numerous degradative physiologic changes in livestock. The usual dietary supplements are choline bitartrate and choline chloride.

References

DuBois, K. P.: "Acetylcholine and Cholinesterase," in "The Encyclopedia of Biochemistry," (R. J. Williams and E. M. Lansford, Jr., Editors), Van Nostrand Reinhold, New York (1967).

Haresign, W., and D. Lewis: "Recent Advances in Animal Nutrition," Butterworth Group, Woburn, Massachusetts (1977).

Hunt, R., and R. Taveau: "On the Physiological Action of Certain Choline Derivatives and New Methods for Detecting Choline," *British Medical Journal*, **2**, 1788–1791 (1906).

Staff: "Food Chemicals Codex," National Academy of Sciences, Washington, D.C. (1972).

Staff: "Nutritional Requirements of Domestic Animals," separate publications on beef cattle, dairy cattle, poultry, swine, and sheep, periodically updated. National Academy of Sciences, Washington, D.C.

Swan, H., and D. Lewis: "Feed Energy Sources for Livestock," Butterworth, London (1976).

CHONDRODITE. Chondrodite, a magnesium fluosilicate mineral $Mg_5(SiO_4)_2(F, OH)_2$, crystallizing in the monoclinic system is a product of metasomatic origin in metamorphosed dolomitic limestones. Crystals are uncommon, usually occurring as discrete grains within the limestone, of light yellow to red color. Vitreous luster, translucent, with hardness of 6–6.5, and specific gravity of 3.1–3.2. This mineral is the most prominent member of minerals falling within the chondrodite group. These are norbergite, chondrodite, humite and clinohumite. Individual members of this group require optical evaluation for positive identification.

Noteworthy world occurrences include Mt. Somma, Italy; Pargas Finland; Kafveltorp, Sweden; and in the Eastern United States.

CHROMATIN. The word was coined by Flemming to denote that substance in cell nuclei, which, in the usual treatment with nuclear dyes, takes up the color. In nondividing nuclei, chromatin is distributed throughout the entire nucleus (euchromatin), but in nuclei which are undergoing cell division, chromatin is confined to the chromosomes (heterochromatin). In Flemming's time (late 1800s), the chemistry of the nucleus was entirely unknown. Even in view of recent knowledge, it is not fully understood what particular substance(s) have the special affinity for the dyes. In chromatin isolated from pea embryos by differential centrifugation and purified by sucrose gradient centrifugation, the composition of chromatin was found to be deoxyribonucleic acid (DNA), 31%; ribonucleic acid (RNA), 17.5%; histone protein, 33%; and nonhistone protein, 18%, accounting for all but 0.5%.

CHROMATOGRAPHY. This is an instrumental procedure based upon physical adsorption principles for separating various components from a mixture of chemical substances. In its broad application, chromatography is a combination of separation, identification, and quantitative measurement. The procedure is applied to mixtures of organic and inorganic materials and is particularly useful with mixtures of compounds whose chemical characteristics (composition and molecular structure) and whose physical properties (boiling point, density, etc.) are so nearly identical as to make other separation and analytical techniques difficult or impractical. Some forms of chromatography also offer the distinct advantage of requiring only very small samples—in terms of micrograms. Consequently, chromatography is used widely in the laboratory for organic chemical, biological, medical, and pollution studies—and in industry, notably in chemical, petroleum, and petrochemical plants, for quality and process control.

The recovery of pure compounds from mixtures in the laboratory by means of chromatographic techniques often is as important or more important than making quantitative determinations—even though chromatography is frequently thought of principally as an analytical technique. Chromatographic separations are often followed by other analysis techniques.

For industrial applications, a chromatograph may be regarded as a very complex transducer which not only puts out a signal that identifies the types and amounts of a given substance, but also must first separate the target substance from a stream that may contain numerous other substances. The transducer output can be used for off-line quality control, or on-line as part of a total control loop.

Background. In 1903, a Russian botanist, Mikhail Tsvet, used the chromatographic principle to separate plant pigments. Tsvet filled a vertical glass tube with an adsorbent. As a sample of the pigments was washed through the tube with a solvent, a series of colored adsorption bands was produced. This, in essence, was a graphic presentation of colors. Thus, the name *chromatograph*, or literally 'color writing,' the term first used by Tsvet to describe the procedure. See Fig. 1.

Principles

Chromatography depends upon selective retardation and separation of substances by a stationary bed of porous sorptive media as they are transported through the bed by a moving fluid. The degree of retardation, and hence the rate of migration, of each substance is determined by its relative affinity for the sorbent. The sorbent bed is called the *stationary phase*; the moving fluid is called the *moving phase*. The moving phase may be a liquid or a gas. The stationary phase may be a liquid

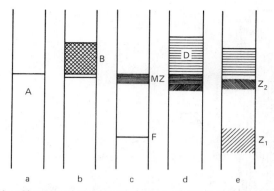

Fig. 1. Chromatography of a binary mixture. (A) Column of adsorbent in the chromatography tube. (B) Solution of mixture to be separated. MZ, mixed zone, applied to the column. F, front of empty solvent. D, developer, applied to the column. Z_1, Z_2, zones of separated components. To the packed column (a) the solution to be analyzed is applied (b). As it passes into the bed of adsorbent the components are retarded to form a mixed zone, while the solvent (also to some extent adsorbed) runs ahead (c). Developer is then applied (d), and as the development occurs the zones of separated material draw apart to produce the developed chromatogram (e).

dispersed on and supported by a porous, inert solid in which the sample components are soluble, and are partitioned in equilibrium with the moving phase. Hence, the solvent or partition liquid are terms used. The term *solvent* is used in both gas chromatography in reference to the stationary-liquid phase and in liquid chromatography to refer to the moving-liquid phase. Alternatively, the stationary phase may be a solid which sorbs sample components on its surface and inner structure by various reversible physical or chemical mechanisms, i.e., *adsorption*, *ion exchange*, or *exclusion* by molecular size.

Nomenclature. Chromatography may be classified in several ways:

(1) By the form of technique: (a) frontal analysis, (b) displacement, (c) development, and (d) elution.

(2) By the nature of the moving and stationary phases, as shown in Fig. 2, such as gas-liquid, or liquid-liquid chromatography.

(3) By the nature of the interaction—whether physical or chemical—between the components and the stationary phase, such as *partition*, *adsorption*, and *exclusion* (permeation) chromatography.

(4) By some distinctive feature of the apparatus or method

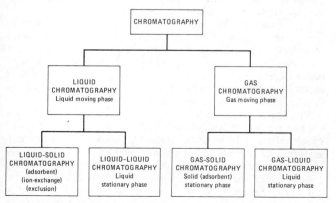

Fig. 2. Chromatography techniques as classified according to nature of moving and stationary phases.

used, such as high-performance liquid chromatography (HPLC), column, paper, or thin layer (TLC) chromatography.

Frontal analysis and displacement chromatography are of limited value and use, particularly in chemical process control applications and are not described here.

Most procedures and techniques are often and variously referred to by two or more of the foregoing nomenclature systems, as shown by Fig. 3. Hence, the widely used gas chromatography can, for different purposes, be referred to as gas, solic (or liquid), column, or elution chromatography.

Development Chromatography. This technique is used only with liquid moving phases. The sample is introduced onto a dry column or sorbent bed and washed through the bed with a solvent that is less strongly sorbed than the sample components. The solvent-washing process is continued until the solvent reaches a point just short of the exit or opposite end of the bed. The migration rate of each sample component is dependent upon its *partition coefficient* or distribution between sorbent and solvent. The most useful forms of development chromatography are paper and thin-layer chromatography (TLC). The use of columns for development chromatography has declined because of difficulties in visualizing and documenting the developed column and in recovering the pure compounds from the sorbent bed.

Elution Chromatography. In this method, the moving phase

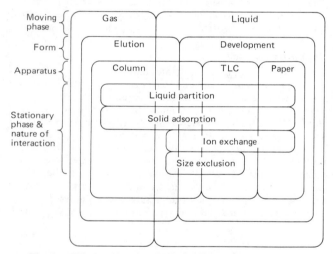

Fig. 3. Relationship of various chromatography techniques.

is passed through the sorbent bed until all sample components have been washed or eluted from the bed. In contrast to development chromatography, elution chromatography uses columns exclusively. In gas-elution systems and in most types of liquid-elution chromatography the column can be reused many times and without interruption of flow of the moving phase. Components move through the column at rates dependent upon the partition coefficient and are separated into bands which elute at characteristic times.

A principal value in elution systems is the precision with which elution times can be reproduced from one analysis to another; an important element in identification and quantitation.

A detector at the end of the column can generate an analog signal that is proportional to the concentration of the sample component in the moving phase. A time record of the detector signal is called a chromatogram and, as shown in Fig. 4(a), contains the characteristic gaussian peaks.

Gas and liquid chromatography are the most popular forms

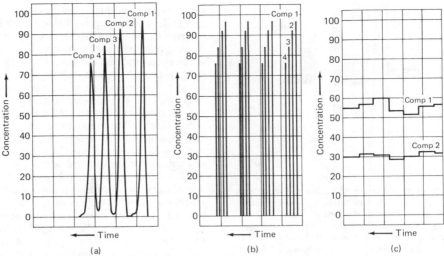

Fig. 4.   Typical chromatograph readouts: (a) chromatogram; (b) bar graph; and
(c) trend. (*Beckman Instruments, Inc.*)

of column elution chromatography. The theory, technique, and apparatus for both methods have been highly developed. Completely automated versions have been developed and widely applied to the continuous, on-line analysis of chemical and petroleum-refining process streams.

Gas Chromatography

Almost any organic or inorganic compound that can be vaporized can be separated and analyzed with a gas chromatography. As shown in Fig. 5, the minimum requirements for a system include (1) a column which contains the *substrate* or stationary phase, (2) a supply of inert *carrier gas* (moving phase) which is continually passed through the columns, with (3) a means for maintaining pressure and flow constant, (4) a means of *admitting* or *injecting* the sample into the carrier-gas stream, (5) a *detector* which senses the sample components as they elute, and (6) a *recorder*. The carrier gas may be any gas that does not react with the sample or adversely affect the detector. Helium, hydrogen, nitrogen and argon are most commonly used.

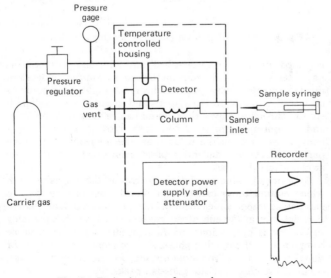

Fig. 5.   Basic elements of a gas chromatograph.

Gas Chromatography in the Laboratory. A typical application is illustrated and described in Fig. 6. Gas chromatography is so flexible that it has many end uses and apparatus configurations in the laboratory, some of which include:

1. *Programming of column temperature* to permit analysis of wide boiling-range fractions. A low starting temperature (sometimes subambient) affords separation of low-boiling components, whereas a later, higher column temperature reduces the elution time of high-boiling components and, in the overall, reduces analysis time. This technique is used widely for complex samples such as essential oils and flavor and aroma concentrates, and, in the petroleum industry for simulated distillation analysis of crude oil and petroleum fractions.

2. *Multiple columns and column switching* to effect separations where a single column would be impractical and also to reduce analysis time.

3. *Fraction collecting or preparative chromatography* where separated compounds are collected for analysis by other methods.

4. *Pyrolysis chromatography* wherein the thermal degradation of high-molecular-weight materials which yield characteristic fragment patterns can be studied.

5. *Reaction chromatography* where compounds may be converted to other related compounds in a reaction zone in the column system to facilitate detection or further separation.

6. *Subtraction chromatography* wherein a reaction zone in the system preferentially reacts with and removes compounds of a class of compounds.

7. *Physiochemical measurements*, such as determination of activity coefficients, partition coefficients, K values, second virial coefficients, heat and entropy of solutions, adsorption, vaporization, and other physical and thermodynamic properties.

8. *Hyperpressure (supercritical) chromatography* in which a fluid above the critical point is used as the moving phase and which enhances the volatility of sample components by as much as 10^4. This permits the analysis of compounds of extremely high molecular weight (mol. wt. = 1,000) and of high boiling point (1000°C) at column temperatures below 300°C. A large number of detectors has been used, many of which are highly specific and of limited application. The most widely used are the thermal conductivity (TCD), hydrogen flame ionization

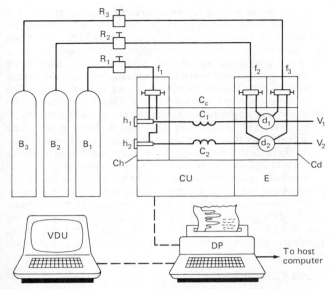

Fig. 6. Basic elements of a gas chromatograph for research and laboratory purposes. Dual column/dual detector gas chromatograph shown is suitable for analysis of complex organic mixtures. Columns C_1 and C_2, with different characteristics, are enclosed in temperature-controlled column compartment C_c, which may be isothermal and/or temperature-programmed during analysis for rapid elution of high-boiling compounds. Individual flash vaporization of liquid samples are enclosed in temperature-controlled inlet compartment Ch. Carrier gas, supplied from high-pressure cylinder B_1, is maintained constant by pressure regulator R_1 and flow controller f_1. Sample components are detected by hydrogen-flame ionization detectors d_1 and d_2 in constant-temperature detector compartment C_d. Hydrogen fuel B_2 and combustion air B_3 are supplied to detectors through pressure regulators R_2 and R_3 and flow controllers f_2 and f_3. Combustion products are vented through V_1 and V_2. Control CU gives individual temperature control of sample inlets and detectors, and programs temperature of column compartment at rate and over range required. Amplifier E, amplifies detector signal and transmits to microprocessor-based Data Processor and Programmer, DP, with keyboard and built-in printer. Data Processor digitizes signal, processes data and calculates composition in accordance with analytical program stored in microprocessor RAM, and prints out chromatogram, and analytical data on printer plotter. Analytical method parameters entered by operator through keyboard and stored in methods file in RAM. All temprature settings and programmed temperature profile set in CU by Data Processor. Auxiliary video display unit (VDU) displays chromatogram, and allows post-run data manipulation for methods development simultaneously with data reduction by DP. Data and programs can be transmitted via RS232 port and serial link to host computer.

(FID), argon ionization, electron-capture (ECD) flame photometric (FPD) and conductivity detectors.

Microprocessor Based Automation. Microprocessor based systems are now widely used to perform the functions previously done manually by the operator. These systems may take many forms:

1. Integral to and dedicated to a single chromatograph.
2. A separate unit which controls one or several chromatographs.

The latter are also available for upgrading older manually operated chromatographs to provide automated data reduction and output.

Systems are equipped with Read Only Memory (ROM) for the operating system and Random Access Memory (RAM) for storage of user specific methods, operating instructions and analytical data acquisition and manipulation.

Depending on user requirements, one or more of the following functions can be provided:

1. Individual temperature control and indication of column ovens, detectors and sample inlets.
2. Dynamic control of programmed temperature profiles; starting and ending temperature, rate of temperature increase with one or more time controlled segments of isothermal temperature; and cooling down cycle.
3. Automatic integration of peak areas or peak height measurements, with choice of several calibration methods for calibration of results.
4. Raw data storage with post-run calculation and manipulation of data for optimizing variables and methods development.
5. Integral printer/plotter for hard copy of chromatograms, and analytical results.
6. Keyboard for input of analytical parameters and operator selected variables.
7. Video terminal for display of chromatograms and analytical results.
8. Cassette or floppy disk extended memory for storage of files, methods and user-specific application programs.
9. High level language for user programming of special calculations such as physical properties, comparison with alarm levels, etc.
10. Serial communications via RS-232 ports with a central computer or other peripheral devices.
11. Timed output control signals to peripheral and ancillary devices such as sample valves, column switching valves and other external functions which are a part of the automated procedure.
12. On-board diagnostics to aid in troubleshooting and isolating and identifying failures.

Chromatograph Data Reduction. Qualitative Analysis. For a given substrate under given conditions, each compound has a characteristic retention time which can be used for tentative identification. However, two or more compounds may have the same elution time on a particular column. In such cases, the compound may be rerun on a different column with other characteristics to reduce ambiguity. Extensive compilations of individual compound-retention times on different substrates are available for reference. Positive identification can be made only by collecting the compound or transferring it as it elutes directly into another apparatus for analysis by other means, such as infrared or ultraviolet spectroscopy, mass spectrometry, nuclear magnetic resonance, etc. Commercially available apparatus are available which combine in a single unit both a gas chromatograph and an infrared, ultraviolet or mass spectrometer for routine separation and identification. The ancillary system may also be microprocessor based with extensive memory for storing libraries of known infrared spectra or fragmentation patterns as in the case of mass spectrometers; with microprocessor controlled comparison and identification of detected compounds.

Quantitative Analysis. Quantitative analysis is based on the proportionality of detector response to the amount of component in the elution band. The most widely used measure of detector response is the area under the chromatogram peak. However, the peak height (amplitude of detector signal at peak maxima) also may be used. Early methods for measuring peak area included direct measurement on the chart with a planimeter, or calculating from dimensions of the height and the width of the peak, or with a ball-and-disk integrator attached directly to the recorder. Electronic integrators have also been used to sense the detector signal directly or after being amplified, and to provide digital printout of peak area and time of peak maxima.

These methods of quantitation have been largely replaced by microprocessor-based data systems. The detector signal is directly coupled to a high speed analog-to-digital converter which samples the detector signal at predetermined rates as high as 40 Hertz. The digitized values are stored in memory for subsequent manipulation. Slope-sensing algorithms find the beginning and end of peaks, peak maxima, and the valley points between incompletely resolved peaks, and can differentiate between peaks and baseline (absence of peaks) to correct the digitized signal for baseline drift. Tangent skimming algorithms can locate and digitize small "rider" peaks on tailing edges or larger peaks.

The microprocessor sums the digitized values for a given peak to obtain the peak area. Results are then calculated by various methods:

- *Area % normalized*. The area of each peak is determined as a percent of the total of the areas of all peaks. Results must add up to 100%, therefore, all components are detected.
- *Concentration by relative response factors, normalized*. Similar to Area %, but each area is corrected by a relative response factor which is characteristic for that compound.
- *Internal standard*. Results are calculated relative to a standard added to the sample in a known amount. Results are independent of sample size. The internal standard must be a compound which is not normally in the sample, and is well separated from other components in the sample.
- *Standard addition*. The sample is analyzed with and without the addition of a known amount of a compound which is also in the sample (spiking). Concentration is calculated from the observed increase in area.
- *External standard*. The area of each component is compared to the area of a separately run standard with known concentration of each component.
- *Special calculations*. Analytical results can be used to calculate other properties or characteristics of the sample, such as: average molecular weight, specific gravity, heating value, octane number, Reid vapor pressure, and boiling point distribution (simulated distribution).

Process Gas Chromatography

A system for continuous, repetitive, and fully automatic on-line analysis of process streams similar in all essential elements of the basic technique to the laboratory chromatograph, but different in design and appearance. Factors affecting design include: (1) the need to comply with the National Electrical Code for operation in hazardous atmospheres; (2) the need to automate the procedure; and (3) the need for ready adaptability to closed-loop process control and communication with process control systems and computers. Principal uses of Process Gas Chromatographs are listed in Table 1. Demand for maximum reliability and minimum maintenance has emphasized simplicity of hardware and methodology. Emphasis is placed on analyzing for a few rather than a number of components and on minimizing analysis time. These design targets have resulted in extensive use of multicolumn techniques for rapid separation of selected components, with large portions of the sample being discarded. As shown in Fig. 7, the major components of a microprocessor-based process-gas chromatograph system are (1) the *analyzer*, (2) the data processor, (3) the sample conditioner system, (4) one or more recorders, and (5) analog and serial outputs to peripheral devices and/or process control systems and computers.

Liquid Chromatography

This method is particularly useful for the separation and analysis of high-molecular weight compounds which are beyond the range of gas chromatography. Most widely used forms are paper and thin layer chromatography, forms of development chromatography, and liquid column chromatography form of elution chromatography. The latter frequently referred to as High Performance Liquid Chromatography (HPLC).

Paper and Thin-layer Chromatography. Both methods are similar in apparatus and technique. The sorbent bed is in the form of a thin sheet of paper or a thin layer of a finely divided sorbent material deposited on a supporting metal, glass, or plastic plate. The sample is spotted near one end of the bend, which is then brought in contact with a source of solvent. As the solvent moves through the bed by capillary action, the sample components are washed through the bed at different rates and are separated into spots of pure compound, which can be recovered after the solvent is allowed to evaporate. Spots can be detected visually if the substances are colored, or with ultraviolet light if they are fluorescent. Components can be reacted to give colored or fluorescent derivatives by spraying with reagents. Conventional quantitative determinations can be made after recovery of the substance of interest with a solvent, or roughly estimated by comparison with spots produced with known quantities and concentrations. Recording densitometers are used for quantitative measurement *in situ*. The paper is drawn automatically past a slit in front of a photocell, and the transmitted or reflected light is recorded. Radioactive compounds are detected by contact exposure with X-ray film, which after development,

TABLE 1. PRINCIPAL USES OF PROCESS GAS CHROMATOGRAPHS

Process Control—Use information to adjust process through open- or closed-loop control.

Process Study—Obtain information about process to improve yield or throughput. Correlate process variables with reaction products and yields.

Process Development—Obtain information about process characteristics, as in pilot plants. Correlate process variables with reaction products and yields.

Material Balance—Use information to calculate material balance for process units.

Product Quality Specification Monitor—Monitor impurities in outgoing or incoming product for conformance to specifications.

Waste Disposal Monitoring—Monitor liquid or gas effluent wastes for loss of valuable product or for presence of toxic compounds.

Personnel Safety-Area Monitoring—Monitor ambient air for presence of toxic compounds.

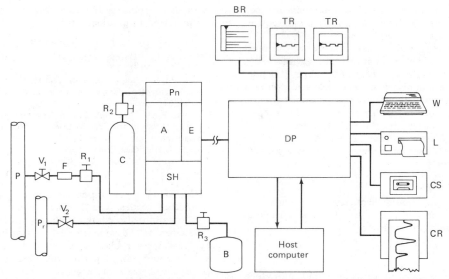

Fig. 7. Basic elements of a process-gas chromatograph. Vapor sample is continuously withdrawn at high rate from process line P, circulated through sample conditioner SH, and returned to lower pressure point P_r, through shutoff valves V_1 and V_2. Particulate matter is removed by filter F, and pressure reduced to constant low level by regulator R_1. Sample conditioner contains flow control and other conditioning components and valve for switching to synthetic calibration blend B through pressure regulator R_3. Sample slipstream is circulated to sample valve in analyzer A, which also contains columns, detectors, and temperature-control system. Carrier gas C is controlled by regulator R_2, and pneumatics control section P. Microprocessor in electronics module E stores analytical program in RAM and controls analyzer functions, and data acquisition and reduction. Analytical results are transmitted over serial link to Data Processor, DP. Data Processor converts results to analog signal for presentation or bar graph recorder, BR, and as many as 30–40 trend recorders, TR. Real time reconstructed chromatogram for maintenance on Recorder, CR. Serial outputs (RS232) to writer or panel mounted line printer, L for data logging, and to cassette recorder, CS, for storing application programs. Results and alarm messages to host computer via serial link. Applications program entered via Data Processor and downloaded to analyzer RAM for execution in analyzer. Processor controls several analyzers.

can be measured with a densitometer in which a radiation detector is used in place of the photocell.

These have gained wide acceptance because of the ease and convenience of the method, and the modest cost and availability of materials. Moreover TLC affords a rapid and inexpensive means for preliminary screening and selecting solvent/stationary phase systems for liquid chromatographs.

Liquid (Column) Chromatography. This method is generally classified according to the stationary phase or the natures of its interaction with sample components:

1. *Liquid-liquid partition chromatography,* wherein the sample components are partitioned between the moving liquid phase and a stationary liquid phase deposited on an inert solid. The two solvent phases must be immiscible. The stationary phase may be a large molecule, chemically bonded to the surface of a solid (bonded liquid phase) to prevent loss by solubility in the moving phase. This method can also be subdivided into *normal phase* systems, in which the moving phase is less polar than the stationary phase, and *reverse-phase*, in which it is more polar.

2. *Liquid-solid or adsorption chromatography,* in which sample components are adsorbed on the surface of an adsorbent such as silica gel.

3. *Ion exchange,* in which ionic sample components interact with functional groups on a permeable ionic resin.

4. *Exclusion or gel permeation,* in which compounds are separated by molecular size by the range of pore sizes in a polymeric gel. This method is useful for measuring molecular weight distribution of polymers.

Isocratic elution uses a solvent of constant composition throughout the analysis.

Gradient elution is a modification of the technique in which the solvent is a mixture of two solvents which differ in solvent strength. The composition or ratio of the two is changed during the analysis in accordance with a predetermined program. The change may be continuous, linear or nonlinear, or stepwise, or a combination.

Apparatus for Liquid Chromatography. The need for more efficient columns and faster separation has led to the development of stationary phase packings with particles as small as 2 micrometers, operating at pressures as high as 10,000 psig (69 MPa), with solvent flow rates as low as 1 ml/minute or less. This has, in turn, led to development of detectors with internal volumes of only a few microliters, and special fittings and connectors with minimum dead volume to prevent band spreading and loss of resolution. These improvements in apparatus and technique have resulted in achieving complex separation with speeds comparable to gas chromatography.

The solvent is moved through the system by constant flow or constant pressure pumps, which are driven mechanically (screwdriven syringe or reciprocating) or by gas pressure with pneumatic amplifiers. For gradient elution two pumps may be synchronized and programmed to provide a controlled and reproducible composition change.

Samples may be introduced by syringe directly into the column through a septum or by means of valves with a fixed volume which has been prefilled with the sample. Valves may be rotary, sliding plug or diaphragm design and of stainless steel and fluoro-plastic construction for inertness. Auto samplers are used for unattended injection of samples loaded into vials and sequentially rotated into the injection mechanism.

The differential refractive index (RI) is probably the most widely used detector; other detectors are based on photometry. Fixed-wavelength and variable-wavelength ultraviolet (UV) detectors are the most commonly used. The delectric constant (DC) detector and electrical conductivity (EC) detector are also used and are especially suitable for Process Liquid Chromatography. Transport detectors transfer the elute to a moving wire or belt through a solvent evaporator and into a pyrolyzer which fragments and transfers the sample components into a hydrogen flame ionization detector.

Process Liquid Chromatography

Like gas chromatography, liquid chromatography lends itself well to automation for on-line process analysis. See Table 2. The constraints on design are the same:

1. Operation in hazardous atmospheres,
2. Need for automation, and
3. Compatibility with process control systems and computers.

Except for the elements of the analytical hardware, a process liquid chromatograph is virtually identical to the process GC in (1) appearance, (2) data acquisition and reduction techniques, and (3) information and data output. Except for the analyzer, other system components such as data processor, communications and output peripherals, etc. are the same and may be shared simultaneously with process gas chromatographs. Data output is also as shown previously in Fig. 4. Emphasis of the technique is placed upon analyzing for a few rather than for all components and minimizing analysis time. This has resulted in extensive use of column switching and backflushing. Isocratic

TABLE 2. REPRESENTATIVE PROCESS LIQUID CHROMATOGRAPHY (PLC) APPLICATIONS

1. Food Industry
   a. Additives
   b. Preservatives
   c. Flavor components
   d. Starches
   e. Heat exchanger fluids
   f. Triglycerides
   g. Cooking oils
2. Beverage Industries
   a. Carbohydrates
   b. Sugars
   c. Wines
3. Agricultural Products
   a. Pesticides
   b. Herbicides
   c. Insecticides
   d. Wood preservatives
   e. Water analysis
4. Inorganic Ions
5. Pharmaceuticals
6. Water Pollution
   a. Phenols
   b. Petrochemicals
   c. Plasticizers
   d. Nitroaromatics
   e. Organic acids
   f. Polyaromatics
   g. Chlorinated organics
7. Macromolecule Industries
   a. Polymers and copolymers
   b. Rubbers and resins
   c. Silicones
8. Dyes and Dye Intermediates
   a. Anthraquinones
   b. Benzidine
   c. Naphthols
   d. Water analysis
9. Petrochemicals and Monomers
   a. Acrylic acid
   b. Acrylic esters
   c. Methacrylic acid
   d. Terephthalic acid
   e. Acrylonitrile
10. Petroleum Products
    a. Water analysis
    b. Fuels and hydrocarbon analysis

methods to optimize the separation of the target components are preferred to gradient elution.

References

Heftmann, E.: "Chromatography," 3rd Edition, Van Nostrand Reinhold, New York, 1975.
Kobayashi, R., Chappelear, P. S., and H. A. Deans: "Physico-Chemical Measurements in Gas Chromatography," *Ind. Eng. Chem.*, **59**(10), 63 (1967).
Mowery, R. A.: "Process Liquid Chromatography," in "Automated Stream Analysis for Process Control," Vol. 1 (D. P. Manka, editor), Academic Press, New York, 1982.
Purnell, J. H.: "Gas Chromatography," Wiley, New York, 1962.
Villalobos, R.: "Process Gas Chromatography," *Analyt. Chem.* **47**, 983A (1975).

CHROMITE. An important mineral in the chromite series of multiple oxides. The dominant compound is $FeCr_2O_4$, but most chromite also contains magnesium Mg and aluminum Al. Crystallizes in the isometric system. Hardness 5.5; specific gravity, 4.5–4.8; color, black. Associated with peridotite, and serpentine (metamorphized peridotite). Commercial amounts occur as placer deposits in serpentine areas. In the United States chromite has been mined in Maryland, Pennsylvania, California, Montana, Oregon, Wyoming and North Carolina. Important deposits occur in Quebec, Canada. Valuable deposits occur in Asia Minor, Rhodesia, New Caledonia, Cuba, India, and the Philippines; also in New South Wales, and in the Urals associated with platinum.

CHROMIUM. Chemical element symbol Cr, at. no. 24, at. wt. 51.996, periodic table group 6b, mp 1837–1877°C, bp, 2671–2673°C, density, 7.2 g/cm³. Elemental chromium has a body-centered cubic crystal structure. The metal is silvery-white with a slight gray-blue tinge, very hard (9.0 on the Mhos scale), capable of taking a brilliant polish, not appreciably ductile or malleable. The element is not affected by air or H_2O at ordinary temperatures, but when heated above 200°C, chromic oxide Cr_2O_3 is formed. There are four stable isotopes, ^{50}Cr, and ^{52}Cr through ^{54}Cr. Four radioactive isotopes have been identified, all with comparatively short half-lives, ^{48}Cr, ^{49}Cr, ^{51}Cr, and ^{55}Cr. The element was first identified by Vauquelin in 1797.

Ionization potential 6.76 eV; second, 16.6 eV. Oxidation potentials $Cr \rightarrow Cr^{3+} + 3e^-$, 0.71 V; $Cr^{2+} \rightarrow Cr^{3+} + e^-$, 0.41 V; $2Cr^{3+} + 7H_2O \rightarrow Cr_2O_7^{2-} + 14H^+ + 6e^-$, −1.33 V; $Cr + 3OH^- \rightarrow Cr(OH)_3 + 3e^-$, 1.3 V; $Cr + 4OH^- \rightarrow CrO_2^- + 2H_2O + 3e^-$, 1.2 V, $Cr(OH)_3 + 5OH^- \rightarrow CrO_4^{2-} + 4H_2O + 3e^-$, 0.12 V.

Other important physical properties of chromium are given under **Chemical Elements.**

Chromium occurs chiefly as chromite (ferrous chromite) $Fe(CrO_2)_2$ in Zimbabwe, the Republic of South Africa, the U.S.S.R., New Caledonia, India, Philippine Islands, Japan, Turkey, Greece, Cuba, and California. (1) Heating chromite in the electric furnace with carbon yields ferrochrome for alloys, and (2) when chromite is heated with sodium carbonate and nitrate, sodium chromate is formed, which is then extracted with H_2O. This is the substance from which chromium compounds are obtained. See also **Chromite.**

Chromium is used extensively for (1) decorative and wear-resistant electroplating, (2) many important alloys, and (3) the manufacture of numerous chemicals and refractory materials.

Alloys. In constructional steels, chromium imparts hardness by improving hardenability and promoting the formation of carbides. These steels have exceptional wear resistance and are relatively stable at elevated temperatures. In stainless and heat-resisting steels, chromium improves corrosion and heat resis-

REPRESENTATIVE STAINLESS STEELS

Type	Composition and Characteristics
Austenitic Stainless Steels	
302	Basic alloy of the group: Cr, 17–19%; Ni, 6–8%
302B	Silicon (2–3%) added to increase scaling resistance
303	Sulfur (0.02–0.05%) added to improve machinability
303Se	Selenium (0.15%) added to improve machinability
304L	Extra low carbon for improved weldability
304	Lower carbon content to improve weldability and inhibit carbide formation: Cr, 18–20%; Ni, 10–12%
305	Nickel increased to lower work-hardening: Cr, 17–19%; Ni, 10–13%
308	Chromium and nickel increased to increase corrosion and scaling resistance: Cr, 19–21%; Ni, 10–12%
309S	Lower carbon content (0.08% max) for improved weldability
314	Silicon added for increased scaling resistance at high temperatures: Cr, 23–26%; Ni, 19–22%; Si, 2% max
316	Molybdenum added to improve resistance to pitting corrosion and strength at high temperatures: Cr, 16–18%; Ni, 10–14%; Mo, 2–3%
316L	Extra low carbon content for improved weldability
317	Additional molybdenum to further improve resistance to pitting corrosion: Cr, 18–20%; Ni, 11–15%; Mo, 3–4%
321	Titanium added to prevent chromium carbide precipitation: Cr, 17–19% Ni, 9–12%; C, 0.08% max; Ti, 5 × C content
347	Columbium (niobium) or tantalum added to prevent chromium carbide precipitation
Martensitic Stainless Steels	
410	Basic alloy of the group: Cr, 11.5–13.5%; Ni, 0.5% max
405	Aluminum added to prevent weld hardening
414	Nickel (2%) added to improve corrosion resistance
416	Sulfur added to improve machinability
416Se	Selenium added to improve machinability
420	Carbon increased for higher hardness: Cr, 12–14%; Ni, 0.5% max; C, over 0.15%
431	Chromium increased to further improve corrosion resistance: Cr, 15–17%; Ni, 1.25–2.50%
440A	Carbon slightly decreased to improve toughness: Cr, 10–18%; Ni, 0.5% max; C, 0.6–0.75%
440B	Carbon decreased slightly to improve toughness even more
440C	Carbon increased to increase hardness; chromium increased to make up loss in corrosion resistance: Cr, 16–18%; Ni, 0.5% max; C, 0.95–1.2%
Ferritic Stainless Steels	
430	Basic alloy of the group: Cr, 14–18%; Ni, 0.5% max
430FSe	Selenium added to improve machinability
442	Chromium increased to improve resistance to scaling and corrosion: Cr, 23–27%; Ni, 0.5% max

tance. As shown by the accompanying table, stainless steels may be grouped into three principal classes (1) austenitic, (2) martensitic, and (3) ferritic stainless steels. The fundamentals of stainless steels are further described under **Iron Metals, Alloys, and Steels.** Stainless steels, as a group of major ferrous alloys, are characterized by their high degree of resistance to chemical attack. They possess a property commonly referred to as passivity, a property eminently displayed by elemental chromium. This property is manifested in steels when the chromium content exceeds about 11%. A very marked improvement in corrosion and heat resistance is achieved when chromium is added to this or a greater extent to low-carbon steels. The addition of nickel, along with chromium, enhances these properties even more. Although a steel that contains 12% chromium will stain, it will not undergo progressive rusting in normal atmospheres. When the chromium content is increased to 18%, staining will not occur in normal atmospheres, but may stain to a limited extent in particularly bad industrial environments. However, the addition of 8% nickel, along with the 18% chromium, will make the steel stain-resistant in all but the most

rigorous industrial atmospheres. Additional resistance to corrosion and heat resistance may be obtained by the presence of molybdenum and other elements in smaller quantities.

An example of the effect of chromium content on the corrosion resistance of steel is demonstrated as follows:

A low-carbon steel placed in boiling 65% HNO_3 for one month:

4.5% chromium	corrosion rate, 12.9 in.
8.0% chromium	0.14
12.0% chromium	0.01
18.0% chromium	0.003
25.0% chromium	0.0006

The corrosion resistance of iron-chromium alloys was known in England and France in the early 1800s, but passivity was not clearly recognized and reasonably understood until 1910 as the result of studies by Borchers and Monnartz in Germany. Commercial stainless steels were introduced shortly thereafter in Germany, France, England, and a bit later in the United States. Worldwide production of stainless steels now is measured in terms of millions of tons annually.

In addition to ferrous alloys, chromium also is added to copper, vanadium, zirconium, and other metals to form several hundred chromium-bearing alloys. Nickel-chromium-iron alloys have high electrical resistance and are used widely as electrical heating elements. *Nichrome* and *Chromel* are examples.

Chromium Plating. Although the tonnage of chromium used for electroplating is far below its consumption for alloys, plating represents a major market for the element. In terms of special protective coatings, chromium is behind only copper, lead, and zinc in consumption. Generally, chromium plating is used for two purposes: (1) wear-resistance and (2) decorative effect, taking on polish and a much brighter surface than the other electroplated metals. The "bright work" of automotive hardware, plumbing fixtures, and electrical appliances are examples. Normally, chromium is plated over nickel where the nickel is about 100 × thicker than the chromium. In decorative plating, the thickness of the chromium plate ranges only from 0.00001 to 0.00002 inch (0.00025 to 0.00051 millimeter). Hexavalent chromium also forms protective chromate coatings over aluminum, cadmium, copper, magnesium, and zinc. These chromate conversion coatings are used on hot-dipped or electrogalvanized parts, on zinc die castings, and extensively on aluminum parts for aircraft.

In 1979, scientists at the General Motors Research Laboratories (Warren, Michigan) announced some interesting findings concerning chromium plating. The conventional process uses a bath containing Cr^{6+} ions. During plating, the Cr^{6+} ions re-

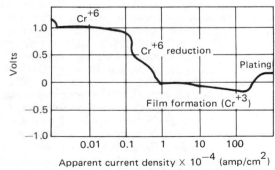

Typical polarization curve made during investigation of chromium plating process. (*General Motors Research Laboratories.*)

duce to Cr^{3+} and then to metallic Cr. Investigators have wondered for years why the process will not succeed if one commences with Cr^{3+} ions, thus avoiding the double step. The GM scientists found that starting with Cr^{3+} fails because it immediately forms a stable complex with water molecules from which Cr cannot be deposited. Commencing with Cr^{6+} succeeds because during reduction a chemical film forms around the cathode (part being plated). Since Cr^{3+} is bound in that film, it does not react with water, but, instead, plates out as chromium metal. The researchers determined all 10 steps that occur as Cr^{6+} reduces to bright Cr, pinpointed the step at which catalysis begins, and also identified the active catalyst, which turned out to be the bisulfate ion, not the sulfate ion as might be expected. See accompanying figure.

Chemistry and Compounds. Chromium metal is soluble in dilute HCl or H_2SO_4 and made passive by dilute or concentrated HNO_3 or concentrated H_2SO_4.

In keeping with the $3d^5 4s^1$ electron configuration of the atom, chromium forms compounds in which it is in the stable 6+ state. The compounds of 2+ and 3+ chromium are also quite numerous. The compounds with chromium oxidation numbers of 4+ and 5+ are unstable except under extremely alkaline conditions. The one known chromium 1+ complex is also unstable.

As is evident from the oxidation potential for Cr^{2+} to Cr^{3+}, the Cr^{2+} ion in aqueous solution is an extremely strong reducing agent, readily reacting with atmospheric oxygen. It is therefore stable in aqueous solution only in a complex or a slightly ionized salt. It may be obtained by reducing hexavalent or trivalent chromium in acid solution with one of the active metals, such as zinc. The four halogens form halides of divalent chromium, CrX_2, which dissolve in water with the evolution of heat, although the fluoride is less soluble than the other halides. The ammoniacal solution of chromium(II) sulfate, $CrSO_4$, is particularly reactive with gases, not only with oxygen like the other Cr^{2+} compounds, but also acetylene. Chromium(II) acetate is only slightly soluble in H_2O.

The Cr^{3+} ion readily forms complexes; it exists in aqueous solution as $Cr(H_2O)_6^{3+}$, and forms other complexes with anions, such as $Cr(H_2O)_5Cl^{2+}$, $Cr(H_2O)_4 Cl_2^+$, etc. Thus chromium(III) halides may be crystallized in a number of forms differing in color and other properties due to variation in the bonding. Compounds of the trichloride have been reported with the following arrangements: $Cr(H_2O)_6Cl_3$, $[Cr(H_2O)_5Cl]Cl_2 \cdot H_2O$, and

$$[Cr(H_2O)_4Cl_2]Cl \cdot 2H_2O.$$

Trivalent chromium also forms double salts, notably the chromium alums, hydrated double salts of Cr(III) sulfate and the alkali metal (or thallium or ammonium) sulfates.

The three oxides of chromium, CrO, Cr_2O_3, and CrO_3, are of interest in exhibiting the transition in properties from basicity to acidity and from electrovalence to covalence, with increasing oxygen content, and the consequent increase in charge upon the Cr atom. Thus Cr_2O_3, the middle oxide, is amphiprotic, dissolving either in strong alkali hydroxide solutions or in acids.

One of the most stable of the compounds of Cr(IV) is the tetrafluoride, CrF_4, brown solid, steel blue vapor at 150°C, prepared by direct reaction of the elements; even this compound readily undergoes hydrolysis. Chromium tetrachloride, $CrCl_4$, may be prepared as a gas by reaction of Cl_2 and $CrCl_3$ at elevated temperature, but decomposes at room temperature. Chromium(V) occurs in CrF_5, fire red, volatile, and in the hypochromates, M_3CrO_4, green, which may be prepared by fusion of alkali chromate and alkali hydroxide at high temperature.

CrO_3 is the anhydride of chromic and dichromic acids, H_2CrO_4 and $H_2Cr_2O_7$, which have not been isolated, but whose anions are found in many salts. pK_{A1} and $pK_{A2} = 0.745$ and 6.49, and -1.4 and 1.64 respectively. They are strong oxidants in acid solution, and are readily obtained in basic solution by oxidation of Cr(III) compounds. The oxyhalogen compounds of chromium are chiefly the CrO_2X_2 (chromyl) compounds, which are acid halides of chromic acid. Chromyl chloride CrO_2Cl_2 deep red, liquid, is prepared by heating sodium dichromate and sodium chloride with H_2SO_4, while chromyl fluoride undergoes polymerization to a white solid. Chlorochromates, e.g., $KCrO_3Cl$, and fluorochromates, e.g., $KCrO_3F$, are also known, but are hydrolyzed in water.

Chromates and Dichromates. Sodium chromate Na_2CrO_4, potassium chromate K_2CrO_4, ammonium chromate $(NH_4)_2CrO_4$, calcium chromate $CaCrO_4$, are yellow soluble solids; barium chromate $BaCrO_4$, pale yellow, strontium chromate $SrCrO_4$, pale yellow, lead chromate $PbCrO_4$, yellow (used as a pigment, "chrome yellow"), zinc chromate $ZnCrO_4$, yellow, used as a pigment, mercurous chromate Hg_2CrO_4, yellow to red to brown, silver chromate Ag_2CrO_4, reddish-brown, are insoluble solids. Sodium dichromate $Na_2Cr_2O_7$, potassium dichromate $K_2Cr_2O_7$, readily crystallized, ammonium dichromate $(NH_4)_2Cr_2O_7$ (forming nitrogen gas and green chromic oxide solid upon heating), are red soluble solids, of important application as oxidizing agents, e.g., sulfurous acid causes reduction to chromic; silver dichromate $Ag_2Cr_2O_7$, red insoluble solid, changing to silver chromate Ag_2CrO_4 upon boiling with H_2O. Solutions of chromate in the presence of acid are changed to the corresponding dichromate.

A number of peroxychromates are known, including the deep blue, organic soluble reaction product of H_2O_2 and acid $Cr_2O_7^{2-}$ solutions, which is H_4CrO_7, i.e., $CrO_3 \cdot 2H_2O_2$.

Trivalent chromium forms many complexes. The large number is due in part to the many possible arrangements in which the same element may be present as part of a complex anion or as the cation, as was indicated for the chloride complexes of the hydrated Cr^{3+} ion. The ligands may also be present in various proportions. As in the case with Co^{3+}, Cr^{3+} forms complexes with ammonia in the various color series, e.g., a violeo-chloride, $[Cr(NH_3)_4Cl_2]Cl$, a purpureo chloride, $[Cr(NH_3)_5Cl]Cl_2$ and a luteo chloride, $[Cr(NH_3)_6]Cl_3$. In addition to NH_3 many amines, especially ethylenediamine and its derivatives, form stable complexes with Cr^{3+}. In the oxalato complexes, the tripositive chromium ion coordinates with oxalate groups to form such anions as $[Cr(C_2O_4)_3]^{3-}$ and $[Cr(C_2O_4)_2(H_2O)_2]^-$. There are many cyano, thiocyano, nitrato, fluoro, bromo, iodo, azido, acetylido and nitro complexes.

Organometallic compounds. These include a large number of cyclopentadienyl compounds. In addition to dicyclopentadienyl chromium itself, $(C_5H_5)_2Cr$, there are many derivatives of it with additional radicals or molecules such as $(C_5H_5)_2CrBr$, $(C_5H_5)_2CrNH_3$, $C_5H_5Cr(NO)_2CH_2Cl$, $H(C_5H_5)Cr(CO)_3$ and $C_5H_5(CO)_3CrCr(CO)_3C_5H_5$. The compounds of Cr with other hydrocarbon radicals are fewer ($Cr(C_6H_6)_2$ is known) such as $(C_6H_5)_5CrOH \cdot 4H_2O$ and

$$(C_6H_5)_3Cr \cdot 3THF \cdot (THF = Tetrahydrofuran).$$

Besides $Cr(C_6H_6)_2$, chromium hexacarbonyl, $Cr(CO)_6$, white, solid, bp (extrap.) 145.7°C, represents chromium in the zero oxidation state. This is a nonpolar compound, insoluble in H_2O, freely soluble in organic solvents. The zero-valent compound $K_6Cr(CN)_6$ has also been made by the reaction of $K_3Cr(CN)_6$ and potassium metal in liquid NH_3.

Cermets. The term *cermet* derives from the combination of ceramic and metal. Cermets are produced by powder metallurgy techniques and represent the bonding of two or more metals. They are particularly useful at high temperatures (850–1250°C). Chromium is used in several cermet combinations, including chromium-bonded aluminum oxide, metal-bonded chromium carbide, and metal-bonded chromium boride.

Biological Aspects. The first indication for a biological role of chromium in metabolism was derived from the enzymatic studies (Horecker et al., 1939). The succinate-cytochrome dehydrogenase system, an important enzyme system for the production of energy, requires certain inorganic cofactors. Of various elements tested, chromium produced the greatest increase of enzyme activity. A significant stimulation of the activity of phosphoglucomutase by chromium was described by Strickland in 1949. This system which has an important function in the early steps of carbohydrate metabolism requires magnesium and one other metal for optimal activity. Chromium was outstanding as a "second metal" because it produced the highest enzyme activation and supported a measured amount of activity even when given alone.

Chromium also stimulates fatty acid and cholesterol synthesis from acetate in liver. That chromium is an essential cofactor for the action of insulin on the rat lens was shown by Farkas in 1964. In the absence of the element, no significant insulin effect on glucose utilization of lens can be demonstrated. Chromium supplementation to the donor animals results in a significant response of lens tissue to the hormone. Numerous other findings indicate that chromium may play several vital roles in biological systems.

Chromium levels in biological matter have been studied extensively. In contrast to findings with other metals, chromium concentrations in the United States population are highest at the time of birth, with a pronounced decline during the lifetime, whereas they appear to remain high in some other countries, such as Thailand and the Philippines. These findings suggest the possibility of a relative chromium deficiency in the United States. The relationship of this to disturbances of carbohydrate metabolism in humans, while under study, remains to be established.

Not all of the various chemical forms of chromium are effective in improving sugar metabolism, and the exact nature of the compound or compounds involved in activating insulin is not established. Some of the chromium in plants may not be present in nutritionally effective forms.

It has not been established that chromium is essential to plants, but high concentrations of the metal are toxic. Most agricultural crops, especially their seeds, contain only low levels of chromium.

References

Allaway, W. H.: "The Effect of Soils and Fertilizers on Human and Animal Nutrition," *Agricultural Information Bulletin 378*, Cornell University Agricultural Experiment Station, and U.S. Department of Agriculture, Washington, D.C. (1975).

Krichgessner, M., Editor: "Trace Element Metabolism in Man and Animals," Institut für Ernährungsphysiologie, Technische Universität München, Freising-Weihenstephan, Germany (1978).

Underwood, E. J.: "Trace Elements in Human and Animal Nutrition," 4th Edition, Academic, New York (1977).

CHROMIZING.

Production of a high-chromium-content surface layer on iron and steel by heating at high temperatures in a solid packing material containing chromium powder, or in an atmosphere containing chromium chloride. The surface layer is formed by diffusion of chromium into the iron in the same manner as carbon diffuses into iron in carburizing; however, the process is much slower and requires higher temperatures than carburizing. A similar result can be obtained by high-temperature diffusion of electrodeposited chromium. Chromized coatings have corrosion resistance and elevated temperature oxidation resistance similar to the high-chromium types of stainless steel.

CHROMOGENIC COUPLERS.

Couplers are used in secondary color development. See **Dyes (Textile)**. The word *coupler* is applied to a large number of organic compounds which combine with a limited number of chromogenic developers to produce dye images with a wide range of color and intensity. Couplers are dye intermediates, but differ from chromogenic developing agents in that they do not, as a rule, have the ability to develop a silver image.

Couplers may be hydroxy or amine derivatives of aromatic compounds, such as benzene, naphthalene, or anthracene. Phenol, naphthol, aniline, cresol, para-aminophenol and dimethyl-paraphenylenediamine, are examples. With these compounds coupling takes place with the hydrogen atom which is in the ortho- or para-position to the hydroxy or amino group on the coupler.

Compounds having active methylene $=CH_2$ groups, with strong polar groups for other valences as the cyano $—CN$, the carbonyl $=CO$, the aceto $CH_3CO—$, the acid ester $—COOC_2H_5$, and the phenyl, $—C_6H_5$ groups will couple. Acetoacetic ester and paranitrophenylacetonitrile are illustrations.

The methylene group may be part of a ring structure as in coumarine and indoxyl, or may be part of a heterocyclic ring attached to a phenyl group as in 1-phenyl-3-methyl-5-pyrazolone.

Compounds having a N in the ring will couple if there is a methyl $—CH_3$ group, attached to the ring in the alpha position to the nitrogen. Two illustrations for this type are picoline and 2-methyl-thiazole.

CHROMOPHORE.

Certain groups of atoms in an organic compound cause characteristic absorption of radiation irrespective of the nature of the rest of the compound. Such groups are called *chromophores* or *color carriers*.

CHROMOPHORIC ELECTRONS.

Electrons in the double bonds of the chromophore groups. Such electrons are not bound as tightly as those of single bonds and can thus be transferred into higher energy levels with less expenditure of energy. Their electronic spectra appear at frequencies in the visible or near-ultraviolet region of the spectrum.

CHROMOPLASTS. Plastids.

CHRYSOBERYL.

The mineral chrysoberyl, an aluminate of beryllium, corresponds to the formula $BeAl_2O_4$, crystallizes in the orthorhombic system with both contact and penetration twins common often repeated resulting in rosetted structures. Hardness, 8.5; specific gravity, 3.75; luster vitreous; color various shades of green, sometimes yellow. A variety which is red by transmitted light is known as alexandrite. Streak colorless; transparent to translucent occasionally opalescent. Chrysoberyl also is known as *cymophane* and *golden beryl*.

Chrysoberyl occurs in granitic rocks, pegmatites and mica schists; often is found in alluvial deposits. The Ural Mountains yield alexandrite. Other localities for chryosberyl are Czechoslovakia; Ceylon; Rhodesia; Brazil; and the Malagasy Republic, where it occurs of gem quality in the pegmatites of that island. In the United States it is found in Maine, Connecticut, and New York. The word chrysoberyl is derived from the Greek words meaning golden and beryl. *Cymophane* has its derivation also from the Greek words meaning *wave* and *appearance*, in reference to the opalescence exhibited at times.

CHRYSOCOLLA. This mineral, a hydrous silicate of copper probably corresponding to the formula $Cu_2H_2Si_2O_5(OH)_4$, is perhaps a mineral gel, for it usually appears as an amorphous mass, in veins, or as incrustations. Common occurrence as massive cryptocrystalline character, possibly orthorhombic; extremely rare as small acicular crystals.

Chrysocolla is generally some shade of blue or green, but if impure may be brown or black. It has a characteristic conchoidal fracture; hardness, 2–4; specific gravity, 2.24; vitreous to dull luster; translucent to opaque.

Chrysocolla is a secondary mineral and associated commonly with other copper minerals of similar origin. It is one of the less important ores of copper and has a minor use as a gem stone. Among the localities for excellent specimens may be mentioned Cornwall and Cumberland, England; Congo; Chile; Lebanon and Berks Counties, Pennsylvania; the Clifton-Morenci Globe and Bisbee districts in Arizona; Dona Ana County, New Mexico, and the Tintic district, Utah.

The word chrysocolla is derived from the Greek words meaning gold and glue, formerly the name for gold solder.

CHRYSOTILE. A delicately fibrous variety of serpentine which separates easily into silky flexible fibers of greenish or yellowish color, with formula $Mg_3Si_2O_5(OH)_4$. It crystallizes in the monoclinic system; hardness, 2.5; specific gravity, 2.55. Its name is derived from the Greek words meaning gold and fibrous. Most of the common asbestos of commerce is chrysotile. It is mined in Thetford, Province of Quebec, and in the Republic of South Africa. See also **Serpentine.**

CINNABAR. The mineral cinnabar, mercuric sulfide (HgS) occurs in small and often highly modified hexagonal crystals, usually of rhombohedral or tabular habit. It is found chiefly in crystalline crusts, granular or simply massive. The fracture of cinnabar is subconchoidal; hardness, 2–2.5; specific gravity, 8–8.2; luster, adamantine tending toward metallic, sometimes dull. This mineral has a characteristic cochineal-red color which, however, may be brownish at times, occasionally dull lead gray. The streak is scarlet; it is transparent to opaque.

Cinnabar occurs in veins or may be in masses in shales, slates, limestones and similar rocks due to the impregnation by mineral-bearing solutions or as replacements. The U.S.S.R., Czechoslovakia, Bohemia, Bavaria, Italy and Spain have furnished excellent specimens. The most important of the world's mercury deposits is at Almaden in Spain. Italy, Peru, Surinam, China and Mexico have commercially valuable occurrences of cinnabar. In the United States this mineral is found in California (most important deposit), Nevada, Utah, Texas and Oregon. Cinnabar is the chief ore of mercury. Its name is supposed to be of Hindu origin.

CIS-COMPOUND. Isomerism.

CITRIC ACID. $C_3H_4(OH)(COOH)_3$, formula weight 192.12, white crystalline solid, mp 153°C, decomposes at higher temperatures, sp gr 1.542. Citric acid is soluble in H_2O or alcohol and slightly soluble in ether. The compound is a tribasic acid, forming mono-, di-, and tri-series of salts and esters. Citric acid may be obtained (1) from some natural products, e.g., the free acid in the juice of citrus and acidic fruits, often in conjunction with malic or tartaric acid, and the juice of unripe lemons (approximately 6% citric acid) is a commercial source; (2) by fermentation of glucose (blackstrap molasses is a major source); and (3) by synthesis.

Citric acid and sodium citrate have found use as additives in effervescent beverages and medicinal salts, although excessive quantities are considered toxic. Citric acid can be used as an effective antioxidant. Because the acid is not readily soluble in fats, it is added to formulations which improve solubility, such as propylene glycol and butylated hydroxy anisol, and thus can be used as a stabilizer for tallow, fats, and greases. Citric acid also is used for adjusting pH in certain electroplating baths. It finds a number of miscellaneous uses in etching, textile dyeing and printing operations.

Citrates (like tartrates) in solution change silver of ammonio-silver nitrate into metallic silver. Calcium citrate on account of its solubility characteristics, is of importance in the separation and recovery of citric acid. Calcium citrate plus dilute H_2SO_4 yields citric acid plus calcium sulfate, and the latter may be separated by filtration. Citric acid may be obtained by evaporation of the filtrate.

In most living organisms, the citric acid cycle constitutes the final common pathway in the degradation of foodstuffs and cell constituents to carbon dioxide and water. This cycle is described in the entry on **Carbohydrates.**

CITRIC ACID CYCLE. Carbohydrates.

CITRINE. The mineral citrine is a yellow variety of quartz sometimes used as a gem. It is often marketed under the name topaz and may mislead the unwary. Brazil and the Malagasy Republic have furnished material of excellent quality. See also **Quartz.**

CLARIFYING AGENTS. Chemical substances used in connection with the purification of various solutions and liquors that occur during the processing of raw materials to final end-products. These agents operate in connection with mechanical equipment to bring about the removal of suspended particles that represent product impurities. Allowing such particles to settle by gravity alone would require very long periods. Clarification is part of the total process of sedimentation, which may be defined as the removal of solid particles from a liquid stream by gravitational force. The operation is effected by slowing the velocity of a feed stream in a large-volume tank so that gravitation settling can occur. Sedimentation is divided into two functions: (1) In *thickening* the primary purpose is to increase the concentration of suspended solids of the feed stream, i.e., to remove liquids. This is largely a mechanical operation, assisted sometimes by clarifying agents. (2) In *clarification* the purpose is to remove fine-sized particles and produce a clear effluent, i.e., to remove solids. Some equipment does both, and the dividing line between thickening and clarification is not always sharp. See also **Classifying (Process).** A clarifying agent to assist in this operation must possess certain properties for acting on the suspended particles—chemical precipitation, attraction via ionic forces, absorption qualities (large surface areas plus weak forces).

In the sugar refining industry, for example, various soluble, nonsugar compounds are present in sugar juices as the result of rupturing plant cells (either from pressing sugar cane stalks or by extraction of the sliced root of sugar beets). Lime is commonly used to precipitate impurities, following by carbonation of the solution with carbon dioxide to remove residual lime as calcium carbonate. Where filtration is used, various filter aids, such as fuller's earth, may be used. Filtering-type centrifuges also may be used. Phosphates, frequently in the form of orthophosphoric acid, may be used to precipitate the calcium. Some authorities suggest the use of polyphosphates, such as superphosphate and pyrophosphate, along with lime in the clarification operation. Phosphates assist in regulating the pH for optimal precipitation of calcium.

Tannin is used as a clarifying agent in wine making. Polyvinylpyrrolidone (PVP) has been used as a clarifying agent in the food industry. Ion exchange processes are also used in various processes, along with, or in lieu of conventional clarification. For example, in ion exchange, the demineralization of sugar solutions can be effected; iron can be removed from wine by substitution with hydrogen ions.

Generally, wherever practical and economical, food processors prefer to accomplish clarification without the aid of chemicals—because all or all but a trace of these chemicals must be removed so that they do not reappear in the final food product.

CLASSIFYING (Process).

An operation or series of operations designed to separate a mixture of substances of various particle sizes and specific gravities into two or more categories. These operations find application in mining, metallurgy, water and sewage treatment, and in other chemical processes.

Flotation. This is a means of separating a relatively small particle from a liquid medium. The particle may have a sp gr greater than, less than, or equal to that of the liquid from which it is floated. There are two main requirements: (1) a gas bubble and particle must come in contact with each other; and (2) the particle should have an *affinity* for attaching itself to the bubble.

Frothers. These are chemicals whose molecules contain both a polar and a nonpolar group. The purpose of a froth is to carry mineral-laden bubbles for a period of time until the froth can be removed from the flotation machine for recovery of its mineral content. Typical frothers include alcohols, cresylic acids, eucalyptus oil, camphor oil, and pine oil, all of which are slightly soluble in water. Soluble frothers in common use include alkyl ethers and phenyl ethers of propylene and polypropylene glycols.

Collectors. These are chemical reagents which selectively coat the particles to be floated with a water-repellent surface which will adhere to air bubbles. Collectors include xanthates, dithiophosphates, thiocarbonilides, and thionocarbonates. Fatty acids and soaps are anionic collectors and serve for nonsulfides. Amine salts are cationic collectors for nonsulfides.

Depressants are mainly inorganic salts, which compete with the collector for position on the sulfide surface. This permits the separation of one sulfide mineral from another. In one case, for example, in an alkaline solution, the addition of sodium cyanide prevents flotation of sphalerite and pyrite by xanthates, but not of galena, thus producing a higher grade of galena concentrates. The cyanide solution does not permanently affect the floatability of sphalerite as it can be floated by adding cupric sulfate and xanthate.

Activators are chemical reagents which alter the surface of a sulfide so that it can absorb a collector and float. Cupric sulfate is the most widely used activator. For example, xanthate as a collector will not readily float sphalerite, but the addition of cupric sulfate to the pulp changes the surface of the sphalerite particles to copper sulfide. Xanthate then will readily float the activated sphalerite as it behaves similarly to copper sulfide.

Although flotation was developed as a separation process for mineral processing and applies to the sulfides of copper, lead, zinc, iron, molybdenum, cobalt, nickel, and arsenic and to non-

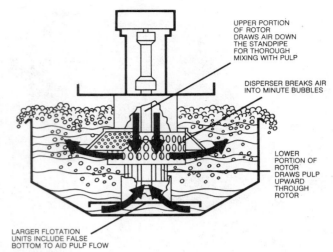

Fig. 1. Flotation cell. Upper portion of rotor draws air down the standpipe for thorough mixing with pulp. Lower portion of rotor draws pulp upward through rotor. Disperser breaks air into minute bubbles. Larger flotation units include false bottom to aid pulp flow. (*Envirotech.*)

sulfides, such as phosphates, sodium chloride, potassium chloride, iron oxides, limestone, feldspar, fluorite, chromite, tungstates, silica, coal, and rhodochrosite, flotation also applies to nonmineral separations. See Fig. 1.

Dense-Media Separation. This operation is useful for the separation of solid particles of different densities. A liquid suspension of finely-divided high-gravity solids is prepared. Ores of different densities, when exposed to such a suspension will tend to separate by rising or settling in the liquid suspension. Numerous types of solids have been used to obtain a high-gravity medium, but the magnetic solids (ferrosilicon and magnetite) are most frequently used. These solids, alone or in combination, can provide a suitable dense medium over a gravity range of 1.25 to 3.40. Dense-media separation is applicable to any ore in which the valuable component has an appreciable gravity difference from the gangue components.

Jigging. In this operation, a pulsating stream of liquid flows through a bed of materials of different specific gravities, causing the heavy material to work down to the bottom of the bed and the lighter material to rise to the top.

Tabling. In this concentration process, a separation between two or more minerals is effected by flowing a pulp across a riffled plane surface inclined slightly from the horizontal, differentially shaken in the direction of the long axis, and washed with an even flow of water at right angles to the direction of motion. A separation between two or more minerals depends mainly on the difference in specific gravity between the effective specific gravity (sp gr of mineral minus sp gr of water) of the valuable and the waste material.

Sedimentation. This is a general term for an operation wherein suspended solids are removed from a liquid by gravitational settling. The two major forms of sedimentation equipment are (1) thickeners, and (2) clarifiers. The term *decanting* also is sometimes used to designate sedimentation.

Thickeners. The primary objective of thickening is to increase the concentration of the feedstream. The mechanical continuous thickener, equipped with sludge-raking arms, is the most common type. Usually the operation is performed in cylindrical tanks. The sludge collection system and the removal system are designed to move the settled material continuously across the tank floor to a discharge point. Feed enters through a central

Fig. 2.   Center-pier type thickener, showing pumps and access. (*Dorr-Oliver.*)

feed well designed to distribute around the periphery. Thickened sludge, raked toward the center by a slowly revolving mechanism, enters a central collecting trough or cone and is discharged through a spigot or removed by a sludge pump. See Fig. 2.

Clarifiers. The primary objective of clarifying is to free solids from a relatively dilute stream. These units operate on the basis of gravity sedimentation and utilize a raking mechanism, as in a thickener. Frequently, clarifiers are operated in conjunction with flocculation equipment, which employs chemical coagulants, such as alum, iron salts, lime, polyelectrolytes, activated silica sol, and other chemical reagents. Clarification essentially expedites the natural gravity settling process.

See also **Clarifying Agents.**

CLATHRATE.   Compound (Chemical).

CLAUDE PROCESS.   Ammonia.

CLAY. A very fine-grained, unconsolidated rock material which normally is plastic when wet, but becomes hard and stony when dried. Common clay essentially consists of hydrous silicates of aluminum, together with a large variety of impurities, such as hematite and limonite, which usually impart color to the clay. Geologically, clay may be defined as a rock or mineral fragment, having a diameter less than $\frac{1}{256}$ millimeter (0.00016 inch), which is about the upper limit of size of a particle that can exhibit colloidal properties. Clays are widely used in the manufacture of tile, porcelain, earthenware, as filtering aids in oil and other processing, and as coatings for paper.

CLINOZOISITE. Clinozoisite crystallizing in the monoclinic system is a hydrous calcium aluminium silicate $Ca_2Al_3Si_3O_{12}(OH)$. Crystals are usually of prismatic habit with striations parallel to the *b*-axis. May occur as granular or columnar masses. Hardness 6.5, and specific gravity 3.21–3.38; vitreous luster, and transparent to translucent. Color gradations from gray through green to pink.

Clinozoisite occurs in crystalline schists which are themselves products of metamorphism from calcic feldspar-rich dark igneous rocks. Zoisite (orthorhombic) represents its dimorphous counterpart.

COACERVATION. An important equilibrium state of colloidal or macromolecular systems. It may be defined as the partial miscibility of two or more optically isotropic liquids, at least one of which is in the colloidal state. For example, gum arabic shows the phenomenon of coacervation when mixed with gelatin. It also may be defined as the production, by coagulation of a hydrophilic sol, of a liquid phase, which often appears as viscous drops, instead of forming a continuous liquid phase. See also **Colloid System.**

COAGULATION.   Blood; Colloid Systems.

COAL. Containing more than 50% (wt) and 70% (vol) of carbonaceous material, including inherent moisture, coal is a readily combustible rock. Coal was formed from the compaction and induration of various altered plant remains similar to those found in peat. Coal was formed during earlier geological periods by a very slow process acting over long time periods. Coal is not a uniform substance, but reflects the conditions of its formation. These include:

1. *Differences in the kinds of plant materials* from which the coal was derived account for *different types of coal.*

2. *Differences in the degree of metamorphism* occurring during the formation of coal determine the *different ranks of coal.*

3. *Differences in the range of impurity* in coal account for the *different grades of coal.*

The fermentation of vegetable matter under conditions of no air and abundant moisture where volatiles are retained, resulting in the formation of bitumens, such as peat and coal, is known as *bituminous fermentation.* The metamorphic transformation of bituminous coal into anthracite is known as *anthracitization. Coalification* is the alteration or metamorphism of plant material into coal; the biochemical process of diagenesis and the geochemical process of metamorphism in the formation of coal. The peat-to-anthracite theory of coal formation is described as a process in which the progressive ranks of coal are indicative of the degree of coalification and, by inference, of the relative geologic age of the deposit. Peat, as the initial stage of coalification, is of recent geological age. Lignite, as an intermediate stage, is usually Tertiary or Mesozoic, and bituminous coal and anthracite, as the more advanced stages of coalification, are usually Carboniferous.

The major coals may be defined as follows:

(*Anthracite Coal*)—Coal of the highest metamorphic rank, in which the fixed carbon content is between 92 and 98%. It is hard, black, and has a semimetallic luster and semiconchoidal fracture. Anthracite ignites with difficulty and burns with a short, blue flame and without smoke. Anthracite coal is also known as hard coal, stone coal, kilkenny coal, and black coal.

(*Semianthracite Coal*)—Coal having a fixed-carbon content of between 86 and 92%. It is between bituminous coal and anthracite coal in metamorphic rank, although its physical properties more closely resemble those of anthracite.

(*Semibituminous Coal*)—Coal that ranks between bituminous coal and semianthracite. It is harder and more brittle than bituminous coal, has a high fuel ratio and burns without smoke. Semibituminous coal is also known as *metabituminous coal* which is defined as containing 89–91.2% carbon, analyzed on a dry, ash-free basis. The term *smokeless coal* also is used.

(*Bituminous Coal*)—Coal that ranks between subbituminous coal and semibituminous coal and that contains 15–20% volatile matter. It is dark brown-to-black in color and burns with a smoky flame. Bituminous coal is the most abundant rank of coal and is commonly Carboniferous in age. The most common synonym is *soft coal.*

(*Subbituminous Coal*)—A black coal intermediate in rank between lignite and bituminous coals, or in some classifications, the equivalent of *black lignite*. It is distinguished from lignite by higher carbon content and lower moisture content.

Subbituminous C Coal—A type of subbituminous coal having 8,300 or more, but less than 9,500 Btu per pound (4615–5282 Calories/kg).

(*Lignite Coal*)—A brownish-black coal that is intermediate in coalification between peat and subbituminous coal; consolidated coal with a calorific value less than 8,300 Btu per pound (4615 Calories/kg), on a moist, mineral-matter-free basis. Synonyms include *brown lignite* and *brown coal*. Further classifications of lignite are made on the basis of calorific value:

Lignite A Coal—A lignite that contains 6,300 or more Btu per pound, but less than 8,300 Btu per pound (3503–4615 Calories/kg). Also known as *black lignite*.
Lignite B Coal—A lignite that contains less than 6,300 Btu per pound (3503 Calories/kg). Also known as brown lignite or brown coal.

(*Peat*)—This is an unconsolidated deposit of semicarbonized plant remains of a water-saturated environment, such as a bog or fen, and of persistently high moisture content (minimum of 75%). It is considered an early stage or rank in the development of coal. The carbon content is about 60%; oxygen content is about 30%. Structures of the vegetal matter can be seen. When dried, peat burns freely.
(*Peat Coal*)—This refers to two materials: (a) a coal transitional between peat and brown coal or lignite; and (b) an artificially carbonized peat that is used as a fuel.
(*Cannel Coal*)—A compact, tough *sapropelic coal* that contains spores and that is characterized by a dull-to-waxy luster, conchoidal fracture, and massiveness. It is attrital and high in volatiles. By American standards, it must contain less than 5% anthraxylon. Synonyms include *candle coal*, *kennel coal*, *cannel*, *cannelite*, *parrot coal*, and *curley cannel*. A sapropelic coal is derived from organic residues (finely divided plant material, spores, algae, etc.) in stagnant or standing bodies of water. Putrifaction is under anaerobic conditions rather than by peatification.
Ranks of Coal. Coals are classified in order to identify end-use and also to provide data useful in specifying and selecting burning and handling equipment and in the design and arrangement of heat-transfer surfaces. One classification of coal is by rank, that is, according to the degree of metamorphism, or progressive alteration, in the natural series from lignite to anthracite. Volatile matter, fixed carbon, inherent or bed moisture (equilibrated moisture at 30°C and 97% humidity), and oxygen are all indicative of rank, but no one item completely defines it. The classification of the American Society for Testing and Materials (ASTM) uses fixed carbon and calorific values, calculated on a mineral-matter-free basis, as the classifying criteria.

In establishing the rank of coals, it is necessary to use information showing an appreciable and systematic variation with age. For the older coals, a good criterion is the "dry, mineral-matter-free fixed carbon or volatile." However, this value is not suitable for designating the rank of the more recent, younger coals. A dependable means of classifying the latter is the "moist, mineral-matter-free Btu" which varies little for the older coals, but appreciably and systematically for younger coals.

The subbituminous coals are further classified in terms of their calorific value:

Subbituminous A Coal—A type of subbituminous coal having 10,500 or more, but less than 13,000 Btu per pound (5838–7228 Calories/kg).
Subbituminous B Coal—A type of subbituminous coal having 9,500 or more, but less than 10,500 Btu per pound (5282–5838 Calories/kg).

Classification of major coals according to rank or age is given in Table 1. The criteria given in the prior paragraph are used in classifying the older and younger coals. Seventeen United States coals are arranged in order of the classification of Table 1 and presented in Table 2.

Classification of coals in Europe and other parts of the world differs somewhat from the American system. European classifications include: (1) the *International Classification of Hard Coals by Type*; and (2) the *International Classification of Brown Coals*. These systems were developed by a Classification Working Party established in 1949 by the Coal Committee of the Economic Commission for Europe. The term "hard coal" is defined as a coal with a caloric value of more than 10,260 Btu per pound (5705 Calories/kg) on the moist, ash-free basis. The term "brown coal" refers to a coal containing less than 10,260 Btu per pound (5705 Calories/kg). In European terminology, the term "type" is equivalent to rank in American coal classification terminology and the term "class" approximates the ASTM rank.

Testing of Coal

Proximate Analysis. This includes the determination of total moisture, volatile matter, and ash; and the calculation of fixed carbon for coals and cokes. The term "Proximate" should not be confused with the word "approximate," since all Proximate Analysis tests are performed according to rigid specifications and tolerances. Proximate Analysis results may be used to establish the rank of coals; to show the ratio of combustible to incombustible constituents, to provide the basis for buying and selling coal, and to evaluate for beneficiation, or other purposes.

Moisture in coal takes three forms: (1) free or adherent moisture, essentially surface water; (2) physically bound or inherent moisture (that moisture held by vapor pressure and other physical processes); and (3) chemically bound water (water of hydration or "combined" water). The ASTM defines total moisture as a loss in weight in an air atmosphere under rigidly controlled conditions of temperature, time, and air flow. *Total moisture* represents a measurement of all water not chemically combined. Total moisture is determined by a two-step procedure, involving air-drying for removal of surface moisture from the gross sample, division and reduction of the gross sample, and determination of residual moisture in the prepared sample. An algebraic calculation is used to obtain the total moisture value.

Ash is the noncombustible mineral matter left behind when coal is burned under rigidly controlled conditions of temperature, time, and atmosphere.

Total nitrogen is determined by chemical digestion (Kjeldahl-Gunning) methods.

Oxygen content is determined by calculations, subtracting total carbon, hydrogen, sulfur, nitrogen, and ash from 100%.

Chlorine is commonly included as part of the ultimate analysis.

Other important chemical and physical tests performed to characterize coal include: (1) Heating value (Btu content); (2) sulfur forms; (3) ash fusibility temperatures; (4) ash analysis; (5) trace elements; (6) free swelling index; and (7) Hardgrove grindability.

Heating value is determined by burning a coal sample in

TABLE 1. CLASSIFICATION OF COALS BY RANK (ASTM)

CLASS	GROUP	FIXED CARBON LIMITS, % (Dry, Mineral-Matter-Free Basis)		VOLATILE MATTER LIMITS, % (Dry, Mineral-Matter-Free Basis)		CALORIFIC VALUE LIMITS, Btu/Pound (Moist*, Mineral-Matter-Free Basis)		AGGLOMERATING CHARACTER
		Equal or Greater Than	Less Than	Greater Than	Equal or Greater Than	Equal or Greater Than	Less Than	
I Anthracite	1. Meta-anthracite	98	—	—	2	—	—	Nonagglomerating
	2. Anthracite	92	98	2	8	—	—	Nonagglomerating
	3. Semianthracite[c]	86	92	8	14	—	—	Nonagglomerating
II Bituminous	1. Low volatile bituminous	78	86	14	22	—	—	Commonly Agglomerating[b]
	2. Medium volatile bituminous	69	78	22	31	—	—	
	3. High volatile A bituminous	—	69	31	—	14,000[a]	—	
	4. High volatile B bituminous	—	—	—	—	13,000[a]	14,000	
	5. High volatile C bituminous	—	—	—	—	11,500	13,000	
						10,500[b]	11,500	Agglomerating
III Subbituminous	1. Subbituminous A	—	—	—	—	10,500	11,500	Nonagglomerating
	2. Subbituminous B	—	—	—	—	9,500	10,500	
	3. Subbituminous C	—	—	—	—	8,300	9,500	
IV Lignitic	1. Lignite A	—	—	—	—	6,300	8,300	Nonagglomerating
	2. Lignite B	—	—	—	—	—	6,300	

* Moist refers to coal containing its natural inherent moisture, but not including visible water on the surface of the coal.

[a] Coals having 69% or more fixed carbon on the dry, mineral-matter-free basis are classified according to fixed carbon, regardless of calorific value.

[b] It is recognized that there may be nonagglomerating varieties in these groups of the bituminous class, and there are notable exceptions in high-volatile C bituminous group.

[c] If agglomerating, the coal is classified in the low-volatile group of the bituminous class.

The terms, *mineral-matter-free fixed carbon*; and *mineral-matter-free Btu* are defined by the following formulas:

Parr formulas

$$\text{Dry, Mm-free FC} = \frac{FC - 0.15S}{100 - (M + 1.08A + 0.55S)} \times 100, \%$$

$$\text{Dry, Mm-free VM} = 100 \times \text{Dry, Mm-free FC}, \%$$

$$\text{Moist, Mm-free Btu} = \frac{Btu - 50S}{100 - (1.08A + 0.55S)} \times 100, \text{per pound}$$

Approximation formulas

$$\text{Dry, Mm-free FC} = \frac{FC}{100 - (M + 1.1A + 0.1S)} \times 100, \%$$

$$\text{Dry, Mm-free VM} = 100 - \text{Dry, Mm-free FC}, \%$$

$$\text{Moist, Mm-free Btu} = \frac{Btu}{100 - (1.1A + 0.1S)} \times 100, \text{per pound}$$

Symbols Used:

Mm = mineral matter; Btu = heating value per pound; FC = fixed carbon, %; VM = volatile matter, %; M = bed moisture, %; A = ash, %; S = sulfur, %. All for coal on a moist basis.

Conversion Factor: 1 Btu/pound = 0.556 Calories/kg).

an oxygen bomb and measuring the temperature rise. See also **Calorimetry.**

Three *sulfur forms* recognized by ASTM are: (1) Sulfate sulfur, which may be in the form of calcium or iron sulfate; (2) pyritic sulfur, which is sulfur combined with iron in the form of minerals pyrite and/or marcasite; and (3) organic sulfur, which is bonded to the carbon structure. Sulfate sulfur is extracted from the coal with dilute hydrochloric acid, precipitated as barium sulfate, ignited and weighed. After sulfate removal, pyritic sulfur is extracted with nitric acid and the extracted iron is measured by redox titration. Organic sulfur is calculated by subtracting sulfate and pyritic sulfur from total sulfur.

Ash fusibility can be defined broadly as the melting temperature of the ash. Small triangular pyramides (cones) prepared from coal or coke ash pass through certain defined stages of fusing, and flow when heated at a specified rate.

Ash analysis is the term used to designate analysis of the major elements commonly found in coal and coke ash. The elements, expressed as oxides, are SiO_2, Al_2O_3, Fe_2O_3, TiO_2, CaO, MgO, Na_2O, K_2O, P_2O_5, and SO_3. Phosphorus, a trace element, has been included historically because of its importance in the subsequent steel making processes where coke is used.

Interest in *trace element analysis* has increased by environmental concerns.

Volatile matter is defined as the gaseous products, exclusive of moisture vapor, driven off during standardized test conditions. The combustible gases are carbon monoxide, hydrogen, methane, and other organic hydrocarbons. Those generally classified as noncombustible are carbon dioxide, ammonia, hydrogen sulfide, and some chlorides. Volatile matter tests are used to establish the rank of coals, to indicate coke yield upon carbonization, and to establish burning characteristics.

Fixed carbon is the solid residue, other than ash, resulting from the volatile matter test. The value is calculated by subtracting moisture, volatile matter, and ash from 100%.

Total carbon is determined by catalytic burning of the sample in oxygen to form carbon dioxide which can be readily measured.

Total hydrogen also is determined by catalytic burning of a sample in oxygen to form water. The water is absorbed in a desiccant and weighed directly. Hydrogen results as determined include the hydrogen present in both the sample moisture and water of hydration.

Importance of Testing. As previously mentioned, heating value, ash, fusibility, and other parameters are extremely important to the ultimate consumer. Advance coal testing is also very important to the layout and operation of coal preparation plants. A new dimension to coal testing has been added in recent years as various coals are considered for gasification and liquefaction processes. As will be obvious from several of the processes to be described shortly, the physical as well as the chemi-

TABLE 2. REPRESENTATIVE UNITED STATES COALS ARRANGED IN ORDER OF ASTM CLASSIFICATION

COAL RANK				COAL ANALYSIS, BED MOISTURE BASIS						RANK FC	RANK Btu
Class	Group	STATE	COUNTY	M	VM	FC	A	S	Btu		
I	1	Pennsylvania	Schuylkill	4.5	1.7	84.1	9.7	0.77	12,745	99.2	14,280
I	2	Pennsylvania	Lackawanna	2.5	6.2	79.4	11.9	0.60	12,925	94.1	14,880
I	3	Virginia	Montgomery	2.0	10.6	67.2	20.2	0.62	11,925	88.7	15,340
II	1	West Virginia	McDowell	1.0	16.6	77.3	5.1	0.74	14,715	82.8	15,600
II	1	Pennsylvania	Cambria	1.3	17.5	70.9	10.3	1.68	13,800	81.3	15,595
II	2	Pennsylvania	Somerset	1.5	20.8	67.5	10.2	1.68	13,720	77.5	15,485
II	2	Pennsylvania	Indiana	1.5	23.4	64.9	10.2	2.20	13,800	74.5	15,580
II	3	Pennsylvania	Westmoreland	1.5	30.7	56.6	11.2	1.82	13,325	65.8	15,230
II	3	Kentucky	Pike	2.5	36.7	57.5	3.3	0.70	14,480	61.3	15,040
II	3	Ohio	Belmont	3.6	40.0	47.3	9.1	4.00	12,850	55.4	14,380
II	4	Illinois	Williamson	5.8	36.2	46.3	11.7	2.70	11,910	57.3	13,710
II	4	Utah	Emery	5.2	38.2	50.2	6.4	0.90	12,600	57.3	13,560
II	5	Illinois	Vermilion	12.2	38.8	40.0	9.0	3.20	11,340	51.8	12,630
III	1	Montana	Musselshell	14.1	32.2	46.7	7.0	0.43	11,140	59.0	12,075
III	2	Wyoming	Sheridan	25.0	30.5	40.8	3.7	0.30	9,345	57.5	9,745
III	3	Wyoming	Campell	31.0	31.4	32.8	4.8	0.55	8,320	51.5	8,790
IV	1	North Dakota	Mercer	37.0	26.6	32.2	4.2	0.40	7,255	55.2	7,610

NOTE: Definition of coal rank is given in Table 1.

M = equilibrium moisture, %; VM = volatile matter, %; FC = fixed carbon, %; A = ash, %; S = sulfur, %; Btu = high heating value, Btu per pound; Rank FC = dry, mineral-matter-free fixed carbon, %; Rank Btu = moist, mineral-matter-free Btu per pound.

All calculations are per the Parr formulas defined in Table 1.

Conversion Factor: 1 Btu = 0.2520 Calorie.

cal properties of coal need vary but a very small amount to foul up some of the gasification and liquefaction processes. Tolerances are narrow; consistency of raw material is extremely important; very tight control over raw materials is mandatory. Variations from rigid specifications not only can adversely affect the yields and hence economic viability of coal conversion processes, but they can cause equipment damage through fouling, sometimes involving conditions that can be corrected only by shutting down operation for from a few hours to several days. The coal raw material used also can adversely affect plant effluents, upsetting delicate operating balances maintained in gas treating and recovery units, adding to the difficulty of close environmental controls.

Coal Conversion Processes

Particularly during the 1970s, much research was directed toward processes for converting coal into practical liquid and gaseous fuels that would be devoid of serious threats to the environment. Well over a score of major processes were partially or fully developed through operating-bench scale and pilot-plant size. These plants did not rival in practical acceptance, however, the numerous "artificial" gas plants, based upon coal technology, that existed from the late 1800s through the first quarter of the present century. These plants furnished gas to many cities in Europe, the United States, and other industrial nations prior to the coming of the age of natural gas. Natural gas as a major and widely distributed fuel really did not get underway on a large scale until the 1930s. With natural gas in abundance, emphasis on synthetic fuels waned for about 40 years. In the more recent era of emphasis on coal conversion the term "artificial gas" has largely been replaced by SNG (substitute natural gas or synthetic natural gas).

Although there are numerous design variations, there are three fundamental ways currently considered for converting coal into desirable liquid and gaseous fuels: (1) Removal of carbon to alter the hydrogen-to-carbon ratio[1]—known as *pyrolysis*; (2) addition of hydrogen to alter the ratio—known as *hydroliquefaction*; and (3) the *synthesis of hydrocarbons* from carbon monoxide and hydrogen, which basically uses Fischer-Tropsch technology. All of these methods are based upon very old chemistry and scientists continue to seek breakthroughs that may make coal conversion more attractive, technically and economically.

Pyrolysis. Essentially straight pyrolysis is a destructive distillation process, the products of which are volatile liquid and gaseous products, plus solid, carbonaceous char.

Hydroliquefaction. This process commences with the same initial step as in the case of pyrolysis, but differs in that hydrogen from an outside source is added to the radicals that have been generated thermally in the initial step. In straight pyrolysis, the radicals are capped by hydrogen originally in the coal, or the radicals combine with carbon, resulting in the formation of high-molecular-weight char. Although the latter two reactions also occur in hydroliquefaction, the net result is production of larger quantities of liquid and gaseous products essentially in proportion to the amount of extra hydrogen added. The process designer has a number of options, accounting in part for the proliferation of differing versions of hydroliquefaction that have been proposed. The hydrogen can be furnished in molecular form, or by way of an organic donor. Catalysts may or may not be used.

Synthesis of Hydrocarbons. Involving Fischer-Tropsch technology, coal is converted to synthesis gas (carbon monoxide and molecular hydrogen) for later conversion to fuels and various chemicals. This approach has been quite successful in processes used in the Republic of South Africa (*South Africa, Coal, Oil, and Gas Corporation Limited—SASOL*) since 1955, during

[1] In raw coal, the hydrogen-to-carbon ratio is less than unity. A ratio of 1.5–2 or more is needed to produce desirable liquid fuels.

which period synthetic motor fuels, pipeline gas, ammonia, and chemicals have been manufactured from coal.

REFERENCES

Considine, D. M., Editor: "Coal Technology," Section 1 (296 pp) in "Energy Technology Handbook," McGraw-Hill, New York (1977).

Goldberg, A. J., Hoover, L. J., and T. Surles: "Consequences of Increased Coal Utilization," *Chem. Eng. Progress*, **76**, 3, 61–64 (1980).

Gorbaty, M. L., et al.: "Coal Science: Basic Research Opportunities," *Science*, **206**, 1029–1034 (1979).

Gray, J. A.: "Gasification of Carbonaceous Feedstocks," *Chem. Eng. Progress*, **76**, 3, 73 (1980).

Moll, N. G., and G. J. Quarderer: "The Dow Coal Liquefaction Process," *Chem. Eng. Progress*, **75**, 11, 46–50 (1979).

Morris, S. C., et al.: "Coal Conversion Technologies: Some Health and Environmental Effects," *Science*, **206**, 654–662 (1979).

Patterson, R. C., and S. L. Darling: "A Low-Btu Coal Gasification Scheme," *Chem. Eng. Progress*, **76**, 3, 55–60 (1980).

Quig, R. H.: "Coal as an Energy Source: Its Impact on Engineering Planning," *Chem. Eng. Progress*, **76**, 3, 47–54 (1980).

Quillman, B.: "Present and Future Coal Utilization at Du Pont," *Chem. Eng. Progress*, **76**, 3, 43–46 (1980).

Raymond, W. J., and A. G. Sliger: "The Kellogg/Weir Scrubbing System," *Chem. Eng. Progress*, **74**, 2, 75–80 (1978).

Smoot, L. D., and D. T. Pratt: "Pulverized Coal Combustion and Gasification for Continuous Flow Processes," Plenum, New York, 1979.

Squires, A. M.: "Chemicals from Coal," *Science*, **191**, 689–700 (1976).

Staff: (For background reading): Coal Research I, II, III, and IV in *Science*, **193**, 665; **193**, 750; **193**, 873; and **194**, 172 (1976).

Staff: "Sasol-II Coal Liquefaction Plant," *Chemical Engineering Progress*, **76**, 3, 85–88 (1980).

Staff: "Chemicals from Coal," *Chemical Engineering Progress*, **76**, 3, 89 (1980).

Stephens, D. R., Brandenburg, C. F., and E. L. Burwell: "Underground Coal Gasification," *Chem. Eng. Progress*, **76**, 4, 89–94 (1980).

Swabb, L. E., Jr.: "Liquid Fuels from Coal: From R & D to an Industry," *Science*, **199**, 619–622 (1978).

Thurlow, G. G.: "Producing Fuels and Chemical Feedstocks from Coal," *Chemical Engineering Progress*, **76**, 3, 81–84 (1980).

Vogt, E. V., and M. J. van der Burgt: "Status of the Shell-Koppers Process," *Chemical Engineering Progress*, **76**, 3, 65–72 (1980).

COAL TAR AND DERIVATIVES. Coal tar constitutes the major part of the liquid condensate obtained from the "dry" distillation or carbonization of coal (mostly bituminous) to coke. The three major products of this distillation are (1) metallurgical coke, (2) gas which is suitable as a fuel after appropriate chemical treatment, and (3) vaporized materials which leave the coke oven along with the gas and which are constituted principally of ammonia liquor and coal tar. The condensables and gas impurities are separated from gas in the condensation and purification train of the coke oven plant. The purified coke oven gas is used as fuel to heat the coke ovens and steel-producing furnaces. Prior to the widespread use of natural gas as a domestic fuel, coke oven gas was widely used for this purpose after additional purification.

Since metallurgical coke for use in blast furnaces is the prime product of coal carbonization, coal tar production is tied closely to the demand for metallurgical coke. Although steel production has increased progressively over many years, the demand for metallurgical coke has remained reasonably steady, for several reasons. Large improvements in blast-furnace efficiency have occurred. The amount of coke required to produce one ton of pig iron has dropped from 1760 pounds coke/ton of pig iron (878 kilograms coke/metric ton of pig iron) to 1250 pounds coke/ton of pig iron (624 kilograms coke/metric ton of pig iron) and in some modern blast furnaces, the rate is below 1000 pounds coke/ton of pig iron (500 kilograms/metric ton of pig iron). Further, coke has been partially replaced by lower-cost carbonaceous materials, such as petroleum oils, powder coal, and tar. Fundamental changes in the production of steel are expected to further reduce the need for coke.

A number of years ago, coal tar was the primary, if not the sole, source for hundreds of important organic chemicals and derivatives, notably the phenols, cresols, naphthalene, and anthracene, as well as other important coal tar end-products, such as solvent naphtha and pitch. In recent years, synthetic processes for the production of phenol, the cresols, and later the xylenols have been developed and thus, to a large extent, have pushed coal tar into the background as a source of materials for which it was at one time extremely important.

Carbonization Process. Present coking processes generally are of two types: (1) *high-temperature* (900–1200°C) carbonization for producing metallurgical coke and practically the only process practiced in the United States, and (2) *low-temperature* (500–750°C) carbonization, still practiced in some countries where there is a market for "semicoke" as a smokeless home fuel.

Currently, in the United States, slot ovens with byproduct recovery systems are used almost exclusively. These ovens are built in the shape of narrow chambers placed side by side with interspaced flues for heating. Usually, up to 90 chambers are placed together to form a battery. The chambers are charged individually with coal from the top. After carbonization is complete (14–20 hours), the chambers are discharged on one side by pushing with a ram from the opposite side. Each chamber is connected at the top to one or two collecting ducts, or *mains*, which carry the gas evolved and the distillate (tar) to suitable coolers and receivers. By high-temperature carbonization, one obtains generally:

1550 pounds coke, 11,000 standard cubic feet of coke oven gas and 10 gallons of tar from 1 ton of coal;

748 kilograms coke, 343 cubic meters of coke oven gas, and 37.9 liters of tar from 1 metric ton of coal.

Processing of Crude Tar. With reference to the accompanying diagram, the crude tar, after being separated from ammonia and other gases, is subjected to an initial distillation (called *topping*) which separates the desired chemical constituents from the higher-boiling, more viscous tar constituents. In a typical case, the distillate from this operation (sometimes referred to as *chemical oil*) has an upper boiling point of about 250°C and contains (1) phenols (*tar acids*), (2) naphthalene, which is the most prevalent single constituent of coal tar (6–10%), (3) pyridine-type bases (*tar bases*), and (4) neutral oils. The tar acids constitute about 1.5–3% of the coal tar.

Tar Acids. These materials are recovered by extraction of

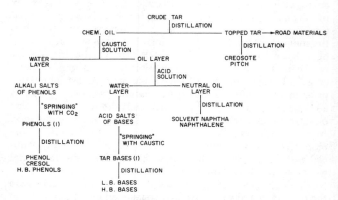

Bulk fractions from crude coal tar.

the chemical oil with aqueous alkali, usually caustic solution. The aqueous layer is separated from the dephenolized (*acid-free*) oil. The phenols then are recovered in crude form by acidification (*springing*) of the aqueous solution, usually by injecting carbon dioxide, followed by gravity settling. The crude phenols then are fractionated to obtain phenol, cresols, and the higher-boiling phenols (mostly xylenols). See also **Phenol.**

Tar Bases. These materials are extracted from the dephenolized oil with aqueous solutions of mineral acids. This operation may be carried out on the entire neutral oil, or it may be done on the solvent naphtha fraction. In the latter case, only the lowest-boiling bases (picolines and lutidines) are recovered, and the higher-boiling bases (mostly quinoline and isoquinoline) can be recovered from postnapthalene fractions or left in the residue for disposal. In European practice, the topping is carried out so that several fractions are obtained: *carbolic oil*, which yields the phenols and lower-boiling bases; and *naphthalene oil*, from which naphthalene is recovered by crystallization.

The tar bases form water-soluble salts with mineral acids which are separated from the oil. They are recovered from their salts by contacting with aqueous alkali (*springing*) and separating the crude bases from the salt solution. The lutidines constitute the major part of the lower-boiling bases. See **Pyridine and Derivatives.**

Solvent Naphtha. The lower-boiling fraction of the neutral oil is a very powerful solvent, particularly for coatings containing coal tar and pitch. The material also is a source of unsaturated compounds, such as indene and, in a lesser amount, coumarone and homologues of these compounds. Resins are formed in situ from these compounds when solvent naphtha is treated with Friedel-Crafts type catalysts. See **Friedel-Crafts Reaction.** These resins are useful in the manufacture of inexpensive floor tiles and coatings. The remaining solvent is recovered by distillation and used as a solvent.

Naphthalene. This compound finds a ready market principally for the production of phthalic anhydride. There is a great variety of processes to isolate napthalene from the acid-free or neutral oils. Frequently, the naphthalene is first concentrated by distillation, and the enriched oil then is worked up by crystallation. This process is prevalent in Europe. It is also possible to isolate the naphthalene by careful fractionation. Depending on the purity desired, additional chemical treatments may be required. Naphthalene usually is traded with freezing point as a measure of purity (80.3°C) for pure naphthalene. A good quality commonly used is "78°C" naphthalene, which is about 96% pure. See also **Naphthalene.**

Topped Tar. With reference to the accompanying diagram, it will be noted that topped tar is the residue remaining from the topping operation where the chemicals are separated as the distillate. The principal use of topped tar is in road materials. A number of standard grades (RT-1 to RT-12) are available, the grade depending on the "consistency" or viscosity of the tar. Road tar has excellent weather and skid resistance, but its use is limited by availability and price as compared with asphalt. This is borne out by the respective amounts used for road building (United States) with about 90% using asphalt.

Creosote. Chemically, creosote is a mixture of a great number of compounds, almost exclusively of cyclic structure. Individual compounds present in creosote in concentrations of 2–4% are acenapththene, fluorene, diphenylene oxide, anthracene, and carbazole. Only one compound, phenanthrene, is present in a larger concentration (12–14%). For many years, chemists in many countries have tried to isolate individual compounds and to find profitable uses for them. Most of these attempts have failed with exception of those involving anthracene. See also

Anthracene. The principal use of creosote is for preservation of wood. Railroad ties, poles, fence posts, marine pilings, and lumber for outdoor use are impregnated with creosote in large cylindrical vessels. If properly treated, the life of the wood is greatly extended. Materials that are competitive with creosote for wood-preservation purposes include various petroleum oils, and pentachlorophenol. Pentachlorophenol is used in solutions of creosote or of petroleum oils. Blends of creosote with petroleum oils also are used for economic reasons.

Pitch. This is the residue from the processing of coal tar. Since pitch constitutes over 50% of the crude tar, its utilization has a major effect on the economics of tar processing. Coal tar contains an estimated 5,000–10,000 compounds, it is reasonable to assume that about half of this number is contained in the pitch. Of the roughly 300 compounds identified in coal tar, about one-half must be in the pitch on the basis of their boiling points. It is probable that none of these compounds is present in pitch in concentrations of more than a fraction of 1%. Coal tar pitch is a black, shiny material which is solid and brittle at low temperatures and liquid at high temperatures. Since it is composed of a great number of different compounds, many of which interact to form eutectic mixtures, it does not show a distinct melting or crystallizing point. Pitch usually is characterized by the *softening point*, which can be measured in several ways.

Because of the importance of pitch in various industries, many studies have been made to elucidate its composition. Solvent fractionation has been used to subdivide pitch into fractions by molecular weight, the higher-molecular-weight fractions requiring more powerful solvents. However, even the highest-molecular-weight compounds are only of moderate molecular weights, ranging up to 6 or 7 condensed rings with molecular weights in the range of 350–400. Constituents isolated from pitch have been identified and appear to be crystalline, well-defined substances. It is generally believed that the glasslike state of pitch is caused by association forces and by the mutual melting-point depression exerted by a large number of multiring compounds that tend to form eutectics. The uses of pitch fall in two general classes: (1) applications based upon the binder properties of carbonized pitch (*pitch coke* or *binder coke*); and (2) uses based upon the other physical qualities of pitch.

Carbon pitch is used for carbon electrodes in electrolytic reduction processes, such as aluminum reduction. Refractory pitch is used in the manufacture of refractory brick, usually burned magnesite or dolomite, the pores of which are filled with pitch by hot impregnation. Upon firing the pitch in the brick is carbonized. The remaining pitch coke retards penetration of molten metals and slags, thus prolonging the life of the brick furnace lining. Core pitch is used in the production of foundry cores.

Roofing Pitch. A substantial amount of pitch is used as covering membrane on flat roofs on industrial plants, large office and apartment buildings, parking garages, and similar structures. Pitches of 50–60°C softening point generally are used.

—George R. Romovacek, Koppers Company, Inc., Monroeville, Pennsylvania; and G. G. Lauer (Retired), Pittsburgh, Pennsylvania.

COATING AGENTS (Foods). Substances that are used to protect the surface (and penetration through the surface) of food materials that are being processed or final products, including fresh produce that requires very little processing per se. Wax-coated cheese is possibly the simplest example of a coating application. Depending upon the materials involved, coating agents may be dusted or sprayed onto the food substance, or the latter may be dipped into a solution of the protective material. Only a few representative applications of coating agents are given here to illustrate the variety of uses. As with so many other food additive or process-assisting chemicals, coating agents frequently serve a number of functions.

Some fruits may be dipped in 1–2% citric acid prior to freezing. This removes any residual lye from the peeling process and any further destruction of ascorbate is prevented. However, citric acid alone is not always sufficient to prevent deteriorative effects during freezing and thus a sequestrant/antioxidant may be added. A solution of 0.5% citric acid with 0.02% D-erythro-ascorbic acid may be used to prevent browning of some fruits during freezing and defrosting. The procedure is also useful with some vegetables.

Dried fruits, such as raisins, prunes, and figs, require a residual moisture content because most consumers do not like a thoroughly dry or crispy product in this category. Residual moisture, of course, encourages mold and yeast growth. Protection can be gained by dipping the fruit in solutions of 2–7% potassium sorbate, this leaving a fine coating of antimicrobial agent on the fruit pieces.

Acetostearin products (di- and triglycerides) solidify into wax-like solids and are used in some protective coatings for food products. Researchers have shown these products to be effective against moisture penetration, as well as against atmospheric gases. The coating is applied by dipping, or in the case of nuts, by spraying.

For tabletlike confections, coating techniques from the pharmaceutical industry have been used. Possibly the most common coating material used is sucrose, but a number of transparent or opaque materials also make excellent film coatings. These include various gums and resins. See also **Gums and Mucilages**. Some products may be multicoated, and through the use of approved colorants, interesting shades and coloring effects can be obtained. The film-type coatings generally use an organic solvent containing a gum or resin plus plasticizers.

The film-forming characteristics of starch have been known and exploited by the food industry for many years. Numerous foods can be protectively and decoratively coated with starch. Starch can be added to sugar solutions used for coatings, providing a less brittle and moisture-sensitive surface. Starch coatings also have the advantage of being oil and grease resistant.

Methylcellulose when incorporated into sweet dough products performs a number of functions, one of which is improving glaze adherence.

Red meats, poultry, and fish can be coated with a protective film prior to freezing by dipping into successive solutions of 10–15% sodium alginate, 3–5% calcium chloride, and 10–20% glycerol, the latter used as a plasticizer. This coating procedure improves retention of juices on freezing and thawing. Sodium alginate coating for sausages, alone or with ethylcellulose, prevents salt rust and increases storage stability. Researchers also have studied the effectiveness of carrageenan as a way to prevent oxidative rancidity after freezing. In a Norwegian process, fish are block-frozen in alginate jelly, forming an air-tight coating, preventing oxidative rancidity.

COATINGS. Conversion Coatings; Paint and Finishes.

COBALAMIN. Vitamin B_{12}.

COBALT. Chemical element symbol Co, at. no. 27, at. wt. 58.9332, periodic table group 8, mp 1495°C, bp 2869–2871°C, density, 8.832 g/cm^3. There are two allotropic modifications of cobalt, a close-packed hexagonal form (ϵ) with space group $P6_3/mmc$, stable at temperature below 417°C; and a face-centered cubic form (α) with space group $Fm3m$, stable at higher temperatures—up to the melting point. The metal is a silvery gray color. The only naturally occurring isotope, ^{59}Co, is stable, but the other twelve known isotopes are radioactive, their mass numbers ranging from 54 to 64. Half-lives range from 0.2 second for ^{54}Co to 5.3 years for the industrially and medically important ^{60}Co. See **Radioactivity**.

Cobalt was identified and described by Georg Brandt in 1735, but had to wait until the last decade of the nineteenth century before the new sources of metal supply from New Caledonia and Canada stimulated its metallurgical use.

First and second ionization potentials are 7.86 eV and 17.05 eV, respectively. Oxidation potential E° is −0.277 V. Further general specifications are given under **Chemical Elements**.

Occurrence. The cobalt content of the earth's crust is estimated to be within the range 10 to 40 ppm. Economic concentrations of the element are the exception so that supply is governed by its by-product output from ores mined for the recovery of other elements, particularly copper and nickel. For technical reasons, the sulfides, arsenides and oxidized minerals form almost the entire economic source of the metal. Its production is restricted to a relatively few countries, the most important being the Republic of Zaire and Zambia in Africa, Canada, Finland, Morrocco, and the United States. The cobaltiferous deposits in the copper belt of central southern Africa vary in content from 1 to 30 parts of cobalt to 100 parts of copper, with an estimated hundreds of millions of tons containing 2 to 15 parts cobalt to 100 parts copper. Whereas sulfide nickel ores the world over have a cobalt content in the range 2 to 5 (very occasionally as high as 10) parts to 100 parts of nickel. Oxide nickel ores vary in cobalt content from 1 to 30 parts per 100 parts nickel with estimated thousands of millions of tons containing 5 to 15 parts cobalt per 100 nickel. The nickel lateritic deposits in the tropical and subtropical areas of the world represent a large potential future source of both nickel and cobalt.

Mining and Recovery. The economically important cobalt-containing minerals at present exploited are listed in the accompanying table. The metal extraction processes, following the usual pretreatment of the ore, are varied and complicated because the metallurgical properties of cobalt differ insufficiently from those of the associated metals and because the cobaltiferrous raw materials comprise the sulfide, arsenide, and oxide, or a mixture of these. The final refining stage invariably involves electrolysis.

Uses of Metal and Alloys. The metallurgical applications of cobalt consume approximately 70–75% of the world production; the remainder goes into chemicals. Relatively little use has been made of the pure metal, the most important being as the radioisotope ^{60}Co for teletherapy and industrial radiation processing and gamma radiography. The increasing availability of the metal in various wrought forms will stimulate its applications in those areas where the intrinsic properties of cobalt are advantageous, e.g., magnetic devices, wear resistance and bearing properties at elevated temperatures, and the manufacture of heterogeneous welding rods for hardfacing. Major uses of the metal may be classified as follows:

High-Temperature Materials. To meet the demands of the gas turbine industry, alloys capable of reliable performance at high temperature and loads have been developed based on cobalt and cobalt-containing nickel and iron-based alloys. The improvement in properties at these very high temperatures (~750–1200°C) has been brought about by the addition to the oxidation and sulfidation resistant cobalt-chromium matrix of varying

PRINCIPAL ECONOMICALLY IMPORTANT COBALT MINERALS

GROUP AND NAME	IDEAL FORMULA	COBALT CONTENT %	SOURCE AREA
Sulfides			
Linnaeite	$(Co, Cu, Ni, Fe)_3S_4$	up to ~ 48	U.S.A., Zaire, Finland
Carrolite	$CuCo_2S_4$	up to ~ 38	Zambia, Zaire, U.S.A.
Pentlandite	$(Fe, Ni, Co)_9S_8$	up to ~ 2	Canada
Cobaltiferous Pyrite	$(Fe, Co)S_2$	up to ~ 2	Canada, Finland, Zambia
Arsenides			
Skutterudite	$(Co, Ni, Fe)As_3$	up to ~ 28	Canada, Morrocco, U.S.A.
Gersdorffite	$(Ni, Co, Fe)AsS$	up to ~ 12	Canada, Zaire, U.S.A.
Oxides			
Heterogenite	$Co_2O_3H_2O$	up to ~ 57	Zaire, Zambia
Asbolite	$(Co, Ni)O2MnO_24H_2O$	up to ~ 27	New Caledonia, Canada

Note: See also **Pentlandite**; **Skutterudite**.

amounts of refractory metals, mainly tungsten, molybdenum, tantalum, and columbium (niobium) to strengthen the matrix further and promote the formation of stable carbides which make a substantial contribution to the high-temperature strength. The application of current techniques of vacuum melting and casting optimum material properties and improved component reliability. A new family of alloys based on Co-Fe-Cr is being successfully applied in the exacting conditions which exist in metallurgical furnaces and petrochemical plants.

Magnetic Materials. For the past 50 years, apart from iron, cobalt has been the most important constituent of permanent magnet materials. This is because cobalt additions raise the saturation magnetization and the Curie temperature to higher values than are obtained from pure iron. Present advanced Alnico-type cast alloys are the results of 40 years of development and are the most widely used permanent magnet materials. More recently, permanent magnet materials based on cobalt-rare-earth (e.g., samarium, praseodymium) compounds have emerged as the most powerful magnets available.

Hardfacing and Wear-Resistant Alloys. These materials, essentially quaternary alloys of cobalt; chromium, tungsten (or molybdenum); and carbon, are widely used for industrial hardfacing purposes. They can be deposited by welding techniques, sprayed on as powders, or produced as separate castings. By using the weld deposition technique, a highly alloyed heat, wear, and corrosion-resistant surface can be applied to a much cheaper substrate, e.g., mild steel. Used originally as a component reclamation process, these alloys and techniques are now primary design requirements for many engineering items. The feasibility of electrodepositing composite cobalt-carbide coatings from an agitated slurry of solid carbide particles in a conventional plating bath has been demonstrated. These coatings provide excellent wear protection and are now established in the aerospace industry, while other engineering applications are being evaluated.

The demand for wear resistance allied with a low coefficient of friction has been successfully met by cobalt-molybdenum-silicon alloys. Structurally they conform to the well proven bearing concept of hard intermetallic phases dispersed through a softer cobalt matrix. They offer outstanding bearing performance in conditions of poor lubrication.

Alloy Steels. The addition of cobalt to high-speed tool steel was one of the earliest uses of cobalt. This application is represented by the super high-speed tool steel grades which contain 5–12% cobalt. Two other cutting tool materials, the nonferrous cobalt-chromium tungsten-carbon cast alloy and the cemented

carbides also provide a steadily growing outlet for cobalt. Hot work die steels are another group of alloy steels which benefit from the effect cobalt has on the tempering characteristics and consequent hot strength retention. The most significant development in recent years, however, is the advent of the Ni-Co-Mo miraging family of steels which combine very high strength and toughness properties to an unusual degree. This is still an area of active development and growing engineering utilization.

Toxicity. Cobalt, like most other metals, is not entirely harmless, although it is not in any way comparable to the known toxic metals, such as mercury, cadmium, and lead. Inhalation of fine cobalt dust over long periods can cause an irritation of the respiratory organs which may result in chronic bronchitis. Complete recovery is usually achieved upon removal from the contaminated atmosphere. Cobalt salts can cause benign dermatoses, either in people new to handling them, or after prolonged exposure, usually several years.

Chemistry of Cobalt. The metal in the massive form is not attacked by air or water at temperatures below approximately 300°C; above this temperature it is oxidized in air. The metal combines readily with the halogens to form the respective halides. It combines with most of the other metalloids, when heated or in the molten state. It does not combine directly with nitrogen but decomposes ammonia at elevated temperature to form a nitride. It reacts with carbon monoxide at 225–250°C to form the carbide Co_2C. Cobalt also forms intermetallic compounds with many metals, e.g., Al, Cr, Mo, Sn, Ti, V, W and Zr. Metallic cobalt is readily dissolved in dilute sulfuric, hydrochloric or nitric acids to form cobaltous salts. Like iron, cobalt is passivated by strong oxidizing agents, such as the dichromates. It is slowly attacked by ammonium hydroxide and sodium hydroxide.

Cobalt Compounds. In general the chemical properties are intermediate between those of iron and nickel. The predominant oxidation states of cobalt compounds, except for a large class of organometallic compounds, are 2+ and 3+. Common usage assigns the terms *cobaltous* and *cobaltic*, respectively, to these.

In aqueous solutions and in the absence of complexing agents, cobalt compounds are stable only in the 2+ oxidation (cobaltous) state. In the complexed state the cobaltous ion is relatively unstable being readily oxidized to the 3+ oxidation (cobaltic) state. An extremely large number of 3+ complex ions have been identified most of which are quite stable in aqueous media.

Cobalt has an electronic configuration $1s^22s^22p^63s^23p^63d^74s^2$. The two 4s electrons are readily removed producing the ordinary Co^{2+} ion. In principle, if the odd $3d$ electron were removed the simple cobaltic ion Co^{3+}, would be formed. This however does not occur, the ion exists only in complex ions or crystal lattices in which cases additional electron orbitals are filled.

Cobalt and oxygen form two stable oxides, cobaltous oxide CoO, stable below 200°C and above 900°C; and cobalto-cobaltic oxide Co_3O_4 which is stable below 900°C. Between 200 and 900°C the CoO oxidizes partially, or completely, to Co_3O_4.

Cobaltous oxide is usually prepared by heating the carbonate. It is insoluble in H_2O, NH_4OH, and alcohol, but dissolves in cold strong acids and in weak acids on heating. Commercial gray oxide which may contain up to 40% Co_3O_4 is used in the ceramic, glass and enamel industries and also in the production of catalysts. It is also used in the preparation of cobalt metal powder. The Co(III) oxide can also be prepared by calcining oxides, hydroxide and salts. The oxide is insoluble in H_2O and only slightly soluble in acids. The commercial black oxide consists essentially of Co_3O_4 with possibly up to 20% of CoO.

Cobalt(II) hydroxide exists in two allotropic forms, a blue α-$Co(OH)_2$ and a pink β-$Co(OH)_2$. The hydroxide is prepared by precipitation from a cobaltous salt solution by an alkali hydroxide. When the alkali is in excess the pink β-form is produced—the blue α-form is produced when the cobalt salt is in excess. The salt slowly oxidizes in air at room temperature and changes to hydrated cobaltic oxide $Co_2O_3 \cdot nH_2O$. The hydroxide is practically insoluble in H_2O and in bases, but highly soluble in mineral and organic acids. The commercial salt is used as the starting material in the preparation of drying agents.

Cobaltous halides are formed with the halogen group, but only fluorine forms a stable cobaltic compound. The other cobaltic halides are stable only in complex ions.

Commercial cobaltic fluoride is formed by the reaction of cobalt, its oxides or simple salts with ClF_3 or BrF_5 to yield a brown anhydrous salt. This is a powerful fluorinating agent readily replacing hydrogen in aliphatic and aromatic hydrocarbons. Cobaltous chloride is a pale blue compound and very hygroscopic. A series of hydrates is known; the hexahydrate is pink but becomes blue on warming. It is extremely soluble in H_2O, and in numerous organic solvents. The commercial salt is used as a starting material in the manufacture of catalysts, in electroplating, agricultural chemicals, and pharmaceuticals. Cobaltous bromide is highly hygroscopic and transforms gradually to the red hexahydrate. The salt is mainly used in the catalysts industry.

The iodide exists in two forms, the α-form, a black graphite-like solid; and iodine-free β-form, which is yellow-brown in color. The iodide is used as a catalyst in organic reactions.

A number of sulfides have been reported, the best characterized being Co_4S_3, Co_9S_8, CoS, Co_3S_4, and CoS_2. They are prepared either by metal or salt solution reaction with S or H_2S. The mixed sulfides of cobalt and molybdenum have catalytic properties of hydrogenation and isomerization.

The sulfate, one of the more important industrial salts of cobalt, is usually available as the red heptahydrate, an efflorescent substance. Upon heating, it converts to the dark blue monohydrate, and after prolonged heating above 250°C, the red anhydrous salt. It is widely used in the electroplating and ceramic industries, and in the preparation of drying agents and agricultural pasture top-dressing.

Cobalt and nitrogen form three nitrides Co_3N_3, CO_2N, and CoN, products of metal/ammonia reaction and compound decomposition. All are gray-black or black in color.

In its commercial form cobalt nitrate appears as the red hexahydrate $Co(No_3)_2 \cdot 6H_2O$ formed by dissolving the metal oxide or carbonate in dilute HNO_3 and concentrating the solution. The compound deliquesces in moist air and effloresces in dry air. The pink anhydrous cobaltous nitrate cannot be formed by dehydrating the hexahydrate but by treating the salt with nitrogen pentoxide gas (or in solution in concentrated HNO_3). The salt is used mainly in the preparation of catalysts.

Two discrete carbides have been characterized—Co_3C and Co_2C. The former, which has the same structure as Fe_3C, has been prepared by reacting cobalt with coal gas at temperatures between 500–800°C. Co_2C is formed by the reaction of carbon monoxide at atmospheric pressure on cobalt powder at 225–230°C.

Cobaltous carbonate $CoCo_3$ is found almost pure in the mineral sphaerocobaltite in the Republic of Zaire and less extensively in Zambia. The pale-red anhydrous salt is obtained by reaction in solution of an alkaline carbonate and a cobaltous salt under a slight pressure of carbon dioxide (up to 1 atmosphere) and subsequent heating at 140°C. The commercial salt is violet-red in color, partially hydrolized with an indeterminate composition. It is insoluble in H_2O and alcohol, but dissolves easily in inorganic and organic acids, and is often used for the preparation of other salts. According to the thermal conditions it decomposes to the different types of oxides.

Cobaltous acetate $Co(CH_3CO_2)_2$ is obtained as a pink salt by the dehydration of the tetrahydrate which is prepared by dissolving the hydroxide or carbonate in acetic acid. The tetrahydrate which is the commercial form is widely used in the preparation of catalysts, e.g., OXO synthesis, and driers for inks and varnishes.

Cobaltous oxalate dihydrate $CoC_2O_42H_2O$ (pink) is obtained by adding oxalic acid or an alkaline oxalate to a cobaltous salt solution. This is the commercial form of the salt and is important as the starting material in the preparation of cobalt metal powders.

Cobalt(III) coordination compounds are the classics of coordination chemistry, many of the cobalt(III) amines having been prepared in the nineteenth century. They are invariably colored and undergo reaction slowly. Because of this many isomers have been isolated and studied. Much of the knowledge of isomerism, mechanisms of reaction, and general properties of octahedral species are based on cobalt-(III) compounds. The important donor atoms (in order of decreasing tendency to complex) are nitrogen, carbon in the cyanides, oxygen, sulfur, and the halogens. The coordination number is invariably six. An extensive class of amines and compounds of the amines is known ranging from haxamine $[CoN_6]^{3+}$ through to monoamines $[CoNX_5]^{2-}$. The compounds are made in several stages by first oxidizing the cobalt(II) species and then various different substituted species are prepared by substitution reactions on the primary cobalt(III) product. When air is drawn through on aqueous solution containing cobalt(II) and ammonia the solution turns brown. From this mixture can be obtained a variety of products which depend on the initial concentrations of reactants, the pH of the solution, the anion present and the presence of heterogeneous catalysts, such as charcoal. The products are invariably colored ranging from blue-violets and green through various shades of red and brown to yellow. The absorption spectra are characteristic of the various structures. Complex cyanides of the types $[Co(CN)_6]^{3-}$ and $[Co(CN)_5X]^{3-}$ are stable and diamagnetic. The anion $[Co(CN)_6]^{3-}$ is pale yellow and is the ultimate product of the reaction when a solution of cobalt(III) cyanide in aqueous KCN is boiled in air. It is very unreactive, being untouched by chlorine, peroxide, alkali, aqueous HCl and H_2S, although it gives CO when treated with concentrated H_2SO_4.

The only cobalt(IV) compound representing this oxidation state appears to be $Cs_2[CoF_6]$ which is prepared as a yellow powder by the fluorination of Cs_2CoCl_4.

Evidence for cobalt(I) was first obtained from the electrolytic reduction of cyano-compounds and some of the reduced species have been isolated. There are also many cobalt(I) coordination compounds of the organometallic class carbonyl, isonitriles, and unsaturated hydrocarbon derivatives. The oxidation state cobalt (O) may be represented in the cyano-compound which has been formulated as $K_8[Co_2(CN)_8]$. It has been prepared as an air-sensitive brown-violet compound by reducing a liquid ammonia solution of $K_3[Co(CN)_6]$ with an excess of K metal. The only other known cobalt (O) species are organometallic compounds.

For the role of cobalt in biological systems, see **Cobalt (In Biological Systems)**.

References

Bailar, J. E., et al., Editors: "Comprehensive Inorganic Chemistry," Vol. 3, Pergamon Press, Elmsford, New York, 1973.
Holden, C.: "Strategic Minerals," *Science*, **212**, 305 (1981).
Maykuth, D. J.: "Cobalt" in "Metals Handbook," 9th Edition, Vol. 2, American Society for Metals, Metals Park, Ohio, 1979.
Smithells, C. J., Editor: "Metals Reference Book," Vol. III, Plenum, New York, 1967.

—E. Williams, Cobalt Information Centre, London.

COBALT (In Biological Systems). Although cobalt is regarded as an essential element for animals, including humans, the element can perform its essential functions only after it has been incorporated into the vitamin B_{12} molecule. The microorganisms living in ruminants are the major producers of vitamin B_{12} in the food chain. Green plants do not synthesize the vitamin. The normal intake of this vitamin is by way of milk, cheese, meat, and eggs. Persons who follow a strictly vegetarian diet may become deficient in vitamin B_{12} unless supplementary sources are used. Single-stomached domestic and wild animals receive their vitamin B_{12} from animal flesh or from animal fecal material. See also **Vitamin B_{12}.**

Cobalt is present in vitamin B_{12} to the extent of about 4%. Lack of cobalt in the soil and feedstuffs prevents ruminants from synthesizing all of the vitamin B_{12} for their needs. Thus, cobalt can be added to feedstuffs as the chloride, sulfate, oxide, or carbonate. Excessive cobalt intakes are toxic, causing a reduction in feed intake and body weight, accompanied by emaciation, anemia, debility, and elevated levels of cobalt in the liver. It is of interest to note that clinical cobalt toxicity closely resembles clinical cobalt deficiency.

Cobalt is required by the microorganisms that live in nodules on the roots of legumes, such as bean and clover. They convert nitrogen from the air into chemical forms that can be used by higher plants. This is possibly the only well established and understood function of cobalt in plant growth. Legumes may grow normally and the microorganisms on their roots fix atmospheric nitrogen, even though the forage does not contain sufficient cobalt to meet the requirements of ruminants.

Areas of low cobalt content in the United States soils, where clovers and alfalfa are too low in cobalt content to meet requirements of cattle and sheep, include northeastern Maine, all of New Hampshire, Vermont, Massachusetts, Connecticut, and Rhode Island, much of New York with exception of the central portion, the northwestern portion of the lower peninsula of Michigan, a small area in Illinois with Peoria at its approximate center, all but eastern Iowa, and southwestern Minnesota. The low-cobalt soils of New England are primarily sandy and were formed from glacial deposits near and to the south of the White Mountains of New Hampshire. Along the south Atlantic Coastal Plain, legumes with very low concentrations of cobalt are primarily on the sandy soils formed in naturally wet areas. These soils, which are called spodosols, have light-colored subsurface layers overlying a dark-brown or dark-gray hardpan layer.

Grasses and cereal grains generally contain less than the 0.07 to 0.10 parts per million of cobalt required by ruminants. Cattle and sheep that are not fed any legumes nearly always require cobalt supplementation.

Adding cobalt to soils, either as cobalt sulfate, or as cobaltized superphosphate, can be used to increase the level of cobalt in plants and prevent cobalt deficiency in cattle and sheep. Cobalt fertilization may not be effective in preventing cobalt deficiency on alkaline soils because in these soils, the added cobalt quickly reverts to forms that are not taken up by plants. Cobalt fertilization is more common in Australia than in the United States. In the United States, cobalt is usually added to mixed feeds, mineral mixes, or salt licks.

Still another method is to place heavy ceramic "bullets" containing cobalt in the animal's rumen. These bullets remain in the rumen and slowly release cobalt to meet the animal's needs for a long period. The diets of hogs and chickens are often supplemented with concentrated forms of vitamin B_{12}.

The relationship of the levels of cobalt in soils and plants to the health of ruminants is one of the striking examples of the importance of a soil and plant relationship to animal health. When some Australian scientists discovered this relationship, new areas in several parts of the world became usable for animal production. The vitamin B_{12} formed within cattle and sheep in these new areas contributed to the vitamin B_{12} nutrition of people, even though adding cobalt to soils does not directly affect human nutrition in the absence of the production of ruminants. Cobalt-deficient grazing soils are found in Australia, New Zealand, and, in the United States, mainly in Florida, although deficient regions are found elsewhere as previously mentioned. Cobalt deficiency can result in a condition known as "pining disease," where the affected animals are quire listless. The disease is also known as "bush sickness" and "salt sickness."

References

Allaway, W. H.: "The Effect of Soils and Fertilizers on Human and Animal Nutrition," Agricultural Information Bulletin 378, Cornell University Agricultural Experiment Station and U.S. Department of Agriculture, Washington, D.C., 1975.
Kirchgessner, M. (editor): "Trace Element Metabolism in Man and Animals," Institut für Ernahrungsphysiologie, Technische Universität München, Freising-Weihenstephan, Germany (1978).
Underwood, E. J.: "Trace Elements in Human and Animal Nutrition," 4th edition, Academic, New York, 1977.

COBALTITE. The mineral cobaltite is a sulfarsenide (see **Arsenic** and **Sulfur**) of cobalt, corresponds to the formula CoAsS, crystallizing in the isometric system as cubes or pyritohedrons, also may be massive. Cobaltite has a very good cleavage parallel to the cube faces; uneven fracture; brittle; hardness, 5.5; specific gravity 6.33; metallic luster; color, silvery-white to reddish, sometimes steel gray or violet to grayish-black; streak, grayish-black. Cobaltite is found with cobalt and nickel minerals deposited commonly by metasomatic processes. It is found in Sweden, Norway, England and the Province of Ontario. It is an ore of cobalt.

COCAINE. Alkaloids.

COCONUT OIL. Vegetable Oils (Edible).

CODEINE. Alkaloids.

COENZYMES. A nonprotein substance that is closely associated with or bound to the protein component (*apoenzyme*) of an enzyme. Together, the coenzyme and apoenzyme form the complete enzyme known as the *holenzyme*. The presence of a coenzyme is necessary for enzyme activity. Coenzymes are organic molecules of a size intermediate between the small-molecule intermediary metabolites, which serve as the substrates of enzymatic reactions, and the macromolecular proteins. Each coenzyme acts usually as acceptor or donor of some specific type of atom or group of atoms to be removed from or added to a small-molecule substrate in a reaction catalyzed by the holoenzyme.

Coenzyme A (CoA). Pantothenic acid is a constituent of coenzyme A, which participates in numerous enzyme reactions. CoA (Fig. 1) was discovered as an essential cofactor for the acetylation of sulfanilamide in the liver and of choline in the brain. It has been established that CoA is involved in many biochemical reactions in the body as an "activator" of normally less reactive carbon fragments and a "transferer" of these fragments to different molecules. CoA is particularly important in the initial reaction of the citric acid cycle of carbohydrate metabolism and energy production. After oxidative decarboxylation of pyruvic acid, CoA combines with the two-carbon acetate fragment to form acetyl-CoA or "active" acetate.

$$CH_3-\overset{\overset{\displaystyle O}{\|}}{C}-COOH + CoA \rightarrow CH_3-\overset{\overset{\displaystyle O}{\|}}{C}-CoA + CO_2$$

(Pyruvic acid)   Acetyl-CoA
("active acetate")

Coenzyme A is necessary for the activation, synthesis, and degradation of fatty acids. Synthesis of cholesterol and ultimately

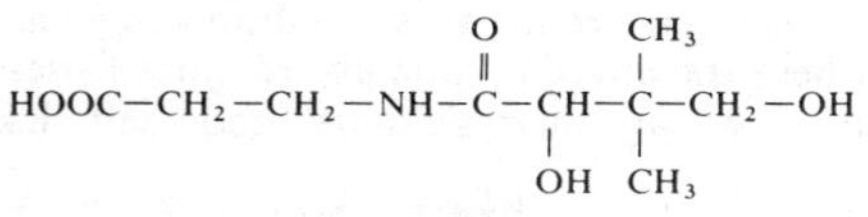

$$HOOC-CH_2-CH_2-NH-\overset{\overset{\displaystyle O}{\|}}{C}-\overset{\overset{\displaystyle OH}{|}}{C}H-\overset{\overset{\displaystyle CH_3}{|}}{\underset{\underset{\displaystyle CH_3}{|}}{C}}-CH_2-OH$$

(Pantothenic acid)

3'-Adenosine-5'-phosphate   Phosphopantatheine   Cysteamine

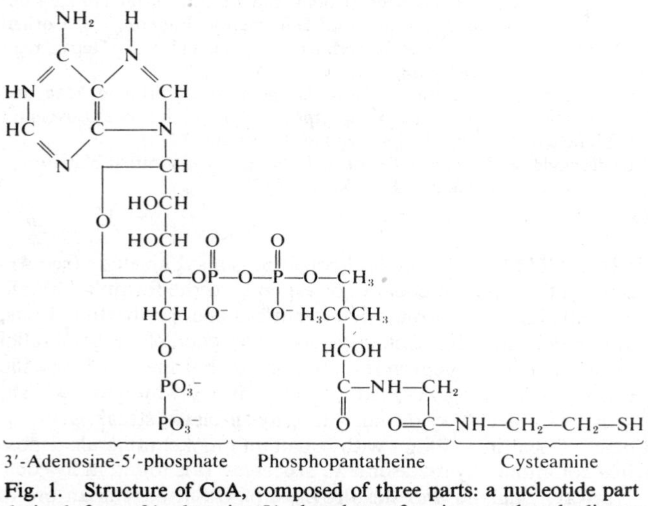

Fig. 1. Structure of CoA, composed of three parts: a nucleotide part derived from 3'-adenosine-5'-phosphate, forming a phosphodiester bond with a 4-phospho derivative of pantothenic acid, and a third part derived from the amino acid, *cysteine*. The side chain —SH group of the latter is free in this compound and is readily acylated, and thus able to act as a carrier for acyl groups in biochemical reactions in which it transfers that group between two substrates.

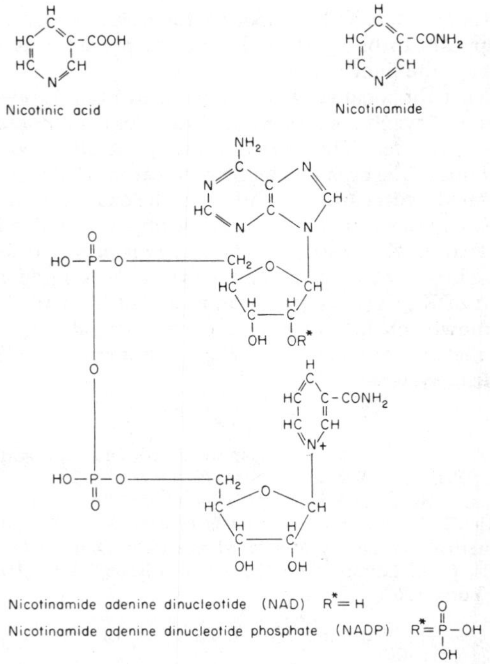

Fig. 2. Structures of nicotinic acid, nicotinamide, and nicotinamide coenzymes.

the production of steroid hormones are also coenzyme A dependent.

Nicotinic Acid Coenzymes. Nicotinic acid can be converted to nicotinamide in the body and, in this form, is found as a component of two oxidation-reduction coenzymes (Fig. 2): *nicotinamide adenine dinucleotide* (NAD); and *nicotinamide adenine dinucleotide phosphate* (NADP). The nicotinamide portion of the coenzyme transfers hydrogens by alternating between an oxidized quaternary nitrogen and a reduced tertiary nitrogen. See Fig. 3.

Enzymes that contain NAD or NADP are usually called *dehydrogenases*. In excess of fifty NAD-dependent enzyme systems are known to exist. They participate in many biochemical reactions of lipid, carbohydrate, and protein metabolism. An example of an NAD-requiring enzyme is lactic dehydrogenase, which catalyzes the conversion of lactic acid to pyruvic acid. NADP is an essential coenzyme for glucose-6-phosphate dehydrogenase which catalyzes the oxidation of glucose-6-phosphate to 6-phosphogluconic acid. This reaction initiates metabolism of glucose by a pathway other than the citric acid cycle. The alternate route is known as the phosphogluconate oxidative pathway, or the hexose monophosphate shunt. The first step is:

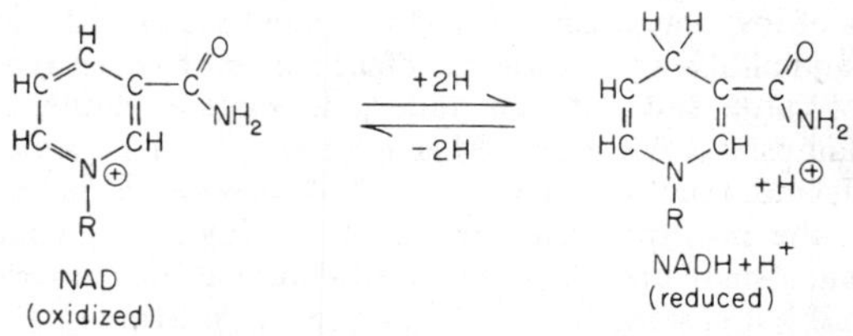

Fig. 3. Oxidized and reduced states of nicotinamide coenzymes as shown in Fig. 2.

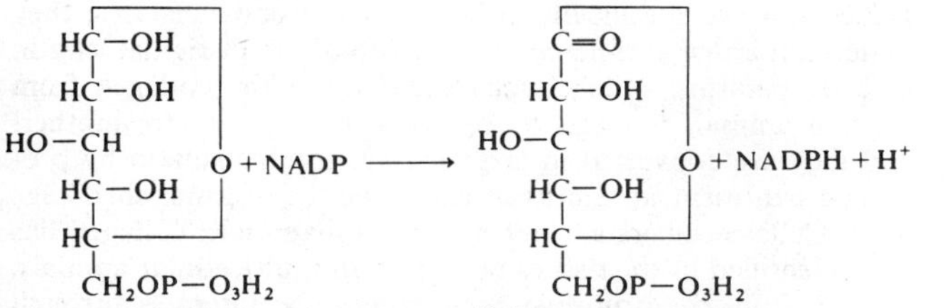

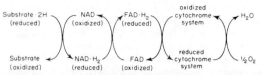

Fig. 5. Simplified representation of the biological oxidation-reduction system.

In the biological oxidation-reduction system, reduced NAD (i.e., NADH) is reoxidized to NAD by the riboflavin-containing coenzyme FAD (*flavin-adenine dinucleotide*).

Riboflavin Coenzymes. Riboflavin has been shown to be a constituent of two coenzymes: *flavin mononucleotide* (FMN) and *flavin adenine dinucleotide* (FAD). See Fig. 4. FMN was originally discovered as the coenzyme of an enzyme system that catalyzes the oxidation of the reduced nicotinamide coenzyme, NADPH, to NADP. Most of the many other riboflavin-containing enzymes contain FAD. FAD is an integral part of the biological oxidation-reduction system, where it mediates the transfer of hydrogen ions from NADH to the oxidized cytochrome system. This is illustrated in Fig. 5. FAD can also accept hydrogen ions directly from a metabolite and transfer them to either NAD, a metal ion, a heme derivative, or molecular oxygen. The various mechanisms of action of FAD are probably due to differences in the protein apoenzymes to which it is bound. The oxidized and reduced states of the flavin portion of FAD are shown in Fig. 6.

Decarboxylation Coenzymes. Thiamine, biotin and pyridoxine (vitamin B) coenzymes are grouped together because they catalyze similar phenomena, i.e., the removal of a carboxyl group,—COOH, from a metabolite. However, each requires different specific circumstances. Thiamine coenzyme decarboxylates only alpha-keto acids, is frequently accompanied by dehydrogenation, and is mainly associated with carbohydrate metabolism. Biotin enzymes do not require the alpha-keto configuration, are readily reversible, and are concerned primarily with lipid metabolism. Pyridoxine coenzymes perform nonoxidative decarboxylation and are closely allied with amino acid metabolism.

Folic Acid Coenzymes. The coenzyme forms of folic acid are derivatives of tetrahydrofolic acid, FH_4. See Fig. 7. Folic acid functions as a coenzyme in enzyme reactions which involve the transfer of one-carbon fragments at various levels of oxidation. Vitamin B_2 (*cobalamin*) may be interrelated with folic acid in these reactions. Folic acid and vitamin B_{12} are also considered together since certain clinical anemias can be corrected by administration of either of the two vitamins.

Coenzyme Q. A series of quinones which are widely distributed in animals, plants, and microorganisms, these quinones have been shown to function in biological electron transport systems which are responsible for energy conversion with living cells. The nature and significance of coenzyme Q was first recognized in 1957. In structure, the coenzyme Q group closely resembles the members of the vitamin K group and the tocopherylquinones, which are derived from tocopherols (vitamin E), in that they all possess a quinone ring attached to a long hydrocarbon tail. The quinones of the coenzyme Q series which are found in various biological species differ only slightly in chemical struc-

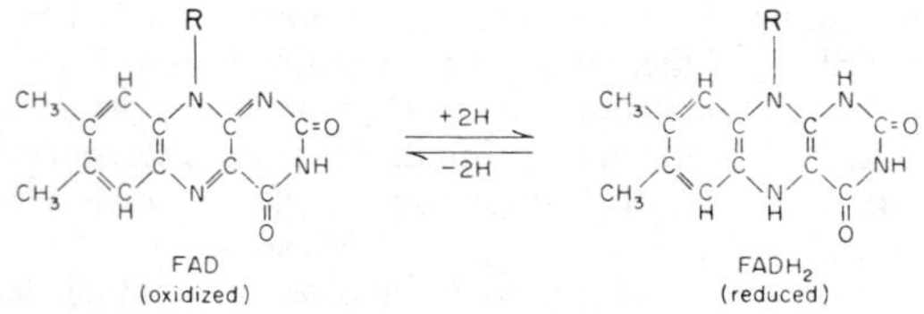

Fig. 6. Oxidized and reduced states of flavin coenzymes. R represents the remainder of the coenzyme as given in Fig. 4.

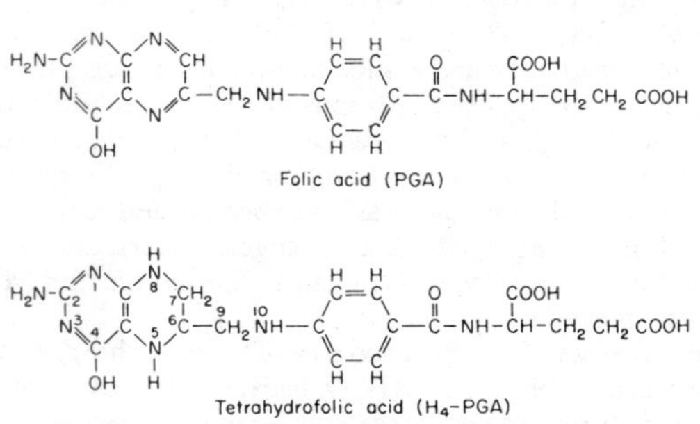

Fig. 7. Structures of folic acid and tetrahydrofolic acid.

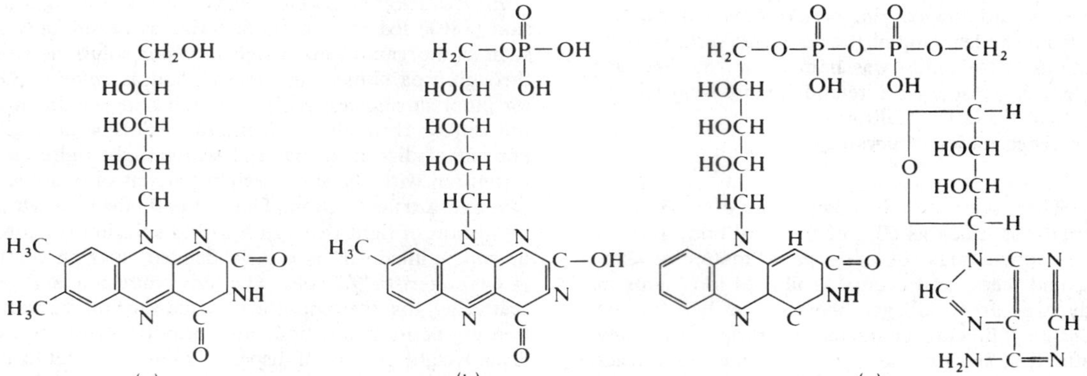

Fig. 4. (a) Riboflavin; (b) flavin mononucleotide (FMN); (c) flavin-adenine dinucleotide (FAD).

ture and form a group of related, 2,3-dimethoxy-5-methyl-benzoquinones with a polyisoprenoid side chain in the 6-position which varies in length from 30 to 50 carbon atoms. Since each isoprenoid unit in the chain contains five carbon atoms, the number of isoprenoid units in the side chain varies from 6 to 10. The different members of the group have been designated by a subscript following the Q to denote the number of isoprenoid units in the side chain, as in coenzyme Q_{10}. The members of the group known to occur naturally are Q_6 through Q_{10}.

Coenzyme Q functions as an agent for carrying out oxidation and reduction within cells. Its primary site of function is in the terminal electron transport system where it acts as an electron or hydrogen carrier between the flavoproteins (which catalyze the oxidation of succinate and reduced pyridine nucleotides) and the cytochromes. This process is carried out in the mitochondria of cells of higher organisms. Certain bacteria and other lower organisms do not contain any coenzyme Q. It has been shown that many of these organisms contain vitamin K_2 instead and that this quinone functions in electron transport in much the same way as coenzyme Q. Similarly, plant chloroplasts do not contain coenzyme Q, but do contain *plastoquinones* which are structurally related to coenzyme Q. Plastoquinone functions in the electron transport processes involved in photosynthesis. In some organisms, coenzyme Q is present together with other quinones, such as vitamin K, tocopherylquinones, and plastoquinones; and each type of quinone can carry out different parts of the electron transport functions.

COLCHICINE. Alkaloids.

COLEMANITE.
The mineral colemanite is a borate of calcium corresponding to a formula which is perhaps best represented as $Ca_2B_6O_{11} \cdot 5H_2O$. It occurs either as massive deposits or in monoclinic crystals. It has a subconchoidal fracture; hardness, 4–4.5; specific gravity, 2.42; vitreous to adamantine luster, may be colorless to milky white, grayish or yellowish; transparent to translucent. Colemanite was found originally in Death Valley, Inyo County, California, and has since been found rather widely distributed in San Bernardino, Los Angeles, Kern and Ventura Counties, California, as well as in Clark, Esmeralda and Mineral Counties in Nevada.

Colemanite was, until the discovery of kernite, the chief source of borax. Kernite, $Na_2B_4O_7 \cdot 4H_2O$, because of its easy solubility in water, has displaced very largely other boron-bearing minerals as a source of borax. Colemanite, kernite and inyoite (probably $Ca_2B_6O_{11} \cdot 5H_2O$) are lake deposits associated with other and rarer boron minerals, laid down during periods of volcanic activity or resulting from the leaching of the adjacent Tertiary sedimentary formations. Colemanite was named for Mr. William T. Coleman of San Francisco; Kernite and Inyoite were named from Kern and Inyo Counties, California.

See also terms listed under **Mineralogy.**

COLLAGEN.
The major protein component of connective tissue. In mammals, as much as 60% of the total body protein is collagen. It comprises most of the organic matter of skin, tendons, bones, and teeth, and occurs as fibrous inclusions in most other body structures. Collagen fibers are easily identified on the basis of the following characteristic properties: They are quite inelastic; they swell markedly when immersed in acid, alkali, or concentrated solutions of certain neutral salts and nonelectrolytes; they are quite resistant to most proteolytic enzymes, but are specifically attacked by the collangenases; they undergo thermal shrinkage to a fraction of their original length, at a temperature which is characteristic of the collagen from a given animal, but this varies from one species to another; and they are converted in large part to soluble gelatin by prolonged treatment at temperatures above the thermal shrinkage level. Collagen fibers are not unique to mammals; collagen has been identified in the tissues of almost all multicellular animals, ranging from the primitive porifera and coelenterates through the annelids and echinoderms, and up to the vertebrates.

As a protein, collagen is unusual in both chemistry and structure. Nearly one-third of its residues are glycine, and an additional 20–25% are imino acids (proline and hydroxyproline). In terms of sequence, glycine occurs regularly in essentially every third position, following as a steric requirement of the secondary-tertiary structure.

It appears that specific side-chain interactions between polar residues on adjacent collagen macromolecules are largely responsible for ordering the macromolecules into fibers. Specific cooperative interactions between functional groups on appropriately oriented macromolecules seem to be involved in the heterogeneous nucleation of hydroxyapatite crystals, and thus the initiation and control of mineralization in bones and teeth. As collagen fibers age, *in vivo*, they seem to become progressively more intermolecularly cross-linked, perhaps by the "esterlike" bonds formed. Little or no soluble collagen can be extracted from most mature connective tissue because of this extensive cross-linking, although the material can be converted into soluble gelatin by drastic thermal treatment.

Collagen and gelatin are of commercial importance. As insoluble collagen, this material may be cross-linked further by tanning and thus converted to leather. The soluble gelatins are used in the manufacture of foodstuffs, film emulsions, and glue.

COLLOID SYSTEMS.
Colloids are usually defined as disperse systems with at least one characteristic dimension in the range $10^{-7}-10^{-4}$ centimeter. Examples include *sols* (dispersions of solid in liquid); *emulsions* (dispersion of liquids in liquids); *aerosols* (dispersions of liquids or solids in gases); *foams* (dispersion of gases in liquids or solids); and *gels* (systems, such as common jelly, in which one component provides a sufficient structural framework for rigidity and other components fill the space between the structural units or spaces). All forms of colloid systems are encountered in nature. Products of a colloidal nature are commonly found in industry and are notably extensive in the food field. Foams, widely used in industrial products, but also the causes of processing problems are described in the article on **Foam.**

Early Background. Thomas Graham's investigations of diffusion (1861) led him to characterize as *crystalloids* substances, such as inorganic salts which in water solutions would diffuse through a parchment membrane; and as *colloids* (Greek word for glue) substances, such as starch and gelatin, which would not diffuse through the membrane. Sols with a given weight percent of dispersed material scatter light more strongly than a solution with the same weight percent of dissolved inorganic salt, i.e., a true solution. The Tyndall effect, in which the path of a beam of light through a turbid solution (or through dusty or smoke-filled air) is clearly defined through scattered light, is characteristic of sols. The slow diffusion and strong light scattering, together with the fact that the boiling-point elevation, freezing-point depression, and osmotic pressure caused by a given weight percent of dispersed material in sol form are much less than the corresponding magnitudes caused by the same weight percent of common inorganic salts—all of these observations indicated to early investigators that the particles dispersed in the sol must be larger than those resulting from dissolving inorganic salts in water.

Development of the ultramicroscope (Siedentopf and Zsigmondy, 1903) permitted particles substantially smaller than the wavelength of light to be observed in scattered light and were thus capable of counting. Invention of the ultracentrifuge by Svedberg (1924) made it possible to cause particles in sols to sediment at observable rates, to measures these rates with reasonable precision, and to infer particle sizes from these rate measurements. The ultramicroscope and ultracentrifuge permitted validation of the early conclusions that colloidal particles are much larger than ions resulting from dissolving metal salts in water.

Svedberg found that in some sols the particles were highly uniform in size. For example, he found that the gram particle weight of insulin (a protein) was 40,900 and that apparently all insulin particles had this gram particle weight. This made it extremely likely that the insulin particles were either single molecules (albeit giant ones), or aggregates of a very definite number of smaller (but still quite large molecules by ordinary standards) molecules. The research of Staudinger (commencing about 1920) and of Carothers (1929) opened up the field of macromolecular chemistry, leading to the recognition that giant molecules were not only abundant in nature, but could be prepared by established principles of chemistry. See also **Macromolecule.**

It is interesting to note that Wolfgang Ostwald, in the late 1800s, stated, "There are no sharp differences between mechanical suspensions, colloidal solutions, and molecular (true) solutions. There is a gradual and continuous transition from the first through the second to the third."

Some colloidal systems are *thixotropic*, i.e., they differ in their fluid behavior from *pseudoplastic* substances in that the flow rate increases with increasing duration of agitation as well as with increased shear stress. When agitation is stopped, internal shear stress exhibits hysteresis. Upon reagitation, generally less force is required to create a given flow than is required for the first agitation. Examples of thixotropic materials include silica gel, most paints, glue, molasses, lard, fruit juice concentrates, and asphalts. By rhythmically shaking or tapping certain thixotropic suspensions, the suspensions will "set" or build up very rapidly. This type of non-Newtonian substance is said to be *rheopectic*. Bentonite sols and suspensions of gypsum in water are rheopectic. *Dilatant fluids* often are termed *inverted plastics* or inverted pseudoplastics. Initial flow under a low shear stress is at a high rate; further increases in shear stress, however, result in lower flow rate. Some liquids may change from thixotropic to dilatant or vice versa as the temperature or concentration changes. Examples of dilatant materials include quicksand, peanut (groundnut) butter, and many candy compounds. For comparison, it should be recalled that a Newtonian substance is a liquid or suspension which, when subjected to a shear stress, undergoes deformation wherein the ratio of shear rate (flow) to shear stress (force) is constant. These varying behavioral patterns of colloidal materials become important considerations in specifying pumps and other process handling equipment. See also **Gold Number.**

Sols. It is convenient to classify sols into three types: (1) *lyophilic* (solvent-loving) colloids. Examples are solutions of gelatin or starch in water. (2) *Association* colloids, of which a solution of soap in water at moderate concentration is an example. (3) *Lyophobic* (solvent-repelling) colloids. An example is sulfur in water. Both lyophilic and association colloids can be prepared in thermodynamic equilibrium, so that when solvent

is removed and then returned to the system, the original properties of the system are regained.

Lyophobic colloids are not (or at most, rarely) an equilibrium system. When solvent is removed and then returned to the system, the original dispersed material fails to redisperse, and it is usually convenient to regard such a system as one in which the dispersed particles are continuously aggregating. A lyophobic sol thus appears to be stable if the aggregation rate is slow; and unstable if it is fast. The terms lyophilic and lyophobic entered the literature before the characteristics of these systems were well understood and thus are somewhat anachronistic, but nonetheless well-established.

Lyophilic sols are true solutions of large molecules in a solvent. Solutions of starch, proteins, or polyvinyl alcohol in water are representative of numerous examples. Properties of these solutions at equilibrium (for example, density and viscosity) are regular functions of concentration and temperature, independent of the method of preparation. The solvent-macromolecule compound system may consist of more than one phase, each phase in general containing both components. Thus, if a solid polymer is added to a solvent in an amount exceeding the solubility limit, the system will consist of a liquid phase (solvent with dissolved polymer) and a solid phase (polymer swollen with solvent), i.e., a polymer with dissolved solvent).

The foregoing characteristics also are found with solutions of small molecules. But properties of solutions, one of whose components is macromolecular, differ in quite understandable ways from those of solutions having only small molecular components. For example, where small molecules are involved, molecular distortion is minor. Quite generally, the shapes of small molecules are little affected by environment unless the small molecules react chemically. In contrast with small molecules, there is a considerable variation in polymer conformation with environment. See also **Molecule;** and **Polymer.**

A polymer dissolved in a good solvent will tend to stretch out, and the resulting entanglement of polymer chains and interference with solvent movement will lead to a high viscosity. If the solvent is a poor one, the polymer molecule will tend to form a small ball, and the viscosity for a given weight percent will be much less. A side group of a polymer may be ionizable. Ionization of this group distributes a charge along the backbone and charge repulsion causes the macromolecule to tend toward a rod shape. If there is a moderate salt concentration in the solution, the backbone charge will be partly shielded by ions of opposite charge from the salt. Thus, the tendency toward rod formation will be less pronounced. The tendency of oppositely charged macromolecules to aggregate is much greater than the tendency of oppositely charged small ions to pair simply because the charges involved are greater in the former case.

The foregoing special properties of solutions of large molecules are relatively easy to describe in qualitative terms, but a difference of a more subtle nature occurs when the system forms two liquid phases. In a macromolecular solution, both phases tend to be rich in the (small molecule) solvent, whereas in systems formed from two molecules of comparable size, one phase is rich in one component, while the other phase is rich in the second component. The formation of two liquids phases from a solvent-macromolecule system is sometimes called *coacervation*; and the phase with the higher percentage of macromolecule is sometimes called the *coacervate*. See also **Coacervation.**

Association Colloids. These are generally encountered in solutions of soaps and detergents in water. These matters become important, of course, to procedures for cleaning and sterilizing equipment (food processors, biochemical manufacturers, hospitals, etc.), but the principles also apply to other association colloids also encountered industrially, notably in the food processing field. A typical soap, such as sodium stearate, $C_{17}H_{35}COONa$, or a detergent, such as sodium dodecyl benzene sulfonate, $C_{12}H_{25}C_6H_4 \cdot SO_3Na$, consists of a long hydrocarbon tail and a polar (in the examples cited, ionizable) head group. The solubility of the soap in water is largely conferred by the head group. As the soap concentration is increased, the soap molecules tend to cluster in aggregates called *micelles*, with hydrocarbon tails in the interior of the micelles and the polar groups in contact with water. The formation of micelles is favored by the interaction between hydrocarbon tails and is opposed by charge repulsion of the polar group which are placed close together at the micelle surface. See also **Surfactants.**

Micelle formation becomes pronounced at soap concentrations exceeding the critical micelle concentration. As hydrocarbon tail length is increased, the interaction of tails is increased, and as salt concentration is increased, the repulsion of head groups is reduced because their charges are partly shielded by ions of the salt. Both of these factors favor micelle formation, causing micelles to be larger and the critical micelle concentration to be smaller. Typically, a micelle might contain about 50 soap molecules. The micelle interior is a hydrocarbon, and as such is receptive to other molecules soluble in hydrocarbons. Hence, a soap solution can 'dissolve' such molecules (taking them up in micelle interiors) even if the molecules are quite insoluble in water. This phenomenon is called *solubilization* and is a factor in detergency.

Lyophobic Sols and Aerosols. These products can be viewed most simply and, in most cases, with sufficient accuracy as two-phase systems in which the dispersed particles are steadily and irreversibly aggregating according to a second-order rate law. Thus, where C is the number of particles per cubic centimeter (an aggregate of many primary particles being counted as one particle) at time t, and where C_0 is the number of particles per cubic centimeter at zero time, and K is a constant, C depends on t according to

$$\frac{C_0}{C} - 1 = KC_0t$$

and will be one-half its value at zero time when $KC_0t = 1$. The time required for this is longer, the smaller K and the smaller C_0. If the time required is weeks, the sol will appear quite stable over a period of days. If there is no barrier to aggregation so that the particles aggregate as fast as diffusion brings them in contact, the rate constant K can be calculated approximately from diffusion theory and is $8kT/3\eta$, where k is Boltzmann's constant, T is the absolute temperature, and η the viscosity of the medium. Initial sol concentration (particles per cubic centimeter) giving one-minute half-lives at room temperature are 1.4×10^9 in water; and 2.7×10^7 in air in the absence of aggregation barriers.

Although the number may appear large, they correspond to quite small volume percentages of dispersed particles. A particle of radius 5×10^{-5} centimeter is at the upper limit of the colloidal range; and 1.4×10^9 such particles occupy 0.07% of space. The behavior of smokes, fogs, and many dispersions of uncharged particles in water accords well with the rate equation and theoretical rate constant given. In contrast, the dispersed particles in many sols are electrically charged, manifesting this charge through electrophoresis (motion of colloidal particles under the influence of an electric field). In fact, Tiselius developed electrophoresis to a high degree, successfully fractionating and classifying proteins thereby. Evidently, like charges on two colloidal particles will contribute to a repulsion between them, which will be greater, the greater the charge on each particle

and the smaller the concentration of salts in solution (since ions from the salt will tend to mask the charges on the particle).

The theory of interaction between colloidal particles with a surface electrostatic potential (due to surface charges) surrounded by an electrical double layer (one layer of which is the layer of surface charges; the other a diffuse cloud of charges of opposite sign due to ions from salts in the solution) was developed by Derjaguin and Landau and independently by Verwey and Overbeek, and is generally known as the DLVO theory after the first letters in the names of these scientists. The DLVO theory shows how a barrier sufficient to reduce the rate constant K (and so to increase the half-life at a given initial concentration) by many powers of 10 may arise from the interaction of charged particles in a solvent, and the magnitudes calculated agree rather well with experiment.

It is evident why the properties of lyophobic sols depend so critically on the chemistry of the interface between dispersed particle and solvent, for this chemistry establishes the means by which the surface charge can be established or altered. Particles of silver halide dispersed in water will acquite a positive charge if silver ion is in slight excess in the water, because the silver ion can readily add to the silver halide lattice. A negative charge is similarly acquired if the halide ion is in slight excess. The silver ions and halide ions are called *potential-determining ions* for the silver halide sol. They can lose their waters of hydration and adsorb on the particle side of the electrical double layer, conferring a charge on the particle. Hydrogen ion and hydroxyl ion are similarly potential-determining ions for many oxide sols, such as silica and alumina, including particles such as carbon and many metals which are ostensibly not oxides, but in fact usually have oxidized surfaces. Finally, charge can be conferred by the adsorption of charged macromolecules, such as gelatin, which are called *protective colloids*. Salts added to the sol form ions which tend to mask the particle charges and so tend to promote flocculation. The ion whose charge is opposite to the particle charge (the counter ion) is of particular importance, and the greater its charge the lower the concentration at which its flocculating effect is evident.

Emulsions. These are dispersions of one liquid in another. Most commonly, one phase is an oil which is at most slightly miscible with water. The disperse phase can either be oil (an oil-in-water emulsion) or water (a water-in-oil emulsion). For apparent stability, an emulsifying agent is almost invariably required. The emulsifying agent has an oil-soluble tail and a polar head. The emulsifying agent concentrates (adsorbs) at the interface between oil and water, lowering the interfacial tension and frequently conferring a charge on the dispersed droplets. The film of emulsificant thus formed is usually only one molecule thick, but it is essential to emulsification. Mixed emulsificants, such as a mixture of sodium stearate and octadecyl alcohol, may be more effective in emulsification than either component alone, and there is a great deal of art and experience required in the formulation of emulsions. Lecithins and some proteins are effective natural emulsificants, and a mixture of lecithin and cholesterol is an effective natural mixed emulsificant. It should be noted that the difference between solubilization (described in connection with association colloids) and emulsification is not sharp, particularly insofar as large, extensively swollen micelles and ultrafine emulsions are concerned.

Gels. These substances involve the formation of a three-dimensional structure. A gel is a colloidal disperse system in which is contained a dispersed component and a dispersion medium, both extending continuously throughout the system. Further, the system has equilibrium-elastic (time-dependent) deformation. Thus, since they have a shear modulus of rigidity, gels are like solids, but in most other physical respects they behave like liquids. It is conceived that the three-dimensional network is kept together by bonds or junction points which essentially have an unlimited lifetime. Junction points may be described as primary valence bonds, attractive forces of long range, or secondary valence bonds which maintain an association between parts of polymer chains or form submicroscopic crystalline regions. A gel may be defined as a flocculant and gelatinous precipitate. A jelly is a transparent elastic mass. Upon standing, a gel may shrink—a process known as *syneresis*.

In 1861, Thomas Graham first used the term syneresis to describe the phenomenon of exuding small quantities of liquid by gels. By definition, syneresis is the spontaneous separation of an initially homogeneous colloid system into two phases—a coherent gel and a liquid. The liquid is actually a dilute solution whose composition depends upon the original gel. When the liquid appears, the gel contracts, but there is no net volume change. Syneresis is reversible if the colloid particles do not become too coagulated immediately after their formation.

In 1937, Heller classified three types of syneresis as to cause: (1) syneresis of desorption, caused by the particle becoming less hydrophilic with time; (2) syneresis of aggregation, whereby discrete gel particles may unite into a denser gel portion; and (3) syneresis of contraction, where a gel with fibrillar structure contracts and squeezes out the intermicellar liquid. Most commonly, syneresis is the visible manifestation of further slow coagulation which follows the initial setting of the gel, the gel-forming process itself being an enmeshing of the hydrous particles into a network. It may be further explained as the exudation of liquid held by capillary forces between the heavily hydrated particles constituting the framework of the gel. Ostwald noted that the phenomenon is one of the most characteristic of the properties of gels.

A common example of syneresis is found when a mold of gelatin remains under refrigeration for a period. A general shrinkage of the body of the gel occurs and a liquid collects around the edge of the mold. The liquid is a dilute solution of the original composition. Since the total volume of the system remains the same, syneresis should not be considered simply as the opposite of imbibition (absorption). Extending the onset of syneresis in various products, notably foods, is of obvious importance.

Dispersions Processes. 1. The simplest method of accomplishing dispersion is by grinding the solid (or liquid) material with the liquid medium until particles of the required size are ultimately obtained. The colloid mill (Plauson, 1921) is used for such purpose, as in mixing paints and pastes, regenerating milk from milk powder, dispersing cellulose in sodium hydroxide and carbon disulfide for the production of xanthates for viscose, and in emulsifying fats and waxes. See accompanying figure.

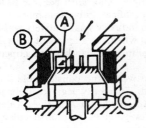

One type of colloid mill. Rotor blades *A* break up slurry. Serrations in rotor and stator provide mechanical shear and force material into adjustable gap (0.0005–0.125 inch; 0.013–3.2 millimeters) between rotor and stator *B* for intense hydraulic shear. Lower part of rotor *C* adds further whirling action.

2. Zinc sulfide, cupric hexacyanoferrate(II), stannic acid, silver chloride are examples of precipitates which, when washed on the filter paper until the accompanying soluble electrolyte has been removed, form colloidal solutions and pass through the pores of the paper. Since this is usually to be avoided in practice, the washing is then done with an electrolyte which does not conflict with the treatment to follow. Frequently ammonium nitrate solution is used. 3. A peptizing agent is frequently employed. Tannin is peptized by water, and by glacial acetic acid. Soaps are peptized by water. Gelatin swells in cold water but is not peptized, but is peptized in warm water. Starch, although insoluble in cold water, behaves similarly to gelatin with warm water (63 to 74°C, depending upon the kind of starch). Cellulose nitrate swells in ethyl alcohol and not in ether, but is peptized in ethyl alcohol-ether mixture. Clay is peptized by ammonium hydroxide, and it is held by some that the action of sodium hydroxide on zinc, aluminum, and chromium hydroxides is one of peptization. 4. Water-peptizable colloidal substances such as gelatin, dextrin, gum arabic, and soap peptize many precipitates, and are often called protective colloids. Gelatin in the solution prevents the precipitation of silver dichromate upon mixing silver nitrate and potassium dichromate solutions. (See Condensation Processes, below.) 5. When dilute silver nitrate and dilute potassium bromide solutions are mixed so that there is a slight excess of either solution, silver bromide is peptized. Acheson's oil-dag and aqua-dag are suspensoids of graphite in oil or water containing a protective colloid, tannin. Oil-dag contains about 15% of a "deflocculated graphite," and is used in dilute solution in lubricating oil (about 0.1% graphite). Bearings gradually become coated with a thin layer of graphite.

Sonic methods also are used to create emulsions. Liquids are pumped under pressure through an orifice of special design and impinge on the edge of a blade causing it to vibrate at ultrasonic frequencies. Cavitation takes place continuously in the stream, causing violet pressure changes to be generated locally. The result is a uniform and stable emulsion and a dispersion of a very high order.

Condensation Processes. 1. When a solution of ferric chloride is poured into a relatively large volume of boiling water, colloidal ferric hydroxide is formed. The ferric hydroxide sol does not react with hydrogen sulfide nor with potassium hexacyanoferrate(II), and like all colloidal substances does not pass readily through animal membranes or parchment. 2. When hydrogen sulfide is passed into a solution of arsenious oxide, arsenious sulfide sol is formed which in the absence of an electrolyte may be made of the high concentration of 60 grams of arsenious sulfide per 100 grams of water. Upon addition of hydrochloric acid, arsenious sulfide coagulates and is precipitated. 3. When hydrochloric acid is added to sodium silicate solution either silicic acid sol or silicic acid gel is formed. 4. When hydrogen sulfide solution is treated with an oxidizing agent, for example, the proper concentration of nitric acid, sulfur sol is formed. 5. When gold chloride very dilute solution (0.01 to 0.001% of gold chloride) is made slightly alkaline (say by the addition of magnesium oxide) and then treated with a reducing agent, for example, formaldehyde or sodium hydrosulfite $Na_2S_2O_4$, red gold sol is formed. 6. Use of a protective colloid in solution prevents the formation of the ordinary and expected precipitate in many cases and causes the formation of the expected substance as colloidal sol. Silver nitrate (0.6 gram per liter) and potassium dichromate (0.5 gram per liter) to one of which is added 0.1 volume of hot gelatin solution (2 grams per 100 milliliters of water) are mixed with stirring silver dichromate sol is formed. 7. When an electric arc is formed under water between two metallic rods, particles of the metal of colloidal size are formed along with more or less separation of free metal. A protective colloid increases the stability. If the metal vaporizes and then condenses to the colloidal state this is strictly speaking a condensation process, if otherwise, a dispersion process.

The disappearance of the colloidal state of a substance may be accomplished in either of two directions, namely, by the colloid passing into solution or into suspension. Practically, the latter is the more important method. Coagulation, agglomeration or precipitation is readily brought about by discharge of the electric charge on the particles. Ions carrying a charge of opposite sign to that carried by the colloidal particles are active precipitants, and the higher the valency of the ion the more effective (Linder-Picton-Hardy). When the colloidal particles are made neutral the conditions are least favorable to their stability. For colloidal arsenious sulfide, which is negatively charged in water, the coagulating power of potassium iodide K^+I, calcium chloride $Ca^{2+}Cl_2$, aluminum chloride $Al^{3+}Cl_3$ is in the ratio of $1:80:1500$ (Svedberg); and for colloidal ferric hydroxide, which is positively charged in water, the coagulating power of potassium chloride KCl^-, potassium sulfate $K_2SO_4^-$ is in the ratio of $1:45$. The active ion is carried down with precipitated particles. Oppositely charged colloids, e.g., arsenious sulfide and ferric hydroxide, when mixed, precipitate each other. Other methods of coagulation are by migration of colloidal particles to and their discharge at electrodes, and by heating, as in the case of egg albumin. Coagulation is usually irreversible, especially when caused by electrolytes.

An interesting case, operating on a large scale in nature, of the precipitation of a colloidal system by an electrolyte is that of the action of sea water on the mud and silt of river water entering the ocean. When river water flows into the ocean the former, on account of its lower specific gravity, tends to flow over the latter and spread out in widening range. As the current diminishes some of the suspended mud and silt settles out, but the finer colloidal particles are coagulated by the electrolyte of the sea water and form deltas at the mouths of rivers.

Importance of Colloidal State. All living matter, whether animal or plant, is made up of many colloidal materials and is largely sustained by colloidal processes. Of similar importance is colloidal chemistry in everyday living, in almost all of our foods, such as proteins and starches, in our clothing, whether of natural or synthetic origin, and in our shelter materials, such as wood, bricks, concrete. When there is added to these, other common things and operations of everyday life, such as pottery and porcelain, paper, rubber and leather, and cooking and washing, where colloidal matter and processes operate, it is evident how broad is the scope and how great is the importance of the field. To these there must also be added other applications in the realm of industry, such as dyeing, printing, photography, water purification, smoke prevention, ore flotation, sewage disposal and soil preparation, paints, varnishes and lacquers, plastics, adhesives, and innumerable other operations and materials.

References

Andres, C.: "Antifoaming Agent Increases Fermentation Capacity 20%," *Food Processing*, **38**, 5, 58–59 (1977).
Graham, H. D.: "Food Colloids," AVI, Westport, Connecticut, 1977.
Matijevic, E.: "Surface and Colloid Science," Vols. 1–5, Wiley, New York, 1969–1972.
Somorjai, G. A.: "Surface Science," *Science*, **201**, 489–497 (1978).
Staff: "Food Chemicals Codex," 2nd edition, National Academy of Sciences, Washington, D.C., 1972 (with numerous subsequent supplements).
Weiss, T. J.: "Food Oils and Their Uses," AVI, Westport, Connecticut, 1970.

COLORANTS (Foods). Color, as a component of appearance, is important in the sensory evaluation of a food substance, including beverages. Color affects the degree of acceptability of a food product in the marketplace. Color also is a frequently useful indicator of the degree of wholesomeness of a foodstuff. Many foodstuffs are attractively colored as the result of their naturally occurring pigments. In these instances, a major objective of the food grower and processor is to protect and preserve the natural colors as long as may be required by the distribution network, considering such factors as temperature and humidity to which the food substance may be subjected before reaching the consumer.

Considering the expectations of consumers, particularly in countries that have an advanced food technology, colorants, along with flavorings and texture modifiers, are important. These factors are particularly stressed in connection with modern fabricated foods, substitute foods, and food analogues. Many years have passed since final approval was given to margarine producers to use colorants and artificial flavorings. Even though artificial flavors are used in some products, ice creams and ices, as well as soft drinks, are color matched to their fruit flavors. Yellow colorings provide a note of richness in cake mixes, eggnogs, and other products associated with their content of eggs; cheese snacks are both colored and flavored to simulate cheese; popcorn oil is usually colored; iron oxide can be added to pet foods to simulate the color of meat; the red coloration of cherries is intensified by using red colorants in the processing of maraschino cherries. Caramel colorings are widely used in both alcoholic and nonalcoholic beverages.

Particularly during the past couple of decades, regulatory agencies in various countries have scrutinized colorants (along with other food additives) against a backdrop of consumer health. Since the early 1900s, several countries have approved colorants for use in foods only after thorough physiological testing. In recent years, analytical instrumentation and research methodologies have become sophisicated and measurements of minute quantities are now practical. With improved analytical tools and a heightened awareness of the effects of foods upon health, a number of colorants that once were considered perfectly safe have come under question. In some countries, most or all synthetic colorants have been banned. Generally, the limitations on the use of colorants have become much more stringent.

In the United States, the first rather complete legislation involving such matters was the Food and Drug Act of 1906. As a result of that legislation, the list of colorants permitted was reduced to only 7 dyes. Because the remaining seven dyes did not provide sufficient flexibility in the formulation of food products, considerable research went forth to find additional colors, not only with more desirable hues, but easier to use (solubility in oil/water, less temperature sensitivity, etc.). During the 66-year period 1906–1971, several additional colors were added. In 1938, the Food, Drug, and Cosmetic Act was passed. The common names of dyes previously used were given color prefixes and numbers. For example, Amaranth became FD&C Red No. 2. Under the new act, certification became mandatory.

The color situation in the food industry was relatively without incident until the early 1950s, when, as the result of a few cases of excessive levels of usage in some candies and popcorn, two colors (FD&C Orange No. 1 and FD&C Red No. 32) were delisted. After much controversy and considerable litigation, more colors were delisted. To rectify legal complexities and unworkability of the 1938 Act, the Color Additives Amendments of 1960 were passed. Nevertheless, as pharmacological studies continued, a number of other colors were delisted during the interim. An excellent summary of the situation up to the early 1970s is contained in the Noonan reference listed. In 1973, Violet No. 1 was delisted; in 1976, Red No. 2 and Red No. 4 were delisted. As of the early 1980s, Red No. 40 is under serious scrutiny.

Current reports are available from the U.S. Food and Drug Administration. The publication "Food Colors" is released periodically by the National Academy of Sciences, Washington, D.C. Reports are issued by the Cosmetic, Toiletries, and Fragrance Association; the Pharmaceutical Manufacturers' Association; the United Kingdom Department of Health; and the Food and Agriculture Organization (United Nations)—all organizations that update information as regards color additive regulations.

Natural Colorants

The use of naturally-derived colorants for foods dates back many centuries. The principal concerns expressed in the foregoing paragraphs relate largely to the use of synthetic colorants. Natural colorants include the anthocyanins, annatto colors, the betalaines, the carotenoids, cochineal, saffron, turmeric, and titanium dioxide. Caramel coloring also falls into this category. Paprika is used in some foods for its coloring attributes.

Anthocyanins are water-soluble pigments which account for many of the red, pink, purple, and blue colors found in higher plants. Most plants contain more than one of these pigments and they occur most prevalently as glycosides. Several hundred different anthocyanins are known. These compounds are most atable at a pH range of 1–4, thus limiting the spectrum of usage. As compared with synthetic colorants, the anthocyanins produced to date generally are less stable, have less tinctorial potency, and lack some color uniformity. They are degraded by light, heat, enzymes, and interact with ascorbic acid. They also tend to form complexes with metal ions to produce off-colors. Their main advantage stems from the fact that they are naturally derived and thus not regarded with suspicion as health deterrents, as are many synthetic materials. On the other hand, attempts to alter and modify anthocyanins to make up for their fundamental disadvantages could also move them toward a suspicious category—a factor which researchers on anthocyanins are taking into consideration.

Numerous commercial sources for the anthocyanins have been and are continuing to be investigated. These sources include grape anthocyanins, apparently with good potential in the carbonated beverage field. Red cabbage also has been seriously considered. Roselle plants, native to the West Indies, have been studied and appear to have potential for use in apple and pectin jellies, but not for carbonated beverages, such as ginger ale. Cranberry pomace and blueberries have been investigated. Considerable interest has been shown in the red anthocyanin pigments of miracle fruit (*Synsepalum dulcificum*, Schum), a tropical plant that produces a red berry.

Betalaines are sometimes referred to as beetroot pigments. They are made up of two main groups: (1) *Betacyanins*, the principle component of which is betanin; and (2) *betaxanthins*, the principal component of which is vulgaxanthin-I. The betacyanins contribute a red color, whereas the betaxanthins are yellowish, Another yellow pigment, betalamic acid derives directly from cleavage of betanin and is probably the key intermediate in the biogenesis of all betalaines.

A factor of concern in connection with the betalaines is the earthy flavor associated with beets. To date, beets have been regarded as the primary source for these substances.

Carotenoids. These yellow-orange colorants are described in the entry on **Carotenoids.**

The other natural colorants previously mentioned have been used for many years and are familiar to nearly everyone. Annatto colors are described in a separate entry, **Annatto Food Colors.**

Synthetic Colorants

Perkin, in 1856, synthesized the first synthetic dye, *mauve* or *mauveine*, by the oxidation of crude aniline. In that time and for about 80 years, coal tar was the principal source of

SULFONATED INDIGO COLORANT
FD&C Blue No. 2

FLUORESCEIN TYPE COLORANT
FD&C Red No. 3

AZO COLORANTS
FD&C Red No. 40

FD&C Yellow No. 5
FD&C Yellow No. 6

TRIPHENYLMETHANE COLORANTS
FD&C Blue No. 1

FD&C Green No. 3

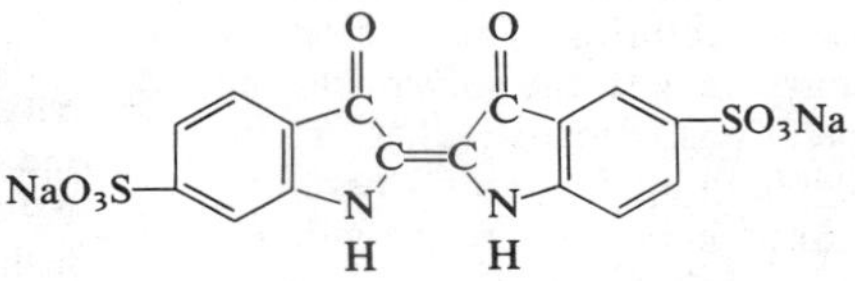

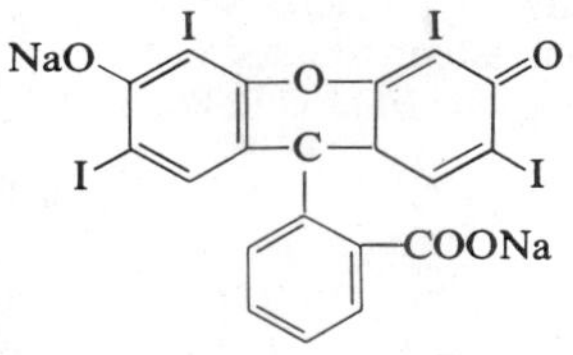

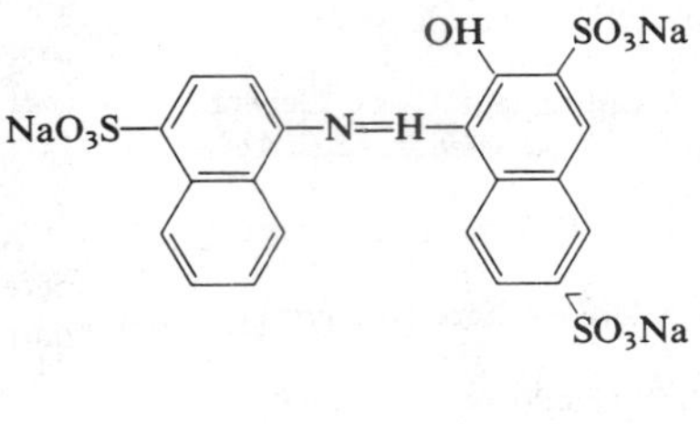

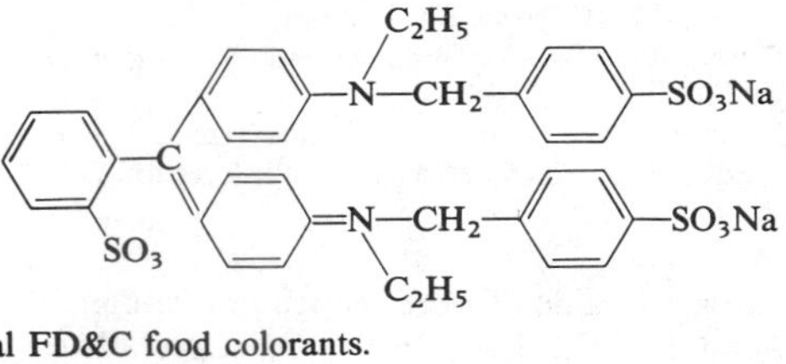

Structural formulas of several FD&C food colorants.

aromatic compounds, which, in turn, were the sources of numerous synthesized dyes, which were used primarily in textiles. Some of these dies were also found to be adapted to other uses, including the coloring of food substances. This generally gave rise to the term "coal tar color," used commonly in the food and cosmetics industries for many years. Of course, with the development of more sophisticated organic syntheses and the petrochemical field, the association with coal tar no longer had a direct meaning. The term was finally eliminated from legislation in connection with the Color Additives Amendments of 1960. It is interesting to note, however, that prior to the first Act of 1906, it is estimated that some 80 such dyes were being used in a large number of food products, at a time when there were no regulations regarding the nature and purity of colorants used in foods.

Structural formulas of synthetic colorants still listed and used as of the early 1980s are shown in the preceding diagram. (Note that the safety of Red No. 40 is under study.)

Lakes

In the United States, FD&C lakes were accepted for the approved list of certified color additives for the first time in 1959. As defined by the FDA, a lake is an "Extension on a substratum of alumina, of a salt prepared from one of the water-soluble straight colors by combining such a color with the basic radical aluminum or calcium." Because the substratum of alumina hydrate or aluminum hydroxide is insoluble, the lake provides an insoluble form of the dye, i.e., a pigment. Colors from dyes result from solution in a solvent; whereas colors from pigments result from dispersion of that pigment throughout the food substance. Prior to the acceptance of lakes, insoluble colorants were formed by absorbing them on insoluble materials, such as cellulose, flour, and starch. Generally, these forms were inadequate because of relatively low coloring power.

When utilized in solid or semisolid vehicles, dyes must be added in solution to achieve effective coloring. Thus, color migration is a problem. Dyes can migrate with the solvent during various drying or processing operations. Because lakes are insoluble, color migration is negligible in applications where distinct interfaces are required. Striped candy pieces provide an example. Where opacity is required, titanium dioxide can be added to lakes. In high-quality lakes, nearly all particles will pass through a 325-mesh screen when wet-tested. Shades of coloration can be produced by blending the various FD&C Lake Colors.

See also **Polymeric Food Additives.**

References

Clydesdale, F. M., and F. J. Francis, Co-chairpersons: "Food Colorants," Institute of Food Technologists 39th Annual Meeting, Saint Louis, Missouri (1979).
Eagerman, B. A.: "Orange Juice Color Measurement Using General Purpose Tristimulus Colorimeters," *Jrnl. of Food Science*, **43**, 2, 428–430 (1978).
Furia, T. E.: "Nonabsorbable, Polymeric Food Colors," *Food Technology*, **31**, 5, 26–33 (1977).
Kramer, A.: "Benefits and Risks of Color Additives," *Food Technology*, **32**, 8, 65–67 (1978).
Noonan, J.: "Color Additives in Food," in "Handbook of Food Additives" (T. E. Furia, Editor), CRC Press, Boca Raton, Florida (1972).
Riboh, M.: "Natural Color: What Works; What Doesn't?" *Food Engineering*, **49**, 5, 66 (1977).
Staff: "HT Lakes," Colorcon, Inc., West Point, Pennsylvania (1979).
Staff: "Food Colors," National Academy of Sciences, Washington, D.C. (Issued periodically).

COLUMBITE. A mineral oxide of iron, manganese, niobium (columbium) and tantalum, $(Fe,Mn)(Nb,Ta)_2O_6$. Crystallizes in the orthorhombic system. Hardness, 6; specific gravity, 5.20; color, red to brown.

COMBUSTION (Fuels). The rapid chemical combination of oxygen with the combustible elements of a fuel. There are three combustible chemical elements of major significance—carbon, hydrogen, and sulfur. However, as a source of heat, sulfur is of minor concern. Sulfur is of particular importance in the combustion of several fuels because of the corrosion and pollution problems which its presence creates.

Carbon and hydrogen when burned to completion with oxygen unite according to:

$$C + O_2 = CO_2$$
$$+ 14{,}100 \text{ Btu/pound (7840 Calories/kilogram) of carbon}$$

$$2H_2 + O_2 = 2H_2O$$
$$+ 61{,}100 \text{ Btu/pound (33,972 Calories/kilogram) of hydrogen}$$

Air is the usual source of oxygen for boiler furnaces. These combustion reactions are exothermic as indicated by the foregoing equations.

The objective of good combustion is to release all of the indicated heat while minimizing losses from combustion imperfections and superfluous air. The combination of the combustible elements and compounds of a fuel with all the oxygen requires *temperature* high enough to ignite the constituents, mixing or *turbulence*, and sufficient *time* for complete combustion.

Table 1 lists the chemical elements and compounds found in fuels generally used in the commercial generation of heat with their molecular weights, heats of combustion, and other combustion constants. The term, "100% total air" used in Table 1 and figures and examples which appear elsewhere in this entry means 100% of the air theoretically required for combustion without excess. Higher percentages indicate the theoretical plus excess air, e.g., 125% total air means 100% theoretical air plus 25% excess air.

Table 2 summarizes the molecular and weight relationships between fuel and oxygen and lists the heat of combustion for the substances commonly involved in combustion.

In power plant practice, the practical source of oxygen is primarily air, which includes, along with the oxygen, a mixture of nitrogen, water vapor, and small amounts of inert gases, such as argon, neon, and helium. Data on the composition of air are given in Table 3.

Water vapor is one of the products of combustion for all fuels which contain hydrogen. The heat content of a fuel depends on whether this water vapor is allowed to remain in the vapor state or condensed to liquid. In the bomb calorimeter, the products of combustion are cooled to the initial temperature and all of the water vapor formed during combustion is condensed to liquid. This gives the high, or gross, heat content of the fuel with the heat of vaporization included in the reported value. For the low, or net heat of combustion, it is assumed that all products of combustion remain in the gaseous state.

While the high, or gross, heat of combustion can be accurately determined by established (ASTM) procedures, direct determination of the low heat of combustion is difficult. Therefore, it is usually calculated using the following formula:

$$Q_L = Q_H - 1040\,w$$

where:

Q_L = low heat of combustion of fuel, Btu/pound
Q_H = high heat of combustion of fuel, Btu/pound
w = pound of water formed per pound of fuel
1040 = factor to reduce high heat of combustion at constant volume
  to low heat of combustion at constant pressure

TABLE 1. COMBUSTION CONSTANTS OF CHEMICAL ELEMENTS AND COMPOUNDS GENERALLY FOUND IN FUELS

No.	Substance	Formula	Molecular Weight	Lb per Cu Ft	Cu Ft per Lb	Sp Gr Air = 1.0000	Heat of Combustion — Btu per Cu Ft Gross (High)	Net (Low)	Btu per Lb Gross (High)	Net (Low)	Vol — Required O₂	N₂	Air	Vol — Flue CO₂	H₂O	N₂	Wt — Required O₂	N₂	Air	Wt — Flue CO₂	H₂O	N₂
1	Carbon[a]	C	12.01	—	—	—	—	—	14,093	14,093	1.0	3.76	4.76	1.0	—	3.76	2.66	8.86	11.53	3.66	—	8.86
2	Hydrogen	H₂	2.016	0.0053	187.723	0.0696	325	275	61,095	51,623	0.5	1.88	2.38	—	1.0	1.88	7.94	26.41	34.34	—	8.94	26.41
3	Oxygen	O₂	32.00	0.0846	11.819	1.1053	—	—	—	—	—	—	—	—	—	—	—	—	—	—	—	—
4	Nitrogen (atm)	N₂	28.01	0.0744	13.443	0.9718	—	—	—	—	—	—	—	—	—	—	—	—	—	—	—	—
5	Carbon monoxide	CO	28.01	0.0740	13.506	0.9672	321	321	4,347	4,347	0.5	1.88	2.38	1.0	—	1.88	0.57	1.90	2.47	1.57	—	1.90
6	Carbon dioxide	CO₂	44.01	0.1170	8.548	1.5282	—	—	—	—	—	—	—	—	—	—	—	—	—	—	—	—
Paraffin series																						
7	Methane	CH₄	16.04	0.0425	23.552	0.5543	1012	911	23,875	21,495	2.0	7.53	9.53	1.0	2.0	7.53	3.99	13.28	17.27	2.74	2.25	13.28
8	Ethane	C₂H₆	30.07	0.0803	12.455	1.0488	1773	1622	22,323	20,418	3.5	13.18	16.68	2.0	3.0	13.18	3.73	12.39	16.12	2.93	1.80	12.39
9	Propane	C₃H₈	44.09	0.1196	8.365	1.5617	2524	2322	21,669	19,937	5.0	18.82	23.82	3.0	4.0	18.82	3.63	12.07	15.70	2.99	1.63	12.07
10	n-Butane	C₄H₁₀	58.12	0.1582	6.321	2.0665	3271	3018	21,321	19,678	6.5	24.47	30.97	4.0	5.0	24.47	3.58	11.91	15.49	3.03	1.55	11.91
11	Isobutane	C₄H₁₀	58.12	0.1582	6.321	2.0665	3261	3009	21,271	19,628	6.5	24.47	30.97	4.0	5.0	24.47	3.58	11.91	15.49	3.03	1.55	11.91
12	n-Pentane	C₅H₁₂	72.15	0.1904	5.252	2.4872	4020	3717	21,095	19,507	8.0	30.11	38.11	5.0	6.0	30.11	3.55	11.81	15.35	3.05	1.50	11.81
13	Isopentane	C₅H₁₂	72.15	0.1904	5.252	2.4872	4011	3708	21,047	19,459	8.0	30.11	38.11	5.0	6.0	30.11	3.55	11.81	15.35	3.05	1.50	11.81
14	Neopentane	C₅H₁₂	72.15	0.1904	5.252	2.4872	3994	3692	20,978	19,390	8.0	30.11	38.11	5.0	6.0	30.11	3.55	11.81	15.35	3.05	1.50	11.81
15	n-Hexane	C₆H₁₄	86.17	0.2274	4.398	2.9704	4768	4415	20,966	19,415	9.5	35.76	45.26	6.0	7.0	35.76	3.53	11.74	15.27	3.06	1.46	11.74
Olefin series																						
16	Ethylene	C₂H₄	28.05	0.0742	13.475	0.9740	1604	1503	21,636	20,275	3.0	11.29	14.29	2.0	2.0	11.29	3.42	11.39	14.81	3.14	1.29	11.39
17	Propylene	C₃H₆	42.08	0.1110	9.007	1.4504	2340	2188	21,048	19,687	4.5	16.94	21.44	3.0	3.0	16.94	3.42	11.39	14.81	3.14	1.29	11.39
18	n-Butene	C₄H₈	56.10	0.1480	6.756	1.9336	3084	2885	20,854	19,493	6.0	22.59	28.59	4.0	4.0	22.59	3.42	11.39	14.81	3.14	1.29	11.39
19	Isobutene	C₄H₈	56.10	0.1480	6.756	1.9336	3069	2868	20,737	19,376	6.0	22.59	28.59	4.0	4.0	22.59	3.42	11.39	14.81	3.14	1.29	11.39
20	n-Pentene	C₅H₁₀	70.13	0.1852	5.400	2.4190	3837	3585	20,720	19,359	7.5	28.23	35.73	5.0	5.0	28.23	3.42	11.39	14.81	3.14	1.29	11.39
Aromatic series																						
21	Benzene	C₆H₆	78.11	0.2060	4.852	2.6920	3752	3601	18,184	17,451	7.5	28.23	35.73	6.0	3.0	28.23	3.07	10.22	13.30	3.38	0.69	10.22
22	Toluene	C₇H₈	92.13	0.2431	4.113	3.1760	4486	4285	18,501	17,672	9.0	33.88	42.88	7.0	4.0	33.88	3.13	10.40	13.53	3.34	0.78	10.40
23	Xylene	C₈H₁₀	106.16	0.2803	3.567	3.6618	5230	4980	18,650	17,760	10.5	39.52	50.02	8.0	5.0	39.52	3.17	10.53	13.70	3.32	0.85	10.53
Miscellaneous gases																						
24	Acetylene	C₂H₂	26.04	0.0697	14.344	0.9107	1477	1426	21,502	20,769	2.5	9.41	11.91	2.0	1.0	9.41	3.07	10.22	13.30	3.38	0.69	10.22
25	Naphthalene	C₁₀H₈	128.16	0.3384	2.955	4.4208	5854	5654	17,303	16,708	12.0	45.17	57.17	10.0	4.0	45.17	3.00	9.97	12.96	3.43	0.56	9.97
26	Methyl alcohol	CH₃OH	32.04	0.0846	11.820	1.1052	868	767	10,258	9,066	1.5	5.65	7.15	1.0	2.0	5.65	1.50	4.98	6.48	1.37	1.13	4.98
27	Ethyl alcohol	C₂H₅OH	46.07	0.1216	8.221	1.5890	1600	1449	13,161	11,917	3.0	11.29	14.29	2.0	3.0	11.29	2.08	6.93	9.02	1.92	1.17	6.93
28	Ammonia	NH₃	17.03	0.0456	21.914	0.5961	441	364	9,667	7,985	0.75	2.82	3.57	—	1.5	3.32	1.41	4.69	6.10	—	1.59	5.51
29	Sulfur[a]	S	32.06	—	—	—	—	—	3,980	3,980	1.0	3.76	4.76	SO₂ 1.0	—	3.76	1.00	3.29	4.29	SO₂ 2.00	—	3.29
30	Hydrogen sulfide	H₂S	34.08	0.0911	10.979	1.1898	646	595	7,097	6,537	1.5	5.65	7.15	1.0	1.0	5.65	1.41	4.69	6.10	1.88	0.53	4.69
31	Sulfur dioxide	SO₂	64.06	0.1733	5.770	2.2640	—	—	—	—	—	—	—	—	—	—	—	—	—	—	—	—
32	Water vapor	H₂O	18.02	0.0476	21.017	0.6215	—	—	—	—	—	—	—	—	—	—	—	—	—	—	—	—
33	Air	—	28.9	0.0766	13.063	1.0000	—	—	—	—	—	—	—	—	—	—	—	—	—	—	—	—

Column group headings:
- **For 100% Total Air — Moles per mole of Combustible or Volume Units of Air per Volume Units of Combustible:** Required for Combustion (O₂, N₂, Air); Flue Products (CO₂, H₂O, N₂)
- **For 100% Total Air — Weight Units of Air per Weight Units of Combustible:** Required for Combustion (O₂, N₂, Air); Flue Products (CO₂, H₂O, N₂)

[a]Carbon and sulfur are considered as gases for molal calculations only.
All gas volumes corrected to 60 F and 30 in. Hg dry. (15.6°C and 101.6 kilopascals).
SOURCE: American Gas Association, Arlington, Virginia.

To convert:
lb/cu ft to kg/cu meter, multiply by 16.026
cu ft/lb to cu meters/kg, multiply by 0.0624
Btu to Calories 0.2520
Btu/cu ft to Calories/cu meter 8.898

279

TABLE 2.   COMMON CHEMICAL REACTIONS OF COMBUSTION

COMBUSTIBLE	REACTION	MOLES	POUNDS	HEAT OF COMBUSTION (HIGH) BTU/POUND OF FUEL
Carbon (to CO)	$2C + O_2 = 2CO$	$2 + 1 = 2$	$24 + 32 = 56$	4,000
Carbon (to CO_2)	$C + O_2 = CO_2$	$1 + 1 = 1$	$12 + 32 = 44$	14,100
Carbon Monoxide	$2CO + O_2 = 2CO_2$	$2 + 1 = 2$	$56 + 32 = 88$	4,345
Hydrogen	$2H_2 + O_2 = 2H_2O$	$2 + 1 = 2$	$4 + 32 = 36$	61,100
Sulfur (to SO_2)	$S + O_2 = SO_2$	$1 + 1 = 1$	$32 + 32 = 64$	3,980
Methane	$CH_4 + 2O_2 = CO_2 + 2H_2O$	$1 + 2 = 1 + 2$	$16 + 64 = 80$	23,875
Acetylene	$2C_2H_2 + 5O_2 = 4CO_2 + 2H_2O$	$2 + 5 = 4 + 2$	$52 + 160 = 212$	21,500
Ethylene	$C_2H_4 + 3O_2 = 2CO_2 + 2H_2O$	$1 + 3 = 2 + 2$	$28 + 96 = 124$	21,635
Ethane	$2C_2H_6 + 7O_2 = 4CO_2 + 6H_2O$	$2 + 7 = 4 + 6$	$60 + 224 = 284$	22,325
Hydrogen Sulfide	$2H_2S + 3O_2 = 2SO_2 + 2H_2O$	$2 + 3 = 2 + 2$	$68 + 96 = 164$	7,100

SOURCE: "Steam—Its Generation and Use," 39th edition, The Babcock and Wilcox Company, New York, 1978.

TABLE 3.   COMPOSITION OF AIR

	COMPOSITION OF DRY AIR	
	% by Volume	% by Weight
Oxygen, O_2	20.99	23.15
Nitrogen, N_2	78.03	76.85[a]
Inerts	0.98	

Equivalent molecular weight of air = 29.0[a]

% Moisture = 1.3% by weight. Standard for boiler industry (ABMA)[b]

$$\frac{\text{Moles air/mole oxygen}}{\text{Cubic feet air/cubic feet oxygen}} = \frac{100}{20.99} = 4.76$$

$$\text{Moles } N_2/\text{mole oxygen} = \frac{79.01}{20.99} = 3.76$$

$$\text{Pounds air (dry)/pound oxygen} = \frac{100}{23.15} = 4.32$$

$$\text{Pounds nitrogen/pound oxygen} = \frac{76.85}{23.15} = 3.32$$

[a] It is convenient in combustion calculations to account for inerts as equivalent nitrogen. The equivalent weight percentage of 76.85 and the equivalent molecular weight of 29.0 have been corrected to account for the extra weight of the inerts.
[b] Air containing 0.013 pound water/pound dry air is often referred to as standard air.
SOURCE: "Steam—Its Generation and Use," 39th edition, The Babcock and Wilcox Company, New York, 1978.

In the United States the practice is to use the high heat of combustion in boiler combustion calculations. In Europe the low heat value is used.

Ignition temperature may be defined as the temperature at which more heat is generated by combustion than is lost to the surroundings so that the combustion process becomes self-sustaining. The term usually applies to rapid combustion in air of atmospheric pressure.

Ignition temperatures of combustion substances vary greatly as indicated in Table 4, which lists minimum temperatures and temperature ranges in air for fuels and for the combustible constituents of fuels commonly used in the commercial generation of heat. Many factors influence ignition temperature so that any tabulation can be used only as a guide. Pressure, velocity, enclosure configuration, catalytic materials, air-fuel-mixture uniformity, and ignition source are only a few of the variables.

Ignition temperature usually decreases with rising pressure and increases with increasing moisture content in the air.

The ignition temperature of the gases of a coal vary considerably and are appreciably higher than the ignition temperatures of the fixed carbon of the coal. However, the ignition temperature of coal may be considered as the ignition temperature of its fixed carbon content, since the gaseous constituents are usually distilled off but not ignited before this temperature is attained.

The adiabatic flame temperature is the maximum theoretical temperature which can be reached by the products of combustion of a specific fuel and air (or oxygen) combination assuming no loss of heat to the surroundings until combustion is complete. This theoretical temperature also assumes no dissociation, a phenomenon discussed later under this heading. The heat of combustion of the fuel is the major factor in the flame temperature, but increasing the temperature of the air or of the fuel will also have the effect of raising the flame temperature. As would be expected, this adiabatic temperature is a maximum with zero excess air (only enough air chemically required to combine with the fuel), since any excess is not involved in the combustion process and only dilutes the temperature of the products of combustion.

The adiabatic temperature is determined from the adiabatic enthalpy of the flue gas as follows:

$$h_g = \frac{\binom{\text{heat of}}{\text{combustion}} + \binom{\text{sensible heat}}{\text{in fuel}} + \binom{\text{sensible heat}}{\text{in air}}}{\text{weight of products of combustion}}$$

where:

h_g = adiabatic enthalpy (adiabatic heat content of the products of combustion), Btu/pound

Knowing the moisture content of the products of combustion and its enthalpy, the theoretical flame or gas temperature can be obtained from published graphs.*

The adiabatic temperature is a fictitiously high temperature which does not exist in fact. Actual flame temperatures are lower for two main reasons:

1. Combustion is not instantaneous. Some heat is lost to the surroundings as combustion takes place. The faster the combustion occurs the less heat is lost before combustion is complete. If combustion is slow enough, the gases may be cooled

* Series of 8 graphs in "Steam—Its Generation and Use," pages 6–8 and 6–9, published by the Babcock and Wilcox Company, New York (39th edition, 1978).

TABLE 4. IGNITION TEMPERATURES OF FUELS IN AIR
(Approximate Values and Ranges at Atmospheric Pressure)

COMBUSTIBLE	FORMULA	TEMPERATURE: °F	°C
Sulfur	S	470	243
Charcoal	C	650	343
Fixed carbon (bituminous coal)	C	765	407
Fixed carbon (semibituminous coal)	C	870	466
Fixed carbon (anthracite)	C	840–1115	449–602
Acetylene	C_2H_2	580–825	304–441
Ethane	C_2H_6	880–1165	471–630
Ethylene	C_2H_4	900–1020	482–549
Hydrogen	H_2	1065–1095	574–590
Methane	CH_4	1170–1380	632–749
Carbon Monoxide	CO	1130–1215	610–657
Kerosine	—	490–560	254–293
Gasoline	—	500–800	260–427

sufficiently for combustion to be incomplete with some of the fuel unburned.

2. At temperatures above 3,000°F (1,649°C), some of the CO_2 and H_2O in the flue gases dissociates, absorbing heat in the process. At 3,500°F (1,926°C), about 10% of the CO_2 in a typical flue gas dissociates to CO and O_2 with a heat absorption of 4,345 Btu/pound of CO formed, and about 3% of the H_2O dissociates to H_2 and O_2, with a heat absorption of 61,100 Btu/pound of H_2 formed. As the gas cools, the CO and H_2 dissociated recombine with the O_2 and liberate the heat absorbed in dissociation, so the heat is not lost. However, the effect is to lower the maximum actual flame temperature.

References

Baumeister, T. (editor): "Mark's Standard Handbook for Mechanical Engineers," 8th edition, McGraw-Hill, New York, 1978.
Beer, J., and N. Chigier: "Combustion Aerodynamics," Fuel and Energy Science Monographs Series, Wiley, New York, 1972.
Grey, J., Sutton, G. W., and M. Zlotnick: "Fuel Conservation and Applied Research," Science, 200, 135–142 (1978).
Johnson, S. A., and A. H. Rawdon: "NO_x Control by Furnance and Burner Design," in "Energy Technology Handbook," (D. M. Considine, editor), McGraw-Hill, New York, 1976.
Smith, R. B.: "Heat Generation," in "Chemical Engineers' Handbook," 5th edition, (R. H. Perry and C. H. Chilton, editors), McGraw-Hill, New York, 1973.
Smoot, L. D., and D. T. Pratt: "Pulverized Coal Combustion and Gasification for Continuous Flow Processes," Plenum, New York, 1979.
Staff: "Steam—Its Generation and Use," 39th edition, The Babcock & Wilcox Co., New York, 1978.

COMMON ION EFFECT. The reversal of ionization which occurs when a compound is added to a solution of a second compound with which it has a common ion, the volume being kept constant. The degree of ionization of the second compound then is lowered, i.e., it retrogresses. The common ion effect also can markedly affect solubility.

COMPLEX COMPOUND. Ligand.

COMPOUND (Chemical). A homogeneous, pure substance, composed of two or more essentially different chemical elements, which are present in definite proportions; compounds usually possess properties differing from those of the constituent elements.

An *addition compound* is one that is formed by the junction or union of two simpler compounds. Effectively the same as a molecular compound (see later definition).

An *additive compound* is formed by an addition reaction, or by the saturation of a double bond, triple bond, or more than one of them.

An *alicyclic compound* is an organic compound containing a saturated ring of carbon atoms, such as a cycloparaffin or other hydroaromatic compound.

An *aliphatic compound* is an organic compound without ring structures, i.e., with straight chain arrangement of carbon and, possibly other, atoms. In the narrower sense, an aliphatic compound is a member of the paraffin series of hydrocarbons, or one of their derivatives.

An *aromatic compound* is an organic compound containing a ring of carbon atoms, usually unsaturated, such as a benzene, naphthalene, anthracene, and acenaphthylene ring.

An *associated compound* is a compound formed by the union of two or more molecules, usually of the same or similar chemical composition, to form a single complex molecule.

Berthollide compound. See nonstoichiometric compound in this entry.

A *binary compound* is made up of two elements in a definite molecular ratio.

A *catenation compound* has a molecular configuration resembling a linked chain, in which the atoms forming one ring pass through, but are not joined by valence forces to, the ring formed by another group of atoms. Since the two rings, while spatially interlocked, are not joined by valence forces, the application of the word compound to such aggregates may be questioned.

H. L. Frisch of Bell Laboratories has made extensive calculations of ring sizes necessary to permit the formation of various catenation compounds. He found that 20 is the minimum number of —CH_2— groups in an alicyclic ring through which another ring can be catenated (threaded) without encountering excessively great repulsive forces from the alicyclic ring atoms. For threadings of two rings through a third ring, the probable minimum alicyclic ring size of the latter is 33 —CH_2— groups; Borromean rings, formed by the interlocking of three rings (with no two of them locked separately), require a minimum of 30 —CH_2— groups.

A laboratory preparation of the simplest of these catenation compounds, two interlocking rings, has been carried out at Bell Laboratories. They started with the dimethyl ester of a 34-carbon paraffinic dicarboxylic acid, CH_3OOC—$(CH_2)_{32}$—$COOCH_3$, which was reacted in a suspension of metallic sodium in xylene with acetic acid to condense the terminal ester groups to form an aceloin ring compound $O{=}\overset{\frown}{C}$—$(CH_2)_{32}$—$CHOH$. Treatment of the latter with deuterated hydrochloric acid reduced it to a 34-carbon alicyclic (ring) compound,

$$\overset{\frown}{HDC}{-}CHD{-}CD_2{-}CHD{-}(CH_2)_{30}$$

containing five deuterium atoms. This hydrocarbon was added to the suspension of metallic sodium in xylene, and then more of the 34-carbon dimethyl ester was added. Ring formation of the latter compound to form the aceloin occurred as before, and a small percentage of the aceloin rings were found to be threaded through the deuterated rings in the solvent, yielding a catenation compound consisting of the 34-carbon aceloin ring and the 34-carbon deuterated alicyclic hydrocarbon threaded together, but without any atoms in one ring being joined by valence bonds to those in the other.

Separation and identification of this catenation compound was effected by chromatography and infrared spectroscopy, the latter being the reason why the hydrocarbon portion of the catenation compound was deuterated.

Chelation compound. See entry on **Chelation Compounds.**

A *clathrate compound* means, literally, an enclosed compound, a term applied to a solid molecular compound in which a molecule of one component is physically enclosed in the crystal structure of a second compound, so that the properties of the aggregate are essentially those of the enclosing compound. Examples of such "cage compounds" are those of the small molecules of SO_2, CO_2, CO and the noble gases with ice and hydroquinone, which have very open crystal structures. Another example is the clathrate of benzene with nickel cyanide.

A *complex compound* is made up structurally of two or more compounds or ions. See **Ligand.**

A *condensation compound* is formed by a reaction in which the largest parts, constituting the essential structural elements, of two or more molecules combine to form a new molecule, with elimination of minor elements, such as those of water.

Coordination compound. See that entry.

A *covalent compound* is formed by the sharing of electrons between atoms; as distinguished from electrovalent compounds, in which there occurs a transfer of electrons.

A *cyclic compound* has some or all of its atoms arranged in a ring structure.

An *electrovalent compound* is formed by ions, or by atoms which become ions by transfer of electrons between them. (See ionic compound below.)

An *endothermic compound* is a compound whose formation is accompanied by a positive change in heat content, i.e., by the absorption of heat.

An *epoxy compound* contains an oxygen bridge, as

$$\underset{CH_2-CH_2-CH_2-CH_2}{\overset{\overset{\displaystyle O}{\frown}}{}}$$

which is 1,4 epoxy butane.

An *exothermic compound* is a compound whose formation is accompanied by a negative change in heat content, i.e., with the liberation of heat.

A *heterocyclic compound* contains one or more rings composed of atoms some of which are of dissimilar elements. A few inorganic substances fall into this classification, but by far the majority of them are carbon compounds. In organic chemistry substances of cyclic structure, as acid anhydrides, lactides, lactams, lactones, cyclic ethers, and cyclic derivatives of dicarboxylic acids which are formed by the elimination of water from aliphatic compounds, are not considered among the heterocyclic substances. Derivatives of pyridine, quinoline, thiophene, thiazole, pyrone, etc., which contain heterocyclic rings that persist in the compound through chemical reactions, are considered the true members of this class. Heterocyclic rings are known that contain nitrogen, sulfur, and oxygen members. The noncarbon members of the ring are termed "heteroatoms," and their number is indicated by the prefixes mono, di, tri, tetra, etc. The number of members in the ring may reach as high as sixteen, as in tetrasalicylide.

A *homocyclic compound* contains a homocyclic ring, i.e., a ring composed of atoms of the same element.

The term *inclusion compound* was once used for the clathrate compounds described in this entry.

In an *inner compound* an additional valence bond has been formed between two atoms of an already existing structure, usually by loss of the elements of water or other simple substance. Inner compound formation commonly results in the formation of a ring. The inner esters, inner anhydrides, and inner coordination compounds are well-known classes of inner compounds.

An *inorganic compound* means, in general, a compound that does not contain carbon atoms. Some very simple carbon compounds, such as carbon monoxide and dioxide, binary metallic carbon compounds (carbides) and carbonates, are also included in the group of inorganic compounds.

An *intermetallic compound* consists of metallic atoms only, which are joined by metallic bonds. Such compounds may be made semiconducting if the two metals between them contribute just sufficient electrons to fill the valence band, e.g., InAs. (See also **Alloys.**)

An *interstitial compound* consists of a metal or metals and certain metalloid elements, in which the metalloid atoms occupy the interstices between the atoms of the metal lattice. Compounds of this type are, for example, TaC, TiC, ZrC, NbC, and similar compounds of carbon, nitrogen, boron, and hydrogen with metals.

An *ionic compound* is one of a class of compounds which are formed when atoms combine to produce molecules having stable configurations by the transfer of one or more electrons within the molecule. This type of combination is illustrated by the combination of sodium atoms and chlorine atoms to form sodium chloride. The sodium atom loses the single elctron in its outer shell, and thus is left with the stable configuration of eight electrons; the chlorine atom acquires an electron to increase the number of electrons in its outer shell from seven to eight; as a result of the loss and gain of the electrons, the atoms have acquired positive and negative charges, respectively, which constitute an electrovalent bond.

A *molecular compound* is formed by the union of two or more already saturated molecules apparently in defiance of the ordinary rules of valence. The class includes double salts, salts with water of crystallization, and metal ammonium derivatives. These salts are usually formed by van der Waals attraction between the constituent molecules. They do not differ in any characteristic manner from compounds formed in strict accordance with the concept of valence. They are also called addition compounds.

A *nonpolar compound* is a compound in which the centers of positive and negative charge almost coincide, so that no permanent dipole moments are produced. The term nonpolar also applies to compounds in which the effect of oppositely directed diple moments cancel. Nonpolar compounds may contain polar bonds, if their effect is cancelled by opposing bonds, as may occur in a perfectly symmetrical molecule. Nonpolar compounds do not ionize or conduct electricity. Most organic compounds are to be classed as nonpolar compounds.

A *nonstoichiometric compound* has a composition not in accord with the law of definite proportions, which is therefore also called a berthollide compound. Non stoichiometric compounds occur among the binary compounds of Group 6b, as exemplified by $TiO_{1\cdot8}$, $Cu_{1\cdot7}S$ and $Cu_{1\cdot6}Se$; among the hydrides, e.g., $CeH_{2\cdot7}$ and especially among the intermetallic compounds.

An *organic compound* is one of the great number of compounds consisting of carbon linked in chains or rings; such compounds usually also contain hydrogen and may contain elements such as oxygen, nitrogen, sulfur, chlorine, etc. Some of the simpler carbon compounds are classified as inorganic compounds.

An *organometallic (or metal-organic) compound* is an organic compound in which one or more hydrogen atoms have been replaced by a metallic atom or atoms, usually with the establishment of a valence bond between the metal atom and a carbon atom. A metallic salt of an organic acid, in which the hydrogen atom of a —COOH group is replaced by a metal atom, is not classified as an organometallic compound.

A *polar compound* is, in general, a compound that exhibits polarity, or local differences in electrical properties, and has a diple moment associated with one or more of its interatomic valence bonds. Polar compounds have relatively high dielectric constants, associate readily in most cases, and include the substances that exhibit tautomerism. In the most general use of the term, polar compounds include all electrolytes, most inorganic substances, and many organic ones. Specifically, the term polar compound is frequently applied to the extreme type of polarity which arises in the presence of an electrovalent bond or, in wave-mechanical terms, to cases in which one ionic term dominates in the orbital function of the molecule. Such compounds are exemplified by the inorganic acids, bases, and salts which possess, to a greater or lesser degree the power to conduct electricity, associate, form double molecules and complex ions, etc.

In a *saturated compound* the valence of all the atoms is completely satisfied without linking any two atoms by more than one valence bond.

A *spiro-compound* contains two ring structures having one common carbon atom.

A *tracer compound* is a compound which by its ease of detection enables a reaction or process to be studied conveniently. Wide use has been made of isotopes, including radioactive isotopes of common elements, which are added in small quantities, in the form of the proper compound, to follow the course of an atom or a compound through a complicated series of reactions; or conversely to determine the properties of a tracer—that is available only in quantities too small to handle alone—by adding it to a system containing chemically related elements, and then following its course throughout a given series of reactions. Considerable use of tracer compounds is made in the study of physiological reactions.

Unsaturated compound is a term specifically applied to a carbon compound containing one or more double bonds or triple bonds. One consequence of the presence of these double bonds or triple bonds, from which a broader concept of unsaturation stems, is the relative ease with which such bonds are split, and other constituents linked to them.

See also **Organic Chemistry.**

COMPRESSION (GAS).

The compressibility of a gas is defined as the rate of volume decrease with increasing pressure, per unit volume of the gas. The compressibility depends not only on the state of the gas, but also on the conditions under which the compression is achieved. Thus, if the temperature is kept constant during compression, the compressibility so defined is called the isothermal compressibility β_T:

$$\beta_T = -\frac{1}{V}\left(\frac{\partial V}{\partial P}\right)_T = \frac{1}{\rho}\left(\frac{\partial \rho}{\partial P}\right)_T \tag{1}$$

If the compression is carried out reversibly without heat exchange with the surroundings, the *adiabatic compressibility* at constant entropy, β_S, is obtained:

$$\beta_S = -\frac{1}{V}\left(\frac{\partial V}{\partial P}\right)_S = \frac{1}{\rho}\left(\frac{\partial \rho}{\partial P}\right)_S \tag{2}$$

Here P is the pressure, V the volume, ρ the density, T the temperature, and S the entropy.

In adiabatic compression, the temperature rises, thus the pressure increases more sharply than in isothermal compression. Therefore β_S is always smaller than β_T.

The *compressibility factor* of a gas is the ratio PV/RT. This name is not well chosen since the value of the compressibility factor by itself does not indicate the compressibility of the gas.

Experimental values for the compressibility of gases can be obtained in several ways, most of which are indirect.

Since the compressibility is proportional to the pressure derivative of the volume, any experiment that establishes the P-V-T relation of a gas with sufficient accuracy also yields data for the isothermal compressibility. For obtaining the adiabatic compressibility from the P-V-T relation, some additional information is necessary [see section (c)], for instance specific heat data in the perfect gas state of the substance considered. A more direct way of determining the adiabatic compressibility is by measuring the speed of sound v, the two quantities being related by

$$v^2 = \frac{1}{\rho \beta_S} \tag{3}$$

This relation is valid only when the compressions and expansions of the sound wave are truly reversible and adiabatic. This is the case if the frequency is fairly low and the amplitude small.

Dilute gases obey the laws of Boyle and Gay-Lussac, $PV = RT$, to a good approximation. Thus, it can readily be shown that the following relations hold for the compressibility:

$$\beta_T = 1/P = V/RT$$
$$\beta_S = 1/\gamma P = V/\gamma RT \tag{4}$$

where $\gamma = c_P/c_V$, the ratio of the specific heats at constant volume and at constant pressure, respectively, and R is the gas constant.

Compressed gases show large deviations from the behavior predicted by equation (4). This is demonstrated by the accompanying diagram, where the isothermal compressibility of argon, divided by the corresponding value for a perfect gas at the same density, is pictured as a function of density for various temperatures. It is seen, first of all, that at all temperatures the compressibility at high densities falls to a small fraction of the value for a perfect gas, and secondly, that super-critical isotherms show a maximum in the ratio $\beta/\beta_{perfect}$ as a function of density, which maximum is the more pronounced the closer the critical temperature. It occurs roughly at the critical density ρ_c. Since at the critical point $(\partial P/\partial V)_T$ equals zero, the isothermal compressibility becomes infinite at this point. The adiabatic compressibility, however, remains finite. Qualitatively, all gases show the same behavior as pictured for argon in the diagram.

The molecular theory can explain the general features of the

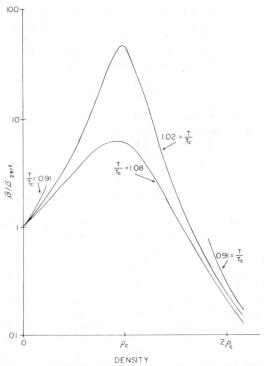

The ratio $\beta/\beta_{perf.}$ of the isothermal compressibility of argon to that of a perfect gas at the same density, as a function of the density, at 0.91, 1.02 and 1.08 times the critical temperature. The critical density is indicated by ρ_c.

compressibility in its temperature and density dependence. The pressure of the gas is caused by the impact of the molecules on the wall. If the volume is decreased at constant temperature, the average molecular speed and force of impact remain constant, but the number of collisions per unit area increases and thus the pressure rises. If the gas is compressed adiabatically, the heat of compression cannot flow off, thus the average molecular speed and force of impact increase as well, giving rise to an extra increase of pressure. Therefore $\beta_S < \beta_T$. The actual magnitude of the temperature rise depends on the internal state of the molecules: the more internal degrees of freedom available, the more energy can be taken up inside the molecule and the smaller the temperature rise on adiabatic compression. Thus for gases consisting of molecules with many internal degrees of freedom, adiabatic and isothermal compressibilities differ but little.

If the gas is assumed to consist of molecules of negligible size and without interaction, then the gas can be shown to follow the laws of Boyle and Gay-Lussac, therefore its isothermal and adiabatic compressibilities must be given by Equation (4). For a perfect gas, the percentage pressure rise is proportional to the percentage volume decrease if the change is small; thus the compressibility is inversely proportional to the pressure.

To explain the very different behavior of real gases, the model must be modified. Suppose the molecular volume is small but not negligible. In states of high compression where the total molecular volume becomes of the order of the volume available to the gas, the free space available to the molecules is only a fraction of what it would be in a perfect gas, and thus the real gas is much harder to compress than the perfect gas. This explains the low compressibility of dense gases and liquids (diagram).

Furthermore, one assumes that molecules, on approaching each other, experience a mutual attraction before they collide; this mutual attraction makes it easier to compress a real gas than a perfect gas. This explains the initial rise of the compressibility of a real gas over that of a perfect gas at temperatures not too far above the critical.

When compressed at subcritical temperatures, the gas condenses; that is, macroscopic clusters or droplets are formed under the influence of the attractive forces. During condensation, the pressure remains constant while the volume decreases, giving rise to an infinite compressibility in the two-phase region. At the critical point the system is on the verge of condensation and the compressibility is also infinite.

Theoretical predictions for the isothermal compressibility can obviously be obtained from any theory of the equation of state. If, in addition, data for the specific heat are supplied, the adiabatic compressibility can be derived in the same way. Thus the compressibility can be derived from the viral expansion of the equation of state which expresses the ratio PV/RT in a power series in the density, the coefficients being related to the interactions of groups of two, three, etc., particles.

In the dense system the convergence of the virial expansion is doubtful. In any case the higher coefficients are hard to calculate; here approximate theories have been developed, of which the cell model[5] is an example.

Many semiempirical equations of state with varying degree of theoretical foundations are in use. The van der Walls equation, a two-parameter equation which gives a qualitatively correct picture of the P-V-T relations of a gas and of the gas-liquid transition, is an example.

Modern developments are centered around the calculations of the radial distribution function $g(r)$, which is the ratio of the density of molecules at a distance r from a given molecule, to the average density in the gas. The compressibility can be expressed straightforwardly in terms of $g(r)$ as follows.

$$KT\beta_T = 1/\rho + \int_0^\infty [g(r) - 1]4\pi r^2 dr \qquad (5)$$

Approximate evaluations of the radial distribution function in dense systems are being obtained as solutions to integral equations derived from first principles under well-defined approximations.

CONCENTRATION (Chemical). Demal Solution; Gram-Equivalent; Gram-Molecular Weight; Mole Fraction; Mole Volume; Molal Concentration; Molar Concentration; Mole (Stochiometry); Normal Concentration.

CONCRETE. A mixture of fine and coarse aggregates firmly bound into a monolithic mass by a cementing agent. The cement ordinarily used is standard Portland cement. Aggregates are usually sand and crushed stone or gravel. Crushed slag or cinders are used in special concretes. Formation of concrete can be considered as a process in which the voids between the particles of coarse aggregate are filled by the fine aggregate, and the whole cemented together by the binding action of the cement. See **Cement.**

CONDUCTION (Heat). Heat Transfer.

CONSISTENCY. Rheology.

CONSOLUTE LIQUIDS. Liquids which are miscible in all proportions, i.e., mutually completely soluble, under specified conditions. The term is not applied to gases because all gases are miscible. The *upper consolute temperature* is the critical temperature above which two liquids are miscible in all proportions. In some systems, the mutual solubility decreases with increasing temperature over a certain temperature range. The *lower consolute temperature* is the critical temperature below which two liquids are miscible in all proportions. Some systems, such as methylethyl ketone and water, have both upper and lower consolute temperatures.

CONSTITUENT. 1. In general, one of the elements or parts of a compound. 2. An identifiable component in the microstructure of an alloy. It may be a phase or a characteristic configuration of several phases.

CONTRACTILE PROTEINS. The fundamental property of living matter on which its power of movement depends is termed *contractility*. In the simplest forms of living things, it is evident in the flowing movement of the material of the cell. In more complex forms, the property is centralized in muscle tissues. Muscle cells are elongate and are so arranged that necessary movement results from their shortening when stimulated.

The study of the fibrillar proteins of muscle invites interest for two reasons. On the one hand, they form the prime ingredients of the mechanism that performs a typical vital activity, movement, in its most specialized form; thus, the elucidation of their functions, and of the physical and chemical properties basic to it, represents one of the cardinal parts of molecular biology. On the other hand, the isolated proteins display such striking physical behavior that to the macromolecular physicist, they are among the most fascinating materials. They are difficult to obtain and exceedingly changeable and labile. Thus, working with them is somewhat difficult.

The major fibrous proteins are myosin and actin. Other proteins that may be involved in myofibrillar structure are tropomyosin and paramyosin. All four proteins share certain chemical properties; among others, they are all exceptionally rich in ionizing amino acids, thus they are highly charged molecules. Myosin is an adenosinetriphosphatase. The molecular weight is not accurately known; a number of determinations cluster around 500,000. Light scattering dissymetry suggests a mean molecular length of about 1600 micrometers. Electron microscopy suggests similar lengths.

CONVECTION (Heat). Heat Transfer.

CONVERSION COATINGS. The industrial application of organic finishes to metals almost always requires the use of an intermediate conversion coating, particularly when the performance demands are high. Conversion coatings are formed chemically by causing the surface of the metal to be "converted" into a tightly adherent amorphous or crystalline coating, part or all of which consists of an oxidized form of the substrate metal. Conversion coatings can provide high corrosion resistance as well as strong affinity for organic coatings. They are also useful as lubricants for the drawing and forming of metals and sometimes are used for decorative purposes. The most important and widespread use of conversion coatings is on steel, zinc or galvanized steel, and aluminum alloys. The most widely used classes of conversion coatings for use on these metals are the phosphates and chromates. Depending on size, shape, volume of production, and other factors, the coating may be applied by spray, immersion, roll coat, or brush. A typical metal pretreatment sequence is: (1) cleaning; (2) rinsing; (3) conversion coating; (4) rinsing; and (5) final rinsing.

Phosphating of Steel. When a ferrous alloy is immersed in phosphoric acid, it initially forms a soluble phosphate. As the pH rises at the metal/solution interface, the phosphate becomes insoluble and crystallizes epitaxially on the substrate metal. The phosphate coating thus produced consists of a nonconductive layer of crystals that insulates the metal from any subsequently applied film and provides a topography with enhanced "tooth" for increased adhesion. The crystals insulate microanode and microcathode centers caused by stress or imperfections in the metal surface. This greatly reduces the severity of electrochemical corrosion.

Practically all phosphating processes involve patented proprietary solutions which produce superior coatings in shorter times and at lower temperatures than are obtainable with phosphoric acid alone. Treating times have been reduced from the earlier 1–2-hour to 1–5-minute periods, while temperatures have been reduced from 98 to 20–30°C.

Increasing energy costs have led to more widespread use of low-temperature phosphatizing baths. These comprise a lower free acid content which results in higher saturation of the primary zinc phosphate and thus provides a greater tendency to deposit at temperatures at or close to ambient temperature. In many cases, depending on the type of soil, it is difficult to reduce the cleaner temperature and, as a result, the cleaned parts tend to heat the phosphatizing stage to approximately 50°C even when no external heat is used.

Iron Phosphate Coatings. These coatings, weighing about 40–60 milligrams per square foot (430–646 milligrams per square meter), provide good paint adhesion, but inferior heat and corrosion resistance. They are used when coating performance is not very demanding. Iron phosphate coatings appear as a very thin blue or brown film to the naked eye.

Crystalline Zinc Phosphate Coatings. These coatings, weighing about 200 milligrams per square foot (2152 milligrams per square meter) when applied by spray, or as much as 3000 milligrams per square foot (32,280 milligrams per square meter) when applied by immersion, are of a medium-gray color and are used when higher quality is mandatory. For even greater corrosion resistance and paint adhesion, *microcrystalline zinc phosphate* coatings are used. They are dark in color and give coating weights of about 150–200 milligrams per square foot (1614–2152 milligrams per square meter) when sprayed and up to 1000 milligrams per square foot (10,760 milligrams per square meter) when applied by immersion. They are used as a paint base for products which are expected to last for years under varying environmental conditions.

Phosphating Chemistry. Iron and iron oxide react with phosphoric acid to form soluble primary iron phosphate, $Fe(H_2PO_4)_2$, liberating H_2 and H_2O, respectively. Due to the consumption of acid at the metal interface, there is a local rise in pH, which causes insoluble secondary iron phosphate, $FeHPO_4$, to coat the metal. The iron in the coating is supplied by the substrate.

In zinc phosphating, a small amount of iron phosphate is formed initially, but the bath contains primary zinc phosphate, $Zn(H_2PO_4)_2$, which crystallizes on the metal surface as secondary and tertiary zinc phosphates, $ZnHPO_4$ and $Zn_3(PO_4)_2$, respectively, when the pH rises at the metal/solution interface. The most frequently used baths contain accelerators, preferably nitrates and nitrites, which oxidize the hydrogen formed by the pickling reactions. The fundamental zinc phosphate reactions occur in three steps, all in the same bath:

(Pickling)

$$Fe^0 + 2H^+ \rightarrow Fe^{++} + H_2$$
$$3Fe^0 + 2NO_3^- + 8H^+ \rightarrow 3Fe^{++} + 2NO + 4H_2O$$

(Coating)

$$3Fe^{++} + 2H_2PO_4^- \rightarrow 4H^+ + Fe_3(PO_4)_2$$
$$3Zn^{++} + 2H_2PO_4^- \rightarrow 4H^+ + Zn_3(PO_4)_2$$

(Iron Removal)

$$4Fe(H_2PO_4)_2 + O_2 \rightarrow 4FePO_4 + 4H_3PO_4 + 2H_2O$$
(sludge)
$$Fe(H_2PO_4)_2 + NaNO_2 \rightarrow FePO_4 + NO + H_2O + NaH_2PO_4$$
(sludge)

In the coating reaction, each 3 moles of iron or zinc liberates 4 moles of hydrogen ion. However, in the pickling reaction, 8 moles of hydrogen ion are consumed. Thus, the pH at the metal interface rises, and insoluble tertiary ferrous phosphate and zinc phosphate crystallize on the iron surface. The coating closest to the metal interface is largely iron phosphate, while that farther away is rich in zinc phosphate.

Iron buildup in the bath is objectionable; in the iron-removal equations above, it is seen that dissolved Fe^{++} can be removed by oxidation, slowly in air or more rapidly by peroxides or nitrite, as shown in the final equation. The iron removed becomes ferric phosphate, while iron in the coating is ferrous phosphate.

Accelerators speed phosphating reactions by reacting with hydrogen liberated at the metal. Were hydrogen not removed, it would form gas bubbles which would interfere with metal/solution contact. Strong oxidizer accelerators also serve to precipitate dissolved iron and, to a degree, act as metal cleaners by oxidizing residual organic soils. In spray application, mild accelerators are usually adequate for maintaining dissolved iron at safe levels because atomization of the solution permits it to absorb from air the oxygen needed for precipitation of iron.

Zinc phosphate coatings consist of varying ratios of hopeite, $Zn_3(PO_4)_2 \cdot 2H_2O$, and phosphophyllite, $Zn_2Fe(PO_4)_2 \cdot 4H_2O$, with hopeite usually predominant. Hopeite and phosphophyllite grow epitaxially on alpha-iron crystallites. Only a slight adaptation deformation is necessary for the lattice planes of both foreign phases compared with the alpha-iron lattice of the substrate. Good adhesive strength can be expected from such a bond.

The smaller the crystals, the better the adhesion: crystal-to-metal, crystal-to-crystal, and crystal-to-final finish. Also, the smaller the crystals, the tighter the packing, the denser the coating, the less total porosity area for corrosive reactions to take place.

Crystal size depends upon such factors as growth rate, agitation, and the effects of nucleating agents and foreign atoms in the crystal lattice. Spray application provides agitation which reduces crystal size. Accelerators increase the number of nucleation sites on the substrate and result in smaller crystals. The most refined technique for producing very small crystals is the introduction of foreign elements of different atomic radii into the crystal lattice. A calcium additive in a zinc phosphating solution produces scholzite, $CaZn_2(PO_4)_2 \cdot 2H_2O$, in which the crystals may average one-twentieth the size of the finest crystals produced by other methods. It is believed that the foreign elements cause uneven growth along one crystal face, creating stresses that either stunt growth at an early stage, or rupture the crystal.

The relatively recent surge into the use of cathodic electrodeposition, especially in the automotive industry, has necessitated change in the metal pretreatment. The generation of hydroxyl ions at the cathode where the organic coating is deposited tends to dissolve zinc from the conventional zinc phosphate conversion coating with consequent deleterious coating performance. By increasing the concentration of phosphoric acid in the zinc phosphatizing stage, the ratio of iron to zinc in the conversion coating is increased. Such conversion coatings show superior performance under cathodically electrodeposited coatings.

The following sequence is typical of a production spray phosphate system: (1) cleaning for 60 seconds at 71–77°C; (2) rinsing for 15 to 30 seconds (hot or cold); (3) phosphating for 60 seconds at 54–60°C; (4) rinsing for 15–30 seconds (cold); and (5) chromate-rinsing for 30–45 seconds at 27–60°C. The final stage, an acid chromate rinse, is extremely important to the overall adhesion and corrosion resistance of the finish. One of its functions is to seal the pores in the phosphate coating. The effect can be demonstrated by placing a drop of chromic acid in the center of a phosphatized panel and subjecting it to corrosive exposure. Corrosion in the chromated area will be strongly inhibited.

Phosphating of Zinc. The chemistry in the phosphating of zinc alloys is similar, with the exception that iron does not play an important role in the coating or in the bath. The only cation involved is Zn. Zinc phosphate coatings are used widely in the treatment of galvanized steel for refrigerators, air conditioners, kitchen cabinets, and house and building siding.

Aluminum Pretreatment. Although Al protects itself against corrosion by forming a natural oxide, the protection is not complete. In the presence of moisture and electrolytes, Al alloys, particularly the high-copper alloys, corrode much more rapidly than pure Al.

Chemically produced oxides can be formed by treatment in 2–3% sodium carbonate containing 0.1% sodium dichromate for 10–20 minutes at 66°C, followed by "sealing" in 5% sodium dichromate at 82–88°C for 10 minutes. Such coatings are softer, more porous, and not as effective as those produced by chromic acid anodizing.

Electrically produced anodic coatings from chromic, sulfuric, or oxalic acid electrolytes are more dense and less porous. Their corrosion resistance is improved by hot-water sealing, which is even more effective with the inclusion of dichromate.

Generally, the higher the alloy content of the base aluminum, the heavier the oxide present and the more difficult it is to remove prior to effective chemical pretreatment. A properly formulated deoxidizer will remove the oxide only to the desired degree, with minimum attack on the base metal. The baths usually consist of fairly high concentrations (5–10%) of either nitric or sulfuric acid, along with chromates and either free or complex fluorides. A deoxidizer will also remove the smut that forms on etching in strong alkali. Smut consists of the alkali-insoluble alloying elements and their oxides.

Amorphous Phosphate Coatings for Aluminium. These coatings were introduced in 1945. Their simplicity of operation, speed, and economy have resulted in wide commercial acceptance. They provide a continuous uniform green coating with excellent paint-bonding properties and underfilm corrosion protection. The coatings consist of varying ratios of chromic phosphate and hydrated aluminum oxide. The bath contains hydrofluoric acid, which removes the natural oxide to permit contact of the coating-forming chemicals with the metal. The complexity of the reactions involved makes it difficult to present a simplified chemistry, but the results of many tests and analyses give the following coating composition: $xCrPO_4 \cdot yAl_2O_3 \cdot zH_2O$. The phosphate coatings vary from 10 to 300 milligrams per square foot (108 to 3228 milligrams per square meter), depending on the end use. The lower coating weights are used for paint bonding; the higher range is used for decorative purposes.

Gold-colored conversion coatings are formed in baths containing hydrofluoric acid to remove the natural oxide, and chromic acid. The coating composition is chromic chromate plus varying amounts of hydrated aluminum oxide. Some baths also contain ferricyanide iron, which greatly accelerates the coating action and forms some chromic ferricyanide in the coating. This constitutes one of the most widely used conversion coatings on aluminum because of its high speed, excellent corrosion resistance, and high affinity for organic finishes. The hexavalent chromium content permits these coatings to withstand somewhat more severe corrosive environments than do the amorphous phosphate coatings. The baths have a pH of about 1.2–1.9 and can be applied by dip, brush, spray, or reverse roll coater.

Proprietary chromate rinses are frequently used over conversion coatings on aluminum for increased corrosion resistance. See also **Autodeposition.**

—Wilbur S. Hall, Amchem Products, Inc., Ambler, Pennsylvania.

COOLER (Thermoelectric). **Thermoelectric Cooling.**

COORDINATION COMPOUNDS.

One of a number of types of complex compounds, usually derived by addition from simpler inorganic substances. Coordination compounds are essentially compounds to which atoms or groups have been added beyond the number of possible in terms of available electrovalent linkages; or the usual covalent linkages to which each of the two atoms linked donates one electron to form the duplet. The coordinate groups are linked to the atom of the compound usually by *coordinate valences*, in which both the electrons in the bond are furnished by the linked atom of the coordinated group. The ammines and complex cyanides are representative of coordination compounds.

In attempting to classify coordination compounds, Sidgwick noted that the number of molecules or atoms coordinated with a metallic atom (which he called the *coordination number*) is 2, 3, 4, 5, 6, or 8, and that 2, 4, and 6 are the most common. In forming such compounds, each molecule or atom donates a pair of electrons to the metallic atom, forming a semi-covalent type of bond. Thus, in the nitrocobaltates, six nitro groups each donate a pair of electrons to a cobalt(III) ion, forming the complex ion.

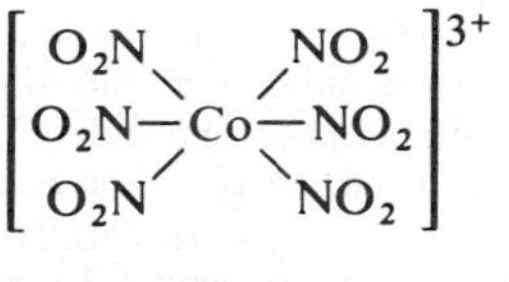

$$\begin{bmatrix} O_2N & & NO_2 \\ O_2N-Co-NO_2 \\ O_2N & & NO_2 \end{bmatrix}^{3+}$$

In the positive triammine cobalt(III) ion, three ammonia molecules donate one pair of electrons each to the cobalt(III) ion, to form the complex ion.

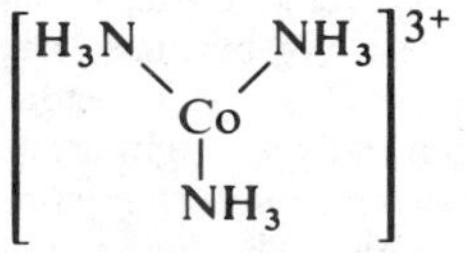

$$\begin{bmatrix} H_3N & & NH_3 \\ & Co & \\ & NH_3 & \end{bmatrix}^{3+}$$

The covalent character of the coordination bond is evident from the fact that both the ions and the molecules which form it fail to exhibit their characteristic reactions after coordination.

Included in the coordination compounds are the double salts, the complex salts, the oxysalts, and the hydrates.

See also **Chelation Compounds; Cobalt; Copper; Gold; Hydrate; Iron; Manganese;** and **Molybdenum.**

COORDINATION NUMBER.

The number of nearest neighbors of a given atom in a crystal structure. In covalent crystals, only those neighbors to which the atom is directly bonded are counted, and the number is usually 4 or less. In metals, the coordination number may be as high as 12, as in the close-packed structures.

COORDINATION POLYHEDRA.

The arrangement of oxygen ions about the cation to which they are closely bonded in an ionic crystal as, for example, the group SiO_4, which forms a tetrahedron. Such polyhedra pack as units in the crystal structure.

COPALITE (or Copaline).

The mineral copalite or "Highgate resin" is a fossil resin found in irregular fragments in the blue clay of London, England. It resembles copal, the resin of certain modern tropical trees. Copalite is pale yellow to greenish or brownish and emits an aromatic odor when broken. It has a hardness of 1.5; a specific gravity of 1.046; burns with a very smoky yellow flame.

COPOLYMER. **Elastomers; Polymerization.**

COPPER.

Chemical element symbol Cu, at. no. 29, at. wt. 63.546, periodic table group 1b, mp 1083°C, bp 2566 ± 0.5°C, density, 8.92 g/cm³. Elemental copper is a yellowish-red, soft, very malleable and ductile metal. Very thin sheet copper is translucent and transmits greenish-blue light. The element is unattacked by dry air, but in moist air containing CO_2, a protective greenish film of basic carbonate is formed. There are two natural isotopes, ^{63}Cu and ^{65}Cu. Seven radioactive isotopes have been identified, all with comparatively short half-lives: ^{58}Cu through ^{62}Cu, and ^{64}Cu, ^{66}Cu, and ^{67}Cu. Copper may have been the first metal to be used by people and today it ranks second, exceeded only by iron in annual consumption.

First ionization potential 7.723 eV; second, 20.29 eV; third, 29.5 eV. Oxidation potentials: $2Cu + 2OH^- \rightarrow Cu_2O + H_2O + 2e^-$, 0.361 V; $Cu + 2OH^- \rightarrow Cu(OH)_2 + 2e^-$, 0.224 V; $Cu^+ \rightarrow Cu^{2+} + e^-$, −0.153 V; $Cu \rightarrow Cu^{2+} + 2e^-$, −0.344 V; $Cu \rightarrow Cu^+ + e^-$, −0.522 V. Other important physical properties of copper are given under **Chemical Elements.**

Copper occurs as native copper particularly in the region south of Lake Superior (often 99.9% Cu), as sulfides (chalcocite, copper glance, cuprous sulfide, Cu_2S; chalcopyrite, $CuFeS$), as oxide (cuprite, cuprous oxide, Cu_2O, red); as basic carbonates (malachite, $CuCO_3 \cdot Cu(OH)_2$, green; azurite, $2CuCO_3 \cdot Cu(OH)_2$, blue). The copper content of its ores varies from 0.3 to 8% Cu and the average is of the order of 2.5%. The value depends largely upon the content of silver and gold. The area of production is widely distributed: in the United States, Montana, Utah, New Mexico, Arizona, Michigan, and Tennessee.

Copper frequently is detected in atmospheric particles, even in those collected at locations far removed from anthropogenic sources. Cattell and Scott (1978) reported similar enrichments found over the north Atlantic and the South Pole. Duce et al. (1975) proposed that measured atmospheric concentrations are larger than those predicted for unenriched crustal weathering or oceanic production. It was proposed that the enrichment may result from natural processes of anomalously enriched elements in aerosol particles. These may derive from low-temperature volatilization processes, such as biological methylation, or volcanism, or direct sublimation from the earth's crust, or emissions from plants or fractionating at the air-sea interface which enriches elements in particles produced from the oceans. Atmospheric studies conducted by Cattell and Scott near the island of Tasmania in 1977 led to the conclusion that a biogenic agent may be responsible for the approximately 20,000-fold enrichment of copper during aerosol production from the ocean.

Native copper ore is crushed, concentrated by washing with water, smelted, and cast into bars, Oxide and carbonate ores are treated with carbon in a smelter. Sulfide ore treatment is complex, but in brief, consists of smelting to a matte of cuprous sulfide, ferrous sulfide, and silica, which molten matte is treated in a converter by the addition of lime and air is forced under pressure through the mass. The products are blister copper, ferrous calcium silicate slag, and SO_2. Refining is conducted by electrolysis, and the anode mud is treated to obtain the gold and silver.

See also **Azurite; Chalcocite; Chalcopyrite; Cuprite; Malachite; Mineralogy.**

Leading world producers of copper include: United States (23.9%); Chile (11.8%); Canada (11.8%); Zambia (10.9%); U.S.S.R. (10.6%); Zaire (7.0%); Peru (3.7%); Philippines (3.0%); South Africa (2.9%); Australia (2.6%); Japan (2.1%); and China (1.7%).

Copper is distinguished by several properties which contribute to its extensive use: (1) a combination of mechanical workability with corrosion resistance to many substances, (2) excellent electrical conductivity, (3) superior thermal conductivity, (4) effect as an ingredient of alloys to improve their physical and chemical properties, (5) efficiency of copper and some of its compounds as catalysts for several kinds of chemical reaction, (6) nonmagnetic characteristics, advantageous in electrical and magnetic apparatus, and (7) non-sparking characteristics, mandatory for tools for use in explosive atmospheres. There are additional attractions of copper for many other applications. The metal would be used even more widely, but for some uses, even though superior, copper cannot compete with substitute materials because of cost.

Unalloyed Copper: In the United States, the term *copper* signifies copper that contains less than 0.5% impurities or alloying elements. Copper-base alloys are those that contain no less than 40% copper. Additionally, copper appears as a minor, but important ingredient of several alloys. There are six major types of commercial, unalloyed copper. These are described briefly in Table 1.

Very-High Copper Alloys: Although not meeting the foregoing definition of copper precisely, there is a group of copper alloys which contain only a few percent of other ingredients and commonly these are also referred to as coppers, usually with the name of the other element preceding copper in the name—as chromium copper or beryllium copper. These very-high copper alloys are described briefly in Table 2.

The Brasses. There are eight principal categories of brasses, not including the leaded and alloy brasses. Brass essentially is an alloy of copper and zinc. Several of the brasses contain other ingredients, such as lead and iron. When zinc is added to copper, there is a progressive alteration of color and lowering of melting point. When the zinc content is about 10%, the metal is a bronze color; with 15% zinc, the color may be described as golden; from 20–40% zinc, there is a range of yellow colors; over 45% zinc, the color is silver-white. The melting point of a 95% copper–5% zinc brass is about 1,065°C, whereas the melting point of 50% copper–50% zinc brazing metal drops to about 880°C. The ratio of copper to zinc also progressively affects mechanical and corrosion-resistance properties. Maximum tensile strength, for example, is attained with a 55% copper content, whereas maximum ductility is attained with a 70% copper content. This exceptional range of properties accounts for the availability and

TABLE 1. COMMERCIAL UNALLOYED COPPERS

Electrolytic Tough-Pitch Copper
Cu, 99.90%; O, 0.04% nominal; density 8.89–8.94 g/cm³, EC, 101%
Architecture: downspouts, flashing, building fronts, gutters, screening, roofing
Automotive: radiators and gaskets
Electrical: conductive-wire contacts, terminals, switch parts, bus bars
Hardware: cotter pins, nails, rivets, soldering copper, ball floats
Other: anodes, chemical process equipment, kettles, pans, printing rolls, expansion plates, rotation bands, die-pressed forgings

Deoxidized Copper
Cu, 99.90% minimum; P, 0.025% nominal; density 8.94 g/cm³, EC, 80–90%
Industrial: condensers, evaporators, heat exchangers, dairy tubes, fractionating columns, kettles, pulp and paper piping, steam and water piping, tanks
Transportation: gasoline, oil, air, and hydraulic fluid lines, oil coolers
Other: shell rotation bands, die-pressed forgings, gauge lines

Oxygen-free Copper
Cu, 99.92% minimum; no residual oxidants; density 8.89–8.94 g/cm³, EC, 101%
Electrical: conductors, electron tubes, bus bars, waveguides (for operation at high temperatures in presence of reducing gases)
Industrial: heaters, oil coolers, gasoline supply lines, radiators, refrigeration lines, water piping

Silver-bearing Copper
Cu, 99.90% minimum; 8–25 ounce (226–708 grams) Ag/ton; density 8.91 g/cm³, EC, 100–101%.
Electrical: commutator bars, heavy-duty motor windings (particularly for retention of strength at elevated temperatures)
Other: brazing solders, die-pressed forgings

Arsenical Copper
Cu, 99.68% nominal; P, 0.025% nominal; As, 0.30% nominal; density 8.94 g/cm³, EC, 90%
Industrial: heat-exchangers, boilers, radiators, condenser tubes

Free-cutting Copper
Cu, 99.4–99.5%; Te, 0.5–0.6%; density 8.94 g/cm³, EC, 90%
Industrial: electrical connectors, motor and switch parts, soldering coppers, screw-machine parts, forgings, welding-torch tips

---

EC, electrical conductivity (International Annealed Copper Standard).

demand for a wide variety of brasses. Metallurgically, brasses may be classified as (1) *alpha brass,* in which the content of zinc is less than 36% and in which the zinc is dissolved in the copper, imparting to the alloy the basic structure of copper; (2) *beta brass,* in which the content of zinc ranges between 36% and 45%. This alloy contains the CuZn as a compound and enhances the hot workability of the alloy; and (3) *gamma brass,* in which the zinc content exceeds 45% and where there are Cu_2Zn_3 crystals in the alloy. This combination does not lend itself to either hot or cold workability. Some of the important commercial brasses are described briefly in Table 3.

The Bronzes. Classically, a bronze is defined as an alloy of copper and tin, but over the years, the term has taken on a much broader meaning. The term may apply to numerous copper alloys that possess a crystalline, bronzelike structure, are of a bronze color, or simply because they may contain some tin. Further, bronze generally is considered a casting metal. In contrast, brass is generally wrought. Some alloys are commercially named bronze even though they contain no tin whatever.

Copper Wire and Cable. *The International Annealed Copper Standard* (IACS), which sets annealed copper as having 100% electrical conductivity as a basis against which to compare other metals, alloys, and materials, is accepted internationally. Using this standard for comparison, the conductivity of copper is exceeded only by silver for which the IACS figures is 108.4%. This comparison is on the basis of conductivity per gram, pound, or other mass unit. Aluminum, although widely used as an

TABLE 2. VERY HIGH COPPER ALLOYS

Cadmium Copper
Cu, 99.00-plus %; Cd, 0.6–1.0%
Cadmium toughens copper and increases resistance to fatigue; also increases softening temperature.
Electrical conductivity (fully annealed) is about 95% (IACS).
Essentially free of oxygen; not susceptible to gassing.
Uses: contact wires used in electrical transportation, notably long-span overhead electric transmission lines.

Chromium Copper
Cu, 99.50%; Cr, 0.5%
Chromium improves mechanical properties while retaining high thermal and electrical conductivities.
Strength and hardness depend on heat treatment and not cold working—hence alloy can be used up to temperature of about 450°C without danger of softening.

Tellurium Copper
Cu, 99.50%; Te, 0.5%
Tellurium increases softening temperature of work-hardened copper.
Alloy is excellent where combination of good machinability and electrical conductivity is required.
Uses: motor and switch parts, electrical connectors, screw-machine parts, electrical instrument parts.

Beryllium Copper
Type 1: Cu, 98%; Be, 2%
Type 2: Cu, 97%; Be, 0.4%; Co, 2.6%
Cobalt is added as a lower-cost substitute for beryllium.
Uses: instrument springs, bellows, diaphragms, bourdon tubes, non-sparking tools for hazardous locations.
Alloy permits springs to be shaped while soft, followed by hardening.

Selenium copper also available for combining high electrical conductivity with free-machining and hot-working properties. Alloy makes excellent copper-to-glass seals.

---

electrical conductor for selected applications, has a rating on this schedule of approximately 61%, steel a rating of 11%, and nickel-chromium alloy (valued because of its high electrical resistance rather than conductivity) has a rating of 1.5%. Thus, the value of copper for electrical conductors, considering its availability and economics, is self-evident.

Processing Equipment. In terms of thermal conductivity, assigning a value of 100 to copper, the metal is exceeded only by silver which has a value of 108. Copper is followed by gold (76), aluminum (56), magnesium (41), zinc (29), nickel (15), iron (15), steel (13–17), lead (9), and antimony (5). Thermal conductivity means good heat transfer and this is extremely important in most industrial processing equipment where heating and cooling cycles are involved. This property, when combined with corrosion resistance, makes copper attractive for the construction and lining of process vessels. Deoxidized copper, admiralty brass, and arsenical copper are effective in condenser tubes operating with fresh water. Copper is less suitable for seawater because of its inability to form a protective film. Aluminum brass and 70–30 cupronickel alloy are favored for severe seawater service. Copper and copper alloys are not suited for use in oxidizing acidic solutions, in mercury, or in the presence of free NH_3. Copper vessels are used extensively in food processing and for numerous organic materials, particularly distillation columns and hardware. The relative high cost of copper as compared with other metals, however, is always an important factor.

Piping. Copper and copper alloys are used in a wide variety of pipes and tubes, both for industrial and domestic systems. In many areas, galvanized water pipe has been almost completely replaced by copper piping in new construction. Advantages include corrosion resistance and ease of installation, which offset higher costs. Because of excellent thermal conductivity,

copper and brass fittings are used widely in hot water and steam-heating systems. However, the greater conductivity of bare copper pipe in long runs requires more attention to insulation covering.

Chemistry and Compounds. Copper is dissolved best by HNO_3; not attacked by cold dilute HCl or H_2SO_4, but in hot HCl dissolves to yield cuprous chloride, in hot concentrated H_2SO_4 to yield copper sulfate; attacked by chlorine, especially when heated, to form cuprous and cupric chlorides; only slight action by H_2S or SO_2 at ordinary temperatures in the absence of air.

In view of its $3d^{10}4s^1$ electron configuration and the relatively small energy difference between the two levels, copper forms dipositive ions as well as monopositive ones. In fact, the former are the more stable in aqueous solution, due primarily to the larger heat of hydration of Cu^{2+} than Cu^+. Moreover, the d-electrons may participate in bonding, and tripositive copper, Cu(III), appears in complexes. In addition, copper forms a number of compounds essentially covalent in character, such as copper(I) oxide, Cu_2O.

This compound is less stable at room temperature than copper(II) oxide, CuO, although Cu_2O occurs in nature (as cuprite). It is the stable oxide above 1,026°C. It is prepared by fusion of copper(I) chloride, CuCl with sodium carbonate, Na_2CO_3. In its crystal, each copper atom has two colinear bonds, and each oxygen atom four tetrahedral ones; two such interpenetrating lattices constitute the structure. Copper(I) hydroxide, CuOH, is relatively stable, and is produced by electrolysis of a sodium chloride solution between copper electrodes (by action of NaOH on the cathode). Copper(II) oxide, produced by heating copper in air, is also essentially covalent (Cu—O, 1.95 Å); it has tetrahedral bonding of the oxygen atoms, and coplanar bonding of the copper atoms. Copper(II) hydroxide, $Cu(OH)_2$, is precipitated by alkali hydroxides from Cu^{2+} solutions. It is gelatinous, and its composition and solubility vary somewhat with the alkali concentration. It is thermodynamically unstable even in contact with liquid water with respect to dehydration to CuO, but this occurs only very slowly, except upon heating or when catalyzed by hypochlorite, hydrogen peroxide, etc.

Copper(I) halides are formed with chlorine, bromine and iodine, the chloride and bromide by reduction of the copper(II) halides with copper powder, and the iodide by reduction of copper(II) sulfate, $CuSO_4$, solution with potassium iodide. The fluoride appears never to have been made, despite reports to the contrary. All are insoluble in H_2O. Copper(II) fluoride, CuF_2 may be made from CuO and hydrofluoric acid at 400°C, copper(II) chloride, $CuCl_2$ by dissolving the oxide or carbonate in HCl, and copper(II) bromide, $CuBr_2$, from copper and bromine water; copper(II) iodide, CuI_2, is unstable at room temperature with respect to decomposition into CuI and iodine. The chloride and bromide are water-soluble, and ionic. The fluoride is only slightly water-soluble. Anhydrous copper(II) chloride, $CuCl_2$, is monoclinic and its structure contains infinite chain molecules formed by $CuCl_4$ groups that share opposite edges. $CuBr_2$ has a similar structure.

Complex halides of both monovalent and divalent copper are known. The monovalent complexes are primarily of the composition $MCuX_2$, where X is a halogen atom and M usually an alkali metal, although $CuCl_3^{2-}$ ions are also known, being found in infinite chain $(CuCl_3^{2-})_n$ structures, as in crystals of Cs_2CuCl_3. The ion $CuCl_4^{3-}$ is also known. The composition of the copper(II) complex halides is primarily in terms of $CuCl_3^-$, $CuCl_4^{2-}$, or $CuBr_4^{2-}$ ions, although the corresponding complex fluoride has trivalent copper, as in K_3CuF_6. Its paramagnetic moment indicates two unpaired electrons.

TABLE 3. REPRESENTATIVE BRASSES AND BRONZES

Gilding Brass
Cu, 95%; Zn, 5%; Pb, 0.03% maximum; Fe, 0.05% maximum; density 8.86 g/cm^3; mp 1066°C; AT, 427–788°C; HWT, 760–871°C
Coinage: coins, metals, tokens
Munitions: firing-pin support shells, bullet jackets, fuse caps, primers
Novelties: emblems, plaques, jewelry
Other: base for gold plate and for vitreous enamel

Commercial Bronze
Cu, 90%; Zn, 10%; Pb, 0.05% maximum; Fe, 0.05% maximum; density 8.80 g/cm^3; mp 1043°C; AT, 427–788°C; HWT, 760–871°C
Architectural: grillwork, etching bronze, screen cloth, weather stripping
Cosmetics: lipstick cases, compacts
Hardware: kickplates, line clamps, marine hardware, escutcheons, rivets, screws
Munitions: rotating bands, primer caps
Other: costume jewelry, screen wire, ornamental trim, vitreous enamel base

Red Brass
Cu, 85%; Zn, 15%; Pb, 0.06% maximum, Fe, 0.05% maximum; density 8.75 g/cm^3; mp 1027°C; AT, 427–732°C; HWT, 788–900°C
Architectural: trim, etching parts, weather stripping
Electrical: screw shells, sockets, conduit
Hardware: fasteners, fire extinguishers, eyelets
Industrial: heat-exchanger tubes, condensers, flexible hose, piping, pumps, radiator cores, pickling crates
Other: compacts, costume jewelry, dials, badges, etched articles, lipstick cases

Jewelry Bronze
Cu, 87.5%; Zn, 12.5%; Pb, 0.05% maximum; Fe, 0.10% maximum; density 8.78 g/cm^3; mp 1035°C; AT, 427–760°C; HWT, 760–900°C
Architectural: angles, channels
Hardware: chains, fasteners, slide fasteners, eyelets
Novelties: costume jewelry, emblems, compacts, etched articles, lipstick cases, plaques
Other: base for gold plate

Low Brass
Cu, 80%; Zn, 20%; Pb, 0.05% maximum; Fe, 0.05% maximum; density 8.67 g/cm^3; mp 999°C; AT, 427–704°C; HWT, 816–900°C
Architectural: medallions, spandrels, ornamental metalwork
Electrical: battery caps
Instruments: bellows and muscal instruments
Hardware: flexible hose, pump lines, tokens, clock dials

Cartridge Brass
Cu, 70%; Zn, 29-plus %; P, 0.07% maximum; Fe, 0.05% maximum; density 8.53 g/cm^3; mp 954°C; AT, 427–760°C; HWT, 732–843°C
Automotive: radiator cores and tanks, reflectors
Electrical: flashlight shells, lamp fixtures, socket shells, screw shells, bead chain
Hardware: fasteners, pins, rivets, eyelets, springs, tubes, stampings
Munitions: various components. Note: Admiralty brass is similar: Cu, 71%; Zn, 28%; Sn, 1%

Yellow Brass
Cu, 65%; Zn, 34-plus %; Pb, 0.15% maximum; Fe, 0.05% maximum; density 8.47 g/cm^3; mp 932°C; AT, 427–704°C
Architectural: grillwork

Automotive: reflectors, radiator cores and tanks
Electrical: lamp fixtures, flashlight shells, screw shells, socket shells, bead chain
Hardware: kick plates, push plates, locks, hinges, grommets, fasteners, eyelets, stencils, plumbing accessories, pins, rivets, screws, springs

Muntz Metal
Cu, 60%; Zn, 39-plus %; Pb, 0.30% maximum; Fe, 0.07% maximum; density 8.39 g/cm^3; mp 904°C; AT, 427–593°C; HWT, 621–788°C
Hardware: large nuts and bolts, brazing rod, condenser plates, valve stems, hot forgings

Leaded Brasses
When lead is added to brass up to about 4%, improved machinability results. The lead has practically no effect on tensile strength or hardness. However, for cold-worked materials, lead does lower ductility and shear strength.

Phosphor Bronzes
Although tin is the primary alloying element in these alloys, their name derives from the addition of small quantities of phosphorus used as a deoxidizing agent in casting the alloys. Tensile strength ranges from moderate to very high, decreasing with amount of tin added. Tin percentage will range from 1.25 to 10%. Of the copper alloys, the phosphor bronzes are best suited for sea duty and where acid reagents may be present.

Silicon Bronzes
Most of these alloys are of proprietary compositions and are known by a variety of trade names. Silicon content ranges from 1.5 to 3.5%; usually less than 1.5% zinc content. Tin, manganese, and iron also may be added in small quantities. Because of their excellent strength, ease of welding, and corrosion resistance, the alloys have become important construction materials. As the silicon content increases, the alloys become more subject to fire cracking.

Aluminum Bronzes
The aluminum content of these alloys ranges from 4 to 10%. They are moderately hard, very ductile, and tough. The alloys resist scaling and oxidation at high temperatures because of the aluminum content. They perform well in both acids and alkalis. The alloys are good for sea duty, particularly in contact with turbulent seawater.

Nickel Silvers
Nickel essentially is added to copper-zinc alloys to enhance color. With a nickel content of about 18%, the alloy is silver-white. Also, most of the mechanical properties and corrosion resistance are improved. The alloys find wide application for operations that require ductility in the cold condition, as in stamping, spinning, deep drawing, and for articles to be plated. An alloy widely used as a spring material because of its high tensile and fatigue properties has the composition: Cu, 55%; Zn, 27%; Ni, 18%. German silver contains: Cu, 50%; Ni, 30%; Zn, 20%. It is interesting to note that the nickel silvers do not contain silver.

Cupronickels
The nickel silvers generally are classified as brasses. Cupronickels fall more into basic copper–nickel alloys. Possible minor ingredients are manganese, iron, and zinc. These alloys can be used for severe drawing, spinning, and stamping operations because they do not work harden readily. They also are extensively used for condenser tubes and plates, heat exchangers, and other process equipment.

AT, annealing temperature range.
HWT, hot-working temperature range.

Copper oxyhalides of a number of different compositions have been reported, but the most definitely established compositions are $Cu(OH)Cl$, $Cu_2(OH)_2Cl_2$, and $CuBr_2 \cdot 2Cu(OH)_2$. The property of forming basic salts is not limited to the halides of copper. Basic sulfates, such as $CuSO_4 \cdot 2Cu(OH)_2$, $CuSO_4 \cdot 3Cu(OH)_2$, $CuSO_4 \cdot 4Cu(OH)_2$, and $CuSO_4 \cdot 5Cu(OH)_2$ have been prepared, more or less hydrated. In copper(II) carbonate, the stable forms are oxycarbonates, $xCuCO_3 \cdot yCu(OH_2)$, where the $x:y$ ratios may be 2:1, 1:1, 2:3, 1:9, and still other values. Many of these compositions occur in minerals, such as malachite and azurite. The halogen complexes were discussed above. In general, copper tends to be 6-coordinate, as in the complex ion $[Cu(H_2O)_2(ethylenediamine)_2]^{2+}$. With NH_3 and many amines, stable complexes are formed, both of Cu^+, such as $[Cu(NH_3)_2]^+$ and Cu_2^+, such as $[Cu(NH_3)_4]^{2+}$ which add halogen, pseudohalogen, and many other anions to form compounds of the composition $Cu(NH_3)_2X_2$ and $Cu(NH_3)_4X_2$.

Among the other copper compounds, copper(II) acetate is used as a pigment and fungicide; in its basic form it is the familiar verdigris that forms on copper surfaces in the presence of moisture and organic matter. The arsenic compounds of copper are used as insecticides and wood preservatives: copper(II) arsenite is called "Scheele's green" and copper(II) acetoarsenite $Cu(AsO_2)_2 \cdot Cu(C_2H_3O_2)_2$ is "Paris green." Copper(II) hexacyanoferrate(II), brown, is precipitated from copper(II) solutions by soluble hexacyanoferrates(II), even from very dilute solutions, and copper(II) sulfide, black, by H_2S or other soluble sulfides. Copper(II) sulfate, when hydrated, forms its characteristic blue crystals (the hydrated Cu^{2+} ion is blue).

Copper(I) cyanide dissolves an alkali cyanide solution to form cyanocuprates(I) of the general formula $M_n[Cu(CN)_{n+1}]$ where M is an alkali metal, and n ranges in value from 1 to 5. Not all of the values, of course, are found for a particular alkali or in the presence of particular anions. With sodium cyanide, NaCN, most of the complex present has an n value of 2, but if the original solute was CuCl, more of the $n = 3$ cyanocuprate(I) is present. At low temperatures, values of n of 4 and 5 are found. Copper(II) appears to coordinate four cyanide ions to form the unstable tetracyanocuprate(II) ion which decomposes at once to the copper(I) complex and cyanogen. In fact, for Cu(II) the chelated complexes are more common than the simple ones, examples being those formed with ethylenediamine and its derivatives, oxalates, catechol, and the β-diketones.

A solution of CuCl in HCl absorbs carbon monoxide, forming copper(I) carbonyl chloride, $Cu(CO)Cl \cdot H_2O$. This reaction, which is used in gas analysis, is indicative of the ability of copper to combine with carbon monoxide. Evidence for a true carbonyl is limited to the observation that if hot carbon monoxide is passed over hot copper, a metallic mirror is produced in the hotter parts of the tube. Other organometallic compounds include the very unstable methyl copper, CH_3Cu, phenyl copper, C_6H_5CU, and bischlorocopper acetylene, $C_2H_2(CuCl)_2$.

Copper Industrial Chemicals. Copper oxides, salts, and organocopper compounds find extensive use in industry and commerce. Some of the more important compounds are:

Cupric Acetate, $Cu(C_2H_3O_2)_2 \cdot H_2O$, sp gr 1.88, mp 115°C, decomposes at 240°C, dark brown powder, slightly soluble in cold H_2O and alcohol; moderately soluble in hot H_2O and ether. Used as a fungicide, insecticide, as a catalyst, and in pigments.

Cupric Acetoarsenite (Paris Green), $(CuOAs_2O_3)_3 \cdot Cu(C_2H_3O_2)_2$, emerald green powder, very slightly soluble in cold H_2O, soluble in alcohol and potassium cyanide. Used as an insecticide, wood preservative, and paint pigment.

Cupric Acid Orthoarsenite (Scheele's Green), $CuHAsO_2$, green powder, insoluble in H_2O, soluble in alcohol, acids, and NH_4OH. Used as an insecticide and wood preservative.

Copper Carbonate (Basic), $CuCO_3 \cdot Cu(OH)_2$, dark green monoclinic crystals, insoluble in cold H_2O, decomposes in hot H_2O, soluble in potassium cyanide. Malachite, a copper ore, is of this composition. Refined compound is used as a pigment.

Cupric Hydroxide, $Cu(OH)_2$, blue, gelatinous compound, insoluble in cold H_2O, decomposes in hot H_2O, soluble in alcohol, NH_4OH, and potassium cyanide. Used as a pigment.

Cuprous Cyanide, $Cu_2(CN)_2$, white monoclinic crystals, insoluble in H_2O, soluble in HCl, NH_4OH, and potassium cyanide. Used in Sandmeyer's reaction to synthesize aryl cyanides.

Cuprous Iodide, Cu_2I_2, cubic white crystals, practically insoluble in H_2O or alcohol, soluble in NH_4OH, potassium iodide, or potassium cyanide. Used in Sandmeyer's reaction to synthesize aryl chlorides.

Cupric Oxide, CuO, black cubic crystals, insoluble in H_2O, soluble in HCl, NH_4OH, or ammonium chloride. Used as a green and blue colorant in ceramics.

Cuprous Oxide, Cu_2O, red cubic crystals, insoluble in H_2O, soluble in HCl, NH_4OH, or ammonium chloride. Cuprite, a copper ore, is of this composition. Refined compound is used in electrical rectifiers.

Cupric Sulfate, $CuSO_4 \cdot 5H_2O$, blue triclinic crystals, moderately soluble in cold H_2O, quite soluble in hot H_2O, very slightly soluble in alcohol. Used in copper plating, dyestuff manufacture, water treatment, germicides, and coppering of steels.

Cupric Chloride, $CuCl_2$, brown-yellow powder, quite soluble in cold H_2O or alcohol, very soluble in hot H_2O. Catalyst for several organic syntheses, including production of vinyl chloride monomer.

References

Blackwood, A. W., and J. E. Casteras: "Copper" in "Metals Handbook," 9th edition, Vol. 2., American Society for Metals, Metals Park, Ohio, 1979.

Cattell, F. C. R., and W. D. Scott: "Copper in Aerosol Particles Produced by the Ocean," *Science*, **202**, 429–430 (1978).

Duce, R. A., Hoffman, G. L., and W. H. Zoller: *Science*, **183**, 198 (1974).

Gale, N. H., and Z. A. Stos-Gale: "Bronze Age Copper Sources in the Mediterranean: A New Approach," *Science*, **216**, 11–19 (1982).

Jovanović, B.: "The Origins of Copper Mining in Europe," *Sci. Amer.*, **242**, 5, 152–167 (1980).

Marchant, G. R., et al.: "Digital Controls for Continuous Copper Smelting," *Instrumentation Technology*, **25**, 6, 51–57 (1978).

Rosenbaum, J. B.: "Minerals Extraction and Processing," *Science*, **191**, 720–723 (1976).

Staff: "Introduction to Copper and Copper Alloys," in "Metals Handbook," 9th edition, Vol. 2, American Society for Metals, Metals Park, Ohio, 1979.

COPPER (In Biological Systems). The activity of copper in plant metabolism manifests itself in two forms: (1) synthesis of chlorophyll, and (2) activity of enzymes. In leaves, most of the copper occurs in close association with chlorophyll, but little is known of its role in chlorophyll synthesis, other than the fact that the presence of copper is required. Copper is a definite constituent of several enzymes catalyzing oxidation-reduction reactions (oxidases), in which the activity is believed to be due to the shuttling of copper between the +1 and +2 oxidation states.

Traces of copper are required for the growth and reproduction of lower plant forms, such as algae and fungi, although larger amounts are toxic.

The effects of copper deficiency in plants are varied and include: die-back, inability to produce seed, chlorosis, and reduced photosynthetic activity. In contrast, excesses of copper in the soil are toxic, as in the application of soluble copper salts to foliage. For this reason, copper fungicides are formulated with a relatively insoluble copper compound. Their toxicity to fungi arises from the fact that the latter produce compounds, primarily hydroxy and amino acids, which can dissolve the copper compounds from the fungicide.

Copper is a necessary trace element in animal metabolism. The human adult requirement is slight. The adult human body contains 100–150 milligrams of copper, the greatest concentrations existing in the liver and bones. Blood contains a number of copper proteins, and copper is known to be necessary for the synthesis of homoglobin, although there is no copper in the hemoglobin molecule.

Anemia can be induced in animals on a low copper diet,

such as milk, and appears to be due to an impaired ability of the body to absorb iron. This anemia, however, is rare, because of the widespread occurrence of copper in foods. In locations such as Australia and the Netherlands, diseases of cattle and sheep, involving diarrhea, anemia and nervous disorders, can be traced either to a lack of copper in the diet, or to excessive amounts of molybdenum, which inhibits the storage of copper in the liver.

Ingestion of copper sulfate by humans causes vomiting, cramps, convulsions, and as little as 27 grams of the compound may cause death. An important part of the toxicity of copper to both plants and animals is probably due to its combination with thiol groups of certain enzymes, thereby inactivating them. The effects of chronic exposure to copper in animals are cirrhosis of the liver, failure of growth, and jaundice.

Copper deficiency in plants is most frequent on organic soils, such as newly drained bogs, and on very sandy soils. The severe copper deficiency often found when bogs and marshes are first used for crop production is called *reclamation disease* in some parts of the world.

Ruminants are sensitive to copper deficiency. The symptoms of copper deficiency in animals vary with the species and age, but often the fading of brown or black hair is evident. On some acidic soils, the use of copper in fertilizers increases crop and pasture production, and the increase in level of copper in the plants helps to prevent copper deficiency in the cattle and sheep. In parts of Australia, livestock production was impossible until copper fertilizers were used on the pastures. Application of copper fertilizers to alkaline soils generally does not increase the copper level in the crop. Farm animals are often supplied with copper in the form of dietary mineral supplements. Compounds used include copper gluconate, copper oxide, and copper sulfate.

Although copper fertilizers will sometimes increase crop yields and improve the nutritional quality of the crops, this practice must be used with caution and only on copper-deficient soils. Both plants and animals are subject to toxicity from excessive levels of copper. Ruminants, especially sheep, are sensitive to copper toxicity as well as to copper deficiency. Adding a copper fertilizer to a soil that naturally contains rather high levels of available copper may increase levels of the metal in the forage to the point of causing copper toxicity in grazing sheep. Copper toxicity from soils naturally high in copper occurs in Australia, but is uncommon in the United States. There are soils in the United States, however, that produce forages with levels of copper close to toxicity limits, and if copper-bearing mineral supplements are inadvertently used with these forages, copper toxicity to sheep may result.

It is not always possible to set a definite limit, in terms of the copper concentration in the diet, that will permit accurate predictions of the danger of copper deficiency, or of copper toxicity in cattle and sheep. In particular, if the molybdenum concentration in the forage is high, extra amounts of copper are needed to prevent deficiency. Also, higher copper levels can be tolerated without danger of toxicity.

Monogastric animals, including humans, are less sensitive than ruminants to either copper deficiency or toxicity. Copper deficiency in people has been found only when other complications, such as excessive bleeding, general starvation, and iron deficiency, are also present. Wilson's disease, an inherited disease of humans, prevents the loss of excess copper from the body and brings on copper toxicity. No direct relationships have been found between levels of available copper in the soil and the copper status of humans.

A number of copper-containing protein compounds are enzymes with an oxidase function (ascorbic acid oxidase, urease, etc.) and these play an important role in the biological oxidation-reduction system. There is a definite relationship of copper with iron in connection with utilization of iron in hemoglobin function.

Copper absorption is depressed by ascorbic acid, dietary phytates, cadmium, mercury, silver, and zinc. It appears that metals impede copper absorption through competition for metal-binding sites. Dietary copper, molybdenum, and sulfur are closely interrelated in optimum copper and molybdenum nutrition of ruminants. Increased pasture molybdenum content and low-pasture copper result in a condition known as "peat scours."

Copper toxicity tends to accumulate in the liver. The capacity to tolerate copper varies considerably with the species. Sheep are most susceptible. Swine have a much greater tolerance and copper may be added to the swine diet for pharmacological reasons, for example, use as an antihelmintic (internal parasite).

Continuing research is providing a better understanding of the biological role of copper. The prooxidant and antioxidant effects of ascorbic acid and metal salts, including copper, in a β-carotene-linoleate model system were studied by Israeli scientists Kanner, Mendel, and Budowski (1977). The interacting effects of ascorbic acid and metal ions on carotene oxidation were studied in an aqueous carotene-linoleate solution at pH 7. Ascorbic acid at concentrations up to 10^{-3}M was a prooxidant. Fe^{3+} and, to a lesser extent Co^{2+}, acted synergistically with ascorbic acid, the prooxidant effect increasing with metal concentration. Cu^{2+} formed a prooxidant system with ascorbic acid only at low metal concentration, but as the copper concentration was raised, inversion of activity occurred and the copper-ascorbic acid system exerted a stabilizing action on carotene. Prooxidant effects were enhanced and antioxidant effects weakened in the presence of added linoleate hydroperoxides. The latter were unstable in the presence of ascorbic acid and especially ascorbic acid $+ Cu^{2+}$. Ascorbic acid itself became unstable in the presence of Cu^{2+}. Oxygen depletion, brought about by the rapid oxidation of ascorbic acid, may be partly responsible for the carotene-stabilizing effect of the Cu^{2+} couple. The investigators postulated that additional stabilization results from the radical-scavenging properties of copper or of a copper chelate formed by ascorbic and/or dehydroascorbic acid.

Y. C. Lee and a team of investigators (Department of Food Science and Human Nutrition, Michigan State University, East Lansing, Michigan) studied the kinetics of ascorbic acid stability of tomato juice as functions of temperature, pH, and metal catalyst, including copper. The rate of copper-catalyzed destruction of ascorbic acid increased as copper concentration in tomato juice increased, and was affected by pH.

Relatively recent hypotheses concerning the effect of zinc-to-copper ratios in the diet as a determining factor of plasma cholesterol levels have been made (Klevay, 1973). In 1978, L. R. Helwig, Jr. and a team (Department of Poultry Science, Cornell University, Ithaca, New York) undertook a study to establish if this hypothesis could be demonstrated in an animal system (other than rat). They selected the White Leghorn laying hen. The chicken was selected as the model system for a number of reasons, including: (1) Any resulting alteration in plasma cholesterol levels in a species other than rat would further support the proposal that the zinc-to-copper ratio could be involved in human cholesterol metabolism; and (2) an alteration in the cholesterol metabolism of the chicken may result in lowered egg cholesterol levels, which would be beneficial to persons wishing to restrict dietary cholesterol without eliminating egg intake. As reported (Helwig et al., 1978), the researchers were unable to demonstrate any effect of the zinc-to-copper ratio upon cholesterol metabolism in the White Leghorn laying hen. The researchers suggested that this inability to support the hypothesis in a species other than rat indicates that further studies, possibly with human subjects, should be undertaken prior to accepting the concept for humans.

A study by Zenoble and Bowers (Department of Foods and Nutrition, Kansas State University, Manhattan, Kansas) undertaken in 1977 is exemplary of the much needed further research in determining the properties of certain elements, including copper, when contained in various food substances. Part of the study was directed at determining the effects of cooking on copper content of turkey muscle. The researchers found that copper was significantly lower in cooked than in raw breast turkey muscle, but similar in raw and cooked thigh muscle.

References

Campbell, J. K., and C. F. Mills: "Effects of Dietary Cadmium and Zinc on Rats Maintained on Diets Low in Copper," *Proceedings Nutr. Society*, 33, 1, 15A (Abstract) (1974).

Cort, W. M., Mergens, W., and A. Greene: "Stability of Alpha- and Gamma-Tocopherol: Fe^{3+} and Cu^{2+} Interactions," *Journal of Food Science*, 43, 3, 797–800 (1978).

Hamilton, R. P., et al.: "Zinc Interference with Copper, Iron and Manganese in Young Japanese Quail," *Journal of Food Science*, 44, 3, 738–741 (1979).

Helwing, L. R., Jr., et al.: "Effects of Varied Zinc/Copper Ratios on Egg and Plasma Cholesterol Level in White Leghorn Hens," *J. Food Sci.*, 43, 666–669 (1978).

Hill, C. H., and G. Matrone: "A Study of Copper and Zinc Interrelationships," *Proc. 12th World's Poultry Congress*, 219 (1962).

Kenner, J., Mendel, H., and P. Budowski: "Proxidant and Antioxidant Effects of Ascorbic Acid and Metal Salts in a Beta-Carotene-Linoleate Model System," *J. Food Sci.*, 42, 1, 60–64 (1977).

Kirchgessner, M. (editor): "Trace Element Metabolism in Man and Animals," Institut für Ernahrungsphysiologie, Technische Universität München, Freising-Weihenstephan, Germany, 1978.

Lee, Y. C., et al.: "Kinetics and Computer Simulation of Ascorbic Acid Stability of Tomato Juice as Functions of Temperature, pH and Metal Catalyst," *J. Food Sci.*, 42, 3, 640–644 (1977).

Sax, N. I.: "Dangerous Properties of Industrial Materials," Van Nostrand Reinhold, New York, 1979.

Zenoble, O. C., and J. A. Bowers: "Copper, Zinc and Iron Content of Turkey Muscles," *J. Food Sci.*, 42, 5, 1408–1412 (1977).

CORDIERITE. The mineral cordierite, composition (Mg, Fe)$_2$Al$_4$Si$_5$O$_{18}$ is an orthorhombic mineral frequently seen, however, in pseudo-hexagonal forms, as well as massive. It is brittle, with a subconchoidal fracture; hardness, 7–7.5; specific gravity, 2.53–2.78; luster, vitreous; color, blue of varying shades; translucent to transparent. Cordierite exhibits pleochroism (or dichroism) being dark blue, light blue and light yellow when examined by transmitted light in different directions. Hence, it is frequently called *dichroite*, and less frequently, *iolite*. It is occasionally used as a gem.

Cordierite is found as a primary mineral in the igneous rocks. It is, however, found ordinarily in gneisses, schists and in areas of contact metamorphism. Localities for good specimens are numerous in Europe, including Bavaria, Finland, Norway. It is found in Greenland, the Malagasy Republic and Ceylon, from which later place come the rolled pebbles of a rich blue color known as saphir d'eau, prized as a gem. In the United States, it is found principally in Connecticut.

Named for the French geologist, Pierre Louis Antoine Cordier, this mineral has also been called iolite from the Greek word meaning violet, and stone, as well as dichroite from the Greek meaning *two-colored*.

See also terms listed under **Mineralogy.**

CORROSION. (1) The electrochemical degradation of metals or alloys due to reaction with their environment; it is accelerated by the presence of acids or bases. In general, the corrodability of a metal or alloy depends upon its position in the activity series (electromotive force series). Corrosion products often take the form of metallic oxides; in the case of aluminum and stainless steel, this is actually beneficial, for the oxide forms a strongly adherent coating which effectively prevents further degradation. Hence, these metals are widely used for structural purposes. Probably the most familiar kind of corrosion is that of *rusting*. This is but a special case of a general classification known as *atmospheric corrosion*, wherein the oxygen of the atmosphere reacts with the material in question. Most metals, with exception of the noble metals, such as gold, can be oxidized by atmospheric oxygen. In the usual case, however, water vapor must be present before any appreciable oxidation can take place. With iron, for example, about 40% relative humidity is needed at ordinary temperatures before rusting will occur.

Acidic soils are highly corrosive. Sulfur is a corrosive agent in automotive fuels and in the atmosphere (SO_2) as well and is frequently mentioned in connection with so-called acid rains. Sodium chloride in the air at locations near the sea is strongly corrosive, especially at temperatures above 70°F (21.1°C). Copper, nickel, chromium, and zinc are among the more corrosion-resistant metals and are widely used as protective coatings for other metals.

(2) The term *corrosion* is also sometimes used in connection with the destruction of body tissues by strong acids and bases.

In a restricted sense, corrosion is considered to consist of the slow chemical and electrochemical reactions between metals and their environments. From a broader point of view corrosion is the slow destruction of any material by chemical agents and electrochemical reactions. This contrasts with *erosion* which is the slow destruction of materials by mechanical agents. The character of the atmospheres to which materials are exposed may be classified as: rural, urban, industrial, urban-marine, industrial-marine, marine, tropical, and tropical-marine. In addition to these general kinds of environments, corrosion is of particular concern in the environments of chemical, petrochemical, and other processing and manufacturing environments where extremely corrosive substances may be encountered.

Metals Corrosion. The relationship between metals and hydrogen in the activity series are important because in the electrochemical processes of corrosion the discharge of hydrogen ions and the evolution of hydrogen as a gas is one of the principal cathodic reactions. The facility with which this can occur is determined by such factors as the hydrogen ion concentration (pH) of the electrolyte, the electrical potential of the corrosion cell, and the other-voltage characteristics of the cathodic surface. Sometimes called *concentration cell corrosion*, two solutions of different concentrations will set up an electrical potential between them similar to that produced by a battery. If oxygen is present in the liquid and is continually being replenished by contact with the air, then the oxygen concentration in the liquid will remain substantially constant. Any liquid that is contained in small holes or cracks on a metal surface will not be able to obtain oxygen from the main bulk of the solution, so when the supply in the holes and cracks is exhausted, no more oxygen can get in to replace it. Therefore, the oxygen concentration in the cracks is different from that of the main bulk of the solution and a concentration cell is set up. This minute electrical effect is sufficient to make corrosion proceed quite rapidly. A similar cell type of corrosion is that called galvanic or two-metal corrosion. Two different metals in contact will set up an electrical potential between them. If the two metals are surrounded by an electrolyte so that a closed circuit can be obtained, corrosion takes place. The magnitude of the electrical potential and, therefore, the speed and extent of the corrosion will depend upon the types of metals in each pair. In general, pairs farther apart in the activity series will corrode faster than those close together. See also **Electrochemistry.**

The electrochemical reactions in corrosion of a divalent metal may be written as follows:

Anodic reaction: $M^0 \rightarrow M^{++} + 2$ electrons
At the cathode: (1) $2H^- + 2$ electrons $\rightarrow H_2$ gas
(2) $\frac{1}{2}O_2 + 2H^- + 2$ electrons $\rightarrow H_2O$
(3) $O_2 + 2H_2O + 2$ electrons $\rightarrow H_2O_2 + 2OH^-$
(4) $\frac{1}{2}O_2 + H_2O + 2$ electrons $\rightarrow 2(OH)^-$

It is evident that oxygen as well as hydrogen plays an important part in metal corrosion. It can accelerate corrosion by participating in cathodic reactions, or it can retard corrosion by forming protective oxides or passive films. The dual effect of oxygen is one of the factors that complicates corrosion processes, including the interpretation of observations of the process and the steps to be taken to avoid corrosion damage.

Forms of Corrosion. (1) *Pitting* results from local action currents, as at discontinuities in protective or passive films or under or around deposits that set up concentration cells. (2) *Stress corrosion cracking* results from the combined effects of corrosion by a specific environment and either applied or internal static tensile stresses. Depending upon the metal and the environment, the cracks may be either intercrystalline or transcrystalline. (3) *Corrosion fatigue* results from the combined effects of corrosion and cyclic stresses. Cracks of this type are characteristically transcrystalline. (4) *Intergranular corrosion* results from preferential attack on, or around, a phase or compound that occupies grain boundaries. (5) *Corrasion-corrosion* results from the combined effects of corrosion and either abrasion or attrition. The mechanism usually involves local or general removal of otherwise protective corrosion product films. One particular form is *impingement attack* due to the effects of high-velocity or turbulence in flowing liquids, e.g., salt water in steam condensers, or other heat exchangers, in piping systems, valves, pumps, among others. A particularly aggressive form is associated with the severe mechanical forces that are characteristic of *cavitation phenomena.* (6) Uniform attack or general wastage may be caused, for example, by the action of strong acids as used for pickling (scale removal) or etching. This is also characteristic of the slow corrosion of durable materials in appropriate environments, such as copper roofing in suburban atmospheres, cupronickel tubes in ships' condensers, Monel-nickel copper alloy racks for pickling steel in sulfuric acid, or stainless steel columns for handling nitric acid.

Metal Corrosion Minimization. In the most recent official assessment of the economic costs of equipment damage arising from corrosion, prepared by the National Commission on Materials Policy (U.S., 1973), it was stated that annual losses in the United States alone are on the order of $15 billion annually. Since that time and going into the 1980s, these losses have increased.

Some of the means used to combat corrosion losses are described below:

(1) Use of the right metal in the proper way and in the correct place. Planners tend to look too closely at first costs and not closely enough at maintenance costs; consequently, there are many applications of materials that are of lower first cost, but of severely limited life. For example, in some applications, inexpensive fasteners that will obviously corrode in a few years are used in place of stainless steel or hardened aluminum fasten-

ers that would have cost only a few dollars more at the time of installation. When discussing applications of possible corrosion-resistant materials, it is important to define clearly the parameters of the environment of usage, such as temperature, pressure, humidity, presence of specific chemical agents, presence of living or dead organic materials, and the characteristics of associated electrical magnetic, light, and other radiation fields. Far more needs to be done on the microclimatology of environments and the specifics needed to ameliorate corrosion problems (Morgan, 1978).

(2) Protective coatings—paints, enamels, other metals, oils, greases, among others. One of the more common methods used is zinc-coating, i.e., galvanizing. Galvanized iron wire with a thin layer of zinc applied to it by dipping in molten zinc or by electrical means usually will resist corrosion for an extended period. Cadmium, nickel, tin, and chromium are metals often used as protective coatings, generally applied by electroplating. See also **Autodeposition; Conversion Coatings; Electrochemistry; Electroplating; Galvanizing;** and **Paint** and **Finishes.**

(3) Inhibitors and neutralizers, i.e., compounds added to the environment in small concentration to form protective films which increase anodic or cathodic polarization or both, or to neutralize some corrosive constituents. For example, it is possible for corrosion to occur at many places in the piping leading to boilers or heaters, but usually it occurs in the boiler itself. The trouble is ordinarily found to be due to an acid condition of the boiler feedwater, or to dissolved oxygen contained in it. The raw water used may be acid from surface pollution or from subsurface drains. Usually this can be detected and readily remedied. A more serious factor is the oxygen dissolved in water. Under the high-temperature conditions existing inside the boiler, this oxygen becomes extremely active in attacking metal surfaces. The operators of large, high-pressure boilers well know the necessity of removing oxygen from feedwater through the use of deactivators or deaerators. Corrosion protection of boilers in power plants is effected by installation of ion exchange resin and other purifiers in the feedwater cycle, and by monitoring them by automatic analysis. In still other installations where acidity and oxygen are a problem, neutralizing chemicals with the dosage governed by automated pH and oxygen control systems are effective. Systems of this type are also effectively used to neutralize plant effluents to streams so that water users downstream have a reasonably neutral and clean supply of water. Clean water programs not only are desirable from an environmental standpoint, but also can help in reducing the costs of corrosion.

(4) Drying of air and other gases to keep humidity below the level where corrosion becomes serious.

(5) Design of hydraulic systems to avoid excessive velocities or localized turbulence or to maintain a velocity high enough to prevent the accumulation of corrosion products or other deposits that will promote localized corrosion.

(6) Various features of design and operation of structures or equipment to favor rapid drainage and drying, prevent accumulation or concentration of corrosive chemicals in crevices or low spots, hold operating stresses and temperatures within desired limits, eliminate fabricating stresses by appropriate metallurgical treatment, avoid galvanically unfavorable combinations of different metals, provide protection against stray electrical currents by appropriate insulation and electrical bonding.

(7) Heat-treating metals to leave them in optimum condition to resist corrosion.

(8) Applying protective electrical currents (*cathodic protection*) from sacrificial metals (galvanic anodes) such as zinc, magnesium or aluminum or from some external source through a

graphite, platinum, or other appropriate anode receiving current from a rectifier, generator, or battery. The location of the anodes, the magnitude of the current, and the applied voltage must be engineered so that without wasting current all surfaces that require protection will receive sufficient current to achieve this effect. Too much current may cause damage by the alkali generated by a cathodic reaction or by hydrogen evolved at the cathode which can destroy protective films or embrittle metals. A current of 1–15 milliamperes per square foot (929 square centimeters) is usually required to protect bare steel areas; for design purposes, the range is generally narrowed to 3–5 milliamperes. The potential required to produce this current flow depends, of course, on the resistivity of the electrolytic path between the electrodes.

Corrosion can also be suppressed by the controlled application of current to the metal as an anode. This is called *anodic protection*. Passivity is induced and preserved by maintaining the potential of the alloy at or above a critical potential in what is called the *range of passivity* in a potentiostatic diagram. Such diagrams are based on the relationship between anodic current density and the corresponding potential in the environment of interest.

When pipes and cables carrying an electric current are underground, they are commonly corroded by electrolytic action from unidirectional electric currents in the ground. Stray current from electric traction equipment is retarded by increasing the resistance of the ground circuit and by reducing the electric resistance of the track. Also, cathodic protection is widely used. An external source of dc voltage is applied so that the protected equipment (pipeline or cable, for example) becomes lower in potential than the soil that surrounds it. Thus, the buried material is the cathode rather than the anode. Usual forms of corrosion can be prevented when the preventive system causes the pipe or metal structure to be 0.25–0.30 volt negative with reference to the soil or liquid that may be surrounding it. The negative lead from a small generator, battery, or rectifier is connected to the metal structure; the positive lead to the ground at some distance, or sacrificial magnesium or zinc rods sunk in the ground may be externally connected to the structure.

Atmospheric Corrosion of Metals and Nonmetals

Taking into consideration the relative order of corrodibility, it is preferable to describe corrosive damage as attributable to certain agents rather than to the indefinite characterization "smoke." Corrosive agents can be placed into four major groups, namely, oxygen and oxidants, acidic materials, salts, and alkalis.

Corrosion attributable to oxygen is deemed to result from the solution of oxygen by a thin film of liquid adjacent to the metallic surface, the transportation of the oxygen through the film, and the subsequent reaction at the surface of the metal. This explains why there is corrosive action even in relatively arid land. In a very dry atmosphere corrosion is, however, markedly reduced.

There are three principal categories of oxidizing agents which occur as air pollutants. These are ozone, nitrogen oxides and nitric acid, and organic peroxide. Many materials which are relatively resistant to attack by the free oxygen of the air are far less resistant to attack by such oxidants and peroxides. These dissolve in the surface film and thus convert metals to their oxides which react readily even with such relatively weak acids as carbonic acid and sulfurous acid. For instance, copper tarnishes rapidly forming the oxide which dissolves readily in dilute acids.

The acid components given off to the air by the various processes of combustion are sulfur dioxide and sulfurous acid, sulfuric acid, hydrogen sulfide, hydrochloric acid, carbon dioxide and carbonic acid, and tar acids. There is little doubt that the material of greatest importance in respect to atmospheric corrosion in this group is sulfur dioxide. Generally, the total acidity of the atmosphere is closely related to the sulfur dioxide content.

Other soluble acidic components such as sulfuric acid, hydrogen sulfide, hydrochloric acid, nitric acid and the like are all of minor importance. Carbon dioxide and carbonic acid play a significant role in acid decomposition.

One aspect of tar and tar acids should be noted, namely, that these are sticky and cling to the surfaces with which they come in contact. This enables the acids which such tar contains to have a prolonged corrosive action. It also increases the difficulty of removal by rain or wind or other action.

It is common to consider that certain salts have a very corrosive action. This is true in the respect that the corrodibility of marine atmospheres has been shown to be greater than rural, tropical, and urban atmospheres. For example, ammonium sulfate and ammonium chloride being salts of strong acids and a weak base, that is ammonium hydroxide, hydrolyze in water to yield the respective acids. These salts then have a corrosive action which is due actually to the acid produced in hydrolysis.

Alkalis seldom occur as air pollutants except under industrial conditions. Nevertheless, the corrosive action of alkalis should not be completely ignored. While a number of metals are relatively resistant to acid attack, they have an amphoteric action and can react with alkalies. For instance, aluminum and zinc are in this category and they are subject to corrosive attack by relatively weak alkalis.

Metals exposed to air pollution can be placed into three groups:
1. Metals that corrode rapidly because they do not form completely protective corrosion products. The major metal in this group is iron. It should be noted, however, that iron oxide Fe_2O_3 does have some protective action.
2. Metals which are initially attacked somewhat readily but subsequently become resistant to attack because of the formation of a corrosion-resisting film which hinders further attack. Among the metals in this group are aluminum, lead, zinc, brass, copper, nickel and magnesium.
3. Metals that are almost completely corrosion resistant, as for instance stainless steel of the 18/8 type, chromium plate products, monel, and gold.

Mention has already been made of the action of oxygen and oxidants on metal. It should be noted that metals react with sulfides, such as hydrogen sulfide, and are subsequently subject to additional slow attack by oxygen and oxidants. Thus, copper reacts to form sulfide and then the basic copper sulfate.

Generally, metals are resistant to attack in dry air; even in pure humid air, corrosion is slight; when, however, air pollutants are present the rate of corrosion will increase measurably, the increase being dependent upon the humidity and the character of the pollutant. Such action may be grouped as follows:

RELATIVE HUMIDITY		DEGREE OF CORROSION
Less than	60	None
More than	60	Slow but definite
	80	Decided increase
Greater than	80	Very high

A factor of note is the settling and adherence of particles on metals. Particles of carbon, ammonium sulfate, and silica cause a marked increase in corrosion, and this is accentuated in atmospheres containing sulfur dioxide. The presence of such hygroscopic particles enhances the adherence of liquid and thus provides for electrochemical attack.

Stone building materials may be placed into relatively resistant and non-resistant categories. The acids of the air, such as sulfuric acid, attack carbonate-bearing stone, such as limestone, converting the calcium carbonate to calcium sulfate. The gypsum formed is dissolved by rainwater, causing pitting. Incrustations may be formed because of the crystallization of soluble salts. These break away in time and leave pitted surfaces. Other types of damage, such as porosity of the stone, are caused by analogous reactions.

See also **Pollution (Air)**.

Corrosion-Resistant Metals

Some concept of the economic importance of corrosion to the production and consumption of various metals can be gleaned from the accompanying table. Many mineral commodities are used in more or less direct proportion to steel production. In the case of the United States and many other countries, a large number of the mineral materials important to combatting corrosion come from distant sources. That is why many countries maintain a stockpile of such materials, particularly those that are regarded as critical or strategic. In the United States, 93 materials are officially classified for defense purposes as basic stockpile commodities. Seventy-nine of these are metals and minerals, including nearly every one of the metals with important corrosion-resistant properties.

The possible use of low-grade, presently noncommercial, mineral deposits requires constant consideration. For example, chromium has been recognized as an important strategic material ever since World War I. Over the years, the U.S. government, through the U.S. Geological Survey and the Bureau of Mines, has discovered and carefully defined numerous domestic chromium deposits. The Bureau of Mines in its metallurgical laboratories has produced acceptable chrome concentrates from these deposits as well as acceptable ferrochromium chemicals. Current Bureau of Mines research includes recovering chromium, nickel, and cobalt from laterite deposits, both domestic and from other countries, and also from flue dusts, plating wastes, and other residues (Morgan, 1978).

Corrosion Monitoring. Combatting corrosion in continuous processing plants which may be scheduled for quite infrequent, but thorough, equipment checking and maintenance is particularly difficult. Process downtime costs for a large unit may be several hundred thousand dollars per day in terms of lost production. To avoid excessive downtime for checking and still

APPLICATIONS OF CORROSION-RESISTANT METALS

METAL	CORROSION-RESISTANCE USE	CONSUMPTION	
		For Corrosion Applications	Percent of All Uses of Metal
Nickel	Alloying, 34%; High-temperature oxidation resistance, 26%; Plating, 13%	14,000 MT	73
Chromium	Alloying, 57%; Coatings and Plating, 7%	306,000 MT	64
Titanium	Coatings, 51%; Alloying, 1%	243,000 MT	52
Cadmium	Plating	2,520 MT	45
Gold	Plating and alloying	50,000 kg	35
Zinc	Galvanizing, 32%; Coatings and sacrificial anodes, 3%	432,000 MT	35
Tin	Plating, Tinning	19,800 MT	34
Tantalum	Alloying, 16%; Cladding, 12%; High-temperature oxidation resistance, 5%	1,935 MT	33
Rare Earths	Alloying	3,600 MT	30
Platinum	Resistance to chemical attack	13,375 kg	27
Silver	Alloying	1,370 kg	26
Columbium (Niobium)	Alloying; High-temperature oxidation resistance	630 MT	25
Iron Oxide Pigments	Coatings	30,600 MT	25
Copper	Alloying and plumbing	370,000 MT	18
Molybdenum	Alloying; Coating	4,860 MT	18
Cobalt	Alloying; High-temperature oxidation resistance	1,530 MT	17
Magnesium	Alloying; Sacrificial anodes	14,400 MT	15
Zirconium	Alloying; Chemical resistance	340 MT	15
Thorium (ThO$_2$)	Alloying	31 MT	12
Hafnium	Alloying	3 MT	11
Beryllium	Alloying	4 MT	10
Lead	Pigments and plating, 8%; Cable covering, 1%	108,000 MT	9
Indium	Coatings	1,555 kg	8
Aluminum	Alloying; Coatings; Cladding	225,000 MT	5
Manganese	Alloying; Cladding	45,000 MT	4

SOURCE: U.S. Bureau of Mines (1978).
MT = metric ton; kg = kilogram

control the effects of corrosion (personnel and equipment safety, product quality and throughput rates, etc.) requires means to measure the status and rate of corrosion that may be taking place within the equipment. The design of corrosion monitors is among the most recent developments in overall process instrumentation. For obvious reasons, such on-line corrosion testing must be of a nondestructive nature.

In one type of monitor, changes in electrical resistance of a measuring element or probe relate to corrosion rate. The measuring element may be a wire, tube, or strip that can be inserted in a tee or an elbow in the process piping. As the measuring element corrodes, the cross-sectional area reduces and the electrical resistance increases. The thickness of the measuring element is directly proportional to a corrosion dial reading. The difference in dial readings is plotted over a period of time and from these data corrosion rate can be determined. Probes are available in a number of different metals for different temperature ranges and corrosion conditions. In another variation, three-electrode probes are used. The corrosion rate is determined by measuring electrical current flow between the test and auxiliary electrodes. That current either cathodically protects or anodically accelerates the corrosion rate of the test electrode, depending upon the flow. The current is measured on a microammeter that has been converted to read directly the corrosion rate in mils (1 mil = 0.001 inch = 25.4 micrometers) per year of the test electrode.

In another instrumental approach, a hydrogen test probe operates on the principle that hydrogen will diffuse through the thin wall of a test probe and set up a pressure within the tube. The rate at which the pressure increases is measured by a pressure gage. The rate at which hydrogen is penetrating per unit area can then be determined, using the exposed surface area, and the internal volume of the probe. Beyond a certain rate, severe hydrogen damage can be anticipated.

Measurement of wall thickness can be determined using ultrasonic nondestructive testing methodology. This method can be used to determine wall thinning, pitting, erosion, and flaws in metals, plastics, and various rubbers and polymers. Some disadvantages of the method include the need to take many readings over a period of time to determine corrosion rate and high-temperature measurements tend to be inaccurate.

Infrared thermographic techniques can be used to identify hot spots on process equipment. The camera works on the theory that the hotter the object, the higher the frequency of radiation. For off-line corrosion monitoring, borescopes for inspecting tubes, pumps, compressors and other equipment may be used. Spot chemical testing can indicate the presence of alloy constitutents of unknown materials. Television camera and holographic techniques also have been used. The monitoring of pH is an invaluable indicator of possible corrosion problems, particularly in cooling water systems. Monitoring is usually done continuously because pH shifts can take place rapidly in many systems, particularly as a result of a process leak.

Probably the weight-loss coupon approach is one of the most reliable methods and is widely used. The accuracy of the data is highly dependent upon good techniques and on the statistical significance of the tests. The engineering quality data produced requires the efforts of many people over a period of several weeks or months, which makes this information quite costly. However, it is the technique of first choice of many processors.

References

Arnold, C. G.: "Using Real Time Corrosion Monitors in Chemical Plants," Chemical Engineering Progress, 74, 3, 26–31 (1978).
Cangi, J. W.: "Characteristics and Corrosion Properties of Cast Alloys," Chemical Engineering Progress, 74, 3, 61–66 (1978).
Covington, L. C., Schutz, R. W., and I. A. Franson: "Titanium Alloys for Corrosion Resistance," Chemical Engineering Progress, 74, 3, 67–69 (1978).
Draley, J. E., and J. R. Weeks: "Corrosion by Liquid Metals," Plenum, New York (1969).
Fontana, M. G. and R. W. Staehle: "Advances in Corrosion Science and Technology," Plenum, New York (1980).
Fontana, M. G., and N. D. Greene: "Corrosion Engineering," McGraw-Hill, New York (1978).
Gasper, K. E.: "Non-Chromate Methods of Cooling Water Treatment," Chemical Engineering Progress, 74, 3, 52–56 (1978).
Harrell, J. B.: "Corrosion Monitoring in the Chemical Process Industries," Chemical Engineering Progress, 74, 3, 57–60 (1978).
Leidheiser, H.: "Corrosion of Copper, Tin and Their Alloys," Wiley, New York (1971).
Morgan, J. D., Jr.: "Supply and Demand of Corrosion-resistant Materials," Chemical Engineering Progress, 74, 3, 26–31 (1978).
Rak, G. R.: "Corrosion Monitoring Equipment," Chemical Engineering Progress, 74, 3, 46–51 (1978).
Schmeal, W. R., MacNab, A. J., and P. R. Rhodes: "Corrosion in Amine/Sour Gas Treating Contactors," Chemical Engineering Progress, 74, 3, 37–42 (1978).
Sedricks, A. J.: "Corrosion of Stainless Steels," Wiley, New York (1979).

CORROSION INHIBITION. Autodeposition; Conversion Coatings; Paints and Finishes; Petroleum.

CORSITE. An orbicular diorite resulting from the segregation, in rounded concentric forms, of ferro-magnesian minerals. It derives its name from its occurrence on the Island of Corsica, and is also sometimes called *Napoleonite*.

CORTICOSTERONE. Hormones.

CORTISONE. Steroids.

CORUNDUM. The mineral corundum, Al_2O_3, aluminum oxide, occurs as well-developed hexagonal crystals which may display prismatic, rhombohedral, pyramidal or tabular habits. The larger crystals are often rounded or barrel shaped. Corundum shows both basal and rhombohedral partings; the fracture is conchoidal; hardness, 9; specific gravity, 4.0–4.1; luster, vitreous to adamantine, may be pearly on base; transparent to translucent. Common corundum is gray, grayish-blue or brown, but may be red, yellow or whitish; it is sometimes called adamantine spar. Transparent corundum may be colorless or of various tints. The highly prized ruby is deep red; the sapphire, blue. Transparent yellow corundum is known as oriental topaz; if vilet, oriental amethyst; if green, oriental emerald.

Emery is a mixture of granular corundum of dark color, magnetite and hematite, sometimes with spinel. Quartz may be present. For a long time emery was supposed to be an ore of iron. Until the introduction of artificial abrasives, emery was much used for such purposes.

Corundum is found as an accessory mineral in the crystalline rocks such as crystalline limestones and dolomites, gneisses, schists as well as in the igneous rock types granite and syenite. Corundum syenites are found in Canada, especially in the Province of Ontario. Rubies have long been mined in Upper Burma; both rubies and sapphires are found near Bangkok, Thailand. Numerous localities in India furnish gem stones of high quality.

In the United States, common corundum is found in New York, New Jersey, Pennsylvania, Virginia, North Carolina, South Carolina and Georgia; sapphires of gem quality near Helena, Montana, associated with alluvial gold in the Missouri River. From the crystalline limestones and schists of the islands of Naxos and Samos in the Grecian archipelago most of the emery of commerce comes. Other deposits are near Ephesus in Asia Minor, and in the town of Chester in Massachusetts. The word corundum comes from the Hindu, *kurand*; emery is derived from the Greek name for this substance.

See also **Bauxite**; and terms listed under **Mineralogy.**

COTTONSEED OIL. Vegetable Oils (Edible).

COUMARIN. Anticoagulants; Furan and Related Compounds.

COVELLITE. The mineral covellite, cupric sulfide, CuS, is hexagonal, usually in thin platey crystals, but may be massive. It has a hardness of 1.5–2; specific gravity, 4.6; luster, submetallic to resinous; color, dark indigo blue, sometimes showing a purplish tarnish, or if moistened may appear purple in color. Its streak is dark gray to black; it is opaque. Covellite is found associated with chalcopyrite, bornite, and chalcocite, and is believed to be chiefly of secondary origin. Covellite occurs in Yugoslavia, Saxony, Sardinia, Argentina, Chile, Bolivia and Peru, and in the United States at Butte, Montana, and in Colorado, Wyoming, and Utah. This mineral was named for Covelli, who discovered it in the lavas of Mt. Vesuvius.

CRACKING PROCESS. A reaction in which a hydrocarbon molecule is broken or fractured into two or more smaller fragments. Sometimes the term *pyrolysis* is used for this reaction. Possibilities for cleavage of a molecule include (1) a carbon-hydrogen bond; (2) a bond between an inorganic atom and a carbon or hydrogen atom; (3) a carbon-carbon bond. Usually the objective of cracking is that of reducing the size of hydrocarbon molecules; hence the target is to fracture the carbon-carbon bonds. The main cracking processes are: (1) thermal cracking; (2) fluid catalytic cracking; and (3) hydrocracking.

Thermal Cracking. Of the thermal cracking processes, two are of major importance: (1) coking, and (2) visbreaking (viscosity breaking). Both of these processes convert nondistillable residues into more valuable products. Thermal cracking was the first of the principal cracking processes used in the petroleum industry. For increasing gasoline production and improving quality, fluid catalytic cracking has essentially replaced thermal cracking.

In *thermal coking*, heavy residual stocks are converted into gas, gasoline, distillates, and coke. Generally, the objective is that of maximizing the yield of distillates; and minimizing the production of gas, gasoline, and coke. Light distillates are used for both domestic and industrial heating oils. There are two types of thermal coking processes: (1) cyclic, semicontinuous process, sometimes referred to as *delayed coking*, decarbonizing, or low-pressure coking; and (2) a continuous fluid coking process. About 70% of the installed capacity in the United States is the delayed coking type.

As shown by Fig. 1, a delayed coking unit is comprised of three sections—a furnace, coke drums, a fractionating unit, plus coke removal and handling equipment. Usually the feedstock is charged to the lower part of the fractionator. Here the feedstock is contacted by hot vapors from the coke drum, causing any light components to be flashed from the feed before the feed joins with the recycle and charged (from the bottom of the fractionator) to the furnace. The charge in the furnace is heated to about 480°C (896°F). The heated effluent from the furnace is introduced into the bottom of one of two or more insulated vessels (coke drums) where, as the result of its contained heat, the material cracks to form a solid coke residue. At the same time, lighter cracked products are evolved and

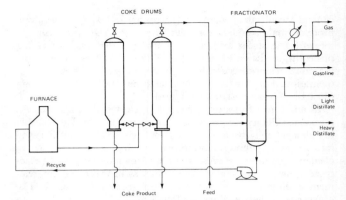

Fig. 1. Delayed coking unit.

proceed from the top of the coke drums to the fractionator. The reaction in the coke drum is endothermic and thus the temperature of the material drops to about 425°C (797°F). The cracked products leave as vapors; the coke remains in the drum. The fractionator separates the cracked vapors into several side streams as shown on the diagram. When accumulated coke reaches a certain level in the drum, that drum is temporarily taken off-stream and the flow is switched to the second drum. Prior to removal of coke, the drum is steamed to remove vapors. Water is also added to cool the coke.

The fluid coking process accomplishes the coking operation in a continuous manner. Feed is sprayed into a fluid bed of hot coke in a coking reactor. Steam introduced into the bottom of the reactor provides the fluidization energy. The cracked products are quenched in an overhead scrubber and then go to the fractionator. The coke is deposited on the particles in the reactor which commute with a heater vessel in which a portion of the coke is burned to heat up the returning coke particles to supply the energy for the coking reaction.

Fluid Catalytic Cracking. In operation, preheated feedstock meets a controlled stream of hot, regenerated catalyst. Vaporized oil and catalyst ascend in the riser, such that the catalyst particles are suspended in a dilute phase. Essentially all of the cracking occurs in the riser. The catalyst particles are separated from the cracked vapors at the end of the riser and the catalyst containing a coke deposit is returned to the regenerator. The cracked vapors pass through one or more cyclones located in the upper portion of the reactor and proceed to the fractionator (main column) that produces the side streams indicated. See Fig. 2.

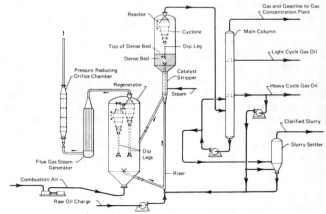

Fig. 2. Fluid catalytic cracking process. (*UOP Inc.*)

Hydrocracking. Processes in this category produce gasoline and light distillates from feed distillates that are higher-boiling than the products. Hydrocracked products are not olefinic. The light gaseous hydrocarbons produced by hydrocracking are entirely paraffinic. The processes operate at elevated pressures in the presence of hydrogen and catalysts. Temperatures are usually lower than 482°C (900°F). Pressures run from 800 to 2,500 psig. Both fixed-bed and ebullating-bed configurations are used. Because carbonaceous deposits accumulate very slowly on the catalyst, the on-line periods for these units is quite long, ranging from several months to over a year. Somewhat more costly to build than fluid catalytic cracking, the hydrocracking process has the advantage that it can handle heavier and dirtier feedstocks and also may be adapted to varying product ratios of gasoline to middle distillate.

—Technical Staff, UOP Inc., Des Plaines, Illinois.

CREOSOTE.   Coal Tar and Derivatives.

CRITICAL DENSITY.   The density of a substance which is at its critical temperature and pressure.

CRITICAL MASS.   Nuclear Reactor.

CRITICAL POINT.   1. A point where two phases, which are continually approximating each other, become identical and form but one phase. With a liquid in equilibrium with its vapor, the critical point is such as combination of temperature and pressure that the specific volumes of the liquid and its vapor are identical, and there is no distinction between the two states. 2. The critical solution point is such a combination of temperature and pressure that two otherwise partially miscible liquids become consolute. See **Consolute Liquids.**

To consider in detail the critical point as defined in (1), examine the accompanying figure, which shows the family of isotherms of a pure substance in the fluid range (liquid or gas) such, for example, as shown in the graph for carbon dioxide.

At sufficiently high temperatures, each isotherm is a continuous curve, but at low temperatures, the isotherm consists of

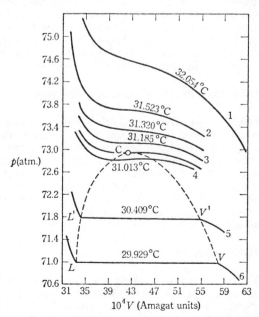

Isotherms of carbon dioxide in the neighborhood of the critical point.

three portions. The first section of the curve at high pressure corresponds to the liquid state, while that at low pressures refers to the gaseous state. These two curves are joined by a horizontal line corresponding to the simultaneous presence of two phases, liquid and gas.

The isotherm between those numbered 3 and 4 in the figure represent the transition between isotherms corresponding to the gas phase only, and those including a horizontal portion corresponding to a liquid-gas equilibrium. In this isotherm the horizontal line has contracted to a single point of inflection C. This is the critical point characterized by the relations

$$\left(\frac{\partial p}{\partial V}\right)_{T_c} = 0; \quad \left(\frac{\partial^2 p}{\partial V^2}\right)_{T_c} = 0; \quad \left(\frac{\partial^3 p}{\partial V^3}\right)_{T_c} < 0.$$

The curve $LL'C$ gives the molar volume of the liquid. Similarly $VV'C$ gives the molar volume of the gas.

At the critical point the molar volumes of the liquid and of the gas become equal. In general a critical state is characterized by the fact that the two coexistent phases (here the liquid and the vapor) are identical.

The curve $VV'CL'L$ is called the *saturation curve.*

The experimental data do not indicate the existence of a critical point for the liquid-solid transition.

Above the critical point the substance can no longer exist in the liquid state. The critical temperature is thus the highest temperature at which the liquid and vapor can coexist.

Ternary Critical Point. The point where, upon adding a mutual solvent to two partially miscible liquids (as adding alcohol to ether and water), the two solutions become consolute and one phase results.

CRITICAL TEMPERATURE.   1. This term is most commonly used to denote the maximum temperature at which a gas (or vapor) may be liquefied by application of pressure alone. Above this temperature, the substance exists only as a gas. 2. The critical temperature of a superconducting transition takes place in zero magnetic field.

CRITICAL VOLUME.   The volume occupied by unit mass, commonly one mole, of a substance at its critical temperature and critical pressure.

CROCIDOLITE (Blue Asbestos).   The mineral crocidolite may be considered as a fibrous variety of the monoclinic amphibole, riebeckite. It is also known as a massive mineral. Its hardness is 4; specific gravity, 3.2–3.3; luster, silky to dull; color, blue or bluish-green. It is found in Austria, France, Bolivia, the Republic of South Africa (the variety known as tiger's-eye); and in the United States, in Massachusetts and Rhode Island. The name *crocidolite* is derived from the Greek meaning *woof,* in reference to its fibrous appearance. See also **Cat's-Eye.**

CROCOITE.   The mineral crocoite, lead chromate, corresponds to the formula $PbCrO_4$, and forms prismatic monoclinic crystals, often acicular. It is also found in columnar or granular masses. It has a rather distinct cleavage parallel to the prism, and a less distinct cleavage parallel to the base. It has a conchoidal fracture, is sectile; hardness, 2.5–3; specific gravity, 5.9–6.1; luster, adamantine to vitreous; color, red; streak, orange-yellow, translucent. Crocoite is a secondary mineral believed to be formed by waters containing chromic acid acting upon lead minerals like galena, with which it is associated. It is found in the U.S.S.R., Rumania, Tasmania, Brazil, the Philippines, and Arizona. It is not of commercial importance. The name *crocoite* is derived from the Greek word for *saffron,* in reference to the color of the powdered mineral.

CROSS LINKING. Rubber (Natural).

CRYOGENICS. The production and study of phenomena which occur at very low temperatures, i.e., below about 80 K. The first step in attaining the required temperature generally involves the liquefaction of a gas or gases. Liquids can exist over a range of temperatures limited by the critical point at the higher end and the triple point at the low-temperature end. It is thus possible to compress a gas to the liquid phase at the critical point and to cool it by boiling under reduced pressure to its triple point. A series of gases having their critical and triple points overlapping can thus be used in a cascade process, each being used as the refrigerant for the next in the series. Pictet used this method to liquefy oxygen, using methyl chloride and ethylene as refrigerants. There are, however, no liquids which cover the range from 77 K to the critical point of hydrogen, or from 14 K to the critical point of helium (5.2 K). Thus, liquid hydrogen and helium cannot be produced by the cascade method.

A gas may also be cooled by making it do work in the course of an expansion. When an ideal gas is expanded through an aperture into a constant volume, no work is done, since there are no interactions between the molecules and the molecules themselves occupy no fixed volume. When a nonideal gas is so expanded, however, an amount of internal work ($W = (PV)_{final} - (PV)_{initial}$) is done against the intermolecular forces. This work may be positive or negative, resulting in a cooling or heating of the gas. Air is cooled by this Joule-Thomson expansion at room temperature, but hydrogen and helium must be precooled to 90 and 15 K, respectively, to obtain further cooling upon expansion. Using this method, Kamerlingh Onnes first succeeded in liquefying helium in 1908. Compressed gases may also be made to do external work, for example, by expansion against a movable piston. In this case, the work is always positive and helium may be cooled and liquefied without any precooling by liquid hydrogen. See **Ammonia**; and **Helium**.

With liquid helium readily available in the laboratory, research in the temperature range 5–0.8 K has become commonplace. By using the isotope of helium ^{3}He, it is possible to attain temperatures down to about 0.3 K, since the isotope has a lower boiling point than ^{4}He. This is about the lowest temperature practically attainable by boiling liquids at reduced pressure. To reach lower temperatures, it is necessary to use magnetic phenomena.

By taking advantage of the fact that the entropy of a superconducting metal is less than that of the metal in its normal state, temperatures below 1 K can be attained. Quenching of a superconductor by the application of a magnetic field can cause a cooling to about 0.1 K. However, since the specific heat of metals is very small at these temperatures, they are not very suitable for cooling other substances.

One of the most interesting phenomena of cryogenics is that of superconductivity, which was discovered by Onnes. When metals are cooled from room temperature, their resistivities decrease and at low temperature, they attain low values which are fairly independent of temperature. Such a metal is known as a *superconductor*. See **Superconductors**.

Another phenomenon of cryogenics is superfluidity, which is described in the article on **Superfluidity**.

CRYOHYDRATE. A eutectic system consisting of a salt and water, having a concentration at which complete fusion or solidification occurs at a definite temperature (eutectic temperature) as if only one substance were present.

CRYOLITE. Cryolite, sodium aluminum fluoride, Na_3AlF_6, crystalizes in the monoclinic system but in forms that closely approach cubes and isometric octahedrons. It is usually found massive. Cryolite has an uneven fracture, is brittle; hardness, 2.5; specific gravity, 2.97; luster, vitreous to greasy; color, snow-white but may be colorless, reddish or brownish; translucent to transparent. The only considerable occurrence of cryolite is at Ivigtut, Greenland, where veins of this mineral are associated with granites and gniesses. Small occurrences of cryolite have been noted in the Ilmen Mountains, U.S.S.R., and at Pikes Peak, Colorado. Cryolite has its chief use in the electrolytic production of aluminum, but small amounts are employed in the manufacture of opalescent glass.

The name cryolite is derived from the Greek words meaning frost (ice) and stone in reference to its translucency.

The aluminum industry no longer uses the natural mineral for operation of electrolytic cells except infrequently for start-up operations. The synthetic cryolite used is produced by two main processes: (1) the sodium aluminate from the Bayer process (see also **Bauxite**) is reacted with hydrofluoric acid: $NaAlO_2 + 2NaOH + 6HF \rightarrow 3NaF \cdot AlF_3 + 4H_2O$; or (2) sodium carbonate may be used instead of NaOH:

$$NaAlO_2 + Na_2CO_3 + 6HF \rightarrow 3NaF \cdot AlF_3 + CO_2 + 3H_2O.$$

The quality hydrofluoric acid required for these reactions may be prepared from fluorspar: $CaF_2 + H_2SO_4 \rightarrow 2HF + CaSO_4$. Normally, cryolite is not considered toxic. A continued exposure to finely divide cryolite in the air may lead to *fluorosis*. However, the compound is toxic to insects.

See also terms listed under **Mineralogy**.

CRYOSCOPIC CONSTANT. A quantity calculated to represent the molal depression of the freezing point of a solution, by the relationship

$$K = \frac{RT_0^2}{10000 l_f}$$

in which K is the cryoscopic constant, R is the gas constant, T_0 is the freezing point of the pure solvent, and l_f is the latent heat of fusion per gram. The product of the cryoscopic constant and the molality of the solution gives the actual depression of the freezing point for the range of values for which this relationship applies. Unfortunately, this range is limited to very dilute solutions, usually up to molalities of $\frac{1}{100}$, and must be modified for many solutes. See also **Freezing-Point Depression**.

CRYPTATE COMPLEXES. Chemical Elements.

CRYPTOCRYSTALLINE. When the texture of a rock is so finely crystalline (that is, made up of such minute crystals) that its crystalline nature is but vaguely revealed even in a thin section by transmitted polarized light, the rock is said to be cryptocrystalline. Among the sedimentary rocks, chert and flint are cryptocrystalline. Lava flows, especially of the acidic type such as felsites and rhyolites, may have a cryptocrystalline ground mass as distinguished from pure obsidian (acidic), or tachylite (basic), which are natural rock glasses.

CRYPTOCRYSTALLINITY. Mineralogy.

CRYSTAL. A macroscopic sample of a solid substance exhibiting some degree of geometrical regularity, or symmetry, or capable of showing these properties after suitable treatment (e.g., cleavage, etching, etc.). Almost all pure elements and compounds are capable of forming crystals.

A perfect crystal is one in which the crystal structure would be that of an ideal space lattice. No such crystals exist, all real crystals containing imperfections which have a strong influence on the physical properties of the crystal.

Structure of Crystals. Early investigators suggested that the regular structure of crystals, embodied in the laws of crystallography, could be explained if they were thought of as built up by the repetition of equal polyhedral cells, fitting together to fill space, each cell representing a characteristic group of particles, perhaps the atoms and molecules of the compound. A rough calculation showed that the spacing of these units in many ionic crystals might be of the same order of magnitude as the wavelengths of x-rays, as deduced from quantum theory. Von Laue suggested, and verified, that diffraction of the x-rays occurs when they are passed through a crystal, suitably oriented. He knew from the density and atomic weights, that the number of atoms in a cubic centimeter of rock salt, for example, is about 4.488×10^{22}, and that therefore, if they are equally spaced in all three directions, their distance apart is 2.814×10^{-8} centimeters or 2.814 Å. Certain quantum theory calculations had already indicated that x-rays have wavelengths of this order. It occured to von Laue that if a beam of x-rays were directed upon a crystal and the crystal turned into a suitable position, one might observe interference maxima analogous to those produced with light by a diffraction grating. This proved to be the case, and the result verified beyond question the existence of regular spacings between reflecting planes of some sort, presumably plane arrays of atoms or ions. The kind of pattern obtainable is demonstrated by Fig. 1.

Subsequent analysis of the problem by Bragg resulted in a formula analogous to that for interference of light reflected by thin plates. If the reflecting layers are spaced at equal distances d, and if the wavelength of the incident x-rays is λ, the angle θ between rays and layers necessary for an interference reflection maximum is given by Bragg's law, viz., $\sin \theta = n\lambda/2d$; in which n is an integer. By slowly turning the crystal, the various plane-families are brought into suitable orientations for the production of maxima. The result is a Laue pattern of black spots on the photographic plate placed beyond the crystal to catch the reflections. Another method, developed by Hull and by Debye and Scherrer, secures the necessary angular variation by crushing the crystal to powder and relying upon the fortuitous orientation of the fragments; the pattern in this case being a system of concentric rings, as exemplified by Fig. 2.

While it is very easy, when one knows the structure of the crystal and the wavelength of the rays, to predict the diffraction pattern, it is quite another matter to deduce the crystal structure in all its details from the observed pattern and the known wavelength. The first step is to determine the spacing of the atomic planes from the Bragg equation, and hence the dimensions of the unit cell. Any special symmetry of the space group of the structure will be apparent from space group extinction. A trial analysis may then solve the structure, or it may be necessary to measure the structure factors and try to find the phases or a Fourier synthesis. Various techniques can be used, such as the F^2 series, the heavy atom, the isomorphous series, anomalous atomic scattering, expansion of the crystal and other methods.

By such methods, the structures of crystals have been determined and all of them can be shown to possess space structures corresponding to one or another of the 14 Bravais lattices.

Crystal Systems and Crystallography. The field which deals with the geometrical relations between the atomic planes within

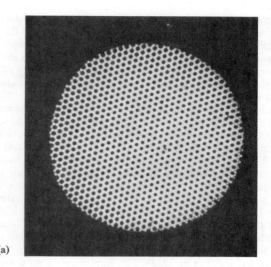

(a)

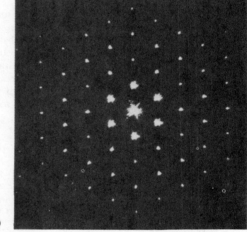

(b)

Fig. 1. Types of x-ray patterns of crystals: (a) steel balls in a crystalline lattice; (b) Bragg reflections with shape S^2.

them is *crystallography*. *Some* minerals, if broken, will separate along given "cleavage planes" into polyhedral fragments. Even when powdered, the minute grains will show this characteristic and measurements will indicate the planes to belong to one or another plane family, the members of any one of which are all parallel. It should be stressed, however, that all minerals do not possess cleavage planes for such experimentation.

In most crystal systems each of the more prominent crystal faces belongs to one of three plane-families intersecting along what are called the crystal axes. In the hexagonal system there are four. These may be conveniently used as coordinate axes, S, Y, Z, though they are not generally at right angles. Haüy

Fig. 2. Concentric ring pattern of crystalline materials.

discovered that if the ratio of the intercepts of two crystal planes on one of these axes is a simple fraction, such as $\frac{3}{5}$, the ratios of the intercepts on the other axes are likewise simple. This suggests that the two intercepts on any one axis are multiples of a common unit. The units are, however, generally different for the different axes, bearing to each other ratios called the axial ratios.

It is more convenient to use the reciprocals of the intercepts. For example, a plane might have intercepts equal to 10,000, 15,000, and 6,000 of the respective units. The reciprocals have the ratios 1/10,000 : 1/15,000 : 1.6,000, which in lowest terms are 3 : 2 : 5. These smallest integers are the Miller indices of the family to which this plane belongs, and the family is thus designated (325). The family (201) is parallel to the Y axis but intersects the X and Z axes. (The hexagonal system has four Bravais-Miller indices for each plane-family.)

If the intercept of any plane has a negative value, that is, if it cuts the axes when extended in a direction opposite to that in the standard arrangement, the fact is shown by a bar over the Miller index, e.g., ($2\bar{2}2$). The eight faces of an octahedron are (111), ($1\bar{1}1$), ($\bar{1}11$), ($\bar{1}\bar{1}1$), ($11\bar{1}$), ($1\bar{1}\bar{1}$), ($\bar{1}1\bar{1}$), and ($\bar{1}\bar{1}\bar{1}$), that is they all belong to the form (111).

Close study of the angles, indices, and axial ratios long since made it clear that every crystalline substance has a structure built upon a space "lattice" characteristic of the substance. It has been established that this is due to the regular arrangement of the atoms, molecules, or ions composing the substance. As shown by the accompanying table, the lattice structures of crystals may be classified into 32 symmetry classes (point groups), which are further divided into seven systems. This topic also is discussed under **Mineralogy**.

ELEMENTS OF CRYSTAL SYSTEMS

System	Crystallographic Elements	Essential Symmetry	Number of Point Groups
Cubic, or regular	Three axes at right angles: all equal.	4 triad axes; 3 diad, or 3 tetrad axes	5
Tetragonal	Three axes at right angles: two equal.	1 tetrad axis	7
Orthorhombic or rhombic	Three axes at right angles: unequal.	3 diad axes, or 1 diad axis and 2 perpendicular planes intersecting in a diad axis	3
Monoclinic	Three axes, one pair not at right angles: unequal.	1 diad axis, or 1 plane	3
Triclinic or anorthic	Three axes not at right angles: unequal.	No axes or planes	2
Hexagonal	Three axes coplanar at 60°: equal. Fourth axis at right angles to other three.	1 hexad axis	7
Rhombohedral or trigonal	Three axes equally inclined, not at right angles: all equal.	1 triad axis	5

Practically all minerals are crystalline, although perfect natural crystals are seldom, if ever, found. Because of the laws of crystallography, however, a crystallographer can usually determine the crystal form of a known species from a fragment of the original crystal, provided that at least two of the crystal faces are visible. Crystalline aggregates are said to be cryptocrystalline when the individual particles are proved to have crystalline structure but their crystal faces are exceedingly small or indistinguishable. A mineral is pseudocrystalline if its external form does not correspond with its crystalline structure.

Determination of Crystal Structure. The object of a crystal-structure determination is to ascertain the position of all of the atoms in the unit cell, or translational building block, of a presumed completely ordered three-dimensional structure. In some cases, additional quantities of physical interest, e.g., the amplitudes of thermal motion, may also be derived from the experiment. The processes involved in such crystal-structure determinations may be divided conveniently into (1) collection of the data, (2) solution of the phase relations among the scattered x-rays (phase problem)—determination of a correct trial structure, and (3) refinement of this structure.

The data consist of intensities $I(hkl)$, where h, k, and l (the Miller indices represent a vector triplet which conveniently identifies the beam diffracted from a single crystal. In a typical determination, there may be one to two thousand such $I(hkl)$. The intensity is related to the structure factor $F(hkl)$ by the relation.

$$I(hkl) = KF(hkl)F^*(hkl) \qquad (1)$$

where K is a known relative factor, and where F^* is the complex conjugate of F. The structure factor itself is related to the scattering by the j atoms in the unit cell by the relation,

$$F(hkl) = \sum_j f_j T_j \exp[2\pi i(hx_j + ky_j + lz_j)] \qquad (2)$$

where f_j are the individual atomic scattering factors, T_j are the individual modifications of the scattering as a result of thermal motion, and x_j, y_j, z_j are the fractional positions of atom j along the three crystallographic axes. In a typical determination, j may be between 10 and 60. The *scattering density* $\rho(xyz)$ is derivable from the relation,

$$\rho(xyz) = V^{-1} \sum_{-\infty h,k,l}^{\infty} F(hkl)\exp[-2\pi i(hx + ky + lz)] \qquad (3)$$

where V is the volume of the unit cell.

The "phase problem" in crystallography arises because in the usual experiment (Eq. 1) the magnitudes of the complex structure factors are obtained, but not the phases. Yet in order to obtain the scattering density, and hence the positions of the atoms, the phases as well as the magnitudes of the structure factors are necessary (Eq. 3).

Once the phase problem is solved, then the positions of the atoms may be refined by successive structure-factor calculations (Eq. 2) and Fourier summations (Eq. 3) or by a nonlinear least-squares procedure in which one minimizes, for example, $\Sigma w(|F_{obs}| - |F_{calc}|)^2$, with weights w taken in a manner appropriate to the experiment. Such a least-squares refinement procedure presupposes that a suitable calculational model is known.

It is perhaps useful to indicate how the attention of crystallographers to these three steps in the solution of a structure has changed in recent years. In 1954, the time involved in the arduous task of collecting the three-dimensional data—step (1)—needed for the solution of a complex problem was generally short in comparison with the time needed to solve the phase problem—step (2). This time involved in step (2) of course depended (and still depends) upon the complexity of the problem, and on the ingenuity, luck, and perseverance of the investigator, but it was true in many cases that step (2) was the rate-determining step in the entire process. This in part was because little attention was paid to detailed refinements—step (3); in 1954, three-dimensional least-squares refinements of complex structures were out of the question computationally, and even Fourier refinements were rare, for on computing systems advanced for those days (e.g., punched-card tabulators, sorters, and primitive electronic computers), a three-dimensional Fourier summation might require forty man-hours. In fact, in 1954 it was usual for the crystallographer to examine the unit cells of a number of related substances and to pick the probelm that was crystallographically most favorable (and perhaps soluble from two-dimensional data), even though this problem might not be the one of greatest chemical or physical interest. Ten years later the situation had changed markedly, mainly because of the availability of high-speed computers. It is still true that there are classes of problems where step (2) is rate-determining, but these problems are far more complex than those attempted in 1954. Yet there is an extensive class of problems in which today the solution of the phase problem is straightforward and rapid. The crystallographer is thus often working on the problem of greatest chemical or physical interest, and is able to obtain a solution in feasible time. Relatively complete refinement of structures is now the rule, since it is a reasonably fast and effortless procedure. Thus it turns out that in many crystallographic problems the rate-determining step is data collection. For this reason, there has been a dramatic increase in interest in ways of making data collection less tedious, more rapid, and more accurate.

Although in the early days the Braggs and others used ionization chambers for the collection of x-ray intensities, these methods were gradually abandoned in favor of photographic film techniques. Up until a few years ago the great majority of structure determinations were based on photographically recorded intensities, usually visually estimated. This process is a slow one: the typical time involved in the collection and estimation of a data set of two thousand intensities is perhaps six to eight weeks. Collection of intensity data from protein crystals is far more challenging and time-consuming, both because the number of data to be collected is far greater and because the crystals are unstable and rapid collection is thus desirable. For these reasons, Harker and his co-workers, particularly Furnas, then at Brooklyn Polytechnic Institute, were among those instrumental in developing scintillation-counter methods for collecting three-dimensional x-ray data. Diffractometers with single-crystal orienters, based on the so-called Eulerian geometry developed by Harker and Furnas, as well as on the more conventional Weissenberg geometry are available commercially and have engendered widespread interest in counter techniques. Data collection by counter techniques, as practiced by most workers, is still an arduous task, since the setting of a number of orientation angles is involved. Program or computer control of such setting operations is an obvious extension. Especially for neutron diffraction studies, such programmed control of diffractometers has been the rule for some time.

Nevertheless, a programmed unit will do only what it was designed to do, whereas a computer can be programmed to perform new tasks or operations as they seem necessary. There are several installations of computer-controlled diffractometers. Counter methods, particularly when semiautomatic or completely automatic, enable more rapid data collection than is possible photographically. What is equally important is that they should also enable more accurate data to be collected. The general level of accuracy of intensities obtained photograph-

ically is perhaps 15 to 20%. Such a level has proved sufficient for the solution of conformational or stereochemical problems, but not necessarily for the determination of meaningful descriptions of thermal motion or bonding.

There are two approaches to the solution of the phase problem that have remained in favor. The first is based on the tremendously important discovery of Patterson in the 1930's that the Fourier summation of Eq. 3, with the experimentally known quantities $F^2(hkl)$ replacing $F(hkl)$ leads not to a map of scattering density, but to a map of all interatomic vectors. The second approach involves the use of so-called direct methods developed principally by Karle and Hauptman of the U.S. Naval Research Laboratory. The direct method of phase determination makes use of probability theory to give probable relations between phases of different structure factors.

The *Patterson function* has been the most useful and generally applicable approach to the solution of the phase problem, and over the years a number of ingenious methods of unraveling the Patterson function have been proposed. Many of these methods involve multiple superpositions of parts of the map, or "image-seeking" with known vectors. Such processes are ideally suited to machine computation. Whereas the great increase in the power of x-ray methods of structure determination in the past few years has come simply from our ability to compute a three-dimensional Patterson function, it is reasonable to expect that as machine methods of unraveling the Patterson function are developed, this power will increase many fold.

Step (3), the refinement of crystal structure, continues to enjoy a considerable amount of interest. Reasonably complete refinement is routine these days, owing in large measure to the availability of suitable computers. For reasons that are both practical and mathematically sound, the least-squares approach to refinement has gained favor over the successive structure-factor-Fourier approach. Yet the computational problems often tax this generation of computers. If one assigns a single isotropic thermal parameter to each atom, then there are four parameters, three positional and one thermal, to be determined for each atom. In the least-squares procedure, if one stores the upper right triangle of the normal-equations' matrix, then $\frac{1}{2}N(N + 1)$ elements are required, where N is the number of variables. In a machine with a memory of 32,000 words, a practical limit is reached at about $N = 200$, if one wishes to keep the rest of the program in core. Thus refinement of a 50-atom problem often taxes the memory capacity of the machine, and for larger problems special computational or mathematical tricks are needed. One of these tricks is to make use of known features of the structure or the thermal motion to reduce the number of parameters. But, even with increased numbers and more rapid availability of data, computerized solutions to crystallographic problems still suffer from a degree of uniqueness. Programs written to operate on one machine must often be extensively revised to be used on another. An international group of workers at the National Resource for Computation in Chemistry at the Lawrence Berkeley Laboratory is trying to overcome this problem of interchangeability by writing a program designed to run on any medium or large sized computer. When completed, the program will allow a crystallographer, armed only with experimental data and a computer, to determine the structure of any crystal without having to write special programs.

Crystal Growth. The direct growth of an ideal and perfect crystal is difficult except at very high supersaturations because of the difficulty of nucleating a new surface on a completed surface of the crystal. But, if there is a screw dislocation present, it is not necessary to start a new surface, and growth proceeds in a spiral fashion by the accretion of atoms at the edge of

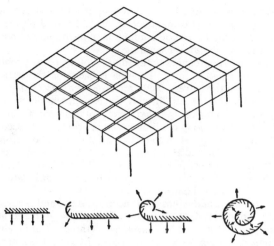

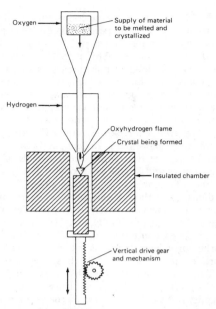

Fig. 4. Verneuil technique for growing crystals.

Fig. 3. Highly schematic representation of crystal growth. (*F. C. Frank.*)

growth steps. The resultant growth spirals have been observed, and it is believed that most cyrstals grow in this manner. See Fig. 3. But spiral growth is not the only mechanism which enables crystal growth at fast rates. Gilmer, in particular, has employed computer simulation models in studies of crystal growth and has demonstrated that, among other influences, temperature and impurity levels have decisive effects upon growth rates. Again, since different crystal faces have different kinetic properties, the particular crystal plane exposed to growth will also be a partial determinant of growth rate.

In modern technology based upon solid state chemistry and physics, much emphasis is placed upon the availability of elements and compounds in single-crystal form. Over the past twenty-five years a highly sophisticated technology has developed in this area. From the relatively simplistic growth of ammonium and potassium dihydrogen phosphates (ADP and KDP) from saturated solutions for reansducer elements, a level has been obtained at which pure metals (Cu, Pb, Al, Ag, Fe etc.), semi-metals (Si, Ge, As, etc.) and compounds (GaAs, InAs, InSb, InP, etc.) are available and even essential in large single crystal form to the electronics industries. Synthetic gems (rubies, spinels, sapphires, emeralds, and zircons) are single crystals of aluminum, beryllium, and zirconium silicates or oxides with controlled impurity levels of transition elements.

Growth of such single crystals can follow several techniques, with thermodynamic constraints dictating the technique for any particular material: crystallization by cooling a supersaturated solution of a compound in a high-temperature flux; crystallization by dropping powder through an intense flame onto a seed pedestal, known as the Verneuil technique (see Fig. 4); crystallization by pulling a "seed" crystal from the surface of a liquid melt, known as the Czochralski method (see Fig. 5); crystallization by lowering a melt through a small, controlled thermal gradient, known as the Stockbarger technique (see Fig. 6); crystallization by zone-melting, known as the Pfann method (see Fig. 7); and crystallization by the vapor-phase approach (see Fig. 8). All of these procedures, however, require three essential ingredients: (1) A good "seed" crystal from which spiral and sometimes oriented growth can occur and develop; (2) highly precise operational conditions—movement of fractions of a millimeter, or temperature variations of 0.5°C per hour; and (3), as previously indicated, materials of a specific impurity level. Given these conditions, single crystals can be grown in large

quantities—ranging from the multimillion carat operations of Linde (United States) an Djevaherdijian (Switzerland), using the Verneuil technique, to the multikilo manufacture of single crystal silicon by Texas Instruments Incorporated (United States), employing the Czochralski and zone melt methods, and including a 27-kilogram single crystal of dislocation-free silicon by the Kayex Corporation (United States) grown by a modified Czochralski approach.

See also **Semiconductor**.

Dislocation in Crystalline Solids. This type of imperfection in a crystalline solid is generated as follows: A closed curve is drawn within the solid, and a cut made along any simple surface which has this curve as boundary. The material on one side of this surface is displaced by a fixed amount called the Burgers vector relative to the other side. Any gap or overlap is made good by the addition or removal of material, and the two sides are then rejoined, leaving the strain displacement intact at the

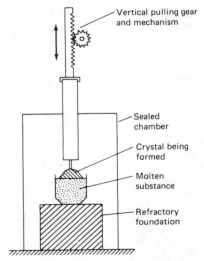

Fig. 5. Czochralski method for growing crystals.

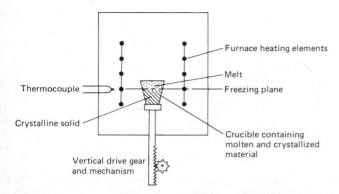

Fig. 6. Stockbarger technique for growing crystals.

moment of rewelding, but afterwards allowing the medium to come to internal equilibrium. If the Burgers vector represents a translation vector of the lattice, the weld is invisible, and the dislocation is characterized only by the original curve, or dislocation line and by the Burgers vector.

A dislocation line may only terminate at the surface of the crystal. The energy of a dislocation is largely stored as strain in the surrounding lattice. The important property of a dislocation is its ability to move quite easily through the lattice, and hence to allow the rapid propagation of slip. The general dislocation defined above usually separates into its components, edge and screw dislocations, which may be treated as rather stable entities in the theory. Direct evidence for the existence of dislocations is the observation of dislocation networks in crystals of silver bromide, and, for their motion, from the spiral growth patterns in crystals.

In recent years high resolution electron microscopes have been used to study dislocations and their movements in thin crystals. This is possible in both metals and nonmetals. Other techniques which have been applied successfully are field ion microscopy and x-ray microscopy. Dislocations are important in determining the mechanical and electrical properties of solids, and play an important part in solid state physics. The density of dislocations (i.e., the concentration of dislocation lines), for example, is believed to vary from about 100,000,000 per square centimeter in good natural crystals, through 1,000,000,000 in good artificial crystals up to about 1,000,000,000,000 in cold-worked specimens. These estimates are based on the energy stored by cold work, on x-ray analysis, and on measurements of electrical resistivity. Among the various types of dislocation are the *edge dislocation*, which is defined as having its Burgers vector (line of displacement) normal to the line of the dislocation. An edge dislocation may be thought of as caused by inserting an extra plane of atoms terminating along the line of the dislocation (Fig. 9). For example, if the dislocation were along the Z-axis and its Burgers vector along the X-axis, then one might think of an extra half plane of atoms being inserted at the surface $x = 0$, $y > 0$. Such a dislocation would be of positive sign. An edge dislocation may move easily only parallel to its Burgers vector, i.e., in its slip plane.

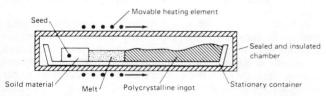

Fig. 7. Zone-melting or Pfann method for growing crystals.

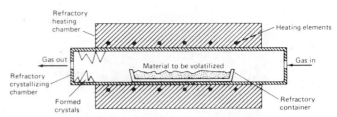

Fig. 8. Crystallization by the vapor-phase approach.

The screw *dislocation* has its Burgers vector parallel to the line of the dislocation. In a screw dislocation, the atomic planes are joined together in such a way as to form a spiral staircase, winding round the line of the dislocation (Fig. 10). A screw dislocation is capable of easy movement in any direction normal to itself. The growth spirals formed in crystal growth appear where such dislocations intersect the surface. Edge dislocations of the same sign repel each other along the line between them, but are most stable when arranged vertically above each other. Edge dislocations of opposite signs attract one another, but otherwise prefer to lie so that the line between them makes an angle of 45° with their slip planes. Screw dislocations of opposite sign attract, of like sign repel.

Dislocation Line. The curve separating displaced and undisplaced positions of a crystal, and thus at the center of a dislocation, is termed the dislocation line. Within a distance of one or two lattice constants of the dislocation line the atoms are displaced by an amount more than can be represented fairly as a strain. A screw location may have a substantial hole down the dislocation line, through which impurity atoms may diffuse.

Dislocation Climb. This is a type of dislocation motion, differing fundamentally from slip, that is associated with the edge components of dislocations. In climb, an edge dislocation moves in a direction perpendicular to its slip plane as atoms are either added to or taken away from the extra plane of the dislocation. The motion of atoms to and from the dislocation is accomplished by vacancy movements. If an atom in the plane next to the edge jumps out of its position and attaches itself to the extra plane, as indicated in the accompanying figure, a vacancy is created which can then diffuse off into the lattice. Repetition of this process over and over will cause the extra plane to grow in size, and, if this occurs, the process is said to be negative climb. On the other hand, if vacancies diffuse up to the extra plane of the dislocation and remove atoms from it, the plane grows smaller in size and positive climb is said to occur.

Dislocation climb only becomes of practical significance at elevated temperatures because of its dependence upon vacancies whose number and mobility depend very strongly on the temperature. Dislocation climb is important in high temperature creep and recovery phenomena. See Fig. 11.

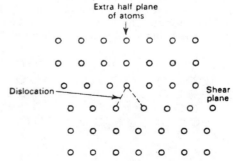

Fig. 9. Atomic arrangement in an edge dislocation.

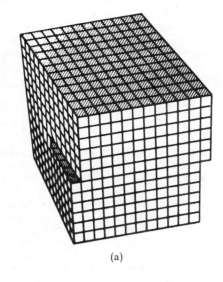

(a)

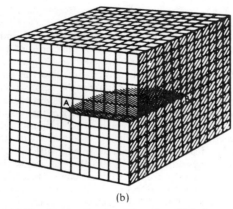

(b)

Fig. 10.   (a) Simple type of screw dislocation; (b) both screw and edge dislocation along the arc from A to B.

Crystal Slip. This is the process by which a crystal undergoes plastic deformation, as a result of which one atomic plane moves over another. Slip is believed to occur through the movement of dislocations. The total deformation of a given crystal is the sum of many small lateral displacements in parallel crystallographic planes of a given family. Moreover, each slip plane becomes more resistant to further deformation than the remaining potential slip planes.

Cross-Slip. This is slip that occurs simultaneously on several slip planes having the same slip direction. See Fig. 12. This type of plastic deformation is normally associated with the movement of screw dislocations. Screw dislocations can move on any slip plane that passes through the dislocation. This is a result of the fact that the slip plane of a dislocation is that plane which contains both the dislocation and its Burgers vector,

Fig. 11.   Negative climb of an edge dislocation.

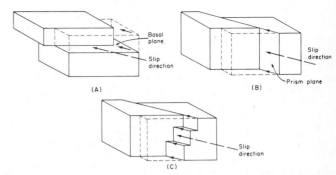

Fig. 12.   Schematic representation of cross-slip in a hexagonal metal: (A) slip on basal plane; (B) slip on prism plane; (C) cross-slip on basal and prism planes.

and the fact that the Burgers vector of a screw dislocation lies parallel to the dislocation itself.

See also **Liquid Crystals.**

References

Bardsley, W., Hurle, D. T. J., and J. B. Mullin: "Crystal Growth," Elsevier-North Holland, New York, 1979.
Buckley, H. E.: "Crystal Growth," Wiley, New York, 1964.
Considine, D. M. (editor): Crystallization and Crystallizers in "Chemical and Process Technology Encyclopedia," pp. 326–336, McGraw-Hill, New York, 1974.
Gilman, J. J. (editor): "The Art and Science of Growing Crystals," Wiley, New York, 1964.
Gilmer, G. H.: "Computer Models of Crystal Growth," *Science*, **208**, 355–363 (1980).
Gruber, Boris: "Theory of Crystal Defects," Academic, New York, 1964.
Hoseman, R.: Diffraction by Matter and Diffraction Gratings in "The Encyclopedia of Physics," (R. M. Besancon, editor), Van Nostrand Reinhold, New York, 1974.
Kroger, F. A.: "Chemistry of Imperfect Crystals," Wiley, New York, 1964.
Lawson, W. D., and S. Neilsen: "Preparation of Single Crystals," Butterworth, London, 1958.
Nowick, A. S. and B. S. Berry: "Anelastic Relaxation Crystalline Solids," Academic, New York, 1971.
Ramachandran, G. N. (editor): "Crystallography and Crystal Perfection," Academic, New York, 1963.
Randolph, A. D. and M. A. Larson: "Theory of Particulate Process—Analysis and Techniques of Continuous Crystallization," Academic, New York, 1971.
Strickland-Constable, R. F.: "Kinetics and Mechanism of Crystallization," Academic, New York, 1967.
Ubbelohde, A. R.: "The Molten State of Matter," Wiley, New York, 1979.
van Gool, W.: "Principles of Defect Chemistry of Crystalline Solids," Academic, New York, 1966.

—R. C. Vickery, Hudson Laboratories, Hudson, Florida.

CRYSTAL FIELD THEORY. A theory which was developed in the early 1930's in research on magnetism by Bethe, Van Vleck and others. It applied particularly to the transition metal ions, and is therefore conveniently treated by reference to those ions.

A transition metal ion in a complex or compound is considered to be subject to the electrostatic field of the molecules and ions in its neighborhood, particularly by those constituting its nearest neighbors. In the compounds to which the theory applies, the only nearest neighbors which are important to the theory are either negative ions, or molecules such as NH_3 which have unshared electron pairs and which orient themselves so that the negative end of the electron-pair dipole is directed toward the transition metal ion. The effect of these negative ions or negatively-oriented dipoles (which we shall hereafter call ligands, with the understanding that ligand field theory is a later development of crystal field theory) is to produce a negative field about the positive transition metal ion.

In the absence of this negative field, the d-electrons of the central ion have orbitals of equal energy, i.e., degenerate orbitals, but the field of the ligands affects the energies of these orbitals to different degrees. To show how this effect arises, consider the example chosen by Griffith and Orgel. This is the regular octahedral complex MX_6, where M is a metal ion of the first transition series of the elements (see **Periodic Table of the Elements**), and X is a ligand such as H_2O, NH_3 or Cl^-.

The five $3d$ orbitals of M have the forms indicated in the figure, in which the coordinate axes lie along the MX bond

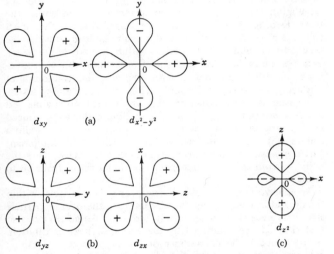

Cross sections of the five d orbitals, chosen in real form.

directions. It is clear from the figure that the d_{z^2} and $d_{x^2-y^2}$ orbitals have substantial amplitudes in the directions of the ligands but that the orbitals d_{xy}, d_{yz}, and d_{zx} tend to avoid them. Hence the energy of an electron in the d_{z^2} or $d_{x^2-y^2}$ orbitals will be substantially raised by the repulsive field of the ligands, whereas the energy of an electron in the d_{xy}, d_{yz}, or d_{zx} orbitals will be comparatively little affected. Furthermore, it is obvious from symmetry that the degeneracy of the last three orbitals is maintained in the octahedral complex and it can be shown by group theory that the d_{z^2} and the $d_{x^2-y^2}$ orbitals also remain degenerate. Consequently the five d orbitals split into a lower group of three and an upper group of two, the two groups being usually designated as t_{2g} and e_g respectively (or sometimes as γ_5 and γ_3, and as $d\epsilon$ and $d\gamma$ respectively).

In tetrahedral complexes it can be shown similarly that the d orbitals are again split into groups of three and two, respectively, but now the doubly degenerate orbital is lower. In all other important cases the degeneracy of the d orbitals is reduced even further.

In order to understand the electronic structure of an octahedral complex and the optical transitions which it can undergo, it is necessary to appreciate the principles determining the distribution of the d electrons among the t_{2g} and e_g orbitals. Let us begin by considering the ground state. Two separate tendencies are at work. The first is the tendency for the electrons to

occupy, as far as possible, the orbitals of lowest energy in the ligand field. The second is for the electrons to go into different orbitals with their spins parallel, since this gives a lower electrostatic repulsive energy and a more favorable exchange energy. Let us now see how these ideas apply to an octahedral complex containing n $3d$ electrons.

The ion $[Ti(H_2O)_6]^{3+}$ has one d electron. In the ground state this will obviously occupy one of the t_{2g} orbitals. A transition is possible in which this electron is transferred to one of the e_g orbitals, and this occurs at 20,400 cm^{-1}. (The intensities of such transitions are low—$\epsilon_{max} \approx 10$—as they are symmetry-forbidden.) The converse situation arises in the hydrated copper(II) ion which has nine d electrons. In this ion the vacancy in the d shell is one of the e_g orbitals. This vacancy can be filled by exciting an electron from one of the t_{2g} orbitals, giving rise to a transition at about 12,500 cm^{-1}. However, the ion $[Cu(H_2O)_6]^{2+}$ is strongly distorted in its ground state so that most of the degeneracy of the t_{2g} and e_g orbitals is removed and there is more than one transition in this region. From these two examples we see that the splitting between the t_{2g} and the e_g orbitals may be quite large, being usually in the range 20–50 kcal $mole^{-1}$.

It is only when we come to consider complexes with several d electrons that complications arise. If there are only two or three d electrons, both the above-mentioned tendencies can be satisfied simultaneously by placing the electrons in different t_{2g} orbitals with their spins parallel. However, when there are more than three d electrons this is no longer possible. If there are 4–7d electrons we then have the choice either of putting as many as possible into the low-energy t_{2g} orbital or distributing them so as to maintain a maximum number of parallel spins. This is illustrated in the accompanying table.

The former choice will be favored if the orbital separation Δ is large, and the latter will be realized if Δ is small. The value of Δ depends primarily on the nature of the ligand and the charge on the ion, and the following generalizations can be made for the first transition series: (1) For hydrated bivalent ions Δ falls in the range 7,500–12,500 cm^{-1}; (2) For hydrated tervalent ions Δ falls in the range 13,500–21,000 cm^{-1}; (3) The common ligands can be arranged in a sequence so that Δ for their complexes with any given metal increases along the sequence. A shortened series is I^-, Br^-, Cl^-, F^-, H_2O, oxalate, pyridine, NH_3, ethylenediamine, NO_2^-, CN^-. Lastly, Δ for the compounds of the second and third series 40–80% larger than for corresponding compounds of the first series.

With these considerations in mind let us consider in turn the two extreme possibilities, known respectively as the "strong-field" and the "weak-field" case. If Δ is very large the tendency for electrons to go into separate orbitals will be outweighed by their tendency to occupy the t_{2g} orbitals (as against the e_g orbitals) in circumstances where these two tendencies conflict. In the strong-field case, therefore, a complex with up to six d electrons will have all these in t_{2g} orbitals with the maximum number of unpaired spins consistent with the restriction to the t_{2g} orbitals. As examples we may take the hexacyanoferrates(II) and (III) which possess six and five d electrons, respectively, all in t_{2g} orbitals with the maximum number of unpaired spins consistent with the restriction to the t_{2g} orbitals. As examples we may take the hexacyanoferrates(II) and hexacyanofer-rates(III) which possess six and five d electrons respectively all in t_{2g} orbitals. The former has no unpaired electrons and the latter one. The next four electrons will then enter the e_g orbitals; the first two will go into different e_g orbitals with their spins parallel as in the octahedral complexes of Ni^{2+}. In Co^{2+} there is just one e_g electron.

d-ELECTRON ARRANGEMENTS IN OCTAHEDRAL COMPLEXES

NUMBER OF d ELECTRONS	ARRANGEMENT IN WEAK LIGAND FIELD		N	ARRANGEMENT IN STRONG LIGAND FIELD		N	GAIN IN ORBITAL ENERGY IN STRONG FIELD
	t_{2g}	e_g		t_{2g}	e_g		
1	↑	—	0	↑	—	0	0
2	↑↑	—	1	↑↑	—	1	0
3	↑↑↑	—	3	↑↑↑	—	3	0
4	↑↑↑	↑	6	↑↓↑↑	—	3	Δ
5	↑↑↑	↑↑	10	↑↓↑↓↑	—	4	2Δ
6	↑↓↑↑	↑↑	10	↑↓↑↓↑↓	—	6	2Δ
7	↑↓↑↓↑	↑↑	11	↑↓↑↓↑↓	↑	9	Δ
8	↑↓↑↓↑↓	↑↑	13	↑↓↑↓↑↓	↑↑	13	0
9	↑↓↑↓↑↓	↑↓↑	16	↑↓↑↓↑↓	↑↓↑	16	0

N Number of distinct pairs of electrons with parallel spin.

The complex $[Co(NH_3)_6]^{3+}$ provides a good example of the strong-field case. The ground state is $(t_{2g})^6$ and the transitions observed at the longest wavelengths involve taking one of these electrons and putting it in an e_g orbital. According to the choice of the orbitals involved the final state may be one of the two triply degenerate states, and the bands associated with the two transitions have been observed for a number of d^6 complexes in each of the three transition series. (The separation between the bands is not in very good agreement with theory, however, so that it is difficult to obtain a reliable value of Δ for such cobalt(III) complexes. The reasons for this are not fully understood at present.)

The weak-field case is that in which the separation Δ is not large enough to overcome the tendency of the d electrons to go into different orbitals with their spins parallel. For example, in the hydrated manganese(II) ion the five d electrons each occupy one of the five d orbitals; this is because the separation Δ is insufficient to break up the highly stable half-filled shell in which all the electron spins are parallel. The same is true of $[Fe(H_2O)_6]^{3+}$ and of the hydrated iron(II) ion, which has six d electrons, the extra one being in one of the t_{2g} orbitals. It may be noted again that only in those complexes containing four, five, six, or seven d electrons is there an important distinction between the strong- and the weak-field cases; if there are one, two, or three d electrons they will necessarily occupy t_{2g} orbitals, while if there are eight or nine the vacancies in the d shell will occur in the e_g orbitals in both the strong- and the weak-field case. (For further discussion, see J. H. Van Vleck, *Phys. Review*, 1932, **41**, 208.)

CRYSTALLIZATION. Crystals are formed (1) from solution, (2) from fusion, and (3) by sublimation.

The formation of crystals from solution, starting with an unsaturated solution, takes place when a solution is evaporated or cooled below the saturation point, except as retarded by supersaturation. Supersaturation is prevented by the addition of seed crystals of the substance. Since, in the case of the majority of soluble substances, solubility increases with increase of temperature cooling below the saturation point favors the formation of crystals. In very few cases, such as sodium sulfate above 32.4°C, calcium sulfate (slightly soluble), calcium hydroxide (slightly soluble), solubility decreases with increase of temperature, and the above statement would not apply. A case of wide scope and great importance is that of crystal formation by precipitation upon mixing two solutions. Actually, this is the same as for substances of greater solubility, since the substance precipitated is first formed in solution and the excess above the satura-

tion point separates as precipitate. As a rule the crystals are larger and more perfect the slower their growth. Conversely, when small crystals are desired, rapid stirring and quick cooling are practical. The smaller the crystals of a given substance, the purer the material generally is. Small crystals may be increased in size by allowing them to stand in the mother liquor before separation.

An industrial forced-circulation evaporative crystallizer is shown in Fig. 1. Sizes range from 18 inches (~ 46 centimeters) to over 42 feet (12.6 meters) in diameter, with no inherent limit to the size of the vessel. Slurry is moved by the circulation pump through the heat exchanger, where it is subjected to a temperature rise of 2 to 10°F. The heated liquor is discharged tangentially into the body at a point sufficiently far beneath the surface so that the liquor entering tangentially is just at the boiling point for the liquid depth at which it is submerged. As the liquid rotates around the body and rises toward the surface, it starts to boil, and this boiling induces a secondary circulation which creates a spinning toroid of fluid within the body. Depending on the location of tangent inlet with respect to the cone, this toroidal circulation can result in considerable secondary circulation and agitation within the body of the vessel. When properly designed, this type of vessel is capable of producing a smooth boiling action with relatively small amounts of salt being deposited on the walls, while still maintaining a suitable suspension of product crystals within the boiling zone and in the lower part of the vessel. Crystalline materials produced in this type of equipment include sodium carbonate, sodium sulfate, and sodium chloride. Operating cycles of this equipment between washouts to remove salt growth from the walls of the body or from the heat exchangers normally range from 30 to 90 days.

To achieve some control of the number of fine particles within the crystallizer body, and thereby increase the overall particle size, it is necessary to selectively remove the fine particles so that they can be destroyed by the action of heat or dilution. A draft-tube baffle crystallizer of the type shown in Fig. 2 achieves these objectives. A body of growing crystals is sus-

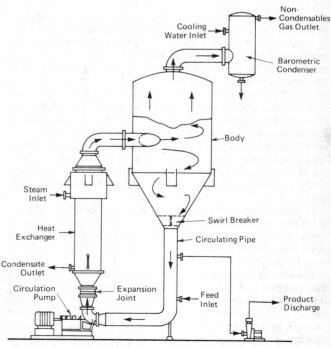

Fig. 1.   Forced-circulation evaporative crystallizer.

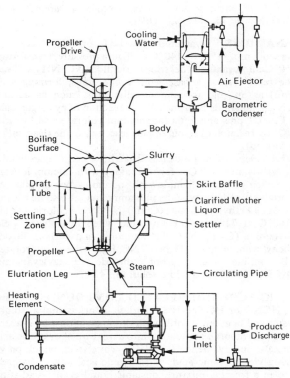

Fig. 2.   Draft-tube baffle crystallizer.

pended by the circulation flowing up the draft tube from the propeller shown close to the bottom of the vessel. From the areas surrounding this body of circulated slurry, a stream is removed at relatively low velocity so that gravitational settling will produce a separation between the product-size crystals and relatively fine crystals which are removed with the clarified mother liquor leaving by the circulating pipe. In the draft-tube baffle crystallizer, both the velocity of the liquor in the settling zone and the quantity of liquor removed by the circulating pump are important to insure that the proper end-product size is achieved and that reasonable stability in particle size is obtained. The equipment is used to crystallize potassium chloride, ammonium sulfate, and other relatively fast-growing inorganic salts.

The foregoing descriptions cover only two of several industrial crystallizer configurations.

The formation of crystals from fusion takes place when the melted substance is cooled sufficiently slowly near and below the fusion point. If the cooling is rapid the fusion may result in the formation of an undercooled liquid of rigidity corresponding to a solid. Glasses, whether artificial, such as glass, vitreous enamels, and slags, or natural, such as vitreous rocks and minerals, e.g., obsidianite, are undercooled liquids. Rocks and minerals which have cooled sufficiently slowly from fusion form crystals, for example, granite. An important method of forming pure crystals is zone melting.

The formation of crystals by sublimation takes place when the vapor of a substance is condensed as a solid without passing through the liquid phase in so doing. This occurs when the temperature of the condenser is below that of the melting point of the substance.

The heat of crystallization is in amount the same as the heat of solution of a given substance but of opposite sign.

For references see entry on **Crystal.**

CRYSTALLOID.   Colloid System.

CRYSTAL PHASES (α-, β-, γ-, ϵ-, η-, etc.).   Certain alloy systems may form different crystal structures, according to the relative proportions of the constituents, e.g., Cu-Zn, for which no less than five different phases are known. In many cases, the same crystal structure occurs with quite different constituent metals, so that it is often possible to use the one expression such, for example, as β-phase, to cover a wide variety of compounds all having the same basic structure. This effect is explained by the Hume-Rother rules. Pure substances, as well as alloys, may exhibit more than one crystal structure, depending on temperature and past history, e.g., cobalt, iron, titanium.

CUMMINGTONITE.   The mineral cummingtonite is a variety of amphibole which is essentially $(Mg, Fe, Mn)_7Si_8O_{22}(OH)_2$, the amounts of magnesium and iron varying as they replace one another. Cummingtonite is generally restricted to material containing from 50 to 70% $MgSiO_3$. The name *grunerite* has been applied to cummingtonite which contains more than 50% of the $FeSiO_3$ molecule. Cummingtonite usually occurs as a dark green to brown fibrous to lamellar mineral. It derives its name from Cummington, Massachusetts.

See also **Amphibole.**

CUPRITE.   The mineral cuprite, cuprous oxide, Cu_2O, occurs as isometric crystals, usually octahedrons, but may be cubes, dodecahedrons or modified combinations. It also is found as a massive, earthy material. Its fracture is cochoidal to uneven; brittle; hardness, 3.5–4; specific gravity, 6.14; luster, submetallic to earthy; color, red; nearly transparent to nearly opaque. Its streak is shining brownish-red. Cuprite is a secondary mineral resulting doubtless from the oxidation of copper sulfides. It is often found associated with native copper, malachite and azurite.

Cuprite is a fairly common mineral, and of the many localities in which it occurs may be mentioned the Province of Perm, in the U.S.S.R.; Chessy, France; Broken Hill, New South Wales; Corocoro, Bolivia; Andacollo, Chile; Bisbee, Arizona; and Del Norte County, California. Magnificent large transparent red gem crystals, some with a coating of malachite, have been found at Ojunga, S.W. Africa. The name cuprite is derived from the Latin *cuprum*, copper.

CURIE POINT (or Curie Temperature).   Ferromagnetic materials lose their permanent or spontaneous magnetization above a critical temperature (different for each substance). This critical temperature is called the Curie point. Similarly ferroelectric materials lose their spontaneous polarization above a critical temperature. For some materials, this is called the *upper Curie point*, for there is also a *lower Curie point*, below which the ferroelectric property disappears.

CURING (Rubber).   Rubber (Natural).

CURIUM.   Chemical element symbol Cm, at. no. 96, at. wt. 247 (mass number of the most stable isotope), radioactive metal of the actinide series; also one of the transuranium elements, mp estimated 1350 ± 50°C. The isotope ^{247}Cm has a half-life of 1.64×10^7 years. Other long-lived isotopes are ^{245}Cm ($t_{1/2} = 9230$ years). ^{246}Cm ($t_{1/2} = 5480$ years), ^{248}Cm ($t_{1/2} = 4.7 \times 10^5$ years), and ^{250}Cm ($t_{1/2} = 2 \times 10^4$ years). Other known isotopes are ^{238}Cm, ^{242}Cm, ^{244}Cm, and ^{249}Cm. Electronic configuration is $1s^22s^22p^63s^23p^63d^{10}4s^24p^64d^{10}4f^{14}5s^25p^65d^{10}5f^76s^26p^66d^17s^2$. Ionic radius: Cm^{3+} 0.98 Å.

First identified in 1944 by G. T. Seaborg, R. A. James, and A. Ghiorso, who found ^{242}Cm in the product obtained by bombarding ^{239}Pu with alpha particles of resonance energies. Later

L. B. Werner and I. Perlman produced and isolated the same isotope by the action of neutrons on ^{241}Am.

In experiments, the concentration of curium must be kept low in order to avoid the formation of a reducing medium due to the action of the ^{242}Cm alpha particles on water. At a concentration of 10^{-5} molar in curium and under conditions where americium(III) is oxidized to americium(VI) in the same solution, the curium is not oxidized above the (III) state with ammonium peroxydisulfate.

The solubility properties of curium(III) compounds are in every way similar to those of the other tripositive Actinide elements and the tripositive Lanthanide elements. Thus the fluoride and oxalate are insoluble in acid solution, while the nitrate, halides, sulfate, perchlorate, and sulfide are all soluble.

Solid curium trifluoride has been prepared by drying the fluoride, which precipitates from dilute HNO_3 upon the addition of hydrofluoric acid. Curium trifluoride can be reduced to the metal by heating at 1275°C in a beryllia crucible with barium vapor. The metal is silvery in color and has the properties of an electropositive element in common with the other Actinide elements.

The ion Cm^{3+} is colorless, as are its compounds generally. CmF_3 is hexagonal, Cm_2O_3 is white and CmO_2 is black and hexogonal. $CmCl_3$ is light yellow and hexagonal. Cm^{4+} is known in solution only as the complex fluoride.

In research at the Institute of Radiochemistry, Karlsruhe, West Germany during the early 1970s, investigators prepared alloys of curium with iridium, palladium, platinum, and rhodium. These alloys were prepared by hydrogen reduction of the curium oxide or fluoride in the presence of finely divided noble metals. The reaction is called a *coupled reaction* because the reduction of the metal oxide can be done in the presence of noble metals. The hydrogen must be extremely pure, with an oxygen content of less than 10^{-25} torr.

Curium was first isolated in the form of a pure compound, the hydroxide, of curium-242 (produced by the neutron irradiation of americium-241) by Werner and Perlman at the University of California in the autumn of 1947. Much of the earlier work with curium used the isotopes ^{242}Cm and ^{244}Cm, but the heavier isotopes offer greater advantages mainly because of their longer half-lives. The isotope ^{248}Cm, obtainable in relatively high isotopic purity as the alpha particle decay daughter of ^{252}Cf, is the most practical for chemical studies. See also **Radioactivity**.

References

Asprey, L. B., Ellinger, F. H., Fried, S., and W. H. Zachariasen: "Evidence for Quadrivalent Curium: X-Ray Data on Curium Oxides," *Amer. Chem. Soc. J.*, 77, 1707–1708 (1955).

Asprey, L. B., Ellinger, F. H., Fried, S., and W. H. Zachariasen: "Evidence for Quadrivalent Curium. II. Curium Tetrafluoride," *Amer. Chem. Soc. J.*, 79, 5825 (1957).

Kanellakopulos, B., et al.: "The Magnetic Susceptibility of Americium and Curium Metal," *Solid State Commun.*, 17, 6, 713–715 (1975).

Keller, C., and B. Erdmann: "Preparation and Properties of Transuranium Element-Noble Metal Alloy Phases," *Proc. 1972 Moscow Symp. Chemistry of Transuranium Elements* (1976).

Samhoun, K., and F. David: "Radiopolarography of Am, Cm, Bk, Cf, Es, and Fm," *Proc. 4th International Transplutonium Element Symp.*, Baden Baden, W. Germany, 1975.

Seaborg, G. T.: "The Chemical and Radioactive Properties of the Heavy Elements," *Chem. Eng. News*, 23, 2190–2193 (1945).

Seaborg, G. T. (editor): "Transuranium Elements," Dowden, Hutchinson & Ross, Stroudsburg, Pennsylvania, 1978.

Stevens, C. M., et al.: "Curium Isotopes 246 and 247 from Pile-Irradiated Plutonium," *Phys. Rev.*, 94, 4, 974 (1954).

Street, K., Jr., and G. T. Seaborg: "The Separation of Americium and Curium from the Rare Earth Elements," *Amer. Chem. Soc. J.*, 72, 2790–2792 (1950).

Werner, L. B., and I. Perlman: "First Isolation of Curium," *Amer. Chem. Soc. J.*, 73, 5215–5217 (1951).

CYANAMIDES.

Cyanamide $NC \cdot NH_2$ or $HN:C:NH$ is a white solid, mp. 44°C, bp 140°C at 20 mm pressure, transformed at 150°C into cyanuramide, tricyantriamide $(NC \cdot NH_2)_3$. Cyanamide reacts (1) as a base with strong acids, forming salts; and (2) as an acid forming metallic salts, such as calcium $CaCN_2$. Cyanamide is formed (1) by reaction of cyanogen chloride $CN \cdot Cl$ plus ammonia (ammonium chloride also formed), and (2) by reaction of thiourea plus lead hydroxide (lead sulfide also formed).

When calcium cyanamide is boiled with water, dicyandiamide $(NC \cdot NH_2)_2$, mp 207°C is formed (along with calcium hydroxide). Fusion of dicyandiamide with sodium carbonate plus carbon produces sodium cyanide, plus ammonia (also some tricyantriamide). Diethylcyanamide $(C_2H_5)_2N \cdot CN$ is a colorless liquid, bp 189°C at 748 mm pressure, and when hydrolyzed yields diethylamine $(C_2H_5)_2NH$ plus ammonia and carbon dioxide. Diphenylcyanamide $C_6H_5N:C:NC_6H_5$, when hydrolyzed, yields aniline plus carbon dioxide. Benzylcyanamide $C_6H_5CH_2NH \cdot CN$ is a white solid, mp 43°C.

CYANIC ACID AND RELATED COMPOUNDS.

Cyanic acid, $HCNO$ or $HOCN$, is a colorless, odorless liquid, soluble in water and in ether, volatile with decomposition when heated, passing at ordinary temperature into a mixture of cyanuric acid, $(HNCO)_3$, and cyamelide $(CONH)_2$, white solid, which on vaporizing yields cyanic acid; when cyanic acid vapor is rapidly cooled in a freezing mixture, unstable liquid cyanic acid is obtained, and when the vapor is condensed above 105°C, cyanuric acid

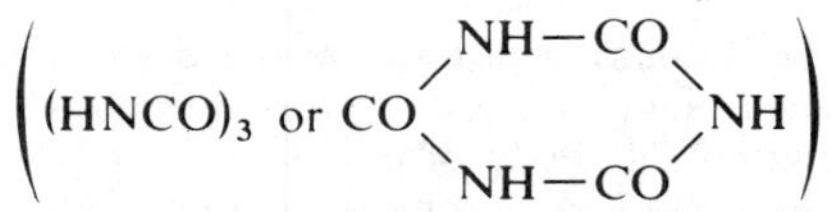

$$(HNCO)_3 \text{ or } CO \left\langle \begin{array}{c} NH-CO \\ NH-CO \end{array} \right\rangle NH$$

is obtained. Cyamelide dissolves in H_2SO_4 unchanged and addition of water causes a precipitation of cyamelide; passes into cyanuric acid when warmed with concentrated H_2SO_4, finally into carbon dioxide plus ammonia; dissolves in sodium hydroxide solution, forming sodium cyanate. Sodium cyanate is prepared by heating sodium cyanide and an oxide, such as lead monoxide PbO, trilead tetroxide Pb_3O_4, or lead dioxide PbO_2, addition of water, and separation of the sodium cyanate solution from the lead oxide by filtration. Sodium cyanate solution upon boiling changes into sodium carbonate plus urea $CO(NH_2)_2$.

Ammonium cyanate, $CNONH_4$, white solid, formed by reaction of sodium cyanate and ammonium sulfate solutions, is transformed to urea upon being heated at 100°C. This reaction was carried out in 1828 by Wöhler, and is the first record of a so-called inorganic substance being transformed outside a living organism into a so-called organic substance. The following esters are known:

Methyl isocyanate, CH_3NCO, boiling point 44°C
Ethyl cyuanate, C_2H_5OCN, decomposes on heating
Ethyl isocyanate, C_2H_5NCO, boiling point 60°C
Phenyl isocyanate, C_6H_5NCO, boiling point 166°C

Ethyl cyanurate $\quad C_2H_5O \cdot C \left\langle \begin{array}{c} N=C(OC_2H_5) \\ N=C(OC_2H_5) \end{array} \right\rangle N$

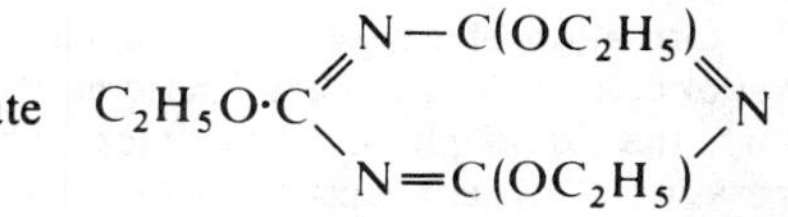

Ethyl isocyanurate $\quad CO \left\langle \begin{array}{c} N(C_2H_5)-CO \\ N(C_2H_5)-CO \end{array} \right\rangle N(C_2H_5)$

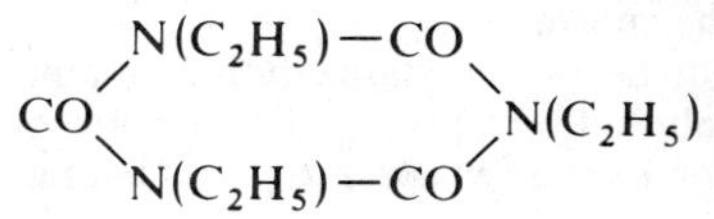

The extensive use of organic isocyanates in various industrial processes for production of high polymers has brought about tonnage production. Toluene diisocyanate is made by nitrating toluene to the dinitro compound, which is then reduced to the diamine, and treated with phosgene to obtain the diisocyanate:

$$C_6H_5CH_3 \xrightarrow{HNO_3} C_6H_3(NO_2)_2CH_3 \xrightarrow{H}$$

$$C_6H_3(NH_2)_2CH_3 \xrightarrow{COCl_2} C_6H_3(NCO)_2CH_3$$

Toluene diisocyanate is widely used in the manufacture of urethane plastics, particularly the urethane foamed plastics. Another isocyanate, diphenylmethane 4,4'-diisocyanate, is produced by reaction of aniline and formaldehyde, followed by reaction with phosgene:

$$2C_6H_5NH_2 \xrightarrow{HCHO} CH_2(C_6H_4NH_2)_2 \xrightarrow{COCl_2} CH_2(C_6H_4NCO)_2$$

The diphenylmethane 4,4'-diisocyanate is used in the manufacture of solid urethane elastomers (primarily for heavy duty tires) and chemically resistant coatings.

Fulminic acid, HONC, and the fulminates are violently explosive. Utilizing this property, mercuric fulminate $Hg(ONC)_2 \cdot \frac{1}{2}H_2O$, is used as a detonator for other explosives. Mercury fulminate is made by the reaction of ethyl alcohol and mercuric nitrate in excess of nitric acid, from which insoluble mercuric fulminate separates. Silver fulminate, $Ag(ONC)$ is more explosive than mercuric fulminate, and is used in the manufacture of firecrackers. Free fulminic acid may be obtained by reaction of potassium fulminate and excess of ether. It volatilizes with the ether upon distilling, and changes rapidly to metafulminic acid. Related to fulminic acid, is fulminuric acid, $(HONC)_3$, or $NO_2 \cdot CH(CN) \cdot CONH_2$.

CYANOGEN. Cyanogen $(CN)_2$ is a colorless gas of marked characteristic odor, very poisonous, density 1.8 (air equal to 1.0), melting point $-28°C$, boiling point $-20°C$, soluble. When passed into water at $0°C$, cyanogen forms hydrocyanic acid plus cyanic acid, but at ordinary temperatures the reaction is complex. With sodium hydroxide solution, there is formed with cyanogen sodium cyanide plus sodium cyanate, with dilute sulfuric acid oxamic acid $COOH \cdot CONH_2$, oxalic acid $COOH \cdot COOH$. By reaction with tin and hydrochloric acid, cyanogen is reduced to ethylene diamine $CH_2 \cdot NH_2 \cdot CH_2 \cdot NH_2$. Cyanogen reacts with hydrogen to form hydrocyanic acid, and with metals, e.g., zinc, copper, lead, mercury, silver, to form cyanides. Cyanogen, (1) when burned in air produces a violet flame forming carbon dioxide and nitrogen in the outer part and carbon monoxide and nitrogen in the inner part, (2) when exploded with oxygen produces carbon dioxide or carbon monoxide and nitrogen depending upon the ratio of oxygen to cyanogen (2 volumes oxygen plus 1 volume cyanogen yields 2 volumes carbon dioxide plus 1 volume nitrogen; 1 volume oxygen plus 1 volume cyanogen yields 2 volumes carbon monoxide plus 1 volume nitrogen). The flame spectrum contains characteristic bands in the blue and violet. By means of the electric spark, the electric arc or a red hot tube, cyanogen is decomposed into carbon plus nitrogen. When heated at ordinary pressure at about 300°C, or under 300 atmospheres pressure at about 225°, cyanogen is converted into paracyanogen, a brown powder, also formed when mercuric cyanide is heated. Cyanogen is prepared (1) by reaction of sodium cyanide and copper sulfate solutions, whereby one half the cyanogen is evolved as cyanogen gas and one half remains as cuprous cyanide. From the filtered cuprous cyanide, by treatment with ferric chloride solution, cyanogen is evolved with accompanying formation of ferrous chloride, (2) by heating mercuric cyanide solid, or a mixture of mercuric chloride and sodium cyanide solutions, mercury

and mercurous, respectively, being formed, (3) by heating ammonium oxalate $COONH_4 \cdot COONH_4$ with phosphorus pentoxide, water being abstracted. Small amounts of cyanogen are present in blast furance gas and raw coal gas.

CYANOHYDRINS. The products of the reaction between an aldehyde or a ketone with hydrogen cyanide HCN are called *cyanohydrins*. Sometimes the compounds are referred to as *hydroxycyanides*.

$$\underset{\text{(acetaldehyde)}}{CH_3CHO} + HCN \rightarrow \underset{\substack{\text{(hydroxyethyl cyanide or} \\ \text{aldehyde cyanohydrin)}}}{CH_3 \cdot CH(OH) \cdot CN}$$

$$\underset{\text{(acetone)}}{(CH_3)_2CO} + HCN \rightarrow \underset{\substack{\text{(hydroxyisopropyl cyanide or} \\ \text{acetone cyanohydrin)}}}{(CH_3)_2C(OH) \cdot CN}$$

CYBOTAXIS. A condition in which certain liquids, under x-ray examination, give evidence of structure resembling that of crystals. By passing a beam of x-rays through various alcohols and other organic liquids, G. W. Stewart and collaborators obtained one, two, or even three diffraction maxima or halos somewhat like the diffraction rings produced by powdered crystals. These suggest that molecules are temporarily arranged in rows, layers, or stacks like bricks in a pile and that they have one, two, or even three different dimensions or spacings, corresponding, in accordance with Bragg's law, to the different angles of diffraction observed.

A closely related property is exhibited by certain substances known as liquid crystals which appear to be intermediate between merely cybotactic liquids and true crystals. In these, there appear to be large groups of molecules which, though able to move and turn about, retain their structural arrangement. Such mesomorphic substances manifest even some of the optical properties of crystals, which the former type do not. See also **Liquid Crystals.**

CYCLOHEXANOL-CYCLOHEXANONE. KA oil is comprised of cyclohexanol, an alcohol, and cyclhexanone, a ketone, and is a principal raw material in the manufacture of nylon 6 and nylon 66 fibers. The KA stands for ketone-alcohol oil. At one time, KA oil was derived principally from phenol, but the majority is currently manufactured from the oxidation of cyclohexane. The KA oil is converted to adipic acid in the production of nylon 66, but in the production of nylon 6 (polycarprolactam), the KA oil is converted into the monomer, caprolactam.

Cyclohexanol, $CH_2[(CH_2CH_2)_2]$ CHOH, formula wt 100.16, mp 23.9°C, bp 160–161°C, sp gr 0.962 (20°C referred to water at 4°C), slightly soluble in H_2O, soluble in alcohol and ether. Cyclohexanone, $CH_2(CH_2CH_2)_2CO$, formula wt 98.14, mp $-45°C$, bp 155–156°C, sp gr 0.947, soluble in H_2O, alcohol and ether.

CYSTEINE. Amino Acids.

CYSTINE. Amino Acids.

CYTOCHROMES. The cytochrome *c*-cytochrome oxidase system represents the terminal segment of the respiratory chain common to the vast majority of organisms utilizing oxygen as the terminal oxidant in tissue respiration. The complete respiratory chain consists of a number of electron carriers, both protein and nonprotein in nature, organized in a definite sequence within the walls and internal partitions of subcellular organelles known as mitochondria. These structures carry the electrons which come from the substrates being oxidized and eventually react with oxygen. The energy released in several of the many steps of this series of reactions is utilized to make the high-energy compound adenosine triphosphate, a process known as *oxidative phosphorylation*. The high-energy compound is, in turn, employed to drive the many reactions of metabolism which require chemical energy. Every component of the terminal respiratory chain is reduced by the component immediately preceeding it and then reduces the component immediately following it in the chain, itself becoming reoxidized. The *c*-cytochrome oxidase system is common to all vertebrates and invertebrates, plants, as well as numerous microorganisms, and must be distinguished from systems having similar functions, but very different properties, which occur in numerous bacteria.

The cytochromes were first observed by MacMunn as early as 1886. He described their spectral absorption bands in a large variety of organisms and tissues. His discovery was, however, forgotten after a controversy with Hoppe-Seyler had raised doubts as to the validity of some of his conclusions, and it was not until 1925 that Keilin independently, rediscovered the remarkable cytochrome spectrum in the flight muscles of a living insect.

Keilin's observations came at a time when the understanding of tissue respiration had advanced to the point of providing the foundations necessary for the unraveling of the physiological role and chemical nature of cytochromes. The first step had indeed been taken some 40 years earlier by Ehrlich when he found that a variety of animal tissues could transform a mixture of α-napththol and dimethyl-*p*-phenylenediamine to indophenol, in the presence of oxygen. A decade later, the enzyme responsible for this effect had been named indophenol oxidase, and it was shown that its activity was inhibited by cyanide. In the first decades of this century, Warburg, from studies of the catalysis of the oxidation of cysteine by iron-charcoal, considered as a "model" of cellular respiration, concluded that oxygen activation was the all-important process in cellular respiration and that an iron-containing enzyme, the "respiratory enzyme" or *atmungsferment*, is solely responsible for the transport of the oxidizing equivalents of oxygen to the substrates. An opposing view was taken by Thunberg, who had detected a large variety of dehydrogenases in tissues, and by Wieland who used palladium-hydrogen as a "model" of tissue respiration and believed that substrate-specific hydrogen activations were characteristic of all biological oxidation processes, the reaction with oxygen being nonspecific and relatively unimportant.

The controversy as to the respective roles and importance of hydrogen and oxygen activation faded into the background when, following his initial observations, Keilin demonstrated that the four-banded spectrum of cytochrome, observed in a large variety of tissues and organisms, was in fact the spectrum of the ferrous or reduced forms of three distinct cytochromes, cytochrome *a*, cytochrome *b* and cytochrome *c*. Keilin obtained a soluble preparation of cytochrome *c* from baker's yeast, and together with Hartree, in 1938–1939, showed that the indophenol oxidase activity of particulate tissue preparations was simply the result of a nonenzymic reduction of cytochrome *c* by dimethyl-*p*-phenylenediamine, the reduced heme protein being oxidized by indophenol oxidase in the presence of oxygen. Having established the nature of the final steps of tissue respiration, they renamed the enzyme "cytochrome oxidase," since its only function appeared to be the oxidation of cytochrome *c*. There had been no doubt of the overwhelming physiological importance of the system ever since 1934 when Haas found that in

a number of tissues the rate of oxygen uptake was identical to that of cytochrome *c* reduction, demonstrating that nearly all of the oxidizing equivalents of oxygen were transmitted by the cytochrome *c*-cytochrome oxidase system.

That the material in tissues reacting directly with oxygen was in fact a heme compound had been shown by the experiments of Warburg and collaborators on the effect of carbon monoxide on tissue respiration, carried out in the late 1920s. Warburg observed that carbon monoxide inhibits the uptake of oxygen by tissues and that this inhibition is reversed in bright light. Using this phenomenon, he succeeded in measuring the abosrption spectrum of the carbon monoxide complex of the respiratory enzyme, a spectrum which turned out to be clearly that of a heme compound. Thus, when Keilin and Hartree in 1939 found that in the presence of carbon monoxide, cytochrome *a* showed up as two spectroscopic components, they were able to demonstrate that the new cytochrome, cytochrome a_3 was the substance responsible for the photochemical action spectrum of Warburg. Cytochrome a_3 was thus identified with the respiratory enzyme reacting directly with oxygen, and the system was considered to be composed of three entities, cytochromes *c*, *a*, and a_3, reacting consecutively, like all the other components of the respiratory chain.

Cytochrome *c* consists of a polypeptide chain, from 104 to 108 amino acid residues in length. A single heme prosthetic group is attached by thioether bonds formed between the sulfhydryl side chains of two cysteine residues in the protein and the vinyl side chains of the porphyrin ring as shown by

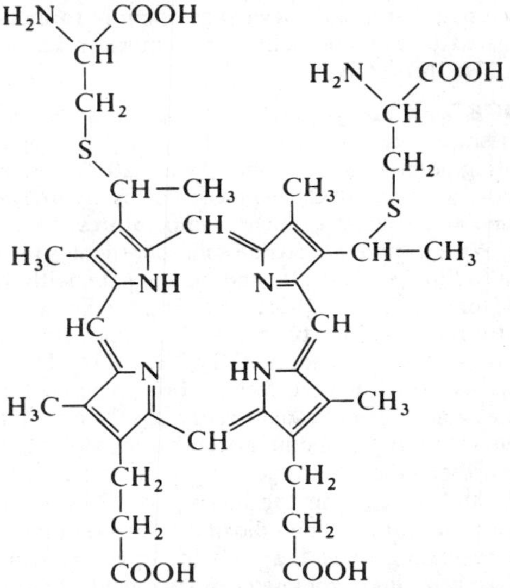

—Ann C. Vickery, Ph.D., University of South Florida, College of Medicine, Tampa, Florida.

CYTOTOXIC CHEMICALS. Chemical agents that damage cells to which they are applied. They are poisons, to which cells respond with injury, disease, or death. There are multitudes of cytotoxic chemicals; they act by a variety of mechanisms; and they have many different kinds of effects. See also **Carcinogens.** An excellent reference on poisonous substances is "Dangerous Properties of Industrial Materials," 5th edition, N. I. Sax, Editor, Van Nostrand Reinhold, New York, 1979.

D

DALTON'S LAW. The law of partial pressures in mixed gases and vapors. If several gases not reacting chemically with each other are introduced into the same container, the pressure of the resulting mixture is equal to the sum of the pressures which would be observed if each gas were separately enclosed in that chamber. Atmospheric pressure, for example, is the sum of a nitrogen pressure, an oxygen pressure, an argon pressure, a carbon dioxide pressure, a water-vapor pressure, etc. The same principle holds for mixtures of the saturated vapors of two or more liquids evaporating in the same closed space, provided one liquid does not dissolve the vapor from the other (as water dissolves ammonia). The law is approximately valid only within limits.

DANBURITE. The mineral dandurite, $CaB_2Si_2O_8$, calcium-boron silicate, crystallizes in the orthorhombic system in prismatic forms somewhat resembling the mineral topaz. Its fracture is subconchoidal; brittle; hardness, 7; specific gravity, 2.97–3.02; color, colorless, yellowish-white, dark wine yellow and brownish-yellow; luster vitreous to greasy; translucent to transparent. It is found at Danbury, Connecticut, from whence its name was derived; Saint Lawrence County, New York; Switzerland; Japan; and the Malagasy Republic.

DATOLITE. Datolite, basic calcium boron silicate, $CaBSiO_4(OH)$, occurs in monoclinic crystals of varied habit, mostly short stout prisms, but often in highly modified forms. Datolite reveals no cleavage, its fracture is conchoidal to uneven; brittle; hardness, 5–5.5; specific gravity, 2.9–3.0; luster, vitreous to dull; color, white to gray or may be greenish, yellowish, or brownish. It has a white streak and is transparent to translucent usually, but has been observed opaque.

Datolite is a secondary mineral, being found in veins and cavities associated with zeolites and calcite, particularly in the basic igneous rocks. It has been found in the Harz Mountains, Germany; in the Trentino district, Italy; in Norway and Tasmania. In the United States it has been found in the Triassic traps of the Connecticut River Valley in Massachusetts and Connecticut, and from similar rocks in New Jersey. In Michigan, datolite has been found associated with the copper-bearing rocks of Keweenaw County. This mineral derives its name if an observation lies more than, say, three times the standard deviation of a whole sample of observations away from the mean, it is unlikely to belong to that sample in the sense of having been obtained under similar conditions.

DEAERATION. A process for removing dissolved air from a liquid and usually refers to the removal of dissolved oxygen and carbon dioxide from boiler feedwater. In the case of high-pressure steam boilers, a small amount of oxygen dissolved in the feedwater may become quite active in attacking the boiler metal. Deaerators are of the deactivating or heating type.

DEBYE-HÜCKEL LIMITING LAW. The departure from ideal behavior in a given solvent is governed by the ionic strength of the medium and the valences of the ions of the electrolyte, but is independent of their chemical nature. For dilute solutions, the logarithm of the mean activity is proportional to the product of the cation valence, anion valence, and square root of ionic strength giving the equation

$$-\log f_{\pm} = Az + z - \sqrt{\mu}$$

See **Electrochemistry.**

DEBYE THEORY OF SPECIFIC HEAT. The specific heat of solids is attributed to the excitation of thermal vibrations of the lattice, whose spectrum is taken to be similar to that of an elastic continuum, except that it is cut off at a maximum frequency in such a way that the total number of vibrational modes is equal to the total number of degrees of freedom of the lattice.

The Debye temperature is defined by the relation

$$\Theta = \frac{h v}{k}$$

where v is the maximum frequency of the thermal vibrations of the lattice, h is Planck's constant and k is the Boltzmann constant.

DECARBURIZATION. This is the reduction in carbon content at the surface of steel or cast iron. In heating metal for hot rolling, forging, or heat treatment, decarburization is usually objectionable and specially prepared neutral furnace atmospheres may be used to reduce or eliminate it. Molten salt or lead baths are also effective in protecting the surface during heat treatment. Decarburization reduces fatigue strength and lowers wear-resistance of bearing surfaces. On the other hand, decarburization is intentionally carried out in the processing of low-carbon sheet steels for electrical applications. In the production of malleable cast iron by annealing white cast iron, decarburization is beneficial.

DECOLORIZING AGENTS. A substance that removes color by a physical or chemical action. Charcoals, carbon blacks, clays, earths, activated alumina or bauxite, or other materials of highly adsorbent character are used to remove undesirable colors (and often odors) from sugar and vegetable and animal fats and oils, among other substances. In a broad sense, decolorizing agents also embrace bleaches, which usually involve a chemical reaction for removing color.

Activated carbon is one of the most widely used of the adsorbents. It is an amorphous form of carbon characterized by high adsorptivity. The carbon is obtained by the destructive distillation of wood, nut shells, animal bones, or other carbonaceous material. It is "activated" by heating to 800–900°C with steam or carbon dioxide, which results in a porous internal structure. The internal surface area of activated carbon averages about 10,000 square feet (929 square meters) per gram. Numerous uses include applications in the brewing and sugar refining industries.

Diatomaceous earth also finds numerous adsorbant applications in food and chemical processing, not only in decolorizing, but as a filter aid and clarifying agent as well. See also **Diatomaceous Earth.**

Fuller's earth, also used as an adsorbant, is a porous colloidal aluminum silicate (clay) which has a high natural adsorptive power. See also **Fuller's Earth.**

Silica gel is a regenerative adsorbent consisting of amorphous silica derived from sodium silicate and sulfuric acid. In addition to color adsorbing and bleaching powers, silica gel is used as a dehumidifying and dehydrating agent and as an anticaking agent.

Prior to crystallization in the refining of sugar, bleaching of the syrup is required. This is sometimes effected through treatment of the solution with calcium hypochlorite, usually in the presence of calcium phosphate, which serves as a buffer and aids in the final precipitation of calcium from the bleached solution.

The physical properties of representative adsorbents and decolorizers are given in the table in the entry on **Adsorption Operations.**

DECOMPOSITION (Chemical).

A chemical change in which a single chemical substance is broken up into two or more other substances, which differ from each other and from the parent substance in chemical identity. Complete decomposition refers to such a condition of the products that they are not readily decomposed further, e.g., such decomposition products as ammonia and carbon dioxide.

Degradation refers to gradual decomposition in which the molecule is diminished in size in small steps.

The *heat of decomposition* is the change of heat content when one mole of a compound is decomposed into its elements. This is equal in quantity, but opposite in sign, to the *heat for formation*.

Sensitized decomposition is a chemical decomposition that is brought about by the presence of a second substance which absorbs an exciting radiation. The essential mechanism of the reaction is the excitation of particles of the second substance by the radiation, followed by collisions between these excited particles and molecules to the decomposed. The process proceeds most effectively if the energy difference between the ground state and excited state of the sensitizer is nearly equal to the energy of the decomposition reaction.

Double decomposition is a term used to express the interaction of molecules which exchange one or more of their constituent atoms or radicals.

DEFOAMING AGENTS.

Film breakers or defoaming agents are substances used to reduce foaming caused by proteins, gases, or nitrogenous materials, among others, which may interfere with processing or the desired characteristics of the end-products. Processes particularly prone to foaming conditions include the Kraft process for papermaking, where a very foamy pulp slurry is formed; phosphoric acid production from phosphate rock; beet sugar processing; several fermentation processes; and, in terms of the end-product, latex paints. See also **Paints.**

The terms *defoamer* and *antifoam* or *anti-foaming agent* are frequently used interchangeably. *Defoamer* best describes a substance that kills the foam from above, once it exists. An *antifoaming agent* stops the foam from forming in the first place. Often a defoamer will be a poor antifoaming agent, and vice versa. Defoaming agents, when used in small concentrations,

can be quite effective. Where a suitable chemical substance cannot be found, physical means may be required. These may be mechanical, electrical, or thermal in nature. The mechanical devices are fundamentally simple, usually taking the form of rotating breaker arms. The presence of a hot surface near a foam tends to destroy the foam: essentially, a portion of the foam is evaporated, causing the acceleration of its breakdown. It has also been established that electrical discharges tend to weaken or destroy foams.

Mechanisms of Defoaming Agents. These agents may operate via a number of mechanisms, but the most common ones appear to be those of entry and/or spreading. The defoamer must be insoluble in the foaming liquid for these mechanisms to function. Second, the surface tension of the defoamer must be as low as possible. The interfacial tension between defoamer and foamer should be low, but not so low that emulsification of the defoamer may occur. Third, the defoamer should be dispersible in the foaming liquid. It was first shown thermodynamically in 1948 that the entry of the defoamer droplet into a bubble surface occurs when the entering coefficient has a positive value. The physics of bubbles is described in the entry of **Foam.**

A type of defoamer may consist of a dispersion in oil of fine particles of silica coated with silicone; the silicone surface of the particles causing them to be hydrophobic. The defoaming action of such a formulation can be explained on the basis of the entry mechanism. Hydrophobic particles can act as an emulsifying agent where the defoamer oil constitutes the continuous phase and the foam constitutes the dispersed phase. In experimental trials, it has been found that excessively hydrophobic particles, such as powdered Teflon, do not function as well as silicone-coated particles. An emulsifier particle must be wetted to some extent by the dispersed phase in order to function as an emulsifier.

The more efficient defoaming mechanism of spreading involves transport of underlying liquid so that the liquid is replaced by a film of defoamer which does not support foam. A drop of oleic acid added to water spreads at a velocity of 30 miles (48.2 kilometers) per hour. The mechanical shock to a film by such a defoamer may be considerable. In addition to the foam-destroying aspect, spreading is also of value as a defoamer-dispersion method, particularly in viscous or poorly stirred systems.

Defoaming agents are in three principal categories, but sometimes are used in combination: (1) solubilized surfactants, (2) dispersions of hard particles, and (3) dispersions of soft particles. The fatty acid-fatty alcohol combination in hydrocarbon oil is an example of a solubilized surfactant defoaming formulation. Paraffinic waxes and fatty amides may be used in soft-particle formulations. The most common of the hard-particle formulations is silica or a mineral coated with silicone dispersed in a vehicle. A particle size as small as 0.02 micrometer may be optimal.

The choice of defoaming agents in the food field is somewhat restricted because substances used obviously must be nontoxic and not produce off-odors, off-colors, or off-tastes. Chemical defoaming agents commonly used in food processing include decanoic acid, dimethylpolysiloxane, lauric acid, mineral oil (white), myristic acid, octanoic acid, oleic acid, oxystearin, palmitic acid, petrolatum, petroleum wax (synthetic), silicon dioxide, sorbitan monostearate, and stearic acid. See also **Foam.**

An example from the brewing industry points out the importance of defoaming agents. Advantages of their use include: (1) higher production through increased fermentation capacity—up to 20% more throughput; (2) the lid of the fermentation tank can be left on, reducing oxidation and improving sanitation; (3) lower oxidation rate, which gives a better physical and chemical stability to the beer, as the denatured or partially denatured protein levels remain low, so that turbidity is reduced; and (4) less yeast build-up on the sides of the tank, leading to reduced cleaning requirements. Inasmuch as foam is a consideration in the final quality of the brewed product, effective foam control during processing can later affect the "head" of the final product, providing for a stable, long-lived creamy foam. Thus, a defoaming agent for process use must be insoluble in the beer and capable of removal so that it does not detract from the "head" of the final product.

References

Andres, C.: "Anti-foaming Agent Increases Fermentation Capacity 20%," *Food Processing*, **38**, 5, 58–59 (1977).
Considine, D. M. (Editor): "The Encyclopedia of Chemistry," 4th Edition, Van Nostrand Reinhold, New York (1982).
Robinson, J. V., and W. W. Woods: *Journal of Society of Chemists*, **67**, 361–365 (1948).
Staff: "Food Chemicals Codex," National Academy of Sciences, Washington, D.C. (Revised periodically).

DEGASIFICATION. Removal of gas, as applied particularly to the removal of the last traces of gas from wires used in vacuum apparatus, from metals to be plated, and from substances to be used in other specialized applications. Untreated glass contains water, carbon dioxide, oxygen, and traces of other gases within it and on its surface, and these are ordinarily in a state of equilibrium with the surroundings. When the pressure is reduced, however, the equilibrium is upset, and these gases are gradually released from solution in and adsorption on the glass, spoiling the vacuum. It is usual to drive the gases out of glass by baking at a temperature of 350–500°C, while on the vacuum pump. Degassing of metal is necessary for the same reason, but because of the larger quantities of gas that may be present in metals, more complex methods must be used. These include baking at elevated temperatures, eddy-current heating, and electron bombardment.

Getters are also used to insure vacuum integrity. A getter film is a metallic deposit in a vacuum system with the function of absorbing residual gas. Electropositive metals, such as sodium, potassium, magnesium, calcium, strontium, and barium have been used as getters. The process of gettering can be done in various ways. In the distillation method, the metal to be deposited is volatilized into the vacuum system from a side tube provided with means for sealing off when the process is complete. The electrolytic method is applicable where the metal to be deposited is sodium, and where the system is made of soda-lime glass. It is well known that sodium may be electrolyzed through soda-lime glass. If, therefore, a thermionic source of electrons is provided inside an evacuated sealed-off vessel, part of which is dipped into a suitable liquid kept at a high potential relative to the source of electrons, a current will pass, carried by electrons between the thermionic cathode and the inner surface of the glass, and by ions within the glass. The only ions in the glass that are mobile are sodium ions, and thus pure sodium is released at the inner surface of the envelope, available to absorb any residual gases.

Other getter materials include cesium-rubidium alloys, tantalum, titanium, zirconium, and several of the rare-earth elements, such as hafnium.

DEHUMIDIFICATION. Removal of water vapor from air and other gases, usually accomplished in one of two ways: (1) cooling below dew point so that part of the water vapor is condensed; and (2) adsorption of moisture by various chemical desiccants; and (3) a combination of both actions. Compressed air used for pneumatic tools and instruments; dry inert gases for blanketing processing operations in the coatings industry; dry gases, such as nitrogen or hydrogen, for metal-annealing operations; blanketing with dry inert gases for semiconductor device manufacture—these are examples of the industrial needs for dehumidification. Dehumidifiers also find wide application for comfort-conditioning of building and residence environments.

Desiccants commonly used are silica gel, activated alumina, and molecular sieves. Desiccant systems permit efficient dew-point performance in the range of −40 to −100°F (−40 to −73°C) and are available in a wide range of capacities and pressure ratings (1–340 atm). To meet tight drying specifications, a solid, regenerable desiccant is often preferred. Normally, dual drying towers are used for continuous service, allowing one tower to be on line, while the other tower is being regenerated.

DEHYDRATION (Chemical). Removal of water from a substance or system or chemical compound, or removal of the elements of water, in correct proportion, from a chemical compound or compounds. The elements of water may be removed from a single molecule or from more than one molecule, as in the dehydration of alcohol, which may yield ethylene by loss of the elements of water from each molecule, or ethyl ether by loss of the elements of water from two molecules, which then join to form a new compound:

$$H\!-\!\overset{\overset{\displaystyle H}{|}}{\underset{\underset{\displaystyle H}{|}}{C}}\!-\!\overset{\overset{\displaystyle H}{|}}{\underset{\underset{\displaystyle H}{|}}{C}}\!-\!OH \xrightarrow{-H_2O} \overset{\overset{\displaystyle H\ \ H}{|\ \ |}}{\underset{\underset{\displaystyle H\ \ H}{|\ \ |}}{C\!=\!C}}$$

$$2H\!-\!\overset{\overset{\displaystyle H}{|}}{\underset{\underset{\displaystyle H}{|}}{C}}\!-\!\overset{\overset{\displaystyle H}{|}}{\underset{\underset{\displaystyle H}{|}}{C}}\!-\!OH \xrightarrow{-H_2O} H\!-\!C\!-\!C\!-\!O\!-\!C\!-\!C\!-\!H$$

Many reactions known in chemistry under special names, such as neutralization, esterification and etherification are dehydration reactions.

In the food processing field, dehydration is sometimes described as the removal of 95% or more of the water from a food substance, by exposure to thermal energy by various means. The aims of dehydration are reduction in volume of the product, increase in shelf-life, and lower transportation costs, among other factors. There is no clearly defined line of demarcation between drying and dehydrating, the latter sometimes being considered as a supplement of drying. Usually, the direct use of solar energy, as in the drying of raisins, hay, etc., is not lumped together with dehydrating. The term *dehydration* also is not generally applied to situations where there is a loss of water as the result of evaporation. *Rehydration* or *reconstitution* is the restoration of a dehydrated food product to essentially its original edible condition by the simple addition of water, usually just prior to consumption or further processing. The distinction between the terms *drying* and *dehydrating* may be somewhat clarified by the fact that most substances can be dried beyond their capability of restoration. Important food products that are dehydrated include animal feedstuffs, hops, malt, oat, peanut (groundnut), potato, rice, and sweet potato.

DEHYDROGENATION. A reaction which results in the removal of hydrogen from an organic compound or compounds. This process is brought about in several ways. Simple heating of hydrocarbons to high temperature, as in thermal cracking, causes some dehydrogenation, indicated by the presence of unsaturated compounds and free hydrogen. Catalytic processes often produce commercially-practicable yields of selected dehydrogenated products. The enzyme dehydrogenase is a selective catalyst of this character. There is considerable evidence to indicate that many reactions commonly classed as oxidations, e.g., the oxidation of methanol to formaldehyde are actually dehydrogenations, i.e.,

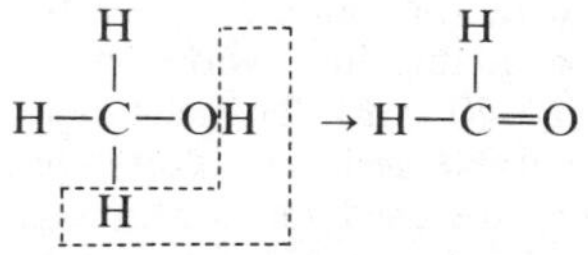

$$\begin{array}{ccc} H & & H \\ | & & | \\ H-C-OH & \rightarrow & H-C=O \\ | & & \\ H & & \end{array}$$

In the chemical process industries, nickel, cobalt, platinum, palladium, and mixtures containing potassium, chromium, copper, aluminum, and other metals are used in very large-scale dehydrogenation processes. For example, acetone (6 billion pounds per year) is made from isopropyl alcohol; styrene (over 2 billion pounds per year) is made from ethylbenzene. The dehydrogenation of *n*-paraffins yields detergent alkylates and *n*-olefins. The catalytic use of rhenium for selective dehydrogenation has increased in recent years. Dehydrogenation is one of the most commonly practiced of the chemical unit processes.

DELIQUESCENCE. When a substance absorbs moisture upon exposure to the atmosphere, the substance is said to be *deliquescent*. At ordinary temperatures, the vapor pressure of water varies. If the solution of a substance in water has a lower water vapor pressure than that of the atmosphere at the given temperature, water vapor condenses in the solution from the atmosphere until the water vapor pressure of the solution equals the water vapor pressure of the surrounding atmosphere.

Substances that are ordinarily deliquescent are sulfuric acid (concentrated), glycerol, calcium chloride crystals, sodium hydroxide (solid), and 100% ethyl alcohol. In an enclosed space, these substances deplete the water vapor present to a definite degree. Other substances can be used to accomplish this end by chemical reaction, e.g., phosphorus pentoxide (forming phosphoric acid), and boron trioxide (forming boric acid). Water is absorbed from nonmiscible liquids by addition of such substances as anhydrous sodium sulfate, potassium carbonate, anhydrous calcium chloride, and sodium hydroxide. The converse phenomenon is known as *efflorescence*.

See **Dehumidification; and Efflorescence.**

DEMAL SOLUTION. A solution which contains one gram-equivalent of solute per cubic decimeter of solution. It is slightly weaker than a normal solution, in the ratio of the magnitude of the liter to the cubic decimeter.

DENDRITE. A treelike crystal formed during solidification of metals or alloys. Dendrites generally grow inward from the surface of the mold (casting), extending branches from a central trunk in a manner resembling a fir tree. In alloys, the central portions of a dendritic crystal are richer in higher-melting-point constituents, while the outer portions consist of lower-melting-point material which is last to solidify. This form of segregation can be eliminated by diffusion during subsequent mechanical working and heat treatment. See accompanying figure.

DENSITY. Mass of substance per unit volume, usually expressed in grams per cubic centimeter. Specific gravity (sp gr) is the ratio of density to that of water, usually at 4 or 20°C, or 60°F in same units, and thus sp gr is an abstract number independent of units. The word *density* is also applied in length-

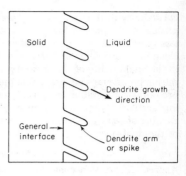

(a) Schematic representation of the first stage of dendritic growth. A temperature inversion is assumed to exist at the interface i.e., the temperature in the liquid drops in advance of the interface.

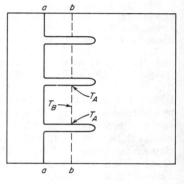

(b) Secondary dendrite arms form because there is a falling temperature gradient starting at a point close to a primary arm and moving to a point midway between the primary arms.

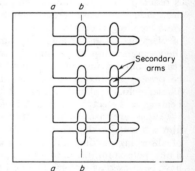

(c) In a cubic crystal, primary and secondary arms are normal to each other.

Formation of dendritic crystals.

force-time systems of units, to the weight per unit volume. Other uses of the word are blackness of an image on a photographic plate and the concentration of other phenomena, such as *luminous density* (luminous energy found in a unit volume of space), *specular density* (logarithm of reciprocal of specular transmittance), etc.

DEOXIDIZING AGENT. A compound that has an affinity for oxygen—hence, chemically removes oxygen from many substances. In essence, a deoxidizing agent plays the role that is reverse that of an oxidizing agent. Thus, a deoxidizing agent is a reducing agent. Of course, at one time, oxidation meant simply a combination with oxygen and reduction meant a loss of oxygen. In their broader interpretations, oxidation now refers to the loss of one or more electrons from the outershell of an atom, and the reverse for reduction. Although the broader interpretation also could apply to deoxidation, the term still is interpreted generally as removal of oxygen.

Oxygen frequently is an impurity in various metallurgical processes, particularly melting and refining processes. Deoxidizing agents are commonly used to reduce or remove oxygen from molten metals. Lithium metal, for example, will preferably absorb oxygen (by combination) from copper and copper alloys. Boron carbide also is used as a deoxidizing agent for casting copper. Silicon usually in the form of ferrosilicon, and manganese in the form of ferromanganeses are widely used in the production of steel and iron alloys. Aluminum, titanium, zirconium, and vanadium also play a deoxidizing part in the production of iron alloys. Magnesium is also a powerful deoxidizer and desulfurizer and is used in the production of such metals as beryllium, hafnium, titanium, uranium, yttrium, and zirconium. Zinc is used as a deoxidant in the refinement of silver and gold.

DEOXYRIBONUCLEIC ACID (DNA). A complex sugar-protein polymer of nucleoprotein which contains the genetic code for enzymes in the cell. It occurs as a major component of the genes, which are located on the chromosomes in the cell nucleus. The DNA molecule is a unique and vastly intricate structure; it is comprised of from 3000 to several million nucleotide units arranged in a double helix containing phosphoric acid, 2-deoxyribose, and the nitrogenous bases adenine, guanine, cytosine, and thymine. The spiral consists of two chains of alternating phosphate and deoxyribose units in continuous linkages. The nitrogenous bases project toward the axis of the spiral and are joined to the chains by hydrogen bonds. Adenine always unites with thymine and cytosine with guanine. The complementarity of the bases on the joined chains allows each chain to act as a template for replication of the other when the chains are separated, thus producing two new strands of DNA. The sequence of the bases on the chains varies with the individual, and it is this sequence that governs the genetic code. DNA works in conjunction with ribonucleic acid (RNA).

The foregoing is a highly generalized definition. Within the last few years, considerable new knowledge has been gained concerning DNA and its genetic function. The early studies of genes concentrated on bacterial genes. In these, the bacterial genes are not spread out in pieces. More recent studies in the mid- and late 1970s concentrated on animal viruses, animals, and humans. The genes in these cases, with the possible exception of the histone genes, are found in pieces that are spread out along DNA. Thus, between gene fragments, there are long stretches of DNA, the functions of which are poorly understood as of the present. This discovery of fragmented or spaced out genes in animals raised the question as to whether such structures are exceptional or the rule. Subsequent research has indicated that they are the norm.

This discovery, while creating several new and fundamental questions, has provided at least partial answers to some former questions. For example, it has been known for some years that there are large quantities of DNA in the cells of higher organisms, that is, DNA in an amount far in excess of the DNA required if the genes were not in pieces. The spacing out of the genes into fragments accounts for all or part of the excess DNA previously noted.

As of the early 1980s, numerous hypotheses have been formulated by molecular biologists and this fundamental discovery has stimulated a whole new line of research in many laboratories throughout the world. Some scientists have observed that the extra DNA cannot be accounted for simply upon the basis of evolutionary theories. The extra DNA may play a role in controlling gene expression. The complexity and current uncertainty of these hypotheses are beyond the scope of this book at this juncture in the research program. Perhaps the topic will be better clarified at the time of the next edition. Several of the references listed shed further insights.

Studies of DNA in the human cell have suggested that the intricate DNA structure may be as much as 1.2 meters (4 feet) long—and yet it is contained within a nucleus which is less than 0.025 millimeters ($\frac{1}{1000}$ inch) in diameter. That the DNA structure is in the form of a complex tangle under these conditions is not difficult to imagine. Upon division of the cell, the DNA condenses into strands in duplicate copies, one set for each daughter cell. It follows that the copying must take place while the DNA is in the form of a tangle, but in such a manner to allow easy separation of the duplicate copies. The mechanics and time required for the copying process still remain undetermined. As early as 1974, Berezney and Coffey (Johns Hopkins University School of Medicine) estimated from electron micrographs that about 5% of the protein in a nucleus appears to be a rigid skeleton or scaffolding, a structure they termed the *nuclear matrix*. These researchers have proposed that the DNA in the nucleus is attached onto the nuclear matrix at thousands of sites, with the genetic material arrayed in thousands of loops. The loops pass through enzyme complexes located at the matrix sites and in so doing are copied. In the process, the loops are preserved and remain attached to the matrix. It is further postulated that when a cell divides, duplicate copies of loops already joined to a scaffolding are drawn apart to the daughter cells. It has been estimated that genetic material in rat liver cells consists of from 10,000 to 15,000 loops. Thus, the nuclear matrix would have a corresponding number of enzyme complexes.

See also **Nucleic Acids and Nucleoproteins.**

References

Bauer, W. R., Crick, F. H. C., and J. H. White: "Supercoiled DNA," *Sci. Amer.*, **243**, 118–133 (1980).
Brown, P. O., and N. R. Cozzarelli: "A Sign Inversion Mechanism for Enzymatic Supercoiling of DNA," *Science*, **206**, 1081–1083 (1979).
Cozzarelli, N. R.: "DNA Gyrase and the Supercoiling of DNA," *Science*, **207**, 953–960 (1980).
Griffith, J. D.: "DNA Structure: Evidence from Electron Microscopy," *Science*, **201**, 525–527 (1978).
Kolata, G. B.: "DNA Sequencing: A New Era in Molecular Biology," *Science*, **192**, 645–647 (1976).
Kolata, G. B.: "Genes in Pieces," *Science*, **207**, 392–393 (1980).
Kolber, A. R., and M. Kohiyama (editors): "Mechanism and Regulation of DNA Replication," Plenum, New York, 1977.
Lehman, I. R., and D. G. Uyemura: "DNA Polymerase I: Essential Replication Enzyme," *Science*, **193**, 963–969 (1976).
Maugh, T. H., II: "Phylogeny: Are Methanogens a Third Class of Life?" *Science*, **198**, 812 (1977).
Mitchell, R. M., et al.: "DNA Organization of *Methanobacterium thermoautotrophicum*," *Science*, **204**, 1083–1084 (1979).
Portugal, F. H., and J. S. Cohen: "A Century of DNA," MIT Press, Cambridge, Massachusetts, 1978.
Razin, A., and A. D. Riggs: "DNA Methylation and Gene Function," *Science*, **210**, 604–610 (1980).
Smith, G. P.: "Evolution of Repeated DNA Sequences by Unequal Crossover," *Science*, **191**, 528–535 (1976).
Sparrow, A. H., and A. F. Nauman: "Evolution of Genome Size by DNA Doublings," *Science*, **192**, 524–529 (1976).
Staff: "Undogmatic Toad," in "Science and the Citizen," page 66–69, *Sci. Amer.*, **242**, 4 (1980).
Temin, H. M.: "The DNA Provirus Hypothesis," *Science*, **192**, 1075–1080 (1976).
Timberlake, W. E.: "Low Repetitive DNA Content in *Aspergillus nidulans*," *Science*, **202**, 973–975 (1978).
Yang, R. C. A., and R. Wu: "BK Virus DNA: Complete Nucleotide Sequence of a Human Tumor Virus," *Science*, **206**, 456–459 (1979).

DEPHLEGMATION. Distillation.

DEPOLARIZATION. Battery.

DESALINATION. Removal of dissolved salts, notably sodium chloride, from seawater and brackish waters to yield potable water for human consumption, process, and irrigation purposes. As of the mid-1980s, there are about 500 desalination plants in operation (or under construction). These plants have a total capacity of some two billion tons of fresh water daily. In addition to direct ocean sources of saline water, a number of smaller desalination plants produce fresh water at coastal and island locations, on drilling structures, and aboard ships. A majority of plants are located in arid regions and in selected locations of high population where demand exceeds supply. Two of the largest plants are located in Hong Kong and Dubai (United Arab Emirates), with other large plants in Kuwait, Saudi Arabia, Qatar, Israel. There are also relatively large plants in Italy, Mexico, and the Netherlands.

The principal desalination processes used (or considered seriously in the past) are described briefly in the following paragraphs. The major plants built to date use distillation. A majority of these plants employ multistage flash distillation. There are also several plants that use multieffect evaporation with falling films, and single- or multiple-stage vapor compression. A plant in Israel uses the freezing process. Electrodialysis and reverse osmosis (membrane plants) made considerably progress during the 1970s and are now seriously considered when new desalination plants are planned.

As established by the U.S. public health authorities, water for human consumption should contain no more than 500 ppm of dissolved solids. Seawater contains about 35,000 ppm; brackish water usually contains about 1000 ppm. The purity of irrigation water varies with crop and soil conditions, but sometimes may contain up to 1200 ppm dissolved salts.

Multistage Flash Distillation (MSF). Saline solution is vaporized and pure water is obtained by condensation. The seawater is heated progressively to a maximum of about 250°F (121°C) and then flashed into a number of successive stages, in which operation is under progressively lower pressures. Incoming seawater is heated by condensing vapors. Much experience has been gained from numerous installations. Scale formation tends to limit the top operating temperatures and thus limits efficiency.

Multiple-Effect Multistage Flash Distillation (MEMS). Fundamentally, this process is similar to the MSF. The stages, however, are divided among two or more effects in which the brine is recycled. Thus, operation at higher temperature is possible. The process also allows higher blowdown concentrations without scaling. Although efficiency is generally higher than the MSF, capital costs tend to be higher. Top operating temperatures still are limited by scaling and the operation tends to be more complex than the MSF.

Vapor Reheat Flash Distillation. This process is similar to the MSF distillation with the exception that fresh water is recycled and used to condense vapors. Heat is recovered from the fresh water recycle stream by use of heat exchangers. If liquid–liquid heat transfer is used, metallic metal transfer surfaces (as in the distillation process) are eliminated, reducing the scale problem. However, the large volume of liquid handling involved increases the complexity and cost of the process. Capital costs are higher than for the MSF process for similar size plants.

Multiple-Effect Evaporation (MEE) (Vertical Tube Evaporator Process). Three or more evaporators are used and steam is used to evaporate portion of the seawater in the first effect. Vapor from the first effect condenses in the second effect and evaporates additional water. This process has been used successfully in numerous industrial evaporation processes for years. The process can operate at higher blowdown concentrations than the MSF. However, more effective antiscaling techniques are required. Initial capital investment is higher than for the MSF, but may be quite competitive for plants exceeding 10 million gpd.

Vapor Compression Distillation. The vapor from boiling brine is compressed mechanically, thus increasing the vapor temperature and pressure. The compressed vapor is then fed back into the evaporator to distill more seawater feed. The multiple-effect approach can be applied to this concept. Attractive for small size units.

Solar Distillation. A large basin is fitted with sloping transparent glass or plastic covers. These serve as condensing surfaces. Saline water fed to the enclosed chamber is heated by solar radiation. Process limited to areas of maximum sunshine. A large land area is required. Advantages include simplicity and minimal fuel costs.

Humidification. Either by solar or fossil fuel energy, the incoming saline water is heated, after which it is sprayed in a tower with dry air flowing countercurrently. The saturated air leaving a first tower is cooled in a second tower where fresh water is condensed and collected. The process is simple and may be competitive in small size units up to about 0.5 million gpd.

Solvent Extraction. A solvent is used to remove either the salts from water or water from the salts, followed by separation of solvent and salts, or fresh water by change in temperature of mixture. Appropriate solvents are costly and removal of traces of solvent (possibly toxic or otherwise objectionable) may pose problems. Large volumes of fluids must be handled.

Electrodialysis. Electrodes impress an electrical field across a cell through which saline water flows. Cations migrate to the cathode; anions to the anode. The positive ions pass through cation-permeable membranes; negative ions through anion-permeable membranes. Between alternate membranes, the water becomes enriched with or depleted of salts. The energy requirements are in proportion to saline content, but increase rapidly with attainment of high purity. Suited to brackish waters that contain up to 5,000 ppm dissolved solids.

Reverse Osmosis. The saline water is pressurized above its osmotic pressure—then flows over a semipermeable membrane. The membrane is permeable to water, but not to dissolved solids. The fresh water is transported through the membranes. Energy requirements increase less rapidly with increasing saline content than with electrodialysis process. Tailoring of membranes to different saline concentrations is possible. High-pressure equipment is required.

Although high pressures are needed to overcome the osmotic pressure of seawater, membranes have been developed that allow single-pass separation of water containing up to 500 milligrams/kilogram of salts from a 3.5% (wt) seawater solution. Available are cellulose acetate and substituted polyamide membranes both in hollow fine-fiber and sheet configurations.

Hydrate Process. A gas, such as propane, is mixed with saline water. Insoluble crystals of a solid hydrate are formed. A slurry containing crystals and brine goes to a wash column where crystals are washed and transferred to a decomposition chamber. Here the hydrate crystals are decomposed by altering temperature and/or pressure. The process can operate at near-ambient conditions.

Direct Freezing-Vapor Compression. The saline water is sprayed into a vacuum chamber where part of the water is evaporated. This produces a cooling effect, causing formation of ice crystals. A slurry of ice and brine goes to a washer-melter where brine is washed from ice after which ice is melted to yield fresh water. The energy requirements are lower than for distillation and corrosion and scaling problems are minimized. However, large volumes of vapor have to be handled and compressed. High capital costs.

Direct Freezing-Secondary Refrigerant. A hydrocarbon refrigerant immiscible with water, such as butane, is vaporized in direct contact with the saline water. This produces an ice–brine mixture. The ice crystals are washed to remove the brine. The refrigerant vapor is compressed and condenses as it melts the ice. Some cost advantages over other freezing processes, and energy requirements are lower than for distillation. Removal of hydrocarbon refrigerant from product water poses problem.

Ion Exchange. Cation and anion ion exchange resins are used to remove salts. The exhausted resins are regenerated with acids, ammonia, or lime, depending upon the particulars of the process. Basic process well-developed for other industrial uses. Foreign matter can foul ion exchange resins. Pretreatment of some saline waters may be in order. Production costs are about proportional to total dissolved solids removed.

References

McIlhenny, W. F.: "Extraction of Inorganic Materials from Seawater," in "Chemical Oceanography," (J. P. Riley and G. Skirrow, editors), Vol. 4, Chap. 10, Academic Press, London, 1975.

McIlhenny, W. F.: "Ocean Raw Materials," in "Encyclopedia of Chemical Technology," 3rd edition, Wiley, New York, 1981.

Multhaupf, R. P.: "Neptune's Gift—A History of Common Salt," Johns Hopkins Univ. Press, Baltimore, Maryland, 1978.

Spiegler, K. S.: "Principles of Desalination," Academic Press, New York, 1966.

Staff: "Desalting Plants Inventory," Office of Water Research and Technology, U.S. Department of the Interior, Washington, D.C. (Revised periodically).

DESULFURIZATION (Steel). Calcium.

DETERGENTS. Complete washing or cleansing products which, among other ingredients, contain an organic surface-active compound known as a *surfactant*, the latter possessing marked soil-removal characteristics. Tonnage detergents are synthetic products sometimes referred to as *snydets*. Soaps are alkali salts of long-chain fatty acids and differ significantly in certain important performance properties and are still widely used. In addition to a surfactant, a detergent usually will contain (1) a *builder*, normally an ingredient which will chelate (sequester) or precipitate polyvalent metal ions present in the cleaning solution, particularly calcium and magnesium ions which are present in substantial quantities in hard water supplies; (2) *bleaches*; (3) *corrosion inhibitors*; (4) *sudsing modifiers*; (5) *fluorescent whitening agents* (FWA); (6) *enzymes*; (7) *antiredeposition agents*; and (8) *various additives* to enhance color, scent, and general consumer acceptability.

Surfactants. A surfactant is an organic compound consisting of two parts: (1) a hydrophobic portion, usually including a long hydrocarbon chain; and (2) a hydrophilic portion which renders the compound sufficiently soluble or dispersible in water or another polar solvent. The combined hydrophobic and hydrophilic moieties render the compound surface-active and thus able to concentrate at the interface between a surfactant solution and another phase, such as air, soil, and textile or other substrate to be cleaned.

Surfactants function to penetrate and wet soiled surfaces, to displace, solubilize, or emulsify various soils, notably oils and greases, and to disperse or suspend certain soils in solution to prevent their redeposition. Surfactants usually are classified into: (1) *anionics*, where the hydrophilic portion of the molecule carries a negative charge; (2) *cationics*, where this portion of the molecule carries a positive charge; and (3) *nonionics*, which do not dissociate, but commonly derive their hydrophilic portion from polyhydroxy or polyethoxy structures. Ampholytic and zwitterionic surfactants are not presently of major commercial importance.

The anionic surfactants are the most commonly used and are linear sodium alkyl benzene sulfonate (LAS), linear alkyl sulfates, and linear alkyl ethoxy sulfates. To obtain improved biodegradability, manufacturers converted a few years ago to linear alkyl chains. Formulas of the major surfactants are:

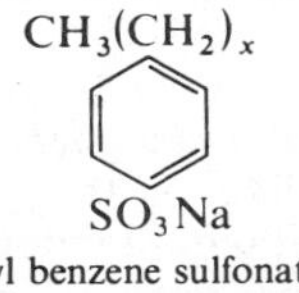

$$CH_3(CH_2)_x$$

(*x* ranges from 9 to 15)

$$SO_3Na$$

Linear alkyl benzene sulfonate (LAS)

$$CH_3(CH_2)_xOSO_3Na \quad (x \text{ ranges from 9 to 17})$$

Alkyl sulfate

$$CH_3(CH_2)_x(O-CH_2-CH_2)_yOSO_3Na$$

(*x* ranges from 7 to 15) (*y* ranges from 0 to 6)

Alkyl ethoxy sulfate

For automatic dishwashing detergents and laundry detergents, non-ionic surfactants also are used, particularly because of their lower sudsing characteristics. Commercially important nonionics include the alkyl ethoxylates, the ethoxylated alkyl phenols, the fatty acid ethanol amides, and complex polymers of ethylene oxide, propylene oxide, and alcohols. Some of these formulas are:

$$CH_3(CH_2)_x(O-CH_2-CH_2)_yOH$$

(*x* ranges from 9 to 15) (*y* ranges from 4 to 20)

Alkyl ethoxylates

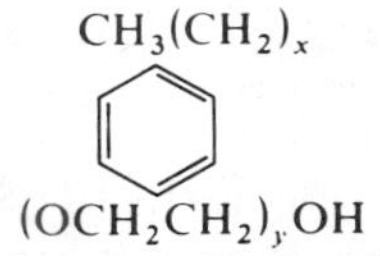

$$CH_3(CH_2)_x$$

$$(OCH_2CH_2)_yOH$$

(*x* ranges from 7 to 13) (*y* ranges from 3 to 12)

Ethoxylated alkyl phenols

$$CH_3(CH_2)_xCON(CH_2CH_2OH)_y(H)_z$$

(*x* ranges from 9 to 17) (*y* ranges from 1 to 2) (*z* is 1 or 0)

Fatty acid ethanol amide

For specialty detergents, such as metal cleaners for electroplating and in connection with accessory laundering products, such as fabric softeners, antistatic, and germicidal preparations, cationic surfactants are used in fairly limited quantities. Tallow trimethylammonium chloride, $CH_3(CH_2)_{13\text{-}17}N^+(CH_3)_3CL^-$, is a representative cationic surfactant used in this manner.

Unlike soaps, synthetic anionic and nonionic surfactants do not form visible insolubles with the calcium and magnesium ions present in hard water. These compounds, along with certain phosphate chelating agents, revolutionized the laundry-product field a number of years ago.

Builders. When a laundry product contains a large quantity of builders it may be termed *heavy-duty* or *built* detergent. By sequestering the calcium and magnesium ions of the water, builders perform the following major functions: (1) polyvalent metal ions are prevented from combining with surfactant (in most cases) to form an adduct which would be less effective than the modified surfactant in cleansing characteristics; (2) polyvalent metal ions are prevented from combining with various soils, such as lipid residues and clays, to form less dispersible residues which adhere tenaciously to the surface to be cleaned; (3) soils removed from a surface are prevented from redepositing back onto the surface because of the dispersing action associated with chelating and charge-distribution effect; (4) additional buffered alkalinity is provided to the wash solution; and (5) destruction and removal of microorganisms is enhanced, particularly important for maintaining commercial sanitation.

The most effective builders appear to be condensed polyphosphates, mainly pentasodium tripolyphosphate (STP) and to a somewhat lesser extent, tetrasodium pyrophosphate. These polyphosphates chelate the polyvalent metal ions to form a soluble complex. Precipitating builders also are used, such as sodium carbonate. Application of this compound is limited because the precipitate formed can deposit on the surface to be cleaned and thus requires special washing procedures. Nonphosphate chelating builders include trisodium nitrilotriacetate (NTA), tetrasodium ethylenediamine tetraacetate (EDTA), and certain other polycarboxylates. Formulas for the condensed polyphosphates are:

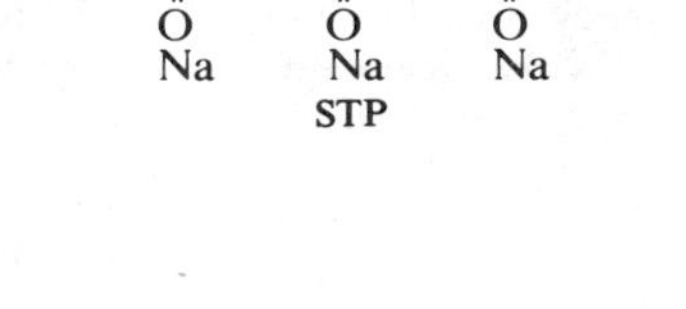

$$
\begin{array}{ccccccc}
& O & & O & & O & \\
& \| & & \| & & \| & \\
NaO- & P- & O- & P- & O- & P- & ONa \\
& \| & & \| & & \| & \\
& O & & O & & O & \\
& | & & | & & | & \\
& Na & & Na & & Na & \\
\end{array}
$$

STP

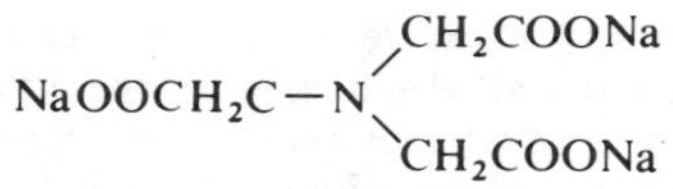

$$NaOOCH_2C-N\underset{CH_2COONa}{\overset{CH_2COONa}{<}}$$

NTA

$$NaO-\overset{\overset{O}{\|}}{P}-O-\overset{\overset{O}{\|}}{P}-ONa$$
$$\underset{Na}{\overset{|}{O}}\quad\underset{Na}{\overset{|}{O}}$$

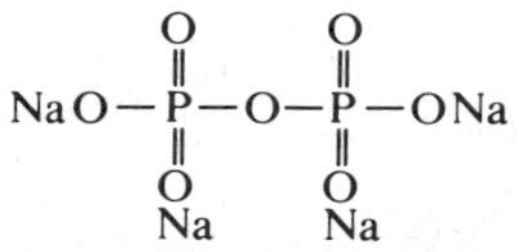

Tetrasodium pyrophosphate

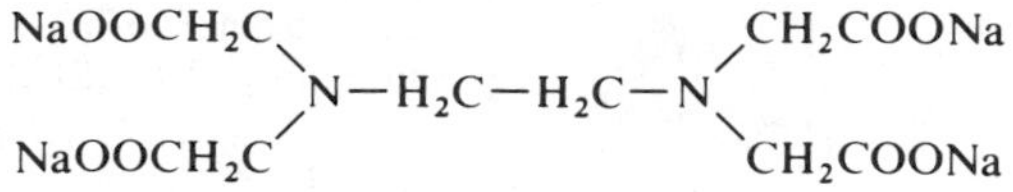

$$\underset{NaOOCH_2C}{\overset{NaOOCH_2C}{>}}N-H_2C-H_2C-N\underset{CH_2COONa}{\overset{CH_2COONa}{<}}$$

EDTA

Bleaches. Two families of bleaching agents may be used: (1) *hypochlorite* or chlorine-type; and (2) *peroxygen* compounds. The hypochlorite compounds tend to be more powerful in their oxidizing action. Commonly used chlorine-type compounds include potassium dichlorisocyanurate (KDCC), and chlorinated trisodium phosphate. The latter is a physical mixture of NaOCl, H_2O, and Na_3PO_4. Of the peroxygen bleaches, sodium perborate, $NaBO_3 \cdot 4H_2O$ is the most common. Because aqueous bleach solutions have not been found to be sufficiently stable in the presence of other detergent ingredients, it is common practice to include a solid bleaching agent in detergents.

Corrosion Inhibitors. The unmodified alkaline detergent can be corrosive to aluminum, porcelain, and the overglaze on fine china. This type of corrosion is essentially prevented by adding soluble silicates into the detergent mix. The soluble silicates contain varying ratios of SiO_2 and Na_2O.

Sudsing Modifiers. For certain types of detergent products, a sudsing action presents an aesthetic appeal even though it may not add essentially to the functioning of the product. Sudsing can be increased by adding small quantities of anionic surfactants, such as mono- and diethanol amides of C_{10-16} fatty acids. In other instances, sudsing depression is desired. This can be accomplished by the addition simply of the C_{10-16} fatty acids.

Fluorescent Whitening Agents (*FWA*). These agents also are termed brighteners and optical bleaches. They are organic chromophores which absorb incident light in the ultraviolet region and reemit part of the absorbed energy as visible light, usually in the blue region of the visible spectrum. For use in detergents, the chromophore is modified with organic substituents to make it substantive to one or more textile substrates from a laundry-wash solution. Thus, there is enhancement of the brightness and whiteness of the fabrics onto which FWA deposit has been made. The result is that an added portion of incident light is reflected by the fabric. Low levels of sulfonated trazinylstilbenes are used as FWAs in detergents for cellulosic fibers. The trend is toward incorporating a brightening agent into other synthetic fibers during their manufacture.

Enzymes. Low levels of enzymes may be added to detergents and pre-soak products. Proteolytic and anylolytic enzymes attack and loosen soils and stains with protein and carbohydrate substituents (including body soils, numerous food stains, grass stains, blood, and others). The enzymes perform catalytically and quite specifically and thus can be used effectively at low levels, assuring safety for fibers and preservation of textile colors. The essential step needed to make enzymatic action available in textile detergency was the discovery and production of the *B. subtilis* and *B. licheniformis* mutants and their metabolites which remain active under laundering conditions.

Antiredeposition Agents. Substances which tend to prevent redeposition of soils once removed from a fabric include carboxymethyl cellulose and polyvinyl alcohol. The actual mechanisms involved still are under investigation.

Although detergents commonly are supplied as powders, they are available in liquid and paste form. In some instances, the liquid products are essentially the same as the dry detergents without going through the spray-drying operation.

DEUTERIUM. The isotope of hydrogen with mass number 2 is called *deuterium*. The symbol D is sometimes used. Using ocean water as a reference, the atomic abundance of deuterium in natural hydrogen is 0.0149%. Deuterium oxide D_2O is known as *heavy water* and was first identified by Urey in 1932. Urey noted a slight shift in the spectrum of deuterium and tritium as compared with protium. The diameter of the electron orbit for deuterium is slightly greater than for ordinary hydrogen, and still greater for tritium. Deuterium and deuterium oxide gained prominence largely because of their excellent properties as moderators in nuclear reactors.

DETINNING. Goldschmidt Detinning Process.

DEUTERON. The nucleus of deuterium (heavy hydrogen) is known as deuteron. A particle that contains one proton and one neutron also is termed a deuteron.

DEVITRIFICATION. Vitreous State.

DEXTROROTATORY COMPOUND. Asymmetry (Chemical); Isomerism.

DEZINCIFICATION. A form of electrolytic corrosion observed in some brasses where the copper-zinc alloy goes into solution with subsequent redeposition of the copper. The small red copper plugs thus formed in the brass are usually porous and of low strength. In recent years, the word *dezincification* has also been applied in a more general sense to signify any metallic corrosion process that dissolves one of the components from an alloy.

DIALLAGE. The mineral term for a calcium-iron pyroxene, similar in chemical composition to diopside but richer in iron oxide. In addition to the typical prismatic cleavage of the pyroxene group, diallage has a marked "cleavage" parallel to the vertical pinacoids, known as *diallage parting*. Diallage is a common constituent of gabbros. The term *diallagite* was proposed by Cloiseaux in 1845 for rocks particularly rich in diallage. The term *diallage* is derived from the Greek meaning *difference*, and referring to the peculiar cleavages of this variety of monoclinic pyroxene. See also **Pyroxene.**

DIALYSIS. The process of separating compounds or materials by the difference in their rates of diffusion through a colloidal semipermeable membrane. Thus, sodium chloride diffuses eleven times as fast as tannin and twenty-one times as fast as albumin. When the process is conducted under the influence of a difference in electrical potential, as from electrodes on opposite sides of the membrane, it is called *electrodialysis*.

An apparatus for carrying out a dialysis usually consists of two chambers separated by a semipermeable membrane (parchment paper, latex, animal tissue, other colloidal substances). In one chamber the solution is placed, and in the other the pure solvent. Crystalline substances diffuse from the solution through the membrane and into the solvent much more rapidly than amorphous substances, colloids, or large molecules.

DIAMOND. An allotropic form of carbon, diamond occurs in nature and, in comparatively small sizes, is produced synthetically. Diamond crystallizes in the cubic system, is the hardest of known substances (10 on the Mohs scale; 5500–7000 on the Knoop scale); sp gr 3.51–3.521 (20°C); dielectric constant at 10^4 Hz, 16.5; at 10^8 Hz, 5.5; index of refraction 2.417–2.4195. Classically, a diamond crystal may be pictured as a huge polymeric molecule, very tightly packed, with a density about 1.6 times greater than that of the other allotropic form of carbon, graphite. The normal C—C single bond distances in the atomic lattice of diamond are all 1.54 Å, whereas in graphite the C—C bond distances are 1.42 Å. The tight packing of diamond accounts for its relatively high density and for its extreme hardness. Graphite, on the other hand, is essentially composed of very laminar, two-dimensional molecules which tend to slide and thus impart lubricity to the substance. A rough comparison of the structures of these two forms of carbon is shown in Fig. 1. Because diamond is relatively rare and beautiful, it is

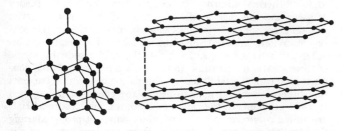

Fig. 1. A gross conceptualization of the space lattices of (a) diamond and (b) graphite. See the new spatial concept for graphite in the article on **Carbon.**

a gem. But, because of its hardness, diamond is also an important industrial abrasive; this aspect of diamond is described in the article on **Grinding and Polishing Agents.**

Diamonds were discovered in 1867 along the Orange River in South Africa, and since that time the region has been preeminent in production of diamond. Diamonds also have been found in Australia, Borneo, British Guiana, and Arkansas. See **Carbonado;** and **Kimberlite.**

Synthetic Diamonds. The high pressures and temperatures shown in Fig. 2 favor the stable form of carbon as diamond

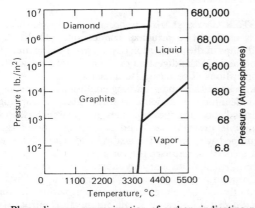

Fig. 2. Phase diagram approximation of carbon, indicating pressure-temperature parameters favoring yield of graphite and diamond. See also the phase diagram in the article on **Carbon.**

rather than graphite. When graphite is heated and compressed into this region, its layer space lattice undergoes a transformation into the much more closely knit lattice of diamond. In synthesizing diamond, graphite is introduced in solution in iron, after which high temperatures and pressures are applied to the sample. After the sample has cooled, the iron matrix is dissolved by acid, leaving a diamond residue. The individual pieces in the residue are of industrial grade, containing various impurities. The size usually is on the order of 0.1 mm, suitable for grinding abrasives. Industrial-grade diamonds as large as 1 carat have been produced synthetically, but costs rise proportionately with the size of diamond yielded. An interesting treatise is "Man-Made Gemstones," by D. Elwell (Wiley, New York, 1980).

Diamond, particularly of gem stone quality, is marketed by the carat. A carat equals about 3.086 grains (troy) or 0.2 gram.

DIASPORE. The mineral diaspore is a hydrous oxide of aluminum corresponding to the formula AlO(OH) occurring in prismatic orthorhombic crystals, usually somewhat flattened, or massive. It displays good cleavage; conchoidal fracture; is brittle; hardness, 6.5–7; specific gravity, 3.3–3.5; luster, vitreous to pearly; color, white, grayish, greenish, yellowish, brownish or colorless; transparent to translucent. Diaspore is found associated with corundum, emery and bauxite, being probably an alteration product of the oxide. It has been made artificially. Diaspore has been found associated with emery in the Ural Mountains, in Asia Minor, in the Island of Naxos, Greece, and in the United States at Chester, Massachusetts. Its name is derived from the Greek word meaning to scatter, because of its decrepitation upon heating.

DIATOMACEOUS EARTH. A soft, bulky, solid material (88% silica) composed of skeletons of small prehistoric aquatic plants related to algae (diatoms). The material is insoluble in acids except hydrofluoric; soluble in strong alkalis. It absorbs 1.5–4 times its weight of water, and also has a high oil-absorption capacity. Diatomaceous earth is a poor conductor of sound, heat, and electricity. It is noncombustible. The material is used in filtration and clarifying and as a decolorizing agent. Sometimes it is used as a mild abrasive. It has been used as a drilling mud thickener; as an extender in paints, rubber, and plastic products; as an anticaking agent in fertilizers; and in asphalt compositions. The material is found in the western United States, Europe, the U.S.S.R., and Algeria.

DIATOMITE. A rather dense, chertlike, condensed version of diatomaceous earth. Also called *indurated diatom ooze.* The term frequently is used synonymously with diatomaceous earth. Diatomite is composed principally of the opaline frustules of diatoms.

DIAZO COMPOUNDS AND DIAZONIUM SALTS. Azo and Diazo Compounds.

DICHLOROBENZENES. Chlorinated Organics.

DIECASTING ALLOYS. Aluminum; Antimony; Zinc.

DIELECTRIC. A substance that has very low electrical conductivity, i.e., an insulator. Such substances have electrical conductivities of less than 1 millionth mho per cm. Those with a somewhat higher conductivity (10^{-6} to 10^{-3} mho per cm) are called *semiconductors.* See **Semiconductor.** Among the common solid dielectrics are glass, rubber, and similar elastomers; wood and other cellulosics. Liquid dielectrics include hydrocarbon and silicone oils.

DIELS-ALDER REACTION. Organic Chemistry.

DIESEL FUELS. Petroleum.

DIFFUSION. This term denotes the process by which molecules or other particles intermingle as a result of their random thermal motion. The molecules of a gas or of a liquid wander about rapidly, colliding frequently and exchanging kinetic energy, but maintaining a certain aimless progress. If an enclosure contains two gases, the lighter initially above and the heavier below, the gases at once begin to mingle because of their molecular motion. The same is true of a dense solution (as of sugar) and pure water; both the sugar and the water molecules wander across the boundary, so that in the course of time the whole body of liquid attains nearly uniform concentration. The process whereby this is effected is called diffusion. In the case of fluids of different color, its progress may be easily watched.

The rates at which different gases diffuse at a given temperature are inversely proportional to the square roots of their molecular weights. Thus, hydrogen diffuses four times as fast as oxygen. This follows, according to the kinetic theory, from the fact that the molecules of various kinds have the same mean kinetic energy and hence their mean square speeds are in the inverse ratio of their masses. In the case of a solution of non-uniform concentration, the diffusion of the solute from the more to the less concentrated regions takes place in accordance with *Fick's law*, expressed by the equation

$$\frac{dm}{dt} = - DS \frac{dc}{dx}$$

This gives the mass of solute diffused per unit time through a cross-section S, in terms of the concentration gradient dc/dx in the direction x perpendicular to the cross section. D is a constant for the given solute and solvent at a given temperature, and is called the diffusion coefficient. For any one pair of substances, D is found to be proportional to the absolute temperature. It should be stated that these statements apply only to nonelectrolytic solutions.

Diffusion in solids is a phenomenon which occurs rather slowly, but can be observed. Three basic processes may be responsible: (a) direct exchange of atoms on neighboring sites; (b) migration of interstitial atoms; (c) diffusion of vacancies. The first process requires very large energy. The energy to make an interstitial migration is rather large, but many atoms migrate easily. Vacancies are fairly readily formed, and diffuse fairly easily. From the Kirkendall effect it appears that (b) and (c) are the usual processes. The diffusion coefficient is related to the ionic mobility by the Einstein relation.

Another use of the term diffusion is to denote the passage of particles through matter in such circumstances that the probability of scattering is large compared with that of leakage or absorption. It is often limited to phenomena described by a member of the class of differential equations known as diffusion equations.

Diffusion operations are of large importance in chemical and process engineering. Both gaseous and thermal diffusion are used to separate one gas from another. In the case of *gaseous diffusion*, if a binary gaseous mixture at a high pressure is passed over a microporous barrier, a fraction of the gas will diffuse through the barrier into a low-pressure discharge chamber and will be found to be richer in the content of one gas than of the other gas. This is termed *Knudsen diffusion*. The passage of gas mixtures through the barrier is governed by the unequal collision frequency of each molecular species upon the walls of the pores. Fast, so-called light molecules separate from slower, heavier molecules within the barrier. The Oak Ridge, Tennessee plant designed for the enrichment of $^{235}UF_6$ from the naturally occurring uranium hexafluoride that contained 99.3% $^{238}UF_6$ represented the first major application of gaseous diffusion on a large scale. The molecular weight of the hexafluoride of ^{235}U is 349, whereas that of the hexafluoride of ^{238}U is 352. Inasmuch as the rate of diffusion of a gas is inversely proportional to the square root of density, the greatest separation factor for one stage of separation is the square root of 352/349, or 1.0043. Inasmuch as only part of the gas can diffuse through a given barrier, the separation factor is less. Thus, the number of diffusion stages for the Oak Ridge plant was approximately 4,000, requiring a plant that covered several acres of ground. Polymeric barriers also are under study and with scientific improvements, gaseous diffusion may become a widely used means for the recovery of carbon dioxide, helium, and nitrogen from natural gas.

In *thermal diffusion*, a thermal gradient is applied to a homogeneous solution (gas or liquid). This causes a concentration gradient and thus affords a means of separating materials. The logic of thermal diffusion is derived from the kinetic theory of gases and the cage model of liquids. If there is no marked size difference, heavier species tend to concentrate in the cold region. Where materials of identical molecular weight are involved, the larger molecules go to the cold region by virtue of their greater momentum. In the static mode, differential concentration can be established by eliminating convection currents that otherwise would tend to negate the effects of the applied thermal gradient. In the reflux method, hot and cold materials are flowed countercurrently. The reflux usually is provided using the density gradient that results from the imposition of the temperature gradient. Equipment of this latter type usually is referred to as a *thermogravitational column*, or a *Clusius-Dickel column*. Limited applications of thermal diffusion separations include those for concentrating dilute mixtures of isotopic gases. However, equipment costs tend to be high and efficiencies low.

References

McGabe, W. L., and J. C. Smith: "Unit Operations in Chemical Engineering," McGraw-Hill, New York, 1967.
Perry, R. H., and C. H. Chilton: "Chemical Engineers' Handbook," McGraw-Hill, New York, 1979.
Slattery, J. C.: "Momentum, Energy and Mass Transfer in Continua," McGraw-Hill, New York, 1972.
Weast, R. C. (editor): "Diffusion Equations and Coefficients," in "Handbook of Chemistry and Physics," 60th edition, CRC Press, Boca Raton, Florida, 1979.

DIGESTER (Process). In the process industries, the term *digester* is used in two principal connections: (1) the digestion of wood chips in the production of pulp prior to the manufacture of paper, and (2) the digestion of sewage sludge in waste-treatment operations. The term also appears in a number of other operations operating under varying conditions and hence a generalized definition is difficult to formulate. In chip digestion (also termed cooking), the chip digester is a large vessel provided with suitable raw-chip feed and cooked-chip discharge ports and equipped with means for heating and maintaining its contents at a specified temperature for a specific time. Batch digesters are vertical, stationary cylindrical pressure vessels into which chips and cooking liquor are charged and in which liquor is constantly moved, either by percolation within the digesters aided by direct addition of steam for heating purposes, or by continual withdrawal of liquor through screened ports and rein-troduction of the liquor, after further heating. Modern batch digesters are typically 4,000 to 6,000 cubic feet (113.3 to 170 cubic meters) in volume with pulp capacities of 10 to 20 tons (9 to 18 metric tons).

By contrast in terms of operating parameters, sewage sludge is digested by aerating a lagoon or pond under normal outdoor temperatures, except that below about 40°F (4.05°C), the activity of the bioorganisms which aid in the digestion falls off considerably.

Autoclaves used in the chemical industry also are sometimes referred to as digesters.

DIGESTER (Pulp).   Pulp (Wood) Production and Processing.

DIGITALIS. A drug prepared mostly from leaves of the foxglove plant (*Digitalis purpurea*). The drug is used in certain kinds of heart disease. Its chief effects are regulation of heart rate, rhythm, tone, contraction, and conduction of impulses.

DILATANCY. The property of certain colloidal solutions of becoming solid, or setting, under pressure. Also known as *inverse plasticity*, since there is an increase in the resistance to deformation with increase in rate of shear.

DILATANT SUBSTANCES.   Rheology.

DILATION NUMBER. Ratio of the volume of a liquid to the volume of a solid of the same composition at the same temperature.

DIOPSIDE. The mineral diopside is a monoclinic pyroxene corresponding to the chemical formula, $CaMgSi_2O_6$, calcium magnesium silicate. Its crystals, like other pyroxenes, tend to be short stout prisms of square or octagonal cross-section. Compact, granular, lamellar and fibrous varieties are often found. The prismatic cleavage is characteristic, cleavage planes intersecting at angles of 87% and 93%. A basal parting is often noted, but should not be confused with the cleavage. The hardness of diopside is 5–6; specific gravity, 3.2–3.3; uneven fracture tending toward conchoidal; luster, vitreous to dull; sometimes pearly on the base; color, light or dark greens, but may be colorless, gray, yellow or blue, although the latter color is rare.

Diopside is a primary mineral in rocks like diorites, gabbros and the like, but is also found in schists, and, as the result of contact metamorphism, in such rocks as crystalline limestones and dolomites. Diopside is found in association with vesuvianite, garnet, spinel, scapolite, tremolite, tourmaline and similar minerals. It is a rather widespread mineral, important localities being found in the following European countries: Finland, Sweden, Switzerland, Italy; it is found in eastern Siberia near Lake Baikal. In Canada diopside localities are in Lanark and Hastings Counties, Province of Ontario, and in the United States in Lewis and St. Lawrence Counties, New York, and in Maine.

See also **Pyroxene;** and terms listed under **Mineralogy.**

—Elmer B. Rowley, Union College, Schenectady, New York.

DIOPTASE. The mineral dioptase is a rather rare copper silicate corresponding to the formula $CuSiO_2(OH)_2$, occurring in prismatic crystals of the hexagonal system, tri-rhombohedral in form. It may be found in crystalline aggregates or simply massive. Dioptase displays a conchoidal to uneven fracture; hardness, 5; specific gravity, 3.28–3.35; luster, vitreous; color,

a beautiful emerald green. It has been found in the U.S.S.R., Congo, Central African Republic, South West Africa, Chile, and in the United States in Arizona. The name is derived from the Greek words meaning *through* and *to see*, because cleavage was observed by looking through the crystals.

DIOXIN. Also called *dimethoxane* (2,3,7,8-tetrachlorodibenzo-*p*-dioxin; 2,6,-dimethyl-*m*-dioxan-4-yl-acetate; TCDD), this is a toxic chlorinated hydrocarbon best known as an impurity in the herbicide 2,4,5-T. Some authorities estimate that the compound has a half-life in soil of about one year. Sax (1979) reports this as "one of the most toxic materials known. Damaging to guinea pigs at 0.6 part per billion." Experimental laboratory tests show the compound to be a carcinogen. The possible hazard to the atmosphere of dioxin has been a subject of debate among authorities since the late 1970s. For an interesting exchange of viewpoints on this topic, reference is made to letters by Westing and Blair, *Science,* **206,** pages 1135–1136 (December 7, 1979). For an earlier review, see "Dioxin: Toxicological and Chemical Aspects" (Wiley, New York, 1978, edited by F. Cattabeni, A. Cavallaro, and G. Galli). See also "Trace Chemistries of Fire: A Source of Chlorinated Dioxins," by R. R. Bumb, et al., *Science,* **210,** 385–390 (1980).

DIPOLE MOMENT. In the simplest case, let two electric charges $+q$ and $-q$ be separated by the distance **d.** Then the permanent electric dipole moment is the vector $\mathbf{p} = q\mathbf{d}$. More generally, if discrete charges q_i are located at points x_i, y_i, z_i the magnitude of dipole moment is given by $p_a = \Sigma q_i \alpha_i$, $\alpha = x$, y, z. If the charge distribution is continuous, the summations are replaced by integrals. An induced dipole moment can be produced by an electric or magnetic field. Atomic or molecular dipole moments, permanent or induced, are of considerable value in the study of atomic or molecular structure. The magnitude of such moments is usually reported in Debye units. The magnetic dipole moment produced by a current i flowing in a loop area A has magnitude $m = iA$. It is a vector with directional normal to the plane of the loop and sense taken as the direction of progression of a right-handed screw rotating with the current.

DISACCHARIDES.   Carbohydrates.

DISPERSION.   Colloid System.

DISPOSAL (Radioactive Wastes).   Nuclear Reactor.

DISSOCIATION. This can be broadly defined as the separation from union or as the process of disuniting. In chemistry, dissociation is the process by which a chemical combination breaks up into simpler constituents due, for example, to added energy as in the case of the dissociation of gaseous molecules by heat, or to the effect of a solvent upon a dissolved substance, as in the action of water upon dissolved hydrogen chloride. Dissociation may occur in the gaseous, liquid or solid state or in solution.

Elementary substances, if polyatomic in the molecule, will dissociate under conditions of sufficient energy. Chlorine and iodine, which are diatomic, are half dissociated at 1700°C and 1200°C, respectively. Just above the boiling point the molecule of sulfur is S_8. Its molecular weight decreases from 250 at 450°C to 50 at 2070°C. Thus there are some monatomic sulfur molecules at 2070°C. The dissociation probably takes place in reversible steps and can be represented by the equation:

$$S_8 \rightleftharpoons 4S_2 \rightleftharpoons 8S.$$

(1)

Many chemical compounds dissociate readily upon heating or otherwise supplying them with energy. Acetic acid vapor consists of double molecules just above the normal boiling point, but dissociates completely into single molecules at 250°C. Nitrogen tetroxide (N_2O_4) is a pale reddish brown gas at temperatures near its normal boiling point of 21.3°C. On heating the density of the gas becomes less and the color becomes darker until it is almost black. At 140°C the molecular weight is 46 which is that of NO_2 molecules. The dissociation can be written:

$$N_2O_4 \rightleftharpoons 2NO_2. \tag{2}$$

If one mole of gas yields ν moles of gaseous products, and α is the fraction of the one mole which dissociates, then the total number of moles present is:

$$1 - \alpha + \nu\alpha = 1 + \alpha(\nu - 1). \tag{3}$$

Now the density of a given weight of gas at constant pressure is inversely proportional to the number of moles, and if d_1 is taken as the density of the undissociated gas and d_2 that of the partially dissociated gas, then:

$$\frac{d_1}{d_2} = \frac{1 + \alpha(\nu - 1)}{1} \tag{3a}$$

or

$$\alpha = \frac{d_1 - d_2}{d_2(\nu - 1)} \tag{4}$$

Therefore the *degree of dissociation* of a substance can be found by measuring the densities of the undissociated and partially (or completely) dissociated substance in the gaseous state. Molecular weights may be substituted for densities giving

$$\alpha = \frac{M_1 - M_2}{M_2(\nu - 1)} \tag{5}$$

The degree of dissociation can be used to calculate the *equilibrium constant* for dissociation. The equilibrium constant may be expressed in terms of concentrations, for example, moles per liter (K_c), or in terms of partial pressures (K_p). The degree of dissociation and equilibrium constants are important theoretically and practically, e.g., the latter can be used to ascertain the extent of a chemical process.

The temperature dependence of dissociation is expressed in terms of the equilibrium constant and is

$$\frac{d \ln K_p}{dT} = \frac{\Delta H}{RT^2} \text{ or } \frac{d \ln K_c}{dT} = \frac{\Delta H}{RT^2} \tag{6}$$

where ΔH is the heat of dissociation. Integrating between the limits T_1 and T_2 one obtains

$$\ln \frac{K_{p_2}}{K_{p_1}} = \frac{\Delta H}{R}\left(\frac{T_2 - T_1}{T_1 T_2}\right)$$

$$\ln \frac{K_{c_2}}{K_{c_1}} = \frac{\Delta H}{R}\left(\frac{T_2 - T_1}{T_1 T_2}\right)$$

Electrolytes, depending upon their strength, dissociate to a greater or less extent in polar solvents. The extent to which a weak electrolyte dissociates may be determined by electrical conductance, electromotive force, and freezing point depression methods. The electrical conductance method is the most used because of its accuracy and simplicity. Arrhenius proposed that the degree of dissociation, α, of a weak electrolyte at any concentration in solution could be found from the ratio of the equivalent conductance, Λ, of the electrolyte at the concentration in question to the equivalent conductance at infinite dilution Λ_0 of the electrolyte. Thus

$$\alpha = \frac{\Lambda}{\Lambda_0} \tag{8}$$

This equation involves the assumption that mobilities of the ions coming from the electrolyte are constant from infinite dilution to the concentration in question. From the degree of dissociation and the concentration, the ionization constant or protolysis constant of a weak electrolyte can be obtained.

Water is a weak electrolyte, ionizing according to the equation:

$$H_2O + H_2O \rightleftharpoons H_3O^+ + OH^- \tag{9}$$

The specific conductance L, of water at 25° is 5.5×10^{-8} mho cm^{-1}, and the equivalent conductance of water at infinite dilution is found from the equivalent conductance of its constituent ions (H_3O^+ and OH^-) to be 547.8 mhos. The equivalent conductance Λ of water at 25°C is LV, where V is the volume of water (18 ml) containing 1 gram equivalent of water. Hence $\Lambda = LV = 5.5 \times 10^{-8} \times 18 = 9.9 \times 10^{-7}$. Therefore $\alpha = \Lambda/\Lambda_0 = 9.9 \times 10^{-7}/547.8 = 1.81 \times 10^{-9}$. Now $C_{H_3O^+} = C_{OH^-} = 55.5 \times 1.81 \times 10^{-9} = 1.00 \times 10^{-7}$ and

$$K = \frac{C_{H_2O^+} \times C_{OH^-}}{C_{H_2O}^2} \tag{10}$$

but C_{H_2O} is a constant, namely 55.5 moles/1 and therefore

$$K_\omega = (55.5)^2 K = C_{H_3O^+} \times C_{OH^-}$$
$$= 1.00 \times 10^{-7} \times 1.00 \times 10^{-7}$$
$$= 1.00 \times 10^{-14}$$

The ionization constant of pure water varies with temperature as shown below.

Temperature °C	0	10	25	40	50
$K_\omega \times 10^{14}$	0.113	0.292	1.008	2.917	5.474

Inserting corresponding values of K_ω and absolute temperature into Eq. (7) and solving for ΔH one finds the heat of ionization per mole of water to be 13.8 kilocalories.

Ionization or dissociation in general can be repressed by adding an excess of a product of the dissociation process.

The acid formed when a base accepts a proton is called the conjugate acid of the base and the base formed when an acid donates a proton is the conjugate base of the acid. Thus in the reaction

$$HA + H_2O \rightleftharpoons H_3O^+ + A^- \tag{12}$$

HA and A^- are conjugate acid and base and H_2O and H_3O^+ are conjugate base and acid, respectively.

The common ion effect then can be found as the following example shows. When using ammonium hydroxide to which the common ammonium ion in the form of ammonium chloride has been added, the ionization can be represented by the equation

$$NH_3 + HOH \rightleftharpoons NH_4^+ + OH^- \tag{13}$$

Ammonium ion NH_4^+ is the conjugate acid of the ammonia molecule. The ionization constant can be written

$$K = \frac{C_{NH_4^+} \times C_{OH^-}}{C_{NH_3}} \tag{14}$$

and

$$C_{OH^-} = K \frac{C_{NH_3}}{C_{NH_4^+}} \qquad (15)$$

Now the base NH_3 is such a weak base that in the presence of NH_4Cl the concentration of unionized base, C_{NH_3}, is equal to the total concentration of base represented by C_{base}, and $C_{NH_4^+}$ coming from the weak base is so small as to be negligible. Hence the NH_4^+ ions can be considered as coming exclusively from the NH_4Cl. Therefore Eq. (15) can be written

$$C_{OH^-} = K \frac{C_{base}}{C_{salt}} \qquad (16)$$

or

$$pOH = pK + \log \frac{C_{salt}}{C_{base}} \qquad (17)$$

Thus C_{OH^-} and hence the degree of ionization of the base NH_3 is decreased with increasing concentration of salt. The salt effect of adding electrolytes with no common ion to a solution of incompletely ionizable substance can be seen from the following considerations and using the equilibriums represented by Eq. (14) which in terms of activities becomes:

$$K = \frac{a_{NH_4^+} \times a_{OH^-}}{a_{NH_3}}$$
$$= \frac{C_{NH_4^+} \times C_{OH^-}}{C_{NH_3}} \cdot \frac{f_{H_3O^+} \times f_{OH^-}}{f_{NH_3}} \qquad (18)$$

This ionization constant in terms of activities is called the true or thermodynamic ionization constant. It does not differ too much from the K in Eq. (14) for sufficiently low ionic strengths. The two differ more markedly for appreciable ionic strengths. Now suppose a salt with no common ion is added to the solution. The ionic strength of the solution will be increased. This increase in ionic strength causes a decrease in the activity coefficients of the ions except in very concentrated solutions. Thus for K of Eq. (18) to stay constant the concentrations of the ions must increase to offset the decrease in their activity coefficients. The ammonia must therefore increase in ionization and K as defined by Eq. (14) must increase. This is known as a salt effect.

Ampholytes in solution give equal concentrations of a weak acid and a non-cojugate weak base. The amino acids are ampholytes which contain within their molecules equal amounts of a weak acid, the COOH group and a weak non-conjugate base, the NH_2 group.

According to Arrhenius those substances which yield the hydrogen ion in solution are acids, whereas bases produce the hydroxyl ion. As long as water was considered the only "ionizing" solvent these definitions were relatively simple. In the case of nonaqueous solvent chemistry at least three other concepts have been advanced. These are: (1) Franklin's *solvent system concept*, first limited to water and ammonia but since extended to nonprotonic media and defining an acid as a substance yielding a positive ion identical with that coming from auto-ionization of the solvent and a base as a substance yielding a negative ion identical with that coming from auto-ionization of the solvent; (2) the *protonic concept* of acids as proton donors and bases as proton acceptors advanced by Brønsted and by Lowry; and (3) Lewis' electronic theory according to which an acid is a molecule, radical or ion which can accept a pair of electrons from some other atom or group to complete its stable quota of electrons, usually an octet, and forming a covalent bond, and a base is a substance which donates a pair of electrons for the formation of such a bond.

In liquid ammonia as in water auto-ionization takes place. Ammonium and amide ions are formed by the dissociation or protolysis according to the following equation

$$2NH_3 \rightleftharpoons NH_4^+ + NH_2^- \qquad (19)$$

The acid and base analogs of ammonia as a solvent is specified by this equilibrium as NH_4^+ and NH_2^- ions. All substances which undergo ammonolysis and hence bring about an increase in the ammonium ion concentration yield acid solutions. Thus P_2S_5 dissolves in liquid ammonia to give an acid solution as follows.

$$P_2O_5 + 12NH_3 \rightarrow 2PS(NH_2)_3 + 3(NH_4)_2S. \qquad (20)$$

The solution is acid since an ammonium salt is formed and also because a solvo acid is obtained.

Many substances dissolve in liquid sulfur dioxide to yield ionic, conducting solutions. It has been found that such conductance data extrapolated to very high dilution yield the limiting conductance of sulfur dioxide. Both the Ostwald dilution law and the law of independent mobility of ions hold for "strong" electrolytes in highly dilute solutions.

The order of increasing dissociation and conductivity of salts in liquid sulfur dioxide apparently parallel the order of increasing cationic size. Probably because of solvation effects a similar relationship does not hold with respect to anion size. The mobilities of various ions in liquid sulfur dioxide have been studied. The van't Hoff i factors or mole numbers have been obtained by the ebullioscopic method for a wide variety of solutes in liquid sulfur dioxide. For non-electrolytes the mole number is one within experimental error. In liquid sulfur dioxide, univalent electrolytes give mole numbers which indicate large effects of ion-association of some kind. As would be expected the mole numbers of these electrolytes approach two in very dilute solutions. See Jander and Mesech, Z. physik. Chem., *A183*, 277 (1939). The mole number in general can be found from the ratio of the value of a colligative property of the solute is solution to the value of the same colligative property for a normal solute such as sugar, both solutes being at the same molal concentration.

The protonic concept of acid and bases is applicable to many of these high temperatures solvent systems such as the fused ammonium salts which possess the "onium" ion or solvated proton, and the fused anionic acids which are salts possessing a metallic ion and a hydrogen containing anion. One of the most useful of the anionic acids is KHF which is used to dissolve ore minerals containing silica, titania and other refractory oxides.

In many high temperature reactions there is an absence of hydrogen-containing ions. The Lewis electron pair concept of acid and bases can be used to advantage in such systems. In such systems strong anion bases such as the $O^=$ ion coming from basic compounds such as metallic oxides, hydroxides, carbonates or sulfates react with acidic oxides such as silica through the intermediate formation of polyanionic silicate complexes. The average ionic size of these complexes depend no doubt upon the temperature and the amount of added base.

Anion bases include the sulfide and fluoride ions coming from the corresponding alkali metal compounds. Likewise, metaphosphate and metaborate melts are acid in nature. Also proton-like character can be ascribed to any positive ion. The smaller the positive particle and the higher its charge, the greater is its polarizing tendency in bringing about deformation of negative ions, and the more reasonably can such an ion be looked upon as an acid analog.

When the potential energy of a diatomic molecule is plotted

versus the distance separating the nuclei in the molecule, the potential-energy curve shows a minimum of zero in the energy at the distance separating the nuclei where the molecule is most stable, that is, where the nuclei are at the equilibrium internuclear separation. Energy is required to force them closer together or to pull them farther apart. The energy required to separate the nuclei to an infinite distance is D', the dissociation energy measured from the minimum of the potential energy curve. The spectroscopic dissociation energy D is smaller than D' by the zero point energy $\frac{1}{2} h\nu_0$. This results in the relationship,

$$D' = D + \frac{1}{2} h\nu_0 \tag{21}$$

The spectroscopic dissociation energy D is the dissociation energy of an ideal gas molecule at absolute zero, where all the gas molecules are in the zero potential energy level, h is Planck's constant (6.62×10^{-27} erg second), and ν_0 is the frequency of vibration of the nuclei at the lowest vibrational level, which is above the point of zero potential energy at the equilibrium internuclear separation. Thus, for the hydrogen molecule, $D = 4.476$ electron volts, $\nu_0 = 1.3185 \times 10^{14}$ sec^{-1}, and since 1 electron volt $= 23.06$ kilocalories per mole we calculate D' using Eq. (21) as follows:

$$D' = (4.47\text{eV}) (23.06 \text{ kcal mole}^{-1}\text{eV}^{-1}) +$$

$$\frac{6.023 \times 10^{23} \text{ mole}^{-1} \times 6.62 \times 10^{-27} \text{ erg sec} \times 1.3185 \times 10^{14} \text{ sec}^{-1}}{(2) (4.184 \times 10^{10} \text{ ergs kcal}^{-1}).}$$

$$= 109.5 \text{ kcal mole}^{-1}$$

—E. S. Amis, University of Arkansas, Fayetteville, Arkansas.

DISTILLATION. This is one of the most important and widely used of the chemical unit operations, both in the laboratory and on a large industrial scale, for separating the components of a liquid mixture. Distillation provides a means for partially vaporizing the mixture and separately recovering the vapor and residue. Consequently, the method is dependent upon the vapor pressures of the components making up the mixture. The vapor pressure of a pure substance is a constant, but varying with temperature. In distillation, the lighter, more volatile components of the original mixture (*distilland*) concentrate in the vapor when heat is applied. Advantage is taken of the fact that the ratios of the component substances in the vapor and liquid phases, except for special situations, are different. The less volatile components concentrate in the liquid residue (*bottoms*). The vapors evolved by distillation are condensed and are termed the *distillate*.

Distillation should be contrasted with evaporation wherein the vapor (frequently water) is not usually condensed (except where it is desired to conserve water). The principal product desired in evaporation is the solid material which remains in the evaporator vessel. Thus distillation would be used to separate two or more miscible liquids, such as glycol and water, whereas evaporation would be used to separate solid sodium chloride from brine.

The effectiveness of separation by distillation is largely determined by differences in the boiling points of the starting components. Where these are widely separated, as in the case of water (100°C) and ethylene glycol (197.6°C), the separation is relatively easy and redistillation is not required. Closer-boiling mixtures, such as the isomers of xylene, are much more difficult to separate by distillation. As the boiling points of the components approach each other, effective separation by distillation becomes more difficult.

Boiling-point diagrams or what also are termed boiling and condensation curves, usually experimentally determined, are useful guides in the design of distillation equipment. Fig. 1 is

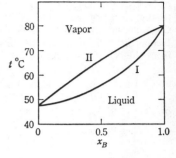

Fig. 1. Temperature-composition of a liquid-vapor system at constant pressure.

the phase diagram of a binary system forming a liquid and a vapor phase at constant pressure. Curve I is the boiling curve, which gives the coexistence temperature as a function of liquid composition; and curve II is the condensation curve, which gives the coexistence temperature as a function of the composition of the vapor phase. If the temperature is increased, vaporization begins when the boiling curve is crossed. Inversely, condensation begins when the temperature is decreased below the condensation curve. The use of boiling-point diagrams in still design will be discussed later.

Major Types of Distillation

Distillation can be a batch or a continuous operation. Batch distillation is frequently used in the laboratory for determining the chemical composition of mixed liquids, such as hydrocarbons. In the majority of industrial processes, distillation is continuous. In a batch operation, the charge material is boiled and vapors are removed continuously, condensed, and collected until such point is reached where there is the desired average composition. The separation is not sharp. This type of operation is also termed *simple* or *differential distillation*. In another approach, the mixture may be heated until a definite fraction of the liquid batch is vaporized, during which time the liquid and vapor are kept in intimate contact, i.e., with no vapors being removed. At the prescribed temperature and after the liquid and vapor have had opportunity to reach full equilibrium, the vapor is suddenly withdrawn and condensed. This approach is known as *equilibrium* or *flash distillation*. The method finds wider application in connection with multicomponent systems than with simple binary systems.

In the majority of industrial distillation systems, some of the distillate will be continuously returned to the distillation column. The returned condensate is contacted countercurrently with the rising vapors, thus bringing about an enrichment of the vapor in the more volatile components than otherwise would be accomplished with a single distillation and most often obviates the need for one or more redistillations to obtain the degree of purity desired. This approach is known as *rectification* or *fractional distillation*. The material returned is termed *reflux*. In most rectifying columns, the raw feed to the column is introduced at about the mid-level of the tower or column. The portion of the column above the point of feed is called the *rectifying section*; the portion below, the *stripping section*. Where the feed may be introduced at the top of the column, the entire column then is usually referred to as a *stripping column*, with no reflux used.

Dephlegmation is a means for increasing the efficiency of fractional distillation by forcing the vapors from the still to bubble through shallow layers of condensate in a column or dephelgmator whereby the amount of low-boiling component in the vapor is increased and a substantial portion of the higher-boiling components is retained in the condensate.

Steam distillation is a process whereby compounds which are sparingly soluble in water may be distilled by heating with water or by blowing steam through the mixture. Compounds of relatively high boiling point may be distilled at lower temperatures by this method and thus prevent degradation.

A representative fractional distillation column of which there are thousands in use in the process industries, notably in the petroleum and petrochemical industries, is shown in Fig. 2. The material balance of the column is:

$$F = W + D$$
$$Fx_F = Wx_F + Dx_D$$

where

F = feed rate, weight-moles/unit of time
W = bottom product, weight-moles/unit of time
D = distillate, weight-moles/unit of time
x_F = mole fraction of low boiler in feed
x_D = mole fraction of low boiler in distillate
x_W = mole fraction of low boiler in bottom product

In this balance, the assumption is made that the molar heat capacities and the latent heats of vaporization of all components are identical. It is also assumed that heat losses from the column and heats of mixing are negligible. With these assumptions, the upward vapor flow and the downward liquid flow in both the rectifying and stripping sections will be invariant within the sections. It is also assumed that accounting for the column heat balance is independent of the compositions of the product streams. Within these qualifications, the internal material balance is:

$$L_n = (1 + b)R$$
$$V_n = D + (1 + b)R$$
$$L_m = L_n + qF$$
$$V_m = L_m - W$$

$$x_W = f\left(\frac{L_m}{V_m}\right)$$
$$x_D = g\left(\frac{L_n}{V_n}\right)$$

where

V_n = vapor rate in rectifying section, weight-moles/unit of time
V_m = vapor rate in stripping section, weight-moles/unit of time
L_n = liquid rate in rectifying section, weight-moles/unit of time
D = distillate rate, weight-moles/unit of time
R = external reflux, weight-moles/unit of time
b = a numerical factor, depending upon the reflux enthalpy or temperature. (It should be noted that b is greater than zero whenever the reflux temperature is below that at the top of the column.)
q = a numerical factor, depending on the feed enthalpy whose value satisfies certain constraints:

$q < 0$, when feed temperature is below feed plate temperature.
$q = 1$, when feed temperature and composition are identical with those of feed plate.
$1 > q > 0$, when feed enters column partially vaporized.
$q = 0$, when feed is fully vaporized and is at saturated temperature.
$q < 0$, when feed is superheated vapor.

f and g are factors which account for several functional relationships which depend upon such column design criteria as the number of plates in column, location of control plates, location of feed plate, and temperature and other conditions specified for control plates.

The foregoing type of material balance is of large value in determining the best form of automatic control to apply to the column in order to maximize yields. This procedure is explained in detail in Reference 1 given at the end of this entry.

Distillation Calculations. In determining the type of packing or trays to be used in a fractionating column, the diameter, height, location of feed, location of reflux return, vapor and liquid rates, and all other specifications for a column to effect a given separation, equilibrium diagrams of the type shown in Fig. 3 are important. Prior to the availability of high-speed computers, distillation column designers depended heavily upon graphical solutions, notably McCabe-Thiele diagrams, named

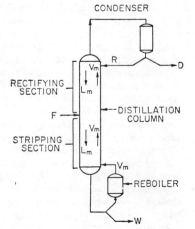

Fig. 2. Typical distillation column.

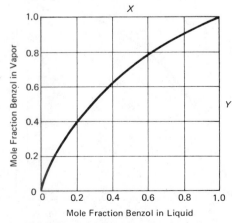

Fig. 3. Equilibrium diagram for the benzol-toluol system.

after the early developers of this concept. A typical diagram of this type for a simple binary distillation is shown in Fig. 4. Oversimplifying the method, first an equilibrium curve of the type of Fig. 3 is constructed. Next a 45° diagonal line is drawn. With knowledge of desired final composition, i.e., composition of the liquid received by the top plate from the condenser, X_p, calculate the intercept of an *operating line* with the Y-axis of the chart. This is indicated as #1 on Fig. 4. The term $X_p/O + 1$ is this Y-intercept of the operating line. O is the reflux ratio. Next, the intersection of the diagonal line with the ordinate X_p is marked. This is indicated as #2 on Fig. 4. The operating line is then drawn in by joining #1 and #2. Now, commencing at #2, rectangular steps between the operating line and the equilibrium curve are drawn in until it crosses the line $X = X_s$. X_s is the starting composition of the mixture. The number of horizontal steps counted (in this case, five) indicates the number of *theoretical plates* required to accomplish the separation desired. A theoretical plate may be defined as a plate wherein complete equilibrium is reached between the vapor rising from the plate and passing to the plate above—with the liquid leaving the plate and passing to the plate below. An actual plate, of course, will not perform with this efficiency. Thus, the designer, depending upon past experience with plates of certain designs, will include an appropriate margin in specifying the number of actual plates needed.

Azeotropic Systems. An azeotropic system is one wherein two or more components have a constant boiling point at a particular composition. Such mixtures cannot be separated by conventional distillation methods. If the constant boiling point is a minimum, the system is said to exhibit *negative azeotropy*; if it is a maximum, *positive azeotropy*. Consider a mixture of water and alcohol in the presence of the vapor. This system of two phases and two components is divariant. Now choose some fixed pressure and study the composition of the system at equilibrium as a function of temperature. The experimental results are shown schematically in Fig. 5.

The vapor curve *KLMNP* gives the composition of the vapor as a function of the temperature T, and the liquid curve *KRMSP* gives the composition of the liquid as a function of temperature. These two curves have a common point M, where the curves are tangent. Because of the special properties associated with systems in this state, the point M is called an *azeotropic point*. In an azeotropic system, one phase may be transformed to the

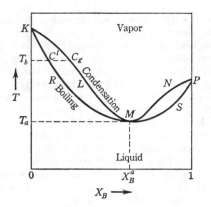

Azeotropic system.

Fig. 5. Boiling-point diagram of an azeotropic system exhibiting negative azeotropy.

other at constant temperature, pressure and composition without affecting the equilibrium state. This property justifies the name azeotropy, which means a system which boils unchanged.

Because a number of industrially important liquid mixtures are azeotropic systems, means had to be found whereby they may be separated by distillation. Two approaches, *extractive distillation* and *azeotropic distillation*, are used. In either case, a *separating agent* is added to the column so as to alter favorably the relative volatilities of the feed components. Usually, water or polar organic compounds are found most effective. They increase the liquid-hase non-ideality of one feed component more than another.

In extractive distillation, the agent (sometimes termed solvent) is significantly less volatile than the regular feed components. The agent will be added near the top of the column. The agent behaves as a *heavier-than-heavy* key component. It is also conveniently separated from the product streams. Because the agent usually must be added in fairly substantial amounts, this means that column diameters and heat loads are increased, while plate efficiencies are lowered.

In azeotropic distillation, an agent is selected that will form an azeotrope with one of the feed components. In essence, separation is accomplished between this "new" azeotrope (as an overhead product) and the other feed component as bottoms product. An agent will be selected preferably that will permit easy separation after distillation.

Instrumentation of Distillation Columns. There are numerous ways in which a distillation column can be controlled. Inasmuch as the usual object of distillation is to achieve one or more products of a specified composition, composition control of either the distillate or bottoms product, or both, is one logical approach. However, because of time lags in such a system and because of the difficulty of finding fully applicable composition sensors, other means of control are frequently used.

Trays and Packing. Trays with bubble caps or other suitable configurations for enhancing a maximum intermingling of rising vapors with falling liquid in a column are usually used where efficiency and close separations are major considerations. Packed columns, filled with ceramic shapes of various types, such as Berl saddles and Raschig rings, are used primarily where cost and acid-resistance are factors. A tunnel cap is much more shallow and replete with peripheral perforations through which vapor and liquid flow. In some very tight separations, several trays separated a few feet apart up the vertical length

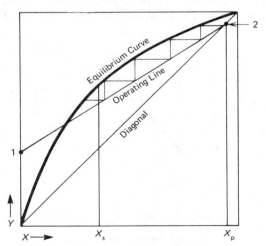

Fig. 4. Representative solution for theoretical plates in rectifying part of a distillation column.

of the column may be required.

A petroleum crude atmospheric distillation unit separates the crude into three major fractions which then are subjected to later separations: (1) a light straight-run fraction, consisting primarily of C_5 and C_6 hydrocarbons, but also containing any C_4 and lighter gaseous hydrocarbons dissolved in the crude; (2) a naphtha fraction having a nominal boiling range of 200–400°F (93–204°C); and (3) a light distillate with boiling range of 400–650°F (204–343°C).

See also **Molecular Distillation.**

References

Buckley, P. S.: "Distillation Column Design Using Multivariable Control," *Instrumentation Technology,* **25,** 9, 115–122 (1978); **25,** 10, 49–53 (1978).
Castellano, E. N., McCain, C. A., and F. W. Nobles: "Digital Control of a Distillation System," *Chem. Eng. Progress,* **74,** 4, 56–60 (1978).
Considine, D. M. (editor): "Chemical and Process Technology Encyclopedia," pp. 354–356, McGraw-Hill, New York, 1974.
Danziger, R.: "Distillation Columns with Vapor Recompression," *Chem. Eng. Progress,* **75,** 9, 58–64 (1979).
Hughart, C. L., and K. W. Kominek: "Designing Distillation Units for Controllability," *Instrumentation Technology,* **24,** 5, 71–75 (1977).
Latour, P. R.: "Composition Control of Distillation Columns," *Instrumentation Technology,* **25,** 7, 67–74 (1978).
Lubowicz, R. E., and P. Reich: "High Vacuum Distillation Design," *Chem. Eng. Progress,* **67,** 3, 59–63 (1971).
Perry, R. H., and C. H. Chilton (editors): "Chemical Engineers' Handbook," McGraw-Hill, New York, 1973.
Shinskey, F. G.: "Distillation Control for Productivity and Energy Conservation," McGraw-Hill, New York, 1977.
Wang, J. C.: "Computer Relative Gain Matrices for Better Distillation Control," *Instrumentation Technology,* **27,** 3, 40–44 (1980).
Watkins, R. N.: "Petroleum Refinery Distillation," 2nd edition, Gulf Publishing, Houston, Texas, 1979.
Weisenfelder, A. J., and R. E. Olson: "Solving Recycle Streams in Multicolumn Distillation," *Chem. Eng. Progress,* **76,** 1, 40–43 (1980).

DIURETICS. These are substances which increase the volume of urine excreted, causing a condition known as *diuresis.* Some natural drugs, such as caffeine, act as diuretics. Some other substances, known as "saline diuretics," when filtered through the renal capsule are incapable of being reabsorbed by the tubules. These substances thus increase the concentration of salts within the tubule so that little of the urine can diffuse back into the blood stream. Some drugs, such as mercurial diuretics, act at the tubules, thus preventing reabsorption. Other drugs, such as digitalis, cause diuresis because of their specific actions on the circulatory system.

A modern classification of diuretics places them in four categories:

(1) *Thiazides* were discovered during the synthesis of carbonic anhydrase-inhibiting analogues of sulfanilamide. The thiazides inhibit reabsorption of sodium and chloride in the distal convoluted tubules of the kidney. These drugs also increase secretion of potassium in the distal convoluted tubule and collecting ducts and thus may cause a depletion of potassium. The thiazides have relatively few other side effects. Where hypokalemia (deficiency of potassium) is noted, this can be corrected through the use of potassium supplements, usually potassium chloride. Examples of the thiazides include chlorothiazide (*Diuril*®), hydrochlorothiazide (*Hydro-Diuril*®), trichlormethiazide (*Metahydrin*®; *Naqua*®), and chlorthalidone (*Hygroton*®). The latter drug differs chemically from the thiazides, but is pharmacologically similar.

(2) *Mercurial diuretics,* which are stronger in their diuretic effects, require parenteral administration. The organic mercurials, such as meralluride (*Mercuhydrin*®) and mercaptomerin sodium (*Thiomerin*®) work at the ascending limb of the loop of Henle to inhibit active chloride transport and reabsorption of sodium. Chloride excretion can be extensive, causing hypochloremic alkalosis. Because mercury, a heavy metal, is involved, the mercurials have lost considerable favor.

(3) *Loop diuretics,* such as ethacrynic acid and furosemide, are the strongest diuretics available. These compounds have largely replaced the mercurials. Loop diuretics inhibit tubular reabsorption of sodium chloride in the ascending loop of Henle. These drugs also enhance potassium excretion and thus cause the same kinds of problems as the thiazides. There are a number of side effects, including hearing loss in the case of ethacrynic acid. Because they are potent, the loop diuretics must be prescribed with considerable discretion. The principal examples are furosemide (*Lasix*®) and ethacrynic acid (*Edecrin*®).

(4) Frequently *potassium-sparing agents,* although in themselves weak diuretics, are combined with other diuretics. In particular, these agents enhance the actions of the thiazides and loop diuretics, increasing the urinary loss of sodium while decreasing the excretion of potassium. Representative of these agents are spironolactone (*Aldactone*®) and triamterene (*Dyrenium*®).

The carbonic anhydrase inhibitors, such as acetazolamide (*Diamox*®) also decrease the absorption of sodium, bicarbonate, and chloride, but are too weak in their effect to be useful as single agents.

In addition to their use in congestive heart failure and angina pectoris, diuretics are widely prescribed in the treatment of hypertension (high blood pressure), acute respiratory failure, hypercalcemia (in cancer patients), hypertrophic cardiomyopathy, and the syndrome of inappropriate ADH secretion, among others.

DNA (Recombinant). **Recombinant DNA.**

DOLOMITE. The mineral dolomite, the carbonate of calcium and magnesium, corresponds to the formula $CaMg(CO_3)_2$ and closely resembles calcite. Its crystals, rhombohedral in habit, fall in the hexagonal system. Like calcite, it may be massive or granular, some marbles being dolomite rather than calcite. It displays a perfect cleavage parallel to the rhombohedron; subconchoidal fracture, brittle; hardness, 3.5–4; specific gravity, 2.85; luster vitreous to pearly; color varies widely, white, reds, greens, black, browns, yellows or colorless; transparent to translucent. Unlike calcite, dolomite dissolves very slowly if at all in dilute cold hydrochloric acid; powdered dolomite will dissolve in warm acid. This is the common test for the two minerals.

Much dolomite occurs as stratified rocks where it is believed to have been formed by a secondary process, probably by the action of waters charged with magnesium compounds. Dolomite also is found as a vein mineral, as is calcite. Iron or manganese, rarely zinc or cobalt, may replace some of the magnesium. Ankerite is the name given to a mineral whose composition is essentially a calcium-magnesium-iron carbonate. Among the many noted localities for dolomite are Saxony, Switzerland, Italy, France, Spain, Brazil, Mexico; in the United States, Roxbury, Vermont; Lockport, New York; Phoenixville, Pennsylvania; Alexander County, North Carolina; Hancock County, Illinois, and the Joplin District, Missouri, Dolomite was named for Deodat de-Dolomieu, who first described its characteristics. See also **Limestone;** and terms listed under **Mineralogy.**

DRYING (Process). Frequently in the process industries, materials must be dried, i.e., liquid must be removed from a solid or gaseous phase. In most instances, the liquid to be removed is water, although in solvent recovery systems, for example, the liquid may be an organic solvent. See **Dehumidification** for a discussion of the drying of gases. Some materials which must be dried include: (1) *solutions*, colloidal suspensions, and emulsions, such as extracts, milk, blood, waste liquors, rubber latex, and inorganic salt solutions; (2) *slurries*, which are pumpable suspensions, as found in calcium carbonate, bentonite, clay slip, and lead concentrates; (3) *sludges and pastes*, such as centrifuged solids, starch, filter-press cakes, and sedimentation sludges; (4) *powders* that may be relatively free-flowing when wet, but very dusty when dry, including pigments, cement, clay, and centrifuged precipitates; (5) *fibrous solids* and granular and crystalline solids, such as sand, ores, rayons staple, salt crystals, and synthetic rubber; (6) *formed and shapes solids*, such as pottery, rayon cakes, shotgun shells, brick, rayon skeins, lumber, and objects that have been painted or otherwise coated; (7) *sheeted materials*, such as impregnated fabrics, paper, plastic, and fiberboard—in the continuous form; or veneers, wallboards, foam-rubber sheets, and photographic prints—in the individual-piece configuration. Because of this very wide variety of drying requirements, it is obvious that there does not exist what might be termed a universal dryer. Further, universal dryer design criteria are difficult to develop and summarize. See accompanying table.

Basic Concepts of Drying. Two criteria hold for a large number of drying situations, namely, that of breaking the drying process down into two main periods: (1) the *constant-rate period*, the rate of removal of liquid per unit of drying surface is essentially steady, but to qualify the surface of the material must remain fully wet (saturated) during this period; and (2) the *falling-off period*, during which time the rate of drying decreases as it becomes increasingly difficult to move moisture from the capillaries and interstices of the material to the surface where the moisture can be taken up and moved away. With temperature and air flow (if air is the absorbing and moisture conveying medium to be used) well established, the constant-rate period is relatively easy to forecast and to design because this situation is somewhat analogous to that of drying a shallow container of water. Knowing the foregoing conditions as well as the wetted area involved, drying time periods can be predicted. But, in the case of the falling-off period, the dryer designer must be intimately familiar with the mechanism whereby moisture is held to the material and the mechanics involved in moving the moisture to the surface where it can be picked up.

Thus, the falling-off period has been broken down into two phases: (1) the *unsaturated-surface drying period*; and (2) the *internal-moisture movement period*. This first period does not impose unusually difficult analysis because essentially the condition is analogous to a partially-filled shallow vessel whose surface is decreasing as drying proceeds. Essentially, this is an extension of the analysis of the constant-rate period, still leaving the internal-movement situation to be predicted. Formulas are available to assist the designer, based upon past experience with specific drying situations and materials. This is a process which is difficult to assess theoretically, and almost always requires experimental runs or comparisons with other at least somewhat similar materials that have been dried successfully. One can generally state that where materials are to be dried to a low-moisture content, the internal-moisture movement period will require the largest portion of the total drying time.

Classification of Dryers. In reviewing the wide variety of drying equipment available, there are a few major classifications that are helpful. There is the distinction between *continuous*

MAJOR TYPES OF PROCESS DRYERS

Continuous Dryers

DIRECT HEATING	INDIRECT HEATING
Tunnel Dryers. Material to be dried is placed on trucks or carts which are moved through a tunnel in which there is a flow of hot gases. Temperature of the tunnel may be zone controlled.	*Drum Dryers.* For materials in liquid and slurry form. One or several drums dip into the liquid or slurry and thus coat the heated drums. The drum temperature is controlled to effect drying during a part of the rotation of the drum from which the dried material is removed by knife prior to dipping one again into the liquid material.
Through-circulation Dryers. Material is supported on a conveying screen that moves continuously. Hot gases from below or above conveyor pass through the material and pick up moisture.	*Cylinder Dryers.* Material in the form of a continuous sheet passes over and around cylinders which rotate and which are heated, usually by steam or hot water.
Rotary Dryers. Material (liquid) is pumped to and showered within a rotating cylinder through which hot gases flow.	*Screw-conveyor Dryers.* A conveyor is housed within a closed, heated housing. This operation may proceed at atmospheric pressure or under vacuum.
Tray Dryers. Material is placed on vibrating trays over or under which hot gases flow.	*Vibrating-tray Dryers.* Similar to the directly heated tray dryer except that the heat is conducted to the trays indirectly (as by electrical heating) rather than by hot gases.
Sheeting Dryers. Material in sheet form passes continuously through a hot chamber. Depending upon product, sheet may be taut (as pinned to a frame), or it may pass through dryer in festoon manner.	*Steam-tube Rotary Dryers.* Material is passed through a long, rotating cylinder. A shell around the cylinder contains steam, hot water, or other heating medium.
Pneumatic Conveyor Dryers. Material is moved in a stream of gas at higher-velocity and high temperature and finally collected by a cyclone separator.	

Batch Dryers

Through-circulation Dryers. Material is placed on trays with screen bottoms. Hot gases are blown from below through material.	*Vacuum Rotary Dryers.* Material is subjected to agitation within a stationary, horizontal shell under vacuum. The agitator may be heated to increase drying effectiveness.
Tray and Compartment Dryers. Material is placed on trays which then may be placed on trucks or on permanent shelves within dryer. Hot gases are blown across the trays.	*Agitated-pan Dryers.* Material is placed in covered shallow pans. Pans are jacketed for heating. An agitator stirs the material constantly. This design may be operated at atmospheric pressure or under vacuum.
	Vacuum Tray Dryers. Trays in which material is placed are heated by conduction from supporting shelves. The whole compartment may be under a relatively high vacuum. The material is not agitated.

and *batch* operations. If a continuous drying operation is desired because it will fit into the overall manufacturing operations best, then this decision will rule out those drying concepts that can be applied only in a batch manner. In some instances, it may turn out that in an otherwise fully continuous manufacturing process, the drying portion will have to be handled by batches simply because, for the particular product, batch drying offers the greatest efficiency. There is also the distinction between *direct heating* and *indirect heating* methods used in dryers. In direct heating, the heat needed is applied by way of immediate contact between the wet material and hot gases. In indirect heating, the heat needed is transferred to the wet material through an intervening medium, commonly pipes or a retaining wall. There are numerous instances, for example, where a product may be too sensitive to withstand exposure to a moving hot gas.

Since a large majority of drying equipment involves the use of hot gases (usually air), other means of heating can be overlooked. Dielectric heating and freeze-drying, for example, are other means where practical.

Drying equipment also can be made available with means to accelerate the drying process if these means are acceptable to product quality. For example, agitation, stirring, and otherwise keeping the material to be dried in constant motion naturally assists the amount of exposure of surfaces to the drying medium. Numbers of materials, however, cannot be handled in this fashion and require relatively conservative, still conditions.

Principal design configurations of drum-type dryers are illustrated and described in Figs. 1 through 4; conveyor-type dryers in Figs. 5 through 7.

Dehydration. This operation is sometimes described as the removal of 95% or more of the water from a substance, by exposure to thermal energy by various means. Frequently used in food processing, the aims of dehydration are reduction in

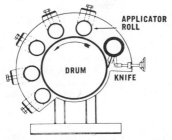

Fig. 1. Single-drum dryer (atmospheric). Dryers of this type may be dip or splash fed (not shown), or, as shown, equipped with applicator rolls. The latter is particularly effective for drying high-viscosity liquids or pasty materials, such as mashed potatoes, applesauce, fruit-starch mixtures, gelatin, dextrine-type adhesives, and various starches. The applicator rolls eliminate void areas, permit drying between successive layers of fresh material and form the product sheet gradually. While single applications may dry to a lacy sheet or flake, the multiple layers generally result in a product of uniform thickness and density with minimum dusting tendencies. (*Buflovak Division, Blaw-Knox Food & Chemical Equipment, Inc.*)

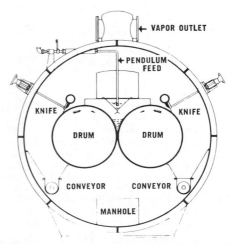

Fig. 3. Single- and double-drum dryer (vacuum). Design configurations like this are applicable whenever products must be dried without exposure to high temperatures or reactive atmospheres. Vacuum operation is particularly suited for vitamin extracts, protein hydrolyzates, soluble coffee, and malt products. Continuous drying without breaking vacuum is achieved by a special conveyance system that utilizes two receivers and air locks. Single-drum feed is usually the pan type with pump and spreading device, or a spray film for materials that are repelled by contact with heated surfaces. Double-drum utilizes pendulum or perforated tube feed. Specially designed dispersion devices for feed also can be used. (*Buflovak Division, Blaw-Knox Food & Chemical Equipment, Inc.*)

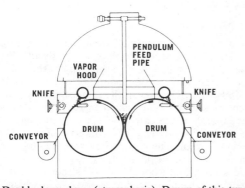

Fig. 2. Double-drum dryer (atmospheric). Dryers of this type handle a variety of food products of widely varying densities and viscosities—dilute solutions, heavy liquids, or pasty materials. A number of products can be dried successfully with this kind of configuration, inasmuch as exposure to temperature above the boiling point is restricted to just a few seconds. The movable drum permits effective control over product film thickness. Feed may be by perforated tube trough, pendulum, or various special configurations. (*Buflovak Division, Blaw-Knox Food & Chemical Equipment, Inc.*)

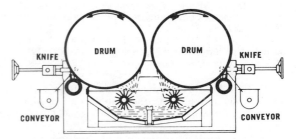

Fig. 4. Twin-drum dryers are designed for handling slurries, corrosive solutions, crystal-bearing or crystal-forming liquids and delicate, heat-sensitive products which may be exposed to high temperatures only for a very limited time. These drums may utilize pendulum or perforated tube for top feed, or splash or spray for bottom feed. Heat-sensitive materials are not in direct contact with the drums. The temperature remains fairly constant and preconcentration is minimized. Cooling and agitation of the material in the pan may also be provided when necessary. (*Buflovak Division, Blaw-Knox Food & Chemical Equipment, Inc.*)

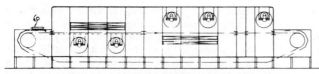

IN ♦ CROSS-APRON WIPER-FEEDER • LEVELER • UP-THRU &
DOWN-THRU ZONES • AIR LOCK • COOLING ZONE ♦ OUT

Fig. 5. Single-stage, single-pass conveyor dryer of convection type commonly used to dry, cool, toast, roast, bake, and heat materials. Such units can be zoned for various product drying temperatures, cooling, humidifying, and conditioning requirements. The air flow through the product can be upward or downward and the heat source can be steam, gas, oil, electric, or waste heat. Dryers with this design configuration are applicable to the processing of breakfast cereals, pet foods, fresh vegetables, fruits, and nuts, among numerous other food substances. (*The National Drying Machinery Company.*)

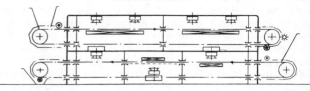

♦ FIRST PASS ♦
IN ♦ HOPPER-FEEDER • LEVELER • AIR LOCK • 2 DOWN-THRU ZONES • AIR LOCK
DOFFER • BRUSH ➥

♦ SECOND PASS ♦
COOLING ZONE • UP-THRU & DOWN-THRU ZONES • AIR LOCK • LEVELER
OUT ♦ BRUSH • AIR LOCK

Fig. 6. A multi-tier, multi-pass conveyor dryer. Due to space limitations it is sometimes necessary to tier dryers and use gravity or transfer conveyors to carry the product from one tier to the other. The location of the discharge end will vary with the number of tiers. (*The National Drying Machinery Company.*)

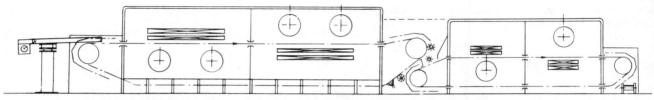

IN ♦ OSCILLATING FEEDER • UP-THRU & DOWN-THRU ZONES • TRANSFER & REORIENT SECTION (UNINSULATED) • UP-THRU & DOWN-THRU ZONES • CROSS CONVEYOR DELIVERY BELT ♦ OUT

Fig. 7. A multi-stage, 4-zone drying range consisting of a series of single-stage conveyor dryers designed so that the individual stages present a new set of conditions to the product being processed. These stages, in turn, can be zoned to give maximum processing flexibility. Transfer devices between stages reorient the product to provide effective processing uniformity. (*The National Drying Machinery Company.*)

volume of the product, increase in shelf life, and lower transportation costs, among other factors. There is no clearly defined line of demarcation between drying and dehydrating, the latter sometimes being considered as a subelement of drying.

References

Flink, J. M.: "Energy Analysis in Dehydrations Processes," *Food Technology*, **31**, 3, 77–84 (1977).
Keey, R. B.: "Introduction to Industrial Drying Operations," Pergamon, Elmsford, New York, 1978.
King, C. J., and J. P. Clark: "Water Removal Processes: Drying and Concentration of Foods and Other Materials," Amer. Inst. of Chem. Engr., New York, 1977.
Leed, D. A.: "Five Steps Simplify Selecting a Dryer," *Food Engineering*, **47**, 3, 63–66 (1975).
Mujumdar, A. S. (editor): "Proceedings of the First International Symposium on Drying," Science Press, Princeton, New Jersey, 1978.
Rippen, A. L.: "Energy Conservation in the Food Processing Industry," *J. Milk Food Technol.*, **38**, 11, 715 (1975).
Sapakie, S. F., Mihalik, D. R., and C. H. Hallstrom: "Drying in the Food Industry," *Chem. Eng. Progress*, **75**, 4, 44–49 (1979).
Staff: "Drying Techniques," *Chem. Eng. Prog.*, **75**, 4, 37–60 (1979).

DUMORTIERITE. This mineral found within schists and gneisses is valuable for use in the manufacture of high-grade porcelain. It is a borosilicate of aluminum $Al_7(BO)_3(SiO_4)O_3$, crystallizing in the orthorhombic system. Crystals are rare; commonly occurs as fibrous aggregates of blue to pink color, with vitreous luster. Transparent to translucent with hardness of 8.5, and specific gravity of 3.41.

World occurrences include France, the Malagasy Republic, Brazil, and Mexico, with California and Nevada the principal United States localities.

DYEING. The technology of applying coloring matter to a substrate in such a way that the color appears to be a property of the substrate not readily removed by rubbing, washing or exposure to light. The term dyeing is usually understood to be confined to the application of colors to textiles and such related materials as furs, leather, plastics and paper, and is used to differentiate between other coloration proccesses such as printing, painting or staining. Printing and painting involve the application of colored pigments in a resinous vehicle to a textile or paper surface, while in staining small amounts of color are held on the surface of the substrate. True dyeing implies an interaction of the dye with the substrate on a molecular basis and involves chemical and physical principles which are responsible for bringing about a permanent union between the material to be dyed and the coloring matter applied. See the following article on **Dyes.**

General Principles

There is no uniform theory of dyeing, and different chemical and physical principles are involved in the dyeing of different substrates. In cotton dyeing with direct dyes, the nature of the dye-fiber interaction is predominantly physical, while the application of fiber-reactive dyes to the same substrate involves primarily chemical and to only a small degree physical phenomena. Another example is the case in which the same dye (magenta) may be used for the dyeing of two very different fibers such as cotton and wool. The cotton, however, has to be prepared prior to the application of the dye (i.e., with a mordant such as tannic acid), while wool acts itself as the mordant.

The simplest dyeing procedure is the one in which the fiber or fabric is placed in a solution or dispersion of the dye which is then absorbed upon increasing the temperature of the system. The ease with which the coloring matter is taken up by the substrate varies widely, and has made it necessary to devise dyeing processes which accelerate the penetration of the substrate by the color by controlling temperature, pH, salt concentration and other variables of the system. The trend in modern industry is to develop more rapid and economical dyeing processes, characterized by a greater emphasis on the utilization of scientific methods. Improved tools such as instruments for color measuring and matching, are now widely used.

Dyeing Processes

Different classifications have been suggested for dyeing processes referring either to the type of substrate or to the method of application of the dye. The classification of dyeing processes chosen in this article is based on the method of application.

Direct Dyes on Cellulosic Fibers. Dyeing of cotton, rayon and other cellulosic fibers with direct dyes is carried out in a neutral or mildly alkaline bath with additions of common salt or sodium sulfate at temperatures near the boil. The dyeing is facilitated by the use of surface-active compounds, which are used extensively in the trade for this purpose. Direct dyes are generally linear azo and polyazo dyes, but dioxazine, phthalocyanine and other structures are also encountered. The common features of these dyes are good water solubility, relatively high molecular weight and fiber affinity, i.e., they transfer directly (hence the name) from the aqueous solution to the cellulosic medium. They differ from acid dyes in that they are substantive and are fixed on the fiber by the addition of only NaCl. Best affinity is shown by dye molecules of linear and planar configuration. A full spectrum of shades is available from direct dyes, varying from dull to relatively bright. Wash and light fastness of direct dyes are not very good. However, the wash fastness can be improved by after-treatments. These treatments involve cationic agents which form water-insoluble complexes, or crosslinking agents such as formaldehyde which often adversely affect the light fastness. Other after-treatments such as diazotizing and coupling on the cloth aim at increasing the molecular weight of the dye without adding further solubilizing groups, thus decreasing the diffusion rates of the dye from the fiber during wet treatments.

The light fastness of o,o'-dihydroxyazo derivatives can be improved markedly by metallization with copper or nickel. This treatment often causes a broadening of the peaks in the absorption spectrum with resulting dullness, and is therefore restricted in use.

Fiber-Reactive Dyes on Cellulosic and Polyamide Fibers. Dyeing with reactive dyes is accomplished from aqueous solution by controlling pH and temperature. These dyes represent one of the most important developments in recent years. Among the properties which have been responsible for their outstanding commercial success are ease of application, brilliance of shade, and fastness to wet treatment. The common feature of all fiber-reactive dyes is a reactive group capable of forming covalent bonds with the substrate under the dyeing conditions. Many such reactive groups have been suggested and patented throughout the world over the past 17 years. Among them are the mono- and dichlorotriazinyl groups, the di- and trichloropyrimidyl groups, the 2,3-dichloro-6-quinoxalinecarbonyl and the 2-sulfatoethylsulfonyl group. A wide range of shades is offered in this class and different lines have been designed for cellulosic fibers, for wool and for polyamides. One of the still unsolved problems with reactive dyes is the accomplishment of quantitative fixation. Usually only 80–85% of the dye is chemically linked to the substrate, while the remainder is lost due to premature hydrolysis or to side reactions in the dyeing step. An interesting compromise between the fiber-reactive concept and mordanting is the fixation of dyes to the substrate in the presence of crosslinking agents. In this case no premature hydrolysis can occur since the reactive group is not covalently bound to the dye molecule. Fiber-reactive dyes readily lend themselves to continuous application.

Acid Dyes on Wool, Silk and Polyamide Fibers. The dyeing of natural and synthetic polyamides is achieved by salt formation between the basic fiber (free —NH$_2$ groups) and the acid groups of the dye. It is accomplished from aqueous medium by control of pH and temperature. Chemically, acid dyes are mainly sodium salts of sulfonic acids. Except for the common feature of the sulfonic acid, acid dyes vary widely in their chemical structure and derive from azo, anthraquinone, quinophthalone, nitroarylamine and triarylmethane chemistry. They vary widely in fastness to light, brilliance and tinctorial properties. They are usually applied to the fiber from an aqueous solution containing some acetic, formic or sulfuric acid; a few neutral-dyeing acid dyes are also known. These differences are reflected by the common grouping into:

(a) Level-dyeing types, which require sulfuric acid to exhaust.
(b) Neutral-dyeing or milling colors, which exhaust with acetic or formic acid.
(c) Metal-complex colors derived from o,o'-dihydroxyazo compounds and chromium or cobalt, which require relatively high amounts of sulfuric acid.

The higher concentration of acid for leveling the metallized dyes on wool is necessary to avoid demetallization of the dye through fiber-metal interaction. On nylon metallized acid dyes tend to accentuate barré and their exhaust rate, therefore, must be controlled carefully.

Nonionic surfactants facilitate level dyeing of many acid dyes by forming a complex (aggregate) with them, which releases the dye slowly. Anionic surfactants also assist in leveling acid dyes, but they function differently; they rapidly form salts with the basic sites in the fiber and are replaced only gradually by the dye molecules.

Disperse Dyes on Acetate Rayon, Polyester, Polyamide and Acrylic Fibers. Disperse dyes are generally applied in the form of dispersions from an aqueous medium and are "developed" by thermal means. The mechanism of dyeing is considered to involve a solid solution of the dye in the hydrophobic fiber.

Disperse dyes are compounds of low water solubility and are nonionic in nature. As a group they are largely anthraquinone, mono- and disazo derivatives; they provide a complete shade range for most of the hydrophobic fibers. At first restricted to the dyeing of cellulose acetate, their use has grown steadily. They now find extensive use on nylon, particularly since the advent of tufted carpets. Nearly all polyester fabrics are dyed with disperse dyes by wet and "Thermosol" application methods. The use of relatively high temperatures in the dyeing and durable-press application steps required the development of sublimation-fast disperse dyes.

The large number of end uses, the introduction of convenience finishes, and extreme processing conditions have imposed stringent requirements on these dyes. In automotive fabrics they have to show resistance to heat and light, and in the dyeing of polyester they must not be affected by the "carriers" (used in exhaust dyeing) or by heat applied in pleating, heat setting, or dyeing. They have to resist high pH conditions in carpet dyeing and must be fast to gas-fume fading and ozone bleaching. The latter requirement is especially important in hot and humid climates.

Disperse dyes are useful in light and medium shades on acrylic and modacrylic fibers. Even polypropylene is sometimes dyed with disperse dyes, for instance when used as primary carpet backing in place of jute. Elastomeric fibers are also dyed with disperse dyes.

With the continuing introduction of new antisoil, antistatic and fire-retardant finishes, the number of requirements will increase and many new dyes will have to be synthesized by the chemist.

Basic Dyes on Acrylic, Modified Polyamide and Polyester Fibers. Most manufacturers of acrylic fibers offer types that will dye with basic or cationic dyes. The mechanism by which these dyes color the substrate is believed to entail the adsorption of the dye on the fiber surface, neutralizing the negative potential of the fiber. Upon raising the temperature the dye diffuses into the fiber and combines with acid groups in the fiber. Basic dyes exhaust quickly within narrow limits of temperature and it is normally necessary to use cationic retarders to achieve level dyeings, particularly for light and medium shades. A wide range of shades can be dyed with basic dyes and in general the brilliance and fastness of these colors are outstanding. Chemically the basic dyes belong to two main types, (1) those in which cationic site and chromophore are insulated from each other, and (2) those in which the cationic site is a part of the chromophore.

Important members of this class are the triarylmethanes and amino salts of azo dyes, but xanthenes, acridines, azines, oxazines and thiazines also have found commercial uses. Cyanine derivatives are employed extensively as photographic sensitizers.

Basic dyes may also be applied to acid-modified polyester and polyamide fibers. These fiber types have found great acceptance in the carpet industry, where they permit the production of multicolored effects on tufted carpets.

Reduced Dyes

Vat Dyes. Vat dyes are applied by the so-called vatting procedure, in which they are first reduced in alkaline solution with sodium hydrosulfite to form a water-soluble "leuco" derivative, then applied to the fiber (the "leuco" form shows substantivity) and oxidized with air or other mild oxidizing agents to precipitate the colored insoluble species in the fiber. Vat dyes of the anthraquinone series generally exhibit outstanding fastness to washing, light and bleaching agents; their only drawbacks are a relatively dull shade and high cost. Vat dyes of the indigoid and thioindigoid type do not afford the outstanding fastness properties of the anthraquinone derivatives, but they do find extensive use in textile application where price is as important as dye performance. Vat dyes are readily applied by continuous dyeing methods and are used extensively in the dyeing of blend fabrics such as polyester/cotton. Vat dyes are primarily used for the coloration of cotton and viscose rayon, but since the indigoid dyes require only a weakly alkaline vat, they can also be used for the dyeing of wool and silk. A versatile subgroup are the soluble vat dyes, which are stabilized leuco salts (sodium salts of sulfuric acid esters of leuco vat dyes). In addition to their water solubility these dyes exhibit practically no color and possess a low affinity for cellulosic fiber but high affinity for wool and silk. They are generally applied by padding and developed by acid oxidation, using sulfuric acid in combination with mild oxidizing agents such as sodium nitrite or potassium dichromate.

Sulfur Dyes. Sulfur dyes are water-insoluble pigments which are applied by a vatting technique. By alkaline reduction with sodium sulfide the dye is dissolved and applied to the fiber at elevated temperatures. The affinity of the reduced dye is not great and salt has to be added to obtain acceptable exhaustion. The leuco dyes have good leveling properties but care has to be taken that they do not oxidize prematurely. After the dyeing is complete, the insoluble original form of the sulfur dye is then regenerated in the fiber by oxidation. Sulfur colors are made by treating a wide variety of organic ompounds with sulfur or sulfur derivatives. The commercially important products are made from aminophenols, diphenylamines, indophenols, azines, carbazoles, and aminoaphthalenesulfonic acids. Structures of these dyes remain largely unidentified, but structural features such as mercapto, sulfide, disulfide and polysulfide groups have been identified and heterocyclic systems such as thiazole, thiadiazole and thiazine are often formed. The sulfur dyes tend to afford dull shades; because of their low cost and good fastness properties (except to chlorine) they have penetrated the market for dyeing cellulosic substrates and are produced in large amounts. Important technical advances have been reported in the production of sulfur dyes in the form of: (a) dispersions, (b) stable, clarified, reduced solutions, and (c) nonsubstantive thiosulfonic acid derivatives. These new types of sulfur dyes simplify the dyeing process and make it amenable to continuous application.

Wool and synthetic polyamides are not dyed with sulfur dyes, because the strong alkaline dyeing conditions would produce extensive chemical damage in the fiber by hydrolyzing amide linkages.

Ingrain Dyes. Ingrain dyes are all of those dyes which are formed *in situ* in the substrate by the development or coupling of one or more intermediate compounds which in themselves are not dyes. The principal subclasses are the azoic dyes and the oxidation dyes to be discussed below and the phthalocyanines, which are developed on the fiber by special treatments.

Azoic Dyes. The azoic dyes are water-insoluble mono- or disazo dyes produced *in situ* on the fiber. Because of their pigment-like structure they exhibit good washfastness and often very satisfactory light fastness. The azoic dyes are used primarily on cellulosic fiber but have found limited use also on wool, fur, acetate rayon, nylon and polyesters. A complete range of shades is available but the colors of greatest utility are the bright red, scarlet and orange shades. Commercial dyeing with azoic dyes is a two-step process. First a hydroxy coupling component is brought onto the fiber from an aqueous bath. The fiber subsequently is dried and then treated with a diazonium salt solution. In this latter step the azo pigment is formed *in situ* in the fiber. The coupling components primarily are naphthol derivatives. Because of the ease with which diazonium salts degrade they have to be applied at low temperature (ice colors is another name for azoics) or have to be employed in a "stabilized" form. These stabilized diazonium salts are easy to handle and have been used extensively in dyeing and printing.

The azoic dyes are less expensive than the quality vat dyes but their use is often difficult and this has limited their acceptance. Reproducibility of shade depth is a well-known problem with them and it is not practical to use mixtures of azoic combinations, since each diazonium salt in a mixture can react with every coupling component.

Oxidation Dyes. The principal dye of this class is aniline black, which is produced on the fiber by a two-step acid oxidation of aniline. In the first step emeraldine is generated by ageing or steaming, in the second step an ungreenable black is produced by further oxidation. The color has been used primarily on cotton and rayon and rarely on wool and silk.

Mordants

Basic Dyes on Cellulosic Fibers. Basic dyes are applied to cotton by a two-step process. The cloth is first mordanted with a solution of tannin, then dried and fixed (precipitation of antimony tannate). The basic color is now dyed on the fiber, the fiber's negative potential is neutralized on the surface allowing diffusion into the fiber and combination with the mordant. The common structural feature of the dyes is one or several positive charges, which are not necessarily localized on any specific site in the molecule (distributed charge).

The application of basic dyes to mordanted cotton today is essentially obsolete, since dyeings with equal brilliance and superior fastness can be obtained with fiber-reactive dyes.

Acid Dyes on Cotton. Acid dyes have little or no affinity for substrates such as cotton, yet may still be fixed if a mordant has been applied first. The mordanting (complex formation) is achieved with salts of metals such as aluminum and chromium which form a complex with the dye on the substrate. Often the self shade of the dye is changed slightly by the treatment and properties such as fastness to light and to washing are greatly improved.

The cheaper dyes of this type find use in paper dyeing, where they are applied to the surface of the finished sheet or by precipitation in the pulp with alum.

Mordant Acid Dyes on Wool. Mordant acid dyes (chrome dyes) provide excellent fastness on wool. They are applied by several methods. By one technique the wool is mordanted with sodium or potassium dichromate by boiling in the presence of the salt. After washing the wool, it is then dyed with the appropriate dye under acid conditions (bottom-chrome method). Another technique permits the simultaneous application of dye and mordant (autochrome method).

—Hermann Pohland, E. I. DuPont de Nemours & Co., Inc., Wilmington, Delaware.

DYES. These are intensely colored chemical compounds which when applied to a substrate impart color to this substrate. Retention of color as well as stability are required functional properties and are accomplished by chemical and physical forces such as chemical bonding, hydrogen bonding, Van der Waals forces, adsorption, solution, electrostatic interaction and others. The color of dyes is due to the interaction of visible light with the electron system of the dye molecule. Several hundred thousand known compounds qualify as dyes based on their light absorption in the visible region of the electromagnetic spectrum. However, of these only about 1500 have proved to be of practical value and are being manufactured. Commercial uses of dyes include the coloration of textiles, paper, leather, wood, inks, fuels, food items, and metals. Dyes are used also in photographic paper, as indicators in analysis, and as biological stains. Colors used in food products are described in the article on **Colorants (Foods)**. See immediately preceding article on **Dyeing**.

The functional properties required of commercially useful dyes vary markedly and depend to a great extent on the end use. For certain uses such as printing inks fastness properties are of secondary importance and the dyes are largely selected on the basis of economy. For other uses, e.g., carpets, resistance to light fading is more important than washfastness; the opposite is true for apparel, which is washed more frequently than it is exposed to an intensive source of light. A number of the synthetic fibers are dyed and finished at elevated temperatures (200°C) and all dyes present have to exhibit sublimation fastness. Since many of the fastness properties of the dyes are dependent upon the environment in which the dye molecule finds itself, dyes have to be carefully screened under many different dyeing conditions before they can be commercialized. With textile technology advancing towards more complicated blends of different fiber types, greater demands are made on new dyes. Often it is advantageous to dye each type of fiber in the fabric with a different kind of dye. It is of great importance in such a system to avoid cross staining (a fiber dyed with a dye not intended for it) and to work with dyes essentially free of impurities.

Dyes may be classified in various ways, according to color (yellow, red and blue, etc.), origin (natural or synthetic), chemical structure, substrates to which they are applied, and methods of application.

Classification of dyes based on their chemical structure is the most precise. However, for the user of coloring matters this is not always the best arrangement, since it makes no provision for dyes whose constitutions are unknown or have not been disclosed. An excellent reference is the "Colour Index," published periodically by AATCC and SDC (United Kingdom). This classifies dyes according to both structure and usage.

Nitroso Dyes (Quinone Oximes). The chromophore —N=O characterizes nitroso dyes. They are classified as mordant dyes and, except for the green iron lakes, have found very limited application in dyeing and printing. Usually they are prepared by reacting phenols or naphthols with nitrous acid. Other synthetic methods have also been reported, such as the simultaneous introduction of the nitroso and hydroxyl groups by the reaction of nitrosyl radical and an oxidizing agent.

Nitro Dyes. o- and *p*-Nitrophenols and *o-* and *p*-nitroanilines and their derivatives make up this class. Picric acid, one of the first synthetic dyes, is no longer employed as a textile colorant, but Naphthol Yellow S in form of its sodium salt is still used as a cheap dye for wool and silk giving pure yellow shades.

A general method of preparation of these dyes is by nitration of phenols, naphthols and diphenylamines. Many dyes have another chromophore in addition to the nitro group, but they are generally classed with the other dyes containing the second chromophore.

Wool and silk have been dyed with water-soluble acid nitro dyes in yellow and orange shades. Nitro dyes in the form of insoluble disperse dyes have served for the coloration of acetate, polyamide, polyester and similar fibers.

Azo Dyes. Azo dyes are characterized by the common chromophore the azo group —N=N— and generally have such auxochromic groups as hydroxyl, amine, and substituted amino groups. Azo dyes to a large extent are manufactured by reacting primary arylamines with nitrous acid in mineral acids (diazotization), yielding diazonium salts which are reacted, usually without isolation, with aromatic amines, phenols or enolizable ketones (coupling). The coupling with amines generally proceeds under acidic conditions, while phenols require alkaline media.

The process of diazotization and coupling can be repreated giving rise to disazo and polyazo dyes. The number of possible synthetic alternatives is so great that azo dyes easily form the largest class of synthetic dyes. Azo dyes have been developed for coloring every fiber, natural and synthetic, and for the coloration of solvents and a wide range of nontextile substrates.

The shade range of monoazo dyes covers all colors from greenish yellow to blue. However, there are only a few blue structures with the exception of metallized dyes, which generally show a bathochromic shift compared to the corresponding unmetallized dyes. Additional azo groups in the molecule produce a change to predominantly dark colors such as browns, navies and blacks.

Azoic Dyes. There is no fundamental chromophoric difference between azo and azoic dyes. The differentiation is made to characterize a group of azo pigments which are precipitated within the cellulosic fiber by carrying out the dye coupling on the fiber. Good functional properties and a wide range of brilliant shades can be attained with these dyes, but with the advent of the equally brilliant but more easily applied fiber-reactive dyes the azoics have lost some of their importance. A few of the naphthol azoics still find use in discharge and resist printing.

Stilbene Dyes. This group of dyes has a limited range of shades and is essentially confined to yellows, oranges and browns; they result from the condensation of 5-nitro-*o*-toluenesulfonic acid in alkaline medium with other aromatic compounds, generally arylamines. Most of the structures have not been elucidated and most commercial stilbene dyes are mixtures rather than single compounds.

The stilbene dyes are essentially all direct dyes with a few exceptions where these colors find use as acid and solvent dyes.

Diphenylmethane Dyes. Diphenylmethane dyes are basic colors which exhibit relatively poor fastness properties. The common feature is the chromophore $\diagdown C = NH$, which shows sensitivity to hydrolysis. The best known members of this group are the Auramines, used on wool, cotton, paper, leather, silk and jute.

Triarylmethane Dyes. The triarylmethane dyes are among the oldest synthetic dyes. The chromophore of this class is the quinonoid grouping, which may appear as

where Ar symbolizes a substituted aryl group. The color and properties depend on the kind and numbers of the substituents. The presence of the sulfonic acid group confers water solubility and such dyes are applied as acid dyes. In the absence of acidic groups the dyes are called cationic or basic dyes and, where hydroxyl groups are present as auxochromes, adjacent carboxy groups confer mordant dyeing properties. Thus the class includes basic, direct, acid, mordant and solvent dyes. Nearly all are of brilliant hue, the range running from reds and violets to blues and greens. Fastness to light is generally poor to fair, except on polyacrylonitrile, where it is outstanding.

Xanthene Dyes. Xanthene dyes have the heterocyclic ring system of xanthene in common. The properties of the individual members depend largely on the substituents on the ring system and the class is subdivided into amino, aminohydroxy, and hydroxy derivatives. Some of the xanthene dyes resemble triarylmethane dyes in that they exhibit unusual brilliance and basic dye properties such as the following CI Basic Red.

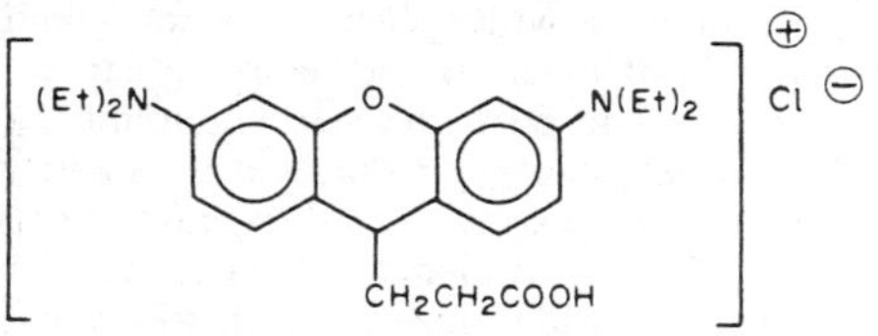

Several xanthene dyes are of commercial importance as cationics for polyacrylonitrile. Their use in the dyeing of wool and silk from weak acid baths and cotton on a tannin mordant is very limited because fastness properties on these substrates are generally poor, especially fastness to light.

Acridine Dyes. Acridine Orange R is a typical member of this class of dyes which incorporates mostly basic dyes of yellow, orange, red and brown shade. The dyes have found extensive use on leather. Their use on silk, wool or cotton, however, is restricted due to their lack of adequate fastness.

Quinoline Dyes. The condensation of methylquinoline and its derivatives with phthalic anhydride affords quinophthalones of the following structure:

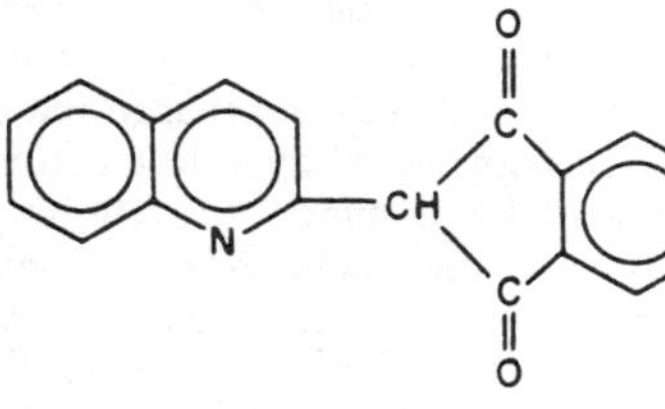

2 – (2 – Quinolyl) –1, 3 – Indanedione

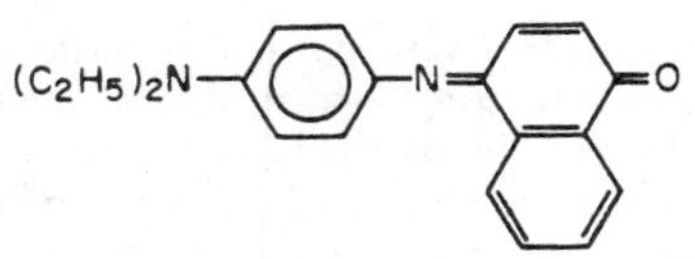

Indophenol
(CI Solvent Blue 22)

These dyes, mainly yellow or red compounds, may be applied as paper, food, solvent, basic and disperse dyes; when sulfonated they give excellent acid dyes for wool.

Methine Dyes. Methine dyes, also known as cyanines, are characterized by the presence of the methine group —CH= or a conjugate chain of such groups. The majority of the dyes contain heterocyclic systems such as quinoline, benzothiazole or trimethylindoline, and are formed by linking these heterocycles together by means of a chain of methine groups.

The main use of these dyes is as photographic sensitizers. In general they have poor fastness to light and therefore have limited use as basic dyes on textile fibers. One notable exception is CI Disperse Yellow 31, a dye of exceptional fastness to light and gas fume fading.

Thiazole Dyes. The parent substance of this group of dyes is dehydrothio-*p*-toluidine, which upon sulfurization at higher temperatures gives rise to the so-called Primuline Bases, which can be sulfonated to afford direct cotton dyes. They also can be diazotized and coupled with amines, phenols, and naphthols to produce azo dyes of yellow, orange and red shades. Primuline is used to color cotton where washfastness but not fastness to light is the primary consideration. Hypochlorite oxidation of the dehydrothio-p-toluidine, on the other hand, produces dyes of relatively high light fastness.

Indamine and Indophenol Dyes. Oxidation of mixtures of an aromatic *p*-diamine with an aromatic monoamine affords the indamines, while indophenols result from oxidation of mixtures of *p*-amino-phenol and phenol.

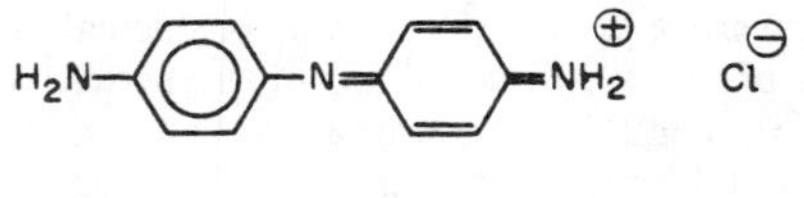

Indamine
(Phenylene Blue)

The functional properties of these dyes make application to fibers impractical. However, they have found use in photography as well as in the preparation of sulfur dyes.

Azine Dyes. Azine dyes are derivatives of the heterocyclic phenazine system. They are generally basic dyes, have been used on wool and silk, and dye cotton on a tannin mordant. Sulfonated derivatives are acid dyes and nonsulfonated compounds are used to color fats, lacquers, and oils. The most famous but now obsolete member of this dye class is Mauveine.

Oxazine and Thiazine Dyes. Oxazine dyes are derived from the heterocyclic system of the same name and are manufactured by condensing *p*-nitroso-N,N-dimethylaniline with suitable phenols. Most oxazine dyes are classified as basic colors for polyacrylonitrile, wool and cotton (on a tannin mordant). These dyes have also been used on leather and, after sulfonation, as direct dyes. Among the latter group are several brilliant blue dyes with excellent fastness to light.

The thiazine dyes are closely related, their structure including the thiazine ring. Several members of this group are used as stains for tissues in biology and medicine. As coloring matters for textile fibers they are of less importance.

Sulfur Dyes. Sulfur dyes are complex mixtures of uncertain composition and, while certain structural features such as the thioketone, the disulfide, the thiazole group have been identified, it has not been possible to assign precise chromophores. Sulfur dyes are used extensively on cellulosic fibers because of their low price and relatively good fastness. They are applied by a method similar to the vatting process. Since they are water-insoluble compounds, they are rendered water-soluble by reduction with sodium sulfide. In this form they are applied to the cotton and then are reoxidized (in air) to the original dye. Sulfur dyes are produced by heating relatively common aromatic compounds with sulfur or sulfur derivatives. The usual classification of sulfur dyes is based on the type of starting material such as amines, phenols, nitro compounds, and carbazoles.

The shade range provided by sulfur colors spans the spectrum from yellow to black. However, most of them are fairly dull and their great asset is their cheapness.

Lactone Dyes. The group of lactone dyes is a very small one. They are prepared by oxidation or hydrolysis of polyhydroxy aromatic compounds, such as gallic acid to produce Alizarine Yellow (MLB).

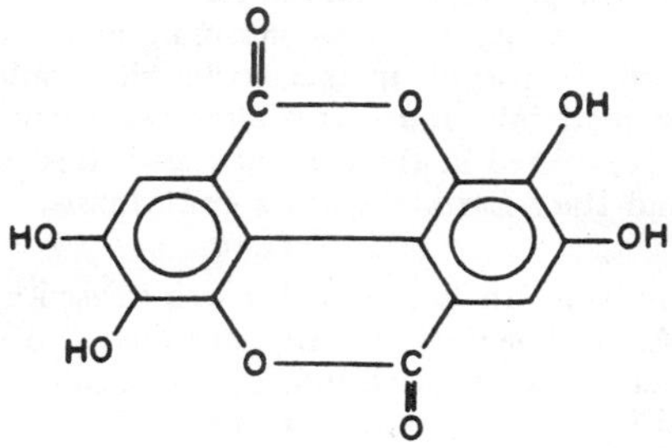

CI 55005 Alizarine Yellow (MLB)

The shade range of these colors is limited to yellows and olives and their application is restricted to use on chrome-mordanted wool.

Aminoketone and Hydroxyketone Dyes. These two groups of dyes derive their color from the carbonyl chromophore and amino and hydroxy groups as auxochromes. The amino-ketone dyes are generally arylaminoquinones and are applied to wool by a vatting procedure. The hydroxyketone dyes are mostly hydroxyquinones and hydroxy derivatives of aromatic ketones and are applied by mordanting techniques.

Anthraquinone Dyes. Anthraquinone dyes derive their color from one or more carbonyl groups in association with a conjugated system. Auxochromes include amine, hydroxyl, alkylamino, arylamine, and acylamino groups, as well as complex heterocyclic system. Anthraquinone dyes are found in many different usage groups, the more important ones being the acid, direct, disperse, mordant, solvent, vat, pigment and reactive dyes. The hydroxy derivatives of anthraquinone, such as alizarin (the coloring matter of Madder), exist naturally.

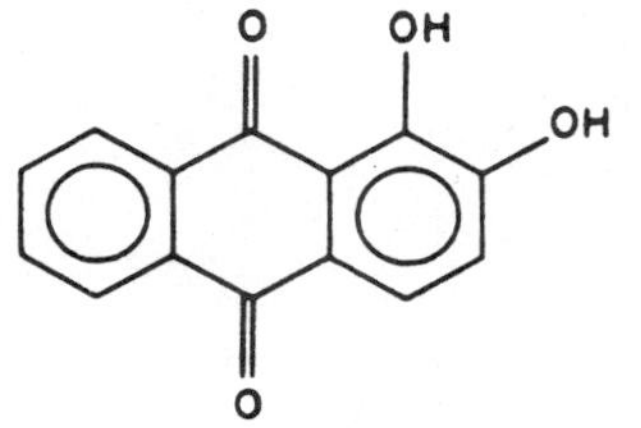

Alizorin

After the discovery of its constitution many other hydroxyanthraquinones and their derivatives were introduced as commer-

cial mordant dyes. Anthraquinone acid dyes are usually sulfonated arylamino-anthraquinones. As a group they exhibit good fastness to light and wet treatment and provide bright shades on wool and synthetic polyamides. An important widely used example is CI Acid Blue 47.

Of ever-increasing importance are the anthraquinone disperse dyes. Structurally they are fairly simple, solubilizing groups such as the sulfonic acid group being absent. Initially these dyes were limited to use on acetate fiber; with the spectacular growth of synthetic fibers after World War II, however, production of disperse dyes increased rapidly. A well-known member of this group is CI Disperse Red 11.

The first of the anthraquinone vat dyes was indanthrone, discovered in 1901. A wide range of vat dyes followed over the years, highlighted by the introduction of a bright green (Caledon Jade Green) in 1920 and the commercialization of stable leuco dye solutions in 1921. As a class, vat dyes exhibit excellent fastness to light, washing and bleaching. They are applied in the reduced form which is soluble under strong alkaline conditions and then are reoxidized to their water-insoluble colored pigment form by use of air or other mild oxidizing agents. The alkaline dyeing conditions have confined the use of vat dyes essentially to cellulosic fibers. However, the development of stable, neutral-dyeing leuco esters has broadened the applicability of the vat dyes to include wool, silk and nylon.

Indigoid Dyes. This class of dye is characterized by the chromophore

$$-\overset{\displaystyle O}{\overset{\displaystyle \|}{C}}-\underset{\displaystyle |}{C}=\underset{\displaystyle |}{C}-\overset{\displaystyle O}{\overset{\displaystyle \|}{C}}-. $$ The auxochromes of these dyes

are —NH— groups or —S— atoms. The dyes may be symmetrical or unsymmetrical. The symmetrical dyes are generally produced by oxidative coupling, while the unsymmetrical ones result from condensation reactions of suitable molecules. Most indigo dyes are readily reduced in mildly alkaline medium and this makes them applicable not only to cellulose but also to wool and silk. The introduction of halogen, alkyl and alkoxy substituents gives rise to numerous derivatives of indigo and its analogs, many of which have gained commercial importance.

Phthalocyanine Dyes. The tetrabenzoporphyrazine nucleus is one of the more recently discovered chromophores and represents the common feature of the phthalocyanines. This group of dyes has produced the most brilliant known blues and greens. Many phthalocyanines are metallized such as in CI Pigment Blue 15 or copper phthalocyanine

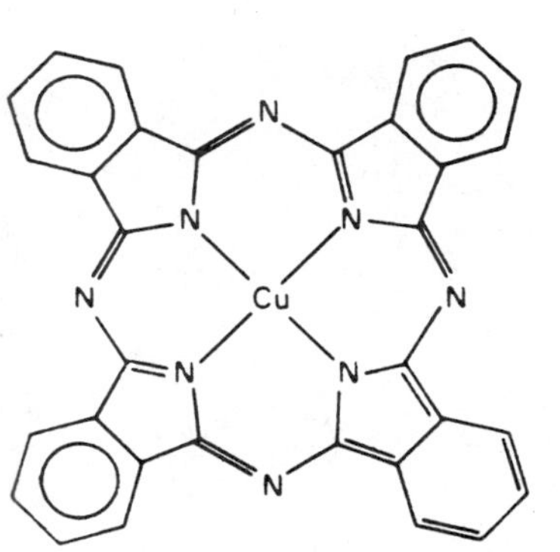

Copper Phthalocyanine

and have set new standards for brilliance and fastness properties. While most of the phthalocyanines are pigments in respect to their application properties, dye chemists have been successful in synthesizing a few vat, direct, solvent and fiber-reactive dyes based on the same chromophore.

The parallel development and interdependence of synthetic dyes and organic chemistry are well known. Many of the reactions pioneered in dye chemistry have proved fruitful in such areas as heterocyclic and steroid chemistry. Today interest in dyes and their synthesis is almost entirely confined to the research departments of large dye manufacturers.

Advances in organic synthesis and reaction mechanisms, as much as development of new or modified fibers, will determine the direction the research on synthetic dyes will take in the future.

—Hermann Pohland, E. I. DuPont de Nemours & Co., Inc., Wilmington, Delaware.

DYSPROSIUM. Chemical element symbol Dy, at. no. 66, at. wt. 162.50, ninth in the Lanthanide Series in the periodic table, mp 1412°C, bp 2567°C, density 8.551 g/cm³ (²⁰°C). Elemental dysprosium has a close-packed hexagonal crystal structure at 25°C. The pure metallic dysprosium is silver-gray in color and retains its luster at room temperature. Although stable up to approximately 400°C, the metal then oxidizes at a slow rate up to 600°C. Because of its comparative softness, the metal can be worked by conventional equipment to form rod, foil, and ribbon configurations. There are seven natural isotopes ^{156}Dy, ^{158}Dy, ^{160}Dy through ^{164}Dy. Twelve artificial isotopes have been identified. In terms of abundance, dysprosium is present on the average of 3 ppm in the earth's crust and is ranked 42nd in terms of total abundance, thus making it potentially more available than tin or beryllium. The element first was identified by Lecoq de Boisbaudran in 1886. The metal has a low acute-toxicity rating. Electronic configuration

$$1s^2 2s^2 p^6 3s^2 3p^6 3d^{10} 4s^2 4p^6 4d^9 4f^{10} 5s^2 5p^6 5d^1 6s^2$$

Ionic radius Dy³⁺ 0.908 Å. Metallic radius 1.775 Å. First ionization potential 5.93 eV; second 11.67 eV. Dysprosium appears to be exclusively trivalent. Other important physical properties of dysprosium are given under **Rare-Earth Elements and Metals.**

Dysprosium occurs in apatite, euxenite, gadolinite, and xenotime. All of these minerals also are processed for their yttrium content. With liquid-liquid organic and solid-resin organic ion-exchange techniques, the separation of dysprosium from yttrium is favorable.

Because dysprosium has a high neutron-absorbing capability, the metal, usually in the form of foil, is used for detecting and measuring nuclear reactions and exposures. Dy_2O_3 also is used in dosimeters. Dysprosium does not emit harmful decay radiations when it is used as a neutron sponge. Further, helium-generated alpha particles which may ultimately crack structural parts for containing nuclear fuel are not generated. Thus, stainless steel containing about 3% Dy_2O_3 sometimes is used in control rod hardware for high-flux-beam reactors. Although Dy_2O_3 catalyzes the polymerization of ethylene and other synthetic reactions, the cost to date has limited these uses. Since dysprosium oxide fluoresces yellow in glass under ultraviolet radiation, it has been considered as an activator for the yellow component of the phosphors used in black-and-white television picture tubes. Investigations are continuing in connection with future uses of dysprosium in thermoelectric, semiconducting, photoelectric materials, and garnet microwave devices.

See references listed at the ends of the entries on **Chemical Elements;** and **Rare-Earth Elements and Metals.**

NOTE: This 4th Edition entry was revised and updated by K. A. Gschneidner, Jr., Director, and B. Evans, Assistant Chemist, Rare-Earth Information Center, Energy and Mineral Resources Research Institute, Iowa State University, Ames, Iowa.

E

EBULLIOMETER. An instrument, sometimes referred to as an *ebullioscope*, that measures the property of a substance by noting a deviation from a normal known boiling point. The term applies to apparatus for estimating the percentage of alcohol in a mixture through observation of the boiling point. Beckmann's apparatus for molecular weight determination is an ebullioscope. The *ebullioscopic constant* is a quantity calculated to represent the molal elevation of the boiling point of a solution by the relationship:

$$K = \frac{RT_0^2}{1000 \, l_e}$$

in which K is the ebullioscopic constant, R is the gas constant, T_0 is the boiling point of the pure solvent, and l_e is the latent heat of evaporation per gram. The product of the ebullioscopic constant and the molality of the solution gives the actual elevation of the boiling point for the range of values for which this relationship applies. The range is limited to very dilute solutions, not extending to solutions of unit molality.

EFFLORESCENCE. When a substance evolves moisture upon exposure to the atomosphere, the substance is said to be *efflorescent*, and the phenomenon is known as *efflorescence*. If a substance has a higher water vapor pressure than that corresponding to the atmosphere at the given temperature, water vapor is evolved from the substance until the water vapor pressure of the substance equals the water vapor pressure of the surrounding atmosphere. Substances that are ordinarily efflorescent are sodium sulfate decahydrate, sodium carbonate decahydrate, magnesium sulfate heptahydrate, and ferrous sulfate heptahydrate. When the saturated solution of a substance in water has a water vapor pressure greater than that of the surrounding atmosphere, evaporation of the water from solution takes place. See **Deliquescence** for converse phenomenon.

EINSTEINIUM. Chemical element symbol Es, at. no. 99, at. wt. 254 (mass number of the most stable isotope), radioactive metal of the actinide series; also one of the transuranium elements. Both einsteinium and fermium were formed in a thermonuclear explosion which occurred in the South Pacific in 1952. The elements were identified by scientists from the University of California's Radiation Laboratory, The Argonne National Laboratory, and the Los Alamos Scientific Laboratory. It was observed that very heavy uranium isotopes which resulted from the action of the instantaneous neutron flux on uranium (contained in the explosive device) decayed to form Es and Fm. The probable electronic configuration of Es is $1s^2 2s^2 2p^6 3s^2 3p^6 3d^{10} 4s^2 4p^6 4d \ 10 4f^{14} 5s^2 5p^6 5d^{10} 5f^{11} 6s^2 6p^6 7s^2$. Ionic radius: Es^{3+} 0.97 Å.

All known isotopes of einsteinium are radioactive. The first evidence of their existence was obtained by ion-exchange methods applied to coral rocks obtained from Eniwetok Atoll after the thermonuclear explosion. The first pure isotope found was ^{253}Es, produced by prolonged treatment of plutonium-239 with neutrons in the Arco, Idaho, Testing Reactor. The most stable is ^{254}Es, half-life 270 days, and therefore the mass number 254 is carried in the atomic weight table. Others include ^{245}Es–^{246}Es, ^{248}Es–^{252}Es, and ^{253}Es, ^{255}Es.

The ion Es^{3+} is stable. The isotopes of mass numbers 245, 252, 253 and 254 decay by alpha-particle emission; that of mass number 250 by electron capture, those of mass numbers 246, 248, 249 and 251 by both of these processes, while those of mass numbers 255 and 256 emit electrons to form the corresponding fermium isotopes.

Sufficient einsteinium, produced through intense neutron bombardment of plutonium-239 in a reactor, was not available until 1961 to allow separation of a macroscopic amount. Cunningham, Wallmann, L. Phillips, and Gatti worked with submicrogram quantities to separate a small fraction of pure einsteinium-235 compound and measure its magnetic susceptibility. Only a few hundredths of a microgram were available at that time.

References

Cunningham, B. B., Peterson, J. R., Baybarz, R. D., and T. C. Parsons: "The Absorption Spectrum of Es^{3+} in Hydrochloric Acid Solutions," *Inorg. Nucl. Chem. Lett.*, 3, 519–523 (1967).

Cunningham, B. B., and T. C. Parsons: "Preparation and Determination of the Crystal Structure of Californium and Einsteinium Metals," *Lawrence Berkeley Laboratory Nuclear Chemistry Annual Report*, UCRL–20426, University of California, Berkeley, California (1970).

Fields, P. R., et al.: "Additional Properties of Isotopes of Elements 99 and 100," *Phys. Rev.*, 94(1), 209–210 (1954).

Fujita, D. K., Cunningham, B. B., and T. C. Parsons: "Crystal Structures and Lattice Parameters of Einsteinium Trichloride and Einsteinium Oxychloride," *Inorg. Nucl. Chem. Lett.*, 5(4), 307–313 (1969).

Ghiorso, A., et al.: "New Elements Einsteinium and Fermium, Atomic Numbers 99 and 100," *Phys. Rev.*, 99(3), 1048–1049 (1955).

Seaborg, G. T., Editor: "Transuranium Elements," Dowden, Hutchinson & Ross, Stroudsburg, Pennsylvania (1978).

Studier, M. M., et al.: "Elements 99 and 100 from Pile-Irradiated Plutonium," *Phys. Rev.*, 93(6), 1428 (1954).

Thompson, S. G., Harvey, B. G., Choppin, G. R., and G. T. Seaborg: "Chemical Properties of Elements 99 and 100," *American Chemical Society Journal*, 76, 6229–6236 (1954).

ELASTOMERS. Of natural or synthetic origin, an elastomer is a polymer possessing elastic (rubbery) properties. A polymer is a substance consisting of molecules which are, in the most part, multiples of low-molecular-weight units, or monomers. As an example, isoprene (2-methylbutadiene-1,3) is C_5H_8, whereas polyisoprene is $(C_5H_8)_x$, where $x \geq 2$ and normally is from 1,000 to 10,000 for rubbers. Although they differ in composition from natural rubber, many of these high-molecular-weight materials are termed *synthetic rubbers*. See also **Rubber (Natural)**.

Synthetic rubbers fall into two major classifications; (1) general-purpose rubbers, but nevertheless the major volume of which is used for tire production; and (2) specialty rubbers that essentially find little use in tires, but that are important for a number of other uses. Synthetic rubbers have not replaced natural rubber for numerous uses. For large, heavy-duty truck and bus tires, natural rubber tends to run considerably cooler and wears better than a blend of natural and synthetic rubbers. On the other hand, a tire tread made of a blend of styrene-butadiene (SBR) and butadiene rubber (polybutadiene) wears longer than natural rubber in conventional automobile usage, where lower temperatures can be maintained.

Styrene-Butadiene (SBR) Rubbers. This series of rubbers includes monomer ratios up to about 50% styrene. The addition of more than 50% styrene makes the materials more like plastic than rubber. The most commonly used SBR rubbers contain about 25% styrene, which is polymerized in emulsion systems at 5–10°C. Most SBR goes into tires, but the type for the tread differs from that of the sidewall or carcass. SBRs for adhesives, shoe soles, and other products also differ. The formulation permits vast varieties of end-products. Among the processing variables that can be manipulated to provide different end-characteristics are temperature, viscosity, use of different emulsifiers and solvents, use of different antioxidants for stabilization, different oils, carbon blacks, and coagulation techniques.

Initial processes for emulsion (E-SBR) called for polymerization at 50°C. However, it was found later that cold processing produced a rubber particularly good for tires. This is sometimes referred to as *cold SBR.* The overall process of making SBR is shown in the accompanying figure. Typical formulations are:

| | Parts by Weight | |
	Hot SBR	*Cold SBR*
Deionized water	180	200
Fatty acid soap	5	4.5
Styrene	25	28
t-Dodecyl-mercaptan	0.35	0.2
Butadiene	75	72
Potassium persulfate	0.3	
Redox initiator	——	small quantity
Temperature	50°C	5°C
Time	12 hours	8 hours
Conversion	75%	60%

Polymerization of emulsion SBR is started by free radicals generated by the redox system in cold SBR and by persulfate or other initiator in hot SBR. The initiators are not involved in the molecular structure of the polymers. Almost all molecules are terminated by fragments of the chain transfer agent (a mercaptan). Schematically, the molecules are RSM_nH, where RS is the $C_{12}H_{25}S$ part of a dodecyl mercaptan molecule; M is the monomer involved; n is the degree of polymerization, and H is a hydrogen atom formerly attached to the sulfur of a mercaptan. In the case of free-radical-initiated polymerization of butadiene, by itself to form homopolymers or with other monomers for form copolymers, the butadiene will be about 18%; 16% *cis*-1,4; and 66% *trans*-1,4.

Considerable quantities of SBR latex are used in the manufacture of foam rubber, adhesives, fabric treating, and paints. The solids content of latices runs from 50% to 65–70%.

Solution (S-SBR) consists of styrene butadiene copolymers prepared in solution. A wide range of styrene-butadiene ratios and molecular structures is possible. Copolymers with no chemically detectable blocks of polystyrene constitute a distinct class of solution SBRs and are most like styrene-butadiene copolymers made by emulsion processes. Solution SBRs with terminal blocks of polystyrene (S-B-S) have the properties of self-cured elasto-

mers. They are processed like thermoplastics and do not require vulcanization. Lithium alkyls are used as the catalyst.

Stereospecific solution polymerization has been emphasized since the discovery of the complex coordination catalysts that yield polymers of butadiene and isoprene having highly ordered microstructures. The catalysts used are usually mixtures of organometallic and transition-metal compounds. An example of one of these polymers is *cis*-1,4-polybutadiene, the bulk of which is used in tires. However, it must be blended with other materials because of its poor processability and traction.

Butyl Rubber. Known as IIR, butyl rubber is a copolymer of isobutylene and isoprene. The elastomers contain only 0.5–2.5 mole % of isoprene. This is introduced to effect sufficient unsaturation to make the rubber vulcanizable. Polymerizations are usually carried out at low temperature (−80 to 100°C) with methyl chloride as solvent. Anhydrous aluminum chloride and a trace of water serve as catalyst.

Butyl is one of the earlier synthetic rubbers. However, it lost favor after the development of SBR. Butyl rubber is incompatible with natural rubber and is difficult to cure. Chlorobutyl rubbers, a much more recent development and containing up to 1.3% chlorine, apparently do not exhibit these former problems. Major uses for butyl rubber have been inner tubes for tires. The appearance of tubeless tires, however, depressed this market. Butyl rubber makes excellent motor mounts because of its high energy absorption and low rebound. Essentially free of double-bonds, butyl rubber has a high resistance to aging, attractive properties for use in curing bags for tire production and in outside coating materials.

Acrylonitrile-Butadiene Rubers (NBR). Except for the monomers used, the production of NBRs is quite similar to that described for the SBRs. The NBR family is sometimes referred to as the *nitrile rubbers.* The acrylonitrile-butadiene ratios cover a wide range from 15:85 to 50:50. NBRs are noted for their solvent resistance, increasing with the acrylonitrile content. Thus, they are used for gaskets and oil and gasoline hoses, solvent-resistant electrical insulation, and food-wrapping films. Nitrile latices also are used in treating fabrics for dry-cleaning durability. Because the NBRs become quite inflexible (stiff) at low temperatures (actually brittle at about −20°C), they are blended with polyvinyl chloride for some applications.

Neoprene. This family of dry rubbers and latices was introduced in 1932, called *Duprene* by Du Pont at that time. The material is made by the free-radical-initiated polymerization of chloroprene in emulsion systems. As with most synthetic rubbers, a variety of neoprenes is made possible by variation of the polymerization conditions and ingredients. Neoprene is particularly good for its fire-retardant, solvent-resistant, and high-temperature stability properties. The chlorine in each segment deactivates the adjoining carbon-carbon double-bond, thus making it less sensitive to oxidative attack. Metal oxides, such as zinc oxide and magnesium, serve as curatives rather than sulfur.

Polyurethanes. Polyethylene in solution is treated with chlorine and sulfur dioxide to introduce approximately 1.3% sulfur and 29% chlorine into the polymer. Most of the chlorine is attached directly to the carbon atoms in the backbone of the polymer. The remainder is in the form of sulfuryl chloride groups, $\cdot SO_2Cl$, through which cross-linking occurs in the curing step with metal oxides. The material has good oxidation and ozone resistance and thus overall excellent weather resistance. Calendered stocks are used for lining ditches and ponds, for example.

Thiokol® Rubbers. These are polysulfide rubbers and are prepared by the condensation polymerization of sodium polysulfides with a dichloro (sometimes blended with a trichloro) or-

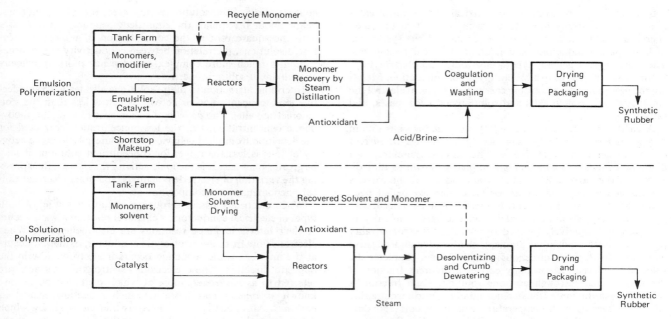

Synthetic rubber production process.

ganic compound. Type A, the first family of rubbers, was made from Na_2S_4 and ethylene dichloride. Thiokols are known for high resistance to organic solvents.

Polyacrylate Elastomers. These are made in emulsion or suspension systems involving the copolymerization of ethyl acrylate with the acrylate esters of higher-molecular-weight alcohols. These materials have excellent solvent-resistant properties and stability at elevated temperatures. A major use is for automatic-transmission gaskets for automobiles.

Silicone Elastomers. These materials have alternating Si and O atoms for a backbone and the members differ mainly in the nature of the organic substituents on the Si atoms and the degree of polymerization. Because of the absence of double-bonds in the backbone, the numerous forms of stereoisomers found in unsaturated hydrocarbon rubbers do not have counterparts in the silicone rubbers. The chemical combination of organic and inorganic materials gives the silicone rubbers useful properties over a wide temperature range (-70 to $+225°C$). These rubbers are well known for excellent dielectric stability and high resistance to weathering, oils and chemicals.

Fluoroelastomers. These materials are prepared by emulsion co-polymerization of perfluoropropylene and vinylidene fluoride; or of chlorotrifluoroethylene and vinylidene fluoride. Also there are fluorosilicones in this family. Useful at temperatures up to and over $300°C$, the fluoroelastomers have excellent resistance to aromatic solvents, acids, and alkalies. They are also among the more costly of the available commercial elastomers.

Ethylene-Propylene Elastomers. Known as EPR, this material is of limited use because it cannot be vulcanized in readily available systems. However, the rubbers are made from low-cost monomers, have good mechanical and elastic properties, and outstanding resistance to ozone, heat, and chemical attack. They remain flexible to very low temperatures (brittle point about $-95°C$). They are superior to butyl rubber in dynamic resilience.

ELECTROCAPILLARITY. The surface tension between two conducting liquids in contact, wuch as mercury and a dilute acid, is sensibly altered when an electric current passes across the interface. As a result, when the contact is in a capillary tube, the pressure difference on the opposite sides of the meniscus is affected by a current traversing the capillary column, to an extent dependent upon the direction of the current across the boundary.

ELECTROCHEMISTRY. That branch of science which deals with the interconversion of chemical and electrical energies, i.e., with chemical changes produced by electricity as in electrolysis or with the production of electricity by chemical action as in electric cells or batteries. The science of electrochemistry began about the turn of the eighteenth century.

Background. In 1796, Alessandro Volta observed that an electric current was produced if unlike metals separated by paper or hide moistened with water or a salt solution were brought into contact. Volta used the sensation of pain to detect the electric current. His observation was similar to that observed ten years earlier by Luigi Galvani who noted that a frog's leg could be made to twitch if copper and iron, attached respectively to a nerve and a muscle, were brought into contact.

In his original design Volta stacked couples of unlike metals one upon another in order to increase the intensity of the current. This arrangement became known as the "voltaic pile." He studied many metallic combinations and was able to arrange the metals in an "electromotive series" in which each metal was positive when connected to the one below it in the series. Volta's pile was the precursor of modern batteries.

In 1800, William Nicholson and Anthony Carlisle decomposed water into hydrogen and oxygen by an electric current supplied by a voltaic pile. Whereas Volta had produced electricity from chemical action these experimenters reversed the process and utilized electricity to produce chemical changes. In 1807, Sir Humphry Davy discovered two new elements, potassium and sodium, by the electrolysis of the respective solid hydroxides, utilizing a voltaic pile as the source of electric power. These electrolytic processes were the forerunners of the many industrial electrolytic processes used today to obtain aluminum, chlorine, hydrogen, or oxygen, for example, or in the electroplating of metals such as silver or chromium.

Since in the interconversion of electrical and chemical energies, electrical energy flows to or from the system in which chemical changes take place, it is essential that the system be, in large part, conducting or consist of electrical conductors. These are of two general types—electronic and electrolytic—though some materials exhibit both types of conduction. Metals are the most common electronic conductors. Typical electrolytic conductors are molten salts and solutions of acids, bases, and salts.

A current of electricity in an electronic conductor is due to a stream of electrons, particles of subatomic size, and the current causes no net transfer of matter. The flow is, therefore, in a direction contrary to what is conventionally known as the "direction of the current." In electrolytic conductors, the carriers are charged particles of atomic or molecular size called *ions*, and under a potential gradient, a transfer of matter occurs.

An electrolytic solution contains an equivalent quantity of positively and negatively charged ions whereby electroneutrality prevails. Under a potential gradient, the positive and negative ions move in opposite directions with their own characteristic velocities and each accordingly carries a different fraction of the total current through any one solution. Each fraction is referred to as the ionic transference number. Furthermore, the velocity increases with temperature causing a corresponding increase in electrolytic conductivity. This characteristic is opposite to that observed for most electronic conductors which show less conductivity as their temperature is increased.

The concept that charged particles are responsible for the transport of electric charges through electrolytic solutions was accepted early in the history of electrochemistry. The existence of ions was first postulated by Michael Faraday in 1834; he called negative ions "anions" and positive ones "cations." In 1853, Hittorf showed that ions move with different velocities and exist as separate entities and not momentarily as believed by Faraday. In 1887, Svante Arrhenius postulated that solute molecules dissociated spontaneously into *free ions* having no influence on each other. However, it is known that ions are subject to coulombic forces, and only at infinite dilution do ions behave ideally, i.e., independently of other ions in the solution. Ionization is influenced by the nature of the solvent and solute, the ion size, and solute-solvent interaction. The dielectric constant and viscosity of the solvent play dominant roles in conductivity. The higher the dielectric constant, the less are the electrostatic forces between ions and the greater is the conductivity. The higher the viscosity of the solvent, the greater are the fractional forces between ions and solvent molecules and the lower is the electrolytic conductivity.

In 1923, Debye and Hückel presented a theory which took into account the effect of coulombic forces between ions. They introduced the concept of the ion atmosphere, in which at some radial distance r from a central ion, there is, on a time average, an ionic cloud of opposite charge which sets up a potential field whose magnitude depends on the magnitude of r. This interionic attraction leads to two effects on the electrolytic conductivity. Under a potential gradient, an ion moves in a certain direction. However, the ion cloud, being of opposite sign will tend to move in the opposite direction, and because of its attraction for the central ion, will have a retarding effect on the ion velocity and thereby lead to a lowering in the electrolytic conductivity. On the other hand, the central ion will tend to pull the ion cloud with it to a new location. The ion atmosphere will adjust to its new location in time, but not instantaneously, and the delay results in a dissymmetry in the potential field around the ion. This also causes a lowering in the conductance of the solution. These effects become more pronounced as the concentration of the solution is increased; for dilute solutions, below about 0.1 molal, the equivalent conductance decreases with the square root of the concentration. For more concentrated solutions, the relation between conductivity and concentration is much more complex and depends more specifically on individual solute properties.

Interionic attraction in dilute solutions also leads to an effective ionic concentration or activity which is less than the stoichiometric value. The *activity* of an ion species is its thermodynamic concentration, i.e., the ion concentration corrected for the deviation from ideal behavior. For dilute solutions the activity of ions is less than one, for concentrated solutions it may be greater than one. It is the ionic activity that is used in expressing the variation of electrode potentials, and other electrochemical phenomena, with composition.

When electricity passes through a circuit consisting of both types of electrical conductors, a chemical reaction always occurs at their interface. These reactions are electrochemical. When electrons flow from the electrolytic conductor, oxidation occurs at the interface while reduction occurs if electrons flow in the opposite direction. These electronic-electrolytic interfaces are referred to as *electrodes*; those at which oxidation occurs are known as *anodes* and those at which reduction occurs, as *cathodes*. An anode is also defined as that electrode by which "conventional" current enters an electrolytic solution, a cathode as that electrode by which "conventional" current leaves. Positive ions, for example, ions of hydrogen and the metals, are called *cations* while negative ions, for example, acid radicals and ions of nonmetals are called *anions*.

Laws of Electrolysis. In 1833, Faraday enunciated two laws of electrolysis which give the relation between chemical changes and the product of the current and time, i.e., the total charge (coulombs passed through a solution. These laws are: (1) the amount of chemical change, e.g., chemical decomposition, dissolution, deposition, oxidation, or reduction, produced by an electric current is directly proportional to the quantity of electricity passed through the solution; (2) the amounts of different substances decomposed, dissolved, deposited, oxidized, or reduced are proportional to their chemical equivalent weights. A chemical equivalent weight of an element or a radical is given by the atomic or molecular weight of the element or radical divided by its valence; the valence used depends on the electrochemical reaction involved. The electric charge on an ion is equal to the electronic charge or some integral multiple of it. Accordingly, a univalent negative ion has a charge equal in magnitude and of the same sign as a single electron, and its chemical equivalent weight is equal to its atomic weight, if an element, or to its molecular weight, if a radical. A trivalent ion has $+3$ or -3 electronic charges, depending on whether it is a positive or a negative trivalent ion. For trivalent ions, then, the equivalent weight would be equal to its atomic weight, if an element, or to its molecular weight, if a radical, divided by three.

The quantity of electricity required to produce a gram-equivalent weight of chemical change is known as the *faraday*. A faraday corresponds, then, to an *Avogadro number of charges*. The most accurate determination of the faraday has been made by a silver-perchloric acid coulometer in which the amount of silver electrolytically dissolved in an aqueous solution of perchloric acid is measured. This method gives 96,487 coulombs (or ampere-seconds) per gram-equivalent for the faraday on the unified ^{12}C scale of atomic weights adopted in 1961 by the International Commission on Atomic Weights.

Electrochemical Equivalent. Preferably termed *coulomb equivalent* of an element or radical, this is the weight in grams which is equivalent to 1 coulomb of electricity and is given by the gram-equivalent weight divided by the faraday (96,487 coulombs per gram-equivalent); for example, the electrochemical equivalent of silver is given by 107.870/96,487 or 0.00111797 grams/coulomb where 107.870 is the atomic weight of silver based on the unified ^{12}C scale adopted in 1961. The electrochemical equivalents of other elements may be calculated in like fashion.

In electrolysis and in any electric cell or battery, there is an electromotive force (emf) or voltage across the terminals. This emf is expressed in the practical unit, the volt, which is equal to the electromagnetic unit in the meter-kilogram-second system, in any one cell, the emf is the sum of the potentials of the two electrodes and of any liquid-junction potentials that may be present. Neither of the individual electrode potentials can be evaluated without reference to a chosen reference electrode of assigned value. For this purpose, the hydrogen electrode has been universally adopted and is arbitrarily assigned a zero potential for all temperatures when the hydrogen ion is at unit activity and the hydrogen gas is at atmospheric pressure. A hydrogen electrode consists of a stream of hydrogen gas bubbling over platinized platinum or gold foil and immersed in a solution containing hydrogen ions; the electrochemical reaction is: $\frac{1}{2}H_2(gas) = H^+$ (solution) $+ \epsilon$, where ϵ represents the electron. The potential of the hydrogen electrode, E_H, as a function of hydrogen ion concentration and hydrogen-gas pressure is given by

$$E_H = E_H^0 - (RT/nF)\ln(a_{H^+}/p_{H_2}^{1/2})$$
$$= E_H^0 - (RT/nF)\ln(c_{H^+}f_{H^+}/p_{H_2}^{1/2}),$$

where E_H^0 is the standard quantity assigned a value of zero, R is the gas constant, T the absolute temperature, n the number of equivalents, F the faraday, p_{H_2} the pressure of hydrogen and a_{H^+}, c_{H^+} and f_{H^+}, respectively, the activity, concentration, and activity coefficient of hydrogen ions. When a_{H^+} and $p_{H_2}^{1/2}$) equal one, $E_H = E_H^0$. For very dilute solutions below 0.01 molal f_{H^+} may be taken as unity without appreciable error.

The standard potentials, E^0, of other electrodes are obtained by direct or indirect comparison with the hydrogen electrode. Values are determined at 25°C. The values for several metals and other elements are given in entry on **Activity Series.** The reducing power of the elements decreased on going down the column from those elements with negative standard electrode potentials to those with positive potentials. These values are for the ions at unit activity, and reversible or thermodynamic values as a function of metal or radical concentration are given by equations similar to the one above. For the general reaction: $M = M^{n+} + n\epsilon$, the potential is given by $E_m = E_M^0 - (RT/nF)\ln a_{M^{n+}}$.

In electrolysis, at very low current densities, the potentials of the electrodes approximate in magnitude their reversible values and deviate somewhat from these values because of an *IR* drop in the solution and possible concentration polarization (the concentration at the electrode surface may differ from that in the bulk of the solution). Also for high current densities, especially for the generation of gases such as hydrogen, oxygen or chlorine, the voltage required exceeds the reversible voltage; the excess voltage is known as overvoltage, or overpotential for a single electrode, and arises from energy barriers at the electrode. Overpotential, in general, increases logarithmically with an increase in current density.

Scope of Electrochemistry. In addition to what has previously been described, it is customary to include under electrochemistry: (1) processes for which the net reaction is physical transfer, e.g., concentration cells; (2) electrokinetic phenomena, e.g., electrophoresis, electro-osmosis, streaming potential; (3) properties of electrolytic solutions if determined by electrochemical or other means, e.g., activity coefficients and hydrogen ion concentration; (4) processes in which electrical energy is first converted to heat which in turn causes a chemical reaction to occur that would not do so spontaneously at ordinary temperature. The first three are frequently considered a portion of physical chemistry, and the last one is a part of electrothermics or electrometallurgy.

The passage of electricity through gases is sometimes included under electrochemistry. However, in electrical discharges in gases, the principles are entirely different from what they are in the electrolysis of electrolytic solutions. Whereas in the latter, ionic dissociation occurs spontaneously as a result of forces between solvent and solute and without the application of an external field, for gases relatively high voltages must be applied to accelerate the electrons from the electrode to a velocity at which they can ionize the gas molecules they strike. In this case, the resulting chemical reaction taking place between ions, free radicals, and molecules occurs in the gas phase and not at the electrodes as in the electrolysis of solutions. Studies of the electrical conduction of gases, accordingly, are generally considered under the physics of gases.

Electrochemistry finds wide application. In addition to industrial electrolytic processes, electroplating, and the manufacture and use of batteries already mentioned, the principles of electrochemistry are used in chemical analysis, e.g., polarography, and electrometric or conductometric titrations; in chemical synthesis, e.g., dyestuffs, fertilizers, plastics, insecticides; in biology and medicine, e.g., electrophoretic separation of proteins, membrane potentials; in metallurgy, e.g., corrosion prevention, electrorefining; and in electricity, e.g., electrolytic rectifiers, electrolytic capacitors.

See also **Battery; Corrosion; Electrophoresis;** and **pH (Hydrogen Ion Concentration).**

References

Delahay, P., and C. Tobias: "Advances in Electrochemistry and Electrochemical Engineering," Volumes 1 through 8, Wiley, New York, 1971.

Heise, G. S., and N. C. Cahoon: "The Primary Battery," Wiley, New York, 1971.

Salamon, M. B. (editor): Physics of Superionic Conductors," Springer-Verlag, New York, 1979.

Yeager, E., and A. J. Salkind (editors): "Techniques of Electrochemistry," Wiley, New York, 1972.

ELECTRODE POTENTIALS. Activity Series.

ELECTRODIALYSIS. The removal of electrolytes from a colloid solution by a combination of electrolysis and dialysis. Usually the colloidal solution is placed in a vessel with two dialyzing membranes with pure water in compartments on the other side of the membranes. Two electrodes are inserted in the pure water compartments and an applied emf causes the ions to migrate from the colloidal solution. See also **Desalination.**

ELECTROLYSIS. Electrochemistry.

ELECTROLUMINESCENCE. Luminescence.

ELECTROMOTIVE SERIES. Activity Series.

ELECTRON. An elementary particle with a mass of $(9.109558 \pm 0.000054) \times 10^{-28}$ gram and a negative charge of $(1.6021917$

± 0.0000070 × 10^{-19} coulomb [(4.803251 ± 0.000036) × 10^{-10} electrostatic unit]. The name electron (from the Greek, elektron, amber) was proposed in 1891 by G. Johnstone Stoney for the unit of electricity which ruptures each chemical bond within an electrolyte solution during electrolysis. Hence one mole of electrons would be one faraday of electrical charge. However, the electron as a fundamental particle was not discovered until 1897 by J. J. Thomson in his studies of the effect of the passage of x-rays through gases. Thomson also measured the ratio of charge to mass (e/m) for the electron by observing the deflection of the path of a beam of electrons in an electric field. He found the value to be of the order of 10^{11} coulombs per kilogram; the presently accepted value of e/m for the electron is 1.7589 × 10^{11} coulomb/kilogram (5.273 × 10^{17} esu/gram). By observing the rate of fall of a cloud of negatively charged water droplets in air Thomson and Townsend obtained a preliminary value for e, the charge on the electron, of 6.5 × 10^{-10} esu. A better value of e was obtained by Robert A. Millikan, who in 1909 observed the rate of fall of charged oil droplets under the combined influence of gravity and an electric field. His result of 4.774 × 10^{-10} esu was slightly low, owing to his use of an erroneous value for the viscosity of air.

In 1921 Arthur H. Compton expressed the idea that an electron might possess an intrinsic angular momentum or spin and thus act as a magnet. In 1925 Wolfgang Pauli investigated the problem of why lines in the spectra of the alkali metals are not single, as predicted by the Bohr theory of the atom, but actually made up of two closely spaced components. He showed that the doublet in the fine structure could be explained if the electron could exist in two distinct states. G. E. Uhlenbeck and S. Goudsmit identified these states as two states of different angular momentum. They showed that the spectral multiplets could be explained by introducing a new quantum number s which would have either of two values, $+\frac{1}{2}$ or $-\frac{1}{2}$, so that the intrinsic angular momentum of the electron could be $\frac{1}{2}(h/2\pi)$ or $-\frac{1}{2}(h/2\pi)$. The spinning electron thus behaves as a tiny magnet with a magnetic moment of $eh/4\pi m_e c$, where h is Planck's constant, m_e is the mass of the electron, and c is the velocity of light. Electronic spin was confirmed experimentally by the famous Stern-Gerlach experiment. Later, when P. Dirac worked out a relativistic form of wave mechanics for the electron it was found that the property of spin fell out naturally from the theory.

The allowed stationary states of the electrons in even the most complex atoms can be referred to the same four quantum numbers n, l, m, and s, i.e., the principal (or shell), the azimuthal (or angular momentum), the magnetic and the spin quantum numbers, respectively. However, an important rule governs the quantum numbers allowed for an electron in an atom. This rule is known as the Pauli Exclusion Principle, which states that no two electrons in an atom can exist in the same quantum state, i.e., no two electrons in a given atom can have all four quantum numbers the same. A general statement of the Pauli Exclusion Principle is: a wave function for a system of electrons must be *antisymmetric* for the exchange of the spatial and the spin coordinates of any pair of electrons.

The Pauli Exclusion Principle provides the key to the explanation of the structure of the Periodic Table of the Elements. The most stable state, or *ground state*, of an atom is that in which all the electrons are in the lowest possible energy levels that are consistent with the exclusion principle. To explain the structure of the table arrange the different orbitals, each specified by its unique set of four quantum numbers, in the order of decreasing energy and then place the electrons, one by one, into the lowest open orbitals until all the electrons, equal in number to the nuclear charge Z of the atom, are safely accommodated. This process was called by Pauli the Aufbauprinzip. When this building process is carried out it is found that the number of electrons that can be placed in successive shells is *two* in the first, *eight* in the second, *eighteen* in the third, and *thirty-two* in the fourth. In general the number of electrons that can be placed in the nth shell is $2n^2$.

The tendency for atoms to form complete electron shells accounts for their valence states. For example, sodium, which has atomic number eleven, in the ground state has two electrons in the first shell, eight in the second, and one in the third. When electrons are removed from an atom, the atom is said to be ionized. Removal of the electron from the outermost shell of sodium, so that it assumes a configuration with the other two shells complete, accounts for its observed valence of + 1. Similarly, fluorine has its nine electrons arranged with two in the first shell and seven in the second. The tendency is for the fluorine atom to accept one more electron to complete the second shell, forming a fluoride *ion* with a charge of −1.

Although the Bohr theory of the atom, which treated electrons as particles, had many successes, it also had troublesome failures. For example, the theory could not explain the spectra of helium and other more complex atoms. In 1923 it was suggested by Louis de Broglie that electrons, and in fact all material particles, must possess wavelike properties. For example, an electron that has been accelerated through a potential of ten kilovolts, according to de Broglie, should have a wavelength of 0.12 Å or about the same as that for a rather hard x-ray. The experimental demonstration that a beam of electrons may be diffracted and hence that electrons do indeed have wave properties was accomplished by C. Davisson and L. H. Germer in the U.S. and by G. P. Thomson and A. Reid in Scotland. The fact that electron beams may be diffracted is the basis for the electron microscope.

Free electrons for study can be produced from metal or other surfaces in various ways: heating the surface (thermionic emission), allowing light of the appropriate wavelength to strike the surface (photoelectric effect), placing the surface in a strong electric field (field emission) or bombarding the surface with charged particles (secondary electron emission).

—M. H. Leitzke and R. W. Stoughton, Oak Ridge National Laboratory, Oak Ridge, Tennessee.

ELECTRONEGATIVITY. That property of each atom that determines the direction and extent of polarity of the covalent bond that holds two atoms together. A difference in electronegativity between the two atoms causes the valence electrons to spend more than half time more closely associated with that atom which was originally of higher electronegativity and therefore attracted them more strongly. This increases the total average electron population around this atom, which therefore imparts a partial negative charge. The other atom is left with an average deficiency of electrons, corresponding to a partial positive charge since each atom was initially neutral. The magnitude of these partial charges depends on the initial electronegativity difference between the two atoms and also in the nature of other atoms in the compound.

Unfortunately, a precise, quantitative definition of absolute electronegativity is thus far lacking. However, the property certainly must originate with the nuclear charge and the negative charge on a valence electron. It may be regarded as proportional to the coulombic force between a valence electron and that fraction of the total nuclear charge which, despite the screening effect of the underlying electronic cloud, remains effective at that point. The problem of evaluating the effective nuclear charge as it would be felt by a valence electron is far too difficult for exact solution at present. However, approximate evaluation has been achieved by recognizing that although underlying electrons are quite effective in cancelling positive charge of the nucleus, outermost shell electrons are only about one-third efficient in this respect. In other words, one outermost shell electron cannot very effectively interfere with the coulombic attraction between the nucleus and another outermost shell electron. Consequently, in the building up of elements across a period of the Periodic Table, each successive proton added to the nucleus is only about one-third neutralized or cancelled by the corresponding successive electron added to the outermost shell. Thus with each unit increment in atomic number, the effective nuclear charge increases by about two-thirds of a unit protonic charge.

This progressive increase in the effective nuclear charge across a period has two highly significant results. One is a reduction of the atomic radius as the electronic cloud is more effectively pulled together by the increased nuclear charge. The other is an increase in the electronegativity because of the combined effects of increasing the effective nuclear charge and decreasing the average distance between it and the valence electrons. This corresponds to a steady increase in electronegativity from one element to the next across the Periodic Table from left to right. Within each period, therefore, the lowest electronegativity is that of the alkali metal and the highest, that of the halogen. (The concept of electronegativity has only a very special and different application to the helium family elements that end each period.)

Relative values of electronegativity have been determined by a wide variety of methods. For example, electronegativities have been derived from ionization energies and electron affinities, from heats of reaction and bond energies, from calculations of the relative compactness of electronic clouds, from calculations of the effective nuclear charge at the surface of an atom, from work functions of metals, from force constants determined by infrared spectroscopy, and by miscellaneous other methods. When adjusted to the same arbitrary scale, that established by Pauling and ranging approximately from 1 to 4, the different methods show surprisingly good agreement, with only a few minor discrepancies that are still controversial. The general order of increasing electronegativities is Cs, Rb, K, Na, Li, Ba, Sr, Ca, Mg, Al, Be, Cd, Si and In, B and Hg, Zn, Tl, Pb, Sn, Bi, Sb, P, H, Ge, and Te, C, I, As, S, Se, N, Br, Cl, O, and F. Reliable values for the transition elements are not yet available but they would probably be placed near to magnesium.

If it is assumed that the homonuclear single covalent bond energy of an element is proportional to the coulombic energy of the interaction between the valence electrons and the effective nuclear charge, then it follows that homonuclear single covalent bond energy E is quantitatively related to the electronegativity by the simple equation, $E = CrS$, where r is the nonpolar covalent radius, S the electronegativity, and C an empirical constant which can be derived from the electronic structure type. Future development of these concepts gives promise of the possibility of determining absolute electronegativity values from experimentally determined homonuclear single covalent bond energies.

The most useful practical application of electronegativities is to a quantitative evaluation of heteronuclear bond energies and heats of formation, as described under **Bonding (Chemical)**. In such calculations, electronegatives afford the basis for apportioning the total bond energy between covalent and ionic contributions. Also extremely useful is the quantitative estimation of the relative condition of combined atoms made possible

through the use of electronegativities. Much of chemistry becomes intelligible if one recognizes that the contribution made by an atom to the properties of its compound depends at least as much on the condition of that atom in the compound as it does on which element it is. The best available index of the condition of a combined atom is its partial charge, which results from the initial atomic electronegativities in the following way.

The electrons involved in a covalent bond must, in effect, be equally attracted to both nuclei. When these two atoms are initially different in electronegativity, their bonding orbitals must of necessity be different in energy. Therefore the process of forming the bond must provide some mechanism by which these energies can be equalized. Such a mechanism can be based on the fact that the electronegativity of an atom must decrease as the atom begins to acquire an electron and increase as it begins to lose an electron. An atom of fluorine has a very high electronegativity but a fluoride ion has none. An atom of calcium has relatively little attraction for electrons but a calcium ion attracts them strongly. Consequently, the energies of the bonding orbitals can be equalized through an equalization of electronegativity, which in turn can result from uneven sharing of the bonding electrons. When the bonding electrons spend more than half time more closely associated with the nucleus of the atom that initially attracted them more, they reduce its electronegativity, at the same time imparting a partial negative charge. By spending less than half time more closely associated with the initially less electronegative atom, they leave it with net partial pasitive charge and correspondingly a higher electronegativity. The final state is one of even attraction through uneven sharing.

Such adjustment is postulated by the *Principle of Electronegativity Equalization*, which may be stated: When two or more atoms initially different in electronegativity unite, their electronegativities become equalized in the compound. The intermediate electronegativity in the compound is taken as the geometric mean of the electronegativities of all the atoms before combination. An estimate of the relative condition of a combined atom with respect to partial charge can be based on the assumption that the charge changes linearly with electronegativity. Assignment of a particular ionicity to a particular bond (75% to Na-F) then permits calculation of the change in electronegativity per unit charge for any element of known electronegativity. The relative partial charge on any combined atom is then defined and estimated as the ratio of the electronegativity change undergone in forming the compound to the electronegativity change that would correspond to the gain or loss of one electron. The sign of the charge is negative if the electronegativity of the element has decreased on combination, and positive if it has increased.

The partial charge distribution in any compound composed of elements of known electronegativity can thus easily be estimated. Such information provides many valuable insights that contribute to a fundamental understanding of chemistry.

Theoretical chemists have recently found it advantageous to consider electronegativity as the property of a particular orbital rather than of the atom as a whole. For the purposes of practical interpretations of chemistry through knowledge of partial charge distribution and calculation of bond energy, however, the additional complication of this refinement has thus far seemed unnecessary.

—R. T. Sanderson, Arizona State University, Tempe, Arizona.

ELECTROPHORESIS. A procedure that involves the migration of charged particles, either colloidally dispersed substances or ions, through conducting liquid solutions, under the influence of an applied electric field.

Prior to about 1950, two methods were generally employed for studying the electrophoretic behavior of charged particles in a liquid. In the microscopic method, the migration of individual particles is observed in a solution contained in a glass tube placed horizontally on the stage of a microscope. The method is suitable for the study of relatively large particles such as bacteria, blood cells or droplets of oil. Its usefulness was extended somewhat by the finding that various finely divided inert materials, such as tiny spheres of glass, quartz or plastic can, in some instances, be so completely covered with adsorbed protein that they act as if they were large protein particles and respond to an electrical field in terms of the charge on the protein. The method is mainly of historical interest.

In the moving boundary technique of electrophoresis, the movement of a mass of particles is measured, thus obviating the necessity of observing individual particles. The displacement of the particles in an electric field is recorded photographically as the movement of a boundary between a solution of a colloidal electrolyte, such as a protein, and the buffer against which it was dialyzed. The material to be studied is poured into the bottom of a U-tube and on top of it, in each arm of the U-tube, a buffer solution is carefully layered in order to produce sharp boundaries between the two solutions. Electrodes inserted in the top of each arm of the tube are attached to a DC electric source. If the material under study is a protein bearing an excess of negative charges on its molecular surface, the boundary will move toward the positive electrode. Since the net electric charge on the protein molecule varies with the acidity of the buffer solution, the charge on the molecule, and hence its velocity, may be varied by changing the acidity of the buffer. As the hydrogen-ion concentration is increased or the pH lowered, the velocity of the protein is reduced until a point on the pH scale is reached at which it fails to move (isoelectric point or pI). If the pH of the buffer is further reduced, the protein will acquire a net positive charge and will move toward the negative electrode. The migration velocity of any particular migrant is, of course, directly proportional to the applied voltage gradient. Other important factors which may affect the observed velocity include the molecular shape and structure of the specific substance under study as well as the concentration of the buffer solution or, more specifically, its ionic strength, the temperature and electroosmosis. Electroosmosis refers to the constant flow of liquid relative to a stationary charged surface, for example, a strip of filter paper, under the influence of an applied electric field. The net movement in the case of aqueous solutions is generally toward the negatively-charged electrode. Owing to its experimental complexity and cost, this technique has been largely replaced as an analytical tool by the simpler method of electrophoresis described below.

The third technique for carrying out electrophoretic separations has been named variously as ionography, zone electrophoresis, electrochromatography, etc. Although the rootlets of the technique are discernible in the publications of Lodge dating back to 1886, the modern era began in 1950, when several papers on electromigration in peper-stabilized electrolytes were published from a number of countries. From that time on, the number of reports on various applications, modifications and limitations of the technique has grown phenomenally, and this procedure has become one of the important tools in biochemical research and in routine clinical chemical analyses. It lends itself equally well to the study of the electromigration of ionic substances of low molecular weight, such as amino acids, peptides, nucleotides and inorganic ions, and of colloidal materials such as proteins and lipoproteins. Under favorable conditions, substances are not only separated totally from mixtures but may be recovered almost completely. This aspect of the procedure is particularly valuable in work with radioactive materials. Rigid restrictions on the temperature, current and composition of the solutions, in order to minimize convectional disturbances, are largely removed when electrophoresis is carried out in a solution stablized with a relatively inert material, originally paper but later on principally cellulose acetate, starch gel, polyacrylamide gel or agar, which permit sharper separations. The high resolving power of gel media is to a large extent a consequence of molecular sieving acting as an additional separative factor.

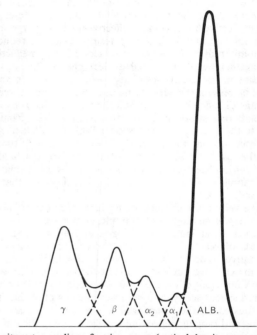

A densitometer reading of an ionogram (optical density versus distance of migration along the ionogram) for normal human plasma. The major peak represents the albumin fraction; the lesser peaks represent the alpha 1, alpha 2, beta, and gamma globulin fractions.

As an example, blood serum can be separated into about 25 components in polyacrylamide gel but only into five components on filter paper or by moving-boundary electrophoresis. Only minute amounts of material are required, the equipment is relatively simple and inexpensive, and the method can be utilized over a wide range of temperature. Some of the many uses to which electrophoresis in stabilized media has been applied are identification of the individual components of a mixture, establishing homogeneity, concentration and purification. In conjunction with other microanalytical techniques such as immunology and polarography, it may often be the means by which identification is ultimately established and quantitative assessment made. Adaptations of the technique, e.g., curtain and planar electrophoresis, provide for continuous operation thus permitting the separation of relatively large quantities of substances.

Normally, a micro amount of the mixture to be separated is streaked across the midpoint of a horizontal column or strip of stabilizing agent, e.g., a narrow, paper-thin cellulose acetate strip saturated with a buffer solution of known pH and ionic strength; a controlled DC source of electric potential is then applied to the ends of the column or strip. The substances under study, the migrants, begin to move and each rapidly reaches a constant velocity of electromigration through the stabilizing structure. The velocity depends, among other factors, upon the potential gradient, the charge on the substance, the ionic strength of the solution, the temperature, the barrier effect interposed by the stabilizing agent, the wetness, electroosmosis and hydrodynamic movement of the solvent through the stabilizing structure. In general, as the electromigration proceeds, the original zone separates into several discrete zones having different specific electromigration velocities. The distance of each separated spot, zone or band on the ionogram from the point of application provides a measure for determining the mobility of the particles making up the zone and the density and area of the spot provide an index to the quantity of material in the mixture. If the substances separated are colorless, the bands may be developed by the use of suitable dyeing reagents, e.g., bromophenol blue for blood serum proteins.

Several methods have been utilized for the quantitative determination of the dyed zones, e.g., protein fractions, on the ionogram. The strip may be cut into sections, the colored material eluted with suitable solvents and its concentration determined in a spectrophotometer fitted with small cuvettes. The most common method in use today involves direct determination by a transmission densitometer.

The charge on a particle may arise from charged atoms or groups of atoms that are part of its structure, from ions which are adsorbed from the liquid medium, and from other causes. It is evident that the behavior of colloidal particles, as compared to that of ions, in an electric field differs only in degree rather than in kind. Although a colloidal particle is much larger than an ion, it may also bear a much greater electrical charge, with the result that the velocity in an electric field may be about the same, varying roughly from $0–20 \times 10^{-4}$ cm/sec per second in a potential gradient of 1 volt per cm.

To understand the phenomenon of electrophoresis, let us suppose, for simplicity, that a nonconducting particle, spherical in shape, of radius r, and bearing a net charge of Q coulombs is immersed in a conducting fluid of dielectric constant D and a viscosity of η poises. Suppose, further, that the particle moves with a velocity of v cm/sec under the influence of an electric field having a potential gradient of x volts per cm. The force causing the particle to move, namely, $Qx \times 10^7$ dynes, is opposed by the frictional resistance offered to its movement by the liquid medium. From Stokes' law, the latter is given by $6\pi\eta rv$. Under steady-state conditions, and introducing the electrophoretic mobility, $u = v/x$; rearrangement yields the expression $u = Q \times 10^7/6\pi\eta r$. It is evident that if the electrophoretic mobility of a particle can be computed, it should be possible to determine Q, the net charge on the particle.

Electrophoresis in stabilized media is used mainly as an analytical technique and, to a lesser extent, for small-scale preparative separations. The most important applications are to the analysis of naturally occurring mixtures of colloids, often of plant or animal origin, such as various proteins, lipoproteins, polysaccharides, nucleic acids, carbohydrates, enzymes, hormones and vitamins. Electrophoresis often offers the only available method for the quantitative analysis and recovery of physiologically active substances in a relatively pure state. It provides a most convenient and dependable means of analyzing the protein content of body fluids and tissues and provides an important tool in most hospital laboratories. Like chromatography, electrophoresis in stabilized media is mainly a practical technique and the most important advances have involved improvements in experimental procedures and the introduction and development

of a wide range of suitable stabilizing media. The marked differences between normal and pathological serum samples are useful in the diagnosis and understanding of disease. Such changes in the electrophoretic pattern of blood serum are evident in diseases characterized by marked protein abnormalities such as multiple myeloma, nephrosis, liver cirrhosis and various parasitic disorders. Because of the small amount of fluid required, the method is applicable to the study of spinal fluids.

The electrophoretic pattern is not to be interpreted as specific for a given disease but rather as an index to the physiological and nutritional condition of the patient, to be combined with other information for a more complex diagnosis. If the electrophoretic pattern of a patient's blood plasma shows an excess of gamma globulin, the inference is that the body may be suffering from an infection, since most of the antibodies evoked by the presence of infectious microbes are gamma-globulin-like proteins. An increase in the alpha-globulin, a result of the breakdown of tissue proteins, may herald a fever-producing disease, such as pneumonia or tuberculosis. When the blood shows a decrease in albumin, the clinician looks to the liver as a possible seat of the disease because it is the main site of albumin production. When the liver fails, other tissues will often react to the lower albumin level by producing an excess of globulins.

Because fractionation of mixtures by ionography and quantitative estimation of the components has been of such great practical value, the potentiality of the technique for theoretical electrochemistry and colloid chemistry is often overlooked. Since substances can be characterized by their electrophoretic mobilities, phenomena such as the binding of various ions to proteins, which produce changes in the net charge of a molecule, can, in principle, be evaluated from the changes in the mobility. It has been possible to arrive at some conclusions as to relative binding strength and the charge on the chemical site involved. The isoelectric points of ampholytes, such as proteins and lipoproteins, can be determined and progress has been made toward determining such factors as ionization constants and associated thermodynamic quantities. Since these determinations can be made with less than a milligram of substance, the technique is often of considerable value for establishing the nature of functional groups.

—Hugh J. McDonald, Loyola University, Maywood, Illinois.

ELECTROPLATING. The deposition of a layer, usually thin, of some material on an object by passing an electric current through a solution containing the material in which the object has been immersed, the object being one of the electrodes. The purpose of electroplating is usually to protect and/or to decorate. Also by suitable procedures, shapes such as that of printer's type are reproduced.

Usually the object being plated is a metal; the thin coating deposited is a different metal, and the solution is an aqueous solution of a salt containing the element being deposited. The object being plated is the cathode, that is, it is the electrode to which an outside electric source delivers electrons. The anode often is composed of the metal being deposited, and is the electrode from which the outside electrical source accepts electrons. Ideally it dissolves as the process proceeds.

The current supplied is almost always steady D.C. but sometimes is made a pulsating one by superimposing an A.C. upon the D.C. The A.C. may be of the sine wave type, square wave type, etc.

The thin layer being deposited sometimes is composed of two or more metallic elements, in which case it is an alloy.

The solution, or plating bath as it is called, contains dissolved salts of all the metals which are being deposited. It often also contains an appropriate acid, base and/or salt added for the puspose of holding the pH at a desired level.

Other substances, called addition agents, are often added to the plating bath for the purpose of giving the plate a desired texture, such as one which is strong, adherent and mirror smooth rather than rough, granular, loose and mechanically weak. Boric acid, glue, gelatin, urea, glycine are examples of such substances. Often, if not always, these are substances which when added to the solvent of the bath alone will form a solution having a higher dielectric constant than that of the pure solvent. Small amounts of these addition agents are likely to find their way into the plate itself which may or may not be desirable.

While the object undergoing plating usually is a good electrical conductor, it may be a nonconductor which has previously received a thin coating of some conductor. Graphite is often used for this purpose, but other materials, such as the gold paint normally used on chinaware will serve, or a thin silver film produced in a manner similar to that on the back of an ordinary mirror.

The anode must be an electrical conductor but may or may not be of the same chemical composition as the plate being deposited and may or may not dissolve during the electroplating process.

In addition to the plating of metals, colloids are sometimes electroplated. A positively charged colloid in a plating bath will plate or tend to plate on the cathode and a negatively charged colloid on an anode. Rubber is electroplated from latex by this means. This process is called *electrodeposition*; it permits rubber coating of complicated shapes and thin films such as surgeon's gloves.

There is nearly always a temperature range within which it is desirable to hold the plating bath and a range of current density within which it is desirable to hold the current. By current density is meant the current per unit area of the object being plated. This may be expressed in amperes per square centimeter, per square decimeter, per square foot, etc.

Ideally to plate monovalent ions such as Ag^+ from the plating bath would require one electron per ion, divalent ions such as Cu^{++} two electrons per ion, etc. Thus ideally one faraday of electricity would deposit one equivalent weight of any metal, i.e., 107.880 g of silver, $63.54/2 = 31.77$ g of divalent copper, etc. Actually this ideal is almost never realized. The term "current efficiency" (of an electroplating process) is defined as the mass of a metal actually deposited divided by the mass of the same metal which would ideally have been deposited by the passage of the same quantity of electricity.

Objects to be plated are often irregular in shape and thus have recessed and remote areas. With some plating baths such areas may receive very little deposition while with other baths considerable deposition. The term "throwing power" is used to describe the ability of a plating bath to reach these remote or recessed parts of the object; the thicker they are plated, compared with the more exposed parts, the better the throwing power of the particular plating bath.

The voltage required for electrodeposition under ideal conditions would be given by the following equation:

$$E = E^0 + \frac{RT}{VF} \ln A$$

where E is the required voltage relative to the solution as measured by a hydrogen electrode in a unit molal activity hydrogen ion solution. E^0 is the electrolytic potential, in volts, of the metal being plated when immersed in a solution containing its ions at unit molal activity (approximately unit molal concentration). See **Electrode Potential.** $R = 8.31$ joules per degree mole; T is the Kelvin temperature; $F = 96,500$ joules per gram-equivalent; V is the valence of the ions which are depositing out; $\ln A$ is the natural logarithm of the activity of these ions (approximately the natural logarithm of their molality).

In actual practice the concentration and therefore the activity of the ions soon after the electroplating process starts is different in the solution just next to the cathode, called the cathode "film," than in the main body of the bath. The foregoing equation must in practice be modified to read:

$$E = E^0 + \frac{RT}{VF} \ln A - P,$$

where A is the molal activity in the cathode film of the ions being electrodeposited and P is the extra potential required to keep the plating going. A and P depend on temperature, current, density, concentration, valence, pH, and ion mobility.

Some of the metallic ions that are commonly electroplated and their electrolytic potential, E^0, are as follows:

Ion	E^0	Ion	E^0
Zn^{+2}	−0.762	Cu^{+2}	+0.345
Cr^{+3}	−0.71	Cu^{+1}	+0.522
Cd^{+2}	−0.402	Ag^{+1}	+0.800
Ni^{+2}	−0.250	Au^{+3}	+1.42
Sn^{+2}	−0.136	Au^{+1}	+1.68
(H^{+1}	0.000)		

As will be seen, in order to plate out the metals above hydrogen from an aqueous solution, the concentration of the hydrogen in the cathode film must be low. Hence a basic solution such as provided by a cyanide bath may be resorted to. A cyanide bath may also be used to give a smooth adherent plate.

To plate out an alloy, that is, to codeposit at least two kinds of metals, the plating bath must be so contrived that the electrodeposition potentials of the two metals in the cathode film are equal or nearly so.

Four typical electroplating baths are as follows:

Copper: CuCN 26 g/l, NaCN 35 g/l,Na_2CO_3 30 g/l $KNaC_4H_4O_6 \cdot 4H_2O$ 45 g/l, NaOH to give pH of 12.6
Copper: $CuSO_4 \cdot 5H_2O$ 188 g/i, H_2SO_4 74 g/l
Tin: Tin (as tin fluoborate concentrate) 60 g/l, free fluoboric acid 100 g/l, free boric acid 15 g/l.
Zinc: $Zn(CN)_2$ 60 g/l, NaCN 23 g/l, NaOH 53 g/l

The bright, hard, ornamental chrome so popular on automobiles and household and office equipage is produced by electroplating. In this electroplating process the chromium is not present in the bath as a positive metal ion but rather as part of the anion of chromic acid, H_2CrO_4. The object being plated is made the cathode. Usually it has already been electroplated first with copper and then with nickel. The pleasing to the eye, long lasting surface is the result of electroplating a coating of chromium only 0.00001 to 0.00005 inch thick on the nickel.

The plating bath is an aqueous solution of chromic trioxide, CrO_3, and sulfuric acid, H_2SO_4, with a ratio of approximately 100 to 1 by weight. The sulfuric acid acts only as a catalyst. The total concentrations may vary widely in different baths. Fluosilicate catalysts are also sometimes added to the bath.

As an example a chromium electroplating bath might contain 300 grams of CrO_3 and 3.0 grams of H_2SO_4 per liter, operate at 120°F with a current density of 2.6 amperes per square inch requiring a potential of 10 volts or more, with a current efficiency of around 20 percent. It would have a nondissolving anode, lead, which would also serve to oxidize trivalent chromium, produced at the cathode, back to chromic acid.

Thicker coatings of chromium are electroplated onto wearing surfaces such as cutting tools and the walls of cylinders of internal combustion engines, with or without a previous plating with copper and nickel.

—Penrose S. Albright, Wichita, Kansas.

ELECTRUM. A native alloy of gold and silver in which the latter metal may be present in quantities up to 40%. Electrum from the Urals is said to carry 20% copper. The color of electrum is pale yellow or yellowish-white. The name is derived from the Greek word mentioned in the "Odyssey," meaning a metallic substance consisting of gold alloyed with silver. This same word was also used for the substance amber, doubtless because of the pale yellow color of certain varieties.

ELEMENTARY PARTICLES. Particles (Subatomic).

ELEMENTS (Chemical). Chemical Elements.

ELUTRIATION. A process of washing, decantation, and settling which separates a suspension of a finely divided solid into parts according to their weight. It is especially useful for very fine particles below the usual screen sizes and is used for pigments, clay dressing, and ore flotation.

EMBDEN-MEYERHOF PATHWAY. Carbohydrates.

EMBRITTLEMENT. A lowering of the ductility of a metal as a result of physical or chemical changes. Metals may be embrittled under many different conditions. Ordinary steel, wrought iron, and body-centered cubic metals generally, as well as zinc alloys and magnesium alloys, suffer a reduction in impact toughness at subnormal temperatures. The effect is only temporary, full recovery of toughness occurring upon return to normal temperatures. Austenitic stainless steels, brasses and bronzes, nickel alloys, aluminum alloys, and lead alloys are not subject to severe embrittlement at low temperatures. Nickel additions to ordinary steels have a favorable effect in decreasing embrittlement.

Hydrogen embrittlement of iron and steel may be caused by absorption of atomic hydrogen in electroplating processes or in pickling baths. After such exposure, the normal toughness can usually be restored by prolonged aging or by subjection to a short period of heating at slightly elevated temperature, as in a steam bath.

Season cracking of high-zinc brasses is a severe form of embrittlement, resulting in cracking or disintegration. Somewhat similar forms of stress-corrosion cracking occur in many other metals and alloys. Embrittlement of boiler plate may be considered a special case. This is *caustic embrittlement*, which occurs in steels and ferrous alloys after prolonged exposure to alkaline substances, like caustic soda, in solution. Failures and explosions in boilers and evaporators have been caused by this action. This condition can be corrected through effective boiler water treatment.

EMERALD. This beautiful green variety of the mineral beryl has been known since ancient times and always prized as a gem, both because of its color and relative rarity. It is frequently cloudy or flawed, hence the expression "rare as an emerald without a flaw." The original source of emeralds seems to be the so-called Cleopatra's mines in Egypt, where in a range of low mountains about 15 miles from the Red Sea, they are found in schists. The quality of these emeralds is not high, but there is much evidence of considerable workings in a former period. See also **Beryl**.

Although emeralds are found in the Urals and to some extent elsewhere the most important locality for emerald is at Muso, Colombia, South America, about 75 miles northwest of Bogotá. These mines are believed to be in part at least the source of the emeralds which Cortez and the Spanish conquistadores ruthlessly seized and which were believed for a long time to have come from Peru.

The word *emerald* is probably derived from the Persian.

EMULSION. Colloid System.

ENANTIOTROPY. The property possessed by a substance of existing in two crystal forms, one stable below, and the other stable above, a certain temperature called the transition point.

ENARGITE. A mineral sulfosalt composed of copper, arsenic and antimony, Cu_3AsS_4. Crystallizes in the orthorhombic system. Hardness, 3; specific gravity, 4.451; color, gray to black with metallic luster. From the Greek, meaning distinct cleavage.

ENERGY LEVEL. A stationary state of energy of any physical system. The existence of many stable, or quasi-stable, states,

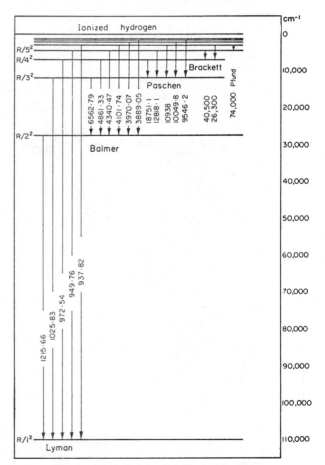

Fig. 1. Energy levels of the hydrogen atom.

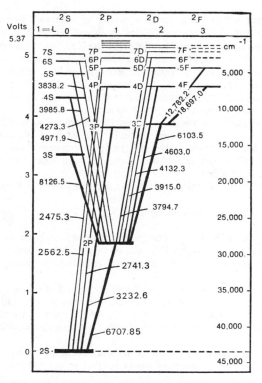

Fig. 2.   Energy level diagram of the lithium atom. (*After Grotian*).

in which the energy of the system stays constant for some reasonable length of time, is an essential characteristic of quantum-mechanical systems, and is the basis of large areas of modern physics.

1. The motions of electrons within an atom may be described by various orbitals, which are wave functions corresponding to various quantized orbits. The characteristic spectra of atoms are emitted when an electron shifts from an orbital of higher energy to one of lower energy. Such transitions are shown on an energy level diagram. Figure 1 shows the energy level diagram for hydrogen, on which the various spectral series are designated. The atomic energy level diagrams for other elements having only one electron in their outer "shells" (that is, those orbitals in an atom which are designated by the highest principal quantum number possessed by any of the electrons in that atom) are quite similar to those of hydrogen, as is apparent in Fig. 2, which is the atomic energy level diagram for lithium.

On the other hand, the atomic energy diagram for elements having more than one electron in their outer "shells" is more complex, as is seen from the energy level diagram for calcium (Fig. 3), which has two electrons in its outer "shell."

2. The fine structure of x-ray lines may be described by an energy level diagram that is closely similar in appearance to those of optical spectra (Fig. 4). However, the selection rules permit fewer transitions of the kind that produce x-rays (that is, between "shells") so that the x-ray energy levels (and the x-ray spectra they represent) are simpler than the atomic energy levels (and their corresponding optical spectra).

3. Even more complex optical and x-ray energy diagrams are found in heavy elements, which have multiplet levels (that

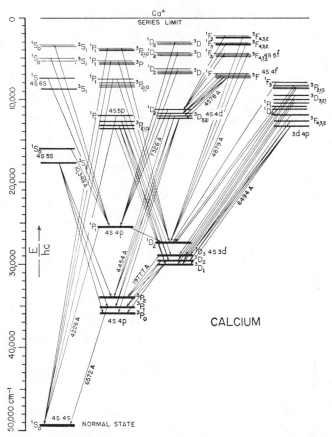

Fig. 3.   Energy level diagram for neutral calcium atoms, showing electron configurations and some of the more prominent spectral transitions.

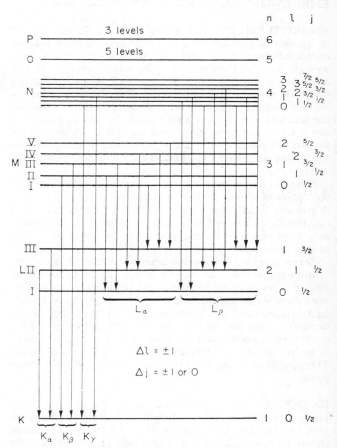

Fig. 4.   Fine structure of x-ray energy levels.

is, groups of spectra lines very close in frequency and their corresponding levels). The transition elements exhibit particularly high multiplicities (e.g., chromium and iron have them as high as seven and manganese as high as eight). Moreover, in magnetic and electric fields the Zeeman and Stark effects cause splitting of the energy levels and further complexity.

4. As might be expected, the molecular energy level diagrams are far more complex than the atomic ones. Figure 5 shows some of the transitions that may occur between only three levels of one molecular energy state in the ultraviolet band system of the nitrogen molecule.

5. Just as electronic transitions of the orbital electrons of the atom are represented by energy level diagrams, so are the nuclear changes in radioactivity. Figure 6 shows the decay process of Pb-210, in which this lead isotope emits an electon (and a neutron) to produce a bismuth-210 nucleus in an excited state. Emission of a gamma ray effects deexcitation, yielding a bis-

muth-210 nucleus at ground level. The latter in turn emits another electron to become polonium-210 (not shown in figure).

6. Of course, energy level diagrams are not limited to single atoms, molecules, or nuclei. If a block of stone, or any other object, is represented at different heights, the change in potential energy may be shown on a diagram. Moreover, a system of entities may also be represented on an energy level diagram.

ENDORPHINS. Enkephalins and Endorphins.

ENKEPHALINS AND ENDORPHINS. During the past decade of brain research, the number of chemical-messenger systems identified has increased dramatically. This advance is highlighted by the discovery of a large family of brain chemicals known as *neuropeptides*. As shown by Fig. 1, these molecules are made up of long chains of amino acids, ranging from a few to as many as 39 amino acids. Research indicates that these substances are resident within the neutrons. A few of these substances, such as corticotropin (ACTH) and vasopressin, have been known for many years and identified with the hypothalamus and pituitary gland. Probably of greatest interest to researchers have been the *enkephalins* and *endorphins*. These chemicals are strikingly similar to the opiate morphine.

Identification of these substances followed the finding that specific regions of the brain possess receptor sites that bind opiates with a high affinity. These sites were revealed through the use of radioactively labeled opiate compounds. Further research showed that opiate receptors are located in those regions of the brain and spinal cord which are known to be associated with emotion and pain. Pioneering research in the field included the work of Snyder and Pert (Johns Hopkins University School of Medicine and Terenius (University of Uppsala). The enkephalins (Met- and Leu-), as shown in Fig. 1, were first isolated by Hughes and Kosterlitz (University of Aberdeen) in 1975. Each enkephalin contains five amino acids, one of these being methionine in one case, and leucine in the other case. Shortly after the isolation of these compounds, the endorphins, also morphinelike compounds, were isolated from the pituitary gland. Some researchers have suggested that some of the non-traditional methods used for relieving chronic pain, such as acupuncture, direct electrical stimulation of the brain, and possibly hypnosis, may be effective because these procedures may cause enkephalins or endorphins or both to be released to the brain and spinal cord. The drug naloxone (*Narcan®*) blocks the binding of morphine and experiments have shown that naloxone also blocks the effects of the aforementioned pain-relieving procedures; hence the tentative hypothesis.

Early research findings indicate that neuropeptides are released from axon terminals through the presence of calcium ions known to be the releasing mechanism in connection with established transmitters. Thus, some investigators believe that the neuropeptides also may be transmitters. This is particularly true of a compound simply identified as *substance P*. See Fig. 1. Some investigators have found that substance P is associated with spinal neurons involved in pain stimuli.

One of the surprising findings of recent brain and central nervous system research is that chemical substances previously considered to be exclusive to the province of the brain have been found in organs outside the nervous system (e.g., somatostatin, neurotensin, and enkephalins have been found in the gut), and conversely, that other substances active in other organs, but not previously associated with the central nervous system, have been found in the latter (e.g., gastrin, vasoactive intestinal polypeptide (VIP), and cholecystokinin, traditionally associated with the gastrointestinal tract, but now found in the central nervous system).

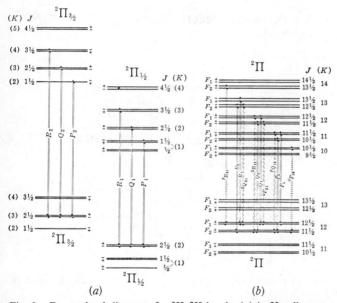

Fig. 5. Energy level diagrams for $^2\Pi$–$^2\Pi$ bands: (*a*) in Hund's case, (a) [$^2\Pi$(a)–$^2\Pi$(a)]; (*b*) in Hund's case, (b) [$^2\Pi$(b)–$^2\Pi$(b)]. Only one line of each branch is given. In the designation of the branches, components of the same Λ-type doublet are not distinguished. In (*a*), the $^2\Pi_{1/2}$ and $^2\Pi_{3/2}$ levels form the $F_1(J)$ and $F_2(J)$ series, respectively. The dotted branches in (*b*) do not appear when both $^2\Pi$ states belong strictly to Hund's case (*b*).

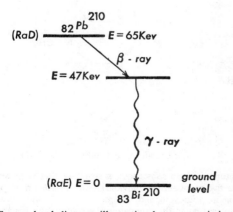

Fig. 6. Energy level diagram illustrating beta-ray emission followed by gamma ray emission.

Tyr-Gly-Gly-Phe-Met   Tyr-Gly-Gly-Phe-Leu   Arg-Pro-Lys-Pro-Gln-Gln-Phe-Phe-Gly-Leu-Met-NH₂
(Met-ENKEPHALIN)   (Leu-ENKEPHALIN)   (SUBSTANCE P)

p-Glu-Leu-Tyr-Glu-Asn-Lys-Pro-Arg-Arg-Pro-Tyr-Ile-Leu   Asp-Arg-Val-Tyr-Ile-His-Pro-Phe-NH₂
(NEUROTENSIN)   (ANGIOTENSIN II)

Tyr-Gly-Gly-Phe-Met-Thr-Ser-Glu-Lys-Ser-Gln-Thr-Pro-Leu-Val-Thr-Leu-Phe-Lys-Asn-Ala-Ile-Val-Lys-Asn-Ala-His-Lys-Lys-Gly-Gln
(Beta-ENDORPHIN)

Ser-Tyr-Ser-Met-Glu-His-Phe-Arg-Tyr-Gly-Lys-Pro-Val-Gly-Lys-Lys-Arg-Arg-Pro-Val-Lys-Val-Tyr-⟩
⟨Pro-Asp-Gly-Ala-Glu-Asp-Glu-Leu-Ala-Glu-Ala-Phe-Pro-Leu-Glu-Phe   ←   Asp-Tyr-Met-Gly-Trp-Met-Asp-Phe-NH₂
(ACTH, CORTICOTROPIN)   (CHOLECYSTOKININ-LIKE PEPTIDE)

His-Ser-Asp-Ala-Val-Phe-Thr-Asp-Asn-Tyr-Thr-Arg-Leu-Arg-Lys-Gln-Met-Ala-Val-Lys-Lys-Tyr-Leu-Asn-Ser-Ile-Leu-Asn-NH₂
(VASOACTIVE INTESTINAL POLYPEPTIDE, VIP)

p-Glu-His-Pro-NH₂   p-Glu-His-Trp-Ser-Tyr-Gly-Leu-Arg-Pro-Gly-NH₂
(THYROTROPIN RELEASING HORMONE, TRH)   (LUTEINIZING-HORMONE RELEASING HORMONE, LHRH)

Ala-His   p-Glu-Gln-Arg-Leu-Gly-Asn-Gln-Trp-Ala-Val-Gly-His-Leu-Met-NH₂
(CARNOSINE)   (BOMBESIN)

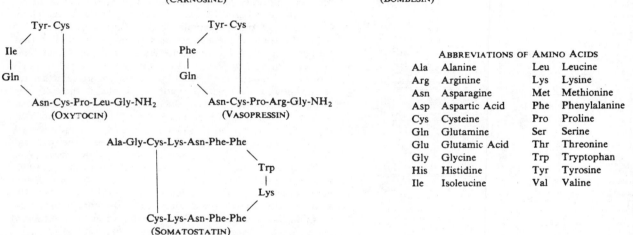

ABBREVIATIONS OF AMINO ACIDS

Ala	Alanine	Leu	Leucine
Arg	Arginine	Lys	Lysine
Asn	Asparagine	Met	Methionine
Asp	Aspartic Acid	Phe	Phenylalanine
Cys	Cysteine	Pro	Proline
Gln	Glutamine	Ser	Serine
Glu	Glutamic Acid	Thr	Threonine
Gly	Glycine	Trp	Tryptophan
His	Histidine	Tyr	Tyrosine
Ile	Isoleucine	Val	Valine

Fig. 1.   Now believed to be transmitters, neuropeptides, which are short chains of amino acids found in brain tissue and notably localized in axon terminals, participate in complex mental activity, such as thirst, memory, and sexual behavior.

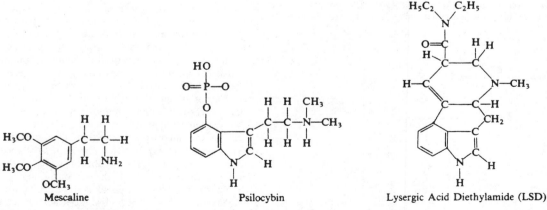

Fig. 2.   Representative hallucinogenic drugs which bear structural similarities to some of the monoamine transmitters. It has been hypothesized that these similarities may cause the hallucinogens to mimic natural transmitters at synaptic receptors in the brain. Note presence of benzene- ring structure in these substances, a structure that is present in four out of the five monoamines previously shown. Also note presence of indole ring in psilocybin and lysergic acid diethylamide, a structure that is also present in the monoamines serotonin and histamine.

Also among recent discoveries in brain chemistry are the so-called *trophic substances*, which are believed to be secreted from nerve terminals. One of these is *nerve growth factor* (NGF). It has been established that this protein is required for the differentiation and survival of peripheral sensory and sympathetic neurons. Another benefit of recent research is a better understanding of how psychoactive drugs interact with the brain and central nervous system. See Fig. 2.

References

Barchas, J. D., et al.: "Behavioral Neurochemistry: Neuroregulators and Behavioral States," *Science*, **200**, 964–973 (1978).
Costa, E., and M. Trabucchi, Editors: "The Endorphins," Raven, New York (1978).
Iversen, L. L.: "The Chemistry of the Brain," *Sci. Amer.*, **241**(3), 134–149 (1979).
Snyder, S. H.: "Brain Peptides as Neurotransmitters," *Science*, **209**, 976–983 (1980).

ENSTATITE. The mineral enstatite is an orthorhombic pyroxene, rarely in distinct crystals, usually found as fibrous or lamellar masses or perhaps compact. It has one easy cleavage parallel to the prism; brittle with uneven fracture; hardness 5–6; specific gravity 3.2–3.4; luster pearly to vitreous, sometimes somewhat metallic in bronzite, a variety of enstatite carrying up to 15% ferrous oxide. FeO. Color grayish to greenish or yellowish-white, green and brown. Chemically, enstatite is a silicate of magnesium, $MgSiO_3$. It occurs in igneous rocks which are high in magnesium content, like gabbros, diorites, and pyroxenites, and less commonly in metamorphic rocks. Meteorites of both the stony and metallic types have been shown to contain enstatite. It has been found at many places in Europe, Czechoslavakia, Austria, Bavaira, Germany, Norway, and the Republic of South Africa. In the United States it occurs in Putnam and St. Lawrence Counties, New York; Lancaster County, Pennsylvania; Jackson County, North Carolina, and near Baltimore, Maryland. The name enstatite is derived from the Greek word meaning *opponent*, in reference to its refractory nature; it is almost infusible. See also **Pyroxene.**

ENTEROVIRUSES.   Virus.

ENTHALPY.   Thermochemistry.

ENTNER-DOUDOROFF PATHWAY.   Carbohydrates.

ENTROPY (Kinetics).   Kinetic Theory.

ENZYME. Collectively, the term *enzymes* refers to a group of proteins which catalyze a variety of chemical reactions. Over a thousand enzymes have been identified. In 1964, the International Union of Biochemistry officially named and listed nearly 900 enzymes. Since that time, progress has continued toward developing a consistent and standardized procedure for naming the enzymes. The existence of enzymes has been known since the early 1600s, mainly from the observation of their role in digestion and fermentation processes used to make alcohols and other allied products. Only in later years were the simpler enzymes isolated. Urease was produced in crystalline form in 1926. Other enzymes that were later isolated include amylase, carboxypeptidase, chymopapain, papain, pepsin, and starch phosphorylase. See also **Enzyme Preparations.**

Enzyme complexes are generated by living cells. They function as catalysts in reactions that involve the metabolism of living organisms and thus play vital roles at practically all levels of food involvement—production, processing, and consuming, whether by fish, bird, insect, primate, etc. Enzymes and the reactions in which they participate thus are ever present during the entirely of the food chain—from start to finish. Investigations in botany, pursuits of agronomy, studies of nutrition, inquiries into plant and animal pathology, and the numerous other aspects of the sciences that are involved in life processes, when probed in depth, ultimately encounter the vital roles played by enzymes.

Expanding the knowledge of the life processes depends critically upon furthering the investigation of enzymes, their constitution, structure synthesis, and behavior. Common properties of enzymes include: (1) their predominant, established role as catalysts for several types of chemical reactions, often providing the means of effecting chemical conversions otherwise difficult and at lower rates of energy expenditure; (2) their structure which suggests that most enzymes are simple or conjugated proteins; (3) their relatively high sensitivity to environmental conditions, including temperature and pH, and the presence of certain organic and inorganic materials; and (4) their origin from living cells. The environmental tolerance of enzymes closely parallels other substances associated with life processes. They tolerate a relatively narrow temperature span with denaturation (deactivation) occurring at temperatures generally above 50°C (122°F) and greatly reduced activity often occurs well above the freezing point of water. Enzymes have a low tolerance to a pH below 4, a minimal to no tolerance of certain organic solvents such as alcohol and acetone, and destruction by numerous organic and inorganic substances.

Unlike most inorganic catalysts, enzymes are very specific for the reactions which they catalyze. As a case in point, an acid catalyst will yield glucose, fructose, and galactose in the hydrolysis of raffinose (a trisaccharide). But, diastase will yield melibose and fructose; emulsin will yield sucrose and galactose. The glucosidic linkages are hydrolyzed at about equal rates with an acid catalyst, whereas the enzyme catalysts act on just one kind of linkage even though the difference in linkages is small. Whereas acids may catalyze numerous compounds, including amides, acetals, and esters, a given enzyme will confine its action to a very specific compound or related group. Because of this behavior, mixtures of enzymes often can be effective.

In addition to the very large role that enzymes play in life processes and medicine and in industrial fermentation and related processes, enzymes are finding a growing role in industrial products, such as detergents, where enzymes tend to break down proteins to water-soluble proteoses or peptones. Obviously for purposes of this type, enzymes must be selected that can remain active at relatively high pH values (8.5 to 9.5) and remain stable for long periods of storage.

The great number of reactions catalyzed by the enzymes in living organisms can be indicated by mentioning some of the major types. They include all the oxidation processes by which the organism obtains its energy—mechanical and thermal; the hydrolysis processes by which food carbohydrates, proteins, and fats are broken down into simpler molecules capable of direct oxidation or of use by the organism in constructing its own structure; and all the detoxification reactions by which many harmful substances that may be absorbed by the organism, as well as its normal waste products, are converted into forms suitable for excretion.

Activators. Most enzymes can function only with the assistance of certain other substances. These are broadly designated as *activators*, and are commonly grouped into two classes. The first is that of the nonspecific activators, which take no part in the reaction and appear to act by their effect upon the enzyme itself. The most important of these are the metallic ions K^+, Na^+, Rb^+, Cs^+, Mg^+, Ca^{2+}, Al^{3+}, Zn^{2+}, Cd^{2+}, Cr^{2+}, Mn^{2+}, Fe^{2+}, Co^{2+}, Ni^{2+}, Cu^{2+}. The second class of activators, mentioned earlier in this entry, are organic molecules, which enter into the reaction itself, often as carriers of a particular group. These substances are the group discussed in the entry on **Coenzymes**. In general, they are regenerated in their original form by other processes, so that they are not strictly substrates. On the other hand, the nicotinamideadenine nucleotides, which act as hydrogen carriers for various oxidoreductase reactions, may well be regarded as substrates.

A *substrate* may be defined as a substance modified by the action of an enzyme, or by the growing upon it of microorganisms. A *coenzyme* is a low-molecular-weight organic substance which can attach itself and thus supplement specific proteins to form active enzyme systems.

Inhibitors and Primers. Two other adjunct substances are the *inhibitors*, which retard or block enzyme action; and the *primers*, which enhance, or in some cases, are essential to it. An example is the priming of polyribonucleotide phosphotransferase by short ribonucleotide polymers.

Structure of Enzymes. The knowledge of the structure of enzymes is growing at a rapid rate, but much research remains before a high confidence level can be established pertaining to even the fundamentals of certain basic reactions.

One of the intensively investigated enzymes is ribonuclease (trivial name), which has the systematic name, polyribonucleoside 2-olionucleotide-transferase. This enzyme transfers a phosphate group from one position to another within a polynucleotide, forming a cyclic compound, and the pancreatic form of this enzyme can also catalyze the transfer of the phosphate group to water, which is a step in the depolymerization of RNA.

The molecular weight of this enzyme was found (by analysis of its constituent amino acids) to be 13,700. It was found to consist of a single polypeptide chain, internally cross-linked by four cystine residues, as evidenced by the failure of any drop in molecular weight to accompany the oxidation of all four cystines to cysteic acid, and also by the occurrence of only one terminal $-NH_2$ group and one terminal $-COOH$ group.

From this point, the primary structure was fully determined. The *primary structure* of an enzyme, or other protein, is the number, length, and composition of the polypeptide chains, the linear arrangement of their amino acids, and the number and position of the cross-links between chains. (The geometrical configuration of the molecule, which is usually a three-dimensional coiled and folded structure, and the side chains and their interactions were not determined.)

To determine the primary structure, after oxidation of the disulfide bridges between the cysteine molecules, the enzyme was cleaved into a series of linear polypeptides by the enzymatic action of trypsin. The fragments were separated by chromatography, and their individual amino acids were split off by acid hydrolysis. Then by repeating this process with another enzyme, a different series of polypeptide fragments were obtained, because the chains split at different points. By studying the overlaps of the two series, the fragments could be arranged in linear order. Combining the peptide structure so determined with the amino acids found gave the provisional primary structure of the enzyme shown in the accompanying figure.

The final step in complete elucidation of the three-dimensional structure of ribonuclease was made by researchers at the Roswell Park Memorial Institute, Buffalo, New York. This group, headed by Dr. David Harker, employed x-ray diffraction techniques. Roughly 500,000 diffraction points were recorded and the data so obtained was fed to a computer.

The elucidation of the structure of ribonuclease follows that of lycozyme by a group at London's Royal Institute headed by Dr. David C. Phillips, and that of the other protein, myoglobin, for which Dr. Max F. Perutz and Dr. John C. Kendrew of Cambridge University received the Nobel Prize in chemistry in 1962.

Note in this structure of enzyme that specific points are marked as those at which other enzymes act. This feature of "active sites" is characteristic of enzyme behavior. In the case of the enzymes trypsin or chymotrypsin, the active sites for peptide or ester hydrolysis contain the functional groups of two histidine residues and of a serine residue. The ester enters the active region, forming temporary bonds with the enzyme at that point, the $-OR$ group of the ester becoming bonded to a hydrogen atom of the enzyme. Then the bond between the hydrogen atom and the enzyme breaks, releasing the alcohol of the hydrogen atom and the enzyme breaks, releasing the alcohol of the ester. As a next step, H_2O adds from the solution to the complex, and by another bond rupture, the acid part of the ester is released, leaving the enzyme in its original condition. The overall reaction is a simple hydrolysis of the ester,

$$R'COOR + H_2O \rightarrow ROH + R'COOH$$

but a large number of steps may be involved.

The determination of sites of active centers is effected not only by splitting of enzymes, but also by treating them with temporary or permanent inhibitors, and determining their points of attachment. Other methods of studying enzymes are by means of enzyme induction and enzyme repression.

Enzyme Induction. An example of *enzyme induction* is the growth of the bacteria *Escherichia coli* in a suitable culture medium. If no beta-galactoside is added to the medium, the bacteria form scarcely any of the enzyme that hydrolyzes that sugar. The addition of the sugar to the medium increases the production of the enzyme by the cell by as much as 10,000 times. On the other hand, the same bacteria will produce the enzyme tryptophan synthase only if tryptophan is absent from the culture medium. These observations are useful, not only in interpreting enzyme action, but also in determining its relationship to genetics, for in this case, the genes determining the ability to synthesize both enzymes are present on the chromosome map of the organism.

There are methods used to study enzymes other than those of chemical instrumental analysis, such as chromatography, that have already been mentioned. Many enzymes can be crystallized, and their structure investigated by x-ray or electron diffraction methods. Studies of the kinetics of enzyme-catalyzed reactions often yield useful data, much of this work being based on the Michaelis-Menten treatment. Basic to this approach is the concept that the action of enzymes depends upon the formation by the enzyme and substrate molecules of a complex, which has a definite, though transient, existence, and then decomposes into the products of the reaction. Note that this point of view was the basis of the discussion of the specificity of the active sites discussed above.

A simple enzyme reaction may thus be written as

$$E + S \leftrightarrows ES \rightarrow \text{products}$$

In the Michaelis-Menten treatment, this equation can be regarded as the result of the three processes:

rate of formation of $ES = k_1[E][S]$
rate of decomposition of ES into products $= k_2[ES]$
rate of decomposition of ES into original reactants $= k_3[ES]$

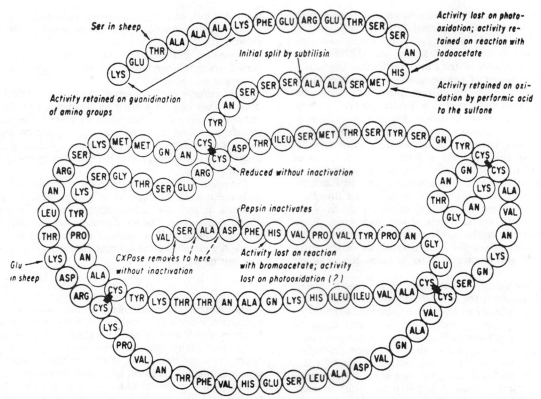

Ribonuclease. Abbreviations used: alanine (ALA), adenine (AN), arginine, ARG), aspartic acid (ASP), cysteine (CYS), cytosine (CN), glutamic acid (GLU), glycine (GLY), guanine (GN), histidine (HIS), isoleucine (ILEU), leucine (LEU), lysine (LYS), methionine (MET), phenylalanine (PHE), proline (PRO), serine (SER), threonine (THR), thymine (TN), tryptophan (TRY), tyrosine (TYR), uracil (UN), and valine (VAL).

where the terms in square brackets denote concentrations of E, S, and ES, and the k's are rate constants. By representing the ratio $(k_2 + k_3)/k_1$ by K_m, the *Michaelis constant*, we can obtain a form of the Michaelis equation

$$\frac{[E_t]}{v} = \frac{K_m}{k_3 + [S]} + \frac{1}{k_3}$$

where E_t is the total concentration of enzyme (as distinguished from $[E]$, the concentration of free enzyme), and v is the velocity of the reaction. By plotting $[E_t]/v$ against $1/[S]$, we can obtain, at the axis-intercepts, values of $1/K_m$ and $1/v$.

This approach has been successfully extended to the more complex enzymatic reactions involving inhibitors, activators, and even multiple reactions in which the successive action of more than one enzyme is involved.

The observation follows that in addition to the specificity of the enzymes with respect to reaction type and to structure of the substrate, with respect to reaction type and to structure of the substrate, the action is also confined to a single configuration of the substrate. If the molecular structure of the substrate is unsymmetrical (asymmetric) and therefore two compounds exist—one the mirror image of the other—a specific enzyme will act upon only one of the stereoisomers. This specificity is undoubtedly due to the fact that the region of interaction on the enzyme is also asymmetric and exists in only one form or configuration. Racemic mixtures or certain substrates may sometimes be separated by making use of the fact that enzymic action will affect only one of the two forms.

Enzyme Repression. Repression as applied to biochemical reactions is a process of feedback control whereby a cell limits its production of the substances produced within it. An example that has been investigated shows the nature and mechanism by which this limitation is effected. It has been found that production of the amino acid L-isoleucine by cells of the bacterium *Escherichia coli* is repressed in the presence of an excess of the product. This excess is obtained experimentally by adding the substance to the culture medium in which the bacterium is grown. A form of the L-isoleucine is used that has been labeled with a radioactive isotope, so that the mechanism of the repression can be followed.

By this means, it has been found that the excess of L-isoleucine has two distinct effects—one that is relatively slow, and another that is rapid. The slower effect is to repress production by the cell of all the enzymes required to catalyze the series of biochemical reactions in the metabolic pathway by which the cell synthesizes L-isoleucine. The fast effect is to inhibit production of the enzyme for the first reaction in the series. This enzyme is L-threonine deaminase, which removes the amino group from L-threonine as a preliminary step to its oxidation and reamination in order to product L-isoleucine from it.

The independent existence of these two effects was demonstrated by the discovery among mutations of *E. coli*, a mutation that exhibited only one of the effects, and of another mutation that exhibited only the other, leading to the conclusion that two distinct genes are involved in the control system.

An even more striking instance of feedback control is found

in the synthesis of DNA (see **Nucleic Acids and Nucleotides**). As pointed out in that entry, normal DNA is composed of the nucleotides deoxyguanosine, deoxycytidine, deoxyadenosine, and thymidine, and the amounts of the first and second of these are the same, as are those of the third and fourth. Obviously, close control is required of the amounts of these nucleotides that are synthesized by the cell, if they are to be made in the quantities required for DNA synthesis. Evidence has been found that the enzyme carbamoylphosphate: L-aspartate carbamoyl transferase, which catalyzes the reaction between asparatic acid and carbamoyl phosphate (which has deoxycytidine triphosphate, CTP, as its final product), is inhibited by an excess of the CTP, and is also initiated (or activated) by an excess of deoxyadenosine triphosphate, which requires an equal amount of CTP to react with it in forming DNA. There are thus both positive and negative feedback controls on the synthesis of the enzymes that catalyze the synthesis of the nucleotides. Since all enzymes are proteins, the mechanism of this control is believed to be that suggested for protein synthesis in the entry on **Nucleic Acids and Nucleoproteins**. See also **Recombinant DNA**.

An aspect of the control of enzymatic action that is related to the effect of initiation (or activation) just discussed is the effect of *induction*, which can readily be illustrated experimentally. Many years ago, it was found that the yeast, *Saccharomyces ludwigii*, although able to ferment many sugars, was ineffective on lactose (milk sugar), because it did not synthesize the necessary enzyme, lactase (β-D-galactoside galactohydrolase). However, if this yeast was grown for several generations on a medium containing lactose, it acquired the ability to make lactase, and its subsequent generations retained that ability. In the years since this discovery, so many instances of induction have been discovered that they are regularly cited in discussions of the properties of those enzymes for which they are known, as are also the repressing and blocking substances.

Experimental evidence is accumulating to lead to an explanation of the precise genetic and molecular mechanism of the action of both repression and induction.

Since all enzymes are proteins, and since, as explained in the entry on nucleic acids, proteins are synthesized by a process of replication, starting from the DNA of the genes, it is considered that there are "repressor" and "inductor" molecules, which, respectively, either block or play a necessary part in that process for the particular enzyme.

See also **Bile; Carbohydrates; Detergents; Starch**.

As shown by the accompanying table, enzymes are classified into six groups: (1) oxidoreductases, (2) transferases, (3) hydrolases, (4) lyases, (5) isomerases, and (6) ligases or synthetases. The main group to which an enzyme belongs is indicated by the first figure of the code number. The second figure indicates the subclass; for the oxidoreductases, it shows the type of group in the *donors* which undergoes oxidation; for the transferases, it indicates the nature of the group which is transferred; for the hydrolases, it shows the type of bond hydrolyzed; for the lyases, the type of link which is broken between the group removed and the remainder; for the isomerases, the type of isomerization involved; and for ligases, the type of bond formed.

The third figure of the code number, indicating the sub-sub class, shows for the oxidoreductases the type of acceptor involved; for the transferases and hydrolases, it shows more precisely the type of group transferred or bond hydrolyzed; for the lyases, it shows the nature of the group removed; for the isomerases, it indicates in more detail the nature of the isomerization; and for the ligases, it shows the nature of the substance

formed. Thus, an enzyme number, commonly indicated by the prefix EC, provides fairly detailed information about a specific enzyme.

Categories of Reactions. The comparative simplicity of the classification scheme bears testimony to the underlying unity of enzymatic catalysis.

Oxidoreductases. The overall reaction catalyzed by the oxidoreductases can be written as hydrogen transfer, and these enzymes might be considered to be merely one section of the transferases. The oxidoreductases are classified separately because of their large number and because of their great biological importance in bringing about the main energy-yielding reactions of living tissues.

Transferases. The main groups of transferases are concerned with the transfer of *one-carbon* groups, acyl groups, glycosyl residues, amino- and other nitrogen-containing groups, phosphate, and sulfate. Oxidoreductases and transferases together represent about half or more of the enzymes presently recognized. A general reaction for both oxidoreductases and transferases can be written:

$$AX + B \rightleftharpoons A + BX$$

Hydrolases. These enzymes include esterases, glycosidases, peptidases, deaminases, and enzymes which hydrolyze acid anhydrides (such as the pyrophosphate group in adenosinetriphosphate). Many hydrolases have been shown to be able, under appropriate conditions, to catalyze transfer reactions; a high concentration of acceptor is usually necessary, since there is competition between the added acceptor and water for the group transferred. The detailed mechanism in these cases probably involves transfer of a part of the substrate onto a group on the enzyme, with subsequent transfer to an acceptor or hydrolysis, e.g., for a hydrolase acting on a substrate AB to produce AOH and BH:

$$EH + AB \rightarrow E—A + BH$$

and

$$E—A + X \rightarrow E + AX$$

or

$$E—A + H_2O \rightarrow EH + AOH$$

These hydrolases, if not all, can therefore be regarded as transfereases which include H_2O among their possible acceptors. Under normal conditions, in aqueous solution, hydrolysis will be the dominant reaction.

Lyases. Enzymes in this grouping catalyze reactions of the type:

$$AX—BY \rightleftharpoons A = B + X—Y$$

Molecules, such as H_2O, H_2S, NH_3, or aldehydes, are added across the double bond of a second unsaturated molecule. Decarboxylases, such as those acting on amino acids can be regarded as lyases (carboxy-lyases), assuming CO_2 and not H_2CO_3 to be the immediate product of decarboxylation. Over one hundred lyases are known.

Isomerases. These include enzymes which bring about reactions similar to those in several other groups, but distinguished in that the reaction takes place entirely within one molecule, which is not cleaved, so that the overall reaction is

$$A \rightleftharpoons B$$

1. OXIDOREDUCTASES
  1.1 *Acting on the CH—OH group of donors*
    1.1.1 With NAD or NADP as acceptor
    1.1.2 With cytochrome as an acceptor
    1.1.3 With O_2 as acceptor
    1.1.99 With other acceptors
  1.2 *Acting on the aldehyde or keto group of donors*
    1.2.1 With NAD or NADP as acceptor
    1.2.2 With a cytochrome as an acceptor
    1.2.3 With O_2 as acceptor
    1.2.4 With lipoate as acceptor
    1.2.99 With other acceptors
  1.3 *Acting on the CH—CH group of donors*
    1.3.1 With NAD or NADP as acceptor
    1.3.2 With a cytochrome as an acceptor
    1.3.3 With O_2 as acceptor
    1.3.99 With other acceptors
  1.4 *Acting on the CH—NH$_2$ groups of donors*
    1.4.1 With NAD or NADP as acceptor
    1.4.3 With O_2 as acceptor
  1.5 *Acting on the C—NH group of donors*
    1.5.1 With NAD or NADP as acceptor
    1.5.3 With O_2 as acceptor
  1.6 *Acting on reduced NAD or NADP as donor*
    1.6.1 With NAD or NADP as acceptor
    1.6.2 With a cytochrome as an acceptor
    1.6.4 With a disulfide compound as acceptor
    1.6.5 With a quinone or related compound as acceptor
    1.6.6 With a nitrogenous group as acceptor
    1.6.99 With other acceptors
  1.7 *Acting on other nitrogenous compounds as donors*
    1.7.3 With O_2 as acceptors
    1.7.99 With other acceptors
  1.8 *Acting on sulfur groups of donors*
    1.8.1 With NAD or NADP as acceptor
    1.8.3 With O_2 as acceptor
    1.8.4 With a disulfide compound as acceptor
    1.8.5 With a quinone or related compound as acceptor
    1.8.6 With a nitrogenous group as acceptor
  1.9 *Acting on heme groups of donors*
    1.9.3 With O_2 as acceptor
    1.9.6 With a nitrogenous group as acceptor
  1.10 *Acting on diphenols and related substances as donors*
    1.10.3 With O_2 as acceptor
  1.11 *Acting on H_2O_2 as acceptor*
  1.12 *Acting on hydrogen as donor*
  1.13 *Acting on single donors with incorporation of oxygen (oxygenases)*
  1.14 *Acting on paired donors with incorporation of oxygen into one donor (hydroxylases)*
    1.14.1 Using reduced NAD or NADP as one donor
    1.14.2 Using ascorbate as one donor
    1.14.3 Using reduced pteridine as one donor
2. TRANSFERASES
  2.1 *Transferring one-carbon groups*
    2.1.1 Methyltransferases
    2.1.2 Hydroxymethyl-, formyl-, and related transferases
    2.1.3 Carboxyl- and carbamoyltransferases
    2.1.4 Amidinotransferases
  2.2 *Transferring aldehydic or ketonic residues*
  2.3 *Acyltransferases*
    2.3.1 Acyltransferases
    2.3.2 Aminoacyltransferases
  2.4 *Glycosyltransferases*
    2.4.1 Hexosyltransferases
    2.4.2 Pentosyltransferases
  2.5 *Transferring alkyl or related groups*
  2.6 *Transferring nitrogenous groups*
    2.6.1 Aminotransferases
    2.6.3 Oximinotransferases
  2.7 *Transferring phosphorus-containing groups*
    2.7.1 Phosphotransferases with an alcohol group as acceptor
    2.7.2 Phosphotransferases with a carboxyl group as acceptor
    2.7.3 Phosphotransferases with a nitrogenous group as acceptor
    2.7.4 Phosphotransferases with a phospho-group as acceptor
    2.7.5 Phosphotransferases, apparently intramolecular
    2.7.6 Pyrophosphotransferases
    2.7.7 Nucleotidyltransferases
    2.7.8 Transferases for other substituted phospho-groups
  2.8 *Transferring sulfur-containing groups*
    2.8.1 Sulfurtransferases
    2.8.2 Sulfotransferases
    2.8.3 CoA-transferases

3. HYDROLASES
  3.1 *Acting on ester bonds*
    3.1.1 Carboxylic ester hydrolases
    3.1.2 Thiolester hydrolases
    3.1.3 Phosphoric monoester hydrolases
    3.1.4 Phosphoric diester hydrolases
    3.1.5 Triphosphoric monoester hydrolases
    3.1.6 Sulfuric ester hydrolases
  3.2 *Acting on glycosyl compounds*
    3.2.1 Glycoside hydrolases
    3.2.2 Hydrolyzing N-glycosyl compounds
    3.2.3 Hydrolyzing S-glycosyl compounds
  3.3 *Acting on ether bonds*
    3.3.1 Thioether hydrolases
  3.4 *Acting on peptide bonds (peptide hydrolases)*
    3.4.1 α-Aminoacyl-peptide hydrolases
    3.4.2 Peptidyl-amino acid hydrolases
    3.4.3 Dipeptide hydrolases
    3.4.4 Peptidyl-peptide hydrolases
  3.5 *Acting on C—N bonds other than peptide bonds*
    3.5.1 In linear amides
    3.5.2 In cyclic amides
    3.5.3 In linear amidines
    3.5.4 In cyclic amidines
    3.5.5 In cyanides
    3.5.99 In other compounds
  3.6 *Acting on acid-anhydride bonds*
    3.6.1 **In phosphoryl-containing anhydrides**
  3.7 *Acting on C—C bonds*
    3.7.1 In ketonic substances
  3.8 *Acting on halide bonds*
    3.8.1 In C-halide compounds
    3.8.2 In P-halide compounds
  3.9 *Acting on P—N bonds*
4. LYASES
  4.1 *Carbon-carbon lyases*
    4.1.1 Carboxy-lyases
    4.1.2 Aldehyde-lyases
    4.1.3 Ketoacid-lyases
  4.2 *Carbon-oxygen lyases*
    4.2.1 Hydro-lyases
    4.2.99 Other carbon-oxygen lyases
  4.3 *Carbon-nitrogen lyases*
    4.3.1 Ammonia-lyases
    4.3.2 Amidine-lyases
  4.4 *Carbon-sulfur lyases*
  4.5 *Carbon-halide lyases*
  4.99 *Other lyases*
5. ISOMERASES
  5.1 *Racemases and epimerases*
    5.1.1 Acting on amino acids and derivatives
    5.1.2 Acting on hydroxyacids and derivatives
    5.1.3 Acting on carbohydrates and derivatives
    5.1.99 Acting on other compounds
  5.2 *Cis-trans isomerases*
  5.3 *Intramolecular oxidoreductases*
    5.3.1 Interconverting aldoses and ketoses
    5.3.2 Interconverting keto- and enol-groups
    5.3.3 Transposing C=C bonds
  5.4 *Intramolecular transferases*
    5.4.1 Transferring acyl groups
    5.4.2 Transferring phosphoryl groups
    5.4.99 Transferring other groups
  5.5 *Intramolecular lyases*
  5.99 *Other isomerases*
6. LIGASES OR SYNTHETASES
  6.1 *Forming C—O bonds*
    6.1.1 Aminoacid-RNA ligases
  6.2 *Forming C—S bonds*
    6.2.1 Acid-thiol ligases
  6.3 *Forming C—N bonds*
    6.3.1 Acid-ammonia ligases (amide synthetases)
    6.3.2 Acid-amino acid ligases (peptide synthetases)
    6.3.3 Cyclo-ligases
    6.3.4 Other C—N ligases
    6.3.5 C—N ligases with glutamine as N-donor
  6.4 *Forming C—C bonds*

Thus, there are intramolecular oxidoreductases (e.g., ketolisomerases), intramolecular transferases (e.g., phosphomutases), and intramolecular lyases. About fifty isomerases are known.

Ligases. These enzymes catalyze reactions which are more complex than those of the other groups and must involve at least two separate stages in the reaction. The overall result is the synthesis of a molecule from two components with a coupled breakdown of adenosine triphosphate, or some other nucleoside triphosphate. In general, this may be written:

$$X + Y + ATP \rightarrow XY + AMP$$

$$+ \text{ Pyrophosphate (or ADP + Phosphate)}$$

These enzymes, of which about fifty are known, are of great importance in the conservation of chemical energy within the cell and in the coupling of synthetic processes with energy-yielding breakdown reactions.

References

Baker, T. S., Eisenberg, D., and F. Eiserling: "Ribulose Bisphosphate Carboxylase: A Two-layered, Square-shaped Molecule of Symmetry 422," *Science*, **196**, 293–295 (1977).

Brattsen, L. B., Wilkinson, C. F., and T. Eisner: "Herbivore-Plant Interactions: Mixed-Function Oxidases and Secondary Plant Substances," *Science*, **196**, 1349–1352 (1977).

Cornforth, J. W.: "Asymmetry and Enzyme Action," *Science*, **193**, 121–125 (1976).

Dixon, N. E., et al.: "Metal Ions in Enzymes using Ammonia or Amides," *Science*, **191**, 1144–1150 (1976).

Eisenberg, D.: "Enzyme Structure Control," Academic, New York (1970).

Gray, J. C.: "Enzyme Catalysed Reactions," Van Nostrand Reinhold, New York (1971).

Gutfreund, H.: "Enzymes: Physical Principles," Wiley, New York (1972).

Jakoby, W. B.: "Enzyme Purification and Related Techniques," Academic, New York (1971).

Meister, A.: "On the Enzymology of Amino Acid Transport," *Science*, **180**, 33–39 (1973).

Ory, R. L., and A. J. St. Angelo, Editors: "Enzymes in Food and Beverage Processing," American Chemical Society, Washington, D.C. (1977).

Perlman, G., and L. Lorand: "Proteolytic Enzymes," Academic, New York (1970).

Plowman, K. M.: "Enzyme Kinetics," McGraw-Hill, New York (1972).

Segal, H. L.: "Enzymatic Interconversion of Active and Inactive Forms of Enzymes," *Science*, **180**, 25–32 (1973).

Wang, D. C., et al.: "Fermentation and Enzyme Technology," Wiley, New York (1979).

ENZYME PREPARATIONS.

During the past several years, a number of commercially prepared enzyme formulations have been available to processors, notably for use in the food industry. These preparations fall into three basic categories: (1) Animal-derived preparations; (2) plant-derived preparations; and (3) microbially derived preparations.

Fruit juices, jams, and jellies, corn (maize) syrups and sweeteners, structured protein foods, and tenderized meats are exemplary of products, the quality of which has been improved through the use of enzyme preparations. Principal areas of development in food-grade enzymes research have been toward upgrading quality and byproduct utilization, higher rates and levels of extractions, synthetic food developments, sweetener development, improving flavor of foods, and the stabilization of food quality and nutrition. Enzyme preparations also are used in the detergent and pharmaceutical fields.

Animal-derived enzyme preparations include catalase (bovine liver), lipase, pepsin, rennet, and trypsin. Plant-derived preparations include bromelain, cellulase, ficin, malt, papain, and pectinase. Microbially derived preparations include amylases, carbohydrase, catalase, glucose oxidase, lipase, protease, and zymase.

EOSINOPHILS. Blood.

EPHEDRINE. Alkaloids.

EPICHLOROHYDRIN.

A highly reactive and industrially important compound with the structural formula

$$\underset{\underset{H}{|}}{Cl}-\underset{\underset{H}{|}}{\overset{\overset{H}{|}}{C}}-\underset{\underset{H}{|}}{\overset{\overset{O}{\diagup\!\diagdown}}{C}}-\underset{\underset{H}{|}}{\overset{}{C}}-H$$

Epichlorohydrin is also called 1-chloro-2,3-epoxypropane and is classified as an organic epoxide. The compound is a colorless, clear, mobile liquid with an odor something like chloroform. Molecular wt 92.53, mp −57.1°C, bp 116.07°C, sp gr 1.1750 (25°C). Solubility is 6.53 g/100 g H_2O. Epichlorohydrin is made by the chlorohydrination of allyl chloride, in which 1,2-dichlorohydrin and 1,3-dichlorohydrin are produced as intermediates.

One of the most common epoxy resins is produced by the reaction between epichlorohydrin and bisphenol A. See also **Epoxy Resins.** The compound also is used in the production of epichlorohydrin-based rubbers which have good aging, high resiliency, and flexibility at low temperatures, advantage of which is taken in automotive and aircraft parts, seals, gaskets, hose, belting, wire, and cable jackets. These rubbers also have good resistance to solvents, fuels, oils, and ozone. A number of wet-strength resins for use in the paper industry also are derived from epichlorohydrin, including (a) epichlorohydrin-modified polyamides; and (b) the addition of epichlorohydrin to high-molecular-weight polyalkylene polyamines. The advantages of these resins is that no alum or acid medium is required for incorporating the resin into the cellulose pulp. During the drying process, the resin crosslinks and thus yields a paper with permanent wet-strength properties. Ion-exchange resins also can be prepared by reacting epichlorohydrin with ethylenediamine or a similar amine. The resulting material is a stable, water-insoluble anion-exchange resin.

In addition to its use in the production of epoxy resins, epichlorohydrin is used in large quantities in the manufacture of glycerin. Other uses include textile applications, where it is used to modify the carboxy groups of wool, thus increasing durability and improving moth resistance; in the synthesis of antistatic agents, wrinkle-resistance agents, and coating sizings. Effective against the larvae of certain insects, the compound is used in control chemicals for agriculture where permitted.

EPIDOTE.

This mineral is a hydrous silicate of calcium, aluminum, and iron with the formula, $Ca_2(Al, Fe)_3Si_3O_{12}(OH)$. The ratio of aluminum to iron ranges from 6:1 to 3:2. Epidote is found in prismatic monoclinic crystals, which may be acicular to fibrous. Fine granular and compact masses are common. The mineral displays one good cleavage, an uneven fracture; is brittle; hardness, 6–7; specific gravity, 3.25–3.5; luster, vitreous to resinous; typical color, pistachio green, but may be yellowish- to brownish-green, sometimes red, yellow, gray, white or colorless. Colorless to grayish streak; transparent to opaque. The characteristic color of ordinary epidote makes it usually an easily identified mineral.

It occurs commonly in metamorphic rocks as gneisses and

schists; however, it seems probable that under certain conditions it may appear as a primary mineral, for example in granitic rocks. The Urals, Austria, Switzerland, Italy, France and Norway are known for their occurrences of fine epidote crystals. In the United States epidote has been found in excellent specimens at Franconia and Warren, New Hampshire; Huntington, Massachusetts; Willimantic and Haddam, Connecticut; Chaffee County, Colorado, and Riverside County, California. The word epidote is derived from the Greek. The name pistacite, from the Greek word meaning the pistachio nut, has been occasionally applied to this mineral. It has been used as a gemstone but is in little demand for this purpose.

See also terms listed under **Mineralogy.**

EPINEPHRINE. Alkaloids; Hormones.

EPOXY RESINS. A family of thermosetting resins known for their excellent mechanical and electrical properties, dimensional stability, resistance to high temperatures and numerous chemicals, and for their strong adhesion to glass, metal, fibers, and numerous other materials. Structurally, the epoxy groups are three-membered rings with one oxygen and two carbon atoms. The most common epoxy resins are made by reacting epichlorohydrin with a polyhydroxy compound, such as bisphenol A, in the presence of a catalyst. Epoxy resins produced in this fashion are known as diglycidyl ethers of bisphenol A (bis-A). The structural formula is shown below.

By changing the ratio of epichlorohydrin to bis-A, resins range from low-viscosity liquids to high-melting solids. The structure shown represents a solid epoxy novolak resin. The epoxy phenol novolak resins are the most important. Basically they are novolak resins whose phenolic hydroxyl groups have been converted to glycidyl ethers. Epoxidized novolaks are used principally in solid single-stage molding compounds and high-temperature laminating systems.

The liquid cycloaliphatic epoxies normally are produced by the peracetic acid expoxidation of cyclic olefins where the epoxide groups are attached directly to the cycloaliphatic ring. These materials have better weatherability, arc and tracking resistance and dielectric strength over conventional epoxies.

Usually the major makers of resins, hardeners and other chemicals for epoxy systems do not supply finished compounds. The compounding is done by specialized firms and by some large epoxy users. The epoxy resins per se are not finished products, but are reactive chemicals to be combined with other chemicals to yield systems capable of conversion to a predetermined thermoset structure.

Epoxy resins are cured by cross-linking agents known as hardeners or by catalysts which promote self-polymerization. Some of the cross-linking agents used include the primary and secondary aliphatic polyamines, such as diethylenetriamine, triethylenetetramine, tetraethylenepentamine, diethylaminopropylamine, and piperazines. Most of these materials are liquids of moderately low viscosity that can be blended with the resins at room temperature.

Because of excellent electrical properties, epoxies are used in casting, potting, and encapsulation of electrical/electronic parts. Advantage is taken of the low shrinkage of the epoxies, with absence of cracking or separation of the resins from the parts during cure. Encapsulated parts vary from miniature coils and switches that may weigh but a few grams to large motors and insulators that may weigh several pounds.

Epoxy resins also find use in making chemical-loaded molecular sieves, adhesives, and protective coatings. Epoxy-based adhesives are widely used for bonding dissimilar materials like plastics and metal, wood and metal, and ceramics and rubber. Minimum pressure is required to obtain a satisfactory bond. For such applications, epoxy adhesives are available as one- or two-part systems. The one-part systems require curing at elevated temperatures. The two-part systems can be cured at room temperature, but both have better resultant properties if cured under heat. Epoxy resin-based coatings provide outstanding chemical resistance, toughness, flexibility, and adhesion to most substrates.

Epoxy resins are used in the chemical industry because of their excellent resistance to attack by many corrosive chemicals. One application is a protective coating for container, pipe, and tank liners, floors, and walls. High-pressure vessels are made by filament winding with cycloalphatic epoxies.

EPSOMITE. Epsomite is normally found as an efflorescence on mine and cave walls. It belongs to the orthorhombic crystal system, being a hydrous sulfate of magnesium, $MgSO_4 \cdot 7H_2O$, with vitreous to earthy luster and colorless to white, of transparent/translucent quality. Hardness of 2–2.5, and specific gravity of 1.68. Very bitter to the taste. Found with the soluble salt lake deposits in Strassfurt, Germany, and in limestone caves in Kentucky, Tennessee and Indiana; also in several California and Colorado abandoned mines.

EPSOM SALTS. Magnesium.

EPSTEIN-BARR VIRUSES. Virus.

EQUILIBRIUM (Chemical). Chemical Equilibrium.

EQUILIBRIUM DISTILLATION. Distillation.

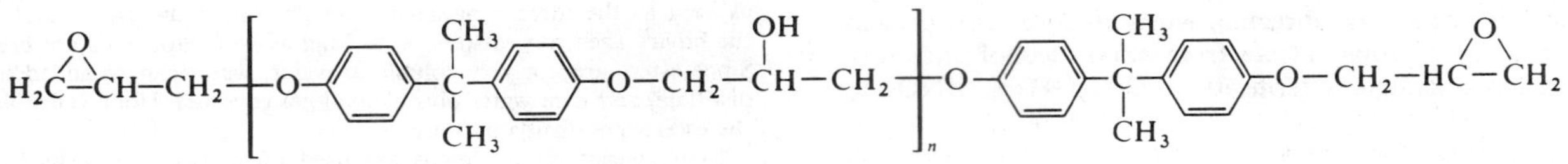

ERBIUM. Chemical element symbol Er, at. no. 68, at. wt. 167.26, eleventh in the Lanthanide Series in the periodic table, mp 1529°C, bp 2868°C, density 9.066 g/cm³ (20°C). Elemental erbium has a close-packed hexagonal crystal structure at 25°C. The pure metallic erbium is silver-gray in color and retains its luster at room temperature, not affected by moisture or normal atmospheric gases. Large pieces of the metal do not oxidize readily even when heated. Fine chips and powder, however, will ignite and burn. Because of its comparative softness, the metal can be worked by conventional equipment. The metal should be annealed after size-reduction. There are six natural isotopes ^{162}Er, ^{164}Er, ^{166}Er through ^{168}Er and ^{170}Er. Twelve artificial isotopes have been prepared. The natural isotopes are not radioactive. In terms of abundance, erbium is present on the average of 2.8 ppm in the earth's crust, making its potential availability about equal with uranium. The element was first identified by C. G. Mosander in 1843. The thermal-neutron-absorption cross section of erbium is 160 barns per atom, relatively high and tenth among the natural elements. The metal has a low acute-toxicity rating. Electronic configuration $1s^2 2s^2 2p^6 3s^2 3p^6 3d^{10} 4s^2 4p^6 4d^{10} 4f^{11} 5s^2 5p^6 5d^1 6s^2$. First ionization potential 6.10 eV; second 11.93 eV. Ionic radius Er^{3+} 0.881 Å. Metallic radius 1.758 Å. Other important physical properties of erbium are given under **Rare-Earth Elements and Metals.**

Erbium occurs in certain types of apatites, xenotime, and gadolinite. These minerals also are processed for their yttrium content as well as for other heavy *Lanthanide* elements. With liquid-liquid organic and solid-resin organic ion-exchange techniques, the separation of erbium from the other elements is favorable.

Because of the metal's high thermal-neutron-absorption cross section, it has been of much interest in terms of use in nuclear reactor hardware. When an erbium-activated phosphor is coated onto a gallium-arsenide diode, the latter emits infrared radiation, which is converted to visible light by the phosphor. Through variation of the energizing power and by use of a combination of rare-earth-activated phosphors, the primary colors of light can be produced. Thus, erbium holds promise for use in display panels and color-television picture tubes. An erbium hydride-hydrogen system at a fixed temperature creates an extreme vacuum and when used for comparative purposes makes it possible to measure vacuums in the range of 10^{-4} to 10^{-11} torr with much precision. The system has been used for the calibration of ionization gauges used for very high vacuums as found in outer space. Erbium is in an early stage of investigation for application to lasers, semiconductor devices, garnet microwave devices, ferrite bubble devices, and catalysts.

See references listed at the ends of the entries on **Chemical Elements;** and **Rare-Earth Elements and Metals.**

NOTE: This 4th Edition entry was revised and updated by K. A. Gschneidner, Jr., Director, and B. Evans, Assistant Chemist, Rare-Earth Information Center, Energy and Mineral Resources Research Institute, Iowa State University, Ames, Iowa.

ERYTHRITE. A mineral of the composition, $Co_3(AsO_4)_2 \cdot 8H_2O$, isomorphous with annabergite. The color ranges from rose to crimson. The mineral sometimes contains nickel and occurs in monoclinic crystals, in earthy forms (as a weathering product of cobalt ores) in the oxidized portions of the veins, or in globular and reniform masses. The mineral sometimes is referred to as *erythrine, cobalt bloom,* and *peachblossom ore.*

ERYTHROCYTES. Blood.

ERGOSTEROL. Photochemistry and Photolysis.

ESSENTIAL AMINO ACIDS. Amino Acids.

ESTERS. The compound resulting from the reaction of an alcohol with an acid is termed an *ester.* The reaction is termed *esterification* and is accompanied by the yield of H_2O along with the ester. The reaction is highly reversible and hydrolysis will occur in the reverse direction when H_2O remains present. The formation of ethyl nitrate from ethyl alcohol and nitric acid typifies a simple esterification: $C_2H_5OH + HNO_3 \rightleftharpoons C_2H_5NO_3 + H_2O$.

Under normal conditions, esterification occurs slowly and inasmuch as the reaction is fully reversible, an equilibrium is reached which tends to withhold completion of the reaction in either direction.

The esterification reaction can be speeded up by the use of a catalyst. Such a catalyst is hydrogen ion, as from HCl or H_2SO_4. Side reactions may occur, HCl furnishing some organic chloride, and H_2SO_4 causing dehydration of the alcohol. Phosphoric acid generally avoids both these results. Salts that hydrolyze to furnish hydrogen ions are also used, e.g., zinc chloride, aluminum sulfate, ferric chloride, sometimes by the addition of acid to these salts.

The equilibrium point can be displaced to produce more ester by increasing the relative amounts of either the acid or the alcohol as desired.

A complicating factor of considerable significance when recovery of the ester is to be made by distillation is the existence of 2-component (binary) azeotropes of constant boiling points with (1) ester and H_2O and (2) ester and alcohol, and 3-component (ternary) azeotropes with (3) ester and H_2O and alcohol, selected from the accompanying table.

Esters are high-tonnage chemicals. Among the more important esters are normal, secondary, and isobutyl acetates, ethyl acetate, normal, secondary, and isoamyl acetates, and methyl acetate. These acetates are used primarily in the lacquer industry. Cellulose nitrate is used in the plastics, lacquer, and explosives industries; cellulose acetate in the plastics and lacquer industries; glyceryl trinitrate in the explosives industry; while cellulose xanthate (viscose) is an important synthetic product for textiles. In the specialized field of plasticizers for the plastics and lacquer industries, numerous synthetic esters are used; as examples, butyl stearate, diamyl phthalate, dibutyl oxalate, dibu-

AZEOTROPES ENCOUNTERED IN ESTERIFICATION PROCESSES

ESTER [Alcohol]	bp °C		AZEOTROPE bp °C	Ester %	Other Component %
Ethyl acetate	77.1	a	70.4	93.9	6.1 water
[Ethyl alcohol	78.3]	b	71.8	69.2	30.8 alcohol
		c	70.3	83.2	{ 7.8 water { 9.0 alcohol
n-Propyl acetate	101.6	a	82.4	86.0	14.0 water
[*n*-Propyl alcohol	97.2]	b	94.7	49.0	51.0 alcohol
		c	82.2	59.5	{ 21.0 water { 19.5 alcohol
iso-Propyl acetate	91.0	a	77.4	93.8	6.2 water
[*iso*-Propyl alcohol	82.5]	b	81.3	40.0	60.0 alcohol
		c			
n-Butyl acetate	126.2	a	90.2	71.3	28.7 water
[*n*-Butyl alcohol	117.8]	b	117.2	53.0	47.0 alcohol
		c	89.4	35.3	{ 37.3 water { 27.4 alcohol
n-Amyl acetate	148.8	a	95.2	59.0	41.0 water
[*n*-Amyl alcohol	137.8]	b			
		c	94.8	10.5	{ 56.2 water { 33.3 alcohol

The ternary azeotrope has the lowest boiling point and in distillation, as long as the three components are present, it distills off first, then the binary azeotrope distills off as long as its components are present. Since esters are not very soluble in water, water can be added to the distillate, and then water plus alcohol are removed. Upon redistillation the ester soon distills off pure.

Many variations in details are used in applying the principles of esterification.

tyl phthalate (also for smokeless powder), dibutyl sebacate, dibutyl tartrate, diethylene glycol monostearate, diethylene glycol distearate, diethyl phthalate, dimethyl phthalate, diphenyl phthalate, glyceryl tripropionate, isobutyl phthalate, tributyl borate, tributyl citrate, tributyl phosphate, tricresyl phosphate, triethylene glycol dihexoate, triethylene glycol dioctoate, triethyl citrate, triethyl phosphate, triphenyl phosphate. Methyl methacrylate ester is an important plastic. Ethyl silicate is used to cover concrete, brick and stone with a coating of silicic acid to resist water penetration. Diocytl phthalate is used as a plasticizer in cable and wire insulation, and dimethyl phthalate as an insect repellent.

See also **Organic Chemistry**.

ESTROGEN. Hormones; Steroids.

ETHANE. C_2H_6, formula wt 30.07, colorless, odorless gas, mp $-172°C$, bp 88.6°C, sp gr 1.05 (air = 1), practically insoluble in H_2O, moderately soluble in alcohol. The compound burns when ignited in air with a pale, faintly luminous flame. Forms an explosive mixture with air over a moderate range. With excess air, products of combustion are CO_2 and H_2O. Ethane is among the chemically less reactive organic substances. However, ethane reacts with chlorine and bromine to form substitution compounds. Ethane occurs, usually in small amounts, in natural gas. The fuel value of ethane is high, 1,730 Btu/cu ft (15,397 Cal/cu meter). Ethane may be prepared by reaction of magnesium ethyl iodide in anhydrous ether (Grignard's reagent) with H_2O or alcohols. Ethyl iodide, bromide, or chloride are preferably made by reaction with ethyl alcohol and the appropriate phosphorus halide. Important ethane derivatives, by successive oxidation, are ethyl alcohol, acetaldehyde, and acetic acid.

ETHANOL (Ethyl Alcohol). C_2H_5OH, formula wt 46.07, colorless liquid with mild characteristic odor, mp $-$ 114.1°C, bp 78.32°C, sp gr 0.789. Also known as *ethanol*, the compound is miscible in all proportions with H_2O or ether. When ignited, ethyl alcohol burns in air with a pale blue, transparent flame, producing H_2O and CO_2. The vapor forms an explosive mixture with air and is used in some internal combustion engines under compression as a fuel. Such mixtures are frequently referred to as *gasohol*.

Anhydrous ethyl alcohol is made from the constant boiling mixture with H_2O (95.6% ethyl alcohol by weight)—(1) by heating with a substance such as calcium oxide, which reacts with H_2O and not with alcohol, and then distilling, or (2) by distilling with a volatile liquid, such as benzene (bp 79.6°C) which forms a constant low-boiling mixture with H_2O and alcohol (bp 64.9°C), so that H_2O is removed from the main portion of the alcohol; after which alcohol plus benzene distills over (bp 78.5°C). Anhydrous ethyl alcohol is required for certain purposes as a solvent and reagent and fuel applications.

Commercially, ethyl alcohol is marketed by the proof gallon, 200 proof on the scale representing pure alcohol (100%). When the term *alcohol* alone is used, it refers to a liquid that ranges from 188 to 192 proof (94% to 96% ethyl alcohol). When the terms *grain alcohol*, *high-purity* alcohol, or *pure* ethyl alcohol are used, these usually refer to a liquid that is 190 proof. In most countries, beverage alcohol is highly taxed and to make the product available for nonbeverage purposes, denaturants will be added. Denaturants include methyl alcohol, pyridine, benzene, kerosene, pine oil, mixtures of primary and secondary aliphatic higher alcohols, and hydrogenated organic compounds. Thousands of nonbeverage industrial and commercial products,

notably food extracts, toiletries, pharmaceuticals, solvents, and cleaning products, contain denatured ethyl alcohol.

Worldwide, ethyl alcohol is the basis for a huge alcoholic beverage industry, offering a wide range of products wherein the alcoholic content varies from a few to over 50% (100 proof). Industrially, ethyl alcohol is very important high-tonnage raw and intermediate material for numerous processes, and is used extensively in solvents, antiseptics, antifreeze compounds, and fuels.

Production. Natural fermentation is the oldest process for making ethyl alcohol and still constitutes the principal means for creating the alcoholic content of beverages. Except in connection with other alcohol-containing products, industrial producers of ethyl alcohol use processes other than fermentation. For fermentation, almost any agricultural raw material with a carbohydrate content in the form of sugars or starches that are easily converted to sugars can be used. Once the raw materials are in the form of sugars, yeast enzymes are added to commence natural fermentation. Traditionally, in the United States, industrial alcohol prepared by fermentation has used blackstrap molasses, which contains up to 50% sugars and can be easily fermented. The starting mash is prepared by diluting the molasses with H_2O to bring the sugar content down to about 15% (weight). The mash is slightly acidified, after which invertase (enzyme to convert sucrose) and zymase (enzyme to convert glucose and fructose) are added. The products are ethyl alcohol and CO_2. Yeast activity is sustained by the addition of nutrients. With careful control of temperature and acidity, the fermentation process can be completed in about two days. The resulting mash (beer) usually contains about 12% ethyl alcohol which is recovered from the beer by distillation. See **Fermentation.**

In modern industrial ethyl alcohol plants, the compound is produced in two principal ways: (1) by *direct hydration of ethylene*, or (2) *by indirect hydration of ethylene*. In the direct hydration process, H_2O is added to ethylene in the vapor phase in the presence of a catalyst: $CH_2:CH_2 + H_2O \rightleftharpoons CH_3CH_2OH$. A supported acid catalyst usually is used. Important factors affecting the conversion include temperature, pressure, the $H_2O/CH_2:CH_2$ ratio, and the purity of the ethylene. Further, some by-products are formed by other reactions taking place, a primary side reaction being the dehydration of ethyl alcohol into diethyl ether: $2C_2H_5OH \rightleftharpoons (C_2H_5)_2O + H_2O$. To overcome these problems, a large recycle volume of unconverted ethylene usually is required. The process usually consists of a reaction section in which crude ethyl alcohol is formed, a purification section with a product of 95% (volume) ethyl alcohol, and a dehydration section which produces high-purity ethyl alcohol free of H_2O. For many industrial uses, the 95%-purity product from the purification section suffices.

In the indirect hydration process, ethylene first is absorbed in concentrated H_2SO_4 to form mono- and diethyl sulfates:

$$CH_2:CH_2 + H_2SO_4 \rightarrow CH_3CH_2OSO_3H; \text{ and}$$

$$2CH_2:CH_2 + H_2SO_4 \rightarrow (CH_3CH_2)_2SO_4.$$

The ethyl sulfates then are hydrolyzed to ethyl alcohol:

$$CH_3CH_2OSO_3H + H_2O \rightarrow CH_3CH_2OH + H_2SO_4; \text{ and}$$

$$(CH_3CH_2)_2SO_4 + 2H_2O \rightarrow 2CH_3CH_2OH + H_2SO_4.$$

Remaining steps in the process include recovery and purification of the crude ethyl alcohol and reconcentration of the dilute H_2SO_4. The crude ethyl alcohol is steam-stripped from the dilute acid solution, followed by distillation for purification.

Azeotropes. The physical properties of ethyl alcohol are influenced by the hydroxyl group that imparts hydrogen-bonding characteristics and polarity to the substance that are analogous to water. Ethyl alcohol displays a highly nonideal behavior in numerous solutions, forming several azeotropes. The list of binary azeotropes of ethanol is long, including acetonitrile, benzene, carbon disulfide, chloroform, ethyl acetate, hexane, toluene, and water. See also **Azeotropic System.**

Chemistry. Ethyl alcohol reacts (1) with sodium metal, forming sodium ethoxide C_2H_5ONa plus hydrogen gas, (2) with phosphorus chloride, bromide, iodide, forming ethyl chloride, bromide, iodide, respectively, (3) with H_2SO_4 concentrated, forming at 100°C ethyl hydrogen sulfate $C_2H_5OSO_2OH$, at 140°C diethyl ether $(C_2H_5)_2O$, at 200°C ethylene $CH_2:CH_2$, (4) with organic acids, warmed in the presence of H_2SO_4, forming esters, e.g., ethyl acetate $CH_3COOC_2H_5$, ethyl benzoate $C_2H_5COOC_2H_5$ (see various individual acids), (5) with magnesium methyl iodide in anhydrous ether (Grignard's solution), forming methane as in the case of primary alcohols, (6) with calcium chloride to form a solid addition compound $4C_2H_5OH \cdot CaCl_2$, which is decomposed by H_2O, (7) with oxygen, using sodium dichromate solution and H_2SO_4, to form acetaldehyde (and acetic acid), using air, in the presence of acetic bacteria, to form vinegar (dilute acetic acid along with the substances present in the alcohol used, e.g., wine, cider), (8) with HNO_3 (a) concentrated, free from nitrogen tetroxide, to form ethyl nitrate, (b) dilute to form glycollic acid, (c) concentrated acid containing nitrogen tetroxide (fuming HNO_3) explosive reaction, (9) with chlorine (or bromine) to form chloral CCl_3CHO (or bromal).

See **Organic Chemistry.**

References

Batter, T. R., and E. R. Wilke: "A Study of the Fermentation of Xylose to Ethanol by *Fusarium Oxysporum*," NTIS, LBL-635 (June 1977).

Benson, W. R.: "Biomass Potential from Agricultural Production," in "Proc. Biomass—A Cash Crop of the Future?," Midwest Research Inst. and Battelle Columbus Laboratories, Kansas City, Missour, March 1977.

Sutton, O. C., et al.: "Ethanol from Agricultural Residues," *Chem. Eng. Progress,* **75,** 12, 52–58 (1979).

ETHANOLAMINES. There are three ethanolamines, all hydroxyamines and all high-production industrial chemicals: (1) monoethanolamine, $NH_2CH_2CH_2OH$, industrial symbol (MEA), formula wt 61.08, mp 10.5°C, bp 171°C, sp gr 1.018; (2) diethanolamine, $NH(CH_2CH_2OH)_2$, industrial symbol (DEA), formula wt 105.14, mp 28.0°C, bp 270°C, sp gr 1.019; (3) triethanolamine, $N(CH_2CH_2OH)_3$, industrial symbol (TEA), formula wt 149.19, mp 21.2°C, bp 360°C, sp gr 1.126. Mono- and triethanolamine are miscible with H_2O or alcohol in all proportions and are only slightly soluble in ether. Diethanolamine will dissolve in H_2O up to 96.4% at 20°C, is very soluble in alcohol, and only slightly soluble in ether.

Wurtz first reported the ethanolamines in 1860, but they were not used commercially on any significant scale until the late 1920s. All of the compounds are clear, viscous liquids at standard conditions and white crystalline solids when frozen. They have a relatively low toxicity. Industrially, the compounds are important (1) because they form numerous derivatives, notably with fatty acids, soaps, esters, amides, and esteramides; and (2) for their exceptional ability for scrubbing acidic compounds out of gases. MEA, for example, will effectively remove H_2S from hydrocarbon gases. The compounds also remove CO_2 from process streams and, where desired, the CO_2 may easily be recovered by heating the absorptive solutions. The soaps of the ethanolamines are extensively used in textile treating agents, in shampoos, and emulsifiers. The fatty acid amides of DEA

are applied as builders in heavy-duty detergents, particularly those in which alkylarylsulfonates are the surfactant ingredients. The use of TEA in photographic developing baths promotes fine grain structure in the film when developed. Ethanolamines are also used as humectants and plasticizing agents for textiles, glues, and leather coatings; and as softening agents for numerous materials. Morpholine is an important derivative.

Current ethanolamine production processes react ethylene oxide, $(CH_2)_2O$ with NH_3, usually in aqueous solution. The ratio of MEA, DEA, and TEA varies in accordance with the amount of NH_3 present. This is controlled by the quantities of MEA and DEA recycled. Higher NH_3-ethylene oxide ratios favor high DEA and TEA yields, whereas lower ratios are used where maximum production of MEA is desired. The reaction is noncatalytic. The pressure is moderate, just sufficient to prevent vaporization of components in the reactor. The bulk of the H_2O produced is removed by subsequent evaporation. The dehydrated ethanolamines then proceed to a further drying column, after which they are separated in a series of fractionation columns, not difficult because of the comparatively wide separation of their boiling points.

ETHER (Chlorinated). Chlorinated Organics.

ETHERS. The homologous series of ethers has the formula $C_nH_{2n+2}O$. Structurally, the ethers have an oxygen linkage between two radicals (R—O—R'). R and R' may be the same, as in dimethyl ether, or they may differ, as in ethylisopropyl ether. The latter may be referred to as a *mixed ether.* Mixed ethers frequently are made from mixed alcohols. Where R and R' are alkyls, the ether may be called an *alkyl ether* or an *alkyl oxide.* They may be considered to be derivatives of the monohydric alcohols. Each ether is isomeric with a saturated alcohol. Both diethyl ether and butyl alcohol are $C_4H_{10}O$. Also, there are many isomeric ethers, starting with $C_4H_{10}O$. Methylpropyl ether and diethyl ether are isomeric. Where compounds such as these have the same general formula, are members of the same family, and differ only by the alkyl group present, they are called *metameric.*

Since they are similar structurally to the alcohols, phenols also form ethers. An example of an aromatic ether is methylphenyl ether (anisole), C_6H_5—O—CH_3. There are few ethers where both R and R' are aryls. The structure of thioethers is similar to the other ethers, but with a sulfur atom in the link instead of an oxygen atom, as R—S—R'. Examples of thioethers include diethyl sulfide C_2H_5—S—C_2H_5 and methylethyl sulfide CH_3—S—C_2H_5, which is a mixed thioether.

The properties of ethers may be summarized: (1) with the exception of dimethyl ether, which is a gas, the ethers are volatile, mobile, inflammable liquids that are lighter than H_2O; (2) they are relatively inert chemically, not being acted on by alkali metals or alkalis and not reacting with dilute acids; (3) they form substitution products when reacted with chlorine and bromine; and (4) they are decomposed when heated with strong acids, yielding esters:

$$(C_2H_5)_2O + 2H_2SO_4 \rightarrow 2C_2H_5 \cdot HSO_4 + H_2O$$
(diethyl ether)   (ethyl hydrogen sulfate)

$$CH_3 \cdot O \cdot C_2H_5 + 2HBr \rightarrow CH_3Br + C_2H_5Br + H_2O$$
(methylethyl ether)   (methyl bromide)   (ethyl bromide)

Ether. $(C_2H_5)_2O$, formula weight 74.12, mp − 116.3°C, bp 34.6°C, sp gr 0.708. Probably the best known of the ethers, diethyl ether, commonly called simply *ether,* is slightly soluble in H_2O (1 volume in 10 volumes H_2O) and is miscible with alcohol in all proportions. Ether dissolves iodine and many organic substances, e.g., oils and fats, waxes, resins, and alkaloids and hence is widely used as a solvent for these substances in the preparation of numerous products, including explosives and collodion. Ether explodes in oxygen in the presence of a flame or spark, yielding H_2O and CO_2. When heated with an acid, such as H_2SO_4, ether yields ethyl alcohol. With phosphorus halides, ethyl halide (2 moles) is formed, Ether reacts with HNO_3 to form ethyl oxide.

Although still used medically, at one time ether was the major anesthetic, for which it must be scrupulously pure. In addition to various side effects which may result from the use of ether as an anesthetic, it is a definite hazard in the operating room because of its explosive properties, particularly in enriched oxygen atmospheres.

See also **Organic Chemistry.**

ETHYL ALCOHOL. Ethanol.

ETHYL CELLULOSE.
A versatile thermoplastic cellulose ether that is compatible with a wide variety of solvent systems, resins, oils, and plasticizers. This versatility permits a wide diversity of end-product properties. It is an excellent film former—as from a wide range of neat, lacquer, or dispersion formulations. Molded ethyl cellulose has excellent toughness, flexibility, and shock resistance. Useful temperature range is from about −40 to +100°C. In the preparation of ethyl cellulose, wood pulp or cotton linters with a high alpha-cellulose content are reacted with ethyl chloride and sodium hydroxide. Structural formulas of cellulose and ethyl cellulose with complete (54.9%) ethoxyl substitution are:

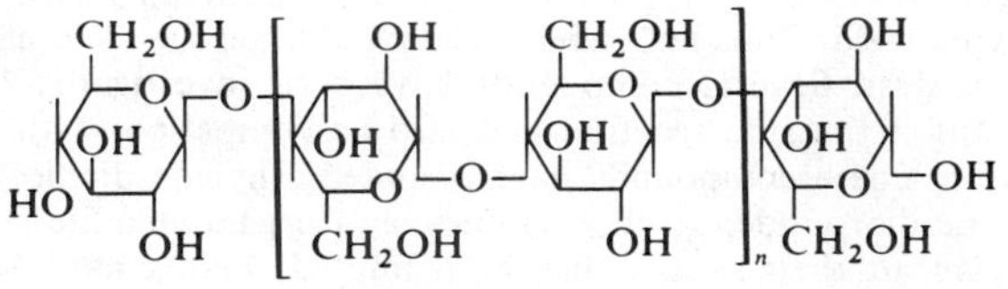

Cellulose, $n > 500$

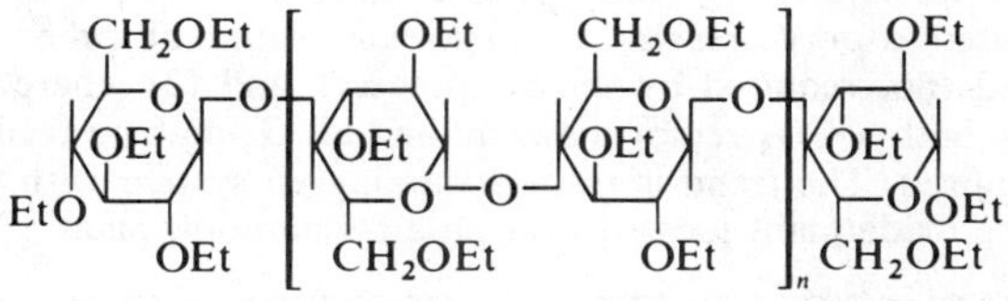

Tri-O-ethyl cellulose, $n = 50$–150; Et $= CH_3CH_2$

The natural color of ethyl cellulose is colorless to light amber, but it can be formulated into a wide range of transparent, translucent, and opaque colors. The material should be dried before molding because it is slightly hygroscopic. Compression molding temperatures range from 121–200°C and pressures from 500–5,000 psi. Injection molding temperatures range from 175–260°C and pressures from 8,000–32,000 psi. Strong acids decompose the material, but weak acids and strong alkalis have only a slight effect. Weak alkalis do not attack the material. Ethyl cellulose is soluble in a large number of organic solvents.

Among the commercial uses for ethyl cellulose are strippable coating for metal parts, paper coatings, and in medicinal tablets. An interesting application is coating for bowling pins. Ethyl cellulose sheeting is tough, flexible, and transparent, yet sufficiently rigid to withstand rough handling.

ETHYL CHLORIDE. Chlorinated Organics.

ETHYLENE.
C_2H_4, formula wt 28.03, colorless gas with slight odor, normal bp −103.7°C, critical pressure of 49.98 atm, critical temperature of 9.5°C, density 1.26 g/liter (0°C and 760 mm), sp gr 0.97 (air = 1.0), very slightly soluble in H_2O. Ethylene burns with a luminous flame when ignited in air. The presence of ethylene in coal gas is chiefly responsible for the luminosity of that gas. Ethylene forms an explosive mixture with air and has a high fuel value, 1,615 Btu/cu ft (14,374 Cal/cu meter).

Even though there are few direct end-uses for ethylene, it is probably the most important petrochemical feedstock, both in terms of quantities used and economic value. Ethylene is the feedstock for ethylene oxide, ethylbenzene, ethyl chloride, ethylene dichloride, ethyl alcohol, and polyethylene, most of which, in turn, are used to produce hundreds of other end-products. Most ethylene is produced by steam cracking of ethane, or propane.

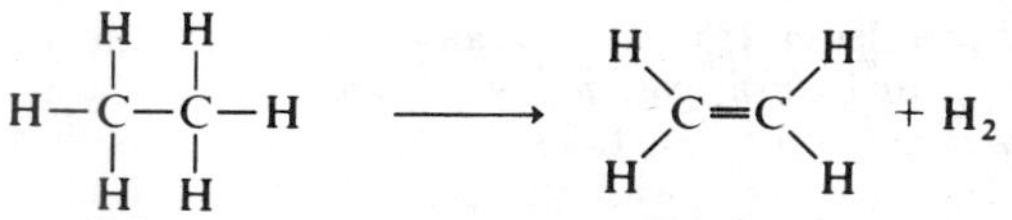

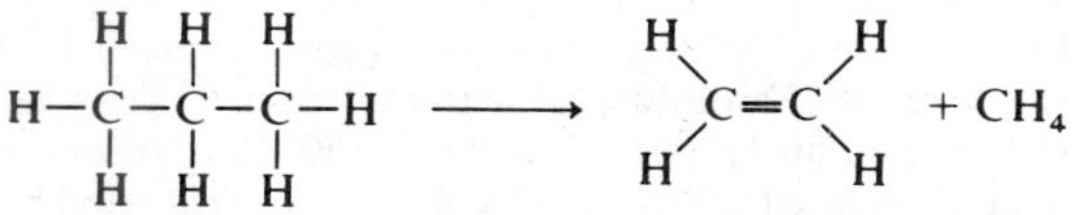

Ethylene also may be produced from other paraffinic or naphthenic hydrocarbons. The reactions are highly endothermic (34,400 kcal/kg mole of ethane cracked at approximately 900°C) and proceed in the direction indicated at temperature exceeding approximately 620°C without a catalyst.

Ethylene is of importance as a petrochemical feedstock because of its great versatility in reacting to form several chemical intermediates. The double-bond provides reactivity; the compound also has the ability to homopolymerize and copolymerize with other monomers. Some of the important reactions involving ethylene include:

Chlorination

$$CH_2{=}CH_2 + HCl \xrightarrow[\text{cat.}]{\text{acidic}} CH_3{-}CH_2Cl$$
Ethyl chloride

Oxidation

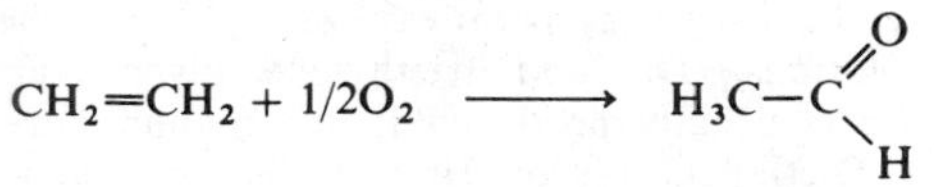

Acetaldehyde

Hydration

$$CH_2{=}CH_2 + H_2O \longrightarrow C_2H_5OH$$
Ethyl alcohol

Oxychlorination

$$CH_2{=}CH_2 + 2HCl + 1/2O_2 \longrightarrow C_2H_4Cl_2 + H_2O$$
Ethylene
dichloride

Ethylene dichloride is used for the production of vinyl chloride.

Alkylation

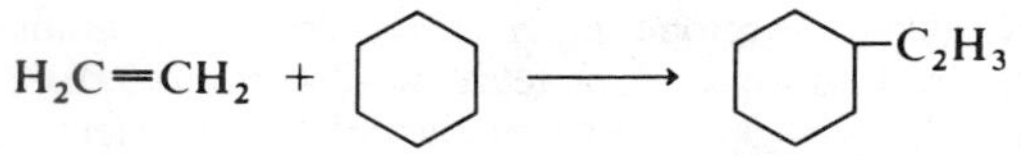

Ethyl benzene

Ethyl benzene is used for the production of styrene.

Polymerization

$$n\text{CH}_2\!\!=\!\!\text{CH}_2 \longrightarrow \begin{array}{c} (-\text{CH}_2-\text{CH}_2)_n \\ \text{High- and low-density polyethylene} \end{array}$$

Ethylene is also oxidized in large quantities to ethylene oxide:

$$\text{H}_2\text{C} \overset{\text{O}}{-\!\!\!-} \text{CH}_2$$

At one time, ethylene was produced by the dehydration of ethyl alcohol over alumina.

Almost any naphthenic or paraffinic hydrocarbon heavier then methane can be steam-cracked to yield ethylene. The preferred feedstock in the United States has been ethane and/or propane recovered from natural gas, or from the volatile fractions of petroleum. However, because of long term uncertainties pertaining to natural gas, many producers have been turning to heavier petroleum fractions, such as gas oils, as feedstocks.

Ethylene reacts (1) with the halogens to form substitution halides; (2) with hypochlorous and hypobromous acid to form ethylene chlorohydrin or ethylene bromohydrin, respectively; (3) with hydrogen iodide or bromide (not chloride) to form ethyl iodide or ethyl bromide; (4) with hydrogen, in the presence of a catalyst, e.g., finely divided nickel at 150°C, to form ethane; (5) with concentrated sulfuric acid at 160°C to form ethyl hydrogen sulfate; and (6) with potassium permanganate to form ethylene glycol, although glycol is preferably made from ethylene dichloride or chlorohydrin.

In addition to its uses in the preparation of intermediates for a large variety of petrochemical reactions, ethylene is used as an anesthetic, as a fuel with oxygen for high-temperature flames, and as a coloring and ripening agent for citrus fruits and tomatoes. Ethylene chlorohydrin is used as an agent for decreasing the dormant period of seeds. See also **Polyethylene.**

ETHYLENE DICHLORIDE.   Chlorinated Organics.

ETHYLENE GLYCOL.   This compound, $HOCH_2CH_2OH$, is traditionally associated with its use as a permanent-type antifreeze for engine cooling systems. However, since the early 1960s, large amounts of ethylene glycol have been used in the production of polyesters for fibers, films, and coatings. The compound also finds important uses in hydraulic fluids, in the manufacture of low-freezing-point explosives, glycol ethers, and deicing solutions. Di- and triethylene glycols are important coproducts usually produced in the manufacture of ethylene glycol. Diethylene glycol, $HOCH_2CH_2OCH_2CH_2OH$, is used in the production of unsaturated polyester resins and polester polyols for polyurethane-resin manufacture, as well as in the textile industry as a conditioning agent and lubricant for numerous synthetic and natural fibers. It is also used as an extraction solvent in petroleum processing, as a desiccant in natural gas processing, and in the manufacture of some plasticizers and surfactants. Triethylene glycol, $HOCH_2CH_2OCH_2CH_2OCH_2-CH_2OH$, finds principal use in the dehydration of natural gas and as a humectant.

Ethylene glycol (1,2-ethanediol) is a clear, colorless, syrupy liquid; sweet taste; hygroscopic; odorless; soluble in H_2O, alcohol, and ether, sp gr 1.1155 (20°C), bp 197.2°C, fp −13.5°C (∼8°F), combustible. Diethylene glycol (dihydroxydiethyl ether) is a colorless, practically odorless, syrupy liquid; sweet taste, extremely hygroscopic; noncorrosive; miscible with water, ethyl alcohol, acetone, ether, ethylene glycol; immiscible with benzene, toluene, carbon tetrachloride; bp 245.0°C, fp −8°C (∼17.6°F), sp gr 1.1184, combustible. Triethylene glycol is a

colorless, hygroscopic, practically odorless liquid, bp 287.4°C, fp −7.2°C (19°F); soluble in water, immiscible with benzene, toluene, and gasoline, sp gr 1.1254, combustible.

In one process for the manufacture of the aforementioned glycols, ethylene oxide is formed by direct oxidation of ethylene with oxygen over a silver catalyst. After purification, the stabilized ethylene oxide is mixed with a large excess of water, preheated, and fed to an ethylene oxide reactor. Here the ethylene oxide and water react under high temperature and high pressure to form principally ethylene glycol, with the other aforementioned glycols as coproducts. The crude glycols are dehydrated and then recovered individually as highly pure overhead streams from a series of vacuum-operated purification columns.

ETHYLENE OXIDE. $\overline{CH,CH,O}$, formula wt 44.05, liquid, mp −111.3°C, bp 13.5°C, sp gr 0.887. The compound is miscible in all proportions with H_2O or alcohol and is very soluble in ether. Ethylene oxide is slowly decomposed by H_2O at standard conditions, converting into glycol $CH_2OH \cdot CH_2OH$. Ethylene oxide is a very high production chemical. In terms of consumption (1) 60% for manufacture of ethylene glycol, (2) 12% for preparation of surfactants, (3) 8% for manufacture of ethanolamines, (4) 10% for production of plasticizers, solvents, and lubricants, and (5) 10% for making glycol ethers which are used as jet-fuel additives and solvents.

Direct oxidation of ethylene in the presence of a silver catalyst is the predominant large-scale process used: $CH_2 \colon CH_2 + \frac{1}{2}O_2 \rightarrow \langle\!\langle (CH_2)_2 \rangle\!\rangle O$. The yield is approximately 70% of the theoretical. For maximum yield, very careful temperature control is required, the yield dropping as the temperature climbs. The side reaction: $CH_2 \colon CH_2 \rightarrow 3CO_2 + 2H_2O$ is the main factor for reducing yield. Thus far, silver has proved to be the most effective catalyst. Several compounds have been investigated that can inhibit the side reaction and also be compatible with the catalyst. These compounds have included ethylene dichloride, ethylene dibromide, alcohol, amines, and organometallic compounds, but their success has been limited. Plants have been designed to use either air or pure oxygen for oxidation. Selection presents an interesting study in economics because (1) where air is used, a purge reactor and associated purge absorber are required (not required by the O_2 process), and (2) where O_2 is used, both a CO_2 removal system and an O_2-making facility are required. The trend is toward the oxygen system with the ethylene oxide plant located near an air-separation plant.

ETHYLENE-PROPYLENE ELASTOMERS.   Elastomers.

ETHYLENE-VINYL ACETATE COPOLYMERS.   Known as EVA copolymers, these materials are polyolefins which can be processed like other thermoplastics, but which approach rubbery materials in softness and elasticity. The resins meet regulatory requirements for use in direct contact with food in food-processing machinery and in packaging applications. They are used in a number of applications to replace plasticized polyvinyl chloride and rubber. EVA copolymers require no curing or plasticizer. Parts made from EVA have little or no odor. Their elasticity is permanent. The copolymers can be injection-, blow-, compression-, transfer- and rotationally molded or extruded into film, sheeting, pipe, and profiles. EVA copolymers offer advantages over polyvinyl chloride and rubber in that they have good clarity and gloss, stress-crack resistance, good barrier properties, low-temperature flexibility and toughness, good adhesive properties, and good resistance to ultraviolet radiation. Their main limitation is a comparatively low resistance to heat and solvents. The resins soften at a temperature of about 70° C. EVA copolymers are not attacked by alcohols, glycols, or weak organic

acids. However, to a varying degree, the materials are attacked by chlorinated hydrocarbons, straight-chain paraffinic solvents, and benzene and its derivatives.

EUCLASE. The mineral euclase is a silicate of beryllium and aluminum corresponding to the formula $BeAlSiO_4(OH)$, which crystallizes in the monoclinic system. It has a perfect prismatic cleavage; hardness, 7.5; specific gravity, 3.1; luster, vitreous; is colorless to sea-green or blue. It has been used to a very slight extent for jewelry, as its transparent crystals somewhat resemble aquamarine. Euclase occurs in the Minas Geraes region, Brazil, associated with topaz and beryl, and also in the Ural Mountains, where it is found in gold-bearing sands. The name *euclase* is derived from the Greek, meaning *easiness* and *fracture*, in reference to its easily cleaved crystals.

EUDIOMETER. A graduated tube closed at one end in one form of which two platinum wires are sealed so that a spark may be passed through the contents of the tube; used to measure volume changes in combustion of gases.

EUROPIUM. Chemical element symbol Eu, at. no. 63, at. wt. 151.96, sixth in the lanthanide series in the periodic table, mp 822°C, bp 1529°C, density, 5.245 g/cm³ (20°C). Elemental europium has a body-centered cubic crystal structure at 25°C. Pure metallic europium is silvery-gray in color under vacuum, but oxidizes readily in air and must be handled in an inert atmosphere. Europium is very soft as compared with the other rare-earth metals. Two stable isotopes of the element occur naturally, ^{151}Eu and ^{153}Eu. Upon absorption of thermal neutrons, ^{151}Eu forms ^{152}Eu with a half-life of 13 years. The isotope ^{153}Eu forms ^{154}Eu with a half-life of 16 years. The latter further decays to ^{155}Eu with a half-life of 1.7 years. In terms of abundance, europium is present on the average of 1.2 ppm in the earth's crust, making its potential availability greater than those of antimony, bismuth, or cadmium. The element was first identified by Sir William Crookes in 1889. Europium dissolves readily in dilute mineral acids and reacts with water at room temperature. The metal is not known to be toxic, but because of its high reactivity in air, great care in handling it is required. The electronic configuration is $1s^22s^22p^63s^23p^63d^{10}4s^24p^64d^{10}$-$4f^65s^25p^65d^16s^1$. Ionic radius Eu²⁺ 1.09 Å, Eu³⁺ 0.950 Å. Metallic radius 1.995 Å. First ionization potential 5.67 eV; second 11.25 eV. Oxidation potential Eu²⁺ → Eu³⁺ + e⁻, 0.43 V. Other important physical properties of europium are given under **Rare-Earth Elements and Metals.**

Europium occurs in the rare-earth fluocarbonate mineral bastanite, mainly found in southern California. The mineral contains between 0.09 and 0.11% Eu_2O_3. Other minerals, such as xenotime and monazite, also contain europium compounds and sometimes are used as sources of the element.

Because of the desirable nuclear properties of the element, europium has received serious consideration for the construction of nuclear reactor hardware. Earlier commercial unavailability of the element, however, favored the use of other materials. Some small reactors have been constructed in which europium molybdate has been the major control-rod component. With much increased availability of the metal in recent years, the prospects of further usage of europium in reactor design are good. A europium-activated yttrium orthovanadate Eu:YVO₄ has shown promise as a red phosphor for commercial television. An increase of 40% in light output has been claimed. With this system, the average color television set would require about ½ g of Eu_2O_3 and 6 g of Y_2O_3. The stimulus resulting from this discovery resulted in the development of other new phos-

phors involving europium in various host matrices. These new materials have been used in high-intensity mercury-vapor lamps, general-purpose fluorescent lamps, x-ray screens, charged-particle detectors, and neutron scintillators. In some optically-read memory systems, ferromagnetic europium chalcogenides (sulfides, selenides, and tellurides) have been used. Other electronic and semiconductor uses of europium are under serious investigation.

See references listed at the ends of the entries on **Chemical Elements;** and **Rare-Earth Elements and Metals.**

NOTE: This 4th Edition entry was revised and updated by K. A. Gschneidner, Jr., Director, and B. Evans, Assistant Chemist, Rare-Earth Information Center, Energy and Mineral Resources Research Institute, Iowa State University, Ames, Iowa.

EUTECTIC. An eutectic reaction is a reversible isothermal transformation in which, during cooling, a single liquid phase is transformed into two or more solid phases, the number of solid phases being equal to the number of components. In a given alloy system, at a fixed pressure, all phases will have fixed compositions during the isothermal transformation. The temperature at which freezing occurs is known as the *eutectic temperature,* while the composition of the liquid phase is called the *eutectic composition.* On a temperature-composition binary phase diagram, the eutectic point is determined by the eutectic composition and the eutectic temperature. In general, an alloy of the eutectic composition freezes at a minimum temperature. For this reason, eutectic compositions, or compositions close to the eutectic, are frequently used in low-melting-point solders.

By a similar usage in petrology, a eutectic is a discrete mixture of two or more minerals, in definite proportions, which have simultaneously crystallized from the mutual solution of their constituents. The eutectic point is the lowest temperature at any given pressure at which the foregoing physical-chemical process may take place. The eutectic ratio is the ratio by weight of two minerals which originate by the aforementioned process.

EUTECTOID. This is a phase transformation analogous to an eutectic where a single solid phase, instead of a liquid phase, is transformed into two or more different solid phases. The number of solid phases in the resulting eutectoid structure is equal to the number of components in the system. Under very slow cooling, the eutectoid transformation should occur at the eutectoid temperature. However, due to the sluggishness of solid state transformations, there is usually some hysteresis with the transformation temperature depressed on cooling and raised on heating. Under equilibrium conditions, the compositions of the various phases are fixed in an eutectoid reaction just as they are in an eutectic transformation.

The best known eutectoid reaction is that which occurs in steel where the austenite phase, stable at high temperatures, transforms into the eutectoid structure known as pearlite. In this transformation, the austenite phase, containing 0.8% carbon in solid solution, transforms to a mixture of ferrite (nearly pure body-centered cubic iron) and iron-carbide (Fe_3C). At atmospheric pressure, the equilibrium temperature for this reaction is 723°C. This temperature is the eutectoid temperature.

In binary alloy systems, a eutectoid alloy is a mechanical mixture of two phases which form simultaneously from a solid solution when it cools through the eutectoid temperature. Alloys leaner or richer in one of the metals undergo transformation from the solid solution phase over a range of temperatures beginning above and ending at the eutectoid temperature. The structure of such alloys will consist of primary particles of one of the stable phases in addition to the eutectoid, for example ferrite and pearlite in low-carbon steel. See also **Iron Metals, Alloys, and Steels.**

EVAPORATION. The evaporation of a liquid consists of the escape from the main body of a liquid those molecules which, in their thermal agitation, are moving with sufficient speed to break through the surface tension, i.e., whose kinetic energy exceeds the work function of cohesion at the surface. Since only a small proportion of the molecules are at any instant located near enough to the surface and are moving in the proper direction to escape, the rate of the evaporation is limited. It is evident why evaporation proceeds more rapidly with higher temperature and why liquids of low surface tension are relatively volatile. Also, as the faster-moving molecules emerge, those left behind have less average energy, and the temperature of the liquid is thereby lowered. If the evaporation takes place in a closed vessel, the escaping molecules accumulate as a vapor above the liquid. Many of them return to the liquid, such returns being more frequent, the greater the density and pressure of the vapor. Ultimately, the processes of escape and return reach equilibrium; the vapor is then said to be saturated, its density and pressure no longer increasing, and the cooling effect ceases.

Evaporation is a major chemical engineering unit operation for bringing about separations of liquids and solids and, in particular, to recover the solute (such as a dissolved salt) from the solvent (frequently water). Usually, the main object of the separation is the solute. The pulp and paper industry is a large user of evaporation equipment. In pulp mills, after the digestion system, the pulp is leached with water and the chemical solids are dissolved out almost completely by a pulp-washing system. The recovered liquid from these operations is fed to an evaporator, generally at about 15% total dissolved solids content. The evaporator removes much of the water and in so doing concentrates the liquid to 55–65% total dissolved solids, whereupon the solution then can be further processed in a chemical recovery furnace. Other types of pulp processing also involve chemical-containing solutions which must be evaporated for recovery of valuable chemicals. Evaporation also is used extensively in the production of table and industrial salt (sodium chloride) as well as other salts, in caustic-chlorine production, in the phosphate industry, and in food processing. Evaporators can be very large structures.

Evaporation is in principle the same operation as plain distillation, with the modifications in practice that (1) the vapor may or may not be recovered, (2) the residue in the evaporator may or may not contain solids, and (3) vacuum evaporation is frequently used in a single compartment or in multiple stages with each successive stage operated at an increasing vacuum utilizing the heat of condensation of the vapor from the preceding stage. In multiple stage evaporators there is a saving in the cost of heat and an increased expenditure for apparatus. Vacuum evaporation is frequently utilized to lower the temperature to which a substance is subjected and thus avoid decomposition by passing a current of warm dry air over the substance. Combined high-vacuum and very low temperature evaporation or drying is practiced in the final removal of water vapor from frozen penicillin, due to the heat-sensitive nature of this material. Water vapor passes from the place of higher concentration, that is, the substance, to the place of lower concentration, that is, the air, and is thus removed from the substance. If oxygen of the air reacts with the substance, an inert gas such as nitrogen may be substituted for air.

An evaporator system may be single effect (accompanying figure), in which the steam is produced from one evaporator,

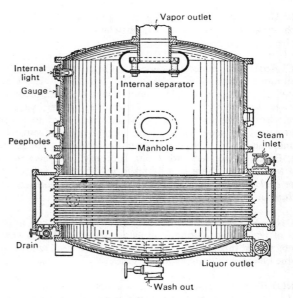

Horizontal tube evaporator.

or multiple effect, in which the steam is produced from several evaporators in series. In a multiple effect system the vapor from one evaporator becomes the heating steam in the succeeding. Unusual conditions met in industrial or steam heating plants may require so large a fraction of make-up as to warrant double, triple, or quadruple effect evaporators. The central generating station ordinarily employs single effect and rarely requires more than a double effect system. The ratio vapor produced/steam used is about 0.8 for the single effect, 1.5 for the double effect, and 2.5 for the triple effect system. Evaporator feed is sometimes preheated to increase evaporator capacity.

Evaporators are classed as film, flash, or submerged-tube types. The first and last are steam-tube types; in the former the raw water trickles over the hot tubes, in the latter the tubes are entirely surrounded by the water being evaporated. The flash type produces steam by dropping the pressure on water at the saturation temperature. The excess heat flashes part of the water into steam, then the remainder is drawn off, reheated, and again flashed.

References

Charm, S. F.: "Fundamentals of Food Engineering," 3rd Edition, AVI, Westport, Connecticut (1978).
Rosenblad, A. E.: "Three Evaporator Flow Diagrams that Adapt to Current Energy Costs and Emission Requirements," Rosenblad Corp., Princeton, New Jersey (1975).
Schwartzberg, H. G.: "Energy Requirements for Liquid Food Concentration," *Food Technology*, **31**, 3, 67–76 (1977).
Staff: "Evaporation Technology," *Chem. Eng. Progress*, **72**, 4, 41–71 (1976).

EVAPORATION (Desalination). Desalination.

EXOSMOSIS. An osmotic process by which a diffusible substance passes from the inner or closed, to the outer parts of a system, as in the loss of substances from a portion of a plant root to water in the surrounding soil.

EXPLOSIVES. Substances that are capable of exerting, by their characteristic high-velocity reactions, sudden high pressures usually generating loud noise and more or less destructive action on surroundings. Explosive classification is based on the types of reactions that produce explosions, namely mechanical, chemical, and nuclear. This article is concerned only with the chemical explosives. They comprise two main types, the "low" or "deflagrating" (sometimes also called "propellant") explosives, and the "high" or "detonating" explosives. The latter are further classified as "primary" and "secondary" detonating explosives.

Pre-Explosion Reactions

Chemical explosives are based on exothermic (heat-liberating) reactions of the type that increase in rate (exponentially) with temperature. At ambient temperatures chemical explosives undergo negligible or no reaction. Indeed the reaction rate must not be appreciable if the explosive is to be considered safe for preparation and storage at temperatures up to 80–100°C, although many show evidence of slow decomposition at the upper limit of this range. At higher temperatures the characteristic reaction of the explosive takes place at rapidly increasing rates, and will transform suddenly into an explosion reaction at the "explosion temperature." Usually self-heating occurs at temperatures considerably below the "explosion temperature." Its extent depends on the heat balance ratio (fraction of the total heat of reaction retained in the explosive).

Reaction rates increase with temperature much more rapidly than heat dissipation to the surroundings (radiative, conductive and convective). While the "explosion temperature" may seem to be a sharply and distinctly defined value in any particular test designed to measure it, this is seldom, if ever the case; it is really a rather ill-defined property which depends critically on the many factors determining the heat balance ratio (confinement, size of charge, surrounding medium, rate of reaction, heat of reaction, and numerous other factors).

Sources of initiation include direct heating, flame, electrical discharge, impact or shock, contamination with incompatible chemicals which promote spontaneous decomposition, instability of the explosive itself, etc. Experimental evidence shows that all these sources act by producing sufficiently high temperature to initiate self-heating. Even those explosives which exhibit maximum shock or impact sensitivity, for example, are probably never exploded directly by shock or impact; shock of sufficient intensity to cause explosion merely generates hot spots or high-temperature regions which initiate self-heating, and the explosion results eventually from the latter.

While the pre-explosion process is characterized by a thermal chain branching or self-accelerating reaction, the explosion itself does not continue to accelerate indefinitely but quickly reaches a maximum rate, which may maintain until the explosive is entirely consumed, as in the detonating explosives, or the reaction may even decrease after a maximum rate is attained, as in some propellants. The conditions determining the limiting rate of reaction in explosion are the physical state of the explosive, the magnitude of exothermicity (or heat) of the reaction, and (in the low explosives) the heat balance ratio. The explosion proper is actually a very high-velocity reaction requiring total reaction times of the order of only miliseconds in the "low" explosives and microseconds in the high explosives. While reported "explosion temperatures" for useful explosives are in the range between about 150 and 350°C, no explosion ever occurs at these low temperatures, but requires a minimum temperature of perhaps more than 1000°C. However, this temperature range is bridged so suddenly that the temperature transient is seldom even observed by conventional experimental methods.

Explosive Deflagration

"Low" explosives are characterized by a reaction rate which increases nearly in direct proportion to the pressure (as a result of the influence of pressure on surface temperature), but always remains one or two orders of magnitude lower than in the detonating type. The limiting rate of reaction and pressure in propellant or granular "low" explosives is determined by the effective burning surface and the upper limit of surface temperature. The pressure-time (or p-t) curve of a propellant tends to exhibit a maximum usually below about 50,000 psi. By careful control of the geometry and size of the propellant grains one may control to a large extent the nature of the p-t curve associated with the explosion of a propellant. Indeed, this is one of the most important problems in propellant technology.

The relatively low rates of pressure development and peak pressures of the "low" explosives, in addition to rendering them useful in guns and rockets as propellants, give them a desirable "blasting action" for coal mining and other commercial blasting operations where "heaving action" (or sustained pressure) and controlled fragmentation are desired. Black powder (an intimate mixture of sodium or potassium nitrate, charcoal and sulfur) for centuries the sole source of explosive powder in both the military and commercial fields, has remained in use as a blasting explosive only because of its superior action for this type of blasting. Unfortunately, it is a very dangerous explosive owing to its extreme sensitivity to primary ignition sources, particularly spark and flame. Moreover, because of its relatively long reaction time, it is a very hazardous explosive for use in an explosive dust or gas environment, e.g., in dusty or gassy coal mines. Its chief present use in the military—as the igniter or fuse element—takes advantage of its ease of ignition and hot flame of relatively long duration.

The smokeless powder propellants, which represent the only other "low" explosives in extensive use today, are indispensable in the military field, but owing to their cost they have little commercial value. However, government-surplus smokeless powder is now being used extensively as a sensitizer for "slurry" explosives used in commercial blasting. Smokeless powders are the basis for most modern artillery, small arms, and rocket ammunition. Three general types are available, the "single base" powders, in which nitrocotton is the basic ingredient, the "double base" powders containing primarily nitrocotton and nitroglycerine, and the "triple base" powders made principally with nitroguanidine, nitroglycerine and nitrocellulose.

Detonation

Detonation is a process in which the (effective part of the) explosive reaction takes place within a high-velocity shock wave known as the "detonation wave" or "reaction shock." This wave generally propagates at a constant velocity from below 800 to above 8500 meters/sec, depending on the explosive, its density and physical state. Detonation velocity may be measured accurately by direct flame photography using "streak" cameras of various types and by electronic chronographs. Since the pressures generated by detonation may be controlled over a broad range as low as the maximum pressures of "low" explosives to about 100 times greater, the detonating explosives are more readily adaptable to most military and commercial demolition and blasting purposes than the low explosives. Even in operations where the low explosives exhibit better blasting action, they are less desirable than the detonating explosives because of their tendency to initiate explosions in explosive gas or dust-air mixtures. The commercial industry is today therefore based on the detonating type.

Owing to their great tendency to transform very suddenly into detonation from a very short duration pre-explosion reaction, the primary explosives are the basis of blasting caps and fuzes (military detonators) which are used to create the detonation wave in the less sensitive secondary explosives. Some of the most important primary explosives are mercury fulminate, lead azide, diazodinitrophenol, nitromannite, and lead styphnate.

In the applications of the secondary explosive one always requires the use of a detonator and in some cases also a booster in order that detonation may be produced easily and reliably. A booster is usually a small, pressed, cast, or low-density (loose-packed) charge of one of the more sensitive secondary explosives capable of being detonated readily by a detonator. It is used to reinforce the detonation wave from the detonator and thus deliver to the secondary explosive a more powerful detonation wave than is possible with a detonator alone. This makes possible the use of many relatively insensitive explosives that otherwise would have no direct application.

Even the least sensitive secondary explosive may explode in essentially the same type process involved in forming the detonation wave in primary explosives, except that the transition time and the amount of explosive required to go from the pre-explosion reaction to detonation may be very much larger in the secondary than the primary type. In some cases tons or even thousands of tons of the explosive are required for explosion to occur spontaneously following initiation. This is strikingly illustrated by the Texas City catastrophe and that of Brest, France, which involved one of the least sensitive of all secondary explosives, namely (fertilizer) ammonium nitrate, which in both instances ignited spontaneously, burned for some time, and then exploded.

Secondary explosives develop detonation pressures from a minimum of about 2,500 to a maximum of about 350,000 atmospheres. Even in a single explosive the detonation pressure may be varied over a wide range simply by varying the apparent density. The lowest detonation pressures are obtained by combining the effects of low density and dilution either by inert additives or incomplete chemical reaction accomplished by particle size control. Both means are used in the design of coal mining explosives ("permissibles") to produce lump coal where the lowest detonation pressures are required. The detonation temperatures of secondary explosives may range from as low as about 1500°C to 5,500°C or higher depending on the nature of the explosive.

Some of the most important military and commercial pure explosive compounds are RDX (cyclotrimethylenetrinitramine; HMX (cyclotetramethylenetetranitramine); PETN (pentaerythritol tetranitrate); NG (nitroglycerine); tetryl (trinitrophenylmethylnitramine); TNT (trinitrotoluene); AN (ammonium nitrate), picric acid and ammonium picrate. While pure AN can be detonated with heavy boostering only in large diameters (e.g., 5″ or greater) and/or under heavy confinement, it is an important source of explosive power in the commercial field because of its low cost. It is used, however, only in mixtures with combustibles and/or other explosives in which its strength is greatly enhanced. A suitable mixture of AN and combustible such as wax or wood pulp has approximately the same explosive potential as cast or pressed TNT. While RDX, HMX, and PETN are among the most powerful chemical explosive compounds known, they are never used in the pure state owing to their excessively high sensitivity, but instead are used in conjunction with other less sensitive explosives or nonexplosive ingredients in desensitized or phlegmatized condition.

In the past liquid NG, comparable in strength to RDX and PETN, was used extensively for oil well shooting, despite its exceedingly high sensitivity, but its use incurred many devastating accidental explosions. Today NG has been largely replaced in oil well shooting by far less sensitive types, e.g., non-NG high-AN explosives or desensitized liquid mixtures of NG, dinitrotoluene, and/or mononitrotoluene. However, oil well shooting itself has been largely replaced by other methods of sand fracturing.

Slurry explosives are replacing the older types of oil well explosives. The coarse TNT type slurry explosives are especially effective under very high hydrostatic pressures, but still better, more powerful slurry explosives capable of functioning at much higher temperatures are coming into use for deep, hot oil well blasting.

NG is also the basic ingredient of dynamites, and in rocket propellants. Dynamites are of two general classes, the "straight" dynamites containing "balanced dopes" consisting of mixtures of sodium nitrate and combustibles (wood pulps, meals, starch, sulfur, etc.), and the "ammonia" dynamites consisting of mixtures of ammonium nitrate and combustibles. These mixtures are carefully "oxygen balanced" (i.e., adjusted in composition to utilize all the available oxygen of the nitrate) for fume control and optimum strength. Calcium carbonate and other "antacids" are used in dynamites to prevent acids from accumulating during storage; dynamites are rendered unstable and dangerous in the presence of acid.

Numerous explosives based on AN (used either as the only explosive ingredient or sensitized by the use of a small amount of another explosive, e.g., TNT) have been used for commercial blasting, notable among which are the duPont "Nitramon" blasting agents. However, today these have been largely replaced by slurry explosives.

A decade ago two new types of commercial "blasting agents" came into extensive commercial use: "ANFO" and "slurry." ANFO, or "prills and oil," is a 94/6 mixture of porous prilled ammonium nitrate and fuel oil of average density about 0.85 g/cc. Owing to low cost and high weight strength, ANFO has moved ahead rapidly since its discovery in 1955 and today comprises about half of the commercial explosives industry. "Slurry Blasting Agents" or SBA are based on thickened or gelatinized aqueous ammonium nitrate slurries sensitized with TNT or other usually coarse solid high explosives, smokeless powder or aluminum. SBA, discovered by Cook and Farnam in 1956, is much higher in density and velocity than ANFO and is water resistant. SBA is intermediate in cost between dynamites and ANFO, but because of its superior properties is today gaining rapidly in commercial blasting even replacing ANFO in large open pit mines where the latter had been used very successfully. In hard rock blasting, the economic advantages of SBA are unsurpassed. Moreover, the aluminized slurry blasting agents are the safest, most powerful and economic modern explosives available, especially when handled by new field-mixing and loading system known as the "pump truck system."

The second most significant development in commercial explosives since the invention of slurry explosives by Cook and Farnam over a decade ago was the on-site mixing and loading system called the "pump truck," also a development of Cook and associates. First introduced in 1963, the "pump truck" has revolutionized open pit commercial blasting. Today commercial explosives are being bulk handled with an efficiency not even visualized a decade ago.

While slurry explosives first came into use in large diameter applications, they have since been perfected to the point that they may replace dynamites not only in large but also in small diameter, underground applications. Some slurry types are cap sensitive, i.e., able to be detonated with a blasting cap alone, without the use of the high pressure "booster" generally required for explosives of minimal sensitivity, the basic characteristic of slurry explosives.

Composition B (60/40/1 RDX-TNT-wax), cast from the slurry produced by melting the TNT, develops a detonation pressure of about 240,000 atm. Other secondary explosives of still higher detonation pressures are now in use. For example, the military explosive "PBX 9404" has a measured detonation pressure of about 350,000 atm. Another very useful military explosive capable of developing detonation pressures in the same range is cast 50/50 pentolite (PETN-TNT). These explosives are most useful where high "brisance" is needed, e.g., in military demolition operations and shaped charges. "Brisance" is identified with detonation pressure; before the development of the thermo-hydrodynamic theory it was an empirical somewhat ill-defined quantity interpreted as "shattering action." The aluminized military explosives, while not highly "brisant," are characterized by high available explosive energy; they develop only moderately high but sustained pressures. Explosives of this type, of which Torpex (RDX, TNT and aluminum) is a notable example, are among the most powerful chemical explosives available.

—M. A. Cook

EXPRESSION (Process). The separation of liquids from solids by compressively squeezing certain liquid-containing substances, such as separating oils from vegetable seeds and nuts. Equipment is designed to permit the liquids to be removed while still retaining the solids between the compressing surfaces.

In a *plate press*, material, such as fruit or seeds, is wrapped in special plate cloths. These then are placed between a series of hydraulically operated plates, which act on the materials much like a vise. The *box press* is similar with exception that shallow boxes enclose the pressed cake on two sides, thus simplifying the folding of press cloths. The *cage press* consists of a cylinder finely perforated with a hydraulically operated ram. Essentially, the material is squeezed at one end of the cylinder by the ram. The expressed oil flows through the perforations. There are numerous other configurations, including the pot press, the curb press, the V-disk press, and the roll press. In recent years, pressure also has been applied by using pneumatically inflated bags which serve the function of the mechanical ram.

EXTRACTION (Liquid–Liquid). Sometimes referred to as solvent extraction, this operation is effected by treating a mixture of different substances with a selective liquid solvent. At least one of the components of the mixture must be immiscible or partly miscible with the treating solvent so that at least two phases can be formed over the entire range of operating conditions. To be effective, one or more of the components must be dissolved from the mixture by the solvent preferential to the other components present. The solvent-rich phase that contains the preferentially dissolved component is termed the *extract layer*. The residual phase formed by the undissolved component (or diluent) and usually containing some solvent is called the *raffinate layer*. Either layer may be at the top or bottom of the separating vessel, depending upon relative densities. Other forms of solvent extraction include leaching, washing, and precipitative extraction (*salting out*).

Liquid–liquid extraction finds application in separating the components of condensed mixtures where vaporization methods, such as distillation or evaporation, may be impractical. This condition may arise because the substances to be separated may have comparable volatilities, are relatively nonvolatile, are heat-sensitive, or have one component present in very small concentration.

Liquid–liquid extraction finds wide application throughout the processing industries, including the manufacture of toluene, uranium vanadium, amino acids, coal-tar products, lube-oil refining, protein processing, and solvent refining of coal and oil shales.

Solvent extraction also is an important laboratory operation as in the recovery of oils from oil-bearing material. The material is placed in a porous container and subjected to treatment with solvent. The solvent containing some dissolved material passes through the porous membrane, leaving the undissolved residue in the container. The principle of countercurrent extraction may be utilized in consecutive containers, or the solvent may be vaporized from the solution, condensed onto the material and, by means of a syphon in the apparatus, withdrawn periodically to the solution compartment below, as in the Soxhlet type of apparatus. When a third substance is of different solubility in two nonmiscible liquids, this substance may be separated from the solution of lower concentration by shaking with the preferential solvent, and then separating the two liquid layers. The desired substance may be recovered from the solution by evaporation of the solvent. The effectiveness of separation is increased by the use of a given amount of extracting solvent in successively smaller portions rather than by a single extraction with the total amount.

Example: upon shaking 1 volume of liquid A plus 1 volume of liquid B, assume a concentration ratio of 1 (concentration of B)/10 (concentration in A) of the third substance C. See Table 1. Upon shaking 1 volume of liquid A plus $\frac{1}{2}$ volume of

TABLE 1. LIQUID–LIQUID EXTRACTION USING EQUAL VOLUMES IN SINGLE EXTRACTION

CONCENTRATION RATIO $\frac{B}{A}$	VOLUME RATIO $\frac{B}{A}$	AMOUNT OF C IN		FRACTION OF C IN	
		B	A	B	A
10	1	$10 \times 1 = 10$	$1 \times 1 = 1$	$\frac{10}{11} = 0.91$	$\frac{1}{11} = 0.09$

liquid B and, after separation, shaking 1 volume of liquid A (containing the residue of C) plus $\frac{1}{2}$ volume of liquid B, the results are indicated by Table 2.

TABLE 2. EFFECT OF SUCCESSIVELY SMALLER PORTIONS ON LIQUID–LIQUID EXTRACTION

	CONCENTRATION RATIO $\frac{B}{A}$	VOLUME RATIO $\frac{B}{A}$	AMOUNT OF C IN		FRACTION OF C IN	
			B	A	B	A
First extraction	10	0.5	10×0.5 $= 5$	1×1 $= 1$	$\frac{5}{6} = 0.83$	$\frac{1}{6} = 0.17$
Second extraction	10	0.5	0.17×0.83	0.03	0.14	0.03
Combined					0.97	0.03

A single equal-volume extraction would, therefore, remove 91% of C from A, whereas a double half-volume extraction would remove 97%.

EXTRACTIVE METALLURGY. That phase of metallurgy dealing with the removal of metals from minerals. Methods are discussed in this Encyclopedia under individual metals articles.

EXTRUSION. A majority of stock plastic shapes (bars, cylinders, special cross sections) are made in this way. Thermoplastic materials are heated in a plasticizing cylinder and by means of a rotating screw are forced through a die to provide the desired cross section. A variation of the process is used for extruding coatings of soft plastic materials over other materials.

Almost any profile can be imparted to the product, but of course, variations in profile are limited to two dimensions. The tooling costs for extrusion are low compared with injection molding. Thickness of the material can be controlled quite precisely. Production rates are high.

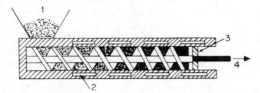

Mixer-type extruder. (1) Charge stock; (2) heating or cooling chambers in extruder jacket; (3) die; (4) extruded product.

Certain metals, including aluminum, and various rubbers are also extruded.

Extrusion is also combined with mixing in some applications. In the type of device shown in the accompanying diagram, the material is fed as a dry solid, is fluxed in the barrel to form a paste, and then resolidified at the discharge. Action in the barrel is one of shearing, rubbing, and kneading. One continuous screw or two screws rotate in the closely-fitting barrel. The work of the screw is augmented by forcing the material through breaker screens and around breaker disks just before the material is forced out through the exit nozzle. Such machines are often used for extruding soft chemical and food mixes which do not require fluxing, as well as for the extrusion of hard plastics, some of which must be fluxed at temperatures above 400°F (204°C). Wires can be covered and shapes of intricate cross section can be produced. Plastic resins also are blended in extruders to form pellets for later press and injection molding.

F

FATTY ACIDS. Carboxylic Acids; Chlorinated Organics; Lipids; Vegetable Oils (Edible).

FELDSPAR. Feldspar is the name of a group which includes the most important of the rock-forming minerals, making up perhaps as much as 60% of the earth's crust.

This group of minerals consists of three silicates: a potassium-aluminum silicate, a sodium-aluminum silicate, and a calcium-aluminum silicate ($KAlSi_3O_8$, $NaAlSi_3O_8$, and $CaAl_2Si_2O_8$) and their isomorphous mixtures.

The various members of the feldspar group show many characteristics in common. Crystallizing in the monoclinic and triclinic systems, they show similarity of crystal habit, cleavage and other physical properties as well as similar chemical relationships.

Orthoclase, $KAlSi_3O_8$, derives its name from the Greek words meaning right or straight, and fracture, because its two cleavages are at right angles to each other. It crystallizes in the monoclinic system and its crystals are usually prismatic; it occurs also in coarsely cleavable masses. Hardness, 6; specific gravity, 2.56–2.58; luster, vitreous to pearly; colorless to white, gray, yellow or red, rarely green. Twin crystals not uncommon.

Orthoclase is a common constituent of many igneous rocks and is often found in huge masses in pegmatite veins. Localities for orthoclase are so numerous as to prohibit a complete list. Adularia (from Adular) is essentially a pure potassium silicate; when pearly and opalescent it is called moonstone and frequently used for jewelry. These opalescent varieties are known to be an intergrowth of orthoclase and albite. A glassy kind of orthoclase, sanidine, is found in the trachytes of the Drachenfels, Germany. Beautiful moonstones come from Ceylon and Switzerland; in the United States, from California and Virginia.

Orthoclase is found in the New England pegmatites, in New York, Pennsylvania, Virginia, North Carolina, Arkansas, Texas, Colorado, California, and elsewhere. Its commercial use is in the manufacture of porcelain and as a constituent in scouring powders.

Microcline, $KAlSi_3O_8$, is chemically the same as orthoclase, but belongs to the triclinic system, the prism angle being slightly less than a right angle ($89°30'$), hence the name microcline from the Greek meaning small, and to slope. Microcline is like orthoclase in all physical properties and can be distinguished from it surely only by optical examination. Under the polarizing microscope microcline displays a minute multiple twinning which results in a grating-like structure that is unmistakable. It is probable that much orthoclase would, upon proper examination, prove to be microcline. Amazon stone or amazonite is a beautiful green microcline occurring in the Ilmen Mountains in the Urals, Italy, Norway, the Malagasy Republic; and in the United States, in the Pikes Peak region, Colorado, Virginia, North Carolina, and sparingly in the pegmatites of New England.

The name amazon stone is derived from the application of this term to some green mineral found by the Spaniards among the aborigines of the Amazon Valley in South America. As no microcline is known to occur in the region there must have been some confusion with another green-colored substance.

A soda microcline, anorthoclase, is known, which is probably an isomorphous mixture of $KAlSi_3O_8$ and $NaAlSi_3O_8$, the sodium-aluminum silicate being in the greater proportion. The soda feldspar albite, $NaAlSi_3O_8$ and the calcium feldspar anorthite, $CaAl_2Si_2O_8$ form an isomorphous series from pure albite at one end to pure anorthite at the other, the two molecules appearing to be completely miscible one with the other. The members of this series are spoken of as the soda-lime (or lime-soda) feldspars, and as a group are called the plagioclase feldspars from the Greek meaning *oblique* and *fracture*, referring to the two cleavages at an angle that differs slightly from a right angle. Nearly always present are the striations, fine parallel lines, resulting from minute multiple twinning, which, never seen on orthoclase or microcline, are therefore an important diagnostic feature.

More or less arbitrarily, four intermediate plagioclase feldspars are recognized between albite and anorthite; these are listed below together with the approximate percentage of each molecule present.

	% of $NaAlSi_3O_8$	% of $CaAl_2Si_2O_8$
Albite	100 to 90	0 to 10
Oligoclase	90 to 70	10 to 30
Andesine	70 to 50	30 to 50
Labradorite	50 to 30	50 to 70
Bytownite	30 to 10	70 to 90
Anorthite	10 to 0	90 to 100

Albite is so called from the Latin, *albus*, in reference to its usual pure white color. It is a sodium aluminum silicate corresponding to the formula $NaAlSi_3O_8$. It crystallizes in the triclinic system commonly in tabular crystals. Twinning is very common, thin twinning lamellae producing a series of fine striations on certain crystal faces. There are two good cleavages at an angle of $86°24'$ to each other. Hardness, 6; specific gravity, 2.62; luster, vitreous to pearly. It may be colorless to white or gray and transparent to opaque.

Albite is a relatively common and important rock-making mineral associated with the more acid rock types and in pegmatite dikes, often with rarer minerals like tourmaline and beryl. There are many famous localities in Europe in the Swiss and Austrian Alps, the Urals, the Harz Mountains, in Italy, France, and Norway. Brazil has yielded fine specimens. In the United States, notable localities are Paris and Auburn, Maine; Chesterfield, Mass.; Haddam, Connecticut; Amelia County, Virginia; and the Pikes Peak region of Colorado. It is used in the ceramic industries and also in the manufacture of artificial teeth.

Anorthite was named by Rose in 1823 from the Greek meaning oblique, referring to its triclinic crystallization. The physical properties are essentially the same as for albite, except that the specific gravity of anorthite is somewhat greater, 2.74–2.76. Anorthite is characteristic of the basic igneous rocks such as gabbro and basalt. Anorthite is found in the lavas of Vesuvius and Monte Somma, Italy; in Finland; Japan; and in the United States, in Sussex County, New Jersey.

The intermediate members of the plagioclase group are all very similar and with the exception of certain labradorites, cannot be distinguished from each other ordinarily save by optical means. Oligoclase is a common mineral in such rocks as granites, syenites, diorites, their extrusive equivalents and many gneisses. It is a frequent associate of orthoclase. The word *oligoclase* is derived from the Greek meaning *little* and *fracture*, in reference to the fact that its cleavage angle differs slightly from 90°. Sunstone is mainly oligoclase (sometimes albite) spangled with flakes of hematite.

Andesine is a characteristic mineral of rocks such as diorites which contain a moderate amount of silica and related extrusives, such as andesites. Because of its occurrence in these latter, andesine derives its name from them as well as from the Andes Mountains.

Labradorite is the characteristic feldspar of the more basic rock types like diorite, gabbro, andesite or basalt and it is usually associated with some one of the pyroxenes or amphiboles. Labradorite frequently shows a beautiful play of iridescent colors due to minute inclusions of another mineral. However, the labradorescent phenomenon has not been fully determined. The classic locality of this mineral is of course Labrador, whence its name. It is a constituent there of the rock anorthosite and is found in the anorthosites of the Provinces of Quebec and Ontario, and in the Adirondack region in New York State.

Bytownite, named from Bytown, the former name for Ottawa, Canada, is a rare mineral occasionally found in the more basic rocks.

The feldspars crystallize from the magma in both extrusive and intrusive rocks; they occur as contact minerals, in veins, and are developed in many sorts of metamorphic rocks, e.g., albite schists. They may also be found as mechanical deposits in various sedimentary rocks. See also terms listed under **Mineralogy.**

—Elmer B. Rowley, Union College, Schenectady, New York.

FERGUSONITE. A mineral multiple oxide containing niobium (columbium), tantalum, and titanium, corresponding to formula $YNBO_4$. Essentially an oxide, or niobate-tantalate of yttrium with varying amounts of erbium, cerium, and iron. Crystallizes in the tetragonal system. Hardness, 5.5–6.5; specific gravity, 5.6–5.8; color, variable. Named after Robert Ferguson (1799–1865), Scotland.

FERMENTATION. A modern definition of fermentation would be those energy-yielding reactions in which organic compounds act as both oxidizable substrates and oxidizing agents. Anaerobic reactions in which inorganic compounds are utilized as electron acceptors may be termed "anaerobic respirations," whereas reactions in which oxygen serves as a terminal electron acceptor are respirations. Almost any organic compound may be fermented provided it is neither too oxidized nor too reduced, since it must function as both electron donor and electron acceptor. In some fermentations, a compound is degraded via a series of reactions in which intermediates in the sequence act as electron donors and acceptors; in others, one molecule of the substrate may be oxidized while another molecule is reduced; or two different organic compounds may be degraded after a coupled oxidation-reduction reaction. These fermentations provide

energy required for the growth of a variety of cells. In addition, many microorganisms can carry out, in appropriate conditions, a number of fermentative reactions (e.g., oxidations, reductions, cleavages) which do not yield useful energy, or do not yield sufficient energy for growth.

In view of the great variety of different compounds that may be fermented and the enzymatic capabilities of different microorganisms, it is not surprising that numerous compounds important in industry (e.g., ethanol, particularly in connection with alcoholic beverages, butyl alcohol, acetone, 2,3-butylene glycol, among many others) are produced by fermentation. It should be pointed out, however, that although fermentation sometimes requires the least investment in processing equipment and starting raw materials, it is not always the most economical route for many end-products. At one time, nearly all industrial alcohol (ethyl) was produced via fermentation. Currently, most ethanol is prepared by the direct hydration of ethylene (vapor phase, catalytic addition of water); or by the indirect hydration of ethanol via the sulfation-hydrolysis process. See **Ethanol.** In terms of long-range conservation of nonrenewable hydrocarbon raw materials, interest has been growing in recent years in the fermentation of agricultural feedstocks to make fuels, as now practiced extensively in Brazil in the production of gasahol from sugarcane.

In food production, fermentative reactions play three fundamental roles: (1) They make it possible for goods that are ingested to be metabolized (used for energy and the building of indispensable substances for growth and maintenance). (2) They make it possible for the conversion of chemicals in natural food substances (such as the conversion of starches into sugars in the ripening of fruit), and for the conversion of chemicals in food processing to create such interesting items as bread, cheese, and wine, among many others. (3) They cause problems in the storage of various food products, not only in terms of economic loss, but also in terms of a number of foodborne diseases.

Alcoholic Fermentation. This process was first identified by Gay-Lussac in 1810, but at that time it was thought that the process resulted from contact catalysis and the decay of animal or vegetable materials. This explanation was refuted by the work of Pasteur (1857) on lactic acid fermentation. Pasteur determined that fermentation was caused by living cells, that different microbial species caused different fermentations, that the nitrogenous materials present served only to support the growth of the cells, that lactic acid was produced when cells (removed from the fermentation mixture) were added to a sugar solution, and that the natural fermentation yielded both alcohol and lactic acid, but that the amount of each could be altered by changes in pH. In later studies, Pasteur showed that the conversion of glucose to alcohol, $C_6H_{12}O_6 \rightarrow 2CO_2 + 2C_2H_5OH$, was caused by yeast cells growing under anaerobic conditions. This lead to the earlier definition that fermentation was "life without air," which has since been displaced by the modern definition given earlier in this entry.

In addition to alcoholic fermentation, there are scores of other types of fermentative processes, including:

Amolytic fermentation—the fermentation of starch, but specifically it is an incomplete fermentation of starch in which simple sugars are not produced.

Butyric Fermentation—in which butyric acid is produced. The organisms producing this type of fermentation are mainly anaerobic like *Clostridium butyricum.* Some organisms, such as *Clostridium tetani*, the organism causing tetanus, and *Clostridium botulinum*, theorganism causing botulism, also produce this type of fermentation.

Lactic Fermentation—in which lactic acid is produced. This is an important fermentation for the preservation of food. *Lacto-*

bacillus bulgaricus, L. casei, and *Streptococcus lactis* are used for the manufacture of dairy products, such as sour cream. *Lactobacillus plantarum* is used in the preservation of certain vegetables, such as the production of pickles and kraut.

Controlled Oxidative Fermentations—by which a number of industrial chemicals are produced. *Citromyces,* for example, can be used for the production of citric acid from sugar. *Aspergillus niger* will yield oxalic acid by partial oxidative fermentation, but if the mold is permitted to remain in contact with the acid, it will convert it to carbon dioxide.

Some sugars such as glucose may be completely oxidized to carbon dioxide by certain bacteria, most molds, and some yeasts. Such microorganisms produce complete oxidation by fermentation. Many bacteria and yeasts are able to produce a gassy fermentation. The gaseous end product in the fermentation of vegetable products with *Leuconostoc mesenteroides* and *Lactobacillus brevis* is carbon dioxide. The gaseous end products of the coliform group are carbon dioxide and hydrogen.

Ropy Fermentation—which causes the spoilage of foods. Ropy milk is caused by *Aerobacter aerogenes, Lactobacillus bulgaricus, L. casei,* and *Alcaligenes viscosus.* Ropy bread is caused by members of the *Bacillus mesentericus group* which are identical with or are strains of *B. subtilis* or *B. pumilus.* Rope in maple syrup is produced by *A. aerogenes.*

A characteristic cultural reaction of *Clostridium perfringens* and many other clostridia is known as a *stormy fermentation.* When the organism is inoculated into milk the lactose is fermented and the casein is coagulated.

The anaerobic respiration which takes place in the muscles of higher animals, when insufficient oxygen is available for a complete breakdown of the food, is also called fermentation. Lactic acid and carbon dioxide are the products of this type of fermentation.

For closely related subject matter, see also **Yeasts and Molds.**

References

Bodyflet, F. W., and H. Yang: "Utilization of Cheese Whey for Wine Production," Proceedings of 70th Annual Meeting of American Dairy Science Association, Manhattan, Kansas, 1975.
Bungay, H. R.: Inexpensive Computers Aid Continuous Fermentation," *Chem. Eng. Progress,* **76,** 4, 53–54 (1980).
Humphrey, A. E., and E. K. Pye: "Enzyme Technology, Past, Present, and Future," Proceedings of the 1st RANN Symposium on Productivity, National Science Foundation, Washington, D.C., 1973.
Humphrey, A. E.: "Fermentation Technology," *Chem. Eng. Progress,* **73,** 5, 85–91 (1977).
Kunkee, R. E.: "Malo-lactic Fermentation," *Adv. Appl. Microbiol.,* **7,** 235 (1967).
Pirt, S. J., "Principles of Microbe and Cell Cultivation," Halsted Press, New York, 1975.
Webb, A. D. (editor): "Chemistry of Winemaking," Advances in Chemistry Series, Vol. 137, American Chemical Society, Washington, D.C., 1974.

FERMIONS. Those elementary particles for which there is antisymmetry under intra-pair production. They obey Fermi-Dirac statistics. Fermionic hadrons are called *baryons;* other hadrons are *mesons.* See also **Baryons;** and **Mesons.** The experimentally observed fermions are the leptons and baryons, each of which has a spin angular momentum of $\sqrt{3}h/4\pi$, where h is Planck's constant; this is equivalent to stating that each fermion has a spin quantum number of magnitude $\frac{1}{2}$. See **Particles (Subatomic).**

FERMI SURFACE. Of a metal, semimetal, or semiconductor, that surface in momentum space which separates the energy states which are filled with free or quasi-free electrons from those which are unfilled. Such a surface exists simply because the electrons obey Fermi-Dirac statistics. It is a surface of constant energy and is sometimes called the *Fermi level.*

If one considers an elementary model of a metal consisting of a lattice of fixed positive ions immersed in a sea of conductance electrons which are free to move through the lattice, every direction of electron motion will be equally probable. Since the electrons fill the available quantized energy states starting with the lowest, a three-dimensional picture in momentum coordinates will show a spherical distribution of electron momenta and, hence, will yield a spherical Fermi surface. In this model, no account is taken of the interaction between the fixed positive ions and the electrons. The only restriction on the movement or "freedom" of the electrons is the physical confines of the metal itself.

For many metals, the "nearly free" electron description corresponds quite closely to the physical situation. The Fermi surface remains nearly spherical in shape. However, it may now be intersected by several Brillouin zone boundaries which break the surface into a number of separate sheets. It becomes useful to describe the Fermi surface in terms not only of zone or sheets filled with electrons, but also of zones or sheets of holes, that is, momentum space volumes which are empty of electrons. A conceptually simple method of constructing these successive sheets, often also referred to as "first zones," "second zones," and so on was demonstrated by Harrison. An example of such construction is shown in the accompanying figure.

This construction works out quite well, for example, for aluminum which has three valence electrons per atom. Experiments and more elegant theoretical calculations show that the fourth zone is totally unoccupied and that the third zone is not multiply connected in the manner shown.

FERMIUM. Chemical element symbol Fm, at. no. 100, at. wt. 257 (mass number of the most stable isotope), radioactive metal of the actinide series; also one of the transuranium elements. Both fermium and einsteinium were formed in a thermonuclear explosion which occurred in the South Pacific in 1952. The elements were identified by scientists from the University of California's Radiation Laboratory, the Argonne National Laboratory, and the Los Alamos Scientific Laboratory. It was observed that very heavy uranium isotopes which resulted from the action of the instantaneous neutron flux on uranium (contained in the explosive device) decayed to form Es and Fm. The probable electronic configuration is $1s^2 2s^2 2p^6 3s^2 3p^6 3d^{10} 4s^2 4p^6 4d^{10} 4f^{14} 5s^2 5p^6 5d^{10} 5f^{12} 6s^2 6p^6 7s^2$. Ionic radius is 0.97 Å.

All known isotopes of fermium are radioactive. They include isotopes of mass numbers $248(t_{1/2} = 0.6\text{m.})$, $249(t_{1/2} = 150\text{s.})$, $250(t_{1/2} = 0.5\text{h.})$, $251(t_{1/2} = 7\text{h.})$, $252(t_{1/2} = 23\text{h.})$, $253(t_{1/2} = 4.5\text{d.})$, $254(t_{1/2} = 3.24\text{h.})$, $255(t_{1/2} = 22\text{h.})$, $256(t_{1/2} = 160\text{m.})$, and an isotope of mass number 257 which has a half-life of about 10 days. The mass number of fermium given in atomic weight tables is 253, the mass number of the isotope of longest half-life (except possibly for the last one cited). The first three decay by alpha-particle emission only; the others, through ^{255}Fm, by that and other modes, including spontaneous fission for ^{254}Fm and ^{255}Fm. Isotopes of 256 and higher decay by this mode only. Fermium isotopes are made (1) by heavy iron bombardment of uranium and plutonium isotopes, (2) by the action of alpha particles on californium isotopes, and (3) from several of the other transuranium elements by multiple neutron capture—or in the case of the heaviest isotopes, from einsteinium by single neutron capture.

The ion Fm^{3+} is stable, and fermium is probably exclusively trivalent.

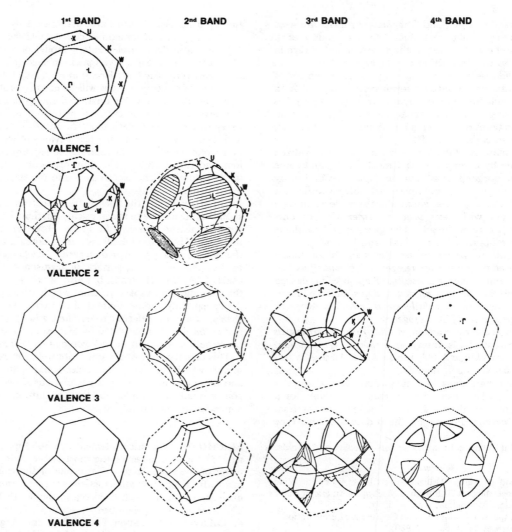

1st BAND   **2nd BAND**   **3rd BAND**   **4th BAND**

VALENCE 1

VALENCE 2

VALENCE 3

VALENCE 4

Fermi surfaces in several zones or bands, for face-centered cubic metals having various numbers of "quasi-free" electrons per atom, as constructed by Harrison.

The discovery of fermium (also einsteinium) was not the result of very carefully planned experiments, as in the cases of the other transuranium elements, but fermium and einsteinium were found in the debris of an atomic weapon test in the Pacific in November 1952. Researchers, using the Oak Ridge High Flux Isotope Reactor (HFIR) which produced 3.2-hour ^{254}Fm, determined the magnetic amount of the atomic ground state of the neutral fermium atom with a modified atomic-beam magnetic resonance apparatus. In essence, this observation represented that of a macroscopic property of the metallic 0-valent state of fermium.

References

Diamond, H., et al.: "Heavy Isotope Abundances in Mike Thermonuclear Device," *Phys. Rev.*, **119**, 6, 2000–2004 (1960).
Fields, P. R., et al.: "Additional Properties of Isotopes of Elements 99 and 100," *Phys. Rev.*, **94**, 1, 209–210 (1954).
Ghiorso, A., et al.: "New Elements Einsteinium and Fermium, Atomic Numbers 99 and 100," *Phys. Rev.*, **99**, 3, 1048–1049 (1955).
Goodman, L. S., et al.: "g_J Value for the Atomic Ground State of Fermium," *Phys. Rev.*, **4**, 2, 473–475 (1971).
Hulet, E. K., Hoff, R. W., Evans, J. E., and R. W. Lougheed: "79-Day Fermium Isotope of Mass 257," *Phys. Rev. Lett.*, **13**, 10, 343–345 (1964).
John, W., Hulet, E. K., Lougheed, R. W., and J. J. Wesolowski: "Symmetric Fision Observed in Thermal-Neutron-Induced and Spontaneous Fission of ^{257}Fm," *Phys. Rev. Lett.*, **27**, 1, 45–48 (1971).
Seaborg, G. T., Editor: "Transuranium Elements," Dowden, Hutchinson & Ross, Stroudsburg, Pennsylvania (1978).
Thompson, S. G., Harvey, B. G., Choppin, G. R., and G. T. Seaborg: "Chemical Properties of Elements 99 and 100," *American Chemical Society Journal*, **76**, 6229–6236 (1954).

FERRITE. The existence of ceramic magnetic materials capable of combining the resistivity of a good insulator (10^{12} ohm-cm) with high permeability was announced by Snoek (1946). Shortly thereafter, Néel (1948) introduced the term *ferrimagnetism* to describe the novel magnetic properties of these materials. A simple ferrite is composed of two interpenetrating ferromagnetic sublattices with magnetizations $M_a(T)$ and $M_b(T)$ which decrease with increasing temperature and vanish at the Curie point, T_c. In a ferromagnetic material, the resulting saturation magnetization, M, would be $M_a + M_b$. However, in a ferrite, strong antiferromagnetic interaction between sublattices results in antiparallel alignment, and $M = M_a - M_b$. In general,

$M_a(T) \neq M_b(T)$, and the material behaves in most respects like a ferromagnet, exhibiting domains, a hysteresis loop, and saturation of the magnetization at relatively low applied magnetic fields. Practical values for saturation magnetization and Curie temperature range from 250 to 5000 oersteds (19,984–397,887 ampere turns/meter) and from 100 to 600°C.

Ferrites resemble ceramic materials in production processes and physical properties. The high resistance of ferrites makes eddy-current losses extremely low at high frequencies. The direct current resistivities correspond to those of semiconductors, being at least one million times those of metals. Magnetic permeabilities may be as high as 5,000 and dielectric constants in excess of 100,000. Ferrites provide design advantages over strip and powder cores for filter cores, deflection transformers, and yokes and in antenna rods, pulse transformers, delay lines, waveguide elements, and a number of other electronic components.

Ferrimagnetic materials have spinel, garnet, and hexagonal structures. A typical spinel ferrite is $NiFe_2O_4$. Other ferrites may be obtained by substituting magnetic (cobalt, nickel, manganese) or nonmagnetic (aluminum, zinc, copper) ions for some of the nickel or iron ions, e.g., $Ni_{1-y}Co_yAl_xFe_{2-x}O_4$, where x and y may be varied to modify M and T_c. Yttrium-iron garnet (YIG), $Y_3Fe_5O_{12}$, is the classical ferrimagnetic garnet which combines very low magnetic loss with high resistivity. Substitution of magnetic rare-earth ions (gadolinium, ytterbium, holmium, etc.) for Y and of nonmagnetic ions (gallium, aluminum) for some of the Fe ions leads to many different compositions with a wide range of M and magnetic loss. The rare-earth ions form a third magnetic sublattice with attendant magnetization M_c antiparallel to the resultant magnetization $M_{a,b}$ of the two Fe sublattices. Since M_c and $M_{a,b}$ exhibit different variations with temperature, the net magnetization may vanish twice, at T_c and at an intermediate temperature called the compensation point, T_{comp}, where $M_c = M_{a,b}$.

A typical hexagonal ferrite is $BaFe_{12}O_{19}$. Again, other magnetic ions, such as manganese, cobalt, and nickel may be introduced to produce wide variations in M and T_c. Hexagonal ferrites are characterized by large anisotropy fields with an axis of symmetry which may be either a direction of hard (planar ferrites) or easy (uniaxial ferrites) magnetization.

To distinguish among major fields of applications, ferrites can be separated into five groups: soft, square-loop, hard, microwave, and single-crystal ferrites.

Soft ferrites have a slender S-shaped hysteresis loop with low remanence and low coercive force permitting easy magnetization and demagnetization with little magnetic loss. These ferrites are uniquely suited to low-loss inductor and transformer cores for radio, television, and carrier telephony.

Square-loop ferrites are materials exhibiting an almost rectangular hysteresis loop with two distinct states of remanence and with a coercive force of a few oersteds. All practical square-loop ferrites have a spinel structure. The Mg-Mn (Zn) system has retained an important position in computer memory applications. Lithium-nickel ferrites and more complex systems containing Li, Mn, and Al have shown stability and fast switching over a wide range of temperatures.

Hard ferrites are characterized by hexagonal structure, a hysteresis loop enclosing a large area, and a coercive force of several thousand oersteds. These ferrites can store a significant amount of magnetic energy and have been used as permanent magnets in loudspeakers, small motors, generators, and measuring instruments.

Microwave ferrites have garnet, spinel, or hexagonal crystal structure and very low electric and magnetic loss factors. In general, the required M increases with the frequency f of applications. Substituted and pure garnets, Mg-Mn-Al ferrites and Mg-Mn ferrites are used at the lower part of the microwave spectrum where $M = 200$–3000 gauss is adequate. In the millimeter wave region, $f = 30$–100 GHz, Ni-Zn ferrites ($M = 5000$ gauss) and hexagonal ferrites of various compositions may be used. Microwave ferrite devices, such as isolators, circulators, switches, phase shifters, limiters, parametric amplifiers, and harmonic generators are based upon interactions of rf signals with the ferrite magnetization.

Single-crystal ferrites of practical importance are rare-earth garnets grown in a flux of molten lead oxide. Some of these are optically transparent, permitting direct observation of magnetic domains. Interactions of infrared and visible light with the electron spins is called the magneto-optic effect. It permits electronic modulation of a beam of light which propagates through a single-crystal garnet. These devices are also of interest in laser technology.

Single-crystal, rare-earth garnet sheets have been grown on a substrate with a preferred direction of magnetization perpendicular to the plate. In these plates, tiny round magnetic domains called "bubbles" can be formed by an applied magnetic field. These bubbles can be propagated, erased, and manipulated to perform binary functions in computers, including logic, memory, counting, and switching. See also **Rare-Earth Elements and Metals.**

FERTILIZER. A substance, but often a combination of substances, of organic composition, natural and/or manufactured, in solid or liquid slurry forms (in some cases in the gaseous phase) made available to plants to promote normal, healthy, and often vigorous growth. Most frequently added to soils, fertilizers also are applied directly to plant parts above ground (foliar sprays), in nutrient fluids as furnished in hydroponic systems and by irrigation systems.

Unless poisoned or severely leached, some soils still may retain some of the nutrients required by growing plants and may support plants of a weak, straggly nature, with submarginal yields and poor quality for a number of years. Some soils may be generally poor, that is, they originally did not contain any of the primary plant nutrients in adequate concentration; or they may be poor soils because they have an imbalance of nutrients. It should be stressed that the proper chemical nutrients must be present along with suitable physical properties of the soil if highest yields and quality are to be achieved. A given soil may be classified as generally rich and yet lack the needed concentration of only one or two micronutrients. Further, considering the soil-plant system, soils should be customized to the needs of specific crops; or, in reverse, certain crops should not be planted on soils that are severely out of balance with their needs.

Because of the intensity of cultivation in many regions of the world, notably among the larger producers of food and fiber crops, where there is repetitive use of the same tracts of land year after year, the uniformity of nutrient content, as may have been present in the virgin soil, cannot be assumed. In fact, experience has shown that the same soil class may vary widely in nutrient content from one field to the next. The ability of soils to provide plant nutrition is a reflection of how the soils have been artificially treated (fertilized and conditioned) over the remote and recent past and what kinds of crops have been grown on them. With increasing use of control chemicals, soils require increasing frequency of testing and analysis, not only for required plant nutrients, but also for the possible presence of chemicals that tend to accumulate rather than dissipate downward to deeper ground levels or to the groundwater. Some substances may not be biodegradeable and hence may be present from one growing season to the next.

Particularly during the last few decades, astute growers have learned to depend upon reliable scientific analysis of their soils, augmented by tissue analyses of plant parts, and sometimes total plant analysis—as a basis for planning an annual fertilizing program. Perhaps not sufficiently stressed are problems that can arise from overfertilization as well as from underfertilization. Crops vary immensely in their ability to tolerate deficiencies and excesses of nutrients. Overfertilization also can cause serious pollution problems. Overfertilization also adds a needless element to the cost of food production.

Basic Fertilizer Functions. Major reasons for adding fertilizer to soils and plants include: (1) replenishment of chemical elements that have been reduced or exhausted by the soils to the crops previously grown or leached from the soils as the result of poor tillage practices, overirrigation, natural flooding, and, in some cases, adding nutrients that are naturally deficient in a given type of soil; and (2) customizing the nutrient content of soils to particular growing objectives.

Fertilization, in combination with irrigation, has been responsible for converting vast semiarid and arid lands with lean soils into land useful for production of a number of important food and fiber crops. This effective combination is particularly important to many of the developing countries that have large holdings of land in these categories.

Although essentially self-evident, the fact that crops exhaust the soil of key chemical ingredients is made even clearer by examination of Table 1. In the analysis, nitrogen is lost and does not appear in the ash, but other methods for analyzing for nitrogen confirm its significant content in crops. Carbon, oxygen, and hydrogen, of course, are made available to crops by way of water and the atmosphere. See Table 1.

Principal Categories of Nutrients. The nutrients required by plants fall into three categories: (1) *primary nutrients or elements*—nitrogen, phosphorus, and potassium, because they are generally required by plants in larger amounts and are often present in more limited amounts in soil; (2) *secondary nutrients or elements*—calcium, magnesium, and sulfur, because they generally are not so limited in soils and they are required in smaller amounts; and (3) *micronutrient elements*, of which there are several and which are required in very small amounts. See Table 2.

Nitrogen Requirements

Although the requirements for nitrogen are well understood today, this was far from self-evident just a few centuries ago.

TABLE 2. PRINCIPAL CATEGORIES OF PLANT NUTRIENTS

PRIMARY NUTRIENTS OR ELEMENTS	SECONDARY NUTRIENTS OR ELEMENTS
Nitrogen (N) Phosphorus (P) Potassium (K)	Calcium (Ca) Magnesium (Mg) Sulfur (S)

MICRONUTRIENTS

| | General Range in Soils | |
Element	Pounds/Acre	Kilograms/Hectare
Boron (B)	20–200	22.4–224
Manganese (Mn)	100–10,000	112–11,200
Zinc (Zn)	10–600	11.2–672
Copper (Cu)	2–400	2.2–448
Iron (Fe)	10,000–200,000	11,200–224,000
Molybdenum (Mo)	1–7	1.1–7.8

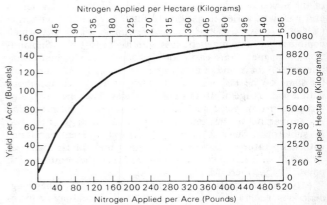

Fig. 1.—Effect of nitrogen fertilizer application on crop of irrigated corn (maize) in state of Washington. It should be pointed out that the very high yields were obtained under experimental, carefully controlled conditions. Although much higher yields are obtained in some regions, the average yield in the United States is 87.3 bushels/acre (5489 kilograms/hectare). The world average is 45 bushels/acre (2829 kilograms/hectare).

See Fig. 1. It seems reasonably safe to postulate that the early growers of plants practiced rudimentary forms of fertilization without understanding what they were doing beyond "some-

TABLE 1. CONSTITUENTS OF ASH OF NORMAL CROPS

| | POUNDS OF CONSTITUENT REMOVED PER ACRE OF GROUND (Kilograms per Hectare given in parentheses) | | | | | | | | |
CROP AND PART	Silica	Potash	Soda	Magnesia	Lime	Ferric Oxide	Chloride	Sulfate	Phosphate
Grain	15 (16.8)	14 (15.7)	7 (7.8)	2 (2.2)	8 (9)	1 (1.1)	0 (0)	0 (0)	36 (40.3)
Straw	233 (261)	33 (37)	1 (1.1)	28 (31.4)	12 (13.4)	6 (6.7)	4 (4.5)	13 (14.6)	11 (12.3)
Roots	27 (30.2)	143 (160)	17 (19)	46 (51.5)	18 (20.2)	4 (4.5)	12 (13.4)	46 (5.5)	26 (29.1)
Tops	3 (3.4)	89 (99.7)	17 (19)	72 (80.6)	10 (11.2)	3 (3.4)	50 (56)	39 (43.7)	29 (32.8)
Hay	78 (87.4)	38 (42.6)	12 (13.4)	45 (50.4)	7 (7.8)	1 (1.1)	4 (4.5)	9 (10.1)	15 (16.8)

thing taken from the soil must be returned." The latter observation in itself was rather profound for ancient peoples. It remained for the awakening of chemistry in the 1600s and notably by the early scientists of the late 1700s and early 1800s to commence investigations of the technical links between plants and their nutrients.

Chilean saltpeter, $NaNo_3$, was the first of the chemical nitrogenous fertilizers. Ammonium sulfate, $(NH_4)_2SO_4$, made available as a byproduct of coal-gas produced in large quantities prior to wide use of natural gas, was also an early fertilizer, but followed Chilean saltpeter by several years. Ammonium sulfate is still an important source of nitrogen, but no longer holds the lead role in most parts of the world.

In the early 1900s, attempts to fix atmospheric nitrogen in a compound that could be applied directly to soil were made. During that period, nitrogen fixation was a much-discussed aspect of industrial chemistry, much as synthetic fuels are a timely topic today. For some years, calcium cyanamide, $CaCN_2$, was produced in the United States, largely as the result, at that time, of newly developed hydroelectric energy of the Tennessee Valley region. A large facility was located at Muscle Shoals, Alabama. The last plant for making $CaCN_2$ in the United States was closed in June 1971. In Norway, also because of low-cost hydroelectric energy once available, production of calcium nitrate by way of first producing nitric acid was pioneered.

The first breakthrough in the large-scale synthesis of ammonia resulted from the work of Fritz Haber (Germany, 1913), who found that ammonia could be produced by the direct combination of nitrogen and hydrogen in the presence of a catalyst at a relatively high temperature and very high pressure. See also **Ammonia.** During the interim, much improved processes for producing synthetic ammonia have been designed.

For many years, the bulk of nitrogen fertilizers has been based upon ammonia, as synthesized. There has been an increasing trend in some major agricultural regions and countries to favor the use of liquid nitrogen for some crops. These products include anhydrous ammonia, aqua ammonia, ammonium salts in solution, and numerous combinations, such as formulations also containing phosphate salts. Because of the relatively easy solubility of these materials, some ecologists have expressed concern over so-called *nitrogen runoff.* Fortunately, however, most soils will fix ammonium ions quite rapidly, thus reducing a pollution threat.

It is interesting to observe that the compound urea, NH_2CONH_2, was discovered (in urine) as early as 1773 and first synthesized by Wöhler in 1828. See also **Urea.** Although there was an obvious connection between urea and life processes, little if any thought was given to its use as a fertilizer until after World War I, when the German firm BASF (Bosch) developed a process for synthesizing urea from carbon dioxide and ammonia. Urea-form fertilizers, now widely used, are the result of reacting urea and formaldehyde to provide a form of controlled-release nitrogen. Urea also can be coated with sulfur (itself required by some soils) to reduce the rate of solution. Also, slowly soluble compounds such as isobutylidene diurea can be made, but because of high cost these are marketed only to horticulturists rather than growers of commercial crops.

Although they are not chemical fertilizers (artificially produced), natural nitrogenous materials returned to the soil in various degrees of *organic farming* are of extreme importance and markedly reduce the total demand for chemical nitrogen fertilizers. Significant nitrogen needs are met by returning *crop residues* to the soil. There are some negative aspects to this practice as well, however, because residues may contain insects, fungi, bacteria, and weed seeds, which generally must be con-

trolled chemically. *Green manure* is a crop purposely grown for plowing under to enrich the soil. Animal manure also is an important contributor to the soil, not only in terms of nitrogen, but phosphate and potash as well. Also, certain crops, notably legumes, create more nitrogen for the soil than they use—this as a result of bacteria that inhabit the root zones of these plants. Collectively, these kinds of materials are sometimes called *organic fertilizer.* (See Fig. 2.)

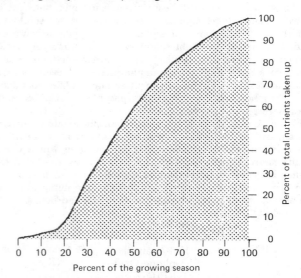

Fig. 2.—Composite nutrient uptake chart representing an average of many plants. Each plant has its own specific nutrient uptake curves. (*Source: Leitch reference listed.*)

Other chemical fertilizers used as sources of nitrogen include ammonium nitrate, which contains approximately 35% nitrogen by weight. Calcium ammonium nitrate contains between 20.5 and 28% nitrogen by weight, depending upon the amount of calcium carbonate added. Ammonium sulfate nitrate contains from 26 to 28% nitrogen by weight. See also **Nitrogen** and specific nitrogen compounds described in this book.

Phosphorus Requirements

Deficiencies of available phosphorus in soils are a major cause of limited crop production. Phosphorus deficiency is regarded by some authorities as the most critical mineral deficiency in grazing livestock.

Liebig and other investigators in the mid-1800s indicated that the much earlier noted fertilizing properties of bones were derived mainly from their phosphate content and that the treatment of bones with sulfuric acid increased their effectiveness in soils. With these advantages proclaimed, a large market for what was then called "chemical manure" resulted in a shortage of bones throughout the European agricultural community. Fortunately, a number of years later, several guano deposits were located in Peru, and quite a bit later phosphate minerals were located in Florida and other parts of the world. Guano fertilizers today are used mainly for special horticultural products, whereas commercial chemical phosphate fertilizers are derived from phosphate rock.

In formulating phosphate fertilizers, solubility is of particular concern. Good solubility assures availability of phosphorus to the plants. For example, in water and in alkaline and neutral soils, the apatites, which are $Ca_5(PO_4)_3R$ (where R is usually but not always fluorine), are quite insoluble and hence of little value to the soil. Tricalcium phosphate is also quite insoluble. On the other hand, these compounds are moderately soluble in acid soils and thus can be used with discretion. Somewhat in contrast, dicalcium phosphate, $CaHPO_4$, is quite soluble in acid soils and only moderately soluble in water and alkaline and neutral soils. Fortunately, monocalcium phosphate, $Ca(H_2PO_4)_2$, is soluble in water and all moist soils. The presence of iron and aluminum phosphates also has an effect on total solubility, these compounds being insoluble in water, but soluble in weak acids. Some countries require total water solubility and thus monocalcium and ammonium phosphates must be used. In other areas, slight solubility in water or appreciable solubility in weak acids is adequate to meet regulations. Some humic soils can assimilate ground phosphate rock even without chemical treatment. Commonly in phosphate fertilizer manufacture, phosphorus is recovered from phosphate rock. See also **Phosphorus;** and **Phosphoric Acid.**

Nomenclature. Because the ammonium phosphate fertilizers contain two of the primary plant nutrients, it is pertinent at this point to briefly comment on the manner in which these fertilizers are named. A series of three numbers, separated by dashes, is used to indicate the primary nutrient content of fertilizer mixtures. In order, from left to right, the numbers show the percentage of nitrogen, phosphoric oxide, and potash:

$$5 = 10 = 5$$

| 5% Total Nitrogen (N) | 10% Available Phosphoric Oxide (P_2O_5) | 5% Soluble Potash (K_2O) |

Single Superphosphate is produced in large quantities and is the oldest of the water-soluble phosphates. The material contains about 20% P_2O_5 by weight. Single superphosphate is made by reacting ground phosphate rock with 70% sulfuric acid. This reaction results in a solid mass of monocalcium phosphate and gypsum. The fluorine and silicon evolved are removed by water scrubbing.

Wet-Process Orthophosphoric Acid contains 30–54% P_2O_5 by weight. When sulfuric acid is added to phosphate rocks in a proportion greater than needed to make single superphosphate, orthophosphoric acid, H_3PO_4, is produced. This acid is used as an intermediate in preparing other phosphates.

Triple Superphosphate is made by acidulating phosphate rock with phosphoric acid. The concentrated triple superphosphate produced is essentially monocalcium phosphate containing very little gypsum. The principle use of triple superphosphate is in mixed fertilizers to make P_2O_5 available in water-soluble form.

Ammonium Phosphates. Although several ammonium phosphates can be prepared, only the mono- and di-compounds are produced for fertilizer use. In some processes, anhydrous ammonia is reacted with phosphoric acid, with the resultant slurry converted to solid form by drying. The ratio of ammonia to phosphoric acid can be varied between 1 and 2, and consequently several product grades can be made.

Nitrophosphates. Phosphate rock is readily dissolved in nitric acid to yield a mixture of calcium nitrate, phosphoric acid, and monocalcium phosphate. When calcium nitrate is converted to solid form, the material is highly hygroscopic and thus not desirable for packaging and storing. This is overcome by forming calcium nitrate tetrahydrate crystals through chilling and later removal by filtering or centrifuging. Ammoniation of the mother liquor produces a mixture of ammonium phosphate, dicalcium phosphate, and ammonium nitrate. These can be concentrated and prilled or granulated. Thus, in terms of the traditional fertilizer nomenclature, that is, %N : %P_2O_5 : %K_2O, numerous product grades are possible according to raw-material ratios and process conditions. By adding potash to 20-20-0, a formulation of 15-15-15 can be obtained, as one of numerous examples.

Nonorthophosphates. If 54% orthophosphoric acid is dehydrated to remove the remaining free water, a pyro acid is yielded. Continued heating removes more whater and various insoluble compounds are formed. Evaporation by submerged combustion or under vacuum yields a eutectic between ortho and pyro acids with a P_2O_5 content of about 72% by weight. This superphosphoric acid can be ammoniated to yield liquid ammonium polyphosphate (APP) fertilizers, such as 10-34-0. Because of their strong sequestering properties, these liquids keep impurities in solution, as well as the salts of micronutrient metals which are insoluble derivatives of orthophosphoric acid.

APP fertilizers also are produced by reacting ammonia directly with wet-process phosphoric acid and dissolving the melt in ammonia solution to produce 10-34-0. By adding potash and a bit of clay, nitrogen-phosphorus-potassium suspensions, such as 13-13-13, are possible. If the melt is granulated, a solid with proportions 12-57-0 will result. Thus, by adding urea and potash, a wide variety of mixes (28-28-0, 19-19-19, and so on) are possible.

Potassium Requirements

Researchers in the early 1800s found that potassium is an essential plant nutrient although the details of its function were unknown. The potassium in wood ashes was first used in Europe, later to be replaced by sylvite, KCl, and carnallite, $KCl \cdot MGCl_2 \cdot 6H_2O$, found in deposits in Germany. During the interim, billions of tons of these materials have been located. A number of sites are continuously mined. Large reserves are found in the U.S.S.R., Canada, and east and west Germany. Significant deposits also are found in Israel and Jordan (Dead Sea region), Spain, France, the United Kingdom, and the United States. Deposits are as water-soluble salts, such as sylvite, and minerals with varying content of magnesium.

The physiological role of potassium in life processes is described in the entry on **Potassium and Sodium (In Biological Systems).** Potassium is a usual, but variable constituent of most soils. However, available potassium may be depleted through loss arising from leaching by rain water, flooding, and overirrigation, or through loss to crops continuously planted on a given tract. Thus, potassium is categorized as a primary nutrient. Potassium differs from most other essential constituents of plant cells in that it is not built into the cell as part of an organic compound, but rather it is an ion from a soluble inorganic or organic salt. Potassium ions may chelate with cellular constituents, such as polyphosphates. The ion is of the correct size to fit into the water lattice adsorbed to the proteins in the cell. In general, potassium ions are attracted to protein or other colloidal or structural units having a negative charge. Mucopolysaccharides within the cell, on the cell surfaces and on the surfaces of the intercellular structures, are of particular importance in holding potassium. Active centers or other configurational features of the proteins in the cell may be affected or altered by the potassium held by electrostatic or covalent binding. There are several enzyme systems which are activated by potassium.

In plants, the meristematic tissues in general are particularly rich in potassium, as are other metabolically active regions, such as buds, young leaves, and root tips. Potassium deficiency may produce both gross and microscopic changes in the structure of plants. Effects of deficiency include leaf damage, high or low water content of leaves, decreased photosynthesis, disturbed carbohydrate metabolism, and low protein content, among other abnormalities. The importance of potassium is also reflected by livestock who consume plants and feedstuffs prepared from plant materials.

Potassium fertilizers are prepared from the potassium minerals previously mentioned. The ores are crushed, beneficiated, crystallized, and dried to commercial *potash* or *muriate* (KCl), in various grades and particles containing from 60 to 62% KCl. Relatively small amounts of other potassium salts are used as fertilizers. Some vegetables and tobacco are adversely affected by high chloride concentrations and some growers prefer to use potassium sulfate, K_2SO_4, or potassium nitrate, KNO_3.

Potash is frequently applied to the soil along with salts containing nitrogen and/or phosphoric oxide, P_2O_5, in amounts varied for different soil and crop requirements. One method used is that of combining crushed muriate with moist nitrogen- and P_2O_5-containing compounds, followed by granulation of the mixture. Another method is to dry-blend materials, such as urea diammonium phosphate, and potash, and apply the mixture to the soil. The total water solubility of potash results in full initial K_2O availability in moist soils. However, in clays, when rainfall or irrigation is limited, excessive chloride build-up can be harmful.

Calcium Requirements

The role of calcium in biosystems is described in the entry on **Calcium (In Biological Systems)**. Fortunately, as the fifth most abundant element in the earth's crust, calcium is usually available to plants in abundance. Nevertheless large amounts of calcium are added to soils by virtue of the use of lime (calcium oxide, CaO) as means to adjust the pH of soils. Among many reasons why soil pH is so important are the effects of acidic soils on the availability of manganese, a micronutrient, and also of aluminum, more recently identified as toxic to some crops. The use of dolomitic lime, $CaMg(CO_3)_2$, also is an effective way to correct magnesium deficiencies. Agricultural lime is not water soluble and therefore will not correct soil acidity immediately after application.

Sulfur Requirements

The role of sulfur in biosystems is described in the entry on **Sulfur (In Biological Systems)**. Sulfur in some form is required by all living organisms. Among important sulfur-containing compounds are the amino acids cysteine, cystine, and methionine; the vitamins thiamine and biotin; and certain complex lipids, such as sulfatides, among others. In the chain from soils to plants to animals, including humans, inorganic sulfur (sulfate ion, SO_4^{2-}) is taken up by plants and converted within the plant to organic compounds (sulfur amino acids). The most important feature of sulfur in the food chain is that plants use inorganic sulfur compounds to make the aforementioned amino acids, whereas animals use the sulfur amino acids for their own processes and excrete inorganic sulfur compounds.

Although sulfur is a widespread element in the earth's crust, ranking as the 14th element in abundance, it is obviously less abundant than calcium and, further, it is not so evenly distributed. Thus some soils show sulfur deficiency. The trend toward high-analysis fertilizers without sulfur can create a need for more deliberate use of sulfur fertilizers.

Sulfur is present in ammonium sulfate, used as a fertilizer, but ammonium sulfate is only one of many fertilizers that may be used. Elemental sulfur and sulfur-containing compounds are also used for controlling various plant pests. The soils also pick up sulfur from air pollution. Nevertheless, soil analysis should be made to provide an accurate diagnosis of whether or not a particular soil may require additional sulfur.

Magnesium Requirements

The role of magnesium in biosystems is described in entry on **Magnesium (In Biological Systems)**. Magnesium is generally abundant in the earth's crust, ranking eighth in abundance. Nevertheless, magnesium deficiencies are quite common. Magnesium is a fairly common cause of poor crop yields, especially among crops produced on sandy soils. Accumulation of magnesium from the soil by plants is strongly affected by the species of plant. Legumes usually contain more magnesium than grasses, tomatoes, corn (maize), regardless of the level of magnesium in the soil. A high level of available potassium in the soil interferes with the uptake of magnesium by plants, and thus magnesium deficiency can occur even though there are adequate amounts of the element in the soil. The role of magnesium highlights the systems aspects of agricultural management, exemplified by need to maintain a proper magnesium intake from pasture and from feedstuffs—because when animals are fed diets primarily of grains, a proper balance among magnesium, calcium, and phosphorus must be maintained to minimize danger of urinary calculi. Magnesium deficiency among cattle (grass tetany or grass staggers) is observed most frequently when animals are first grazed on lush grass or wheat pastures, indirectly indicating the relative low uptake of magnesium by certain crops.

Magnesium deficiencies are easily corrected by applying magnesium minerals, such as kieserite or dolomite.

Micronutrients

Even though the traditional micronutrients may be required only in minute quantities, deficiences can lead to diseased crops and stunted livestock. See also entries on **Boron; Copper (In Biological Systems; Iron; Manganese; Molybdenum (In Biological Systems); and Zinc (In Biological Systems)**.

Much more detail on all aspects of fertilizers, including worldwide consumption, methods of application, etc. can be found in the "Foods and Food Production Encyclopedia," (D. M. Considine, Editor), Van Nostrand Reinhold, New York (1982).

References

Bress, D. F., and M. W. Packbier: "Major Urea Plants," *Chem. Eng. Progress*, **73**, 5, 80–84 (1977).

Considine, D. M. (editor): "Foods and Food Production Encyclopedia," Van Nostrand Reinhold, New York, 1982.

Considine, D. M. (editor): "The Encyclopedia of Chemistry," 4th Edition, Van Nostrand Reinhold, New York, 1983.

Leitch, D.: "Matching Fertilizer Application to Crop Requirements," Chevron Chemical Company, San Francisco, California (Revised periodically).

Scheldrick, W. F.: "The Fertilizer Industry in Developing Countries," *Chem. Eng. Progress*, **75**, 10, 21–27 (1979).

Slack, A. V., and G. R. James (editors): "Ammonia, Part IV (Fertilizer Science and Technology Series), Vol. 2, Marcel Dekker, New York, 1979.

Staff: "Fertilizer Situation," U.S. Department of Agriculture, Washington, D.C. (Issued several times each year).

Staff: "Annual Fertilizer Review," Food and Agriculture Organization (United Nations), Rome (Issued annually).

Staff: "Production Yearbook," Food and Agriculture Organization (United Nations), Rome (Issued annually).

Swenson, R. J.: "The Need for Soil Testing," Chevron Chemical Company, San Francisco, California (Revised periodically).

Trivedi, R. N., Chari, K. S., and V. Pachaiyappen: "Polution Control in Fertilizer Industry," Fertiliser Assn. of India, New Delhi, 1979.

FIBER (Dietary).

Particularly during the past decade, there has been increased interest in the importance of fiber in the human diet. In terms of perspective, Hippocrates, as early as 400 B.C., identified bran as a laxative, but did not attribute this property to the fiber content of bran. As of the mid-1980s, fiber in terms of diet is defined rather loosely. In 1970, the Association of Official Analytical Chemists defined crude fiber as "the residue remaining after treatment with hot sulfuric acid, alkali, and alcohol. It consists primarily of cellulose, lignin, and trace amounts of other polysaccharides." In more recent years, the meaning of the term has been expanded, but not along very definitive lines. Perhaps a modernized definition would be "that part of plant material in the diet which is resistant to digestion by the secretions of the human gastrointestinal tract—consisting of variable proportions of complex carbohydrates, such as celluloses, hemicelluloses, pentosans, and uronic acids, as well as lignin," as proposed by a number of authorities.

A wide variety of foods supply significant amounts of dietary fiber. Total dietary fiber content (weight %) of representative foods is as follows: bran, 48%; wholemeal flour, 11%; brown flour, 8.7%; white flour, 3.5%; flesh of fresh pear, 2.5%; cooked cabbage, 2.8%; raw strawberry, 2.1%; banana, 1.8%; raw plum (flesh and skin), 1.5%; and raw tomato, 1.4%. Processed breads and cereals range from 0.2% (white bread) to 7.8% (all bran cereal).

From the standpoint of application in preventive medicine, fiber has two major functional properties: (1) absorption capacity, and (2) water-binding properties. Cardiovascular diseases relate to the ability of fiber to absorb materials which reduce blood cholesterol. The intestinal diseases appear to be related to fiber's water-binding properties and are influenced by the physical tone of the large intestine. Obesity is indirectly related to both properties. Wheat bran absorbs two times its weight in water.

Many claims for the beneficial effects of high fiber content of the diet remain to be proved scientifically. In addition to cardiovascular and intestinal diseases, some authorities have claimed beneficial effects in connection with colon cancer and diabetes.

References

AOAC: "Official Methods of Analysis," Association of Official Analytical Chemists, Washington, D.C. (1970).

Colmey, J. C.: "High-fiber Foods in the American Diet," *Food Technology*, **32**, 3, 42–47 (1978).

Heaton, K. W.: "Dietary Fiber," Food and Nutrition Press, Westport, Connecticut (1979).

Spiller, G. A., Editor: "Topics in Dietary Fiber Research," Plenum, New York (1978).

Wen-Li, J. J., and M. E. Zabik: "ENDF (Neutral Detergent Fiber Content) Values for Selected Foods," *J. Food Science*, **44**, 3, 924–925 (1979).

FIBER GLASS.

Glass in fibrous form. The material generally has properties similar to the glass from which it is made except that the tensile strength may be increased up to over 100 times that of the base glass. History records the use of strands of glass for decorating vases by the early Egyptians. The famed Venetian craftsmen had a limited knowledge of drawing glass fibers, but it was not until the 1930s and 1940s that glass producers perfected a way to make fibers commercially.

Two forms of glass fibers are produced, a staple or short-length fiber or monofilament and continuous strand composed of many monofilaments bonded together in a threadlike form. The continuous strands are often chopped into short lengths, ranging from $\frac{1}{8}$ inch to 2 inches (3 millimeters to 5 centimeters) or longer, and this product is referred to as *chopped strand*. Staple fibers are used for thermal and acoustical insulation. Continuous strands are used for yarn (as in fabrics), tire cord, and plastic reinforcement. Both thermoset and thermoplastic resins are reinforced by chopped strand. Some varieties of chopped strand are also converted to monofilament paper, which is utilized in roofing shingles, flooring materials, and other products. Products using continuous filaments show high fiber strength (as high as 400,000 pounds per square inch; 2759 megapascals) compared with organic fiber strengths of less than 150,000 psi (1034 megapascals).

The base glass is made by heating raw materials, such as silica sand, limestone, dolomite, clay, boric acid, soda ash, and other minor ingredients, in a high-temperature furnace. See also **Glass**. Typical glass-fiber compositions are given in Table 1.

TABLE 1. FORMULATIONS FOR TYPICAL FIBER GLASS TYPES.

INGREDIENT	TYPE OF GLASS, wt%				
	E	Insulating	A	S	C
SiO_2	54	63	73	64	65
Al_2O_3	14	5	1	24	4
MgO	4	2	2	10	3
CaO	19	6	10	—	14
R_2O	0.5	16	14	—	8
B_2O_3	8	7	—	0–2	6
Fe_2O_3	0.3				
F_2	0.2	1			

NOTE: R = rare-earth element.

Fiber made from electrical-grade glass E is used most commonly for yarn, tire cord, and plastic reinforcement because of its high strength and electrical properties. Specialty glasses, although low in volume of production, fill important needs. S glass is a superior-strength glass primarily for defense applications, such as missile cases. C glass is more chemically resistant than E and is used for battery separator plates and chemical filters. Alkali glass A is used to some extent in the production of plastic reinforcement products.

Continuous-Fiber Products. In the *direct-melt* process, shown in the accompanying figure, raw materials are fed to a tank furnace to convert the mixture to glass. The glass flows to forehearths, which have platinum-alloy bushings or spinnerettes in the bottom. The bushings contain many holes, or orifices, each of which supplies a small stream of molten glass from which monofilaments are drawn. Mechanical attenuation, which produces a forming package, is accomplished by attaching the fibers to a rotating drum which turns up to 20,000 peripheral feet (6000 meters) per minute.

The *marble-melt* process consists of producing 1-inch (2.5-centimeter) marbles by a separate tank furnace. The marbles are then fed to a bushing unit, which is heated by electrical resistance. From this point, the process is identical to the direct-melt process.

Sizing. Because of the basic character of glass, the filaments are somewhat fragile and tend to abrade each other in close

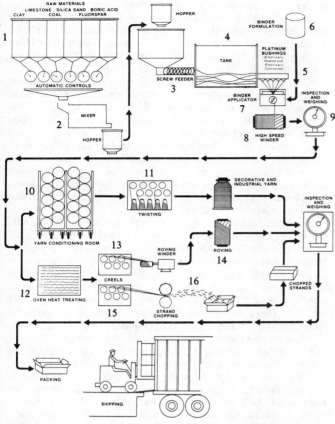

Direct-melt process for producing fiber glass. Raw materials (1) are automatically weighed and batched to mixer (2) prior to passing through screw feeder (3) to the glass melting tank (4). The molten glass flows to forehearths (5), at the bottom of which are platinum-alloy bushings or spinners. The latter are electrically heated and carefully temperature-controlled. Formulated binder material (6) is applied to the newly formed filaments (7) prior to high-speed winding (8). After weighing and inspecting (9), the wound multi-filament (in the form of a strand) follows one of three paths in accordance with desired end product. In making decorative and industrial yarn, the packages are placed in a conditioning room (10) prior to twisting (11). For the production of roving and chopped strand, the material from inspection operation (9) passes to an oven (12), where the filaments are heat-treated. This is followed by creeling (13) and roving winding (14) for production of roving. Following creeling (15), chopped strands (16) may also be made. There are several additional weighing and inspecting stations. (*PPG Industries.*)

contact. A protective coating or sizing is necessary for the production, processing, and end use of all continuous-fiber glass products. Generally, a fiber-glass sizing or binder for textile or reinforcement products may contain (1) a film former, generally resinous in nature, that forms a strand or thread from grouped monofilaments; (2) a lubricant to aid in processing and end use of the fiber-glass product; and (3) additives to accomplish specified purposes, e.g., providing antistatic characteristics.

Sizings for plastic reinforcement also will have a coupling agent, such as a chrome complex, a silane, or combination of these two, to assure an interfacial bond between the glass surface and the resin matrix. Yarns for weaving normally have an oil-

starch sizing. These coatings are applied before winding of the forming package.

Filaments ranging in number from 20 to 2000 are then gathered together as a thread or strand before winding. As shown by Table 2, the filaments are available in many diameters and

TABLE 2. CODING SYSTEM FOR FIBER GLASS DIAMETERS.

DESIGNATION	DIAMETER $\times 10^{-5}$ in.
AAA	<3.0
AA	3.0– 5.9
A	6.0– 9.9
B	10.0– 14.9
C	15.0– 19.9
DE	23.0– 27.9
G	35.0– 39.9
H	40.0– 44.9
K	50.0– 54.9
P	70.0– 74.9
Q	75.0– 80.0
R	81.0– 85.0
S	86.0– 90.0
T	91.0– 95.0
U	96.0–100.0

are letter-designated. In continuous-fiber products, filaments range from designation B to U.

Continuous-filament products are designated by a letter-number system which specifies properties important to end uses. For example, listing a strand as ECK67.5 (200) 630 indicates that the material is made from E glass and is a C continuous fiber of K diameter; the strand contains 200 monofilaments and has a yield of 67.5 × 100 (or 6750 yards; 6172 meters per pound). In the metric system, yield is expressed in Tex or grams per kilometer. The number 486,235 divided by yards per pound is equal to Tex. This product is coated at the bushing with 630 binder, an oil-starch type making it suitable for weaving into fabric.

Forming packages composed of wound strands normally are not supplied to industrial users without further processing. Strands are twisted and plied before being woven into fabric. A plied yarn, for example, is coated with a latex binder before being used as a tire-cord reinforcement.

End products made from continuous strand include fire-resistant curtains, reinforced tires and transmission belts, and many reinforced plastic items, such as boats, auto bodies, corrosion-proof pipe, roofing panels, and missile cases.

Staple-Fiber Products. Monofilament, short-length fibers are used for thermal and acoustical insulation, filtration, and cushioning. These products are made in basically three ways: In the high-temperature blast-jet process, 30-mil-diam. fibers or rods first are produced by a bushing-type process. The coarse primary rods then are filamentized by a high-temperature, high-velocity blast burner. The blown mass of filaments is collected on a conveyor belt and bonded together by an inert thermosetting resin in a manner which creates many tiny air spaces throughout the material. The bonding process may be modified to produce flexible blankets, rigid board, or special molded shapes, such as pipe insulation. Coatings, facings, or jackets usually are applied for reflective, vapor-barrier, or decorative purposes.

In another process (replacing the high-temperature blast-jet process in some areas), a stream of molten glass is directed onto a rapidly rotating wheel which contains holes in its periphery. Centrifugal force directs glass through each hole to create fibers. A third process involves conversion of 2–6-inch (5–15-centimeter) chopped strand and textile-type yarns into separate and random monofilaments by a garnetting machine.

Properties of Staple Fibers. Fiber diameters range from AAA to G (Table 2), with the largest production volume in the range C–G. Thermal conductivity of glass-fiber products is influenced by fiber diameter, density or compactness of the fiber mass, and temperature conditions. Generally, thermal conductivity ranges between 0.20 and 0.80 (Btu)(in.)/(hr)(ft^2)(°F). In metric units, this is 0.029 and 0.115 watt/meter-Kelvin (W/m·K).

Temperature of applications ranges between near absolute zero to 593°C (1100°F), or to the softening point of the glass. Unbonded mat is used at extreme temperature conditions, whereas standard bonded insulation covers the range of −40–232°C (−40–450°F). Fiber-glass acoustical products are particularly good energy absorbers at the frequency levels of 500–2000 Hz. Fiber diameter, density, and method of mounting control the absorbing characteristics.

See also **Optical Fibers.**

—L. Dow Moore, PPG Industries, Pittsburgh, Pennsylvania.

FIBERS. Long, thin, threadlike, strong, and flexible, fibers, both natural and synthetic, are the fundamental structural components of yarn, thread, string, rope, paper, woven, and matted goods. Fibers usually exhibit considerable elasticity, the ability to return to their original dimension without permanent stretching. Most fibers tend to interlock or mechanically bond with other fibers, forming fiber matrices. Natural fibers are usually quite *un*uniform; for example, cotton staple ranges from $\frac{1}{2}$ to $2\frac{1}{2}$ inches (1.3 to 6.4 centimeters) in length, with a diameter of about 1/1,000 inch (0.025 millimeter). Some synthetic fibers are made in the form of very thin filaments and thus may be quite uniform, with fiber length controlled and tailored for specific applications. Many materials required for the production of synthetic fibers are derived from petroleum and, along with synthetic plastics and resins, the synthetic fiber industry contributed importantly to the great growth of the chemical and petrochemical industries. To some extent, the essential raw materials for synthetic fibers are threatened because of competition for the same raw materials that are consumed for power in terms of fuels—both petroleum and natural gas.

The use of fibers dates back to antiquity. Very early uses and still found among primitive peoples are *tying* applications, using easily obtainable fibers from plants. Fibers have been used for centuries for making rough cordage, huts, and rope suspension bridges. Broomcorn and broomroot fibers have long been used for what might be termed *brush* applications. Straw, bamboo, rattan, and palm leaves were among the early *plaiting* and *rough-weaving fibers* for use in furniture making and basketry and, of course, are still extensively used in various parts of the world for these purposes. The use of various reeds, husks, and grasses as *filling fibers* is very old, but large tonnages of such fibers still are used for packing materials, upholstery padding, and like applications.

The use of fibers in nonwoven sheetlike products is quite old, although there has been a resurgence of interest in so-called nonwovens during recent years. Felts, paper, and, more recently, some of the nonwoven disposable garments and products, notably for hospital use, are representative of the use of fibers for what might be termed *matting* and webbing applications. But by far the most important use of fibers, and the application most often visualized in connection with their use, is for the manufacture of knitted and woven textile fabrics and products. Spinning of the fibers to make yarn increases the utility of fibers for uses which far exceeded the imagination of those persons who first applied fibers in their cruder forms.

Although fibers can be classified in numerous ways, in terms of present-day technology, they are fundamentally classified as (1) natural fibers, and (2) synthetic fibers. The principal natural fibers are cotton, wool, and to a much lesser extent, silk, flax, and mohair. Synthetic fibers have made inroads into the use of all natural fibers, but the greatest impact has occurred in connection with the latter three fibers. Cotton continues to be a major textile fiber, measured in terms of billions of pounds used per year. Cotton is one of the most versatile of all fibers and blends well with synthetics. This is also true of wool, but to a somewhat lesser extent.

Synthetic Fibers. Introduced in 1910 as a substitute for silk, rayon was the first artificial or synthetic fiber. Rayon, of course, differs completely in chemical constitution from silk. Rayon typifies most reconstituted or synthetic fibers which perform almost as well and, in a number of respects, far better than their natural "counterparts." Some of the more recently developed synthetic fibers have little if any resemblance to naturally available fibers and thus entirely new types of end-products with previously unobtainable end-qualities are available.

It is interesting to note that some authorities define a synthetic fiber as a "noncellulosic fiber of synthetic origin," a definition which excludes rayon and acetate. Other authorities, however, include rayon and acetate, along with nylons, polyesters, acrylics, and others, in the full spectrum of synthetic fibers. The reasoning behind the fine distinction is that, with cellulose-derived synthetics, one commences with a naturally fibrous material and grossly modifies it, whereas with most other synthetics, the starting materials are strictly chemicals that bear no relationship whatever to a fibrous structure, many of the starting ingredients actually being in the gaseous or liquid phase.

In classifying synthetic fibers, there is also a narrow, twilight zone between fibers and elastomers. There are elastomers with fiberlike qualities; and vice versa. For example, spandex is a fiber with rubberlike qualities. See also **Elastomers.**

Specific Synthetic Fibers

In the following several paragraphs there are profiles of the principal synthetic fibers in high-scale production as of the mid-1980s. Unless otherwise noted, moisture regain is stated for a relative humidity of 65% at 70°F (21.1°C). The following symbols are used: RT = regular tenacity; HT = high tenacity; IT = intermediate tenacity; * = variations of this property are found with particular proprietary brand of fiber.

ACETATE FIBERS

A manufactured fiber in which the fiber-forming substance is cellulose acetate. Where not less than 92% of the hydroxyl groups are acetylated, the term triacetate may be used as a generic description of the fiber.

(acetate)   (triacetate)

Specific gravity: 1.3–1.32
Moisture regain: 3.2% (triacetate); 6.3–6.5% (acetate)
Tensile strength: 18–22 × 10³ psi (124–152 MPa) (triacetate)
 20–24 × 10³ psi (138–166 MPa) (acetate)
Excellent to impervious to aging. Good resistance to mildew discoloration and sunlight (acetate), although there may be some loss of strength from long exposure to sunlight. Triacetate has poor resistance to sunlight. Fair resistance to abrasion.
Attacked by strong oxidizing agents. Resists common solvents, normal hypochlorite, peroxide bleaching. Dissolves or swells in acetone, ketones, trichloroethylene, concentrated and glacial acetic acid, and methylene chloride.
Triacetate does not stick when ironed at cotton setting temperature, 450°F (232°C). Melts at 572°F (301°C). Acetate sticks at 350–375°F (177–191°C). Softens at 400–445°F (204–230°C). Melts at 500°F (260°C).
Dyes: Dispersed and developed dyes are commonly used. Acid dyes are used for printing. Solution dyed available.
Types available:
 Triacetate (filament)
 Acetate (filament and staple)
 See also separate article on Fibers (Acetate).

ACRYLIC FIBERS

A manufactured fiber in which the fiber-forming substance is any longchain synthetic polymer composed of at least 85% by weight of acrylonitrile units:

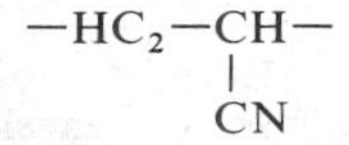

$$-HC_2-CH-$$
$$\quad\quad\quad | $$
$$\quad\quad\quad CN$$

A portion of the molecule may appear as:

$$-C-C-C-C-$$ with H CN H CN / H H H H

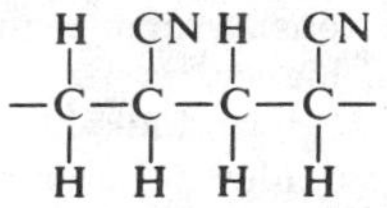

Specific gravity: 1.16–1.18
Moisture regain: 1.0–2.5%
Tensile strength: 30–54 × 10³ psi (207–373 MPa)
Excellent resistance to mildew and aging. Good resistance to sunlight and abrasion.
Good resistance to bleaches and common solvents.
Generally good resistance to mineral acids and weak alkalis.*
Safe ironing temperature up to 300°F (150°C).*
Does not support combustion.
Dyes: Disperse and cationic.*
Principal brands:
 Acrilan®, Monsanto (staple)
 Creslan®, American Cyanamid (staple and tow)
 Orlon®, DuPont (staple and tow)
 Zefran®, Badische (staple)

AREMID FIBERS

A manufactured fiber in which the fiber-forming substance is a longchain synthetic polyamide in which at least 85% of the amide linkages are attached directly to two aromatic rings.
Amide linkage:

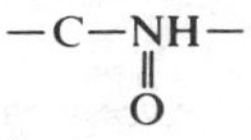

$$-C-NH-$$
$$\ \|$$
$$\ O$$

Specific gravity: 1.38–1.44
Moisture regain: 4.5–7% (at 55% RH)
Tensile strength: 90–400 × 10³ psi (621–2760 MPa)

Excellent resistance to mildew and aging. Prolonged exposure to sunlight causes deterioration, but fibers are self-screening. Good abrasion resistance.*
Some are degraded by bleaching; others are not affected. No degradation in solvents, except slight loss of strength from exposure to sodium chlorite.*
Unaffected by most acids, except some strength loss after long exposure to hydrochloric, hydrobromic, nitric, and sulfuric acid. Generally good resistance to alkalis.*
Difficult to ignite—does not propagate flame—does not melt. Decomposition temperature is from 700 to 930°F (371 to 499°C).*
Dyes: Industrial yarn is nondyeable. Staple is dyeable with cationic dyes.
Principal brands:
 Kevlar®, DuPont (filament)
 Nomex®, DuPont (staple, tow and filament)

FLUOROCARBON FIBERS

Fiber formed of longchain carbon molecules whose available bonds are saturated with fluorine. A portion of the molecule may appear as:

$$-C-C-C-C-C-$$ with F F F F F above and F F F F below

Specific gravity: 0.8–2.2
Moisture regain: 0
Tensile strength: 25–115 × 10³ psi (173–794 MPa)
Good to excellent resistance to mildew, aging, sunlight, and abrasion.
Essentially inert to bleaches and solvents except for alkali metals at high temperature and/or pressure. Fluorine gas and chlorine trifluoride react with fibers at high pressures and temperatures.*
Essentially inert to acids and alkalis.
Very heat resistant. Usually can be safely handled from −350 to +550°F (−212 to +288°C).* Melts between 550 and 620°F (288 and 327°C).*
Dyes: Some cannot be dyed; others can be pigmented and dyed with selected solvent system.
Principal brands:
 Gore-Tex®, W. L. Gore (expanded PTFE staple, filament, tow, and slit film-RT)
 Teflon®, DuPont (TFE multifilament, staple, tow and flock; FEP monofilament)

GLASS FIBERS

A manufactured fiber in which the fiber-forming substance is glass.
Specific gravity: 2.48–2.69
Moisture regain: None
Tensile strength: 313–700 × 10³ psi (2160–4830 MPa)
Not attacked by mildew, although binder may be affected by it. Excellent resistance to aging and sunlight.
Unaffected by bleaches and solvents.
Resists most acids and alkalis.
Nonburning. Generally holds 75% tensility up to 650°F (343°C). Softens between 1560 and 1778°F (843 and 970°C). Melts at 2720°F (1493°C).
Dyes: Resin-bonded pigment systems. Vat, acid, or chrome dyes will tint.
Available from numerous manufacturers.
 See also Optical Fibers.

MODACRYLIC FIBERS

A manufactured fiber in which, when not qualified as rubber or anidex, the fiber-forming substance is any longchain synthetic polymer composed of less than 85%, but at least 35% by weight of acrylonitrile units:

A portion of the molecule may appear as:

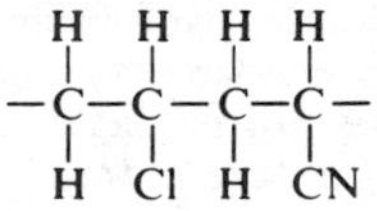

Specific gravity: 1.35–1.37

Moisture regain: 2.5–3.0%

Tensile strength: $29–47 \times 10^3$ psi (200–324 MPa)

Good to excellent resistance to mildew, aging, and sunlight. Good resistance to abrasion.

Good resistance to bleaches, dry-cleaning fluids, and most common solvents. Dissolves in warm acetone and acrylic-type solvents.

Resistant to most acids and good resistance to weak alkalis; some discoloration may result.*

Boiling-water shrinkage, about 1%. Good resistance to shrinkage in dry heat, with about 5% shrinkage at 390°F (200°C). Pressure and heat at 300°F+ (150°C+) may cause stiffening and discoloration.*

Does not support combustion.

Dyes: Neutral-premetalized, cationic (basic), and disperse.*

Principal brands:

Monsanto (staple)

Verel®, Eastman (staple-regular)

NYLON FIBERS

A manufactured fiber in which the fiber-forming substance is a longchain synthetic polyamide in which less than 85% of the amide linkages are attached to two aromatic rings. Amide linkage:

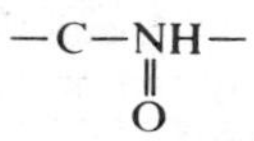

A portion of the Nylon 6,6 molecule, based upon hexamethylene diamine and adipic acid may appear as:

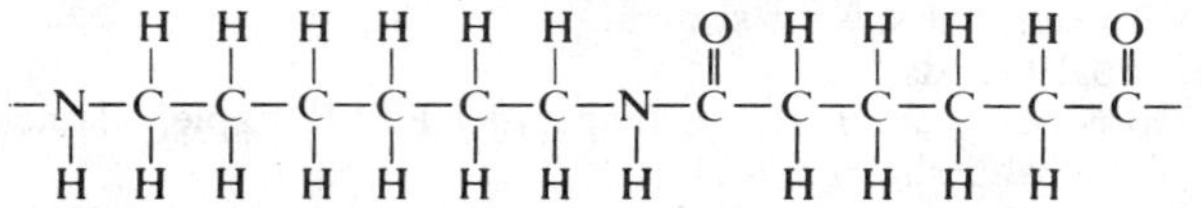

A portion of the Nylon 6 molecule, based upon caprolactam, may appear as:

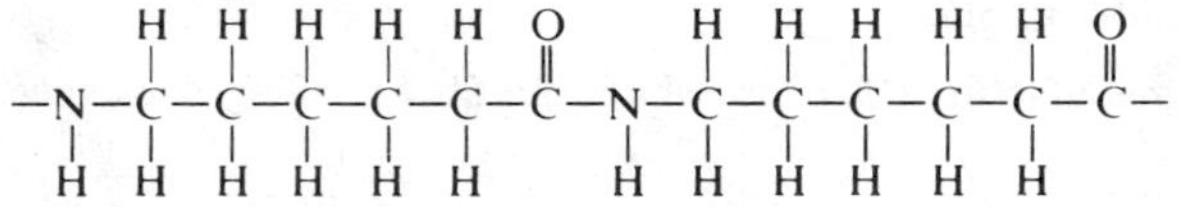

Specific gravity: 1.03–1.14 (most = 1.14)

Moisture regain: 2.8–5%*

Tensile strength: $40–134 \times 10^3$ psi (276–925 MPa)*

Excellent resistance to mildew, aging, and good-to-excellent resistance to abrasion. Prolonged exposure to sunlight causes some deterioration.

Excellent resistance to bleaches and other oxidizing agents. Generally insoluble in most organic solvents except some phenolic compounds.

Strong oxidizing agents and mineral acids may cause degradation of some brands. However, generally unaffected by most mineral acids, except when hot. Dissolves with partial decomposition in concentrated solutions of hydrochloric, sulfuric, and nitric acids.* Substantially inert in alkalis.

Sticking temperature is about 445°F (229°C).* Melts between 480 and 525°F (249 and 274°C).* Some yellow slightly if held at 300°F (150°C) for several hours. Decomposes between 600 and 730°F (316 and 388°C).*

Dyes: Has marked affinity for all types of dyestuffs, including pigment, direct, acid, premetalized acid, disperse, chrome, and vat colors, including complex types.*

Principal brands:

Nylon 6, DuPont and others (staple, monofilament and filament-RT and -HT; staple and tow)

Nylon 6,6, DuPont and others (staple and tow; monofilament and filament-RT; and filament-HT)

Qiana®, DuPont (filament-RT)

OLEFIN FIBERS

A manufactured fiber in which the fiber-forming substance is any longchain synthetic polymer composed of at least 85% by weight of ethylene, propylene, or other olefin units. A portion of the molecule may appear as:

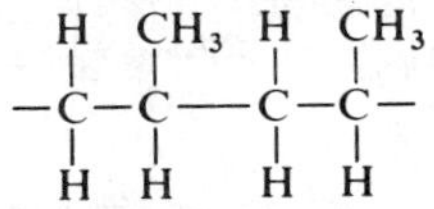

(polypropylene)

Specific gravity: 0.9–0.96

Moisture regain: Negligible (polyethylene); 0.01–0.1% (polypropylene)

Tensile strength: $11–90 \times 10^3$ psi (76–621 MPa)*

Not attacked by mildew. Good to excellent resistance to sunlight, abrasion, and aging.

Resistant to bleaches and most solvents, but some swelling in chlorinated hydrocarbons at room temperature and dissolves at 160°F (71°C) and higher.*

Excellent resistance to acids and alkalis, with exception of oxidizing agents, such as chlorosulfonic acid and concentrated nitric acid.

Softens at 225–235°F (107–113°C) (polyethylene); at 285–330°F (141–166°C) (polypropylene).

Melts at 230–250°F (110–121°C) (polyethylene); at 320–350°F (160–177°C) (polypropylene).

Dyes: Traditionally fibers are pigmented during manufacture, but some can be dyed with disperse, acid, and chelating dyes and certain vats, sulfurs, and azoics.

Types available:

Polyethylene (monofilament—conventional low density; and high density)

Polypropylene (staple and tow-isotactic; mono- and multifilament-isotactic)

POLYESTER FIBERS

A manufactured fiber in which the fiber-forming substance is any longchain synthetic polymer composed of at least 85% by weight of an ester of a substituted aromatic carboxylic acid, including but not restricted to substituted terephthalate units:

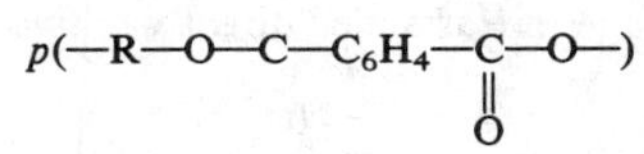

and parasubstituted hydroxybenzoate units:

A portion of the molecule may appear as

$$-O-\underset{\underset{H}{|}}{\overset{\overset{H}{|}}{C}}-\underset{\underset{H}{|}}{\overset{\overset{H}{|}}{C}}-O-\overset{\overset{O}{\|}}{C}-\bigcirc-\overset{O}{\underset{\|}{C}}$$

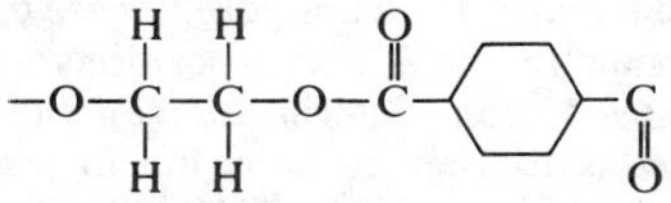

Specific gravity: 1.34–1.39 (most = 1.38)
Moisture regain: 0.4%
Tensile strength: 33–165 × 10³ psi (228–1139 MPa)*
Good to excellent resistance to mildew and sunlight, although prolonged exposure to full sunlight degrades some brands (strength loss). Abrasion resistance ranges from good to excellent.
Good to excellent resistance to bleaches, soaps, synthetic detergents, drycleaning agents, sea water, and perspiration. May be soluble in some phenolic compounds.
Sticking temperature is 440–445°F (227–230°C). Melts between 480–500°F (249–260°C).*
Good resistance to most mineral acids. Dissolves with partial decomposition in concentrated sulfuric acid. Good resistance to weak alkalis. Moderate resistance to strong alkalis.*
Dyes: Disperse, azoic, and cationic dyes.*
Principal brands:
  Avlin®, Avtex (filament-RT)
  Dacron®, DuPont (staple and tow; partially oriented filament; filament-RT; filament-HT)
  Encron®, American Enka (staple-RT; Filament-RT; producer-textured filament)
  Fortrel®, Fiber Industries Div. Celanese (staple, RT AND HT; filament, RT and HT)
  Kodel®, Eastman (staple, RT, IT, and HT; filament-RT)
  Monsanto®, (staple-RT; partially oriented filament; producer-textured filament)
  Trevira®, Hoechst (staple, partially oriented filament; filament-HT)
  A.C.E.®, Allied Chemical (filament-HT)
  S-3, S-3H®, IRC (filament-HT)

RAYON FIBERS

A manufactured fiber composed of regenerated cellulose, as well as manufactured fibers composed of regenerated cellulose in which substituents have replaced not more than 15% of the hydrogens of the hydroxyl groups. A portion of the molecule may appear as:

$$\begin{array}{c} \overset{O}{\|} \\ HC \\ | \\ HCOH \\ | \\ HOCH \qquad O \\ | \\ HCOH \\ | \\ HC \\ | \\ HCH \end{array}$$

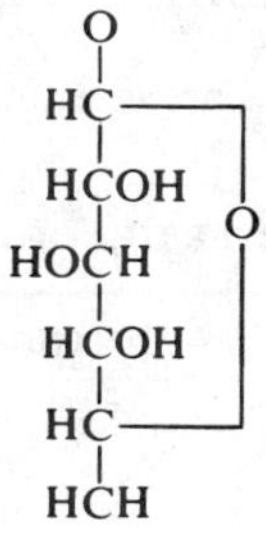

Specific gravity: 1.46–1.54
Moisture regain: 11–13%
Tensile strength: 28–66 × 10³ psi (193–455 MPa)*
Attacked by mildew. Resistant or stable to aging.* Mostly good resistance to sunlight and abrasion, but long exposure may yellow some intermediate rayons.*
Not affected by solvents. Insoluble in common organic solvents. Some attacked by strong oxidizing agents, but generally not damaged by hypochlorite or peroxide.*

Most rayons behave to acids much as cotton. Hot dilute or cold concentrated acids cause disintegration of fibers. Strong alkaline solutions cause swelling and reduce strength.*
Fibers do not melt, but may lose strength above 300°F (150°C) and decompose between 350 and 464°F (177 and 240°C).*
Dyes: Direct, vat, fiber-reactive, sulfur and pigment.*
Principal types and brands:
  Cuprammonium, available from several sources (filament).
  Viscose, available from several sources (filament and staple-RT and IT)
  Avril, Avril Prima, Avril III®, Avtex® (staple-HT)
  Zantrel 700®, American Enka (staple)

SPANDEX FIBERS

A manufactured fiber in which the fiber-forming substance is a longchain synthetic polymer comprised of at least 85% of a segmented polyurethane. A portion of the molecule may appear as

$$-O(CH_2)_4OCONH-\underset{}{\overset{\overset{CH_3}{|}}{\bigcirc}}-NHCOO(CH_2)_4O-$$

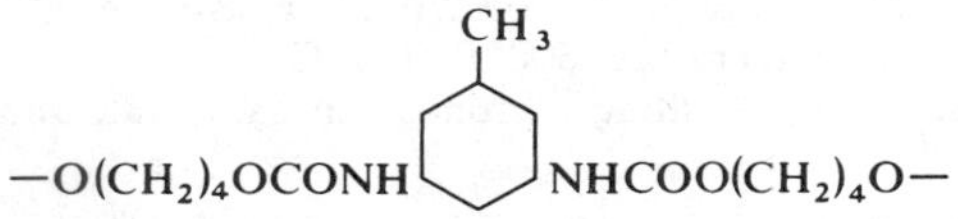

Specific gravity: 1.2
Moisture regain: <1.0–1.3
Tensile strength: 11–15 × 10³ psi (76–104 MPa)
Good to excellent resistance to mildew, aging, sunlight, and abrasion.
Good resistance to deterioration by bleaches, but some discolored slightly by hypochlorite bleaches. Resistant to solvents, including dry-cleaning fluids, and oils, except glycols.*
Good resistance to mild acids and alkalis, but may be degraded by strong acids and alkalis at high temperatures. Some are slightly yellowed by dilute hydrochloric and sulfuric acids.*
Sticking point from 347 to 420°F (75 to 216°C).* Melts at 511–518°F (267–269°C).*
Dyes: Good affinity for most classes of dyes, but disperse, acid, and premetalized dyes are generally preferred.*
Principal brands:
  Glospan/Cleerspan®, Globe (multifilament)
  Lycra®, DuPont (coalesced monofilament)
  Numa®, Ameliotex® (multifilament)

Comparative Profiles of Natural Fibers

COTTON

General formula, $(C_6H_{10}O_5)_x$. Chemical composition is cellulose.
Specific gravity: 1.54
Moisture regain: 7.0–8.5%
Tensile strength: 60–120 × 10³ psi (414–828 MPa)
Cotton fibers are quite resistant to thermal degradation. After about 5 hours at 250°F (121°C), material yellows. Decomposes above 300°F (150°C).
Cotton is attacked by cold concentrated acids and by hot dilute acids. Alkalis cause mercerization, but without danger. Cotton is quite resistant to most solvents.
Cotton fabrics have an excellent hand, good abrasion resistance, excellent pilling resistance, excellent stability to repeated launderings (if preshrunk), fair sunlight resistance, excellent colorfastness, good wash and wear performance (if resin-treated), and good wrinkle resistance (if resin-treated).
Safe ironing temperature: 425°F (219°C)

Dyes used: Direct, vat, azoic, basic, mordant, pigment, sulfur, and fiber-reactive.

WOOL

General formula, $(C_{42}H_{157}O_{15}N_5S)_x$. Chemical composition is keratin.

Specific gravity: 1.32

Moisture regain: 11–17%

Tensile strength: 17–29×10^3 psi (117–200 MPa)

Wool shows marked effects of thermal degradation above 212°F (100°C). Scorches at 400°F (204°C); chars at 570°F (299°C).

Wool is destroyed by hot sulfuric acid; otherwise it is quite resistant to acids. Strong alkalis destroy the material; it is attacked by weak alkalis. Wool is quite resistant to most solvents.

Wool fabrics have an excellent hand, fair abrasion resistance (but good in carpets), good pilling resistance (pills form, but tend to break off), poor stability to repeated launderings, good sunlight resistance, good colorfastness, poor wash and wear performance, and good wrinkle resistance.

Safe ironing temperature: 300°F (149°C)

Dyes used: Acid, milling, chrome, mordant, vat, and indigo.

SILK

Silk is comprised essentially of the protein (fibroin).

Silk fabrics have an excellent hand, fair abrasion resistance, good pilling resistance, good stability to repeated launderings, poor sunlight resistance, good colorfastness, poor wash and wear performance, and good wrinkle resistance.

Safe iron temperature: 300°F (149°C)

FLAX

A bast fiber.

Flax fabrics (linen) have an excellent hand, fair abrasion resistance, fair pilling resistance, good stability to repeated launderings, fair sunlight resistance, excellent color fastness, very poor wash and wear performance, and poor wrinkle resistance.

Safe ironing temperature: 450°F (232°C).

References

Staff: "Properties of the Manmade Fibers," *Textile Industries*, Atlanta, Georgia (Published annually).

Staff: "Manmade Fiber Chart," *Textile World*, McGraw-Hill, Atlanta, Georgia (Published annually).

Staff: "ASTM Standards on Textile Materials," American Society for Testing and Materials, Philadelphia, Pennsylvania (Revised periodically).

FIBERS (Acetate). Cellulose acetate fiber, or *acetate*, is a chemical derivative of the naturally occurring polymer *cellulose*. Two types of acetate fibers are produced:

1. Fibers made from partially hydrolyzed cellulose triacetate, called *secondary acetate* (or simply acetate).
2. Fibers that are fully acetylated cellulose, called *triacetate*.

Chemistry. Highly purified cellulose wood pulp (greater than 95% alpha cellulose) is the basic raw material for making cellulose acetate. The natural polymer, cellulose, in wood pulp has a degree of polymerization of 500 to 1000, the basic repeating unit of which is cellubiose:

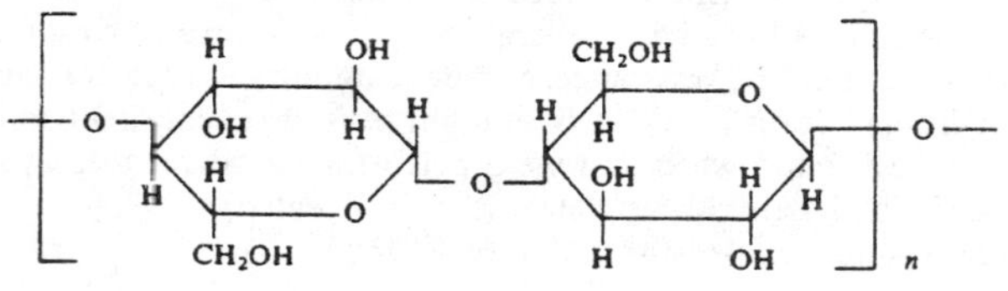

Cellubiose contains two anhydroglucose units connected by a B-1,4-glucoside linkage. Each anhydroglucose unit contains two secondary alcohols in the 2 and 3 positions and one primary alcohol in the 6 position. Esterification of the three hydroxyl groups produces cellulose triacetate, but in practice, the commercial triacetate has a degree of substitution (DS) of 2.80–2.95, and secondary acetate has a DS of 2.35–2.40. The DS number is the average number of hydroxyl groups esterified per anhydroglucose unit. Since cellulose and acetic acid do not react directly to an appreciable extent, the ester is prepared by reacting cellulose with acetic anhydride in a solvent of glacial acetic acid, using sulfuric acid as a catalyst. The chemical reaction for converting cellulose to the triacetate thus is:

$$C_6H_7O_2(HO)_3 + (CH_3CO_2)_2O \rightarrow$$
$$C_6H_7O_2(OCOCH_3)_3 + 3\ CH_3COOH$$

where $C_6H_7O_2(OH)_3$ is one anhydroglucose unit in the cellulose molecule.

The solution of triacetate is converted to secondary acetate through the addition of aqueous acetic acid, which results in hydrolysis of about 20% of the ester groups. Addition of water precipitates the secondary acetate and stops the deesterification. This procedure results in a random distribution of the acetyl and hydroxyl groups on the polymer. This, in turn, enhances the solubility of the secondary acetate in acetone. Acetone is commonly used as the solvent for spinning operations. The triacetate is produced by precipitating the product without hydrolysis.

Properties. Commercial acetate and triacetate yarns are produced in a range of 45–900 denier. Denier is a measure of the fineness of a yarn and is the weight in grams of 9000 meters of yarn. The largest portion is in the range of 55–150 denier. The filament count for these yarns ranges from 14 to 100 and the denier per filament ranges from 2.5 to 9. The yarns are produced in both bright and dull lusters, the latter type containing 1–2% titanium dioxide (TiO_2). Color-pigmented or "solution dyed" acetate yarns are also produced. The cross-sectional shape of acetate fibers is typically crenellated and circularly symmetric. The crenellated cross section results from the formation of a rigid skin during the initial filament formation when the fiber is largely fluid. As the acetone volatilizes through the skin, the fiber solidifies and shrinks. This results in skin folds or crenellations. Shaped fiber cross sections, such as Y shapes, are also produced and provide the capability for increasing the bulk of yarns and fabrics.

Key physical properties are given in the accompanying table.

PHYSICAL PROPERTIES OF ACETATE YARN

PROPERTY	SECONDARY ACETATE	TRIACETATE
Tenacity, g/denier:		
Conditioned	1.2–1.4	1.1–1.3
Wet	0.8–1.0	0.8–1.0
Breaking elongation, %:		
Conditioned	25–45	26–35
Wet	35–50	30–40
Density, g/cm³	1.32	1.30
Percent moisture regain	6.3–6.5	3.2
(65% relative humidity at 22°C)		

Chemical Properties and Dyeing. Acetate and triacetate fibers are readily dyed with disperse and azoic dyestuffs. Acid dyes may be used for print dyeing the triacetate fibers. Acetate and triacetate fibers, being organic esters, are susceptible to hydrolysis in strong alkaline solutions. Weakly basic or acid solutions have little effect on these fibers. Severe degradation occurs through hydrolysis when these fibers are subjected to strong mineral acids. Both the secondary acetate and triacetate show strong resistance to hypochlorite and peroxide bleaches. The fibers are not affected by normal dry-cleaning solvents.

Thermal Properties. Secondary acetate, having a random substitution of the hydroxyl groups, shows very little crystallinity and cannot be induced to crystallize or heat set. On the other hand, triacetate is a stereospecific molecule capable of crystallizing when heated above 205°C, the glass transition temperature, for a short period. Hence, triacetate offers heat-setting advantages in fabric finishing processes. Triacetate fabrics, heated at 240°C for 30 seconds, in a state of tension, exhibit good retention of the flat geometry and resist wrinkling at lower temperatures. Pleats and creases applied to triacetate garments, which are then heated above the glass transition temperature, will be retained over the normal period of usage. The melting points for secondary acetate and triacetate are 260 and 300°C respectively. Secondary acetate softens in the range of 205–230°C, which necessitates a moderate ironing temperature. Triacetate does not soften to the sticking temperature of 232°C and hence can accept ironing temperatures usually used for cotton.

Manufacture of Acetate Fibers. Acetic anhydride is produced by some of the acetate fiber producers because cellulose acetate is the major end user for this chemical. The anhydride is the product of reaction with ketene and acetic acid. Ketene is made by the catalytic pyrolysis of either acetic acid or acetone. The manufacturer of acetate fibers proceeds through a number of steps as delineated below.

(1) Wood pulp (high alpha type) in roll or sheet form is shredded to permit rapid penetration of the pulp by the reactants.

(2) The shredded pulp is treated with glacial acetic acid which may contain part or all of the sulfuric acid catalyst. This "pretreatment" swells the cellulose fibers and increases the accessibility of the acetylating agent. Ratios of acetic acid to cellulose may be 1:1 to 3:1 and the treatment time may vary from one-half to several hours, depending upon temperature.

(3) The activated pulp is then treated with the acetylation solution, a mixture of acetic anhydride and acetic acid in about 2:3 ratio and containing the remainder of the sulfuric acid catalyst (5–20% based on cellulose). Since the reaction is exothermic and it is necessary to keep the reaction temperature low (<50°C), precooling of the acetylation mixture and/or external cooling of the reaction vessel are employed. Reaction times of 5 to 10 hours, under controlled temperature, result in the pulp mass dissolving in the acetylating mixture. At this point, the cellulose is completely acetylated in the triacetate form. If triacetate is the desired product, aqueous acetic acid is applied to destroy the anhydride and the product is precipitated, washed, and dried.

(4) In producing the secondary acetate, the excess anhydride is reacted with aqueous acetic acid under careful temperature control. When all the anhydride is reacted, excess water is added to the extent of 10–30% of the mixture. The acetate is allowed to hydrolyze until a DS of 2.35–2.40 is obtained. Modern practice permits hydrolysis to occur in 4 to 8 hours at temperatures of 70–80°C. Sulfur esters are also hydrolyzed and sodium or magnesium acetate may be introduced with the hydrolysis water to neutralize the acid produced. Additional water precipitates

the secondary acetate and control is exercised to obtain an open "flake" rather than a pelletized material. The open flake makes more efficient the removal of acetic acid in subsequent washing. The flake is dried and conveyed to storage bins.

Cellulose acetate is converted into fibers by solvating the polymer in an acetone solution which is then spun by a process called *dry spinning*. In the spinning process, the polymer solution is metered through a spinneret into a column containing warm air which evaporates the acetone. The solidified filaments are oiled and package collected at the base of the spinning tube.

(5) Preparation of the polymer spinning solution or "dope" begins with blending of flake batches to ensure uniformity of chemical composition, particularly acetyl content, which is important for uniform dyeing properties of the spun yarn. Either batch or continuous mixers are used for dissolving the blended flake in acetone-water solution. The final composition of the "dope" is about 26% acetate, 72% acetone, and 2% water. The water serves as a cosolvent with the acetone, leading to a marked reduction in the viscosity of the solution, which is approximately 900–1000 P (poises). A delustrant, such as titanium dioxide (TiO_2) may be added to the dope in concentrations of 1–2%, based on the acetate content. Colored pigments may also be added to the dope to produce mass-dyed fibers.

(6) The viscous "dope" solution is filtered several times to eliminate insoluble fiber and debris. Plate-and-frame as well as continuous filtration may be used. After filtration, the dope is transported to storage tanks where it is permitted to deaerate just prior to spinning.

(7) The deaerated dope is pumped to the spinning machine headers which distribute the dope to the individual spinning positions. Each machine has a hundred or more spinning positions where a single yarn of acetate yarn is produced. A metering pump provides accurate control of the quantity of dope pumped through a small line filter and heat exchanger which raises the temperature to 50–70°C just prior to reaching the spinneret. Spinnerets are made of stainless steel and incorporate 14–100 or more holes. The holes are 0.0015–0.0025 inch (0.04–0.06 millimeter) in diameter, having a depth slightly less than their diameter.

(8) The nascent acetate filaments leaving the jet are carried through the vertical spinning tube which may be 10–20 feet (3–6 meters) long and from 6 to 12 inches (15 to 30 centimeters) in diameter. Hot air moving concurrent or countercurrent to the path of the filaments evaporates most of the acetone and solidifies the liquid stream into the acetate filament.

(9) The acetone-laden air removed from the spinning tubes is stripped of acetone as it passes through activated carbon absorbers. The absorbers are steam-stripped and the acetone recovered by distillation. An acetone recovery efficiency of 96% overall is commonly experienced in the industry.

(10) As the yarn leaves the spinning tube, it is passed over an oiling device which applies 2–3% of lubricant of the weight of the yarn. The lubricant serves to reduce friction in subsequent textile processing and to diminish static electrification. From the oiler, the yarn passes over a rotating *godet* or wheel which supplies the force to transport the yarn from the jet to the take-up mechanism which packages the product.

Spinning speeds for acetate yarn range from 492 to 2297 feet (150 to 700 meters) per minute. Since acetone must be removed from the yarn in the spinning tube, the spinning speed depends upon the denier of the yarn, air velocity and temperature, dope temperature, and composition and spinning tube length.

Spinning of cellulose triacetate yarns follows the procedures used for the secondary acetate except that a different solvent is used. This is generally a 90:10 mixture of methylene chloride and methanol.

The manufacture of acetate staple and tow follow the same scheme as employed for continuous filament yarns, except that the yarns are combined as they leave the spinning tube to form a tow which is mechanically crimped and cut into staple lengths. Acetate and triacetate are packaged in bales of about 400 pounds (181 kilograms).

—Joseph W. Schappel, Avtex Fibers, Inc., Front Royal, Virginia.

FILM. In general usage, the word *film* means any thin sheet of material used for covering, coating or wrapping, or any thin layer that enters into a structure, usually on or near the surface. Film also denotes the *monomolecular layer* which is formed on the surface of a solution or at an interface between two immiscible liquids. The adsorption is of such a nature that the free surface energy is minimum. Insoluble and nonvolatile substances placed on the surface of a liquid such as water may also under certain conditions spread out on the water surface to give a monomolecular film. Adsorbed films of gases or liquids are also formed on solids, such as mica, sodium chloride, glass, or metals. In some cases, such as the adsorption of vapors on solids at relatively high pressures, the films may be thicker, attaining a thickness of three or four molecules.

Condensed Film. A surface film in which the molecules are closely packed and steeply oriented to the surface. The molecular packing approaches that observed in the crystalline state.

Expanded Film. A state of film intermediate in area and other properties between gaseous and condensed films.

Gaseous Film. A film in which molecules move about independently on the surface and their lateral adhesion for each other is very small. At low surface pressures (π) and large area (A), a gaseous film obeys the relation $\pi A = kT$. At higher pressures an equation of the form ($\pi A - A_0$) $= xkT$ holds, where x is a constant.

Liquid-Expanded Film. This film occupies a much larger area than a condensed film, but is still a coherent film. It can form a separate phase from a gaseous film with which it is in equilibrium, and obeys the relation

$$(\pi - \pi_0)(A - A_0) = C$$

where π is the surface pressure, A the surface area, and A_0 the coarea of the molecule.

FILM (Bubble). Foam.

FILTRATION. A very common requirement of several industries, such as chemical and biologicals manufacturing, food processing, ore processing, and water and waste treatment, is the separation of solids that are suspended in liquids. Filtration is a principal means for effecting such separation. Other means include centrifuging, clarifying, and sedimentation, described elsewhere in this volume.

In filtration, the suspension containing the solids is caused to pass through a porous medium. Numerous filtering media are used, including paper, cloth, and wire cloth. Filtration may be conducted under positive pressure or vacuum.

Electrokinetics and Filtration. Process engineers in many in-

dustries have long regarded filtration as a very efficient and relatively low-cost means of product recovery, clarification, stabilization, and sterilization. For many decades, a primary component of many filters has been asbestos. See also **Asbestos.** Occupational hazards associated with asbestos were reported as early as 1964. The negative findings on asbestos stimulated the search for substitute filtering media. However, the search has been proceeding slowly; this has led process engineers to restudy the fundamentals of filtration mechanisms. With a better understanding of these fundamentals, the search for substitute media may be accelerated. Until recently, most research leading to development of models of the filtration process concentrated essentially on physical mechanisms and provided little, if any, attention to the electrochemical phenomena involved. Several researchers during the past decade have demonstrated a resemblance of filtering to coagulation and have noted that electrochemical phenomena can be controlling factors.

Because asbestos has been such a successful medium, many researchers decided to restudy the mechanical and electrochemical characteristics of this medium. Yada's research in 1967 demonstrated that chrysotile fibers (asbestos) have (1) a hollow cylindrical form with an average outer diameter of 50–80 $\times$ 10^{-10} meter (1 angstrom $= 10^{-10}$ meter); (2) that all such tubes are not simple cylinders, but that some may be spirally wound layers; and (3) that the distance between spiral layers may range between 4 and 7 $\times$ 10^{-10} meter. In short, chrysotile fibers possess an exceptionally large surface area per unit weight. See also **Chrysotile.** Also, in 1968, Riddick showed that asbestos has an isoelectric or zero point charge at a pH of 8.3 and thus has a positive charge when in neutral, aqueous solutions. It is to be noted that, with few exceptions, the majority of natural substances are negatively charged under these conditions. This observation led to the conclusion that a suitable substitute for asbestos fibers should have an isoelectric point about pH 7 when in neutral, aqueous solutions.

Wnek (1974) described the operation of a filter bed in this way: "Various physical transport mechanisms convey the particles to the surface of the medium. If the medium and particles are of opposite charge, electrokinetic attraction will deposit particles on the bed. If the medium and particle charges are of the same sign, repulsion will occur and deposition will be hindered, if not prevented. As the particles accumulate, the charge on the medium decreases, diminishing its ability to remove particles. Loss of efficiency starts at the top of the medium bed,

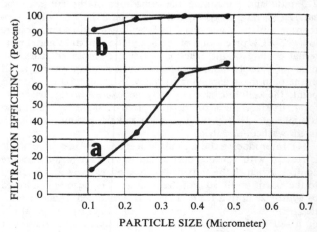

Fig. 1. Effect on filtration efficiency of modification of surface charge of diatomite: (a) untreated diatomite; (b) diatomite treated with melamine-formaldehyde colloid. (*After Fiore/Babineau.*)

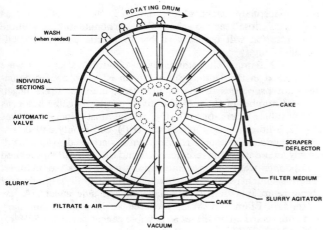

Fig. 2. Sectional view of rotary drum filter. This design dates back to 1908 (E. L. Oliver) and continues to be one of the most versatile and widely used continuous filters in the process industries. A horizontal drum is partially submerged in a vat that contains the slurry to be filtered. A vacuum is applied through a central valve on the drum shaft to individual compartments that provide support and drainage for the filter medium. The filter cake is formed while the sections are immersed. When the sections emerge (because of the continuous rotation of the drum), additional dewatering takes place as air passes through the cake, thus displacing a significant portion of the mother liquor. Before final dewatering, wash water may be applied to remove any remaining soluble solids. Discharge of the dewatered cake is effected by cutting off the vacuum (by individual sections) and applying a reverse air blow. As the cake separates from the filter cloth, a scraper blade deflects it whereupon it is dropped to a conveyor or discharge trough below.

and eventually electrokinetic removal of particles ceases, although deposited particles may have sufficiently reduced the pore size of the medium to allow further particle retention by straining. As the top of the medium bed becomes saturated, the lower parts remove more and more particles, until they too become saturated. At this point, if the medium bed has not become mechanically plugged, electrokinetic breakthrough will occur, i.e., charged particles will pass through."

Recent research has been directed toward chemically treating otherwise suitable filter media (sand, perlite, diatomite, etc.) so as to impart a positive charge (as found on chrysotile fibrils) instead of the negative charge of untreated materials. One example of this modification research is shown in Fig. 1.

Types of Industrial Filters. Filters may be continuous or intermittent and operate under positive pressure or vacuum. Numerous combinations and configurations are available. One of the most common styles is the rotary drum filter, cross section

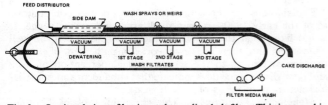

Fig. 3. Sectional view of horizontal traveling belt filter. This is a combination of a filter medium and a drainage belt which travels over a series of fixed vacuum boxes. Feed is supplied by a distributor. The slurry is contained by side dams along the filter belt until dewatering and washing are completed. The cake discharge is by gravity. The filter medium, on the return side, is washed by sprays applied to both sides.

of which is given in Fig. 2. A horizontal traveling belt filter is shown in Fig. 3. Ultrafilters for separating macromolecules are described in the separate article on **Ultrafiltration.**

References

Fiore, J. V., and R. A. Babineau: "Filtration," *Food Technology*, **33**, 4, 67–72 (1979).
Kirjassoff, D., Pinto, S., and C. Hoffman: "Ultrafiltration of Waste Latex Solutions," *Chem. Eng. Progress*, **76**, 2, 58–61 (1980).
Miles, H. V.: "Filtration," in the "Chemical and Process Technology Encyclopedia," (D. M. Considine, editor), McGraw-Hill, New York, 1974.
Perry, R., and C. Chilton: "Chemical Engineers' Handbook," 5th edition, McGraw-Hill, New York, 1973.
Severeson, S. D., et al.: "The Economics of Fabric Filters and Precipitators," *Chem. Eng. Progress*, **76**, 1, 68–73 (1980).
Smith, C. V.: "Electrokinetic Phenomena in Particulate Removal from Water by Rapid Sand Filtration," Ph.D. thesis, Johns Hopkins University, Baltimore, Maryland, 1966.
Staff: "Filtration and Separation," *Chem. Eng. Progress*, **73**, 4, 57–91 (1977).
Wnek, W.: "Electrokinetic and Chemical Aspects of Water Filtration," *Filt. Separ.*, **11**, 237 (1974).
Yada, K.: "Study of Chrysotile Asbestos by a High Resolution Electron Microscope," *Acta Cryst.*, **23**, 704 (1967).
Yao, Y. M., Habibian, M. T., and C. R. O'Melia: "Water and Waste Water Filtration: Concepts and Applications," *Environ. Sci. Tech.*, **5**, 11, 1105 (1971).

FIRE-RETARDANT PAINTS. Paints and Finishes.

FIRMING AGENTS (Foods). Certain foodstuffs, such as apples, potatoes, and beans, tend to be rather fragile when subjected to processing operations prior to packaging (canning, freezing, etc.) and, if not treated in some way to retain their natural firmness to a relatively high degree, the mouthfeel of the final product will be disappointing (mushiness versus slight chewiness or crispness). Certain chemical substances can be added prior to or during processing to protect and retain natural firmness. These substances include: aluminum potassium sulfate, aluminum sodium sulfate, aluminum sulfate, calcium carbonate, calcium chloride, calcium citrate, calcium gluconate, calcium hydroxide, calcium lactobionate, calcium phosphate (monobasic), calcium sulfate, and magnesium chloride, among others.

Researchers have found that calcium lactate, in particular, can be an effective agent for preserving the firmness of apple slices during processing and prior to canning or freezing. Studies have shown that calcium salts participate in firming the tissues of various fruits and vegetables by forming calcium pectates. Calcium citrate is quite useful for firming peppers, potatoes, tomatoes, lima and snap beans. Manufacturers of pet foods have found that from 1 to 2.5% monoglyceride contributes to the firming of pet foods, as well as aiding in the prevention of fat separation.

FISSION (Nuclear). Nuclear Fission.

FITTIG REACTION. The formation of aromatic hydrocarbons from aryl or aryl and alkyl bromides by use of sodium, e.g., bromobenzene plus ethyl bromide plus sodium forms ethylbenzene plus sodium bromide: $C_6H_5Br + C_2H_5Br \rightarrow C_6H_5 \cdot C_2H_5 + 2NaBr$.

FIXED BEDS. In chemical processing terminology, a fixed-bed installation (usually a reactor) requires that materials in the solid phase that are to be reacted with gases and vapors remain in a fixed location. In other words, the flow in such equipment is confined to materials in the gaseous or vapor phase. The solid materials require careful preparation to permit a maximum of surface to be exposed to the gases which pass through them and to avoid the formation of channels through the gases would pass without contacting the bulk of the solids. Beds of this type are often used in connection with various catalytic operations and are used, for example, in the production of benzene, in catalytic reforming, hydrocracking, hydrotreating, vinyl chloride monomer production, and in ion-exchange operations. A later reaction development, but also one that is several decades old, is the *fluidized bed* reactor in which the solids to be reacted are essentially "fluidized" and intermix with other solids or gases in a rapidly moving turbulent stream—in contrast with the solids remaining in a fixed bed.

Whether or not a fixed bed is used instead of a fluidized bed is determined by numerous factors, notably time and cost. Fixed-bed reactors are simpler and less costly, but often require cyclic operation because of the need to replenish the solids (catalysts, etc.). Thus to achieve continuous production, two or more beds are required (one on stream; the other regenerating).

FIXED OILS. These are fats, compounds of glycerin and various complex fatty acids. Fixed oils are often called the nonvolatile oils, in distinction to the essential or volatile oils, which are readily vaporized by heat. It is characteristic of the fixed oils to leave a spot when dropped on paper. Many of them remain liquid at common room temperatures; others are solid at ambient temperatures. Solid forms are usually called *fats*, a purely arbitrary distinction, since slight changes in temperature will cause many of them to change from liquid to solid or vice versa.

Fixed oils, especially those of economic importance, are largely obtained from the seeds of plants. They have a high energy value and thus can be valuable food substances if they are palatable.

Fixed oils usually are classified into three groups: (1) drying, (2) semidrying, and (3) nondrying. Sometimes a fourth group is made up of those substances which are usually seen in solid form, although they differ but little otherwise from the other groups.

Drying oils are those which upon exposure to air form a tough elastic film. Linseed oil from flax seeds is one of the most important and is largely used in paints and varnishes. Tung oil, also important, is used in the production of waterproof varnishes and quick-drying enamels.

Nondrying oils are those which remain permanently greasy or sticky, becoming rancid after a time. Among these oils are olive oil, castor oil, rape seed oil, peanut oil, and almond oil.

Semidrying oils are intermediate in nature. In this category, the major oils are cottonseed oil, soybean oil, corn or maize oil, and sesame oil. These are frequently used as food oils and for cooking. See also **Paint and Finishes.**

FLAME-RETARDING AGENTS. A material used as a coating on or a component of a combustible product to raise its ignition point. The protection provided is usually only partial, and most materials so treated will burn when exposed to sufficiently high temperatures. The three principal types of agents are: (1) *nondurable*, consisting of water-soluble inorganic salts, which are easily removed by washing or accidental exposure to water; (2) *semi-durable* (removed by repeated laundering or dry-cleaning); and (3) *durable* (not affected by laundering and dry-cleaning). The latter type includes or has included in the past organic compounds of bromine and chlorine, and insoluble metal salts. Antimony trioxide, tricresyl phosphate and

other phosphate esters, chlorendic acid, etc. are effective, as well as cellulose-reactive agents. Zinc carbonate in high volume concentration will render a rubber or plastic compound self-extinguishing.

In 1972, flammability standards for children's sleepwear were established in the United States. In an effort to confer flame-resistant properties to the fabrics used, manufacturers began to use a number of chemical additives, notably organic halogens or phosphate esters, or both. One of the most widely used was *tris*-(2,3-dibromopropyl)phosphate (BP), commonly called tris-BP. Other closely associated compounds were used. At a considerably later date, some researchers found that tris-BP and related compounds were carcinogenic, among other negative qualities. There is much room for further research into finding effective flame retardants that do not have adverse side effects. A few years ago, Blum and Ames prepared an interesting summary of flame-retardant additives as possible cancer hazards (*Science*, **195**, 17–23, 1977).

FLASH DISTILLIATION. Distillation; Desalination.

FLAVIN ADENINE DINUCLEOTIDE (FAD). Coenzymes; Vitamins.

FLAVONOID. A group of aromatic, oxygen-containing heterocyclic pigments widely distributed among higher plants. They constitute most of the yellow, red, and blue colors in flowers and fruits. Exceptions are the carotenoids. See also **Carotenoids.** The flavonoids include: (1) catechins; (2) leucoanthocyanidins and flavonones; (3) flavanols, flavones, and anthocyanins; and (4) flavonols. See also **Anthocyanins;** and **Colorants (Foods).**

FLAVOR ENHANCERS AND POTENTIATORS. A *flavor enhancer* is a substance which when present in a food accentuates the taste of the food without contributing any flavor of its own. This is reminiscent of the role of a catalyst in a chemical reaction which promotes a reaction without chemically participating in the reaction. Although not usually regarded as a flavor enhancer, common salt, if not used excessively, enhances the taste of food substances. Salt does not fully meet the definition of an enhancer, however, because the salt is detectable as such.

Monosodium glutamate for many years has been the best known and most widely used of the flavor enhancers. MSG is normally effective in terms of a relatively few parts per thousand, but far less powerful than the newer flavor potentiators. Like enhancers, potentiators do not add any taste of their own to food substances, but intensify the taste response to the flavorings already present in the food. Because a potentiator is more powerful, smaller quantities of the substances are required than in the case of the enhancers. Generally, the available potentiators are from about 15 to nearly 100 times more effective than the enhancer.

Explanations for the actions of enhancers and potentiators, as of the early 1980s, remain qualitative and rather vague. It is not likely that the actions of these substances will be well understood until there are new theories or refinement of existing theories pertaining to the sensations and perception of taste and odor. Experience does indicate that enhancers and potentiators act more in terms of taste than odor.

Monosodium Glutamate. The chronology of MSG commenced centuries ago when certain seaweeds were used in the Far East to improve the flavor of soups and certain other foods. It was not until 1908, however, when the curiosity of K. Ikeda (university of Tokyo) caused him to study the seaweed *Laminaria japonica*, traditionally used by Japanese cooks to enhance food flavoring. After much research on the seaweed, MSG was isolated and identified as an excellent flavor enhancer, particularly for high-protein foods. As an aside, it is interesting to note that Ritthausen in Germany had isolated glutamic acid as early as 1866 and his associates had prepared the sodium salt of the acid, namely monosodium glutamate. But the path of research in Germany was targeted in other directions and the flavor enhancing qualities of MSG were left to Ikeda to determine.

The Japanese throughout the first half of the century produced glutamic acid by extraction from natural materials, a slow and costly method. Nevertheless, the demand for MSG grew rapidly and cost tended to be a secondary factor. It was not until 1956 that Japanese microbiologists succeeded in developing the first industrial production of L-glutamic acid by means of fermentation. See Fig. 1. The problem of producing glutamic acid, as

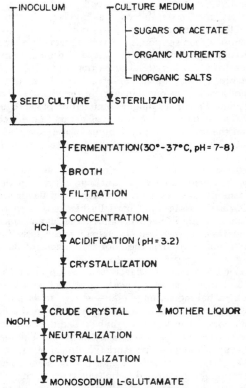

Fig. 1. Preparation of monosodium L-glutamate by fermentation. Sugar beets, corn (maize), and wheat gluten have been used in the process.

well as a number of other important amino acids, by fermentation was the lack of suitable strains of microorganisms for starting the cultures. Initially, the Japanese researchers were successful in isolating microbial strains from natural sources that possessed good abilities to excrete and accumulate a large amount of the amino acid in the cultural broth, but only under very carefully controlled conditions. For example, S. Kinoshia found that a high yield of glutamate could be attained only when the level of biotin (a vitamin required by glutamate-producing bacteria) was held within certain limits. An excess of biotin killed the microorganisms. Both antibiotics and detergents were used to control the biotin levels. Later, work was conducted with an artificial mutant by way of investigating genetic tech-

niques. Ultimately, large-scale MSG production was achieved by the fermentation route, sugar beets commonly used as a raw material. More detail on the fermentation process can be found in the Oeda (1974) reference listed. MSG also can be produced by chemical synthesis, as shown in Fig. 2.

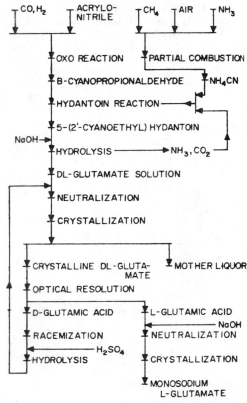

Fig. 2. Preparation of monosodium L-glutamate by chemical synthesis.

Although listed as a GRAS substance (generally regarded as safe) for many years, questions concerning its safe usage have arisen from time to time and, as of the mid-1980s, MSG still remains somewhat controversial. It is known that overconsumption of MSG can produce an illness, usually of just a few hours duration, in some persons. This is commonly referred to as the "Chinese Restaurant Syndrome." Apparently even when usage is somewhat excessive, only a relatively few people exhibit the symptoms of the syndrome. The possible seriousness of any deleterious effects of MSG tend to be countered by the many years the substance has been used by literally thousands of food processors and many millions of chefs and household food preparers. Some authorities also have a comfortable regard for MSG because of occurrence of the substance in many natural foods, notably mushrooms, tomatoes, and human milk. The topic is summarized by Krueger (1979).

The 5'-Nucleotides. Also dating back many years in the Far East was the knowledge that bonita tuna possesses a substance that very effectively enhances the flavor of foods. However, it was not until 1913 that S. Kodama (Tokyo University) commenced a serious investigation directed toward identifying and isolating the substance from tuna. Initially, Kodama believed that the substance was the histidine salt of 5'-inosinic acid, but later found that the substance was actually 5'-inosinic acid itself. This nucleotide was found to be many more times as effective as MSG. Further research by Kodama and others has shown that these nucleotides are present in many natural foods.

The nucleotides, in addition to their effectiveness at much lower concentrations, have been found to be superior to MSG for certain types of foods (in addition to high-protein foods). It also has been observed that the nucleotides tend to create a sense of increased viscosity, providing more body, for example, to soups. One manufacturer (Takeda) produces a series of the nucleotides by the enzymatic hydrolysis of ribonucleic acid. See formulas of Fig. 3. As of the early 1980s, these compounds

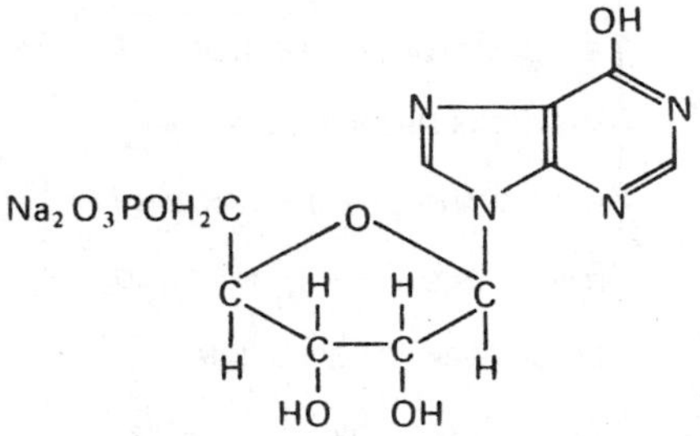

Disodium 5′-inosinate
$C_{10}H_{11}N_4Na_2O_8P \cdot H_2O$
Molecular Weight (anhydrous): 392.17

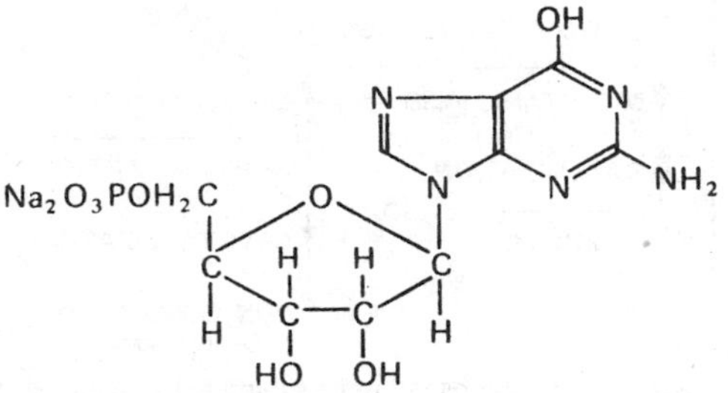

Disodium 5′-guanylate
$C_{10}H_{12}N_5Na_2O_8P \cdot xH_2O$
Molecular Weight (anhydrous): 407.19

Fig. 3. Salts of ribonucleotides. (*Takeda, Tokyo.*)

are enjoying a high volume of production and usage. It should be pointed out that the nucelotides are commonly used together with MSG. Some researchers point out that while the nucelotides and MSG have a lot in common, there is a considerable difference in their use. The nucelotides are up to 100 times more effective than MSG on a weight basis, and whereas MSG has been a favorite of processors for the enhancement of "meaty flavor," the range of the nucleotides is broader, modifying salty or sweet flavors and suppressing many undesirable flavors. The nucleotides are not a replacement for MSG. The substances do have a synergistic effect when used together. Generally, 1 kilogram of nucelotide used with 50 kilograms of MSG will have the same flavor intensifying result as 100 kilograms of MSG alone.

Other Potentiators. Because the market is so large, research continues at a good pace in seeking other potentiators. Established since the early 1940s, *maltol* is effectively used in foods that are high in carbohydrates, such as beverages, jams, and gelatins. Claims of reducing sugar content by 15% in products

using maltol have been made. Other potentiators used or proposed include dioctyl sodium sulfosuccinate, N,N′-di-o-toly-ethylenediamine, and cyclamic acid.

References

Furia, T. E., and N. Bellanca: "Fenaroli's Handbook of Flavor Ingredients," CRS Press, Cleveland, Ohio (1971).

Furia, T. E.: "Handbook of Food Additives," 2nd Edition, CRC Press, Cleveland, Ohio (1972).

Krueger, J.: "MSG: One of the Food Industry's Most Studied Ingredients," *Processed Prepared Foods*, **148**, 1, 128–140 (1979).

Oeda, H.: "Amino Acids," in "Chemical and Process Technology Encyclopedia," (D. M. Considine, Editor), McGraw-Hill, New York (1974).

Staff: "An Introduction to Nucleotide Seasonings," Ajinomoto, Tokyo (1980).

Staff: "Ribonucleotide Flavor Enhancers," Takeda, Tokyo (1980).

FLAVORINGS. The *flavor* of a food substance is the combined sensation of *taste* and *odor* as perceived by the eater/drinker of that substance. Although the components (*flavorings*) are present in a food substance, the full aspects of flavor require intimate contact between substance and consumer. The odors emanating from a bakery tend to be richer and more pleasant than the bread itself; the flavor of coffee seldom attains the richness of aroma that one perceives in the vicinity of a coffee roasting plant. Flavor is a unique combination of nerve impulses on the brain centers as the result of actions upon receptors located on the tongue and in the lining of the nose and is thus the result of interaction between the food substance and the consumer.

In terms of total flavor sensation, many authorities agree that odor is usually more important than taste. Experience, of course, demonstrates the marked reduction of flavor sensation when the nasal passages are partially blocked, as in the case of a common cold. In such instances, the layman may refer to the flat taste of the food. In actuality, the taste buds are functioning normally; it is the odor component of flavor that is missing.

The odor component of flavor is made up of at least two vectors. Sniffing of a substance without contact with the tongue provides a partial indication of odor—that is, molecular vapors or gases pass directly to the olfactory sensors in the nose via the nasal cavities. This vector might be called the *absolute external odor* or *fundamental odor* of a substance. This vector is dependent upon the vapor pressure (volatility) of the food substance itself. The other vector of odor is what some researchers call *internal odor* because the molecules reach the olfactory sensors by way of the pharynx, a flattened tubular passage that connects the back of the mouth with the nasal cavities. In the mouth, the food substance is wetted by saliva, altering not only the vapor pressure of the flavorings present, but sometimes exposing more and different flavorings, thus affecting flavor intensity and quality. It is well known, of course, that exceedingly dry substances tend to be odorless or nearly so. The odor of a polished metallic surface, for example, is difficult for most persons to detect. The addition of only modest amounts of moisture to most substances significantly increases their fundamental odor—by increasing vapor pressure and by activating the flavoring substances present. The effect of moisture on odor is dramatically illustrated by the dog at the fireside and the dog that has just come in out of the rain.

The Technology of Flavorings

To say that the overall topic of flavorings is intricate and complex is indeed an understatement. Far from exhaustive in its coverage, the "Food Chemicals Codex" describes well over 300 flavoring agents. Fenaroli's "Handbook of Flavor Ingredients" (T. E. Furia and N. Bellanca, Editors) describes in detail nearly 200 natural flavorings and nearly 750 synthetic flavorings—and it also does not fully cover the field.

Natural flavorings come from a number of *plant sources*— bushes, herbs, shrubs, trees, weeds, specific parts of which are used as flavoring sources. These include arils, balsams, barks, beans, berries, blossoms, branches, buds, bulbs, calyxes, capsules, catkins, cones, exudates, flowering tops, flowers, fronds, fruits, gums, hips, husks, juices, kernels, leaves, needles, nuts, oils, oleoresins, peels, pits, pulps, resins, rhizomes, rinds, roots, seeds, shoots, stalks, stigmas, stolons, thallus, twigs, wood, and wood sawdust—as well as some entire plants. Exploitable plants are found growing throughout the world.

Natural flavorings are prepared for commerce in various ways, as will be explained shortly, but it should be pointed out that adding to the complexities of raw natural flavorants are factors of timing and maturity. Some plant parts are only suitable when green (unripe); others must be fully ripe or nearly so. The timing of harvest can be critical. For example, jasmine flowers must be harvested before dawn. The roots of the orris plant must be aged 2 years before they are ready for the commercial market. Several of the natural flavorings in any given category also are obtainable in two or more quality classes. For example, there are at least four classes of crude camphor oil, in addition to the true or distilled camphor oil. There is camphor safrole (Hon-sho variety—Japan); camphor safrole (Taiwan); camphor cineol (Yu-sho variety—China); and camphor linalool (Ho-sho variety—Taiwan). And, there is Ceylon cinnamon bark; Ceylon cinnamon leaf oil; Seychelles cinnamon leaf oil; and Chinese cinnamon bark oil, among several others. There are also several classes of nutmeg essential oil, including the subvarieties of mace oil.

Only a relatively few *animal sources* of fundamental flavoring substances are used. There is the musk deer (*Moschus moschiferus* L.) found in the Himalayan highlands. The reddish-brown secretion of the male is the odorous principle identified as 3-methylcyclopentadecanonone-one, the principal use of which is in perfumery as a fixative, but which has been reported as an additive in certain food products. Levels of usage are low (3 parts per million in some syrups). There is the civet (*Viverra civetta* Schreber), a cat that lives in Africa and southeastern Asia. The glandular secretion is of main interest in perfumery as a fixative, but it has been reported in foodstuffs at low levels (about 4 ppm). There is the beaver (genus *Castor*) of the northern climes of Alaska, Canada, and Siberia, the dried and ground glandular secretion of which is also used in perfumery, but which has been reported in chewing gum up to concentrations of 400 ppm. The flavoring additive is known as *castoreum*. And there is beeswax, a crude yellow wax that represents a secondary secretion of the honeybee. In addition to its use as a modifier in perfumery, the substance is used up to levels of 5 ppm to enhance the flavor and textural qualities of honey. There always has been a close link between the technology of flavorings for the food field and of fragrances used in perfumes, cosmetics, and related products.

Synthetic flavorings are frequently prepared from natural raw materials that bear a close relationship with the end-product. Fractions and isolates made from refined essential oils are frequently starting points for synthetic flavorings. On the other hand, a number of synthetic flavorings are made from chemical industry raw materials, such as petrochemicals, via the route of organic synthesis. Major unit operations often part of the synthesis include addition, condensation, cyclization, dehydro-

genation, esterification, hydrogenation, oxidation, pyrolysis, reduction, and saponification. Examples of the two approaches include synthetic eucalyptol which commences with purified and concentrated eucalyptol obtained by fractionation of the essential oil. In contrast, cinnamaldehyde ethylene glycol acetate, which possesses a soft, warm, spicy odor reminiscent of cinnamon, is prepared by reacting cinnamaldehyde with ethylene glycol. Cinnamaldehyde is prepared by the condensation of benzaldehyde with acetaldehyde in the presence of sodium or calcium hydroxide.

Although a number of synthetics represent essentially identical replacement of natural flavorings, in many instances the synthetics are an approximation of the natural materials; or represent entirely new flavorings which have significantly broadened the spectrum of available flavorings. The flavorist has many more flavoring substances to call on and many more options than were available several decades ago. Frequently, there are cost advantages for the synthetics, but cost is not always the predominating advantage for synthetics. The flavorist determines the most suited natural or synthetic flavoring, with physical, chemical, and organoleptic qualities considered along with cost.

The synthetic equivalents (approximations or analogues) of the common natural flavors are given in the following abridged listing:

Almond	Tolualdehyde (o, m, p)
Apple	Allyl butyrate, cyclohexylvalerate, isovalerate, propionate; Benzyl isovalerate; Butyl isovalerate, valerate; Cinnamyl formate, isobutyrate, isovalerate; Citronellyl isovalerate; Cyclohexyl acetate, butyrate, isovalerate; Ethyl isovalerate, valerate; Isopropyl acetate, valerate; 2-Methylallyl butyrate; Methyl butyrate; Terpenyl isovalerate.
Apricot	Allyl butyrate, cyclohexylcaproate, cyclohexylvalerate, propionate; Amyl phenylacetate; Benzyl formate, propionate; Butyl propionate; Cinnamic acid; Citronellyl acetate; gamma-Decalactone; gamma-Dodecalactone; Ethyl cinnamate; Geranyl butyrate, isobutyrate, isovalerate, Heptyl acetate, propionate; Methyl ionone; Phenylethyl dimethyl carbinol; Phenylpropyl alcohol; Propyl cinnamate; Santalyl acetate; Tetrahydrofurfuryl propionate; gamma-Undecalactone.
Banana	Cyclohexyl acetate, butyrate, propionate; Ethyl valerate.
Butter	Diacetyl.
Caramel-Butterscotch	Maltol.
Caraway	D-Carvone.
Cheese	Hexanoic acid; Isovaleric acid (rancid).
Cherry	Allyl benzoate; Anisyl butyrate, propionate; Cyclohexyl cinnamate, formate; Methyl anthranilate; Rhodinyl formate; Tetrahydrogeraniol; Tolualdehyde (o, m, p).
Chocolate	Tetrahydrofurfuryl propionate.
Cinnamon	Cinnamaldehyde; Alpha-methylcinnamaldehyde.
Citrus	Decanal dimethyl acetal.
Cloves	Methyl cinnamate; isoeugenol.
Cocoa	Neryl butyrate; Phenylpropyl cinnamate.
Coconut	Allyl undecylate; Ethyl undecylate; Methyl undecyl ketone; gamma-Nonalactone; gamma-Octalactone.

Cognac	Allyl pelargonate; Cyclohexyl caproate.
Cola	2-Ethyl-3-furylacrolein.
Currant	Cyclohexyl butyrate; Guaiol acetate (black currant); Linalyl acetate, isobutyrate, propionate (black currant); Methyl ionone; Methyl propionate (black currant).
Fatty	Decanal; Ethyl nonanoate; Heptyl alcohol; Lauryl alcohol, aldehyde, Nonanal; Octanal; 1-Octanol; Undecanal; 10-Undecenal.
Flowery	Anisyl alcohol; Benzyl acetate, phenylacetate; Cinnamic acid; Cinnamyl acetate; Citronellyl formate; Cresyl acetate; Decanal; Dimethyl benzyl carbinol; Dimethyl benzyl carbinyl acetate; Ethyl anthranilate; Geranyl acetate; Hydroxycitronellal dimethyl acetate; Linalool; Linalyl acetate; Methyl benzoate; Penethyl acetate; 2-Phenylpropionaldehyde; 3-Phenylpropionaldehyde.
Flowery/Fruity	Anisyl acetate; Cinnamyl isovalerate; Citronellyl; Ethyl laurate, octanoate; Geranyl butyrate, propionate; Nonyl acetate.
Fruity	Benzyl propionate; Butyl acetate; Cinnamyl anthranilate, formate; Citronellyl acetate, butyrate, isobutyrate; propionate; Delta-decalactone; Diethyl malonate; Dimethylbenzyl carbinyl acetate; Delta-dodecalactone; Ethyl p-anisate, benzoate, butyrate, heptanoate, hexanoate, maltol, nonanoate; Isoamyl butyrate, hexanoate, isovalerate, cinnamate; Linalyl isobutyrate, propionate; Maltol; Methyl benzoate, cinnamate; 2-Methylundecanal; Nerolidol; Octanol; Octyl formate; Penethyl isobutyrate, isovalerate; gamma-Undecalactone.
Grape	Allyl salicylate; Cinnamyl anthranilate; Guaiol acetate; Isobutyl anthranilate; Isovalerophenone; Octyl isobutyrate; Phenylpropyl acetate, ether.
Grapefruit	Styralyl acetate.
Green Leaves	Allyl anthranilate
Hawthorne	Acetanisole; p-Methoxybenzaldehyde.
Heliotrope	Piperonal
Honey	Allyl phenoxyacetate, phenylacetate; Benzyl cinnamate; Carvacryl acetate; Cinnamyl butyrate; p-Cresyl acetate; p-Cresyl ethyl ether; m-Cresyl phenylacetate; p-Cresyl phenylacetate; Cyclohexyl phenylacetate; Ethyl phenoxyacetate, phenylacetate; Guaiol phenylacetate; Isobutyl phenylacetate; Linalyl butyrate; Methyl phenylacetate; Phenethyl acetate, butyrate, phenylacetate; Phenylacetic acid; Propyl phenylacetate; Santalyl phenylacetate.
Lemon	Citral; Citronellal.
Licorice	Methylcyclopenteneolone.
Maple	Methylcyclopenteneolone.
Melon	Cinnamaldehyde; Ethyl hexadienoate; Methyl amyl ketone; Octyl butyrate.
Menthol	3-p-Methanol.
Mushroom	Hexyl furan carboxylate.
Mustard	Allyl formate.
Orange	Linalyl anthranilate.
Peach	Allyl cyclohexylcaproate, cyclohexylvalerate, undecylate; Amyl phenylacetate; Anisyl alcohol, butyrate; L-Citronellol; Cyclhexyl caproate, cinnamate;

	gamma-Dodecalactone; Ethyl cinnamate; Isopropyl benzyl carbinol; Methyl methylanthranilate; Methyl nonyl ketone; Methyl octine carbonate; gamma-Octalactone; Oxyl acetate; Phenethyl alcohol, isovalerate, salicylate; Phenylallyl alcohol; Phenylpropyl isobutyrate; Propyl cinnamate; Rhodinyl formate; gamma-Undecalactone.
Pear	Benzyl butyrate; 2-Ethylbutyl acetate; Ethyl heptylate; Hexyl acetate; Hexyl furan carboxylate; Isoamyl acetate; 2-Methylallyl caproate; Methylheptenone; Propyl acetate.
Pineapple	Allyl caproate, cyclohexylacetate, cyclohexylbutyrate, cyclohexylpropionate, 2-nonylenate, phenoxyacetate; Benzyl formate; Bornyl acetate; n-Butyl acetate; Butyl isobutyrate; Cinnamyl acetate; Decanal dimethyl acetal; Ethyl butyrate, hexadienoate, phenoxyacetate; Hexyl butyrate; 2-Methylallyl caproate; Methyl beta-methylpropionate; Methyl undecylate; Propyl isobutyrate.
Plum	Butyl formate; Citronellyl butyrate, formate, propionate; gamma-Decalactone; Guaiol butyrate; Heptyl formate; Hexyl formate; Isoamyl formate; Isopropyl formate, propionate; Linalool; Neryl propionate; Phenethyl formate (green plum); Phenethyl isobutyrate (green plum); Phenylallyl alcohol; Phenylpropyl butyrate; Propyl formate; Terpenyl butyrate.
Raspberry	Benzyl salicylate; alpha-Ionone; Isobutyl cinnamate; Methyl ionone; Neryl acetate; Santalol.
Rose	Phenethyl alcohol; Phenethyl dimethyl carbinyl isovalerate; Rhodinol.
Rum	Ethyl formate; isobutyl formate.
Sassafrass	p-Propyl anisole.
Spearmint	1-Carvone.
"Spice"	Eugenol; Isoeugenyl acetate; 3-Phenylpropyl acetate.
Strawberry	Anisyl formate; Benzyl isobutyrate; Cuminic alcohol; Ethyl methylphenylglycidate; Ethyl phenylglycidate; Isoamyl salicylate; Isobutyl anthranilate; Methylacetophenone; Methyl cinnamate; Methyl naphthyl ketone; Nerolin; Neryl isobutyrate; Phenylglycidate.
Vanilla	Propenyl guaiethol; Vanillydene acetone.
Violet	alpha-Ionone; beta-Ionone; Methyl-2-octynoate.
Walnut	gamma-Octalactone.
Wine	Ethyl acetate; heptylate.
Wintergreen	Allyl salicylate; Methyl salicylate.

Physiological Aspects of Flavors

Receptor cells especially sensitive to chemicals are found in virtually all animals. By convention, those receptors normally excited by contact with chemicals in liquid phase at relatively high concentrations are termed taste or *gustatory receptors*, although the distinctions between taste and smell are not critical at cellular or molecular levels.

Chemical aspects of taste receptor functions can be studied by recording the patterns of electrical potentials in receptor cells while the cells are being stimulated with pure chemicals of known structures and properties. Since the mid-1950s, when this method was first successfully applied to single taste receptor cells, using receptors on the mouth parts of a fly, many earlier theories of taste stimulation have been revised.

Electrolyte stimulation is chiefly a function of monovalent cations in all animals which have been studied. Consequently, the receptor sites are thought to be anionic. The pH relationships of stimulation also indicate that strongly acidic (e.g., PO_4^{2-} or SO_4^{2-}) receptor groups are involved. Calculations of free energy changes of the reaction between salt and receptor site give values between 0 and -1 kcal/mole; and low ΔF values suggest that the reaction involves only weak physical forces. The reaction occurs extremely rapidly, since typical nerve impulses can be recorded within 1 millisecond after stimulating electrolytes are applied. In blow-flies, $0.004M$ NaCl, which produces 1 impulse per second, represents the threshold for behavior response. These thresholds appear to be somewhat higher in humans.

No one type of receptor site or reaction can account for the extreme structural specificities observed. A curious assortment of molecules can elicit "sweet" sensations. Early studies on a variety of organisms demonstrated that ring structures and D-isomers were more stimulating in polyol compounds than straight-chain and L-isomers. See accompanying figure. Inositol

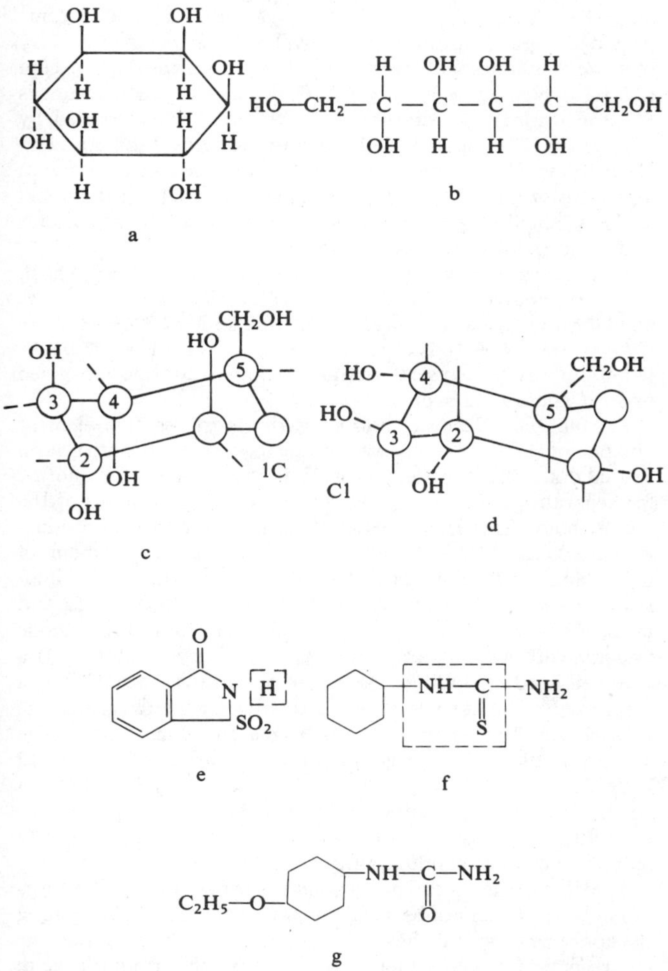

Molecules illustrating relationships between structure and effects on taste receptors. See explanation in text.

(a), with its ring structure, was found to stimulate. The straight-chain polyhydric alcohols sorbitol, dulcitol (b), and mannitol did not stimulate. Possession of an alpha-D-glucopyrano side linkage was found to generally increase the stimulating capacity of sugars. Maltose, with a 1,4-linkage; turanose, with a 1,3-linkage; and the nonreducing sugars stimulate. Lactose, with its 1,4-linkage, and melibiose, with a 1,6-linkage, both lack the alpha link and are relatively nonstimulating.

Conformation, as well as configuration, is important in determining the stimulating power of sugar molecules. Glucose, which exists in solution almost entirely in an aldopyranose "chair" formation has derivatives of both 1C and C1 conformations (c and d). Those of the C1 type are considerably the more stimulating. The hydroxyl groups attached to C3 and C4, inclined 19 degrees above and 19 degrees below the adjacent plane of the molecule, appear to be necessary for the critical linkage at the receptor site. Lack of effects by metabolic inhibitors (azide, fluoride, ioodoacetate, etc.) or of temperature effects upon the initial excitatory process, suggests that this step depends upon specific physical rather than chemical reactions.

The nature of other polyol receptor sites and the molecular basis for genetic and species differences in taste capabilities remain largely unknown. Saccharin (o-sulfobenzime) (e) exemplifies both puzzles, since its molecule does not fit any known sugar receptor site, yet it is confused with sugar stimuli by humans and other primates, but probably not by nonprimate animals. The substitution of other groups for one hydrogen (dotted lines in (e)) renders saccharin tasteless. The genetic basis of taste has been studied with phenylthiocarbamide (PTC). A strong bitter taste of PTC depends upon the chemical components indicated by dotted lines in (f), and upon possession of a dominant "taster" gene in humans. Curiously, a small change in the molecule (g) yields a product 250 to 300 times sweeter than sugar.

Taste receptors for water have been reported to occur on mouth parts of mammals and invertebrates. Specialized amino acid and amine receptors are found on the legs of many anthropods. The mechanisms by which adequate stimuli initiate nerve impulses in these cells offers a rich field for further investigation. Some stimuli, especially longchain hydrocarbons, are known to act in an opposing manner, i.e., by decreasing, rather than increasing, the output of receptor impulses. Their effects resemble the actions of narcotics. Some authorities suggest that taste sensation, as ultimately perceived, probably results from a complex coded pattern of augmented or depressed frequencies of nerve impulses, originating in the different cells of a heterogeneous population of taste receptors.

Sense of Smell. For many years, physiologists have explained that the sense of smell is located in the mucus lining inside the nose. Traditionally, it has been observed that a person with a "dry" nose has little if any sense of smell. Molecules characteristic of certain flavorings must be moistened by the mucus before they can be detected. Traditionally, it also has been observed that these nerve cells (estimated to be a million or more) tend to become blocked (refuse further transmission of signals) upon prolonged exposure to any given odor. This blocking phenomenon can occur within just a few minutes after exposure to certain, usually powerful odors. When all is sorted out concerning the mechanics of tasting and smelling, it is highly likely that operationally the cells in the nose lining and the papillae of the tongue will be highly similar, if not identical—simply because many other interrelationships have been shown to be similar. Of course, over the years, some researchers have approached the phenomena of taste and odor separately. Some of the odor detection theories proposed have included: (1) the *vibrational theory* (Demerdach; Dyson; Wright); (2) the *steteochemical theory* (Amoore, Johnston, Naves); (3) the *theory of interfacial adsorption* (Beck; Davies); and the *profile functional group theory* (Beets).

Considerably more detail on flavorings can be found in the "Foods and Food Production Encyclopedia" (D. M. Considine, editor), Van Nostrand Reinhold, New York, 1982.

References

Amoore, J. E., Johnston, J. W., Jr., and M. Rubin: "The Stereochemical Theory of Odor," *Sci. Amer.* (February 1964).

Arvidson, K., and U. Friberg: "Human Taste: Response and Taste Bud Number in Fungiform Papillae," *Science*, **209**, 807–808 (1980).

Beck, L. H.: "A Quantitative Theory of the Olfactory Threshold Based upon the Amount of the Sensory Cell Covered by an Adsorbed Film," *Proc. New York Academy of Sciences*, **116**, 448 (1964).

Beets, M. G. J.: "Structure and Odor," in "Molecular Structure and Organoleptic Quality," Soc. of Chem. Ind., London, 1961.

Beets, M. G. J.: "Odor and Molecular Constitution," *Amer. Perf.*, **76**, 54 (1961).

Beets, M. G. J.: "Some Aspects on the Problems of Odor," *Parf. Cosm. Savons*, **5**, 4 (1962).

Davies, J. T., and F. H. Taylor: "Olfactory Thresholds: A Test of a New Theory," *Perfum. Essent. Oil Rec.*, **46**, 1, 15 (1955).

Demerdache, A., and R. H. Wright: "Low-Frequency Molecular Vibration in Relation to Odor," in "Olfaction and Taste" (T. Hayaski, editor), Vol. 2, Pergamon, Elmsford, New York, 1979.

Dethmers, A. E.: "Utilizing Sensory Evaluation to Determine Product Shelf Life," *Food Technology*, **33**, 9, 40–42 (1979).

Dorland, W. E., and J. A. Rogers, Jr.: "The Fragrance and Flavor Industry," Dorland, Mendham, New Jersey, 1977.

Dravnieks, A.: "Comparison of Theories on Relations between Odor Parameters and Other Properties of Odorants," NATO Advanced Study Institute on Odor Theories and Odor Measurements, Robert College, Blebes, Istanbul, Turkey, 1966.

Dyson, G. M.: "Raman Effect and Concept of Odor," *Perfum. Essent. Oil Rec.*, **28**, 13 (1937).

Furia, T. E., and N. Bellanca: "Fenaroli's Handbook of Flavor Ingredients," CRC Press, Boca Raton, Florida, 1971.

Furia, T. E.: "Handbook of Food Additives," 2nd endition, CRC Press, Boca Raton, Florida, 1972.

Heath, H. B.: "Flavor Technology," AVI, Westport, Connecticut, 1978.

Hodgson, E. S.: "Taste Receptors," in "The Encyclopedia of Biochemistry," (R. J. Williams and E. M. Lansford, Jr., editors), Van Nostrand Reinhold, New York, 1967.

Johnson, J. W., and A. Sandoval: "Organoleptic Qualities and the Stereochemical Theory of Olfaction," *Proc. Sci. Sect. Toilet Goods Assoc.*, **34** (1960).

Katz, M. H., and G. Jacobson: "Flavor Chemistry," Symposium, 39th Annual Institute of Food Technologists, Saint Louis, Missouri, 1979.

Kazeniac, S. J.: "Flavor Trends in New Foods," *Food Technology*, **31**, 1, 26–28 (1977).

Naves, Y. R.: "The Relationship between the Stereochemistry and Odorous Properties of Organic Substances," in "Molecular Structure and Organoleptic Quality," Soc. of Chem. Ind., London, 1957.

Staff: "Food Chemicals Codex," National Academy of Sciences, Washington, D.C. (Revised periodically).

Staff: "Essential Oils," Foreign Agriculture Circular, U.S. Department of Agriculture, Washington, D.C. (Issued periodically).

Wolfe, K.: "Use of Reference Standards for Sensory Evaluation of Product Quality," *Food Technology*, **33**, 9, 43–44 (1979).

FLINT. Flint is a rock composed essentially of a cryptocrystalline form of silica. It is very dense and tough, breaking with a conchoidal fracture; colors, usually dark grays, blues, or browns, often black. It occurs chiefly as nodules and masses in chalks and limestones. Flint is particularly interesting because it was used by primitive man for making instruments (artifacts) for thousands of years before he learned to use bone and metal. Flint still remained an essential mineral resource for making fire, including the flint locks on guns, until the close of the eighteenth century. From the dawn of civilization the best flint has come from Belgium and the coastal chalks of the English Channel and the Paris Basin. See also **Chert.**

FLOTATION. Classifying (Process).

FLUORESCENCE. This term has three common usages: 1. The process of emission of electromagnetic radiation by a substance as a consequence of the absorption of energy from radiation, which may be either electromagnetic or particulate, provided that the emission continues only as long as the stimulus producing it is maintained. That is, fluorescence is a luminescence which ceases within about 10^{-8} second after excitation stops; this period of time being the lifetime of an atomic state for a normal allowed transition. 2. The term fluorescence may also be applied to the radiation emitted, as well as to the emission process. 3. In x-ray terminology, the term fluorescence may be used in the more specific sense (than given in the general definition above) to denote the characteristic x-rays emitted as a result of the absorption of x-rays of higher frequency.

FLUORINE. Chemical element symbol F, at. no. 9, at. wt. 18.9984, periodic table group 7a (halogens), mp $-219.62°C$, bp $-188.1°C$, density, 1.696 g/liter (gas at $0°C$), 1.108 g/cm³ (liquid at bp). Fluorine is a pale-yellow gas, poisonous, very reactive, combines with most other elements in the dark, except it does not combine readily with oxygen. Critical pressure is 55 atm; critical temperature is $-129.2°C$. First identified by Scheele in 1771, but not isolated until 1886 by Moissan, who electrolyzed fused potassium hydrogen fluoride in a platinum apparatus. Fluorine is a high-tonnage chemical, used mainly in the production of fluorides, in the synthesis of fluorocarbons, and as an oxidizer for rocket fuel.

First ionization potential 17.42 eV; second, 34.6 eV; third, 58.02 eV; fourth, 84.88 eV; fifth, 113.0 eV; sixth, 152.9 eV. Oxidation potential $F^- \rightarrow \frac{1}{2}F_2 + e^-$, -2.85 V; $2F^- + H_2O \rightarrow F_2O + 2H^+ + 4e^-$, 2.1 V; $HF \rightarrow \frac{1}{2}F_2 + H^+ + e^-$, 3.03 V. Other important physical characteristics of fluorine are given under **Chemical Elements.**

Production. Because fluorine is the most reactive element and one of the strongest oxidizing agents known, its preparation caused difficulties for many years. The requirements for fluorine for separating ^{235}U from ^{238}U during the development of the atomic bomb accelerated research on finding improved production methods. Much research went into the development of compounds and materials that would resist the actions of fluorine for use in diffusion plants. Materials finally selected and in use in modern fluorine production plants include (1) a cathode integral with a mild-steel cell body, (2) a carbon anode, (3) a steel cell head, including a Monel skirt, and (4) a Monel screen diaphragm. The skirt is required to prevent admixture of the fluorine gas formed with the hydrogen gas also formed. The elctrolyte consists of a fused mixture of potassium fluoride and hydrofluoric acid. The overall reaction: $2HF \rightarrow H_2 + F_2$. Hydrogen is liberated at the cathode; fluorine at the anode. The potassium fluoride is required because hydrofluoric acid of the purity required does not conduct an electric current.

Fluorine causes both chemical and thermal burns and, unfortunately, they may not be detected immediately, depending upon the concentration. Further, upon contact with the skin, fluorine gas reacts with water in the skin to form hydrofluoric acid, an excellent solvent for protein. Personnel directly involved in the handling of fluorine must be equipped with gauntlet gloves, neoprene rubber apron, chemical goggles and when in atmospheres above the TLV, gas masks with cannisters must be used. In emergency situations where there is a very high concentration of fluorine, the area must be evacuated, directing personnel upwind.

All containers, processing equipment, and piping to be used in fluorine service first must be passivated before use and thereafter designated for fluorine service. These requirements result from the severe oxidizing characteristics of fluorine gas. Passivation removes any easily oxidized materials, such as paint, pipe dopes, metal oxides, grease, and metal filings. During the procedure, a metal fluoride film will form on metal surfaces, thus minimizing further corrosion of the metal by fluorine.

Special permits are required to ship fluorine. Generally, fluorine is transported as a nonliquefied compressed gas in seamless steel or nickel cylinders. Upon receipt, multijacketed dewars are frequently used to contain the product.

Chemistry and Compounds. Fluorine exhibits in common with the other halogen elements a marked readiness to form singly charged negative ions, as would be expected from the fact that these atoms need only one electron to acquire an inert gas configuration. However, the electron affinity of fluorine (3.74 eV) is not the highest of the four common halogens, but is less than that of chlorine and bromine (4.02 eV and 3.78 eV respectively). The greater reactivity of fluorine in aqueous solution is due to the fact that its lower electron affinity is more than offset by its lower energy of dissociation (38 kcal against 58 kcal for Cl_2 and 46 kcal for Br_2) and the higher energy of hydration (122 kcal for F^- against 89 kcal for Cl^- and 81 kcal for Br^-). The overall result is to give the system $F^- \rightarrow \frac{1}{2}F_2 + e^-$, the largest negative oxidation potential (-2.85 V) of any simple ion to its element.

The reactions of fluorine have, in general, high temperature coefficients. At low temperatures its reactivity with hydrogen is very slight, but becomes rapid and even violent at higher temperatures and in the presence of impurities. Fluorine reacts with all metals, the vigor of the reaction and the composition of the resulting fluoride depending upon the temperature and the reactivity of the metal. Sulfur, silicon, carbon, and antimony ignite in fluorine; cesium, rubidium, and potassium form trifluorides, which, however, do not contain trivalent cations, while the noble metals react only at very high temperatures. Unlike the three alkali metals mentioned, however, most metals do not form fluorides in exceptional oxidation states, and, in fact, many elements form oxyanions of higher valence than they do fluorocomplexes. Thus manganese forms two fluorides, MnF_2 and MnF_3, and its fluorocomplex ions of highest valence are MnF_6^{2-} and MnF_5^-; chromium forms four stable fluorides, CrF_2, CrF_3, and CrF_4, and CrF_5 (and possibly CrF_6).

Four binary compounds of fluorine and oxygen have been reported, O_4F_2, O_3F_2, O_2F_2, and OF_2. The polyoxygen difluorides are produced from the elements by action of the silent electric discharge at low pressures, lower temperature favoring higher oxygen content. Dioxygen difluoride is a yellow to orange solid, melting at about $-160°C$. It decomposes rapidly at temperatures above $-25°C$. The others are even less stable. Oxygen difluoride, OF_2, is prepared by passing fluorine rapidly through weak NaOH solutions or by electrolysis of liquid HF solutions of H_2O or other oxygen compounds. It has an O—F bond distance of 1.4Å, F—F, 2.22Å, and FOF angle, about 105°. It reacts with metals to form fluorides and oxygen, with other halides to form fluorides, oxygen, and the other halogens, and with other compounds usually to yield fluorides.

Hydrogen fluoride is the most stable of the hydrogen halides (heat of formation 64 kcal). Its bond moment shows the compound to have a marked covalent character. In the pure state, liquid hydrogen fluoride is slightly more conducting than pure water, and in the anhydrous state it reacts only with the alkali metals, alkaline earth metals (excluding beryllium and magnesium) and with thallium. Liquid HF is an extremely strong acid, having a Hammett acidity function of 10.2 (compared to H_2SO_4 11.3), HCl, hydrobromic and hydriodic acids being

essentially unionized in it, although the dielectric constant is comparable to that of water. However, in 0.1 N aqueous solution, hydrogen fluoride is only about 15% ionized. A correlative property is the extensive polymerization through hydrogen bonding, various polymers being present. In the vapor state, the degree of polymerization depends upon temperature and pressure, varying from mostly monomer to linear hexamer or even higher polymers, with the ring hexamer being particularly favored. The liquid likewise contains monomer and polymers, the average degree of polymerization being three of four. The units of the polymers undergo very rapid exchange. In aqueous solution there is a strong tendency for fluoride ion to associate with hydrogen fluoride molecules, forming the symmetrical HF_2^- ion. It reacts with metal oxides and hydroxides to produce the fluorides, and with metal ions to produce complex ions, or their salts. It reacts with phosphorus pentoxide to produce complex fluorides, oxyfluorides, and, on continued action, mono-, di- or hexafluorophosphates.

Due to the high oxidation potential of fluorine, and the small size of the fluoride ion, the element enters into many compounds with the other halogens. Diatomic compounds of this type include ClF, BrF, and IF (the latter two having been identified but not isolated in a pure state), tetratomic compounds include ClF_3, BrF_3, and IF_3 hexatomic ones include BrF_5, and IF_5, while the octatomic type is limited to one member, IF_7. These compounds are discussed under the entry for the halogen forming the donor atom in the compound.

Fluorine forms polyhalide anionic complexes including $IFBr^-$, $IFCl_3^-$, BrF_4^-, IF_6^-, which occur in salts of the higher alkalies, as well as cations such as BrF_2^+, which occurs in such salts as BrF_2SbF_6, BrF_2AuF_4, BrF_2SO_3F, etc. The difluoroiodate ion, $IO_2F_2^-$, is also known in such salts as KIO_2F_2.

Organic Compounds. Organic fluorine compounds are made by reaction of the corresponding alkane chloro-compounds with silver fluoride, mercurous fluoride, antimony trifluoride, titanium tetrafluoride, and the arene fluoro-compounds by the diazo-reaction using hydrogen fluoride, and otherwise. The effect of the continued replacement of hydrogen atoms by fluorine atoms is an initial increase in reactivity, followed by a reversal of this effect, so that the highly substituted compounds are relatively inert. See also **Fluorocarbon.**

References

O'Keeffe, M., and J. O. Bovin: "Solid Electrolyte Behavior of $NaMgF_3$: Geophysical Implications," *Science,* **206,** 599–600 (1979).

Quentin, K. E., et al.: "Analytical Determination of Small Amounts of Fluorine in Foods and Waters. 4. Fluorine Studies in Foods (in German) (in *Chem. Abstr.* 1960, **54**(6):5967).

Rudge, A. J.: "The Manufacture and Use of Fluorine and Its Compounds," Oxford, London, 1962.

Slesser, C., and S. R. Schram: "Preparation, Properties, and Technology of Fluorine and Organic Fluoro Compounds," McGraw-Hill, New York, 1951.

Simons, J. H. (editor): "Fluorine Chemistry," vols. 1–5, Academic Press, New York, 1950–1964.

Venkateswarlu, A., and P. Sita: "A New Approach to the Microdetermination of Fluoride Adsorption-Diffusion Technique," *Analytical Chemistry,* **43,** 78 (1971).

NOTE: See also references listed at end of entry on **Fluorine (In Biological Systems).**

FLUORINE (In Biological Systems). Fluorides are not required for plant growth, but in animals, including humans, low levels of fluorides have been shown to have beneficial effects on teeth and on bone structure. Growth increases in experimental animals have been reported when low levels of fluorides have been added to purified diets. However, fluoride substances show toxicity in both animals and plants when encountered in fumes and dusts from industrial facilities as well as natural emissions from the eruption of volcanoes. Abnormally high levels of fluoride in water also have caused fluorine toxicity in animals and mottled teeth in humans.

Fluorides do not usually move from the soil to plants and on to livestock feedstuffs and human foodstuffs in amounts that are toxic. Injury to plants from fluoride in the soil has been noted on soils that are too acid for the satisfactory growth of most plants. On limed soils or soils with sufficient calcium for optimum growth, any fluorine added to the soil reacts with the calcium and other soil constituents to form insoluble compounds, which are not taken up by the plants. Rock phosphate and some kinds of superphosphate fertilizers contain large amounts of calcium fluoride, but the fluorine content of the plants grown on soils that have been heavily fertilized with these phosphates is not appreciably increased. Tea and some other members of the Theaceae family are the only plants that take up very much fluorine from the soil.

The soil-to-plant segment of the food chain contains some built-in safeguards against fluorine toxicity. This toxicity has been due to the deposition of airborne fumes and dusts on the above-ground parts of plants, followed by the consumption of these contaminated plants by animals, including humans. Also, fluorine toxicity has been caused by direct inhalation of the fumes and dusts, or by drinking water with abnormally high fluorine levels. If the fumes and dusts are mixed into the soil, they will be inactivated and will not find their way into the food chain in toxic amounts.

The safeguards against toxicity provided by the chemistry of fluorine in soils make it unlikely that applying fluorine-containing compounds to soils will be a useful way to insure that plants will contain sufficient fluorine to prevent dental caries. However, tea and mechanically deboned meats may contribute to these needs. When increased fluoride intake is desirable, carefully controlled direct additions to drinking water, to dentifrices, or to specific foods are more promising than adding fluorides to soils that produce food crops.

Fluoridation. In a broad sense, this term would signify the addition of fluorine to a substance much as chlorination means the addition of chlorine. In a more specific, but commonly used sense, fluoridation means the addition of very small amounts of a fluoride-containing compound to water supplies for the purpose of preventing dental caries. It has been shown over a number of years that the introduction of about 1 part per million (ppm) of fluoride to drinking water will reduce the incidence of tooth decay in children by as much as 60% as compared with similar groups of children who consume nonfluoridated water. Because of the striking nature of these findings, a few cities in the United States commenced experimental treatment of water supplies during the mid-1940s. As of the early 1980s, it is estimated that close to 100 million persons in the United States are now supplied with fluoridated water.

The commonly used compound is sodium fluoride or sodium silicofluoride in a dry crystalline or powdered form. Hydrofluosilicic (flusilicic) acid is also used in liquid form. Inasmuch as concentrations of fluoride in excess of 1.5 ppm may cause mottling of tooth enamel, it is mandatory to exercise very careful control to maintain the desired 1.0 ppm dosage. Water supply samples are frequently tested by municipal authorities. Fluoride concentrations may be determined by colorimetric or electrometric methods, and the latter can be adapted to continuous reporting and controlling purposes.

The concept of fluoridation has created numerous controversies among the populace, a situation that occurs frequently when decisions to install fluoridation systems for the first time are under consideration. Until the mid-1970s, it was believed that such practice was unquestionably safe and that arguments against fluoridation were essentially emotionally motivated. However, some second thoughts are now being taken, particularly with reference to possible reactions of fluorine with certain pollutants now found in raw water supplies that once were not present.

Feedstuffs. Excessive amounts of fluoride in the soil can cause tooth and bone damage in livestock. Parts of Arkansas, California, South Carolina, and Texas have soils abnormally high in fluorine content. In serious situations, diarrhea and emaciation will be exhibited by the livestock. The effects depend upon the fluorine source and species of livestock. Exceptionally high fluoride levels can be encountered near smelters where pollution safeguards have not bee installed or are ineffectively maintained. As compared with other livestock, pigs can tolerate much more fluorine (up to nearly 300 ppm of fluorine derived from rock phosphates).

Fluorine in Tea. The majority of foods found in the average diet contain 0.2–0.3 ppm or less fluorine in the food as consumed. Tea and seafoods are notable exceptions (McClure, 1949). Different values are reported for fluorine content of various teas by different investigators (Wang et al., 1949; Fabre and de Campos, 1950; de Campos, 1950; Zimmerman et al., 1957; Quentin et al., 1960; Okada and Furuya, 1969; Cook, 1970; Venkateswarlu and Sita, 1971). The fluorine content of tea depends upon the origin of the plant, the type of soil and fertilizer, the age of the leaves and the time of harvesting (Garber, 1962).

In 1978, investigators at the University of Teheran (Iran) undertook a study to find out the fluorine content of teas consumed in Iran and to evaluate the potentiality of tea as a contributor of fluorine. Tea is an important item in the Iranian diet and drunk mostly by laborers and peasants; furthermore, diluted infused tea is used as a supplement in between breast feedings of infants. The investigators concluded that, considering the optimal intake of fluorine of 1 milligram per day suggested for protection from dental caries, the drinking of tea in Iran provides about half of this amount without considering the fluorine content of water and other sources.

Fluorine Content of Mechanically Deboned Beef and Pork. One question that has arisen from time to time in connection with mechanically deboned meat (MDM) is its possible fluoride content because some microscopie bone particles may be present in the product. Investigators Kruggel and Field, Division of Biochemistry and Division of Animal Science, University of Wyoming, made a study of this in 1977. Samples were collected from regions where high levels of fluoride occurring in the water and vegetation have been reported. Higher magnesium, iron, and fluoride contents were found in beef MDM from the western and midwestern regions of the United States when compared with the southern region. Higher iron and fluoride contents were found in beef MDM than in pork MDM. Among a number of conclusions drawn was that the consumption of fluoride from MDM and other foods combined would be far below the 20–80 milligrams or more of fluoride that must be consumed daily to produce toxicity (Food and Nutrition Board, 1974). Mottling of teeth in children has been observed at fluoride concentrations in the diet and drinking water of 2–8 ppm. A frankfurter containing 10% MDM would contain about 1.7 ppm fluoride. Since the daily fluoride intake in many areas of the United States is not sufficient to afford optimal protection against dental cavities (Food and Nutrition Board, 1974), products which contain MDM may be of value in furnishing needed fluoride and in reducing the incidence of tooth decay (Kruggel/Field, 1977).

Fluorine in Marine Sponge Halichondria moorei. It is well known that many marine organisms accumulate the halogens—iodine, bromine, and chlorine. However, reports of fluorine accumulation have been rare. Thus the report of fndings by Gregson et al. (1979) that the marine sponge *Halichondria moorei* has a fluorine content (dry weight basis) of 10% of the total weight is of interest. In this species, the fluorine occurs as potassium fluorosilicate, which is known to be a powerful antiinflammatory agent. It is of interest that closely related sponge varieties of the same habitat contain little if any fluorine—and, further, the habitat is free of fluorine except for the small amount naturally present in seawater.

References

Allaway, W. H.: "The Effect of Soils and Fertilizers on Human and Animal Nutrition," *Agriculture Information Bulletin 378*, Cornell University Agricultural Experiment Station and U.S. Department of Agriculture, Washington, D.C. (1975).

de Campos, P.: "Fluorine Content of Tea Cultivated in Sao Paulo" (in *Chem. Abstr.* 1952, **44**(22):11498).

Cook, H. A.: "Fluoride Intake through Tea by British Children," *Fluoride Quart. Rept.*, 3, 12 (1970).

Fabre, R., and P. de Campos: "Distribution of Fluorine in Plants—Tea Leaves," *Ann. Pharm. Franc.*, 8, 391 (1950).

Food and Nutrition Board: "Effects of Fluoride in Animals," National Academy of Sciences, Washington, D.C. (1974).

Food and Nutrition Board: "Recommended Daily Allowance," National Academy of Sciences, Washington, D.C. (1979).

Garber, K.: "Plants and Fluorine" (in German), *Qualitas Plant., Mater., Vegetabiles*, 9, 33 (in *Chem. Abstr.* 1962, **52**(2):2589).

Gregson, R. P., et al.: "Fluorine is a Major Constituent of the Marine Sponge Halichondria moorei," *Science*, **206**, 1108–1109 (1979).

Kruggel, W. G., and R. A. Field: "Fluoride Content of Mechanically Deboned Beef and Pork from Commercial Sources in Different Geographical Areas," *Journal of Food Science*, 42, 1, 190–192 (1977).

McClure, F. J.: "Fluoride in Foods," *U.S. Public Health Report 64*, 1061 (1949).

Okada, F., and K. Furuya: "Flouride Content in Tea" (in Japanese), *Chago Gijutsu Kenkyu*, 37: 32 (in *Chem. Abstr.* 1966, **71**(23):245).

Sax, N. I.: "Dangerous Properties of Industrial Materials," 5th Edition, Van Nostrand Reinhold, New York, 1979.

Wang, T. H., et al.: "Flourine Content of Fukein Teas," *Food Research*, **14**, 98 (1949).

Zimmerman, P. W., et al.: "Fluorine in Food with Special Reference to Tea" (in German) (*Chem. Abstr.* 1958, **52**(1):610).

FLUORITE. Fluorite is a calcium fluoride mineral CaF_2 crystallizing in the isometric system, often in superb cubic crystals. Twinned crystals are common, usually as cubic penetration twins. It is found in many diverse geological environments, from vein material associated with metallic ores, especially lead and silver, to sedimentary formations associated with celestite, gypsum, dolomite, and calcite, as a component mineral in high-temperature pneumatolytic deposits with cassiterite, topaz and tourmaline, and in pegmatites. Exceptional crystals are found in Alpine type veins on quartz crystals from Switzerland. Also occurs as massive compact to granular aggregates. Possesses perfect 4-directional cleavage planes, with uneven to splintery fracture. It is a brittle mineral with a hardness of 4 and a specific gravity of 3.180. Vitreous luster when crystallized, dull to glimmering in massive material. Colorless when pure, but shades of blue, green, yellow, brown, white, rarely rose-red and pink, are known, including intermediate color graduations of each

type. Certain colored crystals appear blue by reflected light, green by transmitted light. This phenomenon may be a product of heat, ultraviolet light, pressure, or exposure to radiation, as from x-rays. Varying color zones are commonly observed in areas parallel to the crystal faces. Massive varieties may also exhibit parallel zones of varying color.

Phosphorescence is not uncommon when certain fluorites are exposed to sunlight, ultraviolet rays, or are heated. Vivid fluorescence is a common attribute of many fluorites, with blue to violet fluorescence predominant. The word fluorescence is derived from the mineral name, fluorite, owing to its strong fluorescent character.

Certain dark blue fluorite from Bavaria known as *antozonite* contains free fluorine and calcium, which when released either by grinding or exposure to cathode rays produce a distinctive odor, caused by the reaction of the fluorine with water.

Fluorite is a ubiquitous mineral and is so widespread in its occurrence that only the most noteworthy can be mentioned. The English localities at Cumberland, Durham, and Weardale are world famous. Exceptionally beautiful banded material of blue fibrous character from Derbyshire, known as Blue-John, has been much used for decorative carved peices, such as vases and other ornamental objects. Norway has produced exceptional specimens from the famous Kongsberg silver veins, as well as yttrium-rich fluorite from northern Norway associated with rare-earth minerals. Fine material has been obtained from the Transvaal in the Republic of South Africa, Tasmania, and Australia. Large quantities of fluorite is mined in Mexico at Guadalcazar and Guanajuato.

Notable United States localities include Hardin and Pope Counties in Illinois, and also adjacent Kentucky areas where it is intimately associated with calcite, barite, quartz, with minor galena and sphalerite in sedimentary rock veins. Large deep green masses yielding exceptional cleavage octahedrons were obtained from Westmoreland, New Hampshire. Macomb, New York produced large sea-green crystallized cubes. Various sedimentary formations in Ohio have yielded fine brown crystals associated with celestite. Many occurrences are known throughout Colorado and Idaho. Optical-quality crystals have been obtained from Madoc, Ontario, Canada in association with barite and calcite; also in British Columbia near Grand Forks, and at several localities in Mexico.

Fluorite is highly valued as a flux in the manufacture of steel; also as a raw material for hydrofluoric acid. When of optical quality, the mineral is used for lens and prisms in scientific instruments. See also terms listed under **Mineralogy.**

—Elmer B. Rowley, Union College, Schenectady, New York.

FLUOROCARBON. A number of organic compounds analogous to hydrocarbons, in which the hydrogen atoms have been replaced by fluorine. The term is loosely used to include fluorocarbons that contain chlorine; these should properly be called chlorofluorocarbons or fluorocarbon chlorides. The distinction is important, because it is these latter compounds which are thought to deplete the ozone layer of the upper atmosphere. Fluorocarbons are chemically inert, nonflammable, and stable to heat up to 260–316°C. They are denser and more volatile than the corresponding hydrocarbons, and have low refractive indices, low dielectric constants, low solubilities, low surface tensions, and viscosities comparable to hydrocarbons. Some are compressed gases; others are liquids. These compounds were once used extensively in aerosol packages. They are used as refrigerants, solvents, blowing agents, fire extinguishers, lubricants, and hydraulic fluids—as components of complete systems.

Fluorocarbon polymers include polytetrafluoroethylene, polymers of chlorotrifluoroethylene, fluorinated ethylene-propylene polymers, polyvinylidene fluoride, and hexafluoropropylene, among others. These are thermoplastic substances, resistant to chemicals and oxidation; noncombustible; with broad useful temperature range (up to 285°C); of high dielectric constant; resistant to moisture, weathering, ozone, and ultraviolet radiation. Their structure comprises a straight backbone of carbon atoms symmetrically surrounded by fluorine atoms. These materials are available as powders and dispersions for further processing, as films, sheets, tubes, rods, tapes, and fibers. They find use in high-temperature wire and cable insulation, other electrical equipment, chemical processing equipment, coatings for cooking utensils, piping, gaskets. Among the fluorocarbon polymers are a number of fluoroelastomers. These polymers are amorphous, thermally stable, noncombustible, have low glass transition temperature (−77°C), and are generally resistant to attack by solvents and chemicals.

FOAM. A tightly packed aggregation of gas bubbles, separated from each other by thin films of liquid. If foams were not so common, their existence would cause some wonderment. None of the obvious properties of a liquid would lead one to suppose that thin liquid films could sustain themselves for any appreciable time against the effect of gravity. The existence and stability of a foam depend, in fact, on a surface layer of solute molecules, which form a structure quite different from that of the underlying liquid inside the interbubble film.

At the surface of a liquid, molecules are in a state of dynamic equilibrium, in which the net attractive forces exerted by the bulk of the fluid cause molecules to move out of the surface; this motion is counterbalanced by ordinary diffusion back into the diluted surface layer. The equilibrium results in the surface layer being constantly less dense than the bulk fluid, which creates a state of tension at the surface. The tension can be somewhat relieved by adsorption of foreign molecules either out of the bulk solution, or out of the vapor phase. Soluble substances that have a strong tendency to concentrate in the surface layer are collectively known as surface-active agents; examples are soap, synthetic detergents, and proteins. The excess concentration of solute at the surface reduces the surface tension of water. The general relation, in the form of a differential equation, was first deduced thermodynamically by Gibbs. It is called the *Gibbs adsorption theorem*, and is

$$\mu = -\frac{c}{RT}\frac{d\gamma}{dc}$$

where μ is the excess concentration at the surface, c is the bulk concentration, and $d\gamma/dc$ is the change of surface tension with concentration of solute.

An excess of solute at the surface, as measured by $+\mu$, can be termed *positive adsorption* to distinguish it from an excess of solvent at the surface $(-\mu)$, or *negative adsorption*. According to the Gibbs equation, positive adsorption and the lowering of surface tension always appear simultaneously. When a fresh liquid surface is newly created, however, and before the excess solute molecules have had time to diffuse to the surface, the surface tension must remain high.

Lord Rayleigh showed, by means of a vibrating jet experiment, that about 5 milliseconds are required for the surface tension of a fresh surface to reach equilibrium. During this time, the tension continuously declines from a high initial value of about 70 dynes/centimeter to a final equilibrium value of about 35 dynes/centimeter. The cause of the stability of a foam film resides in this effect. Should, for any reason, the equilibrium surface layer be disturbed, fresh surface is created and the tension immediately increases; a difference in surface tension cannot, however, be sustained for long because of the mobility of the liquid, which flows in response to the higher tension toward the area in which it has appeared. The first response of the liquid is no doubt just at the surface, but a considerable quantity of the bulk liquid is dragged along to the area of high tension. The following simple experiment illustrates the effect. Pour a layer of water on to a thin metal plate, and touch the underside of the plate with a piece of ice. A high surface tension is created in the cold water just above the ice, and the motion of the surrounding liquid toward the colder area is noted immediately.

On a foam film, the stress that creates regions of higher surface tension is always present. The liquid film is flat at one place and curved convexly at another, where the liquid accumulates in the interstices between the bubbles. The convex curvature creates a capillary force that sucks liquid out of the connected foam films (Laplace effect), so that internal liquid flows constantly from the flatter to the more curved parts of the films. As the liquid flows, the films are stretched, new surface of higher tension is created, and a counter-flow across the surface is generated to restore the thinned-out parts of the films (Marangoni effect). In this way, the foam films are in a constant state of flow and counterflow, one effect creating the conditions for its reversal by the other. Pure liquids do not foam because of the absence of a Marangoni effect. See accompanying figure.

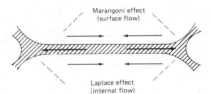

Dynamic equilibrium in a stable foam film. The Marangoni effect reverses the destructive action of the Laplace effect.

The Marangoni effect maintains the stability of the foam films even against other disruptive actions, such as hydrodynamic drainage, that, like the Laplace effect, cause stretching of the films. To the Marangoni effect can be traced all the resilient ability of foam films for elastic recovery after external mechanical shock. Foam films sometimes have this property to a remarkable degree. Lead shot, cork balls, mercury drops, and jets of water can be dropped through some foam films without causing rupture. In the fragility and brittleness of aged foams are seen the effects of impaired resilience, probably due to the extreme depletion of solution from old films by prolonged drainage.

While the primary stabilizing factor in foam is the resilience of the film, provided by the Marangoni effect, in special cases additional surface-layer phenomena are significant. These include gelatinous surface layers and low gas permeability. Such effects can add significantly to the stability of the foam, resulting in such relatively stable structures as meringue, whipped cream, fire-fighting foams, and shaving foams.

When a foam is first produced, as by bubbling a gas through a liquid, each bubble is a little sphere, separated from its neighbors by thick liquid partitions. But soon after the foam is formed, a large amount of liquid drains away by gravity and the spheres of gas become closer together. At this stage, there is a passage of gas from one bubble to another, through the curved liquid convexities that separate them.

Bubble Pressure. The pressure within a bubble of gas in liquid is greater than the pressure in the surrounding liquid by $2\gamma/R$, where γ is the surface tension and R is the bubble radius. If, as in a soap bubble, a bubble of gas is separated from the surrounding gas by a thin film of liquid, the pressure difference is twice this value.

In a foam, where γ is the same for every bubble, the pressure inside each bubble is inversely proportional to its size, i.e., the gas inside the smaller bubbles is at a higher pressure than the gas inside the larger bubbles. When, through drainage, the bubble wall becomes thin enough to be permeable, the gas in the smaller bubbles diffuses into adjacent larger bubbles to equalize the pressure. This spontaneous process increases the average bubble size without any coalescence of bubbles taking place by film rupture. The final, stable equilibrium product is a fragile, honeycomb structure, in which the separating films have plane surfaces. At this stage of the life-history of a foam, it is particularly vulnerable to external mechanical or thermal shocks, or air-borne contamination. Sir James Dewar kept plane soap films in a horizontal position inside a closed bottle for several months. In the open air, the bubbles would have ruptured almost instantly. An interesting treatise is "Science of Soap Films and Soap Bubbles," by C. Isenberg, Tieto Ltd., Clevedon, Avon, England (1979).

When comparing the foam stability of one solution with that of another, it is necessary to attend to the means by which foam is made, inasmuch as foams of quite different characteristics can be obtained from the same solution by different treatments. The area of interfacial surface, and the mechanical efficiency with which it is created, are the determining factors. The smaller the bubble, the more persistent the foam; and more foam can be produced if excessive agitation is avoided.

Different physical properties of foams are utilized in various industrial applications. In ore flotation, advantage is taken of the presence of an air-liquid interface and of the buoyancy of the bubble. Finely divided solid particles that are not wetted by aqueous solutions serves as the partitions between bubbles, and so rise with the foam. Thus, hydrophobic sulfide particles can be separated from hydrophilic silica. In fire-fighting foams, use is made of the ability of a foam to retain a noncombustible gas (such as carbon dioxide), thus preventing air from reaching the fire. Foams in which the liquid phase is solidified are useful because of their insulating property and the low density conferred by the gaseous phase. Solid foams can be flexible, as in sponges, because the foam cells have all been ruptured in the course of production. Cellulose sponges, foam rubber, and polyurethane foam are examples of this type. See also **Foamed Plastics.** Solid foams with closed cells are also made to form rigid structures.

In many industrial processes, excessive foaming of liquid causes waste and delays. Excessive foaming is particularly troublesome in the paper industry, the refining of beet sugar, the manufacture and use of glue, and numerous other examples are found in the food processing industry. The foaming can often be inhibited by the addition of an insoluble liquid that is able to spread spontaneously, by virtue of surface-tension forces, over the surface of the foam films even as they are being formed. The spreading of the insoluble droplet is so violent, and the spreading liquid drags along with it so much of the underlying film, that a hole is gouged in the film, which is thus destroyed.

Defoaming agents may be classified as solubilized surfactants, dispersions of hard particles, and dispersions of soft particles.

The classifications more often than not overlap. In all cases, a liquid nonaqueous vehicle is present, even where the defoamer is represented as a solid formulation. Water may also be present, particularly in emulsified silicone formulations.

FOAMED PLASTICS.   The great variety of uses for foamed rubber has in some measure brought about, and in some measure been accompanied, by an even greater development of foamed plastics. In fact, one of the most striking features of this development in plastics has been the wide variety of processes, and the consequent great range of products.

Foamed plastics range in density anywhere from $\frac{1}{10}$ to 65 pounds per cubic foot (1.6 to 1041 kilograms per cubic meter); they range in consistency from rigid materials suitable for structural use to flexible substances for soft cushions; they range in cellular formation from the open- or interconnecting-cell type to the closed- or unicell type. Their electrical, thermal, mechanical and chemical properties show a similar variation. Many types of foamed plastics may be produced "on the job"; this procedure often offers important production or construction advantages.

Many methods have been developed for the manufacture of foamed plastics, and void-containing plastics in general. For convenient discussion, they may be classified in three groups: (1) Methods for adding gas to the plastic mass during processing; (2) methods for producing gas in the plastic mass during processing; and (3) methods for forming a plastic mass from granules, and thus obtaining a cellular structure.

1. An obvious method of forming a foamed plastic is to whip air into the plastic mass before it sets. This method is used for unreaformaldehyde and polyvinyl-formaldehyde plastic foams. A related method is to introduce air (or some other gas or volatile solvent) into the plastic mass, and then to form pores of the desired size by expanding the gas bubbles by application of heat, or reduced pressure.

2. Methods for producing gas chemically during the reaction or reactions that produce the plastic are as almost varied as these reactions themseves. For example, a polyester resin and an aromatic diisocyanate react to form a resin prepolymer, which then reacts with water to form a urethane polymer (plastic). Since carbon dioxide gas is also formed in this reaction (which is a condensation), its presence causes the urethane resin to be cellular. Another example is a condensation which eliminates water from the reacting molecules. This process is used in making phenolformaldehyde resin foams, where the reaction is so exothermic (yields so much heat), that the water is produced as steam bubbles that expand the plastic. However, some reactions for forming plastics do not yield gaseous by-products; in such cases gases may be produced by adding to the materials a "blowing agent," that is, a substance which decomposes to form a gas, at a temperature below the gel temperature of the plastic. Substances so used include dinitroso compounds, such as dinitroso pentamethylenetetramine, and hydrazides, such as benzene sulfonyl hydrazide. These compounds evolve nitrogen gas, but inorganic blowing agents, such as bicarbonates which evolve carbon dioxide, have also been used. This technique is applied with many kinds of plastics, including polyethylene, silicone, epoxy, and vinyl resins.

3. An example of the granular type of foam-forming plastics is the polystyrene product furnished as small granules for this use. They expand on heating, and fuse together to form rigid unicellular materials.

FOLIC ACID.   Sometimes referred to as the antianemia factor or folacin and earlier called vitamin B_c, vitamin M, and the

L. casei factor, the chemical name for folic acid is pteroylglutamic acid. Most animals require folic acid. The substance is synthesized by bacteria in some vertebrates, including human, rat, dog, pig, and rabbit. Exogenous sources are required by most other vertebrates and invertebrates. In ruminants, synthesis of folic acid occurs in the rumen, but some researchers believe that newborn lambs require a dietary supplement. The most common manifestation of a deficiency in livestock is development of a characteristic macrocytic, hyperchromic anemia (also called megaloblastic anemia). Bone marrow changes, red cells are large and immature, usually with an accompanying reduction of white cell numbers. Folic acid deficiency in poultry retards growth. Other disorders include glossitis, diarrhea, gastrointestinal lesions, intestinal malabsorption, and sprue.

In 1931, Wills demonstrated a factor from yeast active in treating anemia. In 1938, Day et al. found yeast or liver extracts active in treating anemia in monkeys. Hogan and Parrot, in 1939, showed how anemia in chicks could be prevented by using liver extract. The *L. casei* growth factor was isolated from liver and yeast by Snell and Peterson in 1940. Hutchings et al., in 1941, found the *L. casei* factor also essential for chicks. Also, in 1941 Mitchell, Snell, and Williams isolated bacterial (*S. lactis R*) growth factor similar to *L. casei* factor from yeast and named the substance folic acid. Stokstad, in 1943, reported *L. casei* factor from liver more active than that from yeast; and provided evidence of multiple factors. Pterolymonoglutamic acid was finally isolated, the structure proved, and the substance synthesized by Angier et al. in 1946. Commercial production of folic acid is either by extraction from yeast or liver, or by synthesis wherein 2,3-dibromopropanol, 2,4,5-triamino-6-hydroxypyrimidine, and para-aminobenzoyl glutamic acid are reacted.

Folic acid and derivatives are involved biologically in the synthesis of nucleic acid, coenzyme in purine-pyrimidine metabolism, serine-glycine conversion, intermediate in metabolism of purines and pyrimidines, differentiation of embryonic nervous system, one-carbon transfer mechanisms, metabolism of tyrosine and histidine, formation of active formate and methionine, and the synthesis of choline. Antagonists of folic acid include aminopterin (4-aminopteroylglutamic acid), methotrexate (amethopterin), pyrimethamine, and 4-aminopteroylaspartic acid. Synergists include biotin, pantothenic acid, niacin, vitamins B_1, B_2, B_6, B_{12}, C, and E, somtatotrophin (growth hormone), and testosterone.

Folic acid coenzymes are derivatives of tetrahydrofolic acid. See also **Coenzyme.** Structurally, these are:

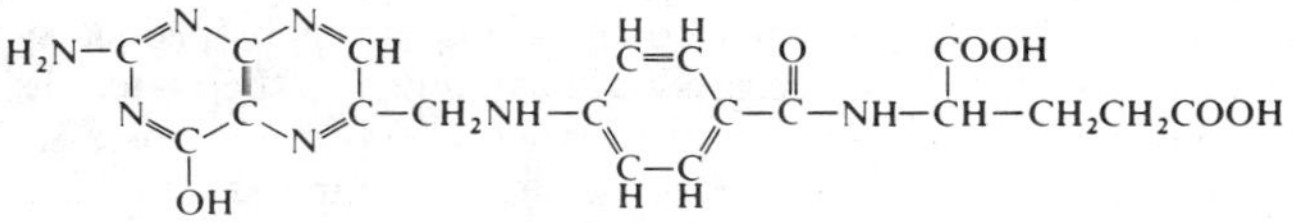

Folic acid

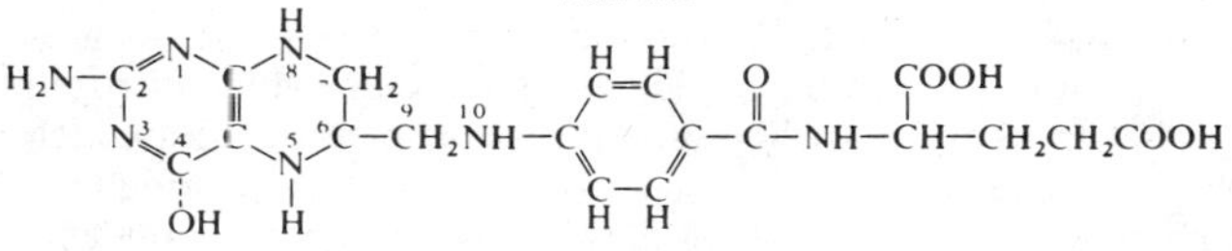

Tetrahydrofolic acid

One-carbon fragments in various oxidation states are: (1) formyl ($-CHO$); (2) hydroxy-methyl ($-CH_2OH$); and (3) methyl ($-CH_3$). The coenzyme forms of folic acid have one of these groups attached to either the N—5 or N—10 of tetrahydrofolic acid. One folic acid coenzyme, methyltetrahydrofolate (CH_3-FH_4) transfers its methyl group to homocysteine to yield methionine, in a reaction which also requires a vitamin B_{12} coenzyme:

$$HS-CH_2CH_2\overset{\overset{\displaystyle NH_2}{|}}{C}HCOOH + CH_3-FH_4$$
Homocysteine

$$\xrightarrow{B_{12}\ coenzyme} CH_3-S-CH_2CH_2\overset{\overset{\displaystyle NH_2}{|}}{C}HCOOH + FH_4$$
Methionine

As pointed out by Chen and Cooper (Division of Food Science and Nutrition, California State University, Northridge, California), the group of compounds denoted by the term *folacin* is a heterogeneous group of derivatives with a similar basic structure and biological function. Folic acid is the basic structural unit in these compounds. Other monoglutamate folates are formed when the pteridine moiety of this basic molecule is reduced or substituted in the 5-nitrogen or 10-nitrogen position. In addition, all of these monoglutamate folates may be transformed into polyglutamates of various length by the addition of glutamic acid residues to the basic molecule.

Distribution and Sources. In the biosynthesis of folic acid, paraminobenzoic acid, glutamic acid, and some substances yet to be identified serve as precursors. Paraminobenzoylglutamic acid is an intermediate in the synthesis. In plants, folic acid is produced within the leaves, seeds, cereal germ. Production also occurs in algae, fungi, and by bacteria, as in the intestines of species previously mentioned. The liver is the primary storage site.

Bioavailability of Folic Acid. Factors which cause a decrease in bioavailability include (1) high urinary excretion; (2) destruction by certain intestinal bacteria; (3) increased urinary excretion caused by vitamin C; (4) presence of sulfonamides which block intestinal synthesis; and (5) a decrease in absorption mechanisms. Increase in bioavailability can be provided by stimulating intestinal bacterial synthesis in certain species. No toxicity due to folic acid has been reported in humans.

Some of the unusual features of folic acid noted by investigators include: (1) Folic acid antagonists have been used in cancer therapy with temporary remissions; (2) folic acid occurs in chromosomes; (3) folic acid is distributed throughout cells; (4) folic acid is needed for the mitotic step metaphase → anaphase; (5) antibody formation decreased in folic acid deficiency; (6) folic acid shows choline-sparing effects; (7) folic acid is analgesic in humans—the pain threshold is increased; (8) it shows antisulfonamide effects; (9) enterohepatic circulation of folate has been observed; (10) folic acid is synthesized by psittacosis virus; (11) it is concentrated in spinal fluid.

References

Brown, G. M.: "Folic Acid" in "The Encyclopedia of Biochemistry" (R. J. Williams and E. M. Lansford, Jr., editors), Van Nostrand Reinhold, New York, 1967.

Chan, C., Shin, Y. S., and E. L. R. Stokstad: "Studies of Folic Acid Compounds in Nature. 3: Folic Acid Compounds in Cabbage," *Can. J. Biochem.*, **51**, 1617 (1973).

Chen, T. S. and R. G. Cooper: "Thermal Destruction of Folacin: Effect of Ascorbic Acid, Oxygen and Temperature," *J. Food Sci.*, **44**, 3, 713–716 (1979).

Considine, D. M., Editor: "Foods and Food Production Encyclopedia," Van Nostrand Reinhold, New York, 1982.

Cooper, R. G., Chen, T. S., and M. A. King: "Thermal Destruction of Folacin in Microwave and Conventional Heating," *J. Amer. Dietet. Ass.*, **73**, 406 (1978).

Jaenicke, L.: "Folic Acid Coenzymes," in "The Encyclopedia of Biochemistry," (R. J. Williams and E. M. Lansford, Jr., editors), Van Nostrand Reinhold, New York, 1967.

Huskisson, Y. J., and F. P. Retief: "Folate Content of Foods," *South Afr. Med. J.*, **44**, 12, 362 (1970).

Kutsky, R. J.: "Handbook of Vitamins and Hormones," Van Nostrand Reinhold, New York, 1973.

O'Broin, J. E., et al.: "Nutritional Stability of Various Naturally Occurring Monoglutamate Derivatives of Folic Acid," *Amer. J. Clin. Nutr.*, **28**, 5, 438 (1975).

Paine-Wilson, B., and T. S. Chen: "Thermal Destruction of Folacin: Effect of pH and Buffer Ions," *J. Food Sci.*, **44**, 717–722 (1979).

FORGING.   Iron Metals, Alloys, and Steels.

FORMALDEHYDE.

HCHO, formula weight 30.03, colorless gas with pungent odor, mp $-92°C$, bp $-21°C$, sp gr 0.815 (at $-20°C$). The gas is very soluble in H_2O, alcohol, and ether. Formaldehyde usually is produced and marketed as a 37% (weight) solution in water. From 3 to 15% methyl alcohol normally is added as a stabilizer to prevent paraformaldehyde formation. The commercial trend is to furnish a more concentrated product (up to 50% HCHO by weight) which contain as little as 0.5 to 1% methyl alcohol. The addition of special stabilizing agents and storage at elevated temperatures reduces the formation of paraformaldehyde.

Polymerized formaldehyde (trioxane) is a ring compound of anhydrous formaldehyde with the formula $(HCHO)_3$. See also **Acetal Resins.** Trioxane is a colorless crystalline solid with a pleasant odor, mp 62°C, bp 115°C, sp gr 1.17. This compound is used as a tanning agent and solvent and as a source of dry HCHO gas. Because trioxane ignites readily at 113°C and burns with an odorless, hot flame, it has been furnished in tablet form as a replacement for solidified alcohol in portable heating applications.

Paraformaldehyde $(CH_2O)_x$, sometimes called paraform, is an amorphous white powder and may be used for applications where an aqueous solution of HCHO may not be desirable. It finds use as an antiseptic and as a catalyst and hardener for certain synthetic resins. Formaldehyde gas also may be dissolved in methyl or butyl alcohol for applications where H_2O is undesirable.

Formaldehyde is a high-tonnage chemical and, in addition to the uses already mentioned, finds wide application in the manufacture of urea-formaldehyde resins (growing annually at a rate of about 6%), melamine-formaldehyde resins (growth rate of about 5%), and acetal resins (growth rate of about 10%). Formaldehyde also is used in the production of pentaerythritol which, in turn, is used in the manufacture of lubricant additives, resin esters, pentaerythritol tetranitrate, and alkyd resins. Formaldehyde also is required in the manufacture of hexamethylene tetramine, a compound important in explosives manufacture and as a resin-curing agent. Other uses for formaldehyde include the production of ethylene glycol, acrylic esters, ureaformaldehyde fertilizers, textile-treating agents, tetrahydrofuran for elastomeric fibers, trimethylol propane for urethanes, and as a solvent for synthetic and natural resins. The growing need for nitrilotriacetic acid (NTA) and isoprene, for which formaldehyde is required, will account for additional tonnage production in the near future.

Production. All major commercial processes for making formaldehyde initially yield an aqueous solution of HCHO. In over 90% of the installations, methyl alcohol is the chargestock. Other feedstocks uncommonly used include methane, hydrocarbon gases, and dimethyl ether. Those processes commencing with methyl alcohol are of two types: (1) the *silver-catalyzed* process in which formaldehyde results from a combination dehydrogenation–oxidation reaction: $CH_3OH \rightarrow HCHO + H_2 + \frac{1}{2}O_2 \rightarrow H_2O$. The first part of the two-step reaction is endothermic, the second part is exothermic; and (2) the *oxide-catalyzed* process in which methyl alcohol is directly oxidized: $CH_3OH + \frac{1}{2}O_2 \rightarrow HCHO + H_2O$. The latter is an exothermic reaction. A process representing the silver technology is shown in the accompanying flowsheet. The process essentially consists of two sections: (1) the synthesis portion which yields products containing from 3 to 15% methyl alcohol, and (2) the distillation portion which is required only where the methyl alcohol content of the final product must be low. In the oxide-catalyzed process, catalysts used generally are mixtures of oxides of molybdenum, iron, and vanadium.

Formaldehyde reacts with many chemicals in a marked manner, (1) with ammonio-silver nitrate (Tollen's solution), to form metallic silver, either as a black precipitate or as an adherentmirror film on glass, (2) with alkaline cupric solution (Fehling's solution), to form cuprous oxide, red to yellow precipitate, (3) with rosaniline (fuchsine, magenta) which has been decolorized by sulfurous acid (Schiff's solution), the pink color of rosaniline is restored, (4) with NaOH, yields methyl alcohol plus sodium formate, (5) with NH_4OH, when evaporated, yields hexamethylene tetramine "urotropine" $(CH_2)_6N_4$, white solid, mp 263°C, (6) with sodium or hydrogen peroxide in sodium hydroxide, yields sodium formate, (7) with manganese dioxide and H_2SO_4, forms methylal, dimethoxymethane $CH_2(OCH_3)_2$, colorless liquid, bp 42°C.

Formaldehyde gas, when cooled under certain conditions, yields trioxymethylene, metaformaldehyde $(CH_2O)_3$; formaldehyde solution, when evaporated, upon standing, or upon being subjected to low temperatures, yields paraformaldehyde $(CH_2O)_x$, white solid, from which formaldehyde is regenerated upon heating; dilute formaldehyde, in the presence of calcium hydroxide solution, yields a mixture of sugars called formose from which fructose $C_6H_{12}O_6$ has been prepared, suggesting the intermediate formation in nature of formaldehyde in the photosynthetic process of the conversion of carbon dioxide to sugars. Formaldehyde stands chemically between methyl alcohol on the one hand—to which it can be reduced—and formic acid on the other—to which it can be oxidized.

Formaldehyde is commonly detected by the Schiff test (above), and confirmed by the formation of a dimethyl derivative with a mp of 189°C.

—D. G. Sleeman, Davy McKee (Oil & Chemicals) Ltd., London, England.

FORMIC ACID.

HCOOH, formula weight 46.03, colorless liquid, mp 8.6°C, bp 100.8°C, sp gr 1.220. Sometimes referred to as methanoic acid or hydrogen carboxylic acid, this compound is miscible with H_2O, alcohol, or ether in all proportions. Formic acid occurs in some living organisms, such as nettles, ants, and caterpillars. Commercially, the compound is obtained from the black liquor of sulfite paper mills where it is present as sodium formate. In the laboratory, it may be prepared by reacting oxalic acid and glycerol at about 110°C, or by reacting solid lead formate with H_2S gas at about 100°C, yielding anhydrous formic acid. In the textile and leather industries, formic acid is used as a reducing agent, particularly in connection with the chrome dyeing of wool. Formic acid, in a reaction with glycerol at 220°C, is a source of allyl alcohol. Miscellaneous uses for formic acid have included brewing (acts as a fermentation assistant), electroplating for pH and redox adjustment, food preservative, germicide, and coagulant for rubber latex. The compound also is useful in the preparation of metallic formates and esters. The most important ester in methyl formate $HCOOCH_3$, a solvent for acrylic resins and cellulose esters as well as an intermediate for certain organic syntheses.

Formic acid solution reacts (1) with hydroxides, oxides, carbonates, to form formates, e.g., sodium formate, calcium formate, and with alcohols to form esters, (2) with silver of ammonio-silver nitrate to form metallic silver, (3) with ferric formate solution, upon heating, to form red precipitate of basic ferric formate, (4) with mercuric chloride solution to form mercurous chloride, white precipitate, (5) with permanganate (in the presence of dilute H_2SO_4) to form CO_2 and manganous salt solution. Formic acid causes painful wounds when it comes in contact with the skin. At 160°C, formic acid yields CO_2 plus H_2. When sodium formate is heated in vacuum at 300°C, H_2 and sodium oxalate are formed. With concentrated H_2SO_4 heated, sodium formate, or other formate, or formic acid, yields carbon monoxide gas plus water. Sodium formate is made by heating NaOH and carbon monoxide under pressure at 210°C.

FORMULA (Chemical). Chemical Composition; Chemical Formula.

FRACTIONAL DISTILLATION. Distillation.

FRANCIUM. Chemical element symbol Fr, at. no. 87, at. wt. 223 (mass number of the most stable isotope), periodic table group la, mp 26–28°C, bp 676–678°C, density, 2.4 g/cm³. To date, 22 isotopes of francium, with mass numbers ranging from 203 to 224, have been identified. All are radioactive. See **Radioactivity.** Although Mendeleev visualized that element 87 would occupy the bottom position among the alkali metals in his periodic classification, the discovery of francium did not occur until 1939 when it was confirmed by Marguerite Perey, a collaborator of Marie Curie. Many earlier attempts had been made, including the observations of J. A. Cranston in 1913, the search for the element in radioactive ores by O. Hahn and G. Hevesy in 1926, and efforts by M. C. Guében in 1932. Also, earlier investigations by S. J. Meyer, V. Hess, and F. Paneth in 1914, when they were studying the emissions by ^{227}Ac, contributed to the network of information that finally led to the firm identification of francium. See also **Chemical Elements.**

Studies indicate that no further isotopes of francium with half-lives longer than those already known should exist. Thus, among the first 101 elements of the periodic chart, francium is the most unstable. The short half-lives of the isotopes explain the difficulties in isolating and confirming the element and of learning more of the detailed chemistry of the element.

The electronic structure of francium consists of closed K, L, M, and N shells plus $5s^25p^65d^{10}6s^26p^67s^1$. Atomic radius: 2.8 Å.

The isotopes of francium that occur in nature are found in thorium and uranium ores, in which they are continually formed by disintegration chains. These start with ^{232}Th ($4n$ family), ^{237}Np ($4n + 1$ family), and ^{235}U ($4n + 3$ family). It is estimated that 1 ton of natural uranium contains about 3.8×10^{-3} g of ^{223}Fr and 10^{-17} g of ^{221}Fr. The separation of these isotopes from natural sources involves long and complex chemical procedures. Like the other alkaline elements, francium remains in solution when other elements are precipitated as carbonates, hydroxides, fluorides, chromates, sulfates, and sulfides. However, at a pH of 9, francium can be extracted from solution by nitrobenzene in the presence of sodium tetraphenylborate. Extraction from a very dilute sodium solution can be effected with dipicrylamine in nitrobenzene solution, the separation be-

ing much easier to effect than in the case of cesium. Francium can be separated from rubidium and cesium by chromatography on cation-exchange resins or on mineral exchangers.

The heavy isotopes of francium can be formed by irradiation of uranium or thorium by protons of high energy; the lighter isotopes can be obtained by nuclear reactions induced in gold, tellurium, or lead targets by heavy ions.

Because of the difficulties in obtaining any significant quantities of francium, use of the element is confined to scientific investigations. ^{223}Fr is used for the measurement of ^{227}Ac. Studies have shown that francium fixes itself in induced sarcomas in rats. Because of the short half-lives of ^{223}Fr and ^{212}Fr, which would cause no radiation risk to organisms, the property could become useful for the early diagnosis of certain kinds of cancers.

For references, see **Chemical Elements.**

FRANKLINITE. The mineral franklinite is a zinic-iron-manganese mineral whose formula may be written (Zn, Mn^{2+}, Fe^{2+}) $(Fe^{3+}, Mn^{3+})_2O_4$, but the composition varies considerably, in respect to the amounts of the several metals that may be present. Its isometric crystals have an octahedral habit; it may be coarse or finely granular or compact. It shows a parting parallel to the octahedron; fracture uneven; brittle; hardness, 5.5–6.5; specific gravity, 5–5.2; luster, usually metallic, occasionally dull; color, black; streak, brown to black; opaque; may be slightly magnetic. Only in one place in the world does franklinite occur in quantity: at Franklin Furnace, New Jersey, from whence it was named. Here there are two bodies of this mineral, which is used as a zinc ore, about 3 miles distant from each other. The franklinite is found in pre-Cambrian limestones that are associated with gneisses believed to be of igneous origin and responsible for the mineralization. Associated minerals are willemite, zinc silicate, and zincite, zinc oxide, manganoan calcite.

FREE ENERGY AND FREE ENERGY FUNCTIONS. The *free energy* of a system is a precise thermodynamic quantity which is used to predict the maximum work obtainable from a spontaneous transformation of the system. Furthermore it provides a criterion for spontaneity of a transformation (reaction) occurring and predicts the maximum extent to which the transformation can occur (maximum yield).

In order to understand the nature of the "free energy" it is useful to derive it from the internal energy and the entropy of a system. Transformation of a system can be brought about by thermal input (temperature change) or by mechanical work (volume-pressure change). In a cycle such that the system is alternately subject to a thermal change, $\bar{d}Q$, and a mechanical change, $\bar{d}W$, whereby after each cycle the initial state of the system is restored, the first law of thermodynamics postulates that:

$$\oint(\bar{d}Q - \bar{d}W) = 0 \qquad (1)$$

The quantity under the cyclic integral must be the differential of some property of state of the system. This property is called the *Energy, E,* (sometimes also internal energy) of the system. Hence

$$dE = \bar{d}Q - \bar{d}W \qquad (2)$$

which integrates to

$$\Delta E = Q - W \qquad (3)$$

Only a difference in energy has been defined by Eq. 3 since one cannot assign an absolute value of energy to any state. It should be pointed out that dE is an exact differential, in contrast to $\bar{d}Q$ and $\bar{d}W$, which are path dependent.

If one considers, for example, the expansion of a gas at constant pressure, one finds readily that the work gained can be expressed as:

$$\Delta W = p_2 V_2 - p_1 V_1 \qquad (4)$$

Since a temperature change during the expansion has not been excluded Eq. 3 can be writeen as

$$(E_2 + p_2 V_2) - (E_1 + p_1 V_1) = Q_p \qquad (5)$$

The function $E + pV$ is again a state variable and by definition called the enthalpy of the system:

$$H = E + pV \qquad (6)$$

Both E and H have important properties a discussion of which would exceed the scope of this paragraph. Suffice it to note here that:

$$\left(\frac{dE}{dT}\right)_{v\,=\,const} = C_v \quad \text{and} \quad \left(\frac{dH}{dT}\right)_p = C_p \qquad (7)$$

where C_v and C_p are the heat capacities for constant volume and constant pressure processes, respectively.

Again as in the case of the internal energy of a system it is not possible to assign an absolute value to the enthalpy. Hence it is customary to make a conventional assignment to the values of the enthalpy of the elements at a reference state (see later).

Just as the first law of thermodynamics leads to the definition of the energy of a system, the second law of thermodynamics leads to the definition of the entropy. It can be shown that for a Carnot cycle, or in fact for any cyclic engine, the cycle integral

$$\int \frac{\bar{d}Q}{T} = 0 \qquad (8)$$

provided all transformations were carried out reversibly. Since the sum over the cycle of the quantity $\bar{d}Q/T$ is zero, this quantity is the differential of some property of state. This property is called the *entropy* and is given the symbol S. Again, discussing the many properties of S would exceed the scope of this paragraph. Suffice it to say that in any real transformation in an isolated system S always increases, or:

$$dS > \frac{dQ_{\text{irreversible}}}{T} \qquad (9)$$

This prompted Clausius' famous aphorism that "the energy of the Universe is constant, the entropy strives to reach a maximum." Therefore the condition of equilibrium in an isolated system is that the entropy have a maximum value. On this basis then it is possible to formulate the various free energy relationships and to state the conditions for equilibrium or spontaneity of a reaction.

The basic condition for equilbrium or spontaneity is

$$TdS \geqq \bar{d}Q \qquad (10)$$

where the equal sign stands for equilibrium condition. Using the above equations one can write:

$$-dE - P_{op}dV - \bar{d}U + TdS \geqq 0 \qquad (11)$$

where $P_{op}dV + \bar{d}U$ stands for the entire work ($\bar{d}W$) the system is capable of delivering. This fundamental relationship permits one to very rapidly formulate the three major energy relationships one unusually deals with in practical situations.

In an *isolated system* $dE = O$, $\bar{d}W = O$, $\bar{d}Q = O$ per definition, hence

$$dS \geqq 0 \qquad (12)$$

This means that a reaction in an isolated system can take place only if the entropy change of the system is positive. Equilibrium is attained when the entropy has reached a maximum.

If the *temperature* of the system remains *constant* during the transformation, then $\bar{d}Q = O$ and:

$$-d(E - TS) \geqq \bar{d}W \qquad (14)$$

The combination of variables E-TS is given the symbol A and is often called the "work function" of the system. However it is really the maximum amount of work (free energy) obtainable from the system under isothermal, reversible conditions.

Under constant pressure conditions Eq. 11 converts into:

$$-d(E + pV - TS) \geqq \bar{d}U \qquad (15)$$

The combination $E + pV - TS$ is given the symbol G and is called the Gibbs free energy. (Often G is given other names and other symbols such as F. Therefore, when looking up free energy values in tables, one should always reassure oneself of the exact definition of this value.) It is good to remember, that in a certain sense S/T, A and G are all free energy terms although only G is specifically given this name. Finally by way of summarizing, the general relationship between G and A can be established:

$$G = E + pV - TS = H - TS = A - pV \qquad (16)$$

Free Energy Functions

All the above correlations would merely be of only theoretical interest if the various properties of state could not be evaluated numerically. Much of the research in the field of thermodynamics has been devoted to this problem. Originally properties of state such as the enthalpy were linked to easily measurable properties such as the heat capacity. Later however, an effort was made to arrive at more fundamental relationships.

The differential enthalpy can be written in the following manner:

$$dH = \left(\frac{\partial H}{\partial T}\right)_p dT + \left(\frac{\partial H}{\partial T}\right)_T dp \qquad (17)$$

and per definition

$$dH = C_p dT + \left(\frac{\partial H}{\partial T}\right)_T dp \qquad (18)$$

from where it follows that for any constant-pressure transformation the enthalpy differential is simply:

$$dH = C_p dT \qquad (19)$$

Integrated for a transformation from condition 1 to condition 2 we obtain:

$$\Delta H = C_p \Delta T \qquad (20)$$

Eq. 20 permits the calculation of enthalpy differences between two states by means of experimentally readily available data. In order to facilitate this job, an enthalpy standard was defined. The enthalpy at 298.15 °K, $H°$, for elements is zero. For compounds $H°$ then becomes the "heat of formation," and is tabulated in many reference books and tables. In order to calculate the enthalpy at a given temperature Eq. 20 can now be written as:

$$H_T = H° + C_p(T - 298.15) \qquad (21)$$

This equation is correct only if C_p is not a function of temperature, which is rarely the case. Early attempts to express C_p and C_v as temperature functions took the forms of power series:

$$C_p = A + bT - cT^2 + dT^3 \ldots \ldots \quad (22)$$

with the coefficients A, b, c and d having been determined experimentally, and tabulated in standard reference works.

Similarly, an energy function for the entropy can be written and evaluated:

$$S_T = S° + \int_0^T \frac{C_p}{T} dT \quad (23)$$

Being able to evaluate both the enthalpy and the entropy at any desired temperature enables one automatically also to evaluate the Gibbs free energy, G.

More modern approaches have been opened by statistical thermodynamics. If the energy levels of molecules composing a system can be obtained by solving the Schroedinger equation, the partition functions can be evaluated. This in turn permits one to arrive at any thermodynamic property.

Solutions of the Schroedinger equation are difficult and involve a number of approximations which may cast doubt on the validity of the result. The actual values of the energy levels may, however, be obtained experimentally from an analysis of the spectrum of the molecule. Again, inserting these experimental energy values into the partition function provides an approach to calculate any thermodynamic property of interest. This approach is particularly useful for high temperatures, since it may be much easier to determine a spectrum at 3000 or 4000°F than measure the heat capacity at these elevated temperatures. Actually a large amount of the thermodynamic data available now has been determined from spectral information.

—R. H. Hausler, UOP Inc., Des Plaines, Illinois.

FREE RADICAL. Although free radicals have been defined as highly reactive groups of atoms containing unpaired electrons, such a definition is imprecise, as it would include ions, such as those of the lanthanide and actinide series, which not only possess such unpaired electrons, but also—because of this—exhibit the color, magnetic and other characteristics of free radicals so defined. The term is, therefore, to be reserved for short-lived alkyl radicals possessing a magnetically noncompensated electron, or somewhat longer-lived, larger organic molecules of the aryl-alkyl type similarly possessing unpaired electrons in their valence shell.

A few very reactive inorganic radicals, such as $NO\cdot$, $ClO_2\cdot$, or $NO_2\cdot$, may also be construed as "free," but knowledge of the chemistry of free radicals is best exemplified by unsaturated organic fragments, such as $CH_3\cdot$, $C_6H_5\cdot$, among others. These free radical fragments are formed by breaking one or more bonds in a stable molecule by photolysis, electrolysis, pyrolysis, and some other processes. The first free radical to be synthesized was that of triphenylmethyl by Moses Gomberg in 1900. Paneth and Hofeditz, in 1926, found free radicals in the pyrolysis of lead tetramethyl, and Rice subsequently demonstrated their existence in the breakdown products of many organic compounds.

Because of the affinity of their unpaired electrons, free radicals have short lives, tend to dimerize and thus lose their reactivity. Because of their generally short half-lives (1–100 milliseconds), detection and identification of these entities is essentially through spectrophotometric methods. However, in solid systems, free radicals can be trapped for appreciable lengths of time and at least one of these, 2,2-diphenyl-1-picrylhydrazyl, has such a long half-life that it is sold as such for the photometric determination of tocopherol.

Free radicals formed by photolysis play significant roles in the chemistry of the earth's atmosphere, not the least of which is the destruction of ultraviolet protective ozone by chlorofluorocarbons from anthropogenic sources. But, the most important reactions involving free radicals as intermediates in organic chemistry are polymerizations, such as may develop from a free radical and ethylene. These may also be ensured through use of substances known to produce free radicals, e.g., benzoyl peroxide, of which the benzoyl free radical can initiate a chain reaction which may continue indefinitely; or, it may lose CO_2 itself to give a new free radical—phenyl—which is itself capable of polymerization.

A free radical chain reaction proceeds through a succession of free radicals. In the photochemical chlorination of an alkane, the initiating step is the homolytic fission of chlorine molecules to produce chloroalkane molecules and chlorine free radicals. These two reactions constitute the propagating step. However, the chlorine free radicals may also combine to form chlorine molecules or react with the alkane free radicals to form chloroalkane molecules. Both of these reactions constitute terminating steps of the chain reaction. It should be noted, however, that the foregoing sequence cannot take place in the dark. Exposure to light allows the series of reactions then to proceed rather violently.

—R. C. Vickery, Hudson Laboratories, Hudson, Florida.
EDITOR'S NOTE: The biological roles of free radicals are described in a series of books: Pryor, W. A., Editor: "Free Radicals in Biology," Vol. 4 (latest), Academic, New York (1980).

FREE VOLUME. A liquid differs from a solid having the same type of packing of the molecule in having a certain additional volume, the *free volume*, which provides the necessary looseness in the structure to permit free movement of the molecules. The concept of free volume is used in several theories of the liquid state.

FREEZE-CONCENTRATING. In lieu of evaporation and other means of concentrating liquid substances, notably foods, freeze-concentrating may be used, particularly where it is desirable to retain volatile constituents, as in the instances of increasing the alcohol content of wines and to prepare and preserve flavor, as in the case of orange juice or coffee extract concentrates. From an energy standpoint, the energy required by freeze-concentrating is considerably less than that used by evaporative systems.

There are difficulties associated with the achievement of high concentrations through freezing. The liquid viscosity may increase so much as concentration increases and freezing point drops that difficulty in handling the ice-concentrate mixture and of separating the concentrate from the ice will arise. Improvements in recent years have been made by the use of ripening-induced growth of large ice crystals through sacrificial melting of small, easily-formed subcritical ice crystals; and the development of wash columns which provide efficient solute recovery from the ice-concentrate mixture. Even then, it is estimated that 35–50% dissolved solids represents the maximum concentration that can be practically achieved by freeze-concentration. By comparison, in two other attractive concentrating methods the liquid food concentrations are even less: 20–35% by reverse osmosis, and 20–30% by ultrafiltration. However, these latter processes, when considered in conjunction with (prior to) evaporation may be attractive from an energy-expenditure standpoint. A proprietary freeze-concentration system is depicted in accompanying diagram.

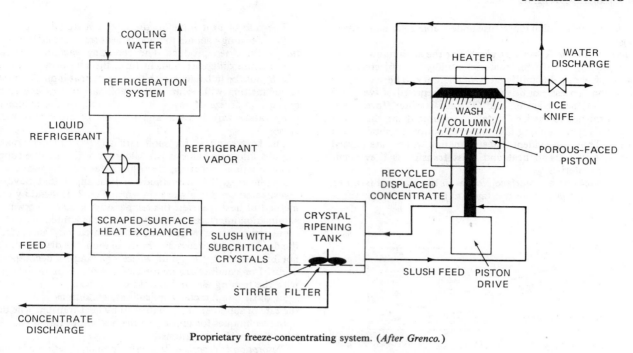

Proprietary freeze-concentrating system. (*After Grenco.*)

References

Leninger, H. A., and W. A. Beverloo: "Food Process Engineering," Reidel, Boston (1975).

Schwartzberg, H. G.: "Energy Requirements for Liquid Food Concentration," *Food Technology*, **32**, 3, 67–76 (1977).

Tjijssen, H. A. C.: "Freeze Concentration," in "Advances in Preconcentration and Dehydration of Foods," (A. Spicer, Editor), Wiley, New York (1974).

FREEZE-DRYING. A process for removing moisture from a wet material by bringing the material to the solid state and subsequently subliming it. This process is used for drying and preserving a number of products, notably food products, including instant coffee, vegetables, fruit juices, and meats. The needs of the food industry, coupled with those of pharmaceutical manufacturers, accelerated research into this process several years ago and commercial applications of freeze drying are now commonplace in these industries.

The wet material in the form of a wet solid or in the form of a suspension or solution is frozen under vacuum or at atmospheric pressure, followed by transforming the ice into vapor and removing it. In the usual case, the dried material remaining will be a spongy mass of about the same size and shape as the original frozen mass, and frequently will be found to have excellent stability and convenient reconstitution when placed in cold water, and will maintain flavor and texture sometimes indistinguishable from the original materials. These properties differ markedly with various materials. Some products are much better adapted to the process than others.

Usually materials to be freeze-dried are complex mixtures of water and several other substances. When such materials are cooled below 0°C (32°F), pure ice crystals will separate out first. With further cooling, the mass will become rigid as the result of formation of eutectics. (A eutectic is that particular mixture out of a possible combination of two or more mixtures of materials that has the lowest melting point.) See Fig. 1. Most

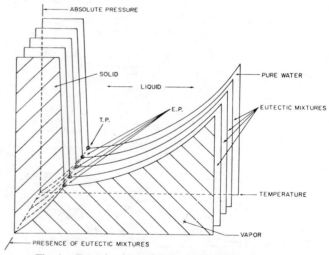

Fig. 1. Eutectic phase diagram in freeze-drying process.

food products and biologicals solidify completely at a temperature in the range of −15 to −73°C (−5 to −100°F). At solidification of the entire mass, all of the free water has been transformed into ice. Only a small quantity of the original water, the bound water, remains fixed in the internal structure of the material.

The quality of the finished product as well as the rate of drying will be affected by the size, shape, and size distribution of the ice crystals which form during freezing. These properties also will be affected by the homogeneity of the frozen mass. Thus, freezing must be effected under carefully controlled conditions (time, pressure, and temperature). Large ice crystals result from slow freezing rates. These may be injurious to certain substances. On the other hand, too rapid freezing results in small ice crystals, which may cause undesirable color and texture changes.

Sublimation (or Primary Drying). For the sublimation phase of the process, the frozen material usually is subjected to a vacuum of about 4.6 millimeters of mercury. The ice-crystal sublimation process can be regarded as comprised of two basic processes: (1) heat transfer, and (2) mass transfer. In essence, heat is furnished to the ice crystals to sublime them; the generated water vapor resulting is transferred out of the sublimation interface. Thus, it is evident that sublimation will be rate-limited by both resistances to heat and mass transfer as they occur within the material.

As the sublimation interface recedes in the material (see Fig. 2), the dry layer presents a resistance to the flow of water vapor

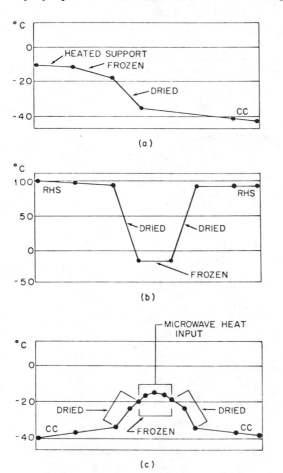

Fig. 2. Heat-input methods for freeze-drying processes: (a) conduction; (b) radiation; (c) microwave. CC = cold condenser; RHS = radiant-heat device.

and a pressure difference must exist between the ice interface and the surface of the dry layer. A large pressure difference will facilitate high mass-transfer rates; however, the maximum allowable sublimation temperature at which no melting will occur and the cost of the vacuum equipment restrict this driving force to a limited range. In practice, the maximum allowable temperature and corresponding presure at the sublimation interface is in the range of +15 to −40°F (−9.4 to −40°C) and 2000–100 micrometers, respectively.

The rate of heat input to the frozen material is a function of the operating-vacuum method of heat transfer and the properties of the dried product. The operating vacuum determines the pressure difference and, in turn, the rate of mass transfer, which must be in balance with the rate of heat input. Otherwise, either melting will occur at the sublimation interface and the purpose of freeze-drying will be defeated; or the sublimation temperature will decrease and the cost of processing will increase.

The heat required for sublimation (1200 Btu per pound of ice; 664 kilogram-calories per kilogram of ice) can be supplied by conduction, radiation, electric resistance, microwave, or infrared heating. Three methods of heat input that have been investigated extensively are shown in Fig. 2. Depending on the method of heat transfer, the temperature gradient between the sublimation interface and the heat source is limited by the maximum temperature which can be tolerated on the surface of the dry layer or frozen mass. For radiation, the dry layer should not be heated to the point where charring or decomposition occur. For conduction, melting of the frozen mass in contact with the heating element should be avoided.

In most commercial applications, conditions are such that the rate of sublimation is controlled by heat transfer. The developing techniques for improving the heat-input rate is the objective of many investigations.

Desorption (Secondary Drying). Upon completion of sublimation of the ice crystals, final dehydration is carried out to remove the bound water which did not crystallize out during freezing and is bound by adsorption phenomena to the dried product. The product temperature is increased to 80–120°F (26.7–49°C), and under high vacuum, the bound water and oxygen are removed from the dried product. The rate of desorption is considerably slower than sublimation. Although the bound water is only 5–10% of the total water in many substances, the secondary drying may require up to 35% of the total drying time.

Drying Rates. Drying a frozen material proceeds initially at a constant rate with rapid evolution of water vapor. As the sublimation interface recedes within the product, water-vapor evolution decreases. This is the start of the falling-rate period. When only bound water remains within the cellular structure of the product, the desorption period begins. During the constant-rate period, the sublimation rate can be expressed in terms of the heat of sublimation of ice and the heat-rate equation:

$$\text{Rate of sublimation} = \frac{UA\Delta T}{\Delta H_{\text{ice}}}.$$

The overall heat-transfer coefficient U depends upon the properties of the dry product and the method of heat transfer. The heat-transfer rate A is influenced by the mechanical design of the heating elements and the conditioning of the frozen mass. The temperature gradient ΔT is limited by the maximum allowable temperatures at the sublimation interface and dry-layer surface. In the constant-rate period, the first $\frac{1}{2}$ to $\frac{2}{3}$ of the drying cycle, about 80% of the water is removed.

Processes and Equipment

In addition to the three fundamental operations just described, the freeze-drying process involves several other operations necessary to achieve an economically feasible system for large-scale production. The general commercial process comprises: (1) preparation of the material; (2) freezing; (3) conditioning of the frozen mass; (4) drying, that is, sublimation and desorption; and (5) conditioning the product.

References

Ang, T. K., Ford, J. D., and D. C. T. Pei: "Microwave Freeze-drying of Food—A Theoretical Investigation," *Int. J. Heat Mass Transfer*, 20, 517 (1977).

Ang, T. K., Ford, J. D., and D. C. T. Pei: "Optimal Modes of Operation for Microwave Freeze-drying of Food," *J. Food Sci.*, 43, 2, 648–649 (1978).

Considine, D. M., Editor: "Foods and Food Production Encyclopedia," Van Nostrand Reinhold, New York, 1982.

Flink, J. M.: "Energy Analysis in Dehydration Processes," *Food Technology*, 37, 3, 77–84 (1977).

Flink, J. J.: "A Simplified Cost Comparison of Freeze-Dried Food with Its Canned and Frozen Counterparts," *Food Technology*, 31, 4, 50–56 (1977).

Porter, W. L., Levasseur, L. A., and A. S. Henick: "Evaluation of Some Natural and Synthetic Phenolic Antioxidants in Linoleic Acid Monolayers on Silica" (related to rancidity of whole tissue, freeze-dried foods), *J. Food Sci.*, 42, 1533–1535 (1977).

Zarkarian, J. A., and C. J. King: "Asymmetry in Freeze-Drying," *J. Food Sci.*, 43, 3, 992–997 (1978); "Acceleration of Limited Freeze-Drying in Conventional Dryers," ibid., pages 998–1001.

FREEZE-PRESERVING.

Knowledge of the fact that food substances remain edible for longer periods of time when cooled probably dates back to antiquity, centuries before the process for making ice was developed. Ice-making led to cold storage, a practice that persisted for several decades and which, of course, remains useful for a number of products today, especially in cerregions. Cold storage was first limited by the minimum achievable temperature, dictated by the melting point of ice. Chemicals to depress the freezing point were an additional step toward cold-preservation. Generally accredited with the initial breakthrough from cold-storage practices to present freezing technology was the step to quick-frozen foods, pioneered by Clarence Birdseye, among others, in the late 1920s. Consumers began to accept the fact that fresh, high-quality food when frozen quickly and retained at a temperature of about $-17.8°C$ ($0°F$) was a good substitute for fresh produce out of season. Frozen and stored in this way, these foods represented a marked improvement over the earlier available, slowly cooled food products.

Freeze-preserving is an across-the-board operation in the food industry of countries with advanced technology. There are relatively few foods—vegetables, fruits, fish, poultry, and meat—that cannot be frozen with reasonable success.

Fundamentals of Freezing

In food materials, water is the major component.[1] Thus, when foods are cooled below $0°C$, ice formation occurs, starting at a temperature between 0 and $-3°C$ (32 and $26.6°F$), which depends upon the molar concentration of soluble cell components. As the temperature is progressively reduced, more and more water is turned into ice and the latent heat of ice formation adds to the sensible heat involved in cooling both ice and the unfrozen portion. This leads to large variations in heat capacities while thermal conductivities also change considerably, mainly because the thermal conductivity coefficient of ice is nearly 4 times greater than that of water. For most biological materials, the largest part of the freezing process takes place in a temperature interval between -1 and $-8°C$ (30.2 and $17.6°F$), while the largest variations of heat capacity occur between -1 and $-3°C$ (30.2 and $26.6°F$). Only at temperatures ranging from

[1] Paragraph summary prepared by L. Rebellato, S. Del Giudice, and G. Comini (see reference list).

-20 to $-40°C$ (-4 to $-40°F$) and below, there is no more measurable change with temperature in the amount of ice present, and the remaining water, if any, can be considered as nonfreezable. However, for practical purposes, a lower limit to the phase-change interval can be defined on the basis of a ratio of ice to total water content of, say, 90%. This choice, in addition to providing an easily applicable criterion, allows one to approximate heat capacity and thermal conductivity curves, above and below the phase-change zone, by means of constant values. These techniques are described by Bonacina et al. (1974).

Rebellato et al. (1978) have developed a finite-element-analysis approach to freezing processes in foodstuffs and apply this method to the computation of temperature distributions in foodstuffs of irregular shape during freezing in an air-blast tunnel.

Wide Range of Freezing Configurations

The food processor has several options available when selecting the best freezing format for a given set of product characteristics and marketing objectives. Methods can be classified in several ways, as for example the medium used to contact and extract heat from the food substance—air or other gases, liquids, or mechanical contact.

Air-blast systems commonly take the form of large rooms, tunnels, or cells. In a room, the air velocity may be low or range up to 1500 feet (457 meters) per minute. The temperature for air-blast freezing usually ranges from -29 to $-40°C$ (-20 to $-40°F$). Blast freezing requires longer than other available methods and product quality cannot be assured unless very efficient insulation is used.

Plate freezers generally are limited in application to prepackaged products. In this system, the food substance is placed in direct contact with refrigerated metal plates (usually steel or aluminum). Cooling coils are located within the interior of the metal plates. The required contact refrigeration time ranges from about 30 to 90 minutes, depending upon size and nature of food substances.

Liquid-immersion freezing has grown in acceptance during the past few years. This system requires placing the product in a bath of cooling liquid, which must be nontoxic, noncorrosive, and have a low freezing point, low viscosity, and high thermal conductivity. Wrapping of the product is required in many cases. Salt solutions and propylene glycol are frequently used. Advantages over air-blast freezing include operational energy savings that result from the high heat-transfer coefficients and high heat capacities.

Cryogenic freezing has gained wide acceptance during recent years. Because cryogenic freezing is very fast and accomplished at extremely low temperatures (down to $-196°C$; $-320°F$), less dehydration occurs. Problems of cell damage, caused by sharp ice crystals formed during slower freezing processes, are largely overcome with short freezing times. Also, the sooner a product is deeply frozen, the sooner will be the halting of bacterial and enzyme degradation.

For cryogenic freezing, nitrogen is used in several forms—as a shower of liquid droplets, as a liquid bath for direct immersion, or as a cold gas. Carbon dioxide is used as a liquid or in solid "snow" form. When used in a tunnel for IQF applications, liquid carbon dioxide can freeze product at a temperature from -62 to $-78°C$ (-80 to $-109°F$). Fluorocarbons and halocarbons also have been used in conjunction with tunnel and spiral-type freezers that are used in IQF methods. The advantages of rapid freezing by cryogenic systems, including reduced tissue damage and improved quality, have been summarized many times, including reports by Meryman (1956).

References

Bonacina, C., et al.: "On the Estimation of Thermophysical Properties in Nonlinear Heat-Conduction Problems with Special Reference to Phase Change," *International Jrnl. of Heat Mass Transfer*, **17**, 861 (1974).

Rebellato, L., et al.: "Finite Element Analysis of Freezing Processes in Foodstuffs," *Jrnl. of Food Sci.*, **43**, 1, 239–243 (1978).

FREEZING-POINT DEPRESSION. The freezing point of a solution is, in general, lower than that of the pure solvent and the depression is proportional to the active mass of the solute. For dilute (ideal) solutions

$$\Delta T = Km$$

where ΔT is the lowering of the freezing point, K, the *cryoscopic constant* for the given solvent, and *m*, the molality of the solution.

There are several methods for measuring this depression: In the *Beckmann Method* the freezing point of pure solvent and that of solution is measured by a special type of thermometer. See BECKMANN METHOD.

In the *equilibrium method* a relatively large amount of solvent crystals are allowed to form and the system allowed to come to equilibrium. The temperature is recorded and some of the solution withdrawn and analyzed.

See also **Cryoscopic Constant.**

FRIEDEL-CRAFTS REACTION. Aluminum chloride anhydrous, introduced by Friedel and Crafts, is used as reagent, generally in CS_2 solution to avoid rise in temperature, for the preparation of (1) arylalkyl hydrocarbons, (2) di- and triphenyl-methane and derivatives, and (3) aryl-alkyl and diaryl ketones. Other chlorides, such as those of zinc, iron(III), and tin(IV), are often effective in certain cases.

1. Aryl-alkyl hydrocarbons. The reaction takes place between benzene or its homologues and the alkyl haloid, thus:

$$C_6H_5 \cdot H + Cl \cdot CH_3 \longrightarrow C_6H_5 \cdot CH_3 + HCl$$

Benzene Methyl chloride   Toluene Hydrogen chloride gas evolved

$$C_6H_4 \Big\langle {}^{H}_{H} + {}^{Cl \cdot CH_3}_{Cl \cdot CH_3} \Big\} \longrightarrow C_6H_4(CH_3)_2 + HCl$$

Benzene Methyl chloride   Xylene

2. Di- and triphenylmethane, derivatives. The reaction takes place between benzene and benzyl haloid or methylene haloid in the case of diphenylmethane, and between benzene and benzal haloid or chloroform in the case of triphenylmethane, thus:

$$C_6H_5CH_2Cl + HC_6H_5 \longrightarrow C_6H_5CH_2C_6H_5 + HCl$$

Benzyl chloride Benzene   Diphenylmethane

$$H_2CCl_2 + {}^{HC_6H_5}_{HC_6H_5} \Big\} \longrightarrow C_6H_5CH_2C_6H_5 + {}^{HCl}_{HCl}$$

Methylene chloride Benzene   Diphenylmethane

$$C_6H_5CHCl_2 + {}^{HC_6H_5}_{HC_6H_5} \Big\} \longrightarrow C_6H_5CH \Big\langle {}^{C_6H_5}_{C_6H_5} + {}^{HCl}_{HCl}$$

Benzal chloride Benzene   Triphenylmethane

$$HCCl_3 + HC_6H_5 \cdot {}^{HC_6H_5}_{HC_6H_5} \Big\} \longrightarrow C_6H_5CH \Big\langle {}^{C_6H_5}_{C_6H_5} + HCl$$

Chloroform Benzene   Triphenylmethane

3. Ketones. The reaction takes place between benzene and paraffin or benzenoid acyl haloid thus:

$$CH_3COCl + HC_6H_5 \longrightarrow C_6H_5COCH_3 + HCl$$

Acetyl chloride Benzene   Acetophenone

$$C_6H_5COCl + HC_6H_5 \longrightarrow C_6H_5COC_6H_5 + HCl$$

Benzoyl chloride Benzene   Benzophenone

The keto-group occupies the position para to alkyl already present. Two acyl groups have been placed in mesitylene to form diacetylmesitylene:

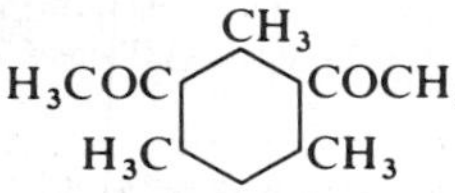

Summarizing: benzene or its homologues plus paraffin-substituted haloid in the presence of aluminum chloride anhydrous react with the elimination of hydrogen chloride. In several cases an intermediate compound of the reactants with aluminum chloride has been identified.

Other reactions involving aluminum chloride anhydrous are:

1. Xylene plus benzene to yield toluene, and the reverse, namely, toluene to yield xylene plus benzene. Boiling temperature.

2. Benzene, toluene and homologs chlorinated by reaction with chlorine gas.

3. Benzene sulfinated by reaction with SO_2. Benzene sulfinic acid $C_6H_5 \cdot SOOH$ formed.

FROTHING AGENTS. Classifying (Process).

FRUCTOSE. Carbohydrates; Sweeteners.

FUEL CELLS. The fuel cell is an electrochemical device which directly combines hydrogen and oxygen from air to produce electricity and water. With prior processing, a wide range of fuels, including natural gas and coal-derived synthetic fuels, can be converted to electric power. The basic process is highly efficient, pollution-free, and since single fuel cells can be assembled in stacks of varying sizes, systems can be designed to produce a wide range of output levels and thus satisfy numerous kinds of applications.

The concept of the fuel cell has been known for well over a century, but was not utilized in a practical sense until advent of the various space-exploration programs. Cells developed by Bacon (1960) in the late 1930s provided the basis for the Apollo power plant and led to a sophisticated technology of cells using alkaline electrolyte and pure hydrogen and oxygen as reactants.

The adaptation of this technology to the seemingly endless possibilities in terrestrial applications proved to be difficult because operation on hydrogen of lesser purity and air rather than oxygen presented many problems. These shortcomings of the early alkaline cell were overcome by new cell types of which the phosphoric acid cell is the most advanced.

Operating Principles of Fuel Cells

As an energy conversion device, the fuel cell is distinguished from a conventional battery by the fact that the electrodes are invariable and catalytically active. Current is generated by reaction on the electrode surfaces which are in contact with an electrolyte. As a rule, fuel and oxidant are supplied as required by the current load; water is continuously removed.

Single Cell. Under load, the voltage of one individual fuel cell element is less than one volt. Therefore, the assembly of many cells, connected in series as a stack, is necessary. Each

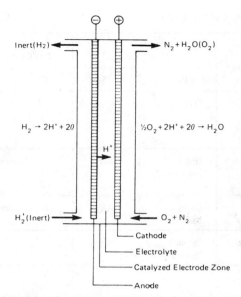

Fig. 1.   Principles of operation of the hydrogen-air cell with acid electrolyte. Product water is removed by the flowing air.

individual cell contains the elements needed for feeding reactants to the electrode surface, and removal of water from the cell, as shown schematically in Fig. 1 for a hydrogen-air cell with acid electrolyte.

The electrode reactions are comprised of the oxidation of hydrogen on the anode (the negative electrode) to hydrated protons with the release of electrons; and on the cathode the reaction of oxygen with protons to form water vapor with the consumption of electrons. Electrons flow from the anode through the external load to the cathode and the circuit is closed by ionic current transport through the electrolyte. In an acid cell, the current is carried by protons.

Reactants in this cell need not be pure. Hydrogen may be extracted from fuel mixtures and oxygen from air. Since product moisture is formed in an acid cell on the cathode, the air depleted in oxygen can be used for water removal if the cell is operated at a sufficiently high temperature to vaporize the water as it is formed.

The electrode has a central function in cell operation. In its catalyzed layer, it provides a large number of sites where gases and electrolyte can react. By virtue of a porous configuration, fast reactant transport and removal of inerts and product moisture is possible. The electrode also provides a path for current to flow to the terminals and serves to contain the electrolyte. The latter not only provides ionic conduction, but also assures separation of the reactants.

Cell Voltage. The cell voltage and the free energy of the underlying reaction are defined by

$$U = \frac{\Delta F}{nF}$$

where U = theoretical cell voltage; n = number of electrons transferred in the reaction; and F = Faraday constant.

Since $\Delta F = \Delta H - T\Delta S$, it follows that, depending upon the value of ΔS, the electrical energy to be derived from the cell can be larger or smaller than the energy ΔH obtained by direct combustion of the fuel. ΔH = reaction enthalpy for the current generating reaction; ΔS = entropy change; and T = absolute temperature.

For the hydrogen-oxygen couple, the corresponding theoretical cell voltage at 25°C is 1.23 volt if liquid coater is formed; or 1.18 volt if the product water is vaporized.

The thermodynamically possible conversion efficiency, however, is only partly realized in a practical fuel cell. Two basic losses are encountered: (1) the ohmic loss and (2) the electrode polarization, that is, the deviation of the actual from the thermodynamic electrode potential. The polarization is the result of the irreversibility of the electrode process, that is, the activation polarization and the voltage loss which develops from concentration gradients of the reactants. This leads to the current-voltage characteristics as shown in Fig. 2.

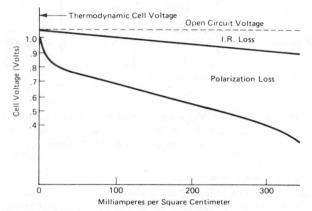

Fig. 2.   Current-voltage characteristic of hydrogen-air cell with phosphoric acid electrolyte. Operating temperature is 125°C (257°F).

Cell Technology

Fuel cells currently in use or in an advanced state of development in the early 1980s are based upon the electrochemical systems summarized in the accompanying table. The selection resulting from electrochemical as well as systems and cost considerations include cells with aqueous and fused salt electrolyte and high-temperature cells in which ionic transport is provided by oxygen mobility in the solid state. The alkaline cell with platinum-activated electrodes, a highly efficient cell, but sensitive to reactant impurities, is used in spacecraft power plants. The remaining cells are considered for commercial power generation and are air-breathing.

Commercial cells operate at elevated temperatures and, in larger power plants, also at elevated pressures. This leads to an improvement in cell performance and also provides reject heat at a useful temperature level.

In molten carbonate cells and solid-electrolyte cells, the temperature of the reject heat is sufficiently high to permit integration of these cells with coal-gasification systems (Dawes/Peterson, 1981).

Historically, the fuel cells listed in the table were not the first to be perfected. The first practical fuel cell was built in connection with the National Aeronautics and Space Administration's Gemini Program by the General Electric Company and relied on an unconventional electrolyte, namely, a solid polymer electrolyte membrane. Referred to as an *ion exchange membrane*, it consisted of a lacelike organic structure with anionic groups firmly bonded and hydrogen ions loosely held in the polymer chain. This provided sufficient mobility for ionic transport. A main advantage of the ion-exchange membrane

MAJOR ELECTROCHEMICAL SYSTEMS FOR FUEL CELLS

Electrolyte	Current Transport	Operating Temperature		Electrode Catalyst	Reactants		Operating Pressure	
		°C	°F		Fuel	Oxidant	PSIG	Atmosphere
Aqueous potassium hydroxide (KOH)	OH⁻	20–90	68–194	Nickel, silver, or platinum metals	Hydrogen	Oxygen	15	1.02
Concentrated phosphoric acid (H_3PO_4)	H⁺	190	374	Platinum metals	Impure hydrogen*	Air	Atmospheric to	
							120	8.16
Fused alkali carbonate	CO_3^{2-}	600–800	1112–1472	Nickel, silver	Impure hydrogen*	Air	Elevated	
Stabilized zirconium oxide	O^{2-}	700–1000	1292–1832	Base metal oxides	Impure hydrogen*	Air		

* Hydrogen-containing mixtures, such as steam reformate.

was the elimination of the need for electrolyte containment because of its well-defined boundaries. Separate electrodes were not required since a layer of platinum black bonded to the surface served that function.

Phosphoric Acid Matrix Cell

Among air-breathing cells, those with phosphoric acid electrolyte are the most advanced. The matrix-type cell construction is used. In this configuration, a limited amount of electrolyte is trapped in a microporous structure by capillary forces. As a result, thin, highly porous, and comparatively low-cost electrodes can be employed inasmuch as the electrodes are not required to contain the electrolyte. In practice, for electrolyte absorption, a thin layer of Teflon-bonded silicon carbide powder is directly applied to the electrode surface. The electrodes are made up of 0.3 millimeter-thick Teflon-impregnated carbon fiber sheets, which are activated on one surface with a layer of platinum dispersed on carbon black. See Fig. 3.

Single cells are sandwiched between carbon plates with a suitable pattern for current collection and grooved for reactant distribution. The direction of flow channels for air and hydrogen fuel are perpendicular to each other. Air is used to remove the product water as it is generated and may also serve to remove reject heat (Adlhart, 1972).

In large cells, in order to minimize temperature gradients, heat is removed through cooling plates or coils located in the stack either by recirculation of liquid coolants (Johnson/Kaufman, 1981) or by generation of steam.

Fuel Sources for Fuel Cells

The reactants converted in the fuel cell to electric power are hydrogen and oxygen from air. Hydrogen, however, is not the primary fuel source in many modern cells. It is obtained from selected fossil fuels to be replaced eventually by coal-derived synthetic fuels. These fuels are converted by suitable processing, such as steam reforming into a hydrogen-containing gas stream from which hydrogen is extracted in the fuel cell for power generation. See Fig. 4.

As of the mid-1980s, the preferred fuel for major applications is methane derived from natural gas—substitute natural gas (Rigney, 1981). The cost is favorable, a distribution system is in place, and availability in quantity appears assured for the remainder of the century.

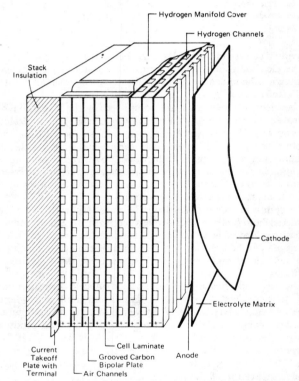

Fig. 3. Stack section of phosphoric acid matrix cell. Operation and design of cell are exceedingly simple. Product water is removed with air.

Alternate fuels under consideration are light distillates, coal gas, and fuel-grade methanol. The latter is a potentially coal-derived liquid fuel. Methanol can be steam reformed at low temperatures and is considered partly for this reason as well as for its adaptability to smaller, transportable fuel-cell power plants (Smith, 1981).

As shown by Fig. 5, a commercial fuel cell system is comprised of three subsystems: (1) the reformer section, (2) the fuel-cell power section, and (3) in most instances, an inverter to generate utility-grade alternating current. To achieve high fuel conversion

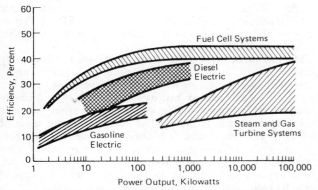

Fig. 4. Thermal efficiency of fossil fuel-operated fuel-cell power plants compares favorably with conventional means of energy conversion. The efficiency is reduced in small units because of losses in the fuel processing. (*United Technology Corp.*)

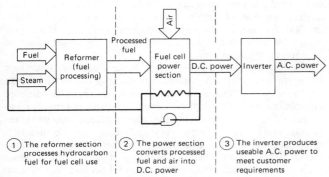

Fig. 5. Three principal subsystems required to convert hydrocarbon fuel to electric power.

efficiency, close thermal integration of the fuel cell and the fuel-processing section is required. In particular, part of the reject heat in the fuel cell is used to generate steam for the reforming of the fuel.

References

Adlhart, O. J.: "The Air-Cooled Matrix Type Phosphoric Acid Cell," in "From Electrocatalysis to Fuel Cells," Univ. Washington Press, Seattle, Washington, 1972.

Bacon, F. T.: "High Pressure Hydrogen-Oxygen Fuel Cells," in "Fuel Cells" (F. J. Young, editor), Van Nostrand Reinhold, New York, 1960.

Berger, C.: "Handbook of Fuel Cell Technology," Prentice-Hall, Englewood Cliffs, New Jersey, 1968.

Bockris, J. O'M., and T. Srinivasan: "Fuel Cells: Their Electrochemistry," McGraw-Hill, New York, 1969.

Breiter, M. W.: "Electrochemical Processes in Fuel Cells," Springer-Verlag, Berlin, 1969.

Dawes, M. H., and J. R. Peterson: "Carbonate Fuel Cell Technology Progress," Intersociety Energy Conversion Engineering Conference Proceedings, Atlanta, Georgia, 1981.

Gitlow, B.: "Strip Cell Test and Evaluation Program," Final Report, Contract NAS3-20042, National Aeronautics and Space Administration, Washington, D.C., 1980.

Grevstad, P. E.: "40 Kilowatt On-Site/Integrated Energy System," National Fuel Cell Seminar, Norfolk, Virginia, 1981.

Huff, J. R.: "Progress in Fuel Cells for Transportation," National Fuel Cell Seminar, Norfolk, Virginia, 1981.

Johnson, G. K., and A. Kaufman: "Phosphoric Acid Technology Progress," National Fuel Cell Seminar, Norfolk, Virginia, 1981.

Kurpit, S. S., and E. A. Gillis: "1.5 Kilowatt Indirect Methanol-Air Fuel Cell Power Plant," Engineering Foundation Conference on Methanol Fuel, New England College, Henniker, New Hampshire, 1974.

Liebhafsky, H. A., and E. J. Cairns: "Fuel Cells and Fuel Batteries: A Guide to Their Research and Development," Wiley, New York, 1968.

Rigney, D. M.: "Fuel Cell Power Plants—A User's Group Evaluation," National Fuel Cell Seminar, Norfolk, Virginia, 1981.

Smith, G. A.: "Methanol-Based Fuel Cell Systems," National Fuel Cell Seminar, Norfolk, Virginia, 1981.

Sperberg, R. T., and R. R. Woods: "Cogeneration with Fuel Cells," Intersociety Energy Conversion Engineering Conference Proceedings," Atlanta, Georgia, 1981.

Williams, K. R.: "Introduction to Fuel Cells," Elsevier, Amsterdam, 1966.

—Otto J. Adlhart, Senior Research Associate, Engelhard Industries Division, Engelhard Minerals and Chemicals Corporation, Iselin, New Jersey.

FUEL CONVERSION FACTOR. Nuclear Reactor.

FUEL CYCLE. Nuclear Reactor.

FUEL (Hydrogen). Hydrogen (Fuel).

FUEL (Nuclear). Nuclear Reactor.

FUELS. Coal; Natural Gas; Petroleum.

FUGACITY. Only perfect gases obey exactly the ideal gas law, which is the basis for the derivation of many other equations for the properties of gases. Therefore, we cannot substitute the measured pressure of real gases for the p in term in such equations without more or less inaccuracy. Since, however, calculations are simplified by using ideal equations for real gases, the quantity fugacity is defined as the equivalent pressure of a real gas for which the ideal gas equations are valid, so that by tabulating calculated values of fugacity corresponding to measured pressures for real gases, we can use the relatively simple equations derived for real gases.

For example, the chemical potential for a mixture of ideal gases can be written in the form

$$\mu_i = \mu_i^*(T) + RT \ln p_i \qquad (1)$$

where p_i is the partial pressure of component i. By analogy one may write for a mixture of real gases

$$\mu_i = \mu_i^*(T) + RT \ln p_i^* \qquad (2)$$

where $\mu_i^*(T)$ is the same function as for the ideal gas, while all the effects of molecular interactions (that is, of the departure of the real gas from ideality) are included in the p_i^*. This function $p_i^*(T, p, n_1, \cdots n_c)$ is called the fugacity of component i. This definition, due to G. N. Lewis, permits the preservation for real gases of the general form of the equations for ideal gases, with the fugacities replacing partial pressures.

In the lower pressure limit, p_i^* reduces to p_i.

FULLER'S EARTH. A fine-grained earthy substance similar to clay, both in appearance and composition, but lacking the usual plasticity, possessing a higher water content, and usually high in magnesia. The material consists mainly of hydrated aluminum silicates, such as the clay minerals, montmorillonite and palygorskite. Generally, it is believed that fuller's earth was formed as a residual deposit as the result of decomposition of rock in place, perhaps by the devitrification of volcanic glass. The color of the material ranges from light brown through yellow and white to light and dark green. Fuller's earth is used for decolorizing oils, degreasing raw wool, and as a natural bleaching agent.

FUNGICIDES. The fungicides are a heterogeneous group of chemicals that mitigate, inhibit or destroy fungi. Most chemicals, except the very inert ones, are fungitoxic if present in sufficient quantity, but they usually are not designated as fungicides unless they are effective at nominal dosages of 1000 ppm or less in aqueous suspensions. Those chemicals that inhibit spore germination or mycelial growth without destroying the fungous body are properly known as fungistats, although in common usage they are referred to as fungicides.

The commercial fungicides are indispensible to the welfare of people in preventing or curing the diseases of plants, man and animals and in suppressing the deterioration of stored agricultural produce, material and structures made of cellulosic, lignified or plastic materials. Such diseases as athlete's foot, skin mycosis and ringworm of the human scalp, pulmonary infection of fowl, and moist eczema of dogs are amenable to control by fungicides. The major use of fungicides, however, is in preventing plant diseases such as the leaf blights, powdery mildews, downy mildews, rusts, anthracnoses, fruit rots, fruit scab and stem cankers by spray or dust application of protective or eradicant fungicides. They are also used as soil fumigants and ground sprays to destroy spores and mycelium in their natural habitat before they can attack plants, to disinfest seed known to bear spores or mycelium of smut and other types of fungi, and to protect seed from decay and damping-off organisms in the soil. Other uses include impregnation of fabrics to prevent mildewing and decomposition when in contact with moist substrates, impregnation of wallpaper, paints and leather goods to suppress mildew, treatment of structural timbers, piles, fenceposts, etc. to prevent dry rot and decay, and incorporation into bread to suppress mold growth.

About 100 fungicides are required for these various uses in the United States. The principal ones are sulfur; lime-sulfur (polysulfides of calcium); copper sulfate (or its equivalent in the oxides, basic sulfates, oxychloride and other relatively insoluble copper compounds); creosote products and zinc chloride, both used as wood preservatives; and a wide variety of organic compounds. Among the latter are several dithiocarbamates, such as ferbam and zineb, and other thio compounds, like N-(trichloromethylthio)-phthalimide (folpet); cis-N((trichloromethyl)thio)-4-cyclohexane-1,2-dicarboxyimide (captan); and 8-hydroxyquinoline.

Prior to 1939, the inorganic sulfur and copper compounds were used almost exclusively as spray and dust materials and the copper and organic mercury compounds as seed treatments. Sulfur had been used from before the time of the Romans in various plant prescriptions and had been used for powdery mildews since the beginning of the 19th century. Copper sprays were introduced in 1882 as Bordeaux mixture for control of downy mildew on grapes, and as a copper carbonate seed treatment in 1917. Lime-sulfur was developed as an apple spray in 1906. A new era in fungicides was initiated in the period 1934–1939 with the announcement of the dithiocarbamate and quinone fungicides which indicated the potentialities of organic compounds.

Mercurial fungicides were abandoned in 1971 because of environmental pollution and hazard of conversion into poisonous methyl mercury.

The intensive search for new organic fungicides has continued for over 30 years. The technique employed in most laboratories is to make an extensive survey of various structures by empirical methods to locate materials that suppress spore germination on glass slides or prevent mycelial growth on nutrient agar plates or rolled tubes. Those materials having an ED_{50} (effective dose for 50% inhibition) in the order of 10 ppm are further tested in use applications.

The mechanism of action of the fungicides has been only partially solved. Sulfur deposits volatilize and within two to three minutes after coming into contact with a spore the sulfur begins to be released as hydrogen sulfide. It was formerly thought that hydrogen sulfide was the lethal agent, but it is now evident that it is the product rather than the cause of fungus destruction, probably because the hydrogen transport system of the sulfur-sensitive spore becomes overtaxed after about 10,000 ppm of sulfur has been reduced. In any event, sulfur appears unique in that its lethal effect does not depend upon its accumulation inside or on the spore.

The organic fungicides have tremendous affinity for spores. For example, within 30 seconds after spores are placed in suspension containing 2 ppm of 2-heptadecyl-2-imidazoline they will remove up to 6,000 ppm of their body weight. There is reason to believe that this attribute may very well depend upon the compound's lipoid solubility induced by the 17-carbon chain in the 2-position. Homologs containing longer or shorter chains are less fungitoxic and fungitoxicity in this series is directly correlated with ability to fractionate into the lipoid phase of an aqueous-lipoid system.

Once the fungicides penetrate to the cell membrane or into the cytoplasm they may operate by devious means to disrupt vital functions. There is substantial evidence that the quinones immobilize the sulfhydryl and imino prosthetic group of enzymes. The 8-hydroxyquinoline and dithiocarbamate compounds are active against copper and other metallic members of an enzyme system, presumably by their ability to chelate metals. Heavy metals such as mercury affect certain enzymes such as amylases and may serve as general protein precipitants.

The best evidence available indicates that most, if not all, fungicides are not particularly specific in their action on vital cell systems and may react with nonvital molecules; thus a large percentage of them is detoxified before an essential biochemical process can be effected. The actual dosage required on a spore weight or mycelial basis for a lethal effect has been determined for very few compounds but most of those investigated by Miller and McCallan, who used radioactively labelled molecules, were not lethal until 5,000 to 20,000 ppm were accumulated by the spores.

There is appreciable specificity in the action of fungitoxicants on different types of fungi. Ferric dimethyldithiocarbamate, for example, is much more effective than sulfur against the cedar-apple rust fungus (*Gymnosporangium juniperi-virginian-nae*) but has no particular advantage in control of apple scab (*Venturia inaqualis*). Ferbam is also superior to copper sprays against anthracnose of tomato (*Colletotrichum phomoides*) but is totally inadequate against *Alternaria* and *Phytophthora* leaf blights of this crop. Insofar as is known at present, specificity is only qualitative and not absolute so fungicides may be detected in a general group of candidate compounds with reasonable accuracy by measuring their effects on two or three indicator species of fungi.

See also **Pyridine and Derivatives.**

—George L. McNew, Boyce Thompson Laboratories, Yonkers, New York.

FUNGICIDE (Pyridine). Pyridine and Derivatives.

FURAN AND RELATED COMPOUNDS. Furan C_4H_4O contains a ring of one oxygen and four carbons, with one hydrogen attached to each carbon:

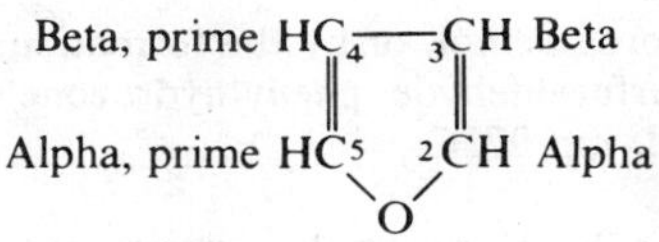

Beta, prime HC$_4$———$_3$CH Beta

Alpha, prime HC$_5$ $_2$CH Alpha

O

Furan is a colorless liquid, bp 32°C, insoluble in H$_2$O, soluble in alcohol and ether. Furan vapor produces a green coloration on pine wood moistened with hydrochloric acid. Furan may be made from mucic acid, COOH(CHOH)$_4$COOH, by dry distillation into pyromucic acid, C$_4$H$_3$O·COOH, and then heating the latter under pressure at 270°C. Furan derivatives are known, such as methyl, primary alcohol, aldehyde, carboxylic acid, in which the group attachment is at carbon number 2.

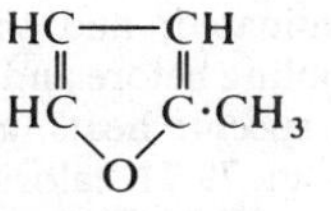

α-Methyl furan
bp 65°C

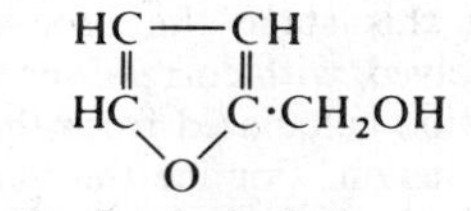

Furfuryl alcohol
bp 170°C
(750 mm)

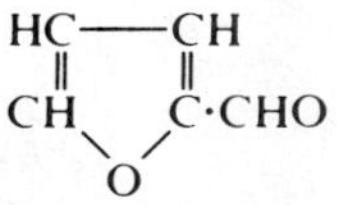

α-Furfuraldehyde
bp 160°C
(740 mm)

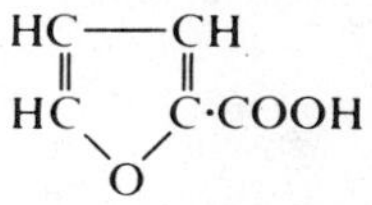

Pyromucic acid
mp 133°C
bp 230°C

See also **Furfuraldehyde.**
Coumarone is benzofuran, C$_8$H$_6$O, colorless liquid, bp 173°C.

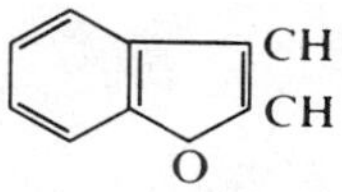

Diphenylene oxide is dibenzofuran, C$_{12}$H$_8$O, white solid, mp 81°C, bp 288°C.

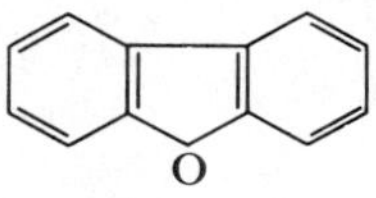

Coumarin is 1,2-benzopyrone, C$_9$H$_6$O$_2$, white solid, mp 67–68°C, bp 301°C.

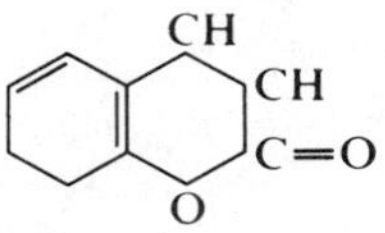

Gamma-pyrone, C$_5$H$_4$O:O(4), is a gamma-ketone (4) containing a ring of 1 oxygen and 5 carbons with 1 hydrogen attached to each of 4 carbons, namely, 2,3,5,6.

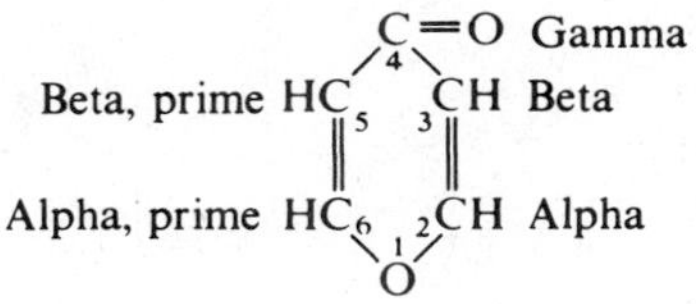

C=O Gamma

Beta, prime HC$_5$ $_4$ $_3$CH Beta

Alpha, prime HC$_6$ $_1$ $_2$CH Alpha

O

Gamma-pyrone is a colorless liquid, mp 32°C bp 218°C. Pyrone derivatives are known, e.g.,

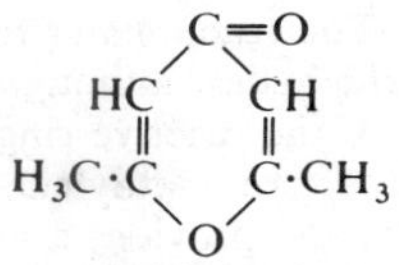

Alpha, alpha prime
dimethyl-gamma-pyrone

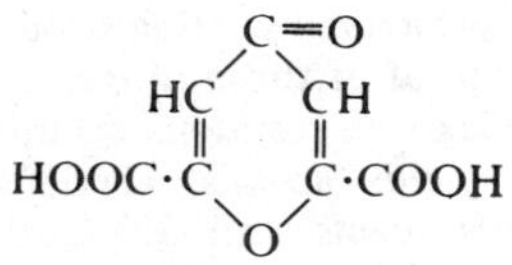

Chelidonic acid
gamma-pyrone-alpha,
alpha-prime-dicarboxylic acid

Chromone is benzo-pyrone C$_9$H$_6$O$_2$, a white solid, mp 59°C

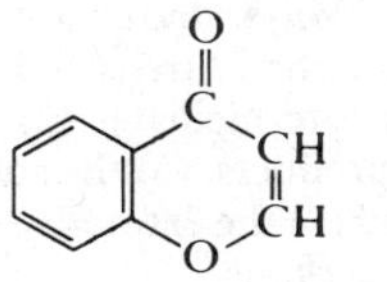

and chromane is colorless liquid, bp 214°C, 750 mm.

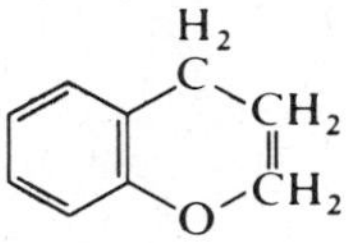

Flavone is phenyl chromone: a white solid mp 97°C.

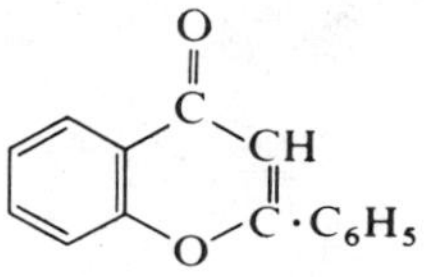

Xanthone is dibenzo-pyrone, C$_{13}$H$_8$O$_2$ or C$_6$H$_4$ O   C$_6$H$_4$
                                                          C
                                                          O

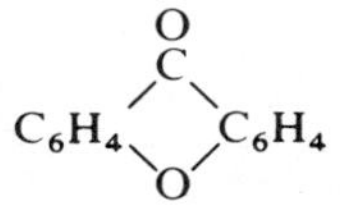

or 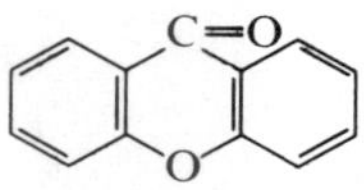   white solid, mp 174°C, bp351°C, and

xanthene is 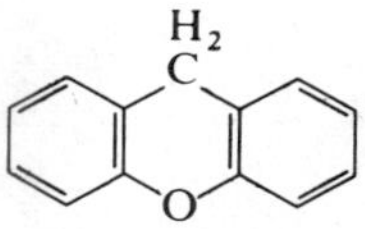   white solid, mp 100°C, bp 315°C.

From chromone and xanthone a number of yellow dyes are made, which dyes also occur in nature. Such dyes are chrysin, fisetin, buteolin, morin, quercetin, rhamnetin.

Where oxygen of furane is replaced by sulfur, thiophene is the compound, and of coumarone, benzothiophene; and where oxygen of furane is replaced by nitrogen (group=NH), pyrrole, and of coumarone, indole.

FURFURALDEHYDE.   2-C$_4$H$_3$O·CHO, formula weight 192.16, colorless, odorous (pungent, almond-like) liquid aldehyde, mp −38.7°C, bp 161.7°C, sp gr 1.159. Also known as 2-furaldehyde or 2-furancarboxaldehyde, this compound becomes brown in color when in contact with air. Furfural is modestly soluble in H$_2$O (up to 8% by weight at 20°C) and is miscible in all proportions with alcohol and ether. At atmospheric pressure, a mixture of furfural and H$_2$O (65%) forms a minimum-boiling azeotrope when a distillation temperature of 97.9°C is reached.

Aside from a darkening in color, furfural is relatively stable thermally and does not exhibit changes in physical properties after prolonged heating up to 230°C. The reactions of furfural are typical of those of the aromatic aldehydes, although some complex side reactions occur because of the reactive ring. Furfural yields acetals, condenses with active methylene compounds, reacts with Grignard reagents, and provides a bisulfite complex. Upon reduction, furfural yields furfuryl alcohol; upon oxidation, it yields furoic acid. It can be decarbonylated to furan.

Furfural is obtained commercially by treating pentosan-rich agricultural residues (corncobs, oat hulls, cottonseed hulls, bagasse, rice hulls) with a dilute acid and removing the furfural by steam distillation. Major industrial uses of furfuraldehyde include (1) the production of furans and tetrahydrofurans where the compound is an intermediate, (2) the solvent refining of petroleum and rosin products, (3) the solvent binding of bonded phenolic products, and (4) the extractive distillation of butadiene from other C_4 hydrocarbons.

When pentoses, e.g., arabinose, xylose, are heated with dilute HCl, furfuraldehyde is formed, recognizable by deep red col-

oration with phloroglucinol, or by the formation, with phenylhydrazine, of furfuraldehyde phenylhydrazone $C_4H_3O \cdot CH : NNHC_6H_5$, solid, mp 97°C.

FUSION (Heat of). Very simple experiments show that the fusion of a given mass of any crystalline substance requires a definite quantity of heat. The quantity required per unit mass, without any change of temperature, is called the heat of fusion of the substance. It may be measured by means of a calorimeter. The fused substance is introduced into the calorimeter at a temperature somewhat above its melting point and allowed to cool, the heat evolved being measured. At the melting point it ceases to cool for a time, but continues to give out heat as it solidifies; and when all congealed, it begins to cool again. At this stage the process is terminated; and the total heat evolved, with corrections for the cooling before and after solidification calculated from the known specific heats, gives the heat of fusion. For ice the value is about 79.71 calories per gram.

G

GADOLINIUM. Chemical element symbol Gd, at. no. 64, at. wt. 157.25, seventh in the lanthanide series of the periodic table, mp 1312°C, bp 3273°C, density, 7.901 g/cm³ (20°C). Elemental gadolinium has a close-packed hexagonal crystal structure at 25°C. Pure metallic gadolinium is silvery-gray in color, slow to tarnish in normal atmospheres. The metal is soft, malleable, and easy to fabricate with normal tools provided that processing temperatures are maintained below 150°C. The turnings and chips of gadolinium are mildly pyrophoric and care must be exercised in their handling. There are seven natural isotopes of gadolinium, ^{152}Gd, ^{154}Gd through ^{158}Gd, and ^{160}Gd. Eleven artificial isotopes have been prepared. The natural isotopes are not radioactive. In terms of abundance, gadolinium is present on the average of 5.4 ppm in the earth's crust, making it potentially more available than tantalum, tin, or tungsten. The element was first identified by J. C. G. Marignac in 1880. The natural isotopic mixture of gadolinium has the greatest thermal-neutron-absorption cross section of all elements, 40,000 barns. This is approximately 10 times greater than the next two elements, samarium (5800 barns) and europium (4300 barns). However, gadolinium is limited to nuclear applications mainly as a start-up and shutdown material because only two of the natural isotopes, ^{155}Gd and ^{157}Gd behave in this manner. These are separated by isotopes which do not so react—hence, no chain relationship exists. The isotopes ^{155}Gd and ^{157}Gd make up 31% of the total weight of elemental gadolinium. The metal has a low acute-toxicity rating. Electronic configuration is $1s^22s^22p^63s^23p^63d^{10}4s^24p^64d^{10}4f^75s^25p^65d^16s^2$. Ionic radius: Gd^{3+}, 0.938 Å. Metallic radius, 1.801 Å. First ionization potential, 6.16 eV; second, 12.1 eV. Other important physical properties of gadolinium are given under **Rare-Earth Elements and Metals.**

Gadolinium reacts vigorously with dilute mineral acids, but is practically inert to strong bases and boiling H_2O. Gadolinium is an active reducing agent for metals, including iron, chromium, manganese, tin, lead, and zinc. The major sources of gadolinium are xenotime, monazite, gadolinite, and residues from uranium mining.

Although the nuclear properties of the element are attractive, gadolinium has enjoyed rather limited applications in reactor technology. An important discovery in the 1960s showed that gadolinium iron garnets (called GIGs) $Gd_6Fe_5O_{12}$ possess a crystalline structure which finds useful application in microwave frequency control, circulators, isolators, and bandpass filters in electronic circuitry. Gadolinium oxide also is used as the host matrix in the red phosphor for color television picture tubes, where it is activated by europium. Gadolinium oxysulfide Gd_2O_2S is used as an x-ray image intensifier making possible less x-ray dosage for medical explorations. Along with yttrium and lanthanum activated by cerium, gadolinium is used in a phosphor for single-gun beam-indexing flying-spot scanning cathode ray tubes. Gadolinium also provides magnetic properties when alloyed with cobalt, cerium, iron, and copper ($Co_{3.5}CuFe_{0.5}Ce$) in permanent magnets, imparting a desirable negative temperature coefficient of magnetic saturation. A glass with magnetic properties (5% wt Gd_2O_3) has been produced. Gadolinium metal and several of its salts are under consideration for use in a magnetic heat pump device.

See references listed at ends of entries on **Chemical Elements; and Rare-Earth Elements and Metals.**

NOTE: This 6th Edition entry was revised and updated by K. A. Gschneidner, Jr., Director, and B. Evans, Assistant Chemist, Rare-Earth Information Center, Energy and Mineral Resources Research Institute, Iowa State University, Ames, Iowa.

GAHNITE—ZINC-SPINEL. The mineral gahnite is isometric with an octahedral habit but may appear as dodecahedrons or modified cubes. Chemically it is zinc aluminate corresponding to the formula $ZnAl_2O_4$. There is a tendency for cleavage parallel to the octahedron, fracture varies from conchoidal to uneven; brittle; hardness 7.5–8; specific gravity 4.6; luster, vitreous; color ranges from dark green through various shades of greenish or bluish-black, yellowish-black or grayish, subtransparent to almost opaque. Gahnite is found in association with other zinc minerals at several European localities, notably in Bavaria and Sweden. In the United States it is found at Franklin and Sterling Hill, New Jersey; at Rowe, Massachusetts; and in Maryland, North Carolina, Georgia, and Colorado. Gahnite was named in honor of the Swedish chemist, J. G. Gahn.

GALENA. The mineral galena, lead sulfide, PbS, crystallizes in the isometric system, usually in cubes or cube-octahedron combinations, less frequently in octahedrons. It is often found in cleavable masses, but may be granular or fibrous. The highly perfect cubic cleavage is an important characteristic of this mineral: it may, however, sometimes show an octahedral parting. Its hardness is 2.5; specific gravity, 7.58; luster, metallic; color, lead gray; streak, grayish-black; opaque. Galena is the most important ore of lead and in addition often carries values of silver; it is then known as argentiferous galena. It occasionally is actually mined as a silver ore. Sometimes galena contains small amounts of zinc, cadmium, antimony, bismuth, and copper as sulfides.

Galena is a very common and widely spread mineral, it occurs in veins and beds in various rocks, both crystalline and sedimentary. Some of these deposits are doubtless replacements, others seem to show a close connection with intrusive igneous rocks. Of the many European localities, the classics are Freiberg, Saxony, and the silver mines of the Harz Mountains. This mineral has been found in the lavas of Vesuvius, in Italy, and fine specimens came from Cornwall and Cumberland, England. Australia, South America, Chile, and Peru produce galena. In the United States, Missouri, Illinois, Iowa, and Wisconsin contain large and important galena deposits. In Colorado and Idaho it has been mined for its silver content. Galena is usually associated with sphalerite, smithsonite, and at Phoenixville, Pennsylvania, with beautiful pyromorphite crystals. The name is derived from the Latin *galena*, a term which was applied both to the lead ore and slag from refining. See also terms listed under **Mineralogy.**

—Elmer B. Rowley, Union College, Schenectady, New York.

GALLIUM. Chemical element symbol Ga, at. no. 31, at. wt. 69.72, periodic table group 3a, mp 29.78°C, bp 2403 $\pm$ 0.5°C, density, 5.90 g/cm^3 (solid at 20°C), 6.095 (liquid at 29.8°C), 5.445 (liquid at 1100°C). Elemental gallium has a one-face-centered orthorhombic crystal structure. Among the elements, gallium (like mercury) is liquid at ordinary temperatures. Gallium is a white, tough metal, but so soft that it can be cut with a knife. A freshly-exposed surface soon oxidizes superficially to a bluish-gray color. When heated to about 500°C, the metal burns in air. Gallium is only slightly affected by water at room temperature, but reacts vigorously in boiling water. The metal is only slowly attacked by concentrated acids, but does dissolve readily in aqua regia. The two stable isotopes of gallium are ^{69}Ga and ^{71}Ga. The eight radioactive isotopes include ^{64}Ga through ^{68}Ga, ^{70}Ga, ^{72}Ga, and ^{73}Ga. All have a relatively short half-life, the longest, ^{67}Ga with a half-life of 78 hours. See **Radioactivity.** Gallium was one of the elements predicted by Mendeleev from his early periodic arrangement of the chemical elements. The element was first identified by Francois Lecoz de Boisbaudran in 1875 from observations in a spectroscopic study of zinc blende. In terms of abundance, gallium ranks 31st among the elements, with about 15 ppm in the earth's crust.

First ionization potential 6.00 eV; second, 20.43 eV; third, 30.6 eV. Oxidation potentials Ga $\rightarrow$ Ga^{3+} + 3e$^-$, -0.52 V; Ga + 4OH$^-$ $\rightarrow$ H$_2$GaO$_3^-$ + H$_2$O + 3e$^-$, 1.22 V.

Other important physical characteristics of gallium are given under **Chemical Elements.**

Gallium occurs in very small amount in zinc blende, magnetite, pyrite, bauxite, and kaolin of certain localities. A few parts per million is present in Oklahoma zinc ores. The recovery of gallium from zinc flue dust is effected by solution of the dust in excess of HCl, addition of potassium chlorate, and distillation to remove germanium. When the residue is converted into sulfate, fractional electrolysis of the slightly acid solution removes zinc, and the gallium is obtained almost free from indium. The only known deposit of gallite, CuGaS$_2$, is in southwest Africa. The mineral contains about 1% gallium. The most important commercial source of gallium is bauxite which contains up to 0.01% gallium. The metal is recovered from the sodium aluminate used in the extraction of aluminum from bauxite. In one process, calcium hydroxide is mixed with the sodium aluminate solution. At this juncture the ratio of gallium to aluminum is about 1 to 3,000. By precipitating and filtering out calcium aluminate, a gallium-rich solution remains. The filtrate then is agitated with CO$_2$ which precipitates more aluminum out as aluminum hydroxide. At this point, the enriched gallate-in-caustic solution containsapproximately 0.2 grams of gallium per liter. This solution is used as an electrolyte in a mercury cathode cell. The gallium amalgamates with the mercury. It is dissolved out of the mercury with boiling NaOH in the presence of iron which serves as a catalyst. At this point, the concentration is approximately 80 g of gallium per liter. The process is repeated several times, after which the gallium concentrate is electrolyzed, using a stainless steel cathode on which the gallium plates out. The gallium is easily removed from the cathode by raising the temperature above the melting point. For highly-pure metal, subsequent purification processes are required, including (1) crystallization as monocrystals, (2) chemical treatment with acids or oxygen at high temperatures, or

(3) repeat resolution in pure boiling NaOH and reelectrolyzing. A metal of 99.99999% purity thus can be obtained.

Uses. The availability of gallium in very high purity is important to its use as a semiconductor in various electronic devices, such as diodes, laser diodes, and electroluminescent diodes. The compound usually used in these applications is gallium arsenide GaAs which is prepared by reacting hydrogen and arsenic vapor with gallium oxide Ga$_2$O$_3$ (prepared from very pure metal) at a temperature of about 600°C. Properties of the GaAs so produced include: intrinsic electron concentration, 10^7; energy gap, 1.38 eV at 20°C; electron mobility, 8,800 cm^2/V-s.

Gallium arsenide also is used in solar batteries. Gallium metal is used as an activator in luminous paints and phosphors, as well as in arc rectifiers, dental amalgams, as a sealant in vacuum systems, in transistors, and in some organic syntheses. Because the metal expands upon solidifying (3.1%), it should not be stored in fragile containers. Although potentially useful in high-temperature thermometers because of its liquidity over a wide temperature range, these applications have been limited, partially because of the high cost of the element.

Chemistry and Compounds. Gallium metal is quite corrosive to most other metals because of the rapidity with which it diffuses into the crystal lattices of metals. For example, only a very small amount of gallium in contact with an aluminum plate or sheet will result in immediate embrittlement as the result of the diffusion of gallium through the grain boundaries separating them. Gallium readily forms alloys with most metals over 600°C, including barium, copper, gold, iron, lead, lithium, magnesium, manganese, nickel, platinum, silver, sodium, titanium, vanadium, zirconium, and zinc. The few metals that tend to resist attack by gallium are molybdenum, niobium, tantalum, and tungsten.

Gallium trihalides include the trifluoride, tribromide, triiodide, and the trichloride. The trichloride is readily formed by heating the metal with chlorine or HCl, is soluble in ether, and like aluminum chloride, is effective as a catalyst in various organic reactions. Both the trichloride and the tribromide are dimeric in the vapor state. Other known trivalent gallium compounds are the sesquisulfide, sesquisulfate (which forms double salts analogous to the alums), trinitrate, nitride, sesquioxide (which is polymorphic like alumina), and trihydroxide, which is, however, of variable composition, and which forms salts, the gallates, in alkaline solution.

Known gallium(II) compounds include the sulfide, selenide, telluride, dichloride, and dibromide. The last two are unstable, reacting vigorously with water to give hydrogen, and also undergoing oxidation, or disproportionation to the metal and the gallium(III) compound. They also are diamagnetic and their structure is Ga$^+$[GaX$_4$]$^-$.

Simple gallium(I) compounds are also unstable, but Ga$^+$ may be stabilized in the presence of large anions, e.g., in Ga[AlCl$_4$]. The sulfur and selenium compounds Ga$_2$S and Ga$_2$Se have been shown to exist, but the oxide is uncertain.

Triethylgallium and trimethylgallium have been prepared, but are extremely reactive, even with air and H$_2$O. Like aluminum and indium, gallium forms a number of chelated oxy compounds, almost all of which are of 6-coordinate type. They include the stable crystalline inner complexes of which the β-diketones coordinate in the proportion of 3 molecules of diketone per atom of gallium. Trioxalato as well as dioxalato salts are known, and compounds such as 8-quinolinol and substituted 8-quinolinols form trimolecular chelate rings involving nitrogen and donor oxygen.

Gallium, like boron, forms a dimeric hydride, Ga$_2$H$_6$, from which a series of tetrahydrogallates, containing the GaH$_4^-$ ion, is derived.

Gallium and most of its compounds are not highly toxic. For rats and rabbits, the LD_{100} has been established at approximately 100 mg of gallium per kilogram.

See list of references at end of entry on **Chemical Elements.**

GALVANIZING. The application of a layer of zinc to the surface of iron or steel for protection from corrosion. Galvanizing is one of the most effective means of preventing rusting because of its relatively low cost and ease of application. See **Zinc.**

GAMMA GLOBULIN. The fraction of the protein globulins of the blood plasma in which are found the antibodies. See **Blood.**

GANGLIOSIDES. Identified by Kleng in 1935, the gangliosides are a family of acidic glycolipids that are characterized by the presence of sialic acid. The compounds bear a strong negative charge and are unusual in that they contain both hydrophobic and hydrophilic regions. These compounds are membrane components. Plasma cell membranes are rich with gangliosides. It has been suggested that gangliosides participate in the transmission of membrane-mediated information in living systems. As described by Fishman and Brady, "the carbohydrate portion of gangliosides is made up of molecules of sialic acid, hexoses, and *N*-acetylated hexosamines. The hydrophobic moiety is called ceramide, and it consists of a long-chain fatty acid linked through an amide bond to the nitrogen atom on carbon 2 (C-2) of the amino alcohol, sphingosine. Oligosaccharides are linked through a glycosidic bond to C-1 of the sphingosine portion of ceramide." Svennerholm (1963) suggested the configuration given by the accompanying diagram. The role of gangliosides is still rather obscure, but Fishman and Brady (1976) have studied in some detail the interaction of cholera toxin with ganglioside-deficient cells, as well as their interaction with glycoprotein hormones and their effect on the action of these hormones.

References

Fishman, P. H., and R. O. Brady: "Biosynthesis and Function of Gangliosides," *Science,* **194,** 906–915 (1976).
Klenk, E.: *Z. Physiol. Chem.,* **235,** 24 (1935).
Svennerholm, L.: *J. Neurochem.,* **10,** 613 (1963).

GARNET. The name *garnet* is now applied to a group of very important minerals crystallizing in the isometric system and showing the same habit of dodecahedrons and trapezohedrons. Garnets belong to the nesosilicate group of silicate minerals and conform to the general formula $A_3B_2(SiO_4)_3$. The elements represented by A and B, respectively, may include calcium, magnesium, or maganese, and ferrous iron, aluminum, ferric iron, chromium, or titanium. While garnets show no cleavage a dodecahedral parting is rarely noted; fracture conchoidal to uneven; some varieties very tough and valuable for abrasive purposes and for polishing eye-glass lenses. The hardness of garnet varies between the different varieties from 6.5 to 7.5; and the specific gravity from 3.4 to 4.3. Luster, vitreous to resinous; colors, red, yellow, brown, black, green or colorless; transparent to opaque. The word *garnet* is derived from the Latin *granatus,* meaning *a grain*.

In general six varieties of garnet are recognized, based on their chemical composition: grossularite (which is also called hessonite and cinnamon-stone); pyrope; almandine or carbuncle; spessartine; uvarovite; and andradite. Grossularite is a calcium-aluminum garnet which corresponds to the formula $Ca_3Al_2(SiO_4)_3$; the calcium may, however, be in part replaced by ferrous iron and the aluminum by ferric iron. The name *grossularite* is derived from the botanical name for the gooseberry, *grossularia,* in reference to the green garnet of this composition found in Siberia. Other shades are the well-known cinnamon brown, reds, and yellows. Because of its inferior hardness to zircon, which mineral the yellow crystals resemble, they have been termed *hessonite,* from the Greek meaning *inferior.* Curiously, in the gem-bearing gravels of Ceylon, both zircon and hessonite are found and indiscriminately called *hyacinth.* This term, from the Greek, was apparently a general term used by Pliny for the transparent varieties of corundum; later it was used for yellow zircons. Grossularite is found in crystalline limestones with vesuvianite, diopside, wollastonite and wernerite. Among the many localities are the Urals, Italy, Switzerland, Mexico, and, in the United States, Maine and New Hampshire. Fine specimens are obtained from the Jeffrey Mine, Asbestos, Quebec, Canada.

Pyrope, sometimes called *Cape ruby,* is ruby-red in color and chemically a magnesium aluminum silicate with the formula $(Mg, Fe)_3Al_2(SiO)_3$; the magnesium may be replaced in part by calcium and ferrous iron. The color of pyrope varies from deep red to almost black. The transparent pyropes are used as gems, but some have a slight tinge of yellow. The name pyrope is derived from the Greek word meaning *fire-like.* A sub-variety of pyrope from Macon County, North Carolina, is of a violet-red shade and has been called *rhodolite,* from the Greek meaning *a rose.* In chemical composition it may be considered as essentially an isomorphous mixture of pyrope and almandine, in the proportion of two molecules of pyrope to one molecule of almandine. Pyrope is found at Teplitz and

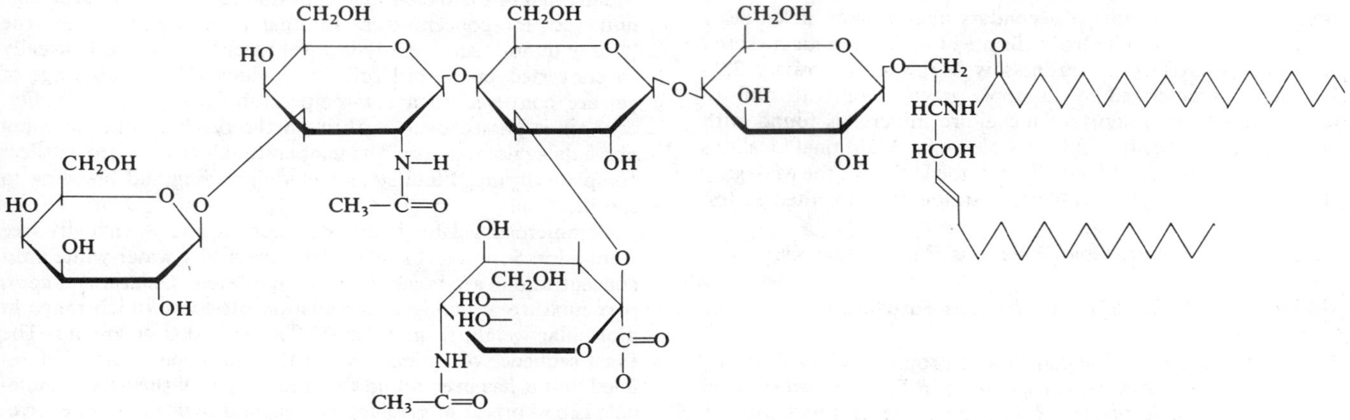

Configuration of monosialoganglioside G_{MI} as suggested by Svennerholm (1963).

Aussig, Bohemia; in the Kimberley diamond mines in the Republic of South Africa; in Australia and elsewhere. In the United States, important localities are in Arizona, New Mexico, and Utah.

Almandine is the modern gem the carbuncle, although in Pliny's time this term was used for almost any red stone. The term *carbuncle* is derived from the Latin *carbunculus*, meaning *a little spark*. The name almandine is a corruption of Alabanda, a locality in Asia Minor where, in ancient times, these red stones were cut. Chemically alamandine is an iron-aluminum garnet corresponding to the formula $Fe_3Al_2(SiO_4)_3$. The deep red transparent stones are often called precious garnet and used for gems. Almandine occurs in metamorphic rocks like mica schists usually associated with typically metamorphic minerals such as staurolite, kyanite, and andalusite. Good gem material comes from India and Brazil. Almandine is also found in Australia, Alaska, Africa, Norway, Sweden, the Malagasy Republic, and Japan. In the United States almandine with 11.48% MgO pyrope content is found in the gneisses of the Adirondack region of New York, sometimes of very large size, in New England, and elsewhere.

Spessartine is manganese aluminum garnet, $Mn_3Al_2(SiO_4)_3$. The name of this mineral is derived from Spessart in Bavaria, a well-known European locality. Spessartine of a beautiful orange-yellow comes from the Malagasy Republic. Violet-red spessartine has occured in rhyolites in Colorado and Maine. Uvarovite is a calcium chromium silicate the formula being $Ca_3Cr_2(SiO_4)_3$. It is a rather rare garnet, bright green in color, usually in small crystals associated with chromite in serpentines, sometimes in crystalline limestones or schists. It is found in the Urals, the Republic of South Africa, Canada, and, in the United States, in California and Pennsylvania. Andradite, calcium-iron garnet, $Ca_3Fe_2(SiO_4)_3$, is of variable composition and may be red, yellow, brown, green, or black, or of intermediate shades. The subvarieties topazolite, yellow or green, demantoid, green, and melanite, a black sort, are recognized. Andradite is found both in deep-seated igneous rocks like syenite as well as in serpentines, schists, and crystalline limestones. Demantoid has been called the "emerald of the Urals" from its occurrence there. Varieties of andradite are found in many localities in Europe: Italy, Switzerland, Norway, and Saxony. In the United States it is found at Franklin, New Jersey; Magnet Cove, Arkansas; and elsewhere.

See also terms listed under **Mineralogy.**

—Elmer B. Rowley, Union College, Schenectady, New York.

GARNIERITE. This mineral occurs at amorphous masses, presumably as a product of secondary alteration of nickel-bearing peridotites. It is a hydrous silicate of nickel and magnesium, $(Ni, Mg)_3Si_2O_5(OH)_4$. Hardness is 2–3; specific gravity 2.2–2.8, and characterized by its apple green color with dull-to-earthy luster. An important nickel-ore mineral is found with chromite and serpentine in New Caledonia. Additional localities include the Republic of South Africa, the U.S.S.R., the Malagasy Republic, and Oregon and North Carolina in the United States.

GASAHOL. Biomass and Wastes as Energy Sources.

GAS-COOLED REACTOR. Nuclear Reactor.

GAS CONSTANT. The constant of proportionality R in the equation of state of a perfect gas $pv = RT$, when referring to one gram molecule of gas. R has the value of 1.985 calories per mole degree (°C).

GASEOUS DIFFUSION. Diffusion; Graham Law.

GASIFICATION PROCESSES (Coal). Coal; Substitute Natural Gas (SNG).

GASOLINE. Petroleum.

GEL AND GELATION. Colloid System.

GELATIN. Widely used in food processing, gelatin is a mixture of simple proteins derived from the collagen of animal connective tissues through a series of degradation or hydrolytic steps. Gelatin is frequently encountered in home-made soup stock, to which it imparts the tendency to form a gel under refrigeration. Collagen, one of the most abundant proteins found in the animal body, is the major component of the white fibers of connective tissue and is present in skin, sinews, hides, and ossein, the connective tissue protein of bones.

Through a number of hydrolytic steps effected by heat and/or acids and alkalis, collagen is cured and plumped into a translucent mass of swollen fibers from which the gelatin can be extracted with hot water. When raw materials, equipment, and procedures are maintained in accordance with strict regulations, this product may be marketed as gelatin. Crude, less-refined products are commonly called technical gelatins (glues).

The raw material is frequently pork skins, calf skins (tannery byproducts), or ossein from the demineralization of bones. Dependent upon the raw material, acids or alkalis are used to cure the collagen preparatory to extraction of the gelatin. Pork skins obtained from the packing industry are treated with dilute hydrochloric acid. Calf stock from tanneries usually is pretreated (lime-cured), but if not so, it is treated at the gelatin facility. Ossein, as the residue remaining from the acid leaching of the mineral calcium phosphate from bone, is acid-cured in the process of demineralization.

Irrespective of the cure used, the collagen is thus prepared for hydrolytic cleavage with hot water after exhaustive washing to remove excess mineral acid or alkali. Since conversion of the hot-water-insoluble collagen into soluble gelatin is a time-temperature reaction, a series of extractions are made by soaking in hot water until maximum solids content is reached in the liquor. The average molecular weight will decrease with each extraction and will be manifested by a gradual decrease in jelly strength and viscosity. Fresh from the extraction kettles, these "light liquors" are quite dilute and lend themselves readily to subsequent steps of settling, centrifugation, filtration, and deionization.

The liquors are then concentrated by vacuum evaporation, usually first in a triple-effect evaporator, followed by a turbulent-film vacuum concentrator. In final concentrated form, the "heavy liquor" sets rapidly to a gel when cooled and is usually so converted on a chill roll from which sheets or strings of gel are conveyed to a low-temperature drier. Low-humidity, cool air is passed over or through the product until skinning prevents agglomeration. The temperature is then raised to effect complete drying. Final steps include grinding and blending to specification.

Commercial gelatin is almost colorless and is virtually free from odor. Solutions of gelatin are practically water-white until concentrations are raised to very high levels. Gelatin is a complex mixture of collagen degradation products which range in molecular weight from about 30,000 to 80,000 or greater. The exact sequence of amino acids in the molecular chain is ill defined, but it has been found that nearly half of the gelatin molecule is comprised of glycine, proline, and hydroxyproline. Irrespective of the degree of degradation found in a gelatin sample, it will contain about 45 milliequivalents of amino functions and about 70 milliequivalents of carboxyl functions per 100 grams. These values remain substantially unchanged unless gross hydrolysis is effected by acids, alkalis, or proteolytic enzymes.

The three outstanding properties of gelatin are: (1) ability to form reversible gels; (2) viscosity; and (3) high strength of films. These properties account for the many uses of gelatin. As a dry, granulated product, gelatin appears amorphous. It is not soluble in cold water, but swells rapidly until it has imbibed about 6 or 8 times its weight of water. The swollen gelled particles then melt to a viscous solution when warmed above the melting point, which lies between 104 and 113°F (40 and 45°C).

In many respects, gelatin resembles numerous other dry proteins in that it does not melt and functions as an amphoteric electrolyte. Gelatin from acid-cured stock has an isoelectric point of about 8.3 to 8.5 (Type A); and alkali-cured gelatin an isoelectric point of about 4.75 to 5.0 (Type B). Either type is most difficult to precipitate at its isoelectric point unless powerful protein precipitants are used. Various aldehydes and heavy metal salts exert a tanning action on gelatin so as to produce insoluble material. The tendency to gel can be sharply decreased by a number of peptizing agents in much the same way as salts depress the freezing point of water.

Commercial gelatins are graded according to the viscosity and gel rigidity of their aqueous solutions. Since these easily measured constants describe the potential suitability of gelatin for a given use, the commercial product is sold on the basis of its viscosity and Bloom test (jelly strength). These two physical constants are reproducible within 1% and indicate not only the suitability of the product for a given use, but also the type of gelatin being tested. An acid-cured gelatin will have a jelly strength-to-viscosity ratio of 4 or 5 to 1, whereas an alkali-cured gelatin will have a ratio of 2.5 or 3 to 1.

Large amounts of gelatin are used in the production of flavored powders for the preparation of desserts and salads. Gelatin also finds wide use in ice cream production, where the substance functions to prevent or reduce the growth of ice crystals. In the confectionery and baking industry, gelatin finds use as an essential ingredient of marshmallow. Gelatin is used in connection with the packing of a wide variety of meat-base products (jellied meats). Gelatin also finds wide application in hard and soft capsules for packaging drugs and vitamin preparations. An important nonfood use is found in photography.

GEM STONES.

A gem stone is a mineral substance which because of its beauty or rarity is in demand for ornamental purposes, chiefly personal adornment. The origin of such use for what we now call gem minerals is lost in the dim vistas of early human history. Ancient records describe the various gem stones, and archeologists find them in their investigations of bygone peoples. When we look at a collection of minerals with their bright colors and varying degrees of transparency or light-reflecting power, we cannot doubt that primitive man was much attracted by them and valued them greatly. We may imagine, too, that the occasionally found crystals with their regular geometric forms were more highly prized than broken fragments of the same minerals. Later they learned to polish them. Apparently the oldest form into which stones were shaped is that known as *cabochon*, a French term derived from the Latin word for *head* and referring to its rounded shape. The forms were either hemispherical or hemiellipsoidal. The Emperor Nero is supposed to have had a large emerald cut en cabochon, and indeed, for several centuries after this time this seems to have been the only sort of cutting employed. The supposedly accidental discovery in 1475 that diamonds would scratch each other began the era of modern gem cutting. Previously it had been believed that diamonds were so hard that they could not be artificially shaped. At first, however, little progress was made in fashioning gems other than polishing a number of facets without any definite arrangement.

We owe to Vicenzio Peruzzi, a Venetian, the credit for devising the so-called "brilliant cut," the style of the modern diamond cutting, which, except for certain refinements due to a more thorough understanding of the behavior of minerals toward light, remains the same as in Peruzzi's day. At the present time, transparent stones of all sorts are usually "brilliant cut," whereas translucent or opaque stones are cut en cabochon.

Since time immemorial, dealers in gems have used as the unit of weight the carat, undoubtedly introduced from the east. The word is derived from the Greek meaning a small horn, referring to the pods of the locust tree, *Ceratonia siliqua*, a common Mediterranean tree whose seeds were said to have been taken as the unit of weight in buying and selling gems. In the nineteenth century the actual weight of the carat differed slightly in different countries of Europe, from a little under to somewhat over $\frac{1}{5}$ of a gram. The metric carat is exactly $\frac{1}{5}$ of a gram.

In recent years, synthetic stones have made large inroads in both the jewelry and industrial fields. Although numerous gem materials have been manufactured synthetically, only a few are cut as gemstones for the jewelry trade, notably corundum, spinel, emerald, rutile, garnet, sphene, and strontium titanate. Synthetic diamond for industrial use is produced by subjecting a carbonaceous material to very high temperature and pressure. One of the processes in use for creating synthetic crystals of corundum and spinel was developed by Auguste V. L. Verneuil (1856–1913), a French mineralogist and chemist. The starting composition is an alumina powder which then is melted in an oxyhydrogen flame, whereupon a series of drops are formed that build up the boules of the synthetic gems. See also **Beryl, Corundum, Diamond.**

GENETIC CODE.

Cellular information that is transferred to the daughter cells in the process of mitosis, or, by way of the fertilized ovum, to an entirely new organism. This information makes it possible for the new cell or the new organism to construct itself in a fashion identical, or nearly identical, with that of the parent cell or the parent organisms.

That such information must exist is obvious; the location of the information within the cell is less so. By 1904, Sutton specifically named the chromosomes as the site. The basic reasoning behind this lay in the fact that the chromosomes are carefully replicated in the process of cell division and as carefully shared among the daughter cells so that each has its full complement of different pairs. Furthermore, the sperm cell, carrying all the information contributed by the male parent, is a tiny bag containing virtually nothing but chromosome material.

The next problem lay in locating the information within the chromosome and here there seemed no doubt. The chromosomes contained complex protein molecules and protein molecules were the largest and most complex molecules in living tissue. Each protein molecule is made up of twenty different but related units, the amino acids, arranged in a chain, or in several connected chains. Each different order of amino acids produces a different protein molecule (with even trivial differences sometimes proving important) and the number of different orders possible is formidable. A medium-sized protein molecule, such as hemoglobin, can have its amino acids arranged in any of about 10^{600} ways.

The chromosomes also contained nucleic acids but through the first quarter of the twentieth century, these were viewed as relatively small molecules of relatively simple structure and were dismissed out of hand as possible carriers of genetic information. The fact that Kossel found, in 1896, that fish sperm contained only extraordinarily simple proteins, yet possessed a full complement of nucleic acid was puzzling but was, on the whole, ignored.

The turning point came in connection with two strains of pneumococci, the S strain ("smooth," because it formed a smooth pellicle) and the R strain ("rough," because it did not.) If dead S strain, or even an extract of the strain, were added to living R strain, the R strain became capable of forming the pellicle and changed over into S strain. Apparently, something in the R strain contained the information required to oversee the formation of the pellicle.

In 1944, Avery, MacLeod and McCarty produced the active extract in great purity and were able to show that it contained nucleic acid *only*, with no protein at all. For the first time, genetic information had been pinpointed and it was found in the nucleic acid component of chromosomes, not in the protein.

By then, nucleic acids were recognized as complex molecules after all, as large as, or larger than, protein molecules, and differing characteristically from species to species. In 1953, Watson and Crick had determined the structure of the variety of nucleic acid in chromosomes ("deoxyribonucleic acid" or DNA) and had worked out the manner by which it formed a replica of itself ("replication") in the course of cell division (see **Nucleic Acids and Nucleoproteins.**

But with the site of the information located, another problem was raised. How could the nucleic acids carry the necessary information?

The working of the cell depends on the nature and relative quantities of the thousands of different enzymes it contains. Each enzyme catalyzes a particular reaction and it is the network of reactions within the cell or cells of an organism that is responsible for all its characteristics. It is the nature of the enzymes, their relative quantities, the manner in which the working of one stimulates or inhibits the working of another, that is all the difference between a human and a mandrill, a cat and a catfish, a lion and a dandelion.

The essence of what now came to be called the genetic code, then, was the manner in which the structure of specific DNA molecules guided the production of specific enzyme molecules. The enzyme molecules were all protein in nature, made up of arrangements of twenty different amino acids. The DNA molecules were made up of arrangements of only four different nucleotides, and for all the great size of the molecules, that paucity in number of different units made DNA seem much simpler than proteins and too simple, perhaps, to carry the necessary information to guide protein-manufacture. (The four nucleotides, by the way, are adenylic acid, guanylic acid, cytidylic acid, and thymidylic acid, usually symbolized as A, G, C, and T, respectively. In another variety of nucleic acid, ribonucleic acid or RNA, thymidylic acid is replaced by the very similar uridylic acid, symbolized as U.)

In 1954, Gamow pointed out that the presence of only four different nucleotides did not matter. The code might involve groups of nucleotides. If a long chain, made up of four different nucleotides, is taken two at a time the number of different combinations is 4^2 or 16; if three at a time, the number is 4^3 or 64.

Investigation showed that it was indeed a chain of three adjacent nucleotides ("triplets" or "codons") in the nucleic acid molecule that represented a specific amino acid. The fact that there were 64 codons and only 20 amino acids, meant that several closely allied codons might all stand for the same amino acid, thus allowing redundancy, so that small errors in the repli-

cation of a nucleic acid often do not affect its working. One way of indicating that more than one triplet stands for a particular amino acid is to say that the genetic code is "degenerate." It is also "universal" for, as far as biochemists have been able to tell, the same triplet stands for the same amino acid in all organisms.

The next problem is to work out the code "dictionary," by determining which particular codon stands for which particular amino acid. The first breakthrough in this direction came in 1961 when Nirenberg and Matthaei made use of a synthetic RNA molecule made up of a long chain of a single nucleotide, uridylic acid. All the codons were therefore UUU. By adding to a solution of this RNA a supply of amino acids and all the cellular paraphernalia required for protein manufacture, a chain made up of a single amino acid, phenylalanine, was produced. Thus, the first item of the dictionary was determined: UUU = phenylalanine.

By 1967, the dictionary was completed. It was determined, for instance, that both UUU and UUC stood for phenylalanine. Similarly, GGU, GGC, GGA, and GGG all stood for the amino acid, glycine. The codons, AUG and GUG, were "capital letters" serving to indicate the beginning of a chain, while UAA was a "period" serving to end a chain.

There was also the problem of how the information originally contained in the DNA molecule, which existed in the chromosomes and never left the nucleus, reached the site of protein manufacture, which was in the cytoplasm. (In 1956 Palade had demonstrated the site of protein manufacture to be on tiny cytoplasmic particles rich in RNA, which he called "ribosomes." There are as many as 150,000 ribosomes to the cell.)

Clearly, the information had to pass from chromosomes to ribosomes and suspicion fell on RNA, which was similar enough in structure to DNA to carry the imprint of DNA information, and which existed both in nucleus and in cytoplasm. Thus, a section of DNA, using the process which ordinarily builds up another section of DNA of complementary structure (where A = T or U, and vice versa, and where G = C, and vice versa (see **Nucleic Acids**), builds up, instead, a section of RNA of that structure. The RNA molecule leaves the nucleus for the cytoplasm, carrying the message (so that it is called "messenger-RNA") to the ribosomes.

What is then needed is some sort of translating device to convert nucleotide codons into amino acids. The necessary device was located by Hoagland in the form of RNA molecules, small enough to be freely soluble in the cell plasma. One end of this small RNA molecule possessed a three-nucleotide combination that would attach itself only to a particular complementary codon on the messenger-RNA chain. At the other end of the small RNA molecule is a section that can combine only with a particular amino acid.

There are twenty different RNA molecules of this sort, one for each of the twenty different amino acids. Each one has a characteristic codon at the opposite end. Therefore, the small RNA molecules line up on the messenger-RNA according to the places into which their codons, at one end, fit; and on the other end their amino acids line up as well and are bound together to form an enzyme molecule. The structure of the enzyme molecule has thus been dictated by the structure of the messenger-RNA, the structure of which was in turn dictated by the structure of the DNA in the chromosomes.

Because the small RNA molecules transfer information from the messenger-RNA to the protein molecule, they are called "transfer-RNA."

In 1964, Holly and co-workers, determined the detailed structure of "alanine transfer-RNA," the one which combines with the amino acid, alanine. It was found to be made up of a chain of 77 nucleotides, bound together to form a three-leaf-clover structure. The structures of other transfer-RNAs were worked out in subsequent years and this three-leaf-clover effect seems common to all of them.

—Isaac Asimov

GENETIC ENGINEERING. Recombinant DNA

GEOCRONITE. A mineral sulfide of lead, antimony and arsenic, Pb_5SbAsS_8. Crystallizes in the monoclinic system. Hardness, 2.5; specific gravity, $6.4\pm$; color, gray to blue with metallic luster; opaque.

GERMANIUM. Chemical element symbol Ge, at. no. 32, at. wt. 72.59, periodic table group 4a, mp 937°C, bp 2830°C, density, 5.36 g/cm³ (20°C). Elemental germanium has a diamond cubic crystal structure. Germanium is a silvery-white, lustrous, hard, brittle metal. When heated in oxygen to 730°C, the metal is partially oxidized to dioxide. The element is unaffected by solutions of acids and bases, but is soluble in fused NaOH. In the form of powder (dull gray), combines readily with chlorine to form the volatile tetrachloride. Although predicted by Mendeleev as early as 1871, the element was not fully identified until 1886 by Winkler. Mendeleev had previously termed the missing element *eka-silicon*. There are five natural isotopes, ^{70}Ge, ^{72}Ge through ^{74}Ge, and ^{76}Ge. Seven radioactive isotopes include ^{67}Ge through ^{69}Ge, ^{71}Ge, ^{75}Ge, ^{77}Ge, and ^{78}Ge. All have a relatively short half-life, the longest, ^{68}Ge, with a half-life of 275 days. In terms of abundance, germanium ranks 32nd among the elements in the earth's crust and thus is about as abundant as gallium, selenium, arsenic, and bromine. First ionization potential, 8.13 eV; second, 15.86 eV; third, 31.97 eV; fourth, 45.5 eV. Other important physical characteristics of the element are given under **Chemical Elements.**

Germanium occurs in very small amounts in many sulfide ores, such as American zinc ores (0.25% GeO_2), and the rare mineral argyrodite (silver germanium sulfide) of Saxony and Bolivia. The primary source is flue dust from the zinc industry. Also, it may be obtained from the reduction of oxide and sulfide ores. A major ore is germanite, a copper ore found in southwest Africa. The ore is quite complex, containing some 20 different elements. The copper content ranges as high as 45%, sulfur up to 30%, whereas the germanium content is from 6 to 9%. The ore also contains up to 1% gallium. A major sulfide ore is renierite which contains up to about 8% germanium. Small quantities of germanium are found in lepidolite, sphalerite, and spodumene. Some English coals contain as much as 1.6% germanium oxide. Germanium metal of 99.99 + % purity is obtained by zone melting. In this system, electric heating coils are moved slowly along the length of an ingot. Impurities in the metal tend to raise or lower the freezing point of the molten alloy. By progressively melting the metal along the length of the ingot, the impurities which tend to lower the melting point will be swept to the last portion of the ingot to freeze, whereas the impurities which tend to raise the melting point will concentrate in the first region to freeze.

Uses. The principal uses of germanium have been in solid-state electronic devices, notably transistors, which can be used as amplifiers and oscillators. The electrical properties of germanium metal which have brought about this wide use in semiconductors are its high specific resistance at ordinary temperatures and the narrow gap between its filled energy band and its conduction band. Thus, germanium is an intrinsic semiconductor, wherein an increase of temperature or the addition of very small amounts of group 3 or group 5 elements can cause electrons to move readily to the conductive band to form "holes," thus making the material conductive. A key to the manufacture of semiconductor devices is making materials of high purity, great uniformity, and in sufficient quantity.

The addition of as little as 0.35% germanium to tin doubles the hardness of tin. Similarly, germanium improves the strength and hardness of aluminum and magnesium alloys. These applications are limited, however, because of the current high costs of germanium. Germanium-silicon alloys are under intensive study for use in thermoelectric generators. Advantages claimed for these metals include better thermoelectric qualities above 600°C, an improved efficiency per unit weight factor, and virtually no corrosion or decomposition.

Chemistry and Compounds. Germanium forms compounds in which the oxidation states are (II) and (IV). The divalent ones are unstable. Thus the monoxide is readily oxidized by air when hydrated. However, when completely dehydrated it resists the action of H_2SO_4 and potassium hydroxide, and reacts only slowly with fuming HNO_3. On heating in an inert atmosphere it disproportionates to the element and germanium dioxide, GeO_2. The latter resembles silicon dioxide in existing in more than one form, with a difference in chemical properties. The stable form at room temperature has the rutile structure, but just below the melting point the stable form has the cristobalite structure. Germanium(IV) oxide, GeO_2, prepared by hydrolysis of germanium(IV) chloride, $GeCl_4$, is somewhat soluble in water, acids, and alkalis, but GeO_2 from heating of germanic acid is insoluble. Like silicon dioxide, GeO_2 forms gels readily.

Germanium(II) hydroxide, $Ge(OH)_2$, is obtained by action of alkali hydroxides upon germanium(II) chloride, $GeCl_2$, solutions; it is amphiprotic, dissolving in excess of the alkali. Moreover, the acid form, sometimes called germanous acid, is obtained upon heating the hydroxide: $Ge(OH)_2 \rightarrow HGe(O)H$. GeO_2 is slightly acid in solution and when freshly precipitated ($pK_A = 9.4$). There is no experimental evidence for the existence of a definite hydrate, although melting point diagrams of germanate salts have indicated the existence of ortho($\equiv GeO_4$), meta($= GeO_3$), and tetra($= Ge_4O_9$) compounds.

Germanium forms dihalides and tetrahalides with all four of the common halogens. In general, the dihalides readily react with halogens or other oxidizing agents to form tetravalent germanium compounds, and some, e.g., the iodide, disproportionate to the metal and tetravalent compound.

Suggestive of carbon and silicon is the existence of hydrides of germanium, though they are much fewer in number. The compound GeH_4 is called germane (mp -165°C, bp -90°C). Compounds having the general formula Ge_nH_{2n+2} ($n = 2$, 3, etc.) are called digermane, trigermane, etc., according to the number of germanium atoms present. The first three compounds in this series have been obtained by treatment of magnesium germanide with ammonium bromide in liquid ammonia. Compounds such as $GeHCl_3$ and alkylgermanes are also known. Germane and the alkyl- and aryl-substituted germanes retaining at least one hydrogen atom are somewhat more acidic than the corresponding silanes in nonaqueous media, easily forming alkali salts, R_3GeM and even dialkali salts R_2GeM_2 under some circumstances. Germane, GeH_4, appears to be thermodynamically stable, although no quantitative data are available on its heat of formation. It decomposes at about 285°C.

Germanium resembles the other main group IV elements in forming organometallic compounds. Over two hundred have been reported, from chloromethyl trichlorogermane,

$ClCH_2GeCl_3$ to cyclotetrakis (diphenyl germanoxane), $[(C_6H_5)_2GeO]_4$.

See list of references at the end of the entry on **Chemical Elements.**

GERSDORFFITE. A mineral related to cobaltite and ullmannite in the cobaltite group. A sulfide-arsenide of nickel, NiAsS. Crystallizes in the isometric system. Hardness, 5.5; specific gravity, 5.9; color, white to gray with metallic luster; opaque. Named after von Gersdorf (1842) at Schladming.

GETTERING. Degasification.

GIBBERELLIC ACID AND GIBBERELLIN PLANT GROWTH HORMONES. These organic chemical compounds, first isolated from the parasitic fungus *Gibberella fujikuori* in Japan in the late 1930s, produce unusual results when applied to plants, including various food crops. The results can be advantageous or disadvantageous. The phenomena of the gibberellins were uncovered as the result of studying the excessive leaf elongation in rice plants. This fungus disease of rice is sometimes referred to as the "foolish seedling" disease in rice. When infected with this fungus, the rice plants grow ridiculously tall and the stems break before the plants can flower and produce seed. When experimentally applied to higher plants, the gibberellins have varied effects. The most common reaction is the rapid lengthening of the stems. The stems of citrus trees, for example have been stimulated to grow at a rate six times greater than normal. When applied to the young fruit of seedless grapes, the gibberellins cause the fruit to grow much larger and to stay on the vine longer. Although some results can be predicted from experience with other species, generally results must be observed through long trial-and-error experimentation with many plants and many different concentrations and forms of the chemical growth hormones. The gibberellins are but one category of several kinds of plant hormones which affect food crop production. See also **Plant Growth Modification and Regulation.**

Since the 1960s, commercial gibberellin formulations have been available. These take several forms, ranging from liquid concentrates through tablets and powders. In some countries, registration is required of these compounds. The following practical results, among others, have been achieved when gibberellins are used properly on certain food plants:

Artichoke: prolongs picking period

Barley: enzyme content increased

Bean: more rapid emergence of plant

Blueberry: better fruit set

Celery: extends winter crop

Cherry (sour): combats cherry yellow virus

Cucumber: produces staminate flowers

Grape: loosens and elongates clusters; increases grape size

Hops: increases yields; aids harvesting

Lemon: delays yellow color development

Oats: promotes more rapid emergence of plant

Orange (navel): retards aging of rind

Lettuce: increases seed production; effects uniform bolting

Potato: stimulates sprouting

Prune (Italian): increases yield; reduces internal browning

Rhubarb: for forced crops, increases yield

Rye: promotes more rapid emergence of plant

Soybean: Promotes more rapid emergence of plant

Sugarcane: increases sucrose yield

Tangerine: increases yield and fruit set

Wheat: promotes more rapid emergence of plant

The gibberellins are actually a family of closely related substances. To date, structures have been determined for well over a dozen of these and a number have been isolated from higher plants. See accompanying diagrams. The structure of three fused saturated or nearly saturated rings, with two additional rings perpendicular to them, suggests relationship to the diterpens for which there is strong isotopic evidence. For example, C^{14}-kaurene is readily converted to gibberellic acid (GA_3) by *Gibberella* cultures. The biosynthesis is apparently inhibited by chlorocholine, which is suspected as the basis for the dwarfing action of this compound. GA_7 to date has had the highest activity in most tests.

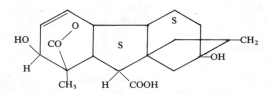

Gibberellic acid (GA_3)

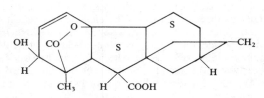

Gibberellic acid (GA_7)

Gibberellins cause rapid elongation of shoots; many of the dwarf forms of maize (corn), bean, pea, and morning glory (closely allied to sweet potato) are caused to grow into tall forms indistinguishable from their tall genetic relatives. Many long-day plants are brought into flower in short days by gibberellin, and some biennials, including *Hyoscyamus* (henbane), are made to flower in one year. This process depends on the activation of cell divisions in the shoot apex. Like auxins (other plant hormones), gibberellins produce parthenocarpic fruits, especially on tomato, but unlike auxins, they do not inhibit lateral bud development, but they inhibit rooting of cuttings and promote the germination of many seeds. Their transport shows no polarity. They are active at concentrations comparable to those of the auxins. There is good evidence that the gibberellins act only when auxin is present.

In their biological function, it is believed that the gibberellins destroy or bypass naturally occurring inhibitors which normally prevent premature germination. However, high concentrations of the gibberellins and like substances actually prevent germination in certain varieties of seed.

GIBBS-DUHEM EQUATION. In a system of two or more components at constant temperature and pressure, the sum of the changes for the various components, of any partial molar quantity, each multiplied by the number of moles of the component present, is zero. The special case of two components is the basis of the Gibbs-Duhem equation of the form:

$$n_1 \, d\overline{X}_1 = -n_2 \, d\overline{X}_2$$

in which n_1 and n_2 are the number of moles of the respective components and $\overline{X}_1$ and $\overline{X}_2$ are the partial molar values of any extensive property of the components.

GLASS. Glass is an inorganic product of fusion which has cooled to a rigid solid without undergoing crystallization. It is a solid. It may be transparent, translucent, or opaque, and it may be colored. The chemical composition and corresponding properties may vary over a wide range. Glass will support a load, and may be shaped, broken, or cut. It is much like other solid materials, and yet it is unique.

Its uniqueness becomes obvious when it is examined on a submicroscopic level. Most solids have regular, orderly patterns for the arrangement of atoms, molecules, and ions, but glassy materials are highly disordered. There is some short-range order in glass, but beyond one or two atoms or ions the ordering may be described as random. Thus, on a submicroscopic level, glassy solids look more like liquids than solids.

Since glasses do not have ordered structures with correspondingly specific bonding energies between rows, stacks, planes, or discrete ions, they do not have definite melting points. When a glassy material is heated, it softens slowly and transforms to the liquid state. Crystalline solids generally transform from a solid to a liquid at a single specific temperature, the melting point. On cooling, a material that has a tendency to crystallize to solid will do so at the same temperature at which it transformed to a liquid. When a glass is cooled from a high temperature, it becomes increasingly viscous in a manner which is related to the inverse of the temperature until it becomes a rigid solid again. Thus, a specific temperature where melting or freezing takes place cannot be found for glass; i.e., glass does not have a melting point.

Most glasses can be made to crystallize if they are subjected to the right conditions of temperature and rate of cooling, which suggests that the glassy state is like a supercooled liquid. This is not borne out by measurements of density and other volume properties, which do not decrease in a linear manner as glass is cooled below its crystallization temperature.

Why is it that some melts when cooled through a crystallization temperature form glasses while others do not? It is simply a question of whether the melt can be cooled through the temperature range of maximum crystal growth rate faster than the crystals can grow. Thus table salt cannot be formed as a glass, but sand, or SiO_2, can be. The maximum crystal growth rate is normally just below the melting point of the material, but materials that tend to form glasses easily are much more viscous at these temperatures. For example, in the extreme cases of salt and sand, the differences in viscosities at their respective melting points is about eight orders of magnitude.

The two-dimensional drawing in Fig. 1 shows SiO_2 in the ordered, or crystalline, and in the random, or glassy, state to illustrate the difference on a submicroscopic scale. Figure 2 shows how the volume properties of a material would respond to temperature if they could be prepared as a glass, a supercooled liquid, or crystalline material.

Most glasses are composed of inorganic oxides, and most commercial glasses contain SiO_2 as their major constituent, but there are organic glasses and elemental metallic glasses. Glass is typically hard and brittle, and exhibits a conchoidal fracture. Most commercial glasses are transparent or translucent in the visible portion of the spectrum.

Types of Glasses

A wide range of glass products exists, each type having special properties. The properties of glass are determined primarily by chemical composition, and since the composition may be varied almost infinitely, there are many thousands of different glasses. However, they may be generally classified into soda-lime-silica glasses; lead glasses; borosilicate glasses; and a number of special glasses, including solder glasses, laser glasses,

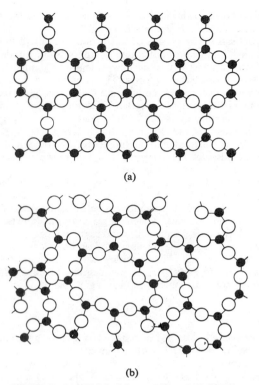

Fig. 1.   Silicon dioxide (SiO_2): (a) crystalline; and (b) glassy state. ● = silicon; ○ = oxygen.

silica glass, glass-ceramics, and colored glasses. These types essentially bracket the commercial glasses.

Soda-Lime-Silica Glasses. This is the most important group in terms of tonnage melted and variety of use. The combination of silica sand, soda ash, and limestone produces a glass that is easily melted and shaped and has good chemical durability. The raw materials are indigenous to most areas of the world and inexpensive. Soda-lime glasses are particularly suited to automatic machine-forming methods and are the basis for most of the bottle-, sheet-, and window-glass industry. Very small amounts (often less than 1% of the total batch) of alumina,

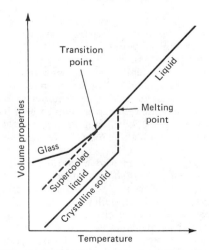

Fig. 2.   Volume properties of glass in contrast with crystalline solids as a function of temperature.

magnesia, boric oxide, and other chemicals are added to act as stabilizers and to increase durability.

Lead Glasses. The glasses of this group, composed basically of silica sand and lead oxide, have a high refractive index and high electrical resistivity. Potash is present as a significant constituent in most of these glasses. The slow rate of increase in viscosity with decrease in temperature makes lead glass particularly suitable to hand fabrication. The amount of lead may vary considerably, even up to 92% lead oxide; it is a more expensive glass, as the raw materials are relatively expensive and special care is needed in melting to avoid bubbles and seeds. Glasses of this type are used in high-quality art and tableware and for special electrical applications.

Borosilicate Glasses. This group of glasses is basically a combination of silica sand with boric oxide and soda ash. The glasses have excellent chemical durability and electrical properties, and their low thermal expansion yields a glass with a high resistance to thermal shock. High durability makes them ideal for demanding industrial and domestic use, such as chemical laboratory ware, cook ware, and pharmaceutical ware. These glasses were developed in the early part of this century to cope with the problem of cold rain on hot railway-signal lights.

Special Glasses

Solder Glasses. These glasses have low softening and annealing temperatures together with expansion characteristics which permit them to be used as intermediate glasses in making seals between two glass surfaces, between a glass and a metal, or between two ceramic surfaces. In fact, solder glass might be described as a high-grade glass glue. Normally, sealing temperatures are well below the annealing temperature of the glass being sealed, and there is little permanent effect on the glass parts being joined. The major constituents of these glasses include lead oxide, boric oxide, and zinc oxide.

Laser Glasses. Glass has various characteristics which make it an ideal laser host material. Its random structure permits broad emission and absorption bands, which provide higher efficiency, more energy storage, and greater energy per pulse than any other material. In addition, most lasing ions are easily soluble in the glass, and rods, fibers, or disks of any size and of high optical quality are easily fabricated. Of the several rare-earth ions which have been made to lase in a glass host, only neodymium has received commercial application. When a neodymium glass lases, it emits light at a rather fixed wavelength of 1.06 nm. Neodymium-doped silica and phosphate glasses have been used to provide the energy source for laser fusion research throughout the world.

Silica Glass. A glass composed of silicon dioxide as the only constituent has a very high softening temperature and a very low thermal expansion. It is costly to make and fabricate because temperature in excess of 1800°C is required to manufacture it. However, its refractory character coupled with its very high resistance to thermal shock makes it ideal for special laboratory equipment, windows in high-temperature environments, and instruments.

Glass-Ceramics. These materials are formed in the same manner as conventional glasses and then subjected to heat treatments which caused controlled nucleation and crystallization. Although nearly completely crystalline, their properties can range from transparent to opaque; electrically insulating to weakly conducting; hard to machineable; and with zero or negative thermal expansions depending upon the composition and heat treatment. This family of materials is based on glasses whose major constituents are magnesium oxide, lithium oxide, alumi-

TABLE 1. COMMONLY USED INGREDIENTS FOR COLORING GLASS

GLASS COLOR	COLORING AGENT	STATE
Red	Cadmium sulfide, cadmium selenide	Reduced
	Cuprous oxide	Reduced
	Gold (metal)	
Yellow	Cerium oxide with titanium oxide	
Yellow-green ..	Chromic oxide	Oxidized
Blue-green	Iron chromite	Reduced
Blue	Cobalt oxide	
Purple	Neodymium oxide	
Gray	Nickel oxide with titanium oxide	
Black	Copper, cobalt, nickel, and iron oxides in combinations of two or more	
Amber	Iron sulfide	Reduced
Flint (or colorless)	Selenium and cobalt oxide*	Oxidized

* Selenium and cobalt are used in flint glass to add red and blue hues in amounts only sufficient to balance the green hue resulting from iron oxide present as impurity in most naturally occurring raw materials. The intended result is an even light transmission over the whole visible spectrum.

num oxide, and silicon dioxide. The crystalline phase or phases and their morphology control the properties of the materials, but the starting chemical composition and the heat treatment determine which crystalline phases will result. Glass-ceramics, which are the result of recent research efforts, have found applications as household cooking ware, reflective optics substrates, chemical processing components, and cooking-stove tops.

Colored Glasses. Nearly all glasses can be colored by adding one or more colorants to the batch in correct amounts. Production of some colors requires, or is enhanced by, the state of oxidation of the coloring agents and the atmospheres in which the glasses are melted. Table 1 indicates the colors obtainable, colorants used, and chemical states required or utilized.

While the preceding paragraphs describe several classes of glass, within each class there can be infinite composition variations to fit the exact requirements of the user. Table 2 shows typical composition ranges for commercial glasses.

See also **Optical Fibers.**

Manufacturing Processes

Glass products are many and varied, and glass compositions range rather widely, depending on the desired products. Figure 3 shows a typical cross section of a glass manufacturing facility. Raw-materials weighing, mixing, charging, and melting are common requirements regardless of the forming operation that is to follow. Most melting furnaces have a primary melting area, followed by a refining or homogenizing section, which is connected to the forming operation by channels called *feeders*. Although fiber glass is not passed through an annealing furnace after it is formed, most other glass products are annealed to relieve stresses caused by uneven cooling during and immediately after forming.

It is apparent that although there are several similar steps in all glass-manufacturing processes, the forming operations are the most diverse.

TABLE 2. COMPOSITION OF COMMERCIAL GLASSES (Weight Percent)

| | Soda-Lime-Silica Glass | | | | Borosilicate Glass | Laser Glass | Solder Glass | Lead Glass | Glass-Ceramics |
	Containers	Plate and Window Glass	Tableware	Fiber Glass Fabrics and Insulation					
SiO_2	70–74	71–74	71–74	65–74	70–82	61–69	0.5–16	35–70	62–70
Al_2O_3	1.5–2.5	1–2	0.5–2	2–4.5	2–7.5	0–5	0.1–4	0.5–2.0	17–22
B_2O_3				3–5.5	9–14		7–20		
Li_2O									3–5
Na_2O	13–16	12–15	13–15	8–16	3–8	12–24		4–8	
K_2O				0–1				5–10	
CaO	10–14	8–12	5.5–7.5	5–16	0.1–1.2	3–10			0–5
MgO			4.0–6.5	3–5.5					0–7
BaO					0–2.5		0–4		
ZnO							7–62		
PbO							4–77	12–60	
CuO							0–10		
Nd_2O_3						1–6			
CeO_2						0.1–1			
F_2							0–2		
ZrO_2 and TiO_2									3–10

Batch Preparation. This begins with the selection, procurement, and storage of an adequate quantity of the raw materials. Selection is made on the basis of the oxides which each material contains and will provide to a glass and on the basis of purity and grain size. Naturally occurring raw materials are used wherever possible for economy, e.g., silica sand, limestones, feldspars, borates, soda ash, boric acid, potash, and barium carbonate. The prescribed quantities of these raw materials, depending on their chemical composition, are measured carefully and mixed together to provide a homogeneous batch. Such mixing is done on an intermittent or a continuous basis, depending on the volume of batch needed to charge the furnaces. The batch is conveyed by a variety of means to the furnaces but always in such a way that segregation is avoided. The importance of grain size of the various raw materials becomes evident in preventing dusting and/or segregation.

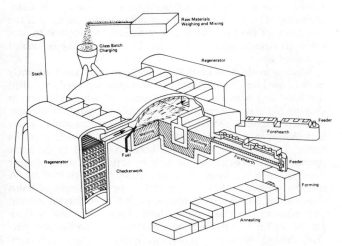

Fig. 3. Representative glass-producing facility.

Furnaces. A variety of furnaces are used in the industry to melt the batch to produce glass. They must all accomplish the two purposes of confining the heat to the necessary area and containing the melted glass within the furnace. Crucibles or pots are sometimes used to contain the batch and the melted glass, in which cases the furnace merely retains heat; however, tank furnaces (Fig. 3) are far more common. They are so constructed that the lower portion contains the glass and the superstructure retains the heat and provides combustion space for the fuels used. "Day" tanks are used in some instances where the operation is intermittent and the quantity of glass is small. The great majority of glass produced is melted in continuous furnaces, which are charged initially with batch and cullet (broken-up pieces of previously melted glass) which are melted, filling the tank to a specified depth, sometimes up to 66 inches (~168 centimeters). Thereafter, batch and cullet are charged continuously at a rate equal to that at which the molten glass is withdrawn from the working end.

Continuous tank furnaces are designed to provide for a separate melter section and a refiner or conditioning section. The melting end is maintained at the necessary high temperatures to accomplish the melting and chemical reactions of the batch materials. The refining, or conditioning, section retains the glass long enough for it to cool to the necessary lower working temperatures.

Glass-melting furnaces are built of refractory materials of various types which will withstand the severe conditions to which they are exposed. The lower portion of the melter section, for instance, must be of the highest quality to withstand the corrosive action of the glass as well as the high temperatures used. Some sections may use lower-quality refractories because the temperature or corrosion conditions are not as severe.

Fuels used in today's furnaces in the United States are natural gas or oil. The fuel is fed to burners that project flames over the surface of the glass. Nearly all continuous furnaces utilize regenerators, which reclaim a portion of the heat from the exhausting combustion gases. Although some glass is melted entirely by the use of electric power, it is generally too expensive to use as a source of energy. When electric power is used to augment the fossil fuels, it is called *electric boosting*.

For the areas that do have sufficiently low-cost electric power, the furnaces are constructed with conventional bottoms but with superstructure only adequate for initial heat-up. They depend on a blanket of batch floating on the surface of the glass to retain the heat within the tank that is provided by the submerged electrodes. Fresh batch is added to the blanket at a rate equal to the rate of melted glass withdrawn.

Melting. This provides the mutual solution of the oxide materials at high temperatures to yield a homogeneous liquid. Temperatures may range from 1427°C to over 1593°C, depending on the glass composition. Water vapor, entrapped air, and CO_2 are given off, some of which become entrapped in the glass, resulting, initially, in a foamy mass. As the melt moves to the higher-temperature regions, the viscosity is lowered and the gases escape. Deliberate hot spots enhance the natural convection currents, promoting homogeneity. More modern furnaces utilize bubblers, which introduce controlled pulses of air through the furnace bottom, further enhancing convection. This is particularly valuable for increasing temperatures near the tank bottom in melting those glasses which are more opaque to infrared radiation.

The glass is essentially free from bubbles (or seeds) when it reaches the end of the melting chamber. It then passes under floaters in some furnaces, or through submerged throats in most, to the so-called refining section (more properly, the conditioning section). Here the refining or conditioning consists of allowing the glass to cool to a more usable viscosity level by uniformly lowering the temperature; this step also allows the remaining tiny seeds or gaseous inclusions to dissolve.

Furnaces supply glass to up to eight forming machines. Forehearths or alcoves serve to channel the glass to the individual machines or machine locations and to further change the temperature and viscosity.

Forming Operations. These are many and varied, involving two, three, or four major steps. The first is a further temperature conditioning to place the glass in the exact viscosity range, sometimes wide but often quite narrow, suitable for the selected primary forming operation. The second step is the primary forming itself, followed usually, but not always, by an annealing step. Single or multiple secondary operations may ensue. Only the major forming processes of drawing, pressing, blowing, and casting will be discussed.

Drawing is one of the simpler forming methods by which thousands of tons of window glass and millions of feet of rod and tubing are produced annually. Drawing window glass frequently utilizes a rectangular refractory frame, called a *debiteuse*, placed on the surface of the conditioned glass. It has a slot roughly 4–8 in. (10–20 cm) wide and 8 ft. (2.4 m) or more long through which the glass is pulled vertically. The width and length of the slot in the debiteuse, together with the drawing speed, aid materially in controlling the width and thickness of the sheet. The upward draw may continue until the sheet is nearly cold, when it can be stored and cracked off in suitable lengths, or it may be bent over a large roller at nearly the last moment it will withstand bending and conveyed horizontally into the annealing lehr.

Glass tubing may be drawn vertically in a manner similar to that for window glass. Another common method is the Danner process, in which a suitable stream of glass is flowed onto a conical rotating mandrel supported with its small end downward and its axis at a suitable angle to the horizontal. The tubing is drawn from the small end, through which sufficient

air is blown to retain the desired cross section of the tubing. Drawing continues horizontally over rollers until the tubing can be cracked off in lengths at the cold end.

Plate glass may be formed by flowing the molten glass over the lip of the discharge end of the furnace between a set of large water-cooled rollers and then pulling it away by means of driven rollers. The resulting sheet is up to 1 in. (2.54 cm) or more thick and 10–12 ft (3–3.7 m) wide. However, most flat glass made throughout the world today is made by the recently developed *float-glass* process. In this process the molten glass is formed into a sheet by floating it on a bath of molten metal such as tin. The glass flowing onto the bath of tin is pulled across the surface and cooled to the temperature at which it is rigid while still on the molten metal. The outstanding advantage of this process is that it produces a plate of glass both surfaces of which require no further polishing.

Modern methods of pressing, blowing, and casting usually involve an intermediate step, the formation of a suitable charge of glass, or gob, for the ensuing operation. The most common method involves a gob feeder located at the end of the forehearth. This consists of a bowl, or spout, kept full of glass by flow from the forehearth and having an orifice in its bottom and a refractory tube suspended in the bowl over the spout. The tube may be lowered to shut off the flow of glass or raised to permit flow at a selected rate. A refractory plunger operates vertically inside the tube. It provides a pumping action on its upstroke, momentarily restraining the flow of the glass. Its downstroke forces the accumulated glass out of the orifice, where it is sheared off. The results is a charge of glass, called a gob, of controlled size, which is delivered to the forming machine by gravity.

Pressing, or press-forming, operations normally are used for relatively shallow, heavy-walled products. Pressing is accomplished by means of a metal mold (usually iron or steel), a ring which is centered on top of the mold, and a plunger which is forced into the mold through the ring. The mold shapes the exterior of the product, the ring the top, and the plunger the interior. A pressing machine may have many molds mounted on its circular rotating table, a ring for each mold or, more commonly, a single ring mounted on the same mechanism as the plunger, and a single plunger. After a gob is charged into the mold, the machine indexes one station under the plunger and the plunger moves down into the mold, dwells momentarily, then retracts. It is noteworthy that the plunger action flows the glass into the mold cavity rather than stamping out the product by a quick movement. Since considerable heat is removed from the glass by the plunger, it is cooled with water internally. The product remains in the mold for about half the revolution of the press table before removal to allow it to cool below its deformation temperature. The molds may be cooled by forced air.

Blowing methods work best for deep products and frequently must be used for thin-walled items. A common procedure, called the *blow and blow*, involves two steps, of which the first is shaping the glass charge into a form called a *blank* or *parison*. Gob-fed machines receive the gob in the parison mold, where it is shaped into a cylinder about two-thirds the height of the bottle. The finish, or top, of the bottle is formed in the same operation at the bottom of the mold by action of a small plunger entering the mold from below and delivering a puff of air. A transfer mechanism holding the parison by the completed finish then swings and inverts it into a second mold for the second step, blowing the glass into its final shape. The most modern machinery for rapidly forming containers and bottles commercially are individual section (IS) machines. Each section is capable of forming up to three gobs at the same time and there are as many as ten sections per machine. The individual sections can be sequenced electronically to produce 300 or more bottles per minute on a ten section machine.

The *Owens process* employs vacuum to charge the glass into the blank or parison mold. Here, a blank mold dips into a shallow pot of molten glass, a vacuum is applied, and a charge of viscous glass is pulled into the blank mold. The finish is formed simultaneously at the top of the blank. This blank or parison is subsequently transferred into the blow mold, where the bottle is blown into its final form. See Fig. 4.

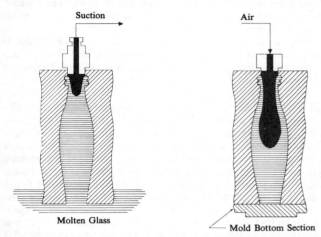

Fig. 4. Owens process. *Left:* Blank mold is dipped into the surface of molten glass, where it is filled by a vacuum suction. As the mold is lifted from the glass, a knife cuts off the glass and closes the mold. *Right:* The blank mold opens, and a puff of air is introduced to shape the parison before transferring it to the blow mold, where it is blown to its final shape.

In another modern machine, the glass flows downward from an orifice in a continuous stream which passes between rollers that flatten it into a ribbon with alternate thick and thin spots. The ribbon is picked up by a horizontally moving support in which voids coincide with the thick portions of the ribbon. Blow heads on an endless belt operating from above the ribbon provide puffs of air to aid in producing a bulbous sagging in the thick portion of the ribbon. After sufficient sagging, molds on an endless belt close around the sagging glass from below, and air from the blow heads blows the glass into the shape of the mold. After the molds open, the product, frequently light bulbs or Christmas ornaments, can be cracked off the ribbon.

Casting is usually restricted to two types of operations. The first involves the simple pouring of molten glass into molds. Examples include such massive shapes as the borosilicate mirror blank for the Mt. Palomar telescope and the large glass-ceramic mirror blanks for observatories in Australia and South America. The molds are specially constructed of refractory materials.

The second type of casting is spin casting, in which a gob from a gob feeder is fed into the bottom of a metal mold supported so that it can be rotated rapidly or spun on its vertical axis. The centrifugal force thus generated causes the glass to flow up the inclined sides of the mold, producing a conical shape. The initial movement of the glass is aided by insertion of a conical plunger into the glass at the bottom of the mold when spinning is begun. Mold speeds of up to 1,600 rpm are attained within one second. The funnel portions of television tubes are sometimes produced by this method.

Annealing. As with most substances on cooling, the temperature differential between the surface and interior layers of a piece of glass establishes temporary stresses, and the higher this differential the greater the stresses. Fracturing can occur when the stresses exceed the tensile strength of the glass. Permanent stresses can be avoided by carefully controlled cooling from a little below the annealing point to the strain point. This is the annealing range. Thereafter, the rate of cooling need only be such that the temporary stresses do not exceed the tensile strength of the glass. Glass manufacturers have learned to take advantage of these phenomena.

Annealing immediately follows glass-forming operations. In continuous processes, the ware is placed on an endless belt, which carries it through the lehr, a tunnel in which the temperature is carefully controlled. Temperature of the ware is raised initially to near the softening point, then lowered slowly through the annealing range and thereafter at a more rapid rate to the point where it can be packed or stored. The process is designed to result in the degree of permanent stresses desired. Optical glass must be annealed very thoroughly to produce an essentially distortion- and strain-free lens; however, some stresses can be tolerated or become beneficial to most other products. Small rods and tubing, for instance, are strong enough because of their regular cross section to require no annealing, while tempered glass has uniformly controlled stresses to increase its mechanical performance.

Secondary Operations. Lampworking is one of the many and varied operations utilized to produce glassware following the initial forming. The materials used are rod and tubing, which are softened in the flame of burners and shaped or blown as desired.

Grinding and polishing are important steps in many glass-manufacturing processes. Use of a sequence of increasingly finer gradations of abrasives, usually ending with jewelers' rouge or cerium oxide powder for polishing, produces the desired results. Optical lenses, prisms, and reflective optics parts are prominent examples. The plate-glass industry has used long lines of grinding and polishing equipment, but the glass produced by the float process has replaced much ground and polished plate glass.

Bending procedures are utilized to produce shapes otherwise difficult to fabricate, e.g., automotive windshields. They are produced by placing the flat pieces of proper shape and size on molds and exposing them to temperatures above the softening point. The glass takes the shape of the mold by sagging or slumping with or without assistance from mold parts contacting the glass from above. Temperatures are maintained sufficiently low and the mold material is such that the surface of the glass is unaffected.

Laminating to produce safety-glass parts, as for automotive windows, is a common practice. A sheet of resin such as polyvinyl butyral is placed between properly sized sheets of glass and the whole exposed to slightly elevated temperatures and pressures to bond the glass tightly to the resin.

Coating of glass products such as containers is quite common, the objective being to protect the container from abuse to which it is subjected in handling during filling and shipping. A coating which is not visible, can be labeled, protects the surface, and provides lubricity is required; this usually calls for a two-layer coating, such as tin or titanium oxide followed by a lubricious coating such as polyethylene. The oxide coatings are obtained by subjecting the hot container to a vapor of chloride which oxidizes to the oxide. Thick opaque or translucent oxide and metallic coatings are sometimes used to provide attractive color effects or light protection. Many precision optical lenses are coated with thin, vapor-deposited layers which reduce the light losses by reflection from the surface, and some architectural glass is coated to provide attractive colors and reflect undesirable infrared radiation.

Decorating glass or glassware is an old art that takes many and varied forms. Cutting, grinding, and mechanical or chemical polishing or etching are well known. Opaque, translucent, and transparent enamels can be applied by silk screens or other means in multiple colors and in almost any pattern. Low-melting vitreous enamels have been used for many years, and when properly fired, they provide good durability. More recently, organic polymers have been substituted for the vitreous enamel. They are not quite as durable as vitreous enamels, but they do not require high curing temperatures.

Tempering is the direct reverse of annealing; i.e., high permanent stress is induced in the glass. Rapid cooling or quenching is applied to the glass surfaces at a temperature slightly below the softening point, placing the surfaces in a high degree of compression while the balancing tensile forces are confined to the interior. Since glass always breaks in tension, very considerable strength is incorporated. Typical products are glass doors, windows, goggles, spectacles, and tableware. Tempering must be the final step in the production line. Other products can be strengthened by judicious control of the degree of annealing if their shapes permit it.

Sealing glasses to each other or to other materials must take into account the thermal expansion-and-contraction characteristics. Many glasses have thermal-expansion properties which allow them to be sealed to metals, but each metal usually requires a different glass composition. Solder glasses are used to seal two pieces of glass to each other, two pieces of metal, or a piece of metal and a piece of glass. The glass seals on light bulbs and vacuum tubes are examples of commercial glass-metal seals, while color TV tubes are sealed together with solder glass at a temperature at which the phosphors are not degraded. See also **Ceramics.**

References

Babcock, C. L.: "Silicate Glass Technology Methods," Wiley, New York (1977).
Chaudhari, P., and D. Turnbull: "Structure and Properties of Metallic Glasses," *Science*, **199,** 11–21 (1978).
McMillan, P. W.: "Glass Ceramics," 2nd Edition, Academic, New York (1979).
Tooley, F. V., Editor: "The Handbook of Glass Manufacturing," Volumes 1 and 2, Books for Industry, Inc. and *The Glass Industry Magazine*, New York (1974).
Uhlman, D. R., and N. J. Kreidl, Editors: "Glass: Science and Technology," Vol. 5: "Elasticity and Strength in Glasses," Academic, New York (1980).

—Earl D. Dietz, Owens-Illinois, Inc., Toledo, Ohio.

GLAUBERITE. This anhydrous sulfate of sodium and calcium mineral, $Na_2Ca(SO_4)_2$, crystallizes in the monoclinic system. Hardness of 2.5–3; specific gravity of 2.8; with vitreous luster and pale yellow to gray in color. Grades from transparent to translucent. Perfect basal pinacoidal cleavage with conchoidal fracture. Glauberite is a product of salt lake evaporation. World occurrences include the German saline deposits, and Borax Lake in San Bernardino County, California.

GLAUCONITE. Glauconite is a hydrous silicate of potassium, iron and aluminum with considerable ionic substitution, crystallizing in the monoclinic system. A general formula is $(K,NA)(Al,Fe^{3+},Mg)_2(Al,Si)_4O_{10}(OH)_2$. It possesses perfect basal cleavage; hardness, 2; specific gravity, 2.4–2.95; color dull green to blue-green. It is often a constituent of marine deposits forming so-called green sands. It is believed to have been produced through the alteration of iron-bearing silicates, chiefly biotite and possibly augite and hornblende. It occurs along the Atlantic Coastal Plain of the United States. Frequently found filling the interiors of the shells of *Globigerina*, a common genus of the foraminfera (Protozoa). Since *Globigerina* occurs as a deep-sea deposit, some European geologists have claimed that glauconite is only found in deep water. On the other hand, typical green sands occur associated with sand and clays which are certainly of shallow marine origin. Glauconite derives its name from the Greek word meaning *bluish-green*.

Glauconite, being of sedimentary origin, can be used to determine the age of those sediments by evaluating its $^{40}K/^{40}A$ ratio (potassium-argon isotope ratio). See also **Ocean Resources (Mineral).**

GLAUCOPHANE. Galucophane, essentially a complex silicate of sodium and iron or aluminum, $Na_2(Mg, Fe)_3Al_2Si_8O_{22}(OH)_2$, is a rather rare mineral, although it has been noted from widely separated occurrences. It is monoclinic and ordinarily is fibrous or granular. It is brittle; hardness, 6; specific gravity, 3–3.1; color, azure blue, blackish-blue or gray; luster, vitreous to pearly; translucent to opaque. Glaucophane is found only in the metamorphic rocks sometimes forming glaucophane schists. It is found in Switzerland, Italy, Siberia, and Japan; in the United States it is found chiefly in the rocks of the Coast Ranges in California and Oregon. The name *glaucophane* is derived from the Greek words meaning *bluish-green*, and *appear*.

GLUCOSE. **Carbohydrates; Starches; Sweeteners.**

GLUTAMINE. **Amino Acids.N. Starch.**

GLYCEROL. Glycerol, propanetriol, glycyl alcohol, "glycerine" $CH_2OH \cdot CHOH \cdot CH_2OH$ is a colorless, viscous liquid, of sweetish taste, odorless, boiling point 290°C, or at 12 mm pressure 170°C, gradually solidifies at 0°C to solid, melting point 18°C, miscible in all proportions with water or alcohol, insoluble in ether or chloroform, absorbs water on exposure to the atmosphere. Glycerol reacts (1) with phosphorus pentachloride to form glyceryl trichloride $CH_2Cl \cdot CHCl \cdot CH_2Cl$, (2) with acids to form esters, e.g., glycerol monoacetate $CH_2OH \cdot CHOH \cdot CH_2OOCCH_3$, glycerol diacetate $CH_3H_5(OH)(OCOCH_3)_2$, glycerol triacetate, triacetin $CH_2OOCCH_3 \cdot CHOOCCH_3 \cdot CH_2OOCCH_3$, glycerol mononitrates (alpha, $CH_2OH \cdot CHOH \cdot CH_2ONO_2$; beta, $CH_2OH \cdot CHONO_2 \cdot CH_2OH$), glyceryl dinitrates (1,2,$CH_2OH \cdot CHONO_2 \cdot CH_2ONO_2$; 1,3, $CH_2ONO_2 \cdot CHOH \cdot CH_2ONO_2$), glyceryl trinitrate, "nitroglycerine" $CH_2ONO_2 \cdot CHONO_2 \cdot CH_2ONO_2$, glyceryl tristearate, tristearin $CH_2OOCC_{17}H_{35} \cdot CHOOCC_{17}H_{35} \cdot CH_2OOCC_{17}H_{35}$, indirectly, glycerol monophosphates (alpha, $CH_2OH \cdot CHOH \cdot CH_2OPO(OH)_2$, beta, $CH_2OH \cdot CHOPO(OH)_2 \cdot CH_2OH$, (3) with oxidizing agents, e.g., dilute nitric acid, to form glyceric acid $CH_2OH \cdot CHOH \cdot COOH$, tartronic acid $COOH \cdot CHOH \cdot COOH$, mexoxalic acid $COOH \cdot CO \cdot COOH$, (4) with phosphorus plus iodine, to form allyl iodide $CH_2 : CHCH_2I$, which with hydrogen iodide yields propylene $CH_2 : CHCH_3$ and then iso-propyl iodide CH_3CHICH_3, (5) with sodium or sodium hydroxide to form alcoholates, (6) with sodium hydrogen sulfate or phosphorous pentoxide heated, to form acrolein $CH_2:CHCHO$. Glycide alcohol

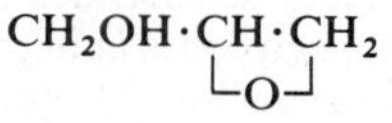

$$CH_2OH \cdot CH \cdot CH_2$$
$$\underset{O}{\llcorner \quad \lrcorner}$$

is obtained by treatment of glycerol alphamonochlorohydrin $CH_2OH \cdot CHOH \cdot CH_2Cl$, which is made by reaction of hypochlorous acid and allyl alcohol with barium hydroxide. With hydrogen chloride, glycide alcohol yields epichlorohydrin

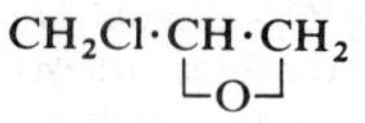

$$CH_2Cl \cdot CH \cdot CH_2$$
$$\underset{O}{\llcorner \quad \lrcorner}$$

Glycerol is obtained (1) from vegetable and animal oils and fats, most of which are mainly glycerol esters of stearic, palmitic and oleic acids by treatment with alkali (sodium hydroxide commonly used), acid (sulfobenzone—or naphthalene—stearic acid, "Twitchell's reagent"), superheated steam, or an enzyme (lipase of castor beans). Glycerol is recovered from the water solution by evaporation under diminished pressure, and purified by treatment with decolorizing carbon followed by filtration, (2) by fermentation of glucose in the presence of yeast and sodium sulfite.

Glycerol may be detected by the characteristic odor of acrolein, found on heating with potassium bisulfate.

Glycerol is used (1) in the manufacture of high explosives, e.g., glyceryl trinitrate ("nitroglycerin"), which is the main component of dynamite, (2) in antifreeze solutions, especially for automobile radiators, (3) to maintain a moist condition in fruits and tobacco, (4) in cosmetics and skin preparations, (5) to prepare glycerol phosphoric acid, used in medicine, and "boroglyceride" used as a preservative. See accompanying table.

CHARACTERISTICS OF WATER SOLUTIONS OF GLYCEROL

% GLYCEROL BY WEIGHT	SPECIFIC GRAVITY (15.6°C/60°F)	FREEZING POINT, °C
20	1.049	−5.0
40	1.103	−15.6
60	1.158	−34.0

GLYCOGEN. Carbohydrates.

GLYCOL.
A dihydric alcohol (i.e., a compound containing two alcoholic hydroxyl groups). The chemical properties are represented by those of the simplest members of the class, ethylene glycol, 1,2-ethanediol, $CH_2OH \cdot CH_2OH$, which is a colorless, viscous liquid, of sweetish taste, odorless, boiling point 197°C, miscible in all proportions with water or alcohol, slightly soluble in ether. Like ethyl alcohol, ethylene glycol is often called by the class name.

Glycol reacts (1) with sodium to form sodium glycol $CH_2OH \cdot CH_2ONa$ and disodium glycol $CH_2ONa \cdot CH_2ONa$; (2) with phosphorus pentachloride to form ethylene dichloride $CH_2Cl \cdot CH_2Cl$; (3) with carboxy acids to form mono- and disubstituted esters, e.g., glycol monoacetate $CH_2OH \cdot CH_2OOCCH_3$, glycol diacetate $CH_3COOCH_2 \cdot CH_2OOCCH_3$; (4) with nitric acid (with sulfuric acid), glycol mononitrate $CH_2OH \cdot CH_2ONO_2$, glycol dinitrate $CH_2ONO_2 \cdot CH_2ONO_2$; (5) with hydrogen chloride, heated, to form glycol chlorohydrin (ethylene chlorohydrin, $CH_2OH \cdot CHCl$); (6) upon regulated oxidation to form glycollic aldehyde $CH_2OH \cdot CHO$, glyoxal

$CHO \cdot CHO$, glycollic acid $CH_2OH \cdot COOH$, glyoxalic acid $CHO \cdot COOH$, oxalic acid $COOH \cdot COOH$.

In the preparation of glycol derivatives, important substances are: (a) the sodium glycols; (b) ethylene dichloride 1,2-dichloroethane, best prepared by reaction of ethylene and chlorine; (c) ethylene chlorohydrin 1-hydroxy-2-chloroethane, best prepared by reaction of ethylene and hypochlorous acid. From these are readily made, respectively: (a) ethers, e.g., glycol monoethyl ether $CH_2OH \cdot CH_2OC_2H_5$, glycol diethyl ether $CH_2OC_2H_5 \cdot CH_2OC_2H_5$; (b) ethylene diamine $CH_2NH_2 \cdot CH_2NH_2$, ethylene dicyanide $CH_2CN \cdot CH_2CN$, glycol itself; (c) hydroxyethylamine $CH_2OH \cdot CH_2NH_2$, ethylene cyanhydrin $CH_2OH \cdot CH_2CN$, ethylene oxide by sodium hydroxide.

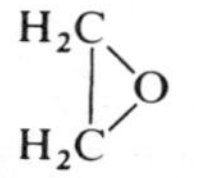

$$\underset{H_2C}{\overset{H_2C}{\Big\rangle}}O$$

Glycol is made by reaction of ethylene and chlorine or hypochlorous acid to form ethylene dichloride or ethylene chlorohydrin, respectively, followed by treatment of either of these with sodium carbonate solution heated under pressure. Glycol is also formed when ethylene is treated with potassium permanganate.

Glycol is used (1) in antifreeze solutions, especially for automobile radiators; (2) in the preparation of ethers and esters, especially nitrate for explosive; (3) as a solvent substitute for glycerol.

See accompanying table.

CHARACTERISTICS OF WATER SOLUTIONS OF GLYCOL

% GLYCOL BY VOLUME	SPECIFIC GRAVITY (15.6°C/60°F)	FREEZING POINT, °C
17	1.026	−6.7
32.5	1.048	−17.8
44	1.063	−28.9

GLYCOLYSIS.
A series of about 10 enzyme-stimulated reactions in which glucose is broken down into pyruvic acid in cell respiration. No oxygen is needed for glycolysis and it is used as the sole energy source for anaerobic organisms. In aerobic metabolism, however, the pyruvic acid is then taken through the tricarboxylic acid cycle (see **TCA Cycle**) and the balance of the energy is extracted. It appears that glycolysis can take place free in the cytoplasm, but the tricarboxylic acid cycle must take place within the mitochondria of the cell.

Glycolysis was defined in the late 1920s by Otto Warburg as "the splitting of carbohydrate into lactic acid." This type of lactic acid fermentation was well known to Berzelius, Liebig, Pasteur, and Claude Bernard in the mid-1800s, as was also alcoholic fermentation. Various kinds of carbohydrates may serve as substrates for glycolysis. It is remarkable that although glycolysis is the sum of a very large number of consecutive intermediate compounds, enzymes, and coenzymes, knowledge of these components and their sequences was acquired many years ago. For most animal cells studied, the biochemical sequence from glucose may be summarized as shown in the accompanying table.

The splitting of sugar to lactic acid is thus, briefly, the shifting of hydrogen by means of the nicotinamide moeity of diphosphopyridine nucleotide (also termed nicotinamide adenine dinucleotide). Nicotinamide in DPN takes away two atoms of H from

EMBDEN-MEYERHOF PATHWAY

Step	Product	By way of
	Glucose (start)	
1	D-Glucose	Glucokinase, ATP, Mg^{2+}, insulin: anti-insulin regulators
2	D-Glucose-6-phosphate	Phosphoglucoisomerase
3	D-Fructose-6-phosphate	Phosphofructokinase, ATP, Mg^{2+}
4	D-Fructose-1,6-diphosphate	Fructaldose
5	D-Glyceraldehyde-3-phosphate	Glyceraldehyde-3-phosphate dehydrogenase, DPN, $HOPO_3^{2-}$
6	1,3-Diphospho-D-glycerate	3-Phosphoglycerate kinase, ADP, Mg^{2+}
7	3-Phospho-D-glycerate	Phosphoglycerate mutase, Mg^{2+}
8	2-Phospho-D-glycerate	Enolase, Mg^{2+}
9	Phosphoenolpyruvate	Pyruvate kinase, ADP, Mg^{2+}
10	Pyruvate	Pyruvate reductase = lactate dehydrogenase, $DPNH_2$
11	L-Lactate	

phosphorylated carbohydrate, and after dephosphorylation gives back 2H (in $DPNH_2$) to pyruvic acid.

The biochemical importance of the foregoing sequence in glycolysis is at least twofold: (1) Each one of the intermediate compounds formed leads to one or more important possible side reactions also, and these, in turn, lead to innumerable reactions that are indispensable to life processes, including respiration; and (2) in the entire sequence, and also in some of its parts, comparatively large amounts of free energy are made available, up to a maximum of some 28,000 cal/mole lactate formed under common *in vivo* conditions from one-half mole of glucose. This free energy available is considerably larger than the some 9000 calories free energy available from hydrolysis of the high-energy ATP to ADP and inorganic phosphate, although much smaller than the free energy of combustion of a mole of lactate to carbon dioxide and water, some 332,000 calories. Whereas the free and heat energies of combustion of lactate are nearly equal, lactic acid fermentation from glucose represents an instance of the relatively rare situation in which the free energy liberated is considerably greater (about 50%) than the heat energy liberated, owing to the large entropy change involved in the formation of the additional carbonyl ($=C=O$) bond in two lactates derived from one glucose molecule.

The foregoing reaction sequence, commonly called the Embden-Meyerhof pathway after its initial investigators, was in due course worked out in greatest part by Warburg. This pathway is also common to ethyl alcohol fermentation down to the pyruvate stage, which then branches off (via carboxylase) to form acetaldehyde and finally (via alcohol dehydrogenase, $DPNH_2$) ethanol. Alcoholic fermentation is sometimes erroneously referred to as glycolysis. Ordinary respiration, by this same reasoning, could be called glycolysis, since it too shares the common pathway down to pyruvate. Just as lactate fermentation is the most common fermentation met with in animal cells, so alcoholic fermentation is the most common fermentation met with in plant cells, a distinction most easily observed under anaerobic conditions.

GLYCOSIDES (Steroid). In plants, steroids occur as glycosides, as acyl glycosides, as esters, and in the free form. Many of the steroid glycosides are important drugs or starting materials for the partial synthesis of drugs.

Sterolins. Although the sterols are largely in the free and esterified form, plants contain significant quantities of sterolins (steryl glycosides and acyl glycosides). As a rule, the 3-hydroxyl

group of the sterol is linked to glucose or some other common sugar to form a heteroside, but other hydroxyl groups in the sterol molecule and higher saccharides may be involved. The two most common sterol aglycones in higher plants are sitosterol (previously called β-sitosterol) and stigmasterol, shown in Fig. 1. Cholesterol is usually present in very small amounts. (See **Cholesterol.**) The biosynthesis of sitosterol differs from that of cholesterol in the alkylation of a precursor at C-24 and involves the successive introduction of two methyl groups. Stigmasterol is formed by the dehydrogenation of sitosterol.

Cholesterol and the C_{29} sterols are used by plants as the starting materials for the biosynthesis of other steroids with the same number of carbons, such as the insect-molting hormones and the steroidal sapogenins and alkaloids or they are converted to progesterone and other steroids with a lower number of carbons. Fig. 1 shows the structure of one representative of the insect-molting hormones, ponasterone A, which has been isolated from plants in the form of its 3-glucoside, ponasteroside A.

Steroidal Saponins and Glycoalkaloids. Certain plants have the ability to hydroxylate cholesterol stereospecifically at C-26 or C-27. A glycoside of cholesterol with a glucose residue attached to this terminal hydroxyl group and chacotriose (2L-rhamnoses + D-glucose) attached to its 3-hydroxyl group is a biogenetic precursor of the saponin, dioscin, shown in Fig. 1. Steroidal saponins are glycosides of spiroketals, which form spontaneously, when the terminal sugar in an analogous 16-hydroxy-22-keto-steroid is enzymatically removed. While the configuration at C-22 is the same in all natural sapogenins, the orientation of the methyl group at C-25 depends on the position of the terminal hydroxyl group in the sterol precursor. In the D- or isosapogenin series the methyl group is α-oriented (equatorial), as in diosgenin (see Fig. 1), whereas in the L-, normal, or neosapogenin series it is β-oriented (axial).

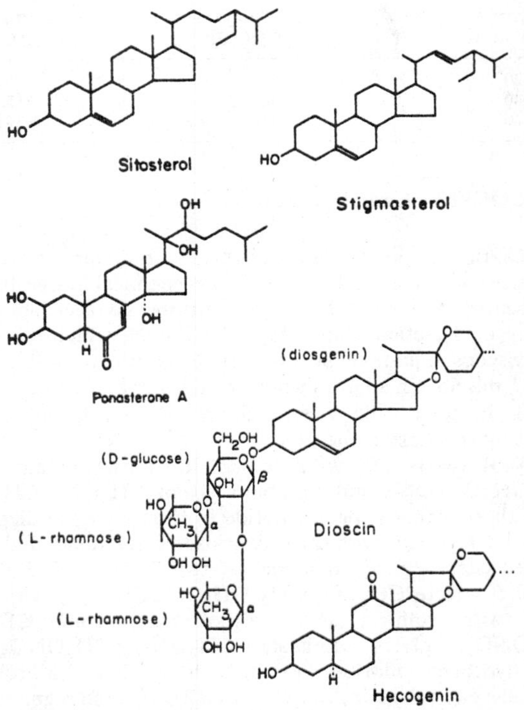

Fig. 1. Steroidal saponins and related compounds.

Sapogenins are widely distributed in monocots belonging to the genera *Yucca, Trillium, Chlorogalum, Smilax, Nolina, Agapanthus, Agave, Manfreda,* and *Dioscorea,* and in dicots belonging to the genera *Digitalis, Solanum, Lycopersicon,* and *Cestrum.* Diosgenin (Fig. 1) from Mexican barbasco root (*Dioscorea* tubers) and hecogenin (Fig. 1) from wastes of African sisal fibers (*Agave sisalana*) are important starting materials for the commercial preparation of synthetic hormones.

The glycoalkaloids are nitrogen analogs of the saponins, occurring in the Solanaceae. Two of their aglycones are shown in Fig. 2. Tomatidine occurs in the form of a glycoside, tomatine, in tomato vines and may also be used for the partial synthesis of steroid hormones. Solanidine is found in potatoes and other *Solanum* species in the form of various glycosides, the solanines and chaconines. For instance α-chaconine differs from dioscin (Fig. 1) only in the nature of the aglycone. The glycoalkaloids are likewise synthesized by plants from cholesterol.

Saponins and glycoalkaloids are characterized by their surface activity and hemolytic effect as well as their ability to form complexes with cholesterol and similar sterols. The best-known cholesterol-precipitating agent is digitonin, a saponin in *Digitalis.* While ingested saponins are nontoxic to warmblooded animals, the glycoalkaloids are toxic. Tomatine and other glycoalkaloids have antifungal and cytostatic activity.

Pregnane Derivatives. The degradation of cholesterol in plants and animals produces pregnenolone (Δ^5-pregnen-3β-ol-20-one), which is oxidized to progesterone (Δ^4-pregnene-3,20-dione). (See **Steroids.**) Various plants contain neutral pregnane derivatives with C and D rings in cis-fusion in the form of glycosides. They have been called digitanol glycosides, because they were first isolated from *Digitalis* plants, and contain the rare hexoses otherwise only found in the cardiac glycosides. Fig. 2 shows an example of the digitanols, digipurpurogenin I, which occurs in *Digitalis purpurea* as digipurpurin, a glycoside in which three molecules of D-digitoxose are attached to the 3-hydroxyl group.

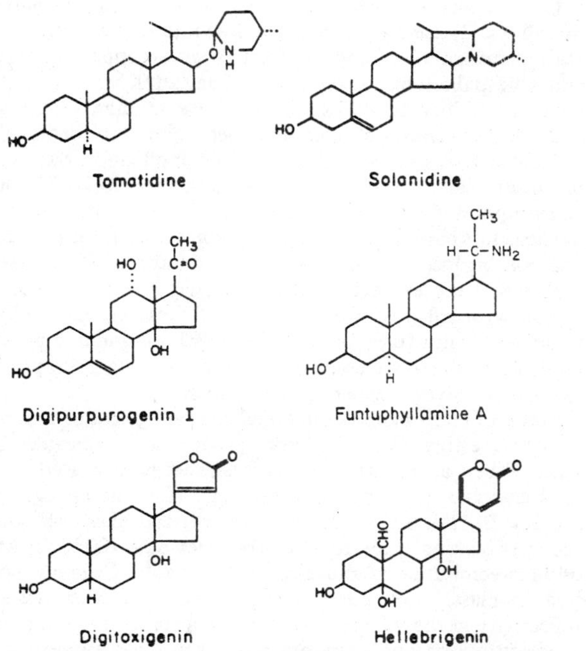

Plants belonging to the Apocynaceae and Buxaceae aminate the pregnane derivatives to produce alkaloids, which may be present as either glycosides or esters. Only one example of Apocynaceae alkaloid is shown in Fig. 2, funtuphyllamine A, which was isolated from *Funtumia africana.* The alkaloids of kurchi bark (*Holarrhena antidysenterica*) have many interesting pharmacological properties and one of them, conessine, is used as an amebicide and as a starting material for the partial synthesis of aldosterone.

Cardiac Glycosides. About eleven plant families are known to elaborate cardiac glycosides. Their genins have either 23 (cardenolides) or 24 (bufadienolides) carbon atoms and their sugars are not found elsewhere in nature. Cardenolides and bufadienolides have not been found together in the same genus. Both types of genins are synthesized in plants from a C$_{21}$ steroid, usually progesterone, and contain a 14β-hydroxyl group. Fig. 2 gives one example of a cardenolide, digitoxigenin, and one representative of the bufadienolides, hellebrigenin. *Digitalis* plants contain three cardenolides: digitoxigenin, gitoxigenin (16β-hydroxy digitoxigenin), and digoxigenin (12β-hydroxydigitoxigenin). These genins are combined with 2 molecules of digitoxose, 1 molecule of acetyldigitoxose, and 1 molecule of glucose to form the lanatosides (digilanides) A, B, and C, respectively. When the acetyl group is removed by mild alkaline hydrolysis, one obtains the corresponding purpurea glycosides (desacetyllanatosides or desacetyldigilanides). Enzymatic removal of the glucose unit, on the other hand, gives the acetyl derivatives of the digitoxose triosides. Combination of both hydrolytic procedures yields the digitoxose triosides digitoxin, gitoxin, and digoxin.

The bufadienolides occur in plants as well as animals, but only in plants are they in the form of glycosides. Their 3-hydroxyl group is attached to glucose, rhamnose, or thevetose. Hellebrigenin (Fig. 2), which is also known as bufotalidin, occurs in the rhizomes of the Christmas rose and other *Helleborus* species in the form of a rhamnoside. Bufadienolides have so far been found in plants in only two families, the buttercup and the lily family.

Crude leaf preparations of *Digitalis* have been in medical use since 1785. Pure cardiac glycosides are now available. These preparations in injectable tinctures or powdered leaf tablets are used extensively for the treatment of congestive heart failure. They increase the force of the heart muscle and the power of systolic contraction, apparently by inhibiting the active transport of K$^+$ and Na$^+$ ions through cell membranes.

—Erich Heftmann

GLYCYRRHIZINS. Sweeteners.

GLYOXYLATE SHUNT PATHWAY. Carbohydrates.

GOETHITE. The mineral goethite is a hydroxide of iron corresponding to the formula FeO(OH) crystallizing in the orthorhombic system. It occurs in prisms, but is often found in foliated or other massive forms. When observable it shows one good cleavage parallel to the prism; fracture, uneven; hardness, 5–5.5; specific gravity, 3.3–4.3; luster, adamantine to dull; color; yellowish, reddish, brownish to nearly black; translucent to opaque. It is found associated with hematite and limonite, being perhaps in part an alteration product of the latter mineral. Goethite is used as an ore of iron. There are many European localities, including Bohemia, Saxony, Westphalia, and Cornwall. In the United States it is found in the hematite mines of the Lake

Fig. 2. Representative glycoalkaloids occurring in Solanaceae.

Superior region and in Colorado. This mineral was named in honor of the German poet Johannes Wolfgang von Goethe.

GOLD. Chemical element symbol Au (from Latin *aurum*), at. no. 79, at. wt. 196.967, periodic table group lb (transition metals), mp 1064.43°C, bp 2805–2809°C, density, 19.32 g/cm³ (20°C). Elemental gold has a face-centered cubic crystal structure. Gold is a yellow metal, soft, and extremely malleable. The purity of gold (sometimes referred to as *fineness*) is expressed in *karats*. Pure gold is 24 karat. Gold has 19 isotopes, ^{185}Au through ^{203}Au. Only one of these, ^{100}Au is stable. See **Radioactivity**. In terms of cosmic abundance, the estimate of Harold C. Urey (1952), using silicon as a base with a figure of 10,000, gold was ranked number 79 among the elements, with an abundance figure of 0.0015. In terms of abundance in seawater, gold is ranked number 59 among the elements, with an estimated content of 38 lb/cu mi (4 kg/cu km) of seawater.

Electronic configuration is

$$1s^2 2s^2 2p^6 3s^2 3p^6 3d^{10} 4s^2 4p^6 4d^{10} 4f^{14} 5s^2 5p^6 5d^{10} 6s^1.$$

First ionization potential is 9.223 eV; second 19.95 eV. Oxidation potentials: $Au \rightarrow Au^{1+}$, $E° = -1.68$ V; $Au \rightarrow Au^{3+}$, $E° = -1.50$ V. Other important physical properties of gold are given under **Chemical Elements**.

Gold is one of the most ancient metals. Gold jewelry and ornaments made as early as 3,500 B.C. have been discovered at Ur in Mesopotamia. During the period from 3,000 to 2,000 B.C., lead cupellation was used to purify gold and most modern jewelry techniques were developed during that time.

Occurrence and Processing. Gold is found chiefly as the free metal scattered through gravel (*placer gold*) or disseminated in veins of quartz (*vein gold*). Small quantities also are found in lead and copper sulfide ores. Nuggets of native gold, varying in size from that of a tiny pebble to a mass weighing as much as 248 pounds (112.5 kg) have been found. In a combined state, gold occurs in sylvanite, a telluride of gold and silver, (Au, Ag)Te₂, a rich ore found in Colorado. The bulk of the gold ores contain very little gold (about 5 to 15 grams/metric ton). Some of the richest ores found in Africa contain from 20 to 30 grams/metric ton. Almost all countries produce some gold. The leader, by far, is the Republic of South Africa, followed by the U.S.S.R. and Canada. Far behind, other producers include the United States, Australia, Ghana, and Zimbabwe. See also **Mineralogy**.

The treatment of gold ores involves: (1) grinding, amalgamation, and/or cyanidation of those ores containing coarse free gold, and (2) the very fine grinding, flotation, roasting, and amalgamation and/or cyanidation of those ores containing gold telluride or sulfide. These processes produce an impure gold metal containing considerable silver and some copper plus other base metals. The impure gold is purified by melting and oxidizing the base metals or by melting and chlorinating (Miller process) which removes the base metals and silver. The silver-containing oxidized gold is purified by the electrolysis of gold chloride solutions containing an HCl solution (Wohlwill process). In the latter process, the anode is the alloy (gold-silver) and the cathode is pure gold. The gold deposits then on the cathode and the silver forms silver chloride and remains as a deposit about the anode.

Uses of Gold. The monetary aspects of gold have long dominated commercial interest in the metal. Gold through history has provided a common base from which the value of materials and services can be measured. Gold probably became a medium of exchange as early as 3400 B.C.

Jewlery is the largest commercial user of gold, accounting for nearly 65% of the total consumption. Most jewelry is made by the lost-wax process, a casting method that dates to 3000 B.C. or earlier. Usually these jewelry products use *karat golds* which contain 10 and 14 karats, and less commonly, 18 karats, of gold (41.7, 58.3, and 75.0 wt percent of gold, respectively). These gold alloys are of two general types. Red, yellow, and green golds are basically alloys of gold, copper, and silver. A wide variety of color shades can be produced by varying composition within this ternary alloy system—with reddish hues provided by high copper-to-silver ratios, and a pale-green tint when silver is predominant. These alloys almost always contain minor amounts of zinc and deoxidizers or grain refiners to facilitate fabrication.

The second widely used class is the *white karat golds*, which are produced in two basic alloy types. These are the original gold-nickel-zinc-copper (18 karat) and the gold-copper-nickel-zinc (10 and 14 karats) alloys, and the more recent gold-palladium-silver-copper and the gold-copper-nickel-palladium-silver alloys which are usually 14 and 10 karat alloys. The *pink golds* are derived from the system gold-silver-copper-nickel-zinc. These are essentially red golds, which are "whitened" by the addition of silver, nickel, and zinc.

Considerable brazing is done by jewelry manufacturers and the solders that are used may be of a lower karat content than the alloy being brazed. Usually they contain much more silver and zinc than the alloys themselves.

The use of gold in the electrical, electronic, and other industrial fields has grown considerably in recent years, estimated at about 25%. The electrical and thermal conductivity, resistance to oxidation, and ease of being electroplated make gold an excellent coating for electrical contacts. This has been particularly true in metallized ceramics for use in microelectronics and other electronic components. Here gold does not migrate into the ceramic as does silver. Gold is widely used as a conductor in thin and thick film circuitry. It is also useful as bonding wire for integrated circuit electrical connections and mechanical packaging of semiconductor chips (die bonding).

Gold is used extensively in many industrial solders and brazing alloys. These range from the low-melting eutectics of gold with germanium, silicon, and tin to gold-copper, gold-nickel, and gold-palladium-nickel alloys. The latter brazing materials have the ability to withstand long use at high temperatures and are particularly applicable to jet engine fabrication.

Gold is also used in dentistry. This application has declined in recent years; however, it still accounts for about 7% of gold consumption. Gold alloys, such as gold-silver-copper with varying amounts of platinum and palladium, are used for restorations and for bridges, inlays, and partial dentures. These are cast with much more precision than jewelry, and have in fact, replaced wrought gold wire in many of these dental appliances. Gold wire is now used principally in orthodontic and prosthetic appliances. These are complex alloys containing gold, platinum, palladium, silver, copper, nickel, and zinc.

Some of the minor commercial uses of gold are among the most interesting. Gold is used to produce a very beautiful ruby glass. When an oxidizing glass is melted with a gold salt, the gold dissolves forming colorless ions. If reducing agents like Sn, Sb, Bi, Pb, Se, or Te are present, the glass will become red after heating at temperatures between 600–700°C, as a result of the precipitation of minute particles of gold. Gold films deposited on glass by evaporation are superior to other metals for reflectivity in the infrared. Mirrors thus coated have application in spectroscopy and space science. Thin films applied to plate glass give adequate transmission of light combined with good infrared reflectivity, reducing the overheating of office windows during hot weather. Gold is extremely malleable. It can be rolled and beaten into foil less than 5 millionths of an inch (0.00013 mm) thick. Such foil has been used for indoor and outdoor decoration for centuries. One of the most conspicuous examples is the gold leaf dome, an architectural highlight in many important structures.

Chemistry of Gold. Gold has a $5d^{10}6s^1$ electron configuration, like the similar ones at lower levels of copper and silver, and thus the d electrons can take part in bonding. However, for gold the +3 oxidation state is the most stable, and the +1 state next to it in stability, so that Au^{3+} as well as Au^+ are found both in simple compounds and in complexes. As with copper and silver, the bonds in most gold compounds, including the oxides, are largely covalent. In most of its compounds gold is univalent or trivalent. While a few compounds are known in which it is divalent, some of these are considered to consist of Au(I) and Au(III), rather than Au(II). Thus, the compound with cesium and chlorine, $CsAuCl_3$, is black and diamagnetic, and so contains both Au(I) and Au(III). A similar compound with cesium, silver, and chlorine, $Cs_2AuAgCl_6$, yields $[AuCl_4]$ and $[AgCl_2]^-$ ions on hydrolysis. However, the sulfide, AuS, probably contains divalent gold.

Gold does not combine directly with oxygen. Gold(I) oxide, Au_2O, formed by heating AuOH to 200°C, is very easily reduced to gold. It is essentially covalent. Gold(I) hydroxide, AuOH, is prepared from a gold(I) solution by the addition of potassium hydroxide solution in theoretical amounts. It forms a deep blue "solution" believed to be a colloidal sol. It dissolves in excess alkali to form aurates(I), such as $KAu(OH)_2$. Gold(III) oxide, Au_2O_3, is formed by heating $Au(OH)_3$ at 100°C in the presence of a dehydrating agent. Like Au_2O, it is easily reduced to gold. It dissolves in hydrochloric, hydrobromic, and hydriodic acids, forming the haloauric acids, $HAuX_4$. It also dissolves in excess of alkali hydroxide, forming an aurate, containing the ion $[Au(OH)_4]^-$. Gold(III) hydroxide, $Au(OH)_3$, is precipitated by the addition of potassium hydroxide solution in equivalent amount, to a solution of chloroauric acid (obtained by dissolution of gold in aqua regia). It is insoluble in H_2O, gives many of the reactions of Au_2O_3, and may be a hydrous form of that compound. Gold(II) oxide, AuO, formed by the action of potassium bicarbonate upon solutions of chloroauric acid, is believed, as stated above, to consist of gold(I) and gold(III), based on properties of other divalent gold compounds.

Gold does not react directly with fluorine, but dissolves in bromine trifluoride BrF_3, to form BrF_2AuF_4, which loses BrF_3 at a 120°C to give gold(III) trifluoride, AuF_3, which decomposes into the elements at about 500°C. Water decomposes AuF_3 into hydrogen fluoride and $Au(OH)_3$. The chlorides, on the other hand, are the most important of the gold salts. Gold(I) chloride, AuCl, may be produced by heating gold(III) chloride, $AuCl_3$, in air at 170°C; it is hydrolyzed by H_2O to $AuCl_3$ and gold. Gold(III) chloride, $AuCl_3$, is formed directly from the elements at 200°C; unlike AuCl, it is soluble in H_2O, forming initially $H[AuCl_3(OH)]$, which then undergoes further hydrolysis. With hydrochloric acid, $AuCl_3$ forms tetrachloroauric(III) acid, $H[AuCl_4]$, of which many salts are known. Gold(I) bromide, AuBr, is formed by continued heating of bromoauric(III) acid above 100°C. Like the AuCl, it readily undergoes hydrolysis. Gold(III) bromide is formed by the action of bromine water upon gold. The equivalence of its three Au–Br bonds have been proved by a tracer technique with radioactive bromine. With hydrobromic acid it forms $H[AuBr_4]$. Gold(I) iodide is prepared from the elements at 50°C, or by the slow decomposition of AuI_3 at room temperature. It decomposes on heating above 120°C. It dissolves in potassium iodide, KI, solution, forming

$KAuI_2$, which then decomposes to gold and $KAuI_4$. Gold(III) iodide, obtained by evaporation of an 1:1 hydriodic acid solution of $AuCl_3$ is unstable, decomposing when dry or when heated with H_2O, into the elements. It dissolves in hydriodic acid as $H[AuI_4]$. The gold(I) halides are the least soluble of the univalent halides except for silver iodide. The solubility product constants are AuI, 1.6×10^{-23}; AuBr, 5.0×10^{-17}; AgI, 8.30×10^{-17}; AuCl, 2.0×10^{-13}; and AgBr, 4.27×10^{-13}.

There are many gold complexes. The gold(I) and gold(III) halocomplexes, involving the groups $[AuX_2]^-$ and $[AuX_4]^-$ have already been described. Apparently, there are no other gold(I) halocomplexes than the chloro-compound. There are also fluorocomplexes of the form $M[AuF_4]$ formed by fluorination of $M[AuCl_2]$, where M is an alkali metal or ammonium. Due to the polar character of the AuF bond, they are readily hydrolyzed. Many hexachloroaurates, such as $Cs_2M[AuCl_6]$ are known.

Gold forms complexes with ammonia much less readily than do copper and silver. A few ammonia complexes of gold(III), such as $KAuCl_4 \cdot 3NH_3$ have been prepared. Gold(I) halides react more readily. AuCl forms $[Au(NH_3)_2]Cl$, while AuBr and AuI react, but only with anhydrous ammonia, to form $[Au(NH_3)_2]Br$ and $[Au(NH_3)_6]$. Gold(I) cyanide dissolves in excess cyanide to form the very stable ion $[Au(CN)_2]^- K_{inst} = 10^{-38.3}$. (This complex is so stable that gold metal dissolves in potassium cyanide solution in the presence of air. This is of importance in the separation of gold from its ores); while $Au(CN)_3$ reacts to form $[Au(CN)_4]^- K_{inst} = 10^{-56}$. Treatment of salts of this ion with sulfites, gold(III) forms such complexes as $K_5[Au(SO_3)_4] \cdot 5H_2O$ and $Na_5[Au(SO_3)_4] \cdot 14H_2O$. In these complexes, the sulfito group is monodentate and is attached to the gold atom through the sulfur atom (really an aurisulfonate ion); however, a bidentate compound is also known.

Gold(III) chloride or tetrachloroaurates(III) also form thiosulfate complexes, especially in the presence of NaI, of the form $Na_3[Au(S_2O_3)_2]$, in which the gold is monovalent.

Gold forms thiocyanate complexes $M[Au(SCN)_2]$ and $M[Au(SCN)_4]$.

A striking difference between gold and copper or silver, is the fact that its oxyacid compounds do not exist in stable form, and few have been isolated. Among the few that are known are the gold(III) orthoarsenite, $AuAsO_3 \cdot H_2O$, the gold(III) selenate, $Au_2(SeO_4)_3$, and the gold(III) iodate, $Au(IO_3)_3$. Nevertheless a number of complexes of oxyacids are known, including $M[Au(NO_3)_4] \cdot 2H_2O (M = H_3O^+, NH_4^+, K^+, Rb^+)$, $Mg[Au(CH_3CO_2)_4]$.

Gold is unique among the coinage metals in forming true (i.e., sigma-bonded) stable organometallics. The action of methyllithium on $AuBr_3$ in ether at −65°C, produces a solution of $(CH_3)_3Au$, which begins to decompose at −35°C into gold, ethane, and methane. The presence of benzylamine or ethylenediamine, however, stabilizes the solution up to room temperature. Triethylgold is less stable than trimethylgold. The action of a hydrogen halide on a trialkylgold or the action of an alkyl Grignard reagent in pyridine on gold(III) halides produces dialkylgold halides, which are much more stable. Appropriate metathetical reactions of these produce the corresponding cyanides, sulfates, etc. These are all covalent compounds, as attested by the solubility of the sulfates, $(R_2Au)_2SO_4$, in benzene and chloroform. The melting points of a few dialkylgold compounds are: $(CH_3)_2AuBr$, 68°C; $(C_2H_5)_2AuCl$, 48°C; $(C_2H_5)_2AuBr$, 58°C; $(C_2H_5)_2AuCN$, 103–105°C; $(n\text{-}C_3H_7)_2AuCN$, 94–95°C; $(i\text{-}C_3H_7)_2 AuCN$, 88–90°C; $(i\text{-}C_5H_{11})_2AuCN$, 70°C; $(C_6H_5CH_2)_2AuCl$, 100°C decomposes; $(C_6H_5CH_2CH_2)_2AuBr$, 112.5°C. The n-propyl chloride and bromide, and the n-butyl, *i*-butyl, and *i*-amyl bromides are liquid at room temperature.

The dialkylgold halides are dimeric, having the planar structure

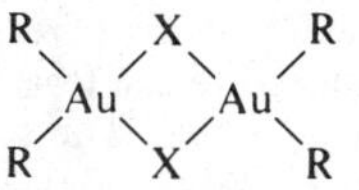

$$
\begin{array}{ccccc}
R & & X & & R \\
 & \diagdown & | & \diagdown & \\
 & Au & & Au & \\
 & \diagup & | & \diagup & \\
R & & X & & R
\end{array}
$$

The cyanides, on the other hand are tetrameric, having the structure shown below.

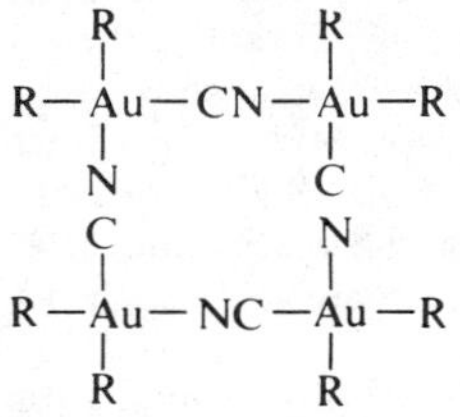

$$
\begin{array}{ccc}
R & & R \\
| & & | \\
R-Au-CN-Au-R \\
| & & | \\
N & & C \\
| & & | \\
C & & N \\
| & & | \\
R-Au-NC-Au-R \\
| & & | \\
R & & R
\end{array}
$$

References

Chynoweth, A. G.: "Electronic Materials: Functional Substitutions," *Science*, **191**, 725–732 (1976).

Pudde-Phatt, R. J.: "The Chemistry of Gold," Elsevier, New York (1978).

Smithells, C. J.: "Metal Reference Book," Vol. 3, Plenum, New York (1967).

Wise, E. M. (Editor): "Gold: Recovery, Properties and Applications," Van Nostrand Reinhold, New York (1964).

—Donald A. Corrigan, Handy & Harman, Fairfield, Connecticut.

GOLD NUMBER. When certain hydrophilic colloids, such as gelatine, are added to a gold sol, the gold sol is strongly protected against the flocculating action of electrolytes. This protective action on red gold sols may be measured by utilizing the color change red to blue which indicates the first stage of coagulation. The "gold number" as defined by Zsigmondy is the weight in milligrams of protective colloid which is just sufficient to prevent the change from red to blue in 10 cm³ of a standard gold sol (0.0053 to 0.0058% Au) after the addition of 1 cm³ of a 10% solution of sodium chloride.

GOLDSCHMIDT DETINNING PROCESS. A method for the recovery of tin, based upon the action of dry chlorine gas on scrap tinplate. The tin reacts readily to form stannic chloride, and the iron reacts only slightly. By fractional distillation, the stannic chloride is separated from the small amount of ferric chloride that is formed.

GOLDSCHMIDT REDUCTION PROCESS. 1. Reaction of oxides of various metals with aluminum to yield aluminum oxide and the free metal. This reaction has been used to produce certain metals, e.g., chromium and zirconium, from oxide ores; and it is also used in welding (iron oxide plus aluminum giving metallic iron and aluminum oxide, plus considerable heat—known as the *thermite process*). 2. A method of producing formates by heating sodium hydroxide with carbon monoxide under pressure. 3. A process for recovery of tin, as described in entry on **Goldschmidt Detinning Process.**

GRAHAM LAW. The rates of diffusion of two gases are inversely proportional to the square roots of their densities.

GRAIN BOUNDARY. The surface separating two regions of a solid in which the crystal axes are differently oriented. It has been shown that such a boundary may be thought of as built up of an array, or network of dislocations, whose spacing depends on the tilt θ of the axes across the surface. The energy (per unit area) of a grain boundary is given by

$$E/E_m = (\theta/\theta_m)\{1 - \ln(\theta/\theta_m)\}$$

where E_m and θ_m are parameters depending on the material.

Grain boundary relaxation is a source of internal friction in solids due to the motion of grain boundaries under stress.

GRAIN REFINER. An additive agent used to obtain finer grains in a casting. The material is added to the molten metal prior to casting.

GRAIN SIZE. In metallurgy, it is common practice to call the crystals of a polycrystalline metal its grains. The grain or crystal size of metals is determined by microscopic examination of a suitably prepared section. There are two principal standards of grain size in use in the United States. Both are standards of the American Society for Testing and Materials.

For most non-ferrous alloys, particularly brass and bronze and other alloys having homogeneous grain structures with twin bands, a set of ten photomicrographs having average grain diameters ranging from 0.010 to 0.200 millimeter are used for direct comparison with microstructures at a magnification of 75 times.

The A.S.T.M. standard grain size chart for steels covers about the same range of average grain diameters but the comparison is made at 100 times magnification and the grain size is expressed by numbers from 1 to 8. The following single equation relates the grain size number to the grain sizes:

$$n = 2^{N-1}$$

where N is the grain size number and n the numer of grains per square inch. In general, grain sizes 1 to 3 are considered coarse, 4 to 6 intermediate, and 7 to 8 fine. The grain size of steel can also be judged from a clean fracture if the steel can be fractured without appreciable plastic deformation because the fracture surface mirrors the grain structure. This is possible with most heat-treated machine steels and tool steels, but low-carbon steels are often too tough to break with a crystalline fracture. A series of standard fractures is available for direct visual comparison, and the numbering system for these standards coincides with that of the charts used for microscopic determination of grain size.

The grain size of metals is related to many important properties. In general, fine grain size is an indication of relatively high strength, hardness, and toughness while coarse grain indicates softness and plasticity. However, the hardenability of steels by heat treatment is highest for coarse grain steel. Coarse grain size is usually desirable for creep strength at elevated temperatures.

In the case of sheet and strip for drawing or stamping, coarse grain may give a rough surface. On the other hand, metal with too fine a grain size may lack plasticity and crack in the dies; therefore, a compromise must be reached.

The grain size of castings is generally much coarser than that of wrought products such as rod or sheet. In the case of steel castings the original coarse structure may be refined by heat treatment. This is not possible in the case of most nonferrous alloys because they do not undergo a change in type of crystal structure on heating or cooling.

In the case of hot-rolled or forged metals, the finishing temperature has an important influence on grain size. A high finish-forging temperature, for example, will permit grain growth after recrystallization. In the case of metals finished by cold-working processes, the final annealing temperature establishes the grain size. A high annealing temperature results in coarse grain size.

GRAM-ATOM. That quantity of an element having a mass in grams numerically equal to the atomic weight. One gram atom contains the Avogadro number of atoms.

GRAM-EQUIVALENT. The gram-atomic weight of an element (or formula weight of a radical) divided by its valence. In the case of multivalent substances, there will be more than one value for the gram-equivalent (e.g., Fe(II) = 27.92 grams, Fe(III) = 18.61 grams) and the proper value for the particular reaction must be selected.

GRAM-MOLECULAR WEIGHT. That amount of a pure substance having a weight in grams numerically equal to the molecular weight. One gram-molecular weight contains the Avogadro number of molecules. It is also designated as the *mole* or *mol*.

GRAM STAIN. A method of staining microorganisms which enables such organisms to be classified into two main groups, those which retain the stain being described as *Gram-positive*; and those from which the stain is decolorized being described as *Gram-negative*. The organisms are first stained with either gentian violet or its analogue, crystal violet, and then treated with a solution of iodine. An organic solvent, usually alcohol, is then applied which washes out the stain from the Gram-negative organisms, leaving Gram-positive organisms with the violet stain unaffected. A counterstain of some contrasting color is then applied to demonstrate the Gram-negative organisms. Gram-positive organisms include staphylococci, streptococci, pneumococci; among the Gram-negative organisms are gonococci, meningococci, *Bacillus coli*, and the salmonellae.

GRAPHITE. An allotropic form of carbon, graphite occurs in nature and also is produced artificially. Graphite crystallizes in the hexagonal system, often in the form of scales or plates, or in large foliated masses. Graphite has a perfect basal cleavage, is soft (hardness between 0.5–1 on the Mohs scale—similar to talc), and feels greasy to the touch. Specific gravity 2–2.2, black to steel gray, lusterous metallic appearance, very opaque. Graphite finds many uses: (1) In the manufacture of "lead" pencils, graphite (the marking medium) is mixed with clay as a bonder, the amount of clay used determining the hardness of the pencil lead. (2) In the manufacture of self-lubricative metals graphite is mixed with copper, lead, and tin, after which the mix is sintered and subjected to powder metallurgy techniques to form alloys which will hold relatively large volumes of lubricating oil over long periods of use. (3) In the construction of heat-resistant structures, such as rocket casings and chemical process equipment, graphite allows operating temperatures up to 3,000°C and greater. (4) Graphite is used in the manufacture of corrosion-resistant apparatus for chemical processing. (5) In the manufacture of packings the lubricative and corrosion-resistant characteristics of graphite are advantageous. (6) It is used in the production of electrodes for electric furnaces and electrolysis equipment. (7) A special pyrolytic graphite, with excellent electrical and thermal conductivity properties, good tensile strength at temperatures up to about 2,800°C, and impervious to gases and liquids, finds use in various electrical apparatus and, when mixed with boron, makes an effective nuclear radiation shield. Graphite slows the flow of neutrons without capturing them. (8) Graphite epoxy composite materials are now widely used in aircraft as a means of retaining strength while cutting down on weight. In the F/A-18 strike fighter, the wing skins, trailing edge flaps, stabilators, vertical tails and rudders, speedbrake, and many access doors are made of graphite composite material, comprising about 9% of the aircraft structural weight.

Graphite is formed during the metallurgical operations of producing pig iron, cast iron, malleable iron, and some special die steels and has a marked effect upon the characteristics of these materials. See also **Iron Metals, Alloys, and Steels.** The effects may be positive or negative. When present in cast iron in excessive amounts, or in the form of large interlocking flakes or films, graphite reduces the tensile strength.

Graphite is a rather widely distributed mineral and is found in a variety of rocks. It occurs in marbles, gneisses or schists; granites and other igneous rocks often carry graphite. It has been noted in pegmatites. It is likely that graphite has been formed by different processes, by magmatic separation of the graphite as an original constituent or as the result of assimilation of carbonacous rocks, by pneumatolytic action, or by the metamorphism of sedimentary rocks that contained original carbonacous matter. Well-known localities are in Siberia, on the Island of Ceylon, which is the chief producing district at present; England, Malagasy Republic, Mexico, and Canada. In the United States it is found in the Adirondack region of New York State, in Massachusetts, Rhode Island, Pennsylvania, Alabama, New Mexico, and Montana. Natural graphite sometimes is referred to as plumbago, black lead, and Flanders stone.

Graphite is made artificially by heating coke to a very high temperature, usually in an electric furnace. To prevent oxidation, the coke is covered with a layer of sand.

The German mineralogist, A. G. Werner, devised the name graphite from the Greek meaning *to write*, with reference to its use in pencils.

For a comparison of the characteristics and crystalline structure of graphite and diamond, see **Carbon;** and **Diamond.**

GREASE. A lubricating agent of higher viscosity than oils, consisting originally of calcium or sodium soap jelly emulsified with mineral oil. Greases are used where heavy pressures exist, where oil drip from the bearings is undesirable, and where the motion of the contacting surfaces is discontinuous so that it is difficult to maintain a separating film in the bearing. Grease-lubricated bearings have greater frictional characteristics at the beginning of operation, causing a temperature rise which tends to melt the grease and give the effect of an oil-lubricated bearing.

The principal categories of greases are: (1) calcium soap greases; (2) sodium soap greases; (3) complex soap greases—combinations of soaps and fatty acids used to impart high-temperature properties and moisture resistance, in which a low-molecular-weight soap can be used as a binding agent between the oil and soap in place of water; (4) lithium soap greases—excellent as multipurpose greases; (5) extreme-pressure greases, usually containing some form of sulfur, phosphorus, or other reactive agent—particularly suited to uses where there are sudden shock loads or continuous high pressures, as in steel rolling mill bearings; (6) nonsoap greases—exemplified by organically modified clays which hold the lubricating oil both by absorption and adsorption, often used in high-temperature applications because they actually have no fixed melting point; (7) asphalt-base greases—blends of asphaltic materials with lubricating oil, enabling a wide range of consistencies; (8) filler-type greases—frequently calcium-base greases that contain solid materials having unctuous properties, where the filler essentially serves as a cushion for absorbing impacts.

Calcium and sodium base greases are most commonly used.

Sodium-base greases have higher melting point than calcium greases, but are not resistant to the action of water. Graphite, either alone or mixed with grease, is also employed as a lubricant. Gear greases consist of rosin oil, thickened with lime and mixed with mineral oil, with a percentage of water. The special-purpose greases often contain glycerol and sorbitan esters. They are used, for example, for low-temperature conditions. See also **Lubricating Agents.**

Standard methods for testing greases are published by the American Society for Testing and Materials, Philadelphia, Pennsylvania.

GREENOCKITE. The mineral greenockite is cadmium sulfide, CdS, and is used as an ore of that metal. It is found rarely in hexagonal crystals, sometimes as earthy coating on other minerals. Its hardness is 3–3.5; specific gravity, 4.9–5.0; luster adamantine to earthy; color, yellow to yellowish-orange; sub-transparent. It is found in Scotland, Bohemia, and France; also, in the United States, at Franklin Furnace, New Jersey, and Marion County, Arkansas, where it occurs as a yellow coloring matter in smithsonite, and in Mono County, California. It was named for Lord Greenock.

GRIGNARD REACTIONS. Very important to the synthesis of numerous organic compounds, both in the laboratory and on a large scale in industry, is a two-step reaction involving the use of organomagnesium halides. These reactions were studied intensively by Victor Grignard during the early 1900s and for this work he was awarded the Nobel Prize in Chemistry in 1912. The reactions are referred to universally as Grignard reactions and the many magnesium compounds required by the reactions are known as Grignard reagents. Grignard's work stemmed from a discovery by Barbier in 1899 that dimethylheptenol could be prepared by reacting methyl iodide, dimethylheptenone, and magnesium in ethyl ether. In studying the mechanics of Barbier's reaction, Grignard found that the reaction proceeds in two steps: (1) the reaction of magnesium and an alkyl halide to form the corresponding alkyl magnesium halide; and (2) the reaction of the alkyl magnesium halide with a compound containing a carbonyl group to form a new carbon-carbon bond. Through subsequent years of experience, researchers have learned that nearly all alkyl and aryl halides react with magnesium to form Grignard reagents. However, the aryl and vinyl derivatives are more difficultly achieved. In the mid-1950s, Normant and Ramsden showed that some of the less reactive halides, such as vinyl chloride and chlorobenzene will form a Grignard reagent with comparative ease if tetrahydrofuran is used as the solvent. See accompanying table.

Because of the importance of the Grignard reaction techniques, they have received much study and numerous proposals have been made concerning the detailed mechanics involved. Originally, Grignard represented a Grignard reagent by RMgX, where R is the alkyl or aryl radical and X is the halide. Thus, magnesium ethyl bromide, a Grignard reagent, would appear in Grignard's symbolism as C_2H_5MgBr. Two of the main factors which make Grignard reagents so important are: (1) the many kinds of reagents that can be formulated, considering the substitution possibilites of the R and the X in the formula; and (2) the variety of reactions in which the Grignard reagents participate to yield numerous kinds of compounds. This versatility is demonstrated partially by the accompanying table.

In addition to the mono-Grignard reagent RMgX, di-Grignard reagents have proved valuable in organic synthesis. These may be symbolized by XMgRMgX. Most important of these for the synthesis of heterocyclic compounds have been

REACTIONS OF GRIGNARD REAGENTS

GRIGNARD REAGENTS REACT WITH	TO YIELD
H_2O, alcohols, primary or secondary amines	Hydrocarbons
Oxygen	Alcohols and phenols
CO_2	Carboxylic acids
Nitriles	Ketones
Metal halides	Organometallic compounds
NH_3	Hydrocarbons
γ-Lactones	Glycols
Acid esters	Tertiary alcohols (except formic acid which yields secondary alcohols or aldehydes)
Aldehydes	Secondary alcohols (except formaldehyde which yields primary alcohols)
Carboxylic acids	Tertiary alcohols
Acid halides	Tertiary alcohols or ketones
Ketones	Tertiary alcohols
Hydrogen halides	Hydrocarbons
Sulfur	Mercaptans

$BrMg(CH_2)_4MgBr$ and $BrMg(CH_2)_5MgBr$. The di-Grignard reagents of o-bromoiodobenzene also have been used in the synthesis of o-phenylene tertiary diphosphines.

Among industrial and commercial products that involve Grignard reactions in their synthesis are certain vitamins, pharmaceuticals, hormones, motor fuel additives, insecticides, organometallic compounds, and synthetic perfumes.

GRINDING AND POLISHING AGENTS. These materials are comprised of abrasives in some form. An abrasive is a hard substance that, in particulate form, is capable of effecting a physical change in a surface, ranging from the removal of a thin film of tarnish to the cutting of heavy metal cross sections and cutting stone. Abrasive action can be negative, as in the case of grit in lubrication oil that will cause engine wear; or it can be positive, as used in scores of different abrasive products, ranging from sandpaper to grinding wheels.

The two principal categories of abrasives are: (1) natural abrasives, such as quartz, emery, corundum, garnet, tripoli, diatomaceous earth (diatomite), pumice, and diamond; and (2) synthetic abrasives, such as fused alumina, silicon carbide, boron nitride, metallic abrasives, and synthetic diamond. Quartz, emery, garnet, and corundum were used in prehistoric times. Natural diamond was first used in India as an abrasive in about 800 B.C. The first grinding machines were developed in France in about 1300 A.D. Early shellac-bonded abrasives were developed in India about 1825. The first cylindrical grinding machine was made in the United States in 1860. Vitrified bonded abrasives were developed in the United States in 1872. Fused alumina and silicon were developed in 1901. Resinoid bonded abrasives were developed by L. Baekeland in Belgium in 1923. Metal bonded diamond wheels appeared in 1936, and synthetic diamond abrasives and cubic boron nitride were developed in the United States during 1955 and 1957, respectively.

The general properties of natural and synthetic abrasives are given in Table 1. Both natural and synthetic abrasives must be crushed to small particle size before bonding to cloth or paper for mounting on various tools. See Table 2.

The Nature of the Grinding Process has been the object of much research. It appears to be a mixture of chemical and physical processes. Chips are cut from the metal by the sharp abrasive points which are heated in the process to the melting

TABLE 1. GENERAL PROPERTIES OF ABRASIVES

Abrasive	Composition	Trade Names and Synonyms	Knoop Hardness	Melting Point C	Specific Gravity
Quartz	SiO_2	Sand, flint	820	1.700	2.65
Emery	imp. Al_2O_3		2000	1.900	4.00
Corundum	imp. Al_2O_3		2000	2.050	3.95
Garnet			1360	1.200[a]	4.25
Tripoli	98% SiO_2	Rottenstone	820	1.700	2.50
Diatomite	89% SiO_2	Diatomaceous earth Kieselguhr	820	1.700	2.50
Pumice	70% SiO_2 + Oxides	Pumicite, black ash			2.50
Diamond	C'		6500	1.000[a]	3.51
Fused alumina	93–97% Al_2O_3	Alundum, Aloxite, Lionite	2000	2.050	3.95
Silicon carbide	SiC	Carborundum, Crystolon	2450	2.400?	3.20
Cubic boron nitride	BN	Borazon	4700	2.000[a]	

[a] decomposes

SOURCE: Norton Company.

temperature of the metal. At this temperature chemical reactions take place which involve the abrasive, the metal, and the surrounding atmosphere. These reactions cause a dulling of the abrasive points necessitating a wearing of the wheel structure to expose new ones. The reactions also have a beneficial effect in preventing the rewelding of the chips to the base metal and their adhesion to the wheel structure. The latter is termed "loading." The detrimental reactions involving the abrasive are minimized by the control of its purity. The beneficial reactions can be augmented by the incorporation of chemical aids either into the wheel structure or into a fluid applied to the point of contact. Substances commonly used for this purpose are organic and inorganic sulfides and chlorides.

Polishing appears to be a quasi-chemical process in which the metal (or other material) is removed in particles approaching molecular size.

GROWTH REGULATOR (Plant). Plant Growth Modification and Regulation.

GUANIDINE. Guanidine, or carbamidine or iminourea $(NH_2)_2C{=}NH$, is formed (1) by heating ammonium thiocyanate to 180°C, (2) by ammonolysis of orthocarbonates, $C(OC_2H_5)_4 + 3NH_3 \rightarrow (NH_2)_2C{=}NH + 4C_2H_5OH$, (3) by ammonolysis of chloropicrin, $Cl_3CNO_2 + 7NH_3 \rightarrow (NH_2)_2C{=}NH + 3NH_4Cl + N_2 + 3H_2O$. (4) by ammonolysis of cyanogen chloride, $ClCN + NH_3 \rightarrow ClC(NH_2){=}NH \rightarrow HN{=}C{=}NH \rightarrow (NH_2)_2C{=}NH$.

Guanidine forms salts with acids, e.g., guanidine nitrate, $HNC(NH_2)_2 \cdot HNO_3$. By heating at 120°C for several hours, a mixture of ammonium thiocyanate and dicyanodiamide, guanidine thiocyanate solution is obtained by extracting with water. Treating guanidine with a mixture of nitric and sulfuric acids forms nitroguanidine

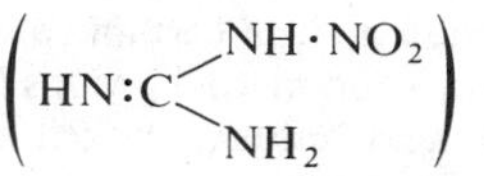

$$\left(HN{:}C{\Large<}_{NH_2}^{NH \cdot NO_2} \right)$$

which is reduced by zinc and acetic acid to aminoguanidine

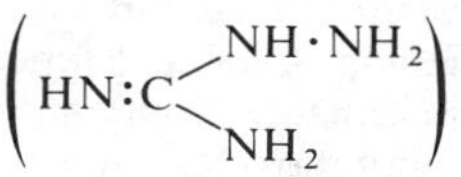

$$\left(HN{:}C{\Large<}_{NH_2}^{NH \cdot NH_2} \right)$$

By treating aminoguanidine (1) with dilute acid or alkali, there is obtained first, semicarbazide, finally hydrazine; (2) with nitrous acid, diazoguanidine

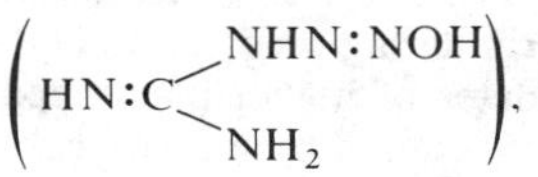

$$\left(HN{:}C{\Large<}_{NH_2}^{NHN{:}NOH} \right).$$

TABLE 2. AVERAGE PARTICLE SIZE OF ABRASIVE GRAIN USED IN GRINDING WHEELS

Grit Size	Inches	Micrometers
8	.1817	4,620
10	.1366	3,460
12	.1003	2,550
14	.0830	2,100
16	.0655	1,660
20	.0528	1,340
24	.0408	1,035
30	.0365	930
36	.0280	710
46	.0200	508
54	.0170	430
60	.0160	406
70	.0131	328
80	.0105	266
90	.0085	216
100	.0068	173
120	.0056	142
150	.0048	122
180	.0034	86
220	.0026	66
240	.00248	63
280	.00175	44
320	.00128	32
400	.00090	23
500	.00065	16
600	.00033	8

GUANIDINES

GUANIDINE	FORMULA	MELTING POINT C.	BOILING POINT °C.
1. Guanidine	$HN:C\underset{NH_2}{\overset{NH_2}{<}}$		
2. 1,3-diphenylguanidine	$HN:C\underset{NHC_6H_5}{\overset{NHC_6H_5}{<}}$	147	
3. 1,1,3,3-tetraphenylguanidine	$HN:C\underset{N(C_6H_5)_2}{\overset{N(C_6H_5)_2}{<}}$	130	
4. 1,2,3-triphenylguanidine	$C_6H_5N:C\underset{NHC_6H_5}{\overset{NHC_6H_5}{<}}$	144	
5. 1,1,3-triphenylguanidine	$HN:C\underset{N(C_6H_5)_2}{\overset{NHC_6H_5}{<}}$	131	
6. Guanylurea	$HN:C\underset{NHCONH_2}{\overset{NH_2}{<}}$	105	160 decomposes
7. Aminoguanidine	$HN:C\underset{NH_2}{\overset{NHNH_2}{<}}$	decomposes	

which is decomposed by alkali into alkali azide (e.g., NaN_3) plus cyanamide ($H_2N \cdot CN$) plus water.

In the Pauling theory of its structure, guanidine is a resonance compound of the molecular structure cited $[(NH_2)_2C=NH]$ and two ionic structures in which the nitrogen of the imino group gains an electron lost by one of the amino groups.

The monoalkyl- and N,N-dialkyl-guanidines are somewhat weaker bases than guanidine, because resonance of the double bond to the substituted —NH_2 group is restricted by the fact that carbon is more electronegative than hydrogen, and renders more difficult the acquisition of a positive charge by an adjacent nitrogen atom. This effect is still more marked with the N,N'-dialkyl guanidines, while in contrast, the N,N',N"-trialkyl guanidines are essentially as strong bases as guanidine.

The accompanying table lists seven representative substituted guanidines.

GUMS AND MUCILAGES.

Natural gums and mucilages are carbohydrate polymers of high molecular weight obtained from plants. They can be dispersed in cold water to give viscous or mucilaginous solutions which normally do not gel. They are composed of acidic and/or neutral monosaccharide building units joined by glycosidic bonds. The acid groups (—CO_2H, —SO_3H) are usually present as salts of calcium, magnesium, sodium, and potassium and, in certain cases, substituents, such as acetyl (karaya gum) and methyl groups (mesquite gum), may be present. Pyruvic acid residues, linked as ketals, are present in several cases (such as agar).

Gums are of particular importance in the food processing field where they perform at least three functions—emulsifying, stabilizing, and thickening. A few also function as gelling agents, bodying agents, foam enhancers, and suspension agents. Gum guiac also serves as an antioxidant and preservative.

Sources of Gums. Gums and mucilages may be found either in the *intracellular parts* of plants or as *extracellular exudates*. Those found within plant cells represent storage material in seeds and roots. They also serve as a water reservoir and as protection for germinating seed. The polysaccharides found as extracellular exudates of higher plants appear to be produced as a result of injury caused by mechanical means or by insects.

It has not been well established whether the exudates are formed at the site of the injury, or whether they are generated elsewhere and then transported to the injured area.

The true exudates, such as gum arabic and the East African and Indian gums are picked by hand. Seldom are commercial samples pure. This is a serious disadvantage in product control. They are classified according to grade, which, in turn, depends upon color and contamination with foreign bodies, such as wood and bark. The exudates are processed simply by grinding, their only prior treatment being sorting and sometimes bleaching under the sun. In some cases, they are purified by extraction with water and precipitated by alcohol.

Gums and mucilages present in roots, tubers, and seaweeds are usually extracted with hot water, dried, and marketed as a powder. Those gums found on the inner side of the seed coat as vitreous layers (e.g., locust bean, guar bean, etc.) are best obtained by a suitable milling process which first removes the seed coat and then makes use of the fact that the gum layer is very hard and tough as compared with the seed endosperm. The intracellular gums and mucilages can be purified by precipitation with alcohol from aqueous solution as in the case of the plant gum exudates, or by a process, such as acetylation. In a similar way, the bacterial polysaccharides can be precipiated from the cell-free culture fluid with alcohol, or as the salt of a quaternary ammonium compound where acidic groups are present.

Characteristics of Gums. The extracellular plant gums and mucilages (gum arabic, karaya gum, and tragacanth, for example) generally have a more complex structure than the intracellular types. They are made up of a number of different sugar-building units linked together by a variety of glycosidic bonds. They possess a central core or nucleus composed mainly of D-galactose and D-glucoronic acid units joined by glycosidic bonds which are relatively stable to hydrolysis by acids. To this central nucleus, there are attached as side chains those sugar units which are removed by mild acid hydrolysis. Thus, in the case of gum arabic, the acid-resistant portion of the molecule is composed of D-glucuronic acid and D-galactose, and to this nucleus are attached units of L-arabinose, L-rhamnose, and D-galactopyranosyl(1 → 3)L-arabinose.

The neutral mucilages and gums, such as mannans, glacto-mannans, and glucomannans extracted from seed and roots, have a relatively simple structure. The kinds of building units are fewer and the molecules are much less branched. The galactomannans are usually composed of a backbone of linear chains of D-mannose units jointed by 1,6-glycosidic bonds, to which are attached at regular intervals side chains of D-galactose residues. The glucomannans are essentially linear polymers united by 1,4-linkages.

The algal polysaccharides resemble the relatively simplified structures of the neutral mucilages, as in the case of carrageenan. A wider spectrum of structures is found in the bacterial gums, which are generally of the highly branched type exuded by higher plants.

Food processing and other industrial applications of gums and mucilages take advantage of their physical properties, especially the viscosity and colloidal nature. They are substances of high molecular weight. For example, gum arabic has a molecular weight of 250,000 to 300,000. The gums and mucilages which possess relatively linear molecules, such as gum tragacanth, form more viscous solutions than the more spherically shaped gums, such as gum arabic, when at the same concentration. Consequently, for some applications, the gums with linear molecules are more economical to use. Due also to the elongated molecular shape of the seed gums and mucilages, the viscosity of their aqueous solutions varies widely with concentration. They exhibit structure viscosity. In contrast, the gums and mucilages of more spherical shape, i.e., the exudates, give solutions whose viscosities do not depend so much upon concentration.

Gums and mucilages influence each other. Mixing of two gums of the same viscosity may result in a mixture with a different viscosity. The viscosity of solutions of gums and the mucilages is dependent upon the pH, especially for those containing acid groups. In certain cases, the viscosity decreases upon standing as the result of enzymatic breakdown of the molecules. The molecules can undergo large changes in shape and size under the osmotic influence of opposing ions. Some of them, such as carrageenan from Irish Moss, can be fractionated by dilute salt solutions (potassium chloride) and the poly-β-glucosan from barley grain may be precipitated with ammonium sulfate. Gum arabic shows the phenomenon of coacervation when mixed with gelatin. See **Coacervation.**

The specific uses of gums are wide and diverse. By way of example, seaweed gums (e.g., carrageenan) and seed mucilages (guar gum) are used as stabilizers in dairy products, such as ice cream and certain cheeses. They are used in confectionery, in making jams and jellies, and in stabilizing citrus oil emulsions and salad dressings. They have been used as fixatives for 2,3-butanedione in the baking industry. Outside the food field, the gums and mucilages find scores of applications.

In 1974, the Northern Regional Research Center (Peoria, Illinois) of the U.S. Department of Agriculture and the Kelco Company were joint recipients of the Institute of Food Technologists award for the development and commercialization of xanthan gum. See also **Xanthan Gum.** This gum differs by virtue of its production by pure-culture fermentation of a carbohydrate, as contrasted with refining a naturally occurring substance.

The principal gums used today are described in the following several paragraphs.

ACACIA GUM (arabic gum)
The dried water-soluble exudate from stems of *Acacia senegal* or related species. Thin flakes, powder, granules, or angular fragments; color white to yellowish white; almost odorless; mucilaginous taste. Completely soluble in hot and cold water, yielding a viscous solution of mucilage; insoluble in alcohol. Aqueous solution is acid to litmus. Produced in the Sudan, Nigeria, and other parts of west Africa. Used in adhesives, inks, textile printing, cosmetics; as a thickening agent and colloidal stabilizer in confectionery and other food products.

AGAR
Thin, translucent, membranous pieces or pale bluff powder. Strongly hydrophilic—absorbs 20 times its weight of cold water with swelling; forms strong gels at about 40°C. Agar (sometimes called agar-agar) is a phycocolloid derived from red algae, such as *Gelidium* and *Gracilaria*. It is a polysaccharide mixture of agarose and agaropectin. Agar is used as a culture medium in microbiology and bacteriology; as an antistaling agent in bakery products; in confectionery; in meats and poultry; as a gelation agent; in desserts and beverages; as a protective colloid in ice cream; in pet foods, health foods; as a laxative; in pharmaceuticals; for making dental impressions; as a laboratory reagent; in photographic emulsions.

ALGINIC ACID $(C_6H_8O_6)_n$
White to yellow powder, possessing marked hydrophilic colloidal properties for suspending, thickening, emulsifying, and stabilizing. Insoluble in organic solvents; slowly soluble in alkaline solutions. Used in food industry as thickener and emulsifier; as a protective colloid; in tooth paste, cosmetics, pharmaceuticals, textile sizing, coatings; as a waterproofing agent for concrete; in boiler water treatment; in oil-well drilling muds; in storage of gasoline as a solid.

CALCIUM ALGINATE
White or cream-colored powder, or filaments, grains, or granules. Slight odor and taste. Insoluble in water; insoluble in acids, but soluble in alkaline solutions. It is used in pharmaceutical products; as a food additive; as a thickening agent and stabilizer in ice cream, cheese products, canned fruits, and sausage casings also used in synthetic fibers.

CARRAGEENAN
A yellowish to colorless, coarse to fine powder, practically odorless, but with a mucilaginous taste. Moderately soluble (1 gram in 100 milliliters of water at 27°C), forming a viscous, clear, or slightly opalescent solution which flows readily. Carrageenan disperses in water more readily if first moistened with alcohol, glycerin, or a saturated solution of sucrose in water. Carrageenan is a hydrocolloid consisting mainly of a sulfated polysaccharide, the dominant hexose units of which are galactose and anhydrogalactose. It is a two-component, polyanionic colloid. The *kappa* and *lambda* components occur in varying proportions and degrees of polymerization and are associated with ammonium, calcium, potassium, or sodium ions, or with a combination of these four. Varying proportions alter the physical qualities of the substance. Carrageenan is obtained by extraction with water of members of the *Gigartinaceae* and *Solieriaceae* families of the class *Rhodophyceae* (red seaweed). The seaweed is also called Irish Moss and is prevalent off the coasts of Canada, New England, and New Jersey, but is found in other parts of the world. Carageenan is used as an emulsifier in food products, especially chocolate milk; in toothpastes, cosmetics, pharmaceuticals; as a protective colloid; and as a stabilizing aid in ice cream (0.02%).

GUAR GUM
Yellowish-white powder. Dispersible in hot or cold water. It possesses 5–8 times the thickening power of starch. Reduces friction drag of water on metals. Guar gum is obtained from the ground endosperms of *Cyanopsis tetragonoloba*, which is cultivated in Pakistan and used there as a livestock feed. The water-soluble portion of the flour (85%) is called *guaran* and consists of 35% galactose, 63% mannose, probably combined in a polysaccharide, and $5\frac{7}{8}\%$ protein. Guar gum is used in paper manufacture; cosmetics; pharmaceuticals; as an interior coating of fire-hose nozzles; as a fracturing aid in oil wells; in textiles, printing, polishing; as a thickener and emulsifier in food products.

GUIAC GUM
Moderate yellow-brown powder, becoming olive brown upon exposure to air. Odor is balsamic. Taste is slightly acrid. Dissolves incompletely but readily in alcohol, ether, chloroform, and in solutions of alkalies. Slightly soluble in carbon disulfide and benzene. Occurs as irregular masses enclosing fragments of vegetable tissues, or in large, nearly homogenous masses. Source is resin of the wood of *Guajacum officinale*, principally found in Central America.

KARAYA GUM

A pale yellow to pinkish brown, translucent, and horny gum with a slightly acetous odor and a mucilaginous and slightly acetous taste. In powdered form it is light gray to pinkish gray. Karaya gum is insoluble in alcohol, but swells in water to form a gel. Karaya gum is obtained as a dried gummy exudate from *Sterculia urens* and other species of *Sterculiaceae* family, or from *Cochlospermum gossypium*. It occurs in tears of variable size or in broken irregular pieces having a somewhat crystalline appearance. The properties depend upon freshness and time of storage. Viscosity greatly decreases over a 6-month period. The gum is used in pharmaceuticals, textile coatings, ice cream and other food products, adhesives; as a protective colloid, stabilizer, thickener, and emulsifier.

LOCUST BEAN GUM (carob-bean gum)

White to yellowish-white, nearly odorless powder. It is dispersible in either hot or cold water, forming a sol, having a pH between 5.4 and 7.0, which may be converted to a gel by the addition of small amounts of sodium borate. It has a molecular weight of about 310,000. The gum swells in water, but viscosity increases when heated. Insoluble in organic solvents. The gum is extracted from the ground endosperms of *Ceratonia siliqua* of the *Leguminosae* family. The gum is used in foods as a stablizer, thickener, and emulsifier; in packaging material, cosmetics, sizing and finishes for textiles, pharmaceuticals, paints.

POTASSIUM ALGINATE

Occurs in filamentous, grainy, granular, and powdered forms. It is colorless or slightly yellow and may have a slight characteristic odor and taste. Slowly soluble in water, forming a viscous solution; insoluble in alcohol. The gum is used as a thickening agent and stabilizer in dairy products, canned fruits, and sausage casings. It is variously used as an emulsifier.

SODIUM ALGINATE

A colorless or slightly yellow solid occurring in filamentous, granular, and powdered form. Forms a viscous colloidal solution with water; insoluble in alcohol, ether, and chloroform. It is extracted from brown seaweeds. The gum is used as a thickener, stabilizer, and emulsifier in foods, especially ice cream. Also used in boiler compounds, pharamaceuticals, textile printing, cement compositions, paper coatings, and in some water-base paints.

TRAGACANTH GUM

Dull white, translucent plates or yellowish powder. Soluble in alkaline solutions, aqueous hydrogen peroxide solution; strongly hydrophilic; insoluble in alcohol. One gram in 50 milliliters of water swells to form a smooth, stiff, opalescent mucilage free from cellular fragments. It is obtained as a dried gummy exudate from *Astragalus gummifer*, or other Asiatic species of *Astragalus* (*Leguminosae* family). The gum is used in pharmaceutical emulsions, adhesives, leather dressings, textile printing and sizing, dyes, food products (notably ice cream and desserts), toothpastes; for coating soap chips and powders; and in hair wave preparations.

XANTHAN GUM

See separate entry on **Xanthan Gum.**

See list of references under **Colloidal Systems.** A particularly good reference covering the physical properties and procedures for testing various gums and mucilages is 'Food Chemicals Codex,' published by the National Academy of Sciences, Washington, D.C. (Revised periodically).

GYPSUM. The mineral gypsum is hydrous calcium sulfate, $CaSO_4 \cdot 2H_2O$. It occurs as flattened monoclinic crystals, often twinned; transparent cleavable masses, called selenite; or silky and fibrous, called satin spar; it may also be granular or quite compact. It is a soft mineral, hardness 2; has two good cleavages which yield rhombic plates whose angles are 66° and 114°. Its specific gravity is 2.31–2.33; luster, vitreous to silky or pearly; color, colorless to white and gray, may be tinted red, yellow, blue, brown, etc., by impurities; transparent to opaque. A very fine-grained white or lightly tinted variety of gypsum is called *alabaster*, and prized for ornamental work of various sorts.

Gypsum is a very common mineral, thick and extensive beds of which are associated with sedimentary rocks. The largest deposits known occur in strata of Permian age. Besides being a result of deposition in sea and lake waters, gypsum has been deposited by hot springs, from volcanic vapors, and by sulfate solutions in veins. Notable localities for gypsum are in Greece, Czechoslovakia, Austria, Saxony, Bavaria, Italy, France, Spain, England and Mexico. In the United States, well-known localities are at Lockport, New York; the Mammoth Cave, Kentucky; Ellsworth, Ohio; Grand Rapids, Michigan; Hermosa, South Dakota; Wayne County, Utah; and San Bernardino County, California. In Canada, the Provinces of New Brunswick and Nova Scotia have large gypsum deposits. Because the gypsum from the quarries of the Montmartre district of Paris has long furnished burnt gypsum used for various purposes, this material has been called plaster of Paris. See also classified index under **Mineralogy.**

Often, there is confusion between the mineral gypsum, $CaSO_4 \cdot 2H_2O$, and the useful product of partial dehydration, $CaSO_4 \cdot \frac{1}{2}2H_2O$. See accompanying table. There are numerous commercial products based upon gypsum. *Plaster*, made from gypsum, is widely used for the economical fabrication of building products. Importantly, the setting time of gypsum plaster can be carefully controlled through the addition of fractional percentages of *accelerators* (typically water-soluble salts, such as K_2SO_4, or finely-ground gypsum) and *retarders*, which frequently are modified organic substances, such as glue, casein, blood, hair, and hoof meal; or citric, boric, and phosphoric acids and their salts. Accelerators are believed to function by providing additional nuclei for crystallization, whereas retarders are believed to provide protective colloids or insoluble salts which block water access to the plaster particle. A controlled rate of reaction can be obtained by incorporating a combination of retarders and accelerators in the gypsum plaster mix.

Wallboard is the largest single user of gypsum. The product usually consists of a core of gypsum sandwiched between two layers of paper. Characteristics of the product include fire resistance, dimensional stability, low cost, and easy workability. Wallboard conventionally measures $\frac{1}{2}$ inch (1.3 centimeters) thick, 48 inches (1.2 meters) wide, and 8 to 20 feet (2.4 to 6 meters) in length. The average weight is 1.8 pounds per square feet. In manufacture, foamed plaster slurry is mixed and discharged on a moving web of paper. The edges of the bottom paper are scored and folded so that the slurry is completely contained between that sheet and the top paper, which is laid on the slurry. The paper surfaces not only provide strength and paintability to the finished board, but also form a continuous mold within which the gypsum is cast. The board machine operates continuously. Within five minutes after forming, the gypsum is sufficiently hard to be cut, after which the sheets are dried further before storage and shipment. Fibers may be added to provide crack resistance and additional fire resistance. Water-repellent chemicals may be added to the board core or to the paper surface. Also, decorative and functional finishes may be factory-applied.

Industrial plasters of a gypsum base include dental plasters, used in making tooth impressions, orthopedic plasters for immobilizing broken bones, pottery plasters, oil-well cements, permeable plasters for casting nonferrous metals, art and statuary casting, lamp bases, patching and grouting compounds, insulating-brick production, and pattern and model making for the aircraft and automotive industries. Water-reducing additives

TERMINOLOGY AND PROPERTIES OF CALCIUM SULFATE-WATER COMPOUNDS

CHEMICAL FORMULA	DESIGNATIONS COMMONLY USED	PROPERTIES
$CaSO_4 \cdot 2H_2O$	Calcium sulfate dihydrate; rock gypsum; chemical gypsum; alabaster (white fine-grained); selenite (translucent, platey); satin spar (fibrous); land plaster (pulverized gypsum).	All forms (natural, synthetic, and recrystallized) are thermodynamically and crystallographically equivalent. Habit may be needles, plates, or prisms.
$CaSO_4 \cdot 1/2H_2O$	Calcium sulfate hemihydrate; calcined gypsum; stucco; plaster of Paris; molding plaster; gypsum plaster; chemical hemihydrate.	Alpha and beta types exist, depending upon conditions of calcination. Alpha type is more stable, crystalline, of lower energy. Beta type is less stable, disordered, of higher energy.
$CaSO_4$	Anhydrite	
I	Anhydrite I; high-temperature anhydrite.	Produced by high-temperature ($> 1,000\,°C$) calcining. Contains free CaO.
II	Anhydrite II; insoluble anhydrite; inactive anhydrite; dead-burned gypsum; chemical anhydrite; mineral anhydrite.	Produced by calcining at $250-1,000\,°C$. Relatively inert. Reactivity depends upon calcining-time-temperature relationship and particle size.
III	Anhydrite III; soluble anhydrite; active anhydrite; dehydrated hemihydrate.	Produced by low-temperature ($175-250\,°C$) dehydration of hemihydrate. Reacts vigorously with water and moist air to form hemihydrate.

SOURCE: United States Gypsum Company, Des Plaines, Illinois.

and reinforcing resins and cements may be added to achieve a compressive strength of over 15,000 pounds per square inch (1021 atmospheres).

Portland cement also consumes large quantities of gypsum. About 5% of gypsum is added to the cement clinker before grinding. Addition of gypsum aids in increasing the early strength of the cement and prevents undesirable false set.

Agriculturally, gypsum serves as a soil conditioner, providing a source of available calcium and sulfate, assisting the retention of organic nitrogen, without the addition of acidity or alkalinity to the soil. Gypsum is widely used in areas where the soils are deficient in sulfur. Gypsum also has been used in mixed fertilizers and animal feeds.

Terra alba or dead-burned, fine white gypsum is used as a paper filler, in plastics, and as an extender for titanium dioxide. Pharmaceutically-pure gypsum can be added to bread and other bakery products, finds use in beer production, and as a pharmaceutical-tablet diluent. In Japan, calcium sulfate is used in making *tofu*, a soybean curd.

Gypsum may be a potential source of sulfur and sulfuric acid. Some European plants make portland cement and sulfuric acid from gypsum or anhydrite. In the Muller-Kuhne process, gypsum is mixed with clay and silica in quantities necessary to make cement, along with coke to reduce $CaSO_4$ to CaO. In equipment similar to that for portland-cement manufacture, the SO_2 is driven off and converted to sulfuric acid by the contact process.

References

Edinger, S. E.: "The Chemistry of Gypsum and Its Dehydration Products," *U.S. Dept. Commerce Bull.* NTIS, PB-203, September 1971.

Kelly, K. K., Southard, J. C., and C. T. Anderson: Thermodynamic Properties of Gypsum and Its Dehydration Products, *Bur. Mines Tech. Paper* (U.S.), **625**, 1941.

Hansen, W. C. and J. S. Offutt: "Gypsum and Anhydrite in Portland Cement," 2nd edition, U.S. Gypsum Co., Chicago, Illinois, 1969.

Lane, M. K.: Disintegration of Plaster Particles in Water, *Rock Prod.*, **71** (3), 60 and (4), 73, 1968.

Schroeder, H. J.: "Mineral Facts and Problems (Gypsum)," *Bur. Mines Bull.* (U.S.), **650**, 1970.

H

HADRONS. These are subatomic particles, the strong interactions of which are manifested by the forces that hold neutrons and protons together in the atomic nucleus. Hadrons include the proton, the neutron, and the pion, among others. These particles show signs of an inner structure, i.e., they are made up of other particles, which has led over a period of the last several years to consider the hadrons as combinations of constituents known as quarks. See **Particles (Subatomic);** and **Quarks.**

HAFNIUM. Chemical element symbol Hf, at. no. 72, at. wt. 178.49, periodic table group 4b, mp 2207–2247°C, bp 4601–4603°C, density, 13.3 g/cm³. The alpha form of elemental hafnium has a close-packed hexagonal crystal structure; the beta form, a body-centered cubic structure. Metallic hafnium, like zirconium, exhibits passivity in air due to formation of adherent coatings of oxide or nitride. Urbain reported evidence of the element in 1911, but hafnium was not fully identified until 1923 by D. Coster and G. C. de Hevesy. The remarkable similarity between hafnium and zirconium accounts mainly for its late isolation, as compared with the majority of elements. In terms of abundance, there is an average of about 4 ppm hafnium in the earth's crust. The element occurs with zirconium in certain varieties of zircon, including malacon, cyrtolite, and alvite. One mineral found in Scandanavia, thortveitite, contains more hafnium than zirconium. Pegmatite, monazite, baddeleyite, and zerkelite also contain hafnium. First ionization potential, 5.5 eV. Oxidation potentials Hf + H_2O → HfO^{2+} + $2H^+$ + $4e^-$, 1.68 V; Hf + $4OH^-$ → $Hf(OH)_2$ + H_2O + $4e^-$, 2.60 V. Electron configuration $1s^22s^22p^63s^24d^{10}4s^24p^64d^{10}4f^{14}5s^25p^6$ $5d^26s^2$. Ionic radius Hf^{+4}, 0.75Å. Other important physical properties of hafnium are given under **Chemical Elements.**

Hafnium usually is extracted from ores along with zirconium. In one process, zircon sand is broken down by carbiding or carbonitriding, followed by chlorination. The mixture formed is dissolved with a complexing agent, after which it is introduced into a liquid-liquid extraction process. The final product is $HfCl_4$. Fractional crystallization of the fluorides of hafnium and zirconium also is practiced. Metallic hafnium is made by the Kroll process in which the $HfCl_4$ is reduced in an inert atmosphere by magnesium. The hafnium sponge and magnesium chloride resulting is vacuum-distilled to accomplish the final separation. In a modified Kroll process, sodium or sodium amalgam may be used. The latter requires less rigid temperature and pressure control during processing, costs less, and introduces fewer impurities into the process. For further purification of hafnium metal, a number of methods have been used, including electrorefining, arc and induction melting, zone refining, and the hot-wire or van Arkelde Boer process.

Uses. Compared with most metals, the annual production of hafnium is low. Mainly produced in the United States, France, and the Soviet Union, the combined production is in the range of 100 metric tons annually, or less. Several uses have been found for hafnium: (1) as a control material in water-cooled nuclear reactors. Also hafnium is an effective flux-depressor in a reactor for absorbing neutrons to decrease the peaks in neutron flux; (2) as a filament in gas-filled incandescent light blubs; (3) as an alloying ingredient to add strength to tungsten and molybdenum filaments and electrodes used in high-pressure discharge tubes; (4) as a cathode in x-ray tubes; (5) as a getter material in vacuum tubes and systems; (6) as a minor alloying ingredient in nichrome heating elements where hafnium appears to significantly increase the lifespan of the elements; and (7) usually with zirconium, as an ingredient of several alloys.

Chemistry and Compounds. Hafnium metal dissolves in HCl (warm) and slowly in H_2SO_4, more rapidly if fluoride ion F^- is present, forming compounds of HfO^{2+}, or fluoro complexes in the latter case. The metal resists the attack of weak acids and their salts.

Due to its $5d^26s^2$ electron configuration, hafnium forms tetravalent compounds readily, although the Hf^{4+} ion does not exist as such in aqueous solution except at very low pH values, the common cation being HfO^{2+} (or $Hf(OH)_2^{2+}$) and many of the tetravalent compounds are partly covalent. There are also less stable Hf(III) compounds. There is close similarity in chemical properties to those of zirconium due to the similar outer electron configuration ($4d^25s^2$ for zirconium) and the almost identical ionic radii (Zr^{4+} is 0.80Å) the relatively low value for Hf^{4+} being due to the Lanthanide contraction.

With improved means to separate the compounds of these two elements, future research will yield more details of specific hafnium compounds. The methods of separation used effectively include ion exchange techniques, a particularly successful one using a column of silica gel, with a solution of the tetrachlorides in methanol as feed and a 1.9N HCl solution as eluant for zirconium. Separations also have been accomplished through the distillation of the phosphorus oxychloride addition products.

See references listed at the end of the entry on **Chemical Elements.**

HALF-LIFE (Elements). Chemical Elements.

HALIDES. A compound made up of a halogen (astatine, bromine, chlorine, fluorine, or iodine) and another element or radical may be termed a *halide*. Fundamentally, there are three classes: (1) the *ionic* (saline) halides, (2) the *covalent* (acid) halides, and (3) the *complex* halides. The ionic halides are most sharply characterized by the halides of the alkali and alkaline earth metals, plus those of certain lanthanide and actinide metals. They form ionic or semi-ionic crystals in the solid state, have high boiling points and melting points, and are soluble in polar solvents. Their bonding is electrovalent, varying in degree with the difference between the electronegativities of the halogen and the metal. Potassium iodide and silver fluoride are ionic, but silver iodide is essentially covalent. The fluorides exhibit a primarily ionic character for most of the metals, but the other halogens form fewer ionic compounds. The degree of ionicity varies down as well as across the priodic table.

The covalent (acid) halides have low boiling and melting

points, are soluble in nonpolar solvents and insoluble in polar solvents, although they often react with the latter. The degree of covalence generally is greatest for the nonmetals. For a given nonmetal, the boiling point depends upon both the number of atoms of the halogen with which it is combined and the symmetry of the molecule. For example, the boiling points of bromine(I) fluoride, bromine(III) trifluoride, and bromine(V) pentafluoride, BrF, BrF_3 and BrF_5, are 20, 135, and 40.5°C, respectively.

The complex halides are very numerous, because of the readiness with which halide ions form coordination compounds with metals. In general, stability of these complexes depends upon the size and electronic structure of the metal ion—the smaller cations form their more stable compounds with the smaller halide ions, notably with fluoride, while with larger cations the order of stability is that of polarizability of the halide, i.e., decreasing from iodide to fluoride. The more electronegative transition elements form especially stable complexes; e.g., those of palladium, platinum, etc., $PdCl_4^{2-}$, PtF_6^{2-}, etc. The most common halo complexes have four or six halogen ions coordinated with the cation, although such complexes as those of copper, gold and mercury, e.g., CuI_2^-, $AuCl_2^-$, $HgCl_3^-$, etc., are notable exceptions.

See also **Bromine; Carbon; Chlorine; Chlorinated Organics; Fluorine; and Iodine.**

HALITE (Rock Salt). The mineral halite (rock salt) is naturally occurring sodium chloride, NaCl, common salt. It is isometric with cubic habit and cleavage. It is brittle; hardness, 2.5; specific gravity, 2.168; luster, vitreous; colorless when pure, but usually white, yellow, red, or blue. It is soluble in water. Halite occurs in association with anhydrite and gypsum. In the United States "salt beds" of this type have been exploited in Michigan, New York, Ohio, and Pennsylvania. Louisiana produces salt from great subsurface dome-shaped masses, often 2,000–4,000 feet (600–1,200 meters) thick. The salt domes of the Gulf Coastal Plain are particularly important as subsurface structures, on the flanks of which are apt to occur large and important pools of petroleum. Poland, Saxony, Austria, and France possess well-known deposits of salt, as do the U.S.S.R., England, Algeria, India, and China. Salt is chiefly used in cooking and as a preservative; in the manufacture of soda ash for the glass industry; and as a source of many sodium compounds. It derives its name from the halogen group of elements to which chlorine belongs.

See also **Sodium Chloride.**

HALL EFFECT. This is one of several galvanomagnetic effects first described by E. H. Hall in 1879. If a bar of electrically conducting material is placed in a magnetic field and current is passed through the bar in a direction normal to the magnetic field, a potential is created in a direction perpendicular to both the current and magnetic field by deflection of the charged current carriers. The Hall effect is important because it provides a means for a direct estimate of the density of the charge carriers in the material and their sign, and the Hall coefficient can be combined with the result of a measurement of the material conductivity to yield a value for the mobility (velocity per unit applied electric field) of the charge carriers. In addition to its use for characterizing the electrical properties of a material, the Hall effect principle is applied in several practical devices.

Consider the bar shown in the accompanying figure. The current density I (amps/cm²) is proportional to the applied field **E** (volts/cm), $I = \sigma E$. The proportionality factor σ is called the conductivity; σ can be expressed in terms of the

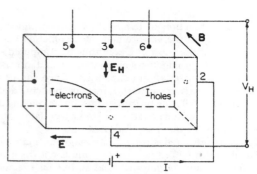

Sample with electrical contacts arranged for measurement of Hall effect and conductivity. The Hall voltage V_H appears between contacts 3 and 4. The voltage drop between contacts 5 and 6 can be used to determine the conductivity of the material.

density of charge carriers n (cm^{-3}) in the material, the charge on each carrier e (coulombs), and the carrier mobility μ (cm²/volt-sec), by the relationship $\sigma = ne\mu$. The carriers may be either negatively charged electrons or positively charged holes.

The magnetic field causes the charge carriers to be deflected by Lorentz force $\mathbf{F} = e\mathbf{v} \times \mathbf{B}$ toward one side of the bar. In this expression **B** is the magnetic field strength and **v** is the velocity of the charge carriers, $\mathbf{v} = \mu\mathbf{E}$. Note that carriers of either sign are deflected to the same side of the bar. The charge buildup creates a transverse field $\mathbf{E}_H$ which acts on each charge carrier in the steady state to balance the Lorentz force, $e\mathbf{E}_H = e\mathbf{v} \times \mathbf{B}$ or $\mathbf{E}_H = \mu\mathbf{E} \times \mathbf{B}$. The Hall coefficient R is defined by the relation $E_H = RIB$, so that $R = \mu E/I = \mu E/\sigma E = 1/ne$(cm³/coulomb). Thus, the magnitude of the Hall coefficient is inversely proportional to the product of the density of charge carriers and the magnitude of charge on each. The polarity of the charge carriers is determined by the direction of the Hall field created in the sample. The sign of R is the same as the polarity of the charge carriers. If the conductivity σ of the material is known, the mobility of the charge carriers can be determined, since $R\sigma = \mu$.

In monovalent metals, the electron density calculated from the Hall coefficient agrees quite well with the density of atoms. In polyvalent metals, the results are more erratic. In semiconductors, the Hall coefficient for a single carrier must be modified to $R = r/ne$, where r is a correction factor dependent on the type of scattering the charge carriers encounter during conduction and the details of the energy band structure. The magnitude of r is rarely far from unity with the usual band structure; the value ranges from $3\pi/8$ for predominant lattice scattering to $315\pi/512$ for ionized impurity scattering. More complicated formulae must be used for the Hall coefficient when more than one type of charge carrier is participating in the conduction process.

At very low temperature, the conductivity of a semiconductor may be determined by "hopping" or tunneling of charge carriers from one atom to another. As the temperature is increased, impurities which may be present in the semiconductor become ionized and contribute charge carriers to the conduction process (extrinsic activation). The slope of the Hall coefficient measured as a function of temperature is in this case determined by the activation energy of the impurities. When the temperature is increased still further, conduction is enhanced by ionization of the semiconductor atoms themselves to produce pairs of holes and electrons (intrinsic activation). In the simple case of only one type of intrinsically activated carrier being mobile and the impurity contribution being negligible, the slope of the Hall coefficient as a function of temperature is determined by the energy gap between the valence and the conduction band, and this parameter of the substance may thus be computed simply. In general, since hopping, tunneling, extrinsic, and intrinsic activation effects may overlap in temperature, since any of several kinds of charge carrier scattering may be present, and since charge carriers of differing properties may be conducting current simultaneously, the interpretation of Hall effect data from semiconductors must be quite sophisticated.

In ferromagnetic and strongly paramagnetic materials, the magnetic field introduced into the formula is not the externally applied field H nor the internal field B; but an effective field must be used, $B_{eff} = H + 4\pi M\alpha$, where α is in general temperature and material dependent and can exceed unity, and M is the sample magnetization. In this case, $E_H = RIH = D\pi\alpha RMI$. R is the ordinary Hall coefficient and $4\pi\alpha R$ is called the extraordinary Hall coefficient. Other variations of the Hall effect include, for example, the photo-Hall effect, in which the density of carriers liberated by light may be measured in a photoconducting medium.

The Hall effect is used in some gaussmeters. A known current is passed through a small strip of a semiconductor which has a Hall constant large in magnitude and independent of temperature. The Hall voltage generated when the strip is inserted in a magnetic field is an indication of the magnetic field intensity.

Because the Hall voltage is proportional to the product of I and B, a Hall device can yield a voltage proportional to the product of the current in the device and the current producing a magnetic field surrounding the device. Such units are called Hall multipliers.

—D. Warschauer, Naval Weapons Center, China Lake, California.

HALLUCINOGENS. There are many substances which will, if taken in appropriate quantities, produce distortion of perception, vivid images, or hallucinations. Most of these substances will produce powerful peripheral as well as the central effects. Some few agents are characterized by the predominance of their actions on mental and psychic functions. This group of drugs has been called hallucinogens, psychotomimetics, psycholytics, and psychodelics, among several ambiguous terms. None of these names is adequately descriptive of these compounds.

Hallucinogens may be classified into five groups of chemically distinct compounds: (1) lysergic acid derivatives of which lysergic acid diethylamide (LSD-25) is the prototype; (2) phenylethylamines, such as mescaline; (3) indolealkylamines, which include psilocybin, psilocin, and bufotenin; (4) piperidyl benzilate esters, typified by ditran (a 70:30 mixture of N-ethyl-2-pyrrolidymethyl phenylcyclopentylglcolate and N-ethyl-3-piperidyl phenylcyclopentylglycolate), and (5) phenylcyclohexyl piperidines (sernyl). The chemical structures of these compounds is shown in the accompanying figure.

Drugs from the first three groups have been isolated from naturally-occurring sources. LSD-25 is a molecular component of ergot, a fungus which infects cereal grains. Mescaline, historically the oldest hallucinogen, was isolated from a Mexican peyote cactus. Psilocybin and psilocin were isolated from the Mexican mushroom, *Psilocybe mexicana*. Bufotenin is found in some varieties of toadstools. The indole derivatives are chemically closely related to serotonin (5-hydroxytryptamine), a compound which plays an important, yet unknown role in the central nervous system.

The piperidyl benzilate esters and phenylcyclohexyl piperidines are synthetic compounds, and have not been shown to

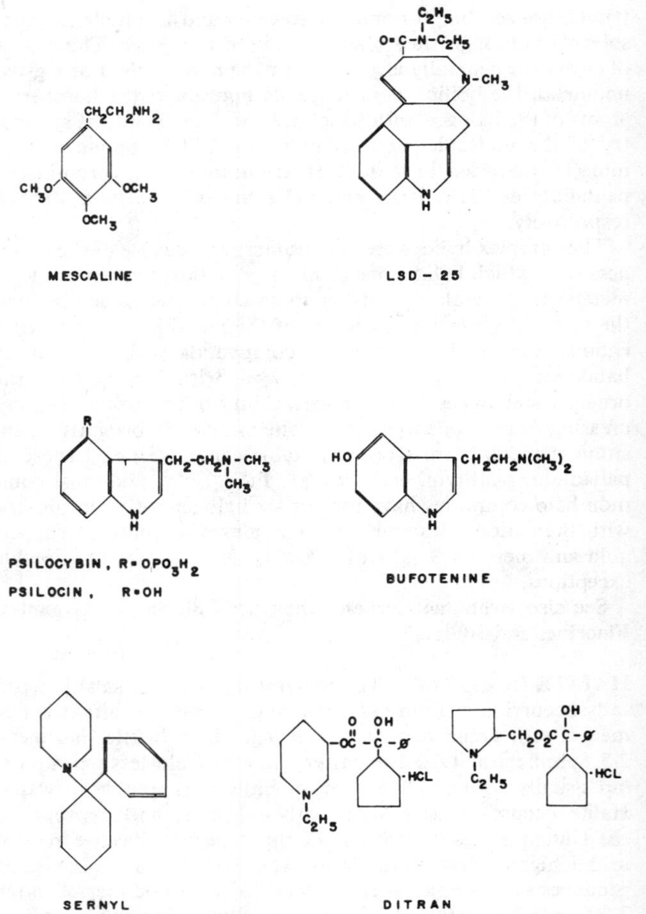

Structures of some hallucinogenic drugs.

occur naturally. Some authorities do not consider them to be hallucinogens, but active researchers in the field include them among the most active psychotomimetics.

Clinical syndromes from LSD-25, mescaline, and the indoleamines are similar. Somatic symptoms are nausea, dizziness, loss of appetite, blurred vision, paresthesia, weakness, drowsiness, and trembling. These result frequently and are usually associated with sympathomimetic effects, such as increased pulse rate and slight temperature elevation. Perceptual and psychic changes are marked. Visual illusions and vivid hallucinations, decreased concentration, slow thinking, depersonalization, dreamy states, changes in mood, and often anxiety are commonly found.

The clinical syndromes from ditran are different from those produced by the aforementioned drugs in some respects. Disorganization of thought, disorientation, confusion, mood changes, and visual and auditory hallucinations are observed. The piperidyl benzilate esters are central anticholinergics, and mental states produced by them are reminiscent of those from other anticholinergics, such as scopolamine.

The effects of phenylcyclohexyl derivatives are also distinctive. Comparatively minor somatic symptoms are evoked. Psychic effects predominate, being typically characterized by feelings of unreality, depression, anxiety, and delusional or illusional experiences. The effects of these drugs are said to be more analogous to natural psychoses than those of the other drugs; however, the same claim has been made for ditran.

When LSD-25 was discovered, it was believed that the drug would provide an extremely useful tool in the investigation of psychoses and mental illness. However, therapeutically, the hallucinogens, including LSD-25, have been of little value to psychiatrists.

References

Lincoff, G., and D. H. Mitchel: "Toxic and Hallucinogenic Mushroom Poisoning," Van Nostrand Reinhold, New York, 1977.
Seiden, L. S., and L. A. Dykstra: "Psychopharmacology," Van Nostrand Reinhold, New York, 1977.

HALOGENATED COMPOUNDS. Chlorinated Organics; Organic Chemistry.

HALOGEN GROUP (The).

The elements of group 7a of the periodic classification are sometimes referred to as the Halogen Group. The individual elements commonly are referred to as *halogens*. In order of increasing atomic number, they are fluorine, chlorine, bromine, iodine, and astatine. The elements of this group are characterized by the presence of seven electrons in an outer shell, and hence have the ability to gain an electron to form negative ions with a completed octet of valence electrons. The halogens present striking similarities of chemical behavior, all being very reactive and, in particular, readily form substitution compounds with numerous organic compounds. Although these elements also have other valences, all have a -1 valence in common.

HARDENING (Metals).

There are three principal methods for hardening metals and alloys: cold working by plastic deformation, precipitation hardening, and quench hardening as applied to steel. The last two methods involve heating and cooling operations. A pure metal may also be hardened through the addition of alloying elements. When a solid solution is formed, it is normally harder than the pure metal. If additional phases are formed by alloying, these may also be harder than the pure metal and contribute to the hardness of the metal.

The hardenability of steel refers to the ease with which it can be hardened rather than the maximum hardness value attainable. For example, a 1-inch (2.5-cm) bar of a certain 0.20% carbon alloy steel can be hardened to a 50 Rockwell C in the center by quenching in oil. A similar bar of plain carbon steel requires a drastic quench in brine to attain the same hardness, and therefore has a lower hardenability. Neither bar can be quenched to a greater hardness because 50 Rockwell C is the maximum attainable for a 0.20% carbon steel. A 0.40% carbon steel can be hardened to a maximum of about 60 Rockwell C, and the maximum for high-carbon steel is about 65 Rockwell C.

HARDNESS.

The meaning of this terms as applied to solids has various interpretations. Commonly, it refers to the resistance of the substance to surface abrasion, so that of two solids, the one that will scratch the other, as diamond scratches glass, is the harder. Or hardness may denote rigidity or lack of plasticity—or even strength; in some cases, it denotes a combination of several such properties. The original Mohs' scale of hardness is given in Table 1.

In metallurgy and engineering, hardness is determined by methods based upon resistance to penetration by an indenter of greater hardness than the material being tested. Indenters commonly used are hardened steel balls for testing the softer metals, such as aluminum, copper, lead, magnesium, tin, and their alloys. Testing methods and scales commonly used include

Fig. 1. Magness Taylor puncture test for evaluation of apples. (*Instron Corporation.*)

Brinell, Rockwell, Vickers, Tukon, Eberbach, and Scleroscope (based upon the rebound of a diamond-tipped body falling under the force of gravity from a fixed height). Hardness values of various metals are given in Table 2.

Hardness and related variables are also of much interest in the food processing field. For example, a puncture test is used for evaluating fruits, such as apple, to determine the effect of storage conditions on the raw fruit. The Magness Taylor puncture test has been developed to stimulate the pressing of the raw fruit with the thumb. See Fig. 1. The probe has a sphrical tip surface with rounded edges and a line which can be used to visually determine when a fixed penetration has been achieved. Performing the Magness Taylor test in an instrument with constant-speed control and graphic readout allows accurate and meaningful data to be obtained. See Fig. 2. The test probe is driven into the surface of the product at a fixed speed to a preset depth of penetration. Similar tests are applied to other fruits, vegetables and processed foods, such as cheeses.

HARDNESS (Mineral). Mineralogy.

HARMOTOME.

The mineral harmotome is a zeolite, composition approximately (Ba, K)(Al, Si)$_2$SI$_6$O$_{16}$ · 6H$_2$O; it is monoclinic but often forms double twins giving the effect of a square prism. It is a brittle mineral; hardness, 4.5; specific gravity 2.41–2.50; luster, vitreous; color, white to gray or perhaps yellow,

TABLE 1. HARDNESS SCALES

MOHS' SCALE	RIDGWAY'S EXTENSION OF MOHS' SCALE	METAL EQUIVALENT	OTHERS
1. Talc			
2. Gypsum			
			2.5. Finger Nail
3. Calcite			
4. Fluorite			
5. Apatite			
			5.5. Window Glass
6. Feldspar (Orthoclase)	6. Orthoclase or Periclase		
			6.5. Steel (Knife Blade; File)
7. Quartz	7. Vitreous Pure Silica		
8. Topaz	8. Quartz	8. Stellite	
9. Corundum or Sapphire	9. Garnet		
	10. Topaz		
	11. Fused Zirconia	11. Tantalum Carbide	
	12. Fused Alumina	12. Tungsten Carbide	
	13. Silicon Carbide		
	14. Boron Carbide		
10. Diamond	15. Diamond		

1. In the above scales each abrasive is capable of scratching all others above it in each scale and may be scratched by all abrasives below it.

2. The gap between 9 and 10 in the original Mohs' scale is much greater than that between 1 and 9 in the same scale.

3. Various additional hardness scales have been devised by different investigators; in general, different materials maintain the same order of hardness in all these scales.

TABLE 2. TYPICAL HARDNESS VALUES

MATERIAL	BRINELL		ROCKWELL	VICKERS 50 kg
	500 kg	3000 kg		
Aluminum, annealed	23		H 45	25
Magnesium alloy	63		B 21	63
Armco iron	66	73	B 31	71
Yellow brass, annealed	72	82	B 40	77
Copper, cold rolled	99	83	B 55	110
Mild steel, annealed	107	117	B 70	123
Aluminum alloy, 24st	130	144	B 78	146
Stainless steel, annealed	121	145	B 80	153
Yellow brass, cold rolled	174	178	B 91	189
Ni-Moly steel, quenched in water, tempered at 1200°F (649°C)		241	C 23	255
Same, 1000°F (538°C)		293	C 31	310
Same, 800°F (427°C)		363	C 38	380
High-speed tool steel		684	C 62	740

red or brown; white streak; translucent. Harmotome like other zeolites is found in cavities in basalts and similar rocks, sometimes in trachytes or in gneisses, occasionally as a gangue mineral in veins of metallic minerals. Some well-known localities are in Bavaria; the Harz Mountains; Norway; and Scotland. Harmotome occurs in the United States with stilbite, near Port Arthur, Lake Superior. The name *harmotome* comes from the

Greek meaning *joint* and *to cut*, referring to the division of the pyramid formed by the prismatic faces of the mineral when in the twinned position.

HEAT TRANSFER. Although there are three generally accepted methods for transferring heat from one medium to another, or from one locale to another within a given medium,

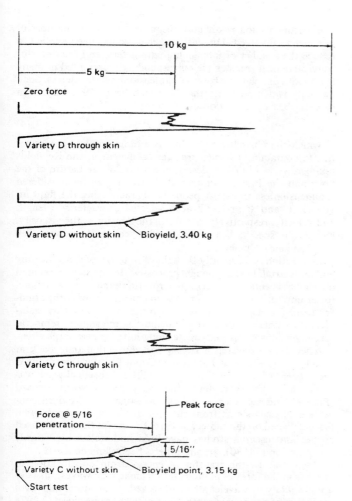

Fig. 2.   Curve obtained with Magness Taylor puncture testing of apple texture. (*Instron Corporation.*)

it is uncommon for one method to act unilaterally. Particularly where convection may predominate, some conduction of heat will be involved. In conduction, heat must diffuse through material substances; in convection, heat is essentially carried from one locale to another by actual movement of the transport medium; in radiation, heat transfer involves radiant wave energy.

Conduction. From a microscopic standpoint, thermal conduction refers to energy being handed down from one atom or molecule to the next one. In a liquid or gas, these particles change their position continuously even without visible movement and they transport energy also in this way. From a macroscopic or continuum viewpoint, thermal conduction is quantitatively described by Fourier's equation, which states that the heat flux q per unit time and unit area through an area element arbitrarily located in the medium is proportional to the drop in temperature, -grad T, per unit length in the direction normal to the area and to a transport property k characteristic of the medium and called *thermal conductivity*:

$$q = -k \ \text{grad} \ T. \tag{1}$$

Predictions for the value of the thermal conductivity k can be made from considerations of the atomic structure. Accurate values, however, require experimentation in which the heat flux q and the temperature gradient, grad T, are measured and these values are inserted into Fourier's equation. Thermal con-

ductivity values for a number of media over a large temperature range are shown in Fig. 1. Metals have the largest conductivities and, among these, pure metals have larger values than alloys. Gases, in contrast, have very low heat conductivity values. Electrically nonconducting solids and liquids are arranged in between. The low thermal conductivity of air is utilized in the development of thermally insulating materials. Such materials, like cork or glass fiber, consist of a solid substance with a very large number of small spaces filled by air. The thermal transport occurs then essentially through the air spaces, and the solid structure only supplies the framework which prevents convective currents. It will be noted that the thermal conductivities indicated in Fig. 1 (at ambient temperature) extends through 5 powers of 10. This range is still small when compared with the range for the electric conductivity of various substances, where electric conductors have values which are larger by 25 powers of 10 than those of electric insulators. As a consequence, it is much easier to channel electricity along a desired path than to do so with heat, a fact which accounts for the difficulty in accurate experimentation in the field of heat transfer.

Fourier's equation can be used together with a statement on energy conservation to derive a differential equation describing the temperature field in a medium. Fourier was the first person to develop this equation and to device means for its solution. In vector notation, this equation is:

$$\rho c = \frac{\partial T}{\partial t} = \nabla(k \nabla T) \tag{2}$$

where ρ is the density, c is the specific heat, t is time, and ∇ is the Nabla operator. The temperature field in a substance can either change in time (unsteady state), or it can be independent of time (steady state $\partial T/\partial t = 0$). For a steady-state situation, the temperature field depends primarily on the geometry of the body involved and on the boundary conditions. The simplest case of a steady-state temperature field is the one in a plane wall with temperatures which are uniform on each surface, however different at the two surfaces. The temperature in the wall then changes linearly in the direction of the surface normal as long as the variation of the thermal conductivity in the temperature range involved can be neglected. For an unsteady process, the capacity of the medium to store energy enters the energy conservation equation; correspondingly, the specific heat of the material and its density become factors for the conduction process, as well as the thermal conductivity. A combination of these properties, defined as the ratio of the thermal conductivity to the product of specific heat and density, called *thermal diffusivity* ($k/\rho c$), then determines how fast existing temperature differences equalize in time. It is found that metals and gases have thermal diffusivity values which are approximately equal in magnitude and are considerably higher than thermal diffusivities of liquid and solid nonconductors. This means that temperature differences equalize much faster in metals and gases than in other substances.

Various other physical processes lead in their mathematical description to equations of the same form as Eq. (2), especially in its steady-state form. Such processes include the conduction of electricity in a conductor, or the shape of a thin membrane stretched over a curved boundary. This situation has led to the development of analogies (electric analogy, soap film analogy) to heat conduction processes which are useful because they often offer the advantages of simpler experimentation.

Convection. When energy is transported by convection in fluids, conduction usually takes care of the transport of heat from one stream tube to another and is the dominating mode of transfer near solid walls. Convection transports heat along the stream lines and is dominating in the main body of the

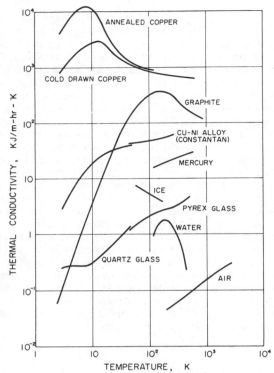

Fig. 1. Thermal conductivity values for a wide range of substances over a temperature range of 1–10⁴ K.

fluid where the velocities are large. In many situations, the flow is turbulent; this means that unsteady mixing motions are superimposed on the mean flow. These mixing motions contribute also to a transport of heat between stream tubes, a process which can be described by an "effective" conductivity which often has values by several powers of ten larger than the actual conductivity of the fluid.

Movement of the fluid may be generated by means external to the heat transfer process, as by fans, blowers, or pumps. It may also be created by density differences connected with the heat transfer process itself. The first mode is called *forced convection*; the second one *natural* or *free convection*. Convective heat transfer may also be classified as heat transfer in *duct flow*, or in *external flow* (over cylinders, spheres, air foils, and similar objects). In the case of external flow, the heat transfer process is essentially concentrated in a thin fluid layer surrounding the object (boundary layer).

Of special interest in such heat transfer processes is the knowledge of the heat flux from the surface of a solid object exposed to the flow. This heat flux q_w per unit area and time is conventionally described by Newton's equation:

$$q_w = h(T_w - T_f) \qquad (3)$$

where T_w is the surface temperature and T_f is a characteristic temperature in the fluid. This equation defining the heat transfer coefficient h is convenient because in many situations the heat flux is at least approximately proportional to the temperature difference $T_w - T_f$. Information on the heat transfer coefficients can be obtained by a solution of the Navier-Stokes equation describing the flow of a viscous fluid and the related energy equation, or they are found by experimentation. Computers enhance the ability to study heat transfer analytically, at least for laminar flow, whereas in turbulent flow the bulk of the information is determined experimentally.

Experimentation is difficult because of the large number of parameters involved. Dimensional analysis has been applied to reduce the number of influencing parameters, and relations for convective heat transfer are correspondingly presented in many handbooks as relations between dimensionless parameters. Such an analysis demonstrates that heat transfer in forced flow can be described by a relation of the form:

$$Nu = f(Re, Pr) \qquad (4)$$

in which the Nusselt number Nu is a dimensionless parameter hL/k, containing the heat transfer coefficient h; the Reynolds number $Re = \rho(VL/\mu)$ describes essentially the nature of the flow; and the Prandtl number $Pr = c_p\mu/k$ can be considered a dimensionless transport property characterizing the fluid involved. L and V are arbitrarily selected characteristic length and velocity, respectively; ρ denotes the density, μ the viscosity, and c_p the specific heat of the fluid at constant pressure. See also **Reynolds Number.**

Convection is frequently thought of in terms of space heating and industrial heat-exchange processes. It should be pointed out that convection plays a cosmic role (in the sun's photosphere, for example), and a very large role in connection with the atmosphere of the earth and some other planetary bodies. For example, when normal convective transport is inadequate temperature inversions occur and create smog hazards over large cities.

Attempts to develop a theory for convection date back at least to the 1790s when Thompson (Count Rumford) introduced the concept of heat convection. Very little theoretical work was undertaken, however, until the early 1900s when Bénard (France) undertook experimental investigations. Modern convection physics stems from the work of Lord Rayleigh, who first published on the subject in 1916. In current times, advanced convection research studies have been undertaken by Velarde and Normand (1980), among others. See reference listed.

Radiation. In the transfer of energy from one location to another in the form of photons (electromagnetic waves), usually a multiplicity of wavelengths is involved. In vacuum, all waves regardless of their wavelength move with the same speed (2.9977 $\times 10^8$ meters per second). In various substances, the wave velocity c changes somewhat with wavelength, and the ratio of the wave velocity in vacuum to the velocity in a substance is equal to the optical refraction index. Air and generally all gases have refraction indices which differ from unity only in the fourth decimal. Their wave velocity, is therefore, practically equal to that in vacuum.

Prévost's principle states that the amount of energy emitted by a volume element within a radiating substance is completely independent of its surroundings. Whether the volume element increases or decreases its temperature by the process of radiation depends upon whether it absorbs more foreign radiation than it emits, or vice versa. One refers to thermal radiation when the emission of photons is thermally excited, i.e., when the substance within the volume element is nearly in thermodynamic equilibrium. For such radiation, Kirchhoff was able to derive a number of relations by consideration of a system of media in thermodynamic equilibrium. If $j\nu$ indicates the coefficient of emission, i.e., the radiative flux at the frequency ν^* emitted per unit volume into a unit solid angle, and η is the coefficient of absorption at the same frequency, i.e., the fraction of the intensity of a radiant beam which is absorbed per unit path length, then one of these relations states:

$$c^2 \frac{j_\nu}{\eta_\nu} = f(T, \nu) \qquad (5)$$

with c denoting the wave velocity. According to this relation, the combination of parameters on the left-hand side of Eq. (5)

is a function of temperature T and frequency ν of the radiation only, but does not depend upon the substance under consideration. Kirchhoff's law can also be expressed in parameters which refer to the interface of two media, 1 and 2. It then takes the form:

$$c^2 \frac{i_\nu}{\alpha_\nu} = f(T, \nu) \qquad (6)$$

in which i_ν is the monochromatic intensity of the radiative flux at frequency ν originating in medium 2 and traveling through the interface into medium 1 per unit solid angle and area normal to the direction of the radiant beam; α_ν is the monochromatic absorptance or absorptivity, i.e., that fraction of a radiant beam approaching the interface in the medium 1 in the opposite direction that is absorbed in medium 2. The wave velocity in medium 1 is c. Kirchhoff's law states that the combination of the parameters on the left-hand side of Eq. (6) is again a function of temperature and frequency only, but does not depend upon the nature of the medium. A medium which absorbs all the radiation traveling into it through an interface ($\alpha_\nu = 1$) is called a *blackbody*. The intensity of radiation emitted by an arbitrary medium is, according to Eq. (6), in the following way related to the intensity of radiation $i_{b\nu}$ emitted by a black body at the same temperature and frequency:

$$\frac{i_\nu}{\nu} = i_{b\nu} \qquad (7)$$

The amount of heat transferred by radiation can be determined by use of the *Stefan-Boltzmann law*:

$$Q = bA(T_1^4 - T_2^4) \qquad (8)$$

where Q is the amount of heat transferred per unit time, b is a constant, A is the area of the radiating surface, T_1 is the absolute temperature of the radiating body and T_2 is the absolute temperature of the receiving body. Various correction factors are introduced into the formula to account for the shape of the bodies, their thermal radiation characteristics and the properties of the media through which the radiant rays must pass while traveling from radiator to absorber. The thermal radiation characteristics are its emissivity, a measure of its ability to radiate at a given temperature, its absorptivity, a measure of its ability to absorb heat and its reflectivity, which measures its ability to reflect without absorbing.

Radiant energy travels in a straight line. Therefore to transmit it to an object out of sight of the radiator requires a reflector, such as a furnace wall, to deflect the rays to their objective.

It is possible to set up controlled laboratory radiation between simple plane surfaces and determine therefrom accurate coefficients to incorporate into radiation equations. However, the radiation of heat from furnace gases, consisting of non-luminous gases, luminous carbon particles in flame, ash globules, etc., to the walls and tubes of a steam generator in commercial operation at variable load, is another matter. Here, empirical data which are gathered and interpreted from field tests on similar equipment, must still be resorted to however great the designer's urge to go back to basic laws of heat transfer.

Radiant heat transfer in furnaces is roughly proportioned to the difference in the fourth power of the absolute temperatures of the radiating and receiving surfaces. The water wall surface is approximately at boiler saturation temperature, while the superheater surface varies from this to somewhat about the temperature of the steam at the superheater outlet. However, the mean radiating temperature of the furnace gases is usually over 1204°C. The fourth power of the receiving surface temperature is thus seen to be small compared to the fourth power of the

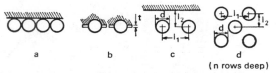

Fig. 2. Arrangements of radiant heat-absorbing surface.

transmitting surface temperature; consequently the latter controls the transmittance, and boiler tube temperature does not need to be considered a variable to be accounted for.

Figure 2 shows some of the arrangements in which radiant heat-absorbing surface is disposed. It may be used to illustrate another of the difficulties which beset the designer in following a rational or semi-rational form of radiation analysis. Projected radiant surface is one thing; actual radiant energy receiving surface may be quite a different area. For example, suppose the tubes of case (a) to be separated and spaced l_1 inches on centers. The *projected* areas of cases (a) and (c) would then be the same, but it seems obvious that re-radiation from the wall causes more of a (c) tube to receive radiant energy than is the case with an (a) tube. Also, if δ is a factor correcting projected area to *equivalent* absorbing surface, what value should be assigned to it in the case of a bank of tubes which may receive by re-radiation some radiant energy deep in the tube bank? Here δ has a minimum value of 1, but some investigators have derived expressions which indicate that δ may have a magnitude of 3 or more.

Industrial Heat Transfer. Plate-type heat exchangers are widely used in the process industries. For example, in the dairy industry alone, they serve such purposes as heating milk with hot water or steam (especially for high temperature-short time pasteurization), heat recovery by milk to milk heat transfer and milk cooling with water or brine. In the modern plate heat exchanger, the plates are so shaped that at frequent intervals the passage of milk is retarded. This produces a pulsating flow, which gives high heat transfer with low pumping power.

Some of the more common cases of industrial heat transfer are:

1. Radiation from fuel beds and luminous gases to absorptive surfaces such as boilers, cylinder walls, etc.
2. Radiation from heat generators such as drying lamps.
3. Convection of heat out of combustion regions.
4. Convection of heat from hot surfaces under either free or forced convection.
5. Conduction of heat through the tubes of boilers, heaters, heat exchangers, condensers, etc.
6. Conduction in walls, pipe covering, and other so-called "heat insulators."
7. Conduction of heat through the plates of plate-type heat exchangers and regenerators.

Types of Heat Exchangers. Heat exchangers perform many functions within a manufacturing plant. See Fig. 3. Often they are given special names even though they remain fundamentally heat exchangers. These include:

Chiller—a device which cools fluids to temperature below those obtainable with ordinary cooling water by using the vaporization of a refrigerant. The fluid to be cooled is routed through the tubes while the low-boiling refrigerant vaporizes from a pool of liquid in the shell.

Partial Condensers—Many overhead vapors from distillation columns in petroleum-refinery services are a mixture of light and heavy hydrocarbons and noncondensable gases, i.e., gases

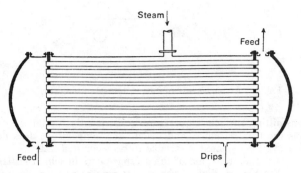

Fig. 3.   Cross section of one type of heat exchanger.

that are not condensed at the outlet temperature and pressure of the condenser (air, hydrogen sulfide, methane, and other light ends). These vapors are routed through the shell side while water is used as the cooling medium on the tube side of the unit. Condensation on the shell side begins at the saturation temperature of the heavy components and continues over a decreasing temperature range until part of the lighter components are condensed. Part of the existing liquid is sent back to the tower as reflux, while the remainder is further refined or passes to the trim cooler and storage.

Trim Cooler—This unit condenses the last remaining light-end vapors and cools the liquid to the ultimate storage temperature (often about 100°F; 38°C) by using cooling water. This cooling usually is not conducted in the main condenser because it would reduce column pressure.

Thermosiphon Reboiler—Flow of the vaporizing fluid depends upon the difference in static head between the column of liquid flowing from the tower to the reboiler and the partially vaporized column of liquid returning from the exchanger to the tower.

Reboilers—These exchangers operate in conjunction with a distillation tower to vaporize enough liquid to assure vaporization of the overhead product. A hot process stream of steam may be used as the heating medium. Most reboilers are shell-and-tube exchangers located at the base of the tower. The vaporizing fluid is routed through the shell side of the exchanger.

Forced-circulation Reboiler—A pump is used to provide more positive circulation than available with the thermosiphon effect, e.g., in the vaporization of viscous fluids.

Vapor Heat Exchanger—Units of this type preheat a cool stream of process fluid by using heat from partially condensing vapor. The objective is to conserve heat and eliminate the requirement for a separate preheater.

Air-cooled Exchanger—As used in the petroleum industry, air-cooled exchangers normally comprise two headers joined by a horizontal bank of finned tubes. Usually two motor-driven fans located above (induced draft) or below (forced draft) the tubes are used to circulate the air over the finned surface.

Superheater—A unit of this type heats vapor above the saturation temperature.

Waste-heat Boiler—A unit of this type generates steam and is similar to a regular steam generator except that hot gas or liquid produced by a chemical reaction (often combustion) is the heating medium.

Heat Storage

It is often necessary to store heat in rather large quantities in specially designed apparatus. Hot water, of course, is one of the easiest forms in which to store thermal energy that is immediately available. As contrasted with hot water, electric energy and steam have to be generated on an as-needed basis. The blast furnace poses a difficult heat-storage problem which obviously cannot be handled by storing heat in water. Great amounts of hot gas are required on a cyclic basis. To heat such quantities of air on a continuous, as-required, basis would be quite impractical with the present stage of the art. The solution used involves several stoves which are quite large, often over 100 feet (30 meters) in height and about 25 feet (7.5 meters) in diameter. The blast temperature of approximately 1,000°F (528°C) is accomplished by preheating the stove checkerwork to a much higher temperature. Checkerwork is comprised of refractory material forms constructed in high walls in checkerboard fashion to permit free passage of air through the interstices when under pressure. The gas passing through the stove exhausts initially at 2000°F (1093°C). Mixing this with unheated air produces the required blast temperature for the blast furnace. The stoves usually are heated for a period of three hours and exhaust (termed "on wind") for a period of about one hour. A similar system of checkerwork regenerators is used in connection with glass-tank heat-storage systems.

Flowing streams of pebbles also have been used in the chemical industry for removing heat from gases. Pebbles and stones are also used in some solar energy storage systems.

References

Chandrasekhar, S.: Hydrodynamic and Hydromagnetic Stability," Oxford Univ. Press, New York, 1961.
Kittel, C.: "Thermal Physics," 2nd edition, Freeman, San Francisco, 1980.
Normand, C., Pomeau, Y., and M. G. Velarde: "Convective Instability: A Physicist's Approach," *Rev. Mod. Phys.*, **49**, 3, 581–624 (1977).
Prigogine, I.: "Time, Structure, and Fluctuations," *Science*, **201**, 777–785 (1978).
Staff: "Heat Transfer," Special issues of *Chem. Eng. Progress*: **74**, 7, 41–46 (1978); **75**, 7, 41–91 (1979).
Staff: "Practical Aspects of Heat Transfer," Amer. Inst. of Chem. Engineers, New York, 1978.
Turner, J. S.: "Buoyancy Effects in Fluids," Cambridge Univ. Press, Cambridge, England, 1973.
Velarde, M. G., and C. Normand: "Convection," *Sci. Amer.*, **243**, 1, 92–108 (1980).

HEAT TREATING. A term usually confined to metals. Where heat is applied over a period of time to other materials, the word *curing* is usually used. Heating and cooling of metals to effect changes in properties is the objective of heat-treating technology. *Annealing* and *normalizing* are generally for the purpose of softening or improving the grain structure. *Patenting* is also a softening process in which cold-drawn carbon-steel wire, for example, is heated above its critical temperature range, followed by cooling to below this range in a molten lead or molten salt bath, with subsequent cooling to room temperature.

While heat treating includes the aforementioned softening procedures, it most often implies hardening and strengthening. In the case of steels, this requires heating to above the critical temperature range, followed by rapid cooling (*quenching*) in oil, water, or brine, except in the case of special grades which harden upon cooling in air. This is followed by *tempering*, a low-temperature reheating treatment which reduces the internal stresses caused by the hardening treatment. Tempering may be carried to a sufficiently high temperature to reduce somewhat the extreme hardness of the as-quenched steel and increase the toughness and ductility, depending upon the requirements of the part.

HEAVY HYDROGEN.   Deuteron.

HEAVY WATER REACTOR.   Nuclear Reactor.

HELIUM. Chemical element symbol He, at. no. 2, at. wt. 4.0026, periodic table group 0 (inert or noble gases), mp $-272.2°C$ (20 atm), bp $-268.93°C$ (4.2144 K), specific gravity, 0.124 at 4.2144 K. The element has no triple point and can be solidified only by applying high pressure to the liquid phase. Described later, liquid helium undergoes a change in its physical properties at 2.178 K, known as the *lambda point*. Solid helium has a close-packed hexagonal crystal structure (subject to further study and confirmation). At standard conditions, helium is a colorless, tasteless, odorless gas. There are two natural isotopes, ^3He and ^4He, with ^4He being slightly less than 100% abundant. The boiling point is 3.2 K for ^3He. Radioactive ^5He and ^6He have extremely short half-lives. See **Radioactivity.** The first ionization potential is 24.58 eV; second, 54.14 eV. Other physical properties of helium are described under **Chemical Elements.**

Like the other rare gases, helium exhibits negative chemical properties with ordinary materials under normal conditions. Under the influence of electric glow discharge or electron bombardment, helium forms compounds with tungsten and other metals, as well as with iodine, sulfur, and phosphorus. In a vacuum electric discharge tube shows green to canary-yellow glow. Discovered first in the vapors surrounding the sun by Lockyer in 1868, through the yellow spectral line near the two yellow lines of sodium, then by Ramsay in 1895 in the mineral clevite.

Helium occurs (1) in minerals of uranium and thorium, such as clevites, pitchblende, carnotite, monazite, and also in beryl, (2) in mineral waters (1 part He per thousand of water, in some Iceland waters), (3) in volcanic gases, (4) especially in certain natural gases of the United States. The first discovery of this kind was made in Kansas. The richest helium wells are in Utah. In northeastern Texas, four wells have produced 55 million ft³ of helium. The fields in which are located the wells having the greatest percentage (1.3–8.0%) of helium are now held as government reserves. (5) in ordinary air, about 1 part in 200,000.

Uses. Industrially, helium is used to provide an inert gaseous shield for arc welding, for growing transistor crystals, in the production of titanium and zirconium, to fill the space where optical lenses in instruments, as the carrier gas in some chromatographic apparatus, as a liquid bath for masers and cryotrons, as a refrigerant for furnishing the low temperature required for superconducting electrical equipment, in lasers, as a diluent gas in deep-sea diving applications, as a heat-transfer medium in gas-cooled nuclear reactors, and as a leak-detecting medium for testing pressure and vacuum equipment. Now, to a rather limited extent, helium is used as a lifting gas for airships and for balloons used in meteorological investigations. Helium is used in aerospace programs in several ways, including its use in propellant tanks as a compressed gas which expands and takes the place of fuel as the fuel is consumed, in ground-support equipment, and in communication satellites for providing the low temperature required for sensitive electronic systems. In medicine, helium sometimes is mixed with oxygen for patients with certain respiratory ailments and also it is mixed with certain anesthetics to reduce the hazards of forming an explosive mixture with air.

The liquefaction of helium was accomplished by Onnes in 1908 in Leiden, and Keosom in 1926 succeeded in solidifying helium in the same laboratory.

Relatively recently, helium has been solidified at room temperature. The melting pressure at 24°C is 115 kilobars, in complete agreement with the Simon equation. Besson and Pinceaux (1979) developed an original apparatus for the experiment, which allowed loading of the cell at room temperature. Diamond anvil cells were used in the procedure.

Liquid Helium II. Upon cooling, ^4He liquefies at atmospheric pressure at 4.216 K to form an essentially normal liquid, liquid helium I. On further cooling to the lambda-point, 2.178 K at one atmosphere, a change occurs to liquid helium II. The latter has a very low viscosity (hence the name "superfluid") and a very high thermal conductivity, which produce such phenomena as the creeping of a film over the edge of the container, and the fountain effect, in which the liquid sprays out of a capillary. Superfluidity is commonly explained in terms of a two-fluid theory. Thus, London and Tirza attribute the properties of helium II to a mathematical peculiarity in the distribution function of Bose-Einstein statistics, whereby below the λ-point, a finite fraction of the atoms fall into a ground state of zero thermal energy. In this state they would have the properties of a superfluid. However, this theory has not yielded good quantitative predictions of the properties of the aggregate liquid helium. ^3He, which follows Fermi-Dirac statistics, does not have a superfluid state.

Landau treats liquid helium by an approach similar to that of the Debye theory of solids. The longitudinal and transverse sound waves, which are the elementary excitations of that theory of solids, correspond in the case of liquid helium to phonons and rotons. The *phonons* are the longitudinal sound waves, while the *rotons* are another type of elementary excitation postulated by Landau to represent the rotational motion of the liquid, because a liquid cannot support transverse waves. The specific heat can be expressed as the sum of contributions from phonons and rotons. Landau derived expressions for these which fit the data and experiments quite closely up to 1.6 K.

Feynman developed wave functions to provide an atomistic interpretation of Landau's spectrum of elementary excitations.

The complexity of the helium II problem is apparent at once when one attempts to extend the equations of classical hydrodynamics to this two-component system, in which each component has its own density and velocity. Khalatnikov derived such equations by ignoring terms of second order.

Still another area of investigation has been that of the properties of ^3He-^4He mixtures. As stated above, ^3He exhibits no λ-transition and no superfluidity. It has a critical temperature of 3.35 K and a boiling point of 3.2 K, against values of 5.2 K and 4.216 K for ^4He.

The most abundant helium atoms, ^4He, are bosons, but the ^3He atoms are fermions. This has as a consequence that liquid ^3He does not show superfluidity—a property very probably connected with the Bose-Einstein statistics obeyed by the ^4He atoms.

Chemistry. The most striking properties of helium are its emission as the positively charged (+2) alpha particles in radioactive changes, its formation in radioactive change by uranium-radium and thorium-containing substances, emitting alpha particles, later losing the charge to become helium, and its production artificially by bombardment of lithium or boron with high-velocity protons or alpha rays.

Unlike the other inert gases, helium gives little evidence of compound formation with organic substances. Like neon, but unlike the others, if forms no hydrate. However, if forms compounds much more readily under excitation, due apparently to unpairing of its 1s electrons and promoting of one of them to the 2s state. The 460 kcal/g-atom of energy is readily obtained by electric discharge or electron bombardment. Under such conditions the helium molecule-ion, He_2^+, with a pair of bonding electrons (1s) and a single antibonding electron (1s), is formed, as are combinations of the type of HeH$^+$ and HeH$_2^+$. In a mercury discharge tube, the compound HgHe$_{10}$ has been found, and with various metallic electrodes corresponding helides, such as the compounds of tungsten, platinum, iron, palladium, bismuth, etc., e.g., WHe$_2$, Pt$_3$He, FeHe, PdHe, BiHe, etc. have been formed.

Helium in Resources Planning. The Tip Top natural gas field in Sublette County, Wyoming contains the highest known helium content in a natural gas, namely, 0.8%. There are a few helium-rich field in Alaska and Canada. The percentage of helium in natural gas drops off fast for most fields. The very large gas fields of the world have a helium content of below 0.10%. Most gas fields with an average content as high as 0.3% are found in Kansas, Colorado, Texas, and Oklahoma. Natural gas fields in coastal zones contain as little as 0.007% helium. Outside North America, helium is found in West Germany and South Africa, Algeria, Poland, and the British sector of the North Sea, the content ranging between 0.10 and 0.3%.

As pointed out by Hurley (1954), helium has a geologic occurrence and distribution unique among the elements. It is a product of radioactive disintegration of uranium and thorium within the earth's mantle and crust, but flows to the surface at a rate less than that of its generation, because most of it is driven into crystal structures of rock minerals until released by alpha radiation damage near radioactive concentrations. Mobile helium rising through the crust may then be trapped, along with other gases, beneath relatively impermeable barriers. Nitrogen is almost always associated with helium in natural gases, although this has not been fully explained. Also, carbon dioxide is abundant in some helium-rich gas mixtures.

As the helium-rich gas is burned toward exhaustion, most of its contained helium is dissipated into the atmosphere. It has been estimated that the recovery of helium from the atmosphere would incur 800 times (with 1980 technology) the cost of separating it from natural gas, prior to releasing the natural gas to pipelines. But, as pointed out by Cook (1979), controversy over the need for a government-directed helium-conservation program reflects fundamental differences in viewpoints on the economic future of industrial society, on the limits of substitution and labor and capital for a depleting resource, and on intergenerational equity and risk-bearing. Cook (1979) stresses that future demand for helium on a much larger scale than the demands of the past will depend upon development and deployment of technologies either in their infancy, or not yet even conceived. Among the sources of future demand are magnetic containment systems for fusion reactors, breeder and high-temperature gas reactors, high-temperature gas turbines, laser-based missile-defense systems, magnetic propulsion units for new transport systems, helium refrigeration systems for military aircraft, advanced energy conversion cycles (mainly magnetohydrodynamic systems), and low-temperature energy transmission, distribution, and storage.

As summarized by Cook, helium conservation is a national issue in which thermodynamic certainty collides with economic uncertainty. Low-entropy helium is wasting into the atmosphere at least in part because there is no way to prove that future generations will be better off if it is saved for them. A helium conservation program was started, then aborted. A decision to resume storing helium will be a political decision based more upon prevailing ideas of fairness in intergenerational risk-bearing and equality, and on current view of the qualitative impact on future society of materials scarcities, than on any quantitative forecasts of future needs and costs.

References

Besson, J. M., and J. P. Pinceaux: "Melting of Helium at Room Temperature and High Pressure," *Science,* **206,** 1073–1075 (1979).

Cohen, E. G. D.: "Quantum Statistics and Liquid Helium-3–Helium-4 Mixtures," *Science,* **197,** 11–16 (1977).

Cook, E.: "The Helium Question," *Science,* **206,** 1141–1148 (1979).

Ebisch, R.: "Helium," *Science News,* **116,** 3, 50 (1979).

Lupton, J. E., and H. Craig: "A Major Helium-3 Source at 15°S on the East Pacific Rise," *Science,* **214,** 13–18 (1981).

Staff: "Superfluid Helium," *Science News,* **116,** 8, 135 (1979).

HEMATITE. The mineral hermatite, ferric oxide, Fe$_2$O$_3$, occurs as thick or thin tabular rhombohedral forms, sometimes in pyramids but rarely in hexagonal prisms. It also assumes botryoidal, columnar and lamellar shapes, and may be granular or compact. Its hardness is 5.6; specific gravity, 5.26; luster, metallic to earthy or dull; color, dark gray to black; earthy forms may be different shades of red; streak, red to red-brown; translucent (in very thin flakes) to opaque. Hematite with a metallic luster is called specular iron.

It is a widely distributed and common mineral, found in igneous, sedimentary and metamorphic rocks as beds and veins, having probably been formed in many different ways under very different conditions. Beautifully crystalized hematite has been found in the Urals of the U.S.S.R.; Rumania; Switzerland; the Island of Elba; Alsace, France; Cumberland, England. Extremely rich, large hematite ore bodies have been found and are being worked in Minas Gerais, Brazil; Cerro de Mercado, Durango, Mexico; Quebec and Labrador in Canada. The hematite ore deposits which lay along the southern and northwestern sides of Lake Superior in Michigan, Wisconsin and Minnesota have been worked to near depletion. Extensive beds of hematite are found throughout the Appalachian region from New York to Alabama, being mined near Birmingham in the latter state. Hematite occurs in quantity in Nova Scotia and Newfoundland. It is the most important ore of iron, and has other industrial uses in paint manufacture and polishing compounds. The name hematite is derived from the Greek word meaning blood.

See also terms listed under **Mineralogy.**

HEMIMORPHITE. This mineral is zinc silicate, Zn$_4$Si$_2$O$_7$(OH)$_2 \cdot$2H$_2$O, occurring in tabular and prismatic orthorhombic crystals, although often in massive and fibrous forms. There is a perfect cleavage parallel to the prism; it is brittle with a subconchoidal fracture; hardness, 4.5–5; specific gravity, 3.40–3.50; luster, vitreous; color, white, tending to translucent. Hemimorphite differs from willemite, also a zinc silicate, in that the former contains considerable water which may be driven off when heated to a high temperature.

There are many localities for hemimorphite in Europe, fine specimens having come from Saxony, Sardinia; Cumberland, Alston Moor and Derbyshire, England. It is found in Siberia, Algeria, and Mexico. In the United States, hemimorphite has been found at Sterling Hill, New Jersey; in Lehigh County, Pennsylvania, and in Virginia, Missouri, Montana, Colorado, Utah, New Mexico, and Nevada.

The mineral is so named because of the tendency to form doubly terminated crystals showing a different grouping of faces at either end. The name is derived from the Greek meaning half and form.

HEMOGLOBIN. The main function of the hemoglobin molecule is oxygen transport. The hemoglobin molecules from each species of organism which has been examined differ in the sequence of amino acids in their polypeptide chains unless they are very closely related. Chimpanzee and human hemoglobins are apparently identical. Sometimes two or more different kinds of hemoglobin are found simultaneously in the same organism. These structural variations may give rise to differences in the physiological properties which help to determine the efficiency of oxygen transport by the blood from lungs or gills to the tissues. Hemoglobin also plays an important role in carbon dioxide transport. See also **Blood.**

Vertebrate hemoglobins are usually composed of four polypeptide chains of two types, called α and β. The molecules can, therefore, be described as $\alpha_2\beta_2$. An iron porphyrin moiety, *heme*, is associated with each chain. Evidence indicates that combination of the heme with oxygen results in structural changes in the protein to which it is bound. Studies of single crystals of horse and human hemoglobins by x-ray diffraction show that removal of oxygen from the iron atoms of the four hemes results in a separation of the β-chains from one another; the relative positions of the α-chains do not appear to change. Although the molecular basis is not fully understood, the consequences are important. It is certain that any change in the mutual relationships of the polypeptide chains will alter the environment of many amino acid residues. These environmental changes are probably responsible for the degree of oxygenation to the oxygen pressure; and the dependence of the oxygenation upon pH and upon carbon dioxide concentration.

Mutations which alter the amino acid sequence can occur in either the α- or the β-chain of the adult. However, most mutations are deleterious and changes in the α-chain would be more severely selected against in a process of natural selection because any change in the α-chain would affect the sensitive fetus, whereas changes in the β-chain would affect only the adult. This means that evolution tends to favor changes in the β-chain over changes in the α-chain. These considerations indicate that molecular adaptation of hemoglobin, at least in mammals, may involve changes more in the β-chain than in the α-chain.

Hemoglobins can be dissociated into their α- and β-subunits. Not only are hemoglobins capable of dissociating into their polypeptide subunits, but certain hemoglobins are also capable of polymerization. Many reptiles and amphibians and certain mice possess hemoglobins which polymerize to form double molecules $(\alpha_2\beta_2)_2$ and sometimes triple or quadruple molecules. Many hemoglobins from invertebrate animals have very large molecular weights and are composed of a large number of subunits—as many as 180 in some species. The nature of the forces holding these large aggregates together is under study.

The amino acid sequences of hemoglobins have been extensively altered by mutation during evolution. Data on the amino acid sequences of the chains from a variety of mammalian and other vertebrate hemoglobins show that the sequence can be varied extensively without drastic change in function. There appears to exist a hierarchy in the functional importance of different parts of a protein. Substitutions in different segments of a polypeptide chain may, according to the type and position of the substitution, exhibit a spectrum of effects, ranging from detectable to catastrophic. For example, the single substitution of valine for glutamic acid in the 6th position of the β-chain in human sickle cell hemoglobin results in a large decrease in the solubility of deoxygenated hemoglobin within the red cells. The hemoglobin, by forming a gel, distorts the red cell shape ("sickle") in such a way that flow through the capillaries is retarded. Such drastic consequences do not result if the substitution is lysine rather than glutamic acid (hemoglobin C). Histidine in position 63 of the human β-chain has an essential role stabilizing the ferrous state of the heme iron. Substitution by tyrosine (in hemoglobins "M") results in the loss of this stability because the ferric iron can form a strong linkage with the —OH group of tyrosine. Such a substitution results in a complete loss of capacity to combine reversibly with oxygen.

The foregoing are radical substitutions. Most effective substitutions appear to be relatively conservative and do not drastically affect the oxygen transport function. Therefore, the number of differences between homologous chains appears to be related not to functional differences, but to the time which has elapsed since the chains diverged from a hypothetical polypeptide ancestor. The mean number of differences between the hemoglobin chains of man, horse, pig, rabbit, and cattle is approximately 11. The common ancestor of these mammals may have existed some 80 million years ago. Thus, approximately 11 effective mutations per chain occurred in 80 million years, or 1 substitution per chain in 7 million years. Zuckerkandl and Pauling, using standard probability theory, have used this figure to estimate the time at which the different human hemoglobin chains (α, β, γ, and δ) are believed to have arisen by gene duplication. These estimates are shown in the accompanying table.

DIVERGENCE OF HEMOGLOBIN CHAINS WITH TIME

TYPE OF CHAIN DIVERGENCE	NUMBER OF DIFFERENCES	ESTIMATED TIME SINCE DIVERGENCE
β-δ	10	35 million years
β-γ	37	150 million years
β-α	76	380 million years
(α-β)-myoglobin	~ 135	650 million years

Estimates like these indicate that hemoglobins are very old and that it may be possible to find relatives of vertebrate hemoglobins in invertebrate animals. They also suggest that the gene duplication believed to be responsible for the divergence of the α- and β-chains took place in the Devonian period at the time of the appearance of early amphibians and the dominance of fish.

The suggested relationship between numbers of differences and evolutionary time is not wholly secure. It assumes uniformity in the rate of effective amino acid substitution, but this rate may be neither uniform with time, nor uniform in different parts of the polypeptide chain. Differences in the rate of effective substitution along the polypeptide chain may be due not only to restrictions imposed by the required tertiary structure, but also to differences in the rate at which various parts of the DNA or the gene mutate. The evolution of hemoglobin may be contrasted with that of cytochrome c in which approximately 50% of the molecule appears to have remained invariant during the time yeast and man have evolved.

HEPARIN. A complex organic acid (mucopolysaccharide) present in mammalian tissue; a strong inhibitor of blood coagulation. Precise chemical formula has not been fully established, but the formula, $(C_{12}H_{16}NS_2Na_3)_x$, has been suggested for sodium heparinate. The drug is derived from animal livers or lungs. Heparin is used in deep venous thrombosis therapy. It is also used in rodenticides which cause internal hemorrhaging. Pets exposed to such poisons must receive immediate treatment with the administration of vitamin K, also sometimes called the antihemmorhagic vitamin. See **Vitamin K.**

HERBICIDE. A substance that kills or interferes markedly in the life cycle of certain plants and is used with other control chemicals, such as fungicides and insecticides, to increase the yield and quality of crops. In addition to eliminating or greatly stunting the growth of those plants (weeds) that compete with crops for water and soil nutrients, herbicides achieve a number of objectives. Usually lumped under the phrase *weed control*, the advantages of herbicides include: (1) They eliminate weeds that serve as harboring places for insects which attack crop plants. (2) They eliminate perennial plants that may serve as hosts for survival and build-up of virus diseases. An example is the corn stunt virus, which overwinters in johnson-grass rhizomes. Insects able to carry virus diseases may feed on weeds and move to crop plants, causing infection by damaging virus diseases. (3) They eliminate weeds that serve as traps for moisture. Easy availability of moisture encourages fungus diseases that can be spread easily from weeds to crop plants by wind movement of fungus spores. (4) They eliminate honeysuckle, kudzu, and other plants that grow on fences and that are severely damage to fencing and other minor structures because of the sheer weight of their foliage. Damaged fences adversely affect livestock production. (5) In so-called no-till planting, herbicides are used exclusively, eliminating the mechanical removal of weeds by cultivating equipment. The advantages of herbicides in this regard are time and labor savings.

Classification of Herbicides

There are several ways in which herbicides can be grouped:

Target plant selectivity is a measure of the effectiveness of a herbicide against a range of plants to be destroyed. *Nonselective* herbicides are not difficult to create. There are hundreds of chemicals that will kill just about any living plant within range. These are extremely *wide-spectrum* substances, not only destroying or stunting both broadleaf and grasslike weeds, but woody plants as well. Of course, some control over these very powerful chemical substances can be exerted by regulating concentration. Dilute applications may result in desired defoliating, for example, without fully destroying a stand of plants, such as trees. Broad-spectrum and nonselective herbicides are sometimes regarded as the same, but more generally, broad-spectrum refers to a compound that does not differentiate between broad- and narrowleaf plants. A *selective* or *narrow-spectrum* herbicide is customized to make this selection and, in fact, to differentiate even more closely. In operating with crop-rotation programs, it is usually advantageous to select herbicides with relatively narrow spectrums of effectiveness so that later crops may not be adversely affected by any residues from a prior crop. Herbicide manufacturers continue in their research toward the development of crop-specific control chemicals. Just one example— a herbicide to control wild oats in connection with wheat production.

Timing. Herbicides are usually designated for a *pre-plant* or *pre-emergence* use or for *post-emergence* application. The terms are roughly self-explanatory—pre-emergence signifying the use of the herbicide on the land prior to the cracking stage or emergence of weeds or desired crop above the soil line. The herbicide, possibly in granule or liquid form, may be incorporated into the soil a number of weeks before planting, in which case, the term *pre-plant* is used. For effective control over the growing season, some land areas or crops may have to be treated a number of times between emergence and harvest. Thus, the term *post-emergence*. Several days must elapse between application and harvesting to avoid contamination of the crop when gathered.

Stability. A number of factors determine the stability of a herbicide. For example, some of the control chemicals decompose (and thus become ineffective) when applied at temperatures in excess of about 90°F (32.2°C), or during long periods of intense sunlight. Most herbicides are more efficient when applied on cool, partly cloudy days. In the case of many other herbicides, the presence of moisture (after an irrigation, rain, or during a generally wet period) greatly reduces or destroys their effectiveness. The presence of certain chemicals also affects application success. Some control chemicals are adversely affected by any mixture with acidic materials; others by the presence of alkaline materials, or certain metals, such as iron or copper. Essentially, these materials interact with the original chemical composition and alter it so that, instead of an effective herbicide, for this purpose it may be essentially inert. Careful preparation, particularly of emulsions for spraying, cannot be overemphasized. If the stability and effectiveness of a herbicide is long term, it can be designated as a *persistent herbicide*, i.e., one that is effective over a period of several months. Those substances that break down within several days to a few weeks (biodegradable, in a sense) are *nonpersistent herbicides*. This is an important factor in selecting a herbicide. Some control chemicals can essentially sterilize a plot of land for a period of years, and should one's objectives change after such an application, neutralization or removal of the substance from the affect area can be costly and quite difficult.

All control chemicals can be categorized as either *contact-type* or *systemic* substances. Although these designations are more commonly applied to insecticides, they also are operable in terms of herbicides. In a contact-type substance, the killing action is largely limited to the area of actual contact between chemical and plant (for example, a defoliant that damages a bush or plant without completely destroying the whole plant). In a systemic substance, contact of the substance with plant tissue is progressively spread throughout the plant as, for example, by the plant's vascular system.

Physical Form. A large number of herbicides are available in several forms, including granules, powders and dusts, wettable powders, emulsifiable concentrates, slurries, etc. Some of these are factory-prepared; others can be prepared locally by the user. Sprays and dusts are widely used for foliar applications, whereas dusts and granules may be preferred for soil applications.

Chemical Structure. As with fungicides, insecticides, and other pesticide chemicals, herbicides are usually complex, often synthetic organic chemicals, and of a widely-varying composition, ranging from carbamates, to anilides, to organic acids, salts, etc. Chemical make-up is discussed further a bit later in this entry.

Nomenclature of Herbicides. There are well over 100,000 pesticides and agricultural control chemical formulations; perhaps 25% of these fall into the sphere of herbicides. As with insecticides, although the basic chemicals used in the formulation of herbicides may number in the several hundreds, many thousands of possible formulations arise from the various physical formats offered, as well as minor differences provided by many manufactuers in brand name products. Each manufacturer markets products under tradenames—names that are essentially coined for their marketing charisma and only infrequently connote much about the content or purpose of the product. Thus, there are scores of equivalent (or essentially equivalent) products, adding to the difficulty of selecting these chemicals. Unfortunately, from this standpoint, the generic chemical names of the majority of herbicide chemicals are long and complex and essentially meaningless to persons who are not well versed in organic and biochemistry. Helpful listings of this type can be found in the "Foods and Food Production Encyclopedia" (D. M. Considine,

Editor), Van Nostrand Reinhold, New York (1981). There are also a number of frequently revised directories of control chemicals and considerable information available from various government agencies and universities. See the list of references at the end of this entry. This situation of nomenclature is quite similar to that which applies to generic and tradename drugs and pharmaceuticals.

Chemistry of Herbicides

Aromatic Carboxylic Acids. Considerable research has gone into investigations of the physiological activity of these acids on plants, including benzoic, phenylacetic, and naphthoic acids. Among the benzoic acid derivatives, the greatest activity is shown by those compounds containing substituents in the 2, 3, and 6 positions; and only to a slightly lesser degree, by those substituents in other positions. Included among commercial herbicides in this category are: 2,3,6-trichlorobenzoic acid; 2-methoxy-3,6-dichlorobenzoic acid; 2,5-dichloro-3-nitrobenzoic acid; and 2,5-dichloro-3-aminobenzoic acids. Slightly less active are: 2-bromo-3,5-dichlorobenzoic acid; and 2,3,5-triiodobenzoic acid.

Substituted phenylacetic acids have a high activity. Considerable activity is shown by monohalogen-substituted acids. Introduction of a second halogen does not markedly affect the degree of activity.

Aryloxyalkylcarboxylic Acids and Derivatives. Research has shown that the physiological activity of phenoxyacetic acid toward plants increases when a halogen atom is incorporated into the molecule. The strongest effects are displayed by fluorine and chlorine. The position of the substituent also affects physiological power, with the 4-halophenoxyacetic acids displaying the greatest activity. It is interesting to note that the activity of this compound is about ten times greater than the case of the 2-isomer. Activity is further reduced in the 3-chlorophenoxyacetic acid. There are numerous herbicides in this chemical structural category, including 2,4-D, 3,4-D, and MCPA (4-chloro-2-methylphenoxyacetic acid). It is also interesting to note that while MCPA is effective as an agricultural control chemical, the very closely related compound, 4-chloro-2-chloromethylphenoxyacetic acid is not of great value. See Dioxin.

Derivatives of Carbamic Acid. Whereas the aryl esters of *N*-methylcarbamic acid find wide application as insecticides, the alkyl esters are strong herbicides against monocotyledonous weeds. Their actions against dicotyledonous plants is much weaker. Because of these differences, these herbicides are effective in controlling monocotyledonous weeds in such crops as carrot, cotton, and sugar beet. Research has shown that the esters of naphthylcarbamic, diphenylcarbamic, and other polycyclocarbamic acids are not effective herbicides. It has been found that the arylcarbamic acid ester derivatives of unsaturated alcohols are stronger herbicides than the corresponding esters of saturated alcohols. The carbamates have an ability to form hydrogen bonds with the chlorophyll molecule or proteins of plants, accounting for their effective herbicidal activity.

Derivatives of Thio- and Dithiocarbamic Acids. Research has indicated that the derivatives of the thiocarbamic acids are good penetrants of plants, moving easily through the xylem. Among this structural class of herbicides, the *S*-alkyl-*N*, *N*-dialkylthiocarbamates are the most effective. Most of these compounds are selective herbicides against annual grasses and a few dicotyledons. They have been applied successfully in connection with such crops as bean, beet, other vegetables, and sugarcane. The thiocarbamates are usually mixed with the soil as pre-emergence herbicides. In terms of effectiveness, this usually decreases as the number of carbon atoms in the ester radical increases, particularly in excess of five carbon atoms. The activity also decreases when the total carbon atoms in the alkyl radicals on the nitrogen atom is greater than six.

Derivatives of Urea and Thiourea. A great deal of investigation has gone into the effectiveness of these compounds and, as the result, a number of urea derivatives have found use as effective herbicides as well as growth regulators. This is particularly true among the trialkylureas that contain simple and complex hydrocarbon radicals. Examples include: 3-(3,4-dichlorophenyl)-1,1-dimethylurea (*Diuron*); 3-(3,4-dichlorophenyl)-1-methoxy-1-methylurea (*Linuron*); 1,3-dimethyl-3-3(2 benzothiazoyl) urea (*Methabenzthiazuron*); 3-(4-bromophenyl)-1-methoxy-1-methylurea (*Metobromuron*); and *N*-benzyl-*N*-(dichloro-3,4-phenyl)-*N*,*N*-dimethylurea (*Phenobenzuron*).

The salts of aryldialkylureas tend to be more active than the ureas.

Thiocyanates and Isothiocyanates. At one time, ammonium thiocyanate was a commonly applied nonselective, contact-type herbicide and desiccant. The compound is less important now because of the development of other organic herbicides that do not decompose so readily.

Sulfuric and Sulfurous Acid Derivatives. Earlier, sulfuric acid was applied as a herbicide and still finds some use as a desiccant for potato plant tops prior to mechanical harvesting. The primary drawback of sulfuric acid, not experienced with more recently developed herbicides, is the large amount of acidity which it adds to the soil. Ammonium sulfamate continues to find use as an effective herbicide in some areas, both for the elimination of weeds and as a sterilant for soil. This compound hydrolyzes in soil to form ammonium sulfate, a source of ammonia and nitrogen.

Several other classes of organic chemicals, including heterocyclic compounds, are represented among the scores of herbicides commercially available. As of the late 1970s, in terms of tonnage usage, the Food and Agriculture Organization (United Nations) listed the following major categories: MCPA, 2,4-D, 2,4,5-T, triazines, carbamates, and urea derivatives. It should be stressed here that regulations pertaining to the use of herbicides vary widely from one country to another—and from one time period to another. The proliferation of new products tends to offset those prior compounds that have been banned in some countries, or where usage has been severely curtailed.

References

Andrilenas, P. A.: "Farmers' Use of Pesticides," U.S. Department of Agriculture, Washington, D.C. (Revised periodically)

Crafts, A. S.: "Modern Weed Control," University of California Press, Berkeley, California (1977).

Considine, D. M., Editor: "Foods and Food Production Encyclopedia," Van Nostrand Reinhold, New York (1981).

Melnikov, N. N. (F. A. Gunther and J. D. Gunther, Editors): "Chemistry of Pesticides," Springer-Verlag, New York (1971).

Staff: "2,4-D Increases Insect and Pathogen Pests on Corn," *Science*, **192**, 239–240 (1976).

Thomson, W. T.: "Agricultural Chemicals—Herbicides," Thomson Publications, Fresno, California (1980).

HEROIN. Alkaloids.

HESSITE.

A mineral telluride of silver, Ag_2Te, with some gold, crystallizing in the monoclinic system at normal temperatures; isometric system above 149.5°F (65.3°C). Crystalline form not obvious at normal temperatures. Hardness, 2–3; specific gravity, 8.24–8.45; color, gray with metallic luster; opaque. Named after G. H. Hess (1802–1850).

HETEROCYCLIC COMPOUNDS. Compound (Chemical); Organic Chemistry.

HETEROPOLYACIDS. Acids derived from two or more other acids, under such conditions that the negative radicals of the individual acids retain their structural identity within the complex radical or molecule formed. The term **heteropolyacids** is usually restricted to complex acids in which both radicals are derived from oxides, such as phosphomolybdic acid.

HEULANDITE. The mineral heulandite is a monoclinic zeolite whose crystals are often quite suggestive of orthorhombic forms. Its chemical composition is probably $(Na, Ca)_{4-6}Al_6(Al, Si)_4Si_{26}O_{72}\cdot24H_2O$; strontium may be present. Heulandite has one good cleavage; is brittle with a conchoidal fracture; hardness, 3.5–4; specific gravity, 2.18–2.22; luster, vitreous to pearly; color, white to gray, red or brown; streak, white; transparent to translucent. Occurs chiefly in cavities in basaltic rocks with other zeolites, but may be found in granites, pegmatites, gneisses, and schists. Famous localities are in Iceland, India, the Harz Mountains, Italy, Switzerland, Scotland, Nova Scotia; and in the United States at Bergen Hill and West Paterson, New Jersey. This mineral was named for the English mineralogist Heuland.

HEXAMINE. $(CH_2)_6N_4$, formula wt 140.19, white crystalline solid, mp 280°C, decomposed at higher temperatures. Also known as hexamethylenetetramine, methenamine, or urotropine, the compound is soluble in H_2O and only very slightly soluble in alcohol or ether. Although used to some extent in medicine as an internal antiseptic, the primary use of hexamine is in the manufacture of synthetic resins where the compound is a substitute for formalin (aqueous solution of paraformaldehyde) and its NaOH catalyst. Hexamine also is used as an accelerator for rubber.

On a commercial scale, hexamine is manufactured from anhydrous NH_3 and a 45% solution of methanol-free formaldehyde. These raw materials, plus recycle mother liquor, are charged continuously at carefully controlled rates to a high-velocity reactor. The reaction is exothermic. The reactor effluent is discharged into a vacuum evaporator which also serves as a crystallizer. The hexamine crystals then are washed, dried, and screened. Average yield of the process is about 96% conversion of ingredients to produce hexamine.

HEXOSE MONOPHOSPHATE OXIDATIVE PATHWAY. Carbohydrates.

HISTAMINE. A powerful vasodilator which is released in anaphylactic (hypersensitivity to protein) shock and occurs in blood and tissues in minute amounts.

The injection of 1 microgram intravenously in humans is said to bring about a sharp drop in blood pressure. Its close relationship to histidine is emphasized by the fact that the amino acid can be decarboxylated by certain intestinal bacteria to produce it.

Histamine is a product of the degradation of histidine and is liberated by injury to the tissue, or whenever a protein is decomposed by putrefactive bacteria. Histamine's biological role is both positive and negative. The production of excessive histamine gave rise to the formulation of drugs for countering such excesses. See also **Antihistamine.** Histamine can cause pulmonary edema of noncardiac etiology. Histamine causes both constriction of bronchial smooth muscle and edema of bronchial mucosa by increasing the permeability of small bronchial veins. Histamine is one of several humoral mediators that affect bronchial tissue. Mast cells contain and release histamine. Histamine stimulates connective tissue regeneration by producing edema.

HISTONES. Basic proteins which occur in the nuclei of both plant and animal cells. They are less basic than the protamines, having isolectric points at about pH 11. Some investigators restrict the term *histone* to only those basic proteins anatomically and chemically associated with DNA (deoxyribonucleic acids). The close associations of histones with DNA led to the hypothesis that histones might play a role in the control of genetic expression at the cellular level. Advances in molecular biology have permitted more detailed mechanisms for such control to be proposed. Histones, by blocking some areas of the DNA molecule, may permit only part of the DNA base sequences to act as templates for the formation of messenger RNA. Thus histones, by controlling messenger RNA formation, may ultimately control protein biosynthesis within the cell. Or, the primary role of histone may be structural, histone being essential for stabilizing the DNA helix, for the integration of DNA strands into more complex chromosomal structures, and for fixing and maintaining during cell division chromosomal changes occurring during differentiation and development. The foregoing two concepts are not mutually exclusive, i.e., histone may fix chromosomal structure in a specific configuration in which the position of the histone molecules also limit RNA formation. The possibility that histones play a role in genetic mechanisms suggests that possibility that histone changes may initiate or accompany early cellular changes, leading to the formation of tumors. Further investigation is needed to ascertain whether or not tumor histones differ from those of corresponding normal tissues.

HOLMIUM. Chemical element symbol Ho, at. no. 67, at. wt. 164.93, tenth in the lanthanide series in the periodic table, mp 1472°C, bp 2700°C, density, 0.795 g/cm³ (20°C). Elemental holmium has a close-packed hexagonal crystal structure at 25°C. The pure holmium is a silvery-gray color, slow to tarnish or oxidize at room temperature in normal atmospheres. Even at relatively high temperatures, the metal is slow to oxidize. Under a vacuum of about 10 torr, holmium will react when hot with water vapor, CO_2, NH_3, and hydrocarbons. Holmium is soft and can be worked by conventional equipment. There is one natural isotope, ^{165}Ho, and 18 artificial isotopes have been produced. The natural isotope is not radioactive. In terms of abundance, holmium is present in the earth's crust on the average of 1.2 ppm, ranking ahead of bismuth, antimony, cadmium, and mercury in potential availability. The element was first identified by P. T. Cleve and J. L. Soret in 1879. The metal has a low acute-toxicity rating. Electronic configuration is

$$1s^22s^22p^63s^23p^63d^{10}4s^24p^64d^{10}4f^{10}5s^25p^65d^16s^2.$$

Ionic radius: Ho^{3+}, 0.894 Å; metallic radius, 1.766 Å.

Holmium occurs in apatite, xenotime, and yttrium- and heavy rare-earth minerals. The element of a purity of 99.9% can be obtained through organic ion-exchange techniques. Supplies of holmium are available commercially as the result of yttrium production. To date, the applications for holmium have been quite limited. When added to orthoferrites, holmium has shown promise for use in electric circuits. Uses in semiconductors, lasers, thermoelectric devices, phosphors, and ferrite bubble devices currently are being studied.

See references listed at the ends of the entries on **Chemical Elements**; and **Rare-Earth Elements and Metals**.

NOTE: This 4th Edition entry was revised and updated by K. A. Gschneidner, Jr., Director, and B. Evans, Assistant Chemist, Rare-Earth Information Center, Energy and Mineral Resources Research Institute, Iowa State University, Ames, Iowa.

HOMOGENIZING. A process for reducing the size of particles in a liquid and useful in the preparation of numerous food substances, including milk, ice cream, salad dressings, various fruit juices, flavor concentrates, and infant foods, among others.

A reduction of particle or globule size in a mixture of two immiscible liquids makes an emulsion possible. If an emulsifying agent is present, a more stable emulsion can be produced and coalescence of the dispersed phase is prevented. The homogenizer is also used to produce dispersions by reducing the particle size with solid-in-liquid mixtures. As in the preparation of an emulsion, a dispersing agent is needed to maintain a homogeneous mixture.

Typically, a homogenizer consists of a high-pressure, positive-displacement pump and an adjustable orifice. The pump is a piston or plunger type, usually consisting of three plungers, although some homogenizers are made with five or even seven plungers. The cylinder for each plunger has an inlet and discharge valve. The plunger pump must push the product through the homogenizing valve (adjustable orifice). For two-stage homogenization, two valves are arranged in series.

A typical homogenizing valve consists of a seat and plug of very hard abrasion-resistant materials (alloys such as *Stellite* are used). The seating surfaces must be lapped smooth and be parallel. In operation, the plug is spring-loaded against the seat. Spring compression is adjusted so that when the product flows, energy in the form of pressure is required to lift the plug. Although many products can be homogenized at pressures below 3000 pounds per square inch (204 atmospheres), machines are made to develop pressures in excess of 8000 pounds per square inch (544 atmospheres). In another design, a valve uses a compressed cone of stainless-steel wire inserted into a socket, the product being homogenized by flowing between the wires.

A number of theories have been proposed as to what actually breaks up the particles in the homogenizer: (1) As the product enters the area between the lapped surfaces, it is suddenly accelerated to velocities as high as 30,000 feet per minute (9,144 meters per minute) at a pressure of 5,000 pounds per square inch (340 atmospheres). When acceleration is this sudden, the particle (especially the liquid particle) is stretched or elongated to the point of breaking. (2) At this high velocity, there are shear forces between layers of liquids under flow that break up particles. (3) Cavitation may be the major cause of homogenization. When the pressure energy is converted into velocity energy, the vapor pressure of the product exceeds product pressure, resulting in the formation of vapor cavities which collapse upon leaving the valve at higher pressures. This collapsing, or implosion, of cavitation exerts tremendous force, breaking up the particles. Most homogenizers are designed to incorporate one or more of the foregoing principles.

References

Becher, P.: "Emulsions: Theory and Practice," Van Nostrand Reinhold, New York (1963).
Farral, A. W.: "Food Engineering Systems," AVI, Westport, Connecticut (1976).
Harper, W. J., and C. W. Hall: "Dairy Technology and Engineering," AVI, Westport, Connecticut (1976).
Loo, C. C., Slatter, W. L., and R. W. Powell: "Study of the Cavitation Effect in the Homogenization of Dairy Products," *Journal of Dairy Science*, **33**, 672 (1950).
Selitzer, R., Editor: "The Dairy Industry in America," Dairy and Ice Cream Field, New York (1977).

HOMOLOGOUS SERIES. Two organic compounds are said to be homologous if their molecular formulas differ by CH_2, or a multiple of CH_2. For example, the alkane series has the general formula, C_nH_{2n+2}, its first three members being methane CH_4, ethane, C_2H_6, and propane, C_3H_8.

HORMONES. In animals, hormones are organic compounds, usually of considerable complexity and even after years of research not fully understood, that are secreted by endocrine (ductless) glands, such as the adrenal gland, the thyroid and parathyroid glands, the pituitary gland, and the gonads, among others. Hormones are sometimes commonly called by the names of the gland which secrete them. Thus, there are adrenal cortical hormones, thyroid and parathyroid hormones, etc. Hormones are regulators of physiological processes within the body, exerting control over such processes as metabolism, growth, reproduction, molting, pigmentation, and electrolytic and osmotic balance, among other processes. Apparently, hormones achieve these objectives chemically and electrically, although the mechanisms are not fully understood and, in fact, the mechanisms may vary from one situation to the next. At one time, hormones were loosely called "chemical messengers" because they are transported from point to point within the organism and thus effect actions at distances from the region where they are made. If one visualizes secreting glands as sensors of a type detecting need for correction of some physiological process, then the hormones might be visualized as both the transmitters or carriers of this information and the initiators of actions as well. The conventional concept is that cells have receptors on their surface which sense the presence of specific hormones. At one time, it was firmly believed that hormones, particularly polypeptide hormones, such as insulin, prolactin, and growth hormone, all of which are large charged molecules, could not penetrate through the cell's membrane and actually enter the cell. This belief has since been altered because researchers have shown that insulin, for example, can enter into the cell. Referring to this process as "internalization," one investigator in 1978 suggested that the internalization of polypeptide hormones will be one of the most active topics in cell biology for a number of years.

Research during the late 1970s and continuing into the 1980s has shown too that hormones and/or their receptors may be degraded. As gross examples of this type of situation, it is known that many obese people with high concentrations of insulin in their blood also have normal concentrations of blood sugar. Why doesn't the insulin decrease the blood sugar concentration in these cases? And it is well known that pregnant women produce much angiotensin II, which normally increases blood pressure. But, these women usually do not have hypertension. What alterations in the hormone-cell mechanism provide this result? And there are instances where males have tumors which secrete large quantities of a hormone that stimulates the production of testosterone and yet there is no evidence of abnormal amounts of testosterone. At least two questions are posed: Do certain hormones lose their effectiveness with time? Or are there changes in target receptor cells? Research has indicated that there may be a relationship between concentration of hormones and the surface receptors which bind them, such receptors being inactive for a time or possibly disappearing from the cell surface all together. A number of investigators have observed that a better understanding of the manner in which hormones affect their own and other receptors possibly may result in new ways to treat certain diseases, including insulin-resistant diabetes.

In recent years, it has been shown that a wide variety of receptors are regulated by hormones. Some receptors are sensitive to but one hormone, which appears to be the case of insulin. Others appear to be regulated not only by one hormone, but others as well. For example, it has been shown that the receptors for TRH (thyrotropin-releasing hormone) are not exclusively regulated by TRH, but also by other homones. Receptors for gonadotropins (pituitary hormones that act on the gonads, appear to be regulated by hormones in addition to gonadotropin. There are numerous other instances of this kind.

Although of an exploratory nature, Marx (1979) reports on consideration being given to the overall process of aging in terms of hormones and their receptors by endocrinologists. Details are given by Korenman (1979).

Hormones display not only great variations in function, but also in their chemical nature. They vary widely in chemical nature. Some are steroids, such as estrogen, progesterone, cortisone, etc., while others are amino acids (thyroxine), polypeptides (vasopressin), low-molecular-weight proteins, and conjugated proteins. Amino acid and steroid hormones have been isolated and many, including insulin, have been sythesized. Other types are prepared directly from the endocrine organs of animals.

Hormones produced by one species usually show similar activity in other species. The hormones showing greatest species specificity are proteins or conjugated proteins.

Hormones are markedly affected by deficiencies or excesses of the various vitamins and other dietary essentials.

Because of the great complexity of a number of the natural hormones, conventional approaches of organic synthesis which have been used so successfully over the years in connection with many drugs have not proved viable to date with some of the hormones. Insulin is an example. Presently, millions of diabetics still depend upon animal insulin as extracted from the pancreatic glands of slaughtered pigs. If diabetes mellitus continues to become more prevalent, as it has over the past several years, natural sources may not be ample. Further, means to lower the cost of hormones such as insulin would be beneficial. In the late 1970s and continuing into the 1980s, researchers at several institutions have been probing the possibilities of applying genetic engineering to this problem. One group has been successful in inducing the bacterium *Escherichia coli* to manufacture and secrete rat proinsulin, an immediate precursor of rat insulin that incorporates insulin itself. Research like this is an important step toward the objective of developing bacterium-based industrial systems that can replace animal and human tissues as the source of medically useful proteins, such as insulin, growth hormone, and clotting factor.

Classes of Hormones

Hormones may be grouped into two distinct types: (1) *direct-acting*; and (2) *stimulating*—substances that stimulate other organs to produce their own characteristic hormones. The latter group is sometimes called the *tropic hormones*. See accompanying tabular summary of hormones.

Thyroid Hormones. These are compounds of the amino acid *thyronine*. They are present in the free form only to a slight extent, existing chiefly as constituents of the protein thyroglobulin. The most important of these acids in terms of hormone action are the 3,5,3'-tri, and the 3,5,3',5'-tetraiodocompounds,

triiodothyronine and *thyroxin*, the structures of which are given in the accompanying table. The action of thyroid hormones is to accelerate cellular reactions and to increase the metabolic rate and oxygen consumption of tissues. They effect this action by stimulating many of the enzyme systems, not only the glucose oxidation system and the cytochrome chain for dehydrogenating the coenzyme NADPH, but other processes, such as the synthesis of proteins from amino acids. Their effects are clearly apparent in the pathological changes in the organism caused by their excess or deficiency. The thyrotrophic hormone and other biochemical interactions with the thyroid gland are discussed later in this entry.

Parathyroid Hormones. The influence of the parathyroid glands on the regulation of calcium concentrations in the blood of mammals was first recognized by MacCullum and Voegtlin in 1909.

More recently, several groups of investigators have succeeded in purifying and partially identifying the structure of the hormone, variously called *parathormone* and *parathyroid hormone*. This is a single-chain peptide hormone with a molecular weight of about 8,000. A second parathyroid hormone, *calcitonin*, was postulated by Copp (1961). Subsequent research has indicated that this hormone is actually the hormone which is now known to be produced by the thyroid gland. However, a parathyroid calcitonin may exist in certain species.

The more classical function of parathyroid hormone is concerned with its control of the maintenance of constant circulating calcium levels. Its action is on (1) the kidney, where it increases the phosphate in the urine, (2) the skeletal system, where it causes calcium resorption from bone, and (3) the digestive system, where it accelerates (stimulates) calcium absorption into the blood. The hormone and gland exhibit characteristics of feedback control; when the concentration of calcium ions in the blood falls, the secretion of the hormone increases, and when their concentration rises, the secretion of hormone decreases.

Adrenal Cortical Hormones. The adrenal gland is made up characteristic hormones. The hormones of the adrenal medulla are the catecholamines, epinephrine (adrenalin) and norepinephrine (noradrenalin), which are closely related chemically, differing only in that epinephrine has an added methyl group. See accompanying table. In fact, animal experiments have established a metabolic pathway for the biosynthesis of both compounds from the amino acid phenylalanine, which involves enzymatic oxidation and decarboxylation reactions. It is also to be noted that the isomeric form of norepinephrine is most important; the natural D-form (which incidentally, is levorotatory) has many times the activity of the synthetic isomer. Epinephrine has a pronounced action upon the circulatory system, increasing both blood pressure and pulse rate, and hence the cardiac output by its direct action upon the heart muscle, and especially because it causes constriction of the arterioles. However, its effects upon smooth muscles vary; it relaxes the muscles of the digestive system, but contracts the pyloric sphincter.

Norepinephrine does not affect the cardiac output, although it does raise the blood pressure by constricting the arterioles. Its muscular effects are less pronounced. Both epinephrine and norepinephrine release free fatty acids from adipose tissue, so raising its level in the blood. This effect is due to the action of the hormones in accelerating enzymatic reactions whereby the esters of the fatty acids are hydrolyzed. The third type of action of epinephrine is its effect upon the carbohydrate metabolism, notably the acceleration of the hydrolysis of glycogen in muscular tissue and the liver, and so raising the glucose level in the blood, and the rate of glucose oxidation, with resulting

HORMONE Common Names, (Synonyms), Structure and Production Site	PRINCIPAL PHYSIOLOGIC FUNCTIONS	INTERRELATIONSHIPS WITH VITAMINS
Adrenocorticotropic Hormone (ACTH) (Adrenocorticotrophin; corticotrophic hormone) Straight-chain, simple polypeptide, 39 amino acids, no S—S bridges. (See text, Fig. 1.) Molecular Weight ~4500 Production Site: Anterior pituitary	Maintenance of adrenal cortex Promotes secretion of steroids, oxidative phosphorylation in adrenal cortex Mobilizes and increases oxidation of free fatty acids in adipose tissue Increases gluconeogenesis in liver; increases cyclic adenosine monophosphate (AMP) in adrenal cortex Decreases urea formation in liver	Ascorbic acid: depleted in adrenal cortex on stimulation by ACTH Biotin and vitamin A: adrenocortical insufficiency noted in biotin and vitamin A deficiency Niacin: production of reduced nicotinamide adenine dinucleotide (phosphate) (NADPH) by ACTH via cyclic adenosine monophosphate (AMP) Niacin and pantothenic acid: synergistic with ACTH in steroid hormone synthesis Vitamin D: antagonized directly by ACTH via cortisol action
Aldosterone (Aldocortin; electrocortin; mineralocorticoid; 18-oxo-corticosterone) 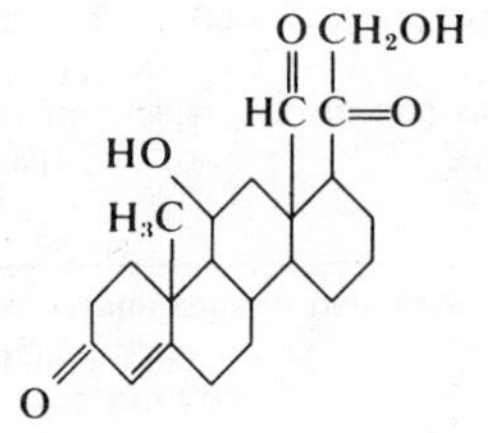 Molecular Weight 360.4 Production Site: Adrenal cortex	Maintenance of normal electrolyte blood balances Prolongs survival of adrenalectomized animals Accelerates gluconeogenesis Regulates kidney function	Ascorbic acid: adrenal cortex depleted of ascorbic acid on production of aldosterone Biotin: prolongs life in adrenalectomized rats Niacin: nicotinamide adenine dinucleotide (phosphate) (NADPH) involved in synthesis of aldosterone
Cortisol (Hydrocortisone, 17-hydroxycorticosterone) 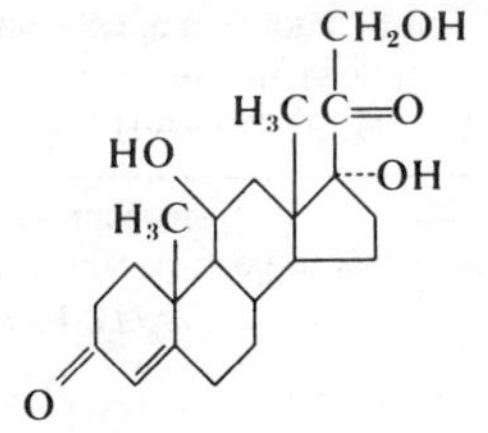 Molecular Weight 362.5 Production Site: Adrenal cortex	Increases (1) protein catabolism (excepting liver) gluconeogenesis; (2) carbohydrate anabolism (liver); (3) blood sugar; (4) glucose absorption; (5) brain excitation; (6) spread of infections; (7) urinary glucose and nitrogen; (8) stress tolerance; (9) lactation; (10) water diuresis Decreases (1) fat anabolism; (2) growth rate; (3) inflammation; (4) eosinophils; (5) lymphocytes; (6) antigen sensitivity; (7) respiratory quotient; (8) ketosis; (9) wound healing; (10) skin pigmentation; (11) RBC hemolysis. Regulates general adaptation syndrome, water balance, blood pressure, and hormone release.	Ascorbic acid: may be required for steroid hormone biosynthesis; depleted from adrenal cortex on cortical secretion Biotin: adrenocortical insufficiency noted in biotin deficiency Folic acid and pantothenic acids maintain secretions of steroids by adrenal cortex Niacin: nicotinamide adenine dinucleotide (phosphate) (NADPH) required for steroid hormone biosynthesis Vitamin A: deficiency causes cortical necrosis Vitamin D: action antagonized by cortisol by reducing calcium absorption in intestine
Epinephrine (Adrenaline, adrenin, suprarenin, vasotonin, vasoconstrictine, adrenamine; levorenine) 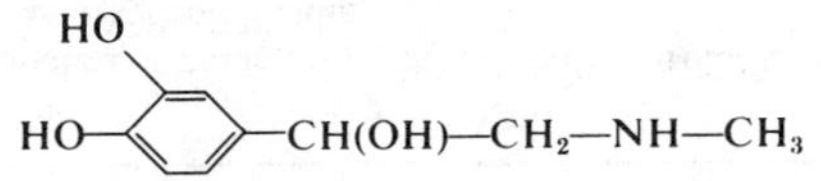 Molecular Weight 183.2 Production Site: Adrenal medulla and chromaffin cells in gut	Blood circulation: increases blood pressure; peripheral vasodilator; increases heart output and rate; flow increased in brain, liver, and skeletal muscle Central nervous system: causes restlessness, anxiety Kidney: reduces glomerular filtration rate Lung, intestine, genital system: inhibited motility Metabolic effects: increases oxygen consumption, temperature, basal metabolic rate, gluconeogenesis Pituitary effects: stimulates production and release of ACTH and corticoids	Ascorbic acid: maintains reduced state of epinephrine Ascorbic acid, folic acid, and vitamins B_6 and B_{12} are cofactors in synthesis of epinephrine from phenylalanine

461

REPRESENTATIVE HUMAN HORMONES (continued)

Hormone / Structure	Physiological Action	Vitamin Relationships
Estradiol (Female hormone; dihydrotheelin; dihydrofollicular hormone; dihydrofolliculin) 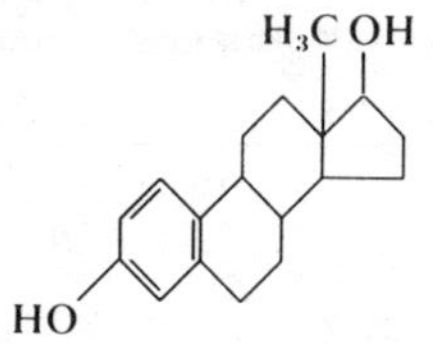 (H$_3$C OH … HO) Molecular Weight 272.4 Production Sites: Ovarian follicles; testes; corpus luteum; adrenal cortex; placenta	Regulates menstrual cycle, female sex behavior Maintains secondary sex characteristics Affects antibody properties Induces estrus, uterine hypertrophy, vaginal cornification; potentiates and stimulates calcitonin secretion	Folic acid: involved in mitotic effect of estradiol Niacin, diphosphopyridine nucleotide (DPN), triphosphopyridine nucleotide (TPN): involved in increased respiration and in cholesterol precursor synthesis Pyridoxine: competes as cofactor with estrogen sulfate in kynurenine aminotransferase activity Vitamin D: synergistic in calcium metabolism with estradiol Vitamin E: involved in follotropin production or release
Follicle-Stimulating Hormone (FSH) (Follotropin, luteoantine, thylakentrin, Prolan A, gonadotropin 1, gametogenic hormone, follicle ripening hormone, gametokinetic hormone) Structure: Not fully definitized. Production Site: Anterior pituitary.	Female: stimulates ovarian follicles to grow and to develop, forming multiple layers and antra Male: stimulates seminiferous tubules; stimulates spermatogenesis	Ascorbic acid: depletion in ovary due to follicle-stimulating hormone and luteinizing hormone action Vitamin E: required to maintenance of membranes in sex organs
Glucagon (HGF) (Hyperglycemic-glycogenolytic factor; glucagon; HG-factor) Structure: Polypeptide, 29 amino acids (structure determined). No S—S bridges. Molecular Weight ~3500 Production Site: Alpha cells in pancreas.	Increases: blood sugar; blood K$^+$, oxygen consumption, liver glycogenolysis, gluconeogenesis, nitrogen and salt excretion Decreases: liver glycogen, protein formation, gastric juice, fatty acid synthesis	Ascorbic acid: depletion of adrenal ascorbic acid by glucagon
Insulin (no synonyms) Structure: 51 amino acids. Known and synthesized. 3 S—S bridges. (See text, Fig. 4) Molecular Weight 5,734 (monomer); 12,000–48,000 (polymer), depending upon pH. Production Site: Beta cells of islets of pancreas.	Regulates carbohydrate and fat metabolism, especially glucose and fat oxidations Stimulates amino acid and glucose transport into cells and protein synthesis	Ascorbic acid: acts similarly to alloxan (i.e., antangonist)
Luteinizing Hormone (LH) (Luteotrophin, ISCH) Structure: Globular glycoprotein with S—S bridges. Molecular Weight 26,000 Production Site: Anterior pituitary.	Female: promotes estrogen and progesterone secretion, ovulation; maintains ovarian tissues Male: stimulates Leydig cells to secrete testosterone; gametogenic with follotropin (FSH)	Ascorbic acid: ovarian depletion on LH stimulation Vitamin E: involved in spermatogenesis
Melanocyte-stimulating Hormone (MSH) (Melanotrophin, chromatophorotropic hormone; pigmentation hormone) Structure: Polypeptide; purified, synthesized; alpha and beta forms; straight chains. Molecular Weight: 1500 (alpha) 2100–2600 (beta) Production Site: Intermediate lobe of pituitary.	Mammals: exerts small effect on skin pigmentation (protection from sunlight not fully proved) Expands or contracts pigments in various chromatophores Expands melanophore pigments with color changes in amphibia (adaptation to environment) Lower vertebrates: increases sensitivity to light; decreases dark adaptation time	Ascorbic acid: adrenal cortex depleted on ACTH and MSH activity Vitamin A: MSH decreases dark adaptation time

Hormone	Physiological Effects	Vitamin/Other Interrelationships
(Arterenol, noradrenaline; levarterenol) 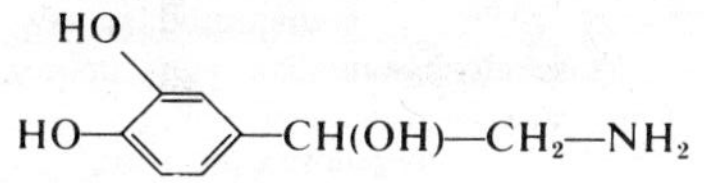 HO—⟨ring⟩—CH(OH)—CH₂—NH₂ (HO substituent) Molecular Weight 169.2 Production Site: Adrenal medulla; adrenergic nerve endings; chromaffin cells.	constrictor without change or slight decrease in output and heart rate. No flow increase in brain, liver, or muscle Central nervous system effects: adrenergic transmitter agent at synapses; no brain excitation Kidney: decreases glomerular filtration rate Lung, intestine, genital system: inhibited Metabolic effects: weak epinephrine effect	Ascorbic acid, folic acid, and vitamin B_6 are cofactors in synthesis of norepinephrine from phenylalanine
Oxytocin (Oxytocic hormone; pitocin; uteracon; α-hypophamine) Cys—Tyr—Ile—Gln—Asn—Cys—Pro—Leu—Gly NH₂ (Cys–Cys joined by bridge) Molecular Weight 1007 Production Site: Hypothalamus.	Uterine contraction, milk ejection, facilitates sperm ascent in female tract Decreases membrane potential of myometrium, basic metabolic rate, and liver glycogen Stimulates oviposition in hen, releases luteinizing hormone (LH) Increases blood sugar and urinary sodium and potassium	Findings on interrelationships with vitamins are not extensive
Parathyroid Hormone (PTH) (Parathormone) Structure: Simple polypeptide (83 amino acids), sequence determined; straight chain; No S—S bridges. Production Site: Parathyroid glands.	Increases blood calcium, kidney calcium reabsorption, phosphate excretion, and blood citrate level Mobilizes calcium and phosphate from bone Activates calcium and phosphate absorption from the gastrointestinal tract (for which vitamin D is required) Increases osteoclast formation	Vitamin D: synergistic with PTH in maintenance of serum calcium
Progesterone (Progestin, luteosterone) 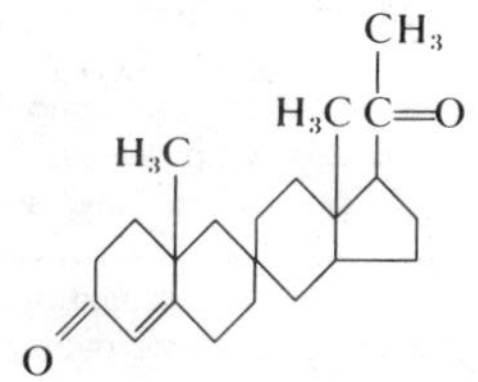 [steroid structure: CH₃—C=O; H₃C; H₃C; O] Molecular Weight 314.5 Production Sites: Ovary (follicles, corpus luteum); testicles; adrenal cortex; placenta	In low concentrations: prepares uterus for blastocyst implantation; promotes ovulation and mammary gland development; regulates female sex accessory organs; weak corticosteroid properties; precursor to sex hormones In high concentrations: maintains pregnancy; represses ovulation and sex activity; inhibits vaginal cornification and parturition; decreases myometrial excitation	Ascorbic acid: depleted from adrenal cortex or ovary on progesterone formation Niacin: diphosphopyridine nucleotide (DPN) involved in progesterone synthesis
Prolactin LTH (Lactogenic hormone; lactogen; galactin; mammotropin) Structure: Single-chain protein, 205 amino acids Molecular Weight 23,000–25,000 Production Site: Anterior pituitary	Initiates lactation Develops mammary glands in female Increases weight and growth (similar to somatotrophin in some species) Participates in nidation of zygote Protein anabolism (some species) Growth and secretion of crop gland (birds) Luteotropic (only in mouse, rat) Promotes maternal behavior	Not fully determined. Generally participates with other substances having growth action
Relaxin (Releasin, cervilaxin) Structure: Polypeptide (4 peptides with activity have been isolated); about 30–40 amino acids in each peptide Molecular Weight 4000–5000 Production Site: Corpus luteum in pregnancy	Enlarges birth canal in preparation for parturition Separation of symphysis pubis, loss of rigidity in pelvic bones Decreases uterine motility Maintains pregnancy Increases sensitivity to oxytocin; releases oxytocin Stimulates mammary gland Stimulates imbibition of water in uterus Inhibits uterine contraction	Ascorbic acid: maintains mucoprotein ground substance in connective tissue, affected by relaxin

463

REPRESENTATIVE HUMAN HORMONES (continued)

Somatotrophin (STH)
(Growth hormone, GH; somatotrophic hormone; hypophyseal growth hormone)
Structure: Known and synthesized; coiled, unbranched protein; 188 amino acid residues; 2 S—S bridges
Molecular Weight 21,500
Production Site: Anterior pituitary

Promotes general growth of organism
Promotes skeletal growth, protein anabolism, fat metabolism, carbohydrate metabolism, water, and salt metabolism

Relates with all vitamins in connection with growth actions

Testosterone
(17 beta-hydroxy-4-androsten-3-one)

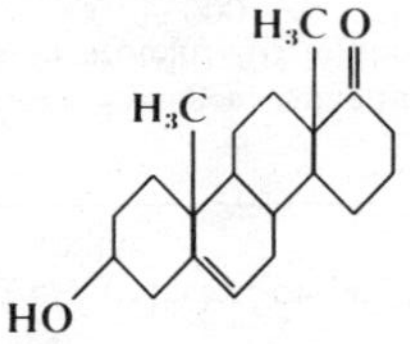

Molecular Weight 288.4
Production Sites: Interstitial cells of ovary and testis; adrenal cortex; embryonic placenta

Controls secondary male sex characteristics
Maintains functional competence of male reproductive ducts and glands
Increases protein anabolism; maintains spermatogenesis; inhibits follotropin
Increases male sex behavior; increases closure of epiphyseal plates

Ascorbic acid, folic acid, vitamins A and E are synergists with testosterone for maturation of germ cells and increased anabolic activity

---

Thyroid-stimulating Hormone (TSH)
(Thyrotrophic hormone, thyrotrophin)
Structure: Glycoprotein (300 amino acids)
Molecular Weight 26,000–30,000
Production Site: S^2 type cell, anterior pituitary

Regulates body temperature via thyroxine
Maintains thyroid gland and its secretory activity (colloid discharge)
Maintains iodine uptake by thyroid gland
Promotes differentiation in embryo during development via thyroxine
Stimulates coupling of diiodotyrosine to form thyroxine

Ascorbic acid, thiamine, riboflavin, and vitamin B_{12}: requirements increase in hyperthyroidism; tissue concentrations reduced
Vitamin A: massive doses of vitamin A inhibit secretion of TSH; thyroid hormones required for carotene and retinene conversions
Vitamins A, D, E, and K: requirements increased in hyperthyroidism; tissue concentrations reduced in hyperthyroidism
Vitamin B_6, niacin: conversion to phosphorylated reactive forms impaired in hyperthyroidism

---

Thyroxine (T_4)
(3,5,3',5' tetraiodothyronine)

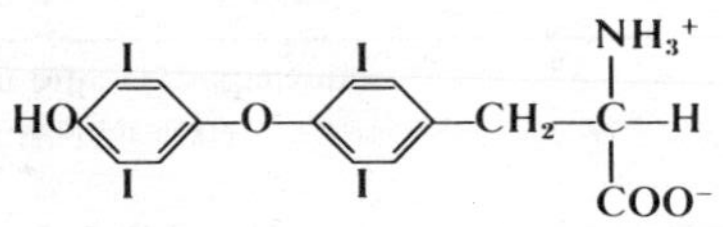

Molecular Weight 776.9
Production Site: Thyroid gland

Regulates growth, differentiation, oxidative metabolism, electrolytic balance
Increases carbohydrate metabolism, calorigenesis, protein anabolism, basal metabolic rate, oxygen consumption, fat catabolism, fertility
Sensitizes nervous system

Ascorbic acid: synergist in cold survival
Niacin: synergist in mitochondrial metabolism
Vitamin A: T_4 is required for vitamin A synthesis in liver
Vitamin B_{12}: T_4 aids in B_{12} absorption
B complex vitamins: deficiencies develop in hyperthyroidism

Vasopressin
(Arginine vasopressin; antidiuretic hormone; ADH; pitressin; tonephin; vasophysin)

Cys—Tyr—Phe—Glu NH_2—Asp NH_2—Cys—Pro—Leu—Gly NH_2
Vasopressin

Molecular Weight 1084 (arginine-vasopressin)
Production Site: Hypothalamus

Elevates blood pressure (mammals) (reverse effect in birds)
Decreases kidney blood flow
Antidiuretic; releases ACTH
Increases sodium chloride and urea excretion
Regulates water balance
Stimulates contraction of smooth muscles
Increases renal tubular water reabsorption
Releases anterior pituitary hormones

Not fully determined

increase in oxygen utilization, carbon dioxide production, and body temperature.

The hormones of the adrenal cortex are steroids. See also **Steroid.** Among them there are a number of hormones with androgenic activity, such as adrenosterone and 17α-hydroxy-progesterone, which are discussed under the sex hormones later in this entry. In all, over ten steroids have been identified in the adrenal cortex, including seven of characteristic cortical activity. These are corticosterone, from which the others are named, 17α-hydroxyl-11-dehydrocorticosterone (cortisone), 17α-hydroxycorticosterone (cortisol or hydrocortisone, and 18-oxocorticosterone (aldosterone). Only two hormones, cortisol and corticosterone, are normally released in fairly large quantities, and another, aldosterone, deserves mention because of its somewhat different effects, even though it is released to a far lesser extent.

All of these hormones are synthesized from cholesterol in the adrenal cortex, by an extended series of reactions which include many related compounds. Although these hormones have widespread effects throughout the organism, their primary mechanism is not known, so that many of the effects may be indirect. Much of the knowledge of their action arises from studies of insufficiency or hyperactivity of the adrenal cortex, which produces a wide variety of pathological conditions. See accompanying table.

It is generally considered that aldosterone, and to some extent the other hormones, have a regulatory effect upon the metabolism of electrolytes and water, particularly upon the concentration of the ions of the alkali metals in intracellular fluids. Administration of steroids also increases the concentration of calcium ions in those fluids. However, all three of these hormones have a number of other effects, roughly in the order of potency—cortisol, corticosterone, aldosterone. (Cortisone would be placed in the second position.) They produce changes in the metabolism of carbohydrates, proteins, and fats.

For the carbohydrates alone, three major effects are evident—increase in the rate of formation of glucose, increase in the rate of release of glucose from the liver, and increase in the rate of utilization of glucose. These hormones affect the digestive system, increasing the secretion of hydrochloric acid, pepsinogen, and trypsinogen. They prevent inflammatory responses to bacterial or even chemical stimuli; they counteract anaphylactic shock, and other effects of hypersensitivity. Obviously, these properties have led to their widespread therapeutic use.

There are relationships between the adrenal cortical hormones and the thyroid and pituitary glands. Depression of the function of the adrenals produces thyroid deficiency, whereas administration of thyroxine stimulates the ACTH-adrenal cortical mechanism.

Pituitary Hormones. The hormones of the hypophysis (pituitary gland) are quite numerous, being secreted variously in three parts of the gland—the neurohypophysis (posterior lobe), the adenohypophysis (anterior lobe), and the *pars intermedia*, which connects the other two.

The chief hormones of the neurohypophysis are the polypeptides oxytocin and vasopressin. The hormone characteristic of the *pars intermedia* is the melanocyte-stimulating hormone. It is usually spoken of in the plural, since in most mammals both alpha and beta forms are known. The structures of the first two are shown in the accompanying table.

The most prominent effect of oxytocin is the contraction of smooth muscle, especially of the uterus. It also has a major effect upon the muscles about the breast, and so stimulates the ejection of milk in lactating animals. It has a definite stimulating effect upon the muscles of the ureter, urinary bladder, intestine, and gall bladder.

The most prominent effect of vasopressin is upon the kidneys, where it stimulates the resorption of water in the tubules (which by repeated release and absorption concentrate the urine). It also constricts the coronary arteries, raises the blood pressure, and exhibits the effect of oxytocin upon smooth muscles, but generally to a lesser degree.

The action of the melanocyte-stimulating hormones has been established by studies of animals, in which they cause dispersal of certain black pigments from the cells that contain them, with resulting darkening of the skin.

The adenohypophysis is the part of the gland in which the tropic hormones are secreted. They include the adrenocorticotropic hormone (ACTH), the thyrotropic hormone (TSH), and somatotropin, as well as three hormones with pronounced effects upon the gonads: the hormone prolactin, the follicle-stimulating hormone (FSH) and the luteinizing or interstitial cell stimulating hormone (LH or ISCH).

ACTH. Adrenocorticotropin (ACTH) in humans is a polypeptide containing a sequence of 39 amino acids, although work with animal forms of it and with degradation products of the human form have shown that not all of them are essential to the activity of the hormone. This sequence for the human ACTH is shown in Fig. 1.

The primary function of ACTH is the stimulation of the adrenal cortex to produce its hormones, which have already been discussed. This is evident from the therapeutic effect of administration of ACTH, which is closely similar to that of these hormones, so that if the action of only one of them is sought, its administration is preferable. Moreover, ACTH stimulates secretion of the androgenic substances mentioned as produced by the adrenal cortex.

Thyrotrophic Hormone. This hormone (TSH) stimulates the development of the thyroid and controls its secretion. Although purified preparations of it have been obtained, they consist of a mixture of proteins of high mean molecular weight (about 30,000). Some of their amino acids have been determined, as well as their carbohydrates, but the structures have not been elucidated.

Growth Hormone. Somatotropin is the growth hormone. Purified preparations of extracts of it from the human adenohypophysis have been crystalized. They are known to be proteins, of mean molecular weight 21,000, and containing a single polypeptide chain. This hormone differs from the others of its group in not acting primarily upon the other endocrine glands, but in controlling the gain in body weight and the rate of skeletal growth. The growth abnormalities, such as dwarfism and giantism, have been shown to result from its hypo- and hypersecretion. In addition to its effect upon growth and anabolism generally, it has been found to affect the kidneys and pancreas, and to influence glucose, galactose, and lipid metabolism.

Gonadotrophic Hormones. These include follicle stimulating hormones (FSH), luteinizing or interstitial cell stimulating hormone (LH or ISCH), and prolactin. Their structures are not known; the molecular weight of human LH is about 26,000, that of human FSH is about 30,000, and that of human prolactin is uncertain. They are proteins, with variable amounts of carbohydrates. FSH induces the growth of Graafian follicles in the

Ser—Tyr—Ser—Met—Glu—His—Phe—Arg—Tyr—Gly—Lys—Pro—

Val—Gly—Lys—Lys—Arg—Arg—Pro—Val—Lys—Val—Tyr—Pro—

NH$_2$
|
Asp—Ala—Gly—Glu—Asp—Glu—Ser—Ala—Glu—Ala—Phe—Pro—

Leu—Glu—Phe

Fig. 1.   Amino acid sequence of human adrenocorticotropin (ACTH).

ovary and the production of spermatozoa in the testis. LH stimulates the final development of the ovarian follicles, the appearance of estrus, and the change of the follicles to corpora lutea. In the male, it stimulates the secretion of testosterone. Since these effects are due to the effect of this hormone upon interstitial cells, it is also called ISCH. Prolactin stimulates lactation after birth, acts with estrogen to promote the growth of the mammary gland, and influences the activity of the corpus lutea.

Male Hormones. The androgenic hormones produced in the testes (and adrenal gland) have a widespread effect upon the development of secondary sexual characteristics (musculature, facial hair, larynx, etc.), as well as upon the sexual organs and responses themselves. They also promote anabolism to a marked degree by their effect upon nitrogen and calcium metabolism. The structure of testosterone is shown in the accompanying table.

Female Hormones. Closely related to the male androgenic hormones, and probably synthesized from them in the female organism, are the estrogenic hormones which are produced principally in the ovary. Although β-estradiol is the normally secreted ovarian hormone, a number of other estrogenic substances have been isolated from urine and from animal studies. They include α-estradiol, estriol, and estrone. The structures of these hormones are given in Fig. 2.

These hormones are important in both the menstrual cycle and the reproductive cycle, and of course play an important role in oral contraceptives (the "pill"). They induce growth of the vaginal epithelium, secretion of mucus by the glands of the cervix, and initiate the growth of the endometrium, which is taken over by progesterone (from the corpus luteum) later in the cycle. They activate the proliferation of the mammary gland during pregnancy. As the androgens do for the male, the estrogens bring about the secondary sexual characteristics of the female. They have a number of effects upon metabolism, notably that of calcium and phosphorus, and of lipids and proteins. A number of other estrogens, some made synthetically and others obtained from animals, are known.

The corpus luteum produces two hormones, progesterone and relaxin. The structures of these hormones are shown in the accompanying table.

Progesterone acts to complete the proliferation of the endometrium, which was initiated by the estrogenic hormones, and to prepare it for the ovum. In pregnancy the continued action of progesterone is necessary. It aids the growth of the breasts and has a definite effect against ovulation. It is also the biosyn-

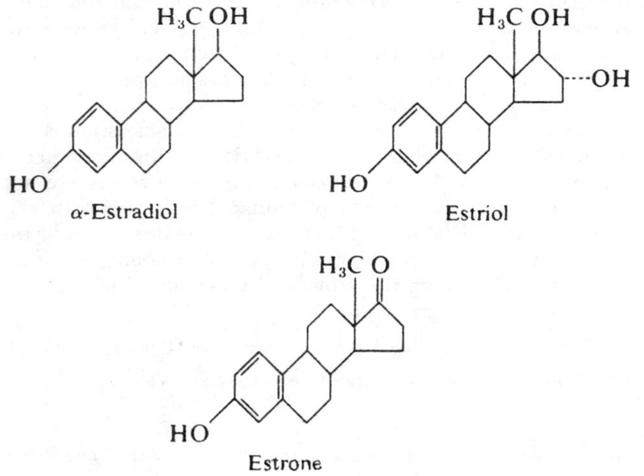

$H_3C\ OH$

α-Estradiol

$H_3C\ OH$ ----OH

Estriol

$H_3C\ O$

Estrone

Fig. 2. Major ovarian hormones.

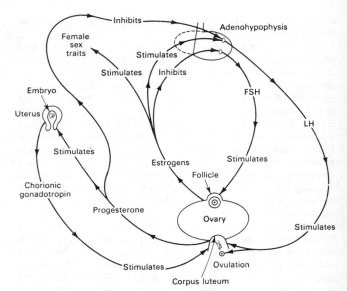

Fig. 3. Cycle of hormone adjustment in human female.

thetic precursor of some of the estrogenic hormones. Relaxin has been shown to have a relaxing effect on the cartilaginous junction of the pubic bones in preparation for parturition.

Feedback in Hormone Control Systems. Not only do the hormones initiate or stimulate biological processes, both directly and by bringing about production of other hormones in other glands, but they also act to maintain the organism in a steady state, or *homeostasis*. Thus the gonadotropic hormones from the hypophysis stimulate the testes, but the resulting production there of androgens like testosterone, inhibits the action of the hypophysis in producing the gonadotropic hormones. The complicated cycle of adjustment in the human female involves both positive and negative feedback, as shown in the cycle illustrated in Fig. 3.

As shown in the figure, the regulation of the ovarian hormones in the human female involves both positive and negative feedback. The follicle-stimulating hormone (FSH) from the adenohypophysis stimulates the Graafian follicles, which thus produce estrogens. These not only inhibit FSH production through negative feedback, but also stimulate the adenohypophysis to increase its production of luteinizing hormone (LH) through positive feedback. This hormone in turn brings about ovulation from the Graafian follicle. After the ova are discharged, the LH stimulates the empty follicle, now the corpus luteum, to produce progesterone.

This hormone brings about the changes in the reproductive organs required for the development of the embryo. Then the progesterone partly inhibits the adenohypophysis from producing further LH, an example of negative feedback; as a result, there is no further ovulation. The progesterone also acts as a positive feedback and stimulates the production of FSH.

When pregnancy intervenes, a new feedback mechanism must be introduced, or the embryo would be expelled by the shedding of the lining of the uterus in menstruation. Here the placenta (chorion) of the embryo itself produces hormones, as already noted. Its LH stimulates continuing production of progesterone from the corpus luteum, thus preventing menstruation and stimulating the continuing development of the uterus as needed by the growing embryo. The extra progesterone also inhibits further ovulation in spite of the presence of the gonadotropin from the placenta (chorion).

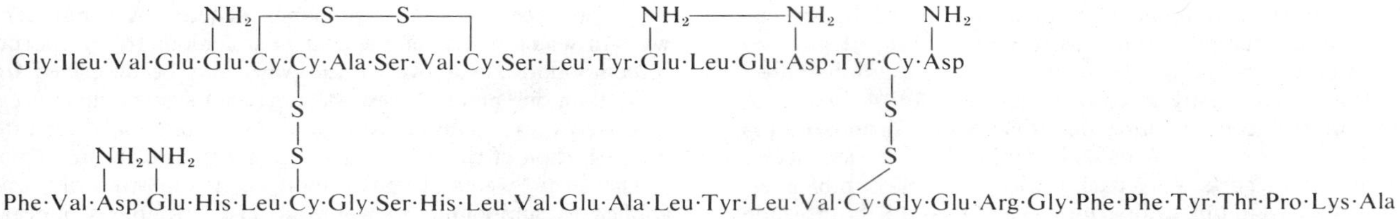

$$\text{Gly·Ileu·Val·Glu·Glu·Cy·Cy·Ala·Ser·Val·Cy·Ser·Leu·Tyr·Glu·Leu·Glu·Asp·Tyr·Cy·Asp}$$

with NH_2 groups and S—S disulfide bridges shown above, and the lower chain:

$$\text{Phe·Val·Asp·Glu·His·Leu·Cy·Gly·Ser·His·Leu·Val·Glu·Ala·Leu·Tyr·Leu·Val·Cy·Gly·Glu·Arg·Gly·Phe·Phe·Tyr·Thr·Pro·Lys·Ala}$$

Fig. 4.   Primary structure of bovine insulin. (Abbreviations for amino acids will be found under **Amino Acids.**)

$$\text{His·Ser·Glu·Gly·Thr·Phe·Thr·Ser·Asp·Tyr·Ser·Lys·Tyr·Leu·Asp·Ser·Arg·Arg·Ala·Glu·Asp·Phe·Val·Glu·Tyr·Leu·Met·Asp·Thr}$$

Fig. 5.   Glucagon. (Abbreviations for amino acids will be found under **Amino Acids.**)

Pancreas and Nonendocrine Hormone Sources. In addition to producing hormones, the pancreas also generates digestive fluids (*pancreatic juice*). It is the hormone function which makes the pancreas a part of the endocrine system. The pancreas secretes *insulin* and *glucagon*, both hormones. The primary structure of bovine insulin is shown in Fig. 4. Abbreviations of the amino acids will be found under **Amino Acids.** The upper chain is termed the A chain; the lower chain, the B chain. In the spatial configuration of the molecule, these chains occur as coiled helices. The structure of glucagon is considerably simpler inasmuch as it consists of a single chain of amino acids. See Fig. 5.

These two hormones have two opposing effects. That of insulin is *hypoglycemic,* i.e., it increases the rate of utilization of glucose, the probable process being an effect of insulin to increase the penetration of glucose through the cell walls as well as increased phosphorylation. The overall result of action of insulin in its relation to glucose is to increase the rate of the reactions by which glucose is oxidized, but also its transformation to glycogen. The enzyme glucagon raises blood glucose levels by increasing the rate of hydrolysis of glycogen (in the liver) to increase the formation ultimately of glucose. Insulin increases the rate of entry of amino acids into cells and their rate of protein biosynthesis. Insulin also accelerates the formation of lipids from carbohydrates, whereas glucagon stimulates the formulation of keto compounds from lipids, inhibits the synthesis of fatty acids, and accelerates the breakdown of various phosphorus and nitrogen compounds. The primary result of insulin deficiency is diabetes mellitus.

Hormones may be produced by organs other than the endrocrine glands. Conspicuous among such organs is the placenta, the organ on the wall of the uterus to which the umbilical cord is attached. It has been found to produce the same estrogenic hormones as the ovary, the same hormones (progesterone and relaxin) as does the corpus luteum, and gonadotropic hormones (and luteinizing hormones) similar to, but not identical with, those produced by the adenohypophysis.

Other hormones which do not originate in endocrine glands are the cholecystokinin of the intestine, and the enterogastrone and gastrin of the stomach. The first is produced by the upper intestinal mucosa and causes the gall bladder to contract; the enterogastrone is produced in the same tissue and inhibits gastric motility and secretion; it also excites secretion of digestive fluids, principally hydrochloric acid.

In plants, a *plant hormone* or *phytohormone* is an organic compound produced by the plant, controlling growth and other functions at sites remote from where the hormone is produced. Plant hormones also act in very minute amounts. Plant hormones include the auxins, gibberellins, and kinetins.

References

Barrington, E. J. W. (editor): "Trends in Comparative Endocrinology," American Society of Zoologists, Thousand Oaks, California, 1975.

Brownstein, M. J., Russell, J. T., and H. Gainer: "Synthesis, Transport, and Release of Posterior Pituitary Hormones," *Science,* **207,** 373–378 (1980).

Crews, D.: "The Hormonal Control of Behavior in a Lizard," *Sci. Amer.,* **242,** 2, 180–187 (1979).

Kolata, G. B.: "Hormone Receptors," *Science,* **196,** 747–748 (1977).

Kolata, G. B.: "Polypeptide Hormones," *Science,* **201,** 895–897 (1978).

Kolata, G. B.: "Sex Hormones and Brain Development," *Science,* **205,** 985–987 (1979).

Korenman, S.: "Conference on the Endocrine Aspects of Aging," National Institute of Aging, the Endocrine Society, and the Veterans Administration, Bethesda, Maryland (October 18–20, 1979). Veterans Administration Hospital, Sepulveda, California, 1979.

Krieger, D. T., and A. S. Liotta: "Pituitary Hormones in Brain," *Science,* **205,** 366–372 (1979).

Kutsky, J. F.: "Handbook of Vitamins and Hormones," Van Nostrand Reinhold, New York, 1973.

Martial, J. A., et al.: "Human Growth Hormone," *Science,* **205,** 602–607 (1979).

Marx, J. L.: "Hormones and Their Effects in the Aging Body," *Science,* **206,** 805–806 (1979).

Mettes, J.: "The 1977 Nobel Prize (Hormone Research)," *Science,* **198,** 594–596 (1977).

Meites, J., Donovan, B. T., and S. M. McCann (editors): "Pioneers in Neuroendrocrinology," Plenum, New York, 1978.

Oppenheimer, J. H.: "Thyroid Hormone Action at the Cellular Level," *Science,* **203,** 971–979 (1979).

Staff: "Insulin by Bacterium," *Sci. Amer.,* **239,** 4, 85–90 (1978).

Turakulov, Y. K. (editor): "Thyroid Hormones," Plenum, New York, 1975.

White, B. A., and C. S. Nicoll: "Prolactin Receptors in *Rana catesbeiana* during Development and Metamorphosis," *Science,* **204,** 851–853 (1979).

HORNBLENDE. The mineral hornblende is a complex silicate which is probably an isomorphous mixture of three molecules, a calcium-iron-magnesium silicate, an aluminum-iron-magnesium silicate and an iron-magnesium silicate. A general formula is

$$(Ca,Na,K)_{2-3}(Mg,Fe^{2+},Fe^{3+},Al)_5(Al,Si)_8O_{22}(OH)_2.$$

Manganese and alkalis are sometimes present as is also titanium. It is monoclinic, with prismatic crystals, often pseudo-hexagonal. Bladed, fibrous, columnar, granular and compact massive varieties also are common. It has a perfect prismatic cleavage; hardness, 5–6; specific gravity, 3.02–3.27; color, green, greenish-brown, brown and black; luster, vitreous to silky; transparent to opaque.

Hornblende is a common constituent of many of the igneous rocks such as granite, syenite, diorite, or gabbro, of gneisses and schists and is the principal mineral of the amphibolites. Hornblende alters easily to chlorite and epidote. A variety of hornblende that contains little (less than 5%) of iron oxides is gray to white in color and named edenite, from its locality in Edenville, New York. Very dark brown to black hornblendes, which contain titanium, ordinarily are called basaltic hornblende from the fact that they are usually a constituent of basalts and similar rocks.

Well-known localities for hornblende are in Czechoslovakia, Mount Vesuvius, Italy, Norway, Sweden, and, in the United States, in Massachusetts, New Hampshire, and New York. Black hornblende is found in Renfrew County, Canada. The word hornblende is derived from the German *horn*, and *blende*, to blind or dazzle. The term blende was often used to refer to a brilliant nonmetallic luster, i.e., zinc blende.

See also terms listed under **Mineralogy.**

—Elmer B. Rowley, Union College, Schenectady, New York.

HUMECTANTS AND MOISTURE-RETAINING AGENTS. Substances that have affinity for water, with stabilizing action on the water content of a material, are called *humectants* or moisture-retaining agents. Ideally, a humectant maintains within a rather narrow range the moisture content caused by humidity fluctuations. These materials are widely used in certain food products, as well as tobacco, and in recent years have taken on increasing importance in the case of intermediate-moisture foods. Traditionally, humectants have been used to retain moisture in foods like coconut and marshmallows which otherwise would quickly dry and become tasteless. For example, flaked coconut is kept moist in the container by adding glycerine and glycerol monostearate.

Among the most commonly used hemectants are glycerin, potassium polymetaphosphate, propylene glycol, sodium chloride, sorbitol, sucrose, and triacetin. Also, phosphates are added to the pickling solutions used to treat cured meats, such as ham, bacon, corned beef, etc., by soaking or injection. Their principal purpose is for moisture binding to reduce the loss of fluids during curing and cooking.

During the last few years, important research has gone into the addition of multiple humectants and water to food systems. Studies have shown that a hysteresis effect may occur with certain humectants, i.e., a different rate of moisture absorption than the rate for moisture desorption. Multiple humectants tend to compensate these hysteresis effects, giving uniform rates in both directions.

References

Gee, M., Farkas, D., and A. R. Rahman: "Some Concepts for the Development of Intermediate-Moisture Foods," *Food Technology*, **31**, 58–64 (1977).

Labuza, T. P., et al.: "Water Activity Determination," *Journal of Food Science*, **41**, 910 (1976).

Staff: "Food Chemicals Codex," National Academy of Sciences, Washington, D.C. (Issued periodically).

HUME-ROTHERY RULES. Intermetallic Compound.

HYDANTOIN PROCESS. Amino Acids.

HYDRATE. Exluding the loose usages in which the term hydrate indicates merely the presence of water or of its elements in 2:1 ratio, as in carbohydrate, the term hydrate denotes the appearance of water in compounds. There are a number of ways in which water may appear in stoichiometric proportions in compounds. Moreover, these ways may be described from more than one point of view. A somewhat systematic approach is to view these compounds from the point of view of the extent of integration of the water, or its elements, into the compound.

The term "water of constitution" is a somewhat old usage, applied to compounds in which no H_2O groupings appear in the structure of the compound, but the compound may undergo reaction, usually reversible, in which water is one of the products. Magnesium hydroxide and sulfuric acid could thus be said to have "water of constitution," even though it appears in their structure as hydroxyl groups, or hydroxyl groups and hydrogen atoms (protons).

The term "cationic water" may be used to describe the situation in which water appears in coordination compounds apparently joined to cations by covalent bonds. However, the fact that a number of such compounds exhibit "hydrate isomerism" is evidence for cationic bonding, as well as it is for the existence of other forms of these compounds in which the presence of water is due to electrostatic attractions or crystal stability requirements.

The term "anionic water" describes the situation in which water is joined to anions through covalent bonds, or more frequently, through hydrogen bonds. The type case is copper(II) sulfate pentahydrate, where the cation has a coordination number of four and presumably the fifth molecule of H_2O is bound to the sulfate ion (as well as to other H_2O molecules) by hydrogen bonds.

The term "lattice water" is commonly applied to cases in which the water molecules are occupying definite positions in the crystal lattice but are apparently not coordinated with either cations or anions. Again, clear-cut cases are those in which the compound is so highly hydrated that both lattice water and "ion water" are present.

The water in crystals may, however, be present in other than definite lattice positions. For example, the water molecules may be found in holes in the lattices, or they may occupy random positions in the lattices. The latter situation is often found in ion exchange resins where loss of water, up to a certain point, does not materially change the lattice structure.

Finally, in essentially noncrystalline materials, such as hydrous precipitates and colloidal gels, the water present is at the limiting case of being a hydrate, in which virtually no bonding, in the chemical sense, exists.

HYDRAZINE. $H_2N \cdot NH_2$, formula weight 32.04, colorless, fuming liquid, mp 1°C, bp 113°C, sp gr 1.011, decomposes when heated above 350°C at atmospheric pressure into N_2 and NH_2, also decomposes in presence of a catalyst (e.g., platinum) into N_2 and NH_3. Hydrazine burns when ignited in air with a violet-colored flame. The compound is soluble in all proportions with H_2O and is soluble in alcohol. Hydrazine forms a hydrate with one molecule of H_2O. Upon moderate heating or in a vacuum, the hydrate yields hydrazine and H_2O. Hydrazine is a base slightly weaker than NH_4OH.

Hydrazine is a tonnage chemical with numerous uses, including that of a propellant for rockets, yielding exhaust products at a high temperature and of a low-molecular weight; use as a strong reducing agent in the manufacture of various chemicals; and as a blowing agent for foamed rubber. The compound reacts with citric acid to form *Continazin*, an antituberculan drug.

Although the earlier processes for the commercial production of hydrazine used urea as a raw material, modern processes employ direct ammonia oxidation. In one such process, reactions occur in two steps:

(1) $NH_3 + NaOCl \rightarrow NH_2Cl + NaOH$,
(2) $NH_3 + NH_2Cl + NaOH \rightarrow H_2N \cdot NH_2 + NaCl + H_2O$.

Highgrade hypochlorite is required for Step 1. Special agents, such as gelatin, ethylenediamine tetracetic acid, glue, high alcohols, or formaldehyde are required to inhibit undesirable side reactions that would reduce the hydrazine yield through formation of ammonium chloride and N_2. In another hydrazine process, chlorine, NH_3, and H_2SO_4, along with methylethyl ketone, are used as the charge. The products of this process include hydrazine hydrate, hydrazine sulfate, ketazine, and dialkyldiazacyclopropane. Hydrazine also is used as a start-up ingredient in the preparation of cooling water for nuclear reactors where it is desired to keep the oxygen content of the water to an absolute minimum and thus decrease corrosion. Oxygen reacts with hydrazine. $H_2H \cdot NH_2 + O_2 \rightarrow N_2 + 2H_2O$. When no oxygen is present in the water, the hydrazine acts as a sink for dissolved oxygen that may enter later, by maintaining metal oxides at their lower oxidation states.

Hydrazine forms two series of salts (1) hydrazinium (1+) chloride, $H_2NNH_3^+Cl^-$, nitrate $H_2NNH_3^+NO_3^-$, hemisulfate $(H_2NNH_3^+)_2SO_4^{2-}$, (2) hydrazinium (2+) chloride $H_3NNH_3^{2+}$ $(Cl^-)_2$, dinitrate $H_3NNH_3^{2+}(NO_3^-)_2$, hydrogen sulfate $H_3NNH_3^{2+}(HSO_4^-)_2$, all soluble in H_2O. This last is produced when hydrogen azide reacts with concentrated H_2SO_4. It is very hygroscopic and decomposes in aqueous solution to give the slightly soluble monosulfate and H_2SO_4. The monosulfate and difluoride, which have been thought to have the structures $N_2H_5^+HSO_4^-$ and $N_2H_5^+HF_2^-$ in the solids, have been shown in fact to be $N_2H_6^{2+}SO_4^{2-}$ and $N_2H_6^{2+}(F^-)_2$. Hydrazinium azide $N_2H_5^+N_3^-$ is a soluble solid.

In the laboratory, hydrazine can be prepared by converting one-half of a given amount of NH_3 into chloramine NH_2Cl by sodium hypochlorite solution in the presence of a colloid and heating. The remaining one-half of the NH_3 reacts with chloramine to form hydrazine. The product is then cooled to $0°C$ and H_2SO_4 added in amount to react with the hydrazine to form hydrazine sulfate $N_2H_6SO_4$, insoluble solid. Hydrazine hemisulfate $(N_2H_5)_2SO_4$ is soluble in H_2O. It can also be made by the reaction of NH_3 and hydroxylamine-O-sulfonic acid.

Phenylhydrazine is a colorless liquid, slightly soluble in H_2O, miscible in all proportions with alcohol or ether, forms salts with acids, e.g., phenylhydrazine hydrochloride or phenylhydrazinium chloride $C_6H_5NHNH_3Cl$, is a powerful reducing agent, with alkaline copper(II) salt solution (Fehling's solution) yields copper(I) oxide precipitate, reacts with carbonyl group of aldehydes or ketones yielding phenylhydrazones, white solids, of definite melting point and utilized in identification of aldehydes and ketones, e.g., acetaldehyde phenylhydrazone $CH_3CH:NNHC_6H_5$.

Phenylhydrazine, as hydrochloride solution plus sodium acetate react with polyhydroxy aldehydes or ketones yielding *osazones* or diphenylhydrazones, yellow solids, of definite melting point and utilized in identification of sugars, e.g., phenyl-d-glucosazone $CH_2OH(CHOH)_3C:(NNHC_6H_5)CH:(NNHC_6H_5)$ plus aniline $C_6H_5NH_2$ plus NH_3.

Attention should be given to the difference between osazones and osones. An *osone* is formed by reaction of an osazone with HCl, e.g., glucosone $CH_2OH(CHOH)_3COCHO$.

1,1-Diphenylhydrazine is made by reduction of diphenylnitrosamine $(C_6H_5)_2N \cdot NO$ by zinc plus acetic acid, the nitrosamine being formed by reaction of diphenylamine $(C_6H_5)_2NH$ and nitrous acid.

Tetraphenylhydrazine is a white solid, soluble in chloroform, acetone, benzene, or toluene, and upon standing is changed into triphenylamine plus azobenzene. In solution, tetraphenylhydrazine dissociates into nitrogen diphenyl $(C_6H_5)_2N \cdot$, free radical, which in toluene at $90°C$ reacts with nitric oxide NO. Tetraphenylhydrazine is formed by oxidation of diphenylamine $(C_6H_5)_2NH$ by lead dioxide.

Hydrazine reacts with ketones to form *azines*.

HYDRAZOIC ACID. HN_3, formula wt 43.03, colorless, odorous, poisonous liquid, mp $-80°C$, bp $37°C$, explodes with marked violence. Also known as azoimide and hydronitric acid, the compound is miscible in all proportions with H_2O, alcohol, and ether. Hydrazoic acid reacts (1) with metals, e.g., magnesium, aluminum, zinc, iron, to form azides or hydrazoates (or trinitrides); (2) with heavy metal salt solutions to form insoluble azides, e.g., silver azide AgN_3, mercury(I) azide HgN_3, etc., which decompose in the light to form nitrogen plus the metal; (3) with NH_4OH to form ammonium azide $NH_4 \cdot N_3$, (4) with hydrazine to form hydrazine azide $N_2H_4 \cdot HN_3$; (5) with sodium hypochlorite plus acetic acid to form chlorazide ClN_3, explosive; (6) with sodium amalgam to form NH_3 with some hydrazine; (7) with potassium permanganate to form nitrogen and H_2O.

Hydrazoic acid is formed (1) by reaction of sodium nitrate with molten sodamide; (2) by reaction of nitrous oxide with molten sodamide; (3) by reaction of nitrous acid and hydrazonium ion $(N_2H_5^+)$; (4) by oxidation of hydrazonium salts; (5) by reaction of ethyl nitrite with NaOH solution and acidifying. See also **Azides.**

HYDRAZONES. The products of the reaction between an aldehyde or a ketone with phenylhydrazine are termed *hydrazones*. Sometimes the compounds are referred to as phenylhydrazones.

$CH_3 \cdot CHO$ (acetaldehyde) $+ C_6H_5 \cdot NH \cdot NH_2 \rightarrow CH_3 \cdot CH:N \cdot NH \cdot C_6H_5 + H_2O$ (phenylhydrazine) (acetaldehyde hydrazone)

C_6H_5CHO (benzaldehyde) $+ C_6H_5 \cdot NH \cdot NH_2 \rightarrow C_6H_5 \cdot CH:N \cdot NH \cdot C_6H_5 + H_2O$ (benzylidenehydrazone)

$(CH_3)_2CO$ (acetone) $+ C_6H_5 \cdot NH \cdot NH_2 \rightarrow (CH_3)_2C:N \cdot NH \cdot C_6H_5 + H_2O$ (acetone hydrazone)

$C_6H_5 \cdot CO \cdot CH_3$ (acetophenone) $+ C_6H_5 \cdot NH \cdot NH_2 \rightarrow (C_6H_5)(CH_3)C:N \cdot NH \cdot C_6H_5 + H_2O$ (acetophenonehydrazone)

Several of the hydrazones may be decomposed by strong acids, whereupon the original aldehyde or ketone is regenerated, along with the formation of a phenylhydrazine salt. When reduced, hydrazones yield primary amines.

HYDRIDE. A binary compound of hydrogen. Hydrides traditionally have been classified into three groups. In modern terminology, these are conveniently designated as covalent, electrovalent and metallic, although reference to the entries for hydrogen and the various hydrogen halides shows that a number of binary hydrogen compounds are partly ionic and partly covalent. See also **Hydrogen.**

Covalent hydrides are formed by the nonmetals. In general, the elements of main groups III and VII form single compounds consisting of a single atom of the element combined with a number of hydrogen atoms equal to the number of electrons which the element needs to complete its octet. Exceptions are beryllium, aluminum, and indium, which have polymeric hydrides, and boron and gallium, which have dimeric hydrides. Then also the elements of lower atomic number in main groups IV, V and VI (carbon, silicon, germanium, nitrogen, phosphorus, oxygen, and sulfur) and boron form more than one hydride. The covalent hydrides are volatile with low melting points and low boiling points (except as those properties are modified, as in the case of hydrogen fluoride, water and ammonia, by hydrogen bonding). They are nonconductors of electricity in the liquid state or when dissolved in nonpolar solvents.

Complex hydrides are formed by some elements (particularly

in main group III) having too few electrons to attain an octet in the neutral hydrides. These are structurally similar to the corresponding complex chlorides and are all excellent reducing agents. The most important are the tetrahydroborate, BH_4^-, -aluminate (frequently called alanate in the European literature), AlH_4^-, -gallate, GaH_4^-, and -indate, InH_4^-. There is evidence for polymeric ions, such as $B_2H_7^-$. Anions derived from higher hydrides are also known, e.g., $B_4H_{11}^-$.

Only the strongly electropositive elements, the alkali metals, the alkaline earth metals, and certain lanthanide and actinide metals, form electrovalent hydrides. The compounds are definitely crystalline, the alkali hydrides being cubic, but the structure increasing in complexity in going from main group 1 to main group 2 and to the lanthanides and actinides. In fact, hydrides of the last two groups, while approaching the alkali and alkaline earth hydrides in electropositive character, and while also giving evidence of the presence of H^- ions in their structures, are usually non-stoichiometric, compositions such as $CeH_{2.70}$, $PrH_{2.85}$ and $ThH_{3.07}$ being found. In this respect, those compounds approach in character the metallic hydrides.

This gradation in properties extends to the metallic hydrides themselves, some of which, such as copper hydride, approaches closely, but never quite reaches, a 1:1 atomic ratio of hydrogen to copper. In the case of palladium, the pressure-composition graph at temperatures below 200°C indicates a wide range of composition at little or no increase in pressure. At higher temperatures the flat portion shortens, and two breaks develop before and after it, indicating solid solutions, one of which approaches a 2:1 atomic ratio of H to Pd.

A group of complex metal hydrides have been used successfully for the preparation on an industrial scale of many organic and metallorganic compounds. Among these complex hydrides are highly reactive lithium aluminum hydride and the related sodium aluminum hydride and magnesium aluminum hydride. A more selectively reactive group of complex hydrides are the lithium, sodium, and potassium borohydrides. These compounds also have properties which make them useful as high energy fuels and rocket propellants.

Lithium aluminum hydride, $LiAlH_4$, also known as lithium aluminohydride and often abbreviated as LAH is prepared by the reaction of lithium hydride and aluminum chloride in ether solution

$$4LiH + AlCl_3 \rightarrow LiAlH_4 + 3LiCl$$

with some prior prepared complex hydride used as a seeding material to control the reaction rate. Lithium aluminum hydride forms a microcrystalline powder which is stable in dry air but decomposes above 125°C. It is soluble in many organic compounds like ether, dimethyl Cellosolve, tetrahydrofuran but is only slightly soluble in dioxane. It reacts vigorously with water, yielding hydrogen in a manner similar to the reaction of the simple hydrides:

$$LiAlH_4 + 4H_2O \rightarrow 4H_2 + LiOH + Al(OH)_3$$

It reacts with carbon dioxide to form methyl alcohol, or formaldehyde, or formic acid. It is a powerful reducing agent and reduces aldehydes, ketones, quinones, acids, esters, anhydrides, lactones, epoxides, and acid chlorides to the corresponding alcohols; amides, lactams, imides, nitriles, isocyanides, oximes, hydroxylamines, and related compounds to amines; dithiols, disulfides, polysulfides, sulfoxides, sulfones, and related compounds to the mercaptan or sulfide (see **Sulfur**); and aromatic nitro compounds to azo compounds. Olefinic bonds are not attacked unless conjugated with a nitrile, phenyl, or carbonyl group.

Sodium aluminum hydride can be prepared like the lithium

analogue but tetrahydrofuran is used as the solvent because the sodium complex is insoluble in ether. The sodium compound produces virtually the same reductions as the lithium compound.

Magnesium aluminum hydride may be prepared by treating an etherate of magnesium bromide with an ether solution of lithium aluminum hydride

$$2LiAlH_4 + MgBr_2 \rightarrow Mg(AlH_4)_2 + 2LiBr$$

or by use of an excess of magnesium hydride in ether solution with an ether solution of aluminum chloride. While the reducing activity of magnesium aluminum hydride is similar to that of the lithium complex in that polar double and triple bonds such as carbonyl and nitrile groups are reduced whereas nonpolar groups are not attacked, the magnesium complex, however, does not reduce the triple bond of propargyl aldehyde nor the double bond of cinnamic acid in contrast to the lithium complex.

Lithium borohydride, $LiBH_4$, may be prepared by the reaction of aluminum borohydride on ethyllithium or by the action of diborane on ethyllithium. It forms orthorhombic crystals which decompose at 250 to 272°C and while it is stable under usual conditions it is decomposed by humid air. It reacts readily with water and is a strong reducing agent.

Aluminum borohydride, AlB_3H_{12}, is a liquid which boils at about 44.5°C. It ignites in air. It can be prepared by the reaction of diborane with trimethylaluminum. It reacts readily with hydrogen chloride and water to yield hydrogen. It can be used in organic syntheses.

Metals like palladium and platinum absorb hydrogen forming mixtures which may be considered as alloys. Such mixtures may be placed into two groups: those in which the absorption takes place with decrease in temperature like those mixtures of hydrogen with palladium and tantalum; and those which absorb hydrogen with an increase in temperature like those with calcium, iron, nickel, and platinum.

HYDROCARBONS. Organic Chemistry.

HYDROCHLORIC ACID.

HCl (hydrogen chloride gas) in aqueous solution, colorless when pure. Commercial grades of HCl (also known as muriatic acid) generally are marketed in three concentrations: (1) 18° Bé (sp gr 1.1417 at 15.6°C, 27.92% HCl); (2) 20° Bé (sp gr 1.160, 31.45% HCl); and (3) 22° Bé (sp gr 1.1789, 35.21% HCl). Frequently the commercial grades are slightly yellow because of impurities, notably dissolved iron. Fuming hydrochloric acid contains about 37% HCl, with a sp gr 1.194. Reagent grade hydrochloric acid usually is of this latter high strength, is perfectly clear and colorless. The maximum limits set on impurities commonly are: NH_4 0.003%; arsenic 0.000001%; free chlorine 0.0001%; heavy metals, such as lead 0.001%; iron 0.00002%; sulfates 0.0001%; sulfites 0.0001%; and residue after ignition 0.0005%. A mixture of three parts HCl and one part HNO_3 is known as *aqua regia*, a powerful solvent and oxidizing agent which will dissolve materials that may be unaffected by either acid alone. Gold and platinum are soluble in aqua regia.

Hydrochloric acid is a very high-tonnage chemical, finding major uses in (1) the cleaning and preparation of metals prior to application of coatings, (2) the recovery of zinc from galvanized iron scrap, (3) the production of numerous chlorides, and (4) production of chlorine. At one time, HCl was extensively used as a source of both hydrogen and chlorine by way of electrolysis. This process was made obsolete many years ago when the chlor-alkali process (electrolysis of sodium chloride brines) was introduced for the production of chlorine. In recent years, however, the production of by-product HCl, resulting from chlorination of numerous organic compounds, has increased. In some of these instances, the installation of a HCl electrolysis plant may be economically feasible. For industrial consumption anhydrous HCl gas also is available in steel cylinders under a pressure of 1,000 psi (68 atmospheres). Hydrochloric acid forms a constant-boiled solution with H_2O (20.22% HCl) which has a bp 108.58°C (760 mm Hg).

Dilute HCl reacts (1) with many hydroxides, e.g., NaOH, to yield the corresponding chloride, e.g., sodium chloride, solution, (2) with many ordinary oxides, e.g., magnesium oxide, to yield the corresponding chloride, e.g., magnesium chloride, solution, (3) with many carbonates, e.g., calcium carbonate, to yield the corresponding chloride, e.g., calcium chloride solution plus CO_2, (4) with many sulfides, e.g., ferrous sulfide, to yield the corresponding chloride, e.g., ferrous chloride, solution plus H_2S, (5) with many metals, e.g., zinc (but not copper) to yield the corresponding chloride, e.g., zinc chloride, solution plus hydrogen gas, (6) with some special oxides, e.g., lead or manganese dioxide, to yield lead or manganese chloride plus chlorine gas, (7) with solution of some salts, e.g., silver nitrate, to yield the corresponding chloride, silver chloride, precipitate. Higher strengths of hydrochloric acid usually react similarly to the dilute. Hydrochloric acid sometimes reacts as a reducing acid, e.g., (6) above.

All metallic chlorides, except silver chloride and mercurous chloride, are soluble in H_2O, but lead chloride, cuprous chloride and thallium chloride are only slightly soluble. Metallic chlorides when heated melt, and volatilize or decompose, e.g., sodium chloride, mp 804°C; calcium, strontium, barium chloride volatilize at red heat; magnesium chloride crystals yield magnesium oxide residue and hydrogen chloride; cupric chloride yields cuprous chloride and chlorine. See also **Chlorine; Chlorinated Organics; Halides; Hypochlorites;** and **Sodium Chloride.**

Hydrogen Chloride. This is a colorless gas, heavier than air, density 1.639 g/l at standard conditions. The gas is poisonous and quickly causes suffocation. Formula weight 36.47, mp −111°C, bp −85°C, critical pressure 83 atm, critical temperature 51.3°C. The gas is very soluble in H_2O, accounting for the high concentrations of hydrochloric acid obtainable. Although hydrogen chloride gas may be used directly in some industrial operations, normally it is generated for the purpose of dissolving in H_2O to form hydrochloric acid. The most common route to HCl is by reacting sodium chloride with H_2SO_4. This is a two-step, exothermic reaction: (1) $NaCl + H_2SO_4 \rightarrow NaHSO_4 + HCl$, and (2) $NaCl + NaHSO_4 \rightarrow Na_2SO_4 + HCl$. Preparation of hydrochloric acid from the gas involves an absorption tower where the gas meets a fine spray of H_2O. Ratio controllers are used to assure maximum yield of the acid of desired concentration. These controls are easily adjusted for obtaining different concentrations. In most chlorinations of organic compounds, only half of the chlorine is used to substitute for hydrogen atoms, the remaining chlorine forming HCl. Frequently, this by-product HCl is recycled or recovered.

HYDROGEN. Chemical element symbol H, at. no. 1, at. wt. 1.0080, periodic table group 1a, mp −259.14°C, bp −252.87°C, density, 0.089 (solid at 4.2 K), 0.071 (liquid at 20.4 K), sp gr 0.0696 (air = 1.0000). Solid hydrogen has a hexagonal crystal structure. Hydrogen at standard conditions is a colorless, odorless, tasteless gas, suffocating, but not toxic. Hydrogen occurs chiefly combined with oxygen in H_2O, with carbon in hydrocarbons, with carbon and oxygen, and with carbon and several other elements, including oxygen, nitrogen, sulfur, phosphorus, and most metals in a vast variety of hundreds of thousands of organic compounds. See **Organic Chemistry.** Hydrogen is considered by some scientists as the primordial substance from which all other elements in the universe were developed. In terms of cosmic abundance, with a rating of silicon = 10,000, it has been estimated that the figure for hydrogen is about 3.5×10^8, this figure compared with that of carbon =80,000, nitrogen = 160,000, and oxygen =220,000. For further comparison, the figure for gold is 0.0015 and for uranium it is 0.0002. In terms of abundance of the chemical elements in seawater, hydrogen ranks second (behind oxygen), with an estimated 510 million tons/cubic mile (~109 million metric tons/cubic kilometer). Hydrogen ranks eleventh in terms of content in igneous rocks in the earth's crust, the estimate of average content being 0.13%. Although free hydrogen escaped from the earth's lower atmosphere, some of the planets appear to have significant amounts, including the atmospheres of Jupiter, Saturn, and Uranus. At an altitude of 1000 miles (1609 kilometers) above the surface of the earth, there is a greater abundance of hydrogen atoms than of nitrogen or oxygen atoms.

Hydrogen was first identified by Cavendish in 1766. The element was named by Lavoisier in 1783. However, it was not until 1931 that a second isotope of hydrogen (deuterium) with a mass number 2 was discovered by Urey. In 1934, Rutherford, Oliphant, and Harteck prepared a third isotope (tritium) with a mass number 3. Normal hydrogen (protium) and deuterium are stable, whereas tritium is radioactive, with a half-life of 12.26 years. Tritium emits a negative electron to form ^3He. It is estimated that the isotopic abundance of ^1H (protium) in natural occurring hydrogen is 99.9851% and on the basis of carbon = 12 (atomic weight scale), protium has a mass of 1.007825 amu. The isotopic abundance of ^2H (deuterium) is estimated at 0.0149% with a mass 2.014101 amu. The artificially-prepared ^3H (tritium), $^9Be + {}^2H \rightarrow 2\,{}^4He + {}^3H$, has a mass of 3.01605 amu. Heavy water is deuterium oxide $2H_2O$, usually written D_2O. Deuterium and deuterium oxide gained prominence largely because of their excellent properties as moderators in nuclear reactors. The ionization potential of hydrogen is 13.59765 ± 0.00022 eV. Other physical properties of hydrogen are given under **Chemical Elements.** See also **Deuteron; Deuterium.**

When ignited, hydrogen burns in air with a pale blue to colorless, nonluminous flame, yielding H_2O. When mixed with air, the flammability limit is 4–74% hydrogen. When mixed with oxygen, the flammability limit is 4–94% hydrogen. Care always must be exercised where there may be hydrogen mixtures with air or oxygen because violent explosions may occur. In sunlight or magnesium light, hydrogen combines with chlorine with violent release of energy, forming hydrogen chloride HCl. When hydrogen is heated with sodium, calcium, and several other metals, the corresponding hydride is formed. In the presence of a catalyst, hydrogen reacts with nitrogen to form ammonia NH_3. Upon heating sulfur in the presence of hydrogen, hydrogen sulfide, H_2S, is formed. At elevated temperatures, hydrogen will reduce many of the metal oxides to the metal, notably copper, iron, nickel, tin, and lead. The oxides of zinc, aluminum, and magnesium are not so reduced. Hydrogen reacts with unsaturated organic compounds in most cases to form saturated compounds. For example, in the presence of a catalyst, hydrogen will add to oleic acid $C_{17}H_{33}COOH$ to form stearic acid $C_{17}H_{35}COOH$. See also **Hydrogenation.**

Production of Hydrogen. For chemical and petroleum processes, hydrogen is a very high-tonnage and one of the most fundamental raw materials. Sources of hydrogen and processes for producing it are described in the entry on Hydrogen (Fuel).

Uses. In terms of consumption, NH_3 is by far the largest user of hydrogen. Petroleum refining processes and methanol synthesis are the next largest consumers. Hydrogen needs for these uses are almost always fulfilled by hydrogen-generation capacity on the premises. What might be termed commodity hydrogen is shipped from hydrogen plants to various users. Some of the more important uses include the hydrogenation of numerous organic compounds, such as vegetable and animal oils, the oxyhydrogen and atomic-hydrogen welding applications, the reduction of several metallic oxides, such as iron, copper, nickel, cobalt, tungsten, and molybdenum, and the use of liquid hydrogen as a rocket fuel. See also **Ammonia; Hydrogenation; Methanol (Methyl Alcohol); Petrochemicals;** and **Synthesis Gas.** For the potential role as a fuel, see **Hydrogen (Fuel).**

Ortho- and Para-Hydrogen. On the basis of nuclear spin, two forms of hydrogen are known: *ortho-hydrogen*, in which the two nuclei in the H_2 molecule have parallel spins, and *para-hydrogen*, in which the nuclear spins are anti-parallel. At ordinary temperatures (and above) ortho-hydrogen is present to the extent of above 75%; at lower temperatures, the ortho changes to para-hydrogen, until at very low temperatures, as that of liquid hydrogen, the para form is present to the extent of 99.7%. There is some difference in properties between the two, notably in thermal conductivity.

The transition from ortho- to para-hydrogen releases heat in amount of 168 cal/g. The heat of vaporization of liquid hydrogen is 107 cal/g. Thus, more than ample heat is released to revaporize liquid hydrogen. Knowledge of the existence of the ortho-para transition and the development of catalysts to equilibriate the liquid during liquefaction essentially have made possible the very large-scale manufacture, use, and storage of liquid hydrogen.

Below $-220°C$ the specific heat of hydrogen is that of a monatomic gas like helium (He). Practically pure para-hydrogen may be obtained by adsorption of ordinary hydrogen, which is three-fourths ortho and one-fourth para, on charcoal at about $-225°C$. The mp of para-hydrogen is 0.13°C lower (ortho-hydrogen 0.04°C higher) than ordinary hydrogen, and the bp at 60 mm pressure is 0.13°C lower (ortho-hydrogen 0.04°C higher) than ordinary hydrogen. Parahydrogen reverts slowly to ordinary hydrogen, but immediately in the presence of platinized asbestos.

Atomic Hydrogen. At high temperatures, the loss of heat from a glowing wire in hydrogen is larger than expected on regular assumptions. This is believed to be due to dissociation of ordinary hydrogen into atomic hydrogen (H). See accompanying table.

When hydrogen is passed through an electric arc between tungsten poles, a considerable transformation into atomic hydrogen occurs, and when a stream of this gas strikes a surface a large evolution of heat takes place through recombination to ordinary hydrogen. This atomic hydrogen flame is of temperature sufficiently high to melt tungsten (mp 3,370°C). The half-life of the hydrogen atom is one-third second at 0.5 mm pressure. This reaction is endothermic, values of 98–105 kcal per mole having been reported for it. It is an active reducing agent, reducing many metallic oxides and halides to the free metals, and

DISSOCIATION OF HYDROGEN

TEMPERATURE, °C	PRESSURE	
	At 760 mm	At 1 mm
1730	0.33%	8.7%
2230	3.1	57.5
2730	34	99.3

forming hydrides with many nonmetals. The energy of its exothermic recombination is utilized, in combination with the energy released by the oxidation of the H_2 formed, by atmospheric oxygen, in the oxyhydrogen welding process.

Ionization. The ionization potential of hydrogen is 13.59765 ± 0.00022 eV, and the ionization process (in the case of protium) yields an electron and a free proton. The electric field of the proton is strong, due to its small radius, so that it readily combines with polarizable atoms. Thus, in aqueous solution, it shares an unshared pair of electrons of the oxygen atom of H_2O to form H_3O^+, the hydronium ion; with NH_3 it forms NH_4^+, the ammonium ion; with phosphine it forms the phosphonium ion, PH_4^+, etc. The hydrogen atom can also add an electron, to form the hydride anion, H^-, this potential (electron affinity) being only about 0.7 eV. Hydride ions have been shown (by electrolysis, crystal structure, etc.) to exist in the hydrides which hydrogen forms with the alkali metals and some of the other metals on the left side of the periodic table. While most other hydrogen compounds are essentially covalent, the binary compounds with the halogens and some of the other elements on the right side of the periodic table exhibit a considerable degree of ionicity, varying considerably in the same group.

The hydrogen atoms in many compounds tend to be shared between the electronegative atom or group to which they are attached and similar groups on other molecules. These hydrogen bonds increase the intermolecular forces and boiling points of hydrogen fluoride, water, organic acids and alcohols, etc. A descriptive explanation of the process is the positive polarity of the H atom that is attached to the electronegative atom or group, which gives it an effective coordination number of 2, so that it can attract an unshared electron pair of a fluorine, oxygen, nitrogen, atom of another molecule. The atom having the unshared pair must be negatively polarized or easily polarizable. For example, tertiary arsines form stronger hydrogen bonds with phenols than do tertiary phosphines.

A number of hydrogen compounds ionize to yield solvated protons, i.e., $2H_2O \leftrightarrows OH_3^+ + OH^-$, and $2NH_3$ (liq.) $\leftrightarrows NH_4^+ + NH_2^-$. Moreover, many hydrogen compounds, when dissolved in such solvents, ionize more or less completely to give solvated protons and anions. In the case of polybasic acids, ionization constants are reported for each step in this dissociation.

Hydrides. See the section on hydrides in next entry on **Hydrogen (Fuel);** and also the separate entry on **Hydride.**

Water and Acids. The properties of the most prevailing hydrogen-bearing compound, water, are given in the entry on **Water.** The characteristics of acids are attributed essentially to the presence of hydrogen ions. These topics are treated in the entries on **Acids and Bases;** and **pH (Hydrogen Ion Concentration).**

See the list of references at end of the next entry.

HYDROGENATION (Vegetable Oils). Vegetable Oils (Edible).

HYDROGEN FLUORIDE. Fluorine.

HYDROGEN (Fuel). Because of the wide use of hydrogen in the processing industries and for the hydrogenation of various oils and fats in the food and related fields, hydrogen has become much better understood during the past several decades. For many years, hydrogen has served as a specialized fuel for certain applications, such as oxyhydrogen cutting and welding torches, but generally, until the 1960s, the possible role of hydrogen as a major energy source fuel was rarely discussed. The word *hydrogen* took on a negative connotation with the development of the hydrogen bomb, as it also did some years ago when the hydrogen-filled dirigible Hindenburg exploded as it moved to its mooring mast in Lakewood, New Jersey in 1937.

The probable future of hydrogen in the world's energy system was the subject of prophecy over one hundred years ago. In 1874, Jules Verne wrote: "I believe that water will one day be employed as a fuel; that hydrogen or oxygen, which constitute it, used singly or together, will furnish an inexhaustible source of heat and light." In the early 1900s, Britain's Lord Haldane said: "It is axiomatic that the exhaustion of our coal and oil fields is a matter of centuries only ... As it has often been assumed that their exhaustion would lead to the collapse of industrial civilization, I may perhaps be pardoned if I give some of the reasons which led me to doubt this proposition." Haldane envisioned networks of windmills generating the electricity needed to separate hydrogen from water. The hydrogen would then be liquefied and stored underground.

Some readily apparent advantages of hydrogen, both as a direct and indirect fuel (discussed later) have been extrapolated into terms of a future hydrogen economy. As the result of continuing and concentrated research and development in the energy field, many experts see a hydrogen energy economy gradually emerging.

As with other energy proposals, and there have been many in the past decade or two, three factors will likely determine the pace of hydrogen energy technology: (1) the manner in which, step-by-step, hydrogen-oriented systems and subsystems will compete economically and environmentally with other energy source, conservation, and utilization proposals; (2) the pace of technological advancement in related fields, such as nuclear engineering, upon which hydrogen systems may depend; and (3) the pace of unilateral efforts on behalf of hydrogen-oriented systems, including the refinement of current planning-purpose data and opinions into actual operating information relating to hydrogen generation, transportation, conversion and/or end-utilization, and safety. Without the funding of a series of "crash programs," unilateral developments probably will be relatively slow. Most likely, the information bank for hydrogen systems will stem from an increasing awareness of the energy characteristics of hydrogen and the progressive use of hydrogen subsystems in situations where they are eminently superior.

The present concept of a hydrogen fuel economy includes a primary energy source, such as a nuclear fission or fusion reactor, a geothermal source, or a solar-powered source, with hydrogen being produced as the portable energy carrier. See accompanying figure. Thermal energy from nuclear sources would be used to generate electricity that would then be used to electrolyze water for the production of hydrogen and oxygen. The hydrogen would be distributed by pipeline to distant points of use, with storage provided by underground gas storage, or by liquefaction and refrigerated storage.

Fuel-Related Background of Hydrogen. Although the abundant hydrogen isotope *protium* is the simplest known atom, it forms two diatomic molecules, namely, *ortho-hydrogen*, in which the two atomic nuclei spin in the same direction; and *para-hydrogen*, in which the nuclei spin in opposite directions. While the equilibrium composition of hydrogen gas is 75% ortho at ambient temperature, it changes to 99% para in the liquid state. The transition from ortho- to para-hydrogen is exothermic (168 cal/gram), so that the heat released is more than enough to revaporize liquid hydrogen (heat of vaporization 107 cal/gram). Recognition of the existence of the ortho-para transition and the development of catalysts to equilibrate the liquid during liquefaction have made possible the large-scale production, use, and storage of liquid hydrogen.

Hydrogen molecules dissociate to atoms endothermally at high temperatures (heat of dissociation about 103 cal/gram mole), in an electric arc, or by irradiation. This property is used to effect atomic-hydrogen arc welding, in which hydrogen gas is dissociated by an ac electric arc between two tungsten electrodes, the hydrogen atoms recombining at the metal surface to provide the heat required for welding.

Pertinent properties of hydrogen are given in Table 1.

Actual and potential uses for hydrogen can be predicted by inspection of its properties. Its low density, 7% that of air, plus its high thermal conductivity, 6.7 times that of air, have led to its use as a coolant in large rotating electrical equipment. The low density reduces windage friction losses to less than 10% those with air, while its high thermal conductivity and heat capacity permit more efficient heat transfer, the result being an overall increase in generator efficiency of as much as 1%.

The high heats of reaction of hydrogen with oxygen or fluo-

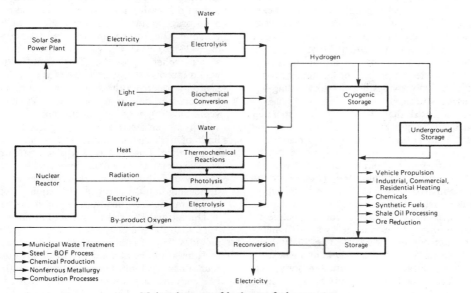

Major elements of hydrogen fuel economy.

TABLE 1. FUEL PROPERTIES OF HYDROGEN

Melting point, K	13.96
Heat of fusion at 14.0 K, calories/gram	14.0
Boiling point at 1 atmosphere, K	20.39
Heat of vaporization at 20.4K, calories/gram	107
Density, grams/cubic centimeter	
Solid at 4.2K	0.089
Liquid at 20.4K	0.071
Critical temperature K	33.3
Critical pressure, atmospheres absolute	12.8
Critical volume, cubic centimeters/mole	65.0
Critical density, grams/cubic centimeter	0.031
Heat of transition, ortho to para at 20.4K	
calories/gram	168
Specific heat (At constant pressure C_p,	
calories/gram)	
Liquid at 17.2K	1.93
Solid at 13.4K	0.63
0–200°C	3.44
Specific heat (At constant volume C_v (0–200°C)	
calories/gram	2.46
Specific heat: Ratio C_p/C_v (0–200°C)	1.40
Gas density, 0°C and 1 atmosphere, grams/liter	0.0899
Gas specific gravity (Air = 1.0)	0.0695
Gas thermal conductivity, 25°C (cal)(cm)/(s)	
(cm²)(°C)	0.00044
Gas viscosity, 25°C and 1 atmosphere, centipoise	0.0089
Coefficient of thermal expansion per °C	0.00356
Heat of combustion at 25°C, kcal/gram mole	
Gross	63.3174
Net	57.7976
Energy release upon combustion, calories/gram	29,000
calories/cubic centimeter	2,050
joule/gram	1.21×10^5
Flame temperature, K	2,483
Autoignition temperature, K	858
Heat of formation of HF at 25°C, kcal/gram	
mole ΔH	−64.2
Flammability limit, percent	
In oxygen	4 to 94
In air	4 to 74

rine, plus the low molecular weights of the product gases, have made hydrogen a prime fuel for rocket propulsion, since rocket thrust increases directly with the temperature and inversely with the molecular weight of the exhaust gases. Liquid hydrogen and oxygen were used in the second- and third-stage Saturn engines in the Apollo moon flights. The low atomic weight of hydrogen has made it the preferred propellant for nuclear rockets, in which nuclear emission provides heat for exhausting hydrogen gas at high temperatures.

Some studies have indicated that the cost of transporting and distributing hydrogen by pipeline may be less than the cost of transporting and distributing electric power. Presumably existing natural gas pipelines and distribution systems can be adapted to the use of hydrogen. Although hydrogen has a net heating value of only 275 Btus per cubic foot (2448 Calories per cubic meter), as compared with 913 Btus per cubic food (8126 Calories per cubic meter) for methane, the lower density and viscosity of hydrogen make it possible for a pipeline to deliver about the same amount of thermal energy as with methane, at a somewhat greater compression cost. The thermal energy in hydrogen can be utilized more efficiently in home heating than natural gas, because hydrogen can be burned in nonvented heaters, with no loss of heat, since its only primary combustion product is water. By using flameless catalytic heaters, nitrogen

oxide formation can be eliminated. However, oxygen depletion of closed spaces still presents a serious hazard.

One advantage of hydrogen as a source of thermal energy, as compared with electricity, is that it can be stored for later use—hydrogen is a commodity with weight and volume. Electricity, although it can be converted into chemical energy in batteries, essentially is a form of energy that must be consumed at the same rate it is generated. Hydrogen, like natural gas or substitute natural gases, may be stored and transported as a refrigerated liquid, or stored as a gas under pressure in an underground system. Hydrogen also may be stored as a metallic hydride.

Categories of Energy-Related Hydrogen Uses. The probable functions of hydrogen in future energy technology may be put into two major categories: (1) *direct functions*, in which hydrogen serves as a fuel, i.e., as the source of heat, power, and light without prior conversion to some other energy form; and (2) *indirect functions*, in which hydrogen is an important component of the total energy system, but before the end-use of that energy, the hydrogen is involved in some conversion, possibly chemical, used in the creation of a synthetic fuel, such as substitute natural gas, or possibly converted into electrical energy which becomes the final end-energy used. One of the major indirect or secondary roles proposed for hydrogen is that of an energy transporter, where in one scheme, other forms of energy would be consumed to generate hydrogen, which then would be pipelined and stored at distant points available for another conversion step—for example, converted into electrical energy as needed.

Hydrogen as Energy Source for Motive Power

Aside from their relatively low costs until the mid-1970s and continuing into the 1980s, the hydrocarbon fuels, notably gasoline and kerosine, have offered convenience in handling and transportability for use in connection with powered vehicles. During the past decade, the political factors that arise from the striking geographic imbalance between petroleum resources and petroleum consumption in most regions of the world have provided ample incentives to strike out for alternative sources of vehicular power.

Hydrogen, when cost competitive, can provide many of the advantages of petroleum liquids and offer the additional attraction of decreasing air pollution. Because of its low density, the net storage volume required would be at least as much as for gasoline. The storage tank must be maintained at a temperature of −423°F (−253°C), which is the boiling point of hydrogen at atmospheric pressure. This would require insulation that would increase the overall size of the storage container. Vaporization losses from the storage tank, amounting to perhaps 2% or more per day, must be vented so that no ignition of the vented hydrogen gas can occur, and no accumulation of explosive hydrogen-air mixtures are possible. The lower explosive limit of hydrogen in air is 4%, so adequate ventilation must be provided. Fortunately, hydrogen gas, being the lightest gas with a specific gravity of 0.07 referred to air, will rise and diffuse rapidly and thus can be easily dispersed. Service stations for dispensing liquid hydrogen will require more expensive storage and pumping facilities than required for gasoline.

The estimated weights and volumes expressed in Table 2 are relative to the same energy content of gasoline. Relative weight includes that of containers. The data indicate that magnesium hydride would be at a 4.6 weight disadvantage and thus require four times the tankage in comparison with the use of gasoline in a conventional automobile. New hydrides, as described later, may change this. This would be equivalent to 450 pounds (204

TABLE 2. SOME HYDROGEN STORAGE OPTIONS FOR VEHICLES

STORAGE SYSTEM	RELATIVE SYSTEM WEIGHT[a]	RELATIVE CONTAINED VOLUME[a]
Gaseous phase, 2,000 psi (136 atmospheres)	~30.0	~24.0
Solid (as magnesium hydride with 40% porosity)	4.6	4.0
Liquid phase at 37°R	2.4	3.8

[a] Relative to gasoline, as unity for same energy content.

kilograms) of added vehicle weight and 60 more gallons (227 liters), requiring $2 \times 2 \times 2$ feet or 0.2 cubic meter volume over that required for a vehicle with a 20-gallon (76-liter) tank. Bursting upon collision for liquid storage can be overcome by using containers capable of withstanding 30 Gs, which are presently available.

If a designer were to elect the option of using hydrogen in the gaseous phase at 2000 psi (136 atm), this would require a metal container weighing some 30 times and requiring a volume of some 24 times that required for an energy-equivalent volume of a hydrocarbon fuel. Also important in the total energy equation is the additional energy required to compress hydrogen (gaseous phase) or to liquefy it.

It is most likely that the first major use of liquid hydrogen as an energy source for motive power will be jet aircraft, largely because of the excellent weight advantage and the less serious nature of the boil-off loss and distribution problems as compared with other forms of transportation. City buses and long-haul motor trucks, already equipped mainly with hydride hydrogen power, have been tested in the United States, West Germany, among other countries. In the future, refueling may be effected through replacement of entire storage tanks (dewars). Because the private motorcar presents the most crucial logistics, including the small capacity of the fuel system, concern with safety, boil-off loss of fuel even when vehicle is not in use, and the education and acceptance involving millions of users, it probably will follow rather than lead the use of hydrogen in other modes of transportation. However, during the last few years, a few firms have offered hydrogen-powered private motor vehicles, set up to switch from hydrocarbon fuel to hydrogen, and vice versa, but at a cost that is not competitive with mass-marketed vehicles.

Metal Hydrides. For a number of years among many scientists and advanced planners there have been those who have considered the use of metal hydrides to store hydrogen at atmospheric or reasonable pressures and at relatively low temperatures (comparable to current metal temperatures in some conventional engines). The fact that hydrogen will form hydrides with most metals has been known for many years, during which time, a number of these hydrides have been formed and tested. Until comparatively recently, magnesium hydride and a hydride of a rare-earth metal plus nickel, such as $LaNi_5$, appeared to be best suited as hydrogen storage media.

The hydride-forming reaction is exothermic and reversible: Metal + Hydrogen $\rightleftarrows$ Metal Hydride + Heat. Thus, when it is desired to call for the separation of hydrogen from the hydride, heat (of decomposition) is required. As may be expected, the heat of decomposition is roughly proportional to the stability of the particular hydride. It is thus evident that for a metal hydride to serve as an efficient and viable means for storing hydrogen, it should be capable of decomposition at a relatively low temperature—say 300°C or lower. At the same time, the hydride must be reasonably stable and, of course, not require a high hydrogen pressure to manufacture it. It is further evident

that the metal portion of the hydride must be comparatively inexpensive and thus common and readily available in the quantities that may be required. Metal hydride storage, operating as it does in terms of hydrogen as a battery operates in terms of electricity, must be capable of easy and efficient replenishing or "recharging" cycles. In every respect, the hydride storage element must be as safe as current vehicular fuel systems.

Researchers have investigated a large number of known binary hydrides, i.e., compounds which contain one metal and hydrogen. Investigators now regard magnesium hydride (MgH_2) as a borderline possibility. This binary hydride evolves hydrogen at a pressure of one atmosphere and requires a decomposition temperature of 289°C.

In comparatively recent research, much has been learned concerning the manner in which hydride compounds hold hydrogen. It has been known for a long time, of course, that the metal portion of the hydride should be comprised of tiny particles so that there is a large surface area available for reaction. In searching for reasons why hydrides permit such a high density of hydrogen, Reilly/Sandrock (1980) have observed that it is possible to pack more hydrogen into a metal hydride than into the same volume of liquid hydrogen. When the subject metal is first exposed to diatomic hydrogen (H_2), the hydrogen atoms are absorbed onto the surface of the metal. Immediately, some of the hydrogen is dissociated into monoatomic hydrogen (H). This permits the monoatomic hydrogen to penetrate deeply into the crystal lattice of the metal and to occupy what are known as *interstitial sites*. Investigators have found that these sites must have a critical minimum volume if they are to easily receive the hydrogen atom. Upon increasing the pressure of the hydrogen applied, the metal reaches a saturated phase—the metal hydride phase. It has been found that under certain conditions and with certain metals, the number of hydrogen atoms contained in the crystal will range from 2 to 3 times the number of metal atoms.

The most recent experimentation with metal hydrides has involved multiple-metal hydrides. It has been known for some time that hydrogen reacts with alloy metal combinations. Considerable research has gone forth in connection with ternary hydrides (2 metals + hydrogen) and one of the most promising of these compounds, as of the early 1980s, is iron-titanium hydride ($FeTiH_x$), where x may range from 1 to 2. Reilly/Sandrock report that the hydrogen storage capacity by weight percent of this ternary hydride is 1.75 and by volume (grams per milliliter) is 0.096. The energy density by weight is 593 calories per gram; and the energy density by volume is 3254 calories per milliliter. Thus, this ternary hydride has a higher hydrogen-storage capacity than an equal volume of liquid or gaseous hydrogen (at 100 atmospheres). Another promising intermetallic hydride is lanthanum-pentanickel hydride ($LaNi_5H_x$), although it is more costly to produce. In this hydride, x may range from 1 to 6. Both the iron titanium hydride and the lanthanum-pentanickel hydride have lower temperatures of formation and decomposition, contributing to easier charging and discharging at ambient temperature.

In connection with hydrogen engine design, it is assumed that the heat of decomposition required by the hydride can be furnished from the inevitable waste heat generated by any engine. See also **Hydride.**

Hydrogen as a Heating Fuel

The routine use of hydrogen as a heating fuel for industry and commercial-residential installations entails even greater complications and would appear to be much more dependent upon the overall economic and technical aspects of a so-called hydrogen fuel economy. From many standpoints, assuming availability, hydrogen can be an excellent fuel for almost any heating application. Hydrogen can be used in the home for cooking and heating (and even lighting) and likewise in commerce and industry. Compared with natural gas, hydrogen burns with a faster, hotter flame. Hydrogen-air mixtures are flammable over wider limits of mixtures. Hydrogen burns without producing noxious exhaust products, allowing unvented appliances except where water vapor and resulting increased humidity may be objectionable. In winter, the additional humidity can, in fact, be highly desirable. But, in humid locations in summer, the water vapor produced could be objectionable. Adequate ventilation must be provided to prevent depletion of oxygen in closed spaces.

But, generally because of the absence of hazards from carbon monoxide and other fumes, large savings could be achieved from the elimination or at least simplification of flues. Some experts suggest that not only construction costs could be lowered as the result of clean burning, but that an increase of some 30% in the efficiency of a gas-fired home heating system could be achieved. The concept of peripherally placed unflued devices, particularly through the use of catalytic "flameless" heaters, could ultimately lead to a serious revision of the widely accepted central heating concept. By maintaining the temperature of a catalytic bed as low as 100°C, the production of nitrogen oxides would be virtually eliminated.

Because hydrogen burns with a hotter flame, some design features of heating apparatus would require change. The energy content per unit mass of liquid hydrogen is about 2.75 times greater than that of hydrocarbon fuels. On the other hand, there are only 325 Btus per standard cubic foot (2893 Calories per cubic meter) of hydrogen as compared with about 1000 Btus per standard cubic foot (8900 Calories per cubic meter) of natural gas, thus dictating further design changes. The ignition energy of hydrogen is about 0.02 millijoule, which is less than 7% that of natural gas, a major factor in making low-temperature catalytic burners possible; also a major factor in designing for safe operation.

Despite the numerous advantages of hydrogen as a direct heating fuel, particularly in the home, the application of hydrogen must be viewed in terms of the total energy concept of an exclusively hydrogen-supplied (all-hydrogen home) installation. Where the direct use of hydrogen for heating is large, the economy will be most favorable. If a substantial amount of the hydrogen must be converted into electrical energy, as by a fuel cell, then economic justification becomes more difficult.

Lighting in the all-hydrogen home may be accomplished by condoluminescence, a cold process. A phosphor is spread on the inside of a tube similar to the conventional fluorescent lamp. Upon coming in contact with the phosphor, small amounts of hydrogen combine with the oxygen in the air to excite bright luminescence in the phosphor.

Conversion of burners and other design aspects of heating systems and appliances to pure hydrogen, or to a hydrogen-enriched natural or substitute natural gas supply, while costly and inconvenient, is certainly not in the economically insurmountable category. Similar alterations over the years were made in the United States when communities switched from manufactured gas (about 50% hydrogen) to natural gas. Such switchovers are even more recent in European communities.

As more hydrogen becomes available for transportation use and as more hydrogen is pipelined regionally or transcontinentally, depending largely on the demand placed upon the supply of hydrogen for industrial uses, it may be that the hydrogen content of community gas supplies will be progressively enriched with hydrogen (in a periodic, stepwise manner because of switchover problems) and thus contribute in a gradual manner to less pollution and to the conservation of natural gas.

Sources of Hydrogen

The major source of chemical hydrogen over the past several decades has been natural gas. In strictly terms of chemical needs, where economic factors are favorable, natural gas has served this need well. Obviously, in terms of total energy conservation, where hydrogen is looked to as a means of conserving fossil-fuel sources, a much less costly and much more abundant hydrogen-containing raw material must be sought. The logical candidate is water. Particularly in areas of the world where hydrocarbons are not readily available, reasonably large water electrolysis installations have been made, notably in locations with low electricity costs.

In addition to electrolysis, the principal means under consideration for deriving hydrogen from water is that of thermochemical splitting. The waste heat and high temperature available from certain types of nuclear reactors would effect a series of chemical reactions, still much in the research phase, to free hydrogen and oxygen from water. Additional proposals have included the use of ultraviolet radiation from the plasma of a fusion reactor for the direct photolysis of water vapor and the use of some forms of algae, under the stimulation of light, to convert hydrogen ions to hydrogen gas by a complex chain of biochemical reactions (L.O. Krampitz, Case Western Reserve University).

Electrolysis. Because of years of operating experience, electrolysis is possibly an order of magnitude ahead of other proposals from a technological standpoint. Although simple in concept, electrolysis is costly—hence the research efforts to find other ways of splitting water carry a high incentive. Nevertheless, this side of one or more breakthroughs in other areas, most likely electrolysis operations will continue to serve as the basis for costs in extending the use of hydrogen in the relatively near term.

As of 1983, industrial electrolyzers range in size from 500 standard cubic feet (14.2 cubic meters) of hydrogen production per day, consuming 3 kilowatts of electricity, to more than 40 million standard cubic feet (~1.1 million cubic meters) of hydrogen per day, consuming 240,000 kilowatts. Most common installations are from 10,000 to 500,000 standard cubic feet (283–14,160 cubic meters) of hydrogen per day. Two factors generally characterize an electrolyzer installation: (1) access to comparatively low-cost electricity, as found in some areas served by hydroelectric installations; and (2) need for the oxygen which accompanies the production of the hydrogen. Industrial electrolyzers usually operate at efficiencies of about 60% to 70%. Some high-pressure prototype models have reached 85%. It has been pointed out (D. P. Gregory, Institute of Gas Technology) that, in theory, electrolyzers can approach a maximum electrical efficiency of nearly 120% as the result of the ideal unit absorbing ambient heat and also converting this energy into hydrogen. A reasonable, practical target for an improved electrolyzer appears to be around 100%. Thus, the production of electrolytic hydrogen would be limited only by the efficiency of electric current generation, namely, between 35% and 45%. An estimate has been made (E. C. Tanner, Princeton University; R. Huse, Public Service Electric & Gas Co.) that the overall conversion efficiency of electricity-to-hydrogen-to-electricity will approximate 38%. The theoretical power required to produce hydrogen from water is 79 kilowatts per 1000 cubic feet (~28 cubic meters) of hydrogen gas. One of the largest electrolyzers operating commercially is that of Cominco, Ltd. (British Columbia). This is a 90-megawatt installation that produces approximately 36 tons (32.4 metric tons) of hydrogen gas per day for use in ammonia synthesis. Other large plants are located in Norway and Egypt.

Thermochemical Splitting. The major objective is to find one or more series of chemical reactions that will result in the satisfactory separation of hydrogen (and oxygen) from water. Considerable work has been going forth at the Nuclear Research Center, Julich, Federal Republic of Germany, where much attention has been given to sulfur- and chlorine-base thermochemical cycles. Other researchers (Institute of Gas Technology; General Electric Co.; European Atomic Energy Community) have been probing various combinations of at least 56 chemical elements, including over 700 different compounds, that may show promise in various schemes for a closed-water-splitting cycle. It is understood that approximately 20 promising schemes have emerged, mainly centered in chlorine compounds. The most frequent flaw encountered among prospective reactions is the large amount of free energy required to force one or possibly two of the series of reactions; and the appearance of reactions that produce stable compounds incapable of regeneration.

Some of these reactions would rely upon a nuclear reactor as a heat source and would not have to await the emergence of a practical, operating fusion reactor. One sequence of reactions, in particular, is of interest:

$$CaBr_2 + 2H_2O \rightarrow Ca(OH)_2 + 2HBr$$
$$Hg + 2HBr \rightarrow HgBr_2 + H_2$$
$$HgBr_2 + Ca(OH)_2 \rightarrow CaBr_2 + HgO + H_2O$$
$$HgO \rightarrow Hg + \tfrac{1}{2}O_2$$

A drawback of this sequence is its use of highly corrosive hydrogen bromide. The scheme also requires a large inventory of mercury.

Of major concern to investigators in the thermochemical splitting schemes is the availability of appropriate materials of construction. Heat exchangers between the nuclear side and the chemical side must withstand both corrosion and radioactive contamination. The conventional nickel-chromium alloys are capable up to about 1050K; exotic, but available alloys, up to about 1400K. Above these temperatures, ceramics and new alloys may have to be used. Considerable materials research along these lines is going forth at the Los Alamos Scientific Laboratory.

Conventional Hydrogen Uses. Even before serious consideration was given to a hydrogen energy economy, the demand for hydrogen grew at a rate of about 15% annually since World War II. About 3 trillion standard cubic feet (85 million cubic meters) of hydrogen (8 million tons; 7.2 million metric tons) were produced in the United States during the 1970s. Not including energy needs, the chemical requirements for hydrogen are expected to increase by about 7% per year to the year 2000.

References

Bassett, L. C., and R. S. Natarajan: "Hydrogen—Buy It or Make It?" *Chem. Eng. Progress*, **76**, 3, 93–98 (1980).
Considine, D. M., Editor: "Energy Technology Handbook," McGraw-Hill, New York (1977).
Corniel, H. G., Heinzelmann, F. J., and E. W. S. Nicholson: "Production Economics for Hydrogen, Ammonia, and Methanol During the 1980–2000 Period," Exxon Research and Engineering, New York (1977).
Cox, K. E., and K. D. Williamson: "Hydrogen: Its Technology and Implications," Vol. 2: *Transmission and Storage of Hydrogen*, CRC Press, Boca Raton, Florida (1977).
Gregory, D. P.: "The Hydrogen Economy," *Sci. Amer.*, **228**, 1, 13–21 (1973).
Hänsch, T. W., Schawlow, A. L., and G. W. Series: "The Spectrum of Atomic Hydrogen," *Sci. Amer.*, **240**, 3, 94–110 (1979).
Mao, H. K., and P. M. Bell: "Observations of Hydrogen at Room Temperature (25°C) and High Pressure (to 500 kilobars)," *Science*, **203**, 1004–1006 (1979).
Reilly, J. J., and G. D. Sandrock: "Hydrogen Storage in Metal Hydrides," *Sci. Amer.*, **242**, 2, 118–129 (1980).
Ross, M., and C. Shishkevish: "Molecular and Metallic Hydrogen," *Paper R-2056-ARPA*, Rand Corporation, Santa Monica, California (1977).
Sandrock, G. D., Reilly, J. J., and J. R. Johnson: "Metallurgical Considerations in the Production and Use of FeTi Alloys for Hydrogen Storage," *11th Intersociety Energy Conversion Engineering Conference Proceedings*, Vol. 1, pages 965–971 (September 1976).

HYDROGENATION. In its simplest interpretation, to hydrogenate is to add hydrogen. There are scores of examples where hydrogenation is used as a unit process through the chemical, food, and process industries. Generally, the process is associated with relatively high pressure, elevated temperature, and the presence of a catalyst.

Nickel, prepared in finely divided form by reduction of nickel oxide in a stream of hydrogen gas at about 300°C, was introduced by Sabatier (1897) as a catalyst for the reaction of hydrogen with unsaturated organic substances conducted at about 175°C. Nickel proved to be one of the most successful catalysts for such reactions. The unsaturated organic substances that are hydrogenated are usually those containing a double bond, but those containing a triple bond also may be hydrogenated. Platinum black, palladium black, copper metal, copper oxide (Adkin catalyst), nickel oxide, aluminum, and other materials have subsequently been developed as hydrogenation catalysts. Temperatures and pressures have been increased in many instances to improve yields of desired product. The hydrogenation of methyl ester to fatty alcohol and methanol, for example, occurs at about 3000 psig (204 atm) and 290–315°C. In the hydrotreating of liquid hydrocarbon fuels to improve quality, the reaction may take place in fixed-bed reactors at pressures ranging from 100 to 3000 psig (7 to 205 atm). Many hydrogenation processes are of a proprietary nature, with numerous combinations of catalysts, temperatures, and pressures.

Among the better known products of hydrogenation are hydrogenated vegetable and fish oils which may be hardened or solidified by catalytic hydrogenation. Some of these oils can be partially hydrogenated to clarify and deodorize them. Fatty oils, such as oleic acid, may be converted into stearic acid by hydrogenation. Through hydrogenation, peanut oil, cottonseed oil, and coconut oil can be converted to materials that taste, appear, and smell like lard; or by varying the process, they can be made to resemble tallow. Most synthetic shortenings are comprised of hydrogenated oils. Usually, hydrogenated oils will have higher melting points and lower iodine values than the natural untreated oils.

Hydrogenation of Coal and Crudes. The interest in hydrogenation has been greatly intensified since the early 1970s in connection with the synthesis of new types of fuels to augment energy supplies. Basically, however, the hydrogenation of coal is not a new concept, but dates back at least a half-century to the time when manufactured gas (artificial, illuminating, producer, water gas, etc.) was used prior to the more general availability of low-cost, cleaner natural gas. In 1927, a serious paper was published discussing the processes then available for production of oil from coal. One of the first large-scale applications of the Fischer-Tropsch process for the production of oil from coal was that of the South African Coal, Oil and Gas Corporation's plant in Sasolburg, Republic of South Africa, constructed in the mid-1950s and since expanded many times.

Similarly, sour crudes, heavy residuums, and other petroleum-base starting materials can be hydrogenated, sometimes coupled with other processes to sweeten (remove sulfur), reduce viscosity, and otherwise improve the materials for better use as fuels. See also **Coal; Hydrotreating;** and **Petroleum.**

HYDROGEN CYANIDE. HCN, formula weight 27.03, colorless gas with characteristic odor, very poisonous, mp $-14°C$, bp $26°C$, critical temperature $183.5°C$, critical pressure 50 atmospheres, density 0.20 g/cm^3, sp gr 0.697 ($18°C$). There are two isomeric forms: (1) HCN which forms cyanides, (2) HNC (inferred from its derivatives) which forms isocyanides. Hydrogen cyanide is soluble in H_2O, or alcohol, or ether in all proportions. The compound usually is marketed as an aqueous solution containing 2–10% (weight) HCN. For many process uses, it is frequently more convenient to generate HCN as needed and thus avoid storage and handling problems. HCN burns with a red-blue flame, yielding CO_2, nitrogen, and H_2O. Aqueous solutions of HCN decompose slowly, yielding ammonium formate: $HCN + 2H_2O \rightarrow HCOONH_4$. Decomposition is slowed by storage in dark locations. Peaches, apricots, bitter almonds, cherries, and plums contain some HCN derivatives in their kernels, frequently in combination with glucose and benzaldehyde as a glucoside (amylgdalin). The bitter almond fragrance of HCN and its derivatives sometimes can be detected in such kernels.

Production. Hydrogen cyanide can be prepared from a mixture of NH_3, methane, and air by partial combustion in the presence of a platinum catalyst:

$$NH_3 + CH_4 + 1.5 \, O_2 + 6N_2 \rightarrow HCN + 3H_2O + 6N_2.$$

The process is carried out at about 900–1,000°C; yield ranges from 55–60%. In another process, methane (contained in natural gas) is reacted with NH_3 over a platinum catalyst at from 1,200–1,300°C, the reaction requiring considerable heat input. In still another process, a mixture of methane and propane is reacted with NH_3; $C_3H_8 + 3NH_3 \rightarrow 3HCN + 7H_2$; or $CH_4 + NH_3 \rightarrow HCN + 3H_2$. An electrically-heated fluidized bed reactor is used. Reaction temperature is approximately 1,510°C.

The high-tonnage uses of HCN are in the preparation of numerous chemical products and intermediates for organic syntheses. As a gas, HCN sometimes is applied as a disinfectant; or cellulosic disks impregnated with HCN may be used. In ore processing and metal treating, cyanides are widely used.

Hydrogen cyanide reacts with hydrogen at 140°C in the presence of a catalyst, e.g., platinum black, to form methyl amine CH_3NH_2; when burned in air, produces a pale violet flame; when heated with dilute sulfuric acid forms formamide $HCONH_2$ and ammonium formate $HCOONH_4$; when exposed to sunlight with chlorine forms cyanogen chloride CNCl, plus hydrogen chloride. An important reaction of hydrogen cyanide is that with aldehydes or ketones, whereby cyanhydrins are formed, e.g., acetaldehyde cyanhydrin $CH_3CHOH \cdot CH$, and the resulting cyanhydrins are readily converted into alpha-hydroxy acids, e.g., alpha-hydroxypropionic acid $CH_3 \cdot CHOH \cdot COOH$.

Metallic cyanides are (1) soluble, e.g., sodium cyanide NaCN, potassium cyanide KCN, calcium cyanide $Ca(CN)_2$, mercuric cyanide $Hg(CN)_2$, aurous cyanide AuCN, (2) insoluble, e.g., silver cyanide AgCN, cuprous cyanide CuCN, (3) complex, (a) decomposed by dilute H_2SO_4 and not affected by dilute NaOH, e.g., sodium silver cyanide $NaAg(CN)_2$ solution, sodium cuprous cyanide $NaCu(CN)_2$ colorless solution, (b) changed only to acid by dilute H_2SO_4 and reactive with dilute NaOH, e.g., potassium hexacyanoferrate(II) $K_4Fe(CN)_6$ yields, with dilute H_2SO_4, hexacyanoferric(II) acid, cupric hexacyanoferrate(II) $Cu_2Fe(CN)_6$ yields, with dilute NaOH, cupric hydroxide.

Sodium cyanide solution dissolves certain metals (1) with absorption of oxygen, e.g., gold, silver, mercury, lead, (2) with evolution of hydrogen, e.g., copper, nickel, iron, zinc, aluminum, magnesium; and solid sodium cyanide, when heated with certain oxides, e.g., lead monoxide PbO, stannic oxide SnO_2, yields the metal of the oxide, e.g., lead, tin, respectively, and sodium cyanate NaCNO. Two classes of esters are known, cyanides or nitriles, and isocyanides, isonitriles or carbylamines, the latter being very poisonous and of marked nauseating odor.

Methyl cyanide CH_3CN, bp 82°C, formed by reaction of (1) methyl iodide and potassium cyanide, (2) acetamide and phosphorus pentoxide. Methyl isocyanide CH_3NC, bp 60°C, formed by reaction (1) of methyl iodide and silver cyanide, (2) of methylamine, chloroform and NaOH solution warmed. Ethyl isocyanide C_2H_5NC, bp 78°C. Phenyl isocyanide C_6H_5NC, bp 78°C at 40 torr pressure.

Oxidation of cyanide ion (e.g., by copper(II)) gives cyanogen or oxalonitrile NCCN, poisonous colorless gas, bp $-21°C$. This reacts with organic compounds and bases like a halogen, for example, disproportionating in aqueous alkali to cyanide and cyanate. In aqueous acid, hydrolysis to oxalamide and ultimately oxalic acid takes place. Oxidation of cyanides by oxygen donors (e.g., lead monoxide or dioxide, manganese dioxide or dichromate) a little below red heat produces cyanates.

HYDROGEN PEROXIDE. H_2O_2, formula weight 34.02, in pure, anhydrous form is a viscous, colorless liquid, sp gr 1.44, mp $-0.89°C$, bp 151.4°C. Hydrogen peroxide is soluble in H_2O in all proportions, soluble in alcohol, or ether, but not in hydrocarbons. Reagent, chemically-pure (CP) grade H_2O_2 is a solution of 90% H_2O_2 and 10% H_2O, sp gr 1.39. This concentration contains 42% active oxygen by weight. One volume yields 410 volumes of oxygen. Hydrogen peroxide solutions are high-tonnage chemicals and are supplied commercially in several strengths, ranging from 3–35% H_2O_2 by weight. Commercial grades for oxidation and bleaching normally contain 27.5–35% H_2O_2.

To reduce the tendency of H_2O_2 solutions to decompose, storage must be at comparatively low temperatures and in light-tight containers. Often, an organic material, such as acetanalide, will retard degradation. H_2O_2 has been used as an oxidizer in liquid bipropellant systems, or as a monopropellant through controlled catalytic decomposition, in supplying oxygen to various fuel mixtures for rockets and torpedoes. Low-concentration (normally 3% H_2O_2) solutions have been used for many years as antiseptics in medical applications. Bleaching is a primary outlet for H_2O_2, particularly in connection with cotton, wool, groundwood pulp—as well as hair-bleaching formulations. The compound is used as a source of gas in foaming rubber plastics. The highly reactive H_2O_2 molecule readily participates in oxidation, epoxidation, and hydroxylation reactions and is frequently used in an intermediate capacity in chemical syntheses. In restoring old paintings, H_2O_2 has been used to convert black PbS tarnish into the original white lead sulfate.

Industrial Production of Hydrogen Peroxide. The traditional process for manufacturing H_2O_2 has been the electrolysis of aqueous solutions of $KHSO_4$, H_2SO_4, or NH_4HSO_4. In recent years, chemical autoxidation processes have grown in favor, largely because of energy costs. In these processes, the feedstock may be an alkylated quinone, alkylated anthraquinone, and hydroquinone solvents, together with hydrogen, air or oxygen, H_2O, and a nickel, palladium, or platinum catalyst. The process yields a 15–75% solution of H_2O_2 in H_2O, depending upon adjustment of process concentrations and conditions to provide desired concentration. The yield for this type of process is about

90% of theoretical. The process proceeds essentially in two steps. In the first step, anthraquinone contained in a solvent is hydrogenated at a temperature of about 40°C and a pressure of 1–3 atmospheres. The anthraquinone is reduced to hydroquinone (*p*-dihydroxybenzene):

$$C_6H_4:(CO)_2:C_6H_3 + H_2 \rightarrow C_6H_4:(COH)_2:C_6H_3R.$$

R is a radical such as ethyl or tertiary butyl. In the second step, the hydroquinone solution is oxidized with air or oxygen: $C_6H_4:(COH)_2:C_6H_3R + O_2 \rightarrow C_6H_4:(CO)_2:C_6H_3R + H_2O_2$. In theory, the process consumes only hydrogen, atmospheric oxygen, and H_2O. A solvent must be used that will minimize side reactions during hydrogenation while also dissolving both the hydrogenated and oxidized forms of the organic compound. Solvents referred to in this connection are benzene-methyl-cyclohexanol mixtures and primary and secondary nonyl alcohols. Very tight purity precautions are required because any impurities in the H_2O_2 cause spontaneous catalytic decomposition of the product. As the result of these necessary precautions, the resulting H_2O_2 is one of the purest of commercial chemicals.

The process is highly corrosive. At one time, enameled steel vessels were standard for H_2O_2 processing. Aluminum, once properly passified through pickling and treatment after fabrication, has been found satisfactory.

Hydrogen peroxide reacts (1) with alkalis to form peroxides, (2) with potassium iodide solution, in presence of ferrous sulfate, to liberate iodine. This reaction serves to indicate the presence of as small an amount as 1 part by weight of hydrogen peroxide in 25,000,000 parts of H_2O, (3) with lead sulfide PbS, brown solid, to form lead sulfate $PbSO_4$, white solid, and sometimes used to brighten the lead pigment of darkened oil paintings, (4) with lead dioxide to form lead oxide, (5) with sulfites, especially in alkaline solution, to form sulfates, (6) with nitrites to form nitrates, (7) with arsenites for form arsenates, (8) with ferrous compounds to form ferric, (9) with chromic compounds to form chromates (see **Chromium**), (10) with permanganates in acid solution to form manganous compounds plus oxygen of twice the volume available from the hydrogen peroxide, (11) with dichromates in acid solution cold to form perchromic acid, blue solution, more soluble in ether than in acid, (12) with titanic salt solutions to form pertitanic acid, yellow solution, (13) with colored organic materials, e.g., litmus, indigo, to destroy the color, and thus used for bleaching hair, silk, feathers, straw, ivory, teeth, bones, gelatin, flour. When hydrogen peroxide solution is treated with finely divided platinum or other substances, or comes in contact with rough surfaces, e.g., ground glass, oxygen is evolved (water also formed).

In the laboratory, hydrogen peroxide is prepared from barium peroxide by treatment with ice-cold dilute acid; when H_2SO_4 is used barium sulfate (insoluble) may be separated by filtration. Other peroxides, e.g., sodium peroxide, react similarly with acids to form hydrogen peroxide plus the salt corresponding to the peroxide and acid used. Hydrogen peroxide is formed when ether is exposed to sunlight, when a hydrogen-oxygen flame impinges on ice, and when H_2O in a quartz vessel is exposed to ultraviolet light.

HYDROGEN SCALE. Since there is no reliable method for determining the absolute potential of a single electrode, electrode potentials are measured against a reference electrode whose potential is arbitrarily taken as zero. The arbitrary zero in general use is the potential of a reversible hydrogen electrode, with gas at one atm pressure, in a solution of hydrogen ions of unit activity, or other electrodes calibrated against the hydrogen electrode. The hydrogen scale is also the name given to a thermometric scale.

HYDROGEN SULFIDE. H_2S, formula weight 34.08, colorless, odorous gas, mp −82.9°C, bp −59.6°C, sp gr 1.1895 (air = 1). The gas must be handled carefully because of (1) its toxic properties (particularly dangerous because it may paralyze the olfactory nerves), and (2) its explosive tendencies (low ignition temperature of 260°C and wide flammability range from 4.3 to 44% by volume in air). Hydrogen sulfide liberates considerable heat upon burning (6,230 calories/liter at 15.6°C). The gas is produced by acid hydrolysis of many sulfides and by water hydrolysis of those elements higher in the hydrogen scale.

An aqueous solution of hydrogen sulfide is termed hydrosulfuric acid which undergoes slow atmospheric oxidation to sulfur. The acid is a strong reducing agent, usually with the separation of sulfur, e.g., with nitric acid (nitric oxide formed), with concentrated H_2SO_4 (SO_2 is formed), with permanganate (manganous ion formed in the presence of acid), dichromate (chromic ion formed in the presence of acid).

Fluorine, chlorine, bromine, and iodine react with H_2S to form the corresponding halogen acid. Metal sulfides ire formed when H_2S is passed into solutions of the heavy metals, such as Ag, Pb, Cu, and Mn. This reaction is responsible for the tarnishing of Ag and is the basis for the separation of these metals in classical wet qualitative analytical methods. Hydrogen sulfide reacts with many organic compounds.

The gas results from the decomposition of metal sulfides and albuminous matter and is found in the areas of mineral springs, sewers, and in some mines where it is referred to as "stink damp." H_2S also is a by-product of several industrial processes, including synthetic rubber, viscose rayon, petroleum refining, dyeing, and leather-treating operations. In the laboratory, H_2S usually is prepared by treating a sulfide with an acid, such as iron pyrites and HCl, or by heating thioacetamide $CH_3C(:S)NH_2$. Three processes are used industrially to produce H_2S in large quantities: (1) treating a sulfide with an acid, $2NaHS + H_2SO_4 \rightarrow 2H_2S + Na_2SO_4$, (2) reacting sulfur with an alkali, $4S + 2NaOH + 2H_2O \rightarrow 2H_2S + Na_2S_2O_3$, and (3) directly reacting sulfur with hydrogen, $S + H_2 \rightarrow H_2S$. Large quantities of by-product H_2S usually are converted into elemental sulfur or H_2SO_4.

Industrial uses for H_2S include (1) the preparation of sulfides, such as sodium sulfide and sodium hydrosulfide, (2) the production of sulfur-bearing organic compounds, such as thiophenes, mercaptans, and organic sulfides, (3) the removal of Cu, Cd, and Ti from spent catalysts where the gas acts as a precipitant, (4) the formulation of extreme-pressure lubricants, and (5) the preparation of rare-earth phosphors used in color TV tubes.

HYDROLYSIS. A chemical reaction in which water reacts with another substance to form two or more substances. This involves ionization of the water molecules as well as splitting of the compound hydrolyzed, e.g., $CH_3COOC_2H_5 + H \cdot OH \rightarrow CH_3COOH + C_2H_5CH$. Examples are conversion of starch to glucose by water in the presence of suitable catalysts; or the conversion of sucrose (cane sugar) to glucose and fructose by reaction with water in the presence of an enzyme or acid catalyst; or conversion of natural fats into fatty acids and glycerin by reaction with water, as occurs in one stage of soap manufacturing; or the reaction of the ions of a dissolved salt to form various products, such as acids, complex ions, etc.

HYDRONIUM ION. An ion (H_3O^+) formed by the transfer of a proton (hydrogen nucleus) from one molecule of H_2O to another; a companion ion (OH⁻) is also formed; the reaction is $2H_2O \rightarrow H_3O^+ + OH^-$. Formation of such ions is statistically rare, resulting from the interaction of water molecules in a ratio of 1 to 556 million.

HYDROPHILIC. Colloid Systems.

HYDROQUINONES. These are dihydroxy aromatic compounds with the two groups in positions corresponding to *ortho* or *para* substitution in the benzene ring. They are closely related to the quinones from which they can be obtained by reduction. Thus *o*-dihydroxybenzene (catechol) can be obtained from *o*-benzoquinone, and hydroquinone (*p*-dihydroxybenzene or quinol) from *p*-benzoquinone. Resorcinol (*m-d*ihydroxybenzene is not properly a hydroquinone since the corresponding *meta* quinone is not known to exist. Homologs of hydroquinone are usually named after the parent hydrocarbon. Thus toluhydroquinone is 2,5-dihydroxy-1- methylbenzene and naphthohydroquinone is 1,4-dihydroxy-naphthylene. Unlike many of the quinones, the ring systems are fully aromatic and undergo substitution reactions common to phenols and other benzene derivatives. However they are easily oxidized by some reagents to the less stable quinones and degradation products frequently result. Thus treatment of hydroquinone with nitric acid yields oxalic acid while halogenation with sulfuryl chloride results in a mixture of chlorohydroquinones, quinone, quinone chlorides, and tetrachloro-*p*-benzoquinone. The formation of side and degradation products can be minimized if the molecule is protected against oxidation by acetylating or benzoylating at least one of the hydroxyl groups. For example 2-nitro-hydroquinone can be prepared in good yield by the nitration of monobenzoyl hydroquinone followed by hydrolysis. Concentrated sulfuric acid gives hydroquinone-2.5-disulfonic acid directly, and tertiary amyl groups can be introduced into the ring in the 2 and 5 positions by treatment with amylene in the presence of sulfuric acid. The hydroxyl groups are weakly acidic and can readily be converted to ethers by treatment with alkyl halides or sulfates in the presence of alkali. A diacetate is formed on treatment with acetic anhydride. The most characteristic reaction of hydroquinones is their reversible oxidation to quinones.

Catechol, or 1,2-dihydroxybenzene, was first prepared by the dry distillation of catechin obtained from *Mimosa catechu*. It can also be formed by the hydrolysis of its methyl ether, guaiacol, which is a constituent of beechwood tar. It is prepared synthetically by fusing phenol-*o*-sulfonic acid with sodium hydroxide, or treating *o*-chlorophenol with aqueous alkali in the presence of copper at a high temperature and pressure. It crystallizes from benzene in colorless monoclinic plates which melt at 105°C. The lead salt can be oxidized to *o*-benzoquinone by a solution of iodine in chloroform. The ethers of catechol are of considerable importance and can be derived from a number of naturally occurring substances. The methylene ether of protocatechualdehyde is known as piperonal, and is closely related to various natural products including piperine, safrole, and isosafrole, from which it can be derived. These compounds have been used for the synthesis of pyrethrin synergists. *Vanillin*, the principal flavoring constituent of vanilla, is the 3-methyl ether of protocatechualdehyde.

Hydroquinone is found in nature combined in the glycoside arbutin, from which it can be released by hydrolysis with emulsin or dilute sulfuric acid. It is prepared commercially from *p*-benzoquinone by reduction with sulfur dioxide. It is a dimorphic solid with the stable form melting at 170.5°C. Hydroquinone is one of a number of compounds that possess the property of forming molecular compounds with gases such as hydrogen

sulfide, sulfur dioxide, krypton, xenon, etc. These are known as clathrate compounds, and their existence is due to the entrapment of atoms or molecules of the gas in the crystal lattice of the hydroquinone. Three moles of hydroquinone can entrap one mole of gas, which is firmly held but which is liberated when the clathrate is dissolved in water. The most important commercial use of hydroquinone is for the development of photographic film. Its effectiveness is dependent on its ability to reduce the silver subhalide formed on exposure of the film to light to metallic silver. It gives films of high density and it is often necessary to reduce the harshness of contrast by using it in combination with other developers such as metol or paramidophenol. Hydroquinone and its derivatives are effective antioxidants for the preservation of fats, oils, and rubber. It has also been used as a short-stopping agent for controlling polymerization in the production of synthetic rubber of the butadiene-styrene type.

—H. P. Burchfield, Gulf South Research Institute, New Iberia, Louisiana.

HYDROTREATING. A specialized kind of hydrogenation in which the quality of liquid hydrocarbon streams is improved by subjecting them to mild or severe conditions of hydrogen pressure in the presence of a catalyst. The objective is to convert undesirable material in the feedstock to either desired materials or easily disposed byproducts, on a highly selective basis. As of the early 1980s about 45% of the crude oil refined in the United States is hydrotreated. Some applications of hydrotreating include: (1) improvement of the burning quality of jet fuels, kerosines, and diesel fuels; (2) purification of light aromatic by-products from pyrolysis operations; (3) pretreatment of naphtha feeds for catalytic reforming units; (4) reduction in sulfur content of residual fuel oils; (5) pretreatment of catalytic cracking feeds and cycle oils by removal of metals, sulfur, nitrogen, and reduction of polycyclic aromatics; (6) desulfurization of distillate fuels; (7) upgrading of lubricating oil quality; and (8) improvement of color, odor, and storage stability of various fuels.

Some of the specific reactions involved include: (1) hydrogenation of monoaromatics to naphthenes to improve burning quality of certain fuels; (2) removal of nitrogen as ammonia from its organic combinations; (3) removal of oxygen from its organic combinations as water; (4) hydrogenation of polycyclic aromatics so that only one aromatic ring remains in the molecule; (5) hydrogenation of diolefins and olefins to paraffins or naphthenes; (6) removal of sulfur from its organic combinations in various types of sulfur compounds by hydrodesulfurization to form hydrogen sulfide; and (7) decomposition and removal of organometals, such as arsenic compounds in naphthas, by retention of these metals on the catalyst. Vanadium and nickel also can be removed.

In the hydrotreating process shown by the accompanying diagram, the liquid feed is preheated by exchange with the reactor effluent. It is then heated to the desired reactor-inlet temperature in a fired heater. At this point, recycle hydrogen joins the feedstock. An excess of hydrogen is used to suppress accumulation of deactivating carbonaceous deposits on the catalyst. Fresh makeup hydrogen enters the process to maintain a sufficient supply and also pressure on the system. Cooled effluent from the reactor goes to a separator vessel at which point the recycle or net hydrogen is removed. The liquid then goes to a stripper or stabilizer where hydrogen, hydrogen sulfide, ammonia, water, and light hydrocarbons dissolved in the separator liquid are removed. The stabilized hydrotreated liquid, free of

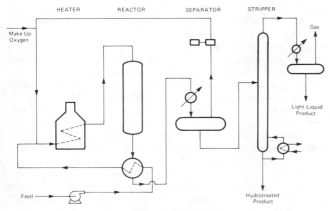

Representative hydrotreating unit. (*UOP Inc.*)

dissolved, unwanted contaminants, is routed to subsequent processing or to product fuel blending.

It is interesting to note that there are over 25 proprietary versions of this basic process. Numerous modifications are required, depending upon the nature of the feedstock and desired end-products.

—Technical Staff, UOP Inc., Des Plaines, Illinois.

HYDROXYLAMINE. H_2NOH, formula weight 33.02, white, odorless solid, mp 33°C, bp 56°C (22 mm pressure), explosive, soluble in all proportions in H_2O or alcohol. Hydroxylamine is: (1) A weak base forming with acids soluble salts that decompose more or less violently when heated, e.g., hydroxylamine hydrochloride (hydroxylammonium chloride, $H_2NON \cdot HCl$), mp 151°C, nitrate $H_2NOH \cdot NHO_3$, hemisulfate $H_2NOH \cdot \frac{1}{2}H_2SO_4$. Dihydroxylamine oxalate and trihydroxylamine phosphate are insoluble in H_2O. Hydroxylamine hydrochloride is soluble in alcohol. (2) A weak acid forming with bases soluble salts, e.g., sodium hydroxylamite H_2NONa. Hydroxylamine salt solution is a powerful reducing agent, more especially in alkaline than in acid solution, for example, cupric salt solutions changed to cuprous oxide, silver salt solutions to silver, mercuric chloride solution to mercurous chloride, ferric salt solutions (in acid) to ferrous. Ferrous hydroxide in sodium hydroxide is, however, oxidized by hydroxylamine to ferric hydroxide plus NH_3.

Hydroxylamine reacts with carbonyl group $=CO$ of aldehydes, ketones or quionones, yielding *oximes*, white solids, of definite melting point and used in identification of aldehydes and ketones, e.g., acetaldehyde oxime $CH_3CH:NOH$.

Beta-phenylhydroxylamine, N-phenylhydroxylamine, is a white solid, slightly soluble in water, very soluble in alcohol or ether, forms salts with acids, e.g., beta-phenylhydroxylamine hydrochloride $C_6H_5NHOH \cdot HCl$, upon exposure to air the water solution forms azobenzene $C_6H_5N:NC_6H_5$. Beta-phenylhydroxylamine reacts (1) with oxidizing agents, such as chromic acid or ferric chloride, to form nitrosobenzene C_6H_5NO, (2) with reducing agents, such as tin plus hydrochloric acid, to form aniline $C_6H_5NH_2$, (3) with alkaline cupric salt solution (Fehling's solution) at room temperature to form cuprous oxide, (4) with ammonio-silver salt solution (Tollen's solution) at room temperature to form silver, (5) in the presence of hydrochloric acid to form paraminophenol $HO \cdot C_6H_4 \cdot NH_2(1,4)$.

Beta-phenylhydroxylamine is formed by reduction of nitrobenzene (1) by zinc and calcium chloride or ammonium chloride solution, (2) by electrolysis in acetic acid plus sodium acetate solution.

Diphenylhydroxylamine is prepared by reaction of nitroso-benzene and phenylmagnesium bromide in anhydrous ether, followed by treatment with H_2O (magnesium hydroxybromide also formed).

When hydroxylamine reacts with aldehydes, the resulting compounds are termed *aldoximes* as, for example, acetaldoxime. $CH_3 \cdot CHO + H_2NOH \rightarrow CH_3 \cdot CH:N \cdot OH$ (acetaldoxime) + H_2O. Hydroxylamine reactions with ketones produce *ketoximes*. $(CH_3)_2CO + H_2NOH \rightarrow (CH_3)_2C:N \cdot OH$ (dimethylketoxime) + H_2O.

The lower aldoximes are essentially odorless, volatile liquids, and miscible with H_2O in all proportions. The higher members are only slightly soluble. Ketoximes have similar properties.

HYDROXYPROLINE.   Amino Acids.

HYGROSCOPIC.   1. Pertaining to a marked ability to accelerate the condensation of water vapor. In meteorology, this term is applied principally to those condensation nuclei composed of salts that yield aqueous solutions of a very low equilibrium vapor pressure compared with that of pure water at the same temperature. **2.** Descriptive of a substance, the physical characteristics of which are appreciably altered by effects of water vapor. The hygroscopicity of certain materials has been advantageously utilized in humidity control.

HYPERFINE STRUCTURE. In general, a set of very closely spaced lines in atomic spectra or other kinds of spectra. There may be many causes of hyperfine structure: (1) for a single atomic species or nuclide, the occurrence of spectral lines as doublets, triplets, etc., due to the interaction, or coupling, of the total angular momentum of the orbital electrons with the nuclear spin and associated magnetic moment; (2) for an element consisting of several isotopes, the occurrence of components for each spectral line that is observable under high resolution, each isotope contributing one or more components. This type of hyperfine structure is often called *isotope structure* to differentiate it from the type described under (1).

HYPERONS. These are subatomic particles that are more massive than nucleons (protons and neutrons). *Strangeness* is a property of elementary particles found useful in classifying hyperons. Each particle is assigned a strangeness quantum number S which is related to the electric charge Q, the isospin number T, and the baryon number B by the formula $Q = T + (S + B)/2$. ($T + \frac{1}{2}$ for a proton and $-\frac{1}{2}$ for a neutron; other particles may have $T = 0$ or $T = 1$, depending on the type.) Strangenessis conserved in reactions involving the strong interaction. The selection rules resulting from strangeness conservation are important in understanding why some reactions take place much more slowly than others. See **Particles (Subatomic).**

HYPERSTHENE. The mineral hypersthene is an orthorhombic pyroxene, chemically a ferro-magnesian silicate, differing from enstatite in that the iron content is considerable (FeO being greater than 15%). A general formula is $(Mg, Fe)SiO_3$. It is usually found as a massive mineral, whose crystals tend to be prismatic or tabular in habit. It has a distinct prismatic cleavage; fracture, uneven; brittle; hardness, 5–6; specific gravity, 3.42–3.84; luster, pearly to somewhat metallic; color, brownish-green, brown, greenish-black to grayish-black; streak, grayish-brown; translucent to opaque. Hypersthene is often associated with labradorite in gabbro and norite and in extrusive rocks like andesite. It is occasionally encountered in meteorites. Hypersthene is associated with pyrrhotite in Bavaria, with labradorite on the Isle St. Paul, Labrador. It is also found in Montmorency County, Quebec; and in the United States in the rocks of the Cortlandt series in the Hudson River Valley, and the andesites of Colorado and northern California. Superb crystals of exceptional size and quality have been found growing into and within the almandine-pyrope garnets at Gore Mountain, North River, New York. The rarity of hypersthene in crystal form makes this occurrence noteworthy. The word hypersthene comes from the Greek words meaning *strong* or *tough*.

See also **Pyroxene**.

HYPOBARIC (Controlled-Atmosphere) SYSTEMS. Sensitive materials, notably fresh foods, normally cannot withstand long periods of transportation and storage prior to their consumption. Over the years, much of the effort extended toward offering produce in marketplaces far distant from the source was concentrated on reducing the time for delivery. Thus, the extensive use of air express and air freight. Conventional refrigeration systems for trucks and railway cars also were specially adapted for use during transport. But, even with all of these improvements in technology, certain transporting feats (as, for example, shipping Midwestern pork to the California and even Hawaii markets) were difficult to achieve. The controlled-atmosphere hypobaric concept has greatly extended the potential for far distant shipping of delicate, perishable materials (not necessarily limited to foodstuffs).

In 1964, the Institute of Food Technologists' annual award was given in recognition of the development of a controlled-atmosphere storage system. Essentially, the process was designed to reduce the rate of deterioration of certain fruits and vegetables in refrigerated storage by reducing the oxygen level, increasing the carbon dioxide level, and maintaining the relative humidity close to 100%. In an initial design, the conditions were created by using a home-furnace size catalytic generator which burned natural gas or propane gas to create the atmosphere that essentially halts the natural respiration of the stored products. Later, the gas generator was replaced by cryogenic liquefied gases, allowing additional flexibility and the creation of any desired gas mixture. Atmospheres can be tailored to particular perishables. For example, an atmosphere of 15–20% carbon dioxide and 80–85% nitrogen is optimal for strawberries. For iceberg lettuce, an atmosphere of 8–10% oxygen, less than 10% carbon dioxide, with the remainder nitrogen is used. As of the early 1980s, the system has been installed on over ten thousand rail cars and 7000 sea vans.

In 1979, another IFT award was given in recognition of a hypobaric transport and storage system for fresh meats and meat products. Hypobarics is defined as a precisely controlled combination of low pressure, low temperature, high humidity, and ventilation which, when properly applied, extends up to six times the length of time a perishable commodity remains fresh. This makes possible the shipment of perishable items by way of relatively low-cost surface transportation to distant points. In developing the concept, it was observed that refrigerated storage of fruits in closed containers will result in accumulation of gases generated by the fruit, i.e., ethylene and carbon dioxide, an atmosphere which hastens ripening and spoilage. Although ventilation of fruit containers can prevent accumulation of the gases, the gases are not removed from within the product itself—with no prevention of accumulation of gases within the cells of the fruit. The researchers made the supposition that by drawing a partial vacuum on a closed vessel containing the fruit, the low pressure would increase the diffusivity of the gases, thus promoting release and removal of the gases. At the same time, a reduction of pressure would reduce the oxygen concentration, thus retarding respiration and attendant spoilage. Combined with refrigeration, this would decelerate the metabolic processes, not only of the fruit, but also of any bacteria present. Humidification of the chamber would prevent any drying of the fruit. After testing the concept on bananas and other perishables, the system was patented.

The capability and flexibility of the system has been illustrated a number of times. Early in 1976, a shipment of 14,000 pounds (6350 kilograms) of fresh chicken was shipped in a hypobaric container from Arkansas to Arizona, a trip requiring 48 hours. The shelf life of the chicken was extended by 10 to 14 days over that attainable by conventional refrigeration. In the fall of 1977, 21,000 pounds (9525 kilograms) of fresh pork cuts were shipped by truck from South Dakota to California, and then by ship to Hawaii, requiring a total of 7 days. Official inspectors judged the meat to be in excellent condition and superior to that shipped by conventional means. In the spring of 1978, 14,000 pounds (6350 kilograms) of fresh lamb carcasses, 1400 pounds (635 kilograms) of fresh beef, and 900 pounds (408 kilograms) of fresh veal were shipped by truck from Texas to Maryland and then by ship to Iran, requiring a total of 42 days. The load was judged by Iranian officials to be in excellent condition upon receipt.

In addition to the foregoing examples, the hypobaric concept has been shown to be practical for various seafoods, avocados, cherries, limes, mangos, papayas, mushrooms, tomatoes, and peppers. Generally the storage temperature for meats is about $-1°C$, and up to 10 or 12°C for various fruits and vegetables. In all cases, the relative humidity is controlled at about 95%. Pressure ranges between 10 and 80 millimeters of mercury. Lower pressures are maintained for meats and seafoods; somewhat higher pressures for fruits and vegetables.

HYPOCHLORITES. When chlorine is reacted with an alkali, a hypochlorite is formed. These compounds are very high-tonnage chemicals for sanitizing and bleaching purposes. Commercial sodium hypochlorite NaClO usually is available in two strengths (1) the familiar household liquid bleach which contains about 5.25% (weight) NaClO, and (2) commercial bleach which contains about 13% (weight) NaClO. The latter compound sometimes is referred to as 15% bleach because the chlorine content is approximately 150 grams/liter of available chlorine. The term "liquid chlorine" usually refers to a solution of NaClO (up to 10%) used in the swimming-pool trade. "Dry chlorine" is part of the registered trade-mark of a proprietary calcium hypochlorite product containing 70% available chlorine. See also **Bleaching Agents**.

Sodium hypochlorite normally is manufactured in batches by diluting caustic soda to the proper starting concentration. This is approximately 6.8% NaOH for the 5.25% bleach; and about 18.5% NaOH for the 15% bleach. After cooling the caustic soda solution, chlorine gas is added through a sparger pipe until the desired concentration is reached. This usually is determined by making a series of titration analyses. Bleaching powder $CaOCl_2$ is made by passing chlorine gas over slaked lime. This was the first type of chlorine bleaching agent made and dates back to 1799. The product usually contains about 30% available chlorine. Over the years, it was used extensively in the bleaching of textiles and for sanitizing even though the compound is unstable and difficult to use. The original bleaching powder largely has been replaced by an improved calcium hypochlorite product which contains about 70% available chlorine. The compound essentially is a calcium hypochlorite dihydrate and, in one process, is made by chlorinating a slurry of lime and caustic soda. The crystals which precipitate out are mixed with calcium chloride and chlorinated lime. When warmed, the calcium hypochlorite dihydrate precipitates, with sodium chloride remaining in solution. After filtering, the cake is dried, granulated, sized, and packaged. In addition to use in swimming pools, products of this type are used widely for water purification, algae control, and sanitation. On a very high-tonnage basis, calcium hypochlorite $Ca(ClO)_2 \cdot 4H_2O$ is used for pulp bleaching in the paper industry. Bleach liquor containing from 20–40% available chlorine may be produced in batches or continuously. In a continuous system, the flow of chlorine is controlled by making frequent (or continuous) measurements of oxidation-reduction potential.

A common means of detecting hypochlorites is the production of a blue color (caused by free iodine) with starch iodide paper by hypochlorites in weakly alkaline solution. Silver nitrate also precipitates part of the hypochlorite in solutions as white silver chloride.

Hypochlorous Acid. This compound, HOCl, is prepared by the reaction of (1) chlorine monoxide Cl_2O with H_2O, (2) sodium hypochlorite and an acid, excess acid yielding chlorine and oxygen, and (3) chlorine with mercuric oxide suspended in water, mercuric chloride being formed simultaneously. Hypochlorous acid is a yellow solution of characteristic odor. It decomposes upon standing, the rate depending upon (1) concentration, (2) exposure to light, (3) presence of a catalyst (cobaltous hydroxide, for example, promotes the evolution of oxygen), and (4) acidity or alkalinity. Hypochlorous acid is a powerful oxidizing agent and sometimes used as a bleaching agent for organic colors.

Perchloric Acid. This compound, $HClO_4$, is a colorless, fuming, oily liquid, miscible with H_2O, volatile under diminished pressure. A maximum constant-boiling solution (203°C, 760 millimeters Hg) results when the concentration of $HClO_4$ reaches 73% in H_2O. Cold dilute perchloric acid reacts with such metals as zinc and iron, yielding hydrogen gas and the corresponding perchlorate in solution; is stable from the point of view of oxidation and reduction (except that iodine is oxidized to periodic acid, with liberation of chlorine, ferrous salt solutions to ferric, titanous salt solutions to titanic). Concentrated hot perchloric acid, on the other hand, is a powerful oxidizing agent, exploding violently in contact with charcoal, paper, alcohol; causes serious wounds in contact with the skin.

Prepared by distilling ammonium perchlorate with HNO_3 and HCl.

Metallic perchlorates are soluble in water, except that potassium perchlorate is slightly soluble. Potassium perchlorate is, however, insoluble in alcohol containing perchloric acid, a property made use of in the qualitative recognition and quantitative estimation of potassium in salt solutions. Perchlorates, when heated, evolve oxygen and leave the chloride as a residue. Potassium perchlorate decomposes at 400°C.

HYPOFLUORITE. Any compound containing the group —OF. The simple anion FO^- is unknown. A number of covalent hypofluorites are known, including such compounds with carbon, oxygen, nitrogen, sulfur, chlorine and arsenic (uncertain), CF_3OF, CF_3COOF, C_2F_5COOF, NO_2OF, OF_2, O_2F_2, O_3F_2, SF_5OF, FSO_2OF, ClO_3OF and possibly AsF_4OF. These are all powerful fluorinating agents. They react violently with water yielding OF_2 as one product. The oxygen fluorides O_3F_2 and O_2F_2 decompose about −158°C and −100°C, respectively, the former into the latter and the latter into the elements. Nitryl and perchloryl hypofluorites (fluorine nitrate and fluorine perchlorate) easily detonate. The perfluoracyl hypofluorites are much more stable but may also decompose violently. The others appear to be stable.

HYPOIODOUS ACID AND HYPOIODITES. Hypoiodous acid (HOI) is a greenish-yellow solution, of characteristic odor. It is unstable, and cannot be distilled unchanged.

Prepared by reaction (1) of iodine and mercuric oxide (see **Mercury**) suspension in water, mercuric iodide being simultaneously formed, (2) of sodium hypoiodite and an acid, excess acid yielding iodine.

Sodium hydroxide solution reacts with iodine to form iodide and hypoiodite, the latter decomposing in a few hours at ordinary temperatures to form iodide and iodate.

HYPONITROUS ACID AND HYPONITRITES. Hyponitrous acid $H_2N_2O_2$ is a white solid, explosive even at as low a temperature as 0°C, soluble in water, more soluble in ether, can thus be extracted from water solution by ether and the latter evaporated; water solution decomposes quickly into nitrous oxide plus water. Hyponitrous acid is nonreactive with hydriodic acid (a strong reducing agent), but reactive with permanganic acid (a strong oxidizing agent) to form nitrous or nitric acid.

Prepared (1) by reaction of silver hyponitrite $Ag_2N_2O_2$ and hydrogen chloride in anhydrous ether, an evaporation of the resulting solution, (2) by reaction of hydroxylamine H_2NOH plus nitrous acid HONO.

Sodium hyponitrite $Na_2N_2O_2$ is formed (1) by reaction of sodium nitrate or nitrite solution with sodium amalgam (sodium dissolved in mercury), after which acetic acid is added to neutralize the alkali. Sodium stannite ferrous hydroxide, or electrolytic reduction with mercury cathode may also be utilized, (2) by reaction of hydroxylamine sulfonic acid and sodium hydroxide. Silver hyponitrite is formed by reaction of silver nitrate solution and sodium hyponitrite.

HYPOPHOSPHORIC ACID AND HYPOPHOSPHATES. Hypophosphoric acid (H_2PO_3 or $H_4P_2O_6$) is a solid, melting point 55°C, decomposing in solution to form phosphorous plus phosphoric acids. Hypophosphoric acid is used in solution and is a reducing agent, but only with strong oxidizing agents, such as potassium permanganate; and the acid is unaffected by zinc and dilute sulfuric acid (distinction from phosphorous acid). Dehydration of hypophosphoric acid does not yield phosphorus tetroxide; hydration of phosphorus tetroxide does not yield hypophosphoric acid but phosphorous plus phosphoric acids.

Hypophosphoric acid is formed by reaction (1) of yellow phosphorous and potassium permanganate in sodium hydroxide medium, (2) of red phosphorus and calcium hypochlorite solution, (3) also one of the products of slow oxidation at ordinary temperatures of phosphorus in moist air.

There are recorded the following sodium hypophosphates: Na_2PO_3 (or $Na_4P_2O_6$), $NaHPO_3$ (or $Na_2H_2P_2O_6$), $Na_3H(PO_3)_2$ (or $Na_3HP_2O_6$), and $(NaH_3PO_3)_2$ (or $NaH_3P_2O_6$). There is evidence in support of each of the formulas H_2PO_3, $H_4P_2O_6$ for hypophosphoric acid.

Ester: Dimethyl hypophosphate $(CH_3)_2PO_3$ or $(CH_3O)_2PO$. See also **Phosphorus**.

HYPOPHOSPHOROUS ACID AND HYPOPHOSPHITES. Hypophosphorous acid (H_3PO_2, or $H \cdot PO_2H_2$) is a colorless liquid, melting point 26.5°C, density 1.493.

Hypophosphorous acid is miscible with water in all proportions and a commercial strength is 30% H_3PO_2. Hypophosphites are used in medicine.

Hypophosphorous acid is a powerful reducing agent, e.g., with copper sulfate forms cuprous hydride Cu_2H_2, brown precipitate, which evolves hydrogen gas and leaves copper on warming; with silver nitrate yields finely divided silver; with sulfurous acid yields sulfur and some hydrogen sulfide; with sulfuric acid yields sulfurous acid, which reacts as above; forms manganous immediately with permanganate.

Hypophosphorous acid is formed by reaction of barium hypophosphite and sulfuric acid, and filtering off barium sulfate. By evaporation of the solution in vacuum at 80°C, and then cooling to 0°C, hypophosphorous acid crystallizes.

Sodium hypophosphite $NaPO_2H_2$, the only sodium hypophosphite, is formed (1) by reaction of yellow phosphorus and sodium hydroxide solution (phosphine simultaneously formed), (2) by reaction of hypophosphorous acid and sodium hydroxide, and evaporating. Sodium hypophosphite, upon heating, yields sodium phosphate and sodium phosphide. Common tests for the hypophosphites are as follows:

1) Zinc reduces dilute sulfuric acid solution of hypophosphites to phosphine recognizable by odor (difference from phosphates).

2) Barium chloride produces no precipitate (difference from phosphites). See also **Phosphorus.**

HYPOSULFUROUS ACID AND HYPOSULFITES. Hyposulfurous acid $H_2S_2O_4$ is a yellow solution rapidly oxidized in air to sulfurous acid and then to sulfuric acid. Commercially known as hydrosulfurous acid and its salts as hydrosulfites (but not to be confused with "hypo" which is sodium thiosulfate).

Hyposulfurous acid is a powerful reducing agent, e.g., with copper sulfate forms cuprous hydride Cu_2H_2, brown precipitate, which evolves hydrogen gas and leaves copper on warming, with silver nitrate yields finely divided silver, with permanganate yields manganous compounds. Hyposulfurous acid is formed by reaction of sodium hyposulfite and an acid.

Sodium hyposulfite, sodium hydrosulfite $Na_2S_2O_4 \cdot 2H_2O$ is formed (1) by reaction of zinc and sulfurous acid (or sodium hydrogen sulfite), yielding zinc hyposulfite and then converted by sodium chloride into sodium hyposulfite, (2) by electrolysis of sodium hydrogen sulfite and then addition of sodium chloride.

Sodium hyposulfite is used to bleach sugar, indigo, wood pulp. With moist hydrogen sulfide, sulfur is precipitated and sodium thiosulfate simultaneously formed.

I

IDEAL GAS LAW. An ideal gas would, if kept at a constant temperature, behave with respect to volume and pressure in strict accord with Boyle's law. If the temperature is also allowed to vary, we must combine the law of Charles (or of Gay Lussac) with Boyle's law, yielding the Boyle-Charles law:

$$pv = p_0 v_0 (1 + at), \qquad (1)$$

in which $p_0 v_0$ is the value of the pressure-volume product pv when the temperature t is zero, a is the coefficient of expansion of the gas, practically the same for all gases, and in the ideal case equal to the reciprocal of the absolute temperature of the scale zero. If the centigrade scale is used, the value of a is approximately $1/273.2$ per degree. Substituting this, Equation (1) may be written

$$pv = \frac{p_0 v_0}{273.2°} (t + 273.2°) \qquad (2)$$

which is one expression for the ideal gas law.

The factor $t + 273.2°$ will be recognized as the absolute temperature T of the gas. And since the ideal gas obeys Boyle's law, the product $p_0 v_0$ is constant however p_0 and v_0 may vary between themselves. We may thus denote the coefficient $p_0 v_0 / 273.2°$ by a single constant symbol, say R, and the ideal gas equation then takes the usual form

$$pv = RT \qquad (3)$$

The value of R depends, of course, upon the quantity of gas used, since at any pressure p_0 it is proportional to the volume v_0. For 1 gram of air, R equals about 2,868,000 g cm^2/sec^2 deg. At the zero of temperature and at any given pressure p_0, the gram molecular weights, or moles, of all pure gases have equal volumes. (This follows from Avogadro's law.) Hence if one mole of any pure gas is used, R will always have the same value, in c.g.s. units about 8.316×10^7 g cm^2/sec^2 deg; which is called the "ideal gas constant." Many physical formulas involve a quantity which may be regarded as the ideal gas constant per molecule, that is, the above molar gas constant divided by the number of molecules in a mole, 6.025×10^{23}, giving 1.3803×10^{-16} g cm^2/sec^2 deg. This is the "Boltzmann constant."

Since actual gases, even those with the smallest molecules, hydrogen and helium, do not obey the ideal gas law exactly, various empirical characteristic equations have been devised to represent their behavior.

ILMENITE. A mineral oxide of iron and titanium, $FeTiO_3$. Magnesium and manganous manganese may replace ferrous iron to form a complete isomorphous series between ilmenite, and its magnesium-manganese end members, geikielite and pyrophanite. It crystallizes in the rhombohedral division of the hexagonal system; hardness, 5–6; specific gravity, 4.72; brittle, with uneven to conchoidal fracture. Crystals tabular, rarely rhombohedral, also massive, lamellar, granular. Color, iron black; opaque, with metallic to dull luster.

Ilmenite occurs as a common accessory mineral in both igneous and metamorphic rocks, and as heavy concentrations in certain black beach sands with magnetite, rutile, and zircon. Also found in pegmatites and as vein deposits. Valuable deposits are found in Norway, Sweden, Mexico, Finland, Ilmen Mountains, U.S.S.R., Canada, England, Brazil, and Italy. Brazil and India are rich in beach sand deposits. United States localities include California, Idaho, Colorado, Wyoming, Arkansas, Kentucky, Pennsylvania, Massachusetts, Connecticut, Orange County and the Adirondack Mountain Deposits in New York, and as beach sands in Florida north of St. Augustine.

Named after the Ilmen Mountains, U.S.S.R.

IMIDES. An imide may be defined as a compound that has the divalent radical NH combined with two acid radicals. The definition implies that the acid from which an imide is derived must be dibasic, such as oxalic acid $HOOC \cdot COOH$ or succinic acid $HOOC \cdot CH_2 \cdot CH_2 \cdot COOH$. The derivatives of these two acids illustrate the relationship between amides and imides:

COOH	CO·NH₂	CO·NH₂	CO
|	|	|	| NH
COOH	CO·NH₂	CO·OH	CO
(oxalic acid)	(oxamide)	(oxamic acid)	(oximide)

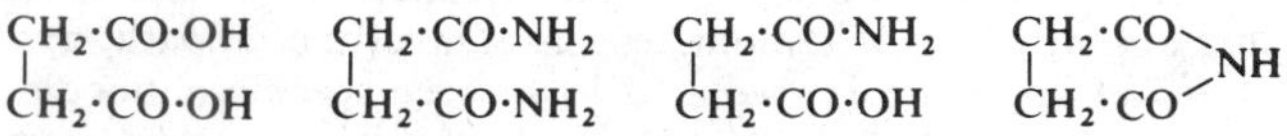

CH₂·CO·OH	CH₂·CO·NH₂	CH₂·CO·NH₂	CH₂·CO
|	|	|	| NH
CH₂·CO·OH	CH₂·CO·NH₂	CH₂·CO·OH	CH₂·CO
(succinic acid)	(succinamide)	(succinamic acid)	(succinimide)

Phthalimide $C_6H_4 \cdot (CO_2 \cdot NH$ is an imide of commercial and industrial importance, forming a number of interesting derivatives. With alcoholic potash, phthalimide forms a potassium derivative $C_6H_4 \cdot (CO)_2 \cdot NK$ which, when reacted with ethyl iodide (or other alkyl halides), yields ethylphthalimide $C_6H_4 \cdot (CO)_2 \cdot N \cdot C_2H_5$. The latter product, when hydrolyzed with an acid or alkali, further yields ethylamine. Such reaction chains are useful in the preparation of certain primary amines and their derivatives.

IMINO COMPOUNDS. Imino compounds are organic compounds containing the imino group $> NH$, e.g., dimethylamine $(CH_3)_2NH$, dibenzamide $(C_6H_5CO)_2NH$, succinimide

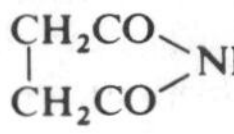

$$\begin{matrix} CH_2CO \\ | \quad\quad\;\; NH \\ CH_2CO \end{matrix}, \text{ pyrrole } (C_4H_4NH), \text{ uric acid}$$

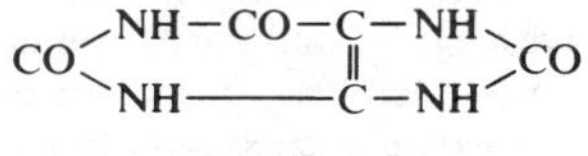

$$\begin{matrix} CO & \overset{NH-CO-C-NH}{\underset{NH\quad\quad\;\; C-NH}{}} & CO \end{matrix}$$

IMMUNE SYSTEM AND IMMUNOCHEMISTRY. Immunity is from the Latin *immunis* (*free of*). The term originally referred to the ability of the body to resist invasion by pathogenic organisms, but has now been expanded to include specific reactions to antigens (Ags) in general and to include reactions observed in the emerging field of tumor immunology. See **Antigen.**

Immunity is derived from the *immune system*, whose function is the preservation of the body's integrity against antigens, which are agents recognized by the host as foreign, e.g., surface structures of microorganisms, tissue transplants, or a wide variety of chemicals. Specifically, antigens include such structurally diverse substances as proteins, polysaccharides, nucleic acids, and lipids. Large rigid proteins are the most antigenic, and the more insoluble the foreign material, the more antigenic it appears to be. Thymus-dependent antigens are those in which antibody production requires thymus-derived cell participation, i.e., serum proteins. Thymus-independent antigens do not require this participation, i.e., polysaccharides, such as endotoxin lipopolysaccharides.

Antigens invoke immune responses by the host which include production of *antibodies* (Abs) possessing a specificity for the antigen which is determined by the latter's structure. See **Antibodies.** Antibodies belong to a group of serum proteins called *immunoglobulins* (Igs) which bind with the antigens to form complexes in which the two components are held together by weak hydrogen bonds, Van der Waals forces or ionic bonds, but not by covalent bonding. An antibody which binds a given antigen will also bind antigens having similar structural configurations. This is referred to as *cross-reactivity*. The extent to which this binding occurs indicates the measure of the similarity of the two antigens. Most antigens have several antigenic determinants or Ab binding sites and the antibody response to any antigen is thus the sum of responses to each individual determinant. In addition to antigens *per se*, two other types of substances are recognized by the immune system; (1) *haptens* which are molecules capable of reacting with antibodies, but which are not able to stimulate their production unless coupled to a carrier; and (2) *adjuvants*, which enhance the immune response to an antigen.

The immune response to antigens can proceed through several paths, the two major categories of which are *humoral* and *cellular*. In general, most immune responses involve both pathways. Humoral immunity is mediated by antigen-specific antibodies which circulate through the body and act at a site distant from that of their production. Humoral immunity can be transferred from one person to another by serum. Cell-mediated immunity is mediated by specifically sensitized cells which release mediators in the vicinity of the antigen. This form of immunity can be transferred from one person to another by cell transfer. Cell-mediated immune responses usually take longer to occur and are responsible for resistance to many infectious agents and tumors as well as for some drug allergies, rejection of foreign grafts, and some autoimmune diseases.

The level of immune response to antigens varies. A *primary response* is seen to antigens which the body has never before encountered. In this, the first antibody to be produced is IgM, described a bit later, the serum concentration of which peaks after a lag of several days and then decreases. IgG production shows a longer lag time, but its concentration remains elevated for a more extended period. *Secondary responses* are more rapid and of greater intensity; they differ markedly from primary responses in having a shorter lag period, higher antibody level with earlier and more pronounced emphasis on IgG production. These secondary responses occur when the immune system faces a previously encountered antigen and illustrates a basic characteristic of immune response—*memory*. For this reason, secondary responses are also known as *anamnestic*. These memorized responses are derived from two of the major cell types which work together to generate immune responses—*T* and *B* lymphocytes. The former are dependent upon the thymus and are responsible for cell-mediated immune responses and for providing help for most antibody responses. *B* cells depend upon another central lymphoid organ (in birds, this has been identified as the Bursa of Fabricus). When stimulated by antigens, *B* cells become *plasma cells* and produce antibodies specific for that antigen. *T* and *B* cells cannot be distinguished morphologically so, in order to differentiate them, one must use surface markers, e.g., theta antigen on *T* cells and surface immunoglobulin on *B* cells.

The spleen, lymph nodes, tonsils, and gut-associated lymphoid tissue make up the secondary or peripheral system where *T* and *B* lymphocytes undergo terminal differentiation in response to antigen stimulation.

The third major type of cell involved in the immune process is the *macrophage*. The *B* and *T* lymphocytes are rarely phagocytic—this is the function of the macrophage. After specific recognition of the invading antigen by the lymphocytes, the macrophage acts nonspecifically and migrates by *chemotaxis* (migration along an increasing chemical concentration gradient) to the site of the immune response; the macrophage ingests and eventually digests the antigenic inert particles or living or dead microorganisms responsible for engendering the immune response.

It is possible to divide these major cell types into subpopulations which interact by means of soluble mediators called *lymphokines* and *monokines*. Lymphokines are products of activated lymphocytes which exert regulatory effects upon other cells of the immune system. Monokines are products of activated macrophages. The soluble mediators help activities such as suppression or cytotoxicity to be manifested by the target cells.

A given antigen induces proliferation and differentiation of clones of cells capable of producing antibodies in response to that antigen—a process called *clonal selection*. Each stimulated cell produces antibodies of only one specificity. Thus, the immense heterogenicity of antibodies to a given antigen results from a great diversity of responsive cells.

As previously indicated, antibodies are immunoglobulins (Igs) produced by *B* cell-derived plasma cells during a humoral response to antigen. They are synthesized on polyribosomes attached to the rough endoplasmic reticulum of plasma cells upon inoculation of antigen into the host and are specific for that antigen. They have two functions—specific recognition of the antigen, and effector functions, such as agglutination or lysis of bacteria, complement fixation, or opsonization.

Immunoglobulins are high-molecular-weight glycoproteins having symmetrical, four-polypeptide-chain structures composed of two heavy and two light chains held in configuration by disulfide linkages. Based upon serological characteristics of the heavy chains, five distinct classes or isotypes have been found—IgG, IgA, IgM, IgD, and IgE. Subclasses of these immunoglobulins have also been found. Bonds between chains—heavy-heavy or heavy-light—are by disulfide bridges, the number and positions of these bridges being characteristic of different classes and subclasses of immunoglobulins. See Fig. 1. In a given immunoglobulin molecule, both heavy chains are identical and, although there are two types of light chain—kappa and lambda—each of which can be associated with any heavy chain, both the light chains in a given immunoglobulin molecule are identical. Polymerization of the basic immunoglobulin configurations can occur, but IgG, IgD, and IgE occur only as monomers. IgA may be either monomeric or dimeric; IgM is pentameric. See Fig. 2.

In addition to the basic configurations of the immunoglobulin molecule, the light and heavy chains have variable and constant regions where more or less variability in amino acid sequences occurs. The variable regions of the antibody molecule contain the structures responsible for the antigenic specificity of the immunoglobulin. These are found at the amino terminal of the

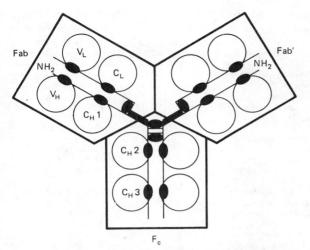

Fig. 1. Hypothetical model of human IgG along the lines proposed by Poljak and others. Note heavy solid bars which indicate hinge region (at center of configuration). Solid ellipses indicate S—S linkages.

heavy and light polypeptide chains—the F_{ab} portion obtained as cleavage fragment following treatment with pepsin or papain. Since each antigen binding region is composed of the variable region of one light and one heavy chain, an immunoglobulin

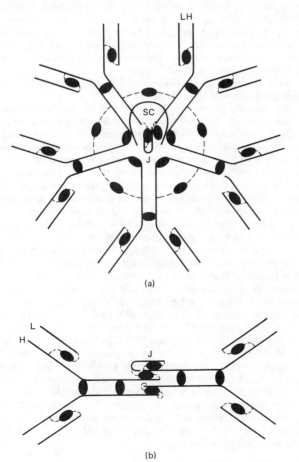

Fig. 2. Hypothetical models utilizing the clasp configuration proposed by Koshland et al. (1975). (a) pentameric IgM; (b) dimeric IgA. L = light chain; H = heavy chain; J = J chain. Solid ellipses indicate S—S linkages.

molecule has two identical antigen binding sites. The antigen binding site of every immunoglobulin molecule is structurally unique. Therefore, each immunoglobulin has unique antigenic determinants not shared by other immunoglobulin molecules. These unique determinants are called the *idiotypes* of the particular immunoglobulin. In contrast, an *allotype* is an antigenic specificity representing any one of a number of possible allelic markers present on a significant percentage of immunoglobulin molecules in an individual which are recognizable as foreign by another individual lacking that allotype.

The F_c portion of the immunoglobulin molecule identified by papain cleavage is responsible for biological activity other than antigen binding, i.e., complement fixation, transplacental transfer, binding to cells such as macrophages and granulocytes, and the rate of synthesis and catabolism of the immunoglobulin molecule.

IgG and IgM are present in the highest concentrations throughout the body and compose the major portion of a systemic humoral response to antigenic challenge. IgM exists as a pentamer of five immunoglobulin molecules bound together by a protein molecule called a *J* chain. IgM has a molecular weight of approximately 900,000 and a sedimentation coefficient of 19S and is thus referred to as 19S Ig. It is the main immunoglobulin of early humoral response, it will fix complement, and it is not cytophilic. IgG is the major component of secondary humoral response, and its ability to diffuse through tissues makes it indispensable for host defense. It has a molecular weight of 150,000 and a sedimentation coefficient of 7S. Eighty percent of serum immunoglobulin is IgG and several subclasses of this immunoglobulin exist which have different functions, i.e., complement fixation, binding to macrophages, or passage across the placenta.

IgA is the primary secretory immunoglobulin and occurs in tears, nasal and intestinal secretions, saliva, and bile. It is present as a 7S monomer in serum, where it occurs in low concentration. As secretory IgA, it is an 11S dimer with the monomeric structures joined by a *J* chain and linked to a glycoprotein called *secretory component*. Secretory IgA plays a role in initial protection against external pathogens, aggregates bacteria, and neutralizes viruses.

IgE, with a molecular weight of 190,000, has a short half-life and occurs at low concentrations in the serum. It is cytophilic and binds strongly to most cells and basophils. Plasma cells producing IgE are primarily found in gastrointestinal mucosa and along the respiratory tract. This immunoglobulin is responsible for *immediate hypersensitivity* reactions, since crosslinking by antigens of two IgE molecules causes the release of mediators, such as histamine, SRS-A (Slow Reacting Substance of Anaphylaxis), and ECF-A (Eosinophil Chemotactic Factor) which are responsible for the immediate hypersensitivity reactions. Hence, IgE plays a definitive role in allergies and may have a part in providing resistance to parasites and in protection of mucosal surfaces.

IgD occurs in extremely low concentrations in serum and has a short serum half-life. It has a molecular weight of 170,000–200,000 and is monomeric and fixes complement. Otherwise its role is unknown, although it may act as a cell surface receptor on *B* lymphocytes.

Associated with humoral immune responses involving antibodies is *complement*. The *complement system* is a complex series of proteins circulating in inactive form in the extracellular fluid. Activation of complement involves a series of steps, each one generating a new active component that, in turn, activates the next component, and so on. There are two major pathways of complement activation—the classical pathway and the alternate or properdin pathway. The classical pathway is initiated by the binding of antigen to antibody (two molecules of IgG or one molecule of IgM). The alternate pathway does not require the presence of antigen-specific antibody, but can be activated by endotoxin, bacterial cell wall polysaccharides, and aggregated immunoglobulins. The ability to be initiated before an immune response can occur makes the alternate pathway a first line of defense against microorganisms. Although the ultimate result of complement activation is cell lysis, mediators are produced during the activation sequence which play major roles by allowing *immunoadherence*, *opsonization*, or *chemotaxis* to occur, others such as *anaphylatoxins* may be harmful to the host. The importance of complement to normal host defense is illustrated by many pathologic conditions and high susceptibility to infections seen in persons with congenital complement deficiencies.

References

Bach, J. F., Editor: "Immunology," Wiley, New York (1978).

Bach, F. H., Bonavida, B., Vitetta, E. S., and C. F. Fox, Editors: "*T* and *B* Lymphocytes: Recognition and Function," Academic, New York (1979).

Pernis, B., and H. J. Vogel, Editors: "Cells of Immunoglobulin Synthesis," Academic, New York (1979).

Rose, N. R., Milgrom, F., and C. J. van Oss, Editors: "Principles of Immunology," 2nd Edition, Macmillan, New York (1979).

Thaler, M. S., Klausner, R. D., and H. J. Cohen: "Medical Immunology," Lippincott, Philadelphia (1977).

—Ann C. Vickery, Assistant Professor, Department of Comprehensive Medicine, Division of Tropical Health, College of Medicine, University of South Florida, Tampa, Florida.

INDICATOR (Chemical). A substance which shows by a color change, or other visible manifestation, some change in, or particular condition of the chemical nature of a system. Thus acid-base indicators may be used to indicate the end point of a particular neutralization reaction, or they may be used to indicate the pH value of a system. There are over 50 useful indicators for determining pH, covering the full range from 0 to 14. See accompanying table. Although indicators still are used in connection with colorimetric pH determinations, they are not nearly so important as they were several decades ago prior to the availability of electronic methods. Indicators also are useful in following oxidation-reduction reactions, precipitation reactions and, in general, throughout most volumetric analyses.

INDIUM. Chemical element symbol In, at. no. 49, at. wt. 114.82, periodic table group 3a, mp 156.6°C, bp 2078–2082°C, density, 7.31 g/cm³ (20°C). Elemental indium has a face-centered tetragonal crystal structure. Indium is a silvery-white metal, softer than lead, malleable, ductile, and crystalline. It is stable in dry air, but upon heating in air burns with a blue flame to form indium trioxide, In_2O_3. Up to a temperature of 100°C, the element does not decompose water. Indium becomes a superconductor at 3.37 K. The element dissolves in HCl, H_2SO_4, or HNO_3, but not in NaOH. Metallic indium combines readily with chlorine and sulfur. The isotope ^{113}In is the only nonradioactive isotope and is in isotopic abundance of 4.23%. Isotope ^{115}In, with an extremely long half-life of 6×10^{14} years, accounts for the other 95.77% of naturally occurring indium. Other radioactive isotopes include ^{107}In through ^{112}In, ^{114}In, ^{116}In, and ^{117}In. The half-lives of these isotopes are relatively short, measured in minutes, hours, or days. First ionization potential, 5.785 eV; second, 18.86 eV; third, 28.03 eV. Oxidation potentials: In → In^{3+} + $3e^-$, 0.34 V; In → In^+ + e^-, −0.25

pH RANGES AND COLOR CHANGES OF SELECTED INDICATORS

INDICATOR	pH RANGE OF COLOR CHANGE	COLOR CHANGE WITH INCREASING pH
Alpha naphtholbenzein	0– 0.8	Colorless to yellow
Methyl violet	0.2– 1.9	Yellow to blue-violet
Para methyl red	1.0– 3.0	Red to yellow
Thymolsulfonphthalein (Thymol Blue)	1.2– 2.8	Red to yellow
	8.0– 9.6	Yellow to blue
Methyl orange	3.3– 4.5	Red to yellow
Methyl red	4.2– 6.2	Red to yellow
Aurin (rosolic acid)	6.2– 7.2	Amber to pink
Phenolsulfonphthalein (Phenol Red)	6.8– 8.5	Yellow to red
Phenolphthalein	8.3–10.2	Colorless to purple
Thymolphthalein	9.4–10.7	Colorless to blue
Sodium nitrobenzeneazo-salicylate (Alizarin Yellow R)	10.1–12.0	Yellow to red
Malachite green	11.4–13.0	Blue-green to colorless
1,3,5-Trinitrobenzene	12.0–14.0	Colorless to orange

V. Other important physical properties of indium are given under **Chemical Elements.**

Indium occurs in very small amount in zinc blende, tungsten, tin and iron ores of certain localities. The recovery of indium from zinc flue dust (sometimes, 1 part per thousand) is effected by treating with a slight deficiency of HCl and allowing to stand. The residue is subjected to a series of treatments until finally pure indium sulfate is obtained, a solution of which when electrolyzed yields compact indium metal. A thin surface layer of indium is used on some bearings.

As the result of spectroscopic studies, the element was discovered by Reich and Richter in 1863.

On the scale of nonferrous metals, the production of indium is quite limited, annual production probably not exceeding 1.3 million troy ounces (40.4 million grams). The availability of the element is affected by zinc production because it is a minor coproduct in the refining of zinc ores. The metal, in the form of an electroplate over lead and silver, has been used in aircraft bearings, the primary benefit being improved corrosion resistance. Indium also has been used as a dopant for germanium diodes and transistors. Several significant semiconductor compounds have been formulated, including InAs, InSb, and InP. The oxide has been used in electroluminescent panels. Indium alloys readily with several metals and it has been found particularly effective as a low-melting point fusible alloy when alloyed with bismuth, lead, tin, and cadmium. The eutectic alloy of indium-tin is an effective solder for glass-to-glass or glass-to-metal seals. With a melting range of 700–800°C, copper-gold-indium and copper-silver-indium alloys are used as brazing materials. The eutectic alloy of mercury, thallium, and indium has a solidifying temperature of −63°C, considerably below the mp of mercury, a feature which makes the alloy attractive for seals, switches, and thermometers for low-temperature applications. Control rods for nuclear reactors sometimes are produced from an alloy of silver, indium, and cadmium. It is also used in the manufacture of low-pressure sodium lamps. Indium for electroplating generally is furnished as the normal sulfate, $In_2(SO_4)_3 \cdot H_2O$, the acid salt, $In_2(SO_4)_3 \cdot H_2SO_4$, or the basic salt, $In_2O(SO_4)_2 \cdot 6 H_2O$.

Chemistry and Compounds

Since indium has only three electrons in its valence shell, it is an electron acceptor. Indium trihalides include the trifluoride, trichloride, tribromide, and triiodide. They can be prepared by heating the metal or oxide in the halogen acid, or in the case of the trichloride and tribromide, by use of the halogen acid, or in the case of the trichloride and tribromide, by use of the halogen itself. Indium sesquisulfate forms double salts like the alums with alkali metal sulfates. A monohydrogen sulfate, $HIn(SO_4)_2 3\frac{1}{2}H_2O$, is known. Other compounds of indium(III) include the oxide (and its gelatinous hydrate), nitride, the nitrate, and the sulfide, selenide, and telluride.

Indium(II) compounds include the oxide, sulfide, fluoride, and chloride. They are prepared either by reduction of the corresponding trivalent compounds or, in the case of the chloride, by heating the metal in hydrogen chloride. They disproportionate, under suitable conditions, to give as end products the metal and the stable trivalent compound. Like gallium, indium(II) chloride is diamagnetic, having the structure $In^+[InCl_4]^-$.

Indium(I) compounds are formed by reduction of the corresponding In(III) compounds with hydrogen (on heating) or with indium metal, as in the case of the chloride. They are reactive compounds, the chloride disproportionating with water to form the metal and $InCl_3$; the oxide being oxidized on heating in air to the sesquioxide, and the sulfide reacting in dilute acids to form the sesquisulfide.

A number of indium trialkyls have been prepared, starting from the trimethyl, and some diaryl compounds are known, such as the diphenyl bromide. The lower trialkyls are tetramers. Like aluminum, indium forms a polymeric hydride, $(InH_3)_n$, from which tetrahydroindates, such as the lithium compound, $LiInH_4$, can be derived.

Like aluminum and gallium, indium forms a number of chelated oxycompounds, almost all of which are of 6-coordinate type. They include the stable crystalline inner complexes of which the β-diketones coordinate in the proportion of 3 molecules of diketone per atom of indium. Trioxalato as well as dioxalato salts are known, and compounds such as 8-quinolinol and substituted 8-quinolinols form trimolecular chelate rings involving nitrogen and donor oxygen.

References

Carapella, S. C., Jr.: "Indium" in "Metals Handbook," Vol. 2, 9th Edition, 739–740, American Society for Metals, Metals Park, Ohio (1979).
Hampel, C. A., Editor: "Rare Metals Handbook," 2nd Edition, Van Nostrand Reinhold, New York (1961).
Ludwick, M. T.: "Indium Corporation of America (1959, out of print).

—S. C. Carapella, Jr., ASARCO Incorporated, South Plainfield, New Jersey.

INDOLES. These are weakly basic heterocyclic nitrogeneous compounds in which a benzene ring and a pyrrole nucleus are fused in the 2,3 positions of the pyrrole ring. The structure of indole was first elucidated by Adolf von Bayer while carrying out investigations on indigo.

Indole is a colorless crystalline solid, mp 52°C, bp 254°C, soluble in hot H_2O, stable under neutral and basic conditions, but polymerizes appreciably in acids. Chlorine and bromine react violently with indole, forming tars. Indole is a chemical reagent in various syntheses and has been produced for use in the manufacture of tryptophan and indole-3-acetic acid. Perhaps the most broadly applicable synthesis of indole derivatives is the Fischer indole synthesis, in which aryl hydrazones of aldehydes, ketones, aldehydic acids, keto acids, or esters of such acids are converted into indole derivatives by heating with an acid catalyst. Substituted indoles are prepared by applying the Reissert synthesis to substituted ortho-nitrotoluenes. Indole derivatives include skatole (3-methylindole); serotonin (5-hydroxy-tryptamine), a human hormone; reserpine, an indole-containing alkaloid; and a number of hallucinogenic drugs, including lysergic acid diethylamide (LSD), psilocybin (orthophosphoryl-4-hydroxy-N,N-dimethyltryptamine), and lysergic acid, which occurs esterified with alkaloids in the rye mold ergot. Indole chemistry owes much of its development to earlier interests in indigo and allied dye chemistry.

INERT GASES (The). The elements of group 0 of the periodic classification sometimes are referred to as the *inert gases*, or the *noble gases*. In order of increasing atomic number, they are helium, neon, argon, krypton, xenon, and radon. The elements of this group are characterized by their closed shells or subshells of electrons. They generally are considered as having zero valence. The name of the group derives from the lack of chemical activity which the elements display, forming compounds only under abnormal conditions (high pressures, strong electrical fields, etc.). See specific articles on each of the aforementioned elements.

INOSITOL. A constituent of body tissue. In purified form it is used as a nutrient and dietary supplement in some foods and feedstuffs. The chemical name of inositol is hexahydroxycyclohexane, $C_6H_6(OH)_6 \cdot 2H_2O$. There are nine isomeric forms of inositol. Myoinositol or meso-inositol (*cis*-1,2,3,5-*trans*-4,6-hexahydroxycyclohexane) is the isomer that possesses essential nutrient activity. The substances, often identified as a vitamin, is found in small amounts in many vegetables, citrus fruits, cereal grains, liver, kidney, heart and other meat. The commercial source is corn (maize) steep liquor. In addition to its use in nutrition, inositol finds use in medicine and as an intermediate for organic syntheses.

INSECTICIDE. A substance that kills or interferes in the life cycle of certain insects and is thus useful for reducing and controlling insect populations. The reduction or elimination of such populations is desirable for several reasons, including: (1) Preventing the spread of certain diseases by insects which serve as transmitters or carriers of infective organisms. Thus, insecticides are used widely as a public health measure. (2) Preventing or reducing the damage caused by insects eating and habitating food plants, trees, and other crops. Such damage, without control, sometimes approaches 100% and averages 25% or more in some regions of the world. (3) Preventing or reducing physical property damage, notably of wood, cloth, and other materials of an organic nature that attracts certain insects as a source of food and place of habitation. Termites and moths are examples. (4) Reducing or eliminating discomfort, annoyance, and sometimes injury that results from the immediate presence of insects on or near people and domestic and farm animals. Ticks and face flies are examples.

Classification of Insecticides

Insecticides may be classified and characterized in many ways: (1) By their *selectivity*, i.e., their ability to control or not to control different forms, varieties, and species of insect, ranging from compounds which have a rather *narrow control spectrum*, which enable the user to eradicate or reduce selected target insects—all the way to the very *wide spectrum* insecticides that will destroy practically all insects, including those of a beneficial nature that are of positive economic importance. (2) By the manner in which insecticides *interact* with the insect enemy or target, i.e., (a) whether or not a chemical requires overwhelming contact with the insect; or where the chemical acts systemically within and throughout the insect once local contact is made; (b) how the chemical interferes with the life process of the insect as, for example, is it a stomach poison? or does it interfere with the insect's nervous or respiratory system, etc.? (c) how the insecticide enters the body of the insect, i.e., via the alimentary or respiratory system, etc.; and (d) whether or not the primary function of the chemical is to kill the insect, or essentially to sterilize the insect sexually and thus reduce insect population in this manner. (3) By application—is the insecticide a solid, liquid, or gas? Can it be sprayed, dusted, or incorporated into bait? Can it be worked into the soil? Is it miscible with oil, water, or both? Can it be applied by aircraft? (4) By the useful life or persistence of the chemical, i.e., a few days, weeks, or months. (5) In terms of safety to humans and livestock and pets. Insecticide dangers range from highly toxic, to moderate, light, and low toxicity. Quite important, is the insecticide harmful to bees? fishes?, other wildlife?—or to adjacent crops and orchards should a spray or dust of the material drift away in the wind from the immediate point of application? (6) By the chemical structure of the insecticide, i.e., chlorinated hydrocarbon, organic phosphate, etc., and how produced—extracted from natural bacterial and biological materials? or synthesized from basic raw materials and chemical intermediates? (7) By cost—a very important and practical consideration for large food producers, health authorities, and other users.

Selectivity and Spectral Range of Insecticides. There is no practical universal insecticide because such a chemical, to successfully eradicate all forms of insect life, would be dangerous to other life forms. There are, however, multipurpose pesticides, such as hydrogen cyanide gas, which are used *most carefully and with the ultimate of safety provisions* to kill not only all insects within a given area or space, but all other animal life forms as well. Some insects are much easier to destroy than others. Thus, numerous compounds are available to control aphid, fly, leafhopper, and thrip, whereas only a few chemicals are effective against fire ant, and certain beetle and weevil species.

Action of Insecticides. The average user of insecticides is essentially interested in the results of insecticide application rather than the exact manner in which these chemicals act upon the life and life cycle of the insect. But there are some important distinctions which govern both the timing and selection of a given insecticide. For example, some chemicals are much more effective in destroying larvae or nymphs than when applied to insects in the adult stage, or vice versa. Also, some perennial crop plants and orchards are undamaged by certain chemicals (dormant sprays, for example) during the winter inactivity, whereas the same chemicals would cause severe damage if applied during springtime budding. Also, the food producer, by taking advantage of multifunction compounds, can reduce the number of control chemicals required and hence the number of applications.

Control chemical manufacturers frequently offer combined formulations. In addition to blending acaricides with insecticides, molluskicides, etc., it is not uncommon to blend in herbicidal compounds, but usually as the blending becomes more complex, the selection becomes more difficult—because great care must be taken to assure that (1) the chemicals will mix well without destroying any of their intended effectiveness (many control chemicals are weakened or destroyed, for example, if mixed with strongly alkaline materials, such as lime, or with sulfur-bearing compounds); (2) the mixture is truly customized to the crop at hand and that the combination is not used on other crops without specifically checking; and (3) there will be no unexpected environmental damage. A combination of chemicals may be entirely effective and safe when applied to a specific crop or area, whereas it may be inappropriate if used on other crops and other areas that may be adjacent to pastures and streams or lakes and thus be dangerously polluting. For crops of large economic value and commonly grown, manufacturers offer numerous crop-specific formulations.

Nomenclature of Insecticides. It is estimated that there are well over 100,000 pesticide formulations and perhaps half of these would be basically classified as insecticides. Although the basic chemicals used may number in the hundreds, variations arise from the many thousands of possible formulations, not only combinations of materials, but formats (sprays, wettable powders, emulsifiable concentrates, dusts, baits, etc.). Added to this are the scores of control chemical manufacturers worldwide. Each manufacturer markets products under tradenames—names that are essentially coined for their marketing charisma and infrequently connoting the content or purpose of the product. Thus, there are scores of equivalent (or essentially equivalent) products, adding to the difficulty of selecting these chemicals. Unfortunately, the generic chemical names of the majority of insecticide chemicals are long and complex and essentially meaningless to persons who are not well versed in organic and biochemistry. Helpful listings of this type can be found in the "Foods and Food Production Encyclopedia" (D. M. Considine, Editor), Van Nostrand Reinhold, New York (1982). There are also a number of frequently revised directories of control chemicals, and considerable information is available from various government agencies and universities. See the list of references at the end of this entry. This situation of nomenclature is quite similar to that which applies to generic and tradename drugs and pharmaceuticals.

Toxicity of Insecticides. Even though insecticides and other pesticides vary greatly in toxicity, all may be considered hazardous if they are not handled properly and with precautions. The following elementary rules are always worthy of repetition: (1) Observe all directions, restrictions, and precautions on pesticide labels. It is dangerous, wasteful, and, in some regions and countries, illegal to do otherwise. (2) Store all pesticides behind locked doors in the original containers, with labels intact. (3) Use pesticides at correct dosage and intervals to avoid excessive residues and injury to plants and animals. (4) Apply pesticides carefully to avoid drifting of the compound to nearby fields, lakes, and streams. (5) Bury surplus pesticides and destroy used containers so that contamination of water and other hazards will not result. (6) Certain pesticides must not be used during a specified period just prior to harvest because of the danger to workers and pickers who will be handling the product and because of the danger of inadequately washing away or otherwise removing residues prior to releasing the commodity to the consumer. (7) Do not mix two or more compounds without prior knowledge of their compatibility. (8) When in doubt concerning the applicability of a given compound to a given situation, such as a specific crop, seek advice from local sources of expertise, such as extension service representatives, reliable suppliers, neighboring users who have faced similar problems, regional colleges and universities, and agricultural experiment stations. Above all, the final responsibility for safe usage rests with the person who ultimately applies the chemicals.

In the United States, when registering insecticides, pesticides, and other control chemicals, regulatory agencies have been using acute LD_{50} values to determine the toxicity category and the words or symbols that must be placed on labels and containers. For this purpose, the test animals usually are rats, mice, or rabbits, but other mammals are sometimes used. The LD_{50} value is the dosage of the chemical at which 50% of the test animals are killed. It is based on the body weight of the animal and is expressed in milligrams of the chemical per kilogram of animal (mg/kg). One mg/kg = 1 part per million (ppm). Thus, the lower the LD_{50} value, the higher the toxicity of the chemical. The usual way of administering chemicals to test animals is by mouth, application to the skin, and in some cases, by inhalation. Toxicity may be either acute or chronic. Acute refers to rather quick action from a single exposure, whereas chronic refers to the toxic effect of many exposures over a period of time.

Chemistry of Insecticides

In classifying compounds used as insecticides, from the standpoint of chemical characteristics and structure, the most fundamental division separates the inorganic from the organic chemicals. The latter are, by far, the most widely used. Acaricides, bactericides, fungicides, and nematocides, along with insecticides, are included in the following descriptions.

Inorganic Compounds. These control chemicals are comparatively simple compounds and include calcium and lead arsenates, elementary sulfur and inorganic sulfur compounds, such as calcium polysulfide (lime-sulfur) and sodium thiosulfate. Because of their effectiveness against certain fungus infections, these compounds are more frequently considered as fungicides than as insecticides. This is also true of a number of copper, zinc, and other metal inorganics, such as copper carbonate, copper oxychloride, copper sulfate, and copper zinc sulfate. It should be pointed out that Paris Green (cupric acetoarsenite), one of the older and once widely used inorganic insecticides, essentially has been phased out in most regions of the world because of the effects of chronic arsenic poisoning of workers and users who come in contact with the substance. This was particularly true in the case of vineyard workers a number of years ago.

A number of other metals, such as iron and tin, enter into insecticide and pesticide compounds, but as part of an organic chemical structure, as exemplified by triphenyltin hydroxide. Such compounds are sometimes referred to as organometallics (or, specifically in the case of tin, as organotins). Mercury compounds are rapidly being phased out because of their long-term toxic residual effects as pollutants, particularly of fresh and saline waters. Regulations vary from one country to another.

A few other inorganic chemicals, more frequently identified as multipurpose pesticides than as insecticides, do have very strong insecticidal properties. These compounds are used sparingly, often requiring special permits in some places, and with the greatest of safety cautions observed. Such compounds would include calcium cyanide, carbon bisulfide, carbon tetrachloride, hydrogen cyanide, paraformaldehyde, and phosphine.

Organic Compounds. A listing of the categories of organic chemicals used reads like the table of contents of an organic chemistry text, with but relatively few subfamilies of organic compounds not represented in one way or other.

Alcohols. The open straight or branched chain (alkyl; aliphatic) saturated alcohols, such as methanol, ethanol, up through tetradecanol, etc., are not powerful insecticides, although they are somewhat more effective than the related hydrocarbons (methane, ethane, etc.). As is evidenced in other areas of activity, these alcohols increase in insecticidal effectiveness roughly in proportion to their molecular weight (number of carbon atoms in compound). This is an effect, however, which levels off when 9–12 carbons (molecular weight of 144–186) are present in the chain. For example, although this is not a direct measure of insecticidal activity, nonyl (9 carbons) and decyl (10 carbons) alcohols are most effective in reducing the sprouting of stored potatoes. Beyond this point, a greater number of carbon atoms does not increase the effectiveness.

Unsaturated and Cyclic Alcohols. Although these compounds show somewhat stronger insecticidal effectiveness, as compared with the saturated aliphatic alcohols, this added strength is still not sufficient to warrant their serious consideration as insecticides. Some of these compounds, however, do make effective herbicides.

Aldehydes. The aldehydes possess greater insecticidal effectiveness than the alcohols, formaldehyde, for example, serving as a stomach poison. The compound also exhibits strong bactericidal and fungicidal activity. Formaldehyde tends to polymerize into paraformaldehyde, the pesticidal properties of which are considerably weaker than those of formaldehyde. The same increasing effectiveness with increasing molecular weight exhibited by the alcohols also applies to the aldehydes.

Metaldehyde, the polymer of acetaldehyde, is a widely used molluskicide and is effective in the control of snails. Some of the unsaturated aldehydes are more potent in their pesticidal and herbicidal effectiveness. However, only a few compounds are of commercial significance, notably acrolein and related compounds, which are used as aquatic herbicides in connection with water reservoirs and systems. Some aldehydes also have growth-regulating properties. See **Plant Growth Modification and Regulation.**

Amines. This class of organic compounds also exhibits the same relationship between molecular weight and insecticidal activity as previously described. In the case of aliphatic amines, studies of toxic potency against house fly larvae have shown that the most effective compound is di-*n*-octylamine, with the compounds of higher or lower molecular weight in the series proving less toxic. Effectiveness also improves as one proceeds from the aliphatic amines to the aromatic amines; note, for example, the greater toxicity of aniline as compared with hexylamine. Although *o*-iodoaniline and 2,5-dichloroaniline demonstrate some toxicity for caterpillar and louse, respectively, generally and perhaps surprisingly, inclusion of halogen atoms within the aromatic amine nucleus does not promote greater toxicity. Some toxicity increase is shown, however, when nitro groups are introduced into the nucleus. Diphenylamine, a diarylamine, is effective against louse and at one time enjoyed wide use for troops during wartime.

The insecticidal usefulness of the amines is hindered by their tendency to severely injure plants (phytotoxic effects) to which they may be applied. Advantage of this property, however, is taken by using some amine compounds as herbicides. Examples include benefin, nitralin, and trifluralin.

Carbonic Acid Derivatives. Mixed esters of carbonic acid display toxic potency as acaricides and fungicides, including their inhibiting action against powdery mildew. When sulfur is introduced into the structure, as in the case of the mixed esters of thio- and dithiocarbonic acids, the acaricidal and fungicidal toxicity is further increased. Derivatives of thio and dithiocarbonic acids used commercially include carbon bisulfide, CS_2, and 6-methylquinoxaline-2,3-dithiocyclocarbonate (*Morestan*), an effective acaricide, fungicide, and insecticide.

Carbamic Acid Derivatives. A rather large number of commercially available food crop control chemicals fall into this category of organic compounds. Carbamic acid (or aminoformic acid) is NH_2COOH, but is best known in the form of its salts and esters. Several insecticides are found among the aryl esters of *N*-methylcarbamic acid, whereas the alkyl esters of *N*-arylcarbamic acids possess strong herbicidal powers, particularly in connection with undesired monocotyledonous plants.

The biological and physiological actions involved in the toxicity of the carbamates to animal and plant life processes are quite complex and not fully understood. It has been established that the esters of *N*-alkylcarbamic acids with insecticidal properties inhibit cholinesterase. Of the *N*-methylcarbamic acid ester series, the most powerful insecticidal compound is 1-naphthyl-*N*-methylcarbamate, the basis for such commercially produced compounds as carbaryl, naphthylcarbamate, and *Sevin*. Other important carbmate-type acaricides, fungicides, insecticides, and nematicides include *Aldicarb, Allyxycarb, Aminocarb, Bassa, Benomyl, Buffencarb, Carbendazim, Carbofuran, Ethiofencarb, Formetanate, Knockbal, Landrin, Mancozeb, Maneb, Meobal, Metacrate (Tsumacide), Metiram, Mexacarbate, Pirimcarb, Promecarb,* and *Propoxur (Baygon)*. Most of these names are proprietary.

Thio- and Dithiocarbamic Acid Derivatives. The derivatives of thiocarbamic acid are essentially herbicidal in nature. See **Herbicide.** However, excellent nematocidal effectiveness is illustrated by the dithiocarbamic acid derivatives, notably by sodium *N*-methyldithiocarbamate (*Vapam*), which serves the multiple purposes of eradicating not only nematodes, but many insects and weeds as well. It is frequently used as a soil sterilant.

In the case of the alkali metal salts of alkyldithiocarbamic acids, studies have shown that the fungicidal, nematocidal, and herbicidal effectiveness decreases as the length of the alkyl radical is increased. The nematocidal activity of the esters of alkyl- and dialkylcarbamic acids is greater than that of the salts. Toxicity is greatest in the methyl and ethyl esters, and decreases as the number of carbons in the ester radical increases.

Commonly available (check regulations) bactericides, fungicides, insecticides, and nematocides that fall into this category include: *Carbothion, Eptam, Ferbam, Nabam, Propineb* (zinc bearing), *TEC, Thiram, Zineb,* and *Ziram*. Most of the foregoing names are proprietary.

Aliphatic Carboxylic Acids. Studies indicate a very low pesticidal activity for the aliphatic monobasic acids (acetic, propionic, etc.) and the dibasic carboxylic acids (oxalic, fumaric, maleic, etc.). As is usually expected, the pesticidal activity of the acids increase when halogen atoms are introduced to displace hydrogens in the alkyl radicals. Examples of this effect include the monohaloacetic acids, which are sometimes used commercially. The salt *calcium propionate* is used in bread- and cheese-making as a preservative for its mild antibactericidal and fungicidal effects. Generally, the fluorine-containing derivatives are more toxic than those compounds with chlorine atoms. As can be expected, the unsaturated compounds exhibit greater toxicity than their saturated counterparts.

Alicyclic Carboxylic Acids. With exception of copper-bearing compounds, such as copper naphthenate, which is strongly fungicidal, the free alicyclic acids are not important as control chemicals. This is not the case, however, for a number of their derivatives. The natural pyrethrins and their synthetic analogs are included in this category. Among them are *Allethrin, Barthrin, Bioallethrin, Cinerin I, Cyclethrin, Dimethrin, Furethrin*; as well as *Neopyanimin, Pyresy, Pyrexcel,* and *Pyrocide*. Most of the foregoing names are proprietary.

The alicyclic carboxylic acid derivatives also include a family of growth-regulating compounds known as the *gibberellins*. See also **Gibberellic Acid and Gibberellin Plant Growth Hormones; and Plant Growth Modification and Regulation.**

The alicyclic carboxylic acid derivatives have been studied extensively and, to date, with the exception of the compounds already mentioned, relatively few have been found to possess commercially important potential as pesticides. An exception is dimethylcarbate (*Dimelton*), which finds use as a repellent for certain blood-sucking diptera. The compound frequently is mixed with other pesticides.

Aromatic Carboxylic Acids. As in the case of aliphatic and the alicyclic carboxylic acids, the free aromatic acids (benzoic,

naphthenic, etc.), their halogen and nitro derivatives, as well as their alkali metal salts, possess a low insecticidal effectiveness. However, for some species of mite, the benzyl ester of benzoic acid is very effective. As a general rule, the incorporation of chlorine or other halogen atoms in the benzoic acid and benzyl alcohol configurations accentuate the biological power. Introduction of chlorine into the para position of the benzyl radical appears to provide a maximum effect against both egg and adult mite. Further enhancement of acaricidal potency is obtained by the presence of amino, hydroxy, and nitro groups. For codling moth and body louse, the aliphatic esters of anisic, anthranoic, and salicylic acids have toxic effectiveness. Chlorobenzilate is an effective selective acaricide, useful against numerous species of mite, including the tracheal mite which is parasitic to honeybee.

Numerous other aromatic carboxylic acids possess some bactericidal, fungicidal, and herbicidal characteristics, as well as growth-regulating properties. Salicylanilide, the amide of salicylic acid, is effective against leaf mold and tomato brown spot.

Heterocyclic Compounds. Quite a large number of insecticides and other pesticides as well as herbicides are heterocyclic compounds. The variations pertaining to structure and composition are so many that this is strictly a generalized, umbrellalike classification. For the biochemist, organic chemist, entomologist, or other professional who is concerned with the development and theoretical aspects of control compound chemistry, other more detailed classifications are required. A first step in this direction was undertaken by Melnikov. (See the list of references at the end of this article.) Among important fungicides and insecticides in this classification are copper quinolate and *Phenazim* fungicide.

Aliphatic Hydrocarbons. After extensive research into the biological activity of the aliphatic hydrocarbons, relatively few of the pure (nonderivative) compounds have been found worthy of commercial attention. Popular for use in orchard spraying are the petroleum derivative oil sprays, which possess a good combination of acaricidal and insecticidal activity with low phytotoxicity. These sprays are effective against San Jose scale and mite.

Ethylene, for a number of years, has been used to hasten the ripening of some fruits. Although difficult to apply, ethylene is also an excellent defoliant. The insecticidal and nematocidal effectiveness of the halogenated aliphatic hydrocarbons is in proportion with the general chemical activity of these compounds. A number of these compounds have been used in the form of fumigants, notably in treating stored commodities and storage areas. Some of these include methyl bromide and *DD* pesticide.

Aromatic Hydrocarbons. Compounds in this category (such as benzene, naphthalene, xylene, etc.) have undergone extensive investigation. Although many of their derivatives are important, the pure compounds find little if any pesticidal use. Some of them, however, and in particular the xylenes, are used as solvents for carrying other control chemicals. For a number of years, naphthalene did enjoy widespread usage as a control agent against moth species. This has largely been replaced by synthetic compounds.

Halogen derivatives of benzene vary considerably in acaricidal and insecticidal toxicity, depending upon type, location, and number of halogen atoms introduced. Bromine appears to impart maximum effectiveness, followed by chlorine and fluorine. Biological activity increases with halogen atom introduction up to a total of three such atoms. A larger number tends to decrease effectiveness. Dichlorobenzene is more powerful than hexachlorobenzene. Loading a compound with bromine exhibits a greater effect than chlorine in reducing effectiveness.

Although there are eight stereoisomers of benzene hexachloride, only one of these (1,2,3,4,5,6-hexachlorocyclohexane) is important commercially.

DDT (1,1,1,-trichloro-2,2-*bis*(*p*-chlorophenyl)ethane, now banned in many countries, is a derivative of an unsymmetrical diarylethane and an effective insecticide. Paradichlorobenzene is an effective multipurpose pesticide. The compound is useful against sugarbeet weevil and in the control of phylloxera.

Ketones. Because of their rather weak insecticidal effectiveness, the ketones are used mainly as solvents for control chemical formulations.

Mercaptans. The aliphatic mercaptans with four or fewer carbon atoms have rather powerful insecticidal properties and can be used as fumigants against certain insect species. This is not true of those compounds containing over four carbon atoms; and also not true of aromatic mercaptans.

The control chemical interest of the mercaptans essentially is in the derivatives that incorporate chlorine or bromine. Methyl mercaptan rivals hydrogen cyanide as a powerful and useful fumigant. This compound is also an important intermediate in the synthesis of *Captan* and *Folpet* fungicides.

Some of the closely-related organic sulfides and thioacetals have found commercial pesticidal use. These formulations include *Mikazin, Fluoroparacide,* and *Fluorosulfacide.* Names are proprietary in most cases.

Nitro Compounds. Biological activity of organic nitro compounds in general compares favorably against that of the corresponding pure (nonderivative) hydrocarbon compounds. This activity is considerably enhanced in those nitro compounds that contain one or more halogen atoms. Bromine imparts a greater toxic power than chlorine. Many of the halonitro compounds command a wide spectrum of functionality, including acaricidal, bactericidal, fungicidal, herbicidal, insecticidal, and nematicidal attributes. Some of the action of these compounds is ascribed to their strong oxidizing properties. Examples of effective nitro compounds include chloropicrin and several related compounds, such as dichloronitroethane and chloronitropropane, as well as *Binapapacryl* acaricide-fungicide, *Dicloran* fungicide, *Dinobuton* acaricide, *Dinocap* acaricide fungicide, and *Dinoterb* acetate pesticide. Most of these names are proprietary.

Mercury Compounds. The powerful biological activity of mercury in simple compounds, such as the inorganic mercuric chloride, particularly against molds and other bacterial and fungus infections, has been recognized for generations. Also, the toxicity not only to microorganisms, but to higher animal forms also has been known and of major concern for a long time. Because of emphasis on environmental factors and safety during the past few decades, mercury-containing chemicals are being phased out in many countries. In connection with mercury compounds, it is interesting to note that many of these compounds have a good chemotherapeutic rating (or index), i.e., the dosage required to control a plant disease organism is *many, many* times smaller than the dosage that would be harmful to the plant. Some mercury compounds stimulate plant growth and yield.

Tin Compounds. Several organotin compounds are quite biologically active. Some of the simple inorganic tin salts, such as stannous or stannic chloride, have little if any pesticidal value. The fungicidal effectiveness is achieved by substituting alkyl or aryl groups for the chlorine atoms. A peak of insecticidal activity is achieved with the trialkyl- and triaryltins. It is interesting to note that the tetraalkyl- and tetraaryltins are essentially ineffective. Research has indicated that tributyltin chloride and tributyltin fluoride are the most active of the possible combina-

tions. Most popularly used (check regulations) commercially are triphenyltin hydroxide and triphenyltin acetate.

Copper Compounds. Principally used as fungicides, copper inorganic compounds are widely applied. Of the organic copper compounds, the most commonly used are copper linoleate, copper naphthanate, and copper quinolate.

Zinc Compounds. Zinc is associated in a number of organic pesticides, notably Propineb fungicide, *Zineb* fungicide, and *Ziram* fungicide. Names are proprietary.

Phenols. As compared with the aliphatic alcohols, the phenols are more active biologically, but even with this greater activity, most of these compounds are not of practical commercial importance. Introduction of halogen, nitro, thiocyano, and some other groups increase their activity, the nitro group appearing to have the greatest insecticidal power. Included among these compounds are some which date back 50 years or more—dinocap (*Karathane*) acaricide-fungicide and dinobuton (*Dinoseb*) acaricide. The phenols tend to severely burn plants, a property that led to their use, commencing in the late-1930s, as contact, selective-type herbicides and desiccants.

Phosphorus Organics

Possibly the most extensive of all categories of organic compounds used as control chemicals, the organophosphorus compounds, are derived from the inorganic acids of phosphorus. It is estimated that the very fundamental compounds in this category number well over one hundred and that the commercial formulations resulting may number into the several hundreds. For study it is sometimes convenient to classify these compounds as derivatives of (1) phosphorous acid, H_3PO_3; (2) phosphoric acid, H_3PO_4; (3) thiophosphoric acid, $PS(OH)_3$; (4) pyrophosphoric acid, $H_4P_2O_7$; and (5) phosphonic acids (phosphine = PH_3).

Phosphorous Acid Derivatives. The principal control chemical potential of these organic phosphate compounds lies with their abilities as herbicides. Acaricidal, fungicidal, insecticidal, and nematocidal powers are comparatively weak. But, as is nearly always the case, the toxic potential increases with many different kinds of derivatives that can be prepared. Commercial products based upon derivatives include: DDVP (*Dichlorvos*) acaricide-insecticide; *tris*(2,4-dichlorophenoxyethyl)phosphite (*Falone*); *Gestid* (*Mevinphos, Phosdrin*) acaricide-insecticide; *Naled* (*Dibrom*) acaricide-insecticide; and *Phosphamidon* acaricide-insecticide. Most of these names are proprietary.

Phosphoric Acid Derivatives. The biological activity of the phosphites is considerably greater than that of the phosphates, and notably among the mixed esters of phosphoric acid where one of the estre radicals is acidic. The toxicity of the resulting derivative is roughly proportional to the dissociation constant of the parent alcohol, phenol, or acid, the toxicity decreasing as the dissociation constant decreases. Research on the phosphoric acid mixed esters shows that the methyl derivatives are the most toxic. The toxicity of these types of compounds is believed to be the result of (1) high alkylating potential as regards certain biological nitrogen and sulfur constituents; and (2) elevated rates of hydrolysis.

Numerous proprietary examples of the derivatives of phosphoric acid include: *Bromophos* insecticide; *Chlorpyrifos* (*Dursban*) insecticide; *Demeton* (*Mercaptophos*) acaricide-insecticide; *Diazinon* acaricide-insecticide; *Fenitrothion* insecticide; *Fensulfothion* insecticide-nematicide; *Fenthion* acaricide-insecticide; *Kitazin* fungicide; *Methyl Parathion* (*Metafos*) acaricide-insecticide; *Oxydemeton-Methyl* acaricide-insecticide; *Parathion* (*Thiophos*) acaricide-insecticide; *Vamidothion* acaricide-insecticide; and *Zytron* insecticide. Most of these names are proprietary. Some of the compounds are banned in some countries.

Thiophosphoric Acid Derivatives. A fortunate combination of characteristics occurs in the phosphoric acid derivatives upon substitution of a sulfur atom for one of the oxygens of the parent compound, namely, the toxicity of the derivatives to higher forms of life is substantially diminished, while at the same time, the acaricidal and insecticidal powers, with few exceptions, remain strong. Derivatives of thiophosphoric acids may feature a thiono or a thiolo (most toxic) structure. Commercial preparations are usually mixed esters of thiophosphoric acid, as well as of dithio- and trithiophosphoric acids. The trithio compounds usually are markedly less effective than the dithio compounds.

Some commercial formulations based upon derivatives of the dithiophosphoric acids include: *Azinphos-Methyl* (*Guthion*) acaricide-insecticide; *Carbophenothion* (*Trithion*) pesticide; *Dimethoate* acaricide-insecticide; *Disulfoton* acaricide-insecticide; *Malathion* (*Carophos*) acaricide-insecticide; *Mecarbam* acaricide-insecticide; *Menazon* acaricide-insecticide; *Phorate* acaricide-insecticide; *Phosalone* acaricide-insecticide; and *Phosmet* (*Imidan*, *Phthalodophos*) acaricide-insecticide. Most of these names are proprietary. Some of these compounds are banned in some countries.

Pyrophosphoric Acid Derivatives. A pioneer among the phosphorus-containing organic control chemicals, tetraethyl pyrophosphate (*Bladen*; *TEPP*), was developed by Bayer AG in Germany in the early 1940s. The pyrophosphates are powerful contact-type acaricide-insecticides that have little or no tendency to function systemically. In addition to *TEPP*, some of the commercial formulations in this category include *NPD*; *Pirophos*; *Schraden*; and *Sulfotepp*. Most of these names are proprietary. Some of these compounds have been banned in some countries.

Other Organic Bases

Phosphonic Acid Derivatives. Very important in this category of proprietary compounds are *Trichlorfon* (*Chlorophos*, *Dipterex*, *Dylox*); and *Trichloronate*.

Quinones. As compared with the alcohols and aldehydes, the biological activity of the quinones is greater, notably in their actions as fungicides. Although benzoquinones exhibit relatively low activity, the presence of halogen and hydrocarbon radicals in the ring structure, as is true of many other organic compounds, significantly increases the effectiveness. Some of the derivatives of the quinones used commercially include tetrachlorobenzoquinone (*Chloranil*), which is particularly effective for disinfecting seed; and 2,3-dichloronaphthoquinone-1,4 (*Dichlone*; *Phygon*).

Sulfonic Acid Derivatives. The sulfonic acids, including their salts, have found value as agents for treating wool fabrics against species of moth for a number of years. One of these products is *Eulan*, a number of variations of which have been produced. *Mitin-FF*, a derivative of urea, also has been used in this way. To date, the free sulfonic acids have not played an important role as insecticides for crops.

Impressive activity against mite, larva and egg, has been shown by some of the aromatic esters of arylsulfonic acids. Some of the commercial formulations in this category include *CPCBS* (*Chlorofenson*; *Ester Sulfonate*; *Ovex*; *Ovotran*) acaricide.

Thiocyanates and Isocyanates. The intense biological activity of hydrogen cyanide has been known for scores of years. Similarly, derivative compounds have been known and used for many years. The derivatives of thiocyanic acid are particularly powerful fungicides and pesticides. Research has indicated that the straight-chain thiocyanates are more effective than those with branched chains and, as can be expected, introduction of halogen atoms into the compounds increases their toxicity. Some of the commercial formulations that fall into this overall category include *Thiophanate* fungicide; and *Thiophanate Methyl* fungicide. Regulations over usage must be checked.

Urea and Thiourea Derivatives. The value of urea as a fertilizer is well known. See **Fertilizers**. However, the most elementary derivatives of urea show strong phytotoxic effects. Thus urea derivatives are widely used as herbicides. A number of derivatives of thiourea also show strong bactericidal and fungicidal activity. In this category are found *Dodine* fungicide and *Guazatine* fungicide.

Bacterial and Botanical Compounds as Insecticides and Control Chemicals

One of the long established and better known of the botanical compounds is pyrethrum or pyrethrin insecticide. Known since the early 1800s, the active ingredient of this formulation is obtained from pyrethrum plants found in Africa and South America. Pyrethrin is considered to be one of the safest insecticides, but it is costly. The first synthesis of allerthrin, the analog of pyrethrin, was developed in the late 1940s and is now widely used.

A much more recent natural insecticide, developed in the early 1960s, is *Bacillus thuringiensis* Berliner, from living spores of a bacterial strain that causes disease among certain types of insects.

Other naturally derived commercial insecticides include *Evisect*, *Hellebore*, nicotine sulfate, and *Sabadilla insecticide*.

Pheromones. Some entomologists believe that insect pests may be controlled economically with a minimum of environmental disruption by exploiting the hormones and pheromones by which an insect regulates its growth, development, and behavior. Peromones may be defined as chemicals which are secreted by one insect that affect the behavior of other individuals of the same species. Pheromones evoke several behavioral responses, but the sex-attractant pheromones are those most frequently mentioned by entomologists. It is believed that inasmuch as pheromones are natural substances the insects may be less likely to develop a resistance to them than to some synthetic organic insecticides. However, entomologists point out that insects are quite adaptable and that a change in their pheromones is a possibility.

Three approaches in the use of pheromones have appeared in the literature: (1) Use of traps baited with sexual attractant material as a means for monitoring the infestation of areas with select insects; (2) similar use of traps except on a massive scale to attract males (female sex pheromone used as bait); and (3) "male confusion" technique in which female sex pheromone is permeated in the air, frustrating the attempts of males to locate females. The pheromone developed in connection with the gypsy moth is called disparlure and has been used for monitoring. More recently (late 1978), the Environmental Protection Agency (United States) released a compound known as Glossyplure H.F., a chemical duplicate of the natural mating scent emitted by female pink bollworm moths to attract male moths. When Glossyplure is applied to a cotton field, it so saturates the air with the artificial sex scent that male moths become confused are rarely find females with which to mate. This particular compound is the first to be approved in the United States for commercial use.

The use of pheromones is particularly attractive because of the high selectivity of the method, enabling the destruction of pests by way of large reductions in future populations and doing this without interfering with the normal life and habits of beneficial insects.

Juvenile Hormones. These are organic chemicals that are present in insects during the greater part of the insect's development. It is only during metamorphosis (period when a larva changes into an adult) that these chemicals are absent. When juvenile hormones are applied to insects during metamorphosis, the adults produced are deformed and lack the capacity for further development and soon die. Because juvenile hormones are relatively simple compounds, synthetic analogs are not too difficult to prepare and thus can be used as effective insecticides. However, timing is very critical because effectiveness is limited to the relatively short period of metamorphosis. If applied before or after this period, the compounds are essentially ineffective. Wide use of these chemicals could become practical if tied into a computerized pest management system control center as previously described.

Antiallatotropins. As part of their defense mechanism, some plants contain chemicals with juvenile hormone activity. Plants also have been found that contain chemicals with antijuvenile hormone activity, known as antiallatotropins. Although still not fully understood, it is assumed that the biological control system of the plant distinguishes between the metamorphosis period of an insect (during which time the juvenile hormones would be used as weapons) and the other periods of the insect's life cycle (during which time juvenile hormones are required by the insect, hence use by the plant of the antijuvenile or antiallatotropin compounds). Two antiallatotropins have been isolated from a common bedding plant (*Ageratum houstoniatum*). Chemically, these are 7-methoxy-2,2-dimethylchromene and 6,7-dimethoxy-2,2-dimethylchromene. W. S. Bowers of the New York State Agricultural Experiment Station, where these substances were first isolated, has named the compounds Precocene I and Precocene II because the compounds cause precocious metamorphosis in insects. Studies have shown that contact with the precocenes causes premature metamorphosis in milkweed bug nymphs, cotton stainers, and other *Hemiptera* species. Apparently, the precocenes cause the nymphs to skip one or more of their 5 instars (stages of life between molts) to become imperfect, miniature adults, the ovaries of which do not undergo development or yolking, and the females among which are sexually unreceptive. Exposure of mated milkweed bugs causes the eggs to be resorbed and the ovaries to regress to the underdeveloped condition. When mature eggs are present at the time of exposure, the eggs are deposited and hatch normally, but undergo precocious metamorphosis.

The precocenes cause other effects which are detrimental to the adult insect. For example, in cockroach, the chemicals block the production of sexual pheromones. In the Colorado potato beetle, the precocenes induce diapause, that is, after exposure, the beetle simply burrows into the ground and sleeps for extended periods. Different insects respond to the precocenes in different ways. Few generalities can be drawn at this time, but one common result of exposure is that female insects almost always become sterile before completion of their sexual development. Some scientists suggest that the precocenes interfere with secretion of juvenile hormones because the effects of the precocenes are much reduced when they are administered along with juvenile hormones.

Sometimes juvenile hormones and their analogs are referred to as the third generation of insecticides. It has been suggested that precocenes may become the fourth generation of insecticides.

Phytoalexins. First reported by K. Müller in Germany in 1940, phytoalexins are lipidlike chemicals that are synthesized by some plants. Research indicates that these compounds are toxic to fungi and bacteria, as well as some other pests. It has been found that the chemicals are produced as the result of an attack upon the plant and an analogy between these compounds and inteferon (an antiviral substance produced in humans in response to a viral infection) has been suggested. To date, about 100 phytoalexins have been isolated. One of their roles is believed to be prevention of germination of fungus spores.

Research by K. Uehara at the University of Osaka, Japan, in 1958 produced the concept of elicitors. Since then, at the University of Colorado (Boulder), the first phytoalexin elicitor was isolated. This was obtained from filtrates of cultures of *Phytophthora megasperma* var. *sojae*, a fungus that attacks soybean. This compound stimulates the accumulation of the phytoalexin glycerollin, previously characterized by research at the University of London.

Viruses. There are epidemics caused by viruses which occur periodically in the insect populations and thus naturally help to check their spread. Entomologists would like to find ways of infecting pest insects before they can cause serious damage. Most insect viruses are specific for a few closely related hosts. The general structure of the viral particles include DNA plus protein, all imbedded in a protein matrix. They are termed nuclear polyhedrosis virus (NPV). The viruses spread when larvae eat contaminated foliage.

Apparently viral infections in the past have helped to control the population of the Douglas-fir tussock moth. This insect undergoes population explosions at intervals of about 10 years. The outbreaks may last up to 3 years, during which time severe damage results. DDT helped to control the tussock moth population before it was banned. Attempts are now being made to combat a current outbreak of tussock moths with virus. Preliminary experiments in spraying virus on trees have been encouraging.

A virus called NPV is commercially produced and is used for the control of the cotton bollworm, a species closely related to the tobacco budworm. With present technology, the viruses can reproduce only in living cells. The insect viruses are usually grown in the appropriate hosts. Investigators are attempting to develop cell culture systems for propagating insect viruses that will eliminate the inconvenience of contamination of large numbers of insects or insect larvae for virus production.

The possible impact of virus insecticides on other forms of life, including humans, in the long term will obviously require some years of actual experience at a high level of usage. Regulatory agencies at this juncture are understandably extremely cautious in approving more than limited use. There appears to be evidence in support of the safety of the viruses and the negative aspects of this juncture appear to be in the category of theoretical possibilities in the absence of any substantive evidence. One scientist at the U.S. Department of Agriculture, with reference to the case of the cabbage looper caterpillar that succumbs to a virus attack, refers to the so-called coleslaw example. The insect body dissolves and sheds onto the leaf of cabbage large quantities of virus which are not killed in the preparation of coleslaw. In mid-October, when mortality of the loopers is at the highest level, the average bowl of coleslaw will contain about 4 billion live particles of cabbage looper nuclear polyhedrosis virus. The scientist reasons that if the virus were harmful to people, this would have been long evident.

However, there are unconfirmed reports from Japan that polyhedrosis virus of the silkworm can infect human amnion cells. The cotton bollworm nuclear polyhedrosis virus has passed a number of severe tests which have included attempts to infect animals, such as rhesus monkeys and tissue cultures of humans.

The possibilities of long-term effects on humans, such as the production of cancer or birth defects, are difficult to test, but it is reasoned that where a virus in question does not multiply in human or laboratory animal cells at body temperature, the risk is minimal.

Genetic Controls. A number of genetic methods are under consideration for regulation of pest populations. Wallace and Dobzhansky described conditions that led to genetic collapse and extinction of a population in 1959. The simplest cases considered were induced recessive lethal mutations and dominant lethal mutations. As reported by Smith and von Borstel in *Science*, **178**, 4066, 1166–1174, more insidious genetic mechanisms that could cause population collapse and obliteration have been suggested during the last decade and include: (1) Meiotic drive inseparably associated with genes for female sterility; (2) conditional lethal mutations; (3) unstable genetic equilibrium caused by compound chromosomes or translocations.

Early research by Müller indicated that after irradiation of males, events were induced in *Drosophila* sperm that resulted in dead embryos in the first generation. Müller termed these dominant lethal mutations. The details of this and later investigations are reported by Smith and von Borstel. As early as 1908, Boveri showed that multiple fertilizations and an egg cause chromosomes to be unequally distributed in cells during early cleavage, because the multiple centrioles set up multiple spindle orientation sites and chromosomes proceed to these sites at random. Boveri's study demonstrated that the well-being of the organism depends upon a full complement of chromosomes with corresponding total genic balance.

The subject is beyond the scope of this encyclopedia but to summarize, four of the major areas for insect population control through genetic means are: (1) By induced dominant lethality; (2) by contrived dominant lethality; (3) by induced inherited sterility; and (4) by contrived inherited partial sterility.

References

Burges, H. D., and N. W. Hussey: "Microbial Control of Insects," Academic, New York, 1971.
Carter, L. J.: "Eradicating the Boll Weevil," *Science*, **183**, 4124, 494–499 (1974).
Carter, Luther J.: "Pest Control: NAS Panel Warns of Possible Technological Breakdown," *Science*, **191**, 4229, 836–837 (February 27, 1976).
Djerassi, Carl, et. al.: "Insect Control of the Future: Operational and Policy Aspects," *Science*, **186**, 4164, 596–607 (November 15, 1974).
Giese, Ronald L., Peart, Robert M., and Roger T. Huber: "Pest Management," in "Food: Politics, Economics, Nutrition, and Research," American Association for the Advancement of Science, Washington, D.C., 1975.
Klassen, W., Creech, J. F., and R. A. Bell: "The Potential for Genetic Suppression of Insect Populations by Their Adaptations to Climate," Agricultural Research Service, Miscellaneous Publication No. 1178, United States Department of Agriculture, Washington, D.C., October 1970.
Marx, J. L.: "Insect Control: Use of Pheromones," *Science*, **181**, 4101, 736–737 (1973).
Marx, J. L.: "Insect Control: Hormones and Viruses," *Science*, **181**, 4102, 833–835 (1973).
Maugh, Thomas H., II: "Plant Biochemistry: Two New Ways to Fight Pests," *Science*, **192**, 4242, 874–876 (May 28, 1976).
Melnikov, N. N. (Gunter, F. A. and J. D., editors): "Chemistry of Pesticides," Springer-Verlag, New York, 1971.
Nogge, G., and M. Giannetti: "Specific Antibodies: A Potential Insecticide," *Science*, **209**, 1028–1029 (1980).
Rockstein, M.: "The Physiology of Insects," Vol. 1, Academic, New York, 1973.
Smith, Roger H. and R. C. von Borstel: "Genetic Control of Insect Populations," *Science*, **178**, 4066, 1164–1174 (1972).
Thomson, W. T.: "Agricultural Chemicals—Insecticides," Thomas Publications, Fresno, California, 1980.
Wade, N.: "Insect Viruses: A New Class of Pesticides," *Science*, **181**, 4103, 925–928 (1973).
Ware, G. W.: "Pesticides," Freeman, San Francisco, 1975.
Watson, T. F., Moore, L., and G. W. Ware: "Practical Insect Pest Management," Freeman, San Francisco, 1976.

INSTRUMENTATION (Analytical).   Analysis (Chemical).

INSULATION (Thermal). Thermal insulation is any substance or configuration of materials that resists the flow of heat. Thermal insulation does not stop heat flow, but retards it to rates that suit particular requirements. For example, in the case of buildings or residences in climates or during seasons when the ambient temperature of the atmosphere is uncomfortably hot or cold, thermal insulation will be used to retain heat within a structure during cold weather and to shield or insulate the structure from the penetration of external heat during hot weather. In the one case, by reducing the flow of heat from structure to atmosphere and near outer space during winter, less energy is required to maintain the desired temperature within the structure. In the other case, by reducing the flow of heat from atmosphere and sun to structure during summer, less energy is required to artificially cool the inside temperature. Many parallel instances occur in industry. By thermally insulating processing vessels, piping, etc., where it is desired to maintain warm or hot conditions, energy is not lost to ambient surroundings. The efficient maintenance of temperature is critical to many industrial situations because temperature affects the physical and chemical properties of materials, such as viscosity, and determines the rate at which chemical reactions occur, among numerous other temperature-sensitive properties. In cryoprocessing, as in the liquefaction of gases, the freezing of foods, etc., the objective is the maintenance of low temperatures and thermal insulation in such cases restricts the flow of heat from ambient surroundings and thus reduces the amount of energy required to maintain desirable low temperatures.

Although the principles of heat flow have been understood and treated mathematically since the early 19th century (Fourier, LaPlace, Poisson, Peclet, Lord Kelvin, Riemann, and many others), it was not until nearly the beginning of the 20th century that major developments of commercial thermal insulating materials and systems were undertaken.

In addition to increasing thermal efficiency (conservation of energy), thermal insulation is frequently used to protect personnel from injury by burns and to shield adjacent structures from overheating, thus assisting in protection against fire. Thermal insulations are not usually suited to fire protection per se once a fire is in progress, but by restricting heat flow in the first place, insulation plays an important role in preventing some kinds of fires from starting. The behavior of thermal insulation once a fire has started is not necessarily positive in all situations and thus requires the attention of equipment and structure designers. A major concern in fires is thermal diffusivity rather than thermal resistance. Thermal insulation may not be suited for protection against high-velocity radiation.

Although not the primary function, the retardation of moisture migration into insulated spaces, where condensation may occur, is an important engineering consideration in the design of thermal insulation systems.

Insulation and Heat-Flow Principles. Heat flows from places of higher temperature to those of lower temperature by one or more of three modes: (1) Conductance through solids; (2) convection by induced motion of fluids carrying heat; and (3) radiation by heat waves emitted from a surface. The rate of heat flow in solids depends upon temperature difference $T_2 - T_1$ and the resistances encountered. The heat flow, under steady state, is expressed by:

$$Q_{\text{heat flow}} = \frac{T_2 - T_1}{R\text{-value}}$$

R-value is a measure of thermal resistance and varies from one insulating material to the next.* Q can be expressed Btu/square foot/hour or as Cal/square meter/hour, depending upon

TABLE 1. CONVERSION FROM U.S. CUSTOMARY UNITS TO METRIC UNITS[1]
(Data are for thermochemical values unless noted. International Table values differ slightly.)

W = watt; m = metre; J = joule; kg = kilogram;
C° = temperature difference Celsius

MULTIPLY	BY	TO OBTAIN
Btu (mean) British thermal unit	$1.055\,870 \times 10^3$	J
Btu ft/h ft² F°	1.729 577	W/m C°
(k-factor, thermal conductivity)		
Btu in/h ft² F°	$1.441\,314 \times 10^{-1}$	W/m C°
(k-factor, thermal conductivity)		
Btu in/s ft² F°	$5.188\,732 \times 10^2$	W/m C°
Btu/h	$2.928\,751 \times 10^{-1}$	W
Btu/ft² h	3.152 481	W/m²
Btu/ft² min	$1.891\,489 \times 10^2$	W/m²
Btu/ft² s	$1.134\,893 \times 10^4$	W/m²
Btu/h ft² F°	5.674 466	W/m² C°
(C-factor, thermal conductance)		
(U-factor, overall thermal conductance)		
Btu/s ft² F°	$2.042\,808 \times 10^4$	W/m² C°
Btu/lb	$2.324\,444 \times 10^3$	J/kg
(Heat capacity)		
Btu/lb F°	$4.184\,000 \times 10^3$	J/kg C°
(Specific heat capacity)		
Btu/ft³	$3.723\,402 \times 10^4$	J/m³
Calorie (mean)	4.190 020	J
Calorie (kilogram)	$4.184\,000 \times 10^3$	J
(Kilocalories)		
Calorie/cm²	$4.184\,000 \times 10^4$	J/m²
Calorie/g	$4.184\,000 \times 10^3$	J/kg
Calorie/g C°	$4.184\,000 \times 10^3$	J/kg C°
Calorie/min	$6.973\,333 \times 10^{-2}$	W
Calorie/s	4.184 000	W
Calorie/cm² min	$6.973\,333 \times 10^2$	W/m²
Calorie/cm² s	$4.184\,000 \times 10^4$	W/m²
Calorie/cm s C°	$4.184\,000 \times 10^2$	W/m² C°
F° h ft²/Btu	$1.762\,280 \times 10^{-1}$	C° m²/W
(R-value, thermal resistance)		
F° h ft²/Btu in.	6.928 113	C° m/W
(ru-value, thermal resistivity)		
Therm (100,000 Btu)	$1.055\,056 \times 10^8$	J

[1] Most metric units shown are SI (the universally adopted designation for Le Système International d'Unités), except SI uses K (kelvin) for both absolute temperature and for temperature differences even though temperatures are determined on the Celsius scale. Since temperatures will usually be measured on the Celsius scale, the symbols used here are °C for temperature Celsius, as in the past, while C° is Celsius degrees difference. Similarly, °F is temperature Fahrenheit and F° is Fahrenheit degrees difference.

the units used in the equation. Helpful relationships between English and metric units are given in Table 1. See also entry on **Heat Transfer.**

While convection may be a significant factor within processes,

* Resistance, thermal, R-value—the mean temperature difference at equilibrium between two defined surfaces of material, or a construction, that induces unit heat flow rate through unit area. (From ASTM Std. C168-80a.)

in general, convection affects thermal insulation primarily at the surface of the insulation or its jacket, where the air film is a resistance to heat flow from (or to) that surface. Wind reduces the air film resistance. To a lesser convection can occur within some low-density fibrous insulations, especially in walls, and to a greater extent within cavities and unfilled spaces within constructions. In walls, it is important that insulation that does not fill the space completely be installed so that remaining spaces are uniform, not skewed. Radiant heat flows through space, either vacuum or gaseous, from a higher-temperature surface toward a lower-temperature substance by the difference in absolute temperatures to the fourth power and the surface characteristics called emittance, e, as shown by the Stefan-Boltzman relation:

$$Q_{\text{rad}} = 0.174e \left[\left(\frac{T_2}{100} \right)^4 - \left(\frac{T_1}{100} \right)^4 \right] \text{Btu/ft}^2 \text{ hr}$$

(on Rankine scale)

$$Q_{\text{rad}} = 5.670e \left[\left(\frac{T_2}{100} \right)^4 - \left(\frac{T_1}{100} \right)^4 \right] \text{W/m}^2 \quad \text{(on Kelvin scale)}$$

A common use of low convection with high-reflectance/low-emittance surfaces is in the food and liquid containers (Dewar flask, Thermos™ bottle, etc.). The double glass-wall space is under high vacuum so convection is virtually eliminated, and the surfaces are coated with silver to reduce heat transfer by the low emittance on the outside of the inner wall and the high reflectance on the inside of the outer wall.

Low emittance can be observed if the hand is held close to a very hot silver teapot without feeling much heat despite the high temperature of the surface of the teapot.

In high-temperature process plants, men have been burned on low-emittance hot metal jackets on insulation because the low emittance did not give them a sense of heat. Yet the thermal resistance from the heat conservation standpoint was excellent.

In some materials, especially foams, the spaces may contain gases other than air and the performance in convective and radiative heat transfer in the spaces affects the overall performance of the material. Also, the emittance/reflectance performances of the walls of the spaces affect performance overall.

Technically, the performance of materials and systems depends upon all three modes of heat transfer to varying degrees in different materials, so it is the effective or apparent conductance that is to be evaluated. Although such terminology may be correct technically, and is appearing again in the literature, it was discussed many years ago and abandoned because it aroused too many questions by users of insulations with limited technical knowledge. It was felt that those with necessary technical competence would understand that a multimode heat transfer was involved, and the simple thermal conductance would satisfy users so long as the data were correct. R-values are even more readily understood by users, and technical analyses can be made by identifying by subscript that phase of the analysis being evaluated.

Thermal Insulation Systems and Materials

ASTM* Committee C-16 on Thermal Insulating Materials defines thermal insulation as a material or assembly of materials used primarily to resist heat flow. The reference to assembly of materials indicates that the concern is thermal insulating systems, because it is not until materials have been designed into systems that performance can be estimated. Thermal insu-

* American Society for Testing and Materials, 1916 Race St., Philadelphia, Pa. 19103.

lating systems include not only the basic materials, but also the auxiliary materials and the methods of application and protection in service.

Time of exposure differentiates the needs for insulation performance when used in relatively continuous exposures, in cyclic increases and decreases of temperature, in processes with wide ranges of temperature in the various phases, especially in pipelines carrying fluids that must not fall below critical temperatures lest they solidify and necessitate dismantling and replacement of the lines.

A special short-time performance of thermal insulation is the ablative protection on the bottom of astronautical capsules returning from outer space when they are heated to sudden high temperatures by impact with the atmosphere. As principles stated below indicate, ablation is the process of resisting heat flow by using the heat absorptance in changes of state from solid to liquid and to vapor of the ablative insulation, which is thereby lost, so that a one-time or at most a few times of exposure is practicable.

Classes of Insulating Materials. See Fig. 1. While glass has

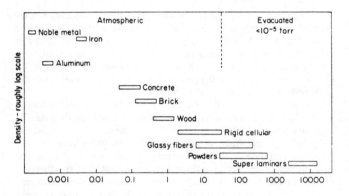

Fig. 1. Thermal resistivity of materials in insulation systems.

a high conductance, if it is fiberized and formed into wool-like masses, the high conductance of the fibers is counteracted by the still air that is held within the mass. Still air (no motion) has high thermal resistance, and at one time it was presumed that still air was the best insulator. A few other materials have been found with somewhat greater thermal resistance than still air, but they are so costly that they are suited only to very special applications. Other fiberized materials perform similarly, and rock, slag, and glass wools are collectively called mineral wools, but each has its own temperature limits.

All mass-type thermal insulations rely for their thermal resistance upon dispersion of the solid phase with air, or sometimes with gases. Plastics are reduced in density by foaming them into low-density material (0.5 to 2 lb/ft³; 8 to 32 kg/m³). Molten glass is foamed so that it performs as thermal insulation; its advantages are high compressive strength and virtually imperviousness to moisture, although thermal resistance is not as high as in some other materials.

Reacted materials, such as hydrous calcium silicate, are made with the solids dispersed to create density on the order of 12 lb/ft³ (192 kg/m³) and still provide high compressive strength. While the reflectance and emittance of the solid phase have an effect on heat transfer, it is the still air within the mass that gives it significant thermal resistance.

Still air is the important factor in dispersed solids, such as wood fibers, exfoliated mica, powdered diatomite, and expanded perlite.

The earliest reacted insulation was so-called 85% magnesium that had wide acceptance for many years, but since it was suited to less than 500°F (260°C), it has been replaced by other insulations with higher R-values or more desirable physical properties.

Although metals are good conductors of heat, they can also perform as good thermal insulators when their surface properties are used to advantage. While solid metal conducts heat readily, the emittance and reflectance of some metals are used to provide high overall R-values, especially when they are used in multiple sheets with spacers. Moreover, metals have relatively high temperature tolerances but have no absorption, so that all-metal thermal insulation may be suited to higher temperature services than some other kinds of materials. While silver and gold are the best metals for high reflectance service, their use is limited to special applications where cost is not governing. For long-time exposure, gold surfaces would maintain high reflectance longest. However, aluminum sheets are used effectively when all-metal insulation is required. The first sheet of aluminum spaced about 12 mm ($\frac{1}{2}$ in.) from the hot surface reflects a large percentage of heat striking it. The unreflected heat passes through the metal rapidly by conductance, but the low emittance of the reverse surface prevents much of the heat leaving to strike the next sheet of aluminum where, again, the reflectance prevents a large portion of the emitted heat to enter that sheet. The number of reflective sheets is designed for the R-value desired. If the service temperature is above the working temperature of aluminum, about 1000°F (540°C), the first one or two sheets can be made of polished stainless steel.

Since aluminum is also used for jackets, it should be noted that although the thermal performance may be acceptable, the low emittance of an aluminum surface on the outside may introduce a personnel hazard, mentioned above.

An opposite effect of emittance occurs when an attempt is made to use aluminum jackets on very cold (cryogenic) piping. In this case, heat reaching the surface from surroundings is reflected away so that the metal surface becomes so cold that it will condense and freeze moisture from the air. To overcome this surface condensation and freezing by use of thicker thermal insulation on the lines would require a very great increase in thickness over that needed if the jacket had been made of a heat-absorbing material that would keep the surface above the dewpoint.

In the high-temperature case, low emittance is desirable from an operating standpoint, whereas in the cryogenic case low emittance is undesirable.

Two general types of heat flow systems exist: one in which it is desired that heat flow as rapidly as practicable, and another in which heat flow must be resisted as much as practicable. The former is a high thermal conductance type of system, whereas the latter is a high thermal resistance, high R-value, type. Consequently, in heat conserving systems it is simpler to think in terms of thermal resistances, because resistances are additive whereas conductances are not.

However, sometimes both high and low and low R-value materials are needed simultaneously, as in traced lines (Fig. 2). Although systems are designed with limitation on overall heat loss, high-conductance cements are used for process safety on pipelines carrying hot fluids that would solidify if flow was interrupted or fluids become so viscous that the pumps could not handle the material. In such cases, one or more small pipes called tracers carrying hot fluids are enclosed with the thermal insulation envelope, and high conductance cements are used to improve heat flow from tracers to the main pipe. The size of the pipe insulation must be large enough to enclose the main pipe and the tracers, and also the insulation on fittings.

Importance of Moisture Migration. Heat conservation has

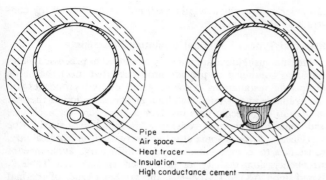

Fig. 2. Pipe insulation over pipe and heat tracer line, with and without high-conductance cement.

been treated from an energy standpoint as if it were an independent subject, whereas great costs and energy losses have been incurred by premature failures because it was not realized that in many cases thermal insulations can not be installed without inducing an effect on the migration of moisture. The problems are usually not great from the standpoint of solutions, but are great in getting people to realize that they have created problems that have been overlooked. In most constructions, heat and moisture performance must be considered jointly, because even high-temperature systems are shut down for alterations, maintenance, or repair. For example, in a house wall, adding thermal insulation makes the indoor wall warmer, as desired, but at the same time it makes the outdoor wall colder, and it is well known that when surfaces are colder than the dewpoint, condensation occurs. The old log cabin with its loose construction had no moisture problems, but it was often only the side of the body that faced the fireplace that felt warm while the rest did its best to accommodate the facts.

Economic Considerations. For many years, cost appraisal of insulation was based on a publication by L. B. McMillan in *ASME Proceedings,* December 1926, in which several factors enter into the analysis, as shown in Fig. 3. At present, the

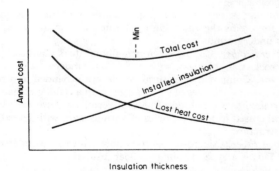

Fig. 3. Relation of incremental cost of additional thickness of insulation to the resultant heat savings and total cost. (*After L. B. McMillam.*)

cost of money and the costs of materials and labor are so unstable, that such analyses are of little value except that they do indicate the factors that enter into actual costs to plant management.

To presume that thermal insulation is always necessary is false. For example, when cryogenic fluids are transported from a supply source to a vessel, even in bright sunshine, it is usually most economical not to insulate the line at all. The reason for this is twofold: (1) for the short time of transport the area of

the pipe exposed to the sun is much smaller than the area that would be exposed if it were insulated; and (2) the heat that would be in the insulation when the transport starts would have to be removed by the cryogenic fluid. The combination of these two factors often makes the use of insulation undesirable for this type of cold fluid transport system. Moreover, the rapid formation of ice crystals from moisture in the air constitutes a thermal insulation.

While thermal insulations are selected first for their resistance to heat flow, their other properties need evaluation for each application. Hence there is no "best" insulation because a material well suited to one service may be poorly suited to another. Economics must be studied in detail because in some services the cost of a highly efficient material per unit of thickness may be overcome by additional thickness of a less efficient material, provided there is room for the greater thickness. All properties of materials must be considered for each exposure, even within the same system. Recognize that when different materials are used on different parts of insulation systems that are close together, the probability of using the wrong material on a particular surface is increased appreciably. Unless there is specific reason for wide use of multiple types of materials, it may be prudent to accept some compromises of properties.

In general, it is desirable to place thermal insulation on or near the outside of constructions, including basement walls, because this location reduces appreciably the temperature stresses in the structure induced by the changes in exposure.

Selection of Thermal Insulation

Test methods, specifications, and some recommended practices are in ASTM Book of Standards, Part 18. Producer's literature gives forms, properties and design data.

1. *Thermal Resistance—R-value—*Thermal data in general tables, as Table 2, are for dry materials, and for one or two mean temperatures, but most materials are not linear with mean temperature, hence, for specific designs, the whole range of resistance should be obtained.

A factor often overlooked is that space to be occupied by the insulation must be made available; a 3-inch (76-millimeter) iron pipe with 3 inches (76 millimeters) of insulation covered with a protective jacket would have a diameter of about 10 inches (254 millimeters).

2. *Temperature Limit—*Usually the high temperature limit governs because shrinkage then becomes excessive. Generally, high-temperature materials are physically stable at low temperatures but another material may be preferable.

Note that shrinkage data are usually from tests in soaking heat, whereas the field condition is for heat on only one surface. Hence, such data should not be presumed to mean that the insulation will shrink that amount in service; a 1% shrinkage would imply a change of almost $\frac{3}{8}$ inch (9.5 millimeters) in a standard 36-inch (914-millimeter) length and leave a wide crack. However, in service such a large shrinkage does not occur; shrinkage would be $\frac{1}{16}$–$\frac{1}{8}$ inch (1.6–3.2 millimeters). Moreover, hot metals expand, and it is this expansion that must be considered in compensating for openings between adjacent pieces of insulation, usually by use of double-layer insulation with staggered joints.

For cold-temperature service, moisture ingress is a problem that must be designed against and the material selected for its resistance to moisture and potential freezing.

3. *Corrosion—*While corrosion is usually not a concern on hot surfaces, recognize that all systems have shut-down periods, with the probability that moisture will find ingress and condense. Many insulations are alkaline and have little adverse effect on iron and copper, but aluminum is affected adversely. A major

concern is stress-corrosion of stainless steel induced by even trace amounts of soluble chlorides; an ASTM Test Method may be used.

4. *Density*—While the density of thermal insulation is low enough not to be a problem in most cases, density (or mass-weight) must be considered in airplanes, balloons and other antiterrestrial constructions.

5. *Moisture-Wetting*—If moisture can enter an insulation, the high conductance of water (or ice) reduces the thermal resistance. However, all materials that admit moisture are not affected adversely to the same degree. Some fibrous insulations have a threshold moisture content below which the adverse effect on resistance is not significant; thresholds may be on the order of 8% by weight. The reason is that small droplets adhere to the contact points of one fiber against another so that the volume of relatively still air that provides the thermal resistance is not reduced significantly. Moreover, the temperature difference within the insulation drives the moisture to the cold side, and if it condenses there, the thickness of the wet insulation with decreased resistance is still a small portion of the total resistance.

In materials that readily absorb water, a decrease in thermal resistance will occur, but consideration must be given to the performance after the material is redried.

Caution is needed in interpreting absorption tests by immersion because the internal structure may resist displacement of air or gas so that a low absorption is indicated. However, if moisture is induced to flow through the specimen by vapor pressure differential and a condensing temperature is reached, high absorption may occur in service.

Materials with very high internal surface areas may absorb a measurable amount of moisture; even a monomolecular thickness of moisture on a large area becomes readily measurable. However, exposure to high relative humidity will not "saturate" the material, and as relative humidity changes so will the absorbed moisture. Some materials are tested wet to indicate wet to dry strength ratio.

6. *Handleability*—In the field, materials must have strength properties that enable them to be handled in application by the usual procedures of the industry without excessive breakage. ASTM tests for flexure, impact, and friability are aimed at indicating handleability potential. Sometimes vibration tests are indicated if an unusual vibrating condition is to be encountered, although this test is receiving less attention than in the past because vibrations are usually designed against.

7. *Reflectances*—When an all-metal system is desired for temperatures to 1000°F (540°C) usually to avoid absorptions in case of leaks, multilayer sheets of aluminum are spaced on the order of $\frac{1}{2}$ inch (13 millimeters). For temperatures to 1400°F (760°C) the first sheets are made of stainless steel.

8. *Fire Behavior*—While thermal insulations are not intended for fire protections (treated elsewhere) their behavior in fire is important, especially from the standpoint of contribution of combustible matter to a fire that has started at the site. Material behavior may be complex, e.g., an absorptive material that would hold a combustible fluid (say, kerosene) would not be a major contribution to fire intensity because the fluid would not flow to the surface to burn as rapidly as it would from a pool of the fluid. Materials that contain organic binders may not be a serious contribution in an open fire, but if they are totally enclosed they may contribute to persistence of fire by smoldering.

While some thermal insulation may not constitute a significant fire hazard, consideration must be given to jackets, coatings, or coverings, and to the methods of attachment, so that even in moderate fires the insulation will not readily fall off the construction they insulate.

Principal Types of Thermal Insulating Materials

Thermal insulations are made from natural or processed materials and combined to provide properties that meet the needs of specific installations. Obviously, all desired properties are not available in any one insulation. Hence, selection of thermal insulation for specific uses involves comparisons for each use, and some high thermal resistance (*R*-value) may need to be sacrificed in favor of some other property, such as handleability, resistance to compression, thermal diffusivity, avoidance of stress-corrosion, toxicity, or in-fire performance. See Table 2. Based upon ASTM C 168–80a, the principal types of thermal insulation are:

Ablative. Heavy density combination of materials that change state from solid to liquid or vapor in high temperature so that heat absorbed through their change of state reduces substantially the heat transfer rate through the material. Suited to one-time or very few times use.

Calcium Silicate. Composed principally of hydrous calcium silicate, usually containing reinforcing fibers.

Cellular Elastomeric. Composed principally of natural or synthetic elastomers, or both, processed to form flexible, semirigid, or rigid foams which have a predominantly closed-cell structure.

Cellular Glass. Composed of glass processed to form a rigid foam usually having a predominantly closed-cell structure.

Cellular Polystyrene. Composed principally of polymerized styrene resin processed to form rigid foam having a predominantly closed-cell structure.

Cellular Polyurethane. Composed principally of the catalyzed reaction product of polyisocyanate and polyhydroxy compounds, processed usually with fluorocarbon gas to form a rigid foam having a predominantly closed-cell structure. Under investigation by regulators.

Cellulosic Fiber. Composed principally of cellulose fibers usually derived from paper, paperboard stock, or wood, with or without binders.

Diatomaceous Silica. Composed principally of diatomite (diatomaceous earth) with or without binders, usually containing reinforcing fibers.

Gilsonite. A pure form of asphalt processed into powdered form for use as the enclosing insulating mass around pipes or tanks underground.

Mineral Fiber. Composed principally of fibers manufactured from rock, slab, or glass, with or without binders.

Perlite. Composed of natural perlite ore expanded and processed to form particles of various sizes with a cellular structure.

Vermiculite. Composed of natural vermiculite ore expanded and processed to form particles of various sizes with an exfoliated structure.

Wood Fiber. Composed of wood fibers, with or without binders. This is a type of cellulosic fiber insulation.

Principal Forms of Thermal Insulation

Also based upon ASTM C 168–80a, the principal forms of thermal insulation are:

Blanket Insulation. A relatively flat and flexible insulation in coherent form furnished in units of substantial area. Some forms are called batts.

Blanket Insulation, Metal Mesh. Blanket insulation covered by flexible metal-mesh facings attached on one or both sides.

Block Insulation. Rigid insulation preformed into rectangular units.

Board Insulation. Semirigid insulation preformed into rectan-

TABLE 2. REPRESENTATIVE THERMAL INSULATIONS

Temperature Limit °F	Temperature Limit °C	Material and Usual Forms	Inorganic (I) or Organic (O)	Density lb/ft³	Density kg/m³	Thermal Resistivity At mean temp., °F	Thermal Resistivity F° ft² h / Btu in	Thermal Resistivity At mean temp., °C	Thermal Resistivity C° m / W
		HIGH-TEMPERATURE SERVICE							
2300	1265	Alumina-silica ceramic fiber, soft mass	I	3–12	48–192	300	3.2	160	22.2
						1000	1.2	540	8.3
2200	1200	Potassium titanate fiber, soft mass	I	15–18	240–288	300	3.0	160	20.8
1900	1040	Diatomaceous silica, bonded, semirigid, preformed block and pipe	I	23–25	368–400	300	1.5	160	10.4
						1000	1.3	540	9.0
1800	1000	Mineral fiber, rock, and slag, loose fill, preformed block and pipe	I	16–24	256–384	400	1.7	204	11.8
						1000	1.3	540	9.0
1600	875	Perlite, expanded, loose granules	I	4–10	64–160	0	3.0	−18	20.9
						1000	0.9	540	6.2
1200	650	Hydrous calcium silicate, may contain unexposed organic reinforcing fibers, preformed rigid block and pipe, compression over 100 psi (0.69 Mpa)	I/O	11–14	176–224	300	2.5	160	17.3
						700	1.7		11.8
1000	540	Glass fiber, no binder, loose mass	. I	3–5	48–80	300	2.9	160	20.1
						800	1.4	430	9.7
500	275	Gilsonite, processed pure asphalt powder for underground fill, compacted, impervious	O	35–48	560–640	50	1.9	10	13.2
						3300	1.6	1820	11.1
800	425	Glass, cellular, preformed block, impervious, compression over 100 psi (0.69 Mpa)	I	10–18	160–288	300	1.8	160	12.5
						600	1.0	320	7.0
450	225	Glass fiber, organic binder, loose fill, blankets, batts, preformed block and pipe	I/O	0.5–3	8–24	75	3.8	25	26.4
						300	2.8	160	19.4
200	95	Cellulosic fibers of wood, cane, reused paper, as loose fill	O	0.3–5	4.8–80	75	3.8	25	26.4
		LOW-TEMPERATURE SERVICE							
−225	−140	Plastic Foams Polyurethane (under investigation)	O	1.8–2.2	28.8–35.2	−200	11.0	95	76.3
						100	6.0	40	41.6
−200	−130	Polystyrene	O	1.0–4.0	16–64	40	3.9	4.4	27.0
						75	2.6	25	18.0
−40	−40	Polyvinyl chloride	O	4–25	64–400	75	3.9	25	57.0
−40	−40	Rubber, cellular	O	3–20	48–320	25	4.3	−3.9	29.8
						75	3.3	25	22.9
−400	−245	Glass, cellular, preformed block	I	10–18	160–288	25	2.8	−3.9	19.4
						100	2.4	40	16.7
−450	−270	Mineral Fibers	I	0.5–10	8–160	25	3.7	−3.9	25.7
						100	3.3	40	22.9
−459	−273	Evacuated multilayer foil and fiber mats	I	Various		On the order of 25–100+		On the order of 175–700+	

1. Most high-temperature insulations are usable also at low temperatures, but other materials may be preferable.
2. Maximum temperatures apply to surface service, not soaking heat.
3. Thermal resistivity data are approximate and vary widely with density (for specific designs and temperatures, consult current manufacturers' data).
4. Celsius temperatures are rounded. °C is temperature; C° is temperature difference.

gular units having a degree of suppleness particularly related to their geometrical dimensions.

Cement, Finishing. A mixture of dry fibrous or powdery materials, or both, that when mixed with water develops a plastic consistency, and when dried in place forms a relatively hard, protective surface.

Cement, Insulating. A mixture of dry granular, flaky, fibrous or powdery materials that when mixed with water develops a plastic consistency, and when dried in place forms a coherent covering that affords substantial resistance to heat transmission.

Fitting Covers. Manufactured or assembled segments of insulation to form covers for various pipe and vessel fittings such as elbows, tees, crosses, valves, etc. see ASTM Standard C450-76 (or later) for dimensions.

Loose-Fill Insulation. Insulation in granular, nodular, fibrous, powdery, or similar form designed to be installed by pouring, blowing, or hand placement.

Pipe Insulation. Insulation in a form suitable for application to cylindrical surfaces.

Reflective Insulation. Insulation depending for its performance upon reduction of radiant heat transfer across spaces by use of one or more surfaces of high reflective and low emittance.

Roof Insulation. Rectangular boards or blocks of various thicknesses with properties for use beneath the roofing membrane protected from the weather, or with properties for use above the roofing membrane exposed to the weather.

Underground Systems. Systems that enclose insulated piping in small tunnels that include expansion arrangements and provide drainage. Since accidental general flooding may occur, the insulations should be capable of withstanding "boiling water" effects so that when the system has been dewatered the redried insulation will perform thermally essentially as it did prior to flooding.

Super Insulations. Several insulation systems have been developed that have very higher thermal resistances, such as multilayer radiation shields, specially selected and matted fibers with small interfiber distances, special powders, ceramic foams, honeycomb composites, often highly evacuated, but they are too costly for usual services with which the general public is familiar.

References

ASHRAE: "Handbook of Fundamentals," American Society of Heating, Refrigerating, and Air Conditioning Engineers, Inc., New York, N.Y. (Published every 5 years).

ASTM: "Book of Standards," Part 18, American Society for Testing and Materials, Philadelphia, Pennsylvania (Issued annually).

ASTM: "Thermal Insulation Performance," STP718, American Society for Testing and Materials, Philadelphia, Pennsylvania (1981).

Glaser, P. E., et al.: "Thermal Insulation Systems," National Aeronautics and Space Administration, NASA SP-5027, Washington, D.C., 1967.

Staff: "How to Determine Economic Thickness of Insulation," Thermal Insulation Manufacturers Association, Mt. Kisco, New York (Revised periodically).

Turner, W. C., and J. F. Malloy: "Thermal Insulation," Krieger, Melbourne, Florida, 1981.

Tye, R. P.: "Thermal Conductivity," Vol. I and II, Academic, New York, 1969.

Wilson, A. C.: "Industrial Thermal Insulation," McGraw-Hill, New York, 1959.

—E. C. Shuman, Consulting Engineer, State College, Pennsylvania.

INTERMEDIATE (Chemical). An intermediate generally is considered to be a material (usually a chemical compound) that occurs somewhere in a chemical manufacturing process between the introduction of the basic raw materials and the creation of the final end products. When two or more separate chemical reactions are involved, the intermediate may be the product of one of the *between reactions* and serve as a charge material for a subsequent reaction. For example, in the manufacture of aromatic polyester, several materials and reactions are required. The fundamental raw materials are nitric acid, xylene, methanol, and ethylene glycol. In one reaction, *p*-xylene and nitric acid yield terephthalic acid. The terephthalic acid then is esterified with methanol using sulfuric acid as a catalyst to yield dimethyl terephthalate. The dimethyl terephthalate then undergoes an ester interchange with ethylene glycol which yields *bis*(β-hydroxyethyl)terephthalate, later condensed to polyethylene terephthalate. This low-molecular-weight polymer then is polymerized to a high-molecular-weight polyethylene terephthalate. In this operation, terephthalic acid and dimethyl terephthalate can be regarded as intermediates. In some instances, a producer will procure intermediate materials from the outside rather than produce them in house, particularly in the pharmaceutical and dye industries. Thus, a number of intermediates are high-tonnage items of commerce. Some intermediates are of low-tonnage requirements and sometimes the economics is in favor of one producer who supplies a number of using firms. A representative list of intermediates would include: *o*-aminophenol-*p*-sulfonic acid, 2,6-dichloro-4-nitroaniline, 4-sulfophthalic acid, *o*-tolidine dihydrochloride, diphenylmethane, diphenylacetaldehyde, methyl cyclopentylphenylglycolate, and 2,3-dichloro-5,6-dicyanobenzoquinone. See also **Synthesis (Chemical).**

INTERMEDIATE-MOISTURE FOODS. Authorities have various defined intermediate-moisture foods (IMF) as having from 15 to 40% water at a water activity ($a_w = 0.6$–0.8). Interest in intermediate-moisture foods for the human diet largely stemmed from the introduction and acceptance of "soft-moist" pet foods which were first marketed in the 1970s. Research continues at a good pace and considerable progress has been made as, for example, in processed cheese foods with high moisture contents.

More important than moisture content per se are the objective of creating a preserved food substance that is stable and can be eaten directly. An idealized IMF system will have (1) microbial stability at reduced water activity, (2) storage stability without special conditions, (3) reduction of weight and more compactness of product, and (4) the ability to be consumed from the package without rehydration. Some authorities believe that IMF systems represent excellent potential in the development of the snack market.

Two factors have largely delayed expansion of IMF systems: (1) technology and (2) consumer acceptance. Some of the technical problems involved in formulating IMF systems include: (1) rates of lipid oxidation, (2) enzymatic deterioration, and (3) nonenzymatic deterioration. With some products, there are also problems associated with the desired texture.

References

Davies, R., Birch, G. G., and K. J. Parker, Editors: "Intermediate Moisture Foods," Applied Science Publishers, Barking, Essex, England (1976).

Gee, M., Farkas, D., and A. R. Rahman: "Some Concepts for the Development of Intermediate Moisture Foods," *Food Technology*, **31**, 4, 58–64 (1977).

Lee, B. B., and A. A. Kraft: "Microbiology of Intermediate-Moisture Pork," *Journal of Food Science*, **42**, 3, 735–737 (1977).

INTERMETALLIC COMPOUND. In certain alloy systems, distinct intermediate phases occur where the constituent atoms are in fixed integral ratios, e.g., CuZn (beta-brass). Such a compound is held together by metallic bonding and may form a very complex crystal structure. The constitution of such an alloy is often governed by the Hume-Rothery rules. In some cases, if the electron concentration is such as to just fill a band, the material may even be semiconducting (e.g., InAs). See also Compound (Chemical).

Hume-Rothery Rules. When alloy systems form distinct phases, it is found that the ratio of the number of valence electrons to the number of atoms is characteristic of the phase (e.g., β, γ, ϵ) whatever the actual elements making up the alloy. Thus, both $Na_{31}Pb_8$ and Ni_5Zn_{21} are γ structures, with the electron-atom ratio 21:13. The rules are explained by the tendency to form a structure in which all the Brillouin zones are nearly full, or else entirely empty.

IN VITRO. An event or process occurring outside a living organism—in an unnatural environment, as in a test tube or petri dish.

IN VIVO. An event or process occurring naturally or spontaneously within a living organism.

IODINE. Chemical element symbol I, at. no. 53, at. wt. 126.9045, periodic table group 7a (halogens), mp 113.5°C, bp 184.35°C, density, 4.94 g/cm³ (20°C). Iodine has an orthorhombic crystal structure. Solid iodine is a violet-to-black color; the vapor is a beautiful violet color. The element sublimes readily and is easily purified in that way. Iodine is insoluble in water, soluble in alcohol, ether, CS_2, or carbon tetrachloride. The element was first identified by Courtois in 1812 when making a study of kelp. There is one stable isotope, ^{127}I, and fourteen radioactive isotopes, ^{122}I through ^{126}I, and ^{128}I through ^{136}I. The lengths of the half-lives of the isotopes vary widely, the shortest, ^{136}I, with a half-life of 86 seconds; the longest, ^{129}I, with a half-life of 1.72×10^7 years. See Radioactivity. In terms of abundance in the crust of the earth, the element ranks 53rd and is about as plentiful as tin, antimony, cesium, and barium. Considerable quantities of iodine have concentrated in the oceans. The average iodine content of a cubic mile of seawater is 230 tons (50 metric tons per cubic kilometer).

First ionization potential 10.44 eV; second, 19.4 eV. Oxidation potentials $I^- \rightarrow \frac{1}{2}I_2 + e^-$, −0.535 V; $I^- + H_2O \rightarrow HIO + H^+ + 2e^-$, −0.99 V; $I^- + 3H_2O \rightarrow IO_3^- + 6H^+ + 6e^-$, −1.085 V; $\frac{1}{2}I_2 + 3H_2O \rightarrow IO_3^- + 6H^+ + 5e^-$, −1.195 V; $\frac{1}{2}I_2 + H_2O \rightarrow HIO + H^+ + e^-$, −1.45 V; $IO_3^- + 3H_2O \rightarrow H_5IO_6 + H^+ + 2e^-$, ca. −1.7 V; $I^- + 6OH^- \rightarrow IO_3^- + 3H_2O + 6e^-$, −0.26 V; $I^- + 2OH^- \rightarrow IO^- + H_2O + 2e^-$, −0.49 V; $IO + 4OH^- \rightarrow IO_3^- + 2H_2O + 4e^-$, −0.56 V; $IO_3^- + 3OH^- \rightarrow H_3IO_6^{2-} + 2e^-$, ca. −0.70 V. Other important physical characteristics of iodine are given under Chemical Elements.

Sea plants, particularly kelp found in the waters around California and the Bay of Biscay, have been a source of iodine. Because of pollution, the kelp beds in California are no longer a major source. Iodine also is found in the petroleum oil well brine of California and, in small percentages, in sodium nitrate of Chile. The latter was once the primary source of the element. Brines now are the major source.

Uses. For many years, iodine tincture (3% to 7% dissolved in ethyl alcohol) has been an important antiseptic. The commercial tinctures also usually contain 5% potassium iodide to provide stability. This form produces a mild burning of the skin and strains both skin and fabrics. A milder preparation is available in which about 2% iodine is contained in an oil-water emulsion which also contains lecithin. The burning effect of the compound is greatly reduced and the mild stains produced usually wash off easily. There are a number of prepared medicines that contain iodine, although some of these have been removed from the market in recent years. At one time, iodoform CHI_3, a yellow, insoluble, crystalline powder with a very penetrating odor, was a very popular antiseptic and used widely in the preparation of gauzes and packings for infected cavities. Because of possible toxic effects with some individuals, the compound largely has been replaced by other less objectionable, less odorous materials.

The medical use of iodine compounds, particularly organoiodine substances, has been researched with resultant limited applications. The dietary requirement for iodine was established many years ago as needed for the maintenance of cell growth in humans and animals. The largest concentration of iodine occurs in the thyroid gland where the hormone thyroxine $C_{15}H_{11}O_4I_4N$ is present. The waters and soils of some areas, as in the inland areas of the United States, do not contain the minimal trace quantities of iodine required by the normal diet. Thus for many years so-called iodized table salt with a content of about 0.01% potassium iodide has been available. This preventive practice has been accredited with forestalling goiter and associated glandular disturbances in untold thousands of instances. For health reasons, iodine supplements also are added to cattle feeds.

Iodine tablets provide an easy means for sterilizing drinking water in small portions, usually resulting in less odor and taste objections than chlorine compounds for the same purpose. Iodine chemicals are widely used in photography and printing reproduction processes.

The production of vanadium metal essentially is the calcium reduction of vanadium pentoxide in the presence of iodine and is known as the McKechnie-Seybolt process. The reaction is carried out in a steel bomb at about 700°C. The end-products are vanadium metal, lime, and calcium iodide. A similar iodide process is used in the production of high-purity zirconium.

Chemistry and Compounds. Iodine exhibits in common with the other halogens a marked readiness to form singly charged negative ions, as would be expected from the fact that these atoms need only one electron to acquire an inert gas configuration. However, of the four common halogens, iodine has the lowest electron affinity (3.2 eV) due to more effective screening of the nucleus. The iodides range in character from ionic to covalent compounds, many of them, such as hydrogen iodide, having bonds of intermediate nature. Iodine is the most electropositive of the common halogens, functioning as the positive univalent "ion" I^+, as in the compound iodine perchlorate, which, however, is not a salt, as well as in forming trivalent complex radicals, such as IO^+. Iodine also forms essentially covalent linkages with negative elements, in which it has positive valences 1, 3, 5, or 7.

In binary combination with oxygen, however, only one simple compound has been isolated, iodine pentoxide, I_2O_5, a white compound. The yellow I_2O_4, prepared from sulfuric and iodic acids, is considered to be made up of 3+ and 5+ iodine, and the structure iodyl iodate, $(IO^+)(IO_3^-)$, is assigned to it, in which the trivalent state is stabilized by the acid radical. Similarly, yellow tetraiodine enneaoxide, I_4O_9, is considered to have the structure $(I^{3+})(IO_3^-)_3$, in which again the trivalent state is stabilized by the acid radicals.

Hydrogen iodide, HI, is the least stable of the four common hydrogen halides, and correlatively, the best reducing agent, readily reducing vanadic acid, nitrous oxide to ammonium, nitrous acid to nitric oxide, and HNO_3 to nitrous acid. Because it is so readily oxidized, it cannot be prepared by action of

H_2SO_4 on an iodide, but can be made by the action of weak acids, e.g., H_2S, upon iodine, or by hydrolysis of certain iodides. It can be prepared by direct combination of hydrogen and iodine vapor on a platinum catalyst. It is also liberated in many organic iodination reactions, such as the reaction of iodine and refluxing tetralin. The H—I bond is considered to be partly covalent. HI is a monoprotic acid, and is stronger than hydrogen chloride and hydrogen bromide.

The oxyacids of iodine are essentially covalent compounds, the known acids being hypoiodous acid, HIO, iodic acid, HIO_3, and the various periodic acids. Hypoiodous acid is formed, along with iodide ion, by dissolving iodine in dilute alkali, or by action of mercuric oxide upon iodine and water. On standing hypoiodous acid disproportionates to iodic acid and hydriodic acid. Hypoiodous acid is a powerful oxidant. It is also amphiprotic: $H^+ + IO^- \leftrightarrows HIO \leftrightarrows I^+ + OH^-$, $pK_A = 10.4$, $pK_B = 9.49$. The evidence for the formation of the I^+ ion is the existence of a number of compounds of composition $I(r)_n X$, where r is pyridine or some other nitrogen organic base, n is 1 or 2, and X is hydroxyl, nitrate, or chlorate ion. The conductivity of liquid iodine also indicates ionization into solvated I^+ and I^-.

Hydriodic acid (HI) is a colorless solution formed when hydrogen iodide gas is dissolved in water, commercially of strength 10% HI, frequently colored brown by iodine. There is a maximum constant boiling point 127°C (774 mm) at 57% HI (distillate) for mixtures of hydriodic acid and water. Hydriodic acid is used in the preparation of iodides, and as an important reagent in organic chemistry.

All metallic iodides except silver iodide, mercurous iodide, mercuric iodide, lead iodide, cuprous iodide, thallium iodide, and palladium iodide, are soluble. The iodides of antimony, bismuth, tin require a little free acid to keep them in solution.

Dilute hydriodic acid reacts with hydroxides, oxides, carbonates, sulfides, metals in a manner chemically analogous to dilute HCl; with solutions of some salts, e.g., silver nitrate, to yield the corresponding iodide, e.g., silver iodide, precipitate. Higher strengths of hydriodic acid react with oxygen of the air upon standing to yield free iodine, which imparts a brown color to the solution, thus indicating the reducing character of the acid.

Hydroidic acid is made by the reaction (1) of iodine and hydrosulfuric acid (or sulfurous acid), (2) of phosphorus plus iodine plus water, with subsequent distillation in all cases.

Iodic acid, HIO_3, is commonly prepared by oxidizing iodine with HNO_3. It is a strongly oxidizing acid, oxidizing iodide to iodine, sulfite to sulfate, and H_2S to sulfur. It reacts vigorously even with dry carbon, phosphorus, or organic matter.

Iodate ion in aqueous solution appears to be $I(OH)_6^-$, and the iodine atom in crystalline periodates always is coordinated to six oxygen atoms, three nearest neighbors on one side and three next nearest neighbors on the other side. Thus, potassium iodate, KIO_3, has distorted perovskite structure.

In the broad picture, the iodides, like the other halides, range in character from completely ionic structures to covalent ones. The transition is clearly exhibited by the iodides of the first four groups in the (extended) periodic table, potassium iodide, KI, being ionic and titanium(IV) tetraiodide, TiI_4, essentially covalent. On the righthand side of the periodic table, even the group I elements form bonds with iodine that exhibit a considerable degree of covalence. In general, the ionic iodides are the most soluble of the halides of the given element, e.g., sodium iodide, NaI, is the most soluble sodium halide, while the covalent (or partly covalent) iodides are the least soluble, e.g., silver iodide, AgI. These effects are correlated with the size and polarizing power of the cation and the increasing size and polarizability of the halogen ion, as is increase of reducing character and

stability of coordination complexes. A related fact is the readiness of formation of the complex ion, I_3^-, upon dissolving iodine in an aqueous solution of KI. The variation in character of iodine solutions in various organic solvents as well as water, is also attributed to complex formation.

Because iodine is the least electronegative of the four common halogens, it forms a relatively larger number of compounds with the other three. They include iodine trichloride, iodine chloride, iodine bromide, iodine fluoride, iodine trifluoride, iodine pentafluoride, and iodine heptafluoride. These are reactive compounds, especially the lower ones, which enter into reactions with many organic and some inorganic substances. With ICl and organic substances, the end product is usually an iodine or chlorine substitution product, the solvent being important in determining which is formed. With inorganic compounds, the addition product is often the final product, thus ICl and antimony(V) pentaiodide $SbCl_5$ give a product of composition $ISbCl_6$, which ionizes to I^+ and $SbCl_6^-$.

Several oxyfluorides exist: IOF_3 (iodyl hexafluoriodate, $[IO_2][IF_6]$), iodyl fluoride, IO_2F and periodyl fluoride, IO_3F.

Related to the complex ions, as well as the interhalogen molecules, are their association products, the polyhalide complexes. For iodine they include the alkali (and ammonium) triiodides, along with higher compounds such as that with cesium, Cs_2I_8, ammonium, NH_4I_5, tetraethylammonium iodide, $(C_2H_5)_4NI_7$, tetramethylammonium iodide, $(CH_3)_4NI_9$, and benzene, $KI_9 \cdot 3C_6H_6$. They also include compounds of iodine with alkali metals and one or two other halogens, such as NH_4IBr_2, $KICl_2$, $RbICl_2$, $CsICl_2$, $HICl_4 \cdot 4H_2O$, $CsFIBr$, $RbClIBr$, $CsClIBr$, and KIF_6. Most of these compounds hydrolyze readily, decompose on heating to give the metal halide of greatest lattice energy, and ionize to give ions such as $ClICl^-$, $BrIBr^-$, and $BrII^-$.

Structural studies show that the complex $HICl_4 \cdot 4H_2O$ contains a positive trivalent iodine atom, and the planar ion ICl_4^- is one of the most stable of the polyhalide anions, as might be expected from consideration of the large size of the iodine atom and the small size of the chlorine atoms.

The difluoroiodate ion, $IO_2F_2^-$, is a trigonal bipyramid with two apical fluorine atoms, two equatorial oxygen atoms and one equatorial nonbonding electron pair surrounding the central iodine atom.

Iodine forms many organic compounds, chiefly by replacement of hydrogen or by addition reactions at double bonds.

Periodic acid H_5IO_6 has been isolated as a colorless solid, mp about 130°C, and at 138°C begins to decompose, metaperiodic acid HIO_4 being formed and at higher temperatures iodine pentoxide plus oxygen plus water. $H_4I_2O_9$ and H_3IO_5 have been reported as fairly well established in identity; in solution the evidence points to the presence of HIO_4. Prepared by reaction of iodine and perchloric acid.

Paraperiodic acid H_5IO_6, is obtained from sodium paraperiodate, formed by action of chlorine upon a NaOH solution containing I_2. On vacuum drying, paraperiodic acid yields metaperiodic acid, HIO_4, and dimesoperiodic acid, $H_4I_2O_9$; which form heteropoly acids with a number of oxides and acids. The periodic acids and their salts are strong oxidizing agents both with inorganic and organic compounds. Although, in general, the iodates do not disproportionate to give periodates, barium iodate, $Ba(IO_3)_2$, on ignition gives barium paraperiodate, $Ba_5(IO_6)_2$, iodine and oxygen.

Sodium periodate $Na_2H_3IO_6$ is formed by reaction of sodium iodate plus sodium hydroxide plus chlorine (sodium chloride also formed), and the periodate separates as crystals from the medium. In solution, it is stated, periodate gradually forms ozone and iodate at the ordinary temperatures.

Metallic periodates are solids, slightly soluble in water. Periodates, when heated, evolve oxygen with simultaneous formation of iodate, which is decomposed at higher temperatures. Periodate in acid solution oxidizes hydrosulfuric acid or sulfurous acid to H_2SO_4, oxalic acid to CO_2, manganous to manganate, and with hydrogen peroxide yields oxygen and iodate.

See the list of references at the end of the entry on **Chemical Elements.** The biological aspects of iodine are covered in **Iodine (In Biological Systems).**

IODINE (In Biological Systems). Iodine is not required by plants, but if iodine is present in the soil, it is taken up by most plants and moves on into the diets in forms that are effective in preventing goiter. In areas where the soils are high in iodine, ground water is also high in iodine, but the food supply is still the major source of iodine for people in these areas. Seafoods are good sources of dietary iodine.

Many of the iodine-deficient regions of the world have been identified. They are generally either mountainous or in the centers of continents, and distant from the oceans in the prevailing wind directions. Studies of the geochemistry of iodine indicate that this element is volatilized from oceans, carried overland by winds, and deposited on the soil by rain. The mountainous areas are low in iodine because little of that volatilized from the seas reaches sufficient altitude to be deposited in high altitudes. In some areas, the younger soils have less iodine than the older ones because of less time for the geochemical processes to build up the iodine level.

Although the amount of iodine in the soil is the primary factor determining iodine levels in food crops from various regions, the level of iodine in plants and the dietary requirements for iodine are modified to some extent by the plants themselves. There are important differences among plant species, and even among varieties of the same species, in their tendency to take up iodine from the soil. Certain plants, especially some of the genus *Brassica*, such as cabbage, contain compounds called *goitrogens*, which interfere with the effect of iodine on the thyroid gland. The amount of iodine required to protect animals, including humans, against goiter depends upon the kinds of plants in the diet as well as the iodine level characteristic of the soils of the region.

Iodine in food crops can be increased by adding iodine compounds to soil, but that is a very inefficient way of insuring adequate dietary levels of the element. Much of the iodine added to the soil would be leached out and returned to the seas before it could be taken up by the crop plants. The use of iodized table salt is such an effective way of supplying this element that there is little need to include iodine in fertilizers.

As a historical fact, 3,5-diiodotyrosine (iodogorgoic acid) has been discovered in sponges and corals long before there was any hint concerning thyroid hormone structure even in mammalian forms. Since this time, iodotyrosines have been demonstrated in algae as well as in many animals possessing a horny skeleton. It has been suggested that iodide in the water is activated by peroxidases due to the presence of oxygen in the water and the resulting iodine is accepted by tyrosine in the protein molecule, similarly to the process in vertebrates. In animals possessing an exoskeleton, the presence of benzoquinones is part of the formation of scleroproteins by a quinone tanning process, and the iodination of tyrosine may be related to this in some way. In rotochordates, the endostyle, a structure secreting mucus, has been found capable of iodination of protein present in the mucus which is then secreted into the alimentary canal.

The next phylogenetic development takes place in the vertebrates, in the most primitive members of which there is a structure located in the hypopharynx similar to a thyroid capable of collecting iodine and of forming iodinated protein, which is broken down by a protease, liberating iodinated amino acids. In some of these forms, small amounts of thyroxine are actually found in addition to the iodinated tyrosines. In other vertebrates, culminating in the amphibia and higher vertebreas, a thyroid gland is present in which the iodinated protein is held in a storage form known as "colloid." In some, but not all, of these forms, hormonal material liberated by action of proteolytic enzymes is secreted into the bloodstream and plays an essential role in the development of the young animal as well as in the behavior and metabolic activity of the adult. Because of the process known as *ontogeny*, it is not surprising that the development of the thyroid gland in the human embryo commences with a structure located near the alimentary canal, which then separates and develops its peculiar follicular structure. Not until this type of development has occurred can a genuine function for the iodinated substances be demonstrated.

In the thyroid gland, a series of specific chemical reactions can be demonstrated. See accompanying figure. Some of these reactions take place in the absence of any apparently specific synthesis of iodotyrosines, and it has not been fully explained how many of these steps require enzymes, even in mammalian

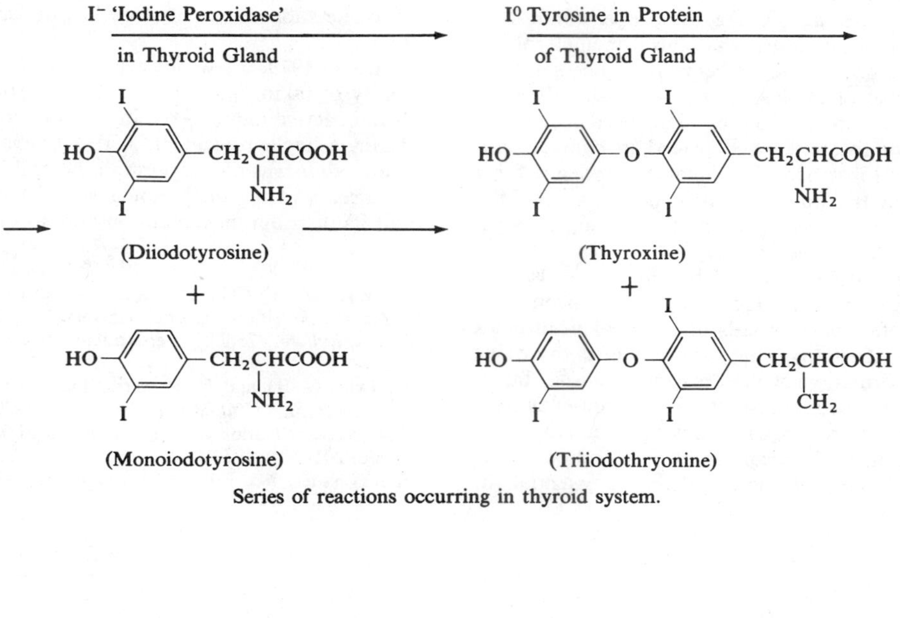

Series of reactions occurring in thyroid system.

forms, although these processes are usually considered as enzymatic. It is possible to iodinate tyrosine in soluble proteins *in vitro* by addition of elemental iodine, and under these circumstances thyroxine and triiodothyroxine will also be formed, in company with small amounts of iodohistidine.

In amphibian vertebrates, thyroxine is known to be essential for metamorphosis from immature to mature forms. There is considerable evidence that after metamorphosis has occurred, the thyroid gland is no longer essential, although it may be involved in seasonal changes, such as molting.

There is no such clear-cut differentiation as metamorphosis in the mammal, but development is an extremely complex process and has been shown to depend upon the presence of adequate amounts of thyroid hormones. Deficient development, especially of the central nervous system, is marked in children suffering from thyroid deficiency early in life, and this inadequacy cannot be overcome completely by medication commenced after the first few weeks. In the adult, thyroxine is important in the maintenance of energy turnover in most of the tissues of the body, such as the heart, skeletal muscle, liver, and kidney. Other physiological functions, most notably brain activity and reproduction, are also dependent upon thyroxine, although the metabolic rates of the tissues concerned in these functions do not seem to be altered.

A great deal of work has been done on determining the portions of the thyroxine molecule which are essential to biological activity. The fact that the hormone is an amino acid is almost certainly due to the widespread existence of the excellent iodine acceptor, tyrosine. Thus, the deaminated and decarboxylated metabolic product of thyroxine, tetraiodothyroacetic acid, has been shown to have appreciable biological activity, although quantitatively less than that of thyroxine itself. As for the halogens present on the diphenylether portion of the molecule, bromines and chlorines are also active, although diminishing considerably in that order from iodine.

Assuming there are some quite definite structural requirements for thyroid hormone activity, it is important to inquire into the specific actions of this material. It seems clear that thyroxine does not participate directly in any enzyme system, but rather affects the function of many systems, presumably by some far more general process. One of the earliest of such demonstrated actions was the uncoupling of oxidation from formation of high-energy phosphate compounds, such as adenosine triphosphate. However, such uncoupling is also produced by many other substances not showing thyroid hormone-like effects, and it has been shown that thyroxine is actually capable of accelerating coupled reactions under the proper conditions. From this evidence, it is suggested that the principal role of thyroxine may be the acceleration of enzyme processes ordinarily limiting the level of metabolic turnover. Mitochondria isolated from broken cell preparations by high-speed centrifugation have been shown to swell when placed in contact with thyroxine and similar substances. This may be evidence for a membrane function of the hormone, although it is not fully understood as to how this may alter cellular function in such specific manner as the hormone does *in vivo*.

An acceleration of protein turnover by thyroxine also has been shown, implying that the hormone may alter various processes by a specific effect on synthesis of certain key proteins involved in enzymatic reactions. Thus, not only does thyroxine increase the rate of formation of new protein material, but it also may be responsible for the transformation of non-enzymatically active protein into protein with enzymatic activity. The hormone has been shown to be capable of acceleration of the synthesis of urea cycle enzymes and probably is essential for

the production of a sodium ion transporting mechanism, both of which are essential in the metamorphic transformation of larval forms into mature amphibia.

Steps in the synthesis of thyroid hormone include: (1) Active concentration of inorganic iodides within the thyroid epithelial cells. Concentration achieved is approximately 30 times that of plasma concentration. The so-called trapping of iodide is stimulated by thyroid-stimulating hormone (released by the pituitary gland). When present, this step is competitively opposed by thiocyanate and perchlorate ions. (2) Next, the inorganic iodide is oxidized to an organic form, in which peroxidase participates. The iodine becomes part of tyrosine residues in the thyroglobulin molecule. In this way, monoiodotyrosine (MIT) and diiodotyrosine (DIT) are formed. (3) The MIT and DIT are coupled by way of an ether linkage to form tetraiodothyronine (T_4) and triiodothyronine (T_3).

It has been observed that non-iodinated thyronine is not found in the thyroid gland.

The storage capacity and slow release mechanism of thyroid hormone appear to be unique among the endocrine glands. Usually a reserve of 100 days' needs (about 80,000 micrograms) are stored in the gland. For diseases related to malfunction of the thyroid gland.

Sometimes iodides in drugs can cause a condition known as *eosinophilia*, in which there are reduced counts of eosinophil in the plasma. However, there are many other possible causes of this condition. Eosinophilic granulocytes are components of the blood in which the cytoplasm is filled with coarse acidophilic granules which may be spherical or rod-shaped, and the nucleus is bilobed and stains deeply.

Iodine Deficiency and Supplementation. It is estimated that the mature mammal contains only 0.0004% iodine, most of which is present in the thyroid gland. This micro amount is extremely important and must be established and maintained throughout the life of the animal as the result of ingesting small quantities of the element from various feed inputs.

Iodine deficiency is manifested by impairment of thyroid gland function. Early symptoms of iodine deficiency are the appearance of a swelling or slight goiter, depressed appetite, listlessness, difficulty in swallowing, a hacking cough, and a tendency for the eyes to be tear-filled. Since corrective measures usually are simple, iodine toxicity does not usually occur, but overtreatment (particularly as the result of careless formulation of feedstuffs for livestock) can occur. Generally, through effective homeostatic mechanisms, livestock can tolerate levels of iodine far greater than their minimum requirements.

Iodine supplements include kelp (source), potassium iodide, etc.

In late 1979, a few physicists brought attention to the desirability of taking measures to protect populations from uptake of radioactive iodine, ^{131}I, in case of exposure to the substance during a nuclear accident. A readily available supply of potassium iodide was one suggestion. Not all authorities agreed with the need and the methodologies and formats of the substance and its distribution remain obscure as of the early 1980s.

References

Allaway, W. H.: "The Effects of Soils and Fertilizers on Human and Animal Nutrition," Cornell University Agricultural Experiment Station, *Bulletin 378*, U.S. Department of Agriculture, Washington, D.C. (1975).

Baldwin, D. B., and D. Rowett: "Incidence of Thyroid Disorders in Connecticut," *Jrnl. Amer. Med. Assn.*, **239**, 742 (1978).

Carter, L. J.: "Nationwide Protection from Iodine-131 Urged," *Science*, **206**, 201–204 (1979).

Kirchgessner, M., Editor: "Trace Element Metabolism in Man and

Animals," Institut für Ernährungsphysiologie, Technische Universität München, Freising-Weihenstephan, Germany (1978).

Kutsky, R. J.: "Handbook of Vitamins and Hormones," Food and Nutrition Press, Westport, Connecticut (1973).

Sterling, K.: "Thyroid Hormone Action at the Cell Level," Parts 1 and 2, *New England Jrnl. Med.*, **300**, 117, 173 (1979).

Thijs, L. G., and J. D. Wiener: "Ultrasonic Examination of the Thyroid Gland: Possibilities and Limitations," *Amer. Jrnl. of Med.*, **60**, 96, (1976).

Underwood, E. J.: "Trace Elements in Human and Animal Nutrition," 4th Edition, Academic, New York (1977).

Welby, M. L.: "Laboratory Diagnosis of Thyroid Disorders," *Adv. Clin. Chem.*, **18**, 103 (1976).

Werner, S. C., and S. H. Ingbar, Editors: "The Thyroid: A Fundamental and Clinical Text," 2nd Edition, Harper & Row, Hagerstown, Maryland (1978).

IODINE VALUE (Unsaturates).   Vegetable Oils (Edible).

ION.   An atom or molecularly bound group of atoms which has gained or lost one or more electrons, and which has thus a negative or positive electric charge, and sometimes a free electron or other charged subatomic particle. Ions may be produced in gases by the action of radiation of sufficient energy; ionic solids are built up of ions bound together by their electrostatic forces, and when dissolved in a polar liquid, such as water, the salt dissociates into its ions, which have an independent existence.

Ions may be characterized in various ways. When they are described by the sign of their electric charge, they are identified as positive, negative, or amphoteric (or zwitter), the latter being an ion which carries both a positive and a negative charge, commonly at opposite ends of a long, or fairly long, chain, as in the case of ions of amino acids. See **Amino Acids.**

Ions may also be described by their atomic structure, when they consist of more than one atom. Thus, a *complex ion* is a complex electrically charged radical or group of atoms such as $Ag(CN)_2^-$ or $Cu(NH_3)_2^+$, which may be formed by the addition to an ion of another ion or ions, or of an electrically neutral radical or molecule. When the particle combined with the ion is a large molecule, and the attachment is essentially by an adsorption process, the complex ion is called a *heteroion*; when the complex particle consists of a simpler ion combined with one or more molecules of water, it is known as an *aquoion* or *hydrated ion*. On the other hand, charged molecules, commonly produced by electrical discharges through gases, are often called *molecular ions*.

In meteorology, there are two special types of "ions" that enter into atmospheric processes: small "ions" and large "ions."

A *small ion* (also called a "light ion" or "fast ion") is the type that has the greatest mobility; hence, collectively, it is the principal agent of atmospheric conduction. The exact physical nature of the small ion has never been fully clarified, but much evidence indicates that each is a singly charged atmospheric molecule (or, rarely, an atom) about which a few other neutral molecules are held by the electrical attraction of the central ionized molecule. Estimates of the number of satellite molecules range as high as twelve. When freshly formed, by any of several atmospheric ionization processes, small ions are probably singly charged molecules; but after a number of collisions with neutral molecules, they acquire (actually, in a fraction of a second) their cluster of satellites.

A *large ion* (also called a "slow ion" or a "heavy ion") is an ion of relatively large mass and low mobility, produced by the attachment of a small ion to an Aitken nucleus.

ION-EXCHANGE RESINS.   Ion exchange has been defined as the reversible exchange of ions between a liquid phase and a solid phase which is not accompanied by any radical change in the solid structure. Ion exchangers are insoluble acids or bases and consequently they are capable of reacting as ordinary acids and bases with the particular exception that one of the reaction products is removed from solution as an insoluble salt.

The only commercial inorganic exchangers used in any appreciable quantity are the natural and synthetic aluminosilicate cation exchanger gels. The original greensands and zeolites can be used only to soften water by exchanging sodium for the hardness-forming calcium and magnesium ions in the water. They lose some of their capacity in acid or alkaline waters and are objectionable because, in waters of low desired silica content, the effluent from them contains silica. These substances are inorganic.

From the point of view of its synthesis, there are two methods of forming an organic ion exchange resin. One consists in building the ionic groups into a resin structure prior to polymerization; i.e., the ionic groups are integral parts of the monomer. The second method consists in first forming a polymer and subsequently introducing the ionic groups into the polymer structure. The first method has the advantage that the resulting product is a homogeneous mass of good mechanical strength. In either case, the ionic groups cause the linear polymers to which they are attached to be soluble and chemically unstable. These undesirable properties are overcome by cross-linking the polymers. The higher the degree of cross-linking, the less the solubility, the less the amount of swelling in solution, and the more impermeable the gel formed. In practice, the resins are made to have the densest structure compatible with a satisfactory rate of ionic diffusion.

It may thus be seen that ion-exchange resins are a special type of polyelectrolyte, i.e., a cross-linked polyelectrolyte. Their properties are determined to a great extent by the nature and number of the fixed and mobile charges and of the cross-linking bonds. The fixed charges or ionic groups determine the acidity

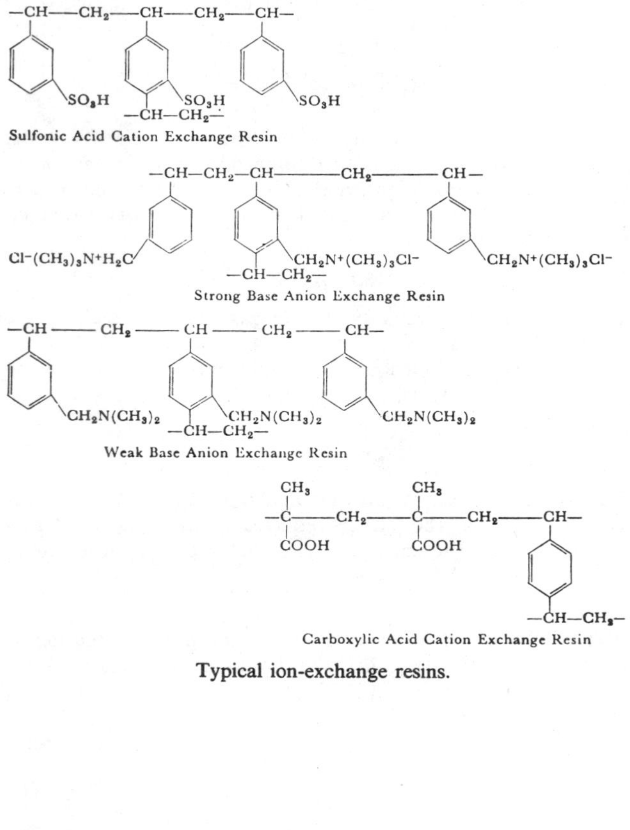

Typical ion-exchange resins.

or basicity of the resins, their capacity to exchange ions, and the amount of hydration of the resins. The mobile or exchangeable ions determine the type of exchange and also the amount of hydration of the resins. The degree of cross-linking determines the degree of swelling of the resin, the rate of diffusion of ions through the resin, and, to some extent, the exchange equilibrium constant. The degree of porosity of the vinyl-type polymers is controlled by regulating the amount of cross-linking (polyfunctional) agent used; with the (usually phenolic) condensation polymers, the amount of formaldehyde used controls the degree of cross-linking. The skeletal structure, although inert with respect to ion exchange behavior, determines the stability and life of the resin in actual use.

Considerable effort has been devoted to the synthesis of resins showing high specificity for particular ionic species and to the synthesis of resins of special physical form and shape. Most attempts to synthesize highly selective resins have been based upon the incorporation of functional groups usually found in the selective organic analytical reagents. For chemical structures of some resins, see the accompanying figure.

Cation-exchange resins usually contain a sulfonic acid, carboxylic acid, or phosphonic acid group. Anion-exchange resins usually contain a quaternary ammonium, secondary amine, or tertiary amine active group. In addition to use in water purification, ion-exchange resins are widely used for the purification of chemical materials, such as formaldehyde, methanol, glycerin, sorbitol, gelatin, sucrose, dextrose, lactose, citric acid, uranium, chromic acid, and copper.

IONIC CRYSTAL. A crystal which consists effectively of ions bound together by their electrostatic attraction. Examples of such crystals are the alkali halides, including potassium fluoride, potassium chloride, potassium bromide, potassium iodide, sodium fluoride, and the other combinations of sodium, cesium, rubidium or lithium ions with fluoride, chloride, bromide or iodide ions. Many other types of ionic crystals are known.

IONIC MOBILITY. 1. The ratio of the average drift velocity of an ion in solution to the electric field. It is expressed by the relationship

$$\mu_+ \text{ or } \mu_- = \frac{\lambda_+ \text{ or } \lambda_-}{F}$$

in which μ_+ or μ_- is the mobility of the ion, λ_+ or λ_- is the ion conductance, i.e., the contribution of the particular ion to the equivalent conductance, and F is the Faraday constant.

2. For gaseous ions in an electric field, the quantity k defined by the relationship

$$k = vp / E$$

where v is the drift velocity, p, the gas pressure, and E, the electric field strength.

3. Conduction of electricity in ionic crystals is due to the motion of lattice defects, either of the Schottky or Frenkel type. The mobility is given by

$$\mu = (eD_0 / kT)e^{-E / kT}$$

where D_0 is a numerical constant, and E is an activation energy, which depends on the energy required to make a defect and on the height of the energy barrier that must be surmounted in order that the defect may move.

IONIZATION. A process which results in the formation of ions. Such processes occur in water, liquid ammonia, and certain other solvents when polar compounds (such as acids, bases, or salts) are dissolved in them. Dissociation of the compounds occurs, with the formation of positively- and negatively-charged ions, the charges on the individual ions being due to the gain or loss of one or more electrons from the outermost orbits of one or more of their atoms. The ionization of gases is a process by which atoms in gases similarly gain or lose electrons, usually through the agency of an electrical discharge, or passage of radiation, through the gas.

Ionization by collision is an ionization process occurring by removal of an electron or electrons from an atom as the result of the energy gained in a collision with a particle (or quantum of radiation) possessing sufficient energy.

Specific ionization is the number of ion pairs formed per unit distance along the track of an ion passing through matter. This is sometimes called the total specific ionization to distinguish it from the primary specific ionization, which is the number of ion clusters produced per unit track length. The relative specific ionization is the specific ionization for a particle of a given medium relative either to that for (1) the same particle and energy in a standard medium, such as air at 15°C and 1 atmosphere, or (2) the same particle and medium at a specified energy, such as the energy for which the specific ionization is a maximum.

Total ionization is a term used to denote either the total specific ionization (defined above); or the total electric charge on the ions of one sign when the energetic particle that has produced these ions has lost all of its kinetic energy. For a given gas the total ionization is closely proportional to the initial energy and is nearly independent of the nature of the ionizing particle. It is frequently used as a measure of particle energy.

Minimum ionization is the smallest possible value of the specific ionization that a charged particle can produce in passing through a particular substance. When the specific ionization produced along the path of a charged particle is plotted as a function of the particle energy, minimum ionization appears as a broad dip, bound on one side by a rather sharp rise for decreasing particle energy, and on the other side by a gradual rise for increasing particle energy. For singly charged particles in ordinary air, the minimum ionization is about 50 ion pairs per centimeter of path. In general, it is proportional to the density of the medium and the square of the charge of the particle. It occurs for particles having velocities of 95% of the velocity of light, which corresponds to a kinetic energy of 1 MeV for an electron, 2 BeV for a proton and 8 BeV for an alpha-particle.

Ionization potential is the energy per unit charge, for a particular kind of atom, necessary to remove an electron from the atom to infinite distance. The ionization potential is usually expressed in volts, and is numerically equal to the work done in removing the electron from the atom, expressed in electron-volts. See also **Chemical Elements.**

IONIZED GASES. Various agencies, such as fast-moving electrons, alpha particles, various forms of radiation, and high temperature, are capable of dislodging electrons from atoms or molecules of a gas and thereby leaving them positively charged. Some of the dislodged electrons may attach themselves to other molecules and render them negatively charged. In some cases, two or more electrons may be removed from the same molecule, or a molecule with a double positive charge may unite with a singly charged negative molecule, forming a singly charged complex, etc. Such charged atoms, molecules or molecular groups are called ions, and their production from neutral molecules is called ionization. The complete separation of an electron from a molecule or an atom requires a definite amount of energy. This may be expressed in ergs, but is more commonly given in electron volts (1.59×10^{12} erg), its value being the ionization potential. A lesser amount of energy may excite the atom or molecule to emit radiation, but will not ionize it.

If an ionized gas is left to itself, the ions soon recombine and become neutral. But if it is subjected to an electric field, as in an ionization chamber, the ions pass to the electrodes, such migration being an "ionization current." Such currents, commonly called electric discharges, are attended by diverse phenomena and vary widely in character from the silent glow discharge to the lightning stroke.

At ordinary pressures, discharges may be classified into four types: (1) If the voltage between two electrodes in open air is gradually increased, the electrodes become surrounded with a luminosity. This "glow" or "corona" gives way, at the negative electrode first, to (2), a "brush," composed of hair-like branches. (3) Finally, the disruptive spark passes. (4) Under other conditions, an arc may be formed. If, however, the electrodes are enclosed in a tube and the pressure reduced, a point is reached at which the tube becomes filled with a beautiful luminosity. Close examination shows this to have structure. Very close to and surrounding the cathode is a thin, luminous layer c, the cathode glow (see figure); and outside this, the Crookes dark

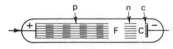

Elementary gas-discharge tube.

space C. Next, extending toward the anode, is the short negative glow n, then the Faraday dark space F. From this to the anode extends the long positive column p, with its regular, transverse striations. As the pressure is further reduced, the cathode dark space enlarges and the other features dwindle toward the electrodes until they finally disappear at about 0.001 mm pressure. From this point on, the cathode rays are the predominant feature.

Upon exploring the discharge in a Crookes tube with suitable probes, it is found that in certain regions the positive and negative ions are so nearly equal in number as to neutralize each other's effect. Such a region is called a "plasma." The plasma may be surrounded by a "sheath" of ionized gas in which ions of one sign greatly predominate, the effect being that of a space charge.

IONIZING PARTICLE. A particle that produces ion pairs in its passage through a substance. Ionizing particles may be divided into two groups: (1) *directly ionizing particles*—charged particles (electrons, protons, alpha particles, and so on) having sufficient kinetic energy to produce ionization by collision; and (2) *indirectly ionizing particles*—uncharged particles, such as neutrons and photons, which can liberate directly ionizing particles or can initiate a nuclear reaction.

Ionizing radiation is any radiation consisting of directly or indirectly ionizing particles, or a mixture of both. Ionizing radiation, unless controlled, poses a biological and environmental hazard.

IRIDESCENCE. The display of colors of the rainbow, commonly by interference of light of various wavelengths reflected from superficial layers in the surface of a substance.

IRIDIUM. Chemical element symbol Ir, at. no. 77, at. wt. 192.20, periodic table group 8 (transition metals), mp 2410°C,

bp 4130°C, density, 22.42 g/cm³ (solid at 17°C), 22.8 (single crystal at 20°C). Elemental iridium has a face-centered cubic crystal structure. The two stable isotopes of iridium are ^{191}Ir and ^{193}Ir. The ten unstable isotopes are ^{187}Ir through ^{190}Ir, ^{192}Ir, and ^{194}Ir through ^{198}Ir. In terms of earthly abundance, iridium is one of the scarce elements. In terms of cosmic abundance, the investigation by Harold C. Urey (1952), using a figure of 10,000 for silicon, estimated the figure for iridium at 0.0025. No notable presence of iridium in seawater has been found. The element was identified and named by Tennant (England) in 1804. The electronic configuration is $1s^2 2s^2 2p^6 3s^2 3p^6 3d^{10} 4s^2 4p^6 4d^{10} 4f^{14} 5s^2 5p^6 5d^9$.

Metallic Ir is not attacked by any mineral acid unless it is very finely divided. It can be brought into solution by fusion with indium at 800–1000°C to yield a soluble alloy.

When fused with Na_2O_2, or an alkaline oxidizing flux, water-soluble iridates(IV) are formed. The finely divided metal is oxidized by air or O_2 at red heat to the dioxide, which decomposes into its elements at higher temperature. The valences of Ir are 1–6, the 3 and 4 valences being most common.

Iridium black is only slightly soluble in aqua regia. When fused with alkalies and alkaline nitrates or Na_2O_2, the metal is converted to an acid-soluble form. The metal at red heat reacts to a small extent with O_2, S, and P. At elevated temperature, the metal is attacked by Cl_2 and F_2. When fused with NaCl and treated with Cl_2, the water-soluble sodium hexachloroiridate(IV), Na_2IrCl_6, is formed.

Iridium(III) hydroxide is a yellow-green or blue-black compound soluble in alkali and insoluble in water. It is made by adding KOH to a solution of potassium hexachloroiridate(III), K_3IrCl_6, in an inert atmosphere. When the trihydroxide is heated, a mixture of iridium(IV) oxide and the metal is formed. Iridium(III) oxide, Ir_2O_3, is made by fusing potassium hexachloroiridate(IV) with Na_2CO_3 and then leaching the mixture with water. At about 1100°C, both the oxide and the hydroxide decompose into the metal and O_2. When a solution of Ir is heated with $NaBrO_3$ at a pH of approximately 6, the darkblue precipitate $Ir(OH)_4$ or $IrO_2 \cdot H_2O$ is formed. This water-insoluble compound, when heated to 350°C in N_2, loses its H_2O and is converted to the black oxide, IrO_2.

When iridium(III) chloride is heated in Cl_2 at 773–798°C, iridium(I) chloride is formed. The copper-red crystals are insoluble in acids and alkalies. The compound sublimes in Cl_2 at 790°C and decomposes into Cl_2 and metallic Ir. Iridium(II) chloride is stable from 763 to 773°C. The brown crystals are insoluble in H_2O, acids, and alkalies. Iridium(III) chloride is an insoluble green compound made by reacting the elements at about 600°C. The reaction is catalyzed by CO. Iridium(III) chlorides are prepared by reducing the corresponding iridium-(IV) chlorides with oxalate or SO_2. Iridium(III) bromide is made by dissolving iridium(III) hydroxide in HBr. The blue solution yields olive-green crystals of $IrBr_3 \cdot 4H_2O$. When heated, the anhydride is formed. The triiodide is formed in analogous fashion as a trihydrate. Iridium(IV) chloride can be made in solution by the action of Cl_2 or aqua regia on ammonium hexachloroiridate(III). The relative insolubility of ammonium hexachloroiridate(IV), $(NH_4)_2IrCl_6$, makes it useful in the purification of Ir. The compound may be reduced in H_2 to the metal. The analogous sodium salt is very soluble, and the potassium salt is relatively insoluble.

Iridium(VI) fluoride is made from the element at 300–400°C. The bright yellow solid melts at 44°C and boils at 53°C. Potassium hexafluoroiridate(V) also has been prepared. Iridium fluoride can be made by heating the hexafluoride with the metal in a sealed tube at 150°C or heating it with glass above 200°C, at which temperature the glass reduces it to the tetrafluoride. This yellow solid melts at 106–107°C and boils above 300°C. Iridium(III) fluoride is formed by reducing the tetrafluoride with glass for 12–18 hr at 430–450°C.

Iridium(II) sulfide is formed by burning the metal in sulfur or by heating a higher sulfide at 700°C in N_2. This black solid is insoluble in H_2O, acids, and aqua regia. Iridium(III) sulfide, Ir_2S_3, is formed as a brown-black insoluble precipitate by passing H_2S through a hot acidic solution of an iridium(III) chloride. The precipitation is usually not quantitative. This amorphous black solid is not attacked by HNO_3 but is slowly dissolved by aqua regia or fuming HNO_3. The brown insoluble iridium-(IV) sulfide, IrS_2, is partly formed by treating a tetravalent Ir solution with H_2S; when it is prepared in this way, some iridium-(III) sulfide also is formed by reduction. Iridium(III) sulfate can be formed by dissolving iridium(III) hydroxide in H_2SO_4 in the absence of air. Trivalent iridium forms numerous cationic and anionic complexes in which it has a coordination number of 6. The ammines are extremely stable and, once formed, difficult to destroy. Tetravalent Ir also forms complex ions, but to a lesser extent.

See also **Chemical Elements; and Platinum Group.** More detail on physical, mechanical, and fabricating properties of iridium can be found in the "Metals Handbook," 9th Edition, Vol. 2, article on "Iridium" by L. Bozza, American Society for Metals, Metals Park, Ohio (1979).

—Linton Libby, Simmons Refining Company, Chicago, Illinois.

IRON. Chemical element symbol Fe, at. no. 26, at. wt. 55.847, periodic table group 8 (transition metals), mp 1536°C, bp approximately 3000°C, density, 7.874 g/cm³ for the pure solid (20°C); 7.92 for a single crystal of α-iron. Iron has a body-centered cubic crystal structure (α-iron).

Iron is a silvery-white metal, capable of taking a high polish; ductile; malleable; can be welded when hot. Pure iron is attracted by a magnet, but does not retain the magnetism. Silicon steel is preferred for electromagnets because it retains magnetism even less than pure iron. The discovery of iron was prehistoric. There are nine isotopes of iron, ^{52}Fe through ^{60}Fe. Isotopes 54, 56, 57, and 58 are fully stable, whereas four others have fairly short half-lives, ranging from 8.9 minutes for ^{53}Fe to 2.94 years for ^{55}Fe. The half-life of ^{60}Fe is approximately 3 × 10⁵ years. Iron has valence numbers of 2₊ (ferrous) and 3₊ (ferric). The hardest of the ductile metals, iron is surpassed only by cobalt and nickel in tenacity. Iron is an extremely versatile construction and engineering material and serves both in relatively pure forms, such as malleable and wrought iron, and in many hundreds of iron-base alloys of major importance, including the numerous types of steel.

Electronic configuration is $1s^2 2s^2 2p^6 3s^2 3p^6 3d^6 4s^2$. First ionization potential is 7.896 eV; second 16.5 eV. Oxidation potentials: Fe → Fe^{2+} + 2e⁻, 0.441 V; Fe → Fe^{3+} + 3e⁻, 0.036 V; Fe^{2+} → Fe^{3+} + e⁻, −0.771 V; Fe + 2OH⁻ → $Fe(OH)_2$ + 2e⁻, 0.877 V; $Fe(OH)_2$ + OH⁻ → $Fe(OH)_3$ + e⁻, 0.56 V; $Fe(OH)_4^-$ + 4OH⁻ → FeO_4^{2-} + 4H_2O + 3e⁻, E = 0.55 V (80°C; 40% NaOH). Metallic radius, 1.2412Å; ionic radius Fe^{2+}, 0.80Å, Fe^{3+}, 0.67Å. Other important physical properties of iron are given under **Chemical Elements.** See also Table 1.

Three allotropic forms of iron are known: (1) *alpha iron*, which is present below 769°C; (2) *gamma iron*, which exists between 906° and 1,404°C, and (3) *delta iron*, which occurs between 1,404° and 1,536°C. On slow cooling, the reverse changes occur, but may be slowed or partly or entirely prevented in the presence of alloying elements.

TABLE 1. SELECTED PHYSICAL PROPERTIES OF IRON

Electrical Properties

Electrical conductivity, volume, % of annealed copper at 20°C	17.75
Electrical resistivity, microohm-cm,	
at 0°C	8.9
at 20°C	9.7
Temperature coefficient of electrical resistance (0–100°C)	0.65×10^{-4}
Electrode potential (standard hydrogen scale), at 25°C, volts	−0.44
Electrochemical equivalent, milligrams/second-absolute amperes	
Fe^{2+} − Fe	0.1929
Fe^{3+} − Fe	0.2893
Magnetic susceptibility, at 820°C	$1,000 \times 10^6$

Thermal Properties

Emissivity, at 0.65 micrometers, %	40
Heat of combustion, cal/gram atom	88,355
Latent heat of fusion, cal/gram	65.5
Latent heat of vaporization, cal/gram	1,598
Specific heat, at 25°C, cal/(°C)(gram atom)	6.55
Thermal conductivity, at 0°C, (cal)(cm)/(second)(cm²)(°C)	0.18
Thermal expansion, linear coefficient,	
cm/(cm)(°C)	11.76×10^{-6}
in/(in)(°F)	6.53×10^{-6}
Average coefficient of linear expansion, at 77°F,	
cm/(cm)(°C)	12.3×10^{-6}
in/(in)(°F)	6.83×10^{-6}

Mechanical Properties

Brinell hardness, at 25°C, 99.9% Fe	82–100
Percent elongation, at 25°C, 99.9% Fe	30–40
Yield strength, psi, at 25°C, 99.9% Fe	10,000–20,000
Tensile strength, psi, at 25°C, 99.9% Fe	30,000–40,000
Modulus of elasticity, psi	28.5×10^6
Modulus of rigidity, psi	11.64×10^6
Poisson's ratio	0.28

Other Properties

Density, liquid, at 1564°C,	
g/cm³	7.00
lb/in³	0.253
Density, solid, at 20°C,	
g/cm³	7.874
lb/in³	0.284
Reflectivity (light from tungsten filament),	
2,500Å	38
10,000Å	65
Surface tension, at 1,550°C, dynes/cm	1,835–1,865
Thermal-neutron-absorption cross section, barns	2.53
Viscosity, cP,	
at 1,743°C	4.45
at 1,390°C	7.85

In terms of abundance in the earth's crust, iron ranks fourth, being estimated as comprising about 5% of the weight of igneous rocks. Of course, only a very small portion of this large amount of the element in the earth's crust is obtainable as iron ore. In terms of cosmic abundance, the estimate of Harold C. Urey made in 1952 put iron as number 9 among the elements, having a figure of 7,250 related to a base for silicon of 10,000. Iron is ranked number 23 among the elements in terms of its presence in seawater, an estimated 47–48 tons per cubic mile (10.2–10.4 metric tons per cubic kilometer) of seawater. In this regard, it is approximately equal with aluminum, molybdenum, and zinc.

Iron Ores

Iron occurs abundantly in several materials, mainly in the form of oxides, carbonate, silicates, and sulfides. These are shown in Table 2. Most of the ores shown in the table are described under separate alphabetical listings in this volume. Briefly, the major iron-bearing materials are:

Magnetite, Fe_3O_4, corresponding to 72.4% Fe and 27.6% O_2, dark gray to black, sp gr 5.16–5.18, strongly magnetic, permitting magnetic exploration methods. Some ores contain small amounts of titanium whereupon they are referred to as titaniferous magnetite.

Hematite, Fe_2O_3, corresponding to 69.94% Fe and 30.06% O_2, steel gray to dull red or bright red, earthy to compact

TABLE 2. PRINCIPAL IRON-BEARING MINERALS

CLASS AND MINERALOGICAL NAME	CHEMICAL COMPOSITION OF PURE MINERAL	COMMON DESIGNATION
Oxides		
Magnetite	Fe_3O_4	Ferrous-ferric oxide
Hematite	Fe_2O_3	Ferric oxide
Ilmenite	$FeTiO_3$	Iron-titanium oxide
(Goethite)	$HFeO_2$	
Limonite		Hydrous iron oxides
(Lepidocrocite)	$FeO(OH)$	
Carbonate		
Siderite	$FeCO_3$	Iron carbonate
Silicates		
Chamosite		
Stilpnomelane	Various;	
Greenalite	often	Iron silicates
Minnesotaite	complex	
Grunerite		
Sulfides		
Pyrite (iron pyrites)	FeS_2	
Marcasite (white iron pyrites)	FeS_2	Iron sulfides
Pyrrhotite (magnetic iron pyrites)	FeS	

or crystalline, sp gr 5.26, most important of iron ores, occurs widely in many types of rocks of varying origin.

Ilmenite, $FeTiO_3$, corresponding to 36.8% Fe, 31.6% Ti, and 31.6% O_2, iron-black, opaque, generally mined for titanium with iron as a by-product, also called iron titanate.

Limonite, mineralogically composed of various mixtures of the minerals *goethite* and *lepidocrocite*, $HFeO_2$ and $FeO(OH)$, respectively. Goethite contains 62.9% Fe, 27% O_2, and 10.1% H_2O, sp gr 3.6–4.0, commonly yellow or brown to nearly black, compact to earthy and ocherous. Limonites are important sources of iron throughout the world.

Siderite, $FeCO_3$, corresponding to 48.2% Fe, 51.8% CO_2, sp gr 3.83–3.88, white to greenish-gray and brown, contains variable amounts of calcium, magnesium, and manganese, varies from dense, fine-grained and compact to crystalline, sometimes referred to as spathic iron ore, or black-band ore. The carbonate ores are calcined before they are charged into the blast furnace; frequently contain sufficient lime and magnesite to be self-fluxing.

Silicate Group Ores. There are comparatively few silicates with iron as the principal base. Often they have a rather complex chemical formula, with sp gr higher than 2.8, occurring in various shades of green or black tones. Important iron-silicate minerals are *chamosite*, *stilpnomelane*, *greenalite*, *minnesotaite*, and *grunerite*. Presently of minor importance as a source of iron.

Sulfide Group Ores. The principal materials in this group are *pyrite*, *pyrrhotite*, and *marcasite*. Pyrite, FeS_2, corresponding to 46.6% Fe, 53.4% S, sp gr 4.95–5.10, is pale brass yellow, and the most widespread of the iron sulfides. Pyrrhotite (magnetic pyrite), varies in composition from FeS to FeS + S, typically contains 59.4% Fe, 40.6% S, bronze yellow to copper red, frequently tarnished, and is often considered an indicator of nickel deposits because of common association with pentlandite. Marcasite (white iron pyrites), FeS_2, corresponding to 46.6% Fe, 53.4% S, is pale brass yellow, commonly associated with limestones, clays, and lignite deposits. It differs from pyrite only in its crystal structure and greater chemical instability. Iron sulfides sometimes are mined for their sulfur content; more commonly because of their association with other valuable metallic elements, such as copper, nickel, zinc, gold, and silver. Iron sometimes is recovered as a by-product.

Geology and Genesis of Iron Deposits. Iron ores have a wide range of formation in geologic time as well as a wide geographic distribution. They are found in the oldest known rocks of the earth's crust, with an age in excess of 2.5 billion years, as well as in rock units formed in various subsequent ages. Iron ores are forming presently where iron oxides are being precipitated in marshy areas, and where magnetite placers are being formed on certain beaches. Thousands of iron deposits are known throughout the world. The deposits range in size from a few tons to many hundreds of millions of tons. Many of the world's largest deposits of iron ore are located in the oldest geologic series, the Pre-Cambrian.

Iron ores occur in igneous, metamorphic or sedimentary rocks, or as weathering products of various primary iron-bearing materials. For convenience of analysis, iron ores are grouped into (1) igneous ores, (2) contact ores, (3) hydrothermal ores, (4) sedimentary ores, with several subclassification of the latter ores. Brief definitions follow:

Igneous Ores. These were formed by crystallization from liquid rock materials, either as layered-type deposits that possibly are the result of crystals of heavy iron-bearing minerals settling as they crystallize to form iron-rich concentrations, or as bodies which show intrusive relationship with their wall rocks. These ore bodies may be tabular or irregular and are composed largely of magnetite with varying amounts of hematite.

Contact Ores. Iron-ore deposits formed at or near the contact between igneous rocks and sedimentary rocks, the latter usually limestones, are commonly composed of magnetite and hematite with associated carbonates and pyrite. The ore deposits are commonly in the sedimentary rocks as irregular or tabular replacement bodies.

Hydrothermal Ores. Iron-ore deposits formed by hot solutions which transported iron and replaced rocks of favorable chemical composition with iron minerals to form irregular ore bodies, commonly in limestones, are termed hydrothermal deposits. The iron often occurs as siderite, or sometimes as oxides.

Sedimentary Ores. There are six subclassifications:

Bedded ores. These often are composed of oölites of hematite, siderite, iron silicate, or less commonly, limonite in a matrix of siderite, calcite, or silicate. They have a wide geographic distribution associated with other sedimentary rocks. They sometimes contain fossils and fine grains of sand. They often have a fairly high phosphorus content and may be self-fluxing.

Placer ores. Iron oxides, when compact, are rather resistant to weathering and erosion, and under favorable conditions may form placer deposits which, in relatively few instances, constitute iron ores. Generally they are of rather minor importance as sources of iron.

Bog Iron ores. Bog ores occur in many swampy areas, particularly in glaciated areas in Europe, Asia, and North America. They occur commonly as dark-brown, cellular masses, or granular or fine particles of limonite. Once important when iron furnaces were local and small, they have ceased to be of commercial importance.

Metamorphic ores. These ores include sedimentary iron-ore deposits which have been metamorphosed as well as ores associated with metamorphic rocks in which the origin of the ore is obscured by recrystallization. Essentially all of the Pre-Cambrian sedimentary iron formations are of this type.

Residual ores. These ores are commonly products of the surficial weathering of rocks, but may include ores formed by hydrothermal oxidation and leaching. Ores of this kind were formed extensively in Pre-Cambrian iron formations by leaching of silica, which commonly constituted in excess of 50% of the rock. Oxidation changes iron carbonate, silicate minerals, and magnetite to hematite or limonite.

Siderite ores. These ores consist of beds of siderite or siderite nodules associated with shales. They are common in coalassociated beds and commonly contain associated sulfides, with a fairly high sulfur and phosphorus content.

Principal Iron Deposits. The main iron deposits used as commercial sources of the metal include

Kerch oölitic limonite	Crimea, Russia
Salzgitter limonite and hematite	Germany
Minette limonite and hematite	France, Germany, Luxembourg
Blackband ironstones	British Isles
Siegerland siderite	Germany
Clinton hematites	Alabama (U.S.A.)
Wabana oölitic hematites	Newfoundland
Minas Gerais hematite	Brazil
Krivoi Rog hematites	Ukraine, Russia
Bihar, Orissa, and Bastar hematites	India
Labrador hematite	Quebec, Labrador (Canada)
Lake Superior taconites, jaspilites, hematites, and magnetites	Michigan, Wisconsin, Minnesota (U.S.A.) Ontario (Canada)
Cerro Bolivar and El Pao hematites	Venezuela
Kirunavaara magnetite	Sweden
Hematites	Australia

The world iron-ore production per annum is well in excess of 700 million tons, a gain of nearly 200 million tons during the past couple of decades. The dependence of major steel-producing nations on imported ore include: France, 15%; United States, 36%; United Kingdom, 56%; West Germany, 84%; Japan, 92%; and Italy, 93%.

World reserves of iron total about 250 billion tons. An additional 500+ billion tons are present, but under existing economic conditions they are considered commercially nonexploitable. Because of very large ore carriers, accompanied by lowered transportation costs, some iron reserves previously passed over have been undergoing exploitation. Such comparatively new sources

of iron include locations in Brazil, Chile, Labrador, Liberia, western Australia, and the U.S.S.R.

Iron Ore Processing. Two major developments have occurred during recent years because of the increasing needs for iron: (1) increased search for new supplies of high-grade iron ores, and (2) expansion of iron-ore pellet-plant production, particularly in those industrial nations where supplies of high-grade ores have diminished. More recently, iron ore has been shipped in slurry form in ocean-going tankers. Once the slurry, containing about 75% solids, settles, the excess water is pumped off, leaving a nonshifting cargo in the hold of about 92% solids. Upon arrival at the receiving port, the cargo is reslurried with high-pressure water jets. Considerable savings in dockside loading and unloading expense are thus effected.

Beneficiation. This is a term which describes all processes used to improve the chemical and physical characteristics of ore (not limited to iron ore) for later use. In the case of iron ore, beneficiation makes the ore better for handling by the blast furnace. The principal methods include crushing, screening, blending, grinding, concentrating, classifying, and agglomerating. Concentration operations include jigging, flotation, and magnetic separation. A blast furnace operates best with a permeable burden which permits not only a high rate of gas flow, but also a uniform gas flow with minimum channeling of the gas. Agglomeration improves burden permeability and thus the gas-solid contact in the furnace. This reduces blast-furnace coke rates and increases the rate of reduction. Agglomeration also decreases the amount of fines blown out of the blast furnace, thus reducing the load on the gas-recovery system.

Practice has shown that the best agglomerate for a blast furnace will contain about 60% or more of iron, a very minimum of undesirable constituents, and a minimum of material larger than 1 inch (2.54 cm). However, the agglomerate must have sufficient strength to withstand degradation while in stockpiles and during transportation and handling. The target is to have the material arrive at the blast furnace, after prior handling, with about 85–90% of the material over $\frac{1}{4}$ inch (6.3 mm). The agglomerate must be able to withstand the high temperature and the degradation forces within the furnace without slumping or decrepitating. Further, the agglomerate should be reasonably reducible so that a satisfactory reduction rate can be maintained in the furnace.

There are four major types of agglomerating processes: (1) sintering, (2) pelletizing, (3) briquetting, and (4) nodulizing. The first two processes have been most popular.

Sinter consists of small particles of iron-bearing materials which are fused or fritted together at high temperature. The latter is achieved by burning carbon in the form of coke breeze in a sintering-machine feed mix. Optionally, fluxing material may be added to eliminate later additions to the blast furnace. A number of materials can be converted by sintering, such as flue dust, naturally fine ores, ore fines from screening operations, and other iron-bearing material of small particle size.

In the pelletizing process, the agglomeration of material is effected prior to heat treatment. A green, unbaked pellet or ball (glomerule) is formed and hardened by heating. The iron ores to be pelletized are ground finely to present an adequate surface area for the formation of green balls and mixed with water and a binding agent, such as bentonite clay. A small amount of fine solid fuel may be added to the pellet mix or coated on the pellets to furnish part of the heat required. Oxidation of a pelletized magnetite concentrate to hematite during the firing step may also furnish a substantial portion of the heat requirements. The optimum moisture content for pellets is a function of fineness and use of additives, but usually ranges from 9–12%. To improve pellet strength, soda ash, limestone, or dolomite may be added.

Chemistry of Iron

Reduction of iron ores in preparation of iron and steelmaking is described under **Iron Metals, Alloys, and Steels.**

With its $3d^6 4s^2$ electron configuration, iron forms Fe^{2+} and Fe^{3+} ions, the latter involving the removal of one $3d$ electron. The ferrate ion, FeO_4^{2-}, containing hexavalent iron, is unstable in acidic solution, being a very strong oxidizing agent.

Three oxides of iron are known, FeO, Fe_3O_4 and Fe_2O_3, although pure FeO does not exist. The actual composition of iron(II) oxide may be approximated by replacement of a small proportion of the Fe(II) atoms by two-thirds their number of Fe(III) atoms. If the operation is continued until three-quarters of the Fe(II) atoms have been replaced, then the composition Fe_3O_4 is reached, which may thus be described by the formula $Fe^{2+}(Fe^{3+}O^{2-}O^{2-})_2$. Continuation of the replacement until all the Fe(II) atoms have been replaced yields $\alpha\text{-}Fe_2O_3$. The γ-allotrope of Fe_2O_3 and the compound Fe_3O_4 are both ferromagnetic.

Iron(II) sulfide, FeS, also may show considerable departure from stoichiometric proportions, exhibiting an electrical conductivity when in large crystals that resembles an alloy rather than a sulfide. Iron(III) sulfide, Fe_2S_3, cannot be prepared in solution in pure form, because of the oxidizing action of Fe^{3+} upon H_2S, and even upon S^{2-} ions in alkaline solution, and when Fe_2S_3 is prepared by reaction of dry H_2S and the hydrated Fe_2O_3, it breaks up into FeS and FeS_2. The latter is made up of Fe^{2+} and S_2^{2-} ions, and in the mineral, pyrites, it has a cubical structure composed of these ions.

Iron forms dihalides with all four of the common halogens, and trihalides with all but iodine. FeF_2 and $FeCl_2$ are readily formed in anhydrous state by action of the hydrogen halide upon the heated metal, and the others can be made directly from the elements. The iron(III) halides, like iron(III) salts generally, are more readily hydrolyzed than the corresponding ferrous compounds, due to the smaller size and greater change of the Fe^{3+} than the Fe^{2+} ion.

Other elements with which iron forms binary compounds, especially at higher temperatures, are boron, carbon, nitrogen, silicon, and phosphorus. Like FeO, these compounds often depart slightly or even considerably from daltonide composition, frequently being interstitial compounds, and in higher elements of groups VB and VIB, merging into the interstitial compound-solid solution picture which iron exhibits with the transition metals.

The oxyacid salts of iron(III) are more numerous than those of iron(II). Among the former, the sulfates are of interest because of the readiness with which iron(III) sulfate replaces aluminum sulfate in the alums, which are hydrated double sulfates formed by certain trivalent and alkali metal (and other monovalent) sulfates. Iron(III) sulfate, $Fe_2(SO_4)_3$, is isomorphous with aluminum sulfate, $Al_2(SO_4)_3$, because the radius of the Fe^{3+} ion is so close to that of the Al^{3+} ion (0.57Å). For that reason, the isomorphous relationship extends to other salts, i.e., the fluorides and some of the nitrates.

Like the neighboring elements of group 8, iron forms large numbers of complexes. This is due to the availability of two $3d$ orbitals in Fe^{2+} and Fe^{3+} to form hybrid orbitals with the $4s$ and $4p$ orbitals to yield spin-paired complexes (the so-called "covalent" or "inner" complexes).

Many ions contain iron combined with oxygen atoms or hydroxyl groups and are known, respectively, as ferrates or hydroxyferrates. Those of iron(II) include FeO_2^{2-}, $Fe(OH)_4^{2-}$, and $Fe(OH)_6^{4-}$, while those of iron(III) include FeO_2^-, $Fe_2O_4^{4-}$, and $Fe(OH)_8^{5-}$. Their compounds are also called ferrates, except that the name ferrite is used for such compounds of iron(III) as MFe_2O_4, in which M is a divalent metal, this last type of compound being used to form magnetic cores because of their low core losses when properly fabricated. See also **Ferrite.**

The complexes of iron(II) are commonly octahedral (formed by d^2sp^3 hybridization), and include chelate and other cyclic compounds as well as monocyclic ones. The most stable of the latter are the ferrocyanides, or hexacyanoferrates(II), containing the $Fe(CN)_6^{4-}$ ion, which is a spin-paired, diamagnetic complex. It is produced by reaction of cyanides with Fe^{2+} solutions. The Fe^{2+} diammine ion, $Fe(H_2O)_4(NH_3)_2^{2+}$ is a spin-free complex, and is paramagnetic. Divalent iron forms several pentacyano complexes, which contain an ion (such as NO_2^- or Cl^-) or a molecule (NH_3, CO, or H_2O) besides the five cyano groups. Examples are $[Fe(CN)_5NO_2]^{4-}$ and $[Fe(NH_3)_5Cl]^+$. A complex with a single nitroso group in addition to H_2O occurs in $Fe(NO)^{2+}$, formed by reaction of Fe^{2+} and NO.

The ferric ion also forms many octahedral complexes. A spin-paired type is the ferricyanide (hexacyanoferrate(III)) ion, $Fe(CN_6)^{3-}$, while a spin-free type is the hexafluoroferrate ion, FeF_6^{3-}.

The effect of coordination in stabilizing higher states of oxidation is seen in the occurrence of iron(IV) in certain cationic complexes, such as $[Fe(Cl)_2 \cdot 2C_6H_4(As(CH_3)_2)_2](FeCl_4)_2$ (cf. the corresponding nickel(IV), Ni(IV), complex).

Iron forms polydentate chelate compounds with a number of organic substances, including the oxalates, dipyridyl and orthophenanthroline. Such a structure is found in hemoglobin.

Iron forms a number of carbonyls in which its atomic charge number is zero, $Fe(CO)_5$ being produced by heating iron powder with carbon monoxide under pressure, and $Fe_2(CO)_9$ and $Fe(CO)_{12}$ being prepared from it. These iron carbonyls are reactive, yielding hydrogen compounds, e.g., $H_2Fe(CO)_4$ with alcoholic potassium hydroxide; halogen compounds, $Fe(CO)_4X_2$, with halogens; amino or substituted-amino compounds $Fe(CO_2)_n(Am)_{5-n}$, where Am is an amino group, with ammonia, pyridine or ethylenediamine. Iron carbonyls form compounds with R_3P, R_3As, R_3Sb, or diphosphines, etc., such as o-phenylenediarsine. They also form nitroso derivatives, such as $Fe(CO)_2(NO)_2$, and mercaptides, such as $Fe_2(CO)_6(SC_2H_5)_2$. In strong acids (e.g., liquid hydrofluoric acid), $Fe(CO)_5$ gives $Fe(CO)_5H^+$ with the proton attached directly to the iron atom.

Iron also forms a series of similar nitrosyls, such as $Fe(NO)_4$, which is probably $[NO^+][Fe(NO)_3^-]$, $Fe_2(NO)_4I_3$, $Fe(NO)_2I$, Fe-(NO)I, $Fe(NO)_3Cl$, $Fe_2(NO)_4(SA)_2$, where A = H, a metal, a sulfonate group, an alkyl or aryl group, $M^I[Fe(NO)_2S]$ (Roussin's red salt), $M^I[Fe_4(NO)_7S_3]$ (Roussin's black salt), obtained by treating the red salts with alkali, $Fe(NO)_2SR$, where R = C_2H_5 or C_6H_5, $M^I[Fe(NO)SSO_3]$ and others described under hexacyanoferrates.

$$CH{=}CH{-}CH{=}CH{-}CH_2$$

By heating powdered iron with cyclopentadiene the compound ferrocene, insoluble in water, soluble in organic solvents, consisting of two cyclopentadienyl radicals connected to an iron atom, $C_5H_5{-}Fe{-}C_5H_5$ is prepared. It is of great interest because the iron atom is sandwiched between the two parallel and symmetrical rings with delocalized bonding. Alkali metals produce salts of the type $M^I[Fe(C_6H_5)_2]$, powerful reductants. Halogens produce water-soluble ferrocenium salts, such as $[Fe(C_5H_5)_2]Cl$. From both of these, sigma-bonded alkyl and aryl derivatives, such as $Fe(C_2H_5)2CH_3$, can be obtained in which the alkyl group is attached directly to the iron atom.

See the reference list at the end of the entry on **Iron Metals, Alloys, and Steels.**

IRON (In Biological Systems). Iron deficiency is a serious problem in crop production in certain regions of the world and some nutritionists consider iron deficiency anemia to be one of the most frequently observed mineral element deficiency conditions in humans. Yet iron fertilization of soils is not likely to be effective in decreasing the incidence of this deficiency. The reasons for this apparent contradiction are based upon the behavior of iron at several stages in the food chain.

In the United States, severe iron deficiency in crop plants occurs most frequently on the alkaline soils of the western states and on very sandy soils, although some plants, especially broad-leaved evergreens, are sometimes iron deficient on many other kinds of soils. Iron deficiency is rarely due to a total lack of iron in the soil. It is nearly always due to the low solubility of the iron that is present. For example, some soils that are red from iron compounds may contain too little available iron for normal plant growth. The relative susceptibilities of cultivated crops to iron deficiency are listed in the accompanying table.

RELATIVE SUSCEPTIBILITY OF CULTIVATED CROPS TO IRON
DEFICIENCY

Crop	Highly Susceptible	Moderately Susceptible	Relatively Tolerant
Alfalfa		x	x
Barley		x	x
Berries	x		
Citrus	x		
Corn		x	x
Cotton		x	x
Field beans	x	x	
Flax	x	x	x
Forage sorghum	x		
Grain sorghum	x	x	
Grapes	x		
Grasses		x	
Groundnuts (Peanuts)	x		
Millet			x
Mint	x		
Oats		x	x
Potatoes			x
Rice		x	x
Soybeans	x	x	x
Sudangrass	x		
Sugar beets			x
Tree fruits	x	x	
Vegetables	x	x	x
Walnuts	x		
Wheat		x	x

NOTE: Crops listed in more than one category have a wide range of tolerance, depending upon variations in soils, crop varieties, and growing conditions.
SOURCE: Martvedt/Wallace/Curley (1977).

To correct iron deficiency in plants, it is usually necessary to add a soluble form of iron to the soil or to spray the foliage. Since soluble iron added generally will revert to insoluble forms, these procedures for plants are only temporarily effective. Soil treatments that make alkaline soils more acid, such as incorporating large amounts of sulfur, may offer a more lasting correction. Incorporating large amounts of manure into soil makes the iron more soluble and may be effective in correcting iron deficiency, particularly in fine-textured alkaline soils.

Iron-deficient plants are generally stunted and chlorotic, i.e., normally green leaves are yellow or streaked with yellow. When the iron deficiency is treated by adding soluble iron to the soil, the plants turn green, grow larger and yield more, but sometimes the concentration of iron per unit weight of plant material may be no higher than in the stunted iron-deficient plants. Thus, in terms of forage plants, correction of iron deficiency in the plant does not necessarily improve the plant as a source of dietary iron. The iron-treated plants, however, may contain a higher concentration of carotene or provatimin A than the yellow, stunted, deficient plants. Thus, iron fertilization can be more useful in improving the vitamin A than the iron level in diets.

In livestock, iron deficiency is most common in young pigs raised in confinement on concrete floors. Injecting iron compounds and painting the sow's udder with iron compounds are measures used to prevent this deficiency. Grazing animals seldom suffer from iron deficiency unless they are heavily parasitized.

Synthetic chelates and some natural organic complexes tend to resist the adverse effects of soil reactions, climate, and management practices which change available iron into unavailable forms. Natural organic complexes, including some lignin sulfonates and polyflavonoids, or synthetic chelates of iron tend to remain in an available form during most of the growing season. Synthetic chelates are mobile, however, and may be lost from the root zone if subjected to excessive leaching due to overirrigation or high rainfall. Not all iron chelates behave alike in soils, since some are differentially fixed onto clay particles. Fixed chelates do not function in delivering iron to plants. Stability or resistance to decomposition of metal chelates is related to soil pH. Some chelates tend to release their iron more readily than others. For application to calcareous soils of pH 7.5 or higher, the principal iron chelate which will correct iron deficiency for most dicotyledonous plants is FeEDDHA or other similar compounds which have high metal chelate stability. Commercial sources of iron for agricultural application include:

Inorganic Sources
  Ferrous, sulfate; ferric sulfate; ferrous carbonate; ferrous ammonium sulfate; iron frits
Chelates
  FeEDTA (ethylenediaminetetraacetic acid)
  FeHEDTA (hydroxyethylethylenediaminetriacetic acid)
Organic Complexes
  Lignin sulfonate; methoxyphenylpropane complex; polyflavonoid

During recent years, more emphasis has been placed on testing soils for iron deficiency. A commonly used test, developed at Colorado State University, employs a solution of the chelating agent DTPA (diethylenetriaminepentaacetate) and calcium chloride buffered at pH 7.3 for extracting the soil under test. For testing larger areas, a new method of detection and visual assessment of iron chlorosis or deficiency symptoms in growing crops has been developed. This involves aerial infrared photography, which records distinctive color differences between chlorotic and normal green plants. Assessment of a spotty chlorotic field by proportion (percent) and 3-dimensional projection makes possible economic evaluation of the iron deficiency problem existing in any given field on a large-scale basis. See also **Fertilizers.** An excellent summary of iron deficiency and correction in plants and soils is given in the Mortvedt/Wallace/Curley (1977) reference.

Availability of Iron in Foods

The biological availability of iron has been researched extensively. Investigators have found this to be quite variable because numerous factors influence absorption of iron, including the consumer's needs and the composition of the diet. Chemical factors which affect iron availability include the valence, solubility, and degree of chelation or complex formation of the iron. Researchers have shown that the ferrous valence is considerably more available than the ferric valence. Others have shown that prior to absorption in the gut, iron must be in solution. Further, it has been demonstrated that chelation may augment iron absorption by maintaining the iron in solution under conditions where it would otherwise be insoluble. Because of more ready availability, ferrous sulfate is used in bread and flour enrichment even though it has a greater reactivity with foods. Ascorbic acid has been shown to increase iron availability when in the diet.

Iron Absorption

The iron content of the normal adult human is dependent on the size of the individual and the hemoglobin concentration. The distribution of iron in a male weighing 155 pounds ($\sim$70 kilograms) has been estimated at just under 3.5 grams. About 64% of this iron is in hemoglobin as part of the peripheral blood; and 2.5% as hemoglobin in the bone marrow. Another 4% is present as myoglobin which also participates in oxygen transport and storage. Another 13% is present as ferritin and 16% as hemosiderin, which are storage forms. Extremely small amounts are found in cellular cytochrome and in the enzyme catalase.

It is generally agreed that iron is absorbed for the most part in its ferrous form directly into the bloodstream. Radioactive iron has been shown to be absorbed from any portion of the intestinal tract, but its uptake appears greatest in the duodenum. On the basis of experiments done on the absorption of iron from the intestinal tract of guinea pigs, it was earlier postulated that iron is taken into the mucosa cells and ferritin is formed by a combination of a protein, *apoferritin*, with iron. After the cell is saturated with ferritin, absorption no longer takes place until the iron of ferritin is transferred to plasma. For a number of years, this concept of a mucosal block was the accepted explanation for iron absorption. However, later research showed that there is no absolute block to iron absorption. It is found that the absorption of iron in patients and in experimental animals is greater than normal in iron deficiency and in cases where erythropoiesis is accelerated, even when the body iron reserves are elevated. Later evidence indicated that the ferritin concentration in the intestinal mucosa neither controls nor blocks absorption. An active transport mechanism requiring energy is concerned with iron transfer across the intestinal mucosa.

The factors involved in the absorption of iron in food are more complex than those involving inorganic iron. To obtain ^{59}Fe-marked foods, radioactive iron has been injected into hens to obtain labeled eggs and meat; plants have been grown in media containing ^{59}Fe, and ^{59}Fe-enriched bread has been prepared. It has been shown that iron-deficient subjects absorb more food iron than normal subjects. Absorption from liver, hemoglobin, muscle, and "enriched" bread is greater than from eggs or plants. Most probably, the low absorption from egg yolk derives from the presence of a ferric iron-phosphate complex. In such research, large variations in results have been obtained.

In the presence of a large amount of ascorbic acid (vitamin C), the absorption of iron is appreciably enhanced, due to the reduction of Fe^{3+} to the Fe^{2+} form. In the presence of phosphates, carbonates, and phytates, insoluble iron compounds are formed, thus reducing absorption.

It has been estimated that normal subjects ingesting a mixed diet containing 12–15 milligrams of iron retain 5–10% (0.6–1.5 milligram); whereas iron-deficient patients retain 10–20% (1.2–3 milligrams) iron.

Iron Transportation. After iron enters the bloodstream, it is immediately bound by a specific plasma protein which is a β_1-globulin. This protein, *transferrin* (siderophilin) has a molecular weight of about 90,000 and binds 2 atoms of ferric iron. About 0.25 gram of transferrin in 100 milliliters of plasma is capable of binding about 300 micrograms Fe^{3+}, but normally it is only one-third saturated while the remaining two-thirds are unbound reserve. If a small amount of ionized iron is injected intravenously, it is bound by the transferrin which may be completely saturated. If the binding limit is exceeded, ionized iron exhibits toxic effects. The transferrin concentration is increased in iron deficiency and during the latter half of pregnancy; it is decreased during infection and a variety of other disorders.

Electrophoretic studies show there are at several genetically controlled variants of human transferrins. They all deliver iron in an equivalent manner for utilization and storage. Evidence indicates that iron may be transferred directly to the developing erythroblast. It has been demonstrated that transferrin-bound iron is utilized by reticulocytes for hemoglobin formation. The transfer of iron is not maximum until 25% of the transferrin is saturated.

Excretion. The total loss of iron from an adult is about 1 milligram daily and is distributed in sweat, feces, hair, and urine. Since approximately 1 milligram of iron is normally absorbed daily, the organism is in iron balance. The loss of red cells from the body in normal menstruation would account for 16–32 milligrams of iron, which would amount to an average daily loss of 0.5–1.0 milligram during the 28-day menstrual cycle. Pregnancy would also represent a loss of iron from the body, but this is compensated by the absence of menstruation. During normal hemoglobin catabolism, about 20–25 milligrams of iron are released per day. The excretion of minute amounts of iron allows the body to conserve and reutilize the iron for the synthesis of hemoglobin. This tenacious conservation has been demonstrated repeatedly by radioactive techniques.

Enzymes. Heme serves as the prosthetic group for catalase, peroxidase, cytochrome oxidase, and the related cytochromes. Catalase and peroxidase iron are presumably present in the ferric form while the iron of the cytochromes may exist in the reduced or oxidized form. A number of flavoproteins, including succinic dehydrogenase, contain iron in the molecule. Iron appears to act as a coenzyme for aconitase. A number of other enzymes require the presence of iron for their activities.

Storage Iron. Ferritin and hemosiderin represent practically all the iron which is present in the reticulo-endothelial cells of the liver, spleen, and bone marrow and in the parenchymal cells of the liver. Ferritin is an iron protein complex containing up to 23% iron. It is composed of a protein which has a molecular weight of 450,000 and a colloidal ferric-hydroxide-phosphate complex. Preparations of hemosiderin granules contain up to 40% iron and are insoluble in water. They appear to be iron-loaded organelles, similar to mitochondria. The granules contain a small amount of ferritin, but the remaining material is composed of heterogeneous proteins.

Hemoglobin. The approximate formula (molecular weight

65,000) is $(C_{738}H_{1166}FeN_{208}S_2)_4$. Hemoglobin is the respiratory protein of the red blood cells. It transfers oxygen from the lungs to the tissues and carbon dioxide from the tissues to the lungs. Its affinity for carbon monoxide is over 200 times that for oxygen. Hemoglobin is a conjugated protein consisting of approximately 94% globin (protein portion) and 6% heme. Each molecule can combine with one molecule of oxygen to form oxyhemoglobin. The iron (in the heme portion) must be in the reduced (ferrous) state to enable the hemoglobin to combine with oxygen. Heme $(C_{34}H_{32}FeN_4O_4)$ is the nonprotein portion of hemoglobin and myoglobin, consisting of reduced (ferrous) iron bound to protoporphyrin. See also **Hemoglobin.**

Iron Deficiency and Dietary Supplementation

Iron deficiencies and corrections in plants have been previously discussed. Iron deficiency in humans is usually associated with loss of blood or the inefficient utilization of dietary iron.

Nutritional anemias may result from nutritional deficiencies, or decreased bone marrow function, both of which cause defective blood formation. The least severe, but most common of these anemias results from an inadequate amount of iron required for red cell formation. The result is microcytic hypochromic anemia. About 100 milligrams per day of iron is needed for hemoglobin manufacture. About 85% of this iron may be obtained from the iron released by breakdown of older red cells. However, some iron is always lost in the excretions and thus must be made up by the diet. Where there is chronic blood loss, as in cases of ulcers or hemorrhoids, or where the iron may not be properly absorbed from foods, the need for iron may be greater. Milk, cereals, and many refined foods, unless artificially supplemented, do not contain much iron. Better sources of iron include meat and leafy vegetables. Iron deficiency is not uncommon.

A common form of *iron-deficiency anemia* frequently seen in young women during the last century was sometimes called chlorosis, or "green sickness" because of the peculiar hue of the skin. With the discovery that iron salts can effect a cure, the disease almost completely disappeared. *Idiopathic hypochromic anemia* is another iron-deficiency anemia associated with a lack of proper stomach acidity. When hydrochloric acid in the stomach is lacking, iron cannot be liberated from foods and converted into a form that can be absorbed. Administration of iron in proper form also alleviates this condition.

During pregnancy, the mother must furnish greater volumes of blood to support herself and the developing baby. Blood volume is increased and causes dilution unless sufficient iron is available. Vomiting in early pregnancy may increase the danger of an iron deficiency. Usually, babies are born with adequate supplies of iron in their tissues to last several months. However, infants born of a mother with an iron deficiency have low reserve stores of iron and will require a diet that is supplemented with the proper amounts of iron. Milk is a poor source of iron, and infants strictly on a diet of milk almost invariably develop hypochromic anemias. Anemic babies are much more subject to infections, which may in turn further increase the anemia. Thus, such children should be treated early.

A protein deficiency also may contribute to iron-deficiency anemias, as well as to the anemia of pregnancy. Minute amounts of copper are required in order for the body machinery to use iron properly and where copper is lacking from the diet, a hypochromic anemia can result. Vitamin C and certain other vitamins also are thought to be involved in the process of iron utilization and hemoglobin construction. These facts emphasize the importance of diet in maintaining proper health of the blood.

The macrocytic hyperchromic anemias resulting from impaired bone marrow function are generally of a more serious nature. The process of red cell formation in the bone marrow is complex. In order for the red cell to mature properly, the bone marrow must have adequate amounts of what may be termed the *growth or maturation factor*, which is obtained from the liver. Most of the macrocytic anemias are caused by the inability of the bone marrow to obtain the proper supplies of this substance, generally believed to be closely related to or identical with vitamin B_{12}. Previously, this vitamin was referred to as the *extrinsic factor* because it came from outside the body. An *intrinsic factor* occurs in the normal gastric juice and is necessary for the absorption of vitamin B_{12}. Some of the macrocytic anemias are caused by the absence of the intrinsic factor in the gastic juice and others by inadequate supplies of the maturation factor. Other macrocytic anemias arise from the inability of the bone marrow to make use of the maturation factor. In all but the last type, injection of vitamin B_{12} or a liver extract which contains the vitamin normally causes prompt disappearance of the anemia.

Iron Supplements. Compounds commonly used to supplement and fortify various food products in terms of iron content include: ferric ammonium citrate, ferric phosphate, ferric pyrophosphate, ferrous gluconate, ferrous sulfate, iron (electrolytic and reduced), and sodium ferric pyrophosphate, among others.

References

Brise, H., and L. Hallberg: "A Method for Comparative Studies on Iron Absorption Analysis of Fe, Mn, Zn, and Cu in Plant Tissue," *Journal of Agricultural and Food Chemistry*, **22**, 103 (1962a).

Brise, H., and L. Hallberg: "Effect of Ascorbic Acid on Iron Absorption," *Acta Medica Scand.*, **171** (Supplement 376), 51 (1962b).

Cort, W. M., Mergens, W., and A. Greene: "Stability of Alpha- and Gamma-Tocopherol: Fe^{3+} and Cu^{2+} Interactions," *Journal of Food Science*, **43**, 3, 797–800 (1978).

Farmer, B. R. et al.: "Iron Bioavailability of Hand-Deboned and Mechanically-Deboned Beef," *Journal of Food Science*, **42**, 6, 1630–1632 (1977).

Forth, W., and W. Rummel: "Iron Absorption," *Physiol. Rev.*, **53**, 724 (1973).

Fritz, J. C., et al.: "Estimation of the Bioavailability of Iron," *JAOAC*, **58**, 902 (1975).

Hamilton, R. P., et al.: "Zinc Interference with Copper, Iron and Manganese in Young Japanese Quail," *Journal of Food Science*, **44**, 3, 738–741 (1979).

Harrison, B. N., et al.: "Selection of Iron Sources for Cereal Enrichment," *Cereal Chemist*, **53**, 78 (1976).

Jacobs, A., and D. A. Greenman: "Availability of Food Iron," *British Medical Journal*, **1**, 673 (1969).

Kanner, J., and H. Mendel: "Prooxidant and Antioxidant Effects of Ascorbic Acid at Metal Salts (including Iron) in a β-Carotene-Linoleate Model System," *Journal of Food Science*, **42**, 1, 60–64 (1977).

Kirchgessner, M., Editor: "Trace Element Metabolism in Man and Animals—3," Institut für Ernährungsphysiologie, Technische Universität München, Freising-Weihenstephan, Germany (1978).

Lee, K., and F. M. Clydesdale: "Quantitative Determination of the Elemental, Ferrous, Ferric, Soluble, and Complexed Iron in Foods," *Journal of Food Science*, **44**, 2, 549–554 (1979).

Lee, L., and F. M. Clydesdale: "Iron Sources for Food Fortification and Their Changes Due to Food Processing," *CRC Crit. Rev. Food Sci. Nutr.* (1979).

Mahoney, A. W., and D. G. Hendricks: "Some Effects of Different Phosphate Compounds on Iron and Calcium Absorption," *Journal of Food Science*, **43**, 5, 1473–1475 (1978).

Mortvedt, J. J., Anderson, F. N., and P. Grabouski: "Placement Effect of Ferrous Sulfate-Ammonium Polyphosphate Suspension on Crop Response to Iron," *Fertilizer Solutions*, **18**, 6, 78–86 (1974).

Mortvedt, J. J., Wallace, A., and R. D. Curley: "Iron—The Elusive Micronutrient," *Fertilizer Solutions*, **21**, 1, 26–36 (1977).

Saltman, P., et al.: "Tired Blood and Rusty Livers," *Annals Clin. Lab. Sci.*, **6**, 167 (1976).

Sayers, M. H., et al.: "Iron Absorption from Rice Meals Cooked with Fortified Salt Containing Ferrous Sulphate and Ascorbic Acid," *British Journal of Nutrition*, **31**, 367 (1974).

Theuer, R. C., et al.: "Effect of Processing on Availability of Iron Salts in Liquid Infant Formula Products," *Journal of Agricultural and Food Chemistry*, **19**, 555 (1973).

Thompson, S. A., and C. W. Weber: "Effect of pH on Minerals (Iron, etc.) in Fibers," *Journal of Food Science*, **44**, 3, 752–754 (1979).

Underwood, E. J.: "Trace Elements in Human and Animal Nutrition," 4th Edition, Academic, New York (1977).

Zenoble, O. C., and J. A. Bowers: "Copper, Zinc, and Iron Content of Turkey Muscles," *Journal of Food Science*, **42**, 5, 1408–1409 (1977).

IRON METALS, ALLOYS, AND STEELS.

Chemically pure iron is used essentially in powder metallurgy and for chemical applications where the element serves as a catalyst, or as a base ingredient for ferrous and ferric chemicals. Iron is principally used as the dominant ingredient of cast irons and steels. Iron-base alloys notably are known for their physical strength and toughness and, when compared with most other metals for similar applications, reasonable cost. The following properties depend upon the nature and extent of the ingredients, such as carbon, present in the alloys and also upon the mechanical and heat treatments given to the formed metals: (1) impact strength or brittleness, (2) cohesive strength, (3) compressive strength, (4) creep, (5) fatigue, (6) ductility, (7) hardness, (8) malleability, (9) shear strength, (10) yield strength, (11) torsional strength, (12) electrical conductivity, (13) thermal conductivity, (14) thermal stability, (15) thermal expansion, (16) corrosion resistance, (17) magnetic properties, and (18) heat treatability.

The iron metals family of products may be classified into (1) the *pure irons*, such as ingot iron and wrought iron, which have only traces of carbon (see accompanying table) and other

TYPICAL ANALYSES OF PURE IRONS

INGREDIENT	INGOT IRON	ELECTROLYTIC IRON	CARBONYL IRON	HYDROGEN-PURIFIED IRON
	Percent of Total			
Carbon	< 0.020	0.006	0.0004	0.005
Manganese	< 0.020	—	—	0.028
Phosphorus	0.005 ±	0.005	—	0.004
Sulfur	0.020 ±	0.004	—	0.003
Silicon	Trace	0.005	—	0.001
Copper	0.04 ±	—	—	—
Oxygen	Some	Some	< 0.01	0.003
Nitrogen	0.004 ±	—	—	0.0001

elements and are very ductile; (2) *cast irons*, which are alloys of iron and carbon, with or without other elements, normally containing 2.4–4.5% carbon; (3) *steels*, which are alloys of iron and carbon, with or without other elements, in which the carbon content seldom exceeds 1.7%; and (4) *alloy steels* whose properties mainly are attributed to the presence of one or more elements other than carbon. There are other groups and numerous subgroups in the total iron metals family.

Ironmaking

The starting ingredients for the numerous iron metals are obtained in three major ways: (1) by smelting run-of-mine or beneficiated iron ore in a blast furnace, low-shaft furnace, or electric smelter to yield a liquid, molten product; (2) by reducing run-of-mine or beneficiated iron ore via direct reduction processes to produce sponge iron; and (3) by melting ferrous scrap in a cupola, electric furnace, or fuel-fired furnace. Inasmuch

as the majority of iron ores are in the form of oxides of iron, the iron is obtained by employing suitable reducing agents to reduce the oxides. Reducing agents most often used are carbon, carbon monoxide, hydrogen, and hydrocarbons, such as methane. With carbon monoxide, the reduction is exothermic; with the other reductants, it is endothermic.

Blast Furnace. For many years, the blast furnace was the traditional way to produce pig iron, the raw material for the steel industry. The furnace is a tall, refractory-lined vessel. Raw materials, including iron ore (sinter or pellet), coke (the reducing and thermal agent), and limestone (for fluxing the gangue material) are charged into the top of the furnace. A blast of hot air is introduced at the bottom of the furnace to burn the coke and thus to heat, reduce, and melt the charge as it descends toward the bottom of the furnace. Liquid iron and slag collect in the furnace hearth. These materials are tapped at regular intervals. Although the furnace can be damped down for short periods, the process essentially is continuous. The waste gas contains about 28% carbon monoxide with a calorific value of about 90 Btu/cubic foot (800 kcal/cubic meter). After collection at the top of the furnace, dust is removed, and the gas is used as a fuel for heating the hot-blast stoves. Blast furnaces range from 100 to 10,000 tons/day in capacity; the hearth diameter may range from 9 to 46 feet (2.7 to 13.8 meters); the height from 50 to 150 feet (15 to 45 meters). Improvements in blast furnace operations over the last several years have resulted from better preparation of the charge, fuel injection through tuyeres, reducing-gas injection in bosh, and oxygen enrichment of blast.

Low-Shaft Furnace. These furnaces are circular or oval in cross section. The oval shape permits greater hearth area without increasing the required depth of penetration of the blast supplied through the tuyeres. Such furnaces are designed for finer raw materials and low-grade coke or lignite. Once considered ideal for small-scale iron production, only a limited number have been installed. The operating principles are essentially similar to those of a blast furnace.

Electric Smelter. Electric energy provides the heat in these designs. Low-grade coke can be used as the reducing agent. Electric smelters are generally limited to areas of low-cost electric power. The most commonly used furnaces of this type employ a submerged arc, using the Söderberg continuous self-baking electrodes. Developed by Tysland and Hole in Norway, the furnace is circular or rectangular in cross section with transformer ratings of up to 60,000 kVA. Production capacity can be increased and power and coke requirements lowered by preheating and prereducing the iron-ore charge.

Direct Reduction Process. Numerous schemes over the years have been attempted as an alternative to the blast furnace. These include rotary and stationary kilns and furnaces, reverberatory furnaces, retorts, fluid-bed reactors, pot furnaces, and jet smelting. Similarly a variety of reductants have been used, including lignite, coal, char, fuel oil, tar, and various gases. The direct reduction approach is essentially useful for producing a highly reduced product containing mostly metallic iron and little gangue material, thus providing a substitute for ferrous scrap for steelmaking operations.

SL/RN Process. In this process, using a rotary kiln and solid reductant, high-grade pellets or lump ores and anthracite are used. The ore or pellets, anthracite, dolomite or limestone, and return coal are fed to the rotary kiln. The temperature is controlled by means of shell burners which are furnished with air and gas or oil. A uniform temperature of about 1.100°C is maintained over about 60% of the kiln length. After leaving the reduction kiln, the charge is passed through a gastight seal into a water-cooled drum, whereupon it is cooled to a temperature below 100°C to prevent reoxidation of the sponge iron.

Most of the sponge iron can be separated by screening, augmented by magnetic separators.

HyL Process. This batch-cyclic process reduces rich lump-iron ores by flowing a reducing gas in a fixed-bed reactor. The reducing gas may be prepared by steam reforming of natural gas or other hydrocarbons. A typical reducing gas may contain 74% hydrogen, 13% carbon monoxide, 8% carbon dioxide, and 5% methane (all volume percent). The process requires four reactors, each reactor following four steps in a 12-hour cycle: (1) removal of cold sponge iron and loading with fresh iron ore or pellets; (2) preheating and secondary reduction with partially spent reducing gas from another reactor; (3) primary reduction to sponge iron; and (4) cooling the sponge iron with fresh, cool reducing gas and controlled deposition of carbon where required.

Purofer, Midrex, and Armco Processes. These are continuous processes in which shaft furnaces are used. Iron ore or pellets are charged from the top and the reduced product is withdrawn from the bottom. The reducing gas may be generated by reforming natural gas, or methane rich gas from naphtha may be used. The hot reducing gas flows countercurrent to the descending charge.

Fior and U.S. Steel Processes. Similar reducing gases are used in connection with a fluid-bed reactor. The iron ore may require further grinding and drying before introduction into the fluid bed. The reduced ore may be briquetted or used as fines.

Cupola. This type of furnace is widely used for making iron for casting in foundries and sometimes may be used to augment the iron required by steelmaking operations. A cupola may be operated for just a few hours/day, or up to 2 to 3 months continuously. A cupola is a vertical cylindrical shaft furnace that uses the countercurrent-flow principle to heat and melt the charge as it descends. Cupola capacity may range from 2 to 75 tons/hour. Unlike a blast furnace, a cupola is not a reducing unit, but is essentially used to melt ferrous scrap and cold pig iron or previously reduced sponge iron. The heat required is supplied by nearly complete combustion of coke. Air is injected through tuyeres near the hearth zone. Some cupolas are equipped for hot-blast supply through recuperators. The iron raw material is charged in alternate layers with coke. Limestone is added to flux the ash from the coke and form slag. The molten iron collects in the hearth and may be removed continuously or intermittently through a taphole.

Steelmaking

The major raw materials used for steelmaking are: (1) liquid iron, (2) steel scrap, (3) prereduced sponge iron, and (4) a mixture of liquid iron, scrap, and sponge iron. Major processes include the open-hearth process, rapidly becoming of historical interest; the Bessemer process; the basic oxygen processes; and several additional specialized processes.

Open-Hearth Process. For well over 100 years, this was the principal commercial process for making steel. The major feature of the open-hearth process is its versatility in handling a variety of raw materials to make most grades of steel. Charges may be 100% scrap, 100% hot metal, or scrap and hot metal in intermediate ratios. The acid open-hearth process imposes limitations on the sulfur and phosphorus contents of the metallic charge inasmuch as these elements can be removed during refining. The basic open-hearth process imposes no such limitations. The open-hearth process employs a refractory-lined furnace (stationary or tilting). It essentially is a shallow hearth enclosed by walls. There is a charging door on one side; a taphole on the other side. Regenerative checkers are installed at either end to heat fuel and/or air before burning over the hearth. The checkers are cycled, thus providing the temperature required for steelmaking. In the Ajax process, a modified tilting-type open-hearth furnace is used wherein a 100% hot metal charge can be refined with oxygen.

Bessemer Process. In this process, liquid iron is refined in a bottom-air-blown converter, a refractory-lined pear-shaped cylindrical vessel open at the top to permit charging of materials and to allow escape of gases. The acid-Bessemer process demands iron that is low in phosphorus and high in silicon. The basic-Bessemer process (also known as Thomas process) can handle high-phosphorus iron. Since a considerable amount of heat is wasted in heating the nitrogen in the air, the Bessemer process can melt only 5 to 10% scrap. It follows that the nitrogen content of the steel also is high. Oxygen-steam or oxygen-carbon dioxide mixture may be used instead of air to produce low-nitrogen steels.

Basic Oxygen Process. This process enables the production of high-grade steels at rapid rates, thereby reducing operating and capital costs as compared with other steelmaking processes. In the process, molten pig iron is refined to steel by top-blowing oxygen at high pressure onto the surface of the metal through a water-cooled lance contained in a tilting furnace. The oxidation of carbon, silicon, manganese, and phosphorus provides sufficient heat for converting the molten iron into steel. Because of excess heat generated, up to 30% scrap can be charged. A conventional basic oxygen process (BOP) can refine iron containing up to 0.3% phosphorus into most grades of steel. Where the phosphorus content is higher, a modified process uses injection of powdered lime with the oxygen stream; or double slagging is required.

Kaldo and Rotor Processes. Developed in Sweden, the refining of molten iron to steel is carried out in a tilted pear-shaped basic-lined converter. Oxygen is blown at an oblique angle to the metal bath through a water-cooled lance. Much of the carbon monoxide produced by the carbon/oxygen reaction is burned inside the converter. The heat generated is absorbed by the rotating vessel and transferred to the bath, thus providing a high thermal efficiency and allowing up to 40% scrap in the charge. Both low- and high-phosphorus irons can be handled. Refractory consumption is high, tending to reduce the availability of the furnace.

Electric Furnace Process. Direct-arc electric furnaces are widely used. Essentially, the furnace is a tilting cylindrical bowl-shaped hearth with three graphite electrodes inserted vertically through the roof. The electrodes are supplied with three-phase current via a transformer. Heat is supplied by the arc struck between the charge and the electrodes. The arc temperature is approximately 3,400°C. The furnace is highly versatile, in that operation may be under oxidizing, reducing, or neutral conditions. The versatility is comparable to that of the open-hearth process. Some electric furnaces operate with liquid iron in the charge, but the majority use steel scrap and pre-reduced pellets.

With very high power input operation and transformer ratings of up to 100,000 kVA, common grades of steel, requiring single slag practice only, can be produced in up to 300-ton heats in less than 3 hours. Special steels also are made in induction furnaces where a current of high or medium frequency is passed through a coil surrounding a refractory crucible containing the charge.

Metallurgy of Iron and Steel

The physical properties of ferrous products can be altered by cold working or heat treatment, both processes which affect the microstructure. The role of carbon in ferrous products is explained to some degree by the *iron-carbon diagram*. See accompanying figure. This diagram shows the relationship between carbon content and temperature and includes key information on microstructure and heat treatment.

When cooled, pure iron solidifies at about 1,536°C as delta iron, having a body-centered cubic lattice structure. This form changes allotropically to gamma iron with a face-centered cubic lattice structure below 1,404°C, and is nonmagnetic. When cooled further, gamma iron changes to alpha iron, with a body-centered cubic lattice structure. At a temperature of 768°C, a nonallotropic change occurs, making the alpha iron strongly magnetic, accompanied by marked changes in electrical resistance, rate of thermal expansion, and specific heat. The foregoing changes occur in reverse order if pure iron is heated instead of cooled.

Carbon dissolves in molten iron to form iron carbide. When the carbon content is increased, the liquidus temperature (melting point of iron) is lowered. The eutectic point (lowest melting temperature) is 1,130°C when the carbon content is 4.3%. Similarly, the solidus temperature is lowered to 1,130°C up to a carbon content of 1.7%. The resulting transformations and the products formed after cooling the iron-carbon alloy below the solidus temperature depend on the carbon content, the temperature, and the rate of cooling. Some of these substances are defined briefly below:

Austenite. This is an allotropic form of gamma iron with carbon in solid solution. Austenite transforms to other products on cooling below 723°C. The products depend on the rate of cooling. At ordinary temperatures, austenite containing only carbide is not stable and thus cannot be completely retained by quenching. The stability can be increased by adding certain alloying elements.

Ferrite. This is practically pure iron and can exist in magnetic alpha-iron form in iron, with up to 0.83% carbon. Ferrite exists at room temperature and up to about 910°C in the absence of carbon. Its upper limit of existence is lowered progressively to about 723°C as the carbon content increases up to 0.83%. Ferrite cannot dissolve carbon, is soft and ductile, and has poor abrasive resistance.

Cementite. This is iron carbide, Fe₃C, containing 6.67% carbon. The substance is hard, brittle, and crystalline. Cementite is precipitated when austenite cools.

Pearlite. This is a eutectoid comprised of a laminated structure of ferrite and cementite. Pearlite is formed by transformation of austenite upon cooling. The fineness or coarseness of the laminated structure is determined by the rate of cooling. The lamellar arrangement of ferrite and cementite produces a very tough structure. It is responsible for the mechanical properties of steels.

Graphite. This is the free or uncombined carbon usually found in cast irons. Because graphite occurs as flakes, cast irons are easily machinable even though they have a high resistance to abrasion.

Effects of Ingredients on Steels

The iron-carbon diagram shows the effects of carbon in steel. Steels containing less than 0.83% carbon are known as *hypoeutectoid steels*. When cooled slowly, the microstructure of these steels consists of pearlite and ferrite. *Eutectoid steel* contains 0.83% carbon and consists entirely of pearlite. Steels containing more than 0.83% carbon are known as *hypereutectoid steels*. When cooled slowly, their microstructure is comprised of pearlite and cementite. Each increase in the carbon content of the

steel increases the hardness and tensile strength of the steel in the as-rolled or normalized condition up to 0.83% carbon. The effect is less pronounced above this figure.

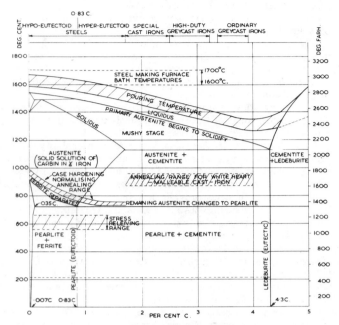

Iron-carbon equilibrium diagram.

Manganese is normally added to the extent of 0.5–1.0%. Normally some manganese is present in many iron ores. Manganese contributes to strength and hardness, but lowers ductility. Manganese is of major importance in increasing hardenability. Steel containing 13% manganese is widely used as a wear-resistant steel.

Phosphorus in steel is undesirable because it decreases ductility and impact toughness. However, in lower-carbon steels, phosphorus promotes machinability and, with copper, improves resistance to atmospheric corrosion.

Sulfur in steel is detrimental to surface quality, but beneficial to machinability, particularly in low-carbon and low-manganese steels. Sulfur is added to the extent of 0.2–0.4% in free-cutting steels to improve machinability.

Silicon, when specified within the limits of 0.6–5.0% qualifies a steel as an alloy steel. Silicon increases the resiliency of steel for spring applications and raises the critical temperature for heat treatment. Silicon promotes the susceptibility of steel to decarburization. Silicon also promotes the adherence of zinc coating on hot-dipped galvanized wire and, because silicon steels have a low hysteresis loss and a high electrical resistance, low-carbon steels with 0.6–5.0% silicon are used as transformer steels.

Aluminum is used mainly to deoxidize steels and to obtain a fine grain size. Aluminum in amounts of about 1% also promotes nitriding properties.

Copper is beneficial to atmospheric corrosion resistance if present in steels in excess of 0.2%. Because copper is not removed in conventional steelmaking processes, it is becoming increasingly difficult to control copper within low limits where a lot of scrap is used. Copper is detrimental to surface quality and adversely affects forge welding.

Nickel, aside from manganese, is the most common alloying element for steel. Nickel is used up to 5% to increase strength and improve shock resistance. When nickel is used in quantities greater than 5%, the steels fall into the stainless and heat-resistant steel categories.

Niobium, in amount of up to 1%, stabilizes chromium and stainless steels. Additions of only about 0.02% increase the yield point of medium-carbon steels by about 50% without any loss of weldability.

Tungsten, when added to steels in amounts up to 20%, greatly improves hardness, producing a hardness that is retained at high temperature and thus is important to high-speed tool steels.

Zirconium, in small amounts, improves the machinability of high-chromium steels.

Cobalt provides cutting efficiency to high-speed steels and also is a constituent of heat-resisting steels because it conveys a resistance to creep and scaling.

Chromium increases hardness, improves hardenability, and promotes the formation of carbides. Chrome steels are relatively stable at elevated temperatures and have outstanding wear resistance. Chromium is an important ingredient in stainless steels.

Molybdenum in steels provides less susceptibility to temper brittleness. Molybdenum also increases hardenability and to some degree increases the high-temperature tensile and creep strengths of alloy steels.

Vanadium, as a strong deoxidizing agent, promotes a fine austenitic grain size. Construction steels contain about 0.03–0.25% vanadium; larger quantities are used in tool steels. These additions increase the hardenability of medium-carbon steels with minimum effect on grain size.

Titanium acts as a deoxidizer in pearlitic steels. The element increases the yield point of plain-carbon steels when added between 0.02 and 0.05%. Titanium promotes weldability without the need for normalizing.

Boron may be added to increase hardenability. The element intensifies the hardenability characteristics of elements already present in the steel.

Industrial Classification of Steels

Numerous systems are used for classifying steels—some based on composition, others on physical properties, special properties, and so on. For convenience, there are two broad categories: (1) plain carbon steels; and (2) alloy steels. Plain carbon steels account for about 95% of all steel production. As described earlier, plain carbon steels are classed as hypoeutectoid or hypereutectoid steels, depending on whether the carbon content is above or below 0.83% (the eutectoid composition). *Low-carbon* steels have a carbon content below 0.20%. *Medium-carbon* steels have a carbon content in the range between 0.20 and 0.50%. *High-carbon* steels have a carbon content in excess of 0.50%.

Stainless steels have a large degree of resistance to chemical attack. This property sometimes is referred to as *passivity*. This property results when iron is alloyed with at least 11% chromium. The corrosion resistance is further enhanced by higher chromium additions and the addition of nickel. A steel with 12% chromium will stain, but will not exhibit progressive rusting in normal atmospheres. Under normal circumstances, a steel with 18% chromium will not stain, but may discolor, particularly in heavy industrial areas. When 8% nickel is added to an 18% chromium steel, the metal will be stain-resistant in all but the very worst of atmospheres. Even further enhancement of corrosion and heat resistance results with the addition of molybdenum.

Iron-chromium alloys and their general corrosion-resistant properties were known in England and France nearly 150 years ago, but the phenomenon of passivity was not formally recognized until 1910 (Borchers and Monnartz in Germany). This discovery led to rapid development of a series of commercial

stainless steels. Stainless steels fall into three broad categories: (1) *martensitic types*—chromium-iron alloys with chromium in the lower range (12 to 17%) and with a wide range of carbon. A main characteristic is an ability to harden by heat treatment in a manner similar to carbon steels. Tensile strengths range from 70,000 to 105,000 psi (483 to 725 MPa) for hardened steels. They are particularly well suited for hot working and forging; the lower-carbon types can be cold-worked. (2) *ferritic types*—chromium-iron alloys with higher chromium in a range of 18 to 30% and with a lower carbon content. They have a microstructure that is predominantly ferritic. The steels are not hardenable by heat treatment. They are ferromagnetic. They have a relatively low coefficient of thermal expansion. These steels exhibit good resistance to oxidation and corrosion; they are frequently selected for high-temperature service, notably for applications involving intermittent heating and cooling because of their ability to retain the oxide scale which has formed. (3) *austenitic types*—iron-chromium-nickel alloys, with a chromium content ranging from 8 to 30% and a nickel content ranging from 6 to 20%. They retain austenite at room temperature. They are characterized by high ductility of the austenite, work-hardening ability, good corrosion resistance, and superior high-temperature properties. Austenitic stainless steels are inherently tough; well adapted for fabrication by deep drawing. They are easily welded and soldered. Their tensile strength (annealed) approximates 90,000 psi (621 MPa) with a yield strength of about 35,000 psi (242 MPa).

References

Bardes, B. P., Editor: "Metals Handbook," 9th Edition, Vol. 1, "Properties and Selection of Irons and Steels," American Society for Metals, Metals Park, Ohio (1979).

Chandler, H. E.: "Trends in Selecting Steels for Economies of Manufacture and Application in the '80s," *Metal Progress*, **118**, 5, 24–34 (1980).

Chandler, H. E.: "Economical Alternative Steels," *Metal Progress*, **118**, 5, 54–61 (1980).

Chandler, H. E.: "Steels that Conserve Materials/Energy," *Metal Progress*, **118**, 5, 64–84 (1980).

Kakela, P. J.: "Iron Ore: Energy, Labor, and Capital Changes with Technology," *Science*, **202**, 1151–1157 (1978).

Schmidt, P., and D. H. Avery: "Complex Iron Smelting and Prehistoric Culture in Tanzania," *Science*, **201**, 1085–1089 (1978).

Staff: "Almost Near Net Shape Steels," *Metal Progress*, **118**, 5, 37–41 (1980).

Staff: "The Making, Shaping, and Treating of Steel," United States Steel Corporation, Pittsburgh, Pennsylvania (Revised periodically).

ISODESMIC STRUCTURE. An ionic crystal structure in which there are no distinct groups formed within the structure, i.e., where no bond is stronger than any of the others.

ISODIAPHERE. One of two or more nuclides having the same difference between the number of neutrons and protons in their nuclei. In alpha-particle decay, for example, the parent and daughter nuclides are isodiaphers.

ISOELECTRIC POINT. Amino Acids.

ISOLEUCINE. Amino Acids.

ISOMER (Nuclear). One of two or more nuclides that are both isotopes (same atomic number) and isobars (same mass number) of each other, but which have some measurably different physical property, such as half-life. Of any two isomeric states, one must be an excited metastable state of the other. Ultimately, the nuclide in the excited state decays with a measurable lifetime to a lower energy state, usually its ground state.

At present about 200 nuclear isomeric states with half-lives longer than 10^{-6} second are known. Metastable isomeric states are denoted by adding the letter m to the mass number where it appears in the nuclidic symbolism, as for example ^{80m}Br. In this particular case, ^{80m}Br and ^{80}Br are nuclear isomers. Ordinarily excited states with lifetimes too short to be measured are generally not considered to be isomeric states, but this is only a matter of convention. On rare occasions, such as for ^{124}Sb, more than two isomeric states may exist for a single atomic number and a single mass number (isotopic isobar).

ISOMERISM. If two chemical compounds incorporate the same elements in exactly the same numbers, the compounds are referred to as *isomers* or *isomerides*. An excellent example of relatively simple isomers is the case of normal butane, $CH_3CH_2CH_2CH_3$, which is an open, straight chain of four carbon atoms, and of isobutane, $(CH_3)_2CHCH_3$, wherein one of the carbons lies in a short branch from the main chain of three carbon atoms. Obviously, as the number of carbon atoms in a compound increases, the possibility of branches and subbranches increases. Normally, then, compounds with high carbon counts, at least theoretically, are capable of numerous isomers.

In *geometric isomerism*, the isomeric relationship can be explained in terms of two dimensions—as shown by the relationship of the two isomers, maleic acid and fumaric acid:

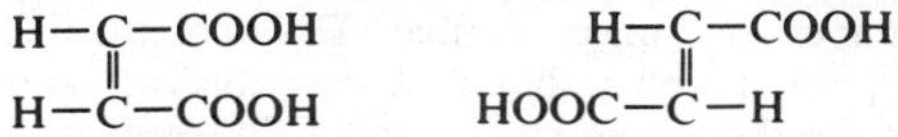

$$H-\underset{\parallel}{C}-COOH \qquad H-\underset{\parallel}{C}-COOH$$
$$H-C-COOH \qquad HOOC-C-H$$
(Maleic acid) (Fumaric acid)

Where the identical atoms or groups are in juxtaposition, as in maleic acid, the compound is designated as the *cis* form. (*Cis* = "on this side" in Latin.) Where the identical atoms or groups are on the opposite sides, as in fumaric acid, the compound is designated as the *trans* form.

In *stereoisomerism*, three dimensions must be considered. In steroeisomerism (also termed optical isomerism), there is no plane of symmetry in the molecule, so that the two forms are mirror-images, and thus cannot be turned into a position of coincidence. Thus, compounds containing a carbon atom (or other tetravalent atom) to which four different atoms or radicals are bonded are optical isomers. They receive this name from the fact that one isomer rotates the plane of polarized light to the right (*dextro form*); the other rotates it to the left (*levo form*). Lactic acid is an example. See also **Lactic Acid,** and formulas below:

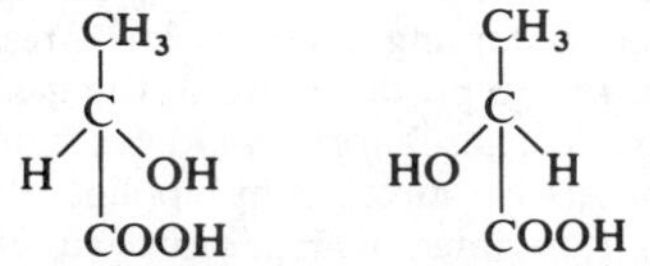

$$\underset{(d\text{-lactic acid})}{\underset{COOH}{\overset{CH_3}{\overset{|}{\underset{|}{C}}}}} \qquad \underset{(l\text{-lactic acid})}{\underset{COOH}{\overset{CH_3}{\overset{|}{\underset{|}{C}}}}}$$

The carbon atom, to which are attached the four different groups to produce stereoisomerism, is known as *asymmetric*, and when written (not shown structurally), that carbon may be printed more prominently than the nonasymmetric carbon atoms.

The projection formulas of the four forms of tartaric acids are shown below. Note that the arrows indicate the direction of rotation of light by the asymmetric carbon atoms.

$$\begin{array}{c} COOH \\ H-C-OH \\ H-C-OH \\ COOH \end{array}$$

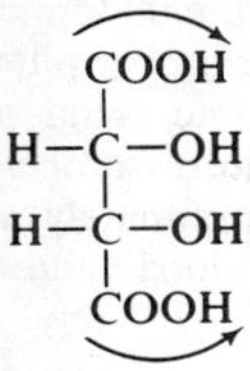

inactive or *meso*-tartaric acid (internally compensated; possesses a plane of symmetry; optically inactive).

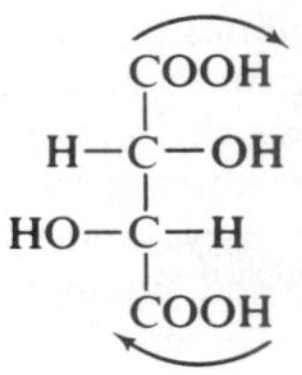

$$\begin{array}{c} COOH \\ H-C-OH \\ HO-C-H \\ COOH \end{array}$$

dextro-tartaric acid (arrangement of groups around each asymmetric carbon atom is cumulative; optically active; dextrorotatory).

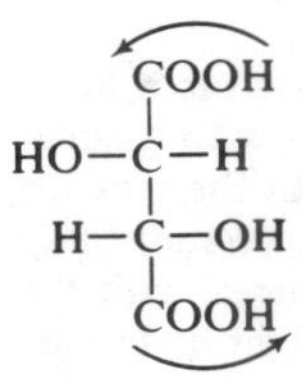

$$\begin{array}{c} COOH \\ HO-C-H \\ H-C-OH \\ COOH \end{array}$$

levo-tartaric acid (arrangement of groups around each asymmetric carbon atom is cumulative; optically active; levo-rotatory).

$$\left\{ \begin{array}{c} d\text{-tartaric acid} \\ l\text{-tartaric acid} \end{array} \right\}$$

Dextrolevo-tartaric acid, racemic tartaric acid (externally compensated; optically inactive; can be resolved into d and l components).

The two optically active tartaric acids when crystallized differ in the arrangement of the faces—one is the mirror image of the other. Pasteur (1848), observing this difference, was able to separate the two optically active forms of ammonium sodium tartrate crystals made from racemic tartaric acid.

Tautomerism is a form of isomerism in which a substance exists in two forms which are in equilibrium and exhibit characteristic reactions; either one may predominate, depending upon the conditions. Thus acetoacetic ester may react as a ketone or an enol (a compound containing a carbon atom having both an alcoholic hydroxyl group and a double bond) depending upon the conditions:

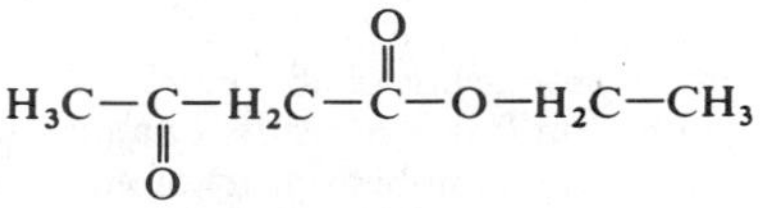

$$H_3C-\underset{\underset{O}{\parallel}}{C}-H_2C-\overset{O}{\overset{\parallel}{C}}-O-H_2C-CH_3$$
Ketone form

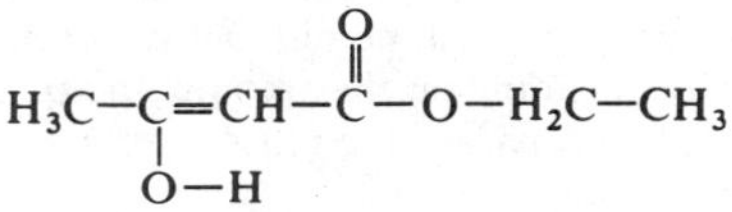

$$H_3C-\underset{\underset{O-H}{|}}{C}=CH-\overset{O}{\overset{\parallel}{C}}-O-H_2C-CH_3$$
Enol form
Acetoacetic ester

Chirality. In chemistry, *chiral* is a term used to describe asymmetric molecules that are mirror-images of each other, i.e., they are related to each other optically as right and left hands. Such molecules are also called *enantiomers* and are characterized by optical activity. An excellent summary of chirality in chemistry is given by Prelog, *Science,* **193,** 17–24 (1976).

ISOMERIZATION. The rearrangement of the structural configuration of a molecule without changing its molecular weight. Although structural changes of this type occur in other processes, e.g., catalytic reforming and cracking, isomerization can be the principal reaction desired in some processes. In petroleum refining, isomerization processes are used to change the structural configuration of C_4 paraffins (alkanes), such as normal butane into isobutane, in order to supplement other sources to provide enough butane for alkylation with olefins (alkenes) in the production of motor fuel. C_5 and C_6 paraffins are isomerized to the more highly branched structures to improve their antiknock ratings. Isomerization is also applied to a lesser extent in C_8 aromatic hydrocarbons.

In one isomerization process (UOP), the unit is arranged

to process a C_5/C_6 mixture with fractionating facilities to provide for the recycling of both n-pentane and n-hexane. A desulfurized C_5/C_6 blend first is fractionated to remove the native isopentane as a net product. The deisopentanizer bottoms are desiccant-dried before being joined by n-hexane recycle and brought to reaction temperature by heat exchange and suitable preheating. Before entering the reactor, the combined feed stream is joined by hydrogen recycle gas, which functions to suppress catalyst-deposit formation.

The fixed-bed reactor effluent is cooled and passed to a high-pressure separator. Gas from the separator, along with a small quantity of dried make-up hydrogen, is recycled to the reactor. The separator liquid is stabilized as a next step to remove any C_4 and lighter hydrocarbons that may be introduced with the make-up hydrogen, plus a very minor amount of light hydrocarbons formed by hydrocracking in the reactor. Hydrogen dissolved in the separator liquid is also removed by the stabilizer.

The next fractionator in series receives the stabilized liquid, from which it separates an equilibrium isopentane-n-pentane mixture that is routed back to the deisopentanizer for separating the isopentane as a net product. Thus, the n-pentane content of the feed is converted entirely to isopentane.

As a final step in the fractionation sequence, the hexane fraction is separated into a dimethylbutanes concentrate as a net overhead product and a n-hexane-rich bottoms stream to be recycled for the further isomerization of the n-hexane and methylpentanes. With economically practical fractionation, the methylpentanes split between the overhead and bottoms of the deisohexanizer column. For the C_5 fraction, the boiling points of the two isomers are far enough apart to make a relatively clean split economically feasible. For the C_6 fraction, the greater number of isomers and the bunching of some of their boiling points preclude precision separation in columns having a reasonable number of plates.

Once-through processing of a typical C_5/C_6 (68–70 octane number) straight-run fraction results in a product having a Research Method octane number (clear) of about 83. By recycling the unconverted n-pentane, n-hexane, and most of the methylpentanes, a product having an octane number of about 93 would result. Obviously, any octane number between 83 and 93 could be produced, depending on the amount and quality of the equipment installed to separate the reactor effluent into net product and recycle streams.

—M. J. Sterba, formerly with UOP Process Division, Universal Oil Products Company, Des Plaines, Illinois.

ISOPROPYL ALCOHOL. Also called dimethylcarbinol or secondary propyl alcohol, formula $(CH_3)_2CHOH$, *isopropyl alcohol* is a colorless liquid at room temperature. Pleasant odor, bp 82.4°C, specific gravity 0.7863 (20/20°C), autoignition temperature, 400°C. The compound is soluble in water, ethyl alcohol, and ether.

Two basic methods of production are in commercial use: (1) absorption of propylene in sulfuric acid to form alkyl hydrogen sulfate, followed by the hydrolysis of the ester; and (2) by direct hydration with water, using a catalyst. An inherent disadvantage in the first process is the need to handle sulfuric acid. Further, the first process yields little more than 70% isopropanol as compared with the second process in which liquid propylene is used as the charge stock. All direct-hydration processes can be represented by: $C_3H_6 + H_2O \rightarrow C_3H_7OH +$ heat.

Isopropyl alcohol is a widely used chemical, finding use as a starting material for making acetone and its derivatives; the manufacture of glycerol and isopropyl acetate; as a solvent for essential and other oils, alkaloids, gums, and resins; as a latent solvent for cellulose derivatives. It is used as a deicing agent for liquid fuels; in pharmaceuticals, perfumes, lacquers, in extraction processes, as a dehydrating agent and as a preservative.

See also **Alcohols;** and **Organic Chemistry.**

ISOTHERMAL. Of constant temperature, with respect to either space or time. Isothermal processes are those conducted without temperature change.

ISOTONE. One of two or more nuclides having the same number of neutrons in their nuclei.

ISOTOPE. An isotope is one of two or more nuclides that have the same number of protons in their nuclei. Any two isotopes have the same atomic number, Z. However, their mass numbers, A, differ. Isotope is a term that stems from the Greek words, *isos* (same) and *topos* (place), to designate substances having different atomic weights and yet having chemical properties so much alike that in the early days of research it was not possible to perform a chemical separation of the isotopes of a given element.

Sometimes, the term *nuclide* is confused with isotope. A nuclide may be defined as a species of atoms, with a specified atomic number and mass number. Different nuclides having the same atomic number should be described as isotopes. This is evident from the accompanying table. Different nuclides having the same mass number are termed *isobars*.

The existence of isotopes first became evident in the early years of this century, from the investigation of natural radioactivity. Then it was found that the natural radio elements underwent successive nuclear disintegrations, that they could be arranged in radioactive series according to these changes, and that in these series there were several instances in which atoms of the same atomic number (or as then stated, atoms occupying the same place in the periodic table) differed widely in their radioactive behavior. For example, it was found that radium C, radium E, thorium C and actinium C were all identical in their chemical properties with bismuth (atomic number 83) but differed in their radioactive properties and origins.

In the long course of research that led to the conclusion that more than one stable isotope of an element may exist, an important milestone was the method of positive-ray analysis. As applied by J. J. Thomson an electric discharge was passed through a vessel containing a gas at low pressure. The effect of the discharge was to produce ions in the gas, and these ions, because of their electric charge, could be formed into beams, deflected and otherwise directed by applied electric and magnetic fields. An experimental apparatus was designed so that the amount of this deflection would depend upon the masses of the particles of the gas. By using neon gas (atomic weight 20.183) in the tube, Thomson obtained photographs showing two beams of particles, one of them in a position calculated for particles having a mass of about 20, and the other for a mass of about 22. Although the conclusion was not immediately reached, it was later concluded (from the work of Aston) that neon (and other elements) consisted of atoms of more than one mass. Aston expressed this conclusion in the whole number rule, according to which all atomic weights of individual atomic species are close to whole numbers, and the whole number plus decimal values calculated chemically for the atomic weights of elements are due to the presence of two or more isotopes each of which has an atomic weight that is approximately a whole number. The fact that the chemically determined values

RELATIONSHIP OF ATOMIC AND MASS NUMBERS IN DESIGNATING ISOTOPES

Element	Mass Number A	Atomic Number Z	Atomic Number Z	Atomic Number Z	Atomic Number Z	
Hydrogen	1	1				
Hydrogen	2	1				
Hydrogen	3	1				
Helium	3		2			
Helium	4		2			
Helium	5		2			
Lithium	5			3		All nuclides
Helium	6		2			
Lithium	6			3		
Lithium	7			3		
Beryllium	7				4	
Lithium	8			3		
Beryllium	8				4	
Beryllium	9				4	
Beryllium	10				4	
	Isobars	– – – – – – – Isotopes – – – – – – –				

Table indicates that hydrogen has three isotopes, each with same atomic number, but with differing mass numbers, and designated as ^{1}H, ^{2}H, and ^{3}H (thus using the mass number to designate a particular isotope. Similarly, the four lithium isotopes (all with atomic number 4) are designated by their mass numbers, ^{5}Li, ^{6}Li, ^{7}Li, and ^{8}Li. Although redundant, because the element symbol implies the atomic number, symbols sometimes are written to indicate both mass and atomic numbers, as $_3^7$Li.

for the elements of naturally occurring materials from different sources are the same is because the isotopic composition of naturally occurring materials (except those of radioactive origin) is essentially the same.

Aston's work was founded upon accurate measurements of the deflections of charged particles. These measurements were made in an instrument he devised, the mass spectrograph. Many later instruments were developed following Aston's work, or following the Dempster instrument, which was built before Aston's. The direction-focusing mass-spectrographs and the later velocity-focusing instruments and composite instruments facilitated the determination, not only of the masses (and hence mass numbers) of the isotopes of an element, but their quantities as well. As a result of the immense amount of research in this field, the isotopic composition of the stable elements has been closely determined, and can now be said to be subject to only slight revision as more refined methods, and the possible discovery of stable isotopes present in very small quantities, are found.

It will be seen from that table that naturally occurring elements differ widely in the number of isotopes they contain. Some (usually of odd atomic number) are composed entirely of atoms of one mass number. Others have many stable isotopes. For example, the element of atomic number 50, which is tin, has at least ten stable isotopes and many radioactive ones.

The great importance of isotopes is due to two facts: (1)

Since the atomic number of an atom determines its chemical properties, all the isotopes of a given element exhibit essentially the same chemical behavior; that is, all atoms of the same number undergo essentially the same reactions with atoms of other atomic numbers. Thus, all three of the isotopes of hydrogen (protium, deuterium and tritium) undergo essentially the same reactions with oxygen, carbon and all the other elements. Therefore, if we add to the ordinary form of an element (which has a known isotope composition) a measured amount of an isotope of that element, we can follow the course of our sample through chemical reactions, especially those in which the same element enters at other points. This instance cited is representative of the many applications of isotopes which will be discussed in this article. (2) The other fact that accounts for the great present-day importance of a knowledge of isotopes is that, while the isotopes of an element exhibit similar chemical behavior, the nuclear characteristics of these isotopes often differ greatly. A most important example of this difference is that between ^{235}U and ^{238}U. See **Uranium**, where the separation of these and other isotopes is described.

The isotopes of all elements are tabulated under **Chemical Elements**. Also, there are descriptions of isotopes under the alphabetical entries for each element. The nature and importance of radioisotopes are described under **Radioactivity**.

J

JADE. Jade is a general term for a compact green mineral substance much prized for ornamental purposes. Jade is either a compact actinolite called nephrite, a variety of amphibole, or jadeite, a monoclinic pyroxene. It is easily worked, and many prehistoric implements have been found of this material in Mexico, Switzerland, France, Greece, and Egypt. The word jade is derived from the Spanish *pietra di hijada*, kidney stone, because it was supposed to be beneficial to diseases of the kidneys. Nephrite is derived from the Greek word for kidney, the allusion being the same as in the case of jade.

See also **Amphibole; Pyroxene.**

JADEITE. The mineral jadeite, essentially sodium-aluminum silicate, $Na(Al, Fe^{2+})Si_2O_6$, is a monoclinic pyroxene usually appearing in crystalline masses, or may be granular, fibrous, or compact. It has a prismatic cleavage; splintery fracture; hardness, 6 in crystals, 6.5–7 massive variety; specific gravity, 3.24–3.43; luster, vitreous to pearly; color, various shades of green, bluish-green, greenish-white or almost white; translucent to opaque. The processes that have acted to form this mineral are little understood both because of the confusion that exists between jadeite and nephrite, and the fact that the localities are not well known. Jadeite is found in Burma and China and has been reported from Mexico. It has probably resulted from the metamorphism, at great depths, of rocks rich in soda and aluminum, such as nephelite syenites. Its association in Burma with serpentine suggests its origin in more basic igneous rocks. See also **Pyroxene.**

Jadeite is a tough and yet rather easily worked substance, and has long been used for ornamental purposes. Evidence has been found in Europe, Mexico, Egypt, and elsewhere that it was used in prehistoric times for both ornaments and implements. The word jadeite is the general term used for all green-colored tough compact stones that have been used as indicated above.

See also terms listed under **Mineralogy.**

JAMESONITE. This mineral is commonly called *brittle featherore* from its brittle character and usual habit in acicular mats of crystals. It crystallizes in the monoclinic system and is of metallic luster, opaque with gray-black color and streak. It has a hardness of 2.5 and a specific gravity of 5.63–5.67. It is a sulfide of lead and iron antimony, $Pb_4FeSb_6S_{14}$.

Jamesonite occurs in low- to moderate-temperature hydrothermal veins with other lead sulfosalt minerals. Exceptional specimens are found in Bolivia, at Potosi and Oruro, and as choice felted masses on pyrite at Zacatecas in Mexico. Jamesonite is also found in Arkansas, Idaho, and Utah in the United States and in Ontario and British Columbia, Canada. It is named after the English mineralogist, Robert Jameson, from specimens obtained from Cornwall, England. It is a minor ore of lead.

JAROSITE. The mineral jarosite is a basic hydrous sulfate of potassium and iron corresponding to the formula $KFe_3(SO_4)_2(OH)_6$. It is formed in the outcrops of ore deposits during oxidation of iron sulfides. It is a hexagonal mineral with basal cleavage; is brittle; hardness 2.5–3.5; specific gravity 2.9–3.26; luster vitreous to dull; color, dark yellow to yellowish-brown; pale yellow streak; translucent to opaque. Jarosite was originally reported from and named for Barranco Jaroso in the Sierra Almagrera, Spain. It has been found in Bohemia, France, the Island of Elba, Siberia and Bolivia. In the United States it is found in Arizona, Colorado, Texas, New Mexico, Utah, Nevada, and South Dakota.

JASPER. A variety of chert, always quartz, that is associated with iron ores and thus contains iron-oxide impurities which give the rock a variety of colorations, often red, but ranging through yellow, green, grayish-blue, brown, and even black. The term also has been used with reference to any red chert or chalcedony. The material is dense, cryptocrystalline, usually opaque, although it may be slightly translucent.

JELLY. Colloid System.

JOSEPHSON TUNNEL-JUNCTION. As early as 1962, B. Josephson recognized the implications of the complex order parameter for the dynamics of the superconductor, in particular when one considers a system consisting of two bulk superconductors connected by a "weak link." The basic requirement for the weak link is that the amplitude of the order parameter at the link should be substantially smaller than in the bulk regions. In early experimentation, such a situation was realized in a variety of ways—two evaporated films separated by a thin (less than 20 angstroms) oxide layer; a light point contact between two bulk superconductors; a single hourglass-shaped evaporated film, with the constriction of dimensions small compared to the coherence length; or even a bare niobium wire with a pendant frozen blob of soft solder, where the weak links, indeterminate in number, are formed by solder bridges through pinholes in the surface oxide. Collectively, all such weak link junctions are referred to as *Josephson junctions*.

Both the dc and ac Josephson effects have found interesting and novel applications. The high sensitivity to magnetic field of the dc Josephson current in certain circuit configurations has been used to develop a family of devices called squids (*super*conducting *qu*antum *i*nterferometric *d*evices) which can be used to measure extremely small currents, voltages, and magnetic fields. The ac effect has been useful in making very precise measurements in connection with research toward improving the maintenance of the U.S. legal volt.

Much interest is presently exhibited in the use of Josephson superconducting devices of the tunnel-junction type (operated at near absolute zero) in connection with building superfast computers. In such devices, the electrical signals have only a millimeter or two to travel. Switching time is about 10^{-11} second. The power dissipation of Josephson devices permits high circuit density. It is estimated that the power dissipation of transistors is about 100 times that of Josephson devices. Different materials are being studied to improve the original lead alloy thin-film materials. These include lead-indium-gold alloys and niobium-tin alloys.

References

Anacker, W.: *IEEE Spectrum*, **16**, 26 (1979).
Brown, A. V.: *IBM J. Res. Dev.*, **24**, 167 (1980).
Howard, R. E., Rudman, D. A., and M. R. Beasley: *Appl. Phys. Lett.*, **33**, 671 (1978).
Kircher, C. J., and M. Murakami: "Josephson Tunnel-Junction Electrode Materials," *Science*, **208**, 944–950 (1980).
Matisoo, J.: *IBM J. Res. Dev.*, **24**, 133 (1980).

JOULE-THOMSON EFFECT. In passing a gas at high pressure through a porous plug or small aperture, a difference of temperature between the compressed and released gas usually occurs. This phenomenon is called the Joule-Thomson effect. The equation for this effect contains two partial derivatives and is

$$\left(\frac{\partial T}{\partial P}\right)_H = \frac{T\left(\frac{\partial V}{\partial T}\right)_P - V}{C_P}$$

where the expression on the left is the rate of change of temperature with pressure at constant enthalpy (heat content) (since no heat is supplied to, or removed from the system). The expression on the right has in its numerator the difference between the product of the temperature and the rate of change of volume with temperature at constant pressure, from which the volume is subtracted; the denominator contains the molar specific heat at constant pressure. The term on the left of the equality sign is called the *Joule-Thomson Coefficient*. It varies with the temperature and pressure of the gas, passing from positive values through zero to negative values. The temperature at which it is zero is called the *Joule-Thomson Inversion Temperature* and varies, of course, with the particular gas. It is to be noted, however, that for hydrogen, and also for helium, the temperature is low, far below 0°C. For other gases, however, much higher values are found, the maximum value for oxygen being 1,058K (785°C). At temperatures above their inversion temperature, gases are warmed, while at temperatures below it, they are cooled by the effect. For that reason, this type of expansion is often used in industrial processes for cooling gases. An interesting application is the process for producing solid carbon dioxide by expanding carbon dioxide through an aperture.

For an ideal gas, $PV = RT$; thus, $(\partial V/\partial T)$ is equal to R/P, that is, to V/T, so that the numerator of the right-hand term in the equation above becomes 0; thus an ideal gas shows no Joule-Thomson effect.

JUVENILE HORMONES. Insecticide.

K

KAOLINITE. The most common mineral of a group of hydrous aluminum silicates, which result from the breaking down of aluminum-rich silicate rocks, such as the feldspars and nepheline syenites, either through weathering or hydrothermal activity. Kaolinite, when pure, corresponds to the formula $Al_2Si_2O_5(OH)_4$, and occurs in white, claylike masses. Impurities may cause various colors or tints. X-ray powder photographs by one investigator (J.W. Grunner, 1932) revealed a two-layered monoclinic cell symmetry; two later investigators (G. W. Brindley and K Robinson 1946) found only a single-layer triclinic cell symmetry; it is also found, very rarely, in hexagonal scales. It has a perfect basal cleavage; is flexible but not elastic; hardness, 2–2.5; specific gravity, 2.6–2.63; luster, pearly to dull; color, white when pure, as described above, but may be yellow, red, blue, or brown; translucent to opaque.

Kaolinite is a mineral of widespread occurrence, well distributed throughout the world. The finest kaolinite locality in Europe is said to be in France, from whence the clay is obtained for porcelain ware. Cornwall and Devonshire in England supply large quantities of this mineral. In the United States, Pennsylvania, Virginia, Colorado, Georgia, and South Carolina contain deposits of kaolinite. The word kaolin or kaolinite is said to be a corruption of a Chinese word *kauling*, the name of a locality where this mineral is found. Kaolinite is very important commercially in the manufacture of china and pottery. See also terms listed under **Mineralogy.**

KERATIN. A chemically complex material (scleroprotein) of which horns, nails, claws, hoofs, and the scales of reptiles, birds, and mammals are formed. Hair and feathers also contain much keratin. It is present in the external layers of the skin, where it develops by the transformation of clear granules of keratohyalin of lower layers.

KERMA. Of ionizing particles, the kerma is $\Delta E_K/\Delta m$, where ΔE_K is the sum of the initial kinetic energies of all the charged particles liberated by indirectly ionizing particles in a volume element of the specified material, and Δm is the mass of the matter in that volume element. In these definitions, the symbol Δ precedes the letters E_K and m to denote that these letters represent quantities that can be deduced only from multiple measurements that involve extrapolation or averaging procedures. Since ΔE_K is the sum of the initial kinetic energies of the charged particles liberated by the indirectly ionizing particles, it includes not only the kinetic energy which these charged particles expend in collisions, but also the energy they radiate in bremsstrahlung. The energy of any charged particles is also included when these are produced in secondary processes occurring within the volume element. Thus the energy of Auger electrons is part of ΔE_K. In actual measurements, Δm should be so small that its introduction does not appreciably disturb the radiation field. This is particularly necessary if the medium for which kerma is determined is different from the ambient medium; if the disturbance is appreciable, an appropriate correction must be applied.

KERNITE. This hydrated sodium borate mineral, $Na_2B_4O_7 \cdot 4H_2O$, occurs in a single known world locality at Kramer, Kern County, California. It is found here as veins and interbedded masses in clay, hundreds of feet thick, thought to be a product of heating and dehydration of a large buried body of borax by intrusive igneous rocks. Crystals are monoclinic and attain individual size to 3 by 8 feet. Cleavage is very perfect to both the macro and basal pinacoids, with hardness, 2.5–3, and specific gravity of 1.908; possesses vitreous luster grading to satiny on cleavage surfaces; colorless to white; transparent. Surface alters readily to tincalconite, a white powder. It is a major source of borax and boron compounds. See also **Borax.**

KEROGEN. Oil Shale.

KEROSINE. Petroleum.

KETENES. Members of a class of compounds which contain the functional group $=C=C=O$. Examples are ketene itself (CH_2CO), methylketene $CH_3CH=C=O$, dimethylketene $(CH_3)_2C=C=O$, diphenylketene $(C_6H_5)_2C=C=O$, and carbon suboxide $O=C=C=C=O$. Of these, ketene is by far the most important. It is a reactive, colorless gas of considerable industrial importance. Physiologically, it is extremely poisonous and care must be taken to avoid breathing it.

The availability of ketone by pyrolysis of acetone (or acetic acid) is the reason for the attention it has received, contrasted to other ketenes which are relatively unavailable. Ketene may be prepared also by pyrolysis of acetic anhydride or phenyl acetate or diketene. Other sources are quite unsatisfactory from a standpoint of yield. Small quantities may be made conveniently by heating acetone in a "ketene lamp." This is a glass apparatus containing a Nichrome filament, heated electrically to red heat. Larger amounts are made by passing acetone or acetic acid through a tube at 700°C. A very brief contact time is required, so that much of the acetone is undecomposed and has to be condensed and recycled. Also, it is imperative that the reaction tube be of inert material such as porcelain, glass, quartz, copper or stainless steel. A copper tube, if used, should be protected from oxidation by an iron sheath. Inert packing may be used (glass, vanadium pentoxide, porcelain), but just as good yields are obtained with empty tubes. No catalyst is known which accelerates this decomposition at significantly lower temperatures.

Methyl ethyl ketone is totally unsatisfactory as a source of methylketene by pyrolysis, but pyrolysis of propionic anhydride in a quartz tube at 400–600°C and low pressures does produce it in a stated yield of about 90%. Another synthetic approach is to prepare methylketene dimer by allowing a mixture of propionyl chloride and triethylamine to stand at 25° for 24 hours and then to pyrolyze the dimer.

The known disubstituted ketenes include dialkyl-ketenes, diarylketenes, and the ester analogs. Dimethylketene may be made from α-bromoisobutyryl bromide by reaction with zinc in boiling ether. Diphenylketene may be made similarly, but the usual way to prepare it is to oxidize benzil hydrazone with yellow mercuric oxide to benzoylphenyldiazomethane which, on heating in benzene solution, decomposes into the ketene.

The best way to make carbon suboxide is a pyrolytic method, starting with tartaric acid. The latter is converted into diacetyltartaric anhydride and then pyrolyzed at 625–650°C (either in an empty tube or in a ketene lamp) into acetic acid and carbon suboxide, the latter in 35–50% yields.

Some recent synthesis of ketenes involve interesting chemistry. A butadienylketene is obtainable at −100 to −150°C by photolysis (mercury lamp) of the appropriate cyclohexadienone:

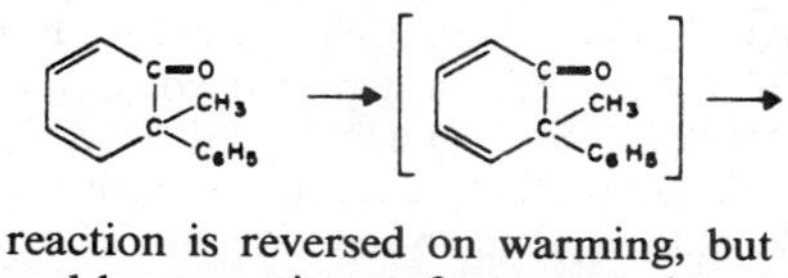

The reaction is reversed on warming, but the ketene may be captured by an amine to form an amide.

1-Ethoxy-1-alkyne, $R—C{\equiv}C—OC_2H_5$, pyrolyzes at 120° into ethylene and alkylketene, but the latter is consumed by the original alkynyl ether to form

2,4-dialkyl-3-ethoxy-2-cyclobutenone,

$$\begin{matrix} R—CH—C{=}O \\ | \qquad | \\ C_2H_5O—C{=}C—R \end{matrix},$$

in high yield.

Ketene and diimide are formed during the alkaline decomposition of chloroacetic hydrazide. The diimide, however, spontaneously changes into hydrazine plus nitrogen, and the hydrazine consumes the ketene to form acetohydrazide, $CH_3CONHNH_2$. If an olefin is present in the reaction mixture it is reduced to a paraffin by the diimide, thus preventing hydrazine formation.

The formulas of acetic acid, acetic anhydride and ketene show that the anhydride differs from the acid by 0.5 mole of water, and that ketene differs by 1.0 mole. Ketenes, therefore, may be regarded as super acid anhydrides, and this viewpoint leads to a good appreciation of their reactions. Because of an original lack of understanding of this relationship, the monosubstituted ketenes were once classed as "aldoketenes," and the disubstituted ketenes as "ketoketenes." Such terms are quite misleading, however, and should be abandoned.

Ketene is absorbed in sodium hydroxide solution, yielding sodium acetate. Aniline adds to ketene to form acetanilide. Both of these reactions are quantitative and are used to assay the ketene in a gas stream.

Primary alcohols react readily with ketene to form acetic esters but tertiary alcohols require the catalytic help of sulfuric acid. Even with primary alcohols, as 1-butanol, it has been established that addition of ketene ceases at about the 75% conversion point unless a little sulfuric acid is present as catalyst. Phenol, which is inert toward ketene at ordinary temperature, may be converted into phenyl acetate by reaction at the boiling point of phenol or by reaction at room temperature if a trace of sulfuric acid is present.

An important industrial synthesis of acetic anhydride is via acetic acid and ketone. Mixed acetic anhydrides are made similarly by passing ketene into the acid in question: RCOOH + $CH_2CO \rightarrow$ RCO—O—COCH$_3$. This is the basis of a good method of synthesizing symmetrical anhydrides in view of their formation from the mixed anhydrides on heating:

$$2RCOOCOCH_3 \rightarrow (RCO)_2O + (CH_3CO)_2O.$$

Comparable reactions of ketene are those with mercaptans to form thio esters (CH_3COSR), with amino acids (in water) to obtain N-acetyl derivatives ($CH_3CONHCHRCOOH$), with hydroxylamine to yield acetohydroxamic acid ($CH_3CONHOH$), dimethylchloroamine to form chloroacetic dimethylamide ($ClCH_2CON(CH_3)_2$), and Grignard reagents to form ketones ($RCOCH_3$).

Ketene adds to pyridine in a 4:1 ratio to form a yellow, crystalline compound,

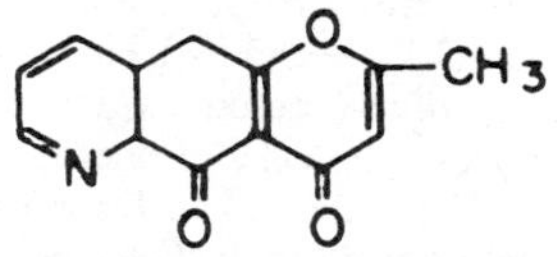

Quinoline, isoquinoline, and phenanthridine resemble pyridine in reacting comparably, and diketene may be substituted for ketene.

Aromatic aldehydes take up ketene in the presence of potassium acetate in the manner of a Perkin reaction. The product is a cinnamic acid:

$$ArCHO + CH_2CO \xrightarrow{AcOK} ArCH{=}CHCOOH.$$

Friedel-Crafts catalysts are effective in converting formaldehyde and ketene into β-propiolactone. This is a process of industrial importance. Similarly, furfural and ketene give rise to 3-(2-furyl)-propionolactone.

When aluminum chloride is used as catalyst, ketene reacts with benzene to form acetophenone: $C_6H_6 + CH_2CO \rightarrow C_6H_5COCH_3$. Also, methyl chloromethyl ether under such conditions reacts to yield 3-methoxypropionyl chloride:

$$CH_3OCH_2Cl + CH_2CO \xrightarrow{AlCl_3} CH_3OCH_2CH_2COCl.$$

Ketones such as acetone, ethyl acetoacetate, ethyl levulinate or acetylacetone react smoothly with ketene if a trace of sulfuric acid is present. The products are enol acetates of the ketones, acetone giving rise to isopropenyl acetate,

$$CH_2{=}C(CH_3)—OCOCH_3,$$

and ethyl acetoacetate changing into ethyl 3-acetoxycrotonate. Cyclobutanone derivatives are made quite easily by addition of ketenes to styrene or to vinyl ethers or to enamines:

$$R_2C{=}CH—NR_2 + O{=}C{=}CH_2 \xrightarrow{0°} \begin{matrix} R_2C——CH—NR_2 \\ | \qquad\quad | \\ OC——CH_2 \end{matrix}$$

One of the most characteristic reactions of ketene or dialkylketenes is that of polymerization into dimers. Diarylketenes do not display this tendency. The dimer from ketene, or "diketene," is a liquid that boils without decomposition at 43° (28 mm), but the compound tends to decompose (into dehydroacetic acid and resinous substances) on distillation (b.p. 127°C) at atmospheric pressure. The structure of diketene was in doubt for many years, but recent critical chemical and physical evidence indicates the structure as 3-buteno-β-lactone. Diketene is an acetoacetylating agent. Thus, with aniline it yields acetoacetanilide, and with methanol (catalyzed by sulfuric acid) it produces methyl acetoacetate. These reactions are useful industrially.

—Charles D. Hurd, Northwestern University, Evanston, Illinois.

KETONES. The homologous series of ketones (like aldehydes) has the formula $C_nH_{2n}O$. Structurally, the ketones consist of a carbonyl group (C:O) linkage between two radicals (R—CO—R′). R and R′ may be the same as in acetone (dimethyl ketone) CH_3—CO—CH_3; or they may differ as in methylethyl ketone CH_3—CO—C_2H_5. The latter may be referred to as a *mixed ketone*. A ketone is isomeric with the aldehyde that contains the same number of carbon atoms. Thus, acetone C_3H_6O is isomeric with propaldehyde C_3H_6O. Where R and R′ are alkyls, the ketone may be called an *alphyl ketone* and may be considered to be derived from the secondary alcohols. Ketones also may be formed from aromatic alcohols as in the case of benzophenone C_6H_5—CO—C_6H_5. The latter compound is a fully aromatic or *diaryl* ketone. Further, there are mixed *aryl-alphyl* ketones as in the case of acetophenone (phenylmethyl ketone) C_6H_5—CO—CH_3. The foregoing examples illustrate the use of trivial names for the ketones. In another system, the ketone may take its name from the alcohol from which it may be derived—thus, propione (from propyl alcohol); or the ketone may be named for the acid to which it may be oxidized—thus, acetone (acetic acid the oxidation product).

Essentially ketones exhibit the following properties: (1) all ketones up to C_{11} are neutral, mobile, volatile liquids. Ketones above C_{11} are solids under usual ambient conditions; (2) all ketones have a reasonably agreeable odor; (3) all ketones except those with a very high carbon count are soluble in H_2O, the solubility decreasing with a rise in formula weight; (4) most ketones are soluble in alcohol or ether, and (5) the specific gravity of ketones rises uniformly to about 0.83 with a rise in formula weight.

The presence of the double bond (carbonyl group C:O) markedly determines the chemical behavior of ketones. The hydrogen atom connected directly to the carbonyl group is not easily displaced. The chemical properties of the ketones may be summarized as follows: (1) They are readily reduced to form *secondary alcohols*, particularly in the presence of a catalyst. This property is used to advantage in the production of numerous organic compounds where ketones serve as a starting material or as an intermediate; (2) unlike the aldehydes, ketones are considerably more stable and do not combine readily with the alcohols, nor generally with NH_3 at ordinary temperatures and pressures; (3) they do not reduce alkaline solutions of metals and also, unlike aldehydes, ketones do not undergo polymerization; (4) they combine with hydroxylamine to yield *ketoximes*, (5) they react with hydrazine to form *hydrazones*; (6) they combine with semicarbazine to form *semicarbazones*; (7) with H_2SO_4 or HCl present, ketones can be induced to undergo a cyclic trimerization with the loss of H_2O; (8) when oxidized, ketones decompose to form two acids. Each acid will contain fewer carbon atoms than the originating ketone; and (9) ketones react with HCN to form *cyanohydrins*.

When ketones are reduced to secondary alcohols, varying amounts of ditertiary alcohols (pinacols) are produced. In a reaction known as the pinacol-pinacoline rearrangement, effective in synthesis of difficult compounds, ketones when treated with magnesium amalgam, after hydrolysis, yield a 1,2-glycol. Important to the purification of methyl and some cyclic ketones is their ability to form crystalline additive compounds with sodium bisulfite solutions. Ketones also react with phosphorus pentachloride and pentabromide to form dihalogen derivatives of the alkyls in which instances the oxygen atom of the carbonyl group is replaced by two hydrogen atoms.

Ketones are widely used as starting and intermediate ingredients in the production of numerous synthetics, such as resins, and they find wide application as solvents. The manufacture of acetone is described under **Acetone.** Other important ketones produced on a tonnage basis include methylethyl ketone (MEK) and methylisobutyl ketone (MIBK). Commercially, MEK may be manufactured by the direct oxidation of butylene in which air is used, along with a catalyst solution comprising cop-

per chloride and palladium chloride. The overall reaction is $C_4H_8 + 1/2)_2 \rightarrow CH_3COC_2H_5$. During the reaction, the palladium chloride is reduced to elemental palladium and HCl. Cupric chloride causes reoxidation. The resulting cuprous chloride is reoxidized to cupric chloride during the catalyst regeneration cycle. The process proceeds under moderate pressures at a temperature of about 100°C. The MEK product must be treated with sodium bisulfite and caustic soda, followed by distillation, to yield pure MEK.

In the production of MIBK, acetone may be the chargestock. In a first step, acetone is converted to diacetone alcohol (DAA) by condensation under pressure with an alkaline catalyst. The latter may be calcium or barium hydroxide, both of which are slightly soluble in H_2O. After cooling, in a second step, mesityl oxide (MSO) is produced by dehydrating the DAA. For this step, an acid catalyst is used and the step proceeds in the temperature range of 100 to 120°C. When the MSO is formed from the DAA, the latter partially decomposes into acetone. Thus, a distilling phase is required to separate and recover the acetone. In a third step, the MSO is hydrogenated and a mixture of MIBK and methylisobutyl carbinol (MIBC) is produced. These must be separated by a further fractionating phase. The hydrogenation reactions are: $(CH_3)_2C{=}CHCOCH_3 + H_2 \rightarrow (CH_3)_2CHCH_2COCH_3$ (MIBK); and $(CH_3)_2CHCH_2COCH_3 + H_2 \rightarrow (CH_3)_2CHCH_2CHOHCH_3$ (MIBC). The temperature of the process and the hydrogen mole ratio used determines the ratio of MIBK/MIBC produced and thus the manufacturer has effective control over the amounts of end-products (MIBK and MIBC) which can be manipulated in accordance with market requirements.

Related Compounds. A polyhydric ketone is referred to as a *ketose.* Monosaccharoses are examples of open-chain polyhydroxyaldehydes or ketones. The aldehyde sugars are termed *aldoses*; the ketone sugars are called *ketoses.* Fructose is a ketose. Dihydroxyacetone $HO \cdot CH_2 \cdot CO \cdot CH_2 \cdot OH$ is a very simple ketose. Compounds that contain both a carbonyl and a carboxylic group are termed *ketonic acids.* They display the reactive properties of both an acid and a ketone. Pyroracemic acid $CH_3 \cdot CO \cdot CO_2H$, acetoacetic acid $CH_3 \cdot CO \cdot CH_2 \cdot CO_2H$, and laevulic acid $CH_3 \cdot CO \cdot CH_2 \cdot CH_2 \cdot CO_2H$ are examples of monobasic ketonic acids. *Ketonic hydrolysis* is a term used to describe such actions as the hydrolysis of the ethyl ester of acetoacetic acid into acetone, CO_2, and ethyl alcohol. Compounds such as ketene $CH_2{:}CO$, methyl ketene $CH_3CH{:}CO$, dimethyl ketene $(CH_3)_2C{:}CO$, and diphenyl ketene $(C_6H_5)_2C{:}CO$ are of the family known as *ketenes.* See immediately preceding article.

KETOSES. Carbohydrates.

KEYES EQUATION. An equation of state for a gas, deduced from the concept of the nuclear atom. This equation is designed to correct the van der Waals equation for the effect upon the term b of the surrounding molecules. The equation is written as

$$P = \frac{RT}{V - Be^{-\alpha/V}} - \frac{A}{(V + l)^2}$$

in which P is pressure, T is absolute temperature, V is volume, R is the gas constant, e is the base of natural logarithms, 2.718 . . ., and A, α, and B, and l are constants for each gas.

KIMBERLITE. The name applied to a mica peridotite which occurs at Kimberley and other places in the Republic of South Africa, the source of rich deposits of diamonds. These valuable gem stones were originally found in the decomposed kimberlite which, being colored yellow by limonite, was termed "yellow ground." Deeper workings disclosed the less altered rock, kimberlite, which miners call "blue ground."

KINETIC THEORY. A theory (proved by experiment) which explains the phenomena of heat and pressure as due to the kinetic motion and elastic collisions of atoms and molecules. The phenomena include gas and vapor pressure, evaporation, and diffusion of fluids.

Gases expand indefinitely when released, not because of repulsion between the molecules as formerly supposed (though the Joule-Thomson effect under certain conditions may involve this), but because the molecules are in rapid motion and do not stop unless they collide with something. Air is not "forced" out through a tire puncture; only those air molecules pass out which, in their aimless wanderings, happen to encounter the opening. Molecules also pass in from the outside; but since there are several times as many per unit volume inside as outside, many more pass out than in. This continues until, a statistical equilibrium being reached, the air inside is no more dense than that outside, and the tire is "flat." The rapidity with which this takes place emphasizes the speed of the molecular motion and the relative insignificance of the "internal friction" opposing it.

What appears to be a steady pressure is due to the incessant impacts of the gas molecules on any surface exposed to them. If n molecules of equal mass m are released in an enclosure of volume v, and if their speeds are $u_1, u_2, \ldots u_n$, it is easy to show that the average pressure set up by these impacts, neglecting the effects of collisons and gravity is

$$p = \frac{m}{3v}(u_1^2 + u_2^2 + \cdots + u_n^2) \qquad (1)$$

This may be written

$$p = \frac{1}{3}\frac{nm}{v}\frac{\Sigma(u^2)}{n}$$

or, since nm/v is the gas density ρ, and $\Sigma(u^2)/n$ is the mean square molecular speed $\bar{u}^2$,

$$p = \tfrac{1}{3}\rho\overline{u^2} \qquad (2)$$

This relation gives the mean square speed as $\overline{u^2} = 3p/\rho$; which is the square of the effect speed, corresponding to average kinetic energy. From this it may be shown that the average speed is

$$\bar{u} = \sqrt{\frac{8p}{\pi\rho}} \qquad (3)$$

easily determined since p and ρ are measurable.

Again, Equation (1) may be written

$$pv = \tfrac{2}{3}(\tfrac{1}{2}mu_1^2 + \tfrac{1}{2}mu_2^2 + \cdots + \tfrac{1}{2}mu_n^2) = \tfrac{2}{3}E \qquad (4)$$

in which E is the total kinetic energy of linear motion of the molecules. From this it follows that the absolute temperature T of the gas bears a constant ratio to this total kinetic energy, and hence to the average translational kinetic energy of the molecules. (See **Ideal Gas Law.**)

Further analysis shows that when gravity is considered, the pressure in an undisturbed pure gas of uniform temperature, at an elevation h, is given by

$$p = p_0 e^{-3gh/\overline{u^2}} \qquad (5)$$

in which p_0 is the pressure at the zero of elevation. This is a form of Laplace's "law of atmospheres," useful in barometric altitude determinations.

A quantity much used in kinetic theory is the "mean free path," which is the average distance traversed by a molecule between collisions. There are ways of calculating this and also the effective diameters of molecules, and these data lead to many conclusions as to frequency of collisions, rate of diffusion.

Irreversibility. The kinetic theory assumes that the velocity of a molecule may depend on the conditions in the region where it has just suffered a collision, but is otherwise random—in other words, independent of its previous history. This assumption permits one to use the methods of probability theory even though, in classical mechanics, the actual motions of the molecules are regarded as completely determined by their initial configurations. As long as one uses the theory only to calculate properties of a gas that can actually be measured during a relatively short time, the assumption of randomness leads to no serious errors. However, it introduces an element of irreversibility which is inconsistent with the reversibility of the laws of classical mechanics. (A reversible process is one that can go equally well forwards or backwards, in contrast to an irreversible process). Lord kelvin pointed out the importance of irreversible processes in 1852. The irreversible aspect of the kinetic theory is shown most clearly by Boltzmann's "H-theorem," which has led to a considerable controversy over the foundations of kinetic theory. Boltzmann showed in 1872 that a certain quantity, later called H, which depends on the velocity distribution, must always decrease with time, unless the velocity distribution in Maxwell's distribution, in which case H remains constant. In the latter case, which corresponds to the equilibrium state, H is proportional to the negative of the entropy. Thus, the H-theorem provides a molecular interpretation of the second law of thermodynamics or, in particular, the principle that the entropy of an isolated system must always increase or remain constant.

Irreversible processes are those in which entropy increases. The entropy itself can be regarded as a measure of the degree of randomness or disorder of the gas, although it must be recognized that disorder really means a lack of knowledge about the details of molecular configurations. The equilibrium state represents the maximum possible disorder; the H-theorem implies that a gas which is initially in a nonequilibrium (partly ordered) state will eventually reach equilibrium and then stay there forever if it is not disturbed.

If the long-term consequences of the H-theorem were applicable to all matter in the universe, one might expect that the universe would eventually "run down"—although the total energy might always remain the same, no useful work could be done with this energy because all matter would be at the same temperature. This final state has been called the "heat death" of the universe.

The contradiction between the H-theorem and the laws of classical mechanics is shown by two well known criticisms of the kinetic theory: (1) the reversibility paradox and (2) the recurrence paradox. The first paradox is based on the fact that Newton's laws of motion are unchanged if one reverses the time direction, so that it would seem to be impossible to deduce from these equations a theorem that predicts irreversible behavior. Kelvin discussed this paradox in 1874, and concluded that while any single sequence of molecular motions could be reversed, leading to an ordered state, the number of disordered states is so much greater than the number of ordered states that it is virtually impossible to stay in an ordered state for any period of time. Thus, irreversibility is a statistical, but not an absolute consequence of kinetic theory. Boltzmann gave a similar answer when the problem was pointed out to him by Loschmidt a few years later.

The second paradox is based on a theorem of Henri Poincaré—if a mechanical system is enclosed in a finite volume, then after a sufficiently long time it will return as closely as one likes to its initial state. Hence *H* must return to its original value; if it has decreased during some period of time, it must increase during some other period. The time between successive recurrences of the same state for the molecules in one cubic centimeter of air is much longer than the present age of the universe, so one does not have to be concerned about recurrences in any actual experiment. In his attempt to resolve the recurrent paradox, Boltzmann was finally led to a remarkable psycho-cosmological speculation; he suggested that the direction of time as perceived by an animate being is determined by the direction of irreversible processes in his environment and in his body. Thus, when the time comes for a recurrence, entropy will decrease, but subjective time will flow in the opposite direction. The concept of alternating time-directions in cosmic history was further explored by Reichenbach in 1956 and has been proposed again in recent theories of the expanding (and contracting) universe.

Updating of Kinetic Theory. Since the late 1940s, there has been a revival of interest in the classical kinetic theory of gases, based on the assumptions of Clausius, Maxwell, and Boltzmann, and ignoring quantum effects except insofar as these may determine the intermolecular force law. In part this interest grew out of applications involving high-speed aerodynamics and plasma physics and, in part, to renewed attempts to construct reliable theories of liquids as well as dense gases. Methods for obtaining accurate solutions of the Boltzmann equation were developed by Grad, Pekeris, Ikenberry, and Truesdell, among others. These solutions were used to describe the behavior of gases in many circumstances more complex than those treated in the 1800s (including the interactions of charge particles and magnetic fields). Problems such as the propagation and dispersion of sound waves were treated by Uhlenbeck and his collaborators.

In 1946, three general formulations of kinetic theory were published (Born and Green; Kirkwood; and Bogoliubov). In each case, the goal was to derive a generalized Boltzmann equation in a form that would be valid when simultaneous interactions among more than two molecules have to be taken into account, and thence to obtain solutions of the equation from which transport properties of dense gases and liquids could be calculated. In each formulation, certain approximations had to be made in order to obtain practical results; because of the difficulty in estimating the error involved in these approximations, and the great complexity of the equations involved, there was no clear evidence that the results for properties such as the viscosity coefficient would be significantly more accurate than those obtained by Enskog from his modified kinetic theory for dense gases, published in 1922. Eventually, in the early 1960s, attention was centered on the systematic derivation of series expansions for the transport coefficients in ascending powers of the density, together with attempts to calculate the first few terms in such series for special molecular models, such as elastic spheres.

In the meantime, an alternative and apparently more rigorous method for deriving theoretical expressions for transport coefficients, based on the fluctuation-dissipation theory introduced in 1928 by H. Nyquist in electrical engineering problems, was developed by Green, Mori, and Kubo. This method had the heuristic advantage of bringing out clearly the connection between transport theory and the description of fluctuations in equilibrium statistical mechanics. Later, it was proved that the Green-Mori-Kubo method gives results precisely equivalent to those that could be obtained from the Born-Green, Kirkwood, and Bogoliubov methods. Thus, just as in the case of quantum mechanics, several alternative approaches are equally valid in modern kinetic theory.

After intensive efforts to calculate terms in the density expansion of transport coefficients, it was finally discovered in 1965 that such a density expansion does not actually exist, for mathematical reasons associated with the persistence of weak correlations between colliding particles over very long times. The divergence of the expansion (and thus the inadequacy of the approximations on which most earlier theories had been based) was established by a number of investigators, including Dorfman and Cohen, Weinstock, and Goldman and Frieman. The result was an increased overall interest in kinetic theory.

KININS. Chemical substances which stimulate differentiation of plant cells that otherwise may have lost permanently the power of differentiation. The kinins contain adenine, which seems to give these substances their biological activity. Small bits of carrot taken from a region containing no cambium can be stimulated to growth and differentiation to produce entire new carrot plants through the addition of kinins and a small amount of indoleacetic acid.

KJELDAHL NITROGEN DETERMINATION. An analytical method for the determination of nitrogen in an organic compound by its reduction to ammonium salts, from which the ammonia is liberated by a nonvolatile alkali, and then distilled into a standard acid. The initial decomposition of the organic material is effected by sulfuric acid digestion, which converts aminoid nitrogen to ammonium sulfate.

KRAFT PULP PROCESS. Pulp (Wood) Production and Processing.

KREBS CITRIC ACID CYCLE. Carbohydrates.

KRYPTON. Chemical element symbol Kr, at. no. 36, at. wt. 83.80, periodic table group 0 (inert or noble gases), mp $-156.6°C$, bp $-152.3°C$, density, 3.4 g/cm^3 (solid at $-273°C$). Solid krypton has a face-centered cubic crystal structure. At standard conditions, krypton is a colorless, odorless gas and does not form stable compounds with other elements. Due to its low valence forces, krypton does not form diatomic molecules, except in discharge tubes. It does form compounds under highly favorable conditions, as excitation in discharge tubes, or pressure in the presence of a powerful dipole. Krypton forms a hydrate at 14.5 atm pressure and 0°C. The element also forms addition compounds with a number of organic substances, such as $Kr \cdot 2C_6H_5OH$ with phenol, which has a dissociation pressure of 6–10 atm at 0°C. Krypton also forms compounds, possibly clathrates, with certain substances in nonstoichiometric proportions. Examples are its crystalline compounds with benzene and the compounds formed in aqueous solutions of hydroquinone, under 40 atm pressure of krypton, which contain 15.8% krypton by weight. First ionization potential, 13.996 eV; second, 26.4 eV; third, 36.8 eV.

Krypton occurs in the atmosphere to the extent of approximately 0.000114% and thus is the second least abundant of the rare gases in ordinary air. In terms of abundance, krypton does not appear on lists of elements in the earth's crust because it does not exist in stable compounds. However, because of its limited solubility in H_2O, krypton is found in seawater to the extent of approximately 1.4 tons per cubic mile. Commercial krypton is derived from air by liquefaction and fractional distillation. With exception of very special applications, krypton usually is not prepared in pure form, but supplied along with other rare gases, such as argon and neon, for filling fluorescent and incandescent lights. As a filler in lamps, the gas assists in reducing filament evaporation and enables higher operating temperatures for lamps. Very high-candlepower aircraft-approach lamps contain krypton. When contained in an electric-discharge tube, krypton, when pure, emits a characteristic pale-violet light; when impure, it emits a brilliant red color characteristic of so-called neon tubes. Krypton also has been used in lasers. There are five natural isotopes ^{78}Kr, ^{80}Kr, and ^{82}Kr through ^{84}Kr, and five radioactive isotopes ^{76}Kr, ^{77}Kr, ^{79}Kr, ^{81}Kr, and ^{85}Kr. The latter isotope is generated in atomic reactors, is a beta emitter with a half-life of approximately 10.6 years. This isotope, in solid form combined with a hydroquinone, has been used for activating phosphors and in luminous paints. ^{81}Kr has a particularly long half-life of 2.1×10^5 years. While investigating the properties of liquid air in 1898, Ramsay and Travers found krypton in the residue remaining after nearly all of the liquid air had boiled away. The element then was identified spectroscopically. The element emits a characteristic brilliant green and yellow line in its spectrum.

By international agreement in 1960, the *fundamental unit of length*, the *meter*, is defined in terms of the orange-red spectral line of ^{86}K. This corresponds to the transition $5p[O_{1/2}]_1 - 6d[O_{1/2}]_1$. One meter = 1,650,763.73 wavelengths (*in vacuo*) of the orange red line of ^{86}K.

Meteorological Effects of Krypton-85 in the Atmosphere. As pointed out by Boeck (1976), projections indicate that ^{85}Kr, a radioactive, chemically inert gas, may be produced and released in such quantities that it will create atmospheric ions at rates comparable to the present ion production rate near the tropical ocean surface. Krypton-85 is a byproduct in nuclear fission reactors and explosions. The ^{85}Kr is sealed in fuel elements of reactors, but during reprocessing of the fuel or to separate plutonium, ^{85}Kr is released in a controlled manner to the atmosphere. The radioactive half-life of ^{85}Kr is 10.76 years. With no apparent natural mechanism to remove it, the gas will accumulate in the atmosphere until a balance between the rate of release and decay is reached. Ultimately this could produce a unique form of atmospheric radiation as contrasted with such natural radioactivity as emanates from uranium, thorium, etc., which are limited to producing ions near ground level. In the atmosphere, of course, are several radioactive isotopes (half-lives are given in parentheses): ^{22}Na (2.6 years); ^{32}P (14.22 days); ^{7}Be (53.6 days); ^{33}P (24.4 days); ^{35}S (87.1 days), among others. Not much concern has been given to these isotopes because of their comparatively short half-lives. Boeck (1976) makes a case for the need for more fundamental research into the possible effects of ^{85}Kr in the atmosphere, including possible climatic alterations.

Krypton in Meteorites. Along with other rare gases, krypton has been found in meteorites, notably the Murchison carbonaceous chondrite. Some scientists have been predicting for years that meteorites may entrap materials which date back to the beginning of the universe. For many years, the field was dominated by the dogma of an isotopically and chemically uniform early solar system (Srinivassan and Anders, 1978). However, careful analyses of the Murchison meteorite have produced anomalies in the form of unexpected differences in the ratios of the various rare element isotopes present.

References

Alexander, E. C., Jr., and M. Ozima, Editors: "Terrestrial Rare Gases," Japan Scientific Societies Press, Tokyo (1978).
Boeck, W. L.: "Meteorological Consequences of Atmospheric Krypton-85," *Science*, **193**, 195–198 (1976).
Srinivassan, B., and E. Anders: "Noble Gases in Murchison Meteorite," *Science*, **201**, 51–55 (1978).

KYANITE. The mineral kyanite is an aluminum silicate, corresponding to the formula Al_2SiO_5. It is triclinic, and has a good cleavage parallel to the macropinacoid. Its hardness varies considerably, depending on the crystallographic direction from 5 to 7.5; specific gravity, 3.56–3.67; luster, vitreous to pearly; color, commonly blue to white, but sometimes gray to green or nearly black; transparent to translucent; usually found in long-bladed crystals or columnar to fibrous structures. Kyanite is found in some metamorphic rocks as gneisses or mica schists. Of the many European localities for fine specimens might be mentioned the Ural Mountains of the U.S.S.R.; Czechoslovakia; Austria; Trentino, Italy; the St. Gotthard region of Switzerland; and France. In the United States, Chesterfield, Massachusetts; Litchfield, Connecticut; and Gaston County, North Carolina, have furnished fine specimens. Kyanite derives its name from the Greek word meaning *blue*, in reference to the delicate blue of the inner portions of the bladed crystals.

L

LACTIC ACID. Alpha-hydroxypropionic acid, H—$C_3H_5O_3$, formula wt 90.05, colorless liquid, mp 18°C, bp 122°C, sp gr 1.248, miscible with H_2O, alcohol, or ether in all proportions. The substance exists in two forms: (1) *dextro* lactic acid, which rotates the plane of polarized light to the right; and (2) *levo* lactic acid, which rotates the plane of polarized light to the left. A mixture of these two forms is ordinary lactic acid, which does not rotate the plane of polarized light. Ordinary lactic acid is termed *dextrolevo* lactic acid. It is a product of corn refining. Lactic acid was one of the first biological substances to be investigated from the standpoint of the existence of two optically active forms.

Lactic Acidosis. Lactic acid is the cause of one of many possible disorders in human acid-base metabolism. Lactic acidosis represents an accumulation of lactic acid in the blood and tissues. This condition gradually depletes the natural buffers in the body and there is a consequent lowering of pH. As described in the entry on **Glycolysis**, lactic acid is the end product of that process. Lactic acid blood levels are determined by at least four factors. The rate of generation of lactic acid; the rate of transport from tissues to plasma and from plasma to the liver (point of utilization of lactic acid); the rate of utilization; and excretion of lactic acid by the kidneys. Normally, all of these functions are maintained in balance to give a normal blood lactate concentration of about 1 mEq/l.

On the generation side, three factors are involved. The availability of oxygen—because, as adenosine triphosphate (ATP) generation from oxidative phosphorylation diminishes, the cells naturally respond with a greater rate of glycolysis. This increases tissue lactate levels and ultimately lactate blood levels. See also **Phosphorylation (Oxidative)**. If, as may caused by various factors, there are increases in pH, the activity of phosphofructokinase will increase (this is the rate-limiting enzyme of glycolysis). With increases in pH, the enzyme is more active and more lactate is formed. Factors which affect the biological oxidation-reduction potentials also are a determininant that influence the rate at which glucose is metabolized to lactate.

Fundamental predisposing conditions causing an increased generation of lactate include decreased tissue perfusion. This is associated with shock and may occur in cardiac arrest. Another factor is increased skeletal muscle activity. The rate of glycolysis increases with exercise. This also may be associated with convulsive states that may follow severe exercise—brought about by increased blood lactate concentrations. Large tumors may cause an increase in lactate generation. Tumors (leukemias, lymphomas, etc.) may have an increased rate of glycolysis even in the presence of a sufficient supply of oxygen. Both cyanide and carbon monoxide poisoning can increase lactate levels because of insufficient oxygen supply.

On the utilization side, there are a number of influencing factors. In liver failure, the principal lactic acid utilization center, a surplus of lactate builds up. A condition like this may be associated with reduced hepatic perfusion, hepatocyte failure, and hepatocytes replaced by tumor. Blood lactate concentrations

are elevated in persons with diabetic ketoacidosis. Although the effects of alcoholism on lactic acid levels are not fully understood and may be contribute by increasing the generation or by decreasing the utilization, the latter is now believed by many authorities as the cause. This is brought about by the fact that ethanol completes for electrons in the liver, thus decreasing utilization of lactic acid in that organ.

LACTOSE. Carbohydrates.

LACTULOSE. Sweeteners.

LAKES (Colors). Colorants (Foods); Dyes.

LAMBERT'S LAW. Bouger and Lambert Law.

LANTHANIDE CONTRACTION. The decreasing sequence of crystal radii of the tripositive rare-earth ions with increasing atomic number in the group of elements (57) lanthanum through (71) lutetium of the lanthanide series in the periodic table. See also **Actinide Contraction.**

LANTHANIDES. The chemical elements with atomic numbers 58 to 71 inclusive, commencing with cerium (58) and through lutetium (71) frequently are termed, collectively, the *lanthanide series*. Lanthanum is the anchor element of the series, but also appears in group 3b of the periodic table. Some authorities consider lanthanum a part of the series per se. These elements are described under **Rare-Earth Elements and Metals.** An analogous series is described under **Actinides.**

LANTHANUM. Chemical element symbol La, at. no. 57, at. wt. 138.91, periodic table group 3b, homolog of the lanthanide series of elements, mp 918°C, bp 3464°C, density, 6.146 g/cm^3 (20°C). Elemental lanthanum has a double close-packed hexagonal crystal structure at 25°C. Pure metallic lanthanum is silvery-gray in color, but with a luster that remains only briefly upon exposure to air, rapidly oxidizing to a white powder. The oxide is hygroscopic and tends to spall, thus exposing fresh surfaces of the metal for oxidation. Thus the metal must be handled in an inert atmosphere. Chips and powdered lanthanum are quite pyrophoric. Under required inert atmospheric conditions, the metal is easy to work with normal tools, paralleling tin in its workability. There are two natural isotopes, ^{139}La and ^{138}La. The latter is mildly radioactive with a half-life of 10^{10}–10^{15} years. The element becomes a superconductor below 6 K. There are 19 known artificial isotopes, all radioactive. Of the light (or cerium group) rare-earth metals, lanthanum is the second most plentiful and ranks 57th in abundance of elements in the earth's crust, exceeding gold, tantalum, platinum, mercury, bismuth, and several other commonly known elements. The element was first identified by C. G. Mosander in 1839. The electronic configuration is $1s^22s^22p^63s^23p^63d^{10}4s^24p^64d^{10}5s^25p^65d^16s^2$. Ionic radius of

La^{3+}, 1.061; metallic radius, 1.879. First ionization potential, 5.571 eV; second, 11.06 eV. Oxidation potentials $La \rightarrow La^{3+} + 3e^-$, 237 V; $La + 3OH \rightarrow La(OH)_3 + 3e^-$, 2.76 V. Other important properties of lanthanum are given under **Rare-Earth Elements and Metals.**

Much of the commercial lanthanum production uses bastnasite, a rare-earth fluorocarbonate found in Southern California, as the source. See also **Bastnasite.** The element is separated from other rare-earth elements in an ion-exchange process after acid leaching of bastnasite (or monazite) minerals. Pure lanthanum is obtained by (1) electrowinning from the oxide La_2O_3 in a molten fluoride electrolyte, (2) electrolysis of fused anhydrous $LaCl_3$, or (3) metallothermic reduction of LaF_3 by calcium in a reactor under an inert atmosphere.

Lanthanum metal dissolves readily in dilute mineral acids. The oxide is dissolved by concentrated mineral acids and acetic and formic acids. The metal is a component of mischmetal used for lighter "flints" and in the cores of carbon electrodes for high-intensity lighting. See also **Cerium.** Several of the best grades of optical glass require pure lanthanum oxide as an ingredient for lowering the dispersion of light and for improving the index of refraction. The oxide melts at 2,310°C. The oxide ranks eleventh among the most refractory metal oxides, but finds limited use because of its highly hygroscopic nature. The oxide also is used as a host matrix for fluorescent phosphors and in thermistors and capacitors and other elements of electronic circuitry. By far, the largest use of lanthanum (mixed with other rare-earths) is for molecular-sieve catalysts for cracking crude petroleum. This use has required over 10 million pounds of lanthanum rare-earth chloride annually.

As an alloying metal, lanthanum finds broad use. Although lacking mechanical strength, lanthanum has a high affinity for oxygen, sulfur, nitrogen, and hydrogen and thus makes an effective component for scavenging gases from molten metals. Cobalt-base alloys containing lanthanum have shown increased resistance to oxidation and hot corrosion. One of the intermetallic compounds of lanthanum $LaCo_5$ possesses excellent magnetic properties—well in excess of those of alnico and platinum cobalt permanent magnets. The intermetallic compound $LaNI_5$ shows exceptional properties for absorbing and desorbing large amounts of hydrogen at room temperature.

See references listed at the ends of the entries on **Chemical Elements;** and **Rare-Earth Elements and Metals.**

NOTE: This 4th Edition entry was revised and updated by K. A. Gschneidner, Jr., Director, and B. Evans, Assistant Chemist, Rare-Earth Information Center, Energy and Mineral Resources Research Institute, Iowa State University, Ames, Iowa.

LASER-INDUCED REACTION. Photochemistry.

LASER-ENRICHMENT PROCESS (Uranium). Uranium.

LASER-INDUCED FUSION. Nuclear Reactor.

LATENT HEAT. Heat which is gained by a substance or system without an accompanying rise in temperature during a change of state. As examples, the latent heat of fusion is the amount of heat necessary to convert a unit mass of a substance from the solid state to the liquid state at the same temperature, the pressure being that to allow coexistence of the two phases. A considerable part of the latent heat arises from the entropy increase consequent on the greater disorder of the liquid state. The latent heat of sublimation is the amount of heat necessary

to convert a unit mass of a substance from the solid state to the gaseous state at the same temperature, the pressure being that to allow coexistence of the two phases.

LATEX. A milky substance found in many plants. It is a complex emulsion in which such substances as proteins, alkaloids, starches, sugars, oils, tannins, resins, and gums are found. In most plants, the latex is white, but in some it is yellow; in others, orange or scarlet.

The cells or vessels in which latex is found make up the *laticiferous system.* There are two very different ways in which this system may be formed. In many plants, the laticiferous system is formed from cells laid down in the meristematic region of the stem or root. Rows of these cells are formed. The cell walls separating them are dissolved, so that continuous tubes, called *latex vessels,* are formed. This method of formation is found in the poppy family; in the rubber plant, *Hevea brasiliensis;* and in the *Cichoriae,* a section of the composite family distinguished by the presence of latex in its members. Dandelion, lettuce, hawkweed, and salsify are members of the *Cichoriae.* See also **Rubber Natural.**

Synthetic latexes are now produced. See **Elastomers.**

LATTICE COMPOUNDS. Chemical compounds formed between definite stoichiometric amounts of two molecular species which owe their stability to packing in the crystal lattice, and not to ordinary valence forces.

LATTICE ENERGY OF CRYSTAL. The decrease in energy accompanying the process of bringing the ions, when separated from each other by an infinite distance, to the positions they occupy in the stable lattice. It is made up of contributions from the electrostatic forces between the ions, from the repulsive forces associated with the overlap of electron shells, from the van der Waals forces, and from the zero-point energy.

LAURIC ACID. Also called dodecanoic acid, formula $CH_3(CH_2)_{10}COOH$. A fatty acid that occurs in many vegetable oils and fats as the glyceride, especially in coconut oil and laurel oil. See also **Vegetable Oils (Edible).** Combustible. It takes the form of colorless needles at room temperature. Specific gravity 0.833; mp 44°C; bp 225°C (100 millimeters pressure). Insoluble in water; soluble in alcohol and ether. It is derived by the fractional distillation of coconut oil. Lauric acid is used in alkyd resins; wetting agents; soaps; detergents; cosmetics; insecticides; food additives.

LAVA. Molten material which has poured out on the surface of the earth and, due to relief of pressure, may have lost much of its original gas and water content during its relatively rapid consolidation. The term is used for both the liquid and the consolidated state of the igneous material. Lava may be erupted either by volcanoes or from fissures. The most extensive lava flows are fissure eruptions, such as the Columbia Plateau basalts in Oregon or the plateau basalts of the Deccan, India, which are derived from basic magma. Had this magma, either basic or acid, cooled slowly beneath the surface of the earth under great pressure and with all its original gases, the resulting rock would have had a coarser texture and somewhat different mineral content.

LAWRENCIUM. Chemical element symbol Lw, at. no. 103, at. wt. 257 (mass number of known isotope), radioactive metal of the actinide series; also one of the transuranium elements. ^{103}Lw was identified in 1961 by A. Ghiorso, T. Sikkeland, A.

Larsh, and R. Latimer at the University of California at Berkeley.

The method used to produce and identify lawrencium was similar to that used in the later, direct-counting experiments performed in connection with the production of nobelium at Berkeley. About 3 micrograms of a mixture of californium isotopes were bombarded with boron ions accelerated in the heavy-ion linear accelerator. The atoms of lawrencium recoiled from the target into an atmosphere of helium, where they were electrostatically collected on a copper conveyor tape. This tape was then periodically pulled into place before radiation detectors to measure the emission rate and the energy of the alpha particles being emitted. By this means, it was possible to identify the lawrencium isotope ^{257}Lw with a half-life of 8 seconds. At present, because of the short half-life and the lack of a suitable daughter isotope, available in the case of nobelium, it has not been possible to perform a chemical identification.

Another isotope, ^{256}Lw, half-life of about 45 seconds, was reported by the U.S.S.R. in 1965. It was said to have been produced by impact of oxygen atoms (^{18}O) on americium (^{243}Am). It decayed by alpha particle emission and electron capture to form ^{252}Fm.

References

Eskola, K., Eskola, P., Nurmia, M., and A. Ghiorso: "Studies of Lawrencium Isotopes with Mass Numbers 255 through 260," *Phys. Rev.*, **4**, 2, 632–642 (1971).

Ghiorso, A., Sikkeland, T., Larsh, A. E., and R. M. Latimer: "New Element, Lawrencium, Atomic Number 103," *Phys. Rev., Lett.*, **6**, 9, 473–475 (1961).

Seaborg, G. T., Editor: "Transuranium Elements," Dowden, Hutchinson & Ross, Stroudsburg, Pennsylvania (1978).

Silva, R., Sikkeland, T., Nurmia, and A. Ghiorso: "Tracer Chemical Studies of Lawrencium," *Inorg. Nucl., Chem. Lett.*, **6**, 9, 733–739 (1970).

LAWSONITE. This calcium aluminum silicate mineral $CaAl_2(Si_2O_7)(OH)_2 \cdot H_2O$, is found as grains and veins within the metamorphic rocks, gneisses, and schists. It was found originally on the Tiburon Peninsula, San Francisco Bay, California, but also occurs in schistose rocks in France and New Caledonia. The mineral has a hardness of 7; specific gravity 3.05–3.10. It is colorless, pale blue to bluish gray, translucent, with vitreous to greasy luster. The mineral crystallizes in the orthorhombic system with perfect prismatic cleavage.

LAZULITE. The mineral crystallizes within the monoclinic system, a basic phosphate of magnesium and aluminum, $MgAl_2(OH)_2(PO_4)_2$. Ferrous iron can substitute for the magnesium and the isomorphous mineral scorzalite is the product. Usually occurs massive but acute pyramidal crystals are not uncommon. Color is azure-blue to bluish-green, usually translucent (rarely transparent), with vitreous luster. It has a hardness of 5.5–6, with specific gravity of 3–3.1.

Lazulite is a rare mineral found principally within high-grade metamorphic rocks. Notable world crystal occurrences are Salzburg, Austria; Styria; Hörnsjöberg, Sweden; Malagasy Republic; Brazil; and Graves Mountain, Georgia. When transparent, the mineral can be cut into gem stones.

LAZURITE. The mineral lazurite or lapis lazuli has been used since ancient times for jewelry and other ornamental purposes. Ground to powder it forms the pigment ultramarine, now, however, largely superseded by artifical preparations. Lapis lazuli is a mixture of minerals, lazurite being the chief component. The mineral is isometric, and chemically a sodium, calcium, aluminum sulfochlorosilicate. A general formula is (Na, Ca)$_8$(Al, Si)$_{12}$O$_{24}$(S, SO$_4$). Lapis lazuli has a hardness of 5–5.5; specific gravity, 2.4; color, various shades of blue; luster, vitreous to greasy; translucent to opaque. Localities are Afghanistan, Siberia, Chile, and California.

LEAD. Chemical element symbol Pb, at. no. 82, at. wt. 207.19, periodic table group 4a, mp 327.5°C, bp 1740°C, density, 11.35 g/cm^3 (20°C). Elemental lead has a face-centered cubic structure with an edge length of 4.950 Å. Lead is a white-to-bluish-gray metal, soft, malleable, and slightly ductile. Lead tarnishes in air, forming a film of oxide, forms oxide scum upon heating in air; soluble in dilute HNO$_3$; HCl or H$_2$SO$_4$ attack lead only slightly, the extent depending markedly upon the concentration and temperature; slowly dissolves in water and consequently the use of lead constitutes a health hazard because of its toxicity; attacked by solutions of organic acids or sodium hydroxide. Lead is one of the four most utilized metals and considerable scrap metal is recovered. Used (1) in construction and apparatus where workability is demanded, and definite resistance to corrosion is supplied by the metal; (2) as a constituent of various alloys, especially solder, type metal, pewter, and fusible alloys; (3) for storage battery plates; (4) for shot and bullets; (5) as a protective coating for iron and steel.

Lead has four naturally occurring isotopes. In order of abundance, these are ^{208}Pb, ^{206}Pb, ^{207}Pb, and ^{204}Pb. There are ten unstable isotopes, 200–203, 205, and 209–214. See **Radioactivity.** In terms of abundance, lead is scarcely represented in the earth's crust, the average composition of igneous rocks containing only 0.002% Pb by weight. In terms of cosmic abundance, an estimate made by Harold C. Urey (1952), using silicon as a basis with the figure of 10,000, lead had an abundance figure of less than 0.02. In terms of presence in seawater, lead is 27th among the elements, with an estimated 14 tons per cubic mile (3 metric tons per cubic kilometer) of seawater. In this regard, it is comparable to tin, copper, arsenic, protactinium, and selenium.

The atomic weight varies because of natural variations in the isotopic composition of the element caused by the various isotopes having different origins: ^{208}Pb is the end-product of the thorium decay series, while ^{207}Pb and ^{206}Pb arise from uranium as end-products of the actinium and radium series, respectively. Lead-204 has no existing natural radioactive precursors. Electronic configuration is $1s^22s^22p^63s^23p^63d^{10}4s^24p^64d^{10}4f^{14}5s^25p^65d^{10}6s^26p^2$. Ionic radius Pb^{2+} 1.18Å, Pb^{4+} 0.70Å. Metallic radius 1.7502Å. Covalent radius (sp^3) 1.44Å. First ionization potential 7.415 eV; second, 14.97 eV. Oxidation potentials Pb → Pb^{2+} + 2e$^-$, 0.126 V; Pb^{2+} + 2H$_2$O → PbO$_2$ + 4H$^+$ + 2e$^-$, − 1.456 V; Pb + 2OH$^-$ → PbO + H$_2$O + 2e$^-$, 0.576 V; Pb + 3OH$^-$ → HPbO$_2^-$ + H$_2$O + 2e$^-$, 0.54 V. Other physical properties are given under **Chemical Elements.**

Lead is of interest as being the terminal product of radioactive decay. Thus while ordinary lead has the atomic weight 207.19 (being composed of 1.37% ^{204}Pb, 26.26% ^{206}Pb, 20.8% ^{207}Pb and 51.55% ^{208}Pb), the isotopic composition, and hence the atomic weight, varies somewhat in lead from meteorites, from deep-seated rocks and from uranium ores (the last being somewhat less dense, as would be expected from the fact that ^{206}Pb is the end-product of the uranium series). These variations in isotopic composition of lead permit of calculations of the age of the earth (and the meteorites).

Occurrence and Processing. Galena, PbS, is the source of over 95% of the lead currently produced. Bodies containing galena range from 3% to 30% lead. One of the most widely distributed sulfide minerals, galena frequently occurs along with sphalerite, ZnS. The lead-zinc ores processed usually contain recoverable quantities of copper, silver, antimony, and bismuth.

Principal sources being worked are in Australia's Broken Hill area in New South Wales, the western United States, Canada, Mexico, Peru, Yugoslavia, and the Soviet Union. When ground-water reacts with galena, cerussite, $PbCO_3$, is formed; when galena is in contact with sulfate solutions generated by the oxidation of sulfide minerals, anglesite, $PbSO_4$, may be formed. See also **Anglesite; Cerussite; Galena.**

In processing, the ore first is crushed, wet-ground, and classified to a point where it is at least 90% less than 200 mesh. Separation of the sulfide ore from the gangue is aided by flotation agents. The resulting concentrates contain from 45% to 60% lead, from zero to 15% zinc, and often a few ounces (~50 grams) of gold and up to 50 ounces (1.4 kilograms) of silver per ton. Copper content may be as much as 3%, arsenic, 0.4%, and antimony, 2%. The sulfur content (10–30%) is reduced by roasting in a Dwight-Lloyd sintering machine. This sulfur reduction is necessary because PbS is not reduced by carbon or carbon monoxide at blastfurnace temperatures. Once formed, the sinter, together with limestone and coke, is fed into a blast furnace. Further oxidation and electrolytic methods may be used to refine the lead. Lead is commercially produced to standards of very high purity. The minimum lead content permitted by specifications for Pig Lead (7 classifications) is 99.73%. Fully refined lead averaging 99.99% lead is obtainable. Large quantities are used for production of chemicals. At one time, primary uses for lead chemicals were in the production of paint pigments and lead tetraethyl gasoline additive.

Lead Metals and Alloys.   Lead is soft and ductile and is readily worked by common methods, predominantly by rolling and extruding. Lead is easily formed and readily joined by welding (burning), or by soldering and can be bonded to steel, or used as a liner for steel, wood, concrete, and other materials. Lead is widely used in this manner because of its excellent resistance to atmospheric and soil corrosion, and attack by sulfuric and phosphoric acids. Lead generally does not resist the action of the organic acids, nor the oxidizing mineral acids, such as HNO_3. Lead is attacked by alkalies.

Due to its low melting point, pure lead will very gradually flow or creep at room temperature. Thus, lead sheeting used as a roofing material on old buildings will usually be thicker at the lower edge than at the upper edge. Other examples of creep occur under low sustained stresses due to the oil pressure in lead-covered power conducting cable, for example, or due to the weight in the case of a deep tank lined with sheet lead. To counter the effects of creep, lead containing 0.06% copper (*chemical lead* or *acid lead*) is preferred.

The addition of antimony in amounts up to 12% greatly improves the casting properties and increases the hardness very materially. These properties make possible the casting of intricately shaped antimonial lead storage-battery grids which, including the weight of the lead oxide paste applied to them, constitute the largest single use for the metal.

Tin and lead in various proportions form a highly useful series of alloys generally known as the soft solders which are used for joining copper, iron, nickel, lead, zinc and even glass. The solders can be applied by means of a soldering tool, by wiping, by hot-dipping or by special machines as in the tin-can industry. Numerous compositions are used, the most popular of which are listed in the accompanying table.

Further additions of bismuth, cadmium, and antimony to the tin-lead alloys result in the low melting or "fusible" alloys widely used as safety devices, the melting points of which can be varied to suit a wide range of requirements. The type metals of the printing industry are lead-tin-antimony alloys having the requisite hardness and good casting properties needed for high-fidelity reproduction.

Babbitt metals (white-metal bearing alloys) are generally classified as either tin-base or lead-base. The true tin-base Babbits contain only tin, antimony and copper, and have been used for many years. The practice of adding up to 25% lead to the tin Babbitts to reduce their cost is to be avoided since the net result is an expensive series of alloys with inferior properties to the inexpensive lead-base Babbitts. The lead-base bearing alloys of the older type usually contain lead, antimony and tin, and while not considered the equal of the tin-base alloys for severe service have been widely employed due to their low cost. The lead-base alloy containing arsenic has found extensive use and has come to the fore of this group since it has successfully met many automotive and other severe service requirements. All of these alloys render their most efficient service when used in the form of a thin lining bonded to a bronze or steel shell. See accompanying table.

Chemistry of Lead.   A number of oxides of lead are known, but not all are daltonide compounds. Thus, lead(I) oxide, Pb_2O, made by heating lead(II) oxalate, has been shown by x-ray analysis to be a mixture of the metal and lead(II) oxide, PbO. The latter is obtained by heating lead in air, which yields a yellow, rhombic material, which has a peculiar layer structure having each lead atom attached to four oxygen atoms all lying on the same side of it, forming a square pyramid with the lead at the apex. Each oxygen is surrounded tetrahedrally by four lead atoms. Another form of PbO, somewhat more stable and soluble in water, red in color, and tetragonal in structure, may be obtained along with the yellow form by alkaline dehydration of $Pb(OH)_2$. PbO is amphiprotic, but only weakly acidic. Lead(IV) oxide, PbO_2, is obtained by action of chlorine on alkaline solutions of lead(II) oxide or acetate. The reaction is $Pb(OH)_3^- + ClO^- \rightarrow PbO_2 + Cl^- + OH^- + H_2O$. PbO_2 can also be produced on a lead or platinum anode by electrolysis in acidic solution. Like the lower elements of main group 4, lead(IV) forms tetrahedral bonds exhibiting sp^3 hybridization. In its relatively more stable salts, however, the $6s^2$ electrons are unused, and Pb^{2+} ions are formed by loss of the $6p^2$ electrons. These facts explain the marked difference between the essentially covalent character of many of the tetravalent compounds and the essentially electrovalent character of the divalent compounds, as well as the peculiar structure of PbO and many other Pb(II) compounds.

The dioxide, PbO_2, has rutile structure, and the compound is a strong oxidizing agent. It is also amphiprotic, giving unstable lead(IV) salts with acids, and orthoplumbates, $M_4^I PbO_4$, or metaplumbates, $M_2^I PbO_3$, upon fusion with alkalies. Lead dioxide dissolves in aqueous alkali with formation of the ion $Pb(OH)_6^{2-}$, the alkali salts of which are isomorphous with the corresponding stannates and platinates. Lead sesquioxide, Pb_2O_4, has been shown not to exist as a stable phase.

Lead orthoplumbate, Pb_2PbO_4, red lead, is similarly described as a salt, in this case, an orthoplumbate of divalent lead, Pb_2PbO_4, because upon treatment with nitric acid, two-thirds of the lead dissolves and one-third remains as PbO_2. It is prepared in the red form by atmospheric heating of PbO, and in a black form by reaction of PbO with pure oxygen. Red lead is formed of PbO_6 octahedra (with one common edge) linked by lead atoms covalently bonded to three oxygen atoms.

The lead dihalides are known for all four of the common halogens. They are not strictly ionic in the anhydrous state, but they dissolve in (hot) water to give Pb^{2+} ions, more or less hydrated. They are much less soluble in cold water. They also form complex compounds such as M_2PbCl_4, MPb_2Cl_5, M_4PbF_6, and $MPbF_3$, where M is an alkali metal. The compound formed, especially of the fluoroplumbates(II) depends somewhat on the alkali metal, some of which form nondaltonide

REPRESENTATIVE LEAD AND TIN ALLOYS

Name	Pb	Sn	Sb	Cu	Bi	Ag	Cd	Typical Applications
Lead Alloys								
Chemical or acid lead	99.9			.06				Tank linings. coils. etc., power cable sheath.
Cable sheath	98.9		1.0					Telephone cable sheath.
Hard lead	96–92		4–8					Cast shapes, wrought sheet and pipe.
Battery grid metal	92–88	.25	8–12					Cast battery grids.
Solders								
Soft solder	50	50						General purposes, most popular solder.
Wiping solder	60	40						For wiping joints in cables, lead
	60	37.5	2.5					pipes, etc.
"Fine solder"	40	60						For making joints at low temperature.
Solder	95–97.5					5–2.5		High temperature solder.
Fusible Alloys								
Wood's metal	25	12.5			50		12.5	**Melts in hot water at 154°F. (~63°C) Wets glass. Wide range of melting points possible with changes in composition for automatic sprinkler systems and other safety devices.**
Matrix metal	28.5	14.5	9		48			For anchoring punches, etc., in jigs and fixtures. Expands on freezing.
Bending alloy	26.5	13.5			50		10	Filler for tubes, etc., during bending. Melts out in hot water.
Type Metals								
Electrotype	93	3	4					
Linotype	84	4	12					
Stereotype	80.5	5.75	13.75					
Monotype	76	8	16					Single type.
Tin Base Babbitts								
		89	7.5	3.5				General usage.
		83.3	8.3	8.3				Hard Babbitt.
Lead Base Babbitts								
	82.5	1.0	15	.5	1.0 As			General usage.
	80	5	15					General usage.
	75	10	15					General usage.

NOTES: Figures given in percent. Wood's metal melts at ~68°C in water.

(berthollide) compounds. Of the lead tetrahalides, only PbF_4 and $PbCl_4$ are known, the fluoride being prepared by fluorination of PbF_2. The chloride, which easily loses chlorine, is made by careful acidification of a hexachloroplumbate(IV). $PbCl_4$ forms the complex compound ammonium hexachloroplumbate, $(NH_4)_2PbCl_6$ upon addition to its solution of solid ammonium chloride.

Lead(II) inorganic compounds and salts of organic acids are far more numerous than those of lead(IV), as is to be expected from the essentially covalent character of the latter. In addition to the oxides and halides already discussed, there are lead(II) compounds of essentially all of the common anions, including many basic compounds. Thus lead(II) chloride forms such basic compounds as $PbCl_2 \cdot Pb(OH)_2$, $PbCl_2 \cdot PbCl_2 \cdot 2PbO$, $PbCl_2 \cdot 3PbO$, and $PbCl_2 \cdot 7PbO$. In fact, a whole series of lead salts are derived from the hydroxide, some of which are double compounds, such as $PbX_2 \cdot 2Pb(OH)_2$ and some of which, of composition $Pb(OH)X$, have been shown to be dimeric of the general formula

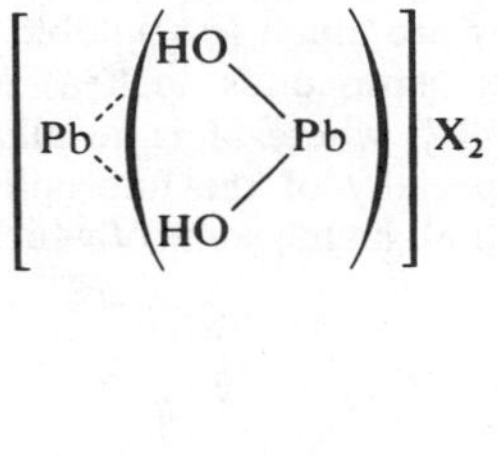

$$\left[Pb \underset{HO}{\overset{HO}{\diagdown\diagup}} Pb \right] X_2$$

Other lead compounds include the following:

Acetates. Lead acetate, "sugar of lead" $Pb(C_2H_3O_2)_2 \cdot 3H_2O$, white crystals, soluble, formed by reaction of lead oxide and acetic acid, and then crystallization. Used (1) to furnish a soluble lead salt, (2) as a mordant in dyeing and printing textiles, (3) as a paint and varnish drier; basic lead acetate, white crystals, soluble, formed by reaction of lead acetate solution and lead oxide, and then crystallization. Used as a coagulating, clarifying, and deacidifying agent for many organic solutions.

Arsenate. lead arsenate, arsenate of lead $Pb_3(AsO_4)_2$, white precipitate, formed by reaction of soluble lead salt solution and sodium arsenate solution. Used as an insecticide.

Azide. Lead azide PbN_6, white precipitate, formed by reaction of soluble lead salt solution and sodium azide solution (white solid, formed by reaction of sodamide $NaNH_2$ upon heating in nitrous oxide N_2O gas). Used as a detonator.

Borate. Lead borate $Pb(BO_2)_2$, white crystals, insoluble, by reaction of lead oxide and boric acid solution. Used in preparing special types of glass.

Carbonates. Lead carbonate $PbCO_3$, white precipitate, formed by reaction of soluble lead salt solution and sodium carbonate solution in the cold; basic lead carbonate, formed by reaction of (1) soluble lead salt solution and hot sodium carbonate solution, (2) lead sheets, carbon dioxide and acetic acid, and pigment, the quality depending largely upon the conditions of the reaction.

Chromates. Lead chromate, "chrome yellow" $PbCrO_4$, yellow precipitate, by reaction of soluble lead salt solution and sodium dichromate or chromate solution, melting point of lead chromate 844°C. Used as a pigment; basic lead chromate, red solid, insoluble, formed by heating lead chromate and sodium hydroxide solution.

Nitrates. Lead nitrate $Pb(NO_3)_2$, white crystals, soluble, formed by reaction of lead oxide and nitric acid, and then crystallization, decomposes on heating leaving lead oxide residue. Used to furnish a soluble lead salt; basic lead nitrate, formed by reaction of lead nitrate solution and lead oxide.

Oxalate. Lead oxalate PbC_2O_4, white precipitate, formed by reaction of soluble lead salt solution and ammonium oxalate solution, yields lead suboxide on heating at 300°C out of contact with air.

Phosphate. Lead phosphate $Pb_3(PO_4)_2$, white precipitate, by reaction of soluble lead salt solution and sodium phosphate solution.

Sulfates. Lead sulfate $PbSO_4$, white precipitate, formed by reaction of soluble lead salt solution and sulfuric acid or sodium sulfate solution; basic lead sulfate, "sublimed white lead," white solid, formed (1) by reaction of lead sulfate and lead hydroxide in water (slow reaction), (2) by roasting galenite in a current of air.

Sulfide. Lead sulfide PbS, brownish-black precipitate, formed by reaction of soluble lead salt solution and hydrogen sulfide or sodium or ammonium sulfide, soluble in dilute nitric acid.

In the great majority of organometallic compounds of lead, the metal is tetravalent and covalently bonded, although the organolead group includes many compounds with both organic radicals and halogen atoms attached to Pb which are not to be described merely as covalent compounds. More than five hundred organometallic compounds of lead have been reported, many of which are named as substituted plumbanes, although PbH_4 is not a starting point in their production. Tetraethyl lead, $Pb(C_2H_5)_4$, is made from a sodium-lead alloy and ethyl chloride.

Like carbon and silicon, and to a lesser extent, germanium and tin, lead forms binary compounds with metals, such as Na_4Pb_7 and Na_4Pb_9. These materials are essentially salt-like, and contain polyplumbide anions. They are of theoretical interest, because they are intermediate in character between stoichiometric compounds (daltonide compounds) and intermediate phases. The two compounds cited dissolve in liquid ammonia, electrolyze in such solutions to give the metals, and apparently form ions such as $[Pb_7]^{4-}$ and $[Pb_9]^{4-}$ which readily form ammine complexes.

Lead Toxicity. As pointed out by Albahara (1972) and White and Selhi (1975), exposure to lead results in a clinical picture of hypertensive encephalopathy, neuropathy, and hemolytic anemia characterized by coarse basophilic stippling in red blood cells. The mechanism of lead's action on human tissue is complex. For one thing, lead blocks heme synthesis. This leads to a buildup of red blood cell protoporphyrin. Lead interferes with cell metabolism by causing a deficiency of pyrimidine 5'-nucleotidase. Lead attacks erythrocyte membrane phospholipids with resultant loss of potassium and interference with the sodium-potassium balance. Diagnosing lead poisoning may involve a determination of the free erythrocyte protoporphyrin level as well as determination of blood and urine levels. Once confirmed, further exposure to lead must be stopped immediately. Chelating compounds, such as $CaNa_2EDTA$, may be administered intravenously over an 8-hour period for several days. This may be followed by treatment with oral penicillamine for several days.

Lead poisoning can lead to chronic renal failure. In its effect on kidney function, leads acts much like cadmium (Maher, 1976). Chronic exposure to or ingestion of practically any heavy metal, such as lead, is the most common path to polyneuropathy. Where effects of heavy metals on the peripheral nervous system are suspect, many physicians will requite a determination of metal presence in hair, fingernails, serum, and urine of the patient. Habitual sniffing of leaded gasolines can lead to lead poisoning (Robinson, 1978; Hansen, 1978).

Settle and Patterson (1980) have compared the lead concentration in the diets of present Americans (0.2 part per million) with the diets of prehistoric peoples (estimated to be less than 0.002 part per million). Some investigators believe that the presence of "natural" lead contamination has been grossly overestimated and that what has appeared to be natural has been the result mainly of a gradual buildup of lead pollution in the air derived from anthropogenic sources. The principal sources of atmospheric lead contamination include (1) natural sources, such as wind-blown volcanic dust, sea spray, forest foliage, and volcanic sulfur compounds; and (2) anthropogenic sources, such as lead alkyls (present in fuels), iron smelting, lead smelting, zinc and copper smelting, and the burning of coal. Some investigators feel that lead-soldered cans should be eliminated because they are a source of lead in foods. Other investigators are not so sure. Much remains by way of research into the sources of lead contamination, including the contributions of atmospheric pollution, of food containers, and of food processing equipment. Current lead toxicity values are given in Sax (1979).

References

Albahary, C.: "Lead and Hemopoiesis: The Mechanics and Consequences of the Erythropathy of Occupational Lead Poisoning," *American Jrnl. of Medicine*, **52**, 367 (1972).
Gale, N. H., and Z. Stos-Gale: "Lead and Silver in the Ancient Aegean," *Sci. Amer.*, **244**(6), 176–192 (1981).
Hansen, K. S., and F. R. Sharp: "Gasoline Sniffing, Lead Poisoning, and Myoclonus," *Jrnl. American Medical Assn.*, **240**, 1375 (1978).
Maher, J. F.: "Toxic Nephropathy. The Kidney," Vol. 2, page 1355 (B. M. Brenner and F. C. Rector, Jr., Editors), Saunders, Philadelphia (1976).
Robinson, R. O.: "Tetraethyl Lead Poisoning from Gasoline Sniffing," *Jrnl. of American Medical Assn.*, **240**, 1373 (1978).
Sax, N. I.: "Dangerous Properties of Industrial Materials," 5th Edition, Van Nostrand Reinhold, New York (1979).
Settle, D. M., and C. C. Patterson: "Lead in Albacore: Guide to Lead Pollution in Americans," *Science*, **207**, 1167–1176 (1980).
Smith, J. F.: "Lead," in "Metals Handbook," 9th Edition, Vol. 2, American Society for Metals, Metals Park, Ohio (1979).
Stukas, V. J., and C. S. Wong: "Stable Lead Isotopes as a Tracer in Coastal Waters," *Science*, **211**, 1424–1427 (1981).
Valciukas, J. A., et al.: "Central Nervous System Dysfunction Due to Lead Exposure,": *Science*, **201**, 465–467 (1978).
Valentine, W. N., Paglia, D. E., K. Fink, et al.: "Lead Poisoning: Association with Hemolytic Anemia, Basophilic Stippling, Erythrocyte Pyrimidine 5'-Nucleotidase Deficiency, and Intraerythrocytic Accumulation of Pyrimidines," *Jrnl. Clin. Invest.*, **58**, 926 (1976).
White, J. M., and H. S. Selhi: "Lead and the Red Cell," *British Jrnl. Haematology*, **30**, 133 (1975).

LEAVENING AGENTS. The generation of carbon dioxide for use as dough leavening is produced by reacting sodium bicarbonate (baking soda) with one of several leavening acids.

Claims for use of acidic phosphate salts, in addition to formation of carbon dioxide, are the buffering effects for providing an optimal pH for the baked product, as well as interactions with protein constituents of flour, with resulting optimal elastic and viscosity properties of the dough batter.

Other leavening acids used in modern bakeries include sodium aluminum sulfite, $Na_2SO_4 \cdot Al_2(SO_4)_3$; sodium aluminum phosphate hydrate (and anhydrous); potassium acid tartrate,

$KHC_4H_4O_6$ (cream of tartar); and glucono-delta-lactone. The baker is concerned with (1) *dough rate of reaction* (DRR), a measure of the rate at which the leavening acid reacts with the baking soda during both the mixing stage and the holding period after mixing (bench action); and (2) *neutralizing value* or neutralizing strength, i.e., the weight of leavening acid required to neutralize a given weight of sodium bicarbonate. This value is used to compute the amount of leavening acid required to yield the needed amounts of leavening gas as well as its effect upon the pH of the baked goods.

Properties of the principal leavening acids are given in the accompanying table, which shows the most appropriate baking applications for each.

Baking powders, as prepared for the home baker and for use in premixes, usually incorporate, along with sodium bicarbonate, one of the following leavening acids: (1) potassium hydrogen tartrate (2 parts for 1 part sodium bicarbonate; (2) tartaric acid (infrequent), 1 part; (3) calcium hydrogen phosphate (crystallized), 1.5 parts; (4) sodium aluminum sulfate or ammonium aluminum sulfate, 1.8 parts. With 7 parts by weight of this finely powdered mixture, there is usually mixed about 3 parts by weight of starch to diminish the effects of moisture in storage. In some cases, dry powdered egg albumin is added to decrease the loss of carbon dioxide upon wetting the flour and baking powder mixture when used. For some purposes, ammonium carbonate can be used alone, since upon heating, this material furnishes both ammonia and carbon dioxide gases to make the product light. These gases escape from the product during the baking process. In selecting a baking powder, one must keep in mind the speed with which they react at room temperature. Thus alum-containing baking powders act slowly; phosphate baking powders have a medium speed; and tartrate baking powders act quickly to produce carbon dioxide. Hence, when using the latter type, it is necessary to bake quickly after mixing to eliminate the loss of too much gas.

Within the last several years, advantage has been taken of mixing different leavening acids in premixes and household baking powders. Because the use of emulsifiers in most cake mixes reduces the need for early leavening action, it is common prac-

tice to use combinations of slow-acting leavening acid which retain much of their leavening reaction for the baking stage. In mixes, the leavening process must be regarded as a system because, in addition to gas generation, the leavening system controls the pH of the finished product and thus affects crumb and crust color, the intensity of flavor, as well as other properties. For various cakes, the optimum pH values are: white cakes, 6.9–7.2; yellow cakes, 7.2–7.5; chocolate or devil's food cakes, 7.1–8.0. Monocalcium phosphate (anhydrous) and sodium aluminum phosphate are frequently used together in white and yellow cake mixes; monocalcium phosphate and sodium acid pyrophosphate or dicalcium phosphate dihydrate are used In chocolate cake mixes. Generally, the combination will comprise 10–20% of fast-acting leavening acid and 80–90% of slow-acting leavening acid.

For pancake and waffle mixes, a common blend of leavening acids is 20–30% monocalcium phosphate monohydrate or monocalcium phosphate (anhydrous), combined with 70–80% sodium aluminum phosphate. A batter of this type can be prepared several hours in advance if retained under refrigeration. It has been observed that such a batter will sour before a serious loss of leavening power occurs.

Prepared biscuit mixes made of flour, shortening, and salt usually contain 30–50% monocalcium phosphate (anhydrous) and 50–70% sodium aluminum phosphate or sodium acid pyrophosphate. Self-rising flours and corn meals usually contain flour or corn meal, salt, soda, and leavening acid. Usually used in these products are combinations of sodium aluminum phosphate and monocalcium phosphate (anhydrous).

Refrigerated doughs available for preparation of biscuits, dinner rolls, and various sweet rolls, usually contain flour, water, shortening, nonfat milk solids (or dried whey solids), sugar (or corn sugar), salt, soda, and a leavening acid. Long-term refrigerated storage requires that only slow-acting leavening acids be used, frequently the sodium acid pyrophosphates. The latter have the disadvantage of possibly producing orthophosphates under certain conditions. The orthophosphates have a rather disagreeable, astringent flavor.

Unleavened Products. The principal unleavened bakery prod-

PROPERTIES OF LEAVENING ACIDS

Chemical Name and Formula	Abbreviation	Relative Speed at Room Temperature	Neutralizing Value[1]
Sodium aluminum phosphate (anhydrous)	SALP	Medium	110
Sodium aluminum sulfate, $Na_2SO_4 \cdot Al_2(SO_4)_3$	SAS	Slow	100
Monocalcium phosphate (anhydrous), $CaH_4(PO_4)$	MCP	Slow	83
Monocalcium phosphate (monohydrate), $CaH_4(PO_4) \cdot H_2O$	$MCP \cdot H_2O$	Quite fast	80
Sodium acid pyrophosphate, $Na_2H_2P_2O_7$	SAPP	Medium	72
Glucono-delta-lactone, $C_6H_{10}O_6$	GDL	Slow	55
Potassium acid tartrate, $KHC_4H_4O_6$	—	Medium to fast	50
Dicalcium phosphate dihydrate, $CaHPO_4 \cdot 2H_2O$	DCP	Very slow	33
Sodium aluminum phosphate hydrate	$SALP \cdot H_2O$	Slow	100

[1] Values in this column indicate the parts of sodium bicarbonate that will be neutralized by 100 parts of the leaivning acid under nominal conditions. Values vary with composition of dough.

uct is pie crust, which is low in moisture and high in fat content. The ingredients and method of preparation prevent the formation of a continuous gluten network through the dough mass. The porosity associated with leavened products is not desirable because the crust literally acts as a container and requires some strength.

LEPIDOLITE. This member of the mica group of minerals is a silicate of potassium, lithium and aluminum, sometimes with sodium, fluorine, or rarely rubidium. A general formula is $K(Li, Al)_3 (SiAl)_4O_{10}(F, OH)_2$. Crystals of lepidolite are monoclinic but often pseudo-hexagonal; cleavage, basal and perfect, being susceptible of splitting into thin laminae; hardness, 2.5–4; specific gravity, 2.8–3.3; luster, pearly; color, reddish to violet, grayish-blue, gray to white. A variety carrying rubidium is yellowish-gray; translucent. It usually is found as granular to scaly masses, in short stocky prisms or less often in easily cleavable sheets. Lepidolite is characteristic of pegmatite veins, frequently being associated with other lithium-bearing minerals such as tourmaline, spodumene, amblygonite, and others. It occurs in the Ural Mountains, Czechoslovakia, the Island of Elba, and the Malagasy Republic, where it is often found in large sheets. In the United States, it is found in the pegmatites of New England, California, South Dakota, and New Mexico. The name lepidolite is derived from the Greek, meaning scale. See also **Lithium**; and **Lithium (For Thermonuclear Fusion Reactors).**

LEPTONS. The electron, muon, and two kinds of neutrino are collectively called *leptons*. The leptons are considered to be pointlike particles without structure and thus truly elementary. Leptons can interact with other particles through the weak interactions. Electrons and muons also can interact through electromagnetic and gravitational forces, but they appear to be without capability of interactions through the strong (nuclear) forces. The neutral members, the electron neutrino, the muon neutrino, and their antiparticles have extremely weak interaction with matter and do not participate in electromagnetic interactions. Leptons make excellent probes in particle physics experiments. The other major family of subatomic particles is referred to as *hadrons*. See also **Electron; Muon; Neutrino;** and **Particles (Subatomic).**

The name *lepton* from its derivation means *light*, referring to the fact that the masses of the leptons are all less than that of the lightest meson. The properties of the electron are discussed in the entry on that topic. Here, it will be noted that the word *electron* is used to denote the negative electron (often called the *negatron* when ambiguity might arise). Its antiparticle is the *positron* (also called *positive electron*).

LEUCITE. The mineral leucite is a metasilicate of potassium and aluminum corresponding to the formula $KAlSi_2O_6$. It is isometric at a temperature of about 600°C (1112°F) and pseudoisometric at lower temperatures, at which leucite is tetragonal but retains an external isometric crystal form usually trapezohedral. It has a conchoidal fracture; is brittle; hardness, 5.5–6; specific gravity, 2.47–2.50, luster, vitreous; color, white or some shade of gray; translucent to opaque. It is commonly found in the more recent lavas of high alkali content. Leucite is seldom reported from plutonic rock types. It is a relatively rare mineral. It is found plentifully at Vesuvius and Monte Somma and elsewhere in Italy, and Germany in the Tertiary volcanic district of the Eifel. In the United States, leucite has been found in the Leucite Hills of Wyoming, the Highwood Mountains of Montana, and as pseudomorphs (pseudoleucites) representing

a mixture of nepheline, orthoclase, analcime, and aegerine from New Jersey, Arkansas and Montana. Its name is derived from the Greek word *leukos*, referring to its white color.

LEVOROTATORY COMPOUNDS. Asymmetry (Chemical); Isomerism.

LIGAND. Any atom, radical, ion, or molecule in a complex (polyatomic group) which is bound to the central atom. Thus, the ammonia molecules in $[Co(NH_3)_6]^{3+}$, and the chlorine atoms in $[PtCl_6]^{2-}$ are ligands. Ligands are also complexing agents, as for example, EDTA, ammonia, etc. See also **Chelation Compounds.** Examples of common configurations of complex ions in various geometric arrangements are shown by the accompanying diagram.

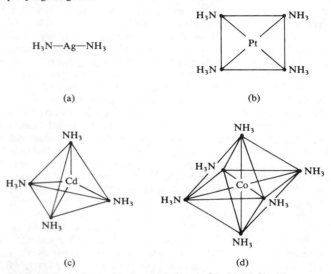

Configurations of complex ions. (a) *linear,* $[Ag(NH_3)_2]^+$; (b) *square-planar,* $[PE(NH_3)_4]^{2+}$; (c) *tetrahedral,* $[CD(NH_3)_4]^{2+}$; (d) *octahedral,* $[Co(NH_3)_6]^{3+}$.

Ligand field theory incorporates elements from the valence bond theory of Pauling and the molecular orbital method of Hund, Mullikan, and others. As pointed out by Mortimer, the chemists of the late- 19th Century had difficulty in understanding how "molecular compounds" or "compounds of higher order" are bonded. The formation of a compound such as $CoCl_3 \cdot 6NH_3$ was baffling, particularly in this case, since simple $CoCl_3$ does not exist. In 1893, Alfred Werner proposed a theory to account for compounds of this type. Werner wrote the formula of the cobalt compound as $[Co(NH_3)_6]Cl_3$. Werner assumed that the 6 ammonia molecules are symmetrically "coordinated" to the central cobalt atom by "subsidiary valences" of cobalt, while the "principal valences" of cobalt are satisfied by the chloride ions. Werner devoted over 20 years to preparing and studying coordination compounds and perfecting and proving his theory. Although modern work has amplified his theory, it has required relatively little modification.

In ligand field theory, one is concerned with the origin and the consequences of splitting the inner orbitals of the central metal by the surrounding ligands. The most satisfactory correlations have been demonstrated with the first transition series, in which the $3d$-orbitals are split into different energy levels. To appreciate the effect of a ligand field, imagine that a symmetrical group of ligands is brought up to a charged ion from a distance. First, the electrostatic repulsions between the ligand electrons and those in the d-orbitals of the metal will raise the energy of all five d-orbitals equally. Then, as the ligands approach to within bonding distances, the repulsion interactions will take on a directional character that will vary with the particular d-orbitals under consideration. This arises because of the different shapes and orientations of the five d-orbitals in splace along a Cartesian coordinate system. The splitting of the orbitals for a given central metal ion is dependent on the set of ligands.

Applications of ligand field theory to many transition metal complexes has played an important role in the interpretation of visible absorption spectra, magnetism, luminescence, and paramagnetic resonance spectra.

LIGNIN. Approximately 25% of the content of most woods is lignin. Lignin concentration in wood substance is greatest in the middle lamella (the zone around each individual fiber cell), decreasing in concentration through the cross section of the fiber, reaching a concentration of about 12% at the inner layer of the fiber adjacent to the fiber cavity, or lumen. Lignin and hemicellulose cement the fiber cells together, providing rigidity to the fibrous wood structure. In the destructive distillation of wood, the methanol produced is derived from the lignin. In the manufacture of paper pulp, it is necessary to remove the lignin, usually accomplished by treatment of the wood fibers with such agents as sulfur dioxide, calcium bisulfite, and sodium sulfate/sodium sulfide solutions. Sodium hydroxide is sometimes used. An important by-product of the paper pulp industry is dimethyl sulfoxide, $(CH_3)_2SO$ which is produced from the lignin released during wood pulping by the Kraft process. Dimethyl sulfoxide has a number of industrial uses—as an intermediate in organic syntheses, as a solvent in spinning synthetic fibers, and in some pharmaceuticals.

The wall material of plant cells is one of their distinguishing characteristics. As a result, lignin, cellulose, and other wall constituents have been studied in many plant tissue cultures. Phenylpropanoids, for example, have been shown to precursors of lignin formation in white pine, *Sequoia*, lilac, rose, carrot, and geranium tissue cultures. Moreover, the biosynthesis of lignin has been shown to be affected by kinetin, boron, and major elements, such as calcium.

Lignin is a major source of vanillin.

LIMONITE. The mineral limonite, hydrated oxide of iron, corresponds to the formula $Fe_2O_3 \cdot nH_2O$, but is often very impure due to the admixture of sand and clay. It is not found crystallized but grades from loose porous material to compact masses. Its hardness is variable but pure material is 5–5.5; specific gravity 3.6–4; usual luster, dull to earthy but may be silky to submetallic; color, various shades of yellowish-brown, sometimes nearly black; streak, yellowish-brown; opaque. Limonite is a secondary mineral from the alteration of various other iron-bearing ores or minerals; it is of widespread occurrence and used both as an ore of iron and as a pigment. Limonite has been formed in marshy and boggy areas and is frequently called bog iron ore. Limonite is an important ore of iron in Lorraine, Luxemburg, Bavaria and Sweden. It is found in Saxony, Austria and England. In the United States, limonite is found particularly in Connecticut, Massachusetts, Pennsylvania, New York, Virginia, Tennessee, Georgia and Alabama, but these deposits are of little economic importance at the present time.

LINOLEIC ACID. Also called linolic acid, formula $CH_3(CH_2)_4HC:CHCH_2CH:CH (CH_2)_7COOH$. This is a polyunsaturated fatty acid (2 double bonds) existing in both conjugated and unconjugated forms. It is a plant glyceride essen-

tial to the human diet. It is found in linseed oil, safflower oil, and tall oil. See also **Vegetable Oils (Edible).** At room temperature linoleic acid is a colorless to straw-colored liquid. Specific gravity 0.905 (15/4°C); mp −5°C; bp 228°C (at 14 millimeters pressure). Insoluble in water; soluble in most organic solvents. Combustible. Sources are the oils previously mentioned. Linoleic acid is used in soaps; special driers for protective coatings; in emulsifying agents; pharmaceuticals; livestock feeds; margarine.

LINOLENIC ACID. Also called 9,12,15-octadecatrienoic acid, formula $CH_3CH_2CH:CHCH_2CH:CHCH_2CH:CH (CH_2)_7COOH$. This is a polyunsaturated fatty acid (3 double bonds). It occurs as the glyceride in many seed fats. It is an essential fatty acid in the diet. See also **Vegetable Oils (Edible).** At room temperature, linolenic acid is a colorless liquid; soluble in most organic solvents; insoluble in water. Specific gravity 0.916 (20/4°C); mp −11°C; bp 230°C. Combustible. Linolenic finds use in various pharmaceuticals; drying oils.

LIPIDS. A heterogenous group of substances which occur ubiquitously in biological materials. They may be categorized as a group by their extractability in nonpolar organic solvents, such as chloroform, carton tetrachloride, benzene, ether, carbon disulfide, and petroleum ether. Structural types within the group range from simple straight-chain hydrocarbon molecules to complex ring structures with varying side chains. A useful classification of the lipids is: (1) fatty acids; (2) neutral fats; (3) phosphatides; (4) glycolipids; (5) aliphatic alcohols and waxes; (6) terpenes; and (7) steroids. See **Steroids.**

Many lipids, especially the phospholipids, have a strong tendency to form complexes with each other and with various substances. Complex formation is due to the electrostatic attraction of polar groups and to the mutual solubility of the long hydrocarbon chains. Thus, the lipoproteins and proteolipids are complexes of proteins and a variety of lipids, such as cholesterol, phospholipids, glycerides, and glycolipids. The lipids are linked to the proteins by several types of forces. Electrostatic forces, van der Waals forces, hydrogen bonding, and hydrophobic bonding hold these complexes together. Because of their attraction for water, the polar groups of the protein and phospholipid arrange themselves on the outside of the complex, while the hydrocarbon groups of the lipids are folded into the center. Thus, there is presented to the aqueous phase those groups which have an affinity for water. This arrangement accounts for the solubility of the complexes in water. The phospholipids, owing to their polar groups, act as water solubilizers for the nonpolar lipids. The arrangement may be different in the proteolipids of the brain and nerves, since they are not soluble in water. In these complexes, the lipids may completely envelop the protein.

Knowledge of lipid metabolism has increased at an accelerated rate during the past few decades. The detailed biochemical reactions whereby the fatty acids are synthesized and oxidized, how phospholipids, glycolipids and cholesterol are synthesized, and how lipids are absorbed and transported have been elucidated. Fatty acids are synthesized from acetyl coenzyme A and malonyl coenzyme A thiol esters. The vitamin biotin plays a vital part in the fixing of carbon dioxide to form malonyl coenzyme A, an important intermediate in fatty acid synthesis. The hormone insulin also favors fatty acid synthesis. The oxidation of fatty acids occurs as their coenzyme A esters in the Krebs cycle of the mitochondria. Cholesterol is biosynthesized from acetyl coenzyme A. Cholesterol in humans is converted to bile acids, fecal sterols, and to steroid hormones. The synthesis of lecithin is mediated via phosphatidic acid and diglyceride precursors. Cytidine nucleotides play a role in the transfer of choline (as phosphorylcholine) to a diglyceride to form lecithin. Uridine nucleotides act to transfer sugar residues in the synthesis of glycolipids.

The transport of lipids in the blood plasma is effected by complex formation with proteins to yield lipoproteins. The liver is the major organ for the synthesis of the lipoproteins. Analysis of serum lipoprotein patterns is important in the understanding of vascular disease (atherosclerosis). The clearing of lipemic blood, such as may occur after a heavy fat meal, is brought about by enzyme known as lipoprotein lipase. This enzyme yields free fatty acids which combine immediately with the plasma albumin to form complexes known as NEFA (nonesterified fatty acids). NEFA act as important transport vehicles for transport of triglycerides and the levels of blood NEFA are very sensitive to hormonal control and neural control. Certain hormones, such as epinephrine, stimulate the membrane-bound adenyl cyclase which converts ATP (adenosine triphosphate) to cyclic AMP (adenosine monophosphate). The latter stimulates adipose tissue lipase and mobilizes depot fat.

The excess utilization of lipids and excess oxidation of fatty acids causes an increase in acetoacetic acid in the body. This condition is known as *ketosis* and can lead to *acidosis*. This situation is common in severe diabetes and can occur whenever carbohydrate utilization is severely decreased.

Research continues in an effort to gain a more thorough understanding of: How are lipoproteins synthesized? How are lipids arranged and combined with proteins to form cell membranes? What specific role do lipids play in transport across cell membranes? How do hormones act to regulate lipid metabolism? What is the biochemical basis of such abnormal lipid metabolic states as Gaucher's disease? Niemann-Pick's disease, etc.? How do lipids per se permeate cell membranes? How many phenotypic lipoproteins occur in serum?

Particularly since the early 1960s, the research into causes of atherosclerosis/coronary heart disease has been rather intensive. *Hyperlipidemia* is considered a major risk factor in the pathogenesis of atherosclerosis (Muller, 1973; Moore et al., 1976). A diet high in fiber is reported to prevent the elevation of serum levels and thus minimize the risk (Trowell, 1972; Kritchevsky et al., 1975). In 1976–1977, Ranhotra et al. demonstrated that bran, under the condition of dietary-induced hyperlipidemia in rat (short-term), appreciably prevented the elevation of serum and liver cholesterol levels. It has been postulated that this occurred, in part, through an increased fecal bile acid and cholesterol loss. In 1978, the team (American Institute of Baking, Manhattan, Kansas) attempted to find the major subfraction(s) of bran which may prevent elevation of tissue lipid levels over an extended period of intake. Their conclusions, in part: "It thus seems that although lipids in bran appear to favorably affect tissue cholesterol and triglyceride levels in hyperlipidemic rats, such an effect seems to be even more pronounced through direct or indirect (effect on caloric density of diet and on nutrient availability) action of fiber in bran." See also **Fiber (Dietary).**

Vitamin D consumption has increased markedly in several regions of the world since the 1920s by way of increased intakes of vitamin supplements and enrichment of products, such as milk and margarine. Myasnikov (1958), Bajiwa et al. (1971), and Kummerow et al. (1976) have observed that inasmuch as vitamin D is a form of cholesterol, it may be a possible precursor of circulatory and deposit forms of cholesterol in the body. Retzlaff and a team at the Department of Food and Nutrition, University of Nebraska, Lincoln, Nebraska conducted a study in 1977 to determine the effect of vitamin D_3 supplementation of diets fed to middle-aged men on their blood serum cholesterol, triglyceride, and phospholipid levels. Their conclusions, in part: "The results of this and earlier studies suggest that the directional influence of vitamin D on blood serum cholesterol levels may be dose and/or species related. Vitamin D supplements for humans at nontoxic levels seemingly present no hazard in relationship to circulating blood lipids and may have a slight beneficial effect. These results should not be interpreted to indicate that no damage from excessive vitamin D intakes exist at higher dose levels or in parameters not measured." See also **Vitamin D.**

Additional aspects of lipids are described in entries on **Antioxidants; Enzyme Preparations;** and **Proteins.**

References

Bajiwa, G. S., Morrison, L. M., and B. H. Ershoff: "Induction of Aortic and Coronary Atherosclerosis in Rats Fed a Hyper-vitaminosis D, Cholesterol-Containing Diet," *Proc. Soc. Exp. Biol. Med.*, **138**, 975 (1971).
Gray, P.: "The Encyclopedia of the Biological Sciences," 2nd edition, Van Nostrand Reinhold, New York, 1970.
Hamilton, R. M. G., and K. K. Carroll: "Plasma Cholesterol Levels in Rabbits Fed Low-fat, Low-cholesterol Diets. Effects of Dietary Proteins, Carbohydrates, and Fiber from Different Sources," *Atherosci*, **24**, 47 (1976).
Heaton, K. W., and E. W. Pomare: "Bran and Blood-Lipids," *Lancet*, **1**, 800 (1975).
Kritchevsky, D., Tepper, S. A., and J. A. Story: "Symposium: Nutritional Perspectives and Atherosclerosis. Nonnutritive Fiber and Lipid Metabolism," *Nutr. Rept. Int.*, **7**, 271 (1975).
Kummerow, F. A.: "Symposium: Nutritional Perspectives and Atherosclerosis. Lipids in Atherosclerosis," *J. Food Sci.*, **40**, 12 (1975).
Moore, M., et al.: "Dietary-Atherosclerosis Study on Decreased Persons," *J. Amer. Diet. Assn.*, **68**, 216 (1976).
Myant, N. B.: "The Influence of Some Dietary Factors on Cholesterol Metabolism," *Proc. Nutr. Soc.*, **34**, 271 (1975).
Ranhotra, G. S., et al.: "Effect of Some Wheat Mill-Fractions on Blood and Liver Lipids in Cholesterol-Fed Rats," *Cereal Chem.*, **53**, 540 (1976).
Ranhotra, G. S., et al.: "Effect of Particle Size of Wheat Bran on Lipid Metabolism in Cholesterol-Fed Rats," *J. Food Sci.*, **42**, 1587 (1977).
Ranhotra, G. S., et al.: "Effect of Wheat Bran and Its Subfractions on Lipid Metabolism in Cholesterol-Fed Rats," *J. Food Sci.*, **43**, 1829 (1978).
Retzlaff, K. A. V., Kies, C., and H. M. Fox: "Blood Lipid Levels of Adult Men on Vitamin D-Supplemented and Unsupplemented Diets," *J. Food Sci.*, **43**, 135 (1978).
Staff: "Lond-term Effects of Diets Prescribed in Coronary Prevention Programs," *Nutr. Review*, **35**, 140 (1977).
Truswell, A. S.: "Food Fiber and Blood Lipids," *Nutr. Review*, **35**, 51 (1977).
Whyte, H. M., and N. Havenstein: "A Perspective View of Dieting to Lower the Blood Cholesterol," *Amer. J. Clin. Nutr.*, **29**, 784 (1976).

LIQUID. Matter in a fluid state but relatively incompressible. An ideal liquid offers no permanent resistance to a shear stress but is incompressible. It has then a constant volume and incompletely fills any container of less than this volume. A real liquid is appreciably compressible, and the liquid state of a substance might be defined as the denser, and less compressible, phase of the two-phase fluid system which can exist in equilibrium at temperatures below the critical temperature. X-ray diffraction experiments show that, near the melting-point, the molecules of a liquid show a considerable degree of short-range order and that, in small volumes, they are arranged much as in a solid crystal. This crystalline structure persists over volumes comparable with the intermolecular distances, but cannot be traced beyond. This local or short-range order means that the average molecule is at any moment surrounded by a number of molecules occupying nearly the same relative positions as they would in the solid state. The degree of short-range order is described by the radial distribution function.

This concept of a liquid as an imperfect crystal requires that the molecules in a liquid are packed sufficiently loosely for comparatively free movement, i.e., the energy required to move a molecule from a lattice site to a vacant space is not large compared with thermal energies. Under these conditions, shear flow of the liquid resembles closely the high temperature creep of crystalline solids. A number of theories of the liquid state have this concept as their starting point.

With a few exceptions, including helium, the accompanying universal phase diagram applies for all pure compounds. The

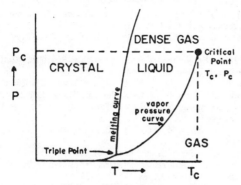

Universal phase diagram.

triple point is the single point at which all three phases (crystal, liquid, and gas) are in equilibrium. The triple point pressure is normally below atmospheric. Those substances, such as carbon dioxide, where $P_t = 3,885$ millimeters, $T_t = -56.6°C$, sublime without melting at atmospheric pressure. From the triple point, the melting curve defines the equilibrium between crystal and liquid, usually rising with small but positive dT/dP, and presumably always with positive dT/dP at sufficiently high P values. The line is believed to extend infinitely without a critical point (it has been followed to $T \cong 16T_c$ for helium, and calculations indicate that hard spheres would show a gas-crystal phase change). The gas-liquid equilibrium line, the vapor pressure curve, has dT/dP always positive and greater than the melting curve. The vapor pressure curve always ends at a critical point, $P = P_c$, $T = T_c$, above which the liquid and gas phase are no longer distinguishable. Since the liquid can be continuously converted into the gas phase without discontinuous change of properties by any path in the P-T diagram passing above the critical point, there is no definite boundary between liquid and gas. Two liquids of similar molecules are usually soluble in all proportions, but very low solubility is sufficiently common to permit the demonstration of as many as seven separate liquid phases in equilibrium at one temperature and pressure (mercury, gallium, phosphorus, perfluoro-kerosene, water, aniline, and heptane at 50°C, 1 atmosphere).

References

Hansen, J. P., and I. R. McDonald: "Theory of Simple Liquids," Academic, New York, 1976.
March, N. H., and M. P. Tost: "Atomic Dynamics in Liquids," Wiley, New York, 1977.
Stephan, K., and K. Lucas: "Viscosity of Dense Fluids," Plenum, New York, 1979.

LIQUID CRYSTALS. Liquids that have the structural character of cybotactic liquids (see **Cybotaxis**), but which are consid-

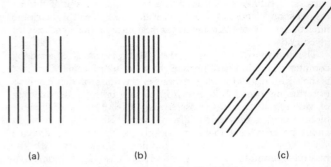

Fig. 1. Smectic liquid crystals.

erably more viscous, with viscosities extending from that of a light glue to that of a glassy solid. They also exhibit much more definite evidences of structure than the cybotactic liquids.

Liquid crystals must be geometrically highly anisotropic—usually long and narrow—and revert to an isotropic liquid through thermal action (thermotropic mesomorphism) or by the influence of a solvent (lyotropic mesomorphism). Several thousand organic compounds are now known which meet these criteria, but significant molecular features found in thermotropic liquid crystals are among the following. The molecule will be elongated and rectilinear; if "flat segments," e.g. benzene rings, are present its liquid crystallinity will be enhanced. The molecule will be rigid along its long axis and double bonds will be common in this direction. The simultaneous existence is seen in the molecule of strong dipoles and easily polarizable groups. Of lesser importance are weak dipolar groups at the extremities of the molecule.

The present-day classification of thermotropic liquid crystals is threefold. *Smectic liquid crystals*, such as p-ethyl azoxybenzoate, have their molecules arranged in definite strata, with a variety of molecular arrangements being possible within each stratification. In *smectic Type A* crystals, the molecules may be considered to "stand on end" with their long axes perpendicular to the plane of the layer but with their centers irregularly spaced. Where the molecular centers adopt hexagonal closest packing, the crystals are considered *smectic Type B*, and when they adopt a tilted form of Type A, they are classified as *smectic Type C*. See Fig. 1.

In *nematic liquid crystals*, the molecular structures possess a high degree of long-range orientation order, but no long-range translational order. The molecules are spontaneously oriented with their long axes approximately parallel, but without the stratification seen in smectic crystals. Nematic liquid crystals like p-azoxyanisole are generally optically uniaxial, positive and strongly birefringent, and some are composed of hundreds of

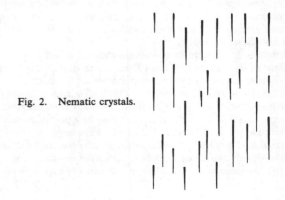

Fig. 2. Nematic crystals.

Fig. 3.   Cholesteric crystals.

molecules (cybotactic groups) with the molecular centers in each group being arranged in layers. See Fig. 2.

Lyotropic liquid crystals possess at least two components. One of these is water and the other is amphible (a polar head group attached to one or more long hydrocarbon chains). In the lamellar form, water molecules are sandwiched between the polar heads of adjacent layers while the hydrocarbon tails lie in a nonpolar environment. Lyotropic liquid crystals have very complex structures, but occur abundantly in nature, particularly in living systems. See Fig. 3.

References

Brown, G. H., Doane, J. W., and V. D. Neff: "Review of Structure and Physical Properties of Liquid Crystals," CRC Press, Boca Raton, Florida (1971).

Brown, G. H.: "Advances in Liquid Crystals," Academic, New York (Issued annually).

Chandrasekar, S.: "Liquid Crystals," Cambridge Univ. Press, New York (1977).

Dennis, D., and L. Richter: "Textures of Liquid Crystals," Verlag Chemie Weinheim, New York (1978).

—R. C. Vickery, Hudson Laboratories, Hudson, Florida.

LITHIUM.   Chemical element symbol Li, at. no. 3, at. wt. 6.939, periodic table group Ia (alkali metals), mp 180.54°C, bp 1340°C (at 760 torr), density, 0.534 g/cm³ (20°C). Lithium is lightest in weight of all the chemical elements that are solid at standard conditions. Elemental lithium in the solid phase has a body-centered cubic crystal structure. In comparison with other members of the alkali metal series, lithium has the smallest ionic radius, the highest ionization potential, the highest electronegativity, and the greatest heat capacity. Generally, lithium is the least reactive of the alkali metals. Lithium is a silvery-white metal, harder than sodium, but softer than lead. It is tough and may be drawn into wire or rolled into sheets. The element tarnishes rapidly in air and often is preserved under naphtha. The reaction with water is vigorous, producing LiOH (lithium hydroxide) and hydrogen. In these respects, lithium parallels the behavior of sodium.

There are two naturally occurring isotopes, ^6Li and ^7Li. They are not radioactive. Two radioactive isotopes have been identified, ^5Li and ^8Li, both with very short half-lives, measured in fractions of a second. Among elements occurring naturally in the earth's crust, lithium ranks 28th, with an estimated average content of about 10–20 ppm. In terms of content in seawater, lithium ranks 17th, with an estimated content of approximately 950 tons of the element per cubic mile (204 metric tons per cubic kilometer). The element was first identified by Johann August Arfvedson in 1817 in the laboratory of Berzelius. The naming of the element is accredited to Berzelius. Electronic configuration is $2s^1s$. First ionization potential, 5.39 eV. Oxidation potential $Li \rightarrow Li^+ + e^-$, 3.02 V. Other physical properties of lithium are given under **Chemical Elements.** The reserves needed and available for possible future use in thermonuclear fusion reactors are described in the following entry.

The main sources of lithium are pegmatites and brines. The most important pegmatite mineral is *spodumene*, $LiAlSi_2O_6$, which contains a theoretical content of 8.03% Li_2O. *Petalite*, $LiAlSi_4O_{10}$, contains between 4 and 4.5% Li_2O. *Lepidolite*, a complex mica, contains between 3 and 4% Li_2O. See also entries on **Lepidolite; Petalite;** and **Spodumene.** Brines contain normally a few hundred to a few thousand parts per million (ppm) of lithium. The only commercial source of spodumene in North America is located in North Carolina. Abundant resources of lithium pegmatites occur in Canada, the African continent, and unconfirmed sources in the U.S.S.R. and China. Significant quantities of lithium (as carbonate) are produced from the brines of Clayton Valley, Nevada. A recently discovered lithium-rich brine deposit has been located in the Atacama desert of Chile. More detail on lithium resources is given in the following entry.

There are three major processes for extracting lithium from pegmatite ores: (1) An acid process, wherein the spodumene concentrate, after calcining at about 1095°C, is reacted with sulfuric acid, followed by water leaching of the resulting lithium sulfate, Li_2SO_4. The sulfate is then converted to the carbonate with soda ash. (2) An alkaline process, wherein the ore is reacted with lime or limestone at high temperatures followed by water leaching of the resulting lithium hydroxide. (3) A base exchange method, whereby the ore is reacted with an alkaline chloride or sulfate at a high temperature in an aqueous phase to yield a soluble lithium salt. The sulfuric acid leaching method is the only commercial process for extraction of lithium from spodumene in practice today.

The first lithium metal was prepared by Sir Humphry Davy in 1818 by electrolyzing lithium oxide. At about that same time, Brande also isolated the metal. In 1855, R. Bunsen and A. Matthiessen prepared gram amounts by electrolyzing fused lithium chloride. Modest commercial quantities were first made in Germany during World War I when the metal was considered as a potential alloying material. Limited production did not commence in the United States until the early 1930s. Present commercial methods were pioneered by Guntz in 1893 and involve electrolyzing a low-melting mixture of LiCl and KCl. Graphite anodes and mild steel cathodes are used. Lithium is formed at the cathode and rises to the surface from which it is skimmed periodically. Pure lithium chloride is added to the bath as required. Chlorine gas is liberated at the anodes. The process yields a lithium metal of about 99.8% purity. The metal normally is cast into ingots of different sizes, but is also available as extruded rod, ribbon, or wire. The metal also is available as "sand"—fine dispersions in the 10–30 μm range.

Uses.   In metallurgy, lithium metal is used as a deoxidizer, desulfurizer, and degasifier in the production of a number of molten metals, notably copper and copper alloys. Lithium also is an ingredient of the aluminum alloy X2020 (1% Li), which is a structural alloy with improved high-temperature strength. About 14% lithium is alloyed with magnesium in the LA 141 alloy, a very lightweight structural alloy. Lithium alloyed with silver has been used for fluxless brazing. Considerable attention has been given to the potential of lithium in lightweight primary and secondary batteries. The metal also has potential as a coolant and heat-transfer medium for nuclear reactors. Very important potentially, lithium can be used as a tritium source for controlled-fusion reactors.

Lithium is attractive for use in batteries for a number of reasons. Lithium is the most reducing of the alkali metals and also has the lowest equivalent weight. Thus, as suggested by Roth and Farrington (1977), a high-conductivity Li^+ solid electrolyte would lend flexibility in the design of energy storage systems. The aforementioned researchers have reported that lithium-sodium beta alumina as having a lithium/sodium ratio greater than about 1 and thus appears to be the first generally useful lithium superionic conductor reported thus far. It is described as exhibiting strikingly nonlinear ion exchange properties and thus may presage the discovery of similar coionic interactions in other superionic conductors. In the reference cited, the investigators discuss the properties of lithium-sodium beta alumina in relation to present concepts of ionic interaction and distribution in the beta alumina conduction plane. The use of lithium in batteries is further discussed in the entry on **Battery**.

The possible physiological role of lithium in brain function has been described in the literature for a number of years.

Chemistry and Compounds

Lithium has the highest ionization potential (i.e., $Li \rightarrow Li^+$, in the vapor) of the alkali metals. However, the measured value of its oxidation potential against a normal aqueous solution of its ion is 3.02 V, which does not differ from those of the other main group I metals by as much as the difference in ionization potentials. That difference, attributed to the high heat of hydration of Li^+, explains why lithium is a vigorous reductant in aqueous systems, but reacts slowly with H_2O, and not at all with dry oxygen except above 100°C.

The single $2s$ electron in the outer shell of lithium is easily removed to form the positive ion, and stability of the remaining $1s^2$ electron pair requires too high a potential (75.62 eV) for any further ionization (by chemical means) so that lithium is exclusively monovalent in its compounds.

Because of the reactivity of lithium with water to form its hydroxide, LiOH, and hydrogen, its properties when dissolved in other solvents have been studied extensively. It does not decompose liquid NH_3, but does form a blue solution, which decomposes to yield its amide, $LiNH_2$, and hydrogen, when catalyzed by metallic salts. With the elements of main groups 2 to 7, lithium in liquid NH_3 reacts to form binary compounds, which may vary from simple halides, as with the halogens, to intermetallic phases, as with cadmium and mercury. Lithium amide in liquid NH_3 as regarded in the same class as a hydroxide in aqueous solution.

Many other lithium compounds not obtainable in aqueous solution can be produced from the solution of lithium in liquid ammonia. Thus the acetylide is obtained by action of acetylene

$$C_2H_2 + LiNH_2 \rightarrow LiC_2H + NH_3$$

$$2LiC_2H \rightarrow Li_2C_2 + C_2H_2$$

The amide, as stated above, is produced by catalyzed decomposition of the liquid NH_3 solution, and the nitride, Li_3N, by heating the amide or by direct combination of the elements.

Lithium salts exhibit general high solubility and a high degree of dissociation in other nonaqueous solvents than liquid ammonia, such as liquid sulfur dioxide and acetic acid.

Like the other alkali metals lithium forms compounds with virtually all of the anions, inorganic as well as organic. The lithium salts are in many instances different in their solubility properties from the corresponding salts of the other alkali metals. Thus lithium fluoride, phosphate, and carbonate are the least soluble alkali metal fluoride, phosphate, and carbonate, the solubilities for the other alkali metals increasing with increasing ionic radius. Lithium chlorate and dichromate are on the other hand, the most soluble alkali chlorate and dichromate, the solubilities for the other alkali metals decreasing with increasing ionic radius. These differences are partly explained, as was that in the oxidation potential, by the considerable hydration of the lithium ion, which also explains the fact that lithium salts generally crystallize as hydrates. Lithium salts, probably because of the small size of the lithium ion, do not form mixed crystals with the other alkali salts, but they do form double

salts, notably the two series of lithium-sodium and lithium-potassium sulfates.

Lithium forms several organic compounds. Most of them are lithium salts or lithium acid salts of organic acids or other oxygen-connected lithium compounds. The number of lithium-carbon bonded compounds that have been reported is very small including, in addition to the carbide, methyllithium, CH_3Li, ethyllithium, C_2H_5Li, n-propyllithium, C_3H_7Li, n-butyllithium, C_4H_9Li, benzyllithium, $C_6H_5 \cdot CH_2Li$, and methylenedilithium $LiCH_2Li$.

The alkyllithium compounds are usually colorless, soluble in organic solvents, and capable of distillation or sublimation. They are nonelectrolytes and are widely used in synthetic organic chemistry, since, like other lithium compounds, they resemble in their properties the corresponding magnesium compounds.

Lithium Carbonate. Li_2CO_3, mp 72.6°C, slightly soluble in H_2O. Used in glass, enamel, and ceramic formulations, in the electrowinning of aluminum, and in the manufacture of other lithium compounds. The compound also has been used in the treatment of manic-depressive psychoses.

Lithium Hydride: LiH, mp 686.4°C, reacts vigorously with H_2O. With NH_3, it forms the amide. The compound is used to produce $LiAlH_4$ and other double hydrides. Lithium hydride is an excellent lightweight source of hydrogen. One pound yields 45 cubic feet of hydrogen (one kilogram yields 2.8 cubic meters of hydrogen) at standard conditions. The compound also can serve as a lightweight shield for thermal neutrons.

Lithium Hydroxide Monohydrate. $LiOH \cdot H_2O$ loses water at 101°C. LiOH melts at 450°C. The compound is soluble in water. The compound is used in the formulation of lithium soaps used in multipurpose greases; also in the manufacture of various lithium salts; as an additive to the electrolyte of alkaline storage batteries. LiOH also is an efficient lightweight absorbent for carbon dioxide.

Lithium Bromide. LiBr, mp 550°C, soluble in H_2O or alcohols. The compound is very hygroscopic and forms four hydrates. Major use has been in absorption-refrigeration air-conditioning systems in which H_2O is the refrigerant—strong LiBr is used to absorb H_2O vapor.

Lithium Chloride. LiCl, mp 608°C, soluble in H_2O or alcohols. Very hygroscopic and forms four hydrates like the bromide. The compound is a component of brazing fluxes for aluminum and magnesium; is used in dehumidification systems, as an additive to the electrolyte of dry cells for low-temperature applications, is used in low-freezing fire-extinguishing systems; as an ingredient of fused-salt baths to lower fusing temperature; and, as a coating, in humidity-sensing instruments.

Lithium Fluoride. LiF, mp 848°C, soluble in H_2O (slight). Used in enamel and glass formulations; as a component of welding and brazing fluxes; in the electrowinning of aluminum; and as an ingredient of molten salts.

References

Bach, R. O., Kamienski, C. W., and R. B. Ellestad: "Lithium and Lithium Compounds," in "Kirk-Othmer Encyclopedia of Chemical Technology," Wiley, New York (Updated periodically).

Jasinksi, D. R., et al.: "Lithium: Effects on Subjective Functioning and Morphine-Induced Euphoria," Science, **195**, 583–584 (1977).

Mulligan, R.: "The Geology of Canadian Lithium Pegmatites," Geol. Surv. Can. Econ. Rep. 21, Ottawa, Canada (1965).

Roth, W. L., and G. C. Farrington: "Lithium-Sodium Beta Alumina: First of a Family of Co-ionic Conductors," Science, **196**, 1332–1334 (1977).

Schreck, A. E.: "Lithium: A Materials Survey," U.S. Bur. Mines Inf.

Circ. 8053, U.S. Government Printing Office, Washington, D.C. (1961).

—Parts of this entry prepared by Patrick M. Brown, Director of Research, and I. A. Kunasz, Chief Geologist, Foote Mineral Company, Exton, Pennsylvania.

LITHIUM (For Thermonuclear Fusion Reactors). Natural lithium occurs as a mixture of two isotopes: (1) the more abundant ^7Li, and (2) ^6Li, which makes up 7.4% of the atomic weight of the natural substance. Although it occurs in some 145 minerals, only spodumene, lepidolite, petalite, amblygonite, eucryptite, and brines are commercial sources of lithium. These sources are described a bit later.

Use in Thermonuclear Fusion Reactors. There is little doubt that lithium will be used in thermonuclear fusion power plants. However, because of the wide diversity of design approaches and the early stage of many of these designs as of the early 1980s, it is impractical to predict with any accuracy how much lithium will be required. Initially, all fusion reactors probably will use deuterium plus tritium in the core (Ref. 1):

$$^2D + {}^3T \rightarrow {}^4HE + n + 17.6 \text{ MeV.}$$

The tritium will be produced from lithium. The quantity of lithium necessary to produce the tritium is relatively small. For example, assuming 100% efficiency, the lithium consumption for a 1000-MW(e) fusion power plant would be approximately 200 kg per year.

Lithium will be used in the blanket surrounding the core. This probably will be liquid lithium, although it is possible it might be in the form of certain lithium compounds. The function of the lithium blanket is:

1. To bring about the breeder reaction

$$^6LI + n \rightarrow {}^4He + {}^3T + 4.8 \text{MeV.}$$

This, of course, is necessary to generate the tritium to be used in the core.

2. The blanket may also be used to bring about the reaction

$$^7LI + n \rightarrow {}^3T + {}^4He + n - 2.5 \text{MeV.}$$

This reaction produces additional tritium and supplementary neutrons so as to make up for inevitable neutron losses in the system. However, the quantity of lithium in the blanket could be reduced by the use of beryllium compounds to supply the supplementary neutrons.

3. The blanket is used for heat absorption to remove energy from the very high energy neutrons coming from the core.

Most designs involve the use of natural lithium metal in the blanket. However, it is possible that the lithium in the blanket might partially be enriched in ^6Li. The quantity of lithium in the blanket might be as small as 50 metric tons for a 100-MW(e) fusion power plant. This assumes the minimum use of lithium and the use of beryllium or some other light element for the production of supplementary neutrons. On the other hand, the quantity of lithium in the blanket might be 500 metric tons or even somewhat higher if it is assumed that ^7Li will be utilized for the production of tritium and supplementary neutrons. This quantity, again, is for a 1000-MW(e) fusion power plant.

It is possible that additional lithium might be employed as the cooling fluid to transfer heat out of the blanket to the steam-producing portion of the overall power plant.

Another uncertainty is how many fusion power plants will be in operation at some future time. If the fusion power system is a technical and economic success, it will certainly be used for an appreciable fraction, but probably not all, of the total United States electric power requirements. Thus, within a few decades of the year 2010 we might expect to have as many as five hundred 1000-MW(e) fusion power plants in operation. This corresponds to the total United States electric power production for the year 1970.

Using this number as a reasonable guess, the quantity of lithium necessary would range from 25,000 to 250,000 metric tons. The lower quantity could be supplied readily by the present lithium industry from its existing reserves. The higher quantity would require the development of additional lithium sources.

Lithium Raw Materials. Although lithium is contained in some 250 minerals there are only a few species that have been mined as commercial ores: spodumene, lepidolite, eucryptite, and amblygonite.

Spodumene, a lithium aluminum silicate ($LiAlSi_2O_6$), is a monoclinic member of the pyroxene group. It has a very pronounced cleavage plane (110) which results in typically lath-shaped particles upon breaking. The color of spodumene is variable, being nearly white in the low-iron variety, and dark green in iron-rich crystals. Sources of spodumene are described later. Spodumene constitutes the most abundant commercial source of lithium. Theoretically, it may contain up to 8.03% Li_2O, but the actual lithia concentrations vary from 2.91 to 7.66%, probably as a result of sodium and potassium substitution for lithium.

Lepidolite is a phyllosilicate with the general formula

$$K_2(Li,AL)_{5-6}Si_{6-7}AL_{2-1}O_{20}(OH,F)_4$$

The lithia concentration in lepidolite varies between 3.3% and a possible theoretical maximum of 7.74%. In commercial deposits, the concentrations are more normally 3–4% Li_2O. In addition to lithium, lepidolites also carry substantial concentrations of rubidium (0.91–3.80% Rb_2O) and cesium (0.16–1.90% Cs_2O).

The major commercial occurrences of lepidolite are located in Zimbabwe (Bikita), Southwest Africa (Karibib), Canada (Bernic Lake), and Brazil.

Petalite, $LiAlSi_4O_{10}$, is a monoclinic mineral with a framework silicate structure. Its color is grayish white and more rarely pinkish. It has two cleavage directions which form an angle of 38.5°. The basal cleavage is perfect.

The theoretical lithia content of petalite is 4.88%. In actual commercial deposits, the concentration varies from 3.0 to 4.7% Li_2O. Sizable deposits of petalite occur with lepidolite in Zimbabwe, Southwest Africa, Brazil (Aracuai), Australia (Londonderry), the U.S.S.R. (eastern Transbaikalia), and Sweden (Utö).

Eucryptite is also a lithium aluminum silicate which is deficient in silica. It has the formula $LiAlSiO_4$ and may contain 11.88% Li_2O. The only large deposit of eucryptite is found in Rhodesia (Bikita), where its occurrence with quartz suggests a spodumene origin. The grade of eucryptite is 5.03% Li_2O.

Amblygonite, with the generalized formula $LiAl(PO_4)$ (F,OH), is the fluorine-rich end member of a phosphate series, while montebrasite represents the hydroxyl-rich end member. It occurs in white to gray masses. Basal cleavage planes are pearly; others are vitreous. Although amblygonite may contain as much as 10.2% Li_2O, commercial ores usually carry 7.5–9% Li_2O. Amblygonite has been mined in Canada, Brazil, Surinam, Rhodesia, Ruanda, Mozambique, Southwest Africa, and the Republic of South Africa.

Brines, Lithium is found in commercial quantities in certain brine deposits. The brines are present in desert areas and occur in playas and saline lakes where solutions have been concentrated by solar evaporation. In Searles Lake (California) where production of dilithium phosphate began in 1938, the lithium concentration is 0.015% Li_2O. In Clayton Valley (Nevada), lithium-bearing brines contain 0.065% Li_2O. Smaller concentrations of lithium (0.009–0.012% Li_2O) are found in the Great Salt Lake (Utah).

Recent work in South America has identified the presence of high lithium concentrations in the brines of the Salar de Atacama in Chile. (Ref. 2) Significant concentrations of lithium have also been found at the Salar de Uyuni in Bolivia. (Ref. 3) Geologists in Argentina have also reported the presence of lithium in the brines of the salares of La Puna. (Ref. 4) In Asia, lithium associated with other chemical components has been reported in the dry lakes of the Tibetan Plateau. (Ref. 5)

The U.S. Bureau of Mines has reported the association of lithium with a number of oil-field brines. (Ref. 6) Lithium is also known to occur in geothermal brines. (Ref. 7)

Lithium Reserves and Resources

In 1975, a lithium subpanel, formed under the auspices of the National Academies of Science and Engineering, evaluated the lithium raw materials availability in the Western world. The results of the study showed that the reserves (adjusted for mining losses) amount to 2.54×10^6 metric tons Li, while total resources were estimated at 10.65×10^6 metric tons of lithium. See accompanying table. These figures do not include

LITHIUM RESERVES AND RESOURCES OF WESTERN WORLD[7]

	METRIC TONS Li[b]	
	United States	Other Western Countries
Class A[a]		
From pegmatites	329,100	366,400
From brines	40,500	1,290,000
From stockpile	6,100	—
	375,700	1,656,400
Class B		
From pegmatites	47,300	456,800
From brines	—	—
	47,300	456,800
Class C		
From pegmatites	2,780,900	1,969,000
From brines	77,300	3,000,000
	2,858,200	4,969,000
Class D		
From brines	283,700	—
Total, Class A	375,700	1,656,400
Total, Classes A and B	423,000	2,113,200
Total, Classes A, B, and C	3,281,200	7,082,200
Total, Classes A, B, C, and D	3,564,900	7,082,200
Combined Totals:		
Class A	2,032,100	
Classes A and B	2,536,200	
Classes A, B, and C	10,363,400	
Classes A, B, C, and D	10,647100	

[a] Classification system:
 A. Reserves proved by systematic exploration
 B. Reserves indicated by limited exposures and/or exploration
 C. Resources inferred on geological evidence
 D. Quantities largely known but economic lithium extraction probably dependent upon marketing of co-products.
[b] In absence of data indicating otherwise, mine losses are assumed at 25% for open pit and 50% for underground operations. Brine totals are for recoverable brine.

potentially large resources contained in: (A) Other South American salares, (B) geothermal brines, (C) oil field brines, and (D) lithium-rich clays.

United States. The bulk of the world lithium production (mineral concentrate, chemicals, and metal) is currently produced in two areas—North Carolina and Nevada.

The tin-spodumene belt of North Carolina supplies the raw materials for the two major producers of lithium chemicals in the world: Foote Mineral Company (a subsidiary of Newmont Mining Corporation) and Lithium Corporation of America (a subsidiary of Gulf Resources and Chemicals Corporation). Foote Mineral Company also produces lithium carbonate from a brine deposit located in Clayton Valley, Nevada—presently the only brine deposit in the world exploited exclusively for lithium.

The two companies control the total lithium resources of 660,000 metric tons, of which 370,000 metric tons are classified as reserves.

Other Sources of Lithium. In addition to the two major producers, lithium ores are produced in Brazil, Portugal, Zimbabwe, the U.S.S.R., and China. Major producers of lithium chemicals from imported raw materials are Metalgesellschaft of Germany (West) and Honjo Zinc (Japan).

If a large demand for lithium arises in the future because of accelerated development in the battery and thermonuclear fusion fields, several important occurrences of lithium raw materials could become active suppliers. (Ref. 8) In Canada, a Preissac-LaCorne area of Quebec and the Bernic Lake area of Manitoba hold combined lithium resources of 156,000 metric tons. In Quebec, the property of the Quebec Lithium Corporation is anactive, while in Manitoba, the Tantalum Mining Corporation produces only tantalum concentratrates from its pegmatite ore. In Zimbabwe, the resources at Bikita are estimated at 114,000 metric tons of recoverable lithium. In Zaire, the largest known source of lithium pegmatites in the world and presently mined for tin and niobium (columbium), have resources estimated at 2.3 million metric tons of recoverable lithium. In Chile, the Salar de Atacama is scheduled to become a new source of lithium chemicals in 1984. Foote Mineral Company, in a joint venture with the Chilean Government, will produce lithium carbonate at an initial rate of 1050 metric tons per year.

The recent increase in interest in lithium has resulted in the identification of a variety of lithium occurrences, such as brines associated with oil fields, geothermal brines, and lithium-rich clays. Studies indicate that these contain very large resources of lithium, which, because of the relatively small demand as compared with existing reserves, have not been quantified.

References

1. Bogart, L. S.: "Potential Lithium Requirements for Fusion Electric Power," *Foote Prints*, **43**, 1, 2–11 (1980).
2. Kunasz, I. A.: "Lithium in Brines," *5th Symposium on Salt*, Vol. 1, 115–117 (1980).
3. Ericksen, G. E., et al.: "Chemical Composition and Distribution of Lithium-Rich Brines in Salar de Uyuni and Nearby Salares in Southwest Bolivia," *Energy*, **3**, 3, 355–363 (1978).
4. Nicolli, H. B., et al.: "Geochemical Characteristics of Brines in Evaporitic Basins, Argentine Puna," *26th International Geologic Congress*, *Abstract* page 786 (1980).
5. *Mining Journal*, page 161 (March 2, 1979).
6. Collins, C. A.: "Lithium Abundance in Oil-Field Brines," *U.S. Geological Survey Prof. Paper 1005*, 116–123, U.S. Geological Survey, Washington, D.C. (1976).
7. Berthold, C. E., et al.: "Lithium Recovery from Geothermal Fluids," *U.S. Geological Survey Prof. Paper 1005*, pages 61–66, U.S. Geological Survey, Washington, D.C. (1976).
8. Kunasz, I. A.: "Lithium—How Much?" *Foote Prints*, **43**, 1, 23–27 (1980).

—Wayne T. Barrett, Consultant, and Ihor A. Kunasz, Chief Geologist, Chemicals & Minerals Division, Foote Mineral Company, Exton, Pennsylvania.

LNG (Liquefied Natural Gas).   Natural Gas; Substitute Natural Gas (SNG).

LOW-TEMPERATURE TECHNOLOGY.   Cryogenics.

LUBRICATING AGENTS.   There are two fundamental classes of lubricating agents—natural and synthetic.

Lubricating oils are fluids whose function is the reduction of friction and wear between solid surfaces (generally metals), in relative motion. This function is accomplished in either of two ways: (1) by formation of adsorbed films on the two opposed surfaces, which can be more easily sheared than the solid substrate, or (2) by interposition of a fluid film between the two opposed surfaces. In the former case the shear strength of the film, and in the latter, the viscosity of the fluid determine the magnitude of the work which must be done to maintain the opposed surfaces in relative motion. In most cases, bearings are designed to operate under fluid film conditions and thus the viscosity of an oil is its most important property in classifying it for lubricating purposes. The range of viscosities required is approximately 1 centistoke at 100°F (38°C) for high-speed, lightly loaded spindle bearings to approximately 4000 centistokes at 100°F (38°C) for some gear units. The great majority of oils used fall within a much narrower viscosity range of about 20–200 centistokes.

Aside from the primary function of friction and wear control, lubricating oils are often called on to serve other purposes, such as corrosion prevention, electrical insulation, power transmission and cooling. This last is particularly important in metal cutting and grinding.

As petroleum consists of a highly complex mixture of hydrocarbons and other organic compounds, the raw material must be carefully processed to separate the fractions which are desired for lubricating purposes. This is generally accomplished by a combination of fractional distillation, solvent extraction and adsorption treatments to obtain various cuts which are blended if necessary, to produce the desired viscosity. A typical sequence of operations would be:

(1) Preliminary distillation of the crude petroleum to strip off volatile matter, i.e., dissolved gases and light hydrocarbons through the fuel oil range.

(2) Secondary vacuum distillation to separate the various lubricating oil fractions, which in some cases, may include the residuum. Since hydrocarbons tend to undergo thermal decomposition above 600°F (316°C), the distillation is generally carried out under vacuum to enable the higher-boiling fractions to be distilled.

(3) Solvent extraction to remove waxes, unsaturates, aromatics, asphalt, and nonhydrocarbon material.

(4) Contact with an adsorbent, e.g., activated clay, for final removal of polar impurities.

(5) Incorporation of special-purpose additives and blending to obtain the desired viscosity.

The lubricating oil fractions thus obtained consist of some of the largest and most complex hydrocarbon molecules to be found in crude petroleum. Their molecular weight ranges from about 250–1000 or more, based on structures containing 20–70 carbon atoms. They may be classified broadly as (1) straight-chain paraffins, (2) branched chain paraffins, (3) naphthenes (one or more saturated 5 or 6 membered rings with paraffin side chains), (4) aromatics (benzene ring structures with paraffinic side chains), and (5) mixed aromatic-naphthene-paraffin.

Of the above classes, the most desirable are the branched chain paraffins and the naphthenes. The refining processes are directed to increasing the concentration of these and removal of the others, as well as of organic compounds containing sulfur, oxygen, and nitrogen which are found in varying proportions in all crude petroleum.

Most hydrocarbons in the lubricating oil range are subject to low-temperature oxidation (180°F; 82°C and above) by atmospheric oxygen, particularly under the influence of catalysts such as metals (copper is particularly active), moisture, and occasionally nonhydrocarbon impurities. The paraffinic types show the greatest resistance to oxidation, followed by the naphthenes and the aromatics. An empirical rule states that the rate of oxidation approximately doubles for each 20°F (11°C) rise in temperature. Thus, in practice where long life of the oil is desired, bearing temperatures are held to around 150°F (66°C) as a maximum. In an internal combustion engine where temperatures at the sliding surfaces of the order of 400–500°F (204–260°C) are encountered, shorter life will result, e.g., 100 hrs. in an engine, compared with several years in a water-wheel generator.

Besides viscosity, the viscosity-temperature variation of lubricating oils is a most important property. It has been found that this behavior can be accurately expressed by the relation: $\log_{10} \log_{10} (v + 0.8) = b \log_{10} (t°F + 460) + c$ where v is the viscosity in centistokes, t the temperature in °F, and b and c are constants. In most cases an oil with the smallest viscosity-temperature variation is considered the more desirable. As a generalization, it may be said that the oils with the highest paraffin-to-naphthene ratios have the smallest change, while low paraffin-to-naphthene ratios correspond to a higher viscosity-temperature change. However, it is possible to dissolve materials in oil (e.g., olefinic polymers) which will effect a decrease in this variation.

While most bearings depend on the viscosity of the oil for the formation of a lubricating film (0.0001″ to 0.002″; 0.0025 to 0.051 mm thick), there are many cases where this is not feasible and successful lubrication depends on the formation of very thin films (10 to 10^3 Å thick). This was generally thought to occur through preferential adsorption by the solid surface of polar compounds (e.g., naphthenic acids, sulfur compounds) present in small amounts in the oil.

Another factor possibly contributing to the lubricating ability of very thin films is the variation of viscosity with pressure. The viscosity-pressure relationship is approximately $\log_{10} v = kP + c$ where P is the pressure and k and c are constants. Thus, at very high unit pressures of the order of 10^5 psi (690 MPa), viscosity increases of the order of 1000–10,000 times are encountered. This can account for a significant increase in load-carrying capacity. More recent work on rolling contact bearings has shown that these extremely high viscosities may be partially responsible for the observed formation of true hydrodynamic lubricating films in the 10^{-6} to 10^{-5} inch (0.00254–0.000254 mm) range.

Though the major portion of industrial lubricants is derived from petroleum, a small but significant portion is being obtained from other sources.

(1) Natural liquid fatty esters, such as lard oil, palm oil, sperm oil, etc. These are good lubricants but have poor chemical stability.

(2) Synthetic hydrocarbons, prepared by polymerization of olefinic hydrocarbons; these have good stability when saturated and good viscosity-temperature coefficients.

(3) Polyalkylene glycol oils, made by reaction of alcohols with polymerized ethylene and propylene glycols. They are either water-soluble or -insoluble. Fair stability (improved with additives), good viscosity-temperature coefficient, and good lubricating qualities.

(4) Synthetic esters, (a) primarily esters of dibasic acids such as adipic and sebacic, though in Europe some monobasic acid

esters have been prepared and used; (b) organic esters of phosphoric and silicic acid, which have some advantage of being more fire-resistant than the other organic compounds but which are subject to hydrolysis on exposure to water.

(5) Silicone oils, which are linear and cyclic siloxane polymers of the formula $(-SiR_2O)_n$. They generally possess good thermal and oxidative stability, and good viscosity-temperature coefficients. When $-R$ is a hydrocarbon substituent their lubricating ability is poor. This can be somewhat improved by the introduction of aromatic halogen in $-R$.

(6) Halogenated hydrocarbons (chlorinated or fluorinated). These have good lubricating properties but very poor viscosity-temperature relations. The fluorinated materials are extremely stable.

(7) Perfluorinated polyalkylene glycols which have better viscosity-temperature behavior and temperature behavior and temperature resistance.

(8) Polyphenyl ethers are very stable organic fluids which can be used in the 500–700°F (260–371°C) temperature range although, like the silicones, they do not have good boundary lubrication properties. These reportedly, however, can be improved with suitable additives.

Generally speaking, these synthetic materials are at present used primarily as specialty lubricants where the need for a particular property outweighs high cost, e.g., diesters as military aircraft engine lubricants. As synthetic processes are improved more extensive application may be expected.

Synthetic Lubricants

Chemistry has been the guiding science in creating the various classes of chemical lubricants. Each class is named in terms of the chemical structure involved and each has at least one property that is unobtainable with naturally occurring materials. In varying degrees and combinations, synthetic lubricants provide excellent lubricity over extremely wide temperature ranges, high thermal and oxidative stability, fire-resistance, and outstanding resistance to nuclear radiation. Other noteworthy advantages of synthetic lubricants are the ease with which physical or performance properties can be altered by chemical modification or the incorporation of additives. The facility with which modifications can be achieved makes it possible for the chemists to tailor individual members of the family to meet exactly the requirements of a particular application. Compromises in properties are likewise made to obtain a satisfactory lubricant at lowest cost.

Polyglycols: $RO(CH_2CHR'O)_xR''$, where R's can be hydrogen and/or organic groups. Terminal groups determine the type of polyglycol such as diol, monoether, diether, ether-ester, etc. Various alkylene oxides including mixtures are the basic raw materials. The major attributes of the polyglycols are viscosity-temperature characteristics, volatility of products of decomposition, relatively low cost, and the wide variety of properties (water-soluble to organic-soluble) obtainable by structural variations. Major uses are as lubricants in automotive hydraulic brake fluids and water-based hydraulic fluids, as compressor lubricants, textile lubricants and rubber lubricants, and in greases where the polyglycol is the carrier for solid lubricants.

Phosphate Esters. $R'OP(O)(OR'')(OR''')$ where at least one R represents an organic group while the remaining represent organic groups or hydrogen. These products are prepared from phosphorus oxychloride or phosphoryl chlorides plus phenols, alcohols or their sodium salts. The tertiary phosphate esters are often classified as triaryl, trialkyl, and alkyl aryl phosphates. The primary and secondary phosphates are used extensively as lubricant additives in various chemical forms. Phosphate esters are best known for their fire-resistant characteristics, and as a result have found extensive application as industrial and aircraft lubricants and hydraulic fluids.

Dibasic Acid Esters. These can include both simple and complex materials. The simple dibasic acid esters, $ROOCR'COOR''$ are made by reacting a dibasic acid, such as sebacic acid, with a primary branched alcohol, such as ethyl hexanol. Complex esters are prepared by reacting a dibasic acid with a polyglycol, such as polyethylene glycol, and capping the chain with a branched primary alcohol or a monobasic acid. The outstanding characteristics of the dibasic acid esters are favorable viscosity-temperature characteristics, excellent lubricating ability, and high stability. Because of this combination of properties, these products are now used as lubricants in almost all aircraft turbine engines.

Chlorofluorocarbons. The polymerization or telomerization of chlorotrifluoroethylene is the route to these relatively low-molecular-weight synthetic lubricants. Manufacture involves elaborate polymerization techniques and stabilization methods to eliminate all of the hydrogen and terminal chlorine introduced into the polymer chain by peroxide fragments or the chain transfer agent. The major characteristics of the chlorofluorocarbons are chemical inertness and thermal stability. Industrial and aerospace applications involving exposure to corrosive or oxidizing atmospheres are the largest uses for these lubricants.

Silicones.

$$\begin{array}{ccc} R & R & R \\ SiO & SiO & SiR \\ R & R & R \end{array}$$

where the R's may be the same or different organic groups. The properties are varied by the use of different types of organic substituents; the most popular are methyl, phenyl, and chlorophenyl groups. Recent advances involve fluorine-containing substituents. Manufacture entails the preparation of organochlorosilane intermediates, hydrolysis and condensation of these intermediates, and polymer finishing. In addition to good stability and low volatility, silicones have the best viscosity-temperature characteristics of any lubricant. Although they perform well under many conditions of lubrication, silicone lubricants are generally unsatisfactory for situations involving sliding contact of steel-on-steel. The many and varied uses include lubricating electric motors, precision equipment, plastic and rubber surfaces, and as greases for antifriction bearings.

Silicate Esters. $ROSi(OR')(OR'')(OR''')$, where the R's may be similar or dissimilar groups. The best-known types are the tetraalkyl, tetraaryl, and mixed alkylaryl orthosilicates. The classic means of preparation is through the reaction of phenol or alcohol with silicon tetrachloride. A closely related group of products, the hexaalkoxy- and hexa-aryloxydisiloxanes, is also generally included in the silicate esters classification. These products, the so-called "dimer silicates," are conveniently made by the reaction of an alcohol or phenol with hexachlorodisiloxane. Notable characteristics of the silicate esters are low volatility, low-temperature fluidity, and thermal stability. The hydrolytic stability varies from poor to good, depending upon chemical structure. The products are used as high-temperature heat-transfer fluids, wide-temperature range hydraulic fluids, electronic coolants, and automatic weapon lubricants.

Neopentyl Polyol Esters. These polyesters are prepared by the esterification of 5-carbon polyfunctional alcohols with monofunctional acids. Because the beta carbon of the starting alcohol does not contain hydrogen, these esters are superior in thermal stability to the diesters. Most of the other characteristics are similar to those of the diesters. As a result of their superior stability, the neopentyl esters are finding increasing use as the lubricant for aircraft turbine engines.

Polyphenyl Ethers. Both alkyl-substituted and unsubstituted polyphenyl ethers are included in this class of synthetic lubricants. General preparation involves the Ullman ether synthesis. The unsubstituted polyphenyl ethers have outstanding thermal, oxidative and radiation resistance, however, poor low-temperature characteristics are a major drawback. Alkyl substitution improves low-temperature viscosity, but detracts from stability. Most lubricant uses are developmental in nature and involve aircraft and aerospace applications.

Lubrication

Lubrication is the process of separating two rubbing solid surfaces by means of a layer which is effectively "softer" than either surface. Depending upon circumstances, the "soft" layer (the lubricant) may be a gas, a liquid, a solid, or a combination of various phases. The many ways of accomplishing lubrication can be conveniently grouped into two categories: hydrodynamic and solid lubrication.

Hydrodynamic Lubrication. The separation of moving surfaces by a "fluid" (gas, liquid, or gel) is accomplished according to the laws of hydrodynamics. One of the surfaces "swims" on the lubricant, i.e., it is lifted through the simultaneous possession of velocity relative to the lubricating fluid and of an acute angle of attack. This so-called glider bearing is the basis of design of nearly all fluid-lubricated bearings, including the journal bearing, which is just a circular glider bearing.

The basic relation which expresses the load-carrying ability of a glider bearing can be given in the form:

$$W/\eta U = a(r/h) \qquad (1)$$

where W is the load on the glider, U the velocity of the glider relative to the stator, h is the minimum distance between the rubbing surfaces, r is a dimension characterizing the angle of attack, a is a numerical coefficient (somewhere between 1 and 10) and η is the viscosity of the lubricating fluid under the temperature and pressure conditions of the application. The energy expenditure required for the service performed by the lubricant film is usually expressed in terms of the friction coefficient, f, defined as the ratio of the frictional force required to move the glider (or the journal) to the load carried by it. For the case under discussion:

$$f = K\sqrt{\eta U/W} \qquad (2)$$

where K is a numerical coefficient which depends upon the geometry of the system and varies between about 2 and 6.

While Equations (1) and (2) provide only a qualitative guide—the detailed calculations for specific bearings are quite complicated—they clearly indicate that the only variable at the disposal of the chemist, the viscosity, will be chosen such as to give the maximum load-carrying ability of the bearing consistent with a reasonable amount of frictional energy lost in the lubricant.

The low friction losses (f is usually of the order 10^{-3}) and the naturally wear-free operation generally make the maintenance of hydrodynamic lubrication the primary aim of bearing design. This goal is attained perfectly only with journal and with glider (pad) bearings and to some extent in roller bearings. It can be achieved by a special mechanism (thermal expansion of the flowing fluid) in parallel thrust bearings. Its attainment is uncertain (and not readily subject to calculation) in ball bearings and in gear lubrication.

Since the viscosity of readily available fluid varies over a 10^{10}-fold range between gases and the thickest liquids, and the temperature and pressure coefficients of viscosity vary similarly over an about 10^4-fold range, the choice of a suitable lubricant is usually determined by the ancillary conditions of temperature, volatility, chemical stability, etc. The use of sulfuric acid as lubricant for oxygen compressors is a typical illustration of a choice dictated by chemical conditions. The evaporation and deterioration of liquid lubricants in high energy particle fluxes and/or in hard vacua ($<10^{-9}$ torr) has led to increased use and rapid development of gas lubricated bearings. These call generally for extremely high and therefore very costly standards of workmanship. Hence the use of gas lubrication is at present largely restricted to precision instruments and to military and space applications.

The most important source of lubricants is petroleum. There is hardly a chemical species (esters, ethers, sulfides, metal-organic compounds, etc.) which has not contributed to the array of synthetic lubricants now available. Some chemicals, such as the perfluorocarbon compounds and the siloxane polymers were originally synthesized for just this service.

Elastohydrodynamic Lubrication. Well designed and equally well built machine elements with very smooth bearing surfaces permit the imposition of very high bearing loads. The resulting elastic deformation of the bearing may then change the geometry of the load-bearing surfaces substantially. Hence the elastic properties of the load-bearing materials enter the bearing calculations. The change of lubricant properties with pressure, temperature, and with the duration of exposure to the pressure and shear regime also enter the bearing calculations. The required viscoelastic properties of lubricants at high-frequency deformations are only beginning to be determined.

Hydrostatic Lubrication. Slow-moving or even static "gliders" can be separated from the bearing surface by a fluid film if the hydrostatic pressure required to carry the load is provided by an external source, as for instance by a pump. The very large journals of turbo generators, or heavy thrust bearing pads are generally lifted by these means before the onset of rotation in order to avoid scoring damage to the valuable bearings. Exceedingly low friction coefficients are obtained in these hydrostatic bearings, the most spectacular being that of the Mt. Palomar 200-inch (508 cm) telescope mirror pad bearings, where $f = 0.000004$, such that the 500-ton (450 metric ton) structure is easily moved by a $\frac{1}{2}$ HP clock motor.

Solid Lubrication. While fluid film separation of rubbing surfaces is the most desirable objective of lubrication, it is often unattainable, especially when bearings are too small or unsuitable for liquid lubricants. Even bearings built for full fluid lubrication during most of their operating periods experience solid-to-solid contact when starting and stopping.

Solid surfaces in rubbing contact are characterized by friction coefficients varying between 0.04 (Teflon on steel) and >100 (pure metals *in vacuo*). Solid lubrication, in contrast to fluid lubrication, is generally accompanied by a certain amount of wear of the rubbing parts. Optical inspection of the surfaces after rubbing reveals macroscopic (i.e., bulk) damage of the metal both when unlubricated and when lubricated. Quantitative differences between the two cases are easily measured radiographically when a radioactive glider has been used. In this manner it is found that effective solid lubrication can reduce the amount of wear—compared to the dry case—by a factor as high as 10^5.

Typical solid lubricants are the soft metals lead, indium, and tin, the layer lattice crystals graphite and molybdenum disulfide, many soft organic solids, such as metallic soaps, and waxes as well as the crystalline polymers Teflon (polytetrafluorethylene), polythene (polyethylene), and nylon. The integral bonding of these solids to the surface of the hard solid to be lubricated is essential for good performance. The bonding is accomplished either by alloying (copper-lead bearing metals), by flash coating, by introduction of the lubricating solid into the interstices of the sintered metal bearing (Teflon emulsion into sintered bronze), by chemical coating (phosphate coatings), or by anchorage to a phosphate or bonded plastic coating.

Special cases of solid lubrication are boundary and EP (extreme pressure) lubrication. In both cases the solid lubricant is formed by chemical reaction of special compounds, usually applied as oil solutions, with the metallic rubbing surfaces. Typical boundary lubricants are the fatty acids which react with the metal surface to form metallic soaps which then carry the load. Strongly adsorbed but nonreacting substances of linear structure, such as long chain fatty alcohols, can also act as boundary lubricants but only under very mild conditions.

Under the very severe conditions, sometimes encountered in automobile transmissions and especially in hypoid gear differentials as well as in machining operations, only those substances act as lubricants which contain chemically active chlorine, sulfur, or phosphorus to form the corresponding iron chloride, sulfide or phosphide by instantaneous attack on the surface hot spots resulting from the collisions of surface asperities. The chemical stability of these so-called E.P. agents is designed to permit activity at the temperature near the rubbing surface, say 200°C and above, but not be corrosive under normal conditions.

Mixed Film Lubrication. Mixed film lubrication is almost invariably the true state of affairs when boundary and EP lubrication are encountered, i.e., an appreciable fraction of the load is carried by the fluid film in the "valleys" of the surface while the asperities in contact are permitted to carry the balance of the load without seizure through the beneficent intervention of the boundary or EP lubricant. The very important break-in process of rubbing surfaces consists in the controlled reduction of the number and the size of the surface asperities so that fluid lubrication will prevail for most of the time.

"Real" Lubrication. In "real" lubrication of machinery, such as automotive engines, turbines, etc., the lubricating oil has to perform many functions besides lubrication. The most important of these is the cooling of the bearings. It also has to keep internal combustion engines clean by dispersing the partial combustion products of the fuel and its own degradation products, it has to carry chemicals to counteract wear, and—in common with many other lubrication applications—it must prevent corrosion of the equipment and be inhibited against its own deterioration in service. A relatively large amount of synthetic organic chemicals is therefore carried by many oils to perform the additional functions which one must expect from a modern lubricant.

Ideally one should always have full fluid separation of rubbing surfaces. But in inaccessible locations, or reactive environments, or under conditions of very slow motion or of intermittent operation, recourse must be had to solid lubrication.

—Henry E. Mahncke, King of Prussia, Pennsylvania. (Lubricating Oils)
—Reigh C. Gunderson, The Dow Chemical Co., Midland, Michigan. (Synthetic Lubricants)
—A. A. Bondi, Shell Development Co., Houston, Texas. (Lubrication)

LUMINESCENCE. An emission of light which is greater than could occur from a temperature radiation alone. Under various circumstances it is known as photo-, thermo-, tribo-, cathodo-, electro-, and chemi-luminescence. In all cases, it depends upon the conversion of a compound into an unstable, excited state; the light emission arises on the return of this phase to the normal one.

Its most noteworthy forms are photoluminescence that occurs under excitation by light of shorter wave lengths, and cathodo- and electro-luminescence that result from excitation by electrons. The light that is confined to the period of excitation is called *fluorescence*; that which persists after excitation has ceased is the *phosphorescence*, or afterglow. *Phosphor* is the name given to those compounds capable of developing fluorescence. Its emission from inorganic compounds is associated with the presence, at a very low concentration, of an impurity feature called an activator, which may be a foreign element; silver or copper at about 0.05% in zinc sulfide, and manganese at 1–2% in the silicates and phosphates of zinc, cadmium, magnesium and calcium, in the borates of zinc and cadmium, and in zinc fluoride. It may also be a foreign condition, such as monovalent zinc in zinc sulfide, or perturbed groups of atoms in the self-activated tungstates. The effectiveness of monovalent activators in zinc sulfide requires the introduction of coactivators to secure charge compensation. Equivalent amounts of chlorine or aluminum serve this purpose.

The color of fluorescence ranges over the entire spectrum, and depends upon the activator and the composition of the compound. In the zinc sulfides, addition of cadmium sulfide shifts the color from blue to red for silver activation, and from green to infrared for copper. The fluorescence due to manganese ranges from green to red in different compounds. Lead, thallium, antimony, bismuth and titanium activators generally give rise to fluorescence in the ultraviolet or blue.

For photoexcitation to occur, there must be an absorption band at the spectral location of the exciting radiation, and this radiation must be of such wave length that its energy equivalent is sufficient to raise the activator ions to their excited state. In the absence of such an absorption band, the phosphor may still be photoexcited if another activator is added, the sensitizer, which introduces the needed absorption. The phosphor then emits two bands, one due to the sensitizer and the other to the activator proper. The latter receives excitation energy from the sensitizer by a resonance step.

Phosphorescence is generally of the same color as fluorescence. It occurs in two stages. The first has an exponential decay, lasting from a few microseconds to a few milliseconds and even to a second for zinc fluoride activated with manganese. It is independent of temperature below 100°C or so. The second stage of decay is bimolecular and its persistence ranges from seconds to hours, being longest for the bismuth-activated sulfides of calcium and strontium. It is strongly temperature dependent. At low temperature, it is frozen in and is released on warming—the phenomenon of thermoluminescence. At higher temperatures it is quenched and dissipated as heat to the lattice.

Phosphors are used for illumination and for television and radar screens. Fluorescent lamps utilize a low-pressure mercury discharge whose predominant emission lies in the 2537 Å line. Phosphors responding to this excitation include calcium and magnesium tungstates for blue, zinc silicate for green, calcium halophosphates for a combined blue and orange emission, calcium silicate for red, and calcium phosphate for a still deeper red. Combinations of these are used to give the color of light desired.

Bioluminescence is also a common and spectacular form of chemiluminescence. It is exhibited by living organisms, such as bacteria, glow worms, fireflies, and luminous fish. It is due to the admixture of two types of substances present in the organism; one of them, *luciferin*, is capable of oxidation in the presence of the second, an enzyme termed luciferase. The reaction produces an excited form of luciferase, and its return to the normal form is accompanied by the emission of light.

Electroluminescence

Electroluminescence is light generated in crystals by conversion of energy supplied by electric contacts, in the absence of incandescence, cathodo- or photoluminescence.

It occurs in several forms. The first observation of the presently most important form, "radiative recombination in p-n junctions," was made in 1907 by Round, more thoroughly by Lossev from 1923 on, when point electrodes were placed on certain silicon carbide crystals and current passed through them. Explanation and improvement of this effect became possible only after the development of modern solid state science since 1947.

If minority carriers are injected into a semiconductor, i.e., electrons are injected into a p-type material, or "holes" into n-type material, they recombine with the majority carriers, either directly via the bandgap, or through exciton states, or via impurity levels within the bandgap, thereby emitting the recombination energy as photons. Part of the recombinations occur nonradiatively, producing only heat.

Exploitation of the effect was strongly dependent on progress in compound semiconductor crystal preparation and solid state electronics, since crystal perfection (absence of defects) is of prime importance. At present, single-crystalline p-n diodes made of gallium arsenide, GaAs, a "III-V compound" (from groups III and V of the Periodic System), yield the highest efficiencies (40% of the electrical power input converted into optical power output) in the near-infrared, and diodes made of gallium phosphide, GaP, yield red light with about 8% efficiency. Very important are alloys such as $In_xGa_{1-x}P$, $Al_xGa_{1-x}As$ and $GaAs_xP_{1-x}$ where the color of luminescence can be changed by changing the composition. Wave lengths of light emitted by III-V crystals range from 6300 Å to 30μ.

An important phenomenon, "injection laser action," was discovered in GaAs diodes in 1962. The crystal faces at the ends of a p-n junction are made optically parallel so as to form a Fabry-Perot optical cavity. Beyond a certain injection current density, (the "threshold current") the individual recombination processes no longer occur randomly and independent of each other, but in phase, so that a near-parallel beam of coherent light (~9000 Å) of enormous intensity (10^7 w/cm² in pulsed operation) is emitted. The efficiency has been improved by using graded bandgap $Al_xGa_{1-x}As$ heterojunctions and special doping profiles so that the lasing region near the p-n junction acts as a "light pipe," preventing light straying out sideways. The current threshold is now reduced to 2000 amp/cm² at room temperature (continuous operation).

These coherent or incoherent electroluminescent p-n diodes are small point sources used for pilot lights, opto-electronic data processing, ranging systems, direct-sight communication, and as IR-lamps for night vision devices.

Another kind of electroluminescence, discovered by Destriau in 1936, uses inexpensive powders consisting of small particles of essentially copper-doped zinc sulfide, ZnS, a II-VI compound, embedded into an insulating resin and formed into a large flat plate capacitor with one plate transparent (e.g., SnO_2-coated glass). If an ac-voltage is applied, light is emitted (blue, green, red, depending on the exact material composition) twice per cycle, with brightnesses up to thousands of foot-lamberts. Brightness increases linearly with drive frequency, and exponentially with voltage, until saturation occurs. The efficiency is about 1% but it decreases with increasing brightness.

Microscopic examination of the interior of an individual particle reveals that the light is emitted inhomogeneously, in the form of two sets of comet-like striations which light up alternatingly, each set once per cycle. These comets coincide with long, thin conducting copper sulfide precipitates which form along crystal imperfections. The applied field relaxes in these needles and concentrates at the tips, so that electrons and holes are alternatingly field-emitted into the surrounding insulating luminescent ZnS. The holes are trapped there until they recombine with the more mobile electrons, emitting the typical luminescent spectra.

Among applications of this large-area, thin light source are safety lights for home use, luminous instrument faces, alphanumeric and other information display panels. The deterioration of these light sources is presently still a problem, but the time to half-brightness has been improved to useable intervals. Compared to p-n junction recombination electroluminescence the main advantage of this type is low cost.

Large-area dc-electroluminescence of polycrystalline, faintly n-type films of ZnS on glass, doped with copper and manganese, has also been achieved. The mechanism involves high-field-aided hole injection.

Still another type of electroluminescence is acceleration-collision electroluminescence. In back-biased p-n junctions, Schottky barriers, and near conducting inclusions, the local field can become high enough (10^5v/cm or more) so that electrons acquire sufficient kinetic energy to impact-ionize luminescent centers or the host lattice, creating secondaries. Electroluminescence occurs upon recombination. The efficiency is very poor, because only a small fraction of the electrons can attain sufficient energy, the others create only heat.

This type of electroluminescence occurs in all materials displaying the earlier described types, and in materials such as ZnO, $BaTiO_3$, and Zn_2SiO_4. It has been optimized in large-area ZnS polycrystalline films, with a Cu_2S cathodic film electrode. Efficiencies are 10^{-2}% or less.

General Luminescent Characteristics of Crystals.

Luminescence is the phenomenon of light emission in excess of thermal radiation. Excitation of the luminescent substance is prerequisite to luminescent emission. Photoluminescence depends upon excitation by photons; cathodoluminescence, by cathode rays; electroluminescence, by an applied voltage; triboluminescence, by mechanical means such as grinding; chemiluminescence, by utilization of the energy of chemical reactions. Luminescent emission occurs with gases, liquids and solids—both organic and inorganic. The emission involves optical transitions between electronic states which are characteristic of the radiating substance. The luminescence of inorganic crystals involves in many cases the electronic states of impurities or imperfections; in some cases, the electronic bands of perfect crystals. Inorganic crystals which luminesce are called phosphors. Impurities responsible for luminescent emission are called activators.

Luminescent excitation and emission are separate processes, for example, photoluminescence is not a scattering phenomenon but consists of absorption of radiation followed by emission. The time delay between excitation and emission is long compared to the period of the radiation λ/c, where λ is the wavelength and c is the velocity of light. This time delay distinguishes luminescence from the Raman effect, from Compton and Rayleigh scattering, and from Cherenkov emission. For luminescent inorganic crystals the radiative lifetimes of the emitting states vary from 10^{-10} to 10^{-1} second, depending on the identity of the crystal and, if impurity luminescence, also on the identity of the activator. At ordinary intensities of excitation, the spontaneous transition probability predominates so that the luminescent radiation is incoherent; with high intensities of excitation, the induced transition probability may predominate for suitable phosphors, the emitted radiation is coherent and laser action is attained.

Excitation and Emission Spectra. For most phosphors the photoluminescent excitation and emission spectra are different. Normally, there is a Stokes' shift so that the emission occurs at a longer wave-length than the excitation. In other words, in contrast to Kirchhoff's law for incandescence, luminescent emission for most crystals occurs in spectral regions where the absorption is low. The spectral distribution of individual luminescent emission bands is normally the same with all types of excitation; for example, with photo-, cathodo-, or electro-luminescent excitation. These effects are qualitatively understandable on the basis of the configuration coordinate model shown in Fig. 1. In this model the Born-Oppenheimer adiabatic approxi-

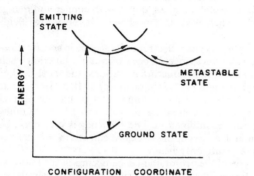

Fig. 1.   Configuration coordinate model for activator system.

mation is used as a basis for plotting the energy of each electronic state of the activator as a function of the coordinates of the nuclei comprising the activator system. Photoluminescent excitation occurs vertically in accordance with the Franck-Condon principle, maintaining the nuclear coordinates of the ground state; emission occurs following a lattice relaxation to the equilibrium coordinates which are characteristic of the emitting state. Both the excitation and emission bands are broadened because of thermal vibrations at ordinary temperatures and the zero point vibrations at low temperatures.

Fluorescence and Phosphorescence. As noted earlier, the initial persistence of luminescent emission following removal of excitation is a matter of the lifetime of the emitting state. This emission decays exponentially and is often called fluorescence. With many luminescent inorganic crystals there is an additional component to. the afterglow which decays more slowly and with more complex kinetics. This component is called phosphorescence. The emission spectra for fluorescence and phosphorescence are the same for most phosphors; the difference in afterglow arises from trapping states from which thermal activation is prerequisite to emission. In some cases the trapping state is a metastable state of the activator; in other cases, is a state of another imperfection. A metastable state of the activator is included in Fig. 1.

Luminescence of Perfect Crystals and of Activator Systems. The luminescence of most phosphors originates from impurities or imperfections, however, there are also phosphors whose luminescence is characteristic of the perfect crystals. The latter include the alkaline earth tungstates whose luminescence is characteristic of the $WO_4^=$ group perturbed by the crystal field, rare

earth salts which emit in narrow bands or lines characteristic of transitions within the 4f shell, and intrinsic semiconductors in which radiative recombination occurs between a conduction electron and a valence band hole coupled to form an exciton.

The impurities and imperfections which form activator systems in inorganic crystals are of diverse atomic and molecular types whose characteristics depend on the structure of the defect, on the electronic states of the defect and on the electronic structure of the pure crystal. Point defects which have been identified as capable of luminescence include lattice vacancies and substitutional impurities. Associated defects capable of luminescence include pairs of oppositely-charged point defects, that is, donors and acceptors. A donor-acceptor pair radiates following capture of a conduction electron and a valence band hole. In some cases the electronic states which participate in the luminescence of inorganic crystals can be described in terms of the energy levels of the impurity ion perturbed by the crystal field; in other cases, in terms of the crystal band structure perturbed by the impurity.

Ionic Crystals. The alkali halides are simple ionic crystals which become luminescent when doped with suitable impurities. Tl^+ substituted at cation sites in alkali halides exhibits characteristic absorption and emission bands. The absorption bands correspond to the $^1S \rightarrow {}^3P$, $^{o1}P^o$ transitions of the free ion perturbed by crystal interactions; the principal emission band, $^3P^o \rightarrow {}^1S$, similarly perturbed. The spectra can be understood qualitatively with the aid of Fig. 1 and with the configuration coordinate interpreted as the displacement of the nearest-neighbor halide ions of the Tl^+. It is the nearest-neighbor interaction which is strongly dependent on the electronic state of the Tl^+ and therefore is primarily responsible for the band widths and the Stokes' shift. In^+, Ga^+, Pb^{2+} and other impurities with electronic structures similar to Tl^+ have been shown to be activators in alkali halide crystals. In addition, alkali halides can be made luminescent by the introduction of imperfections. For example, the F-center which consists of an electron trapped at a halide ion vacancy has a characteristic infrared luminescence at low temperatures. Also, the self-trapped positive hole or V_k center has been shown to luminesce following electron capture.

Many inorganic crystals become luminescent when certain transiton metal ions are dissolved in them. The luminescence involves intercombination transitions within the $3d$ shell, therefore crystal field theory can be used to interpret the absorption and emission spectra. Divalent manganese is a common activator ion. Zn_2SiO_4, ZnS and $3Ca_3(PO_4)_2 \cdot Ca(F,Cl)_2$ are important phosphors activated with Mn^{2+}. The last, activated also with Sb^{3+}, is the principal fluorescent lamp phosphor. The excitation at 254 nm from the Hg discharge occurs at the Sb^{3+}, whose energy level structure is similar to that of Tl^+ modified by local charge compensation by an O^{2-} at a nearest-neighbor halide site; part of the energy is radiated in a blue band due to Sb^{3+}, and part is transferred to the Mn^{2+} which is responsible for an orange emission band. The emission spectrum of this important phosphor is shown in Fig. 2. The ruby laser involves the luminescence of Cr^{3+} in Al_2O_3. Excitation occurs in a broad absorption band, the system relaxes to another excited state from which emission occurs in a narrow band.

Rare-earth ions in solid solution in inorganic crystals frequently exhibit the emission characteristic of transitions within the 4f shell. For examples, samarium, europium, and terbium give visible emission; neodymium, infrared; and gadolinium, ultraviolet. Because of their narrow emission bands, the rare earth activated phosphors are of interest as lasers and photon counters. Crystal field theory can be used to explain the optical absorption and luminescent emission of the 4f transitions of

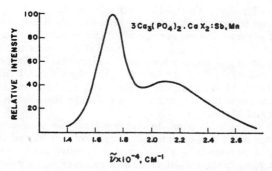

Fig. 2. Emission spectrum of manganese- and antimony-activated calcium halophosphate.

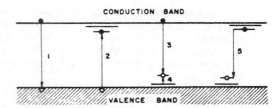

Fig. 3. Energy levels and transistor in semiconducting phosphors.

acceptors and the pairs, emission bands due to transition metals are well-known for zinc sulfide as noted earlier. In zinc sulfide crystals the donors which are unassociated with acceptors serve as electron traps and are responsible for phosphorescence.

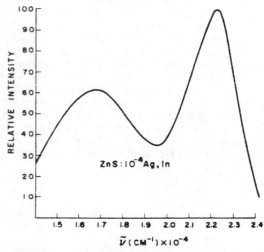

Fig. 4. Emission spectrum of zinc sulfide containing donors and acceptors.

rare earth ions in crystals. The rare earth ion is trivalent in most rare earth laser phosphors. Examples are $Y_2O_3:Eu^{3+}$, $Y_3Al_5O_{12}:Yb^{3+}$ and $CaWO_4:Er^{3+}$. In the last material the trivalent rare earth is charge compensated by an equivalent concentration of monovalent cations such as Na^+. For some materials photo excitation or optical pumping involves transitions within the rare earth, i.e., transitions from the $4f^n$ to the $4f^{n-1}5d$ configurations, and for others, optical absorption by the matrix followed by energy transfer to the rare earth. Sensitized emission may also occur with suitable combinations of rare earths. An example is $CaMoO_4:Er^{3+},Tm^{3+}$ where the Tm^{3+} emission is excited by energy transfer from excited Er^{3+}. In addition to crystals, glasses such as barium crown glass containing Nd^{3+} are capable of coherent luminescent emission and are used as high-power lasers.

Anti-Stokes' emission of visible radiation with infrared excitation occurs with suitable doubly rare earth doped crystals, for example, $GdF_3:Yb^{3+},Tm^{3+}$. Triple, consecutive energy transfer from Yb^{3+} to Tm^{3+} takes place so that the Tm^{3+} is triply excited and emits a single energetic photon in returning to its ground state.

Covalent Phosphors. The zinc sulfide phosphors, which are widely used as cathodoluminescent phosphors and are well-known for their electroluminescence, are large band gap, compound semiconductors. Two impurities or imperfections are essential to the luminescence of many of these phosphors: an activator which determines the emission spectrum, and a coactivator which is essential for the emission but in most cases has no effect on the spectrum. Activator ions such as Cu, Ag and Au substitute at Zn sites and perturb a series of electronic states upward from the valence band edge. In a neutral crystal containing only these activator impurities, the highest state is empty, that is, contains a positive hole, can accept an electron from the valence band, and therefore, in semiconductor notation the activator is an acceptor. In a similar way coactivators such as Ga or In at Zn sites or Cl at S sites are donors in ZnS. The simultaneous introduction of both types of impurities results in electron transfer from donor to acceptor lowering the energy of the crystal and leaving both impurities charged. The coulomb attraction of the donor and acceptor leads to a departure from a random distribution over lattice sites and to pairing. The electronic states, and the band-to-band intrinsic transition and some of the transitions of acceptors, donors and donor-acceptor pairs are shown in Fig. 3. The emission spectrum of ZnS:Ag,In is shown in Fig. 4. The longer wavelength emission band involves the transition from the lowest donor state to highest acceptor state (transition 5) in approximately fifth nearest-neighbor pairs; the shorter wave length emission corresponds more nearly to transition 3 of Fig. 3. In addition to luminescence due to donors,

The luminescence of Group III-Group V semiconductors has been extensively investigated during the past decade. In contrast to the broad gaussian emission of II-VI phosphors such as ZnS, the emission of III-V semiconductors such as GaP may consist of scores of lines. These lines have been identified as donor-acceptor transitions (transition 5 of Fig. 3) in pairs with different donor-acceptor distances. Two types of pair spectra are observed: those arising from donor and acceptor occupying equivalent sites, i.e., Si and S at P sites in GaP, and those arising from donor and acceptor occupying nonequivalent sites, i.e., Zn at Ga sites and S at P sites in GaP. Luminescent emission from the 11th to the 68th nearest-neighbor pairs are observed with theoretically predicted differences in their transition energies. In addition, luminescent emission at isoelectronic dopants (impurities with the same charge as those substituted for) is observed for III-V phosphors, for example GaP:N. The III-V semiconductors can be made either n- or p-type, that is, be doped so that electrical conduction is by electrons in the conduction band or by positive holes (empty electron states) in the valence band and therefore p-n junctions can be formed. With radiative recombination centers such as donor-acceptor pairs in the junction electroluminescence occurs with forward bias. This is a light emitting diode. Both GaP and Ga(P,As) doped with pairs, and possibly isoelectronic dopants as well, are typical light emitting diodes.

Coherent luminescent emission is obtained from appropriate semiconductor phosphors with photo-, cathodo- or electroluminescent excitation. Lasers with electrical injection of charge carriers include In(P,As) and Ga(P,As), whereas cathodoluminescent lasers include CdS, ZnO and ZnS. The latter involve exciton emission (transition 1 of Fig. 3 modified for coupled electron and positive hole). Semiconductor lasers can also be pumped optically.

—G. R. Fonda (Luminescence)

—A. G. Fischer, Pittsburgh, Pennsylvania. (Electroluminescence)

—F. Williams, University of Delaware, Newark, Delaware. (Luminescent Organic Crystals)

LUTETIUM. Chemical element symbol Lu, at. no. 71, at. wt. 174.97, fourteenth element in the lanthanide series in the periodic table, mp 1663°C, bp 3402°C, density, 9.842 g/cm^3 (20°C). Elemental lutetium has a close-packed hexagonal crystal structure at 25°C. Pure metallic lutetium is a silvery-gray color and retains its luster at room temperature indefinitely. Although experimental observations of the element remain limited (workability, alloying behavior, etc.), extrapolations of known data do not forecast any anomalies in the chemical, mechanical, or physical properties of the element as compared with the other elements in the lanthanide series. There are two natural isotopes, ^{175}Lu and ^{176}Lu. The latter isotope is radioactive, with a half-life of 2.2 × 10^{10} years. Fourteen artificial isotopes are known. The element was first identified by G. Urbain in 1907 and independently by C. A. von Welsbach in 1908. Although not investi-gated fully, lutetium is classified with a low acute-toxicity rating. Lutetium is the least abundant of the lanthanide elements, estimated as present on the average of 0.5 ppm in the earth's crust. Potentially, however, it is more plentiful than mercury, cadmium, or any of the precious metals. Electronic configuration is $1s^2 2s^2 2p^6 3s^2 2p^6 3d^{10} 4s^2 4p^6 4d^{10} 4f^{14} 5s^2 5p^6 5d^1 6s^2$. Ionic radius, Lu^{3+}, 0.848 Å; metallic radius, 1.735 Å. Other important physical properties of lutetium are given under **Rare-Earth Elements and Metals.**

The source of lutetium to date has been the processing of the other heavy rare-earth metals. Because of very limited availability, little research was conducted on lutetium until the mid-1960s. Most of these studies now are concentrating on prospective uses in phosphors, semiconductor, and other electronic circuitry components.

See references listed at the ends of the entries on **Chemical Elements;** and **Rare-Earth Elements and Metals.**

NOTE: This 4th Edition entry was revised and updated by K. A. Gschneidner, Jr., Director, and B. Evans, Assistant Chemist, Rare-Earth Information Center, Energy and Mineral Resources Research Center, Iowa State University, Ames, Iowa.

LUTIDINES. Coal Tar and Derivatives; Pyridine and Derivatives.

LYSINE. Amino Acids.

M

MACROMOLECULE. A molecule, usually organic, comprised of an aggregation of hundreds or thousands of atoms. Such giant molecules are in general of two types: (1) Individual entities (compounds) that cannot be subdivided without losing their chemical identity. Typical of these are proteins, many of which have molecular weights running into the millions. (2) Combinations of repeating chemical units (monomers) linked together into chain or network structures called polymers. Each monomer has the same chemical constitution as the polymer, e.g., isoprene (C_5H_8) and polyisoprene ($C_5H_8)_x$. Synthetic elastomers (plastics) are typical of this kind of macromolecule. Cellulose is the most common example found in nature. Most macromolecules are in the colloidal size range. See also **Colloid Systems;** and **Polymerization.**

MAGNESITE. The mineral magnesite is carbonate of magnesium, $MgCO_3$. It is a hexagonal mineral, but usually found massive. It has a rhombohedral cleavage; conchoidal fracture; brittle; hardness, 3.5–4.5; specific gravity, 3.75–4.25; luster, vitreous to dull; color, white, gray, yellow, or brown; transparent to opaque. Most magnesite is believed to have been derived from the action of carbonated waters upon rocks rich in magnesium. Magnesium-bearing waters, on the other hand, may have in some cases acted upon calcite or dolomite. Magnesite deposits are known in Greece, Austria, Norway, India, Australia, and the Republic of South Africa. In the United States, magnesite is found in California and Nevada, some of which deposits seem to be of original sedimentary character. Magnesite is in demand for the manufacture of refractories and various compounds of magnesium.

MAGNESIUM. Chemical element symbol Mg, at. no. 12, at. wt. 24.312, periodic table group 2a (alkaline earths), mp 649°C, bp 1106–1108°C, critical temperature (calculated), 1867°C, density, 1.74 g/cm³ (20°C). Elemental magnesium has a close-packed hexagonal crystal structure, as do the common alloys of magnesium except those that contain lithium in excess of 11%. Magnesium is a silvery-white metal, malleable and ductile when heated; unattacked by dry oxygen, by water or alkalis at room temperature; when heated to about 800°C, magnesium reacts in air or steam and emits a brilliant white light of high actinic power; reactive with acids, including carbonic at room temperature; reactive upon heating with nitrogen, phosphorus, arsenic, sulfur, in some cases with such vigor as to constitute a hazard.

Magnesium occurs extensively in the earth's crust, ranking 8th among the chemical elements in terrestrial abundance. An average composition of igneous rocks contains 2.09% magnesium. Of the elements present in seawater, magnesium ranks 5th, with an estimated 6,125,000 tons of magnesium per cubic mile (1,323,000 metric tons per cubic kilometer) of seawaters. Its content is exceeded only by hydrogen, oxygen, sodium, and chlorine. Magnesium is a constituent of over 150 minerals and also is found in bitterns and subterranean brines and salt beds.

Only a few magnesium minerals are important commercially, notably dolomite $CaO \cdot MgO \cdot 2CO_2$, as a source of magnesium. See also **Dolomite.** More than half of metallic magnesium produced is extracted from seawater. There are three naturally occurring isotopes ^{24}Mg through ^{26}Mg; and three radioactive isotopes have been identified ^{23}Mg, ^{27}Mg, and ^{28}Mg, all with comparatively short half-lives measured in seconds, minutes, or hours. The first known magnesium compound to be isolated was epsom salt $MgSO_4$ which Nehemiah Grew obtained in 1695 by evaporating the mineral waters at Epsom, England. In 1754, Joseph Black demonstrated that magnesia and lime were two different substances, but the exact identity of magnesia was not reported until 1808 by Sir Humphrey Davy who demonstrated that magnesia was an oxide of a heretofore unknown element. He first termed the element, *magnium*. Metallic magnesium was first isolated by A. Bussy in 1828 when he fused magnesium chloride with potassium. Michael Faraday produced the first magnesium metal electrolytically in 1883. First ionization potential 7.64 eV; second, 14.97 eV. Oxidation potential Mg $\rightarrow$ $Mg^{2+} + 2e^-$, 2.375 V; $Mg + 20H^- \rightarrow Mg(OH)_2 + 2e^-$, 2.67 V. Other important physical properties of magnesium are given under **Chemical Elements.**

Production. There are two principal magnesium production processes: (1) electrolytic, and (2) thermal. Electrolytic processes account for 80% of commercial production. In this process, seawater is pumped into large settling tanks where it is treated with lime. Roasted oystershells sometimes are used if a convenient source is nearby. The lime precipitates the magnesium as the insoluble hydroxide. The hydroxide is filtered and then converted into a slurry with fresh H_2O. Subsequent treatment with HCl converts the $Mg(OH)_2$ into $MgCl_2$. The latter compound is dried and then electrolyzed in the fused state to produce molten magnesium and chlorine gas. The latter is recycled. The magnesium is cast into ingots. In the thermal or ferrosilicon process, used in some European countries, a mixture of magnesium oxide and powdered ferrosilicon (an iron-silicon alloy) is fed into a retort and heated under vacuum to about 1,200°C. The magnesium is freed in the form of vapor and condenses into crystals at the cool end of the retort. The crystals then are remelted and cast into pigs.

Uses. Magnesium finds principal uses as a primary metal to which other metals are added in varying amounts, as an alloying ingredient for other metals, as an important metallurgical chemical in the form of a deoxidizer and desulfurizer, and as the constituent of numerous important inorganic and chemical compounds.

In recent years, some technologists have stressed a promising future for magnesium, pointing out that, although magnesium costs more than aluminum per unit weight in the early 1980s, the cost differential may be narrowed considerably or eliminated by the mid-1980s and early 1990s. Aluminum costs will increase because of deficiencies in the world supplies of bauxite, whereas magnesium remains abundant in seawater. Magnesium also has an advantage over aluminum in that it requires only about 71%

of the energy required to produce an equivalent weight of aluminum. Automotive suppliers are continuing their attention of recent years to greater use of magnesium alloy parts. This has been particular true of German car makers, but the weight-savings advantage are now being studied intensively by American and other car makers. For example, it has been estimated that magnesium intake manifolds for a popular American car (V-8 engine) provide a 43-pound (19.5-kilogram) weight savings over iron manifolds. Investigators at the Massachusetts Institute of Technology have calculated that each pound of magnesium substituted for similar steel parts will save about 6.9 gallons of fuel per 100,000 miles (16 liters per 100,000 kilometers) of driving.

Alloys. Lightweight is the key word that has made magnesium alloys so attractive. In many applications where weight reduction is an objective, magnesium alloys are specified. These uses include aircraft and space vehicles, hand trucks, containers, materials-handling equipment, portable electric and pneumatic tools, such as chain saws, hand tools, luggage, sporting goods, dockboards, and tooling jigs and fixtures. It has been found that lighter-weight equipment significantly reduces accidents and lost time due to injuries. Convenience, conservation of power, and safety continue to accentuate the advantages of magnesium alloys for many items of equipment. On an arbitrary scale, where the power required to machine magnesium alloys is 1.0, the figures for other metals are: aluminum alloys, 1.8; brass, 2.3; cast iron, 3.5; mild steel, 6.3; and nickel alloys, 10.0. Some of the principal magnesium alloys are described in the accompanying table.

Magnesium also is an important alloying ingredient in the production of other base metal alloys. When added during metallurgical processing, magnesium in small quantities has a marked effect on final properties of the metals:

Aluminum—Magnesium increases resistance to corrosion, facilitates heat treatment, and increases most mechanical properties. If magnesium-containing aluminum is remelted, the magnesium may be lost and should be replaced by adding pure magnesium to the casting ladle or pot.

Copper—Magnesium improves tensile strength and allows age hardening. Magnesium is used mainly as a deoxidizer, notably in copper-nickel-zinc alloys and in leaded brasses and bronzes. The magnesium is added during melting.

Lead—Magnesium increases hardness, strength, and resistance to creep. Magnesium also is used as a debismuthizer in refining primary lead.

Nickel—Magnesium, in combination with carbon, forms an age-hardenable alloy. The main use of magnesium is to deoxidize and desulfurize the melts, including pure nickel, nickel-chrome, and nickel-copper alloys.

Tin—Magnesium increases hardness and tensile strength. The effect of magnesium on tin can be dramatic. However, too much magnesium will reduce corrosion resistance and ductility.

Zinc—Magnesium improves dimensional stability and reduces the intergranular corrosion of zinc die castings. Magnesium refines the grain and increases hardness and creep strength of zinc sheet. Magnesium also is used in zinc-base bearing metals and in zinc alloy metalworking dies.

Chemistry and Compounds. The behavior of magnesium is intermediate between that of beryllium and the higher alkaline earths. While it reacts readily with halogens, oxygen, and sulfur to form halides, oxide, and sulfide, it reacts with cold water only when the formation of protective oxide is prevented by amalgamation. All its compounds are divalent. Its oxide does not react with water to form the hydroxide, and it does not normally form a peroxide. Its major difference from the higher elements of the group is its much greater number of complexes. Anhydrous magnesium halides, especially, combine easily with many oxygen-functional organic compounds to form addition compounds. These reactions usually suggest covalent or dative bonding (both electrons from oxygen) of the magnesium. Magnesium salts often form ammines and amine complexes, though these are less stable than beryllium complexes. Magnesium also forms some basic salts, and many more of its salts are hydrated than are those of the higher alkaline earths. The metal reacts with alkyl and aryl halides to form the Grignard reagents, through which many organic reactions are conducted. The Grignard reagents themselves form complexes with ethers, tertiary amines, tertiary phosphines, and many other type compounds. See **Grignard Reactions.**

Important compounds of magnesium include the following:

Magnesium acetate, $Mg(C_2H_3O_2)_2 \cdot 4H_2O$, white solid, soluble, formed by reaction of magnesium carbonate and acetic acid.

Magnesium ammonium arsenate, $MgNH_4AsO_4$, white precipitate, solubility 0.0013 molar, formed by reaction of soluble magnesium salt solution and sodium arsenate in the presence of excess ammonium hydroxide, and upon igniting yields magnesium pyroarsenate, $Mg_2As_2O_7$, white solid.

Magnesium borate, $Mg_3(BO_3)_2$, or $Mg(BO_2)_2$, white precipitate, by reaction of soluble magnesium salt solution and sodium borate.

Magnesium boride, Mg_3B_2, brown solid, by reaction of boron oxide and magnesium powder ignited.

Magnesium bromide, $MgBr_2 \cdot 6H_2O$, white solid, soluble, formed by reaction of magnesium carbonate and hydrobromic acid.

Magnesium carbonate, $MgCO_3$, white solid, K_{sp} 4.0 × 10^{-5}, formed by reaction of soluble magnesium salt solution and sodium carbonate or bicarbonate solution. Present in carbonate minerals and rocks, magnesite (more or less pure magnesium carbonate), dolomite (magnesium-calcium carbonate mixtures), dolomitic limestone. When ignited yields magnesium oxide and CO_2; when treated with acids yields the corresponding magnesium salt and CO_2, but with carbonic acid yields soluble magnesium bicarbonate. Magnesium bicarbonate, $Mg(HCO_3)_2$, colorless solution, by reaction of magnesium carbonate and carbonic acid, yields, upon boiling, magnesium carbonate and CO_2; magnesium ammonium carbonate, $(MgCO_3 \cdot NH_4)_2CO_3 \cdot 4H_2O$, white precipitate (soluble in ammonium chloride solution) by reaction of soluble magnesium salt solution and excess ammonium carbonate.

Magnesium chloride, $MgCl_2 \cdot 6H_2O$, white solid, soluble, formed by reaction of magnesium carbonate (or hydroxide, oxide, or metal) and HCl, loses hydrogen chloride when heated, yielding magnesium oxychloride; anhydrous magnesium chloride $MgCl_2$, white solid, soluble, formed (1) by heating hydrated magnesium chloride crystals in a current of dry hydrogen chloride, (2) by heating magnesium ammonium chloride, mp 712°C. Magnesium ammonium chloride, $MgCl_2 \cdot NH_4CL \cdot 6H_2O$, white solid, soluble, when heated yields anhydrous magnesium chloride; magnesium potassium chloride, $MgCl_2 \cdot KCl \cdot 6H_2O$, white solid, soluble, when heated fuses to anhydrous magnesium potassium chloride; magnesium oxychloride, white solid, insoluble, formed (1) by heating hydrated magnesium chloride crystals, (2) by mixing magnesium chloride solution and magnesium oxide.

Magnesium chromate, $MgCrO_4 \cdot 7H_2O$, yellow solid, soluble, formed by reaction of magnesium carbonate and chromic acid solution.

IMPORTANT MAGNESIUM ALLOYS

Alloy Designation	Elements Added	Tensile Strength 1,000 psi	Brinell Hardness	Melting Point °C	Forms Available	Features
AZ31B	3% Al 1% Zn	29	49	627	Sheet, plate, extrusions, forgings.	Moderate strength, good formability, general-purpose alloy. Dent resistant, weldable.
AZ91B	9% Al 0.6% Zn	33	67	596	Die casting alloy.	Good strength and castability. Popular for portable tools, business machines, vehicles.
AZ91C	8.7% Al 0.7% Zn	40	53	596	General-purpose sand and permanent-mold casting alloy.	Good castability, pressure tightness, and weldability. Moderate strength.
HK31A	3% Th 0.7% Zn 0.7% Zr	38	57	649	Sheet and plate for aerospace uses. (200–370°C). Sand and permanent-mold castings.	Good short-time, elevated temperature characteristics. Weldable without stress relief. Low microporosity in cast form.
HM21A	0.6% Mn 2% Th	35	56	650	Sheet, plate, forgings for aerospace uses. (200–425°C)	Very stable at elevated temperatures. Good creep strength and formability. Weldable without stress relief.
HM31A	1.2% Mn (min) 3% Th	44	63	605	Extrusions for aerospace uses. (200–425°C)	Excellent elevated temperature properties. Weldable without stress relief.
QE22A	2% Pr 0.7% Zr 2.5% Ag	40	78	549	Castings for aerospace uses. (up to 260°C)	Superior tensile strength plus excellent creep and fatigue strength.
ZK60A	5.7% Zn 0.5% Zr	47	—	635	Highly stressed parts of aerospace and military uses. Used as a forging alloy.	High strength, good toughness, good spot-weldability. Limited arc-weldability.

NOTE: 1 psi (pounds/square inch) = 0.0069 megapascal.

Magnesium citrate, $Mg_3(C_6H_5O_7)_2 \cdot 4H_2O$, white solid, soluble, formed by reaction of magnesium carbonate and citric acid.

Magnesium fluoride, MgF_2, white precipitate, K_{sp} 6.5×10^{-9}, formed by reaction of soluble magnesium salt solution and sodium fluoride solution.

Magnesium hydroxide, $Mg(OH)_2$, white precipitate. K_{sp} 9.0×10^{-12}, formed by reaction of soluble magnesium salt solution and NaOH solution.

Magnesium hypophosphite, $Mg(H_2PO_2)_2 \cdot 6H_2O$, white solid, soluble, formed by reaction of magnesium carbonate and hypophosphorous acid.

Magnesium iodide, $MgI_2 \cdot 8H_2O$, white solid, soluble, formed (1) by reaction of magnesium carbonate and hydriodic acid, (2) anhydrous, by heating magnesium metal and iodine.

Magnesium lactate, $Mg(C_3H_5O_3)_2 \cdot 3H_2O$, white solid, soluble, formed by reaction of magnesium carbonate and lactic acid.

Magnesium nitrate, $Mg(NO_3)_2 \cdot 6H_2O$, white solid, soluble, formed by reaction of magnesium carbonate and HNO_3.

Magnesium nitride, Mg_3N_2, yellow solid, with moist air or water yields ammonia and magnesium hydroxide, formed by heating magnesium to a high temperature in nitrogen or NH_3 (hydrogen gas evolved).

Magnesium oleate, $Mg(C_{18}H_{33}O_2)_2$, yellow solid, insoluble, formed by reaction of soluble magnesium salt solution and sodium oleate.

Magnesium oxalate, $MgC_2O_4 \cdot 2H_2O$, white solid, insoluble, K_{sp} 8.6×10^{-5}, formed for reaction of soluble magnesium salt solution and ammonium oxalate solution.

Magnesium oxide, MgO, white solid, reacts slowly with H_2O to form magnesium hydroxide, has cubic structure, absorbs CO_2 from the air to form magnesium carbonate, is readily soluble in acids, insoluble in alkalies; formed (1) by heating magnesium carbonate to high temperature (CO_2 gas evolved), (2) by heating magnesium hydroxide, nitrate, sulfate, or oxalate, (3) by burning magnesium metal in air or oxygen.

Magnesium peroxide, MgO_2, white solid, insoluble, formed by reaction of soluble magnesium salt solution and sodium peroxide.

Magnesium ammonium phosphate, $MgNH_4PO_4$, white precipitate, K_{sp} 2.5×10^{-12}, by reaction of soluble magnesium salt solution and sodium phosphate in the presence of excess ammonium hydroxide, upon igniting yields magnesium pyrophosphate, $Mg_2P_2O_7$, white solid.

Magnesium salicylate, $Mg(C_7H_5O_3)_2 \cdot 4H_2O$, white solid, soluble, formed by reaction of magnesium carbonate and salicylic acid in H_2O.

Magnesium sulfate, $MgSO_4 \cdot 7H_2O$, white solid, soluble, formed by reaction of magnesium carbonate and H_2SO_4.

References

Ball, C. J. P.: "The History of Magnesium," *J. Inst. Metals* (London), **84** (Pt. II), 399–411 (1955–1956).

Comstock, H.: "Magnesium and Magnesium Compounds," U.S. Bureau of Mines, Washington, D.C. (Issued periodically).

Emley, E. F.: "Principles of Magnesium Technology," Pergamon, London (1966).

Erickson, S. C.: "Magnesium," in "Metals Handbook," 9th Edition, Vol. 2, American Society for Metals, Metals Park, Ohio (1979).

Mattill, J. I.: "The Coming Age of Magnesium," *Technology Review (MIT)*, **80**, 5, 24–25 (1978).

MAGNESIUM (In Biological Systems). Magnesium is an integral part of the molecule of chlorophyll, the green pigment in plants that absorbs solar energy. See also **Chlorophylls.** Magnesium deficiency is a fairly common cause of poor crop yields, especially among crops produced on sandy soils. Magnesium is a prosthetic ion in enzymes that hydrolyze and transfer phosphate groups. Hence it is essential for energy-requiring biological functions, such as membrane transport, generation and transmission of nerve impulses, contraction of muscles, and oxidative phosphorylation. See also **Phosphorylation (Oxidative).** Magnesium is essential for the maintenance of ribosomal structure and thus protein synthesis. Magnesium may be related to the incidence if ischemic heart disease among Western populations.

The accumulation of magnesium from the soil by plants is strongly affected by the species of plant. The leguminous plants, such as clovers, beans, and peas, usually contain more magnesium than grasses, tomatoes, corn (maize), and other nonleguminous plants, regardless of the level of available magnesium in the soil where they grow.

A very high level of available potassium in the soil interferes with the uptake of magnesium by plants, and magnesium deficiency in plants is often found in soils that are very high in available potassium. High levels of available potassium may occur naturally, especially in soils of subhumid and semiarid regions; or they may be caused by heavy applications of certain commercial fertilizers or animal manure. On sandy and loamy soils, applications of magnesium fertilizers are often effective in increasing crop yields and the concentration of magnesium in the crop, but on fine-textured, clay-containing soils, especially those with substantial reserves of potassium, the application of a magnesium fertilizer may not cause higher magnesium concentration in crops. Since magnesium is not a highly toxic element in either plants or animals, precautions against its overuse are rarely necessary. When animals are fed diets primarily of grains, a proper balance among magnesium, calcium, and phosphorus should be maintained to minimize danger from urinary calculi.

The biological functions of magnesium, such as its essential role as a nutrient, its activation of enzyme systems, and its pharmacological properties, have been widely investigated. Nevertheless, some aspects of its critical physiological role remain obscure.

Distribution in System. Magnesium, primarily an intracellular ion, is distributed among all tissues. It constitutes about 0.05% of the animal body and, of this, 60% occurs in the skeleton and only 1% in extracellular fluids.

Reported serum magnesium values for most species range from 1.0 to 3.5 meq/liter with a mean value of about 2. Between 65 and 80% of the plasma magnesium is ultrafilterable, and most of this exists as the free ion. The nonfilterable portion is reversibly bound to plasma protein. Cerebrospinal fluid contains slightly more than plasma. Interstitial fluid is similar to plasma ultrafiltrate.

The magnesium content of soft tissues varies from 0.06 to 0.13% of dry weight and remains remarkably constant regardless of the magnesium status of the animal. Normally, the intracellular concentration is more than 20 times that of the interstitial fluid and the highest concentration occurs in the cell nucleus. Maintenance of such a large concentration gradient across the cell membrane suggests an active transport mechanism.

The relatively large proportion of magnesium found in the skeleton, which amounts to about 0.6% of dry fat-free bone, serves in part as a body reserve. It occurs largely as Mg^{2+} and $MgOH^+$ ions held by electrostatic attraction to the apatite crystal surface. During deficiency in young animals, 30% or more of bone magnesium can be mobilized for metabolic functions. Calcium ions appear to replace the magnesium which occupied the original adsorption sites.

Metabolism. The rate of absorption from the intestine exerts an important role in magnesium metabolism. Whereas *in vitro* studies show that magnesium absorption is positively correlated with the concentration of magnesium, it does not appear to be a purely passive process. Magnesium absorbed in excess of body needs is excreted primarily by way of the kidney. Urinary excretion is controlled primarily by a filtration-reabsorption mechanism so that magnesium appears in the urine only when glomerular filtration exceeds tubular reabsorption. Acute renal failure is accompanied by hypermagnesemia. In some species, considerable endogenous magnesium is lost by way of the feces, the amount depending upon the magnesium status of the animal and upon other dietary factors, such as the digestibility of the diet. The endogenous fecal magnesium in calves has been estimated at 3.5 milligrams/kilogram of body weight.

In contrast with the metabolism of calcium, no one endocrine gland exerts a primary regulatory function on magnesium. Thyroparathyroidectomy in dogs causes only a temporary lowering of plasma magnesium. Adrenalectomy causes a rise, whereas hyperaldosteronism produces a fall in the plasma level. Administration of deoxycorticosterone or aldosterone to sheep lowers the magnesium concentration in plasma. Magnesium-deficient animals exhibit a higher metabolic rate than normal, and the toxic effect of excess thyroxine is partially overcome by increasing the dietary level of magnesium.

Function. Although magnesium activates isolated enzymes, in most cases an absolute requirement is difficult to establish because the enzymes are partially active without added magnesium. The stimulating effect is not always specific for magnesium. In some cases, manganese or calcium also will activate the system.

Magnesium is particularly concerned with enzyme-catalyzed reactions involving the cleavage of phosphate esters and the transfer of phosphate groups. Magnesium ions activate phosphatases and the phosphorylation reactions involving adenosine triphosphate (ATP). Among the latter group may be mentioned glucokinase, phosphoglucokinase, phosphofructokinase, myokinase, creatine transphosphorylase, arginine transphosphorylase, and flavokinase. It has been suggested that an ATP-Mg complex is the active substrate inasmuch as ATP forms a 1:1 complex with magnesium and maximal activation occurs when the ATP:Magnesium ratio is 1. Alkaline phosphatases, pyrophosphatases and ATPase are activated by magnesium, as are enolase, certain peptidases, and pyruvic oxidase. Since magnesium is tied to ATP utilization, it follows that magnesium plays a role in important metabolic processes, including the synthesis of protein, fat and nucleic acids, and in the trapping and utilization of energy derived from catabolism of carbohydrate and fat.

There is little change in magnesium concentration of soft tissues from deficient animals even at the point of expiration. This does not preclude the possibility that a small component of the cell, such as the nucleus or a cell particulate, is deprived of its critical level, but the dramatic drop in extracellular magnesium suggests that a function outside the cell is of greatest significance. It appears that tetany and convulsions in deficient animals result from a derangement of neuromuscular transmission. Magnesium ion possesses strong pharmacological properties, depressing both the central and peripheral nervous systems. These effects are counteracted by calcium. In the presence of normal calcium levels, a reduction of extracellular magnesium is believed to increase the release of acetylcholine and to decrease the rate of its hydrolysis. Such effects would increase the irritability of the neuromuscular system.

Magnesium generally has not been considered a major factor in bone formation and strength, but recent studies suggest closer attention be given to dietary levels of magnesium in this regard. Because of the close interrelationship with calcium, it is not surprising to see research findings of magnesium interfering with calcium entry into cells of the islets of the pancreas in studies of diabetes. The recognized presence of magnesium as part of numerous enzyme systems has led to observations of the reduction in carbohydrate metabolism associated with a deficiency and to beneficial effects in reducing blood cholesterol and lipids associated with other dietary agents when supplemental magnesium is added to the diet. The relationship to calcium also shows up in a study showing that adding magnesium to the rations of laying hens causes an increase in shell thickness, with a consequent reduction in the number of broken eggs (Davis, 1977).

Pathology of Magnesium Deficiency. Although there are numerous clinical symptoms, two cardinal aspects of pathology have been observed in all species of higher animals. These are hyperirritability and soft tissue calcification. While there are species differences as to the dominating syndrome, this is determined in part by the severity of the deficiency. Metastatic calcification is more likely to occur in a chronic deficiency in which the animal does not succumb at an early age. Hyperirritability, terminating in convulsions and death, has been observed in rat, rabbit, pig, calf, chick, and duck. Magnesium deficiency in humans is characterized by muscle tremors and twitching, often accompanied by delirium and occasionally by convulsions. The guinea pig, calf, dog, and cotton rat are prone to metastatic calcification and develop grossly visible deposits in and around joints, along the muscles of the rib cage, and also in the heart, great vessels, and other critical organs. Most soft tissues show an elevated ash content and marked histopathology. Seelig and Heggtveit (1974) hypothesized that long-term intakes of marginal dietary levels of magnesium may be related to the incidence of ischemic heart disease.

The first clinical symptom of magnesium deficiency is a hypomagnesemia which occurs in cattle and less frequently in sheep and is described by such names as grass tetany, grass staggers, lactination tetany, and wheat pasture poisoning. It is observed most frequently when animals are first grazed on lush grass or wheat pastures. The disease is characterized by irritability, tetany, and convulsions, and all animals have a subnormal plasma magnesium. Symptoms can be relieved by administration of magnesium salts and can be prevented by providing extra magnesium in the diet.

Nutritional Requirements and Dietary Supplementation. As is true of many mineral nutrients, the requirements for magnesium is affected by other dietary constituents, by the age and species of the animal, and by the criterion of adequacy applied. An allowance for magnesium has been included in the Recommended Dietary Allowance since 1968 (Food and Nutrition Board, 1968). Calcium and phosphate have an important effect upon magnesium availability. Either of these ions in excess increases the requirements for magnesium, and their effects are additive. Since calcium is known to compete with magnesium pharmacologically, it is reasonable to believe that it also competes with magnesium for absorption sites in the intestine. It is believed that phosphate decreases magnesium absorption by formation of insoluble magnesium phosphates and excess of calcium aggrevates the effect by creating a more alkaline intestinal medium. Excess magnesium can be considered toxic, but this effect is largely due to the induction of a calcium deficiency. Magnesium deficiency in humans generally has not been fully documented except in cases of predisposing and complicating disease states (Shils, 1976).

J. L. Greger and associates (Purdue University) have pointed out that while there are several compilations of the magnesium content of foodstuffs (Watt and Merrill, 1963; Seelig, 1964; Schlettwein-Gsell/Mommsen-straub, 1973), there is insufficient published information on magnesium content of diets consumed by adolescents who eat a variety of processed and convenience foods. With this in mind, Greger, et al. (1978) studied about 150 commonly consumed foods. The investigators also calculated the ratios of magnesium to calorie content. The researchers generally found the ratios were higher in vegetables than in meat, milk, and cereal products. Distribution of magnesium in food classes is summarized by the accompanying table. The investigators also observed that the distribution of magnesium in foodstuffs is very different than the distribution of zinc (Allen et al., 1977; Haeflien and Rasmussen, 1977; Johnson et al., 1977; Osis et al., 1972).

Commonly used magnesium additives include magnesium phosphate (dibasic and tribasic) and magnesium sulfate.

References

Allaway, W. H.: "The Effect of Soils and Fertilizers on Human and Animal Nutrition," Cornell University Agricultural Experiment Station, *Agriculture Information Bulletin 378*, U.S. Department of Agriculture, Washington, D.C. (1975).

Allen, K. G. D., et al.: "The Zinc and Copper Content of Seeds and Nuts," *Nutr. Rep. Int.*, **26**, 227 (1977).

Davis, G. K.: "Magnesium in Animal Nutrition, *Feedstuffs*, 25–36 (June 21, 1977).

Gregar, J. L., Marhefka, S., and A. H. Geissler: "Magnesium Content of Selected Foods," *Journal of Food Science*, **43**, 5, 1610–1612 (1978).

Haeflein, K. A., and A. I. Rasmussen: "Zinc Content of Selected Foods," *Journal of American Dietet. Assn.*, **70**, 610 (1977).

Johnson, P. E., Straus, C., and G. W. Evans: "Metallocalorie Ratios for Copper, Iron, and Zinc in Fruits and Vegetables," *Nutr. Rept. Int.*, **15**, 469 (1977).

DISTRIBUTION OF MAGNESIUM IN VARIOUS FOOD GROUPS

GROUP (TYPES OF SAMPLES TESTED)	MAGNESIUM CONCENTRATION (Milligrams/100 grams (wet))	MAGNESIUM-TO-CALORIE RATIO (Micrograms/Kilocalorie)
Milk products (cheeses, ice cream, milk, puddings)	6.8–25.7	18–198
Meat and meat alternates (chicken, dried beef, eggs, fish, sausage)	9.8–37.6	20–353
Vegetables (cabbage, carrot, onion, turnip)	6.7–20.6	196–1000
Breads and cereals (buns, cereals, cornbread, crackers, croutons, English muffins, pasta, taco shells)	10.6–126.0	27–325
Bakes desserts (cakes, cookies, doughnuts, pastries, sweet rolls)	4.6–53.2	18–307
Candies	21.8–89.9	63–225

SOURCE: Greger et al. (1978).

Kirchgessner, M., Editor: "Trace Element Metabolism in Man and Animals," Institut für Ernährungsphysiologie, Technische Universität München, Freising-Weihenstephan, Germany (1978).

Schlettwein-Gsell, D., and S. Mommsen-Straub: "Spurenelemente in Lebensmitteln: 7. Magnesium." *Internat. Vit. Nutr. Res.*, **43**(Suppl. 13), 100 (1973).

Seelig, M. S.: "The Requirement of Magnesium by the Normal Adult," *Amer. Jrnl. Clin. Nutr.*, **6**, 342 (1964).

Shils, M. E.: "Magnesium" in "Present Knowledge in Nutrition," page 247, The Nutrition Fndn., New York (1976).

Stewart, A. K., and A. C. Magee: "Effect of Zinc Toxicity on Calcium, Phosphorus and Magnesium Metabolism in Young Rats," *Jrnl. of. Nutrn.*, **82**, 287 (1964).

Underwood, E. J.: "Trace Elements in Human and Animal Nutrition," 4th Edition, Academic, New York (1977).

Watts, B. K., and A. L. Merrill: "Composition of Foods," *Agricultural Handbook* U.S. Department of Agriculture, Washington, D.C. (Revised periodically).

MAGNETITE. The mineral magnetite, ferroferric oxide, Fe_3O_4, is isometric, commonly occuring in octahedrons, dodecahedrons, and massive, granular, and laminated forms. It is brittle with an uneven fracture; cleavage is not distinct, but with pressure an octahedral parting may develop; hardness, 5.5–6.5; specific gravity, 5.18; luster, metallic to dull; color, iron black, streak, black. It is opaque and strongly magnetic; when possessing polarity it is known as lodestone. Important large ore bodies are products of magmatic segregations, with titanium a prominent constituent of such deposits. Magnetite is a common mineral in the igneous rocks, especially those of the ferromagnesian varieties, and is found in many metamorphic types. It is associated with corundum in emery.

In northern Sweden are located what may be the largest magnetic deposits in the world, believed to have been formed by segregation in the magma. Magnetite is also found in Norway, in the Urals, Italy, Switzerland, Australia, and Brazil. In the United States, the Pre-Cambrian rocks of the Adirondacks contain large beds of magnetite, as well as extensive deposits of titaniferous magnetite, and the mineral is found also in Jew Jersey, Arkansas, and Utah. In Canada it is found in Quebec and Ontario. The lodestone or natural magnet is found in Siberia, the Harz Mountains, the Island of Elba, and at Magnet Cove Arkansas. The same magnetite is said to be derived from the district of Magnesia, near Macedonia. There is, however, a fable that it was named for a shepherd, Magnes, whose iron-bound staff and shoes with iron nails stuck to the ground in which magnetite was present.

This mineral is an important ore of iron, 72% being metallic iron. Magnetite sometimes is referred to as magnetic iron ore. See also terms listed under **Mineralogy**; and **Ocean Resources (Mineral)**.

—Elmer B. Rowley, Union College, Schenectady, New York.

MALACHITE. The mineral malachite is a basic carbonate of copper corresponding to the formula $Cu_2(CO_3)(OH)_2$. It is monoclinic, crystals tending to be acicular, but usually found massive. It is a brittle mineral; hardness, 3.5–4; specific gravity, 4.05; vitreous to silky or dull; color, green; streak, green; translucent to opaque. Malachite is an alternation product found associated with other copper-bearing minerals. It is a rather common mineral and is found quite widely distributed. Large quantities have been found in the Ural Mountains; it is also found in Germany, France, England, Zaire, Rhodesia, and Australia. In the United States, beautiful radiated masses of fibrous crystals have been found in Berks County, Pennsylvania, as well as in Tennessee at Ducktown, and in Arizona, Nevada, and Utah. Malachite, besides being an ore of copper, has been used for various ornamental purposes. The word *malachite* is derived from the Greek, meaning a *mallow*, because of its green color.

MALEIC HYDRAZIDE GROWTH INHIBITOR. This compound (1,2-dihydro-3,6-pyridazinedione) is used to inhibit the growth of certain food commodities when in storage, including onion and potato. Maleic hydrazide is also used to promote dormancy in citrus trees as well as increasing protection from frost.

MALTITOL. Sweeteners.

MALTOSE. Carbohydrates.

MANGANESE. Chemical element symbol Mn, at. no. 25, at. wt. 54.9380, periodic table group 7b (transition metals), mp $1244 \pm 3°C$, bp 1962°C, density, 7.3 g/cm³ (solid); 7.21 (single crystal at 20°C). Manganese has a cubic (complex) crystal structure. Manganese is a silvery-white metal, not notably hard (becomes hard upon alloying with carbon), brittle, capable of taking a brilliant polish, but readily oxidized upon heating; reacts with water upon boiling, soluble in dilute acids. Discovered by Scheele in 1774.

In terms of abundance, manganese is present in igneous rocks to an average extent of 0.10% (wt). In terms of cosmic abundance, the estimate by Harold C. Urey (1952), using a base figure of 10,000 for silicon, the figure for manganese is 75. Manganese is estimated as the 34th among the elements in its content in seawater, an estimated 9.5 tons per cubic mile (2 metric tons per cubic kilometer) of seawater. There are eight isotopes of manganese, ^{50}Mn through ^{57}Mn, all radioactive with exception of ^{55}Mn. Half-lives range from a fraction of a second for ^{50}Mn to approximately 140 years for ^{53}Mn. Electronic configuration $1s^2 2s^2 2p^6 3s^2 3p^6 3d^5 4s^2$. Ionic radius Mn^{2+} 0.83Å. Metallic radius 1.365.Å. First ionization potential 7.32 eV; second, 15.7 eV. Oxidation potentials $Mn \rightarrow Mn^{++} + 2e^-$, 1.18 V; $Mn^{2+} + 2H_2O \rightarrow MnO_2 + 4H^+ + 2e^-$, -1.28 V; $Mn^{2+} \rightarrow Mn^{3+} + e^-$, 1.51 V; $Mn^{2+} + 4H_2O \rightarrow MnO_4^- + 8H^+ + 5e^-$, -1.52 V; $MnO_2 + 2H_2O \rightarrow MnO_4^- + 4H^+ + 3e^-$, -0.168 V; $Mn(OH)_2 + OH^- \rightarrow Mn(OH)_3 + e^-$, 0.40 V; $MnO_4^- \rightarrow Mn_4^-$ + e^-, -0.54 V; $MnO_2 + 4OH^- \rightarrow MnO_4^- + 2H_2O + 3e^-$, -0.58 V. Other important physical properties are given under **Chemical Elements.**

Occurrence. The most common manganese ore is pyrolusite, MnO_2. Other commercial ores include braunite, Mn_2O_3; hausmannite, Mn_3O_4; and rhodochrosite, $MnCO_3$. Although not of industrial value, manganese also exists in nature as the silicate, sulfate, sulfite, and tungstate. See also **Pyrolusite; Rhodochrosite.**

Worldwide production of manganese (early 1980s) is estimated at 26.4 million tons (23.8 million metric tons). The U.S.S.R. leads with 34.9% of total production, followed by South Africa (22.7%); Angola (8.7%); Australia (8.3%); Brazil (8.0%); India (6.8%); China (4.2%); Mexico (1.8%); Zaire (0.7%); with a number of other countries accounting for 3.9%. These statistics are from "Metal Statistics 1978" (Fairchild Publications, New York).

Manganese nodules are rocks composed largely of ferromanganese oxides formed by precipitation at the bottom of lakes and the oceans. They range in size from micrometers to meters. Their morphology is highly variable. They contain up to 55% manganese, 35% iron, and 2% (nickel, cobalt, and copper). Manganese nodules were first discovered in the open ocean by Thompson, Murray, and Renard during the *Challenger* expedition (1873–1876). Buchanan reported the occurrence of nodules in the Firth of Clyde, a shallow-water area, and by the end of the century at least five additional occurrences of manganese nodules in shallow marine environments had been discovered. Early workers chemically analyzed about a score of manganese nodules and hypothesized about their mechanism of growth. Two principal concepts emerged: (1) they grow by the slow precipitation of manganese from seawater; and (2) they are formed by the rapid precipitation of manganese released in submarine volcanism.

Until the 1950s, little additional work was done except for some early measurements of manganese nodule growth rates. During recent years, however, there has been a strong revival of interest in manganese nodules, stimulated both by the expansion of oceanographic facilities and the realization of the economic importance of the nodules as ores. It has been found that in large areas of the ocean floor, manganese nodules may be absent. In other areas, they may cover nearly 100% of the area. In all of the Pacific Ocean, nodules have been estimated to cover approximately 10% of the ocean floor. The estimated coverage in the Indian and Atlantic Oceans is less. The local variability in manganese nodule concentration is large. Two ocean bottom photographs only a few meters apart may show very different nodule concentrations. In some locations, the weight concentration of nodules ranges up to 5 g/cm².

Manganese nodules are composed of cryptocrystalline minerals. They are known to consist of three major manganese phases: (1) δ MnO_2 (birnessite); (2) 10-Å manganite; and (3) 7-Å manganite. The first is the most highly oxidized form, and has a chemical composition of about $MnO_{1.9}$. Barnes (1967) examined the depth dependence of the mineralogy in nodules taken from the Pacific and his data indicate that above 3,500 m in depth, the only important manganese phase is δ MnO_2, but below the 3,500 m depth, both 10-Å manganite and 7-Å manganite coexist with the δ MnO_2. The observed phase changes may be pressure induced.

During recent years, the growth rates of manganese nodules have been determined by various methods. Results all indicate that the nodules measured grow at a rate of a few millimeters per million years. This does not exclude the possibility that nodules in certain areas evolve more rapidly, but it appears that most deep-sea nodules grow slowly. There is some belief that the nodules are primarily the result of bacterial fixation of manganese. Other investigators believe that the nodules are formed by inorganic precipitation of metals supersaturated in seawater. There is some experimental and theoretically tenable evidence to support both concepts. Research gathered during the Deep Sea Drilling Project (DSDP) and the International Decade of Ocean Exploration (IDOE) programs (1970s) is currently being analyzed.

Processing. Manganese metal can be obtained from oxide ores by reduction with carbon, aluminum, magnesium, or sodium in an electric furnace. The main form in which manganese is used is *ferromanganese*. This material contains approximately 80% manganese and 20% iron. Ferromanganese is generally made in a blast furnace or an electric-arc furnace. Usually a mixture of ores is used, proportioned to yield the final desired specifications of the alloy. To reduce slag volume, low-silica ores are preferred. It is also desirable to maintain a low phosphorus content in the alloy. The charge to the electric furnace process for making ferromanganese is the manganese ore, coke, and limestone. The loss of metal to the slag is determined by the silica present. Usually about 85% of the metal is recovered. Where high-purity manganese is produced, the ore is first roasted to MnO, then leached with H_2SO_4 to form the sulfate. The solution is then neutralized to precipitate iron and aluminum. Other impurities are removed as the sulfides. Electrolysis of the resulting solution yields a 99.94% pure manganese metal.

The high-purity (electrolytic) manganese is used as a deoxidizing agent and sometimes as a constituent of nonferrous metals where it improves strength, ductility, and hot-rolling properties. Because of their very high temperature thermal coefficient of expansion, manganese-base alloys with 72% manganese (balance is copper and nickel) are used in bimetals for switching applications. Manganese (60–80% Mn) and copper alloys find application because of their vibration-damping properties.

The standard ferromanganese (7% carbon; 74–78% Mn) is used both to produce a manganese alloy steel; or as a deoxidant. As early as 1856, Robert Mushed used *spiegeleisen* (10–23% Mn; 4–5% C) in alloys. Where the carbon content of steel is critical, low-carbon ferromanganese is added. Silicomanganese is used as a blocking agent to stop the reaction of carbon and oxygen in steel. Developed in 1888, Hadfield steel contains about 13% manganese. It finds use where a very hard material is needed and it has the interesting property of increasing its hardness when subject to repeated impacts. In the 200 Series of stainless steels, manganese is replacing nickel in order to achieve more economical austenitic materials.

Manganese Inorganics. A number of chemical processes have been developed to upgrade Mn ore which produces an intermediate Mn compound. These intermediates usually are free of most siliceous matter. Although these processes were designed to convert the compound to an oxide for use in metallurgical applications, the purity of the compounds often renders them suitable for commercial use.

The ammonium carbonate process (developed by Manganese Chemicals Corp.) is the first such upgrading process that has reached commercial application. The high-grade manganese carbonate produced is sold to the chemical industry. The process involves reducing the ore to MnO by roasting with gases rich in CO as the initial step. The calcine is then ground and leached in an aqueous solution containing 18 moles of NH_3 and 3 of CO_2. The resulting product is decomposed to yield $MnCO_3$ and NH_3.

The manganese nitrate process is the second upgrading process which has reached commercial application. The high purity manganese oxides produced are sold to the chemical and ferrite industries. The process involves the reaction between NO_2 and manganese ore to form manganese nitrate solution. The resulting aqueous solution is then thermally decomposed to produce MnO_2 and nitrogen oxides. The nitrogen oxides are recycled to the leaching step, while the MnO_2 is recovered and processed by reduction to Mn_2O_3, Mn_3O_4, or MnO. Processes of lesser importance include the chloride and sulfur oxide processes and bacterial leaching.

Chemistry of Manganese. Manganese has a $3d^54s^2$ electron configuration, and compounds in all oxidation states from 0 to 7+ are known, although those of 1+ and 5+ are uncommon. The reducing power of the manganese atom (Mn → Mn^{2+}, 1.18 V) is less than that of magnesium, although the first and second ionization potentials are closely similar, due to the higher heat of sublimation of manganese. However, manganese is oxidized by the halogens, H^+ or even H_2O to the dipositive state.

Like so many other metals manganese forms compounds with nitrogen, carbon and even oxygen that exhibit unusual valences, or are even of nonstoichiometric character. With nitrogen manganese combines with unusual valence of 5+ to form Mn_3N_5; with carbon it forms Mn_3C, while with free oxygen it forms first MnO, then Mn_3O_4, and finally Mn_2O_3. An exception to this rule is the MnO_2 produced by thermodecomposition of concentrated manganese nitrate solutions where the oxygen-to-manganese ratio is 1.99+.

Manganese (0) compounds are exemplified by the carbonyls discussed below.

Manganese(I) is found chiefly in the few complex ions, such as the hexacyanomanganate(I) ion $[Mn(CN)_6]^{5-}$, produced by vigorous reduction (e.g., by aluminum in alkaline solution) of the corresponding manganese(II) ion $[Mn(CN)_6]^{4-}$, or in isocyanide complexes (formed by reduction of the diiodide with alkyl isocyanides) $[Mn(RNC)_6]^+$ where R is an alkyl radical.

Manganese(II) (manganous) compounds are obtained, as stated above, by action of water, halogens (except fluorine) or acids upon the metal, or by reduction of more highly oxidized compounds in acid solution. Many salts of Mn^{2+} are known, including all four of the common halides, the nitrate, the sulfate, the sulfite, various phosphates, the arsenate, and many salts of organic acids, e.g., the acetate, butyrate, citrate, lactate,

oleate, and tartrate. The manganese(II) compounds are in general relatively resistant to oxidation, due to the stability of the half-filled $3d$ subshell. However, the oxide, MnO, and the hydroxide, $Mn(OH)_2$, are rather easily oxidized by air.

This stability of the Mn(II) state is also reflected in the relatively strong oxidation potential of Mn^{3+} (manganic) ion (the value for Mn^{3+}/Mn^{2+} being −1.51 V), and the readiness with which Mn(III) compounds disproportionate. Manganese(III) fluoride, produced by the action of fluorine on lower compounds, reacts with H_2O to produce the difluoride, hydrogen fluoride, and MnO_2. In general, however, the manganic compounds such as dimanganese trisulfate and manganese triacetate, decompose in H_2O to divalent manganese ions and Mn_2O_3, forming the MnO_2 only if the pH is definitely below 7. The phosphate, $MnPO_4$, is easily formed by action of nitric acid on manganese(II) phosphate in concentrated phosphoric acid. The fluoro salt K_3MnF_6 is formed by reduction of potassium permanganate, $KMnO_4$, in 40% hydrofluoric acid by an excess of diethyl ether or manganese(II) salt, Manganese(III) also forms a variety of complexes with chelating agents, e.g., oxalate, glycine, acetylacetone, and the like. Like other tripositive transition metal ions it forms alums. The cyanide $K_3Mn(CN)_6$ is stable. All Mn(III) compounds undergo hydrolysis except the complexes.

In addition to the dioxide and the manganites, formed by fusion of MnO_2 with alkali, manganese(IV) forms a number of complexes, such as K_2MnF_6 by reduction of potassium permanganate in 40% hydrofluoric acid with a limited amount of diethyl ether or manganese(II) salts; and Cs_2MnCl_6, by action of cold concentrated HCl containing cesium chloride on MnO_2. Complex iodates are known, e.g., $M^I_{\frac{1}{2}}[Mn(IO_3)_6]$, as are cyanides, formed by the action of potassium cyanide on potassium permanganate and said to be $K_4Mn(CN)_8$ (cf. $K_4Mo(CN)_8$ and $K_4Re(CN)_8$).

Manganese rarely occurs with an oxidation number of 5+. In addition to the nitride, there is another compound which is of interest in that it can be formed in solution. It is an oxyanion of pentavalent manganese, MnO_4^{3-}, which occurs in the compound, $Na_3MnO_4 \cdot 7H_2O$, formed by reduction of the manganate in strongly alkaline formate or sulfite solutions or by heating MnO_2 in alkali hydroxide at very high temperature. Upon neutralization, it disproportionates to the manganate (and MnO_2).

The manganates, containing MnO_4^{2-}, and produced by alkaline oxidation of MnO_2, are the principal known compounds of hexavalent manganese. They are unstable in neutral or acidic solution, undergoing disproportionation to permanganate (MnO_4^-) and MnO_2. In basic solution, the reaction is reversible. The equilibrium is displaced toward the MnO_4^- by the action of strong oxidants.

The permanganates are strong oxidizing agents, and are usually reduced down to Mn^{2+} under acidic conditions, but to MnO_2, manganate, MnO_4^{2-}, or even hypomanganate, MnO_4^{3-}, under progressively more alkaline conditions. Permanganic acid, $HMnO_4$, and its anhydride, Mn_2O_7, can be obtained at lower temperatures, but are unstable, decomposing above 0°C. Permanganyl fluoride, MnO_3F, formed by the action of liquid hydrogen fluoride on potassium permanganate, decomposes above 0°C. In strongly acidic media, such as 100% H_2SO_4, manganese(VII) appears to exist as permanganyl ion, MnO_3^+. The sigma bond hybridization in MnO_4^-, MnO_4^{2-} and MnO_4^{3-} is best represented as d^3s.

The complexes of manganese(I) have already been mentioned. Manganese(II) forms a cyano complex ion, $Mn(CN)_6^{4-}$, which undergoes reduction to the Mn(I) state as well as ready oxidation, even with atmospheric oxygen, to the Mn(III) state, especially in acid solution. Mn(III) forms a number of complexes with acid anions, including oxalate, $Mn(C_2O_4)_3^{3-}$ and the acid pyrophosphate, $Mn(H_2P_2O_7)_3^{3-}$, as well as a chlorocomplex, MnC_4^-. The known halogen-complexes, i.e., the complex fluoroion, MnF^{2-} and the chloroion, MnC_6^-, are among the few occurrences of tetravalent manganese, other than the dioxide.

The only compound of manganese with carbon monoxide alone is the decacarbonyl dimanganese, $(CO)_5MnMn(CO)_5$, but several hydrogen-containing carbonyls, such as $HMn(CO)_5$, halogen-containing carbonyls, such as $Mn(CO)_5Br$, alkyl carbonyls, such as $C_2H_5Mn(CO)_5$ and oxygen-function organometallic compounds, such as

$$[CH_3C(=\!\!=O)O]_3Mn$$

are known. With the exception of the dicyclopentadienyl compounds, $C_5H_5MnC_5H_5$ manganese does not combine with unsubstituted hydrocarbons or their radicals.

It is interesting to note that some bacteria found near manganese ore plants have the ability to dissolve manganese oxides in solutions of pH 5–6 by the slow addition of H_2SO_4. The only requirement other than the organisms is a nutrient solution. The extraction of manganese as a sulfate is on the order of 71.7–99%, depending on the ore. The action of the bacteria is not fully understood.

References

Bender, L., Ku, T. L., and W. S. Broecker: "Manganese Nodules: Their Evolution," *Science*, **151**, 325–328 (1966).
Bonatti, E., and Y. R. Nayudu: "The Origin of Manganese Nodules on the Ocean Floor," *Am. J. Sci.*, **263**, 17–39 (1965).
Staff: "Manganese," in "Metals Handbook," 9th Edition, Vol. 2, American Society for Metals, Metals Park, Ohio (1979).
Turner, S., and P. R. Buseck: "Manganese Oxide Tunnel Structures and Their Intergrowths," *Science*, **203**, 456–458 (1979).
Turner, S., and P. R. Buseck: "Todorokites: A New Family of Naturally Occurring Manganese Oxides," *Science*, **212**, 1024–1027 (1981).

—J. Y. Welsh and D. F. DeCraene, Chemetals Corporation, Baltimore, Maryland.

MANGANITE. The mineral manganite is a hydrous oxide of manganese corresponding to the formula $MnO(OH)$, it occurs in prismatic monoclinic crystals, sometimes in massive columnar forms, granular, concretionary, and stalactitic. It is a brittle mineral, perfect prismatic cleavage; hardness, 4; specific gravity, 4.33; luster, submetallic; color, steel gray to iron black; streak, red-brown to almost black; opaque. Manganite is of secondary origin and it may itself alter to pyrolusite. It is usually associated with other manganese minerals. It is found in the Harz Mountains, Germany; Sweden; Cornwall and Cumberland, England; and in the United States in Michigan. It is and ore of manganese.

See also **Pyrolusite.**

MANNANS. Sweeteners.

MANNITOL. Sweeteners.

MARCASITE. The mineral marcasite, sometimes called white iron pyrites, is, like ordinary pyrites, disulfide of iron corresponding to the same formula, FeS_2. Marcasite, however, crystallizes in the orthorhombic system often yielding serrate, spear-shaped twins, hence the name "cockscomb pyrites." It is a brittle mineral; hardness, 6–6.5; specific gravity, 4.92; luster, metallic; color, light bronze-yellow; streak, greenish-black; opaque. Marcasite alters very easily and may disintegrate with the formation of sulfuric acid and iron sulfate. Fossils replaced by marcasite are therefore often destroyed after being placed in collections.

Marcasite is found in numerous places in Europe, notably in Czechoslovakia, France, and England; in Mexico, and in the United States in the lead districts of Illinois, Wisconsin, and Missouri. The name *marcasite* is believed to be of Arabic origin and formerly was applied to common pyrite.

MASS NUMBER. The total number of nucleons in the nucleus of an atomic species is its mass number, which then is numerically equal to the sum of the atomic number and the neutron number of the species. See also **Chemical Elements.**

MASTICATORY SUBSTANCES. The property of chewiness is one of several components that comprise so-called mouth feel experienced by the consumer of a food product. Whereas chewiness may be highly undesirable in a cut of roast beef, this property is the predominant rewarding factor of some food products, notably certain kinds of novelties, such as chewing gum. Biting and deformation resistance can be created or improved by the use of a number of essentially rubberlike substances. These are commonly termed *masticatory substances.* Chewiness is the main advantage contributed by such substances and thus other ingredients, such as sweeteners, flavorings, colorants, etc. are admixed with them to result in an overall attractive product for particular consumers.

Masticatory substances are of (1) vegetable origin, or (2) the products of organic synthesis. In the case of anhydrous lanolin, the source is fat from the wool of sheep.

Vegetable Substances. Masticatory substances derived from vegetables are gums from various plants and trees of the families *Apocynaceae* (dogbane family), *Euphorbiacea* (spurge family), *Moraceae* (mulberry family), and *Sapotaceae* (sapodilla family). Most of the naturally derived substances have unfamiliar names, usually known well only by persons in the trade. For example, from the *Apocynaceae* family are obtained jelutong, leche caspi or sorva, pendare, and perillo. From the *Euphorbiaceae* family, there are candelilla wax, chilte, and natural rubber (latex solids). From the *Moraceae* family, there are Leche de vaca, Niger gutta, and tunu or tuno. From the *Sapotaceae* family, there are chicle, chiquibul, crown gum, gutta hang kang, gutta katiau, massaranduba balata, nispero, rosidinha, and Venezuelan chicle. Other naturally derived substances include lanolin, petroleum wax, rice bran wax, and natural terpene resin.

Synthetic Substances. During the last several decades, naturally derived masticatory substances have been displaced to a considerable degree by synthetic materials—for reasons of availability, economics, and, frequently, better control over purity. These developments essentially paralleled the development of the synthetic rubbers for industrial uses. Some of the synthetic substances now used and listed in the "Food Chemicals Codex," published by the National Academy of Sciences (Washington, D.C.) include: butadiene-styrene 75/25 rubber; butadiene-styrene 50/50 rubber; glycerol ester of partially dimerized rosin; glycerol ester of partially hydrogenated wood rosin; glycerol ester of tall oil rosin; isobutylene-isoprene copolymer (butyl rubber); methyl ester of rosin (partially hydrogenated); paraffin (synthetic by Fischer-Tropsch process); pentaerythritol ester of partially hydrogenated wood rosin; polyethylene; polyisobutylene, polyvinyl acetate; and terpene resin (synthetic).

MEERWEIN-PONNDORF-VERLEY REDUCTION. Organic Chemistry.

MEGALOBLASTS. Blood.

MELAMINE. $(N\equiv\!\!\!C\!-\!NH_2)_3$, formula weight 126.12 white solid, mp 355°C, sp gr 1.56. The compound may be considered the trimer of cyanamide, or as the triamide of cyanuric acid. Melamine resembles an amide more than an amine. Liebig first prepared melamine in 1834. In early production methods, melamine was prepared from calcium cyanamide through conversion to the cyanodiamide and then to the trimer, melamine. The compound now is synthesized from urea. The production of melamine exceeded 450 metric tons annually in the early 1970s and has been growing at a rate of about 5% annually. Most of the melamine made is condensed with formaldehyde or other aldehydes to form resins. These resins possess particularly outstanding resistance to heat, water, and many chemicals. The electrical properties and surface hardness also are rated high. The consumption of melamine for these resins is: (1) protective and decorative laminates, 45%; (2) molding compounds, 30%; (3) textile resins, 9%; (4) coatings, 7%; (5) paper-treating resins and various adhesives, 9%.

Typically, in a modern synthesis process, (1) urea is thermally decomposed into a gas mixture of cyanic acid and NH_3: $H_2N \cdot CO \cdot NH_2 \rightarrow HCNO + NH_3$; (2) cyanic gas is thermally decomposed into a melamine-CO_2 vapor:

$$6HCNO \rightarrow (N \equiv C - NH_2)_3 + 3CO_2.$$

Step 1 is endothermic; step 2 is exothermic; the overall reaction is endothermic. Because of the large quantities of CO_2 and HN_3 generated, the process often is undertaken in connection with urea manufacture which permits the off-gases to be recycled usefully. The melamine synthesis may be carried out at low or medium pressures with the assistance of a catalyst; or at higher pressures without a catalyst.

MELTING POINTS (Chemical Elements). Chemical Elements.

MEMBRANE (Semipermeable). Semipermeable Membrane.

MENDELEVIUM.
Chemical element symbol Md, at. no. 101, at. wt. 256 (mass number of known isotope), radioactive metal of the actinide series; also one of the transuranium elements. The element was produced synthetically and first identified by A. Ghiorso, B. G. Harvey, G. R. Choppin, S. G. Thompson, and G. T. Seaborg at the University of California at Berkeley in 1955. The isotope ^{256}Md was produced by the bombardment of ^{253}Es on gold foil with 48 MeV alpha particles in the 60-in. (152.4-cm) cyclotron at Berkeley. By ion exchange treatment of the dissolved gold foil, only one or two atoms of ^{256}Md were obtained, which decayed ($t_{1/2} = 1.3$ hr) by K-electron capture to ^{256}Fm, which underwent its characteristic spontaneous fission.

Probable electronic configuration $1s^2 2s^2 2p^6 3s^2 3p^6 3d^{10} 4s^2 4p^6 4d^{10} 4f^{14} 5s^2 5p^6 5d^{10} 5f^{13} 6s^2 6p^6 7s^2$. Ionic radius Md^{3+} 0.96Å.

Another isotope, ^{255}Md, is also formed during the bombardment of ^{253}Es by alpha particles. It also decays by electron capture, and has a half-life of 30 minutes.

Regarding the first identification, scientists considered it notable in that only in the order of 1–3 atoms per experiment were produced, thus making Md the first to be discovered on an atom-at-a-time basis. The techniques developed in the search for Md served as a prototype for the discovery of subsequent transuranium elements.

References

Ghiorso, A., Harvey, B. G., Choppin, G. R., Thompson, S. G., and G. T. Seaborg: "New Element Mendelevium, Atomic Number 101," *Phys. Rev.*, **98**, 1518–1519 (1955).

Hulet, E. K., et al.: "Mendelevium: Divalency and Other Chemical Properties," *Science* **158**, 486–488 (1967).

Seaborg, G. T., Editor: "Transuranium Elements," Dowden, Hutchinson & Ross, Stroudsburg, Pennsylvania (1978).

MERCAPTANS.
Hydrogen sulfide yields two classes of organic compounds: (1) hydrosulfides, and (2) sulfides. The hydrosulfides are termed *mercaptans*, a name derived from the Latin phrase *mercurium captans*, because of their ability to react with mercuric oxide to form crystalline compounds. Mercaptans also are termed *thioalcohols* and *sulfur-alcohols*. The more general term *thiols* also is used. This term not only embraces mercaptans, but also covers thioethers, sulfhydrates, and thiophenols.

Ethyl mercaptan C_2H_5SH, one of the better known mercaptans, is an odorous liquid, mp $-121°C$, bp $36–37°C$, sp gr 0.839. The compound is very slightly soluble in H_2O; soluble in alcohol and ether. It is prepared by distilling ethyl potassium sulfate with potassium hydrogen sulfide. Additional mercaptans can be prepared in a similar manner with the corresponding proper ingredients. All mercaptans have unpleasant garlic-type odors; when oxidized with HNO_3, they yield sulfonic acids.

Some formulations for styrene-butadiene rubber (GR-S) contain dodecyl mercaptan which plays the role of a chain-transfer agent used to control the molecular weight of the final synthetic product.

MERCURY.
Chemical element symbol Hg, at. no. 80, at. wt. 200.59, periodic table group 2b, mp $-38.87°C$, bp $356.58°C$, density, 13.546 g/cm^3 (liquid); 14.193 g/cm^3 (solid). Solid mercury has a rhombohedral crystal structure. The element, sometimes referred to as *quicksilver*, is a silvery-white liquid metal at standard conditions. There are seven stable isotopes of mercury, ^{196}Hg, ^{198}Hg through ^{202}Hg, and ^{204}Hg; and seven radioactive isotopes, ^{192}Hg through ^{195}Hg, ^{197}Hg, ^{203}Hg, and ^{205}Hg. With exception of ^{194}Hg ($t_{1/2} =$ approximately 130 days) and ^{203}Hg ($t_{1/2} =$ approximately 46 days), the half-lives of the radioactive isotopes are short, measured in terms of minutes or hours.

First ionization potential 10.434 eV; second, 18.65 eV; third, 34.3 eV. Oxidation potentials $Hg \rightarrow \frac{1}{2}Hg_2^{2+} + e^-$, -0.7986 V; $Hg \rightarrow Hg^{2+} + 2e^-$, -0.852 V; $Hg_2^{2+} \rightarrow 2Hg^{2+} + 2e^-$, -0.905 V. $Hg + 2OH^- \rightarrow HgO + 2H_2O + 2e^-$, -0.098 V; $2Hg + 2OH^- \rightarrow Hg_2O + H_2O + 2e^-$, -0.123 V. Other important physical properties of mercury are given under **Chemical Elements.**

Mercury forms alloys, called amalgams, with most metals, but not with iron or platinum; does not wet glass but forms a convex surface when in a glass container; is slightly volatile at ordinary temperatures and a health hazard due to its poisonous effect; slowly tarnishes in moist air; upon heating in air or oxygen, somewhat below its boiling temperature of 357°C, forms mercuric oxide slowly, as in the classical experiment by Lavoisier on the composition of air; may be purified by distillation and condensation (health hazard); unattacked by dilute HCl or H_2SO_4, but dissolved by dilute or concentrated HNO_3 with the formation of mercurous and mercuric nitrates, respectively, and by hot concentrated H_2SO_4 with the formation of both mercurous and mercuric sulfates; unattacked by alkalis. Discovery ancient.

Mercury was mined as early as 500 B.C. and currently ranks tenth in worldwide production of nonferrous metals. The unusual combination of physical properties possessed by the element give it an importance exceeding its production rating. The chief source of mercury is cinnabar HgS, the red sulfide which contains 86.2% mercury. See also **Cinnabar.** Although the mineral occurs widely throughout the world, relatively few deposits are of commercial importance, notably those in China, Italy, Mexico, the Philippines, Peru, Spain, the Soviet Union, Yugoslavia, and the United States. World reserves of mercury currently are estimated at 4 million flasks, of which sources in the United States account for about 300,000 flasks. A flask contains 76 pounds (34.5 kilograms) of the liquid metal.

The ore is concentrated to about 25–50% mercury by flotation. Beneficiation of mercury ores is not commonly practiced. The concentrate is roasted: $HgS + O_2 \rightarrow Hg + SO_2$. The process is essentially one of distillation because the freed mercury quickly volatilizes, after which it is condensed with a resulting purity of 95% (furnace plans) to 98% (retort plants). The mercury is further refined, through filtering, oxidation, acid leaching of the impurities, or by distillation, to yield prime or virgin mercury with an average purity of 99.9%. This purity is satisfactory for all but the most exacting requirements. Some special mercury chemicals may require that the virgin mercury be triple distilled. Significant quantities of secondary mercury are recovered from waste products, such as dental amalgams, sludges, used batteries, used instruments, and other mercury-bearing materials. Secondary recovery accounts for close to 20% of the total domestic production of mercury in the United States.

Mercury production is dominated by Italy and Spain. About one-half of the mercury imported into the United States is from Spain; 20% from Italy; 17% from Mexico 10% from Yugoslavia, with the remainder made up by imports from Japan and the Philippines. The production and price vacillate widely.

Uses. Total consumption of mercury in the United States has been averaging between 65,000 and 85,000 flasks annually. The uses and percentages of consumption include: (1) electrical apparatus, 28%; (2) electrolytic preparation of chlorine and caustic soda (mercury cells), 20%; (3) antifouling and mildew proofing paints and agents, 18%; (4) industrial thermometers and electric control instruments and switches, 8%; (5) pharmaceutical preparations, 8%; (6) agricultural herbicides, pesticides, and other formulations, 7%; (7) dental preparations, 3%; (8) general laboratory uses, 2%; (9) catalysts, 2%; (10) paper and pulp manufacture, 1.5%; and (11) preparation of amalgams, 1.5%. Miscellaneous applications account for the remaining 1.5%.

The consumption of mercury by the chlor-alkali industry may range from 15% to as high as 35% of the total in any given year, depending upon new construction. Large amounts of mercury are required for the start-up of mercury cell operations, whereas replacement requirements are quite low. Several areas of mercury use are declining gradually, particularly in the pharmaceutical field where sulfa drugs, iodine, and various antiseptics and disinfectants have made inroads on mercury chemicals. Mercury compounds, used for many years in the treatment of syphilis, for example, largely have been displaced by antibiotics and other treatments. Because of the fundamentally toxic nature of mercury and its compounds, the agricultural uses of mercury formulations are being de-emphasized, with constant research for substitute materials. In the dental field, a number of metal powders, porcelain, and plastic materials have displaced mercury amalgams in many dental applications. In the explosives field, several compounds, such as lead azide, diazodinitrophenol, and other organic initiators, are serving the same function as mercury fulminate. The use of mercury as a heat-transfer medium in boilers was essentially abandoned a number of years ago. On the other hand, mercury-base catalysts are increasing in application and, to date, suitable substitutes for mercury in the antifouling and mildew proofing formulations area have not been found. The essential properties of mercury which will be difficult to replace are its high specific gravity, fluidity at room temperatures, and excellent electrical conductivity.

In addition to some diminishment of mercury usage for various products because of increasing awareness of its toxicity potential, conservation-minded technologists also have pointed to the relatively limited world resources for the metal. Considerable ingenuity has been used to replace mercury. For example diaphragm cells can be used in caustic-chlorine production, organic biocides are replacing mercury-containing compounds, gold recovery by the cyanide process versus the amalgamation process, and plastic paints and copper oxide paints can be used in place of biocidal paints. The use of the diaphragm cell for chlorine-alkali production possibly is the most dramatic substitution in terms of mercury conservation. Diaphragm cells require no mercury, whereas the traditional mercury cells once accounted for as much as 35% of the mercury use in some years (when new cells were put on stream).

Chemistry and Compounds

The apparent anomaly between mercury and the lighter elements of transition group 2, in that mercury regularly forms both univalent and divalent compounds, while zinc and cadmium do so very rarely, is partly understood from the observation that mercury(I) salts ionize even in the gaseous state to Hg_2^{2+}, rather than Hg^+. Evidence for this double ion is provided by its Raman spectral line, by the linear Cl—Hg—Hg—Cl units in crystals of mercury(I) chloride, and by the emf of mercury(I) nitrate concentration cells. The anomaly is further removed by the observation that cadmium also forms a (much less stable) diatomic ion Cd_2^{2+}, e.g., in $Cd_2(AlCl_4)_2$.

Oxides. Heating of mercury in air yields the (divalent) oxide HgO, which at higher temperatures decomposes into its elements. Mercury(II) oxide is also precipitated from solutions of mercury(II) salts by alkaline solutions. Alkalies precipitate a yellow form, while alkali carbonates give a red one. The yellow is apparently a finely divided form of the red, since they are crystallographically identical, but differ slightly in certain chemical and physical properties, including solubility. Mercury(II) oxide exhibits solubility in solutions of alkali salts, which is attributed to formation of complex ions such as $[Hg(OH)_2NO_3]^-$ and $[Hg(OH)_2SO_4]^{2-}$.

Halogen Compounds. All eight compounds of univalent and divalent mercury with the single halogens are known, as well as several compounds of mercury(II) with two halogens, such as HgBrI and HgClI. The mercury(I) halides are insoluble in water, with the exception of the fluoride which, like mercury(II) fluoride, is hydrolyzed by water. Like the zinc and cadmium halides, mercury(II) halides behave anomalously in aqueous solution, and for similar reasons, i.e., the presence of complex ions and unionized molecules. In the case of mercury(II) halides, with their more covalent character than zinc or cadmium halides, the ionization is somewhat less, and the concentration of Hg^{2+} relatively low. Thus, in aqueous solution, mercury(II) chloride, $HgCl_2$ is present largely as unionized molecules, but also ionizes to $HgCl^+$ and Cl^-, and only secondarily and to a slight extent to give Hg^{2+}. In the presence of added Cl^-, an $HgCl_2$ solution is a complex system involving equilibria between $HgCl_2$, $HgCl^+$, Cl^-, Hg^{2+}, and the complex ions $HgCl_3^-$ and $HgCl_4^{2-}$. Similarly, the hydrolysis of $HgCl_2$, though slight, involves several equilibria whose relative importance varies with the concentration of the solution. In more concentrated solutions the hydrolysis of $HgCl_2$ to HgOHCl, Cl^- and H^+ is prominent, while in more dilute solutions the most important equilibria involve the ionization of $HgCl_2$ to $HgCl^+$ and Cl^-, and the hydrolysis of $HgCl_2$ to $[HgOHCl_2]^{2-}$ and H^+, and that of $HgCl^+$ to HgOHCl or $[HgOHCl]_2$ and H^+. Finally, oxyhalides, such as $HgBr_2 \cdot 3HgO$, $HgCl_2 \cdot 2HgO$, $HgCl_2 \cdot 3HgO$ and $HgCl_2 \cdot 4HgO$ are also obtainable, usually by action of alkali hydroxides upon mercury halides. Mercury(II) iodide, like the oxide, is polymorphic. It has three forms, yellow, red, and white, the second being the most stable up to $127°C$, where it undergoes a definite transition to the yellow. The colorless HgI_4^{2-} is very stable, especially to alkalies, and is used in Nessler's reagent.

Salts. Mercury forms many salts, both of mercury(I) and mercury(II). In general, action of oxidizing acids upon the metal yields the latter, while the former requires either a limited amount of the oxidant or indirect methods. Mercury(I) salts are made by treating a solution of a soluble mercury(II) salt with metallic mercury. Thus, heating mercury with H_2SO_4 or HNO_3 yields mercury(II) sulfate or nitrate, $HgSO_4$ or $Hg(NO_3)_2$, respectively, crystallizing as hydrates; while mercury(I) nitrate results from the use of cold acid in limited amount and mercury(I) sulfate is produced by the last method as well as from mercury(I) nitrate and sulfuric acid. The other salts of both univalent and divalent mercury include the acetates, antimonates, arsenates, bromates, carbonates, chlorates, chromates, fluorosilicates, iodates, oxalates, perchlorates, periodates, phosphates, tartrates, thiocynates, tungstates, uranates, and vanadates. Also known in both valences are the arsenides (from arsine and the mercury solutions), azides (from hydrozoic acid and the mercury solutions), nitrides and phosphides. Only mercury(II) selenide and telluride exist. Hg_2S_2 has been reported to be obtained as a black powder, but is believed to be a mixture of Hg and HgS. The latter exists in two forms, the black form that is usually precipitated by hydrogen sulfide, and the red, cinnabar, precipitated by H_2S from a solution of mercury(II) acetate and ammonium thiocyanate. The black changes to the red in liquid H_2S, and the red to the black on heating to $386°C$. Cinnabar is the thermodynamically stable form at room temperature.

Compounds with Nitrogen and Sulfur. The reactions of mercury(I) compounds and NH_3 are complex, and published results vary. Recent (x-ray) studies show that this reaction, modified by the presence of ammonium chloride, NH_4Cl, yields three ammonobasic compounds containing divalent mercury; $Hg_2NCl \cdot H_2O$, $HgNH_2Cl$, and $Hg(NH_3)_2Cl_2$. The first of these is the chloride of Millon's base, $Hg_2NOH \cdot 2H_2O$, which is produced by warming HgO with aqueous ammonia.

Mercury is the least active of the elements of its group as an electron acceptor from oxygen; however, with sulfur it is more active, the mercury halides forming dialkyl sulfide addition products, $R_2S \cdot 2HgX_2$, and HgS dissolves in alkali sulfides forming $[HgS_2]^{2-}$ or $Hg(SH)_4^{2-}$. Like zinc and cadmium, mercury forms a series of ammines, which with the mercury halides are principally the diammines, $[Hg(NH_3)_2]X_2$, where X is a covalently-bonded halogen atom, and with more ionic mercury compounds, e.g., the nitrate and sulfate, especially in the presence of high concentrations of ammonium salts, the tetrammines, e.g., $[Hg(NH_3)_4](NO_3)_2$. These complexes also form with amines and diamines, e.g., ethylenediamine, which contributes three molecules per Hg^{2+} ion. Besides forming univalent and divalent cyanides, mercury forms complex cyanide ions, $[Hg(CN)_3]^-$ and $[Hg(CN)_4]^{2-}$, as well as addition compounds with the mercury(II) halides of the structure HgCNX. Mercury forms insoluble thiocyanates and complex ions, e.g., $Hg(SCN)_3^-$, $Hg(SCN)_4^{2-}$. Mercury cyanates and fulminates are insoluble. Mercury(II) mercaptides, $Hg(SR)_2$, decompose on heating to give HgS and R_2S. Mercury(II) hydrogen sulfite, $Hg(HSO_3)_2$, is actually mercuridisulfonic acid, $Hg(SO_3H)_2$, with Hg—S bonds. Like the halides, compounds such as $Hg[C(NO_2)_3]_2$, $Hg[C(CN)_3]_2$, $Hg(NO_2)_2$ (dinitromercury), and $Hg(CF_3)_2$ are noteworthy for their lack of ionic dissociation.

Organometallic Compounds. Mercury also forms a large number of organometallic compounds of the type HgR_2, where R may be not only an alkyl radical, but an alkoxy radical, an acyl radical, a halogenated alkyl radical, an alkylthio radical, an aryl radical or a perfluoroalkyl radical. In addition, mercury also forms numerous organometallic compounds of structure RHgX, where X is a halogen atom, and R one of the foregoing organic radicals.

The organic compounds containing mercury which are used as disinfectants, germicides, and antiseptics are known as mercurials. Among these are Merthiolate, Mercurochrome, and Metaphen. Merthiolate is the sodium salt of ethylmercurithiosalicylic acid, $C_2H_5HgS \cdot C_6H_4 \cdot COONa$. It contains 49.5% of mercury. It is a crystalline, cream-colored powder which is very soluble in water for about 1 gram dissolves in 1 milliliter of water. It is much less soluble in alcohol, 1 gram in 8 milliliters. It is insoluble in organic solvents like benzene and ether. It is used as an antiseptic for tissues in concentrations of the order of $1:1,000$ to $1:30,000$. It is commonly used as an antiseptic in biologics.

Mercurochrome, also known by the name of *Merbromin* and by many other trade names is the disodium salt of 2,7-dibromo-4-hydroxymercurifluorescein, $C_{20}H_8Br_2HgNa_2O_6$. It forms green, iridescent scales or granules which are freely soluble in water yielding a bright red solution with dilute solutions having a yellow-green fluorescence. It is generally used in a 2% aqueous solution as a mild antiseptic. It is nearly insoluble in alcohol and is insoluble in organic solvents like acetone and ether. Most other common mercurials and iodine solution are considered to be better antiseptics.

Mercurophen is sodium hydroxymercuri-o-nitrophenolate, $NaOC_6H_3(NO_2) \cdot HgOH$. It is a brick-red powder which is slightly soluble in water. It contains 55.2% of mercury. It is used as an antiseptic for skin and mucous membranes in $1:2,000$ to $1:15,000$ dilutions with the stronger solutions to be applied for not more than one minute. It is not effective for sporulating pathogens.

Mertoxol is acetoxymercuri-2-ethylhexylphenolsulfonic acid, $CH_3COOHg \cdot C_6H_2OH \cdot (C_8H_{17}) \cdot SO_3H$, though the exact positions of the substituent groups is doubtful. It is used as an antiseptic in $1:1,000$ isotonic solution. It contains 40% mercury.

Meroxyl is a mixture of sodium 2,3-dihydroxy-3,5-(dihydroxymercuri) benzophenone-2'-sulfonate, comprising about 50% of the mixture, and ammonium 2,4-dihydroxybenzophenone-2'-sulfonate, sodium acetate, and water. It contains from 26 to 29% of mercury. It is used as an antiseptic in $1:200$ to $1:1,000$ stabilized aqueous solution.

Mercarbolide is 2-hydroxyphenylmercuric chloride, $C_6H_4OH \cdot HgCl$. It contains about 61% of mercury. It is a solid melting at about 150–152°C. It is slightly soluble in cold water and is freely soluble in alcohol. It is used as an antiseptic in $1:1,000$ aqueous solution and in soaps. Mercresin is a mixture of Mercarbolide and a number of isomeric amyl-o-cresols.

Phenylmercuric nitrate or more particularly basic phenylmercuric nitrate, $C_6H_5HgOH \cdot C_6H_5HgNO_3$ which is also known as Merphenyl nitrate, Merphene, and by a number of other names is a white to grayish white powder, melting in the range of 178 to 184°C. It is only slightly soluble in water and alcohol but is somewhat more soluble in glycerol. It is used as a preservative in some cosmetics and can be used in $1:1,500$ to $1:24,000$ dilutions as an antiseptic. Analogous compounds are phenylmercuric chloride and acetate. The latter has been used as an herbicide.

Toxicity. Mercury and its compounds, with few exceptions, are highly poisonous to living organisms. Particularly, in connection with finely-divided mercury metal, extreme care must be taken to avoid inhalation of and contact with the element. The chances of poisoning are increased because awareness of the presence of the metal is reduced when it is finely divided. The fine gray mercury powder is easily generated when liquid mercury is rubbed against or agitated with grease, chalk, sugar, ether, and numerous other substances.

Mercury can cause acute renal failure and nephrotic syndrome. Chronic exposure to or ingestion of mercury may lead to polyneuropathy. Confirmation of poisoning is sometimes made by analysis for the presence of mercury in hair, fingernails, serum, and urine. Removal of the metal may be hastend by the oral administration several times a day for a limited period of D-penicillamine. As pointed out by Beary (1979), with reference to the relatively high mercury content of whale meat, when humans consume contaminated whale meat, the lipid-soluble methylmercury is concentrated in the cells of the nervous system and very slowly eliminated from the body, even when all intake is stopped.

References

Beary, J. F., III: "Mercury in Sperm Whale Meat," *Science*, **206**, 1266 (1979).
Broad, W. J.: "Sir Isaac Newton: Mad as a Hatter," (mercury poisoning), *Science*, **213**, 1341–1344 (1981).
Cook, E.: "Limits to Exploitation of Nonrenewable Resources," *Science*, **191**, 677–682 (1976).
McAuliffe, C. A., Editor: "The Chemistry of Mercury," Macmillan-Canada/Maclean Hunter Press, Toronto (1977).
Nowak, M.: "Mercury" in "Metals Handbook," 9th Edition, Vol. 2, American Society for Metals, Metals Park, Ohio (1979).
Sax, I.: "Dangerous Properties of Industrial Materials," 5th Edition, Van Nostrand Reinhold, New York (1979).

MERESBURG REACTION. Fertilizers.

MESITYL OXIDE (MSO). Ketones.

MESONS. The *mesons* are subatomic particles of the hadron family. See **Hadrons.** Fermionic hadrons are called *baryons*, the others are called *mesons*. The meson family consists of eight members which fall into a triplet of *pions*, a singlet *eta*, a doublet of *kaons*, and a doublet of *antikaons*. They are all pseudoscalar (spin zero and odd parity) and exhibit strong interactions). The charged particles are coupled to the photon, but even the neutral members can participate in electromagnetic interaction by virtue of the large probability for virtual dissociation into charged particles. They participate in a variety of weak interactions including the nuclear beta decay interaction.

It is found that the kaons, the hyperons (baryons other than the neutron and proton) and their antiparticles, collectively known as *strange particles*, can decay by weak interactions not involving leptons or photons, with a lifetime which is large compared to the natural periods appropriate to strong interactions. On the other hand, these particles are produced copiously in high-energy nuclear collisions. These two circumstances can be understood in terms of the existence of another additive quantum number (*hypercharge*) which is conserved in strong and electromagnetic interactions, but violated in weak interactions.

The meson-baryon system exhibits further regularities as far as strong interactions are concerned. The neutron and the proton have very nearly the same mass and similar nuclear interactions although their electromagnetic properties are quite different. The three *pions* have different electric charges, but again they have approximately equal masses and similar nuclear interactions. This kind of multiplet structure is evident for other strongly interacting particles: the kaons form a doublet, the

sigma hyperons form a triplet, the *xi hyperons* form a doublet, and the *lambda hyperon* remains a singlet. See also **Particles (Subatomic).**

The pion (pi meson) was first recognized in 1947 in photographic films made by C. F. Powell, P. S. Occhialini, and their collaborators of cloud chamber tracks made by cosmic rays high in the Andes. The masses of these pions were greater than those of the previously-discovered muons, corresponding more closely to those predicted by K. Yukawa in his theory of the nuclear structure of the atom. The positive or negative pion has a mass 273 times that of the electron, and a charge equal in magnitude to that of the electron. Both positively charged and negatively charged pions are found in cosmic rays. Neutral pions may also be present in cosmic rays, but are produced in much greater abundance by high-energy particle accelerators and are therefore more easily detected in these laboratories. The first artificially produced pions were made in 1948 by the impact of 380 MeV alpha particles, from the Berkeley synchro-cyclotron, on a target of carbon or certain metals. These were charged pions. Others were produced later by beams of protons and deuterons. The first evidence of neutral pions were the gamma-rays produced in 1950 by the impact of 175 MeV protons upon similar targets (carbon, beryllium, etc.). These gamma-rays had a minimum energy of about 140 MeV, which would be expected from the decay of a (neutral) pion into two gamma-rays. The same method is used today to produce beams of charged pions. The life of the neutral pion is so much shorter (about 10^{-16} seconds against 2.6×10^{-8} seconds) that beams cannot be produced. Unless it is captured by an atom, or reacts with another particle, a charged pion decays into a muon of the same sign and a neutrino or antineutrino. Like the other mesons the pion is a boson. It has zero spin.

Evidence of the *kaon* was found in 1944 by L. Laprince-Ringuet and M. Lhéritier in a cloud chamber photograph of a cosmic ray event. It was found again in 1947 by Rochester and Butler as a V-shaped track in a cloud chamber, the particle forming the other side of the V being probably a pion. For that reason it was first called a charged V-particle, which has been superseded by kaon.

META COMPOUNDS. Organic Chemistry.

METALLOBIOMOLECULES. Natural products, the biologically active forms of which contain one or more metallic elements. Metallobiomolecules may be transport and storage proteins, such as cytochromes (Fe), ferritin (Fe), transferrin (Fe), ceruloplasmin (Cu), myoglobin (Fe), or hemoglobin (Fe); or they may be enzymes, such as carboxypeptidases (Zn), aminopeptidases (Mg, Mn), phosphatases (Mg, Zn, Cu), hydroxylases (Fe, Cu, Mo), or isomerases and synthetases, such as coenzymes (Co). Metallobiomolecules also may be nonproteins, such as siderophores (Fe) or chlorophyll (Mg). Ibers and Holm provide an excellent review of metallobiomolecules in *Science*, **209**, 223–235 (1980).

METALLOCENES. A class of neutral transition metal compounds containing two cyclopentadienyl (C_5H_5) ligands π-bonded to a central metal atom in a "sandwich" structure, as exemplified by ferrocene (see below).

The original metallocene, ferrocene, was first reported simultaneously and independently by two groups of workers in 1951. Miller, Tebboth, and Tremaine at British Oxygen Ltd. passed cyclopentadiene over iron powder and an ammonia catalyst; Kealy and Pauson at Duquesne University treated ferric chloride with cyclopentadienyl magnesium bromide. Both groups obtained the same orange, air-stable, hydrocarbon-soluble crystals (m.p. 173°) and proposed a sigma bonded structure for this first organometallic derivative of iron. Within a year groups at Harvard University including Woodward, Wilkinson, Rosenblum and Whiting revealed that $C_{10}H_{10}Fe$ exhibits unusual thermal stability, has no dipole moment, and undergoes typical aromatic substitution reactions; whereupon, they proposed the novel sandwich structure and advanced the name *ferrocene* to reflect both the iron content of the material and its aromaticity. X-ray crystallographic studies subsequently supported the sandwich structure, showing the preferred staggered, antiprismatic conformation in the solid state. In solution or the vapor phase the barrier to rotation of the rings has been shown to be very small.

One proposed bonding scheme, using a molecular orbital model, calls for utilizing the π orbitals of the $C_5H_5^-$ group and the d_{xz}, d_{yz}, s, p_x, p_y, and p_z iron orbitals to obtain a total of six bonding orbitals with the twelve π-electrons from the rings to fill them.

After the discovery of ferrocene many additional metallocenes and their derivatives were prepared, including titanocene (green), vanadocene (purple), chromocene (scarlet), cobaltocene (purple-black) and nickelocene (green). Dicyclopentadienyl manganese (amber) is ionic, and dicyclopentadienyl mercury, tin and lead are sigma bonded; their derivatives are not properly called metallocenes. In the second and third transition series, only ruthenocene and osmocene are known as simple, neutral metallocenes; the other metallocenes are known as cations (e.g. Cp_2Ti^+) or with additional ligands such as halide or hydride attached to the metal atom, as in Cp_2TiCl_2, Cp_2ReH, Cp_2TaH_3, etc. Though several preparative approaches are known, the most general reaction leading to metallocenes is the treatment of a metal halide with sodium cyclopentadienide in tetrahydrofuran. Ideally a +2 salt of the metal is used, although excess of the cyclopentadienide will often reduce a higher valent metal halide to a lower oxidation state during the reaction.

Chemically, ferrocene undergoes many typical aromatic substitution reactions such as Friedel-Crafts acylation or alkylation, sulfonation, mercuriation, lithiation, the Villsmaier reaction and the Mannich reaction (with dimethylamine, formaldehyde and acetic acid). Its reactivity is very great ("superaromatic") and is comparable in rate to that of phenol. Mono- and disubstitution on one or both rings can be realized, though some measure of control to predominately mono- or disubstitution can be exercised by adjusting conditions.

However, the central metal atom in ferrocene imposes some limitations on the chemistry of the system or provides in some cases additional reaction pathways. For example, the central iron atom is readily and reversibly oxidized from the Fe(II) state to Fe(III) in the form of the water-soluble red-blue dichroic ferrocinium ion. This oxidation occurs with halogens or with nitric acid so that direct aromatic halogenation and nitration cannot be realized; however, halo- and nitroferrocene have been prepared by indirect methods. Aminoferrocene cannot be diazotized, presumably due to oxidative destruction of the system. Coupling with diazonium salts is anomalous; ferrocene reduces diazonium salts to phenyl radicals in aqueous or nonaqueous solution to yield Gomberg (phenylation) products. Condensation with aldehydes in acid solution gives rearranged products due to a role that the iron atom can play. The central iron atom is readily protonated in strong acid media.

In the solid state ruthenocene and osmocene prefer the eclipsed, pentagonal prismatic structure and in solution exhibit a chemistry similar to that of ferrocene. The other metallocenes, on the other hand, are quite different chemically, none of them showing the typical aromatic substitution reactions of ferrocene.

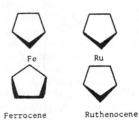

Cobaltocene is rapidly oxidized in air or in solution (it liberates H_2 from water, slowly) to yield the very stable yellow +1 cation, which has been reported to be stable to aqua regia. Neutral cobaltocene reacts with alkyl and acyl halides to give adducts $(C_5H_5)(C_5H_5R)Co^+X^-$ in which the substituted ring is π-bonded to the cobalt as a diene.

Nickelocene, on the other hand, is more slowly oxidized and undergoes addition of suitable activated olefins to one ring converting it to a bicyclic ligand. Some ligand replacement reactions are also known for nickelocene.

Titanocene derivatives undergo substitution reactions at the metal atom:

$$Cp_2TiCl_2 + C_6H_5Li \rightarrow Cp_2Ti(C_6H_5)_2 + LiCl_2$$

$$Cp_2TiCl_2 + 2OCH_3^- \rightarrow Cp_2Ti(OCH_3)_2 + 2Cl^-$$

The utility of these materials is limited and disappointing. Ferrocene has been used to promote the burning of fuel oils and as a catalyst in rocket fuels. Its extreme thermal stability (to over 470°) spurred an interest in its derivatives as high-temperature fluids. Some polymers have been made, but they have found no substantial application. Titanocene dichloride has been shown to react with molecular nitrogen in systems containing added Grignard reagents or similar other materials.

It is generally agreed that the real importance of metallocenes lies in the effect of their discovery twenty years ago along with Ziegler's catalyst on the course and direction of organometallic chemistry. There has been a veritable explosion of interest and effort in this field during the past two decades that is unabated today. Many new materials and new insights into bonding of transition metals have evolved. Great strides in understanding of homogeneous catalysis and the design of new catalysts for hydrogenation, coupling, carbonylation, polymerization, etc., have been made. Even a closer understanding of the action of vitamin B_{12} has resulted from the new brand of organometallic chemistry, and we are moving closer to a practical method for fixation of nitrogen in solution.

See also **Osmium.**

—William F. Little, University of North Carolina, Chapel Hill, North Carolina.

METALLOID. A chemical element which may exhibit physical and chemical properties both of a metal and a nonmetal sometimes is referred to as a *metalloid*. Antimony, arsenic, and tellurium are examples. Less frequently, metalloid refers to elements, such as carbon, silicon, phosphorus, and sulfur, which are added in small amounts in the manufacture of iron and steel.

METALLOPROTEINS. Proteins, especially in solution, readily participate in a greater variety of chemical reactions than any other class of compounds of biological interest. This reactivity is a function primarily of the many polar side chains containing —OH, —COOH, —NH₂, —SH, and other groups, all of which can, to varying extents, interact with metal ions. Proteins can bind metals, some of them very tightly. However, relatively specific and nonspecific binding should be differentiated.

A negatively charged protein molecule exerting a nonspecific electrostatic attraction on metal ions would not qualify as metalloprotein. The term metalloprotein is restricted to compounds in which under natural conditions a metal ion is relatively specifically and strongly bound to a protein molecule in such a way that the compound can be isolated and shown to contain a stoichiometric amount of metal.

A variety of metal ions are found in biologically important metalloproteins. Metalloproteins occur in a wide range of biological systems. The function of the metalloproteins, as indicated in the accompanying table, varies widely from one compound to another.

REPRESENTATIVE METALLOPROTEINS AND THEIR FUNCTIONS

Name	Metal	Source	Function
Hemocuprein	Cu	Erythrocytes	Unknown
Ceruloplasmin	Cu	Serum	Oxidase (probable)
Hepatocuprein	Cu	Liver	Unknown
Polyphenol oxidase	Cu	Mushroom	Enzyme
Hemocyanin	Cu	Mollusks	Respiratory pigment
Tyrosinase	Cu	Mushroom	Enzyme
Metallothionein	Cd + Zn	Kidney	Na Reabsorption (probable)
Xanthine oxidase	Mo	Liver	Enzyme
Carbonic anhydrase	Zn	Erythrocytes	Enzyme
Alcohol dehydrogenase	Zn	Yeast	Enzyme
Ferritin	Fe	Spleen	Fe storage
Transferrin	Fe	Plasma	Fe transport
Conalbumin	Fe	Eggs	Fe storage
Ferredoxin	Fe	Bacteria	Electron transport
DPNH-cytochrome c reductase	Fe	Heart muscle	Electron transport
Hemovanadin	V	Tunicates	Respiratory pigment

The chemical properties of the metal in these compounds may be greatly affected by bonding to a protein ligand. The bound metal can play one of many roles. Thus, in an enzyme, the metal ion may permit the formation of a ternary complex between protein, metal and substrate or coenzyme. An instance of this role is provided by the enzyme enolase, which is unable to catalyze the equilibrium between 2-phosphoglycerate and 2-phosphopyruvate in the absence of Mg ions. In other enzymes, the metal may actually participate in electron transport by cyclic oxidation and reduction. Such is probably the case with the Cu in polyphenol oxidase. The metal may serve primarily for the maintenance of a specific spatial folding of the polypeptide chains in the protein molecule.

The strength of metal-protein bonds in metalloproteins may vary from relatively loose association to very tight binding. When the metal ion is able to dissociate with some ease from the protein, it is usually a single ligand responsible for the metal binding. Such ligand groups are mainly found in the amino acid side chains of the protein molecule (e.g., —NH$_2$ or —OH groups). The interaction of metal and ligand may exhibit strong pH dependence because of competition between metal and hydrogen ions. Of the single ligand groups, by far the strongest is the —SH group in the amino acid cysteine. Even stronger metal bonding to protein may be observed when a divalent or trivalent metal forms chelate complexes with the protein. Chelation is often indicated not only because of the strength of the

bond, but also because of the specificity of the reacting site on the protein molecule for one particular metal. Such a specificity may reflect the coordination requirements of the various metals. The preferred electron donor in the formation of protein-metal coordination compounds is N, such as that of the imidazole nucleus of histidine, but S and O may also participate in this process. If the protein contains carboxyl or phosphoryl groups, strong ionic bonds between metal and protein may be formed. A completely different type of protein-metal interaction is illustrated by the Fe-containing protein ferritin. Basically, this compound consists of a coat of protein (apoferritin) surrounding a micelle of hydrated iron hydroxide. The metal can be readily and reversibly removed from the apoprotein.

See also **Chelation Compounds.**

METALS (The). In terms of classification, several of the chemical elements are referred to as metals, principally because of the metallic qualities which they exhibit. There are several subclassifications:

Group 1b. In order of increasing atomic number, these are copper, gold, and silver. Sometimes, these metals also are referred to as *noble* metals, principally because they sometimes occur in nature in elemental form. Gold and silver also are frequently referred to as "coinage" metals. The elements of this group are characterized by the presence of one electron in an outer shell. Although copper and gold also have other valences, all of the elements in this group have a 1+ valence in common.

Group 2b. In order of increasing atomic number, these are zinc, cadmium, and mercury. The elements of this group are characterized by the presence of two electrons in an outer shell. Although mercury also has a valence of 1+, all of the elements in this group have a 2+ valence in common.

Group 4b. In order of increasing atomic number, these are titanium, zirconium, and hafnium. The elements of this group are characterized by the presence of two electrons in an outer shell. Although titanium and zirconium also have other valences, all of the element in this group have a 4+ valence in common.

Group 5b. In order of increasing atomic number, these are vanadium, niobium (sometimes called columbium), and tantalum. Vanadium and tantalum have two electrons in an outer shell; niobium has one electron in its outer shell. Although niobium and vanadium also have other valences, all of the elements in this group have a 5+ valence in common.

Group 6b. In order of increasing atomic number, these are chromium, molybdenum, and tungsten. Chromium and molybdenum have one electron in their outer shells; tungsten has two electrons in its outer shell. Although chromium and molybdenum also have other valences, all of the elements in this group have a 6+ valence in common.

Group 7b. In order of increasing atomic number, these are manganese, technetium, and rhenium. Manganese and rhenium have two electrons in their outer shells; technetium has one electron in its outer shell. Although manganese and rhenium also have other valences, all of the elements in this group have a 7+ valence in common.

Group 8. In order of increasing atomic number, these are iron, cobalt, nickel, ruthenium, rhodium, palladium, osmium, iridium, and platinum. Ruthenium, rhodium, and platinum have one electron in their outer shells; iron, osmium, cobalt, and nickel have two electrons in their outer shells; iridium has 17 outer electrons and palladium 18 outer electrons. Although all of these elements fall into one group, they appear in the classification in three subgroupings (hence the sometimes-used term *triads*): (1) iron, cobalt, and nickel each have valences of 2+ and 3+; (2) ruthenium, rhodium, and palladium each have valences of 4+, in addition to other valences; (3) osmium, iridium, and palladium each have valences of 4+, in addition to other valances.

In terms of the periodic classification, all elements designated as the metals here fall between highly alkaline elements (alkali metals and alkaline earths) at the left end of the table and the acidic elements, ending with the halogens at the right end of the table. Thus, the term "transition elements" sometimes is used to describe these inbetween elements. Actually, the term transition can be applied to the differences between any series of elements within the overall classification, or between individual elements within a group—because of the gradual alteration in chemical behavior which takes place between groups and between elements.

METASTABLE NUCLEI. Nuclei in excited nuclear states that have measurable lifetimes exceeding 10^{-10} to 10^{-9} second.

METHACRYLATES. Acrylates and Methacrylates.

METHANATION. Coal; Substitute Natural Gas (SNG).

METHANE. CH_4, formula weight 16.04, colorless, odorless (when pure) gas, mp $-182.6°C$, bp $-161.4°C$, sp gr 0.415 (at $-164°C$). Sometimes referred to as *marsh gas* or *fire damp*, methane is practically insoluble in H_2O, and moderately soluble in alcohol or ether. The gas burns when ignited in air with a pale, faintly luminous flame, forming an explosive mixture with air between gas concentrations of 5% and 13%. Methane is the principal constituent of natural gas, averaging 75% by weight. Natural gas from the Pennsylvania fields is almost 99% methane, but some gas from Kentucky fields contains as little as 23% methane. Pipeline gas from several fields typically will contain about 78% methane, 13% ethane, 6% propane, 1.7% butane, and 0.6% pentane. The remaining fraction consists of gases higher in the alkyl series. While generally not referred to as such, methane can be classified as a major fuel. The heating value of pure methane is 995 Btu/ft³ (8856 Calories per cubic meter).

Methane, as the major constituent of natural gas, is an extremely important raw material for numerous synthetic products. For most processes, it is not required to isolate and purify the methane, but the natural gas as received may be used. The high percentage of CH_4 in various feedstocks makes possible the formation of synthesis gas: $CH_4 + H_2O \rightarrow CO + 3H_2$. The percentages of CO and H_2 in synthesis gas vary depending on the end-product to be made. Synthesis gas is used widely in the manufacture of NH_3, oxo-chemicals, and methyl alcohol. See also **Synthesis Gas.**

In addition to the preparation of synthesis gas, which is used so widely in various organic syntheses, methane is reacted with NH_3 in the presence of a platinum catalyst at a temperature of about 1,250°C to form hydrogen cyanide: $CH_4 + NH_3 \rightarrow HCN + 3H_2$. Methane also is used in the production of olefins on a large scale. In a controlled-oxidation process, methane is used as a raw material in the production of acetylene.

Most artificial gases, such as producer gas, coal gas, water gas, manufactured gas, and town gas contain a high content of methane. In addition to its use as a basic chemical and fuel, methane is of notable interest because of its role as the anchor compound of the *alkanes* (paraffin or aliphatic hydrocarbons). All of these compounds may be considered derivatives of methane.

Carbon monoxide and hydrogen react to form CH_4 in the presence of a nickel catalyst. Methane also is formed by reaction of magnesium methyl iodide in anhydrous ether (Grignard's reagent) with substances containing the hydroxyl group. Methyl iodide (bromide, chloride) is preferably made by reaction of methyl alcohol and phosphorus iodide (bromide, chloride).

See also **Biomass and Wastes as Energy Sources.**

METHANOGENS. Cells that resemble bacteria in a superficial way, but that have unique genetic and metabolic characteristics. Methanogens are anaerobic, methane-producing microorganisms that occur in a wide variety of places—the gastrointestinal tract of animals, including humans, in the sediments of natural waters, in sewage treatment plant vessels and piping, and in natural hot springs. As proposed by Woese (University of Illinois) and Fox (University of Houston), the methanogens probably make up a third line of descent of cells in addition to the prokaryotes (bacteria and blue-green algae cells which do not have a well-defined nucleus) and eukaryotes (more complex cells with a nucleus). These researchers also have suggested that there may be still other kinds of cells that do not meet the criteria set down for prokaryotes and eukaryotes.

Methanogens are distinguished from bacteria on at least three counts: (1) The cell walls do not contain muramic acid, the characteristic constituent of the peptidoglycans that form bacterial cell walls. (2) Their metabolism differs markedly from bacteria. A number of coenzymes apparently unique to methanogens have been identified. Some of these enzymes are involved in methyl transfer reactions, including the formation of methane. One of the coenzymes is possibly the smallest coenzyme yet to be discovered. The methanogens also differ in the manner in which carbon dioxide is fixed into cellular carbon. However, the pathway has not been clearly identified. (3) The RNA sequences of methanogens differ from those of other organisms. These observations have indicated to Woese and Fox that although the methanogens share a common ancestor with prokaryotes and eukaryotes, an independent line of descent branched off at possibly about the same time the other cell types diverged.

Although not fully understood, the methanogens place new challenges to the evolutionary biologists for further explanation in terms of the development of early life on earth and may be very valuable toward understanding life on extraterrestrial bodies as these may be explored over future years.

Barker (University of California at Berkeley) and Huntgate (University of California at Davis) as early as the mid-1950s noted that methanogens differ radically from bacteria.

Methanogens take part in the terminal stages of organic matter degradation and survive on carbon dioxide and hydrogen yielded by anaerobic bacteria, converting them to methane.

METHANOL (Methyl Alcohol). CH_3OH, formula wt 32.04, colorless, mobile liquid with mild characteristic odor, mp $-97.6°C$, bp 64.6°C, sp gr 0.792. The compound is miscible in all proportions with H_2O, ethanol, or ether. When ignited, methanol burns in air with a pale blue, transparent flame, producing H_2O and CO_2. The vapor forms an explosive mixture with air. Upper explosive limit (% by volume in air) is 36.5 and lower limit is 6.0.

Methyl alcohol possesses distinct narcotic properties. It is also a slight irritant to the mucous membranes. The principal toxic effect is exerted on the nervous system, particularly the optic nerves and possibly the retinae. The effect upon the eyes has been attributed to optic neuritis, which subsides, but is followed by atrophy of the optic nerve. Once absorbed, methyl alcohol is only very slowly eliminated. Coma resulting from massive exposures may last as long as 2–4 days. In the body, the products formed by the oxidation of methanol are formaldehyde and formic acid, both of which are toxic. The toxicology of methyl alcohol is further described in "Dangerous Properties of Industrial Materials," (N. I. Sax, Editor), Van Nostrand Reinhold, New York (1979).

Chemical Properties. Methyl alcohol is a versatile material, reacting (1) with sodium metal, forming sodium methylate, sodium methoxide CH_3ONa plus hydrogen gas, (2) with phorphorus chloride, bromide, iodide, forming methyl chloride, bromide, iodide, respectively, (3) with H_2SO_4 concentrated, forming dimethyl ether $(CH_3)_2O$, (4) with organic acids, warmed in the presence of H_2SO_4, forming esters, e.g., methyl acetate CH_3COOCH_3, methyl salicylate $C_6H_4(OH) \cdot COOCH_3$, possessing characteristic odors, (5) with magnesium methyl iodide in anhydrous ether (Grignard's solution), forming methane as in the case of primary alcohols, (6) with calcium chloride, forming a solid addition compound $4CH_3OH \cdot CaCl_2$, which is decomposed by H_2O, (7) with oxygen, in the presence of heated smooth copper or silver forming formaldehyde. The density of pure methyl alcohol is 0.792 at 20°C compared with H_2O at 4°C (the corresponding figure for ethyl alcohol is 0.789), and the percentage of methyl alcohol present in a methyl alcohol-water solution may be determined from the density of the sample.

A common test for methyl alcohol is by itx oxidation in air with a hot copper wire to form formaldehyde.

At one time, most methyl alcohol was obtained by the destructive distillation of hardwoods (hence the name *wood alcohol*) at about 350°C, along with a yield of acetic acid and small percentages of acetone in the water condensate. Interest in returning to wood as a source has revived because of fossil fuel shortages. See **Biomass and Wastes as Energy Sources.**

Production of Methyl Alcohol.[1] Synthetic methanol is one of the major raw materials of the organic chemical industry. Methanol has economic stability and a steady growth rate owing to the low costs of production and diversity of applications. Nearly all the methanol producers also make formaldehyde, which is the main end use (more than 50%) of methanol. The other main end uses are dimethyl terephthalate, methacrylates, methylamines (for resins, herbicides, and fungicides), methyl halides (for silicones, tetramethyl lead, butyl rubbers, paint removers, photographic films, aerosol propellents (diminishing use), and degreasing compounds), acetic acid, and solvents.

An important process for production of synthetic protein uses methanol as feedstock. The use of methanol as a fuel, either as pure methanol, as a mixture (approximately 15%) with gasoline, or as a feedstock for synthetic gasoline is envisaged for possible large-scale application; as well as use in gas turbines for electricity generation.

There are three principal commercial grades of methanol (as defined in U.S. Federal Specification 0-M-232 f: June 5, 1975): *Grade A*, synthetic, 99.85% by weight (solvent use); *Grade AA*, synthetic, 99.85% by weight (hydrogen and carbon dioxide generation use); and *Grade C*, wood alcohol (denaturing use).

The most recent advances in methanol synthesis are the low- and intermediate-pressure processes of the type shown in accompanying figure. The synthesis step of this process[2] relies upon a copper-based catalyst, which gives good yields of methanol at pressures of 50 and 100 atmospheres. These pressures are

substantially below those of the 250–350 atmospheres required by earlier processes. The high catalyst activity allows the synthesis reaction to take place at a relatively low temperature of 250–270°C. As a result, methanation is avoided, and byproduct formation is lower, giving increased process efficiency.

The development of this low-pressure technology has caused a major reassessment of the economics of methanol production. The energy required to compress the synthesis gas from its production pressure to the synthesis unit is reduced by a factor between 2 and 3. The lower synthesis pressure allows the exclusive use of centrifugal compressors in plants with capacities as low as 15 million gallons (0.57 million hectoliters) per year. Small producers find attractive the savings in investment, operating, and maintenance costs made possible by low-pressure operation. Plants range in capacity from 15 million gallons (0.57 million hectoliters) to 250 million gallons (9.46 million hectoliters) per year.

Synthesis gas is prepared by the steam reforming or partial oxidation of a liquid or gaseous hydrocarbon feedstock, or by direct combination of carbon dioxide with purified hydrogen-rich gases. Economic considerations usually favor the steam-reforming route for a naphtha or natural-gas feedstock. In this instance, desulfurized feedstock is preheated, mixed with superheated steam, and reacted over a conventional catalyst (normally nickel-based) in a multitubular reformer. The reformer usually is operated at between 15 and 30 atmospheres and at a tube outlet temperature of 840–900°C. The reforming conditions are chosen to give the most economic overall production costs. Methane slip (amount of unconverted methane) usually is greater than for conventional high-pressure synthesis processes, since the cost of compressing the additional methane is less significant with the low-pressure process. With a naphtha feedstock, an almost exact stoichiometric ratio of carbon oxides to hydrogen in the synthesis gas is achieved, but when natural gas is the feedstock, there is an inherent deficiency of carbon. Established practice for many years has been to add carbon dioxide from an external source in preparing a stoichiometric synthesis gas. Development of the low-pressure process has shown that this addition of carbon dioxide is not required and that, depending upon the cost of carbon dioxide production, the production of methanol from natural-gas feedstock alone is economic.

After heat recovery and cooling, the synthesis gas is compressed to the required synthesis pressure and passed into the synthesis loop at the suction of a circulator. The circulator, which boosts the pressure of the circulating gases to make up the total loop pressure drop, also is a centrifugal machine. Feed-gas preheating is carried out by heat exchange with the hot gases leaving the converter. Heat recovery is incorporated into the loop to recover the heat of reaction of methanol synthesis.

Synthesis takes place in a hot-wall converter over the low-pressure methanol-synthesis catalyst at 250–270°C. Temperature control of the converter is effected by injecting cold gas at appropriate levels in the catalyst bed, using specially developed distributors which provide excellent gas mixing while allowing free passage of the catalyst for easy charging and discharging. After leaving the converter and passing through the feed-gas preheater, the converted gases are cooled, and crude methanol is condensed and separated from the uncondensed gases, which are recycled with makeup synthesis gas to the converter. A continuous gas purge is taken from the synthesis loop in order to remove an accumulation of inert gases. This purge is recycled to the synthesis-gas preparation section as reformer fuel. The crude methanol is reduced in pressure before passing forward to the methanol-purification section, where methanol of the required purity is produced by conventional distillation methods.

Economics in fuel gas consumption are achieved by use of recovered heat in reboiling in the distillation columns. In addition, distillation schemes involving 3 or 4 columns have also been developed with reduced reboil heat requirements.

METHIONINE. Amino Acids.

METHYL CHLORIDE. Chlorinated Organics.

METHYLENE CHLORIDE. Chlorinated Organics.

METHYLISOBUTYL CARBINOL (MIBC). Ketones.

METHYLISOBUTYL KETONE (MIBK). Ketones.

METHYLPYRIDINE. Pyridine and Derivatives.

MEVALONIC ACID. Steroids.

MICA. The mica group of minerals includes several closely related species, having a highly perfect basal cleavage; all are monoclinic with a tendency toward pseudohexagonal crystals, and are closely similar in chemical composition. The highly perfect cleavage, the most prominent characteristic of the group, is explained on a basis of x-ray studies of these minerals, which seems to show a sheetlike arrangement of the atomic structure and a hexagonal grouping of atoms; this apparently explains the pseudohexagonal crystals abovementioned. The word *mica* is believed to have been derived from the Latin *micare*, meaning to shine, in reference to the brilliant appearance of this mineral, especially when in small scales.

MICELLE. An electrically charged colloidal particle, usually organic in nature, composed of aggregates of large molecules, such as found in surfactants and soaps. The term is also applied to the casein complex in milk. See also **Colloid Systems.**

MICRONUTRIENTS (Soil). Fertilizers.

MICROSTRUCTURE FABRICATION (and Chemistry). Microstructures are patterns formed on or imbedded in the surface of some substrate material. Microstructure implies that the transverse dimensions of the patterns are in the microscopic range, no greater than a few thousandths of an inch (a few micrometers). The patterns may be formed in layers or insulators deposited on a surface or may consist of chemical or physical modification of shallow regions of the substrate. By far the most important use of microstructure fabrication is in the manufacture of integrated circuits, large numbers of transistors, diodes, resistors and capacitors fabricated with the interconnections that enable them to perform useful electronic functions on a single piece or "chip" of silicon. The art of microstructure fabrication has evolved around this application. Integration and miniaturization are the measures of its progress.

Integration, the fabrication of large numbers of interconnected components on a single substrate, is intimately associated with and dependent on miniaturization, making components smaller. Another substantial contribution to integration has come from increase in the size of the substrate that can be produced with perfection. The rate of advance has been summed up in the observation that the number of components that can be produced on a chip has approximately doubled every year since 1960. Integration and miniaturization have proved to be

the keys to several economically important features of electronic systems. The cost of manufacturing a chip is only weakly dependent on the number of components it contains, so that the cost per component is reduced. Miniaturization reduces the capacitance of interconnections and devices, allowing higher speeds of operation and lower power dissipation per component. Also, it has turned out that the replacement of soldered and pluggable connections by those fabricated by integrated circuit technology has increased the reliability, that is, decrease the probability of failure, of the connections. These economic advantages have led to the growth of a multibillion dollar industry. Low cost, low power dissipation, and high reliability have made very large digital computers and electronic memories possible. High levels of integration have led to the development of powerful, low cost single chip information processing elements.

The pursuit of miniaturization and integration continues. A large body of technology devoted to the fabrication of microstructures has grown up. The technology involves the introduction of impurities into substrates by diffusion or ion implantation, the deposition of many kinds of thin films, the chemical modification of surfaces through such processes as oxidation, and the removal of material by chemical etching. All of these processes are carried out in selected areas, controlled by lithography, the production of masking layers by exposure of radiation-sensitive substances to optical or particulate radiation images.

Miniaturization has advanced to a stage at which the dimensions of structures can be close to the wave length of visible light. Electron microscopes are needed to photograph the structures with adequate resolution. Dimensions are ordinarily measured in micrometers (μm). 1 micrometer = 1/25000 inch. Although silicon microelectronics has been the primary motivation for the development of microstructure technology, it has found many other applications. The methods are used to pattern magnetic materials for magnetic bubble devices and optical materials for integrated optical elements. Fabrication of gallium arsenide transistors and circuits, surface acoustic wave transducers, superconducting electronics, x-ray optical elements, nozzles for ink jet printers, magnetic recording heads, information displays, and impression of video information on recording media also use the techniques. Examples of complex microstructures are shown in Figs. 1–3.

Microstructure fabrication requires mastery of many skills, tools, and processes. All of the process steps interact. Fine details, such as the exact vertical profile of a hole created in a masking resist layer, may determine the success or failure of the next process step.

Pattern Formation and Demagnification. The pattern that is to be produced on a chip must be created first by an experienced designer and on a scale that can be easily resolved and manipulated. One method of doing this involves drawing a large pattern on a rigid transparent sheet covered by a thin opaque film. The opaque film is removed in selected areas to create a mask 100 to 1000 times larger than the desired final image. The large mask is demagnified by photographic processes in one or two steps to form an image five to ten times the final size called a reticle.

Formation of the original image by cutting a film is being replaced by computerized methods. Programs permit one to form an image on a cathode ray tube at a computer terminal. Great flexibility in creating, altering, and reproducing shapes can be provided and a library of useful structures can be maintained and recalled as needed. The relation of masks used for successive operations to one another can be seen by displaying two or more masks simultaneously. The product of the computer-created design is a tape which can be used to control a

Fig. 1. A scanning electron micrograph of a portion of an experimental miniaturized logic chip fabricated by electron beam exposure of resist layers. The bright strips running across the photo are metal lines about 1.5 micrometers wide; the depressions are the sites of field effect transistors. Six separate lithographic processes were used to produce this chip. (*IBM Corporation*.)

light table, an instrument that mechanically moves a light spot over a photosensitive material to form a reticle by a photographic process. A tape that controls a drawing instrument that draws the chip image enlarged several hundred times and uses different colors for different masking steps can also be produced.

Microstructure processing is ordinarily performed with wafers of several inches diameter which are subsequently divided into

Fig. 2. A bubble memory chip that stores 1000 bits of information. The width of the photograph is about $\frac{1}{32}$ inch (~0.8 millimeter) on the chip. (*IBM Corporation*.)

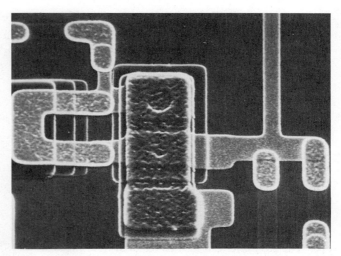

Fig. 3. A superconducting Josephson logic element. The width of the photo represents about 65 micrometers on the chip. The circles in the central rectangle are tunnelling junctions in which layers of metal are separated by an insulator less than 0.01 micrometer thick. (*IBM Corporation*.)

a large number of chips. The reticle, which contains the image of one chip, can be used to form a mask of the size of the wafer by a "step and repeat" process; a photosensitive mask material is exposed to a demagnified image of the reticle and then the reticle (or mask) is moved one chip-to-chip distance and another chip image is transferred to the mask. The process is continued until a mask the size of the wafer has been exposed as a matrix of chip images, which is then developed. The photoresist on the wafer can then be exposed by placing the mask in contact with the wafer.

Several alternative procedures may also be used. The reticle image can be projected and demagnified directly onto the wafer and stepped from chip position to chip position without the intermediary contact masks. The tape produced by a computerized design system can be used to generate a reticle by electron beam rather than optical exposure. Or the tape can control an electron beam that writes directly on the wafer without any kind of masking operation. The choice among the available processing options must be made on an economic basis, which is strongly influenced by the throughput, or rate at which wafers can be exposed.

Microstructure Fabrication. An example will illustrate the process of microstructure fabrication. Consider the structure shown in Fig. 4. Here an *n*-type region has been created by diffusion of a donor impurity into a surface of *p*-type silicon, forming a *p-n* junction diode. There is a metal contact to the *n*-region, and the contact line is insulated from the *p*-type surface by a layer of silicon dioxide. The diameter of the diode is on the order of 10 micrometers (μm).

The fabrication begins with the application of a layer of photoresist to the oxidized surface of a silicon wafer. The photoresist is then exposed to light in the region where the diode is to be formed. Photoresist is a polymeric mixture that is deposited as a thin layer, perhaps 1 μm thick, upon an SiO_2 film on a silicon wafer. Irradiation with light in the near UV region of the spectrum modifies the chemical properties of the photoresist, and, in "positive" photoresist, makes it more soluble in certain developers. Thus, one step frequently employed in microstructure fabrication is the projection of the image of a mask onto the photoresist layer. It becomes possible to remove the exposed

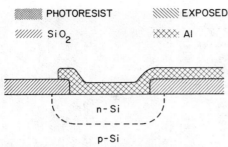

Fig. 4. A diode fabricated on the surface of a wafer of silicon. An *n*-type region has been created by diffusing a donor impurity through an opening in a layer of SiO_2 on the silicon. Electrical contact is made to the *n* region by a deposited aluminum conductor. The SiO_2 insulates the silicon from the aluminum.

region of the photoresist by dissolving it with a suitable developer. The SiO_2 layer can then be removed from the areas that were exposed to light by hydrogen fluoride etches. The photoresist is resistant to HF etches and the SiO_2 in the unexposed areas is not affected by the etch. After etching, the remaining photoresist can be removed by a solvent, leaving a silicon substrate covered with SiO_2 only in the unexposed areas. The SiO_2 film acts as a barrier in the contact of impurities in a gaseous phase with the silicon. Thus, when the silicon wafer covered by the patterned SiO_2 film is exposed, to, for example, a gas containing phosphorus at high temperatures, the phosphorus, being very soluble in silicon, diffuses into the exposed areas rapidly. An idealized description of this sequence of process steps is shown in Fig. 5. The effect of this doping is very important in electronics, since phosphorus is a donor impurity and a region of n-type or electron conductivity is produced where it is present.

Many physical phenomena, however, obstruct the formation of the ideal structure depicted in Fig. 5. The technical literature

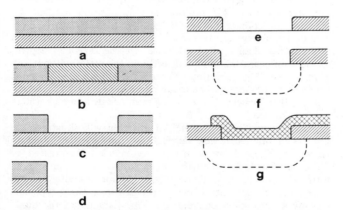

Fig. 5. Process steps used to produce the structure shown in Fig. 4: (a) A film of SiO_2 has been formed by oxidizing the silicon and a layer of photoresist has been deposited on the SiO_2. (b) Shading shows a region of the photoresist that has been exposed to light and thereby made more soluble. (c) The exposed photoresist has been removed. (d) An etchant that reacts with the SiO_2, but not with the photoresist, has been removed. (d) An etchant that reacts with the SiO_2, but not with the photoresist, has been removed. (e) Another solvent has been used to remove the unexposed photoresist. (f) Donor atoms have diffused into the silicon through the opening in the SiO_2 to produce an *n*-type region. (g) Additional masking steps, not shown, have permitted aluminum to be evaporated onto the diode in a pattern that forms a contact to the *n* region of the diode. (See Fig. 4 for legend.)

is well-supplied with papers devoted to each of the steps illustrated in Fig. 5. None is as straightforward as appears at first sight. It is instructive to discuss them further, since much of the essence of microstructure fabrication is revealed by examining them in detail.

Figure 5(a) suggests that the thickness of the photoresist and SiO_2 layers are independent of position. While the layer thickness is not an extremely critical process parameter, its control cannot be entirely neglected, as the time needed for the subsequent developing or etching steps depend on it. The wafers used in modern silicon technology have diameters of three or more inches, and maintaining uniformity of layers and process parameters across a wafer is not a trivial task. Also, very high standards of cleanliness must be maintained, as any particulate contamination will affect the resist adversely.

Figure 5(b) shows a well-defined boundary between the exposed and unexposed areas of the photoresist. In fact, the dimensions of the structures produced in modern microelectronics are comparable to the wavelength of the exposing light, so that diffraction prevents such sharp contrast from being achieved. Furthermore, high-resolution projection exposure schemes require that the light be monochromatic to avoid the problems of chromatic aberration in the lenses. The photoresist must be reasonably transparent to insure that its full thickness is exposed to the light. The silicon surface is, however, reflective, so that the interference of the incident and reflected light produces standing waves in the photoresist and nonuniform exposure of the photoresist in the vertical direction (Dill, 1975; Walker, 1975). Complicated effects of this kind are clearly important to microstructure fabrication. It must also be apparent that, as the amount of exposure received is a continuous function of position, the time of development required to remove a given region of photoresist will also be a continuous function, and that the profile of the developed photoresist will depend on the time of development. In particular, the size of the opening in the photoresist, to which Fig. 5 is oriented, will depend on the time.

Resists can also be exposed with focused electron beams, as used in electron microscopes, instead of light. The electrons, however, pass through the resist layer into the substrate, where they are scattered, and some eventually return from the substrate to the resist, exposing it at a distance from the intended opening. Great care is needed to allow for the backscattering phenomenon in calculating exposures for nearby openings.

Development of the photoresist proceeds somewhat as shown in Fig. 6, with simultaneous lateral and vertical removal of material. The tapered edge of the resist film may be a disadvantage, as the exact point at which the film is thick enough to protect the underlying SiO_2 layer during the succeeding etching step, and thus the size of the hole that will be produced in the SiO_2, is not clearly defined. Prolonging the development beyond the point shown in Fig. 6(b) allows continued lateral development and increase in the size of the opening. Also, however, achieving perfect adhesion of the photoresist film to the SiO_2 is difficult, and the developer may invade the interface between the two layers, producing the undesirable result shown in Fig. 6(d).

The developed photoresist, Fig. 5(c), is then used as a mask to etch the SiO_2 layer. Again, perfection is hard to achieve. Etching for too short a time will leave a certain amount of photoresist in the hole. Etching for too long a time can cause undercutting, as shown in Fig. 7(c). After removal of the photoresist, the wafer is exposed to a diffusant, affording additional opportunities for deviations from idealized behavior. Time and temperature of diffusion are important and can produce results

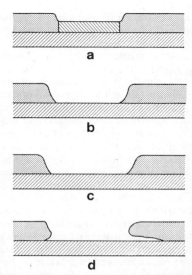

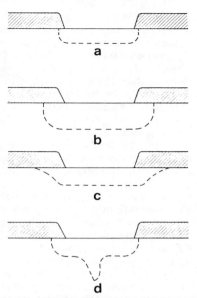

Fig. 6. Development of the photoresist: (a) Although exposure increases the solubility of photoresist in the developer, there is only a finite ratio of dissolution of the exposed photoresist to that of the unexposed region. In addition, the exposure received is not a perfect step function at the boundary of the exposed areas. Thus, dissolution proceeds laterally as well as vertically. (b) The opening in the photoresist has penetrated to the surface of the SiO_2. (c) With continued development the opening continues to enlarge. (d) Poor adhesion of the photoresist to the SiO_2 has allowed the development to penetrate the interface. (See Fig. 4 for legend.)

Fig. 8. A donor impurity is diffused into the silicon from a gaseous phase. (a) A shallow n region has been created. (b) Continued diffusion, longer times, or higher temperatures increase the extent of the n region. (c) Surface diffusion has caused spreading of the n region along the SiO_2-silicon interface. (d) A crystal defect, such as a dislocation, has provided a path for anomalously high diffusion and led to penetration of the junction to unanticipated distance from the surface. (See Fig. 4 for legend.)

resembling Figures 8(a) or 8(b). Diffusion can proceed rapidly along interfaces in certain cases, leading to junction profiles of the kind shown in Fig. 8(c). Preferential diffusion along crystalline defects can give rise to a profile resembling that shown in Fig. 8(d).

Next, a metal connection is to be made to the diffusion-doped region. Aluminum is frequently used for this purpose, as it has high electrical conductivity and does not enter the silicon and alter its properties. The aluminum is also evaporated through a mask that defines the shape and location of the conductor. Examples of the region of contact between the aluminum and the doped semiconductor are shown in Fig. 9. It is seen that the current will be forced to flow through a narrow constric-

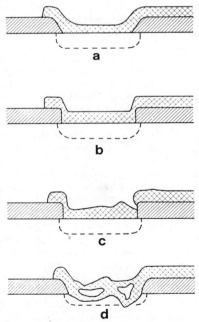

Fig. 9. (a) Masking steps, not shown, have permitted the deposition of aluminum in selected areas to form a contact to the n region. (b) A sharp vertical profile in the SiO_2 opening may cause a reduction of the cross section of the aluminum conductor where it passes from the SiO_2 insulator to the silicon surface. (c) The high current density in the constriction shown in (b) has led to electromigration of aluminum atoms and opening of the conductor. (d) Solution of silicon in the aluminum has resulted in deformation of the metal-semiconductor interface and penetration of the aluminum through the n region. (See Fig. 4 for legend.)

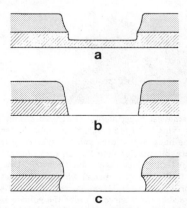

Fig. 7. The opening in the photoresist is used to mask the etching of a hole in the SiO_2: (a) An early stage of development. (b) The opening in the SiO_2 has reached the silicon. (c) Prolonged etching can lead to removal of SiO_2 underneath the photoresist masks. (See Fig. 4 for legend.)

tion if the profiles are as shown in Fig. 9(b). The high current densities may cause electromigration of the aluminum atoms, leading to the open circuit shown in Fig. 9(c).

Also, silicon is somewhat soluble in aluminum. One can thus encounter the situation shown in Fig. 9(d), where enough silicon has been dissolved to allow the metal to completely penetrate the doped region, shorting the junction.

None of the problems illustrated in Figs. 5–9 is insurmountable. A great many ingenious ways to avoid the difficulties described are known. Sometimes these are guided by physical or chemical knowledge; frequently they are empirical fixes.

The microstructure engineer must also be aware of constraints that have little to do with chemistry and materials science, but tend to be more closely related to mechanical technology. One of the most difficult of these is registration or alignment. A substrate is usually passed through several process steps in order to form a desired structure. For example, fabricating a transistor may involve diffusing a base dopant through a window in SiO_2, subsequently diffusing an emitter dopant through a somewhat smaller opening in an SiO_2 layer, and finally, using masking layers to form contacts to these transistor elements. It is necessary to ensure that the emitter region is created in the correct position within the previously diffused base region. The mask that is used to define the emitter region lithographically must be precisely located with respect to the geometrical structure already established on the substrate by previous processing steps. This positioning is known as alignment. It may require that separate structures that can be easily located but have no electronic function be provided on the substrate.

Alignment all over a large substrate is made more difficult by dimensional changes that may take place during processing at high temperatures. Materials soften at high temperatures and can deform under the force of gravity or the stresses that accompany temperature gradients and contacts between different materials.

Further, economic factors also constrain the utilization of microstructure fabrication technology. These are the factors that control the cost of production, such as throughput, the rate at which substrates can be processed by the fabrication tools, capital investment required, and demands on operator time and skill. Electron beam exposure, for example, provides high resolution but uses expensive equipment that works slowly. Naturally, all of the elements of cost must be weighed against the value of the product produced.

Upon observing the practice of microstructure fabrication, one cannot fail to notice a resemblance to certain aspects of modern metallurgy. For example, the precipitates produced by metallurgical processing have a dimensional scale similar to that of electronic microstructures. Inhomogeneities on a scale of 0.01 μm to 10 μm control the desirable properties of a structure. This preoccupation with the properties of solids on a microscopic scale produces a common interest in techniques and in interactions with basic science. Thus, both microstructure fabrication and physical metallurgy: (a) rely on phenomena that take place in the solid state; (b) depend on analytical tools that are capable of chemical analyses with the highest possible spatial resolution; (c) involve the motion of atoms through solids, controlled by diffusion, solution, nucleation, and precipitation; (d) involve interface phenomena at the contact between different solids; and (e) are sensitive to crystal defects.

It must further be noted that both metallurgy and microstructure fabrication are practical disciplines; they are oriented toward the economic production of structures that have a useful role in commerce and industry. In this respect both are engineering rather than scientific disciplines. On the other hand, their

deep probing of phenomena on an atomic scale and under unusual conditions produce new discoveries and lead to new concepts that enhance basic science.

This is not to say that microstructure fabrication is a branch of metallurgy. The detailed motivation of the two disciplines is rather different, metallurgy concentrating on the mechanical properties of solids, while microstructure fabrication controls the electronic properties of structures made from magnetic and optical materials, semiconductors, metals, and insulators. The basic difference, of course, is that microstructure fabrication involves control of the fabrication process in detail at the dimensional level of the structure, while metallurgical processing exercises control at a much grosser level. The application of lithography, with the attendant use of exposure tools, clean rooms, resists, masks, and etchants is the province of microstructure fabrication. Crystalline defects are usually undesirable in microstructures; the metallurgist can frequently use them to advantage. Metallurgy also encompasses its extractive aspects. There is no doubt that microstructure fabrication is a distinct activity.

The technique of microstructure fabrication has grown up as an art in response to a continuous economic and functional motivation to push to the smaller and smaller. Chemistry, physics, and empiricism are combined into the creation of novel physical structures that have enormous economic impact, that, indeed, form the basis of whole new industries. Progress has

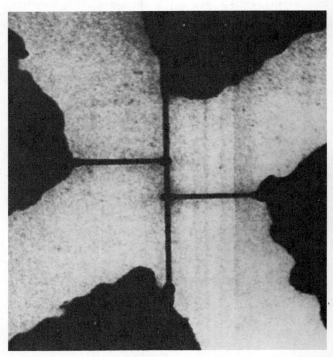

Fig. 10.   The smallest experimental electronic circuit reported as of the early 1980s. This is a superconducting niobium nanobridge which features ultrafine lines with widths and thicknesses of only 100 to 200 atomic diameters, far smaller than human nerve fibers. The three fine lines in the four-terminal configuration shown in this view are only 50 nanometers (billionths of a meter) wide and 30 nanometers thick. The longest line length is 1000 nanometers, and the shortest is 600 nanometers. The short lines serve to probe the voltage along the longer line, which itself functions as a superconducting bridge, or weak link, between the niobium pads in studies of the Josephson effect. Devices of this type and the research leading to them should continue to yield important information about superconductivity and other physics at small dimensions. (*IBM Corporation.*)

been made by adaptation and invention to meet the needs of the moment. By and large, the art has grown rapidly, adapting and improving old methods to new situations and inventing as needed and possible. It is common experience that obtaining reproducible results requires very careful control of all aspects of the fabrication processes, such as temperatures, pressures and time. High standards of cleanliness and reagent purity must be enforced. Seemingly identical apparatuses and starting materials often yield different results. Recipes must be carefully followed, with little understanding of which aspects of a process are critical or what contaminants are important, and why.

The rapid development has outstripped basic understanding of the fundamental mechanisms underlying the techniques. It must be recognized, however, that the optimal exploitation of microstructural technology, the maximization of performance, yields, and utilization of available silicon area will depend on a detailed interpretation of each step in fabrication as a chemical or physical process. Furthermore, unanticipated phenomena are encountered and inject surprises into basic science. A new interdisciplinary field of applied science is emerging. See Fig. 10.

Portions of the text of this article and Figs. 4–9 are reproduced with permission from the paper "Microstructure Fabrication" (R. W. Keyes, *Science*, vol. 196, pp. 945–949, 27 May 1977), copyright 1977 by the American Association for the Advancement of Science.

References

Ballantyne, J. M. (editor): "Proceedings of the National Science Foundation Workshop on Opportunities for Microstructures Science, Engineering, and Technology," National Science Foundation, Washington, D.C., 1978.

Dill, F. H.: *IEEE Trans. Electron Devices*, **ED-22**, 440–444 (1975).

Fogiel, M.: "Modern Microelectronics," Research and Education Association, New York, 1972.

Horowitz, P., and W. Hill: "The Art of Electronics," Cambridge Univ. Press, New York, 1981.

Moore, G. E.: "Progress in Digital Integrated Electronics," in *Technical Digest*, 1975 International Electron Devices Meeting (December 1–3, 1975), IEEE, New York, 1975, pp. 11–13.

Petritz, R. F.: "Current Status of Large-Scale Integration Technology," *IEEE J. Solid State Circuits*, **SC-2**, 130–147 (1967).

Staff: "Scaling the Barriers to VLSI's Fine Lines," *Electronics*, pp. 115–126, (June 19, 1980).

Triebwasser, S.: "Large Scale Integration and the Revolution in Electronics," *Science*, **163**, 429–434 (1969).

Walker, E. J.: *IEEE Trans. Electron Devices*, **ED-22**, 440–444 (1975).

—Robert W. Keyes, IBM Corporation, Thomas J. Watson Research Center, Yorktown Heights, New York.

MILK AND MILK PRODUCTS.

Intricate chemical and microbiological problems arise in the production, processing, and distribution of milk products. Milk is a complex mixture of fat (4%), protein (3.5%), carbohydrate (4.8%) and mineral components (0.7%) and is an excellent bacterial growth medium; hence the need for care and cleanliness in handling.

Milk Fat. Milk fat or butterfat is a mixture of triglycerides of various fatty acids. Milk fat also contains a small percentage of cholesterol (0.37%), a substance characteristic of fats and oils of animal origin in contrast to those of plant origin which contain plant sterols. The phospholipids, lecithin, caphalin, and sphingomyclin are present in milk within the range of 0.03–0.04%. These are fatlike substances containing phosphorus and nitrogen. They have emulsifying properties and are associated with the fat-globule surfaces: hence their tendency to concentrate in butter and buttermilk.

Milk fat is distinguished from all other fats in that it is the only one containing butyric acid (C_4) as a component of glycerides. This acid occurs in the fat in a concentration of approximately 10 mole % and the total of C_6, C_8, and C_{10} fatty acids accounts for another 10 mole % of the component fatty acids. These acids (C_4–C_{10}) are often called the volatile fatty acids of milk fat. In the free form, they have a pungent characteristic flavor which is important in many types of cheese.

The deterioration of milk fat is an important cause of off-flavor development in dairy products and its control requires technical understanding of the processes involved. Three major types of deterioration associated with milk fat are recognized:

(1) *Rancidity*, due to free volatile fatty acids liberated from the glycerides by enzymic (lipase) hydrolysis. Lipases are normal components of raw milk, and are inactivated by the heat of pasteurization.

(2) *Tallowiness or oxidation*, due to autoxidation of unsaturated fatty acids with the production of flavorful unsaturated aldehydes. These reactions are accelerated by oxygen, high storage temperatures, and copper catalysts. Oxidation is usually the primary cause for spoilage of dried whole milk, cream, butter, and butteroil.

(3) *Heat-generated flavors*, due to the formation of lactones and methyl ketones from hydroxy and keto acid precursors, which occur in trace quantities in milk fat. These flavors are considered to be desirable in fried and baked goods and are partly responsible for the unique condiment properties of butter in food preparation. However, they are undesirable in dried whole milk and evaporated milk where the objective is to make a bland product as much like fresh milk as possible.

Proteins. The proteins of milk fall into two groups: casein, precipitable by both acid and proteolytic enzymes such as rennin; and whey proteins, which are acid soluble but heat-denaturable. There is about 3% casein in milk, the removal of which leaves a whey of approximately 1.0% nitrogenous matter. Of this 0.6% is heat-coagulable protein and 0.4% is nonheat-coagulable. The latter fraction is composed of protein-like fragments of a proteose or peptone nature plus other nitrogenous substances. Among these are small percentages of urea, creatin, creatinine, uric acid, and various forms of amino nitrogen.

Casein exists in milk as a calcium caseinate-calcium phosphate complex; the ratio of these components is approximately 95.2 to 4.8. The dispersed casein particles appear to be spherical in shape and of various sizes. The size distribution of the casein micelles is not constant, but varies with aging, heating, concentration, and other processing treatments. Processing alters the water-binding of casein and this in turn affects the apparent viscosity of products that contain casein. Changes in hydration have not been measured quantitatively although the casein particles of raw milk appear to consist of one volume of water-free protein and three volumes of solvate liquid.

The whey or serum proteins have been partially resolved into three relatively homogeneous, crystallizable proteins: (1) β-lactoglobulin (50% of total serum protein), (2) an albumin resembling the albumin of bovine blood (5% of the serum protein), and (3) α-lactalbumin (12% of the serum protein).

On heat denaturation, the serum proteins show decreased solubility at pH 4.7 and in concentrated salt solutions. There is some variability in response toward heat treatment, but complete denaturation will occur during heating within the range of 60 to 80°C for periods of time up to two hours. There is practically no denaturation during normal pasteurization. Heat increases the activity of the sulfhydryl groups and the sulfhydryl titer can be employed as a measure of denaturation. The —SH groups are readily oxidized in liquid systems and consequently they appear to act as antioxidants to protect milk fat in dairy products. The fat of fluid milk, heated to produce a high —SH titer, shows increased resistance to oxidation and this carries through to the dried product which exhibits superior storage stability, if it is made from high heat milk. The high sterilization temperature to which evaporated milk is subjected and the low oxygen content in the can protect this product from development of oxidized and tallowy flavors during storage. See **Protein.**

Milk is widely used as an ingredient for bread and other baked goods to which it adds substantial nutritional value. Milk is heated when used in bread to avoid softening of the dough and reduction of loaf volume. Why heat improves the baking properties of milk is not clear, but good baking properties have long been associated with low whey-protein-nitrogen values.

Lactose. The sugar of milk, lactose ($C_{12}H_{22}O_{11}$), occurs in the milk of all mammals. It is mildly sweet with a final solubility in water of 10.6% at 0°C, 17.8% at 25°, 29.8% at 49°, 58.2% at 89°. Lactose, on hydrolysis by acid or the enzyme lactase, yields a mixture of approximately equal parts of glucose and galactose, together with a small but variable quantity of oligosaccharides. The products of lactose hydrolysis are much more soluble than the original disaccharide. Lactose is a reducing sugar which is converted to lactobionic acid on mild oxidation. Two forms which differ in solubility and optical rotation are known. Alpha-lactose hydrate crystallizes at ordinary temperatures with one molecule of water, but this is lost with the formation of the anhydrous form during heating to a temperature between 149 and 200.3°F. Anhydrous β lactose, more soluble than alpha, crystallizes from supersaturated lactose solutions above 200.3°F. Solid beta-hydrate has never been prepared. The crystalline alpha-hydrate is stable in dry air at room temperatures, but both anhydrous forms readily absorb moisture and change to alpha-hydrate at ordinary temperatures. Alpha-lactose crystallizes out in some dairy products and because of the hardness of its crystals and their slow and limited solubility, "sandy" products may result.

The crystallization of lactose in frozen concentrated milk has been associated with a denaturation of casein which ultimately appears as a gel structure in the thawed product. Gelation in frozen milk can be retarded by enzymatic hydrolysis of part of the lactose before freezing or by addition of a polyphosphate salt.

Lactose, when fermented by lactic bacteria, is the source of the lactic acid formed in sour milk and whey. Lactose is helpful in establishing a slightly acid reaction in the intestine, which assists in calcium assimilation.

Mineral Components. When milk is heated to a temperature high enough to volatilize the water and oxidize the organic constituents, the residue of inorganic oxides that remains is called the milk ash; its major components are: K_2O, CaO, Na_2O, MgO, Fe_2O_3, P_2O_5, Cl, and SO_3. The calcium and phosphorus of the ash are of special interest because of their nutritional importance and because calcium phosphate is part of the casein micelle, influencing its physiochemical behavior toward coagulation with rennin, acid, and heat. Minor inorganic constituents are present in milk in trace amounts, i.e., iron, copper, zinc, aluminum, manganese, iodine, and cobalt.

Miscellaneous Components. The hydrogen-ion concentration of milk increases slightly with age, after milking, as natural carbon dioxide escapes. Most samples of cow's milk vary within the range of pH 6.5–6.7. Titratable acidity of fresh milk which may vary from 0.13 to 0.16%, expressed as lactic acid is an arbitrary measurement influenced by the protein and salt-buffer systems present in the particular sample. Citrates, phosphates, and carbonates are the principal buffers in milk.

Milk contains some important vitamins. The vitamin D content may vary from 30 I.U. per quart in summer to 6 in winter, depending upon the feed and the sunlight which reach the cow. Both pasteurized and evaporated milk are often fortified by the addition, on a fluid basis, of 400 I.U. of vitamin D per quart. Vitamins A, D, and E (alpha-tocopherol) are fat-soluble and stable at the heat treatments used in processing milk and milk products. The remaining vitamins are water-soluble and of varying stability. Vitamins B_1 (thiamine) and C (ascorbic acid) are partially destroyed by heat, while B_6 and B_{12}, are relatively heat-stable. Vitamin B_2 (riboflavin) is heat-stable but it is quickly destroyed by light. In spite of the varying sensitivity of the water-soluble vitamins toward heat, pasteurized milk is a good source of all the milk vitamins except C.

Two types of enzymes in milk are important: those useful as an index of heat treatment and those responsible for bad flavors. Phosphatase is destroyed by the heat treatments used to pasteurize milk; hence its inactivation is an indication of adequate pasteurization. Lipase catalyzes the hydrolysis of milk fat which produces rancid flavors. It must be inactivated by pasteurization or more severe heat treatment to safeguard the product against off-flavor development. Other enzymes reported to have been found in milk include catalase, peroxidase, protease, diastase, amylase, oleinase, reductase, aldehydrase, and lactase.

References

Considine, D. M., Editor: "Foods and Food Production Encyclopedia," Van Nostrand Reinhold, New York, 1982.

Ernstrom, C. A., and N. P. Wong: "Fundamentals of Dairy Chemistry," AVI, Westport, Connecticut, 1974.

Harper, W. J. and C. W. Hall: "Dairy Technology and Engineering," AVI, Westport, Connecticut, 1976.

Marth, E. H., Editor: "Standard Methods for Examination of Dairy Products, American Public Health Association, Washington, D.C. (Revised periodically).

Selitzer, R., Editor: "The Dairy Industry in America," *Dairy and Ice Cream Field* (publishers), New York, 1977.

—Byron H. Webb, U.S. Department of Agriculture, Washington, D.C.

MILLER INDICES. In mineral crystallography the identity of a crystal face consists of a series of whole numbers which are the products of the parameters relating to that face by their inversion, and where required the clearing of fractional values. A parameter is the relative intercept of a crystallographic axis on a given crystal face.

Assuming parameter values on a given crystal face to be $1a$, $1b$, $\frac{1}{2}c$ would on inversion yield $\frac{1}{1},\frac{1}{1},\frac{2}{1}$, parameters $1a$, $1b$, $2c$ would on inversion yield $\frac{1}{1},\frac{1}{1},\frac{1}{2}$, and parameters of $3a$, $2b$, $6c$ would on inversion yield $\frac{1}{3},\frac{1}{2},\frac{1}{6}$. Clearing the fractions in each instance would yield Miller Indices of (112), (221) and (231), respectively.

The three Miller Indices for a crystal face in all systems except the hexagonal, which requires four indices, are always given in the same order as their crystallographic axes, a, b, c respectively; a^1, a^2, a^3 in the isometric system; a^1, a^2, c in the tetragonal system; and a^1, a^2, a^3, c, in the hexagonal.

If the parameter intercepts for a given face are unknown, general indices (hkl) may be used if that face intercepts all three axes; four in the hexagonal with general indices ($hkil$). If a crystal face cuts two axes and parallels the third, general indices would be identified as ($h0l$), ($0kl$), or ($hk0$) as applicable to that face; in the hexagonal system as (*h0hl*) etc.

See also **Crystal** and **Mineralogy**.

MILLERITE.   The mineral millerite is nickel sulfide, NiS, whose slender hexagonal interwoven crystals so suggestive of hairs has led to the application of the name "capillary pyrites." It occurs also as radiated masses and coatings. It is brittle; hardness, 3–3.5; specific gravity, 5.48–5.52; luster, metallic; color, brass-yellow, often with an iridescent tarnish. Millerite is found in association with other nickel-bearing minerals and other sulfides. European localities are Bohemia, Westphalia, Wales, etc.; and in the United States at Antwerp, New York; with pyrrhotite in Lancaster County, Pennsylvania; at St. Louis, Missouri; Keokuk, Iowa, and Milwaukee, Wisconsin. In Canada millerite occurs in Oxford, Quebec, and in the famous Sudbury District, Ontario. It is used as an ore of nickel. Millerite was named for the English mineralogist, W. H. Miller.

MIMETITE.   The mineral mimetite is a chloroarsenate of lead corresponding to the formula $Pb_5(AsO_4)_3Cl$. It is monoclinic (pseudohexagonal); brittle; hardness, 3.5; specific gravity, 7.0–7.25; luster, resinous; color, usually yellow to brown but may be colorless or white; translucent. Mimetite is a rather rare secondary mineral occurring in altered lead deposits. Found in Bohemia; Saxony; Cornwall and Cumberland, England; South West Africa; Mexico; and in the United States, in Pennsylvania and Utah. The name *mimetite* is derived from the Greek word meaning *imitator*, because of the similarity of mimetite and pyromorphite. See also **Pyromorphite**.

MINERALOGY.   The science of mineralogy is concerned with the formation, occurrence, properties, composition, and classification of minerals. Various definitions of a mineral have been proposed. Possibly, the most acceptable may be, "a naturally occurring inorganic substance, usually crystalline, possessing a relatively definite chemical composition and physical characteristics." It should be pointed out that some naturally formed organic substances, particularly of an economic resource nature, are sometimes classified as minerals.

Although in its broadest application, mineralogy is as ancient as human civilization, mineralogy is a modern science, the mineralogist taking full advantage of all modern tools and instruments for exploration, analysis, testing, and study of minerals. Several major scientific advances in the materials field have stemmed from the study of minerals, as will be pointed out shortly.

Presumably in the early ages man used minerals as weapons. Through the passage of time and attainment of knowledge regarding certain mineral characteristics man learned, notably initially by accident but later by design, that the content of minerals provided essential materials for his expanding needs. Very early, the natural form and beauty of certain minerals become objects for personal adornment. Later it was found that the form and innate beauty of minerals were enhanced by cutting and polishing them. Although the science of mineralogy touches the life of every person, a fundamental understanding of minerals is not common.

Modern mineralogy is the product of research and discovery by many persons. Robert Hooke (1665) foretold the atomic theory by constructing models of alum crystals out of leaden musket balls. Nicolaus Steno (1669) discovered the constancy of interfacial angles between corresponding faces of quartz crystals from many localities. This was later formalized by Rome de l'Isle (1764) under his *law of constancy of crystal interfacial angles*. In 1784, René Just Haüy proposed the theory of *integral*

molecules by stacking calcite rhombs to show that structural units could produce exact external facial planes of various forms of calcite crystals. Haüy is known as the "father" of geometrical crystallography. A great advance in mineral studies was made in 1828 when Nicol invented the nicol prism for investigating the behavior of polarized light in crystallized minerals. In 1912, Max von Laue, a student of Roentgen, theorized that the wavelengths of x-rays and atomic spacing of crystals may be of the same magnitude. Laue found that the diffracted rays, when passed through a crystal, substantiated his theory. This discovery opened up an entirely new field of mineral research, i.e., *crystal chemistry*.

Origin of Minerals

Returning to the definition of mineral, minerals are substances which are the products of deposition and formation in an open, natural system as opposed to a closed, controlled laboratory system, and thus vary in many instances from an exact chemical formula and content. Foreign elements within the host open system may, under favorable conditions of chemical affinity, replace certain specific elements named in the given mineral formula. In certain minerals, the formula can be expressed in definite terms, such as SiO_2 for quartz. In other minerals, the formula may vary within restricted limits, such as (Zn, Fe)S for sphalerite, where iron substitutes for zinc within the sphalerite structure. There are two key factors in such chemical substitution within the mineral structure (1) reasonably comparable ionic radii; and (2) maintenance of electrical neutrality of the compound. Basically, a mineral is a homogeneous solid with an ordered atomic arrangment which places it in the category of a crystalline material and possessing a definite, though not a fixed, chemical formula.

Natural systems strive toward a state of equilibrium when all component units attain their lowest energy level. The respective energy level of the elements within any mineral is dependent upon the physical environment, principally the temperature, pressure, and chemical substances present, at the time and place of its formation. Any later change in its environment may cause a change in the mineral's composition and form. Whatever the primary or intermediate environmental conditions may have been, the mineral, as observed, represents its present equilibrium energy state or crystal structure.

Matter exists in three states—gaseous, liquid, and solid.* In the gaseous state, the elements move freely about within their environment, their only contact being haphazard collisions. Elements in the liquid state are in closer contact with each other, but still retain freedom of mobility. The solid state is characterized by the chemical elements combining under atomic bonds of various types and strengths into a structured system. For any element within the structure, there are equivalent elements in a definite three-dimensional crystalline pattern. Under restricted circumstances, two or more minerals may form in contact with one another in such a manner as to preclude their complete development as single crystals. Those minerals which develop fully and which exhibit well-formed polyhedral (many sides) facial planes are referred to as being *euhedral*. Those minerals that exhibit no facial planes are referred to as *anhedral*. Imperfectly formed crystals are referred to as *subhedral*. Crystalline minerals, where the crystallinity can be determined only by aid of a microscope, are said to be *microcrystalline*. Where the materials are so finely divided as to be discernible only by x-ray analysis they are referred to as *cryptocrystalline*. The rate

---

* Traditional concepts. Details on more advanced concepts on the states of matter are given elsewhere in this volume.

of crystallization plays an important part in the resultant crystalline character of minerals.

Crystalline solids are bonded together by electrical forces which originate in the constituent atoms. The position of the respective unit particles within the crystal structure are determined by geometric factors, along with considerations of electrical neutrality and lattice energy. One type of bond is produced by the lending or borrowing of electrons in an attempt to complete the outer electron shell of the atom. A stable compound is formed when the outer shell is complete and the mutual attraction is termed an *ionic bond*. For example, sodium with one electron in its outer electron shell lends that electron to a neutral chlorine atom with seven electrons in its outer electron shell. The result is a stable electron shell for each ion (charged atom) which then join by electrostatic attraction in a crystal structure to form sodium chloride, NaCl. Minerals so bonded are of moderate hardness, readily soluble, and poor conductors of electricity and heat.

The bond produced by the sharing of electrons by which a stable configuration is achieved is termed a *covalent bond*. For example, chlorine with seven electrons in its outer shell needs one additional electron to achieve stability. If its nearest neighbor is another chlorine atom, the two atoms combined in such a way that the one electron serves in the outer shell of each atom, thus achieving a stable configuration. Diamond is another example of this type. Carbon with four electrons in its outer shell mutually shares the four electrons with four adjoining carbon atoms, thus achieving a stable configuration by completing the shell with eight electrons. This type of electron-sharing or covalent bonding is the strongest of all chemical bonds. Minerals so bonded are generally insoluble, possess a high degree of stability, and are nonconducting.

Metallic bonds consist of a structure of positive ions through which free electrons can drift. The structure of metals may be envisioned as a mesh of electrons which surround the atomic nuclei and bind them together. This electron mobility produces minerals of generally a low hardness and of high electrical and thermal conductivity.

The *Van der Waals bond* is a very weak attraction between atoms, usually between essentially neutral atoms or groups of atoms. These neutral and essentially uncharged structure units are held together within the crystal lattice by virtue of small residual charges on their surfaces. This type of bonding is rare in minerals. Minerals so bonded often display ready cleavage and low hardness. For example, the foliated scales of graphite are linked weakly together by Van der Waals bonds.

Every mineral is a product of the redistribution or recombination of its component chemical elements to form a stable substance. The process is known as *crystallization*. The process may involve *precipitation* of chemical elements from aqueous solutions at the earth's surface; or from siliceous melts (magmas) from the earth's interior. In either situation, the process is dependent upon the degree of concentration of the constituent chemical elements present and the temperature/pressure conditions. Precipitation from vapor also is possible. An example is the hot vapor, rich in sulfur dioxide, which is emitted from vents associated with volcanoes. Upon becoming exposed to the cooler atmosphere, crystal sulfur is deposited around those vents. Snow crystals are another example of precipitation from vapor.

Crystal Structure

Associated chemical units become systematically arranged in the crystal structure which is constructed from a single motif that develops repetitively. The resulting three-dimensional array is called the *space lattice* of the crystal. The lattice or framework is defined by three directions and by the distance along those directions where the motif repeats itself. Because the units within the structure adhere to a strict arrangement, the external facial planes of a crystal represent the limiting surfaces of that growth and are an external expression of its internal atomic order. Crystals are formed, therefore, where constituent atoms or ions are free to combine in constant chemical proportions and are an expression of the environmental conditions which promote their formation.

Periodic repetitions of a space lattice cell in three dimensions from the original cell will completely partition space without overlapping or omissions. It is possible to develop a limited number of such three-dimensional patterns. Bravais, in 1848, demonstrated geometrically that there were but fourteen types of space lattice cells possible, and that these fourteen types could be subdivided into six groups called *systems*. Each system may be distinguished by symmetry features which can be related to four symmetry elements:

1. *Symmetry with respect to a point*. If through a central point in a geometric figure, lines are drawn from a point on one side of the figures to a similar point equidistant on the other side, the figure is symmetrical to a point.

2. *Symmetry with respect to a plane*. A geometrical figure is symmetrical with respect to a plane when for each edge, solid angle, or face on one side, there is a corresponding edge, solid angle, or face on the other side of that plane. One side is, in fact, a mirror image of the opposite side. That plane is called a *plane of symmetry*.

3. *Symmetry with respect to a line*. If during a complete revolution of 360° about a given axis, a geometrical figure repeats itself in appearance two or more times, it is said to be symmetrical with respect to a line, or to an *axis of symmetry*. Possible axes of symmetry are twofold, threefold, fourfold, and sixfold. A rotation of onefold or 360° is equivalent to no rotation at all.

4. When a crystal is rotated about an axis and inverted about the central point and at that point repeats itself, it is said to have an *axis of rotary inversion*. It is a twofold axis of rotary inversion if the geometrical figure is rotated 180° and then inverted. Additionally, there are threefold, fourfold, and sixfold axes of rotary inversion possible.

Within each of the six crystal systems, there are specific *crystal classes*. Each class displays distinctive symmetry elements. There are 32 possible classes distributed among the six crystal systems. One of the crystal classes within each system possesses all of the symmetry elements which are characteristics of its space lattice cell. These are called the *holohedral* class of that system. Other classes within each system possess somewhat fewer symmetry elements and are called *merohedral* classes.

It is significant to note that in most geometrical situations regarding crystal study, the distinguishing characteristic is symmetry, not geometry. This is especially true in the case of a mineral such as pyrite (FeS_2) which assumes a cubic shape, but its symmetry, controlled by the three sets of opposing striations on its crystal faces, identify it as belonging to a much lower order symmetry than that of a true cube.

The six crystal systems are identified by hypothetical lines of reference known as *crystallographic axes*, and their angular relationship to each other. Crystal orientation and axial order are given as: (1) front to back; (2) right side to left side; and (3) top side to bottom side. The front, right, and top side axial ends represent the positive ends of those respective axes. The back, left, and bottom side ends are designated as the negative ends. The relative intensity of crystal growth along those axes gives to each crystal face a distinctive identifying character. This character is evidenced by the physical similarity of equivalent facial planes on the crystal and their relationship to the crystallographic axes of that form. The six systems, their crystallographic axes, and interaxial angular relationship are defined and illustrated in the accompanying diagram.

Crystallographic identification of facial planes on a crystal become possible through assignment of numerical values to a face which represents its relationship to the crystal axes. *Parameters*, or relative intercepts, are obtained by plotting coordinates of crystal faces with respect to their crystallographic axes. The actual distances of axial intercepts of the crystal face are determined and expressed as a unit of measurement. The product of these values is known as *Miller indices*.

A Miller indices face on the front face of a cube would be (100), signifying that face intersects the a^1 axis at 1 unit length from the center of the crystal, and is parallel to axes a^2 and a^3, or intersects those axes at infinity. In this system, zero (0) is the numerical substitute for infinity. A (111) Miller indices face identifies that facial plane as intersecting each of the three crystallographic axes of that form at 1 unit length from the crystal center.

The Miller Indices identify the orientation of a face in relationship to its axes of reference regardless of its size and position on the crystal. Haüy first proposed this basic law of crystallography—"crystal faces make simple rational intercepts on suitable crystal axes." Inasmuch as the intercepts are simple, it follows that the Miller indices should likewise be simple whole numbers.

See **Crystal**.

Crystal Forms

Form is used here to designate the general outward appearance of a crystal, specifically to a group of crystal faces which bear identical relationships to crystal's symmetry elements. It is essentially a geometric form with equivalent facial planes in their relationship to the crystal lattice symmetry elements. In this regard an octahedron in the isometric system would be identified as a "closed form," inasmuch as its eight equivalent faces totally enclose the crystal space; its form identification would be [111], enclosed in brackets. Even though the crystal faces may be equivalent they may vary widely in size and distance from the crystal center, owing to irregular or distorted development during their formation. Brackets shown around Miller indices, e.g., [111], signify form identification, as opposed to Miller indices shown in parentheses, e.g., (111), which identify a specific crystal face only. In crystals possessing different lengths of their crystallographic axes the general form [hkl] would be used. The hexagonal system, with four axes, requires a four-unit form identification.

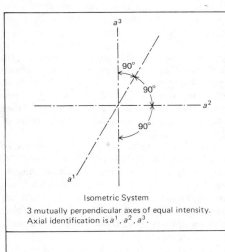

Isometric System

3 mutually perpendicular axes of equal intensity. Axial identification is a^1, a^2, a^3.

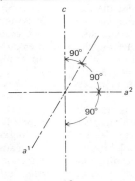

Tetragonal System

3 mutually perpendicular axes; 2 of equal intensity, the third unequal. Axial identification is a^1, a^2, c.

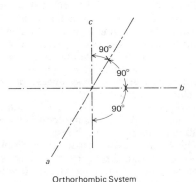

Orthorhombic System

3 mutually perpendicular axes; all of unequal intensity. Axial identification is a, b, c.

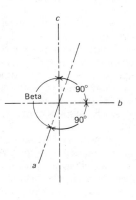

Monoclinic System

2 mutually perpendicular axes, and 1 axis inclined to the plane of the other 2 axes. Axial identification is a, b, c, with axis a as the inclined axis.

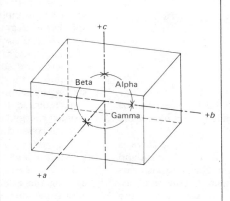

Triclinic System

3 unequal axes, all intersecting at oblique angles. Axial identification is a, b, c. Interaxial angles identified as alpha, beta, and gamma.

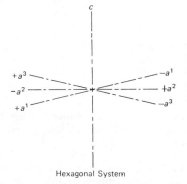

Hexagonal System

4 axes, 3 of equal intensity in a horizontal plane intersecting at angles of 60°, and a fourth vertical axis of unequal intensity perpendicular to the plane of the other three axes. There are two divisions within this system: the *hexagonal* and *rhombohedral*. The first displays hexagonal or 6-fold symmetry; the second displays rhombohedral or 3-fold symmetry.

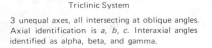

Crystallographic axes and interaxial angular relationship of six crystal systems.

In the isometric system, crystals are recognized by their geometrical forms, e.g., cube, octahedron, trapezohedron, pyriteohedron, and so on. In all other systems, crystal faces are given specific names which refer to their relationship with their respective crystallographic axes. The most common facial forms in these systems are *pedion*, *pinacoid*, *dome*, *prism*, and *pyramid*. Crystal forms bear specific relationships to symmetry axes or planes, and not to the general shape of the crystal.

Physical Mineralogy

This aspect of mineralogy is concerned with several observable physical characteristics, such as color, hardness, lustre, fracture, cleavage, magnetic properties, radioactivity, fluorescence, and specific gravity.

Color. The color of a mineral is a product of selective absorption of certain wave lengths of visible white light by atoms within the mineral structure. A mineral color may be indicative of a species, but more often is of descriptive value only. There are two broad classification categories of color in minerals—*allochromatic* and *idiochromatic*. Allochromatic minerals are those which occur with variable colors, such as quartz, SiO_2; corundum, Al_2O_3; and calcite, $CaCO_3$. Color in such minerals may be a product of included foreign elements, rate of crystallization, or a defective lattice structure. Idiochromatic minerals are those in which the color is a characteristic constant, such as the green of malachite, $Cu_2CO_3(OH)_2$; the blue of azurite, $Cu_3(CO_3)_2(OH)_2$; the black of magnetite, Fe_3O_4; and the red of cinnabar, HgS. The actual color of most minerals can be obtained by rubbing the mineral on an unglazed porcelain plate and observing the powder (streak) of that mineral left on the plate. This test is essential in certain instances inasmuch as many minerals display a physical color foreign to their actual color, e.g., hematite may appear black or blue, but its streak color is always red.

Hardness. The hardness of a mineral is its resistance to scratching. Testing for hardness is based upon the premise that all minerals possess such resistance to a lesser or greater degree. The Mohs scale of hardness has been universally adopted to test this physical property. In the list below, the ascending numeric order of the mineral named will scratch all of those of lower order:

1 Talc
2 Gypsum
3 Calcite
4 Fluorite
5 Apatite
6 Orthoclase
7 Quartz
8 Topaz
9 Corundum
10 Diamond

In a relative sense, minerals from 2 to 2.5 hardness can be scratched by a fingernail; 4 by a penny, and 5 to 6 by a knife blade or piece of glass.

This scale is strictly relative and nonlinear because there is a much wider differential, for example, between corundum and diamond than between topaz and corundum. Hardness also varies according to crystallographic direction in certain minerals. This test should always be made on a freshly broken area because certain minerals are subject to surface alteration. A hardness test on an altered area can be misleading.

The hardness (Mohs scale) of several minerals is given in Table 1.

Lustre. The lustre of a mineral is a product of both light

TABLE 1. HARDNESS OF REPRESENTATIVE MINERALS
(Mohs Scale)

Agate	6–7	Galena	2.5
Alabaster	1.7	Garnet	6.5–7
Andalusite	6.5–7.5	Graphite	0.5–1
Aragonite	3.5	Kaolinite	2.0–2.5
Asbestos	5.0	Magnetite	6
Barite	3–3.5	Marble	3–4
Beryl	7.8	Mica	2.8
Corundum	9	Opal	5.5–6.5
Diatomaceous earth	1–1.5	Pumice	6
Dolomite	3.5–4	Pyrite	6–6.5
Emery	7–9	Serpentine	3–4
Flint	7	Tourmaline	7–7.5

reflection and refraction. Reflection is the governing factor in translucent to opaque minerals; refraction is transparent minerals. There are two broad categories in describing lustre—*metallic* and *nonmetallic*. Minerals of nonmetallic lustre are further subdivided into categories, such as vitreous (glass); adamantine (diamond); resinous (sphalerite); silky (asbestos); waxy (chalcedony and opal); greasy (some quartz and diamonds); and pearly (talc or mica).

Fracture. The product of irregular breaking of a mineral is termed fracture. Categories include conchoidal or conch-like (quartz); uneven (serpentine); and hackly (copper). Fracture in a mineral is unrelated to its crystal structure.

Cleavage. The cleavage of a mineral is the product of regular breaking of the mineral along specific surfaces related to the mineral's internal structure. The cleavage plane is always parallel to a possible crystal face, and, therefore, is a reflection of the atomic structure of the mineral. Three planes of cleavage are present in the cube and rhombohedron. Four cleavage planes produce an octahedron, six planes a dodecahedron, both of which fall within the isometric system.

Magnetic Properties. A few minerals possess the property of being attracted by a magnet and they are known as *ferromagnetic* minerals. The more common ferromagnetic minerals are magnetite, Fe_3O_4, and pyrrhotite, $Fe_{2-x}S$. Those minerals which are natural magnets are known as *lodestones*. Most minerals are affected to some extent in a magnetic field. The minerals which are repelled are known as *diamagnetic*; those which are weakly attracted are known as *paramagnetic*.

Radioactivity. Basically, this involves the spontaneous disintegration of uranium and thorium minerals. Both of these mineral species disintegrate at a steady rate that is completely unaffected by external chemical or mechanical conditions. The end-product of this disintegration is lead within the mineral structure. Geophysicists are able to determine the geologic age of a specific uranium or thorium mineral and its host environment by measuring the amount of lead present and computing the known time required to produce this amount. With the impetus of atomic energy, many previously unknown uranium minerals were uncovered and identified.

Fluorescence. This is the property of certain minerals to absorb ultraviolet radiation and convert that energy into visible light. The rays energize certain elements within the mineral, causing the excitation of electrons in the orbital shell. If the activating source is removed and visible light continues for a period of time following, the phenomenon is known as *phosphorescence*. The visible light continues until the electrons return to their normal orbital electron shell.

Specific Gravity. The density or specific gravity of a substance is the amount by which that substance decreases in apparent weight when first weighed in air, followed by weighing in water. The density is obtained by dividing the weight of the substance in air by the loss of its weight in water.

Isomorphism

Isomorphism is the substitution of an atom, ion, or radical for another within a mineral compound. The degree of substitution is controlled by two factors: (1) temperature, and (2) ionic size. A high temperature allows a greater degree of thermal disorder. Only ions of similar size will readily substitute for one another.

Under certain conditions, isomorphism may be either complete or partial. The spinel and garnet mineral groups represent a complete isomorphous series. Partial isomorphism is exemplified by chemical components displaying extensive ionic substitution at high-temperature levels, which, upon cooling, may segregate according to their respective radii sizes and then group together, forming two separate mineral phases. This phenomenon is known as *exsolution* and is common in the plagioclase feldspar mineral, perthite. Other isomorphous examples include calcite and siderite; and magnesite and siderite.

Polymorphism

Polymorphism is the capability of a substance to exist in more than one crystal form. The basic controlling factors appear to be temperature-pressure conditions at the time of formation which controls the type of atomic packing within the structure. Examples of polymorphism are given in Table 2.

TABLE 2.   REPRESENTATIVE POLYMORPHIC FORMS

Chemical Composition	Mineral	Crystal System	Specific Gravity
Carbon	Diamond	Isometric	3.5
Carbon	Graphite	Hexagonal	2.2
Al_2SiO_5	Sillimanite	Orthorhombic	3.2
Al_2SiO_5	Kyanite	Triclinic	3.7
FeS_2	Pyrite	Isometric	5.0
FeS_2	Marcasite	Orthorhombic	4.9

Pseudomorphism

Pseudomorphism is the change of the original chemical composition of a substance into some other equally definite compound by the action of natural agencies. Pseudomorphism exists when the external crystalline form of a mineral is inconsistent with its internal chemical composition and atomic structural arrangement. It is always a secondary process. The altered substance is known as a *pseudomorph*.

Types of pseudomorphic alteration include:

Substitution. This is a process whereby silica replaces wood fiber to form silicified (petrified) wood. Quartz replacing fluorite is another example. In the latter instance, the original fluorine in the fluorite is removed by silica-rich solutions which first remove the fluorine and then substitute silica in its place.

Incrustation. This is a process whereby one mineral forms a crust over another mineral, e.g., prehnite over anhydrite crystals. Later solutions may remove the anhydrite, but the space occupied by the anhydrite remains as a cast surrounded by the prehnite.

Alteration results from the addition of new material or partial removal of original material from a mineral, e.g., anglesite after galena; or gypsum after anhydrite.

Paramorphism results when the internal crystal structure of a mineral is changed to a polymorphous form, yet retaining the external crystal form of the original mineral, e.g., aragonite after calcite.

Twinning

Twinning in crystals results from the intergrowth of two or more individuals in such a way as to yield parallelism in the case of certain parts of the different individuals and at the same time, other parts of the different individuals are in reverse positions in respect to each other. For example, an octahedral crystal of magnetite is twinned when one-half of the crystal is rotated 180° parallel to an octahedral facial plane. This type of twinning is known as *spinel twinning*, owing to its common occurrence in the spinel group of minerals.

Mineral aggregates are often grouped together to form compound crystal structures. Such grouping may be a product of irregular and accidental growth which do not conform to basic twinning laws. Two or more intergrown crystals should not arbitrarily be labeled as twins unless their twin relationship can be established.

Rock Types, Associated Minerals, and Their Uses

Minerals are the basic building blocks of all rock types found in the earth's crust. Rocks are classified into three broad categories: *igneous*, *sedimentary*, and *metamorphic*.

Igneous rocks have their origin deep within the earth's interior from molten magmas of siliceous (silica-rich) melts. The rocks resulting from deep-seated solidification are known as *plutonic*; those which have been extruded on the surface as lava flows are termed *extrusives*. Plutonic rocks are products of slow cooling and possess well-formed crystalline structure, e.g., granite. Extrusive rocks are products of fast cooling and occur as glassy or microcrystalline masses, e.g., obsidian and basalt.

Minerals occurring in igneous rocks possess crystalline character, but their rate of precipitation from the parent melt prevented their development as euhedral crystals and thus they occur as granular (anhedral) aggregates within these rocks. The more prominent component minerals include quartz, feldspar, and feldspathoid family members, micas, pyroxenes, amphiboles, and olivine. Zircon, magnetite, ilmenite, hematite, apatite, pyrite, and garnet are commonly associated in these rock types.

Pegmatites represent a residual phase of igneous depositions, characterized by extremely coarse crystalline material, that results from the presence of associated volatiles, e.g., water vapor, carbon dioxide, sulfur dioxide, and others, which decrease the viscosity and facilitate crystallization. Quartz, feldspar, and mica are the more common minerals found in this environment, but such bodies are also hosts for many rare minerals and several types of gem stones, e.g., beryl, tourmaline, and topaz.

Sedimentary rocks are products of deposition from either the mechanical or chemical breakdown of all pre-existing rock types. Precipitation from aqueous bodies rich in soluble salts produces economically valuable beds of halite and gypsum. Weathering of iron-bearing rocks has produced extensive deposits of hematite. Limestone, chalk, and diatomite are products of biochemical precipitation. Minerals commonly associated with sedimentary rocks include calcite, galena, sphalerite, pyrite, marcasite, fluorite, barite, celestite, and quartz.

Metamorphic rocks are those that have undergone a reconstitution or redistribution of the chemical elements contained in the original formations to new mineral species. The process of change involves attaining a state of equilibrium of the constituent elements with the newly imposed environment. Chlorite, biotite, garnet, staurolite, andalusite, kyanite, and sillimanite, with ubiquitous quartz, are a few of the more common minerals associated with metamorphic rocks.

Minerals of Primary Industrial and Economic Importance

Economically valuable mineral occurrences are the products of particular types of worldwide geological formations which are, or have been, hosts to such minerals. In each occurrence, the minerals found represent the products of geological processes, which include:

1. The character and concentration of the mineral components within the source magmas from which the minerals were formed.

2. Secondary deposition from either percolating solutions or gaseous emanations from intrusive formations, causing chemical reactions within the intruded formations.

3. Precipitation from chemically supersaturated solutions.

4. Alluvial deposits resulting from the erosion weathering of the original host rocks.

Minerals which are a product of the first type of formation include the basic native metals, gold, silver, and copper.

Gold is primarily a product of deposition from ascending hydrothermal solutions associated genetically with siliceous-rich igneous rocks. Pyrite and other sulfide minerals are common associates within which the gold is often physically admixed. Surficial weathering of such deposits removes the sulfides, leaving free gold as a residual deposit. Erosion of these deposits results in alluvial deposits of placer gold, both as flakes and nuggets. Other characteristics and the uses for gold are described under **Gold.**

Silver occurs both as native ore and in combination with various silver sulfide minerals. Native silver is predominantly a product of primary deposition from hydrothermal solutions. Minor occurrences are products of oxidation of silver sulfide minerals with which the native ore is secondarily associated. See also **Silver.**

Native copper commonly occurs in the oxidized zones of copper deposits in association with cuprite, malachite, and azurite. The native copper deposits on the Michigan Keeweenaw Peninsula represent an exceptional occurrence. The copper occurs there as veins within igneous trap rocks interbedded with conglomerates. See also **Copper.**

Similar and valuable minerals include cinnabar (mercury); antimony (for type metal and battery plates); galena (lead and silver); argentite, pyrargyrite, and proustite (silver); sphalerite (zinc); chalcocite, chalcopyrite, bornite, malachite, azurite, and cuprite (copper); nickeline and pentlandite (nickel); bauxite (aluminum); magnetite, hematite, and goethite (iron).

Diamonds were found originally as loose crystals in geologically ancient alluvial stream beds. Later, their host formations were found to be a basic igneous rock (kimberlite) in the Republic of South Africa. Diamonds are the products of extremely high-temperature, high-pressure environment and are composed of pure carbon. See also **Diamond.**

Trap rocks (basalts) are products of volcanic action, either as extensive lava flows, or as intrusive dikes in pre-existing rocks. Secondary mineralization within such rocks from circulating waters produces interesting suites of zeolitic minerals, such as analcime, heulandite, natrolite, stilbite, mesolite, and others.

These minerals possess the ability to exchange ions contained in the mineral structure for those in solutions. This facility promotes the use of zeolitic minerals (or their synthetic counterparts) as water softeners. Water rich in calcium (hard water), when passed in solution through a tank containing zeolites, loses the calcium ions by absorption in the zeolite structure, with substitution of calcium ions by sodium ions. A reverse process may be initiated with the sodium ions replacing calcium ions in the structure, thereby reconstituting the original zeolite composition.

Halite, gypsum, and anhydrite are products of precipitation from large bodies of supersaturated salt water. The salt and gypsum of commerce are derived from such deposits. See also **Gypsum; Sodium Chloride.**

Landlocked inland seas and lakes become enriched with various soluble elements from waters draining into those basins. Sylvite and carnallite are valuable for their potassium content. They represent the final evaporation products of landlocked bodies of supersaturated sea water. Two famous localities are Stassfurt, Germany and near Carlsbad, New Mexico. Minerals formed from the evaporation of boron-rich waters include borax, kernite, colemanite, and ulexite. See also separate alphabetical entries for these minerals. The only known locality for kernite is the Mohave Desert in California, where a deposit of great extent exists—with potential reserves of millions of tons beneath the desert floor.

Pegmatites are valuable mineral sources. These formations represent the residual phase of igneous crystallization from magmas rich in siliceous content. As crystallization of their component elements proceeds, these magmas become increasingly enriched with volatile substances (mineralizers), such as water vapor, carbon dioxide, chlorine, fluorine, phosphorus, and others. The volatiles reduce the viscosity of the residual magmas and facilitate crystallization, as previously mentioned. When the residual liquids are injected into cooler rocks, they crystallize from their peripheral borders inward. The great mobility of the constituents enhances the growth of large mineral crystals—a characteristic feature of pegmatite bodies. Beryl and spodumene crystals from pegmatites attain sizes in terms of feet and tons. Feldspar, quartz, and mica crystals of comparable character are not uncommon.

There are two genetic pegmatite types—*simple* and *complex*. Simple pegmatites are recognized by their coarse texture and normal granite components, e.g., quartz, feldspar, and mica. Pegmatites produce the feldspar of commerce and mica for industrial and commercial uses. See also **Feldspar; and Mica.** Complex pegmatites are characterized by the presence of rare elements, in addition to the normal feldspar, quartz, and mica. Such bodies are also hosts for many semiprecious gem stones, such as amethyst, rose quartz, topaz, tourmaline, beryl, and chrysoberyl. Many rare-earth minerals obtained from complex pegmatite minerals include columbite/tantalite (columbium or niobium; tantalum), lepidolite, triphylite, spodumene, amblygonite (lithium), zircon (zirconium), and monazite (thorium oxide).

A most unusual pegmatite occurs near Ivigtut, Greenland. This consists of a cryolite with subordinate siderite, chalcopyrite, galena, and sphalerite. Cryolite is a fluoride of sodium and aluminum. For many years, cryolite was mined from this single occurrence for use as a flux in the electrolytic recovery of aluminum from bauxite, the major ore source of aluminum. Synthetic sodium aluminum fluoride essentially has replaced the need for natural cryolite. See also **Aluminum; Bauxite; and Cryolite.**

Quartz and tourmaline crystals once were commercially important for their piezoelectric properties as radio oscillation wafers and other electronic and instrumental uses. Synthesized quartz crystals have largely replaced the need for natural quartz for such applications.

Nuclear fission reactors are supplied with materials from uranium-bearing minerals of primary origin, e.g., uraninite/pitchblende, and other uranium-bearing minerals of secondary origin, e.g., carnotite, tyuyamunite, torbernite, and autunite.

Minerals of economic importance within sedimentary formations include, but are not limited to fluorite, barite, phosphorite, and oolitic hematite. Fluorite is utilized as a flux in steelmaking and when of high quality as lenses and prisms in the optical industry. Barite is an essential mineral used in gas- and oil-well drilling. Phosphorite, a product of chemical precipitation from seawater, when treated with sulfuric acid, produces superphosphate fertilizer. Oolitic hematite deposits of extensive size are important sources of iron ore. Garnet is a common mineral component in metamorphic rocks. A major occurrence of this type is at the summit of Gore Mountain, North River, in Warren County, New York. The garnet is a composit of almandine and pyrope with a hardness exceeding that of most world garnets. Gore Mountain garnet retains sharp cutting edges even when crushed to sub-micron size, making it an outstanding abrasive. It is used extensively as an abrasive (garnet paper) and as a glass-polishing agent in the optical industry.

Titaniferous iron ores, represented by the mineral ilmenite, occur within crystalline metamorphic environments. These ores are the major source of titanium. See also **Titanium.**

United States Imports of Minerals

With exception of the Soviet Union, the United States is more autonomous in nonfuel minerals than any other nation. U.S. Department of the Interior estimates indicates that all nonfuel mineral imports by the United States were approximately 6 billion dollars in 1971 and that these imports will increase to about 20 billion dollars by 1985.

Ocean Sources of Minerals

Mineral requirements for future world needs has focused increased attention to potential ocean resources. Major attention has been directed to petroleum resources. Associated with the petroleum are salt domes or bedded salt deposits, often with anhydrite and sulfur. Their potential is dependent upon development of economically feasible recovery methods. Not enough is known at this time about the origin of sulfur to satisfactorily predict precise occurrences. The Frasch process is presently being utilized in certain offshore deposits of this type. The economics of sulfur also may be affected by availability of large quantities of the element from Claus recovery units used in connection with the desulfurization of flue gases.

Valuable deposits of detrital sands and lime muds occur on the continental shelves of many world areas. These can be recovered by dredging operations. Diamonds are presently being recovered by means of vacuum suction tubes from detrital subsea sands adjacent to the Orange River section off the south-western African coast. Dependent upon the near-shore geology, it is known that iron, copper, and coal deposits extend into the subsea areas. In several world areas (Scotland and Japan for coal; Finland and Canada for iron ore; the English coast for tin and copper) deposits have been mined from underground entrances from the adjacent land areas. Sphene and zircon, plus other heavy minerals, have been noted in Texas offshore sediments.

Phosphorite, a major source of phosphorus, is known to occur both as nodular masses and crusts on rocks in subsea areas. Although enormous amounts of phosphorite are accessible in relatively shallow water, marine phosphorites have not been economically competitive with terrestrial supplies.

Metallic sulfides of copper, zinc, and iron have been found in central oceanic rocks and muds under conditions which indicate their deposition from hydrothermal solutions. Such solutions, rich in carbon dioxide, leached metallic elements from both basic rock masses and sedimentary formations with which they came in contact. When such solutions ascended, with concomitant cooling, the minerals were precipitated in the overlying sediments.

Manganese and iron oxides occur as nodular masses in many subsea world areas. They presently are of more interest for their copper, nickel, and cobalt content than for the manganese. Most extensive occurrences are at great ocean depths, as much as 3,500 to 4,500 meters. Fullest exploitation of these deposits and the metallic sulfides will require not only additional technical knowledge for their initial recovery, but also for their refinement to a marketable form. Again, the economics depends upon future demands for the metals as the continental deposits become depleted. Beyond these considerations, the persistent problem of ownership of oceanic resources must be solved as a condition of large-scale recovery. See also **Manganese.**

Lunar Rocks

Geological specimens collected on the Apollo lunar missions are indicative of an anhydrous igneous origin. There are three major rock types: (1) a potassium-rich basalt; (2) anorthosite; and (3) an iron, titanium-rich basalt. The first two types are prevalent in the highland areas; the latter in the maria terrain. They occur as crystalline vescicular masses, breccias, and regolithic mantle dust. The absence of an atmosphere and weathering processes on the moon has left the rocks and their component minerals unchanged through eons of time since their formation. Secondary mineralization, therefore, is generally absent and the rocks exhibit a rather limited mineralogy.

Lunar rocks differ in their chemical content rather than type of rock from their terrestrial counterparts. They consistently contain more titanium and chromium, less sodium, and most are richer in iron content. Lunar plagioclase, the major mineral component of anorthosite, is almost always the calcium-rich anorthite, $CaAl_2Si_2O_8$, indicating extreme magmatic differentiation. Lunar basalts are olivinerich and have been found to be from 3 to 10 times richer in ilmenite (titaniferous iron) as compared with terrestrial basalts.

Clinopyroxene materials which are common in terrestrial basalts are well represented in lunar rocks. They include diopside, hedenbergite, johannsenite, aegerine-augite, spodumene, jadeite, augite, pigeonite, omphacite, and fassaite. The more prominent lunar mineral species noted include ilmenite, with rutile intergrowths in certain subfloor maria basalts, cristobalite/tridymite, and pyroxferroite, a mineral closely related in both structure and composition to terrestrial pyroxmangite. Accessory minor minerals include troilite, chromite, ulvospinel, apatite-whitlockite, potash feldspar, quartz, hafnium-rich baddeleyite, and perovskite. Two newly classified species have been recorded—armalcolite, an iron-magnesium titanate, named after Apollo astronauts *Arm*strong, *Al*drin, and *Col*lins; and zirkelite, an oxide of calcium, iron, thorium, uranium, with titanium, niobium, and zirconium. Euhedral iron crystals in a pyrozene-rich vug of recrystallized breccia were recovered on the Apollo 15 mission.

Tiny translucent-to-opaque glassy spherules are prominent in the lunar regolith, within which ilmenite (as thin plates) with minor olivine are present.

Lunar mineralogy is generally analogous to that of terrestrial basalts. The major difference being the lack of oxygen during crystallization which has resulted in the presence of free iron and the exotic minerals, such as troilite, pyroxferroite, armalcolite, and zirkelite. The bulk mineralogy, however, is quite similar to terrestrial rocks with pyroxene, plagioclase, ilmenite, and olivine as the dominant minerals.

Phase equilibrium research of mineral solids has revealed vital information regarding their molecular structure. Application of knowledge gained from this research has extended into the fields of metallurgy, glass and ceramics, and a more adequate interpretation of mineral geology.

Mars Surface Geology/Mineralogy

Mariner 9 fly-by of Mars revealed a surface terrain of massive blocks of tumbled character cut by ridges and graben-type troughs. Huge volcanic peaks dominate a pockmarked landscape. Extensive channels characteristic of concentrated erosive powers of torrential floods were also evident, as were braided stream systems emanating from what were resolved to be plateau-type elevations.

Viking spacecraft equipped with a hoe-type scoop and spectrometer analyzed the surface soil to be of a character suggestive of an igneous mafic rock origin, rich in magnesium and iron. Landers' spacecraft analysis of the surface soil chemical composition by X-ray fluorescence revealed a low SiO_2 concentration (~45%) with iron as Fe_2O_3 near 20%. Further analysis revealed that the regolith mantle soil consisted essentially of iron-rich clay mineral with iron hydroxide, and minerals of sulfate and carbonate content with approximately 1% water by weight. Magnets attached to Vikings' scoop attracted magnetic material aggregates. It is quite probable that the magnetic material represents a component part of that regolith and possibly the soil is enriched by both magnetite (Fe_3O_4, color black) and maghemite (γ-Fe_3O_4, color yellowish-brown). The yellowish-brown surface color may be product of thin coating of hydrated iron oxides, with nontronite/montmorillionite as the host soil.

Much remains to be resolved before final definitive answers can be given in this area of planetary investigation and evaluation.

Classification of Minerals

Minerals are classified in groups according to their chemical composition, based upon the dominant anion or anionic group. The system works well in various ways. Generally, the dominant anion or anionic group brackets minerals of corresponding characteristics which tend to occur in quite similar environments. The dominant chemical subdivisions are:

Elements. Minerals composed of uncombined chemical elements, e.g., Au, gold; Ag, silver; Cu, copper; although minor impurities may be present within the structure.

Sulfides. Minerals composed of compounds of metals with sulfur.

Sulfosalts. Minerals composed of compounds of semimetals with sulfur.

Halides. Minerals composed of compounds of metals with flourine, chlorine, bromine, and iodine.

Oxides and Hydroxides. Minerals composed of compounds of the metallic elements with oxygen.

Carbonates. Minerals composed of compounds of a metal with the carbonate radical CO_3.

Borates. Minerals composed of compounds of a metal with the borate radical, BO_3.

Nitrates. Minerals composed of compounds of a metal with the nitrate radical, NO_3.

Sulfates. Minerals containing the sulfate radical, SO_4.

Chromates. Minerals containing the chromate radical, CrO_4.

Molybdates. Minerals containing the molybdate radical, MoO_4.

Tungstates. Minerals containing the tungstate radical, WO_4.

Phosphates. Minerals containing the phosphate radical, PO_4.

Arsenates. Minerals containing the arsenate radical, AsO_4.

Vanadates. Minerals containing the vandate radical, VO_4.

Silicates. This mineral classification encompasses the largest group of mineral species and includes most of the important rock-forming minerals, such as the feldspars, feldspathoids, pyroxenes, amphiboles, micas, olivine, and quartz. Silicon is the basic chemical element, as the name implies. The small silicon cation combines with four oxygens to form an SiO_4 tetrahedral structure. The SiO_4 formula leaves a net negative charge which requires additional combinations with other tetrahedra or anions to effect a neutral balance. The type and degree of such tetrahedral combinations control the final structural character and act as a convenient classification of the silicate mineral family.

Subclassification of silicates are:

Nesosilicates, with each tetrahedron existing within the structure as isolated SiO_4 units.

Sorosilicates involve the pairing of SiO_4 tetrahedra. The shared-oxygen anion represents the link between these tetrahedra.

Cyclosilicates involve two oxygens from each SiO_4 tetrahedron combining with oxygen in adjacent tetrahedral units to form *ring* structures.

Inosilicates are the product of oxygen sharing between adjacent tetrahedra to form *single* or *double* chains. In the single-chain structure, two oxygens from each tetrahedra combine with adjacent tetrahedra. In the double chain, half of the tetrahedra share three oxygens, while the other half share only two.

Phyllosilicates involve the sharing of three oxygens in each tetrahedron with adjacent tetrahedrons to form *sheet* structures. Minerals in this classification are usually flaky in character and relatively soft.

Tectosilicates involve the sharing of all four oxygens in each tetrahedral unit with adjacent tetahedrons to form a three-dimensional *framework* of SiO_4 units linked together. The product is a strongly bonded structure with a silicon-oxygen ratio of 1:2. The greater portion of the earth's crust is composed of minerals found within this classification.

References

Deer, W. A., Howie, R. A., and J. Sussman: "Rock-Forming Minerals," vol. 1 (Ortho- and Ring Silicates), 1962; vol. 2 (Chain Silicates), 1963; vol. 3 (Sheet Silicates), 1962; vol. 4 (Framework Silicates), 1965, and vol. 5 (Non-Silicates), 1964, Wiley, New York: Longmans, Green, London.

Derry, D. R.: "A Concise World Atlas of Geology and Mineral Deposits," Wiley, New York, 1981.

Flint, R. F. and B. J. Skinner: "Physical Geology," Wiley, New York, 1974.

Frondel, C.: "The System of Mineralogy," 7th edition, vol. 3, Wiley, New York, 1972.

Frye, K.: "Modern Mineralogy," Prentice-Hall, Englewood Cliffs, New Jersey, 1974.

Gary, M., McAfee, R., and C. L. Wolf (editors): "Glossary of Geology," American Geological Institute, Washington, D.C., 1972.

Issues of *Geotimes*. "Lunar Samples from the Moon—Preliminary Results," (October 1972); "Apollo 17 Exploration at Taurus-Littrow," (November 1972); "Astrogeology," (January 1973); "4th Lunar Science Conference," (June 1973); "Astrogeology-Geochemistry," (January 1974).

Hurlbut, C. S.: "Dana's Manual of Mineralogy," 18th edition, Wiley, New York, 1971.

Mason, B. and W. G. Melson: "The Lunar Rocks," Wiley, New York, 1970.

Pirie, R. G.: "Oceanograph-Contemporary Readings in Ocean Sciences," Oxford University Press, New York, 1973.

Roberts, W., Rapp, G., and J. Weber: Encyclopedia of Minerals, Van Nostrand Reinhold, New York, 1974.

Staff: "Preliminary Examination of Lunar Samples from Apollo 11," *Science*, (September 19, 1969).

Issues of *Scientific American*: "The Physical Resources of the Ocean," (September 1969); "The Red Sea Hot Brines," (April 1970); "Plate Tectonics and Mineral Resources," (July 1973); "The Lunar Rocks," (October 1971); "The Surface of Mars" (March 1978); "Geology of the Planet Mars," edited by Vivien Cornitz (1979).

—Elmer B. Rowley, Union College, Schenectady, New York.

MINERALS (Dietary). The vertebrate body contains about 4–6% total mineral matter, the major portion of this consisting of calcium and phosphorus in the skeletal tissues. Interspecies variation in mineral composition of the adult whole body is not great except as a result of differences in relative skeletal size. Thus, the average percentage concentration of minerals in the lean body mass of vertebrates is approximately: Calcium, 1.1–2.2; phosphorus, 0.70–1.20; magnesium, 0.045; potassium, 0.30; sodium, 0.15; chloride, 0.15; iron, 0.008; zinc, 0.003; and copper, 0.0003. Sulfur, in some form, is also required by all living organisms.

Minerals and mineral elements that occur in smaller amounts, yet with proven nutritional and biological roles, include chromium, cobalt, iodine, manganese, and molybdenum. Mineral elements suspected of being valuable biologically to animals and humans include boron, cadmium, nickel, and vanadium, but further research is required to identify specific biological roles, if any. A number of the trace element minerals that are required in very small quantities are quite toxic when ingested even in relatively small amounts. Flourine occupies a peculiar position in being nonessential and toxic and yet distinctly beneficial in preventing tooth decay under conditions that otherwise lead to dental caries.

A brief summary of mineral pathways in human and animal biological systems is given here. Each mineral is discussed in detail in a series of separate entries in this volume. See also: **Boron; Cadmium; Calcium; Chloride; Chromium; Cobalt; Copper; Fluorine; Iodine; Iron; Magnesium; Manganese; Molybdenum; Nickel; Phosphorus; Potassium; Selenium; Sodium; Sulfur; Vanadium;** and **Zinc.**

Mineral Absorption. Minerals appear in animal tissues mainly as a result of ingestion with food and water. The primary entry into the body is by way of the small intestine, from which the minerals are absorbed into the bloodstream and then transported through the body. Some of the minerals are absorbed by simple diffusion processes, while others enter into metabolic pathways that enable them to be absorbed against a concentration gradient by energy-requiring processes. Sodium and chlorine are examples of actively absorbed elements. Absorption of calcium is much slower than that of sodium and is at least in part regulated by hormonal and dietary factors. Parathyroid hormone stimulates calcium absorption and thus provides for increased absorption in states of calcium deficiency. Vitamin D likewise stimulates calcium absorption as it does the presence in the diet of relatively slowly absorbed carbohydrate, such as lactose. Evidence for active absorption of other divalent cations is not fully developed, but there is general agreement that iron absorption often increases in case of increased need by the animal. Mineral absorption is inhibited by dietary factors that tend to reduce solubility of minerals at the pH of the intestinal contents. Excesses of phosphate will inhibit absorption of calcium, iron, and magnesium. Excesses of calcium will inhibit absorption of phosphate, magnesium, and, particularly in the presence of phytic acid, zinc.

The interactivity of metals is under considerable study. For example, some investigators have observed that, although cadmium and mercury are both extremely toxic, they may be rendered harmless in the presence of selenium. Others have observed that the amount of copper require varies widely, depending upon the presence of molybdenum and sulfur. Because of possible mineral interactions, some researchers have indicated that it may be difficult to establish quantitative requirements for trace elements unless the composition of the entire diet is known.

Metabolic Roles. The diversity of roles served by minerals in the body defies simple summarization, and indeed they involve the entire field of nutritional biochemistry. Structurla rigidity of the body is largely a function of calcium and phosphorus. Sulfur is an integral part of a number of organic structures, including the amino acids methionine and cysteine, the vitamins biotin and thiamine, and a wide variety of sulfated polysaccharides. Cobalt is found in vitamin B_{12}; iron in hemoglobin; iodine in thyroxine; phosphorus in lecithin, related phospholipids, and proteins; iron, copper, molybdenum, zinc, and magnesium in a diversity of enzymes. In addition to these structural functions, minerals play important parts in a regulatory sense. Sodium, potassium, and chloride are important in pH regulation and in maintenance of osmotic pressure relationships. These minerals, with calcium and magnesium, are much concerned with neuromuscular irritability and the transmission of nerve impulses. Activators of enzymic reactions include calcium for prothrombin synthesis; magnesium for phosphate transfer; magnesium, manganese, copper, and zinc for transfer of acyl groups, hydration-dehydration reactions, and numerous others. The exact mechanism whereby minerals function either as activators of enzyme systems or as constituents of enzymes continues under intensive investigation. It is known that certain metals contribute stability to the configuration of protein molecules and also provide sites for ionic attraction and binding of reactants. Ribonucleic acids also are involved with certain trace elements.

Mineral Excretion. Excretion of absorbed minerals is by way of the intestinal tract, urinary tract, and in sweat. Which excretory pathway will be used depends upon the mineral in question and upon a variety of dietary, hormonal, and environmental factors which act interdependently to exert homeostatic control over the mineral concentrations in the body. Calcium, magnesium, phosphorus, and zinc are primarily excreted by way of the intestinal tract. Only small amounts of calcium and zinc are found in urine. Magnesium and phosphorus excretion may be quantitatively greater in the urine, dependent upon the cation-anion ratio. The monovalent ions, sodium, potassium, and chlorine, are primarily excreted in the urine, a process largely regulated by aldosterone (hormone produced by adrenal gland). Excretion of absorbed iron by way of urine and feces is very slight. In the human, sweat can be an important pathway of mineral loss, especially for sodium, chlorine, and iron.

Excesses. Minerals may be deleterious to animals if ingested in excessive amounts. The toxic mechanisms are as diverse as the functional roles of the minerals and may range from an effect on absorbability of the element to a displacement of one element by another in an enzyme system, or the direct inhibition of a metabolic reaction. Hence the toxicity of mineral elements may be a function not only of the amount present, but also of the supply of other minerals in the diet. The toxicity of copper, for example, is enhanced by lowered levels of molybdenum; and molybdenum toxicity is at least in part a reflection of an induced copper deficiency. Both of these interactions are modified by the level of inorganic sulfate in the diet. Zinc tolerance in animals is enhanced by increasing intakes of copper and iron above normally employed levels. Selenium toxicity may be alleviated by judicious administration of a variety of arsenic compounds. Although effects of these kinds can be observed, a mechanistic explanation of many of them remains for continuing research in this field. An integral aspect of proper mineral nutrition planning is due consideration of the potential hazards of mineral excesses as well as their deficiencies.

Measurements of mineral requirements for the normal growth and maintenance of animals have demonstrated that the quantitative results found are greatly influenced by many factors, such as age of the animal, physiological function and dietary constituents—hence specific recommendations are difficult without having full data concerning the individuals and their diets. Some researchers feel that tables of allowance should be considered nonrigidly and only as estimates and guidelines.

See also **Fertilizer.**

MIRROR NUCLIDES. Pairs of nuclides, having their numbers of protons and neutrons so related that each member of the pair would be transformed into the other by exchanging all neutrons for protons and vice versa.

MISSILE PROPELLANTS. **Rocket Propellants.**

MIXING AND BLENDING. These operations are important to chemical research and processing. In exploratory work in the laboratory the effects of mixing may be very great, and it is essential that the desired type and amount of mixing can be reproduced or can be varied by known amounts. The primary purpose of mixing is to distribute components as uniformly as possible; temperature distribution is frequently a major purpose. These may be followed by a chemical reaction or a transfer of matter between phases, and by a transfer of heat for temperture control. The mixer produces mechanical effects only. Molecules of themselves will diffuse, but mixing impellers produce flow which results in forced convection and mixing. Hence, reactants can be brought to an interface as rapidly as desired by controlling the fluid motion. Most fluid mixing is done by rotating impellers.

Both large scale (mass flow) motion and small scale (turbulent) motion are ordinarily required to bring about rapid mixing. The discharge stream from an impeller initiates the large scale flow pattern. Turbulence is generated mostly by the velocity discontinuities adjacent to the stream of fluid flowing from the impeller, and also by boundary and form separation effects. Turbulence spreads throughout the mass flow and is carried to all parts of the container. Some mixing operations require relatively large mass flows for best results, whereas others require relatively large amounts of turbulence. There is usually an optimum ratio of flow to turbulence for a desired mixing operation, whether it is a simple blending of immiscible liquids or a mass transfer followed by chemical reaction.

In the research laboratory it is important to recognize the effect of mixing on reaction rate or on other performance criteria. Energy must be supplied to produce fluid motion, thus, to compare mixing with different equipment or with different sizes of the same type impeller, it is essential that the comparisons be made on the basis of equal power input.

For the same power, the ratio of flow to turbulence from mixing impellers can be varied by changing the size and speed of the impeller. Figure 1 illustrates the differences in mass flow and turbulence which can be achieved for the same power input

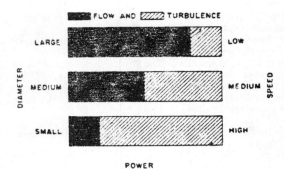

Fig. 1. Constant power, effect of impeller size, and speed on flow and turbulence.

for dimensionally similar impellers. A large-diameter low-speed impeller produces a large ratio of flow to turbulence, whereas a small-diameter high-speed impeller will give a small ratio. Curve A, Figure 2, illustrates a reaction best accomplished by

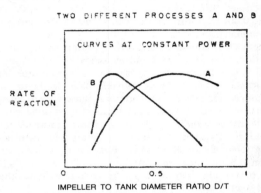

Fig. 2. Effect of impeller size on reaction rate at equal power output.

large flow and small turbulence. This curve, which is typical of blending operations, shows that the rate of blending increases to a maximum with a large impeller as impeller diameter is increased (and impeller speed is decreased) with power input constant.

Curve B of Figure 2 is typical of gas-liquid contacting operations. Here the rate of mass transfer between phases increases to a maximum at small impeller diameter and then decreases as impeller diameter is increased. The significance is that more turbulence is available with the small impeller and that turbulence is more important than flow in this operation.

In all bench-scale and pilot plant work where mixing is important, the effect of the impeller diameter-turbine diameter ratio should be determined so that the type of flow motion best suited to the operation can be found. If an optimum ratio is found, it becomes the basis for larger scale design.

Mixing Vessels, Flow Patterns, and Impellers. Flow motion is dependent upon the shape and fitting of the container, the shape and position of the rotating impeller, and the physical properties of the fluid. The best mixing is usually one which produces lateral and vertical flow currents, and these currents must penetrate to all portions of the fluid; swirling motion should be avoided. Cylindrical vessels provide the best environment for mixing.

The most useful impellers are the simple flat paddle, the marine-type propeller, and the turbine. If any of these are on a vertical shaft rotating on the center line of a cylindrical vessel, the fluid motion will be one of rotation. A vortex forms around which the liquid swirls. A minimum of turbulence and of vertical and lateral flow motion will result. Very little power can be applied.

Rotary motion (and surface vortex) can always be stopped by inserting projections in the body of the fluid; when these are at the side of the tank they are called baffles, and this is the method most commonly used to obtain good mixing in large industrial equipment. The propeller with baffles will produce an axil flow pattern, Fig. 3, and the paddle and turbine will produce radial flow, Fig. 4.

Motionless Mixers.   Mixing of molten polymers has been a problem in the plastics industry for many years mainly because of the high viscosity of the melt. Blending of color concentrations, fillers, stabilizers, and other additives have sometimes been difficult to achieve efficiently with traditional mixing approaches, such as the extruder screw and other rotating devices. These problems have been partially solved through the application of so-called motionless mixers. A stationary baffle installed in a pipe can utilize the energy of the flowing fluid to produce mixing. Turbulent conditions are required for this simplistic approach. In recent years, more sophisticated motionless mixers have been developed. One design consists of a number of short right- and left-hand helices. The opposite hand helices are welded together so that their leading edges are 90 degrees to the trailing edge of the preceeding element. The mixing unit is housed in a tube or pipe for in-line installation. Materials entering the static mixer experiences flow division at the leading edge of each element. As the flow divides, it follows the semicircular channel of each element and repeatedly divides at succeed-

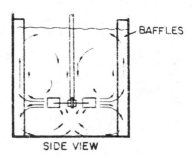

SIDE VIEW

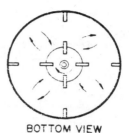

BOTTOM VIEW

Fig. 4.   Radial flow pattern for flat blade turbine positioned on center in baffled tank.

ing junctions of additional elements, resulting in flow division and radial mixing. Other approaches involving complex geometric stationary elements have been developed. Some of these units also can function for heating and cooling as well and mixing. [See "Motionless Mixers in Plastic Processing," Schott, Weinstein, and LaBombard, *Chem. Eng. Progress*, **71** (11), 52–58 (1975)].

Automatic Blending Systems.   For a number of industrial applications, in-line blending of liquids and solids has replaced former batch-type operations. In these systems, all components flow together simultaneously to a central collection point where they combine to form the finished product. A modern in-line blending system is shown in Fig. 5. The blend controller will nominally utilize microporcessor technology with a cathode-ray tube (CRT) display. Each fluid component is pumped from a storage tank, through a strainer, and then through a flowmeter, with the meter and valve carefully selected for prevailing process conditions (viscosity, temperature, pressure, flow rates). The signal from the flowmeter is fed to the blend controller which compares the actual flow rate to the desired flow rate. For minor components, such as dyes or additives, it is sometimes most practical to control the flow rate by means of proportioning pumps which inject a precise amount of the fluid when a pulse signal from the blend controller is received. This type of open-loop control is cost efficient, but some means for assuring flow (detecting any dry line) should be considered inasmuch as there is no direct fluid measurement device used. Other variations of measurement and control involve the use of variable-speed pump motor controllers (SCRs) for flow control; adding a flowmeter in series with an injection pump. Weigh-belt feeders with variable-feed/speed control and tachometer/load-cell outputs are frequently used for blending powders and aggregates.

A blend controller block diagram is shown in Fig. 6. A system for preparing bread and pastry dough is shown in Fig. 7. Applications for continuous blending systems are frequently found in the petroleum, petrochemical, food and beverage, building materials, pharmaceutical, automotive, and chemical industries, among others.

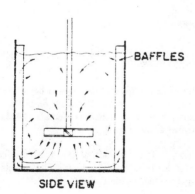

SIDE VIEW

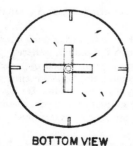

BOTTOM VIEW

Fig. 3.   Typical flow patterns from axial flow impeller in baffled tank.

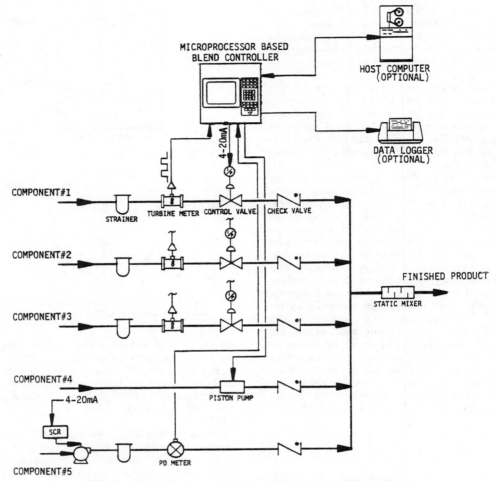

Fig. 5.   Typical blender configuration. (*Waugh Controls Corporation.*)

MODACRYLIC FIBERS.   Acrylic Fibers; Fibers.

MODERATOR.   A substance used to slow down neutrons by means of collisions. Moderators play an important role in the design and operation of nuclear reactors. Moderators *thermalize* neutrons to an energy of about 0.025 eV.

MOIETY.   An indefinite portion of a sample.

MOISTURE-RETAINING SUBSTANCES.   Humecatants and Moisture-Retaining Substances.

MOLAL CONCENTRATION.   A one molal solution contains one mole of a particular substance (the solute) in 1000 grams of solvent. Thus, a 0.5 molal solution of potassium chloride in water contains $0.5 \times$ (gram-molecular weight of KCl = 74.555), or 37.278 grams of the salt in 1000 grams of H_2O.

MOLAR CONCENTRATION.   A one molar solution contains one mole of a particular substance (the solute) in 1000 milliliters of solution. Thus, a 0.5 molar solution of potassium chloride in water will be prepared by placing $0.5 \times$ (gram-molecular weight of KCl = 74.555), or 37.278 grams of the salt in a vessel and then adding H_2O, while thoroughly mixing to assure complete solution of the salt, until a total volume of 1000 milliliters of solution is obtained. Molar is abbreviated M. Thus, the solution in the example given here would be $0.5M$ KCl. Molar solutions sometimes are referred to as *formal* solutions, not to be confused with normal solutions. See also **Molal Concentration;** and **Normal Concentration.**

MOLAR HEAT.   The product of the gram-molecular weight of a compound and its specific heat. The result is the heat capacity per gram-molecular weight.

MOLD.   Yeasts and Molds.

MOLD INHIBITORS.   Antimicrobial Agents (Foods).

MOLECULAR BEAM.   A unidirectional stream of neutral molecules passing through a vacuum, generally with thermal velocity. Such a beam may be produced by emergence from a pinhole in a chamber containing low-pressure gas or vapor, and it may be defined by a system of slits. By passing the beam through known electric or magnetic fields, quantities, such as nuclear magnetic moments, can be determined.

MOLECULAR DISTILLATION.   A special form of distillation conducted at pressures of 1–7 micrometers in the laboratory and 3–30 micrometers in industrial applications. Compared with conventional laboratory vacuum distillations carried out be-

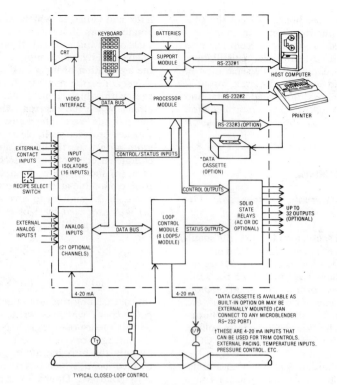

Fig. 6.   Blend controller block diagram.

tween 1 and 10 millimeters of mercury pressure, this is a very high vacuum. One micrometer equals 0.001 millimeter of mercury pressure. The other feature of the molecular still is that the condenser is located within a distance less than the mean free path of the evaporating molecules from the evaporator portion of the apparatus. Thus, although a molecule may return to the distilland many hundred times before reaching the exit of a conventional vacuum still, 50% of the molecules in a properly functioning molecular will reach the exit on their first try. Thus, efficiency is remarkably high.

Because of the absence of convection due to ebullition and because high viscosities and high molecular weights may impede diffusion within the distilland, the surface of the distilland in a molecular still may not always represent the total liquid. Therefore, efficient molecular distillation requires the mechanical renewal of the surface film. This is achieved by vigorous agitation, as in the *stirred-pot still*; by employing a *falling film*; or by using centrifugal force, as in the *centrifugal* molecular still. Commercial installations of falling-film stills achieve

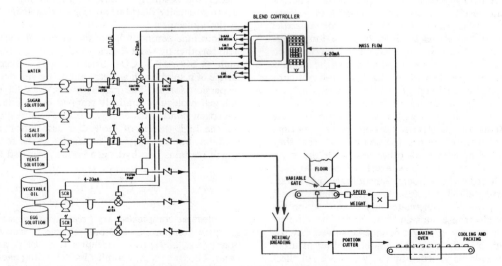

Fig. 7.   Blending system for preparing bread and pastry dough. (*Waugh Controls Corporation*.)

throughputs of many tens of liters per hour, whereas the centrifugal still is capable of several hundred liters per hour. Centrifugal stills usually are arranged in groups of from three to seven. This permits fractionation by multiple redistillation. Among the uses of molecular distillation are the separation of mono- and diglycerides for bread and paraffin wax for milk cartons; the distillation of plasticizers, fatty acid dimers, and synthetics; the distillation of vitamin A esters and intermediates; and the stripping of α-, β-, γ-, and δ-tocopherols and sitosterols from vegetable oils.

See also **Distillation.**

MOLECULAR ORBITALS. Orbitals.

MOLECULAR STRUCTURE (Organic Compounds). Organic Chemistry.

MOLECULE. In the traditional sense, a molecule is the smallest particle of a chemical substance capable of independent existence with retention of all its chemical properties. Molecules comprise one or more atoms which need not be of the same kind. Only the rare, or noble gases form single- atom or monatomic molecules. All other elements form di-, tri-, tetra-, etc. atomic molecules, e.g., hydrogen, H_2; ozone, O_3; phosphorus, P_4; and sulfur, S_8; or hydrogen chloride, HCl; sodium sulfide, Na_2S; aluminum chloride, $AlCl_3$; carbon tetrachloride, CCl_4; and so on.

Structurally, a more specific definition would be that a molecule is a local assembly of atomic nuclei and electrons in a state of dynamic stability. The cohesive forces are electrostatic, but, in addition, relatively small electromagnetic interactions may occur between the spin and orbital motions of the electrons, especially in the neighborhood of heavy nuclei. The internuclear separations are of the order of $1-2 \times 10^{-10}$ meter, and the energies required to dissociate a stable molecule into smaller fragments fall into the $1-5$ eV range. The simplest diatomic species is the dihydrogen ion, H_2^+, with two nuclei and one electron. At the other extreme, the protein ribonuclease contains 1876 nuclei and 7396 electrons per molecule.

Another form of molecule is known, however, and this is formed by atomic nuclei alone. Although these have not yet been found in nature, they may well play a role in stellar evolution. Under special conditions in high-energy interactions, nuclei can be momentarily held together by effective bonds. Whether these bonds are the result of exchange or sharing of valence protons and neutrons is as yet moot. In these *nuclear molecules* a somewhat unstable balance is attained between long-range electrostatic repulsion of positively charged nuclei and the much stronger short-range nuclear force which determines the motions of protons and neutrons. Nuclear molecules are significant entities because they live much longer ($\sim 10^{-21}$ second) than the time usually taken for nuclei to collide ($\sim 10^{-23}$ second).

The molecular, or kinetic, theory of matter assumes that the molecules of which matter is composed are constantly in motion, that their energy is increased by the addition of heat, that they undergo elastic collision with each other and with the walls of a containing vessel, and that they exert forces upon each other. As first developed by Heisenberg, Schrödinger, and Dirac, reduction of these theoretical assumptions to mathematical bases is somewhat inadequate and can relate only to interaction between hypothetical electron clouds and, except for the simplest of systems, the Schrödinger equation cannot be solved exactly. On the other hand, a more accessible understanding of atom interactions in molecules is afforded by the Valence Shell Electron Pair Repulsion Theory, independently enunciated by Ny-

holm and Gillespie, which proposes that both bonding and nonbonding pairs of outer atomic shell electrons in a molecule repel each other and establish themselves as far apart as possible.

Historically, molecules were regarded as being formed by the association of individual atoms. This led to the concept of *valency,* i.e., the number of individual chemical bonds or linkages with which a particular atom can attach itself to other atoms. When the electronic theory of the atom was developed, these bonds were interpreted in terms of the behavior of the valence, or outer shell, electrons of the combining atoms. Each atom with a partly filled valence shell attempts to acquire a completed octet of outer electrons, either by electron transfer, as in (a) shown below, to give an electrovalent bond, resulting from coulombic attraction between the oppositely charged ions; or, as in (b) and (c) to give a covalent bond. The concept of (a) was proposed by Kossel in 1916; that of (b) and (c) by Lewis, also in 1916.

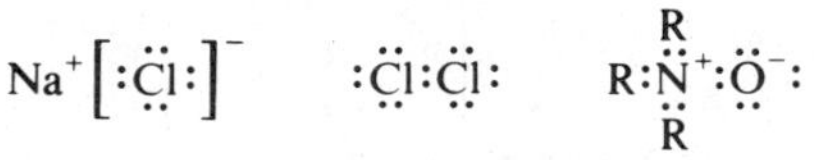

$$Na^+ \left[:\ddot{\underset{..}{C}}l: \right]^- \qquad :\ddot{\underset{..}{C}}l:\ddot{\underset{..}{C}}l: \qquad \underset{\underset{R}{|}}{\overset{\overset{R}{|}}{R:N^+:\ddot{\underset{..}{O}}:^-}} \qquad R = CH_3$$

$$\quad\quad (a) \qquad\qquad\quad (b) \qquad\qquad (c)$$

In (b), each chlorine atom donates one electron to form a *homopolar bond,* which is written Cl—Cl where the bar denotes on this theory one single bond, or shared electron pair. In (c), the nitrogen-oxygen bond is formed by two electrons donated by only the nitrogen atom, giving a *semipolar,* or *coordinate-covalent bond,* which is written $R_3N \rightarrow O$, and which is electrically polarized. Double or triple bonds result from the sharing of 4 to 6 electrons between adjacent atoms. More information on these bonding theories is given in the entries on **Chemical Composition; Chemical Elements;** and **Compound (Chemical).**

However, difficulties arise in describing the structures of many molecules in this manner. For example, in benzene, C_6H_6, a typical aromatic compound, the carbon nuclei form a plane regular hexagon, but the electrons can only be conventionally written as forming alternate single and double bonds between them. Furthermore, an electron cannot be identified as coming specifically from any one of these bonds upon ionization. Such difficulties disappear in the quantum-mechanical theory of a polyatomic molecule, whose electronic wave function can be constructed from nonlocalized electron orbitals extending over all of the nuclei. The concept of valency is not basic to this theory, but is simply a convenient approximation by which the electron density distribution is partitioned in different regions in the molecule.

Molecular compounds consist of two or more stable species held together by weak forces. In *clathrates,* a gaseous substance, such as SO_2, HCl, CO_2, or a rare gas is held in the crystal lattice of a solid, such as beta-quinol, by van der Waals-London dispersion forces. The gas hydrates, e.g., $Cl_2 \cdot 6H_2O$, contain halogen molecules similarly trapped in ice-like structures. The hydrogen bond, with energy ~ 0.25 eV, is responsible not only for the high degree of molecular association in liquids, such as water, but also for such molecules as the formic acid dimer, which contains two hydrogen bonds indicated by dashed lines.

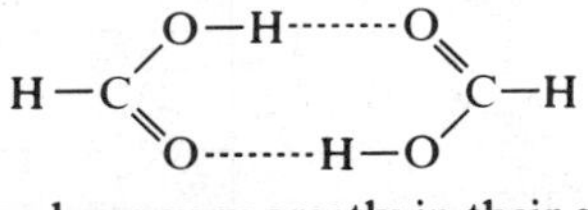

$$\begin{array}{ccc} & O\!-\!H\cdots\cdots O & \\ H\!-\!C & & C\!-\!H \\ & O\cdots\cdots H\!-\!O & \end{array}$$

Molecular complexes vary greatly in their stability; in donor-acceptor complexes, electronic charge is transferred from the donor (e.g., NH_3) to the acceptor (e.g., BF_3), as in a semipolar bond. The $BF_3 \cdot NH_3$ complex has a binding energy with respect to dissociation into NH_3 and BF_3 of 1.8 eV. The bond here is relatively strong; the electron transfer can occur between the components in their electronic ground states. On the other hand, in weaker complexes, such as $C_6H_6 \cdot I_2$, with binding energy of about 0.06 eV, there is only a fractional transfer of charge from benzene to iodine. The actual ionic charge-transfer state lies at much higher energy than the ground state of the complex.

Complete pairing of all electrons present in a molecule and absence of any bonding orbitals was long taken to be a stable, unreactive state exemplified by the inert, or rare gases. In 1962, however, Bartlett unequivocally synthesized $XePtF_6$, and this was rapidly followed by the synthesis of other rare gas compounds whose existence was not predicted by classical valency theories. Compounds such as XeF_2, XeF_4, XeF_6, and $XeOF_4$ are quite stable, the average Xe—F bond energy in the square planar XeF_4 being 1.4 eV.

A molecule is characterized by (1) a stoichiometric formula; (2) the spatial distribution of the nuclei in their mean equilibrium or "rest" positions; and (3) the dynamical state.

The ratio $a:b:c: \ldots$ in a formula $A_aB_bC_c$, where $a,b,c,$ $\ldots$ are the numbers of atoms of elements $A,B,C, \ldots$ that it contains is found by chemical analysis for these elements. The absolute values of $a,b,c, \ldots$ are then fixed by determination of the molecular weight. This principle is further described under the entry on **Compound (Chemical)**.

The spatial distribution of the nuclei in their mean equilibrium positions, at an elementary level, is described in geometrical language. For example, in carbon tetrachloride, CCl_4, the four chlorine nuclei are disposed at the corners of a regular tetrahedron, and the carbon nucleus is at the center. In the $[CoCl_4]^{2-}$ ion, the arrangement of the chlorine nuclei about the central metal nucleus is also tetrahedral, whereas in $[PdCl_2]^{2-}$, it is planar. For example, the pyramidal ammonia molecule NH_3 has a threefold rotation axis C_3 through the nitrogen nucleus and three reflection planes σ_v intersecting at the axis, and belongs to the $C_{3v}(3m)$ point group. Tetrahedral molecules CX_4 belong to the $T_d(43m)$ point group. Linear diatomic and polyatomic molecules belong to either of the continuous point groups $D_{\infty h}$ or $C_{\infty v}$ according to whether a center of symmetry is present or not.

The symmetry classification does not define the geometry of a molecule completely. The values of certain bond lengths or angles must also be described. In carbon tetrachloride, it is sufficient to give the C—Cl distance (1.77×10^{-10} meters), since classification under the T_d point group implies that all four of these bonds have equal length and the angle between them is 109°28″. In ammonia, both the N—H distance (1.015×10^{-10} meters) and the angle HNH(107°) must be specified. In general, the lower the molecular symmetry, the greater is the number of such independent parameters required to characterize the geometry. Information about the symmetry and internal dimensions of a molecule is obtained experimentally by spectroscopy, electron diffraction, neutron diffraction, and x-ray diffraction.

The dynamic state is defined by the values of certain observables associated with orbital and spin motions of the electrons and with vibration and rotation of the nuclei, and also by symmetry properties of the corresponding stationary-state wave functions. Except when heavy nuclei are present, the total electron spin angular momentum of a molecule is separately conserved with magnitude Sh, and molecular states are classified as singlet, doublet, triplet . . . according to the value of the multiplicity $(2S + 1)$. This is shown by a prefix superscript to the term symbol, as in atoms.

The Born-Oppenheimer approximation permits the molecular Hamiltonian H to be separated into a component H_e that depends only on the coordinates of the electrons relative to the nuclei plus a component depending upon the nuclear coordinates, which in turn can be written as a sum $H_v + H_r$ of terms for vibrational and rotational motion of the nuclei, translation being ignored. The eigenfunctions Ψ of H may correspondingly be factorized as the product $\Psi_e\Psi_v\Psi_r$ of eigenfunctions of these three operators, and the eigenvalues E decomposed as the sum $E_e + E_v + E_r$. In general, $E_e > E_v > E_r$.

Molecular Spectra. The spectra of substances in the molecular state, like atomic spectra, are made up of lines, although more complex. The transitions in a molecule which release the most energy (largest quanta) are due to electron changes, as in atoms, and the results of these changes are observed as lines in the ultraviolet region. But there are other ways in which a molecule can release or absorb energy. Thus, the component atoms oscillate with reference to each other within the molecule, and this motion apparently is "quantized," i.e., changes abruptly from one state to another of different energy. But these "vibrational" energy changes are much less than the electronic, so that the resulting quanta and spectrum lines are of much lower frequency, and appear in the extreme red or near-infrared. Again, the molecule rotates, and the quantization of its rotational energy results in the emission of quanta of still lower frequency, appearing as lines in the far-infrared.

Polymerization. This term is used to designate a reaction in which a complex molecule of high molecular weight (or macromolecule) is formed from a number of simpler molecules. Thus the monomer formaldehyde $CHOH$ can form the trimer trioxane $(CHOH)_3$, or the long-chain polymer paraformaldehyde, $HO-(CHOH)_nH$, where $n = 8-100$. But the combining molecules may be of the same or different sorts. An *addition polymerization* is one in which like or unlike molecules combine without the elimination of any atoms or molecules. A *simple polymerization* involves only one species of molecule. *Copolymerization* is an addition polymerization in which two or more distinct molecular species are involved, each one of which is capable of polymerizing by itself. The high polymer formed contains each molecular constituent or an essential portion, as a distinct unit in the structure of the polymer. *Heteropolymerization* is an addition polymerization in which two or more molecular species are involved, one of which species will not polymerize by itself. It does, however, form distinct units in the high polymer.

A condensation polymerization is one in which the molecules undergoing polymerization react with the elimination of simple molecules like water, ammonia, and the like.

Polymerizations are also characterized by the state in which they are carried out such as *gaseous polymerization* or those carried out in the vapor or gaseous phase; *mass polymerizations* in the liquid state; *solution polymerizations* carried out by first disolving the material to be polymerized in an adequate solvent; *emulsion polymerizations* carried out in which one of the components is in an emulsion as in the case of rubber polymerizations; and *bulk polymerizations* in which the polymerization takes place without the use of a solvent or other medium. The wide variety of methods by which the process of polymerization can be used in the manufacture of plastics is exemplified by the production of polystyrene from styrene (vinylbenzene, $CH_2 \cdot CH \cdot C_6H_5$). One factor that conditions all these processes is the highly exothermic (high heat production) nature of this polymerization. This fact, together with the low heat conductivity of polystyrene, determines certain characteristics of an industrial process. This is particularly the case because the extent of the polymerization of styrene, like that of many other plastics, depends on the temperature. When the process is conducted at higher temperatures, the resulting polymer has a low molecular weight and low physical strength properties (i.e., it is weak and brittle). Extremely high molecular weight polymers, although mechanically tough, are more difficult to fabricate. Since the higher polymerization temperatures also result in faster polymerization rates, economic considerations dictate a compromise between faster production and better physical properties. Practical experience indicates a temperature range from 60 to 150°C.

The polymerization of styrene illustrates the application of several of the methods defined above.

In *batch mass polymerization* the reaction vessel is loaded with styrene monomer and heated to a temperature sufficient to initiate polymerization within a reasonable time. As polymerization proceeds, the temperature within the vessel rises, thus increasing the rate.

In *continuous mass polymerization* the reaction vessel contains monomer and polymer. As more polymer is formed it is drawn off the bottom of the reaction vessel while monomer is added to the top of the vessel. The temperature of the reaction is controlled by cooling coils within the vessel.

In *solution polymerization* the styrene monomer is diluted with a solvent. The solvent acts as a diluent which decreases the rate of polymerization and also serves as a heat transfer medium for removing the excess heat developed by the reaction.

In *suspension polymerization* water is used as a diluent and as a heat transfer aid. Suspending agents such as starch and methylcellulose are used to keep the styrene monomer particles in suspension. The more efficient heat transfer of this process also allows for a narrower molecular weight distribution.

In *emulsion polymerization* the styrene monomer is emulsified with water by the addition of certain emulsifying agents. This results in very small particles and rapid polymerization rates. The heat of polymerization is dissipated by the water ingredient.

See **Fibers**; and **Macromolecule**.

References

Alfrey, T., Jr.: "Polymerization," in "Chemical and Process Technology Encyclopedia" (D. M. Considine, Editor), McGraw-Hill, New York (1974).

Braun, D., Cherdron, H., and W. Kern: "Techniques of Polymer Syntheses and Characterization," Wiley, New York (1972).

Bromley, D. A.: "Nuclear Molecules," *Sci. American*, **239**, 58–69 (1978).

Fruton, J. S.: "Molecules and Life," Wiley, New York (1972).

Huber, K. P., and G. Herzberg: "Constants of Diatomic Molecules," Van Nostrand Reinhold, New York (1979).

Knox, J. H.: "Molecular Thermodynamics," Wiley, New York (1971).

Lewis, G. N.: "Valence and the Structure of Atoms and Molecules," Dover, New York (1973).

Prausnitz, J. M.: "Molecular Thermodynamics for Chemical Process Design," *Science*, **205**, 759–766 (1979).

Pullman, B.: "The Modern Theory of Molecular Structure," Dover, New York (1973).

Rao, K. N., and C. W. Mathews, editors: "Molecular Spectroscopy: Modern Research," Academic, New York (1972).

Rice, O. K.: "Electronic Structure and Chemical Binding," Dover, New York (1973).

Syrkin, Y. K., and M. E. Dyatkina: "Structure of Molecules and the Chemical Bond," Dover, New York (1973).

Weissbluth, M.: "Atoms and Molecules," Academic, New York (1978).

—R. C. Vickery, Hudson Laboratories, Hudson, Florida.

MOLE FRACTION. As applied to a system, the *mole fraction* (sometimes spelled *mol fraction*) of a given substance in the system is found by dividing the number of moles of that substance by the total number of moles in the system. For a mixture of ideal gases, the mole fraction is equal numerically to the volume fraction. The *volume fraction* of a component in a mixture is found by dividing the volume of that component at the total pressure and at the temperature of the mixture by the volume of the mixture at the same pressure and temperature.

In a binary solution consisting of components X and Y, the mole fractions, F_X and F_Y; respectively, are:

$$\text{Mole fraction of } X = F_X = \frac{f_X}{f_X + f_Y}$$

$$\text{Mole fraction of } Y = F_Y = \frac{f_Y}{f_X + f_Y}$$

where f = number of moles of specific component present.

It is apparent, of course, that the mole fraction of X plus the mole fraction of Y must equal unity, or if expressed as a percentage (mole percent), must equal 100. In the instance of three or more components, the denominators of the prior expressions must reflect the additional moles present.

In considering a solution containing 50 grams of methyl alcohol in 1,000 grams of H_2O, the mole percent of each component will be:

(In this example, the molecular weights are rounded off: $CH_3OH = 32$; $H_2O = 18$)

Moles of $CH_3OH = 50/32 = 1.562$
Moles of $H_2O = 1,000/18 = 55.556$
Total Moles 57.118

Mole percent $CH_3OH = 1.562/57.118 \times 100 = 2.735\%$
Mole percent $H_2O = 55.556/57.118 \times 100 = 97.265\%$

The same solution, expressed in terms of weight percentage, is:

Weight percent $CH_3OH = 50/1,050 = 4.762\%$
Weight percent $H_2O = 1,000/1,050 = 95.238\%$

In nuclear chemistry, the term mole fraction may be used to indicate the number of atoms of a given isotope in an isotopic mixture as a fraction of the total number of atoms of that element in the mixture.

MOLE (Stoichiometry). Sometimes spelled *mol*, a mole is a quantity of a substance, expressed in specific mass units, that is equal to the molecular weight of the substance. For example, a *gram-mole* or *gram-molecular mass* of hydrogen H_2 will have a mass of 2 × (atomic weight of hydrogen), or 2.016 grams. A gram-mole of carbon dioxide CO_2 will have a mass of 1 × (atomic weight of carbon) plus 2 × (atomic weight of oxygen), or 12.011 plus 31.998 = 44.009 grams. A pound-mole of ammonia gas NH_3 will have a mass of 1 × (atomic weight of nitrogen) plus 3 × (atomic weight of hydrogen), or 17.031 pounds. See also **Avogadro Constant**.

MOLE VOLUME. A mole of gas will occupy a definite volume under definite conditions regardless of the nature of the gas. This definite volume is called the *mole volume*. Under a pressure of 760 torr and at a temperature of 0°C, a gram-mole of gas will occupy 22.41 liters. This situation also applies to a mixture of gases. A pound-mole of gas will occupy 359 cubic feet at a pressure of 760 torr and at a temperature of 32°F (0°C).

Because the volume of one mole of gas at any specific pressure and temperature contains the same number of molecules even though there may be several different gases in the mixture, the percent by volume of any given gas is equal to the percent pressure exerted by that gas and is also equal to the mole percent of that gas. Mole percent equals volume percent equals pressure percent.

MOLYBDENITE. The mineral molybdenite is sulfide of molybdenum, MoS_2. Its hexagonal crystals are usually tabular to short prismatic, but if in massive form it may be foliated or granular. Has a perfect basal cleavage; is sectile; hardness, 1–1.5; specific gravity, 4.52–5.06; luster, metallic; color, very slightly bluish, lead gray; streak, greenish-gray; opaque. Molybdenite is one of the few minerals soft enough to give a distinctly greasy feel. Molybdenite is found as a contact mineral with cassiterite and wolframite, in granite pegmatites and sometimes in granites, syenites, or gneisses. It is found associated with tin ore in Saxony and Bohemia; in Norway; England; Australia; and in the United States in Colorado, Washington County and Oxford County, Maine, in New Hampshire, Connecticut, Pennsylvania, and Washington. Its name, derived from the Greek meaning *lead*, was formerly applied to minerals containing lead, to graphite, and to molybdenite as well. Later the term was restricted to the latter mineral. It is an ore of molybdenum.

MOLYBDENUM. Chemical element symbol Mo, at. no. 42, at. wt. 95.94, periodic table group 6b, mp 2610°C, bp 5560°C, density, 9.01 g/cm^3 (solid, 20°C), 10.2 g/cm^3 single crystal). Molybdenum has a body-centered cubic crystal structure. Molybdenum is silvery-white, tough, malleable, softer than glass, not oxidized by air at ordinary temperatures, but above 600°C burns to form white molybdenum oxide. The metal is dissolved by dilute HNO_3 and aqua regia, but is made passive by concentrated HNO_3. Like chromium, molybdenum exhibits the phenomenon of passivity to a marked degree and even though it shows a strong reducing action when offering a fresh surface or in potentiometric determinations, it may be quite resistant to chemical action. The metal is attacked by fused alkalis. Seven isotopes occur naturally, ^{92}Mo, ^{94}Mo through ^{98}Mo, and ^{100}Mo; and five radioactive isotopes have been identified, ^{90}Mo, ^{91}Mo, ^{93}Mo, ^{99}Mo, and ^{101}Mo. With exception of ^{93}Mo, which has a half-life of approximately 10^4 years, the other radioactive isotopes have short half-lives, measured in terms of minutes and hours. The element does not appear on the list of the first 36 most frequently occurring elements in the earth's crust and thus may be considered a scarce element. Molybdenum is listed as the 42nd element in terms of estimated occurrence in the universe. The element ranks 25th among the elements occurring in seawater, there being an estimated 50 tons of molybdenum per cubic mile (10.8 metric tons per cubic kilometer) of seawater. Molybdenum was one of the first of the uncommon elements to be identified; reported by Carl Wilhelm Scheele in 1778 when he was investigating lead ores.

First ionization potential 7.18 eV. Oxidation potentials Mo → Mo^{3+} + 3e^-, *ca.* 0.2 V; Mo(IV) → Mo(V) + e^-, (0.1 M KCl, pH 3) −0.01 V.

Other physcial properties of molybdenum are given under **Chemical Elements.** See also summary of properties of refractory metals under **Niobium.**

The main sources of molybdenum are MoS_2 (molybdenite), $Ca(MoW)O_4$ (powellite), and $PbMoO_4$ (wulfenite), of which the first is by far the most important. See also **Molybdenite; Wulfenite.** A representative molybdenite ore contains about 0.2% molybdenum. A concentrate containing about 90% MoS_2 is prepared through crushing, grinding and flotation operations. Most of the concentrate so produced is roasted in air to form technical molybdic oxide (about 90% MoO_3), the product used to add molybdenum in most steelmaking processes. The preva-

lent industrial method for producing ferromolybdenum, the material used to add molybdenum to some cast irons and steels, is a thermit process wherein aluminum and silicon reduce a charge of iron oxide and molybdic oxide. Ferromolybdenum contains about 60% molybdenum.

Pure molybdic oxide (99.95% MoO_3) is prepared by sublimation of technical oxide or by calcining ammonium molybdate. Metallic molybdenum powder is prepared commercially by hydrogen reduction of either pure molybdic oxide or ammonium molybdate in a two-step process. The first step, carried out at about 500°C yields MoO_2. The second step, reduction of MoO_2 to Mo, is carried out at about 1100°C.

Uses. Molybdenum is added to a number of alloy steels because its presence increases hardenability, toughness, cold-formability, and weldability. Even though the steels so treated may contain less than 1% Mo, these uses account for about 95 million pounds (~43 million kilograms) of the metal per year. One of the more recent special steels contains manganese, molybdenum, and niobium for use in pipelines under arctic conditions. Stainless steels account for over 40 million pounds (~13.6 million kilograms) of Mo per year. In these steels, molybdenum increases corrosion resistance, elevated-temperature strength, and weldability. However, ferritic stainless steels that contain 18–26% chromium and 1–2% molybdenum are displacing the traditional 18–8 type stainless steels for many applications. For example, a steel containing 18% chromium and 2% molybdenum is finding wide use in solar energy panels.

The use of molybdenum in tool steels accounts for about 18 million pounds (~8.2 million kilograms) of Mo per year. In these steels, molybdenum provides better hot strength and improved resistance to softening and thermal cycling effects. A number of the earlier tungsten-grade tool steels have been replaced by molybdenum steels because of improved performance, lower metal density, and greater price stability. About 15 million pounds (~6.8 million kilograms) of Mo per year are consumed by the foundry industry where the metal improves the strength and abrasion resistance of cast iron. About 6 million pounds (~2.7 million kilograms) of Mo per year go into superalloys for high-temperature environments as required, for example, by jet engines. One of the later-developed superalloys is a nickel-base alloy that contains about 18% Mo and has the advantages of a very high melting point, low density, and low coefficient of thermal expansion. Uses of molybdenum compounds are described below.

Chemistry and Compounds: In keeping with its $4d^5 5s^1$ electron configuration, molybdenum forms many compounds in which its oxidation state is 6+, to an even greater extent than chromium. Also, like chromium, it forms compounds in which it is divalent and those in which it is trivalent; unlike chromium, it forms a number of pentavalent compounds, and a few more tetravalent compounds, especially complexes.

Among the divalent compounds of molybdenum are the dibromide, $MoBr_2$, and the dichloride, $MoCl_2$. There is also a complex ion of divalent molybdenum $Mo_6Cl_8^{4+}$, that is of particular interest because it does not yield Mo^{2+} ions. The chloride salt of the ion, Mo_6CL_{12} has been shown, by precipitation with Ag^+, to have two-thirds of its chlorine content present in a complex, so its structure is established as $[Mo_6Cl_8]Cl_4$. It is obtained by higher temperature dismutation of $MoCl_3$, while the corresponding dibromide can be produced by direct reaction of the elements.

The relatively small number of trivalent molybdenum compounds generally exhibit the marked reducing action of the Cr^{3+} compounds, though not quite so strongly. They include the trichloride $MoCl_3$ and the tribromide $MoBr_3$, as well as the sesquioxide Mo_2O_3.

In addition to Mo_2O_3, the other oxides include MoO_2, Mo_2O_5, MoO_3, the tetravalent and pentavalent oxides being obtainable from the hexoxide by hydrogen reduction, the MoO_3 being formed by direct combination of the elements.

In addition to the dioxide MoO_2 and the disulfide, MoS_2, example of tetravalent molybdenum compounds include the tetrachloride, $MoCl_4$, and tetrabromide, $MoBr_4$, both of which are hydrolyzed in hot H_2O. Other tetravalent compounds include a few of the complexes.

The complexes also exist among the pentavalent molybdenum compounds, which include a number of simple compounds as well. In addition to the pentoxide, Mo_2O_5 and pentasulfide, Mo_2S_5, pentahalides and oxyhalides, such as $MoCl_5$ and $MoOCl_3$ are also known. Direct reaction with fluorine, yields MoF_6; with chlorine, $MoCl_5$, and with bromine, $MoBr_3$.

As stated above, hexavalent compounds of molybdenum constitute the most numerous group. The hexavalent molybdenum oxyhalides include $MoOF_4$, MoO_2F_2, $MoOCl_4$, MoO_2Cl_2, and MoO_2Br_2. $Mo_2O_3Cl_5$ may contain tetravalent and hexavalent molybdenum. Hexavalent molybdenum also exists in the sulfide, MoS_3, and in various oxyacid salts, and molybdates.

Molybdenum trioxide dissolves in alkaline solutions to yield, in more or less hydrated form, the molybdate ion, MoO_4^{2-}. However, the ionic species that exist in solutions of low pH are more complex than hydration of MoO_4^{2-} would indicate, and this is especially true of the compounds obtained from such solutions. Thus from strongly ammoniacal solution, $(NH_4)_2MoO_4$ is obtained, but from nearly neutral solutions containing NH_4^+ and MoO_4^{2-}, the complex $(NH_4)_6Mo_7O_{24} \cdot 4H_2O$ crystallizes. The process of complex anion formation is considered to occur upon neutralization, or particularly upon acidification, of a solution, by the addition of a proton to an MO_4^{2-} ion, forming HMO_4^- ions, which condense with other oxyanions to form complexes, which can be crystallized as salts. Such salts are considered to be derived from "poly" acids. When such acids have one kind only of metal atom (e.g., Mo) in their anions, like the complex ammonium salt cited above, they are called *isopoly acids*; with more than one kind they are called *heteropolyacids*. The latter group comprises an entire field of molybdenum chemistry (as well as that of tungsten and related elements); the molybdophosphates are important in analysis and other applications. Other examples are the heteropolyacid salts formed by molybdates with oxyanions of boron, silicon, germanium, tin, arsenic, titanium, zirconium, and hafnium. In such compounds, one important group is the "12-group," containing 12 atoms of molybdenum, but many other proportions are known. Other interesting compounds are the "molybdenum blues," complex oxides of colloidal nature obtained by reduction of molybdates.

Molybdenum forms many other complexes. Of particular interest are the octacyano complexes, containing eight cyanide ions, CN^-, coordinated to a single tetravalent of pentavalent molybdenum ion, $Mo(CN)_8^{2-}$ and $Mo(CN)MO_8^{3-}$, the latter being exceptionally stable, and both form octacyanomolybdic acids $H_3[Mo(CN)_8] \cdot 3H_2O$ and $H_4[Mo(CN)_8] \cdot 6H_2O$. Molybdenum(III) forms salts of

$$H_3MO(CN)_6$$

A fluoro complex of Mo(VI) has the structure, $[MoF_8]^{2-}$.

In liquid NH_3 solution, potassium amide reacts with MoO_3 to form the salt K_3MoO_3N, completely hydrolyzed by H_2O, in which the molybdenum atom is the center of a monomeric tetrahedral anion, being surrounded by three oxygen and one nitrogen atoms.

Like chromium, molybdenum forms a number of cyclopentadienyl compounds, many of which are carbonyls, e.g., $C_5H_5Mo(CO)_2NO$, $C_5H_5Mo(CO)_3X$ (where X may be Cl, Br, I, H, CH_3, C_2H_5 or C_3H_7). Molybdenum also forms a simple carbonyl, $Mo(CO)_6$.

Molybdates: Ammonium polymolybdate, also referred to as ammonium dimolybdate or molybdic acid (85%), is the highest-purity molybdenum compound commercially available (up to 99.97% purity). The compound is a source of very high-purity MoO_3 which is used as a catalyst in hydrogen-treating and hydrocracking processes. The compound also is used in electroplating baths and as an important laboratory reagent for determinations of phosphates, arsenates, and lead. Sodium molybdate Na_2MoO_4 or $Na_2MoO_4 \cdot 2H_2O$ is made by dissolving MoO_3 in excess NaOH. The compound is widely used in the manufacture of molybdate-chromate orange pigments and several phosphomolybdic acid-organic pigments. The compound also is used as a condensation catalyst for phthalocyanine pigments, as a synthetic cutting fluid, and as a corrosion inhibitor in circulating cooling water systems and in glycol-based antifreeze formulations. Several of the soluble molybdates react with organic intermediates to form dyes that are used for furs and hair and have the advantage of being very colorfast. Zinc molybdate, because of its nontoxicity and excellent corrosion-inhibiting properties, is an excellent white pigment. Lithium molybdate is an additive for porcelain enamel coatings. Iron, cobalt, and nickel molybdates are used as catalysts in hydrogenation, desulfurization, denitrification, and hydrocracking processes. Lead molybdate is used in connection with applying vitreous designs to glass bottles.

Sulfides. In addition to serving as the primary natural source of molybdenum, purified molybdenum disulfide MoS_2 is an excellent lubricant when in the form of a dry film, or as an additive to oil or grease. The compound also is used as a filler in nylons, and as an effective catalyst for hydrogenation-dehydrogenation reactions. Molybdenum also combines with sulfur as the sesquisulfide Mo_2S_3 and the trisulfide MoS_3, uses for which are under study.

Halides. Molybdenum pentachloride $MoCl_5$ is used as a catalyst for several polymerization reactions involving olefins, vinyl monomers, trioxane, ethylene, vinylcyclohexane, cyclopentene, and butadiene. Vapor-phase coatings of molybdenum on metallic or ceramic substrates are prepared, using $MoCl_5$ as the starting compound.

Organomolybdenum Compounds. Soluble molybdates, molybdenum hexacarbonyl $Mo(CO)_6$, and several molybdenum halides form complex compounds with many organic oxygen, nitrogen, and sulfur compounds. Some of the oxygen-coordinated compounds include alkoxides, acetonates, oxalates, carboxylates, phenoxides, and organic chlorides. Nitrogen-coordinated compounds include several organic molybdates and chlorides. Sulfur-coordinated compounds include dialkyldithiophosphates, cysteine complexes, α-diketone complexes, and dialkyldithiocarbamates. Examples of industrial use of organomolybdenum compounds include pyrogalol-molybdate complexes in dyes, molybdenum oxalate in photochemicals, molybdenum dithiocarbamate as a lubricant additive, and molybdenum acetylacetonate as a catalyst for the polymerization of ethylene and the formation of polyurethane foam.

A discussion of the biological aspects of molybdenum is given in the next entry.

References

Barry, H. F., and P. C. H. Mitchell, Editors: "Chemistry and Uses of Molybdenum," *Proceedings of Symposium*, sponsored by Climax

Molybdenum Co., and University of Michigan, Ann Arbor, Michigan (August 1979).

Braithwaite, E. R.: "The Chemical Uses of Molybdenum and Its Compounds," *Chem. and Ind.*, **12**, 405–412 (1978).

Chianelli, R. R., et al.: "Molybdenum Disulfide in the Poorly Crystalline 'Rag' Structure," *Science*, **203**, 1105–1107 (1979).

Climax: Various product data sheets revised periodically, Climax Molybdenum Company, Greenwich, Connecticut 06830.

Lander, H. N.: "Technological Progress with Molybdenum," *Molybdenum Mosaic*, **3**, 1, 2–11 (1978).

Sutulov, A.: "International Molybdenum Encyclopedia," Vol. III, Intermet Publications, Santiago, Chile (1980).

—Robert Q. Barr, Director Technical Information, Climax Molybdenum Company, Greenwich, Connecticut.

MOLYBDENUM (In Biological Systems). Molybdenum is required in very low amounts by both plants and animals. Nutrient imbalances involving molybdenum and copper have caused serious problems in cattle and sheep production.

Molybdenum deficiencies are found in plants grown on certain acid soils, and sometimes the deficiency can be corrected by adding either small quantities of manganese compounds or larger quantities of limestone to the soil. The limestone makes the soil more alkaline and increases the availability of the native molybdenum in the soil. In certain parts of the world (including the eastern United States), small amounts of molybdenum fertilizer are used regularly for producing some vegetables, notably cauliflower. In Australia, large areas have been changed from near-desert conditions to productive agriculture through the application of molybdenized superphosphate.

In alkaline soils, molybdenum is more available to plants. Forage crops growing on some alkaline soils (as in the western United States) may take up high concentrations of molybdenum. The element is not toxic to the plants. They grow normally and may produce excellent yields. But cattle and sheep that eat these forages may suffer from molybdenum toxicity. It is now well established that what appears to be molybdenum toxicity is actually a copper deficiency that is induced by the molybdenum. Thus, the symptoms of molybdenum toxicity are the same as those of copper deficiency and include fading of the hair and diarrhea. The condition may be prevented by supplementing the animal diet with extra copper, or by injecting copper compounds into the animal body, usually by an experienced veterinarian. Cattle are more susceptible to molybdenum-induced copper deficiency than other types of livestock. Horses and pigs are rather tolerant of high levels of dietary molybdenum.

High levels of molybdenum are generally considered to be 20 parts per million (ppm) or more in dry forage. Some symptoms of interference with copper metabolism in cattle may be evident when the forage contains as little as 5 ppm molybdenum if the forage is also low in copper. The effects of high-molybdenum forage in interfering with copper metabolism in animals are generally more severe if the animal diet is also high in sulfates.

In terms of humans, some research in New Zealand and the United Kingdom indicates that diets containing moderately high levels of molybdenum help to prevent dental decay. The high-molybdenum soils in the United States are seldom used for production of food crops and thus the effects of molybdenum toxicity from food substances are not well known.

Restriction of the molybdenum intake by young rats in a synthetic purified casein diet results in a decreased level of tissue, particularly small intestinal, xanthine oxidase. The enzyme levels are restored to normal by the inclusion of sodium molybdate and other molybdate compounds. Sodium tungstate is a competitive inhibitor of molybdate, and dietary intakes of tungstate

greatly reduce the molybdenum and xanthine oxidase concentrations in tissues.

Legumes, cereal grains, and some green leafy vegetables are good sources of molybdenum, whereas fruits, berries, and most root or stem vegetables are poor sources. Vertebrate tissues are generally low in molybdenum with concentrations in liver and kidney being higher than in other organs and cells. Excess molybdenum intake by cattle causes the disease known as "teart," characterized by severe diarrhea and degradation of general health.

References

Allaway, W. H.: "The Effect of Soils and Fertilizers on Human and Animal Nutrition," Cornell University Agricultural Experiment Station, *Agriculture Information Bulletin 378*, U.S. Department of Agriculture, Washington, D.C. (1975).

Kirchgessner, M., Editor: "Trace Element Metabolism in Man and Animals," Institut für Ernährungsphysiologie, Technische Universität München, Freising-Weihenstephan, Germany (1978).

Sax, N. I.: "Dangerous Properties of Industrial Materials," Van Nostrand Reinhold, New York (1979).

Underwood, E. J.: "Trace Elements in Human and Animal Nutrition," 4th Edition, Academic, New York (1977).

MONATOMIC GASES. Chemical Elements.

MONAZITE. The mineral monazite is essentially a phosphate of the rare-earth metal cerium, $(Ce, La, Nd, Th)PO_4$; but other rare-earth metals are usually present. So constant is the presence of thorium that monazite is the chief source of thorium dioxide. It is monoclinic, but found ordinarily as translucent yellow to brown grains with a resinous luster, often as sand. Its hardness is 5.0–5.5; specific gravity, 4.6–5.4. Monazite is found in granites, pegmatites and similar rocks, but rarely in any concentration. The commercial deposits are residual sands. The Ilmen Mountains in the U.S.S.R., Norway, India, Malagasy Republic, the Republic of South Africa, and Brazil are well known for their monazite deposits. In the United States monazite is known in Connecticut, New York, Virginia, North Carolina, and Idaho. Monazite derives its name from the Greek word meaning *solitary*, in reference to the relative rarity of this mineral.

MOND PROCESS. Nickel.

MONEL. Nickel.

MONOMER. A single molecule or a substance consisting of single molecules. The word *monomer* is used in differentiation of dimer, trimer, etc., words designating polymerized or associated molecules, or substances composed of them, in which each free particle is composed of two, three, etc., molecules.

MONOMOLECULAR LAYER. The early work of Rayleigh, Langmuir, Hardy, and others showed that it is possible to deposit on solid or liquid surfaces films which are only one molecule thick. Any such layer is called a *monomolecular layer*, *unilayer*, or *monolayer*.

MONOSODIUM GLUTAMATE (MSG). Flavor Enhancers and Potentiators.

MORPHINE. About 10% of the weight of opium is morphine which was the first of the vegetable alkaloids to be isolated in 1805 by Sertürner. Since the source of the natural alkaloids is opium, all narcotics whose actions resemble those of morphine are sometimes referred to as opiates. Semisynthetic agents are usually made by altering the morphine molecule, and include such agents as diacetylmorphine (heroin), ethylmorphine (*Dionin*), dihydromorphinone (*Dilaudid*), and methyldihydromorphinone (metopon). Synthetic narcotics include agents with a wide variety of chemical structures. Some of the important synthetic agents are meperidine (piperidine type), levorphanol (morphinian type), methadone (aliphatic type), phenaxocine (benzmorphan type), and their derivatives. The structures of the various narcotics are given in Fig. 1.

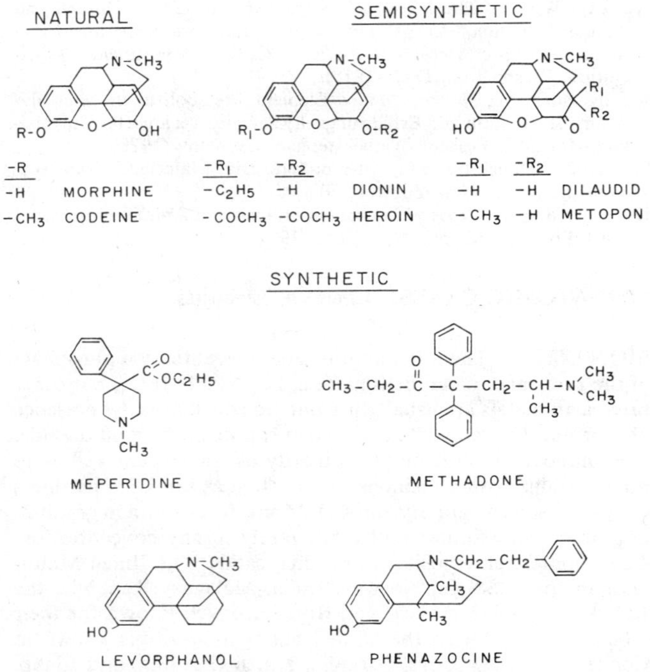

Fig. 1.   Chemical structures of various narcotics.

Since morphine is responsible for the major actions of opium and the actions of all narcotics are qualitatively similar, morphine can be used as a model for discussing narcotic agents. The most prominent effects of morphine in the human body are on the central nervous system and the gastroenteric tract. The principal central action of morphine is the relief of pain, and this occurs in at least three ways: (1) morphine reduces central perception of pain probably at the thalamic level, (2) it alters the reactions to pain probably at the level of the cerebral cortex, and (3) it elevates the pain threshold by inducing sedation or sleep. In the medulla, morphine depresses the respiratory, cough and vasomotor centers and indirectly stimulates the vomiting center. The nuclei of the occulomotor (III) and vagus (X) nerves are stimulated by sufficient doses of morphine causing myosis (constriction of the pupils), bradycardia (slowing of the heart rate), and increased gastroenteric tone. The overall effect of morphine on the gastroenteric tract is spasmogenic and constipative. Morphine causes the constipative action by several means, including increased segmental movement of the large bowels, spastic tonus of the sphincters, decreased defecation reflex, and increased reabsorption of water in the large intestines to cause drying of feces.

The metabolic effects of morphine are not marked and are clinically unimportant. The metabolic rate may be decreased

slightly due to the lowered activity and tone of the skeletal muscles resulting from the central depression. A rise in blood sugar may be observed after the injection of morphine. The hyperglycemia is due to glycogenolysis in the liver resulting from the release of epinephrine from the adrenal medulla. The lowering of urine production noted after the administration of the drug is due mainly to the release of antidiuretic hormone from the posterior pituitary gland.

Morphine is detoxified or biotransformed mainly in the liver by conjugation with glucuronic acid. Morphine is conjugated by a series of reactions involving the formation of uridine diphosphoglucose (UDP-glucose), the oxidation of carbon-6 of glucose to form uridine diphosphoglucuronic acid (UDP-glucuronic acid) and the transfer of glucuronic acid to morphine to form the morphine glucuronide. This reaction is diagramed in Fig. 2. The following enzymes catalyze the sequential reactions: reac-

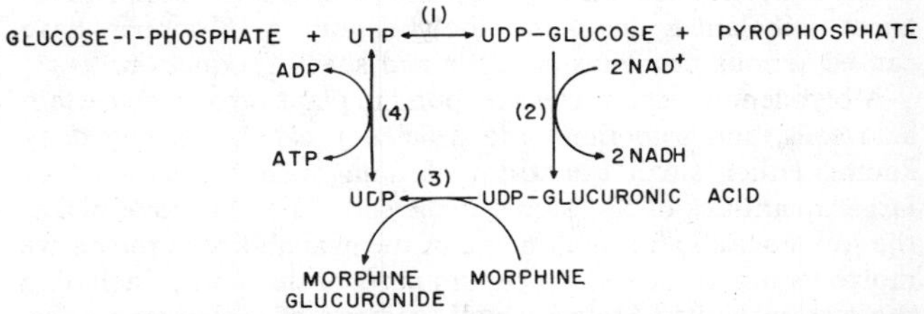

Fig. 2.   Formation of morphine glucuronide. NAD$^+$ = nicotinamide adenine dinucleotide; NADH = reduced NAD$^+$; ATP = adenosine triphosphate; ADP = adenosine diphosphate; UTP = uridine triphosphate; UDP = uridine diphosphate.

tion (1), UDP-glucose pyrophosphorylase; reaction (2), UDP-glucose dehydrogenase; reaction (3), glucuronyl transferase; reaction (4) nucleoside diphosphokinase.

The most serious drawback in the use of morphine and other narcotic analgesics is their addictive potentiality. The characteristics of drug addiction include psychological need or habituation, tolerance and physical dependence. Habituation consists of an emotional and psychic dependence, and in addiction, the habituation becomes an overpowering desire to take the drug. Tolerance is a phenomenon whereby the dosage of the drug must be continually increased to maintain equivalent pharmacologic effects. Physical dependence develops when the tissues of the body become so adapted to the effects of the drug that the cells of the tissues cannot function normally without the drug in the environment. This is the most vicious characteristic of drug addiction.

The mechanisms underlying the development of tolerance are not fully understood. Biochemically, it may be attractive to explain tolerance by decreased absorption, altered distribution, increased biotransformation, and/or increased excretion of the drug. However, these processes have been shown to be unrelated to the development of tolerance. Thus, cellular adaptation offers the greatest likelihood for clarifying the phenomenon. Evidence for cellular adaptation is the finding that the respiration of chemically stimulated cortical slices of brain from normal rats is markedly depressed by morphine, whereas the respiration of those from rats chronically dosed with morphine is unaffected.

Heroin is diacetylmorphine (diamorphine hydrochloride) and is prepared by the action of acetic anhydride on morphine, possessing four times the analgesic affect of morphine, but having a considerably less depressant effect. Addiction is common, the drug being taken in the form of snuff, or by injection.

Nalorphine, the allyl (—CH$_2$—CH=CH$_2$) derivative of morphine (N-allylnormorphine) is remarkable in that it is antagonizing to almost all the effects of narcotics. The antagonizing action is specific for the narcotic analgesics. For instance, nalorphine will antagonize the respiratory depression due to morphine or other narcotics, but not that caused by other depressants, such as hypnotics or anesthetics. This property of nalorphine makes it a particularly useful antidote in cases of acute morphine poisoning. The agent can also precipitate acute withdrawal symptoms if administered to persons addicted to narcotics. The agent has become a useful biochemical tool for studying the mechanism of action of narcotics and tolerance. Since the chemical structures between morphine and nalorphine are so similar, it has been suggested that nalorphine acts by competing with morphine for the receptor site. The antagonistic effect of nalorphine cannot be explained by a simple competitive inhibition if equal affinity for the receptor site with the agonist and antagonist is assumed, because small doses of nalorphine antagonize the effects of much higher doses of the narcotic. Nalorphine also antagonizes the effects of synthetic narcotics of varying chemical structures, such as methadone and meperidine.

Morphine, $C_{17}H_{19}NO_3 \cdot H_2O$, is a white powder melting at 253°C and is derived from opium which is the dried juice obtained from unripe capsules of the poppy plant (*Papaver somniferum*), variously cultivated in the Near East and Far East. The opium poppy is an annual. When the petals drop from the white flowers, the capsules are cut. The juice exudes and hardens, forming a brownish mass which is crude opium. It contains a total of about 20 narcotics, including morphine.

MÖSSBAUER EFFECT.

The phenomenon of recoilless resonance fluorescence of gamma rays from nuclei bound in solids. It was first discovered in 1958 by R. L. Mössbauer. The extreme sharpness of the recoilless gamma transitions and the relative ease and accuracy in observing small energy differences make the effect an important tool in nuclear physics, solid-state physics, and chemistry.

If a gamma ray is emitted by an atomic nucleus, the system to which the emitting atom belongs must recoil, in order to conserve momentum, in a direction opposite to that in which the gamma rays is emitted. Similarly, if an atomic nucleus absorbs a gamma ray, the system must continue to move, following absorption, in such a way that momentum is conserved. If the recoiling system is a single atom, such as in a gas and shown schematically in the accompanying diagram (a), the emitting atom carries away enough energy from the transition for the observed energy $E_0 - R$ of the emitted gamma ray to be measurably less than the energy E_0 of the nuclear transition that caused the gamma ray to be emitted, also indicated in (a) of the diagram. Furthermore, a gamma ray that is absorbed by a single atom must transfer a measurable kinetic energy to that atom, as well as the energy of the nuclear transition. On the other hand, if the emitting or absorbing nucleus belongs to an atom that is bound into a crystalline structure, such that the structure as a whole can recoil, and as indicated schematically in (b) of the diagram, the kinetic energy that must be given to the crystalline system to conserve momentum is greatly reduced, compared to the energy that must be given to a single atom, because of the much larger mass of the system. The recoil energy is then so small that the gamma ray carries away essentially the full energy E_0 of the transition in the case of emission, transferring such a small fraction of its energy to the absorbing system that emission and absorption appears to be recoil-free.

This process is observed, of course, in the analogous case of the resonance radiation in atomic transitions, in which case, the photons have energies in the range of light, commonly visible light. However, the protons of gamma radiation are so much

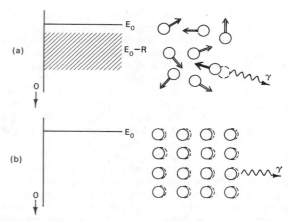

(a) Emission of a gamma-ray by a single atom moving randomly in a gas transfers appreciable energy to the emitting atom in the form of recoil kinetic energy, and reduces the energy of the gamma ray from the transition energy E_0 to some lower energy $E_0 - R$. (b) Emission of a gamma ray by an atom bound into a crystalline structure may sometimes cause recoil of the whole crystal, in which case, the loss of energy in the form of kinetic energy of the crystal is negligibly small and the gamma ray appears to be emitted with an energy E_0.

more energetic that their energy loss by recoil of the nucleus emitting them is great enough, in the case of free atoms, for the resonance effect not to occur.

Mössbauer discovered, however, that in the case of atoms which are not free, but bound in a solid, the effect can often be observed. It is easily demonstrated when the normal, free-atom recoil energy is comparable to the energy of the quantized lattice vibrations. Under these conditions, zero-phonon processes are possible in which the entire energy of the nuclear transition goes into the gamma ray and the recoil momentum is taken by the solid as a whole. The resulting gamma rays then have the proper energy to be resonantly absorbed or scattered in an analogous zero-phonon process.

The Mössbauer effect is useful in determining nuclear level widths and Doppler effects. Another application is based upon the measurement of nuclear hyperfine structure, a measurement which is possible when the line-width of the gamma ray is smaller than the hyperfine interaction (that due to the coupling of the nuclear moments with external fields). In this application, the Mössbauer effect is almost unique because one obtains the splitting of both the ground and the first excited nuclear states. This effect; in turn, makes possible the determination of nuclear moments of the excited states, which can be important tests of nuclear models. Another important feature is the so-called isomer or chemical shift (terms used interchangeably) which measures the simple electrostatic interaction of the nucleus with its own s-electron and has given information about the difference in the nuclear radii of the ground and excited states.

MOTOR OCTANE NUMBER.   Petroleum.

MUCILAGES.   Gums and Mucilages.

MUON.

The *muon* (μ^-) is an elementary particle of the lepton family. Properties include: Spin, $\frac{1}{2}$; mass (MeV), 105.66; lifetime, 2.20×10^{-6} second. The antiparticle is the positive muon (μ^+). The muon neutrino (ν) has spin, $\frac{1}{2}$; 0 mass; and is stable. The muon family appears to be simply a duplicate of the electron family except for a change in the unit of mass. See also **Particles (Subatomic).**

The positive muon was discovered in cloud chamber photographs made by C. D. Anderson and S. H. Neddermyer on Pike's Peak in 1935, and the negative muon almost simultaneously in cloud chamber photographs made by J. C. Street and E. C. Stevenson. These particles have long been called mu-mesons, but since they are fermions (spin $\frac{1}{2}$) while all other mesons are bosons, the name *muon* is preferred, as is their classification with the leptons because of their small rest mass, which is about 206 m_e, where m_e is the mass of the electron. Another reason is their inability to interact with other particles through the nuclear forces.

Their charges are equal in magnitude to that of the electron. They are produced by the decay of pions (pi mesons) and (to a limited extent) by the decay of kaons and hyperons. Positive-negative muon pairs also can be generated by the action on matter of gamma-rays of energy greater than the rest masses of the particles, i.e., exceeding 211 MeV. Their lives are short, about 2.2×10^{-6} seconds in the free state, and the negative muon usually decays into an electron, a neutrino, and an antineutrino, while the positive muon usually gives a positron, as well as a neutrino and antineutrino. As explained in the entry on neutrino, there are two types of neutrinos and antineutrinos (ν_e or $\bar{\nu}_e$) like that produced in the decay of radionuclides, and a muon-associated neutrino or antineutrino (ν_μ or $\bar{\nu}_\mu$) so that these reactions would be written

$$\mu^- \rightarrow e^- + \nu_\mu + \bar{\nu}_e$$

$$\mu^+ \rightarrow e^+ = \bar{\nu}_\mu + \nu_e$$

Muons can easily penetrate many meters of iron and can sometimes cause problems in particle physics research. For example, the upsilon experiment at Fermilab in 1977, conducted by L. M. Lederman and others, required building a simple magnetic system that would remeasure each muon's energy after it emerged from the main detector. See also **Upsilon Particle; and Particles (Subatomic).**

MUONIUM. The atom consisting of a positive muon and an electron. Thus, muonium may be regarded as a light isotope of hydrogen in which the positive muon replaces the proton. When a beam of positive muons is stopped in a gas (argon under such pressures as 50 atmospheres has been used in much of this research), muonium is formed directly in its ground state by the capture of an electron by a positive muon. The reaction is important because of its bearing upon the nature of the muon-electron interaction and the muon itself. The study has included measurement of the hyperfine structure interval in the ground state of muonium, and measurement of muon polarization as a function of time and impurity concentration. By adding such gases as oxygen (O_2) and nitric oxide (NO) as impurities to the argon, data on spin exchange of electron and muon is obtained, while with impurities such as nitrogen dioxide (NO_2) and ethylene (C_2H_2), evidence of such reactions as $NO_2 + M \rightarrow NO + OM$ and $C_2H_4 + M \rightarrow C_2H_4M$ is obtained.

MUTATION. Recombinant DNA.

MYOSIN. Contractility and Contractile Proteins.

MYRISTIC ACID. Also called tetradecanoic acid, formula $CH_3(CH_2)_{12} COOH$. At room temperature, it is an oily, white crystalline solid. Soluble in alcohol and ether; insuluble in water. Specific gravity 0.8739 (80°C); mp 54.4°C; bp 326.2°C. Combustible. The acid is derived by the fractional distillation of coconut oil. Myristic acid is used in soaps; cosmetics; in the synthesis of esters for flavorings and perfumes; and as a component of food-grade additives. Myristic acid is a constituent of several vegetable oils. See also **Vegetable Oils (Edible).**

N

NAPHTHALENE. Coal Tar and Derivatives; Organic Chemistry.

NARCOTICS. Alkaloids; Analgesics; Hallucinogens; Morphine.

NATROLITE. The mineral natrolite, one of the zeolites, is a sodium aluminum silicate corresponding to the formula $Na_2Al_2Si_3O_{10} \cdot 2H_2O$. It is orthorhombic, crystallizing in slender prisms of nearly square cross section which are terminated by relatively flat pyramids. There are also fibrous to compact varieties. Natrolite is a brittle mineral; hardness, 5–5.5; specific gravity, 2.2; luster, vitreous; color red, yellow, white, or colorless; transparent to opaque. Natrolite is found with other zeolites in fissures and cavities in basaltic and related rocks. Czechoslovakia, France, Italy, Norway, Scotland, Ireland, Iceland, Greenland, and South Africa contain well-known localities for natrolite. In the United States it is found in the Triassic traps of New Jersey; also from Oregon, Washington, Montana, Colorado, and as exceptional crystals from San Benito County, California. Superb crystals occur at Mt. St. Hilaire, Quebec, Canada, and from an asbestos mine in Quebec, crystals up to 3 feet (0.9 meters) long and 4 inches (10 centimeters) in diameter have been found. The name *natrolite* refers to its soda content.

NATURAL GAS. The composition of natural gas varies with the source, but essentially it is made up of methane, ethane, propane, and other paraffinic hydrocarbons, along with small amounts of hydrogen sulfide, carbon dioxide, nitrogen, and, in some deposits, helium. Natural gas is found underground at various depths and pressures, as well as in solution with crude oil deposits. Principal gas deposits are found in the United States, Canada, the U.S.S.R., and the Middle East, but there are significant reservoirs in a number of other locations. The analysis of a gas sample taken from the Panhandle natural gas field in Texas is given in Table 1.

Worldwide, there are substantial reserves of natural gas in which the reservoir formation hydrocarbons are contaminated with nonburning components. The presence of helium, nitrogen, or carbon dioxide reduces the heating value of the gas mixture and can result in the gas being unsuitable for existing transmission and distribution systems. Such contaminated mixtures are termed *low-Btu gases* when their heating value falls below the minimum standards.

Cryogenic Processing. Low-Btu gas mixtures can be upgraded through cryogenic processing. This is a physical process in which subambient temperatures are used to bring about a separation between the hydrocarbons and nonhydrocarbons. The reduction of temperature during cryogenic processing produces a two-phase (gas-liquid) mixture. The relative volatilities between the components in the mixture results in selective mass transfer between the two phases. One phase is enriched with hydrocarbons, thus elevating the heating value. The second phase is

TABLE 1. ANALYSIS OF NATURAL GAS FROM NATURAL GAS FIELD IN TEXAS PANHANDLE

COMPONENT	MOLE PERCENT
Methane	76.2
Ethane	6.4
Propane	3.8
Normal butane	1.3
Isobutane	0.8
Normal pentane	0.3
Isopentane	0.3
Cyclopentane	0.1
Hexane plus other hydrocarbons	0.35
Nitrogen	9.8
Oxygen	Trace
Argon	Trace
Hydrogen	0.0
Hydrogen sulfide	0.0
Carbon dioxide	0.2
Helium	0.45

NOTE: Heating value of various natural gases averages between 975 and 1180 Btu/cubic foot (8678–10,502 Calories/cubic meter) at 60°F (15.6°C) and 30 inches (76.2 centimeters) mercury pressure.

denuded of hydrocarbons, with a very marked reduction in heating value.

One of the main considerations in the design and operation of cryogenic upgrading is to identify and remove any component which can adversely affect operation of the cold sections of the plant. Such components are carbon dioxide, water vapor, and heavy hydrocarbons, each of which has a high solidification temperature and low solubility. Such components can be separated through traditional absorption and adsorption processes.

A simplified flowsheet of a cryogenic upgrading process is shown in Fig. 1. The plant stream parameters are given in Table 2. In this example, the low-Btu gas is available at 800 psig (54 atm) and is mainly a nitrogen-methane mixture. The more volatile materials as previously described have been removed. Small quantities of CO_2 remaining are absorbed by a monoethanolamine (MEA) solution, remaining water is taken out on molecular sieves, and any heavy hydrocarbons remaining are adsorbed on activated carbon. The gas is then cooled in plate-fin exchangers against the returning high-Btu product gas and vent gas. The gas is then expanded to 380 psig (25 atm) and a vapor-liquid mixture passes into a high-pressure (H.P.) fractionator. This fractionator brings about an initial separation of the nitrogen-methane and produces a liquid nitrogen reflux for the low-pressure (L.P.) fractionator.

A nitrogen-enriched vapor flows up the H.P. fractionator, while methane is returned to the sump of this column by a nitrogen reflux stream produced in the tubes of the overhead condenser. The refrigeration required to produce the nitrogen

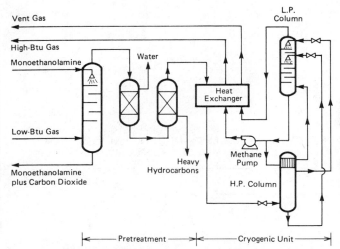

Fig. 1.   Plant for nitrogen removal from natural gas using cryogenic upgrading. (*Petrocarbon Developments, Ltd.*)

reflux is provided by evaporating some of the liquid methane from the L.P. column in the shell of the overhead condenser. Two liquid streams are taken from the H.P. fractionator, and these become the feed and reflux for the L.P. fractionator. The L.P. feed is an enriched methane stream taken from the base of the H.P. fractionator just below the condenser. The upgrading is completed in the L.P. fractionator. The feed stream is stripped to produce a high-Btu liquid containing 4% nitrogen and having a heating value of 980 Btu/standard cubic foot (8722 Calories per cubic meter). The liquid is pumped from the column sump, evaporated, and superheated against the incoming low-Btu gas. By using this arrangement of two distillation columns, the separation of nitrogen and methane can be achieved using only the pressure energy available in the low-Btu gas.

TABLE 2.   CRYOGENIC UPGRADING OF NATURAL GAS—
STREAM PARAMETERS
Composition—Mol. %

	LOW-BTU GAS	HIGH-BTU GAS	VENT GAS	HELIUM
Helium	0.40	—	0.09	100.00
Nitrogen	42.75	4.00	98.95	—
Methane	56.02	95.09	0.96	—
Ethane +	0.53	0.91	—	—
CO_2	0.30	—	—	—
Flow (million standard cubic feet/day)	246	143	100	0.43
Flow (million cubic meters/day)	7	4	2.8	0.012
Heating value (Btu/standard cubic foot)	580	980	—	—
Heating value (Calories/cubic meter)	5162	8722	—	—

Liquefied Natural Gas (*LNG*). The liquefaction of natural gas for storage and transportation and regasification for final distribution dates back several decades. A few major accidents in the handling of LNG thwarted the progress of the field for

a while, but in the early 1970s, LNG was again considered in a major way because of energy concerns. One of the more serious LNG accidents occurred in Cleveland, Ohio on October 20, 1944, when a storage tank developed a leak with spillage and subsequent fires in the surrounding neighborhood in which 135 persons lost their lives. While liquefaction offers marked storage space savings and convenience, the predominant advantage occurs in connection with both pipeline and ship transportation. Energy-short nations, such as Japan, and some of the European nations, have turned in recent years to the concept of shipping LNG in specifically designed ships.

Oil-and-gas-rich nations, which at one time flared to the atmosphere much of the natural gas that accompanied the production of crude oil, have turned toward conservation—either through reinjection of much of the natural gas under ground or through constructing LNG production facilities for shipment of product overseas. Three types of liquefaction processes may be used for production of LNG. The standard cascade process which uses three refrigerants—methane, ethylene, and propane—all circulating in closed cycles, is shown in Fig. 2. There is a separate

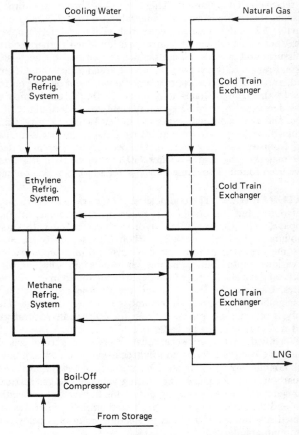

Fig. 2.   Conventional or standard cascade system.

compressor for each of these refrigerants. The methane and propane are available from the feed gas (natural gas). The ethylene must be furnished separately. Ethane may be used in place of ethylene at subatmospheric suction pressure. The cascade process has the highest rank in terms of thermal efficiency. As a possible improvement over the cascade process, the mixed refrigerant cascade (MRC) system shown in Fig. 3. In one plant using this process, a hydrocarbon-plus-nitrogen mixture of rela-

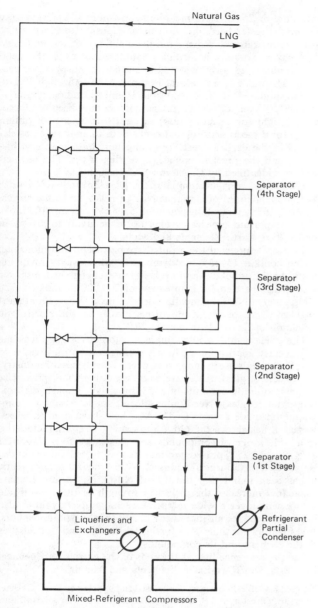

Fig. 3. Single-pressure mixed refrigerant cascade system. (*U.S. Patent No. 3,593,535*)

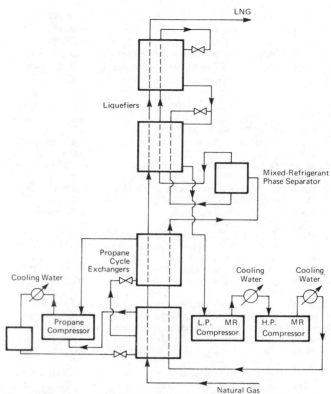

Fig. 4. Propane–mixed-refrigerant liquefaction system. L.P. = low-pressure; H.P. = high-pressure; MR = mixed refrigerant.

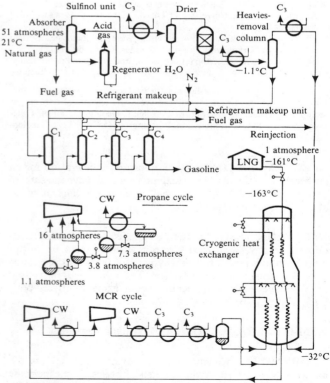

Fig. 5. LNG production plant flow scheme. MCR = mixed-refrigerant cascade; CW = cooling water.

tively wide boiling range (N_2 through C_5) is used as the refrigerant. All of these components can be recovered from natural gas in separate apparatus. In still another system, shown in Fig. 4, a propane and mixed-refrigerant cycle is used. In this process, the cooling load is divided horizontally at about −34.4°C into an upper portion absorbed by propane and a lower portion absorbed by the mixed refrigerant. In essence, the system is a dual refrigerant cascade in which the lower-boiling fluid is a mixed refrigerant. The cascade combination with propane makes it possible to reduce the boiling range of the mixture refrigerant substantially, which improves the thermodynamic efficiency over that of the straight MRC process.

As previously described in connection with cryogenic upgrading of low-Btu natural gas, the liquefaction of natural gas commonly requires pretreatment prior to cryogenic liquefaction. Hydrogen sulfide, if present, must be removed. The absorption process involved introduces water vapor which must be subsequently removed. Carbon dioxide and heavy hydrocarbons must be eliminated. One of five trains of a large LNG plant, showing pretreatment and cryogenic operations, is shown in Fig. 5.

References

Arnoni, Y. G.: "The Marriage of LNG and Offshore Facilities," *Chem. Eng. Progress*, **75**, 10, 60–65 (1979).

Cline, J. D., and M. L. Holmes: "Submarine Seepage of Natural Gas in Norton Sound, Alaska," *Science*, **198**, 1149–1153 (1977).

Considine, D. M., Editor: "Energy Technology Handbook," McGraw-Hill, New York (1977).

Hayes, E. T.: "Energy Resources Available to the United States, 1985–2000," *Science*, **203**, 233–239 (1979).

Hodgson, B.: "Natural Gas: The Search Goes On," *Natl. Geographic*, **154**, 5, 632–651 (1978).

Howard, J. L., and P. G. Andersen: "A Barge-Mounted Gas Liquefaction and Storage Plant," *Chem. Eng. Progress*, **75**, 10, 76–81 (1979).

Keeney, R. L., Kulkarni, R. B., and K. Nair: "Assessing the Risk of an LNG Terminal," *Technology Review (MIT)*, **81**, 1, 64–72 (1978).

Neustadti, S. J.: "LNG Safety," *Technology Review (MIT)*, **80**, 4, 56–57 (1978).

Staff: "Reserves of Crude Oil, Natural Gas Liquids and Natural Gas in the United States and Canada and United States Productivity Capacity," American Gas Association, Arlington, Virginia, (issued annually). Also numerous special publications relating to pipelines, natural gas properties, production, and consumption, listed in the Catalog of Publications.

Staff: Technical and economic reports on natural gas (issued periodically). American Petroleum Institute, Washington, D.C. Listed in the Catalog of Publications.

Staff: "Reserves of Crude Oil, Natural Gas Liquids, and Natural Gas in Canada," Canadian Petroleum Association, Calgary, Alberta, Canada (issued annually).

Staff: "Future Natural Gas Requirements of the United States," Denver Research Institute, University of Denver, Denver, Colorado (issued annually).

Staff: "LNG Materials and Fluids Users Manual" National Bureau of Standards, Cryogenics Division, Boulder, Colorado (1977).

Staff: Technical and economic reports on natural gas issued periodically, National Petroleum Council, Listed in the Catalog of Publications.

NOTE: Other organizations that may be contacted for current information on the various technological and economic aspects of natural gas include: American Association of Petroleum Geologists, Tulsa, Oklahoma; Federal Power Commission, Washington, D.C.; and U.S. Geological Survey, Reston, Virginia.

NÉEL TEMPERATURE. The transition temperature for an antiferromagnetic material. Maximal values of magnetic susceptibility, specific heat, and thermal expansion coefficient occur at the Néel temperature.

NEGATRON. A word sometimes applied to the normally occurring negatively charged electron when it must be distinguished from a positron. In many parts of the world, the name *negaton* is used instead of negatron. The word *negatron* is used in this Encyclopedia wherever distinction is made between positively and negatively charged electrons.

NEMATIC LIQUID CRYSTALS. Liquid Crystals.

NEODYMIUM. Chemical element symbol Nd, at. no. 60, at. wt. 144.24, third in the lanthanide series in the periodic table, mp 1016°C, bp 3068°C, density, 7.004 g/cm³ (20°C). Elemental neodymium has a close-packed hexagonal crystal structure at 25°C. Pure metallic neodymium is a silvery-gray color, the luster becoming dull upon exposure to moist air at room temperatures. When pure, the metal is soft and malleable and may be worked with ordinary equipment. Because the metal is pyrophoric, it must be stored in an inert atmosphere or vacuum. There are seven natural isotopes, ^{142}Nd through ^{146}Nd, ^{148}Nd, and ^{150}Nd. Isotope ^{144}Nd is mildly radioactive ($t_{1/2} = 10^{10}–10^{15}$ years). Seven artificial isotopes have been produced. Of the light (or cerium-group) rare-earth metals, neodymium is the third most plentiful and ranks 60th in abundance of elements in the earth's crust, exceeding tantalum, mercury, bismuth, and the precious metals, excepting silver. The element was first identified by C. A. von Welsbach in 1885.

Electronic configuration: $1s^22s^22p^63s^23p^63d^{10}4s^24p^64d^{10}4f^35s^25p^65d^16s^2$. Ionic radius: Nd^{3+}, 0.995 Å; metallic radius, 1.821 Å. First ionization potential, 5.49 eV; second, 10.72 eV. Other important physical properties of neodymium are given under **Rare-Earth Elements and Metals.**

Primary sources of the element are bastnasite and monazite, which contain 15–25% neodymium. Plant capacity involving liquid-liquid or solid-liquid organic ion-exchange processes for recovering the element is in excess of 200,000 pounds (90,720 kilograms) of Nd_2O_3 annually. Metallic neodymium is obtained by electrolysis of fused anhydrous $NdCl_3$ or the electrolytic reduction of the oxide in molten NdF_3.

Use of elemental neodymium as a colorant for glass was one of the early applications. The color ranges from pure violet to purple and finds use in sunglasses, protective glasses for industry, art objects of glass, tableware, and decorative fiber optics. Use of neodymium in amounts of 3–5% by weight imparts dichroic properties to glass. Neodymium-doped single-crystal yttrium-aluminum oxide garnets (Nd : YAG) have been used in lasers. Research has shown the Nd ion to exhibit laser characteristics in a wide range of compounds and glasses. A formulation of 75% Nd and 25% praseodymium, frequently called *didymium*, is used as a metallurgical additive. Within the last several years, it has been found that the use of Nd_2O_3 in barium titanate capacitors increases the dielectric strength of these electronic components over a wide temperature range. Neodymium also has been used as an ingredient of phosphate-type phosphors. Investigators continue into further electronic and optical uses of the element and its compounds.

See the reference lists at the ends of the entries on **Chemical Elements;** and **Rare-Earth Elements and Metals.**

NOTE: This 4th Edition entry was revised and updated by K. A. Gschneidner, Jr., Director, and B. Evans, Assistant Chemist, Rare-Earth Information Center, Energy and Mineral Resources Research Institute, Iowa State University, Ames, Iowa.

NEON. Chemical element symbol Ne, at. no. 10, at. wt. 20.183, periodic table group 0 (inert or noble gases), mp −248.68°C, bp −246°C, density, 1.204 g/cm³ (liquid). Specific gravity compared with air is 0.674. Solid neon has a face-centered cubic crystal structure. At standard conditions, neon is colorless, odorless and does not form stable compounds with other elements under ordinary conditions. Due to its low valence forces, neon does not form diatomic molecules, except in discharge tubes. It does form compounds under highly favorable conditions, as excitation in discharge tubes, or pressure in the presence of a powerful dipole. However, the compound-forming capabilities of neon, under any circumstances, appear to be far less than those of argon or krypton. No known hydrates have been identified, even at pressures up to 260 atm. First ionization potential, 21.599 eV.

There are three natural isotopes, ^{20}Ne through ^{22}Ne, and four radioactive isotopes, ^{18}Ne, ^{19}Ne, ^{23}Ne, and ^{24}Ne, all with half-lives of less than 5 minutes. Ramsay and Travers first found the element when investigating the properties of liquid air in 1898. The element is easily identified spectroscopically. Neon emits characteristic red and green lines in its spectrum.

Neon occurs in the atmosphere to the extent of approximately 0.00182%. In terms of abundance, neon does not appear on lists of elements in the earth's crust because it does not exist in stable compounds. However, because of its limited solubility in H_2O, neon is found in seawater to the extent of approximately 1.5 tons per cubic mile (0.3 metric ton/cubic kilometer). Commercial neon is derived from air by liquefaction and fractional distillation. For most applications, the gas need not be in a highly pure form, but may be supplied along with small quantities of the other rare gases, such as argon and krypton. The gas finds principal applications in various electronic devices and lamps, but the most familiar application is the neon tubes used mainly in signs. Neon emits the familiar orange light. Neon also has been used in certain lasers.

NEPHELINE. Nepheline, of hexagonal crystallization, is a sodium-potassium aluminum silicate $(Na, K)(AlSiO_4)$. It is found in silica-poor geological environments, where there has been insufficient silica to form feldspar. Nepheline rocks are characterized by the absence of quartz within them. They constitute a mineral family group known as the *feldspathoids*. Crystals are extremely rare; usually occurs massive to compact. Luster is greasy in the massive varieties, vitreous in crystals. Color grades from yellowish to colorless in crystals; gray, green, and reddish in massive material. It ranges from transparent to translucent. Hardness 5.5–6, specific gravity 2.55–2.65.

Immense masses of nepheline-rich rocks occur on the Kola Peninsula, U.S.S.R., in Norway and in the Republic of South Africa; also in the Bancroft, Ontario, Canada region. Smaller deposits are found in Maine and Arkansas in the United States. Fine crystals are found in lavas on Mt. Vesuvius, Italy.

Nepheline is used extensively in the manufacture of glass.

NEPTUNIUM. Chemical element symbol Np, at. no. 93, at. wt. 237.0482 (predominant isotope), radioactive metal of the actinide series; also one of the transuranium elements. Neptunium was the first of the transuranium elements to be discovered and was first produced by McMillan and Abelson (1940) at the University of California at Berkeley. This was accomplished by bombarding uranium with neutrons. Neptunium is produced as a byproduct from nuclear reactors. ^{237}Np is the most stable isotope, with a half-life of 2.20 $\times$ 10^6 years. The only other very long-lived isotope is that of mass number 236, with a half-life of 5 $\times$ 10^3 years. ^{237}Np is the parent of the neptunium $(2n + 1)$ alpha decay series. Other isotopes include those of mass numbers 229–235 and 238–241; metastable forms of ^{236}Np, ^{240}Np and two of ^{237}Np are known. Electronic configuration is $1s^2 2s^2 2p^6 3s^2 3p^6 3d^{10} 4s^2 4p^6 4d^{10} 4f^{14} 5s^2 5p^6 5d^{10} 5f^5 6s^2 6p^6 6d^1 7s^2$. Ionic radii: NP^{4+}, 0.88 Å; Np^{3+}, 1.02 Å (*Zachariasen*). Oxidation potential: Np $\rightarrow$ Np^{3+} + 3e^-, 1.85 V; Np^{3+} $\rightarrow$ Np^{4+} + e^-, −0.155 V; Np^{4+} + 2H_2O $\rightarrow$ NpO$_2^+$ + 4H^+ + e^-, −0.739 V; NpO$_2^+$ $\rightarrow$ NpO$_2^{2+}$ + e^-, −1.137 V.

Neptunium has the oxidation states (VI), (V), (IV), and (III) with a general shift in stability toward the lower oxidation states as compared to uranium. The compounds which are formed are very similar to the corresponding compounds of uranium.

The ionic species corresponding to the oxidation states vary with the acidity of the solution; in acid solution of moderate strength the species are Np^{3+}, Np^{4+}, NpO$_2^+$, and NpO$_2^{2+}$ as in the case of uranium and plutonium. The potential scheme in 1 M HCl is as follows:

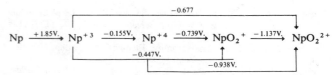

It will be seen that the metal is highly electropositive, in common with the other actinide elements. The Np^{3+} $\rightarrow$ Np^{4+} couple is reversible and this oxidation can be accomplished by the oxygen of the air. The (IV) state is stable, not oxidized by air, and only slowly oxidized to NpO$_2^+$ by nitric acid. The Np^{4+} $\rightarrow$ NpO$_2^+$ couple is not readily reversible, whereas the NpO$_2^+$ $\rightarrow$ NpO$_2^{2+}$ couple is reversible; this is reasonable on the basis that the former involves making or breaking the neptunium-oxygen bonds, whereas the latter does not. The oxidation of NpO$_2^+$ to NpO$_2^{2+}$ requires moderately strong oxidizing agents. Neptunium differs from uranium and plutonium in that its potential relations are such as to render NpO$_2^+$ moderately stable with respect to disproportionating, even in solutions containing moderate concentrations of hydrogen ion.

The potentials are altered extensively by change in the hydrogen ion concentration and by the presence of any of a number of anions capable of forming complex ions.

Neptunium ions in aqueous solution possess characteristic colors: pale purple for Np^{3+}, pale yellow–green for Np^{4+}, green–blue for Np^{5+}, while NpO$_2^{2+}$ varies from colorless to pink or yellow–green depending on the acid present.

The precipitation reactions of Np^{3+} are similar to those of the tripositive rare earths, those of Np^{4+}, to the other tetrapositive actinides and to Ce^{4+}, and those of NpO$_2^{2+}$ to the corresponding ions of uranium and plutonium. All of the simple salts of NpO$_2^+$ appear to be soluble.

The neptunium oxide system exhibits complexity similar to that found in the uranium oxide system. Thus, the important oxide is NpO$_2$ and there exists a range of compositions, depending upon conditions, up to Np$_3$O$_8$.

The important halides of neptunium are the trifluoride, NpF$_3$, purple or black and hexagonal, the hexafluoride, NpF$_6$, brown and orthorhombic, the trichloride, NpCl$_3$, white and hexagonal the tetrachloride, NpCl$_4$, red–brown and tetragonal, and the tribromide, NpBr$_3$, α-form green and hexagonal, β-form green and orthorhombic.

In research at the Institute of Radiochemistry, Karlsruhe, West Germany, during the early 1970s, investigators prepared alloys of neptunium with iridium, palladium, platinum, and rhodium. These alloys were prepared by hydrogen reduction of the neptunium oxide in the presence of finely divided noble metals. The reaction is called a *coupled reaction* because the reduction of the metal oxide can be done only in the presence of noble metals. The hydrogen must be extremely pure, with an oxygen content of less than 10^{-25} torr.

References

Keller, C., and B. Erdmann: "Preparation and Properties of Transuranium Element–Noble Metal Alloy Phases," *Proceedings of the 1972 Moscow Symposium on the Chemistry of Transuranium Elements* (1976).

Krot, N. N., and A. D. Gel'man: "Preparation of Neptunium and Plutonium in the Heptavalent State," *Dokl. Chem.*, **177**, 1–3, 987–989 (1967).

Magnusson, L. B., and T. J. LaChapelle: "The First Isolation of Element 93 in Pure Compounds and a Determination of the Half-life of $_{93}$Np^{237}," *American Chemical Society Journal*, **70**, 3534–3538 (1948).

McMillan, E., and P. H. Abelson: "Radioactive Element 93," *Phys. Rev.*, **57**, 1185–1186 (1940).

Seaborg, G. T.: The Chemical and Radioactive Properties of the Heavy Elements," *Chemical & Engineering News*, **23**, 2190–2193 (1945).

Seaborg, G. T., and A. C. Wahl: "The Chemical Properties of Elements 94 and 93," *American Chemical Society Journal*, **70**, 1128–1134 (1948).

Seaborg, G. T., Editor: "Transuranium Elements," Dowden, Hutchinson & Ross, Stroudsburg, Pennsylvania (1978).

Wahl, A. C., and G. T. Seaborg: "Nuclear Properties of 93^{237}," *Phys. Rev.*, **73**, 9, 940–941 (1948).

NERNST-THOMPSON RULE. A solvent of high dielectric constant favors dissociation by reducing the electrostatic attraction between positive and negative ions, and conversely a solvent of low dielectric constant has small dissociating influence on an electrolyte.

NERVE BLOCK.   Anesthetics.

NEUTRINO. A neutral particle of very small (presumed zero) rest mass and of spin quantum number $\frac{1}{2}$. This particle was initially postulated to account for the continuous energy distribution of beta particles and to conserve angular momentum in the beta-decay process. Experimental evidence indicates that, for the linear momentum to be conserved in the beta process, there must be a contribution from a departing neutrino. Presumably, a neutrino (or antineutrino) is emitted in every beta transition. The energy of a neutrino emitted in a beta disintegration is assumed equal to the difference between the energy of the particular beta particle and the energy corresponding to the upper limit of the continuous spectrum for that beta transition. The neutrino has also been postulated as one of the particles in pion (π) decay and as two of the particles in muon (μ) decay. These processes, however, lead to two types of neutrinos: an electron-associated neutrino v_e and a muon-associated neutrino v_μ. For example, $\pi^+ \rightarrow \mu^+ \rightarrow \bar{v}_\mu + e^+ + v_e$ or $\pi^+ \rightarrow e^+ + v_e$, whereas neutron decay obeys only the process $n \rightarrow p^+ + e^- + \bar{v}_e$, where the bar over v indicates an anti-particle. The difference between neutrinos was established at Brookhaven in 1962 when it was shown that a beam of neutrinos from the process $\pi^+ \rightarrow \mu^+ + v_\mu$ gave rise to the process $v_\mu + n \rightarrow p + \mu^-$ but not to $v_\mu + n \rightarrow p + e^-$. There is also a neutrino associated with the tau particle. Because of its properties, the neutrino has negligible interactions with matter and has proved difficult to detect. It was first positively identified experimentally in 1956 by Reines and Cowan, Jr. See also **Particles (Subatomic).**

The term antineutrino usually denotes an antiparticle whose emission is postulated to accompany radioactive decay by negatron emission, such as, for example, in neutron decay into a proton p^+, negatron e^- and antineutrino $\bar{v}_e$, expressed by the equation $n \rightarrow p^+ + e^- + \bar{v}_e$. Capture of a neutrino by the neutron, $v_e + n \rightarrow p^+ + e^-$ would be an equally good description of the process. Positron emission is accompanied by a neutrino, as in the decay $^{64}Cu \rightarrow {}^{64}Ni + e^+ + v_e$. Orbital electron capture also involves a neutrino, as for example, $e^- + {}^{64}Cu \rightarrow {}^{64}Ni + v_e$. Since there is no possibility of charge differentiation between the antineutrino and the neutrino, differentiation between these two particles can be made only on the basis of such properties as the sign of the ratio of magnetic moment to angular momentum.

References

Bromley, D. A.: "Physics," *Science*, **209**, 110–121 (1980).

Learned, J. G., and D. Eichler: "A Deep-Sea Neutrino Telescope," *Sci. Amer.*, **244**, 2, 138–154 (1981).

Reines, F.: "The Early Days of Experimental Neutrino Physics," *Science*, **203**, 11–16 (1979).

Waldrop, M. M.: "Massive Neutrinos: Masters of the Universe?" *Science*, **211**, 470–471 (1981).

Reines, F.: "Neutrino," in "The Encyclopedia of Physics," (R. M. Besancon, Editor), Van Nostrand Reinhold, New York (1974).

NEUTRON. The discovery of the neutron by Chadwick in 1932 represented a great step forward in the investigation of nuclei of atoms. Chadwick found that a radiation emitted when α-rays from polonium reacted with beryllium could project protons from a thin sheet of paraffin wax. Although the radiation itself produced no observable ionization when passing through a gas, the protons released from the paraffin were detected in an ionization chamber. Inability to produce ionization was interpreted as a lack of electric charge. From measurements of the ionization from the protons, Chadwick deduced that the so-called beryllium radiation must consist of neutral particles with a mass very nearly equal to that of the proton. He announced the discovery of the neutron, a previously unknown particle. It has been confirmed that the neutron has no charge and a mass of 1.088665 atomic mass units. Thus, it is heavier than the proton by 0.00139 mass unit. The introduction of the neutron into nuclear structure produced a sharp change in previously held concepts. Lacking knowledge of the neutron, masses of atomic nuclei had been attributed solely to protons. The number of protons required on this basis for most nuclei greatly exceeded the known charge number. In an attempt to solve this dilemma, a number of electrons were assigned to each nucleus to adjust the charge number to the proper value. This compromise created an even greater problem, that of accommodating so many electrons in the small space occupied by a nucleus. Bringing the neutron into the picture meant that a nucleus contains only protons to equal the charge number, with the rest of the mass contributed by neutrons. No additional electrons were required.

Decay. The neutron in the free state undergoes radioactive decay. Elaborate experiments by Robson were required to identify the products of the decay and to measure the half-life of the neutron. He showed that the neutron emits a β-particle and becomes a proton. The half-life was found to be 12.8 minutes. In stable nuclei, neutrons are stable. In radioactive nuclei, decaying by β-emission, the neutrons decay with a half-life characteristic of the nuclei of which they are a part. See also **Radioactivity.**

Detection. Because it is a neutral particle the neutron is detected by means of a secondary charged particle which it releases in passing through matter or by means of the radioactivity which the neutron can induce in stable elements. Protons may be projected by collisions with neutrons in hydrogenous material and the ionization from the protons can be measured in an ionization chamber, as in the original experiment with neutrons. Secondary charged particles may be the direct result of nuclear disintegration produced by neutrons. Commonly, the radioactivity induced in a stable element by neutron capture serves to detect neutrons, and this technique is known as the *activated foil method.* Also, fission may be utilized for detection of neutrons by placing fissionable material inside an ionization chamber and observing the ionization generated by the fission fragments.

Energies. The kinetic energy of neutrons has an important bearing on the behavior of neutrons when interacting with nuclei. These kinetic energies may range from near zero to as much as 50 MeV. It is therefore natural to classify neutrons in terms of energy according to their properties in each range of energy. For example, energies from zero to about 1000 eV are usually called *slow neutrons*. Because they are more readily captured by nuclei than faster neutrons, slow neutrons are responsible for a large number of nuclear transformations. When slow neutrons have velocities in equilibrium with the velocities of thermal agitation of the molecules of the medium in which they are situated, they are called *thermal neutrons*. The distribution of these velocities approaches the Maxwell distribution

$$dn(\nu) = A\nu^2\, e^{-(M\nu^2/2kT)}\, d\nu$$

where ν is the neutron velocity, M its mass, k is Boltzmann's constant, and T is the absolute temperature. In the slow neutron range of energies, various atomic nuclei show strong absorption (capture) of neutrons at fairly well-defined energies. Neutrons having energies corresponding to those of the absorption bands are called *resonance neutrons*. Frequently, neutrons with energies greater than 1000 eV and less than 0.5 MeV are termed *intermediate neutrons*. In more general terms, all neutrons with energies greater than 0.5 MeV are called *fast neutrons*. The practical upper limit of neutron energy is set by the device thus far developed for accelerating charged particles to extremely high energies.

Magnetic Moment and Spin. Alvarez and Bloch succeeded in measuring the moment of the magnetic dipole associated with the known spin of $\frac{1}{2}$ possessed by the neutron. More refined measurements by Cohen, Corngold, and Ramsey of the magnetic moment μ_n yielded a value of

$$\mu_n = 1.913148 \text{ nuclear magnetons.}$$

Interactions with Nuclei. Neutrons may be scattered or captured by heavy nuclei. Scattering may be elastic, resulting only in the change of direction of the neutrons, or inelastic, in which the neutron loses part of its energy to the scattering nucleus. Collisions with light nuclei, in absence of capture, communicate a considerable fraction of the neutron energy to the target nucleus. A neutron colliding head-on with a proton will give practically all its kinetic energy to the proton. As the mass of the target nucleus increases, the transfer of energy decreases, in accordance with the laws of conservation of energy and momentum. The loss of energy by mechanical impact is utilized in slowing down fast neutrons, a process known as *moderation*. Slow neutrons are most useful, for example, in the production of radioelements from stable elements by neutron capture. A good moderator should have low mass and a small capture cross section. The rate r of capture of neutrons from a neutron flux F (neutrons cm^{-2}sec^{-1}) incident on a layer of matter having N nucleu per square centimeter is given by

$$r = F\sigma N$$

where σ is the complete probability of capture. Replacing r by dN/dt and writing the flux as $n\nu$, where n is the number and ν is the velocity of the neutrons, we have

$$dn/dt = n\nu\sigma N$$

which integrated gives

$$N = N_0\, e^{-n\nu\sigma t}$$

where N is the number of unchanged nuclei in the target area at time t and N_0 is the number at time $t = 0$. The cross section σ is so named because it has the dimensions of an area. The unit for the cross section is the *barn*, equal to 10^{-24} cm^2. When, as is often the case, σ is proportional to $1/\nu$, the advantage of slow neutrons in capture interactions becomes apparent. When the value of σ departs sharply from that predicted by the $1/\nu$ law, it usually increases over a narrow range of energies, and we have what is called a *resonance*. Slow neutron cross sections are customarily quoted for thermal neutrons at 20°C, corresponding to a value of ν of 2200 m/sec. Under these conditions, the thermal neutron capture cross sections

of boron = 759 barns; cobalt, 38 barns; cadmium, 2450 barns; gadolinium, 46,000 barns; gold, 99.8 barns; helium, 0; lead, 0.170 barn; and oxygen, < 0.0002 barn. See also **Nuclear Reactor.**

Additional interactions of neutrons with nuclei include the release of charged particles by neutron-induced nuclear disintegration. Commonly known reactions are n–p, n–d, and n–α. In these cases, the incident neutrons may contribute part of their kinetic energy to the target nucleus to effect the disintegration. Hence, more than mere neutron capture is involved. There is also usually a lower threshold for the neutron energy below which the reaction fails to occur. Another important reaction involving neutrons is fission, which may occur under different conditions for either slow or fast neutrons with appropriate fissionable material.

Sources of Neutrons. Any nuclear reaction in which neutrons are released might serve as a source of neutrons. In the initial experiments on neutrons, an α–n reaction was used. Because of the charge on the α-particle, it must have a high kinetic energy to penetrate a nucleus. Thus, polonium α-particles could release neutrons from beryllium. Such a natural source produces relatively few neutrons. The yield of neutrons from charged particle reactions can be increased manyfold by the use of particle accelerators. Here large numbers of charged particles of high energy can be used in the bombardment of the target to release numerous neutrons. Frequently dueterons or protons are used for the bombardment. A far more prolific source is the nuclear reactor. Fission of uranium is usually the source of the neutrons in this case. A nuclear reactor as usually constructed generates neutrons of different energies in various parts of its structure. Neutrons of suitable energy for a given experiment may be brought outside the reactor through channels into appropriate sections of the reactor. See also **Nuclear Reactor.**

Traditionally, the neutron is regarded as a particle which is a component of nuclei and which exists only briefly in the free state. For many purposes, this view is sufficient. However, it became obvious some years ago from various experiments, for example, in very high energy accelerators, that the neutron must have a complex structure. This view was reinforced by the nature of the decay of the neutron. A β-particle is ejected from the neutron on decay, but it is quite certain that the electron did not exist within the neutron prior to the decay. Rearrangements of an internal structure of the neutron must provide the energy for the formation and ejection of the β-particle. An early theory would have the neutron consist of a proton and a π^- meson bound together so that they oscillate between a completely bound state and a more loosely bound state. This concept might explain the feeble interaction which has also been observed between electrons and neutrons at very short range.

Fermi Age Model. This is a model for the study of the slowing down of neutrons by elastic collisions. It is assumed that the slowing down takes place by a very large number of very small energy changes. Phenomena due to the finite size of the individual losses are ignored. In this model, the word age is somewhat of a misnomer, since its units are those of area rather than time. The name arises because the variable τ, the Fermi age, appears in the Fermi age equation in the same way that time appears in the standard heat-diffusion equation. The equation for the Fermi age in a unit volume of nuclear reactor is $\tau = D\phi/q$, in which D is the diffusion coefficient for fast neutrons, $\phi = n\nu$ is the fast neutron fluence, and q is the number of neutrons thermalized per second per cubic centimeter. For this purpose fast neutrons are inclusively all neutrons with energies between those acquired at fission and that energy at which they are thermalized.

Ultracold Neutrons. As pointed out by King (Massachusetts Institute of Technology), there is probably as much to be learned between the lowest energy yet reached and zero energy as there is between the highest energy attained and infinite energy. Ultracold neutrons may provide an avenue to very-low-energy research. It should be recalled that a neutron, with the energy of 10^{-7} electron volt, is at the low end of the energy scale. Such a neutron has the energy that would be imparted to an electron by a potential difference of one ten-millionth of a volt (0.1 microvolt). As pointed out by Golub et al. (1979), this is the amount of energy of a particle in a gas whose temperature is one millidegree K. Unlike high-energy particles, ultracold neutrons move at a rate measured in a few meters per second. Golub et al. have proposed that inasmuch as ultracold neutrons cannot penetrate a solid surface, they can be confined in a metal bottle and by storing over long periods, it may be possible to measure the fundamental properties of the neutron.

Outside the nucleus, the neutron is an unstable particle. Free neutrons are rare in nature. By the process of beta decay, the neutron breaks down into a proton, an electron, and a neutrino (a massless particle). For probing atomic and molecular structures, thermal neutrons have traditionally been used. As explained by Golub et al., ultracold neutrons can be employed in a similar way, but their low energy and long wavelength adapt them to the examination of materials on a somewhat larger scale. At the Technical University (Munich), Steyeri and associates have developed an ultracold-neutron spectrometer. In conventional neutron spectrometers, particles are analyzed by a magnet that bends their trajectories. In an ultracold-neutron spectrometer, the earth's gravitational field is used. In the device, the neutrons enter the spectrometer in a horizontal movement and are accelerated as they fall a fixed distance to a specimen. Those neutrons that rebound from a target are collected by an exit slit of the instrument. An exchange of energy with the specimen is reflected by the maximum height to which the neutrons rebound.

During the past decade, since their detection, much research has been directed toward methods of extracting, storing, and manipulating ultracold neutrons. It now appears that the next period will be one of investigating the neutron per se and possibly of using this new knowledge for the study of other systems of particles.

See also **Particles (Subatomic).**

References

Alvarez, L. W., and F. Bloch: *Phys. Rev.*, **57**, 111 (1940).
Bromley, D. A.: "Physics," *Science*, **209**, 110–121 (1980).
Chadwick, J.: *Proc. Roy. Soc. London Ser. A*, **136**, 692 (1932).
Cohen, V. W., Corngold, N. R., and N. F. Ramsey: *Phys. Rev.*, **104**, 283 (1956).
Condon, E. U., and H. Odabşi: "Atomic Structure, Cambridge Univ. Press, New York (1980).
Cowen, R. C.: "Particle Physics," *Technology Review (MIT)*, **82**, 3, 10–11 (1980).
Pilkuhn, H. M.: Relativistic Particle Physics," Springer-Verlag, New York (1979).
Reines, F.: "The Early Days of Experimental Neutrino Physics," *Science*, **203**, 11–16 (1979).
Robson, J. M.: *Phys. Rev.*, **83**, 349 (1951).
Weisskopf, V. F.: "Contemporary Frontiers in Physics," *Science*, **203**, 240–244 (1979).

NEUTRON ACTIVATION ANALYSIS. This is a method of elemental analysis based upon the quantitative detection of radioactive species produced in samples via nuclear reactions resulting from neutron bombardment of samples. The neutron-induced reactions are of two main types: (1) those induced by very slow (thermal) neutrons, having energies of about 0.025 eV; and (2) those induced by fast neutrons having energies in the range of MeV. The method is used in two different forms. The purely instrumental form is fast and nondestructive and is based upon the quantitative detection of induced gamma-ray emitters by means of multichannel gamma-ray spectrometry. The amount of the element present is usually computed from the photopeak (total absorption peak) height or area of its gamma ray, or one of its principal gamma rays, compared with that of the standard. Where interferences from other induced activities are serious and cannot be removed by decay, spectrum subtraction, or computer solution, one must turn to the radiochemical separation method. Here the activated sample is put into solution and equilibrated chemically with measured amounts (typically 10 milligrams) of added carrier of each of the elements of interest, before chemical separations are carried out. The element to be detected needs then to be recovered in chemically and radiochemically pure form, but it need not be quantitatively recovered, since the carrier recovery is measured and the counting data are then normalized to 100% recovery. This form is slower, but it applies to pure beta emitters, as well as to gamma emitters, and it does eliminate interfering activities.

NIACIN. Sometimes referred to as nicotinic acid, nicotinamide and earlier called the P-P factor, antipellagra factor, anti-blacktongue factor, and vitamin B_4, niacin is available in several forms (niacin, niacinamide, niacinamide ascorbate, etc.) for use as a nutrient and dietary supplement. Niacin is frequently identified with the B complex vitamin grouping. Early in the research on niacin, nutritional deficiency of this substance was identified as the cause of pellagra in humans, blacktongue in dogs, and certain forms of dermatosis in humans. Niacin deficiency is also associated with perosis in chickens as well as poor feathering of the birds.

Varying in degree in relationship to the length and severity of diet deficiency of niacin, pellagra is clinically manifested by skin, nervous-system, and mental conditions. The disease occurs most frequently among the economically deprived, particularly in areas where the diet may be high in maize (corn) intake. The disease was first described by Gaspar Casal in 1735 and was common in many areas, including Europe, Egypt, Central America, and the southern portion of the United States for many years. The largest outbreak occurred in the United States during the period 1905–1915 and resulted in a high mortality. The medical awareness and understanding of vitamins and dietary deficiencies, coupled with the availability of dietary supplements in staple foods, have resulted in a great lessening in the occurrence of pellagra. Niacin is a specific for the treatment of acute pellagra, where found. Those afflicted are accustomed to a diet low in protein and made up largely of carbohydrates. Predisposing causes are food idiosyncrasies, chronic alcoholism, and diseases which interfere with the assimiliation of a proper diet.

Huber first synthesized nicotinic acid in 1867. In 1914, Funk isolated nicotinic acid from rice polishings. Goldberger, in 1915, demonstrated that pellagra is a nutritional deficiency. In 1917, Chittenden and Underhill demonstrated that canine blacktongue is similar to pellagra. In 1935, Warburg and Christian showed that niacinamide is essential in hydrogen transport as diphosphopyridine nucleotide (DPN). In the following year, Euler et al. isolated DPN and determined its structure. In 1937, Elvhehjem et al. cured blacktongue by administration of niacinamide derived from liver. In the same year, Fouts et al. cured pellagra with niacinamide. In 1947, Handley and Bond established conversion of tryptophan to niacin by animal tissues.

In the physiological system, niacin and related substances maintain nicotinamide adenine dinucleotide (NAD) and nicotinamide adenine dinucleotide (phosphate) (NADP). Niacin also acts as a hydrogen and electron transfer agent in carbohydrate metabolism; and furnishes coenzymes for dehydrogenase systems. A niacin coenzyme participates in lipid catabolism, oxidative deamination, and photosynthesis.

Nicotinic acid can be converted to nicotinamide in the animal body and, in this form, is found as a component of two oxidation-reduction coenzymes, NAD and NADP, as previously mentioned. Structurally, these are:

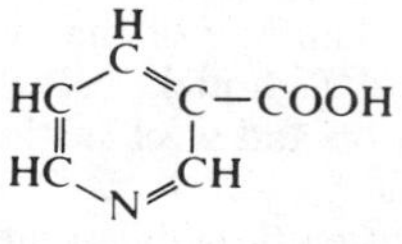

Nicotinic acid 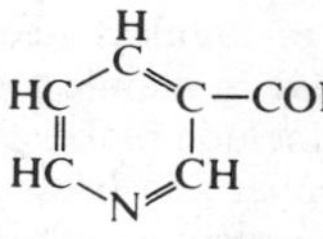   Nicotinamide

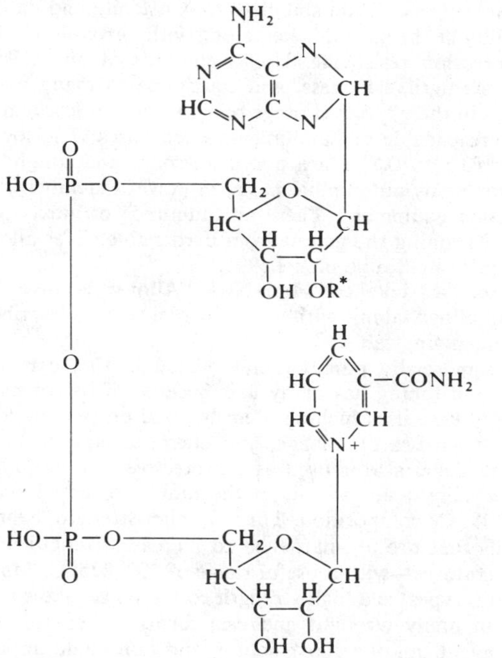

Nicotinamide adenine dinucleotide (NAD) R* = H

Nicotinamide adenine dinucleotide phosphate (NADP) R* = $\overset{\overset{\displaystyle O}{\|}}{\underset{\underset{\displaystyle OH}{|}}{P}}$—OH

The nicotinamide portion of the coenzyme transfers hydrogens by alternating between an oxidized quaternary nitrogen and a reduced tertiary nitrogen as shown by:

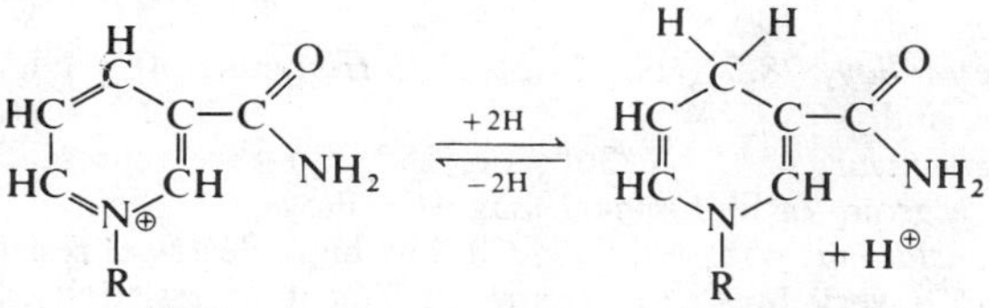

NAD   NADH + H$^\oplus$
(oxidized)   (reduced)

Enzymes that contain NAD or NADP are usually called dehydrogenases. They participate in many biochemical reactions of lipid, carbohydrate, and protein metabolism. An example of an NAD-requiring system is lactic dehydrogenase which catalyzes the conversion of lactic acid to pyruvic acid. Numerous NAD-dependent enzyme systems are known.

$$CH_3-\overset{\overset{\displaystyle OH}{|}}{\underset{\underset{\displaystyle H}{|}}{C}}-COOH + NAD^+ \rightarrow CH_3\overset{\overset{\displaystyle O}{\|}}{C}-COOH + NADH + H^+$$

Lactic acid   Pyruvic acid

Bioavailability of Niacin. In plants, niacin production sites occur in leaves, germinating seeds, and shoots. In humans, niacin is not available from intestinal bacteria, but some conversion is made from tryptophan which occurs in tissues. Foods with high niacin content (10–100 milligrams/100 grams) include chicken (white meat); peanut; halibut; calf heart; beef and pork kidney; beef, calf, chicken, pork, and sheep liver; meat extracts; rabbit (white meat); swordfish; tuna; turkey (white meat); and yeast.

Precursors in the biosynthesis of niacin are: in animals and bacteria, tryptophan; in plants, glycerol and succinic acid. Intermediates in the synthesis include kynurentine, hydroxyanthranilic acid, and quinolinic acid. In animals, the niacin storage sites are liver, heart, and muscle. Niacin supplements are prepared commercially by: (1) hydrolysis of 3-cyanopyridine; or (2) oxidation of nicotine, quinoline, or collidine.

Factors which cause a decrease in niacin availability include: (1) cooking losses; (2) bound form in corn (maize), greens, and seeds is only partially available; (3) presence of oral antibiotics; (4) diseases which may cause decreased absorption; (5) decrease in tryptophan conversion as in a vitamin B_6 deficiency. Factors which increase availability include: (1) alkali treatment of cereals; (2) storage in liver and possibly in muscle and kidney tissue; and (3) increased intestinal synthesis.

Antagonists of niacin include pyridine-3-sulfonic acid (in bacteria); 3-acetylpyridine, 6-aminonicotinamide, and 5-thiazole carboxamide. Synergists include vitamins B_1, B_2, B_6, B_{12}, and D, pantothenic acid, folic acid, and somatotrophin (growth hormone).

In humans, overdosage of niacin causes a limited toxicity (1 to 4 grams/kilogram) with individual variations in sensitivity.

References

Haresign, W., and D. Lewis: "Recent Advances in Animal Nutrition," Butterworth Group, Woburn, Massachusetts (1977).

Krautmann, B. A., and C. R. Zimmerman: "A New Look at the Vitamins in Layer Feeds," *Feedstuffs*, 22–23 (August 21, 1978).

Kutsky, R. J.: "Handbook of Vitamins and Hormones," Van Nostrand Reinhold, New York (1973).

Preston, R. L.: "Typical Composition of Feeds for Cattle and Sheep—1977–1978," Feedstuffs, A-2-A (October 3, 1977); and 3A-10A (August 18, 1978).

Staff: "Food Chemicals Codex," National Academy of Sciences, Washington, D.C. (1972).

Staff: "Nutritional Requirements of Domestic Animals," separate publications on beef cattle, dairy cattle, poultry, swine, and sheep, periodically updated, National Academy of Sciences, Washington, D.C.

Swan, P., and L. M. Henderson: "Nicotinic Acid (Niacin)," in "The Encyclopedia of Biochemistry" (R. J. Williams and E. M. Lansford, Jr., Editors), Van Nostrand Reinhold, New York (1967).

Wosilait, W. D.: "Nicotinamide Adenine Dinucleotides," in "The Encyclopedia of Biochemistry" (R. J. Williams and E. M. Lansford, Jr., Editors), Van Nostrand Reinhold, New York (1967).

NICKEL. Chemical element symbol Ni, at. no. 28, at. wt. 58.71, periodic table group 8, mp 1454–1456°C, bp 2725–2735°C, density, 8.9 g/cm³ (solid, 20°C), 9.04 g/cm³ (single crystal). Elemental nickel has a face-centered cubic crystal structure. Nickel is a silvery-white metal, harder than iron, capable of taking a brilliant polish, malleable and ductile, magnetic below approximately 360°C. When compact, nickel is not oxidized on exposure to air at ordinary temperatures. The metal is soluble in HNO_3 (dilute), but becomes passive in concentrated HNO_3. The metal does not react with alkalis. Finely divided nickel dissolves 17 times its own volume of hydrogen at standard conditions.

There are five naturally occurring stable isotopes ^{58}Ni, ^{60}Ni through ^{62}Ni, and ^{64}Ni. Six radioactive isotopes have been identified ^{56}Ni, ^{57}Ni, ^{59}Ni, ^{63}Ni, ^{65}Ni, and ^{66}Ni. ^{59}Ni has a half-life of 8×10^4 years, and ^{63}Ni has a half-life of 80 years. The half-lives of the remaining radioactive isotopes are relatively short, expressed in hours and days. The element ranks 21st among the elements in terms of abundance in the earth's crust, the estimated average content of igneous rocks being about 0.02%. In terms of cosmic abundance, nickel ranks 28th among the elements. Nickel ranks 40th in terms of concentration in seawater, the estimated content being about 2.5 tons of nickel per cubic mile (540 kilograms per cubic kilometer) of seawater. Awareness of nickel probably dates back to antiquity, but the element was not firmly identified until 1751 when Axel Fredric Cronstedt isolated the metal from the sulfide ore, NiAsS.

First ionization potential 7.33 eV second, 18.13 eV. Oxidation potentials $Ni \rightarrow Ni^{2+} + 2e^-$, 0.230 V; $Ni^{2+} + 2H_2O \rightarrow NiO_2 + 4H^+ + 2e^-$, -1.75 V; $Ni + 2OH^- \rightarrow Ni(OH)_2 + 2e^-$, 0.66 V; $Ni(OH)_2 + 2OH^- \rightarrow NiO_2 + 2H_2O + 2e^-$, -0.49 V.

Other physical properties of nickel are given under **Chemical Elements.**

In the early 1800s, the principal sources of nickel were in Germany and Scandinavia. Very large deposits of lateritic (oxide or silicate) nickel ore were discovered in New Caledonia in 1865. The sulfide ore deposits were discovered in Sudbury, Ontario in 1883 and, since 1905, have been the major source of the element. The most common ore is pentlandite, $(FeNi)_9S_8$, which contains about 34% nickel. Pentlandite usually occurs with pyrrhotite, an iron-sulfide ore, and chalcopyrite, $CuFeS_2$. See also **Chalcopyrite; Pentlandite; Pyrrhotite.** The greatest known reserves of nickel are in Canada and the U.S.S.R., although significant reserves also occur in Australia, Finland, the Republic of South Africa, and Zimbabwe.

Principal producers and/or exporters of nickel include, in diminishing order, Canada, the U.S.S.R., the United Kingdom, Norway, and Indonesia. Main consumers are the United States, Japan, the United Kingdom, Norway, West Germany, Canada, and France.

After beneficiation of the raw ore to form a sulfide concentrate, the latter is roasted to achieve partial oxidation of iron and partial removal of sulfur. The roasted material then is smelted with a flux to eliminate the rock content. At this point, part of the iron goes into the slag. The remaining material is a copper-bearing nickel-iron matte, made up mainly of the sulfides of these metals. The matte is then treated in a Bessemer converter to achieve further removal of iron and sulfur. After controlled cooling, which assists separation, the Bessemer product is finely ground and subjected to magnetic separation and differential flotation. The separated product is an impure nickel sulfide. The sulfide then is sintered to nickel oxide. This product may be marketed for some applications, but the majority of the oxide is cast into anodes for refining into nickel metal by one of two major processes.

In (1) the electrolytic process, a nickel of 99.9% purity is produced, along with slimes which may contain gold, silver, platinum, palladium, rhodium, iridium, ruthenium, and cobalt, which are subject to further refining and recovery. In (2) the Mond process, the nickel oxide is combined with carbon monox-

ide to form nickel carbonyl gas, $Ni(CO)_4$. The impurities, including cobalt, are left as a solid residue. Upon further heating of the gas to about 180°C, the nickel carbonyl is decomposed, the freed nickel condensing on nickel shot and the carbon monoxide recycled. The Mond process also makes a nickel of 99.9% purity.

Uses: The three main commercial forms of primary nickel are: (1) electrolytic sheets, (2) pellets resulting from the decomposition of nickel carbonyl, and (3) ferronickel. Traditionally, pellets are favored in Europe, whereas electrolytic nickel is favored in North America. Additional forms of commercial nickel are powder, ingots, shot, and briquettes. Ferronickel, containing 24–48% nickel with the remainder iron, is used mainly in the production of stainless steel. More than half of the nickel produced is used in stainless steels and high-nickel alloys. Additional uses include nickel plating, iron and steel castings, coinage, and copper and brass products.

The main consumer of nickel is austenitic stainless steel which contains from 3.5 to 22% nickel and 16 to 26% chromium. In these steels, nickel stabilizes the austenite and enhances the ductility of the steel. Nickel, along with chromium, contributes to corrosion resistance. Up to amounts of about 9%, nickel adds strength, hardness, and toughness to many alloy steels. Alloys in the 9% nickel range remain stable at low temperatures and are capable of handling liquefied gases. The lower-nickel steels (0.5 to 0.7%) are ductile, strong, and tough, and find use for many automobile parts, in power machinery, and construction equipment. There are hundreds of nickel-containing alloys, running the gamut from hardenable silver alloy (0.02% Ni) up to malleable nickel (99% Ni).

Wrought Nickel and High-Nickel Alloys: Some of the major nickel alloys, along with wrought nickel, are described in the accompanying table.

Commercially pure wrought nickel in the form of sheets, wire, and tubing has many uses because of its corrosion resistance. These uses include utensils, food-processing equipment, marine hardware, coinage, and chemical equipment. Electroplated nickel also is used as a protective coating on steel. *Nimonic* alloys, not shown in the table, are based on an 80% Ni–20% Cr composition. They are high-strength, heat-resistant metals that are age-hardened to increase strength at elevated temperatures—with a useful range of 700–825°C. *Monel* metal (several types) is a high-strength corrosion resistant alloy available in many wrought and cast forms for use in processing equipment, marine construction, and household appliances. *K Monel* can be heat treated by precipitation hardening to about $2 \times$ the strength of annealed *Monel. Hastelloy*-type alloys are well known for their excellent resistance to HCl, H_2SO_4, and other acids. The *Incoloy*-type alloys (35% Ni approximately) are heat-resistant alloys used mainly as castings for furnace parts. The lower-nickel/higher-chromium alloys generally are classified as stainless steels. See also **Iron Metals, Alloys, and Steels.**

Although not of high-tonnage production, several nickel metals serve important uses, such as:

Permalloy, 78.5% Ni, 21.5% Fe; *Hipernik*, 50% Ni, 50% Fe; and
Perminvar, 45% Ni, 30% Fe, 25% Co—are representative of a group of high-nickel magnetic alloys.
Constantan, 45% Ni, 55% Cu, has high electrical resistivity and a very low temperature coefficient of resistivity. It is extensively used with copper as a thermocouple element.
Nichrome, 80% Ni, 20% Cr (several types with variations of these percentages and additions of other elements, such as silicon in small amounts), is used as resistance wire for

WROUGHT NICKEL AND REPRESENTATIVE NICKEL ALLOYS

	MELTING RANGE °C	POISSON'S RATIO
Wrought nickel 99% Ni, 0.25% Cu, 0.15% C	1,435– 1,445	0.31
Duranickel 301 93.9% Ni, 0.05% Cu, 0.15% C, 0.15% Fe, 0.5% Ti, 4.5% Al	1,400– 1,440	0.31
Monel 400 66.0% Ni, 31.5% Cu, 0.12% C, 1.35% Fe	1,300– 1,350	0.32
Hastelloy B 63.5% Ni, 0.05% C, 5.0% Fe, 2.5% Co, 1.0% Cr, 28.0% Mo, 0.3% V	1,320– 1,460	—
Hastelloy F 45.5% Ni, 0.05% C, 20.5% Fe, 2.5% Co, 22.0% Cr, 6.5% Mo, 1% W, 2% (Nb + Ta)	1,290– 1,295	0.305
Inconel 600 72% Ni, 0.5% Cu, 0.15% C, 8.0% Fe 15.5% Cr	1,370– 1,425	0.29
Incoloy 800 32.5% Ni, 0.75% Cu, 0.10% C, 45.6% Fe, 21.0% Cr	1,355– 1,390	0.30
Illium G 56.0% Ni, 6.5% Cu, 22.5% Cr, 6.5% Mo	1,255– 1,340	0.29

heating elements.

Calorite, 65% Ni, 8% Mn, 12% Cr, 15% Fe, also is used in electric heating elements.

Alumel, 94% nickel, 2.5% Mn, 0.5% Fe plus small amounts of other elements, is used in thermocouples.

Chromel, 35–60% Ni, 16–19% Cr, generally with the balance Fe, also is used as resistance wire and for thermocouples.

Invar, 36% Ni, 64% Fe, has a very low temperature coefficient of expansion and is used for measuring tapes, instruments, and bimetallic thermostats.

Elinvar, 34% Ni, 57% Fe, 4% Cr, 2% W, has a very low temperature coefficient of elasticity which makes it useful for springs in watches and precision instruments.

There are hundreds of special nickel-bearing alloys of proprietary formulations and tradenames. The alloys described here are registered names of the following firms:

Duranickel, Incoloy, Inconel, Invar, Monel—International Nickel Company.
Permalloy—Western Electric Company
Nimonic—Mond Nickel Company
Constantan, Nichrome—Driver-Harris Company
Hipernik—Westinghouse Electric Company
Hastelloy—Haynes Stellite Company
Calorite—General Electric Company
Alumel, Chromel—Hoskins Manufacturing Company

Illium—Burgess-Parr Company
Elinvar, Invar—Hamilton Watch Company

Chemistry and Compounds. With its $3d^8 4s^2$ electron configuration, nickel forms Ni^{2+} ions. Having a nearly complete $3d$ subshell, nickel does not yield a $3d$ electron as readily as iron and cobalt, and trivalent and tetravalent forms are known only in the hydrated oxides, Ni_2O_3 and NiO_2, and a few complexes.

Nickel(II) oxide, NiO, produced by heating the carbonate, is thermally stable. Higher oxides of nickel, including Ni_2O_3 and NiO_2 are known only as hydrates, being prepared by vigorous oxidation of NiO in alkaline solution.

Nickel(II) sulfide, precipitated from Ni^{2+} solutions by ammonium sulfide, may show quite a little departure from stoichiometric composition. Like iron(II) and cobalt(II) FeS and CoS, it has in crystal form an electrical conductivity and other properties similar to a metal or alloy. There is no conclusive evidence that Ni_2S_3 can be prepared, but NiS_2 is known and believed to be, like FeS_2, a compound of Ni^{2+} and the S_2^{2-} ion.

All four dihalides of nickel with the common halogens are known: NiF_2, formed by reaction of hydrofluoric acid or nickel (II) chloride or by thermal decomposition of $[Ni(NH_3)_6][BF_4]_2$, is greenish yellow, while the other three dihalides, formed directly from the elements, are green for the chloride, yellow for the bromide, and black for the iodide. In general, anhydrous Ni^{2+} salts are yellow and the ion $Ni(H_2O)_6^{2+}$ in aqueous solution is green.

Other elements with which nickel forms binary compounds, especially at higher temperature, are boron, carbon, nitrogen, silicon, and phosphorus. Like NiO, these compounds may depart slightly or even considerably from daltonide composition, frequently being interstitial compounds, and with higher elements of transition groups 5 and 6, merging into the interstitial compound-solid solution picture which nickel exhibits with the other transition metals.

Divalent nickel forms two main types of complexes. The first consists of complexes of the spin-free ("ionic" or outer orbital) octahedral type (see **Ligand** for their discussion) in which the ligands are principally H_2O, NH_3, and various amines such as ethylenediamine and its derivatives, e.g., $Ni(H_2O)_6^{2+}$, $Ni(NH_3)_6^{2+}$, $Ni(en)_6^{2+}$. These complexes usually have colors toward the high-frequency side of the spectrum, i.e., violet, blue, and green. The other class consists of tetracovalent square complexes with ligands such as CN^-, the dioximes and their derivatives, and other chelates, which usually have colors on the low frequency side of the spectrum, i.e., red, orange, and yellow. The structure of the nickel-dimethylglyoxime complex is

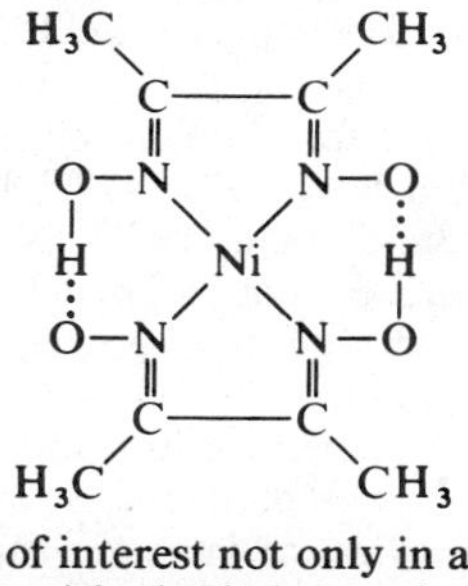

This compound is of interest not only in analysis, but because by limited oxidation with the halogens it yields a unipositive ion containing trivalent nickel and also because the hydrogen bonds formed to the oxygen atoms are among the shortest known. Similarly, the tetracyanide complex of nickel, $Ni(CN)_4^{2-}$, may be reduced by sodium amalgam to give an ion of composition $Ni(CN)_4^{3-}$, or $(NC)_3Ni—Ni(CN)_3^{4-}$ containing Ni(I). This latter ion forms a potassium salt of nickel(I) of the formula

$K_4Ni_2(CN)_6$ which is reduced in liquid NH_3 by metallic potassium to give the compound $K_4Ni(CN)_4$ in which the nickel has an effective valence of zero. Of course, this zero valence also exists in the carbonyls of nickel (and other elements) which, however, are covalent. $Ni(CO)_4$ is prepared by reaction of carbon monoxide with freshly reduced nickel, which occurs at ordinary temperatures and pressures. As with the carbonyls of other metals, the CO groups may be directly or indirectly, partially or completely, replaced by other groups. Derivatives of trivalent phosphorus form many such compounds of general formula $Ni(CO) \times_{4-x}(PR_3)_x$, where R may be one or more of such groups as F, Cl, Br, I, alkyl, aryl, alkoxy, aryloxy, etc.

Like iron, nickel combines with two cyclopentadienyl radicals (see **Iron** for structure) similar to the iron compound $(C_5H_5)_2Ni$. It is not quite so stable as the iron analog.

Toxicity. Nickel contact dermatitis can occur among wearers of nickel-containing jewelry, particularly more common among females than males. This is particularly true of nickel sulfate present in some jewelry. Localization of sites involves the ear lobes, neck, fingers, and wrists. Nickel is a major offender in connection with AECD (allergic eczematous contact dermatitis).

As mentioned earlier, nickel carbonyl is a volatile intermediate in the Mond process for nickel refining. This compound also is used for vapor plating of nickel in the semiconductor industry, and as a catalyst in the chemical and petrochemical industries. The toxicity of the compound has been known for many years. Exposure of laboratory animals to the compound has induced a number of ocular anomolies, including anophthalmia and microphthalmia and has been shown to be a carcinogenic for rat (Sunderman et al., 1979).

Biological Aspects of Nickel. Despite its many pharmacologic and *in vitro* actions, convincing evidence showing that nickel is an essential element for some animal species did not appear until the early 1960s.

Like most trace elements, nickel can activate various enzymes *in vitro*. No enzyme has been shown to require nickel, specifically, to be activated, but urease has been shown to be a nickel metalloenzyme and has been found to contain 6–8 atoms of nickel per mole of enzyme (Fishbein et al., 1976). RNA (ribonucleic acids) preparations from diverse sources consistently contain nickel in concentrations many times higher than those found in native materials from which the RNA is isolated (Wacker-Vallee, 1959; Sunderman, 1965). Nickel may serve to stabilize the ordered structure of RNA (Fuwa et al., 1960). Nickel may have a role in maintaining ribosomal structure (Tal, 1968, 1969). These studies and other information have led to the suggestion that nickel may play a role in nucleic acid and/or protein metabolism.

Nickel also may act to stimulate or inhibit the release of various hormones (Nielsen, 1971, 1972; Dormer et al., 1973; Clay, 1975; Horak-Sunderman, 1975). Nickel has been found to inhibit insulin release from the pancreases (Dormer et al., 1973; Clay, 1975) and stimulates glucagon secretion (Horak-Sunderman, 1975).

Nickel as an essential element in ruminant nutrition has not been proved conclusive as of the early 1980s. However, with nonruminants, some evidence indicates that certain species fed low nickel diets have a greater infant mortality rate and a general degradation of the reproductive process (Nielsen, 1975; Anke et al., 1973).

Zinc and nickel appear to behave similarly at certain sites in the biological system. Both elements are capable of activating certain enzymes; arginase is one enzyme which can be activated by either element (Parisi-Vallee, 1969). Stimulation of enzyme activity is a site at which trace element substitutions or interactions may occur. However, some sacrifice of activity usually results when normally occurring metal is replaced by a trace metal. Nucleic acids as well as the ribosomes are likely sites of interaction between nickel and zinc. Both metals are consistently found in high concentrations firmly bound to RNA. It has been suggested that they function in maintaining the structure of RNA, thus preventing conformational changes. Nickel appears to be as effective as zinc at equal concentrations in this respect. Nickel and zinc are also found in ribosomal ash and studies have indicated that both can contribute to ribosomal conformation. The white blood cell is another possible site at which nickel and zinc may interact. Leukocytes are high in zinc and total leukocyte counts as well as differential white cell counts change drastically during a zinc deficiency. The interrelationship between nickel and zinc has been studied *in vitro* primarily in swine and rats. Their relationship has been studied largely from a substitution standpoint. Nickel appears to substitute for zinc to a certain extent in both species.

Similarly, the relationship between nickel and copper has been under study. One of the major functions of copper is in hemoglobin formation. Hemoglobin and hematocrit values decline rapidly during a copper deficiency. Copper is currently believed to exert its effect on hemoglobin metabolism through ceruloplasmin. Early work also indicated that nickel might be involved in hematopoiesis. Investigators in 1974 found a decreased concentration of copper in the lung and spleen of rats receiving 5 parts per million of nickel in drinking water. High levels of dietary nickel in rats and mice have been reported to decrease the activity of cytochrome oxidase, a copper-containing enzyme.

References

Anke, M., et al.: "Low Nickel Rations for Growth and Reproduction in Pigs," in "Trace Element Metabolism in Animals" (W. G. Hockstra et al., Editors), University Park Press, Baltimore, Maryland (1973).
Clay, J. J.: "Nickel Chloride-induced Metabolic Changes in the Rat and Guinea Pig," *Toxicol. Appl. Pharmacol.*, **31**, 55 (1975).
Cook, E.: "Limits to Exploitation of Nonrenewable Resources," *Science*, **191**, 677–682 (1976).
Classen, R. S.: "Materials for Advanced Energy Technologies," *Science*, **191**, 739–745 (1976).
Crangle, J., and G. M. Goodman: "The Magnetization of Pure Iron and Nickel," *Proceedings of the Royal Society* (Series A), London, Vol. 321 (1971).
Dormer, R. L., et al.: "The Effect of Nickel on Secretory Systems," *Biochem. Jrnl.*, **140**, 135 (1973).
Fishbein, W. N., et al.: "The First Natural Nickel Metalloenzyme: Urease," *Fed. Proc.*, **35**, 1680 (1976).
Horak, E., and F. W. Sunderman, Jr.: "Effects of Ni (II) upon Plasma Glucagon and Glucose in Rats," *Toxicol. Appl. Pharmacol.*, **33**, 388 (1975).
Nielsen, F. H.: "Studies on the Essentiality of Nickel," in "Newer Trace Elements in Nutrition" (W. Mertz and W. E. Cornatzer, Editors), Marcel Dekker, New York (1971).
Parisi, A. F., and B. L. Vallee: "Zinc Metalloenzymes: Characteristics and Significance in Biology and Medicine," *Amer. Jrnl. Clin. Nutr.*, **22**, 1222 (1969).
Pasquin, D. L.: "Nickel" in "Metals Handbook," 9th Edition, Vol. 2, American Society for Metals, Metals Park, Ohio (1979).
Spears, J. W., and E. E. Hatfield: "Role of Nickel in Animal Nutrition," *Feedstuffs*, 24–28 (June 13, 1977).
Sunderman, F. W., Jr.: "Measurements of Nickel in Biological Materials by Atomic Absorption Spectrometry," *Amer. Jrnl. Clin. Path.*, **44**, 182 (1965).
Sunderman, F. W., Jr., et al.: "Eye Malformations in Rats: Induction by Prenatal Exposure to Nickel Carbonyl," *Science*, **203**, 550–552 (1979).

Tal, M.: "On the Role of Zn²⁺ and Ni²⁺ in Ribosome Structure," *Biochem. Biophys. Acta*, **169**, 564 (1968).

Wacker, W. E. C., and B. L. Vallee: "Nucleic Acids and Metals. I. Chromium, Manganese, Nickel, Iron and Other Metals in Ribonucleic Acid from Diverse Biological Sources," *Jrnl. Biol. Chem.*, **234**, 3257 (1959).

NICKELINE. A nickel arsenide mineral, NiAs, crystallizes in the hexagonal system but is usually found massive. Color, light copper-red; hardness, 5.0–5.5; specific gravity, 7.784; luster, metallic; opaque. Found in several European localities and in the Province of Ontario, Canada; in the United States at Franklin, New Jersey, and Silver Cliff, Colorado. It is an ore of nickel.

NICOTINE. Alkaloids.

NINHYDRIN REACTION. Amino Acids.

NIOBIUM. Chemical element symbol Nb, at. no. 41, at. wt. 92.906, periodic table group 5b, mp 2468 ± 10°C, bp 4740–4744°C, density 8.6 g/cm³ (20°C). Elemental niobium has a body-centered cubic crystal structure. The metal has a slightly bluish tinge, is ductile and malleable, and when polished resembles platinum. The metal burns upon being heated in air. There is one natural isotope, ^{93}Nb. Seven radioactive isotopes have been identified, ^{90}Nb through ^{92}Nb and ^{94}Nb through ^{97}Nb, with a wide range of half-lives. The isotope ^{94}Nb has the longest half-life (2×10^4 years). The element was first identified by C. Hatchett in 1801 and was originally called *columbium*, which name persisted for many years and still appears in the literature, particularly in connection with alloys bearing the element, such as columbium steels.

First ionization potential 6.77 eV; second 13.895 eV; third 24.2 eV. Oxidation potential Nb → Nb³⁺ + 3e⁻, ca. 1.1 V; 2Nb + 5H₂O → Nb₂O₅ + 10H⁺ + 10e⁻, 0.62 V.

Other important physical properties of niobium are given in the accompanying table and under **Chemical Elements.**

Niobium occurs, usually with tantalum, in columbite Fe(NbO₃)₂, (80% Nb₂O₅), pyrochlore (50% Nb₂O₅), samarskite (50% Nb₂O₅), chiefly found in western Australia, and South Dakota. Recovered along with tantalum by fusion with potassium bisulfate, and obtained in the residue after subsequent extraction with H₂O. Niobium and tantalum are separated by fractional crystallization of the potassium fluorides, niobium concentrating in the mother liquid and tantalum in the crystals.

The principal uses for the element are in alloys. Niobium also has gained prominence in research as a superconducting material. At the temperatures of liquid helium, niobium becomes a superconductor and, in the form of a fine wire, has been incorporated in a superconducting cell. The element has both size and cost advantages over electronic materials. The alloy Nb₃Sn becomes superconducting at a somewhat higher temperature. Niobium-titanium and niobium-zirconium alloys also have potential as superconductors.

Alloys. Niobium is used in steel, notably stainless steels, to stabilize the carbon present (as carbide) and for preparing niobium carbide, used for dies and cutting tools. Ferroniobium is a strong carbide-forming material and, when added to 18–8 stainless steel, stabilizes areas that may be heat-affected during welding and thus cause subsequent intergranular corrosion. Niobium steels are used for rotors in gas turbines where temperatures up to 700°C must be withstood. Niobium-base alloys find application in fast reactors. Superalloys for very demanding use, as in military applications, contain niobium with cobalt

and zirconium. When alloyed with titanium, molybdenum, and tungsten, the elevated-temperature hardness of niobium is enhanced, whereas when alloyed with vanadium and zirconium, the strength of niobium up to temperatures of 500°C is increased. Metallurgically, niobium is attractive because of its density, good workability, retention of tensile strength at high temperatures, and its high melting point. In the temperature range 920–1,200°C, niobium has been found superior to most other metals on a strength-to-weight basis for aerospace applications. In multi-component alloys, zirconium and hafnium when added with niobium add effectively to strength, even more so than molybdenum or tungsten, but there is some sacrifice in ductility.

In metallurgy, niobium is classified as a refractory metal, along with tungsten, tantalum, and molybdenum. A comparison of the four metals is given in the accompanying table.

Chemistry and Compounds. Elemental niobium is insoluble in HCl or HNO₃, but soluble in hydrofluoric acid or a mixture of hydrofluoric and HNO₃.

As might be expected from its $4d^45s^1$ electron configuration, niobium forms pentavalent compounds. However, the stability of its compounds of lower valence is greater than that of the corresponding tantalum compounds, in keeping with the group 5 position of niobium and tantalum. Nevertheless the similarity of the properties of the compounds of the two metals is so great that special methods are required for their separation, such as solvent extraction of the pentachlorides or chromatographic removal of adsorbed TaF₅ with an ethylmethyl ketone-water system. In addition, divalent and tetravalent compounds are known, and an interstitial, nonstoichiometric hydride.

Niobium forms a divalent oxide, NbO, insoluble in water, but readily soluble in acids or NH₄OH. It also gives by direct combination of the metal on heating with oxygen, the pentoxide, Nb₂O₅, which can be reduced by hydrogen at high temperature to NbO₂, and on heating with magnesium to Nb₂O₃.

Niobium(III) halides are known, notably the chloride, NiCl₃, which is of particular interest because its solution has been shown to contain Nb³⁺ ions (in equilibrium with NbCl₆³⁻ complex ions).

Tetravalent niobium is believed to occur in the form of NbOCl₄²⁻ ions in a solution obtained, with color change, by reduction of HCl solution of NbCl₅, and by inference in similarly reduced solutions of the other pentahalides. Tetravalent niobium also is found in the dioxide (see above) and the carbide, NbC.

Four pentahalides of niobium, NbF₅, NbCl₅, NbBr₅, and NbI₅ have been prepared by heating the pentoxide with carbon in a current of the halogen. They are hydrolyzed in H₂O, and even in concentrated aqueous solution of the respective halogen acids; the Nb⁵⁺ ion is apparently not present, but rather complex ions such as [NbOCl₄]⁻ or [NbOCl₅]²⁻. The products of partial hydrolysis of the pentahalides are oxyhalides, such as NbOF₃, NbOCl₃, and NbOBr₃. They are designated in the older literature as columbyl or columboxy compounds. The more stable oxyhalogen compounds of niobium are complexes such as NbOF₃·3NaF, NbOF₃·ZnO·6H₂O, and NbOF₃·2KF·H₂O.

Further complexes of Nb(V) are formed with oxygen-function compounds, such as *o*-dihydroxybenzene and acetylacetone.

The so-called niobic acid is the hydrated pentoxide, Nb₂O₅·xH₂O, insoluble in H₂O.

The metaniobates of the alkali metals, MNbO₃, the orthoniobates M₃NbO₄ and the pyroniobates, M₄Nb₂O₇, where M is an alkali metal, can be prepared by various alkali carbonate or hydroxide fusion processes.

Niobium forms a nitride, NbN, and a carbide, NbC.

Niobium forms a diamino compound, (NH₂)₂NbCl₃, and an

REPRESENTATIVE PROPERTIES OF REFRACTORY METALS

PROPERTY	TUNGSTEN	TANTALUM	MOLYBDENUM	NIOBIUM
Density, g/cm³	19.3	16.6	10.2	8.7
Melting point, °C	3,410	2,996	2,617	2,468
Boiling point, °C	5,660	5,430	4,620	4,740
Linear coefficient of expansion per °C	4.3×10^{-6}	6.5×10^{-6}	4.9×10^{-6}	7.2×10^{-6}
Thermal conductivity, 20°C (cal/cm²/cm/°C/s)	0.40	0.13	0.35	0.13
Specific heat, 20°C (cal/g/°C)	0.032	0.036	0.061	0.065
Working temperature, °C	1,700	ambient	1,600	ambient
Electrical conductivity, % IACS	31	13	30	12
Nuclear cross section (thermal neutrons, Barns/atom)	19.2	21.3	2.4	1.1
Tensile strength, 1000 psi				
20°C	100–500	100–150	120–200	75–150
500°C	175–200	35–45	35–65	35
1,000°C	50–75	15–20	20–30	13–17
Young's Modulus of Elasticity, psi				
20°C	59×10^6	27×10^6	46×10^6	14×10^6
500°C	55×10^6	25×10^6	41×10^6	7×10^6
1,000°C	50×10^6	22×10^6	39×10^6	—
Poisson's Ratio	0.284	0.35	0.32	0.38
Corrosion resistance, 100°C				
Dilute HNO₃		N	R	N
Dilute H₂SO₄		N	S	VS
Concentrated H₂SO₄		N	S	R
Dilute HCl	See	N	S	—
Concentrated HCl	**Tungsten**	N	SL	SL
Concentrated Hydrofluoric acid		R	SL	R
Phosphoric acid, 85%		N	SL	VS
Concentrated NaOH		R	N	R

N = no appreciable corrosion.
VS = <0.0005 inch (0.013 millimeter) per year.
SL = 0.0005–0.005 inch (0.013–0.13 millimeter) per year.
 S = 0.005–0.01 inch (0.13 0.25 millimeter) per year.
 R = >0.01 inch (0.25 millimeter) per year.

ammine complex, $NbCl_5 \cdot 9NH_3$. It forms two cyclopentadienyl compounds, $(C_5H_5)_2NbBr_3$ and $(C_5H_5)Nb(OH)Br_2$. Its other organometallic compounds are essentially oxygen-functional ones, such as $Nb(OCH_3)_5$, $Nb(OC_2H_5)_5$, $Nb(O)(OC_5H_{11})_3$, and $Nb(OC_5H_{11})_5$. These compounds are named as substituted niobanes (thus, the last is pentabutoxy niobane) or as alkyl niobate esters.

See the list of references at the end of the entry on **Chemical Elements.**

NITER. This potassium nitrate mineral KNO_3, of orthorhombic crystallization, usually occurs as thin crusts, or as silky acicular crystals. It has a hardness of 2, and specific gravity of 2.09–2.14, is of white color, translucent with vitreous luster. It occurs as a surface efflorescence, or in soils rich in organic material in arid regions. World occurrences include Spain, Italy, Egypt, Arabia, India, the U.S.S.R. and the western United States. Also in the Republic of South Africa, and Bolivia, South America. Large quantities were recovered from limestone caves in Tennessee, Kentucky, Alabama, and Ohio during the Civil War for use in the manufacture of gunpowder. It is used as a source of nitrogen compounds, for explosives, and in fertilizers.

NITRATION. The process of adding nitrogen to a carbon compound, generally to create a nitro-derivative (adding a —NO_2 group) is termed nitration. An example is the formulation of nitrobenzene from benzene: $C_6H_6 + HNO_3 \rightarrow C_6H_5NO + H_2O$. In most instances, the —NO_2 group replaces a hydrogen atom. More than one hydrogen atom may be replaced, but each succeeding hydrogen represents a more difficult substitution. The nitrogen-bearing reactant may be: (1) strong HNO_3, (2) mixed HNO_3 and H_2SO_4, (3) a nitrate plus H_2SO_4, (4) nitrogen pentoxide N_2O_5, or (5) a nitrate plus acetic acid. Both straight-chain and ring-type carbon compounds can be nitrated. The alkanes yield nitroparaffins.

Various rules of addition govern the position of the entering nitro group, depending upon the conditions. For example, in the nonsubstituted benzene series, the nitro group can enter in the ortho, meta or para position, but the presence of some other group usually fixes the position of the entering nitro group. For example, it enters meta to a nitro, sulfonic,, or carbonyl group, and ortho and para to a chloro, bromo, or hydroxy group. (These statements apply to the principal product formed, since in most substituted benzene reactions, a limited quantity of all ring positions are entered.) Various other rules govern other conditions in other aromatic series.

One of the great uses of nitration is to break into a pure hydrocarbon, which is usually more difficult to do by other means. The nitro group may then be changed and another group take its place. Typical examples are nitration of ethane to form nitroethane, and of benzene to form nitrobenzene, which is easily changed to aniline.

An important economic consideration in any nitration process is the recovery of the spent acid. Since the nitration reaction forms H_2O, the reagents gradually become diluted to a point where they will not react any more. The water may be taken up during the reaction by removing it with oleum or acetic anhydride, a practice which still leaves large amounts of the reagents at the end of the process.

The HNO_3 is usually concentrated by distilling it from H_2SO_4 solution which retains the H_2O. After the HNO_3 has been driven off, the temperature is raised and the H_2O is driven off the H_2SO_4, thereby concentrating the latter.

As an example of nitration, let us consider the preparation of nitrobenzene. Mixed acid consisting of strong H_2SO_4 plus HNO_3 is slowly added to benzene in a closed iron vessel provided with stirrer and reflux condenser. The acid must be added to the benzene. If it were done the other way, the benzene which was added first would be quickly nitrated all the way to a trinitrobenzene. The temperature is maintained from 45–55°C. After the nitration is finished, the nitro compound is separated from the acid by decantation, since the nitrobenzene is lighter and does not mix with the acid. The nitrobenzene is washed with water and with dilute caustic or sodium carbonate solution and then again with water to give a neutral product. To obtain dinitrobenzene the reaction would be run with stronger acid and at a temperature of about 100°C.

Since the reaction used concentrated H_2SO_4, ordinary iron vessels can be used, but the neutralization process must be carried out in lead-lined tanks. Good agitation and adequate cooling facilities are necessary to avoid any local overheating and the formation of higher nitrated compounds.

NITRIC ACID.

This important industrial chemical has been known for at least 1000 years. The acid was known to alchemists as *aqua fortis* (strong water) or *aqua valens* (powerful water). Nitric acid was of particular interest to the early experimenters because of its ability to dissolve a number of metals, including copper and silver. Early chemists were also fascinated by the fact that addition of sal ammoniac (ammonium chloride) gave *aqua regia* (royal water) which dissolves gold as well as silver.

Nitric acid is a colorless liquid, sp gr 1.503 (25°C), freezing point −41.6°C, and boiling point 86°C. The 100% acid is not entirely stable and must be prepared from its azeotrope (constant-boiling mixture) by distillation with concentrated sulfuric acid. Reagent grade HNO_3 is a water solution containing about 68% HNO_3 (weight). This strength corresponds to the constant-boiling mixture of the acid with water, which is 68.4% HNO_3 (weight) at atmospheric pressure and boils at 121.9°C. Nitric acid is completely miscible with water. It forms two solid hydrates, $HNO_3 \cdot H_2O$ and $HNO_3 \cdot 2H_2O$, with corresponding melting points of approximately −38 and −18.5°C. Nitric acid is a strong acid and a powerful oxidizer. In dilute solutions, it is almost completely ionized to H^+ and NO_3^- ions and behaves like a strong acid.

With organic compounds, HNO_3 may act as a nitrating agent, as an oxidizing agent, or simply as an acid. The classic example of nitration is its reaction with benzene or toluene in the presence of concentrated H_2SO_4 to form nitrobenzene or nitrotoluene (TNT). An example of oxidation properties is in the oxidation of cyclohexanol by HNO_3 to produce adipic acid, an intermediate of nylon. Behaving like an acid, it forms nitroglycerin by esterification of glycerol in the presence of concentrated sulfuric acid.

An interesting property of HNO_3 is its ability to passivate some metals, such as iron and aluminum. This property is of significant industrial importance, since modern processes for producing the acid depend on it. Modern suitably formulated stainless steel alloys are usefully resistant to nitric acid through a wide range of conditions. The acid's passivity or the metal's resistance to attack is attributed to the formation of a protective oxide layer on the surface of the metal.

Nitric acid is a high tonnage industrial chemical. Much of the production is used in the manufacture of agricultural fertilizers, largely in the form of ammonium nitrate, NH_4NO_3. See **Fertilizers.** About 15% of the nitric acid produced is used in explosives (nitrates and nitro compounds), and about 10% is consumed by the chemical industry. As the red fuming acid or as nitrogen tetroxide, HNO_3 is used extensively as the oxidizer in propellants for space rockets and missiles.

Production of Nitric Acid. Three commercial methods have been developed for nitric acid production: (1) the reaction between sulfuric acid and sodium nitrate, (2) the thermal combination of oxygen and nitrogen in the air, and (3) the catalytic oxidation of ammonia and absorption of the gaseous products in waters. There are numerous variations of these fundamentals processes. The principal process used today is based on the catalytic oxidation of ammonia and absorption of the gaseous products in water. This process was developed by Ostwald (Germany) and based on earlier work of Kuhlmann (France). In the Ostwald process, HNO_3 is produced in a three-stage operation: (1) Ammonia is oxidized to nitric oxide; (2) the nitric oxide is further oxidized to nitrogen dioxide; and (3) the gases are absorbed in water to yield HNO_3 according to

$$4NH_3 + 5O_2 \rightarrow 4NO + 6H_2O$$
$$2NO + O_2 \rightarrow 2NO_2$$
$$3NO_2 + H_2O \rightarrow 2HNO_3 + NO.$$

The nitric oxide formed in the last equation returns to the gas phase, is reoxidized to nitrogen dioxide, and reabsorbed. These reactions are highly exothermic. In actuality, numerous complex reactions occur in addition to the main reactions just outlined.

In a manufacturing plant, air is preheated, mixed with superheated ammonia vapor, and reacted catalytically over a gauze composed of 90% platinum and 10% rhodium at a temperature of 800–960°C and operating pressures between atmospheric and 8.2 atmospheres. The reaction produces nitrogen dioxide, NO_2, and nitric oxide, NO. The latter is oxidized to NO_2 in the reaction train. The NO_2 actually exists in equilibrium with its dimer, N_2O_4. This equilibrium mixture, sometimes referred to as nitrogen peroxide, is absorbed in water in a cooled absorber tower to form HNO_3 at a strength of 55–60% HNO_3.

NITRIDING.

Surface hardening of alloy steels by heating the metals to a temperature of 490–650°C in an atmosphere of partially dissociated NH_3 (ammonia). As in cyaniding, hardening results from the formation of nitrides of iron and of certain alloying elements that may be present in the steel. Much longer heating time is required than in carburizing practice, and while the depth of penetration is generally less, the maximum hardness at the surface is higher, 900–1,100 D.P.H. (*Vickers* Brinell) compared to 800–900 D.P.H. for an average carburized case. Nitriding also differs from carburizing in that the parts are fully heat-treated to develop the required core properties before the nitriding treatment. Because of the comparatively low temperature of the process, distortion and dimensional changes are at a minimum. Nitrided steels have good corrosion-resistance when used for valves, pump parts, shafting, and bearing surfaces operating in steam, crude oil, gasolines, and gaseous products of combustion. The fatigue strength is also improved by nitriding.

Other typical applications are piston pins, crankshafts, cylinder liners, timing gears, gauges, and ball and roller bearing parts.

NITRILE RUBBER. Elastomers.

NITRILES. Amines.

NITRO- AND NITROSO-COMPOUNDS.

Nitro-compounds contain the nitro-group ($-NO_2$) attached directly to a carbon atom; nitroso-compounds contain the nitroso-group ($-NO$) similarly attached. A very important member of this group is nitrobenzene, which upon reduction yields a variety of products, important in the synthesis of drugs and dyes.

ALKYLNITRO-COMPOUNDS:

Primary	Secondary	Tertiary
$CH_3CH_2 \cdot NO_2$	$(CH_3)_2CH \cdot NO_2$	$(CH_3)_3C \cdot NO_2$
Nitroethane	Nitrodimethylmethane (2-nitropropane)	Nitrotrimethylmethane

ISOMERIC NITRITES:

$CH_3CH_2 \cdot ONO$	$(CH_3)_2CH \cdot ONO$	$(CH_3)_3 \cdot ONO$
Ethyl nitrite	Isopropyl nitrite	1,1-dimethylethyl nitrite

ALKYLNITROSO-COMPOUNDS:

$(CH_3)_3C \cdot NO$
Nitrosotrimethyl-methane

NITRATES:

$CH_3CH_2 \cdot ONO_2$	$(CH_3)_2CH \cdot ONO_2$	$(CH_3)_3C \cdot ONO_2$
Ethyl nitrate	Isopropylnitrate	1,1-dimethylethylnitrate

NITROSAMINE:

$(C_2H_5)_2N:NO$
Diethylnitrosamine

Upon reduction, nitro- and nitroso-compounds form the corresponding amine; nitrosamines form the corresponding hydrazine. Upon oxidation, nitroso-compounds form the corresponding nitro-compounds. Upon treatment with NaOH solution, nitrites and nitrates form the corresponding alcohol plus sodium nitrite and nitrate, respectively. Primary and secondary nitro-compounds, with sodium methylate $NaOCH_3$ in alcohol form salts of isonitro-compounds, $CH_3CH_2NO_2$ yielding $CH_3CH:NO(ONa)$, and $(CH_3)_2CHNO_2$ yielding $(CH_3)_2C:NO(ONa)$. These salts are derived from an acid form of the pseudo-

$$-CH:N\begin{smallmatrix}O\\ \\OH\end{smallmatrix}$$

acid

true nitro-compounds $-CH_2 \cdot N\begin{smallmatrix}O\\ \\O\end{smallmatrix}$

Upon treatment with nitrous acid, primary nitro-compounds form nitrolic acids, e.g., nitroethane $CH_3CH_2 \cdot NO_2$ yields ethylnitrolic acid

$$CH_3C\begin{smallmatrix}NOH\\ \\NO_2\end{smallmatrix}$$

which dissolves in NaOH to form

$$CH_3C\begin{smallmatrix}NONa\\ \\NO_2\end{smallmatrix}$$

red color; secondary nitro-compounds form pseudo-nitrols, e.g., 2-nitropropane $(CH_3)_2CHNO_2$ yields 2,2-nitrosonitropropane

$$(CH_3)_2C\begin{smallmatrix}NO\\ \\NO_2\end{smallmatrix}$$

colorless, solid but on fusion or in solution changes to blue color; tertiary nitro-compounds are unaffected.

Upon treatment with sodium hypobromite (or hypochlorite) primary and secondary nitro-compounds form bromo- (or chloro-) nitro-compounds, thus, nitroethane $CH_3CH_2 \cdot NO_2$ yields 1-bromo-1-nitroethane $CH_3CHBr \cdot NO_2$, and 1,1-dibromo-2-nitroethane $CHBr_2 \cdot CH_2NO_2$; 2-nitropropane $(CH_3)_2CH \cdot NO_2$ yields 2-bromo-2-nitropropane $(CH_3)_2CBr \cdot NO_2$; tertiary nitro-compounds are unaffected.

Alkylnitro-compounds are made (1) by reaction of the alkyl iodide and silver nitrite. Higher alkyl members yield increasing proportions of nitrite along with the nitro-compound, but these frequently may be separated by fractional distillation. Tertiary alkyl iodides do not behave in this manner; (2) by reaction of alpha-substituted halogen acids and sodium nitrite, followed by loss of CO_2, e.g., chloroacetic acid $CH_2Cl \cdot COOH$ yields nitroacetic acid $CH_2NO_2 \cdot COOH$ and then nitromethane plus CO_2, (3) by reaction of the hydrocarbons with HNO_3.

BENZENOID NITRO- AND NITROSO-COMPOUNDS:

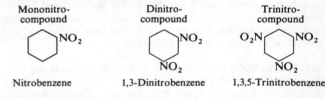

Mononitro-compound	Dinitro-compound	Trinitro-compound
Nitrobenzene	1,3-Dinitrobenzene	1,3,5-Trinitrobenzene

Nitroso-compounds

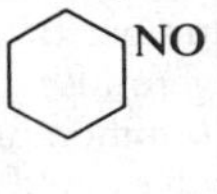

Nitrosobenzene 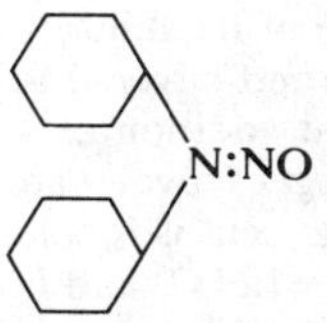   Diphenylnitrosamine

Under the proper conditions of concentration of HNO_3 and of temperature, benzene forms mainly nitrobenzene, nitrobenzene forms mainly 1,3-dinitrobenzene, and 1,3-dinitrobenzene, mainly 1,3,5-trinitrobenzene.

When nitrobenzene is treated (1) with zinc and calcium chloride or ammonium chloride solution, beta-phenylhydroxylamine C_6H_5NHOH is formed, and from this by treatment with chromic acid or ferric chloride nitrosobenzene is formed, (2) with tin or iron and HCl, aniline $C_6H_5NH_2$, is formed and from this by treatment with nitrous acid followed by treatment with stannous chloride plus HCl phenylhydrazine $C_6H_5NH \cdot NH_2$ is formed.

Mono- or poly-substituted nitro-compounds are changed in whole or in part to the corresponding amino-compounds by proper choice of reducing agent and temperature, e.g., in acid medium 1,3-dinitrobenzene yields 1,3-phenylenediamine $C_6H_4(NH_2)_2(1,3)$, and with ammonium sulfide yields 3-nitroaniline $(1)H_2NC_6H_4NO_2(3)$. When diphenylnitrosamine is reduced, 1,1-diphenylhydrazine $(C_6H_5)_2N\cdot NH_2$ is formed.

See also **Nitration**.

NITROCELLULOSE. Cellulose.

NITROGEN.

Chemical element symbol N, at. no. 7, at. wt. 14.0067, periodic table group 5a, mp $-209.86°C$, bp $-195.8°C$, critical temperature $-147.1°C$, critical pressure 33.5 atm, density, 1.14 g/cm^3 (solid), 1.25057 g/liter (0°C, 760 torr), 0.9675 (air = 1.0000). Solid nitrogen has a hexagonal crystal structure. Nitrogen at standard conditions is a colorless, odorless, tasteless gas. The gas is slightly soluble in water (2.35 parts nitrogen in 100 parts water at 0°C), the solubility decreasing with increasing temperature (1.55 parts nitrogen in 100 parts water at 20°C). Nitrogen is slightly soluble in alcohol and is essentially insoluble in most other known liquids. There are two naturally occurring isotopes, ^{14}N and ^{15}N, with ^{14}N by far the most abundant (99.635%). Four radioactive isotopes have been identified, ^{12}N, ^{13}N, ^{16}N, and ^{17}N, all with extremely short half-lives measured in seconds or minutes. In terms of abundance in igneous rocks in the earth's crust, nitrogen does not appear among the first 37 most abundant elements. In terms of abundance in seawater, nitrogen ranks 16th, with an estimated 2300 tons of nitrogen per cubic mile (493 metric tons per cubic kilometer). In terms of cosmic abundance, nitrogen ranks 7th. For comparison, assigning a value of 10,000 to silicon, the figure for nitrogen is 160,000 and that for hydrogen, estimated the most abundant, a figure of 3.5×10^8. Of dry air in the earth's atmosphere, disregarding pollutants, 78.09% is nitrogen by volume and 75.54% by weight. In the atmosphere, the nitrogen is mixed with oxygen, argon, the rare gases, CO_2, and water vapor.

Nitrogen was first identified as an element by Daniel Rutherford in 1772. Lavoisier further confirmed Rutherford's findings in 1776. Like oxygen, nitrogen is essential to practically all forms of life, making some of the compounds of this element extremely important as foods and fertilizers. Nitrogen serves the important function of diluent in the earth's atmosphere, controlling natural burning and respiration rates that otherwise would proceed much faster with higher concentrations of oxygen. Nitrogen is an important ingredient of numerous inorganic and organic compounds, including alkaloids, amides, amines, cyanides, cyanogens, diazo compounds, hydrazines, imides, nitrates, nitrides, nitrites, nitriles, oximes, purines, pyridines, and ureas. In terms of high-tonnage production, the nitrogen compound NH_3 (ammonia) ranks first with worldwide production exceeding 50 million tons annually.

First ionization potential 14.84 eV; second, 29.47 eV; third, 47.17 eV; fourth, 73.5 eV; fifth, 97.4 eV. Oxidation potentials $H_2N_2O_2 + 2H_2O \rightarrow 2HNO_2 + 4H^+ + 4e^-$, -0.80 V; $N_2O_4 + 2H_2O \rightarrow 2NO_3^- + 4H^+ + 2e^-$, -0.81 V; $HNO_2 + H_2O \rightarrow NO_3^- + 3H^+ + 2e^-$, -0.94 V; $NO + 2H_2O \rightarrow NO_3^- + 4H^+ + 3e^-$, -0.96 V; $NO + H_2O \rightarrow HNO_2 + H^+ + e^-$, -0.99 V; $2NO + 2H_2O \rightarrow N_2O_4 + 4H^+ + 4e^-$, -1.03 V; $2HNO_2 \rightarrow N_2O_4 + 2H^+ + 2e^-$, -1.07 V. $N_2O + 3H_2O \rightarrow 2HNO_2 + 4H^+ + 4e^-$, -1.29 V; $N_2O + H_2O \rightarrow 2NO + 2H^+ + 2e^-$, -1.59 V; $N_2 + H_2O \rightarrow N_2O + 2H^+ + 2e^-$, -1.77 V; $N_2 + 4OH^- \rightarrow 2NO_3^- + 2H_2O + 2e^-$, 0.85 V; $NO + 2OH^- \rightarrow NO_2^- + H_2O + e^-$, 0.46 V; $N_2O_2^{2-} + 4OH^- \rightarrow 2NO_2^- + 2H_2O + 4e^-$, 0.18 V; $NO_2^- + 2OH^- \rightarrow NO_3^- + H_2O + 2e^-$, -0.01 V; $N_2O_2^{2-} \rightarrow 2NO + 2e^-$, -0.10 V; $N_2O + 6OH^- \rightarrow 2NO_2^- + 3H_2O + 4e^-$, -0.15 V; $N_2O + 2OH^- \rightarrow 2NO + H_2O + 2e^-$, -0.76 V; $2NO_2^- \rightarrow N_2O_4 + 2e^-$, -0.88 V.

Other physical properties of nitrogen are given under **Chemical Elements**.

Industrial Nitrogen. Like many of the elements, the compounds of nitrogen by far exceed the use of elemental nitrogen (discounting its important role as diluent in the atmosphere). Industrially, nitrogen gas is produced as a by-product in the liquefaction of air to produce pure oxygen. For some applications, nitrogen provides an excellent inert atmosphere for electric furnace operations and for the gaseous insulation of transformers. An inert atmosphere is required where air must be excluded. Nitrogen is one of the three main gases used for such atmospheres, the other two being carbon monoxide and hydrogen. In providing an inert atmosphere, nitrogen reduces the velocities of reactions, lowers the partial pressure and reduces the flammability of any active gases that may be present. Since commercial nitrogen usually contains traces of oxygen, H_2O vapor, and CO_2, sufficent to cause some oxidation at high temperatures, methane may be added to make the gas fully inert.

Nitrogen gas also is required for nitriding certain alloy steels, but pure gas is not required. The nitrogen is provided by dissociating ammonia at the process temperatures ranging from 475–650°C. Metals treated in this manner are hardened by the formation of nitrides on their surface (casehardening). In cyaniding, iron-base alloys simultaneously absorb carbon and nitrogen by heating the metals in a cyanide salt. Again, the nitrogen is not required in initial gaseous form. See also **Nitriding**. Several powder metallurgy techniques also utilize dissociated NH_3 atmospheres.

Environmental Aspects of Nitrogen: The oxides of nitrogen are among the most critical of air pollutants—both in their effects and in their abatement. These aspects of nitrogen are discussed under **Pollution (Air)**.

Chemistry and Compounds: Most of the high-tonnage nitrogen-bearing compounds are described elsewhere in this volume. See also **Ammonia; Ammonium Chloride; Ammonium Hydroxide; Ammonium Nitrate; Ammonium Phosphates; Ammonium Sulfate;** and **Fertilizer**.

In the laboratory, nitrogen, mixed with argon, neon, krypton, and xenon, is obtained from the air by passing it over heated copper to remove the oxygen, or pure by fractional distillation of liquid air, whereby the nitrogen distills off before the oxygen. Pure nitrogen may also be obtained by heating such compounds as ammonium nitrite or ammonium dichromate, and collecting the gas. Mixed with carbon monoxide in producer gas, nitrogen may be utilized without separation by first making methyl alcohol from carbon monoxide and hydrogen and then using hydrogen and nitrogen for ammonia. When nitrogen at low pressure is subject to a silent electric discharge, activated nitrogen is produced. Activated nitrogen displays a golden yellow afterglow upon cessation of the current, increased by cooling and decreased by heating. This form of nitrogen is very active with phosphorus, with alkali metals (forming azides), with many metallic chlorides (forming a green fluorescence), and with hydrocarbons (forming hydrocyanic acid and cyanides). The transformation of nitrogen to activated nitrogen is partial, and its return to ordinary nitrogen takes place rapidly, in about one minute.

The metal amides and imides are important in the nitrogen system. The amides of the active metals are produced by (1) reaction of the metal with NH_3, (2) reaction of the metal hydride with NH_3, (3) reaction of the metal nitride with ammonia, (4) reaction with another amide, as $KNH_2 + NaI \rightarrow NaNH_2 + KI$ (in liquid NH_3). This last method is generally useful for the preparation of the heavy metal amides and imides from halides and binary halogenoids of the heavy metals. Cadmium amide, $Cd(NH_2)_2$ and lead imide, PbNH, for example, are readily prepared in this way. In some cases neither the amide nor the imide is stable, and the reaction proceeds to the nitride.

$$3HgBr_2 + 6KNH_2 \xrightarrow[\text{liq.}]{NH_3} Hg_3N_2 + 6KBr + 4NH_3$$

The metal amides and imides are very reactive with oxygen, and are often unstable or even explosive. Some nitrides (e.g., of silver, gold, and mercury) are explosive, but others are stable. The latter may be obtained, (1) by reaction with the metal with nitrogen or ammonia at higher temperatures, e.g., aluminum nitride and magnesium nitride, AlN and Mg_3N_2, (2) by deamination of the metal amide or azide on heating, e.g., Ba_3N_2. The great thermal stability of certain nitrides, e.g., those of boron, silicon and phosphorus, BN, Si_3N_4 and P_3N_5, is attributed to polymerization. Many of the transition metal nitrides are interstitial compounds and are hard and metal-like in their properties.

In the nitrogen system, hydrazine is analogous to hydrogen peroxide in the oxygen system, its structure being

$$\begin{matrix} & H & \\ H:\!\ddot{N}\!:\!\ddot{N}\!:\!H \\ & H & \end{matrix}$$

It is readily oxidized, even undergoing auto-oxidation under many conditions, and it is a powerful reducing agent. Like hydrogen peroxide it readily disproportionates (e.g., with a platinum catalyst), giving nitrogen and NH_3. Its reactivity (and other properties) makes it, and its derivative, unsymmetrical dimethylhydrazine, important rocket fuels. It forms addition compounds with many substances, including a monohydrate with H_2O. Hydrazine ($pK_{B1} = 6.04$, $pK_{B2} = 14.88$) forms hydrazinium(1+) compounds, containing the $N_2H_5^+$ ion, analogous to ammonium, and hydrazinium(2+) compounds containing the $N_2H_6^{2+}$ ion.

Hydroxylamine is related in its structure both to hydrazine (see formula above) and to hydrogen peroxide.

$$\begin{matrix} & H & & & & H \\ H:\!\ddot{N}\!:\!\ddot{O}\!: & & & :\!\ddot{O}\!:\!\ddot{O}\!: \\ & H & & & & H \end{matrix}$$

The chemical properties of hydroxylamine also suggest a compound intermediate between hydrazine and hydrogen peroxide. Its bond lengths are, N—O, 1.46Å, N—H, 1.01Å, O—H, 0.96Å, and its angles are H—O—N, 103°, H—N—O, 105°, and H—N—H, 107°. It is a base ($pK_B = 9.02$), forming salts containing the hydroxylammonium ion $HONH_3^+$.

Hydrazoic acid, HN_3, $pK_A = 4.72$, and most of its covalent compounds (including its heavy metal salts) are explosive. It is formed (1) in 90% yield by reaction of sodium amide with nitrous oxide, (2) by reaction of hydrazinium ion with nitrous acid, (3) by oxidation of hydrazinium salts, (4) by reaction of hydrazinium hydrate with nitrogen trichloride (in benzene solution). Hydrazoic acid forms metal azides with the corresponding hydroxides and carbonates. It reacts with HCl to give ammonium chloride and nitrogen, with H_2SO_4 to form hydrazinium acid sulfate, with benzene to form aniline, and it enters into a number of oxidation-reduction reactions.

The azides, except those of mercury(I), Hg(I), thallium(I), Tl(I), copper, Cu, silver, Ag, and lead, Pb, are readily prepared from hydrazoic acid and the oxide or carbonate of the metal, or by metathesis of the metal sulfate with barium azide. They are all thermally unstable, giving nitrogen and free metal or

occasionally nitride. The azide ion appears to resonate between four structures:

$$:\!\ddot{N}\!=\!\overset{+}{N}\!=\!\ddot{N}\!:^-, \quad ^-:\!\ddot{N}\!=\!\overset{+}{N}\!=\!\ddot{N}\!:^-$$
$$^=:\!\ddot{N}\!-\!\overset{+}{N}\!\equiv\!N\!:, \quad :\!N\!\equiv\!\overset{+}{N}\!-\!\ddot{N}\!:^=$$

These structures are in accord with a spacing of 1.15 Å and electronic charges of -0.83, 0.66, and 0.83 on the three nitrogen atoms.

N(I) Compounds. Hydration of nitrogen(I) oxide, N_2O, to hyponitrous acid, $H_2N_2O_2$, is not possible. However, the latter decomposes (in three steps) to yield the former, which is thus its anhydride. Spectroscopic studies indicate a linear structure for N_2O, resonating between $^-:\!\ddot{N}\!=\!\overset{+}{N}\!=\!\ddot{O}\!:$ and $\ddot{N}\!\equiv\!\overset{+}{N}\!-\!\ddot{O}\!:^-$. However, heat capacity measurements give a higher entropy at low temperatures than spectroscopic studies do, which is explained by a partial randomness of the structure at low temperatures.

Hyponitrous acid ($pK_{A1} = 7.05$, $pK_{A2} = 11.0$) and its salts are obtained by (1) reduction of sodium nitrite with (a) sodium amalgam, (b) by electrolysis, (c) by stannous or ferrous salts, (2) by reduction of alkyl nitrates, (3) by reduction of hydroxylamine by noble metal oxides, and (4) by reduction of sodium hydroxylamine monosulfonate in alkaline solution.

Explosive salts such as NaNO can be prepared by the reaction of NO and liquid ammonia solutions of alkali metals. The unstable free acid, HNO, is thought to be an intermediate in many redox reactions of nitrogen compounds.

Nitramide, NO_2NH_2, a weak acid ($pK_A = 6.59$) is relatively more stable than its isomer, hyponitrous acid.

N(II) Compounds. Nitrogen(II) oxide is formed in many reductions of nitrous acid, but is best prepared pure by reduction with ferrous ions, Fe^{2+}, or iodide ions, I^-. It undergoes many types of addition reactions, but its very slight tendency to dimerize and its low reactivity under ordinary conditions suggest that its odd electron lies in an antibonding orbital of very low energy; and the molecular orbital formulation is

$$NO[KK(z\sigma)^2(y\sigma^*)^2(x\sigma)^2(w\pi)^4(v\pi^*)]$$

The nitrosyl compounds can be readily classified on the basis of three modes of reaction of the NO molecule in accordance with the above formulation.

1. It can lose (or partly lose) the odd electron to form an ion of the formula $:\!N\!\equiv\!O\!:^+$. This formula gives rise to ONF, ONCl and ONBr by direct reaction of NO and the halogen. These are covalent compounds. Such salts as $NOBF_4$, $NOPF_6$, $NOAuF_4$, $NOSO_3F$, and $NOHSO_4$, on the other hand, are ionic. These may be considered the salts of nitrous acid acting as a base, $ONOH \rightleftharpoons NO^+ + OH^-$, $pK_B = 18.2$.
2. It can gain an electron to form a negative ion of the formula $:\!N\!\equiv\!\ddot{O}\!:^-$. Thus dry NO reacts with sodium in liquid ammonia to form sodium nitrosyl, NaNO (empirical formula).
3. It can share a pair of electrons to form a coordinate link, as it does in coordination compounds. In most of these, it appears to be coordinate as the positive ion, by transfer of an electron to an acceptor metal, which is thereby reduced by 1 unit in oxidation state. This causes, in some cases, the need for placing a negative charge on the metal. To avoid this, Pauling assumed the presence of four bonding electrons, involving structures of the type $M\!=\!\overset{+}{N}\!=\!\ddot{O}\!:$.

Nitrogen(III) Compounds. Nitrogen(II) oxide, NO, readily enters into equilibrium with NO_2 to form N_2O_3, nitrogen sesquioxide. The latter is unstable even at room temperature and consists of an equilibrium mixture of the three compounds. Its structure appears to be $O\!=\!N\!-\!NO_2$. If an equimolar mixture of NO and NO_2 is cooled and condensed, a blue liquid, bp 3.5°C, largely N_2O_3, is obtained. The latter readily combines with H_2O to form nitrous acid, HNO_2 ($pK_A = 3.29$). Nitrous acid is unstable, forming the equilibrium mixture, $3HNO_2 \rightleftharpoons NO_3^- + 2NO + H_3O^+$, which in concentrated solution or on warming is largely displaced to the right ($K = 39.6$ at 30°C). Moreover, the NO undergoes further reactions, so that actual system is complex. One of these reactions is: $NO + OH \rightarrow NO^+ + OH^- \rightleftharpoons NO \cdot OH \rightleftharpoons HNO_2$.

The existence of NO^+ and NO^- helps to explain the kinetics of nitrous acid as an oxidizing agent. It oxidizes I^-, Sn^{2+}, Fe^{2+}, Ti^{3+}, $S_2O_3^{2-}$, SO_2, and H_2S. It reacts with NH_3, urea, sulfonates and some other nitrogen compounds to produce nitrogen. With aromatic amines in the cold, it gives diazo compounds, while with secondary amines it gives nitroso compounds. Nitrous acid also functions as a reducing agent, as in the reactions with permanganate and hydrogen peroxide, in which nitrate ion is formed.

The nitrites vary widely in solubility, those of the alkalies and alkaline earths being very soluble, while those of the heavy metals are only slightly so. Moreover, the latter are relatively unstable, some decomposing at room temperature. The nitrites, like nitrous acid, function either as oxidizing or reducing agents. X-ray and spectroscopic studies give a triangular structure for the nitrite ion, with the N—O bond length 1.13Å and the O—N—O angle 120–130°. Values of 1.23Å and 116° have also been reported. Complex ions containing the NO_2 group may be either nitrito complexes (e.g., $Co(NH_3)ONO^{2+}$) or nitro complexes (e.g., $Co(NH_3)NO_2^{2+}$). The former of these two examples readily isomerizes to the latter.

Nitrosyl fluoride, NOF, and nitrosyl chloride, NOCl, are quite stable, but the bromide decomposes at room temperature. They are prepared by direct union of NO and the halogen, among other methods. Three trihalides NF_3, NCl_3, and NI_3 are known. The first is a colorless stable gas; NCl_3 is a yellow liquid and NI_3, a brown solid; both are explosive. The contrast in stability is attributed to the large amount of ionic resonance energy of the N—F bond, which gives NF_3 a negative heat of formation.

Nitrogen(IV) Compounds. Nitrogen dioxide, NO_2, readily associates to form the tetroxide, N_2O_4, so that at ordinary temperatures and pressures both forms are present in equilibrium. Since nitrogen dioxide has an unpaired electron, it is paramagnetic and colored (red). N_2O_4 is diamagnetic and colorless. As with NO, the odd electron is in an antibonding orbital but of higher energy so that NO_2 is more reactive and more readily undergoes dimerization. The N—O bond length is 1.20Å and the angle is 132° (electron diffraction). The structure of N_2O_4 is, on the basis of spectral and entropy considerations,

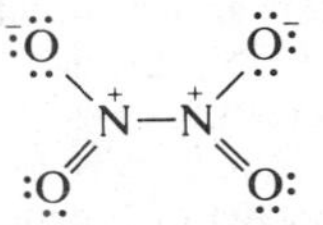

This formula is at variance with Pauling's stability argument, but is supported by Ingold's evidence (*Nature*, **159**, 743, 1947). Longuet-Higgins has proposed the structures

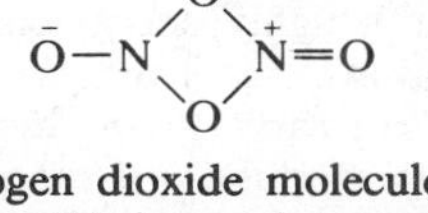

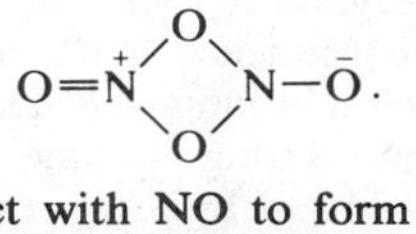

Nitrogen dioxide molecules react with NO to form N_2O_3, in an equilibrium mixture. The equilibrium mixture of NO_2 and N_2O_4 also reacts with water in a series of reactions

$$2NO_2 + H_2O \rightleftharpoons H^+ + NO_3^- + HNO_2$$

$$3HNO_2 \rightleftharpoons H^+ + NO_3^- + 2NO + H_2O$$

In warm solution, at high acidity, the second reaction is very rapid. In basic solutions the simple disporportionation $N_2O_4 + 2OH^- \rightarrow NO_2^- + H_2O$ takes place.

Nitrogen(V) Compounds. Nitrogen(V) oxide, N_2O_5, the anhydride of nitric acid, is a white solid subliming at 32.4°C and 760 mm. It hydrates readily to HNO_3, is a strong oxidizing agent, and decomposes at 20°C slowly into NO_2 and O_2. Its structure in the gas state consists of the molecules

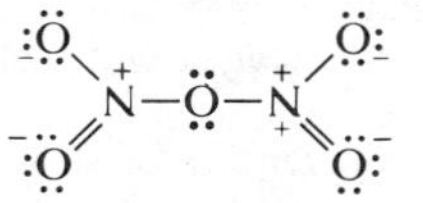

However, x-ray, Raman, and infrared spectra show the crystalline solid to consist of NO_2^+ and NO_3^- ions.

Pure nitric acid, HNO_3, is a colorless liquid boiling with decomposition at 86°C and 760 torr. Upon continued heating it decomposes into NO_2, O_2 and H_2O. It is a fairly strong acid ($K_A + 22$), showing dissociation in concentrated solutions, and the presence of nitryl cation, NO_2^+ (nitronium ion). Solutions of HNO_3 in H_2SO_4 owe many of their properties to ions such as NO_2^+ and NO^+, as well, of course, as to HSO_4^- and oxonium ions.

The properties of HNO_3 are in accordance with resonance between the three electronic structures:

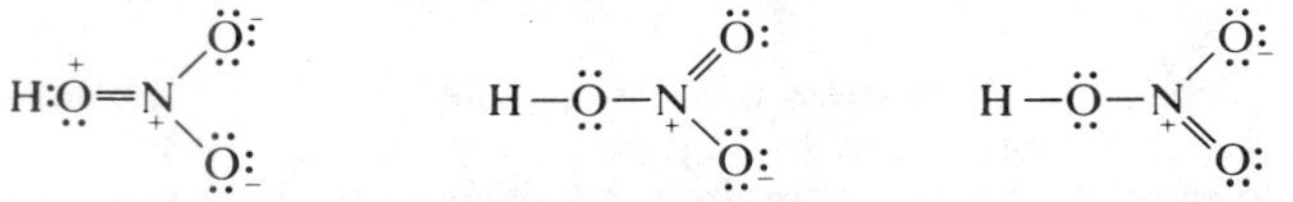

in which the last formula contributes a relatively small proportion to the overall structure. The two N—O bond lengths are 1.22Å, and N—O—H bond lengths 1.41Å and 0.96Å. The N—O—H angle is 90° and the O—N—O angle 130°.

The reaction of nitric acid are of three types: (a) acid-base reactions which are typical of a strong acid; (b) oxidation reactions, such as those with metals and organic materials, the latter often involving carbonization; (c) substitution reactions such as the replacement of —H by —NO_2 in aromatic hydrocarbons, to form nitro compounds, or of hydroxyl hydrogen by —NO_2 to produce esters of HNO_3.

These esters of nitric acid form one of the two groups of nitrates, the covalent group, which are also exemplified by nitryl hypofluorite and hypochlorite ($FONO_2$ and $ClONO_2$), often called fluorine and chlorine nitrate. Most nitrates, however, are ionic, i.e., salts of HNO_3. All metal nitrates are soluble in H_2O. Anhydrous metal nitrates, such as $Cu(NO_3)_2$, $Ti(NO_3)_4$, $VO(NO_3)_3$, $CrO_2(NO_3)_2$, $Si(NO_3)_4$, can be made by the action of liquid N_2O_4 on the metal (e.g., Cu) or of $ClONO_2$ on the corresponding chloride (e.g., the other examples given above).

The nitrate ion is considered to resonate between three equivalent structures of the form:

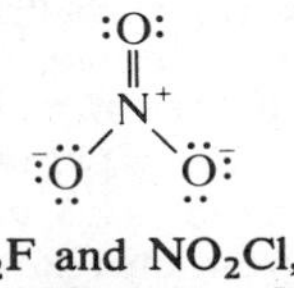

Two nitryl halides, NO_2F and NO_2Cl, are known, as well as nitryl salts, such as NO_2AsF_6, NO_2SbF_6, $(NO_2)_2SiF_6$, NO_2ClO_4, etc. Nitrogen also forms higher oxides, such as NO_3, and possibly NO_4, under action of the electric discharge.

Nitrate Losses from Disturbed Forest Ecosystems

Nutrient losses occur following a forest harvest or other disturbance, whether natural or anthropogenic. Studies have shown a variety of patterns of such losses. Vitousek et al. (1979) reports on a systematic examination of nitrogen cycling in disturbed forest ecosystems and shows that at least eight processes, operating in three stages in the nitrogen cycle, can delay or prevent solution losses of nitrate from disturbed forests. The study involved 19 forest sites in the United States, ranging from Pack Forest, Findley Lake, and Cascade Head in the northwest, Tesuque Watersheds in the southwest, Lake Monroe in southern Indiana, Coweeta in southwestern North Carolina, and Harvard Forest, Mount Mossilauke, and Cape Cod in the northeastern United States. The three stages and eight operative processes identified are:

Stage 1. Processes preventing or delaying ammonium accumulation
    (a) nitrogen immobilization
    (b) ammonium fixation
    (c) ammonia volatilization
    (d) plant nitrogen uptake

Stage 2. Processes preventing or delaying nitrate accumulation
    (e) lag in nitrification
    (f) denitrification to: —N_2, N_2O, or NO_x, —NH_4

Stage 3. Processes preventing or delaying nitrate mobility
    (g) lack of water
    (h) nitrate soprtion
    (i) denitrification at depth

The researchers stress that the net effect of all of these processes, except uptake by regrowing vegetation, is insufficient to prevent or delay losses from relatively fertile sites and thus such sites have the potential for very high nitrate losses following disturbance.

Nitrogen Fixation

A positive balance of usable nitrogen on earth depends upon nitrogen fixation which is the process by which atmospheric nitrogen, N_2, is converted either by biological or chemical means to a form of nitrogen, such as ammonia, NH_3, that can be used by plants and other biological agents. Insofar as the total amount of N_2 fixed, the biological processes for converting from N_2 to NH_3 are the most significant. In biological nitrogen fixation, microorganisms, either free-living or in symbiosis with plants (mainly in root nodules), reduce N_2 to NH_3 at atmospheric pressure and within the temperature range of 20–37°C. This natural process is to be contrasted with industrial chemical conversion processes which may require up to 300 atmospheres of pressure and a reaction temperature range of 200–300°C.

Biological Nitrogen Fixation. The occurrence and the importance to soil fertility of biological nitrogen fixation have been known since the early 1800s. The first major finding did not occur until 1960, however, when it was shown that cell-free extracts of the anaerobic bacterium *Clostridium pasteurianum* could be made to fix nitrogen if molecular oxygen, O_2, were rigorously excluded—and also if pyruvic acid, a source of energy and electrons, was supplied. This finding demonstrated that studies no longer were restricted to whole cells, as previously indicated, but that it should be possible to isolate and chemically identify the components of the nitrogen-fixing system.

The first demonstrable product of cell-free N_2 fixation is NH_3, as had been strongly suggested by previous whole-cell studies. Since the reduction of N_2 to $2NH_3$ requires six electrons and since most electron transfer systems known in biochemical pathways involve either a one- or a two-electron transfer, it could be expected that either six one-electron or three two-electron

transfer steps would be involved in nitrogen fixation. This would also suggest the existence of nitrogen compounds of valence states (reduction states) intermediate between N_2 and NH_3. However, no such intermediates have been found even in systems using cell-free extracts.

Because of failure to detect intermediates, attention was focused on the mechanism in extracts of *Clostridium pasteurianum* through which electrons were transferred from pyruvic acid to the nitrogen-fixing system. These investigations led to the discovery and isolation of the new electron carrier ferredoxin (Fd) which functioned by accepting electrons released during pyruvate oxidation by enzymes present in the clostridial extracts. The electrons from reduced Fd were transferred to a variety of different acceptors as directed by the cell. For example, some of the electrons from reduced Fd were transferred to hydrogenase, an enzyme which combined the electrons with protons (H^+) to produce molecular hydrogen, H_2, a major by-product of this anaerobe. Other electrons from reduced Fd were transferred via a flavoprotein carrier to nicotinamide adenine dinucleotide phosphate ($NADP^+$) to yield NADPH, a reduced electron carrier shown to be important in the metabolism of all biological agents. It was also found that electrons from Fd were required for nitrogen fixation when pyruvate was present as supporting substrate.

A major finding was that H_2, through hydrogenase, would act as an electron source for reducing ferredoxin. Thus, in these extracts, H_2 could be used to reduce $NADP^+$ to NADPH and NO_2^- to NH_3, and Fd was necessary as an intermediary electron carrier. Since Fd is required for pyruvate-supported N_2 fixation, it may be expected that H_2 would support nitrogen fixation, since reduced Fd is readily produced from H_2 in these extracts. Molecular H_2 alone, hoever, did not support N_2 fixation. This suggested either that a component other than reduced Fd was required, or that H_2, although capable of reducing Fd, was inhibitory to H_2 fixation as prior whole-cell studies had indicated. If an additional component were required, it appeared that it was produced from pyruvic acid, since pyruvic acid supported active N_2 fixation.

Several unsuccessful attempts were made to obtain H_2 fixation in extracts to which H_2, N_2, and one of the other products of pyruvate metabolism, ATP, were added. Active N_2 fixation did occur, however, when another product of pyruvate metabolism, acetyl phosphate, was added in addition to H_2 and N_2. When compounds such as ADP were removed from cell extracts by dialysis, no N_2 fixation occurred unless ADP was added together with phosphate, H_2, and N_2. Acetyl phosphate then was acting as a source of ATP. The reason ATP did not work directly was that a continuous supply of ATP was required, and a high concentration of ATP, if added directly to a cell-free extract, was highly inhibitory to N_2 fixation. In whole cells that are fixing N_2, a continuous supply of ATP is made available during sugar metabolism. Further details of this mechanism can be found in "The Encyclopedia of Biochemistry," (see references).

Genetic Manipulation. High on the list of many researcher's agendas for projects using the practical application of recombinant DNA research has been the possible development of a living organism that will produce ammonia—in an effort to lessen dependence upon costly and highly energy-consuming synthetic ammonia fertilizers. However, at symposia held on this topic during the late 1970s and early 1980s, these goals are considered by most researchers as quite long-range. There are fundamental problems difficult to overcome, including (1) the possibility that increasing biological nitrogen fixation, for which the plant furnishes the energy, can cause a net decrease in crop yields by depriving nitrogen to the plant for the production of certain critical growth elements; and (2) the very rapid-acting inactivation by oxygen of nitrogen-fixation mechanisms. Cloning techniques may be a path toward introducing nitrogen-fixation genes into certain bacteria. One objective is that of developing new forms of bacteria that will enter into symbiotic relationships with crop plants, such as corn (maize) and wheat, that in themselves do not possess their own nitrogen fixation symbients.

In addition to recombinant DNA and molecular cloning techniques, some scientists have combined their research with more conventional genetic techniques. An *E. coli* plasmid capable of carrying nitrogen-fixation genes of *K. pneumoniae* has been developed. Some researchers also believe that nitrogen-fixation genes may be introduced directly into plant cells to result in a plant that requires no nitrogen fertilizer.

In research activities such as these, much knowledge has been gained concerning the energy needs for biological nitrogen fixation. More energy is consumed than originally contemplated (20 moles of adenosine triphosphate, ATP, are rquired to fix one mole of nitrogen). This contributes largely to the first problem mentioned a few paragraphs earlier, namely, the great amount of energy required for the plant to fix its own nitrogen, possibly leading to yield reduction.

The well-known nitrogen fixation by rhizobia (see **Legume**) depends on photosynthesis by the plant. Although essentially impractical, photosynthesis can be increased by blanketing the plant with an atmosphere enriched in carbon dioxide. When this is done in the laboratory, yields of legumes do increase. Some investigators postulate that this is the result of decreased photorespiration by the plant, a rather wasteful process in which carbon dioxide gained through photosynthesis is diverted into a series of less productive pathways. Investigators have also found that 30% of the energy used by the nitrogenase of most rhizobial species goes to producing hydrogen rather than ammonia. Research has also shown that the organisms which perform the nitrogen fixation function in plants are indeed quite diverse in themselves. Thus, new combinations of plants and organisms may increase efficiency in some cases.

As pointed out by Evans-Barber (1977), nitrogen is fixed by a variety of microorganisms in addition to legumes. Some of these include bacteria located in soils, in decaying wood, and on the surfaces of plant roots. They also include free-living blue-green algae with fungi, ferns, mosses, liverworts, and higher plants (Hardy-Havelka, 1975). Reviews of numerous nitrogen-fixing organisms are given by Silvester (1976), Dalton (1974), Bond (1974), and Stewart (1974).

Madigan (1979) and associates found that photosynthetic purple bacteria can grow with dinitrogen gas as the only source of nitrogen under anaerobic conditions, with light as the energy source. They also found that *Rhodopseudomonas capsula* can fix nitrogen in darkness with alternative energy conversion systems.

It appears that, as of the early 1980s, researchers are on the eve of capitalizing on the body of research developed during the 1960s and 1970s.

References

Allcock, H. R.: "Phosphorus-Nitrogen Compounds: Cyclic, Linear, and High Polymeric Systems," Academic, New York (1971).
Bond, G.: in "The Biology of Nitrogen Fixation," (A. Quispel, Editor), North-Holland, Amsterdam (1974).
Campbell, J. W., Editor: "Comparative Biochemistry of Nitrogen Metabolism," Academic, New York (1970).
Dalton, H.: *Crit. Rev. Microbiol*, **3**, 183 (1974).
Evans, H. J., and L. E. Barber: "Biological Nitrogen Fixation for Food and Fiber Production," *Science*, **197**, 332–339 (1977).
Hardy, R. W. F., and M. D. Havelka: *Science*, **188**, 633 (1975).
Hollaender, A., et al., Editors: "Genetic Engineering for Nitrogen Fixation," Plenum, New York (1977).
Madigan, M. T., Wall, J. D., and H. Gest: "Dinitrogen Fixation by Photosynthetic Microorganisms," *Science*, **204**, 1429–1430 (1979).
Marx, J. L.: "Nitrogen Fixation: Prospects for Genetic Manipulation," *Science*, **196**, 638–645 (1977).
Silvester, W. B.: in *Proceedings of the 1st International Symposium on Nitrogen Fixation* (W. E. Newton and C. J. Nyman, Editors), Washington State University Press, Pullman, Washington (1976).
Stewart, W. D. P.: in "The Biology of Nitrogen Fixation," (A. Quispel, Editor), North-Holland, Amsterdam (1974).
Streuli, C. A., and P. R. Averell: "Analytical Chemistry of Nitrogen and Its Compounds," Wiley, New York (1971).
Vitousek, P. M., et al.: "Nitrate Losses from Disturbed Ecosystems," *Science*, **204**, 469–474 (1979).
Zafiriou, O. C., McFarland, M., and R. H. Bromund: "Nitric Oxide in Seawater," *Science*, **207**, 637–639 (1980).

NITROGEN GROUP (The). The elements of group 5a of the periodic classification sometimes are referred to as the Nitrogen Group. In order of increasing atomic number, they are nitrogen, phosphorus, arsenic, antimony, and bismuth. The elements of this group are characterized by the presence of five electrons in an outer shell. The similarities of chemical behavior among the elements of this group are less striking than hold for some of the other groups, e.g., the close parallels of the alkali metals or alkaline earths. Although all of the elements of this group have valences in addition to 5+, all do have the 5+ valence in common. Unlike, for example, the alkali metals or alkaline earths, the elements of the nitrogen group are not similar chemically that they comprise a separate group in classical qualitative chemical analysis separations. Three of the five, however (antimony, arsenic, and bismuth) are members of the second group in terms of qualitative chemical analysis.

NITROGLYCERIN. Explosives.

NOBELIUM. Chemical element symbol No, at. no. 102, at. wt. 254 (mass number of ^{254}No), radioactive metal of the actinide series; also one of the transuranium elements. Nobelium has valences of 2+ and 3+. In 1957, a group of American, English, and Swedish scientists bombarded a target of several curium isotopes (largely ^{244}Cm) with a beam of ^{13}C ions from the cyclotron at the Nobel Institute for Physics. They obtained a few alpha particles of 8.5 MeV energy and half-life of 10 minutes. This was considered to indicate the presence of element 102 with a probable mass number of 251 or 253. At that time, the element was named nobelium with assignment of the symbol No. Further experiments at the University of California, however, failed to confirm this discovery. In April 1958, Ghiorso, Sikkeland, Walton, and Seaborg, working with the heavy ion linear accelerator (HILAC) at Berkeley, showed the isotope 102^{254} to be a product of the bombardment of ^{246}Cm with ^{12}C ions. Confirming experiments at Berkeley in 1966 showed the existence of ^{254}No with a 55-second half-life; ^{252}No with a 2.3-second half-life; and ^{257}No with a 23-second half-life. Four other isotopes are now recognized, including ^{255}No with a half-life of 3 minutes.

In 1973, scientists at Oak Ridge National Laboratory and Lawrence Berkeley Laboratory, produced a relatively long-lived isotope of nobelium through the bombardment of ^{248}Cm with ^{18}O ions. A total half-life of 58 ± 5 minutes was computed from the combined data of both laboratories.

References

Ditmer, P. F., et al.: "Identification of the Atomic Number of Nobelium by an X-ray Technique," *Phys. Rev. Lett.*, **26**, 17, 1037–1040 (1971).

Fields, P. R., et al.: "Production of the New Element 102," *Phys. Rev.*, **107**, 5, 1460–1462 (1957).

Flerov, G. N., et al.: "Experiments to Produce Element 102," *Sov. Phys. Dokl.*, **3**, 3, 546–548 (1958).

Ghiorso, A., Sikkeland, T., Walton, J. R., and G. T. Seaborg: "Attempts to Confirm the Existence of the 10-minute Isotope of 102," *Phys. Rev. Lett.*, **1**, 1, 17–18 (1958).

Ghiorso, A., Sikkeland, T., Walton, J. R., and G. T. Seaborg: "Element No. 102," *Phys. Rev. Lett.*, **1**, 1, 18–20 (1958).

Maly, J., Sikkeland, T., Silva, R., and A. Ghiorso: "Nobelium: Tracer Chemistry of the Divalent and Trivalent Ions," *Science*, **160**, 1114–1115 (1968).

Mikheev, V. L., et al.: "Synthesis of Isotopes of Element 102 with Mass Numbers 254, 253, and 252," *Sov. At. Energy*, **22**, 93–100 (1967).

Seaborg, G. T., Editor: "Transuranium Elements," Dowden, Hutchinson & Ross, Stroudsburg, Pennsylvania (1978).

Silva, R. J., et al.: "The New Nuclide Nobelium-259," *Nucl. Phys.*, **A216**, 97–108 (1973).

NOMENCLATURE (Fertilizer). Fertilizer.

NOMENCLATURE (Organic Chemistry). Organic Chemistry.

NONNUTRITIVE SWEETENERS. Sweeteners.

NORMAL CONCENTRATION. A one normal solution (often abbreviated 1 *N*) contains one gram-equivalent weight of a particular substance dissolved in 1 liter of *solution*. The equivalent weight of a substance may be defined as that weight of the substance which will involve, in a chemical reaction, one atomic weight of hydrogen, or the weight of any other element of portion of a substance which, in turn, would involve in reaction one atomic weight of hydrogen.

As an example, the chlorine atom of potassium chloride (KCl) also is found in hydrochloric acid (HCl) in combination with one hydrogen atom. Thus, the gram-equivalent weight of KCl is 74.555, which is the same as its gram-molecular weight. A 1 *N* solution of KCl will contain 74.555 grams of the salt per liter of solution.

For a particular solution, the molar and normal cocentrations are the same only when the gram-molecular and gram-equivalent weights are the same. Sulfuric acid H_2SO_4 represents a case where these values are not the same. This acid contains two active hydrogen ions and, therefore, its gram-equivalent weight is one-half its gram-molecular weight. Phosphoric acid H_3PO_4 contains three active hydrogen ions. Consequently, the gram-equivalent weight for this acid is one-third that of the gram-molecular weight. Calcium hydroxide $Ca(OH)_2$ contains two active hydroxyl ions, each being equivalent to a hydrogen ion. Therefore, the gram-equivalent weight of $Ca(OH)_2$ is one-half its gram-molecular weight.

NUCLEAR FISSION. A type of nuclear reaction in which the compound nucleus splits into two nearly equal parts, rather than ejecting one or a few small nuclear particles, as in most nuclear reactions. Our knowledge of nuclear fission dates back to the mid-1930s when Fermi and his coworkers showed that the number of distinctly different radioactive nuclides that could be induced by neutron bombardment of uranium far exceeded the number expected, unless some previously unknown pattern of isomerism could be found. Furthermore, the radio-chemical properties of many of these radio-elements different quite markedly from expectations. For example, both Hahn and Strassman in Germany and Curie and Savitch in France found that certain unknown activities, thought to be radioactive radium, always followed the chemically separated barium fraction rather than the radium fraction. Hahn and Strassman found several other similar examples and were able to show that uranium, when bombarded by neutrons, undergoes what then appeared to be a very unusual nuclear reaction in that the products are radioelements with about half the atomic number of uranium. These findings were interpreted by Meitner and Frisch as the division of an excited nucleus into nuclei of medium mass, a process that was given the name *nuclear fission*.

The first such process to be extensively studied was fission induced in ^{235}U by thermal neutrons (neutrons with energies of about 0.03 eV). This reaction, symbolically represented by the equation

$$^{235}U + n \longrightarrow {}^{236}U \longrightarrow \text{fission},$$

produces an unstable system which achieves stability by splitting into two large fragments, not by ejecting one or a few small particles.

An individual fission does not procude a unique pair of fragments, but in a large number of such processes, the mass distribution of the fragments can be predicted with reasonable certainty, leading to precitable fission yields. A fission yield, usually expressed as a percentage, describes that fraction of nuclear fission processes that give rise to a specified nuclide or group of isobars. The yields of single nuclides are known as independent yields and those of a set of isobars as mass yields or chain yields. Since two fragments are produced by each fission, the total of all fission yields for a given fission process is 200%. The fission yield curve is different for each mode of induced fission, the most commonly known one being that for thermal neutron induced fission of ^{235}U, shown in Fig. 1. The chemical

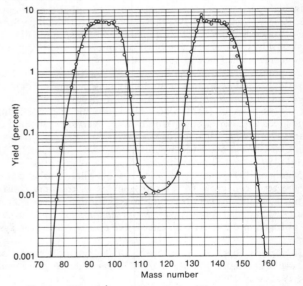

Fig. 1. Mass fission yield curve for $^{235}U + n$ (thermal).

characteristics of the two fragments vary within limits, so that many elements are formed. Analysis of the fission products shows that most of them are in two mass groups, a "light" group consisting of elements having mass numbers between 85 and 104, and a "heavy" group consisting of elements having mass numbers between 130 and 149. Fragment mass numbers that have been detected range from around A = 70 to around

A = 160. The determination of independent yields is made more difficult by the fact that many of the products are highly radioactive and undergo extensive secondary changes, sometimes in extremely short times, a very small fraction of a second.

A most significant aspect of nuclear fission is its great release of energy. The source of this energy is the loss of mass between the initial and final products of the reaction. The total mass of all atoms and nuclear particles produced in a single fission process in less than the original mass of the ^{235}U atom and the neutron that combined with the ^{235}U to induce fission. During fission of ^{235}U, the total energy released because of loss of mass is about 200 MeV. In practical units, the fissioning of 1 gram of ^{235}U yields 24,000 kilowatt-hours of energy.

Another important feature of fission is the presence of neutrons among the reaction products, slightly more than two for each fission of a ^{235}U atom. These neutrons are not an immediate consequence of fission, but are boiled off the original fission products, their release being possible because of the very large amount of available energy. In all these neutrons were captured by other ^{235}U nuclides, the number of available neutrons would multiply by factors of two for every generation of fission processes, a very rapid increase. However, some neutrons escape from the region containing the ^{235}U and others are absorbed in nonfission capture processes. The minimum conditions for a self sustaining chain reaction is that at least one neutron from each nucleus undergoing fission must cause fission of another nucleus, a multiplication factor of one or greater. Maintenance of a chain reaction is essential to the proper functioning of both nuclear weapons and nuclear reactors.

The probability that fission can occur (generally called the cross section for fission) varies widely among different nuclides. Only a few nuclides, such as ^{235}U, have a high probability of undergoing fission when they capture a neutron. In other nuclides, the probability of fission is generally much smaller. As an example, the cross section as a function of incident neutron energy is shown in Fig. 2 for fission of ^{235}U and of ^{238}U. Although fission can be induced in ^{238}U, such a process is possible only if the indicent neutron has an energy greater than 1 MeV, whereas neutrons of any energy can induce fission in ^{235}U. The characteristic double hump yield curve of Fig. 1 (asymmetric fission) is common only for low neutron excitation energy and targets consisting of highly fissile elements. For either higher excitation energies or less fissile elements, such as actinium or radium, symmetric fission becomes much more important, creating a triple humped fission-yield curve, shown in Fig. 3. Slightly fissile elements, such as lead and bismuth, or very high excitation energies further emphasize the symmetric mode of fission, also illustrated in Fig. 3. Nuclear fission may be induced by particles other than neutrons, such as alpha particles and photons. In some nuclides, it also occurs spontaneously, although the probability of such occurrence is so low that it has almost no effect on the radioactive decay characteristic of the nuclide.

Nuclear fission has generally been explained theoretically in terms of the liquid-drop model of the nucleus. In this model, the incident neutron combines with the target nucleus to form a compound nucleus at a high excitation energy. A small part of this excitation energy can be attributed to the kinetic energy of the incident neutron, but most of it usually comes from the binding energy of the incident neutron. This added energy initiates oscillations in the drop, which then sometimes assumes an elongated shape, similar to B in Fig. 4. If oscillations become sufficiently violent that a form similar to D is reached, fissioning (form E) becomes inevitable, since the positive charge at the two ends of the dumbbell-shaped nucleus then produces an elec-

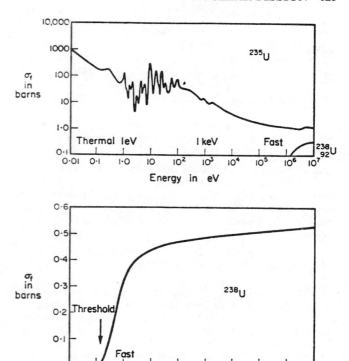

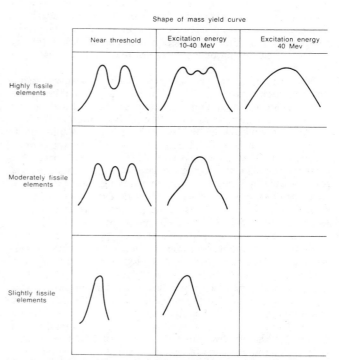

Fig. 2. Fission cross section as function of energy for ^{235}U and ^{238}U.

Fig. 3. Mass fission yield curves as function of excitation energy and degree of fission probability.

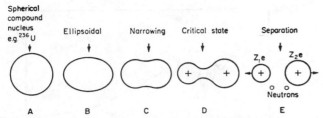

Fig. 4. Fission mechanism according to liquid-drop model of the nucleus.

trostatic repulsive force greater than the attractive nuclear force holding the neck of the dumbbell together. The reason for asymmetric fission is not clearly understood. The liquid drop model predicts symmetric fission. Most people believe that asymmetric fission results because of the effects of the closed shells of the nucleus.

—C. Sharp Cook, Professor of Physics, The University of Texas at El Paso, El Paso, Texas.

NUCLEAR FORCES. Strong, short-range, attractive forces that interact between the individual nucleons of an atomic nucleus. Unfortunately, despite several decades of research, a clear and unambiguous description cannot be given for the forces that hold individual protons and neutrons together in an atomic nucleus. Unlike the electrostatic force that holds electrons in an atom, no equation can be written that completely describes the nature of the force that holds an atomic nucleus together, or the nature of its associated potential energy. A description of the detailed structure of a nucleus cannot, therefore, be derived directly from calculations based on knowledge of nuclear forces. Instead, detailed knowledge of the structure of atomic nuclei has been derived from nuclear models. These models have been constructed by using results from other fields of physical science which display the same or similar characteristics as those observed in nuclear reactions and in radioactive decay. From such analogies, construction of a partial description of nuclear structure and of the nature of nuclear forces has been possible.

Because of the unknown characteristic of nuclear forces, many different suppositions have been made, using available experimental evidence, regarding the nature of the potential energy V of a nuclear particle as a function of its position in the field of a nucleus, or of another nuclear particle. To a first approximation, the nuclear potential is assumed to be spherically symmetric, such as V is a function only of the distance r from the center of the field, thus being the same in all directions, and is representable by a curve as in accompanying curves (a) to (f).

A *potential well* is the name given to a region in which a minimum in the potential is formed; it results from attractive forces. A *potential barrier* is the name given to a region in which there is a maximum in the potential; it results from repulsive forces, either alone or in combination with attractive forces. Some central potentials commonly used as approximations to nuclear potentials are illustrated in the curves. Curve (a) shows a square well potential, which has a constant negative value $-V_0$ for $r \leq r_0$ and zero value for $r \geq r_0$. When this curve represents the potential between two nucleons, r_0 is called the *range of nuclear forces*; when it represents the potential of a nucleus, as this nucleus interacts with an individual nucleon, r_0 is called the nuclear radius. Curve (b) shows a square well potential for $r \leq r_0$ with a Coulomb potential resulting from repulsive electrostatic forces, for $r > r_0$. The resulting barrier is called a *Coulomb barrier*, and the maximum energy b is

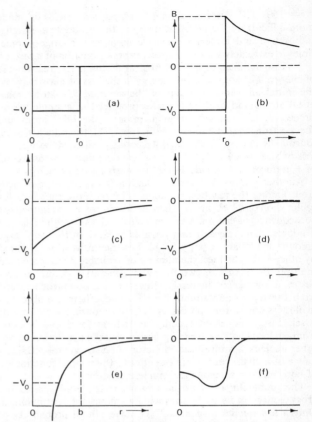

Potential energy of a nuclear particle versus distance from the center of the field.

called the *barrier height*. Such a potential approximates that of a positively charged particle in the field of a nucleus, and is often used in the theory of alpha particle disintegration and nuclear reactions. Curve (c) shows an exponential well, $V = V_0 e^{-r/b}$; curve (d) shows a Gaussian well, $V = V_0 e^{-r/b^2}$. Curve (e) shows a Yukawa potential, $V = -(V_0/r)e^{-r/b}$ used in the meson theory of nuclear forces for the interaction between two nucleons; and (f) shows a wine-bottle potential, characterized by a low central elevation. If a high central elevation is present, the resulting barrier is called a *central barrier*, or *repulsive core*.

Although the predominant part of the nuclear potential is the part described above that is derived from the central force produced by the average effects of all other nucleons in the system on the individual nucleon under observation, evidence exists that nonsymmetric tensor and spin-orbit coupling terms must be included in the description of the nuclear potential. These are derived from a tensor force resulting from a coupling between individual pairs of nucleons and from the coupling between spin and orbital angular moments of the individual nucleus, as described by the shell model of the nucleus.

A considerable amount of evidence indicates that nuclear forces are charge-independent, i.e., the neutron-neutron, neutron-proton, and proton-proton forces are identical. The meson theory of nuclear forces, originated by Yukawa, postulates the atomic nucleus being held together by an exchange force in which particles, now called mesons, are exchanged between individual nucleons within the nucleus.

—C. Sharp Cook, Professor of Physics, The University of Texas at El Paso, El Paso, Texas.

NUCLEAR FUSION. The character of the atomic nucleus is such that the individual nuclear particles are most tightly bound in elements of intermediate atomic number. When energy is sought, attention is focused on the more loosely assembled elements, releasing energy by splitting (*fissioning*) the heavy isotopes, or by joining (*fusing*) the lighter ones. There is less energy release per fusion reaction than there is per fission reaction, but the reactants are more plentiful and, in many respects, easier to handle. A particular fusion reaction is of interest if the power produced can be sufficiently large to offset the power consumed in generating and maintaining the reacting medium, and if the relevant rates can be large enough so that economically interesting regimes are accessible to modern technology. There are over thirty such reactions possible. The most appealing of the fusion reactions as possible routes to fusion energy are (1) those which involve the heavy hydrogen isotopes, deuterium, ^2_1H or D; and (2) those which involve tritium ^3_1H or T. These tend to have the largest fusion reaction probability (cross section) at the lowest energies. Deuterium is abundant, naturally occurring and in wide use now as D_2O in heavy-water-moderated reactors. Tritium is a radioactive isotope with a 12.3-year half-life and does not occur in nature. Tritium emits an electron and decays to stable helium-3.

The deuterium (D-D) reaction chain may be represented by:

$$D + D \rightarrow {}^3He + n + 3.2 \text{ MeV}$$
$$D + D \rightarrow T + p + 4.0 \text{ MeV}$$
$$D + T \rightarrow {}^4He + n + 17.6 \text{ MeV}$$
$$D + {}^3He \rightarrow {}^4He + p + 18.3 \text{ MeV}$$

$$\cdots\cdots\cdots\cdots\cdots\cdots\cdots\cdots\cdots\cdots\cdots$$

$$6D \rightarrow 2{}^4He + 2p + 2n + 43.1 \text{ MeV}$$

The first two equations represent the fact that the D-D reaction can follow either of two paths, producing tritium and one proton; or helium-3 and one neutron, with equal probability. The products of the first two reactions form the fuel for the third and fourth reactions and are burned with additional deuterium. The net reaction consists of the conversion of six deuterium nuclei into two helium nuclei, two hydrogen nuclei, and two neutrons along with a net energy release of 43.1 MeV. The reaction products—helium, hydrogen, and neutrons—are harmless as contrasted with the myriad fission products obtained in a fission reactor. The neutrons produced may be absorbed in sodium to produce an additional 0.25 MeV per cycle. Therefore, the D-D reaction produces at least 7 MeV per deuterium atom (deuteron) and, with absorption in sodium, more than 10 MeV per fuel atom.

The peak reaction rate coefficient of the D-D reaction is considerably less than that of the deuterium-tritium (D-T) reaction occurring within the (D-D) cycle. Thus, attention tends to focus on the latter. Because tritium does not occur naturally, the reaction must be supplemented by one using lithium to reproduce the tritium fuel:

$$D + T \rightarrow {}^4He + n + 17.6 \text{ MeV}$$
$$n + {}^6Li \rightarrow {}^4He + T + 4.8 \text{ MeV}$$

$$\cdots\cdots\cdots\cdots\cdots\cdots\cdots\cdots\cdots\cdots$$

$$D + {}^6Li \rightarrow 2{}^4He + 22.4 \text{ MeV}$$

This reaction is tritium-regenerating and produces only helium as a reaction product.

The D-T reactor is technologically more complex than the D-D reactor because of the need to facilitate the second reaction (which takes place outside the plasma) and because very energetic neutrons must be slowed down to allow the reaction with lithium to take place. However, the conditions needed to achieve net power output are less demanding than for the D-D fuel reactor. The D-T reaction will probably be exploited first, but its ultimate, very long term use may be limited by the availability of lithium. See also **Lithium (For Thermonuclear Fusion Reactors).**

Fusion reactions can take place only when the nuclei of the fuel atoms are brought into close enough conjunction. The nuclei are positively charged and so repel each other. This repulsion is equivalent to an energy barrier which can be penetrated with reasonable efficiency only if the reacting nuclei have kinetic energy comparable to the barrier height. The level of kinetic energy required depends upon the particular reaction and the desired reaction rate, but in general, plasmas of interest have average energy per particle in excess of 5 keV. A collection of particles with average energy 5 keV has an effective temperature of at least 10^8 degrees Kelvin. At these temperatures, the gas is completely dissociated into its constituent positively charged nuclei and free electrons. The density ranges between 10^{13} to 10^{14} cm^{-3}. The electrical charge density is such that the behavior of the collection of particles is completely dominated by electrostatic and electromagnetic phenomena. Such a charge-dominated collection of ionized matter is known as *plasma*. This plasma at such extremely high temperatures cannot be confined by walls made of materials, known or imagined. But confinement, even for a nanosecond or less, is required if fusion reactions are to occur. Ways to confine the plasma have been researched for 20–30 years by scientists in a number of countries.

After nearly three decades of effort, fusion ignition, that is, the efficient burnup of deuterium and tritium has been accomplished only in one way, namely, the thermonuclear or hydrogen bomb. In this instance, obviously tremendously greater amounts of energy are released than are required to trigger the fusion reactions. In the case of the hydrogen bomb, an atomic bomb was used to generate the extremely high temperature and the degree of confinement required for hydrogen nuclei to fuse. The methods researched to date for confining the plasma include containment within magnetic fields; inertial confinement methods in which the fuel is pelletized in a special way and fusion reactions are initiated either by laser beams or beams of particles; and by heating the plasma with high-power microwave radiation.

NUCLEAR POTENTIAL. The potential energy V of a nuclear particle as a function of its position in the field of a nucleus or of another nuclear particle. A central potential is one that is spherically symmetric, so that V is a function only of the distance r of the particle from the center of force. A noncentral potential, on the other hand, is one that is not spherically symmetrical, or that depends upon the relative directions of the angular momenta associated with the particle and the center of force as well as upon the distance r. A negative potential corresponds to an attractive force, while a positive potential corresponds to a repulsive force.

Although the expression can certainly be applied to the problem of nuclear forces, the usual meaning of a nuclear potential refers to the interaction of a nucleon (neutron or proton) with a complex nucleus. Although the potential energy of a single nucleon inside a nucleus is clearly a rapidly varying function of position and time (since it represents the interaction with a large number of closely packed, fast-moving particles), one may nevertheless speak of the average potential energy, and regard this as a smoothly varying function. For a neutron, the nuclear potential is essentially negative inside the nucleus, rising rapidly to zero outside the nuclear radius R. For a proton, the long-range electrostatic repulsion must, of course, be added. Owing to the Pauli exclusion principle, and to the exchange nature of nuclear forces, however, such a potential cannot in general be regarded simply as a function of position, $V = V(r)$; it depends in addition upon the momentum of the particle, which in quantum mechanics does not commute with the position. Hence, the potential must be regarded as a nondiagonal matrix operator $V = \langle r|V|r' \rangle$ in configuration space, or a similar operator in momentum space.

Although the concept of a nuclear potential in this latter sense cannot be defined in a precise way, it has nevertheless been useful, both qualitatively and quantitatively in the investigations of nuclear structure and nuclear reactions. It has been of particular usefulness in the optical model of nuclear reactions.

NUCLEAR REACTOR.   An assembly of equipment (system) in which a nuclear reaction (fission or fusion) can be sustained, producing large quantities of heat which can be converted into other energy forms, commonly steam. The steam is then used to generate electricity in the same manner that steam produced in fossil-fuel power plants is utilized. Thus, only portions of a nuclear power plant are actually nuclear in nature, just as the boiler in a conventional steam plant is the only part of the plant that deals with fuel combustion.

Since the first commercial nuclear reactor produced steam for electric power in 1958, the practical reactors used have based upon *nuclear fission*. Mainly, they have been light-water reactors (predominate in the U.S.), heavywater reactors (Canada), and gas-cooled reactors (pioneered in the UK). The fast-breeder reactor is in an advanced state in a number of countries, including France, the UK, West Germany, and Japan. Less information is known concerning the details of nuclear power in the U.S.S.R. and iron-curtain countries.

Nature of Nuclear Fission Reactions

The energy of a nuclear fission reaction can be computed from the change in mass between reactants and products according to Einstein's law:

$$\Delta E = \Delta m c^2$$

where E is the energy in ergs, m is mass in grams, and c is the velocity of light in centimeters per second. For example, the mass difference in this equation is $\Delta m = 0.2058$ amu (atomic mass units). Therefore, $\Delta E = 931$ MeV/amu $\times 0.2058$ amu $= 191.6$ MeV. The average amount of energy released in the various fissions reactions is about 200 MeV (million electron volts). This energy is distributed in the fission process as:

	MeV
Kinetic energy of fission fragments	165
Radioactive-decay energy	23
Kinetic energy of neutrons	5
Prompt gamma-ray energy	7

The energy of a chemical reaction, approximately 3–4 eV, is dramatically lower than that of a nuclear reaction. Hence, the fission of ^{235}U yields 2.5 million times as much energy as the combustion of the same weight of carbon.

The importance of fission in energy (power) production lies in two facts: (1) an exceedingly large amount of energy is released in the fission reaction; and (2) the production of excess neutrons permits a chain reaction. These two circumstances make it possible to design nuclear reactors in which self-sustain-ing reactions occur with the continuous release of energy. Although described later, it may be pointed out here that nuclear fission is not the only energy-releasing nuclear reaction. The *fusion* of light nuclides, like hydrogen, into heavier elements is also an energy-producing process.

The heat generated in nuclear power plants is transferred to a working fluid and from this point on the nuclear power plant and the conventional fossil-fueled power plant are essentially similar.

Fission Reaction. In nuclear fission, the nucleus of a heavy atom is split into two or more fragments. The reaction is initiated by the absorption of a neutron. A typical reaction is

$$^{235}_{92}U + ^{1}_{0}n \rightarrow ^{137}_{56}Ba + ^{97}_{36}Kr + 2^{1}_{0}n + \Delta E$$

In this reaction, a ^{235}U atom absorbs one neutron, becomes unstable, and subsequently fissions into two fission fragments plus two neutrons. This is just one of the many ways in which ^{235}U might fission. The number of neutrons produced in a fission reaction is usually 2 or 3. The excess neutrons produced by the fission reaction provide the means of self-sustaining the chain reaction. Nuclides including ^{233}U, ^{235}U, and ^{239}Pu, which are fissionable by neutrons of all energies, are termed *fissile* nuclides.

Nuclear Fuels. There are two broad categories of nuclear fuels: (1) the fissile nuclides previously mentioned; and (2) the fissionable nuclides, ^{232}Th (thorium) and ^{238}U. Thermal reactors use fissile nuclides as fuel, while *fast reactors* are designed to burn fissionable materials. In fast reactors, only a small portion of the ^{232}Th and ^{238}U are fissioned directly. A larger portion of these materials is converted into ^{233}U and ^{239}Pu, respectively, through neutron absorption. Thus, this type of reactor not only consumes fuel, but also produces (breeds) new fuel material. Hence, the term *breeder reactor* is used for reactors designed to take advantage of this phenomenon. Breeding is possible in thermal reactors also, but to a lesser extent. The fuel material in a fast reactor must contain a significant amount (about 10%) of one of the fissile materials. The remainder of the fuel must have a high mass number in order to avoid slowing down the neutrons. The natural reserves of fissionable materials are more than 100 times greater than the reserves of fissile materials. Consequently, from the viewpoint of utilization of available energy resources, fast reactors are of great importance. Breeder reactors are described later.

Moderators. The most important slowing-down mechanism is elastic scattering on elements of low mass number. Materials like light and heavy water, beryllium oxide, and graphite are used to slow down, or *thermalize* the neutrons to an energy of about 0.025 eV. As neutrons collide with the nuclei of these atoms, their kinetic energy and speed are gradually reduced until thermal equilibrium is achieved with the reactor structure. The fewer such collisions before deceleration is complete, the less chance of ^{238}U atoms absorbing neutrons.

Critical Mass. Thermal neutrons, which move like atoms in a low-pressure gas, diffuse throughout the reactor. They must be absorbed by a nucleus of the reactor structure, in which case they merely make that nucleus radioactive. Or, they may strike a fissionable atom of ^{235}U, causing fission and, in turn, releasing more neutrons to maintain the reaction. Should the number of neutrons absorbed by the moderator and ^{238}U be greater than about 1.5 excess neutrons emitted from each fission, the chain reaction will not be maintained. Therefore, the reactor core must be designed so that the mass of fuel will be just sufficient to ensure one neutron from each fission causing fission in another atom. A mass and configuration of fissionable material in which this occurs is termed the *critical mass*—or a reactor in which this condition is achieved is said to have "gone critical."

To measure a chain reaction, a multiplication factor k is used to indicate the ratio of neutrons in one generation to those in the preceding generation. Thus, in a constant chain reaction where the total number of neutrons neither increases nor decreases, the heat output is constant and $k = 1$. Should k rise above unity, the rate of fission, and hence the rate of heat productivity, steadily rises. This is so even if k is held constant at its new value. Here lies one *major difference between nuclear reactors and conventional steam generators*. In the latter, heat output is proportional to firing rate. If the firing rate is increased, the steam output is increased; but it remains constant at its new level. In a nuclear reactor, an increase in k results in continuously rising heat output. Only by returning the rate of neutron production to its original ratio can heat output be maintained at its new level.

Reactivity Control. Absorption of excess neutrons, above those needed to maintain a constant reactivity level, provides close control over the degree of reactivity. This is accomplished by inserting materials having a high neutron-capture rate into the core. Control rods of special alloy metals are moved into and out of the cores as required. To start the reactor from shutdown (black start), control rods are partially withdrawn until k becomes greater than one. Neutron flux and heat output grow until the desired level is reached. At this point, control-rod movement is quickly reversed to keep k at unity. The reactor is shut down by inserting the rods to their full extent. In this position, the rods absorb more than 1.5 excess neutrons per fission and the chain reaction quickly stops. Heat production continues for a time, but is usually dissipated by an auxiliary cooling system.

Types of Fission Reactions

Light-Water Reactors (LWR)[1]

These reactors are of two principal designs: (1) *pressurized water reactors* (PWR), and (2) *boiling-water reactors* (BWR). In a PWR, heat generated in the nuclear core is removed by water (reactor coolant) circulating at high pressure through the primary circuit. The water in the primary circuit both cools and moderates the reactor. Heat is transferred from the primary to the secondary system in a heat exchanger, or boiler, thereby generating steam in the secondary system. The BWR differs from the PWR primarily in that boiling takes place in the reactor itself. Comparable steam temperatures are possible at pressures of about 1000 pounds per square inch (6.9 mPa) as contrasted with 2000 psi (13.8 mPa) for pressurized reactors.

Boiling Water Reactor (BWR). Aside from its heat source, the boiling water reactor (BWR) generation cycle is substantially similar to that found in fossil-fueled power plants. One of the first BWRs was the *Vallecitos* BWR, a 1000 psi (6.9 mPa) reactor which powered a 5 MW electric generator and provided power to the Pacific Gas & Electric Company grid through 1963. Power output capabilities have increased many times during the intervening years.

The direct-cycle boiling water reactor nuclear system (Fig. 1) is a steam generating system consisting of a nuclear core and an internal structure assembled within a pressure vessel, auxiliary systems to accommodate the operational and safeguard requirements of the nuclear reactor, and necessary controls and instrumentation. Water is circulated through the reactor core

producing saturated steam which is separated from the recirculation water, dried in the top of the vessel, and directed to the steam turbine-generator. The turbine employs a conventional regenerative cycle with condenser deaeration and condensate demineralization. The direct-cycle system is used because of its inherently simple design, contributing to reliability and availability.

The steam from a BWR is, of course, radioactive. The radioactivity is primarily ^{16}N, a very short-lived nitrogen isotope (7 seconds half-life) so that the radioactivity of the steam system exists only during power generation. Extensive generating experience has demonstrated that shutdown maintenance of a BWR turbine, condensate, and feed-water components can be performed essentially as a fossil-fuel plant.

The reactor core, the source of nuclear heat, consists of fuel assemblies and control rods contained within the reactor vessel and cooled by the recirculating water system. A 1,220-MWe BWR/6 core consists of 732 fuel assemblies and 177 control rods, forming a core array 16 feet (4.8 meters) in diameter and 14 feet (4.2 meters) high. The power level is maintained or adjusted by positioning control rods up and down within the core. The BWR core power level is further adjustable by changing the recirculation flow rate without changing control rod position, a feature which contributes to excellent load-following capability.

The BWR is the only light water reactor system that employs bottom-entry control rods. From the very first BWRs, bottom-entry control rods have been used because reactivity and moderator density is highest in the lower part of the core. They provide optimum power shaping characteristics for the type of core where moderator density is varied as a function of power level. Bottom-entry and bottom-mounted control rod drives also allow refueling without removal of rods and drives, and allow drive testing with an open vessel prior to initial fuel loading, or at each refueling operation. The hydraulic system, using reactor system pressure, provides rod insertion forces that are greater than gravity or mechanical systems.

The BWR requires substantially lower primary coolant flow through the core than pressurized water reactors. The core flow of a BWR is the sum of the feedwater flow and the recirculation flow, which is typical of any boiler. Unique to the BWR is the application of jet pumps inside the reactor vessel. See Fig. 2. The jet pumps deliver their driving force from the external recirculation pumps and generate about two-thirds of the recirculation flow within the reactor vessel. The jet pumps also contribute to the inherent safety of the BWR design under loss-of-coolant emergency conditions because they continue to provide internal circulation with one or both external recirculation loops out of service. The BWR can deliver about one-third power through this natural jet pump circulation mode, a vital capability in effecting a "black start" (a fully fresh start-up of a reactor) of the plant without external power.

The BWR operates at constant pressure and maintains constant steam pressure similar to most fossil-fueled boilers. The BWR primary system operates at pressure about one-half that of a pressurized water reactor primary system, while producing steam of equal pressure and quality.

The integration of the turbine pressure regulator and control system with the reactor water recirculation flow control system permits automated changes in steam flow to accommodate varying load demands on the turbine. Power changes of up to 25% can be accomplished automatically by recirculation flow control alone, at rate of 15% per minute increasing and 60% per minute decreasing. This provides a load-following capability that can track rapid changes in power demand.

---

[1] In nuclear power terminology, ordinary water, in contrast with heavy water, is termed *light water*.

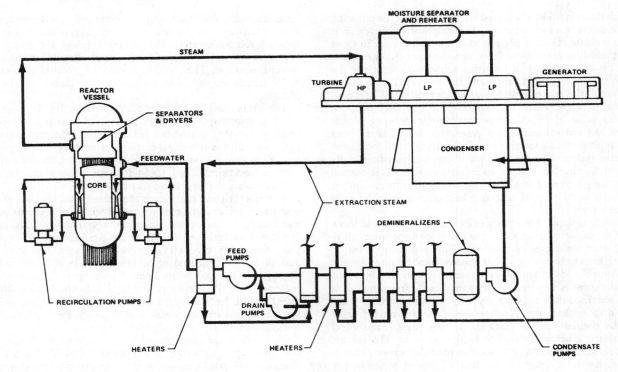

Fig. 1.   Direct-cycle reactor system. (*General Electric Co.*)

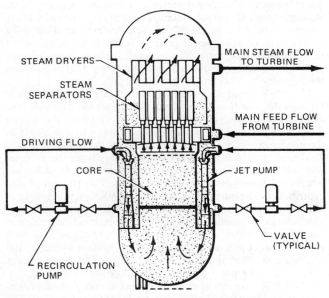

Fig. 2.   Steam and recirculation water flow paths of boiling water reactor. (*General Electric Co.*)

Reactor Assembly. This assembly (Fig. 3) consists of the reactor vessel, its internal components of the core, shroud, top guide assembly, core plate assembly, steam separator and dryer assemblies and jet pumps. Also included in the reactor assembly are the control rods, control rod drive housings and the control rod drives.

Each fuel assembly that makes up the core rests on an orificed fuel support mounted on top of the control rod guide tubes.

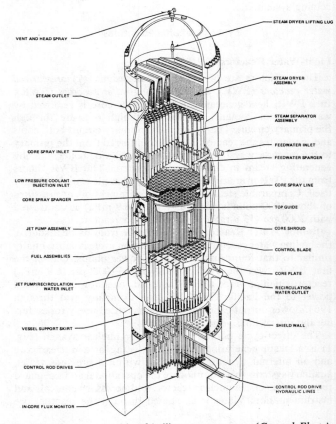

Fig. 3.   Reactor assembly of boiling water reactor. (*General Electric Co.*)

Each guide tube, with its fuel support piece, bears the weight of four assemblies and is supported by a control rod drive penetration nozzle in the bottom head of the reactor vessel. The core plate provides lateral guidance at the top of each control rod guide tube. The top guide provides lateral support for the top of each fuel assembly.

Control rods occupy alternate spaces between fuel assemblies and may be withdrawn into the guide tubes below the core during plant operation. The rods are coupled to control rod drives mounted within housings which are welded to the bottom head of the reactor vessel. The bottom-entry drives do not interfer with refueling operations. A flanged joint is provided at the bottom of each housing for ease of removal and maintenance of the rod drive assembly.

Except for the Zircaloy in the reactor core, the reactor internals are stainless steel or other common corrosion-resistant alloys. The reactor vessel is a pressure vessel with a single full-diameter removable head. The base material of the vessel is low alloy steel which is clad on the interior except for nozzles with stainless steel weld overlay to provide the necessary resistance to corrosion.

Reactivity Control. The movable boron-carbide control rods are sufficient to provide reactivity control from the cold shutdown condition to the full-load condition. Supplementary reactivity control in the form of solid burnable poison is used only to provide reactivity compensation for fuel burnup or depletion effects. The movable control rod system is capable of bringing the reactor to the subcritical when the reactor is an ambient temperature (cold), zero power, zero xenon, and with the strongest control rod fully withdrawn from the core. In order to provide greater assurance that this condition can be met in the operating reactor, the core is designed to obtain a reactivity of less than 0.99, or a 1% margin on the "stuck rod" condition.

Pressurized Water Reactor (PWR). In a typical pressurized water reactor (PWR), heat generated in the nuclear core is removed by water (reactor coolant) circulating at high pressure through the primary circuit. The water in the primary circuit cools and moderates the reactor. The heat is transferred from the primary to the secondary system in a heat exchanger, or boiler, thereby generating steam in the secondary system. The steam produced in the steam generator, a tube-and-shell heat exchanger, is at a lower pressure and temperature than the primary coolant. Therefore, the secondary portion of the cycle is similar to that of the moderate-pressure fossil-fueled plant. In contrast, in boiling-water or direct-cycle systems, steam is generated in the core and is delivered directly to the steam turbine.

The similarities of basic pressurized water reactor design from one manufacturer to the next are more striking than the differences. Therefore, the description of one particular configuration (Combustion Engineering, Inc.) can suffice to convey the general operating principles. The major components of a **PWR** are: (1) The reactor vessel which contains the oxide fuel core, core intervals, control element assemblies, and in-core instruments; (2) the electrically-heated pressurizer; (3) the electric-motor-driven primary coolant pumps; and (4) the U-tube type steam generators. See Fig. 4. The primary coolant system layout can be fitted into a variety of containment types and concepts. A prestressed cylindrical containment is common. Figure 5, shows the arrangement in a spherical containment. This type of building lends itself to separation of safeguards equipment, steam lines, and emergency power supplies.

Steam Generators. The basic geometry is shown in Fig. 6. With the nuclear steam supply system operating at 3,817 MW, two steam generators produce a total of 17.18 × 10⁶ pounds

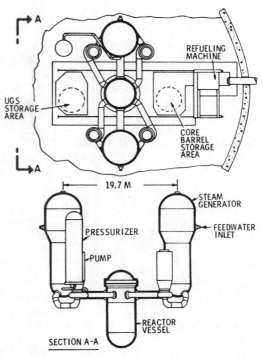

Fig. 4. Nuclear steam supply system for pressurized water reactor. (*Combustion Engineering, Inc.*)

(7.89 × 10⁶ kilograms) of steam per hour at 1,070 psia (72.8 atmospheres). The steam generators are constructed, using carbon steel pressure-containing members and Inconel-600 tubes. The tube-sheet is clad by weld deposit for maximum strength; tongue and groove construction of the divider plate places no stress on the tube-sheet cladding. Fusion welding of the end of each tube to the tube-sheet primary cladding provides an effective seal for leakage control, and "expanding" (explosively expanding) the tubes in the full length of the tube-sheet eliminates corrosion-prone crevices. An economizer section on the units improves heat transfer by preheating the incoming feedwater, using the low (primary side) temperature heat transfer area of the U-tubes. Multiple feed nozzles allow the economizer flow distribution to be optimized for each power level.

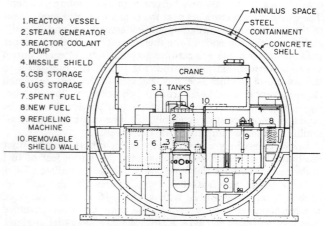

Fig. 5. Spherical containment for pressurized water reactor. (*Combustion Engineering, Inc.*)

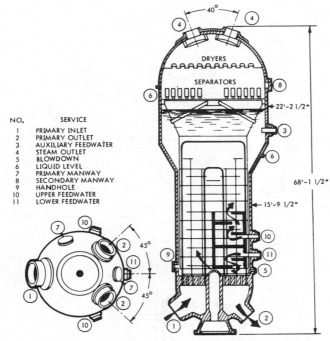

NO.	SERVICE
1	PRIMARY INLET
2	PRIMARY OUTLET
3	AUXILIARY FEEDWATER
4	STEAM OUTLET
5	BLOWDOWN
6	LIQUID LEVEL
7	PRIMARY MANWAY
8	SECONDARY MANWAY
9	HANDHOLE
10	UPPER FEEDWATER
11	LOWER FEEDWATER

Fig. 6. Steam generator for pressurized water reactor. (*Combustion Engineering, Inc.*)

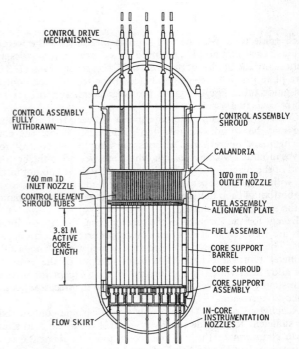

Fig. 7.   Reactor arrangement in pressurized water reactor.

Reactor Coolant Pumps. As indicated by Fig. 4, four reactor coolant pumps are used, two for each steam generator. The pumps are vertical, single-bottom-suction, horizontal-discharge, motor-driven centrifugal units. The pump impeller is keyed and locked to its shaft. A complex system of seals is used to prevent any leakage. The motors are designed to start and accelerate to speed under full load with a drop to 80% of normal rated voltage at the motor terminals. Each motor is provided with an anti-reverse rotation device. Each reactor coolant pump is provided with four vertical support columns, four horizontal support columns, and one vertical snubber. The structural columns provide support for the pumps during normal operation, earthquake conditions, and any hypothetical loss-of-coolant accident in either the pump suction or discharge line.

The reactor arrangement is shown in Fig. 7. The barrel-calandria guide structure is a rugged [3-inches (76.2 mm) thick, barrel section] unit which can withstand and protect all control element fingers from the combined effects of seismic and blowdown loads that may result from a loss-of-cooling accident. The calandria structure fits over the control element guide tubes of the fuel assemblies, aligning all fuel assemblies, and laterally restraining the top ends of the fuel assemblies. With the upper guide structure in place, a continuous guide tube for each control finger is formed, extending from the top of the tube-sheet to the bottom of the fuel assembly. Because of this feature, which isolates every control finger from the coolant crossflow, flexibility is obtained in the number of control fingers that can be attached to one control assembly, i.e., one control element assembly can serve more than one fuel assembly.

Severe emergency core cooling system criteria require that the builders of water reactors increase the linear feet of fuel in the reactor core for the same power in order to reduce LOCA (loss-of-cooling accident) fuel temperatures. In the unit described here, an assembly with a 16 × 16 fuel rod array of smaller diameter rods is used in the same assembly envelope that was occupied by a 14 × 14 assembly in earlier designs.

The results in a maximum linear heat rate decrease in the assembly of about 25%.

Fuel Assembly. As shown in Fig. 8, the active core is made up of 241 fuel assemblies, all of which are mechanically identical. Each fuel assembly contains 236 Zircaloy clad, UO_2 fuel rods retained in a structure consisting of Zircaloy spacer grids welded at about 15-inch (38.1-centimeter) intervals to five Zircaloy control element assembly guide tubes which, in turn, are mechanically fastened at each end to stainless steel end fittings. The overall length of the fuel assembly is about 177 inches (450 centimeters) and the cross section is about 8 inches (20.3 centimeters) by 8 inches (20.3 centimeters). Each fuel assembly weighs about 1,450 pounds (657.7 kilograms). With reference to Fig. 8, fuel rods, consisting of uranium dioxide (UO_2) pellets of low enrichment canned in thin-walled Zircaloy-4 tubing, are

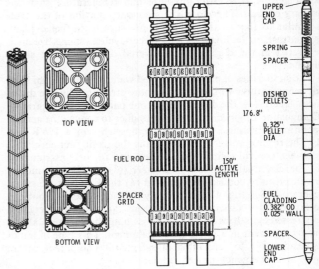

Fig. 8.   Fuel assembly used in pressurized water reactor.

designed to achieve average burnups of about 33,000 MWD/MTU (thermal megawatt days/metric tons of uranium) and peak burnups of about 50,000 MWD/MTU. The design factors limiting burnup of the fuel are the effects on the clad of volumetric changes of the fuel pellet and fission gas release.

High-Temperature Gas-Cooled Reactors (HTGR)

The high-temperature gas-cooled reactor (HTGR) is an advanced thermal reactor that produces modern steam conditions. Helium is used as the coolant. Graphite, with its superior high-temperature properties, is used as the moderator and structural material. The fuel is a mixture of enriched uranium and thorium in the form of carbide particles clad with ceramic coatings.

The high-temperature conditions and high thermal efficiency (approximately 39%) of the HTGR result in high performance through conservation of fuel, lower capital cost, and the use of conventional turbine-generating equipment. The amount of cooling water required to carry away the waste heat is significantly less than in a light-water reactor (LWR). The use of thorium in the fuel cycle decreases fuel cost, improves the conservation of fuel, and adds the large deposits of thorium to available fuel reserves. The HTGR has significant environmental advantages, including: (1) Lower thermal discharge because of its high efficiency; (2) low release of radioactive waste because of the high-integrity fuel and the inert coolant; and (3) low consumption of raw materials because of high efficiency and use of thorium in the fuel cycle.

High operating temperatures at moderate pressures are achieved through the use of helium as the coolant. Helium has the fundamental advantage that it always remains in the gas phase, making complete loss of coolant no longer a problem. Helium is attractive as a coolant because it: (1) is chemically inert; (2) absorbs essentially no neutrons; (3) makes no contributions to the reactivity of the system; and (4) will continue to be in good supply to meet the requirements of the HTGR.

Graphite is used as the moderator and core structural material because of (1) excellent mechanical strength at high temperatures; (2) very low neutron-capture cross section; (3) good thermal conductivity; and (4) high specific heat. Graphite has a long history of use in thermal reactors. Because of low neutron-capture cross section, no neutrons are lost within the core through absorption in metallic fuel cladding or structural supports. Graphite also is well suited to high-temperature operations, increasing in strength with temperature up to a point (2,482°C) well beyond the operating range of the HTGR.

The use of the thorium-uranium fuel cycle in the HTGR provides improved core performance over the plutonium/uranium low-enrichment cycle used in LWRs. The principal reason for this is that fissile ^{233}U produced from neutrons captured in thorium during reactor operation is neutronically a better fuel than ^{239}Pu, produced from ^{238}U in the low-enrichment cycle. The excellent neutronic characteristics of the graphite-moderated thorium/uranium cycle leads directly to high conversion ratios and low fuel inventories. Reduced ^{235}U inventories and make-up requirements spell reduced sensitivity to increasing uranium prices.

A simplified flow diagram of the HTGR installed at Fort St. Vrain (Colorado) is shown in Fig. 9. The prestressed concrete reactor vessel (PCRV) is 31 feet in internal diameter with a 75-foot (23 meters) internal height. The upper and lower heads are nominally 15 feet (4.5 meters) thick, and the walls have a nominal thickness of 9 feet (2.7 meters). Thus, the PCRV provides the dual function of containing the coolant at operating pressure and also providing radiological shielding. The exterior vertical surface of the vessel may be described as a hexagonal

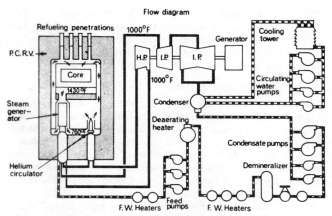

Fig. 9.   Simplified flow diagram of Fort St. Vrain Nuclear Generating Station. (*General Atomic Company*)

prism with vertical pilasters at each corner. It is 61 feet (18.6 meters) across pilasters, 49 feet (14.9 meters) across flats, and 106 feet (32.3 meters) high. The concrete walls and heads of the PCRV are constructed around a carbon steel liner which is nominally $\frac{3}{4}$-inch (1.9 centimeters) thick. The liner is anchored

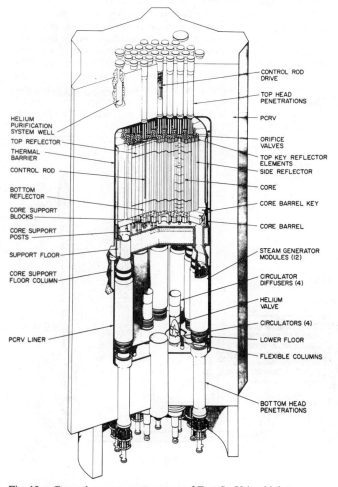

Fig. 10.   General reactor arrangement of Fort St. Vrian high-temperature gas-cooled reactor.

to the concrete and provides a helium-tight membrane. A system of water-cooled tubes welded to the concrete side of the liner provides a heat removal system to control concrete temperature.

The general reactor arrangement of the Fort St. Vrain high-temperature gas-cooled reactor is shown in Fig. 10.

Heavy Water Reactor (HWR)

During the atomic energy developments in the World War II years and for a period thereafter, the United States, the United Kingdom, and Canada cooperated closely and many of the nuclear scientists of these countries appreciated the merits of heavy water as a moderator. Each of these countries pursued some development of HWRs for commercial power generation, but at different paces and dedication. Only Canada took to the HWR for commercial power generation.

One of the first high-priority nuclear applications of the United States was for naval propulsion. Because of a very tight minimal physical size criterion, LWRs offered advantages over the HWR. The United Kingdom placed emphasis on the production of plutonium for weapons programs. Gas graphite reactors were a reasonable early choice. When commercial nuclear power was recognized as a needed source of energy, the reactors developed in the United States and the United Kingdom for military purposes were logical extensions for each country as the result of accumulation of experience in operating them. Longterm savings at that time were not a major criterion.

In the postwar years, hydroelectric power amply met a large portion of Canada's power needs and its abundance made nuclear power quite noncompetitive. Canadian utility operators were used to capital-intensive plants combined with low operating costs. In analyzing the prospects for nuclear power in Canada, utility planners and engineers placed a significant value on low fueling costs and thus neutron economy was paramount. Thus, when commercial nuclear power studies commenced in Canada in the mid-1950s, the choice was the HWR. This was

also bolstered by experience and knowledged gained on heavy water production plants when Canadian scientists were trading experience from the heavy water-moderated NRX research reactor when the United States was developing the Savannah River production reactors.

The first Canadian nuclear power demonstration (NPD) reactor was of 20-MWe capacity and was configured similarly to a light water reactor. Because of limited facilities for making large pressure vessels, a modular pressure-tube design of the configuration shown in Fig. 11(a) was investigated. Zircaloy-2 had become available at that time for fabrication of the pressure tubes. Thus, the NPD was constructed, using Zircaloy as cladding material and uranium dioxide as fuel. The NPD reactor has been in operation since 1962. The CANDU (Canada Deuterium Uranium) power reactors, including the NPD, numbered eleven installations as of the early 1980s, with 4 in Ontario, 2 in Quebec, 1 in New Brunswick, 2 sold to India, 1 to Pakistan, and 1 to Argentina. Scheduled for installation during the mid- and late- 1980s are 3 additional units in Ontario, and 1 for Korea.

As pointed out by Robertson (1978), CANDU power reactors are characterized by the combination of heavy water as moderator and pressure tubes to contain the fuel and coolant. Their excellent neutron economy provides the simplicity and low costs of once-through natural uranium cycling. Future benefits include the prospect of a near-breeder thorium fuel cycle to provide security of fuel supply without the need to develop a new reactor, such as the fast breeder. The CANDU system is appropriate for countries of intermediate economic and industrial capacity, such as Canada. Producing heavy water is fundamentally simpler than enriching uranium and commercial heavy-water plants have been built in smaller sizes than would be possible for uranium enrichment plants. Although Canada has rather generous supplies and reserves of uranium, there is increasing pressure on Canada to export uranium, a pressure that

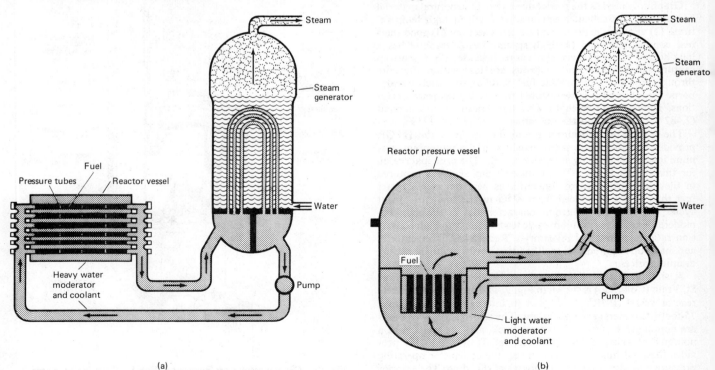

(a)   (b)

Fig. 11.   Comparison of heavy-water reactor (a) with light-water reactor (b). (*After Robertson, Atomic Energy of Canada Limited.*)

will probably increase if the introduction of fast breeder reactors in other countries is delayed. The current simplest possible fuel cycle for the CANDUs, which is not dependent upon fuel reprocessing, will probably be retained in Canada so long as uranium remains plentiful and comparatively economic. However, for future planning, research to date has indicated that a "self-sufficient thorium cycle" may be practicable in the CANDUs with minimum modification. It has been observed that, at equilibrium, the thorium cycle would require no further uranium. Only small quantities of thorium, which is more abundant than uranium, would be required. Also of interest for the future is *electronuclear breeding*, i.e., the use of electric power to convert fertile to fissile material for neutron economy.

Fast Breeder Reactors

The fast breeder reactor derives its name from its ability to breed, that is, to create more fissionable material than it consumes. This ability stems from the fact that neutrons travel faster than they do in a thermal reactor. The breeding process depends, in part, upon the neutrons maintaining a high speed, or high energy. If their speed or energy is allowed to degrade as occurs in thermal reactors, the number of neutrons produced per absorption in uranium or plutonium decreases. Furthermore, at lower velocities, neutrons tend to be captured in various structural materials of the reactor, and this further reduces the breeding potential. It is important, therefore, in fast reactors to keep the velocity of the neutrons high. Water, which is used as a coolant in some thermal reactors, tends to slow the neutrons down and thus prevent efficient breeding. Therefore, it is necessary to use a coolant which does not slow the neutrons or capture them as they travel through the coolant. Liquid sodium and gaseous helium under pressure are the two principal coolants under study and development.

Fuel cycle Considerations. Approximately 99.3% of uranium as it is found in nature is the isotope ^{238}U and 0.7% is ^{235}U. Uranium 235 is a fissile isotope, that is, if it is struck by a neutron it will split, or fission yielding on the average approximately two neutrons and 200 MeV of energy. This amount of energy corresponds to approximately 78 million Btu for every gram of uranium which fissions (3.5×10^{10} Btu/pound) (1.95 Calories/kilogram). Most reactors which exist today are largely dependent upon ^{235}U for their energy. However, some of the neutrons released in fission of ^{235}U also are absorbed in nonfissionable ^{238}U. As the ^{238}U absorbs a neutron, it is transformed into fissionable ^{239}Pu (plutonium). Thus, while the reactor is sustaining the fission process and thereby creating energy, it is also generating fresh fuel which can later be used to create more energy. Unfortunately, this is an inefficient process in present thermal reactors where the neutron velocity is established by the temperature, or thermal energy, so only limited amounts of additional energy are made available by transformation of ^{238}U into ^{239}Pu.

The fast breeder reactor makes possible the recovery of most of the available energy in uranium. This occurs because during fission in the fast breeder nearly three neutrons are released for every neutron absorbed as compared with only approximately two neutrons in a thermal reactor. On the average, between one and two neutrons are necessary for sustaining the fission process, and the extra neutron in a fast reactor can be absorbed in nonfissionable ^{238}U and thereby transformed into fissionable ^{239}Pu. Reactors which have a breeding ratio greater than one create more fuel than they need for their own purposes, and the extra plutonium can be used to fuel new breeder reactors. By this means, 80% or more of the available energy in uranium can be recovered and used in reactors.

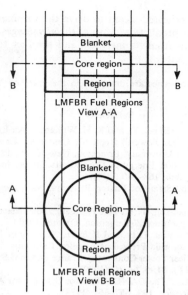

Fig. 12. Liquid-metal fast breeder reactor core and blanket arrangement.

In a typical fast breeder, most of the fuel is ^{238}U (90 to 93%). The remainder of the fuel is in the form of fissile isotopes which sustain the fission process. The majority of these fissile isotopes are in the form of ^{239}Pu and ^{241}Pu, although a small portion of ^{235}U can also be present. Normally, the fissile isotopes are located in a central "core" region which is surrounded by the fertile isotopes in the "blanket" region. This is illustrated in Fig. 12.

There are many design differences among reactor designs, including the primary cooling system arrangement, steam generator type, core support method, structural material choices, and safety features. The flow circuit for a liquid metal fast breeder reactor (LMFBR), where two sodium circuits are included, is shown schematically in Fig. 13.

In another type, the gas-cooled fast breeder reactor (GCFR), the heat generated in the reactor is transferred to pressurized helium as the coolant, thence to the steam generators that, in

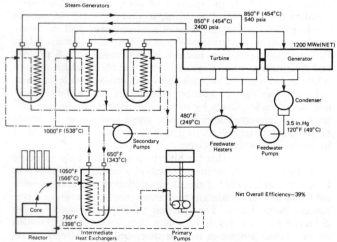

Fig. 13. Liquid-metal fast breeder reactor flow circuit. (*General Electric Co.*)

turn, supply superheated steam to drive the turbine-generator for electric power production. Most of the advantages of helium as a coolant stem from its inertness and the fact that it is a single-phase coolant.

NOTE: Radioactive wastes are discussed in the entry on **Wastes and Water Pollution.**

References

Feiveson, H. A., von Hippel, F., and R. H. Williams: "An Evolutionary Strategy for Nuclear Power," Rept. 67, Center for Environmental Studies, Princeton University, Princeton, New Jersey, 1978.

Holdren, J. P.: "Fusion Energy in Context: Its Fitness for the Long Term," *Science,* **200,** 168–180 (1978).

Lewis, H. W.: "The Safety of Fission Reactors," *Sci. Amer.,* **242,** 3, 53–75 (1980).

Lönnroth, M., Johansson, T. B., and P. Steen: "Solar versus Nuclear: Choosing Energy Futures," Pergamon, Oxford, 1980.

Lönnroth, M., Johansson, T. B., and P. Steen: "Energy in Transition," University of California Press, Los Angeles, California, 1980.

Mattill, J. I.: "Nuclear Power Contrasted: France vs. the U.S.," *Technology Review (MIT),* **81,** 5, 66–67 (1979).

Robertson, J. A. L.: "The CANDU Reactor System: An Appropriate Technology," *Science,* **199,** 657–664 (1978).

Robinson, A. L.: "Fusion Energy in Our Time," *Science,* **207,** 622–624 (1980).

Sefcik, J. A.: "Decommissioning Commercial Nuclear Reactors," *Technol. Rev. (MIT),* **81,** 56–71 (1979).

Staff: "The Strategy of Fusion," *Sci. Amer.,* **240,** 1, 74 (1979).

Zaleski, C. P.: "Breeder Reactors in France," *Science,* **208,** 137–144 (1980).

NUCLEAR STRUCTURE. The nucleus of an atom of atomic number Z and mass number A contains Z protons and A-Z neutrons bound together under the influence of shortrange nuclear forces much as molecules are bound together in a drop of liquid. The strength of binding may be determined by subtracting the actual mass of the atom from the mass of its constituent particles considered as free particles. The binding energy E_B is then related to this mass defect ΔM by Einstein's relation, $E_B = \Delta Mc^2$, where c is the velocity of light. The precise value of E_B depends upon the nucleus concerned, and upon how many neutrons and protons it contains, but it is of the order of 8 MeV per nucleon in most nuclei.

The binding energy determines whether the nucleus is stable or unstable. Among the lighter nuclei, the ones which are stable are those in which the number of protons is approximately equal to the number of neutrons, so that $A \approx 2Z$. In heavier stable nuclei, there is an excess of neutrons over protons owing to the repulsive electrostatic forces between the protons. Thus, the most stable oxygen nucleus is ^{16}O, containing 8 protons and 8 neutrons, while the most nearly stable uranium nucleus is ^{238}U, containing 92 protons and 146 neutrons. Nuclei containing a disproportionate number of neutrons tend to be unstable and decay radioactively by emission of electrons whereby neutrons are converted into protons; those containing an excess of protons similarly tend to decay by emission of positrons or by capture of orbital electrons.

It has been shown that the nucleus is approximately spherical in shape and of volume proportional approximately to its mass. It is, however, capable of executing oscillations about the spherical form, and in certain circumstances may even acquire a permanent deformation. The heaviest nuclei are unstable under deformation, as a result of which they undergo spontaneous fission. These properties may be described qualitatively by regarding the nucleus as an electrically charged drop of liquid possessing volume energy and surface tension.

Although the nucleus is normally found in its lowest energy state, it may be produced as the result of a nuclear reaction, or through radioactivity in a number of excited states whose detailed properties may differ quite markedly from the lowest state. If formed in an excited state, it will decay, normally by the emission of electromagnetic radiation (gamma rays) to the lowest state, or by the emission of particles to another nucleus.

NUCLEIC ACIDS AND NUCLEOPROTEINS. Nucleic acids are compounds in which phosphoric acid is combined with carbohydrates and with bases derived from purine and pyrimidine. Nucleoproteins are conjugated proteins consisting of a protein moiety and a nucleic acid. Originally, nucleoproteins were thought to occur only in the nuclei of cells, but it was later established that they are far more widely distributed, being found in cells of all types, animal and plant. They are found in the chromosomes, in the genes, in viruses, and bacteriophages.

The protein portion of the nucleoproteins is basic in nature and being complex in structure may form several types of linkage, depending upon the type of nucleic acid. In gastric digestion or hydrolysis with weak acid, nucleoproteins yield protein and nuclein. The latter in pancreatic digestion or hydrolysis with weak alkali yields additional protein and nucleic acid.

Upon additional hydrolysis, nucleic acids yield four characteristic constituent groups: (1) heterocyclic nitrogenous bases of the purine type; (2) heterocyclic nitrogenous bases of the pyrimidine type; (3) a carbohydrate, either ribose or deoxyribose; and (4) phosphoric acid.

Hydrolysis of nucleic acid with enzymes belonging to the group of nucleases gives nucleotides, nucleosides, and the constituent groups already mentioned. Thus, polynucleotidase catalyzes the hydrolysis of nucleic acid to give nucleotides which consist of purine or pyrimidine, ribose or deoxyribose, and phosphoric acid. Nucleotidase catalyzes the hydrolysis of nucleotides to nucleosides (which consist of a purine or a pyrimidine and ribose or deoxyribose) and phosphoric acid. Nucleosidase catalyzes the acid hydrolysis of nucleosides to the respective base or carbohydrate. Since the nucleic acids are composed of nucleotides, they may be considered polymers of nucleotides and thus be called polynucleotides.

There are two groups of nucleic acids differentiated by the carbohydrate present. Those which contain D-ribose,

$$CHO \cdot CHOH \cdot CHOH \cdot CHOH \cdot CH_2OH,$$

are generally known as *ribonucleic acids*, usually termed RNA. Those which contain D-2-deoxyribose,

$$CHO \cdot CH_2 \cdot CHOH \cdot CHOH \cdot CH_2OH,$$

are usually known as *deoxyribonucleic acids*, generally termed DNA. Sometimes the latter are also termed DRNA.

Deoxyribonucleic acid is found not only in the nucleus of normal cells but is also present in mitochondria of both plants and animals as well as in plant chloroplasts. As far as is known, the DNA occurs in circular, double-stranded structures, and the presence of DNA polymerase in the mitochondria suggests that the organelles may be able to replicate themselves. Formerly, this type of nucleic acid was thought to be a constituent only of animal cells and was known as thymus nucleic acid. Ribonucleic acid is found principally in the cytoplasm, although small amounts are found in the nucleus, nucleoli, and the chromosomes. RNA at one time was known as yeast nucleic acid and was thought to be the characteristic nucleic acid of plants.

It has been observed that deoxyribonucleic acid is a constituent of the chromosomes of the cell nucleus. Since the number of chromosomes of a cell and its daughter cells are equal, the quantity of DNA in the normal cells of any given species or type should be and is remarkably constant. This quantity is not changed by starvation or other action or form of stress. The quantity in a normal diploid nucleus is twice that of a normal haploid nucleus and in polyploid cells, such as cultivated wheats, the quantity of DNA is some multiple of the quantity in the haploid cell. This quantity is of the order of 6×10^{-9} milligrams per nucleus in the case of mammals. Birds and fishes have lesser amounts.

Until recently, the chemical synthesis of DNA and the resultant ability to construct totally synthetic genes appeared to be impractical. However, under the influence of recombinant DNA technology, the field of synthetic DNA has matured. Two chemical methods—the diester and triester methods—and one enzymatic method—the polynucleotide phosphorylase method—are generally used for the synthesis of oligodeoxyribonucleotide. Once made the resultant gene can be cloned (see **Clone**) and then used to produce useful peptide products. Products such as insulin and somatostatin have been made in *Escherichia coli* following insertion of the chemically synthesized DNA into the bacterial gene.

The quantity of RNA varies in different tissues. It also varies in amount in the cytoplasm and nucleus of the same cells. The amount of RNA is affected by the nutritional state of the cells, the type of tissue, and the metabolic action in which they participate. Thus, there is more ribonucleic acid in growing embryonic tissue, in tumor tissue, in pancreas, salivary glands, and other gland with secretory functions. All of these indications led to the concept that RNA controls the rate of the production of protein by cells, white DNA controls the transmission of hereditary characteristics from one generation to the next.

At one time, it was believed that nucleic acids were tetranucleotides, but it has been established that they are much larger molecules. Ribonucleic acids have been found to have molecular weights as high as 300,000 (implying approximately 1,000 nucleotides per molecule), while deoxyribonucleic acids have molecular weights of the order of 1 or 2 million. It has been shown that in the living organism, RNA and DNA are long double chains, with each chain spiraling around the other, also a characteristic of some proteins.

Each of the two spiral chains that constitute the DNA molecule is composed of alternating sugar (deoxyribose or ribose) and phosphate groups, and attached to each sugar is one of the four bases, usually adenine, guanine, thymine, and cytosine. (Adenine and guanine are pyrimidines, while thymine and cytosine are purines). Linkage between the two chains is effected by hydrogen bonds formed between the adenine groups of each chain and the thymine groups of the other, and between the guanine groups of each chain and the cytosine groups of the other. Variations occur in the sequence in which these four bases appear in each chain. These variations are considered to be a means of transmission of genetic "information," e.g., a basic mechanism of heredity. This view has been supported by experimental evidence that replication of DNA molecules occurs only to give new molecules with identical sequence of bases.

Delineation of DNA as the bearer of genetic information stems from the findings of Chargaff et al. that the base composition of DNA is related to the species of origin. Indeed Chargaff's chromatographic studies enabled the establishment of the following conclusions:

1. The base composition of DNA varies from one species to another.
2. DNA specimens isolated from different tissues of the same species have the same base composition
3. The base composition of DNA in a given species does not change with age, nutritional status, or changes in environment.
4. In nearly all DNAs examined, the number of adenine residues is always equal to the number of thymine residues, i.e., A = T, and the number of guanine residues is always equal to the number of cytosine residues, G = C. As a corollary, it is clear that the sum of purine residues equals the sum of pyrimidine residues, i.e., A + G = C + T.
5. The DNAs extracted from closely related species have similar base composition, whereas those of widely different species are likely to have widely different base composition.

NUCLEONS. Two nuclear particles, the proton and the neutron, and their antiparticles are known as *nucleons*. The rest mass of the proton is 1.0076 amu; that of the neutron, 1.0089 amu. The antiproton bears the same relation to the proton that the electron does to the positron, i.e., its charge is equal and opposite and its mass is the same, the charge being equal in magnitude to the electronic charge. Protons and antiprotons also annihilate each other when they collide, the reaction of a single pair producing positive and negative pions or kaons. If the proton and antiproton do not collide, but experience a "near miss," then an exchange of charge can occur, resulting in the formation of a neutron-antineutron pair. The four nucleon particles are fermions and have a spin angular momentum quantum number of $\frac{1}{2}$. See also **Neutron**; and **Particles (Subatomic)**.

NUCLEOPHILE. An ion or molecule that donates a pair of electrons to an atomic nucleus to form a covalent bond. The nucleus that accepts the electrons is called an *electrophile*. This occurs, for example, in the formation of acids and bases according to the Lewis concept, as well as in covalent bonding in organic compounds.

NUTRIENTS (Soil). Fertilizer.

NYLON. The first nylon developed (type 6/6) was discovered in 1938 by W. H. Carothers. Since that time, nylons have filled an important role for industry and the consumer in various formulations, shapes, and forms, e.g., oriented fibers, which are subsequently processed into fabrics, fishing line, and other monofilament uses; injection-molded nylons, used as bearings, gears, and other parts subjected to wear and impact; extruded nylon tubing and hose, used in large quantity because of its chemical inertness, high strength, and flexibility; oriented nylon strip used as strapping for packaging, displacing traditional steel strapping; and heavy cast-nylon parts, frequently used in the textile, paper-making, and bottle-handling fields.

Most nylons exhibit a combination of high melting point, high strength, impact resistance, wear resistance, chemical inertness, and a low coefficient of friction.

Types of Nylon. Type 6/6 and type 6 nylons are widely used, dominating the field of textile fibers. Nylon 6/10, a lower-strength material produced in less volume, is used for industrial applications requiring improved moisture stability and high dielectric strength. It also has a lower melting point, lower specific gravity, and higher cost than types 6/6 and 6. Nylons 11 and 12 appeared considerably later than the other formulations. Generally, these nylons have a lower order of moisture absorption and are thus preferred where consistent properties are required in the presence of moisture. They are also more chemically inert, flexible, and in certain cases, are transparent.

Formulations. Types 6/6 and 6/10 are formed by the condensation of diamines with dibasic organic acids into linear chains containing amide groups. Types 6, 11, and 12 are self-condensed amino acids.

Type 6/6:

$$NH_2(CH_2)_6NH_2 + HOOC(CH_2)_4COOH \rightarrow$$

Hexamethylenediamine   *Adipic acid*

$$[NH(CH_2)_6NHCO(CH_2)_2CO]_n + H_2O$$

Polyhexamethyleneadipamide

Type 6/10:

$$NH_2(CH_2)_6NH_2 + HOOC(CH_2)_8COOH \rightarrow$$

Hexamethylenediamine   *Sebacic acid*

$$[NH(CH_2)_6NHCO(CH_2)_8CO]_n + H_2O$$

Polyhexamethylenesebacamide

Type 6:

$$NH(CH_2)_5CO \rightarrow [NH(CH_2)_5CO]_n$$

ε-Caprolactam   *Polycaprolactam*

Type 11:

$$NH_2(CH_2)_{10}COOH \rightarrow [NH(CH_2)_{10}CO]_n + H_2O$$

Aminoundecanoic acid   *Polyaminoundecanamide*

Type 12:

$$NH(CH_2)_{11}CO \rightarrow [NH(CH_2)_{11}CO]_n$$

Laurolactam   *Polydodecanolactam*

In addition to the basic nylons, a variety of copolymers can be manufactured, some of which are commercially available. Nylons and nylon copolymers can be blended to form alloys with specific customized properties.

Nylons can be modified by the addition of certain plasticizers, fillers, reinforcements, and stabilizers. Ordinarily, nylons used for injection molding, such as type 6/6, have relatively low molecular weights (on the order of 15,000 to 20,000). High molecular weights are available to provide higher melt viscosity for nylon resins which are to be extruded into tubing or shapes. The molecular weight of nylon generally is determined by the ASTM relative-viscosity test.

Nylon resins usually are supplied in the form of cylindrical or rectangular diced pellets. Most commercial nylon molding resins are nontoxic. If a large amount of residual monomer is retained in the resin, as can occur with certain unextracted formulas, the material should not be in prolonged contact with food because of the possibility of monomer leaching.

Nylons require modification or stabilization to improve their resistance to certain environmental effects. Unstabilized nylon is degraded by ultraviolet light. The most widely used stabilizer has been approximately 2% well-dispersed carbon black, which has proved effective in the absorption of ultraviolet light. The nylons are considered adequate for outdoor applications if they are not exposed to direct sunlight.

See also **Caprolactam;** and **Fibers.**

O

OCEAN MINERALS. Since antiquity, the oceans and seas have been a major source of salt (NaCl) and continue to be so. Today, solar sea salt is produced in about sixty countries. Worldwide, as of the mid-1980s, slightly under 40% of the sodium chloride produced is evaporated from seawater.

Many of the ore deposits found on the continents are the result of ancient oceans. Tin is found in offshore deposits, such as in Indonesia, In Cornwall (Saint Ives Bay), and Phuket Island off the west coast of the Malay Peninsula.

For several decades, seawater has been a significant source of bromine (production commenced by DuPont in 1931), potassium, sulfur, and several other elements and their compounds. Today, over 13% of the requirements for bromine come from seawater, as do over 70% of magnesium metal and 33% of magnesium compounds. Sulfur, associated with the cap rock of salt domes, has been produced from two salt domes off the coast of Louisiana for many years. For the last few decades, the continental shelves have been producing large volumes of natural gas and petroleum. Although not a mineral, fresh water is derived from the seas by over 500 desalination plants in operation or under construction. See **Desalination.**

Within the past 30 years, much interest has been shown in manganese nodules on the sea floor in various locations. More recently, the discovery of hot brines in the Red Sea, "black smokers" on the East Pacific Rise and suspected in many other locations, and ophiolites has excited the scientific community and attracted industrialists because these phenomena are associated with metals such as cadmium, copper, nickel, and zinc. These findings have largely resulted from the funding provided for geological and oceanographic research as part of the Deep Sea Drilling Project (DSDP) and the International Decade of Ocean Exploration (IDOE), projects which were commenced in the late 1960s and some not fully completed as of the mid-1980s.

Beach sands also have received considerable attention in recent years as sources of metals and other materials. Marine beaches may contain gold, silver, platinum, and diamonds in addition to magnetite, cassiterite, chromite, columbite, ilmenite, rutile, scheelite, zircon, monazite, and wolframite. Heavy mineral beach sands are usually commercially worked for the titanium content of the rutile and the ilmenite. The same sands may also be processed to recover thorium from monazite and zircon for use in foundry sands. Currently, marine beaches are mined for heavy mineral production in Australia, Brazil, India, Madagascar, Mozambique, Sierra Leone, South Africa, and Sri Lanka, among other countries.

The resource potential of the oceans awaits further technological development. It is interesting to note that the famous German chemist, Fritz Haber, spent more than eight years after World War I in attempts to recover gold from seawater in order to pay the German war debt. The results were disappointing, but large quantities of gold are indeed in seawaters. Currently, there is considerable interest in attempts to recover uranium from seawater, particularly by nations with no assured supply. Should fusion power come to fruition, the oceans will be considered as a source of lithium to augment land-based sources. See **Lithium (For Thermonuclear Fusion Reactors).**

Diamonds are found in the seafloor sediments on the coast of the Kalahari Desert in southwest Africa.* The origin of the diamonds is obscure, but it is generally believed that basaltic and kimberlite pipes exist on the ocean floor as on the nearby land. There is a relative abundance of gemstones in the marine deposits and a few large stones have been recovered (Webb, 1965). Dredging began in 1961, using suction dredges capable of operating in waters to depths of 50 meters. Because of rough seas on this exposed coast, a number of barges were lost and the operation was concluded (Cruickshank, 1973). However, in recent years a subsidiary of DeBeers is using a dredge protected by a seawall, thus permitting mining offshore about 120 meters at depths of 90 meters. A capacity of 172,000 tons per month of diamondiferous gravel for processing has been reported.

Calcium carbonate often precipitates from tropical or subtropical waters when the water becomes supersaturated due to enrichment of the carbonate content by intense biological photosynthesis and by solar heating of carbon dioxide-rich cooler waters. The aragonite precipitates as single needles in the shallow waters at a rate of about one millimeter of wet sediment per year. Continuing deposition leads to cementation and the formation of successive concentric sheaths known as oöids (Cloud, 1965). The most extensively studied oölithic aragonite deposit is that distributed over the 250,000-square-kilometer (96,525-square-mile) Great Bahama Bank on the continental shelf near islands of the Bahamas. Most of the areas are less than 5 meters deep and are composed of quite pure calcium carbonate containing higher levels of strontium and uranium than are found in limestones of biological origin. Similar deposits occur in the Gulf of Batabanó (Cuba) and in the Mediterranean Sea off Egypt and Tunisia, as well as on the Trucial coast of the Persian Gulf (Calvert, 1965).

Iron is a common constituent of marine sediments. Magnetite is found in beach sands and iron is common in glauconitic marine silicates. Iron oxides and sulfides occur where anaerobic conditions and elevated temperatures are found, as in the hot, salty brines found near rifts. Iron is a major constituent of the ferromanganese nodules (Calvert, 1965).

Magnetite-rich iron sands have been dredged from the ocean floor just off Kagoshima Bay (Japan) in water averaging from 15 to 40 meters in depth. Iron sand concentrates were produced in Japan as recently as 1976, although a major marine iron sand operation in Kyushu ceased operation in 1966.

Marine sand and gravel for fill and for aggregate have been produced on all coasts of the United States, particularly from San Francisco and San Pedro Bays in California and from Long

* Acknowledgment of assistance obtained from W. F. McIlhenny, the Dow Chemical Company, Freeport, Texas in preparation of several of the following paragraphs is hereby made.

Island Sound. Marine sand and gravel are found in significant quantities in the United Kingdom.

Phosphorites (marine apatites) are dense, light-brown-to-black concretions, ranging in size from sands to nodules and irregular masses. Phosphorites have been found off Argentina, Chile, Japan, Mexico, Peru, South Africa, and Spain, and several islands in the Indian Ocean. Some also have been found off the west coast of North America and on the eastern North American continental shelf. These deposits occur where water upwelling transports phosphorus and where the rate of sedimentation is slow. The nodules are usually found as a monolayer on the surface. The mineralogy of the marine phosphorites is similar to western U.S. land deposits which were almost certainly marine in origin. Phosphorites are quite constant in composition, containing 45–47% calcium oxide and 29–30% phosphorus trioxide. Seawater is generally saturated with tricalcium phosphate so that, under the oxidative conditions normally present, the phosphates precipitate in colloidal form and accrete to existing surfaces, rather than forming a phosphorite suspension. Although most of the phosphorite is believed to have formed during the Miocene epoch, it is believed that precipitation is currently taking place. The largest known seafloor phosphorite deposit is off the coast of California from Point Reyes to the Gulf of California along the inner edge of the continental shelf. Additional deposits have been more recently found on the edges of the Blake Plateau east of Florida (Mero, 1965). A recovery project was commenced in 1962–1963, but failed to materialize.

Glauconite or green sands (a hydrated silicate with potassium, iron, and aluminum as cations) is widely distributed on the ocean floor in both ancient and more recent marine sediments. Glauconite is often found with phosphorite and occurs on the tops of banks, submerged hillcrests, and on slopes in water from 50 to 2000 meters in depth. Glauconites are known off the coasts of Africa, Australia, China, Japan, Portugal, South America, the United Kingdom (Scotland), the United States (California and the Atlantic shelf), and New Zealand. A 130-square-kilometer (50-square-mile) deposit has been identified on the Santa Monica shelf off California.

Submarine Hydrothermal Deposits

Discovery of the East Pacific Rise hot springs has created extensive interest and plans are underway to commence a four-year, multi-institutional project to explore the East Pacific Rise for additional areas of hot spring activity and ore deposition. The major objective of the program will be to examine the nature of hydrothermal processes along the mid-ocean ridge system from the slow-spreading to the very fast-spreading segments, such as at 10–30° South. The project will involve the use of surface ships, deeply towed instrument packages, new high-precision multibeam echo sounding for making highly accurate topographic maps of the seafloor, and ultimately manned submersibles, such as *Alvin*.

The knowledge of submarine hydrothermal deposits was advanced by a large measure in 1979 when the hot springs on the East Pacific Rise at 21° North were discovered. Unlike the warm springs discovered on the Galápagos Spreading Center a few years ago, the springs on the East Pacific Rise are hot, with water venting at temperatures as high as 350°C and at velocities of several meters per second. These formations are precipitating large quantities of sulfide ore and minerals rich in copper, zinc, and iron. The precipitates form chimneys around the individual vents that spout black or white smoke composed of precipitated crystals of sulfides and other minerals. The discovery is the most exciting and significant in this field since the discovery of the Red Sea hot brines and metal deposits (Mottl, 1980).

By *hydrothermal* is meant hot water. When deposits are formed by chemical precipitation from hot solutions, they are termed hydrothermal. Hydrothermal deposits on land represent a very important class of economically retrievable ore deposits and provide a significant percentage of various metals, such as copper, zinc, lead, silver, gold, tin, molybdenum, among others. Mottl (1980) suggests that five factors are involved in forming hydrothermal ore deposits: (1) A source of the ore metals; (2) a source of water that dissolves and later precipitates the metals, concentrating them during the total cycle; (3) a source of heat; (4) a pathway between the site where metals are dissolved and precipitated and the site where they are finally deposited which is permeable and permits solution flow; and (5) the ultimate collection or deposition site. For preservation, it is also important that ores be deposited in places where they will not be eroded away, as by weathering. Because so many factors are involved, there is a wide variety of hydrothermal ore deposits.

In terms of submarine hydrothermal ore deposits, the source of heat is the thermal energy associated with the formation of new oceanic lithosphere along the mid-ocean ridge system, where the seafloor is spreading apart and basaltic magma wells up. Because of tensional forces present, the newly formed crust becomes fractured, allowing seawater to percolate down through the fractures. During this percolation, the seawater is heated by contact with hot rock and commences to react with the rock, leaching metals that may be present. Because of the lesser density of the seawater (due to temperature), it rises and ascends to the seafloor and exists at submarine hot springs. Because there are several factors involved, there is, as on land, a wide variety of submarine hydrothermal ore deposits. Much remains to be understood and to confirm some of the early postulates, as given above, pertaining to the actual formation of submarine deposits.

For example, the concentrations of ore metals in most natural waters are quite low, particularly so in "normal" seawater. Measuring these low concentrations has been a problem of marine chemistry for many years. It is interesting to note that when artificial seawater, made up from pure reagent chemicals, is exposed to metallic elements, the ultimate solutions produced will contain from 100 to 1000 times the concentrations of these metals as compared with natural seawater.

As pointed out by Mottl (1980), the first submarine hot springs discovered along a mid-ocean ridge, those at the Galápagos Spreading Center, were emitting water at only 20°C, but the chemistry of this water indicated that it had reacted with basalt at 350–400°C. Then came the discovery of the 350° springs on the East Pacific Rise. Currently, the chemistry of this water is being studied at the Massachusetts Institute of Technology. To date, no submarine hydrothermal deposit has been sufficiently studied that all components contributing to its formation are known. Nevertheless, data at hand as of the early 1980s suggest some intriguing relationships among known deposits along mid-ocean ridges and point out the importance of special situations in producing and preserving large deposits. Considerable detail pertaining to current conceptual thinking is given in Mottl (1980) and in other references listed at the end of this entry.

Offshore Oil and Gas Resources

Although oil and gas exploration and production activities which occur offshore involve an extension of continents (the continental margins), they are nevertheless considered more in

Fig. 1. Frontier Outer Continental Shelf areas. (*Left*) Alaska; (*Right*) Contiguous United States, including Gulf of Mexico. (*After Edgar and Bayer, 1979.*)

the general terms of oceanological rather than continental resources (the latter generally considered land or above-sea-level resources). The centers of off-shore activity are indicated by the map of Fig. 1. Obviously a great deal of study, investigative work, and planning must take place to achieve successful (but often unsuccessful) drilling and production of gas and oil. Full advantage must be taken of all prior knowledge of a given region.

Manganese Nodules

Deep-sea nodules, comprised mainly of manganese and iron oxides, have been found in abundance over large areas of the deep ocean floor that have been examined to date. In some locations, the nodules have been found to contain generous proportions of nickel, copper, cobalt, molybdenum, and vanadium, as well as manganese and iron. From the standpoint of potential commercial exploitation, a deposit is not considered promising unless the nickel-copper content is about 1.8% (weight) or greater. On this basis, one of the most promising areas is the Clarion and Clipperton fracture zone, an immense area some 4400 kilometers (2730 miles) long and 900 kilometers (560 miles) wide at its widest point. This zone is located southeast of Hawaii and southwest of Baja California. See Fig. 2.

An investigation of the origin and distribution of manganese nodules and the processes by which they selectively concentrate copper, nickel, and other metals was one of the first major projects under the Seabeds Assessment project of the IDOE program. A workshop was held in 1972, during which over a hundred scientists from various countries selected the most

likely locations for the exploration and study of manganese nodules. The north central Pacific was identified as the zone where the nodules have the highest metal content. A team of American scientific investigators proposed that a comprehensive field and laboratory program be initiated to relate the high metal content to the local geological conditions. Data gathering was concerned along a transect that both academic and industrial scientists agreed could serve as a potential mining site. In addition to dredge sampling and piston cores, bathymetric measurements, side-scan sonar, and high-resolution television pictures were obtained. See Fig. 3. The results provided a broad-scale picture of the conditions under which nodules form, but the mechanisms for concentrating specific metals are still not well defined.

Although the nodules vary widely in their composition over the world oceans, metals are concentrated in three distinct types. One type comprises the nickel-copper-rich nodules of the Clarion-Clipperton variety, which is mainly formed in the equatorial regions. Another type, high in cobalt (1% or more) and low in nickel and copper, appears to be most commonly formed on sea mounts. The third type is high in manganese (35% or more), but low in other metals; it is known mainly on the eastern side of the Pacific Basin. As of the early 1980s, the most economically attractive are the cobalt-rich nodules.

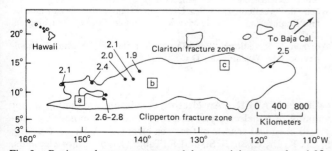

Fig. 2. Regions where manganese nodules containing more than 1.8% nickel-copper occur in the northeastern equatorial Pacific Ocean. Numbers indicate average percent of nickel-copper in one-degree squares. Areas a, b, and c indicate locations of activity carried out by Deepsea Ventures, Inc., or Deep Ocean Mining Environmental Studies program. (*After McKelvey, U.S. Geological Survey.*)

Fig. 3. Type of manganese nodules found in north Central Pacific Ocean zone. (*Woods Hole Oceanographic Institution.*)

The nodules form in a layered structure around a nucleus, which may be almost any material on the ocean floor. Most deep-sea nodules tend to be spherical or oblate in form. Nodules may occur up to 25 centimeters in diameter, but they average about 5 centimeters. The deposits may occur as slabs or agglomerates, or as incrustations on rocks or as pavement in some areas. The nodules are disorderly crystalline materials with layers of MnO_2 (mixed Mn^{2+}–Mn^{4+} oxides) alternating with $Mn(OH)_2$ and $Fe(OH)_3$. Excess iron appears as a mixture of goethite ($Fe_2O_3 \cdot H_2O$) and lepidocrocite.

The nodules are formed by the oxidation and precipitation of iron and manganese. The oxidation of Mn^{2+} is catalyzed by a reaction surface to a tetravalent state which absorbs additional Fe^{2+} or Mn^{2+} which, in turn, becomes oxidized. A surface is required and the initial deposition may be of iron oxide, possibly from volcanic or geothermal sources. Proper conditions of pH, redox potential, and metal ion concentration are found in deep ocean waters. The rate of accumulation appears to be very slow. The growth also may be discontinuous, and is estimated at a faster rate near the continental margins.

Precious Coral

Only a few species of coral have a combination of beauty, hardness, and luster, such the black coral species of Hawaii (*Antipathes dichotoma* and *Antipathes grandis*). These are highly valued by the jewelry trade. There are also a few red, pink, gold, and bamboo varieties that are in demand. Black coral also occurs in the Gulf of California and in the Pacific Ocean off Baja California, plus a few scattered locations in the Pacific Ocean east of Australia and north of New Zealand. Traditional sources of red and pink corals have been the Mediterranean Sea, various locations in the western Pacific Ocean, ranging from the Philippines, the Ryuku and Bonin islands and south of Japan. There is also a string of precious red and pink coral beds northwest of Hawaii and off the Cape Verde islands in the Atlantic Ocean off west Africa. A submersible vessel, the *Star II*, operated by Maui Divers of Hawaii, Ltd., is used to harvest pink coral (*Corallium secundum*) from the Makapuu bed. State regulations permit the collection of only 4400 pounds (1996 kilograms) within a 2-year period.

Legal Aspects of Ocean-Bed Exploitation

Although the political aspects of science and technology are not generally within the sphere of this encyclopedia, the reader may wish to refer to the McKelvey (1980) and Legatski (1980) references listed at the end of this entry. The ultimate Law of the Sea will affect not only commercial operations, but scientific investigations of the sea as well.

References

Bischoff, J. L., and W. E. Seyfried: "Hydrothermal Chemistry of Seawater from 25° to 350°C," *Amer. J. Sci.*, **278**, 838–860 (1978).

Bolger, G. W., Betzer, P. R., and V. V. Gordeev: "Hydrothermally Derived Manganese Suspended Over the Galápagos Spreading Center," *Deep-Sea Res.*, **25**, 721–734 (1978).

Calvert, S. E.: "The Mineralogy and Geochemistry of Near-Shore Sediments," in "Chemical Oceanography," Vol. 6 (J. P. Riley and G. Skirrow, editors), Academic, London, 1965.

Cloud, P. E.: "Carbonate Precipitation and Dissolution in the Marine Environment," Chap. 17 in "Chemical Oceanography," Vol. 1 (J. P. Riley and G. Skirrow, editors), Academic, London, 1965.

Considine, D. M.: "Natural Gas Production and Reserves in the United States," in "Energy Technology Handbook," (D. M. Considine, editor), pages 2–9 to 2–19, McGraw-Hill, New York, 1977.

Considine, D. M.: "Natural Gas Reserves in Canada," in "Energy Technology Handbook," (D. M. Considine, editor), pages 2–34 to 2–37, McGraw-Hill, New York, 1977.

Considine, D. M.: "World Natural Gas Production and Reserves," in "Energy Technology Handbook," (D. M. Considine, editor), pages 2–38 to 2–46, McGraw-Hill, New York, 1977.

Considine, D. M.: "Petroleum Production and Reserves in the United States," in "Energy Technology Handbook," (D. M. Considine, editor), pages 3–45 to 3–77, McGraw-Hill, New York, 1977.

Considine, D. M.: "World Petroleum Production and Reserves," in "Energy Technology Handbook," (D. M. Considine, editor), pages 3–78 to 3–95, McGraw-Hill, New York, 1977.

Cruickshank, M. J.: "Marine Mining," Chap. 20 in "SME Mining Engineering Handbook," Society of Mining Engineers, New York, 1973.

Edgar, N. T., and K. C. Bayer: "Assessing Oil and Gas Resources on the U.S. Continental Margin," *Oceanus*, **22**, 3, 12–22 (1979).

Edmond, J. M., et al.: "On the Formation of Metal-Rich Deposits at Ridge Crests," *Earth Planet. Sci. Lett.*, **46**, 19–30 (1979).

Faulkner, D. J., and W. F. Fenical (editors): "Marine Natural Products Chemistry," Plenum, New York, 1977.

Faulkner, D. J.: "The Search for Drugs from the Sea," *Oceanus*, **22**, 2, 44–50 (1979).

Grigg, W. R.: "Precious Corals: Hawaii's Deep-sea Jewels," *National Geographic*, **155**, 5, 718–732 (1979).

Hahlbrock, U.: "Mining Metalliferous Muds in the Red Sea," *Ocean Industry*, **14**, 5, (1979).

Kaul, P. N. (editor): "Food-drugs from the Sea—Myth or Reality," Univ. Oklahoma Press, Norman, Oklahoma, 1979.

Klinkhammer, G., Bender, M., and R. F. Weiss: "Hydrothermal Manganese in the Galapagos Rift," *Nature*, **269**, 319–320 (1977).

Legatski, R. A.: "Law of the Sea: Are American Interests being Sacrificed to Achieve a Treaty?" *Ocean Industry*, **15**, 4 (1980).

Mattick, R. E., et al.: "Petroleum Potential of the U.S. Atlantic Slope, Rise, and Abyssal Plain," *AAPG*, **62**, 4, 592–608 (1978).

McKelvey, V. E.: "Seabed Minerals and the Law of the Sea," *Science*, **209**, 464–472 (1980).

Mero, J.: "The Mineral Resources of the Sea," Elsevier, New York, 1965.

Mottl, M., Holland, H. D., and R. F. Corr: "Chemical Exchange During Hydrothermal Alteration of Basalt by Seawater," *Geochim. Cosmochim. Acta*, **43**, 869–884 (1980).

Mottl, M. J.: "Submarine Hydrothermal Ore Deposits," *Oceanus*, **23**, 2, 18–27 (1980).

Pearson, J. S.: "Ocean Floor Mining," Noyes Data Corp., 1975.

Rassinier, E. A., and P. N. Glover: "Potential Supply of Natural Gas in the United States," in "Energy Technology Handbook," (D. M. Considine, editor), pages 2–20 to 2–33, McGraw-Hill, New York, 1977.

Ryan, W. B. F., et al.: "Bedrock Geology in New England Submarine Canyons," *Oceanol. Acta*, **1**, 2, 233–254 (1978).

Schlee, J. S., et al.: "Regional Geologic Framework Off Northeastern United States," *AAPG*, **60**, 926–951 (1976).

Staff: "Hot Springs and Geophysical Experiments on the East Pacific Rise," *Science*, **207**, 1421–1433 (1980).

Webb, B.: "Technology of Sea Diamond Mining," in "Ocean Science and Engineering," Vol. 1, Marine Technology Society, Washington, D.C., 1965.

Worthen, L. R. (editor): "Food-Drugs from the Sea," Marine Technology Society, Washington, D.C., 1973.

OCEAN WATER. An electrolyte solution containing minor amounts of nonelectrolytes and composed predominantly of dissolved chemical species of fourteen elements O, H, Cl, Na, Mg, S, Ca, K, Br, C, Sr, B, Si, and F (Table 1). The minor elements, those that occur in concentrations of less than 1 ppm by weight, although unimportant quantitatively in determining the physical properties of sea water, are reactive and are important in organic and biochemical reactions in the oceans.

TABLE 1.   ABUNDANCES OF THE ELEMENTS AND PRINCIPAL DISSOLVED CHEMICAL SPECIES
OF SEAWATER, RESIDENCE TIMES OF THE ELEMENTS

Element	Abundance (mg/l)	Principal species	Residence time (years)
O	857,000	H_2O; $O_2(g)$; SO_4^{2-} and other anions	
H	108,000	H_2O	
Cl	19,000	Cl^-	
Na	10,500	Na^+	2.6×10^8
Mg	1,350	Mg^{2+}; $MgSO_4$	4.5×10^7
S	885	SO_4^{2-}	
Ca	400	Ca^{2+}; $CaSO_4$	8.0×10^6
K	380	K^+	1.1×10^7
Br	65	Br^-	
C	28	HCO_3^-; H_2CO_3; CO_3^{2-}; organic compounds	
Sr	8	Sr^{2+}; $SrSO_4$	1.9×10^7
B	4.6	$B(OH)_3$; $B(OH)_2O^-$	
Si	3	$Si(OH)_4$; $Si(OH)_3O^-$	8.0×10^3
F	1.3	F^-; MgF^+	
A	0.6	$A(g)$	
N	0.5	NO_3^-; NO_2^-; NH_4^+; $N_2(g)$; organic compounds	
Li	0.17	Li^+	2.0×10^7
Rb	0.12	Rb^+	2.7×10^5
P	0.07	HPO_4^{2-}; $H_2PO_4^-$; PO_4^{3-}; H_3PO_4	
I	0.06	IO_3^-; I^-	
Ba	0.03	Ba^{2+}; $BaSO_4$	8.4×10^4
In	<0.02		
Al	0.01	$Al(OH)_4^-$	1.0×10^2
Fe	0.01	$Fe(OH)_3(s)$	1.4×10^2
Zn	0.01	Zn^{2+}; $ZnSO_4$	1.8×10^5
Mo	0.01	MoO_4^{2-}	5.0×10^5
Se	0.004	SeO_4^{2-}	
Cu	0.003	Cu^{2+}; $CuSO_4$	5.0×10^4
Sn	0.003	(OH)?	5.0×10^5
U	0.003	$UO_2(CO_3)_3^{4-}$	5.0×10^5
As	0.003	$HAsO_4^{2-}$; $H_2AsO_4^-$; H_3AsO_4; H_3AsO_3	
Ni	0.002	Ni^{2+}; $NiSO_4$	1.8×10^4
Mn	0.002	Mn^{2+}; $MnSO_4$	1.4×10^3
V	0.002	$VO_2(OH)_3^{2-}$	1.0×10^4
Ti	0.001	$Ti(OH)_4$?	1.6×10^2
Sb	0.0005	$Sb(OH)_6^-$?	3.5×10^5
Co	0.0005	Co^{2+}; $CoSO_4$	1.8×10^4
Cs	0.0005	Cs^+	4.0×10^4
Ce	0.0004	Ce^{3+}	6.1×10^3
Kr	0.0003	$Kr(g)$	
Y	0.0003	(OH)?	7.5×10^3
Ag	0.0003	$AgCl_2^-$; $AgCl_3^{2-}$	2.1×10^6
La	0.0003	La^{3+}; $La(OH)^{2+}$?	1.1×10^4
Cd	0.00011	Cd^{2+}; $CdSO_4$	5.0×10^5
Ne	0.0001	$Ne(g)$	
Xe	0.0001	$Xe(g)$	
W	0.0001	WO_4^{2-}	1.0×10^3
Ge	0.00007	$Ge(OH)_4$; $Ge(OH)_3O^-$	7.0×10^3
Cr	0.00005	(OH)?	3.5×10^2
Th	0.00005	(OH)?	3.5×10^2
Sc	0.00004	(OH)?	5.6×10^3
Ga	0.00003	(OH)?	1.4×10^3
Hg	0.00003	$HgCl_3^-$; $HgCl_4^{2-}$	4.2×10^4
Pb	0.00003	Pb^{2+}; $PbSO_4$	2.0×10^3
Bi	0.00002		4.5×10^5
Nb	0.00001		3.0×10^2
Tl	<0.00001	Tl^+	
He	0.000005	$He(g)$	
Au	0.000004	$AuCl_2^-$	5.6×10^5
Be	0.0000006	(OH)?	1.5×10^2
Pa	2.0×10^{-9}		
Ra	1.0×10^{-10}	Ra^{2+}; $RaSO_4$	
Rn	0.6×10^{-15}	$Rn(g)$	

Adapted from Goldberg, E. D., "Minor elements in sea water," in "Chemical Oceanography," v. 1, pp. 164–165, J. P. Riley and G. Skirrow, Eds., Academic Press, New York, 1965.

Dissolved Species. The form in which chemical analyses of sea water are given records the history of our thought concerning the nature of salt solutions. Early analytical data were reported in terms of individual salts NaCl, CaSO₄, and so forth. After development of the concept of complete dissociation of strong electrolytes, chemical analyses of sea water were given in terms of individual ions Na⁺, Ca⁺⁺, Cl⁻, and so forth, or in terms of *known* undissociated and partly dissociated species, e.g., HCO₃⁻. In recent years there has been an attempt to determine the thermodynamically stable dissolved species in sea water and to evaluate the relative distribution of these species at specified conditions. Table 1 lists the principal dissolved species in sea water deduced from a model of sea water that assumes the dissolved constituents are in homogeneous equilibrium, and (or) in equilibrium, or nearly so, with solid phases.

Both associated and nonassociated electrolytes exist in sea water, the latter (typified by the alkali metal ions Li^+, Na^+, K^+, Rb^+, and Cs^+) predominantly as solvated free cations. The major anions, Cl^- and Br^-, exist as free anions, whereas as much as 20% of the F in sea water may be associated as the ion-pair MgF^+ and IO_3^- may be a more important species of I than I^-. Based on dissociation constants and individual ion activity coefficients the distribution of the major cations in sea water as sulfate, bicarbonate, or carbonate ion-pairs has been evaluated at specified conditions by Garrels and Thompson (1962).

About 10% each of Mg and Ca is tied up as the sulfate ion-pair. It is likely that the other alkaline earth metals, Sr, Ba, and Ra, also exist in sea water partly as undissociated sulfates; about 60% and 21%, respectively, of the total $SO_4^=$ and HCO_3^- are complexed with cations, and two-thirds of the $CO_3^=$ is present as the ion-pair $MgCO_3^0$.

The activities of Mg^{++} and Ca^{++} obtained from the model of sea water proposed by Garrels and Thompson have recently been confirmed by use of specific Ca^{++} and Mg^{++} ion electrodes, and for Mg^{++} by solubility techniques and ultrasonic absorption studies of synthetic and natural sea water. The importance of ion activities to the chemistry of sea water is amply demonstrated by consideration of $CaCO_3$ (calcite) in sea water. The total molality of Ca^{++} in surface sea water is about 10^{-2} and that of $CO_3^=$ is 3.7×10^{-4}; therefore the ion product is 3.7×10^{-6}. This value is nearly 600 times greater than the equilibrium ion activity product of $CaCO_3$ of 4.6×10^{-9} at 25°C and one atmosphere total pressure. However, the activities of the free ions Ca^{++} and $CO_3^=$ in surface sea water are about 2.3×10^{-3} and 7.4×10^{-6}, respectively; thus the ion activity product is 17×10^{-9} which is only 3.7 times greater than the equilibrium ion activity product of calcite. Thus, by considering activities of sea water constituents rather than concentrations, we are better able to evaluate chemical equilibria in sea water; an obvious restatement of simple chemical theory but an often neglected concept in sea water chemistry.

Constancy and Equilibrium. The concept of constancy of the chemical composition of sea water, i.e., that the ratios of the major dissolved constituents of sea water do not vary geographically or vertically in the oceans except in regions of runoff from the land or in semienclosed basins, was first proposed indirectly in 1819 by Marcet and expanded later by Forchammer and Dittmar. The concept was established on a purely empirical basis whereas in actual fact there is a theoretical basis for the concept.

Barth (1952) proposed the concept of residence (passage) time of an element in the oceanic environment and formalized this concept by the equation

$$\lambda = \frac{A}{dA/dt},$$

where λ is the residence time of the element, A is the total amount of the element in the oceans, and dA/dt is the amount of the element introduced or removed per unit time. Sea water is assumed to be a steady-state solution in which the number of moles of each element in any volume of sea water does not change; the net flow into the volume exactly balances the processes that remove the element from it. Complete mixing of the element in the ocean is assumed to take place in a time interval that is short compared to its residence time. Table 1 shows the residence times of the elements, and Table 2 compares the residence times of some elements on the basis of river input and removal by sedimentation. For the major elements the results are strikingly similar and suggest that at least as a first approximation sea water is a steady-state solution with a composition fixed by reaction rates involving the removal of elements from the ocean approximately equalling rates of element inflow into the ocean. Thus, as a first approximation the steady-state oceanic model implies a fixed and constant sea water composition and provides a theoretical basis for the concept of the constancy of the chemical composition of sea water. However, it is possible that at any time, t_0, for example, the present, the ratios of the major dissolved constituents in the open ocean may be nearly invariant simply because the amounts of new materials introduced by streams and other agents to the ocean are small compared to the amounts in the ocean, and these new materials are mixed into the oceanic system relatively rapidly. But over time periods of 1000 to 2000 years or more the major ionic ratios can only remain constant if the ocean is a steady-state solution whose composition is controlled by mechanism(s) other than simple mixing.

TABLE 2. THE RESIDENCE TIMES OF ELEMENTS IN SEAWATER CALCULATED BY RIVER INPUT AND SEDIMENTATION

Element	Amount in ocean (in units of 10^{20} g)	Residence time in millions of years	
		River input	Sedimentation
Na	147.8	210	260
Mg	17.8	22	45
Ca	5.6	1	8
K	5.3	10	11
Sr	0.11	10	19
Si	0.052	0.035	0.01
Li	0.0023	12	19
Rb	0.00165	6.1	0.27
Ba	0.00041	0.05	0.084
Al	0.00014	0.0031	0.0001
Mo	0.00014	2.15	0.5
Cu	0.000041	0.043	0.05
Ni	0.000027	0.015	0.018
Ag	0.0000041	0.25	2.1
Pb	0.00000041	0.00056	0.002

After Goldberg, E. D., "Minor elements in sea water," in "Chemical Oceanography," v. 1, p. 173, J. P. Riley and G. Skirrow, Eds., Academic Press, New York, 1965..

Further insight into the constancy concept can be gained by exploring possible mechanisms governing the steady-state composition of sea water. The steady-state solution could be simply a result of the rates of major element inflow into the oceans being equal to rates of outflow by biologic removal, flux through the atmosphere, adsorption on sediment particles, and removal in the interstitial waters of marine sediments. For exam-

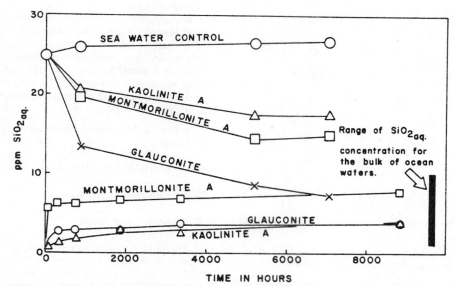

Fig. 1.   Concentration of dissolved silica as a function of time for suspensions of silicate minerals in sea water. Curves are for 1-g ($<62\mu$) mineral samples in silica-deficient (SiO_2 in water was initially 0.03 ppm) and silica-enriched (SiO_2 was initially 25 ppm) sea water at room temperature. Notice that the minerals react rapidly and that the dissolved silica concentration for individual minerals becomes nearly constant at values within or close to the range of silica concentration in the oceans (from Mackenzie, F. T., Garrels, R. M., Bricker, O. P., and Bickley, F., "Silica in sea water: Control by silica minerals" *Science*, **155**, 1404 (1967)).

ple, Ca^{++} carried to the oceans by streams is certainly removed, in part, in sea aerosol generated at the atmosphere-ocean interface and transported into the atmosphere, later to fall as rain or dry fallout on the continents. However, recent theoretical and experimental work suggests that sea water may be modeled as a steady-state solution in equilibrium with the solids that are in contact with it. Sillén has modeled the oceanic system as a near-equilibrium of many solid phases and sea water. Experimental work has shown that aluminosilicate minerals typical of those in the suspended load of streams and in marine sediments react rapidly with sea water containing an excess or deficiency of dissolved silica. Reactions involving these aluminosilicates may control on a long-term basis the activities of H_4SiO_4 and other constituents in sea water. Thus it has begun to emerge that the composition of the oceans represents an approximation of dynamic equilibrium between the water and the solids that are carried into it in suspension or are precipitated from it by the continuous evaporation and renewal by streams. Therefore, if sea water is a solution in equilibrium with solid phases, or even closely approaches such a system, then the *ion activity ratios* of the major dissolved species would be fixed and the chemical composition of the ocean would be "constant." Consequently, the activity of Ca^{++} in the ocean is not simply a result of removal processes involving sea aerosol, adsorption and so forth but is controlled by solid-solution equilibria. A model leading to nearly invariant ion activity ratios geographically and vertically at any time, t_0, in the oceans based on mixing rates alone may be sufficient to explain the constancy of sea water composition but is somewhat misleading and uninformative when considered in light of the recent advances in treating the oceans as an equilibrium system.

Some limitations of the equilibrium model of sea water do exist. Sillén has pointed out that based on equilibrium calculations all the nitrogen in the ocean-atmosphere system should be present as NO_3^- in sea water; however, most of the nitrogen is present as N_2 gas in the atmosphere. Also, the concentrations of the major alkaline earth elements, Mg, Ca, and Sr, in sea water may vary slightly with depth or geographic location.

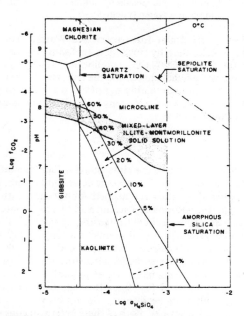

Fig. 2.   Logarithmic activity diagram depicting equilibrium phase relations among aluminosilicates and sea water in an idealized nine-component model of the ocean system at the noted temperatures, one atmosphere total pressure, and unit activity of H_2O. The shaded area represents the composition range of sea water at the specified temperature, and the dot-dash lines indicate the composition of sea water saturated with quartz, amorphous silica, and sepiolite, respectively. The scale to the left of the diagram refers to calcite saturation for different fugacities of CO_2. The dashed contours designate the composition (in % illite) of a mixed-layer illite-montmorillonite solid solution phase in equilibrium with sea water (from Helgeson, H. C. and Mackenzie, F. T., 1970, Silicate-sea water equilibria in the ocean system: Deep Sea Res.).

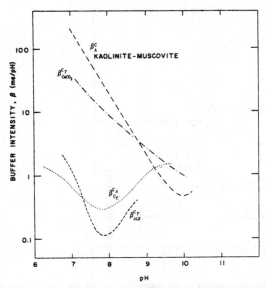

Fig. 3. Buffer intensity as a function of pH for some homogeneous and heterogeneous chemical systems. The buffer intensities are defined for $\beta^{C_A}_{\text{Kaolinite-muscovite}}$, addition of a strong acid (or base) to sea water in equilibrium with kaolinite and muscovite; $\beta^{C_T}_{\text{CaCO}_3}$, addition of CO_2 in a sea water system of zero noncarbonate alkalinity in equilibrium with $CaCO_3$; $\beta^{C_A}_{C_T}$, addition of a strong acid (or base) to a sea water solution of constant total dissolved carbonate; and $\beta^{C_T}_{\text{Alk}}$, addition of total CO_2 to a sea water solution of constant alkalinity (data from Morgan, J. J., preprint, "Applications and limitations of chemical thermodynamics in natural water systems").

Buffering and Buffer Intensity of Sea Water. The view has long been held that hydrogen-ion buffering in the oceans is due to the CO_2–HCO_3^-–$CO_3^=$ equilibrium. Within recent years this view has been challenged, and the importance of aluminosilicate equilibria in maintaining the pH of sea water emphasized. The buffer intensity of a system is of thermodynamic nature and is defined as

$$\beta^{c_i}_{c_j} = \frac{dC_i}{dpH},$$

where $\beta^{c_i}_{c_j}$ is the pH buffer intensity for incremental addition of C_i to a closed system of constant C_j at equilibrium. Homogeneous buffer intensities are defined for systems without solid phases, e.g., the addition of a strong acid to a carbonate solution, whereas heterogeneous buffer intensities are defined for systems with solid phases, e.g., the addition of a strong acid to a solution in equilibrium with calcite, $CaCO_3$, or with kaolinite and muscovite. The homogeneous buffer intensities for the range of sea water and interstitial marine water pH values (7.0 to 8.3) are about 10- to 100-fold less than the heterogeneous intensities involving equilibrium between calcite and sea water or kaolinite, muscovite, and sea water. Both of these heterogeneous equilibria represent large capacities for resistance to sea water pH changes. Unfortunately, the kinetic aspects of these buffer systems have not been investigated quantitatively. However, it is apparent that aluminosilicate equilibria have buffer intensities equal to and perhaps greater than (the buffer intensities of most aluminosilicate equilibria in natural waters have only been qualitatively evaluated) the CO_2—$CaCO_{3(s)}$ equilibria in sea water. Small additions of acid or base to the oceans could be buffered by the homogeneous equilibrium CO_2—HCO_3^-—$CO_3^=$. However, large incremental additions of acid or base or additions over a duration of time would involve the heterogeneous carbonate

and aluminosilicate equilibria; the relative importance of each would depend on the buffer intensities of the various equilibria and the relative rates of aluminosilicate and carbonate reactions.

For geologically short-term processes on the order of a few thousands of years, it is likely that the carbon dioxide-carbonate system regulates oceanic pH. The long-term pH is controlled by an interplay of various near-equilibria involving carbonates and silicates.

—Fred T. Mackenzie, Northwestern University, Evanston, Illinois.

OCTANE NUMBER. Petroleum.

OIL (Petroleum). Petroleum.

OIL SANDS. Tar Sands.

OIL SHALE. A carbonaceous rock that yields petroleumlike materials when heated to pyrolysis temperatures in the range of 427–538°C. The oil precursor in the rock is a high-molecular-weight organic polymer known as *kerogen*, an elemental analysis of which shows: C = 80.5%; H = 10.3%; N = 2.4%; S = 1.0%; O = 5.8% (weight). In the United States, there are vast shale oil deposits in Colorado and Utah, as well as in the east and midwest. Potential oil content of these shales is estimated in the trillions of barrels, or many times greater than traditional petroleum reserves. Interest in shale oil tends to cycle with the intensity of the energy supply and cost situation. The factors that tend to limit research and development are technical, economic, political, and environmental. The two fundamental recovery approaches considered over a period of many years are (1) surface recovery in some form of retort, requiring the mining and handling of materials that rival and exceed current production of coal—with a high ratio of tonnage of overburden and raw materials for tonnage of final product, and (2) in-situ recovery. In terms of the latter approach, many oil shales are quite impermeable to passage of liquids or gases and thus the shale bed has to be fractured hydraulically or with explosives so that a flame front can pass through the bed to convert the kerogen and thus release the oil.

References

Atwood, M. T.: "Oil Shale Retorting," in "Energy Technology Handbook," (D. M. Considine, Editor), McGraw-Hill, New York (1977).
Berry, K. L.: "Combined Retorting Technique for Oil Shale," *Chem. Eng. Progress*, **75**, 9, 72–77 (1979).
McNamara, P. H., and J. P. Humphrey: "Hydrocarbons from Eastern Oil Shale," *Chem. Eng. Progress*, **75**, 9, 87–92 (1979).
Sundquist, E. T., and G. A. Miller: "Oil Shales and Carbon Dioxide," *Science*, **208**, 740–741 (1980).

OIL (Vegetable). Vegetable Oils (Edible).

OLEIC ACID. $CH_3(CH_2)_7CH:CH(CH_2)_7 \cdot COOH$, formula weight 282.45, colorless liquid, mp 14°C, bp 286°C, sp gr 0.854. Sometimes referred to as red oil, elaine oil, or octadecenoic acid, this compound is insoluble in H_2O, but miscible with alcohol or ether in all proportions. Oleic acid solidifies into colorless needle crystals.

Oleic acid differs from stearic acid chemically by possessing 33 instead of 35 hydrogen atoms in the radical $C_{17}H_{33} \cdot COOH$ (oleic acid), $C_{17}H_{35}COOH$ (stearic acid). It is possible to convert oleic acid and oleate esters into stearic acid and stearate esters by treatment with hydrogen gas in the presence of finely divided nickel as a catalyzer at 250°C under pressure as in the hydrogenation of oils and fats. Either by careful oxidation, or by addition of ozone and splitting, oleic acid yields products of 9 carbon atoms, thus leading to the conclusion that the double bond is in the center of the carbon chain. Oleic acid adds bromine or iodine in definite amounts to confirm the conclusion that one double bond is contained. Nitric acid converts oleic acid into elaidic acid $C_{17}H_{33}COOH$, mp 51°C (oleic and elaidic acids are related, cis- and trans-, as maleic and fumaric acids).

Oleic acid may be obtained from glycerol trioleate, present in many liquid vegetable and animal non-drying oils, such as olive, cottonseed, lard, by hydrolysis. The crude oleic acid after separation of the water solution of glycerol is cooled to fractionally crystallize the stearic and palmitic acids, which are then separated by filtration, and fractional distillation under diminished pressure. Oleic acid reacts with lead oxide to form lead oleate, which is soluble in ether, whereas lead stearate or palmitate is insoluble. From lead oleate oleic acid may be obtained by treatment with H_2S (lead sulfide, insoluble solid, formed). With sodium oleate, a soap is formed. Most soaps are mixtures of sodium stearate, palmitate, and oleate.

Representative esters of oleic acid are: methyl oleate $C_{17}H_{33}COOCH_3$ bp 190°C at 10 millimeters pressure; ethyl oleate $C_{17}H_{33}COOC_2H_5$ bp 205°C at 10 millimeters pressure; glyceryl trioleate (triolein) $C_3H_5(COOC_{17}H_{33})_3$ bp 240°C at 18 millimeters pressure.

Oleic acid is used in the preparation of metallic oleates, such as aluminum oleate for thickening lubricating oils, for waterproofing materials, and for varnish driers. The glyceryl ester oleic acid is one of the constituents of many vegetable and animal oils and fats.

See also **Vegetable Oils (Edible).**

OLIVINE. The mineral olivine is a silicate of magnesium and ferrous iron corresponding to the formula $(Mg,Fe)_2(SiO)_4$. Olivine is the group name for the isomorphous series from forsterite, Mg_2SiO_4, to fayalite, Fe_2SiO_4. The ratio of magnesium to iron varies considerably, but the more common olivines are richer in Mg than in Fe. Olivine crystallizes in the orthorhombic system, usually in flattened prismatic forms, also granular and massive. It has a conchoidal fracture and is rather brittle; hardness, 6.5–7; specific gravity, 3.22–4.39; luster, vitreous; color, olive to gray, green; may be yellowish-brown from the oxidation of the iron. It is transparent to translucent. Olivine occurs both in igneous rocks as a primary mineral and in certain rocks of metamorphic origin. It has also been discovered in meteorites.

Olivine crystallizes from magmas that are rich in magnesia and low in silica and which form such rocks as gabbros, norites, peridotites and basalts. The metamorphism of impure dolomites or other sediments in which the magnesia content is high and silica low seems to produce olivine.

Transparent olivines of good color are sometimes used as a gem, often called *peridot*, the French word for olivine; it is also called chrysolite from the Greek meaning gold, and stone. Olivine occurs in the lavas of Vesuvius and Monte Somma and in the Eifel district of Germany. Gem material comes from St. John's Island in the Red Sea, Upper Burma, and from Minas Geraes, Brazil. In the United States olivine localities are Orange County, Vermont; Webster and Jackson counties, North Carolina. Arizona and New Mexico have also furnished some gem material.

See also **Peridotite.**

OPAL. The mineral opal, long classified as an amorphous mineral gel, has been found by x-ray analysis to consist of a microcrystalline aggregate of crystallites of cristobalite. On this basis, opal may be considered as a variety of cristobalite bearing the same relative relationship to that mineral as chalcedony does to quartz. Opal is hydrous silica, $SiO_2 \cdot nH_2O$, with variable water content. It never occurs in crystal form; usually as irregular veins or masses, or as pseudomorphous replacements after wood or fossilized material such as bones and shells. Opaline silica occurs in many forms; geyserite from geyser deposits, siliceous sinter (fiorite) from siliceous waters of hot springs, and diatomite (diatomaceous earth) from siliceous shells of diatoms and comparable microscopic species. It has a conchoidal fracture; hardness 5.5–6.5; specific gravity 2.1–2.3; luster, vitreous or greasy to dull; color, very variable, colorless, white, milky-blue, gray, red, yellow, green, brown, and black. Often a beautiful play of colors may be observed in the gem varieties. The color play in opals is attributed to three different mechanisms: finely divided pigmentation of foreign material; light interference by open spaced grid or cristobalite crystallization; and reflected light. It may well be that two or all three causes contribute to the color effect in any given opal specimen. Until a more complete understanding of opal color is established these phases seem to be of prime significance.

Besides the gem varieties which show the delicate play of colors, there are other kinds of common opal such as the milk opal, a milky bluish to greenish kind; resin opal, which is honey-yellow with a resinous luster; wood opal, resulting from the replacement of the organic matter of wood by opal, and hyalite, a colorless glass-clear opal sometimes called Muller's Glass. Opal is deposited at relatively low temperatures and may occur in the fissures of almost any type of rock. Hungary, Australia, Honduras, Mexico and in the United States Nevada and Idaho, have been the sources of gem opals. Hyalite comes from Czechoslovakia, Mexico, Japan and British Columbia. Other common varieties of opal are widespread in their occurrence. The word opal is derived from the Latin *opallus*.

OPTICAL FIBERS. This entry describes optical fibers and glasses used for fiber waveguide systems, also called fiber optic systems. Optical fibers act as a pipeline for light and are designed in such a way that very little light can leak out of the sidewalls from the transparent materials of which they are made. Because of this, optical fibers find many applications. Glass fibers are the most appropriate medium to guide light from end to end, because glass properly manufactured from very pure raw materials absorbs less of the light passing through it than does any other practical material.

Light Transmission Through Fibers. A light beam (or ray) passes through one common type of fiber, known as the *step index* design (Fig. 1) by traveling through its central glass core and richocheting off the interface of the cladding adhering to and surrounding the core. This core/cladding interface acts as a cylndrical mirror that returns light to the core by a process known as total internal reflection. To insure that total internal reflection occurs, fibers are usually made from two glasses, one for the core which has a relatively higher refractive index, the other a clad glass or possibly a plastic layer surrounding the core that has a somewhat lower refractive index. When the seal interface between core and clad is essentially free of imperfections and the relative refractive indices of the glasses used are proper, many millions of internal reflections are possible and light can travel through many kilometers of fiber and emerge from the far end with only a modest loss in brightness or intensity.

Another type of optical fiber is one having a *graded index profile*, shown in Fig. 2, that guides light through it by means of refraction or bending to refocus it about the center axis of

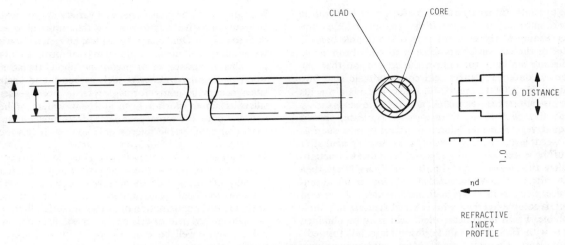

Fig. 1.   Step index optical fiber design.

the fiber core. Here each layer of glass from the center of the fiber to the outside extremity has a slightly decreased refractive index from the layer preceding it. This type fiber construction causes the light ray to move through it in the form of a sinusoidal or snakelike curve rather than the zig-zag fashion in the step index variety. With this type of fiber, when physical design is proper and the glass flaws are limited, light can also be conducted through very long distances without severe loss because the light is trapped inside and guided in an efficient manner.

Characteristics Affecting Optical Fiber Performance

The *fiber core* (Fig. 1) is that portion of an optical fiber that conducts the light from one fiber end to the other. The glass materials used are generally selected fro their transmission capability at the wavelength of light to be transmitted. Fiber core diameters range from 6 to about 250 micrometers, depending upon fiber design.

Fiber Cladding. To help retain the light being conducted within the core, a layer surrounding the core of an optical fiber is required. Glass is the preferred material for the cladding, although plastic clad silica fibers are common in less demanding applications. The cladding thickness may vary from 10 to about 150 micrometers, depending upon design. The cladding layer

is shown in the cross sectional view of an optical fiber (Fig. 1).

Basic fiber Designs. These fall into two catagories:

(1) *Multi-mode optical fibers* are made with a core of large diameter and normally have a relatively high numerical aperture. They have a step index profile, and are able to accept large amounts of light and readily transmit light from noncoherent sources. The disadvantage of this design is the tendency for light pulse spread and resulting low bandwidth.

(2) *Single-mode optical fibers* have a step index profile, but are designed for accepting the light mode that essentially only travels parallel with the fiber core. This is accomplished by making a very small core diameter of approximately 6 micrometers and/or selecting glasses for core and clad that yield a small numerical aperture. These fibers have minimal pulse spread, resulting in a large bandwidth, but require very close dimensional and glass homogeneity control.

Index of Refraction. This is the ratio of the velocity of light passing through a transparent material to the velocity of light passing through a vacuum using light at the sodium *D* line as a reference. The higher the refractive index of a material, the lower the velocity of light through the material, and the more the ray of light is bent on entering it from an air medium.

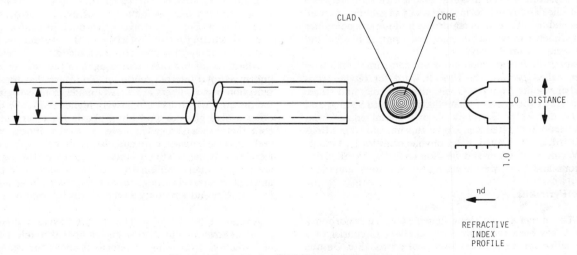

Fig. 2.   Graded index (GRIN) optical fiber design.

Numerical Aperture (*N.A.*). For an optical fiber, this is a measure of the light capture angle and describes the maximum core angle of light rays that will be reflected down the fiber by total reflection. The formula from Snell's law governing the numerical aperture number for a fiber is N.A. $= \sin \theta =$ V $n_1^2 - n_2^2$, where n_1 is the refractive index of the core and n_2 is the refractive index of the clad glass.

Most optical fibers have numerical apertures between 0.15 and 0.4 and these correspond to light acceptance half-angles of about 8 and 23 degrees. Typically fibers having high N.A.'s exhibit greater loss and lower bandwidth capabilities.

Light Loss or Attenuation-Through a Fiber. This is expressed in decibels per kilometer (dB/km). This is a relative power unit according to the formula dB $= 10 \log (I/I_0)$, where (I/I_0) is the ratio of light intensity at the source to that at the extremity of the fiber. A comparison of light transmission to light loss in dB through one kilometer of fiber is as follows:

80% transmission/km $\simeq$ a loss of $\sim$ 1 dB/km

10% transmission/km $\simeq$ a loss of $\sim$ 10 dB/km

1% transmission/km $\simeq$ a loss of $\sim$ 20 dB/km

Bandwidth. This is a rating of the information-carrying capacity of an optical fiber and is given either as pulse dispersion in nonoseconds per kilometer (ns/km) or bandwidth length in megahertz-kilometers (MHz-km). Light pulses spread or broaden as they pass through a fiber, depending on the material used and its design. These factors limit the rate at which light carrier pulses can be transmitted and decoded without error at the terminal end of the optical fiber. In general, large bandwidth and low losses favor those optical fibers with small core diameter and low numerical aperture.

Fiber Optic Systems and Applications

Fiber optic systems refer to those systems that use optical fibers to carry information between a light source and a receiver. The information may be in the form of coded light pulses or as completely coherent images. Systems are made up of several important components that include such items as the light source and a detector, but the common and possibly the most important component is the optical fiber which conducts the light. These fibers may be flexible or rigid. They may comprise a single fiber or, for the special function of imaging, be assembled in groups so that each fiber carries a single element of an image.

Imaging. In imaging both ends of the groups of fibers must maintain an unchanging orientation one to another so that a coherent image is transmitted from the source to the receiver. Flexible coherent bundles of optical fibers having only their terminal ends secured in coherent arrays are used primarily in endoscopes to examine the inside of cavities with limited access, as in such body cavities as the stomach, bowel, urinary tract, etc.

Rigid fiber bundles fused tightly together along their entire length can be made to form a solid glass block of parallel fibers. Slices from the block with polished surfaces are sometimes used as fiber optic faceplates to transmit an image from inside a vacuum to the atmosphere. A typical application is the cathode-ray tube (CRT) used for photorecording. The requirement for this type of application is for both image coherence and vacuum integrity so that when the fiber optic array is sealed to the tube, the vacuum required for tube's operation is maintained. However, any image formed electronically by phosphor films on the inside surface of the fiber optic face is clearly transmitted to the outside surface of the tube's face. High-resolution CRT images can easily be captured on photographic film through fiber optic faceplates.

Telecommunications. One of the most demanding systems using optical fibers is in lightwave telecommunications where application requires only a single fiber as the medium to carry sufficient information for many individual telephone conversations. It is normal however to provide more than one fiber per circuit as a redundant spare in a suitable cable for an unexpected yet possible fiber failure. The most demanding technical requirements for a high performance light-wave telecommunication system relate to trunk phone line (city to city) economics, where 20-kilometer repeater-free circuit links with greater than 200 MHz carrying capacity require the fiber to be made of very pure constituents and constructed with virtually no imperfections. Happily there is also a need for many somewhat less demanding communication links for intercity telecommunication use and in process control equipment requiring less than 1 km line length and lower than a 200 MHz information carrying capacity. Intermediate loss optical fibers of less demanding design and glass purity will function for these types of needs readily and economically.

Figure 3 is a simplified schematic illustration of a lightwave telecommunication system. Key components are the (1) *Transmitter circuitry* that modulates or pulses in code the light of the (2) *light source*, which may be a light emitting diode (L.E.D.) or a solid state laser positioned to launch light into one end of the (3) *optical fiber waveguide*, which conducts the light signal over the prescribed distance because it has particularly good transmission capability at the wavelength of the light source. The terminal end of the waveguide is attached to the (4) *detector*, which may be a PIN or avalanche photodiode; it in turn accepts light and changes the signal into an electromagnetic form for the (5) *receiver circuitry*, which decodes the signal, making it available as useful electronic analog or digital output. When two-way communication is needed, the system is fully duplexed and two circuit links of the type illustrated are required.

Advantages of Fiber Optic Systems

Fundamental advantages offered by fiber optic systems seem to insure wider and wider applications in place of the existing electromagnetic counterparts. One very important advantage is the excellent electrical isolation that can be achieved, since the electronic components used have no electrical interconnection with other components of the system. In lightwave telecommunication systems optical fiber cables are already cost competitive with coaxial copper cables now used on an installed dollar per kilometer basis. Fiber economics will improve greatly as volume usage permits further manufacturing economies, while little if any improvement can be anticipated for copper conductor systems of the type now in place. Signal losses are much lower in fiber waveguide cable than in coaxial cable, and the bandwidth or information carrying capacity is far greater. Additional advantages favoring a telecommunication system using optical fibers include small cable size, virtual elimination of crosstalk between conductors, immunity of the cables to electromagnetic interaction, and increased communication security due to relative difficulty of tapping the line.

Glasses Used for Optical Fibers

Many glass compositions can be used for optical fibers, but for intermediate- and low-loss applications the options become increasingly limited. Multicomponent glasses containing a number of oxides are adequately suited for all but the very low-loss fibers, which are usually made from pure fused silica doped with other minor constituents. Multicomponent glasses are prepared by fairly standard optical melting procedures, with special attention to details for increasing transmission and controlling defects from later fiber drawing steps. In contrast, doped fused

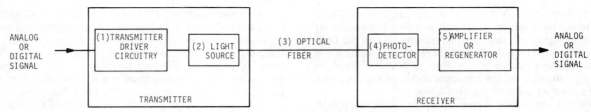

Fig. 3.   Schematic of lightwave telecommunication system.

silica glasses are produced by very special techniques that place them almost directly in a form from which fibers may be drawn. A brief discussion of each approach follows.

Multicomponent glasses are generally characterized by ease of melting ($<2800°F$; $1538°C$) and fewer fiber processing problems when compared to the high-temperature fused silica glasses. Preparation of adequately pure raw materials as batch for these glasses is one of the major areas of concern for the multicomponent approach. The use of metal-organic-derived batch constituents is one means to accomplish an adequate purity. Maintenance of high material purity, once attained, is another though lesser challenge.

Special precautions are routinely taken during the melting operation to insure that multicomponent glasses are stirred until they are well homogenized. This reduces the amount of objectionable cords or stria in the glass. Stirring is followed by a careful "fining" procedure that allows the molten glass to remain relatively motionless and become seed and blister free. These glasses are normally melted using platinum melting utensils without any rhodium constituent, as the latter metal tends to impart an undesirable absorption band in the glass. Atmospheres over the melt range from oxidizing (O_2 gas) to reducing (CO in CO_2 gas mixture) depending on the adjustment needed for control of the redox state of transition metal oxides present as trace impurities. The objective is to place the impurities in the most favorable valence for minimum light absorption. In certain cases dry CO_2 is bubbled through the melts to reduce the OH^- ion content in the glass to enhance transmission at the 1.39 μm wavelength. When full reducing atmospheres are required, a pure silica crucible may be substituted for platinum to avoid crucible damage. If this is done, a new crucible will likely be required for each melt operation because the silica crucible has only a limited life.

For doped fused silica optical fibers, the melting step is usually not a distinct operation, as is the case for multicomponent glasses. The preform layup step, preceding fiber draw for this type fiber, is unique because it simultaneously provides for batch ingredient purification, metering, and a fusion operation (melting) during and after the layer-by-layer construction. These integral fusions essentially avoid the requirement for an initial bulk glass melting step. This fiber-making process, known as Chemical Vapor Deposition (CVD), will be discussed further in a later paragraph. It should be emphasized here, however, that the optical fibers with highest performance are usually produced by this technique.

Preform Preparation and Fiber Draw

Most optical fiber drawing sequences start with the assembly of a preform ($\frac{1}{2}$–1 in. diameter $\times$ 12–36 in. long) (12.7–25.4 mm diameter $\times$ 30.5–91.4 cm long) which in cross section is a large version of the final fiber to be drawn. Only the double-crucible fiber-making process avoids the preform assembly step; in that approach small fibers are drawn directly from molten glass.

There are a number of optical fiber drawing processes, but the three most frequently used are as follows:

Rod-in-Tube Method. A cylinder of glass with a relatively high refractive index is produced from an appropriate melt by casting. It may then either be fire polished or machined and mechanically polished. A glass of lower refractive index is either formed directly into a tube or a tube is machined from a cast cylinder and appropriately polished (particularly well on the inside surface). The core rod is carefully nested in the clad tube as shown in Fig. 4 and together they act as a preform

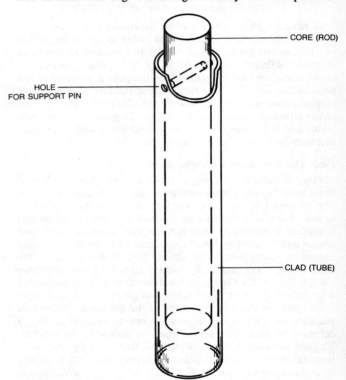

Fig. 4.   Glass core and tubing preform.

for the follow-on optical fiber drawing operation. As the preform is fed into a furnace, the two glasses melt simultaneously, seal to each other in the redraw machine (Fig. 5), and are drawn into the fiber, which is accurately controlled for uniform physical dimensions. To help insure core-clad seal quality, a vacuum may be produced in the space between core and the clad during the redraw operation. The advantage of this approach is its simplicity. Some of the major problems are the difficulties of obtaining perfect mating surfaces and avoiding foreign material entrapment at the core-clad interface.

Double-Crucible Method. Two crucibles are nested together, as in Fig. 6, having concentric orifices exiting their bottoms.

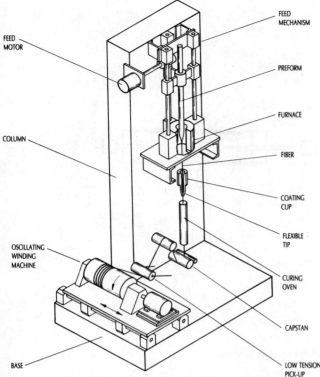

Fig. 5.   Redraw machine schematic.

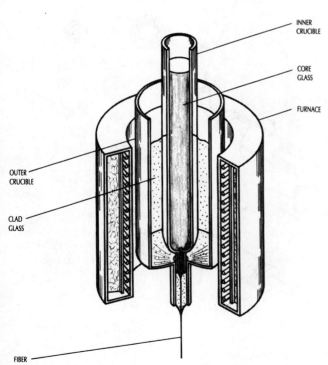

Fig. 6.   Double crucible fiber draw method.

The larger crucible contains the clad glass and the smaller inside crucible the core glass. The crucibles are heated until the glasses flow simultaneously through the double orifice. As in the initial glass melting operation, which precedes crucible draw, various atmospheres may be employed over the melts to optimize fiber performance. Optical fibers are drawn from the composite glass using a drawing mechanism like that shown in the lower part of Fig. 5. This technique has the advantage of yielding superior core-clad glass seal perfection, and when desired, allows for a controlled interdiffusion of this interface to provide a graded index fiber.

Chemical Vapor Deposition (CVD) Method. This approach to optical fibers has many variations. One of these, referred to as inside CVD illustrated in Fig. 7, involves producing a preform using a rotating fused silica tube (with high loss characteristics) and coating the inside surface with successive glassy layers of silica (SiO_2) doped with boron oxide (B_2O_3). The source of the two oxides are silicon tetrachloride ($SiCl_4$) and boron trichloride (BCl_3), whose reactions with oxygen form a vapor that deposits very pure doped silica films on the inside of a heated tube enclosing the reaction. After a suitable thickness of doped silica layers has built up inside the tube wall (to become the cladding), the boron dopant is stopped and germanium oxide from a germanium tetrachloride ($GeCl_4$) reactant is started, which with pure silica is codeposited in a succession of film layers that will become the core. After sufficient core layers have been built up the tube assembly is then overheated as it is rotated to shrink the entire tube, close the center hole, and thus produce a solid preform rod. Figure 8 is an illustration of the preform making sequence steps. The preform is then redrawn with equipment similar to that previously discussed, the resulting fiber being spooled on mandrels for later use. As stated earlier, the highest-performance optical fibers are commercially produced by this process. It has great flexibility and is an easily controlled technique for making the various optical fiber designs of interest.

Typical Optical Fiber Performance

Optical fibers are closely specified by manufacturers for a system designer's selection and optimum use, particularly for lightwave telecommunication system design. Performance data presented here is only an attempt to illustrate existing commercial capabilities for optical fibers and make a crude comparison for the different techniques of optical fiber manufacture.

A typical optical fiber loss spectrum vs wavelength is shown in Fig. 9 for a graded index fiber made by the CVD process. While it is normal for losses in all optical fibers to vary with the wavelength of the transmitted light used, the curve clearly illustrates that our typical fiber meets an objective of achieving relatively low loss at about 0.82 μm where available laser diode emitters function. Even lower losses are possible at longer wavelength if the OH absorption peak is controlled. However adequately reliable emitters have not yet been developed for longer wavelength operation.

An idea of the relative performance of optical fibers made by various processes is given in accompanying table. Since improvements are continually being made in the processes used, a range of values is shown with the best values being those made in controlled laboratory endeavors versus those available from commercial operations.

—L. V. Pfaender, Corporate Technology, Owens-Illinois, Toledo, Ohio.

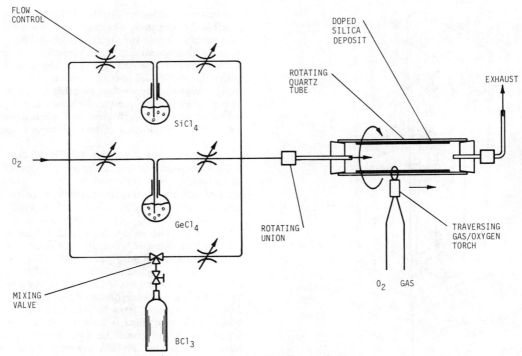

Fig. 7. Schematic CVD preform making process.

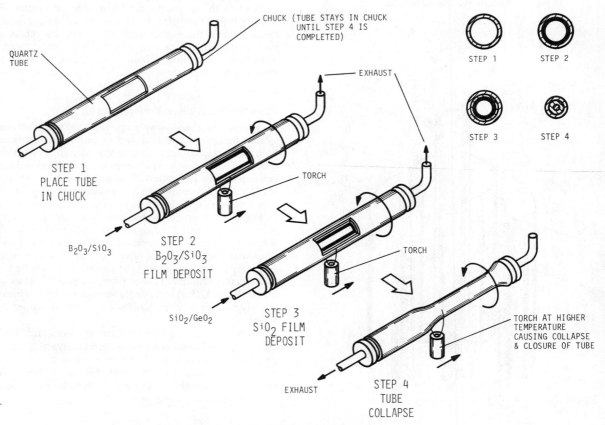

Fig. 8. Inside CVD preform process sequence.

RELATIVE PERFORMANCE OF OPTICAL WAVEGUIDE GLASSES AND FIBERS

	MULTICOMPONENT GLASS CORE/TUBE REDRAW PROCESS		MULTICOMPONENT GLASS DOUBLE CRUCIBLE PROCESS		DOPED FUSED SILICA CVD PROCESS	
	Typical Core (wt%)	Typical Clad (wt%)	Typical Core (wt%)	Typical Clad (wt%)	Typical Core (wt%)	Typical Clad (wt%)
SiO_2	40	51	$\simeq 25$	$\simeq 60$	$\simeq 96$	$\simeq 92$
GeO_2	—	—			$\simeq 3$	—
B_2O_3	12	14	$\simeq 50$	$\simeq 13$	$\simeq 1$	$\simeq 8$
Al_2O_3	9	19				
CaO	17	5				
MgO	—	8				
SrO	10	—				
La_2O_3	9	—				
Na_2O	3	3	$\simeq 25$	$\simeq 27$		
	100%	100%	100%	100%	100%	100%

	STEP INDEX FIBER	GRADED INDEX FIBER	STEP INDEX FIBER
Loss at 820 μm (db/km)	50–1000	5–50	2–5
Bandwidth (MHz-km)	70–200	20–200	500–1500
N.A.	0.3–0.66	0.1–0.3	0.1–0.25
Distance limitations	Short (.01–.1 km)	Medium (.1–1 km)	Long (>1 km)

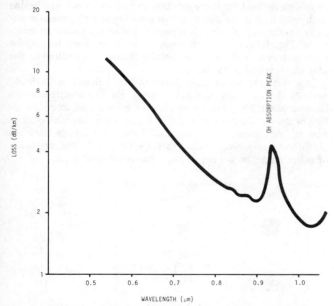

Fig. 9.   Typical optical fiber loss spectrum. (Graded index fibers made by CVD.)

References

Andreiev, N.: "Industrial Fiber Optics," *Control Eng.*, **24**, 3, 36–39 (1977).

Baumbick, R. J., and J. Alexander: "Fiber Optics Sense Process Variables," *Control Eng.*, **27**, 3, 55–75 (1980).

Boraiko, A.: "Harnessing Light by a Thread," *National Geographic*, **156**, 4, 516–534 (1979).

Baues, P.: "The Anatomy of a Fiber Optic Link," *Control Eng.*, **26**, 8, 46–49 (1979).

Elion, G. R., and H. A. Elion: "Fiber Optics in Communications Systems," Marcel Dekker, New York (1978).

Faust, G.: "Programmable Controller Offers Fiber Optic Data Link for Remote I/O," *Control Eng.*, **26**, 10, 53–54 (1979).

Gawlowicz, D. J.: "Asynchronous Transmission in Electronic, Fiber Optic Systems," *Control Eng.*, **25**, 2, 67–68 (1978).

James, K. A., Qukck, W. H., and V. H. Strahan: "Fiber Optics: The Way to True Digital Sensors," *Control Eng.*, **26**, 2, 30–33 (1979).

Klein, R., and P. Onorato: "Glass Fibers for Optical Communication," *GTE Profile*, **4**, 8–12 (1979).

Miller, S. E.: "Photons in Fibers for Telecommunication," *Science*, **195**, 1211–1216, (1977).

Morris, H. M.: "Fiber Optics for Industrial Control," *Control Eng.*, **26**, 10, 49–52 (1979).

Staff: "Turbine Blade Temperature Monitored by Fiber Optic System," *Control Eng.*, **23**, 10, 22 (1976).

Staff: "Voices in the Light," *Sci. Amer.*, **242**, 3, 96 (1980).

Weik, M. H.: "Fiber Optics and Lightwave Communications Standard Dictionary," Van Nostrand Reinhold, New York (1980).

OPTICAL GLASS. Glass to be useful for lenses, prisms and other optical parts through which light passes, as distinguished from mirrors, must be completely homogeneous. This includes freedom from bubbles, striae, seeds, strains, etc. In order to reduce aberrations, the optical designer needs many different kinds of glass. A few typical types are described in the accompanying table. The v-number is the reciprocal of the dispersive power of the glass.

	TYPE	n_D	v-NUMBER
Borosilicate	Crown	1.5170	64.5
Barium	Crown	1.5411	59.5
Spectacle	Crown	1.5230	58.4
Light	Flint	1.5880	53.4
Ordinary	Flint	1.6170	38.5
Dense	Flint	1.6660	32.4
Extra dense	Flint	1.7200	29.3

ORBITALS.   This article embraces both atomic and molecular orbitals.

Atomic. From spectroscopic studies, it is known that when an electron is bound to a positively charged nucleus only certain fixed energy levels are accessible to the electron. Before 1926, the old quantum theory considered that the motion of the electrons could be described by classical Newtonian mechanics in which the electrons move in well defined circular or elliptical orbits around the nucleus. However, the theory encountered numerous difficulties and in many instances there arose serious discrepancies between its predictions and experimental fact.

A new quantum theory called wave mechanics (as formulated by Schrödinger) or quantum mechanics (as formulated by Heisenberg, Born and Dirac) was developed in 1926. This was immediately successful in accounting for a wide variety of experimental observations, and there is little doubt that, in principle, the theory is capable of describing any physical system. A strange feature of the new mechanics, however, is that nowhere does the path or velocity of the electron enter the description. In fact it is often impossible to visualize any classical motion that could be consistent with the quantum mechanical picture of the atom.

In this theory the electron is viewed as a three-dimensional standing wave. The pattern of the wave is described by a wave function ϕ (analogous to the amplitude of a water wave). This one-electron wave function is called an atomic orbital. Since the wave function can be positive or negative (and real or complex) it does not describe an observable property of the electron. However, the square of the wave function (ϕ^2 or ϕ times its complex conjugate) is always positive and real, and can be identified with the probability of finding the electron at any point. This was first suggested by Born, but has now received ample experimental support. Hence, when the wave function is calculated for any electron we can determine the regions in space where the electron is most likely to be found, though we cannot say what type of motion results in that particular probability pattern.

Atomic orbitals are usually labeled by a set of designating numbers called quantum numbers. The one that determines the energy of the resulting state (for hydrogen) is given the symbol n and called the "principal quantum number." It assumes the values 1, 2, 3, 4, 5, . . . to infinity, with increasing electron energies. The second quantum number is given the symbol l. It can be identified with the angular momentum of the electron due to its orbital motion, and assumes values of 0, 1, 2, 3, . . . to $(n-l)$. For historical reasons the orbitals with these values are referred to as s, p, d, f, . . . orbitals respectively. Hence a $3d$ orbital is one for which $n = 3$ and $l = 2$. The third quantum number is usually given the symbol m and is difficult to define in the absence of an external field. However, under all conditions it can assume $2l + 1$ values.

From the rules given in the previous paragraph it can easily be shown that for $n = 1$, 2 and 3 there are a total of 14 allowed orbitals. For an isolated hydrogen atom all orbitals must be spherically symmetrical and have the shapes shown in series (a) of Fig. 1. These are plots of the probability function (ϕ^2) for the orbitals and hence the darkest areas represent regions where the electron is most likely to be found. Though there are actually three $2p$ orbitals, three $3p$ orbitals and five $3d$ orbitals, in the absence of an external field the orbitals within a given set are identical in shape and energy, and are said to be degenerate. If a direction is defined by the presence of a magnetic field or the approach of another atom, the degeneracy is removed. The shapes of the allowed orbitals under these conditions are shown in series (b) of Fig. 1. The quantum number m is well defined for these orbitals and its value is given beside each orbital. For the discussion of bonding in polyatomic molecules another set of orbitals is useful. These are shown in series (c) of Fig. 1. The quantum number m is not well-defined for these orbitals, and they are usually labeled according to the axis along which they lie.

In addition to the orbitals shown in Fig. 1 there are "hybrid" orbitals that are not stationary states for the electron in an isolated atom. They can be obtained by taking a linear combination of the standard orbitals in Fig. 1. Since the electron distribution is "off center" they are useful only for atoms that are perturbed by an electric field (Stark-effect) or by the approach of other atoms as occurs in chemical-bond formation.

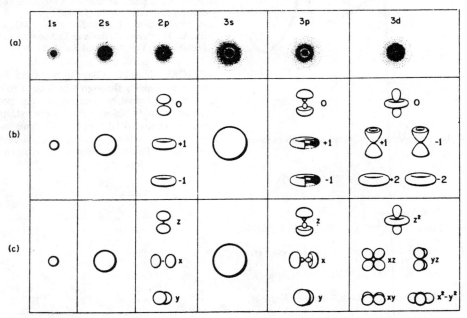

Fig. 1.   Hydrogen atomic orbitals: (a) for isolated atoms; (b) with one direction defined; (c) with three directions defined.

In addition to the three quantum numbers discussed above, experimental evidence requires an additional quantum number m_s, which by analogy to classical mechanics is attributed to an intrinsic (i.e., position-independent) property of the electron called "spin." Unlike the other quantum numbers, however, it can assume only two values ($\pm\frac{1}{2}$). As we shall see, this fact determines the orbital population of the many-electron atom.

It is an unfortunate consequence of the mathematical complexity of the quantum mechanical equations that the hydrogenic atom (i.e., one-electron atom) is the only system for which an exact probability distribution (ϕ^2) can be obtained. Approximate methods must be used to calculate the wave functions for the many electron atoms. We begin with the natural assumption that the wave function for such a more complex atom (ψ) can be obtained by taking a product of the appropriate one electron functions (ϕ). However, even if we ignore for the moment the coulombic repulsion between electrons, there are two fundamental postulates of quantum mechanics that complicate the picture. One is that electrons are indistinguishable. Hence, when their positions are interchanged, the probability function (ψ^2) must remain unchanged. This means that the wave function (ψ) must either remain the same or only change signs when electron positions are interchanged. A second postulate (called the Pauli principle) is that the total wave function must change signs when electron positions are interchanged. From these requirements it can be seen that when two electrons are placed in the same orbital (i.e., assigned the same orbital wave function ϕ) the total wave function will change signs (as required by the Pauli principle) only if the electrons have different spin quantum numbers ($+\frac{1}{2}$ and $-\frac{1}{2}$). It is hence a general rule that hydrogenic orbitals can only be occupied by a maximum of two electrons and these must have opposite spin.

If we begin with the most tightly bound orbital and add two electrons to each orbital the so called "ground state configuration" of the atom is obtained. For example the electronic configuration of the silicon atom is written $1s^2$, $2s^2$, $2p^2$; $3s^2$, $3p^2$. The orbital shapes and energies are, however, considerably altered by electron-electron repulsion, especially between electrons whose orbitals overlap appreciably. This has several marked effects. For a given value of n, all the orbitals no longer have the same energy. The binding energy now decreases with increasing values of the quantum number l. Secondly, because of the coulombic repulsion between electrons, they will tend to occupy separate orbitals whenever feasible (for example the $3p$ orbitals configuration in the silicon atom is actually $3p^1$, $3p^1$). Furthermore, when electrons are forced together into the region of one hydrogenic orbital it is quite likely that electron-electron repulsion (always greater than 25 kcal) leads, in effect, to slightly different orbitals for each electron.

For chemical purposes we are most interested in the shapes of orbitals in which the valence (outermost) electrons reside. By assuming that the inner electrons act only to screen some of the positive charge on the nucleus the valence electrons can be shown to assume the shape of the appropriate hydrogenic orbital, with the insignificant difference that the inner nodes in the orbitals shown in Fig. 1 are drawn in closer to the nucleus because the shielding is poorer in this region. It is worth noting that for the many-electron atom, the indistinguishability of electrons has the effect of making only the total probability function physically meaningful. For this reason and because of the repulsion between electrons, the individual one-electron wave functions are so correlated that, though the "independent orbital concept" remains a very useful approximation, it is not fundamental to the problem.

Molecular. By analog with atomic orbitals, the wave function (ψ) for one electron in a molecule is called a molecular orbital, and the probability of finding the electron at any point is similarly given by the value of ψ^2 at that point. Just as in the case of atomic orbitals, an exact solution of the equations is possible only for the one-electron molecule (H_2^+) and it is only for this species that accurate molecular orbitals can be obtained.

The shapes of 10 of the most tightly bound orbitals of H_2^+ are shown in Fig. 2. Three quantum numbers can be used to label these orbitals. The one that is always well defined is given the symbol λ and can be identified with the component of the orbital angular momentum along the internuclear axis. It can take on values 0, 1, 2, 3 . . . for which the orbitals are called σ, π, δ, . . . respectively. The other two quantum numbers are defined differently depending on the nuclear separation. When the internuclear distance is short, as the case of H_2^+, it is convenient to consider the atomic orbital that would result from a given molecular orbital if the two nuclei were made to coalesce (in our imaginations). In the case of homonuclear diatomic molecules, for example, the molecular orbital designated $3d\ \sigma$ has $\lambda = 0$ and correlates to a $3d$ atomic orbital when the nuclei coalesce (i.e., the atomic quantum numbers n and l become well-defined at short internuclear distance). On the other hand, when the internuclear distance is large, a more useful and significant label is one which identifies the atomic orbitals with which a given molecular orbital correlates when the distance between the two nuclei approaches infinity. For this "separate atom" designation an additional symbol must be used to distinguish between the two molecular states that can arise from a given pair of atomic states. Chemists find it most useful to use the superscript * to indicate the higher energy (antibonding) orbital and the absence of * to indicate the lower energy (bonding) orbital. When the difference in symmetry between the two states is important the symbols g (gerade-symmetric) and u (ungerade-symmetric) are used. Thus, the orbital $\sigma^*\ (2p_x)$ is an antibonding orbital with $\lambda = 1$ that correlates with two $2p_x$ orbitals on the separated atoms. Similarly, a $\sigma_g 2(\text{sp})$ orbital is a bonding orbital with $\lambda = 0$ that correlates with two atomic orbitals on the separated atoms. It is worth noting that the "antibonding" orbitals possess a nodal surface (i.e., a region of zero electron-probability density) between the two nuclei.

Exact solutions such as those given above have not yet been

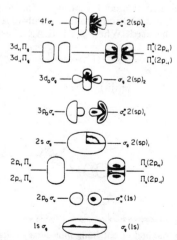

Fig. 2.   Molecular orbitals of H_2^+. The "united atom" designation is given on the left-hand side and the "separate atom" designation on the right-hand side.

obtained for the usual many-electron molecules encountered by chemists. The approximate method which retains the idea of orbitals for individual electrons is called "molecular-orbital theory" (M. O. theory). Its approach to the problem is similar to that used to describe atomic orbitals in the many-electron atom. Electrons are assumed to occupy the lowest energy orbitals with a maximum population of two electrons per orbital (to satisfy the Pauli exclusion principle). Furthermore, just as in the case of atoms, electron-electron repulsion is considered to cause degenerate (of equal energy) orbitals to be singly occupied before pairing occurs.

It has not proved mathematically feasible to calculate the electron-electron repulsion that causes this change in orbital-energies for many-electron molecules. It is even difficult to rationalize the qualitative changes in sequence on the basis of the shapes of the H_2^+ orbitals. Greater success has been achieved by an approximate method which begins with orbitals characteristic of the isolated atoms present in the molecule, and assumes that molecular orbital wave functions can be obtained by taking linear combinations of atomic orbital wave functions (abbreviated L.C.A.O.). For homonuclear diatomic molecules, the atomic orbitals used are normally those with which a given molecular orbital correlates. An example of how molecular orbitals are manufactured in the L.C.A.O. approximation is shown in Fig. 3. According to this theory, the relative molecular-orbital energies will depend on (a) the spread of orbital energies on the individual atoms (since this determines the extent of hybridization) and (b) on the internuclear distance which is largely determined by the relative number of bonding and antibonding orbitals that are filled.

The L.C.A.O. approximation has also proved useful for the description of heteronuclear diatomic molecules, although valence-bond theory has been somewhat more successful in its quantitative calculations of bond energies. For these molecules the selection of appropriate atomic orbitals to be used in linear combination is governed by considerations of symmetry, energetics and overlap. These considerations also apply to the formation of molecular orbitals in polyatomic molecules by the L.C.A.O. approximation. In the case of polyatomic molecules it is always possible to develop either (a) molecular orbitals that extend between only two atoms localized M.O.s) or (b) molecular orbitals that extend over the entire molecule (delocalized M.O.s). In the case of saturated molecules, the two descriptions are practically equivalent. However, for molecules with conjugated double bonds and especially for aromatic compounds, delocalized molecular orbitals which extend over several atoms must be used. When the L.C.A.O. method is used

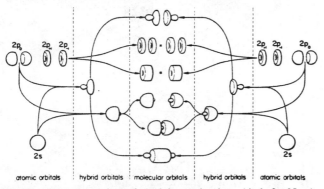

Fig. 3. L.C.A.O. method of obtaining molecular orbitals for N_2. Arrows indicate the combinations of atomic orbitals used to generate molecular orbitals. The stabilities of the orbitals increase from top to bottom.

to develop molecular orbitals in complexes between metal ions (possessing d orbitals) and ligands such as CO, OH^-, NH_3, CN^- etc., the procedure is called Ligand Field Theory.

A thorough test of the L.C.A.O. molecular orbitals has been possible only for H_2^+ where both the wave functions and orbital energies are known accurately. Such a comparison shows that though the shapes of the wave functions are reasonably represented by the approximation, their energies can be appreciably in error, especially for excited state. Nevertheless, the L.C.A.O. approximation has proved the most fruitful method of obtaining molecular orbitals that has been developed to date.

—E. A. Ogryzlo, University of British Columbia, Vancouver, British Columbia, Canada.

ORDER-DISORDER THEORY AND APPLICATIONS. Phase transitions in binary liquid solutions, gas condensations, order-disorder transitions in alloys, ferromagnetism, antiferromagnetism, ferroelectricity, antiferroelectricity, localized absorptions, helix-coil transitions in biological polymers and the one-dimensional growth of linear colloidal aggregates are all examples of transitions between an ordered and a disordered state.

The two quantities which apparently must be used to explain or describe these phenomena are the presence of a potential between the particles or spins and a small volume which must be assigned to each particle so that two particles or spins cannot occupy the same space. All the above phenomena may be semi-quantitatively treated by a single statistical model called the Ising model which consists of a lattice in space, each site of which possesses either a 0 or a 1. Thus two rows of a two dimensional lattice might look like

$$. . . 011000011011110011 . . .$$
$$. . . 110110010100110111 . . .$$

The 0's and 1's represent different particles or spins depending upon the system being studied and their definition will be given below for different systems. The disordered state corresponds to a random array of 0's and 1's. The ordered state is described by an ordered arrangement of the 0's and 1's. The total number of lattice sites equals N. The number of 0's and 1's on the lattice are respectively n_0 and n_1 and $n_1 + n_0 = N$.

The dimensionality of the lattice depends on the physical nature of the phase. For transitions in linear biopolymers and associative colloids, a one-dimensional lattice is used because these materials become ordered in a one-dimensional manner even though they actually exist in three dimensions. For absorption of gas onto a surface, a two-dimensional lattice is sufficient. For bulk phase changes, however, a three-dimensional lattice must be used.

The great advantage of this model is that it gives essentially the same explanation for a large number of seemingly completely diverse physical and chemical phenomena, often with quantitative success. This permits a deeper insight into the statistical thermodynamic behavior of different types of matter.

The calculation of the thermodynamic properties begins by selecting values for w_{00}, w_{11}, w_{10}, which are the potential energies between like and unlike particles or spins occupying nearest neighbor sites. Since two particles cannot occupy one lattice point, the energy of repulsion of two particles on the same site is considered infinite. The energy of a given configuration is then given by

$$E_{Conf} = n_{11}w_{11} + n_{10}w_{10} + n_{00}w_{00} \qquad (1)$$

where n_{11}, n_{00} and n_{10} are the number of nearest neighbor pairs. The probability of a given configuration is given by Boltzmann's theorem as

$$P(E_{\text{Conf}}) = \frac{e^{(-E_{\text{Conf}}/kT)}}{\sum\limits_{\text{Conf}} e^{(-E_{\text{Conf}}/kT)}} \qquad (2)$$

where the summation occurs over all possible configurations, keeping the number of 0's and 1's constant. The restriction of a constant number of 0's and 1's may be removed by letting the lattice interact with its surroundings. The probability that the lattice possesses a given number of 1's and a configurational energy, E_{Conf}, is given by

$$P(E_{\text{Conf}}, n_1) = \frac{e^{-Xn_1/kT} e^{-E_{\text{Conf}}/kT}}{\sum\limits_{n_1=0}^{N} e^{-Xn_1/kT} \sum\limits_{\text{Conf}} e^{-E_{\text{Conf}}/kT}} \qquad (3)$$

where X is a suitable thermodynamic quantity such as the chemical potential or magnetic field, etc.

For a lattice gas, the 0 and 1 stand for an empty and filled site respectively. Consequently, w_{11} is the attractive potential energy between two gas molecules when they occupy adjacent sites on the lattice, and $w_{00} = w_{10} = 0$.

For binary solutions including linear associative colloids, the 0 and 1 represent solvent and solute respectively; w_{11} and w_{00} are the potential energies between like molecules and w_{10} is the potential energy between unlike molecules.

For a ferromagnet the 0 and 1 represent spins of $-\frac{1}{2}$ and $+\frac{1}{2}$ respectively. It is generally assumed that the potential energy between like spins is always the same, i.e., $w_{11} = w_{00}$. To show how the quantity, X of Eq. (3), is obtained, consider a ferromagnet interaction with an external magnetic field. The total energy is

$$E = (n_0 - n_1) H \cdot d + E_{\text{Conf}}$$

where H is the magnetic field strength, and d is the magnetic dipole moment of a spin. Thus for a ferromagnet $X = 2H \cdot d$ since $N - 2n_1 = n_0 - n_1$.

The above described order-disorder transitions are all three-dimensional phase transitions and occur with essentially infinite sharpness unless the condensed phase exists in a colloidally dispersed state. Recently, it has been shown that certain polymers and associative colloids, particularly those of biological interest, have one-dimensional order-disorder transitions which may be explained in exactly the same terms that describe the three-dimensional phase changes discussed above. However, because the condensed state is colloidally dispersed and the ordering occurs only in one dimension, it may be shown that such transitions cannot be infinitely sharp.

The accompanying figure illustrates the transition from an ordered helical to a disordered state which occurs in a large number of colloidal systems. For the helix-coil transition in polymers, the 1 and 0 represent, respectively, a hydrogen bonded turn of the helix and an unhydrogen bonded section of randomly fluctuating polymer. In an associative colloid, the 0 or 1 represent respectively a cell containing a solvent or colloid monomer. In either case, there exists an energy of attraction between 1's which leads to the formation of a one dimensional ordered helix. The fact that the transition tends to be rather sharp comes about because the 01 and 10 configurations which always occur at the beginning and end of a helical sequence are energetically unfavorable. This means that configurations with a lot of ends containing many short segments are suppressed since the configurational energy is large when n_{10} is large and this leads to a small Boltzmann factor for this configuration.

The fact that so many diverse phenomena can be correlated

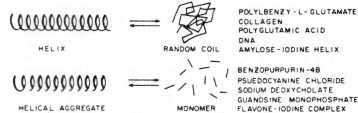

POLYLBENZY-L-GLUTAMATE
COLLAGEN
POLYGLUTAMIC ACID
DNA
AMYLOSE-IODINE HELIX

BENZOPURPURIN-4B
PSUEDOCYANINE CHLORIDE
SODIUM DEOXYCHOLATE
GUANOSINE MONOPHOSPHATE
FLAVONE-IODINE COMPLEX

Diagrammatic illustration of the helix-coil transitions in biopolymers (*top*). Helix-monomer transition in associative biocolloids (*bottom*).

and explained by such a simple model makes it possible to develop tables of equivalent thermodynamic properties. This was first done by Yang and Lee who showed the equivalence between the properties of the lattice gas and Ising ferromagnet. Hill extended their analogies to the case of binary liquid mixtures. After the development of the helix-coil transition theory by Zimm and Bragg and others, Peticolas gave a corresponding table of equivalent thermodynamic properties between polymers and associative colloids. Thus the force-length curve for a helical polymer is the one dimensional analogue of the three dimensional pressure-volume curve for gas condensation and is equivalent to the chemical potential-mole-fraction curve for the associative linear colloid.

—W. L. Peticolas, University of Oregon, Eugene, Oregon.

ORE. A mineral aggregate in which the valuable metalliferous minerals are sufficiently abundant to make the aggregate worth mining. Types and origins of ore deposits are illustrated in the figure.

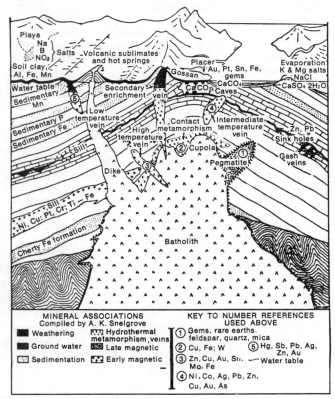

MINERAL ASSOCIATIONS
Compiled by A. K. Snelgrove

- ■ Weathering
- ■ Ground water
- ▢ Sedimentation
- ▨ Hydrothermal metamorphism, veins
- ▨ Late magnetic
- ▨ Early magnetic

KEY TO NUMBER REFERENCES USED ABOVE

① Gems, rare earths, feldspar, quartz, mica
② Cu, Fe; W
③ Zn, Cu, Au, Sn, Mo, Fe
④ Ni, Co, Ag, Pb, Zn, Cu, Au, As
⑤ Hg, Sb, Pb, Ag, Zn, Au
--- Water table

Diagrammatic illustration of the origin of the ore deposits. (*Field, "Outline of Geology," Princeton Univ. Press.*)

ORGANIC CHEMISTRY.* The term *organic*, which means pertaining to plant or animal organisms, was introduced into chemical terminology as a convenient classification of substances derived from plant or animal sources. Early it was believed that organic compounds could arise only through the operation of a vital force inherent in the living cell. However, Wöhler discovered in 1828 that the organic compound urea, identified in urine by Rouelle in 1773, could be produced by heating the inorganic salt ammonium cyanate:

$$NH_4^+ \, OCN^- \longrightarrow H_2NCONH_2$$

Ammonium cyanate   Urea

Subsequently, the association of organic compounds only with living organisms was discontinued.

The term organic has persisted, but the modern definition of organic chemistry has changed to mean the chemistry of carbon compounds. Sometimes a few carbon compounds are excluded from this category, such as carbon dioxide, CO_2; metal carbonates, e.g., Na_2CO_3; carbonyls, e.g., $Ni(CO)_4$; cyanides, e.g., KCN; carbides, e.g., CaC_2; and a few others, but this exclusion is somewhat arbitrary. The designation organic is still pertinent because the chemistry of carbon compounds is more important to everyday life than that of any other element.

The uniqueness of carbon stems from its ability to form strong carbon-carbon bonds that remain strong when the carbon atoms are simultaneously bonded to other elements. Whereas both the carbon-hydrogen and carbon-fluorine compounds, CH_3CH_3 and CF_3CF_3, are highly stable and relatively unreactive, the corresponding compounds in which the carbon atoms are replaced by boron, silicon, phosphorus, and others, either are thermodynamically unstable or highly reactive.

Theoretically, an infinite number of different carbon compounds can exist. Carbon atoms alone or in combination with other atoms, such as oxygen, nitrogen, etc., can join to form linear, branched, and cyclic chains of nearly any length. One, two, or three bonds may be shared between two carbon atoms. In most stable organic compounds, the total number of bonds to each carbon atom is four.

Classification of Organic Compounds

A subject with the wide scope that characterizes organic chemistry requires a logical approach to its organization so that knowledge can be gathered and applied manageably. Molecular structure has become the key method for classifying this subject. Scientists use two-dimensional diagrams or three-dimensional models to depict molecular structures. Although these analogies are sometimes crude representations of actual molecules, they are useful for communicating information about the molecules. Structural diagrams characteristic of some basic types of organic molecules are shown in Table 1. A more detailed discussion on naming organic compounds is given later.

The compounds shown in Table 1 contain only carbon and hyrogen and are called hydrocarbons. Most organic compounds that contain other kinds of atoms in addition to carbon and hydrogen are considered as formally derived from hydrocarbons in which a hydrogen atoms has been replaced by another atom or collection of atoms. However, these derivatives usually are not formed directly from the hydrocarbons. The other atoms usually are referred to as functional groups. A group of compounds having the same functional group is referred to as a family. Table 2 shows a number of such families. The R and

* Publication B-600-375-79 from The Dow Chemical Company.

TABLE 1. EXAMPLES OF STRUCTURAL DIAGRAMS OF HYDROCARBONS

Type of Compound	Formula	Name
Linear alkane	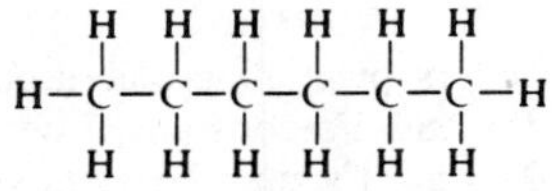 or $CH_3CH_2CH_2CH_2CH_2CH_3$ or $CH_3(CH_2)_4CH_3$ or $n\text{-}C_6H_{14}$ or 	*n*-Hexane
Branched alkane	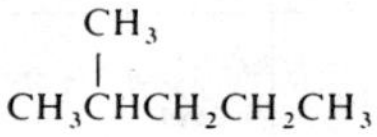 or 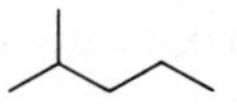	2-Methylpentane
Monocyclic alkane	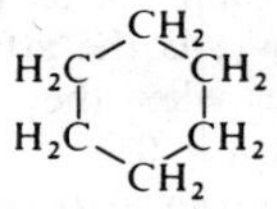 or 	Cyclohexane
Bicyclic alkane	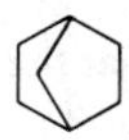	Bicyclo[2.2.1]heptane or Norbornane
Polycyclic alkane	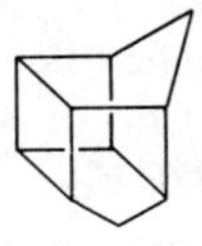	Pentacyclo[5.3.0.0.^{2,5}-0^{3,9}.0^{4,8}]decane or 1,3-Bishomocubane
Alkene	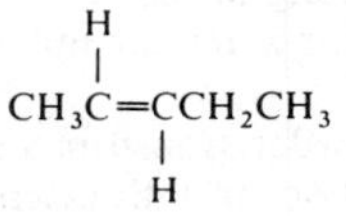 or 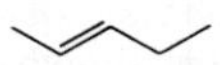	*trans*-2-Pentene
Alkyne	$CH_3C{\equiv}CCH_2CH_3$ or 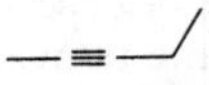	2-Pentyne
Aromatic	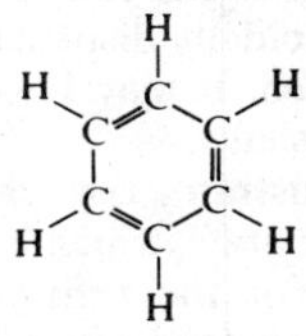 or  or 	Benzene
Polymer	$X(CH_2)_nY$ ($n \geq 1000$, X and Y vary according to how polymer was prepared)	Polymethylene or Polyethylene

R' in the formula of Table 2 represent any hydrocarbon in which a hydrogen atom has been removed from the position to which the functional group is attached.

TABLE 2. EXAMPLES OF SOME MAJOR FAMILIES OF ORGANIC COMPOUNDS

FAMILY	GENERAL STRUCTURE	FAMILY	GENERAL STRUCTURE
Alcohol, phenol	ROH	Isocyanate	$RN{=}C{=}O$
Ether	ROR'	Thiol	RSH
Aldehyde	$\overset{\displaystyle O}{\overset{\|}{RCH}}$	Sulfide	RSR'
Ketone	$\overset{\displaystyle O}{\overset{\|}{RCR'}}$	Sulfoxide	$\overset{\displaystyle O}{\overset{\|}{RSR'}}$
Carboxylic acid	$\overset{\displaystyle O}{\overset{\|}{RCOH}}$	Sulfone	$\overset{\displaystyle O}{\underset{\displaystyle O}{\overset{\|}{\underset{\|}{RSR'}}}}$
Ester	$\overset{\displaystyle O}{\overset{\|}{RCOR'}}$	Sulfonic acid	$\overset{\displaystyle O}{\underset{\displaystyle O}{\overset{\|}{\underset{\|}{RSOH}}}}$
Amine	RNH_2	Chloride	RCl
Amide	$\overset{\displaystyle O}{\overset{\|}{RCNH_2}}$	Bromide	RBr
Nitrile	$RC{\equiv}N$	Iodide	RI
Isonitrile	$R\overset{+}{N}{\equiv}C$	Organolithium	RLi
Nitro compound	$R\overset{+}{N}{\diagdown}{\overset{\diagup O}{\diagdown O^-}}$	Heterocycle	$\overset{\frown}{R\;Y}$
Nitroso compound	$RN{=}O$		
Imine	$\overset{\displaystyle NH}{\overset{\|}{RCR'}}$		
Azo compound	$RN{=}NR'$		
Diazo compound	$RR'C{=}\overset{+}{N}{=}\overset{-}{N}$		
Diazonium salt	$RN{\equiv}N^+X^-$ (anion)		

(Y is an atom other than carbon such as N, O, Si, P, S, etc.; Y is bonded to R at two or more positions.) Examples:

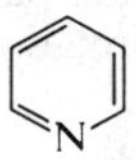

, pyridine;

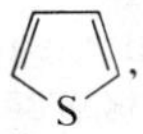

, thiophene.

Molecules with common features also are grouped into more specialized families. The compounds in one such family, carbohydrates, contain only carbon, hydrogen, and oxygen. The hydrogen and oxygen atoms are in the same ratio as in water (2H:1O), hence the suffix -hydrate. Other specialized families include terpenes, alkaloids, steroids, lipids, proteins, enzymes, vitamins, and organometallic compounds.

Although these classifications are important for education and documentation in organic chemistry, research usually is rather specialized. It is often oriented in a practical way, not to the structure of compounds or to the kinds of atoms they contain, but to the manner in which the compounds are used. A partial list of such uses includes plastics, pharmaceuticals, insecticides, fungicides, herbicides, paints, petroleum, fuels, dyes, photography, and adhesives.

An important area of organic chemistry is that which deals with life and living substances. Organized under the title biochemistry, this is a subfield of organic chemistry, since most of the compounds involved contain carbon. Numerous advances have been made recently in biochemistry, and of all the areas of organic chemistry it probably will produce the greatest progress in the next decade. Noteworthy advances can be expected in the areas of biochemistry relating to medicine and human health. Some of the categories along which the study of biochemistry is organized are proteins, peptides, amino acids, nucleoproteins, enzymes, nucleotides, carbohydrates, lipids, steroids, carotenoids, porphyrins, nucleic acids, vitamins, and hormones. These topics are described in detail elsewhere in this volume.

Theoretical organic chemistry is another field that has progressed rapidly in recent years. Chemists have derived molecular orbital symmetry rules that allow understanding and predicting the stereochemistry and relative rates of organic reactions in electronic ground and excited states. In a ground state molecule, all electrons are in their lowest energy levels, whereas, in an excited state molecule, at least one electron is in a higher energy level. For example, the Woodward-Hoffmann orbital symmetry rules for concerted reactions predict that ground state (thermal) cycloaddition reactions involving $4n + 2$ (where n is an integer) π-electrons, and excited state (photochemical) cycloaddition reactions that involve $4n$ π-electrons, may occur via a concerted

SCHEME 1

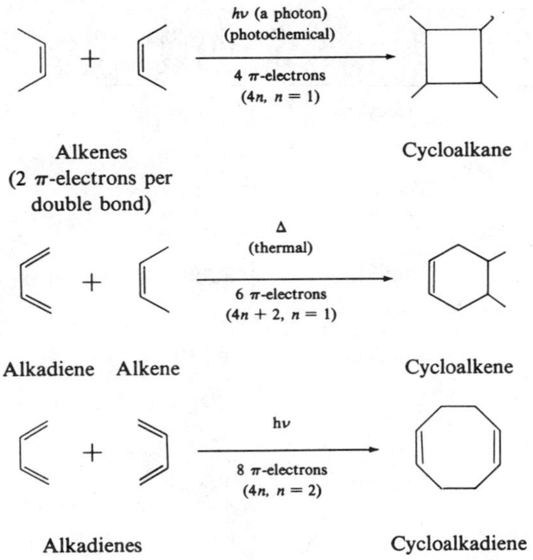

process (Scheme 1). A concerted process or reaction occurs without the involvement of an intermediate. The stereochemistry of the reactants is retained in the product, and the reaction is usually more facile than a comparable nonconcerted reaction. The other combinations of ground state, excited state, $4n +$ 2, $4n$ reactions cannot be concerted reactions.

Another area that has received increased attention is environmental organic chemistry. Reactions that organic compounds undergo when they are released to the environment are becoming as significant as the reactions by which the compounds are prepared or the reactions that take place in the use of the compounds. Some environmentally important types of reactions are hydrolysis, oxidation, sunlight initiated photochemical decomposition, and biodegradation by microbes.

Nomenclature of Organic Compounds

With the foregoing review in mind, the reader will appreciate the difficulties associated with naming organic molecules. Originally, chemical names were indicative of the sources of compounds. For example, catechol was the name given to a compound isolated from the natural product *gum catechu*. Chemists have coined other nonsystematic names such as cubane or basketene to pictorially describe molecules. These nonsystematic names are called common or trivial names. The names phenol, acetic acid, and styrene are also nonsystematic but are widely understood in chemistry.

As the number of known organic molecules increased, a systematic approach to nomenclature was required. To minimize confusion in communicating chemical information, a name should be consistent with other systems in use and should clearly define the structure of a molecule. Specialists in organic chemistry have developed nomenclatures that are logical for their disciplines, thus devising systems for naming alcohols, antibiotics, carboxylic acids, etc.

Systematic nomenclature on a worldwide scale began in 1892 when a committee of the International Chemical Congress established a set of standards known as the Geneva Rules for naming organic compounds. The International Union of Pure and Applied Chemistry (IUPAC) was formed in 1919 and further developed this nomenclature system. In 1886 in the United States,

the American Chemical Society (ACS) established a Committee on Nomenclature. The ACS and IUPAC have developed parallel rules for naming organic compounds.

Alternative rules within the latter system may allow assigning more than one unambiguous name to a compound. The controlled alphabetic listing in the *Chemical Substances Index* of Chemical Abstracts Service (CAS), a part of the ACS, requires that each compound have a unique name. This convention ensures that all information for a single compound such as H_2N—CH_2—CH_2—OH, which can be unambiguously named 2-aminoethanol, 2-aminoethyl alcohol, 2-hydroxyethylamine, etc., will appear in the index under one name: ethanol, 2-amino-. Because universal nomenclature systems are complex and sometimes inconvenient, some chemists have retained the older methods of naming molecules. These older names are used in the parts of this section that do not deal specifically with nomenclature.

The following paragraphs introduce some basic areas of organic chemical nomenclature. References at the end of this section can be consulted for further details.

In addition to trivial names, some of the other categories of organic compound names are:

Generic name: one that indicates a class of compounds; e.g., alkanes, esters.

Parent name: a base from which other names are derived; e.g., ethanol, from ethane; butanoic acid, from butane.

Systematic name: a name composed of syllables defining the structure of a compound; e.g., chlorobenzene, 2-methylhexane.

Substitutive name: one describing replacement of hydrogen by a group or element; e.g., 1-methylnaphthalene, 2-chloropropane.

Replacement name: a name describing compounds that have carbon replaced by a hetero atom; also called "a" nomenclature; e.g., 2-azaphenanthrene.

Subtractive name: one that indicates removal of specific atoms; e.g., in the aliphatic series names ending in -ene or -yne, such as ethene or ethyne, and names involving anhydro-, dehydro-, deoxy-, nor-, etc.

Additive name: one that signifies addition between molecules and/or atoms without replacement of atoms; e.g., styrene oxide.

Conjunctive name: a combination of two names, one of which represents a cyclic structure and the other an acyclic chain, with one hydrogen atom removed from each; e.g., benzenemethanol.

Fusion name: a combination that results from linking with an "o" two names of cyclic systems fused by two or more common atoms; e.g., benzofuran.

Multiplicative name: nomenclature describing the symmetrical repetition of radicals about a central unit; e.g., 2,2'-oxybis(ethanol).

Hantzch-Widman name: a name devised by Hantzsch and Widman for describing heterocyclic systems, in which the prefix denotes a hetero atom(s) and the suffix denotes the ring size and degree of saturation; e.g., oxirene, aziridine.

Von Baeyer name: a name that describes alicyclic bridged systems; e.g., bicyclo[2.2.1]heptane.

The procedure for naming a compound involves some or all of the following steps, depending on the structure of the molecule under study: (1) the type of nomenclature to be used (conjunctive, multiplicative, etc.) is chosen; (2) the parent structure is named; (3) the prefixes, suffixes, and names of functional and substituent groups that were not included in (2) are attached; (4) the numbering is completed.

Hydrocarbons

Acyclic Hydrocarbons. A knowledge of the structural features of hydrocarbon skeletons is basic to the understanding of organic chemical nomenclature. The generic name of saturated acyclic hydrocarbons, branched or unbranched, is *alkane.* The term *saturated* is applied to hydrocarbons containing no double or triple bonds.

The simplest saturated acyclic hydrocarbon is called methane. Names of the higher, straight-chain (*normal*) homologs of this series contain the termination "-ane," as shown in Table 3. The structures of the first four members of the series in

TABLE 3.   EXAMPLES OF SATURATED ACYCLIC HYDRO-CARBONS

MOLECULAR FORMULA	NAME	MOLECULAR FORMULA	NAME
CH_4	methane	$C_{18}H_{38}$	octadecane
C_2H_6	ethane	$C_{19}H_{40}$	nonadecane
C_3H_8	propane	$C_{20}H_{42}$	icosane
C_4H_{10}	butane	$C_{21}H_{44}$	henicosane
C_5H_{12}	pentane	$C_{22}H_{46}$	docosane
C_6H_{14}	hexane	$C_{23}H_{48}$	tricosane
C_7H_{16}	heptane	$C_{30}H_{62}$	triacontane
C_8H_{18}	octane	$C_{31}H_{64}$	hentriacontane
C_9H_{20}	nonane	$C_{32}H_{66}$	dotriacontane
$C_{10}H_{22}$	decane	$C_{40}H_{82}$	tetracontane
$C_{11}H_{24}$	undecane	$C_{50}H_{102}$	pentacontane
$C_{12}H_{26}$	dodecane	$C_{100}H_{202}$	hectane
$C_{13}H_{28}$	tridecane	$C_{101}H_{204}$	henhectane
$C_{14}H_{30}$	tetradecane	$C_{102}H_{206}$	dohectane
$C_{15}H_{32}$	pentadecane	$C_{110}H_{222}$	decahectane
$C_{16}H_{34}$	hexadecane	$C_{120}H_{242}$	icosahectane
$C_{17}H_{36}$	heptadecane	$C_{132}H_{266}$	dotriacontahectane
		$C_{200}H_{402}$	dictane

Table 3 are: CH_4, CH_3—CH_3, CH_3—CH_2—CH_3, and CH_3—CH_2—CH_2—CH_3. The structures of subsequent members of this series are formed by inserting additional CH_2 units.

Univalent groups derived from the preceding acyclic hydrocarbons by removal of one hydrogen atom from a terminal carbon atom are named by replacing the ending "-ane" with "-yl."

EXAMPLES: ethyl CH_3—CH_2—

butyl CH_3—CH_2—CH_2—CH_2—

A saturated branched acyclic hydrocarbon is named by numbering the longest chain from one end to the other, and the positions of the side chains are indicated by the lowest possible numbers. The numbers precede the groups, and are separated from them by a hyphen.

EXAMPLES:

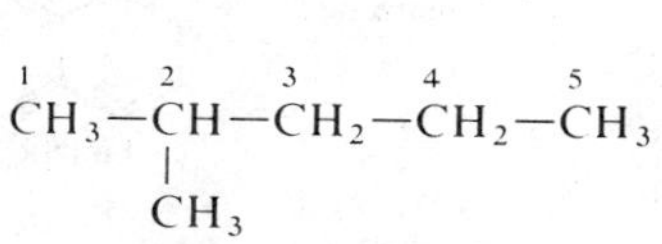

$$\overset{1}{C}H_3-\overset{2}{C}H-\overset{3}{C}H_2-\overset{4}{C}H_2-\overset{5}{C}H_3$$
$$|$$
$$CH_3$$

2-methylpentane (not 4-methylpentane)

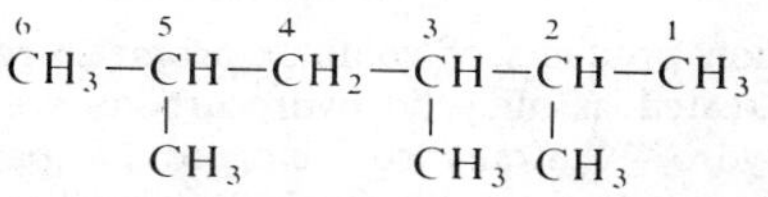

$$\overset{6}{C}H_3-\overset{5}{C}H-\overset{4}{C}H_2-\overset{3}{C}H-\overset{2}{C}H-\overset{1}{C}H_3$$
$$| \qquad\qquad | \quad |$$
$$CH_3 \qquad\quad CH_3\ CH_3$$

2.3.5-trimethylhexane (not 2.4.5-trimethylhexane)

If two groups are attached to the same carbon atom the number is repeated.

EXAMPLES:

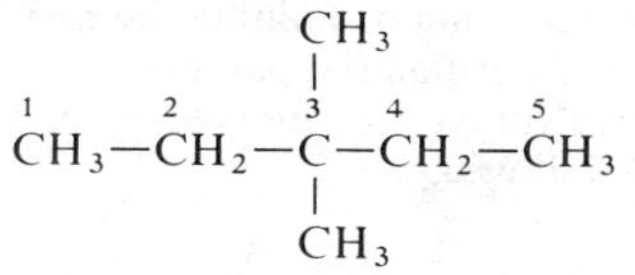

$$CH_3$$
$$|$$
$$\overset{1}{C}H_3-\overset{2}{C}H_2-\overset{3}{C}-\overset{4}{C}H_2-\overset{5}{C}H_3$$
$$|$$
$$CH_3$$

3.3-dimethylpentane (not 3-dimethylpentane)

For some purposes, such as alphabetical listing of basic skeleton names, inverted word orders are used:

EXAMPLES: pentane, 2-methyl-benzene, chloro-

The two propyl groups are distinguished by calling them *normal*-propyl or *n*-propyl, CH_3—CH_2—CH_2—, and isopropyl or *i*-propyl, $[(CH_3)_2$—CH—$]$, in common usage. The latter is called 1-methylethyl in systematic nomenclature. The butyl groups are named as follows:

Structure	*Systematic Name*	*Trivial name*
CH_3—CH_2—CH_2—CH_2—	butyl	*n*-butyl
CH_3—CH—CH_2— $\quad\ \ |$ $\quad\ \ CH_3$	2-methylpropyl	*i*-butyl
CH_3—CH_2—CH— $\qquad\qquad |$ $\qquad\qquad CH_3$	1-methylpropyl	*s*-butyl (*s* = *secondary*)
$\quad\ \ CH_3$ $\quad\ \ |$ CH_3—C— $\quad\ \ |$ $\quad\ \ CH_3$	1,1-dimethylethyl	*t*-butyl (*t* = *tertiary*)

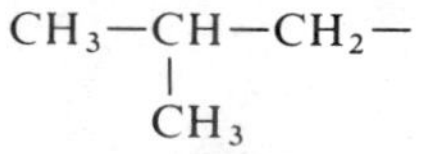

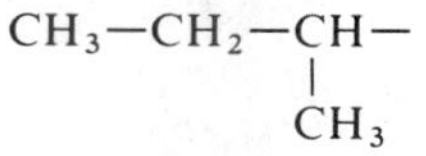

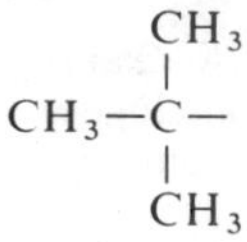

Hydrocarbons that contain one or more double bonds are called "unsaturated," and are named by replacing the ending "-ane" of the corresponding saturated hydrocarbon with the ending "-ene," "-adiene," or "-atriene," etc. The generic names of unsaturated hydrocarbons are alkene, alkadiene, alkatriene, etc. The double bonds receive the lowest possible numbers.

In the following examples, the names listed in the *Chemical Substances Index* of the CAS system are given first. Other names are given in parentheses.

EXAMPLES: 2-butene
(2-butylene)

$$\overset{1}{C}H_3-\overset{2}{C}H=\overset{3}{C}H-\overset{4}{C}H_3$$

1,4-hexadiene

$$\overset{1}{C}H_2=\overset{2}{C}H-\overset{3}{C}H_2-\overset{4}{C}H=\overset{5}{C}H-\overset{6}{C}H_3$$

ethene
(ethylene)

$$CH_2=CH_2$$

2-methyl-1,3-butadiene
(isoprene)

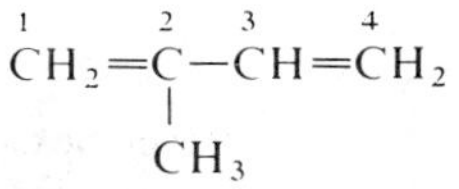

$$\overset{1}{C}H_2=\overset{2}{C}-\overset{3}{C}H=\overset{4}{C}H_2$$
$$|$$
$$CH_3$$

Hydrocarbons containing one or more triple bonds are named by replacing the ending "-ane" of the corresponding saturated hydrocarbon with the ending "-yne," "-adiyne," "-atriyne," etc. The triple bonds receive the lowest possible numbers. Double bonds take precedence over triple bonds when there is a choice in numbering.

EXAMPLES: 1-butyne

$$\overset{1}{C}H\equiv\overset{2}{C}-\overset{3}{C}H_2-\overset{4}{C}H_3$$

1-hexene-3,5-diyne

$$\overset{1}{C}H_2=\overset{2}{C}H-\overset{3}{C}\equiv\overset{4}{C}-\overset{5}{C}\equiv\overset{6}{C}H$$

Unsaturated branched acyclic hydrocarbons are numbered in the same manner as alkanes. The longest chain is chosen as the parent. If the alkene or alkyne contains two or more chains of equal length, the chain containing the maximum number of double bonds is chosen as the parent.

Univalent or multivalent groups derived from alkenes and alkynes are named as follows:

EXAMPLES: ethenyl (vinyl) $\overset{2}{C}H_2=\overset{1}{C}H-$

ethynyl $\overset{2}{C}H\equiv\overset{1}{C}-$

2-propynyl $\overset{3}{C}H\equiv\overset{2}{C}-\overset{1}{C}H_2-$

methylidyne $CH\equiv$

ethylidene $CH_3-CH=$

ethylidyne $CH_3-C\equiv$

Alicyclic Hydrocarbons. Saturated monocyclic hydrocarbons, or "cycloalkanes," are named by attaching the prefix "cyclo" to the name of the acyclic unbranched alkane.

EXAMPLES: cyclopropane

cyclohexane

Univalent groups derived from unsubstituted cycloalkanes are named cyclopropyl, cyclohexyl, etc., in a manner analogous to that used for naming acyclic alkanes. The carbon atom with the free valence is numbered as 1.

EXAMPLES: cyclopropyl

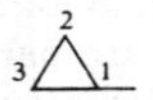

cyclohexyl

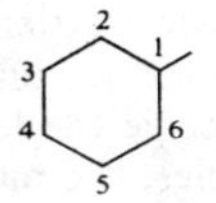

Unsaturated monocyclic hydrocarbons are named by substituting "-ene," "-adiene," "-atriene," "-yne," "-adiyne," etc., in the name of the corresponding cycloalkane. Double and triple bonds are given numbers as low as possible.

EXAMPLES: cyclohexene

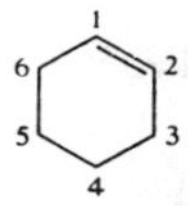

1,3-cyclohexadiene

1-cyclodecen-4-yne $\overset{7}{C}H_2\overset{8}{C}H_2-\overset{9}{C}H_2-\overset{10}{C}H_2-\overset{1}{C}H$
$\overset{6}{C}H_2-\overset{5}{C}\equiv\overset{4}{C}-\overset{3}{C}H_2$ $\overset{2}{C}H$

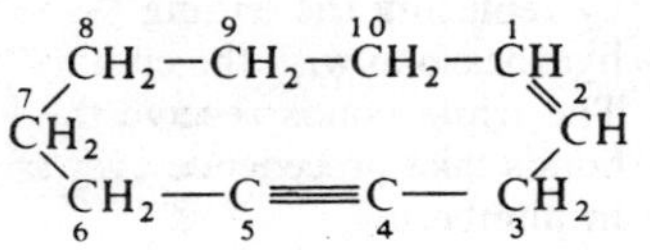

Aromatic Hydrocarbons. Aromatic hydrocarbons generally are considered those which have the characteristic chemical properties of benzene. Many such compounds are known more commonly by their trivial names than by their systematic names.

EXAMPLES: benzene

methylbenzene (toluene)

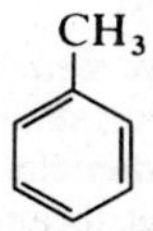

1,2-dimethylbenzene (o-xylene)

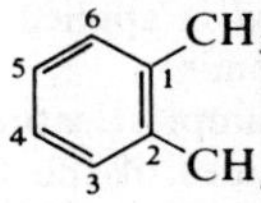

ethenylbenzene (styrene)

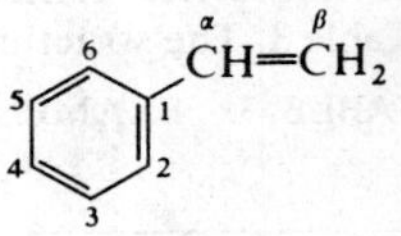

(1-methylethyl)benzene (cumene)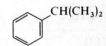

The terms *ortho*, *meta*, and *para* (abbreviated *o*, *m*, and *p*) refer to the location of substituents on the benzene ring and are equivalent to 1,2-, 1,3-, and 1,4-substitution in systematic nomenclature, respectively. The lowest numbers possible are given to substituents.

Fused aromatic systems are named by prefixing the largest parent trivial names with combining forms such as benz(o)- and napthth(o)-. Hydrocarbons that contain five or more fused benzene rings in a linear arrangement are named from a numerical prefix followed by "-acene."

EXAMPLES: naphthalene

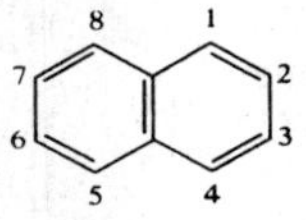

anthracene

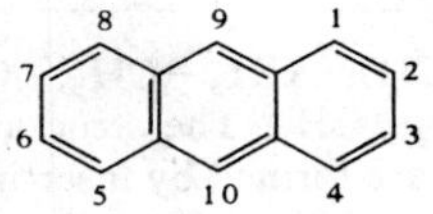

hexacene

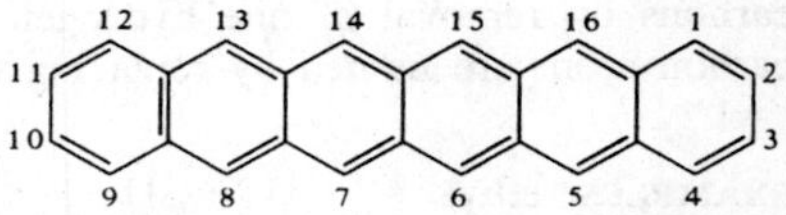

Fusion prefixes are used to designate to which side of the parent hydrocarbon a substituent ring is attached.

EXAMPLE:

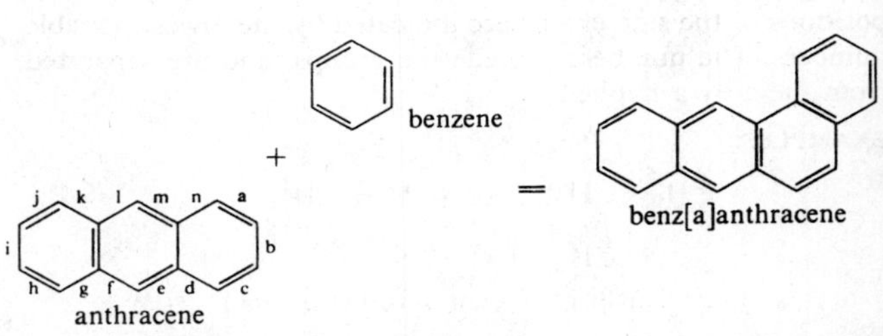

benzene

+

anthracene

=

benz[a]anthracene

Hydrogenation products of complex aromatic ring systems that are not treated as alicyclic hydrocarbons are named by prefixing "dihydro," "tetrahydro," etc., to the parent name. The lowest locants are used. "Perhydro" is used in trivial nomenclature to indicate a fully hydrogenated compound.

EXAMPLES: 1,2-dihydronaphthacene

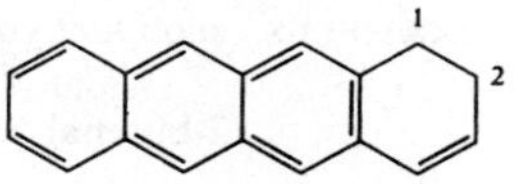

docosahydropentacene
(perhydropentacene)

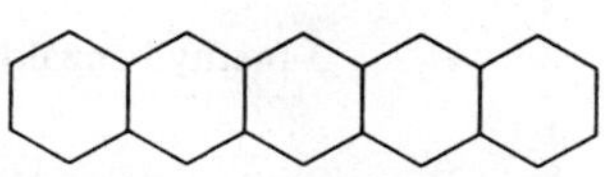

Multiple unsubstituted assemblies of benzene rings are named by using the appropriate prefix with the radical name "phenyl."

EXAMPLES: 1,1'-biphenyl

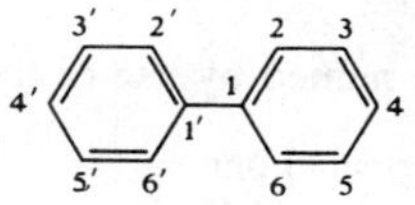

1,1':4',1''-terphenyl
(p-terphenyl)

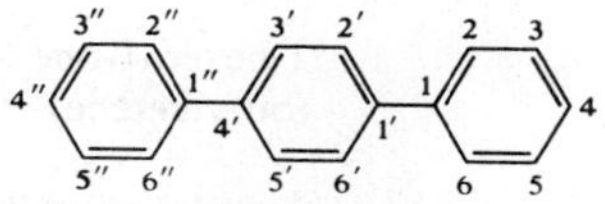

Indicated hydrogen in aromatic systems is assigned to angular or non-angular positions when needed to accommodate structural features in systematic nomenclature. The lowest locants are used.

EXAMPLE: 1H-indene
(not 3H-)

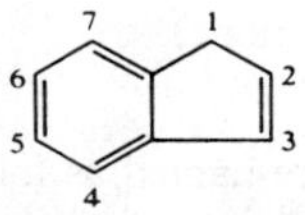

Bridged Hydrocarbon Ring Systems. These compounds are named by prefixing the parent ring with "bicyclo," "tricyclo," etc., or in some complex cases by using prefixes that denote the nature of the bridges. The three numbers in brackets denote the number of carbon atoms in each of the three bridges in descending order.

EXAMPLES:
bicyclo[3.1.0]hexane

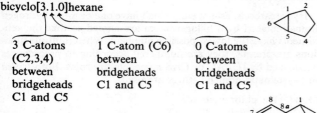

3 C-atoms (C2,3,4) between bridgeheads C1 and C5	1 C-atom (C6) between bridgeheads C1 and C5	0 C-atoms between bridgeheads C1 and C5

1,2,3,4-tetrahydro-1,4-methanonaphthalene

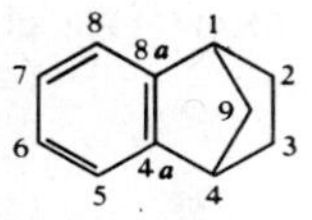

Spiro Hydrocarbon Ring Systems. Spiro systems contain pairs of rings or ring systems that have only one atom (a "spiro atom") in common. The name of the simplest monospiro system is formed by prefixing the acyclic hydrocarbon name with "spiro" and numerals separated by periods. The numerals are given in ascending order to define the number of atoms in each ring linked to the spiro atom. Numbering begins at the atom next to the spiro atom in the smallest ring for monospiro systems.

EXAMPLES: spiro[3.4]octane

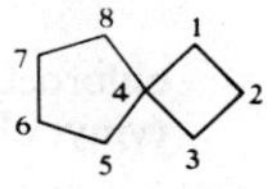

dispiro[5.1.7.2]heptadecane

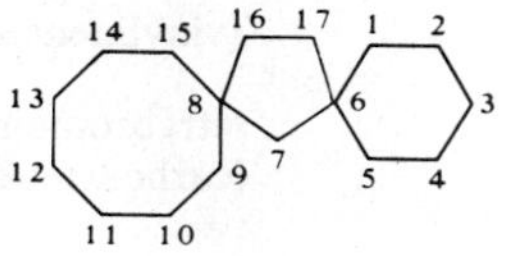

Carboxylic Acids and Their Anhydrides

Acids are named according to the Geneva ("-oic") or "-carboxylic" system. They are regarded as derived from parent hydrocarbons having the same number of carbon atoms so that CH_3 is replaced by COOH. The carbon atom of the carboxyl group is assigned number one in aliphatic monocarboxylic acids. In an alternative numbering system, the Greek letters *alpha*, *beta*, etc., are assigned to the second, third, etc., carbon atoms, respectively, leading away from the —COOH group. When chain branching is present, the longest chain containing the carboxylic acid group at one end is chosen for naming the molecule. Unsaturated aliphatic acids are named so the longest chain includes the maximum number of unsaturated linkages. Double bonds are given preference over triple bonds. Trivial names are retained for some common molecules.

EXAMPLES:

formic acid — HCOOH

acetic acid — $\overset{2}{C}H_3\overset{1}{C}OOH$

hexanoic acid (caproic acid) — $CH_3(CH_2)_4COOH$

octadecanoic acid (stearic acid) — $CH_3(CH_2)_{16}COOH$

2-propenoic acid (acrylic acid) — $\overset{3}{C}H_2{=}\overset{2}{C}H\overset{1}{C}OOH$

2-methylbutanoic acid — $\overset{4}{C}H_3\overset{3}{C}H_2\overset{2}{C}H\overset{1}{C}OOH$ with CH_3

cyclobutanecarboxylic acid — ⬜—COOH

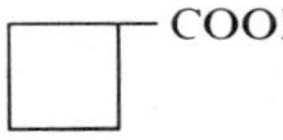

benzoic acid — ⬡—COOH

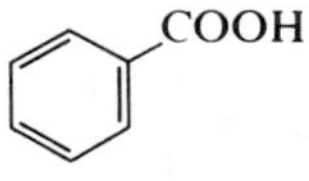

ethanedioic acid (oxalic acid) — HOOC—COOH

butanedioic acid (succinic acid) — $HOOC{-}CH_2CH_2{-}COOH$

1,2-benzenedicarboxylic acid (phthalic acid) — ⬡ with COOH, COOH

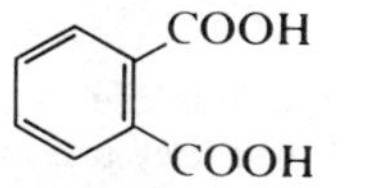

Conjunctive nomenclature may be used for naming cyclic acids. It is applied to any ring system attached by a single bond to one or more acyclic hydrocarbon chains, each of which bears only one principal functional group.

EXAMPLE: cyclohexaneacetic acid — ⬡—CH_2COOH

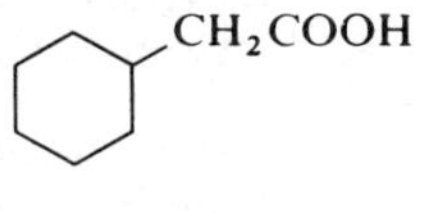

Acid anhydride names are formed from systematic, Geneva, trivial, conjunctive, or other types of acid names.

EXAMPLES: propanoic acid anhydride (propionic anhydride) — CH_3CH_2CO — O — CH_3CH_2CO

benzoic acid anhydride
(benzoic anhydride)

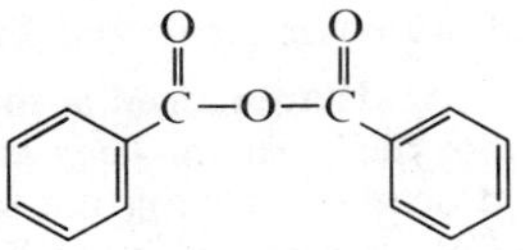

See also **Carboxylic Acids.**

Alcohols

Monohydric alcohols are named by adding "-ol" to a molecular skeleton name. Carbon chains, unsaturation, etc., are numbered in a manner analogous to that used for carboxylic acids. (See preceding section.)

EXAMPLES: methanol CH_3OH
(methyl alcohol)

cyclohexanol
(cyclohexyl alcohol)

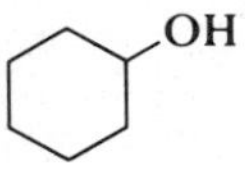

2-propen-1-ol $\overset{3}{C}H_2=\overset{2}{C}H\overset{1}{C}H_2OH$
(allyl alcohol)

See also **Alcohols.**

The simplest aromatic hydroxy compound is called phenol:

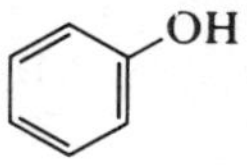

Esters

Simple esters are named on the basis of their alcohol and acid functions. Carbon chains, unsaturation, etc., are numbered in a manner analogous to that used for carboxylic acids.

EXAMPLES:
ethyl acetate $CH_3COOC_2H_5$
(inverted name:
acetic acid,
ethyl ester)

propyl 2-butenoate $\overset{4}{C}H_3\overset{3}{C}H=\overset{2}{C}H\overset{1}{C}OOCH_2CH_2CH_3$

methyl benzoate

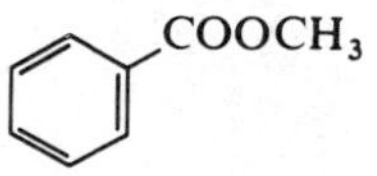

See also **Esters.**

Ethers

Alkoxy compounds are commonly called ethers. In current CAS nomenclature they are named as derivatives of functional parent compounds, hydrocarbons, etc., by use of "oxy" radicals.

EXAMPLES: 1,1'-oxybis(ethane)
(ethyl ether or $C_2H_5OC_2H_5$
diethyl ether)

(hexyloxy)cyclopropane $\triangleright$—$O(CH_2)_5CH_3$
(cyclopropyl hexyl ether)

methoxybenzene
(anisole)

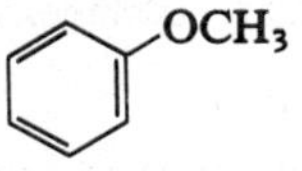

See also **Ethers.**

Aldehydes

Aldehydes are named from the corresponding acids by use of "-carboxaldehyde" and "-al" suffixes.

EXAMPLES: acetaldehyde CH_3CHO

2-butenal $\overset{4}{C}H_3\overset{3}{C}H=\overset{2}{C}H\overset{1}{C}HO$

3-methylbenzaldehyde

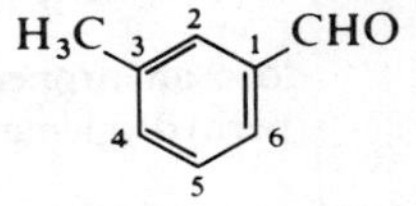

See also **Aldehydes.**

Ketones

Ketones are named by use of the characteristic suffix "-one."

EXAMPLES: 2-propanone $\overset{1}{C}H_3\overset{2}{C}O\overset{3}{C}H_3$
(acetone)

1-hexen-1-one $CH_3(CH_2)_3\overset{2}{C}H=\overset{1}{C}O$
(butylketene)

4-cyclopentyl-2-butanone $\overset{4}{C}H_2\overset{3}{C}H_2\overset{2}{C}O\overset{1}{C}H_3$

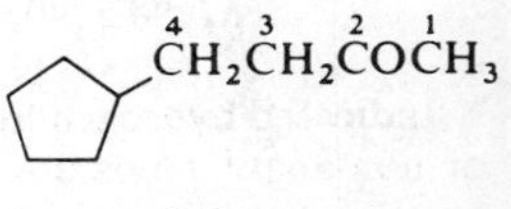

diphenylmethanone
(benzophenone)

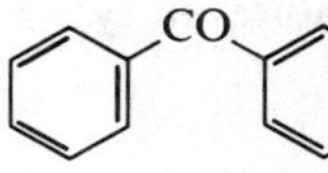

Peroxides

Simple peroxides are named as follows:

EXAMPLES:
ethyl methyl peroxide $C_2H_5OOCH_3$

benzoyl peroxide

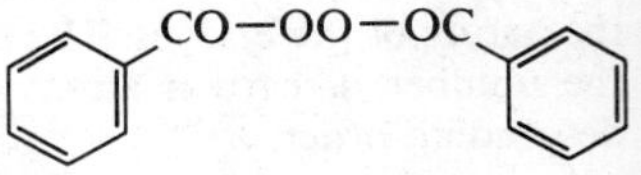

Halogenated Compounds

Hydrocarbons, esters, etc., which have one or more hydrogen atoms replaced by a halogen atom are named so that the substituents have the lowest possible numbers. When multiple functions are present, they are numbered according to precedences established by IUPAC or CAS nomenclature rules. Both trivial and systematic nomenclatures are used.

EXAMPLES: chloromethane CH_3Cl
(methyl chloride)

2-iodobutane $\overset{1}{C}H_3\overset{2}{C}H\overset{3}{C}H_2\overset{4}{C}H_3$
(sec-butyl iodide) $|$
I

1-bromo-2-fluorobenzene

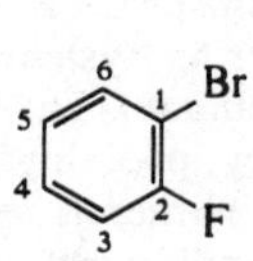

3-chloropentanoic acid $\overset{5}{C}H_3\overset{4}{C}H_2\overset{3}{C}H\overset{2}{C}H_2\overset{1}{C}OOH$
(β-chlorovaleric acid) $|$
Cl

chloroethene $CH_2=CHCl$
(vinyl chloride)

1,1-dichloroethene $CH_2=CCl_2$
(vinylidene chloride)

tetrabromomethane CBr_4
(carbon tetrabromide)

Acid halides are named as follows:

EXAMPLES: acetyl chloride CH_3COCl

6-heptenoyl chloride $\overset{7}{CH_2}=\overset{6}{CH}(CH_2)_4\overset{1}{COCl}$

cyclohexanecarbonyl bromide

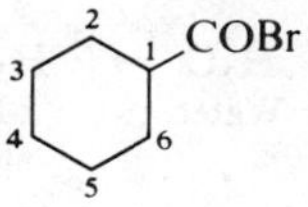

benzoyl fluoride

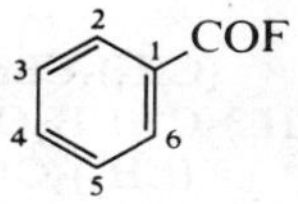

See also **Chlorinated Organics.**

Nonheterocyclic Nitrogen Compounds

Amines are named by adding the suffix "-amine" either to the name of the hydrocarbon or to the hydrocarbon radical. A second system names all amines as derivatives of primary amines.

EXAMPLES: methanamine (methylamine) CH_3NH_2

N,N-dipropyl-1-propanamine (tripropylamine) $(CH_3CH_2CH_2)_3N$

benzenamine (aniline)

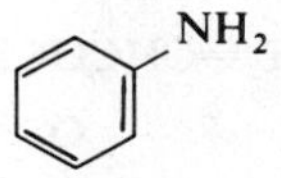

N-ethylcyclohexanamine (N-ethylcyclohexylamine)

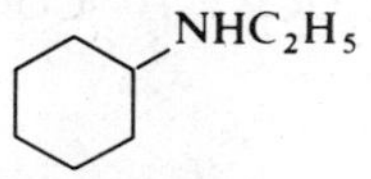

Imines are named from the hydrocarbon by the addition of the suffix "-imine."

EXAMPLES: ethanimine $\overset{2}{CH_3}\overset{1}{CH}=NH$

2,4-cyclopentadien-1-imine

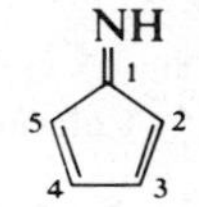

See also **Amines.**

Names of amides are based on the corresponding acids. Thus, "-oic acid" becomes "-amide," and "-carboxylic acid" becomes "-carboxamide."

EXAMPLES:
acetamide $\overset{2}{CH_3}\overset{1}{CONH_2}$

benzamide

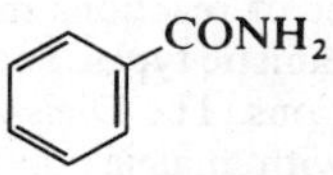

cyclohexanecarboxamide

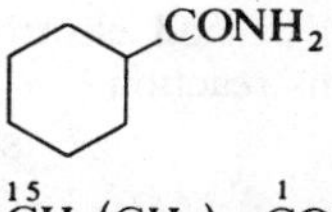

N-methylpentadecanamide $\overset{15}{CH_3}(CH_2)_{13}\overset{1}{CONHCH_3}$

N,N-dimethyl-2,4-pentadienamide $\overset{5}{CH_2}=\overset{4}{CH}\overset{3}{CH}=\overset{2}{CH}\overset{1}{CON}(CH_3)_2$

See also **Amides.**

Nitro and nitroso compounds are named so that the substituents have the lowest possible numbers.

EXAMPLES: 2-nitrobutane $\overset{1}{CH_3}\overset{2}{CH}\overset{3}{CH_2}\overset{4}{CH_3}$
$\quad\quad\quad\quad\quad\quad NO_2$

4-nitrosobenzoic acid

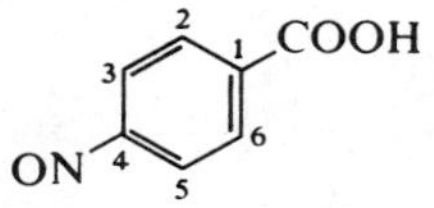

See also **Nitro-** and **Nitroso- Compounds.**

Nitrile names are formed from common names of carboxylic acids.

EXAMPLES: acetonitrile CH_3CN

benzonitrile (phenyl cyanide)

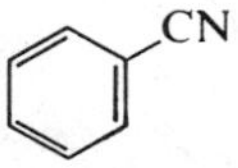

3-butenenitrile $\overset{4}{CH_2}=\overset{3}{CH}\overset{2}{CH_2}\overset{1}{CN}$

ethenetetracarbonitrile (tetracyanoethylene)

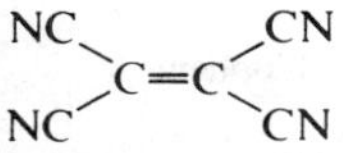

In the presence of more senior functional groups, the nitrile function is expressed by the prefix "cyano."

EXAMPLES:
4-cyanobenzamide

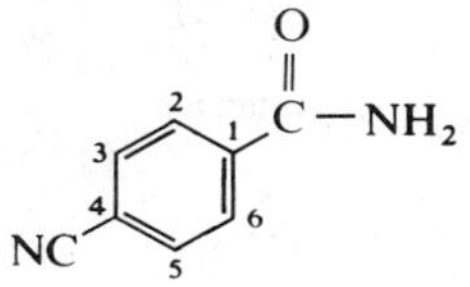

3-cyanobutanoic acid $\overset{4}{CH_3}\overset{3}{CH}\overset{2}{CH_2}\overset{1}{COOH}$
$\quad\quad\quad\quad\quad CN$

Nonheterocyclic Sulfur Compounds

Sulfur compounds are named similarly to oxygen compounds.

EXAMPLES: (methylthio)benzene (methyl phenyl sulfide)

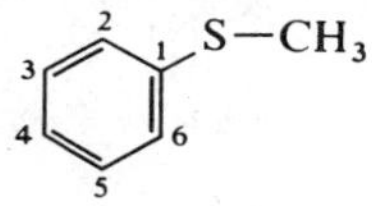

1,1'-sulfinylbis(benzene) (diphenyl sulfoxide)

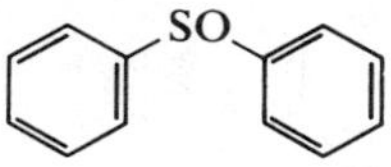

2-(propylsulfonyl)naphthalene (2-naphthyl propyl sulfone)

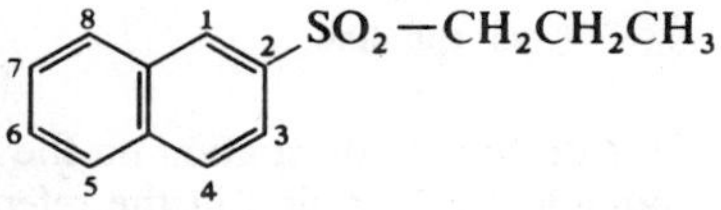

2-butanethiol (2-mercaptobutane) $\overset{1}{CH_3}\overset{2}{CH}\overset{3}{CH_2}\overset{4}{CH_3}$
$\quad\quad\quad\quad SH$

2-propanethione (thioacetone) $CH_3\overset{S}{\overset{\|}{C}}CH_3$

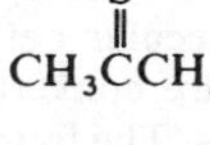

N-methylbenzenesulfonamide

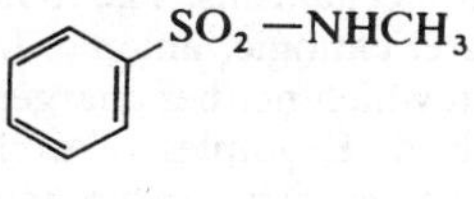

benzenesulfonic acid

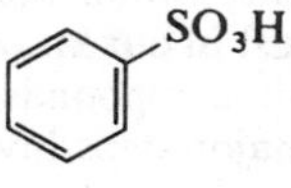

Heterocyclic Compounds

Some of the common heterocyclic nitrogen, oxygen, and sulfur compounds and their numbering systems are shown:

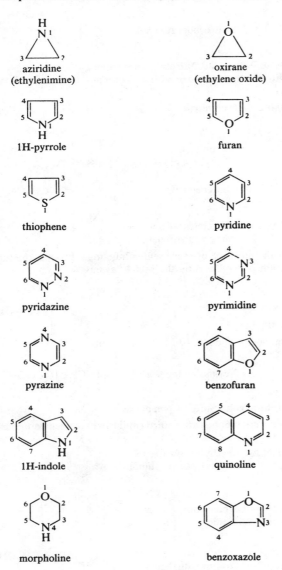

aziridine
(ethylenimine)

oxirane
(ethylene oxide)

1H-pyrrole

furan

thiophene

pyridine

pyridazine

pyrimidine

pyrazine

benzofuran

1H-indole

quinoline

morpholine

benzoxazole

Additional information on the nomenclature of organic compounds can be found in the references listed at the end of this entry.

Organic Reaction Mechanisms and Processes

There are many reactions by which one organic compound is converted into another, and new reactions are constantly being developed. The molecular details of the intermediate steps by which compounds are converted into new products are called reaction mechanisms. The four broad classes of reaction mechanisms are: cationic, anionic, free radical, and multicenter processes in which neither charged species nor odd electron species is involved. Examples of each type will be given, but many variations can exist within each type. Also, varying degrees of sophistication exist in our knowledge of the exact reaction pathways that organic compounds follow. The examples discussed show only the major steps involved.

Cationic and anionic mechanisms involve species that have either positive or negative charges, respectively (heterolytic reactions). An example of a reaction that proceeds via a cationic mechanism is the hydrolysis of t-bromide (Scheme 2).

SCHEME 2

$$(CH_3)_3CBr + H_2O \rightarrow (CH_3)_3COH + (CH_3)_2C{=}CH_2 + HBr$$

t-Butyl   Water t-Butyl   Isobutylene   Hydrogen
bromide   alcohol   bromide

Mechanism:

$$(CH_3)_3CBr \rightarrow (CH_3)_3C^+ + Br^-$$
$$(CH_3)_3C^+ + H_2O \rightarrow (CH_3)_3COH + H^+$$
$$(CH_3)_3C^+ \rightarrow (CH_3)_2C{=}CH_2 + H^+$$

The addition of methanol to methyl acrylate in the presence of sodium methoxide is an example of a reaction that proceeds by an anionic mechanism (Scheme 3).

SCHEME 3

$$CH_2{=}CHCOCH_3 + CH_3OH \xrightarrow{CH_3O^-Na^+} CH_3OCH_2CH_2COCH_3$$

Methyl acrylate   Methanol   Sodium   Methyl
                               methoxide β-methoxypropionate

Mechanism:

$$CH_2{=}CHCOCH_3 + CH_3O^- \rightarrow CH_3OCH_2CH{=}COCH_3$$

$$CH_3OCH_2CH{=}COCH_3 + CH_3OH \rightarrow$$

$$CH_3OCH_2CH_2COCH_3 + CH_3O^-$$

Free radical (homolytic) reactions involve species with an unpaired electron. The ultraviolet light initiated reaction of methane with chlorine is an example (Scheme 4).

SCHEME 4

$$CH_4 + Cl_2 \xrightarrow{h\nu} CH_3Cl + HCl$$

Methane   Chlorine   Methyl   Hydrogen
                       chloride   chloride

Mechanism:

$$Cl_2 \xrightarrow{h\nu} 2Cl\cdot$$

$$Cl\cdot + CH_4 \rightarrow CH_3\cdot + HCl$$

$$CH_3\cdot + Cl_2 \rightarrow CH_3Cl + Cl\cdot$$

A number of reactions do not seem to belong to any of the above mechanistic types. Such processes are referred to as multicenter reactions. The Diels-Alder cycloaddition reaction of 1,3-butadiene with maleic anhydride is an example (Scheme 5). No charged or odd electron intermediates seemingly are involved in this reaction.

SCHEME 5

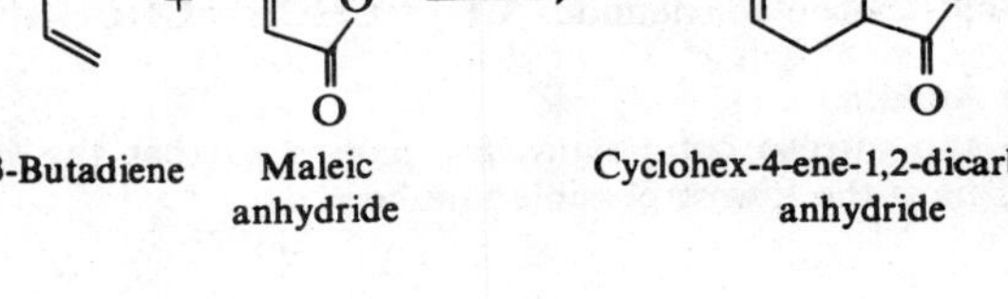

1,3-Butadiene   Maleic
                  anhydride

Cyclohex-4-ene-1,2-dicarboxylic
anhydride

The reactions shown in Schemes 2–5, with the exception of the photodissociation of Cl_2 to Cl atoms (Scheme 4), occur in molecules that are in electronic ground states. Reactions also can occur in molecules existing in excited electronic states. Commonly, these excited states are produced by irradiating the reactants with ultraviolet or visible light, hence the term photochemistry. When a molecule is in an excited state, its reactions are often different from those it normally exhibits in its ground state. An organic compound often can exist in more than one excited state, as shown in Scheme 6.

SCHEME 6

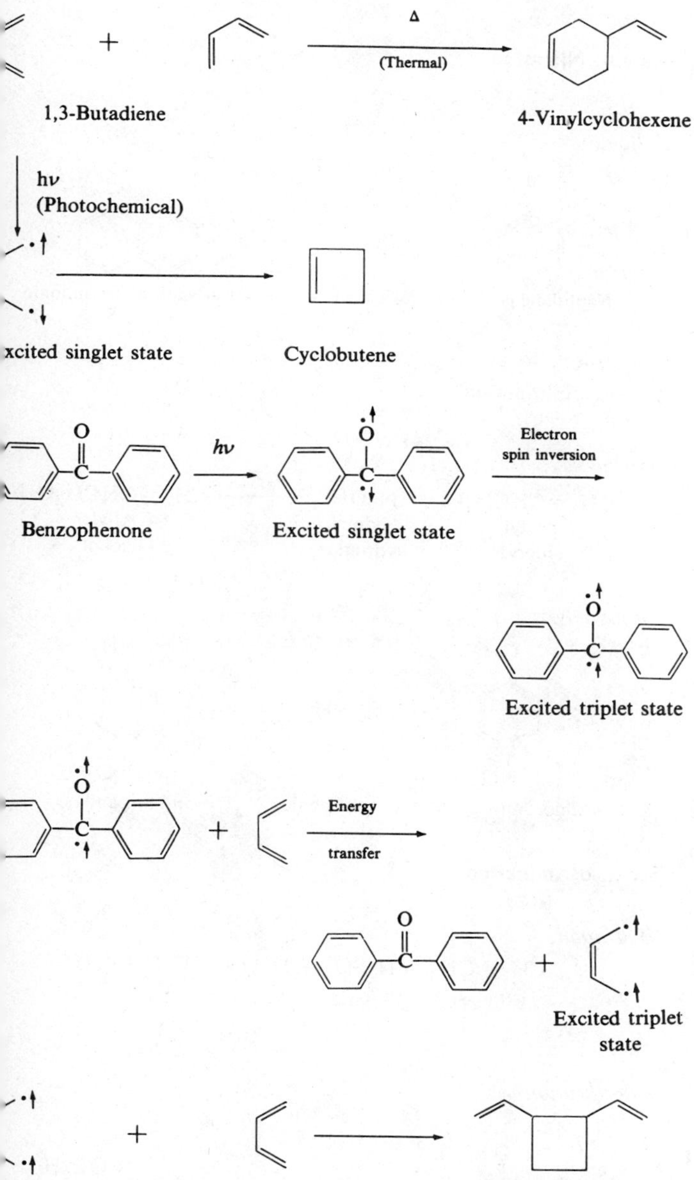
Δ
(Thermal)
1,3-Butadiene
4-Vinylcyclohexene
hν
(Photochemical)
xcited singlet state
Cyclobutene
hν
Electron
spin inversion
Benzophenone
Excited singlet state
Excited triplet state
Energy
transfer
Excited triplet
state
1,2-Divinylcyclobutane

Excited state molecules usually differ from their ground state counterparts by having dissimilar electronic and geometric configurations, and shorter lifetimes, e.g., 10^{-2} to 10^{-10} second.

Many organic reactions are referred to by the inventor's or discoverer's name. A few name reactions are shown in Scheme 7.

SCHEME 7. Examples of Organic Name Reactions.

Friedel-Crafts Alkylation

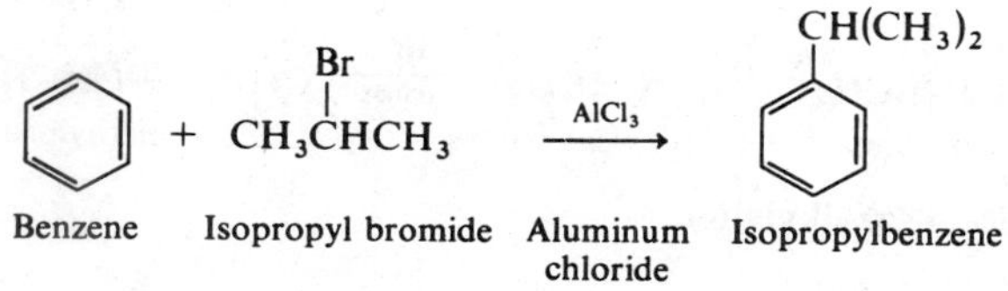
Br
CH₃CHCH₃
AlCl₃
CH(CH₃)₂
Benzene
Isopropyl bromide
Aluminum
chloride
Isopropylbenzene

See also **Friedel-Crafts Reaction.**

Grignard Reaction

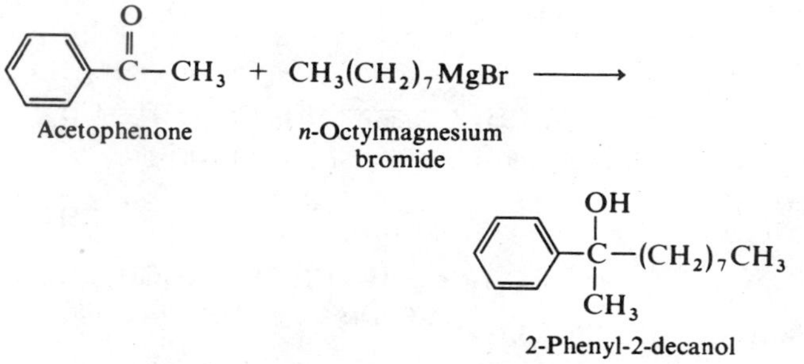
C—CH₃ + CH₃(CH₂)₇MgBr
Acetophenone
n-Octylmagnesium
bromide
OH
C—(CH₂)₇CH₃
CH₃
2-Phenyl-2-decanol

Baeyer-Villiger Oxidation

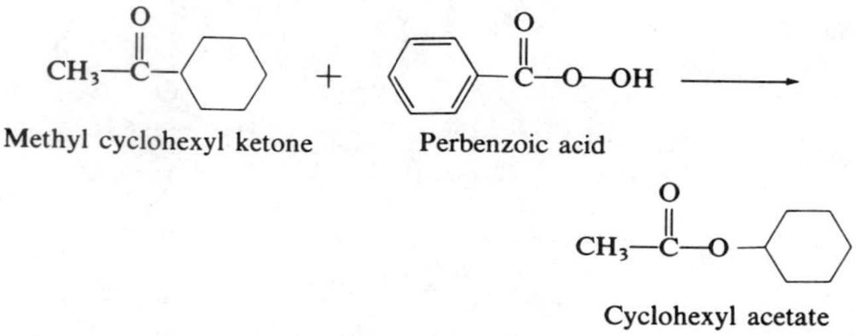
CH₃—C—
C—O—OH
Methyl cyclohexyl ketone
Perbenzoic acid
CH₃—C—O
Cyclohexyl acetate

Meerwein-Ponndorf-Verley Reduction

$$CH_3CH{=}CHCH{=}O + Al[OCH(CH_3)_2]_3 \rightarrow CH_3CH{=}CHCH_2OH$$

Crotonaldehyde — Aluminum isopropoxide — But-2-en-1-ol

Skraup Quinoline Synthesis

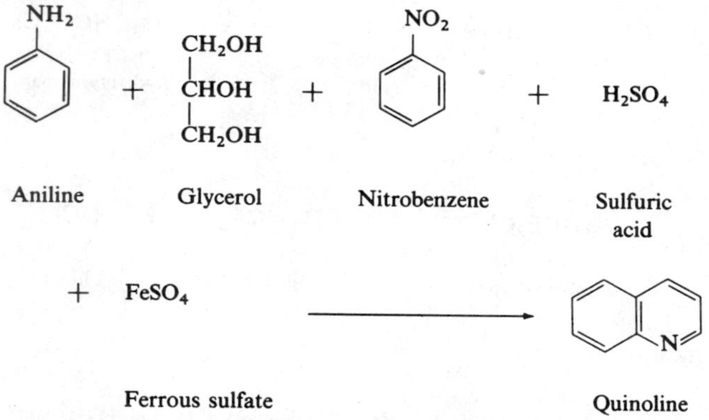
NH₂
CH₂OH
CHOH
CH₂OH
NO₂
H₂SO₄
Aniline
Glycerol
Nitrobenzene
Sulfuric
acid
FeSO₄
N
Ferrous sulfate
Quinoline

Some important industrial organic processes are shown in Scheme 8. Although most of these reactions involve mixtures of isomers or homologs as reactants and products, for simplicity only the major components are shown.

SCHEME 8. Examples of Industrial Organic Processes.

Alkylation

$(CH_3)_3CH + CH_2=CHCH_3 \xrightarrow[\text{catalyst}]{HF} CH_3CH-CHCH_2CH_3$

Isobutane   Propylene   2,3-Dimethylpentane

See also **Alkylation.**

Isomerization

$CH_3(CH_2)_2CH_3 \xrightarrow[\text{catalyst}]{AlCl_3} CH_3CHCH_3$

n-Butane   *i*-Butane

Cracking

$CH_3(CH_2)_{14}CH_3 \xrightarrow[\text{catalyst}]{\text{Silica alumina}} CH_3(CH_2)_4CH=CH_2$

n-Hexadecane   1-Heptene

$+ CH_3CH_2CH=CH_2 + CH_3CH_2CHCH_3$

1-Butene   Isopentane

See also **Cracking Process.**

Oxidation

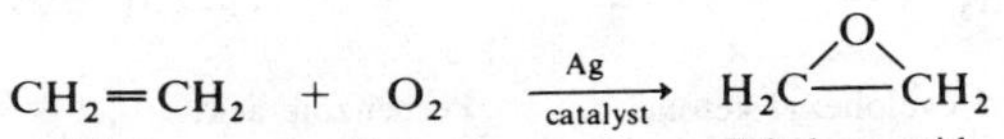

$CH_2=CH_2 + O_2 \xrightarrow[\text{catalyst}]{Ag} H_2C\overset{O}{-}CH_2$

Ethylene   Oxygen   Ethylene oxide

Chlorination

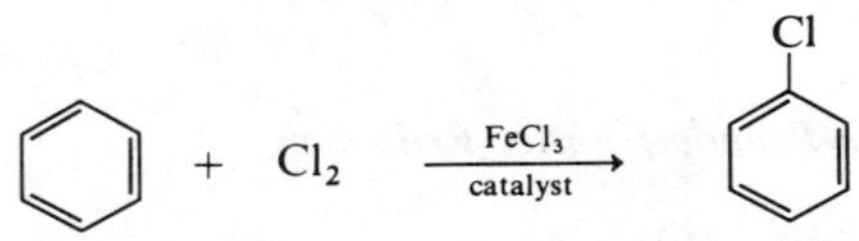

$\text{Benzene} + Cl_2 \xrightarrow[\text{catalyst}]{FeCl_3} \text{Chlorobenzene}$

Benzene   Chlorine   Chlorobenzene

Hypochlorination

$CH_3CH=CH_2 + HOCl \longrightarrow CH_3CHCH_2Cl$

Propylene   Hypochlorous   1-Chloro-
           acid   2-hydroxypropane

Dehydrochlorination

$CH_3CF_2Cl \xrightarrow{\text{Pyrolysis}} CH_2=CF_2 + HCl$

1-Chloro-1,1-   Vinylidene   Hydrogen
difluoroethane   fluoride   chloride

Bromination

$CH_2=CH_2 + Br_2 \longrightarrow CH_2BrCH_2Br$

Ethylene   Bromine   Ethylene
                  dibromide

Nitration

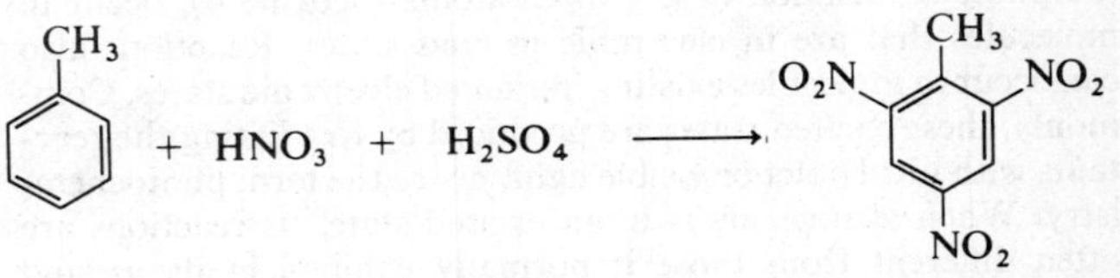

$\text{Toluene} + HNO_3 + H_2SO_4 \longrightarrow \text{2,4,6-Trinitrotoluene}$

Toluene   Nitric acid   Sulfuric acid   2,4,6-Trinitrotoluene
                              (TNT)

See also **Nitration.**

Sulfonation

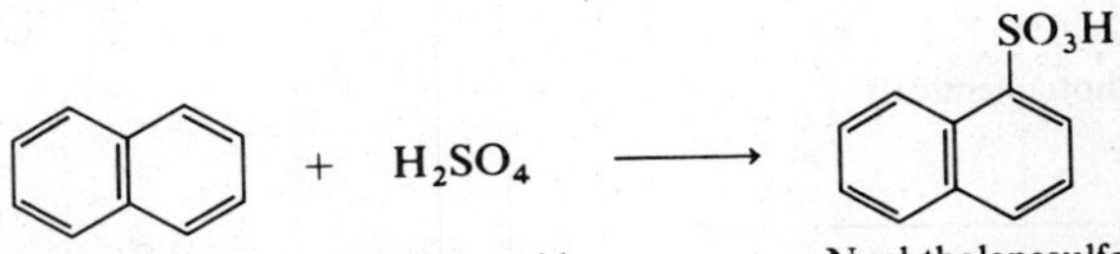

$\text{Naphthalene} + H_2SO_4 \longrightarrow \alpha\text{-Naphthalenesulfonic acid}$

Naphthalene   Sulfuric acid   α-Naphthalenesulfonic
                             acid

See also **Sulfonation.**

Hydrolysis

$CH_2=CHCH_2Cl + NaOH \longrightarrow CH_2=CHCH_2OH$

Allyl   Sodium   Allyl
chloride   hydroxide   alcohol

Ammonolysis

$\text{p-Chloronitrobenzene} + NH_3 \longrightarrow \text{p-Nitroaniline}$

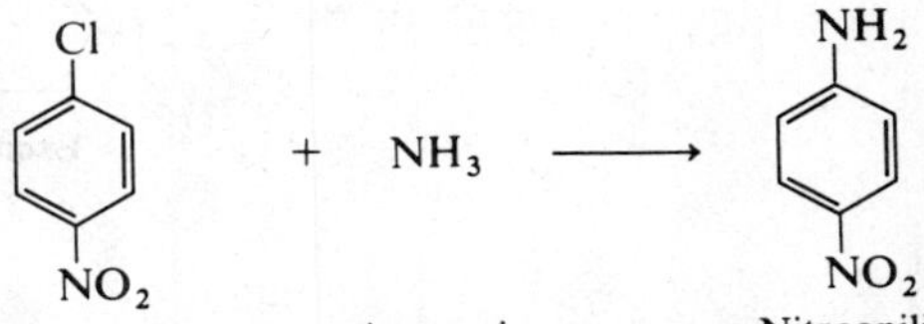

p-Chloronitrobenzene   Ammonia   *p*-Nitroaniline

See also **Amination.**

Hydration

$CH_2=CH_2 + H_2SO_4 + H_2O \rightarrow CH_3CH_2OH$

Ethylene   Sulfuric   Water   Ethanol
           Acid

Esterification

$\text{Phthalic anhydride} + C_8H_{17}OH \xrightarrow[\text{catalyst}]{H_2SO_4} \text{Dioctyl phthalate}$

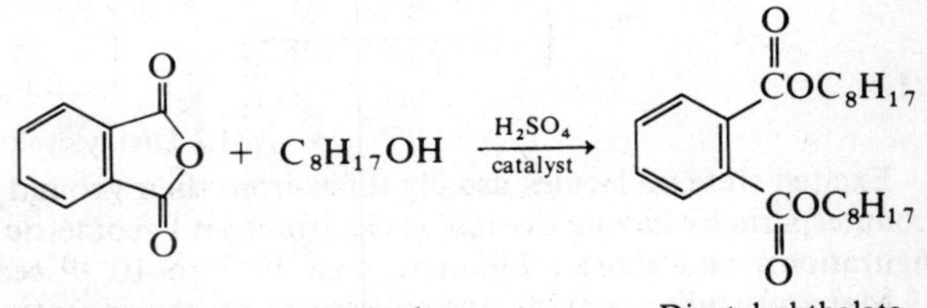

Phthalic anhydride   Octanol   Dioctyl phthalate

Hydrogenation

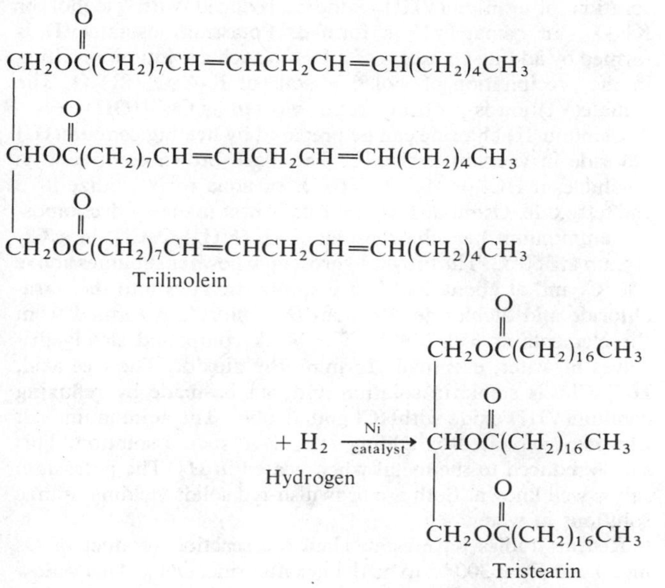

$$CH_2OC(CH_2)_7CH=CHCH_2CH=CH(CH_2)_4CH_3$$
$$\overset{O}{\underset{|}{CHOC(CH_2)_7CH=CHCH_2CH=CH(CH_2)_4CH_3}}$$
$$CH_2OC(CH_2)_7CH=CHCH_2CH=CH(CH_2)_4CH_3$$

Trilinolein

$$+ H_2 \xrightarrow[\text{catalyst}]{\text{Ni}} \begin{array}{c} CH_2OC(CH_2)_{16}CH_3 \\ CHOC(CH_2)_{16}CH_3 \\ CH_2OC(CH_2)_{16}CH_3 \end{array}$$

Hydrogen   Tristearin

Hydrogenolysis

$$CH_3(CH_2)_{10}COCH_3 + H_2 \xrightarrow[\text{catalyst}]{\text{Copper-chromic oxide}} CH_3(CH_2)_{11}OH$$
Methyl   Hydrogen   Lauryl
laurate   alcohol

Dehydrogenation

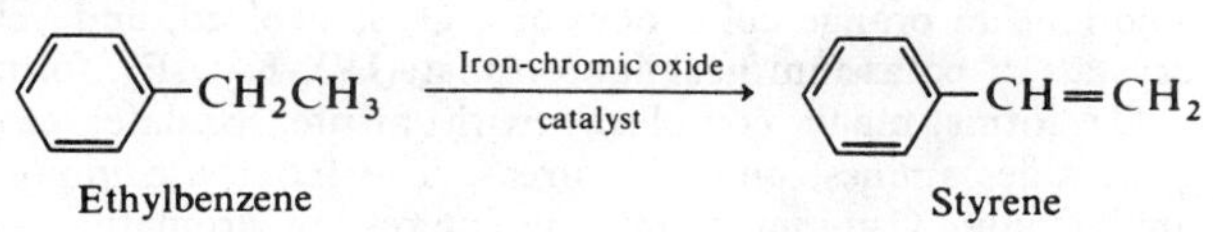

Ethylbenzene $\xrightarrow[\text{catalyst}]{\text{Iron-chromic oxide}}$ Styrene

Coupling

p-Sulfobenzenediazonium   Sodium
chloride   2-naphtholate

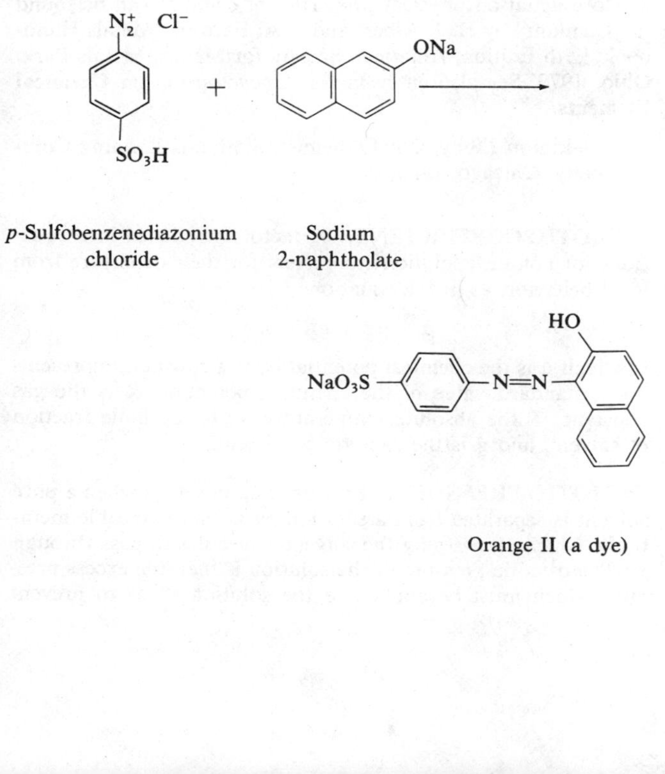

Orange II (a dye)

Hydroformylation

$$CH_3CH=CH_2 + CO + H_2 \longrightarrow CH_3(CH_2)_2\overset{O}{C}H$$
Propylene   Carbon   Hydrogen   *n*-Butyraldehyde
            monoxide

Fermentation

Sucrose $\xrightarrow[\text{(a microorganism)}]{\textit{Aspergillus niger}}$

Citric acid

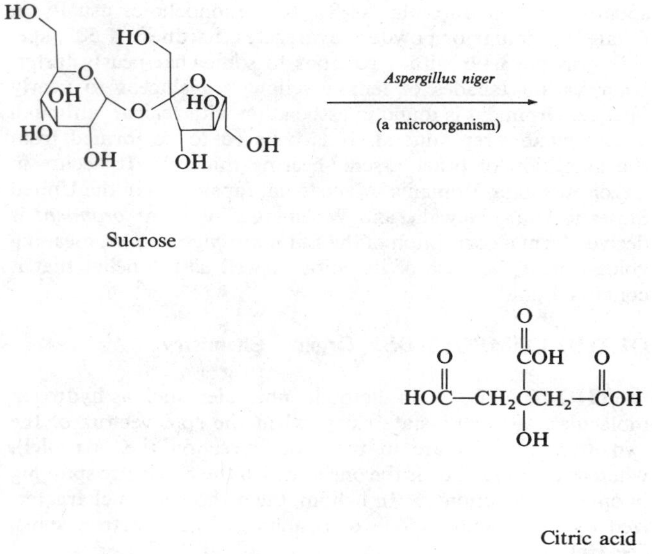

See also **Fermentation.**

Vinyl Polymerization

$$CH_2=CHCl \xrightarrow[\text{initiator}]{K_2S_2O_8} (CH_2CHCl)_{\sim 1,000-10,000}$$
Vinyl chloride   Polyvinyl chloride

Condensation Polymerization

$$HOC(CH_2)_4COH + H_2N(CH_2)_6NH_2$$
Adipic acid   Hexamethylenediamine

$$\xrightarrow{\Delta \text{ (heat)}} HO[C(CH_2)_4CNH(CH_2)_6NH]_{\sim 50-90}H$$
Poly(hexamethyleneadipamide)
[Nylon 66]

References

1. *Naming and Indexing of Chemical Substances for CHEMICAL AB-STRACTS during the Ninth Collective Period (1972–1976)*; Section IV (Selection of Names for Chemical Substances) from the *CHEMICAL ABSTRACTS* volume 76 Index Guide; published by the American Chemical Society, Columbus, Ohio 43210; 1973 (and references cited therein).
2. *Nomenclature of Organic Chemistry*; Sections A, B, C, D, E, F and H combined; Pergamon Press, Oxford, 1979. (IUPAC nomenclature).

—Wendell L. Dilling and Marcia L. Dilling, The Dow Chemical Company, Midland, Michigan.

ORGANIC SYNTHESIS.   Synthesis (Chemical).

ORGANOMETALLIC COMPOUND. Compound (Chemical).

ORNITHINE. Amino Acids.

ORNITHINE CYCLE. Urea.

ORPIMENT. This mineral, like realgar, its frequent associate, is an arsenic sulfide. Orpiment, however, is the trisulfide corresponding to the formula As_2S_3. It is monoclinic, usually in foliated, granular or powdery aggregates; hardness, 1.502; specific gravity, 3.49; with a resinous to somewhat pearly luster; color various shades of lemon yellow; translucent to nearly opaque. Orpiment is found in association with realgar, although a somewhat rarer mineral. It is believed to be formed from the alteration of other arsenic-bearing minerals. It occurs in Czechoslovakia, Romania, Macedonia, Japan, and in the United States in Utah, Nevada, and Wyoming. The name *orpiment* is derived from a corruption of the Latin *auripigmentum*, meaning golden paint, because of its color as well as the belief that it contained gold.

ORTHO COMPOUNDS. Organic Chemistry.

ORTHO-STATE. 1. In diatomic molecules, such as hydrogen molecules, the ortho-state exists when the spin vectors of the two atomic nuclei are in the same direction (i.e., parallel), whereas the para-state is the one in which the nuclei are spinning in opposite directions. 2. In helium, the ortho-state is characterized by a particular mode of coupling of the electron spins. See **Helium.**

OSMIUM. Chemical element symbol Os, at. no. 76, at. wt. 190.2, periodic table group 8, mp 3015°C, bp 4930–5130°C, density, 22.5 g/cm^3 (solid), 22.8 g/cm^3 (single crystal at 20°C). Elementary osmium has a close-packed hexagonal crystal structure. Compact osmium is a bluish-white metal and is not attacked by acids. Discovered by Tennant in 1804. The seven stable isotopes of osmium are ^{184}Os, ^{186}Os through ^{190}Os, and ^{192}Os. The six unstable isotopes are ^{182}Os, ^{183}Os, ^{185}Os, ^{191}Os, ^{193}Os, and ^{194}Os. Electronic configuration: $1s^22s^22p^63s^23p^63d^{10}4s^24p^64d^{10}4f^{14}5s^25p^65d^66s^2$. Ionic radius Os^{4+}, 0.65 Å; metallic radius, 1.337 Å. Oxidation potential: $Os + 4H_2O \rightarrow OsO_4 + 8H + 8e^-$, −0.85 V.

Finely divided Os oxidizes in air, producing the poisonous and volatile tetroxide. The compact metal is not attacked by nonoxidizing acids. The finely divided metal dissolves in fuming HNO_3, aqua regia, and alkaline hypochlorite solutions. When fused with Na_2O_2 or KNO_3 and KOH, the metal is converted to the corresponding water-soluble osmate, K_2OsO_4. The brown or black insoluble osmium(IV) oxide, OsO_2, can be made by heating Os with a limited amount of O_2 or with osmium(VIII) oxide. This compound forms a brown to black-blue dihydrate that can be prepared by reducing a solution of the tetroxide or by hydrolyzing a solution of sodium hexachloroosmate, Na_2OsCl_6.

Osmium(VIII) oxide, the most important compound, is formed in one of the reactions unique to the platinum metals. Its ease of formation and volatility make it useful in a purification step for the refining or analysis of Os. The tetroxide is readily formed by heating the metal in air or distilling an osmium-containing solution from HNO_3. Although an aqueous solution of osmium(VIII) oxide is neutral to litmus, it is a weak acid with first dissociation constant of about 8×10^{-13}. Osmium(VIII) oxide is soluble in water, alcohol, and ether. The com-

pound is widely used as a stain for tissues. When an alkaline solution of osmium(VIII) oxide is reduced with alcohol or KNO_2, an osmate(VI) is formed. Potassium osmate(VI) is formed by adding an excess of KOH to such a solution, resulting in the precipitation of violet crystals of $K_2OsO_4 \cdot 2H_2O$. The osmate(VI) ion is probably better written as $OsO_2(OH)_2{-}$.

Osmium(II) chloride can be prepared by heating osmium(III) chloride in vacuum at 500°C. This dark brown compound is insoluble in HCl or H_2SO_4. HNO_3 or aqua regia oxidize it to the tetroxide. Osmium(III) chloride is best made by decomposing ammonium hexachloroosmate(IV), $(NH_4)_2OsCl_6$, in a Cl_2 stream at 350°C. The brown hygroscopic powder sublimes above 350°C, and at about 560°C it disproportionates into the tetrachloride and dichloride. Osmium(IV) chloride is formed from the elements at 650–700°C. The black compound slowly dissolves in water, eventually forming the dioxide. The free acid, H_2OsCl_6, is stable in solution and can be made by refluxing osmium(VIII) oxide with HCl and alcohol. The ammonium salt can be precipitated by adding NH_4Cl to such a solution. This salt is reduced to the metal when heated in H_2. The potassium salt is well known. Both are brownish-red solids yielding orange solutions in water.

Recent studies have established the reaction product of Os metal and F_2 at 300°C to be the hexafluoride, OsF_6. This yellow volatile solid had previously been described as an octafluoride. Osmium(VI) fluoride melts at 33.4°C and boils at 47.5°C. OsF_6 can be reduced to a pentafluoride and a tetrafluoride. The pentafluoride is a blue-gray crystalline solid that melts at 70°C to a green viscous liquid and boils at 226°C. The tetrafluoride distills at about 290°C. Potassium hexafluoroosmate(V), $KOsF_6$, can be made by reacting KBr, osmium(IV) bromide, and bromine trifluoride. The white powder dissolves in water to form a colorless solution that hydrolyzes to yield some osmium(VIII) oxide. On addition of 1 equiv of KOH to a fresh solution, an orange color develops, O_2 is evolved, and yellow crystals of potassium hexafluoroosmate(IV), K_2OsF_6, form.

Os forms many complexes with nitrite, oxalate, carbon monoxide, amines, and thio ureas. The latter are important analytically. Osmium forms the interesting aromatic "sandwich" compound, osmocene. A *metallocene* is described under **Ruthenium.**

More detail on physical properties of osmium can be found in "Osmium" by H. J. Albert and J. A. Bard in "Metals Handbook," 9th Edition, American Society for Metals, Metals Park, Ohio, 1979. See also entry in this Encyclopedia on **Chemical Elements.**

—Linton Libby, Chief Chemist, Simmons Refining Company, Chicago, Illinois.

OSMOTIC COEFFICIENT. A factor introduced into equations for nonideal solutions to correct for their departure from ideal behavior, as in the equation:

$$\mu = \mu_x^0 + gRT \ln x_1$$

in which μ is the chemical potential μ_x^0 is a constant, representing a standard value of the chemical potential, R is the gas constant, T the absolute temperature x_1 is the mole fraction of solvent, and g is the osmotic coefficient.

OSMOTIC PRESSURE. Pressure that develops when a pure solvent is separated from a solution by a semipermeable membrane which allows only the solvent molecules to pass through it. The osmotic pressure of the solution is then the excess pressure which must be applied to the solution so as to prevent the passage into it of the solvent through the semipermeable membrane.

Because of the similarity in the relations for osmotic pressure in dilute solutions and the equation for an ideal gas, van't Hoff proposed his bombardment theory in which osmotic pressure is considered in terms of collisions of solute molecules on the semipermeable membrane. This theory has a number of objections and has now been discarded. Other theories have also been put forward involving solvent bombardment on the semipermeable membrane, and vapor pressure effects. For example, osmotic pressure has been considered as the negative pressure which must be applied to the solvent to reduce its vapor pressure to that of the solution. It is, however, more profitable to interpret osmotic pressure using thermodynamic relations, such as the entropy of dilution.

A number of methods have been developed for measurement of osmotic pressure.

In the Berkeley and Hartley method, a porous tube with a semipermeable membrane such as copper ferrocyanide deposited near the outer wall, and a capillary tube attached to one end contains the pure solvent. The solution surrounds the tube and is enclosed in a metal vessel to which a pressure may be applied which is just sufficient to prevent the flow of solvent into the solution. Berkeley and Hartley also developed a dynamic method for measuring osmotic pressure.

Simple osmometers have also been developed by Adair particularly for aqueous colloidal solutions. A thimble-type collodion membrane is attached to a capillary tube and contains the solution. When equilibrium is established the difference in level inside and outside the capillary is measured. Capillary corrections are made. For organic solvents a dynamic type osmometer may be used. A membrane of large surface area is clamped between two half cells and attached to each half cell a fine capillary observation tube. With such an apparatus, equilibrium is rapidly established between solution and solvent contained in the half cells. The volume of the half cell may be small (about 20 cubic centimeters). The level of the solvent is usually arranged to be a little below the equilibrium position, and the height of the solvent in the capillary as a function of time is measured. This procedure is repeated with the level of the solvent just above the equilibrium position. A plot is then made of the half sum of these readings.

Since the osmotic pressure is related to the concentration of dissolved solute particles, it is related to the lowering of the freezing point and elevation of the boiling point.

The relation between osmotic pressure and lowering of the freezing point and the elevation of the boiling point may be expressed by the relation:

$$\Pi = \frac{LT}{\bar{\nu} T_0^2} \Delta T$$

where L is the molar heat of fusion or of vaporization of the solvent, T the temperature at which the osmotic pressure is measured, T_0 the freezing point or the boiling point of the solvent, $\bar{\nu}$, the partial molar volume of the solvent, and Π, the osmotic pressure.

Moreover, the relation to lowering of the vapor pressure is

$$\Pi = \frac{RT}{\bar{\nu}} \ln \frac{p_0}{p}$$

or

$$\Pi = -\frac{RT}{\bar{\nu}} \ln x_0 = -\frac{RT}{\bar{\nu}} \ln(1 - x)$$

For dilute solutions,

$$\Pi = cRT$$

where Π is the osmotic pressure, $\bar{\nu}$ the partial molar volume of the solvent in the solution, p_0 the vapor pressure of pure solvent, p the partial vapor pressure of the solvent in equilibrium with the solution, x_0 the mole fraction of the solvent, x the mole fraction of the solute, c the concentration of the solution in moles per liter, R, the gas constant, and T, the absolute temperature.

OXALIC ACID. $H_2C_2O_4 \cdot 2H_2O$, formula weight 126.07, white solid, mp 101°C (crystals), 180°C (anhydrous), sublimes at 150°C, soluble in H_2O, or alcohol, or ether.

Oxalic acid may be obtained (1) from some natural products, e.g., wood sorrel, and other members of the oxalis family, as potassium hydrogen oxalate, the bark of certain species of eucalyptus (sometimes containing 20% calcium oxalate in the cells and cell walls), (2) by reaction of acid with an oxalate, e.g., calcium oxalate plus H_2SO_4 as above. Sodium oxalate is made (1) by heating sodium formate at 300°C in a vacuum, with the evolution of hydrogen gas, (2) by reaction of CO_2 plus metallic sodium at 360°C, (3) by heating wood powder ("sawdust"), particularly of coniferous woods, with NaOH (addition of potassium hydroxide enables one to use a moderate temperature, i.e., 220°C), (4) by oxidation of sucrose or starch with HNO_3. Oxalic acid is a dibasic acid, that is, two series of salts are known, a third series is also known, thus, sodium oxalate $Na_2C_2O_4$, sodium binoxalate $NaHC_2O_4$, sodium tetroxalate $NaH_3(C_2O_4)_2$.

Oxalic acid is used (1) in the preparation of oxalates, e.g., titanium potassium oxlate, and esters, (2) in the purification of certain chemicals, e.g., glycerol, stearates, (3) in bleaching straw, (4) in ink and rust remover, (5) in the leather and textile industries, (6) in the manufacture of dyes, (7) in the preparation of glyoxalic and glycolic acids by regulated reduction.

A common test for oxalic acid is as follows: Heat solution with resorcinol in a test tube and on cooling add carefully a layer of H_2SO_4. A blue ring at the junction of the two layers indicates oxalic acid.

Representative esters of oxalic acid are: Methyl oxalic acid $COOH \cdot COOCH_3$, mp 37°C, bp 108°C at 12 mm pressure; dimethyl oxalate $(COOCH_3)_2$, mp 54°C, bp 163°C; diethyl oxalate $(COOC_2H_5)_2$, mp −41°C, bp 186°C.

OXIDATION AND OXIDIZING AGENTS. Many years ago, the term *oxidation* signified a reaction in which oxygen combines chemically with another substance. The term now has a much broader meaning and includes any reactions in which electrons are transferred. Oxidation and reduction always occur simultaneously (redox reactions), and the substance which gains electrons is termed the *oxidizing agent*. For example, cupric ion is the oxidizing agent in the reaction: Fe (metal) + $Cu^{++} \rightarrow Fe^{++}$ + Cu (metal). Here, two electrons (negative charges) are transferred from the Fe atom to the Cu atom. Thus the Fe becomes positively charged (is oxidized) by loss of two electrons, while the Cu receives the two electrons and becomes neutral (is reduced). Electrons may also be displaced within the molecule without being completely transferred away from it. Such partial loss of electrons likewise constitutes oxidation in its broader sense and leads to the application of the term to a large number of processes which at first inspection might not be considered to be oxidations. Reaction of a hydrocarbon with a halogen, for example: $CH_4 + Cl_2 \rightarrow CH_3Cl + HCl$, involves partial oxidation of the methane. Also, when a halogen addition to a double bond is made, this is regarded as an oxidation.

Dehydrogenation is also a form of oxidation, when two hydrogen atoms, each having one electron, are removed from a hydrogen-containing organic compound by a catalytic reaction with air or oxygen, as in oxidation of alcohols to aldehydes. See also **Dehydrogenation.**

Oxidizing agents are widely used throughout the chemical and petrochemical industries. It is also interesting to note that, while a primary thrust in food processing is to prevent oxidation (associated with rancidity and spoilage in foods), powerful oxidizing agents are required by some processes to perform the function of bleaching. See also **Bleaching Agents.** As one example, after a crude fat or oil is refined to remove its impurities, it must be further treated by bleaching to remove coloring materials that are typically present in crude fats and oils. Sulfuric and metaphosphoric acids and hydrogen peroxide have been used for this purpose. Calcium hypochlorite is used in bleaching sugar syrup prior to crystallization. Not all effective bleaching agents can be used because of their toxicity and resistance to complete removal. The aim is to remove all traces of the bleaching agent during subsequent processing so that they do not occur in the final product. Calcium peroxide, acetone peroxide, and benzoyl peroxides are other bleaching agents sometimes used in the food industry.

The use of oxidizers such as potassium bromate, potassium iodate, and calcium peroxide as dough modifiers in the baking industry dates back many years. However, their mechanism has not been fully explained. Among authorities, there are at least two major viewpoints. It has been proposed that oxidizers inhibit proteolytic enzymes present in flour. It has also been proposed that the number of —S—S— bonds between protein chains is increased, forming a tenacious network of molecules. This action leads to a tougher, drier, and more extensible dough. The need for oxidizers is less with well-aged flours and in connection with supplemented flours that have been effectively brominated at the mills. When used properly, oxidizers contribute to improved appearance, brighter crumb, and better texture of breads. Oxidizers do not appear to interfere with the generation of gas by yeast or leavening chemicals. However, oxidizers used to excess can destroy the desirable properties of the dough and end-products.

OXIDATION NUMBER. In its original and restrictive sense, the number of electrons which must be added to a cation to neutralize the charge. The concept has been extended to anions by assignment of negative oxidation numbers. Moreover, it has been further extended, first to all atoms or radicals joined by electrovalent bonds, and then to covalent compounds in which the shared electrons are distributed equally. For the broadest use of the concept, the expression *oxidation state* is often used.

OXIDATION POTENTIAL. The potential drop involved in the oxidation (i.e., ionization) of a neutral atom to a cation, of an anion to a neutral atom, or of an ion to a more highly charged state (e.g., ferrous to ferric).

OXIDATION-REDUCTION. Phosphorylation (Oxidative); Phosphorylation (Photosynthetic).

OXIDATION-REDUCTION EQUATIONS. Chemical Equation.

OXIMES. One of a number of compounds that result from the interaction of aldehydes, ketones, and other carbonyl-containing substances with hydroxylame, e.g., acetone yields acetoxime

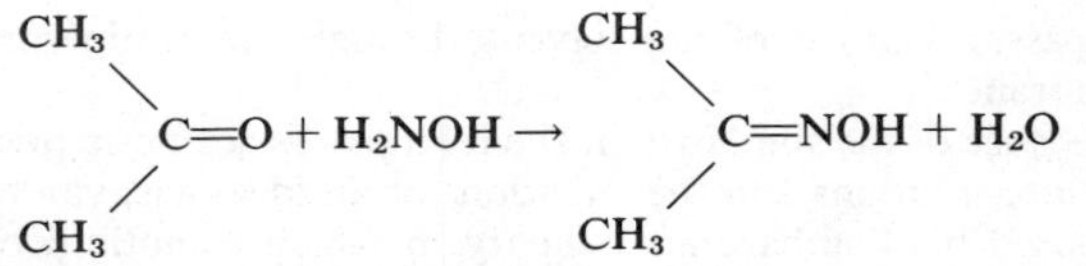

$$\begin{matrix} CH_3 & & CH_3 \\ \diagdown & & \diagdown \\ C{=}O + H_2NOH \rightarrow & & C{=}NOH + H_2O \\ \diagup & & \diagup \\ CH_3 & & CH_3 \end{matrix}$$

OXIME TEST (Rheinboldt). An ethereal solution of an aldoxime gives a blue color with aqua regia.

OXONIUM COMPOUNDS. Coordination compounds, commonly of certain oxygen-containing organic substances, with mineral acids, of the general type $[R_2O]HCl$. These compounds bear a strong resemblance to the oxonium (hydronium) ion, which is a proton in combination with a water molecule, and is the form in which protons commonly exist in aqueous solutions.

OXO PROCESS. The general name for a process in which an unsaturated hydrocarbon is reacted with carbon monoxide and hydrogen to form oxygen-function compounds, such as aldehydes and alcohols. In a typical process for the production of oxo alcohols, the chargestock comprises an olefin stream, carbon monoxide, and hydrogen. In a first step, the olefin reacts with CO and H_2 in the presence of a catalyst (often cobalt) to produce an aldehyde which has one more carbon atom than the originating olefin: $R \cdot CH{:}CH_2 + CO + H_2 \rightarrow R \cdot CH_2 \cdot CH_2 \cdot CHO$. This step is exothermic and requires a cooling cycle. The raw aldehyde exiting from the oxo reactor then is subjected to a higher temperature to convert the catalyst to a form for easy separation from the reaction products. The subsequent treatment also decomposes unwanted by-products. The raw aldehyde then is hydrogenated in the presence of a catalyst (usually nickel) to form the desired alcohol: $R \cdot CH_2 \cdot CH_2 \cdot CHO + H_2 \rightarrow R \cdot CH_2 \cdot CH_2 \cdot CH_2OH$. The raw alcohol then is purified in a fractionating column. In addition to the purified alcohol, by-products include a light hydrocarbon stream and a heavy oil. The hydrogenation step takes place at about 150°C under a pressure of about 100 atmospheres. The olefin conversion usually is about 95%.

Among important products manufactured in this manner are substituted propionaldehydes from corresponding substituted ethylenes, normal and iso-butyraldehyde from propylene, iso-octyl alcohol from heptenes, and trimethylhexyl alcohol from di-isobutylene.

OXY. The radical —O— in organiz compounds, performing in a manner similar to the oxo radical (O=) and the epoxy radical (—O—).

OXYACID. An acid that contains oxygen, such as chloric acid ($HClO_3$).

OXYAZO COMPOUNDS. Compounds of the type $RN{=}NC_6H_4OH$, containing both the azo group —N=N—, and a hydroxyl group —OH, both attached to carbon atoms in the same ring. These compounds are commonly produced by the action of diazo compounds upon phenols in alkaline solution. They constitute a class of dyes. See DYES.

OXYCHLORINATION. Chlorinated Organics.

OXYGEN. Chemical element symbol O, at. no. 8, at. wt. 15.9994, periodic table group 6a, mp −218.4°C, bp −182.96°C, critical temperature −118.8°C, critical pressure 49.7 atm, density, 1.568 g/cm³ (solid), 1.429 g/liter (0°C). Solid oxygen has a cubic crystal structure. Oxygen is slightly soluble in water (4.89 parts oxygen in 100 parts water at 0°C), the solubility decreasing with increasing temperature (2.6 parts oxygen in

100 parts water at 30°C); 1.7 parts oxygen in 100 parts water at 100°C). Oxygen is slightly soluble in alcohol. Molten silver dissolves up to 10 times its volume of oxygen, but easily gives up the gas upon cooling. There are three stable isotopes, ^{16}O through ^{18}O. Three radioactive isotopes have been identified, ^{14}O, ^{15}O, and ^{19}O, with short half-lives measured in seconds and minutes. In terms of abundance in igneous rocks in the earth's crust, oxygen ranks first, with an average composition by weight of 46.6%. In terms of abundance in seawater, oxygen also ranks first, with an estimated 4 billion tons of oxygen per cubic mile (0.86 billion metric tons per cubic kilometer) of seawater. In terms of cosmic abundance, oxygen ranks eighth. For comparison, assigning a value of 10,000 to silicon, the figure for oxygen is 220,000 and that for hydrogen, estimated the most abundant, a figure of 3.5×10^8. Of dry air in the earth's atmosphere, 23.15% is oxygen by weight; 20.98% by volume. In the atmosphere, the oxygen is mixed with nitrogen, argon, the rare gases, CO_2, and water vapor.

Oxygen first was identified by Priestly in 1774 when he was experimenting with mercuric oxide. In the same year, Scheele also identified the element. Oxygen is required for burning and combustion, although the conditions of combustion vary widely. For example, phosphorus burns in air at the low temperature of 34°C when ignited. The temperature of ignition for ether in air is 340°C, for ethyl alcohol in air, 560°C, kerosene in air, about 300°C, and hydrogen in air, about 600°C. The oxidation process may occur with the rapidity and violence of an explosion, or may be as slow as the rusting of iron. Nearly all known species of living things require oxygen in some form, either free or chemically bound. First ionization potential 13.614 eV; second, 34.93 eV; third, 54.87 eV. Oxidation potentials $H_2O_2 \rightarrow O_2 + 2H^+ + 2e^-$, $- 0.68$ V; $3H_2O \rightarrow \frac{1}{2}O_2 + 2H_3O^+$ $(10^{-7}M) + 2e^-$, -0.815 V; $3H_2O \rightarrow \frac{1}{2}O_2 + 2H^+ + 2e^-$, 1.229 V; $4H_2O \rightarrow H_2O_2 + 2H_3O^+ + 2e^-$, -1.77 V; $3H_2O \rightarrow O(g) + 2H^+ + 2e^-$, $- 2.42$ V; $HO_2^- + OH^- \rightarrow O_2 + H_2O + 2e^-$, 0.075 V; $4OH^- \rightarrow O_2 + 2H_2O + 4e^-$, -0.401 V; $3OH^- \rightarrow HO_2^- + H_2O + 2e^-$, -0.87 V; $OH^- \rightarrow OH + e^-$, -1.4 V. Other physical properties of oxygen are given under **Chemical Elements.**

Allotropic Forms. The three known allotropic forms of oxygen are (1) the ordinary oxygen in the air, with two atoms per molecule O_2, (2) ozone O_3, with three atoms per molecule, and (3) the rare, very unstable, nonmagnetic, pale-blue O_4. The latter breaks down readily into two molecules of O_2.

When oxygen is subjected to the silent electric discharge, activated atomic oxygen is produced. Atomic oxygen displays an afterglow upon cessation of the current, and the oxygen is notably active with hydrogen bromide, forming bromine; with H_2S, forming sulfur, SO_2, sulfur trioxide, and H_2SO_4; with CS_2, forming carbon monoxide, CO_2, and SO_2; and, strangely, reduces molybdenum trioxide to a white oxide not reducible with hydrogen. The concentration of atomic oxygen obtainable by the silent electric discharge through oxygen is estimated at 20%.

The normal electron distribution of the electrons of the oxygen atom is $1s^2 2s^2 2p_x^2 2p_y^1 2p_z^1$, with 2 unpaired electrons in the $2p$ orbitals. The covalent or partly covalent compounds of oxygen would be expected to have 90° bonding angles. But in many cases they have values significantly greater (ca. 104° for R_2O and 105° for H_2O). This suggests the promotion of a $2s$ electron to a $2p$ orbital (i.e., $2p_y$ orbital) still leaving two unpaired electrons (a $2s$ and a $2p_z$ electron), and permitting partial sp^3 hybridization (which is incomplete because sufficient energy is not available) but producing bond angles between 90° and 109° for the sp^3 tetrahedral structure, with covalent-polar bonds.

The oxygen molecule is paramagnetic with a moment in accord with two unpaired electrons. In molecular orbital terms, the configuration is written

$$O_2[KK(z\sigma)^2(y\sigma*)^2(x\sigma)^2(w\pi)^4(v\pi*)^2]$$

in which KK designates the complete $1s$ shells of the two atoms, which are non-bonding, the term $(z\sigma)^2$ denotes the bonding effect of one pair of $2s$ electrons, one from each of the O atoms, $(y\sigma*)^2$ denotes the antibonding effect of the second pair, the $(x\sigma)^2$ term represents the σ-bond formed by one pair of p-electrons, $(w\pi)^4$ represents the 2 π-bonds formed by the other two pairs of p-electrons, while the $(v\pi*)^2$ term denotes the last pair of p-electrons, which go into the next π subshell (two orbitals) with unpaired spins, and are hence antibonding.

Ozone, O_3, obtained by electrical discharge through oxygen or high-current electrolysis of sulfuric acid, is considered on the basis of electron diffraction studies to have an O—O—O bond angle of $127 \pm 3°$ O—O bond length of 1.26 ± 0.02Å. Its structure is considered to resonate among several forms, chiefly

Ozone O_3, is a blue gas, of characteristic odor, formed when ordinary oxygen is subjected to electrostatic discharge, density 1.5 times that of oxygen gas, mp −251.4°C, bp −111.5°C. Explosive by percussion or under variations of pressure. Ozone reacts (1) with potassium iodide, to liberate iodine, (2) with colored organic materials, e.g., litmus, indigo, to destroy the color, (3) with mercury, to form a thin skin of mercurous oxide causing the mercury to cling to the containing vessel, (4) with silver film, to form silver peroxide Ag_2O_2 black, produced most readily at about 250°C, (5) with tetramethyldiaminodiphenylmethane $(CH_3)_2N \cdot C_6H_4 \cdot CH_2 \cdot C_6H_4 \cdot N(CH_3)_2$ in alcohol solution with a trace of acetic acid to form violet color (hydrogen peroxide, colorless; chlorine or bromine, blue; nitrogen tetroxide, yellow). In contrast to hydrogen peroxide, ozone does not react with dichromate, permanganate, or titanic salt solutions. Ozone reacts with olefin compounds to form ozonide addition compounds. Ozonides are readily split at the olefin-ozone position upon warming alone, or upon warming their solutions in glacial acetic acid, with the formation of aldehyde and acid compounds which can be readily identified, thus serving to locate the olefin position in oleic acid $C_{17}H_{33} \cdot COOH$ as midway in the chain $(CH_3(CH_2)_7CH:CH(CH_2)_7COOH)$. Ozone is used (1) as a bleaching agent, e.g., for fatty oils, (2) as a disinfectant for air and H_2O, (3) as an oxidizing agent. See also **Aerosol.**

Some researchers have reported that the growth of human cancer cells from lung, breast, and uterine tumors may be selectively inhibited by enriching ambient air with less than one part per million of ozone over a period of several days. A tentative conclusion is that the mechanisms for defense against ozone damage are impaired in human cancer cells. (See Sweet et al. reference listed)

Role of Oxygen in Water. The solvent properties of H_2O are due in great part to the dipole moment of its molecules (1.8 debye units) and its high dielectric constant (ca. 78). Its hydrogen atoms form hydrogen bonds with electronegative atoms such as fluorine, nitrogen, or oxygen. In fact, the H_2O molecules associate in H_2O by this mechanism. Also, the oxygen atoms of H_2O because of their residual negative charges are electrically attracted by cations, so that the H_2O molecules arrange themselves around cations, facilitating solution and ionization. In the same way, H_2O molecules surround anions by attraction of the positive ends of the dipoles. By these two processes, as well as the dissociation of water into oxonium and hydroxide ions, it forms hydrates with many compounds. Moreover, H_2O readily reacts with large numbers of compounds because of these properties. Thus the hydrolysis of covalent halides which have at least one lone pair of electrons is initiated by the donation of a proton by the H_2O, followed by splitting off of hydrogen chloride.

Oxides. Oxygen forms oxides with all the elements except some inert gases. Oxides are said to be normal when they contain no oxygen atoms that are bonded to each other as in the peroxides. The normal oxides may be divided into three groups, basic, acidic, and neutral. The basic oxides, which react with or dissolve in H_2O to produce alkaline solutions, are formed by the alkali and alkaline earth elements (except beryllium) by the lighter Lanthanides and actinium, by silver(I), thallium(I) and lead(II). The oxides of the nonmetals and of the transition metals in their higher oxidation states are in general acidic. The oxides lying in the positions between the two groups exhibit both basic and acidic properties (amphiprotic or amphoteric) such as those of aluminum, tin(II) and iron(III), Al_2O_3, SnO, and Fe_2O_3.

The known facts about the structure of hydrogen peroxide, H_2O_2, are that the O—H distances are 0.97 Å, the O—O distances 1.47 Å, the HOO angles 94°, and the dihedral angle between the planes of the two O—H radicals 97°. The O—O bond is essentially a single one. In the liquid, H_2O_2 is somewhat more self-ionized than water. In water $pK_A = 11.75$, $pK_B = 17$. Its reactions may be oxidizing or reducing. Thus, it oxidizes Fe(II) to Fe(III), Ti(III) to Ti(IV) and SO_3^{2-} to SO_4^{2-}; but it reduces MnO_4^- (acid solution) to Mn^{2+}. Peroxides are known for the alkali and alkaline earth metals, as well as zinc, cadmium, mercury, thorium, uranium, plutonium, etc. However, not all compounds of formula MO_2 (where M is a metal atom) are peroxides; some are merely dioxides, as MnO_2, PbO_2, etc., others are superoxides, such as NaO_2, KO_2, RbO_2, CsO_2, CaO_4, SrO_4, and BaO_4. These last compounds contain the group O_2^-, as evident from their paramagnetism and crystal structure. Perhydroxyl, the free acid corresponding to the superoxides, is unstable ($H_2O_2 \rightarrow HO_2 + H^+ + e^-$, $E° = 1.5$ V; $HO_2 \rightarrow O_2 + H^+ + e^-$, $E° = +0.13$ V). It is a moderately strong acid, $pK_A = 2.2$.

The peroxyacids containing —O—OH groups, are formed with all the transition elements in groups 4, 5, 6 of the periodic table, with main group elements 4 and 5 as well as elements of atomic numbers from boron to sulfur, inclusive. Representa-

tive peroxyacids are peroxymonosulfuric acid, $H : \overset{\cdot\cdot}{\underset{\cdot\cdot}{O}} : \overset{\overset{\cdot\cdot}{O}\cdot\cdot}{\underset{\cdot\cdot}{\underset{\cdot\cdot}{O}}} : \overset{\cdot\cdot}{\underset{\cdot\cdot}{O}} : \overset{\cdot\cdot}{\underset{\cdot\cdot}{O}} : H$

and peroxychromic acid $H : \overset{\cdot\cdot}{\underset{\cdot\cdot}{O}} : \overset{\cdot\cdot}{\underset{\cdot\cdot}{O}} : \overset{\overset{\cdot\cdot}{O}\cdot\cdot}{\underset{\cdot\cdot}{\underset{\cdot\cdot}{Cr}}} : \overset{\cdot\cdot}{\underset{\cdot\cdot}{O}} : \overset{\cdot\cdot}{\underset{\cdot\cdot}{O}} : H$. The only peroxydiacids are formed by sulfur, phosphorus, carbon and boron, of which the most important is peroxydisulfuric acid

$H : \overset{\cdot\cdot}{\underset{\cdot\cdot}{O}} : \overset{\overset{\cdot\cdot}{O}\cdot\cdot}{\underset{\cdot\cdot}{\underset{\cdot\cdot}{S}}} : \overset{\cdot\cdot}{\underset{\cdot\cdot}{O}} : \overset{\cdot\cdot}{\underset{\cdot\cdot}{O}} : \overset{\overset{\cdot\cdot}{O}\cdot\cdot}{\underset{\cdot\cdot}{\underset{\cdot\cdot}{S}}} : \overset{\cdot\cdot}{\underset{\cdot\cdot}{O}} : H$, although peroxy bridge compounds are

also formed by certain transition element complexes, e.g.,

$[Co(NH_3)_5OOCo(NH_3)_5]^{4+}$ and $[Co(NH_3)_5OOCo(NH_3)_5]^{5+}$.

Industrial Oxygen: As with hydrogen, the electrolysis of water offers one approach to the production of pure oxygen. However, the economics are as unfavorable for oxygen production in this manner as for hydrogen. See also **Hydrogen.** For industrial oxygen production, air is the raw material. Using air, processes are of two major types: (1) liquid oxygen processes wherein the oxygen is fractionally distilled from liquid air, and (2) gaseous-oxygen processes. See also **Cryogenics.**

Because of the relatively high energy costs of compressing and refrigerating involved in oxygen production, many process have been developed and tested over the years, a high percentage of these later abandoned. An idea of the alternatives which face the process designer can be gathered from scanning the methods available specifically in the area of producing refrigeration for these processes: (1) Joule-Thomson effect only; (2) Joule-Thomson effect plus auxiliary refrigeration with an ordinary liquid-vapor cycle at moderate or high-temperature levels, i.e., relative to liquid-air temperature; (3) Joule-Thomson effect plus approximately reversible expansion of the air or products in an expander; (4) refrigeration essentially due only to approximately reversible expansions of auxiliary fluid or fluids operating in liquid-vapor cycles, i.e., the cascade process; and (7) processes using an auxiliary nitrogen-liquefaction cycle.

Designers also face the choice of capacity of an oxygen plant. Costs per unit weight of oxygen made are lowered as the capacity of the plant goes up. For example, a plant with a capacity of 2000 tons (1800 metric tons) per day will produce oxygen at approximately 50% the cost per unit weight as a plant with a 200-ton (180-metric ton) capacity per day.

The demand for industrial oxygen has created the need for several new plants during the past 15 years. Capacities of most recent plants range from about 1100 tons (9.9 metric tons) per day to 2500 tons (2250 metric tons) per day. In one installation (SASOL II in South Africa), six of the high-capacity units were installed in 1979. The SASOL II complex feeds oxygen to the largest industrial complex in the world for the production of synthetic natural gas. As of 1981, the plant provides about 30% of the country's needs for hydrocarbons (from coal).

An oxygen pipeline system was established in western Europe in the late 1970s that is 592 miles (956 kilometers) long. The eastern network of this system serves 30 consumers in France, Luxembourg, and West Germany; the northern network serves some 40 additional users in France, Belgium, and the Netherlands.

Uses. In addition to the requirements by the chemical industry for oxygen as a reactant, either directly from the air or in purer, more concentrated form as from a separation plant, significant quantities of purified oxygen are used for welding and cutting metals. Oxygen of a purity of 99.5% is required for oxyacetylene and oxyhydrogen torches. When combined in proper proportions, acetylene and oxygen yield a flame with a temperature of about 3480°C. Oxyhydrogen flames are somewhat lower in temperature, but they are particularly useful for welding light-gage aluminum and magnesium alloys and for underwater cutting. In welding, a reduction in purity of oxygen used from 99.5% to 99.0% will cut welding efficiency by over 10%. During the past several years, basic oxygen steelmaking has increased requirements for pure oxygen. In this process, nearly pure oxygen is introduced by means of a lance into molten iron and scrap. The oxygen combines with carbon and other unwanted elements and refines raw steel in much less time than the older open-hearth furnaces. The basic oxygen process exceeded the open hearth process in terms of output in the United States in 1970 for the first time. On the total scale of consumption, relatively limited amounts of oxygen go into medical and life-support applications, as required for emergency situations in aircraft at high altitudes.

Role of Oxygen in Corrosion. Oxygen and oxidizing agents exert both a positive and negative influence on corrosion of metals. On the one hand, an oxidizing agent may form a protective oxide film on the surface of certain metals, aluminum being an excellent example, which essentially arrests corrosion by many external agents. On the other hand, the presence of oxidants may increase the rate of corrosion by supporting cathode reactions. As an example, Monel metal fully resists attack by oxygen-free 5% H_2SO_4 at room temperature. The corrosion rate rises, however, in almost direct proportion to oxygen content. A 20% oxygen content will cause a corrosion rate of about 150 mdd (milligrams of metal corroded per square decimeter per day). A concentration of 40% will increase the rate to about 250 mdd; a concentration of 80% to about 450 mdd. The oxygen need not be present in all of the acid contained in the metal vessel, but simply present in that concentration at the interface of metal, acid, and surrounding atmosphere. The effect of oxidizing salts on corrosion can be dramatic. For example, the rate for Monel metal exposed to 1.5% H_2SO_4 at room temperature is 140 mdd. If ferric sulfate is added to the acid to the extent of 0.18%, the corrosion rate is increased to 1,215 mdd. Several factors, in addition to oxygen, affect corrosion, including the presence of other metals (electromotive-force displacements of one metal by another), temperature, acidity, and velocity. These factors are discussed further under **Corrosion.**

Oxygen Toxicity of Plants. As early as 1801, Huber and Senebier observed that grains develop more satisfactorily in an atmosphere containing a mixture of 3 parts nitrogen and 1 part oxygen than in an atmosphere containing 3 parts of oxygen and 1 part nitrogen. Considerably later, in 1878, Bert noted that the earlier observations also apply to the development of many plant species and are not peculiar to grains. Bert further suggested that excessive oxygen may slow down various reactions involving fermentation. It was not until much later, in the mid-1940s, that scientists (Dickens, Haugaard, and Stadie) further confirmed that enzymes are inactivated by oxygen excesses. They particularly stressed this fact in connection with enzymes that contain a sulfhydryl group in the active site. Michaelis (1946), Barron (1946), and Gilber (1963) later pointed out that molecular oxygen alone acts in a rather sluggish manner in this regard and that, therefore, a special process or phenomenon must be involved. Molecular oxygen can be reduced only be accepting one electron at a time.

A number of scientists in the late 1960s through the mid-1970s pointed out that may sources in biological systems produce oxygen *free radicals*. For example, some oxidative enzymes which contain flavin as a prosthetic group proceed by a radical mechanism. When illuminated, chloroplasts produce superoxide ions and singlet oxygen. Because of its singlet configuration, the latter is not hindered in its interactions with biological materials. As pointed out by Griffiths and Hawkins, singlet oxygen can be formed from the ground state when energy, usually in the form of light, is supplied in the presence of a photosensitizer. The compounds that are photosensitized include many dyes and pigments, such as chlorophyll, flavins, and hematoporphyrins. The interaction between the sensitizer and oxygen results in the transfer of electrons, with the formation of superoxide ion. McCord and Fridovich (1969) discovered the enzyme superoxide dismutase. Their later findings show that aerobic organisms contain it, giving further credence to the proposal that all oxygen-metabolizing organisms form superoxide free radicals as a result of a univalent reduction of oxygen. As pointed out

by Kon (1978), those free radicals that are toxic to the organism, by themselves or through interaction with other active forms of oxygen, are dismutated by the action of this enzyme.

There is a close relationship between oxygen toxicity and radiation on enzymes, DNA, and fats. Gerschman et al. (1954) showed that the same substances that afford protection against oxygen poisoning also increase resistance to radiation. Their results were further strengthened by experiments that demonstrated the additive nature of the two effects. Work on the effects that free radicals have on some of the polysaccharides used in food processing was commenced by Kon and Schwimmer and reported in 1977. An excellent summary of the status of research in this area, as of 1978, is given by Kon.

Environmental Aspects of Oxygen. Gaseous oxides, notably those of carbon, nitrogen, and sulfur which result from the combustion of fossil fuels and numerous industrial processes comprise a large portion of the air pollution problem. These compounds are discussed under the specific elements and, in particular, are described under **Pollution (Air).** In connection with the pollution of water in streams, lakes, ponds, rivers, etc., the content of dissolved oxygen in water is of prime concern. Dissolved oxygen must be available to support fish and other desirable living species in natural waters, and sufficient additional oxygen must be available in the water to effect biological degradation of both natural and manufactured materials which reach the water. The overuse of streams for disposal purposes in many instances has almost fully depleted the dissolved oxygen available for life support and hence has given rise to the term "dead" lakes or streams. Two terms are widely used: (1) BOD (biological oxygen demand) which is the requirement for dissolved oxygen in water to degrade or decompose organic matter within a measured time period at a given temperature, and (2) COD (chemical oxygen demand) which is the requirement for dissolved oxygen in water to combine with chemicals, essentially of an inorganic nature, which are introduced into a stream as the result of disposal operations. These aspects of oxygen are discussed under **Water Pollution.**

Earth's Oxygen Supply

The manner in which the earth's present oxygen system and reserves were formed has been the subject of much postulation for many years. Many of the details remain unclear and unconfirmed. In a theory proposed by Berkner-Marshall (1964, 1965), as the earth's atmosphere evolved, there was a slow buildup of the concentration of oxygen—proceeding from a trace to the present content of 23.15% (weight). This theory also proposes that the oxygen content of the atmosphere fluctuated from time to time in a major and relatively rapid manner. There is speculation that these major alterations may have accounted for the extinctions of animal and life forms which took place at the ends of the Paleozoic and Mesozoic eras. For example, there was a great reduction in life in the latter part of the Permian period (Paleozoic era) when many kinds of strange reptiles and trilobites disappeared and seem to have left no descendents. Plant life declined greatly too during the late Paleozoic. From thousands of species in the Pennsylvania period, there remained only a few hundred during the late Permian. Numerous explanations, particularly of a climatic nature, have been offered for these periods of reduction in life.

As pointed out by Van Valen (1971), photosynthesis does not produce a net change in oxidation. Except in bacterial photosynthesis, oxygen production is accompanied by a stoichiometrically equal quantity of reduced carbon. Thus, almost all of the oxygen is eventually used to oxidize reduced carbon. Predominantly, this oxidation occurs as the result of respiration in animals and plants. Further oxidation occurs as the result of forest fires. As observed by Borchert (1951), the only net gain in oxygen equals the amount of reduced carbon buried, as in the form of peat, black mud, and similar sediments. It has been estimated that most individual molecules of carbon remain reduced only for relatively short periods (months or years) because animals and plants have geologically very short lives. Plants respire and so oxidize some reduced carbon almost immediately. Other net sources of oxygen include nitrogen fixation and the photolysis of water in the upper atmosphere. Some investigators have considered these sources quantitatively unimportant, although Brinkmann (1969) suggests that this process would produce, over the earth's history (4.5×10^9 years), about seven times the present mass of oxygen in the atmosphere.

Numerous ways have been proposed to explain a net loss of molecular oxygen. Oxidation of volcanic gases, production of ferrous iron, sulfur, sulfide, and manganese, and the accretion of hydrogen from the solar wind are among these. Such processes are sometimes referred to as *oxygen sinks*. Estimates by Holland (1964) indicate that the net gain and net loss over geologic time are essentially in balance.

Van Valen has posed the question, "What can happen if photosynthesis is suddenly and drastically reduced?" Under such conditions, at a new steady state, production of oxygen and its consumption in the oxidation of carbon would be equal. But, before the new steady state occurs, would animals and decomposers use up much of the previously stored carbon in plants, thus creating a new loss of oxygen? Several investigators have observed that even if all the carbon in all organisms now alive were oxidized, this would decrease the atmospheric concentration of oxygen by less than 0.1% of its present value. And, further, still less than 1% of the present oxygen concentration would be used if all the reduced carbon available in soils and the like were reduced.

Much more detailed explanation of the stability of atmospheric oxygen is contained in the excellent review by Van Valen (1971).

As pointed out by Broecker (1970), the earth's oxygen supply is frequently included in lists of concerns over alterations in the environment, particularly as brought about by anthropogenic activities. Several investigators have made a number of observations which tend to invalidate any claims that oxygen is in danger of serious depletion. Broecker observes that each square meter of the earth's surface is covered by 60,000 moles of oxygen gas. Further, plants living in the ocean and on land produce about 8 moles of oxygen per square meter of surface each year. It is also observed that animals and bacteria destroy nearly all of the products of this photosynthetic activity—thus they use an amount of oxygen nearly equal to that generated by plants. Using the rate at which organic carbon enters the sediments of the ocean as a measure of the amount of photosynthetic product preserved each year, Broecker estimates this to be about 3×10^{-3} mole of carbon per square meter per year. This corresponds to approximately 1 part in 15 million of the oxygen present in the atmosphere. It is estimated, however, that this small amount of oxygen is probably being destroyed by a number of processes, including oxidation of reduced carbon, iron, and sulfur (weathering mechanisms). Broecker points out that the oxygen content of the atmosphere is thus well buffered, particularly in terms of relatively short time spans (100–1000 years).

Over a period of time, people have recovered about 10^{16} moles of fossil carbon and the fuels containing this carbon have been oxidized as sources of energy. Byproduct carbon dioxide from this combustion represents about 18% of the carbon dioxide content of the atmosphere. Two moles of atmospheric oxygen are used to liberate each mole of carbon dioxide from fossil fuel sources. Broecker points out that this process uses up only 7 out of every 10,000 available oxygen molecules. It is estimated that if these fuels are burned at an accelerating rate (5% per year), by the end of this century, only about 0.2% of available oxygen (20 molecules in every 10,000) will be used. It is estimated that if all known fossil fuels were ultimately burned, only 3% of available oxygen would be consumed. In terms of urban oxygen needs, particularly for automotive combustion needs, it is estimated that carbon monoxide levels in the atmosphere (in terms of physiological damage) would reach intolerable levels before the oxygen content of the atmosphere would have decreased by 2%.

The case of anthropogenic alterations of photosynthetic rates and its possible effects on oxygen supply has been covered previously by the observation that stoppage of all photosynthetic activity would require less than 1% of the present oxygen concentration.

Sverdrup et al. (1942) estimated that the oxygen content of deep sea water averages about 2.5 cubic centimeters at standard temperatures and pressure per liter (0.1 mole per cubic meter). Thus, there are about 250 moles of oxygen gas in the deep sea for each square meter of earth surface. The oxygen content of the deep sea waters is renewed about ever 1000 years. The magnitude of this oxygen reservoir is tremendous. Broeker emphasizes this by observing that if the entire terrestrial photosynthetic production were dumped each year into the deep sea, the supply of deep-sea oxygen would last 50 years. But, if the waste products of 1 billion people were limited to 100 kilograms of dry organic waste per year, this would consume 0.01 mole of oxygen per square meter of earth surface and the deep-sea oxygen supply would last some 25,000 years.

In the summary of his report, Van Valen (1971) states, "There are three processes weakly concentration-dependent that keep changes in concentration of atmospheric pressure from being a random walk—inhibition of net photosynthesis by oxygen, the passage of hydrogen through the oxidizing part of the atmosphere before it escapes from the earth, and burial of reduced carbon in anaerobic water. A stronger regulator seems desirable but remains to be found. The cause of the initial rise in oxygen concentration presents a serious and unresolved quantitative problem."

And, in the summary of his report, Broeker (1970) states, in part, "It can be stated with some confidence that the molecular oxygen supply in the atmosphere and in the broad expanse of open ocean are not threatened by man's activities in the foreseeable future. Molecular oxygen is one resource that is virtually unlimited."

Ozone Layer

It has been established for many years that there is a ozone layer in the upper atmosphere of the earth. This ozone is formed by the absorption of ultraviolet radiation by oxygen. At from 50,000 feet (15,000 meters) to 120,000 feet 36,000 meters), ozone can be found in concentrations of a few parts per million, variously estimated between 5 and 10 ppm. In the formation of ozone, ultraviolet radiation of wavelengths between 290 and 320 nanometers are absorbed and thus not allowed to reach the surface of the earth. In recent years, this particular ultraviolet radiation has sometimes been called *damaging ultraviolet radiation* (DUV) because the radiation is associated with a number of damaging human health effects, principally skin cancers. However, the radiation also has been found damaging to other animals and some plants. Some fishes and other forms of sea life are particularly susceptible to the DUV. Thus, any phenomena that may alter the ozone layer (concentration) can greatly reduce the shielding effects of DUV which the ozone provides.

Several years ago, when seeking a means to propel ingredients from spray cans by filling the cans under pressure with a substance that would not interfere with the product (paint, insecticide, deodorants, etc.), scientists found that the inert chlorofluorocarbons (also called chlorofluoromethanes, CFMs) were well tailored for use as propellants. They apparently were no threat to health and they were inert. Because of the good match between the job to be done and the CFM propellants, millions upon millions of aerosol cans were made and marketed. The principal CFMs used were trichlorofluoromethane (CCl_3F) and dichlorodifluoromethane (CCl_2F_2). It was considerably after the CFMs were well-established and widely used that Rowland and Molino (University of California, Irvine) first suggested that potential of these materials for harming the stratosphere. These researchers reasoned that inasmuch as the CFMs are essentially inert in the lower atmosphere, they would eventually work their way to the stratosphere, where they would be dissociated into their constituent atoms by sunlight. Thus released, chlorine atoms could serve as a catalyst in a complex series of reactions, part of the net effect being that of converting ozone into oxygen. This line of thought remains to be confirmed by actual measurement of ozone depletion in the atmosphere, but many people in the scientific community believe the hypothesis to be essentially correct. An abridged version of the series of steps involved in converting ozone to oxygen: $Cl + O_3 \rightarrow ClO + O_2$; $ClO + O \rightarrow Cl + O_2$.

A report issued by the National Academy of Sciences (1979) states in effect that the potential for depletion of ozone in the stratosphere caused by the release of chlorofluorocarbons is greater than had previously been predicted. The report projects that continued production and release of the chemicals at the 1977 rate will lead to an eventual 16.5% depletion of stratospheric ozone. This turns out to be just a little more than double the depletion projected in a 1976 report. The panel responsible for the report also predict that continued growth in production of certain other halocarbons will aggrevate the problem, even though a significant portion of these materials is destroyed in the lower atmosphere. The United States took the lead in banning these chemicals in aerosol packages, but most other countries of the world have not done so (early 1980s).

Increased depletion projections in the 1979 report resulted from direct measurement of several key rate constants. These had previously been obtained only indirectly. The rate constant for the reaction of nitric oxide with peroxide ($NO + HO_2 \rightleftharpoons NO_2 + OH$) was found to be about 40 times larger than previously estimated. Using this and additional rate-constant data, atmospheric models were constructed. As a result of these studies, it is estimated that an eventual ozone depletion of 18.6% if CFMs continue to be released at the 1977 rates. At the current rate, a 5% depletion is not expected until about the year 2000.

Actually since aerosol sprays were banned in the United States, the worldwide production has been increased.

As of the mid-1980s, it is now believed by some researchers that the injection of hydrogen chloride (HCl) directly into the stratosphere (and thus not absorbed by rain) may be a major, if not the predominating, cause for degradation of the ozone layer. It has been estimated that the Augustine Volcano (Alaska), which erupted in 1976, contributed some 570 times the amount of chlorine to the stratosphere as was produced worldwide in 1975 as the result of the manufacture of chlorine and fluorocarbons. Thus, some scientists are having second thoughts and it would appear in order to reassess ozone layer phenomena against volcanic eruption records.

When the filtering of DUV that reaches the surface of the earth is reduced, the induction of skin cancers increases. Two types of skin cancers are produced. Nonmelanoma skin cancer, of which some 300,000 to 600,000 cases are reported annually in the United States, is the most common form of cancer. These cancers generally can be removed by simple surgery and are rarely fatal. These cancers usually develop on parts of the body that are exposed to direct sunlight. Statistics indicate that there is a correlation between cumulative lifetime exposure to DUV radiation and incidence of the disease. Melanoma, the other form of cancer, is much more severe, with fatalities in about one-third of the cases. The total number of melanoma cases reported in the United States in 1979 was 13,600, with 4,000 fatalities. The incidence is greatest among individuals who do not tan readily and who are exposed to direct sunlight for appreciable periods only intermittently. It is estimated that both kinds of cancer will increase markedly if additional DUV reaches the earth.

Investigators at the University of Kentucky and at the National Marine Fisheries Service in La Jolla, California, among others, have researched the impact of greater DUV on aquatic ecosystems. It has been estimated, for example, that currently the anchovy exists in environments very near to its top DUV tolerance. Crab and shrimp larvae also have been reported as being close to their DUV tolerance. Some aquatic forms, of course, can occupy lower depths of the sea (DUV is important down to about 10 meters), but great damage could be done before such an ecological adjustment could be made.

CFMs also have been implicated in contributing to the so-called "greenhouse effect" that could lead to atmospheric warming and major alterations in climate. See also **Climate**.

CFMs also are used industrially, but industry already has made a good start in finding substances or methods that can replace the CFMs. An important use of CFMs, for example, is as a refrigerant. Proposals have been made to recycle or recover the CFMs from abandoned and obsolete refrigeration equipment, much as aluminum cans are collected and recycled today.

References

Berkner, L. V., and L. C. Marshall: in "The Origin and Evolution of Atmospheres and Oceans" (P. J. Brancazio and A. G. W. Cameron, editors), pages 102–126, Wiley, New York (1964).

Berkner, L. V., and L. C. Marshall: *Jrnl. Atmos. Sci.*, **22**, 225 (1965).

Bert, P.: "Barometric Pressure Researchers in Experimental Physiology" (translation of original, 1878), College Book Company, Columbus, Ohio (1943).

Borcher, H.: *Geochim. Cosmochim. Acta*, **2**, 62 (1951).

Brinkman, R. T.: *J. Geophys. Res.*, **74**, 5355 (1969).

Broeker, W. S.: "Man's Oxygen Reserves," *Science*, **168**, 1537–1538 (1970).

Caughey, W. S., Editor: "Biochemical and Clinical Aspects of Oxygen," Academic, New York (1979).

Dickens, F.: "The Toxic Effects of Oxygen on Brain Metabolism and on Tissue Enzymes," *Biochem. Jrnl.*, **40**, 145, 170 (1946).

Fridovich, I.: "The Biology of Oxygen Radicals," *Science*, **201**, 875–880 (1978).

Gerschman, R., et al.: "Oxygen Poisoning and X-irradiation: A Mechanism in Common," *Science*, **119**, 623 (1954).

Gilbert, D. L.: "The Role of Pro-Oxidants and Anti-Oxidants in Oxygen Toxicity," *Radiation Research Supplements*, **3**, 44 (1963).

Griffiths, J., and C. Hawkins: "Mechanistic Aspects of the Photochemistry of Dyes and Their Intermediates," *Jrnl. of Soc. of Dyers and Colourists*, **89**, 173 (1973).

Haugaard, N.: "Oxygen Poisoning. XI. The Relation Between Inactivation of Enzymes by Oxygen and Essential Sulfhydryl Groups," *Jrnl. Biol. Chem.*, **164**, 265 (1946).

Holland, H. D.: in "Petrologic Studies: A Volume in Honor of A. F. Buddington," (A. E. J. Engel, H. L. James, and B. F. Leonard, Editors), pages 447–477, Geological Society of America, Washington, D.C. (1962).

Huber, F., and J. Senebier: "Mémoirs sur l'influence de l'air et de divers substances gaseuses dans la germination de différentes graines," page 38, Geneva, Switzerland.

Kon, S., and S. Schwimmer: "Depolymerization of Polysaccharides by Active Oxygen Species Derived from Xanthine Oxidase Systems," *Food Biochem.*, **1**, 141 (1977).

Kon, S.: "Effects of Oxygen Free Radicals on Plant Polysaccharides," *Food Technology*, **32**, 5, 84–94 (1978).

McCord, J. M., and I. Fridovich: "Superoxide Dismutase: An Enzymatic Function for Erythrocuprein," *Jrnl. Biol. Chem.*, **244**, 6046 (1969).

Michaelis, L.: "Fundamentals of Oxidation and Reduction," in *Currents in Biochemical Research* (D. E. Green, Editor), page 207, Wiley, New York (1946).

Schiff, H. I., Chairman: "Stratospheric Ozone Depletion by Halocarbons: Chemistry and Transport," National Academy of Sciences, Washington, D.C. (1979).

Smith, R. C., and K. S. Baker: "Stratospheric Ozone, Middle Ultraviolet Radiation, and Carbon-14 Measurements of Marine Productivity," *Science*, **208**, 592–593 (1980).

Stadie, W. C., and N. Haugaard: "Oxygen Poisoning," a series of articles in *Jrnl. of Biol. Chem.*, **160**, 191 (1945); **161**, 175 (1945); **161**, 181 (1945); and **164**, 257 (1946).

Staff: "Protection Against Depletion of Stratospheric Ozone by Chlorofluorocarbons," National Academy of Sciences, Washington, D.C. (1979).

Sverdrup, H. U., Johnson, M. W., and R. H. Fleming: "The Oceans, Their Physics, Chemistry and General Biology," Prentice-Hall, Englewood Cliffs, New Jersey (1942).

Sweet, F., et al.: "Ozone Selectivity Inhibits Growth of Human Cancer Cells," *Science*, **209**, 931–932 (1980).

Van Valen, L.: "The History and Stability of Atmospheric Oxygen," *Science*, **171**, 439–443 (1971).

OXYGEN DEBT. A term used to refer to the build-up of a need for oxygen through anaerobic respiration of muscle cells in a higher vertebrate animal during violent exercise. When the energy demands are too great to be satisfied by the aerobic respiration, the cells turn to anaerobic respiration. Lactic acid is an end product of such respiration; this acid tends to accumulate in the muscles and some of it diffuses into the blood and accumulates in the liver. When the activity ceases, deep breathing continues and the extra oxygen is used to reconvert the lactic acid back to pyruvic acid and to carry the pyruvic acid through the tricarbocyclic acid cycle. It may also be reconverted back to glucose and glycogen.

OXYGEN GROUP (The). The elements of group 6a of the periodic classification sometimes are referred to as the Oxygen Group. In order of increasing atomic number, they are oxygen, sulfur, selenium, tellurium, and polonium. The elements of this group are characterized by the presence of six electrons in an outer shell. The similarities of chemical behavior among the elements of this group are less striking than hold for some of the other groups, e.g., the close parallels of the alkali metals or alkaline earths. With exception of oxygen, all elements of the group have a valence of 4+, in addition to other valences. All of the elements with the exception of polonium also have a valence of 2−. Unlike the alkali metals or alkaline earths, for example, the elements of the oxygen group are not so similar chemically that they comprise a separate group in classical qualitative chemical analysis separations. Tellurium and selenium do appear together among the rarer metals of the second group in terms of qualitative chemical analysis.

OXYTOCIC HORMONE. Hormones.

OXYTOCIN. A polypeptide hormone which is secreted by the posterior lobe of the pituitary gland of mammals and other vertebrates. Oxytocin exerts a stimulating effect upon the muscles of the breast (milk-ejection) and those of the uterus of mammals. It is sometimes used medically to stimulate labor in cases of difficult childbirth and to time the onset of labor.

OZOCERITE. Sometimes spelled ozokerite, this is a natural brown to jet black mineral (paraffin) wax comprised mainly of hydrocarbons. The melting point is variable. The material is soluble in chloroform. When heated with sulfuric acid (20–30%) from 120 to 200°C, ozocerite yields ceresine. Sometimes called earth wax, fossil wax, mineral wax, and native paraffin.

OZONE; OXYGEN. Photochemistry and Photolysis.

P

PAINT AND FINISHES. Paint is a mixture composed of solid coloring matter suspended in a liquid medium and applied as a coating to various types of surfaces—for decorative, protective, or other functional purposes. Decorative effects may be produced by color, gloss, or texture. A secondary decorative function of paint is lighting, as the color of the surface affects the reflectance. The proportion of light reflected by a surface is expressed as a percentage of complete reflectance and, for various colors is: White 90–80%; very light tints, 90–70%; light tints, 70–60%; medium-to-dark tints, 60–20%; aluminum paint, 45–35%; deep colors, 20–3%; black, 2–1%. Examples of the functional properties of paint include a great variety of signs, traffic markings on highways, and color coding for piping in process plants, among others. The protective properties of paints include defense against air, water, sunlight, and chemicals (such as acids, alkalis, and atmospheric pollutants), as well as mechanical properties, such as hardness and abrasion resistance. Numerous special paints are made, including those for inhibiting molds or rotting of woods, as coatings on electrical parts to provide a degree of insulation, fire retardant paints, and coatings on plaster, wood, and concrete for ease of cleaning.

Classification. The type of binder, such as alkyds, vinyls, or epoxy, may be the basis for classifying paints. Since most binders are used in several different kinds of paint, however, paints also may be classified according to the properties of the product or end use. Alkyd enamels, for example, are gloss paints with good abrasion resistance and good cleanability, while alkyd flat wall paints are characterized by a very low sheen and inferior abrasion resistance and cleanability.

Paints used as the final coat on a surface are referred to as *finish coats* or *topcoats*. Paint applied before the topcoat is called an *undercoat*. Undercoats are often classified according to use. *Fillers* are undercoats used to fill holes, pores, or irregularities to provide a uniform surface for the topcoat. *Primers* are used to aid the adhesion of the topcoat to a surface and to prevent absorption of the topcoat into a porous surface. Primers can also be used to prevent corrosion of metals that are to be painted. *Surfacers* are highly pigmented undercoaters used to make a surface more uniform and give adhesion to the final coat. Surfacers often are formulated so that they can be sanded smooth before the topcoat is applied. *Sealers* are clear or pigmented materials applied to a surface to prevent some materials in the surface to be painted, e.g., a dye, from migrating into the top coat.

Paints also may be classified into two very broad groups: (1) those which use water as the primary liquid of the paint; and (2) those which use other liquids, such as hydrocarbons or aliphatic or oxygenated compounds. Many of the same polymers and pigments can be used in both types of paint. The primary difference is that most non-water dilutable paints are *solution paints*, where the liquid is a solvent in which the polymer is dissolved. *Water-based paints*, however, are primarily latex paints where the polymer particles are a discontinuous phase and water is the continuous phase. *Organosols*, however,

consist of a resin dispersed in a liquid other than water. There are also water-based paints which have a water-soluble binder instead of a latex dispersion.

In addition to the pigment, binder, and liquid, a paint also may contain many additives, such as defoamers, thickeners, flow agents, catalysts, wetting agents, and plasticizers to improve various properties of the paint.

Solvent Paints. Because of its high refractive index, titanium is a widely used pigment in solvent paints. This material provides opacity to white and light-colored paints. Among other white pigments used is zinc oxide for hardness, drying, and mildew resistance of exterior paints.

Extender pigments, such as talc, silica, mica, clay, and calcium carbonate can be used to control many different properties of a paint (hardness, gloss, setting, rheological properties, and rust prevention). Since the solvent evaporates and contributes nothing to the dried film, the least expensive combination of solvents can be used which will dissolve the polymer, give the desired viscosity, and conform to air-pollution standards.

Antiskinning agents are often used to prevent oxidation and surface hardening of the paint resin in the can. Oxime compounds, e.g., methyl ethyl ketoxime, and substituted phenols, e.g., guaicol, can be used.

Scratch- and marproof additives, such as amino resins, wax, polyethylene, and cellulose derivatives, are sometimes added to harden the paint film or develop a lubricated surface.

Flooding and floating is the separation of one or more pigments from the rest of the paint at the surface of the film, usually due to difference in densities of the pigments. Antiflood and antifloating agents, such as silicones, can be used to correct this problem.

Additives which will attach to the surface of the pigments and form a pigment-vehicle bond for better grinding or dispersing are often used. Compounds which contain both hydrophilic and lipophilic groups, e.g., morpholine, methyl ethyl ketoxime, and soya lecithin, are used so that the hydrophilic part will attach to the pigment and the lipophilic part will attach to the vehicle.

Flow or leveling agents are used to correct irregularities or defects in applied paint films. These defects can be caused by differences in concentration between the surfaces and the interior of the paint during evaporation of the solvent, overpolymerized particles in the vehicle, an unclean surface, pigment flocculation, or solvent mixtures with a wide range of evaporation rates. Some of the materials used are silicone resins and unsaturated organic acids.

Drying-Oil Paints. Drying oils, which have been used for years in paint binders or vehicles, are naturally occurring materials, usually of vegetable origin, i.e., liquid triglycerides with three molecules of long-chain fatty acids to each glycerin molecule. The majority of these acids have 18 carbon atoms, many of which are unsaturated. The oils differ greatly in their drying properties, which depend upon the degree of unsaturation of the acids. Linseed, safflower, soya, tall oil, cottonseed, tung, and oiticia contain fairly large percentages of unsaturated acids, e.g., oleic, linoleic, linolenic, eleostearic, and licanic acids.

Drying oils are often used in exterior architectural paints because of their durability, ease of application, and moderate coat. Disadvantages are slow drying and poor chemical resistance.

Alkyd Paints. Alkyds are polyester resins made from polybasic acids and polyhydric alcohols. Glycerol and pentaerythritol are often used for the polyalcohol, and phthalic anhydride and maleic acid are often used for the polycarboxylic acids. All the oils previously mentioned are also used in alkyds by converting the fatty acid oils into monoglycerides and then reacting with a dibasic acid, such as phthalic anhydride. See also **Alkyd Resins.**

Alkyd resins vary greatly in their properties because of the many different oils, alcohols, and acids that can be used to make them. Alkyds are faster drying, have better gloss retention, and better color than oils. Most unmodified alkyds have low chemical and alkali resistance. Alkyds can be modified with rosin esterified in place of some oil acids. Phenolic resins, such as *o*- or *p*-phenylphenol, can also be used in order to produce greater hardness and better chemical resistance.

Styrene and vinyl toluene are also used to modify alkyds for faster drying, improved hardness, and toughness. Copolymers of silicones and alkyds are often used for stove or heater finishes or for coating articles which may be subjected to heat up to about 450°F (232°C). They have good adhesion, hardness, flexibility, toughness, exterior durability, and resistance to solvents, acids, and alkalis.

Acrylic monomers can be copolymerized with oils to modify alkyd resins for fast drying, good initial gloss, adhesion, and exterior durability. Aromatic acids, e.g., benzoic or butylbenzoic, may be used to replace part of the fatty acids for faster air drying, high gloss, hardness, chemical resistance, and adhesion.

Vinyl Paints. Vinyl chloride-vinyl acetate copolymers in solution form can be used in paints. These paints are resistant to alkali and organic acids, alcohols, oils, and aliphatic hydrocarbons. They can be dissolved in ketones, esters, and chlorinated hydrocarbons. They have good water resistance, toughness, and flexibility and are nonflammable. They have good clarity and good exterior durability. Vinyl copolymers can be used in paints for such difficult applications as railway cars. See also **Vinyl Ester Resins.**

Epoxy Paints. These are prepared from epichlorohydrin and a dihydroxy compound, usually a biphenol. Two reactions are involved in the polymerization—condensation to eliminate hydrochloric acid and addition reactions to open epoxide rings along the chain to produce hydroxyl groups. The polymer has epoxide rings at each end and hydroxyl groups along the chain, which ensure good adhesion to polar surfaces, such as metals. There are two types of epoxy resins—the *catalyzed types* and the *epoxy esters.* See also **Epoxy Resins.**

Catalyzed epoxies must be converted to useful products by reaction with curing agents, e.g., amines, polyamide resins, polysulfide resins, anhydrides, metallic hydroxides, or Lewis acids. Most of these materials are supplied as two-package systems to separate the materials until just before application. Polymer curing takes place by reaction of a curing agent with epoxide rings to crosslink the polymer. Paints made from these polymers have excellent chemical resistance and hardness and can be used for maintenance coatings and industrial finishes. Chalking with exterior exposure and the two-package format limits their use.

The epoxy resin can also be reacted with drying oils or fatty acids to produce epoxy esters, which cure by air drying or heat. Paints made with epoxy esters do not have as good chemical and solvent resistance as catalyzed epoxies, but they are superior to oils and alkyds in this respect.

Acrylic Paints. Acrylic resins can be divided into thermoplastic and thermosetting types. Acrylic resins used in paints are mono- or copolymers of acrylic acid or methacrylic acid esters. Some of the common monomers are methyl methacrylate, butyl methacrylate, methyl acrylate, butyl acrylate, ethyl acrylate, and 2-ethylhexyl acrylate. See also **Acrylates and Methacrylates.**

Thermoplastic resins become soft when heated and reharden when cooled. Paints made with these resins have excellent exterior durability and excellent gloss and color retention. They are nonyellowing and can have good resistance to stain. Heating after application allows these paints to reflow, producing excellent appearance. Thermoplastic acrylic finishes are primarily used for automotive and product finishes. They have the disadvantages of poor adhesion and high price for both polymer and solvents.

Thermosetting acrylic resins have at least one monomer belonging to the acrylic family which will react with itself or other resins at elevated temperatures to crosslink in order to cure. In addition to the acrylic monomers previously mentioned, acrylonitrile, acrylamide, styrene, and vinyl toluene are often used in these polymers. Polymers which react to crosslink primarily because of hydroxyl groups are usually combined with an epoxy resin; those which react mainly with carboxyl groups usually are combined with an amine resin. Thermosetting acrylic paints, which are hard and stain-resistant and have high gloss, are often used for appliance finishes. Tough, flexible finishes can be formulated for coil coatings.

Cellulosic Paints. Cellulosic polymers, such as nitrocellulose, ethyl cellulose, ethyl hydroxyethyl cellulose, cellulose acetate, and hydroxyethyl cellulose, can be dissolved in solvent mixtures which are primarily oxygenated liquids, such as esters, ketones, ethers, and alcohols. Coatings made with these polymers dry only by loss of solvent, which is often referred to as *lacquer drying.* Cellulosic lacquers must often be plasticized with compounds, such as dibutyl phthalate or butylbenzyl phthalate. Compatible resins, e.g., ester gum or acrylic or maleic resins, are often used to improve gloss, adhesion, and elongation. Cellulosic lacquers are fast drying, but low film builders. They are used extensively for furniture finishing and automobile refinishing. See also **Cellulose Ester Plastics.**

Chlorinated-Rubber Paints. Chlorinated rubber, made by chlorinating natural rubber, is soluble in aromatic hydrocarbons. It is used to make paints with excellent chemical resistance, water resistance, and fast-drying qualities. The polymer is rather brittle and must be plasticized with materials such as drying oils and chlorinated biphenyl. The disadvantage of these paints is poor resistance to organic solvents and oils; exterior exposure is fairly good. They are used primarily where good chemical or water resistance is required, as in certain types of industrial maintenance.

Polyester Paints. Polyesters are unsaturated thermosetting polyester resins similar to those used for reinforced plastic. Although alkyds can be considered unsaturated polyesters, this term has been reserved for resins which have unsaturated compounds in the backbone of the polymer. These resins are made by reacting unsaturated dibasic acids, e.g., maleic anhydride, citraconic anhydride, fumaric acid, itaconic acid, phthalic anhydride, and adipic acid, with polyhydric alcohols, e.g., propylene glycol. Styrene or some other aromatic vinyl monomer is added to the polyester resin, which is then solubilized and made into a paint. Inhibitors, e.g., hydroquinone, are added to prevent premature polymerization in the can; organic peroxides or some other catalyst must be added to initiate polymerization of the styrene monomer and the polyester resin for curing, which is often carried out at elevated temperatures. See also **Polystyrenes.**

Polyester finishes are very hard, tough, resistant to solvents, and fairly heat-resistant. They are often used for furniture finishes. Adhesion of these paints is often poor.

Phenoloic Paints. Phenolic resins as used in coatings are primarily made from phenol and para-substituted phenols reacted with formaldehyde to form methylol groups on the phenol ring. Condensation polymers are then produced by reacting these groups with phenol. Phenoloic coatings are fast drying and have high build and good resistance to moisture and chemicals. Their poor initial color and tendency to yellow after application limit their use. Phenolic coatings are sometimes used for baked can coatings, and oil-modified phenolaldehyde finishes are sometimes used for marine finishes and aluminum paints.

Polyurethane Paints. Polyurethanes are based upon reactions of isocyanates. Urethane coatings have excellent solvent and chemical resistance, abrasion resistance, hardness, flexibility, gloss, and electrical properties. They are somewhat expensive and the aromatic isocyanates yellow after application.

Coumarone-Indene Paints. Coumarone-indene resins which are derived from coal tar are used widely to make aluminum paints since they aid leafing of the aluminum and minimize gas formation. They have a yellow color, however, and only fair durability except in aluminum paints.

Bitumen Paints. Bitumen resins are also used for aluminum paints as well as for maintenance and insulating coatings because of their relatively low cost. They are dark in color and have low solvent resistance and only fair durability except with aluminum.

Latex Paints. Rutile (titanium dioxide) is the primary pigment used in latex paints to obtain opacity, or hiding power, in white or pastel paints. Semichalking grades are used for interior paints; chalk-resistant grades for exterior paints. Zinc oxide is sometimes used for exterior paints to help prevent mildew, but care must be taken in using this pigment because of its chemical reactivity.

A number of pigments, such as calcined clay and deaminated clay, can be used to advantage in latex paints as titanium dioxide extenders to increase opacity. These pigments have large surface areas due to irregular surfaces or fine particle size, and the latex vehicle will not cover all the pigment surface when the paint film is dry, leaving entrapped air in the film. The interfaces of air with pigment and vehicle increase the light refraction of the film and thus the opacity. Good hiding power can be obtained with these pigments at a relatively low cost, but the paint film often becomes porous and difficult to clean.

Most of the common extenders, e.g., mica, calcium carbonate, clay, talc, silica, and wollastonite, can be used in latex paints. Since these pigments vary in particle size, shape, hardness, color, surface treatment, and water demand, they can affect viscosity, flow, gloss, color, cleanability, scrubbability, enamel holdout, uniformity of appearance, and even opacity to some extent. Extender pigments are selected to obtain the desired properties for each type of paint. Slightly soluble ammonium phosphate compounds are used as the primary pigment in intumescent fire-retardant paints.

Surfactants are used in latex paints to help wet and disperse pigments, emulsify liquids, and function as defoamers. These materials have a balanced polar-nonpolar structure, which in water-base paints is usually referred to as a *hydrophile-lipophile balance.* The chemical composition of surfactants can vary greatly, and they are usually only classified into anionic, cationic, and nonionic types.

Thickeners or protective colloids are used in latex paints to produce the desired viscosity and help stabilize emulsions and pigment dispersions. Water-soluble protein or casein dispersions and cellulosic polymers are the most commonly used. Soluble polyacrylates, starches, natural gums, and inorganic colloidal materials have also been used. Protective colloids can affect many properties of paint, such as washability, brushability, rheological properties, and color acceptance. Since latex paints are susceptible to bacterial attack, they should contain preservatives. Several types of preservatives can be used: phenolic, mercuric, arsenic, or copper compounds (although some of these are now under strict regulations in many countries), formaldehyde, and certain quaternary chlorinated compounds. Some of these compounds are chemically active, and some are toxic, facts which must be considered when selecting a preservative.

Many of the polymers used in the plastics industry and in solution coatings previously described also can be obtained in latex form. An advantage of using polymers in this form is that high-molecular-weight fully cured polymers can be made to flow well whereas in solution form they would have high solution viscosities.

Latex paints dry fast and are easy to apply; equipment can be cleaned with water; and there is no fire hazard or atmospheric pollution during application. A latex paint does not lose its liquid gradually with a gradual increase in viscosity during drying as is the case of a solution paint. Among the limitations of latex paints are inability to hold more than 55% solids and their possible coalescence when frozen. The main types of latex polymers used in latex paints are styrene-butadiene, vinyl homo- or copolymers, and acrylic polymers or copolymers.

Water-Soluble Paints. Water-soluble paint binders, such as egg albumin, gum arabic, and casein, have been used for many years. Most of these materials have serious limitations, including water sensitivity and poor durability. Many of the synthetic polymers now used in solvent or latex paints can be solubilized in water. Carboxylic, hydroxyl, epoxy, or amine groups on a polymer in conjunction with coupling solvents, such as alcohols, alcohol ethers, or glycol ethers, are the primary mechanisms by which resins are solubilized. Maleic or formaric acids can be reacted with drying oils to produce resins with some carboxy groups which can be solubilized with ammonia or amines. Alkyds can be solubilized by leaving a reactive carboxylic group on the resin instead of terminating the reaction with a monobasic acid or drying-oil acid.

Water-soluble resins have been used for air-dry and low-bake industrial primers and finishes, coatings applied by electrodeposition, and architectural semigloss paints. Contamination and mechanical and chemical stability can be a problem with these resins.

Drying Mechanisms

There are two main mechanisms by which a wet paint applied to a surface becomes a dry solid coating. Some paints, e.g., lacquers and most latex paints, dry only by the evaporation of the liquids in the paint. The polymer binder is completely cured when the liquids evaporate, and no chemical change is required to harden the polymer. In latex paints, the latex binder consists of very small particles of solid polymer separated by water, which is the continuous phase. When the water evaporates, the polymer particles touch each other and fuse together, or coalesce, into a continuous paint film. Pigment particles are also dispersed in the water phase, and the dry paint film consists of a mixture of pigment and polymer particles fused together. If the latex particles are so hard that they will not fuse together when the water evaporates, coaslescents must be added, e.g., carbitol acetate or dibutyl phthalate.

The other fundamental method of drying takes place by a chemical reaction to cross-link a soft polymer after the liquids have evaporated to produce a hard paint film. Some paint binders contain unsaturated compounds, such as linseed or soya drying oils, which react with oxygen in the air to form solid polymers. Other materials, such as isocyanates, react with water vapor in the air in order to polymerize. Some materials, e.g., epoxy resins, must be cured by reaction with curing agents, such as amines and polysulfide resins. These generally are two-package systems.

Application of Paint

In addition to the familiar brush and roller methods, paint can be applied with air or airless spray equipment; electrostatic, hot, or steam spraying; use of aerosol packaging; dip, flow, and electrodeposition coating; roller coating machines; and powder coating.

With electrostatic spraying, the atomized paint is attracted to the conductive object to be painted by an electrostatic potential between the paint and the object. Very little paint is lost with this process, and irregular objects can be coated uniformly. Heat spray application consists of heating the paint so that it is more fluid and higher-solids paints can be applied. With steam spraying, steam is used to atomize the paint. Two-component spray equipment consists of two material lines to the spray gun so that two materials, e.g., an epoxy and a catalyst, can be mixed in the gun just before application. In flow coating, the paint is allowed to flow over the object to be painted, which is usually suspended from a conveyor. The process is similar to dip coating, but is used where the object, e.g., a bed spring, is too large for a tank.

Electrodeposition consists of depositing a paint on a conductive surface from a water bath containing the paint. The negatively charged paint particles are attracted to the object to be coated, which is the anode when an electric potential is applied. Paint can be applied to very irregular surfaces at very uniform thickness with little loss of paint. This system is limited to one coat of limited film thickness and equipment costs are high.

Roller coating machines are used to apply paint to one or both sides of flat surfaces, e.g., fiberboard or tin plate. The thickness of the coating can be controlled by the clearance between a doctor blade and the applicator rolls. Decorative effects, such as wood-grain patterns, can be applied with these machines. In powder coating, paint in a dry powder form is applied on the surface of a heated or electrostatically grounded object to be coated. Following powder application, the object is heated to fuse and cure the coating.

—Ralph S. Armstrong, The Sherwin-Williams Company, Cleveland, Ohio.

PALLADIUM.   Chemical element symbol Pd, at. no. 46, at. wt. 106.4, periodic table group 8 (transition metals), mp 1550–1552°C, bp 3139–3141°C, density, 12.16 g/cm³ (solid), 12.25 g/cm³ (single crystal at 20°C). Elemental palladium has a face-centered cubic crystal structure. The six stable isotopes of palladium are ^{102}Pd, ^{104}Pd through ^{106}Pd, ^{108}Pd, and ^{110}Pd. The seven unstable isotopes are ^{100}Pd, ^{101}Pd, ^{103}Pd, ^{107}Pd, ^{109}Pd,

^{111}Pd, and ^{112}Pd. In terms of earthly abundance, palladium is one of the scarce elements. Also, in terms of cosmic abundance, the investigation by Harold C. Urey (1952), using a figure of 10,000 for silicon, estimated the figure for palladium at 0.0091. No notable presence of palladium in seawater has been detected. The element was discovered by Wollaston in 1803.

Electronic configuration is $1s^22s^22p^63s^23p^63d^{10}4s^24p^64d^{10}$. Ionic radius, Pd²⁺, 1.3755 Å. First ionization potential, 8.33 eV; second, 19.8 eV. Oxidation potentials Pd → Pd²⁺ + 2e⁻ (4M HClO₄), −0.83 V; Pd + 1OH⁻ → Pd(OH)₂ + 2e⁻, −0.1 V. Further physical properties are given under **Platinum Group.**

Palladium has some similarities with both nickel and silver and many with platinum. Palladium dissolves more readily in acids than any other members of the platinum group of metals. In aqua regia, the metal dissolves quickly. Even the compact metal dissolves slowly in HCl. In finely divided form, it is quite soluble in all acids. When heated in air at red heat, the monoxide, PdO is formed. Palladium is similarly converted to the dihalides under the same conditions when it is exposed to F₂ or Cl₂. The metal is not affected by hydrogen sulfide.

The black compound palladium(II) oxide is formed by fusing palladium(II) chloride with NaNO₃ at 600°C and then leaching out the salts with water. This strong oxidizing agent is easily reduced to the metal by hydrogen. The compound is insoluble in water and acids, including aqua regia. The hydroxide, Pd(OH)₂, is made by the hydrolysis of palladium(II) nitrate. The compound is soluble in acids, and water is evolved on heating, but even at 500–600°C some water still remains. At this temperature, the compound starts to lose oxygen.

Palladium(III) oxide, P₂O₃, is made as a hydrate by careful oxidation of a solution of palladium(II) nitrate either by anodic oxidation or ozone treatment at −8°C. This unstable brown powder reverts to the monoxide in about four days. When heated, the compound loses water and may explode as it changes to the monoxide.

Palladium(II) chloride is formed by direct combination of the elements at 500°C. It is the only stable solid chloride over 500–1500°C. The red crystals are partly soluble in water and completely soluble in HCl. The fraction insoluble in water is probably a polymer. Palladium(II) chloride also is the product obtained by evaporation of a solution of Pd in HCl. Palladium(II) bromide can be made from the elements.

When potassium iodide is added to a solution of palladium(II) chloride, an insoluble diiodide is precipitated. The dark red-black crystals are soluble in excess iodide with formation of the tetraiodide complex ion. Palladium(II) iodide evolves iodine at 100°C, the decomposition to the elements being complete at 330–360°C. The black compound palladium(III) fluoride is made by direct combination of the elements. On reduction, the brown difluoride is formed.

Divalent palladium forms many planar complexes with a coordination number of 4. The tetrachlorides are quite soluble. When a solution of palladium(II) chloride is oxidized with chlorite or chlorate ion, Pd(IV) is formed, which has a coordination number of 8. The addition of ammonium chloride to such a solution precipitates ammonium hexachloropalladate(IV) as a red compound. It is somewhat less stable than the platinum analog.

The soluble yellow-brown palladium(II) nitrate is formed by dissolving finely divided palladium in warm HNO₃ and then crystallizing the compound from this solution. The analogous sulfate is similarly formed from H₂SO₄. It crystallizes as a red-brown dihydrate. Both of these compounds easily hydrolyze.

Palladium(II) sulfide is precipitated as a brown powder by adding H₂S to a solution of palladium (II) ion. When this sulfide is heated with sulfur at 400°C, the insoluble disulfide is formed. The excess sulfur can be extracted with carbon disulfide to yield the gray-black crystalline palladium(IV) sulfide. This compound is not soluble in single acids, but is soluble in aqua regia.

Some palladium complexes are important analytically or in the refining of palladium. The yellow dimethylglyoxime compound is quantitatively precipitated from a HCl solution of palladium (II) chloride by the addition of an alcoholic solution of dimethylglyoxime. Palladium (II) has a great affinity for nitrogen-containing ligands. The di- and tetramine find use in refining.

Palladium, as with other members of the platinum group, exhibits catalytic activity for various reactions. One of its best known uses is in conjunction with other platinum metals in the catalytic converters of present automobiles.

As reported by Chung-Chium Liu et al. (*Science*, **207**, 188–189, 1980), a palladium-palladium oxide miniature pH electrode has been developed. The miniature wire-form electrode exhibits a super-Nernstian behavior and gives a mean pH response of 71.4 mV per pH unit. The electrode may find application in biological, medical, and clinical studies.

See also **Chemical Elements; and Platinum Group.**

—Linton Libby, Chief Chemist, Simmons Refining Company, Chicago, Illinois.

PALMITIC ACID. $CH_3(CH_2)_{14}COOH$, formula weight 256.42, white crystalline powder, mp 64°C, bp 271.5°C, sp gr 0.849. The acid is insoluble in H_2O, moderately soluble in alcohol, soluble in ether. About 60% of the content of palm oil is palmitic acid.

Palmitic acid is present as cetyl ester in spermaceti from which, by hydrolysis, the acid may be obtained; it is present in bee's wax as the melissic ester; and in most vegetable and animal oils and fats, in greater or less amounts, as glyceryl tripalmitate or as mixed esters, along with stearic and oleic acids. Palmitic acid is separated from stearic and oleic acids by fractional vacuum distillation and by fractional crystallization. With NaOH, palmitic acid forms sodium palmitate, a soap. Most soaps are mixtures of sodium stearate, palmitate, and oleate.

Representative esters of palmitic acid are: methyl palmitate $C_{15}H_{31}COOCH_3$, mp 30°C, bp 195°C at 15 mm pressure; ethyl palmitate $C_{15}H_{31}COOC_2H_5$, mp 24°C, bp 185°C at 10 mm pressure; cetyl palmitate $C_{15}H_{31}COOC_{16}H_{33}$, mp 54°C; glyceryl tripalmitate (tripalmitin) $C_3H_5(COOC_{15}H_{31})_3$, mp 65°C, bp 310°C, approximately.

As the glyceryl ester, palmitic acid is one of the constituents of many vegetable and animal oils and fats.

Palmitic acid finds use in the production of cosmetics, food emulsifiers, pharmaceuticals, plastics, and soaps. One commercial formulation contains 95% palmitic acid, 4% stearic acid, and 1% myristic acid; another preparation contains 50% palmitic acid and 50% stearic acid.

See also **Vegetable Oils (Edible).**

PALMITOLEIC ACID. Also called *cis*-9-hexadeconoic acid, formula $CH_3(CH_2)_5CH:CH(CH_2)_7COOH$. This is an unsaturated fatty acid found in nearly every fat, especially in marine oils (15–20%). At room temperature, it is a colorless liquid. Insoluble in water; soluble in alcohol and ether; mp 1.0°C; bp 140–141°C (5 millimeters pressure). Insoluble in water; soluble in alcohol and ether. Combustible. Palmitoleic acid is used in organic synthesis; and as a standard in chromatographic analysis. See also **Vegetable Oils (Edible).**

PALM OIL.   Vegetable Oils (Edible).

PANTOTHENIC ACID. The designation vitamin B_3 for this essential substance is now used only infrequently, as are other, earlier terms, such as chick antidermatitis factor, Bios IIa, and anti-gray-hair factor. Pantothenic acid is a constituent of coenzyme A, which participates in numerous enzyme reactions. CoA was discovered as an essential cofactor for the acetylation of sulfanilamide in the liver and of choline in the brain. CoA is particularly important in the initial reaction of the TCA cycle (citric acid cycle) of carbohydrate metabolism and energy production. These factors are described in greater detail in the entries on **CARBOHYDRATES;** and **COENZYMES.** Pantothenic acid is unique among the vitamin group, in that it was one of the first to be isolated, using as a basis a microbiological assay method. Even more unique is the fact that its structure was largely determined, using a highly quantitative biological yeast test, long before it was isolated or obtained in concentrated form. R. J. Williams and coworkers described it as an acid with an ionization constant lower than that of an alpha-hydroxyacid, but about right for a hydroxyacid in which the hydroxyl group was farther removed from the carboxyl group.

In 1901, Wildiers described *bios,* an essential for yeast growth. In 1933, Williams isolated crystalline bios from yeast and named it *pantothenic acid*. In 1938, Williams isolated pantothenic acid from liver; and, in 1939, Jukes determined liver antidermatitis factor (chick) to be identical with yeast factor. Also, in 1939, Woolley et al. demonstrated beta-alanine as a vital part of pantothenic acid (see diagram below). In 1940, Harris, Folkers, et al. reported structure determination, synthesis, and crystallization of pantothenic acid. In 1950, Lipmann et al. discovered coenzyme A; and, in 1951, Lynen characterized the coenzyme A structure.

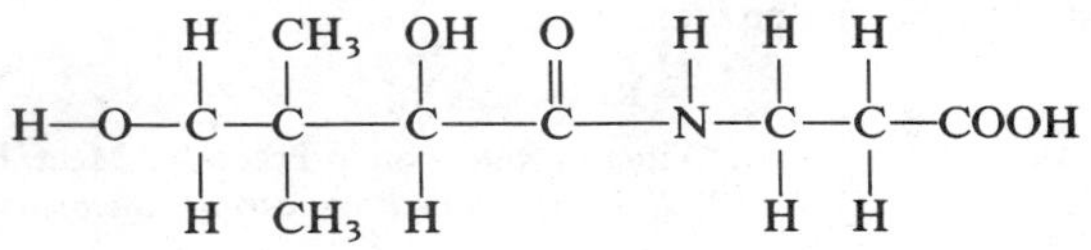

(Pantoic Acid)   (Beta-Alanine)

d (+) Pantothenic acid ($C_9H_{17}O_5N$)

Physiological functions in which pantothenic acid participates include: it serves as part of coenzyme A in carbohydrate metabolism (2-carbon transfer-acetate, or pyruvate) and participates in lipid metabolism (biosynthesis and catabolism of fatty acids, sterols, + phospholipids), protein metabolism (acetylations of amines and amino acids), porphyrin metabolism, acetylcholine production, and isoprene production.

Particularly high in pantothenic acid content are yeasts, animal glands, and organs. Fruits have a low content.

Pantothenic acid is produced commercially by synthesis involving the condensation of *d*-pantolactone with salt of beta-alanine. Some of the dietary supplement forms include calcium pantothenate, dexpanthenol, and panthenol.

Antagonists of pantothenic acid include pantoyltaurine, ω-methylpantothenic acid, *bis*-(β-pantoylaminoethyl)disulfide, 6-mercaptopurine, and pantoylamino ethanethiol. Synergists include biotin, folic acid, niacin, somatotrophin (growth hormone), and vitamins B_1, B_2, B_{12}, and C.

Precursors in the biosynthesis of pantothenic acid include α-ketoisovaleric acid (pantoic acid), uracil-(β-alanine), and aspartic acid. Intermediates in the synthesis include ketopantoic acid, pantoic acid, and β-alanine.

Some of the unusual features of pantothenic acid noted by some investigators include: (1) It promotes amino acid uptake; (2) it is potentiated by zinc in preventing graying of hair in rats; (3) it promotes resistance to stress of cold immersion; (4) there is a deficiency of pantothenic acid in tumors; (5) it is required for chick hatchability; (6) it is useful in treating vertigo, postoperative shock, poisoning with isoniazid and curare, and in accelerating wound healing; and (7) it is useful in treating Addison's disease, liver cirrhosis, and diabetes.

Thermal Deactivation of Pantothenic Acid. Investigators have shown how thermal processing inactivates pantothenic acid (Hellendoorn, 1971; Greenwood, 1944; Wituszynska, 1973). In 1971, Schroeder summarized existing information and suggested that up to 78% of pantothenic acid is lost during canning of vegetables. Hamm and Lund (1978) noted that, although evidence would indicate that pantothenic acid is very heat sensitive, few studies have been done to determine the kinetic parameters which describe reaction rate and its dependence upon temperature. The investigators prepared a stock solution of calcium pantothenate for addition to samples of pea and meat purees. It was noted that, as pH values of samples increased from 4 to 7, activation energies also increased. The effects of buffering were also studied. They concluded that generally the large D values calculated indicated that pantothenic acid is quite heat stable in tested food and model systems, leading to observation that one would not expect much pantothenic acid to be destroyed during thermal processing of food. These conclusions are contrary with earlier findings, and the investigators stressed that further research is needed in this area.

Determination of Pantothenic Acid. Bioassays are made by determining the growth rate of chicks, or microbiologically by determining the growth of *L. casei*. Physicochemical methods include estimating β-alanine after hydrolysis; or estimating CoA by citrate cleavage enzyme.

References

Greenwood, D. A., et al.: "Vitamin Retention in Processed Meat. Effect of Thermal Processing," *Industrial and Engineering Chemistry*, **36**, 922 (1944).

Hamm, D. J., and D. B. Lund: "Kinetic Parameters for Thermal Inactivation of Pantothenic Acid," *Journal of Food Science*, **43**, 2, 631–633 (1978).

Haresign, W., and D. Lewis: "Recent Advances in Animal Nutrition," Butterworth Group, Woburn, Massachusetts (1977).

Hellendoorn, E. W., et al.: "Nutritive Value of Canned Meats," *Journal of American Diet. Association*, **58**, 434 (1971).

Kutsky, J. F.: "Handbook of Vitamins and Hormones," Van Nostrand Reinhold, New York (1973).

Preston, R. L.: "Typical Composition of Feeds for Cattle and Sheep—1977–1978," *Feedstuffs*, A–2A (October 3, 1977); and 3A–10A (August 18, 1978).

Staff: "Food Chemicals Codex," National Academy of Sciences, Washington, D.C. (1972).

Staff: "Nutritional Requirements of Domestic Animals," separate publications on beef cattle, dairy cattle, poultry, swine, and sheep, periodically updated. National Academy of Sciences, Washington, D.C.

Williams, R. J.: "Pantothenic Acid," in "The Encyclopedia of Biochemistry," (R. J. Williams and E. M. Lansford, Jr., Editors), Van Nostrand Reinhold, New York (1967).

Wituszynska, B.: "Determination of Some B Vitamins in Fresh and Canned Fish," *Bromatol. Chem. Toksykol.*, **6**, 13 (1973).

PAPERMAKING AND FINISHING. One of the most important factors in the progress of civilization has been paper, a thin flat tissue composed of closely matted fibers obtained almost entirely from plant sources. In modern life paper finds a variety of uses, for writing, for containers, wrappers, wall covering, and—perhaps most important—in all forms of printing: newspapers, magazines, books.

The art of making paper seems to have been discovered first by the Chinese, who were making paper as early as the beginning of the Christian era. From China the process was carried to Arabia and thence to Europe. Paper was not an important article at first and, since it is not a very durable substance under ordinary conditions, could not compete with parchment or vellum as a medium for the written word. In the fifteenth century writing became more general and the demand for a cheaper material increased. Paper became an important product. At this time paper was made largely from vegetable fibers reclaimed from cloth (especially linen), as had been done since the invention of paper in China. This paper was made entirely by hand, as is done even today in the manufacture of certain expensive types of paper. In making hand-made paper, a pulp is formed by soaking the vegetable fibers in water in a vat. From this vat the pulp is dipped out in a mold, the bottom of which is a fine screen. By a deft motion of this mold the soft pulp is spread over the screen in a thin layer of matted fibers. The water in the pulp drains off, leaving a rather firm mass which is turned out on a piece of felt. More pieces of half-dried pulp spread on felt are added. The whole pile is then pressed to squeeze out more of the water, press the fibers closer together and form a firm sheet. These are then removed from between the felts, pressed again, and dried. During the final treatment surface sizing is added to render a surface more suitable to receive ink. Sheets of hand-made paper are naturally of limited size and expensive.

To meet the great demand for paper, machine methods were developed. This increased demand for paper also led to the utilization of material which could be obtained in quantities much greater than rags. Out of this developed the vast pulp industry which today converts vegetable material, mostly soft woods such as spruce and fir, as well as poplar, into a white felt-like mass of fibrous substance, known as pulp. See **Pulp (Wood) Production and Processing.**

Wet End. A modern paper machine begins with a flow spreader or distributor, conveying a dilute fiber suspension (0.1–1% fibers) to a headbox which delivers a jet of the suspension or slurry through a slice (sluice) across the full width of the machine, almost 400 inches (~8.6 meters) in some large machines. In the headbox, the fibers are dispersed, and the flow is rectified as well as possible so that the jet is delivered onto a moving, endless, fine-mesh wire screen with uniform composition, flow rate, and velocity. The pressure in the headbox and its slice opening are adjusted so that the jet velocity matches the speed of the wire screen, which may be up to 4000 feet (~1220 meters) per minute for newsprint. The proper stock flow per unit width corresponds to the desired *basis weight* of the paper. (Basis weight is weight per unit area and varies with grades and sizes of papers.)

The dispersion of fibers in the headbox is brought about by subjecting the slurry or suspension to shear stresses, usually with turbulence. Various designs have been developed to accomplish this.

As shown in Fig. 1, the most common type of paper machine is the Fourdrinier, in which the moving wire screen is in the form of an endless conveyor belt stretched between two large rolls. The roll situated under the headbox slice is called the *breast roll*. The roll located generally at the end of the straight wire run is the *couch roll*. Drainage of the slurry through the wire screen is induced by several types of driving forces. In the early, slow-speed machines, the principal force was gravity. Later, the hydrodynamic action of table rolls, which support the wire and rotate with it, began to play an important part in drainage as speed increased. More recently, foils came into use, i.e., rigid, stationary, hydrodynamically shaped elements

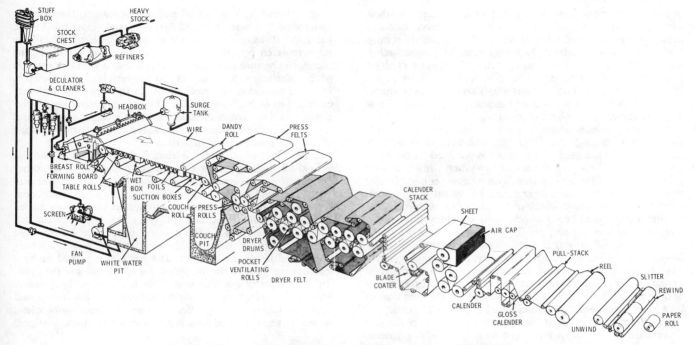

Fig. 1.   Fourdrinier machine for producing printing-grade paper. (*Beloit Corporation.*)

which support the wire and exert a pumping action through the wire screen. Other means are perforated or slotted boxes with vacuum over which the wire runs. When only water is drained, they are called *wet boxes*. When applied toward the dry end of the wire screen, they also draw air through the wet paper mat and are called *suction boxes*. Other equipment configurations have been developed to meet these objectives. On all modern Fourdriniers, a forming board located close to the breast roll is used to scrape off the water, drained initially by gravity, from the bottom of the wire.

A relatively recent and important development in paper forming is the *twin-wire former (Beloit Bel Baie Former)* shown in Fig. 2. In this type of machine, the fiber suspension is confined

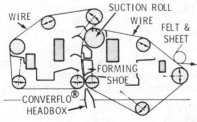

Fig. 2.   Twin-wire former. (*Bel Baie former, Beloit Corporation.*)

between two wire screens, and water is removed through both wires either simultaneously or alternately. This two-sided drainage leads to greater symmetry of distribution of fines and other nonfibrous particles through the thickness of the sheet. A significant feature of twin-wire forming is the elimination of the free surface of the fiber-water suspension while the sheet is being formed. This greatly reduces the larger-scale disturbances (waves, streaks, and jumps) which occur at higher speeds on Fourdrinier wires.

Not all the fiber and other solid materials are retained by the forming wire. For this reason and because so much water is used in the papermaking process, the *white water* removed

in the sheet-forming process is recirculated in the overall system. A large part of it is added directly to the highconsistency stock and fed back to the headbox, while a small portion goes into a *save-all* device, which recovers much of the solids from the white water. These extracted fibers and other solids are returned and added to the suspension. The clarified water is used in showers for cleaning wires and felts and other purposes so that only a small amount of the reused water eventually is discharged.

Press Section. At the end of the forming system, the *paper web* is transferred from the wire to a *press felt*, a fine-textured, usually synthetic fabric. At this point, the web contains about 4 or 5 parts water to 1 part solids. The wet paper web and one or more press felts pass through two or more press-roll nips, where water is squeezed out. Pressing also compacts the paper mat. This increases the potential interfiber contact areas where bonds will be formed.

The early *plain press* used a pair of metal and rubber-covered solid rolls. The expressed water had to flow out of the nip in the upstream direction, parallel to the paper web, as in an old washing-machine wringer. Nip pressures were then limited by the damage to the wet web (crushing) caused by this lateral flow. Although the plain press was improved in many ways, later development work led to the *fabric press* in which the felt contacting rolls are wrapped with a relatively coarse and incompressible mesh fabric. In another development (Beloit *Ventanip* press), the felt contacting rolls have narrow, closely spaced circumferential grooves. In both types, the lateral flow is virtually eliminated.

While the development of the modern presses has achieved high performance with simple constructions, the remaining problems of flow resistance and web rewetting leave room for improvement. It is generally recognized that mechanical removal of water is much less costly than drying.

Dryers. After water removal by pressing has been done to the extent which is practical with present technology, the paper web leaves the press section with 1.5–2 parts of water to 1 part fibers. Most of this remaining water, down to 5–10%, must be removed by evaporative drying. In the most common method, the paper web is passed over a series of staggered cast-iron drums internally heated by condensing steam at pressures ranging up to approximately 10.2 atmospheres. The paper web is held in contact with the rotating drums by means of dryer felts or fabrics under tension. The diameter of the dryer drums is typically 5–6 feet (1.5–1.8 meters). There may be as many as 100 of them in heavyweight paperboard machines. These dryer drums are shown in the panoramic view (Fig. 1).

Ventilating devices which blow air of controlled temperature and humidity through the dryer felts into the spaces between adjacent dryers are used. Here the air is confined by the sheet and felt runs. These pocket ventilating systems, together with greater control of the flow patterns within the dryer hood (which usually encloses the entire drier section) have led to significant improvements in cross-machine uniformity of paper drying. This results in paper and board of improved suitablity for modern high-speed converting and printing operations.

Other types of dryers, including radiant heating, dielectric and microwave heating, and high-velocity, hot air impingement, have been developed. These devices are generally applied to drying coated paper where sheet contact to a solid surface may be detrimental during drying. Wider application has been limited because of low thermal efficiencies and high capital costs.

Size Press and Coaters. Many printing grades of paper and paperboard are coated with an aqueous suspension of pigments (such as clay) in adhesives (such as starch) to provide a smoother surface, control the penetration of inks, and improve the pick resistance, appearance, brightness, and opacity. These and other materials are also applied, such as *functional coatings*, to provide such features as water resistance, pressure sensitivity for carbonless copying, and a wide variety of other properties. The appropriate materials may be added to the papermaking furnish during some stage of stock preparation (called *internal sizing*). Application of sizing or coating to one or both surfaces of the formed and dried sheet, rather than as internal sizing, simplifies the sheet-forming process and provides better control of surface properties.

The principal methods of surface coating may be classified as roll, blade, and air-knife coating, according to the method used to apply and control the final coating-layer thickness and smoothness. A recently developed coater (Beloit *Billblade*) simultaneously coats and smooths both surfaces of the paper web by running it down through the nip between a blade and a roll while maintaining two puddles, one between the web and the roll and other between the flexible blade and the web, thus eliminating the necessity for two coating stations.

After sizing or coating, the solvent, usually water, must be removed from the coating by evaporative drying. With some coating formulations and paper grades, drying can be done on ordinary steam-heated drums without damage to the coated surface, particularly if the surface of the first drum is smooth (sometimes chrome-plated). However, it is often desirable to do the initial drying with air impingement or radiant heating. Surface coating can be done on the machine as a step in the paper-machine operation, as shown in Fig. 1.

Calenders and Winders. Nearly all paper grades are calendered after they have been dried to the desired final moisture content. Ordinary calendering involves passing the paper web through one or more nips between metal rolls with high linear pressures. The calendering process flattens out the paper structure by virtue of the high pressure and "irons" the sheet. Calendering causes bulk reduction, which often is not desired, and surface smoothing, which is desired. The results strongly depend upon moisture content, calender-roll temperature, roll pressure, and speed.

In *supercalendering*, an off-machine operation, the calender rolls consist of alternating chilled-steel and paper-filled rolls, i.e., paper disks clamped on a steel shaft. These roll fillers have to be replaced periodically. Very high pressures are used. The increased pressure and shear forces associated with deformation of the relatively soft paper roll and the very high roll pressures impart a smoother, glossier surface to the web than ordinary calendering with all-metal rolls. This type of calendering is frequently used on coated sheets to provide a glossy coated surface.

There are other process configurations for the various coating effects and specifications desired.

Other Types of Machines. Although the Fourdrinier machine is used for making almost all grades of paper and board, other designs are sometimes more advantageous. The *cylinder machine*, invented at about the same time as the Fourdrinier, consists of a rotating cylindrical mold covered with a wire screen and partially submerged in a vat. The stock flows into the vat, and a mat is formed on the cylinder under a hydraulic head difference between the stock level in the vat and the white-water level inside the cylinder. The wet mat is picked up by a felt running through the nip between a couch roll and the cylinder. The cylinder machine is used for making multi-ply board, employing several vats in series. Because of slow speed and other limitations, the cylinder machine is becoming obsolete. In recent years, several new types of machines have emerged.

References

Ainsworth, J. H.: "Paper, the Fifth Wonder," 2nd Edition, Thomas Printing and Publishing Co., Kaukauna, Wisconsin (1959).

Britt, K. W.: "Handbook of Pulp and Paper Technology," Van Nostrand Reinhold, New York (1964).

Casey, J. P.: "Pulp and Paper Chemistry and Chemical Technology," Vol. 2, "Papermaking," Wiley, New York (1960).

Clark, J.: "Pulp Technology and Treatment for Paper," Freeman, San Francisco (1978).

Libby, C. E., Editor: "Pulp and Paper Science and Technology," Vol. 2, McGraw-Hill, New York (1962).

Hagemeyer, R. W.: "Future Technical Needs and Trends in the Paper Industry, III," Tech. Assn. of the Pulp and Paper Ind., TAPPI Press, Atlanta, Georgia (1979).

Hanna, W. T., and W. J. Frederick, Jr.: "Energy Conservation in the Pulp and Paper Industries," *Chem. Eng. Progress*, **74**, 5, 71–77 (1978).

Parker, J. D.: "The Sheet-Forming Process," *STAP No. 9*, Tech. Assn. of the Pulp and Paper Ind., TAPPI Press, Atlanta, Georgia (1972).

Staff: "Pulp and Paper Technology," *Chem. Eng. Progress*, **72**, 6, 45–69 (1976).

—Robert A. Daene (original preparer and formerly Beloit Corporation); revised and updated by L. H. Busker, Research and Development, Beloit Corporation, Beloit, Wisconsin.

PARABENS. Antimicrobial Agents (Foods).

PARA COMPOUNDS. Organic Chemistry.

PARAFFINS. Organic Chemistry.

PARAFFINS (Chlorinated). Chlorinated Organics.

PARTIAL PRESSURE. The pressure exerted by each component in a mixture of gases. In a mixture of perfect gases

$$p_i = \frac{n_i RT}{V}$$

The partial pressure of i is then the same as if component i occupies the same volume at the same temperature in the absence of the other gases. This is Dalton law, which is treated more fully under that heading.

PARTIAL PRESSURE. Dalton's Law.

PARTICLES (Subatomic).

For many years, the atom was traditionally described as having a central positively charged nucleus possessing considerable mass, but of minute dimension—this nucleus surrounded by a number of electrons in orbits at a relatively great distance from the nucleus. The number of electrons and their orbital arrangement determined the chemical properties of the atom, with the atoms of each chemical element possessing their own unique configuration. Recognition of the electron, the first elementary (presumably indivisible) particle, by J. J. Thomson and his associates in the 1890s ushered in an era of interest in *subatomic particles*.

In the light of much research that has yielded knowledge of scores of additional subatomic particles, these particles may be classified today into two broad categories: (1) The *leptons* include the electron, the muon, the tau particle, and the neutrinos associated with each of them. Leptons do not interact strongly, and research to date has not revealed that they contain any constituent parts, i.e., they are *elementary particles*. (2) The *hadrons* include the proton, the neutron, and the pion, among others. More than 200 kinds of hadron have been identified over the past few decades. These particles do interact strongly and do show signs of an *inner structure*, made up of other particles. The strong interaction between hadrons is manifested by the force that holds neutrons and protons together in the atomic nucleus. Hadrons, in turn, are divided into two categories, based upon how the particles decay: (a) *baryons*, which ultimately decay into the proton, and (2) *mesons*, which decay into leptons and photons, or into proton-antiproton pairs. Relatively recent research has led to the postulate that hadrons are combinations of constituents known as *quarks*.

Background.[1] A better understanding of the building blocks of nature has been a goal for many centuries, extending back to the period in Greek history of Anaxagoras of Ionia (500–428 B.C.) who held that "there was an infinite number of different kinds of elementary atoms, and that these, in themselves motionless and originally existing in a state of chaos, were put in motion by an eternal, immaterial, spiritual, elementary being, from which motion the world was produced." The concept of atoms appeared from time to time in medieval works, although the concepts expressed now seem vague. However, they seem to have been based on the idea that there could be a limit to the divisibility of matter and, consequently, the idea of a final indivisible particle out of which large pieces of matter could be built.

In the early 1800s, it became clear that chemical reactions could be most simply explained if each chemical element was thought of as composed of very small, identical elements characteristic of the element. Thus there arose a rather well-defined idea of a chemical element composed of identical atoms, as distinguished from a compound composed of groups of different atoms combined into molecules. During the later part of the 1800s, the kinetic theory of gases made use of the idea of atoms and molecules in explaining the behavior of gases. During this period, few scientists still doubted the actual material existence and "reality" of atoms.

It is perhaps rather curious that the idea of atoms became

[1] In this short historical review, some of the concepts mentioned have long since been abandoned or altered.

really well-established only after it became clear that the atoms were not in any true sense indivisible, but that instead they probably had a complex structure that should be investigated. Since these investigations required equipment and methods which had been developed by physicists rather than chemists, the physicists took the lead and the work became known over a long period as *atomic physics*, or the physics of atomic structure. As mentioned previously, this era was inaugurated by Thomson, who first isolated and established the existence of electrons. He showed that electrons have only about 1/2000 the mass of the lightest known atom, hydrogen. He also showed that these particles, as indicated by their name, carry negative electrical charges. It was later shown by Millikan that all the electronic charges are the same. Thus, the identification of electrons, as small electrically charged pieces of matter, and as constituents of all matter, became firmly established.

Since it was clear that normal matter is electrically neutral, it had to be assumed that each atom contained a positive electrical charge, as well as negative electrons. J. J. Thomson developed the picture of a somewhat spherical, jelly-like mass of positive electricity, in which electrons are located at various positions, and bound to them by "quasi-elastic" forces.

A principal means of investigating the structure of atoms was the examination of light emitted by the material in the gaseous state. This light was found to consist of a number of discrete wavelengths, or colors. Each of these wavelengths was associated, in the early days of the present century, with a mode of vibration of the electrons in the positive jelly. In particular, Lorentz (University of Leiden) was able to show that such electrons, when placed in a magnetic field, would have their modes of vibration changed in a way that explained the findings of Zeeman, who had made early observations of the wavelengths of the light emitted by a radiating gas in a magnetic field.

During 1910–1911, Sir Ernest Rutherford suggested an experiment, carried out by Geiger and Marsden, in which alpha particles from a radioactive source were scattered from thin foils. The angles at which the alpha particles were scattered were found to be such as could best be described by the close approach of a heavy positively charged particle, the alpha particle, to another heavier and more highly positively charged particle, representing the scattering atom.

From the results of these experiments, Rutherford concluded that the mass in the positive charge of an atom, instead of being distributed throughout the volume of a sphere of the order of 10^{-8} centimeter in radius, was concentrated in a very small volume of the order of 10^{-12} centimeter in radius. He thus developed the idea of a nuclear atom. The atom was pictured as a small solar system with the very heavy and highly charged nucleus occupying the position of the sun, and with electrons moving around it, as planets in their respective orbits.

Although this picture of nuclear atoms served to describe the alpha-particle scattering experiments, it still left many questions unsolved. One of these questions referred to the apparent stability of the atoms. An electron moving around the nucleus would tend to emit radiation, to lose its energy, and thereby to spiral into the nucleus. Why did it not do so? Why did the atoms all seem to be quite stable, and all to be of approximately the same size, even though some contain 90 or more electrons, while hydrogen contains only one?

The first approach to a treatment of these problems was made by Niels Bohr in 1913 when he formulated and applied rules for quantization of electron motion around the nucleus. Bohr postulated states of motion of the electron, satisfying these quantum rules, as peculiarly stable. In fact, one of them would be really permanently stable and would represent the ground state of the atom. The others would be only approximately stable.

Occasionally an atom would leave one such state for another and, in the process, would radiate light of a frequency proportional to the difference in energy between the two states. By this means, Bohr was able to account for the spectrum of atomic hydrogen in a spectacular way. Bohr's paper in 1913 may well be said to have set the course of atomic physics on its latest path.

Out of the experimental work on the scattering of alpha particles and the theoretical work of Bohr, there grew a fairly definite picture of an atom which could be correlated with its chemical properties. The chemical properties were determined in the first place by the nuclear charge. The nucleus contained most of the atomic mass and carried an electric charge equal to an integral number of positive charges, each of the same magnitude as an electronic charge. This positive nucleus then accumulated around itself a number of electrons just sufficient to neutralize its positive charge and form a neutral atom.

The number of positive charges, or the number of negative electrons around the nucleus was designated as the atomic number of the atom. These showed a close parallelism with the arrangement of atoms in the periodic system. Through the formulation of a number of rules based upon Bohr's picture of quantized orbits, the periodic system of the elements could be understood. Hydrogen was given one electron, and helium two. The two electrons in helium constituted a "closed shell" which exhibited almost perfect spherical symmetry and chemical inactivity.

Thus, during the years after 1913, the feeling grew that the chemical properties of atoms could be pretty well understood. The idea that there were undiscovered elements, as indicated by gaps in the periodic system, was reinforced. These elements and more have since been discovered.

However, it was not until 1925 that Bohr's ideas were developed into a mathematical form complete enough and precise enough to permit their general application, under the name *quantum mechanics*. This development associated with the names of Dirac, Heisenberg, and Schrödinger, provided the basic laws which permit, in principle, the complete and quantitative description of an atom consisting of a heavy, positively charged nucleus, and surrounded by enough electrons to make the whole system electrically neutral.

One of the properties of electrons that became evident during the study of optical spectra of atoms was that of *electron spin*. The suggestion was made by Uhlenbeck and Goudsmit in 1925 that one of the features of such spectra could be understood if each electron had associated with it a quantity called *spin*, which is similar in many ways to angular momentum. Each electron also has a certain magnetic moment which affects the energy in the presence of a magnetic field. This property also has been incorporated into the wave ideas of quantum mechanics.

By 1932, it was known that nuclei are made of comparatively small numbers of neutrons and protons.

A quantum theory of nuclei was made possible by the discovery of the proton and the neutron. The nuclear interaction which was responsible for holding the nucleus together (against disruptive electrostatic repulsion of the protons) was found to be of an entirely new kind, much stronger than the electric interaction at short distances, but decreasing very much more rapidly with distance. The various complex nuclei differ in the number of protons and neutrons they contain.

By that time, the theory of the interactions between electrons and photons had developed to the point where the electrostatic repulsion or attraction between electrically charged particles could be understood in terms of the exchange of photons between them. In the lowest nontrivial approximation, it gave the Coulomb law for small velocities. The basic interaction was the emission and absorption of "virtual" photons by charged particles. A similar mechanism could be invoked to explain the short-range nuclear interaction, i.e., it is due to the exchange of particles which have nonzero masses which are a fraction of nuclear mass. These theoretical considerations predicted the existence of a set of three particles called *pions*, which were ultimately discovered.

Another kind of particle and another kind of interaction were discovered from a detailed study of beta radioactivity in which electrons with a continuous spectrum of energies are emitted by an unstable nucleus. The corresponding interactions could be viewed as being due to the virtual transmutation of a neutron into a proton, an electron, and a new neutral particle of vanishing mass called the *neutrino*. The theory provided such a successful systematization of beta decay rate data for several nuclei that the existence of the neutrino was well established more than 20 years before its experimental discovery. The beta decay interaction was very weak even compared to the electron-photon interaction.

Meanwhile, the electron was found to have a positively charged counterpart called the *positron*; the electron and positron could annihilate each other, with the emission of light quanta. The theory of the electron did in fact predict the existence of such a particle. It was later found that the existence of such "opposite" particles (antiparticles) was a much more general phenomenon than once surmised.

With intensification of particle physics research, many more particles were discovered and a classification of these particles into five families was proposed—the photon family, electron family, muon family, meson family, and baryon family. Most of these particles are unstable and decay within a time which is often very small by normal standards, but which is many orders of magnitude larger than the time required for any of these particles to traverse a typical nuclear dimension. There is a wide variety of reactions between them, but they could be understood in terms of three basic interactions—the *strong* (or nuclear), the *electromagnetic*, and the *weak* interactions. The nuclear forces and the interactions between pions and nucleons are strong; the electron-electron and electron-photon interactions are electromagnetic; the beta decay interactions are weak.

As mentioned previously, by 1932 it was known that nuclei are made of comparatively small numbers of neutrons and protons. A new force was discovered (in addition to the electromagnetic and gravitational forces) that held the positive protons and electrically uncharged neutrons together in the nucleus. This nuclear force was very strong, but of limited range. Its quantum, the particle analogous to the photon in the electromagnetic field, was of nonzero rest mass. This particle, later called the π-*meson* or *pion*, was predicted by Yukawa in 1936 and discovered by Lattes, Occhialini, and Powell in 1947. For a short time, it appeared that physicists had achieved a clear, simple, and correct theory of the fundamental constitution of matter. However, shortly thereafter, two new and unpredicted particles were reported. The first of these was another meson, somewhat like the pion but more massive. The second was a *hyperon*, i.e., a strongly interacting particle heavier than the neutron.

With the continuing discovery of more particles, investigators began to suspect that these particles were not in themselves fundamental or elementary, but that they had an internal structure. This paralleled the experiences of the 1800s when the large number of different types of atoms discovered suggested that atoms must have structure. Properties of particles also suggested an internal structure. For example, the neutron has a total electric charge which is indistinguishable from zero down to very fine limits, yet the neutron has a sizeable magnetic moment.

In 1964, M. Gell-Mann and G. Zweig (California Institute of Technology) independently pointed out that all the known hadrons (i.e., particles that interact via the strong nuclear force) could be constructed out of simple combinations of three particles (and their antiparticles). These hypothetical particles had to have slightly peculiar properties (the most peculiar being a fractional electric charge). Gell-Mann called these hypothetical particles, *quarks* (referring to a sentence in James Joyce's work *Finnegan's Wake*, "Three quarks for Muster Mark"). The theory proposed postulated that three quarks bind together to form a baryon, while a quark and an antiquark bind together to form a meson. With supposition that the binding is such that the internal motion of the quarks is nonrelativistic (which requires the quarks be massive and sit in a broad potential well), then many quite detailed properties of the hadrons could be explained.

The purpose of the quark model was to explain the diversity of the hadrons, not to deal with the internal structure of any particle. But awareness of the model created a natural tendency among investigators to associate newly observed particles (among the poorly understood debris from particle experiments) with the hypothetical quarks. A number of properties of *partons* (a name given by Feynman, California Institute of Technology) were measured, including intrinsic spin angular momentum, and these were found to be consistent with the predictions of the quark model. Such observations, of course, added credence to the quark model.

In the 1960s, the quest for a grand unification theory—a theory that would explain all elementary particles and all forces acting between them—grew in intensity among most investigators who had the good fortune of discovering so many new particles, but accompanied with the realization that the ultimate structure of matter was more complex than envisioned in the earlier years. The instrumental means for research (accelerators with higher and higher energies) were getting ahead of the theoretical aspects of the topic. Many particles resulting from collisions were found in the debris of experiments—their presence without plausible explanations. Many questions were posed—why four kinds of force?—each force with its own characteristic strength with the strengths differing by nearly 40 orders of magnitude, electromagnetism with its infinite range, the weak force extending for all practical purposes only 10^{-15} centimeter. For a while, prospects of a unified theory were dim, but a number of theories were proposed and given sufficient serious attention to warrant planning of experimental tests. As pointed out by Glashow (Nobel Prize, Physics, 1979 shared with Salam and Weinberg), in his Nobel Lecture, "In 1956, when I began doing theoretical physics, the study of elementary particles was like a patchwork quilt. Electrodynamics, weak interactions, and strong interactions were clearly separate disciplines, taught and separately studied. There was no coherent theory that described them all. Developments such as the observation of parity violation, the successes of quantum electrodynamics, the discovery of hadron resonances, and the appearance of strangeness were well-defined parts of the picture, but they could not be easily fitted together."

In the early years of investigation, the weak force and the electromagnetic force were regarded as indistinguishable—they were of the same strength and possessed the same infinite range—they were transmitted by four bosons, all of which were massless. The forces manifested a symmetry—that is, they could be interchanged freely. It was believed that no matter which force was applied, the net effect was the same. These views were later to be altered in the light of the process called *spontaneous symmetry breaking*.[2]

The first direct evidence that the proton has not only size but structure was provided by an experiment at the Stanford

[2] Prior to 1956, it was believed that all reactions in nature obeyed the law of conservation of parity, so that there was no fundamental distinction between left and right in nature. However, Yang and Lee pointed out that in reactions involving the weak interaction between particles, parity was not conserved, and that experiments could be devised that would absolutely distinguish between right and left. This was the first example of a situation where a spatial symmetry was found to be broken by one of the fundamental interactions.

The principle of charge conjugation symmetry states that if each particle in a given system is replaced by its corresponding antiparticle, then it would not be possible to tell the difference. For example, if in a hydrogen atom the proton is replaced by an antiproton and the electron is replaced by a positron, then this antimatter atom will behave exactly like an ordinary atom—if observed by "persons also made of antimatter." In an antimatter universe, the laws of nature could not be distinguished from the laws of an ordinary matter universe.

However, it turns out that there are certain types of reactions where this rule does not hold, and these are just the types of reactions where conservation of parity breaks down. For example, consider a piece of radioactive material emitting electrons by beta decay. The radioactive nuclei are lined up in a magnetic field which is produced by electrons traveling clockwise in a coil of wire, as seen by an observer looking down on the coil. Because of the asymmetry of the radioactive nuclei, most of the emitted electrons travel in the downward direction. If the same experiment were done with similar nuclei composed of antiparticles and the magnetic field were produced by positron current rather than an electron current, then the emitted positrons would be found to travel in the upward, rather than in the downward, direction. Interchanging each particle with its antimatter particle has produced a change in the experiment.

However, the symmetry of the situation can be restored if we interchange the words *right* and *left* in the description of the experiment at the same time that we exchange each particle with its antiparticle. In the above experiment, this is equivalent to replacing the word *clockwise* with *counterclockwise*. When this is done, the positrons are emitted in the downward direction, just as the electrons in the original experiment. The laws of nature are thus found to be invariant to the simultaneous application of charge conjugation and mirror inversion.

Time reversal invariance describes the fact that in reactions between elementary particles, it does not make any difference if the direction of the time coordinate is reversed. Since all reactions are invariant to simultaneous application of mirror inversion, charge conjugation, and time reversal, the combination of all three is called *CPT* symmetry and is considered to be a very fundamental symmetry of nature.

A relatively recent type of space-time symmetry has been introduced to explain the results of certain high-energy scattering experiments. This is *scale symmetry* and it pertains to the rescaling or "dilation" of the space-time coordinates of a system without changing the physics of the system. Other symmetries, such as chirality, are more of an abstract nature, but aid the theorist in his effort to bring order into the vast array of possible elementary particle reactions.

A feature of quantum field theory is that the quanta of the fields are initially massless. Spontaneous symmetry breaking offers a mechanism by which weak-field quanta, for example, can acquire mass. Unification of weak and electromagnetic forces may be viewed as follows: at short distances (high energies), the masses of the weak-field quanta became unimportant and thus original symmetry is restored. Symmetry in this context refers to the properties of the equations of motion of particles in the field theories. Spontaneous symmetry breaking occurs when solutions of the equations do not display full symmetry. Some physicists have likened this to a ball moving on a roulette wheel, whose equations of motion are symmetrical about the axis of rotation even though it always stops in an asymmetric position.

Linear Accelerator Center (SLAC) in 1970. Previously it has been established that the proton is not a pointlike particle, but has a finite size—a diameter of about 10^{-13} centimeter. Although is only about 1/100,000 the size of an atom, it is still measurable. In this it is unlike certain other particles, notably the electron, for which no extension has been noted, so that it can be regarded as a mathematical point. In the experiment, electrons were raised to an energy of some 20 billion electron volts and struck protons and neutrons in the atoms of a stationary target. The angular distribution of the scattered electrons and of other particles created in the collisions were carefully monitored. Most of the electrons, as expected, passed through the target with little change in direction. An unexpected excess of widely scattered particles was produced—much greater than if the proton were diffuse and homogeneous. The excess of the widely scattered particles was attributed to a mass embedded within the proton, estimated at no more than $\frac{1}{50}$ the diameter of the proton. In later experiments, a target was illuminated by means of muons (like electrons but with a mass 200 times greater); and by a beam of neutrinos (lack mass and electric charge). The results of the original and later experiments were consistent and the deep scattering of particles was attributed to collisions between the incident leptons and some hard constituent of the proton.

Before discovery of this hadron particle (designated *psi* or *J*), and after much experimental and theoretical effort, physicists had about concluded that three massive, fractionally charged quarks were the primary building blocks of the universe. However, discovery of the psi particles in 1974 indicated a fourth quark was required. Previously, in the three-quark model, all mesons were made up of one quark and one antiquark; baryons of three quarks; and antibaryons of three antiquarks. Prior to 1974, all the known hadrons could be accommodated within this basic scheme. Three of the possible quark-antiquark meson combinations could have the same quantum numbers as the photon, and hence could be produced abundantly in e^+e^- annihilation. These three predicted states had all been found.

As pointed out by Richter, the first publications of a theory based on four quarks rather than three in number was proposed in 1964 by Amati and others. The motivation at that time was more esthetic than practical, and these models gradually expired for want of an experimental fact that called for more than a 3-quark explanation. In 1970, Glashow explained in a paper that the fourth quark (called *charmed* by Glashow) was required to explain the non-occurrence of certain weak decays. The fourth or *c* quark was assumed to have a charge of $+\frac{2}{3}$, like the *u* quark, and also to carry $+1$ unit of a previously unknown quantum number called *charm*, which was conserved in both the strong and electromagnetic interactions, but not in the weak interactions. Discovery of the psi particles demonstrated a more compelling need for the fourth quark. Richter observes that the four-quark model of hadrons seemed to account, in at least a qualitative fashion, for all the main experimental information that had been gathered about the psions, and by the early part of 1976, the consensus for charm had become quite strong. See also **Quarks.**

In 1977, the *upsilon particle* was found as the result of energetic collisions between protons and copper nuclei. The upsilon particle has a mass three times greater than any other subatomic entity yet detected (early 1980s). Researchers on this experiment from Columbia University, the State University of New York (Stony Brook) and the Fermi National Accelerator Laboratory reported that with a mass at its lower energy state equivalent to 9.0 GeV and masses in excited states equivalent to 10 and

10.4 GeV, the upsilon particle has been interpreted as consisting of a massive new quark (the fifth) bound to its antiquark. Confirming experiments were also conducted at the Deutsches Elektronen Synchrotron (DESY) located near Hamburg, Germany. The quantum attribute of the fifth quark was named *bottom*. With a fifth quark reported, many physicists felt that finding a sixth quark (*top*) was highly probable.

In 1979, the Nobel-Prize (Physics) was awarded to Glashow and Weinberg (both of Harvard) and Salam, a Pakistani physicist, in recognition of the significance of a theory which unites the weak force with the electromagnetic force. But most scientists recognize the Nobel action as only a milestone in a series of predictions that include the existence of new particles so massive that they cannot be expected to appear at the energies thus far available to physicists.

The chemists of the 19th Century once thought that all material substances were comprised of only 36 elements.[3] Over the years, the list expanded to over 100 elements. Fifty years ago, it was proclaimed that the elements were made of electrons, protons, and neutrons. Then, commencing in the 1940s, lots of other particles were found, as previously described. Then, for a while it seemed that elementary matter could be reduced to three particles—the quarks. But, quarks multiplied in number, with a sixth quark now seriously proposed. Will there be too many quarks? Perhaps hypothetical particles will be proposed of which the quarks are comprised. Possibly the ultimate answer will lie with the "mathematical groups that order the particles rather than in truly elementary objects."

Particle Accelerators

Subatomic particles, such as electrons, positrons, and protons, can be accelerated to high velocities and energies, usually expressed in terms of center-of-mass energy, by machines which impart energy to the particles in small stages or nudges, ultimately achieving in this way very high-energy beams, measured in terms of billions and even trillions of electron volts. Thus, in terms of their scale, particles can be made to perform as powerful missiles for bombarding other particles in a target substance or for colliding with each other as they assume intersecting orbits. Because the particles are empowered with high energy, their smashing encounters are conducive to breaking them into their constituents. Instruments or machines used to arrange these particle encounters are known as *particle accelerators* and are very large, their dimensions frequently measured in terms of a few miles or kilometers. Inasmuch as the technology involved in creating particles is much more within the bounds of physics than chemistry, this aspect of subatomic particles is not covered here.

There are additional entries in this *Encyclopedia* on specific particles. Consult alphabetical index.

References

Balian, R., Rho, M., and G. Ripka, Editors: "Ions Lourds et Mésons en Physique Nucléaire," North-Holland, Amsterdam (1978).
Bromley, D. A.: "Physics," *Science*, **209**, 110–121 (1980).
Condon, E. U. and H. Odabşi: "Atomic Structure," Cambridge Univ. Press, New York (1980).
Cowen, R. C.: "Particle Physics," *Technology Review (MIT)*, **82**, 3, 10–11 (1980).
Dremin, I. M., and C. Quigg: "The Cluster Concept in Multiple Hadron Production," *Science*, **199**, 937–941 (1978).

[3] Line of thought suggested by L. M. Lederman (Columbia University) in a paper on "The Upsilon Particle," *Sci. Amer.*, **239**, 4, 80 (1978).

Ekstrom, P., and D. Wineland: "The Isolated Electron," *Sci. Amer.*, **243**, 2, 104–121 (1980).

Fulcher, L. P., Rafelski, J., and A. Klein: "The Decay of the Vacuum," *Sci. Amer.*, **241**, 6, 150–159 (1979).

Glashow, S. L.: "Toward a Unified Theory: Threads in a Tapestry," in "Nobel Lectures," Elsevier, Amsterdam and New York (1981).

Goldhaber, M., Langacker, P., and R. Slansky: "Is the Proton Stable?" *Science*, **210**, 851–860 (1980).

Golub, R., et al.: "Ultracold Neutrons," *Sci. Amer.*, **240**, 6, 134–154 (1979).

't Hooft, G.: "Gauge Theories of the Forces between Elementary Particles," *Sci. Amer.*, **242**, 6, 104–138 (1980).

Hung, P. Q., and C. Quigg: "Intermediate Bosons: Weak Interaction Couriers," *Science*, **210**, 1205–1211 (1980).

Jacob, M., and P. Landshoff: "The Inner Structure of the Proton," *Sci. Amer.*, **242**, 3, 66–75 (1980).

Lederman, L. M.: "The Upsilon Particle," *Sci. Amer.*, **239**, 4, 72–80 (1978).

Krisch, A. D.: "The Spin of the Proton," *Sci. Amer.*, **240**, 5, 68–80 (1979).

Martinis, M., Pallua, S., and N. Zovko, Editors: "Particle Physics," North-Holland/American Elsevier, Amsterdam/New York (1974).

Pilkuhn, H. M.: "Relativistic Particle Physics," Springer-Verlag, New York (1979).

Rebbi, C.: "Solitons," *Sci. Amer.*, **240**, 2, 92–116 (1979).

Rho, M., and D. Wilkinson, Editors: "Mesons in Nuclei," North-Holland, Amsterdam (1979).

Salam, A.: "Gauge Unification of Fundamental Forces," in "Nobel Lectures," Elsevier, Amsterdam and New York (1981).

PASSIVITY.

When iron is immersed in nitric acid (concentrated), there is no visible reaction, although dilute nitric acid results in a marked reaction with iron. Upon removal of the iron from the nitric acid (concentrated) and immersed in copper sulfate solution, the iron is not plated by copper, although this occurs with ordinary iron. Iron in such a condition is described as passive iron, and the phenomenon is known as *passivity*. This example may be considered to be wholly chemical, but there are instances of electrochemical passivity. Copper anodes, for example, in a copper cyanide plating bath will, under certain conditions, become insoluble, at least at applied voltages within the capacity of dc sources usually used. Removal of the anodes and scraping of their surfaces destroys the passivity, indicating that the condition is caused by a surface film. Oxidizing conditions tend to promote passivity, whereas reducing conditions lead to elimination of passivity.

PASTEURIZING.

A thermal process that provides a partial sterilization of food substances by inactivating some of the microorganisms present, notably vegetative cells of bacteria, yeast, or molds. The effectiveness with which such organisms are deactivated is a measure of the adequacy of pasteurization. Pasteurization is usually carried out at a temperature of 100°C or lower. Pasteurization of milk and dairy products, including use of HTST (high-temperature, short-time) and UHT (ultrahigh-temperature), is extensively practiced in advanced and industrial countries throughout the world. In the older method of pasteurization, milk was held at a temperature of 63°C for 30 minutes. In the HTST process, the milk is held at 71.7°C for 15 seconds. This process results in greater nutrient retention. As of the mid-1980s, a major objective of the dairy industry is to extend the technology of UHT to a majority of dairy products, including "fresh" milk, and thus extend the shelf life from days to weeks and months.

Pasteurization is also important in the processing of fruit juices, fermented products, and liquid-egg products, among others. Pasteurization is frequently used with other preserving techniques, including refrigeration in the present handling of fresh milk; fermentation in the case of pickles; and the maintenance of an anaerobic condition as in the case of beer.

One configuration of a HTST pasteurizing system is shown on next page.

PCB. Biphenyls.

PECTINS.

Pectin substances are those complex carbohydrate derivatives which occur in or are prepared from plants and contain a large proportion of anhydrogalacturonic acid units, which are thought to exist in a chainlike combination. The carboxyl groups of polyglacturonic acids may be partially esterified by methyl groups and partly or completely neutralized by one or more bases. The general term *pectin* (or *pectins*) designates those water-soluble pectinic acids of varying methyl ester content and degree of neutralization which are capable of forming gels with sugar and acid under suitable conditions. The term *protopectin* is applied to the water-insoluble parent pectic substances which occur in plants and which upon restricted hydrolysis yield pectin or pectinic acids. *Pectic acids* is a term that is applied to pectic substances mostly composed of colloidal polygalacturonic acids and essentially free from methyl ester groups. The salts of pectic acids are either normal or acid *pectates*. The term *pectinic acids* is used for colloidal polygalacturonic acids containing more than a negligible proportion of methyl ester groups. Pectinic acids, under suitable conditions, are capable of forming gels with sugar and acid, or, if suitably low in methoxyl content, with certain metallic ions. The salts of pectinic acids are either normal or acid *pectinates*.

Pectins occur commonly in plants, particularly in succulent tissues, and are characterized by the polygalacturonic acids which are fundamental to their structure. The pectins are important emulsifying, gelling, stabilizing, and thickening agents used in the preparation of numerous food products. About 75% of the pectins produced are used in making fruit jams, jellies, marmalades, and similar products. Additional uses include the preparation of mayonnaise, salad dressings, malted milk beverages, frozen dessert mixes, frozen fruits and berries (to prevent leakage upon thawing), among others. The addition of a dilute pectin solution to milk coagulates the casein. In many food products, the use of pectins as stabilizers is preferred, since they blend better into the flavor complex than do many gums, starches, or a number of carbohydrate derivatives. Pectin jellies do not melt at temperatures below 49°C, a distinct advantage over gelatin gels that require refrigeration. Pectins also have a number of nonfood uses, including pharmaceuticals and cosmetics.

The location of various pectic substances in plant tissues is well established. Pectins make up most of the middle lamella in unripe fruit and are to be found in the cell walls and in small proportion in all plant tissues. The genesis and fate of pectins in plant tissues have not been fully determined.

Citrus peel, apple pomace from juice manufacture, and beet pulp left over from the manufacture of sucrose are common commercial sources of pectins. After some preliminary purification of the raw material, the extraction is usually performed with hot dilute acid (pH = 1.0 to 3.5 in a temperature range of 70–90°C). The pectin is then precipitated from the extract with ethanol or isopropanol, or with metal salts (copper or aluminum). The metal ions have to be subsequently removed by washing with water or acid ethanol. Specific formulas for denatured ethanol for use in pectin manufacture are used. The precipitates are purified, dried, and pulverized to form the yellowish-white powder of commerce.

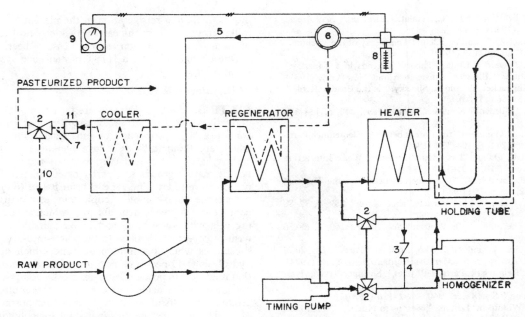

One configuration of a HTST (high-temperature, short-time) pasteurization system with homogenizer of larger capacity than timing pump. (1) Raw product constant-level tank; (2) three-way bypass valves; (3) sanitary check valve; (4) recirculating line; (5) diversion line; (6) diversion valve; (7) horizontal line that must be at least 12 inches (30.5 centimeters) above any raw product in the system; (8) indicating thermometer; (9) recorder-controller; (10) bypass line; (11) vacuum breaker. (*Cherry-Burell Corporation*)

Pectin substances in solution behave as typical colloids. See also **Colloid Systems.** Dry, purified pectins are light in color and soluble in hot water to the extent of 2–3%. The pH of pectin solutions is usually 2–3.5

The proportion of sugar which pectin will form into a firm jelly determines the *jelly grade* of the pectin. In a jelly, jam, or marmalade, the proportions of total solids of sugars, the pH, and the proportion and nature of the pectin used will determine the extent of jellification obtained. The use of added pectin in fruit jams and related products is approved by most food regulators because the addition is believed to compensate for an incidental natural deficiency.

In pectic acids, all carboxyl groups are free, or at least not present as the methyl ester. Under suitable conditions, pectins will form jellies with sugar and acid, whereas the low-ester pectins will form *gels* with traces of polyvalent ions. The general structure of pectin is:

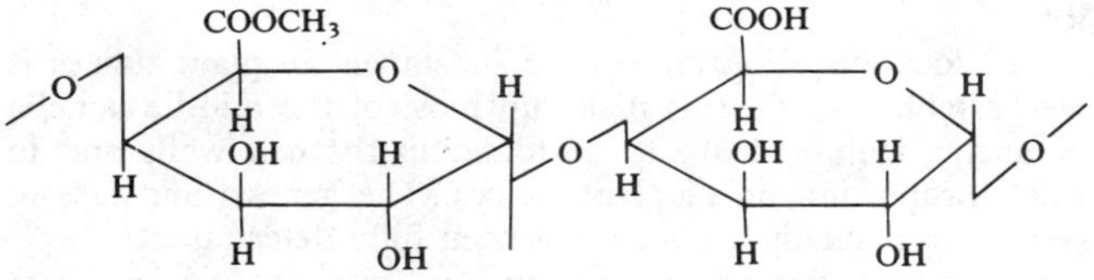

References

Graham, H. D., Editor: "Food Colloids," AVI, Westport, Connecticut (1977).
Kim, W. J., Smit, C. J. B., and V. N. M. Rao: "Demethylation of Pectin Using Acid and Ammonia," *Jrnl. of Food Science*, **43**, 1, 74–78 (1978).
Kim, W. J., Sosulski, F., and S. J. Campbell: "Formulation and Characteristics of Low-Ester Gels from Sunflower Pectin," *Jrnl. of Food Science*, **42**, 3, 746–749 (1978).
Staff: "Food Chemicals Codex," National Academy of Sciences, Washington, D.C. (Revised periodically).

PEGMATITE. The term *pegmatite*, derived from the Greek word meaning "joined together," was first applied by Haüy in 1822 to a peculiar interpenetrating growth of quartz and feldspar sometimes called graphic granite from its resemblance to written characters, particularly those of the Hebrew language. Pegmatite is also used to designate those coarse-grained dikes and sheets, chiefly of granite or syenite, that are apophyses of stocks or batholiths, or of the residual magma, during their congelation. The individual minerals may often reach great size. Granite pegmatites are chiefly composed of alkali feldspar and quartz with some muscovite or biotite but may carry such minerals as tourmaline, topaz, beryl, fluorite, apatite, garnet, lepidolite, etc. See also **Mineralogy.**

PEGMATITES. Mineralogy.

PENICILLIN. Antibiotic.

PENTAERYTHRITOL TETRANITRATE (PETN). Explosive.

PENTLANDITE. The mineral sulfide of iron and nickel corresponding to the formula $(Fe, Ni)_9S_8$. It is isometric, appears in granular masses; hardness, 3.5–4; specific gravity, 5.0; color, bronze-yellow; opaque. Occurs with pyrrhotite, millerite, and nickeline. The best known deposit of pentlandite is at Sudbury, Ontario, Canada, where it is associated with a nickel-bearing pyrrhotite.

PENTOSANS. Carbohydrates.

PENTOSE PHOSPHATE CYCLE. Carbohydrates.

PENTOSES. Carbohydrates.

PEPTONE. A secondary protein derivative that is water-soluble, not coagulated by heat, and not precipitated on saturation of its solutions with ammonium sulfate.

PERFECT GAS. A perfect gas may be defined by the following two laws: The Joule law: the energy per mole, U, depends only on the temperature; the Boyle law: at constant temperature, the volume V occupied by a given number of moles of gas varies in inverse proportion to the pressure.

By combination of these two laws we obtain the equation of state for perfect gas,

$$pV = nRT \qquad (1)$$

where R is the gas constant, T, the absolute temperature. (It is also called the *perfect gas law*.)

The perfect gas is an abstraction to which any real gas approximates according to the nature of the gas and the conditions. For a given temperature and composition, the perfect gas condition is approached when the density tends to zero. From a molecular point of view, the perfect gas laws correspond to the behavior of a system of molecules whose interactions may be neglected in expressing the thermodynamic equilibrium properties. However, even at a low density, the transport properties depend essentially on the interactions.

The thermodynamic properties of a perfect gas are, of course, especially simple. For example, the difference between the molar heat capacities at constant pressure and constant volume is equal to the gas constant R,

$$C_p - C_v = R \qquad (2)$$

The value of R is 0.08205 liter-atm. degree^{-1} mole^{-1}, which in cgs units is equal to 8.314×10^7 g cm2 sec^{-2} degree^{-1} mole^{-1}. This relationship, Formula (2), applies only approximately to real gases.

However, the way in which either C_p or C_v depends on the temperature can only be calculated from statistical mechanics.

PERIDOTITE. The term peridotite is derived from peridot, the French word for olivine.

It is a coarse-grained igneous rock related to gabbro, which consists of olivine and proxene in varying proportions. Certain peridotites contain spinel, chromite, or mica as accessories.

Rocks consisting essentially of olivine alone are known as dunities, the name coming from the occurrence of this rock in the Dun mountains of New Zealand. In the United States, this mineral is found in North Carolina, South Carolina, and Georgia, where corundum is associated with the dunite in commercial quantities. The olivine of peridotites alters readily to the mineral serpentine, often to such an extent that the rock itself is called a serpentine. As mentioned above, the peridotites may contain chromite or other valuable minerals, often to such an extent that they may be commercially exploited, for nickel, platinum, and precious garnet.

Kimberlite from which diamonds are secured is commonly called a mica peridotite but is more closely related to the lamprophyres. See also **Kimberlite.**

PERIODIC TABLE OF THE ELEMENTS. When the chemical elements are arranged in a matrix on the basis of increasing atomic numbers, a pattern of periodicity among the physical and chemical characteristics emerges. By no means is the resulting matrix perfect, but the resemblance of characteristics among groups of elements arranged in this manner is indeed both striking and illuminating. Attempts to classify the elements date back to the early work of deChancourtois (1862) and Newlands (1863), but the discovery of the relationship between atomic-number groupings and characteristics was made by Dimitri Mendeleev in 1869. One year later, Lothar Meyer independently showed the periodicity of the elements in terms of atomic vol-

umes. Meyer defined the latter characteristic as the atomic weight divided by the specific gravity of the element in the solid state.

Although there have been numerous refinements to Mendeleev's early tabulation, fortified by the discovery and isolation of several elements then unknown, the fundamental principles of the matrix are the same. The conventional table is shown in the upper right in Fig. 1. The information also can be presented in polar fashion as shown. It is interesting to note that as one proceeds clockwise around the circle the atomic numbers appear consecutively and that 18 sectors of the circle become the bases for families or groups of elements. Thus, the members of the alkali metals (Group 1a), alkaline earths (Group 2a), halogens (Group 7a), and so on, all bear resemblance, one element to the other, within any given group. There are two significant breakpoints in any representation of periodicity, namely, commencing with atomic number 57 (lanthanum) and atomic number 89 (actinium). Attempts to place the elements which follow—in the one case, atomic numbers 59 through 71 and, in the other case, atomic numbers 90 through 103—in the underlying geometric matrix (whether tabular or circular) do not succeed. These separate groups are known as the lanthanides (rare earths) and the actinides, respectively. Upon completion of the lanthanide series (with lutetium, atomic number 71), the orderly geometry assumes with hafnium, atomic number 72 (Group 4b) and continues through actinium, atomic number 89. The probable positions of elements 104 and 105 are indicated in Group 4b and 5b, respectively.

An amazing result of Mendeleev's pioneering classification was the prediction of elements yet to be discovered. Mendeleev found that he could maintain geometric logic of his table only if he allowed for some blank spaces in the table. He further reasoned that elements later would be discovered that would occupy these vacant positions and, thus, Mendeleev predicted the existence of gallium, scandium, and germanium. In fact, Mendeleev gave a preliminary name to scandium, calling it *eka-boron*, and predicted the probable properties of the element. The element was later isolated by Lars Fredrik Nilson in 1879. Mendeleev lived to see his prediction confirmed.

In retrospect, with a much fuller understanding of the underlying electronic and particle structure of the elements, most aspects of the periodicity of the elements come as no surprise, but the fact remains that Mendeleev, Meyer, and others made these striking observations without benefit of over 100 years of additional knowledge. The periodicity of the elements is demonstrated in Fig. 2, which plots atomic weights along the abscissa versus an arbitrary ordinate for various observed physical characteristics. See also **Chemical Elements.**

PERLITE (or Pearlstone). An unusual form of siliceous lava composed of small spherules of about the size of bird shot or peas. It is grayish in color with a soft pearly luster. The spherules often show a concentric structure and are believed to be formed as a result of a peculiar spherical cracking developed while cooling. They may be confused with oölites, which are classified as concretions.

PEROVSKITE. The mineral perovskite is calcium titanate, essentially $CaTiO_3$, with rare earths, principally cerium proxying for Ca, as does both ferrous iron and sodium, and with niobium substituting for titanium. It crystallizes in the orthorhombic system, but with pseudo-isometric character; subconchoidal to uneven fracture; is brittle; hardness, 5.5; specific gravity, 4; luster, adamantine; color, various shades of yellow to reddish-brown or nearly black; transparent to opaque. It is found

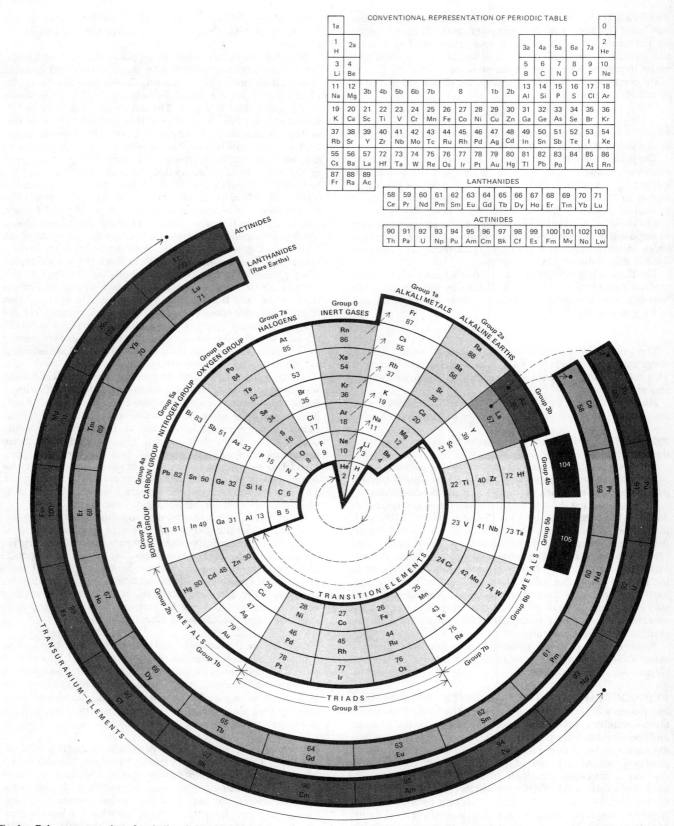

Fig. 1.   Polar representation of periodic relationships of the elements. (*Source*: *Omnibix, U.S.A.*) At upper right is shown the conventional representation.

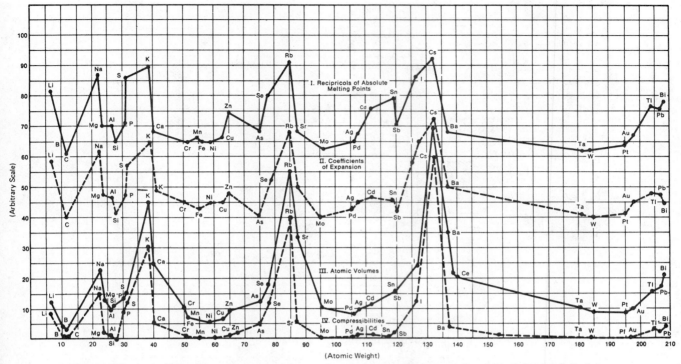

Fig. 2. Pattern obtained when various parameters are plotted against increasing atomic weight of the chemical elements.

associated with chlorite or serpentine rocks occurring in the Urals, Baden, Switzerland, and Italy. It was named for Von Perovski.

PERTHITE. An alkali feldspar comprising parallel or subparallel intergrowths. The potassium-rich phase, usually microcline, seems to be the host from which the sodium-rich phase, usually albite inclusions, exsolved. The exsolved areas typically form blebs, films, lamellae, small strings, or irregular veinlets, and usually are visible to the naked eye.

PETALITE. The mineral petalite, lithium aluminum silicate, $LiAlSi_4O_{10}$ is monoclinic, although crystals are rare, this mineral usually occurring in cleavable, foliated masses, whence the name petalite from the Greek meaning a *leaf*. Its hardness is 6–6.5; specific gravity 2.39–2.46; brittle with subconchoidal fracture; perfect basal cleavage; luster, vitreous, colorless to white or gray but may be greenish or reddish; is transparent to translucent. Petalite occurs in granite pegmatites with sodium-rich feldspar, quartz and lepidolite; has been found in Sweden; U.S.S.R.; on the Island of Elba; and in the United States, at Bolton, Massachusetts, and Peru, Maine. It is interesting to note that lithium was first discovered in this mineral.

PETROCHEMICALS. Chemicals derived from petroleum and, more specifically, substances manufactured from a component of crude oil or natural gas. In this sense, ammonia and synthetic rubber made from natural gas components are petrochemicals. See diagram on next page.

PETROLEUM. A natural oil, ranging in color through black, brown, and green, to a light amber shade. It is often termed *crude oil* and consists principally of hydrocarbons with varying amounts of oxygen-, nitrogen-, and sulfur-bearing compounds almost always present. The term *mineral oil*, which is sometimes used as a synonym for petroleum, is an inadequate description because most geologists believe that petroleum was derived from organic material resulting from reactions of organic materials, such as plants and animals buried in sedimentary rocks. The more important of these geologic formations in which petroleum is found are the Tertiary period of the Cenezoic era (50% of the world's oil production comes from these rocks, including regions in California and the Gulf Coast of the United States, the U.S.S.R., Venezuela, Malaysia, Iran, and Iraq); the Cretaceous period of the Mesozoic era (including the East Texas, Kuwait, and Bahrain fields); the Jurassic period of the Mesozoic era (including the Arkansas and Rocky Mountain regions of the United States, and Saudi Arabia); and the Mississippian period of the Paleozoic era (including the West Texas, Pennsylvania, and Mid-Continent regions of the United States, and the Alberta, Canada fields).

It is evident, therefore, that petroleum oils vary considerably in composition, even when closely associated geographically. Analysis of crude oils found in representative areas of the United States are given in Table 1. It may be generalized that crudes found in the eastern and midwestern sections of the United States are predominantly sweet and paraffinic; those found along the Gulf Coast usually are naphthenic; those occurring in the inland southwest are sour and naphthenic; and those found along the west coast are asphaltic. Analysis of some crude petroleums found outside the United States are given in Table 2. This illustrates the variety of crudes existent, but is not intended to provide a full representation of worldwide petroleum source compositions.

API Gravity. This parameter (API stands for American Petroleum Institute), expressed in "degrees," is mathematically related to specific gravity and can be determined with a hydrometer. The specific gravity of water (arbitrarily defined as unity) is 10.00 when expressed as degrees API. API gravity usually, although not infallibly, indicates the gasoline and kerosine con-

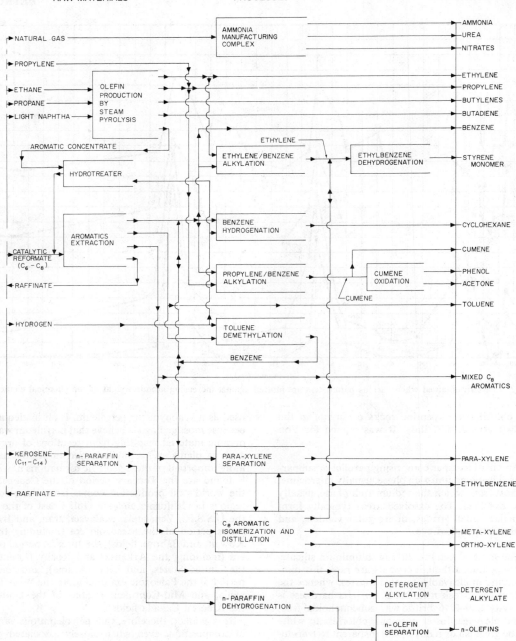

Interlocking processes and flow of materials in a petrochemical complex. (*Universal Oil Products.*)

tents of the crude. As an example, the Mississippi, Texas, New Mexico, and Louisiana crudes have API gravities between approximately 35 to 40; as do the Arabian, Iranian, and Colombian crudes. The gasoline content (that fraction boiling below about 400°F (204°C) of these crudes ranges from about 25% to over 35% by volume. The kerosine portions of such "light" crudes also are usually high. In contrast, Wyoming sour crude with an API gravity of 17.9 contains but 6% gasoline and about 40% asphalt. California crude has an even greater content of residuum and almost no gasoline.

Sulfur Content. The amount of sulfur in crude is important in terms of handling the crude within the refinery and the undesirable effects of sulfur in finished products. High-sulfur crudes require special materials of construction for refinery equipment because of their corrosiveness. Certain refinery processes require desulfurization of sour charge stocks prior to use as a feedstock, not only because of their corrosiveness, but also because of the effect of sulfur-bearing compounds on expensive catalysts. From the standpoint of the consumer, sulfurous gasoline has an unforgettably offensive odor unless specially sweetened and it may corrode the fuel system and engine parts, as well as pollute the atmsphere after it has been burned.

Other factors indicated in the data of Tables 1 and 2 include: *Pour Point*—defined as the lowest temperature at which the material will pour and a function of the composition of the oil in terms of waxiness and bitumen content; *Salt Content*—which is not confined to sodium chloride, but usually is interpreted in terms of NaCl. Salt is undesirable because of the tendency to obstruct fluid flow, to accumulate as an undesirable constituent of residual oils and asphalts, and a tendency of certain salt compounds to decompose when heated, causing corrosion of refining equipment; *Metals Content*—heavy metals, such as vanadium, nickel, and iron, tend to accumulate in the heavier gas oil and residuum fractions where the metals may interfere with refining operations, particularly by poisoning catalysts. The heavy metals also contribute to the formation of deposits on heated surfaces in furnaces and boiler fireboxes, leading to permanent failure of equipment, interference with heat-transfer efficiency, and increased maintenance.

Natural Gas, Oil Shales, and Tar Sands. Natural gas is not formally defined as a component of crude petroleum, although natural gas commonly exists in the same geological formations, often directly in contact with crude petroleum. However, a large percentage of natural gas wells are not associated with producing oil wells. See also **Natural Gas.**

The oils derived from oil shales are not true petroleum, although they are petroleumlike products after being subjected to specialized chemical processing. Shales are sedimentary rocks which have a relatively high content of a bituminous substance called *kerogen* and 30–60% organic matter and fixed carbon. Kerogen, although not a definite chemical compound, yields an oily substance when heated (retorted) in the absence of air. Extraction of oil shale with ordinary solvents produces no oil, and their solubility in solvents is low. This evidence supports the conclusion that the "oil" is the result of a chemical change, i.e., the thermal cracking or fragmenting (pyrolysis) of the molecule that make up kerogen. See also **Oil Shale.**

Tar sands is an expression commonly used in the petroleum industry to describe sandstone reservoirs impregnated with a very heavy viscous crude oil which cannot be produced through a well by conventional production techniques. Two other terms, *bituminous sands* and *oil sands*, are gaining favor. The heavy viscous petroleum substances impregnating the "tar sands" are called asphaltic oils. See also **Tar Sands.**

Origin and Geology of Petroleum. Among the general theories for explaining the origin of petroleum, the most widely accepted is the *organic theory*, which can be quickly summarized. Over millions of years, rivers flowed to the seas, carrying large volumes of mud and sand to be spread out by currents and tides over the sea bottoms near the gradually changing shorelines. New deposits were distributed, layer upon layer, over the floors of the seas. Because of the increasing weight of these accumulations, the sea floors slowly sank, building up a thick series of mud and sand layers. High pressure and chemical forces ultimately converted these layers into sedimentary rocks of the type that often contain petroleum—the sandstones, shales, limestones, and dolomites. The organic theory further stipulates most importantly that tiny marine organisms were buried with the silt. In an airless environment and under high pressures and elevated temperatures, these miniscule carbon- and hydrogen-containing life-forms were converted over an extremely long time span into hydrocarbons. This theory, of course, requires acceptance of the concept of drastically altered shorelines, because obviously oil deposits are found in many parts of the world long distances from the present coastlines.

Geologists find it particularly difficult to trace the history of a given hydrocarbon deposit because the oil and gas may have moved as the result of numerous seismic events, again

TABLE 1. ANALYSIS OF REPRESENTATIVE U.S. CRUDE OILS

PROPERTY	McCOMB, MISSISSIPPI	SOUTHWEST TEXAS	EAST TEXAS	WYOMING (SOUR)	NEW MEXICO	N. KENIA PENINSULA, ALASKA	SAN ARDO, CALIF.	OSPELOUSAS, LOUISIANA	VELMA, OKLA.
Total sulfur, wt %	0.07	0.45	0.2	3.33	1.0	1.04	1.93	0.08	1.13
Pour point, °C	15.6	−1.1	12.8	−20	−3.9			4.4	
°F	60	30	55	−5	25			40	<−30
Gasoline, vol %	35.5	32.0	29.0	6.3	37.8	14.4	1.9	26.1	22.3
Kerosine, vol %	18.1	12.1	10.1	9.1		18.0	16.1	18.9	17.3
Diesel fuel, vol %	14.6	38.0	13.8	14.0		18.4	10.6	22.9	8.5
Gas oil, vol %	28.1	12.6		30.7	41.2	22.3	23.3	27.9	31.9
Asphalt bottoms, vol %	3.7	5.3	47.1	39.9	20.8	26.9	48.1	4.2	20.0
Metals in gas oils, ppm									
Nickel	0.06						0.15		
Vanadium	0.08						<0.1		
Salt, lb/1000 bbl	4	<0.5	31	0.6	14	76		5	78

TABLE 2. ANALYSES OF REPRESENTATIVE WORLD CRUDE OILS

PROPERTY	ARABIAN	MINAS, CENTRAL SUMATRA, TOPPED	PUTOMAYO, COLOMBIA	GULF NIGERIA	ZULIA, VENEZUELA	IRAN	KUWAIT
Total sulfur, wt %	3.05	0.2	0.49	0.16	1.69	1.12	2.62
Pour point, °C	−36.1	−17.8	7.2	−6.7	<−15	15	<−15
°F	−33	0	45	20	<5	5	<5
Gasoline, vol %	29.1	11	34.1	24.9	18.9	32.2	25.5
Kerosine, vol %	16.0	16	9.3	26.5	14.1	18.3	13.7
Gas oils, vol %	12.5	14	40.7	19.3			
Residuum, vol %	42.4	59	15.9	29.3			
Metals in gas oils, ppm							
Vanadium	0		25	7			
Nickel	0		11	5			
Iron	3						
Salt, lb/1000 lb	12		trace	5			

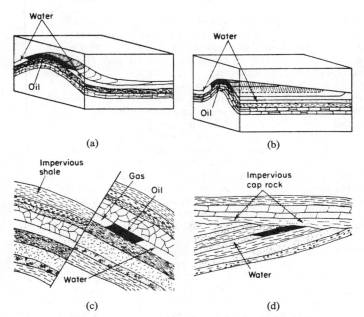

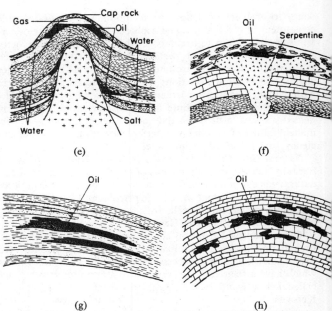

Fig. 1.   Geological types of oil and gas reservoirs: (a) Accumulation of oil in a dome-shaped structure. The dome is circular in outline. (b) Anticlinal types of folded structure. This differs from a dome in being long and narrow. Reservoirs formed by folding of the rock layers or strata usually have the shapes indicated in (a) and (b). These traps were filled by upward migration or movement of oil and/or gas through the porous strata or beds to the location of the trap. Further movement was arrested by a combination of the forms of the structure and the seal or cap rock provided by the formation covering the structure. Examples of domal structures are the Conroe Oil Field in Montgomery County, Texas, and the Old Ocean Gas Field in Brazoria County, Texas. An example of a reservoir formed by an anticlinal structure is the Ventura Oil Field in California. (c) A fault trap. Reservoirs are formed by breaking or shearing and offsetting of strata (faulting). The escape of oil from such a trap is prevented by nonporous rocks that have moved into a position opposite the porous petroleum-bearing formation. The oil is confined in traps like this because of the tilt of the rock layers and faulting. Examples of fields of this type exist along the Mexia fault zone of East-Central Texas. (d) Unconformity. Upward movement of oil has been halted by the impermeable cap rock laid down across the cutoff (possibly by water or wind erosion) surfaces of the lower beds. An example of this type of reservoir is the great East Texas Field. (e) Salt dome. These often deform overlying rocks to form traps like this. An example of a salt-dome field is the Sugarland Oil Field in Fort Bend County, Texas. (f) Serpentine plug. These sometimes form reservoirs similar to the one shown: a porous serpentine plug has formed a reservoir within itself by intruding into nonporous surrounding formations. An example of a serpentine plug field is the Hilbig Field in Bastrop County, Texas. (g) Lens-type trap, formed in sand. An example is the Burbank Field in Osage County, Oklahoma. (h) Lens-type trap formed in limestone. Examples of limestone reservoirs of this type are found in the limestone fields of west Texas. In the lens-type trap, the reservoir is sealed in its upper regions by abrupt changes in the amount of connected pore space within a formation. This may be caused in sandstones by irregular depositing of sand and shale at the time the formation was laid down. In these cases, oil is confined within porous parts of the rock by the nonporous parts of rock surrounding it. In the limestone formations, there are frequent areas of high porosity with a tendency to form traps. (*Sketches by Exxon.*)

occurring over a very long time span. A past requisite for commercially exploitable hydrocarbon deposits has been prior movement and concentration of large quantities of hydrocarbons in various forms of *traps*. In contrast with oil shales and tar sands, natural gas and petroleum flow relatively easily in permeable underground structures and, consequently, tend to concentrate, greatly assisting the economic exploitation of these materials.

There is a broad range of shapes, sizes, and types of traps, of which six principal types are shown in Fig. 1.

Production Processes. It is convenient to classify oil and gas reservoirs in terms of the type of natural energy and forces available to produce the oil and gas. At the time oil was forming and accumulating in reservoirs, pressure and energy in the gas and salt water associated with the oil were also being stored, which would later be available to assist in producing the oil and gas from the underground reservoir to the surface. Obviously, since oil cannot lift itself from reservoirs to the surface, it is largely the energy in the gas or the salt water (or both) occurring under high pressures with the oil that furnishes the force to drive or displace the oil through and from the pores of the reservoir into the wells.

In nearly all cases, oil in an underground reservoir has dissolved in it varying quantities of gas that emerges and expands as the pressure in the reservoir is reduced. As the gas escapes from the oil and expands, it drives oil through the reservoir toward the wells and assists in lifting it to the surface. Reservoirs in which the oil is produced by dissolved gas escaping and expanding from within the oil are called *dissolved-gas-drive reservoirs*.

Often more gas exists with the oil in a reservoir than the oil can hold dissolved in it, under the existing conditions of pressure and temperature in the reservoir. This extra gas, being lighter than the oil, occurs in the form of a cap of gas over the oil. This condition is illustrated by Fig. 1(c) and Fig. 1(e). Such a gas cap is an important additional source of energy because, as production of oil and gas proceeds and as the reservoir pressure is lowered, the gas cap expands to help fill the pore spaces formerly occupied by the oil. Where conditions are favorable, some of the gas coming out of the oil is conserved by moving upward into the gas cap to further enlarge the gas cap. As compared with the dissolved-gas drive, the *gas-cap drive* is more effective, yielding greater recovery of oil. The gas-drive process is typically found with the discontinuous, limited, or essentially closed reservoirs of the types shown in Fig. 1(f) and Fig. 1(g).

Where the formation containing an oil reservoir is quite uniformly porous and continuous over a large area, as compared with the size of the oil reservoir per se, very large quantities of salt water exist in surrounding parts of the same formation, often directly in contact with the oil and gas reservoir. This condition is demonstrated by Figs. 1(a) through (e). These large quantities of salt water occur under pressure and provide a large additional store of energy to assist in producing oil and gas. A situation like this is termed *water-drive* reservoir. The energy supplied by the salt water comes from expansion of the water as pressure in the petroleum reservoir is reduced by production of oil and gas. Water will compress, or expand, to the extent of about one part in 2500 per 100 psi (6.8 atmospheres) change in pressure. Although this effect is slight with reference to small quantities, the phenomenon becomes of importance when changes in reservoir pressure affect large volumes of salt water that are often contained in the same porous formation adjoining or surrounding a petroleum reservoir.

The expanding water moves into the regions of lowered pressure in the oil- and gas-saturated portions of the reservoir caused by production of oil and gas, and retards the decline in pressure. In this way, the expansive energy in the oil and gas is conserved. The expanding water also moves and displaces the oil and gas in an upward direction out of the lower parts of the reservoir. By this natural process, the pore spaces vacated by oil and gas produced are filled with water, and oil and gas are progressively moved toward the wells.

The water drive is generally the most efficient oil-production process. Oil fields in which water drive is effective are capable of yielding recoveries ranging up to 50% of the oil originally in place, if (1) the physical nature of the reservoir rock and of the oil are conducive to the process, (2) care is exercised in completing and producing the wells, and (3) the rate of withdrawal of products is optimal.

When pressure in an oil reservoir have fallen to the point where a well will not produce by natural energy, some method of artificial lift must be used. Oil-well pumps are of three general types: (1) pumpos located at the bottom of the hole run by a string of rods, (2) pumps at the bottom of the hole run by high-pressure liquids, and (3) bottom-hole centrifugal pumps. Another method involves the use of high-pressure gas to lift the oil from the reservoir.

The details of drilling platforms and derricks are beyond the scope of this encyclopedia. A simplified sketch of an undersea drilling operation is given in Fig. 2.

Petroleum Fuels

Inasmuch as petroleum fuels essentially are mixtures of various hydrocarbons, plus various additives and minor impurities, their performance characteristics are not evaluated on the basis of chemical analysis alone, but, rather, a large number of physical tests and properties, such as octane numbers, Reid vapor pressure, distillation temperature, kinematic viscosity, specific gravity, pour point, flash point, color, and corrosiveness, are required to fully describe what might be termed effective performance of the fuels. Further, there are specific chemical data of importance, notably the content of sulfur, lead, phosphorus, water, sediment, and gum-forming materials.

Because petroleum fuels are refined from a rather broad range of feedstocks, and because petroleum fuels are refined and distributed by a number of suppliers, large and small, and because efforts are made by suppliers to adapt their products to the specific needs both of season and of local geography and terrain, a national average of the properties of a given fuel category can be obtained only by periodically conducting rather massive samplings of fuel products, in which all consuming areas and user segments are represented in a proportionate manner. This

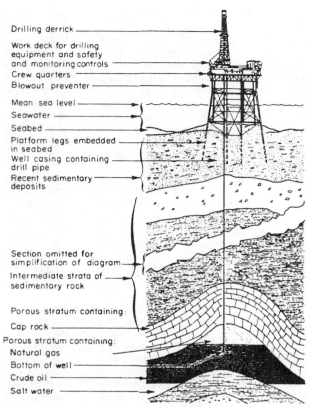

Fig. 2. Simplified sketch of an undersea drilling platform. (*American Gas Association.*)

sampling technique has been used in the United States for many years by the Bureau of Mines, U.S. Department of the Interior, and the American Petroleum Institute as a joint, effort. Annual reports are available. Other private surveys are made periodically.

For these purposes, the four major categories of petroleum fuels are: (1) motor gasoline, (2) diesel fuels, (3) aviation fuels, and (4) burner fuel oils.

Motor Gasolines. In the early days of automotive transportation, gasoline was a relatively simple mixture of petroleum fractions derived from straight-run and thermally-cracked stocks. In contrast, modern fuels are a complex mixture of blends derived from catalytic cracking, alkylation, catalytic reforming, polymerizations, isomerizations, and hydrocracking, plus small amounts of additives designed to further improve the overall efficiency and reliability of the internal combustion engine. From 1925 to 1950, there was a gradual, but steady increase in compression ratio. During the 1950s, this increase was more rapid, with a final leveling off in the late 1960s. This increase in compression ratio was aimed at improving overall engine performance and efficiency. Its effect on fuel composition was direct since increased compression ratio requires an increase in fuel octane number to prevent knocking. The octane increase was achieved partly by the addition of lead alkyls, but mostly via the changes in petroleum refining processes. Without these changes, the amount of gasoline that could be obtained from a barrel of crude would have been greatly reduced.

During the 1950s and 1960s, other changes also were made in automotive gasoline to help performance objectives. Additives were developed that helped to minimize such problems as carburetor icing and fouling, valve and engine deposits, spark plug fouling, fuel system corrosion, and poor fuel distribution. All of these factors have contributed to modern highly complex gasoline formulations. Commencing in the early 1960s, ecological problems also entered into the gasoline picture. Small amounts of carbon monoxide and nitrogen oxides are formed and emitted to the atmosphere, along with some unburned hydrocarbons, when gasoline is burned in an internal combustion engine. The predominant products of combustion, of course, are carbon dioxide and water. Stress on atmospheric pollution coupled with increased demand and lower fuel efficiencies arrived at about the same time that new conventional oil sources were being discovered at a lessened rate. The introduction of unleaded gasolines required a very large capital investment by the petroleum industry.

Antiknock Quality. The longterm trend toward more efficient utilization of fuels in spark ignition, reciprocating engines reached a peak in the 1965–1970 period. In 1971 and 1972, a sharp decrease in compression ratio occurred in passenger car engines in the United States. This decrease was made to permit the engines to operate on unleaded gasolines of an octane quality which could be produced by petroleum refineries in large volumes. This quality was calculated to be 91 Research Octane Number (RON) and 83 Motor Octane Number (MON). Octane number is defined later.

In addition to the need to furnish unleaded gasolines for cars having lead-sensitive catalytic reactors, there is an objective to reduce lead particulate emissions from automotive engines. Scientifically, however, this objective remains controversial. Two additional elements which are believed to poison catalysts are phosphorus and sulfur. Phosphorus additives in the past have been beneficial to engine performance. Sulfur levels in gasoline are low and further declining. See **Catalytic Converter (Internal Combustion Engine).**

The progressively lower limits set on hydrocarbon, carbon monoxide, and nitrogen oxide emissions from automobile exhausts have a significant effect on gasoline mileage relative to older cars (1970) for a number of reasons: (1) reduced compression ratios reduce gasoline mileage and power output; (2) increased engine size or increased gear ratios (to regain power lost by reduced compression ratios) lowers gasoline mileage; (3) use of exhaust gas recirculation to reduce nitrogen oxide emissions lowers gasoline mileage; and (4) operation at nonoptimum spark advance settings to reduce emissions lowers gasoline mileage.

The overall effect of changing gasoline specifications required the raising of the unleaded pool octane number from a value of 88–89 to about 92–94. RON levels are influenced by several factors, including: (1) Removal of lead reduced the freedom once used in blending the various refinery streams because an overall increase in average RON is required to maintain adequate quality control at any given RON level; (2) the average RON requirement of used automobiles is generally somewhat higher than the automobile manufacturer's target for new cars because of the influence of deposit accumulation, average timing maladjustments, and overall engine conditions; (3) the design of engines having higher octane requirements in order to regain efficiencies lost becuase of emission controls.

Volatility. Optimizing volatility for peak engine performance has become increasingly difficult because of seemingly incompatible performance goals. Minimizing automotive emissions while maintaining satisfactory drivability, octane quality, and volume of gasoline production requires fuels having divergent volatility characteristics. It has been suggested that refiners decrease frontend volatility during much of the year to reduce the loss of vapors during handling and to avoid overloading of the vapor recovery systems on emission-system equipped cars. This can

have an adverse effect on starting and warm-up, and in some cars, can cause a definite reduction in the amount of high octane, clean-burning butane which can be blended into automotive fuels. In the mid- and upper-boiling ranges, on the other hand, it has been suggested that refiners increase volatility to reduce final boiling points. These changes make it easier to design cars to give reduced emissions during the warm-up period and have a beneficial effect on the overall drivability of the cars. From the refinery viewpoint, however, these changes reduced the amount of high octane components which can be blended into gasolines and caused a marked reduction in the volume of gasoline which can be produced by mid-1970 vintage refineries. Additionally, these volatility changes involve large expenditures for refinery processing equipment and increased refinery operating costs.

Additives. These are ingredients that are added at concentrations ranging from a few to several hundred parts per million (ppm), and many have become essential ingredients of modern gasolines. Gasoline additives now must be registered with air pollution authorities. Manufacturers and users of gasoline additives must provide the authorities with information on the function, chemical composition, recommended dosage, and emission products of the additives that are registered.

Gasoline additives are categorized as to type and function in Table 3. The number, type, and quantity of each additive will vary among individual marketers of gasoline. In some instances, several functions can be combined in one chemical compound to provide "multifunctional" additives. The increasingly stringent control of automotive emissions has stimulated the development and use of "extended-range" detergents which are designed to promote peak engine performance by maintaining engine cleanliness.

Diesel Fuels. In a cooperative effort similar to the gasoline survey, the Bureau of Mines has published the results of a diesel fuel survey annually since 1950. In a typical recent survey, some 250 samples of diesel fuel, manufactured by over 30 refiners in over 100 refineries, large and small and widely distributed throughout the United States, were reported. Trends dating back to 1960 for four categories of diesel fuels are reported. The categories are: (1) diesel fuel oils for city-bus and similar operations; (2) fuels for diesel engines in trucks, tractors, and similar service; (3) fuel for railroad diesel engines; and (4) heavy-distil-

TABLE 3. AUTOMOTIVE GASOLINE ADDITIVES AND FUNCTIONS

ADDITIVE	FUNCTION
Antiknock compounds	Increase octane number
Scavengers	Remove combustion products of antiknock compounds
Combustion chamber deposit modifiers	Suppress surface ignition and spark plug fouling
Antioxidants	Provide storage stability
Metal deactivators	Supplement storage stability
Antirust agents	Prevent rusting in gasoline-handling systems
Anti-icing agents	Suppress carburetor icing and fuel system freezing
Detergents	Control carburetor and induction system cleanliness
Upper cylinder lubricants	Lubricate upper cylinder areas and control intake system deposits
Dyes	Indicate presence of antiknock compounds and identify makes and grades of gasoline

late and residual fuels for large stationary and marine diesel engines.

Recent changes in diesel fuel properties have resulted from competition for available distillate blending stocks for use in other fuels or as process materials. For example, the rapid increase in commercial jet fuel consumption has significantly reduced availability of the more volatile straight-run components. Thus, the fraction of cracked stocks in diesel fuel continues to increase. Hydrogenation of cracked stocks reduces sulfur content, improves stability, and upgrades rather poor performers to satisfactory diesel components. Most properties have leveled off, but cetane number is slowly declining. This trend is expected to continue as the demand for energy progresses. Cetane number is defined later.

Most diesel engines in truck service can operate satisfactorily on truck and tractor fuels available in the mid-1980s. However, there is a wide range of fuels possible under this classification, and variation within the classification can have a pronounced effect on a given engine's performance. Emphasis on improving air quality is bringing about engine design changes to reduce exhaust emissions. Previously preeminent factors, such as increased power output and improved fuel economy are yielding in importance to reduction of smoke, nitrogen oxides, carbon monoxide, and hydrocarbon emissions. Engine modifications designed to reduce exhaust emissions probably will result in a narrowing of the range of the fuel specifications in regard to cetane number and volatility.

Of the total diesel fuel oil sales in the United States, about 75% is for use in transportation. Trucks, autos, and buses consume about 45%, railroads about 25%, and marine uses amount to about 5%. Industrial plants, utilities, and the military consume the remainder. In metropolitan areas, electric power companies are installing gas turbines that burn diesel fuel for power generation. Gas turbines can be installed quickly at relatively low initial costs and are reasonably free of undesirable emissions.

The quality of railroad fuels has not changed significantly over the years. Some railroads operate on special economy grade fuels that have much broader volatility and lower cetane numbers and always contain large percentages of cracked stock. The large diesel engines in railroad service are less sensitive to fuel properties than their smaller truck or tractor counterparts and can operate satisfactorily on fuels with less exacting specifications.

As in the case of gasolines, the use of additives has become much more common in diesel fuels. Cetane improvers, largely alkyl nitrates, provide ignition quality improvement. Ignition quality influences ease of starting and smoothness of operation. A variety of additives is used to improve storage stability and permit the use of otherwise unstable stocks. Polymeric and other types of additives have been used as detergents and dispersants. The detergents have the ability to maintain fuel injection nozzle cleanliness and will markedly increase fuel-filter life, and many diesel fuels also contain rust inhibitors.

The high demand for diesel and jet fuels has made it difficult to obtain appropriate low-temperature flow characteristics by base stock selection. This has led to increased use of pour-point depressant additives. Several of these materials, both polymeric and nonpolymeric, have the ability to reduce pour points, resulting in substantial improvement in flow through distribution and truck piping systems. However, their effect on cloud point (the first appearance of wax crystals which can cause filter plugging under cold conditions) is small. Fuel system modifications, such as mounting filters in warm locations or installing filter heaters, have been used to provide for the use of pour-point depressants.

Smoke emission laws have led truck operators to investigate antismoking additives. The most functional of these are barium-based organic compounds which are effective at concentrations of about 1,000 ppm. Effective concentrations of the barium additive are costly, and the ash contributed by the barium salts has caused a problem in some engines. Thus, antismoke additives generally are not used in diesel fuels by manufacturers. Two other types of additives are emerging from the development stage, namely, biocides to prevent bacterial attack on the fuel and masking agents to improve the odor of exhaust fumes in city bus service.

Aviation Fuels. National aviation-fuel survey reports have been issued by the U.S. Bureau of Mines since March 1951, in a cooperative effort with the American Petroleum Institute. Quality control of aviation gasoline is even more critical than in motor gasoline for obvious reasons of safety. Antiknock control is especially critical because, unlike the motorist, the pilot of a piston engine plane is unable to hear an engine knock over the high ambient noise level. Volatility, freezing point, heat of combustion, and oxidation stability are all very important to the aviation gasoline user. Aviation gasolines contain up to 4.6 milliliters of tetraethyl lead per gallon. Ethylene dibromide is added to scavenge the lead and has been found more effective under high-load aircraft conditions than the chloride/bromine scavenger mixtures used in motor gasolines. Other alkyl leads, tetramethyl lead, or methylethyl lead compounds, are not used in aviation gasolines. Some hydrocarbon constituents of aviation gasoline tend to oxidize during storage at ambient temperatures. The products of oxidation, fuel-soluble and fuel-insoluble gums, interfere with metering of the fuel to the engine and must be controlled. Certain amine and phenolic chemical compounds have been found effective in this service.

Jet Fuels. Commercial kerosine was used as a fuel in early developmental work on jet aircraft in the United States. The choice of kerosine over gasoline was based on its low volatility to avoid occurrence of vapor lock under certain flight conditions and on its availability as a commercial product with uniform characteristics. JP-1, the first military jet fuel, was highly refined kerosine having a very low freezing point ($-76°F$; $-60°C$). Kerosine from selected crudes high in naphtha was the only fuel having this low freezing point. As the demand for the fuel increased, the Military Petroleum Advisory Board recommended development of a military jet fuel having greater availability in wartime than either JP-1 or aviation gasoline. The second candidate jet fuel was JP-2, but it did not have the desired availability. JP-3 fuel was another possibility which included the total boiling range of kerosine and gasoline. A cooperative program of testing by the Coordinating Research Council demonstrated that the high vapor pressure of JP-3 (Reid vapor pressure of 5 to 7 pounds) resulted in vaporization of the fuel during climb to altitude. In addition, some fuels foamed excessively during vaporization so that very large losses of liquid could occur along with the vented vapors.

To overcome the disadvantages of JP-3, the Reid vapor pressure was reduced to 2 to 3 pounds, and JP-4 was developed in 1951. This fuel is a blend of 25 to 35% kerosine and 65 to 75% gasoline components and has proved satisfactory for military requirements. An important Navy turbine fuel, developed for carrier operation during the Korean War, was a mixture of a special kerosine and aviation gasoline. The latter was stored in tanks in the central zones of carriers to minimize the possibility of hazardous fuel leaks in event of battle damage. But, space was limited for such storage. Thus, JP-5 fuel was developed for aircraft carriers. This was a special 140°F (60°C) flash point kerosine. Because of its low volatility, it could be stored safely in outer tanks of carriers. When mixed with aviation gasoline,

JP-5 produced a fuel similar to JP-4. Later, the Navy eliminated the use of the aviation gasoline mixture and used JP-5 alone.

Commercial airline jet fuels in the United States fall within the general framework of ASTM Jet A, A-1, and B fuels. Jet A and A-1 are of the kerosine type. Jet B corresponds to the military JP-4 fuel. Volume demands for Jet A and A-1 are large, but small for Jet B. Jet fuel fulfills a dual purpose in the aircraft. It provides the energy and serves as a coolant for lubricating oil and other aircraft components. Exposure of the fuel to high temperatures may cause the formation of oxidation materials (gums) which reduce the efficiency of heat exchangers and clog filters and valves in aircraft fuel-handling systems. Thermal stability is the resistance to formation of gums at high temperature. The JP-4, JP-5, and equivalent commercial fuels have satisfactory thermal stability for aircraft operating at speeds up to about Mach 2.0. Future jet aircraft operating at higher speeds, e.g., Mach 3, may expose the fuel to greater thermal stresses and, therefore, may require a more stable fuel. The development of Mach 3–4 turbojets, Mach 6-plus ramjets, and rockets using hydrocarbon fuels will pose additional problems on fuel stability characteristics.

Additives provide an attractive method of improving jet fuel quality. Common additives include:

(*Antioxidants*)—Some jet fuels may oxidize during storage at ambient temperature to form fuel-soluble and fuel-insoluble gums. The oxidation products may cause clogging of filters or coking of engine burner nozzles. The same antioxidants as approved for aviation gasoline are used in jet fuels.

(*Copper Deactivator*)—Traces of dissolved copper in jet fuel accelerate oxidation. A copper chelating agent, N,N′ disalicylidene-1, 20 propanediamine, has been approved for addition to jet fuel to deactivate the prooxidant effects of copper.

(*Corrosion Inhibitors*)—Fuel-soluble corrosion inhibitors have limited specific approval for jet fuels. The inhibitors are added to protect pipelines against corrosion by occluded water and indirectly to reduce contamination of the fuel with rust. Several commercial inhibitors of this type are available.

(*Anti-Icing Additive*)—The major hazard of free water in jet fuels is plugging of fuel lines and filters by ice during flight. Fuel heaters are used in civilian aircraft to prevent ice formation. The military adopted the approach of adding 0.1 to 0.15% of an anti-icing inhibitor comprising a mixture of 99.6% (weight) ethylene glycol monomethyl ether and 0.4% (weight) glycerol. The additive has the extra feature of having biocidal properties. At above freezing temperatures, liquid water can cause erratic operation of electrical fuel gages and provides a suitable environment for growth of bacteria and fungi. More recently, the anti-icing additive has been changed to eliminate the glycerol.

(*Anti-Static Additive*)—High-speed flow (600 to 800 gallons 22.7 to 30.3 hectoliters per minute) through fill lines, filters, and valves generates electrostatic charges in jet fuels. Electrostatic potential differences between the fuel and the fuel tank walls may cause sparking in the vapor spaces of fuel tanks. Intensive investigations have shown that additives which decrease resistivity of the fuel can reduce electrostatic hazards.

(*Contaminants*)—In handling and delivery of jet fuels, there are several places where traces of contaminants can enter the fuel. Particulate matter and free water can be removed by bulk settling or filtering. Removal of solids down to 5 micrometers is generally desired to avoid plugging of filters on fuel control valves and burner nozzles. Surface-active agents (surfactants) in jet fuels are highly undesirable because they promote the formation of water-in-fuel emulsions and reduce the efficiency of filter/coalescers. Surfactants produced by refining operations are removed by neutralization, water washing, clay treating, filtration, or settling. Polar additives that are added to jet fuel are chosen on the basis of having minimum surfactant properties.

(*Future Fuels*)—By proper selection of structural materials and aircraft design, it may be feasible to use conventional hydrocarbon fuels up to flight speeds approaching Mach 4. Flight speeds higher than Mach 4 will require more heat-sink capacity than normally provided by fuels in the liquid state. Vaporization of the fuel can increase the overall heat-sink capacity by about 25%. A promising alternate is to allow endothermic fuels to undergo mild thermal cracking which can absorb several times the heat picked up in the liquid state alone. Cryogenic fuels (liquified hydrogen, methane, and propane) also offer an attractive way of cooling aircraft, although these fuels have the disadvantage of a low volumetric energy content.

Low-altitude, high-speed, ramjet-powered aircraft require a fuel with a high volumetric heat content. The U.S. Air Force has been pursuing the development of slurry fuels in which metals are burned to take advantage of the high heat of formation of metal oxides. Research efforts are continuing to develop fuels having the greatest possible metal content and to overcome problems, such as poor pumpability, abrasiveness, and low combustion efficiency. Another promising area includes the high-density fuels (aromatic and condensed polycyclic hydrocarbons). A desirable material would have a heating value of about 150,000 Btus per gallon or higher, with a freezing point of −50°F (−45.6°C) or lower.

Burner-Fuel Oils. The first annual survey of burner-fuel oils was made and reported by the U.S. Bureau of Mines and the API in 1955. The work was initiated at the request of the National Oil Fuel Institute, Inc., to provide information on the characteristics of burner-fuel oil marketed in the United States. The analytical data are used by manufacturers of heating appliances, consumers, and various governmental agencies. In a typical survey, over 300 samples of burner-fuel oil are represented, covering the products of about 30 petroleum refiners from over 100 domestic refineries located throughout the United States. This survey involves six grades of fuels. Data collection and reporting are done on a regional basis.

Home Heating Oil. Domestic heating oil must be a clean product, and it should not form sediment in storage nor leave a measurable quantity of ash or other deposit upon burning. Since it may be stored at low temperatures, it should be fluid at storage conditions encountered outdoors during winter months. The chemical composition of the product must be controlled to assist in reducing smoke emission. Sulfur content, at one time not considered a major problem, is now important.

Grade No. 2 fuel oil is the designation given to the heating oil most commonly used for domestic and small commercial space heating. This product is a distillate product, normally fractionated to a boiling range of 350° to 650°F (177° to 343°C). Since about 1950, significant improvements have been made in both the quality of home heating oils and in the manufacturing techniques used in producing them. Originally, No. 2 heating oil was composed of selected refinery straight-run stocks, blended to meet product standards. The resultant product has a good stability and was very satisfactory in performance. As the refining industry was required to make larger amounts of motor gasoline at higher octane levels, cracking processes were developed to convert virgin gas oils to lighter boiling products. This necessitated the use of increasing amounts of both catalytically and thermally cracked gas oils in finished heating oil blends. Heating oil blending became more complex in order to maintain a satisfactory quality level without excessive treating expense.

The industry worked in several directions to correct the quality problems associated with the extensive use of cracked distillates. New treating processes, burner alterations, and the development of additives all progressed together. Additives to improve stability also were developed. Caustic washing processes were developed as an improvement over acid washing. In the early 1950s, reforming of straight-run gasoline became widespread. This process made available large volumes of hydrogen, which previously had been costly to produce. With hydrogen available, catalytic hydrogen treating was used to further improve fuel oil quality. The primary objective in hydrogen treatment is enhancement of quality through a reduction in sulfur and removal of small, but objectionable, amounts of nitrogen compounds. The treatment also reduces carbon residue, improves burning characteristics, and reduces sludging tendencies. Sulfur reduction is generally 70 to 80% complete, but can be as high as 90 to 95% if needed. Carbon residue on 10% distillation bottoms is reduced to less than 0.10%.

Residual Fuels. Grades 4, 5, and 6 are the designations given to the fuels most commonly used for commercial, industrial, marine, and other uses involving larger installations than those used for domestic and small commercial establishments. Typically, these fuels are used to provide steam and heat for industry and large buildings, to generate electricity in competition with gas and coal, and to power ships. Most users of residual fuels have converted their equipment to handle higher viscosity Grade 6, which is less costly because it requires less of the distillate "cutter" stocks which can be converted more readily into gasoline. In the shipping industry, heavy bunker fuels are known as Bunker C and generally correspond to Grade 6 fuel oil.

The largest single user of residual fuels is the electrical power generating industry which consumes about 40% of the available residual fuel. Because of air pollution measures which tend to favor residual fuels over coal, this segment of the market demand has been the most rapidly growing segment in recent years. Growth has been to the extent of creating a degree of disequilibrium in the supply/demand picture. Future growing demand for low-sulfur fuel oils can be met in theory, but each strategy has its own particular drawbacks. Adding desulfurization capacity to the existing array of refining equipment involves a substantial time requirement, and desulfurization of fuel oils, especially the heavier grades, is an expensive process. The alternatives of selecting low-sulfur crudes or of importing large amounts of low-sulfur residual fuels are no longer practical because of limited supplies.

Residual fuels are, by their nature, high-boiling and contain stocks which are difficult to burn quickly under "cold" conditions. Accordingly, such fuels are generally burned in equipment which permits relatively steady operation in an environment where firebox temperatures can be high.

The steady increase in the use of catalytic cracking since the mid-1960s has had the effect of decreasing the percentage yield of residual fuels as well as changing their makeup. As more high-boiling materials were charged to catalytic cracking, the remaining oil which was sold as residual fuel became heavier. Common industry practice was to blend these heavy stocks with a distillate to lower their viscosity. Continued work in this field led to the use of milk thermal cracking of the vacuum still bottoms which yielded a small additional amounts of distillate product and reduced the viscosity of the remaining bottoms. Such bottoms required less distillate cutter stock to produce a salable residual fuel oil. This modest advance was developed before World War II and was known as visbreaking.

Most advances in residual fuel technology since World War II have led to improvements in its use rather than in oil quality.

The oil industry and boiler manufacturers have increased their efforts in the areas of desulfurizing fuel oil and flue gas, and reducing fuel oil metals content. A large number of additives have been developed for reducing residual fuel oil sluge, tube deposit formation, and corrosion, and for increasing combustion efficiency.

Basic Physical Characteristics of Petroleum Fuels

In testing various petroleum fuels, the following parameters are of major significance:

(*Octane Number*)—The octane rating of a motor fuel is defined in terms of its knocking characteristics relative to those of blends of isooctane (2,3,4-trimethylpentane) and n-heptane. Arbitrarily, an octane number of zero has been assigned to n-heptane, and a rating of 100 to isooctane. The octane number of an unknown fuel is numerically equal to the volume percent of isooctane in a blend with n-heptane which has the same knocking tendency as the unknown fuel when both the unknown and the reference blend are run in a standard single-cylinder engine operated at specified conditions. Motor Method octane numbers are measured at more severe engine conditions and are numerically lower than those determined by the milder Research Method. The difference between the two numbers is termed *sensitivity*.

(*Equilibrium Volatility of a Gasoline*)—The volatility of a gasoline is determined by the Reid vapor pressure and the ASTM distillation data. The Reid vapor pressure is the vapor pressure of a gasoline at 100°F (37.8°C) under specified conditions. The distillation curve of a fuel indicates the temperatures at which the various amounts of a given sample are distilled under specified test conditions. However, gasoline will completely evaporate in the presence of air at a temperature much lower than the end-point of the distillation curve. According to O. C. Bridgeman (U.S. National Bureau of Standards, Research Paper 694), the volatility of a gasoline is the temperature at which a given air-vapor mixture is formed under equilibrium conditions at a pressure of one atmosphere, when a given percentage is evaporated. According to this definition, one gasoline is more volatile than another for any given percentage evaporated if it forms the given air-vapor mixture at a lower temperature. Distillation temperature curves, for a given test sample, plot amount of sample distilled over (percentage of sample) at the time a given temperature has been reached.

(*Specific Gravity*)—The specific gravity of a petroleum fuel is the ratio of the weight of a given volume of the product at 60°F to the weight of an equal volume of distilled water at the same temperature, both weights corrected for air buoyancy. The relation between API gravity scale and specific gravity is: °APi = 141.5/(sp gr 60/60°F) − 131.5.

(*Pour Point*)—This property is defined as the lowest temperature at which the fuel will pour and is a function of the composition of the fuel. Normally, the pour point of a fuel should be at least 10 to 15 degrees below the anticipated minimum use temperature.

(*Cloud Point*)—The ainline cloud point is a measure of the paraffinicity of a fuel oil, a high value indicating a straight-run paraffinic oil and a low value indicating an aromatic, a naphthenic, or a highly cracked oil.

(*Flash Point*)—The lowest temperature at which a flash appears on the fuel surface when a test flame is applied under specified test conditions. This property is an approximate indication of the tendency of the fuel to vaporize.

(*Fire Point*)—The lowest temperature at which a fuel ignites and burns for at least 5 seconds under specified test conditions.

(*Smoke Point*)—The smoking tendency of a fuel is indicated

by this value, which is the maximum height of a specified type of flame in a given wick lamp that results in no visible smoke.

(*Viscosity*)—This is generally expressed in terms of the time required for a given quantity of fuel to flow through a capillary tube under specified conditions. Kinematic viscosity v is viscosity divided by mass density, or $v = \mu/\rho$. The unit in cgs units is called the *stoke*; a customary unit is the *centistoke* ($\frac{1}{100}$ of a stoke). The value of the kinematic viscosity in (cm²/seconds) can be obtained from the indications in seconds t of various viscometers by:

Saybolt Universal, when $32 < t < 100 = 0.00226t - 1.95/t$
Saybolt Universal, when $t > 100$ $= 0.00220t - 1.35/t$
Saybolt Furol, when $25 < t < 40$ $= 0.0224t - 1.84/t$
Saybolt Furol, when $t > 40$ $= 0.0215t - 0.50/t$

(*Cetane Number*)—The cetane number (C.N.) of a fuel is the percentage by volume of normal cetane in a mixture of cetane and alpha-methylnaphthalene which matches the unknown fuel in ignition quality when compared with a standard diesel engine under specified conditions. The C.N. scale ranges from 0 to 100 C.N. for fuels equivalent in ignition quality to alpha-methylnaphthalene and cetane, respectively. For routine testing, secondary reference fuels having cetane values of about 25 and 74 are blended in any desired proportion.

Refining and Processing of Petroleum

Because of the variety of crude oils (exhibiting important differences from the standpoint of processing), the mix of refined products which vary from one geographic location to the next, and because of the effects of seasonal changes on the demand for certain refined products, there are no two petroleum refineries that are exactly alike. Some of the units, such as catalytic crackers, reformers, hydrotreaters, and the like may be quite similar from one refinery to the next, but the manner in which they are arranged and combined with one another will vary for each local refinery. The diagram of Fig. 3 is representative of an integrated refinery and, of course, is extremely oversimplified.

The major processing units fundamental to the manufacture of fuel products from crude oil include: (1) crude distillation; (2) catalytic reforming; (3) catalytic cracking; (4) catalytic hydrocracking; (5) alkylation; (6) thermal cracking; (7) hydrotreating; and (8) gas concentration. Refineries also will use numerous auxiliary processes, such as treating units to purify both liquid and gas streams, waste-management and pollution-control systems, cooling-water systems, units to recover hydrogen sulfide (or elemental sulfur) from gas streams, desalters, electric-power stations, steam-producing facilities, and provisions for storage of crude oil and products.

Petroleum refining and petrochemical production is a 24-hour, 365-day operation with a very minimum of time planned for downtime. Unless one has visited a refinery firsthand, it is very difficult to comprehend the size and complexity of the equipment used.

Crude Distillation. To minimize corrosion of refining equip-

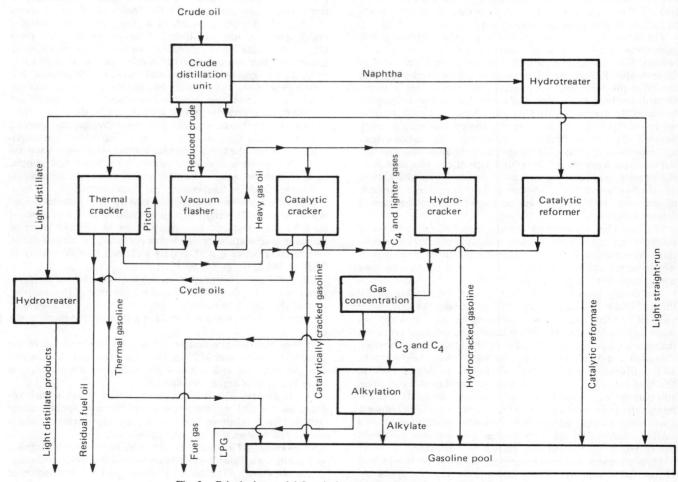

Fig. 3. Principal material flow in integrated refinery for producing fuels.

ment, a crude-oil distillation unit generally is preceded by a *desalter*, which reduces the inorganic salt content of raw crudes. Salt concentrations vary widely (from nearly zero to several hundred pounds, expressed as NaCl per 1,000 barrels). The crude unit functions simply to separate the crude oil physically, by fractional distillation, into components of such boiling ranges that they can be processed by appropriately selected equipment in a long train of processing operations which follow. Although the boiling ranges of components (or fractions) vary between refineries, a typical crude distillation until will resolve the crude into the following fractions:

By distillation at atmospheric pressure,

1. A light straight-run fraction, consisting primarily of C_5 and C_6 hydrocarbons. These also will contain any C_4 and lighter gaseous hydrocarbons that are dissolved in the crude.
2. A naphtha fraction having a nominal boiling range of $200°-400°F$ ($93°-204°C$).
3. A light distillate with boiling range of $400°-540°F$ ($204°-343°C$).

By vacuum flashing,

1. Heavy gas oil having a boiling range of $650°-1,050°F$ ($343°-566°C$).
2. A nondistillable residual pitch.

In the atmospheric-pressure distillation section of the unit, the crude oil is heated to a temperature at which it is partially vaporized and then introduced near, but at some distance above, the bottom of a distillation column. This cylindrical vessel is equipped with numerous trays through which hydrocarbon vapors can pass in an upward direction. Each tray contains a layer of liquid through which the vapors can bubble, and the liquid can flow continuously by gravity in a downward direction from one tray to the next one below. As the vapors pass upward through the succession of trays, they become lighter (lower in molecular weight and more volatile with lower boiling temperature). The liquid flowing downward becomes progressively heavier (higher in molecular weight and less volatile with higher boiling temperature). The countercurrent action results in fractional distillation or separation of hydrocarbons based upon their boiling points. A liquid can be withdrawn from any preselected tray as a net product. Thus, the lighter liquids, such as naphtha, exist from trays near the top of the column, whereas heavier liquids, such as diesel oil, exist from trays near the bottom of the column. Thus, the boiling range of the net product liquid depends upon the tray from which it is taken. The vapors containing the C_6 and lighter hydrocarbons are withdrawn from the top, while a liquid stream boiling at about $650°F$ ($343°C$) is taken from the bottom. The portion taken from the bottom is called *atmospheric residue*.

This residue is further heated and introduced into a vacuum column operated at an absolute pressure of about 50 millimeters of mercury, a vacuum maintained by the use of steam ejectors. A flash separation is made to produce heavy gas oil and nondistillable pitch.

The crude oil and atmospheric residue are heated in tubular heaters. Oil is pumped through the inside of the tubes contained in a refractory combustion chamber fired with oil or fuel gas in such manner that heat is transferred through the tube wall in part by convection from hot combustion gases and in part by radiation from the incandescent refractory surfaces.

The light straight-run gasoline contains all hydrocarbons lighter than C_7 in the crude and consists primarily of the native C_5 and C_6 families. After stabilization to remove the C_4 and lighter hydrocarbons (which are routed to a central gas-concentration unit), the stabilized C_5/C_6 blend is treated to remove odorous mercaptans and passed to the refinery gasoline pool for final product blending. The unleaded octane number (Research Method Number) is less than 70, and thus blending or further processing is required to improve its antiknock qualities. Isomerization can be used to improve octane rating, as well as the addition of lead alkyls.

Naphtha to become a suitable component for blending into finished gasoline pools must be further processed. The octane number will range from 40 to 50. Prior to introduction into a catalytic-reforming unit, most naphtha feedstocks are hydrotreated in the interest of prolonging the life of the reforming catalyst.

Gas oil separated from the crude by vacuum distillation, plus portions of light distillates, is the feedstock to catalytic cracking units. The main function of catalytic cracking is to convert into gasoline those fractions having boiling ranges higher than that of gasoline. Remaining uncracked distillates (cycle oils) are used as components for domestic heating fuels (generally after hydrotreating) and to blend with residual fractions to reduce their viscosity to make acceptable heavy fuel oils. In some refineries, cycle oils are hydrocracked to complete their conversion to gasoline.

The aforementioned processes are described in more detail in separate entries: **Alkylation; Cracking Process; Distillation; Hydrotreating.**

Petroleum Terminology. Some terms commonly used in petroleum processing technology are briefly defined below. However, many of these terms are described in more detail in this entry or in the entries just mentioned.

(Additive)—Any materials incorporated in finished petroleum products for the purpose of improving their performance in existing applications or for broadening the areas of their utility.

(Alkylation)—A refinery process for chemically combining isoparaffin with olefin hydrocarbons. The product, *alkylate*, has a high octane value and is blended with motor and aviation gasoline to improve the antiknock value of the fuel.

(Base Oil)—A refined or untreated oil used in combination with other oils and additives to produce lubricants.

(Blending)—The process of mixing two or more oils having different properties to obtain a final blend having the desired characteristics. This can be accomplished by off-line batch processes or by in-line operations as part of continuous-flow operations.

(Bright Stock)—High-viscosity, fully-refined and dewaxed lubricating oils produced by the treatment of residual stocks and used to compound motor oils.

(Catalytic Cracking)—A refinery process that converts a high-boiling range fraction of petroleum (gas oil) to gasoline, olefin feed for alkylation, distillate, fuel oil, and fuel gas by use of a catalyst and heat.

(Catalytic Reforming)—A catalytic process to improve the antiknock quality of low-grade naphthas and virgin gasolines by the conversion of naphthenes (such as cyclohexane) and paraffins into higher-octane aromatics (such as benzene, toluene, and xylenes). There are approximately ten commercially licensed catalytic reforming processes, including fully regenerative and continuously regenerative designs.

(Clear Octane)—The octane number of a gasoline before the addition of antiknock additives.

(Coking)—Distillation to dryness of a product containing complex hydrocarbons, which break down in structure during distillation, such as tar or crude petroleum. The residue is called coke.

(Cracking)—A process carried out in a refinery reactor in which the large molecules in the charge stock are broken up into smaller, lower-boiling, stable hydrocarbon molecules, which leave the vessel as overhead (unfinished cracked gasoline, kero-

sines, and gas oils). At the same time, certain of the unstable or reactive molecules in the charge stock combine to form tar or coke bottoms. The cracking reaction may be carried out with heat and pressure (thermal cracking) or in the presence of a catalyst (catalytic cracking).

(*Cycle Stock*)—Unfinished product taken from a stage of a refinery process and recharged to the process at an earlier period in the operation.

(*Deasphalting*)—Process for removing asphalt from petroleum fractions, such as reduced crude. A common deasphalting process introduces liquid propane, in which the nonasphaltic compounds are soluble while the asphalt settles out.

(*Desulfurization*)—The removal of sulfur or sulfur-bearing compounds from a hydrocarbon by any one of a number of processes, such as hydrotreating.

(*Distillate*)—That portion of a liquid which is removed as a vapor and condensed during a distillation process. As fuel, distillates are generally within the 400° to 650°F (204° to 343°C) boiling range and include Nos. 1 and 2 fuel, diesel, and kerosine.

(*End Point*)—The temperature at which the last portion of oil has been vaporized in ASTM or Engler distillation. Also called the final boiling point.

(*Flare*)—A device for disposing of gases by burning.

(*Flue Gas Expander*)—A turbine used to recover energy where combustion gases are discharged under pressure to the atmosphere. The pressure reduction drives the impeller of the turbine.

(*Fractions*)—Refiner's term for the portions of oils containing a number of hydrocarbon compounds but within certain boiling ranges, separated from other portions in fractional distillation. They are distinguished from pure compounds which have specified boiling temperatures, not a range.

(*Fuel Oils*)—Any liquid or liquefiable petroleum product burned for the generation of heat in a furnace or firebox or for the generation of power in an engine. Typical fuels include clean distillate fuel for home heating and higher-viscosity residual fuels for industrial furnaces.

(*Gas Oil*)—A fraction derived in refining petroleum with a boiling range between kerosine and lubricating oil.

(*Heating Oils*)—A trade term for the group of distillate fuel oils used in heating homes and buildings as distinguished from residual fuel oils used in heating and power installations. Both are burner-fuel oils.

(*Heavy Ends*)—The highest-boiling portion of a gasoline or other petroleum oil.

(*Hydrocracking*)—The cracking of a distillate or gas oil in the presence of catalyst and hydrogen to form high-octane gasoline blending stock.

(*Hydrogenation*)—A refinery process in which hydrogen is added to the molecules of unsaturated (hydrogen-deficient) hydrocarbon fractions. It plays an important part in the manufacture of high-octane blending stocks for aviation gasoline and in the quality improvement of various petroleum products.

(*Hydrotreating*)—A treating process for the removal of sulfur and nitrogen from feedstocks by replacement with hydrogen.

(*Isomerization*)—A refining process which alters the fundamental arrangement of atoms in the molecule. Used to convert normal butane into isobutane, an alkylation process feedstock, and normal pentane and hexane into isopentane and isohexane, high-octane gasoline components.

(*Kinematic Viscosity*)—The absolute viscosity of a liquid (in centipoises) divided by its specific gravity at the temperature at which the viscosity is measured.

(*Knock*)—The sound or "ping" associated with the autoignition in the combustion chamber of an automobile engine of a portion of the fuel-air mixture ahead of the advancing flame front.

(*Lead Susceptibility*)—The increase in octane number of gasoline imparted by the addition of a specified amount of tetraethyl lead.

(*Low-Sulfur Crude Oil*)—Crude oil containing low concentrations of sulfur-bearing compounds. Crude is usually considered to be in the low-sulfur category if it contains less than 0.5% (weight) sulfur. Examples of low-sulfur crudes are offshore Louisiana, Libyan, and Nigerian crudes.

(*Lube Stock*)—Refinery term for fraction of crude petroleum suitable in terms of boiling range and viscosity to yield lubricating oils when further processed and treated.

(*Mercaptans*)—Compounds of sulfur having a strong, repulsive, garliclike odor. A contaminant of "sour" crude oil and products.

(*Presulfide*)—A step in the catalyst regeneration procedure which treats the catalyst with a sulfur-bearing material such as hydrogen sulfide or carbon bisulfide to convert the metallic constituents of the catalyst to the sulfide form in order to enhance its catalytic activity and stability.

(*Process Unit*)—A separate facility within a refinery, consisting of many types of equipment, such as heaters, fractionating columns, heat exchangers, vessels, and pumps, designed to accomplish a particular function within the refinery complex. For example, the crude processing unit is designed to separate the crude into several fractions, while the catalytic reforming unit is designed to convert a specific crude fraction into a usable gasoline blending stock.

(*Polymer*)—A product of polymerization of normally gaseous olefin hydrocarbons to form high-octane hydrocarbons in the gasoline boiling range. Polymerization is the process of combining two or more simple molecules of the same type, called monomers, to form a single molecule having the same elements in the same proportion as in the original molecule, but having different molecular weights. The combination of two or more dissimilar molecules is known as copolymerization—and the product is called a *copolymer*.

(*Raffinate*)—In solvent refining, that portion of the oil which remains undissolved and is not removed by the selective solvent.

(*Refinery Pool*)—An expression for the mixture obtained if all blending stocks for a given type of product were blended together in production ratio. Usually used in reference to motor gasoline octane rating.

(*Refluxing*)—In fractional distillation, the return of part of the condensed vapor to the fractionating column to assist in making a more complete separation of the desired fractions. The material returned is called *reflux*.

(*Residual Fuel Oils*)—Topped crude petroleum or viscous residuums obtained in refinery operations. Commercial grades of burner-fuel oils Nos. 5 and 6 are residual oils and include Bunker fuels.

(*Riser Cracking*)—Applied to fluid catalytic cracking units where the mixture of feed oil and hot catalyst is continuously fed into one end of a pipe (riser) and discharges at the other end where catalyst separation is accomplished after the discharge from the pipe. There is no dense phase bed through which the oil must pass because all the cracking occurs in the inlet pipe (riser).

(*Road Octane*)—A numerical value based upon the relative antiknock performance of an automobile with a test gasoline as compared with specified reference fuels. Road octanes are determined by operating a car over a stretch of level road or on a chassis dynamometer under conditions simulating those encountered on the highway.

(*SAE Numbers*)—A classification of motor, transmission, and differential lubricants to indicate viscosities, standardized by the Society of Automotive Engineers. They do not connote quality of the lubricant.

(*Solvent Extraction*)—The process of mixing a petroleum stock with a selected solvent, which preferentially dissolves undesired constituents, separating the resulting two layers, and recovering the solvent from the raffinate (the purified fraction) and from the extract by distillation.

(*Sour Crude*)—Crude oil which (1) is corrosive when heated, (2) evolves significant amounts of hydrogen sulfide on distillation, or (3) produces light fractions which require sweetening. Sour crudes usually, but not necessarily, have high sulfur content. Examples are most West Texas and Middle East crudes.

(*Stability*)—In petroleum products, the resistance to chemical change. Gum stability in gasoline means resistance to gum formation while in storage. Oxidation stability in lubricating oils and other products means resistance to oxidation to form sludge or gum in use.

(*Sweet Crude*)—Crude oil which (1) is not corrosive when heated, (2) does not evolve significant amounts of hydrogen sulfide on distillation, and (3) produces light fractions which do not require sweetening. Examples are offshore Louisiana, Libyan, and Nigerian crudes.

(*Tricresyl Phosphate*) (*TCP*)—Colorless to yellow liquid used as a gasoline and lubricant additive and plasticizer. Formula, $PO(OC_6H_4CH_3)_3$.

(*Tetraethyl Lead*) (*TEL*)—A volatile lead compound which is added in concentrations up to 3 milliliters per gallon to motor and aviation gasoline to increase the antiknock properties of the fuel. $Pb(C_2H_5)_4$.

(*Tetramethyl Lead*) (*TML*)—A highly volatile lead compound added to motor gasoline to reduce knock. May be used alone or in mixtures with TEL. $Pb(CH_3)_4$.

(*Thermal Cracking*)—A refining process which decomposes, rearranges, or combines hydrocarbon molecules by the application of heat without the aid of a catalyst.

(*Topped Crude*)—A residual product remaining after the removal, by distillation or other processing means, of an appreciable quantity of the more volatile components of crude petroleum.

(*Unsaturates*)—Hydrocarbon compounds of such molecular structure that they readily pick up additional hydrogen atoms. Olefins and diolefins, which occur in cracking, are of this type.

(*Vacuum Distillation*)—Distillation under reduced pressure, which reduces the boiling temperature of the material being distilled sufficiently to prevent decomposition or cracking.

(*Vapor Lock*)—The displacement of liquid fuel in the feed line and the interruption of normal motor operation, caused by vaporization of light ends in the gasoline. Vaporization occurs when the temperature at some point in the fuel system exceeds the boiling points of the volatile light ends.

(*Virgin Stock*)—Oil processed from crude oil which contains no cracked material. Also called straight-run stock.

(*Visbreaking*)—Lowering or breaking the viscosity of residuum by cracking at relatively low temperatures.

(*Yield*)—In petroleum refining, the percentage of product or intermediate fractions based on the amount charged to the processing operation.

(*Zeolitic Catalyst*)—Since the early 1960s, modern cracking catalysts contain a silica-alumina crystalline structured material called zeolite. This zeolite is commonly called a molecular sieve. The admixture of a molecular sieve in with the base clay matrix imports desirable cracking selectivities.

References

American Society for Testing and Materials: "Annual Book of ASTM Standards, Part 17, Petroleum Products," Philadelphia, Pennsylvania (Revised annually).

Considine, D. M., Editor: "Energy Technology Handbook," McGraw-Hill, New York (1977).

Dick, R. A., and S. P. Wimpfen: "Oil Mining," *Sci. Amer.*, **243**, 4, 182–188 (1980).

Evans, F. L., Jr.: "Equipment Design Handbook for Refineries and Chemical Plants," Gulf Publishing, Houston, Texas (1979).

Maisel, D. S.: "Trends in the Petrochemical Industry," *Chem. Eng. Progress*, **76**, 1, 17–23 (1980).

Moureau, M., and G. Brace: Dictionary of Petroleum Technology," Editions Technip, Paris, France (1979).

Watkins, R. N.: "Petroleum Refinery Distillation," 2nd Edition, Gulf Publishing, Houston, Texas (1979).

—The cooperation of the UOP Process Division (Des Plaines, Illinois); of the American Petroleum Institute; and of the National Petroleum Council in providing source material for parts of this entry is much appreciated.

PHASE RULE. The phase rule, due to Gibbs, gives the number F of intensive variables which can be fixed arbitrary in a system in equlibrium. This number is also called the variance or the number of degrees of freedom of the system. It is given by

$$F = 2 + (C' - R) - P$$

where C' is the number of components, R, the number of independent chemical reactions, and P, the number of phases.

In terms of the number of independent components, C, Equation (1) may also be written

$$F = 2 + C - P$$

If $F = 0$ the system is invariant. We cannot fix either temperature or pressure arbitrarily. Equilibrium can only be established at isolated points. An example is the *triple point* at which pure solid, liquid and vapor are in equilibrium ($F = 2 \times 1 - 3 = 0$).

If $F = 1$ the system is *monovariant*. We can, for example, fix the temperature, but the equilibrium pressure is then fixed. This is the situation for a system containing one component and two phases.

If $F = 2$ the system is *bivariant*. Within certain limits both pressure and temperature can be given arbitrarily. This is the situation for $C = 1$ and $P = 1$, or $C = 2$ and $P = 2$.

pH (Abrasion). The term *abrasion pH* was originated by Sevens and Carron in 1948 to designate the pH values obtained by grinding materials in water as a useful aid in the field identification of minerals. The pH value ranges from 1 for ferric sulfate minerals, such as coquimbite, konelite, and rhomboclase, to 12 for calcium-sodium carbonates, such as gaylussite, pirssonite, and shortite.

The recommended technique for determination of abrasion pH is to grind, in a nonreactive mortar, a small amount of the mineral in a few drops of water for about one minute. Usually, a pH test paper is used. Values obtained in this manner are given in the left-hand column of the accompanying table. Another method, proposed by Keller et al. in 1963 involves the grinding of 10 grams of crushed mineral in 100 milliliters of water and noting the pH of the resulting slurry electronically. Values obtained in this manner are given in the righthand column of the accompanying table.

pH (Hydrogen Ion Concentration). A measure of the effective acidity or alkalinity of a solution. It is expressed as the negative logarithm of the hydrogen-ion concentration. Pure water has a hydrogen ion concentration equal to 10^{-7} moles per liter at standard conditions. The negative logarithm of this quantity is 7. Thus, pure water has a pH value of 7. The pH scale usually is considered as extending from 0 to 14. When a strong acid

ABRASION pH VALUES OF REPRESENTATIVE MINERALS

Mineral	pH by Stevens-Carron Method	pH by Keller et al. Method*
Coquimbite	1	
Melanterite	2	
Alum	3	
Glauconite	5	5.5*
Kaolinite	5, 6, 7	5.5*
Anhydrite	6	
Barite	6	
Gypsum	6	
Quartz	6, 7	6.5
Muscovite	7, 8	8.0
Calcite	8	8.4
Biotite	8, 9	8.5
Microcline	8, 9	8.0 9.0*
Labradorite		8.0 9.2*
Albite	9, 10	
Dolomite	9, 10	8.5
Hornblende	10	8.9
Leucite	10	
Diopside	10, 11	9.9
Olivine	10, 11	9.6*
Magnesite	10, 11	

* indicates more recent values published in literature.

fully dissociates (or ionizes) in water, a 1 N solution of this acid will have a pH value of 0.0. Conversely, a 1 N base fully ionized in water will have a pH value of 14. Both hydrochloric acid and sodium hydroxide come close to meeting these stipulations. Because of the logarithmic nature of the pH scale, there is a tenfold change in hydrogen- and hydroxyl-ion concentration per unit change of pH. Thus, a slightly acidic solution having a pH of 6 will contain ten times as many active hydrogen ions as a solution of pH 7. See also **pK**.

Effective acidity or alkalinity is stressed in pH measurement—not the total hydrogen present. Sulfuric acid and boric acid both contain significant amounts of hydrogen. Nearly all the hydrogen in sulfuric acid dissociates in the presence of sufficient water to become free hydrogen ions. On the other hand, when boric acid is added to water, it dissociates very little into free hydrogen ions. The pH of a 0.1 N sulfuric acid solution will be about 1.3 whereas for the same concentration, boric acid will have a pH of about 5.3. Thus, sulfuric acid is called a strong acid; boric acid a weak acid. In all materials, of course, dissociation increases with temperature, thus the same solution will have a somewhat different pH at a lower temperature than at a higher temperature. Pure water is neutral at a temperature of 25°C, having a concentration of 1×10^{-7} hydrogen ions and 1×10^{-7} hydroxyl ions and, consequently, a pH of 7. Dissociation is less at 0°C, at which temperature the hydrogen-ion concentration is 0.34×10^{-7}, or a pH of 7.47 (slightly basic rather than neutral). But, at a temperature of 100°C, dissociation is greater. The hydrogen ion concentration is 8×10^{-7} and the pH is 6.10 (or slightly acid). The pH of various substances is given in accompanying table.

Buffer solutions can be added to resist changes in pH despite the addition of acid or base to the solution. This is explained under **Buffer (Chemical)**.

pH is measured in two basic ways: (1) colorimetrically, usually where high accuracy is not required and manual methods suffice; and (2) electrometrically. Color changes are based upon various

pH VALUES OF VARIOUS SUBSTANCES
(at 25°C)

MATERIAL	pH
Seawater	7.75 to 8.25
Soils	3 to 10
Plant tissues and fluids	About 5.2
Animal tissues and fluids	About 7.0 to 7.5
Blood	7.35–7.5
Urine	5.0–7.0
Milk	6.5–7.0
Gastric juice	1.7
Pancreatic juice	7.8
Intestinal juice	7.7
Internal tissue fluids:	
Minimum, below which acidosis ensues	7.0
Maximum, above which tetany ensues	7.8
Hydrochloric acid (1N)	0.1
Hydrochloric acid (0.1N)	1.08
Hydrochloric acid (0.001N)	3.00
Sulfuric acid (1.0N)	0.32
Sulfuric acid (0.1N)	1.17
Acetic acid (1N)	2.37
Lemon juice	2.0–2.2
Acid fruits	3.0–4.5
Fruit jellies	3.0–3.5
Sodium hydroxide (1N)	13.73
Sodium hydroxide (0.1N)	12.84
Ammonia (10% NH_3)	11.8
Limewater, $Ca(OH)_2$ saturated	12.4
Trisodium phosphate, 2%	11.95

organic dyes which alter their color within a relatively narrow range of pH values. Numerous dyes are required to cover the full pH range. Electrometric methods are used both in the laboratory and on-line for process control. They are continuous and easily adapted to automatic control systems. The possibility that a thin glass membrane of special composition could develop a potential in relation to hydrogen ion concentration was described as early as 1909 by the German chemist, Fritz Haber. Little progress was made until the middle-1920s. Glass electrodes are now the standard approach to electrometric pH measurement, after periods of trial with quinhydrone and antimony electrodes. The glass electrode responds in a predictable fashion throughout the 0 to 14 pH range, developing 59.2 millivolts per pH unit at 25°C, values which are consistent with the classical Nernst equation. Contrary to earlier pH electrodes, the glass electrode is not influenced by oxidants or reductants in solution. With suitable temperature compensation, pH measurements can be made up to 100°C and higher. In pH measurement, a second or reference electrode is required to complete the circuit. After trials with numerous electrodes (the hydrogen electrode is the standard) for practical plant and laboratory applications, the mercury-mercurous chloride (calomel) electrode is widely used. There is also some use of the silver-silver chloride reference electrode.

pH control systems are widely used in waste control and neutralization systems, in pulp and paper manufacture, in food processing, and in the manufacture of numerous organic chemicals. pH measurement is very important in the medical field.

PHENACITE. The mineral phenacite is a beryllium silicate corresponding to the formula Be_2SiO_4. It is hexagonal but the crystals are usually rhombohedral in habit. It has a conchoidal fracture; is brittle, hardness, 7.5–8; specific gravity, 3; luster, vitreous; colorless to yellowish or reddish, sometimes brown; transparent to translucent. Phenacite is found in pegmatites with topaz, quartz and microcline, and occurs also in emerald-bearing mica schists of the Ural Mountains. It is found also in France, Norway, Switzerland, Africa, Brazil, and Mexico; and in the United States, in Oxford County, Maine; Carroll County, New Hampshire; and in Chaffee and El Paso Counties in Colorado. It derives its name from the Greek meaning *deceiver*, as it resembles quartz and topaz with which it is associated. It is sometimes spelled *phenakite*. It has been used as a gem.

PHENOL. (1) A class of aromatic organic compounds in which one or more hydroxy groups are attached directly to the benzene ring. Examples are phenol itself (benzophenol), the cresols, xylenols, resorcinol, naphthols. Although technically alcohols, their properties are quite different. (2) Phenol (carbolic acid; phenylic acid; benzophenol; hydroxybenzene), C_6H_5OH.

Phenol is a white, crystalline substance which turns pink or red if not perfectly pure, or if under influence of light; absorbs water from the air and liquefies; distinctive odor; sharp burning taste. Toxic by ingestion, inhalation, and skin absorption. Strong irritant to tissue. When in a very weak solution, phenol has a sweetish taste; specific gravity 1.07; mp 42.5–43°C; bp 182°C; flash point 77+°C. Soluble in alcohol, water, ether, chloroform, fixed or volatile oils, and alkalies.

Most of the phenol used in the United States is made by the oxidation of cumene, yielding acetone as a byproduct. The first step in the reaction yields cumene hydroperoxide, which decomposes with dilute sulfuric acid to the primary products, plus acetophenone and phenyl dimethyl carbinol. Other processes include sulfonation, chlorination of benzene, and oxidation of benzene. The compound is purified by rectification.

Major uses of phenol include phenolic resins, epoxy resins, 2,4-D (regulated in many countries); as a selective solvent for refining lubricating oils; in the manufacture of adipic acid, salicylic acid, phenolphthalein, pentachlorophenol, acetophenetidine; picric acid; germicidal paints, and pharmaceuticals, as well as use as a laboratory reagent. Special uses include dyes and indicators, and slimicides.

High-boiling phenols are mixtures containing predominantly meta-substituted alkyl phenols. Their boiling point ranges from 238 to 288°C; they set to a glass below −30°C. They are used in phenolic resins, as fuel-oil sludge inhibitors, as solvents, and as rubber chemicals.

PHENOLIC RESINS. These resins have been known since the 1870s when Baeyer first investigated the reactions of phenols and aldehydes. However, his findings were not commercially utilized until Dr. L. H. Baekeland disclosed his classic work in 1907. Through his use of high-pressure molding, he provided a solution to the problem of making quick-curing moldings which did not blister or crack. Contemporary with Dr. Baekeland was the work of Lebech and Aylsworth who provided the key to the application of large commercial quantities of phenolic Novolacs by suggesting the use of hexamethylenetetramine as a curing agent. Since that time, and because of their desirable price-to-property relationship, phenolic resins have enjoyed steady growth despite the encroachment into their areas of application by a few thermoplastics and other thermosetting materials.

Although in the pure state phenolic resins are quite weak and brittle, they are highly regarded among the plastic materials as being capable of producing very strong physical bonds with a large variety of materials at very low concentration. Consequently, phenolics have found use in many applications as binders. In addition to the strength they impart as bonding agents in matrixes, phenolic resins also possess resistance to chemical attack by all but the most polar organic solvents. The only

inorganic reagents that have a deleterious effect on them are the strongest and most oxidizing of the acids and the strongest bases.

For many years prior to the development of high-temperature thermoplastics and thermosets, such as the polyimides, polysulfones, and epoxies, phenolic molding material dominated the high temperature-resistant market. This emphasizes their ability to resist temperature degradation in the 400–500°F (204–260°C) range. Because phenolics were found to possess excellent ablative properties, it has been reported that both the American and Soviet space efforts used them in combination with certain other polymeric compounds in heat shield materials. Phenolics, like other aromatic hydrocarbon-based resins, possess excellent resistance to high-energy radiation degradation.

Unlike most thermoplastics, phenolic resins and moldings are characterized by high flexural modulus and good tensile strength while having relatively low impact resistance. In addition to their good physical strength properties, phenolic resins are used in the manufacture of many electrical devices where high dielectric breakdown strength and electrical resistance, in combination with excellent dimensional stability, are required.

Chemistry of Phenolics. Phenolic compounds are capable of chemically combining with a large number of aldehydes and other compounds to yield an almost infinite spectrum of modified polymers. However, the reaction of a phenol with an alde-

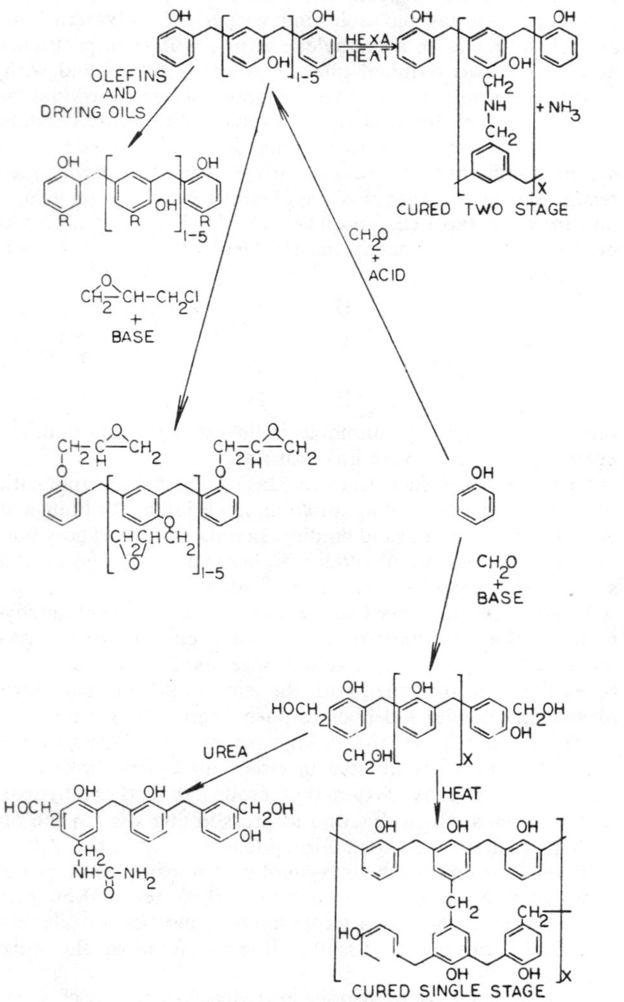

Cured single-stage and two-stage resins.

hyde (most commonly encountered is that between phenol and formaldehyde) leads to the formation of only two classes of phenolic resins. These are Novolacs and resols. In general, these two classes of resins may be differentiated by the fact that Novolacs are prepared with an acid catalyst and substantially less than one mole of aldehyde per mole of phenol and require the addition of a curing catalyst to become thermosetting; while resols, or single-stage resins as they are commonly called, are prepared with from 1 to 3 moles of aldehyde per mole of phenol and employ a basic condensation catalyst, and are inherently thermosetting.

Novolacs. The aldehyde content of Novolac resins is insufficient to render the resin thermosetting, hence, they are true thermoplastics provided that a curing agent is not added. Novolacs may be stored indefinitely in the pulverized state at moderate temperatures even mixed with curing agent.

The final cure speed of simple Novolacs may be accurately controlled by the use of the proper condensation catalyst in the initial phase of the reaction. Structurally, a Novolac consists of a series of phenol nuclei joined by methylene

$$
\begin{array}{c}
H \\
| \\
(-C-) \\
| \\
H
\end{array}
$$

links at the *o* and *p* positions. Only two of the three possible *o* and *p* positions on each ring within the polymeric chain are substituted with a methylene group. Only one position on each of the two terminal phenol groups is substituted with a methylene group, hence, two positions on each terminal ring and one on each internal ring is available for future reactions, including the curing reactions. When the unsubstituted positions are predominantly the para positions, very fast-curing resins result. Slower-setting resins are obtained when the unoccupied positions are the ortho positions. When hexa is used as the hardener at approximately the 10% level, the reactive sites are joined by a

$$
\begin{array}{ccc}
H & & H \\
| & & | \\
-C-&N-&C- \\
| & & | \\
H & & H
\end{array}
$$

linkage. One mole of ammonia is liberated for approximately every three of the above links formed.

Unaltered Novolacs and two-stage resins* find application in grinding wheel bonding, molding material, brake linings and clutch faces, foundry sand binding, premix and wood fiber bonding, thermal insulation. Modified phenolics are found in adhesives, coatings, and aerospace applications.

Resols. On the other hand, resols contain sufficient aldehyde to make them thermosetting without a curing agent. Consequently, they have only finite storage stability and care must be exercised to minimize both the length and the temperature of storage. The high aldehyde-to-phenol ratios used in the preparation of resols insure that a high percentage of the reactive *o* and *p* positions are utilized in either methylene links or are substituted by a hydroxymethyl group. It is these hydroxymethyl groups which function as crosslinking sites in the final curing reaction. In phenolformaldehyde resols, the ratio of methylene groups to hydroxymethyl groups is an important factor in determining the solubility of these resins. Low ratios insure water solubility in almost infinite proportions, while resins with high ratios can be dissolved in only low molecular weight

* By convention, two-stage resins are defined as a mixture of Novolac and curing agent, which is capable of thermosetting.

alcohols, ketones, ethers, and esters. Solubility of resols in nonpolar solvents, such as hydrocarbons, is always very low.

In addition to acting as a crosslinking site, the hydroxymethyl group and unsubstituted *o* and *p* positions may be used as reactive sites to join numerous other compounds to the phenolic polymer. These modified resins often possess many properties normally not attributable to phenolics in the unaltered state.

Resols find application as impregnating resins, in laminating paper, cloth, glass and asbestos; as a pickup agent in grinding wheels, exterior and marine plywood, premix and granular molding material, adhesives, wood waste and particle board manufacture, and in coatings.

The final curing step of both classes of phenolics is accomplished by exposing the resin or the resin containing matrix to temperatures in the 190–450°F (88–232°C) range for an appropriate length of time to render the resin infusible. In most applications high pressure is also applied concurrently with the heating cycle to eliminate blistering, which would normally occur if the trapped gases (ammonia in the case of Novolacs, and water in the case of resols) generated in the cure were allowed to escape unrestrained.

Comparative infrared analysis of the two classes of resins in their pure state show that only resols have strong absorptions in the 1,000 and 880 cm^{-1} range. As Novolacs require a curing agent, their presence may be inferred by a strong, sharp absorption at 510 cm^{-1} which is indicative of the most common curing agent used, hexamethylenetetramine. Unfortunately, hexa also has strong absorptions at or near 1,000 and 880 cm^{-1} which makes differentiation between resols and two-stage resins, by infrared spectroscopy, possible only for the experienced. In the cured state, it is very difficult to determine the class identity of an unknown sample.

—Phillip A. Waitkus, Plastics Engineering Company, Sheboygan, Wisconsin.

PHENYLALANINE. Amino Acids.

PHENYLBUTAZONE. Analgesics.

PHILLIPSITE. The mineral phillipsite is a zeolite, a hydrous silicate of potassium, calcium, and aluminum, corresponding to formula $(K,Na_2Ca)(Al_2Si_4)O_{12} \cdot 4-5H_2O$. It is monoclinic, forming pentration twins, and sometimes crosses resembling orthorhombic or tetragonal forms. It also may occur in radial groups. Phillipsite is a brittle mineral; hardness, 4–4.5; specific gravity, 2.2; luster, vitreous; color, white to light red; translucent to opaque. Like other zeolites, it is found in veins and cavities in basalts, and sometimes in more acidic rocks. It is believed to be a low-temperature mineral. Phillipsite is found in Italy, especially in the lavas of Vesuvius and Monte Somma, and in the basalts of Germany, Ireland, and Australia. It has been reported from Greenland. This mineral was named in honor of the British mineralogist William Phillips.

PHLOGOPITE. The mineral phlogopite is a magnesium-bearing mica, with but little iron, corresponding essentially to the formula $K(Mg, Fe)_3(AlSi_3)O_{10}(F, OH)_2$. Fluorine is sometimes present. This mica is monoclinic like muscovite, biotite and lepidolite, forming prismatic crystals, occasionally very large, and occurring also in scales and plates. Its cleavage is basal and highly perfect with elastic laminae; hardness, 2–2.5; specific gravity, 2.76–2.90; luster, pearly to submetallic; color, yellowish-brown, green, white and colorless; transparent to translucent; may exhibit asterism, probably due to minute inclusions. Phlogopite is more nearly a characteristic of metamorphic than igneous rocks although occasionally occurring in the latter if they are rich in magnesia and with but little iron. Phlogopite is found especially in Rumania, Switzerland, Italy, Finland, Sweden and the Malagasy Republic, where it occurs in the crystalline limestones in huge crystals. In the United States it occurs in New York State at Edwards, Hammond, DeKalb, Monroe and, in New Jersey, at Franklin. In Canada it is found at many places in Ontario and Quebec.

The name phlogopite comes from the Greek word meaning like fire, referring to the copper-like reflections often observed in the reddish-brown varieties.

Phlogopite is in demand commercially by the electrical industry for use as an insulator.

PHOSGENE. Carbonyls; Chlorinated Organics.

PHOSGENITE.
This mineral is a chlorocarbonate of lead $Pb_2(CO_3)Cl_3$, crystallizing in the tetragonal system, associated with other lead minerals of secondary origin, e.g. cerussite and anglesite. Hardness, 2–3; specific gravity 6.133; prismatic to tabular crystals, also massive and granular; adamantine luster; color, white, gray, brown, green, or pink; transparent to translucent. Some specimens show yellowish fluorescence under ultraviolet light.

Found in the United States in California, Colorado, Arizona, and New Mexico. Magnificent crystals up to 5 inches (12.5 centimeters) in diameter have been found at Monte Poni, Sicily; as fine crystals in England at Derbyshire and Matlock, and in Poland, U.S.S.R., Tasmania, Australia, and Tsumeb, Southwest Africa.

PHOSPHATE FERTILIZERS. Fertilizer.

PHOSPHATES. Phosphorus.

PHOSPHATING. Conversion Coatings.

PHOSPHINE. Phosphorus.

PHOSPHOLIPIDS.
These compounds belong to a group of fatty acid compounds sometimes referred to as *complex lipids*. The simplest are esters of fatty acids with glycerol phosphate and are called *phosphatidic acids*. There are also phosphatidylcholines or lecithins, phosphatidylethanolamines, phosphatidylserines, and phosphatidylinositols. The latter may have one or more additional phosphate groups attached to the inositol. A similar series also exists containing an aldehyde attached to the 1-position of the glycerol, in the form of an alpha, beta-unsaturated ether. These are commonly referred to as *plasmalogens*.

The percentage of phospholipid content of tissues varies but little under normal physiological conditions, thus giving rise to the term *element constant*, in contrast to the triglycerides, which have been called the *element variable*.

Phospholipids are considered to be involved in the transport of triglycerides through the liver, especially during mobilization from adipose tissue. Conditions which could be interpreted as interfering with phosphatidylcholine formation, such as deficiency of choline or its precursors, result in a pronounced increase in liver triglycerides.

Mitochondrial phospholipids play a role in electron transport and oxidative phosphorylation, two mechanisms by which the cell accomplishes the final oxidation of the metabolites to produce energy. Phospholipids also are linked in the transport of ions, especially sodium, across membranes.

In summary, phospholipids (phosphatides) comprise a group of lipid compounds that yield, upon hydrolysis, phosphoric acid, an alcohol, fatty acid, and a nitrogenous base. They are widely distributed throughout nature.

PHOSPHOR BRONZE. Copper.

PHOSPHORESCENCE. Luminescence.

PHOSPHORIC ACID.
Generally the term *phosphoric acid* refers to orthophosphoric acid, H_3PO_4. Anhydrous orthophosphoric acid is a white, crystalline solid, which melts at 42.35°C. It forms a hemihydrate, $2H_3PO_4 \cdot H_2O$, which melts at 29.32°C. Although it is possible to produce almost any desired concentration, it is common practice to supply the material as a solution containing from 75% H_3PO_4 (melting point = 17.5°C) to 85% H_3PO_4 (melting point = 21.1°C). When phosphoric acid is heated to temperatures above about 200°C, water of constitution is lost. A series of acids is formed by the dehydration, ranging from pyrophosphoric acid, $H_4P_2O_7$, to metaphosphoric acid, $(H_3PO_4)_n$. Salts of the dehydrated acids are used for the preparation of certain types of liquid fertilizers and have been used in some detergents. However, to counter the effects of "phosphate pollution," there has been a serious cutback in this latter use of the phosphates. See also **Fertilizers**.

One, two, or three of the hydrogens in phosphoric acid may be neutralized, leading to a series of products which range widely in their hydrogen ion concentration (pH): Monosodium phosphate, NaH_2PO_4 with a pH of 4.0; disodium phosphate, Na_2HPO_4 with a pH of 8.3 (approximate); and trisodium phosphate, Na_3PO_4 with a pH of 12.0. Other phosphorous acids of little commercial importance are hypophosphorous acid, H_3PO_2; orthophosphorous acid, H_3PO_3; and pyrophosphorous acid, $H_4P_2O_5$.

Manufacture of Phosphoric Acid. The major sources of H_3PO_4 traditionally have been mineral deposits of phosphate rock. Mining operations are extensive in a number of locations, including the United States (Florida), the Mediterranean area, and the U.S.S.R., among others. The major constituent of most phosphate rocks is fluorapatite, $3Ca_3(PO_4)_2 \cdot CaF_2$. The supply of high-grade phosphates, the raw material of choice for producing high-purity phosphoric acid by the wet process, is rapidly decreasing in some areas.

Two major methods are utilized for the production of phosphoric acid from phosphate rock. The *wet process* involves the reaction of phosphate rock with sulfuric acid to produce phosphoric acid and insoluble calcium sulfates. Many of the impurities present in the phosphate rock are also solubilized and retained in the acid so produced. While they are of no serious disadvantage when the acid is to be used for fertilizer manufacture, their presence makes the product unsuitable for the preparation of phosphatic chemicals.

In the other method, the *furnace process*, phosphate rock is combined with coke and silica and reduced at high temperature in an electric furnace, followed by condensation of elemental phosphorus. Phosphoric acid is produced by burning the elemental phosphorus with air and absorbing the P_2O_5 in water. The acid produced by this method is of high purity and suitable for nearly all uses with little or no further treatment.

Basic reactions of the wet process are

$$3Ca_3(PO_4)_2 \cdot CaF + 10H_2SO_4 + 20H_2O \rightarrow$$
$$10CaSO_4 \cdot 2H_2O + 6H_3PO_4 + 2HF.$$

Numerous side reactions also occur. Phosphate rock and sulfuric acid, together with recycled weak liquors, are carefully metered to a large, stirred reactor, providing retention for 4–8 hours. Conditions in the reaction are carefully controlled to maintain preselected conditions. Temperatures (77–83°C) are controlled by removing excess heat of reaction with a vacuum cooler, or by blowing air through the phosphoric acid slurry. The slurry contains precipitated gypsum and is sent to a filter. The gypsum is washed with water in several countercurrent steps, and weak liquor is returned to the reaction stage. For most uses, the acid requires further concentration, normally done in vacuum evaporators. Merchant-grade acid is generally concentrated to about 54% P_2O_5 (75% H_3PO_4).

Effluents and gypsum disposal pose problems. Fluorine is evolved at various steps in the process and scrubbers are required to reduce release to the atmosphere. Gypsum is frequently piled in diked areas or dumped into abandoned mines. Waste water from these plants is heavily contaminated with fluorine, phosphates, sulfates, and other compounds. It is commonly impounded in large ponds, where a portion of the contaminants may precipitate or be lost by other processes. The cooled effluent from the ponds is recycled to the production unit. Any excess water must be treated with lime before it can be allowed to enter streams.

Developments of recent years include plants designed to precipitate the calcium sulfate in the form of the hemihydrate instead of gypsum. In special cases, hydrochloric acid is used instead of sulfuric acid for rock digestion, the phosphoric acid being recovered in quite pure form by solvent extraction. Solvent-extraction methods have also been developed for the purification of merchant-grade acid, which normally contains impurities amounting to 12–18% of the phosphoric acid content. Processes for recovering part of the fluorine in the phosphate rock are in commercial use.

Although more costly to operate, the electric-furnace process produces phosphoric acid of high purity. A mixture of coke, silica, and phosphate rock is formed into nodules by heating in a nodulizing kiln, and the resulting lump material is transferred to the electric furnace, where it is heated with an electric current introduced by means of graphite electrodes. The entire charge is melted, and elemental phosphorus is volatilized. The slag is tapped off intermittently while the phosphorus vapor is condensed. The phosphorus is then burned in air and the P_2O_5 is absorbed in water. Reactions are

$$2Ca_3(PO_4)_2 + 6SiO_2 + 10C \rightarrow P_4 + 10CO + 6CaSiO_3$$
$$P_4 + 5O_2 \rightarrow 2P_2O_5$$
$$P_2O_5 + 3H_2PO \rightarrow 2H_3PO_4.$$

References

Bergdorf, J., and R. Fischer: "Extractive Phosphoric Acid Purification," *Chem. Eng. Progress*, **74**, 11, 41–45 (1978).

Blumrich, W. E., Koening, H. J., and E. W. Schwehr: "The Fisons HDH Phosphoric Acid Process," *Chem. Eng. Progress*, **74**, 11, 58–60 (1978).

Bridger, G. L., and A. H. Roy: "Producing Acid from Low-Grade Phosphatic Materials," *Chem. Eng. Progress*, **74**, 11, 62–65 (1978).

Djololian, C., and D. Billaud: "Absorbing Fluorine Compounds from Waste Gases," *Chem. Eng. Progress*, **74**, 11, 46–51 (1978).

Kleinman, G.: "Double Effect Evaporation of Crude Phosphoric Acid," *Chem. Eng. Progress*, **74**, 11, 37–40 (1978).

Olivier, P. L., Jr.: "Wet Process Acid Production," *Chem. Eng. Progress*, **74**, 11, 55–57 (1978).

Rushton, W. E.: "Defluorination of Wet Process Acid," *Chem. Eng. Progress*, **74**, 11, 52–54 (1978).

PHOSPHORIC ACID (Fuel Cell). Fuel Cells.

PHOSPHORS AND PHOSPHORESCENCE. A large variety of substances become luminescent when stimulated or excited by suitable radiation, or by emissions, such as cathode rays or beta-rays. This phenomenon is complex and exhibited in various aspects. In some cases, the light is emitted only so long as the exciting emission is maintained, in which case it is called *fluorescence*.

In other cases, the luminescence persists after the excitation is removed and it is then called *phosphorescence*. It has long been known, for example, that zinc sulfide, under certain conditions, glows brightly for a time after exposure to daylight or lamplight, but the luminosity decays rapidly and disappears, usually within a few minutes. The electroluminescent phosphor of zinc sulfide-zinc selenide-copper has the property that the wavelength of the emitted radiation increases with increasing selenium content. The white luminescence of some television tubes is obtained from a combination of cadmium-zinc sulfide phosphors, one that is blue-emitting and the other yellow-emitting. Also, in some color television tubes, the blue-emitting and green-emitting phosphors are of the sulfide type, but earlier use of sulfides for the red-emitting "dots" on the tube surface were replaced by rare-earth red-emitting phosphors. One composition used is prepared by combining about 4% europium oxide and 65% yttrium oxide, with various vanadium compounds and calcining the mixture. Rare-earths also have been used in producing phosphors for high-pressure mercury-arc lamps. These phosphors increase the proportion of red light emitted by reducing the green, blue, and ultraviolet portions.

Quantitatively, phosphorescence may be defined as luminescence that is delayed by more than 10^{-8} seconds after excitation. It may be associated with transitions from a higher excited state to a lower one, the energy going into a radiationless rearrangement of the system. If the lower state is metastable, its lifetime may be considerable before it finally decays by a highly forbidden radiative transition to the ground state. In the case of zinc sulfide, the process depends upon the ionization of activator atoms, the freed electrons being trapped and only released slowly for recombination.

See also **Luminescence.**

PHOSPHORUS. Chemical element symbol P, at. no. 15, at. wt. 30.9738, periodic table group 5a, mp 44.1°C (α-white), bp 280°C (α-white), sp gr 1.82 (white), 2.20 (red). Four allotropes of phosphorus are known, the hexagonal β-white, stable only below −77°C, the cubic α-white, the violet, and the black (which is thermodynamically the most stable). The α-white form is usually taken as the standard state. The violet is obtained by continued heating at 500°C of a solution of phosphorus in lead. When α-white phosphorus is heated to 250°C in the absence of air, a red variety (mp 590°C) is obtained. This is believed to consist of a mixture of the α-white and the violet allotropes, although the studies of the violet component in the mixture have shown that at least four polymorphic forms of red (violet) phosphorus exist.

White phosphorus is considered to be made up largely of P_4 molecules, as are the liquid and vapor up to 800°C, where dissociation becomes appreciable. The P_4 molecule is a tetrahedron, with single covalent bonds between the P atoms, and each having an unshared pair of electrons. White phosphorus is much more reactive than red or violet. Black phosphorus has a graphitelike structure and has a similar electrical conductivity.

There is one stable nuclide ^{31}P. Six radioactive isotopes have been identified ^{28}P through ^{30}P and ^{32}P through ^{34}P, all with short half-lives, measured in terms of seconds, minutes, or days. See also **Radioactivity.** In terms of terrestrial abundance, phosphorus ranks 10th with an estimated average content of igneous rocks being 0.13% phosphorus. The element ranks 19th in abundance in seawater, there being an estimated 325 tons of phosphorus per cubic mile (70 metric tons per cubic kilometer) of seawater. In terms of cosmic abundance, phosphorus is ranked 15th among the elements. The element was first identified by Hennig Brandt in Germany in 1669 during an experiment in which he was distilling urine with sand and coal. White phosphorus is very toxic.

First ionization potential 11.0 eV; second, 19.81 eV; third, 30.04 eV; fourth, 51.1 eV; fifth, 64.698 eV. Oxidation potentials $H_3PO_2 + H_2O \rightarrow H_3PO_3 + 2H^+ + 2e^-$, 0.59 V; $P + 3H_2O \rightarrow H_3PO_3 + 3H^+ + 3e^-$, 0.49 V; $P + 2H_2O \rightarrow H_3PO_2 + H^+ + e^-$, 0.29 V; $H_3PO_3 + H_2O \rightarrow H_3PO_4 + 2H^+ + 2e^-$, 0.20 V; $PH_3(g) \rightarrow P + 3H^+ + 3e^-$, 0.04 V; $P + 2OH^- \rightarrow H_2PO_2^- + e^-$, 1.82 V; $P + 5OH^- \rightarrow HPO_3^{2-} + 2H_2O + 3e^-$, 1.71 V; $H_2PO_2^- + 3OH^- \rightarrow HPO_3^{2-} + 2H_2O + 2e^-$, 1.65 V; $HPO_3^{2-} + 3OH^- \rightarrow PO_4^{3-} + 2H_2O + 2e^-$, 1.05 V; $PH_3(g) + 3OH^- \rightarrow P + 3H_2O + 3e^-$, 0.87 V.

Other physical characteristics of phosphorus are given under **Chemical Elements.**

Because of its reactivity, phosphorus does not occur in nature in the elemental form. Phosphate rock is the principal source of phosphorus and phosphorus compounds. Very large deposits of phosphate rock occur and are worked in the Bone Valley area of Florida, as well as deposits in Tennessee, Idaho, and South Carolina. Large deposits are mined in Northern Africa (Morocco and Tunisia). Very significant reserves have been found on several of the Pacific Islands, the reserves on Christmas Island estimated at some 30 million tons (270 million metric tons) and those on Nauru Island in excess of 100 million tons (90 million metric tons). There also are large active mining operations in the Mediterranean area as well as in the Soviet Union. Known reserves assure a supply for several centuries. The mineral apatite $Ca_3(PO_4)_2 \cdot CaCl_2$ or CaF_2 found in Quebec, Virginia, Brazil, and the South Pacific also contains high percentages of phosphorus, up to 20% P_2O_5. The main constituents of most phosphate rocks is fluorapatite $3Ca_3(PO_4)_2 \cdot CaF_2$. These rocks contain from 30–37% P_2O_5.

Most phosphorus raw materials are converted into phosphorus and phosphorus compounds, such as phosphoric acid on an extremely high-tonnage basis. Percentagewise, relatively little elemental phosphorus is produced for consumption as an end-product. See also **Fertilizer.** Phosphorus is important both to plant and animal nutrition. Traditionally, phosphorus compounds have been key components of cleaning compounds and detergents although there have been trends to reduce or eliminate phosphates from high-consumption items. See also **Detergents.**

Production of Elemental Phosphorus. The tricalcium phosphate in phosphate rock, mixed with coke and silica, is thermally reduced to yield P_2 vapor. The phosphorus vapors condense to a liquid and the carbon monoxide produced is returned for burning in the furnace. The process requires much heat and, in addition to the heat provided by the combustion of the coke and the heating value of the recycled carbon monoxide, an electric arc also is used. The reaction takes place in very large furnaces at a temperature of 1,300–1,500°C and at atmospheric pressure. A 70-MW furnace will produce 44,000 short tons 37,600 metric tons) of P_4 per year, equivalent to 100,000 tons (90,000 metric tons of P_2O_5 (if converted to acid). Although there are numerous intermediate and side reactions, the overall reaction is: $Ca_3(PO_4)_2 + 5C + 3SiO_2 \rightarrow P_2 + 5CO + 3Ca \cdot SiO_3$. Byproduct ferrophosphorus alloy and calcium silicate slag are tapped from the furnace periodically. Maximum furnace efficiency occurs when the SiO_2/CaO weight ratio is about 0.8. This ratio also assures a minimum melting-point eutectic for the melt and thus lengthens furnace life. This process was originally developed by Readman (England) in 1888. The first 1500-

kW furnace in the United States was installed in Niagara Falls, New York in 1896 because of the availability of low-cost hydro-power at that site. For many years, the proximity of Tennessee brown stone (a phosphate rock) to the low-cost power of the Tennessee Valley Authority made a good economic combination. Worldwide production of phosphorus by this process is about 0.75 billion short tons (0.675 billion metric tons) per year (installed capacity). In the United States, about 80% of the phosphorus produced is immediately converted to the oxide and thence to phosphoric acid. The remaining 20% has gone into alloys, organic intermediates for oil and fuel additives, pesticides, and plasticizers, and pyrotechnics. In addition to use in detergents, cleaning compounds, and degreasing formulations, phosphoric acid has been consumed in the preparation of liquid fertilizers, water-treatment, pharmaceutical, and chemical products. Phosphorus-containing fertilizers, such as single superphosphate, wet-process orthophosphoric acid, triple superphosphate, ammonium phosphate, and nitrophosphates do not require elemental phosphorus (or the resulting pure P_2O_5) in their preparation, but are manufactured by directly reacting phosphate rock with requisite chemicals, such as H_2SO_4 or HNO_3. See also **Fertilizer.**

Chemistry and Compounds. Like carbon, phosphorus is covalently bound to its neighboring atoms in all of its compounds, except perhaps for some metallic phosphides. Indeed, the chemistry of carbon and that of phosphorus are somewhat similar as might be expected from the diagonal relationship of these elements in the periodic table.

Probably the major difference between carbon and phosphorus is that the former element is quite closely restricted to the use of s- and p-orbitals, because of the relatively high energy of d-orbitals in the caseof second period elements; whereas, phosphorus, being a third period element, can use d-orbitals in bonding. For both carbon and phosphorus, the most common hybridization for σ-bonding is approximately the tetrahedral sp^3. However, in order to form π-bonds, carbon must go to lower hybrids: sp^2 and sp. Phosphorus, on the other hand, does not do this but can employ d-orbitals for π-bonding. This difference between carbon and phosphorus in sigma bond strength and the ease with which phosphorus uses its d-orbitals for attachment of attacking nucleophilic groups, can be used to explain why catenation is common in carbon compounds, while at the same time phosphorus compounds containing long chains of connected phosphorus atoms have not yet been synthesized.

The known coordination numbers exhibited by phosphorus within the molecule-ions containing this element are 1, 3, 4, 5, and 6 which, to at least a first approximation, exhibit the symmetry of p, p^3, sp^3, sp^3d and sp^3d^2 hybridization, respectively. A very large number (several thousand in each case) of triply- and quadruply-connected phosphorus compounds are known; but there are only a few compounds of higher coordination number in which d-orbitals are involved in the σ-bond base structure. These are the halogen compounds PF_5, PCl_5, PBr_5, PCl_2F_3, PBr_2F_3 and the pentaphenyl compound $(C_6H_5)_5P$, in which the phosphorus is quintuply-connected to its neighboring atoms, and the PF_6^- and PCl_6^- anions, with which the phosphorus has a six-fold coordination. The singly-connected phosphorus atoms appear only in compounds occurring at very high temperatures. Although singly-connected phosphorus is not known under ordinary conditions, interpretation of diatomic spectra has given considerable information.

Several generalities can be stated concerning the phosphorus compounds which are stable under normal conditions:

1. In those compounds in which phosphorus shares electrons

with three neighboring atoms, there are three σ-bonds, with little or no π-character, from the phosphorus.

2. In those compounds in which phosphorus shares electrons with four neighboring atoms there are four σ-bonds, with an average of about one π-bond per P atom.

3. When electrons are shared with five or six neighboring atoms, there is less than one full σ-bond for each connection between the phosphorus and a neighboring atom with apparently very little π-bonding.

These generalities are obviously dependent to a considerable extent upon the specific atoms connected to the phosphorus and, indeed it is possible that the observed differences between the triply and quadruply connected phosphorus atoms may be attributed primarily to the individual ligands. Fluorine appears to contribute nearly as much shortening (assuming that the tabulated values for the fluorine bond length are correct) to the P—F connection in the triply connected as in the quadruply-connected phosphorus compounds. On the other hand, chlorine shows essentially no shortening, whether attached to either triply- or quadruply-connected phosphorus.

Phosphine. PH_3, and its substitution products, have a pyramidal structure. The P-H bond length is 1.42Å, and the H-P-H angle is 93°. It is discussed under phosphines and related compounds. Hypophosphites, containing the radical

$$\left[\begin{array}{c} H \quad\quad O \\ \diagup \; P \; \diagdown \\ H \quad\quad O \end{array}\right]^{-}$$

are produced by alkaline hydrolysis of white phosphorus. The barium salt yields hypophosphorous acid H_3PO_2, upon acidification with sulfuric acid. In this acid, only one H is capable of ionization, suggesting the experimentally confirmed formula $H_2P(O)OH$. Hypophosphorous acid and its salts are reducing agents, although their reaction rates are somewhat low, which is usually explained by an equilibrium between H_3PO_2 and its hydrate form, H_5PO_3.

Phosphorus Halides. PX_3, P_2X_4 and PX_5 are formed by direct reaction of the elements, though the pure substances require special methods. Mixed halides are also known. The trihalides are covalent pyramidal compounds, the X—P—X bond angles being generally between 98° and 104°. They all undergo hydrolysis, the rate being roughly inversely proportional to the sum of the atomic numbers of the halogen atoms.

The halogen derivatives of pentavalent phosphorus may be grouped on the basis of their structure into three classes, the pentahalides, oxyhalides and related compounds, and the fluorophosphoric acids. The pentahalides (except the pentaiodide, which is unknown) are produced by reaction of the elements, or in the case of mixed halides, by reaction of a halogen and a phosphorus trihalide in correct proportions. Their structure in the vapor state has been determined to be a trigonal bipyramid and their bonding is covalent. In the solid, however, phosphorus pentachloride, PCl_5, is $[PCl_4^+][PCl_6^-]$ and phosphorus pentabromide, PBr_5, is $[PBr_4^+]Br^-$. Various mixed halides are known. The five halogen atoms are not in equivalent positions. One may be ionized, with the other four forming sp^3d orbitals or there may be a transition state, to explain the nonequivalence of the exchange between the three equatorial chlorine atoms and the two apical ones in PCl_5 in carbon tetrachloride solution. The pentahalides react with excess water to yield phosphoric acid and hydrohalic acids, but with less water to form phosphorous oxyhalides instead of phosphoric acid.

Phosphorus oxyhalides have the tetrahedral structure

These compounds, particularly $POCl_3$ and $POFCl_2$, readily form complexes with metal halides. Closely analogous to the oxyhalides are the thiohalides of general formula PSX_3 and the phosphorus nitrilic halides, $(PNX_2)_n$, the chloride of the latter being obtained by partial ammonolysis of PCl_5, and existing as cyclic or polymeric structures of alternate nitrogen and phosphorus atoms.

Phosphorus Oxides and Oxyacids. The principal oxides are related to the acids which they yield when dissolved in H_2O in the following manner:

Trioxide, P_2O_3	Hypophosphorus acid, H_3PO_2
Tetroxide, P_2O_4	Phosphorus acid, H_3PO_3
Pentoxide, P_2O_5	Hypophosphoric acid, $H_4P_2O_6$
plus $3H_2O$	Orthophosphoric acid, $2H_3PO_4$
plus $2H_2O$	Pyrophosphoric acid, $H_4P_2O_7$
plus $1H_2O$	Metaphosphoric acid, $2HPO_3$

Normally when the term "phosphoric acid" is used, it is with reference to orthophosphoric acid H_3PO_4. Anhydrous orthophosphoric acid is a white crystalline solid that melts at 42.35°C. It forms a hemihydrate $2H_3PO_4 \cdot H_2O$ which melts at 29.32°C. Although practically any desired concentration can be produced, it is common to supply the material as a solution containing from 75% H_3PO_4 (mp −17.5°C) to 85% H_3PO_4 (mp 21.1°C). When phosphoric acid is heated to above 200°C, the water of constitution is lost. Thus, a series of acids is formed by the dehydration, ranging from pyrophosphoric acid $H_4P_2O_7$ to metaphosphoric acid $(HPO_3)_n$. Salts of the dehydrated acids are used for the preparation of certain kinds of liquid fertilizers and are present in numerous cleaning compounds. The dehydrated acids can form water-soluble complexes with many metals, such as calcium. Either one, two, or three of the hydrogens of phosphoric acid may be neutralized. When one hydrogen is replaced with sodium, for example, the product is slightly acidic; while replacement of all three hydrogens yields a highly alkaline product. The acidity of the solutions is: NaH_2PO_4, a pH of 4.0; Na_2HPO_4, a pH of about 8.3; Na_3PO_4, a pH of 12.0. Although of interest scientifically, the other acids of phosphorus, hypophosphorous acid H_3PO_2, orthophosphorous acid H_3PO_3, and pyrophosphorous acid $H_4P_2O_5$ are not important commercially.

Phosphoric acid is a polyprotic acid and ionizes in a stepwise fashion. There is an ionization constant for each step.

(1) $\quad H_3PO_4 \rightleftharpoons H^+ + H_2PO_4^- \qquad \dfrac{[H^+][H_2PO_4^-]}{[H_3PO_4]}$
$$= K_1 = 7.5 \times 10^{-3}$$

(2) $\quad H_2PO_4^- \rightleftharpoons H^+ + HPO_4^{2-} \qquad \dfrac{[H^+][HPO_4^{2-}]}{[H_2PO_4^-]}$
$$= K_2 = 6.2 \times 10^{-8}$$

(3) $\quad HPO_4^{2-} \rightleftharpoons H^+ + PO_4^{3-} \qquad \dfrac{[H^+][PO_4^{3-}]}{[HPO_4^{2-}]}$
$$= K_3 = 1 \times 10^{-12}$$

Phosphorus sesquioxide, P_4O_6, produced by controlled oxidation of white phosphorus, is hydrolyzed in the cold to produce phosphorous acid, H_3PO_3, a colorless solid (mp 73°C). Only two of its H atoms are capable of ionization, thus compounds such as M_3PO_3 do not exist. This fact leads to the formula $HP(O)(OH)_2$. Phosphorous acid is a somewhat stronger acid than phosphoric, and both it and the phosphite ion (HPO_3^{2-}) are strong reducing agents.

Hypophosphorous acid was discussed earlier in this entry.

Metaphosphorous acid, HPO_2, is produced by atmospheric combustion of PH_3, but in aqueous solution it hydrates to H_3PO_3.

Phosphorous acid is used in solution, and is usually a reducing agent, e.g., in air changes to phosphoric acid, with hot concentrated sulfuric acid yields phosphoric acid plus SO_2, with copper sulfate yields finely divided copper metal, with silver nitrate yields finely divided silver metal, with permanganate after some time yields manganous, but occasionally is an oxidizing agent, e.g., with zinc plus dilute H_2SO_4 it yields phosphine.

Phosphorus tetroxide, P_2O_4, is obtained along with red phosphorus by heating P_4O_6 at 290°C in a closed tube. It is believed to have the formula, P_8O_{16}, and to consist of trivalent and pentavalent phosphorus. Hypophosphoric acid, $H_4P_2O_6 \cdot 2H_2O$ cannot be produced directly from the tetroxide. The acid decomposes into phosphorous and phosphoric acids on heating, and must be prepared by indirect methods, such as treatment of white phosphorus with an HNO_3 solution of $Cu(NO_3)_2$.

Phosphorus(V) oxide is the chief product of atmospheric oxidation of phosphorus, whence it is obtained as the β-allotrope, of formula P_4O_{10}. Several other allotropes are obtained by various thermal treatments of the β-form, differing in structure and physical properties. The compound hydrates rapidly to form various phosphoric acids. With an excess of H_2O, orthophosphoric acid, $(HO)_3PO$, is formed, mp 42.3°C. It is a triprotic acid, yielding $H_2PO_4^-$, HPO_4^{2-} and PO_4^{3-} ions. The other crystalline phosphoric acid, pyrophosphoric acid, is believed to have the formula, $(HO)_2P(O)$—O—$P(O)(OH)_2$. Its acid solution undergoes hydrolysis to the orthoacid. It is tetraprotic, yielding the ions $H_3P_2O_7^-$, $H_2P_2O_7^{2-}$, $HP_2O_7^{3-}$ and $P_2O_7^{4-}$.

Classification of Phosphates.   Audrieth and Hill have proposed a classification on the basis of structure, that is, to divide them into glassy phosphates and crystalline phosphates, the latter class being subdivided into (1) linear phosphates and polyphosphates, and (2) cyclic phosphates. Three important members of class (1) are the orthophosphates containing the PO_4^{3-} ion, the pyrophosphates, having the $P_2O_7^{4-}$ ion, and the triphosphates, having the $P_3O_{10}^{5-}$ ion. The general structural unit of the linear phosphates is the tetrahedron, containing a phosphorus atom surrounded by four oxygen atoms covalently linked to it. Such tetrahedra are linked through a common oxygen atom to form linear polyphosphates, the $P_2O_7^{4-}$ ion having two such tetrahedra, and the $P_3O_{10}^{5-}$ ion, three. Moreover, the tetrahedra may also be double linked through oxygen atoms to form cyclic structures, as in the trimetaphosphate ion, $P_3O_9^{3-}$, which has three such tetrahedra and the tetrametaphosphate ion, $P_4O_{12}^{4-}$, which has four tetrahedra. In general, heating acid salts of simpler phosphates produces polyphosphates (loss of H_2O), while alkaline hydrolysis reverses the process. The glassy phosphates, produced by fusion and rapid cooling of metaphosphates, appear to be true glasses, containing anions with molecular weights well into the thousands.

The widely diverse functionality of phosphates makes them of exceptional importance to technologists, particularly in the food processing field. Phosphoric acid finds many direct uses as an acidulant. It has three available hydrogens which can be replaced one by one with alkali metals, forming a series of *orthophosphate salts* with pH levels ranging from moderately acid (pH = 4) to strongly alkaline (pH = 12). This wide pH range makes phosphates very useful for adjusting the pH of food and chemical systems to almost any desirable level. Heating orthophosphates converts them to condensed phosphates containing two, three, or more phosphorus atoms per molecule. The *condensed phosphates*, or *polyphosphates*, have many properties which the orthophosphates do not enjoy. They are polyelectrolytes, and have dispersing or emulsifying properties. They can sequester or chelate metals, such as calcium, magnesium, iron, and copper, rendering these metals nonreactive. This functionality is useful for controlling oxidative rancidity and color formation, as both are catalyzed by metal ions.

Condensed phosphates containing two atoms of phosphorus are *pyrophosphates*. Sodium acid pyrophosphate (SAPP) is used as a leavening acid in baking, and is particularly useful because of the way it can be modified to give different rates of reaction. Pyrophosphates are good sequestrants for iron and copper, which often catalyze oxidation in fruits and vegetables. Thus, the use of a pyrophosphate effectively prevents the discoloration of such foods during preparation and storage.

Condensed phosphates containing three atoms of phosphorus are *tripolyphosphates*, the most important of which is sodium tripolyphosphate (STPP). This compound reacts with the protein in meat, fish, and poultry to prevent denaturing or loss of fluids. This property is sometimes called *moisture binding*. STPP also solubilizes protein which aids in binding diced cured meat, fish, and poultry. It also emulsifies fat to prevent separation.

The chain length of phosphates can be increased further by melting and chilling to form a glass. Glassy sodium phosphates are generally called *sodium hexametaphosphates* (SHMP). SHMP has excellent sequestering power toward calcium and magnesium. It is used in meat treatment as a partial replacement for STPP to improve solubility in strong pickling brine or to prevent hardness precipitation in very hard water. Considerably more information on the use of phosphates in food processing will be found in the "Foods and Food Production Encyclopedia," D. M. Considine (editor), Van Nostrand Reinhold, New York (1982).

Peroxyphosphoric Acids.   Two are known—peroxymonophosphoric acid, $HOOP(O)(OH)_2$, prepared by treatment of P_4O_{10} with hydrogen peroxide, and peroxydiphosphoric acid, $(HO)_2(O)POOP(O)(OH)_2$, prepared from metaphosphoric acid and peroxide or by electrolysis of an alkali hydrogen phosphate solution (cf. preparation of peroxydisulfuric acid). They and their salts are strong oxidants.

Fluorophosphoric Acids. H_2PO_3F, HPO_2F_2, (POF_3 is not a protic acid), "HPF_6," are obtained by replacing one or more hydroxyl groups with fluorine. They are strong fuming acids, like H_2SO_4 in most properties except its oxidizing power. Their salts are also known, extending as far as completely fluorinated MPF_6 where M is an alkali metal. The solubilities of the monofluorophosphates parallel those of the sulfates; while those of the di- and hexafluorophosphates parallel those of the perchlorates.

Upon acidification or neutralization of solutions containing phosphate anions with other anions, such as those of molybdenum and tungsten, complexes are formed which can readily be crystallized as salts, called phosphomolybdates, phosphotungstates, etc.

Oxygen-Nitrogen Compounds of Phosphorus.   These may be classified as aquo, aquo ammono, or ammono derivatives. The first group includes the various acids and oxides already discussed. The second group includes the amido phosphoric acids in which one or more of the hydroxy groups of the acid are substituted by amino groups. Thus there is amidophosphoric acid (orthophosphoric acid is understood when the substituted acid is not specified), amidopyrophosphoric acid, diamidophosphoric acid, and triamidophosphoric acid. The substitution of all —OH groups of phosphoric acid gives phosphoryl triamide, $OP(NH_2)_3$. Related compounds are phosphoryl amide imide, $OP(NH_2):NH$ and phosphoryl nitride $(OPN)_x$. The second group also includes imidodiphosphoric acid

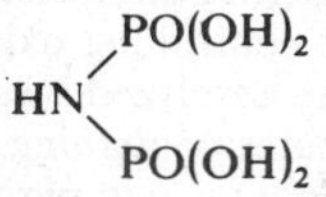

$$HN \begin{matrix} PO(OH)_2 \\ \\ PO(OH)_2 \end{matrix}$$

diimidotriphosphoric acid,

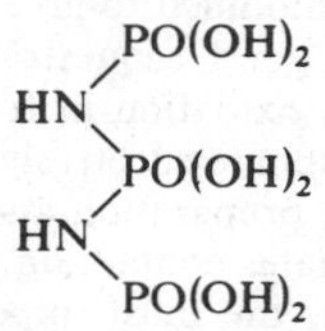

$$HN \begin{matrix} PO(OH)_2 \\ \\ HN \end{matrix} \begin{matrix} PO(OH)_2 \\ \\ PO(OH)_2 \end{matrix}$$

and still longer chain acids. Finally, many other derivatives are possible because of the stability of the $N\equiv P<$ arrangement. This gives rise to the phosphonitrilic acids, $[NP(OH)_2]_x$, as well as to many ammono derivatives containing only the two elements. The latter include phosphonitrilamide $[NP(NH_2)]_x$, phospham $[NP=NH]_x$, and phosphoric nitride $[P_3N_5]_x$. The phosphonitrilic chlorides, already discussed, are derivatives of phosphonitrilamide.

Organophosphorus Compounds. Most of the industrially important organic compounds of phosphorus commence with one of the basic inorganic phosphorus compounds, such as PCl_3, $POCl_3$, P_2S_5, and P_2O_5 reacted with an appropriate organic intermediate. Ester intermediates, such as alkyl phosphoryl chlorides are made by the addition of primary alcohols to $POCl_3$. Triaryl phosphate plasticizers and gasoline additives, such as tricresyl phosphate (TCP) can be prepared from PCl_5 and an appropriate phenolic compound. Alkyl diaryl phosphates can be made from $POCl_3$ and corresponding phenols. A number of thiophosphate esters contain PS plus an ethyl or methyl group and a substituted aryl group and are based on $PSCl_3$ or P_2S_5. These compounds are finding use for pesticide control. Dialkyl dithiophosphates may be prepared from P_2S_5 and appropriate intermediates. They are finding application as flotation-agents, oil-additives, and insecticides. It is much more difficult to prepare organophosphorus compounds containing a C—P bond than it is to form the esters. Numerous organic phosphorus compounds are found in nearly all life processes and remain to be better understood before they can be synthesized. Classes of compounds of this type include the phosphoglycerides required for fermentation, the adenosine phosphates needed in photosynthesis and muscle activity, and the very complex phosphorus-containing groups identified in the nucleotides. Of structural interest are the catenation compounds, as just illustrated, which contain many cyclic phosphates and oxygen-linked chains. Compounds of this type include tetrachlorodiphosphine Cl_2PPCl_2, tetraphenyldiphosphine $(C_6H_5)_2PP(C_6H_5)_2$, diphosphobenzene $C_6H_5PPC_6H_5$, and tetramethyl hypophosphate $(CH_3O)_2(O)PP(O)(OCH_3)_2$.

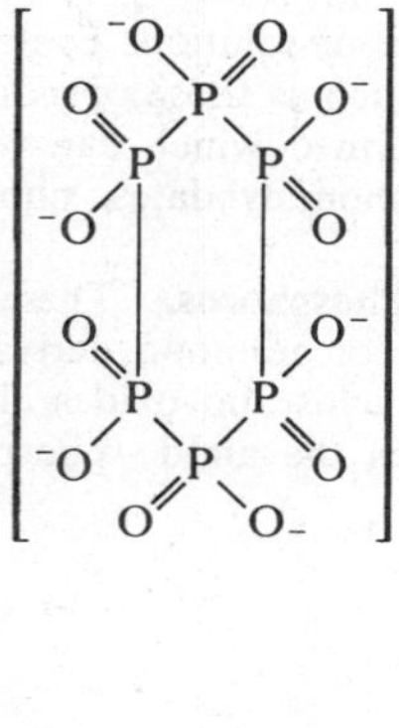

Inorganic Macromolecules. After the accelerated activity in the development of new polymers that took place during the past 30 or 40 years, some researchers observed some lessening of polymer research and polymer achievements during the 1970s. Not all authorities agreed, but most agreed that the time had arrived when polymer chemistry and applications deserved evaluation both in terms of the past and the future. Synthetic polymers generally have a number of relatively negative features—flammability (derived from their organic nature); tendency to melt, oxidize and char at high temperatures in regular atmospheric conditions (again, a result of their organic nature); and a tendency to become stiff and brittle at low temperatures. Many also have a tendency to soften, swell and dissolve in a number of common substances, such as gasoline, jet fuel, hot oil, and numerous other hydrocarbons. In terms of medical applications, most organic polymers tend to initiate a clotting reaction of the blood and many tend to cause toxic, irritant, and sometimes carcinogenic responses. Observers also noted that most polymer research in some way initiated with petrochemicals.

In the early 1970s, a number of investigators decided to shift emphasis and to look at a number of inorganic elements including silicon, phosphorus, sulfur, boron, and some metal atoms that might make up the backbone of a polymer. It was reasoned that the presence of some of the materials in the backbone might remedy some of the aforementioned shortcomings. It should be pointed out that as early as the 1940s, silicone, or poly(organosiloxane) polymers were developed and have proved highly satisfactory in many applications. Peters et al. (1976) reported on a new class of thermally stable polymers which are based upon alternating siloxane and carborane units. In 1965, Allcock et al. (Pennsylvania State University) synthesized the first poly(organophosphazenes). Since that time, well over 60 new polymers have been made and presently constitute a substantial class of new elastomers. They appear to solve some of the biomedical problems previously mentioned.

As pointed out by Allcock (1976), all linear, high polymeric polyphosphazenes have the general structure shown by (a) below. It is interesting to note that over a century ago researchers in Germany and Britain found that phosphorus pentachloride will react with ammonia or ammonium chloride to yield a volatile, white solid. It is now known that this product has the form of (b) below. Later experimentation showed that the compound would melt under strong heating to form a transparent rubbery material. Involving a ring-opening polymerization of the cyclic trimer, a poly(dichlorophosphazene), shown in (c) below, is formed. Mainly for the reason that the compound hydrolyzes slowly in the presence of atmospheric moisture to form a crusty mixture of ammonium phosphate and phosphoric acid, the substance was not given serious thought for many years.

Allcock and associates, during the 1960s, subjected the cyclic trimer with new procedures and, after considerable research, were successful in developing polymers that have molecular weights up to and sometimes exceeding 3 to 4 million. The investigators found that the introduction of different substituent groups had a marked effect on the properties of the polymers. See (d) below. The further detailed research is beyond the scope of this book, but interesting details can be found in the Allcock reference. In summarizing one of their reports, the researchers stated: "Polyphosphazenes are emerging as a new class of macromolecules that have an obvious future as technological elastomers, films, fibers, and textile treatment agents. However, they also possess almost unique attributes for use in biomedicine as reconstructive plastics or as drug-carrier molecules. Moreover, their possible value as 'pseudo-protein' model polymers is an exciting prospect."

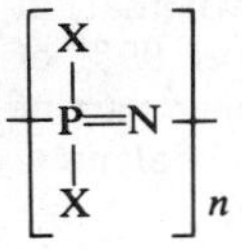

(a) 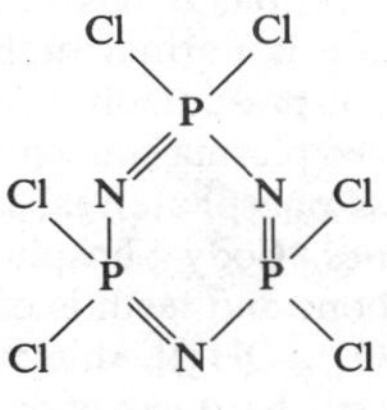   (b)

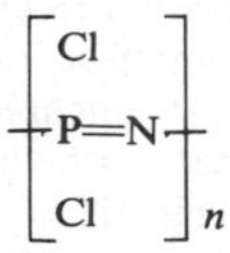

(c) 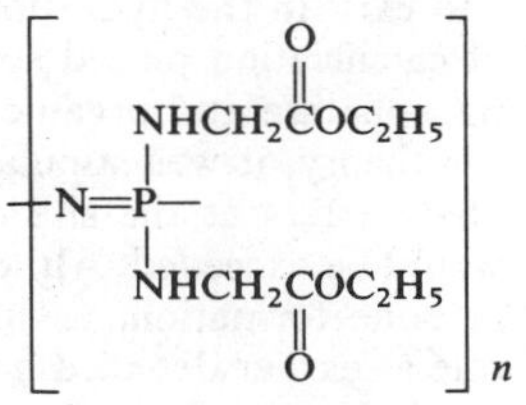   (d)

Phosphorus Ylides. In 1979, G. Wittig received a share of the Nobel Prize for Chemistry in recognition of his development of the use of phosphorus-containing compounds as important reagents used in organic synthesis. According to the selection committee, Wittig's most important achievement was, "the discovery of the rearrangement reaction that bears his name. In the Wittig reaction an organic phosphorus compound with a formal double bond between phosphorus and carbon is reacted with a carbonyl compound. The oxygen of the carbonyl compound is exchanged for carbon, the product being an olefin. This method of making olefins has opened up new possibilities, not the least of which is the synthesis of biologically active substances containing carbon-to-carbon double bonds. For example, vitamin A is synthesized industrially using the Wittig reaction." As early as 1919, some 30 years prior to Wittig's work, the first phosphorus ylide was described by Staudinger. This was diphenylmethylenetriphenylphosphorane, formed by pyrolysis of a phosphazine precursor. During that period, Staudinger conceived the possibility of an olefin synthesis by condensation of this ylide with a carbonyl compound and visualized a four-membered phosphorus-oxygen heterocyclic (oxaphosphetane) as the intermediate. Staudiner's work was accomplished during a period when practical application of the concept was in doubt. At that time, the Lewis theory of electronic structure was new. The exact bonding of phosphonium salts was somewhat veiled and controversial. Attempts at olefin synthesis were put aside. For more detail, see the entry on **Wittig Reaction.**

References

Allcock, H. R.: "Polyphosphazenes: New Polymers with Inorganic Backbone Atoms," *Science,* **193,** 1214–1219 (1976).
Considine, D. M., Editor: "Foods and Food Production Encyclopedia," Van Nostrand Reinhold, New York (1981).
Ellinger, R. H.: "Phosphates in Food Processing," in "Handbook of Food Additives," (T. E. Furia, Editor), Chap. 15, CRC Press, Boca Raton, Florida (1972).
Peters, E. N., et al.: *Rubber Chemistry* (1976).
Staff: "Functions of Phosphates in Foods," FMC Corporation, Philadelphia, Pennsylvania (1977).
Staff: "The Nobel Prizes," *Sci. American,* **241,** 6, 86–88 (1979).
Vedejs, E.: "1979 Nobel Prize for Chemistry," *Science,* **207,** 42–44 (1980).

PHOSPHORUS (In Biological Systems). Phosphorus is required by every living plant and animal cell. Deficiencies of available phosphorus in soils are a major cause of limited crop production. Phosphorus deficiency is probably the most critical mineral deficiency in grazing livestock. Phosphorus, as orthophosphate or as the phosphoric acid ester of organic compounds, has many functions in the animal body. As such, phosphorus is an essential dietary nutrient.

The biological roles of phosphorus include: (1) anabolic and catabolic reactions as exemplified by its essentiality in high-energy bond formation, e.g., ATP (adenosine triphosphate), ADP (adenosine diphosphate), etc., and the formation of phosphorylated intermediates in carbohydrate metabolism; (2) the formation of other biologically significant compounds, such as the phospholipids, important in the synthesis of cell membranes; (3) the synthesis of genetically significant substances, such as DNA (deoxyribonucleic acid) and RNA (ribonucleic acid); (4) contributing to the buffering capacity of body fluids, cells, and urine; and (5) the formation of bones and teeth. Like calcium, the majority of the phosphorus in the vertebrate body is contained in the hard tissues; in the adult, approximately 80–86% of the total body phosphorus is contained in the bones and teeth, with the balance found in the soft tissues and body fluids.

Phosphorus deficiencies are not common in humans and most species, but they have been observed in ruminants. Symptoms of the deficiency are loss of appetite and a depraved appetite (termed *pica*) where the animal chews and consumes extraneous items, such as wood, clothing, bones, etc. Vitamin D deficiency may accentuate a marginal lack of phosphorus in the diet.

Cereals and meats are the major sources of phosphorus in human diets. Phosphorus deficiences in most regions have not been a serious problem in human nutrition. Insofar as human food is concerned, the primary value of phosphorus fertilizers is that they generally increase the total food production, not the content of phosphorus in the food per se.

Experimental phosphorus deficiency can be induced by feeding diets low in this element and by including excesses of calcium, strontium, barium, beryllium, and other cations that precipitate phosphates in the intestinal tract. In this situation, bone formation ceases, and the following histological bone changes have been noted in experimental animals: (1) A thickening of the epiphyseal plate and the formation of a typical rachitic metaphysis; (2) wide osteoid borders of trabecular bone and a considerable rarefaction of the shaft; and (3) irregular or complete cessation of calcification of the zone of provisional calcification of the cartilage matrix. Rickets can be produced in the laboratory by feeding a diet high in calcium and low in phosphorus, and containing little or no vitamin D.

The nutrient requirement for phosphorus depends upon the particular species and the physiological status of the animal. During growth, lactation, gestation, and egg-laying, a higher phosphorus content of the diet is generally required in poultry than for the maintained adult. The availability of phosphorus in the diet varies with its chemical form and the animal species in question. Diets high in foods of plant origin may contain a considerable portion of phosphorus in the form of phytic acid, which is the hexaphorphoric acid ester of inositol. When the acid occurs as salts of calcium, magnesium, sodium, etc., it is referred to as phytin. Phytate phosphorus usually is less available than inorganic phosphate in such species as rat, chicken, dog, pig, and human. However, a phytase has been shown to be present in the intestine and intestinal secretions of some animals, and the formation of this enzyme is dependent, in part, on the presence of vitamin D. Through the action of phytase, some of the phytate phosphorus would be made available for absorption.

Experimentation has indicated that, under normal dietary conditions and calcium intake, food phytate is of no nutritional concern in humans. The microbial population of the ruminant also elaborates a phytase enzyme which makes phytate phosphorus readily available in this class of animals. Phytates may be of nutritional consequence for another reason—dietary calcium can be bound in an unavailable, insoluble complex, thereby decreasing the absorption of this element.

Many studies have involved determination of the availability of phosphorus from other organic and inorganic sources. In chicks, orthophosphates, superphosphates, and phosphate rock products are good sources of phosphorus, whereas metaphosphate and pyrophosphate are relatively unavailable to the species. Most organic phosphorus sources, such as casein, pork liver, and egg phospholipid are found to be equally available as inorganic phosphorus. Commonly used phosphorus supplements in human or animal nutrition or both are steamed bone meal, ground limestone, dicalcium phosphate, and defluorinated rock phosphates. Phosphorus dietary supplements include magnesium phosphate (dibasic and tribasic), manganese glycerophosphate and manganese hypophosphate, potassium glycerophosphate, sodium ferric pyrophosphate, sodium phosphate (mono-, di-, and tri-), and sodium pyrophosphate. Of course, phosphate compounds are not always added in the interest of augmenting phosphorus, but for the other elements which may be contained in the compound.

Absorption of Phosphate. The phosphate ion readily passes across the gastrointestinal membrane. The rate of absorption of phosphate at various intestinal sites in rat has been observed to be most rapid in the duodenu, followed in decreasing order by the jejumen, ileum, colon, and stomach. When transit time is considered, most of the phosphorus is absorbed by the ileum.

The triangular relationship between calcium, phosphorus, and vitamin D is described briefly in the entry on **Calcium (In Biological Systems)**.

Plasma Phosphate. Once absorbed, phosphorus enters the blood and the majority is present therein as orthophosphate ions. About 12% of the phosphorus present is bound to proteins. During egg-laying in birds, the concentration of unionized phosphorus compounds in plasma is greatly increased. The administration of diethylstilbestrol (regulated in some countries) to cockerels results in the formation of a plasma phosphoprotein which forms relatively firm complexes with calcium. The function of the phosphoprotein appears to be one of phosphorus transport; in laying birds, the phosphoprotein is incorporated in egg yolk.

The approximate average plasma phosphorus levels for several species, in milligrams per 100 milliliters of plasma, are: pigs, 8.0; sheep, cattle, and goats, 6.0; horse, 2.3. Erythrocytes contain considerably more phosphorus than plasma, mostly in the form of organic esters. Some of the latter are acid soluble and hydrolyzable by intracellular enzymes.

Plasma phosphate appears to be homeostatically controlled. The primary organ concerned appears to be the kidney, although the skeleton also may play a role. Parathyroid hormone, by way of its direct action on the kidney and bone, is a significant hormonal factor.

Phosphate Excretion. The excretion of body phosphorus occurs via the kidney and intestinal tract, the distribution between these pathways varying with species. For example, relatively small amounts of phosphorus are endogenously excreted into the feces of rat, pig, and human, but in the bovine, perhaps 50% or more of the fecal phosphorus may be from endogenous sources.

The amount of phosphorus excreted in the urine varies with the level of ingested phosphorus and factors influencing phosphorus availability and utilization. It has been shown that, in dog, when plasma phosphate is normal or low, over 99% of the filtered ion is reabsorbed, presumably in the upper part of the proximal tubule. Increased plasma concentrations of alanine, glycine, and glucose depress phosphate reabsorption.

Phosphate of Hard Tissues. Body phosphorus contained in the intracellular matrix of bone and teeth is of the general form of hydroxyapatite $[Ca_{10}(PO_4)_6(OH_2)]$, this calcium phosphate salt providing the characteristic hardness of ossified tissue. Phosphate ions are also adsorbed onto the surface of bone crystals and exist in the hydration layers of the crystals. Early theories of calcification placed special emphasis on the role of alkaline phosphatase and organic esters of phosphoric acid. As part of the theory, it was stipulated that, with the hydrolysis of phosphate esters at the site of calcification, the K_{sp} for bone salt would be exceeded. Although phosphatase may have a function in bone formation, as in the synthesis of organic matrix, its role as earlier depicted has been revised. Later research emphasized the specific and characteristic properties of collagen and other substances, such as chondroitin sulfate. This is related to the local mechanism of calcification; the other component of calcification is the humoral mechanism whereby an adequate supply of calcium, phosphate, and other ions is made available to the calcifying site. A later theory proposed that either the functional groups on collagen are anionic, initially binding Ca^{2+}, or that the first reaction is with phosphate or phosphorylated intermediates. The first-held moiety of bone salt (Ca^{2+} or phosphate) subsequently attracts or binds the other component, providing the aggregation or "seed" for subsequent crystal growth. Since an ATPase-type enzyme has been demonstrated in cartilage, suggesting that ATP may be intimately involved in the calcification mechanism, another proposal is along the line that pyrophosphate is transferred from ATP to free amino groups of collagen, leading to nucleation and followed by combination with calcium and bone salt formation. Or, the ATP provides energy which increases the calcification mechanism.

Dietary inorganic phosphates have been shown to protect experimental animals against dental caries. Orthophosphates were effective cariostats, but $Na_4P_2O_7$ and $Na_5P_3O_{10}$ were not. Dicalcium phosphate, $CaHPO_4$, did not decrease dental caries unless a high level of NaCl was also included in the diet.

Toxicity. Although many phosphorus-containing compounds are vital to life processes, as previously described, there are also many phosphorus compounds that are quite toxic. Elemental phosphorus, for example, is dangerous because of its low combustion temperature, but the absorption of phosphorus also has an acute effect on the liver. The long and continued absorption of small amounts of phosphorus can result in necrosis of the mandible or jaw bone (sometimes called "phossy-jaw"). Chronic phosphorus poisoning may result from long and continued absorption, particularly through the lungs and gastrointestinal tract. The most common symptom is necrosis of the jaw, but this is also usually accompanied by anemia, loss of appetite, gastrointestinal weakness, and pallor. Other bones and teeth may be adversely affected. Phosphine is a very toxic gas. Inhalation of phosphine causes restlessness, followed by tremors, fatigue, slight drowsiness, nausea, vomiting, and, frequently severe gastric pain and diarrhea. Although most cases recover without after-effects, in some cases coma or convulsions may precede death. Phosphorus-halogen compounds are quite toxic. Details of toxicological data can be found in the Sax reference.

Phosphorus in Soils. When phosphorus fertilizers are added to soils deficient in available forms of the element, increased crop and pasture yields ordinarily follow. Sometimes the phosphorus concentration in the crop is increased, and this increase may help to prevent phosphorus deficiency in the animals consuming the crop, but this is not always so. Some soils convert phosphorus added in fertilizers to forms that are not available to plants. On these soils, very heavy applications of phosphorus fertilizer may be required. Some plants always contain low concentrations of phosphorus even though phosphorus availability from the soil may be minimal. See also **Fertilizers.**

References

Dymsza, H. A., et al: "Effect of Normal and High Intakes of Orthophosphate and Metaphosphate in Rats," *Jrnl. of Nutrition*, **69**, 419 (1959).

Kirchgessner, M., Editor: "Trace Element Metabolism in Man and Animals," Institut für Ernährungsphysiologie, Technische Universität Müchen, Freising-Weihenstephan, Germany (1978).

Mahoney, A. W., and D. G. Hendricks: "Some Effects of Different Phosphate Compounds on Iron and Calcium Absorption," *Journal of Food Science*, **43**, 5, 1473–1475 (1978).

Sax, N. I.: "Dangerous Properties of Industrial Materials," 5th Edition, Van Nostrand Reinhold, New York (1979).

Underwood, E. J.: "Trace Elements in Human and Animal Nutrition," 4th Edition, Academic, New York (1977).

PHOSPHORYLATION (Oxidative).

This is an enzymic process whereby energy, released from oxidation-reduction reactions during the passage of electrons from substrate to oxygen over the electron transfer chain, is conserved by the synthesis of adenosine triphosphate (ATP) from adenosine diphosphate (ADP) and inorganic orthophosphate. Since ATP is the major source of energy for biological work, and since most of the net gain of ATP in the animal cell derives from oxidative phosphorylation, research in the area has been intensive.

Oxidative phosphorylation was discovered simultaneously and independently in 1939 by Kalckar (Denmark) and by Belitzer (U.S.S.R.). It was recognized by these workers that aerobic phosphorylation was different from and independent of phosphorylation supported by glycolysis. In addition, they found that the stoichiometry of phosphate esterification (ATP synthesis) and oxygen utilized was 2 or more, or that the reduction of 1 atom of oxygen to form water may be accompanied by the "activation" of 2 or more molecules of phosphorus (P_i), thus leading to an expression of the efficiency of the energy-conserving system. The efficiency expression is known as the P/O ratio, i.e., the ratio of molecules of P_i esterified per atom of oxygen utilized.

The quantitative importance of the ATP synthesized at the expense of energy liberated during electron transfer in the mitochondrion is realized when one follows the conservation of energy during the metabolism of a molecule such as glucose in the cell. The oxidation of 1 mole of glucose to carbon dioxide and water is accompanied by the release of 673,000 calories. In order to degrade the glucose molecule to a form which can be metabolized further by mitochondrial enzymes, the glycolytic enzymes consume 2 molecules of ATP and also synthesize 2 molecules of ATP in the presence of oxygen, a net energy conservation of zero. The mitochondrion may then degrade the pyruvate supplied by glycolysis to carbon dioxide and water, yielding

a net total of 38 molecules of ATP, mostly at the level of the electron transfer process. Thirty-eight molecules of ATP per molecule of glucose results in between 260,000 and 380,000 calories conserved, between 39% and 56% of the total energy released in the complete oxidation of glucose, the remainder being released directly as heat. Inasmuch as the mitochondrion is approximately 50% effective in conserving energy from its major substrate, it is indeed an efficient machine.

PHOSPHORYLATION (Photosynthetic).

Photosynthetic conversion of light energy into the potential energy of chemical bonds involves an electron transport chain, and the phosphorylation of ADP (adenosine diphosphate)

$$ADP + P \xrightarrow[\text{Chlorophyll}]{+\text{Light}} ATP$$

as intermediate stages. The process of phosphorylation, defined by the foregoing equation, was discovered simultaneously by Arnon and coworkers for green plant chloroplasts and by Frenkel, working with Geller and Lipmann for bacteria in 1954. For both systems, the heart of the mechanism is the creation of a very oxidizing and a very reducing component, utilizing the energy of the photoexcited stage of one of the pigment (chlorophyll) molecules. This process will be designated a *photoact.* The redox components are both members of a photosynthetic electron transport chain, bound to the membranes of the *chloroplasts* (for green plants) or *chromatophores* (for bacteria). The photoact can be considered as electron transport against the thermochemical gradient, i.e., away from the member which is a better electron acceptor (high oxidation-reduction potential), through the excited chlorophyll, then to the member which is a better electron donor (low oxidation-reduction potential). Subsequent steps consist of ordinary, dark electron transport with the thermochemical gradient. The energy in at least one of these redox reactions is converted as ATP by a phosphorylation reaction analogous to that found in oxidative phosphorylation by mitochondria. See also **Phosphorylation (Oxidative).**

In bacteria, the photoact proper is accomplished by a special kind of bacteriochlorophyll, amounting to only 3% of the total present. It is unique in having a peak in absorption at 870–890 micrometers, or further into the infrared than the remaining 97% of the chlorophyll molecules. It is unique not by virtue of a difference in structure, but because of its "environment"— most probably a close association or complexing with cytochrome molecules. Since its absorption extends to longer wavelengths, it is an energy trap, and the function of the bulk of the bacteriochlorophyll is that of capturing light and transmitting it to this active center.

Components of the electron transport chain in bacteria have been shown to include b- and c-type cytochromes, ubiquinone (fat-soluble substitute quinone, also found in mitochondria), ferredox (an enzyme containing nonheme iron, bound to sulfide, and having the lowest potential of any known electron-carrying enzyme) and one or more flavin enzymes. Of these, cytochrome (in some bacteria, with absorption maximum at 423.5 micrometers, probably c_2) has been shown to be closely associated with the initial photoact. Some investigators were able to demonstrate, in chromatium, the oxidation of the cytochrome at liquid nitrogen temperatures, due to illumination of the chlorophyll. At the very least, this implies that the two are bound very closely and no collisions are needed for electron transfer to occur.

In both bacterial chromatophores and green plant chloroplasts, the existence of photo-induced high-energy immediates or states leads to reversible conformational changes in the structures of the membranes, and to gross swelling and shrinking. These are observed by changes in light scattering, viscosity, and sedimentation properties, and by electron microscope studies. The mechanisms may include ion transport, followed by water diffusion, internal pH changes leading to conformation changes of proteins, or possibly something resembling a contractile protein.

PHOTOCHEMISTRY AND PHOTOLYSIS. When certain substances are subjected to light, a chemical change results. Such reactions comprise *photochemistry*. The production of an image on a photographic plate is an example. Photosynthesis in the green leaf of a plant is another. Where the change involves chemical decomposition of the radiated material, the process is termed *photolysis*. As used in this context, the term light may include visible light and ultraviolet radiation. One of the better known and most extensive examples of photolysis is the production of ozone, O_3, in the upper atmosphere, a reaction critical to life on earth because ozone acts as a filter of the middle- and far-ultraviolet radiations which destroy living organisms. Regular oxygen, O_2, absorbs solar ultraviolet radiation with a wavelength of 190 nanometers. The released oxygen atoms may combine with oxygen molecules present to form ozone, or the freed oxygen atoms may recombine to form O_2. Thus, there is a continuing combination of processes in dynamic equilibrium, that is, the synthesis and the photolysis of ozone.

Similarly, oceanic nitrite (NO_2^-) photolysis by natural light produces detectable concentrations of nitric oxide (NO). This latter forms in the oceans during daylight and disappears rapidly at sunset when recombination occurs: $NO_2 \rightarrow NO + O \rightarrow NO_2$.

Isomerism can also be induced photochemically although such processes are less well understood and probably require the presence of additional free radicals. The cytotoxic metabolite bilirubin can cause brain damage in infants with neonatal jaundice: this is prevented by exposing the child to intense blue light. The bilirubin is photochemically converted in the skin to metastable geometric isomers which can be transported in the blood and excreted in bile.

Two major instances of photochemical reactions which have reached deeply into modern civilization are the photosensitive silver and uranium salts and dyes which are the basis of photography and the manufacture of Vitamin D by the ultraviolet irradiation of ergosterol.

Photochemical reactions are highly specific and their products quite different from those of thermochemical reaction processes.

Sunlight in the near infrared, visible, and near ultraviolet regions possesses considerable energy; utilization of this through photochemical reactions could make a considerable contribution to energy resources. Since biosynthesis itself is relatively inefficient in conversion of solar energy, emphasis has been placed upon the fabrication of *artificial* photochemical systems. One of the more promising approaches has involved application of photoelectric chemical cells or catalysts of semiconductor materials.

The absorption of light by semiconductors creates electron-hole pairs ($e^- h^+$) which can be separated because their components diffuse in different directions. The energies of these moieties can be stored by several mechanisms or used in photocatalysis or photosynthesis for nitrogen fixation, formation of amino acids, methanol, etc. The efficiencies of such conversions depends almost entirely upon the semiconductor material and as yet these efficiencies are too low for significant application. Currently the most promise is demonstrated by the use of titania

on a platinum substrate or single crystals of strontium titanate.

Fundamental Considerations. In photochemical reactions, light supplies the energy necessary for the activation of the reacting molecules (Grotthus, 1818, and Draper, 1839). Sometimes the light waves which are absorbed by a body produce only an increase in temperature, sometimes fluorescence as in the cases of eosin and fluorescein, and sometimes chemical change. The reaction of hydrogen and chlorine in light was studied by Bunsen and Roscoe (1862), and they discovered that the amount of chemical change is proportional to the intensity of the light and to the length of time of exposure to the light. The first law of photochemistry (Draper-Grotthus) states that light that is absorbed causes chemical change. The energy of light is measured in quanta, and according to the Stark-Einstein law,

$$E = Nhc/\lambda$$

where N is Avogadro's constant, h is Planck's constant, c is velocity of light, λ is wavelength of light; that is, each molecule that takes part in a chemical reaction induced by exposure to light absorbs one quantum of radiation causing the reaction. Photochemical processes are of two kinds: primary and secondary. The primary process in a photochemical reaction is limited by the Einstein law to the absorption of one quantum by a molecule or atom. A knowledge of the spectrum of the reactants is necessary to determine what happens in this process. The molecule may be disrupted into fragments or an electron may be excited from a lower orbit to a higher one. Which of these events takes place can often be determined by spectroscopic studies. The secondary process deals with the fate of the molecular fragments or of the excited molecules. The excited molecule may emit its extra energy as light, causing fluorescence; it may lose it by transferring it to other molecules as thermal energy; or it may cause a chemical reaction. On the other hand, the molecular fragments may either recombine to give the original reactant or cause further chemical reactions. The study of the quantum yield (which is the number of molecules reacting divided by the number of quanta absorbed), is used as a means of formulating the secondary processes. If the quantum yield is less than one, fluorescence, deactivation or recombination of fragments must take place. If the quantum yield is unity every photon absorbed decomposes one molecule. When the quantum yield is greater than unity (and in some reactions it may be as high as a million) chain reactions are involved. The classical example of such a reaction is the combination of hydrogen and chlorine. The primary reaction is Cl_2 and light $\rightarrow$ 2Cl. The chain propagation reactions are

$$Cl_2 + h\nu \rightarrow 2Cl$$
$$Cl + H_2 \rightarrow HCl + H$$
$$H + Cl_2 \rightarrow HCl + Cl$$

creating a cycle which is only stopped by

$$Cl + Cl \rightarrow Cl_2$$
$$H + H \rightarrow H_2$$

Since the last two processes are slow compared to the two before them, one quantum of light can bring about a combination of a million molecules of hydrogen and chlorine.

See also **Photosynthesis.**

Laser Chemistry. Lasers generate a high intensity output of monochromatic photon energy and studies of the photochemical reactions induced by this have created a virtual subdivision of photochemistry known as laser chemistry. While the output of a laser can heat, anneal, burn, cut or be used instrumentally as a spectral source, we are concerned here only with those chemical effects attributable to the photon output at wavelengths between near infrared and near ultraviolet, i.e., between about 12 and 0.2 microns.

When atoms or molecules are excited conventionally by elevated temperatures or pressure, they can follow several reaction paths yielding a variety of byproducts in addition to the desired substance. Since the basis of a chemical reaction is to weaken or break or make specific chemical bonds to yield the final product, energy ideally should be selectively introduced at the particular level necessary to accomplish this. The high energy and monochromaticity of laser output are ideal for imposition of the specific energy changes which induce or catalyze chemical changes.

The absorption of a quantum of energy by an atom or a molecule takes it from a low energy state to a higher one, and the jump will affect the different properties of the atom or molecule depending upon the amount of energy in the quantum. When absorbed, a quantum of visible or ultraviolet radiation raises an electron to a higher orbit; on the other hand a quantum of infrared radiation will alter energy levels on an atomic basis.

A laser can supply a precise amount of energy to an atom or molecule thus effecting a transition from one excited state to a higher one. Once knowledge of the energy level displacement required to effect a chemical reaction is available, the laser can provide the specific energy for the specific excitation required. However, the energy input must be related to the total energy dissipation for, if excess energy leads to ionization or dissociation, a continuum of allowed energy levels will be developed rather than the required discrete levels. If the energy is thus fragmented, the required reaction will proceed only weakly, if at all. Excess energy may be redistributed in two ways. It is either transferred from the excited vibrational state to one or more other vibrational states of the molecule or, it is transferred directly into rotational and translational states. The first mode of energy translation proceeds appreciably more rapidly than the second. Time is a further controlling factor in laser chemistry. The reaction must proceed in time either shorter than, or equal to, that required for transfer of vibrational energy from one state to another in the same molecule; molecule dissociation or atom ionization must take place before there is any depletion of energy by molecular or atomic collisions. Where a reaction proceeds within the lengthy period required for transfer of energy from the initial vibrational state to the much lower rotational and translational states, one cannot hope for laser action to effect a significant degree of reaction specificity since the effect is basically a thermal one.

Operation of visible light and ultraviolet lasers costs more per photon produced than does operations of infrared lasers. Partly because of this appreciable interest has centered over the past several years on unimolecular reactions driven by infrared lasers. But absorption of a single infrared photon will raise a molecule only one step in the energy ladder and, to be dissociated, the molecule will require the absorption of many infrared photons in sequence. The carbon dioxide laser can supply this requirement cheaply and efficiently. A mole of photons (6.02×10^{23}) costs only a few cents in the infrared, but several dollars in the visible and near ultraviolet ranges. This has aided continued study of multiple photon infrared laser excitation. Much study has gone into an exciting and fundamental reaction and its implication. When sulfur hexafluoride (SF_6) is irradiated by infrared laser light, it decomposes to the pentafluoride (SF_5) and fluorine (F). When the laser is tuned to the vibrational absorption of $^{32}SF_6$ in a mixture with $^{34}SF_6$, only the $^{32}SF_6$ decomposes leaving the residual gas enriched some 3000-fold in $^{34}SF_6$. Changing the frequency of the irradiating light slightly

from emission at $10.61\ \mu$ to emission at $10.82\ \mu$ selectively decomposes the $^{34}SF_6$ molecule. This method of isotope separation by lasers is being extensively studied for the separation of fissionable ^{235}U from nonfissionable ^{238}U.

The foregoing is but one example of the application of laser chemistry. Decomposition of chromyl chloride (CrO_2Cl_2) to chromium dioxide (CrO_2) in the two-step reaction:

$$CrO_2Cl_2 \rightarrow CrO_2Cl + Cl \rightarrow CrO_2 + Cl_2$$

is a classic experiment with the CO_2 laser emitting at $10\ \mu$, and

$$NO + O_3 \rightarrow NO_2 + O_2$$

proceeds twenty times as rapidly under laser excitation than normally at room temperature.

Laser induced processes are expected to increase in number and expand in application, but the principal obstacle to large scale introduction of the laser into chemical industry is an economic one. Laser photons are still much more expensive than those from thermal sources and the initial application will undoubtedly be directed to those specialty chemicals and isotopes whose current cost far exceeds that of large volume chemicals.

The increasing number of known photochemical reactions is still very small in comparison with those in ground state chemistry and our understanding of all the factors controlling photochemical reactions is quite primitive. In some cases it is the ease of conversion to the ground state which is significant while in others it is the energy hypersurface surrounding the excited state which dictates the energy pathways through which the free electrons will move back towards the ground state and hence the nature of the photochemical products.

References

Bard, A. J.: "Photoelectrochemistry," *Science*, **207**, 139–144 (1980).

Barltrop, J. A., and J. D. Cole: "Principles of Photochemistry," Wiley, New York (1978).

Green, A. E. S., Editor: "The Middle Ultraviolet: Its Science and Technology," Wiley, New York (1966).

Grunwald, E., Dever, D. F., and P. M. Keehn: "Megawatt Infrared Laser Chemistry," Wiley, New York (1979).

Kosar, J.: "Light-Sensitive Systems: Chemistry and Application of Non-silver Halide Photographic Processes," Wiley, New York (1965).

Kimel, S., and S. Speiser: *Chemical Reviews*, **77**, 437–472 (1977).

Moore, C. B., Editor: "Chemical and Biological Applications of Lasers," Vol. 3, Academic, New York (1977).

Oster, G. K., and H. Kallman: "Energy Transfer from High-Lying Excited States," *J. de Chim. Phys.*, **64**, 1, 28–32 (1967).

Oster, G., and N. Yang: "Photopolymerization of Vinyl Monomers," *Chem. Rev.*, **68**, 2, 125–151 (1968).

Pitts, J. N., Jr., et al.: "Advances in Photochemistry," Vol. 11, Wiley, New York (1979).

Porter, G.: "Flash Photolysis and Some of Its Applications," *Science*, **160**, 1299–1307 (1968).

Ronn, A. M.: "Laser Chemistry," *Scientific American*, **240**, 114 (1979).

Salem, L.: *Science*, **191**, 822 (1976).

Zimmerman, H. E.: *Science*, **191**, 523 (1976).

—R. C. Vickery, Hudson Laboratories, Hudson, Florida.

PHOTOELECTRIC EFFECT. Changes in electrical characteristics of substances due to radiation, generally in the form of light. Radiation of sufficiently high frequency (short wavelength), impinging on certain substances, particularly, but not exclusively, metals, causes bound electrons to be given off with a maximum velocity proportional to the frequency of the radiation, i.e., to the entire energy of the photon. The Einstein photoelectric law, first verified by Millikan, states:

$$E_k = h\nu - \omega$$

where E_k is the maximum kinetic energy of an emitted electron, h is the Planck constant, v is the frequency of the radiation (frequency associated with the absorbed photon), and ω is the energy necessary to remove the electron from the system, i.e., the photoelectric work function for the surface of the emitting substance. An inverse photoelectric effect results from the transfer of energy from electrons to radiation. For example, in an x-ray tube, there is observed the transfer of energy from electrons accelerated by the anode voltage to radiation emitted by the target. This radiation exhibits a continuous spectrum at lower voltages, upon which are superimposed, at higher voltages, intense lines characteristic of the anode material.

Two principal aspects of the photoelectric effect are described here: (1) Photoconductivity; and (2) photovoltage.

Photoconductivity is the phenomenon evidenced by the increase in electrical conductivity of a material by the absorption of light or other electromagnetic radiation. Although insulating or semiconducting materials that give sufficiently large changes of conductivity with illumination for application of the principle in useful devices. The principle can be explained briefly by using a cadmium sulfide photoconductor as an example. As in the case of luminescence, the band-type of energy level diagram is useful. See Fig. 1. Transition 1 represents absorption of a

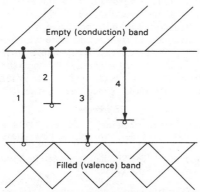

Fig. 1.   Simplified band model for photoconduction processes.

photon of energy at least equal to that of the band gap, giving rise to a free electron and a free hole. Transition 2 represents absorption at a local crystalline imperfection (defect or impurity), also producing a free electron, but with a hole trapped in the vicinity of the imperfection. While these carriers are "free" in the crystal, the conductivity can be greatly enhanced, so that the conductivity in the light can be a million times that in the dark. Recombination of the carriers may occur via transition (3), which is a "direct" electron-hole recombination across the band-gap, or via step 4, an electron recombining with a center containing a hole, so that they no longer contribute to the conductivity.

For a material in which one type of carrier predominates (i.e., electrons), the change in conductivity with illumination can be given as:

$$\Delta\sigma = \Delta n e\mu + en\Delta\mu \qquad (1)$$

where $\Delta\sigma$ = conductivity change, Δn is the change in free carrier density, e is the electronic charge, μ is the carrier mobility, and $\Delta\sigma$ is the change in carrier mobility. Usually, the first term in Equation (1) predominates.

Photoconductivity gain, G, may be defined as the number of interelectrode transits that can be made by an electron until the photo-generated hole is eliminated by recombination. For

the case treated here, namely where one type of carrier predominates, the gain, G, can be stated as:

$$G = \tau\mu^{VL-2} \qquad (2)$$

where τ is the carrier lifetime, μ is the mobility, V is the applied voltage, and L is the spacing between electrodes. Since the specific sensitivity, S, varies as the product of carrier lifetime and mobility, $S\alpha\mu\tau$, we can state that

$$G\alpha(VL^{-2})S \qquad (3)$$

Commercially, photoconductive devices are used as (1) Detectors of radiation; (2) switches which are sensitive to light and which can actuate relays; and (3) in combination with other photoelectronic materials, such as electroluminescent materials, as image intensifiers. Germanium and silicon devices of the *p-n* junction phototransistor type have long been used in computer detectors; lead sulfide has been used in photocells for infrared detection; cadmium sulfide or cadmium selenide have been used in photocells for detection of light in the visible range; zinc oxide and selenium devices have been used in photocopying machines; antimony sulfide has been used in television pickup tubes. There are numerous other applications.

Photovoltage or the *photovoltaic effect* may be defined as the conversion of light photons to electrical voltage by a material. Becquerel, in 1839, was the first to discover that a photovoltage was developed when light was shining on an electrode in an electrolyte solution. Nearly half a century elapsed before this effect was observed in a solid, namely, selenium. Again, many years passed before successful devices, such as the photoelectric exposure meter, were developed. Radiation is absorbed in the neighborhood of a potential barrier, usually a *p-n* junction, or a metal-semiconductor contact, giving rise to separated electron-hole pairs which create a potential. An equivalent circuit for a photovoltaic cell is shown in Fig. 2, where R_{SH} and R_S

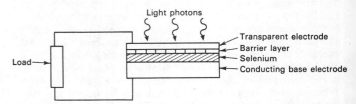

Fig. 2.   Diagram of photovoltaic cell.

are the internal shunting and series impedances; I_J is the junction current; R_L is the load resistance; I_S is a constant-current generator; and V_L and I_L are the voltage and current developed across the load. With R_L optimum, the maximum conversion efficiency, η_{max} can be given by:

$$\eta_{max} = \frac{100 \, V_{mp}I_{mp}}{P_{in}}$$

in which case, V_{mp} and I_{mp} are the voltage and current across R_L, and P_{in} is the radiant input power.

Photovolatic cells have found numerous applications in electronic and aerospace applications, notably in satellites (solar cells) for instrument power. Materials used, in order of decreasing theoretical efficiency, include gallium arsenide (24%); indium phosphide (23%); cadmium telluride (21%); silicon (20%); gallium phosphide (17%); and cadmium sulfide (16%). See Fig. 3. In the past, disadvantages of photovoltaic cells have included: (1) High susceptibility to radiation damage; (2) high cost; and (3) requirement for auxiliary battery power when a source of radiation for the cells is not available.

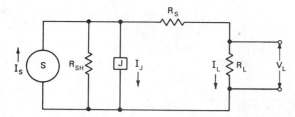

Fig. 3. Photovoltaic cell equivalent circuit.

PHOTOGRAPHIC EMULSIONS.

PHOTOGRAPHIC EMULSIONS. Imagery is the representation (pictorial, graphical, etc.) of a subject by sensing quantitatively the patterns of electromagnetic radiation emitted by, reflected from, or transmitted through a subject of interest (object, body, scene, etc.). Imagery is not wavelength-limited, but is achievable (theoretically if not practically) with all bands of the electromagnetic spectrum—gamma rays, x-rays, ultraviolet radiation, visible light, infrared radiation, radar and radio waves.

Chemical imagery or traditional *photography*, as initially conceived and as commonly practiced, depends upon visible light and uses an optical light-gathering and focusing system (camera) and a light-sensitive medium (film emulsion) to record (store) the image—a *photo-image*. The subsequent availability of infrared, ultraviolet, and x-ray sensitive films extended the capabilities of traditional photography well beyond its dependence upon visible light. The word *photography* derives from the Greek roots *photos* (light) and *graphos* (to draw). Coining of the term is usually attributed to Herschel, although this has not been proved conclusively. Herschel did use the term in a memo dated January 17, 1839 and in a technical paper given on March 14, 1839.

Electronic imagery, instead of using chemical means (emulsions), takes advantage of the sensitivity of various electronic detectors to different bands of the electromagnetic spectrum. The energy received is transduced by these sensors into an electronic or electrical effect (change of resistance, current, emf, the emission of electrons, etc.), from which effects an option of ways to process and display the information is available. The most common form of electronic imagery is found in television. Image orthicons, vidcons, and the more recent TV cameras using charge-coupled devices are among recent developments in electronic imagery.

Electronic imagery is particularly attractive for situations where image information must be transmitted over long distances where digitized signals offer greater accuracy and reliability—and where the incoming information is immediately compatible with digital data processing and computing equipment. Electronic imagery also has made certain imaging tasks possible, such as radar imaging, where traditional photographic means do not suffice. Nevertheless, as of the mid-1980s, advancements in traditional photography, depending heavily upon chemical phenomena, continue at a good pace and film sales are at an historic peak.

Daguerreotype. The first practical process of photography was invented by Louis J. M. Daguerre of Paris in 1837, although the details of the process were not published until 1839. The process was used chiefly for portraiture and became obsolete within a few years after the introduction of the wet collodion process in 1851. Although the daguerreotype process was the original, modern photography is based on the negative-positive methods introduced the same year by William H. Fox-Talbot of England. This was known as the *calotype* process. In the daguerreotype process a light-sensitive layer of silver iodide is formed on a silver plate by contact with iodine. After exposure in the camera, a positive image is produced when the image is exposed to mercury and heated. The mercury, by attaching itself to the unexposed portions, forms a positive image. The silver iodide remaining was removed at first with a solution of sodium chloride (salt) which was soon replaced, however, with sodium thiosulfate (hypo), the properties of which had been discovered by Herschel in 1819. The daguerreotype image so produced is very weak. In 1840 Fizeau described a process of toning with gold which greatly increased the strength of the image and was generally adopted.

At first, from 5 to 10 minutes' exposure was required on open landscapes and street scenes. The invention of a fast, large-aperture portrait lens by Petzval in 1841 and the discovery by Goddard in London (1840) of the superior sensitivity of silver bromide reduced the time of exposure to a few seconds.

Problems were encountered in preparing positive prints from the calotype negatives because of reproduction of the grain of the paper which contained the negative. Attempts were made to wax or oil paper negatives, but these were essentially unsuccessful. De Saint-Victor attempted to coat plates with albumin (egg white). Upon hardening of the albumin, the plates were bathed in silver nitrate, causing precipitation of silver iodide within the film of albumin. This was not successful because the sensitivity of the plates was greatly lowered.

Early Emulsions. The use of collodion in photographic emulsions dates from 1851 when Frederick Scott Archer published details of his wet collodion process. Although this process is no longer in general use, it can be used in making the half-tone negatives required in photoengraving. In the collodion process, a clean glass plate is first coated with collodion containing potassium iodide and potassium bromide. It is next sensitized by immersion in a solution of silver nitrate. It is then placed in a plate holder—specially designed for the handling of the wet plate—and the exposure made. After exposure it is developed in a solution of ferrous sulfate and fixed in potassium cyanide, or in hypo, washed and dried. The wet collodion process, as it is used by the photoengraver, results in a negative of high density and extreme contrast, high resolution and with an extremely fine grain. These characteristics render wet collodion well adapted to the requirements of photoengraving. Much later, the wet collodion process was essentially replaced by the gelatino-bromide emulsions of similar characteristics.

Collodion printing-out paper was introduced by Obernetter of Munich in 1867 and was for many years the favorite printing process of the portrait and professional photographer. It was in general use until the early years of the present century when it was gradually replaced by developing-out paper.

Gelatin Emulsions

These are not true emulsions, but suspensions of minute silver halide crystals dispersed in a protective colloid medium (gelatin). The suggestion of replacing collodion with gelatin was first made by R. L. Maddox in 1871. The first plates made by Maddox were not very sensitive, but their advantages far outweighed their defects, leading to further developments by Charles Bennett in England in the late 1870s, and the first mass production attempts by George Eastman in 1880. One of the several contributions of Eastman to photography was his early recognition of making and marketing gelatin dry plates on a large scale, eliminating the need for the photographer to prepare his own plates, as well as the need for developing and fixing the plates immediately thereafter. There soon followed the concept of strip film, making it unnecessary to change plates after each exposure. Eastman avoided the grain problem by using a coating that

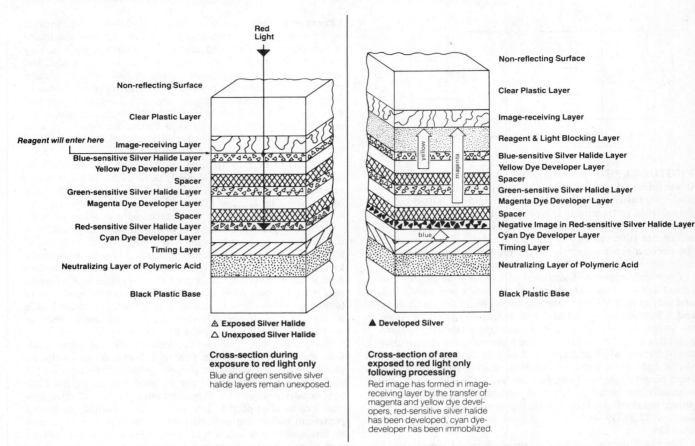

Cross section of Land film used with Polaroid Supercolor Time-Zero SX-70 Land camera. (*Polaroid.*)

enabled the stripping of the thin layer of gelatin from the paper support. Later, in 1889, he replaced the paper support with a transparent plastic support (nitrocellulous), thus making it possible to produce prints without the need of stripping the gelatin layer from the support. Eastman's goals were to make it easy for the masses of people to take photographs in a simplified manner and, through mass production, market equipment at a price within grasp of the public.

Gelatin is a preferred photograpic colloid because the sensitizing bodies in the gelatin make possible emulsions with great sensitivity and speed. Gelatin is an excellent emulsifying agent and is readily transformed, from gel to a liquid or the reverse, by changes in temperature. The latter property makes coating of supports and emulsion processing and working feasible. The strong protective action of gelatin lowers the rate or reduction of unexposed silver halide crystals in developers so that image formation is readily obtained.

Silver halides employed in emulsions are the chloride, the bromide and the iodide. Negative emulsions are composed of silver bromide with a small amount of silver iodide. Positive emulsions for films and paper contain silver chloride, or mixtures of silver chloride and silver bromide in varying amounts, according to the tone, speed, and contrast desired.

In photomicrographs of negative emulsions, the crystals of silver bromide appear as flat triangular or nexagonal plates with rounded corners. Some globular and needle-shaped crystals are also observed. The thickness of the flat plates is approximately $\frac{1}{10}$ of their diameter. The size of silver bromide crystals range from less than 1 to 4 micrometers. Crystals of silver chloride, or mixtures of silver chloride and silver bromide, as used in positive emulsions, are quite uniform and seldom exceed 0.5 micrometer in diameter. Multi-layered emulsions contain approximately 1 billion, 10^9, crystals per square centimeter. The areas of individual crystals range from 0.1×10^{-8} centimeters for low-speed emulsions to 1.0×10^{-8} centimeters for high-speed negative emulsions.

The characteristics of an individual emulsion are primarily dependent on two factors, the size-frequency distribution of the crystals and the composition of the silver halide crystals. The chief problems of the emulsion-maker are the production of uniform suspensions of silver halide crystals with proper-size frequency distribution and correct composition in gelatin, and the ability to reproduce results.

Classification of Emulsions

1. *Printing-Out Emulsions*. These emulsions produce images on exposure without development. They are used largely for making portrait proofs which are distinguished by their red or purplish color. Emulsions of this type differ from others in that they usually contain silver nitrate, some free silver, silver salt of an organic acid and a weak free acid. These are known as P.O.P. Proof Papers.

2. *Developing-Out Emulsions*. Emulsions for development have an excess of alkaline halides. By varying the composition of the silver halide and treatment, developing-out emulsions may be prepared which are suitable for either negative or positive purposes.

a. Negative emulsions. Negative emulsions are prepared by adding a small amount of a soluble iodide to the bromide used in making the silver halide. The mixed crystals of silver-bromiodide formed are more sensitive to light and produce emulsions with greater speed than silver bromide alone. Negative emulsions are referred to as neutral emulsions if precipitation of the silver halide is carried out in a gelatin solution with an excess of soluble bromide, and ammonia emulsions if the precipitation takes place in a gelatin solution with an excess of soluble bromide in the presence of ammonia or ammoniacal silver. The latter method produces emulsions with coarser grains which have the highest sensitivity.

b. Positive emulsions. Positive emulsions are prepared by precipitating silver halides containing chloride or mixtures of chloride and bromide in gelatin. The size of the crystals formed are smaller than those of negative emulsions and have a lower sensitivity. Positive emulsions are divided into four classes, according to the composition of the silver halides and their properties.

Chloride emulsions. Because of their slow speed chloride emulsions are used largely for contact printing.

Bromide emulsions. Bromide emulsions are very sensitive and fast. They are used for projection printing exclusively.

Chlor-bromide emulsions. In chlor-bromide emulsions the amount of silver chloride is greater than that of silver bromide. These emulsions are somewhat faster than chloride emulsions and used for contact or slow projection printing. Chlor-bromide emulsions produce warm-toned silver images with a brown or brown-black color.

Brom-chloride emulsions. Brom-chloride emulsions contain more silver bromide than silver chloride. They are faster than chlor-bromide emulsions and used for projection printing where black images and speed printing are desired. Image tones of brom-chloride emulsions are not as warm as chlor-bromide images nor as cold as bromide images.

Manufacture of Commercial Emulsions

Although the details are proprietary, the basic procedures of manufacturing commercial emulsions are known. A portion of the gelatin in the formula is swelled by soaking in water and later dissolved with heat. Mixtures of soluble bromides and iodides, or chlorides, are placed in water solution and added to the gelatin solution. Precipitation of silver halides is accomplished by slowly adding a solution of silver nitrate, while stirring, to the mixture. The relative concentration of the solutions, the rate of addition and temperature during mixing, are factors which control the formation, size and dispersion of the crystals in gelatin. The emulsion is then heated or "ripened" at 40–80°C to recrystallize the silver halides and readjust the size-frequency distribution. Following ripening, more gelatin is added and the emulsion is chilled so it will set quickly. The emulsion is then placed in a press and forced through a screen to break it into shreds or noodles, which are washed, in cold running water to remove the potassium nitrate formed, the excess soluble halides, and certain soluble by-products of the reaction. Chloride emulsions are often prepared without washing or with only a limited washing. After washing, the emulsion is drained, remelted, and additional gelatin and certain agents, such as fog preventatives, are added. The emulsion is then heated, or "after-ripened," to form sensitizing nuclei on the silver halide crystals. This operation increases the sensitivity and contrast of the emulsion and is necessary for the preparation of high-speed negative emulsions. Certain preservatives, or stabilizers, are added so the emulsion can be stored in refrigerated rooms until needed. Before coating the emulsion is melted and sensitizing dyes, hardening agents, wetting agents, etc., are added. After thorough mixing, filtering and heating to coating temperature, it is placed in a coating machine. Supports, as film, paper, or glass, with substratum coatings are fed through machines at proper rates so they become coated with emulsions in uniform layers of desired thickness. The coated supports pass over chill boxes to set the emulsion and then through a series of drying compartments where the rate of drying is carefully controlled so as not to change the sensitivity on the surface. Following drying, the coatings are inspected under proper safelights and the film or paper is cut to desired size and packaged.

Numerous variations in the manufacturing process make possible a wide range of film characteristics, including film speed and spectral sensitivity. Film, unlike the human eye, can extend beyond the visible region of the spectrum. High-speed film can capture the details of a fast-moving object, seen only as a blur by the eye. By extended exposure, film can capture images entirely too faint to be seen by the eye. The three main types of film emulsions for black-and-white photography are: (1) *ordinary* (color-blind; sensitive to blue light only); (2) *orthochromatic* (sensitive to all but red light); and (3) *panchromatic* (sensitive to light of all colors). Ordinary and orthochromatic films generally offer greater contrast than most panchromatic emulsions. However, the response of panchromatic emulsions can be modified by use of color filters. Film is available in several sizes and formats. Obviously, a delineation of film specifications is beyond the scope of this volume.

Color Films

The trichromatic theory of vision was first proposed by Thomas Young, a British physicist in 1801. He was the first to propose that the retina of the eye incorporates three different types of receptors, responding to blue, green, and red light, respectively. The theory was elaborated upon to the extent that color perception is based upon the stimulation of two receptors, with light stimulating both red and green receptors seen as yellow light; light equally stimulating all three types of receptors seen as white, etc. Young concluded that it should be feasible to match any color of the spectrum through the proper mixing of blue, green, and red light. Although not essentially interested in color photography, Maxwell effectively demonstrated the principle by way of specially-prepared lantern slides before the Royal Institution in London in 1861. Maxwell had demonstrated the *additive color principle* (mixing of blue, green, and red light).

Practical color photography on a massive amateur scale, of course, could not depend upon the preparation of three separate photographs and the use of three projectors, but rather dictated a process that would combine the three records on one plate. In 1907, the Lumièe Autochrome plate was developed. This was comprised of a very coarse mosaic of potato starch grains, one third of which was dyed blue; another third, green, and the remaining third, red. An emulsion layer was exposed, with the light first passing through the mosaic. In the *Kodacolor* system of 1928, filters in the camera were used instead of color mosaics. A major problem of the mosaic and filter approaches was that of loss of light as it passed through one or the other media, greatly reducing sensitivity and loss of brightness of a projected image.

In the *subtractive color system*, the phenomenon of absorption is involved. A dye that will absorb red light will, in turn, reflect green and blue light, thus appears a greenish-blue (cyan); a dye that will absorb green light appears a bluish-red (magenta); and a dye that absorbs blue light appears yellow. Thus, cyan, magneta, and yellow are the three primary subtractive colors. A mixture of all three dyes in proper proportion will absorb all primary light and thus appear black. Most processes of color photography make use of a subtractive synthesis to yield prints or transparencies.

Color-separation negatives are photographic negatives which record the relative intensities of the primary colors used in the analysis necessary to reproduce a subject by means of color photography. In three-color photography, for example, the separation negatives are records, in terms of silver densities, of the amounts of red, green and blue light received at the camera from the subject.

A set of color-separation negatives may be prepared by photographing the subject three times on separate color-sensitive emulsions so that each is a record of one of the primary colors. A panchromatic emulsion is generally employed with a set of tricolor filters, the colors of the primaries. It is only necessary, however, to obtain the color records on separate negatives so it is also possible to use for each record any combination of color filter and emulsion sensitivity that will record one of the primary colors. A set of color-separation negatives may be made by exposing (1) each one in turn in a camera, (2) by the use of a color camera which will expose them simultaneously, or (3) in a tripack.

It is common practice to balance a set of color-separation negatives, by altering the exposure and development times, so that a gray scale will be recorded equally on each negative. The particular densities desired are dependent on the method of color synthesis to be employed.

The majority of color is by use of integral tripacks. There are three layers of photographic emulsion in the tripack, one layer sensitive to red light, another layer to green light, and another to blue light. They are coated, one on top of the other. Since silver iodobromide emulsions usually selected for film emulsions are sensitive to blue light, sensitivity to the green and red light must be conferred by sensitizing dyes. Although this sensitivity can be obtained, the dyes do not negate the emulsion's natural sensitivity to blue light. Thus, those layers that are sensitive to green and red light must be protected from blue light. This is accomplished by inserting a yellow filter layer that will absorb the blue light. Chloride emulsions on the other hand are sensitivity only to ultraviolet light. Whereas they do not require a yellow-filter layer, they have to incorporate a filter for exclusion of untraviolet light. There are a number of dyes that may be used in dye-transfer systems, but for tripacks it is necessary to select only those dyes that will be formed during the development process. In 1912, the German scientist, Rudolf Fischer, discovered the role of couplers. In his early version of a color film, he placed three layers of emulsion one atop another as previously described and he also incorporated a coupler in each layer to cause the development of a particular color. Fischer's concept was brilliant, but the actual process failed because the couplers and sensitizers tended to wander from layer to layer.

In 1931, Leopold Godowsky, a violinist, and Leopold Mannes, a pianist, and both avid amateur photographers made crude experiments in a home laboratory on a type of color film that ultimately became *Kodachrome*, released by Eastman in 1935. In the *Kodachrome* process, the couplers are laced in the developers instead of in the emulsions. Phenols are usually the couplers that form cyan dyes; nitriles or pyrazolones form magenta dyes; and esters, ketones, or amides form yellow dyes. There are many hundreds of couplers and, consequently, there is continuing improvement in color film. Space here does not permit a detailed description of such important matters as the negative-positive system, reversal systems for transparencies, color corrections, etc., but these areas are well covered in some of the listed references.

Direct Positive Images. Even in the early days of black-and-white photography and the early work of Daguerre and Fox-Talbot, it was realized that there would be a great advantage gained from a system that would initially produce a positive rather than a negative image. As early as the late 1830s, Hippolyte Bayard and Robert Hunt proposed systems, but these did not produce satisfactory results. The *chemical transfer* process was developed by A. Rott in Belgium in 1939 and found application in the document copying field. In 1947, E. H. Land demonstrated a camera which produced a finished black-and-white print without need for a negative and one that was available to the photographer within a very short period, approximately one minute. This was the first model of the *Polaroid* camera. In chemical transfer, a normal emulsion is used. Immediately after exposure and while within the camera, it is developed in a solution containing combined developer-fixer agents. The emusion is in contact with a special positive white paper, not light sensitive, on which the finished image is printed. The developing reagent is of a jellylike consistency and in early models was contained in pouches or pods, one for each picture. The exposed grains develop in the normal fashion. The unexposed grains are dissolved by the fixing agent. Thus, in the unexposed areas, the dissolved halide is silver which forms on the nuclei in the receiving sheet. In connection with partially-exposed areas, the developing grains and the receiving sheet nuclei compete for the silver. Thus, a negative image is formed on the original film or paper, whereas a positive image appears on the receiving sheet. Subsequent to the first Polaroid camera, models were developed to provide a permanent negative as well as print, with the processing time reduced to seconds.

A cross section of the film used in the *Polaroid Supercolor Time-Zero SX-70 Land* camera, introduced in 1980, is shown in the accompanying figure.

Descriptions of special films, such as infrared- and ultraviolet-sensitive types, are beyond the scope of this volume. This is also true of cameras other photographic equipment and of the many thousands of applications for photography.

PHOTOIONIZATION. This process, which is also called the *atomic photoelectric effect*, is the ejection of a bound electron from an atom by an incident photon whose entire energy is absorbed by the ejected electron. This statement means that photoionization cannot occur unless the energy of the photon is at least equal at the ionization energy of the particular electron in the particular atom; any excess of energy in the photon above this value appears as kinetic energy of the ejected electron.

PHOTOLUMINESCENCE. Luminescence.

PHOTOLYSIS. Free Radical; Photochemistry and Photolysis.

PHOTONUCLEAR REACTION. A nuclear reaction induced by a photon. In some cases the reaction probably takes place via a compound nucleus formed by absorption of the photon followed by distribution of its energy among the nuclear constituents. One or more nuclear particles then "evaporate" from the nuclear surface, or occasionally the nucleus undergoes photofission. In other cases the photon apparently interacts directly with a single nucleon, which is ejected as a photoneutron or photoproton without appreciable excitation of the rest of the nucleus.

PHOTOSYNTHESIS. This is the most important of all biological processes. With negligible exceptions the existence of the entire biological world hinges upon this process. From a few simple inorganic compounds and from the sugar made in photosynthesis are erected all of the complex kinds of molecules essential in the construction of the bodies of plants and animals or to maintenance of their existence. Some of these subsequent synthetic processes occur in the plant body; others in the bodies of animals after they have ingested plant materials as foods. Likewise, the energy used by plants and animals represents sunlight energy which was entrapped in sugar molecules during photosynthesis. The entire organic world runs by the gradual expenditure of the energy capital accumulated in photosynthesis.

Under suitable conditions of temperature and water supply, the green parts of plants, when exposed to light, abstract and use carbon dioxide from the atmosphere and release oxygen to it. These gaseous exchanges are the opposite of those occurring in respiration and are the external manifestation of the process of photosynthesis by which carbohydrates are synthesized from carbon dioxide and water by the chloroplasts of the living plant cells in the presence of light. For each molecule of carbon dioxide used, one molecule of oxygen is released. A summary chemical equation for photosynthesis is:

$$6CO_2 + 6H_2O \xrightarrow{\text{light}} C_6H_{12}O_6 + 6O_2.$$

In this process, the radiant energy of sunlight is stored as chemical energy in the molecules of carbohydrates and other compounds which are derived from them.

All photosynthetic organisms except *bacteria* use water as the electron or hydrogen donor to reduce various electron acceptors and from the water they evolve molecular oxygen. Anaerobic bacteria cannot endure such oxygen, but derive their sustenance through slightly different photosynthetic routes:

$$2H_2S + CO_2 \xrightarrow{\text{light}} (CH_2O) + H_2O + 2S$$

or

$$2CH_3CHOHCH_3 + CO_2 \xrightarrow{\text{light}} (CH_2O) + CH_3COCH_3 + H_2O$$

Photosynthesis takes place in chlorophyll-containing cells only when carbon dioxide, water and light are available, and when a suitable temperature prevails. Although carbon dioxide constitutes, on the average, only 0.03% of the atmosphere, land plants are entirely dependent upon this source for the carbon dioxide used in photosynthesis. It has been shown experimentally that an increase in the carbon dioxide concentration of the atmosphere results in an increased rate of photosynthesis. On the other hand, a deficiency of water results in a reduced rate of photosynthesis. In nature, sunlight is the source of radiant energy used in photosynthesis, although plants will also photosynthesize under artifical light sources of suitable quality and intensity.

The total radiant energy received at the earth's surface is 1–2 gram calories/square centimeter/minute, depending upon altitude, or approximately 1 hp/10–20 square feet. For crop plants in the field, a maximum of 2–3% of this energy remains stored in the plants at the end of the growing season. During that time about 20% more is actually used in photosynthesis and lost by respiration of the plant, the remainder of the energy being dissipated by radiation, transmission through the leaves, and evaporation of water from the plant.

The intensity, quality and daily duration of illumination all have influence on the amount of photosynthesis accomplished per day. Clearly, the longer the daily period of illumination, the more photosynthesis will be accomplished by a plant in the course of a day. The minimum light intensity at which a measurable rate of photosynthesis occurs varies according to species, but is seldom less than 1% of full mid-day summer sunlight. Under natural conditions, maximum rates of photosynthesis are attained in single leaves of many species at 25–35% of full sunlight intensity and in some shade species at even lower intensities. For equal intensities, more photosynthesis appears to occur in the orange–short red and blue parts of the spectum than in the green and yellow. This is because the chlorophyll pigments of the leaves absorb light energy at wavelengths of 6600 and 4250 micrometers. Radiation is most intense in the green and, if this radiation were absorbed, the plant could not utilize it and would overheat.

The range of temperatures most suitable for relatively rapid rates of photosynthesis is not the same for all kinds of plants. In general, it is higher in tropical than in temperate species, and higher in temperate species than in those of subarctic regions. Increase in temperature results in an increase in the rate of photosynthesis up to an optimum which varies with the variety of plant, but which, for most temperate zone species, lies within the range of 20–30°C. With increase in temperature above the optimum, the rate of photosynthesis progressively decreases.

In the vascular plants, photosynthesis occurs chiefly in the leaves. Carbon dioxide diffuses into the intercellular spaces of the leaf from the atmosphere via the stomates, and then dissolves in the moist walls of the mesophyll cells. In solution, the carbon dioxide diffuses to the surface of the chloroplasts, which are the actual seat of the photosynthetic process. The first major step in photosynthesis is the absorption of radiant energy by the plant pigments in the chloroplasts, with the generation of electrons. The plant pigment consists of two closely similar pigments, chlorophyll and chlorophyll *b*, which are porphyrin-derived complexes of magnesium and which, upon excitation by radiant energy, become electron donors. See also **Chlorophylls.** The chloroplast is a complex, self-replicating organelle that possesses its own DNA and is able to synthesize at least a few of the proteins needed for its own functioning. It is filled with membranous thylakoid sacs which are specifically designed to harness the energy available in the excited electron and to carry out the light phase of photosynthesis. In this, the light energy captured is converted into the chemical energy of adenosine triphosphate (ATP) and nicotinamide-adenine dinucleotide phosphate (NADPH). See also **Adenosine Phosphates;** and **Coenzymes.** Hydrogen atoms are removed from water and used to reduce NADP, leaving behind molecular oxygen. Simultaneously, adenosine diphosphate (ADP) is phosphorylated to ATP:

$$\text{Water} + NADP^+ + PO_4 + ADP \xrightarrow{\text{light}} \text{Oxygen} + NADPH + H^+ + ATP$$

In the second, or dark, reaction phase, NADPH and ATP provide the energy to reduce carbon dioxide to glucose and are themselves oxidized or decomposed:

$$CO_2 + NADPH + H^+ + ATP \rightarrow \text{Glucose} + NADP^+ + ADP + PO_4$$

Peter Mitchell (Nobel Prize 1978) of Great Britain was the first to realize and to propose in his chemiosmotic theory that the energy required for the ADP-ATP reaction could be derived by an accretion of protons in the thylakoid sac to the point at which the electrochemical gradient across the membrane could effect the proton transport required as the driving force for this reaction.

See also **Phosphorylation (Photosynthetic)**.

In most plants, the water used in photosynthesis is absorbed by the roots from the soil whence it is translocated to the leaves and, except for a small portion used in respiration, the oxygen liberated in the process diffuses out of the leaf into the atmosphere, mostly through the stomates.

Carbohydrates other than hexoses are synthesized in the leaves, apparently as a result of secondary reactions following photosynthesis. Sucrose invariably accumulates in actively photosynthesizing leaf cells. This more complex sugar is built up from the molecules of the simpler hexoses. In most plants, insoluble starch also accumulates in leaf cells during photosynthesis. This carbohydrate is synthesized by the condensation of numerous glucose molecules. The sucrose and starch contents of leaves decrease at night as a result of the continued translocation from the leaves to other parts of the plant. The sucrose is probably translocated as such, but the starch must first be converted into simpler, soluble sugars before it can move out of the leaves. Synthesis of starch is not restricted to the green parts of plants; a familiar example of this is the accumulation of starch in potato tubers. Starch in the nongreen cells is made from glucose which comes from the leaves or other photosynthetic organs. Starch occurs in cells in the form of small grains, the type of grain being formed in each kind of plant being more or less characteristic of that species.

Finally, it must be realized that photosynthesis is not the sole prerogative of the higher plants. More than half the photosynthesis on the earth's surface is carried out in the oceans by phytoplankton.

References

Arntzen, C. J.: "Dynamic Structural Features of Chloroplast Lamellae," *Current Topics in Bioenergetics*, **8**, 111–160 (1978).
Barber, J., Editor: "Primary Processes of Photosynthesis," Elsevier, New York (1977).
Bassham, J. A.: "Increasing Crop Production through More Controlled Photosynthesis," *Science*, **197**, 630–638 (1977).
Govindjee, A.: "Bioenergetics of Photosynthesis," Academic, New York (1974).
Miller, K. R.: "A Particle Spanning the Photosynthetic Membrane," *Jrnl. of Ultrastructure Research*, **54**, 1, 159–167 (1976).
Miller, K. R.: "The Photosynthetic Membrane," *Sci. American*, **241**, 4, 102–113 (1979).
Staehelin, L. A., Armond, P. A., and K. R. Miller: "Chloroplast Membrane Organization at the Supramolecular Level and Its Functional Implications," *Brookhaven Symposia in Biology*, **28**, 278–315 (1976).

—R. C. Vickery, Hudson Laboratories, Hudson, Florida.

PHTHALIC ACID. $C_6H_4(COOH)_2$, formula weight 166.13, mp 208°C (ortho), 330°C (meta and iso), the ortho form sublimes and the meta and iso forms decompose with heat, sp gr 1.593 (ortho). Phthalic acid is very slightly soluble in H_2O, soluble in alcohol, and slightly soluble in ether. The solid form is colorless, crystalline. Because of their chemical reactivity and versatility, phthalic acid derivatives find wide use as starting and intermediate materials in important industrial organic syntheses. A common starting material is phthalic anhydride which is formed when phthalic acid loses water upon heating. See also **Phthalic Anhydride**; and **Terephthalic Acid**.

Orthophthalic acid is made by the oxidation of naphthalene (1) with H_2SO_4 fuming heated, in the presence of mercuric sulfate—SO_2 is also formed and recovered; (2) with air in the presence of vanadium pentoxide at 450 to 520°C. Orthophthalic acid also is formed when benzene compounds containing carbon ortho-substituted groups are oxidized. Orthophthalic acid is used in the manufacture of indigo and other dyes.

PHTHALIC ANHYDRIDE. $C_6H_4(CO)_2O$, formula weight 148.11, mp 130.8°C, bp 284.5°C, sp gr 1.527. Phthalic anhydride is very slightly soluble in H_2O, soluble in alcohol, and slightly soluble in ether. The compound is a high-tonnage chemical and is widely used in a variety of industrial organic syntheses. Although phthalic anhydride may be derived directly from phthalic acid by heating and dehydration, it usually is prepared on a large scale by (1) oxidizing naphthalene, or (2) from the petroleum derivative, orthoxylene. Phthalic anhydride, in addition to its use as a raw and intermediate material for syntheses, finds wide application in the chlorinated form as a compounding ingredient for plastics. The chlorine content is approximately 50%. The compound provides increased stability and improved resistance of plastics to high temperatures.

Representative reactions of phthalic anhydride include: (1) phthalic anhydride reacts with phosphorus pentachloride to form phthalyl chloride which, upon rearrangement, can be transformed to unsymmetrical phthalyl chloride; (2) both forms of phthalyl chloride react with zinc plus acetic acid to form unsymmetrical phthalide, or with benzene plus aluminum chloride to form unsymmetrical-diphenylphthalide (phthalophenone); (3) phthalic anhydride reacts with NH_3 to form phthalimide $C_6H_4(CO)_2NH$; (4) phthalimide reacts with KOH in alcohol to form potassium phthalimide; (5) treatment of potassium phthalimide with an alkyl halide (e.g., ethyl chloride) forms an alkyl phthalimide (e.g., ethyl phthalimide); (6) ethyl phthalimide, when heated with fuming HCl, yields the primary amine $C_2H_5NH_2$ (ethyl amine) in a reaction used for the production of many primary amines and known as Gabriel's synthesis; (7) ethyl phthalimide, when treated with sodium hypochlorite, forms sodium anthranilate which upon treatment with an acid yields anthranilic acid; (8) phthalic anhydride reacts with phenol to form phthaleins, such as phenolphthalein, when in the presence of concentrated H_2SO_4; (9) phthalic anhydride reacts with resorcinol to form resorcinolphthalein (fluorescein); (10) fluorescein reacts with bromine to form tetrabromofluorescein, the potassium salt of which is eosin (a red dye for wool and silk); (11) phthalic anhydride reacts with N-diethyl-meta-aminophenol to form N-diethyl-meta-aminophenolphthalein (rhodamine) which is a red dye. See also **Phthalic Acid**; and **Terephthalic Acid**.

PICOLINES. Pyridine and Derivatives.

PIEZOELECTRIC EFFECT. The interaction of mechanical and electrical stress-strain variables in a medium. Thus, compression of a crystal of quartz or Rochelle salt generates an electrostatic voltage across it, and conversely, application of an electric field may cause the crystal to expand or contract in certain directions. Piezoelectricity is only possible in crystal classes which do not possess a center of symmetry. Unlike electrostriction, the effect is linear in the field strength.

The directions in which tension or compression develop polarization parallel to the strain are called the piezoelectric axes of the crystal. Thus the axis of a hexagonal quartz crystal indicated by the arrows in Fig. 1 is known as an "X-axis," and a plate cut, as shown, with its faces perpendicular to this direction

Fig. 1. Hexagonal quartz crystal showing X axis.

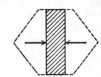

is an "X-cut"; while one cut with its faces parallel to the lateral faces of the crystal is a "Y-cut."

The magnitude of the piezoelectric polarization is proportional to the strain and to the corresponding stress, and its direction is reversed when the strain changes from compression to tension. The principal piezoelectric constants of a crystal are the polarizations per unit stress along the piezoelectric axes. While these constants are much greater for Rochelle salt than for quartz, the latter is better adapted to some purposes because of its greater mechanical strength. It is also stable at temperatures over 100°C.

If a quartz plate is subjected to a rapidly alternating electric field, the inverse piezoelectric property causes it to expand and contract alternately. As an elastic body, the plate has a certain natural frequency of expansion and contraction in the direction of the field, and if the field is made to alternate with the same frequency, the plate responds with a vigorous resonant vibration. This reacts, through the direct piezoelectric property, to augment the electric oscillations. A circuit arranged for this purpose, as in Fig. 2, is known as a piezoelectric or crystal oscillator,

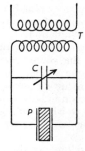

Fig. 2. Piezoelectric oscillator circuit.

the crystal itself, P, being the peizoelectric resonator; T is the oscillation transformer, and C a variable condenser. This device has been much used as a frequency control in radio transmitters. Both X-cut and Y-cut quartz plates are subject to changes of frequency with temperature, due to change of elastic modulus; but certain planes in the crystal have been found, oblique to both X and Y, such that plates cut parallel to them are nearly free from the temperature effect.

In addition to natural quartz, Rochelle salts, and tourmaline, synthetic crystals, such as ethylenediamine tartrate (EDT), dipotassium tartrate (DKT), and ammonium dihydrogen phosphate (ADP) have varying suitability as piezoelectric elements. While Rochelle salt has a greater piezoelectric effect than any other crystal, it has the disadvantage of a greater sensitivity to temperature change than quartz. EDT has an advantage over quartz when used in frequency-modulated oscillators because of the wide gap between its resonant and antiresonant frequencies. See **Quartz.**

PIGMENTATION (Plants).
The distinctive green color of leaves and other plant organs results from the presence in such organs of two pigments called chlorophyll *a* and chlorophyll *b*. In the higher plants these pigments occur only in the chloroplasts. These pigments play so important a role in the fundamental process of plant life, photosynthesis, that their chemical reactions are discussed at length in that entry. The chlorophylls are not water-soluble but can be readily dissolved out of leaf tissues with alcohol, acetone, ether, or other organic solvents. The resulting solutions exhibit the phenomenon of fluorescence; they are deep green when held between an observer and the light, but deep red when viewed in reflected light. By suitable treatments it is possible to obtain pure cyrstals of chlorophyll

from such solutions. Most leaves contain considerably more chlorophyll *a* than chlorophyll *b*, often two to three times as much. In the organs of the higher seed plants, with rare exceptions, chlorophyll is synthesized only upon exposure to light. Leaves of grass which develop under a board, for example, contain no chlorophyll. In the leaves of mosses, ferns, and gymnosperms, however, chlorophyll develops in the dark as well as in the light.

Invariably associated with the chlorophylls in the chloroplasts are the yellow pigments, the carotenes and the xanthophylls. These pigments are not, however, restricted in their occurrence to the chloroplasts, but may also be present in nongreen parts of the plant where they commonly occur in chromoplasts. Collectively, these pigments, together with certain others which are closely related chemically, are called the carotinoids. Carotene refers to a class of orange-yellow pigments. They are especially abundant in the roots of carrots. These compounds are of considerable importance because they are the precursors of vitamin A, one molecule of β-carotene being split into two molecules of vitamin A by a simple hydrolytic reaction.

Lycopene, a red pigment of this class, is responsible for the red color of the fruits of tomato, pepper, rose, and some other species. The commonest xanthophylls found in leaves are lutein and zeaxanthin, although others also occur. Another xanthophyll is fucoxanthin, which imparts to brown algae their distinctive color. None of the carotenoids is water soluble, but all of them can be extracted from plant tissues with suitable organic solvents.

Most of the red, blue, and purple pigments of plants belong to the group of anthocyanins. In general, the anthocyanins are red in an acid solution and change in color through purple to blue as the solution becomes more alkaline. Red pigmentation resulting from the presence of the anthocyanins is found in flowers, fruits, bud scales, young leaves and stems, and sometimes even mature leaves as in those of the red cabbage. Blue and purple pigmentation due to the presence of anthocyanins occur principally in flowers and fruits. The anthocyanins are diglucosides of the compounds pelargonidin, cyanidin, delphinidin and apigenidin. These compounds are closely similar in structure, all having the double ring benzopyrylium.

Another group of cell sap water-soluble pigments is the *anthoxanthins*. These pigments are also chemically related to the glucosides. Anthoxanthins often occur in the plant in a colorless form but under suitable conditions their typical yellow or orange color becomes apparent. Some yellow flowers, such as yellow snapdragons, owe their color to the presence of anthoxanthins, but the color of the majority of kinds of yellow or orange flowers is due to carotenoid pigments.

The autumnal coloration of leaves in temperate regions is one of the most spectacular accompaniments of the march of the seasons. Both carotenoid and anthocyanin pigments play an important role in autumnal leaf coloration which is not, contrary to popular opinion, a result of the action of frost. Brilliant development of the anthocyanin pigments in the fall is, however, favored by dry weather during which cool, but not frosty, nights alternate with clear days. During the late summer and early fall the chlorophyll in the leaves gradually decomposes. In many species this simply results in unmasking the yellow carotenoid pigments already present, accounting for the yellow autumnal pigmentation of such species as birch, sycamore, aspen, and tulip trees. In other species synthesis of anthocyanins occurs more or less concomitantly with the disintegration of the chlorophyll; this accounts for the reds or purplish reds characteristic in the autumnal coloration of such species as many oaks, maples, sumacs, and dogwood.

Except in flowers, white is an uncommon color in the externally visible parts of plants, and results from the complete absence of pigments. In some species white streaks or other markings are of common or regular occurrence in leaves, and in the leaves of some species such as roses completely white leaves or even entire branches bearing only white leaves sometimes occur. Such branches cannot be propagated because no photosynthesis can take place in the absence of chlorophyll. As long as such branches remain attached to a plant bearing green leaves they can obtain necessary food from the branches bearing normally pigmented leaves.

See also **Annatto Colors; Carotenoids; Chlorophylls; Colorants (Food);** and **Photosynthesis.**

PINACOL REARRANGEMENT. Rearrangement (Organic Chemistry).

PITCH. Coal Tar and Derivatives.

pK. A measurement of the completeness of an incomplete chemical reaction. It is defined as the negative logarithm (to the base 10) of the equilibrium constant K for the reaction in question. The pK is most frequently used to express the extent of dissociation or the strength of weak acids, particularly fatty acids, amino acids, and also complex ions, and similar substances. The weaker an electrolyte, the larger its pK. Thus, at 25°C for sulfuric acid (strong acid), pK is about −3.0; acetic acid (weak acid), pK = 4.6; boric acid (very weak acid), pK = 9.24. In a solution of a weak acid, if the concentration of undissociated acid is equal to the concentration of the anion of the acid, the pK will be equal to the pH.

PLANT GROWTH MODIFICATION AND REGULATION. Those chemical substances having the most to do with plant growth and form are given the general term, *plant hormones.* A plant hormone or *phytohormone* may be defined as an organic compound produced naturally in plants, which controls growth or other functions at a site remote from its place of production, and which is very active in minute amounts. Three chemically quite different types of compounds apparently act as plant hormones: the *auxins*, the *gibberellins*, and the *kinetins*. In addition, the growth of roots is dependent upon vitamins of the B group which are synthesized in leaves and transported thence to the roots, thus qualifying as hormones.

Auxins. The best studied hormones are those belonging to the class of auxins. These are defined as organic substances which promote growth along the longitudinal axis, when applied in low concentrations to shoots of plants freed as far as practical from their own inherent growth-promoting substances. Auxins generally have additional properties, but this one is critical.

Natural auxins have been identified in a number of instances. Indole-3-acetaldehyde occurs in a number of etiolated seedlings and in pineapple leaves; indole-3-acetonitrile has been isolated from cabbage and its presence indicated in a number of plants. One of the most widely occurring auxins is indole-3-acetic acid, which has been isolated in pure form from fungi and from corn (maize) grains. Its presence has been conclusively demonstrated by biochemical and chromatographic tests in a wide variety of flowering plants, including both mono- and dicotyledons.

Many synthetic auxins have been produced, including 2,4-dichlorophenoxyacetic acid or 2,4-D; napththalene-1-acetic acid; and 2,3,6-trichlorobenzoic acid, among others. Used as a herbicide, 2,4-D is described in the entry on **Herbicide.** By definition, these synthetic compounds are not hormones, al-

though they are sometimes loosely referred to as hormone-type compounds.

An auxin is formed in fruits, seeds, pollen, root tips, coleoptile tips, young leaves, and especially in developing buds. The auxin travels away from the site of production in shoots by a special transporting system, depending on oxygen, which moves it in a predominantly polar direction from apex toward base. Movement in the opposite direction, i.e., from base toward apex, takes place to a variable extent depending upon the tissue and the plant. In the course of the polar transport, a large part of the auxin becomes bound and is no longer transportable. The transport is rather specifically inhibited by related compounds, particularly 2,3,5-triiodobenzoic acid, 2,4-D, and other synthetic auxins which are transported either more slowly or to a much lesser extent in the polar system. Auxin applied artificially to intact plants can travel rapidly upward by penetrating into the conducting tissues of the wood, where it is carried upward in the transpiration stream.

In its normal polar, downward movement, the auxin stimulates the cells below the tip to elongate and sometimes to divide. Specific tissues, notably the cambium, are caused to divide laterally by auxin coming from the developing buds, which accounts for the wave of cell division occurring in tree trunks in the spring. Stimulation of other stem cells to divide leads to the production of root initials, which grow out as lateral roots. Cells of the young ovary are commonly caused to multiply and enlarge so that an apparently normal fruit is produced without requiring pollination (*parthenocarpic fruit*). This latter phenomenon indicates that the growing seeds normally secrete an auxin to which enlargement of the fruit is due, a conclusion which has been directly confirmed by bioassay in several fruit types.

Gibberellins can also cause enlargement of fruit. On reaching the lateral buds, however, auxin inhibits their elongation into shoots, and this accounts for *apical dominance*, i.e., suppression of the growth of lateral buds by the terminal bud of a shoot. Auxin also inhibits the falling off of leaves or fruits, which normally occurs when they are mature or aged, by the formation of an *abscission layer* of special cells whose walls come apart. That the leaves or fruits do not absciss earlier is due to their steady production of auxin, which prevents formation of these cells. In the root, auxin inhibits elongation except in very low concentrations, but its level therein is usually low. Auxin can be transported for a short distance from the root apex toward the base, but the transport is not fully polar and in the more basal parts of the root the transport is slight.

When the shoot is placed horizontal, auxin is transported toward the lower side, cuasing accelerated growth there and hence upward curvature (*geotropism*); in the root, this causes decreased growth on the lower side and hence downward curvature. However, in the downward geotropic curvature of roots, other phenomena appear to enter in, and the complexities are not yet fully resolved. When shoots are illuminated from one side, auxins accumulate on the shaded side and, therefore, the plant curves toward the light (*phototropism*). Both geotropic and phototropic auxin movements have been confirmed with carboxyl labeled ^{14}C compounds. The first observed effect when auxin is applied is the acceleration of the streaming of cytoplasm, but acceleration of growth begins in 7–14 minutes at about 23°C.

In plants which flower on short days, auxin may inhibit flowering; in plants which flower on long days, however, if close to the transition from the vegetative to the flowering state, auxin may promote flowering. In hemp and some of the squashes, auxin modifies the sexuality of the flowers toward femaleness.

In the special case of pineapple, auxin directly causes flowering in an unusually clear-cut and quantitative response.

The principal uses of synthetic auxins are to promote the formation of roots on stem cuttings, to prevent abscission, especially of apples and pears, to induce flowering in pineapples, and occasionally to produce seedless fruits. The largest use, however, is that of weed killing. This action depends upon the fact that, at concentrations from 100 to 1000 times those concentrations occurring naturally, the auxins are highly toxic. Monocotyledonous plants, however, are usually resistant. In years past, 2,4-D has been favored in North and South America, whereas 2-methyl-4-chlorophenoxyacetic acid (*methoxone*) has been popular in Europe. However, as of the early 1980s, the regulatory status of these compounds in various countries is under study and may be subject to change.

Some chemically related compounds antagonize the action of auxins, for example, by relieving the inhibition of root growth (caused by 2,4-D). In contrast, 2,3,5-triiodbenzoic acid synergizes the action.

Gibberellins. These compounds were originally isolated from a parasitic fungus which causes excessive leaf elongation in rice plants. The mechanisms and applications of this group of compounds are described in entry on **Gibberellic Acid and Gibberellin Plant Growth Hormones**.

Kinetins. Considerably less is known about this class of compounds. The first one to be discovered, produced by autoclaving yeast nucleic acid, was 6-furfurylaminopurine. Somewhat later, zeatin was isolated from immature corn (maize) kernels. The kinetins promote cytokinesis and protein synthesis, thus causing amino acids to accumulate where kinetins are synthesized (or externally applied) and maintaining the chlorophyll content of yellowing leaves. The kinetins antagonize auxin in apical dominance, releasing lateral buds from inhibition by a terminal bud or by applied auxin. It is believed that through the same mechanism, the kinetins promote the development of buds and leaves on tissue cultures. Their action is primarily local, and if there is transport *in vivo*, it probably occurs mainly in the transpiration stream (where amino acids are also often found).

Ethylene. The production of ethylene in fruit tissue and in small amounts in leaves may justify its consideration as a hormone, functioning in the gaseous state. Cherimoyas and some varieties of pear produce 1000 times the effective physiological concentration. Ethylene formation is closely linked to oxidation and may be centered in the mitochondria. Its effects are to promote cell-wall softening, starch hydrolysis, and organic acid disappearance in fruits—the syndrome known as *ripening*. Ethylene also decreases the geotropic responses of stems and petioles.

Daminozide Growth Modifier. *Daminozide*, the chemical name of which is 2,2-dimethyldrazide was developed in the early 1960s as a modifying or regulating agent for the growth process of several food plants. The action varies with each plant. For example, on apple, the compound accelerates the start of flower budding, restricts nonproductive vegetative growth, and assist in fruit drop control. It is also claimed that the compound accelerates fruit coloring and helps to retain the firmness of the fruit. For some of these and other similar reasons, the compound has been used effectively for certain varieties of grape (particularly Concord), for peanuts (groundnuts), for tomatoes, nectarines (except Cherokee), and peaches. Other commercial designations are *Alar*, *B-Nine*, *Kylar*, and *Sadh*.

Ethephon Growth Modifier. This compound, (2-chloroethyl)-phosphonic acid was developed in the United States in the mid-1960s and is used effectively on a number of fruit and vegetable crops for controlling a variety of factors. These include loosening fruit and causing earlier ripening of the fruit (apple, blackberry, blueberry, cherry, cranberry, filbert, tangerine, and walnuts); for encouraging uniform ripening and increasing yield (pepper and tomato); to improve color as well as accelerate maturity (cranberry); and to decrease time required for degreening in citrus fruits, particularly lemon. Other commercial designations for this compound include *Cepha*, *Ethrel*, and *Florel*.

Maleic Hydrazide. This compound (1,2-dihydro-3,6-pyridazinedione) is also used as a growth regulator, herbicide, and plant modifier. It is used in the treatment of tobacco plants; as a post-harvest sprouting inhibitor; and as a sugar content stabilizer in sugar beets.

PLANT HORMONES (Gibberellins).   Gibberellic Acid and Gibberellin Plant Growth Hormones.

PLASMA (Particle). 1. An assembly of ions, electrons, neutral atoms and molecules in which the motion of the particles is dominated by electromagnetic interactions. This condition occurs when the macroscopic electrostatic shielding distance (Debye length) is small compared to the dimensions of the plasma. Because of the large electrostatic potentials which would result from an inhomogeneous distribution of unlike charges, a plasma is effectively neutral. Thus there are equal numbers of positive and negative charges in every macroscopic volume of a plasma. Also, because a plasma is a conductor, it interacts with electromagnetic fields. The study of these interactions is called *hydromagnetics* or *magnetohydrodynamics*.

2. A collection of electrons and ions, usually at a high enough temperature so that the ionization level is above 5% and at densities such that the Debye shielding distance is much smaller than the macroscopic dimensions of the system.

PLASTIC DEFORMATION. When a metal or other solid is plastically deformed, it suffers a permanent change of shape. The theory of plastic deformation in crystalline solids, such as metals, is complex, but well advanced. Metals are unique among solids in their ability to undergo severe plastic deformation. The observed yield stresses of single crystals are often smaller than the theoretical strengths of perfect crystals by a factor of 10^{-4}. The fact that actual metal crystals are so easily deformed has been attributed to the presence of lattice defects inside the crystals. The most important type of defect is dislocation.

PLASTICIZERS. High-boiling solvents or softening agents, usually liquid, added to a polymer to facilitate processing or to increase flexibility or toughness. (Where these effects are achieved by chemical modification of the polymer molecule, e.g., through copolymerization, the resin is said to be "internally plasticized.")

Thermoplastic polymers are composed of long-chain molecules held together by secondary valence bonds. When incorporated in the polymer with the aid of heat or a volatile solvent, plasticizers replace some of these polymer-to-polymer bonds with plasticizer-to-polymer bonds, thereby facilitating movement of the polymer chain segments and producing the physical changes described. Thermosetting polymers, which consist of three-dimensional networks connected through primary valence bonds, are not usually amenable to such softening by external plasticizers. Not all the thermoplastics can be plasticized satisfactorily. Polyvinyl chloride polymers and copolymers, and cellulose esters respond particularly well to plasticizing and represent the major outlets for plasticizers.

The results obtained by addition of plasticizer vary with different polymers. In polyvinyl chloride, for example, plasticizer concentrations of 30–50% convert the hard, rigid resin to rubber-like products having remarkably high elastic recovery, while similar plasticizer concentrations in cellulose acetate produce tough but essentially rigid products.

The plasticizer field has grown tremendously since camphor was patented as plasticizer for nitrocellulose in 1870. The reported U.S. production of plasticizers in 1968 exceeded 1.3 billion pounds. Close to two-thirds of this production volume is used in vinyl chloride polymers and copolymers. Plasticized polyvinyl chloride is fabricated at elevated temperature into film and sheeting, and into molded and extruded articles; it is also processed in the form of dispersions, which may be applied as coatings and subsequently baked to form continuous films. These dispersions include plastisols, in which the plasticizer is the continuous phase; organosols, in which a volatile solvent is added; and water dispersions. Plastisols are also used in molding.

Compatibility. Where the polymer-to-plasticizer attraction is strong, the plasticizer has high compatibility with the resin and is said to be of the "primary" or "solvent" type. With polyvinyl chloride this attraction is furnished particularly well by ester groups. Where the polymer-to-plasticizer attraction is low the plasticizer is of the "secondary" or "nonsolvent" type; it functions as a spacer between polymer chains but cannot be used alone because of limited compatibility. This is manifested by exudation of the plasticizer from the resin. Secondary plasticizers are often employed to take advantage of other desirable properties which they may impart. Where they are used merely to cheapen the formulation, they are usually referred to as "extenders."

Plasticizing Efficiency. It is customary when evaluating plasticizers in polyvinyl chloride to compare them at concentrations which produce a standard apparent modulus in tension, as measured at room temperature. Since the stress-strain relationship is generally nonlinear it is necessary to specify a given point on the stress-strain curve as well as the rate of loading or straining. The efficiency may be expressed as the concentration of a given plasticizer necessary to produce this standard modulus. Other properties, e.g., indentation hardness, may take the place of tensile modulus.

Such tests, while they constitute an adequate basis for routine evaluation of plasticizers, furnish only a rudimentary picture of the elastic properties of the plasticized resin. More complete studies supply valuable information. For example, tensile creep tests have shown that polyvinyl chloride resin plasticized with trioctyl phosphate will deform more in response to stresses of short duration than will resin plasticized with tricresyl phosphate; the reverse is true for stresses of long duration.

For many applications low-temperature flexibility of the plasticized composition is also important. Plasticizers of low viscosity and low viscosity-temperature gradient are usually effective at low temperature. There is also a close relationship between rate of oil extraction and low-temperature flexibility: plasticizers effective at low temperature are usually rather readily extracted from the resin. Plasticizers containing linear alkyl chains are generally more effective at low temperature than those containing rings. Low-temperature performance is evaluated by measurement of stiffness in flexure or torsion or by measurement of second-order transition point, brittle point or peak dielectric loss factor.

Permanence. Where a plasticizer is used in thin films it is important that it have low vapor pressure. For polyvinyl chloride a plasticizer vapor pressure of no more than 4 mm at 225°C has been suggested as a rough criterion, although much higher vapor pressures are tolerated in plasticizers for cellulose acetate. Volatile losses are determined, not only by the plasticizer vapor pressure, but also by the plasticizer-resin interaction; plasticizers of limited compatibility may exhibit unexpectedly high volatile loss. Comparisons are usually made by heating a plasticized sample in contact with activated carbon and measuring the weight loss of the sample. Resistance of the plasticizer to migration determines whether the plasticized composition will mar or soften varnished surfaces with which it comes in contact. Stability and water-and oil resistance are further factors in plasticizer permanenece. Plasticizers which have poor oil resistance are usually also the worst offenders with respect to marring.

Commercial Plasticizers. *Phthalates.* These esters, prepared from *o*-phthalic anhydride, constitute the most important group of plasticizers from the stand-point of production and sales volume. Among these the dioctyl phthalates are the most widely used in vinyl chloride resins, where they are preferred because they offer a good compromise with respect to a wide range of properties: satisfactory volatility, good compatibility, fair low-temperature flexibility, and moderately low cost. Most popular of the group is the 2-ethylhexyl ester, known as DOP, but the isoctyl, n-octyl, and capryl esters also find use. Increased emphasis on low volatility and plastisol viscosity stability has in recent years led to use of octyl decyl and didecyl phthalates. The lower alkyl phthalates are generally employed in resins other than polyvinyl chloride. Dibutyl phthalate is used chiefly in nitrocellulose lacquers and polyvinyl acetate adhesives; diethyl, dimethyl, and di(methoxyethyl) phthalates are used in cellulose acetate. For low toxicity, methyl and ethyl phthalyl ethyl glycolates are favored for cellulose acetate, and butyl phthalyl butyl glycolate for vinyl chloride polymers and nitrocellulose.

Phosphates. The phosphates, second only to phthalates in production volume, are favored for flame resistance and low volatility. Tricresyl phosphate (mixed meta and para isomers) is the most popular; it is used in polyvinyl chloride and in nitrocellulose lacquers. Resins plasticized with tricresyl phosphate are deficient in low-temperature flexibility. Diphenyl cresyl phosphate and triphenyl phosphate are other examples, the former for polyvinyl chloride, the latter for cellulose acetate. Diphenyl-2-ethylhexylphosphate is preferred to tricresyl phosphate in polyvinyl chloride where its low toxicity and improved low-temperature flexibility are required. Tri(2-ethylhexyl)-phosphate is outstanding among phosphates used in polyvinyl chloride with respect to low-temperature flexibility; in flame-and oil resistance, however, it is inferior to tricresyl phosphate. Tri(butoxyethyl)phosphate finds some use in synthetic rubber.

Esters of Aliphatic Dibasic Acids. This group consists of the adipates, sebacates, and azelates. These esters lead the field for low-temperature flexibility and efficiency in vinyls: they are useful in the preparation of low-viscosity stable plastisols. The principal disadvantages are high cost and poor solvent resistance. They are generally incompatible with cellulose acetate. Cheapest but most volatile are the adipates. Among the sebacates, the 2-ethylehexyl ester is outstanding in polyvinyl chlorides for low-temperature flexibility and low volatility; the butyl ester is used in polyvinyl butyral. The 2-ethylexyl ester of azelaic acid falls between the corresponding adipate and sebacate in price and properties.

Fatty Acid Esters. Monohydric alcohol esters of fatty acids usually have limited compatibility with vinyl chloride polymers. However, small amounts of these esters are useful for imparting low-temperature flexibility and softness and for their lubricant action during processing. Butyl oleate and stearate are leaders in this group. Methyl and butyl acetyl ricinoleates are used in nitrocellulose. Various glycol and glycerol esters of fatty acids have properties similar to the monohydric alcohol esters. Castor oil (glyceryl triricinoleate) and its derivatives are used in nitrocellulose. Triethylene glycol (di(2-ethylbutyrate) is the leading polyvinyl butyral plasticizer.

Epoxidized esters of fatty acids are similar in their plasticizing properties to other fatty acid esters, with the added advantage that they improve heat and light stability of vinyl resins by acting as hydrogen chloride scavengers. Although generally used together with conventional stabilizers, they can also be used as the sole stabilizer.

Polymeric Plasticizers. Polyesters prepared from dicarboxylic acids and diols enjoy some use in polyvinyl chloride resins for specialized applications. Ranging in properties from somewhat viscous liquids to soft resins, these plasticizers have outstanding permanence: they are of very low volatility and generally show good resistance to extraction and migration. The acids used commercially in preparing them include adipic, sebacic and azelaic; among the diols employed are propylene glycol and 2-ethyl-1,3-hexanediol; monobasic acids may be included in the preparation to limit molecular weight. Reported molecular weights of commercial polyester plasticizers range from 850 to 8000. Those of highest molecular weight are the most permanent but are somewhat difficult to process because of high viscosity and slow solvation of the resin. Other disadvantages include poor low-temperature properties and high cost. Most polyesters are rated as secondary plasticizers and are blended with the more efficient monomeric types.

Butadiene-acrylonitrile copolymers may also be used in polyvinyl chlorides and are rated as primary plasticizers for these resins. Like polyesters, they are favored for permanence, particularly oil resistance, and because they permit high loading with filler without serious impairment of physical properties. They are difficult to incorporate in vinyl resin and are often used together with conventional plasticizers. GRS rubber and polyisobutylene are also blended with certain polymers for plasticizer-like action. Other addition polymers have been studied as plasticizers, among them a series of promising liquid polymers of dibutyl itaconate.

Other Plasticizers. Acetyl tributyl citrate is an outstanding nontoxic plasticizer for polyvinyl chloride in food packaging. Plastisol formulations containing this ester have exceptional viscosity stability. Acetyl triethyl citrate is a good plasticizer for the cellulosics and acetyl trioctyl citrate shows promise with vinylidene chloride polymers. Other compounds of diverse nature find application as plasticizers. These include tetra-*n*-butyl thiodisuccinate, camphor, *o*-nitrobiphenyl and partially hydrogenated isomeric terphenyls.

—Charles J. Knuth, Pfizer, Inc., New York.

PLASTICS. These are materials formed from resins through the application of heat, pressure, or both. Most starting materials prior to the final fabrication of plastic products exhibit more or less plasticity—hence the term *plastic*. However, the great majority of plastic end-products are quite nonplastic, i.e., they are nonflowing, relatively stable dimensionally, and are hard. There are scores of different kinds of plastics. They fall into two broad categories: (1) *thermoplastic resins*, which can be heated and softened innumerable times without suffering any basic alteration in characteristics; and (2) *thermosetting resins*, which once set at a temperature critical to a given material cannot be resoftened and reworked. Since most plastic fabrication methods, such as casting, molding, or extruding, involve heat, the thermosetting materials must be properly and accurately formed during any thermal cycling that exceeds the critical temperature.

The principal kinds of thermoplastic resins include: (1) acrylonitrile-butadiene-styrene (ABS) resins; (2) acetals; (3) acrylics; (4) cellulosics; (5) chlorinated polyethers; (6) fluorocarbons, such as polytetrafluoroethylene (TFE), polychlorotrifluoroethylene (CTFE), and fluorinated ethylene propylene (FEP); (7) nylons (polyamides); (8) polycarbonates; (9) polyethylenes (including copolymers); (10) polypropylenes (including copolymers); (11) polystyrenes; and (12) vinyls (polyvinyl chloride).

The principal kinds of thermosetting resins include: (1) alkyds; (2) allylics: (3) the aminos (melamine and urea); (4) epoxies; (5) phenolics; (6) polyesters; (7) silicones; and (8) urethanes.

Numerous plastics are described throughout this volume. Consult alphabetical index.

PLASTIDS. Pigments in plants are often located in special bodies called *plastids*. There are many kinds of plastids: *leucoplasts*, which contain no pigment and which are therefore colorless; *chloroplasts*, which contain chlorophyll (by far the commonest kind); and *chromoplasts*, colored plastids which do not contain chlorophyll.

Leucoplasts occur in parts of stems and roots where light fails to penetrate. They absorb glucose and change it to starch.

Chloroplasts occur in cells exposed to light. They are indispensable to photosynthesis. In the algae, the shapes of these bodies are many; in a large number of cases, the plastid is a thick, cup-shaped body occupying the greater part of the volume of the cell; in other algae, the plastids have a central mass from which radiating plates of arms extend outward to the cell wall; spiral, net-shaped, and ring-shaped plastids are not uncommon in this group of plants. In some algae and in nearly all higher plants, the chloroplasts are small subspherical or lens-shaped bodies, varying in number from one to many in a single cell. Always the chloroplasts are found embedded in the cytoplasm of the cell. In many plants, the continuous movement of the cytoplasm in the cell carries the plastids along with it; in others, these bodies have a fixed position. In certain algae and in many cells of higher plants, as for example in the palisade layer of leaves, the chloroplasts may change their position so that they will receive the most favorable amount of light. If the light intensity is low, they will present their flat surface to it; whereas if the light intensity is high, the plastid rotates so that it is placed edgewise to the light. Chloroplasts contain chlorophyll and other pigments. See also **Pigmentation (Plants)**.

PLATINUM AND PLATINUM GROUP. Chemical element symbol Pt, at. no. 78, at. wt. 195.09, periodic table group 8 (transition metals), mp 1772°C, bp 3725–3925°C, density, 21.37 g/cm³ (solid), 21.5 g/cm³ (single crystal at 20°C). Elemental platinum has a face-centered cubic crystal structure. The five stable isotopes of platinum are ^{192}Pt, ^{194}Pt through ^{196}Pt, and ^{198}Pt. The seven unstable isotopes are ^{188}Pt through ^{191}Pt, ^{193}Pt, ^{197}Pt, and ^{199}Pt. In terms of earthly abundance, platinum is one of the scarce elements. Also, in terms of cosmic abundance, the investigation by Harold C. Urey (1952), using a figure of 10,000 for silicon, estimated the figure for platinum at 0.016. No notable presence of platinum in seawater has been detected.

Electronic configuration: $1s^22s^22p^63s^23p^63d^{10}4s^24p^64d^{10}4f^{14}5s^25p^65d^96s^1$. Ionic radius Pt^{2+} 0.52 Å. Metallic radius 1.3873 Å. First ionization potential 8.96 eV. Oxidation potentials Pt → Pt^{2+} + 2e⁻, ca. —1.2 V; Pt + 2OH⁻ → Pt(OH)$_2$ + 2e⁻, —0.16 V.

Platinum is one member of a family of six elements, called the *platinum metals*, which almost always occur together. Before the discovery of the sister elements, the term *platinum* was applied to an alloy with Pt as the dominant metal, a practice that persists to some degree even today. The major properties of the plantinum metals are given in the accompanying table. See also **Iridium; Osmium; Palladium; Rhodium;** and **Ruthenium.**

Occurrence. These metals occur in both primary and secondary deposits. The primary deposits are generally associated with Ni-Cu sulfide ores. The Sudbury ores of Canada and the deposits of the Bushveld complex of South Africa are of this type. Native platinum occurs as a primary deposit in the Ural Mountains of the U.S.S.R. and also in the Choco district of Columbia. Weathering and erosion of these deposits have resulted in the formation of secondary, or placer, deposits of native Pt in riverbeds and streams. One nugget of Pt found in the Urals weighed over 25 pounds (11.3 kilograms). Most of the world's platinum comes from Canada, the U.S.S.R., and South Africa. Minor amounts have been found in Alaska, Colombia, Ethiopia, Japan, Australia, and Sierra Leone.

Because of their unique properties and in spite of their high initial cost, the platinum metals find many applications in industry. Since used platinum metals retain a large portion of their initial value, many scrap materials are a major source of recoverable platinum metals. Practically every application of platinum generates scrap in some form which is eventually returned to the platinum refiner for recycling. Although there are ample mine reserves, they soon would be depleted without constant scrap recycling.

The procedures used in refining provide a good introduction to the complex chemistry of the platinum metals. Some of these methods remain the best analytical techniques available for the separation of the metals. South African ore is smelted to form a copper-nickel matte containing small amounts of the platinum metals (0.18%). The matte is melted, cast into anodes, and electrolytically dissolved. The contained copper is deposited at the cathode, the nickel remains in the sulfuric acid electrolyte, and the platinum metals are contained in the anode slimes. The resulting copper is refined and the $NiSO_4$ solution purified and crystallized. The anode slimes are treated by roasting to remove sulfur and leached with dilute H_2SO_4 and air to remove copper and nickel. The leached slimes are treated with aqua regia. The aqua regia solution is evaporated to concentrate the solution

and expel the excess HNO_3. The residue from this treatment contains rhodium, iridium, ruthenium, osmium, and silver. The solution contains platinum, palladium, and gold.

Platinum is first removed by precipitating as ammonium hexachloroplatinate (IV) [$(NH_4)_2PCl_6$] by the addition of a saturated solution of NH_4Cl. The precipitate is washed, dried, and calcined to form platinum sponge of about 98% purity. The sponge is purified by redissolving in aqua regia and evaporating the solution to dryness with sodium chloride. The resulting sodium hexachloroplatinate is dissolved in water and boiled with $NaBrO_3$ to convert impurities, such as iridium, rhodium, palladium, and base metals, to valence states which produce readily filterable hydroxides. The platinum left in solution is free of impurities. It is then treated with NH_4Cl and the pure ammonium hexachloroplatinate precipitate is calcined at 1000°C to pure platinum sponge.

The first aqua regia solution is treated with $FeSO_4$ to precipitate the gold. Palladium is precipitated by oxidizing the solution with HNO_3 and adding ammonium chloride. Ammonium hexachloropalladate(IV) is formed (analogous to the Pt compound). This salt is purified by dissolving in NH_4OH, filtering off the impurities, and reprecipitating the palladium by the addition of HCl. The insoluble complex $Pd(NH_3)_2Cl_2$ is formed, which when calcined and reduced in hydrogen yields pure Pd sponge.

The insolubles from the first aqua regia treatment are fused with a flux of litharge, soda ash, borax, and carbon in a gas-fired furnace at 1000°C for one hour. This procedure converts silica, alumina, and some base metals to slag. The precious metals are retained in the lead phase. The lead portion is heated with HNO_3, which dissolves the lead and silver. The lead is precipitated as a sulfate and then the silver as a chloride. The residue is treated with concentrated H_2SO_4 at 300°C. Rhodium will dissolve, leaving iridium, ruthenium, and osmium as insolubles. The rhodium solution is treated with zinc powder, precipitating an impure Rh. The impure Rh is heated in an atmosphere of chlorine. Many impurities form volatile chlorides at this temperature and are expelled. Rh forms a polymeric trichloride, which is insoluble in aqua regia. The rhodium trichloride is digested in aqua regia for several hours, then filtered, dried, and calcined, yielding a commercial grade of Rh sponge.

REPRESENTATIVE PROPERTIES OF PLATINUM GROUP METALS

PROPERTY	IRIDIUM	OSMIUM	PALLADIUM	PLATINUM	RHODIUM	RUTHENIUM
Atomic volume, cm^3/g-atom	8.54	8.43	8.88	9.09	8.27	8.29
Atomic radius	1.355	1.350	1.373	1.335	1.342	1.336
Crystalline form	fcc	hcp	fcc	fcc	fcc	hcp
Lattice parameters, Å, a	3.8389	2.7341	3.8902	3.9310	3.804	2.7041
b	—	4.3197	—	—	—	4.2814
Thermal conductivity at 20°C, (cal)(cm)/(s)(cm^3)(°C)	0.14	—	0.168	0.166	0.21	—
Electrical resistivity at 0°C, micro-ohm-cm	5.3	9.5	10.8	10.6	4.5	7.2
Thermal expansivity, °C × 10^6 at 20°C	6.6	6.6	12.4	9.0	8.3	9.6
Hardness, Mohs scale	6.5	7.0	4.8	4.3	—	6.5
Specific heat, cal/g-atom at 20°C	0.031	0.031	−0.0584	0.031	0.059	0.057
Heat of fusion, kcal/mole	6.3	7.0	4.0	4.7	5.2	6.1
Heat of vaporization, kcal/mole	134.7	150	90	122	118.4	135.7

fcc = face-centered cubic

hcp = hexagonal close-packed

The residue insoluble in H_2SO_4 is fused with Na_2O_2, poured into thin slabs, and cooled. Iridium is oxidized in the fusion to IrO_2, which is insoluble in water. Ruthenium and osmium form soluble sodium salts and are separated from the Ir by filtration. The insoluble IrO_2 is dissolved in aqua regia, and ammonium hexachloroiridate(IV) is precipitated by the addition of NH_4Cl. Calcining yields pure Ir sponge.

The filtrate from the dissolution of the Na_2O_2 fusion contains $NaRuO_4$ and $NaOsO_4$. Ethyl alcohol is added to the solution, causing the precipitation of RuO_2, which is separated by filtration.

The ruthenium is purified by distilling with chlorine. Volatile ruthenium tetroxide is collected. A saturated solution of NH_4Cl is added, causing the precipitation of ammonium hexachlororuthenate(III). The precipitated salt is calcined in hydrogen, yielding commercial Ru sponge.

The filtrate from the alcohol precipitation of Ru contains the Os. The solution is neutralized with HCl and is treated with powdered Zn, reducing the Os to the metallic state. Osmium tetroxide is formed by roasting the impure Zn in a current of O_2. The volatile OsO_4 is trapped in an aqueous solution of KOH. Ethanol is added to the solution, precipitating potassium osmate(VI) which is mixed with an excess of NH_4Cl and calcined in an atmosphere of H_2. The resulting Os sponge is leached to remove KCl, leaving a commercial-grade Os sponge.

The refining of secondary scrap follows much the same procedures with minor variations. For example, solid metallic Pt and especially the Rh and Ir alloys of Pt are very difficult to dissolve in aqua regia. Therefore, the scrap generally is alloyed with Cu, Ni, Pb, or Zn before dissolution with acids.

Uses of Platinum Metals. These metals, in various forms, currently are used as catalysts for a wide variety of reactions. Products include high-octane gasoline, nitric acid, sulfuric acid, hydrogen cyanide, vitamins, antibiotics, hydrogen peroxide, cortisone, alkaloids, and fuel-cell chemicals. These catalysts also are used to remove trace impurities, e.g., actylene in ethylene or oxygen in hydrogen, or noxious constituents of partial combustion, e.g., automobile exhausts. Although substitutes are being sought, platinum is by far the best catalyst for pollution control of auto exhausts. In the future, the catalytic converters currently installed in automobiles will become a significant source of platinum metals. See also **Catalysts.**

The corrosion resistance of the Pt metals has made the Pt crucible and the Pt electrodes commonplace laboratory tools. The glass industry makes use of large amounts of Pt and its alloys for manufacturing very pure glass. Synthetic fibers often are extruded through spinnerettes made of Pt alloys. The large use of Pt metals in dental and medical devices, in jewelry, and for decorative purposes is based on the corrosion resistance and general appearance of these metals.

Because of their high melting points and stability, Pt alloys have found applications in thermocouples, resistance thermometers, potentiometer windings, electrodes, insoluble anodes, high-temperature furnace winding, crucibles that can withstand corrosive materials at high temperature, and generally as materials of construction that will not contaminate products at very high temperatures. Often Pt and Pd are alloyed with Rh, Ir, Ru, or Os to increase their strength, hardness, and corrosion resistance.

Platinum metals, in particular Pd, find extensive use in the electrical industry. Most of these metals are used as contacts, particularly in telephone relays, where their resistance to oxidation and sulfidization results in circuits of reliability and stability. Alloys of Pt find use as grids for electronic tubes, in electrodes for aircraft spark plugs, for contact metal in printed and solid-state circuits, and in pressure-rupture disks.

In the medical field, *cis*-dichlorodiammineplatinum(II) has been available for cancer therapy. (*Science*, 192, 774–775, 1976).

Platinum Compounds

Platinum forms many di- and tetravalent compounds. The latter valence is more common and more stable. Pt in compact form is inert to all mineral acids except aqua regia. Under oxidizing conditions, fused alkalies will attack Pt to some extent. Molten halides, carbonates, and sulfates have little effect on the metal. Concentrated boiling H_2SO_4, fused cyanides, and fused alkaline sulfides will attack the finely divided metal. Pt is vigorously attacked by Cl_2 at elevated temperatures. In hot aqua regia or HCl containing chlorate ion of H_2O_2, the metal slowly dissolves, yielding a solution of hexachloroplatinic acid, H_2PtCl_6.

Platinum(II) hydroxide is made by adding KOH to a solution of platinum(II) chloride. The unstable black powder is easily oxidized by air and must therefore be handled in an inert atmosphere. In hot alkali or HCl, it disproportionates into the platinum(IV) compound and the metal. Very careful dehydration results in the formation of a gray powder that approaches the composition of platinum(II) oxide. Platinum(II) oxide can also be made by combining the elements at 420–440°C at an O_2 pressure of 8 atm.

When a solution of hexachloroplatinic(IV) acid is boiled for some time with NaOH, all the chloride ions are replaced by hydroxide ions. The resulting sodium hexahydroxyplatinate(IV), $Na_2Pt(OH)_6$, is soluble in the basic solution, but it can be precipitated as hexahydroxoplatinic(IV) acid, $H_2Pt(OH)_6$, by the addition of acetic acid. The hydroxide ions of the salt are replaced by the corresponding ions of mineral acids when the compound is dissolved in acid. Hexahydroxoplatinic acid can be dehydrated to yield compounds corresponding to the tri-, di-, and monohydrate of platinum(IV) oxide. The last water molecule cannot be removed without some destruction of the dioxide.

Brown-black, insoluble, anhydrous platinum(IV) oxide is made by fusing hexachloroplatinic(IV) acid with $NaNO_3$ at about 500°C. The alkali salts are washed out with H_2O to free the fine insoluble residue of platinum(IV) oxide. This compound is known as *Adam's catalyst.*

When Pt is heated to 500°C in the presence of Cl_2, yellow-green, insoluble platinum(II) chloride is formed. At a pressure of 1 atm of Cl_2, the compound is stable from 435 to 581°C. It can also be made by heating hexachloroplatinic(IV) acid in Cl_2 at about 500°C. Platinum(II) chloride is soluble in HCl as tetrachloroplatinic(II) acid. It forms many salts that are water-soluble. These salts can be made by reducing a hot solution of the corresponding hexachloroplatinate(IV) with oxalic acid or SO_2. Platinum(III) chloride has a narrow range of stability. It can be made by contacting Pt or a platinum chloride with 1 atm of Cl_2 at 364–374°C. This dark-green to black compound is practically insoluble in cold concentrated HCl but does dissolve on warming, forming a mixture of tetrachloroplatinic(II) and hexachloroplatinic(IV) acids. Anhydrous platinum(IV) chloride is very difficult to prepare. This brown soluble solid can be made by heating hexachloroplatinic(IV) acid in Cl_2 at 360°C. The most common Pt compound, hexachloroplatinic(IV) acid, is readily made by dissolving Pt in aqua regia, followed by several evaporations with additional HCl to destroy nitrosyl compounds. The acid crystallizes as a hexahydrate. It is difficult to stop the evaporation at just this point, and slight local overheating causes excess loss of water. The sodium salt is quite soluble, and the compound is resistant to hydrolysis in basic solution, allowing the bromate hydrolysis to precipitate base metals and other Pt metals as their hydroxides. The Pt remains in solution. The insolubility of ammonium hexachloroplatinate(IV) often is used in refining Pt. Its slight solubility can be overcome sufficiently by mass action to allow its use as a gravimetric procedure for the determination of Pt. This yellow compound decomposes at red heat, yielding pure Pt sponge. The insolubility of the potassium salt is used for the gravimetric determination of potassium.

A series of di-, tri-, and tetrabromides is well known. Platinum(II) iodide is precipitated as a black insoluble compound by the addition of 2 equiv of iodide to a hot solution of platinum(II) chloride. The black, insoluble, graphitelike substance, platinum(III) iodide, is made by combining the elements in a sealed tube at 350°C.

In contrast with Pd, Pt does form a Pt(IV) iodide. When a concentrated solution of hexachloroplatinic(IV) acid is treated with a hot solution of KI, this brown-black substance is precipitated. The compound is somewhat unstable and light-sensitive. It dissolves in excess KI to form the complex salt, also rather unstable.

Pt forms a nonvolatile tetrafluoride, a pentafluoride, and a volatile hexafluoride. The dark red PtF_6 melts at 56.7°C and is very reactive. It even reacts with O_2 at 21°C to form dioxygenyl hexafluoroplatinate(V), O_2PtF_6.

When sulfur and Pt sponge are ignited, some platinum(II) sulfide is formed. The naturally occurring mineral is called *cooperite*. When heated in air or H_2, the products are metallic Pt and S. Platinum(IV) sulfide can be made by heating ammonium hexachloroplatinate(IV) or Pt and S at 650°C. When precipitated by H_2S from chloroplatinic acid, the compound may exist as $PtS_2 \cdot H_2S$.

Divalent and tetravalent Pt probably form as many complexes as any other metal. The platinum(II) complexes are numerous with N_2, S, halogens, and C. The tetranitritoplatinum complexes are soluble in basic solution. Tetranitritoplatinum(II) ion is formed when a solution of platinum(II) chloride is boiled, at about neutral pH, with an excess of $NaNO_3$. The ammonium salt may explode when heated. Generally, platinum-metal nitrites should be destroyed in solution. They never should be heated in the dry form. Platinum(II) complexes most often have a coordination number of 4. Many compounds have been prepared with olefins, cyanides, nitriles, halides, isonitriles, amines, phosphines, arsines, and nitro compounds.

Platinum(IV) has a coordination number of 6. It forms complexes with halides, nitrogen and sulfur compounds, and other donors but to a lesser extent than platinum(II).

References are listed at the end of the entry on **Chemical Elements.** For further details on the mechanical and physical properties of platinum, see article on "Platinum" by E. D. Zysk in the "Metals Handbook," 9th Edition, Vol. 2, American Society for Metals, Metals Park, Ohio, 1979.

—Linton Libby, Chief Chemist, Simmons Refining Company, Chicago, Illinois.

PLUTONIUM. Chemical element symbol Pu, at. no. 94, at. wt. 242 (mass number of second most stable isotope), periodic table group 3, radioactive metal of the actinide series; also one of the transuranium elements. The element does not occur in nature except in minute quantities as a result of the thermal neutron capture and subsequent beta decay of ^{238}U; all isotopes are radioactive. The most stable isotope is ^{244}Pu ($t_{1/2} = 7.6 \times 10^7$ years). The isotope of major importance is ^{239}Pu ($t_{1/2} = 2.44 \times 10^4$ years). This isotope is fissionable with slow neutrons. The first isotope to be produced was ^{238}Pu ($t_{1/2} = 86.4$ years). Other known isotopes include those of mass numbers 232–237, 237m, 239, 239m_1, 239m_2, 240, 241, 243, 245, and 246.

Electronic configuration: $1s^2 2s^2 2p^6 3s^2 3p^6 3d^{10} 4s^2 4p^6 4d^{10} 4f^{14} 5s^2 5p^6 5d^{10} 5f^6 6s^2 6p^6 7s^2$. Ionic radius: Pu^{4+}, 0.86 Å; Pu^{3+}, 1.01 Å (Zachariasen). Oxidation potentials in acid solution $Pu \rightarrow Pu^{3+} + 3e^-$, 2.03 V; $Pu^{3+} \rightarrow Pu^{4+} + e^-$, −0.982 V; $Pu^{4+} + O_2 + 3e^- \rightarrow PuO_2^+$, −1.17 V; $PuO_2^+ \rightarrow PuO_2^{2+} + e^-$, −0.91 V. Oxidation potential in alkaline solution $Pu^{3+} + 4H_2O \rightarrow Pu(OH)_4 + 4H^+ + e^-$, 0.4 V; $Pu(OH)_4 \rightarrow PuO_2^+ + 2H_2O + e^-$, 1.0 V; $PuO_2^+ + 2OH^- \rightarrow PuO_2(OH)_2 + e^-$, −0.8 V.

Plutonium is of major importance because of its successful use as an explosive ingredient in nuclear weapons and the role it plays in the industrial applications of nuclear power. Exemplary of the energy available from plutonium: (1) One pound (0.45 kilogram) $\cong$ 10 million kilowatts; (2) one kilogram (2.2 pounds) = 22 million kilowatts; (3) one kilogram (2.2 pounds) = 20,000 tons of chemical explosive. Plutonium has the important nuclear property of being readily fissionable with neutrons. ^{238}Pu was used in the Apollo lunar missions to power seismic and other experimental instruments placed on the lunar surface. Because comparatively large quantities of plutonium are produced in reactors, the amount available for various applications has increased considerably during recent years. It is estimated that as of the mid-1980s, nuclear reactors throughout the world are producing in excess of 20,000 kilograms of Pu per year. Within a few years, there will be an accumulation of some 300,000 kilograms of Pu or more. The element is available for purchase by qualified potential users. The price during the mid-1970s was approximately $1000/gram (80–99% enriched); the price had been lowered by 20–30% by the early 1980s because of a large inventory.

In a typical fast breeder nuclear reactor, most of the fuel is ^{238}U (90 to 93%). The remainder of the fuel is in the form of fissile isotopes which sustain the fission process. The majority of these fissile isotopes are in the form of ^{239}Pu and ^{241}Pu, although a small portion of ^{235}U can also be present. Because the fast breeder converts the fertile isotope ^{238}U into the fissile isotope ^{239}Pu, no enrichment plant is necessary. The fast breeder serves as its own enrichment plant. The need for electricity for supplemental uses in the fuel cycle process is thus reduced. Several of the early liquid metal cooled fast reactors used plutonium fuels. The reactor "Clementine," first operated in the United States in 1949 utilized plutonium metal, as did the BR-1 and BR-2 reactors in the Soviet Union in 1955 and 1956, respectively. The BR-5 in the Soviet Union, put into operation in 1959, utilized plutonium oxide and carbide. The reactor "Rapsodie" first operated in France in 1967 utilized uranium and plutonium oxides.

Plutonium was the second transuranium element to be discovered. The isotope ^{238}Pu was produced in 1940 by Seaborg, McMillan, Kennedy, and Wahl at Berkeley, California by deuteron bombardment of uranium in a 150-cm cyclotron. Plutonium exists in trace quantities in naturally occurring uranium ores. The metal is silvery in appearance, but tarnishes to a yellow color when only slightly oxidized. A relatively large piece will give off sensible heat as the result of alpha decay. Large pieces are capable of boiling water.

Chemical Properties. Plutonium has the oxidation states (III), (IV), (V), and (VI), and a complex chemistry in aqueous solutions, as can be judged from such a multiplicity of states. A large number of solid compounds corresponding to these states have been made, and they are in general similar in formulas and properties to the corresponding compounds of uranium and neptunium. An important difference, especially as regards ranges of stability of these compounds, arises as a result of the much greater stability of the (III) and (IV) states of plutonium. This also leads to differences in the aqueous solution chemistry of plutonium as compared to uranium and neptunium.

The pentavalent state, like that of uranium, but unlike that of neptunium, is unstable in aqueous solution with respect to disproportionation.

The ionic species corresponding to the four oxidation states of plutonium vary with the acidity of the solution. In moderately strong (one-molar) acid the species are Pu^{3+}, Pu^{4+}, PuO_2^+, and PuO_2^{2+}. The ions are hydrated but it is not possible at present to assign a definite hydration to each ion. The potential scheme of these ions in one molar perchloric acid is the following:

$$\text{Pu} \xrightarrow{+2.03V} \text{Pu}^{3+} \xrightarrow{-0.982V} \text{Pu}^{4+} \xrightarrow{-1.17V} \text{PuO}_2^+ \xrightarrow{-0.91V} \text{PuO}_2^{2+}$$

with overbracket $-1.043V$ connecting Pu^{4+} to PuO_2^{2+} and underbrackets $-1.023V$ connecting Pu^{3+} to PuO_2^+.

The potentials are in volts relative to the hydrogen-hydrogen ion couple as zero.

The values given for the potential scheme in one-molar acid may be altered extensively by a change in hydrogen ion concentration (pH) or as a result of the addition of substances capable of forming complex ions with the plutonium species. Among such substances are sulfate, phosphate, fluoride, and oxalate ions, and various organic compounds, especially those known as chelating agents. The tetrapositive and hexapositive ions are complexed appreciably even by nitrate and chloride ions. The stability of the complex formed with a specified anion increases in the order: PuO_2^+, Pu^{3+}, PuO_2^{2+}, Pu^{4+}.

The hydrolysis of the ions follows a similar order; Pu^{4+} begins to hydrolyze even in tenth-molar acid and in hundredth-molar acid forms partly the hydroxide, $Pu(OH)_4$, and partly a colloidal polymer of variable but approximate composition $Pu(OH)_{3.85}X_{0.15}$, where X is an anion present in the solution. Further reduction of the acidity results in the hydrolysis of PuO_2^{2+} near pH 5, of Pu^{3+} at about pH 7, and of PuO_2^+ at about pH 9.

The plutonium ions in aqueous solution possess characteristic colors: blue-lavender for Pu^{3+}, yellow-brown to green for Pu^{4+}, and pink-orange for PuO_2^{2+}.

Plutonium monoxide occasionally appears on the surface of metal exposed to atmospheric oxidation, but is prepared more conveniently by treating the oxychloride with barium vapor at about 1250°C. The oxide is classified with the interstitial compounds rather than with the typical metal oxides.

The so-called sesquioxide ($PuO_{1.5-1.75}$) is a typical mixed oxidation state oxide, similar to those formed by uranium, praseodymium, terbium, titanium, and many other metals. Its composition shows continuous variation with changes in temperature and pressure of oxygen above the oxide.

Plutonium dioxide (yellow-green to brown, cubic) is the most important oxide of the element. Almost all compounds of plutonium are converted to the dioxide upon ignition in air at about 1,000°C.

The important halides and oxyhalides of plutonium are PuF_3 (purple, hexagonal), PuF_4 (brown, monoclinic), PuF_6 (red-brown, orthorhombic), $PuCl_3$ (green, hexagonal), $PuCl_4$ (green-yellow, tetragonal), $PuBr_3$ (green, orthorhombic), PuI_3, $PuOF$, $PuOCl$, $PuOBr$, $PuOI$.

All of the halides except the hexafluoride and the triiodide may be prepared by the hydrohalogenation of the dioxide or of the oxalate of plutonium(III) at a temperature of about 700°C. With hydrogen fluoride the reaction product in PuF_4, unless hydrogen is added to the gas stream, in which case the trifluoride is produced. With hydrogen iodide the reaction product is $PuOI$, and the other oxyhalides may be formed by the addition of appropriate quantities of water vapor to the hydrogen halide gas. Plutonium triiodide is produced by the reaction of the metal with hydrogen iodide at about 400°C. The hexafluoride is produced by direct combination of the elements or by the reaction $2PuF_4 + O_2 \rightarrow PuF_6 + PuO_2F_2$ at high temperature. The hydrides of plutonium include PuH_2 (black, cubic) and PuH_3 (black, hexagonal).

Plutonium forms several binary compounds which are of interest because of their refractory character and stability at high temperatures. These include the carbide, nitride, silicide, and sulfide of the element.

The monocarbide is formed by reacting the dioxide in intimate mixture with carbon at about 1,600°C, and the mononitride may be obtained by heating the trichloride in a stream of anhydrous ammonia at 900°C; it is prepared more easily, however, by reacting finely divided metal with ammonia at 650°C. Although the lower temperatures are favorable to the production of higher nitrides, none are obtained, in contrast to the uranium-nitrogen system in which compositions up to $UN_{1.75}$ are easily realized.

The disilicide is formed when a slight stoichiometric excess of calcium disilicide is heated with plutonium dioxide in vacuum at about 1,500°C. The disilicide is only moderately stable in air and burns slowly to the dioxide when heated to about 700°C.

Plutonium "sesquisulfide" may be prepared by prolonged treatment of the dioxide in a graphite crucible with anhydrous hydrogen sulfide at 1,340°–1,400°C, or by the reaction of the trichloride with hydrogen sulfide at 900°C.

Handling Precautions. Care must be taken in the handling of plutonium to avoid unintentional formation of a critical mass. Plutonium in liquid solutions is more apt to become critical than solid plutonium. The shape of the mass also determines criticality. Plutonium's chemical properties also increase handling difficulty. Metallic plutonium is pyrophoric, particularly in finely divided form. Because of the high rate of emission of alpha particles, and the physiological fact that the element is specifically absorbed by bone marrow, plutonium, like all of the transuranium elements, is a radiological poison and must be handled with special equipment and precautions. To assure the safety of personnel, plutonium operations are normally handled in an essentially closed system, such as a *glovebox*. In addition, shielding is required when certain isotopes, including plutonium-240 and plutonium-241, are present in appreciable quantity. Because research continues on the hazards and toxicity of plutonium, specific toxicity data should be sought from current authoritative literature, including government (U.S., UK, France, etc.) publications. As of the early 1980s, permissible body burden was established at 0.6 microgram; lung burden at 0.25 microgram. Chemical toxicity is trivial compared with radiation effects. The permissible levels for plutonium are the lowest for any of the radioactive elements.

Practical Utilization. Since the potential reserves of uranium-235 are limited, some point will be reached where this power source no longer will be competitive with fossil fuels, synthetic fuels, solar power plants, etc.—unless the development of means for the practical utilization of plutonium can be achieved. An important element of nuclear fuel cost is the credit received from the sale or future utilization of plutonium after its recovery from spent fuel. The plutonium credit is realistic only if the plutonium is used for power production, since at present, there are few commercial uses envisioned where it would yield a similar economic return.

References

NOTE: References pertaining to plutonium in nuclear wastes are listed at the end of the entry on **Nuclear Wastes**.

Cunningham, B. B., Cefola, M., and L. B. Werner: "Ultra Microchemical Investigations on 94," *Metallurgical Laboratory Report CN-250*, University of Chicago, Chicago, Illinois (1942).

Fried, S., et al.: "The Microscale Preparation and Micrometallurgy of Plutonium Metal," *J. Inorg. Nucl. Chem.*, **5**, 3, 182–189 (1958).

Kennedy, J. W., Seaborg, G. T., Segre, E., and A. C. Wahl: "Properties of 94 (239)," *Phys. Rev.*, **70**, 7/8, 555–556 (1946).

Krot, N. N., and A. D. Gel'man: "Preparation of Neptunium and Plutonium in the Heptavalent State," *Dokl. Chem.*, **177**(1–3), 987–989 (1967).

Penneman, R. A., Sturgeon, G. D., Asprey, L. B., and F. H. Kruse: "Fluoride Complexes of Pentavalent Plutonium," *American Chemical Society Journal*, **87**, 5803 (1965).

Seaborg, G. T.: "The Chemical and Radioactive Properties of the Heavy Elements," *Chemical & Engineering News*, **23**, 2190–2193 (1945).

Seaborg, G. T., McMillan, E. M., Kennedy, J. W., and A. C. Wahl: "Radioactive Element 94 from Deuterons on Uranium," *Phys. Rev.*, **69** (7/8), 366–367 (1946).

Seaborg, G. T., and A. C. Wahl: "The Chemical Properties of Elements 94 and 93," *American Chemical Society Journal*, **70**, 1128–1134 (1948).

Seaborg, G. T., Editor: "Transuranium Elements," Dowden, Hutchinson & Ross, Stroudsburg, Pennsylvania (1978).

Staff: "Plutonium" in section on "The Elements," B-17-B-18, "Handbook of Chemistry and Physics," 60th Edition, CRC Press, Boca Raton, Florida (1979).

Staff: "Nuclear Fuels," in "Steam," 39th Edition, Babcock & Wilcox, New York (1978).

POLAR COMPOUND. Compound (Chemical).

POLLUCITE. The mineral pollucite is rather rare. It contains cesium, aluminum, silcon, and oxygen, its chemical composition being approximately $(Cs,Na)_2(Al_2Si_4)O_{12} \cdot H_2O$. It is isometric, usually in cubic crystals or crystalline masses; conchoidal fracture; brittle; hardness, 6.5–7; specific gravity, 2.9; luster, vitreous on fresh surfaces; colorless and transparent. Found on the Island of Elba and in the pegmatites of Maine, and as masses 3–4 feet (0.9–1.2 meters) thick in South Dakota; at Varutrask, Sweden, Italy and Kazakhastan, U.S.S.R. Pollucite and petalite were found in the granites of Elba and at first named pollux and castorite for the two famous brothers of Roman mythology, Castor and Pollux. Pollucite is derived from the Latin genitive *Pollucis*.

POLLUTION (Air). Alteration of the earth's atmosphere is by no means a recent phenomenon. By many orders of magnitude, the greatest alteration of the atmosphere occurred during the middle Precambrian period, between 2.9 and 1.8 billion years ago. It is generally accepted that prior to that time, the terrestrial atmosphere was chemically of a reducing nature— as contrasted with an oxidative environment of the present general composition required to support humans and other mammals and life forms which abound on the earth today. See **Air.** Brought about by greatly accelerated plant growth, that earlier natural change represented the most dramatic pollution effect ever suffered by the earth's environment. During that period, the liberation of oxygen by plant activity proved to be a very toxic substance for anaerobic life forms and eradicated most of the biotic community existing at that time. New types of life had to develop which were capable of survival in an oxidative environment. Geochemical processes took on new characteristics, based upon the slow oxidative degradation of both organic and inorganic materials.

Alteration of the earth's atmosphere as the result of human (anthropogenic) activities is extremely recent on the life scale of the earth. This altering process was essentially commenced when humans first discovered and started to use fire as a means of heating, cooking, etc. It is the *combustion* of organic fuels today that is the principal contribution to anthropogenic air pollution. For centuries the pollutants added to the atmosphere by humans were essentially insignificant in terms of the mass and the dynamics of the earth's atmosphere. Except on a local and sometimes regional basis, air pollution was no problem prior to the invention of the steam engine. Traditionally, air pollution has a direct relationship with increasing population and the growing sophistication of the population which demands ever increasing quantities of energy and the manufacture of goods by processes which yield byproducts that require removal to some kind of sink, the earth's atmosphere being one of these sinks.

Energy-Environmental Conflict. With exception of some of the nontraditional sources of energy, such as nuclear energy and the more direct utilization of solar energy (as contrasted with combustion), the needs of the earth's population for energy tends to follow a collision course with concerns over the environment. For example, until the nontraditional energy sources can be reduced to practical usage (of which economics is an important, if not scientific factor in the equation), coal, wood, biomass, and other organic fuels when combusted are air polluters unless very costly measures are taken to treat the effluents. Even when numerous chemical and electroprecipitation measures, among others, are taken, there remains the problem of increasing the carbon dioxide content of the atmosphere. See the discussion of ozone in the entry on **Oxygen.** As regards the extra energy investment required to accomplish environmental protection, these are shown by the accompanying table.

Principal Air Pollutants

As might be expected, pollutant composition and concentration varies widely with the character (industrial, residential, agricultural, etc.) of a locale as well as with climate (temperature, wind, inversion, humidity, etc.) and with the topography (mountain, desert, plain, natural flat basin, etc.). The importance of these natural factors to the concentration of pollutants is obvious from day-to-day observations of leading and heavily industrialized centers of the world, which when climatic factors are favorable are essentially pollution free with no reduction of polluting activities, but which, when these factors are unfavorable, are excessively polluted even when emergency regulations to cut back on polluting activities are taken. Basins, such as the Los Angeles basin, for example, are by nature very efficient collectors—hence concentrators of pollutants in the air. Capped by an inversion layer during periods of warm, windless weather, such bowls can become highly polluted and hazardous to life forms, particularly to individuals suffering from a variety of respiratory conditions.

There are thousands of substances which contribute to air pollution, but frequently the majority of these materials are airborne in only small quantities and limited to a comparatively few locales. The major air pollutants as identified by a number of countries in recent years in connection with pollutant regulatory programs are: (1) particulate matter, (2) nitrogen oxides (NO_x) sulfur oxides (SO_x), (4) hydrocarbons, and (5) carbon monoxide (CO).

Particulates and aerosols. These may be comprised of numerous mineral and organic materials and frequently result from such operations as milling, crushing, screening, grinding, and demolition operations—as well as quarries and cement plants. Soot and flyash as well as heavy carbonaceous smoke, arising from fuel-burning operations and smudge pots, also may fall into this category of pollutants. Aerosols generally are considered to be very tiny spherical droplets of a liquid that may be as small as 0.01 micrometer in diameter. These small liquid

ENERGY REQUIREMENTS OF VARIOUS TECHNOLOGIES
(10^6 Btu/ton of Product)*

PROCESS OPTION	PRIMARY ENERGY SOURCE	PROCESS ENERGY (10^6 Btu)	AIR POLLUTION CONTROL ENERGY (10^6 Btu)	PERCENT OF TOTAL FOR AIR POLLUTION CONTROL
Glassmaking				
side port regenerative furnace	natural gas	7.0	0.57	7.5
side port regenerative furnace with preheat of charge	natural gas	5.7	0.37	6.0
electric furnace	electric power	8.2	0.03	0.3
coal gasification	coal	8.6	0.9	9.5
direct coal firing	coal	7.0	0.65	8.5
Cement				
long kiln (conventional)	oil	5.6	0.07	1.2
suspension preheater with long kiln	oil	4.2	0.05	1.2
fluid bed	oil	5.0	0.1	2.0
Copper Production				
roast-reverb smelting (conventional)	gas, oil, or coal	22.0	5.3	19.4
flash smelting (90–95% sulfur recovery)	oil	10.0	7.8	43.8

SOURCE: Basic data developed by A. D. Little, Inc. See Kusick, et al. (1977) reference.

* 10^6 Btu = 252×10^3 Calories

particles and the larger liquid particles, including mists and sprays, along with dusts, permit numerous physical separating and isolating means that do not apply to gases and vapors. Recent investigations of particle size distribution of atmospheric aerosols have revealed a multimodal character, usually with a binodal mass, volume, or surface area distribution and frequently trimodal surface area distribution near sources of fresh combustion aerosols. These modes are attributed to the following factors: (a) the course mode (2 micrometers and greater) is formed by relatively large particles generated mechanically or by evaporation of liquid from droplets containing dissolved substances; (b) the nuclei mode (0.03 micrometer and smaller) is formed by condensation of vapors from high-temperature processes or by gaseous reaction products; and (c) the intermediate or accumulation mode (0.1 to 1.0 micrometer) is formed by coagulation of nuclei. This evidence indicates that atmospheric particles tend to form a stable aerosol having a size distribution ranging from 0.1 to 1.0 micrometer in general. However, larger and smaller particles occur. The larger, settleable particles (greater than 1.0 micrometer) fall out, and the very fine particles (smaller than 0.1 micrometer) tend to agglomerate to form larger particles which remain suspended. The nuclei mode tends to be highly transient and is concentration-limited by coagulation with both other nuclei and also particles in the accumulation mode. Therefore, the particulate content of a source emission and the ambient air can be viewed as composed of two portions, i.e., the settleable and the suspended.

Control of emissions in both size ranges is required because both settleable and suspended atmospheric particulates have deleterious effects upon the environment. Significantly, it is the suspended particles from an upper level of about 2 to 5 micrometers and smaller that health experts consider most harmful to humans because particles of this size have been found to penetrate the body's natural defense mechanisms and reach most deeply into the lungs. Efforts to control particulate emissions to the atmosphere have historically been geared to maximizing the efficiency of control (by weight) of the overall particulate loading emanating from the generating process. This work has led to the empirical understanding that present systems can perform with high control efficiencies down to a particle size of about 2–3 micrometers, but, below this size, the control efficiency appears to decrease with decreasing particle size to a minimum between 1.0 and 0.1 micrometer; and then increases

again. This relationship of control efficiency and particle size is highly significant to any strategy for controlling particulate air pollution, and serves to underscore the need to adequately measure and evaluate both ambient particulate air pollution and source emissions.

Other Pollutants. Some of these are gases; others fall into the particulate category. Of considerable importance are beryllium dust—very toxic and arising from ore preparation and metalworking operations, but of relatively limited extent because this metal is not a common structural material. Other contaminants include fluorides, metal fumes, such as arsenic, lead, and zinc, organic phosphates, notably from crop dusting and spraying; numerous kinds of organic vapors, including chlorinated hydrocarbons and hydrofluorocarbons (used in aerosol containers but suspect in connection with altering the ozone content of the upper atmosphere—see **Aerosol**), radioactive fallout, such as ^{14}C, ^{137}Ce, and ^{90}Sr, arising from nuclear-device testing—no longer of the major concern prior to the nuclear test ban accepted by most nations; and uranium dust.

Nitrogen oxides. These compounds result from all fossil-fuel combustion processes where air is used as the oxidant. Oxygen from the air and nitrogen combine at combustion flame temperatures to form nitric oxide. NO according to $N_2 + O_2 \rightleftharpoons 2NO$. The rate at which NO is formed and decomposed depends largely upon temperature. For the majority of stationary combustion processes, there is too short a residence time for the full oxidation of NO to NO_2, an estimated average of only 5 to 10% of this reaction occurring. Thus, it is important to observe that although NO_x emissions generally are given as "equivalent NO_2," the predominant NO_x in combustion gases is NO. Several factors affect the generation of NO_x pollutants. Factors which tend to decrease NO_x emissions are: (a) decrease in excess air for combustion; (b) decrease in preheat temperature; (c) decrease in the heat-release rate; (d) increase in the heat-removal rate; (e) increase in back-mixing; and (f) decrease in fuel nitrogen content. With exception of very large installations, coal appears to generate more NO_x than oil; and oil generates more NO_x than natural gas. Thus, as with SO_x, natural gas is the preferred fuel when properly combusted to minimize NO_x.

The major sources of NO_x are the large-fuel-burning operations as previously mentioned, automotive vehicles, and certain chemical plants, notably nitric acid manufacturing facilities. Research to date indicates that effective steps toward reducing the overall emission of NO_x can be effected from stationary combustion sources by: (a) using low excess air firing; (b) providing for two-stage combustion; (c) utilizing flue-gas recirculation; and (d) using water injection. These objectives, when reduced to terms of hardware, mean changes in the configuration, location, and spacing of burners, and the kinds of firing and combustion techniques used. Two-stage combustion is defined as firing all fuel below stoichiometric amounts of primary air in a first stage of combustion, followed by injecting air in a second stage, whereupon burnout of the fuel is completed. There is removal of heat between the two stages. The formation of NO in the first stage is limited because the available oxygen for combustion with nitrogen is limited. The removal of heat between stages kinetically limits the formation of NO when excess air is added to the second stage. Experience shows that a 90% reduction in NO_x emission can be achieved in this manner. By recirculating flue gas, both the peak flame temperature and oxygen content are lowered. Injecting low-temperature steam or water also provides a diluting effect. Although probably of limited value for electric utility boilers (because thermal efficiency is lowered), the water-injection technique may be one of the better ways to reduce NO_x emissions in connection with internal-combustion engines of the stationary type. The situation in the case of internal combustion engines for automotive vehicles is considerably more complicated—there is a wide range of loads on such engines, high performance is required at all loading conditions, and the combustion process from fuel to exhaust must be simple and relatively low cost. See **Catalytic Converter.**

Sulfur Oxides. Sulfur dioxide, SO_2, and sulfur trioxide, SO_3. The primary sources of these oxides SO_x are sulfur-bearing fuels—as used for heat and power, both industrially and residentially. Chemical and metallurgical plants of various kinds also emit SO_x as the result of processing activities, such as the manufacture of sulfuric acid, the roasting of ores, etc. In order of decreasing pollution, the fossil fuels contributing to SO_x pollution are (a) Untreated coal; (b) untreated petroleum fuels, particularly those originating from so-called sour crude oils; and (c) natural gas. Thus, the preference for natural gas by many large fuel users, such as power plants. With only small variations in the cost of raw fossil fuels, there was an advantage in burning a naturally low-sulfur fuel as contrasted with installing elaborate SO_x removal or reduction systems. But, with a rapidly lessening natural gas supply and accompanying higher costs, it has become economically attractive to pay more for desulfurized coal and petroleum fuels, as well as to install SO_x abatement equipment. The allowable sulfur content of oil and coal fuels varies from one community to the next, ranging from 0.50% by weight or less up to 4% and slightly higher. Such regulations usually take into consideration new versus old fuel-burning equipment, the incidence of serious pollution in a given area, as well as economic impact and practicability. Logically, for some years to come, such regulations must represent a compromise of social, economic, and technological factors.

The chemical nature of the oxides of sulfur is given in the entry on **Sulfur.**

Acid Rain. Sulfur dioxide and the fine sulfate mist into which SO_2 is readily transformed are regarded by most scientists in the field as the principal precursors of *acid rain.* However, nitrogen oxides also are implicated, in that they also form acidic components in the air and thus also contribute to the lowering of the pH of various waters upon which they may fall. Acidification of aquatic ecosystems has been known for a number of years in the northeastern United States, Scandanavia, Germany, and Canada. There remains considerable disagreement among authorities as regards the extent of the ecological effects that

may be caused by acid rain. Unlike many air pollution problems which usually can be alleviated by installing local or regional control measures, the effects of SO_2 and N_2O on the content of natural precipitation may not be manifested until the aerosols have been transported by the winds for long distances. Thus, some authorities attribute the effects of acid rain in the northeastern United States to large amount of acidic aerosols emitted in the heavily industrialized Ohio River Basin—borne by winds that flow northeastward over Pennsylvania, New York, and into New England. It has been observed that atmospheric inputs of sulfuric acid and nitric acid to noncalcareous higher-elevation watersheds in the White Mountains and Adirondack regions, for example, lead to comparatively high concentrations of dissolved aluminum in surface and ground waters. This phenomenon appears to result from moderate increases in soil aluminum leaching. Transport of this aluminum to acidified lakes can lead to fish mortality. Combined results from areas of silicate bedrock in the United States and Europe suggest that aluminum represents an important biogeochemical linkage between terrestrial and aquatic environments exposed to acid precipitation (Cronan and Schofield, 1979). Some scientists have shown that the effects of acid rain depend largely upon the type of terrain upon which the acid rain falls. Certain generalizations have been drawn by Johnson (1979) and others: (1) Hydrologically, the effect of acid rain on stream water quality will be evident mainly in low-order drainage systems, especially where the bedrock is chemically unreactive, as in an igneous and metamorphic terrane. In the marble belts of New England, acid rain neutralization is rapidly accomplished by the solution of carbonate minerals. Probably no sustained acidifying effect will be manifested on major streams, regardless of bedrock type. (2) Lakes whose watersheds are composed of igneous or metamorphic bedrock and which receive water from low-order streams will tend to be acidified and rich in aluminum. In general, it may be anticipated that alpine and upland lakes will show this effect more than lowland lakes or those that receive mostly aged stream water or groundwater, or both. (3) The participation of soil aluminum in the immediate acid neutralization process probably dislocates or otherwise disturbs the normal order of soil formation (Cronan et al., 1978). The soil and regolith of the New England landscape are presently acting as a large sump for the absorption of excess strong acidity. (4) Geologically, no excessive chemical weathering activity can be attributed to acid rain over the northeastern United States. The contemporary ionic denudation rate of New England is well below the North American average, despite the added component of strong acids washing out over New England.

Hydrocarbons. Extensive pollution of air occurs from the introduction of hydrocarbons either from (a) the incomplete combustion of hydrocarbon fuels in both stationary and vehicular engines; or (b) from paint spraying, solvent cleaning, chemical and metallurgical, and other plants that use various fluids that have a high hydrocarbon content. Engine design and tuning are major factors in abating exhaust hydrocarbons. The intent is to fully combust the hydrocarbon content of the fuel. In a major city, industrial and commercial sources of organic solvent fumes (principally hydrocarbons) may average from 300 to 600 tons/day. For years, without legal restrictions, some operators found it more economical to permit vapor-laden air to escape to the atmosphere rather than to invest in solvent recovery equipment. Regulations coupled with higher costs of solvents have gone a long way toward eliminating this source of industrial pollution. Also, the chemical industry has successfully developed newer solvents which are less volatile, and easier to handle and recover.

Carbon monoxide. This pollutant is also associated with combustion operations, again being a product of incomplete combustion. Over the years, there has been a much greater awareness of carbon monoxide as a pollutant than the aforementioned gases because of its potent toxicity, dramatized by numerous deaths in earlier years as the result of keeping an automobile engine running in an enclosed space. Faulty residential heaters continue to take their toll of life and in recent years an important killer is the outdoor grill or hibachi with glowing coals taken into a camper or cabin as a means to temper the evening chill. Vehicular tunnel and large parking garage designers, of course, have practiced careful control over carbon monoxide concentrations for many years. See also **Carbon Monoxide**. The effects of carbon monoxide on human beings is shown in accompanying diagram.

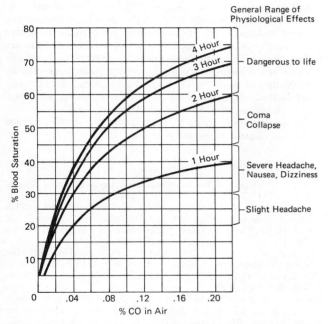

Effects of carbon monoxide on humans. This chart can be considered only as a general guide because the percent of CO blood saturation will vary with exertion, excitement, fear, depth of respiration, anemia, and general physical condition of the individual.

Arctic Haze. For a number of years, pilots flying over the Arctic region in springtime have reported a heavy haze, sometimes reducing visibility aloft from over 100 to less than 10 kilometers. Located long distances from the sources of anthropogenic pollutant sources, it has been tentatively concluded by the *Arctic Air-Sampling Network* organization that two causes may explain this apparently unusual situation: (1) long-distance transport of pollutants from major industrial areas in the northern hemisphere; and (2) natural sources, such as volcanic emissions. A combination of factors is suspected. Composed of scientists from the United States, Canada, Norway, Denmark, and Germany, the aforementioned *Network* monitoring organization has found vanadium, for example, in the atmosphere at Barrow, Alaska, among other locations in the Arctic, which are known to lack sources of this element. Thus, the most likely sources are industrial centers in the middle latitudes. Tentatively, it is believed that at least the vanadium-laden air results from pollution by major European centers, where winds carry the aerosols northeastward into European U.S.S.R. and then northeastward

into Alaska. The study is still in an early phase of sampling and analysis.

Air Treatment Methodologies

Numerous techniques have been used to reduce various forms of air pollution. For example, in connection with particulates, electrostatic precipitators have been quite effective. Most methods, however, fall into what may be called wet-scrubbing processes. As of the mid-1980s, there are nearly one-hundred such processes on the market. Progress in this field has been steady, but not characterized by major breakthroughs. Although the concentration of sulfur dioxide in stack gases emitted by steam generation plants is usually in the range of only 400 to 2000 ppm, the volume of gases produced by the utility industry results, for example, in the liberation of large tonnages of SO_2 into the atmosphere.

Chemical scrubbing systems for SO_2 absorption fall into two broad categories: (a) Disposable systems; and (b) regenerative systems. Typical of systems in use for a number of years are those which use an aqueous slurry of an insoluble calcium compound, which can be discarded after use. Disposable SO_2-removal systems use aqueous slurries of finely ground materials, such as lime, limestone or dolomite, to produce a mixture of insoluble sulfites and sulfates. On passing through the scrubber, SO_2 from the waste gas dissolves to form sulfurous acid: $SO_2 + H_2O \rightarrow H_2SO_3$. The dissolved SO_2 reacts with the lime, $Ca(OH)_2$ or limestone, $CaCO_3$, to form insoluble calcium sulfite, $CaSO_3$: $Ca(OH)_2 + H_2SO_3 \rightarrow CaSO_3 + 2H_2O$; $CaCO_3 + H_2SO_3 \rightarrow CaSO_3 + H_2O + CO_2$. Unfortunately, SO_2 is less soluble (and hence less easily removed by scrubbing) in slightly acid solutions, so that it is extremely difficult in practice to operate a calcium-based system in such a manner that SO_2 removal is maximized while the quantities of calcium chemicals are minimized in order to approach stoichiometric conditions. As calcium-based slurry systems are usually operated at pH 6–10, disposal of the very large masses of used slurry presents a major problem. A typical power station using a calcium-based SO_2-removal slurry system will produce several hundred tons of spent slurry per day. A further disadvantage of lime or limestone systems is their marked tendency to precipitate insoluble calcium salts inside the scrubber. Unless the scale is removed, the scrubber shortly becomes inoperable.

Although chemically analogous to calcium-based systems, magnesium-based scrubbing systems possess several advantages. A slurry of finely divided magnesium hydroxide, $Mg(OH)_2$, is pumped through the scrubber to remove SO_2 from stack gases. Insoluble magnesium sulfite, $MgSO_3$, is formed: $Mg(OH)_2 + SO_2 \rightarrow MgSO_3 + H_2O$. Hydrated magnesium sulfite, $MgSO_3 \cdot 6H_2O$, can be disposed of as such, although it is usually heated to produce a rich stream of SO_2 and regenerate MgO. The SO_2 is compressed, liquefied, and stored in tanks for market; or catalytically oxidized to sulfur trioxide, SO_3, and treated with water to produce sulfuric acid, H_2SO_4. Alternatively, the SO_2 is mixed with hydrogen sulfide, H_2S, to produce elemental sulfur by the Claus process: $SO_2 + 2H_2S \rightarrow 3S + 2H_2O$. Absorption efficiency of SO_2 attainable in a magnesium system is good, and removal efficiencies from 90 to 95% have been claimed without difficulty at reasonable liquor recirculation and MgO feed rates. As with calcium systems, serious scaling occurs due to build-up of insoluble $MgSO_3$.

Scrubbing solutions containing sodium (or other alkali metals) compounds have been extensively studied for removal of SO_2. Justification for the use of sodium compounds includes: (a) complete solubility in water with no formation of scale; and (b) simple reactions with SO_2: $Na_2CO_3 + SO_2 \rightarrow Na_2SO_3 + CO_2$;

$2NaHCO_3 + SO_2 \rightarrow Na_2SO_3 + 2CO_2 + H_2O$; $2NaOH + SO_2 \rightarrow Na_2SO_3 + H_2O$. In one commercial process, a scrubbing solution of sodium sulfite is used, which readily absorbs SO_2 to form the bisulfite: $Na_2SO_3 + H_2O + SO_2 \rightarrow 2NaHSO_3$. In practice, only a portion of the Na_2SO_3 is converted to $NaHSO_3$ because the SO_2 absorption efficiency diminishes as the bisulfite concentration increases. The resulting solution is heated to decompose the bisulfite and thermally regenerate the sulfite. The gaseous SO_2 is compressed, liquefied and handled as previously mentioned under the magnesium system.

Ammonia-based chemicals appear to have some advantages over sodium systems. They are less costly, and regeneration by conventional means is possible, with the byproduct, ammonium sulfate, a marketable commodity for fertilizer.

Solutions containing ammonium sulfate, with or without the addition of ammonium hydroxide, have been widely used. The ammonium system can operate effectively only within a pH range of 4.0 to 7.0. As the pH value increases above 7.0, progressively more gaseous ammonia is liberated and this reacts in the gaseous phase with water vapor and SO_2 to produce a dense aerosol (white plume) which is difficult for scrubbers to remove. In an ammonia system, in order to regenerate the scrubbing solution, the ammonium bisulfite and sulfite mixture is heated to drive off gaseous SO_2: $2NH_4HSO_3 \rightarrow (NH_4)_2SO_3 + H_2O + SO_2$. Alternatively, the ammonium bisulfite/sulfite mixture can be treated with calcium hydroxide. Gaseous ammonia is evolved and trapped in water, which is then recirculated to the scrubber.

Sodium citrate also is used in an SO_2 removal system. The solution is buffered at a pH 3.0–3.7 by the citrate ion, sulfur dioxide is absorbed, and an equilibrium mixture of sodium bisulfite and citric acid is produced.

$$HO-\underset{\underset{CH_2COONa}{|}}{\overset{\overset{CH_2COONa}{|}}{C}}-COONa + 3SO_2 + 3H_2O \rightarrow 3NaHSO_3 + HO-\underset{\underset{CH_2COOH}{|}}{\overset{\overset{CH_2COOH}{|}}{C}}-COOH$$

The bisulfite leaving the scrubber is then reduced with gaseous hydrogen sulfide, which precipitates elemental sulfur by a modified Claus reaction:

$$3NaHSO_3 + HO-\underset{\underset{CH_2COOH}{|}}{\overset{\overset{CH_2COOH}{|}}{C}}-COOH + 6H_2S$$

$$\rightarrow HO-\underset{\underset{C\ H_2COONa}{|}}{\overset{\overset{CH_2COONa}{|}}{C}}-COONa + 9H_2O + 9S$$

A formate system uses two reactions involving potassium formate, HCOOK, which is regenerated after recovery of elemental sulfur. This method has the advantage over other wet scrubbing methods in that no precipitation of insoluble intermediates occurs at any stage of the process. Disadvantages include the need to heat K_2CO_3 solution, at high temperature and pressures, with carbon monoxide to regenerate the potassium formate. The energy requirements thus are high.

While it has been demonstrated that solutions of NaOH, $NaHCO_3$, and Na_2CO_3 are effective for SO_2 removal, these solutions are not effective for removal of mixtures of NO and NO_2, particularly when the gas stream velocities are reasonably high. Under conditions where from 95 to 99% SO_2 may be removed, the solutions may only be effective in removing from 5 to 15% NO_x. The fundamental difference between SO_2 and

NO_x removal is that NO_x gases (mixtures of NO, N_2O_3, NO_2) are approximately 1,000 to 2,000 times less soluble in water than SO_2 at any given temperature. It has been found that conventionally designed wet scrubbers often do not provide sufficient liquid-to-gas contact surface areas or residence times to permit the NO_x to dissolve in the scrubbing solution. Consequently several stages may be required. Concentrations of NO_x in the range of 20,000 to 40,000 ppm require from 6 to 12 stages.

If no SO_2 is present, a sodium-based process may be used to remove NO_x efficiently. In the Neville-Krebs process (patent applied for), removal efficiencies of 60 to 90% have been achieved from gas streams containing up to 1500 to 2000 ppm of NO_x passing through a three-stage scrubber.

The urea system is another system for removing NO_x from low-volume, slow-flowing waste gas streams. The system uses a slightly acid solution of urea, $CO(NH_2)_2$. Unfortunately, the cost of urea is quite high, particularly for a large installation.

Other systems using electron-donor compounds have been tried or are in development. Such compounds include tri-n-butylphosphate, dimethylformamide, triethyleneglycol, dimethylether, dimethylsulfoxide, hexamethylphosphoramide, diethyleneglycoldimethylether, tricresylphosphate, and dioxane. Most of these compounds are expensive compared with inorganic compounds used in most scrubbing systems.

Numerous other entries in this volume pertain to pollution problems. Consult alphabetical index.

References

Carter, L. J.: "Uncontrolled SO_2 Emissions Bring Acid Rain," *Science*, **204**, 1179–1182 (1979).

Claus, G., and G. J. Halasi-Kun: "Environmental Pollution," in "The Encyclopedia of Geochemistry and Environmental Sciences," Vol. IVA, (R. W. Fairbridge, Editor), Van Nostrand Reinhold, New York (1972).

Cleveland, W. S., and T. E. Graedel: "Photochemical Air Pollution in the Northeast United States," *Science*, **204**, 1273–1278 (1979).

Considine, D. M., Editor: "Energy Technology Handbook," McGraw-Hill, New York, 1977.

Cronan, C. S., et al.: "Forest Floor Leaching: Contributions from Mineral, Organic, and Carbonic Acids in New Hampshire Subalpine Forests," *Science*, **200**, 309–311 (1978).

Cronan, C. S., and C. L. Schofield: "Aluminum Leading Response to Acid Precipitation: Effects on High-Elevation Watersheds in the Northeast," *Science*, **204**, 304–305 (1979).

Edmonds, R. L., Editor: "Aerobiology: The Ecological Systems Approach," Academic, New York (1979).

Johnson, N. M.: "Acid Rain: Neutralization within the Hubbard Brook Ecosystem and Regional Implications," *Science*, **204**, 497–499 (1979).

Johnson, A. H.: "Evidence of Acidification of Headwater Streams in the New Jersey Pinelands," *Science*, **206**, 834–835 (1979).

Kenson, R. E., and R. O. Hoffland: "Control of Toxic Air Emissions in Chemical Manufacture," *Chem. Eng. Progress*, **76**, 2, 80–83 (1980).

Kerr, R. A.: "Global Pollution: Is Arctic Haze Industrial Smog?" *Science*, **205**, 290–293 (1979).

Kuski, C. L., et al.: "Energy Use and Air Pollution Control in New Process Technology," *Chem. Eng. Progress*, **73**, 8, 36–44 (1977).

Lawson, D. R., and J. W. Winchester: "Atmospheric Sulfur Aerosol Concentration in South America," *Science*, **205**, 1266–1269 (1979).

Lewis, W. M., Jr., and M. C. Grant: Acid Precipitation in the Western United States," *Science*, **207**, 176–177 (1980).

McBride, J. P., et al.: "Radiological Impact of Airborne Effluents of Coal and Nuclear Plants," *Science*, **202**, 1045–1050 (1978).

Schubel, J. R., and B. C. Marcy, Jr.: "Power Plant Entrainment," Academic, New York (1978).

Singh, H. B., et al.: "Atmospheric Halocarbons, Hydrocarbons, and Sulfur Hexafluoride: Global Distributions, Sources, and Sinks," *Science*, **203**, 899–903 (1979).

Staff: "Air Pollution Control," *Chem. Eng. Progress*, **73**, 8, 31–73 (1977).

Terrill, J. B., Montgomery, R. R., and C. F. Reinhardt: "Toxic Gases from Fires," *Science*, **200**, 1343–1347 (1978).

POLLUTION (Nuclear). Nuclear Reactor.

POLLUTION (Petroleum Fuels). Petroleum.

POLLUTION (Water and Soil). Wastes and Pollution.

POLONIUM. Chemical element symbol Po, at. no. 84, at. wt. 210 (mass number of the most stable isotope), mp 252°C, bp 960°C, sp gr 9.4. The element was first identified as an ingredient of pitchblende by Marie Curie in 1898. The element occurs in nature only as a decay product of thorium and uranium. Because of limited availability and high cost, relatively few practical uses for the element have been found. Meteorological instruments for measuring the electrical potential of air have used small quantities of the metal. It is interesting to note that when Mme. Curie first identified polonium, she found that an electroscope was a far better instrument for detecting the metal than spectroscopic means. Polonium-plated metal rods and strips have been used as static dissipators in textile coating equipment and in various electrical equipment. The alpha particles from the polonium ionize the air, causing it to conduct and draw off accumulations of static electrical charges.

Three isotopes of polonium occur in the uranium $(4n + 2)$ radioactive series: ^{218}Po (radium A), $t_{1/2}$ 3.05 min; ^{214}Po (radium C'), $t_{1/2}$ 1.6 × 10^{-4} s; and ^{210}Po (radium F) $t_{1/2}$ 138.4 days, and the most stable isotope of polonium. It is used as a source of α-radiation. The thorium $(4n)$ series has two isotopes, ^{216}Po (thorium A), $t_{1/2}$ 0.16 s, and ^{212}Po (thorium C'), $t_{1/2}$ 3 × 10^{-7} s. The actinium $(4n + 3)$ series also has two isotopes, ^{215}Po (actinium A), $t_{1/2}$ 1.83 × 10^{-3} s, and ^{211}Po (actinium C'), $t_{1/2}$ 0.52 s, which occurs in a 0.3% branched chain disintegration of ^{211}Bi (actinium C). Several other isotopes of polonium have been prepared, one of which occurs in the neptunium $(4n + 1)$ series as ^{213}Po, $t_{1/2}$ 4.2 × 10^{-6} s.

Polonium exhibits the allotropy of the lower members of the chalcogen group, having a low-temperature, cubic form, α-polonium, and a high-temperature, rhombohedral form, β-polonium.

The tendency of the chalcogens to show increasing metallic character as one moves down the periodic table is quite marked for polonium; in fact, it resembles lead more than tellurium. Its compounds have a more ionic character in its lower oxidation states than do the tellurium compounds. The stability of the 6+ state is low, the existence of polonate(VI) ion being doubtful. The common oxidation states of the element are 2+ and 4+.

The halides, consisting of both dihalides and tetrahalides, are covalent and volatile, and they are not well characterized. The fluorides have not been established. The complex PoCl$_6^{2-}$ is known. Polonium compounds are usually colored, a fact which is useful in following their reactions. Thus, polonium(II) chloride, PoCl$_2$, formed by dissolving polonium(IV) oxide, PoO$_2$, in HCl is pink and on oxidation by heating or treatment with chlorine, yields yellow PoCl$_4$. Polonium(IV) bromide, PoBr$_4$, dark red, gives purple polonium(II) bromide, PoBr$_2$, on heating. PoBr$_4$ also gives ammonium polonium bromide, (NH$_4$)$_2$[PoBr$_6$] with ammonia. Complex iodides M$_2$[PoI$_6$] have been prepared.

Metallic polonium reacts with air readily on heating, to form PoO$_2$, which exists in a yellow face-centered form having fluorite

structure at low temperatures, and a red tetragonal one on heating. Polonium(IV) hydroxide, P(OH)$_4$, precipitated from polonium(IV) solutions by ammonia, exhibits only slight acidity, and is thus not amphiprotic. On reaction of polonium with HNO$_3$, Po(NO$_3$)$_4$ is formed, and on treatment of polonium(IV) chloride, PoCl$_4$ with H$_2$SO$_4$, polonium(IV) sulfate, Po(SO$_4$)$_2$, is formed, both being ionic-type salts, as indeed are other oxyacid compounds. The sulfate, however, is quite reactive, being hydrated in solution, dehydrated on removal from solution, and forming a basic compound 2PoO$_2$·SO$_3$ on heating. H$_2$S precipitates black polonium(II) sulfide, PoS.

See the references at the end of the entry on **Chemical Elements**.

POLYALLOMER RESINS. These are block copolymers prepared by polymerizing monomers in the presence of anionic coordination catalysts. The polymer chains in polyallomers are composed of homopolymerized segments of each of the monomers employed. The structure of a typical polyallomer can be represented as:

$$(PPPPPPPP\cdots)_x(EEEE\cdots)_y$$

where P represents a propylene molecule and E an ethylene molecule. The number and length of the individual segments can be varied within wide limits depending on the process conditions, monomer concentrations, and catalyst systems employed. The copolymers exhibit the crystallinity normally associated only with the stereoregular homopolymers of these monomers.

The word *polyallomer* is derived from the Greek words *allos*, *meros*, and *poly*. *Allos* means "other" and denotes a differentiation from the normal. The word *meros* means "parts" and the prefix *poly* is added to show that these materials are polymeric. Since allomerism is defined as a constancy of crystalline form with a variation in chemical composition, the polyallomers are examples of allomerism in polymer chemistry.

Polyallomers can be synthesized in slurries by contacting the parent monomers with anionic coordination catalysts of the Ziegler-Natta type at temperatures of 70–80°C. Polyallomers are also synthesized in solution at 140–200°C by contacting the parent monomers with hydrogen-reduced alpha-titanium trichloride and a lithium-containing cocatalyst. In both the slurry and solution processes the polyallomers are formed by alternate polymerization of the monomers employed. In the synthesis of polyallomers from monomers which differ widely in polymerization rates it is sometimes desirable to alternately polymerize a single monomer and then a mixture of monomers to obtain the most desirable properties.

Evidence for the crystalline nature of the polyallomers includes x-ray diffraction patterns, infrared spectra, and for olefin polyallomers, low solubility in hydrocarbon solvents.

The physical properties of polyallomers are generally intermediate between those of the homopolymers prepared from the same monomers, but frequently represent a better balance of properties than blends of the homopolymers. This is illustrated by comparing the properties of a propylene-ethylene polyallomer containing 2.5% ethylene with polypropylene, high-density polyethylene, and a blend of 5% high-density polyethylene and 95% polypropylene.

Compared to polypropylene, a propylene-ethylene polyallomer has a lower brittleness temperature, higher impact strength, and less notch sensitivity. Compared to high-density polyethylene, a propylene-ethylene polyallomer is harder, higher-melting, and of higher impact strength. Other advantages of this polyallomer over high-density polyethylene includes its excellent resistance to environmental stress cracking and its low and uniform mold shrinkage, which minimizes sinks and voids in molded parts. A propylene-ethylene polyallomer thus overcomes the most serious property deficiencies of polypropylene (poor impact and low-temperature properties) and of high-density polyethylene (poor stress-crack resistance and excessive mold shrinkage).

Propylene-ethylene polyallomer is used in vacuum forming, blow molding, injection molding, film and sheeting, wire covering and pipe. Propylene-ethylene polyallomer is the easiest of all polyolefins to vacuum form. The excellent melt strength and broad processing range permit a deep draw. Reproduction of mold detail and surface finish is very good. In film and sheeting, propylene-ethylene polyallomer has optical properties equal to polypropylene but much higher impact strength, particularly at low temperatures. In wire covering, blow molding, and pipe extrusion operations, the combination of good processability and excellent environmental stress crack resistance is important.

—H. J. Hagemeyer, Jr. and M. B. Edwards, Longview, Texas.

POLYAMIDE RESINS. These are synthetic polymers that contain an amide group, —CONH—, as a recurring part of the chain. Poly-alpha-aminoacids, i.e., proteins, whether natural or synthetic, are not normally included in this classification.

The polyamides trace their origin to the studies of W. H. Carothers, begun in 1928, on condensation polymerization, a process that involves the repetition many times of a reaction known to the organic chemist as a condensation reaction because it links two molecules together with the loss of a small molecule. Esterification and amidation are examples:

$$CH_3COOH + C_2H_5OH \rightleftharpoons CH_3COOC_2H_5 + H_2O$$

acetic acid   ethyl alcohol   ethyl acetate or ethyl ester of acetic acid

$$CH_3COOCH_3 + C_4H_9NH_2 \rightleftharpoons CH_3CONHC_4H_9 + CH_3OH$$

methyl acetate   butyl amine   N-butylacetamide

Polymerization requires at least two reactive groups per molecule as in the following example:

$$n \; H_2N(CH_2)_6NH_2 + n \; HOOC(CH_2)_4COOH \rightleftharpoons$$

hexamethylenediamine   adipic acid

$$H[NH(CH_2)_6NHCO(CH_2)_4CO]_n OH +$$

poly(hexamethylene adipamide)

$$(2n - 1)H_2O$$

The first truly high-molecular-weight polyamide was made this way in 1935 and lead to the development of the first wholly man-made fiber which became popularly known as *nylon*, a name coined by Du Pont for fiber-forming polyamides. Numerals representing the number of carbon atoms in first the diamine and then the diacid are used to identify the nylon. Thus, poly-(hexamethylene adipamide) is nylon-66 (six-six, not sixty-six), and poly(hexamethylene sebacamide) made from the 10-carbon sebacic acid is nylon-610 (six-ten).

Both reacting species may be present in the same molecule:

$$n \; H_2N(CH_2)_{10}COOH \rightleftharpoons$$

11-aminoundecanoic acid

$$H[NH(CH_2)_{10}CO]_n OH + (n - 1)H_2O$$

poly(11-aminoundecanoic acid)

Here a single number is used to indicate the number of carbon atoms in the original nomomer, i.e., nylon-11 ("eleven" not "one-one"). In some instances the cyclic analogue or lactam is more accessible than the amino acid and is polymerized by a ring-opening rather than condensation mechanism:

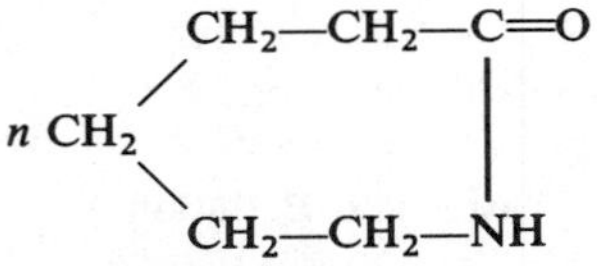

epsilon-caprolactam

$$H[NH(CH_2)_5CO]_n OH$$

polycaprolactam or polycaproamide
nylon-6

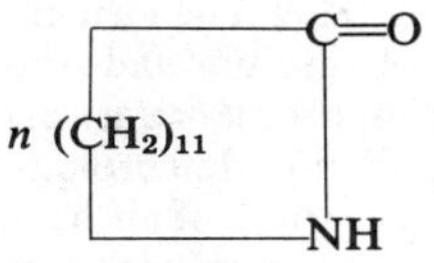

laurolactam

$$+H_2O \rightleftharpoons H[NH(CH_2)_{11}CO]_n OH$$

nylon-12

Some of the monomers commonly used to prepare the nylon resins are shown in the accompanying table. Both petrochemical and vegetable products provide the source materials that are transformed into the reactive intermediates. The table correctly suggests that there is a wider choice in diacids than in diamines. The most important commercial polyamide resins are nylons-66 and -6. Other commercial nylons include 610, 612, 11, and 12.

The above equations illustrate via the double arrows an important facet of polyamides—the equilibrium nature of the polymerization reactions. Achieving and maintaining useful molecular weights (about 10,000 or more) for plastics applications require low moisture contents in order to avoid the reverse reaction of hydrolysis. Most commercial nylons are processed at melt temperatures in excess of 200°C, and molecular weight stability requires a water content below 0.3 weight percent. At a molecular weight of about 11,300, nylon-66 and nylon-6 have, on the average, 99 amide groups linking together 100 monomer units with one unreacted amine group and one unreacted acid group at the ends of the 700-atom long chain. This corresponds to 99% reaction. Because nylon plastics have average molecular weights in the 11,000 to 40,000 range, the need for pure materials and freedom from side reactions is seen to be essential for successful polymerization.

An average molecular weight (number average) is used because any one nylon comprises a broad range of molecular weights. Nylons typically have a "most probable distribution" in which the weight average is twice the number average.

As made, nylon-6 contains up to ten weight percent of lactam monomer that has to be extracted for most applications. Nylons made from larger lactam rings contain only about one percent monomer.

Not all polyamide resins are nylons, and not all nylons are polyamide resins. Some nylons such as nylon-4 or those with a high content of relatively inflexible rings are too unstable or have too high a melt viscosity to be melt processible and are not normally included in the "polyamide resin" category. However, spinning or casting from solution permits some such polymers to be converted into useful fibers or films. A class of polyamide resins distinct from the nylons is based upon polymerization or dimerized vegetable oil acids and polyalkylene polyamines (e.g., ethylenediamine or diethylenetriamine). These are relatively low in molecular weight (2000–10,000), vary from liquids to low melting solids, and are more soluble and flexible than the higher molecular weight, more crystalline nylons.

Nylons are semicrystalline polymers with fairly sharp melting points varying from 180°C for nylon-12 to 270° for nylon-66. The melting point increases in zigzag fashion with increasing concentration of amide groups, being higher where there is an even number of chain atoms between the amide groups. For example, nylon-66 averages five CH_2-groups per CONH but has either four or six chain atoms between amide groups; nylon-6 also has five CH_2/CONH but always has five chain atoms between amides and melts 40°C lower than nylon-66. The degree of crystallinity and the morphology of nylons are more readily controlled by choice of processing conditions than other crystalline polymers such as polyethylene or polyacetal.

The nylons are typically tough and strong. Nylon-66 was the first thermoplastic to provide a combination of stiffness and toughness suitable for mechanical applications. Nylons were therefore the first members of the family of engineering thermoplastics that now include newer materials such as the polyacetals, polysulfones, and polycarbonates. Nylons are outstanding in withstanding repeated impact, in resistance to organic solvents, and in water resistance. A low coefficient of friction, as ASTM self-extinguishing rating, reasonable electrical properties, adequate creep resistance, good fatigue properties, and good barrier properties, particularly to oxygen, are also characteristic of the nylons. Properties change with temperature but are often acceptable in the interval of about −60 to 110°C. Properties may change also with relative humidity because nylons characteristically absorb moisture and water acts as a plasticizer, that is, stiffness decreases and toughness increases. Water absorption decreases as the amide group concentration in the nylon decreases; for example, nylon-6 at saturation contains 9.5% water and nylon-11, 1.9%. Nylons are somewhat notch sensitive, and parts are typically designed to avoid sharp corners. Nylons are attacked by strong acids, oxidizing agents, and a few specific salt solutions such as aqueous potassium thiocyanate or methanolic lithium chloride.

Control of crystallinity and morphology provides one tool for changing properties as desired with any given nylon. But there are many other tools for modification of nylons, and this viability has been an important factor in meeting specific market demands in the face of increased competition from newer materials. Copolymerization and plasticization are alternatives that provide lower melting and tougher compositions with a somewhat different balance of properties. Lubricants as processing aids, nucleating agents to accelerate crystallization and give a little stiffer product, molybdenum disulfide or graphite to improve lubricity in gears or bearings, colorants, fire retardants, carbon black for weather resistance, and antioxidants for better resistance to thermal oxidation are examples of additives employed to achieve specific effects. Combination with other polymers is another technique. Reacting nylon-66 with formaldehyde in alcohol solution has yielded a low melting derivative that can be cross-linked to provide a thermoset. Glass fiber reinforcement has proved to be particularly effective in nylons and has yielded products of exceptional strength, stiffness, and heat resistance.

The nylon resins are converted into useful shapes principally by injection molding or extrusion. These shapes are most often in finished form, but forming is sometimes employed to impart added strength via orientation. Machining of extruded stock shapes is often appropriate where the number of parts does not justify manufacture of a costly mold. Nylon-6 castings are made by the base catalyzed, anhydrous polymerization of monomer in molds that are relatively inexpensive because they do not have to sustain high pressures. Rotational molding, fluidized bed coating, and electrostatic spray coating are also used, especially with low melting nylons such as nylon-11.

Football face guards, hammer handle gears, sprockets, journal

MONOMERS USED TO PREPARE NYLON RESINS

Monomer	Formula	Source(s)
hexamethylene diamine	$H_2N(CH_2)_6NH_2$	butadiene, furfural, or propylene
adipic acid	$HOOC(CH_2)_4COOH$	butadiene or cyclohexane
suberic acid	$HOOC(CH_2)_6COOH$	butadiene or acetylene
azelaic acid	$HOOC(CH_2)_7COOH$	oleic acid
sebacic acid	$HOOC(CH_2)_8COOH$	castor oil
dodecanedioic acid	$HOOC(CH_2)_{10}COOH$	butadiene
"dimer" acid	$HOOC-C_{34}H_{66}-COOH$	oleic and linoleic acids
caprolactam	$(CH_2)_5$ with $\begin{matrix} -CO \\ \quad \\ -NH \end{matrix}$	toluene, benzene, or cyclohexane
7-aminoheptanoic acid	$H_2N(CH_2)_6COOH$	cyclohexane or ethylene
capryllactam	$(CH_2)_7$ with $\begin{matrix} -CO \\ \quad \\ -NH \end{matrix}$	butadiene or acetylene
9-aminononanoic acid	$H_2N(CH_2)_8COOH$	soybean oil
11-aminoundecanoic acid	$H_2N(CH_2)_{10}COOH$	castor oil
laurolactam	$(CH_2)_{11}$ with $\begin{matrix} -CO \\ \quad \\ -NH \end{matrix}$	butadiene

bearings, bristles, filaments, refrigerant tubing, film as a cooking pouch, coil forms, casters, package strapping, loom parts, automobile dome lights, power tool housings, and virtually thousands of other diverse articles attest to the performance of nylon resins in electrical appliances, automotive parts, business equipment, consumer products, and other industrial and home applications.

The non-nylon polyamide resins include relatively low-melting solids that are used with or without modifiers in hot melt cements, heat seal and barrier coatings, inks for flexographic printing of plastic film, and other specialty adhesives and coatings such as varnishes to provide a glossy, transparent, protective layer over print. The large hydrocarbon side chains that are present in the dimerized vegetable oil acids used to make these polyamide resins contribute to their low crystallinity, low water absorption, and flexibility. The fluid, "reactive" polyamides are lowest in molecular weight and contain an excess of amine groups. Monomers with more than two amine groups per molecule such as diethylenetriamine ($H_2NCH_2CH_2NHCH_2CH_2$-NH_2) are used, and care must be taken during polymerization to avoid premature gelation. The excess amine groups permit interaction with epoxy, aldehyde, hydroxmethyl, anhydride, acrylic, and other groups in other resins to produce a variety of thermosetting formulations. Combinations with epoxy or phenol-formaldehyde polymers with or without added fillers are the most common. The epoxy compositions are cured at relatively low temperatures and are useful for bonding aluminum in aircraft, steel in automobiles, wood, leather, glass, ceramics, and other materials. The resin combination broadens the capacity of the adhesive to wet a variety of surfaces and enhances bond strengths. The phenolic-polyamide compositions require higher curing temperatures but withstand higher temperatures in use. They tend to form bonds less tough than the epoxy mixes but bond strongly to copper and other common components of printed circuits. The phenolic blends also produce acid and alkali resistant coatings useful as container linings or wire coatings.

—M. I. Kohan, E. I. DuPont de Nemours & Co., Inc., Wilmington, Delaware.

POLYBASITE.

A mineral antimony sulfide of silver $(Ag,Cu)_{16}Sb_2S_{11}$, in which copper substitutes for silver to approximately 30 atomic percent. Crystallizes in the monoclinic system. Hardness, 2–3; specific gravity, 6.3; color, black, dark ruby red in thin splinters; metallic luster; nearly opaque. From the Greek, meaning many, suggesting the many-metal basis.

Occurs in low-temperature silver deposits commonly associated with silver and lead minerals. Found in various Western States in the United States; as superb crystals at Arizpe and Las Chiapas, Mexico; and in Chile, Peru, Sardinia, Germany, and Australia.

POLYCARBONATE RESINS.

These resins may be defined as linear polyesters of carbonic acid. The carbonate linkage joining the organic units in the polymer gives this thermoplastic resin its name. The carbonate radical is an integral part of the main polymer chain. Both the General Electric Co. and Farbenfabriken Bayer in Germany concurrently and independently conducted research that led to the definition of the properties of polycarbonates based on 4,4'-dihydroxyl-diphenyl alkanes. Procedures for producing polycarbonate thermoplastic resins commercially emerged rapidly from this research.

Processes

Three processes have gained importance in the production of polycarbonates from bisphenol A [2,2'-bis(4-hydroxyphenyl)propane]: (1) transesterification, (2) interfacial or emulsion polymerization, and (3) nonaqueous solution polymerization. In the first method, the reactants are bisphenol A and diphenyl carbonate. In the latter two, the reactants are bisphenol A and phosgene.

Transesterification. In the transesterification preparation of polycarbonates, the reactants are mixed together in a heated vessel under reduced pressure. This procedure is continued until the desired molecular weight has been attained; then the molten polymer is discharged and directly pelletized for use. Specifically, bisphenol A and a slight molar excess of diphenyl carbonate are heated under inert gas to a temperature of 200–300°C at 20–30 mm Hg. During this period, phenol is distilled and the phenyl carbonate of bisphenol is formed. As the rate of phenol distillation slows, the temperature is increased to 290–300°C and the pressure reduced to 0.2 mm Hg. These conditions are maintained for the elimination of diphenyl carbonate until the desired molecular weight has been reached.

The high temperatures required accelerate potential decomposition and rearrangement reactions which primarily affect the color of the product. For this reason, certain conditions are required, e.g., the use of excess diphenyl carbonate, the use of high-purity raw materials and the addition of catalysts to reduce reaction time. Acidic type catalysts are not as effective as basic material. Alkali and alkaline earth metals, oxides, hydrides, and amides have been used; however, these must be neutralized in the final polymer.

The advantages of this process are (a) no solvent is needed, thus eliminating solvent recovery, and (b) no halogen is used directly, thus reducing the polymer purification steps found in the other processes. Disadvantages include (a) a limit on molecular weight due to the high melt viscosity of the resin, (b) color formation from prolonged high temperatures, and (c) mechanical problems associated with high-temperature, high-vacuum technology.

Interfacial or Emulsion Polymerization. The interfacial or emulsion polymerization technique employs the reaction of an aqueous alkaline solution of bisphenol with phosgene in the presence of an immiscible liquid which is a solvent for both phosgene and polymer. The resulting polymer solution is purified and then the polymer is recovered from solution by any of several techniques.

The reaction is carried out by dissolving the bisphenol A in caustic solution, adding methylene chloride [or other chlorinated solvent], then bubbling phosgene into the agitated mixture. The pH of the solution is maintained above 10 by the addition of caustic as necessary. Temperatures below 40°C are desirable to decrease the rate of phosgene hydrolysis, which is an attendant side reaction. The polymerization proceeds in two steps. The first, which is the production of the chlorocarbonate ester of the bisphenol, occurs rapidly. The second step, the reaction of the chloro ester with additional sodium salt [of bisphenol A] is slow. It can be accelerated by the addition of tertiary amines or quarternary ammonium bases.

The resin solution formed is washed with water until the electrolyte [NaCl] is removed. Recovery of the resin can be accomplished in several ways. One method involves polymer precipitation by the addition of a nonsolvent. This yields a powder which is dried and then pelletized. In this method, the solvent and nonsolvent must be separated for reuse. Another method, which avoids this separation is the direct evaporation of the solvent and finally extrusion into pellets.

The advantages of this process are that very high molecular weight polymers can be made and that reaction is simple and at low temperature. The major disadvantage is the difficulty of washing the polymer solution free of electrolyte.

Nonaqueous Solution Polymerization. Polymerization in nonaqueous solution involves the reaction of bisphenol A and phosgene in the presence of an organic base which acts both as acid acceptor and solvent or organic base, plus an inert polymer solvent. The polymer solution is then purified by the removal of the organic base. Recovery of polymer can be effected by precipitation or solvent removal.

In this reaction the bisphenol A is dissolved in dry organic base or organic base plus an inert solvent, then phosgene is bubbled into the solution. The most commonly used base-solvent system is pyridine-methylene chloride. The base acts by removing hydrogen chloride as the salt.

At the end of the reaction, any excess base is converted to its salt and the resulting polymer solution water washed free of the salt. Finally, the polymer is recovered from solution and pelletized. The reaction temperature varies up to the boiling point of the solvent with no adverse effects.

High-molecular weight polymer can be prepared. Control of molecular weight in this process, as with the interfacial polymerization system, is accomplished by the addition of monofunctional materials.

The advantage of this system is that reaction is carried out in a homogeneous solution. The major disadvantage lies in the fact that the organic base must be recovered because of its cost.

Properties and Fabrication

The finished aromatic polycarbonate resins can be processed by virtually every method and technique used for plastics. They are, however, particularly suitable for injection molding and the majority of parts are produced in this manner. The extrusion processes can be used to produce film and sheet, tubing or rods. Film may be obtained from solution casting. Injection molding and extrusion of polycarbonate resins may be done at temperatures ranging from 450°F to 600°F. (232 to 316°C). High pressures for injection molding are suggested and short mold-filling times are preferred. In general, pinpoint gates are not recommended with polycarbonate resins and lands should be as short as possible. Mold temperatures should be around 200°F (93°C). It is important that the resin be as free of moisture as possible to prevent degradation and poor appearance in molded parts.

Polycarbonates offer a combination of very useful properties unmatched by any other thermoplastic material. The outstanding properties are: (1) very high impact strength (16 ft-lb/in. notch) combined with good ductility, (2) excellent dimensional stability combined with low water absorption (0.35% immersed in water at room temperature), (3) high heat distortion temperature of 270°F (132°C), (4) superior heat resistance showing excellent resistance to thermal oxidative degradation, (5) good electrical resistance.

Its excellent toughness properties make it particularly important as a structural and engineering material able to replace hard-to-fabricate metal parts. Advantageous polycarbonate properties plus its ability to be injection molded also dictate its use in many applications where parts consist of several separate sub-assemblies. With polycarbonates, these sub-assemblies can be molded in one shot, thus increasing manufacturing efficiency as well as bettering final assembly properties.

The self-extinguishing properties of polycarbonate combined with its good dielectric and heat resistance make it suitable for use in electrical appliances and tools. Because of its low water absorption, polycarbonate is used as impellers in multistage water pumps.

Its toughness and its transparency enable polycarbonate to be utilized in such applications as lenses, safety shields, instrument windows, glazing applications, and outdoor lighting applications. Its freedom from toxicity has resulted in FDA approval of polycarbonates for food applications and in the medical field in such applications as heart valves, blood oxygenators, etc.

Polycarbonates have good resistance to inorganic and organic acids, solutions of neutral and acid salts, most alcohols, ethers, aliphatic hydrocarbons, oxidizing agents such as hydrogen peroxide, reducing agents and vegetable oils. Alkalies will slowly decompose polycarbonates although they can be used intermittently in dilute solutions of soaps and detergents, for cleaning purposes, for instance. Aromatic hydrocarbons, chlorinated hydrocarbons, ketones and esters will act on polycarbonates as solvents, partial solvents, plasticizers or crystallizing agents.

Although polycarbonates are produced in a great variety of colors, it is entirely possible to use first or second surface finishing by painting, printing, vacuum metallizing, electroplating and other well-known methods of the industry by choosing the recommended materials for this purpose. All normal solvent cementing and bonding methods are used for joining polycarbonate parts for final assembly. Because of their excellent machinability polycarbonates lend themselves to final assembly machining and are used to produce prototype parts for engineering purposes.

POLYELECTROLYTES. These are macromolecules with incorporated ionic constituents. Polyelectrolytes may be cationic or anionic, depending on whether the fixed ionic constituents are positive or negative. Examples of cationic polyelectrolytes are polyvinyl-ammonium chloride and poly-4-vinyl-N-methylpyridinium bromide. Examples of anionic polyelectrolytes are potassium polyacrylate, polyvinylsulfonic acid, and sodium polyphosphate. If a polyelectrolyte contains both fixed positive and negative ionic groups, it is called a polyampholyte. Polyelectrolytes may be synthesized by polymerization of a monomer containing the ionic substituent, as for instance the polymerization of acrylic acid to polyacrylic acid, or by attaching the ionic constituent by chemical means to an already existing macromolecule, as for instance in the quaternization of poly-4-vinyl-pyridine with methyl bromide, or in the preparation of sodium carboxymethylcellulose from natural cellulose. Many macromolecules occurring in nature are polyelectrolytes. Examples are gum arabic, which carries carboxylate groups; carrageenin, which contains sulfate groups; proteins, which carry both negative carboxylate and positive ammonium groups; and nucleic acids, which contain negative phosphate groups and basic purine and pyrimidine groups, which acquire positive charges at low pH. Inorganic long-chain polyphosphates have also been isolated from biological materials.

A solution of a polyelectrolyte in water or other suitable solvent conducts an electric current, indicating that the polyelectrolyte is ionized. Transference and electrophoresis measurements show that both the macroion and the counterions (gegenions) contribute to the conductance. Because the counterions are osmotically active, polyelectrolytes show much higher osmotic pressures and diffusion rates than do nonionogenic macromolecules. The osmotic pressure of a polyelectrolyte solution is greatly reduced by the addition of a simple electrolyte which distributes itself among the two sides of the membrane according to the thermodynamic theory of Donnan equilibrium. Polyelectrolytes are called weak if they carry weakly ionized groups such as —COOH, and strong if they carry strongly ionized groups such as —COONa. On titrating polyacrylic acid with sodium hydroxide, the pH increases much more slowly than it does in a corresponding titration of a monocarboxylic acid, thus indicating a pronounced buffering capacity. Even in the case of strong polyelectrolytes, the full osmotic activity of the counterions is not realized; as counterions leave the macroion, the electrostatic potential on the latter builds up making it increasing difficult for additional counterions to escape. This binding effect becomes especially strong with multivalent counterions, whose effective concentration may be rendered several orders of magnitude smaller by a polyelectrolyte than their stoichiometric concentration.

The electric charge on the macroion has several important secondary effects. If the macroion is a flexible chain, intramolecular repulsion between charged segments will stretch out the macroion from a coiled to a more rod-like structure, resulting in much larger solution viscosities than are usually obtained with uncharged polymers under corresponding conditions. Intermolecular repulsion causes the macroions to arrange themselves so that they are as far from each other as is possible. With this ordering, the light scattering which is characteristic of solutions of ordinary macromolecules is greatly diminished, often to the vanishing point, as a result of destructive interference. These secondary effects of charge may be reduced by the addition of simple electrolytes which screen the charged elements from each other, and in some cases also lower the charge by specific counterion binding. At high enough concentrations of added salt, the light scattering of the polyelectrolyte may become sufficiently pronounced to allow its use for the determination of the molecular weight.

An interesting class of polyelectrolytes, denoted by polysoaps, is obtained by attaching soap-like molecules to the polymer chain. Such a polysoap is for instance produced by the quaternization of polyvinyl-pyridine with n-dodecyl bromide. The polysoap molecules differ from ordinary polyelectrolytes in that they may reach protein-like compactness in solution. They behave like prefabricated soap micelles and solubilize hydrocarbons and other compounds insoluble in water.

While the applications of polyelectrolytes for practical purposes depend on their general ionic properties, nevertheless large differences appear among individual members of the class in their applicability to a specific use. When polyelectrolytes are absorbed at interfaces, they affect the zeta-potential and a suspending action may result. Adsorption at growing crystal surfaces is also believed to be the reason for the high effectiveness of small amounts of certain polyelectrolytes in preventing or retarding the precipitation of calcium carbonate. The dispersion of clays by polyelectrolytes is applied in oil-well drilling. The ability of long-chain polyelectrolytes to bind together small particles has found uses in soil conditioning and in the flocculation of phosphate slimes. Because of their effect on the solution viscosity, certain polyelectrolytes are used as thickening agents. Because of their ability to bind di- and trivalent cations, some anionic polyelectrolytes are used in water softening and as enzyme inhibitors. When polyelectrolytes are adsorbed or otherwise incorporated into membranes, they make the latter permselective, hindering small ions of the same charge as the macroion from passing through the membrane while allowing free passage to small ions of opposite charge. The well-known ion-exchange resins are polyelectrolytes which have been cross-linked to prevent them from dissolving.

The most important and widespread use of polyelectrolytes is to aid in the removal of small suspended solids from waste water in the primary, secondary and dewatering stages of treatment.

—Ulrich P. Strauss, Rutgers University, New Brunswick, New Jersey.

POLYESTER FIBERS. The principal characteristics of these fibers are described in the entry on **Fibers**. Polyester fibers are defined as synthetic fibers containing at least 80% of a long-chain polymer compound of an ester of a dihydric alcohol and terephthalic acid. The first polyester fiber to be commercialized was prepared from the ester in which the dihydric alcohol was ethylene glycol; this fiber is the material used in the largest quantity by the textile industry. For some other commercial uses, the ester 1,4-dimethyldicyclohexyl terephthalate is also good.

The original process, still in use for making the polymer, employs dimethyl terephthalate (DMT) and ethylene glycol as raw materials. A later process, using direct esterification of terephthalic acid (TPA) with ethylene glycol, also gained acceptance after the increased availability of highly purified TPA. With either process, the first step is the preparation of the intermediate diester, bis-hydroxyethyl terephthalate (bisHET), which then is further condensed to the polymer.

The basic process for making polyester fibers from the polymer is called melt spinning, i.e., heating the polymer above its melting point, forcing it through small holes in a metal plate, and then quenching the molten stream as it issues from the holes by means of a current of cool air. The spun yarn is weak and highly extensible because the polymer molecules are randomly oriented. To impart strength and dimensional stability the yarns must be drawn at temperatures above the glass-transition temperature of the material by pulling the yarn between two godet wheels, the second of which is rotating at a speed three to six times as fast as the first. The higher the draw ratio, i.e., the ratio of the two speeds, the more oriented the molecules become and the stronger the yarn.

The two main classes of polyester fibers are continuous-filament yarns and short-cut fibers, called staple. A wide range of deniers is available in continuous-filament yarns, varying from very fine deniers of about 20 up to 2000 for heavy industrial yarns. (Denier is the weight in grams of 9000 meters of yarn.) The number of filaments in these yarns ranges from about 7 for the 20-denier yarns up to 384 for the heavy material. Staple fiber is produced in sizes ranging from 0.5 to 1.5 denier per filament. The finer deniers are used in making blends with cotton and rayon for apparel, while the coarser-denier yarns generally are used for carpets. Staple lengths vary from $1\frac{1}{4}$ to 6 inches (3.1 to 15.2 centimeters).

Fiber with no added delustrant is designated as clear. Bright fiber has about 0.1% titanium dioxide (TiO_2); semidull fiber has about 0.25% TiO_2; and dull fiber has up to 2% TiO_2. Other variations in physical properties and dyeing characteristics include optically brightened, high-modulus, high-shrink, high-tenacity, low-pilling, deep-dyeable, and cationic-dyeable fibers.

POLYETHYLENE. A thermoplastic molding and extrusion material available in a wide range of flow rates (commonly referred to as melt index) and densities. Polyethylene offers useful properties, such as toughness at temperatures ranging from −76 to +93°C, stiffness ranging from flexible to rigid, and excellent chemical resistance. The plastic can be fabricated by all thermoplastic processes.

Polyethylenes are classified primarily on the basis of two characteristics, namely, density and melt index. The former is the criterion used to distinguish the type; and the latter for the designation as to category (ASTM-D-1248). ASTM type I polyethylene (sp gr 0.910–0.925) is commonly referred to as low-density, conventional, or high-pressure polyethylene. ASTM type II polyethylene (sp gr 0.926–0.940) is commonly referred to as medium-density or intermediate-density polyethylene. ASTM type III polyethylene (sp gr 0.941–0.965) is commonly called high-density, linear, or low-pressure polyethylene. High-density type III polyethylene has been divided into two ranges of density: 0.941–0.959 (considered type III); and 0.960 and higher, commonly considered type IV. Within each density classification, products with different melt indexes are categorized numerically as follows: Category 1 has a melt index (MI) greater than 25; category 2 has an MI greater than 10 to 25; category 3, MI > 1.0 to 10; category 4, MI > 0.4 to 1.0; category 5 has a 0.4 maximum.

Chemical Composition. Polyethylene is formed from the polymerization of ethylene under specific conditions of temperature and pressure and in the presence of a catalyst, according to:

$$
\begin{array}{c}
\begin{array}{cc}
\text{H} & \text{H} \\
| & | \\
\text{C}=\text{C}- \\
| & | \\
\text{H} & \text{H}
\end{array}
\xrightarrow[\text{catalyst}]{\text{pressure}}
\left[
\begin{array}{cccc}
\text{H} & \text{H} & \text{H} & \text{H} \\
| & | & | & | \\
-\text{C}-\text{C}-\text{C}-\text{C}- \\
| & | & | & | \\
\text{H} & \text{H} & \text{H} & \text{H}
\end{array}
\right]_n
\end{array}
$$

The reaction is exothermic and may form polymer from a molecular weight of 1000 to well over 1 million. The high-pressure process, which normally produces types I and II, uses oxygen, peroxide, or other strong oxidizers as catalyst. Pressure of reaction ranges from 15,000 to 50,000 psi (~1020–3400 atmospheres). The polymer formed in this process is highly branched, with side branches occurring every 15–40 carbon atoms on the chain backbone. Crystallinity of this polyethylene is approximately 40–60%. Amorphous content of the polymer increases as the density is reduced.

The low-pressure processes, such as slurry, solution, or gas phase, can produce types I, II, III, and IV polyethylenes. Catalysts used in these processes vary widely, but the most frequently used are metal alkyls in combination with metal halides or activated metal oxides. Reaction pressures normally fall within 50 to 500 psi (~3.4–34 atmospheres). Polymer produced by this process is more linear in nature, with branching occurring about every 1000 carbon atoms. Linear polyethylene of types I and II is approximately 50% crystalline and types III and IV are as high as 85% crystalline.

Ethylene has been polymerized with other monomers, e.g., propylene, butene-1, hexene, ethyl acrylate, vinyl acetate, and acrylic acid, to develop such specific properties as environmental stress crack resistance, low-temperature toughness, and improved flexibility and toughness. High-molecular-weight (HDPE) and chlorinated polyethylenes have been developed to extend the property range of polyethylenes from extremely rigid to elastomeric.

Applications. Polyethylene products include extruded films for food packaging (baked goods, frozen goods, produce), nonfood packaging (heavy-duty sacks, industrial liners, shrink and stretch pallet wrap), nonpackaging (agricultural, diaper liners, industrial sheeting, trash bags); extrusion coating of films, foils, paper, and paperboard; blow molding of bottles, drums, tanks, toys, pails; injection molding of industrial containers, closures, housewares, toys; extrusion of electrical cable jacketing, pipe, sheet, and tubing; and rotational molding of tanks, drums, toys, and sporting goods.

Properties. Tensile strength, hardness, chemical resistance, surface appearance, and flexural modulus increase with an increase in density.

Polyethylene is translucent to opaque white in thick sections, opacity increasing with density. Relatively clear film can be extruded from polyethylene, especially if it is quenched rapidly. The plastic accepts pigmentation readily. Most coloring is performed using dry-blend techniques. Color dispersion de-

vices are required to ensure thorough mixing of resin and pigment.

Compression-molded polyethylenes. These materials are available in three ASTM types: I, II, and III. Types I and II are available in two classes: (1) high-pressure and (2) low-pressure. Specific gravity of these materials ranges from 0.915 to 0.930; melt index from 0.8 to 13.5; yield tensile strength from 1300 psi (9.0 MPa) to 2400 psi (16.6 MPa); ultimate tensile strength from 1300 psi (9.0 MPa) to 4000 psi (27.6 MPa); and ultimate elongation from 240 to 1100%. Type III is available only in the low-pressure class, with a specific gravity of 0.960 and melt index that ranges from 0.7 to 5.0; yield tensile strength from 3800 (26.2 MPa) to 4100 psi (28.3 MPa); ultimate tensile strength from 1900 psi (13.1 MPa) to 2200 psi (15.2 MPa); and ultimate elongation from 200 to 400%.

Injection-molded polyethylenes. These materials are also available in three types, with the specific gravity ranging from 0.916 to 0.960, the melt index from 2.0 to 20.0; the yield tensile strength from 1100 psi (7.6 MPa) to 3700 psi (25.5 MPa); the ultimate tensile strength from 1400 psi (9.7 MPa) to 2500 psi (17.3 MPa); and the ultimate elongation from 125 to 500%.

Mechanical properties of polyethylenes vary with density and melt index. Low-density polyethylenes are flexible and tough; high-density products are quite rigid and have creep resistance under load. Toughness is the primary mechanical property affected by melt index, with lower-melt-index polyethylenes having greater toughness. Using loads, polyethylene is subject to creep, stress relaxation, or a combination of both.

Excellent dielectric characteristics at all frequencies and high electrical resistivity have made polyethylene one of the most important insulating materials for wire and cable. Typical electrical properties are shown in accompanying table.

At no-load conditions, polyethylene has good heat resistance. However, small loads can cause distortion at relatively low temperatures. Dimensional stability of polyethylene is fair to good. Dimensional changes caused by crystallization during cooling usually occur in a nonuniform pattern, resulting in warpage. Narrower molecular-weight distribution resins within given families result in less warpage. Types I and II polyethylenes produced by the low-pressure process offer significant improvement in heat distortion temperatures. This property is directly related to melting point; it is much higher for low-pressure low-density resins than for conventional LDPE resins. This allows molded parts to be exposed to significantly higher service temperatures, e.g., dishwasher parts, without undergoing distortion or warpage. Most shrinkage occurs within 48 hours after fabrication and for type I and type II materials is 0.01–0.03 inch/inch (cm/cm).

Rupture of molecular bonds by external and internal stress in the presence of certain compounds is referred to as *environmental stress cracking.* Small molecular fractures in the amorphous regions propagate until visible cracks appear. In time, the part may fail. Chemical agents which accelerate stress cracking in polyethylene include detergents; aliphatic and aromatic hydrocarbons; soaps; animal, vegetable, and mineral oils; ester-type plasticizers; organic acids; and aldehydes, ketones, and alcohols. There is no adequate test for stress cracking.

Deterioration occurs in uncolored polyethylene exposed to weather. Ultraviolet light causes photoactivated oxidation. Satisfactory weathering formulations contain 2–2.5% of well-dispersed carbon black and stabilizers. The carbon black prevents ultraviolet light penetration.

Unmodified polyethylenes are flammable and are classified in the slow-burning category by the National Board of Fire

TYPICAL ELECTRICAL PROPERTIES OF POLYETHYLENE

PROPERTY	TYPE I	TYPE II	TYPE III
Dielectric strength, short-time, $\frac{1}{4}$ in. (~6 millimeter) specimen, V/mil	460–700	460–650	450–500
Volume resistivity, Ω-cm	10^{-17}–10^{-19}	19^{-16}–10^{-18}	10^{-15}–10^{-16}
Dielectric constant at 10^3 and 10^6 Hz	2.25–2.35	2.25–2.35	2.25–2.35
Dissipation factor at 25°C, $\times 10^{-5}$	3–20	3–20	3–20
Arc resistance, s	150	150	150

Underwriters. Burning rate is approximately 1–1.5 inches (2.5–3.8 centimeters) per minute. The flammability of polyethylene may be retarded significantly by the addition of flame retardant compounds, such as antimony trioxide along with halogenated compounds.

At room temperature, polyethylene is insoluble in practically all organic solvents, although softening, swelling, and environmental stress cracking can occur. At high temperatures, some concentrated acids and oxidizing agents chemically attack polyethylene. Above 60°C, the material becomes increasingly soluble in aliphatic and chlorinated hydrocarbons. Chemical resistance increases slightly as density is increased.

Polyethylene is water-resistant and is a good water vapor barrier. Less than 0.1% water is absorbed in a 2-inch (5-centimeter) diameter, $\frac{1}{8}$-inch (3-millimeter) thick disk of polyethylene in 24 hours. Transmission of other gases is high when compared with that of most other plastics. Polyethylene is not satisfactory for retention of vacuum.

Fabrication. Polyethylene is readily fabricated by all methods of thermoplastic processing. The principal methods used are film and sheet extrusion, extrusion coating, injection molding, blow molding, pipe extrusion, wire and cable extrusion coating, rotomolding, and hot melt and powder coatings.

Decorating. Polyethylene parts are decorated by silk screening, hot stamping, or dry offset printing. For satisfactory printing, the surface must be oxidized by hot air, flame, chlorination, sulfuric acid-dichromate solution, or electronic bombardment. Hot air or flame methods are used with molded parts; flame or electronic methods with films. Inks specially made for polyethylene give best results. Roll-leaf hot stamping does not require pretreatment of the surface.

Design. Because of high mold shrinkage, parts must be carefully designed to minimize warpage. Wall cross-sectional thicknesses should be uniform throughout the part. Large flat areas should be avoided. Corners should be curved rather than square. Stiffening ribs should be less than 80% of the thickness of the wall to which they are attached. Thermoformed parts require liberal radii and draft angles. Slight undercuts can be incorporated when a female mold is used. Dimensional variations in a part made of polyethylene are difficult to predict. In general, greater tolerances should be allowed than with more rigid plastics.

—B. W. Heinemeyer, The Dow Chemical Company, Freeport, Texas.

POLYHALITE. Polyhalite, $K_2Ca,Mg(SO_4)_4 \cdot 2H_2O$ is a late evaporate mineral associated with halite, sylvite, and carnalite from the famous oceanic salt deposits at Stassfurt, Germany, and near Carlsbad, New Mexico. It is of triclinic crystallization, with color grading from gray to brick-red; hardness, 3–3.5; specific gravity 2.78; translucent with vitreous luster; very bitter taste. It is a source of potassium.

POLYIMIDES. These are heat-resistant polymers which have an imide group (—CONHCO—) in the polymer chain. Polyimides, poly(amide-imides), and poly(esterimides) are commercially available.

Poly(amide-imides) are prepared by the thermal degradation of a soluble poly[amide-(amic acid)]. The latter may be produced by the condensation of an aliphatic diamine with less than a molar equivalent of pyromellitic dianhydride or with a molar equivalent of a derivative of trimellitic anhydride, such as the acyl chloride in dimethylacetamide as shown in the following equation:

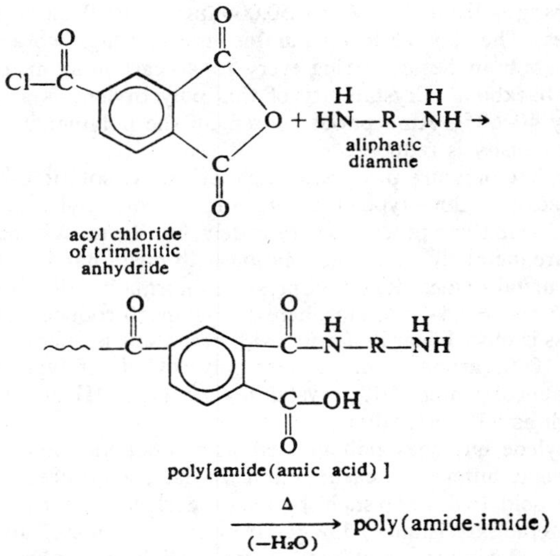

The poly(amide-imides) are soluble in dimethylacetamide but are insoluble in less polar solvents such as toluene and perchloroethylene. They are used for wire enamels, high-temperature adhesives, laminates and molded articles.

The poly(ester-imides) are produced by the thermal decomposition of the soluble poly(amic acids) which are obtained by the condensation of an aromatic diamine and the bis-(ester anhydride) of trimellitic anhydride as shown in the following equation:
These poly(ester-imides) have good electrical properties. Their tensile-modulus is about 400,000 psi (2759 MPa) at 25°C and approximately 50 percent of this modulus is retained at 200°C. Poly(ester-imide) films fail when heated at 240°C for 1000 hrs.

Polyimides are produced by the thermal dehydration of the soluble poly(amic -acid) which is obtained by the condensation of a diamine, such as 4,4'-diaminophenyl ether and a dianhydride, such as pyromellitic dianhydride called PMDA as shown in the following equation:
It is customary to apply these polymers as the poly(amic -acids) and to dehydrate the film, coating, fiber or molded forms by heating to produce the polyimides. Polyimides are insoluble

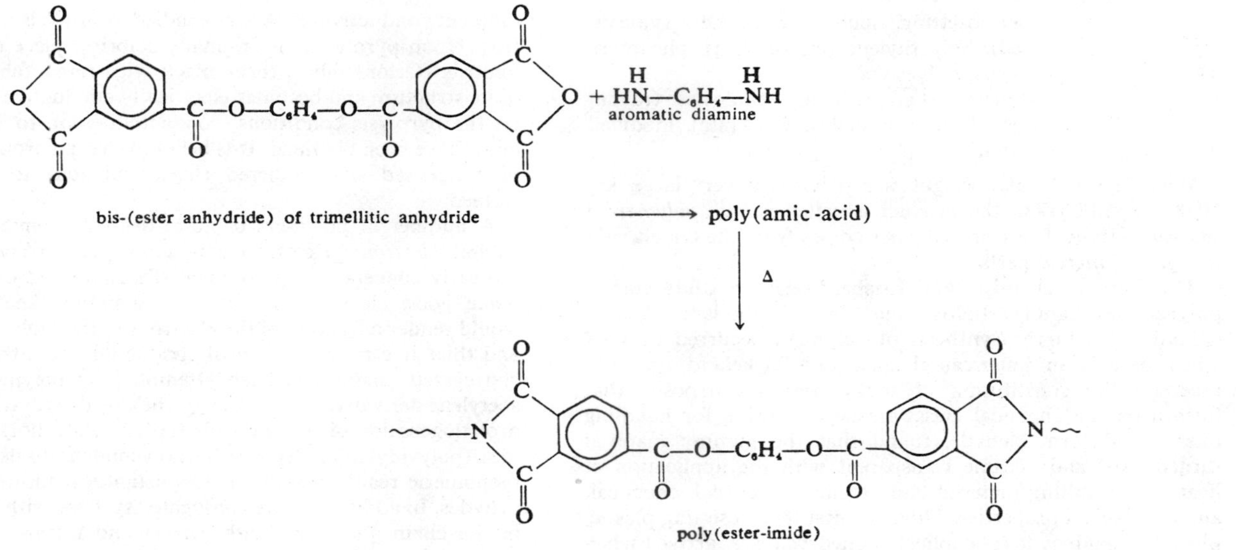

bis-(ester anhydride) of trimellitic anhydride

$+ HN—C_6H_4—NH$
aromatic diamine

⟶ poly(amic-acid)

Δ

poly(ester-imide)

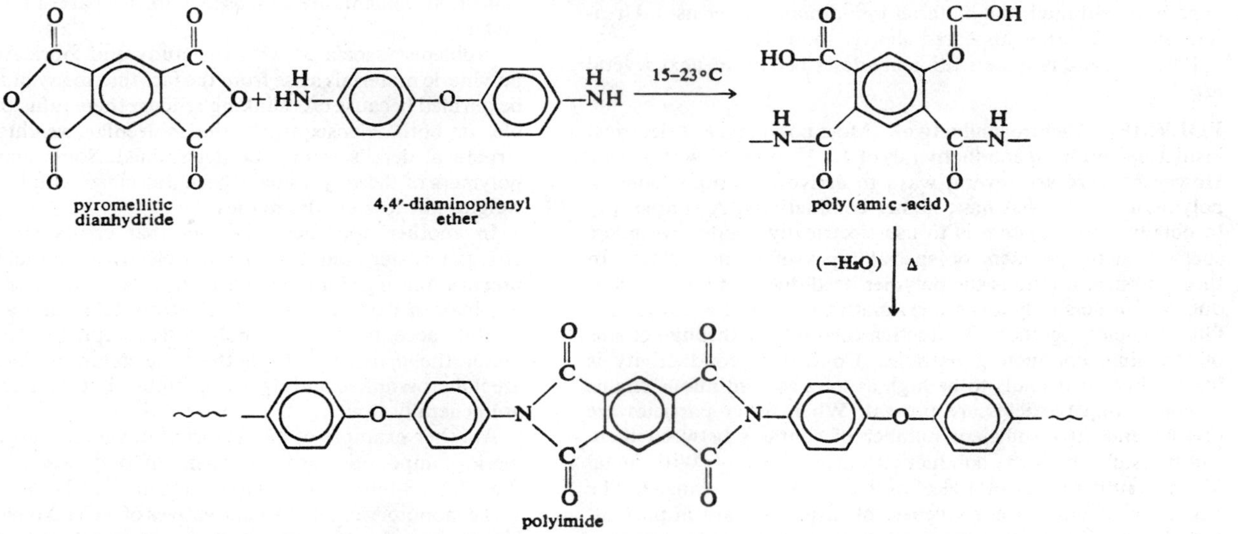

pyromellitic dianhydride

4,4'-diaminophenyl ether

15–23°C

poly(amic-acid)

(−H₂O) Δ

polyimide

in most solvents but are attacked by alkalies, ammonia and amines. These heat resistant polymers are used without fillers and with a graphite filler.

Polyimide films have excellent electrical properties and a tensile modulus of over 400,000 psi at 25°C. Over 60 percent of this modulis is retained at 200°C. Polyimide wire enamels are stable for up to 100 thousand hours at 200°C. Polyimide fibers have a tenacity of 7 g/denier at 25°C and over 1000 hrs at 283°C is required to reduce the value to 1 g/denier.

The coefficient of linear expansion of polyimides is 4.0–5.0 $\times 10^{-5}$ in./in./°C. The heat deflection is 680°F (360°C). Polyimides have been used as binders for abrasive wheels, high-temperature laminates, wire coatings, insulating varnishes and in aerospace applications.

—Raymond B. Seymour, University of Houston, Houston, Texas.

POLYISOPRENE. Rubber (Natural).

POLYMER. A macromolecule formed by the chemical union of five or more identical combining units called *monomers*. In most cases, the number of monomers is quite large (3500 for pure cellulose), and often is not precisely known. In synthetic polymers, this number can be controlled to a predetermined extent, e.g., by shortstopping agents. Combinations of two, three, or four monomers are called, respectively, *dimers, trimers,* and *tetramers,* and are known collectively, as *oligomers.*

Exemplary of well understood polymers are:

(a) *Inorganic*—siloxane; sulfur chains; black phosphorus; boron-nitrogen; silicones
(b) *Organic*
  Natural—polysaccharides, such as starch, cellulose, pectin, seaweed gums, vegetable gums (arabic, etc.)
  —popypeptides (proteins), such as casein, albumin, globulin, keratin, insulin, DNA
  —hydrocarbons, such as rubber (polyisoprene), gutta percha
  Synthetic—thermoplastic, such as elastomers (unvulcanized), nylon, polyvinyl chloride, polyethylene (linear), polystyrene, polypropylene, fluorocarbon resins, polyurethane, acrylate resins

—thermosetting, such as elastomers (vulcanized), polyethylene (crosslinked), phenolics, alkyds, polyesters

—semisynthetic, such as cellulosics (rayon, methylcellulose, cellulose acetate), modified starches

When the molecular weight of a polymer is very large, say 50,000 or 1,000,000, the product is called a *high polymer* or *macromolecule*. The word *polymer* comes from the Greek *poly,* many, and *meros,* parts.

The German chemist Emil Fischer began to study starch, polypeptides, lignin, cellulose, and rubber in the late 1890s. A breakthrough in the synthesis of polymers occurred in 1909 when the Belgian-American chemist, Leo Baekeland, trying to ascertain the constitution of sticky, resinous deposits that formed on the chemical glassware he was using for handling phenol and formaldehyde, found that the gummy material turned hard and became transparent with the application of heat. The resulting material had excellent electrical, chemical, and mechanical properties. Thus, the first thermosetting plastic, phenolformaldehyde (phenolics), named *bakelite,* arrived when there was a great demand for a material to construct electrical apparatus, although the material found many other uses. Urea-formaldehyde resins appeared shortly thereafter.

Polymers and polymerization are described in the next several articles.

POLYMER (Electroconductive). Most polymers are electrical insulators and have conductivities of 10^{-15} ohm^{-1} cm^{-1} or less. However, there are several ways to arrive at compositions of polymeric nature that have higher conductivity. A simple way to obtain such a system is to use electrically conductive fillers such as metal powders or special types of carbon black. In these physical mixtures the polymer itself does not become conductive but acts only as an inert matrix to keep the conducting filler particles together. Conduction then occurs through chains of touching, conducting particles. Control of conductivity is limited in that it tends to be high as long as continuous chains of conducting particles are present. When fewer particles are present and an insufficient number of contacts between them can be established, the conductivity drops sharply. With metal fillers, conductivities of 10^2 ohm^{-1} cm^{-1} can be realized. The particle size and the effectiveness of dispersion are important. In polymer-filler systems the conducting particles may rearrange under the influence of thermal or mechanical cycles, and the bulk conductivity tends to change as a result of such cycles.

Ionic conductivity can be found in polyelectrolytes such as the salts of polyacrylic acid, sulfonated polystyrene or quaternized polyamines (ion-exchange resins). When dry, these materials have low conductivities. However, in the presence of small amounts of polar solvents or water—some of these polyelectrolytes are somewhat hygroscopic—electrical conductivity can be observed. The currents are carried by ions (protons, for instance). Such systems can only be used in cases where very small currents are expected. Large currents would result in observable electrochemical changes of the materials. In applications as antistatic electricity coatings, conductivities of 10^{-8} ohm^{-1} cm^{-1} are sufficient.

Thermal decomposition of a large number of organic solids yield carbonaceous materials which are electrically conductive. It is believed that the conductive pyrolysis products are of polymeric nature, and that at high temperatures a carbon skeleton similar to graphite is formed. Since these products are insoluble, infusible mixtures, very little is known about their structure. Variation of the pyrolysis conditions leads to products with different conductivities. A well-studied example is polyacrylonitrile. Upon pyrolysis, an originally colorless piece of polyacrylonitrile (Orlon) fabric turns black with remarkable retention of its structure and becomes electrically conductive. Depending on the pyrolysis conditions, conductivities up to 10^{-1} ohm^{-1} cm^{-1} have been obtained. It is believed that an aromatic system of condensed six-membered rings analogous to graphite is formed.

A number of polymers of more defined chemical structure exhibit *electronic* electrical conduction. According to one of the early concepts, long conjugated unsaturated chains would make good electronic conductors, assuming that resonance would render a fraction of the electrons in the molecules mobile, and thus give rise to electrical conductivity. Synthesis of long conjugated chains has been attempted by polymerization of acetylene derivatives (phenylacetylene), by dehydration or dehydrohalogenation of polyalcohols (polyvinylalcohol) or polyhalides (polyvinylchloride) and by polycondensations of suitable monomeric reaction partners, for instance, diamines with dialdehydes. In addition to the conjugated systems with only carbon in the chain and those with carbon and nitrogen, polymeric chelates have also been reported. Here the d-orbitals of the transition elements are supposed to form a part of the conjugated system.

Problems associated with the study and fabrication of these polymeric materials arise from the fact that many of them cannot be purified because crosslinking renders them infusible or insoluble, or both. Consequently the molecular weights and other structural details cannot be determined. Some noncrosslinked polymers of these types have been described with low molecular weight and low conductivities.

In another approach, the fact that crystalline monomeric charge transfer complexes exhibit electrical conductivity led to preparation of polymeric charge transfer complexes. These can be obtained from a polymeric electron donor and a monomeric electron acceptor or from a polymeric acceptor and a monomeric donor, the former type being the more common. These polymers are not crosslinked and some are soluble, but their conductivities are generally low.

Another example of an extension of the properties of monomeric compounds into the realm of polymers is the case of the 7,7,8,8-tetracyanoquinodemethan (TCNQ) compounds. Some monomeric, salt-like derivatives of TCNQ have conductivities of the order of 1 ohm^{-1} cm^{-1}. Apparently stacks of TCNQ$^-$ ions and neutral TCNQ are responsible for these high conductivities. The polymeric TCNQ compounds consist of polycations, TCNQ$^-$ ions and neutral TCNQ, and have conductivities ranging from 10^{-10} to 10^{-3} ohm^{-1} cm^{-1}. These polymeric materials are soluble in organic solvents, can have high molecular weights (several million) and can be cast as films from solutions. Although the compounds are polyelectrolytes, they exhibit electronic conduction when dry. Among the many types of electrically conducting polymeric compositions the TCNQ derivatives seem to have an advantage because of an attractive combination of properties, namely controllable molecular weight, solubility, known chemical structure, fair chemical and thermal stability and electronic conduction controllable over several orders of magnitude.

The possibility of synthesizing polymeric superconductors has been proposed, but at the present time these ideas have not been confirmed by successful experiments.

—John H. Lupinski, General Electric Company, Schenectady, New York, and Kenneth D. Kopple, Illinois Institute of Technology, Chicago, Illinois.

POLYMER (Inorganic).

Most inorganic materials can be considered polymeric since they are built up of a relatively simple atomic grouping repeated a very large number of times. Metals and simple ionic materials are easily excluded, but there still remains a large group of covalently bonded, regularly repeating materials. For example, many mineral silicates are based on the monomer $[SiO_4]^{4-}$ which is covalently bonded to form the large, two-dimensional sheets from which these materials are built. Still, we do not normally think of most of these inorganic, covalently bonded polymeric materials as polymers, because their behavior is so different from what we have come to expect of organic polymers. Such properties as high viscosity in the melt and in solution, rubbery elasticity, moldability, ability to form fibers, films, and so on, are not possessed by most of these materials. In a few cases, enough of them are present to suggest the underlying similarity in structure, for example, in the silicate minerals, crysotile asbestos forms fibers of excellent textile quality. Such samples show the possibility of obtaining useful inorganic polymers.

In the light of this discussion inorganic polymers will be considered to be those materials in which the main polymiric chain contains no organic carbon and in which behavior similar to that of organic polymers can be developed.

The question "Why is there such a difference in behavior between the usual inorganic and organic polymeric materials?" is helpful in guiding such a development. The contrast must be due to differences in molecular structure. For example, in the case of quartz the $[SiO_4]^{4-}$ tetrahedra are covalently bonded together. The high regularity of the structure and the large number of cross-links per $[SiO_4]^{4-}$ unit lead to a material which is strong and dimensionally stable, but brittle. The same situation of over-crosslinking can be found with organic polymers. If the number of cross-links in quartz is reduced by substituting organic groups, such as methyl, for some of the oxygen-silicon linkages, the silicone polymers are produced. These polymers, the only commercial inorganic ones, show that inorganic materials which behave as organic polymers can be made. However, a number of obstacles are found which are not as troublesome with organic polymers. For example, six to eight membered rings are more stable than long-chains. In the case of organic materials, if chains can be formed initially, they have considerable stability. With inorganic materials, the bonds are much more labile (constantly forming and breaking) and the long chain may break down to a collection of smaller rings.

Other factors which influence the properties of polymers can be illustrated by examining the bond energies or bond strengths and the ionic character of bonds based on Si as contrasted to similar ones based on C.

Bond Energies

Si—Si	53 kcal	C—C	83 kcal
Si—O	106	C—O	86
Si—N	82	C—N	73
Si—C	78		

Ionic Character

Si—Si	0	C—C	0
Si—O	51%	C—O	22%
Si—N	30	C—N	7
Si—C	12		

From the bond energies we would expect the homo-atomic silane polymers with Si—Si bonds to be considerably less stable than the more familiar C—C chain polymers. This expectation fits the observed facts. On the other hand, one could expect little gain in stability in the carbon series by going to an ether linked chain (—C—O—C—), while in the silicon series a silicon-oxygen linkage is stronger than any of the others. This is reflected in the very good stability of the silicone polymers. From bond energies one might also expect that a chain of alternating Si and N atoms would have good stability.

In addition to pure thermal stability, if the polymer is to be heated in air, one must also consider oxidative stability. In the carbon series oxidation always leads to more stable species and tends to occur, but in the silicon series there is a much higher tendency towards reaction with oxygen. This is the principal reason for the low utility of the silane polymer. Finally, a third factor in polymer stability is the ease of attack by solvents, acids, bases, etc. This is largely determined by the ionic nature of the bonds involved. The silica based polymers should be more susceptible to such attack than carbon, since they have a higher percent of ionic nature.

We do find that acidic or basic water solutions attack silicones when they are heated together under pressure. Their resistance is still high, however, because of other details of the way the polymer molecules are bound together.

The polymers which have been used to illustrate problems of inorganic polymer formation have been heteroatomic, that is, their chains are built from different atoms alternating with each other. The other structure mentioned has been homo-atomic—all the atoms in the chain are the same. There are only a few homoatomic polymers of any promise. Most elements will form only cyclic materials of low molecular weight if they polymerize at all. In addition to the silane polymers, black phosphorus, a high-pressure modification of the element, forms in polymeric sheets.

Boron has similar tendencies in its compounds. The outstanding member of this class is sulfur. A transition from S_8 rings to long sulfur chains takes place over a narrow temperature range around 159°C. An increase in the viscosity of the liquid by 2000 times or more, within a range of 25°, is the tangible evidence of polymerization. The material also forms rubbery, plastic and fibrous forms when chilled to room temperature. However, it has a strong tendency to revert to the cyclic form unless stable groups are placed at the end of the chain or copolymerization hinders the process. Attempts to improve the stability of polymeric sulfur have met with some success. This is the only homoatomic inorganic polymer which appears technically interesting at present.

The class of heteroatomic polymers, besides containing the silicones, offers more promise for useful materials. Most nonmetals and many of the less positive metals form heteroatomic compositions. In many cases they are high polymers. The silicones themselves behave quite the same as organic polymers and are used as oils, rubbers and resins. The rubber is vulcanized either by the reaction of organic peroxides with the methyl groups on the chain or by incorporating groups such as Si—OH or Si—OR which crosslink on exposure to moisture in the atmosphere. The properties in which they excel over organic polymers are high thermal stability, resistance to oxidation and inertness to organic reagents. These are usually the special properties one hopes to get from inorganic polymers. The polymers may be modified by substitution of other groups for the methyl groups on the side chain and by copolymerization with other heteroatoms in which B—O—Si, Al—O—Si, Sn—O—Si, Ti—O—Al and other combinations are produced.

A similar class is the titanates. Three-dimensional Ti—O chains form pigments and pigment binders for paints and waterproofing compounds for use on cloth. Their properties can be modified by substituting monofunctional groups for some of the oxygen, for example, by forming esters to interrupt the chains.

The polyphosphates have also been widely studied. Here the phosphate ion is found as a high polymer. The molecular weight of the polymer ranges from 250,000 to 2,000,000. The polyphosphates are water soluble and form fibers. No uses have been found for this class of materials. They hydrolyze slowly in atmospheric moisture and also embrittle on standing.

Other attempts to base a polymer on B—N heteroatomic chains are being vigorously pursued although the B—O bond with an energy of 130 kilocalories is thermally more stable than the B—N bond at 100 kilocalories. The borates formed with B—O chain links, however, are too hydrolytically unstable and too thoroughly cyclized to be useful. B—N compounds are also plagued with the same weakness. However, the very high thermal stability of low molecular weight materials has encouraged the search for high polymers with the same basic structure. The combination of boron and nitrogen approximates that of carbon with carbon due to its location in the Periodic Table. See **Boron.**

One other area of materials deserves mention here. Coordination polymers are found when metal atoms are joined together by coordinating bonding involving some bridging group, e.g.

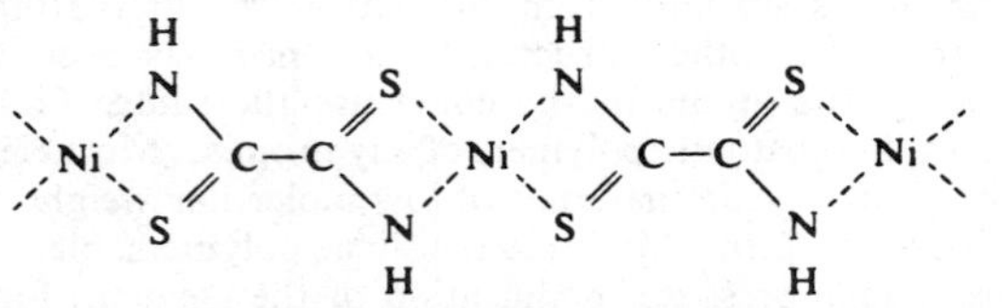

In view of the high thermal stability of monomeric chelation compounds, coordination polymers were expected to be promising for use at high temperatures. This has not proved to be the case. Thermal stabilities are usually lower than for low molecular weight materials. In addition, if the polymerization goes beyond a few monomer units, the materials tend to become insoluble and infusible so they cannot be fabricated into useful items. See **Chelation Compounds.**

—T. E. Ferrington, Clarksville, Maryland.

POLYMER (Organic). Organic high polymers have a great number of different chemical structures, ranging from completely nonpolar to very polar and even ionic materials. They all clearly resemble each other, however. The basis for this resemblance is that many of their properties are governed by their high molecular weights, which range from 5,000 to tens of millions. For example, as the molecular weight increases in a given polymer family, the tensile strength of the polymer increases markedly. In some cases it approaches that of steel on a weight-for-weight basis, especially when oriented fibers are fabricated.

In a similar way, the viscosity of the molten material changes from a free flowing liquid at low molecular weights to the highly viscous polymeric liquid where flow may be observed only over a long period of time or under a considerable applied pressure. A property which shows up only in the case of high polymers is rubbery elasticity. Here again the development of a sufficiently long and flexible molecule is necessary before rubbery behavior develops.

Because of this striking dependence on molecular size, the measurement of molecular weight and dimensions is very important. Some of the most significant early work of Staudinger was the demonstration of the existence of large molecules joined

by covalent bonds. (Others felt that such large molecules were not possible). In order to determine these molecular properties the molecules must be dissolved. Thus each molecule can be separated from its neighbors and its effect measured independently. Solution properties such as osmotic pressure, light scattering and viscosity are used to measure the molecular weight of polymers. In the case of many natural polymers the ultracentrifuge has proved uniquely useful.

The osmotic pressure determination of molecular weights is based on the thermodynamic interaction of solvent and solute to lower the activity of the solvent. Experimentally, the solution is separated from the solvent by a semipermeable membrane. The solvent tends to pass through the membrane to dilute the solution and bring the activity of the solvent in both phases to equilibrium. The quantitative measurement of this tendency is obtained by allowing the liquid solution to rise in a vertical capillary connected to the solution compartment. The equilibrium height it achieves or the rate at which it rises can be measured.

The measurements are converted to effective pressure (π) at zero polymer concentrations (c) and the average molecular weight ($\bar{M}_n$) gotten from the following relation:

$$\lim_{c \to 0} \frac{\pi}{c} = \frac{RT}{\bar{M}_n}$$

(R = gas constant; T = absolute temperature).

The light-scattering method is based on similar thermodynamic interactions. In any solution there are random variations in concentration and refractive index. These scatter some light out of a beam passing through the liquid. In a polymer solution the nature of the fluctuations and thus the amount of scattered light (τ) depend on the attractive forces between polymer and solvent molecules. This, in turn, depends on the polymer molecular weight ($\bar{M}_w$). The following equation describes the behavior:

$$\lim_{c \to 0} \frac{Hc}{\tau} = \frac{l}{\bar{M}_w}$$

(H = a constant)

This method was developed by Peter J. W. Debye in 1944. Its evolution has been one of the most stimulating chapters of polymer physics.

In contrast to these thermodynamic methods, the viscosity molecular weight determination depends on the interference in the flow of the solvent caused by the dissolved molecules. In contrast to osmometry and light scattering, it has not been possible to develop the viscosity effect into an absolute measure of molecular weight. Rather, it must be calibrated, preferably by light scattering measurements.

The relationship between the measured limiting specific viscosity [η] and the molecular weight ($\bar{M}_v$) is as follows:

$$\lim_{c \to 0} \frac{\eta_{sp}}{c} = [\eta] = K\bar{M}_v^{\,a}$$

K and a are determined by calibration for a given polymer-solvent system.

$$\eta_{sp} = \frac{\eta_{\text{solution}}}{\eta_{\text{solvent}}} - 1$$

In each of the equations above a different symbol has been used for the molecular weight. Most polymers are heterodisperse, i.e., have many molecular weight species of the same chemical nature, thus the experiments yield an average molecular weight. Osmotic pressure gives a lower average ($\bar{M}_n$) or number average) because it emphasizes the effect of small molecules, while light scattering emphasizes the larger molecules and gives a higher average, ($\bar{M}_w$) or weight average). The viscosity average ($\bar{M}_v$)is between the two and closer to the weight.

Many natural polymers are monodisperse (all molecules have the same molecular weight). In this case the ultracentrifuge which separates materials according to their effective density in solutions is a most powerful tool for molecular weight determination. With poly-disperse materials, the interpretation of ultracentrifuge results becomes more complex and widespread application of this method to synthetic polymer molecular weight determination has not yet been achieved.

In addition to the primary effect of the great length of the molecule, the details of the distribution of functional groups along the polymer chain modify the behavior of these materials. This leads to differences in their applications. For example, natural rubber exists in the rubbery state at room temperature. If cooled below zero degrees Celsius it becomes a hard, inflexible material, brittle and easily broken. On the other hand, if heated too far above room temperature, it begins to flow quite rapidly and behaves more like a fluid than a rubber. The same pattern is observed with other materials. For example, polystyrene, hard and brittle at room temperature, becomes rubbery when heated up sufficiently. The study of the mechanical behavior of polymers at various temperature is called rheology. The temperature at which the rubbery material becomes glassy is called the glass transition temperature (T_g). This transition temperature depends on the nature of the backbone and the substituent groups on the polymer chain. Rubbery materials, e.g., polyisoprene, polychloroprene, polybutadiene, the copolymer of butadiene and styrene, etc., have molecular chains with considerable flexibility. Usually small side groupings and irregularities in the chain prevent them from coming together in a regular structure. Instead the molecules stay in an amorphous random packing much like a pile of cooked spaghetti. As with cooked spaghetti, there is a tendency for the whole mass to flow, by the movement of chains past one another. With natural rubber this flow at room temperature and above had to be inhibited by tying the chains together with chemical bonds before a useful product was obtained. This cross-linking is called vulcanization in the case of rubber. The process was discovered by Charles Goodyear in 1839. A similar cross-linking to inhibit the motion of the chains is necessary to make useful products from the newer synthetic rubbers also.

Other polymers are not rubbery at room temperature, instead they exist in the glassy state. They are amorphous, but because of more bulky substituents on the polymer chain the molecule is less flexible. There is less ease of molecular motion under applied stress at room temperature. Examples of such polymers are polystyrene and polymethylmethacrylate, which are transparent due to their amorphous, homogeneous nature. When heated they first become rubbery and then, at higher temperatures, show viscous flows so that they may easily be molded. When they are cooled to room temperature the rate of flow is vanishingly small due to the stiff chains; thus items made in this way can be used at ordinary temperatures if they are not required to bear too large a load. These common organic glasses are brittle and easily broken on impact. One of the interesting problems for polymer development is to obtain impact resistance without losing transparency and without increasing the cost by an excessive amount.

A related class of polymers is the crystalline, thermoplastic materials. These also are fabricated by heating to a high temperature so that they flow; but when they are cooled ordered regions develop within them, which makes them translucent. They have much tendency to flow because of these mechanical "crosslinks" and have good dimensional stability. Polyethylene and polypropylene belong to this class. Here the chain is simple and regular so that different polymer molecules, or different parts of the same molecule, can pack next to each other. The same situation exists with "Teflon" (polytetrafluoroethylene).

As might be expected, there are polymers intermediate between the crystalline and glassy ones. For example, polyvinylchloride shows enough order to prevent its classification as glassy, but not enough to be considered crystalline.

Many crystalline polymers form part of another class of materials, the fiber-forming polymers. The formation of fibers of significant strength depends on the growth of ordered structures when the fiber is stretched. Thus crystalline materials, such as polypropylene, whose crystallites can grow on elongation, form strong fibers. In some cases fiber formation is aided by polar groups on the polymer chain. These interact with each other to give strong attractive forces which aid the molecular alignment that is needed. For example, polyacrylonitrile, the base polymer of "Orlon" and "Acrilan," has a —CN dipole in each monomer unit. The cooperative attraction of hundreds of these units along a chain gives a very strong cumulative effect. The fibers are strong even though they are not crystalline. The intermediate degree of order they develop is referred to as paracrystallinity.

In the case of other common fibers the polar groups are found in the polymer chain itself. In nylons the amide group,

$$\begin{array}{ccccc} H & O & & O & H \\ | & \| & & \| & | \\ -R-N-C-R-C-N- \end{array}$$

offers the possibility of both dipolar attraction and hydrogen bonding. The ester group in "Dacron" and similar polyesters contributes the —C=O dipole. See **Fibers.**

The introduction of oxygen and nitrogen into the chain changes its flexibility, stability toward chemical reaction, resistance to solvents, strength and other properties. In this way quite extensive changes in behavior are obtained. Most of these heterochain polymers are prepared by condensation or ring opening reactions. The first important work in this field was the classic investigation of W. H. Carothers in the late 1920's on polyesters and polyamides. See **Polyamides.**

The toughness, high melting points and high tensile strength of many of these polymers have since led to their widespread use of fibers, films and molded objects. In general these polymers are rather high cost materials, which are used, because of their unusual properties, in places where ordinary polymers are inadequate. For example, polysulfide rubbers show outstanding solvent resistance, while the silicones are a unique class of materials inert to many environmental conditions and unwettable by most liquids. Polycarbonates and acetals are so dimensionally stable that they can be used in place of metals in molded items. Many of the interesting developments in polymers over the last few years have involved new syntheses and new variations in structure of such heterochain materials.

The materials described above are thermoplastic resins, i.e., they all melt on being heated to sufficiently high temperatures and can be molded while molten. This characteristic is associated with molecular chains which are long and stringlike with few branches on them. If, however, the polymer chains have many covalent bonds linking them together into a network, a thermosetting resin develops which may flow at an early stage of its history, but is insoluble and infusible after the full crosslinking reaction has taken place. Any of the previous chain compositions can be used in making thermosetting resins if provision for crosslinking is made by using multifunctional monomers. Some are used more commonly, e.g., the epoxy resins, phenolformaldehyde, urea formaldehyde, melamine, etc. Separate articles describing the properties of many of these plastics are found in other parts of this encyclopedia. In general they are useful because of their inertness to solvents, resistance to dimensional change on heating, rigid dimensions, physical strength, chemical resistance and abrasion resistance.

Many of the varieties of polymers which have been discussed have analogs in polymers isolated from natural systems, for example, natural rubber is a polyisoprene with a purely carbon chain. Other natural polymers are based on the C—O—C bond. The cellulose and starch polymers, which are found in plants are composed of chains of six-membered carbon-oxygen rings joined through an oxygen linkage. Cellulose and starches differ from each other in the spatial orientation of the links joining the six membered rings.

Products from different sources in each class differ in degree of branching of the molecule and in amount of crosslinking.

In unmodified cellulose the hydroxyl groups give a large amount of hydrogen bonding which leads to insolubility in most solvents. On the other hand if these are changed by chemical reactions to ether or ester groups a much more tractable material results. Cellulose acetate, butyrate, and nitrate; methyl and ethyl ether and carboxy methyl ether are widely used modified celluloses. Starches also are modified, but much less commercial success has been had with them.

The polymers described above have been chemically pure, although physically heterodisperse. It is often possible to combine two or more of these monomers in the same molecule to form a copolymer. This process produces still further modification of molecular properties and, in turn, modification of the physical properties of the product. Many commercial polymers are copolymers because of the blending of properties achieved in this way. For example, one of the important new polymers of the past ten years has been the family of copolymers of acrylonitrile, butadiene and styrene, commonly called ABS resins. The production of these materials has grown rapidly in a short period of time because of their combination of dimensional stability and high impact resistance. These properties are related to the impact resistance of acrylonitrile-butadiene rubber and the dimensional stability of polystyrene, which are joined in the same molecule.

Since they are organic materials most polymers are not water soluble, however, water solubility can be obtained by substituting the proper side groups on the polymer chain. Such polymers include the nonionic materials polyvinyl alcohol, polyethylene glycol, etc., where the strong dipolar and hydrogen bonding interactions cause the solubility; and the polyelectrolytes where ionizable groups such as the carboxylate, sulfonate, quaternary ammonium, etc., cause the solubility. Many of these water soluble polymers are used to increase the viscosity of water based systems. As little as 0.1 or 0.2% of the polymer is needed to produce a very viscous solution. They also are of wide biological interest.

—T. E. Ferrington, Clarkesville, Maryland.

POLYMERIC FOOD ADDITIVES. Commencing in the early 1970s, developments have been underway in the area of nonabsorbable, polymeric food additives.[1] In brief, functional monomers of suitable chemical and biological stability are augmented in size by way of their incorporation onto a polymer backbone. In this configuration, because of their large molecular dimensions, the functional substances which are a part of the

polymer are rendered virtually nonabsorbable from the gastrointestinal tract of humans. Because they are metabolically inert, polymeric additives offer at least a partial solution to food professionals who desire to improve safety margins by decreasing the potential risk associated with long-term ingestion of food chemicals. See Fig. 1.

It has been known for decades, of course, that substances consisting of large molecules, such as cellulose, are indigestible by humans. In 1972, A. Zaffaroni of Palo Alto, California extended this basic concept to the development of polymeric food additives. Initial concentration of development effort during the interim has been directed (by Dynapol) toward polymeric antioxidants, food colorants, and nonnutritive sweeteners.

As pointed out in the entry on **Additive (Food),** the great majority of additives function prior to consumption of the food product—acting as aids in the manufacture, preservation, coloration, and stabilization of food products, among many other important roles. Nonnutritive sweeteners are added solely to make an initial contact with the taste buds, after which they do not serve (or should not serve) any further function. With the exception of nutrients and dietary supplements (minerals and vitamins), additives should not alter or interfere in any way with the metabolic or other biological processes of the body. Polymeric additives pass virtually untouched through the body and consequently do not participate in body chemistry, relieving concern over possible long-term carcinogenicity or other damaging effects.

Antioxidants. A polymer-leashed antioxidant has been developed that combines the functional groups from three or more of the standard antioxidants used to protect fats and oils. In making this macromolecule, a suitable base compound (divinylbenzene) was used. Various antioxidant groups can be included while building the molecule by condensation of the diolefin radicals. Over 50 different polymers were made and evaluated. The resulting molecule consists of the diolefin branched-chain backbone with numerous antioxidant groups attached, including the functional parts of the hydroxyanisoles and hydroquinones, and the complete TBHQ radical (*t*-butylhydroquinone).

The resulting antioxidant[2] has been tested by food processors. Its antioxidant performance in vegetable oils has been demonstrated to be greater when tested by the AOM (active oxygen method) than either butylated hydroxyanisole (BHA) or butylated hydroxytoluene (BHT) when all are used at concentrations of 200 ppm (parts per million). The oils tested were cottonseed/soy, corn (maize), and palm oils. In potato chips, the polymeric antioxidant demonstrated results comparable with BHA in maintaining shelf life in controlled tests, and proved superior to TBHQ.

The antioxidant also provides better protection against rancidity in ground turkey meat than BHA, based upon TBA (thiobarbituric acid) test values and sniff panel tests. At a concentration of 1000 ppm in a linoleic emulsion system, the polymeric oxidant has been shown to be effective in preventing early onset of subtle, rancidity-caused odors and flavors of the type that occur in breakfast cereals, dehydrated potato flakes, and compounded flavors. In essential oils, it is claimed that the macropolymer antioxidant is more effective at equal-weight concentrations than a number of food-grade and high-performance combination antioxidants.

Polymeric Dye Colors. Borrowing from the fundamentals of color photography, where mixtures of three basic colors (lemon yellow, cyan, and magenta) produce an extensive spectrum of colors, the researchers extended this concept to the development

---

[1] Information for this summary furnished by T. E. Furia, Dynapol, Palo Alto, California.

[2] Named Poly AO®-79 by its makers, Dynapol. U.S. Patent No. 3,996,199 (1976).

Small Molecules

Polymers

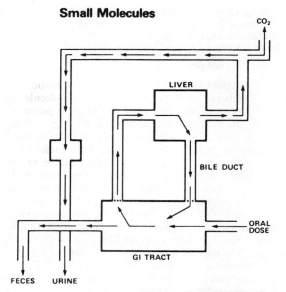

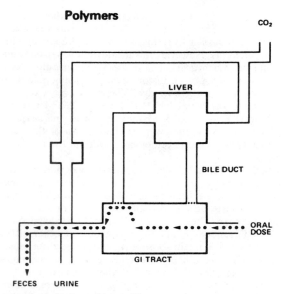

Fig. 1.   Biological fate of monomeric and polymeric dyes. (*Dynapol*.)

of a color system consisting of a yellow, red, and blue polymeric dyes. The schematic representation of the molecular structure of a polymeric dye is shown in Fig. 2. Although the chromo-

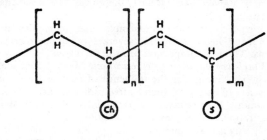

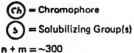

n + m = ~300

Fig. 2.   Molecular structure of a polymeric dye. (*Dynapol*.)

phore per se may or may not be water soluble, the negatively charged solubilizing groups impart to the polymeric dye molecule the desired water solubility and compatibility with other food products. As of the early 1980s, these materials are undergoing final testing prior to obtaining official FDA approval for use in food products. The polymeric materials are spray-dried, free-flowing, stable powders that conform essentially to the characteristics summarized in the accompanying table. The researchers have given particular attention to creating products of reproducible quality and which are essentially free of low-molecular-weight, potentially absorbable impurities, such as oligomers, unreacted monomers, solvents, etc. In solution, the polymeric dyes have shown excellent color stability over a 1.0–8.0 pH range.

The list of approved, traditionally certified food colors has undergone considerable shrinkage during recent years. Should this process of elimination by regulatory bodies (in the United States and some other countries) continue, many existing hues will no longer be available for use in food products. An initial objective of research at Dynapol was that of meeting all or most of the spectral needs of food processors for a large number of colors. It was found that the polymer-leashed colors are feasi-

TYPICAL PHYSICAL AND CHEMICAL CHARACTERISTICS
OF POLYMERIC DYES

	POLY Y(TM)-607	POLY R(TM)-481
Molecular weight	~40,000	~40,000
Particle size, μ	4–18	4–18
Moisture content, %	5.8	9.7
Bulk density, g/ml	0.40	0.17
pH (3% solution)	7.1	6.6
rate of solubility, min.	5	<24
Ca^{++} tolerance (as ppm $CaCO_3$)	>1,250	>1,250
$E_{1cm}^{1\%}$	145	132
λ_{MAX}, nm	429	513

SOURCE: Dynapol, Palo Alto, California.

ble to manufacture within the price range of the premium food-color market. Biological profiles have shown excellent compatibility with the human system.

The polymers used for leashed colors took a different path than for the development of antioxidants. The antioxidant molecule was constructed from scratch, building up the backbone and functional groups from smaller components. It was reasoned that this method would contribute to a complex spectral distribution that would easily dull the desired color. Therefore, the researchers elected to commence with a macromolecule and add complete chromophore groups. This was done to avoid the formation of subsidiary colors and to provide a cleaner spectral distribution.

Because azo dyes are the most widely used and are water soluble, the first approach was to bond azo dyes to selected polymers via a sulfonamide linkage. However, the azo linkages in the dyes themselves were unstable to intestinal microbe action in rat studies, so other dyes were investigated. During the early research phase, an attractive polymer emerged that was uniquely suitable for color leashing. The polymer, PAE (polyaminoethylene), has a high density of nucleophilic sites so that numerous chromophores can be added to give a high color intensity per unit weight.

Having developed a very satisfactory carrying molecule, the researchers moved into the anthraquinone class of chromophores, seeking a red polymer. However, the basic water insolubility of an anthraquinone system kept pushing the research team to consider other options. The backbone was reworked, converting a portion of it to sulfonic acids that impart anionic solubilizing functions. However, this meant that fewer chromophores could be attached and less intense colors would result.

By careful balancing of all factors, two promising colors resulted: a red dye and a yellow dye. These dyes have spectral characteristics of the banned FD&C Red No. 2 and the still available FD&C Yellow No. 5. Manufacturing techniques were developed and test quantities were evaluated. As of the early 1980s, a blue dye is under development to fill out the full color spectrum.

Nonnutritive Sweeteners. The polymeric food additive approach is also being researched in terms of nonnutritive sweeteners. A number of these substances have been developed in recent years, but their taste profiles do not match that of sucrose or corn (maize) syrups. They are derivatives of dihydrochalcones, formed from flavanones found in grapefruit and other citrus peel. These substances retain some of the unwanted taste sensations characteristic of dihydrochalcones, a taste which commences slowly, and lingers longer than that of sucrose.

One explanation of the slow onset/lingering aftertase effect of dihydrochalcone sweeteners apparently relates to a general property of phenols—they bind readily with proteins. Therefore, with their phenolic groups, these substances may bind rapidly and indiscriminately with protein in saliva. In such case, it would require some time for the dihydrochalcones to use up available saliva protein prior to binding to taste receptor sites. This view is supported by the fact that the delayed onset effect does not happen when high concentrations of the dihydrochalcones are tasted, furnishing adequate molecules for all available protein pairing. Another significant finding is that three sites responsible for the sweetness have been located on dihydrochalcone molecules and analogs. These fit the model suggested by Kier, who proposed that surface configuration of a sweetener has three sites at specific angles to one another. These findings (of Dynapol) couple with another observation. The dihydrochalcone molecules have two phenol rings, with part of the three activating sites distributed between them. Saccharin and cyclamates have only one similar site per molecule. Is it possible that both of these compounds must pair up with a single receptor on the tongue to create the sweetness sensation?

At present, the researchers are working with a core concept, in which two or more of the sweetener nuclei are linked together through a common bridge. These sweeteners have been found to have less than 1% absorption in the intestines, as determined by ^{14}C-labeled analogs in rats. See also **Sweeteners.**

References

Bellanca, N.: "Polymeric Dyes: A Novel Color System for the Beverage Industry," *Proceedings of Society of Soft Drink Technologists*, 25th Meeting (April 1978), San Antonio, Texas.

Bellanca, N., and W. J. Leonard, Jr.: "Non-Absorbable Polymeric Dyes for Food," American Chemical Society Food Color Symposium (1977), New Orleans, Louisiana and in "Current Aspects of Food Colorants," (T. E. Furia, editor), CRC Press, Cleveland, Ohio, 1978.

Dodson, A. G., Hammond, G. A., and D. M. Lim: "The Use of Polydyes as Colouring Agents in Confectionary Products," The British Food Manufacturing Industries Research Association, Leatherhead, Surrey, U.K., 1978.

Furia, T. E., and N. Bellanca: "The Properties and Performance of Poly AO⊖-79; A Nonabsorbable, Polymeric Antioxidant Intended for Use in Foods," *J. Amer. Oil Chem. Soc.*, **54**, 6, 239–244 (1977).

Salant, A.: "Non-Nutritive Sweeteners" in "Handbook of Food Additives," (T. E. Furia, editor), Chemical Rubber Company, Cleveland, Ohio, 1968.

Swartz, M. L., and T. E. Furia: "Special Sensory Panels for Screening

New Synthetic Sweeteners," *Food Technol.*, **31**, 11, 51–55, 67 (1977).

Weinshenker, Ned M.: "Polymeric Additives for Food," American Chemical Society Symposium on Polymeric Drugs (1977, New Orleans, Louisiana); and in "Polymeric Drugs" (Donaruma and Vogl, editors), Academic Press, New York, 1978.

POLYMERIZATION. Chemical reaction, particularly in organic chemistry, which provides very large molecules by a process of repetitive addition. These are of great practical importance in the field of rubbers, plastics, coatings, adhesives and synthetic fibers. The initial materials which give rise to such reactions are called *monomers*; they have molecular weights between 50 and 250 and have certain reactive or functional groups which enable them to undergo *polymerization*. The large molecules, which are formed by a polymerization reaction are called *high polymers, macromolecules*, or simply *polymers*; they usually consist of several hundred and in many cases even of several thousand monomeric units and, consequently, have molecular weights of many hundred thousands and even of several millions. The number of monomers contained in a polymer molecule determines its *degree of polymerization* (D.P.).

Polymerization processes never lead to macromolecules of uniform character but always to a more or less broad mixture of species with different molecular weights, which can be described by a *molecular weight distribution function*. The individual macromolecules of such a system belong to a *polymer-homologous series*; the molecular weight and the degree of polymerization of a given material have, therefore, always the character of *average* values.

There exist many ways to assemble small molecules to give large ones and, hence, there exist several types of polymerization reactions. The most important are the following:

(1) *Vinyl-type addition polymerization.* Many olefins and diolefins polymerize under the influence of heat and light or in the presence of catalysts, such as free radicals, carbonium ions or carbanions. Free radicals are particularly efficient in starting polymerization of such important monomers as styrene, vinylchloride, vinylacetate, methylacrylate or acrylonitrile. The first step of this process—the so-called *initiation* step—consists in the thermal or photochemical *dissociation* of the catalyst, and results in the formation of two *free radicals*:

$$\begin{array}{ccc} \text{R—R} & \xrightarrow[\text{or light}]{\text{heat}} & \text{2R} \cdot \\ \text{Catalyst molecule} & & \text{two free radical} \\ & & \text{type fragments} \end{array} \quad (1)$$

The most commonly used catalysts are peroxides, hydroperoxides and aliphatic azocompounds, which need activation energies between 25 and 30 kcal for decomposition.

The free radicals $R \cdot$ attack the monomer and react with its double bond by adding to it on one side and reproducing a new free electron on the other side:

$$R \cdot + CH_2{=}CHX \rightarrow R{-}CH_2{-}CHX \cdot \quad (2)$$

This step is called *propagation* reaction; it adds more and more monomer units to the growing chain and builds up the macromolecules while the free radical character of the chain end is maintained. Each single addition represents the reaction of a free radical with a monomer molecule—a process which requires an activation energy of 8–10 kcal.

Whenever two free radical chain ends collide with each other they can react in such a manner that the resulting products have lost their free radical character and are converted into normal stable molecules. One way is a process of *recombination*:

$$\begin{array}{l} R{-}({-}CH_2{-}CHX{-})_x CH_2{-}CHX \cdot \\ \quad + R{-}({-}CH_2{-}CHX{-})_y CH_2{-}CHX \cdot \end{array}$$

$$\rightarrow R\text{---}(\text{---}CH_2\text{---}CHX)\text{---}CH_2\text{---}CHX$$
$$\text{---}CHX\text{---}CH_2\text{---}(CH_2\text{---}CHX\text{---})\text{---}_yR$$

where *one* macromolecule of the degree of polymerization $(X + y + 2)$ is formed. The other is a process of *disproportionation*:

$$R\text{---}(\text{---}CH_2\text{---}CHX)_x\text{---}CH_2\text{---}CHX \cdot$$
$$+ R\text{---}(\text{---}CH_2CHX\text{---})\text{---}_yCH_2\text{---}CHX \cdot$$
$$\rightarrow R\text{---}(\text{---}CH_2\text{---}CHX\text{---})\text{---}_x\text{---}CH_2\text{---}CHX \cdot \qquad (4)$$
$$+ R\text{---}(\text{---}CH_2\text{---}CHX\text{---})\text{---}_yCH_2\text{=}CH_2X$$

where a hydrogen atom moves from one molecule to the other so that one of the two resulting molecules—the $(x + 1)$ mer—has a double bond at its end, whereas the other one—the $(y + 1)$ mer—has a saturated chain end. Reactions in the course of which free radicals are destroyed are called *termination* or *cessation* steps; they convert the transient reactive intermediates into stable polymer molecules.

Vinyl-type addition polymerization can also be carried out with *acidic catalysts* such as boron trifluoride or tin tetrachloride and with *basic catalysts* such as alkali metals or alkali alkyls. An example of the first case is the low-temperature polymerization of isobutene, which gives "Vistanex" and butyl rubber; an example of the second type is the polymerization of butadiene with sodium, which leads to buna rubber.

(2) Another important kind of addition polymerization is the formation of polyethers by the opening of epoxy ring compounds. Polyoxyethylene ("Carbowax") is produced by a sequence of additions of ethylene oxide to an alcohol or amine, as initiator:

$$CH_3\text{---}CH_2OH + CH_2\text{---}CH_2 \rightarrow$$
$$\underset{O}{\diagdown \diagup}$$
$$CH_3\text{---}CH_2\text{---}O\text{---}CH_2\text{---}CH_2\text{---}OH +$$
$$CH_2\text{---}CH_2 \rightarrow$$
$$\underset{O}{\diagdown \diagup}$$
$$CH_3\text{---}CH_2\text{---}O\text{---}CH_2\text{---}CH_2\text{---}O\text{-}$$
$$\text{---}CH_2\text{---}CH_2OH, \quad \text{and so on}$$

No termination reaction occurs in this case; the reaction proceeds until all the monomer is used. This process is catalytically accelerated by the presence of alkali. A similar addition polymerization involving the opening of a ring compound is the conversion of caprolactam into polycaprolactam ("Perlon" or 6-nylon) under the influence of acidic or basic catalysts. All addition polymerizations are typical *chain reactions* with at least two or three different elementary steps cooperating in building up the resulting macromolecules.

(3) There exist other, different classes of reactions which form large molecules, namely processes in the course of which a *small fragment*, usually H_2O, is *split out* of two reacting monomers and where the monomers are chosen in such a manner that the removal of the fragment can be repeated many times. Multi step reactions of this type are called *polycondensations*; they involve the use of at least a pair of bifunctional monomers and proceed by a sequence of identical condensation steps. One important process of this type is the formation of polyesters from glycols and dicarboxylic acids. Thus the progressive removal of water from ethylene glycol and adipic acid leads to a soft, rubbery polyester ("Paracon")

$$HOCH_2\text{---}CH_2OH$$
$$+ HOOC\text{---}(\text{---}CH_2\text{---})\text{---}_4COOH$$

$$\rightarrow HOCH_2CH_2\text{---}O\text{---}CO\text{---}(\text{---}CH_2\text{---})\text{---}_4COOH$$
$$+ HOCH_2\text{---}CH_2OH \rightarrow HOCH_2\text{---}CH_2OCO$$
$$= \text{---}(\text{---}CH_2\text{---})\text{---}_4COOCH_2\text{---}CH_2OH, \quad \text{and so on.}$$

As long as in processes of this type only *bifunctional* monomers are used, the resulting macromolecules are *linear* and, as a consequence, are of the soluble and fusible type. They can be used as fiber formers, rubbers or thermoplastic resins. If, however, some of the monomers are tri- or tetra-methylolurea, the reaction leads to three-dimensional polymeric networks which are hard and brittle thermosetting resins, such as "Bakelite" or "Glyptal."

The preceeding classification of polymerization reactions concentrates essentially on the organic chemical character of the involved monomers and on the mechanism of their interaction. There exists, however, another classification which is concerned about the manner in which polymerization reactions are carried out in practice and which is of interest and importance whenever industrial application is contemplated. We shall, therefore, briefly enumerate here the most important *polymerization techniques*.

(1) Polymerization in the *gas phase* is usually carried out under pressure (several thousand psi) and at elevated temperatures (around 200°C); the most important example is the polymerization of ethylene to form polythene.

(2) Polymerization in *solution*, essentially under normal pressure and at temperatures from −70°C to 70°C; important examples are the production of butyl rubber with boron trifluoride and the synthesis of the various "Vinylites" with benzoyl peroxide.

(3) Polymerization in *bulk* (or in block) under normal pressure in the temperature range from room temperature to about 150°C. The batch polymerization of methylmethacrylate to give "Lucite" or "Plexiglass" and the continuous polymerization of styrene to give the various types of polystyrene can be quoted as examples.

(4) Polymerization in *suspension* (bead or pearl polymerization) under normal pressure in the range from 60 to 80°C operates with a suspension of globules of an oil-soluble monomer in water and uses a monomer soluble catalyst. Substantial quantities of polystyrene and polyvinyl acetate are made by this method.

(5) Polymerization in *emulsion* under normal pressure and in the temperature range from −20°C to 60°C uses a fine emulsion of oil-soluble monomers in water and initiates the reaction with a system of water-soluble catalysts. This method is probably the most important of all, because it is used in very large scale in the copolymerization of butadiene and styrene and in the polymerization of many other monomers, such as chloroprene and vinyl chloride, to produce latices of the various synthetic rubbers.

See also **Molecule;** and several articles which follow.

(*Reprinted from the 3rd Edition because the article by Herman F. Mark is such an apt review of the fundamentals of polymerization.*)

POLYMERIZATION (Emulsion).

Since an aqueous system provides a medium for dissipation of the heat from exothermic addition polymerization processes, many commercial elastomers and vinyl polymers are produced by the emulsion process. This two-phase (water-hydrophobic monomer) system employs soap or other emulsifiers to reduce the interfacial tension and disperse the monomers in the water phase. Aliphatic alcohols may be used as surface tension regulators.

Formulas for emulsion polymerization also include buffers,

free radical initiators, such as potassium persulfate ($K_2S_2O_8$), chain transfer agents, such as dodecyl mercaptan ($C_{12}H_{25}SH$). The system is agitated continuously at temperatures below 100°C until polymerization is essentially complete or is terminated by the addition of compounds such as dimethyl dithiocarbamate to prevent the formation of undesirable products such as cross-linked polymers. Stabilizers such as phenyl Beta-naphthylamine are added to latices of elastomers.

The final product in latex form may be used for water-type paints or coatings or the water may be removed from the finely divided high-molecular weight polymer. Separation may be brought about by the addition of electorlytes, freezing or spray drying.

It is believed that polymerization of hydrophobic monomers is initiated by free radicals in the aqueous phase and that the surface-active oligomers produced migrate to the interior of the emulsifier micelles where propagation continues. Monomer molecules dispersed in the water phase also solubilize by diffusing to the expanding lamellar micelles. These micelles disappear as the polymerization continues and the rate may be measured by noting the increase in surface tension of the system.

—Raymond B. Seymour, University of Houston, Houston, Texas.

POLYMERIZATION (Oxidative-Coupling).

A technique for preparation of high-molecular-weight linear polymers. Schematically the reaction is represented below and involves the oxidative coupling of certain organic compounds containing two active hydrogen atoms to give a linear polymer. The hydrogens ultimately, with oxygen, form water. High-molecular-weight polymers have been prepared in this manner from phenols, diacetylenes, and dithiols.

$$n\,HRH \xrightarrow[\text{catalyst}]{\frac{n}{2}\,O_2} \text{—(R)—}_n + n\,H_2O$$

When 2,6-dimethylphenol is oxidized with oxygen in the presence of an amine complex of a copper salt as catalyst a high-molecular weight polyether (PPO®) is formed.

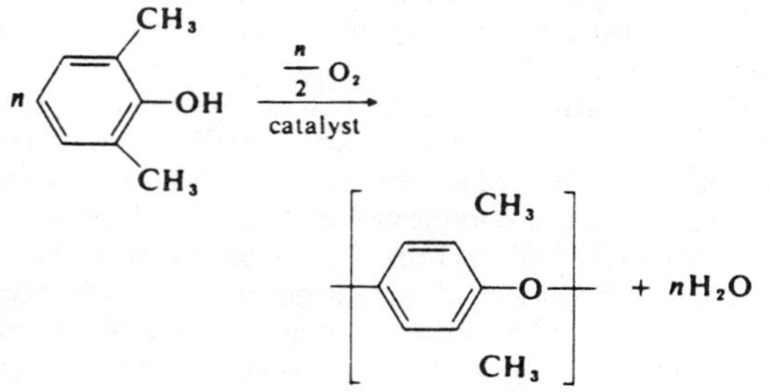

The reaction is exothermic and proceeds rapidly at room temperature. The polymerization is generally performed by passing oxygen or air through a stirred solution of the catalyst and monomer in an appropriate solvent. When the desired molecular weight is attained, the polymer is isolated by dilution of the reaction mixture with a nonsolvent for the polymer. The precipitated polymer is then removed by filtration, washed thoroughly and dried. The polymer is soluble in most aromatic hydrocarbons and chlorinated hydrocarbons and insoluble in alcohols, ketones and aliphatic hydrocarbons.

A large number of other 2,6-disubstituted phenols have been oxidatively coupled. A representative list of the results is presented below.

Polymer formation readily occurs if the substituent groups are relatively small and not too electro-negative. When the substituents are bulky, the predominant product is the diphenoqui-

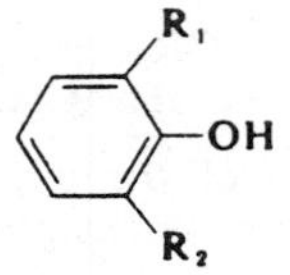

R_1	R_2	Principal Product
methyl	methyl	polymer
methyl	ethyl	polymer
methyl	i-propyl	polymer
methyl	t-butyl	diphenoquinone
methyl	phenyl	polymer
methyl	chloro	polymer
methyl	methoxy	polymer
ethyl	ethyl	polymer
i-propyl	i-propyl	diphenoquinone
t-butyl	t-butyl	diphenoquinone
methoxy	methoxy	diphenoquinone
nitro	nitro	no reaction

none formed by a tail-to-tail coupling. No appreciable reaction occurs when 2,6-dinitrophenol is oxidized even at 100°C.

A family of engineering thermoplastics based on the above technology includes PPO polyphenylene oxide, Noryl® thermoplastic resins (modified phenylene oxide) and glass reinforced varieties of each. The phenylene oxide resins are characterized by: (1) outstanding hydrolytic stability; (2) excellent dielectric properties over a wide range of temperatures and frequencies; and (3) outstanding dimensional stability at elevated temperatures. Because of these properties modified phenylene oxides are finding major application in the areas of business machine housings, appliances, automotive, TV and communications, electrical/electronic, and water distribution.

Oxidative polymerization of 2,6-diphenylphenol yields a crystallizable polymer that is characterized by a very high melting point (~480°C) and excellent electrical properties. It can be spun into a fiber with excellent thermal, oxidative and hydrolytic stability. It is marketed under the trademark Tenax®.

By performing the oxidation at elevated temperatures the phenols which would ordinarily yield polymers are converted instead to diphenoquinones. These quinones are readily reduced to the corresponding hydroquinones, compounds which promise to be useful as antioxidants and polymer intermediates.

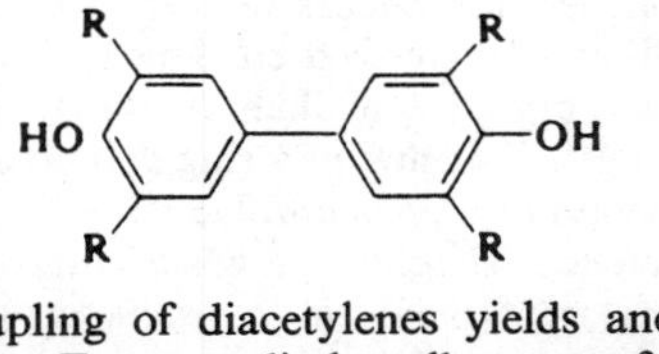

Oxidative coupling of diacetylenes yields another unusual class of polymers. From m-diethynylbenzene, for example, is obtained a high-molecular weight polymer that can be cast into a tough, flexible film.

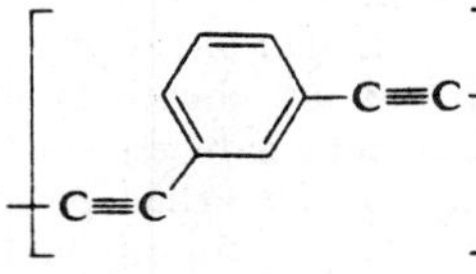

The polymer contains 96.75% carbon and on heating to about 350°F (177°C) or above it spontaneously rearranges to an insoluble and infusible material. When ignited the hydrogen in the polymer burns leaving a carbon residue.

In the same manner dithiols can be converted to polydisulfides.

—Allan S. Hay, General Electric Company, Schenectady, New York.

POLYMERIZATION (Radical). In addition polymerization, polymer is the sole product of the reaction so that the monomer and polymer have essentially the same chemical composition—for example, monomeric styrene and polystyrene. In a polymerization of this type, polymer is formed by a stepwise reaction in which molecules of monomer are added one at a time to a reactive center; the center grows in size while retaining its reactivity.

In a radical polymerization, the reactive centers are free radicals and the process is a typical chain reaction. The monomers in radical polymerizations normally contain carbon-carbon double bonds in their molecules; styrene is typical. Usually radical polymerization is performed in the liquid phase. The chain reaction can be divided into the following steps:

1. *Initiation*. Formation of a reactive free radical and its capture by monomer to form a center
2. *Propagation*. Reaction of a center with a molecule of monomer to form a larger center
3. *Termination*. Deactivation of a center so that it becomes incapable of further growth
4. *Transfer*. Reaction of a center with another molecule so that further growth of that particular center is prevented but a new center, capable of growth, is formed.

Commonly in radical polymerizations, initiation occurs continuously at a steady rate and is balanced by termination so that a steady concentration of growing centers (usually in the region of 10^{-8} mole/1) is established. The number of propagation reactions greatly exceeds the number of reactions of other types so that *macromolecules* are built up. The life-time of an active center is very much less than the duration of the whole process of polymerization and so the macromolecules are produced even in the earliest stages; there is not a continuous rise in the molecular weight of the polymeric product as found in polymerizations of certain other types. It is instructive to consider in some detail the component reactions in the overall process of radical polymerization.

Initiation. In principle, the simplest method for initiation is to add to the purified monomer a small amount of a substance which dissociates to fairly reactive free radicals. This initiator (or sensitizer) is chosen so that its decomposition occurs at a suitable rate at the working temperature; thus azo*iso*-butyronitrile is commonly used at about 60°C dissociating according to the equation:

$$(CH_3)_2C(CH) \cdot N:N \cdot C(CN)(CH_3)_2 \rightarrow 2(CH_3)_2C(CN) \cdot + N_2$$

The radical adds to monomer thus

$$(CH_3)_2C(CN) \cdot + CH_2:CHX \rightarrow (CH_3)_2C(CN) \cdot CH_2 \cdot CHX \cdot$$

forming the starting point of a polymer chain, i.e., an *end-group*; this reaction is the real initiation of polymerization. Initiators of other types are also used, notably peroxides, both organic and inorganic. In some cases, the initiator is chosen to give free radicals under the influence of light; this process can be useful for initiating polymerizations at comparatively low temperatures. Two-component initiating systems are widely used in this connection, an example being

$$H_2O_2 + Fe^{2+} \rightarrow Fe^{3+} + OH^- + \cdot OH$$

which clearly would be selected for aqueous systems. At elevated temperatures or under the influence of external sources of energy (light, high-energy radiations, ultrasonics, mechanical work), many monomers polymerize apparently spontaneously without deliberate addition of sensitizer; the mechanisms of initiation under such circumstances are not completely understood.

Propagation. The propagation reaction in a radical polymerization can be represented by the general equation

$$P \cdot CH_2 \cdot CHX \cdot + CH_2:CHX \rightarrow P \cdot CH_2 \cdot CHX \cdot CH_2 \cdot CHX \cdot$$

which corresponds to the conversion of a carbon-carbon double bond to two carbon-carbon single bonds. The group —$CH_2 \cdot CHX$— in the polymer chain is referred to as the *monomer* unit. If the growing center includes more than a few monomer units, the characteristics of the growth reaction are reasonably supposed to be independent of the size of the center.

The growth reaction is exothermic (in the region of 20 kcal/mole, i.e., about 80 kj/mole); under some circumstances, polymerizations may become self-heating and difficult to control. The growth reaction involves a decrease in entropy since a free molecule of monomer becomes organized in a polymer chain. The opposing effects of changes in enthalpy and entropy indicate that, for every polymerizing system, there is a *ceiling temperature* below which the growth reaction is favored thermodynamically but above which the reverse process is favored; the value of the ceiling temperature depends on the nature of the monomer and on its concentration in the system. Certain monomers, e.g., α-methyl styrene, were once thought not to polymerize by a radical mechanism but it is now clear that they will do so provided that the experiment is performed below the ceiling temperature.

The growth reaction shown above represents *heat-to-tail* addition; the CHX groups occur at alternate sites along the main polymer chain and the unpaired electron is sited on a substituted carbon atom. Head-to-head addition, to give a polymer radical $P \cdot CH_2 \cdot CHX \cdot CHX \cdot CH_2 \cdot$, may occur occasionally but is likely to be followed by tail-to-tail addition to give $P \cdot CH_2 \cdot CHX \cdot CHX \cdot CH_2 \cdot CH_2 \cdot CHX \cdot$ which can be regarded as the normal growing radical. Head-to-head groupings may well be sites of instability in the polymer.

The substituted carbon atoms in the polymer chain are asymmetric. *Stereoregular polymers* are produced if all these carbon atoms have the same configuration (all *d* or all *l*) or if the *d* and *l* configurations occur alternately; pronounced stereo-regularity is seldom achieved in radical polymerizations except perhaps at very low temperatures. When dienes are polymerized by a radical mechanism, the resulting polymers contain several distinct types of monomer unit, thus butadiene can give rise to —$CH_2 \cdot C(CH:CH_2)$—, —$CH_2 \cdot CH:CH \cdot CH_2$— *cis*, and —$CH_2 \cdot CH:CH \cdot CH_2$— *trans*.

Termination. In many radical polymerizations, termination occurs by interaction of pairs of growing radicals, either by combination to give $P \cdot CH_2 \cdot CHX \cdot CHX \cdot CH_2 \cdot P$ or by disproportionation to give ($P \cdot CH:CHX + P \cdot CH_2 \cdot CH_2X$). The relative importances of these alternative processes depend upon the chemical nature of the monomer and, to a lesser extent, upon the temperature in the sense that the chance of disproportionation rises as the temperature is increased. Combination gives rise to a head-to-head grouping in the chain, and disproportionation to some unsaturated end-groups for molecules; both structural features may give rise to instability.

Termination can occur for polymer radicals of any size and so there is inevitably a wide *distribution of sizes* among the final molecules. The distribution can be predicted by application of kinetic principles and can be determined experimentally by fractionation of the whole polymer, e.g., by gel permeation chromatography. It is possible to quote only *average molecular weights* for polymers; they can be determined by several experimental methods, e.g., osmometry and viscometry.

The average *chain length* or *degree of polymerization* (DP) of the molecules in a sample of polymer is the average number of monomer units contained in them. The average *kinetic chain length* (v) in a polymerization is the number of growth reactions which, on average, occur between an initiation step and the corresponding termination process. The relationship between degree of polymerization and kinetic chain length depends on the relative frequencies of combination and disproportionation (for 100% combination DP = $2v$; for 100% disproportionation DP = v) but may also be affected by the occurrence of transfer reactions (see later).

Termination is commonly *diffusion-controlled*, i.e., it is governed by the rate at which the reactive sites in growing radicals can come together rather than by chemical factors. In viscous media, termination may be so seriously impeded that both the overall rate of polymerization and the degree of polymerization increase markedly. In systems where the polymer is insoluble in the reaction medium, polymer radicals may be trapped in the precipitated material and be able to grow but unable to participate in termination processes.

Transfer. The average molecular weight of a polymer produced in a particular system may be substantially reduced by occurrence of some types of transfer reactions. If the system contains certain substances, e.g., mercaptans, a growing polymer radical may abstract hydrogen thus

$$P \cdot + R \cdot SH \rightarrow P \cdot H + RS \cdot$$

giving a dead polymer molecular and a new radical which can react with monomer to reinitiate polymerization. If reinitiation is 100% efficient, the effect of transfer of this type is to reduce the average degree of polymerization without affecting the rate of polymerization or the kinetic chain length. In practice, transfer is commonly accompanied by retardation since some of the new radicals are consumed in side-reactions instead of reacting with monomer; this type of transfer is said to be degradative.

Other components of the polymerization mixture, including monomer and initiator, may engage in transfer reactions. They are particularly significant for allyl monomers for which *degradative transfer* to monomer is of such importance that rates and degrees of polymerization are very low. The radical produced in the reaction

$$P \cdot + CH_2 : CH \cdot CH_2 \cdot O \cdot CO \cdot CH_3$$
$$\rightarrow P \cdot H + CH_2 : CH \cdot CH(O \cdot CH \cdot CH_3) \cdot$$

is so stabilized by resonance that it is not reactive enough to initiate efficiently.

Transfer to polymer, causing reactivation of a polymer molecule at some point along its length, leads to the growth of *branches*. The process can occur intermolecularly and also intramolecularly; the latter process is particularly important in the free radical polymerization of ethylene at high pressure where it leads to the production of numerous short branches which considerably affect the properties of the polymer.

Transfer to polymer, the subsequent growth of branches and termination of their growth by combination lead to *cross-linking* whereby the separate polymer molecules are united to form an insoluble three-dimensional network. Cross-linking is however much more likely to occur during the polymerization of those monomers which contain more than one carbon-carbon double bond per molecule. The monomer unit in the polymer first formed still possesses an unsaturated grouping which can participate in another polymerization chain. Certain monomers of this type however engage in a special type of reaction so that reaction of one double bond in a monomer is immediately followed by reaction of the second double bond; this type of growth is shown by, for example, methacrylic anhydride.

$$P \cdot + CH_2 : C(CH_3) \cdot CO \cdot O \cdot CO \cdot C(CH_3) : CH_2 \rightarrow$$

$$P \cdot CH_2 \cdot \overset{\cdot}{C}(CH_3) \cdot CO \cdot O \cdot CO \cdot C(CH_3) : CH_2 \rightarrow$$

$$P \cdot CH_2 \cdot C(CH_3) \cdot CH_2 \cdot \overset{\cdot}{C}(CH_3) \cdot$$
$$\qquad | \qquad\qquad\qquad |$$
$$\qquad CO\!-\!O\!-\!CO$$

Inhibitors and Retarders. Various substances can reduce the rate at which a monomer is converted to polymer. Inhibitors completely suppress polymerizations whereas retarders only reduce the rate. The former deactivate very readily the primary radicals so that growth of polymer chains cannot begin; the latter deactivate growing polymer radicals so causing premature termination. Inhibitors are commonly used to stabilize monomers during storage. Many nitro compounds and quinones act as inhibitors and retarders.

Copolymerization. A process known as copolymerization can occur if reactive radicals are generated in a mixture of monomers; the resulting polymer molecules contain monomer units of more than one type. Copolymerization is of great significance academically, where it leads to information about the reactivities of monomers and radicals, and also industrially where it is used for the production of materials with special properties. Usually the composition of a copolymer is different from that of the mixture of monomers from which it is derived. For this reason, the average compositions of feed and copolymer drift during the course of a copolymerization. There are useful analogies between copolymerization and fractional distillation; special mixtures of monomers producing copolymers without change of composition are said to give azeotropic copolymerizations.

Extensive tables of so-called *monomer reactivity ratios* are available and make it possible to predict the compositions of copolymers formed from particular mixtures of monomers.

In many binary copolymerizations, there is a pronounced tendency for the two types of monomer unit to alternate along the copolymer chain. In extreme cases, there is almost perfect alteration, notably for pairs of monomers, e.g., maleic anhydride and stilbene, which do not polymerize on their own. Ternary copolymerizations are of practical importance; the kinetic treatments developed for binary copolymerizations can be extended to these systems.

—J. C. Bevington, University of Lancaster, Lancaster, England.

POLYMORPHISM. 1. A phenomenon in which a substance exhibits different forms. Dimorphic substances appear in two solid forms, whereas trimorphic exist in three, as sulfur, carbon, tin, silver iodide, and calcium carbonate. Polymorphism is usually restricted to the solid state. Polymorphs yield identical solutions and vapors (if vaporizable). The relation between them has been termed "physical isomerism." See "Allotropes" under **Chemical Elements.** See also **Mineralogy.**

2. The occurrence of individuals of distinctly different structure or appearance within a species. In many cases two such forms occur and the species is said to be dimorphic rather than polymorphic.

Polymorphism depends upon many different conditions in various groups of animals. The various forms may be adapted for different places in a life cycle, for special parts in a colonial or social organization, or for special stages in a metamorphosis. They may also result from the incidence of different environmental conditions due to seasons or to unusual climatic conditions.

POLYPEPTIDE. A compound composed of two or more amino acids and similar in many properties to the natural peptones. The amino acids are joined by peptide groups

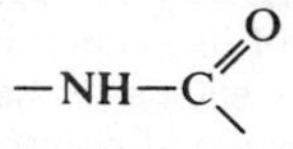

$$-NH-C\diagover{\displaystyle O}$$

formed by the reaction between an —NH_2 group and a

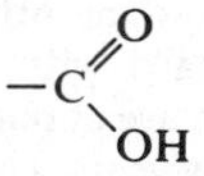

$$-C\diagover{\displaystyle O}{\diagunder{\displaystyle}{OH}}$$

group, whereby there is elimination of a molecule of water, and formation of a valence bond. They may be termed di-, tri-, tetra-, etc., peptide according to the number of amino acids present in the molecule.

The sequence of amino acids in the chain of a protein is of critical importance in the biological functioning of the protein, and its determination is very difficult. The chains may be relatively straight, or they may be coiled or helical. In the case of certain types of polypeptides, such as the keratins, they are crosslinked by the disulfide bonds of cystine. Linear polypeptides can be regarded as proteins. See also **Amino Acids**; and **Protein**.

POLYPROPYLENE. A synthetic crystalline thermoplastic polymer, $(C_3H_5)_n$, with molecular weight of 40,000 or more. Low-molecular-weight polymers are also known which are amorphous in structure and used as gasoline additives, detergent intermediates, greases, sealant, and lubricating oil additives. They are also available as high-melting waxes.

Polypropylenes are derived by the polymerization of propylene with stereospecific catalyst, such as aluminum alkyl. These polymers are translucent, white solids, insoluble in cold organic solvents, softened by hot solvents. They maintain their strength after repeated flexing. They are degraded by heat and light unless protected by antioxidants. Polypropylenes are readily colored, exhibit good electrical resistance, low water absorption and moisture permeability. They have rather poor impact strength below 15°F (−9.5°C). They are not attacked by fungi or bacteria, resist strong acids and alkalis up to about 140°F (60°C), but are attacked by chlorine, fuming nitric acid, and other strong oxidizing agents. They are combustible, but slow burning.

Polypropylenes are available as molding powder, extruded sheet, cast film, textile staple and continuous filament yarn. They find use in packaging film, molded parts for automobiles, appliances, and housewares, wire and cable coating, food container closures, bottles, printing plates, carpet and upholstery fibers, storage battery cases, crates for soft-drink bottles, laboratory ware, trays, fish nets, surgical casts, and a variety of other applications.

POLYSTYRENES. General purpose (or crystal) polystyrene is a clear, water-white, glassy polymer commonly derived from coal tar and petroleum gas. Physical properties of this material can be altered by addition of modifying agents, such as rubber (for increased toughness), methyl or α-methyl styrene (for heat resistance, methyl methacrylate (for improved light stability), and acrylonitrile (for chemical resistance). In general, varying the level of modifying agent (e.g., comonomer) will alter the level of desired property improvement.

Special grades of polystyrene include impact polystyrene modified with ignition-resistant chemical additives. These were de-

veloped because of increased emphasis on product safety and used in many electrical and electronic appliances. The addition of flame-retardant chemicals does not make the polymer non-combustible, but increases its resistance to ignition and decreases the rate of burning when exposed to a minor fire source.

Chemistry. The polymerization of styrene is an exothermic chain reaction which proceeds by all known polymerization techniques. This reaction can be shown schematically as follows:

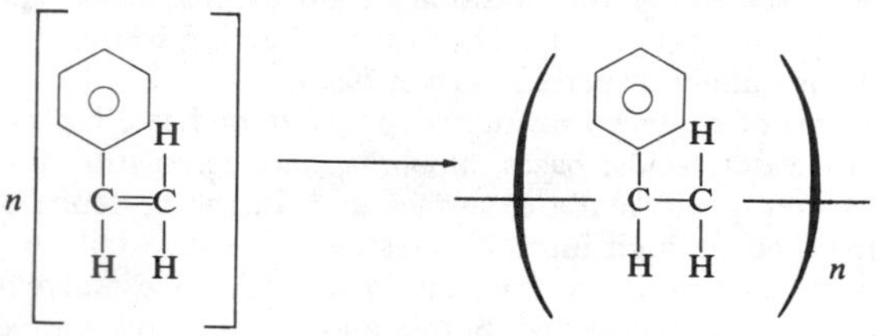

The exact nature of the beginning and end of such a polymer chain is not certain. In general, the polymer can be characterized by its average degree of polymerization, i.e., the value of n, or more precisely by the distribution of n values. The heat of polymerization is 17.4 ± 0.2 kcal/mole at 26.9°C. The reaction may be initiated with heat or by the use of catalysts. Organic peroxides are typical initiators. Styrene also will polymerize in the presence of various inert materials, such as solvents, fillers, dyes, pigments, plasticizers, rubbers, and resins. Moreover, it forms a variety of copolymers with other mono- and polyvinyl monomers.

It is a matter of general observation that with styrene, the polymerization-rate curves will exhibit three distinct phases, the nature of which can be determined by the polymerization conditions and the purity of the monomer: (1) an initial slow period at the beginning of the reaction, known as the *induction period*, which appears to be associated with the presence of an inhibitor or other impurity in the monomer; (2) a period of relatively rapid polymerization, which persists almost to the end of the reaction, and for which the rate is exponentially dependent upon temperature; and (3) a final slowing down in rate as the reaction approaches completion and the monomer becomes exhausted. This effect is particularly apparent at low temperatures with relatively impure monomers.

General Properties. The *specific gravity* of general purpose and impact polystyrene is 1.05. It can vary for copolymers. It is higher for some specialty grades. Density varies slightly with pressure, but for practical purposes, the polymer is non-compressible.

In terms of *heat-resistance*, deflection temperatures range from about 66 to 99°C (170 to 215°F), depending upon the formulation. Continuous resistance to heat for polystyrene is usually 60 to 80°C (140 to 175°F). Time and load have a significant influence on the useful service temperature of a part.

Polystyrene is nontoxic when free from additives and residuals. It has no nutritive value and does not support fungus or bacterial growth.

Dimensional stability of polystyrene resins is excellent. Mold shrinkage is small. The low moisture absorption (about 0.02%) allows fabricated parts to maintain dimensions and strength in humid environments.

General-purpose polystyrene is water white, and transmission of visible light is about 90%. Modifiers reduce this property, and translucence results. The refractive index is about 1.59; critical angle about 39°. Polystyrene molecules do not have the same optical properties in all directions. When molecules become oriented in a given direction during fabrication, a double refraction occurs and a birefringence effect can be observed if the part is examined through a polarized lens under a polarized light source. Injection moldings often exhibit birefringence in a random pattern. This can be beneficial if the birefringence is in the direction of load.

In terms of *weatherability*, polystyrene does not exhibit ultraviolet stability and is not considered weather-resistant as a clear material. Continuous, long-term exposure results in discoloration and reduction of strength. Improvement in weatherability can be obtained by the addition of ultraviolet absorbers, or by incorporating pigments. The best pigmenting results are obtained with finely dispersed carbon black.

In terms of *chemical resistance*, polystyrene has a high resistance to water, acids, bases, alcohols, and detergents. Chlorinated solvents will mar the surface and, in the presence of an external load or high internal stresses, will cause failure. Aliphatic and aromatic hydrocarbons, in general, will dissolve polystyrene. Such foodstuffs as butter and coconut oil should be avoided. The chemical resistance depends upon chemical concentration, time, and stress.

Typical mechanical properties of polystyrene are given in the accompanying table. The long-term load-bearing strength

COMPRESSION MOLDED PROPERTIES OF POLYSTYRENE

PROPERTY	GENERAL PURPOSE psi (MPa)	IMPACT psi (MPa)
Tensile strength	5500–8000 (38–55)	2500–5000 (17–35)
Compressive strength	21000–16000 (145–110)	4500–9000 (31–62)
Flexural strength	9000–15000 (62–104)	5000–10000 (35–69)
Tensile (Young's) modulus	400000–500000 (2760–3450)	200000–400000 (1880–2760)
Impact strength, Izod, foot-pounds/inch	0.3–0.5	1–4
Hardness, Rockwell M	65–80	60
Elongation, %	0.8–2.0	5–50

psi = pounds per square inch; MPa = megaPascals.

of most polystyrene materials is about one-third of the typical tensile strength given in the table.

Uses. Packaging applications are the largest. Meat, poultry, and egg containers are thermoformed from extruded foamed polystyrene sheet. The fast-food market also accounts for a substantial amount of polystyrene for take-out containers where the insulation value of a foamed container is an advantage. Containers, tubs, and trays formed from extruded impact polystyrene sheets are used for packaging a large variety of food. Biaxially oriented polystyrene film is thermoformed into blisher packs, meat trays, container lids, and cookie, candy, pastry, and other food packages where clarity is required.

Housewares is another large segment of the use of polystyrenes. Refrigerator door liners and furniture panels are typical thermoformed impact polystyrene applications. Extruded profiles of solid or foamed impact polystyrene are used for mirror or picture frames, and moldings for construction applications.

General-purpose polystyrene is extruded either clear or embossed for room dividers, shower doors, glazings, and lighting applications. Injection molding of impact polystyrene is used for household items, such as flower pots, personal care products, and toys. General-purpose polystyrene is used for cutlery, bottles, combs, disposable tumblers, dishes, and trays.

Injection blow molding can be used to convert polystyrene into bottles, jars, and other types of open containers.

Impact polystyrene with ignition-resistant additives is used for appliance housings, such as those for television and small appliances. Structural foam impact polystyrene modified with flame-retardant additives is used for business machine housings and in furniture because of its decorability and ease of processing. Consumer electronics, such as cassettes, reels and housings, are a fast growing area for use of polystyrenes. Medical applications include sample collectors, petri dishes, and test tubes.

In an effort to make homes and other buildings more energy efficient, extruded foam board with flame-retardant additives for walls and under slabs the use of polystyrenes has experienced exceptional growth in recent years. Used as a sheeting material, extruded foam board complies with the requirements of the major building codes as well as federal and military specifications.

In general, polystyrene is used in applications where ease of fabrication and decorability are required. Polystyrene has excellent electrical properties, good thermal and dimensional stability, resistance to staining, and low cost. General purpose polystyrene is preferred where clarity is also of prime concern. Impact polystyrene is preferred where toughness is needed.

POLYTROPIC PROCESSES. The expansion or compression of a constant weight of gas may assume a variety of forms, depending on the extent to which heat is added to or rejected from the gas during the process, and also on the work done. There are, theoretically, an infinite number of ways possible in which a gas may expand from an initial pressure p_1, and volume v_1 to a final volume v_2. All these expansions may be grouped generically as polytropic expansions, and all could be represented graphically on the PV plane by the family of curves $pv^n = C$. They are all, in theory, perfectly reversible. n may have any positive value, 0 to ∞, and having been selected numerically it defines the type of expansion. From the infinite number of possible polytropic expansions, it is worth while to isolate four which deserve special attention. When one of the four physical characteristics, to wit, pressure, temperature, entropy, or volume, remains constant, expansions of more than ordinary interest are denoted, since they are frequently employed in a practical way, in situations which can be subjected to thermodynamic analysis. The value of the exponent n of the polytropic family for each of these is:

isobaric	$n = 0$
isothermal	$n = 1$
isentropic	$n = \gamma$ (γ = ratio of specific heat at constant pressure to that at constant volume)
isometric	$n = \infty$

Note, however, that the first and last are limiting cases, since in the first the pressure remains constant, and in the fourth it approaches zero. Note also that the second applies strictly only to ideal gases.

These thermodynamic processes, as they occur in useful machines, are not often of the exact polytropic form desired. For example, an isentropic process which is exemplified, at least theoretically, by expansion of the burned gases after the explosive combustion in the gasoline engine, is modified slightly by the interchanging of heat between gases and cylinder wall, whereas a true isentropic has no heat either added or rejected in this way. The particular polytropic curve which would suit these conditions of expansion would depart somewhat from the adiabatic form.

During a polytropic process conditions of the working medium are constantly varying, and analysis may be aimed at determining one of the following: the work done, the heat added, the variation of temperature, and the change of entropy. Some information may be obtained merely by comparing the value of the exponent n with certain other data. For example, if n lies between 0 and 1, the temperature rises during an expansion and falls during a compression; when n is greater than 1, the temperature falls during expansion and rises during compression. Also, when n is less than γ, heat must be added to obtain an expansion, whereas when it is greater than γ, heat must be expelled. From the above it will be noted that there is a certain range of polytropic expansion in which, although heat is added, the temperature falls. This may seem to some to be paradoxical, but it is readily explained. During these expansions work is being done by the gas at a rate greater than that at which heat is being added, with the result that the deficiency must be made up from within the gas. The only way this may be accomplished is for the gas to cool and give up some of its internal energy.

The equations for work done and for heat added in the case of the general polytropic expansions are:

$$W = \frac{p_1 v_1 - p_2 v_2}{n - 1}$$

$$Q = (p_1 v_1 - p_2 v_2)\left(\frac{1}{n-1} - \frac{1}{\gamma - 1}\right)$$

Both of these are expressed in foot-pounds. Sometimes a substitution of a definite value of n in one or the other of these equations leads to an indeterminate; for example, with the isothermal,

$$W = \frac{p_1 v_1 - p_2 v_2}{1 - 1}$$

But since the equation of the isothermal for an ideal gas is

$$pv = C$$

$$p_1 v_1 = p_2 v_2$$

and the work equation becomes indeterminate,

$$W = \frac{0}{0}$$

By approaching the isothermal from a different angle, however, the equation

$$W = pv \log_e \frac{v_2}{v_1}$$

may be deduced for work done.

POLYURETHANES. These materials comprise a conglomerate family of polymers in which formation of the urethane group,

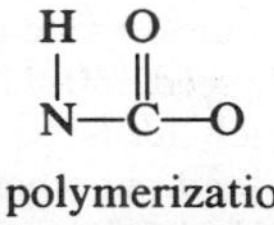

$$\begin{array}{cc} H & O \\ | & \| \\ N & \!-\!C\!-\!O \end{array}$$

is an important step in polymerization. Because the urethane linkage usually is formed by reaction of hydroxyl and isocyanate groups, urethane chemistry is the chemistry of isocyanates. The high reactivity of isocyanates and knowledge of the catalysis of isocyanate reactions have made possible the simple production of diverse polymers from low- to moderate-molecular-weight liquid starting materials. Several isocyanates (tolylene diisocyanate, hexamethylene diisocyanate, dicyclohexylmethane diiso-

cyanate, etc.) are used in preparing polyurethanes. All are low-viscosity liquids at room temperature with the exception of 4,4'-diphenylmethane diisocyanates (MDI), which is a crystalline solid. The aromatic isocyanates are more reactive than the aliphatic isocyanates and are widely used in urethane foams, coatings, and elastomers. The cyclic structure of aromatic and alicyclic isocyanates contributes to molecular stiffness in polyurethanes.

Flexible and rigid urethane foams, probably the most familiar of the polyurethanes, are produced in very large quantities. Foam formulations contain isocyanates and polyols with suitable catalysts, surfactants for stabilization of foam structure, and blowing agents which produce gas for expansion. The largest volume of flexible urethane foam is used as a cushioning material. Expanding uses for flexible foam include carpet underlays and bedding. Weight reduction programs in the transportation field also take advantage of polyurethane foams for seating and trim. Rigid foams find application in insulation for appliances. Thermoplastic urethane elastomers form a widely used family of engineering materials which appear to combine the best properties of elastomers and thermoplastics. They are tough, have high load-bearing capacity, low-temperature flexibility, and resistance to oils, fuels, oxygen, ozone, abrasion, and mechanical abuse. See also **Elastomers.**

POLYVINYL ALKYL ETHERS. These products have properties which range from sticky resins to elastic solids. They are obtained by the low-temperature cationic polymerization of alkyl vinyl ethers having the general formula $ROCH=CH_2$. These monomers are prepared by the addition of the selected alkanol to acetylene in the presence of sodium alkoxide or mercury(II) catalyst. As shown by the following equations, the latter yields an acetal which must be thermally decomposed to produce the alkyl vinyl ether.

$$\underset{\text{acetylene}}{HC\equiv CH} + \underset{\text{alkanol}}{ROH} \xrightarrow[130-180^\circ C]{Na^+, OR^-} \underset{\text{alkyl vinyl ether}}{H_2C=CHOR}$$

$$\underset{\text{acetylene}}{HC\equiv CH} + \underset{\text{alkanol}}{2ROH} \xrightarrow{Hg^{++}}$$

$$\underset{\text{acetal}}{H_3C-CH(OR)_2} \xrightarrow[\substack{200-300^\circ C \\ (-ROH)}]{cat.} \underset{\substack{\text{alkyl vinyl} \\ \text{ether}}}{H_2C=CHOR}$$

These monomers are also produced by an oxidative process in which the alkanols are added directly to ethylene and the alkyl ethers are thermally decomposed to produce hydrogen and the alkyl vinyl ethers.

Commercial polymers have been produced from methyl, ethyl, isopropyl, n-butyl, isotubtyl, t-butyl, stearyl, benzyl and trimethylsilyl vinyl ethers. The poly(methyl vinyl ether) called **PVM** or **Resyn** is produced by the polymerization of the monomer by boron trifluoride in propane at —40°C in the presence of traces of an alkyl phenyl sulfide. The polymer may have isotactic, syndiotactic or stereoblock configurations depending on the solvent and catalyst used.

Nonpolar solvents favor the formation of ion pairs between the polymer cation and the counteranion and favor the production of isotactic polymers. Soluble catalysts, such as diethyl aluminum chloride and ethyl aluminum dichloride, also affect the stereoregularity of the polymer chains. The tendency for the formation of stereoregular polymers is decreased as the size of the alkyl group is increased. Typical structures of these polymers are shown below:

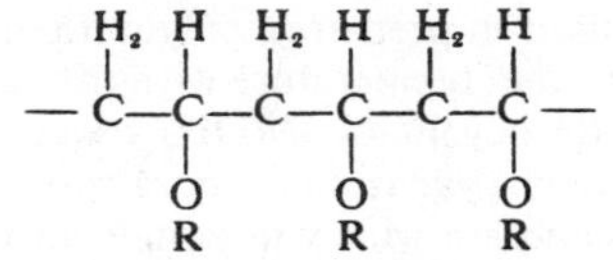

isotactic polymer

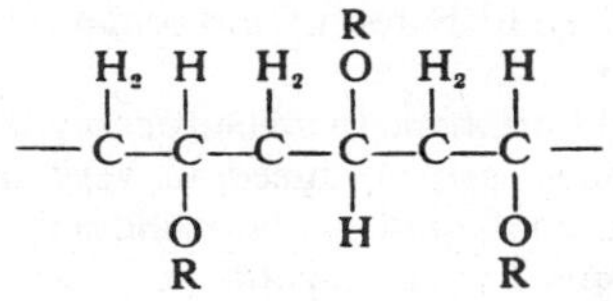

Syndiotactic polymer

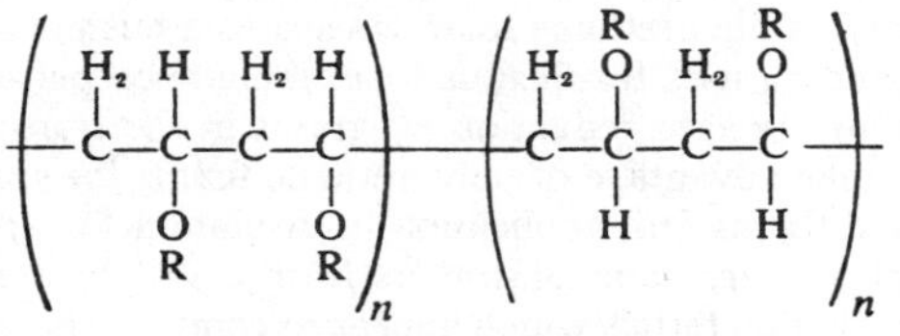

stereoblock polymer

Poly(methyl vinyl ether) is soluble in cold water but becomes insoluble in a reversible process when the temperature is raised to 35°C. This sticky polymer has a glass transition temperature of −20°C. It has been used as an adhesive and as a heat sensitizer for polymer latices.

Poly(vinyl ethyl ether) is soluble in ethanol, acetone and benzene. It is a rubbery product which may be cross-linked by heating with dicumyl peroxide. Poly(vinyl isobutyl ether), has a glass transition temperature of −5°C. It has been used as an adhesive for upholstery, cellophane and adhesive tape.

The processing properties of poly(vinyl chloride) has been improved by copolymerizing vinyl chloride with a small amount of vinyl alkyl ether. Copolymers of vinyl alkyl ethers and maleic anhydride are used as water soluble thickeners, paper additives, textile assistants and in cleaning formulations.

—Raymond B. Seymour, University of Houston, Houston, Texas.

POLYVINYL CHLORIDE (PVC). The manufacture of polyvinyl chloride resins commences with the monomer, vinyl chloride, which is a gas, shipped and stored under pressure to keep it in a liquid state; bp −14°C, fp −160°C, density (20°C), 0.91. The monomer is produced by the reaction of hydrochloric acid with acetylene. This reaction can be carried out in either a liquid or gaseous state. In another technique, ethylene is reacted with chlorine to produce ethylene dichloride. This is then catalytically dehydrohalogenated to produce vinyl chloride. The byproduct is hydrogen chloride. A later process, oxychlorination, permits the regeneration of chlorine from HCl for recycle to the process.

Polymerization may be carried out in any of the following manners:

(1) Suspension: a large particle size dispersion or suspension of vinyl chloride is made in water by addition of a small quantity of emulsifying agent. The product after polymerization and drying consists of granules.

(2) Emulsion: a larger quantity of emulsifier is employed, resulting in a fine particle size emulsion. The polymer after spray drying, is a finely divided powder suitable for use in organosols and plastisols.

(3) Solution: vinyl chloride is dissolved in a suitable solvent for polymerization. The resultant polymer may be sold in solution form, or dried and pelletized.

Emulsions may be polymerized by use of a water-soluble catalyst (initiator), such as potassium persulfate, or a monomer-soluble catalyst, such as benzoyl peroxide, lauroyl peroxide or azobisisobutyronitrile. Suspension and solution polymerizations employ the monomer soluble catalysts only. In addition to the above-mentioned initiators, diisopropyl peroxydi-carbonate may also be employed, where lower-temperature polymerization may be desired, e.g., to reduce branching and minimize degradation.

Because of the low level of emulsifiers and protective colloids, the suspension polymer types are most suitable for electrical applications and end uses requiring clarity. This form is also employed in the bulk of extrusion and molding applications. Cost is lower than for emulsion and solution forms. The emulsion or dispersion resins are employed mainly for organosol and plastisol applications where fast fusion with plasticizer at elevated temperature will occur as a result of the fine particle size of the resin.

Monomers such as vinyl acetate or vinylidene chloride may be copolymerized with vinyl chloride. Up to 15% of the comonomer may be employed. Vinyl acetate increases the solubility, film formation and adhesion. Processing or forming temperatures are generally lowered. Chemical resistance and tensile strength decrease with increasing amount of vinyl acetate.

Rigid Vinyls

These have been separated into two categories according to ASTM:

Type I is rigid PVC with excellent chemical resistance, physical properties and weathering resistance such as obtained from unplasticized high molecular weight PVC.

Type II has the added feature of high impact resistance but with slightly lower chemical and physical requirements.

Perhaps the most important applications for rigid PVC will be in building. This is a rapidly growing market. Fabrication is via extrusion. Examples of applications are pipe, siding, roofing shingles, panels, glazing, window and door frames, rain gutters and downspouts.

Blow-molded bottles, which exhibit excellent product resistance, and good clarity, are also expected to become an important outlet for rigid PVC.

Formulations for extrusion generally include light and heat stabilizers, lubricants, which facilitate molding, and colorants. These materials are generally purchased in a compounded ready to use cube form, in order to minimize irregularities in blending, etc.

The outstanding characteristics of these rigid vinyls are chemical, solvent and water resistance; resistance to weathering when properly stabilized, therefore permitting long-term outdoor exposure; and low cost. Abrasion and impact resistance are satisfactory.

A major deficiency is heat sensitivity. Here, degradation begins with the split-off of HCL. The resultant unsaturation leads to cross-linking and chain cission, causing a degradation of the physical properties. Maximum service temperature for continuous exposure should not exceed 150–175°C. Cold flow or creep is another deficiency, which leads to dimensional changes in materials under constant load, e.g., water pipe under constant service pressures will tend to enlarge in diameter, resulting in decreased strength; long spans of pipe or siding may sag. Temperature accelerates this effect.

PVC has a high coefficient of expansion, one of the highest for all plastic materials, and substantially higher than metals and wood. Therefore, design allowances must be made to provide for movement in order to avoid buckling, breakage, etc.

Flexible PVC

An unusually wide variety of products and usages are possible with plasticized vinyls. Typical applications include floor and wall coverings, boots, rainwear, jackets, upholstery, garden hose, electrical insulation, film and sheeting, foams and many others.

The primary processing techniques are by means of extrusion, calendering and molding. Special techniques involve organosols and plastisols.

Plasticizers used to develop the desired flexibility and performance are selected on the basis of cost and application requirements, e.g., temperature; service life; exposure to solvents, chemicals, water, UV, food; tensile strength; abrasion resistance; flexibility; tear strength, etc.

Plasticizers must be classed as *primary*, where high compatibility is limited, thus restricting the amount that can be tolerated. The addition of secondary plasticizers may import special properties or simply reduce cost (extender plasticizer).

Primary plasticizers may be further subdivided. The *phthalate* types are by far the most popular due to cost and ease of incorporation. Dioctyl phthalate and diisooctyl phthalate are typical of this class. They exhibit good general-purpose properties. *Phosphate* plasticizers are also important for general-purpose use. Typical of these are tritolyl phosphate and trixylenyl phosphate. These plasticizers also impart fire retardant properties. *Low-temperature* plasticizers, such as dibutyl sebacate, are used where good low-temperature flexibility is required. For maximum compatibility and minimum cost, a typical plasticizer combination would be a blend of 50% DOP and 50% dibutyl sebacate.

Polymeric plasticizers are generally polyesters with a relatively low molecular weight. They are used where resistance to high temperatures and freedom from migration and extraction are required. Polymerics are more difficult to incorporate, have poor low-temperature properties, and are expensive.

Epoxy plasticizers are epoxidized oils and esters. These are generally classed with the polymerics. However, molecular weight is lower. Therefore, resistance to extraction and heat are slightly inferior. Low-temperature properties are better and epoxies are more easily incorporated.

Extender plasticizers, which are used mainly to reduce cost, consist of chlorinated waxes, petroleum residues, etc. Incorporation of excessive amounts may result in exudation on aging. The chlorinated types decrease flammability.

Organosols and Plastisols

Plastisols are dispersions of powdered PVC resin in plasticizer. A typical composition would consist of 100 parts of PVC resin dispersed in 50 parts of DOP. The resultant paste when heated to 300°F (149°C) fuses or "fluxes into a solid plastic mass. Stability of this plastisol at room temperature may range from several weeks to several months depending on the plasticizers and resins employed.

An organosol is the same mixture as described above, with the addition of solvent to reduce viscosity. These find their major applications in coatings. The solvent is evaporated before fusion of the film. Various pigments, colorants, stabilizers and fillers may be added, depending on the desired properties. Emulsion polymerization resins are generally employed because of their fast fusion rates. Coarser particle sized PVC resins would require extended time at the elevated temperature.

Plastisols allow the use of inexpensive manufacturing techniques, such as slush and rotational molding, casting, dipping, etc. They are employed for the manufacture of a large variety of parts, e.g., toys, floor mats, handles and many others.

Foams are made by the addition of blowing agents to the plastisol. These may be continuously applied to a moving substrate which includes a pass at an elevated temperature where foaming occurs, followed by fusion of the plastisol.

Organosols find their major application in coatings, which may be applied by spray, dip, knife, roller, etc. Typical products are coated aluminum siding, fabrics, paper, industrial coatings, etc.

An important development was the use of plasticizers which crosslink upon application of heat and thus produce a more rigid end product. This extends the range of products obtainable by plastisol techniques into rigids. By varying the amount of crosslinking plasticizer incorporated, various levels of flexibility are obtained.

—Harold A. Sarvetnick, Westfield, New Jersey.

POLYVINYLIDENE CHLORIDE, a stereoregular, thermoplastic polymer is produced by the free-radical chain polymerization of vinylidene chloride ($H_2C=CCl_2$) using suspension or emulsion techniques. The monomer has a bp of 31.6°C and was first synthesized in 1838 by Regnault, who dehydrochlorinated 1,1,2-trichloroethane which he obtained by the chlorination of ethylene. The copolymer product has been produced under various names, including *Saran*. As shown by the following equation, the product, in production since the late 1930s, is produced by a reaction similar to that used by Regnault nearly a century earlier:

$$H_2ClCCHCl_2 + Ca(OH)_2 \xrightarrow[-2H_2O]{90°C}$$

$$\begin{array}{cc} \text{1,1,2-tri-} & \text{calcium} \\ \text{chloroethane} & \text{hydroxide} \end{array}$$

$$CaCl_2 + 2H_2C=CCl_2$$

$$\begin{array}{cc} \text{calcium} & \text{vinylidene} \\ \text{chloride} & \text{chloride} \end{array}$$

Since this monomer readily forms an explosive

$$\overset{O_2}{\overset{/\backslash}{}}$$

peroxide (H_2C-CCl_2) it must be kept under a nitrogen atmosphere at −10°C in the absence of sunlight.

The copolymers were patented by Wiley, Scott, and Seymour in the early 1940s. A typical formulation for emulsion copolymerization contains vinylidene (85 g), vinyl chloride (15 g), methylhydroxypropylcellulose (0.05 g), lauroyl peroxide (0.3 g) and water (200 g). More than 95 per cent of these monomers are converted to copolymer when this aqueous suspension is agitated in an oxygen-free atmosphere for 40 hrs at 60°C. The glass transition temperature of the homopolymer is −17°C. It has a specific gravity of 1.875 and a solubility parameter of 9.8.

Because of its high crystallinity, the homopolymer (PVDV) is insoluble in most solvents at room temperature. However, since the regularity of repeating units in the chain is decreased by copolymerization, Saran is soluble in cyclic ethers and aromatic ketones. This copolymer (100 g) is plasticized by the addition of α-methyl-benzyl ether (5 g), stabilized against ultraviolet light degradation by 5-chloro-2-hydroxybenzophenone (2.0 g) and heat stabilized by phenoxypropylene oxide (2.0 g).

The poly(vinylidene chloride-co-vinylchloride) may be injection molded and extruded. Extruded pipe and molded fittings which were produced in large quantity in the 1940's have been replaced to some extent by less expensive thermoplastics. A flat extruded filament is used for scouring pads and continuous extruded circular filament is used for the production of insect screening, filter clothes, fishing nets and automotive seat covers.

A large quantity of this copolymer is extruded as a thin tubing which is biaxally stretched by inflating with air at moderate temperatures before slitting. This product, called Saran Wrap, has a tensile strength of 15,000 psi (103 MPa). Since it has a high degree of transparency to light and a high coefficient of static friction (0.95) it is widely used for the protection of foods in the household. It has a low permeability value for gases such as oxygen and nitrogen.

Poly(vinylidene chloride-co-acrylonitrile) is widely used as a latex coating for cellophane, polyethylene and paper. Since this copolymer is soluble in organic solvents, it is also used as a solution coating. The resistance to vapor permeability and the ease of printing on polyethylene and cellophane is increased by coating with this vinylidene chloride copolymer.

The tensile strength of both film and fiber is increased tremendously by cold drawing 400–500 per cent. Thus, tensile strengths as high as 40,000 psi (276 MPa) in the direction of draw have been obtained by cold drawing.

—Raymond B. Seymour, University of Houston, Houston, Texas.

POLYVINYLIDENE FLUORIDE. This product is made by the free-radical chain polymerization of vinylidene fluoride ($H_2C=CF_2$). This odorless gas which has a boiling point of $-82°C$ is produced by the thermal dehydrochlorination of 1,1,1-chlorodifluoroethane or by the dechlorination of 1,2-dichloro-1,1-difluoro-ethane. As shown by the following equations, 1,1,1-chlorodifluoroethane may be obtained by the hydrofluorination and chlorination of acetylene and by the hydrofluorination of vinylidene chloride or of 1,1,1-trichloroethane.

$$HC\equiv CH + 2HF \longrightarrow H_3CCHF_2$$
acetylene   hydrogen fluoride   1,1,-difluoro-ethane

$$\xrightarrow[-HCl]{Cl_2} H_3CCClF_2$$
1,1,1-chloro-difluoroethane

$$H_2C=CCl_2 + 2HF \xrightarrow{-HCl} H_3CCClF_2$$
vinylidene chloride   hydrogen fluoride

$$H_3C-CCl_3 + 2HF \xrightarrow{-2HCl} H_3CCClF_2$$
1,1,1-trichloro-ethane

$$H_3CCClF_2 \xrightarrow[-HCl]{600°} H_2C=CF_2$$
1,1,1-chlorodi-fluoroethane   vinylidene fluoride

$$H_2ClCCClF_2 \xrightarrow[-Cl_2]{\Delta} H_2C=CF_2$$
1,2-dichloro-1,1-difluoroethane

Polyvinylidene fluoride is polymerized under pressure at 25–150°C in an emulsion using a fluorinated surfactant to minimize chain transfer with the emulsifying agent. Ammonium persulfate is used as the initiator. The homopolymer is highly crystalline and melts at 170°C. It can be injection molded to produce articles with a tensile strength of 7000 psi (48 MPa), a modulus of elasticity in tension of 1.2×10^5 psi and a heat deflection of 300°F (149°C).

Poly(vinylidene fluoride) is resistant to most acids and alkalies but it is attacked by fuming sulfuric acid. It is soluble in di-methylacetamide but is insoluble in less polar solvents. Co-polymers have been produced with ethylene, tetrafluoroethylene, chlorotrifluoroethylene and hexafluoroethylene. The latter is an elastomer called Viton or Fluorel.

The homopolymer is used as a chemical resistant coating for steel, for tank linings, hose, and pump impellors. The elastomeric copolymer with hexafluoroethylene when cured with hexamethylenediamine is used as a seal, gasket, o-ring, tubing, coating and lining.

PORPHYRIN. Any of several physiologically active nitrogenous compounds occurring widely in nature. The parent structure is comprised of four pyrrole rings, shown as I, II, III, and IV in the accompanying diagram, together with four nitro-

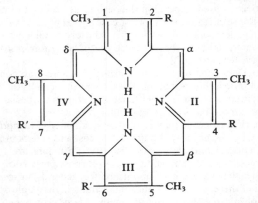

Suggested structure of a metal-free porphyrin molecule.

gen atoms and two replaceable hydrogens, for which various metal atoms can be readily substituted. A metal-free porphyrin molecule has the structure as shown by the diagram. Porphyrins of this type have been made synthetically by passing an electric current through a mixture of ammonia, methane, and water vapor. Some biochemists suggest that this phenomenon may account for the early formation of chlorophyll and other porphyrins which have been essential factors in the development of life.

The most important porphyrin derivatives are characterized by a central metal atom; hemin is the iron-containing porphyrin essential to mammalian blood, and chlorophyll is the magnesium-containing porphyrin that catalyzes photosynthesis. Other derivatives include the cytochromes, which function in cellular metabolism, and the phthalocyanine group of dyes. Porphyrins are described in considerable detail in a 7-volume set of books, "The Porphyrins," Academic, New York (1978).

PORTLAND CEMENT. Cement; Gypsum.

POSITRON. The positron is one of many fundamental bits of matter. Its rest mass (9.109×10^{-31} kilogram) is the same as the mass of the electron, and its charge ($+1.602 \times 10^{-19}$ coulomb) is the same magnitude, but opposite in sign to that of the electron. The positron and electron are antiparticles for each other. The positron has spin $\frac{1}{2}$ and is described by Fermi-Dirac statistics as is the electron.

The positron was discovered in 1932 by C. D. Anderson at the California Institute of Technology while doing cloud chamber experiments on cosmic rays. The cloud chamber tracks of some particles were observed to curve in such a direction in a magnetic field that the charge had to be positive. In all other respects, the tracks resembled those of high-energy electrons. The discovery of the positron was in accord with the theoretical work of Dirac on the negative energy of electrons. These negative energy states were interpreted as predicting the existence of a positively charged particle.

Positrons can be produced by either nuclear decay or the

transformation of the energy of a gamma ray into an electron-positron pair. In nuclei which are proton rich, a mode of decay which permits a reduction in the number of protons with a small expenditure of energy is positron emission. The reaction taking place during decay is

$$p^+ \rightarrow n^0 + e^+ + \mu$$

where p^+ represents the proton; n^0, the neutron; e^+, the positron; and μ a massless, chargeless entity called a neutrino. See also **Neutrino**. The positron and neutrino are emitted from the nucleus while the neutron remains bound within the nucleus. Although none of the naturally occuring radioactive nuclides are positron emitters, many artificial radioisotopes which decay by positron emission have been produced. The first observed case of positron decay of nuclei was also the first observed case of artificial radioactivity. An example of such a nuclear decay is

$$_{11}Na^{22} \rightarrow {}_{10}Ne^{22} + e^+ + \mu \text{ (half-life } \sim 2.6 \text{ years).}$$

This decay provides a practical, usable source of positrons for experimental purposes.

The process of pair production occurs when a high-energy gamma ray interacts in the electromagnetic field of a nucleus to create a pair of particles—a positron and an electron. Pair production is an excellent example of the fact that the rest mass of a particle represents a fixed amount of energy. Since the rest energy ($E_{rest} = m_{rest}c^2$) of the positron plus electron is 1.022 MeV, this energy is the gamma energy threshold and no pair production can take place for lower-energy gammas. In general, the cross section for pair production increases with increasing gamma energy and also with increasing Z number of the nucleus in whose electromagnetic field the interaction takes place.

The positron is a stable particle (i.e., it does not decay itself), but when it is combined with its antiparticle, the electron, the two annihilate each other and the total energy of the particles appears in the form of gamma rays. Before annihilation with an electron, most positrons come to thermal equilibrium with their surroundings. In the process of losing energy and becoming thermalized, a high-energy positron interacts with its surroundings in almost the same way as does the electron. Thus, for positrons, curves of distance traversed in a medium as a function of initial particle energy are almost identical with those of electrons.

It is energetically possible for a positron and an electron to form a bound system similar to the hydrogen atom, with the positron taking the place of the proton. This bound system has been called *positronium* and the chemical symbol Ps has been assigned. Although the possibility of positronium formation was predicted as early as 1934, the first experimental demonstration of its existence came in 1951 during an investigation of positron annihilation rates in gases as a function of pressure.

In principle, positronium can be observed through the emission of its characteristic spectral lines, which should be similar to hydrogen's except that the wavelengths of all corresponding lines are doubled. Positronium is also the ideal system in which the calculations of quantum electrodynamics can be compared with experimental results. Measurement of the fine-structure splitting of the positronium ground state has served as an important confirmation of the theory of quantum electrodynamics.

It is possible for a positron-electron system to annihilate with the emission of one, two, three, or more gamma rays. However, not all processes are equally probable

See also **Particles (Subatomic).**

References

Bromley, D. A.: "Physics," *Science,* **209,** 110–121 (1980).
Cowen, R. C.: "Particle Physics," *Technology Review (MIT),* **82,** 3, 10–11 (1980).
Green, J., and J. Lee: "Positronium Chemistry," Academic, New York (1964).
Reines, F.: "The Early Days of Experimental Neutrino Physics," *Science,* **203,** 11–16 (1979).
Weisskopf, V. G.: "Contemporary Frontiers in Physics," *Science,* **203,** 240–244 (1979).

POSITRONIUM. A quasi-stable system consisting of a positron and a negatron bound together. Its set of energy levels is similar to that of the hydrogen atom (electron and proton). However, because of the different reduced mass, the frequencies associated with the spectral lines are less than half of those of the corresponding hydrogen lines. The mean life of positronium is at most about 10^{-7} seconds, its existence being terminated by negatron-positron annihilation. See also **Positron.**

POTASSIUM. Chemical element symbol K, at. no. 19, at. wt. 39.098, periodic table group Ia (alkali metals), mp 63.7°C, bp 774°C, density, 0.87 g/cm³ (20°C). Elemental potassium has a body-centered cubic crystal structure. Potassium is a silvery-white metal, can be readily molded, and cut by a knife, oxidizes instantly on exposure to air, and reacts violently with water, yielding potassium hydroxide and hydrogen gas, which burns spontaneously in air with a violet flame due to volatilized potassium. In these respects, potassium behaves much like sodium. The element is preserved under kerosene. Discovered by Davy in 1807.

There are three naturally occurring isotopes, ^{39}K through ^{41}K, of which ^{40}K is radioactive ($t_{1/2} = 1.3 \times 10^9$ years). In ordinary potassium, this isotope represents only 0.0119% of the content. There are four other known isotopes, all radioactive, ^{38}K and ^{42}K through ^{44}K, all with relatively short half-lives measured in minutes and hours. In terms of abundance, potassium ranks seventh among the elements occurring in the earth's crust. In terms of content in seawater, the element ranks eighth, with an estimated 1,800,000 tons of potassium per cubic mile (388,800 metric tons per cubic kilometer) of seawater.

First ionization potential, 4.339 eV; second, 31.66 eV. Oxidation potential: $K \rightarrow K^+ + e^-$, 2.924 V. Other important physical properties of potassium are given under **Chemical Elements.**

Potassium does not occur in nature in the free state because of its great chemical reactivity. The major basic potash chemical used as a source of potassium is potassium chloride, KCl. The potassium content of all potash sources generally is given in terms of the oxide, K_2O. The majority of potash produced comes from mineral deposits that were formed by the evaporation of prehistoric lakes and seas which had become enriched in potassium salts leached from the soil.

In addition to natural deposits of potassium salts, large concentrations of potassium also are found in some bodies of water, including the Great Salt Lake and the Salduro Marsh in Utah, the Dead Sea between Israel and Jordan, and Searles Lake in California. All of these brines are used for the commercial production of potash.

The main potassium minerals are sylvite KCl, sylvinite KCl/NaCl, carnallite $KCl \cdot MgCl_2 \cdot 6H_2O$, kainite $MgSO_4 \cdot KCl \cdot 3H_2O$, polyhalite $K_2SO_4 \cdot MgSO_4 \cdot 2CaSO_4 \cdot 2H_2O$, langbeinite $K_2SO_4 \cdot 2MgSO_4$, jarosite $K_2Fe_6(OH)_{12}(SO_4)_4$, leucite $K_2O \cdot Al_2O_3 \cdot 4SiO_2$, alunite $K_2Al_6(OH)_{12}(SO_4)_4$, microcline $K_2O \cdot Al_2O_3 \cdot 6SiO_2$, muscovite $K_2O \cdot 3Al_2O_3 \cdot 6SiO_2 \cdot 2H_2O$, biotite $H_2K(Mg, Fe)_3(Al, Fe)(SiO_4)_3$, and orthoclase $K_2O \cdot Al_2O_3 \cdot 6SiO_2$. See also **Alunite; Biotite; Carnallite, Jaro-**

site; Leucite; Muscovite; Polyhalite. The principal workable mineral deposits are in Stassfurt, Germany, Alsace, New Mexico, Saskatchewan, the Soviet Union, Spain, Poland, Italy, the Atlantic Seaboard of the United States, and Utah. There are significant potassium reserves in many other parts of the world, notably in Canada and the Soviet Union. World consumption of potash is about 18 million tons annually. Potassium metal is obtained by electrolysis of fused potassium hydroxide or chloride fluoride mixture in a specially designed cell.

Uses. Like so many of the chemical elements, the compounds of potassium are far more important than elemental potassium—by several orders of magnitude. The uses for metallic potassium are extremely limited, mainly because metallic sodium serves about the same needs and is much less costly. Sodium production, for example, exceeds potassium production by a factor of at least 1,000. A large amount of elemental potassium is used to produce the superoxide KO_2 which finds application in gas-mask canisters. The compound also goes into the production of a sodium-potassium alloy which is used as a heat-exchange medium. This alloy also has been used in magnetohydrodynamic power generation and as a catalyst for the removal of CO_2, H_2O, and oxygen from inert-gas systems. The handling precautions for potassium metal are similar to those for sodium metal. See **Sodium.**

Chemistry and Compounds. Potassium is more electropositive than sodium in many of its reactions, as is consistent with its position in group 1. Its reaction with water is more vigorous and it reacts violently with liquid bromine, and readily upon heating with solid iodine.

Because of the ease of removal of its single $4s$ electron (4.339 eV) and the difficulty of removing a second electron (31.66 eV) potassium is exclusively monovalent in its compounds, which are electrovalent. (Some experimental work indicates that the potassium alkyls may be covalent, but even then form conducting solutions in other metal alkyls.)

Potassium solutions in liquid NH_3 react readily with the elements on the further right side of the periodic table to produce normal and poly compounds such as potassium sulfide, K_2S, and tetrapotassium plumbide, K_4Pb in the first instance and K_2S_6 and K_4Pb_9 in the second. Ammoniates are not formed by potassium as readily as by sodium or lithium and solubility of salts exhibits a minimum at the cation: anion radius ratio of 0.75 (potassium fluoride, KF, 16 moles per kilogram, potassium chloride, KCl, 0.0177 moles per kilogram, potassium bromide, KBr, 2.26 moles per kilogram, potassium iodide, KI, 11.09 moles per kilogram). Potassium nitrate reacts in liquid ammonia with potassium amide, KNH_2 to form the azide, KN_3.

Like the other alkali metals, potassium forms compounds with virtually all the anions, organic as well as inorganic. Like sodium bicarbonate, the reactivity of potassium bicarbonate with many metallic oxides permits of the preparation of many compounds (such as the meta- and pyroarsenates) which are unstable in aqueous solution. For a general discussion of these reactions, and for a general picture of the inorganic salts of potassium, see the discussion of the compounds of sodium, which differ principally in their greater degree of hydration and greater number of hydrates. However, potassium, rubidium, and cesium coordinate with large organic molecules even though they do not with water. Potassium, like the others, coordinates with salicylaldehyde. It is believed to have two coordination numbers, 4 and 6. The tetracoordinate compounds of potassium (and sodium) are the most stable. The following reasons are given: (1) Increasing atomic number carries with it increasing electropositiveness and ease of ionization, which diminishes the tendency to coordinate. (2) The increasing distance of the nucleus from the coordinating electrons with increasing atomic volume makes it less likely that additional electrons will be held with ease. (3) On the other hand, there is an increase in the maximum coordination number with the elements of higher atomic number. These factors are in keeping with a maximum stability for the tetracoordinate compounds occurring with potassium.

One major difference between potassium and sodium in their salt-forming properties is the much greater ability of potassium to form alums, although potassium does not form quite as many types of these compounds as do the higher alkali metals, or ammonium or monovalent thallium.

Potassium also differs from sodium, and especially from lithium, in the greater stability of its salts of polarizable polyatomic anions, such as peroxide, superoxide, azide, polysulfide, and polyhalides, among others. The corresponding rubidium and cesium salts are even more stable.

Among the other inorganic compounds of potassium are:

Bromide. Potassium bromate $KBrO_3$, white solid, soluble, mp 434°C, upon heating oxygen is evolved and the residue is potassium bromide; formed by electrolysis of potassium bromide solution under proper conditions. Used as a source of bromate and bromic acid.

Carbonate. Potassium carbonate, potash, pearl ash K_2CO_3, white solid, soluble, formed (1) in the ash when plant materials are burned, (2) by reaction of potassium hydroxide solution and the requisite amount of CO_2. Used (1) in making special glasses, (2) in the making of soft soap, (3) in the preparation of other potassium salts (a) in solution, (b) upon fusion; potassium hydrogen carbonate, potassium bicarbonate, potassium acid carbonate $KHCO_3$, white solid, soluble, (4) in vat dyeing and textile printing, (5) in titanium enamels, (6) in boiler water treating compounds, (7) in photographic chemical formulations, (8) in electroplating baths, and (9) as an important absorbent for CO_2 in the process industries.

Chlorate. Potassium chlorate, chlorate of potash $KClO_3$, white solid, soluble, mp about 350°C, powerful oxidizing agent, and consequently a fire hazard with dry organic materials, such as clothes, and with sulfur; upon heating oxygen is liberated and the residue is potassium chloride; formed by electrolysis of potassium chloride solution under proper conditions. Used (1) in matches, (2) in pyrotechnics, (3) as disinfectant, (4) as a source or oxygen upon heating. (Hazardous! Use of potassium perchlorate is recommended instead.)

Chloride. Potassium chloride, KCl, colorless or white crystals, strong saline taste. Occurs naturally as sylvite. Soluble in water; slightly soluble in alcohol. Sp gr, 1.987, mp 772°C, sublimes at 1500°C, noncombustible, low toxicity. Used in fertilizers as a source of potassium salts; pharmaceutical preparations; photography; spectroscopy; plant nutrient; salt substitute; laboratory reagent. See **Fertilizers.**

Chloroplatinate. Potassium chloroplatinate K_2PtCl_6, yellow solid, insoluble, formed by reaction of soluble potassium salt solution and chloroplatinic acid. Used in the quantitative determination of potassium.

Chromate. Potassium chromate K_2CrO_4, yellow solid, soluble, formed by reaction of potassium carbonate and chromite at a high temperature in a current of air, and then extracting with water and evaporating the solution. Used (1) as a source of chromate, (2) in leather tanning, (3) in textile dyeing, (4) in inks.

Cobaltinitrite. Dipotassium sodium cobaltinitrite $K_2NaCo(NO_2)_6 \cdot H_2O$, golden yellow precipitate, formed by reaction of sodium cobaltinitrite solution in acetic acid with soluble potassium salt solution. Used in the detection of potassium.

Cyanate. Potassium cyanate KCNO, white solid, soluble,

formed along with lead metal by reaction of potassium cyanide and lead monoxide solids upon heating. Source of cyanate.

Cyanide. Potassium cyanide, cyanide of potash KCN, white solid, soluble, very poisonous, formed by reaction of calcium cyanamide and potassium chloride at high temperature. Used as a source of cyanide and for hydrocyanic acid, but usually replaced by the cheaper sodium cyanide. Also used in metallurgy, electroplating, extraction of gold from ores, as a pesticide and fumigant, in photography and analytical chemistry. Upon acidification, produces dangerous HCN gas.

Dichromate. Potassium dichromate, chromate of potash $K_2Cr_2O_7$, red solid, soluble, powerful oxidizing agent, formed by acidifying potassium chromate solution and then evaporating. Used (1) in matches, (2) in leather tanning and in the textile industry, (3) as a source of chromate, (4) in pyrotechnics, (5) in colored glass, (6) as an important laboratory reagent, (7) in blueprint developing, and (8) in wood preservation formulations.

Hydroxide. Potassium hydroxide, caustic potash, potassium hydrate KOH, white solid, soluble, mp 380°C, formed (1) by reaction of potassium carbonate and calcium hydroxide in H_2O, and then separation of the solution and evaporation, (2) by electrolysis of potassium chloride under the proper conditions, and evaporation. Used in the preparation of potassium salts (1) in solution, and (2) upon fusion. Also used in the manufacture of (3) soaps, (4) drugs, (5) dyes, (6) alkaline batteries, (7) adhesives, (8) fertilizers. (9) alkylates, (10) for purifying industrial gases, (11) for scrubbing out traces of hydrofluoric acid in processing equipment, (12) as a drain-pipe cleaner, and (13) in asphalt emulsions.

Hypophosphite. Potassium hypophosphite KH_2PO_2, white solid, soluble, formed (1) by reaction of hypophosphorous acid and potassium carbonate solution, and then evaporating, (2) by reaction of potassium hydroxide solution and phosphorus on heating (poisonous phosphine gas evolved).

Iodate. Potassium iodate KIO_3, white solid, soluble, melting point 560°C, formed (1) by electrolysis of potassium iodide under proper conditions, (2)by reaction of iodine and potassium hydroxide solution, andthe fractional crystallization of iodate from iodide. Used as a source of iodate and iodic acid.

Manganate. Potassium manganate K_2MnO_4, green solid, soluble, permanent in alkali, formed by heating to high temperature manganese dioxide and potassium carbonate, and then extracting with water, and evaporating the solution. The first step in the preparation of potassium manganate and permanganate from pyrolusite.

Nitrate. Potassium nitrate, saltpeter, niter, KNO_3, white solid, soluble, mp 333°C, formed by fractional crystallization of sodium nitrate and potassium chloride solutions. Used (1) in matches, explosives, pyrotechnics; (2) in pickling of meat; (3) in glass; (4) in medicines; (5) as a rocket-fuel oxidizer; and (6) in heat treatment of steel. See also **Fertilizers.**

Nitrite. Potassium nitrite, KNO_2, yellowish-white solid, soluble, formed (1) by reaction of nitric oxide plus nitrogen tetroxide and potassium carbonate or hydroxide, and then evaporating; (2) by heating potassium nitrate and lead to a high temperature and then extracting the soluble portion (lead monoxide insoluble) with water, and evaporating. Used as a reagent (diazotizing) in organic chemistry.

Oxides. See later paragraph on "Other Potassium Compounds."

Perchlorate. Potassium perchlorate $KClO_4$, white solid, very slightly soluble, mp 610°C, but above 40°C decomposes with evolution of oxygen gas and formation of potassium chloride residue; formed (1) by electrolysis of potassium chlorate under proper conditions, (2) by heating potassium chlorate at 480°C and then fractional crystallization. Used (1) as a convenient and safe (preferred to use of potassium chlorate) method of preparing oxygen by heating, (2) in the determination of potassium in soluble salt solution.

Periodate. Potassium periodate KIO_4, white solid, very slightly soluble, mp 582°C, formed by electrolysis of potassium iodate under proper conditions.

Permanganate. Potassium permanganate, permanganate of potash $KMnO_4$, purple solid, soluble, formed by oxidation of acidified potassium manganate solution with chlorine, and then evaporating. Used (1) as disinfectant and bactericide, (2) in medicine, (3) as an important oxidizing agent in many chemical reactions.

Persulfate. Potassium persulfate $K_2S_2O_8$, white solid, slightly soluble, formed by electrolysis of potassium sulfate under proper conditions. Used (1) as a bleaching and oxidizing agent, (2) as an antiseptic.

Silicate. Potassium silicate K_2SiO_3, colorless (when pure) glass, soluble, mp 976°C, formed by reaction of silicon oxide and potassium carbonate at high temperature, similar in properties and uses to the more common sodium silicate.

Sulfates. Potassium sulfate, sulfate of potash K_2SO_4, white solid, soluble. Common constituent of potassium salt minerals. Used (1) as an important potassium fertilizer, (2) in the preparation of potassium or potash alums; potassium hydrogen sulfate $KHSO_4$, white solid, soluble; potassium pyrosulfate $K_2S_2O_7$, white solid, soluble, formed by heating potassium hydrogen sulfate to complete loss of H_2O. See **Fertilizers.**

Sulfides. Potassium sulfide K_2S, yellowish to reddish solid, soluble, formed by heating potassium sulfate and carbon to a high temperature; potassium hydrogen sulfide, potassium bisulfide, potassium acid sulfide KHS, formed in solution by reaction of potassium hydroxide or carbonate solution and excess H_2S.

Sulfite. Potassium sulfite $K_2SO_3 \cdot 2H_2O$; potassium hydrogen sulfite $KHSO_3$; white solids, similar in properties and formation to the corresponding sodium sulfites.

Thiocarbonate. Potassium thiocarbonate K_2CS_3, yellow solid, soluble, formed by reaction of potassium sulfide and CS_2.

Thiocyanate. Potassium thiocyanate, potassium sulfocyanide, potassium rhodanate KCNS, white solid, soluble, mp about 170°C, formed by fusing potassium cyanide and sulfur, and then crystallizing. Used as a source of thiocyanate.

Other Potassium Compounds. In addition to the inorganic salts, potassium forms such binary compounds as a phosphide, K_3P, by direct union with phosphorus, a boride, KB_6, electrolysis of fused fluorides and borates in the presence of a metal boride, a nitride, and the oxides. Of the latter, direct reaction of potassium and oxygen yields the superoxide, KO_2, a paramagnetic, orange-colored substance. The likelihood of KO_2 having a monomeric structure is supported by these properties, since the O_2^- ion would have an odd electron which would confer paramagnetism and color upon the compound. The lower oxides of potassium, K_2O and K_2O_2, which are less stable in air than the superoxide, have been prepared, as have their hydrates. K_2O unites explosively with the oxygen of the air. One other oxide, K_2O_3, has been reported, but this appears to be a double salt of KO_2 and K_2O_2. The properties of potassium hydroxide are in keeping with its position in Group 1; thus its heat of solution is somewhat lower than that of rubidium hydroxide, RbOH or cesium hydroxide, CsOH, and much higher than that of lithium hydroxide, LiOH, and NaOH.

The organic compounds of potassium include many oxycompounds, such as salts of organic acids, alcohols and phenols (alkoxides, phenoxides, etc.). A few potassium-carbon linked compounds have been reported, such as, a phenylisopropyl potassium, $C_6H_5C_3H_7K$, and a carbonyl compound of unknown composition $K_x(CO)_x$. The adduct of ethyl potassium and diethyl-zinc is a true salt, $K_2[Zn(C_2H_5)_4]$, potassium tetraethylzincate.

See the list of references at the end of the entry on **Chemical Elements.**

POTASSIUM AND SODIUM (In Biological Systems). Potassium and sodium play major roles in biological processes. Because of the numerous parallels between these two elements in metabolism, they are treated in a single entry, with appropriate distinctions made.

Potassium is required by both plants and animals. Although the total amount of potassium in most soils is usually rather high, the level of available or soluble forms of the element is frequently too low to meet the needs of growing plants. Deficiencies of plant-available potassium are more frequent in the soils of the eastern rather than of the western United States. See also **Soil.** Potassium in the form of soluble potassium salts is a very common constituent of fertilizers. See also **Fertilizers.**

Many plants will not grow at normal rates unless the plant tissues, especially the leaves, contain as much as 1 or 2% potassium and, for some plants, even higher concentrations are required. Therefore, if a plant grows at all, it will nearly always contain sufficient potassium to meet the requirements of the people or animals that consume the plant. Potassium deficiencies do occur in humans and animals, but these are largely due to metabolic upsets and illnesses that interfere with the utilization of potassium in the body, or via excessive losses of potassium from the body, rather than due to inadequate levels of dietary potassium.

The general role of potassium fertilizers in improving human and animal nutrition is to help increase food and feed supplies rather than to improve the nutritional quality of the crops produced. Excessive use of potassium fertilizers may decrease the concentration of magnesium in crops. Sodium is essential to higher animals which regulate the composition of their body fluids and to some marine organisms, but it is dispensable for many bacteria and most plants except for the blue-green algae. Potassium, on the other hand, is essential for all, or nearly all forms of life. The importance of these cations for all forms of life has been related to the predominance of sodium and potassium in the ocean where primitive forms of life are thought to have originated and developed. During most of the period of evolvement of living organisms, there has been little change in the sodium and potassium content of seawater, either as to proportion or total amount. The body fluids of sea animals are, in most instances, similar to seawater in sodium and potassium level and ratio. In freshwater and terrestrial animals, the sodium and potassium level of body fluids is usually somewhat lower, and the ratio is likely to vary from the 40:1 ratio of seawater. Most fresh waters contain small and variable amounts of sodium and potassium, usually in a ratio of from 1:1 to 4:1.

Despite the higher level of sodium in natural water, potassium is universally the characteristic cation found within both plant and animal cells. Although sodium is not an absolute requirement for most plants and bacteria, it is found in these organisms and is essential to higher animals where it is the principal cation of the extracellular fluids. Sodium and potassium are important constituents of both intra- and extracellular fluids. Generally,

the best external and internal medium for function of cells not adjusted to low salt levels is a medium involving a balance of sodium and potassium.

Beyond the osmotic effects depending on the sum of the concentration of the ions in the solution, Ringer found in 1882 that to maintain the contractility of an isolated frog heart, it was necessary to perfuse it with a medium containing sodium, potassium, and calcium ions in the proportion of seawater. It has since been recognized that the normal life activities of tissues and cells may depend on a proper balance among the inorganic cations to which they are exposed. Sodium is required for the sustained contractility of mammalian muscle, while potassium has a paralyzing effect. Thus, a balance is necessary for normal function. Other investigators have found that the antagonism among univalent and divalent cations observed by Ringer is demonstrable with various simpler or more complicated organisms or biological systems.

Excessive salt in soil, such as soils recently soaked with seawater, is toxic to most plants, although there are several plants, e.g., those of the salt marshes and the sea, which are adapted to a high salt concentration. Ingestion of seawater by humans as the only source of water is eventually fatal because of the inability of the body to eliminate salt at a concentration comparable to that of seawater. This results in accumulation of salt.

It is probable that potassium is absorbed by plant roots from the soil by an active transport mechanism which carries it through the cell wall structure. Similarly, potassium and sodium, if required, are accumulated by animals also by active transport. The actual cellular content of potassium and sodium is likewise controlled by transport mechanisms which specifically move potassium in and sodium out of the cell against the concentration gradient. The energy for this is derived from the metabolic processes of the cell. The nature of these transport mechanisms has not been fully established.

Ions and Transport Mechanisms. Potassium differs from most other essential constituents of plant and animal cells in that it is not built into the cell as a part of an organic compound, but is rather an ion from a soluble inorganic or organic salt. Potassium ions may chelate with cellular constituents, such as polyphosphates. The ion is of the correct size to fit into the water lattice adsorbed by the protein in the cell. In general, the potassium and sodium ions are attracted to protein or other colloidal or structural units having a negative charge. Mucopolysaccharides within the cell, on the cell surfaces and of the intercellular structures, are of particular importance in holding cations, such as potassium and sodium. Active centers of other configurational features of the proteins in the cell may be affected or altered by the potassium held by electrostatic or covalent binding. There are several enzyme systems which are activated by potassium.

In general, most of the sodium and potassium in the animal is in a dynamic state, being exchanged between different parts of the cell, between the cell and the extracellular fluid, and intermixing with ingested sodium and potassium in body fluids.

Most cellular constituents do not selectively bind potassium in preference to sodium. Myosin of muscle fibers, for example, will bind either. But, in contrast, the mitochondria and ribosomes are organized cellular organelles able to selectively take up or extrude potassium. This accounts for only a part of the potassium held in the cell.

In blue-green algae and some yeasts, sodium may in part replace cellular potassium. While potassium is usually the principal cation concerned with the maintenance of the osmotic pressure within the cell, sodium contributes appreciably to the total, and amino acids and other organic compounds may help make up any deficit, particularly in marine invertebrates.

The sodium content of the body extracellular fluids of marine invertebrates from the coelenterate through the arthropod phyla is approximately that of seawater. In freshwater and terrestrial invertebrates, the sodium of body fluids varies over a wide range and there is considerable variation among vertebrates. There are both fish and crustaceans which are so highly adaptable that they are able to live in either fresh or salt water.

Osmotic Pressure Regulation. The regulation of osmotic pressure within the cell and the control of the passage of water into or out of the cell is dependent to a considerable extent on the control of the potassium and sodium in the cell by the transport systems of the cell wall. The cell wall itself is of protein-lipid composition and is, in general, impermeable to the passage of water and inorganic salts. Recent studies of the cell walls with electron microscopes and with the use of other investigative techniques indicate that the cell wall contains pores connecting the cell contents with the extracellular fluid, or in some plants, with other cells. In cells having an endoplasmic reticulum, the intracellular vacuolar system may have openings through the cell wall communicating with the extracellular fluid. The ease with which water passes in or out of the cell in response to changes in external or internal osmotic pressure varies over a wide range, from easy passages to rigid control, depending upon the cell and its functions.

Phagocytosis and pinocytosis may bring salts and water, as well as other substances, into the cell.

In some unicellular organisms, osmotic equilibrium may be maintained by a contractile vacuole which collects water; in other organisms, water may be excreted through the cell wall. The kidney and sweat glands of higher animals, gills of fish and salt glands of birds serve to excrete salt. Most animals, through control of sodium and potassium excretion and loss, are able to adapt to a wide range of intake.

The importance of sodium chloride in nutrition has been recognized from the beginning of history. Agricultural populations that lived on cereal grains, nuts, berries, and other vegetable foods poor in sodium, experienced a hunger for salt which led them to go to great lengths to obtain the mineral. This was particularly true if they lived in a hot climate with the attendant increased loss of salt in perspiration. Similarly, herbivorous animals will travel long distances to supply their need for additional salt. In contrast, peoples or animals subsisting on meat, milk and other foods receive quite appreciable amounts of sodium salts in the diet, and experience no special desire or hunger for salt. See **Sodium Chloride.**

In plants, the meristematic tissues in general are particularly rich in potassium, as are other metabolically active regions, such as buds, young leaves, and root tips. Potassium deficiency may produce both gross and microscopic changes in the structure of plants. Effects of deficiency reported include leaf damage, high or low water content of leaves, decreased photosynthesis, disturbed carbohydrate metabolism, low protein content and other abnormalities.

Since potassium is found abundantly in most natural foods consumed by animals, deficiency is ordinarily no problem. With prolonged maintenance through parenteral (intravenous) feeding when normal oral feeding is not possible, potassium must be supplied.

Role of Kidney. Experimental potassium deficiency in rats results in stunted growth, loss of chloride with hypochloremic acidosis, loss of potassium and increase of sodium in muscle. In humans, disease of the gastrointestinal tract, involving loss of secretions through vomiting or diarrhea, may result in serious loss of both sodium and potassium. Trauma, surgery, anoxia, ischemia, shock, and any damage to or wasting away of tissues may result in loss of cellular potassium to the extracellular fluid and plasma, and the loss from the body through kidney excretion. Recovery with rapid uptake of potassium by the tissues may result in low plasma levels. Low extracellular potassium concentration may cause muscular weakness, changes in cardiac and kidney function, lethargy, and even coma in severe cases. There are no reserve stores of either sodium or potassium in the animal body, so any loss beyond the amount of intake comes from the functional supply of cells and tissues.

The kidney is the key regulator of the sodium and potassium content of higher animals and makes possible adaptation to wide variations of intake. In the glomerulus of the kidney nephron (or individual unit), an ultrafiltrate containing the smaller molecules of plasma is normally produced. As this ultrafiltrate passes down the kidney tubule, 97.5% or more of the sodium is actively resorbed, along with nearly all of the potassium. The remaining 2.5% of the sodium is sufficient to account for even the maximum sodium excretion. Potassium is added to the filtrate in the distal tubule through exchange for sodium. Control of this exchange appears to be the principal mode of action of aldosterone, which thus exerts a final control over sodium excretion. Aldosterone is a steroid hormone from the adrenal cortex, secretion of which seems to result from lowering of the Na/K ratio in the blood. Water is passively resorbed with the electrolytes along the length of the tubule.

Water excretion is further controlled by the antidiuretic hormone from the posterior pituitary gland which acts to increase water resorption in the kidney through making the collecting tubule permeable to water for additional resorption beyond what took place in the tubule. The posterior pituitary gland secretes the hormone as a rapid and sensitive response to a rise in the osmotic pressure of the extracellular fluid. The osmotic pressure of the extracellular fluid is, of course, principally due to its sodium chloride content.

With low intake of sodium, excretion is reduced to a very low level to conserve the supply in the body. Potassium is not so efficiently conserved.

The kidney regulates the acid-base balance of the body by control over resorption of sodium ions which may exchange for hydrogen ions in the kidney tubule. Since most dietaries are of acid-ash, the urine is usually more acid than the original plasma filtrate and much of the phosphate excreted is thus changed to the acid monosodium salt. Within the range of normal variability, with an alkaline ash diet, the urine may become alkaline, and in extreme instances, some sodium bicarbonate may be excreted.

The salts of the buffer pairs responsible for control of the pH of plasma and extracellular fluid involve sodium as the principal cation, while the cellular buffers involve potassium salts. See also **Acid-Base Regulation (Blood).**

References

Andreoli, T. E., Grantham, J. J., and F. C. Rector, Jr., Editors: "Disturbances in Body Fluid Osmolality," American Physiological Society, Bethesda, Maryland (1977).

Berl T., Anderson, R. J., and K. M. McDonald, et al.: "Clinical Disorders of Water Metabolism," *Kidney Int.,* **10,** 117 (1976).

Drew, W. L., et al.: "The Clinical Spectrum of Renal Potassium Wasting," *Calif. Med.,* **110,** 493 (1969).

Fitzsimons, J. T.: "The Physiology of Thirst and Sodium Appetite," Cambridge University Press, New York (1979).

Hays, R. M., and S. D. Levine: "Pathophysiology of Water Metabolism. The Kidney," Vol. 1 (B. M. Brenner and F. C. Rector, Editors), W. B. Saunders, Philadelphia, Pennsylvania (1976).

Leaf, A., and R. S. Cotran: "Pathophysiology of Potassium Excess and Deficiency: Renal Pathophysiology," Oxford University Press, New York (1976).

Newmark, D. R., and R. G. Dluhy: "Hyperkalemia and Hypokalemia," *Jrnl. Amer. Med. Assn.*, **231**, 631 (1975).

Rose, B. D.: "Clinical Physiology of Acid-Base and Electrolyte Disorders," McGraw-Hill, New York (1977).

Swales, J. D.: "Sodium Metabolism in Disease," Lloyd-Luke, London (1975).

POTENTIATOR (Flavor).   Flavor Enhancers and Potentiators.

POWDER METALLURGY.   Powder metallurgy embraces the production of finely divided metal powders and their union through the use of pressure and heat into useful articles. The temperatures required are below the fusion point of the principal constituent, and bonding depends on interdiffusion of the metal particles in the solid state. It is necessary to provide intimate contact between particles, hence reducing atmospheres are provided in the sintering process to prevent formation of oxide films. Readily oxidized powders such as aluminum require special technique.

Probably the most important applications of powder metallurgy are those in which a product is made which cannot be duplicated by other methods. There are many examples of this kind. The melting point of tungsten, 6,100°F (3,371°C), is much too high for ordinary melting and casting methods and the only way in which filaments for electric lights can be made is to draw them from rods of compacted and sintered tungsten powder. The cemented carbide cutting tools are another important product of refractory nature which is readily made by powder metallurgy.

Self-lubricating bronze bearings having controlled porosity are products which can be made only by powder metallurgy. The pores are impregnated with oil, and flow to the bearing surface is maintained by capillary action. Graphite is incorporated with the metal powder in one type of oilless bearing. A material made from powdered copper and graphite is used for electric-current collector brushes, and tungsten-copper or tungsten-silver combinations are used for electric contact points. In contrast to these high-conductivity materials, a high-resistance element is produced from a mixture of copper and porcelain powders, combining a metal with a nonmetallic substance.

As of the early 1980s, powder metallurgists are optimistic concerning future developments in this field. It is envisioned that bearing materials will be developed which will run dry or with very little lubrication at temperatures up to 800°C (1470°F). These will be made from powder cermets which will be compacted by hot isostatic pressing. Impregnation of these compacts with metals or oxide spinels that melt at working temperature and provide liquid as well as solid lubrication may allow even higher service temperatures. Before the end of this decade, it is expected that high-strength aluminum alloys will be made by way of powder metallurgy. Connecting rods and pistons may be forged from powdered metal preforms. Cylinder liners for aluminum blocks may be made within the next few years by highly automated isostatic compaction, using iron powder. New developments in powder making by chemical means will allow economical rolling of sheet metal of unprecedented ductility within the next 10–15 years.

Considerable work already has been done on wrought powder metal alloys. The combination of rapid solidification to provide homogeneous structures and extremely fine grain size offers a new range of properties. These include higher strength, better corrosion resistance, higher modulus of elasticity, and improved elevated temperature strength. It is expected that a significant improvement will be made in the cleanliness level coming from the powder-making processes. It is envisioned that the powder process will permit the titanium aluminides to be introduced in jet engine hardware. Elemental titanium powder is expected to be in wide use in aircraft for nonstructural applications and in engines for nonrotating parts.

The average powder metallurgy part weighs less than 0.25 pound (0.11 kilogram). The powder metallurgy industry's goal is to produce large quantities of small, complex-shaped parts that are prohibitively expensive to make by wrought processing techniques. It is expected that demand for such parts, most of them ferrous based, will increase in the future. For more detail on the outlook of powder metallurgy, reference to "Powder Metallurgy," H. E. Chandler, *Metal Progress*, **118**, 4 60–61, is suggested.

References

Bradbury, S., Editor: "Source Book on Powder Metallurgy," American Society for Metals, Metals Park, Ohio (1979).

Hausner, H. H.: "Handbook of Powder Metallurgy," Plenum, New York (1973).

POWER (Nuclear).   Nuclear Reactor.

ppb.   Parts per billion. One part per billion is a frequently used dimension for expressing the composition and analysis of substances—as found in air, water, food substances, etc. Instrument developments and other assay techniques perfected during the past decade or so have made the determination of such minute quantities a practical possibility for many materials. One part per billion is approximately equivalent to 1 drop in a 10,000-gallon (37,850-liter) tank.

ppm.   Parts per million. One part per million is a common dimension for expressing the composition and analysis of substances—as found in air, water, raw materials, food substances, etc. One part per million is approximately equivalent to $\frac{1}{32}$ ounce (1 gram) in 1 ton of substance. One gram is exactly one-millionth of a metric ton.

PRANDTL NUMBER.   A dimensionless number equal to the ratio of the kinematic viscosity to the thermometric conductivity (or thermal diffusivity). For gases, it is rather under one and is nearly independent of pressure and temperature, but for liquids, the variation is rapid. Its significance is as a measure of the relative rates of diffusion of momentum and heat in a flow and it is important in the study of compressible flow and heat convection.

PRASEODYMIUM.   Chemical element symbol Pr, at. no. 59, at. wt. 140.91, second in the lanthanide series in the periodic table, mp 934°C, bp 3512°C, density, 6.769 g/cm³ (0°C). Elemental praseodymium has a close-packed hexagonal crystal structure at 25°C. Pure metallic praseodymium is a silvery-gray color, the luster dulling rapidly upon exposure to air and forming a nonadherent oxide which hastens the process of oxidation. When pure, the metal is soft and workable with ordinary tools. Processing and handling require storage under a nonreactive liquid or inert atmosphere or vacuum. Finely divided praseodymium is pyrophoric, burning at a red heat. There is only one isotope of the element in nature, ^{141}Pr. It is not radioactive and has a low acute-toxicity rating. Fourteen artificial isotopes have been produced. Of the light (or cerium group) rare-earth metals, praseodymium is the fourth most plentiful and ranks 59th in abundance of the earth's crust, exceeding tantalum, mercury, bismuth, and the precious metals, excepting silver. The element was first identified by C. A. von Welsbach in 1885.

Electronic configuration is $1s^22s^22p^63s^23p^63d^{10}4s^24p^64d^{10}4f^25s^25p^65d^16s^2$. Ionic radius Pr^{3+} 1.01 Å, Pr^{4+} 0.90 Å. Metallic radius 1.828 Å. First ionization potential 5.42 eV; second 10.55 eV. Other important physical properties of praseodymium are given under **Rare-Earth Elements and Metals.**

Primary sources of the element are bastnasite and monazite which contain from 4 to 8% praseodymium. Plant capacity involving liquid-liquid or solid-liquid organic ion-exchange processes for recovering the element is in excess of 100,000 pounds Pr_6O_{11} annually. Metallic praseodymium is obtained by electrolysis of Pr_6O_{11} in a molten fluoride electrolyte, or by a calcium reduction of PrF_3 or $PrCl_3$ in a sealed-bomb reaction.

For many years, praseodymium has been a component of light rare-earth mixtures used in mischmetal, a pyrophoric alloy used in cigarette-lighter "flints." Mixtures of cerium, lanthanum, neodymium, and praseodymium, as oxides and fluorides, are used in the cores of arc carbons for the production of light of greater intensity. Similar mixtures of rare-earth oxides, including praseodymium, are used in optical glass polishing formulations. Mixtures of the lanthanide compounds, including about 5% praseodymium, find application as catalysts in petroleum cracking processes. A mixture containing 10% Pr, 30% Nd, and 60% La is used for cracking crude oil and comprises the largest single use of the element as well as of all the other Lanthanide elements. Use of elemental praseodymium as a colorant for glass was one of the early applications. The color ranges from clear yellow to green and finds use in sunglasses, protective glasses for industry, art objects of glass, tableware, and optical filters. In the manufacture of ceramic tile, a praseodymia-zirconia yellow stain is used. Metallurgically, the most important intermetallic compound is $PrCo_5$ which has unsurpassed permanent magnetic properties. The compound has a very high resistance to demagnetization and has a high magnetic saturation value. $PrNi_5$ has been used for adiabatic magnetization cooling of samples down to the milli-Kelvin range for low-temperature research. Investigations continue into further electronic and optical uses of praseodymium and its compounds.

NOTE: This 4th Edition entry was revised and updated by K. A. Gschneidner, Jr., Director, and B. Evans, Assistant Chemist, Rare-Earth Information Center, Energy and Mineral Resources Research Institute, Iowa State University, Ames, Iowa.

PRECIPITATION HARDENING. A large number of alloys are hardenable by a heat treating procedure known as precipitation hardening. Hardening is accomplished by the controlled precipitation of many minute particles of a second crystalline phase (or phases) inside the crystals of the primary metal. In order that the precipitation may be effected, the hardening constituent must be more soluble at higher temperatures than it is at lower temperatures, so that heating of the solid metal at an elevated temperature causes the second phase to dissolve into the matrix. If a precipitation hardening alloy is heated and held at an elevated temperature so as to dissolve the hardening phase and then is quenched to room temperature, a supersaturated solid solution is obtained. This heating and quenching operation is known as the solution treatment. The second phase of precipitation hardening is known as the aging treatment wherein the second phase is precipitated out of the supersaturated solid solution by holding the metal either at room temperature or some intermediate temperature well below the temperature employed in the solution treatment. The various stages involved in the formation of the nuclei of the precipitation particles may be very complex. In general, however, the aim of the aging process is to obtain a distribution of the precipitated particles that produces maximum hardness. This will usually occur when the particles are submicroscopic in size and extremely numerous. Their hardening effect on the crystal lattice of the matrix crystals is believed to result from local strains that they produce in the matrix. These latter hinder the normal easy motion of dislocations, thereby hardening the metal. The term age hardening is synonymous with precipitation hardening, but when so used generally refers to metals aged at room temperature.

PRECURSOR. In biological systems, an intermediate compound or molecular complex present in a living organism, which, when activated physiochemically is converted to a specific functional substance. Sometimes the prefix *pro-* is used to indicate that a compound in question plays the role of a precursor. Examples from the history of vitamin and other essential chemical developments include: Ergosterol (pro-vitamin D_2), which is activated by ultraviolet radiation to form vitamin D; carotene (pro-vitamin A) is a precursor of vitamin A; prothrombin forms thrombin upon activation in the blood-clotting mechanism.

PREGESTOGENS AND PROGESTINS.   Steroids.

PREHNITE.   Prehnite is a hydrous silicate of calcium and aluminum, $Ca_2Al_2Si_3O_{10}(OH)_2$, crystallizing in the orthorhombic system. Usual occurrence as intergrown crystals of reniform, stalactitic character, and as rounded groups of such crystals; hardness, 6–6.5; specific gravity 2.90–2.95; luster, vitreous to pearly; color, various shades of light green to gray or white; translucent. Though not a zeolite it is found associated with them and with datolite and calcite, in veins and cavities of basic rocks, sometimes in granites, syenites, or gneisses. It is found in Austria, Italy, the Harz Mountains, France, Scotland, and the Republic of South Africa, where it was originally discovered. Magnificent crystal casts after an unknown mineral have been found in a single large cavity in the basaltic rocks near Bombay, India. In the United States well-known localities are Somerville, Massachusetts; Farmington, Connecticut; Paterson, New Jersey; and Keweenaw County, Michigan. Named for Colonel Prehn, its discoverer, who was an early Dutch Governor of the Cape of Good Hope colony.

PRENENOLONE.   Steroids.

PRESERVATIVE (Food).   Additives (Food).

PRESSURE.   If a body of fluid is at rest, the forces are in equilibrium or the fluid is in static equilibrium. The types of force which may act on a body are shear or tangential force, tensile force, and compressive force. Fluids move continuously under the action of shear or tangential forces. Thus, a fluid at rest is free in each part from shear forces; one fluid layer does not slide relative to an adjacent layer. Fluids can be subjected to a compressive stress which is commonly called *pressure*. The term may be defined as force per unit area. The pressure units may be dynes per square centimeter, pounds per square foot, torr, mega pascals, etc. Atmospheric pressure is the force acting upon a unit area due to the weight of the atmosphere. Gage pressure is the difference between the pressure of the fluid measured (at some point) and atmospheric pressure. Absolute pressure, which can be measured by a mercury barometer, is the sum of gage pressure plus atmospheric pressure.

Pascal's law states that the pressure in a static fluid is the same in all directions. This condition is different from that for a stressed solid in static equilibrium. In such a solid, the stress on a plane depends upon the orientation of that plane. A liquid in contact with the atmosphere is sometimes called a free surface. A static liquid has a horizontal free surface if gravity is the only type of force acting.

Imagine a body of static fluid in a gravitational field. The mass of the fluid is m (in grams) and the weight of the fluid is mg (as dynes) where g is the local gravitational acceleration. Figure 1 shows a large region of any static fluid with a very small or infinitesimal element. Figure 2 indicates the element in detail. The vertical distance z is measured positively in the direction of decreasing pressure (up); dA is an infinitesimal area; p is the pressure acting on the top surface; and $(p + dp)$ is the pressure acting on the bottom surface. The pressure difference is due only to the weight of the fluid element. Let p represent density, which is mass per unit volume (as grams per cubic centimeter). Thus the weight of the element is $pg\, dz\, dA$. Considering the element as a free body, an accounting of forces in the vertical direction gives:

$$dp\, dA = -pg\, dz\, dA; \qquad dp = -pg\, dz \qquad (1)$$

As z is measured positively upward, the minus sign indicates that the pressure increases with an increase in height. This fundamental equation of fluid statics can be applied to all fluids. In integral form, Equation (1) becomes:

$$\int_1^2 \frac{dp}{g} = \int_1^2 dz = -(z_2 - z_1) \qquad (2)$$

where 1 refers to one level and 2 refers to another level. The functional relation between pressure p and the combination pg must be established before Equation (2) can be integrated. There are two major cases: (a) incompressible fluids, in which the

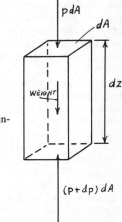

Fig. 2. Vertical forces on infinitesimal element.

density p is a constant; and (b) compressible fluids, in which the density p varies.

Liquids can be considered as incompressible in many cases. For small differences in height, a gas might be regarded as incompressible. For an incompressible fluid, with constant g, Equation (2) becomes:

$$p_2 - p_1 = -pg(z_2 - z_1) \qquad (3)$$

The term $(z_2 - z_1)$ may be called a static "pressure head," and it can be expressed in feet or inches of water, or some height of any liquid. For example, barometric pressure can be expressed in inches of mercury.

A manometer is a device that measures a static pressure by balancing the pressure with a column of liquid in static equilibrium. Many types of manometers are used. See also **Manometer**. The common mercury barometer is essentially a manometer for measuring atmospheric pressure; a mercury column in a glass tube balances the weight of the air above the mercury. Figure 3 illustrates a manometer in which the left leg is open to the atmosphere; the liquid has a specific weight (weight per unit volume) $p_2 g$. In the other leg is a liquid of specific weight $p_1 g$. Starting with the left leg, the gage pressure p_A is:

$$p_A = h_2 p_2 g$$

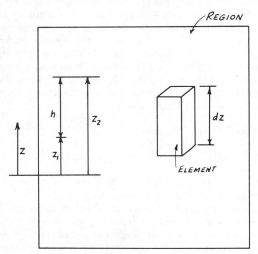

Fig. 1.   Large region of any static fluid.

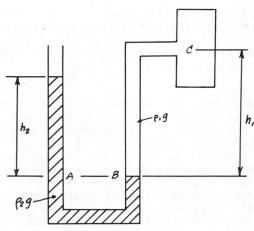

Fig. 3.   Manometer.

Since the fluid is in static equilibrium, the pressure p_B at point B equals the pressure at point A. Thus:

$$p_A = p_B = h_2 p_2 g$$

The pressure p_C at point C is less than that at B. Thus:

$$p_B - p_C = h_1 p_1 g$$

Then the gape pressure at point C is:

$$p_C = g(h_2 p_2 - h_1 p_1)$$

When a body of any kind is partly or fully immersed in a static fluid, every part of the body surface in contact with the fluid is pressed on by the fluid. The pressure is greater on the areas more deeply immersed. The resultant of all these fluid pressure forces is an upward or buoyant force. The pressure on each part of the body is independent of the body material. Archimedes principle states that the buoyant force equals the weight of the displaced fluid.

Equation (3) is for the special case of an incompressible fluid. As an example of a compressible fluid, consider an isothermal or constant-temperature layer of gas. The equation of state for such a gas can be written:

$$p = pRT_1 \qquad (4)$$

where T_1 is the given absolute temperature and R is a gas constant or gas factor depending upon the gas. Assuming a constant g, Equation (2) gives:

$$\frac{RT_1}{g} \int_1^2 \frac{dp}{p} = -(z_2 - z_1)$$

$$z_2 - z_1 = \frac{RT_1}{g} \log_e \frac{p_1}{p_2} \qquad (5)$$

Equation (5) is sometimes called a "barometric height" relation. For an isothermal atmosphere, a measurement of the temperature T_1 and the static pressure (as with a barometer) at two different levels will provide data for the calculation of the height difference.

Other pressure designations include:

Vacuum. A gage pressure below atmospheric.

Hydrostatic Pressure. The pressure at a point below a liquid surface due to the height of fluid above it.

Tons-on-Ram. The force that acts over a given area as in various types of hydraulic machinery.

Partial Pressure. The pressure exerted by one component in a system, usually one gas or vapor in a mixture.

Internal Pressure. The effect of the attractive forces of the molecules of a substance, which is called pressure because its result is the same as that of an added external pressure. In liquids, its effect appears as the ability of liquids to stand substantial negative pressures without rupture.

Cohesion Pressure. A term in van der Waal's equation introduced to take care of the effect of molecular attraction. It is usually expressed as a/V^2, where a is a constant and V is the volume of the gas.

High-Pressure Technology. Until the mid-1970s, the limit to most high-pressure experimentation was confined to about 300 kilobars. As of the early 1980s, much higher pressures have been achieved in the laboratory and scientists are looking to equipment that will achieve pressures in excess of 1.5 megabars (1.5 million atmospheres), which is about the highest limit available with current technology. New high-pressure technology is being directed toward research in two principal areas: (1) the study of rocks at conditions that approach those at the boundary between the earth's mantle and liquid core, an im-

provement over deductions which must be made from seismic wave studies—directed toward a better understanding of mantle structure as well as other geophysical factors; and (2) the conversion of molecular hydrogen into the still hypothetical metallic form (where it is visualized that protons would be embedded in an abundance of electrons)—of possible later use in superconducting electrical generators and controlled nuclear fusion.

The key to the most recent ultrahigh-pressure experiments has been the diamond cell, wherein the experimental sample is held between two diamond anvils. Whereas in earlier high-pressure apparatus, very high forces were applied to relatively large areas, the diamond cell involves concentrating force over very small areas, on the order of a few hundred micrometers in diameter. With this cell, researchers at Carnegie Institution have reported a sustained pressure of 1.7 megabars. At this pressure, one of the diamonds was deformed and this may indicate that the approximate ultimate limit may have been reached with diamond.

In the years when pressures were considered high if they reached the 30 kilobar region, a number of techniques for creating high pressures were developed, and these continue to be used in some research areas (artificial gems, materials testing, etc.). Among the methods used are mechanical, e.g., opposing pistons, belts and girdles, anvils; shock methods, in which the force from a controlled explosion is utilized in specially designed apparatus; and the use of ultrasonic magnetic fields and electrohydraulic methods.

References

Hammond, A. L.: "High-Pressure Geophysics," *Science*, **190**, 967–968 (1975).
Kerr, R. A.: "Ultrahigh Pressure Research," *Science*, **201**, 429–430 (1978).
Mao, H. K., and P. M. Bell: "High-Pressure Physics: The 1-Megabar Mark on the Ruby R_1 Static Pressure Scale," *Science*, **191**, 851–852 (1976).

PROGESTERONE. Hormones; Steroids.

PROMETHIUM. Chemical element symbol Pm, at. no. 61, at. wt. 145 (mass number of the most stable isotope), fourth in the lanthanide series in the periodic table, mp 1042°C, bp 3000°C (estimated), density, 7.26 g/cm³ (20°C). Elemental promethium has a double hexagonal close-packed crystal structure at 25°C. Pure metallic promethium is a silvery-white color, is soft, and can be cast or machined. The naturally occuring isotope, ^{147}Pm, is radioactive ($t_{1/2} = 2.52$ years). Consequently, the element must be handled within a shielded area. Eighteen artificially-produced isotopes, ranging from ^{140}Pm to ^{146}Pm and from ^{148}Pm to ^{158}Pm have been identified, all with very short half-lives. Many of the properties of promethium remain classified, or are known only by proprietary sources. Although first identified as an element by J. A. Marinsky, L. E. Glendenin, and C. D. Coryell in 1947, the element was not available on more than a gram-scale for several years.

Electronic configuration is $1s^2 2s^2 2p^6 3s^2 3p^6 3d^{10} 4s^2 4p^6 4d^{10} 4f^4 5s^2 5p^6 5d^1 6s^2$. Ionic radius, 0.98 Å. Other important physical properties of promethium are given under **Rare-Earth Elements and Metals.**

^{147}Pm is extracted from the wastes of uranium or plutonium reactors, the most important source of the element. ^{146}Pm and ^{148}Pm also are derived from reactor wastes. In 1970, ^{147}Pm became available in kilogram quantities. ^{147}Pm has been under intensive study as a heat and power source; however, before it can be used for this, ^{146}Pm and ^{148}Pm, which produce penetrating gamma radiation, must be eliminated. The desirable property of ^{147}Pm is that it decays by beta emission only, at a low energy level compared with most fission products, and thus requires only light to moderate shielding. ^{147}Pm has been used to activate luminescent phosphors. Beads (*Microspheres®*, 3M Company) containing ^{147}Pm mixed with a phosphor provide a long-lived, reliable green light and were used by astronauts to assist in docking and other maneuvers in outer space. Commercial applications of ^{147}Pm as a power source include betavoltaic cells for surgical implant with heart pumps and pacemakers.

See the references listed at the ends of the entries on **Chemical Elements; and Rare-Earth Elements and Metals.**

NOTE: This 4th Edition entry was revised and updated by K. A. Gschneidner, Jr., Director, and B. Evans, Assistant Chemist, Rare-Earth Information Center, Energy and Mineral Resources Research Institute, Iowa State University, Ames, Iowa.

PROPANE. $CH_3 \cdot CH_2 \cdot CH_3$, formula weight 44.09, colorless gas, mp $-187.1°C$, bp $-42.2°C$, sp gr 0.585 (at $-45°C$). The gas is slightly soluble in H_2O, moderately soluble in alcohol, and very soluble in ether. Although a number of organic compounds which are important industrially may be considered to be derivatives of propane, it is not a common starting ingredient. The content of propane in natural gas varies with the source of the natural gas, but on the average is about 6%. Propane also is obtainable from petroleum sources.

Liquefied propane is marketed as a fuel for outlying areas where other fuels may not be readily available and for portable cook stoves. In this form, the propane may be marketed as LPG (liquefied petroleum gas) or mixed with butane and pentane, the latter also constituents of natural gas (1.7% and 0.6%, respectively). LPG also is transported via pipelines in certain areas. The heating value of pure propane is 2,520 Btu/ft^3 (283 Calories/m^3), butane 3,260 Btu/ft^3 (366 Calories/m^3); and pentane 4,025 Btu/ft^3 (452 Calories/m^3). Propane and the other liquefied gases are clean and appropriate for most heating purposes, making them very attractive where they are competitively priced.

PROPELLANTS (Rocket and Missile). **Rocket Propellants.**

PROPIONATE PLASTICS. **Cellulose Ester Plastics (Organic).**

PROPIONIC ACID AND PROPIONATES. **Antimicrobial Agents (Foods).**

PROSTAGLANDINS. A group of fatty acid derived molecules containing 20 carbon atoms. They have been detected in practically every part and organ in the human body with especially high concentrations in seminal fluid. They are active biologically causing smooth muscle preparations to contract at concentrations of 10^{-9} g/ml and having definite effects on human blood pressure at doses as low as 0.1 microgram per kilogram of body weight. The prostaglandins were first chemically characterized in 1960. The structures of the basic series of compounds are given in the accompanying figure.

The systematic naming of these substances is based on the C_{20} prostanoic acid skeleton. The numbering system starts with the carboxyl carbon designated as number one and proceeds around the chain to the terminal methyl group. This is illustrated

for structure 1. The trivial names for the compounds begin with the letters PG. A third letter distinguishes the four basic series illustrated as 1–4 in the diagram, and a subscript refers to the number of additional double bonds present. The PGE_2 series has another *cis* double bond at carbon 5 and the PGE_3 series has two additional *cis* double bonds at positions 5 and 17.

A description of the biological activity of the prostaglandins dates back to 1930, when it was shown that human seminal plasma could cause either relaxation or excitation of the uterus. It was also noted that the effects could be correlated with a woman's past history of fertility. In 1933–34 two groups independently showed that these effects were caused by a new class of lipid materials different from anything known at that time.

The biological effects of these substances are beyond the scope of this Encyclopedia.

PROTACTINIUM. Chemical element symbol Pa, at. no. 91, at. wt. 231.036, radioactive metal of the actinide series, mp estimated at less than 1600°C. The most stable isotope is ^{231}Pa, with a half-life of 3.43×10^4 years. The latter is a second-generation daughter of ^{235}U and a member of the actinium $(2n + 3)$ decay series. See **Radioactivity.** Electronic configuration is $1s^2 2s^2 2p^6 3s^2 3p^6 3d^{10} 4s^2 4p^6 4d^{10} 4f^{14} 5s^2 5p^6 5d^{10} 5f^2 6s^2 6p^6 6d^1 7s^2$. Ionic radii, Pa^{4+}, 0.91 Å; Pa^{3+}, 1.06 Å.

The probable existence of protactinium was predicted as early as 1871 by Mendeleev to fill up the space on his periodic table between thorium (at. no. 90) and uranium (at. no. 92). Mendeleev termed the uncomfirmed element *ekatantalum*. In 1926, O. Hahn predicted the properties of the element in considerable detail, including descriptions of its compounds. In 1930, Aristid von Grosse isolated 2 milligrams of what then was termed ekatantalum pentoxide and showed that element 91 differed in all reactions from comparable amounts of tantalum compounds, with exception of precipitation by NH_3. However, credit for the discovery of protactinium generally is attributed to Lise Meitner and Otto Hahn in 1917.

Protactinium-231 yields actinium-227 by alpha-particle emission and has a half-life of 3.43×10^4 years. Its other isotopes include two isomers of mass number 234: uranium X_2, with a half-life of 1.17 minutes, and uranium Z, with a half-life of 6.7 hours, the former being an excited state which undergoes deexcitation to give the latter. Other nuclear species have mass numbers 225–230, 232, 233, 235, and 237.

The methods of purification include the use of ion exchange resins, the precipitation of protactinium peroxide and the extraction of aqueous solutions of protactinium salts by various organic solvents.

Protactinium metal is prepared (1) by reducing the tetrafluoride with metallic barium at about 1,500°C; (2) by heating the halide, usually the iodide, under a high vacuum; and (3) by bombardment of the oxide under high vacuum with 35 keV electrons for hours at a current strength of 0.005–0.010 Amperes.

Protactinium (of mass number 231) is found in nature in all uranium ores, since it is a long-lived member of the uranium series. It occurs in such ores to the extent of about $\frac{1}{4}$ part per million parts of uranium. An efficient method for the separation of protactinium is by a carrier technique using zirconium phosphate which, when precipitated from strongly acid solutions, coprecipitates protactinium nearly quantitatively. Then the protactinium is separated from the carrier by fractional crystallization of zirconium oxychloride

Isotopes of protactinium can also be produced artificially, i.e., by the nuclear reactions of other elements with such particles

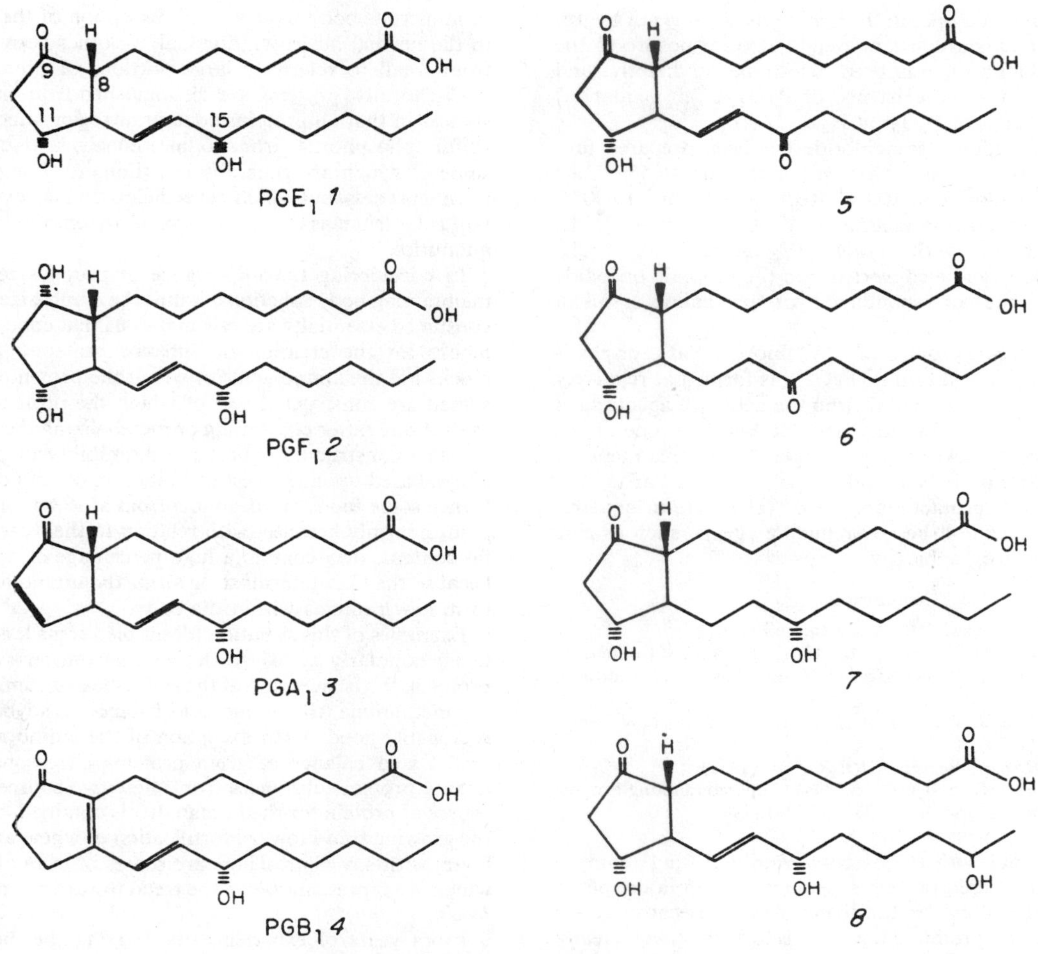

Structures of the basic prostaglandins.

as deuterons, neutrons, and alpha-particles. Thus, when thorium is bombarded with deuterons of various high energies, five of the reactions are: ^{232}Th(d,4n)^{230}Pa, ^{232}Th(d,6n)^{228}Pa, ^{232}Th(d,7n)^{227}Pa, ^{232}Th(d,8n)^{226}Pa, and ^{230}Th(d,3n)^{229}Pa.

Quantitative methods of obtaining protactinium start from the carbonate precipitate from the treatment of the acid extract of certain uranium ores. After this carbonate precipitate is dissolved, the protactinium remains in the silica gel residue, from the solution of which it is obtained on a manganese dioxide carrier. An alternate method effects final separation of the protactinium by formation of a complex compound, protactinium-cupferron, and its extraction with amyl acetate.

As early as 1965, investigators at Los Alamos (Fowler et al., 1965) reported that protactinium metal is superconductive below 1.4 K. In 1972, researchers at Harwell (Mortimer, 1972) reported no superconductivity of the metal down to approximately 0.9 K. An exchange of information to resolve the differences in data was conducted over the next few years (Fowler, 1974; Hall, et al., 1977). Smith, Spirlet, and Müller (1979) reported that differences in experimental research were due to problems with the crystal structure of the metal and sample purity that arise when dealing with radioactive material. These investigators observed very-high-purity protactium, produced by the Van Arkel procedure, and observed an extremely steep superconductivity transition at 0.42 K in protactinium in the presence of rather high self-heating. The superconducting transi-

tion temperature and upper critical magnetic field of protactinium were measured by alternating-current susceptibility techniques. Inasmuch as the superconducting behavior of protactinium is affected by its 5f electron character, it has been further confirmed that protactinium is a true actinide element.

The predominant oxidation state of the element is (V). There is some evidence that the (IV) state is obtained under certain reduction conditions. When the pentapositive form is not in the form of a complex ion it may exist in solution as PaO_2^+. The compounds are very readily hydrolyzed in aqueous solution yielding aggregates of colloidal dimensions, thus showing marked similarity to niobium and tantalum in this respect. These properties play a dominant role in the chemical properties of aqueous solution, because the element is so easily removed from solution by hydrolysis and adsorption. Protactinium coprecipitates with a wide variety of substances, and it seems likely that the explanation for this lies in the hydrolytic and adsorptive behavior.

The element is difficult to maintain in aqueous solution in the form of simple salts. Solubility data seem to indicate that such amounts as can be dissolved probably do so entirely by formation of complex ions. Fluoride ion strongly complexes protactinium, and it is due to this that protactinium compounds are in general soluble in hydrofluoric acid.

Protactinium oxide may be prepared from the hydrated oxide or the oxalate by ignition. The product is a dense white powder with a very high melting point; the ignited material is not hygroscopic and maintains a constant weight upon exposure to the air. The formula Pa_2O_5 has been determined indirectly, and there is evidence for the existence of $PaO_{2.25}$ (air oxidation) and PaO_2 (reduction of P_2O_5 by H_2).

Volatile protactinium pentachloride has been prepared in a vacuum by reaction of the oxide with phosgene at 550°C or with carbon tetrachloride at 200°C. Reduction of this at 600°C with hydrogen leads to protactinium(IV) tetrachloride, $PaCl_4$, which is isostructural with uranium(IV) tetrachloride, UCl_4. The pentachloride can be converted into the bromide or iodide by heating with the corresponding hydrogen halide or alkali halide.

The volatile fluoride protactinium(V) fluoride, PaF_5, or possibly protactinium(V) oxyfluoride, $PaOF_3$, is formed at relatively low temperatures such as 200°C from the action of agents such as bromine tri- or pentafluoride, BrF_3 or BrF_5, on one of the protactinium oxides. At higher temperatures, treatment of Pa_2O_5 with hydrofluoric acid and hydrogen yields PaF_4.

The reduction of protactinium to the (IV) state in aqueous solution can be accomplished by reducing agents, such as zinc amalgam and polarographically.

References

Fowler, R. D., et al.: *Phys. Rev. Lett.*, **15**, 860 (1965).
Fowler, R. D., et al.: "Proceedings of the 13th International Conference on Low Temperature Physics" (K. D. Timmerhaus et al., Editors), Plenum, New York (1974).
Hall, R. O. A., Lee, J. A., and M. J. Mortimer: *Jrnl. Low Temp. Phys.*, **27**, 305 (1977).
Mortimer, J. J.: *Harwell Report AERE-R 7030* (1972).
Smith, J. L., Spirlet, J. C., and W. C. Müller: "Superconducting Properties of Protactinium," *Science*, **205**, 188–190 (1979).

PROTEIN. Along with the carbohydrate and lipid[1] components of the animal diet, protein substances are a major source of nutrition and energy for the living system. Because of his high regard for the proteins, but well before they were really understood, the Dutch Chemist, Gerardus Mulder (1802–1880) pioneered the use of the term *protein*, derived from the Greek word meaning *to come first*. Although proteins furnish energy to the body and thus can be considered as body fuels, as are the carbohydrates and fats, the major nutritional roles of the proteins reside in other functions, usually of a highly specific nature. Thus, there are structural, contractile, processactivating, transport, etc. proteins which essentially are responsible for the chemical workability of the animal system.

In the growing animal body, a significant portion of proteins consumed is required for the creation of new tissue, resulting in an increasing requirement for proteins in the diet of humans, for example, up to about the age of 20 years, at which time the protein requirement tends to level off to a fairly stable figure. After body maturity, the portion of proteins needed for tissue maintenance is greater than the need for new tissue building. It must be emphasized, however, that immediately at the commencement of life both new tissue building and tissue maintenance take place, and even as the body grows older, the two needs continue—only the proportions between the two roles change.

Proteins, on a weight basis, are second only to water in their presence in the human body. If the factor of water is discounted, then about 50% of the body's dry weight is made up of numerous protein substances, distributed about as follows: 33% in muscles; 20% in bones and cartilage; 10% in skin; the remaining 37%

[1] Fats, oils, fatty acids, phospholipids, and sterols.

in numerous body tissues. With exception of the urine and bile in the normal healthy individual, all other body fluids contain from small to relatively large portions of protein substances.

Chemically, proteins are distinguished from other body substances in that all proteins contain nitrogen—and some contain sulfur, phosphorus, iron, iodine, cobalt, and other elements, some of which are generally not thought of as components of the life process, but which nevertheless do play extremely important roles (such as catalysts), even if present only in very minute quantities.

In considering the importance of proteins to building and maintaining body functions, it must be emphasized that proteins consumed essentially are raw materials that contain the building blocks for the creation of different proteins. These building blocks are the amino acids of which the protein molecules consumed are constructed and of which the proteins restructured in the body (after consuming or metabolizing the raw materials) are also constructed. Thus, the desirability of proteins in the diet is based upon the best combination of amino acids present. Hence some foods are desirable from a protein nutrition standpoint not only because, with relation to their carbohydrate and fat content, they contain a high percentage of protein, but also because they contain most or all of the amino acids needed to form new proteins within the body.

Examples of this situation (desirable versus less desirable proteins) popularly cited are the soybean proteins and the grain proteins. With exception of the sulfur-bearing amino acids, notably methionine, the amino acid balance of soybean proteins is reasonably good. With exception of the amino acid lysine, the amino acid balance of grain proteins is reasonably good. By mixing protein substances from these two sources, an excellent source of protein for the human diet is obtained, thus explaining the growing trend toward fortification of wheat and other cereal flours with soy flour. There are scores of examples of this type which are representative of the trend toward so-called *fabricated foods*.

From years of experience in studying the dietary needs of humans, nutritionists and biologists established the hen egg as having the most perfect balance of amino acids in a natural protein substance. Against this standard, other foods can be rated in their performance. In naming the following food substances in order of their diminishing chemical score, it should be stressed that these foods are arranged only in terms of this one nutritional criterion: fish (70), beef (69), cow's milk, whole (60), brown rice (57), polished white rice (56), soybeans (47), green leaves (45), brewer's yeast (44), grandnuts (peanuts) (43), whole grain maize (corn) (41), cassava (manioc) (41), common dry beans (34), white potato (34), white wheat flour (32). The foregoing food items were selected *randomly* to provide a sense of the spectrum of foods from this one particular standpoint. The figures represent only the chemical balance of amino acids present and not the total amount of protein available as a weight percentage of food intake, or from the standpoint of protein utilization, once ingested.

In looking at a number of food substances, again a random selection, from the standpoint of total protein (not necessarily quality of protein) in an average serving, the following amounts of protein (grams) are present: fried chicken breasts (27.8), canned tunafish (24), cooked round roast of beef (24), roasted leg of lamb (22), oven-cooked pork loin (21), dry-cooked soybeans (13), whole milk (1 cup) (9), canned red beans (7.5), cheddar cheese (1 ounce = 28 grams) (7), fresh cooked lima beans (6.5), egg (medium size) (6), vanilla ice cream (6), fried crisp bacon (5), baked potato (3), cooked broccoli (2.5), cooked oatmeal (2.5), enriched white bread (1 slice) (2), cooked green snap beans (1), lettuce ($\frac{1}{4}$ head) (1), and reconstituted frozen orange juice (1).

Consequences of Protein Deficiency. Because proteins are so important to numerous and very complex bodily functions, years of research have just commenced to provide some understanding of most of the mechanisms involved. As would be expected, recognition of the extreme manifestations of protein deficiencies has taken place, at least to the extent of providing new guidelines for assisting millions of inadequately fed people in several regions of the world. As further experience is gained from researching the gross problem, the important subtleties of protein performance within the body will become more apparent.

Exemplary of a better understanding and appreciation of protein nutrition is a comparison of the 1945 report of the Food and Agriculture Organization (United Nations) with more recent findings, recommendations, and nomenclature used. In the first *World Food Survey*, the terms *undernourishment* and *malnourishment* were used throughout the report. The general interpretation of undernourishment was taken to mean an inadequate caloric intake, i.e., insufficient energy input to support normal body functions and activities, with body weight loss the inevitable result. Similarly, *malnourishment* was taken to mean a deficiency of one or all of the protective nutrients, such as proteins, vitamins, and minerals. During the last few years, inasmuch as these two problems are so interrelated, the term *protein-calorie malnutrition* (PCM) has come into wide use. PCM of early childhood, particularly in regions that are a part of some of the less developed countries, is quite widespread. PCM apparently is manifested in minor ways at first, but when prolonged very severe syndromes become evident. These include the conditions known as *kwashiorkor* and *marasmus*.

Kwashiorkor usually occurs in the second or third year in the life of a child. Edema is the principal symptom. The condition arises from a combination of circumstances, but the primary cause appears to be a weaning diet that is both inadequate and indigestible and, notably, is lacking of protein. The principal calories are supplied by carbohydrate. The condition is accelerated by repeated infections of a bacterial, parasitic, or viral nature. Without treatment, the disease is fatal in most cases.

Nutritional marasmus is a severe manifestation of PCM and is a condition that usually occurs during the first year of life. Again, it arises from a combination of conditions, frequently widespread in many regions, of feeding an overly diluted formula of cow's milk, thus reducing the protein input well below minimum needs. The condition is accelerated by filthy surroundings and contaminated bottles. Characteristic of the syndrome are a wasting of muscle and subcutaneous fat, a body weight that may be only 60% of standard, and diarrhea. Children who have access to human milk usually are protected against marasmus and diarrheal disease.

A more recent finding and term now used for a protein deficiency syndrome is *PCM-plus*, or *infantile obesity*. This is a condition that occurs among the more affluent populations where an infant is bottle-fed, where hygiene is adequate, and where funds are adequate. Overfeeding of an improperly-balanced formula can cause the condition. The condition does not occur with breast feeding because the volume of intake is regulated by the infant's appetite and thirst.

Sources of Protein. The two basic categories of protien sources for the animal diet are other *animals* (living or dead) and plants. Thus, in the animal category as a source of human and pet protein foods, there are what might be called terminal sources or nonreplenishing sources, in which the living animal is killed and disassembled into its protein-containing parts, the most common examples including the meaty flesh and organs of beef cattle, pigs, sheep, horses, and goats, as well as the more occasional sources of meat, such as deer, elephant, hippopotamus, etc., depending upon availability and regional eating preferences. To these sources are added the flesh and organs of birds (chickens, ducks, turkeys, pheasants, etc.) and of fish caught in saline and fresh waters. In the overall animal protein category, one also would include those less conventional and essentially unexplored categories, such as earthworms and single-cell proteins (produced by microorganisms) and algae. Renewable or repeating protein sources from living animals, of course, include the milk from dairy cows and buffalos and the eggs from hens, from which hundreds of high-protein foods (cheese, for example) are prepared. And, to this category, must be added the excellent source of protein provided by human milk to the nursing infant.

Plants, of course, also require protein to build and maintain their life processes and, consequently, are protein sources for the animal diet. In the case of herbivores, plants are essentially the exclusive source of proteins, energy, and all other dietary elements.

In terms of percentage of protein content of basic sources, the animal sources far excell the plant sources. For example, the protein content of some typical unfortified foods range from 20–30% for cooked poultry and meats; for cooked or canned fish (19–30%); for cheese (25%); for cottage cheese (13–17%); for nuts (16%); for whole eggs (13%); for dry cereals (7–14%); for white bread (8.5–9%); for cooked legumes (7–8)%; and for cooked cereals (about 2%).

Of course, in achieving the higher protein contents of meat from poultry and cattle, a rather costly two-step production process is involved, wherein the animal first converts plant proteins (as from grasses) into animal protein. In a sense, the animal both converts and concentrates the protein source for humans. Several economic factors enter into the picture—the utilization of land, the costs of labor, the additional costs of feed materials, and the costs related to a greater time span of production, among others. As a case in point, an animal must be fed between 3 and 10 pounds (1.4–4.5 kilograms) of grain for the animal to produce 1 pound (0.45 kilogram) of meat. All of these factors in recent years, particularly in consideration of protein shortages in many regions of the world, have given rise to conflicting opinions pertaining to the ever-increasing production and consumption of meat, not only in several of the western nations of the world, but in the developed nations of the Orient as well. A few authorities have suggested that the western countries should cut back on meat production, thus making more land, skills, etc. available to increasing vegetable protein production to the level where a generous excess supply would be available to underdeveloped countries as well as amply supplying the protein needs of the developed countries. Quickly, these arguments penetrate not only into technological and economic factors, but psychological considerations as well—because any moves of this type necessarily require drastic changes in eating habits and to bring them about successfully would require much more governmental regulation and policing than any system of private enterprise is likely to tolerate. Further, attitudes tend to swing rather widely from times of grain surplus to times of grain shortage.

Fortunately, as of the mid-1980s, it appears that protein processing techniques may be providing a very satisfactory compromise, even though the industry is just getting underway toward a large-scale operation. Protein meat extenders, for example, wherein meat and vegetable protein are blended to produce an edible product that retains much of what is desired of meats, including their good protein content, are finding acceptance. The wide acceptance of vegetable protein in analogue meat products has many hurdles to overcome, but it appears that a solid start has been made. The hurdles not only include acceptability in the marketplace, but also some justifiable resistance on the part of cattle and poultry producers. For many reasons, the transition, if it ultimately takes place, will occur over quite a long period of time. Because of continuing economic inflation, the earlier cost advantages that tended to favor blends of meat and vegetable proteins have become less significant.

An early impetus to soy protein foods was given when the United States introduced soy protein products into its overseas donation program in 1966 as a component of foods formulated to meet special needs of certain population groups. Chief among these were children in developing nations, especially the weanling infant and preschool child whose requirements for growth put special demands on diet composition. Pregnant and lactating mothers also had dietary needs frequently not met in countries where food supplies were marginal. Beyond these needs, there were nutrient deficiencies in large population groups which could be best overcome by enrichment or fortification of commonly eaten foods.

Shortages in the domestic supply of nonfat dry milk, which developed in 1965, stimulated the development of high-protein formulated foods which would serve as supplements in the diets of the children or in the emergency feeding of adults. These formulations had to pass rigid specifications, one of the principal criteria being the recommended daily dietary allowance for protein, vitamins, and minerals. The U.S. Department of Agriculture and the U.S. Agency for International Development (AID) developed the guidelines and designed various formulated foods. Among these formulations were Corn-Soy Milk (CSM), Corn-Soy Blend (CSB), and Wheat-Soy Blend (WSB).

Further impetus was given to protein blends in foods when such products were introduced into the domestic food assistance program in the United States. Soy protein foods were introduced into school lunch and breakfast programs for which federal assistance has been given in the form of a subsidy administered by the federal government. Soy-fortified foods also were distributed to needy families through a family food distribution program.

Textured soy protein products in their use as meat alternatives have become increasingly popular in school lunch programs since their introduction in 1971. A soy-modified macaroni was introduced into the family food assistance program a number of years ago.

Less Conventional Sources of Protein. In addition to the traditional animal sources of protein already described and the very large amounts of vegetable protein derived from the soybean, other sources of protein on a large scale for the future are under intense study. Among these are (1) oilseed crops, such as rapeseed and cottonseed; (2) leaf proteins; (3) algae; and (4) single-cell protein.

Rapeseed, one of the five most widely produced oilseeds, is cultivated mainly in India, Canada, Pakistan, France, Poland, Sweden, and Germany. Past objections to using rapeseed as a source of edible protein has been its content of deleterious glucosinolates. Considerable research has been conducted in Sweden to develop a rapeseed protein concentrate. The first full-scale production plant using a new process was installed in Alberta, Canada. The plant, with a capacity of 5000 tons/year produces a material containing 65% protein. Rapeseed is rich in essential amino acids, with exception of methionine, which soybeans also lack.

Cottonseed offers an attractive source of protein provided that certain objectional ingredients can be removed. One of these is gossypol, a substance in cottonseed gland that is harmful to humans. A process developed by the U.S. Department of Agriculture has been designed to turn out a satisfactory edible cottonseed protein product. Employing silvent-extraction techniques, the first plant was built in Texas. Cottonseed flour extrudes easily and can be water-extracted to produce a nearly 100% protein isolate. The product has been used as a blend extender and fortifier for processed meats, baked goods, candies, and cereals. Research of a different approach has been used in Central America. In this approach, iron compounds are used to tie up the gossypol in nontoxic form without having to remove it.

Leaf Protein Concentrates. Laboratories in Hungary, Japan, the United Kingdom, and the United States, among other countries, have been engaged in perfection of a leaf protein concentrate process, with emphasis upon increasing yields and palatability and reducing flavor problems and cost. To date, alfalfa appears to be most attractive as a source of leaf protein. Alfalfa will produce more protein per unit of land than most other crops—up to 2800–4000 pounds/acre (3136–4480 kilograms/hectare). It has been estimated that the raw material costs for edible protein from alfalfa would be about 50% that for soybean meal. Several processes have been worked out, ranging from a green curd containing 52% protein to a white powder containing about 90% protein.

Single-Cell Protein. The advantages of single-cell protein (SCP) made from growing microorganisms are several: (1) SCP is independent of agricultural or climatic conditions, (2) SCP doubles in mass rapidly for high production rates and fast genetic experimentation, (3) the crop is free of surface-area limitations, and (4) the protein in microbial cells is generally of a high nutritional quality. Many of the processes proposed and tested, some with limited operating experience, commence with hydrocarbon feedstocks—gas oil and normal paraffin substrates. Two objections have been raised. The first is the possibility that carcinogenic polyaromatic materials present in gas oil may be passed along to the final protein product. The second is an adverse public reaction. A more recent, third objection is the proposition that perhaps technology should be concentrating on manufacturing fuels from farm products rather than food from petroleum products.

Some of the more recent SCP process concepts start with other materials, such as ethanol, acetic acid, starches, sugars, and cellulosic products that may be more available and particularly so in the protein-needy developing countries.

Algae have the highest intrinsic rates of photosynthesis and growth found among green plants. Human food and animal feed are being produced from algae. In Japan, a full plant-scale production harvests algae from open ponds to yield green power extract that can be used for animal or human consumption. The genus *Chlorella* has perhaps received the most research to date.

Conservation Sources of Protein. Tightening pollution restrictions have forced cheesemakers in many regions to end a longtime practice of dumping whey (with its high biological oxygen demand) as a liquid waste. Although many of these manufacturers are now evaporating or spray-drying whey to produce a whole-solids product, several fractionation techniques have been devised to separate a concentrated protein. In the United States, whey as a byproduct of chessemaking totals well over 30 billion pounds (13.6 billion kilograms) per year. From 6.5–7% of the whey is solids, of which 0.9% is protein. Some authorities believe that whey and other milk-based protein ingredients offer a high growth potential among all of the nonsoybean sources.

Fish protein concentrate is regarded by some authorities as having a high long-term potential. A major restraint is competition for the whole fish. As fish food sources become increasingly more competitive, fishes currently considered "trash" fishes from a fresh marketing viewpoint may ultimately become more desirable for table use. Animal-feed fish meal also will be a strong contender for available fish. In terms of processes required for preparing fish-protein concentrate, extraction processes using single or mixed solvents of isopropanol, ethylene dichloride, ethanol, and hexane already have been developed. Experiments with enzymatic processing also are underway.

Chemical Nature of Proteins

In defining a protein structurally, it is first necessary to define a peptide. Peptides are compounds made up of two or more amino acids covalently bound in an amide linkage. The characteristic amide linkage, in which the carboxyl group of one amino acid joins with the amino group of the next amino acid, is called a peptide bond. A peptide is a chain of amino acid residues. Provided that the chain is not circular or blocked at either of the ends, the peptide has an N-terminal amino acid, bearing a free amino group, and a C-terminal amino acid, bearing a free carbosyl group. This is illustrated as follows:

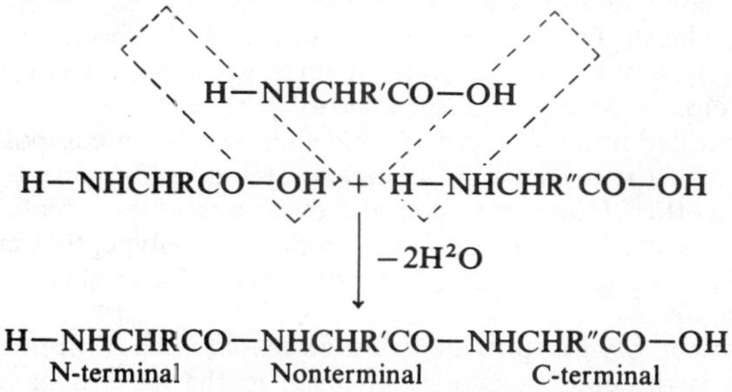

$$H-NHCHR'CO-OH$$
$$H-NHCHRCO-OH + H-NHCHR''CO-OH$$
$$\downarrow -2H^2O$$
$$H-NHCHRCO-NHCHR'CO-NHCHR''CO-OH$$
N-terminal   Nonterminal   C-terminal

Usually a form of shorthand is used to represent the structure of a peptide. For example, H-Val-Gly-Ala-OH, represents a peptide where abbreviation for each amino acid is given in terms of three letters each (Val = valine; Gly = glycine; Ala = alanine). Abbreviations for other amino acids are given in entry on **Amino Acids.** The H denotes the amino terminal (N-terminal) and the suffix OH denotes the carboxyl terminal (C-terminal). Peptides may consist of from two to eight amino acid residues and thus are known as dipeptides, tripeptides, or oligopeptides (eight), depending upon the number of residues contained. A peptide consisting of ten or more amino acid residues and with a molecular weight in the range of $1-5 \times 10^3$ is called a polypeptide. Emil Fischer, father of protein chemistry, proposed early in the twentieth century that proteins are peptide in nature. Actually, no sharp demarcation exists between large polypeptides and small proteins. Examples of small proteins include insulin (hormone protein), protamine, and some components of histone (basic proteins of chromosomes).

Almost all proteins are comprised of amino acid residues, more than 100 in number, and their molecular weight may range from 10^4 to 10^7. A few examples include Insulin (6×10^3); ribonuclease (13×10^3); lysozyme (eggwhite) (15×10^3); chymotrypsinogen (21×10^3); ovalalbumin (43×10^3); serum albumin (66×10^3)—all of the foregoing being single peptide chains. Multiple chains include: Hemoglobin (68×10^3); gamma globulin (IgG) (160×10^3); fibrinogen (340×10^3); urease (460×10^3); thyroglobulin (640×10^3); myosin (850×10^3); hemocyanin (octopus) ($2,800 \times 10^3$); hemocyanine (snail) ($8,900 \times 10^3$); and tobacco mosaic virus ($40,000 \times 10^3$). Proteins of huge molecular weight (millions) are enormous aggregates of protein subunits, each of which may be so large (molecular weight = $1.5-10 \times 10^4$) in most instances. The independent

peptide chains which constitute a protein molecule are often held by the disulfide bridges of cystine residues. From the diagram below, it will be seen that in a single chain the bridges may hold together two quite distant points in terms of the linear amino acid sequence, forming a large loop structure:

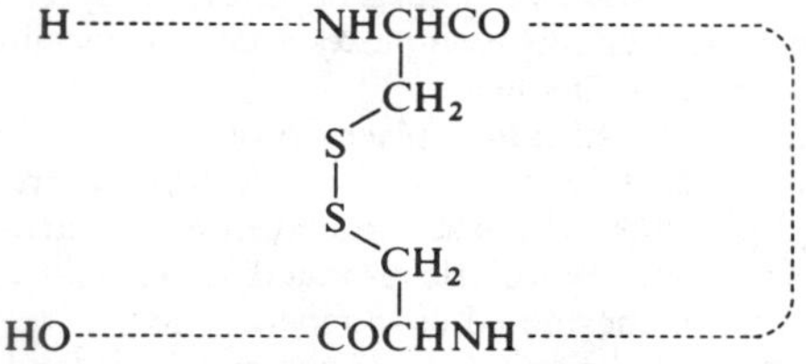

$$H \cdots\cdots NHCHCO \cdots\cdots$$
$$CH_2$$
$$S$$
$$S$$
$$CH_2$$
$$HO \cdots\cdots COCHNH \cdots\cdots$$

Although more than 200 amino acids have been found in living organisms, only 20 alpha-amino acids of the L configuration have been found serving as the building units for proteins and related peptides. These 20 amino acids occur in varying proportions in different proteins. Some proteins are fully lacking in one or more of them. Some amino acids occur only in some of the proteins. For example, hydroxyproline has been found only in collagen and elastin (proteins of animal connective tissue) and in gelatin derived from collagen.

Numerous classifications of proteins have been proposed over the years. In terms of function, there are:

a. *Structural proteins.* Proteins which support the skeletal structures, maintain the form and position of organs, impart the structural rigidity to walls of containers for biological fluids, and often form part of the external tissues. In keeping with their functions, they are insoluble in many liquids, especially body fluids, and are otherwise relatively resistant to biochemical reactions. The proteins of nails, horn, hoofs, and hair are familiar examples.

b. *Contractile proteins.* Those substances which have the property of undergoing a change in configuration which results in a change in length of shape. Thus they give the organism the power to move itself, its parts, or other objects. The proteins of muscles are prominent examples.

c. *Process-activating proteins.* As used here, the term process includes the biochemical reactions, which are catalyzed by enzumes, and in some of which the cytochromes play an intermediate role; it also includes the endocrine reactions activated by the hormones, some of which are proteins.

d. *Transport proteins.* Proteins which transport an essential substance or factor, from that part of the organism where it becomes available from a source external to the organism to the point where it is used. Examples are many of the chromoproteins, such as hemoglobin, or the blue hemocyanins (from mollusks) which contain copper instead of iron as does hemoglobin, or the chlorophyll-protein complexes of plants.

Another basis of classification is that of solubility, which has been applied to proteins from all sources, plant and animal. (a) Thus the albumins were soluble in water and coagulable by heat. They included serum albumin, egg albumin, lactalbumin (from milk), leucosin (from wheat), and legumelin (from legumes, chiefly peas). (b) The globulins are soluble in neutral salt solutions and in strong acids and alkalies. They include blood globulin (which has been separated by electrophoresis into alpha, beta, and gamma fractions, and is further discussed later in this entry), ovoglobulin (from egg yolk), edestin (from hempseed), phaseolin (from beans), arachin (from peanuts), and amandin (from almonds). (c) The glutelins, such as glutenin from wheat, are soluble in dilute acids and alkalies, and insoluble in neutral salt solutions. (d) The scleroproteins are quite insoluble, and the structural proteins (group I mentioned above) belong to this group. All these groups, and several others not included here, are simple proteins, i.e., they consist only of polypeptide chains of amino acids. The many conjugated proteins must then be classified upon the basis of their nonprotein portions: glycoproteins which contain carbohydrate groups, lipoproteins which contain lipid groups, chromoproteins which contain metal-containing complexes that are usually colored, as hemoglobin contains heme.

Still another classification places proteins into three major categories: (a) Simple proteins; (b) conjugated proteins; and (c) derived proteins. The last classification embraces all denatured proteins and hydrolytic products of protein breakdown and no longer is considered as a general class.

A discussion of the structure of proteins is beyond the scope of this volume. However, a relatively simplistic concept of a protein structure is indicated in the accompanying diagram.

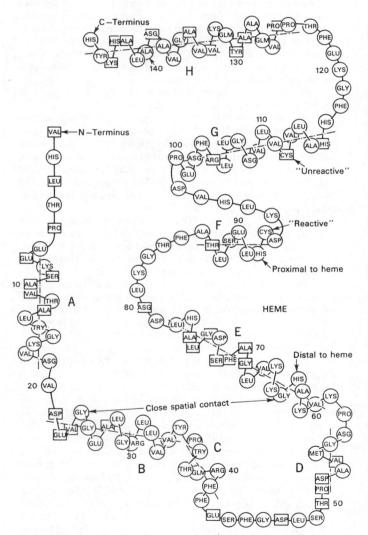

Simplified representation of the beta-chain of human hemoglobin A.

The molecular weight for the hemoglobins is on the order of 68,000. They are conjugated proteins and consist of four heme groups and the globin portion. The heme group is a porphyrin in which the metal ion coordinated is iron, which may be Fe^{3+} or Fe^{2+}, but only in the latter case (ferrohemoglobin) can the molecule bond molecular oxygen and be effective in respiration,

i.e., by forming oxyhemoglobin. The globin portion of the molecule consists of four polypeptide chains. These chains are designated as alpha, beta, gamma, etc. according to their amino acid composition. Normal adult hemoglobin consists of two alpha chains and two beta chains. The composition and conformation of the beta chain is shown in the diagram, together with the point of attachment of the heme groups. Note that they are attached to histidine groups. It has been learned that the central iron atom in heme, which is chelated to the porphyrin ring by four bonds, is attached to the polypeptide chain in adult human hemoglobin by three imidazole ligands of the globin chain, which belong to the histidines at positions 58, 87, and 89 of the alpha chain. See also **Hemoglobin.**

Besides hemoglobin other proteins of blood are of considerable importance. They are the plasma proteins, serum albumin and fibrinogen, and the globulins. Serum albumin is responsible for the major part of the osmotic pressure of human plasma. Its molecular weight is on the order of 68,000. It is a typical globular protein, having nearly one-half helical character. Although not as nearly symmetrical as hemoglobin or myoglobin, it has a symmetry indicated by its molecular dimensions of 150Å long and 38Å wide. It is the smallest, and most abundant, of the plasma proteins; for this reason, and also because of its relatively low isoelectric point, it undergoes migration rapidly in an electric field (see **Electrophoresis**). By this method it may be separated into two types of molecules, similar in composition except for the presence of a single cysteine residue in one and not the other. However, it contains cystine residues, which form seventeen disulfide bridges cross-linking the polypeptide chain, i.e., the molecule of serum albumin consists of a single polypeptide chain.

Another plasma protein to be discussed here is fibrinogen, which is the chief substance involved in the process of blood clotting. Its molecular weight is on the order of 330,000. It contains all twenty of the amino acids described in that entry as the most general in proteins, although it is relatively low in cysteine, and highest in the acidic amino acids (aspartic and glutamic acids). The process of clotting occurs in three major steps. In the first the substance prothrombin, a blood glycoprotein containing about 5% carbohydrate as glucosamine and a hexose sugar, is converted to the clotting enzyme thrombin. (The latter is unstable, and hence must be formed when needed.) The conversion process is catalyzed by the calcium ion and a group of substances known as thromboplastins. In the second step the enzyme thrombin catalyzes the transformation of fibrinogen to an activated form, called profibrin, with an altered pattern of electric charge. This change is considered to be due to the liberation of two short-chain polypeptides (one bearing 18 amino acid residues and the other 20), and a corresponding change in the character of the remainder of the fibrinogen molecule, which collectively constitute the substance profibrin. In the third step, this mixture of substances undergoes spontaneous polymerization to form the substance fibrin, which has been shown in electron microscope photographs to consist of a network of striated fibers. This polymerization occurs in stages, and in some views of the process they are divided into two steps, polymerization and clotting, the former being regarded as the formation of linear polymers, and the latter as their cross-linking by an enzymatic reaction whereby disulfide bonds are formed.

Animal organisms generally require effective assistance of intestinal flora, as in ruminants, to assimilate inorganic nitrogen into very wide variety of foreign substances, called antigens.

The life span of individual proteins in living organisms is relatively short—about 4 months for hemoglobin and but a week or two for serum albumin. The aged proteins are digested by proteolytic enzymes of tissues, such as cathepsin. A significant portion of the recovered amino acids may be available for the biosynthesis of new proteins, but another part is catabolized and the nitrogen is excreted as urea in mammals, uric acid in birds, reptiles, and insects, or ammonia in organisms of lower classes. For the maintenance of nitrogen balance and for growth, the human organism requires a daily intake of from 70 to 80 grams of proteins. The present world population thus requires 3×10^7 tons of animal proteins and 10×10^7 tons of plant proteins annually.

Living organisms can synthesize their own proteins from amino acids. In terms of the ability to carry out the de novo synthesis of amino acids, however, there are wide variations among different organisms. Plants, for example, can synthesize amino acids from nitrogen in the form of ammonium salts or nitrate and other simple compounds. The annual production of cereal and vegetable proteins so assimilated from inorganic nitrogen over the world is estimated in the mid-1970s to be about 10×10^7 tons. Of this, 4×10^7 tons are provided by wheat; 2×10^7 tons by rice; and 1×10^7 tons by corn and other sources. Lactic acid and some other microorganisms require preformed amino acids for growth, lacking some ability to synthesize. However, some microorganisms perform well with ammonium sulfate and carbohydrate as the sole sources of nitrogen, sulfur, and carbon. In some cases, they accumulate particular amino acids in a process referred to as amino acid fermentation.

Animal organisms generally require effective assistance of intestinal flora, as in ruminants, to assimilate inorganic nitrogen into body protein. This accounts for the human needs of a daily requirement of 70–80 grams of protein. However, over half of the protein-constituent amino acids can be derived from other amino acids by their own enzymic reactions. Thus, amino acids are classified as essential or nonessential. Amino acid requirements vary with the physiological state of the animal, age, and possibly with the nature of the intestinal flora.

The Food and Agricultural Organization (FAO) established the following essential amino acids in the ratios indicated:

	Percent of Crude Protein
Isoleucine	4.2
Leucine	6.2
Lysine	4.2
Methionine	2.2
Phenylalanine	2.8
Threonine	2.8
Tryptophan	1.6
Valine	5.0

A distribution of amino acids in dietary proteins can be obtained accordingly by taking both animal and plant proteins at a ratio of 1:3–4. Although plant proteins are lower cost, they are markedly deficient in some essential amino acids. Their protein efficiency is low without addition of deficient amino acids. Enrichment of human and animal diets with free amino acids, such as lysine, methionine, threonine, and tryptophan, as a substitute for animal proteins, has proved successful.

Excesses of amino acids are not harmful—with few exceptions. An imbalance of amino acids can result in a few instances. For example, a rat that feeds on eggwhite proteins with threonine or isoleucine added in high concentrations can produce an undesirable imbalance.

Several industries are based upon proteins as exemplified by the keratins of wool, feather, or horn; the fibroin of silk; the collagenous tissues as leather; protiens in milk, wheat, soybean,

egg, and numerous other natural substances. Making cheese from milk casein and flavor seasonings from plant or fish proteins are old processes. Gelatin derived from collagen has been used widely in processed foods and as an adhesive material and in photography. Glutein in cereal is a protein. Major proteins in meat are *myosin*; in egg, *ovalbumin*; in rice, *oryzenin*; in soybean, *glycinin*; and in corn, *zein*.

Protein Quality and Evaluation

Protein quality relates to the efficiency with which various food proteins are used for synthesis and maintenance of tissue protein. Food industry evaluators of protein nutritional quality must operate on several levels of awareness. In particular, manufacturers of processed foods must measure the biological value of the protein content of a variety of processed foods for several reasons: (1) to comply with various governmental regulations, (2) to satisfy nutrition labeling regulations, and (3) an accurate knowledge of protein effectiveness is required in developing new food products and in controlling sources of protein ingredients. Protein quality is also very important in the formulation of animal feedstuffs.

In a number of countries, including the United States, the stipulated measurement of protein quality is the so-called protein efficiency ratio (PER), which may be defined as the *gain in weight* divided by the *weight of protein consumed* by experimental laboratory animals. As of the early 1980s, the AOAC (Association of Official Analytical Chemists) method, fefined in 1975, is only one of the codifications of PER work since the concept was first proposed in 1919. More specifically, the PER is the ratio of the weight gained by a group of ten weanling rats fed a diet containing about 10% protein, to the weight of protein consumed over a 28-day period. No sample to be studied should contain less than 1.8% nitrogen according to the AOAC method, and the diet should supply 1.6% nitrogen. Since the samples are not analyzed for protein, but rather for nitrogen, and since protein efficiency ratio rather than nitrogen efficiency ratio is reported, it is important to be clear about whether one of the specific nitrogen factors, or the conventional 6.25 figure is used to calculate the protein in the final diet.

Much research is going into new ways of more meaningfully and reliably measuring the performance values of proteins, but discussion of this area is beyond the scope of this book. Several of the listed references are descriptive of this topic.

References

Aman, P.: "Rapesee Protein Isolates," *J. Food Sci*, **42**, 4, 1114–1116 (1978).
Anderson, R. H.: "Protein Quality Testing: Industry Needs," *Food Technol*, **32**, 12, 65–68 (1978).
Anelli, G., et al.: "Leaf Protein Concentrate—the Poly-Protein Process," *J. Food Sci*, **42**, 5, 1401 (1978).
Boyer, R. A.: "Early History of the Plant Protein Industry," *Cereal Foods World*, **21**, 7 (1976).
Brown, H. (editor): "Protein Nutrition," Charles C. Thomas, Springfield, Illinois, 1974.
Considine, D. M. (editor): "Foods and Food Production Encyclopedia," Van Nostrand Reinhold, New York, 1982.
Greengard, P.: "Phosphorylated Proteins as Physiological Effectors," *Science*, **199**, 146–152 (1978).
Hackler, L. R.: "An Overview of the AACC/ASTM Collaborative Study on Protein Quality Evaluation," *Food Technol.*, **32**, 12, 62–64 (1978).
Hurt, H. D., Forsythe, R. H., and C. H. Krieger: "Factors which Influence the Biological Evaluation of Protein Quality by the Protein Efficiency Ratio Methods," in "Protein Nutritional Quality of Foods and Feeds," Part I: Assay Methods—Biological, Biochemical, and Chemical (M. Friedman, editor), Marcel Dekker, New York, 1975.
Inglett, G. E.: "Food Proteins from Unconventional Cereals," *Food Technol.*, **31**, 5, 180–181 (1977).

Kohler, G. O., and B. E. Knuckles: "Edible Protein from Leaves," *Food Technol.*, **31**, 5, 191–195 (1977).

Kolata, G. B.: "Protein Structure: Systematic Alteration of Amino Acid Sequences," *Science*, **191**, 373 (1976).

Kolata, G. B.: "Protein Degradation: Putting the Research Together," *Science*, **198**, 596–598 (1977).

Litchfield, J. H.: "Single-Cell Proteins," *Food Technol.*, **31**, 5, 175–179 (1977).

Marx, J. L.: "Calmodulin: A Protein for All Seasons," *Science*, **208**, 274–208 (1980).

Molina, M. R., et al.: "Nonconventional Legume Grains as Protein Sources," *Food Technol.*, **31**, 5, 188–191 (1977).

Pellet, P. L.: "Protein Quality Evaluation Revisited," *Food Technol.*, **32**, 5, 60–76 (1978).

Perutz, M. F.: "Electrostatic Effects in Proteins," *Science*, **201**, 1187–1191 (1978).

Rosenfield, D.: "Protein Quality Testing," *Food Technol.*, **32**, 51 (1978).

Smith, E. B., and P. M. Pena: "Evaluating Leaf Protein Concentrate," *J. Food Sci.*, **42**, 3, 674–676 (1978).

Staff: "Official Methods of Analysis," 12th edition, Association of Official Analytical Chemists, Washington, D.C., 1975.

Staub, H. W.: "Problems in Evaluating the Protein Nutritional Quality of Complex Foods," *Food Technol.*, **32**, 12, 57–60 (1978).

Uy, R., and F. Wold: "Posttranslational Covalent Modification of Proteins," *Science*, **198**, 890–896 (1977).

Whitaker, J. R.: "Biochemical Changes Occurring During the Fermentation of High-Protein Foods," *Food Technol.*, **32**, 5, 175–180 (1978).

Woon, K. P., and S. Kim: "Protein Methylation," Wiley, New York, 1980.

Yang, H. H., et al.: "Single-Cell Protein Foods from Mesquite," *J. Food Sci.*, **43**, 2, 407–410 (1978).

PROTEIN CALORIE MALNUTRITION (PCM). Protein.

PROTEIN HYDROLYSATE. Solutions of protein hydrolyzed into its constituent amino acids.

PROTHROMBIN. Anticoagulants; Blood.

PROTIUM. The lighter isotope of hydrogen, with a single proton and electron, and constituting 98.51% of ordinary hydrogen.

PROTON. The proton is the atomic nucleus of the element hydrogen, the second most abundant element on earth. Positively charged hydrogen atoms or protons were identified by J. J. Thomson in a series of experiments initiated in 1906. Although the structure of the hydrogen atom was not correctly understood at that time, several properties of the proton were determined. The electric charge on the proton was found to be equal but opposite in sign to that of an electron. The traditionally accepted proton mass is 1836 times the electron rest mass, or 1.672×10^{-24} grams.[1]

An estimate of the size of the proton and an understanding of the structure of the hydrogen atom resulted from two major developments in atomic physics: the Rutherford scattering experiment (1911) and the Bohr model of the atom (1913). Rutherford showed that the nucleus is vanishingly small compared to the size of an atom. The radius of a proton is on the order of 10^{-13} centimeter, as compared with atomic radii of 10^{-8} centimeter. Thus, the size of a hydrogen atom is determined

[1] Particularly since the early 1970s, physicists have been seeking grand unification theory to explain all the elementary particles of matter and all the forces acting between time. Although this goal continues to be elusive, work toward that end is producing many new findings and revised concepts. In the main part of this entry, the traditional viewpoints of the proton are described. Some of the more recent postulations are given toward the end of the entry.

by the radius of the electron orbits, but the mass is essentially that of the proton.

In the Bohr model of the hydrogen atom, the proton is a massive positive point charge about which the electron moves. By placing quantum mechanical conditions upon an otherwise classical planetary motion of the electron, Bohr explained the lines observed in optical spectra as transitions between discrete quantum mechanical energy states. Except for hyperfine splitting, which is a minute decomposition of spectrum lines into a group of closely spaced lines, the proton plays a passive role in the mechanics of the hydrogen atom. It simply provides the attractive central force field for the electron.

The proton is the lightest nucleus with atomic number one. Other singly charged nuclei are the deuteron and the triton which are nearly two and three times heavier than the proton, respectively, and are the nuclei of the hydrogen isotopes deuterium (stable) and tritium (radioactive). The difference in the nuclear masses of the isotopes accounts for a part of the hyperfine structure called the *isotope shift*.

In 1924, difficulties in explaining certain hyperfine structures prompted Pauli to suggest that a nucleus possesses an intrinsic angular momentum or *spin* and an associated magnetic moment. The proton spin quantum number (I) is $\frac{1}{2}$, and the angular momentum is given by $[I(I + 1)h^2/(2\pi)^2]^{1/2}$, where h is Planck's constant. The intrinsic magnetic moment is 2.793 in units of nuclear magnetons (0.50504×10^{-23} erg/gauss), which is about a factor of 660 less than the magnetic moment of the electron.

Two types of hydrogen molecule result from the two possible couplings of the proton spins. At room temperature, hydrogen gas is made up of 75% orthohydrogen (proton spins parallel) and 25% parahydrogen (proton spins antiparallel). Several gross properties, such as specific heat, strongly depend upon the ortho or para character of the gas.

Protons in Nuclei. Protons and neutrons are regarded as *nucleons* or fundamental constituents of nuclei in most traditional theories of nuclear structure and reactions. The nuclear forces operating between them are much stronger than the electrostatic forces which govern atomic and molecular systems, but operate over very short ranges, the order of several times 10^{-13} centimeter. Of particular significance in the structure of nuclei is the apparent charge independence of the forces. That is, the nuclear force between two nucleons may be considered separately from the electrostatic forces due to electric charges the nucleons may carry. In addition to mass and charge, other properties, such as spin and parity, play important roles in determining the mechanics of nuclei.

The mass of a nucleus is the sum of the masses of the nucleons contained, plus a correction due to the total binding energy of the nucleons. This correction is an application of the Einstein mass-energy equivalence ($E = mc^2$). The atomic number or positive charge of a nucleus is given by the number of protons.

A detailed description of the motion of a proton in a nucleus is complicated by the many-body, quantum mechanical nature of the problem, but several simplified theories or models have been helpful in predicting many of the properties of nuclei. One of the best known nuclear structure models is the shell model, in which a nucleon is assumed to move in a central force field. This field represents the average interaction of the proton (or neutron) with all other nucleons. An essential additional assumption is a coupling of the orbital angular momentum of the independent nucleon with its spin.

The advantage of a theory in which a nucleon moves along its close-packed neighbors as if they were not present is due in part to quantum mechanical restrictions, in particular to the Pauli exclusion principle.

The optical model for the scattering of protons by nuclei also rests on the assumption that the interaction with many nucleons may be represented by an average potential well. An imaginary potential term is included which accounts for reactions other than elastic scattering.

Much of nuclear physics may be understood with a picture in which the proton exists as an independent particle in the nucleus. Refinements of an independent particle model to include collective effects and deformation involve a consideration of the residual interaction between nucleons and the details of the individual nucleon interaction with the average potential well.

Certain aspects of the very strong nuclear forces are understood. These forces involve π mesons in somewhat the same way that electrostatic forces responsible for atomic structure involve photons. Yukawa introduced the π meson as a field quantum for nuclear forces in 1935, and the interaction potential derived from this early theory became commonly used in nuclear physics.

Structure of the Proton. A major objective of physics is to identify elementary particles and determine their properties. The proton is important in these investigations in several connections. It is itself an elementary particle. It is used as a projectile in the production and study of other elementary particles. It has been used as a target in the study of nucleon structure.

In the late 1970s, some physicists predicted that all protons and neutrons (and the atoms made from them) will eventually decay into lighter elementary particles—electrons, neutrinos, their antiparticles, and photons. Harvard University theorist Sheldon Glashow is accredited with asking, "What could be more exciting than knowing if a diamond is forever?" Although this concept is important to understanding particle physics, it is essentially academic in practical terms because the probable decay time suggested is in excess of 10^{30} years. This concept would also have a bearing on cosmology. Experiments to determine possible decay of the proton may lead to answers on other fundamental questions. For example, the charge of the electron is quantized and exactly equal in magnitude, but opposite in sign, to that of the proton, but this is difficult to explain.

Some physicists regard studies of the lifetime of the proton as among the most important problems to be tackled in the 1980s. But, as pointed out by other authorities, the concept of proton decay also may prove to be unfounded. It should be pointed out that some 40 years ago, Stuckelberg (Univ. of Geneva) and about 30 years ago, Wigner (Princeton Univ.) questioned the ultimate stability of protons. There is considerable empirical evidence that protons are exceptionally stable. Reines (Univ. of California at Irvine), in the mid-1950s, determined experimentally that protons have a lifetime greater than 10^{22} years. More recently, scientists from Case Western Reserve University, the University of the Witwatersrand (Johannesburg, South Africa), and U. C. Irvine have concluded that the limit on the proton lifetime is probably greater than 10^{30} years.

This area of investigation has taken on added importance in recent years because a number of grand unified theories of relatively recent vintage would require that the life of the proton should be about 10^{32} years. According to the theory developed by Georgi and Glashow in 1974 and, as interpreted by Goldman and Ross (Caltech), the lifetime of the proton should be 10^{33} years at most.

As pointed out by Robinson (1979), the most straightforward and least costly way to search for proton decays is by way of large water detectors. If a proton lives 10^{33} years, a volume of water containing 10^{33} protons should, on the average, produce one decay event per year. This is equivalent to about 1000 tons of water. Two groups of researchers, one at U. C. Irvine, the

Univ. of Michigan, and Brookhaven National Laboratory and the other at the Univ. of Minnesota, Purdue Univ., and the Univ. of Wisconsin, are following this investigative approach. Light is given off by the rapidly moving decay products of the disintegrated proton or neutron (Cerenkov radiation). In a transparent medium, electrically charged particles travel faster than the speed of light in the same medium. Scientists at the European Organization for Nuclear Research (CERN) postulate that about 70% of the time the proton will decay into a positron (anti-electron) and a neutral meson; and in about 60% of the time, the neutron transforms into a positron and a negatively charged meson. Thus, the signature characteristic of the decay of a proton is a set of two light flashes traceable to the two decay particles. The experiment requires an array of some 1500 photomultiplier tubes. The research tanks must be located underground in order to shield against cosmic rays, which otherwise would overwhelm the detectors. The tanks need to be located only a few meters below the earth's surface. The remaining penetrating radiation would comprise cosmic-ray muons and neutrinos which result from interaction of cosmic rays with nuclei of the earth's atmosphere. Fortunately, the muon background can be estimated and taken into consideration. More details along these lines can be found in the Robinson (1979) reference.

Another group from the Univ. of Pennsylvania, since about 1978, has been conducting an experiment in the Homestake gold mine nearly 1500 meters beneath Lead, South Dakota. Instead of one large tank, modules of 200 tons of water each are used, but plans are underway to go to a module of 800 tons.

If, as pointed out by Nanopoulos of CERN, proton decay should be confirmed and support the Georgi-Glashow model, findings could lead to: (1) an explanation of why neutrinos have no mass; (2) why the electron's charge is quantized; (3) why the charges of quarks are either $+\frac{2}{3}$ or $-\frac{1}{3}$; (4) why quarks and leptons (the class of elementary particles not made of quarks) seem to come in families of four particles consisting of quarks with charges $+\frac{2}{3}$ and $-\frac{1}{3}$ and leptons with charges 0 and -1; (5) explain a relation between the masses of some (but not all) of the quarks and leptons; and (6) explain why there may be only six varieties of quarks (i.e., three families of particles).

As pointed out by Jacob and Landshoff (1980), there is something small and hard inside the proton. These objects, which may be the particles called *quarks*, are seen most clearly when a violent collision of particles gives rise to a jet of debris.

See also **Particles (Subatomic); and Quarks.**

References

Bromley, D. A.: "Physics," *Science*, **209**, 110–121 (1980).

Cowen, R. C.: "Particle Physics," *Technology Review (MIT)*, **82**, 3, 10–11 (1980).

Goldhaber, M., Langacker, P., and R. Slansky: "Is the Proton Stable?" *Science*, **210**, 851–860 (1980).

Jacob, M., and P. Landshoff: "The Inner Structure of the Proton," *Sci. Amer.*, **242**, 3, 66–75 (1980).

Krisch, A. D.: "The Spin of the Proton," *Sci. Amer.*, **240**, 5, 68–80 (1979).

Pilkuhn, H. M.: "Relativistic Particle Physics," Springer-Verlag, New York (1979).

Reines, F.: "The Early Days of Experimental Neutrino Physics," *Science*, **203**, 11–16, (1979).

Weisskopf, V. F.: "Contemporary Frontiers in Physics," *Science*, **203**, 240–244 (1979).

PROUSTITE. This ruby-silver mineral crystallizes in the hexagonal system; its name a product of its scarlet-to-vermilion color when first mined. It is a silver, arsenic sulfide, Ag_3AsS, of adamantine luster. Hardness of 2–2.5; specific gravity of 5.55–5.64. Usual crystal habit is prismatic to rhombohedral; more commonly occurs massive. Conchoidal to uneven fracture; transparent to translucent; color, scarlet to vermilion red. Light sensitive; must be kept in dark environment to maintain its primary character. A product of low-temperature formation in most silver deposits. Notable world occurrences include Czechoslovakia, Saxony, Chile and Mexico. Found in minor quantities in the United States; the most exceptional occurrence at the Poorman Mine, Silver City District, Idaho where crystalline mass of some 500 pounds (227 kilograms) recovered in 1865. It was named for the famous French chemist, Louis Joseph Proust.

PSEUDOPLASTIC SUBSTANCES. Rheology.

PSILOMELANE.
Psilomelane is a massive black mineral, essentially a basic oxide of barium with divalent and quadrivalent manganese, corresponding to the formula $BaMn^{2+}Mn^{4+}O_{16}(OH)_4$. It crystallizes in the monoclinic system, but is found only in massive, botryoidal or reniform to earthy habits; hardness, 5–6; less in earthy varieties; specific gravity, 6.45; color, black to gray; opaque; submetallic to dull luster. It is a product of secondary weathering of manganese carbonates and silicates. Of widespread occurrence, usually associated with pyrolusite. Major world occurrences include, Michigan in the United States, Scotland, Sweden, France, Germany, and India. It is a major source of manganese. The word *psilomelane* is derived from the Greek words meaning *smooth* and *black*, in reference to the smooth black surfaces so often exhibited.

PSI PARTICLE.
Discovery of this subatomic particle in 1974 was announced independently by Ting (Brookhave National Laboratory) who named it the *J particle* and by B. D. Richter (Stanford) who named it the *psi particle*. The discovery of this particle resolved a number of important problems in particle physicists. Intensive research on the psi particle was carried out by Richter during 1975 and 1976 by the Stanford group and is reported firsthand by Richter (*Science*, **196**, 1286–1297, 1977). As pointed out by Richter, the four-quark theoretical model became much more compelling with the discovery of the psi particles. The long life of the psi is explained by the fact that the decay of the psi into ordinary hadrons requires the conversion of both c and $\bar{c}$ into other quarks and antiquarks. See also **Particles (Subatomic)**.

PSYCHOACTIVE DRUGS. Enkephalins and Endorphins.

PULP (Wood) PRODUCTION AND PROCESSING.
Pulps can be defined as fibrous products derived from cellulosic fiber-containing materials and used in the production of hardboard, fiberboard, paperboard, paper, and molded-pulp products. With suitable chemical modification, pulps can be used in the manufacture of rayon, cellulose acetate, and other familiar products. Pulps can be produced from any material containing cellulosic fiber; but in North America and several other regions of the world, wood is the predominant source of pulp. This description is confined to the production and processing of wood pulp.

Wood is a cellular substance chemically composed of roughly 70% holocellulose, 25% lignin, and 5% water and ethyl alcohol-benzene soluble extractives. These percentages are based on oven-dry wood.

The chemical composition and physical character of wood vary from species to species, within species grown in different geographical locations, and within a given tree, depending upon the location of the fiber cell in the tree. Both lignin (noncarbohydrate) and holocellulose (carbohydrate) are polymeric substances. Holocellulose is composed of approximately 70% alpha cellulose and 30% hemicellulose, the long-chained alpha cellulose being characterized by nonsolubility in alkali; whereas the shorter-chained hemicellulose is alkali-soluble, the degree depending upon the alkali concentration. Lignin concentration in wood substance is greatest in the middle lamella (the zone around each individual fiber cell), decreasing in concentration through the cross section of the fiber, and reaching a concentration of about 12% at the inner layer of the fiber adjacent to the fiber cavity, or lumen. It is the middle-lamella material (lignin and hemicellulose) that cements the fiber cells together, thus giving rigidity to the fibrous wood structure.

The objective of wood pulping is to separate the cellulose fibers one from another in a manner that preserves the inherent fiber strength while removing as much of the lignin, extractives, and hemicellulose materials as required by pulp end-use considerations. Wood pulp to be used for the manufacture of hardboard, for example, requires only the removal of water-soluble wood sugars and sufficient fiberization, i.e., separation of fibers, to permit effective felting of the fibers in a sheet-forming operation. In a subsequent operation in which the felted fiber sheet is subjected to high pressure and heat, the lignin in the fiber mass softens and flows, ultimately acting as a bonding agent cementing the fibers together into a coherent hardboard. At the other extreme, wood pulp to be used for rayon manufacture must be of high alpha-cellulose content (~88.93%), have extremely low amounts of noncarbohydrate material, and be well fiberized to permit uniform reactions during chemical processing.

Pulping Processes

Wood is converted to pulp by mechanical and chemical actions which constitute the pulping process. Their selection depends upon the type of wood supply available and the pulp qualities desired. Pulps can be characterized on the basis of the unbleached pulp yields achieved by the pulping process used, i.e., the yield of oven-dry (OD) pulp obtained from oven-dry debarked wood.

Five major types, or classes, of pulps, related to pulp yield ranges normally considered to define each class of pulp, are shown in Fig. 1. Pulp yield is a direct indication of degree of chemical action (delignification and chemical attack on carbohydrate and other nonligneous material). Also shown in this figure are the degrees of defibration effected by chemical and mechanical action utilized to produce the pulp, although this representation is not strictly correct. For example, in producing a full chemical pulp, wood chips are subjected to chemical action (digestion or cooking) in a pressure vessel; when digestion is completed, the cooked and softened chips retain the same physical form as the raw chips originally charged to the digester, but separate into essentially discrete fibers as a result of mechanical action occurring upon sudden release of the chips from the pressure vessel into a receiving tank, which ordinarily is at atmospheric pressure.

At the other extreme, no chemicals are used in the production of mechanical pulp, and defibration is effected by subjecting wood to a mechanical grinding or attrition action. In this instance, the defibration is aided by some small degree of chemical

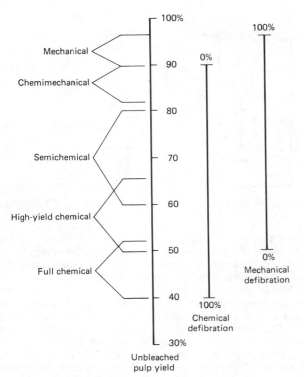

Fig. 1.   Wood pulps characterized on basis of yield.

change and solubilization of wood substance occasioned by heat generated by the grinding operation.

The pulps listed in Fig. 1 are characterized on an unbleached basis as produced by processes conventionally called *pulping processes*. In many instances, these pulps must be further treated chemically to remove residual lignin, hemicellulose, and color bodies before they can be considered suitable for use in specific applications. This further treatment is called *bleaching*, and the bleaching operation is actually an extension of the pulping process.

Customarily, pulping processes and bleaching processes are considered separately, although the choice of bleaching process is highly dependent upon the pulping process used. With this distinction between pulping and bleaching in mind, it will be understood that the pulping processes that are briefly described here pertain only to the production of *unbleached* pulps.

The soda, kraft, and sulfite pulping processes are used to prepare full chemical pulps. The soda process, which uses sodium hydroxide as the cooking chemical for delignification purposes, has largely been superseded by the kraft process, which is characterized by its use of sodium hydroxide and sodium sulfide as active delignification agents in the chip-cooking phase of the process.

Chip-digestion parameters are digester pressure and temperature, digestion time to and at maximum temperature, amount of active alkali used per unit weight of OD wood (percent active alkali), percentage ratio of sulfide to active alkali (percent sulfidity), and weight ratio of cooking liquor (including chip moisture) to OD wood weight. No two kraft pulp mills use the same set of parameter values. Such values must be frequently adjusted, even within a given mill because of variations in incoming wood and pulp-quality requirements.

Kraft processes are applicable to nearly all species of wood,

and effective means of recovering spent cooking chemicals for recycle in the process have been developed. Some sodium and sulfur losses do occur and are replenished in the cooking-liquor system by adding sodium sulfate at the recovery boiler, where it is converted to sodium carbonate and sulfide. In order to maintain a proper sulfur-to-sodium ratio in the recovered chemicals, other chemicals, such as sodium carbonate, sodium sulfite, and sulfur, are sometimes used for chemical makeup. See Fig. 2.

In contrast to the highly alkaline (pH 11–13) kraft processes, sulfite pulping processes are acidic in nature and are of two general types: (1) the *acid sulfate processes* utilize calcium, sodium, magnesium, or ammonium bisulfite in combination with free or excess sulfur dioxide as cooking chemicals (pH 1.7–2.3). (2) the *bisulfite processes* use sodium, magnesium, or ammonium bisulfite (pH 3.5–5.5) for chip digestion.

Several sulfite processes are multistage and use various combinations of acid sulfite and bisulfite cooking stages and can even use the alkaline kraft cook as one of the multistages. Although spent calcium acid sulfite cooking liquor can be incinerated, there is no recovery of calcium or sulfur. The sodium and magnesium bases can be recovered with or without sulfur recovery, and spent ammonium base liquor can be burned with recovery of sulfur as an option.

High-yield chemical pulps can be produced by the soda, kraft, or sulfite processes in which chemical use and digestion time and/or temperature are suitably reduced to effect a milder cook than used for full chemical pulps. Mechanical defibrators are used to complete the separation of wood fibers not accomplished by the chemical action.

Semichemical pulps are usually prepared by the neutral sulfite semichemical (NSSC) process, although modifications of the full chemical processes can be used. Active pulping chemicals are (in the sodium-base NSSC process) sodium sulfite buffered with sodium bicarbonate (pH 7.0–9.0) and (in the ammonium-base NSSC process) ammonium sulfite with ammonium hydroxide used as a buffer. Defiberization is usually accomplished by attrition mills of the disk type.

Mechanical pulps are produced by two basic processes: (1) *stone groundwood pulp* (SGW) is produced by the defibration action of natural or artificial grindstones rotated at moderate speeds (200–300 rpm) against bark-free bolts of roundwood axially aligned across the peripheral face of the stone in the presence of water. By air-pressurization of the grinder, a *pressurized groundwood pulp* (PGW) of improved quality can be produced. (2) *Refiner mechanical pulp* (RMP) is produced by the attrition action upon raw wood chips of an open (atmospheric) discharge disk refiner. By preheating the chips in a pressurized vessel via direct steaming at temperatures of 120°C or higher and fiberizing the heated chips in either a pressurized or atmospheric disk refiner, *thermomechanical pulp* (TMP) is produced.

Chemimechanical pulps (CMP) are produced by processes in which groundwood or chips are treated with weak solutions of pulping chemicals, such as sulfur dioxide, sodium sulfite, sodium bisulfite or sodium hydrosulfite, followed by mechanical defibration. By presteaming chemically treated chips before attrition, *chemithermomechanical pulps* (CTMP) are produced. The mild chemical action, augmented by heat, softens wood lignin and promotes easier defibering with less fiber damage than achieved by the purely mechanical processes.

Batch digesters are vertical, stationary, cylindrical pressure vessels into which chips and cooking liquor are charged under atmospheric conditions. Heating of the digester, after sealing of the feed ports, is effected by direct steam addition or by continual withdrawal of liquor through screened ports and rein-

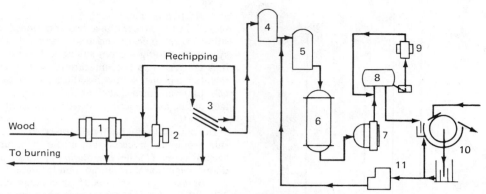

Fig. 2. Flow diagram of kraft pulp mill: (1) debarking, (2) chipping, (3) screening, (4) steaming, (5) impregnating, (6) digesting, (7) fibrilizing, (8) screening, (9) fiberizing, (10) washing, (11) chemical recovery.

troduction of the liquor, after passage through external heat exchangers, onto the top (and sometimes into the bottom) of the chip mass within the vessel. Often, a combination of the direct and indirect heating methods is used. Modern batch digesters are typically 4000–6000 cubic feet (113–170 cubic meters) in volume, with height-to-diameter ratios of 3.5–5.5, and pre-cook pulp capacities of 10–20 tons (9–18 metric tons).

Continuous digesters have been developed as part of the highly successful effort to convert pulp and papermaking from a series of strictly batch operations into an integrated series of continuous operations. A number of successful types of continuous digesters range from horizontal and inclined tube (single or multiple) designs, in which the chip charge is moved through the digester by mechanical screw or bucket conveyors, to vertical digesters, in which chip movement is effected by gravity. See Fig. 3.

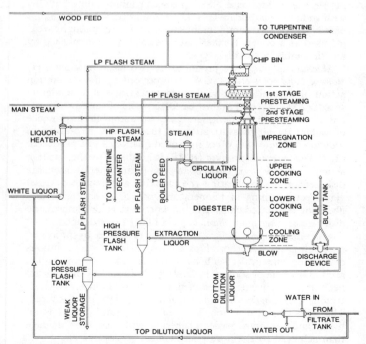

Fig. 3. Continuous digester system. (*Ingersoll-Rand Co., IMPCO Division.*)

Screened chips are conveyed from storage to a chip-supply bin in the digester house. If hardwood and softwood chips are to be cooked together, they are blended by weight proportion during this transfer. Chips feed by gravity from the bin to a chip meter, either a twin-screw or a multipocket rotary feeder, the speed of which determines chip and cooking-liquor flow rate to the digester and pulp discharge rate.

Metered chips drop to a low-pressure rotary feeder valve, through which the chips are introduced into a steaming vessel maintained at a pressure of about 15 pounds/square inch gage (2.2 kPa), where the chips are preheated, air is expelled from the chip interior, and chip moisture content is leveled in preparation for impregnation with cooking liquor. Since cooked chips are continually being removed from the bottom of the digester, chips pass downward in the digester, replacing those discharged; time of passage through the cooking zone is normally 90–120 minutes. As cooked chips reach the bottom zone of the digester, they are plowed to a central well in the bottom of the digester while being mixed with filtrate from the pulp washer for cooling. Mechanical forces exerted in the transfer of chips from the digester to the blow tank effect fiberization of the chips, the degree of which depends upon cooking conditions. The fibrous material collected in the blow tank is called *pulp* and separate blow tanks are normally used to collect the several types of pulp produced alternately in the digester.

Pulp Screening and Washing. Pulp (brown stock) discharged to the blow tank is in admixture with black liquor, a water solution of spent and residual cooking chemicals and dissolved wood substance, and is at a consistency of 10–18%. The term *consistency* has a meaning peculiar to the pulp and paper industry and refers to the percentage ratio of washed, dry (either oven- or air-dried) fiber to total fiber slurry weight. The fiber bundles left in the pulp after blowing must be fiberized, i.e., separated into discrete fibers, and the black liquor removed in order for the pulp to be refined (a conditioning of individual fibers) and formed into a fiber sheet on the linerboard machines.

Pulp is diluted with filtrate from the pulp washer to a consistency of about 4.5% in the lower portion of the blow tank and fed to fibrilizers, which serve the purposes of metal trapping, fiber-bundle breaking, rough screening, and pumping.

Removal of the black liquor from screened brown stock is usually accomplished on rotary-drum vacuum filters, arranged for multistage countercurrent washing.

Refining is accomplished by disk mills, equipped with different plate designs or patterns than those used for defibration. During the refining operation, cellulose fibrils, which wind spirally around the fiber at various positions in its cell wall, are loosened, the cell wall swells due to water absorption, and the fiber is conditioned for sheet formation and interfiber binding in the paper- or board-making operation.

Chemical Recovery. Economic and environmental control factors dictate that chemical and heat values of black liquor solids be carefully conserved, and the recovery system of the modern kraft pulp mill has developed into a highly sophisticated system with still more improvement in efficiency continually being sought. See also **Papermaking and Finishing.**

References

MacDonald, R. G., and J. N. Franklin, Editors: "Pulp and Paper Manufacture," 2nd Edition, 3 Volumes, McGraw-Hill, New York (1969).
Rydholm, S. A.: "Pulping Processes," Wiley, New York (1965).
Rydholm, S. A.: "Continuous Pulping Processes," Tech. Assn. of Pulp and Paper Ind., *Publ. STAP 7, Spec. Tech. Ass.* (1970).
NOTE: See also reference at the end of the entry on **Papermaking and Finishing.**

—Henry F. Szepan and Dunbar G. Terry, Ingersoll-Rand Co., Impco Division, Nashua, New Hampshire.

PUMICE. Rhyolitic lavas with a high gas content, when suddenly discharged by volcanic action, congeal in the form of a highly vesicular natural glass called *pumice*. When ground, mixed with an appropriate binder and pressed into cakes it is the "pumice stone" of commerce, which is used as a light abrasive.

PURINE BASES.   Nucleic Acids and Nucleoporteins.

PURINES. Derivatives of the dicyclodiureide of malonic and oxalic acids. The dicyclodiureide is uric acid and the parent compound is purine:

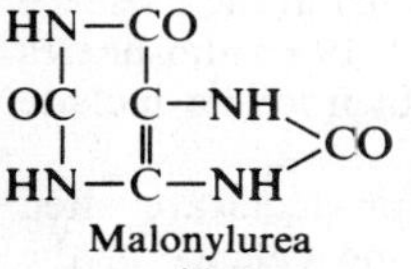

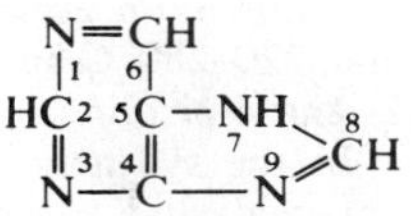

so that uric acid is 2,6,8-trioxypurine or the keto form of 2,6,8-trihydroxypurine. Caffeine, theobromine, and theophylline are other important purine compounds.

Uric acid ($C_5H_4O_3N_4$—formula above) is a white solid, insoluble in cold water, alcohol or ether, sparingly soluble in hot water. Uric acid is a weak dibasic acid thus forming two series of salts, most of which are very slightly soluble in water (lithium urate soluble).

Uric acid is found in the urine, blood, and muscle juices of carnivorous animals (herbivorous animals secrete hippuric acid), in the excrement of birds, serpents and insects, and is an oxidation product of the complex nitrogenous compounds of the animal organism.

Purine Metabolism. Purines are major building blocks for the nucleic acids, DNA and RNA. Adenine, also a purine, plays several important roles—as a cofactor component in energy metabolisms and in enzymatic reactions in which the coenzymes NAD^+ and $NADP^+$ are involved. The end-product of purine metabolism is uric acid. It has been well established for many years that biochemical shortcomings in purine metabolism are the principal cause of gout. An average adult male will excrete between 200 and 600 milligrams of uric acid in the urine per day, representing about two-thirds of the total uric acid production in the body. Less than 10–20% of uric acid can be accounted for as direct dietary intake. When insufficient uric acid is excreted, a condition known as *hyperuricemia* will result. When the concentration of uric acid nears the saturation threshold, precipitation in tissues commences. Increased amounts of uric acid may be produced as the result of faulty enzyme activity or other abnormal factors which may occur in the purine metabolism system. Hyperuricemia may be evidenced by the development of an acute, extremely painful, swollen, inflamed joint, frequently at the base of the great toe (podagra). This condition is most commonly encountered in obese, overindulgent people. Usually this condition persists for several days to several weeks without treatment. The condition may recur periodically. The condition responds well to the administration of colchicine (see also **Alkaloids**). Treatment also includes removal of carbohydrates from the diet for a few days, as well as deprivation of alcohol and certain medications, such as thiazide diuretics.

Abnormalities in purine metabolism also may create a purine nucleoside phosphorylase (PNP) deficiency, which ultimately may surface as *hypoplastic anemia.*

Purine, uric acid, and other associated compounds play a role in organic synthesis of industrial products.

PYRARGYRITE. An antimony-bearing silver mineral corresponding to the formula Ag_3SbS_3. It crystallizes in the hexagonal system, commonly in rhombic prismatic forms. It displays a rhombohedral cleavage; fracture, conchoidal to uneven; brittle; hardness, 2.5; specific gravity, 5.24; luster, adamantine to submetallic; color, deep red, but being light sensitive alters readily to black. In thin fragments deep red by transmitted light, otherwise practically opaque; streak, purplish red. Pyrargyite occurs with proustite, other silver minerals, galena, and sphalerite. It is found in the Harz Mountains, in Czechoslovakia, Bolivia, Chile, Mexico, and in the United States in Colorado, Idaho, and Nevada. In Canada it is found in the Cobalt region of the Province of Ontario. It derives its name from the Greek words meaning *fire* and *silver.*

PYRIDINE AND DERIVATIVES. Pyridine is a slightly yellow or colorless liquid; hygroscopic, bp 115.5°C, fp −41.7°C, unpleasant odor, burning taste, slightly alkaline in reaction, soluble in water, alcohol, ether, benzene, fatty oils. Specific gravity 0.978. Flash point (closed cup) 20°C; autoignition temperature 482°C. Pyridine, a tertiary amine, is a somewhat stronger base than aniline and readily forms quaternary ammonium salts.

Pyridine and derivatives of pyridine occur widely in nature as components of alkaloids, vitamins, and coenzymes. These compounds are of continuing interest to theoretical physical, organic, and biochemistry. They have many uses, e.g., herbicides and pesticides, pharmaceuticals, feed supplements, solvents and reagents, and chemicals for the polymer and textile industries.

Structure and Nomenclature. The pyridine group consists of a six-membered, heterocyclic, aromatic compound with one nitrogen atom in the ring. The parent compound of this group is pyridine (I shown below) with ring positions numbered as shown. Alternate denotations of the 1, 2, and 4 positions in the ring are alpha, beta, and gamma, respectively.

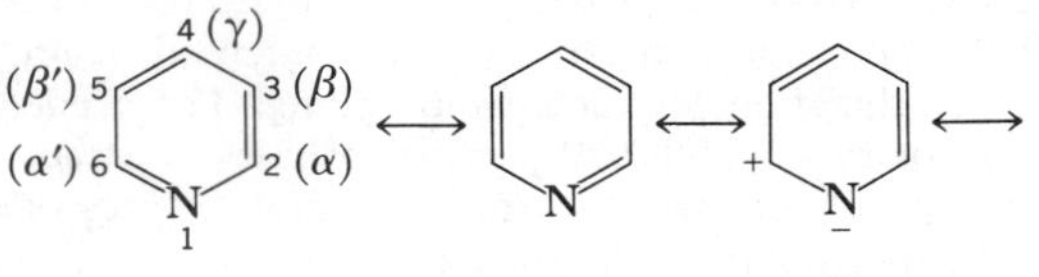

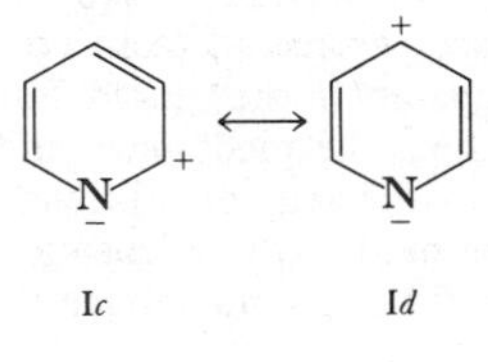

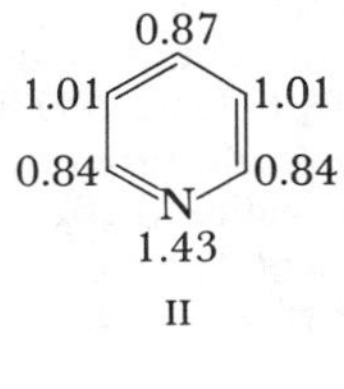

The behavior of pyridine in substitution reactions can be understood on the basis of its resonance structure (I*d*) and on the basis of the electron-density distribution at the various ring positions as derived from molecular-orbital-theoretical calculations. An example of the published pi-electron density distribution is shown in II (above). The resonance energy of pyridine is 35 kcal/mole (versus 39 kcal/mole for benzene).

Electrophilic substitution occurs at the 3 and 5 positions, but usually requires drastic conditions because the species actually being attacked is a pyridinium ion. For example, nitration of pyridine with KNO_3 and concentrated H_2SO_4 at 300°C gives a 15% yield of 3-nitropyridine. Electrophilic substitution in the pyridine ring is facilitated by the presence of electron-donating substituents.

Nucleophilic substitution occurs in the 2, 4, and 6 positions of pyridine under relatively mild conditions. As an example, amination of pyridine with sodium amide in *N,N*-dimethylaniline at 180°C gives 2-aminopyridine in good yield.

Homolytic (free-radical) substitution may occur in any of the 2–6 positions of pyridine. Thus, the reaction of pyridine with benzenediazonium salts gives a mixture of 2-, 3-, and 4-phenylpyridine.

Many pyridine derivatives which are difficult to make directly from pyridine are readily accessible starting from pyridine-N-oxide (below), made by oxidation of pyridine with hydrogen peroxide in acetic acid. As but one example, the nitration of pyridine-*N*-oxide gives 4-nitropyridine-*N*-oxide in high yield.

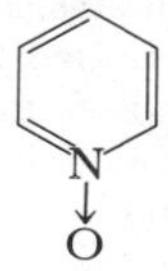

Pyridine *N*-oxide

Trivial names for the methylpyridines are the *picolines*; the dimethylpyridines are the *lutidines*; and the trimethylpyridines (and in older literature the ethyldimethylpyridines) are the *collidines*. The refractive indices for these alkyl pyridines and for pyridine itself fall in the range: $n_D^{20} \sim 1.50$–1.51.

Production of Pyridine and Homologues

Coke Manufacture Byproducts. In United States practice, coking of coal is done almost exclusively by the high-temperature (900–1200°C) process. For many years, the major source of the pyridines was the chemical-recovery coke oven. The volatiles produced in the coke oven are only partially condensed. The noncondensed gases are passed through a scrubber (the ammonia saturator) containing sulfuric acid. After removal of crystals (ammonium sulfate), a solution of ammonium sulfate and pyridinium sulfates is obtained and treated with ammonia to liberate the contained pyridine bases (~70% is pyridine itself). See also **Coal Tar and Derivatives.** The balance of the pyridine bases is extracted from the crude coal tar, i.e., the condensed, main portion of the volatilization products from coking. The crude tar contains approximately 0.1–0.2% pyridine bases. Further separation of the pyridines involves a rather complex series of extractions, distillations and crystallizations.

Synthetic Methods of Manufacture. Due to rising demand, production of the pyridine bases by large-scale synthesis passed the volume of tar bases extracted from coal tar in the 1960s. By the early 1970s, capacity in the United States for the synthetic manufacture of pyridine, the picoolines, and 2-methyl-5-ethyl-pryidine (MEP was in the tens of millions of pounds). All of these products can be made by condensation reactions of aldehydes and ammonia. MEP is no longer made in the United

States. A new synthetic method for preparing 2-methylpyridine has been commercialized by a firm[1] in the Netherlands. This process involves the acid/base-catalyzed condensation of acetone with acrylonitrile to make 4-oxohexanonitrile. Then the nitrile is converted to 2-methylpyridine by catalytic/dehydrogenation:

$$CH_3COCH_3 + CH_2{:}CHCN \rightarrow$$
acetone   acrylonitrile

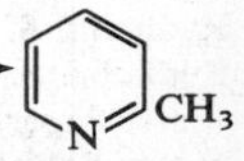

4-oxo-hexanonitrile   2-methylpyridine

When acetaldehyde and ammonia in a 3:1 mole ratio are fed over dehydration-dehydrogenation catalysts (such as PbO or CuO on alumina; ThO_2, ZnO, or CdO on silica-alumina; or CdF_2 on silica-magnesia) at 400–500°C and atmospheric pressure, an equimolar mixture of 2- and 4-picolines can be obtained in 40–60% yields. When a mixture of acetaldehyde, formaldehyde, and ammonia in about 2:1:1 mole ratio is passed over such catalysts, pyridine and 3-picoline are produced; their ratio is usually 1:0.7, but the amounts of pyridine can be increased by changes in the feed.

A major producer[2] of pyridine bases has reported a pyridine–3-picoline synthesis by reacting acetylene, methanol, and ammonia in a fluidized-bed reactor at 400°C, using catalysts consisting of Zn, Cu, or Cd salts on alumina-silica-magnesia supports.

The lowest-cost synthetic pyridine base, 2-methyl-5-ethylpyridine, is made in a liquid-phase process from paraldehyde (derived from acetaldehyde) and aqueous ammonia in the presence of ammonium acetate at approximately 102–190 atmospheres and 220–280°C in 70–80% yield. Minor byproducts include 2- and 4-picoline.

In the synthetic processes, mixtures of products are often obtained, and variation in the supply/demand balance of the alkyl pyridine isomers has led to much research on processes which may alleviate such imbalances, including development of the catalytic hydrodealkylation of alkyl pyridines as well as the alkylation of pyridine.

Major Uses of Pyridine Derivatives

Herbicides. A major outlet for pyridine is in the manufacture of the desiccant herbicides and aquatic weed killers, such as 1,1'-ethylene-2,2'dipyridilium dibromide (below), known as *Diquat*®; and 1,1'-dimethyl-4-4'-dipyridilium chloride (below), known as *Paraquat*®, or dibromide or dimethylsulfate.

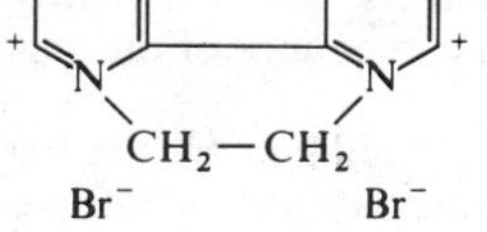

1,1'-ethylene-2,2'-dipyridilium   1,1'-dimethyl-4,4'-dipyridilium
dibromide   dichloride

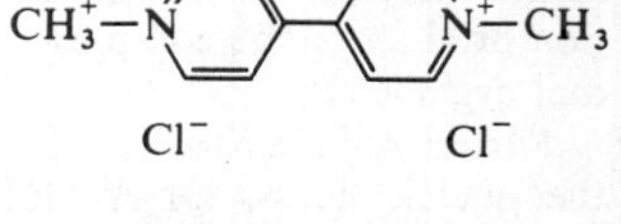

The compound, 2-picoline is the source of 2-chloro-6-trichloromethylpyridine (below), known as *N-Serve*®, which is useful as a fertilizer additive for reduction of nitrogen losses in the soil due to bacterial oxidation. 2-Picoline also is the starting material for the production of 4-amino-3,5,6-trichloropicolinic acid, a powerful broad spectrum herbicide for broad-leaved plants, known as *Tordon*®.

---

[1] Stamicarbon B.V.

[2] Reilly Tar and Chemical Corp.

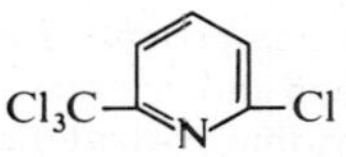

Cl_3C—pyridine—Cl

2-chloro-6-
trichloromethylpyridine

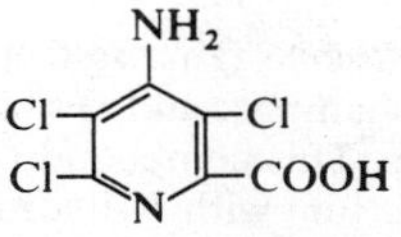

4-amino-3,5,6-
trichloropicolinic acid

Pesticides. The compound, 2-picoline is a component of 1-[(4′amino-2′-*n*-propyl-5′-pyrimidinyl)methyl]-2-picolinium chloride hydrochloride (below), known as *Amprolium*®, a broad-spectrum coccidiostat. A newer coccidiostat is 3,5-dichloro-4-hydroxy-2,6-lutidine (below), known as *Clopidol*®.

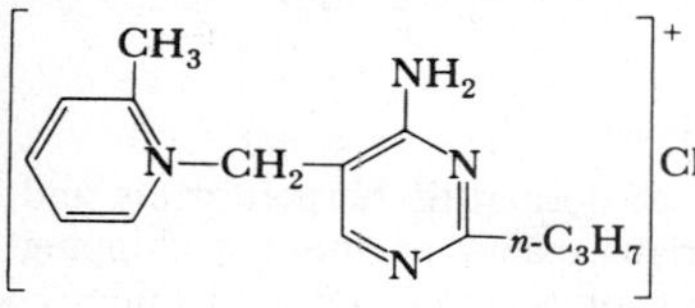

1-[(4′amino-2′-*n*-propyl-5′-pyrimidinyl)methyl]-
2-picolinium chloride hydrochloride

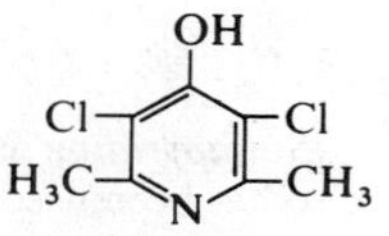

3,5-dichloro-4-
hydroxy-2,6-lutidine

The acaricide *O,O*-diethyl-*O*-(3,5,6-trichloro-2-pyridyl)thiophosphate (below) and known as *Dursban*® is used to control ectoparasites. Di(*n*-propyl)isocinchomerate (below) and known as *MGK Repellent 326*® is used in fly repellents and is made by oxidation of 2-methyl-5-ethylpyridine and esterification of the isocinhomeronic acid obtained. Nicotine (sulfate) (*Black Leaf 40*®) is used as an agricultural insecticide, as an external parasiticide, and as an anthelminthic, and is obtained by extraction of tobacco wastes, not by synthesis.

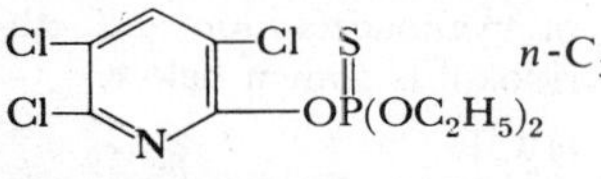

O,O-diethyl-*O*-(3,5,6-trichloro-2-
pyridyl) thiophosphate

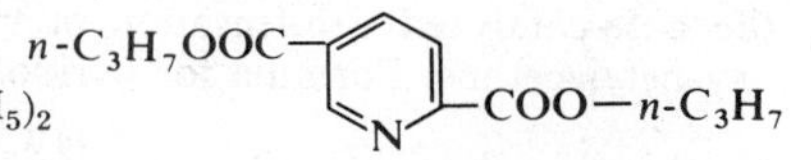

Di (n-propyl) isocinchomerate

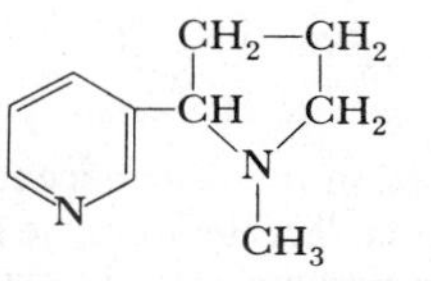

Nicotine

Dimethyl-3,5,6-trichloro-2-pyridyl phosphate (below), known as *Fospirate*® or *Dowco*® 217, is an insecticide useful in antiflea collars for dogs and catas.

The compounds 4-aminopyridine (below), known as *Avitrol*® 100, and 4-nitropyridine-*N*-oxide (below), known as *Avitrol*® 200, are useful as bird repellants.

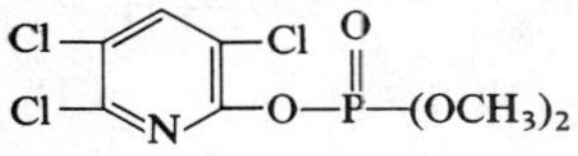

Dimethyl-3,5,6-trichloro-2-pyridyl
phosphate

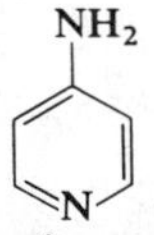

4-aminopyridine

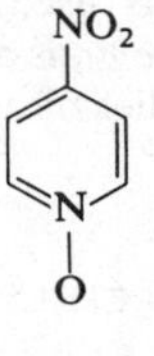

4-nitropyridine-*N*-oxide

Pharmaceuticals. There are several pyridine and piperidine derivatives with varying, often multiple, drug actions. A few examples are given. A number of *antihistamines* contain the pyridine moiety in their structure, as exemplified by chlorpheniramine maleate (2-[*p*-chloro-α-(2-dimethylaminoethyl)-benzyl]pyridine acid maleate), shown below; doxylamine succinate (2-[α-(2-dimethylamino)ethoxy-α-methylbenzyl]pyridine acid succinate), shown below; and pyrilamine maleate (2-(2-dimethylaminoethyl-2-*p*-methoxybenzyl)aminopyridine acid maleate), shown below. These products are synthesized, e.g., from the appropriate benzylpyridines or aminopyridines.

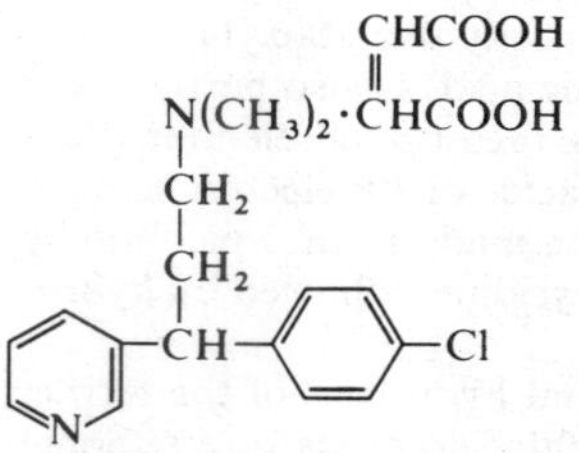

Chloropheniramine maleate

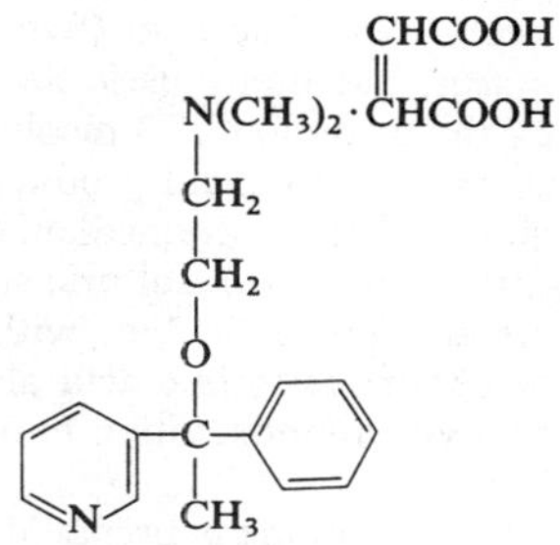

Doxylamine succinate

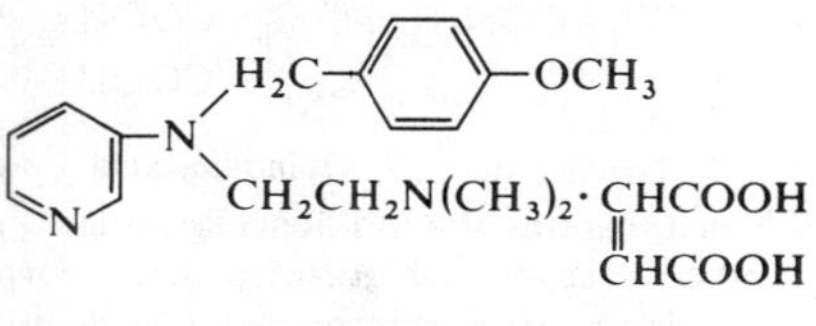

Pyrilamine maleate

Cetylpyridinium chloride (below) is used as a germicide and antiseptic, e.g., in mouthwashes; it is made by quaternization of pyridine with cetyl chloride.

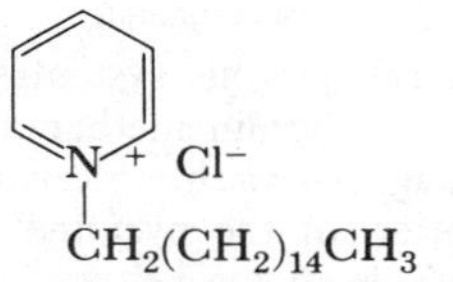

Cetylpyridinium chloride

Isonicotinehydrazide, also known as isoniazid, is an important antitubercular drug made by oxidation of 4-alkylpyridine (or 2,4-lutidine) or by hydrolysis of 4-cyanopyridine to isonicotinic acid (pyridine-4-carboxylic acid) and reaction of an ester or the acid chloride of the latter with hydrazine. The formula is shown below.

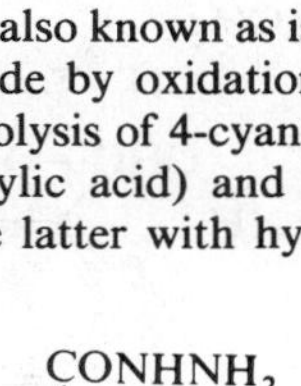

Isonicotinehydrazide

Meperidine hydrochloride (1-methyl-4-carbethoxy-4-phenyl-piperidine), also known as *Demerol*®, is an important narcotic and analgesic. It is not made from piperidine, but rather by ring-closure reactions of appropriate precursors. The formula is shown below.

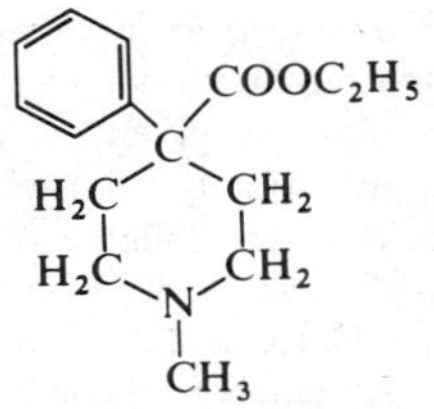

Meperidine hydrochloride

Nicotinic acid and nicotinamide, members of the vitamin B group and used as additives for flour and bread enrichment, among other applications, are made to the extent of 24 million pounds (nearly 11 million kilograms) per year throughout the world. Nicotinic acid (pyridine-3-carboxylic acid), also called *niacin*, has many uses. See **Vitamins.** Nicotinic acid is made by the oxidation of 3-picoline or 2-methyl-5-ethylpyridine (the isocinchomeric acid produced is partially decarboxylated). Or, quinoline (the intermediate quinolinic acid) is also partially decarboxylated with sulfuric acid in the presence of selenium dioxide at about 300°C, or with nitric acid, or by electrochemical oxidation. Nicotinic acid also can be made from 3-picoline by catalytic ammoxidation to 3-cyanopyridine, followed by hydrolysis.

Nicotinamide is prepared by partial hydrolysis of the nitrile, or by amination of nicotinic acid chloride or its esters. Some of the compounds mentioned in the foregoing are shown below:

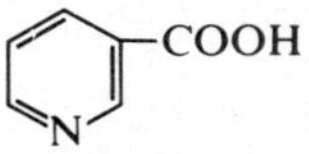

Nicotinic acid 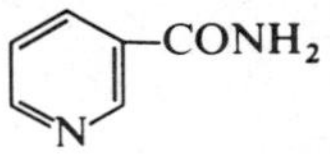   Niacinamide 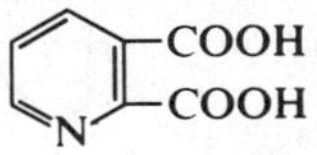   Quinolinic acid 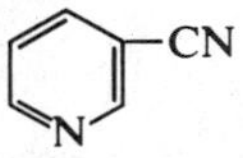   3-cyanopyridine

Nikethamide is a respiratory and heart stimulant, used beneficially against overdoses of barbiturates and morphine. Also known as *Coramine®*, this compound (*N,N*-diethylnicotinamide) is made by reaction of nicotinic acid esters or the acid chloride with diethylamine. The formula is shown below.

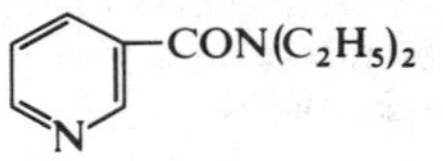

Nikethamide

Pipadrol is a central nervous system stimulant. This compound, α,α-diphenyl-2-piperdinemethanol) is made by condensation of 2-pyridylmagnesium chloride with benzophenone and catalytic hydrogenation of the pyridine ring of the resultant carbinol. The formula is shown below.

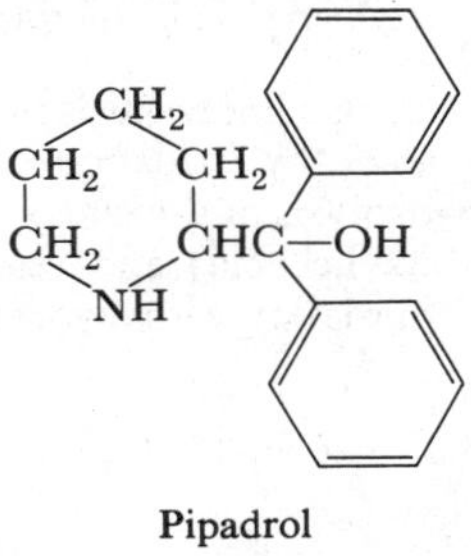

Pipadrol

Piperocaine hydrochloride is used as a local anesthetic. This compound (D,L-(2-methylpiperidino)propylbenzoate hydrochloride) is made by reaction of 2-methylpiperidine with 3-chloropropyl benzoate. Formula is shown below.

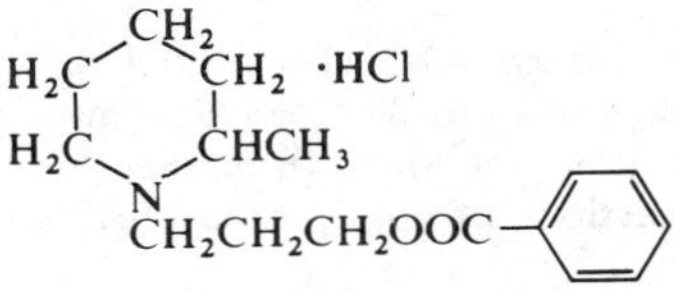

Piperocaine hydrochloride

Pyrithione (zinc salt of) is used as a component of antidandruff shampoos and as a bactericide in soap and detergent formulations. This compound (2-mercaptopyridine-*N*-oxide) exists in equilibrium with *N*-hydroxy-2-pyridinethione and is a fungicide and bactericide, prepared by reaction of 2-chloropyridine *N*-oxide with sodium hydrosulfide and sodium sulfide. This compound is also known as *Omadine®*. The formula is shown below.

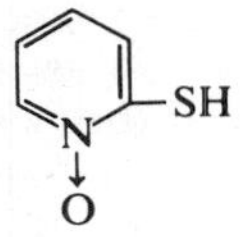

Pyrithione

Sulfapyridine is used to treat dermatitis herpetiformis and also has been used by veterinarians against pneumonia, shipping fever, and foot rot of cattle. This compound (2-sulfanylamidopyridine) is made by condensation of 2-aminopyridine with the appropriate sulfonyl chloride. The formula is shown below.

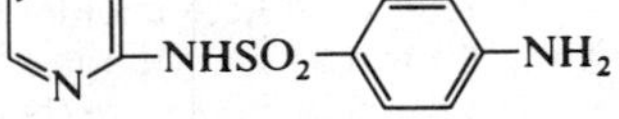

Sulfapyridine

Vitamin B$_6$ is described in detail under **Vitamins.** This is 2-methyl-3-hydroxy-4,5-di(hydroxymethyl)pyridine or pyridoxol. World demand for this compound is estimated at about 5 million pounds (about 2.3 million kilograms) per year. Commercial production is by synthesis starting, for example, with the base-catalyzed condensation of cyanoacetamide and ethoxyacetylacetone. Formula for pyridoxol is shown below.

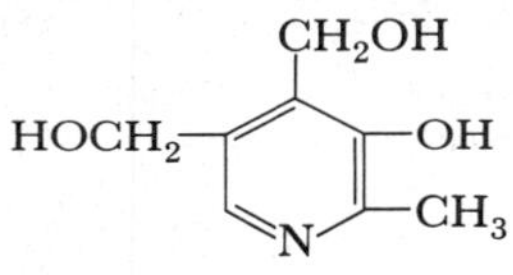

Pyridoxol

Piperidine is used in the pharmaceutical field and for making rubber-vulcanization accelerators, e.g., piperidinium pentamethylenedithiocarbamate (also known as *Accelerator 552®*). On a commercial scale, piperidine (hexahydropyridine) is prepared by the catalytic hydrogenation of pyridine, e.g., with nickel catalysts at 68–136 atmospheres pressure and at 150–200°C, or under milder conditions with noble-metal catalysts. Pyridine derivatives can be similarly reduced to substitute piperidines. See formulas below.

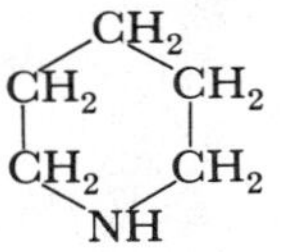

Piperidine 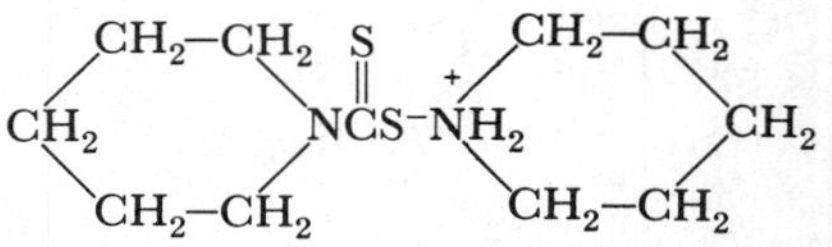   Piperidinium pentamethylendithiocarbamate

Methyridine or 2-(2-methoxyethyl)pyridine, also called *Mintic®*, is used as an anthelmintic. *Piroxicam*, also known as *Feldene®*, is a relatively new anti-inflamatory for the treatment and relief of arthritis. See formulas below.

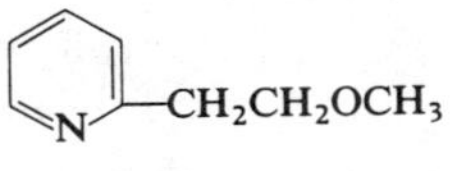

Methyridine 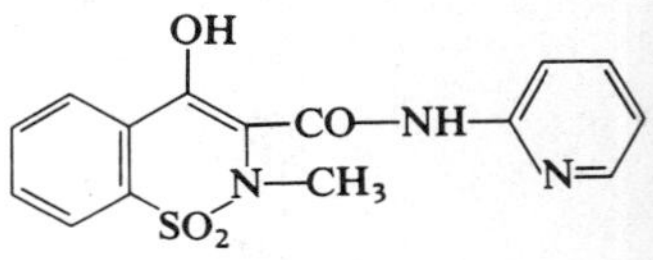   Piroxicam

Pyridinol carbamate has been used as an anti-inflammatory/anti-arteriosclerotic. This compound, 2,6-pyridinedimethanol bis(N-methol-carbamate), is also known as *Anginin®*. See formula below.

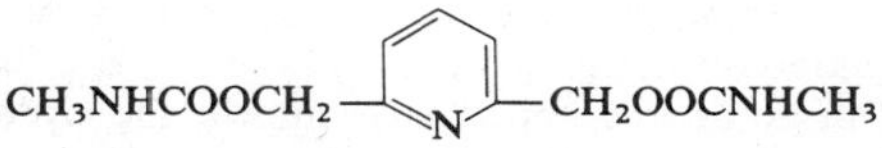

$$CH_3NHCOOCH_2 - \text{(pyridine ring)} - CH_2OOCNHCH_3$$

Pyridinol carbamate

Pyrithioxin is a neurotropic agent which reduces the permeability of the blood-brain barrier to phosphate. This compound, 3,3′-dithiodimethylene-*bis*-(5-hydroxy-6-methyl-4-pyridinemethanol), is also known as *Life®* and *Bonifen®*. The formula is shown below.

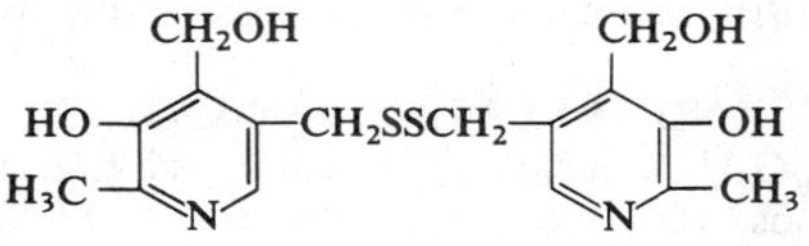

$$HO - \text{(pyridine ring, } CH_2OH, H_3C) - CH_2SSCH_2 - \text{(pyridine ring, } CH_2OH, OH, CH_3)$$

Pyrithioxin

Textile Chemicals. Pyridine derivatives find a number of quite different applications in the textile and related fields.

Stearamidomethylpyridinium chloride is used in waterproofing textiles. It is made by reacting pyridine hydrochloride with stearamide and formaldehyde. *Vinylpyridines* are used as components of acrylonitrile copolymers to improve the dyeability of polyacrylonitrile fibers. The commercially important products are 2-vinylpyridine; 4-vinylpyridine; and 2-methyl-5-vinylpyridine. The formulas are shown below. These compounds are used in the terpolymer latex component of tire cord dips to improve the bonding of textile to rubber. Rubber tires built with steel cord, however, do not require vinylpyridine latex-based adhesives for the steel belt. Therefore, the consumption of vinylpyridines may be affected in the future.

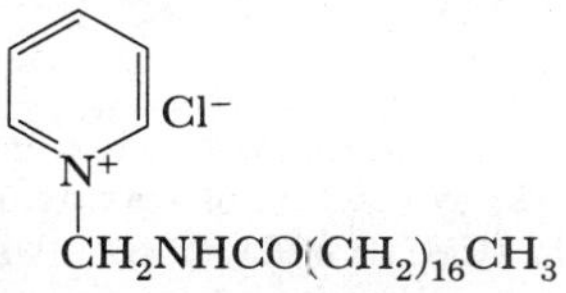

$$\text{(pyridinium ring)} N^+ \; Cl^-$$
$$CH_2NHCO(CH_2)_{16}CH_3$$

Stearamidomethylpyridinium chloride

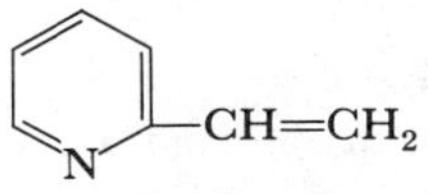

$$\text{(pyridine ring)} - CH=CH_2$$

2-vinylpyridine

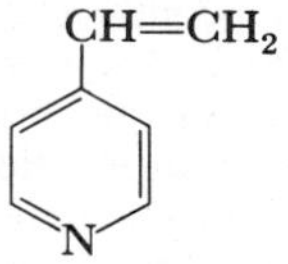

$$CH=CH_2 \; \text{(pyridine ring)}$$

4-vinylpyridine

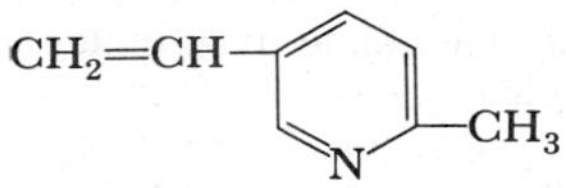

$$CH_2=CH - \text{(pyridine ring)} - CH_3$$

2-methyl-5-vinylpyridine

Pyridine hydrochloride is used in the manufacture of polycarbonates.

—Hans Dressler, Koppers Company, Inc., Monroeville, Pennsylvania.

PYRIDOXINE. Vitamin B_6 (Pyridoxine).

PYRITE. The mineral pyrite or iron pyrites is iron disulfide, FeS_2, its isometric crystals usually appearing as cubes or pyrito-

hedrons. It has a slightly conchoidal to uneven fracture; brittle; hardness, 6–6.5; specific gravity, 5; metallic luster; color, pale to normal brass-yellow; streak, greenish-black; opaque. Arsenic, nickel, cobalt, copper, and gold may be found in small quantities in pyrite, auriferous pyrite being sometimes a very valuable ore. Pyrite is the commonest of the sulfide minerals, and is of world-wide occurrence. It is found associated with other sulfides, or with oxides, in quartz veins, in sedimentary and metamorphic rocks, in coal beds, and as the replacement material in fossils. There are many well-known pyrite localities, among which are the Rio Tinto mines in Spain, where copper-bearing pyrite is obtained from huge deposits. Magnificent crystals and crystal groups occur at Ambasaguas (Logrono) in Spain; Quirivulca, Peru; and from the Island of Elba. In the United States pyrite is found in California, New York, and Virginia in workable deposits. The name pyrite is derived from the Greek word meaning fire, because of the sparks which result when pyrite is struck with steel.

PYROGENETIC MINERALS. A term for the primary magmatic minerals of igneous rocks as distinguished from those minerals which are the result of special and later processes such as come under the head of pneumatolytic, hydrothermal, etc.

PYROLUSITE. The mineral pyrolusite, manganese dioxide, MnO_2, crystallizes in the tetragonal system, but may be only pseudomorphous after manganite. It is found massive or in indistinct crystalline aggregates, often acicular, and as dendritic growths on fractured rock surfaces and inclusions within moss agates and other chalcedony varieties of quartz. Hardness, 6–6.5 (crystals); 2–6 (massive); specific gravity, 5.06; luster, metallic; color, steel gray to black; streak black; opaque. Pyrolusite is found as replacement deposits and as residual and sedimentary masses. Psilomelane is its usual associate. European localities for pyrolusite are in Bohemia, Saxony, the Harz Mountains, England, and elsewhere. Other deposits occur in India and Brazil. In the United States it is found in Arkansas and Michigan. It is an ore of manganese. It is from this latter use that it derives the name *pyrolusite*, from the Greek words meaning *fire* and *to wash*.

PYROLYSIS. Transformation of a compound into one or more other substances by heat alone, i.e., without oxidation. It is thus similar to destructive distillation. Although the term implies decomposition into smaller fragments, pyrolytic change may also involve isomerization and formation of higher-molecular-weight compounds. Hydrocarbons are subject to pyrolysis, e.g., formation of carbon black and hydrogen from methane at 1300°C and the decomposition of gaseous alkanes at 500–600°C. The latter is the basis of thermal cracking (pyrolysis) in the production of gasoline. An application of pyrolysis is the conversion of acetone into ketone by decomposition at about 700°C; the reaction is $CH_3COCH_3 \rightarrow H_2C{=}C{=}O + CH_4$. Pyrolysis of natural gas or methane at about 2000°C and 100 mm mercury pressure produces a unique form of graphite. Synthetic crude oil can be made by pyrolysis of coal, followed by hydrogenation of the resulting tar. Large-scale pyrolysis of solid wastes has been considered in connection with several synfuel projects.

PYROMORPHITE. The mineral pyromorphite is lead chlorophosphate with a formula corresponding to $Pb_5(PO_4)_3Cl$. The phosphorus is sometimes replaced by arsenic and the lead by calcium. It occurs in prismatic, sometimes hollow, hexagonal crystals or may appear in massive forms. It is brittle; hardness, 3.5–4; specific gravity, 7.04; luster, resinous; color, green, yellow-green, yellow, brown, and less often gray or white; translucent to opaque.

Pyromorphite is a secondary mineral associated with other lead minerals, but is seldom found in large quantities. It has probably resulted from the action of waters bearing phosphoric acid upon the pre-existing lead minerals. Localities for pyromorphite are in the Ural Mountains, Saxony, France, Spain, Cornwall and Cumberland, England; in Scotland, Zaire, and Australia. In the United States pyromorphite has been found in Chester and Montgomery Counties, Pennsylvania; in Davidson County, North Carolina, and in the Coeur d'Alene mining district of Idaho. The name is derived from the Greek words meaning fire and form.

PYROPHYLLITE. The mineral pyrophyllite is a hydrous silicate of aluminum corresponding to the formula $Al_2Si_4O_{10}(OH)_2$. Monoclinic with a basal cleavage, it is usually, however, in foliated, radiated, lamellar, or fibrous masses, sometimes compact. It is a soft mineral with a greasy feel; hardness, 1–2; specific gravity, 2.65–2.9; luster, pearly to dull; color, white, greenish, grayish, yellowish, and brownish; translucent to opaque. It is found making up schists or in foliated masses in the Ural Mountains, in Switzerland, Sweden, Brazil, and in the United States in Pennsylvania, North Carolina, Georgia, and California. It is used to some extent and for the same purpose as is the mineral talc, and also for making slate pencils, whence the name *pencil stone* sometimes applied to pyrophyllite.

PYROXENE. This is the name given to a closely related group of minerals, all of which show a distinct cleavage angle of 87° or 93° parallel to the fundamental prism. Chemically the pyroxenes are metasilicates corresponding to the formula $RSiO_3$, where R may be calcium, magnesium, iron, or less commonly manganese, zinc, sodium, or potassium. Rarely titanium, zirconium, or fluorine may be present. A general formula is $ABSi_2O_6$, where A is Ca, Na, Mg, or Fe^{2+}, and B is Mg, Fe^{3+}, or Al. Sometimes the Si is replaced by Al.

The pyroxenes crystallize in the orthorhombic, and monoclinic systems, like the amphiboles, the chief difference between the two groups being the cleavage angles, which for amphibole are 56° and 124°. Pyroxene crystals tend to be short, stout, complex prisms as opposed to the long, slender, and simpler amphiboles.

The pyroxenes are common in the more basic igneous rocks, both intrusive and extrusive, and may be developed by the metamorphic processes in gneisses, schists, and marbles.

For descriptions of members of the pyroxene group, see **Acmite-Aegirite; Augite; Diallage; Diopside; Enstatite; Hypersthene; Jadeite;** and **Spodumene.**

PYRRHOTITE. The mineral pyrrhotite, sometimes called magnetic pyrites, is a sulfide of iron with varying amounts of sulfur. Analyses indicate formulae $Fe_{1-x}S$. Pyrrhotite exists in two modifications: it is monoclinic below, and hexagonal above 138°C (280°F). It is a brittle mineral; hardness, 3.5–4.5; specific gravity, 4.53–4.97; luster, metallic; color, reddish bronze-yellow when fresh, otherwise tarnished; streak, grayish-black; magnetic. It may carry nickel, generally as pentlandite, when it becomes a valuable nickel ore as at Sudbury, Ontario. Pyrrhotite is commonly associated with the basic igneous rock like gabbro and norite, and occurs with chalcopyrite, magnetite, and pyrite. Besides being apparently of magmatic origin, it has been found as contact metamorphic and as vein deposits. Austria, Italy, Saxony, Bavaria, Switzerland, Norway, Sweden, and Brazil have deposits of more or less importance, and in the United States it has been found associated with andalusite crystals at Standish, Maine; also at Brewster, New York; Lancaster County, Pennsylvania; and elsewhere. At Ducktown, Tennessee, it is found together with copper and zinc minerals. It is mined for its nickel content, in the form of admixed pentlandite, in Sudbury, Ontario.

Pyrrhotite derives its name from the Greek word *pyrrhos* meaning *reddish* in reference to the color of the fresh ore.

PYRROLE AND RELATED COMPOUNDS. Pyrrole (monoazole, C_4H_5N or C_4H_4NH), contains a ring of 1 nitrogen and 4 carbons, with 1 hydrogen attached to nitrogen and to each carbon:

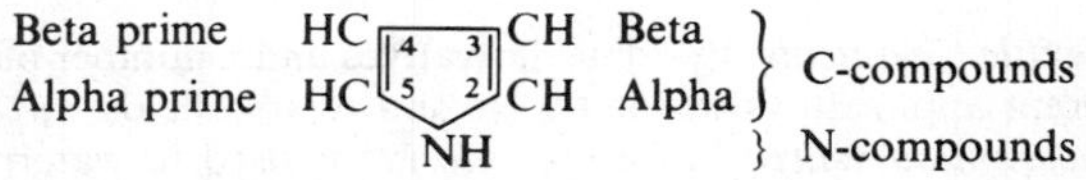

$$
\begin{array}{l}
\text{Beta prime} \quad HC \overset{43}{\underset{52}{\square}} CH \quad \text{Beta} \\
\text{Alpha prime} \quad HC CH \quad \text{Alpha} \\
 NH
\end{array}
\left.\begin{array}{l} \\ \\ \end{array}\right\} \begin{array}{l} \text{C-compounds} \\ \\ \text{N-compounds}\end{array}
$$

Pyrrole is a colorless liquid, boiling point 131°C, insoluble in water, soluble in alcohol or ether. Pyrrole dissolves slowly in dilute acids, being itself a very weak base; resinification takes place readily, especially with more concentrated solutions of acids; and on warming with acid a red precipitate is formed. Pyrrole vapor produces a pale red coloration on pine wood moistened with hydrochloric acid, which color rapidly changes to intense carmine red. Pyrrole may be made (1) by reaction of succinimide

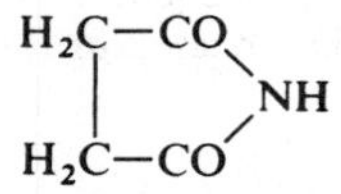

$$
\begin{array}{l}
H_2C-CO \\
\big| NH \\
H_2C-CO
\end{array}
$$

with zinc and acetic acid, or with hydrogen in the presence of finely divided platinum heated, (2) by reaction of ammonium saccharate or mucate $COONH_4 \cdot (CHOH)_4 \cdot COONH_4$ with glycerol at 200°C by loss of carbon dioxide, ammonia, and water.

When pyrrole is treated with potassium (but not with sodium) or boiled with solid potassium hydroxide, potassium pyrrole C_4H_4NK is formed, which is the starting point for *N*-derivatives of pyrrole, since reaction of the potassium with halogen of organic compound and with carbon dioxide, readily occurs. When pyrrole is treated with magnesium metal and ethyl bromide in ether, pyrrole magnesium bromide plus ethane is formed, which may be used as the starting point for *C*-derivatives of pyrrole, since reaction with sodium alcoholates readily occurs (with separation of magnesium oxybromide).

The pyrrole nucleus has been shown to be present in the complex substances chlorophyll (the green coloring matter of plants), hematin (the red coloring matter of blood), and in the coloring matter of bile.

PYRUVIC ACID. Carbohydrates; Coenzymes; Vitamin.

Q

QUARKS. As of the early 1980s, quarks remain hypothetical but nevertheless well accepted particles in the community of particle physicists. The quark hypothesis was first proposed independently in 1964 by M. Gell-Mann and G. Zweig, both at the California Institute of Technology. These researchers pointed out that all the known hadrons (i.e., particles that interact via the strong nuclear force) of that time could be constructed out of simple combinations of three particles (and their three antiparticles). These hypothetical particles had to have slightly peculiar properties (the most peculiar being a fractional electric charge). Gell-Mann called them quarks and they were designated p, n, and λ because they somewhat resemble the proton, neutron, and Λ^0 hyperon. The theory supposed that three quarks bind together to form a baryon, while a quark and an antiquark bind together to form a meson. If it is supposed that the binding is such that the internal motion of the quarks is nonrelativistic (which requires that the quarks be massive and sit in a broad potential well), then many quite detailed properties of the hadrons can be explained. One notable exception—the *omega minus particle* discovered in 1964.

A brief review of the complexities to which the quark theory is addressed is in order. Particles which can interact via the strong nuclear force are called hadrons. Hadrons can be divided into two main classes—the mesons (with baryon number zero) and the baryons (with nonzero baryon number). Within each of the classes there are small subclasses. The subclass of baryons which has been known the longest consists of those particles with spin $\frac{1}{2}$ and even parity. The members of this class are the proton, the neutron, the Λ^0 hyperon, the three Σ hyperons and the two Ξ hyperons. There are no baryons with spin $\frac{1}{2}$ and even parity (or, to the usual notation, $J^P = \frac{1}{2}^+$). The next 'family' of baryons has ten members, each with $J^P = \frac{3}{2}^+$. The mesons can be grouped into similar families. One of the first successes of the quark model was to explain just why there should be eight baryons with $J^P = \frac{1}{2}^+$, ten with $\frac{3}{2}^+$, etc., and why the various members of these families have the particular quantum numbers observed.

The chief drawback of the quark model, as voiced soon after the theory was described, has been the failure to produce a beam of individual quarks, much as one can produce a beam of pions, kaons, negative protons, and so on. As of the early 1980s, this objection remains to be satisfied, but because of much theoretical work and new discoveries since 1964, a majority of the community of particle physicists now accept the theory (with modifications and refinements thereof). Explanations offered for the failure to date to isolate quarks is that quarks exist and have already been seen in numerous experiments, but without recognition. Or, that quarks do not exist, but that the hadrons behave as if they did. A somewhat similar belief was once held by some scientists concerning the neutrino. Some theories have yielded quark-like models in which the particles are not fractionally charged, but they were found to have a number of deficiencies. A compelling argument is that the mass of quarks is greater than the energy available and many scientists are looking forward to more powerful accelerators now underway.

The initial quark model was formulated to explain the diversity of the hadrons and not to explicitly describe the internal structure of any particle. It was inevitable, however, that with further research there was a tendency to identify new findings with the hypothetical quarks. A number of properties of the partons, such as their intrinsic spin angular momentum, have been measured and have proved to be consistent with the predictions of the quark model.

During the course of developing several theories concerning the nature of quarks, physicists required a new nomenclature. To persons outside the field, the words used seem as peculiar as the word quark itself. Some writers have referred to the whimsical character of the names used. The kinds of quarks are called *flavors*.[1] The three kinds of flavors initially proposed are *up*, *down*, and *strangeness*. It was proposed a bit later that a fourth flavor, *charm*, should be added to describe a new property of matter. Discovery of the *psi* or *J particle* in 1974 provided strong evidence for the charmed quark. More recent research has pointed toward the need for a fifth quark, called *bottom*. Inasmuch as the other four quarks appear to be organized in pairs, it is generally assumed that there is also a sixth flavor, named *top*.

Quarks also possess another distinctive property that governs their binding together to form hadrons. This property, called *color*, is of three varieties (*red*, *blue*, and *yellow*) and thus numerous combinations are available, but only certain combinations of them seem possible at this juncture. The property of color[2] plays a role in binding the quarks together in a hadron. Some physicists point out that this is analogous to the role of electric charge in binding together the particles that make up an atom, described by a precise and well-tested theory called quantum electrodynamics. It will be recalled that this theory allows the attraction or repulsion between two charged particles to be communicated by exchange of photons (quanta of electromagnetic radiation).

A theory modeled on the quantum electrodynamics theory has been proposed which describes the interactions between colored quarks and it is known as *quantum chromodynamics*. Mediation of these forces is by hypothetical particles called *gluons*, whose role it is to "glue" the quarks together. Eight kinds of gluon are required. Gluons are themselves colored particles. This means that they are subject to the same forces they transmit. This contrasts with the photon, which communicates electromagnetic forces between charged particles, but in itself carries no electric charge, so that a particle can emit or accept

---

[1] One definition of flavor is: "the characteristic quality of something—distinctive nature."

[2] Color, as used here, has nothing to do with visual color. The manner in which different colored quarks combine in quantum mechanics is suggestive of the way in which visual colors combine. Hadrons are "colorless" since they are averages of the three colors.

a photon without changing its charge. A number of successful predictions have been made by quantum chromodynamics, but as yet they do not compare with quantum electrodynamics in terms of precision.

In the quantum chromodynamics theory, it is suggested that the effective strength of the force between quarks is small when at close range, but much greater when the quarks are separated by a distance comparable to the diameter of a proton. This concept, in itself, is counter to the much better understood forces of electromagnetism and gravitation, where forces become greater as bodies become closer and weaker as bodies recede from one another. It would appear that, within a hadron, the quarks are constrained very little, whereas the energy required to extract them increases at a rapid rate. If it is simply a matter of mounting sufficient energy, then perhaps quarks have not been found in past experiments simply because too little energy was applied. Or, as some physicists have proposed, perhaps the quarks are permanently confined in some manner. For example, K. A. Johnson and a team at the Massachusetts Institute of Technology proposed a *bag model* of quark confinement in 1979. Simply stated, it is hypothesized that quarks are confined in bags analogous to the bubbles in a liquid. This is an interesting concept and is well described by K. A. Johnson in *Sci. Amer.*, **241**, 1, 112–121, 1979. A number of other concepts have been proposed to explain the peculiar behavior of quarks.

Among other experimental objectives in connection with quarks, physicists have been looking for evidence of gluons in the jets of debris which result from particle collisions, notably between electrons and positrons. It is reasoned that some events should produce two oppositely directed jets that become increasingly thin and pencil-like as the collision energy increases. According to some investigators, double-jet events have occurred during the last few years. Some jets have been described as having the shape of a tennis racket. Findings of this kind have been reported by Deutsches Elektronen-Synchrotron (DESY) in Hamburg. In an August 1979 conference held at Fermilab (Fermi National Accelerator Laboratory), researchers from DESY reported events in which three jets form a pattern like the *Mercedes* 3-pronged star. They appeared in early experiments using a new electron-positron colliding beam storage ring called PETRA. For the time being, these results were expressed as "mimicking the expectations of the quantum chromodynamics theory."

NOTE: For references, consult reference list at the end of the entry on **Particles (Subatomic)**.

QUARTZ. The mineral quartz, oxide of the nonmetallic element silicon, is the commonest of minerals, and appears in a greater number of forms than any other. Its formula is SiO_2. Quartz commonly occurs in prismatic hexagonal crystals terminated by a pyramid. This pyramid is due to the equal development of two rhombohedrons, and may be observed in cases where one rhombohedron predominates. Cleavage is not observed; the fracture is typically conchoidal; hardness is 7; specific gravity, 2.65; luster, vitreous to greasy or dull; colorless to white, pink, purple, yellow, blue, green, smoky brown to nearly black; transparent to opaque.

There are two distinct modifications of quartz, depending upon the temperature at which they were formed. The low-temperature variety is formed below 573°C and is the more common sort, being found in veins, geodes, etc. It is called low-quartz. The high-temperature modification is formed between 573°C and 870°C, and is found chiefly in granites and granite or rhyolite porphyries. This is called high-quartz. Above 870°C tridymite is the stable form of SiO_2. The differences between high- and low-quartz are entirely crystallographic, low-quartz having a vertical axis of three-fold symmetry and three horizontal axes of two-fold symmetry, while high-quartz has a vertical axis of six-fold symmetry and six horizontal axes of two-fold symmetry. It is usual to separate the many kinds of quartz into (1) crystalline or vitreous varieties, actual crystals or vitreous crystalline masses, and (2) cryptocrystalline varieties, mostly compact nonvitreous sorts, but which may show a crystalline structure under the microscope.

1. *Crystalline or Vitreous*: Rock crystal, colorless crystals or masses. Amethyst, clear violet or purple, either crystals or masses. Rose quartz, usually massive but rarely in crystals, delicate shades of pink or rose, sometimes red. Citrine or yellow quartz, sometimes called false or Spanish topaz, light to deep yellow. Smoky quartz, smoky brown to almost black, often called cairngorm stone from Cairngorm, Scotland. Milky quartz, often showing delicate opalescence, transparent to nearly opaque, often with a greasy luster. Aventurine quartz incloses glistening scales of mica or hermatite. Rutilated quartz incloses needle-like prisms of rutile called "fleches d'amour." Other acicular minerals such as actinolite, tourmaline, and epidote, may also be thus inclosed; Cat's Eye shows a peculiar opalescence, probably due to inclosed masses of some fibrous mineral. Tiger's Eye is a siliceous pseudomorph after crocidolite of a golden yellow brown color.

2. *Cryptocrystalline*: The following cryptocrystalline varieties of quartz are treated under their own headings: agate, basanite, bloodstone, carnelian, chalcedony, chert, chrysoprase, flint, heliotrope, jasper, moss agate, onyx, plasma, prase, sard, and sardonyx. Quartz readily forms pseudomorphs after various minerals or structures. Silicified wood is a quartz pseudomorph after the organic material of which it originally consisted. Quartz is often pseudomorphic after calcite, barite, and fluorite. Quartz is an essential constituent of many igneous rocks, for example, granites, granite porphyries, and felsites, as well as quartz diorites and their surface equivalents, the dacites. In the metamorphic rocks quartz figures very largely in the gneisses and schists, and, of course, in quartzite. In the sedimentary rocks most sandstones are composed chiefly of grains of quartz, and quartz forms veins and nodules in limestones.

Of the many places that have yielded fine specimens of quartz, a few include: the Swiss Alps, the Piedmont of Italy, the Island of Elba, Dauphiné in France, Cumberland in England, Banffshire in Scotland, the the Malagasy Republic, Uruguay, Mexico and Brazil. Magnificent rose quartz crystals occur at the Arassuahy-Jequitinhonha District, Minas Gerais, Brazil. In the United States the following localities are well known: Paris, Maine, especially for rose quartz; Herkimer County, New York, for small but very brilliant crystals found in the Cambrian dolomites or in the soil. Amethyst County, Virginia, furnishes amethysts, as do Lincoln and Alexander Counties, North Carolina. Other localities for amethyst and smoky quartz are South Dakota, in the Black Hills; the Pikes Peak district; Colorado; Yellowstone National Park, Wyoming; Jefferson County, Montana; and in Canada in the Province of Ontario in the Thunder Bay region. The word *quartz* is believed to have been originally of German origin. Besides the use of the different varieties of quartz for jewelry and other ornamental purposes, this mineral has extensive industrial uses in the ceramic arts, optical and other sorts of scientific instruments, abrasive, scouring, polishing materials, and for refractories.

Certain mineral classes of low symmetry possess no center of symmetry, and their axes, known as polar axes, have different properties at their terminal ends. Quartz belongs to one of those classes. When quartz is exposed to an exerted compressive or mechanical stress along one of these polar axes, electrical charges are developed on that axis; a negative charge is produced at one end, a positive charge at the opposing end. Conversely, when quartz crystals are subjected to an applied electric field along a polar axis, mechanical strains will be developed in those crystals. This phenomenon is known as piezoelectricity. Plates or disks cut perpendicular to such polar axes and properly oriented with established specifications are subject to mechanical vibrations (oscillations) at predetermined frequencies under an applied electric field. Those frequencies are designed to coincide with and stabilize the circuit frequency of radio transmitters and receivers. This property is utilized extensively in the control of frequency oscillations in the field of radio telemetry. Between January 1942 and V-J Day over 70 million such units were manufactured for the United States armed forces, and consumed over 4 million pounds (1.8 million kilograms) of radio grade quartz. The excessive demand for natural quartz of required quality to produce those wafers resulted in the development of a new industry, synthesizing quartz to meet the demand.

See also **Piezoelectricity.**

—Elmer B. Rowley, Union College, Schenectady, New York.

QUARTZITE. A hard, tough, and compact metamorphic rock composed almost wholly of quartz sand grains which have been recrystallized to form a particularly massive siliceous rock. The term is also used for nonmetamorphosed quartzose sandstones and grits whose clastic grains have been firmly cemented by silica which has grown in optical continuity around each grain.

QUINOLINE AND COMPOUNDS. Quinoline compounds are characterized by the structure:

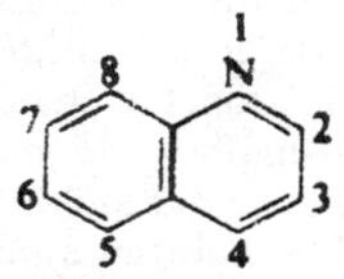

Quinoline, quinaldine(2-methylquinoline), and lepidine (4-methylquinoline), can be isolated from the tar-base fraction obtained from coal-tar distillates. While other quinoline compounds occur naturally (cinchona and angostura alkaloids), most derivatives are synthesized from monocyclic intermediates or, to a lesser extent, from quinoline, quinaldine and lepidine.

Quinoline compounds can be prepared by reaction of an aromatic amine with an appropriately substituted three-carbon fragment. In the Doebner-Miller and Skraup syntheses, this moiety is an α,β-unsaturated carbonyl compound formed *in situ* and produced by an aldol condensation in the former and by the dehydration of glycerol or a substituted glycerol in the latter. Thus, aniline and acetaldehyde yield quinaldine in a Doebner-Miller synthesis and p-toluidine and glycerol afford 6-methylquinoline in the presence of an oxidant in a Skraup synthesis. An α,β-unsaturated carbonyl compound, ethoxymethylenemalonic diethyl ester, is used directly in the Gould-Jacobs synthesis and on treatment with aniline affords 3-carbethoxy-4-hydroxyquinoline.

The three-carbon moiety is a β-dicarbonyl compound in the Combes, Knorr and Conrad-Limpach syntheses. The carbonyl groups can be keto and/or aldehydo groups in the Combes synthesis but β-ketoesters are employed in the latter two syntheses. Thus, aniline and acetylacetone yield 2,4-dimethylquinoline in a Combes synthesis and, depending on the reaction conditions, o-toluidine and acetoacetic ester produce either 2-hydroxy-4,8-dimethylquinoline (Knorr) or 4-hydroxy-2,8-dimethylquinoline (Conrad-Limpach).

Quinoline compounds can be prepared by the Friedlander synthesis, which consists of treating an o-aminoaryl carbonyl compound with an aliphatic carbonyl compound. Hence acetone and anthranilic acid afford 2-methyl-4-hydroxyquinoline. The Pfitzinger synthesis is an analogous reaction, but employs isatin as the nitrogenous reactant. When isatin is treated with acetone, 2-methyl-4-carboxyquinoline is obtained.

Reactions. Quinoline is a weak base ($K_B = 3.2 \times 10^{-10}$) and undergoes many of the reactions of tertiary amines. In general quinoline compounds form quinolinium salts with Lewis acids or with reactive organohalides. When treated with base, N-alkylquinolinium salts carry an alkyl substituent in the 2 or 4 position react with N-alkylquinolinium salts carrying *no* 2 or 4 substituent to form cyanine dyes. Cyanine blue is prepared in this manner.

Electrophilic substitution. Nitration of quinoline affords a mixture of the 8 and 5-nitroquinolines in which the former predominates. Similarly, quinoline-8-sulfonic acid is the main product of sulfonation. Bromination at 300°C yields 3-bromoquinoline but at 500°C 2-bromoquinoline is obtained.

Nucleophilic substitution. High yields of 2-amino or 2-hydroxyquinoline are obtained when quinoline is heated with alkali amides or hydroxides, respectively. When quinoline is heated with Grignard or organolithium reagents, the organic groupings enters the 2 position.

Oxidation and Reduction. The high-temperature oxidation of quinoline by sulfuric acid using a selenium catalyst represents a commercial synthesis of nicotinic acid. Depending on the nature and site of the substituents, substituted quinolines undergo oxidation to yield either pyridine or benzene derivatives or mixtures of both. Perbenzoic acid transforms quinoline into its N-oxide.

The catalytic hydrogenation of quinoline compounds yields a 1,2,3,4-tetrahydroquinoline as the main product. More strenuous conditions produce a decahydroquinoline. Accumulation of substituents in the 2,3 and 4 positions tends to favor the formation of a 5,6,7,8-tetrahydroquinoline.

Reactions of Substituted Quinolines. Except when substituted in the 2 and 4 positions of the quinoline ring, the reactions of substituents situated elsewhere parallel those of the correspondingly substituted naphthalene.

Alkyl derivatives: Quinoline compounds carrying a

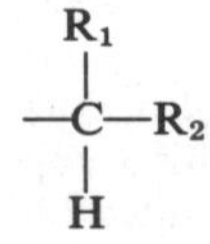

in the 2 or 4 position undergo active methylene reactions, i.e., base-catalyzed condensations with aldehydes, ketones, epoxides, esters, acyl halides, reactive organohalides and carbon dioxide, as well as reactions such as the Michael and the Mannich. Thus quinaldine and ethyl benzoate react in the presence of potassium ethoxide and form the ketone, 2(2-quinolyl) acetophenone.

Halogen derivatives. Halogen atoms substituted in the 2 or 4 positions undergo a facile replacement by hydroxyl, sulfhydryl, alkoxyl, ammono and amino groups as well as by anions like that derived from diethyl malonate. Thus, treatment of 2,7-dichloroquinoline with dilute acid demonstrates the differential reactivity of halogen atoms for 2-hydroxy-7-chloroquinoline is obtained.

Hydroxyl derivatives: The 2 and 4-hydroxyquinolines undergo reactions atypical of phenols in that treating them with phosphorus pentachloride effects conversion to the chloroquinoline, alkylation with alkyl halides or sulfates in an alkaline medium affords an N-alkyl derivative and their ethers can be cleaved with dilute acids. Spectroscopic evidence in the ultraviolet and infrared indicates that they exist in their tautomeric forms. Thus, 2-hydroxyquinoline exists as the quinoline, the latter structure appreciably stabilized through resonance with a dipolar structure. 8-Hydroxyquinoline forms insoluble chelate compounds with many metallic ions, e.g., aluminum, bismuth, cadmium, copper, magnesium, nickel and zinc among others. The use of this reagent as an analytical reagent has been investigated widely.

Amino derivatives. Anomalous results are obtained when a 2 or 4-aminoquinoline is treated with nitrous acid. Instead of the expected formation of the diazonium salt (which is formed from any of the five other aminoquinolines) the corresponding hydroxyquinoline is isolated. When concentrated hydrochloric or hydrobromic acid is used for diazotization, the corresponding halogen derivative is isolated.

Derivatives of 4-amino and 8-aminoquinoline, e.g., chloroquine and pamaquine are invaluable antimalarials.

Quinoline Compounds in Commerce. In addition to quinoline, lepidine, quinaldine, chloroquine, pamaquine, the cinchona alkaloids (antimalerials and cardiac depressants) and the cyanine dyes (photographic sensitizers), the following are marketed commercially: Nupercaine, a local anesthetic; Yatren 105 and Vioform, used as amebicides; chincophen, an analgesic and antipyretic; oxyquinoline sulfate, an antiseptic; and quinoline yellow, a paper dye.

—Ralph Daniels, University of Illinois, Chicago, Illinois.

QUINONES. The quinones are unsaturated cyclic diketones with both the oxygen atoms attached to carbon atoms in simple, fused, or conjugated ring systems. They may be regarded as oxidation products of dihydroxy aromatic compounds with the substituent groups in positions corresponding to the *ortho* or *para* positions in the benzene ring. Thus *o*-benzoquinone can be prepared by the oxidation of catechol with silver oxide and *p*-benzoquinone by the oxidation of hydroquinone. The compound *m*-benzoquinone is unknown, and would not be feasible from structural considerations. In the naphthalene series six quinones are theoretically possible. The 1,2- and 1,4-quinones are similar structurally to the benzoquinones, but in 2,6-naphthoquinone the oxygen atoms are in adjacent rings.

A number of possibilities for quinone formation also exist in the anthracene series; the best known compound in this group is 9,10-anthraquinone, where both oxygens are substituted on the *meso* carbon atoms linking the two benzene rings. Unlike the benzoquinones, the rings are fully aromatic and additions to the double bonds do not take place readily. In the phenanthrene series compounds such as 3,4-phenanthrenequinone are known to exist, but the best known isomer is 9,10-phenanthrenequinone where the carbonyl groups are adjacent and two of the rings are aromatic in character.

A number of quinones are also known which can be regarded as derivatives of more highly condensed ring systems such as naphthacene, chrysene, and pyrene where the carbonyl groups may be adjacent, *para* to one another or contained in different rings. In 4,4'-diphenoquinone the two six-membered rings are not fused, but are linked by a double bond *para* to the carbonyl groups. In stilbenequinone and its derivatives the carbonyl groups are located in different rings which are conjugated through a two-carbon bridge. Acenaphthenequinone is commonly classified with the quinones, although this is not strictly correct since the carbonyl groups are contained in a five-membered ring which cannot be reduced to a fully aromatic ring system.

The most typical reaction of the quinones is their reversible reduction to aromatic dihydroxy compounds by chemical reagents or electrolysis. The oxidation potentials for the quinone-hydroquinone reactions depend upon the degree of conjugation of the reduced and oxidized ring systems. *p*-Benzoquinone has an oxidation potential of +0.715 volt while 1,4-naphthoquinone and 9,10-anthraquinone have potentials of 0.484 and 0.154 volt, respectively. Thus *p*-benzoquinone will readily oxidize 9,10-dihydroxyanthracene to the corresponding anthraquinone. The oxidation potentials of these systems are dependent on oxonium ion concentration; thus the quinhydrone and chloroanil (tetrachloro-*p*-benzoquinone) electrodes have been used for the measurement of pH in neutral or acid solutions. The carbonyl groups of these compounds will undergo typical reactions such as the combination of *p*-benzoquinone with one or two moles of hydroxylamine hydrochloride to yield *p*-benzoquinone monoxime or *p*-benzoquinone dioxime. *p*-Benzoquinone monoxime exists in tautomeric equilibrium with *p*-nitrosophenol.

The quinones will undergo substitution reactions in various ways depending upon the aromatic character of the ring system. *p*-Benzoquinone and 1,4-naphthoquinone can be chlorinated to yield tetrachloro-*p*-benzoquinone and 2,3-dichloro-1,4-naphthoquinone, respectively. The halogen atoms adjacent to the carbonyl group are labile and readily undergo nucleophilic substitution reactions with thiol or amino groups. The halogenation of anthraquinone is very difficult but a dibromoanthraquinone has been prepared. *p*-Benzoquinone is stable in cold concentrated chromic acid. Further oxidation with hydrogen peroxide opens the ring to yield diphenic acid. In the presence of alkaline permanganate one of the carbonyl groups is eliminated and the product is fluorenone. Phenanthrenequinone dissolves readily in sodium bisulfate solution to form an addition product. The double bonds in the ring system are aromatic in character.

—H. P. Burchfield, Gulf South Research Institute, New Iberia, Louisiana, and George L. McNew, Boyce Thompson Laboratories, Yonkers, New York.

R

RACEMIZATION. The conversion of an optically active compound i.e., one that rotates the plane of polarized light, into its racemic or optically inactive form is known as *racemization*. In this process half of the optically active compound is converted into its mirror image. The resultant mixture of equal quantities of the dextro- and levo-rotatory isomers is without effect on plane-polarized light due to external compensation (meso forms are internally compensated). Racemization of compounds possessing more than one asymmetric atom may yield products with residual optical activity due to the presence of unchanged centers of asymmetry. An example of this is found in the mutarotation of either α- or β-D-glucose in which only 1 of the 5 asymmetric centers is affected by the opening and closing of the hemiacetal ring.

Racemization may occur in molecules in which structural changes, such as those due to resonance, enolization, substitution or elimination of groups, temporarily destroy the asymmetry needed to maintain the optical activity. Also, Walden inversion of half of an optically active isomer can yield a racemate without the destruction of the center of asymmetry; this phenomenon is observed in the reaction of D-butanol-2 with $HClO_4$.

In the case of the optically active acids, racemization is postulated as the result of the enolization mechanism:

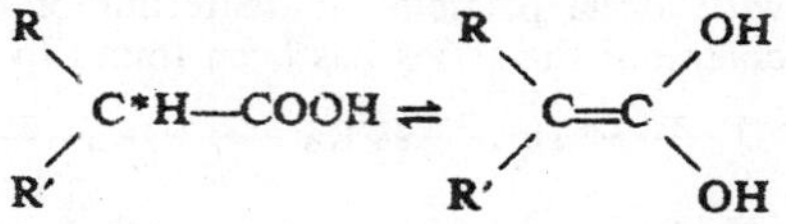

$$
\begin{array}{ccc}
R & & R \qquad OH \\
\backslash & & \backslash \quad / \\
C^*H{-}COOH & \rightleftharpoons & C{=}C \\
/ & & / \quad \backslash \\
R' & & R' \qquad OH
\end{array}
$$

In this case it is seen that the hydrogen attached to the alpha carbon migrates to the oxygen of the carbonyl group forming a carbon-to-carbon double bond and thereby destroying the asymmetry previously found at the alpha carbon atom. Reversion of the enol to the acid reforms the center of asymmetry on a random basis (i.e., giving the same quantities of dextro and levo forms of the acid). Variations of pH, temperature, solvents and catalysts are likely to change the rates of racemization. The progress of the reaction is conveniently followed by examination of the reaction mixture with plane-polarized light in a polarimeter. A similar racemization of disodium L-cysteine in liquid ammonia in the presence of $NaNH_2$ is reported to occur through abstraction of the α-proton to form a carbanion followed by random recombination of the proton.

Although most racemization reactions have been observed to occur in solution, L-leucine on the surface of silicates has been completely racemized by heating to 200°C for 6 hours. Racemization has also been observed in optically active compounds which possess no asymmetric atom but owe their activity to hindered rotation around a single C—C bond. For example, 8'-methyl-1,1'binaphthyl-8-carboxylic acid has been resolved into its optically active forms and racemized by heating in dimethylformamide.

One of the earliest known examples of racemization was described by Pasteur in his studies of tartaric acid. By heating D-tartaric acid to 165°C in water he partially converted it into a mixture of D-, L- and *meso*-tartaric acids. The occurrence of some DL-tartaric acid along with the natural D-tartaric acid in the wine industry is explained on the basis of partial racemization.

The process of racemization has a number of practical applications in the laboratory and in industry. Thus, in the synthesis of an optical isomer it is frequently possible to racemize the unwanted isomer and to separate additional quantities of the desired isomer. By repeating this process a number of times it is theoretically possible to approach a 100% yield of synthetic product consisting of only one optical isomer. An example of the utilization of such a process is found in the production of pantothenic acid and its salts. In this process the mixture of D- and L-2-hydroxy-3,3-butyrolactones are separated. The D-lactone is condensed with the salt of beta-alanine to give the biologically active salt of pantothenic acid. The remaining L-lactone is racemized and recycled.

The process of racemization is important in the survival and growth of living cells and is catalyzed by a group of enzymes called racemases. Alanine racemase, for example, is able to convert D-alanine to DL-alanine if a suitable alpha keto acid is also present. In this reaction the asymmetry of the alpha-carbon atom of alanine is lost as the amino acid is converted to the keto acid and back. This process is analogous to the well-known process of transamination (in which racemization seldom occurs.

See also **Amino Acid;** and **Radioactivity and Other Dating Techniques.**

—Louis H. Goodson, Midwest Research Institute, Kansas City, Missouri.

RADIOACTIVE WASTES. Nuclear Reactor; Wastes and Water Pollution.

RADIOACTIVITY. The spontaneous disintegration of the nucleus of an atom with the emission of radiation is the accepted definition of radioactivity. This phenomenon was discovered by Becquerel in 1896 by the exposure-producing effects on a photographic plate by pitchblende (uranium-containing mineral) while wrapped in black paper in the dark. Soon after this, it was found that uranium minerals and uranium chemicals showed more radioactivity than could be accounted for by the uranium content. About the same time, radioactivity of thorium minerals and thorium chemicals was also discovered.

The excess radioactivity of mineral over chemical uranium led Pierre and Marie Curie to experiment with the mineral. To detect the presence of radioactivity the discharge of a charged gold-leaf electroscope was used. A quantitative estimation of the amount of radioactivity was made by observing the rate of drop of the gold leaf. By chemically separating the uranium mineral into fractions and examining each fraction by the electroscope, they found in the bismuth element fraction the first new radioactive element to be discovered. It was named polonium in 1898. They found that polonium disappeared rapidly, half of its activity vanishing in about 6 months. The fraction containing barium element was also found by them to be radioactive. Repeated fractional crystallizations of the chloride and bromide solutions made possible the recovery by them of practically pure salt of the second new radioactive element. It was named radium in 1898.

Radium is chemically similar to barium; it displays a characteristic optical spectrum; its salts exhibit phosphorescence in the dark, a continual evolution of heat taking place sufficient in amount to raise the temperature of 100 times its own weight of water 1 degree Celsius every hour; and many remarkable physical and physiological changes have been produced. Radium shows radioactivity a million times greater than an equal weight of uranium and, unlike polonium, suffers no measurable loss of radioactivity over a short period of time ($t_{1/2} = 1620$ years). From solutions of radium salts, there is separable a radioactive gas, radon (also sometimes called *radium emanation*), which is a chemically inert gas similar to xenon and disintegrates ($t_{1/2} = 3.82$ days), with the simultaneous formation of another radioactive element, radium A (polonium-218).

Beginning in 1899 and continuing through the next two decades, E. Rutherford and his associates conducted a rather thorough study of the radiations emitted by radioactive substances. During this study the radiations were found to be of three types, called alpha, beta, and gamma radiations. In kind, they resemble anode rays, cathode rays, and x-rays, respectively. In this behavior toward electrical and magnetic fields, the resemblance is qualitatively complete: (1) Alpha rays are positively charged particles of mass number 4 and slightly deflected by electrical and magnetic fields. (2) Beta rays are negatively charged electrons, and strongly deflected by electrical and magnetic fields. (3) Gamma rays are undeflected by electrical and magnetic fields, and of wavelength of the order of 10^{-8} to 10^{-9} centimeters.

Alpha rays have a definite velocity and a definite range for each radioactive nuclide. The velocity is from 5–7% that of light. *Range* is defined as the distance traversed in a homogeneous medium before absorption. The penetrating power of alpha rays is the smallest of the three kinds of rays, the beta rays being of the order of 100 times, and the gamma rays 10,000 more penetrating. The alpha rays are twice-ionized nuclei of helium (He^{2+}). Ramsay and Royds (1909) experimentally demonstrated that accumulated alpha particles, quite independently of the matter from which they have been expelled, consist of helium. They sealed radon in a glass tube with a wall so thin that the alpha particles passed through the wall into a surrounding vessel and after six days the optical spectrum of helium was observed. Helium itself does not diffuse through such a wall. Therefore, alpha particles on losing their positive charge become ordinary helium. This is the first instance of the production of a known element during radioactive transformation. The loss of a single alpha particle by an atom leaves the residual atom four units less in mass number, and two units less in atomic number. The shooting of alpha particles was visibly registered by Crooke's spinthariscope in which the tip of a wire, coated by a tiny amount of radium salts, was placed near a screen coated with zinc blende. Viewed in the dark with a magnifying eyepiece, each alpha particle striking the zinc blende target was observed to produce a visible scintillation. The detection and counting of single alpha particles was accomplished by Rutherford and Geiger (1908), by the deflection of an electrometer needle upon the arrival of each alpha particle in a gas at low pressure in an electric field somewhat below the sparking point.

Beta rays are electrons. They have varying velocities almost up to that of light. The loss of a single negatron by an atom leaves the residual atomic nucleus the same in mass number and one unit greater in atomic number, while the loss of a positron or an orbital electron capture leaves the residual atomic nucleus the same in mass number, and one unit less in atomic number.

Gamma rays are photons of electromagnetic radiation. This radiation is much more penetrating than alpha or beta particles. The presence of gamma rays from 30 milligrams of radium can be observed in an electroscope after passing through 30 centimeters of iron (Rutherford). For the protection of the operator, radium is kept in lead outer containers or screened by lead sheets.

The naturally occurring radioactive elements at the upper end of the periodic table of elements form a number of series, the elements of each series existing in radioactive equilibrium, unless individual elements are separated chemically away from the series. These series include the Uranium Series, the Thorium Series, and the Actinium Series. See Tables 1, 2, and 3. These arrangements are useful in showing the decay-chain (i.e., the parent-daughter) relationships of radioactive elements, including such concepts as radioactive equilibrium. Other naturally occurring radioactive elements are numerous, including, for example, ^{40}K, ^{87}Rb, and ^{148}Sm.

Artificial Radioactivity. In addition to the radionuclides already discussed, there are also the great numbers of artificially produced radioactive elements. They are represented in the Neptunium Series and in various collateral series, because, in addition to the three main natural and the one artificial disintegration series of radioelements, each has been found to have at least one parallel or collateral series. The main series and the collateral series have different parents, but they become identical when, in the course of disintegration, they have a member in common. Collateral with the natural uranium series is an artificial series discovered in the United States by M. H. Studier and E. K. Hyde. Its parent is ^{230}Pa formed by the bombardment of thorium with alpha particles or deuterons of high energy. The decay scheme of the series has been found to be

$$^{230}Pa \xrightarrow{\beta-} {}^{230}U \xrightarrow{\alpha} {}^{226}Th \xrightarrow{\alpha} {}^{222}Ra \xrightarrow{\alpha} {}^{218}Rn \xrightarrow{\alpha} {}^{214}Po \rightarrow$$
Uranium Series

The loss of the alpha particle by the emanation, ^{218}Rn, leads to the formation of ^{214}Po, which is identical with radium C′ of the Uranium Series; the subsequent decay of the collateral series thus becomes identical with that of the main Uranium Series at this point.

Another collateral Uranium Series has for its progenitor ^{226}Pa which is found among the products of bombardment of thorium with 150-MeV deuterons. The decay scheme is represented by:

$$^{226}Pa \xrightarrow{\alpha} {}^{222}Ac \xrightarrow{\alpha} {}^{218}Fr \xrightarrow{\alpha} {}^{214}At \xrightarrow{\alpha} {}^{210}Bi \rightarrow$$
Uranium Series

Still other collateral series are the following:

$$^{228}Pa \xrightarrow{\alpha} {}^{224}Ac \xrightarrow{\alpha} {}^{220}Fr \xrightarrow{\alpha} {}^{216}At \xrightarrow{\alpha} {}^{212}Bi \rightarrow$$
Thorium Series

$$^{232}Pa \xrightarrow{\alpha} {}^{228}U \xrightarrow{\alpha} {}^{224}Th \xrightarrow{\alpha} {}^{220}Ra \xrightarrow{\alpha} {}^{216}Rn \xrightarrow{\alpha} {}^{212}Po \rightarrow$$
Thorium Series

$$^{227}Pa \xrightarrow{\alpha} {}^{223}Ac \xrightarrow{\alpha} {}^{219}Fr \xrightarrow{\alpha} {}^{215}At \xrightarrow{\alpha} {}^{211}Bi \rightarrow$$
Actinium Series

$$^{239}U \xrightarrow{\beta} {}^{239}Np \xrightarrow{\beta} {}^{239}Pu \xrightarrow{\alpha} {}^{235}U \rightarrow \quad \text{Actinium Series}$$

$$^{239}U \xrightarrow{\alpha} {}^{225}Th \xrightarrow{\alpha} {}^{221}Ra \xrightarrow{\alpha} {}^{217}Rn \xrightarrow{\alpha} {}^{213}Po \rightarrow$$
Neptunium Series

See Table 4.

Frederic and Irene Joliot-Curie found in 1933 that boron, magnesium, or aluminum, when bombarded with α-particles from polonium, emit neutrons, protons, and positrons, and that when the source of bombarding particles was removed, the emis-

TABLE 1. THE URANIUM SERIES

RADIOELEMENT	CORRESPONDING ELEMENT (2)	SYMBOL	RADIATION	HALF-LIFE
Uranium I ↓	Uranium (92)	^{238}U	α	4.51×10^9 yr
Uranium X$_1$ ↓	Thorium (90)	^{234}Th	β	24.1 days
Uranium X$_2^*$ 99.87% | 0.13%	Protactinium (91)	^{234}Pa	β and I.T.	1.17 min
Uranium II	Uranium (92)	^{234}U	α	2.48×10^5 yr
Uranium Z	Protactinium (91)	^{234}Pa	β	6.66 hr
Ionium ↓	Thorium (90)	^{230}Th	α	7.5×10^4 yr
Radium ↓	Radium (88)	^{226}Ra	α	1.62×10^3 yr
Ra Emanation ↓	Radon (86)	^{222}Rn	α	3.82 days
Radium A 99.96% | 0.04%	Polonium (84)	^{218}Po	α and β	3.05 min
Radium B	Lead (82)	^{214}Pb	β	26.8 min
Astatine-218	Astatine (85)	^{218}At	α	2 sec
Radium C 99.96% | 0.04%	Bismuth (83)	^{214}Bi	β and α	19.7 min
Radium C'	Polonium (84)	^{214}Po	α	1.5×10^{-4} sec
Radium C''	Thallium (81)	^{210}Tl	β	1.32 min
Radium D ↓	Lead (82)	^{210}Pb	β	19.4 yr
Radium E ~ 100% | ~ 10^{-5}%	Bismuth (83)	^{210}Bi	β and α	2.6×10^6 yr
Radium F	Polonium (84)	^{210}Po	α	138.4 days
Thallium-206	Thallium (81)	^{206}Tl	β	4.23 min
Radium G (end product)	Lead (82)	^{206}Pb	None	Stable

* Undergoes isomeric transition (I.T.) to form uranium Z (^{234}Pa); the latter has a half life of 6.66 hr, emitting β radiation and forming Uranium II ^{234}U.

sion of protons and neutrons ceased, but that of positrons continued. The targets remained radioactive, and the emission of radiation fell off exponentially just as it would for a naturally occurring radioelement. The results of this work may be stated in two equations as follows:

$$^4\text{He} + {}^{27}\text{Al} \longrightarrow {}^{30}\text{P} + n$$
$$^{30}\text{P} \xrightarrow{\beta^+} {}^{30}\text{Si}$$

The first of these equations shows that the result of the nuclear reaction in which aluminum is bombarded with alpha particles is the emission of a neutron and the production of a radioactive isotope of phosphorus. The second equation shows the radioactive disintegration of the latter to yield a stable silicon atom and a positron. Continuation of this line of investigation by several research groups confirmed that radioactive nuclides are formed in many nuclear reactions.

Generally, if any two isobars differ in charge by $\pm e$, one has a higher ground-state energy than the other and is beta radioactive. Any nuclide that can be formed from a nuclear reaction and is not one of the known stable nuclides is radioactive. Nuclides having higher atomic number (Z) than the nearest stable isobar decay to it through positron emission (β^+ decay) or orbital electron capture. Nuclides with lower Z than the nearest stable isobar decay to it through negatron emission (β^- decay). Occasionally, as for ^{64}Cu, a radioactive nuclide is located between two stable isobars and can decay to either of them, in this case either ^{64}Ni or ^{64}Zn. The simplest radioactive nuclide is the neutron, which has a half life of 12 minutes, and decays into a proton, a negatron, and a neutrino.

Energy-level diagrams for nuclear transformations are usually drawn to show the relative energies of levels of an entire neutrally charged atomic system. Since the nuclear charge increases

TABLE 2. THE THORIUM SERIES

RADIOELEMENT	CORRESPONDING ELEMENT	SYMBOL	RADIATION	HALF-LIFE
Thorium	Thorium	^{232}Th	α	1.39×10^{10} yr
↓ Mesothorium I	Radium	^{228}Ra	β	6.7 yr
↓ Mesothorium II	Actinium	^{228}Ac	β	6.13 hr
↓ Radiothorium	Thorium	^{228}Th	α	1.90 yr
↓ Thorium X	Radium	^{224}Ra	α	3.64 days
↓ Th Emanation	Radon	^{220}Rn	α	54.5 sec
↓ Thorium A	Polonium	^{216}Po	α	0.16 sec
~ 100% | 0.014%				
Thorium B	Lead	^{212}Pb	β and α	10.6 hr
Astatine-216	Astatine	^{216}At	α	3×10^{-4} sec
↓ Thorium C	Bismuth	^{212}Bi	β and α	60.5 min
66.3% | 33.7%				
Thorium C′	Polonium	^{212}Po	α	3×10^{-7} sec
Thorium C″	Thallium	^{208}Tl	β	3.1 min
↓ Thorium D (end product)	Lead	^{208}Pb	None	Stable

TABLE 3. THE ACTINIUM SERIES

RADIOELEMENT	CORRESPONDING ELEMENT	SYMBOL	RADIATION	HALF-LIFE
Actinouranium	Uranium	^{225}U	α	7.07×10^{8} yr
↓ Uranium Y	Thorium	^{231}Th	β	25.6 hr
↓ Protactinium	Protactinium	^{231}Pa	α	3.25×10^{4} yr
↓ Actinium	Actinium	^{227}Ac	β and α	21.7 yr
98.8% | 1.2%				
Radioactinium	Thorium	^{227}Th	α	18.2 days
Actinium K	Francium	^{223}Fr	β	21 min
Actinium X	Radium	^{223}Ra	α	11.7 days
Ac Emanation	Radon	^{219}Rn	α	3.92 sec
Actinium A	Polonium	^{215}Po	α and β	1.83×10^{-3} sec
~ 100% | ~ 5×10^{-4}%				
Actinium B	Lead	^{211}Pb	β	36.1 min
Astatine-215	Astatine	^{215}At	α	~ 10^{-4} sec
↓ Actinium C	Bismuth	^{211}Bi	β and α	2.16 min
99.69% | 0.32%				
Actinium C′	Polonium	^{211}Po	α	0.52 sec
Actinium C″	Thallium	^{207}Tl	β	4.76 min
↓ Actinium D (end product)	Lead	^{207}Pb	None	Stable

TABLE 4. THE NEPTUNIUM SERIES

ELEMENT (2)	SYMBOL	RADIATION	HALF-LIFE
Curium (96) ↓	^{245}Cm	α	9300 yr
Plutonium (94) ↓	^{241}Pu	β	13.2 yr
Americium (95) ↓	^{241}Am	α	458 yr
Neptunium (93) ↓	^{237}Np	α	2.20×10^6 yr
Protactinium (91) ↓	^{233}Pa	β	27.4 days
Uranium (92) ↓	^{233}U	α	1.62×10^5 yr
Thorium (90) ↓	^{229}Th	α	7340 yr
Radium (88) ↓	^{225}Ra	β	14.8 days
Actinium (89) ↓	^{225}Ac	α	10.0 days
Francium (87) ↓	^{221}Fr	α	4.8 min
Astatine (85) ↓	^{217}At	α	1.8×10^{-2} sec
Bismuth (83) 96% 4%	^{213}Bi	β and α	47 min
Polonium (84)	^{213}Po	α	4.2×10^{-6} sec
Thallium (81)	^{209}Tl	β	2.2 min
Lead (82) ↓	^{209}Pb	β	3.3 hr
Bismuth (83) (end product)	^{209}Bi	None	Stable

in magnitude by e if a beta radioactive nucleus emits a negatron, one additional external electron must be added to maintain a neutral atom. On the other hand, the nuclear charge changes by $-e$ during positron emission; therefore, in order to maintain a neutral atom, an electron must also be lost by one of the atomic shells. Thus, for negatron decay the total energy difference between initial and final energy states is only the sum of the negatron kinetic energy and the neutrino energy. For positron emission, however, the atom loses a minimum energy equal to twice the rest energy of an electron, $2m_0c^2$. These energy relationships are shown schematically in the accompanying fig-

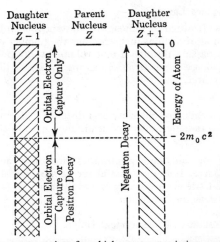

Fig. 1.   The energy regions for which negatron emission, positron emission, and orbital electron capture are energetically possible.

ure. During orbital electron capture the nucleus loses a single positive charge merely by taking an electron from one of its own atomic shells. The only energy loss is that energy emitted as X radiation during rearrangement of the atomic shells following electron capture and the energy carried away by the neutrino. See Fig. 1.

A table of nuclides showing mass number and isotopic abundance is given under **Chemical Elements.**

Analytical Procedures Using Radioisotopes

There is a wide range of applications for methods of analysis that are based upon the energies and intensities of the radiations emitted by radioactive nuclides. These techniques sometimes are termed *radiometric methods of analysis.* The methods are not restricted to the determination of substances initially radioactive, since there is wide use of methods involving the irradiation of stable nuclides to produce radioactive ones, followed by measurement of their radiations, from which the composition of the original stable substance can be inferred. This method is *radioactivation analysis.* Another method for the use of measurements of radioactivity in the analysis of stable substances is that of *tracer techniques,* that is, by the addition to them of radioactive nuclides, which can then be used to follow the course of various reactions or processes. There are various ways of introducing the radioactive nuclides, which are discussed later in this entry.

All methods of radiometric analysis involve, of course, the use of various radiation detection devices. The devices available for measuring radioactivity will vary with the types of radiations emitted by the radioisotope and the kinds of radioactive material. Ionization chambers are used for gases; Geiger-Müller and proportional counters for solids; liquid scintillation counters for liquids and solutions; and solid crystal or semiconductor detector scintillation counters for liquids and solids emitting high-energy radiations. Each device can be adapted to detect and measure radioactive material in another state, e.g., solids can be assayed in an ionization chamber. The radiations interact with the detector to produce a signal.

Since many radionuclides decay with gamma rays, many measurements are being made by gamma-ray scintillation spectrometry. Usually, a crystal detector, such as a sodium iodide crystal, is connected to a spectrometer. As described above, the gamma-rays interact with the crystal to produce light pulses which are converted to electrical pulses by a multiplier phototube. The pulse height analyzer of the spectrometer sorts out the gamma-rays of various energies. From this operation, a spectrum of the radionuclide's gamma-rays can be obtained to the *photopeaks* of full-energy pulses and the continuum of lower-energy pulses associated with the decay of the radionuclide. The photopeak, or photopeaks, in the gamma-ray spectrum can be used to identify and quantitatively measure the radionuclide.

The principle of *radioactivation analysis* is that a stable isotope when irradiated by neutrons, by charged particles such as protons or deuterons or by gamma rays, can undergo a nuclear reaction to produce a radioactive nuclide. After the radionuclide is formed, and its radiations have been characterized by radiation detection devices, calculations can be made of the elements contained in the sample before irradiation.

An important reaction used quite widely for this purpose is irradiation by neutrons and measurement of the energies of radiations emitted. The source of the neutrons may be a nuclear reactor, a particle accelerator, or an isotopic source, that is, a sealed container in which neutrons are produced by alpha rays emitted by a source such as radium, sodium-24(^{24}Na), yttrium-88(^{88}Y), etc., and arranged so that the alpha rays react with a substance such as beryllium which in turn emits neutrons. The neutrons react with stable nuclides in the sample to produce radioactive ones. Thus ordinary sodium undergoes a nuclear reaction with neutrons as follows:

$$^{23}Na + n \longrightarrow {}^{24}Na + \gamma$$

The ^{24}Na decays with a half-life of 15 hours to yield gamma rays and β-particles

$$^{24}Na \longrightarrow {}^{24}Mg + e^- + \gamma$$

Moreover the energies of these β-particles (electrons) are known to be 1.39 MeV and that of the gamma-rays 1.38 MeV so that the measured values of these magnitudes are characteristic of substances containing sodium. (Measurement of the γ-radiation is the usual procedure.) At least 70 of the elements can be activated in this way, by the capture of thermal neutrons, i.e., by *neutron activation analysis*. An activation analysis follows a procedure similar to that shown in Fig. 2. In almost all analy-

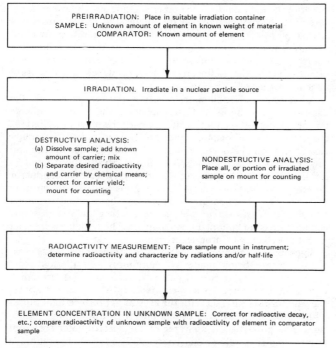

Fig. 2. Representative program followed in activation analysis.

ses, the sample materials are not treated before the bombardment, but are placed directly into the bombardment capsule or container. The length of the bombardment interval is usually determined by the half-life of the radionuclide used for the element of interest and the flux of nuclear particles.

The post-bombardment processing of the activated sample may follow either a nondestructive assay of the radioactivity in the sample (gamma-ray scintillation spectrometry is used most often for this) or a chemical processing of the sample prior to the radioactivity assay. Techniques involving either precipitation, electrodeposition, solvent extraction, and ion exchange or some combination of these form the basis of the radiochemical separation techniques used in activation analysis.

Neutron activation has been successfully applied to a great variety of determinations of small concentrations of elements present in alloys, for example, vanadium and manganese in iron. Other metallurgical applications include the determination of

some 70 elements: including the metals, aluminum, antimony, arsenic, barium, bismuth, cadmium, calcium, cerium, cesium, and so on alphabetically down the list. Minerals and soils have also been extensively analyzed by the method. However, it is not restricted to trace quantities or to inorganic substances, one interesting application being the determination of phosphorus, oxygen and nitrogen in organic phosphorus compounds. Sodium has been determined in blood plasma, and numerous other biochemical determinations have been made accurately. See also **Neutron.**

Tracer analysis is readily performed with radioactive isotopes because their ease of detection by measurement of their radioactivity makes them effective means of "tagging" their stable isotope counterparts (e.g., ^{24}Na(sodium-24) to tag ordinary sodium). Since compounds are usually involved, rather than elements, one merely synthesizes enough of the radioactive compound to tag the compound under analysis. The tagged compound may be followed through any analytical scheme, industrial system, or biological process. It is essential that a compound be tagged with an atom, however, which is not readily exchangeable with similar atoms in other compounds under normal conditions. For example, tritium could not be used to trace an acid if it were inserted on the carboxyl group where it is readily exchanged by ionization with the solvent.

Radiometric methods employing reagent solutions or solids tagged with a radionuclide have been used to determine the solubility of numerous organic and inorganic precipitates, or as a radioreagent for titrations involving the formation of a precipitate. In this type of application it is necessary to establish the ratio between radioactivity and weight of radionuclide plus carrier present. This may be established by evaporating an aliquot to dryness, weighing the residue, and measuring the radioactivity.

Closely related to tracer analysis is the method of *isotopic dilution analysis*. Here, instead of checking the effectiveness of a method from known amounts of an element in the sample, and of its radioactive isotope, one knows only the amount of radioactive isotope added, and by precipitating or otherwise separating the total amount of that element present, and then measuring its radioactivity, one determines its amount, and hence the amount present in the original sample.

Radioactive tracer methods lend themselves well to research applications in studying entire processes in science and industry, and in the biological as well as the physical sciences.

Inverse-Square Law

Radiation emitted by radioactive substances is uniformly distributed in all directions in space; thus the number of particles or quanta passing through a unit volume at any point distant from the source varies inversely as the square of the distance from the source, or

$$I = I_0 r_0^2 / r_i^2$$

where I_0 = radiation intensity at a distance r_0 from a source
I = radiation intensity at distance r_i from the same source.

The foregoing equation neglects absorption effects and assumes a point source. It is most useful for gamma radiation where the source-to-detector distance is usually much larger than any source or detector dimension.

Nuclear Radiation Measurement Units

Curi (Ci). The curie is the SI unit rate of radioactive decay and is defined as the quantity of any radioactive material having

3.7×10^{10} disintegrations per second. The *microcurie*, μCi (3.7×10^4 dps) and the *millicurie*, mCi (3.7×10^7 dps) are commonly used.

Roentgen (R). The roentgen is a measure of the intensity of ionizing radiation in air. It is defined as that quantity of gamma or x-radiation which produces 2.083×10^9 ion pairs (one electrostatic unit of charge, esu) per cubic centimeter of free air at a temperature of 0°C and a pressure of 1 atmosphere. Inasmuch as the same number of roentgens from various types of radiation produce different amounts of body damage, a term reflecting relative biological effectiveness was created. This is described next.

Roentgen Equivalent Man (REM). One REM is that amount of ionizing radiation of any type which produces the same damage to man as 1 roentgen of about 200 kV x-radiation (1 REM = 1 Rad in tissue/RBE. When the physical dose is measured in REP units (see below), the approximate definition, 1 REM ≈ 1 REP/RBE is used. See Table 5.

TABLE 5. VALUES OF RELATIVE BIO-
LOGICAL EFFECTIVENESS

TYPE OF RADIATION	rbe
X- and gamma radiation	1
Beta rays	1
Alpha rays	20
Fast neutrons[a]	10
Thermal or slow neutrons	5

[a] Having energies in the range 0.1 to 10 MeV. Above 10 MeV, the rbe increases rapidly.

Roentgen Equivalent Physical (REP). One REP is that amount of ionizing radiation of any type which results in the absorption of energy at the point in question in soft tissue to the extent of 93 ergs/gram. It is approximately equal to 1 roentgen of about 200-kV x-radiation in soft tissue.

Rad. A rad is an ionizing radiation unit which corresponds to an absorption of energy in any medium of 100 ergs/gram. One rad in tissue = 100/93 REP.

References

Benton, E. V., Henke, R. P., and C. A. Tobias: "Heavy-Particle Radiography," *Science*, **182**, 474–476 (1973).
Heath, R. L.: "Table of the Isotopes," in "Handbook of Chemistry and Physics," 61st Edition, CRC Press, Boca Raton, Florida (1980).
ICRP: "Radiation Protection in Uranium and Other Mines," International Commission on Radiological Protection, *Rept. 24, Annals of the ICRP*, **1** (1), Pergamon Press (1977).
Kathren, R. L., Selby, J. M., and E. J. Vallario: "A Guide to Reducing Radiation Exposure to as Low as Reasonably Achievable (ALARA)," DOE/EV/1830-T5, U.S. Dept. of Energy, Washington, D.C. (April 1980).
Kroger, F. A.: "Some Aspects of the Luminescence of Solids," Elsevier, Amsterdam (1948).
NCRP: "Instrumentation and Monitoring Methods for Radiation Protection," *Rept. 57*, National Council on Radiation Protection and Measurement (March 1978).
NCRP: "A Handbook of Radioactivity Measurements Procedures," *Rept. 58*, National Council on Radiation Protection and Measurement (November 1978).
Penzias, A. A.: "The Origin of the Elements," *Les Prix Nobel en 1978*, Nobel Foundation, Stockholm (1979). Also reprinted in *Science*, **205**, 549–554 (1979).
Sailer, S.: "State of the Industrial Isotope Technology," *Automobilindustrie*, **3**, 45 (1976).
Seaborg, G. T., Editor: "Transuranium Elements: Products of Modern Alchemy," Academic, New York (1979).
Seaborg, G. T., Loveland, W., and D. J. Morrissey: "Superheavy Elements: A Crossroads," *Science*, **203**, 711–717 (1979).
Williams, F.: "Theoretical Basis for Solid-State Luminescence," in "Luminescence of Inorganic Solids" (P. Goldberg, Editor), Academic Press, New York (1966).

RADIOACTIVITY AND OTHER DATING TECHNIQUES. A radioisotope (radionuclide) is an isotopic form (either natural or artificial) of an element that exhibits radioactivity. The Uranium, Thorium, and Actinium Series of elements are naturally occurring radioactive elements. Each isotope has a specific half-life, that is, the time required for an unstable element to lose one-half of its radioactivity intensity in the form of radiation (alpha, beta, and gamma). The half-life is a constant for each unstable element or nuclide. Half-lives range from fractions of a second, as in the cases of many artificially produced radioactive elements, to millions of years. Obviously, radioactive isotopes with the longer half-lives are those of interest for long-term age dating in connection with such determinations as made of archeological objects, fossils, and geological structures and formations.

Age determinations using radioactive nuclides may be looked upon as processes that are the inverse of half-life measurements. If a radionuclide of known half life exists within an object, the age of that object can be determined either by measuring the number of radionuclides that remain or the number of product nuclides of the radioactive decay. In these determinations it is assumed that, if we know the half life of the radionuclide, as elapsed time t, or age, for the object can be found by using the formula $t = (\ln N_1/N_2)/\lambda$, where λ is the decay constant of the radionuclide and N_1 and N_2 are the amounts of the radionuclide present at the beginning and the end of the interval spanning the time t.

In any use of radioactive dating or age determining processes, a basic assumption is, in general, that the concentration of the radioactive element is changed during the life of the sample only by its natural decay process, and that the accuracy of the determination depends primarily, therefore, upon the accuracy with which the half-life of that radionuclide is known.

Ages of specimens may sometimes be determined by other methods than the measurement of radioactivity, as by combination of radioactive measurements with mass spectroscopic determinations. An application of this last type has been made to determine the age of meteorites, which includes a calculation of the ratio of concentrations of the various isotopes of lead, which are of radiogenic and non-radiogenic origin. Comparatively recently, newer measurement techniques have been suggested and used. These will be described a bit later.

Age of Rocks. In the table of nuclides given under **Chemical Elements,** there are listed a number of naturally-occurring radionuclides with long half-lives. From these known half-lives, the geological age of a rock may be calculated. One method of making this estimate is based upon the amount of radionuclide and its daughter nuclide contained in the rock. This method is based upon various assumptions which may be stated as follows:

1. Since the rock was formed, the parent nuclidic content of the mineral has been changed only by radioactive decay.

2. All the decay products produced by the parent nuclide have been retained since the mineral was formed.

3. The geological separation of the parent and daughter elements at the time of formation of the mineral was sufficient to make the determination of the decay products unambiguous.

For example, if a uranium mineral does not exclude all lead at the time it is formed, the isotopic abundance of the lead at the time of formation cannot be calculated with certainty.

4. The radioactive decay scheme of the parent nuclide is well known.

The contributions of modern chemistry, including the availability of separated isotopes, the extension of the range of mass spectrometers, and the developments of new chemical methods, which make possible the determination of microgram quantities, have extended the range of application of radioactive age measurements. This extension has been either to minerals which contain relatively little of the parent element, but maintain a good separation of the parent and daughter elements when they are formed; or to minerals containing radioactive elements that have a very low natural abundance, such as ^{40}K, or a very long half-life, such as ^{87}Rb. Although these extensions have in turn introduced certain new problems and forced some compromises, they have made possible certain conclusions about geological questions and have opened new avenues for research.

A number of possible radioactive dating methods exist, but each method is practical, of course, only if the appropriate radionuclide exists in the mineral. One series of possible dating methods is based on the decay of natural uranium and natural thorium. If the rock has retained the helium produced by the decay of ^{238}U, for example, 8 helium atoms should exist for each nuclide of ^{238}U that has decayed through its complete chain to ^{206}Pb, since 8 alpha particles result from this chain. From a measure of the ratio of the amount of helium to the amount of ^{238}U in the rock, a calculation may then be possible of the age of the rock. In this method, corrections must be made for the decay of ^{235}U and of ^{232}Th, both of which are the initiating nuclides for a natural chain of radioactive nuclides. Because the half lives of ^{232}Th and ^{238}U are different, another method for determining the age of a rock containing both these nuclides is the measurement of the ratio of the amount of ^{206}Pb to the amount of ^{208}Pb, which are the ultimate decay products of the ^{238}U and ^{232}Th chains, provided neither of these isotopes of lead existed in appreciable quantity prior to formation of the rock. A related measurement is the ratio of radiogenic lead (either ^{206}Pb or ^{208}Pb) to nonradiogenic lead (^{204}Pb), which can be assumed to have been of primordial origin. Another correction that may be necessary, especially if the rock comes from a high altitude, is a determination of the amount of helium that has been produced as a result of spallation reactions caused by very high-energy cosmic radiation. Other radioactive age-dating systems are those of potassium-argon (which consists of the decay of ^{40}K to ^{40}Ar, by electron capture, a process with a half-life of 1.27×10^{10} years) and rubidium-strontium which consists of the decay of ^{87}Rb to ^{87}Sr, by electron emission, a process having a half-life of 4.7×10^{10} years.

One conclusions drawn from radioactive measurements is that the pre-Cambrian history of the earth's crust extends beyond 2,700 million years. The pegmatites that have been found to be this old are located in North America and Australia, and they probably exist on all the continents. The oldest rocks in the United States that have been measured are on the south rim of the Bridger Mountains near the Wind River Canyon in Wyoming. These ancient pegmatites intrude geologic formations of sedimentary and volcanic rocks that themselves are the result of even more ancient processes than those in which they were formed. Thus, a period of the order of 3,000 million years or more is available for geologic processes that have formed the crust seen today.

Next, the facility to measure the absolute age of micas in igneous intrusives of pre-Cambrian sediments provides a method of correlating these sediments wherever they occur in much the same fashion that fossil correlation of more recent sedimentary formations is possible. A method that is independent of the lithologic characteristics and the general structure of the sediments will provide a crucial test of the validity of these criteria, which have been all that was available to the geologist. Further, any attempts to look for more subtle evidence of such things as changes in the composition of the atmosphere or origins of life itself must be fitted into a time scale of the pre-Cambrian.

Radiocarbon Dating

This is a method of estimating the age of carbon-containing materials by measuring the radioactivity of the carbon in them. The validity of this method rests upon certain observations and assumptions, of which the following statement is a brief summary. The cosmic rays entering the atmosphere undergo various transformations, one of which results in the formation of neutrons, which in turn induce nuclear reactions in the nuclei of individual atoms of the atmosphere. The dominant reaction is

$$n + {}^{14}N \rightarrow {}^{14}C + p$$

in which the neutrons react with the nuclei of nitrogen atoms of mass number 14 (which make up the nitrogen molecules that constitute nearly $\frac{4}{5}$ of the atmosphere) to form carbon atoms of mass number 14 and protons (p). The ^{14}C atoms are radioactive, having a half life of about 5730 years. The largest rate of formation of ^{14}C atoms from cosmic rays is at 30,000–50,000 feet (9144–15,240 meters) above sea level and at higher geomagnetic latitudes, although formation occurs at varying rates throughout the entire atmosphere. The ^{14}C atoms react with oxygen in the atmosphere to form carbon dioxide, which is mixed with the nonradioactive carbon dioxide in the atmosphere, and with it gains world-wide distribution by various processes. The radioactive $^{14}CO_2$ enters the carbon cycle in which plants take up carbon dioxide from the atmosphere to form carbohydrates, which enter through plant foods into the composition of animals. In another world-wide process, also of exchange nature, carbon dioxide is dissolved in seawater and then, under changing conditions of acidity and temperature, is partially evolved from the seawater again. As a result of these and other processes, the ^{14}C formed in the atmosphere by cosmic rays tends to become distributed throughout all the nonradioactive carbon, not only in the atmosphere, but in the biosphere, the hydrosphere, and even the upper levels of the lithosphere (there are many carbonate-containing minerals).

Obviously this wide distribution of the ^{14}C formed in the atmosphere takes time; it is believed to require a period of 500–1,000 years. This time is not, however, a deterrent to radiocarbon dating because of two factors: the long half life of ^{14}C and the relatively constant rate of cosmic-ray formation of ^{14}C in the earth's atmosphere over the most recent several thousands of years. These considerations lead to the conclusion that the proportion of ^{14}C in the carbon reservoir of the earth is constant, the addition by cosmic ray production being in balance with the loss by radioactive decay. If this conclusion is warranted, then the carbon dioxide on earth many centuries ago had the same content of radioactive carbon as the carbon dioxide on earth today. Thus, radioactive carbon in the wood of a tree growing centuries ago had the same content as that in carbon on earth today. Therefore, if we wish to determine how long ago a tree was cut down to build an ancient fire, all we need to do is to determine the relative ^{14}C content of the carbon in the charcoal remaining, using the value we have determined for the half life of ^{14}C. If the carbon from the charcoal in an ancient cave has only $\frac{1}{2}$ as much ^{14}C radioactivity as does carbon on earth today, then we can conclude that the tree which furnished the firewood grew 5730 ± 30 years ago.

As pointed out by Muller (1977), there are well-documented differences between the ages of materials determined by dating with radioisotopes and the ages determined by other means, such as tree-ring counting. In addition to systematic effects, there are statistical errors due to the limited number of atoms observed. Both types of errors can be considered to be fluctuations in n, the number of atoms observed. A relationship can be derived between the magnitude of these fluctuations and the resulting error in the estimation of age of the sample:

$$n = ke^{-t/\tau} \quad \text{or} \quad t = \tau \ln (n/n)$$

where τ is the mean life of the isotope, t is the age of the sample, and k is the initial number of radioactive atoms in the sample multiplied by the efficiency for detecting them. If n has errors associated with it of $+\delta n_1$ and $-\delta n_2$, then the corresponding values of t will be:

$$t = \tau \ln \frac{k}{n_{-\delta n_2}^{+\delta n_1}} = \tau \ln \left(\frac{k}{n}\right)_{-|\ln [1+(\delta n_1/n)]|}^{+|\ln [1-(\delta n_2/n)]|}$$

Muller has shown that for $n = 1$, inverse Poisson statistics given $n_1 = 1.36$ and $n_2 = 0.62$ and thus the foregoing equation becomes:

$$t = \tau \ln (k)_{-0.86\tau}^{+0.96\tau}$$

Further details of this method can be found in aforementioned reference.

Determination of the ratio of two oxygen isotopes has been effective in fixing the age of fossil sediments and can provide information about ice formation and, possibly, water temperatures. The lighter isotope ^{16}O evaporates preferentially and thus precipitation and hence ice in glaciers and polar caps should be enriched with ^{16}O relative to seawater. Thus, fluctuations in the amount of water locked up as ice can be determined from variations in the oxygen isotope ratio of fossils which have been locked up in deep-sea sediments. And, because this ratio also varies with water temperature, thermal information also can be gleaned. Kennett (University of Rhode Island) has employed this technique in determining when significant amounts of ice first formed at the poles. This research has indicated that the Antarctic ice cap formed only about 16 million years ago, after Australia had split off and moved away from Antarctica, leaving the latter continent isolated at the pole and surrounded by the fast-moving circumpolar current. More details are given by Hammond (1976).

Dating with a Cyclotron. The cyclotron is mainly used as a source of energetic particles. The cyclotron also can be used as a very sensitive mass spectrometer. Alvarez and Cornog (1939) were the first researchers to use a cyclotron in this manner. This was in connection with their discovery of the true nuclear properties of ^3He and tritium. Within the last few years, Muller and associates at the Lawrence Berkeley Laboratory have used this method in a search for integrally charged quarks in terrestrial material. For radioisotope dating, the cyclotron is tuned to accelerate the isotope of interest and the sample is introduced into the ion source, preferably as a gas; the greatest gains over radioactive counting techniques apply to the longer-lived species, which have lower decay rates. It has been estimated that the cyclotron can be used to detect atoms or simple molecules that are present at the 10^{-16} level or greater. For ^{14}C dating, the Berkeley investigators indicate that one should be able to go back 40,000–100,000 years with 1–100-microgram carbon samples; for ^{10}Be dating, 10–30 million years with from

1 cubic millimeter to 10 cubic centimeter rock samples; and for tritium dating, 160 years with a 1-liter water sample. Over 50 cyclotrons are in operation today that could perform radioisotope dating and, although the instruments are costly, the cost for a dating determination experiment may not be much higher than for decay dating technology.

Other isotopes with which an accelerator mass spectrometer may be effective include ^{26}Al, ^{36}Cl, ^{53}Mn, ^{81}K, and ^{129}I. Chlorine-36 has a half-life of 300,000 years and may be used for dating water in underground reservoirs. ^{10}Be is produced in the atmosphere by cosmic rays that break up oxygen and nitrogen nuclei. ^{10}Be has been used in studies of both seafloor spreading and manganese nodule formation. Although tritium (^3H) has a short mean life of 17.8 years, tritium dating has been important in cosmic-ray physics, hydrology, meterology, and oceanography. For example, if one desires to know how long an underground water reservoir may require for refilling, the age of the water can be determined by tritium dating methods.

Conventional ^{14}C dating by means of gas proportional counters has been extended to samples containing as little as 10 milligrams of carbon. The accuracy of the dating procedure has been checked by dating sequoia tree-ring samples of the 1st Century A.D. and B.C. and an oak tree-ring sample of the 19th Century A.D. This work has been conducted at the Brookhaven National Laboratory (Harbottle, Sayre, and Stoenner, 1979).

Non-Radioactivity Dating Techniques

Several methods in addition to those involving radioactivity have been used to estimate the age of various materials and objects.

Obsidian Hydration Rate. Obsidian (rhyolitic volcanic glass) can be used as a key to age determinations for both archeological and geological purposes. As pointed out by Friedman and Long (1976), the method depends upon the fact that obsidian absorbs water from the atmosphere to form a hydrated layer, which thickens with time as the water slowly diffuses into the glass. The hydrated layer can be observed and measured under a microscope on thin sections cut normal to the surface. To convert the measured hydration thickness to an age, the equation relating to time must be known. This requires not only the form of the equation (functional dependence), but also the constants in it. Prior to the early 1960s, age could be related to hydration thickness only if combined with known history of a region or through the use of carbon-14 techniques. In the mid-1960s, Friedman and associates conducted actual experimental hydration experiments on obsidian, exposing the materials (taken from the Valles Mountains in New Mexico) to a temperature of 100°C and steam at a pressure of 1 atmosphere over a 4-year period. An equation of the form, $T = kt^{1/2}$ was developed, where T = thickness of hydration layer, t = time, and k is a constant. Investigators have developed a procedure for calculating hydration rate of a sample from its silica content, refractive index, or chemical index and a knowledge of the effective temperature at which the hydration occurred. The effective hydration temperature (EHT) can either be measured or approximated from weather records. The investigators concluded that if the EHT can be determined and measured for the hydration of a particular obsidian, it should be possible to carry out absolute dating to $\pm 10\%$ of the true age over periods as short as several years and as long as millions of years.

Manufactured Glass Objects. Other investigators (Lanford, 1977) have extended the principles applying to obsidian to manufactured glass, which extends back for thousands of years and thus can be useful to archeologists. However, as observed by Lanford, one cannot use the same optical method for measuring the thickness of hydration layers as used with obsidian. The hydration of the two materials differs. Also, glass that is less than a few hundred years old would generally have hydration layers thinner than the wavelength of visible light. The optical method is destructive in that it requires removal of a slice of glass from an object, something much discouraged by art historians and dealers. The Lanford method involves a resonant nuclear reaction between ^{15}N and ^{1}H for measuring the distribution of hydrogen in solids. With this technique, complete depth profiles of the surface hydration layer can be obtained in a fully nondestructive manner. Lanford summarizes by observing that this method of hydration dating need not be limited to glass. Since most silicates are unstable against slow reactions with atmospheric water, many may develop surface hydration layers suitable for dating and authenticating. The glazes on pottery are chemically similar to glass, and it may be possible that a dating method for glazed pottery based upon these procedures can be developed.

Amino Acid Racemization. This dating method is based upon the incorporation of L-amino acids exclusively into proteins by living organisms. As pointed out by researchers Masters and Zimmerman (1978), given sufficient periods of time over which proteins are preserved after synthesis, a number of spontaneous chemical reactions take place. Among these is racemization, which converts L-amino acids into their enantiomers, the D-amino acids. The different amino acids racemize at various rates, and these rates (true of all chemical reactions) are proportional to temperature. One of the fastest racemization rates known is that of aspartic acid, with a half-life of 15,000 years at 20°C. It follows, then, that the older a fossilized material may be, the higher will be its D-aspartic acid content or D/L Asp ratio. Once the k_{Asp} is known for a given fossil locality, the age of a specimen can be calculated from the D/L ratio.

This method was used in the examination of an Eskimo who died 1600 years ago. The body was discovered in a frozen state on St. Lawrence Island, Alaska in 1972 and remained frozen until it was brought to Fairbanks in 1973. Examination of the female individual revealed that she had a skull fracture, probably resulting from instant burial caused by a landslide. Aspartic acid racemization analysis of a tooth from the mummy yielded an age at death of 53 ± 5 years, which correlated well with earlier estimates based upon morphological features. This method is an example of the need to preserve mummies (Alaskan, Egyptian, and Peruvian, among others) for application of new dating techniques as they develop.

References

Alvarez, L. W., and R. Cornog: *Phys. Rev.*, **56**, 379 (1939).
Bennett, C. L., et al.: "Radiocarbon Dating with Electrostatic Accelerators: Dating of Milligram Samples," *Science*, **201**, 345–347 (1978).
Friedman, I., and W. Long: "Hydration Rate of Obsidian," *Science*, **191**, 347–352 (1976).
Grootes, P. M.: "Carbon-14 Time Scale Extended: Comparison of Chronologies," *Science*, **200**, 11–15 (1978).
Harbottle, G., Sayre, E. V., and R. W. Stoenner: "Carbon-14 Dating of Small Samples by Proportional Counting," *Science*, **206**, 683–684 (1979).
Lanford, W. A.: "Glass Hydration: A Method of Dating Glass Objects," *Science*, **196**, 975–976 (1977).
Levy, P. W.: "Proceedings of the International Seminar on the Application of Science to the Dating of Works of Art" (W. J. Young, Editor), Museum of Fine Arts, Boston, Massachusetts (1978).
Masters, P. M., and M. R. Zimmerman: "Age Determination of an Alaskan Mummy: Morphological and Biochemical Correlation," *Science*, **201**, 811–812 (1978).
Maugh, T. H., II: "Radiodating: Direct Detection Extends Range of the Technique," *Science*, **200**, 636–638 (1978).
Muller, R. A.: "Radioisotope Dating with a Cyclotron," *Science*, **196**, 489–494 (1977).
Muller, R. A., Stephenson, E. J., and T. S. Mast: "Radioisotope Dating with an Accelerator: A Blind Measurement," *Science*, **201**, 347–348 (1978).
Nelson, D. E., Korteling, R. G., and W. R. Stott: "Carbon-14: Direct Detection at Natural Concentrations," *Science*, **198**, 507–508 (1977).
Stuiver, M.: "Carbon-14 Dating: A Comparison of Beta and Ion Counting," *Science*, **202**, 881–883 (1978).

RADIOACTIVITY (Mineral). Mineraology.

RADIOISOTOPES. Radioactivity.

RADIUM. Chemical element symbol Ra, at. no. 88, at. wt. 226.026, periodic table group 2a (alkaline earth), mp 700°C, bp 1140°C, density, 5g/cm³ (20°C). Radium metal is white, rapidly oxidized in air, decomposes water, and evolves heat continuously at the rate of approximately 0.132 calorie per hour per milligram when the decomposition products are retained, and the temperature of radium salts remains above 1.5°C above the surrounding environment. Radium is formed by radioactive transformation of uranium, about 3 million parts of uranium being accompanied in nature by one part radium. Radium spontaneously generates radon gas at approximately the rate of 100 mm³ per day per gram of radium, at standard conditions. Radium usually is handled as the chloride or bromide, either as solid or in solution. The radioactivity of the material decreases at a rate of about 1% each 25 years. All isotopes of radium are radioactive. See **Radioactivity.** The first ionization potential of radium is 5.277 eV; second, 10.099 eV. Other important physical properties of radium are given under **Chemical Elements.**

One year after the discovery of x-rays by Röntgen (1895), Henri Becquerel investigated the relationship between the phosphorescence of various salts after their exposure to sunlight and the fluorescence in an operating x-ray tube. One of the salts under investigation was potassium-uranium sulfate. After exposure to sunlight, Becquerel noted that the salt not only emitted visible light, but also rays similar to x-rays that were able to penetrate the heavy black paper and thin metal foils within which his photographic plates were wrapped. During a period of cloudy weather, Becquerel stored the salt and photographic plates in a closet, awaiting further sunny days. Later, when he inspected the package, he noted that a very intense image had been developed on the photographic plate even though it had not received much prior exposure to sunlight. By further experiments, Becquerel confirmed that the intense image was derived directly from the presence of the salt, regardless of any exposure to sunlight. This constituted the first demonstration of radioactivity. Through further investigations, Ernest Rutherford demonstrated that both alpha and beta radiations were emitted by the salt. Rutherford learned that the alpha rays were easily absorbed by thin sheets of paper, whereas the beta rays acted in the same manner as observed by Becquerel. Later Mdme. Marie Curie found that thorium produced about the same intensity of radioactivity as uranium. Further tests disclosed that the uranium ore with which she was working (pitchblende) exhibited more radioactivity than could be accounted for by its uranium content alone. Subsequently, Mdme. Curie and her husband, Pierre Curie, successfully separated two previously unknown elements, radium and polonium. Thus, both radium and polonium were identified as chemical elements in 1898. It was found that each of these elements was over a million times more radioactive than uranium.

Radium gained prominence not only from its scientific interest, heralding a whole new area of physics and chemistry, but from its wide use in therapeutic medicine, as an ingredient (very dangerous) of luminous paints, and in various instruments for inspecting structures, such as metal castings. Commercially, radium generally is marketed as the bromide or sulfate and is extremely radioactive in these forms. Use of radium in medical technology has largely been replaced by other sources of radioactivity.

Radium occurs in pitchblende, and in carnotite along with uranium. Radium was first obtained from the uranium residues of pitchblende of Joachimsthal, Czechoslovakia, later from carnotite of southwestern Colorado and eastern Utah. Richer ores have been found in Republic of Congo and in the Great Bear region of north-western Canada.

The radium isotope of mass number 226 occurs in the uranium $(2n + 2)$ alpha-decay series. Its half-life is 1,620 years, and it yields radon-222 by α-disintegration. Other naturally occurring isotopes of radium are ^{228}Ra in the thorium series, half-life 6.7 years, producing actinium-228 by β-decay, which yields by β-decay thorium-228, which in turn yields ^{224}Ra, half-life 3.64 days, giving radon-220 by α-decay. Another naturally occuring isotope of radium is found in the actinium series: it is ^{223}Ra, half-life 11.7 days, giving radon-219 by α-decay. In the neptunium series there is ^{225}Ra, half-life 14.8 days, undergoing β-decay to actinium-225. Other isotopes of radium include those of mass numbers 219, 221, 225, 227, 229, and 230.

Chemically related to barium, radium is recovered from its ores by addition of barium salt, followed by treatment as for recovery of barium, usually as the sulfate. The sulfates of barium and of radium are insoluble in most chemicals, so they are transformed into carbonate or sulfide, both of which are readily soluble in HCl. Separation from barium is accomplished by fractional crystallization of the chlorides (or bromides, or hydroxides). Dry, concentrated radium salts are preserved in sealed glass tubes, which are periodically opened by experienced workers to relieve the pressure. The glass tubes are kept in lead shields.

In many of its chemical properties, radium is like the elements of magnesium, calcium, strontium, and barium, and it is placed in group 2, as is consistent with its $6s^26p^67s^2$ electron configuration. Its sulfate ($K_{sp} = 4.2 \times 10^{-15}$) is even more insoluble in water than barium sulfate, with which it is conveniently coprecipitated. Like barium and other alkaline earth metals, it forms a soluble chloride ($K_{sp} = 0.4$) and bromide, which can also be obtained as dihydrates. Radium also resembles the other group 2 elements in forming an insoluble carbonate and a very slightly soluble iodate ($K_{sp} = 8.8 \times 10^{-10}$).

See the list of references at the end of the entry on **Chemical Elements**.

RADON. Chemical element symbol Rn, at. no. 86, at. wt. 222 (mass number of the most stable isotope), periodic table group 0 (inert gases), mp $-71°C$, bp $-61.8°C$. First ionization potential, 10.745 eV. Density, 9.72 g/liter ($0°C$, 760 torr), 7.5 times more dense than air. The gas has been liquefied at $-65°C$ and solidified at $-110°C$. Radon was first isolated by Ramsay and Gray in 1908. Prior to acceptance of the present designation, radon was called *niton* or *radium emanation*. See **Radioactivity**.

Isotope ^{222}Rn is formed by the alpha disintegration of ^{226}Ra. Actinon, its isotope of mass number 219, is produced by alpha disintegration of ^{223}Ra (AcX) and is a member of the actinium series. Similarly, thoron, its isotope of mass number 220, is a member of the thorium series. Since the name "radon" may be considered to be specific for the isotope of mass number 222 (from the radium series), the term "emanation" is sometimes

used for element number 86 in general. Other isotopes of radon include those of mass numbers 209–218 and 221.

A fluorine compound of radon has been formed by reaction of the elements under higher temperature and pressure, similar to the conditions for forming xenon fluorides. Radon forms a hydrate of amospheric pressure at $0°C$. It forms a compound with phenol, $Rn \cdot 2C_6H_5OH$ that is stable enough to give a sharply defined melting point at $50°C$. At low temperatures and pressures, HCl, hydrobromic acid, H_2S, SO_2, and CO_2 all add considerable percentages of radon; the HCl product, although possibly not a compound in the classical sense, being stable enough for its use as a method for separating radon from other gases.

See list of references at end of entry on **Chemical Elements**.

RAOULT'S LAW. The vapor pressure of a substance in solution is proportional to its mole fraction. See also **Vapor Pressure**.

RAPESEED OIL. **Vegetable Oils (Edible).**

RARE EARTH ELEMENTS AND METALS. Sometimes referred to as the "fraternal fifteen," because of similarities in physical and chemical properties, the rare-earth elements actually are not so rare. The term *rare* arises from the fact that these elements were discovered in scarce materials. The term *earth* stems from the fact that the elements were first isolated from their ores in the chemical form of oxides and that the old chemical terminology for oxide is earth. The rare-earth elements, also termed *lanthanides*, are similar in that they share a valence of 3 and are treated as a separate side branch of the periodic table, much like the actinides.

The properties of the lanthanides are given in Tables 1 and 2. Pronunciation of the elements is as follows: Cerium (*sear' ium*), dysprosium (*dis prōz' ium*), erbium (*ur' bium*), europium (*yoo rō' pium*), gadolinium (*gado lin' ium*), holmium (*hol' mium*), lanthanum (*lan' tha num*), lutetium (*loo tee' shium*), neodymium (*neo dim' ium*), praseodymium (*pra zee o dim' ium*), promethium (*pro mee' thium*), samarium (*sa mar' ium*), terbium (*tur' bium*), thulium (*thoo' lium*), ytterbium (*i tur' bium*), and yttrium (*it' rium*).

The lanthanides are further described by individual alphabetical entries for each element.

C. A. Arrhenius, in 1787, noted an unusual black mineral in a quarry near Ytterby, Sweden. This was identified later as containing yttrium and rare-earth oxides. With the exception of promethium, all members of the lanthanide series had been discovered by 1907, when lutetium was isolated. In 1947, scientists at the Oak Ridge National Laboratory (Tennessee) produced atomic number 61 from uranium fission products and named it promethium. No stable isotopes of promethium have been found in the earth's crust.

Natural mixtures of the lanthanides have been used commercially since the early 1900s. *Mischmetal* is the source of the hot spark in cigarette lighter flints. The mixed rare-earth fluorides are burned in the cores of carbon electrodes to create intense sunlike illumination required by motion picture projectors and searchlights. The mixed rare-earth oxides are used to grind and polish almost all optical lenses and television faceplates. In the late 1940s, it was discovered that the rare-earth metals effectively control the shape of carbon in normally brittle cast iron, resulting in ductile or nodular iron. During the 1950s, interest in several of the pure elements (europium, gadolinium, dysprosium, samarium, and erbium) was stimulated because these elements have the highest thermal-neutron-absorption properties among the elements. These elements have found application in the nuclear reactor field as control rods and as

TABLE 1. ATOMIC AND THERMAL PROPERTIES OF RARE-EARTH ELEMENTS

Atomic Number / Symbol / Element	39 Y Yttrium	57 La Lanthanum	58 Ce Cerium	59 Pr Praseodymium	60 Nd Neodymium	61 Pm Promethium	62 Sm Samarium	63 Eu Europium	64 Gd Gadolinium	65 Tb Terbium	66 Dy Dysprosium	67 Ho Holmium	68 Er Erbium	69 Tm Thulium	70 Yb Ytterbium	71 Lu Lutetium
Estimated abundance: ppm	33	30	60	8.2	28	0	6.0	1.2	5.4	0.9	3.0	1.2	2.8	0.5	3.0	0.5
g/ton	28–70	5–18	20–46	3.5–5.5	12–24	10–19	4.5–7	0.14–1.1	4.5–6.4	0.7–1	4.5–7.5	0.7–1.2	2.5–6.5	0.2–1	2.7–8	0.8–1.7
Atomic constants: Atomic weight, (CN = 12)	88.91	138.91	140.12	140.91	144.24	(145)	150.35	151.96	157.25	158.92	162.50	164.93	167.26	168.93	173.04	174.97
Metallic radius, Å, (CN = 12)	1.801	1.879	(+3) 1.846 (+4) 1.672	1.828	1.821	1.811	1.804	(+2) 2.042 (+3) 1.798	1.801	1.783	1.774	1.766	1.757	1.746	(+2) 1.939 (+3) 1.741	1.735
Volume, cm³/g atom	19.89	22.53	(+3) 21.43 (+4) 15.92	20.81	20.60	(20.17)	19.95	(+2) 28.93	19.91	19.30	19.03	18.78	18.49	18.14	(+2) 24.82 (+3) 17.98	17.79
Density, g/cm³	4.469	6.146	6.770	6.773	7.008	7.264	7.520	5.244	7.901	8.230	8.551	8.795	9.066	9.321	6.966	9.841
lb/in.³	0.161	0.222	0.244	0.244	0.253	0.262	0.271	0.189	0.285	0.297	0.308	0.317	0.327	0.336	0.251	0.355
Crystal structure at 25°C	hcp	dhcp	fcc	dhcp	dhcp	dhcp	rhom	bcc	hcp	hcp	hcp	hcp	hcp	hcp	fcc	hcp
Unpaired 4f electrons	0	0	1	2	3	4	5	6	7	6	5	4	3	2	1	0
Number of isotopes: Natural	1	2	4	1	7	0	7	2	7	1	7	1	6	1	7	2
Artificial	14	19	15	14	7	15–18	11	16	11	17	12	18	12	17	10	14
Lattice constants, A: a	3.648	3.774	5.161	3.672	3.658	3.65	3.629	4.583	3.634	3.605	3.592	3.578	3.559	3.538	5.485	3.505
c	5.732	12.171		11.833	11.797	11.656	26.207	—	5.781	5.697	5.650	5.618	5.585	5.554	—	5.549
Ionic radius, Å: +2							1.111	1.09							0.93	
+3	0.893	1.061	1.034	1.013	0.995	0.978	0.964	0.950	0.938	0.923	0.908	0.894	0.881	0.869	0.858	0.848
+4			0.92	0.90							0.84					
Color of 3+ ion (in solution)	Colorless	Colorless	Colorless	Green	Reddish violet	Pink	Yellow	Pale pink	Colorless	Almost colorless	Yellow	Pink	Reddish violet	Green	Colorless	Colorless
Electronegativity	1.177	1.117	(+3) 1.123 (+4) 1.43	1.130	1.134	1.139	1.145	(+2) 0.98 (+3) 1.152	1.160	1.168	1.176	1.184	1.192	1.200	(+2) 1.02 (+3) 1.208	1.216
Absorption bands, 3+ ion, Å	None	None	2105 2220 2380 2520	4445 4690 4822 5885	3540 5218 5745 7395 7420 7975 8030 8680	5485 5680 7025 7355	3625 3745 4020	3755 3941	2729 2733 2754 2756	3694 3780 4875	3504 3650 9100	2870 3611 4508 5370 6404	3642 3792 4870 5228 6525	3600 6825 7800	9750	None
Thermal properties: Melting point: °C	1522	918	798	931	1021	1042	1074	822	1313	1365	1412	1472	1529	1545	819	1663
°F	2772	1684	1468	1708	1868	1908	1965	1512	2395	2489	2574	2685	2784	2813	1506	3025
Boiling point at 1 atm.: °C	3338	3464	3433	3520	3074	3000	1794	1529	3273	3230	2567	2700	2868	1950	1196	3402
°F	6040	6267	6211	6368	5565	5432	3261	2784	5923	5846	4653	4892	5194	3542	2185	6156
Heat of fusion ΔHf, kcal/g atom	2.724	1.482	1.305	1.646	1.705	1.84	2.061	2.202	2.403	2.580	2.643	4.032	4.757	4.025	1.830	4.457
Heat of sublimation ΔHs at 25°C, kcal/g atom	101.287	103.084	101.146	85.286	78.507	64	49.257	42.5	95.347	93.374	70.038	72.330	76.086	55.787	36.473	102.245
Heat capacity ΔCp at 25°c, cal/(g atom)(°C)	6.34	6.48	6.44	6.56	6.56	6.55	7.06	6.61	8.86	6.91	6.72	6.49	6.71	6.46	6.39	6.40
Coefficient of expansion, per °C × 10⁻⁶	10.8	12.1	6.3	6.7	9.6	11.0	12.7	35	9.4	10.3	9.9	11.2	12.2	13.3	26.3	9.9
Nuclear properties: Thermal neutron capture, barns/atom	1.31	8.9	0.73	11.6	50		5,600	4,300	40,000	46	1100	64	170	125	37	108

* Table compiled by Molybdenum Corporation of America, White Plains, N.Y. (Joseph G. Cannon); edited by Rare-Earth Information Center, Energy and Mineral Resources Research Institute, Iowa State University, Ames, Iowa (Karl A. Gschneidner, Jr. and N. Kippenhan). Data from S. R. Taylor, Abundance of Chemical Elements in the Continental Crust: A New Table, *Geochim. Cosmochim. Acta*, vol. 28, pp. 1273–1285, 1964; E. T. Teatum, et al, Compilation of Calculated Data Useful in Predicting Metallurgical Behavior of Elements in Binary Alloy Systems, *Univ. Calif., Los Alamos Sci. Lab. Rep. LA-4003*, pp. 11–12, Dec. 24,

1968; Clifford A. Hampel, "Rare Metals Handbook," 2d ed., chaps. 1 and 35, Van Nostrand Reinhold Company, New York, 1961; O. A. Songina, "Rare Metals: Scandium, Yttrium, Lanthanide and Actinides," chap. 6, trans. from Russian (1970), 3d ed. (1964), U.S. Dept. of Interior and The National Science Foundation, Washington, D.C.; Karl A. Gschneidner, Jr., "Solid State Physics," vol. 16, "Physical Properties and Interrelationships of Metallic and Semimetallic Elements," pp. 275–426, Academic, New York, 1964; Clifford A. Hampel, "The Encyclopedia of the Chemical Elements," Van Nostrand Reinhold Company, New York, 1968,

R. Hultgren, R. L. Orr, and K. K. Kelley, supplement to "Selected Values of Thermodynamic Properties of Metals and Alloys," Wiley, New York, 1963; Data from Department of Mineral Technology and Lawrence Radiation Laboratory, The University of California, Berkeley, Calif. (data and revision published periodically). Data from Karl A. Gschneidner, Jr. and Leroy Eyring, eds., "Handbook on the Physics and Chemistry of the Rare Earths, Vol. 1," North-Holland, Amsterdam, (1979).

TABLE 2. MECHANICAL, ELECTRICAL, AND OXIDE PROPERTIES OF RARE-EARTH ELEMENTS.

Atomic Number / Symbol / Element	39 Y Yttrium	57 La Lanthanum	58 Ce Cerium	59 Pr Praseodymium	60 Nd Neodymium	61 Pm Promethium	62 Sm Samarium	63 Eu Europium	64 Gd Gadolinium	65 Tb Terbium	66 Dy Dysprosium	67 Ho Holmium	68 Er Erbium	69 Tm Thulium	70 Yb Ytterbium	71 Lu Lutetium
Mechanical properties†																
Yield strength:																
kg/mm²	4.3	12.8	2.9	7.4	7.2	N.A.	6.9	N.A.	1.5	N.A.	4.4	22.6	6.1	N.A.	0.7	N.A.
1,000 psi	6.1	18.2	4.1	10.5	10.2	N.A.	9.8	N.A.	2.1	N.A.	6.3	32.1	8.7	N.A.	1.0	N.A.
Elongation, %	34	7.9	22	15.4	25	N.A.	17	N.A.	37	N.A.	30	5	11.5	N.A.	43	N.A.
Tensile strength:																
kg/mm²	13.2	13.3	11.9	15.0	16.7	N.A.	15.9	N.A.	12	N.A.	14.2	26.4	13.9	N.A.	5.9	N.A.
1,000 psi	18.8	18.9	16.9	21.3	23.8	N.A.	22.6	N.A.	17.1	N.A.	20.2	37.5	19.8	N.A.	8.4	N.A.
Vickers hardness, 10-kg load, kg/mm²	41	38	29	37	35	63	40	17	42	38	44	46	42	48	17	44
Elastic properties (values in parentheses estimated):																
Compressibility, cm²/kg × 10⁻⁶	3.98	3.23	4.96	3.39	3.09	(2.96)	2.60	11.76	2.59	2.52	2.44	2.37	2.23	2.21	7.26	2.06
Shear modulus, kg/cm² × 10⁻⁶	0.260	0.152	0.122	0.150	0.169	(0.183)	0.199	(0.079)	0.226	0.232	0.259	0.269	0.289	(0.310)	0.101	0.276
Young's modulus, kg/cm² × 10⁶	0.648	0.392	0.306	0.387	0.431	(0.471)	0.510	0.186	0.569	0.582	0.643	0.665	0.672	0.754	0.314	0.697
Poisson's ratio	0.246	0.288	0.248	0.289	0.279	(0.278)	0.282	0.167	0.254	0.255	0.238	0.237	0.250	0.217	0.207	0.261
Electrical properties at 25°C:																
Resistivity, μΩ-cm	59.6	61.5	74.4	70.0	64.3	75	94.0	90.0	131	115	92.6	81.4	86	67.6	25	58.2
Hall coefficient, V-cm/(A)(Oe) × 10¹²	-0.77	-0.35	+1.81	+0.71	+0.97	N.A.	-0.2	+24.4	-4.48	-4.3	-2.7	-2.3	-0.34	-1.8	+3.77	-0.54
Work function, eV	3.23	3.3	2.84	2.7	3.3	(3.07)	3.2	(2.54)	(3.07)	(3.09)	(3.09)	(3.09)	(3.12)	(3.12)	(2.59)	(3.14)
Magnetic properties:																
Moment, theoretical for 3⁺ ion, Bohr magnetons	0	0	2.5	3.6	3.6	N.A.	1.6	3.5	7.95	9.7	10.6	10.6	9.6	7.6	4.5	0
Susceptibility, emu/g atom × 10⁶	191	101	2430	5320	5650	N.A.	1275	33,100	356,000	193,000	99,800	70,200	44,100	26,100	71	17.9
Curie temperature, °C	None	None	None	None	None	N.A.	None	None	+20	-53	-185	-254	-253	-248	None	None
Néel temperature, °C	None	None	-260.6	None	-253	N.A.	-258	-184	None	-43	-97	-143	-188	-215	None	None
Metal oxide:																
Formula	Y₂O₃	La₂O₃	CeO₂	Pr₆O₁₁	Nd₂O₃	Pm₂O₃	Sm₂O₃	Eu₂O₃	Gd₂O₃	Tb₄O₇	Dy₂O₃	Ho₂O₃	Er₂O₃	Tm₂O₃	Yb₂O₃	Lu₂O₃
Color	White	White	Buff	Black	Light blue	White	Cream	Pale pink	White	Dark brown	Cream	Cream	Rose	Light green	White	White
Molecular weight	225.81	325.82	172.12	1021.79	336.48	342.	348.70	351.92	362.50	747.69	373.00	377.86	382.52	385.87	394.08	397.94
Melting point:																
°C	2410	2300	2210	2183	2233	2320	2269	2291	2339	2303	2228	2330	2344	2341	2355	2427
°F	4370	4172	4010	3961	4051	4208	4116	4156	4242	4117	4042	4226	4251	4246	4271	4401
Density g/cm³	5.03	6.58	7.22	6.83	7.31	7.60	7.11	7.29	7.61	7.87 (Tb₂O₃)	8.16	8.41	8.65	8.90	9.21	9.41

* Table compiled by Molybdenum Corporation of America, White Plains, N.Y. (Joseph G. Cannon); edited by Rare-Earth Information Center, Energy and Mineral Resources Research Institute, Iowa State University, Ames, Iowa (Karl A. Gschneidner, Jr. and N. Kippenhan). Data from S. R. Taylor, Abundance of Chemical Elements in the Continental Crust: A New Table, Geochim. Cosmochim. Acta, vol. 28, pp. 1273–1285, 1964; E. T. Teatum, et al., Compilation of Calculated Data Useful in Predicting Metallurgical Behavior of Elements in Binary Alloy Systems, Univ. Calif., Los Alamos Sci. Lab. Rep. LA-4003, pp. 11–12, Dec. 24, 1968; Clifford A. Hampel, "Rare Metals Handbook," 2d ed., chaps. 1 and 35, Van Nostrand Reinhold Company, New York, 1961; O. A. Songina, "Rare Metals: Scandium, Yttrium, Lanthanide and Actinides," chap. 6, trans. from Russian (1970), 3d ed. (1964), U.S. Dept. of Interior and The National Science Foundation, Washington, D.C.; Karl A. Gschneidner, Jr., "Solid State Physics," vol. 16, "Physical Properties and Interrelationships of Metallic and Semimetallic Elements," pp. 275–426, Academic, New York, 1964; Clifford A. Hampel, "The Encyclopedia of the Chemical Elements," Van Nostrand Reinhold Company, New York, 1968; R. Hultgren, R. L. Orr, and K. K. Kelley, supplement to "Selected Values of Thermodynamic Properties of Metals and Alloys," Wiley, New York, 1963; Data from Department of Mineral Technology and Lawrence Radiation Laboratory, The University of California, Berkeley, Calif. (data and revisions published periodically). Data from Karl A. Gschneidner, Jr. and Leroy Eyring, eds., in "Handbook on the Physics and Chemistry of the Rare Earths, Vol. 1 (metals) & 3 (oxides)," North-Holland, Amsterdam, 1978, 1979.

† Highest reported value for metal at room temperature after 10–50% reduction in area or annealed or as-cast; purity unknown.

N.A.—not available.

burnable poisons. Yttrium metal was fabricated into tubing and mill products because it is almost transparent to thermal neutrons and has a unique stability at high temperature in contact with liquid uranium, potassium, and sodium. Nuclear powered aircraft and submarine propulsion programs were the main motivation for these earlier efforts. Radioactive promethium has been used as a power source for pacemakers.

Early in the 1960s, mixtures of the rare-earth elements were incorporated with synthetic molecular-sieve catalysts, resulting in increased petroleum refining efficiency. Various rare-earth compounds have been found to act as catalysts in several chemical processes, such as hydrogenation. Rare-earth mixed oxides are being considered for possible use in auto exhaust catalysts. In 1964, a new red phosphor for color television was discovered. Relatively large quantities of highly purified europium and yttrium oxides were needed as commercial color television production started. Rare-earth phosphors are now being used in x-ray screens, fluorescent lamps, UV-conversion phosphors, and electro- and thermo-luminescent devices.

Rare-earth permanent magnets having properties several times superior to those made from any other known materials were developed in 1967. In these magnets, praseodymium, yttrium, samarium, lanthanum, and cerium are alloyed with cobalt in the range RCo_5 to R_2Co_{12}, where R = a rare-earth element. This new family of permanent-magnet materials is bringing about improvements in power generation and electronic communications. Conventional applications now include watches, electric motors, computer printers, automotive devices, frictionless bearings, and loudspeakers. Novel applications for the powerful magnets include magnetic earrings and use in medical treatments.

During the 1960s, the rare-earth metals were established as reactive and refining metals in the iron and steel industry. As alloying elements, lanthanum and yttrium improve the high-temperature oxidation and corrosion properties of superalloys. The rare-earth metals are more effective than calcium, magnesium, and aluminum in refining ferrous and nonferrous metals. More recent metallurgical applications of the rare earths include welding solders, brazing alloys, nonferrous alloys, dispersion hardening of complex alloys, explosive shell linings, and transducers.

Several miscellaneous and possible future applications of rare-earth compounds, complexes, and alloys include electronic components, hydrogen storage materials, synthetic jewelry and stones, lasers, magnetic bubble devices, medical uses, NMR (nuclear magnetic resonance) shift reagents, and superconductors.

The Energy and Mineral Resources Research Institute sponsors a Rare-Earth Information Center at Iowa State University, Ames, Iowa, which provides a comprehensive service to science and industry by cataloging the vast amount of data generated about these elements each year.

Occurrence. Rare-earth minerals exist in many parts of the world; the overall potential supply is essentially unlimited. As a group, these elements rank fifteenth in abundance, somewhat more plentiful than zinc. Rare-earth minerals generally are classified as sources for *light* (La through Gd) or *heavy* (Y plus Tb through Lu). Typical mineral distributions are given in Table 3.

Until 1964, monazite, a thorium-rare-earth phosphate, $REPO_4Th_3(PO_4)_4$, was the main source for the rare-earth elements. Australia, India, Brazil, Malaysia, and the United States are active sources. India and Brazil supply a mixed rare-earth chloride compound after thorium is removed chemically from monazite. Bastnasite, a rare-earth fluocarbonate mineral, $REFCO_3$, is a primary source for light rare earths. Since 1965,

TABLE 3.   REPRESENTATIVE   DISTRIBUTION   OF   ACTIVE   MINERAL SOURCES OF RARE EARTHS

REPORTED AS OXIDES	XENOTIME, MALAYSIA, %	U RESIDUES, CANADA, %	MONAZITE, AUSTRALIA, %	BASTNAESITE, CALIFORNIA, %
Lanthanum	0.5	0.8	20.2	32.0
Cerium	5.0	3.7	45.3	49.0
Praseodymium	0.7	1.0	5.4	4.4
Neodymium	2.2	4.1	18.3	13.5
Samarium	1.9	4.5	4.6	0.5
Europium	0.2	0.2	0.05	0.1
Gadolinium	4.0	8.5	2.0	0.3
Terbium	1.0	1.2		
Dysprosium	8.7	11.2		
Holmium	2.1	2.6		
Erbium	5.4	5.5	2.0	0.1
Thulium	0.9	0.9		
Ytterbium	6.2	4.0		
Lutetium	0.4	0.4		
Yttrium	60.8	51.4	2.1	0.1
	100.0	100.0	100.0	100.0

SOURCE: NMAB *Rep.* 266, October 1970.

an open-pit resource at Mountain Pass, California, has furnished about two-thirds of world requirements for rare-earth oxides. The main source for yttrium and heavy rare-earths is a by-product of uranium mining in the Elliot Lake Region, Ontario. Some xenotime, found in Malaysia, is processed in Japan and Europe.

A highly generalized description of the production of some of the rare-earth oxides follows. Crushed and finely ground bastnasite containing about 70% rare-earth oxides is roasted under oxidizing conditions to convert soluble trivalent cerium compounds to insoluble tetravalent CeO_2. The roasted product is leached with HCl, which dissolves the remaining rare earths (La, Pr, Nd, Sm, Eu, Gd), leaving behind a concentrated cerium product. The solution is passed through liquid-liquid organic solvent extraction (SX) cells, resulting in a primary separation of La-Nd-Pr from Sm-Eu-Gd. Further SX separates a pure lanthanum solution and a concentrated Nd-Pr solution, which another SX circuit separates. Europium is reduced to a divalent state in solution and precipitated. A final SX system separates and purifies gadolinium and samarium. Pure elements are usually precipitated as oxalates and calcined to oxides.

In connection with production of the heavy rare earths, monazite, containing about 55% rare-earth oxides and 5% thorium, is treated in one of two ways: (1) finely ground particles are leached with hot H_2SO_4, which dissolves thorium and the rare earths, leaving an insoluble residue; or (2) finely ground particles are reacted with hot caustic (NaOH), which dissolves the phosphate, creating a solution of trisodium phosphate which may be recovered as a by-product. The thorium and rare-earth hydrate cake is then dissolved in H_2SO_4. Thorium sulfate is selectively precipitated by pH adjustment. Separation of the other rare earths in solution is usually completed by selective absorption on ion-exchange resins and elution from ion-exchange columns. After thorium is removed from the H_2SO_4 solution, the rare earths remaining are precipitated, using NaOH, forming a double salt, $NaRESO_4 \cdot xH_2O$, known as pink salt. This salt is dissolved in HCl, treated to remove impurities, and evaporated until the hydrated $RECl_3 \cdot 6H_2O$ can be cast.

In the case of the Canadian yttrium-heavy rare earth concentrate, this is leached with HNO_3, causing all rare earths to go into solution. Solvent extraction separates yttrium from the other heavy rare earths, each of which can eventually be separated by further solvent extraction. In the case of xenotime, this is leached with hot H_2SO_4 and separation of yttrium and the heavy rare earths completed in ion-exchange columns. The liquid-liquid organic solvent extraction cycle is complete within 5–10 days and is a continuous process. The resin ion-exchange cycle requires 60–90 days and is a batch process. Both processes result in pure rare earth oxides and chemicals.

Mischmetal is produced commercially by electrolysis. The usual starting ingredient is the dehydrated rare earth chloride produced from monazite or bastnasite. The mixed rare earth chloride is fused in an iron, graphite, or ceramic crucible with the aid of electrolyte mixtures made up of potassium, barium, sodium, or calcium chlorides. Carbon anodes are immersed in the molten salt. As direct current flows through the cell, molten mischmetal builds up in the bottom of the crucible. A more recent commercial process for making mischmetal and 99.9% pure cerium, lanthanum, praseodymium, and neodymium metals involves electrowinning from the oxide in a molten fluoride bath.

References

Becker, Joseph J.: Permanent Magnets, *Sci. Am.*, pp. 92–100 (December 1970).

Callow, R. I.: "The Industrial Chemistry of the Lanthanons, Yttrium, Thorium, and Uranium," Pergamon, London, (1967).

Chin, G. Y.: "New Magnetic Alloys," *Science*, **208**, 888–894 (1980).

Gschneidner, K. A., Jr.: "Rare Earth Alloys," Van Nostrand Reinhold, New York (1961).

Gschneidner, K. A., Jr.: "Rare Earths: The Fraternal Fifteen," U.S. Atomic Energy Commission, Oak Ridge, Tenn. (December 1964).

Gschneidner, K. A., Jr., and H. Eyring, Editors: "Handbook on the Physics and Chemistry of the Rare Earths," Vols. 1 and 2, North-Holland, Amsterdam (1978–1979).

Molybdenum Corporation of America: *Overview* 1–28, (1968–1971) and published periodically, White Plains, New York.

Spedding, F. H. and A. H. Daane (editor).: "The Rare Earths," Wiley, New York (1961).

Strnat, K. S.: The Recent Development of Permanent Magnet Materials Containing Rare Earth Metals, *Tech. Rep.* AFML-RE69-299, University of Dayton, Dayton, Ohio (June 1970).

Topp, N. E.: "The Chemistry of the Rare-Earth Elements," American Elsevier, New York (1965).

NOTE: This 4th Edition entry was revised and updated by K. A. Gschneidner, Jr., Director, and B. Evans, Assistant Chemist, Rare-Earth Information Center, Energy Mineral Resources Research Institute, Iowa State University, Ames, Iowa.

REACTOR (Nuclear). Nuclear Reactor.

REALGAR. The mineral realgar is a monosulfide of arsenic corresponding to the formula AsS. It is monoclinic, showing short prismatic crystals, or may be in granular or compact masses. It is a soft, sectile mineral; hardness, 1.5–2; specific gravity, 3.5; luster, resinous; color, red to orange-yellow; transparent to translucent. Realgar occurs associated with other arsenic minerals and with gold, silver, and lead ores, although not in great quantities. It has been found as a hot-spring deposit and in volcanic sublimations. Realgar has been found in Macedonia, Japan, Switzerland; and in the United States, in Yellowstone National Park, as a hot-spring deposit, and in Utah and Nevada. The name *realgar* is derived from the Arabic words *rahj al ghar*, which means the powder of the mine.

REARRANGEMENT (Organic Chemistry). These are reactions involving the transfer of an atom or group from one part of the molecule to another. Tautomerism is a special case of rearrangements in which the two forms are in dynamic equilibrium. See "Tautomerism" in entry on **Isomerism.** When such

reactions take place, the establishment of structural formulas becomes complex. Some of the better known rearrangements include:

1. *Allyl Rearrangement.*

$$CH_3CH \colon CHCH_2Br \rightarrow CH_2 \cdot CHCHBrCH_3.$$

2 *Pinacol Rearrangement* takes place when pinacol is heated with dilute acid, pinacolin formed:

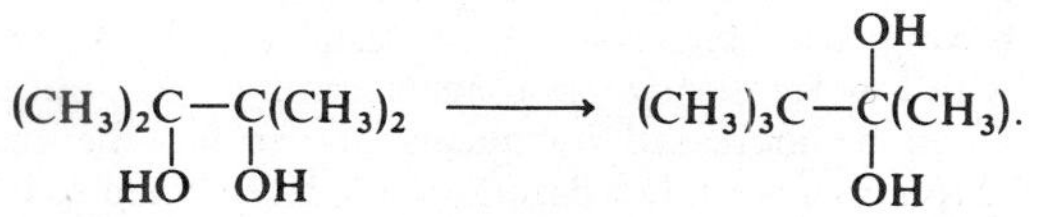

$$(CH_3)_2C-C(CH_3)_2 \longrightarrow (CH_3)_3C-\underset{\underset{OH}{|}}{\overset{\overset{OH}{|}}{C}}(CH_3).$$
$$HOOH$$

3. *Benzil Rearrangement* into benzylic acid:

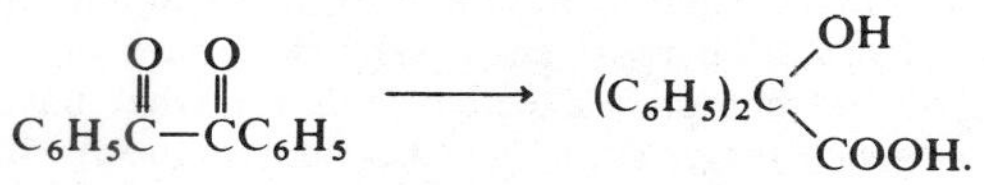

$$C_6H_5\overset{O}{\overset{\|}{C}}-\overset{O}{\overset{\|}{C}}C_6H_5 \longrightarrow (C_6H_5)_2C\overset{OH}{\underset{COOH.}{\diagdown}}$$

4. *Hoffman Rearrangement.*

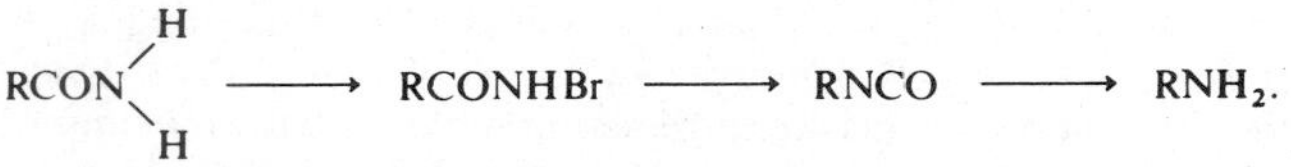

$$RCON\overset{H}{\underset{H}{\diagdown}} \longrightarrow RCONHBr \longrightarrow RNCO \longrightarrow RNH_2.$$

5. *Beckmann Rearrangement* results when oximes of ketones are treated with certain reagents such as phosphorus pentachloride.

$$\underset{NOH}{\overset{R'C-R}{\|}} \longrightarrow \underset{R'N}{\overset{RC-OH}{\|}} \longrightarrow \underset{R\ N-H.}{\overset{R-C=O}{\|}}$$

The exchange is between R' and OH on the opposite sides of the CN bond.

6. *Benzidine Rearrangement.* Treatment of hydrazobenzene with strong acids.

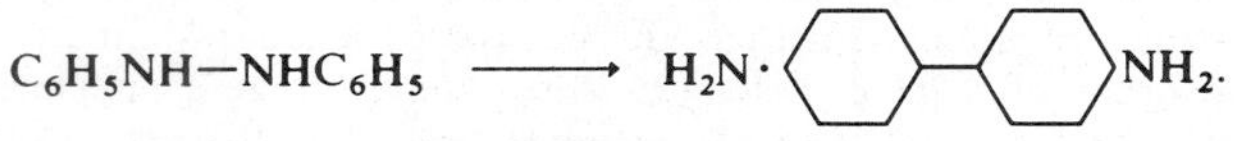

$$C_6H_5NH-NHC_6H_5 \longrightarrow H_2N \cdot \langle\!\!\langle \rangle\!\!\rangle\!\!\langle \rangle\!\!\rangle NH_2.$$

7. *Rearrangement of hydroxylamine derivatives.*

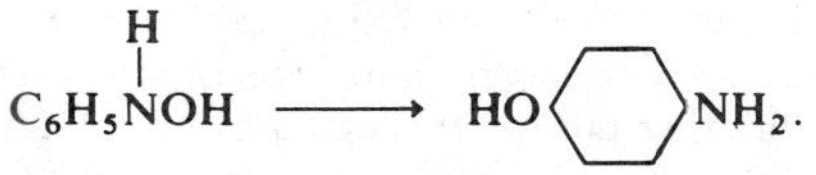

$$C_6H_5\overset{H}{\overset{|}{N}}OH \longrightarrow HO\langle \rangle NH_2.$$

8. *Rearrangement in reaction of nitrous acid and alkyl amines.*

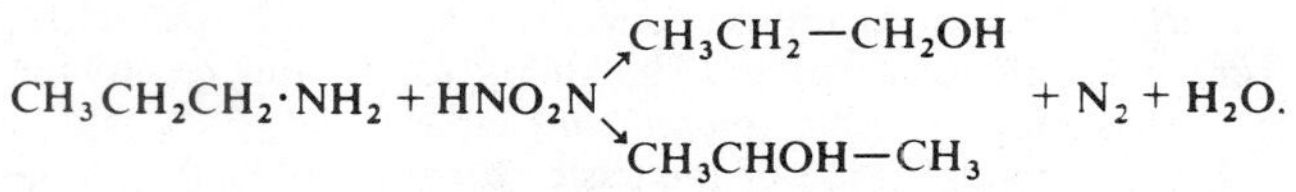

$$CH_3CH_2CH_2 \cdot NH_2 + HNO_2N\overset{CH_3CH_2-CH_2OH}{\underset{CH_3CHOH-CH_3}{\diagup\diagdown}} + N_2 + H_2O.$$

9. *Walden Inversion.* A rearrangement takes place in the reaction of optically active substances. The structural formula does not change but the configuration changes due to the interchange of two groups attached to the asymmetrical carbon atom during the course of the reaction.

RECOMBINANT DNA. The nature of DNA (deoxyribonucleic acid), of RNA (ribonucleic acid), and of mRNA (messenger RNA) molecules and of their respective roles in determining the genetic code are described in entries on **Genetic Code;** and **Nucleic Acids and Nucleoproteins.** New organisms are created in nature by way of very slow processes. A change in the base sequence of the DNA constituting a gene results in an inherited alteration in the code and is called a *gene mutation.* Mutations are genetic changes which occur suddenly and are thereafter heritable.

Mutations arise through three general mechanisms: (1) chemical modification of preformed DNA, such as breakage and aberrant reunion of molecules or the changes elicited by ultraviolet light, for example; (2) errors in incorporation of the purine and pyrimidine bases, or additions and subtractions of bases, during DNA replication; and (3) unequal exchange between two identical or similar DNA molecules ("unequal crossing over" during recombination. These chemical changes normally occur with low frequency (spontaneous mutations), but the frequency can be increased by means of various chemical and physical treatments (induced mutations). Even when so induced, the frequency of bacterial mutants for a particular trait, for example, is low, e.g., one mutant in 10^4–10^{10} bacteria. Thus, any biological evolutionary alterations brought about by the mechanism of mutation represent a very slow pathway. Such procedures do not comprise effective tools for what has been referred to as *genetic engineering* (genetic manipulation) wherein gene structures can be willfully directed under laboratory conditions.

It should be stressed that recombinant DNA methodology is *not* a way of constructing new forms of life *in vitro*. Even the simplest organisms are extremely complex and the maximum alteration of the simplest genome would be of the order of 1%. Also, the genomes of the simplest organisms are highly ordered and the random insertion of a few genes from an unrelated organism is unlikely to create a whole new organism.

In the early 1970s, an interesting observation of large significance to biologists, biochemists, molecular biologists, and related scientists was made, namely, the discovery of certain enzymes that have the ability of cutting and splicing hereditary material of living organisms in an obviously unprecedented and precise manner. Sometimes referred to as *restriction enzymes*, these materials have the ability to cut the extremely long DNA molecules of living organisms into what have been termed manageable fragments. Further, the cut pieces are approximately the order of a gene in length. Also, some of these enzymes have the further ability of cutting a few bases further down than the others, so that what sometimes are known as "sticky ends" are produced. Thus, any species of DNA, if cut by the same enzyme, will possess the same type of sticky ends. Thus, fragments of differing DNAs, through a form of biological scissors-and-paste process, can cause the lower part of one DNA molecule to stick well onto the upper part of another molecule. The result is a hybrid molecule. Theoretically, the technique can cross the boundaries of species by selecting DNA material from fully different sources. The ability to cut and recombine is the basis for the term *recombinant DNA*.

A useful modification of the basic clip-and-paste process involves inserting the DNA fragments into a DNA molecule which has the power of self-replication. Many bacteria contain small, circular, cytoplasmic DNA molecules called *plasmids*, which are capable of self-replication inside the bacterial cell. The characteristics of rapid bacterial growth and multiplication allow quantity replication of the recombinant plasmids in short periods of time. This technique thus offers an obvious advantage over the slow and laborious chemical methods.

However, obtaining sufficient quantities of a specific gene in purified form for insertion into a plasmid is difficult when one considers the genetic complexity of living organisms. An approach to the problem has been through the use of an enzyme known as *reverse transcriptase*. This enzyme synthesizes DNA from RNA. The primary product of genes is mRNA, which possesses base sequences complementary to the genes. The large quantities of specific mRNA available, coded for by the single gene, allow biochemical purification of the mRNA. Thus, if one can isolate the mRNA coded from a particular gene, the corresponding DNA sequence, identical to the gene, can be reconstructed using reverse transcriptase. This synthesized DNA then can be inserted into a plasmid by standard recombinant DNA methods and amplified by growing the plasmid in bacteria.

The advantages of recombinant techniques for increasing knowledge of the genetic construction of any organism are immediately recognized. A number of practical findings from such investigations can be envisaged, such as incorporation of nitrogen-fixing genes in agricultural plants to eliminate the need for nitrogen fertilizers; the bacterial manufacture of large quantities of polypeptide hormones, such as insulin; the bacterial production of vaccines and enzymes as well as the treatment of genetic diseases. Possible production of fermentation products (alcohol, methane, etc.) as fossil fuel substitutes may be aided by this technique.

Although it is now realized that grave consequences from recombinant DNA experimentation are much less likely than originally believed possible in the mid-1970s, nevertheless responsible scientists are aware of the need to continue to exert precautionary measures. In the mid-1970s, a group of scientists who were active isn the field of recombinant RNA established a self-disciplined approach and urged other scientists to do likewise—by restricting or deferring specific types of experiments with recombinant DNA. The concern was possible production of biologically hazardous DNA molecules. The National Institute of Health (United States) published guidelines for this type of research to which most scientists adhered. With the lesser concerns over these problems in the mid-1980s, the guidelines are now being revised. The revisions considered stem essentially from the concept of *biological containment*.

By this means, safety factors may be built into the genetic structure of the organism to be studied—this is of central importance now to the safety of recombinant DNA research. For example, as bases for recombinant experiments, EK2 derivatives of *E. coli* cells are used which are 100 million times less able to survive in nature outside an artificial laboratory environment and thus present no biohazards to normal life. These mutant cell lines are usually constructed by causing a deletion of a portion of DNA in a gene responsible for critical cell characteristics, such as ability to metabolize a certain substrate or to construct a rigid cell wall. Alternatively, defective mutant genes may be inserted into the genome replacing normal genes responsible for properties critical to the survival of the cell.

References

Abelson, J.: "A Revolution in biology," *Science*, **209**, 1319–1321 (1980).

Bukhari, A. I., Shapiro, J. A., and S. L. Adhya, Editors: "DNA Insertion Elements, Plasmids, and Episomes," Cold Spring Harbor Laboratory, Cold Spring Harbor, New York (1977).

Corden, J., et al.: "Promoter Sequences of Eukaryotic Protein-Coding Genes," *Science*, **209**, 1406–1414 (1980).

Davis, M. M., Kim, S. K., and L. E. Hood: "DNA Sequences Mediating Class Switching in α-Immunoglobulins," *Science*, **209**, 1360–1365 (1980).

Gilbert, W., and L. Villa-Komaroff: "Useful Proteins from Recombinant Bacteria," *Sci. Amer.*, **242**, 4, 74–94 (1980).

Gingeras, T. R., and R. J. Roberts: "Steps Toward Computer Analysis of Nucleotide Sequences," *Science*, **209**, 1322–1328 (1980).

Mulligan, R. C., and P. Berg: "Expression of a Bacterial Gene in Mammalian Cells," *Science*, **209**, 1422–1427 (1980).

Scherer, S., and R. W. Davis: "Recombination of Dispersed Repeated DNA Sequences in Yeast," *Science*, **209**, 1380–1384 (1980).

—Ann C. Vickery, Ph.D., University of South Florida, College of Medicine, Tampa, Florida.

REFINING (Petroleum). Petroleum.

REFORMATSKY REACTION. Dating back to 1887, this reaction depends on interaction between a cabonyl compound, an α-halo ester, and activated zinc in the presence of anhydrous ether or ether-benzene, followed by hydrolysis. The halogen component for example ethyl bromoacetate, combines with zinc to form an organozinc bromide that adds to the carbonyl group of the second component to give a complex readily hydrolyzed to carbinol. The reaction

Zn + BrCH₂COOC₂H₅ →

BrZnCH₂COOC₂H₅

is conducted by the usual Grignard technique except that the carbonyl component is added at the start. Magnesium has been used in a few reactions in place of zinc but with poor results, for the more reactive organometallic reagent tends to attack the ester group; with zinc this side reaction is not appreciable, and the reactivity is sufficient for addition to the carbonyl group of aldehydes and ketones of both the aliphatic and aromatic series. The product of the reaction is a β-hydroxy ester and can be dehydrated to the α,β-unsaturated ester; thus the product from benzaldehyde and ethyl bromoacetate yields ethyl cinnamate. α-Bromo esters of the types RCHBrCO₂C₂H₅ and RR′CBrCO₂C₂H₅ react satisfactorily, but

Ethyl β-phenyl-β-hydroxy-propionate (b.p. 130°/6 mm.)

β- and γ-bromo derivatives of saturated esters do not have adequate reactivity. Methyl γ-bromocrotonate (BrCH₂CH=CHCO₂CH₃), however, has a reactive, allylic bromine atom and enters into the Reformatsky reaction.

—L. F. and M. Fieser, Harvard University, Cambridge, Massachusetts.

REFORMING. Decomposition (cracking) of hydrocarbon gases or low-octane petroleum fractions by heat and pressure, either without a catalyst (thermoforming), or with a specific catalyst. The latter is most extensively used. The principal cracking reactions are (1) dehydrogenation of cyclohexanes to aromatic hydrocarbons; (2) dehydrocyclicization of certain paraffins to aromatics; (3) isomerization, i.e., conversion of straight-chain to branched-chain structures, as octane to isooctane. These result in substantial increase in octane number. Steam reforming of natural gas is an important method of producing hydrogen; steam reforming of naphtha is used to produce substitute natural gas. See also **Ammonia; Cracking Process; Isomerization;** and **Petroleum.**

REFRIGERATION. A process of cooling or freezing a substance to a temperature lower than that of its surroundings and maintaining the substance in a cold state. Refrigeration can be accomplished by arranging heat transfer from a warm body to a colder body through processes such as convection or thermal conduction. Other, more exotic methods include the exploitation of thermoelectric properties of semiconductors, the magnetothermoelectric effects in semimetals, or the diffusion of ³He atoms across the interface between distinct phases of liquid helium having high and low concentrations of ³He in ⁴He, among other methods. See **Thermoelectric Cooling.**

Most commercial refrigeration systems operate on a cyclic basis. A refrigerator operating in this manner may be considered a heat pump, for it continuously extracts heat from a low-temperature region and delivers it to a high-temperature region. It is rated by its *coefficient of performance*, which may be defined as the ratio of the heat removed from the cold region per unit of time to the net input power for operating the device, in symbols $K = Q_t/P$. Vapor-absorption and thermoelectric refrigeration systems have lower coefficients of performance than vapor-compression refrigerators, but they have other characteristics that are superior, such as quietness of operation and compactness.

A *vapor-compression refrigerator* consists of a compressor, a condenser, a storage tank, a throttling valve, and an evaporator connected by suitable conduits with intake and outlet valves. See Fig. 1. The refrigerant is a liquid which partly vaporizes

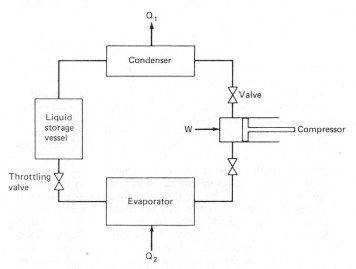

Fig. 1. Vapor-compression refrigeration system.

and cools as it passes through the throttling valve. Among the common refrigerants are ammonia, sulfur dioxide, and various halides of methane and ethane. Commonly used are the *Freons*®, of which *Freon*-12® is dichlorodifluoromethane. Nearly constant pressures are maintained on either side of the throttling valve by means of the compressor. The mixed liquid and vapor entering the evaporator is colder than the near-surround; it absorbs heat from the interior of the refrigerator box or cold room and completely vaporizes. The vapor is then forced into the compressor where its temperature and pressure increase as the result of compression. The compressed vapor then pours into the condenser where it cools down and liquefies as the heat is transferred to cold air, water, or other fluid medium in the cooling coils. Comparative tests have shown that the coefficient of performance of vapor-compression refrigerators depends very little on the nature of the refrigerant. Because of mechanical inefficiencies, the actual value may be well below an ideal value which, ordinarily, lies between 2 and 3.

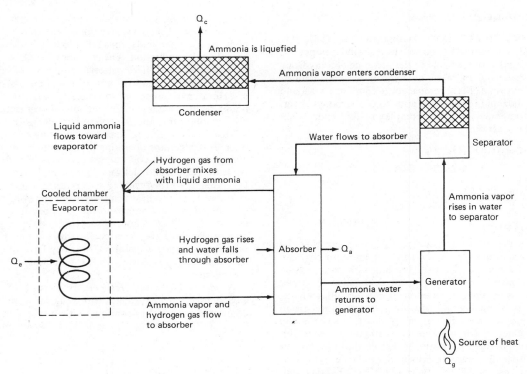

Fig. 2. Vapor-absorption refrigeration system.

In a *vapor-absorption refrigeration system*,[1] there are no moving parts. The added energy comes from a gas or liquid fuel burner or from an electrical heater, as *heat*, rather than from a compressor, as *work*. See Fig. 2. The refrigerant used in this example is ammonia gas, which is liberated from a water solution and transported from one region to another by the aid of hydrogen. The total pressure throughout the system is constant and therefore no valves are needed.

Heat from the external source is supplied to the generator, where a mixture of ammonia and water vapor with drops of ammoniated water is raised to the separator in the same manner as water is raised to the coffee in a percolator. Ammonia vapor escapes from the liquid in the separator and rises to the condenser, where it cools and liquefies. Before the liquefied ammonia enters the evaporator, hydrogen, rising from the absorber, mixes with it and aids in the evaporation process. Finally, the mixture of hydrogen and ammonia vapor enters the absorber, where water from the separator dissolves the ammonia. The ammonia water returns to the generator to complete the cycle. In this cycle, heat enters the system not only at the generator, but also at the evaporator, and heat leaves the system at both the condenser and the absorber to enter the atmosphere by means of radiating fins.

No external work is done, and the change in internal energy of the refrigerant during a complete cycle is zero. The total heat $Q_a + Q_c$ released to the atmosphere per unit of time by the absorber and the condenser equals the total heat $Q_g + Q_e$ absorbed per unit of time from the heater at the generator and from the cold box at the evaporator. Thus $Q_e = Q_a + Q_c - Q_g$, and, therefore, the coefficient of performance is $K = Q_e/Q_g = [(Q_a + Q_c)/Q_g] - 1$.

[1] The following few paragraphs are excerpts from "Refrigeration" by A. L. King in "The Encyclopedia of Physics" (R. M. Besancon, Editor), Van Nostrand Reinhold, New York.

The vapor-absorption refrigerator is free from intermittent noises, but it requires a continuous supply of heat. Once very popular for households, refrigeration systems of this type are now most frequently found in camping facilities, and some rural areas where commercial electric power may not be easily available.

In a *dilution refrigeration system*, the properties of helium are used advantageously for attaining very low temperatures. Below a temperature of 0.87 K, liquid mixtures of ^{3}He and ^{4}He at certain concentrations separate into two distinct phases. One is a concentrated (^{3}He-rich) phase floating on the other, denser (^{4}He-rich) phase with a visible interface between them. The concentrations of ^{3}He in the two phases are functions of temperature, approaching 100% in the concentrated phase and about 6% in the dilute phase at 0 K. The transfer of ^{3}He atoms from the concentrated to the dilute phase, like an evaporation process, entails a latent heat, an increase in entropy, and a lowering of temperature. The main features of a recirculating dilution refrigeration system are shown in Fig. 3. The pump forces helium vapor (primarily ^{3}He) from the still into the condenser, where it is liquefied at a temperature near 1 K in a bath of rapidly evaporating ^{4}He, through a flow controller that consists of a narrow tube of suitable diameter to obtain an optimum rate of flow, and then through the still where its temperature is further reduced to about 0.6 K. The liquefied ^{3}He next passes through a heat exchanger so as to reduce its temperature to nearly that of the dilution chamber, by giving up thermal energy to the counterflowing dilute phase, before entering the concentrated phase therein.

The diffusion of ^{3}He atoms from the concentrated into the dilute phase within this chamber can produce steady temperatures of very low values (0.01 K or less). Liquid ^{3}He from the dilute phase then passes through the heat exchanger to the still where it is warmed to transform the liquid to the vapor phase that goes to the pump, thus completing the cycle. Modified

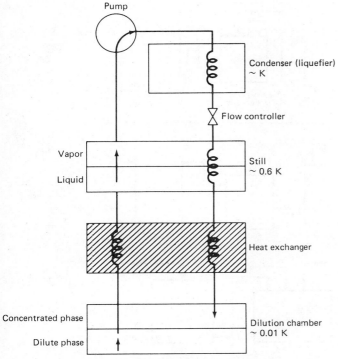

Fig. 3.   Dilution refrigeration system.

versions of this system have been constructed, sometimes with an added single-cycle process for producing temporarily temperatures lower than previously mentioned. The low-temperature limit in any system of this type is governed largely by two important sources of inefficiency that cannot be completely eliminated—heat leakage, especially severe because of the extreme range of temperatues—and recirculation of some ^{4}He with ^{3}He.

See also **Helium**; and **Thermodynamics**.

References

NOTE: See also the list of references at the end of the entry on **Thermodynamics**.

Betts, D. S.: "Helium Isotope Refrigeration," *Contemporary Physics*, **9**, 97–114 (1968).
King, A. L.: "Thermophysics," Freeman, San Francisco (1962).
Zemansky, M. W.: "Heat and Thermodynamics," McGraw-Hill, New York (1968).

RESINS (Acetal).   These are thermoplastic resins, obtainable both as homopolymers and copolymers, and produced principally from formaldehyde or formaldehyde derivative. Acetal resins have the highest fatigue endurance of commercial thermoplastics. A variety of ionic initiators, such as tertiary amines and quaternary ammonium salts, are used to effect polymerization of formaldehyde. Chain transfer, shown by the following reactions, controls the molecular weight of resulting resins:

Step 1, initiation of new chain:

$$-CH_2OCH_2O^- + H_2O \rightarrow -CH_2OCH_2OH + OH^-$$

Step 2, reaction of growing chain with H_2O, releasing hydroxyl ion:

$$OH^- + CH_2O \rightarrow HOCH_2O^-$$

Step 3, end-capping of high-molecular-weight polyoxymethylene glycol to provide stable commercial resin:

$$HOCH_2O^- + nCH_2O \rightarrow HOCH_2O(CH_2O)_{n-1}CH_2O^-$$

Starting ingredients may be formaldehyde or the cyclic trimer trioxane, $CH_2OCH_2OCH_2O$. Both form polymers of similar properties. Boron trifluoride of other Lewis acids are used to promote polymerization where trioxane is the raw material.

Acetals provide excellent resistance to most organic compounds except when exposed for long periods at elevated temperatures. The resins have limited resistance to strong acids and oxidizing agents. The copolymers and some of the homopolymers are resistant to the action of weak bases. Normally, where resistance to burning, weathering, and radiation are required, acetals are not specified. The resins are used for cams, gears, bearings, springs, sprockets, and other mechanical parts, as well as for electrical parts, housings, and hardware.

RESINS (Acrylonitrile-Butadiene-Styrene).   Commonly referred to as ABS resins, these materials are thermoplastic resins which are produced by grafting styrene and acrylonitrile onto a diene-rubber backbone. The usually preferred substrate is polybutadiene because of its low glass-transition temperature (approximately $-80°C$). Where ABS resin is prepared by suspension or mass polymerization methods, stereospecific diene rubber made by solution polymerization is the preferred diene. Otherwise, the diene used is a high-gel or cross-linked latex made by a hot emulsion process.

ABS resins possess an attractive balance of impact resistance, hardness, tensile strength, and elastic modulus properties. The temperature range is wide—from -40 to $107°C$ (-40 to $225°F$). Other advantages include chemical resistance, high gloss, and nonstaining properites. The dimensional stability of ABS is good and creep resistance is excellent. The resins exhibit low water absorption or volume change at varying humidities.

Commercial ABS is in the form of custom color-matched compounded pellets, or granular resin for compounding or alloying with other plastics. A representative alloying ingredient is polyvinyl chloride (PVC). Almost all standard thermoplastic converting processes can be used with ABS plastics. Injection-molded parts include telephone sets, refrigerator parts, plumbing fixtures, fittings, radio, television, and appliance housings, and auto parts. ABS can be extruded into sheet, pipe, and various cross sections.

Thermoforming of large surface areas and deep draws from sheet stock are possible. Examples of parts which involve extrusion and subsequent thermoforming include lawnmower housings, refrigerator liners, pipe and conduit, vehicle bodies, snowmobile shrouds, and camper bodies. The various thermoforming techniques applicable include plug and air assist, vacuum snapback, vacuum-plug forming, and drape forming.

The compatibility of ABS with other plastics makes them useful as impact modifiers and as processing additives with many other polymers to achieve a variety of final product specifications. Substitution of α-methyl styrene for styrene increases heat-distortion temperatures; or of methacrylonitrile for acrylonitrile improves barrier properties to gases such as carbon dioxide in connection with carbonated beverage containers. Over 75 grades of ABS are commercially available, including self-extinguishing, electroplating, antistatic expandable, glass-reinforced, high-heat, cold-forming, and low-gloss sheet grades.

Three polymerization steps are involved in ABS manufacture. The process is shown in block diagram format in the accompanying illustration.

STEP 1—Polybutadiene rubber is formulated by feeding butadiene, water, an emulsifier, and catalyst into a glass-lined reactor. This is an exothermic reaction. About 80% conversion is

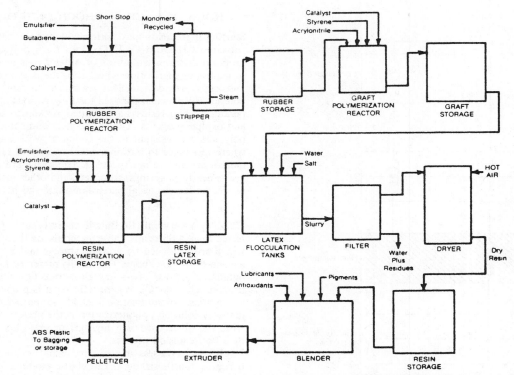

Process for manufacturing acrylonitrile-butadiene-styrene (ABS) resins. (*Marbon Div., Borg-Warner Corp.*)

achieved in a period of about 50 hours. The residual butadiene monomer is recovered by steam-stripping and recycled.

STEP 2—Polybutadiene rubber is further polymerized, but in the presence of styrene and acrylonitrile monomers. This is done in low-pressure reactors under a nitrogen atmosphere. In this operation, the monomers are grafted onto the rubber backbone through the residual unsaturation remaining from the first step.

STEP 3—In a separate step, styrene-acrylonitrile (SAN) resin is prepared by emulsion, suspension, or mass polymerization by free-radical techniques. The operation is carried out in stainless-steel reactors operated at about 75°C (167°F) and moderate pressure for about 7 hours. The final chemical operation is the blending of the ABS graft phase with the SAN resin, plus adding various antioxidants, lubricants, stabilizers, and pigments. Final operations involve preparation of a slurry of fine resin particles (via chemical flocculation), filtering, and drying in a standard fluid-bed dryer at 121–132°C (250–270°F) inlet air temperature.

—Assistance of the Marbon Division, Borg-Warner Corporation in preparation of this entry is appreciated.

RESINS (Natural). Complex compounds composed of carbon, hydrogen, and relatively small amounts of oxygen, which are secreted in various tissues of many plants. In the pine family, where resins are very common, they are secreted as oleoresins in resin canal cells, which break down finally, producing resin canals. These canals appear as longitudinal ducts in the sapwood and inner bark, connected laterally by resin canals in the compound wood rays, thus forming an extensive network. A common name given to the oleoresin in this group is pitch, the sticky juice which exudes from the plant wherever it is wounded. On exposure to the air the volatile oil in this pitch (oil of turpentine) gradually evaporates, leaving a clear hard glassy substance, the resin, which forms a protective coating over the wound.

Most resins have the same physical properties, being clear, translucent, and of a yellow or brownish color. Amber, a fossil resin, is a more or less familiar example. Resins are insoluble in water, but soluble in common organic solvents such as ether and alcohol. All resins burn with a sooty flame. Resins seem to be mainly of value to the plant in that they form protective coverings against the entrance of disease-producing organisms and also prevent excessive loss of water from the thin-walled tissues exposed in the wound.

Resins are separated into several classes. Many of the resins contain almost no volatile oil and are hard, without taste or odor. These are the varnish or hard resins. Other resins, when removed from the plant in which they are formed, and dissolved in volatile oils, form a thick semi-solid mass: these are the oleoresins. In still other cases the resin occurs in combination with a gum, forming a gum resin.

Hard Resins. Several of the hard resins, used mainly for making varnishes, are called copals. Most of them come from Africa and are either found in fossil form or obtained from living plants. Other copals come from Australia, New Zealand and East Indian Islands. The plants which form them are members of the legume and pine families. The African copals are products of several species of *Trachylobium*, fairly large trees growing in east Africa and Madagascar. The best resin from these trees occurs in a fossil form, often deeply buried in the ground—sometimes in regions where the trees no longer grow. These resins dissolve slowly and are used in making varnishes which are very durable. A South American tree of large size, *Hymenaea courbaril*, also of the Leguminoseae, yields a very similar resin, which is also found in lumps in the ground around the trees, and used in varnishes.

Another copal is obtained from *Agathis australis*, a very large coniferous tree native in Australia and New Zealand, where it is known as the Kauri pine. Like the other copals, that from the Kauri pine is found in lumps buried in the ground. Most of these lumps are 1 or 2 inches in diameter, but some are much larger, weighing up to 100 pounds. Nearly all of this resin comes from the northern part of North Island of New Zealand. It is frequently called Kauri gum, though it is not a gum, but a true copal resin. Another group of hard resins, known as dammar resins, is obtained from many different trees growing in southern Asia and the East Indian Islands. These resins dissolve readily in alcohol, forming spirit-varnishes.

One of the commonest and most important of the hard resins is rosin, obtained by distilling the pitch, or turpentine, which is a product of several of the native pines of the southeastern United States. This rosin, also known as colophony, is a very important product of that region. Originally the turpentine was obtained by chopping a deep hollow in the base of the trunk of the tree and allowing it to fill up with the turpentine, which was then scooped out. This method was very destructive and wasteful, since much of the oleoresin, turpentine, was lost during the process. The weakened trees were easily blown down.

Now turpentine is obtained by cutting V-shaped gouges in the bark and inserting metal gutters beneath the gouges. These gutters carry the turpentine to a cup placed underneath. As soon as the cut is made, turpentine beings to flow and continues to do so for two or three days, gradually slowing as the drying turpentine allows resin to accumulate and plug the wounds. A new flow is obtained by cutting off a narrow strip of bark from the upper edge of the cut. The process is continued as long as the pitch will flow, which is usually all summer and well along into the late fall. Each tree may be turpentined for 6 or 7 successive years or even longer before it ceases to be profitable.

The crude turpentine collected in the cups and the product which has dried on the wound of the tree are removed and carried to the still. Here the turpentine, to which a little water is added, is carefully heated to drive off the oil of turpentine present, together with the water added. The distillate is condensed by passing it through a coil around which cold water is flowing, and collected in a barrel or any suitable container. The two substances, water and oil of turpentine, which make up the distillate, are immiscible and soon separate, the lighter oil of turpentine rising to the top and floating on the water, which is drawn off from the bottom. Oil of turpentine is often called spirits of turpentine, or, in the paint trade, turpentine. In medicine the word turpentine is reserved for the oleoresin which upon distillation yields oil of turpentine and rosin.

The residue remaining in the tank at the end of the distillation is skimmed to remove any impurities such as twigs, bits of bark and dirt, and run into vats to cool. Then it is put into barrels and allowed to harden, forming rosin.

Oil of turpentine is used principally as a solvent for paints and varnishes, because it mixes readily with the various substances used and also because it evaporates quickly, causing the paint or varnish to dry. It is also used in making such things as sealing wax and shoe polish. Very pure grades of turpentine (the oleoresin) are used medicinally.

Large quantities of rosin are used in sizing paper, which makes it take ink without spreading or blotting, gives it a smoother surface and makes it heavier. Rosin is also used in cheaper varnishes, in paints, and in soap making. It is furthermore used as an adulterant of the more expensive resins. Linoleum manufacturers use large amounts of rosin.

In early times, large quantities of crude turpentine were used to waterproof the rigging of the sailing vessels and to calk the seams of the hull.

Mastic ia a hard resin exuding from the branches of one of the Pistachio trees, *Pistacia lentiscus*, native of Mediterranean Europe and southern Asia. Formerly, it was extensively used medicinally, for stomach troubles and dysentery, as well as other ailments. Now it is used in making varnishes and in lithographic work. Natives of the region in which it is found chew mastic, which has a pleasant taste.

Since turpentine is a mixture of a volatile substance, spirits of turpentine and a hard resin, it is one of the oleoresins.

Oleoresins. Canada balsam is one of the oleoresins. It is obtained from the bark of *Abies balsamea*, the common balsam fir of northern North America. Canada balsam, because its refractive index is so near that of glass, is much used in optical work and in preparing materials for examination with a microscope.

Little used today is Dragon's blood, an oleoresin obtained from the fruits of *Daemonorops draco*, a native palm of southeastern Asia and the Molucca Islands. The resin exudes from the surface of the ripening fruits. It is removed from them by boiling in water. The resin is then moulded into balls or long sticks. It is sometimes used in making varnishes and lacquers.

True lacquer, obtained from the juice of *Rhus verniciflua*, a sumac tree of southeastern Asia, is another oleoresin. To obtain the juice lateral cuts are made in the bark. The exuded sap is collected not only from these cuts but from small branches which are cut off and soaked in water. The juice is cleaned of any foreign substances by straining it through hemp cloth. By slow heating, either artificial or by the sun, the juice is evaporated and stored until used. Lacquer is a poisonous substance, causing intense irritation of the skin in many people. Others seem to be immune. Lacquer is usually applied over some soft wood, commonly soft pine, the pores of which have first been filled by rubbing in a paste of rice and resin, followed by a paste of soft clay and resin. The surface is then covered with cloth and layer after layer of lacquer put over that. Each layer is allowed to dry and rubbed down very smooth before the next layer is added. Any color which is to be added is mixed with the lacquer, with each colored layer covered by a clear layer before another is put on. The final product is a thick covering composed of many thin layers of lacquers. If this is carved the edges of the carving, on careful examination, will show the fine lines separating the different layers. Lacquering is a very old industry, having been carried on in China since the sixth century. Lacquer work is made in many other oriental countries, and the juice of many other trees used as a source for the lacquer used.

Certain resins occur in combination with fragrant volatile oils. One of these is benzoin, obtained from *Styrax benzoin* by cutting notches in the bark and allowing the resin to collect in them. It is used in making perfumes, in incense, and as a source of benzoic acid, used medicinally.

Another fragrant oleoresin is storax, obtained from *Liquidambar orientalis*, a medium-sized tree growing in southwestern Asia. The resin is obtained by boiling the bark and wood of young branches. It is used medicinally and also in incense.

Gum Resins. Gum resins include myrrh, which exudes from the trunk and branches of *Commiphora myrrha*, a tree growing in the region around the Red Sea. The lumps of resin are used medicinally, and also in making incense. Another gum resin is frankincense, obtained by cutting notches in the stem of *Boswellia carterii*, which grows in northeastern Africa and in Arabia. This resin is used in incense. *Asafoetida* is also a gum resin. See also **Gums and Mucilages.**

REVERSE OSMOSIS. A method for separating inorganic salts and simple organic compounds under pressure. The size of the species usually is considered in terms of molecular weight.

Solvent transport through reverse-osmosis membranes is substantially diffusive in nature. The membranes used are anisotropic (thin-skinned, overlaying a porous substructure). In all probability, the skin contains no pores. If water is the solvent phase, the water passes through the membrane by true diffusion. At some point in its passage through the membrane, the solvent water actually becomes a part of the membrane water structure. As low-molecular-weight solutes exhibit osmotic pressure when concentrated, system pressure in excess of the osmotic pressure of the concentrated solutes must be applied to create a pressure driving force. Depending upon materials being separated and membrane used, reverse osmosis system pressures may range from 200 to nearly 1500 pounds per square inch (13.6 to 102 atmospheres), but the lower values are most frequently encountered.

It has been well appreciated for many years that reverse osmosis concentration has considerable potential in the chemical and food processing field. Reverse osmosis is used for processing whey. A new approach to the production of protein isolates and concentrates from oilseed flour combines ultrafiltration and reverse osmosis. See also **Ultrafiltration**. In 1977, this process was optimized for use with soy flour. The process uses semipermeable membranes to directly process protein extracts from defatted soy flour. Some researchers have found that essentially all the solubilized protein is recovered and the generation of wheylike products is avoided. Other desirable features of the combined ultrafiltration/reverse osmosis process include its shorter time span, increased yield of isolate (since whey proteins are harvested along with proteins normally precipitated), and the enhanced nitrogen solubility of the products. For the ultrafiltration part of the system, second-generation, noncellulosic membranes have been used. The membranes used in reverse osmosis are cellulose-based because noncellulosic reverse osmosis membranes have not been fully developed. Cellulose acetate membranes are generally restricted to operating temperatures below about 50°C and a pH of less than 9.

Japanese researchers have developed a method for washing membranes and overcoming the adhesive effects of mandarin orange juice fouling on reverse osmosis membranes. The importance of keeping fluxes high by controlling concentration polarization and membrane fouling has been observed by several researchers. Other observers have noted that high axial velocities and turbulence promoters minimize the flux decrease caused by concentration polarization and fouling. Considerable research has been undertaken as regards the energy requirements for reverse-osmosis process and these are reviewed in some of the references listed.

Molecular diffusion in reverse osmosis is represented schematically in the accompanying diagram.

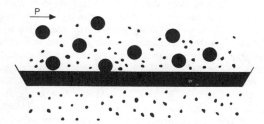

Schematic representation of molecular diffusion in reverse osmosis.

References

Ebara, K. et al.: "The Advanced Waste Water Treatment by Reverse Osmosis Method," *Kagaku Kojo*, **20**, 4, 69 (1976).
Lawhon, J. T., et al.: "Optimization of Protein Isolate Production from Soy Flour Using Industrial Membrane Systems," *Jrnl. of Food Science*, **43**, 2, 361–364 (1978).
Nomura, D., and I. Hayakawa: "Studies on Concentration of Orange Juice by Reverse-osmosis Process," *Jrnl. of Food Science Technology* (Japan), **23**, 404 (1976).
Schwartzberg, H. C.: "Energy Requirements for Liquid Food Concentration," *Food Technology*, **31**, 3, 68–69 (1977).
Watanabe, A., Kimura, S., and S. Kimura: "Flux Restoration of Reverse-Osmosis Membranes by Intermittent Lateral Surface Flushing for Orange Juice Processing," *Jrnl. of Food Science*, **43**, 3, 985–988 (1978).

REYNOLDS NUMBER. A dimensionless number that establishes the proportionality between fluid inertia and the sheer stress due to viscosity. The work of Osborne Reynolds has shown that the flow profile of fluid in a closed conduit depends upon the conduit diameter, the density and viscosity of the flowing fluid, and the flow velocity.

Pipe Reynolds number, R_D, is the dimensionless ratio, $VD\gamma/g\mu_e$ where

V = Velocity in any units consistent with the rest of the equation
D = Inside diameter of the conduit (pipe)
γ = Specific weight in any units consistent with the rest of the equation
g = Acceleration of gravity, feet/second^2
μ_e = Absolute viscosity, pound-second/feet^2

The foregoing equation is inconvenient for commercial use because commonly used units of measurement are rarely consistent. Two alternative equation forms are:

(1) $$R_D = \frac{6.32 \times \text{Rate of Flow (pounds/hour)}}{\text{Pipe Diameter (inches)} \times \text{Absolute Viscosity (centipoise)}}$$

(2) $$R_D = \frac{52.77 \times \text{Rate of Flow (gallons/hour at flowing temperature)}}{\text{Pipe Diameter (inches)} \times \text{Kinetic Viscosity (centistokes)}}$$

Flow generally is considered to be fully laminar at Reynolds numbers below 1500; in transition between 1500 and 4000; and turbulent above 4000. Thus, the Reynolds number is a most useful tool in constructing piping system designs and in sizing flowmeters. Orifice or throat Reynolds number, $R_d = R_D/d/D$, where d = diameter of orifice bore (inches); D = inside diameter of conduit or pipe (inches).

Another general way of expressing Reynolds number is by the ratio

$$\frac{\text{Scale Velocity} \times \text{Scale Length}}{\text{Kinematic Viscosity}}$$

This also is referred to as the Reynolds criterion. Reynolds numbers are only comparable when they refer to geometrically similar flows; and, provided that all the boundary conditions can be described by the scale velocity and scale length, flows of the same Reynolds number are dynamically similar. For an airfoil, it is

$$\frac{\text{Air Velocity} \times \text{Chord of the Airfoil}}{\text{Kinematic Viscosity of the Air}}$$

Reynolds numbers are of value in the various fields because tests of models are directly comparable to full scale results of geometrically similar shapes if the Reynolds ratio for the model equals that of the actual or full-scale project. This has its practical applications in the field of hydrodynamics in the study of water resistance of hulls or floats, and in the study of water velocities, levee problems, etc., of large rivers. It is used also to establish the best proportions of hydraulic turbines through the use of models. Much of the science of aeronautics rests upon experimental data obtained in wind tunnels. Dangerous inaccuracies might exist in drawing conclusions for actual construction from model tests, unless either the model were tested at a Reynolds number equal to that of the completed project, or due corrections and allowances were made for the Reynolds number.

RHENIUM. Chemical element symbol Re, at. no. 75, at. wt. 186.2, periodic table group 7b, mp 3178–3182°C, bp 5900°C, density, 21.04 g/cm³ (20°C). Elemental rhenium has a close-packed hexagonal crystal structure. Rhenium is a platinum-white, very hard metal; stable in air below 600°C; practically insoluble in HCl or hydrofluoric acid, but soluble in HNO_3, with formation of perrhenic acid; forms sodium rhenate when fused with NaOH and nitrate. Discovered by Noddack and Tacke in 1925 in tantalite, wolframite, and columbite by the Mosely x-ray spectrographic method of analysis, and later found present in molybdenite, from which rhenium is obtained. Predicted by Mendeleev in 1871, as an element to be discovered with properties resembling manganese, and named by him *dvi-manganese*. Above a temperature of 600°C, the metal begins to generate a white nonpoisonous vaporous oxide, Re_2O_7.

There are two natural isotopes ^{185}Re and ^{187}Re, of which the latter is radioactive with respect to beta decay, having a half-life of 5×10^{10} years reverting to ^{187}Os. Other radioactive isotopes include ^{177}Re, ^{178}Re, ^{180}Re, ^{183}Re, ^{184}Re, ^{186}Re, ^{188}Re, and ^{189}Re. The latter isotope has a long half-life, something less than 10^3 years; the half-lives of the remaining isotopes are comparatively short, measured in minutes, hours, or days. In terms of abundance, rhenium ranks 75th among the elements, based upon estimated contents of the universe. Rhenium is not plentiful in the earth's crust, being essentially confined to association with the mineral molybdenite. First ionization potential, 7.87 eV. Oxidation potentials $Re + 4H_2O \rightarrow ReO_4^- + 8H^+ + 7e^-$, −0.15 V; $Re + 8OH^- \rightarrow ReO_4^- + 4H_2O + 7e^-$, 0.81 V. Other important physical properties of rhenium are given under **Chemical Elements**.

Rhenium is a minor constituent (100 ppb) of molybdenite-bearing porphyry copper ores, of which there is extensive mining in the United States and South America. Commercial rhenium is recovered from by-product molybdenum. As the result of roasting MoS_2 and MoO_3, the rhenium is concentrated to levels of 300–1,000 ppm. At the high process temperatures, the rhenium is oxidized and volatilized as rhenium heptoxide Re_2O_7. This compound is recovered from flue gases by way of wet scrubbing and chemical separation techniques as the relatively crude ammonium perrhenate. The latter compound is reduced with hydrogen to produce rhenium metal.

Uses. Rhenium finds rather wide use as a catalyst in selective hydrogenation and other chemical reactions. It sometimes is used in conjunction with platinum in reforming operations. Additional processes for which rhenium has been tested and used as a catalyst include alkylation, dealkylation, dehydrochlorination, dehydrogenation, dehydroisomerization, enrichment of water, hydrocracking, and oxidation. The outstanding feature of rhenium catalysts is their high selectivity, very important in hydrogenation reactions. Rhenium catalysts also resist such catalyst poisons as nitrogen, sulfur, and phosphorus. In terms of activity, rhenium commonly surpasses cobalt, molybdenum, and tungsten type catalysts and approximate palladium, nickel, and platinum catalysts.

Because of the very heavy ionic weight (250) of the perrhenate ion, it is one of the heaviest simple anions obtainable in readily soluble salts. It has found use as a precipitant for potassium and some other heavy univalent ions; also as a precipitant for such complex ions as $Co(NH_3)_6^{2+}$, and for the separation of alkaloids and organic bases. Perrhenate also is used in the fractional crystallization of the rare-earth elements.

When rhenium is added to other refractory metals, such as molybdenum and tungsten, ductility and tensile strength are improved. These improvements persist even after heating above the recrystallization temperature. An excellent example is the complete ductility shown by a molybdenum-rhenium fusion weld. Rhenium and rhenium alloys have gained some acceptance in semiconductor, thermocouple, and nuclear reactor applications. The alloys are used in gyroscopes, miniature rockets, electrical contacts, electron-tube components, and thermionic converters.

Because rhenium is very difficult to machine with carbide tools and other conventional methods, electrical-discharge machining (EDM), electrochemical machining (ECM), abrasive cutting, or grinding are usually used.

Rhenium can be consolidated by powder metallurgy techniques, inert-atmosphere arc melting, and thermal decomposition of volatile halides. In the powder metallurgy process, bars are pressed at 200 MPa and this is followed by vacuum sintering at 1200°C and hydrogen sintering at 2700°C. Rhenium is usually fabricated from sintered bar by cold working, following by annealing. Reductions of 10–20% can be taken with intermediate anneals for one to two hours at 1700°C. Primary working is by rolling, swaging, or forging. Wire drawing is possible down to 2 mils for strip and wire. Because of its excellent ductility at room temperature, rhenium is suitable for forming of complex shapes (Knipple, 1979).

Chemistry and Compounds. Rhenium has a $5d^56s^2$ electron configuration and all oxidation states from 0 to 7+ are known, although the heptavalent is the most stable state.

The supposed compounds of Re(−I), formulated as $M^1Re(H_2O)_4$, have been shown actually to be tetrahydrorhenates(III), e.g., potassium tetrahydrorhenate, $KReH_4$. These compounds are obtained by reducing potassium perrhenate, in a solution of water and ethanolamine, with the alkali metal. In addition to these compounds, however, rhenium differs from its congener manganese in that it forms more compounds that are in the higher valence group. Instead of the larger number of divalent compounds formed by manganese, the stable compounds of rhenium begin with the trivalent ones. The oxides include Re_2O_3, ReO_2, ReO_3, and Re_2O_7. The heptoxide differs from the corresponding manganese compound in its stability, as does the acid obtained by its reaction with water, perrhenic acid, $HReO_4$. Perrhenic acid is a strong acid ($pK_A = 1.25$), but not as strong an oxidizing agent as the permanganic acid. Rhenium dioxide is obtained by reduction of perrhenic acid or rhenium heptoxide, and thermal decomposition of the dioxan addition product of the latter yields rhenium trioxide, ReO_3.

Rhenium forms a number of halides and oxyhalides. Direct reaction by heating with chlorine yields $ReCl_5$, but heating that compound in a nitrogen atmosphere yields $ReCl_3$. $ReCl_4$ and $ReCl_6$ are also known. Direct reaction by heating with fluorine carries the halogenation further, to ReF_6 and ReF_7, which may be reduced to ReF_4. Direct reaction by heating with bromine yields $ReBr_3$. Reaction of the halides with oxygen, or the oxides or oxyacid salts with halogen-containing substances yields a considerable number of oxyhalogen compounds, e.g., $ReOF_4$, ReO_2F_2, ReO_2Cl_2, ReO_2Br_2, ReO_3F and ReO_3Cl. ReO_3F may also be prepared by the action of liquid HF on $KReO_4$ and ReO_3Cl may be prepared by reaction of ReO_3 and Cl_2.

The complexes of rhenium have not been studied as extensively as those of manganese. Among the known complex ions are $Re(NH_3)_6^{3+}$, $Re(CN)_6^{3-}$, $Re(CN))_8^{4-}$, $Re(CN)_8^{3-}$, $ReO_2(CN)_4^{2-}$, $ReCl_4^-$, ReF_6^{2-}, ReC_6^{2-}, $ReBr_6^{2-}$ and ReI_6^{2-}.

The sulfides include ReS_2 and Re_2S_7.

Unlike manganese, rhenium is reported to form an alkyl compound, trimethyl rhenium, $Re(CH_3)_3$. Like manganese, it forms a dirhenium compound with carbon monoxide, $(CO)_5ReRe(CO)_5$, as well as hydrogen-halogen and alkyl-carbonyl compounds. It also forms a dicyclopentadienyl compound, $(C_5H_5)ReH$.

References

Davenport, W. H., Dollonitsch, V., and C. H. Kline: "Advances in Rhenium Catalysts," *Ind. Eng. Chem.*, **60** (1968).

Knipple, W. R.: "Rhenium," in "Metals Handbook," 9th Edition, Vol. 2, American Society for Metals, Metals Park, Ohio (1979).

Peters, J. E.: "Recent Developments in Rhenium and Rhenium Alloy Powder Technology," *Jrn. Met.* (April 1961).

Savittskii, E. M., and M. A. Tylkina, Editors: "The Study and Use of Rhenium Alloys," (translated from the Russian), Amerind Publishing Co., Pvt. Ltd. Available through U.S. Bureau of the Interior; or National Science Foundation, Washington, D.C. (1978).

Sutulov, A.: "Molybdenum and Rhenium Recovery from Porphyry Coppers," University of Concepción, Chile (1970).

RHEOLOGY. This topic deals with the flow and deformation of matter. A wide spectrum of flow responses is exhibited by different materials—the mobility of common liquids, the behavior of doughs and batters, the properties of syrups, jellies, and slurries, the extensibility of a rubber band, the flow of paint when brushed on a surface, and the moldability of putty, among many other examples. The study of flow characteristics of various materials and the development of formulas and equations to explain and often to predict the behavior of substances fall within the province of rheology.

In a practical sense, the two rheological variables of key interest to chemistry and process engineering are viscosity and consistency. This article concentrates on these two variable quantities.

Classification of Substances. Fundamentally, liquids or suspensions in liquids when subjected to a shear stress behave in one of two ways and this is the basis for one classification: (1) When a *Newtonian substance* undergoes deformation, the ratio of shear rate (flow) to shear stress (force) is *constant.*

With a *non-Newtonian substance*, this ratio is *not constant.* These two types of behavior are illustrated by Figs. 1 and 2.

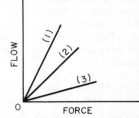

Fig. 1. Behavior of Newtonian substances.

Non-Newtonian substances can be further classified according to the general pattern of functional dependence of rate of shear on shear stress (and, in some cases, time, and occasionally, frequency of vibrational agitation). Five principal classifications are shown in Fig. 2. Examples of specific classes of fluids are given in Table 1. Definitions are given below:

A *true plastic* (Bingham Body) is a substance that when subjected to a constantly increasing shear stress flows only after a definite yield point has been exceeded.

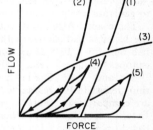

Fig. 2. Behavior of non-Newtonian substances: (1) True plastic (Bingham body), (2) pseudoplastic, (3) dilatant, (4) thixotropic, and (5) rheopectic.

A *pseudoplastic* is a material that appears to have a yield stress beyond which flow commences and increases sharply with increase in stress. In practice, such substances are found to exhibit flow at all shear stresses although the ratio of flow to force increases negligibly until the force exceeds the apparent yield stress.

Dilatant material exhibits an initial flow under a low shear stress at a high rate, but further increases in shear stress result in a lower flow rate. Dilantant substances are sometimes referred to as *inverted plastics* or *inverted pseudoplastics.*

TABLE 1. EXAMPLES OF COMMON FLUIDS EXHIBITING DIVERSE RHEOLOGICAL CHARACTERISTICS

| NEWTONIAN | NON-NEWTONIAN | | | | |
	Pseudo-plastic	Plastic	Thixotropic*	Rheopectic	Dilatant*
Water	Catsup	Chewing gum	Silica gel	Bentonite sols	Quicksand
Most mineral oils		Tar	Most paints	Gypsum in water	Peanut butter
Gasoline	Printers ink	High concentrations of asbestine in oil	Glue		Many candy compounds
	Paper pulp		Molasses		
			Lard		
			Fruit juice concentrates		
Kerosene					
Most salt solutions in water					
Light suspensions of dye stuffs			Asphalts		

* Some liquids may change from thixotropic to dilatant or vice versa as the temperature or concentration changes.

Thixotropic materials experience an increase of flow rate with increasing duration of agitation, as well as with increased shear stress. When agitation is stopped, internal shear stress exhibits hysteresis. Upon reagitation, generally less force is required to create a given flow than is required for the first agitation.

Rheopectic substances. If certain thixotropic suspensions are rhythmically shaken or tapped, they will "set" or build up very rapidly, a phenomenon termed *rheopexy.* Apparent viscosity of a rheopectic substance increases with time (duration of agitation) at any constant shear rate.

Viscosity Relationships

The basic relationship, as given by Newton, is

$$\frac{F}{A} = \mu \frac{V}{L} \tag{1}$$

where F/A = shear stress (force per unit area)
   V/L = shear rate (velocity per unit thickness of layer)
   μ = proportionality constant
Thus, viscosity came to be defined as

$$\mu = \frac{\text{shear stress}}{\text{shear rate}} \tag{2}$$

Hagen-Poiseuille Law. Hagen (Germany) and Poiseuille (France) described viscosity as the ratio of shear stress versus shear rate at the wall of a capillary tube.

$$\mu = \frac{PR/2L \text{ (shear stress)}}{4Q/\pi R^3 \text{ (shear rate)}} = \frac{\pi P R^4}{8QL} \tag{3}$$

where P = pressure differential across liquid in tube
   R = inside radius of tube
   L = length of tube
   Q = volume rate of flow of liquid
   μ = viscosity
This equation is limited to conditions of laminar or viscous (streamline) flow.

Viscosity Units. From Eq. (3) came the unit of viscosity originally in cgs units—a measure of the internal resistance of a material to flow; the ratio of the applied shear stress to the rate of shear. The SI unit of viscosity is the *Pascal second* (Pa·s); the cgs unit is *poise* (1 dyne s/cm^2) and equivalent to 0.1 Pa·s. Values of viscosity are commonly expressed in centipoise (cps); poise ÷ 100, equivalent to millipascal second (mPa·s). The viscosity of water at 68°F (20°C) is 1 cps. A broad classification of fluids by viscosity range is given in Table 2.

Kinematic Viscosity Units. Kinematic viscosity is a measure of the internal resistance to flow of a liquid under gravity; the ratio of the viscosity to the density of a liquid. The SI unit of *kinematic viscosity* is m^2/s. The cgs unit is *stoke* (1 cm^2/s). Stoke ÷ 100 equals centistoke (cks).

$$\text{Kinematic viscosity (stokes)} = \frac{\text{viscosity (poise)}}{\text{density (g/cm}^3)} \tag{4}$$

Viscosity Index. This term has been used with reference to petroleum products. It is an empirical number that indicates the effect of change of temperature on the viscosity of an oil. A low index number signifies an oil that changes viscosity by a substantial amount with a given temperature change.

Viscosity Scales. A few time-based viscosity scales are in common use: (1) Saybolt scales (U.S.), (2) Redwood scales (UK), and (3) Engler scales (Europe). All three scales are based upon the Hagen-Poiseuille law and indicate the time of efflux, under specified conditions, for a fixed volume of fluid through a specific capillary or aperture. Kinematic viscosity (and from it, absolute viscosity) can be determined from such scales by the empirical formula

$$v = At - \frac{B}{t} \tag{5}$$

where A and B = constants applicable to the viscometer
   t = time of efflux, s
   v = kinematic viscosity, centistokes
Commonly accepted values of A and B are given in Table 2. Various viscosity conversion factors are given in Table 4.

Viscosity Measurements

Over many years, numerous ways have been developed to measure viscosity. Some methods are best applied in a laboratory setting; a relatively few methods can be adapted for on-line process control.

Time to Discharge. In this method, the time to discharge a given volume of fluid through an orifice or nozzle is used as a measure of viscosity. This principle is used in the Saybolt, Redwood, Engler, Scott, Ubbelohde, and Zahn viscometers, and in the Parlin cup and the Ford cup. A similar type utilizes a capillary tube in place of the orifice or nozzle. The latter type includes the modified Ostwald, Bingham, and Zeitfuchs.

These are laboratory or batch-type instruments used for many

TABLE 2.   BROAD CLASSIFICATION OF FLUIDS BY VISCOSITY RANGE

Low-Viscosity Fluids (15–2,000,000 centipoises)
Adhesives (Water and Solvent Base)
Chemicals
Cosmetics
Hot Waxes
Inks (Flexo and Roto)
Latex
Oils
Paints and Coatings
Pharmaceuticals
Photo Resist
Rubber Solutions
Solvents
Soups
Textile Fibers (Synthetic)
Water-Based Systems
Medium-Viscosity Fluids (100–8,000,000 centipoises)
Adhesives (Hot Melt)
Creams
Food Products (Dressings, Cheeses, Batters)
Gums
Inks (Screen and Offset)
Organisols
Paints
Paper Coatings
Plastisols
Starches
Surface Coatings
Toothpaste
Varnish
High-Viscosity Fluids (800–64,000,000 centipoises)
Asphalt
Caulking
Chocolate
Epoxies
Gels
Inks (Offset and Thick Film)
Molasses
Pastes
Peanut Butter
Putty
Roofing Compounds
Tars

SOURCE: Brookfield Engineering Laboratories, Inc.

TABLE 3. COMMONLY ACCEPTED VALUES OF *A* AND *B*
(Eq. 5)

Viscometer type	VALUE OF CONSTANTS		
	A	B	Efflux time t, s
Saybolt Universal	0.226	195	32–100
	0.220	135	Above 100
Saybolt Furol	2.24	184	25–40
	2.16	60	Above 40
Redwood No. 1 (Standard)	0.260	179	34–100
	0.247	50	Above 100
Redwood No. 2 (Admiralty)	2.46	100	32–90
	2.45	···	Above 90

years in the petroleum and allied industries. The Zahn, Parlin, and Ford types have been used in the paint and varnish industry. With changes in orifice or capillary dimensions, these instruments are suitable for low-to-moderately-high viscosity ranges and for viscosity approximations. Viscosity is reported in seconds and can be converted to absolute viscosity units only if the fluid is Newtonian.

Timed Fall of Ball or Rise of Bubble. In this method, the time required for a ball to fall or a bubble to rise through a liquid confined in a tube is proportional to absolute viscosity because, in either case, the liquid moves in viscous flow through a restriction. This is primarily a laboratory or batch-type instrument and finds used in the petroleum industry for high-viscosity oils. The ball method can be timed with considerable accuracy by field coils located at the start and finish points.

Drag Torque on Stationary Element in Rotating Cup. In this method, a cup rotates at constant speed. The stationary element, which may be a cylinder or a paddle, is restrained by a calibrated spring. See Fig. 3. The spring deflection indicates torque which can be converted to absolute viscosity units. This is principally a laboratory or batch-type instrument, but may be equipped to record. The bowl may be designed to control sample temperature. The instrument can be used with both Newtonian and non-Newtonian fluids and, by change of rotational speed, a shear stress versus shear rate diagram can be obtained. The instrument is suitable over a wide range of viscosities.

Timed Fall of Piston in Cylinder. In this system, a submerged plunger is raised automatically and then dropped in the cylinder,

TABLE 4. VISCOSITY CONVERSION FACTORS

1 poise = 0.01 centipoise (cps)
1 poise = 0.1 Pascal·second (Pa·s)
1 pound (force)-second/ft² = 1 slug/ft-second
1 pound (mass)/foot-second = 1 poundal second/ft²
μ (stokes or cm²/second) × 0.001076 = μ (ft²/second)

To convert to	MULTIPLY THE NUMBER OF		
	Poise by	Pound (force)-second / Square foot by	Pound (mass) / Foot-second by
Poise	1	478.8	14.88
Pound(force)-seconds per square foot	0.002089	1	0.03109
Pounds(mass) per foot-second	0.06721	32.174	1

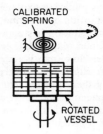

Fig. 3. Viscometer using principle of drag torque on stationary element in rotating cup.

forming an orifice. The piston and cylinder of one type of instrument are shown schematically in Fig. 4. The time of fall may be recorded in terms of absolute viscosity. The instrument is available with a recorder and can be used in automatic control systems as found in the paint, oil, soaps, and plastics industries. The device is suitable for both open and pressurized service and can be designed to handle viscosity ranges from 0.1 to 1,000,000 centipoises with high repeatability at operating temperatures up to 800°F (427°C).

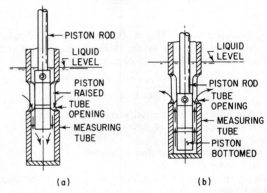

Fig. 4. Viscometer using timed fall of piston in a cylinder. Position of piston during filling phase is shown in (a). The piston is raised, drawing liquid into space formed under the piston. Measuring phase is shown in (b). The piston falls, expelling the liquid through the same path that it entered. Time of fall is directly proportional to viscosity. (*Norcross.*)

Similar types, utilizing a holed disk, or holed cone and designed for laboratory use, report results in seconds, representing the time for the piston to fall a given distance through the sample.

Pressure Drop through a Friction Tube. In this system, the sample is pumped through a friction tube in viscouse flow. Pressure drop across the tube is measured by a differential pressure transmitter. The transmitter output is a direct solution to the Hagen-Poiseuille equation.

Torque Viscometers. In this system, a sensor (spindle) is rotated at some discrete speed through the same couple arrangement which allows the sensor to be displaced to some equilibrium position at which point the calibrated torque spring gives a reference to the viscosity of the fluid being measured. See Fig. 5. Different types of spindles and speeds may be used.

On-Line Process Viscometers. In the instrument shown in Fig. 6, instead of rotating and sensing on the same shaft through the same couple arrangement, the operation has been split into each end of the end cap, i.e., a cylinder is rotated directly from the motor through one end cap and another cylinder is positioned inside the rotating cylinder onto a torque tube which goes through the opposite end cap. With this design, one need be concerned only with pressures and temperatures on seals

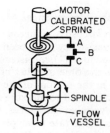

Fig. 5. Viscometer in which torque required to rotate a torque element in a liquid is a measure of viscosity. (*Brookfield.*)

that are rotating and that are not influencing the friction associated with the torque measurement. The nonrotating member is sensed in this design. The instrument shown can measure a wide variety of fluids under various conditions of pressure and flow. Shear rate from 1 to 1200 sec^{-1} can be manually selected in increments of 0.01 to 1.0 sec^{-1}, depending upon range. Shear rate, shear stress (in dynes/cm^2) and viscosity (in poise or centipoise) are continuously displayed by digital LED (light-emitting diode) readouts. These values can also be recorded through back-panel recording outputs. The shear stress range is variable to allow measurement of a wide variety of viscosities at desired shear rates.

The system consists of three basic components: (1) the viscometer, which is mounted in the process pipeline and does the actual measuring. This is the only component that contacts the process fluid; (2) the control unit, which contains most of the system's electronic circuitry and houses the digital readouts and associated switches and controls. An external voltage stabilizer for this unit is supplied; and (3) the frequency generator, which controls the speed of the viscometer drive motor and provides the shear rate signal to the control unit.

Viscosity ranges available: 1–150,000 cps; shear rate ranges: 1–1200 sec^{-1}; temperature range: −40 to 300°F (−40 to 150°C); pressure range: 0–200 psi (1380 kPa); flow rate: 20 gallons (76 liters)/minute maximum; accuracy, ±0.5% of span; repeatability: ±0.5% of span; signal output: 0–1.0 V.

Ultrasonic Probe. Over the years, a number of instruments

have been designed for the measurement of viscosity, utilizing ultrasonic energy.

Representative Viscosity Control Applications

Continuous viscosity measurement and control are widely used (1) in processes involving solvents where viscosity is a measure of solvent concentrations; (2) in certain chemical reactions, such as polymerization, where viscosity is an indication of the *end point* of the reaction; (3) in processes involving starch—as found in the paper and textile industries—where viscosity is a measure of the enzyme actions occurring; and (4) in the blending of petroleum and chemical materials to desired end specifications. Representative applications are shown in Figs. 7 and 8.

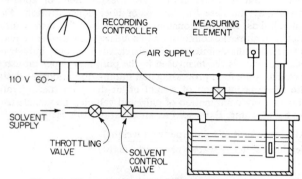

Fig. 7. Viscosity instrumentation system for automatically adding solvent to correct for evaporation losses.

Consistency Measurements

Substances that undergo continuous deformation when subjected to a shear stress are said to exhibit fluid behavior. The resistance offered by the substance to this deformation is termed *consistency*. The consistency is constant for gases and Newtonian fluids, if static temperature and pressure are fixed. When such fluids are referred to, the consistency is called viscosity.

The consistency of non-Newtonian fluids is not constant, but is a function of the applied shear stress and, in some cases, may vary with time. The term *apparent viscosity* is sometimes applied with reference to the consistency of non-Newtonian liquids.

Although consistency is a real property of non-Newtonian fluids, its measurement is usually relative to arbitrary standards.

Fig. 6. Variable-speed rotational viscometer—fully flooded in-line system. (*Brookfield.*)

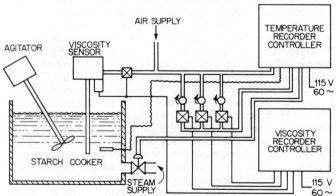

Fig. 8. Automatic batch-type starch conversion system using automatic viscosity measurement and control.

Thus, the measurement cannot be stated in definitive units, such as those of absolute viscosity, temperature, and pressure.

Measurement and control of fluid consistency are of industrial concern because the variable has considerable effect on the efficiency of unit operations and upon the quality of end products from many processes. Consistency control is occasionally a means for reducing the quantity of valuable material lost in a process.

Consistency is important in numerous chemical processes and notably a number of food processes and in the pulp and paper industry. In the latter field, for example, the apparent viscosity of a suspension of certain fibrous materials is related to the percentage of fibers in the suspension. This relationship provides the operating principle for a number of consistency-regulating devices—that is, these devices use measurement of apparent viscosity as the base for control of fiber concentration. This principle has, through constant usage, caused the term consistency to acquire the connotation of solids content and has caused its true scientific definition to be disregarded in some industries.

Consistency is the term used in the pulp and paper industry, as well in some food processing and mining operations, to designate the concentration of weight of air-dry pulp (i.e., of total solids) in any combination of pulp and vehicle, the vehicle usually being water. Expressed as an equation, this is

$$\frac{\text{Air-dry weight of solids (per unit weight)}}{\text{Weight of solids} + \text{water (per unit weight)}} \times 100$$
$$= \text{consistency (\%)}$$

Papermakers occasionally use the terms *bone-dry*, *oven-dry*, and *moisture-free* with reference to pulp fibers or paper from which all water is removed by evaporation. Since such thorough drying is seldom practiced in a paper mill, the term bone-dry is primarily used in the laboratory and in process equipment design. Air-dry pulp is considered to contain 10% moisture so that

Bone-dry consistency (%) = 0.9 × air-dry consistency (%)

Relationship of Consistency to Moisture. Moisture is the virtual reciprocal of consistency. It may be considered as a measure of the liquid water content of the solid medium. This measurement is preferred for materials of higher consistencies, ranging from 12 to 99% consistency. However, it may be applied to consistency measurements outside of this range. For example, a paper web in the fourdrinier section of a paper machine may vary from 0.3% at the breast roll to 20% at the press roll. The drier consistency—entrance to exit—may vary from 10 to 99%.

Relationship of Consistency to Density. Density is the mass per unit volume of a substance, whereas consistency is mass per unit mass of substance. As an example, 3 grams of solids in 100 grams of mixed slurry is a 3% consistency material.

Relationship of Consistency to Freeness. Freeness is the amount of water that leaves a sample of specific volume and consistency. The Canadian Standard Freeness Scale (CSF) is the amount of by-pass water out of a calibrated funnel from a one-liter sample at 0.3% consistency. The freeness is read directly in cubic centimeters. The higher the number, the more readily the slurry will release the water.

Variables Affecting Consistency. Among variables that affect the measurement of consistency are temperature, velocity, freeness, pH, furnish, and inorganic materials content. Temperature changes affect the shear stress of fiber suspensions in that increasing temperature provides decreasing force on a sensor at a fixed consistency. Thus the sensor produces a change in signal, indicating lowering consistency with increasing temperature.

Consistency Measurement Means

As with viscosity, numerous methods for measuring and controlling consistency have been developed over several decades. Methods include (1) consistency cups (where consistency is expressed as the time of efflux of a fixed volume of fluid through an orifice in the bottom of the cup); (2) rotational viscometers, which can be used to measure shear values (apparent viscosity) of non-Newtonian substances; (3) blade-type sensors which operate on a shear rate principle by utilizing a blade sensor suspended in the pipeline. Consistency changes are measured by sensing the change in force required by the sensing blade to shear through the flowing stock. (4) Various styles of probes that sense force within the vessel measured; (5) polarized-light devices which determine the amount of depolarization created by suspended particles; and (6) web-moisture measuring devices which correlate moisture with consistency. Moisture may be measured in several ways, including electrical conductivity, energy (RF, microwave, etc.) absorption, methods which do not require physical contact, for example, with a paper web.

References

Balls, B. W.: "Toward Better Understanding of Consistency Measurements," *Measurement and Control* **1** (9), Institute of Measurement and Control, 20 Peale St., London W8, September, 1968.
Casey, J. P. (editor): "Handbook of Pulp and Paper Technology," 2nd Edition, Van Nostrand Reinhold, New York, 1970.
Considine, D. M., and G. D. Considine (editors): "Foods and Food Production Encyclopedia," pp. 1998–2004, Van Nostrand Reinhold, New York, 1982.
Deman, J. H., et al.: "Rheology and Texture in Food Quality," AVI, Westport, Connecticut, 1976.
Lavigne, J. R.: "An Introduction to Paper Industry Instrumentation," Revised Edition, W. H. Freeman, San Francisco, 1979.
Rha, C. K.: "Viscoelastic Properties of Foods as Related to Micro- and Molecular Structures," *Food Technol.* **33**(10), 71–76 (1979).
Torburg, R. H.: "Fine Tuning of a Consistency Control System for Maximum Performance," *Pulp and Paper*, 134–138, W. H. Freeman, San Francisco, March 1980.

RHEOPECTIC SUBSTANCES. Rheology.

RHODIUM. Chemical element symbol Rh, at. no. 45, at. wt. 102.905, periodic table group 8 (transition metals), mp 1963–1969°C, bp 3625–3825°C, density, 12.44 g/cm³ (solid at 20°C). Elemental rhodium has a face-centered cubic crystal structure. The one stable isotope is ^{103}Rh. The seven unstable isotopes are ^{99}Rh through ^{101}Rh and ^{104}Rh through ^{107}Rh. In terms of earthly abundance, rhodium is one of the scarce elements. Also, in terms of cosmic abundance, the investigation by Harold C. Urey (1952), using a figure of 10,000 for silicon, estimated the figure for rhodium at 0.0067. No notable presence of rhodium in seawater has been detected.

Electronic configuration is $1s^2 2s^2 2p^6 3s^2 3p^6 3d^{10} 4s^2 4p^6 4d^8 5s^1$. Ionic radii, Rh^{3+}, 0.75 Å, Rh^{4+}, 0.65 Å. Metallic radius, 1.345 Å. First ionization potential, 7.7 eV. Other physical properties of rhodium will be found under **Platinum and Platinum Group**.

Rhodium was discovered by Wollaston (England) in 1803. Compact Rh is almost insoluble in all acids at 100°C, including aqua regia. Hot concentrated H_2SO_4 will slowly dissolve the finely divided metal. When alloyed with 90% or more of Pt, it is soluble in aqua regia. The metal is attacked by fused bisulfates. Rh is soluble in molten Pb. This is the basis of the classic separation of Rh and Ir.

Rh compounds exhibit valences of 2, 3, 4, and 6. The trivalent form is by far the most stable. When Rh is heated in air, it becomes coated with a film of oxide. Rhodium(III) oxide,

Rh_2O_3, can be prepared by heating the finely divided metal or its nitrate in air or O_2. Rhodium(IV) oxide is also known. Rhodium trihydroxide may be precipitated as a yellow compound by adding the stoichiometric amount of KOH to a solution of $RhCl_3$. The hydroxide is soluble in acids and excess base. When the freshly precipitated $Rh(OH)_3$ is dissolved in HCl at a controlled pH, a yellow solution is first obtained in which the aquochloro complex of Rh behaves as a cation. The hexachlororhodate(III) anion is formed when the solution is boiled for 1 hr with excess HCl. The solution chemistry of $RhCl_3$ is often very complex. Two trichlorides of Rh are known. The trichloride formed by high-temperature combination of the elements is a red, crystalline, nonvolatile compound, insoluble in all acids. When Rh is heated in molten NaCl and treated with Cl_2, Na_3RhCl_6 is formed, a soluble salt that forms a hydrate in solution. Rhodium(III) iodide is formed by the addition of KI to a hot solution of trivalent Rh.

Rhodium(III) sulfate exists in yellow and red forms. If $Rh(OH)_3$ is dissolved in cold H_2SO_4, the product is the yellow form, in which the sulfate is ionic. If this solution is evaporated in hot H_2SO_4, the product is a red nonionic sulfate. When Rh is treated with F_2 at 500–600°C, RhF_3 is slowly formed. This compound is practically insoluble in water, concentrated HCl, HNO_3, H_2SO_4, H_2F_2, or NaOH.

If a solution of $RhCl_3$ is treated with $NaNO_2$, the very soluble sodium hexanitritorhodate(III), $Na_3Rh(NO_2)_6$, is formed. The solubility of this compound in alkaline solution makes it useful for refining, as many base metals are precipitated as their hydroxides under these conditions. The analogous ammonium and potassium salts are relatively insoluble.

When H_2S is passed into a solution of a trivalent Rh salt at 100°C, the hydrosulfide, $Rh(SH)_3$, is formed. This black precipitate is insoluble in $(NH_4)_2S$. Rh forms many complexes with NH_3, amines, cyanide, chloride, bromide, and numerous polynitrogen and polyoxygen chelating agents.

See also **Chemical Elements.** More detail on the physical properties of rhodium can be found in the article on "Rhodium" by L. Bozza in "Metals Handbook," 9th Edition, American Society for Metals, Metals Park, Ohio, 1979.

—Linton Libby, Chief Chemist, Simmons Refining Company, Chicago, Illinois.

RHODOCHROSITE.
Rhodochrosite, manganese, carbonate, $MnCO_3$, is a rose-pink to red hexagonal mineral, occurring as small crystals, in cleavable masses, granular or compact. It is a brittle mineral; hardness, 3.5–4; specific gravity, 3.7; pure $MnCO_3$ usually 3.4–3.6; luster, vitreous to pearly; color, various shades of pink, red and reddish-brown; transparent to opaque; streak, white. Rhodochrosite has a perfect rhombohedral cleavage. Rhodochrosite is a product of high-temperature metamorphic deposits; as a gangue mineral in ore veins of hydrothermal origin, and as secondary residual deposits from bodies of manganese or iron oxides, and in sedimentary deposits precipitated like siderite by organic matter acting, in the absence of oxygen, upon bicarbonates.

Spectacular, gemmy red rhombohedrons, up to 3 inches (7.5 centimeters) on an edge occur at the Sweet Home Mine, Alma, Colorado, and the John Reed Mine, Alicante, Colorado. Exceptionally fine stalactitic formations are found in Catamarca Province, Argentina. Recently the most beautiful, large transparent gem red scalenohedron crystals ever found have come from Hotazel and the Kalahari manganese field in northern Cape Province, South Africa. This region encompasses one of the largest and richest known manganese deposits in the world.

Other localities for this mineral are in Rumania, Saxony, Westphalia, and Cornwall, England. In the United States rhodochrosite is found at Franklin, New Jersey; Butte Montana; and in various localities in Colorado and Nevada. The name *rhodochrosite* is delivered from the Greek words meaning *rose* and *color.*

RHODONITE.
The mineral rhodonite, manganese metasilicate, $(Mn,Fe,Mg)SiO_3$, crystallizes in the triclinic system forming large, irregular tabular crystals but usually occurring massive. Prismatic and basal cleavages excellent; fracture conchoidal to uneven; hardness, 5.5–5.6; specific gravity, 3.57–3.76; luster, vitreous to pearly on cleavage faces; color, red to brownish-red; rarely yellow to gray; streak, white; transparent to translucent. A variety containing much calcium is called bustamite. Zinc may replace the manganese in rhodonite; it is then known as fowlerite. Rhodonite is found in the Harz Mountains, Germany; in the Urals of the U.S.S.R.; in Hungary, Italy, and Sweden. Bustamite from Mexico, and from Franklin and Sterling Hill, New Jersey, occurs with fowlerite.

Rhodonite has been occasionally used for an ornamental stone. Its name is derived from the Greek meaning *rose*, because of the color.

RIBOFLAVIN.   Vitamin B_2

RIBONUCLEIC ACID (RNA).
Generic term for a group of natural polymers consisting of long chains of alternating phosphate and D-ribose units, with the bases adenine, guanine, cytosine, and uracil bonded to the 1-position of the ribose. Ribonucleic acid is universally present in living cells and has a functional genetic specificity due to the sequence of bases along the polyribonucleotide chain.

Four types are recognized: (1) *Messenger RNA* is synthesized in the living cell by the action of an enzyme that carries out the polymerization of ribonucleotides on a DNA template region which carries the information for the primary sequence of amino acids in a structural protein. It is a ribonucleotide copy of the deoxynucleotide sequences in the primary genetic material. (2) *Ribosomal RNA* exists as a part of a functional unit within living cells called the ribosome, a particle containing protein and ribosomal RNA in roughly 1:2 parts by weight, having a particle weight of about 3 million. Messenger RNA combines with ribosomes to form polysomes containing several ribosome units, usually 5 (e.g., during hemoglobin synthesis), complexed to the messenger RNA molecule. This aggregate structure is the active template for protein biosynthesis. (3) *Transfer RNA* is the smallest and best characterized RNA class. Its molecules contain only about 80 nucleotides per chain. Within the class of transfer RNA molecules, there are probably at least 20 separate kinds, correspondingly related to each of the 20 amino acids naturally occurring in proteins. Transfer RNA must have at least two kinds of specificity: (a) It must recognize (or be recognized by) the proper amino acid activating enzyme so that the proper amino acid will be transferred to its free 2'- or 3'-OH group; (b) it must recognize the proper triplet on the messenger RNA-ribosome aggregate. Having these properties, the transfer RNA accepts or forms an intermediate transfer RNA-amino acid that finds its way to the polysome, complexes at a triplet coding for the activated amino acid, and allows transfer of the amino acid into peptide linkage.

(4) *Viral RNA*, isolated from a number of plant, animal, and bacterial viruses, may be considered as a polycistronic messenger RNA. It has been shown to have molecular weights of 1 or 2 million. Generally speaking, there is one molecule of RNA per infective virus particle. The RNA of RNA virus can

be separated from its protein component and is also infective, bringing about the formation of complete virus.

See also **Deoxyribonucleic Acid (DNA)**; **Nucleic Acids and Nucleoproteins**; and **Recombinant DNA**.

References

Altman, S., Editor: "Transfer RNA," MIT Press, Cambridge, Massachusetts (1978).

Darnell, J. E., Jr.: "Implications of RNA·RNA Splicing in Evolution of Eukaryotic Cells," *Science*, **202**, 1257–1260 (1978).

Kornberg, A., Horecker, B. L., Cornudella, L., and J. Oro, Editors: "Reflections on Biochemistry," Pergamon, New York (1976).

Paigen, K., Labarca, C., and G. Watson: "A Regulatory Locus for Mouse β-Glucuronidase Induction, *Gur*, Controls Messenger RNA Activity," *Science*, **203**, 554–556 (1979).

Quigley, G. J., and A. Rich: "Structural Domains of Transfer RNA Molecules," *Science*, **194**, 796—806 (1976).

Sussman, J. L., and S.-H. Kim: "Three-Dimensional Structure of a Transfer RNA in Two Crystal Forms," *Science*, **192**, 853–858 (1976).

RIEBECKITE. The mineral riebeckite, essentially sodium iron silicate, $Na_2(Fe_3^{2+}, Fe_2^{3+})_5Si_8O_{22}(OH)_2$, is a monoclinic member of the amphibole group, usually in prismatic crystals. It has a prismatic cleavage; hardness, 5; specific gravity, 3.32–3.382; vitreous luster; color dark bluish to black. It occurs in granites and syenites chiefly. It is found in Greenland, Portugal, the Malagasy Republic and South Africa (crocidolite); and in the United States at Quincy, Massachusetts; near Pikes Peak, Colorado; and the San Francisco Mountains, Arizona.

RNA. Nucleic Acids and Nucleoproteins; Ribonucleic Acid.

ROCKET PROPELLANTS. Chemical rocket propellants can be classified in several ways, including liquid or solid, monopropellant, bipropellant or tripropellant, cryogenic or storable, hypergolic or nonhypergolic, double-base, composite or composite double-base. Propellant classification frequently influences the classification of rocket engines—for example, monopropellant rocket, liquid rocket, solid rocket, and hybrid rocket (usually using a liquid oxidizer and a solid fuel). Propellants for chemical rockets serve two primary functions as contrasted to one function for nuclear, solar, electrical or laser heated rockets. In chemical propellant rockets, the propellant is both the energy source and the ejected mass or "working fluid."

Compared with the almost limitless number of chemical compounds that exist or can be formed, the number of chemical propellants in common use are relatively few. This situation arises from criteria including costs, source availability, toxicity, resistance to shock, and other requirements imposed by the vehicle application and the propulsion system design. Another practical reason is that extensive overlap of physical, chemical, and economic properties are displayed by many of the theoretically possible propellants. During the 1960s, the universities, industry, and government in the United States pursued extensive research programs for synthesizing new chemical compounds viewed as candidate propellants which would increase the performance capability of chemical rockets. Although dozens of compounds were synthesized, few results reached the production line. This does not mean, however, that a scientific breakthrough in increasing the molecular energy of a propellant may not be ultimately obtainable.

The characteristics desired of a rocket propellant are several in number and can be divided into economic, safety, materials compatibility, engine-cycle needs and vehicle requirements.

In general terms, the engine-cycle needs ideally are: (1) a propellant or propellant combination that has a high heat of reaction per unit weight (also called heat of combustion). Most vehicles add a requirement for high heat of reaction per unit volume of propellant to minimize the vehicle size. (2) reaction products that are all gaseous, that have a very low molecular weight and that have a very high temperature of dissociation.

In addition to specific impulse, the vehicle requirements usually influence propellant selection in terms of storability, density, toxicity, and other hazards, and other application-sensitive factors, including exhaust plume properties and radar cross section and radiation emissions. Other factors being essentially equal, the higher the heat of reaction of a propellant (or combination), the more attractive the propellant. Sharp exceptions to this rule occur in some missles because of volume limitations, the need for smokeless exhaust or similar restraints.

The heat released by a propellant is the difference in heat between the constituents and the end-products of combustion.

$$\Delta H_r^o = \left[\sum_{k, \text{ products}} \eta_k (\Delta H_f^o)_k - \sum_{j, \text{ reactants}} \eta_j (\Delta H_f^o)_j \right]$$

where ΔH_r^o is the heat generated; ΔH_f^o is the standard heat of formation of the constituent at reference temperature (298K); and η is the number of moles of each j reactant or k product. Large heat release is afforded by reaction products having large negative values, while the reactants should have positive, or at least small negative values, if possible. The heat of reaction is often noted in energy/weight units, such as kilocalories/gram.

Specific impulse, I_{sp}, the universally accepted measure of rocket engine performance, can also be used to indicate the performance of propellants. The most commonly stated expansion ratio is $1,000 \rightarrow 14.7$ giving "sea-level specific impulse at 1,000 psi chamber pressure." Sometimes the expansion ratio is $1,000 \rightarrow 0.2$ to indicate specific impulse for high-altitude or space flight.

By definition, specific impulse I_{sp} is:

$$I_{sp} = \frac{F}{\dot{W}}$$

with the I_{sp} units being seconds; the short designation for units of thrust (force) per units of propellant mass flow per second.

For an ideal rocket with the nozzle exhaust pressure being the ambient pressure, the thrust, recognizing Newton's second law of motion, is:

$$F = \frac{\dot{W}c_e}{g}$$

where $\dot{W}$ is propellant flow rate in pounds per second; c_e is exhaust velocity in feet per second; and g is the gravitational constant in feet/second/second.

In practice, only about 10% of the elements on the periodic chart are adapatable to chemical rocket propellants. Propellants have made little use of elements other than hydrogen, carbon, nitrogen, oxygen, chlorine, fluorine, aluminum, boron, and beryllium.

Liquid propellants fall into two broad classes: (1) earth storable (monopropellants and bipropellants), and (2) cryogenic, depending upon whether they can be kept in the vehicle tankage for months and years, or must be used in a few hours or days. The theoretical performances of storable and cryogenic bipropellants combusting ideally at 1,000 psia chamber pressure and expanding to sea-level pressure without loss, assuming shifting chemical equilibrium of the combustion products during expansion in the engine exhaust nozzle, are listed in Tables 1 and 2. For comparison purposes, a few properties of the more common monopropellants are listed in Table 3. Water (not listed)

TABLE 1. STORABLE LIQUID BIPROPELLANT COMBINATIONS

OXIDIZER	FUEL	OXYGEN-FUEL RATIO BY WEIGHT FOR MAXIMUM I_{sp}	BULK SPECIFIC GRAVITY	THEORETICAL I_{sp}[a]
Nitrogen tetroxide	50/50 Hydrazine/ Unsymmetrical dimethylhydrazine	2.0	1.19	289
Nitrogen tetroxide	Unsymmetrical dimethyl- hydrazine	2.6	1.17	286
Nitrogen tetroxide	Hydrazine	1.3	1.21	292
Red fuming nitric acid (15% NO_2)	Unsymmetrical dimethyl- hydrazine	3.4	1.28	266
Red fuming nitric acid (15% NO_2)	Kerosene-type fuel	5.6	1.37	257
Maximum density red fuming nitric acid (mixture of red fuming nitric acid and N_2O_4)	Unsymmetrical dimethyl- hydrazine	2.9	1.29	278
N_2O_4 with 15% NO	Unsymmetrical dimethyl- hydrazine	2.6	1.15	288
Hydrogen peroxide	Hydrazine	2.0	1.24	287
Hydrogen peroxide	Unsymmetrical dimethyl- hydrazine	4.2	1.22	284
Chlorine trifluoride	Hydrazine	2.8	1.64	295
Chlorine trifluoride	Unsymmetrical dimethyl- hydrazine	3.0	1.39	280
Chlorine pentafluoride	Hydrazine	2.7	1.47	313
Chlorine pentafluoride	Unsymmetrical dimethyl- hydrazine	2.9	1.34	297
Hydrazine	Pentaborane	1.3	0.80	328

[a] 1,000 → 14.7 psia, shifting equilibrium (chemical composition of exhaust gases changes during nozzle flow).

TABLE 2. CRYOGENIC LIQUID BIPROPELLANT COMBINATIONS
(At least one propellant is cryogenic)

OXIDIZER	FUEL	POUNDS OXIDIZER/POUND FUEL STOICHIO- METRIC	POUNDS OXIDIZER/POUND FUEL MAXIMUM I_{sp}	BULK SPECIFIC GRAVITY MAXIMUM I_{sp}	THEORETICAL I_{sp}[a]
Liquid oxygen	Kerosene-type fuel	3.41	2.6	1.02	300
Liquid oxygen	Hydrazine	3.0	0.9	1.07	313
Liquid oxygen	Unsymmetrical dimethylhydrazine	—	1.7	0.98	310
Liquid oxygen	Ammonia	2.37	1.4	0.89	294
Liquid oxygen	Ethyl alcohol	2.09	1.8	0.99	290
Liquid oxygen	Methane	—	3.3	0.82	311
Liquid oxygen	Liquid hydrogen	7.95	4.2	0.29	290
Liquid fluorine	Liquid hydrogen	19.0	8.0	0.46	412
Liquid fluorine	Hydrazine	2.71	2.3	1.31	365
Liquid fluorine	Kerosene-type fuel	4.07	2.6	1.21	322

[a] 1,000 → 14.7 psia, shifting equilibrium (chemical composition of exhaust gases changes during nozzle flow).

TABLE 3. LIQUID MONOPROPELLANTS

PROPELLANT	SPECIFIC GRAVITY AT 68°F	THEORETICAL I_{sp}[a]	EXHAUST (AVERAGE MOLECULAR WEIGHT)
Hydrogen peroxide	1.39	165	22.68
Hydrazine	1.01	199	12.77
Nitromethane	1.12	245	20.34
Ethylene oxide	0.89	199	20.50

[a] 1,000 → 14.7 psia, shifting equilibrium (chemical composition of exhaust gases changes during nozzle flow).

as a source of hydrogen and oxygen via electrolysis has merit as a propellant in long-life satellites (5 years plus), equipped with solar electric cells.

Solid propellants fall into three general types: (1) double-base; (2) composite; and (3) composite double-base. Double-base propellants form a homogeneous cured propellant, usually a nitrocellulose-type of gunpowder dissolved in nitroglycerin plus minor percentages of additives. Both the major ingredients are explosives and both contribute to the functions of fuel, oxidizer, and binder. Composite propellants form a heterogeneous propellant grain with the oxidizer crystals and a powdered fuel (usually aluminum) held together in a matrix of synthetic rubber (or plastic) binder such as polybutadiene. Normally, composite propellants are less hazardous to manufacture and handle than double-base propellants. Composite double-base propellants are a combination of the two aforementioned types—usually a crystalline oxidizer (ammonium perchlorate) and powdered aluminum fuel held together in a matrix of nitrocellulose-nitroglycerin. The hazards of processing and handling this type of propellant are similar to those experienced with the double-base propellants. The characteristics of several common solid propellants are given in Table 4.

Ingredients are generally classified according to their function, e.g., fuel, oxidizer, binder, curing agent, burn-rate catalyst, etc. Ingredients used in small amounts are called additives and usually have functions other than the fuel, oxidizer, or binder. For example, an additive can reduce the viscosity of the propel-

lant during mixing and casting (pouring) of the propellant, increase the burning rate of the propellant, or improve the storage stability. Often an ingredient serves or affects more than one function, the most diffused situation relating to composite double-base ingredients where the binder is a nitrocellulose-nitroglycerine complex with each of these two ingredients having its own fuel and oxidizer chemical elements. The binder contributes also as a fuel and in some propellant formulations, such as asphalt-base nonmetalized propellants, the binder is the fuel.

Ammonium perchlorate, NH_4ClO_4, is the most widely used crystalline oxidizer in solid propellants. Because of its characteristics, including compatibility with other propellant materials, specific impulse performance, quality uniformity and availability, it dominates the solid oxidizer field. Both ammonium and potassium perchlorate are only slightly soluble in water, a favorable trait for propellant use. Nitronium perchlorate is objectionably hygroscopic, is relatively incompatible with available binders, and detonates easily. All of the perchlorate oxidizers produce hydrogen chloride in their reaction with fuels. Their exhaust bases are toxic and corrosive to the extent that care is required in firing rockets, particularly the very large rockets, to safeguard operating personnel and communities in the path of exhaust clouds. Ammonium perchlorate is available in the form of small white crystals and close control of the size range and percentage of several sizes present in a given quantity or batch is required, since particle size influences propellant processing and the physical and ballistic properties of the finished propellant.

Inorganic nitrates are relatively low-performance oxidizers as compared with perchlorates. However, ammonium nitrate is used in some applications for economy and because of its smokeless and relatively nontoxic exhaust. Its main use is in low-burning rate, low-performance applications, such as gas generators for turbine pumps.

One or two crystalline high explosives, such as HMX (cyclotetramethylene tetranitramine) and RDX (cyclotrimethylene trinitramine), are sometimes included in a propellant formulation to achieve a specific performance characteristic. Depending upon the objectives, the percent can range from 5 to 50%.

The one prominent solid fuel is powdered aluminum, and it is used in a wide variety of composite and composite double-

TABLE 4. CHARACTERISTICS OF REPRESENTATIVE SOLID PROPELLANTS

PROPELLANT TYPE	I_{sp} (Seconds)	FLAME TEMPERATURE (°F)	FLAME TEMPERATURE (°C)	DENSITY (Pounds/Cubic Inch)	DENSITY (Grams/Cubic Centimeter)	METAL CONTENT (Weight %)	BURNING RATE (Inches/Second)	BURNING RATE (Centimeters/Second)	HAZARDS CLASS (Military)	STRESS/STRAIN (psi) -60°F (-51°C)	STRESS/STRAIN (%) 150°F (66°C)	PROCESSING METHOD
DB	255	5,340	2449	0.057	1.58	0	0.45	1.1	7	4,600/2	490/60	Extruded
DB/AP/Al[a]	258	6,990	3866	0.069	1.91	25	0.78	2.0	2	2,750/5	120/50	Extruded
DB/AP-HMX/Al[a]	272	6,630	3666	0.067	1.85	20	0.55	1.4	7	2,375/3	50/33	Solvent Cast
PVC/AP[b]	239	4,810	2654	0.065	1.80	0	0.45	1.1	2	369/150	38/220	Solvent Cast
PVC/AP/Al[c]	253	6,120	3382	0.069	1.91	20	0.45	1.1	2	359/150	38/220	Solvent Cast
PBAN/AP/Al[c]	265	5,600	3093	0.063	1.74	19	0.55	1.4	2	520/16 (at -10°F)	71/28	Cast
PU/AP/Al[c]	263	6,000	3316	0.065	1.80	23	0.27	0.7	2	1,170/6	75/33	Cast
CTPB/AP/Al[c]	265	5,540	3060	0.063	1.74	19	0.45	1.1	2	325/26	88/75	Cast
HTPB/AP/Al[c]	264	5,540	3060	0.063	1.74	19	0.40	1.0	2	910/50	90/33	Cast
PBAA/AP/Al[c]	265	5,660	3127	0.063	1.74	20	0.32	0.8	2	500/13	41/31	Cast

Al	Aluminum	HTPB	Hydroxy-terminated polybutadiene
AP	Ammonium perchlorate	PBAA	Polybutadiene-acrylic acid polymer
CTPB	Carboxy-terminated polybutadiene	PBAN	Polybutadiene-acrylic acid-acrylonitrile terpolymer
DB	Double base	PU	Polyurethane
HMX	Cyclotetramethylene tetranitramine	PVC	Polyvinyl chloride

NOTES: [a] AP/Al optimized with 40% DB as binder.
[b] AP/Al optimized with 20% binder.
[c] AP/Al optimized with 15% binder.

base propellant formulations, usually being between 14 and 22% of the propellant by weight.

Boron, even though it appears as one of the high-energy fuels and is lighter than aluminum, has not proven to be a practical fuel because it is so difficult to burn with high efficiency in combustion chambers of reasonable length. Beryllium burns much more easily than boron and improves the specific impulse of a solid propellant motor, usually by about 15 seconds, but as a powder or dust it is highly toxic to animals and humans. The technology with composite propellants using powdered beryllium fuel is sufficiently advanced for vehicle application, with space travel being the most likely application.

Theoretically, both aluminum hydride, AlH_3, and beryllium hydride, BeH_2, are attractive fuels because of their high heat release and gas volume contribution. Both are difficult to manufacture and both deteriorate chemically during storage due to loss of hydrogen. Because of these difficulties, coupled with relatively modest I_{sp} gains, these compounds remain experimental.

Hybrid rocket propellants are various combinations of solid and liquid propellants, usually a solid fuel and a liquid oxidizer. Sometimes, a third propellant, liquid hydrogen, is added, not for energy release, but as a low-molecular-weight working fluid. The main advantages of a hybrid rocket are: (1) use of liquid and solid propellant combinations offering the highest performance attainable with chemical rockets; (2) simplicity of a solid grain (usually fuel); (3) a liquid for nozzle cooling and thrust modulation (compared with a solid rocket); (4) restart capabilities; and (5) good storability and safe storage characteristics.

The chemical bond energy present in propellant molecules is the energy source used by chemical rocket engines to date. This source affords energy densities of approximately 3 kilocalories/gram in the liquid hydrogen/liquid oxygen combination, and up to about 5.7 kilocalories/gram with the lithium/fluorine combination. Theoretically, supplemental energy can be added to molecules or molecular fragments that, upon recombination or relaxation to their normal energy state, release significant amounts of energy. For example, 52 kilocalories/gram is theoretically released when two hydrogen atoms (free radicals) recombine to form hydrogen. Even higher energy densities, as much as 100 kilocalories/gram, are theoretically available from lightweight molecules, such as helium, that are in an excited state.

Metastable, in the sense of propellant ingredients, means that the "energized" molecule, atom, or molecular fragment, tends to promptly return to its normal state. Some molecular species distinctly assume a metastable state upon excitation with the lifetime at room temperature being 10^{-3} to 10^{-2} second as compared with less than 10^{-6} for nonmetastable excited species. Atoms subjected to excitation move into a more energetic state of translational motion of vibration or into a high-energy electron orbital state; diatomic molecules do likewise. Molecules containing more than two atoms can experience higher translational rotational motion, as well as higher electron orbital state.

Most of the research to date on metastable propellants has been with gaseous atoms and molecules. Obviously, energized ingredients in a condensed phase, solid or liquid, would be needed for most rockets. The primary objectives to be reached, if metastable ingredients are to benefit rocket propulsion, are: (1) an efficient process for energizing the ingredients; and (2) a means of storing the ingredients for days at a time without appreciable energy loss. Actual use of metastable ingredients in a rocket is envisioned in the company of liquid hydrogen, or other low-molecular-weight working fluid.

Limited research has been conducted on two approaches to generating and storing (stabilizing) metastable propellant ingredients: (1) free radicals, specifically, atomic hydrogen; and (2) helides which are excited states of helium. In the late-1950s, the U.S. National Bureau of Standards produced low concentrations of free radicals and stored them in inert matrices at very low temperatures. More recently, an approach has been taken to generate hydrogen atoms, immediately condensed at liquid-helium temperature, in the presence of a high density (70 to 100 kilogauss) magnetic field for the purpose of stabilizing the hydrogen atoms. Theoretically, the high-strength magnetic field is capable of aligning the spin of the electron of the hydrogen atom so as to prevent recombination into the hydrogen molecule.

Triplet helium has a theoretical energy level of 114 kilocalories/gram above the ground state. Assuming release of this energy and subsequent expansion through a rocket nozzle gives a specific impulse of 2,800 seconds. Techniques for generating activated helium and other noble gases are well known, but concentrating and storing these metastable species is quite another matter inasmuch as they revert to their ground state by collision processes. Experimental approaches to activating helium and trapping the helium molecules in a hydrocarbon wax have been reported.

The creation and use of metallic hydrogen (hydrogen derived from normal hydrogen subjected to about 2 megabars pressure) should release about 52 kilocalories/gram upon transitioning from the metallic to the normal solid form. The concept dates back to 1935, but interest has been renewed because some scientists believe that metallic hydrogen exists in some large planets.

Antimatter Rockets. Sufficient atomic particle research has been accomplished to warrant discussion of possible methods of applying energy available from particle mass annihilation to rocket propulsion. Complete conversion of matter to energy would allow exhaust velocities near that of light to be obtained from a propulsion device. Antimatter, by definition is matter made up of antiparticles, such as antineutrons, negatrons (antiprotons), and positrons (antielectrons). An annihilation property is known to exist between particles with one particle termed the antiparticle of the other.

Rocket design concepts envisioned for utilizing the reaction between atomic particles and antiparticles (matter and antimatter) are based upon the following postulations: (1) annihilation products can be accelerated using electrical and magnetic forces (consider the annihilation reaction of a neutrino with an antineutrino, yielding a proton and an electron); (2) annihilation products can be used indirectly to heat a working fluid for thermal expansion through a nozzle (consider the annihilation reaction of hydrogen and antihydrogen, leaving high-energy gamma rays); (3) antimatter possesses negative gravitational mass although its inertial mass may be positive. This could give rise to antigravity propulsion; and (4) annihilation products of ordinary quanta give rise to the possibility of a photon-expelling beam for the direct generation of thrust.

Before any form of antimatter rocket can exist, a lightweight method must be developed for producing antiparticles at a flow rate of grams/second in contrast with the few dozen of antiparticles produced in research laboratory generators. Also, a practical storage or containment method must arise inasmuch as antiparticles explode violently upon contact with normal matter. Reference 5 gives a performance estimate of an I_{sp} of 3.06×10^7 seconds for a rocket propelled vehicle with a thrust/weight ratio of 10^{-7}.

Multiple Uses of Propellants. Propellants, both solid and liquid, are used in many secondary propulsion applications, including crew capsule ejection, attitude control and station-keeping of satellites, braking of re-entry vehicles, extra-vehicular space operations—as well as being essential in rocket engine igniters, signal and illumination flares, and fuel-cell type electric generators. New developments, such as the high-powered gas dynamic laser[6] continue to broaden the field of applications.

Acknowledgment. Information for this entry as furnished by Mr. Donald M. Ross, Consulting Engineer, Lancaster, California and for confirmation of tabular performance data by Mr. Curtis C. Selph, Propellant Research Engineer, U.S. Air Force Rocket Propulsion Laboratory, Edwards, California, are gratefully acknowledged.

References

Cohen, W.: "New Horizons in Chemical Propulsion," *Astronautics and Aeronautics,* **2,** 12, 46–51 (1973).

Quinn, L. P., et al.: "High Energy Storage Investigations," Rept. AFRPL TR-71-36, U.S. Air Force Rocket Propulsion Laboratory, Edwards, California (1971).

Ross, D. M.: "Propellants," in "Energy Technology Handbook," (D. M. Considine, Editor), McGraw-Hill, New York (1977).

RUBBER (Natural). Natural rubber is the name applied to the polymer *cis*-polyisoprene obtained chiefly from the *Hevea brasiliensis* tree.[1] Originally, the tree grew wild in the Amazon valley, but during the last part of the 19th Century, it was planted in well-organized plantations in tropical lands of the Far East and later in Africa. See Table 1. The average rubber

TABLE 1.   NATURAL RUBBER EXPORTED

PRODUCING AREA	QUANTITY PRODUCED (Long Tons)				
	1940	1960	1970	1975	1979
Malaysia	547	775	1304	1424	1595
Indonesia	543	587	755	788	866
Other Asian Countries, including Oceania	283	400	484	525	704
Africa	16	149	207	197	141
Tropical America	26	15	10	12	19

NOTE: 1 long ton is slightly more than 1 metric ton (1016 kilograms)

tree stands about 40–50 feet (12–15 meters) high. For optimum growth, a tropical climate having 80 inches (203 centimeters) or more of annual rainfall is required. Estimated worldwide rubber consumption is given in Table 2.

Rubber comes from the tree as a milky white fluid, which is a colloidal suspension of rubber in a liquid consisting mostly of water. The tree is tapped by well-trained workers who use a sharp-edge tool, and the cutting action goes at an angle of 30° from top left to bottom right. It is important that the latex-

TABLE 2.   ESTIMATED WORLD CONSUMP-TION OF NATURAL RUBBER (1978)

CONSUMING AREA	PERCENT OF TOTAL
Total Western Europe	39.8
United Kingdom, 3.7%	
Germany (West), 5.0%	
United States	20.5
Eastern Europe	11.3
Japan	10.0
U.S.S.R.	9.6
Others	8.8

[1] It is alleged that English scientist Joseph Priestly observed that the material could be used for rubbing out lead pencil marks and thus gave the material the name, *rubber.* ("Introduction to the Theory of Perspective," Joseph Priestly, 1835.)

bearing cells be cut, but that the blade not wound the inner cambium layer, as this would harm the tree. A cup is hung below the cut to collect the white, milklike latex, which contains about 35% rubber, the remainder being water, protein, resins, organic materials, and other plant substances.

The yield of the *Hevea* tree can be increased by applying chemicals to the bark. These include 2-chloroethylphosphonic acid, which supplies small quantities of ethylene gas. This type of chemical is applied in a high-viscosity liquid form, usually mixed with palm oil or other diluent. The function of it is to stabilize the latex so that it continues to flow for a longer time and thus increases rubber yield. Because of the higher rubber yield per tapping operation, the cost of tapping labor is reduced. When stimulants are used, the tree is given a longer rest period between tappings to avoid diseases which would eventually kill it. See also **Plant Growth Modification and Regulation.**

As of the early 1980s, the annual yield for Malaysian plantations is about 1010 pounds/acre (1133 kilograms/hectare), which includes high-yielding trees which produce 1500 pounds/acre (1680 kilograms/hectare), as well as older, lower yielding clones.

Properties of Natural Rubber. Chemically, natural rubber or *cis*-polyisoprene, has a broad molecular-weight distribution, ranging from several million to about one-hundred thousand.

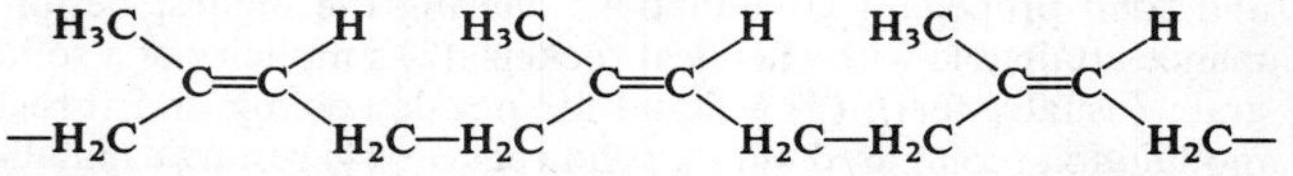

Natural rubber is soluble in practically all aromatic and aliphatic hydrocarbons and particularly in halogenated hydrocarbons. When cements and solvent adhesives are made using natural rubber, methylethylketone (MEK) frequently is used to reduce viscosity. Although MEK is not a solvent, it tends to disperse large molecular particles, resulting in lower-viscosity dilution. Crude rubber is decomposed by heat and can be cyclized at 250°C. It can easily be hydrogenated and reacts readily with halogens. The stress-strain properties of natural rubber are the best of all the elastomeric polymers. In vulcanized films made by the latex process, the tensile strength may exceed 6000 pounds per square inch (41 mPa), and ultimate elongation is as high as 700% or more.

Natural rubber is readily attacked by oxygen. Copper and manganese, if present in amounts greater than the specified 0.001%, greatly accelerate oxidation. There are, however, naturally occurring antioxidants in natural rubber which help preserve it until vulcanization. All vulcanized natural-rubber products contain added antioxidants to ensure satisfactory life.

Rubber burns quite readily and generates more than 10,000 cal/g. The specific gravity of rubber is 0.934, a property utilized in concentrating natural-rubber latex by the centrifuge process. The serum, which is mostly water and has a specific gravity of about 1.0, tends to separate readily from the rubber. The liquid concentrated latex is used in making foam rubber, dipped goods, adhesives, and carpet backing for nonwoven carpets. An industry has developed around this application, which involves spreading foamed latex on the underside of carpeting, making an integral carpet-foam system.

Compounding and Vulcanization. Crude rubber in the raw state has few applications with the exception of crepe soles for shoes. To make commercial rubber products, the material must be mixed with a variety of chemicals and vulcanized into desirable end shapes. Charles Goodyear discovered in 1839 that adding sulfur to rubber and heating the mixture greatly enhances the physical properties of rubber. The material no longer becomes tacky in warm weather and in cold weather it does not become brittle. The material is much tougher, and the quality of products made this way results in service for a much longer period of time. In addition to sulfur, which crosslinks the large rubber molecules and makes it a giant organic molecule, zinc oxide, organic accelerators, antioxidants, reinforcing pigments, and other processing aids are used in compounding rubber for useful vulcanized products.

In the manufacture of rubber products, crude rubber is masticated on a 2-roll mill or in an internal mixing machine (Banbury mixer), where heat and mechanical mixing reduce the rubber to a very viscous plastic mass. Antioxidants, such as an alkylated

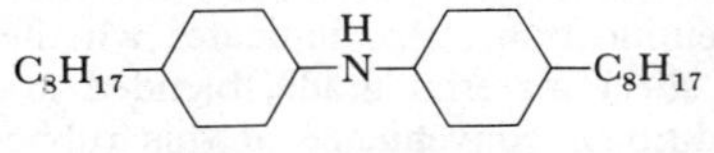

$$C_3H_{17}\!-\!\!\bigcirc\!\!\overset{H}{\underset{N}{}}\!\!\bigcirc\!\!-\!C_8H_{17}$$

Alkylated diphenylamine

diphenylamine (2 parts by weight); two accelerators, one a primary type, such as benzothiazyl disulfide (1 part by weight) and the other a secondary one, such as tetramethylthiuram disulfide (0.1 part by weight); plus activators of cure, such as stearic acid (2 parts by weight) and zinc oxide (5 parts by weight)—all based on 100 parts by weight of rubber—are added to the plastic mass and thoroughly dispersed in it. The compound rubber is passed through the 2-roll mill to make it into $\frac{1}{4}$-inch (~6-millimeter) slabs. These are cut into 8 × 4-inch (~20 × 10-centimeter) dimensions, placed in a flat mold so that there is a slight excess in volume, and subjected to pressure and 140°C temperature for 30 minutes. After this vulcanizing step, the mold is opened and the slab of hot rubber is removed. The next day the sample is tested and normally will have developed a tensile strength of 3850 pounds/square inch (26.6 MPa); elongation of 680%; and stress at 500% elongation of 420 pounds/square inch (2.9 MPa). This is known as *pure gum rubber*. This type of rubber can be used for hot-water bottles, water-hose stock, crude tire tubes, and any application requiring high stress-strain (tensile and elongation) properties.

The function of the antioxidant is to improve service life of the product against such well-known degrading agents as oxygen, light, and nitroso compounds. One theory is that the antioxidant selectively reacts with the degrader, slowing down its reaction with the rubber molecule, which would result in scission and eventually poorer physical properties. During recent years, considerably aggravated by air pollution, degradation of vulcanized rubber by small quantities of ozone (a few parts per million) in the air has become a serious problem. Ozone has little noticeable effect on unstretched rubber, but even under slight stretch it causes cracks in the surface which grow perpendicularly to the direction of extension. Hundreds of different antioxidants and antiozonants are employed, amine and phenol complexes being the basis of most. See also **Antioxidant.**

Accelerators act as catalysts of vulcanization, but unlike most catalysts, they undergo chemical change during the reaction. Benzothiazyl disulfide is one of the oldest types, dating back to 1925, but it still accounts for the greatest use in the industry today. Besides the thiazole types, other popular accelerators are sulfenamides, aryl guanidines, dithiocarbames—extremely fast accelerators used mostly in latex compounds; and thiurams, also very fast and often used as a secondary accelerator to hasten the vulcanization rate. Accelerators also contribute to improved aging properties of the end product.

Stearic acid is an activator of vulcanization, as is zinc oxide, the two substances acting to form zinc stearate, which enhances the activity of the organic accelerators. Zinc stearate is impractical to add directly to the rubber because its slippery, lubricating nature makes it difficult to mix the batch.

With an accelerated system, a simple network structure with dialkenyl mono- and disulfide crosslinks and conjugated triene units as main-chain modifications is obtained.

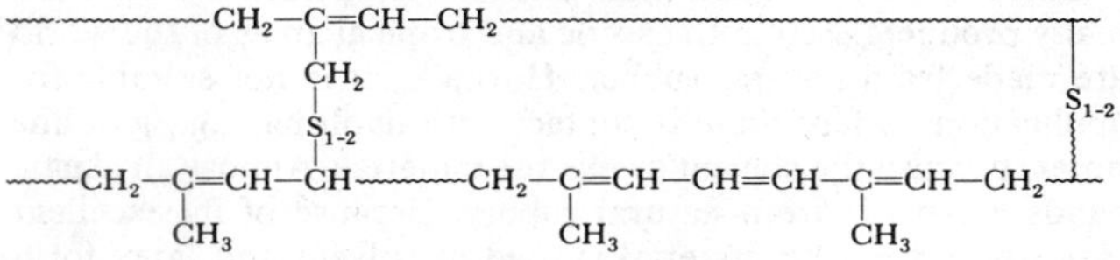

With an unaccelerated sulfur–natural-rubber system, the poor crosslinking efficiency results in sulfur being incorporated into the rubber network as long polysulfide crosslinks, cyclic monosulfides, and vicinal crosslinks, which are very close together and act physically as a single crosslink.

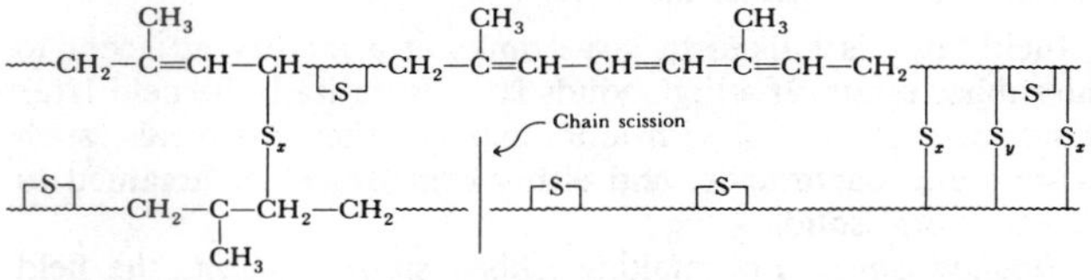

It is theorized that between the complex network structure of the unaccelerated system and the simpler network structure of the accelerated system, structures made up of the two models represent natural-rubber vulcanizates made at various times and temperatures of cures, with different reactant concentrations, and showing the effects of other variants.

At any given degree of crosslinking, the tensile strength is highest with polysulfide bonds. High elongation at break is obtained by slightly decreasing the crosslinking action. If lower elongation is required, slightly excessive crosslinking is used, usually accompanied by higher tensile strength. Vulcanization of rubber decreases its solubility in solvents, and this property frequently is used as a qualitative measure of *cure*.

Vulcanization by sulfur accounts for practically all the commercial products. However, peroxide types of curing systems may be used, especially for some of the synthetic rubbers.

Ultrahigh frequency (UHF) energy may be used for preheating and precuring rubber compounds for continuous vulcanization (CV) of rubber containing carbon black, for such applications as weather stripping, tubing, hose, and, in some instances, tire tread compounds. Advantages of this type of cure are faster throughput and energy savings. Microwave ovens may be used in a line with conventional types of vulcanization equipment.

Carbon black is the major reinforcing pigment used, not only for natural rubber, but for practically all the synthetic rubbers. As much as 40–50 parts by weight, based upon 100 parts of rubber, are used in all tire-tread compounds. Carbon black greatly increases tensile strength at low elongations (modulus) and results in longer-wearing tires. Colloidal silica contributes some reinforcing properties to rubber, but not to the same degree as carbon black.

Uses of Natural Rubber

Thousands of flexible products requiring top performance characteristics are made of natural rubber, e.g., huge earthmover tires, truck tires, tires for large aircraft, bridge supports, and surgeons' gloves. The threads of most passenger-car tires in the United States consist mainly of styrene-butadiene synthetic rubber because of lower cost and lower temperature buildup during use.

The use of natural rubber in passenger car tires has increased in recent years due to the industry going from bias to radial types which, in North America, now account for 75% of the total. Higher degree of tack or cohesive bonding during the building of the radial tire, as compared with that of styrene-butadiene rubber, is largely responsible for this.

Because of its excellent high- and low-temperature properties, many products used in the arctic and tropical areas of the world are made from natural rubber. However, it is not suitable for applications where there is contact with naphtha, e.g., gasoline hoses, because the solvent swells the material. Almost all elastic bands are made from natural rubber. Because of its excellent tack properties, the material is used in solvent and latex form as the base for adhesives.

With the dependence of synthetic rubber on petroleum, natural rubber, which is produced by solar energy, may look increasingly attractive over the years ahead.

Processing Raw Materials

Field latex is bulked in large tanks at a factory adjacent to the rubber estate. If a high-solids latex is desired, the field latex is strained, stabilized with ammonia or other chemicals, such as soap and bactericide, and either centrifuged or creamed to 62–68% total solids.

Smoked Sheet. For making ribbed smoked sheet, the field latex is immediately mixed with dilute formic acid in long horizontal tanks. Because fresh latex is somewhat protected by a protein surface layer, it does not coagulate or gel immediately on addition of the acid. Within a few hours, however, the rubber particles in the latex gel and form a spongy mass, which is then run through a series of smooth metal rolls with clearance decreased from one set to the next, an arrangement that squeezes out the serum and densifies the wet rubber. Water is run over the wet coagulation to wash out nonrubber materials and dirt. The last unit consists of ribbed rolls which imprint ribbed markings on the sheet. After drying in air for a few hours, the sheets are hung in a drying shed at 40–50°C until dry. Modern installations use efficient drying tunnels. Sheets are inspected by holding them over a strong light to determine clarity, color, presence of dirt and other factors. The rubber is classified by various grades. Sheets then are piled up and squeezed in a baling machine to form 250-pound (~113-kilogram) bales that measure $19 \times 19 \times 24$ inches (~$48 \times 48 \times 61$ centimeters).

Crepe. Another popular type of commercial rubber, known as crepe, consists of two major classes—*pale crepe* and *thick blanket crepe*. Pale crepe is made by adding sodium hydrogen sulfite, $NaHSO_3$, to field latex to inhibit discoloration and softening during processing. Formic acid is used as the coagulant. The wet coagulum is passed through rolls with longitudinal grooves which give the rubber a crepelike appearance. Water running over the surface cleans out dirt and other non-rubber ingredients. Sheets are hung up to dry in circulating warm air. The quality of pale crepe is assessed on its whiteness and how well the finished rubber appears.

Blanket crepes are of lower quality and are made from wet slabs obtained usually from small landholders. These are creped, dried, and baled. Other types of crepe are made from coagulum left in collection cups and from dried skin remaining from the tapping incision. In addition to collecting latex, a tapper collects all dried and coagulated rubber that remains from the previous round, usually as skin in the cup or on the tapping panel.

Grading of Rubber. Commercial grades of natural rubber are classified into two main groups: (1) Green Book International Grades, and (2) Technically Specified Forms. The former depends on a visual grading system, the source of the rubber, and the method of preparation. This system, dating back many years and kept current by the International Rubber Quality and Packing Conference Committee, consists of 35 grades under

8 major types, such as Ribbed Smoked Sheets, White and Pale Crepe, Estate Brown Crepes, Compo Crepes, Thin Brown Crepes (Remills), Thick Blanket Crepes, and Pure Smoked Blanket Crepe. Publisher of the "Green Book" is the Rubber Manufacturers Association, Inc., New York.

Technically Specified Rubbers (TSR), originated by the Malaysian Rubber Producers Association, classifies rubber not only by source, but also by physical properties, such as dirt content, ash, and nitrogen content, volatile matter, plasticity; with the higher-quality grades, cure rate and color are standardized. This type of rubber is packaged in 75-pound (34-kilogram) bales and wrapped in transparent plastic; the color of the printing on the bale identification strips indicates whether the source is latex grade, sheet material grade, blended grades, or field grades. An additional convenience of this rubber is that the bales can be charged into the mixing machine (Banbury) without removing the wrapper. This is in contrast with the Green Book grades, which are bonded together by a press, with the outside layer treated with soapstone to keep the bales from sticking together. These larger bales require cutting before they can be charged into the mixer.

Guayule

During the past few years, natural rubber from the desert shrub *Parthenium argentatum* has been under intensive study by scientists in the United States and Mexico as a possible domestic source of natural rubber. This plant grows wild in the arid areas of Mexico and the United States. In 1910, guayule produced 10% of the world's rubber, but lower-cost *Hevea* rubber from the Far East displaced it from the market. Rubber in the guayule plant is present in the roots and branches of the shrub and must be separated and purified by a flotation and solvent system. The purified product is equivalent in chemical properties to the *Hevea* rubber. An advantage of guayule is that it can be grown on semiarid land that is not suitable for other crops. Presently, agricultural experimentation on increasing rubber yield of the plant is underway. The U.S. government has passed legislation providing funds to help in developing an American-based guayule industry. Several large rubber products manufacturers have experimental plots planted with the shrub. The National Research Council, Washington, D.C. published a report, "Guayule: An Alternative Source of Natural Rubber," in 1977.

—Thomas H. Rogers, Consultant (Rubber and Plastics Industries), formerly Research Manager, Goodyear Tire and Rubber Company, Akron, Ohio.

RUBBER (Synthetic). Elastomers.

RUBIDIUM. Chemical element symbol Rb, at. no. 37, at. wt. 85.466, periodic table group la (alkali metals), mp 38.9°C, bp 689°C, density, 1.53 g/cm³ (20°C). Elemental rubidium has a body-centered cubic crystal structure. Rubidium is a silvery-white, very soft metal; tarnishes instantly on exposure to air, soon ignites spontaneously with flame to form oxide; best preserved in an atmosphere of hydrogen rather than in naphtha; reacts vigorously with water, forming rubidium hydroxide solution and hydrogen gas. Discovered by Bunsen and Kirchhoff in 1860 by means of the spectroscope.

There are two naturally occurring isotopes ^{85}Rb and ^{87}Rb, of which the latter is unstable with respect to beta decay ($t_{1/2} = 5 \times 10^{10}$ years) into ^{87}Sr. There are eight other known radioactive isotopes ^{81}Rb through ^{84}Rb, ^{86}Rb, and ^{88}Rb through ^{90}Rb, all with comparatively short half-lives, measured in terms of minutes, hours, or days. In terms of abundance, rubidium ranks 34th among the elements in the earth's crust. In terms of content in seawater, the element ranks higher (18th) with an estimated 570 tons of rubidium per cubic mile (123 metric tons per cubic kilometer) of seawater. First ionization potential 4.176 eV; second, 27.36 eV. Oxidation potential $Rb \rightarrow Rb^+ + e^-$, 2.99 V. Other important physical properties of rubidium are given under **Chemical Elements.**

Rubidium occurs in lepidolite (lithium aluminosilicate, in amount up to 1% Rb), in certain mineral waters and rare minerals. Rubidium salts may be recovered from the mother liquor upon crystallization of (1) lithium salts, (2) potassium salts. Rubidium metal is obtained by electrolysis of the fused chloride out of contact with air.

Uses

The main uses of rubidium are in photocathodes and photoelectric cells. However, rubidium cells are inferior to cesium cells in their sensitivity and range. Although very small quantities are involved, rubidium gas cells now perform as secondary time standards, on the order of quartz crystal oscillators, inasmuch as they must be referenced to more accurate systems. The rubidium systems have a characteristic resonance of 6,835 MHz and, unlike other atomic frequency standards, require little power and are relatively compact. Portable rubidium atomic clocks were introduced by the U.S. Army in 1963. They weigh as little as 44 pounds (20 kilograms) and occupy a volume of only about 1 cubic foot (0.028 cubic meter). The units operate on 110V current, on the 24-V output of military vehicles, or both. Clocks of this type are used to synchronize radar nets, to assist in the accurate tracking of missiles and satellites, and to set precise radio broadcasting frequencies. Rubidium-vapor instruments also were developed as absolute-type magnetometers and introduced in 1958 by U.S. government scientists. The rubidium-vapor magnetometer uses a rubidium lamp, mounted in the tank coil of a radio-frequency oscillator. After collimating and filtering, the rubidium light is circularly polarized and then passed through a rubidium-vapor cell, after which it is focused on a sensitive photocell. Numerous combinations of amplifier parameters and various rubidium isotopes permit considerable range in the measurement of ambient magnetic fields. Inasmuch as the total world range is from 15,000 to 80,000 gammas, a system capable of this span finds use anywhere in the world.

Potential uses of rubidium include use as a fuel for ion propulsion engines and as a heat-transfer medium.

Rubidium alloys easily with potassium, sodium, silver, and gold, and forms amalgams with mercury. Rubidium and potassium are completely miscible in the solid state. Cesium and rubidium form an uninterrupted series of solid solutions. These alloys, in various combinations, are used mainly as getters for removing the last traces of air in high-vacuum devices and systems.

Small quantities of rubidium are found in certain foods, including coffee, tea, tobacco, and several other plants. There is evidence indicating that trace quantities of the element are required by living organisms.

Chemistry and Compounds

Rubidium is more electropositive than potassium (or the lower alkali metals) as is consistent with its position in main group I. It reacts more vigorously with H_2O, and ignites on exposure to oxygen.

Because of the ease of removal of its single $5s$ electron (4.159 eV) and the difficulty (27.36 eV) of removing a second electron, rubidium is exclusively monovalent in its compounds, which are electrovalent.

In its solutions in liquid NH_3, rubidium is, like the other alkali metals, a powerful reducing agent, so that in such solutions titrations of rubidium polysulfide with rubidium are made by electrometric methods. The solubility of rubidium salts in liquid NH_3 increases markedly with the radius of the anion (rubidium chloride, RbCl, 0.024 moles per kilogram, rubidium bromide, RbBr, 1.35 moles per kilogram, and rubidium iodide, RbI, 10.08 moles per kilogram). However, in water they exhibit minimum solubility at cation: anion radius ratio of 0.75 (rubidium fluoride, RbF, 12.5 moles/kilogram, RbCl 6.8 moles/kilogram, RbBr 6.6 moles/kilogram, RbI 7.2 moles/kilogram).

As in the case of the other alkali metals, rubidium forms compounds generally with the inorganic and organic anions; for a general discussion of these compounds, see the entry on **Sodium,** because the sodium compounds differ principally in their greater extent of hydration and greater number of hydrates. However, rubidium coordinates with large organic molecules, such as salicylaldehyde, even though it does not with H_2O.

One respect in which rubidium and cesium are outstanding among the alkali metals is the readiness with which they form alums. Rubidium alums are known for all of the trivalent cations that form alums, Al^{3+}, Cr^{3+}, Fe^{3+}, Mn^{3+}, V^{3+}, Ti^{3+}, Co^{3+}, Ga^{3+}, Rh^{3+}, Ir^{3+}, and In^{3+}.

As in the case of potassium and cesium, rubidium forms a superoxide on reaction of the metal with oxygen. The compound is dark brown in color and paramagnetic, and hence believed to contain the O_2^- ion with an odd electron, and to have the formula RbO_2. On heating, it loses oxygen to form Rb_2O_3. Rubidium also forms a peroxide Rb_2O_2, and a normal oxide, Rb_2O, which is prepared by heating rubidium nitrite with metallic rubidium.

Rubidium hydroxide RbOH, is the strongest, except for cesium hydroxide, CsOH (and francium hydroxide, FrOH), of the alkali hydroxides, as would be expected from its position in the periodic table. For the same reason, it has the next smallest lattice energy (146.4 kilocalories per mole).

The most numerous organic compounds of rubidium are those of oxy compounds, such as the salts of organic acids, the alcohols and phenols (alkoxides, phenoxides, etc.). An ethyl rubidium-zinc diethyl adduct has been reported, $RbZn(C_2H_5)_3$, which is certainly the true salt, rubidium triethylzincate, $Rb[Zn(C_2H_5)_3]$.

References

Kaiser, J. R.: "Rubidium," in "Metals Handbook," 9th Edition, Vol. 2, American Society for Metals, Metals Park, Ohio (1979).

Perel'man, F. M.: "Rubidium and Celsium," (R. G. P. Towndrow and R. W. Clarke, Editors), Pergamon, New York (1965).

Whaley, T. P.: "Sodium, Potassium, Rubidium, Cesium and Francium," in "Comprehensive Inorganic Chemistry," (Trotman-Dickenson et al., Editors), Pergamon, New York (1973).

RUTHENIUM. Chemical element symbol Ru, at. no. 44, at. wt. 101.07, periodic table group 8, mp 2310°C, bp 3900–4000°C, density, 12.1 g/cm³ (solid at 19°C). Elemental ruthenium has a close-packed hexagonal crystal structure. The seven stable isotopes are ^{96}Ru, ^{98}Ru through ^{102}Ru, and ^{104}Ru. The five unstable isotopes are ^{95}Ru, ^{97}Ru, ^{103}Ru, and ^{106}Ru. In terms of earthly abundance, ruthenium is one of the scarce elements. Also, in terms of cosmic abundance, the investigation of Harold C. Urey (1952), using a figure of 10,000 for silicon, estimated the figure for ruthenium at 0.019. No notable presence of ruthenium has been detected in seawater. Ruthenium was discovered by Claus (Germany) in 1844.

Electronic configuration is $1s^22s^22p^63s^23p^63d^{10}4s^24p^64d^7$ $5s^1$. Ionic radius, Ru^{4+}, 0.60 Å. Metallic radius, 1.3251 Å. First ionization potential, 7.5 eV. Other physical properties of ruthenium will be found under **Platinum and Platinum Group.**

The chemistry of ruthenium is still poorly understood. The existence of at least eight valence states, coupled with the tendency to complex with many ions, often results in the presence of several different complexes in a given solution.

Ru metal is quite refractory. It is not significantly soluble in any single acid; even aqua regia has little effect. At room temperature, the metal does not react with O_2, but when heated in air, a film of the dioxide appears. The metal is insoluble in fused sulfates. Molten alkali slowly dissolves the metal. The rate of attack is rapid under oxidizing conditions, and a molten mixture of NaOH and Na_2O_2 will readily dissolve the metal.

The finely divided metal is soluble in hypohalites if an excess of alkali is present. At red heat, the metal combines with Cl_2 to form the dichloride. Ruthenium(VIII) oxide is formed when an alkaline ruthenium solution is treated with a strong oxidant, such as chlorine, or bromate ion when the Ru is in acid solution.

Ruthenium(III) hydroxide is formed by the action of alkali on a solution of ruthenium(III) chloride. It is easily oxidized by air to the tetravalent state. The dioxide, RuO_2, forms when the metal is heated in air. Hydrous ruthenium(VIII) oxide can be precipitated by adding alcohol to a less than 3 M NaOH solution of ruthenium(VIII) oxide, followed by boiling. Above 3 M NaOH, complete reduction is not obtained. The hydrous oxide that is soluble in concentrated HCl tends to occlude impurities.

The only known octavalent Ru compound is the tetroxide, RuO_4, which exists in a yellow and a brown form. The volatile and poisonous tetroxide melts at about 25°C and sublimes readily. It may explode in contact with oxidizable substances or when heated above 100°C. It is formed by distillation from either an alkaline or acid solution under strongly oxidizing conditions. The tetroxide is moderately water-soluble. When dissolved in alkali, it initially forms a green solution of heptavalent perruthenate of the form $MRuO_4$, which further reduces to the orange ruthenate M_2RuO_4. The reduction to the hexavalent state is quicker in strong alkali. The ruthenates also are made by fusing finely divided metal with a mixture of alkali dehydroxide and nitrate or peroxide.

Anhydrous ruthenium(III) chloride, $RuCl_3$, is made by direct chlorination of the metal at 700°C. Two allotropic forms result. The trihydrate is made by evaporating a HCl solution of ruthenium(III) hydroxide to dryness or reducing ruthenium(VIII) oxide in a HCl solution. The trihydrate, $RuCl_3 \cdot 3H_2O$, is the usual commercial form. Aqueous solutions of the trihydrate are a straw color in dilute solution and red-brown in concentrated solution. Ruthenium(III) chloride in solution apparently forms a variety of aquo- and hydroxy complexes. The analogous bromide, $RuBr_3$, is made by the same solution techniques as the chloride using HBr instead of HCl.

Ruthenium(III) iodide, RuI_3, is a black, insoluble compound precipitated by the addition of iodide ion to a solution of $RuCl_3$.

Tetravalent ruthenium chloride, $RuCl_4$, and the hydroxychloride, $Ru(OH)Cl_3$, are intermediate products when $RuCl_3$ is prepared by evaporating the tetroxide in HCl. When the hydroxychloride is hot HCl is treated with Cl_2, it is converted to the tetrachloride. The anhydrous tetrachloride also is known. The tetrabromide and tetraiodide have not been isolated; attempts to prepare these compounds result in the formation of the respective trihalides.

The only pentavalent Ru compounds known are the fluorides; RuF_5 is made by combining the elements. The compound melts

at 107°C and boils at 313°C. The salt, $NaRuF_6$, was recently made by mixing $RuCl_3$ with NaCl and treating the mixture with BrF_3.

Ru forms many complex ions. The nitrosyl compounds are frequently encountered by accident due to the great affinity of Ru for the nitrosyl group. Ruthenium(III) nitrosylchloride, $Ru(NO)Cl_3 \cdot 4H_2O$, is a by-product of most solutions of $RuCl_3$ in aqua regia or solutions containing HNO_3. It also is present in HCl solutions resulting from a KOH and nitrate fusion of the metal. The chloride and bromide are respectively raspberry and violet in solution. Alkaline chlorides form complex salts of the type $M_2Ru(NO)Cl_5$, which can be crystallized from solution. A black gelatinous precipitate of the nitrosylhydroxide, $RuNO(OH)_3$, is slowly formed when a solution of the nitrosylchloride is heated with a strong base. A series of nitrato and nitro derivatives of nitrosylruthenium also have been described and separated.

It is generally accepted that the disulfide is the only certain sulfide of Ru. It is formed by the action of H_2S on a solution of Ru or from the elements at about 1000°C. When ruthenium-(IV) sulfide is treated with HNO_3, the sulfate is formed.

Dichlorodicarbonylruthenium(II), $Ru(CO)_2Cl_2$, is formed when $RuCl_3$ is heated above 210°C in the presence of CO. It is a yellow, insoluble, volatile compound. The bromine and iodine analogs are similarly formed.

When finely divided Ru metal is heated at 180°C under 200 atm of CO, pentacarbonylruthenium(0), $Ru(CO)_5$, is formed.

Ruthenium forms a large number of complex ions with amines.

Recently, a new group of organometallic sandwich compounds, called *metallocenes,* has been discovered. Ruthenocene is made in about 50% yield by reacting $RuCl_3$ with cyclopentadienylsodium in tetrahydrofuran. After refluxing and distilling the solvent, the light yellow crystals of ruthenocene are sublimed. The compound, $Ru(C_5H_5)_2$, undergoes a large number of substitution reactions typical of aromatic systems.

Ruthenium is commonly used with other platinum metals as a catalyst for oxidations, hydrogenations, isomerizations, and reforming reactions. The synergetic effect of mixing ruthenium with catalysts of platinum, palladium, and rhodium has been found for the hydrogenations of aromatic and aliphatic nitro compounds, ketones, pyridine, and nitriles.

References

Atkinson, R. H.: "Ruthenium" in "Metals Handbook," 9th Edition, Vol. 2, American Society for Metals, Metals Park, Ohio, 1979.
Coles, D. G., and L. D. Ramspott: "Migration of Ruthenium-106 in a Nevada Test Site Aquifer," *Science,* **215,** 1235–1237 (1982).
See also references under **Chemical Elements**; and **Platinum and Platinum Group.**

—Linton Libby, Chief Chemist, Simmons Refining Company, Chicago, Illinois.

RUTILE. A mineral composed of titanium dioxide which occurs in three distinct forms: as rutile, a tetragonal mineral usually of prismatic habit, often twinned; as octahedrite (anatase), a tetragonal mineral of pseudo-octahedral habit; and as brookite, an orthorhombic mineral. Both octahedrite (anatase) and brookite as relatively rare minerals.

Rutile has a sub-conchoidal fracture; is brittle; luster, metallic-adamantine; color, commonly reddish-brown but sometimes yellowish, bluish or violet; streak, brown; transparent to opaque. Rutile may contain up to 10% of iron.

Experiments in the artificial preparation of titanium dioxide appear to show that rutile is the most stable form and produced at the highest temperature, brookite at a lower temperature, and octahedrite (anatase) at a still lower temperature.

Rutile is found as an accessory mineral in many kinds of igneous rocks, and to some extent in gneisses and schists. In groups of acicular crystals it is frequently seen penetrating quartz as the "flèches d'amour" from Grisons, Switzerland, and Brazil. Rutile is found also in Austria, Italy, Norway, South Australia, and Brazil. In the United States it occurs in Vermont, Massachusetts, Connecticut, New York, Pennsylvania, Virginia, Georgia, North Carolina, and Arkansas.

Rutile derives its name from the Latin *rutilus*, red, in reference to the deep red color observed in some specimens when viewed by transmitted light.

See also terms listed under **Mineralogy.**

S

SACCHARIDE. Carbohydrates.

SACCHARIN. Sweeteners.

SACCHAROMYCES. Yeasts and Molds.

SACCHAROSE. Carbohydrates.

SAFFLOWER SEED OIL. Vegetable Oils (Edible).

SALICYLIC ACID. Salicylic acid or $C_6H_4(OH)(COOH)$ is a white solid, melting point 159°C, sublimes at 76°C, insoluble in cold water, soluble in hot water, alcohol, or ether. With ferric chloride solution, salicylic acid solutions are colored violet (distinction from benzoic acid).

Salicylic acid may be obtained (1) from oil of wintergreen, which contains methyl salicylate, (2) by heating dry sodium phenate C_6H_5ONa plus carbon dioxide under pressure at 130°C, and recovery from the resulting sodium salicylate by addition of dilute sulfuric acid. Salicylic acid is a mild disinfectant and antiseptic, and has been used as a food preservative. Salicylic acid and certain salicylates are used in medicine as anti-rheumatics.

SALT (Compound). A compound formed by replacement of part or all of the hydrogen of an acid by one (or more) element(s) or radical(s) which are essentially inorganic. Alkaloids, amines, pyridines, and other basic organic substances may be regarded as substituted ammonias in this connection. The characteristic property of salts are the ionic lattice in the solid state and the ability to dissociate completely in solution. The halogen derivatives of hydrocarbon radicals and esters are not regarded as salts in this strict sense of the term.

In the classical concept of the process of neutralization, where an acid and a base in solution react to form a salt, the proton of the acid and hydroxyl ion of the base react to form water, leaving the cation of the base and the anion of the salt by recombination.

Upon evaporation of the solvent, the salt is obtained as such, frequently as crystals, sometimes with, sometimes without water of crystallization. A salt, when dissolved in an ionizing solvent, or fused, e.g., sodium chloride in water, is a good conductor of electricity, and when in the solid state forms a crystal lattice, e.g., sodium chloride crystals possess a definite lattice structure for both sodium cations (Na^+) and chloride anions (Cl^-), determinable by examination with x-rays.

A broader definition than that confined to solutions is demanded in some fields of chemistry, for example, in high-temperature reactions of acids, bases, salts. In the formation of metallurgical slags at furnace temperatures, calcium oxide is used as a base, and silicon oxide and aluminum oxide as acids; calcium aluminosilicate is produced as a fused salt. Sodium carbonate and silicon oxide when fused react to form the salt sodium silicate with the evolution of carbon dioxide.

$$\left[\begin{array}{l}\text{Oxide of any}\\\text{element function-}\\\text{ing as a metal,}\\\text{that is as a base.}\end{array}\right]\text{plus}\left[\begin{array}{l}\text{Oxide of any}\\\text{element function-}\\\text{ing as a nonmetal,}\\\text{that is, as an acid.}\end{array}\right]\text{yields [Salt]}$$

Iron and sulfur when heated react to form the salt ferrous sulfide. In this sense:

metal plus nonmetal yields salt

Salts are, therefore, prepared (1) from solutions of acids and bases by neutralization, and separation by evaporation and crystallization; (2) from solutions of two salts by precipitation where the solubility of the salt formed is slight, e.g., silver nitrate solution plus sodium chloride solution yields silver chloride precipitate (almost all as solid), and sodium nitrate present in solution as sodium cations and nitrate anions (recoverable as sodium nitrate, solid by separation of silver chloride and subsequent evaporation of the solution); (3) from fusion of a basic oxide (or its suitable compound—sodium carbonate above) and an acidic oxide (or its suitable compound—ammonium phosphate), since ammonium and hydroxyl are volatilized as ammonia and water. Thus, sodium ammonium hydrogen phosphate

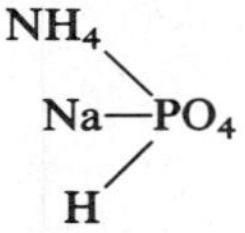

$$\begin{array}{c}NH_4\\|\\Na-PO_4\\|\\H\end{array}$$

yields sodium metaphosphate, $NaPO_3$, upon heating). (4) Salts also are prepared from reaction of a metal and a nonmetal.

Reactions of salts as such in solution, without decomposition of cation or anion, are dependent upon the presence of the cation and the anion of salt.

An *acid salt* is a salt in which all the replaceable hydrogen of the acid has not been substituted by a radical or element. These salts, in ionizing, yield hydrogen ions and react like the acids, e.g., $NaHSO_4$, $KHCO_3$, Na_2HPO_4.

An *amphiprotic* (also called *amphoteric*) *salt* is a salt which may ionize in solution either as an acid or a base, and react either with bases or acids, according to the conditions.

A *basic salt* is a salt which contains combined base as $Pb(OH)_2Pb(C_2H_3O_2)_2$, a basic acetate of lead. These salts may be regarded as formed from the basic hydroxides by partial replacement of hydroxyl, e.g., $HO-Zn-Cl$. They react like bases and, when soluble, ionize to yield hydroxyl ions.

A *complex salt* is a saline compound having the structure of a combination of two or more salts and which is regarded as the normal salt of a complex acid. Complex salts do not split into a mixture of the constituent salts in solution, but furnish a complex ion which contains one of the bases, e.g., potassium molybdophosphate and potassium platinochloride.

A *double salt* is a substance consisting of two simple salts that crystallize together in definite proportions and exist independently in solution (distinction from complex salts). The alums are representative double salts.

An *inner salt* is a member of a special class of internal salts in which an acid group and a neutral group coordinate with metals to form a cyclic complex. These salts occur widely in analytical chemistry, where they are formed between metallic ions and organic reagents, in dyestuffs, in life processes (chlorophyll and hematin belong to this class of compounds), and in many other fields.

An *internal salt* is a compound in which the acidic or basic groups which react to produce the salt linkage (which may or may not entail the formation of water), are in the same molecule. This particular salt linkage may consist of a polar or a nonpolar bond.

A *mixed salt* is a salt of a polybasic acid, in which the hydrogen atoms are replaced by different metallic atoms or positive radicals.

A *pseudo salt* is a compound which has some of the normal characteristics of a salt, but lacks certain others, notably the ionic lattice in the solid state, and the property of ionizing completely in solution. The absence of these properties is due to the fact that the bonds between the metallic and nometalic radicals are covalent or semicovalent, instead of polar. Because these salts do not ionize completely, they are also called *weak salts*.

SAMARIUM. Chemical element symbol Sm, at. no. 62, at. wt. 150.35, fifth in the lanthanide series in the periodic table, mp 1073°F, bp 1791°C, density, 7.520 g/cm³ (20°C). Elemental samarium has a rhombohedral crystal structure (25°C). Pure metallic samarium is a silvery-gray color, retaining a luster in dry air, but is only moderately stable in moist air, with formation of an adherent oxide. When pure, the metal is soft and malleable, but must be worked and fabricated under an inert gas atmosphere. Finely divided samarium as well as chips from working are pyrophoric and ignite spontaneously in air, burning at 150–180°C. There are seven natural isotopes of samarium, ^{144}Sm, ^{147}Sm through ^{150}Sm, ^{152}Sm, and ^{154}Sm. Eleven artificial isotopes have been identified. The natural ^{147}Sm isotope is weakly radioactive ($t_{1/2} = 2.5 \times 10^{11}$ years). The samarium isotope mixture is the second highest (after gadolinium) of all elements in terms of its thermal-neutron-absorption cross section (5800 barns at 0.025 eV). The cross section of ^{149}Sm is about 40,000 barns, but no chain reaction exists because of separation by low-cross-section isotopes. Samarium ranks 62nd in abundance of the elements in the earth's crust, exceeding tantalum, mercury, bismuth, and the precious metals, except silver. The element was first identified by Lecoq de Boisbaudran in 1879.

Electronic configuration is $1s^2 2s^2 2p^6 3s^2 3p^6 3d^{10} 4s^2 4p^6 4d^{10} 4f^5 5s^2 5p^6 5d^1 6s^2$. Ionic radius, Sm^{2+}, 1.11 Å, Sm^{3+}, 0.964 Å. First ionization potential, 5.6 eV; second, 11.1 eV. Other important physical properties of samarium are given under **Rare-Earth Elements and Metals.**

The principal sources of samarium are monazite (4.5% Sm$_2$O$_3$) and bastnasite (0.5% Sm$_2$O$_3$). Current demands for the element are met by the coproduction with europium and gadolinium from these minerals. The residues of uranium mining (Canada) also contain about 4.5% Sm$_2$O$_3$. Unlike the other light rare-earth metals, the salts and oxide of samarium do not reduce to metal using barium, calcium, or lithium, nor can electrolytic processes be used. The most effective reducing agent is lanthanum, which is mixed with Sm$_2$O$_3$ and heated under vacuum in a tantalum crucible. The samarium metal volatilizes and is condensed as powder or sponge on coiled tantalum or copper condenser plates. Subsequently, the samarium must be remelted under an argon or inert atmosphere before it is cast into graphite molds.

Samarium has been alloyed with gadolinium and aluminum to produce nuclear reactor hardware that will absorb neutrons for short periods. The use of samarium in intermetallics, cermets, and other chemical forms for use in nuclear applications holds promise. Small quantities of Sm$_2$O$_3$ are used in optical-glass filters and to encase lanthanum borate glass rods which then are drawn into fine fibers for fiberoptics applications. The element has been used as a coding agent for inks used in data handling systems. Small amounts also have been used for activating phosphate-type phosphors. The addition of samarium oxide produces a strong narrow emission in the near-infrared spectral region. The most significant use of samarium is in the permanent-magnet alloy SmCO$_5$. The strength of these magnets is five times that of other previously developed magnetic materials.

A new co-reduction process has reduced the cost of SmCO$_5$, making it competitive with other magnet alloys. Current and possible applications include electric motors, line printers, frictionless bearings, jewelry, and hospital surgical uses.

See the references listed at the end of the entries on **Chemical Elements;** and **Rare-Earth Elements and Metals.**

NOTE: This 4th Edition entry was revised and updated by K. A. Gschneidner, Jr., Director, and B. Evans, Assistant Chemist, Rare-Earth Information Center, Energy and Mineral Resources Research Institute, Iowa State University, Ames, Iowa.

SAPONIFICATION. A special case of hydrolysis in which an ester is converted into an alcohol and a salt of the appropriate acid by reaction with an alkali. Though the operation has numerous applications throughout the chemical industry, it is noteworthy because some 80% of standard soap is prepared by this method. The esters may be of mono- or polybasic acids and mono- or polyhydric alcohols, the physical conditions under which the reaction occurs being suitably varied to secure an adequate rate. The alkali most commonly used is sodium hydroxide, because of cost and water solubility, but other appropriate alkaline materials are suitable.

Since the preparation of soap is typical of a fairly complex reaction, chemically, and since it is common, it serves as a useful example of the saponification operation. The complication, of course, occurs because the usual esters used for soap are the glycerol esters of fatty acids, saturated and unsaturated. Thus the saponification of stearin (glycerol tristearate) is commonly shown as follows:

$$C_{17}H_{35}COO—CH_2$$
$$C_{17}H_{35}COO—CH + 3NaOH \rightarrow$$
$$C_{17}H_{35}COO—CH_2$$

$$3C_{17}H_{35}COONa + HO—CH_2$$
$$HO—CH$$
$$HO—CH_2$$

Actually, the saponification appears to progress stepwise, the first hydrolytic reaction taking place as follows:

$$C_{17}H_{35}COO—CH_2$$
$$C_{17}H_{35}COO—CH + NaOH \rightarrow$$
$$C_{17}H_{35}COO—CH_2$$

$$C_{17}H_{35}COO—CH_2$$
$$C_{17}H_{35}COO—CH + C_{17}H_{35}COONa$$
$$HO—CH_2$$

The diglyceride formed is subsequently split to the monoglyceride, which finally is converted to glycerol, if sufficient alkali is present. Thus, the reaction is a bimolecular one rather than quadrimolecular, as is commonly indicated. In actual practice, the fats used are complex glycerides of a number of saturated and unsaturated acids, rather than the stearin shown here.

Technologically, the saponification operation varies in degree of difficulty depending on the ester. The reaction rate differs for different esters, for one thing, bu another determining factor is the contact area possible between the alkali and the ester. In the case mentioned above, the fat at the start is insoluble and immiscible in water, so that reaction in a nonagitated vessel would be very slow, occurring only at the limited interface.

Though saponification is the dominant reaction when the techniques described above are used, in many instances side reactions may occur which may profoundly modify the products. Oxidation, of course, is one of the more obvious things to guard against, since many times the esters being treated are unsaturated. Both isomerization and polymerization, however, may occur under the alkaline conditions obtaining, the unconjugated polyethenoid acids becoming conjugated during treatment. This especially appears to be true in the case of highly unsaturated compounds. However, by properly controlling reactants and conditions, saponification remains a very flexible and useful industrial and laboratory operation possible of wide application and at low cost.

—Stanley B. Elliott, Bedford, Ohio.

SAPONIFICATION NUMBER. Vegetable Oils (Edible).

SATURATED COMPOUND. Organic Chemistry.

SATURATED EDIBLE OILS. Vegetable Oils (Edible).

SCANDIUM. Chemical element symbol Sc, at. no. 21, at. wt. 44.956, periodic table group 3b, mp 1540°C, bp 2850°C, density 2.985 g/cm³ (alpha form), 3.19 g/cm³ (beta form). The alpha form is close-packed hexagonal. The face-centered cubic allotrope, although generally accepted, has not received full recognition. Scandium is a relatively soft metal with a silvery luster. The metal oxidizes rapidly in air. Scandium combines readily with water, oxygen, acids, halogens, and chalcogenides. The isotope ^{45}Sc occurs in nature and is not radioactive. Nine radioactive isotopes have been identified, ^{40}Sc through ^{44}Sc, and ^{46}Sc through ^{49}Sc, all with relatively short half-lives, ranging from a fraction of a second up to 84 days. Scandium occurs widely throughout nature, but in reasonably concentrated forms only in a few uncommon minerals. Abundance in the earth's crust is estimated at approximately $5-6 \times 10^{-4}\%$, ranking it ahead of such elements as antimony, bismuth, silver, and gold. It is estimated that a cubic mile of seawater contains about 375 pounds of the element (a cubic kilometer contains about 41 kilograms). Scandium was predicted by Mendeleev in 1869, at which time he called it *ekaboron* and foretold accurately a number of its properties. A small amount of scandium oxide was extracted from euxenite and gadolinite by Nilson in 1879, a material that Nilson called *scandia*. In the same year, Cleve isolated a greater quantity of the oxide, from which several compounds were prepared and favorably compared with Mendeleev's predictions for ekaboron.

Electronic configuration is $1s^22s^22p^63s^23p^63d^14s^2$. First ionization potential, 6.56 eV; second, 12.8 eV; third, 24.64 eV. Other physical properties are given under **Chemical Elements.**

Scandium occurs in some ores with the Lanthanum Series elements. It is easily separated from the Lanthanides, as well as yttrium, by taking advantage of the greater solubility of its thiocyanate in ether. The three recognized scandium minerals are thortveitite, a silicate; and sterrettite and kolbeckite, both phosphates. Wiikite and bazzite, complex niobates and silicates, are known to contain more than 1% scandium. Davidite, with a concentration of 0.02% Sc_2O_3, also is a major source of the element. Scandium has not been found without the Lanthanide elements and an association with yttrium. The element usually is separated from ore extracts and concentrates by precipitation as the oxalate. Scandium with a purity of 99.99% has been produced.

In water solutions, the scandium ion has a triple positive charge. Studies show, however, that the simple Sc^{3+} ion seldom exists. Rather, the form is highly polymerized and hydrolyzed—with hydroxy-bonded structures. In forming compounds, scandium parallels aluminum, yttrium, gallium, indium, and tellurium. Several carbides of scandium have been reported, the most stable being ScC.

Like the hydroxides of the Lanthanides, scandium hydroxide, $Sc(OH)_3$, is precipitated by addition of alkalies to solutions of scandium salts; however, the latter is precipitated at pH 4.9, while the former require pH 6.3 or more, a property which is utilized in one method of separation. Upon heating the hydroxide (or certain oxyacid salts), scandium oxide, Sc_2O_3 is produced. Scandium hydroxide is less acidic than aluminum hydroxide, requiring boiling KOH solution to form the complex potassium compound, $K_2[Sc(OH)_5 \cdot H_2O] \cdot 3H_2O$.

All four trihalides of scandium are known. The trifluoride is very slightly soluble in H_2O, and is precipitated from scandium nitrate, $Sc(NO_3)_3$, solutions by hydrofluoric acid. It dissolves in alkali fluorides to yield the complex ion $[ScF_6]^{3-}$. The chloride is formed in solution by treating the hydroxide or oxide with HCl, yielding hydrated crystals on concentration, which give hydroxychlorides on heating. The bromide is also prepared from the oxide or hydroxide and hydrobromic acid, or in anhydrous form from the oxide, carbon, and bromine, on heating. The iodide is also prepared by the latter method.

The thiocyanate is prepared in solution by adding ammonium thiocyanate, NH_4SCN to HCl solutions of the chloride. Both basic and double carbonates are known. The former is precipitated from Sc^{3+} solutions by adding carbonate solutions, and is probably $Sc(OH)CO_3 \cdot H_2O$. The latter are obtained by the use of an excess of the soluble carbonate. Normal, basic, and double sulfates are known. The first exists in several degrees of hydration; the second is obtained as $Sc(OH)SO_4 \cdot 2H_2O$, by treating the normal sulfate tetrahydrate with the hydroxide. The alkali double sulfates and alums are obtained by treating the sulfate solution with an excess of the alkali (or ammonium) sulfate solution.

The nitrate is readily obtained by action of dilute HNO_3 on the hydroxide. In aqueous solution, the anhydrous nitrate yields a monobasic nitrate on heating.

To date, the applications for scandium and its compounds have been very limited, mainly because of its high reactivity and high cost. In exotic light sources, scandium iodide enhances luminosity. Minor inclusions of scandium are made in substituted yttrium garnets for electronic applications. At one time, it was believed that scandium might serve in a substitute fashion for aluminum, particularly for aircraft applications.

References

Borisenko, L. F.: "Scandium: Its Geochemistry and Mineralogy," Consultants Bureau, New York, 1963.
Vickery, R. C.: "Chemistry of Yttrium and Scandium," Pergamon, London, 1960.

Vickery, R. C.: "Analytical Chemistry of the Rare Earths," Pergamon, London, 1961.

Vickery, R. C.: *Scandium, Yttrium and Lanthanum* in A. F. Trotman-Dickenson (editor): "Comprehensive Inorganic Chemistry," Pergamon, Oxford, 1971.

—R. C. Vickery, Hudson Laboratories, Hudson, Florida.

SCAPOLITE. The mineral scapolite is a silicate of calcium and aluminum which contains also some potassium, sodium, and chlorine. The name identifies all intermediate members of a series with the following end members: marialite $3Na(AlSi_3)O_8 \cdot NaCl$, meinoite $3Ca(Al_2Si_2)O_8 \cdot CaCO_3$. Its tetragonal crystals are coarse and thick, often very large. It occurs also in massive forms. It has a distinct prismatic cleavage; subconchoidal fracture; is brittle; hardness, 5.5–6; specific gravity, marialite, 2.5–2.62, meionite, 2.72–2.78; luster, vitreous to rather dull, color, white to gray, red, green blue, or yellow; translucent to opaque, rarely transparent.

Scapolite is found in metamorphic rocks particularly those rich in calcium, also in contact metamorphic deposits in limestones. It has been found in basic igneous rocks, probably as a secondary mineral. Notable localities are Lake Baikal, Siberia; Arendal, Norway; and the Malagasy Republic. In the United States it is found in Massachusetts, New York, and New Jersey. Grenville, in the Province of Quebec, Canada is an important locality. Superb transparent yellow gem crystals have recently been found in Brazil and Tanzania. Wernerite (scapolite) was named in honor of A. O. Werner, a famous German mineralogist (1749–1817).

See also **Mineralogy.**

SCHEELITE. The mineral scheelite is calcium tungstate, $CaWO_4$, with molybdenum substituting for tungsten up to 24% in the molybdian scheelite variety. It is a tetragonal with an octahedral habit although also at times tabular, and may occur massive. It displays an octahedral cleavage; is bittle; hardness, 4.5–5; specific gravity, 6.1; luster, vitreous; color, white to yellowish, reddish, greenish and brownish; white streak; transparent to translucent. Scheelite is found in pegmatite and ore veins associated with granites, also as a contact metamorphic mineral. It is known from Czechoslovakia; Saxony; Italy; Alsace; Finland; Cumberland and Cornwall in England; and Mexico. Crystals of exceptional length (6–10 inches; 15–25 centimeters) are found at various localities in Korea and Japan; and in the United States, in Connecticut, Colorado, South Dakota, Arizona, Nevada, and California. The Swedish chemist, Karl Wilhelm Scheele, discovered tungsten in this mineral, which later was named for him.

The mineral fluoresces vivid bluish white to white; or yellowish-white with increasing molybdenum content under exposure to short-wave ultraviolet light.

SCHIST. The schists form a great group of metamorphic rocks chiefly notable for the preponderance of the lamellar minerals such as the micas, chlorite, talc, hornblende, graphite, etc. Quartz often occurs in drawn out grains to such an extent that a quartz schist is produced. Most schists have in all probability been derived from clays and muds which have passed through a series of metamorphic processes involving the production of shales, slates and phyllites as intermediate steps. Certain schists have been derived from fine-grained igneous rocks such as lavas and tuffs. Most schists are mica schists, but graphite and chlorite schists are common. Schists are named for the prominent or perhaps unusual mineral constituent, as garnet schist, tourmaline schist, glaucophane schist, etc. The word schist is derived

from the Greek meaning to split, with reference to the easy separation of these rocks in a direction parallel to that in which the platy minerals lie.

SCOLECITE. This mineral is a zeolite, a hydrous calcium-aluminum silicate, $CaAl_2Si_3O_{10} \cdot 3H_2O$. It occurs in slender monoclinic prisms and in fibrous and nodular masses. Hardness is 5; specific gravity, 2.27; luster vitreous to silky; transparent to translucent. When heated, some specimens of scolecite curl up like worms, hence its name, derived from the Greek meaning a *worm*. This mineral occurs with other zeolites, at Baden, Switzerland; Iceland; Greenland; the Deccan region of India; and the United States at Golden, Colorado, and Paterson, New Jersey. Single crystals up to 12 inches (30 centimeters) in length have recently been found in a single large cavity in the basaltic trap rocks near Nasik, India.

SCORODITE. This hydrated arsenate of ferric iron and aluminum $(Fe^{3+}, Mg^{3+})AsO \cdot 2H_2O$, crystallizing in the orthorhombic system, is the iron-rich isomorphous end member of a complete series extending to the aluminum-rich mineral mansfieldite. Crystals usually occur as drusy crusts. Also occurs as massive, compact, and earthy minerals. Hardness of 3.5–4, with specific gravity of 3.278. Vitreous to subadamantine luster, of pale green to liver-brown color.

Scorodite occurs as a secondary alteration mineral in the oxidized zones of metallic arsenic-containing veins. The mineral also may be a product of deposition from certain hot springs. World localities of note include Siberia; Laurium, Greece; Carinthia; Cornwall, England; and Nevada and Utah in the United States. Currently being deposited by hot springs at Yellowstone National Park in Wyoming.

SEAWATER (Desalination). Desalination.

SEAWATER (Elements in). Chemical Elements.

SELENIUM. Chemical element symbol Se, at. no. 34, at. wt. 78.96, periodic table group 6a, mp 217°C, bp 684–686°C, density, 4.82 g/cm³ (solid), 4.86 g/cm³ (single crystal). Selenium has a large number of allotropes, some of which have not been fully investigated. On heating selenium above its melting point and cooling it, a red vitreous mass is formed, probably a mixture of allotropes. A red amorphous allotrope is precipitated by sulfur dioxide from selenious acid solutions. On heating above 150°C, the red vitreous form changes to a gray hexagonal form, the stable form of selenium at ordinary temperatures, with metallic properties, one of which is photoconductivity. By evaporation of a carbon bisulfide solution of the red vitreous form below 72°C, a red α-monoclinic form is obtained; evaporation above 72°C gives β-monoclinic selenium. Black hexagonal selenium, believed to have a ring structure, is produced by heating amorphous selenium to near its melting point. Unlike sulfur, liquid selenium apparently has only one form.

There are six natural occurring isotopes ^{74}Se, ^{76}Se through ^{78}Se, ^{80}Se, and ^{82}Se, and seven known radioactive isotopes ^{72}Se, ^{73}Se, ^{75}Se, ^{79}Se, ^{81}Se, ^{83}Se, and ^{84}Se. With exception of ^{79}Se which has a half-life of something less than 6×10^4 years, the half-lives of the other isotopes are comparatively short, measured in minutes, hours, or days. In terms of abundance, selenium ranks 34th among the elements occurring in the earth's crust. It is estimated that a cubic mile of seawater contains about 14 tons (3 metric tons per cubic kilometer) of selenium. Electronic configuration is $1s^2 2s^2 2p^6 3s^2 3p^6 3d^{10} 4s^2 4p^4$. First ionization potential, 9.75 eV; second, 21.3 eV; third, 33.9

eV; fourth, 42.72 eV, fifth, 72.8 eV. Oxidation potentials $H_2Se(aq) \rightarrow Se + 2H^+ + 2e^-$, 0.36 V; $Se + 3H_2O \rightarrow H_2SeO_3 + 4H^+ + 4e^-$, -0.740 V; $H_2SeO_3 + H_2O \rightarrow SeO_4^{2-} + 4H^+ + 2e^-$, -1.15 V; $Se^{2-} \rightarrow Se + 2e^-$, 0.78 V; $Se + 6OH^- \rightarrow SeO_3^{2-} + 3H_2O + 4e^-$, 0.36 V; $SeO_3^{2-} + 2OH^- \rightarrow SeO_4^{2-} + H_2O + 2e^-$, -0.03 V.

Other physical properties of selenium are given under **Chemical Elements.**

Selenium was first identified by Berzelius in 1817. The element is found associated with volcanic activity, as for example in cavities of Vesuvian lavas and in the volcanic tuff of Wyoming (about 150 parts per million).

Selenium occurs as selenide in many sulfide ores, especiallly those of copper, silver, lead, and iron, and is obtained as a byproduct from the anode mud of copper refineries. The mud is (1) fused with sodium nitrate and silica, or (2) oxidized with HNO_3, and the H_2O extract is then treated with HCl and SO_2, whereupon free selenium is separated.

Uses: Selenium is widely used in photoelectric cells. The element alters its electrical resistance upon exposure to light. The response is proportional to the square root of incident energy. Selenium cells are most sensitive in the red portion of the spectrum. Although an external emf must be applied, the resistance is low and amplification is easy. In the selenium photovoltaic cell configuration, a thin film of vitreous or metallic selenium is coated onto a metal surface. Then, a transparent film of another metal, often platinum, is placed over the selenium. A cell of this type generates its own emf, with a decrease in internal resistance with increasing irradiation. The response essentially is proportional to incident energy. The cells are not importantly sensitive to small temperature changes.

Advantage of the unipolar conduction characteristic of selenium is taken in arc rectifiers. In a typical unit, a nickel or nickel-plated steel or an aluminum disk with a thin layer of selenium applied to one side is used. Selenium also is added to copper alloys and to stainless steel to increase machinability. Advantages claimed for selenium copper are high machinability, combined with hot-working properties and high electrical conductivity. As a decolorizer in glass, selenium counteracts green shades arising from ferrous ingredients. Sodium selenite is used in the production of red enamels and in the manufacture of clear red glass. Addition of from 1 to 3% selenium to vulcanized rubber increases abrasion resistance. The element also is used in photographic and printing reproduction chemicals. Selenium is used as an additive to lead-antimony battery grid metal.

Chemistry and Compounds. Due to its $4s^2p^4$ electron configuration, selenium, like sulfur, forms many divalent compounds with two covalent bonds and two lone pairs; d hybridization is quite common, forming compounds with Se oxidation states of 4+ and 6+.

While selenium dioxide, SeO_2, can be produced by direct reaction of the element with oxygen activated by passage through HNO_3, the compound is easily made by heating selenious acid, H_2SeO_3. Selenium dioxide sublimes at 315–317°C, and is readily reduced by sulfur dioxide to elemental selenium. Selenium trioxide, SeO_3, is not prepared from the dioxide by oxidation, although selenium does react with oxygen to form SeO_3 and SeO_2 in an electric discharge. The preferred method of preparing SeO_3 is by refluxing potassium selenate with sulfur trioxide. The reverse reaction, hydration of SeO_3 to selenic acid, H_2SeO_4, occurs easily. Selenious acid, H_2SeO_3, produced by hydration of SeO_2, is a stronger oxidizing agent than sulfurous acid as judged by its quantitative oxidation of iodide ion in acid solution, but is a weaker acid (ionization constants 2.4×10^{-3} and 4.8×10^{-9} at 25°C). It forms salts, the selenites, many of which, especially those of heavy metals, are reduced

to selenides by hydrazine. Many of the selenites, e.g., those of nickel, mercury, and ferric ion, are very slightly soluble in water. Selenious acid is readily oxidized by halogens in the presence of silver ion or 30% H_2O_2 to selenic acid, H_2SeO_4. Selenic acid is as strong an acid as H_2SO_4, and it is more readily reduced, reacting with hydrobromic acid and hydroiodic acid to form selenious acid or (at high concentration) elemental selenium. Like sulfate ion, SO_4^{2-}, SeO_4^{2-} is tetrahedral in crystals.

Hydrogen selenide, H_2Se, is a stronger acid than H_2S (ionization constants of H_2Se, 1.88×10^{-4} and about 10^{-10}) and is less readily obtained from selenides than H_2S from sulfides (the selenides of aluminum, iron and magnesium, Al_2Se_3, FeSe, and MgSe, require heating with H_2O or dilute acids). In general, the metal selenides are prepared by direct combination of the elements. Those of transition groups 3–8, 1 and 2 and main groups 3 and 4 exhibit many instances of well-defined compounds, berthollide compounds, and substitutional solid solutions. Thus four intermediate phases are found in the palladium-selenium system, Pd_4Se, $Pd_{2.8}Se$, $Pd_{1.1}Se$, and $PdSe_2$.

Selenium hexafluoride, SeF_6, the only clearly defined hexahalide, is formed by reaction of fluorine with molten selenium. It is more reactive than the corresponding sulfur compound, SF_6, undergoing slow hydrolysis. Selenium forms tetrahalides with fluorine, chlorine, and bromine, and dihalides with chlorine and bromine. However, other halides can be found in complexes, e.g., treatment of the pyridine complex of SeF_4 in ether solution with HBr yields $(py)_2SeBr_6$. Selenium tetrafluoride also forms complexes with metal fluorides, giving $MSeF_5$ complexes with the alkali metals.

Selenium forms several oxyhalides, e.g., $SeOF_2$, $SeOCl_2$, and $SeOBr_2$, the first two being liquids and the last a crystalline solid, mp 41.6 C. Selenium also forms tetraselenium tetranitride, Se_4N_4.

Selenocyanates, M^ISeCN, corresponding to the thiocyanates, are prepared by addition of selenium to soluble cyanides. They are similar to the thiocyanates except that HSeCN immediately decomposes in acid to selenium and hydrogen cyanide. The heavy metal selenocyanates are less soluble than the corresponding thiocyanates.

Selenium forms "thio"-type compounds, such as $SeSO_3$ by reaction of selenium and sulfur trioxide, $SeSO_3^{2-}$ (selenosulfates) by reaction of selenium and sulfites, SeS^{2-} (selenosulfides) by reaction of selenium with sulfides, as well as diselenides, Se_2^{2-}, and polyselenides, Se_x^{2-}.

Carbon diselenide is an evil-smelling liquid, and COSe and CSSe are also known.

Biological Role of Selenium. Some very interesting examples of the effect of soils on the nutritional quality of plants are associated with selenium. The element has not been found to be required by plants, but it is required in very small amounts by warm-blooded animals and probably by humans. However, selenium in larger quantities can be very toxic to animals and humans.

In large areas of the world, the soils contain very little selenium in forms that can be taken up by plants. Crops produced in these areas are, therefore, very low in selenium. A selenium deficiency in livestock is a serious problem. The deficiency causes a form of muscular dystrophy in younger animals and poor reproductive qualities in the adult animals. For prevention, sodium selenate or sodium selenite, sometimes augmented with vitamin E, is added in proper proportions to feedstuffs. Some areas, including the Plains and Rocky Mountain states in the United States, have soils that are rich in available selenium. In regions like these, selenium toxicity is a problem.

An interesting feature of selenium is that it occurs naturally in several compounds and these vary greatly in their toxicity and in their value in preventing selenium-deficiency diseases. In its elemental form, selenium is essentially insoluble and biologically inactive. Inorganic selenates or selenites and some of the selenoamino acids in plants are very active biologically, whereas some of their metabolites that are excreted by animals are not biologically active. In well-drained alkaline soils, selenium tends to be oxidized to selenates and these are readily taken up by plants, even to levels that may be toxic to the animals that eat them. In acid and neutral soils, selenium tends to form selenites, which are insoluble and unavailable to plants. Selenium deficiency in livestock is most often found in areas with acid soils and especially soils formed from rocks low in selenium.

In 1934, the mysterious livestock maladies on certain farms and ranches of the Plains and Rocky Mountain states were discovered to be due to plants with so much selenium that they were poisoning grazing animals. Affected animals had sore feet, lost some of their hair, and many died. Over the next 20 years, researchers found that the high levels of selenium occurred only in soils derived from certain geological formations of high selenium content. They also found that a group of plants, called *selenium accumulators*, had an extraordinary ability to extract selenium from the soil. These accumulators were mainly shrubs or weeds native to semiarid and desert rangelands. They usually contained about 50 parts per million (ppm) or more of selenium, whereas range grasses and field crops growing nearby contained less than 5 ppm selenium. These findings helped ranchers to avoid the most dangerous areas when grazing livestock.

In 1957, selenium was found to be essential in preventing liver degeneration of laboratory rats. Since then, research workers have found that certain selenium compounds, either added to the diet or injected into the animal, would prevent some serious diseases of lambs, calves, and chicks. That selenium is an essential nutrient element for birds and animals has been established.

In most diets used in livestock production, from 0.04 to 0.10 ppm of selenium protects the animal from deficiency diseases. If the diet is very high in vitamin E, the required level of selenium may be lower.

In terms of human dietary requirements, much of the wheat for breadmaking in the United States is produced in selenium-adequate sections of the country. Bread is generally a good source of dietary selenium.

Selenomethionine decomposes lipid peroxides and inhibits *in vivo* lipid peroxidation in tissues of vitamin E deficient chicks. Selenocystine catalyzes the decomposition of organic hydroperoxides. Selenoproteins show a high degree of inhibition of lipid peroxidation in livers of sheep, chickens, and rats. Thus, some forms of selenium exhibit *in vivo* antioxidant behavior.

References

Carapella, S. C., Jr.: "Selenium" in "Metals Handbook," 9th Edition, Vol. 2, American Society for Metals, Metals Park, Ohio (1979).

Chen, J. R., and J. M. Anderson: "Legionnaires' Disease: Concentrations of Selenium and Other Elements," *Science*, **206**, 1426–1427 (1979).

McDade, J. E., et al.: *New England Jrnl. Med.*, **297**, 1197 (1977).

Reamer, D. C., and W. H. Zoller: "Selenium Biomethylation Products from Soil and Sewage Sludge," *Science*, **208**, 500–502 (1980).

Sax, N. I.: "Dangerous Properties of Industrial Materials," 5th Edition, Van Nostrand Reinhold, New York (1979).

Serfass, R. E., Hinsdill, R. D., and H. E. Ganther: *Fed. Proc. Fed. Am. Soc. Exp. Biol.*, **33**, 694 (1974).

Serfass, R. E., and H. E. Ganther: *Life Sci.*, **19**, 1139 (1976).

Zingaro, R., and W. C. Cooper: "Selenium," Van Nostrand Reinhold, New York (1974).

SEMICARBAZONES. The products of the reaction between an aldehyde or a ketone with semicarbazide are termed *semicarbazones*.

$$CH_3 \cdot CHO \quad + NH_2 \cdot CO \cdot NH \cdot NH_2 \rightarrow$$
$$\text{(acetaldehyde)} \qquad \text{(semicarbazide)}$$
$$CH_3 \cdot CH : N \cdot NH \cdot CO \cdot NH_2 \; + H_2O$$
$$\text{(acetaldehyde semicarbazone)}$$
$$(CH_3)_2CO + NH_2 \cdot CO \cdot NH \cdot NH_2 \rightarrow$$
$$\text{(acetone)}$$
$$(CH_3)_2C : N \cdot NH \cdot CO \cdot NH_2 + H_2O$$
$$\text{(acetone semicarbazone)}$$

SEMICONDUCTOR. Since the invention of the transistor (contraction for *trans*fer re*sistor*), the transistor and related semiconductor devices have had an unprecedented impact on the electronics industry.

Semiconductors are distinguished from other classes of materials by their characteristic electrical conductivity σ. The electrical conductivities of materials vary by many orders of magnitude, and consequently can be classified as: (1) the perfectly conducting superconductors; (2) the highly conducting metals ($\sigma \sim 10^6$ mho/centimeter; (3) the somewhat less conducting semimetals ($\sigma \sim 10^4$ mho/centimeter; (4) the semiconductors covering a wide range of conductivities ($10^3 \gtrsim \sigma \gtrsim 10^{-7}$ mho/centimeter; and (5) the insulators, also covering a wide range ($10^{-10} \gtrsim \sigma \gtrsim 10^{-20}$ mho/centimeter).

These low-conductivity materials are characterized by the great sensitivity of their electrical conductivities to sample purity, crystal perfection, and external parameters, such as temperature, pressure, and frequency of the applied electric field. For example, the addition of less than 0.01% of a particular type of impurity can increase the electrical conductivity of a typical semiconductor like silicon or germanium by six or seven orders of magnitude. In contrast, the addition of impurities to typical metals and semimetals tends to decrease the electrical conductivity, but this decrease is usually small. Furthermore, the conductivity of semiconductors and insulators characteristically decreases by many orders of magnitude as the temperature is lowered from room temperature to 1 K. On the other hand, the conductivity of metals and semimetals characteristically increases in going to low temperatures, and the relative magnitude of this increase is much smaller than are the characteristic changes for semiconductors. The principal conduction mechanism in metals, semimetals, and semiconductors is electronic, whereas both electrons and the heavier charged ions may participate in the conduction processes of insulators.

Classification. It is customary to classify a semiconductor according to the sign of the majority of its charged carriers, so that a semiconductor with an excess of negatively charged carriers is termed *n*-type. A semiconductor with an excess of positively charged carriers is called *p*-type, while a material with no excess of charged carriers is considered to be perfectly compensated. Many of the important semiconductor devices depend upon fabricating a sharp discontinuity between the *n*- and *p*-type materials, the discontinuity being called a *p-n junction*.

Physical Characteristics*. Most semiconductors exhibit a metallic luster upon visual inspection. Nevertheless, the visual appearance of materials does not provide an adequate criterion for the classification of materials, since the electrical conductivity of all materials is frequency dependent. Visual inspection

---

* The following paragraphs contain several excerpts from "The Encyclopedia of Physics," 2nd Edition (R. M. Besançon, Editor), Van Nostrand Reinhold, New York.

tends to be sensitive to the conductivity properties at visible frequencies ($\sim 10^{15}$ Hz). Although materials with a high optical reflectivity tend also to exhibit high dc conductivity, these two properties are not necessarily correlated in semiconductors and metals. An example of a metal without metallic luster is ReO_3 (rhenium trioxide), a semitransparent, reddish solid. On the other hand, most of the common semiconductors do exhibit metallic luster primarily because electronic excitation across their fundamental energy gaps can be achieved at infrared frequencies. At low frequencies, the principal conduction mechanism is free carrier conduction, which is important in metals and is present to some extent in semiconductors which contain impurities or are found at elevated temperatures. In contrast, interband transitions dominate the conduction process at very high frequencies. Interband transitions contribute to the conductivity by about the same order of magnitude in semiconductors, metals, and insulators.

Since the dc conductivity due to free carriers is characteristically low in semiconductors and insulators, the generation of free carriers by exposure to light at infrared, visible, and ultraviolet frequencies can lead to a large increase in the dc conductivity. This photoconductive effect, which is not observed in metals or semimetals, can be enormous in low-conductivity semiconductors (an increase in the dc conductivity of CdS (cadmium sulfide) by 8 orders of magnitude is observed).

Because of the extreme sensitivity of semiconductors to impurities, temperature, pressure, light exposure, and certain other factors, these materials can be exploited in the fabrication of useful devices, such as the crystal diode, the transistor, integrated circuits, photodetectors, and light switches. Semiconductor devices do, in fact, date back to the infancy of electronics, when crystal sets (usually using galena, lead sulfide) were used for radiowave detection. With the development of reliable and efficient vacuum tubes, the interest in semiconductor devices essentially stagnated for many years. Renewed interest in crystal rectifiers was stimulated by the needs of radar technology during World War II. During this period (for example, at the Massachusetts Institute of Technology's Radiation Laboratory), intensive activity developed in the fabrication of very pure semiconducting materials as well as in the basic understanding of the energy level schemes and of the charged carrier transport in silicon and germanium. This activity culminated in the discovery of the transistor.[1]

Silicon is at present the most commonly used semiconductor, although it was preceded on a large scale by germanium. There are several other elemental semiconductors, including diamond (carbon), gray tin, tellurium, selenium, and boron. Closely related to the group IV semiconductors (diamond, silicon, germanium, and gray tin) are the III–V compounds formed from such elements as indium arsenic, antimony, gallium, cadmium, zinc, tellurium, among others—these various elements used in a variety of combinations in semiconductor devices.

In addition to electrical conductivity, certain classes of semiconductors also possess other interesting properties. For example, the europium chalcogenides form a family of magnetic semiconductors, with EuO, EuS, and EuSe undergoing a ferromagnetic phase transition, while EuTe becomes antiferromagnetic below a Néel temperature of 9.8 K. Magnetic semiconductors have been of particular interest because of the close coupling between the electrical and magnetic properties, such as electrical conductivity and magnetic susceptibility. Some semiconductors also have been found to undergo a superconducting phase transition, as for example GeTe and SnTe for carrier concentra-

[1] Shockley, Bardeen, and Brattain of Bell Laboratories were awarded the Nobel Prize in Physics in 1956 for their 1947 invention of the transistor.

tions of $\sim 10^{21}$/cubic centimer and transition temperatures below 0.3 K.

Although most of the common semiconductor devices utilize crystalline materials, semiconductors are also found in the liquid and amorphous states. Of special interest is the fact that the electrical conductivity of an amorphous semiconductor tends to be much lower than that of its crystalline counterpart. The opposite situation prevails for amorphous and crystalline metals. Common semiconductors like silicon and germanium have been prepared in the amorphous state.

Semiconductors tend to be hard and brittle and become ductile only at high temperatures. The hardness of semiconductors like diamond and SiC is utilized in the manufacture of industrial abrasives. Because of this hardness, high-quality optical surfaces on semiconductors can be achieved, using lapping and etching techniques.

Because of the industrial demands for the fabrication of high-quality, high-purity semiconductors, much attention has been given to the development of a sophisticated semiconductor technology. See also **Crystal; and Microstructure Fabrication (and Chemistry)**. To produce crystals of the highest purity and crystalline perfection, the method of chemical vapor deposition is often favored. Chemical vapor deposition has been used in the fabrication of specific microcircuits in order to exploit the flexibility that this technique provides for varying the type and concentration of dopants which are introduced. The method is also used for the growth of certain mixed compound semiconductors.

Flow of Current in Semiconductors. The flow of electric current depends upon the acceleration of charges by an externally applied electric field. Only those charges that resist collisions or scattering events are effective in the conduction process. Because of collisions, charged particles in a solid are not accelerated indefinitely by the applied field, but rather, after every scattering event, the velocity of a charged particle tends to be randomized. Thus the acceleration process must start anew after each scattering event and charged particles achieve only a finite velocity along the electric field $\mathbf{E}$, the average value of the velocity being denoted by v_D, the drift velocity. The effectiveness of the charge transport by a particular charged particle is expressed by the mobility μ, which is defined as $\mu = v_D/E$. The mobility of a particle with charge e and mass m can be related directly to the mean time between scattering events (also called the relaxation time) by the expressed $\mu = e\,\tau/m$. The electrical conductivity σ depends upon the mobility of the charged carriers as well as on their concentration n, and is simply written as $\sigma = ne\mu$, where e is the charge of the carriers. The advantage of expressing the conductivity in this form is the explicit separation into a factor n which is highly sensitive to external parameters, such as temperature, pressure, optical excitation, irradiation, and into another factor μ, which depends characteristically on scattering mechanisms and on the electronic structure of the semiconductor.

The classical theory for electronic conduction in solids was developed by Drude in 1900. This theory has since been reinterpreted to explain why all contributions to the conductivity are made by electrons which can be excited into unoccupied states (Pauli principle) and why electrons moving through a perfectly periodic lattice are not scattered (wave-particle duality in quantum mechanics). Because of the wavelike character of an electron in quantum mechanics, the electron is subject to diffraction by the periodic array, yielding diffraction maxima in certain crystalline directions and diffraction minima in other directions. Although the periodic lattice does not scatter the electrons, it nevertheless modifies the mobility of the electrons. The cyclotron resonance technique is used in making detailed investigations in this field.

The origin of the *energy barrier*[2] for carrier generation is directly connected with the energy levels for electrons in a solid. Considering electrons in a solid from a tight-binding point of view, the discrete energy levels of the free atom broaden in the solid to form energy bands. For materials which are well described by the tight-binding approximation, the width of the energy bands is sufficiently small that an energy gap between the energy bands is formed; in the forbidden energy gap there are no bound states. Of particular importance to the conduction properties of a solid is the fact that *all* the available states in each band would be filled if each atom were to contribute exactly two electrons, thereby causing every solid with an odd number of electrons per atom to be metallic; while solids with an even number of electrons per atom would be insulating or semiconducting. The occurrence of energy bandgaps is also a consequence of the weak binding approximation, whereby the periodic potential itself is responsible for creating bandgaps through the mixing of states separated by a reciprocal lattice vector.

For semiconductors, the excitation energy lies in the range 0.1 to about 2 eV. Thermal fluctuations are sufficient to excite a small, but significant, fraction of electrons from the occupied levels (the valence band) into the unoccupied levels (the conduction band). Both the excited electrons and the empty states in the valence band (aptly called *holes*) may move under the influence of an electric field, providing a means for conduction of current. A hole acts like an electron with a positive charge. Such electron-hole pairs may be produced not only by thermal energy, but also by incident light, providing photo-effects.

Crystallographic defects, in general, are also electronic defects. In metals, they provide scattering centers for electrons, increasing the resistance to charge flow. The resistance wire in many electric heaters, in fact, consists of an ordinary metal, such as iron, with additional alloying elements, such as nickel or chromium, providing scattering centers for electrons. In semiconductors and insulators, alloying elements and defects provide an even greater variety of effects, since they can change the electron-hole concentrations drastically in addition to providing scattering centers. The semiconductor industry has been built on the alloying of silicon, germanium, and compound semiconductors with *selected impurities* in carefully controlled concentration and geometry.

Most semiconductor materials for electronics are nearly perfect crystals that contain a relatively small fraction of atoms associated with defects. On the other hand, some materials depart greatly from crystallinity, glass and plastics being notable examples. Although their electronic structure is not fully understood, they possess properties of conductivity, dielectric constant, strength, and the like similar to those of crystals. Thus, they are of great practical use.

Of more recent interest is a class of noncrystalline semiconductors called *amorphous semiconductors*. The electronic bands of these materials are not fully understood, but they do provide semiconducting characteristics of value in electronic devices.

Crystalline silicon, used in most present semiconductor devices, is of relatively high cost principally because of the time and care required in growing ultrapure crystals. Amorphous silicon is of lower cost and relatively easy to prepare. But little interest was shown in amorphous silicon until the mid-1970s, when scientists at the University of Dundee (Scotland) found that a certain kind of amorphous silicon containing several percent hydrogen will exhibit attractive electrical properties when small amounts of phosphorus and boron are added as dopants. Solar cells based upon this concept have been experimentally produced by a few manufacturers of semiconductors in the United Kingdom, France, and the United States.

The role of hydrogen atoms in amorphous silicon is now under intensive study. However, as early as the late 1960s, British scientists experimented in this field. Gaseous silane (SiH_4) molecules were disintegrated under vacuum by direct current or a high-frequency alternating current electric field to produce silicon atoms. These atoms built up on a heated substrate to form a layer of amorphous silicon. The process was called *glow discharge*. Investigators found that the electrical conductivity of the amorphous silicon increased by a factor of 600 when 200 parts per million of phosphine (PH_3) were added to the process. It was then assumed that the phosphorus atoms served as dopants and performed in the same manner as they do in crystalline silicon. This path of research was dropped for several years, however, until the findings at the University of Dundee were reported.

Most recent research (such as at Harvard University) indicates that is the hydrogen that really enhances the characteristics of the amorphous silicon. Currently many authorities conclude that Schottky barrier devices show the greatest promise for mass-produced solar cells, but it is also apparent that more will be heard concerning the hydrogen-silicon devices during the early and mid-1980s.

Amorphous *chalcogenide glasses* also have been under investigation since the mid-1950s, and more intensely during the late 1960s and continuing to the present, for possible uses in semiconductor devices. The glasses are named for the chalcogens (Group VI elements in the periodic table). Early in their consideration, these materials created a considerable controversy among solid-state physicists. Claims were made and challenged as regards their possible impact on further revolutionizing the semiconductor industry. It has been shown that chalcogenide glasses can "switch," but some scientists observe that almost any material will switch under the right conditions. Compositions proposed for memory switches are exemplified by $Te_{81}Ge_{15}Sb_2S_2$ and for nonmemory switch materials, $Te_{40}As_{35}Si_{18}Ge_7$. It has also been shown that transitions occur in these glasses when they are exposed to intense light and thus possible photographic uses have been proposed.

Because of the high-volume solar cell demand that may develop at some future date, much research has been directed to this end—as is also the case with the hydrogen-silicon devices previously mentioned. Most scientists consider this to be a possible, rather than a probable achievement. Research in these areas, at a minimum, will continue to increase the knowledge of surface phenomena and solid-state physics.

Representative Semiconductor Devices

Semiconductor Diodes. A diode is a two-terminal device which has the property of permitting current to flow with practically no resistance in one direction and offering nearly infinite resistance to current flow in the opposite direction. Their applications are numerous—in gating circuits used in digital computers, for example.

Silicon, a tetravalent element, i.e., with 4 valence electrons, is a widely used material for diodes. Germanium is also tetravalent. Elements in their pure state are said to be *intrinsic*. Silicon

---

[2] The sets of discrete but closely adjacent energy levels, equal in number to the number of atoms, that arise from each of the quantum states of the atoms of a substance when the atoms condense to a solid from a nondegenerate gaseous condition, make up the energy band, also called the *Bloch band*. For a semiconductor, the highest energy level is the *conduction band*, containing only the excess electrons resulting from crystal impurities. The next highest level is the *valence band*, usually completely filled with electrons. In between these bands is the *forbidden band*, which is wider for an insulating material than for a semiconductor and vanishes in a conducting material.

and germanium are in periodic table Group IV. Elements in periodic table Group III have 3 valence electrons, among them indium and gallium. Elements in periodic table Group V are pentavalent; and these include arsenic and antimony.

The process of introducing one of the Group III or V elements into silicon or germanium is called *doping*. The doped material thus is no longer intrinsic, but is impure and called *extrinsic*. If a trivalent impurity is introduced into silicon or germanium, *holes* are created and the material is said to be *p*-type. Introduction of a pentavalent element into silicon or germanium, on the other hand, creates *free electrons* and the material is said to be *n*-type. Because of thermal effects, free electrons and holes are always being produced in silicon and germanium (intrinsic generation of electron-hole pairs). Consequently, there will be some electrons in *p*-type material and some holes in the *n*-type material. These carriers are referred to as *minority carriers*. Electrons in *n*-type material and holes in *p*-type material are termed *majority carriers*.

The process of placing impurities in the near-surface region of solids is accomplished by a procedure known as *implanting*. A commonly used implanting procedure is to accelerate impurity ions in an electrostatic field with sufficient energy to impinge with the desired force on the solid target. Known as *ion implantation*, this carefully controlled and reproducible procedure has been widely used to dope semiconductors to create *p-n* junction formations. A certain amount of damage, however, occurs in the semiconductor material in some cases. The surface may become amorphous, or because the implanted dopants may not reach substitutional spots in the crystal lattice, the ions may not become electrically active. Thus, it is necessary to anneal the solid for electrical activation of the implanted ions as well as to remove any damage. Normally, the annealing occurs at a temperature of several hundred degrees for periods of a half-hour or longer. The thermal annealing process may be accompanied by some degradation of the device.

Within the last few years, *laser annealing* has attracted much interest. In this procedure, the implanted material is illuminated with laser radiation. Because the laser light does not penetrate far (a few micrometers into the surface layer) and the temperature rise is high but well confined, annealing is not required. Practice with the very rapid procedure has shown that laser annealing can be carried out in an ordinary atmosphere with little introduction of atmospheric impurities.

As may be expected, the laser energy must be directed at just the right location, with tight control over laser radiation wavelength, energy densities, and pulse durations. The technology offers promise for improved fabrication of high-efficiency solar cells and it may prove very valuable for fabricating microelectronic devices in three dimensions. Laser annealing was pioneered by scientists in the U.S.S.R. (1975), and a bit later by italian and American scientists (Oak Ridge National Laboratory).

A widely used semiconductor diode is the *p-n junction diode*. Imagine a crystal (single) of silicon doped so half the material is *p*-type and the other half is *n*-type. The internal boundary

between the two extrinsic regions is a *p-n junction*, and the resulting device is a *diode*. See Fig. 1. Three possible configurations of the *p-n* junction are shown in Fig. 2. The energy diagrams for these three configurations are also shown in Fig. 2.

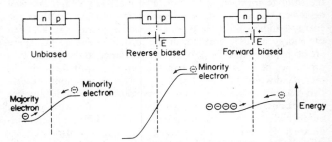

Fig. 2.   Three possible configurations of *p-n* junction diode.

Similar diagrams can be generated for holes. When the diode is unbiased, not net flow of electrons takes place across the junction. Assuming that some electrons on the *n*-side have sufficient energy to overcome the potential hill, electrons on the *p*-side (minority carriers) "slide down" the hill, making the net current flow zero. For the reverse biased example, the potential hill is raised and only the few minority carriers from the *p*-side slide down. This results in a minute reverse saturation current. When the diode is forward biased, the potential hill is lowered. This enables electrons to climb over the hill and current flow occurs. The same considerations apply to holes. In fact, the total diode current is equal to the sum of the electrons and holes flowing across the junction.

The characteristic curve of a semiconductor diode is shown in Fig. 3. An equation for this curve, called the *rectifier equation*, is expressed as:

$$I = I_s \left(e^{-11600\,E/T} - 1 \right)$$

where I = diode current, amperes
I_s = reverse saturated current (temperature dependent), amperes
E = diode biasing voltage ($+E$ for forward bias; $-E$ for reverse bias), volts
T = absolute temperature ($0°C + 273°$), degrees Kelvin.

At room temperature (300 K) and $E = 0.1$ volt,

$$I \cong I_s e^{39E}$$

When E is more negative than 0.1 volt,

$$I \cong -I_s.$$

An example of a simple rectifier employing a *p-n* junction diode is given in Fig. 4. During the positive half-cycle (0–180°) of the ac sinusoidal waveform v_s, the diode is forward-biased and conducts. The voltage v_L across load resistance R_L is, therefore, nearly identical to that of v_s for the positive half-cycle. For the negative half-cycle (180–360°), the diode is reverse-

Fig. 1.   (*Left*) Configuration of *p-n* junction diode. (*Right*) Electrical symbol.

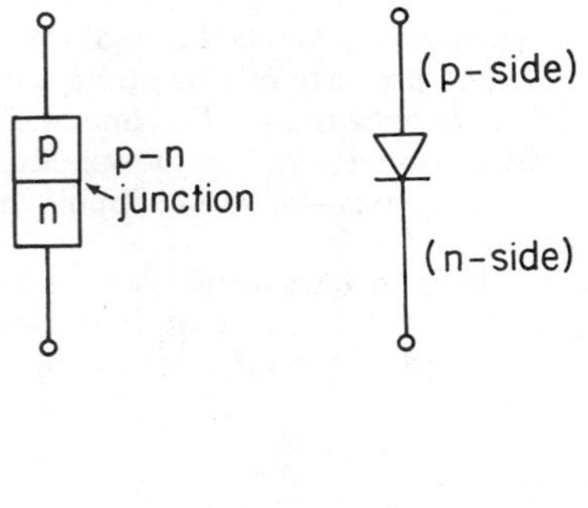

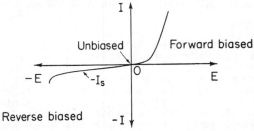

Fig. 3.   Characteristic curve of a semiconductor diode.

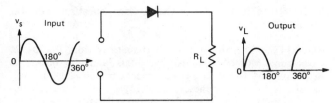

Fig. 4.   Simple rectifier employing a *p-n* junction diode.

biased and does not conduct. No current flows in R_L, and v_L = O during the negative half-cycle. Because the diode conducts for only one-half cycle, the circuit of Fig. 4 is called a *half-wave rectifier*. The waveform of v_L is only unidirectional. To obtain steady dc, like that from a battery, a filter is required. An example of an elementary filter is a large-valued capacitor placed across the load resistor.

The circuit of Fig. 4 can also be used as a detector of amplitude-modulated (AM) radio waves. Figure 5(a) illustrates the components of an AM wave. If this is applied to the input of Fig. 4, the wave is rectified and the output appears as shown in Fig. 5(b). Placing a small-valued capacitor across R_L filters out the carrier frequence and the desired modulating signal is obtained, as shown in Fig. 5(c).

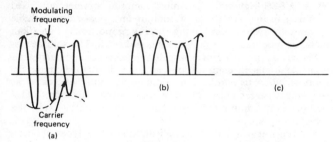

Fig. 5.   Use of *p-n* junction diode as AM radio detector.

Additional types and uses of *p-n* diodes include: (1) *Gunn*—used for the generation of microwave power; (2) *hot carrier* or *Schottky barrier*—used for fast switching of waveforms, such as found in computers; (3) *IMPATT*—used for the generation of microwave power; (4) *injection*—used for the generation of laser frequencies; (5) *light-emitting (LED)*—used for alphanumeric displays; (6) *varactor*—a reverse-biased junction diode that behaves like a variable capacitor, as a function of the applied voltage across the device; (7) *Zener*—a reverse-biased junction diode that exhibits a dc voltage which is nearly independent of a specified range of current flowing in the device. The Zener diode finds wide use as a voltage reference in regulated power supplies.

Silicon-Based Devices Dominate. As of the early 1980s, silicon-based devices continue to be preeminent in the semiconductor industry. During the past few years, there has been much research with Group III–Group V (periodic table) materials, but the most outstanding advances have been made in connection with certain opto-electronic devices, such as lasers, light-emitting diodes (LEDs), photodetectors, and solar cells. However, most authorities currently agree that, because of processing advantages and broadened functions, the future for such devices is quite promising.

References

Abelson, P. H., and A. L. Hammond: "The Electronics Revolution," *Science*, **195**, 1087–1091 (1977).

Harper, C. A.: "Handbook of Components for Electronics," McGraw-Hill, New York (1977).

Harrison, W. A.: "Electronic Structure and the Properties of Solids," Freeman, San Francisco (1980).

Keyes, R. W.: "Physical Limits in Semiconductor Electronics," *Science*, **195**, 1230–1234 (1977).

Kircher, C. J., and M. Murakami: "Josephson Tunnel-Junction Electrode Materials," *Science*, **208**, 944–950 (1980).

Landee, R.: "Electronics Designers' Handbook," 2nd Edition, McGraw-Hill, New York (1977).

Linvill, J. G., and C. L. Hogan: "Intellectual and Economic Fuel for the Electronics Revolution," *Science*, **195**, 1107–1113 (1977).

Mott, N. F., and E. A. Davis: "Electronic Processes in Non-Crystalline Materials," Oxford Univ. Press, New York (1979).

Pierce, J. R.: "Electronics: Past, Present, and Future," *Science*, **195**, 1092–1095 (1977).

Robinson, A. L.: "Amorphous Silicon—New Direction for Semiconductors," *Science*, **197**, 851–853 (1977).

Robinson, A. L.: "Chalcogenide Glasses," *Science*, **197**, 1068–1070 (1977).

Robinson, A. L.: "Laser Annealing—Processing Semiconductors without a Furnace," *Science*, **201**, 333–335 (1978).

Staff: "Making Chips—Ion Implantation," *Sci. Amer.*, **242**, 3, 94–95 (1980).

White, C. W., Narayan, J., and R. T. Young: "Laser Annealing of Ion-Implanted Semiconductors," *Science*, **204**, 461–468 (1979).

Woodall, J. M.: "III-V Compounds and Alloys," *Science*, **208**, 908–915 (1980).

SEMIPERMEABLE MEMBRANE (or Semipermeable Diaphragm).

A membrane or septum through which one (or more) of the substances composing a mixture or solution may pass, but not all.

In osmotic pressure determinations, semipermeable membranes permit the passage of a solvent but not of certain colloidal or dissolved substances. Many natural membranes are semipermeable, e.g., cell walls; other membranes may be made artificially, e.g., by precipitating copper cyanoferrate(II) in the interstices of a porous cup, the cup serving as a frame to give the membrane stability.

Semipermeable membranes are also used in the separation of gases. (See accompanying figure.) When a semipermeable

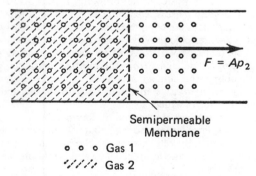

Separation of gases by semipermeable membrane.

membrane is placed in a gas mixture, being impermeable to gas 2 and allowing gas 1 to pass, the force exerted on it will equal the area times the partial pressure of gas 2 only. While there are no ideal semipermeable membranes for gases, there exist in practice reasonable approximations to them, such as incandescent platinum or palladium sheets, which can be penetrated by hydrogen but not by other gases. A film of water also acts as a semipermeable membrane for gases, since it is pervious to NH_3 or SO_2 because of their solubility in water, but gases which are not easily soluble are held back.

SEPIOLITE. The mineral sepiolite or meerschaum is soft, white, light in weight, and occurs in claylike nodular masses. It is a complex, hydrous magnesium silicate corresponding to the formula $Mg_4SI_6O_{15}(OH)_2 \cdot 6H_2O$. It crystallizes in the orthorhombic system; hardness, 2–2.5; specific gravity, 2; color, white grayish white, sometimes a yellowish or bluish-green; opaque. It is capable of floating on water, hence the name *meerschaum* or sea foam. It occurs in Asia Minor associated with serpentine and magnesite, and may be derived from the latter. Other deposits are in Czechoslovakia, Morocco, and Spain; and in the United States in Pennsylvania and New Mexico. The name *meerschaum* is from the German. *Sepiolite* is from the Greek, meaning *cuttlefish*, referring to the similarity of the bone of that animal to the light, porous sepiolite. The material is used in the manufacture of smoking pipes.

SEQUESTERING AGENTS. Chelation Compounds.

SERANDITE. The mineral serandite is a hydrated manganese-sodium silicate corresponding to the formula $Mn_2NaSi_3O_8(OH)$, crystallizing in the triclinic system, of pseudo-monoclinic character. Color, rose-red, pink; transparent; brittle, and uneven fracture. Prominent basal and prismatic cleavage; vitreous to pearly luster. Crystals thick tabular or prismatic, and as intergrown aggregates. Occurs as superb crystals in a carbonatite zone in a host body of nepheline-syenite in association with analcime, aegerine and other rare minerals at Mt. St. Hilaire, Quebec, Canada. Its only other known world occurrence is on the Island of Rouma, Los Islands, Guinea.

SERPENTINE. This is a group name for minerals encompassing two principal polymorphic forms: *chrysotile* and *antigorite*. This monoclinic mineral of hydrous magnesium silicate composition $Mg_3Si_2O_5(OH)_4$ is essentially a product of metamorphic alteration of ultrabasic rocks rich in olivine, pyroxene, and amphibole. Serpentine crystals are unknown except as pseudomorphic replacements of other minerals, e.g., after clinochlore crystals at the Tilly Foster Mine, Brewster, New York, Antigorite occurs as platy masses; *chrysotile* as silky fibers. Most massive serpentine rocks are composed essentially of antigorite. The hardness is 2–5, specific gravity ranges from 2.2 (fibrous varieties) to 2.65 (massive varieties). Color usually mottled green. The name *serpentine* stems from the mottled character, somewhat resembling the skin of a serpent. There is a greasy to wax-like luster in massive material; silky in fibrous material. The minerals are translucent.

Chrysotile fibers are the source of commercial asbestos, although fibrous amphiboles also contribute to similar usage. Asbestos is economically valuable for its incombustibility and low conductivity of heat, thus as fireproofing and insulating material. See also **Asbestos.**

Chrysotile deposits of economic value are found in Quebec, Canada, in the U.S.S.R., and in South Africa. Minor occurrences are found in the United States in Vermont, New York, New Jersey, and Arizona. *Verd antique* marble (serpentine marble) is quarried extensively near West Rutland, Vermont.

SESAME SEED OIL. Vegetable Oils (Edible).

SEWAGE SLUDGE (Energy Source). Biomass and Wastes as Energy Sources; Sludge.

SHIFT REACTION (Water Gas). Coal; Substitute Natural Gas (SNG).

SIDERITE. This mineral is a carbonate of iron, $FeCO_3$. It is hexagonal with rhombohedral crystals, and also occurs in various massive forms. It has a rhombohedral cleavage; uneven fracture; is brittle; hardness, 3.75–4.25; specific gravity, 3.96; luster, vitreous to pearly; color, gray, yellowish or greenish-gray, green, reddish-brown and brown. Siderite is found as concretionary masses in the sedimentary rocks; as a replacement mineral from the action of iron solutions upon limestones; and in metalliferous veins as a gangue mineral. It is relatively common. Siderite is found in Austria, Saxony, Czechoslovakia, France, England, Italy, Greenland, Australia, Brazil, and Bolivia. In the United States important localities are in Connecticut, Pennsylvania, New Jersey, Ohio, and Washington. It is an iron ore. The mineral was at one time called *chalybite*.

SIEVE (Molecular). Materials with special porous and absorbing properties which chemically lock materials into their pores, as used in purification and separation processes, are sometimes called molecular sieves. The sodium aluminum silicates (synthetic zeolites) are examples of materials of this type.

SILICATES (Soluble). The most common commercially used soluble silicates are those of sodium and potassium. Soluble silicates are systems containing varying proportions of an alkali metal or quaternary ammonium ion and silica. The soluble silicates can be produced over a wide range of stoichiometric and nonstoichiometric compositions and are distinguished by the ratio of *silica to alkali*. This ratio is generally expressed as the *weight percent ratio* of silica to alkali-metal oxide (SiO_2/M_2O). With lithium and quaternary ammonium silicates, the *molar ratio* is used.

Sodium silicates find wide application in many types of detergents and cleaning compounds and have been used for many years as adhesives and cements. Both sodium and potassium silicates are important bonding agents in a large variety of ceramic cement and refractory applications, notably because of their heat stability and resistance to chemicals. Alkali-metal silicate bonds are used in high-temperature ceramic products in the fabrication of electrical components. Soluble silicates find wide application for pelletizing, granulating, and briquetting finely divided particles, such as clays, fertilizers, and ores. Sodium silicates also are used as bonding materials for foundry mold and core compositions. Because of their adherence properties, soluble sodium and potassium silidates are widely used as coatings. Frequently, sodium silicates are used to protect against water-line corrosion in tanks. The ability to form sols and gels is an interesting and very useful characteristic of soluble silicates. Silica gels are used in a major way as desiccants and as carriers for the production of petroleum-cracking catalysts, as well as raw materials in the manufacture of zeolites. Activated sols are used in water-clarification.

Generally, sodium and potassium silicates are made by fusion of pure sand with alkali-metal carbonate or alkali-metal sulfate and carbon, this operation carried out in large open-hearth furnaces heated to a temperature range of 1300° to 1500°C. The resulting glasses may be used in this form, or dissolved in water to produce silicate solutions. Sodium and potassium silicate solutions also can be made by dissolving sand in sodium or potassium hydroxide solution at elevated temperatures and pressures. Lithium silicate glasses, although insoluble in water, can be made by dissolving silica gel in, or mixing silica sols with, lithium hydroxide solutions. Anhydrous sodium metasilicate is made from the anhydrous melt. This salt crystallizes rapidly from its aqueous solution at temperatures in the range of 80–85°C.

The most important property of sodium and potassium silicate glasses and hydrated amorphous powders is their solubility in water. The dissolution of vitreous alkali is a two-stage process. In an ion-exchange process between the alkali-metal ions in the glass and the hydrogen ions in the aqueous phase, the aqueous phase becomes alkaline, due to the excess of hydroxyl ions produced, while a protective layer of silanol groups is formed on the surface of the glass. In the second phase, a nucleophilic depolymerization similar to the base-catalyzed depolymerization of silicate micelles in water takes place.

When sodium silicate solutions of intermediate ratios are concentrated to a thick gum, they become very sticky and tacky. This property is important to many of the adhesive applications. It is related to high cohesion and low surface tension rather than primarily to viscosity.

The stability of soluble silicate solutions depends strongly on pH and concentration. The addition of acids and acid-forming compounds gives rise to the formation of silica gels. Soluble alkali-metal silicate solutions are not compatible with most organic water-miscible solvents. The addition of alcohols and ketones causes phase separation into liquid layers. A few organic systems, particularly polyols, such as glycols, glycerins, sugars, and polyethylene glycols, are compatible and miscible with alkali-metal silicate solutions.

References

Schweiker, G. C.: "Sodium Silicates," *JAOCS*, **55**, 1, 36–40 (1978).
Weldes, H. H., and K. R. Lange: "Properties of Soluble Silicates," *Ind. Eng. Chem.*, **61**, 29–44 (1969).
Weldes, H. H.: "Polysilicates as Detergent Builders," *Soap/Cosmetic/Chemical Specialties*, **48**, 4, 72–78 (1972).

SILICON. Chemical element symbol Si, at. no. 14, at. wt. 28.086, periodic table group 4a, mp 1408–1412°C, bp 2355°C, density, 2.242 g/cm^3 (solid crystalline at 20°C), 2.32 g/cm^3 (single crystal at 20°C). Elemental silicon has a face-centered cubic crystal structure (diamond structure). The existence of a hexagonal form of silicon with a wurtzite-type structure and with lattice parameters of $a = 3.80$ Å and $c = 6.28$ Å was established in 1963 (*Wentorf-Kasper*). Claims to different parameters were made by Jennings-Richman (1976).

The common form of silicon is a dark-gray, hard solid. It can be obtained as a brown microcrystalline powder, which is not an allotrope of the gray form. Both forms are unaffected by air at ordinary temperatures, but when heated in air to high temperatures, a protective layer of oxide is formed. Silicon reacts with nitrogen at high temperatures to form the nitride; with chlorine to form the chloride; with several metals to form silicides. Crystalline silicon is unattacked by HCl or HNO$_3$, or H$_2$SO$_4$, but is attacked by hydrofluoric acid to form silicon tetrafluoride gas. Silicon is soluble in NaOH solution, forming sodium silicate and hydrogen gas. Silicon reacts with dry chlorine to form silicon tetrachloride.

There are three naturally occurring isotopes ^{28}Si through ^{30}Si, and three radioactive isotopes have been identified ^{27}Si, ^{31}Si, and ^{32}Si. The latter isotope has a half-life of approximately 700 years, while the half-lives of the other two are short, measured in terms of seconds and hours.

Lavoisier showed in 1787 that SiO$_2$ was not a single element and indicated that it was the oxide of a hitherto unknown element. In the early 1800s, Scheele, Davy, Gay-Lussac, and Thénard attempted to isolate the element, but were not successful. In 1871, Berzelius discovered silicon in a cast-iron melt and, in 1823, succeeded in isolating the element by reduction of potassium fluorosilicate with potassium. Small laboratory amounts were produced by H. E. Sainte-Claire Deville in 1854 and by C. Winkler in 1864. It was not until 1900 that the effective properties of silicon as a deoxidizing agent for steel production were observed. Shortly thereafter, ferrosilicon alloys, using quartzite, coke, and iron pellets, were produced in electric refining furnaces of the type already in use for making calcium carbide. With this technique, it was possible to produce silicon of about 98% purity. It remained for the rigid purity requirements of semiconductors many years later before silicon of higher purities was produced.

Silicon is ranked second in the order of chemical elements appearing in the earth's crust—an average of 27.72% occurring in igneous rocks. It is estimated that a cubic mile of seawater contains about 15,000 tons of silicon (3240 metric tons in a cubic kilometer). In terms of abundance throughout the universe, silicon is ranked seventh. First ionization potential, 8.149 eV; second, 16.27 eV; third, 33.30 eV; fourth, 44.95 eV. Oxidation potentials: $Si + 2H_2O \rightarrow SiO_2 + 4H^+ + 4\bar{e}$, 0.86 V; $Si + 6OH^- \rightarrow SiO_3^{2-} + 3H_2O + 4\bar{e}$, 1.73 V. Other physical properties of silicon are given under **Chemical Elements.**

Because of its chemical reactivity, silicon does not occur in elemental form in nature. The element is present in igneous rocks and clays asalumino-silicate; as the oxide SiO$_2$ in quartz, sand, flint, and the gems amethyst, jasper, chalcedony, agate, onyx, tridymite, opal, crystobalite; as silicates in zircon (zirconium silicate, ZrSiO$_4$), in willemite (zinc silicate, Zn$_2$SiO$_4$), in wollastinite (calcium silicate, CaSiO$_3$), in serpentine (magnesium silicate, Mg$_3$Si$_2$O$_7$). Impure (up to 98% Si) silicon is obtained from the oxide (1) by igniting with aluminum powder, or (2) by reduction with carbon in an electric furnace. See also **Cancrinite.**

Silicon Production for Alloys. The production of raw steel requires about 1.6–1.7 kilograms of silicon per metric ton of steel. The silicon is used in the form of ferrosilicon, which contains about 20% silicon. It is estimated that about 3 million metric tons of ferrosilicon are consumed annually in steel-making. The 20% silicon content ferrosilicon can be made in a conventional blast furnace. Ferrosilicons with higher silicon contents (45, 75, 90, 98%) must be produced in electric furnaces. The raw materials are pure quartzites. The presence of impurities, such as Al$_2$O$_3$ and CaO, interfere with the melting process because of the formation of dross. The reducing agent used is chemical coke. For very high concentrations of silicon (90–98%), ash-free petroleum coke or charcoal is used. Iron is added in the form of small pellets or chips in the production of the 45–75% silicon alloys.

For certain metal alloys, a calcium silicon alloy is required. This alloy also is used as a steel deoxidizer and is favored because it forms a low-melting-point calcium silicate product. A representative composition of the alloy is: 30–33% Ca, 60–64% Si, 3–5% Fe, 1–2% Al, 0.3–0.6% C, and less than 0.15% S and P.

Silicon is used in the primary and secondary aluminum industry. The purity of silicon for metallurgical purposes ranges from 96.7 to 98.5% silicon, 0.10–0.75% aluminum, 0.03–0.04% calcium, with the remainder being principally iron.

Silicon Carbide. This compound is an important industrial abrasive, having a hardness of 9.5 on the Mohs scale. In this compound, each silicon atom is surrounded tetrahedrally by four carbon atoms, and similarly, each carbon atom is surrounded by four silicon atoms. Silicon carbide is made by reducing pure quartz (glass-sand) with petroleum coke in an electric-resistance type furnace, known as the Acheson process. The product is hexagonal crystals ranging from light-green to black. It is used as a ceramic raw material for dross-repellent linings as well as for many abrasive applications.

Super- or Hyperpure Silicon. For semiconductor use, there can be only one atom of impurity for every 100,000 silicon atoms! The starting material for the manufacture of hyperpure silicon is silicon tetrachloride $SiCl_4$ or trichlorosilane $SiCl_3H$. Both of these materials can be reduced with hydrogen to yield a compact deposition of silicon on hot surfaces, ranging from 800–1,200°C. The starting compounds are purified of boron and phosphorus by fractional distillation and absorption techniques. The process hydrogen is purified by passing it through molecular sieves under high pressure, followed by absorption techniques at a low temperature (−190°C). With the highly purified starting ingredients, an excess of hydrogen is circulated through heated quartz tubes. Or, the gas mixture may be blown into quartz bell jars, whereupon the silicon is deposited on filaments of tantalum or tungsten or on thin rods of hyperpure silicon, which may be heated by electrical resistance or radiofrequency energy. This process yields polycrystalline rods of silicon which range up to about one meter in length and 150 millimeters in diameter.

To be used in semiconductor devices, the polycrystalline silicon must be converted to single crystals of a defined, predetermined type of conductivity (n or p type). The crystals must be rigidly controlled as regards their resistivity, and possess the highest degree of crystallographic perfection. The two crystal-growing techniques used are: (1) crucible free vertical float zoning which removes all residual impurities, including phosphorus, arsenic, and oxygen, but boron is essentially irremovable by float zoning, or (2) crucible pulling in which the crystals, particularly those of lower resistivity, are drawn out of a melt in a process known as the Czochralski technique. See also **Semiconductors.** Both processes must be conducted under helium or argon, often under a vacuum of 10^{-5} torr. Approximately 800 tons of polycrystalline hyperpure silicon are required by the semiconductor industry annually.

Chemistry and Compounds. Like carbon, silicon forms chiefly covalent bonds, but its greater atomic radius enables it to form positive ions more readily. Unlike carbon and tin, silicon is not allotropic, having only one elemental form, the diamond structure, in which each atom is surrounded tetrahedrally by four others to which it is covalently bonded. An apparently amorphous brown powder, produced by combusion of silane, SiH_4, has been found to be a microcrystalline variety of this covalently-bonded structure.

Silicon dioxide, SiO_2, exists in at least eleven distinct crystalline forms. Several of them are obtained by heating α-quartz, which has a number of transition points, to produce β-quartz, and to give various forms of tridymite and crystobalite. The unit of structure is the tetrahedron in which each silicon atom is covalently bonded to four oxygen atoms, and the variation is in the ways these tetrahedra are interconnected (by oxygen atoms) to form a three dimensional system.

Silicon dioxide is converted by hydrofluoric acid into silicon tetrafluoride, SiF_4, a gas. SiF_4 can also be produced directly from the elements, as can the other tetrahalides, silicon tetrachloride, $SiCl_4$ (a liquid), silicon tetrabromide, $SiBr_4$ (a liquid), and silicon tetraiodide, SiI_4 (a solid). The silicon halides hydrolyze much more readily than the carbon halides, because the unoccupied silicon 3d orbitals are energetically not far above its 3s and 3p orbitals. This fact also permits the formation of the sp^3d^2 hybrid bonds of the fluorosilicate ion, SiF_6^{2-}, and addition compounds of the halides, e.g., $SiX_4 \cdot 2$ pyridine. Silicon also is intermediate between carbon and the higher members of main group 4 of the periodic table in forming a dichloride,

$SiCl_2$, by strong heating of silicon with silicon tetrachloride.

Quartz and other forms of silica react very slightly with water to form monosilicic acid, $(SiO_2)_n + 2nH_2O \rightarrow nSi(OH)_4$. As shown, this reaction is a depolymerization followed by a hydrolysis, and proceeds rapidly with hot alkalis or fused alkali metal carbonates, yielding soluble silicates containing the SiO_4^{4-} and $(SiO_3^{2-})_n$ ions. The hydrolysis reaction is geologically important, because it is considered to be the starting point in the formation of the innumerable silicate minerals that occur so widely in nature, just as many of the silica minerals may have originated by the reverse reaction. Many of the more complex silicic acids are considered to form by polymerization of $Si(OH)_4$ molecules by sharing of —OH ions between two silicon ions (octahedrally coordinated by six hydroxyl ions) followed by condensation with the loss of water to produce —Si—O—Si— linkages. The polymerization of silicic acid is carried out industrially to produce silica gel, a stable sol of colloidal particles. The various methods involve careful removal of H_2O, the catalytic effect of acid or alkali (or fluoride ion) and controlled pH. Many varieties of silica gel have been made, including the zerogels and aerogels, in which the aqueous phase is displaced by a gaseous one.

The great number of naturally occurring silicates results, as indicated previously, from the polymerization and dehydration of monosilicic acid to form, ultimately, such groups and ions as $(Si_2O_7)^{6-}$, $(Si_3O_9)^{6-}$, $(Si_4O_{12})^{8-}$, and $(Si_6O_{18})^{12-}$. Various cations, such as those of boron, B^{3+}, Aluminum, Al^{3+}, among others, in the structure lie at the centers of anionic polyhedra having as anions the O^{2-} ions of neighboring SiO_4 tetrahedra, in which each Si—O bond has an electrostatic bond strength of one. Cations of lower charge density, on the other hand, like sodium (Na^+), potassium (K^+), calcium (Ca^{2+}), and others, are located interstitially. The great variety of the silicates is due to the considerable degree of isomorphism, exhibited not only by elements of the same group, but by elements of different groups, whereby they partly replace each other in the complex silicates, and by no means necessarily in stoichiometric proportions. Thus troosite may be represented by the formula $(Zn, Mn)_2SiO_4$, chrysolite by $6Mg_2SiO_4 \cdot Fe_2SiO_4$, and vermiculite by $(Mg, Fe)_3(AlSi)_4O_{10} \cdot 4H_2O$, even the silicon in vermiculite being partly replaced (by aluminum). One method of classifying the silicates is based upon the linking of the SiO_4 tetrahedra:

A. Discrete silicate radicals.
   1. Single tetrahedra (SiO_4^{4-}), e.g., phenacite, Be_2SiO_4.
   2. Two tetrahedra $(Si_2O_7)^{6-}$, e.g., hardystonite, $Ca_2ZnSi_2O_7$.
   3. Three tetrahedra $(Si_3O_9)^{6-}$, e.g., benitoite, $BaTiSi_3O_9$.
   4. Four tetrahedra $(Si_4O_{12})^{18-}$, e.g., axinite, (Fe, Mn)Ca_2Al_2 $BO_3Si_4O_{12}$.
   5. Six tetrahedra $(Si_6O_{18})^{12-}$, e.g., beryl, $Be_3Al_2Si_6O_{18}$.
B. Silicon-oxygen chains of indefinite length.
   1. Single chains with one silicon atom to three oxygen atoms, e.g., diopside, $CaMg(SiO_3)_2$.
   2. Double chains (Si:O = 4:11), e.g., tremolite, Ca_2Mg_5 $(Si_4O_{11})_2(OH)_2$.
C. Silicon-oxygen sheets. (Si:O = 2:5), e.g., talc, $Mg_3(SiO_5)_2$ $(OH)_2$.
D. Silicon-oxygen spatial networks.
   1. Composition SiO_2 (composed of interlinked SiO_4 tetrahedral), e.g., quartz, SiO_2.
   2. Composition $M_n(Si, Al)_nO_{2n}$, e.g., feldspar, KSi_3AlO_8. These are probably based upon silicon and aluminum tetrahedra, variously linked.

The increasingly large number of silicon compounds produced by industrial processes may be systematized about the silanes and their substitution products, just as the silicates are about the SiO_4 tetrahedron. Silicon, like carbon, forms a number of hydrides, though their number is much more limited. The silane series, analogous to the paraffin hydrocarbons, has at least six members, silane (SiH_4), disilane (H_3Si—SiH_3) . . . hexasilane

$$(H_3Si$—$SiH_2$—$SiH_2$—$SiH_2$—$SiH_2$—$SiH_3).$$

They are increasingly unstable, hexasilane dissociating at room temperature. They are halogenated with free halogens to form substituted silanes, and catalytically with the hydrogen halides. The halosilanes react with NH_3 to form silylamines or silazanes and are hydrolyzed by water to form siloxanes. Prosiloxane, H_2SiO, polymerizes readily but disiloxane, H_3Si—O—SiO_3, and the higher siloxanes, although they polymerize, can readily be studied. They have properties like the ethers and other analogous carbon compounds. Hydrogen-containing siloxanes, such as HO_2Si—SiO_2H are also known and polymerize readily. There are also ring siloxanes, such as siloxen, which as a polymerized structure of epoxy form (a powerful reducing agent)

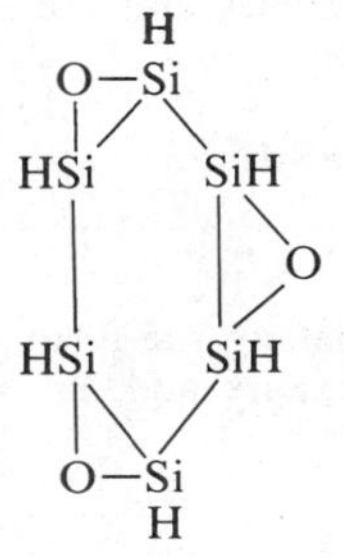

H
O—Si
HSi   SiH
          O
HSi   SiH
O—Si
H

Silyl and polysilyl radicals also combine with nitrogen, arsenic, and other main group 5 elements, as with sulfur and selenium.

The silazanes are such compounds of silicon, nitrogen and hydrogen of the general formula $H_3Si(NHSiH_2)_n NHSiH_3$, being called disilazane, trisilazane, etc., according to the number of silicon atoms present. (In disilazane, n in the above formula has a value of 0, in trisilazane it is 1, etc.)

The silthianes are sulfur compounds having the general formula $H_3Si(SSiH_2)_n SSiH_3$ which are called disilthiane, trisilthiane, etc., according to the number of silicon atoms present. They have the generic name *silthianes*. (In disilthiane, n in the above formula has a value of 0, in trisilthiane, a value of 1, etc.)

Silicones. These are semiorganic polymers with a quartzlike structure in which various organic groups are attached to the silicon atom. By varying the kind and number of organic groups, a variety of materials ranging from liquids through gels and elastomers to rigid solids (resins) can be produced. See also **Silicones.**

The organosilicon compounds may be regarded as substituted silanes, although of course their preparation is not usually in this way. Thus, ethyl silicate, $Si(OC_2H_5)_4$, is prepared from silicon tetrachloride and ethyl alcohol, and tetraethyl silane, $Si(C_2H_5)$ is prepared from silicon tetrachloride and diethylzinc.

The silicon-carbon bond, unlike the carbon-carbon bond, has about 12% of ionic character, varying somewhat with the atoms or groups attached to the two atoms. Other types of organosilicon compounds include the esters, the alkoxyhalosilanes, the higher tetra-alkylsilanes (prepared from silicon tetrachloride and Grignard reagents), the alkylsilanes (H partly replaced by R), the alkylhalosilanes, the alkylalkoxysilanes, the alkylsilylamines, some aryl compounds of the foregoing types, and many related

derivatives of disilane and the polysilanes. Other types of compounds are those having silicon-carbon chains and the organosiloxane compounds, for which the name "silicones" is often used. These are essentially chains or networks of groups,

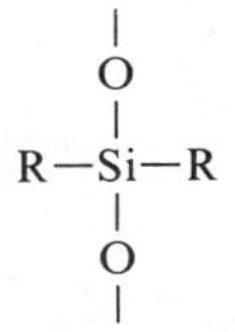

joined by oxygen atoms attached to the silicon atoms as shown. There are many other groups of silicon compounds, as well as individual ones.

Aluminates. Many complex silicoaluminates or aluminosilicates are formed in nature. Of these, clay in more or less pure form (pure clay, kaolinite; kaolin, china clay, $H_4Si_2Al_2O_9$ or $Al_2O_3 \cdot 2SiO_2 \cdot 2H_2O$) is of great importance. Clay is formed by the weathering of igneous rocks, and is used in the manufacture of bricks, pottery, porcelain, Portland cement. Sodium aluminosilicate is used in water purification to remove dissolved calcium compounds.

Fluosilicate. Sodium fluosilicate Na_2SiF_6, white solid slightly soluble; magnesium fluosilicate $MgSiF_6$, white solid, soluble.

Sulfides. Silicon monosulfide SiS, yellow solid, somewhat volatile, formed by heating to redness crystalline silicon in sulfur vapor, reactive with water; silicon disulfide SiS_2, white crystals, formed by heating amorphous silicon and sulfur, and then subliming, reactive with water.

Nitrides. Trisilicon tetranitride, Si_3N_4, by heating silicon oxide plus carbon to 1500°C in a current of nitrogen gas.

Silicates. See **Adhesives;** and **Silicates (soluble).**

Nomenclature of Silicon Compounds. The name of the compound SiH_4 is *silane*. Compounds having the general formula $H_3Si \cdot [SiH_2]_n \cdot SiH_3$ are called *disilane, trisilane*, etc., according to the number of silicon atoms present. Compounds of the general formula Si_nH_{2n-2} have the generic name *silanes*. Example: Trisilane, $H_3Si \cdot SiH_2 \cdot SiH_3$.

Compounds having the formula $H_3Si \cdot [NH \cdot SiH_2]_n \cdot NH \cdot SiH_3$ are called disilazane, trisilazane, etc., according to the number of silicon atoms present; they have the generic name silazanes. Example: Trisilazane $H_3Si \cdot NH \cdot SiH_2 \cdot NH \cdot SiH_3$.

Compounds having the formula $H_3Si \cdot [S \cdot SiH_2]_n \cdot S \cdot SiH_3$ are called disilthiane, trisilthiane, etc., according to the number of silicon atoms present; they have the generic name silthianes. Example: Trisilthiane $H_3Si \cdot S \cdot SiH_2 \cdot S \cdot SiH_3$.

Compounds having the formula $H_3Si \cdot [O \cdot SiH_2]_n \cdot O \cdot SiH_3$ are called disiloxane, trisiloxane, etc., according to the number of silicon atoms present; they have the generic name siloxanes. Example: Trisiloxane $H_3Si \cdot O \cdot SiH_2 \cdot O \cdot SiH_3$.

For designating the positions of substituents on compounds named as silanes, silazanes, silthianes, and siloxanes, each member of the fundamental chain is numbered from one terminal silicon atom to the other. When two or more possibilities for numbering occur, the same principles are followed as for carbon compounds. Examples:

1 Butyl-2,3-dichloro-2-pentyltrisilane
$Cl \cdot SiH_2 \cdot SiCl(C_5H_{11}) \cdot SiH_2 \cdot C_4H_9$
2-Methyl-3-pentyloxytrisilazane
$SiH_3 \cdot N(CH_3) \cdot SiH(OC_5H_{11}) \cdot$
$NH \cdot SiH_3$
1-Methoxytrisiloxane
$CH_3O \cdot SiH_2 \cdot O \cdot SiH_2 \cdot O \cdot SiH_3$

The names of representative radicals containing silicon are shown below. These illustrate the principles on which any further radical names should be formed.

Silicon, hydrogen

silyl	H_3Si-
silylene	$H_3Si=$
silylidyne	$HSi\equiv$
disilanyl	$H_3Si\cdot SiH_2-$
trisilanyl	$H_3Si\cdot SiH_2\cdot SiH_2-$
disilanylene	$-SiH_2\cdot SiH_2-$
trisilanylene	$-SiH_2\cdot SiH_2\cdot SiH_2-$
cyclohexasilanyl	$\begin{array}{c} SiH_2\cdot SiH_2\cdot SiH \\ \mid \qquad\qquad \mid \\ SiH_2\cdot SiH_2\cdot SiH_2 \end{array}$

Silicon, hydrogen, oxygen

siloxy	$H_3Si\cdot O-$
disiloxanyl	$H_3Si\cdot O\cdot SiH_2-$
disilanoxy	$H_3Si\cdot SiH_2\cdot O-$
disiloxanoxy	$H_3Si\cdot O\cdot SiH_2\cdot O-$

Silicon, hydrogen, sulfur

silylthio	$H_3Si\cdot S-$
disilanylthio	$H_3Si\cdot SiH_2\cdot S-$
disilthianyl	$H_3Si\cdot S\cdot SiH_2-$
disilthianylthio	$H_3Si\cdot S\cdot SiH_2\cdot S-$

Silicon, hydrogen, sulfur, oxygen

disilthianoxy	$H_3Si\cdot S\cdot SiH_2\cdot O-$
disiloxanylthio	$H_3Si\cdot O\cdot SiH_2\cdot S-$

Silicon, hydrogen, nitrogen

silylamino	$H_3Si\cdot NH-$
disilanylamino	$H_3Si\cdot SiH_2\cdot NH-$
disilazanyl	$H_3Si\cdot NH\cdot SiH_2-$
disilazanylamino	$H_3Si\cdot NH\cdot SiH_2\cdot NH-$

Silicon, hydrogen, nitrogen, oxygen

disilazanoxy	$H_3Si\cdot NH\cdot SiH_2\cdot O-$
disiloxanylamino	$H_3Si\cdot O\cdot SiH_2\cdot NH-$

Compound radical names may be formed in the usual manner. Examples:

Silyldisilanyl	$(H_2Si)_2SiH-$
Disilyldisilanyl	$(H_3Si)_3Si-$
Triphenylsilyl	$(C_6H_5)_3Si-$

Open-chain compounds which have the requirements for more than one of the structures already defined are named, if possible, in terms of the silane, silazane, silthiane, or siloxane containing the largest number of silicon atoms. Examples:

3-Siloxytrisilthiane
$$H_3Si\cdot S\cdot SiH\cdot S\cdot SiH_3$$
$$\mid$$
$$O\cdot SiH_3$$

1-Siloxy-3-(disilthianoxy)trisilthiane
$$H_3Si\cdot S\cdot SiH\cdot S\cdot SiH_2\cdot O\cdot SiH_3$$
$$\mid$$
$$O\cdot SiH_2\cdot S\cdot SiH_3$$

When there is a choice between two parent compounds possessing the same number of silicon atoms, the order of precedence is siloxanes, silthianes, silazanes, and silanes. Examples:

1-Silylthiodisiloxane
$$SiH_3\cdot O\cdot SiH_2\cdot S\cdot SiH_3$$
1-Silylaminodisilthiane
$$SiH_3\cdot S\cdot SiH_2\cdot NH\cdot SiH_3$$
1-Phenyl-3-silyldisiloxane
$$SiH_3\cdot SiH_2\cdot O\cdot SiH_2\cdot C_6H_5$$

Cyclic silicon compounds having the formula $[SiH_2]_n$ are called cyclotrisilane, cyclotetrasilane, etc., according to the number of members in the ring; they have the generic name cyclosilanes. Example:

Cyclotrisilane
$$\underline{SiH_2\cdot SiH_2\cdot SiH_2}$$

Cyclic compounds having the formula $[SiH_2\cdot NH]_n$ are called cyclodisilazane, cyclotrisilazane, etc., according to the number of silicon atoms in the ring. They have the generic name cyclosilazanes. Example:

Cyclotrisilazane
$$\underline{HN\cdot SiH_2\cdot NH\cdot SiH_2\cdot NH\cdot \qquad\qquad SiH_2}$$

Cyclic compounds having the formula $[SiH_2\cdot S]_n$ have the generic name cyclosilthianes and are named similarly to the cyclosilazanes. Example:

Cyclotrisilthiane
$$\underline{S\cdot SiH_2\cdot S\cdot SiH_2\cdot S\cdot \qquad\qquad SiH_2}$$

Cyclic compounds having the formula $[SiH_2\cdot O]_n$ have the generic name cyclosiloxanes and are named similarly to the cyclosilazanes. Example:

Cyclotrisiloxane
$$\underline{O\cdot SiH_2\cdot O\cdot SiH_2\cdot O\cdot \qquad\qquad SiH_2}$$

Cyclosilanes, cyclosilazanes, cyclosilthianes, and cyclosiloxanes are numbered in the same way as carbon compounds of similar nature. Examples:

2-Methoxycyclotrisilazane
$$\underline{HN\cdot SiH_2\cdot NH\cdot SiH_2\cdot NH\cdot SiH\cdot OCH_3}$$

2-Methoxycyclotrisilthiane
$$\underline{S\cdot SiH_2\cdot S\cdot SiH_2\cdot S\cdot SiH\cdot OCH_3}$$

2-Methoxycyclotrisiloxane
$$\underline{O\cdot SiH_2\cdot O\cdot SiH_2\cdot O\cdot SiH\cdot OCH_3}$$

Polycyclic siloxanes (polycyclic compounds whose members consist entirely of alternating silicon and oxygen atoms) are named as bicyclosiloxanes, tricyclosiloxanes, etc., or as spirosiloxanes, and are numbered according to methods in use for carbon compounds of similar nature. Polycyclic silthianes, silazanes, and silanes are treated similarly. Examples:

3,3,5,5,9,9-Hexamethyl-1,7-diphenylbicyclo[5,3,1]pentasiloxane

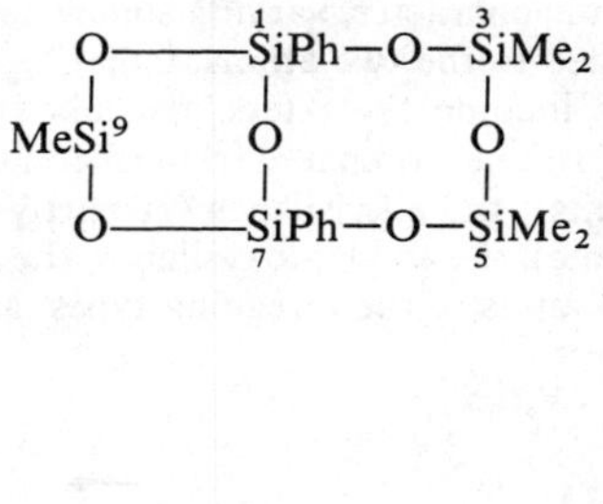

$$
\begin{array}{ccccc}
O & \text{---} & \overset{1}{Si}Ph & \text{---}O\text{---} & \overset{3}{Si}Me_2 \\
\mid & & \mid & & \mid \\
\overset{9}{Me Si} & & O & & O \\
\mid & & \mid & & \mid \\
O & \text{---} & \underset{7}{Si}Ph & \text{---}O\text{---} & \underset{5}{Si}Me_2
\end{array}
$$

Tetramethyltricyclo[3,3,1,1]tetrasiloxane

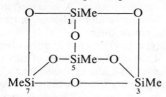

The names of compounds containing silicon atoms as hetero-members (with or without other hetero-members) but not classifiable as (linear or cyclic) silanes, silazanes, silthianes or siloxanes are derived with the aid of the oxa-aza convention. Examples:

2,2,4,4,6,6-Hexamethyl-2,4,6-trisilaheptane
$(CH_3)_3Si \cdot CH_2 \cdot Si(CH_3)_2 \cdot CH_2 \cdot Si(CH_3)_3$
2,4,6,8,-Tetraoxa-5-carbanonasilane
$SiH_3 \cdot O \cdot SiH_2 \cdot O \cdot CH_2 \cdot O \cdot SiH_2 \cdot O \cdot SiH_3$

Octaphenyloxacyclopentasilane

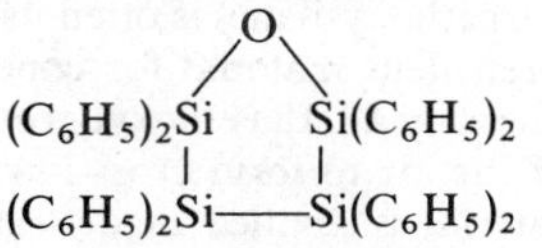

$$
\begin{array}{c}
O \\
(C_6H_5)_2Si \quad Si(C_6H_5)_2 \\
| \qquad\qquad | \\
(C_6H_5)_2Si \!-\!\!-\! Si(C_6H_5)_2
\end{array}
$$

Hydroxy-derivatives in which the hydroxyl groups are attached to a silicon atom are named by adding the suffixes ol, diol, triol, etc., to the name of the parent compound. Examples:

Silanol	$H_3Si \cdot OH$
Silanediol	$H_2Si(OH)_2$
Silanetriol	$HSi(OH)_3$
Disilanehexaol	$(HO)_3Si \cdot Si(OH)_3$
Disiloxanol	$H_3Si \cdot O \cdot SiH_2 \cdot OH$
Cyclohexasilanol	$\begin{array}{c} SiH_2 \cdot SiH_2 \cdot SiH \cdot OH \\ | \qquad\qquad | \\ SiH_2 \cdot SiH_2 \cdot SiH_2 \end{array}$

Polyhydroxy-derivatives in which hydroxyl group is attached to a silicon atom are named wherever possible in accordance with the principle of treating like things alike. Example:

1,1,3,5,5-Pentamethyltrisiloxane-1,3,5-triol

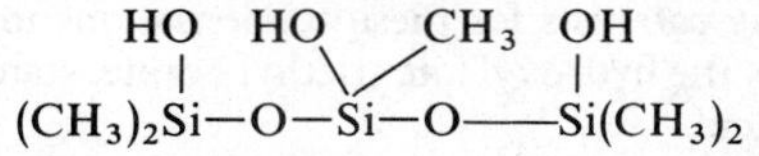

$$
\begin{array}{cccc}
HO & HO & CH_3 & OH \\
| & | & | & | \\
(CH_3)_2Si\!-\!O\!-\!Si\!-\!O\!-\!\!-\!Si(CH_3)_2
\end{array}
$$

Otherwise they are named in accordance with the principle of the largest parent compound. Example:

2-Hydroxysilyltetrasilane-1,4-diol

$$
\begin{array}{c}
SiH_2 \cdot OH \\
| \\
HO \cdot SiH_2 \cdot SiH_2 \cdot SiH \cdot SiH_2 \cdot OH
\end{array}
$$

Substituents other than hydroxyl groups (functional atoms or groups and hydrocarbon radicals) attached to silicon are expressed by appropriate prefixes or suffixes. Examples:

Ethyldisilane
$CH_3 \cdot CH_2 \cdot SiH_2 \cdot SiH_3$
Hexachlorodisiloxane
$Cl_3Si \cdot O \cdot SiCl_3$
Dibutyldichlorosilane
$(CH_3 \cdot CH_2 \cdot CH_2 \cdot CH_2)_2SiCl_2$

Silylamine
$H_3Si \cdot NH_2$
Silanediamine
$H_2Si(NH_2)_2$
Silanetriamine
$HSi(NH_2)_3$
N-Methylsilylamine
$H_3Si \cdot NH \cdot CH_3$
NN-Dimethylsilylamine
$H_3Si \cdot N(CH_3)_2$
NN'-Dimethylsilanediamine
$H_2Si(NH \cdot CH_3)_2$
NN'N''-Trimethylsilanetriamine
$HSi(NH \cdot CH_3)_3$
Acetoxytrimethylsilane
$(CH_3)_3Si \cdot O \cdot OC \cdot CH_3$
Diacetoxydimethylsilane
$(CH_3)_2Si(O \cdot OC \cdot CH_3)_2$

Compounds containing carbon as well as silicon and in which there is a "reactive group" in the carbon-containing portion of the molecule not shared by a silicon atom are named in terms of the organic parent compound wherever feasible. Examples:

α-Trimethylsilylacetanilide
$(CH_3)_3Si \cdot CH_2 \cdot NH \cdot C_6H_5$
1-Trichlorosilylethanol
$Cl_3Si \cdot CH(OH) \cdot CH_3$
2-Trimethylsilylethanol
$(CH_3)_3Si \cdot CH_2 \cdot CH_2 \cdot OH$
(Hydroxydimethylsilyl)methanol

$$
\begin{array}{c}
(CH_3)_2Si \cdot CH_2 \cdot OH \\
| \\
OH
\end{array}
$$

α-(Hydroxydimethylsilyl)acetanilide

$$
\begin{array}{c}
(CH_3)_2Si \cdot CH_2 \cdot CO \cdot NH \cdot C_6H_3 \\
| \\
OH
\end{array}
$$

(Silylmethyl)amine
$H_3Si \cdot CH_2 \cdot NH_2$
But by rules 70.16 and 70.17:
(Methoxymethyl)silanol
$CH_3O \cdot CH_2 \cdot SiH_2 \cdot OH$
N-Methylsilylamine
$H_3Si \cdot NH \cdot CH_3$

Compounds in which metals are combined directly with silicon are, in general, named as derivatives of the metal. Example:

(Triphenylsilyl)lithium
$(C_6H_5)_3SiLi$

However, in exceptional cases, the metal may be named as a substituent. Example:

Sodium p-(sodiosilyl)benzoate
$p\text{-}NaO_2C \cdot C_6H_4 \cdot SiH_2Na$

Metallic salts of hydroxy-derivatives may be named in the customary manner. Example:

Sodium salt of triphenylsilanol
$(C_6H_5)_3Si \cdot ONa$

References

Aufderhaar, H. C.: "Silicon," in "Metals Handbook," 9th Edition, Vol. 2, American Society for Metals, Metals Park, Ohio (1979).

Cotton, F. A., and G. Wilkinson: "Advanced Inorganic Chemistry," 2nd Edition, Wiley, New York (1962).

Glang, R., and E. S. Wajda: "The Art and Science of Growing Crystals," Wiley, New York (1966).

Haberecht, R. R., and E. L. Kern, Editors: "Semiconductor Silicon," Electrochemical Society, New York (1969).

Jennings, H. M., and M. H. Richman: *Science*, **193**, 1242 (1976).

Kasper, J. S., and R. H. Wentorf, Jr.: "Hexagonal (Wurtzite) Silicon," *Science*, **197**, 599 (1977).

Kubaschewski, Evans, and Alcock: "Metallurgical Thermochemistry," 4th Edition, Pergamon, New York (1967).

Wentorf, R. H., Jr., and J. S. Kasper: *Science*, **139**, 338 (1963).

SILICON CHIP (Fabrication).   Microstructure Fabrication (and Chemistry).

SILICONE RESINS.   The chemistry of the silicones is based on the hydrides, or silanes, the halides, the esters, and the alkyls or aryls. The silicon oxides are composed of networks of alternate atoms of silicon and oxygen so arranged that each silicon atom is surrounded by four oxygen atoms and each oxygen atom is attached to two independent silicon atoms:

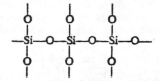

Such a network can be described as a series of spiral silicon-oxygen chains crosslinked with each other by oxygen bonds. If some of the oxygen atoms are replaced with organic substituents, a linear polymer will result:

$$-Si-O-Si-O-Si-O-$$

$$R = CH_3$$

Taking into consideration the stability of structures involving C—Si bonds, it is evident that the basic chain itself must be comparable in its stability to that of silica and the silicate minerals, and if the R substituents contain no carbon-to-carbon bonds, as for example with methyl groups, the combination should have excellent thermal stability and chemical resistance.

Among the most efficient of silicone monomers in early use as building blocks in the preparation of silicone resins are the halogen alkyl or aryl silanes. Compounds of the type represented by the formula R_2SiCl_2 are capable of undergoing hydrolysis to form long-chain polymers of varying consistencies and viscosities with a predetermined number of molecules of R_3SiCl as chain stoppers. If cross links are desired, tri-functional compounds such as $RSiCl_3$ can be used.

Many silicone resins have been prepared through the use of silazine monomers, that is, compounds with amino groups attached directly to silicon. Di-2-pyridyldichlorosilane has also been described as an intermediate in the preparation of oils, emulsifying agents and resins. Fluorinated aromatic rings are found in many silicone resins. The esterification of dimethylbis-(p-carboxylatophenyl)-silane with glycol or glycerol yields thermoplastic materials. Trimethyl-p-hydroxyphenylsilane is acceptable as a monomer in resin formation when compounded with hexamethylenetetramine. Trichlorosilane is often a constituent of cohydrolysis monomeric mixtures.

Emulsifying agents have been prepared from quaternary ammonium salts with silicon in the cation. There is a large number of alkyd-silicone resins.

Water-soluble metal salts of alkyltrisilanols are efficient in the reduction of surface tension. A silicone putty is made by compounding a benzene-soluble silicone polymer with silica powder and an inorganic filler.

Chlorinated alkyl or aryl groups are often found in polysiloxane resins. One of the chief advantages of this type of halogenated product lies in its reduced tendency to burn. Whereas diphenyloxosilane polymer burns readily in a flame, the introduction of one, two or three chlorines in each benzene ring progressively reduces this tendency. Physically, these resins appear in many different forms, from horny through sticky-resinous to rubber-like, depending on the conditions of combination and composition.

Some of the most important types of silicone resins, useful as coating preparations, are analogous to the alkyds. Glycerol, for instance, is allowed to react with trialkylethoxysilanes, by which reaction one or more hydroxyls of the glycerol are replaced by R_3SiO. Polysiloxane resins with terminal ethoxyl groups can be made. Synthesis of silicone resins with terminal halogen or hydroxyl is also possible, though generally the halogen disappears in hydrolysis of further processing.

Ethyl silicate (tetraethoxysilane) is often used without modification as a water-repellent material for concrete and masonry in general. All, or nearly all, the ethoxyl groups are hydrolyzed by the moisture of the air to form cross-linked water-repellent polymers. The material is applied in desirable thickness, dissolved in some volatile solvent which soon evaporates. Silicone resins which are partially condensed before application, or even fully condensed, can also be used here. In the latter case, hardness is achieved on evaporation of the solvent. Certain silicone resins are useful as hydrophobic agents for the impregnation of paper and fabrics.

The simplest silicone resins are formed by the almost simultaneous hydrolysis and condensation (by dehydration) of various mixtures of methylchlorosilanes. Ice water often suffices for the first step, but advanced condensation to resinous materials of satisfactory thermosetting properties generally comes about on heating. As far as solvents are concerned, water alone has its disadvantages in that the organic materials are so slightly soluble therein. Mixed solvents are commonly used, generally water with such compounds as dioxane, one of the amyl alcohols, dibutyl ether or even an aromatic hydrocarbon. Warm water hydrolysis of di-t-butyldiaminosilane forms noncrystallizable liquids or resinous products, and this resinification can be controlled. Among catalysts for these condensations may be found ferric chloride, the hydroxyl ion, triethyl borate, stannic chloride and sulfuric acid.

A water-methylene dichloride mixture is satisfactory as a hydrolyzing agent on groups of compounds such as phenyltrichlorosilane, dimethyldichlorosilane and methyltrichlorosilane. The value of higher-boiling ethers lies in their ability to provide higher-boiling reaction systems.

Ferric chloride is sometimes an important constituent of the hydrolyzing mixture. The formation of gels during polymerization can be controlled. Patents are in existence covering the hydrolysis of chlorosilanes by pouring their solutions onto the surface of a swirling solution of the active electrolyte.

Alkaline hydrolysis of dialkyldialkoxysilanes can sometimes be used for the purpose of preparing silicone resins.

Triethyl orthoborate affects polymerization by dehydration, probably reacting with the water which it abstracts, to form boric acid and alcohol. This principle is commonly used. Antimony pentachloride is used on occasion and also sulfuric acid, but there is always danger that the latter will split off an alkyl or an aryl group. Sulfuric acid also sometimes induces equilibration. This *tendency* on the part of the acid *reacts* sometimes with the opposite effect.

Some polysiloxanes are curable with lead monoxide, with a consequent reduction in both curing time and temperature. High-frequency electrical energy vulcanizes in one case at least. Zirconium naphthenate imparts improved resistance to high temperatures. Barium salts are said to prevent "blooming." Sulfur dichloride is also used. Some resins are solidified by pressure vulcanization, using di-*t*-butyl peroxide. Improvements are to be found in lower condensation temperatures and shorter times of treatment.

Viscosity is often regulated by bubbling air through solutions of polysiloxanes or the liquid material itself. In this manner, alkyl side chains are oxidized and oxygen bridges set up between silicon atoms. Obviously, the greater the number of such cross links, other influences constant, the greater will be the viscosity.

Addition of glycerol, phthalic anhydride and "butylated melamine formaldehyde resins" is sometimes found to improve the thermosetting properties of silicone resins. Methylsilyl triacetate has the same effect in certain cases. Some silicone resins can be advantageously modified by the addition of polyvinyl acetyl resins or nitroparaffins.

A solution of cellulose nitrate in butyl acetate, diluted with toluene, can be plasticized with dibutyl phthalate and tetraethoxysilane. After application to glass, the lacquer cannot be stripped from the surface even after soaking in boiling water. It has been stated, however, that tetraethoxysilane sometimes decreases the tensile strength of a lacquer in spite of its effect in increasing adhesive properties. A uniformly lustrous appearance is imparted to compounds containing plasticized ethylcellulose, cellulose fibers and pigment, by applying liquid polymerized alkyl polysiloxanes with an average radical/silicon ratio of between 1185 and 2.20. Several resinous materials are known which contain sulfur connected to carbon but not directly to silicon.

Inorganic fillers include titanium dioxide, "Celite" and zinc oxide. Lithium or lead salts of acetic acid, stearic acid or phenol are sometimes used as fillers. Silica and alumina are also feasible. Trimethyl-β-hydroxyethylammonium bicarbonate has been used as a curing agent.

There are any number of review articles and patents covering the increase in serviceability of paints and varnishes which are admixed with silicone resins. Products suitable for use as plasticizers, paint vehicles, etc., are sometimes prepared from mixtures which include phthalic anhydride. Silicone paints, in general, show high adhesion and permit greater retention of color and tint. Of special importance here is the absence of color in the resin and the freedom from discoloration during baking and curing. At high temperatures, silicone varnishes show much higher electrical resistance than others. Paints admixed with silicone resins generally show increased resistance to alkalies and to the elevated temperatures involved in baking processes.

Treating spinnarets with silicon resins eliminates much of the plugging.

Aluminum alkoxides are successful as hardening agents.

Copolymers of adipic acid, glycerol, and 1,3-diacetoxymethyltetramethyldisiloxane, or similar compound, are sometimes used. The silicon adaptation of the alkyd resin possesses, in general, increased hardness and flexibility. In addition, there is greater stability at higher temperatures. Copolymers can also be prepared using amyldibutoxyboron. Antiknock properties are claimed for this type of product as well as increased heat resistance.

The most important development in the chemistry of silicone resins embodies the preparation of monomeric silanes with at least one alkenyl group attached to silicon. Hydrolyzable groups are also present, so that polymerization can take place in two ways—by the conventional hydrolytic processes followed by condensation and by addition polymerization on the double bond, usually through the catalytic activity of benzoyl peroxide or similar agent. Some of the products find use as textile finishes, lubricating oils, additives or molding compounds. Vinyl and allyl groups are most common, with an occasional methallyl.

Triethylsilyl acrylate can be induced to undergo hydrolysis of the ethoxyl radicals to a desired extent forming linear or cross linked polymers. Addition polymerization will also take place on the double bond of the acrylate radical. More stable monomers result from the use of allyl or vinyl groups instead of acrylates. The latter contain a silicon-oxygen-carbon linkage which is always more or less susceptible to hydrolysis.

Other copolymers of this type with vinyl acetate or vinyl butyral resins have been found satisfactory for use in the lamination of wood, glass and metals. Glass-resin adhesives are also known.

General resistance to external influences constitutes the most outstanding property of silicone resins. Ultimate failure and breakdown is probably attributable, more than do anything else, to eventual oxidation of the radicals.

Among the more recent developments in this field may be mentioned the use of certain silicone resins, particularly those containing vinyl groups, as adhesives. Others find value as wood-sealing products. A silicone-glycol copolymer has been reported with curing properties, and alkyd resins are now modified with silicones. Combination epoxide-silicone resins have been investigated. A harder type of silicone resins sometimes results from the processing of monomers containing H or olefinic group attached to silicon. Dental impression materials are coming more to the front from this source as well. Finally, more attention is being devoted to studies of the relation between chemical structure and thermodynamic properties.

Resistance to γ-radiation is especially valuable. "Mouth-tissue-simulating molding compositions" for dentures are on the market. Optical lenses prepared from silicone resins are not affected by hot climates. Mold release agents are still the subject of research, as are resins with long storage capabilities. Coating bananas with silicone resins reduces the possibility of bruising. Heat-transfer compositions must have high molecular weights. In oil wells, a new use has come to the fore in the prevention of sand flow. Transparent resins are still in demand, some of which are porous.

Compounding silicone resins with phenyl formaldehyde or melamine polymers seems to produce good results. Sulfur in the resin makes a product suitable for use as fuel hose, gaskets and gas tanks. Phosphorus is found in silicone resin coatings but the product is liable to be toxic. Titanium is found in waterproof coatings. Tin compositions have low toxicity. Some resins of high viscosity contain Si—O—N units. Crosslinking, however, is not recommended. The tendency to become brittle is ever present although a few instances are known to be contrary.

—Howard W. Post, Williamsville, New York.

SILLIMANITE. The mineral sillimanite is an aluminum silicate, the formula Al_2SiO_5 being like that of andalusite and kyanite. It is orthorhombic, usually in slender prisms, but may be fibrous or massive. Its hardness is 6.5–7.5; specific gravity,

3.23–3.27; luster, vitreous to silky; color, various shades of gray, grayish-green, and grayish-brown; transparent to translucent. It occurs in granites and gneisses as tiny prisms and aggregates, and is often associated with andalusite, cordierite, and corundum. Sillimanite has been found in Bavaria, Czechoslovakia, France, India, the Malagasy Republic, Burma, and Ceylon, the latter two localities furnishing transparent sapphire-blue gem stones. In the United States sillimanite has been found in Connecticut, New York, Pennsylvania, Delaware, North Carolina, and in California, where in Inyo County is the largest deposit in the world. This mineral was named in honor of Benjamin Silliman, for many years professor of chemistry and natural science at Yale University. Sillimanite is used in the manufacture of spark plug "porcelains" and laboratory ware.

SILOXANES. Silicon.

SILTHIANES. Silicon.

SILVER. Chemical element symbol Ag (from Latin *argentum*), at. no. 47, at. wt. 107.868, periodic table group lb, mp 961.93°C, bp approximately 2212°C, density, 10.49 g/cm³ (20°C). Elemental silver has a face-centered cubic crystal structure. Silver is a white metal, softer than copper and harder than gold. When molten, silver is luminescent and occludes oxygen, but the oxygen is released upon solidification. As a conductor of heat and electricity, silver is superior to all other metals. Silver is soluble in HNO_3 containing a trace of nitrate; soluble in hot 80% H_2SO_4; insoluble in HCl or acetic acid; tarnished by hydrogen sulfide, soluble sulfides, and many sulfur-containing organic substances (e.g., proteins); not affected by air or water at ordinary temperatures, but at 200°C, a slight film or silver oxide is formed; not affected by alkalis, either in solution or fused.

There are two stable, naturally occurring isotopes, ^{107}Ag and ^{109}Ag. In addition, there are reported to be 25 less-stable isotopes, ranging in half-life from 5 seconds to 253 days. In terms of cosmic abundance, ther estimate of Harold C. Urey (1952), using silicon as a base with a figure of 10,000, silver was asigned an abundance figure of 0.023. In terms of abundance in seawater, silver is ranked number 43 among the elements, with an estimated content of 1.5 tons per cubic mile (0.324 metric tons in the cubic kilometer) of seawater.

Electronic configuration is $1s^2 2s^2 2p^6 3s^2 3p^6 3d^{10} 4s^2 4p^6 4d^{10} 5s^1$. First ionization potential is 7.574 eV; second, 21.4 eV; third, 35.9 eV. Oxidation potentials: $Ag \rightarrow Ag^+ + e^-$, $E^0 = -0.7995$ V; $Ag^+ \rightarrow Ag^{2+} + e^-$, -1.98 V; $2Ag + OH^- \rightarrow Ag_2O + H_2O + 2e^-$, -0.344 V; $Ag_2O + 2OH^- \rightarrow 2AgO + H_2O + 2e^-$, -0.57 V; $2AgO + 2OH^- \rightarrow Ag_2O_3 + H_2O + 2e^-$, -0.74 V. Other important physical properties of silver are given under **Chemical Elements.**

Occurrence and Processing. Silver is widely distributed throughout the world. It rarely occurs in native form, but is found in ore bodies as silver chloride, or more frequently, as simple and complex sulfides. In former years, simple silver and gold-silver ores were processed by amalgamation or cyanidation processes. The availability of ores amenable to treatment by these means has declined. Most silver is now obtained as a by-product or co-product from base metal ores, particularly those of copper, lead, and zinc. Although these ores are different in mineral complexity and grade, processing is similar.

All the ores are concentrated in complex mills by selective froth flotation to produce individual copper, zinc, lead, and, infrequently, silver concentrates. The copper and lead concentrates are smelted to produce lead and copper bullions from which silver is recovered by electrolytic or fine refining. The silver bearing zinc concentrates are commonly processed by leaching and electrolytic methods. Silver is ultimately recovered as a by-product from zinc plant residues.

Canada is the leading silver mining country. Other important sources of the element are Mexico, the United States, Peru, the U.S.S.R., and Australia. A substantial portion of the total world silver supply is obtained from recycled scrap. Much of the scrap comes from photographic film, jewelry, and the electrical field. The high value of the scrap dictates accurate sampling and careful feed preparation. Efficient and fast processing is required to minimize metal losses and a tie-up of high-value materials. The highly complex nature of plant feed, with respect to physical form, chemical composition, and grade, requires use of complex and highly flexible processing procedures.

Uses of Silver. Silver in the 20th century can be classified as an industrial commodity. For most of the 19th century and earlier, silver was a monetary or coinage metal. Industrial consumption of silver is principally in photographic film, electrical contacts, batteries, and brazing alloys. Sterling silver and silver-plated copper alloys are used extensively for tableware and jewelry and other decorative art. In recent years, the field of commemorative and collector arts has become a substantial market for silver alloys, particularly, sterling silver.

The predominant place of silver salts as photographic receptors is not the result of any unusual primary sensitivity to illumination, but is due to the fact that they undergo an unusual secondary amplification process called "development." Silver salts, like the salts of many metals, when immersed in solutions of many reducing agents, are changed to metallic silver. The photographic system depends upon the fact that when certain mild reducing agents (called "photographic developers") are chosen, the rate of reduction is increased many fold if the silver salt crystals carry very small amounts of metallic silver at the developer-crystal interface. The effect produced by the original light exposure is amplified in the development process by a factor of 100 billion. Whereas new photographic or recording devices are being developed not involving silver, none yet approach the packing density of a fine-grained image possible using silver. Thus it appears that silver will be used in photographic recording for many years to come.

Among the electrical uses for silver are electrical contacts, printed circuits, and batteries. By far, the primary use is in electrical contacts where the high electrical and thermal conductivities, as well as corrosion and oxidation resistances, of silver are major reasons for its selection. Although silver has a strong tendency to weld under heavy currents, this is counteracted by alloying or by adding nonmetallic substances (such as cadmium oxide) to the silver matrix. The use of silver-cadmium oxide and silver-tungsten materials in electrical contact applications is widespread. The alloys used to improve the wear resistance and to reduce the sticking tendency of silver include silver-gold, silver-copper, silver-palladium, and silver-platinum. More complex alloys include silver-copper-nickel, silver-magnesium-nickel, silver-gold-cadmium-copper, and silver-cadmium-copper-nickel. Silver-cadmium oxide alloys are unique materials and are prepared either by combining silver and cadmium oxide by powder metal techniques or by the internal oxidation of a silver-cadmium alloy. Electrical alloys, which are impossible to combine by conventional melting, lend themselves to powder metal fabrication. Such composite structures as silver-graphite, silber-iron, and silver-tungsten are good examples of these types of materials.

In silver batteries, the silver oxide-zinc secondary battery has found its place in applications where energy delivered per unit of weight and space is of primary importance. A major disadvantage of these batteries is their high cost of relatively short life. Consequently, a large part of the silver battery market is concerned with defense, space, and some medical applications.

Prior to World War II consumption of silver in silverware and jewelry was the largest industrial use of silver. Competition from stainless steel in flatware and holloware has contributed to a decline in overall use. Most consumption of silver in silverware and jewelry is in the form of sterling silver, an alloy of silver with approximately 7.5 weight percent copper. Silver plate, which is silver electroplated on a base metal, varies widely in specification. The thickness, expressed for example in pennyweights of pure silver per gross of teaspoons can range from a low of 1 to as high as 200.

In the 1920s and 1930s, low-temperature silver-copper brazing alloys were found to be useful in copper and its alloys and iron and its alloys (including stainless steel). Silver and copper form a simple eutectic system with limited solid solubility. This system can absorb elements, such as zinc, cadmium, tin, and indium. These additions lower its melting temperature. It also can absorb higher melting elements, such as nickel or palladium. These raise its melting temperature, but may improve its wetting characteristics, corrosion resistance, and strength at elevated temperatures as well. Silver solders or brazing alloys have the ability of making joints far stronger and more durable than common soft-solder (such as lead-tin) alloys. They are used in most refrigeration systems to join copper tubing. Also, extensive use is found in the assembly of automotive parts, military components, aircraft assemblies, and other hard goods manufacture. The nominal composition of a popular brazing alloy, ASTM Classification BAg-1, is silver 45%, copper 15%, zinc 16%, and cadmium 24%.

One silver alloy containing about 70% silver, 26% tin, 3% copper, and 1% zinc is unique in that it is used extensively by dentists in combination with mercury to fill cavities in teeth. The "amalgam" manufacturers supply dentists with the alloy in the form of powder (filed, or more recently, atomized). This is mixed with mercury, using from 8 to 5 parts of mercury to 5 of alloy, and the cavity is packed. In the cavity, a metallurgical reaction takes place in which the silver-tin compound in the alloy becomes a durable silver-tin mercury compound.

Silver, its oxides, halides and other salts play important roles in chemistry. Silver is an excellent catalyst in oxidizing reactions such as in the production of formaldehyde from methanol and oxygen, ethylene oxide from ethylene and oxygen, and glyoxal from ethylene glycol and oxygen. Silver has oligodynamic properties, that is, the ability of minute amounts of silver in solution to kill bacteria. Modern technology has made use of this property in various ways, mainly as a means of purifying water.

Small amounts of silver are used annually in such diverse applications as a backing for mirrors, and in control rods for pressurized water nuclear reactors. Miscellaneous uses like this account for only a small fraction of total silver consumption.

Chemistry of Silver. Silver(I) oxide, Ag_2O, is made by action of oxygen under pressure on silver at 300°C, or by precipitation of a silver salt with carbonate-free alkali metal hydroxide; it is covalent, each silver atom (in solid Ag_2O) having two collinear bonds and each oxygen atom four tetrahedral ones; two such interpenetrating lattices constitute the structure. Silver(I) oxide is the normal oxide of silver. Silver(II) oxide, AgO, is formed when ozone reacts with silver, and thus was once considered to be a peroxide. Silver(III) oxide, Ag_2O_3, has been obtained in impure state by anodic oxidation of silver.

All of the silver(I) halides of the four common halogens are well known. The fluoride may be prepared from the elements,

the chloride by action of hydrogen chloride gas at 150°C, upon silver, and the bromide and iodide by ionic reaction in solution. The chloride, bromide, and iodide are essentially insoluble in H_2O, but the fluoride is soluble. There is also a subfluoride, Ag_2F, which may be prepared as a cathodic deposit by electrolysis of silver(I) fluoride AgF, or by evaporation of finely divided silver with silver(I) fluoride in dilute hydrofluoric acid. It is an anisotropically conducting solid and is considered to be made up in the solid state of two silver layers, metallic-bonded to each other, and ionic-covalent bonded to a single fluorine layer. It has reverse cadmium iodide structure. Silver subchloride. Ag_2Cl is made by reaction of Ag_2F and phosphorus trichloride. Silver(II) fluoride, AgF_2, made by action of fluorine upon a silver(I) halide, is a fluorinating agent or catalyst for fluorinations. The silver(I) halides vary markedly in ionicity, the values given by Pauling being AgF 70%, $AgCl$ 30%, $AgBr$ 23% and AgI 11%. This is reflected in their crystal structures and in their solubilities in water (or rather, their relative insolubility), the first three having sodium chloride structure, AgI having wurtzite structure; AgF having a molal solubility of 14, and the pK_{sp}'s of the others being 9.75, 12.27 and 16.08, respectively.

Silver differs markedly from copper in forming few oxy compounds. One of these is silver oxynitrate or silver(II, III) nitrate, which has the empirical formula, $AgO_{1.148}(NO_3)_{0.453}$, in which the average oxidation number of silver is 2.448. It is prepared by action of fluorine upon aqueous silver nitrate or is obtained as an anodic deposit by electrolysis of silver nitrate in dilute HNO_3.

Silver enters into complex formation with many ions and molecules. With halogens, the silver complexes are fewer than the copper ones. Silver chloride dissolves in HCl with the formation of such chloroargentate ions as $(AgCl_2)^-$, $(AgCl_3)^{2-}$, and possibly $(AgCl_4)^{3-}$. Complex ions with bromide, $(AgBr_2)^-$ and $(AgBr_3)^{2-}$ are more stable, as are those with iodide, than those with chloride. Complexes of the type Ag_2Cl^+, Ag_3Cl^{2+}, Ag_2Br^+, Ag_3Br^{2+}, Ag_4Br^{3+}, $Ag_2Br_6^{2-}$, Ag_2I^+, Ag_3I^{2+}, Ag_4I^{3+}, $Ag_2I_6^{4-}$, $Ag_2I_7^{5-}$ and $Ag_3I_8^{5-}$ are also known. With ammonia the ions $(Ag(NH_3)_2)^+$ and $(Ag(NH_3)_3)^+$ are definitely known and others may exist. Similar complexes are formed with amines and diamines. With cyanides, silver forms very stable complexes, the number of CN^- ions in the complex depending somewhat upon the excess of cyanide, so that $(Ag(CN)_2)^-$, $(Ag(CN)_3)^{2-}$, and $(Ag(CN)_4)^{3-}$ are definitely known. With thiosulfates, silver forms various complexes. In dilute solution, $(Ag_2(S_2O_3)_2)^{2-}$ exists, while in high concentration of $S_2O_3^{2-}$ ion, the complex $(Ag_2S_2O_3)_6)^{10-}$ has been identified. In HNO_3 solution Ag^+ is easily oxidized to Ag^{2+} by peroxydisulfate. From this solution complex compounds of dipositive silver can be prepared, which are stable because coordination radically alters the oxidation potential of $Ag(I)$ to $Ag(II)$. They include pyridine complexes such as $(Ag(py)_4) \times (NO_3)_2$. 8-Hydroxyquinoline complexes containing the ions $(Ag(oxin)_2)^{2+}$, and o-phenanthroline complexes containing the ion $(Ag(o\text{-phen})_2)^{2+}$. Silver(III) is known in the square, planar complex AgF_4^-, which has been prepared as $KAgF_4$ by direct fluorination of a mixture of potassium chloride and silver chloride. Silver(III), like Cu(III), also occurs in tellurate and periodate complexes.

Other silver compounds include: Silver chromate (Ag_2CrO_4), yellow to red to brown precipitate by reaction of silver nitrate solution and potassium chromate solution.

Silver dichromate ($Ag_2Cr_2O_7$), red precipitate by reaction of silver nitrate solution and potassium dichromate solution, changing to silver chromate upon boiling with H_2O.

Silver phosphate (Ag_3PO_4), yellow precipitate, by reaction of silver nitrate solution and disodium hydrogen phosphate solution, soluble in HNO_3 and in NH_4OH, turns dark on exposure to light.

Silver sulfate (Ag_2SO_4), white precipitate, by the reaction of silver nitrate solution and potassium sodium or ammonium sulfate solution or H_2SO_4, mp of silver sulfate 652°C.

Silver sulfide (Ag_2S), black precipitate, by reaction of silver nitrate solution with hydrogen sulfide.

Silver forms several compounds or complexes with proteins by the action of silver oxide with gelatin in alkali solution, or with albumin, or by suspension in casein solution, among other methods. Such silver-protein complexes containing from 19 to 23% Ag are known as *mild silver protein*, and are sometimes used as antiseptic solutions. They are readily soluble in water.

See the list of references at the end of the entry on **Chemical Elements**.

References

Carter, F. E.: "The Precious Metals," in *Engineering Materials Handbook* (C. L. Mantell, Edit.), 14–4 to 14–12, McGraw Hill, New York, 1958.

Friend, W. Z.: "Corrosion Resistance of Precious Metals," in *Metals Handbook*, Vol. 2, 9th Edit., 668–670, American Society for Metals, Metals Park, Ohio, 1979.

—Donald A. Corrigan, Handy & Harman, Fairfield, Connecticut.

SKUTTERUDITE. This mineral includes an isomorphous series with *smaltite-chloanthite*, essentially cobalt/nickel arsenides, (Co, Ni)As_{2-3}, crystallizing in the isometric system. The usual habit is cubic, octahedral, or cubo-octahedral. The mineral also occurs in massive and granular forms. Skutterudite has a metallic luster; hardness of 5.5–6.0; a specific gravity of 6.5. The mineral is opaque with tin-white to silver-gray color. The nickel-rich material alters surficially to annabergite (green color); the cobalt-rich material to erythrite (rose color). The streak is black. The mineral is an essential ore of cobalt and nickel.

Skutterudite is found in moderate-temperature veins, commonly associated with other cobalt/nickel minerals, e.g., cobaltite and nickeline. The mineral was named for its occurrence at Skutterud, Norway. Important ore sources are Norway, Bohemia, Saxony, Spain, France, and New South Wales, Australia. Notable occurrences are in Ontario, Canada, mainly Sudbury, South Lorrain, and Gowganda.

SLUDGE. When fresh sewage is admitted to settling tanks a certain amount of the solid matter in suspension will settle out, 50% more or less for sedimentation periods of an hour and a half or so. This collection of solids is known as fresh sludge. Such sludge will become actively putrescent in a short time and in modern treatment plants must be passed on from the sedimentation tank before this stage is reached. This may be done in two ways. The fresh sludge may be passed through the slot in an Imhoff tank to the lower-story or digestion chamber. Here, decomposition by anaerobic bacteria takes place with considerable liquefaction and reduction in volume. After the decomposition process has run its course (6–9 months), the resulting sludge is called "digested" sludge and relatively inoffensive in character. It may be disposed by drying on sludge drying beds and spreading on the land. It has little, if any, fertilizing value, being of the nature of humus. The sludge digestion chamber is operated on a periodic schedule of sludge withdrawals. Alternatively, plain sedimentation basins with mechanical equipment for continuous collection of the fresh sludge may be used. The fresh sludge, so collected, is discharged into separate sludge digestion tanks which operate on the principle of the lower story of the Imhoff tank except that by means of higher and better temperature control, the digestion cycle is much more rapid and efficient than for the Imhoff tank.

*Activated Sludge.** This is the biologically active sediment produced by the repeated aeration and settling of sewage and/or organic wastes. The dissolved organic matter acts as food for the growth of an aerobic flora. This flora produces a biologically active sludge which is usually brown in color and which destroys the polluting organic matter in the sewage and waste. The process is known as the activated sludge process.

The activated sludge process, with minor variations, consists of aeration through submerged porous diffusers or by mechanical surface agitation, of either raw or settled sewage for a period of 2–6 hours, followed by settling of the solids for a period of 1–2 hours. These solids, which are made up of the solids in the sewage and the biological growths which develop, are returned to the sewage flowing into the aeration tanks. As this cycle is repeated, the aerobic organisms in the sludge develop until there is 1000–3000 ppm of suspended sludge in the aeration liquor. After a while more of the active sludge is developed than is needed to purify the incoming sewage, and this excess is withdrawn from the process and either dried for fertilizer or digested anaerobically with raw sewage sludge. This anaerobic digestion produces a gas consisting of approximately 65% methane and 35% CO_2, and changes the water-binding properties so that the sludge is easier to filter or dry.

The activated sludge is made up of a mixture of zoogleal bacteria, filamentous bacteria, protozoa, rotifera, and miscellaneous higher forms of life. The types and numbers of the various organisms will vary with the types of food present and with the length of the aeration period. The settled sludge withdrawn from the process contains from 0.6 to 1.5% dry solids, although by further settling it may be concentrated to 3–6% solids. Analysis of the dried sludge for the usual fertilizer constituents show that it contains 5–6% of slowly available N and 2–3% of P. The fertilizing value appears to be greater than the analysis would indicate, thus suggesting that it contains beneficial trace elements and growth-promoting compounds. Recent developments indicate that the sludge is a source of vitamin B_{12}, and has been added to mixed foods for cattle and poultry.

The quality of excess activated sludge produced will vary with the food and the extent of oxidation to which the process is carried. In general, about 1 lb of sludge is produced for each lb of organic matter destroyed. Prolonged or over-aeration will cause the sludge to partially disperse and digest itself. The amount of air or more precisely oxygen that is necessary to keep the sludge in an active and aerobic condition depends on the oxygen demand of the sludge organisms, the quantity of active sludge, and the amount of food to be utilized. Given sufficient food and sufficient organisms to eat the food, the process seems to be limited only by the rate at which oxygen or air can be dissolved into the mixed liquor. This rate depends on the oxygen deficit, turbulence, bubble size, and temperature and at present is restricted by the physical methods of forcing the air through the diffuser tubers and/or mechanical agitation.

In practice, the excess activated sludge is conditioned with 3–6% $FeCl_3$ and filtered on vacuum filters. This reduces the moisture to about 80% and produces a filter cake which is dried in rotary or spray driers to a moisture content of less than 5%. It is bagged and sold direct as a fertilizer, or to fertilizer manufacturers who use it in mixed fertilizer.

The mechanism of purification of sewage by the activated sludge is two-fold i.e., (1) absorption of colloidal and soluble organic matter on the floc with subsequent oxidation by the

---

* By W. D. Hatfield, Decatur, Illinois

organisms, and (2) chemical splitting and oxidation of the soluble carbohydrates and proteins to CO_2, H_2O, NH_3, NO_2, NO_3, SO_4, PO_4 and humus. The process of digestion proceeds by hydrolysis, decarboxylation, deaminization and splitting of S and P from the organic molecules before oxidation.

The process is applicable to the treatment of almost any type of organic waste waters which can serve as food for biological growth. It has been applied to cannery wastes, milk products wastes, corn products wastes, and even phenolic wastes. In the treatment of phenolic wastes a special flora is developed which thrives on phenol as food.

SMECTIC LIQUID CRYSTALS. Liquid Crystals.

SMELTING.
The process of heating ores to a high temperature in the presence of a reducing agent, such as carbon (coke), and of a fluxing agent to remove the accompanying rock gangue is termed *smelting*. Iron ore is the most abundantly smelted ore. It contains about 20% gangue (clay and sand). The ore is heated in an air blast furnace with coke and limestone (fluxing agent) at a temperature above the melting point of iron and slag (fusion mixture of impurities and flux). The molten iron (the more dense material) and molten slag (the less dense material) are removed separately from the furnace. Numerous other metals are smelted, including arsenic, cadmium, cobalt, copper, indium, lead, silver, and tin. See articles on these specific elements.

SMITHSONITE.
Smithsonite is zinc carbonate, $ZnCO_3$, a hexagonal mineral with a rhombohedral cleavage. It is a brittle mineral; hardness, 4–4.5; specific gravity, 4.3–4.5; luster, vitreous to dull; color, usually white, but may be colored yellowish or brownish or perhaps blue or green due to impurities. It is translucent to opaque. Smithsonite is a secondary mineral after sphalerite or may replace limestone or dolomite. It is sometimes called calamine (but true calamine is a zinc silicate) and often associated with it. Smithsonite occurs in Siberia, Greece, Rumania, Austria, Sardinia, Cumberland and Derbyshire, England; New South Wales, Australia; South West Africa, and Mexico. In the United States it is found in Pennsylvania, Wisconsin, Missouri, Arkansas, and Utah. This mineral was named in honor of James Smithson, whose legacy founded the Smithsonian Institution at Washington, D.C.

SNG. Substitute Natural Gas (SNG).

SOAP. Colloid System.

SOAPS.
Salts of long-chain, organic carboxylic acids (R—COO^-X^+). R is usually an unbranched saturated or unsaturated alkyl group, but it may possess branches or even ring groups. The cation (X) present is most usually sodium, although potassium and, to a lesser extent, various ammonium soaps are employed. These soaps are very much more water-soluble than the acids from which they are derived. The most characteristic properties of aqueous soap solutions, their ability to clean and to foam, have been known since ancient times. Soaps are also well recognized by their characteristic waxy feel.

Raw Materials. The most widely used raw materials for soaps are natural fats and oils which may be refined for soap manufacture. These are triglycerides, i.e., fatty acid esters of glycerin, the fatty acids of importance for soaps being essentially in the range 12-carbon to 18-carbon chain length. The chief sources of fatty acids for soap are tallow and coconut oil, the best known soaps being based on a mixture of 80% tallow fat and 20% coconut oil. Tallow, derived from the cattle and sheep meat industry, varies considerably in properties such as color and titer (solidification point of the fatty acids). The titer of tallow used for soap is generally 40°C or higher. The chain length makeup of its fatty acids is typically 2% C_{14}, 32.5% C_{16}, 14.5% C_{18}, 48.3% C_{18}^- (oleic), 2.7% $C_{18}^=$ (linoleic), and of coconut oil 8% C_8, 7% C_{10}, 48% C_{12}, 14.5% C_{14}, 8.8% C_{16}, 2% C_{18}, 6% C_{18}^-, 2.5% $C_{18}^=$. Partial replacements for tallow are lard (hog-fat), hardened marine oils, and, especially abroad, palm oil. Palm kernel oil, which has a fatty acid composition similar to that of coconut oil, is sometimes used to replace the latter.

In certain instances rosin acids (abietic acid, etc.) are used in soap making. Sodium hydroxide, as a concentrated aqueous solution, is the alkali most generally used for fat and oil saponification. Potassium or ammonium hydroxides are used to make more soluble or "soft" soaps; see below. Blends of alkalies are sometimes employed.

Soap Making. The essential process of soap making is the saponification of triglycerides in alkaline media. A substantial fraction of the total production of soap is still carried out by a batch process in large open kettles. Saponification is achieved by a combination of steam and added caustic soda. Thereafter the soap is salted out or "grained" to produce curdy kettle wax soap and lye which is separated for recovery of glycerin. Several washes ensue. The processes of "closing" of the soap, by adding steam and water, and graining are repeated several times and after the final wash-lye phase is removed, the appropriate amounts of water and heat are applied to produce neat soap overlying a lower quality "nigre" soap layer. The pumpable neat soap, containing about 30% water, is then processed. Several continuous processes for soap manufacture have been developed. Some involve preparation of fatty acids from the triglycerides as an intermediate step, whilst others involve direct saponification to the soap stage.

Soap Processing. Neat soap, which contains about 30% water, is processed in several ways. For bar manufacture, chips are formed by cooling the pumpable mass on chill rolls; after drying, mixing and milling operations, screw extrusion of soap in the form of logs follows from a so-called plodder. For preparing spray-dried powders, the neat soap is pumped into large mixing vessels (crutchers), the appropriate additives blended, and the mixture then sprayed from nozzles into heated air for drying.

Properties. An understanding of the preparation and properties of soaps is facilitated by studying the phase diagrams of soap/water mixtures—work pioneered by J. W. McBain. The properties of soap are dictated by its molecular structure, viz., a long hydrophobic hydrocarbon chain coupled to a hydrophilic ionic head group (COO^-X^+). In anhydrous, hydrate and concentrated solution form, soap molecules adlineate into regular, sandwichlike arrays of thickness approximately equal to two soap molecule lengths. This configuration allows hydrocarbon chains and polar groups each to pack side by side. This intermolecular distances in these double layer systems, tilt angle of the chains, and so on, are a sensitive compromise between the way the hydrocarbon chains and the head groups each wish to pack. This is reflected in the fact that, even in anhydrous systems, as many as ten different crystalline phases are encountered on heating the soap from room temperature to the melting point, ~300°C. Hydrated soaps, such as those encountered in commercial bars, also exist in several different phases, the predominant phase(s) present having a marked influence on the properties of these soaps. In hot, concentrated soap/water systems above the Krafft temperature (see below) two important phases are encountered. The first is neat phase, already referred to: x-ray data indicate the basic bimolecular leaflet structure is still present but with several layers of water interposed between the adjacent ionic head group planes. At higher water contents and somewhat lower temperatures, soap exists in "middle phase," a phase avoided during soap manufacture because of its high viscosity. This high viscosity results from the structure of the basic unit aggregates of soap molecules, or "micelles," which have the form of long cylinders of diameter about two soap molecule lengths. At still higher water content, the soap/water system is isotropic and the soap micelles are more or less spherical, with sphere diameter again about two chain lengths. An important phenomenon related to the formation of micelles is the Krafft "point." Below this temperature, soap is rather insoluble above it, the solubility increases rapidly and soap exists in the micellar or liquid crystalline phases described above. For the sample soaps sodium stearate, palmitate and oleate and potassium stearate, the Krafft temperatures are roughly 70°, 60°, 25° and 48°C, respectively.

The dual hydrophobic/hydrophilic nature of soap explains its most characteristic property, viz., surface activity. This is a consequence of the spontaneous attempt of the hydrophobic chains of soap molecules in water to remove themselves from the aqueous environment by a process of adsorption at, for example, the water/air and water/oil interfaces. This results in a lowering of the surface and interfacial tension, and is responsible for the ability of soap to promote foam and emulsion stability. Micelle formation itself, in which the hydrocarbon chains point "in," i.e., away from the water, and the ionic groups "out," i.e., in contact with the water, is a manifestation of the same phenomenon. Below the critical micelle concentration (cmc) the dissolved soap exists as single ions and behaves as a typical electrolyte; above it there is a mixture of micelles and single ions and the solution surface tension becomes insensitive to concentration. The cmc's of typical soaps at 70°C are roughly 0.07, 0.7, 3, $20 \times 10^{-3} M$, for sodium stearate, palmitate, myristate, and laurate, respectively. In a homologous series of fatty acid soaps, characteristic "soap" properties, such as micelle formation, pronounced foaming, etc., first occur at a chain length of 7 carbons.

Because of its tendency to aggregate in solution, soap is known as an association colloid. An aqueous solution of soap micelles can in fact be considered a stabilized suspension of micro droplets of hydrocarbon in water. As such they have the ability to dissolve and suspend oily materials, one of the processes involved in detergency. By adsorption onto oily dirt and the substrate to be cleaned, soaps can lead to a reduction of the contact angle of water at the water/soil/substrate interface so facilitating removal of the dirt by mechanical agitation. By adsorbing onto particulate dirt, and also removing, by precipitation, calcium ions involved in dirt/substrate "bridges," soaps assist detergency of solid soils.

Types. Sodium soaps are by far the most widely used. To offset the limited solubility associated with longer chain-length saturated soaps (e.g., C_{16} and C_{18}), blending is necessary with shorter chainlength soaps, e.g., C_{12} and C_{14}, and unsaturated soaps, such as sodium oleate. Experience has shown the blend of 80 tallow fat and 20 coconut oil, referred to earlier, achieves the desired solubility characteristics. Sodium soaps, furthermore, tend to be "hard," while potassium and ammonium soaps are "soft" and more soluble; thus potassium soaps are often blended with sodium soaps to achieve improved solubility and foaming. For the same reason, potassium, ammonium and triethanolammonium soaps are used in liquid soaps and shampoos.

Soaps are sometimes "superfatted," i.e., contain a small amount of unneutralized free fatty acid. This contributes to the mildness of the soap towards skin, both by reducing the

pH and probably also by deposition on the skin of "acid soaps" which have a relatively low solubility. Acid soaps, e.g., sodium acid stearate, are well-defined complexes of acid and soap with definite stoichiometry. They have been extensively studied.

—E. D. Goddard, Lever Bros. Co. Edgewater, New Jersey.

SODALITE. An isometric mineral, a sodium aluminum silicate containing sodium chloride, with the chemical composition $Na_4Al_3(SiO_4)_3Cl$, potassium sometimes replacing a small amount of sodium. It is commonly found as dodecahedrons or simply massive. When observed sodalite has a dodecahedral cleavage; conchoidal to uneven fracture; brittle; hardness, 5.5–6; specific gravity, 2.14–2.30; luster, vitreous to greasy; color grayish to greenish or yellowish, may be white. It is often a beautiful blue and may sometimes be red. It is transparent to translucent; streak, white. Sodalite is found in igneous rocks of nephelite-syenite type which have been produced from soda rich magmas. Sodalite also has been found in the lavas of Vesuvius. Common minerals associated with it are nephelite and cancrinite. It occurs in the Ilmen Mountains of the U.S.S.R.; at Vesuvius and Monte Somma, Italy; in Norway and Greenland. In Canada, in British Columbia and in Ontario, beautiful blue sodalite is found; and in the United States similar material comes from Kennebec County, Maine. The mineral derives its name from the fact of its soda content.

SODA NITRE. The mineral soda nitre or Chile saltpeter is naturally occurring sodium nitrate, $NaNO_3$. Its hexagonal crystals are rare, this mineral usually being found in crystalline aggregates, crusts or masses. It is soft; hardness, 1.5–2; specific gravity, 2.266; vitreous luster; colorless or white to yellow or gray; transparent to opaque. Soda nitre is a most important mineral commercially, being used in the manufacture of nitric acid, other nitrates and fertilizers. The chief soda nitre deposits of the world are those found in the Atacama and Tarapaca deserts of northern Chile, although others exist in the Argentine and Bolivia. Some small deposits have been found in California, New Mexico and Nevada. The origin of these nitrate deposits is far from being well understood. They have been regarded as nitrates formed originally by oxidation of organic matter and subsequently leached out. Guano, the excrement of birds, might be the original source of the nitrates. Ground water and ancient marine deposits have been suggested as well as the possibility of derivation from nitric acid produced in the atmosphere during electrical storms. Some investigators consider that the nitrates may have come from volcanic sources.

SODA PULP PROCESS. Pulp (Wood) Production and Processing.

SODIUM. Chemical element symbol Na, at. no. 11, at. wt. 22.9898, periodic table group Ia (alkali metals), mp 97.82°C, bp 882.9°C, density, 0.9721 g/cm³ (solid at 0°C), 0.9268 g/cm³ (liquid at mp). Elemental sodium has a face-centered cubic crystal structure. Sodium is a silvery-white metal, can be readily molded and cut by knife, oxidizes instantly upon exposure to air, and reacts with water violently, yielding sodium hydroxide and hydrogen gas, consequently the element is preserved under kerosene. Sodium burns in air at a red heat with yellow flame. Discovered by Davy in 1807.

There is only one naturally occurring isotope, ^{23}Na. There are five known radioactive isotopes, ^{20}Na through ^{22}Na, ^{24}Na, and ^{25}Na, all with short half-lives except ^{22}Na ($t_{1/2} = 2.6$ years). See **Radioactivity.** In terms of abundance, sodium ranks sixth among the elements occurring in the earth's crust, with an average of 2.9% sodium in igneous rocks. In terms of content in seawater, the element ranks fourth (due mainly to excellent solubility of its compounds), with an estimated 50 million tons of sodium per cubic mile (10.8 million metric tons per cubic kilometer) of seawater. First ionization potential, 5.138 eV. Oxidation potential: $Na \rightarrow Na^+ + e^-$, 2.712 V. Other physical properties of sodium are given under **Chemical Elements.**

Sodium does not occur in nature in the free state because of its great chemical reactivity. Sodium occurs as sodium chloride in the ocean (1.14% Na), in salt deposits (salt, halite, NaCl), e.g., in Michigan, New York, Louisiana, in Great Britain, in Germany, in salt lakes, e.g., the Dead Sea (3% Na), Great Salt Lake; in common rocks (average of the solid shell of the earth 2.75% Na) as sodium nitrate (Chile saltpeter, $NaNO_3$) in Chile; as sodium borate (rasorite, kernite, $Na_2B_4O_7 \cdot 4H_2O$, in California; tinkal, $Na_2B_4O_7 \cdot 10H_2O$, in Tibet); as sodium carbonate Na_2CO_3 and sulfate Na_2SO_4 in certain salt lake areas. See also **Sodium Chloride.**

Although sodium metal was isolated in 1807, it remained a laboratory curiosity until Oersted discovered in 1824 that sodium metal will reduce aluminum chloride to produce pure aluminum metal. This discovery led to the development of a commercial process for the manufacture of sodium. The first cell was designed by Castner in 1886 and a plant was built in Niagara Falls, N.Y., because of availability of low-cost electric power, for the electrolysis of fused NaOH. This process was obsoleted in 1921 by introduction of the Downs process in which a mixture of fused sodium chloride and calcium chloride is electrolyzed to produce metallic sodium. The modern cells have four anodes (graphite) surrounded by a steel cathode. Wire mesh diaphragms extend down into the electrolysis zone to prevent recombination of product sodium and chlorine. The use of calcium chloride in the cell significantly lowers the melting point of the mix. Sodium chloride has a mp 800°C, calcium chloride, mp 772°C, the two salt eutectic, mp 505°C. Calcium has limited solubility in sodium. The excess calcium reacts with the sodium chloride present $Ca + 2NaCl \rightarrow 2Na + CaCl_2$ and thus does not contaminate the sodium metal to a large degree. The sodium, which is saturated with calcium, is cooled in a riser pipe. Thus, the solubility of the calcium is reduced and this drops back into the cell. By filtration of the sodium nears its melting point, the calcium content is reduced to about 0.05%. The cells operate at about 8 V, with groups of 25 to 40 cells connected in series.

Uses. Like so many of the chemical elements, the compounds of sodium are far more important than elemental sodium.

Among the attractions of molten sodium metal as a heat-transfer medium are: (1) low density compared with other metals and combinations of salts, contributing to low cost per unit volume and also to relative ease of pumping. (2) relatively low vapor pressure even at temperatures as high as 500°C; (3) greater heat capacity than most common metals in liquid form, the thermal conductivity being 5–10 times greater than the conductivities of lead or mercury and 50 times greater than most organic heat-transfer media; and (4) viscosity of molten sodium is quite low. Despite these qualifications, however, the use of sodium as a heat-transfer medium has enjoyed a mixed reception over the years, partially attributable to a lack of marketing thrust on its behalf.

Sodium is fifth among the metals in terms of electrical conductivity—hence bus bars are constructed from steel pipe filled with sodium. The characteristic yellow sodium light, created by the passage of an electric current through sodium vapor, is used for commercial and industrial lighting. Sodium is used to modify aluminum-silicon alloys. Normally coarse and brittle,

such alloys can be transformed into fine-grained alloys with good casting properties through the addition of a fraction of 1% of sodium. Sodium also has been used as a hardening agent in bearing metals. When added with an alkaline-earth metal, such as calcium, sodium increases the hardness of lead. The German alloy "Bahnmetal" is an alloy of this type.

Generally, plain carbon steel containers are sufficient for handling metallic sodium at temperatures not in excess of the metal's boiling point. All-welded pipeline construction and bellows-sealed packless valves are usually used. Because of the metal's violent reactivity with H_2O, conventional fire extinguishers, including CO_2 and chlorinated hydrocarbons, should not be used. The preferred fire-retarding agents are salt, graphite, or soda ash, but they must be dry. Sand usually is not recommended because it is difficult to obtain perfectly dry sand in an emergency. In manufacturing operations involving sodium, particularly at reasonably high temperatures, an apron, leggings, and a complete face covering should be used. At normal temperatures, or where only small quantities of the metal are required, as may be the case in a research laboratory, conventional protective gear and goggles and gloves usually suffice.

Chemistry and Compounds. Sodium metal is obtained by electrolysis of fused sodium chloride or hydroxide out of contact with air. Its uses are limited in extent, but important in particular cases, as in the liberation of a metal from its chloride by reaction of sodium to form sodium chloride, and in certain reactions of organic chemistry.

The ionization potential of sodium (5.138 eV) is second to that of lithium and higher than that of any other alkali metals. However, the measured value of its oxidation potential against a normal aqueous solution of its ions is 2.712 V, the lowest of the group. Potassium is more electropositive in many of its reactions, even with water; though both react vigorously, the reaction of potassium is the most vigorous. With bromine, sodium reacts only slowly without heating, and with iodine scarcely at all even upon heating—quite in contrast with potassium.

Because of the ease of removal of its single $3s$ electron (5.138 eV) and the great difficulty of removing a second electron (47.29 eV), sodium is exclusively monovalent in its compounds, which are electrovalent. Some experimental work indicates that the sodium alkyls may be covalent, but even they form conducting solutions in other metal alkyls.

Like lithium, sodium and its compounds have been studied extensively in solution in liquid NH_3. Sodium metal in such solutions slowly or with catalysis forms the amide, $NaNH_2$. The solution of the metal is a powerful reducing agent, reacting with metallic salts to free the metal, with which it may form an intermetallic compound

$$Na + AgCl \rightarrow NaCl + Ag$$

$$9Na + 4Zn(CN)_2 \rightarrow 8NaCN + NaZn_4$$

Sodium chloride also forms the amide, or at low temperatures the pentammoniate $NaCl \cdot 5NH_3$.

Like the other alkali metals, sodium forms compounds with virtually all the anions, organic as well as inorganic. These compounds are remarkable for their great variety and for the fact that the reactivity of sodium bicarbonate with many metallic oxides permits of the vity of sodium bicarbonate with many metallic oxides permits of the preparation of many compounds that are unstable in aqueous solution. While other alkali bicarbonates react similarly, the general discussion of these compounds and the inorganic alkali salts generally, is appropriately given in this book under this entry for sodium, of which such a great number of inorganic (as well as organic) salts have been prepared.

Thus, normal (ortho) sodium arsenates $Na_3AsO_4 \cdot xH_2O$ and acid arsenates exist both in solution and in the solid state, whereas the meta- and pyroarsenates exist only as solids, but are readily prepared by heating arsenic pentoxide, As_2O_5, and sodium bicarbonate in correct proportions to produce the primary and secondary sodium arsenates, whence the meta- and pyroarsenates are obtained by heating

$$NaH_2AsO_4 \xrightarrow{\text{Heat}} NaAsO_3 + H_2O$$

$$2Na_2HAsO_4 \xrightarrow{\text{Heat}} Na_4As_2O_7 + H_2O$$

Similarly, the boron salts include metaborates, $NaBO_2 \cdot xH_2O$, tetraborates, $Na_2B_4O_7 \cdot xH_2O$, other polyborates, $Na_2B_{10}O_{16} \cdot xH_2O$, at least one orthoborate, Na_3O_3, and peroxyborates, such as $NaBO_3 \cdot H_2O$. See also **Boron.** Other important sodium salts include the carbonates, cyanides, cyanates, hexacyanoferrates, $Na_4Fe(CN)_6$ and $Na_3Fe(CN)_6$, halides, polyhalides, hypohalites, halites, halates, perhalates, permanganates, ortho-, pyro-, meta-, fluoro-, and peroxyphosphates, hyposulfites, sulfites, sulfates, thiosulfates, peroxysulfates, polythionates, tungstates, vanadates, uranates, etc.

In addition to the simple compounds, sodium forms double salts of various types, although because of the relatively small size of the Na^+ ion, the number of sodium alums (see **Alum**) is relatively small.

Sodium forms binary compounds, such as a phosphide, Na_3P, by direct union with phosphorus; a nitride, Na_3N, by direct union with nitrogen when activated electrically (which decomposes partly to give sodium azide, NaN_3, also obtained by heating sodium nitrate with sodium amide); and the oxides by direct union. Sodium monoxide, Na_2O, is obtained by heating the nitrite with the metal, displacing the nitrogen. Sodium peroxide, Na_2O_2, is the most stable oxide, obtained by reaction of the elements. Sodium superoxide is known, NaO_2, and one other oxide, Na_2O_3, has been reported.

Sodium hydroxide, $NaOH$, is very soluble in water and soluble in alcohol. It is almost completely ionized in water at ordinary concentrations, although its basic character is less than those of the higher elements in the group ($pK_B = -0.70$).

The detailed chemistry and applications of some of the more important compounds, other than those already discussed, follow.

Aluminate. Sodium aluminate $NaAlO_2$, white solid, (1) by reaction of aluminum hydroxide and $NaOH$ solution, (2) by fusion of aluminum oxide and sodium carbonate, the solution reacts with CO_2 to form aluminum hydroxide. Used as a mordant, and in water purification. See also **Aluminum.**

Aluminosilicate. Sodium aluminosilicate is used as a water softener for the removal of dissolved calcium compounds.

Amide. Sodamide, sodamine $NaNH_2$, white solid, formed by reaction of sodium metal and dry NH_3 gas at 350°C, or by solution of the metal in liquid ammonia. Reacts with carbon upon heating, to form sodium cyanide, and with nitrous oxide to form sodium azide NaN_3.

Bromide. Sodium bromide $NaBr$, white solid, soluble, mp 755°C. Used in photography and in medicine. See also **Bromine.**

Carbonates. Sodium carbonate (anhydrous), soda ash Na_2CO_3, sodium carbonate decahydrate, washing soda, sal soda $Na_2CO_3 \cdot 10H_2O$, white solid, soluble, mp 851°C, formed by heating sodium hydrogen carbonate, either dry or in solution. Commonly bought and sold in quantity on the basis of oxide Na_2O determined by analysis (58.5% Na_2O equivalent to 100.0% Na_2CO_3).

Soda ash is a very high-tonnage chemical raw material and approaches a production rate of 10 million tons/year in the United States. About 40% of soda ash is used in glassmaking; approximately 35% goes into the production of sodium chemicals, such as sodium chromates, phosphates, and silicates; nearly 10% is used by the pulp and paper industry; the remainder going into the production of soaps and detergents and in nonferrous metals refining. The first process for preparing soda ash was developed by Leblanc during the first French Revolution. In the Leblanc process, sodium chloride first is converted to sodium sulfate and subsequently the sulfate is heated with limestone and coke: (1) $Na_2SO_4 + 2C \rightarrow Na_2S + 2CO_2$; (2) $Na_2S + CaCO_3 \rightarrow Na_2CO_3 + CaS$. During the mid-1800s, the Solvay process was introduced. In this process, CO_2 is passed through an NH_3-saturated sodium chloride solution to form sodium bicarbonate, then followed by calcination of the bicarbonate: (1) $NH_3 + CO_2 + NaCl + H_2O \rightarrow HNaCO_3 + NH_4Cl$; (2) $2HNaCO_3 + heat \rightarrow Na_2CO_3 + CO_2 + H_2O$. A large proportion of soda ash now is derived from the natural mineral trona, which occurs in great abundance near Green River, Wyoming. Chemically trona is sodium sesquicarbonate $Na_2CO_3 \cdot NaHCO_3 \cdot 2H_2O$. After crushing, the natural ore is dissolved in agitated tanks to form a concentrated solution. Most of the impurities (boron oxides, calcium carbonate, silica, sodium silicate, and shale rock) are insoluble in hot H_2O and separate out upon settling. Upon cooling, the filtered sesquicarbonate solution forms fine needle-like crystals in a vacuum crystallizer. After centrifuging, the sesquicarbonate crystals are heated to about 240°C in rotary calciners whereupon CO_2 and bound H_2O are released to form natural soda ash. The crystals have a purity of 99.88% or more and handle easily without abrading or forming dust and thus assisting glassmakers and other users in obtaining uniform and homogeneous mixes.

Chlorate. Sodium chlorate, chlorate of soda, $NaClO_3$, white solid, soluble, mp 260°C, powerful oxidizing agent and consequently a fire hazard with dry organic materials, such as clothes, and with sulfur; upon heating, oxygen is liberated and the residue is sodium chloride; formed by electrolysis of sodium chloride solution under proper conditions. Used (1) as a weedkiller, (2) in matches and explosives, (3) in the textile and leather industries.

Chloride. Sodium chloride, common salt, rock salt, halite, $NaCl$, white solid, soluble, mp 804°C. See **Sodium Chloride.**

Chromate. Sodium chromate $Na_2CrO_4 \cdot 10H_2O$, yellow solid, soluble, formed by reaction of sodium carbonate and chromite at high temperatures in a current of air, and then extracting with water and evaporating the solution. Used (1) as a source of chromate, (2) in leather tanning, (3) in textile dyeing, (4) in inks.

Citrate. Sodium citrate $Na_3C_6H_5O_7 \cdot 5\frac{1}{2}H_2O$ white solid, soluble, formed (1) by reaction of sodium carbonate or hydroxide and citric acid, (2) by reaction of calcium citrate and sodium sulfate or carbonate solution, and then filtering and evaporating the filtrate. Used in soft drinks and in medicine.

Cyanide. Sodium cyanide $NaCN$, white solid, soluble, very poisonous, formed (1) by reaction of sodamide and carbon at high temperature, (2) by reaction of calcium cyanamide and sodium chloride at high temperature, reacts in dilute solution in air with gold or silver to form soluble sodium gold or silver cyanide, and used for this purpose in the cyanide process for recovery of gold. The percentage of available cyanide is greater than in potassium cyanide previously used. Used as a source of cyanide, and for hydrocyanic acid.

Dichromate. Sodium dichromate $Na_2Cr_2O_7 \cdot 2H_2O$, red solid, soluble, powerful oxidizing agent, and consequently a fire hazard with dry carbonaceous materials. Formed by acidifying sodium chromate solution, and then evaporating. Used (1) in matches and pyrotechnics, (2) in leather tanning and in the textile industry, (3) as a source of chromate, cheaper than potassium dichromate.

Dithionate. Sodium dithionate, "sodium hyposulfate" $Na_2S_2O_6 \cdot 2H_2O$, white solid, soluble, formed from manganese dithionate solution and sodium carbonate solution, and then filtering and evaporating the filtrate.

Fluorides. Sodium fluoride NaF, white solid, soluble, formed by reaction of sodium carbonate and hydrofluoric acid, and then evaporating. Used (1) as an antiseptic and antifermentative in alcohol distilleries, (2) as a food preservative, (3) as a poison for rats and roaches, (4) as a constituent of ceramic enamels and fluxes; sodium hydrogen fluoride, sodium difluoride, sodium acid fluoride $NaHF_2$, white solid, soluble, formed by reaction of sodium carbonate and excess hydrofluoric acid, and then evaporating. Used (1) as an antiseptic, (2) for etching glass, (3) as a food preservative, (4) for preserving zoological specimens.

Fluosilicate. Sodium fluosilicate Na_2SiF_6, white solid, very slightly soluble in cold H_2O, formed by reaction of sodium carbonate and hydrofluosilicic acid. Used (1) in ceramic glazes and opal glass, (2) in laundering, (3) as an antiseptic.

Formate. Sodium formate $NaCHO_2$, white solid, soluble, formed by reaction of NaOH and carbon monoxide under pressure at about 200°C. Used (1) as a source of formate and formic acid, (2) as a reducing agent in organic chemistry, (3) as a mordant in dyeing, (4) in medicine.

Hydride. Sodium hydride NaH, white solid, reactive with water yielding hydrogen gas and NaOH solution, formed by reaction of sodium and hydrogen at about 360°C. Used as a powerful reducing agent.

Hydroxide. Sodium hydroxide, caustic soda, sodium hydrate, lye, NaOH, white solid, soluble, mp 318°C, an important and strong alkali, not as inexpensive as calcium oxide (a strong alkali) or sodium carbonate (a mild alkali), but of wide use. Formed (1) by reaction of sodium carbonate and calcium hydroxide in water, and then separation of the solution and evaporation, (2) by electrolysis of sodium chloride solution under proper conditions and evaporation. Commonly traded in quantity on the basis of oxide Na_2O determined by analysis (77.5% Na_2O equivalent to 100% NaOH). Used (1) in the manufacture of soap, rayon, paper (soda process), (2) in petroleum and vegetable oil refining, (3) in the rubber industry, in the textile and tanning industries, (4) in the preparation of sodium salts. The latter may be prepared from NaOH in solution or as fused NaOH.

Hypochlorite. Sodium hypochlorite NaOCl, commonly in solution by (1) electrolysis of sodium chloride solution under proper conditions, (2) reaction of calcium hypochlorite suspension in water and sodium carbonate solution, and then filtering. Used (1) as a bleaching agent for textiles and paper pulp, (2) as a disinfectant, especially for water, (3) as an oxidizing reagent.

Hypophosphite. Sodium hypophosphite $NaH_2PO_2 \cdot H_2O$, white solid, soluble, formed (1) by reaction of hypophosphorous acid and sodium carbonate solution, and then evaporating, (2) by reaction of NaOH solution and phosphorus on heating (poisonous phosphine gas evolved).

Hyposulfite. Sodium hyposulfite, sodium hydrosulfite (not sodium thiosulfate) $Na_2S_2O_4$, white solid, soluble, formed by reaction of sodium hydrogen sulfite and zinc metal powder, and then precipitating sodium hyposulfite by sodium chloride in concentrated solution. Used as an important reducing agent in the textile industry, e.g., bleaching, color discharge.

Iodide. Sodium iodide NaI, white solid, soluble, mp 651°C, formed by reaction of sodium carbonate or hydroxide and hy-driodic acid, and then evaporating. Used in photography, in medicine and as a source of iodide.

Manganate. Sodium manganate Na_2MnO_4, green solid, soluble, permanent in alkali, formed by heating to high temperature manganese dioxide and sodium carbonate, and then extracting with water and evaporating the solution. The first step in the preparation of sodium permanganate from pyrolusite.

Nitrate. Sodium nitrate, nitrate of soda, Chile saltpeter, "caliche" $NaNO_3$, white solid, soluble, mp 308°C, source in nature is Chile, in the fixation of atmospheric nitrogen HNO_3 is frequently transformed by sodium carbonate into sodium nitrate, and the solution evaporated. Used (1) as an important nitrogenous fertilizer, (2) as a source of nitrate and HNO_3, (3) in pyrotechnics, (4) in fluxes.

Nitrite. Sodium nitrite $NaNO_2$, yellowish-white solid, soluble, formed (1) by reaction of nitric oxide plus nitrogen dioxide and sodium carbonate or hydroxide, and then evaporating, (2) by heating sodium nitrate and lead to a high temperature, and then extracting the soluble portion (lead monoxide insoluble) with H_2O and evaporating. Used as an important reagent (diazotizing) in organic chemistry.

Oleate. Sodium oleate $NaC_{18}H_{33}O_2$, white solid, soluble, froth or foam upon shaking the H_2O solution (soap), formed by reaction of NaOH and oleic acid (in alcoholic solution) and evaporating. Used as a source of oleate.

Oxalates. Sodium oxalate $Na_2C_2O_4$, white solid, moderately soluble, formed (1) by reaction of sodium carbonate or hydroxide and oxalic acid, and then evaporating, (2) by heating sodium formate rapidly, with loss of hydrogen. Used as a source of oxalate; sodium hydrogen oxalate, sodium binoxalate, sodium acid oxalate $NaHC_2O_4 \cdot H_2O$, white solid, moderately soluble.

Palmitate. Sodium palmitate $NaC_{16}H_{31}O_2$, white solid, soluble, froth or foam upon shaking the H_2O solution (soap), formed by reaction of NaOH and palmitic acid (in alcoholic solution) and evaporating. Used as a source of palmitate.

Permanganate. Sodium permanganate, permanganate of soda, $NaMnO_4$, purple solid, soluble, formed by oxidation of acidified sodium manganate solution with chlorine, and then evaporating. Used (1) as disinfectant and bactericide, (2) in medicine.

Phenate. Sodium phenate, sodium phenoxide, sodium phenolate, $NaOC_6H_5$, white solid, soluble, formed by reaction of sodium hydroxide (not carbonate) solution and phenol, and then evaporating. Used in preparation of sodium salicylate.

Phosphates. Trisodium phosphate, tribasic sodium phosphate $Na_3PO_4 \cdot 12H_2O$, white solid, soluble, formed (1) by reaction of sodium hydroxide and the requisite amount of phosphoric acid, and then evaporating, (2) by reaction of disodium hydrogen phosphate plus sodium hydroxide, and then evaporating, Used (1) as a cleansing and laundering agent, (2) as a water softener, (3) in photography, (4) in tanning, (5) in the purification of sugar solutions; disodium hydrogen phosphate, dibasic sodium phosphate $Na_2HPO_4 \cdot 12H_2O$, white solid, soluble, formed (1) by reaction of dicalcium hydrogen phosphate and sodium carbonate solution, and then evaporating the solution, (2) by reaction of sodium carbonate and the requisite amount of phosphoric acid, and then evaporating. Used (1) in weighting silk, (2) in dyeing and printing textiles, (3) in fireproofing wood, paper, fabrics, (4) in ceramic glazes, (5) in baking powders, (6) to prepare sodium pyrophosphate; sodium dihydrogen phosphate, monobasic sodium phosphate $NaH_2PO_4 \cdot H_2O$, white solid, soluble, formed (1) by reaction of sodium carbonate and the requisite amount of phosphoric acid, and then evaporating, (2) by reaction of calcium monohydrogen phosphate and sodium carbonate solution, and then evaporating the solution. Used (1) in baking powders, (2) in medicine, (3) to prepare sodium metaphosphate; sodium pyrophosphate $Na_4P_2O_7 \cdot 10H_2O$, white solid, soluble, mp about 900°C, formed by heating disodium hydrogen phosphate to complete loss of water, followed by crystallization from water solution. Used in electroanalysis; sodium metaphosphate $NaPO_3$, white solid, soluble, mp 617°C, formed by heating sodium dihydrogen phosphate or sodium ammonium phosphate to complete loss of water, is an easily fusible phosphate forming colored phosphates with many metallic oxides, e.g., cobalt oxide. The hexametaphosphate $(NaPO_3)_6$ is an important water-conditioning agent forming soluble complex compounds with many cations, e.g., Ca^{2+}, Mg^{2+}. Many polyphosphate compounds are known; their various uses include water softening and ion exchange. They are widely formulated in detergents, as are several of the simpler phosphates.

A detailed discussion of the many uses of phosphates in food processing will be found in "Foods and Food Production Encyclopedia" (D. M. Considine, Editor), Van Nostrand Reinhold, New York, 1982.

Phosphites. Disodium hydrogen phosphite $Na_2HPO_3 \cdot 5H_2O$, white solid, soluble, formed by reaction of phosphorous acid and sodium carbonate, and then evaporating at a low temperature, mp of anhydrous salt is 53°C, at higher temperatures yields sodium phosphate and phosphine gas; sodium dihydrogen phosphite $NaH_2PO_3 \cdot 2\frac{1}{2}H_2O$, white solid, soluble, formed by reaction of phosphorous acid and NaOH cooled to —23°C when the crystalline salt separates.

Salicylate. Sodium salicylate $NaC_7H_5O_3$, white solid, soluble, formed by reaction of sodium phenate and CO_2 under pressure. Used as a source of salicylate and for salicylic acid.

Silicate. Sodium silicate, sodium metasilicate, "water glass" Na_2SiO_3, colorless (when pure) glass, soluble, mp 1,088°C, formed by reaction of silicon oxide and sodium carbonate at high temperature; solution reacts with CO_2 of the air, or with sodium carbonate solution or ammonium chloride solution, yielding silicic acid, gelatinous precipitate. Sodium silicate solution is used (1) in soaps, (2) for preserving eggs, (3) for treating wood against decay, (4) for rendering cloth, paper, wood noninflammable, (5) in dyeing and printing textiles, (6) as an adhesive (e.g., for paper boxes) and cement. Sold as granular, crystals, or 40° Baumé solution.

Silicoaluminate. (See aluminosilicate, above.)

Silicofluoride. (See fluosilicate, above.)

Stearate. Sodium stearate, $NaC_{18}H_{35}O_2$, white solid, soluble, froth or foam upon shaking the water solution (soap), formed by reaction of NaOH and stearic acid (in alcoholic solution) and evaporating. Used as a source of stearate.

Sulfates. Sodium sulfate (anhydrous), "salt cake" Na_2SO_4, sodium sulfate, decahydrate, "Glauber's salt" $Na_2SO_4 \cdot 10H_2O$, white solid, soluble, formed by reaction of sodium chloride and H_2SO_4 upon heating with evolution of hydrogen chloride gas. Used (1) in dyeing, (2) along with carbon in the manufacture of glass, (3) as a source of sulfate, (4) to prepare sodium sulfide; sodium hydrogen sulfate, sodium bisulfate, sodium acid sulfate, "nitre cake" $NaHSO_4$, white solid, soluble, formed by reaction of sodium nitrate and H_2SO_4, upon heating, with evolution of HNO_3. Used (1) as a cheap substitute for H_2SO_4. (2) in dyeing, (3) as a flux in metallurgy; sodium pyrosulfate $Na_2S_2O_7$, white solid, soluble, formed by heating sodium hydrogen sulfate to complete loss of H_2O.

Sulfides. Sodium sulfide Na_2S, yellowish to reddish solid, soluble, formed (1) by heating sodium sulfate and carbon to a high temperature. Used (1) as the cooking liquor reagent (along with sodium hydroxide) in the "sulfate" or "kraft" process of converting wood into paper pulp, (2) as a depilatory, (3) in

sheep dips, (4) in photography, engraving and lithography, (5) in organic reactions, (6) as a source of sulfide, (7) as a reducing agent; sodium hydrogen sulfide, sodium bisulfide, sodium acid sulfide NaHS, formed in solution by reaction of NaOH or carbonate solution and excess H_2S.

Sulfites. Sodium sulfite Na_2SO_3, white solid, soluble, dilute solution readily oxidized in air, but retarded by mannitol (carbohydrates), formed by reaction of sodium carbonate or hydroxide solution and the requisite amount of SO_2, at high temperature yields sodium sulfate and sodium sulfide. Used (1) as a source of sulfite, (2) as a reducing agent, (3) to prepare sodium thiosulfate, (4) as a food preservative, (5) as a photographic developer, (6) as a bleaching agent and antichlor in the textile industry; sodium hydrogen sulfite, sodium bisulfite, sodium acid sulfite $NaHSO_3$, white solid, soluble, formed by reaction of sodium carbonate solution and excess sulfurous acid. Uses similar to those of sodium sulfite.

Tartrate. Sodium tartrate, $Na_2C_4H_4O_6 \cdot H_2O$, white solid, soluble, formed by reaction of sodium carbonate solution and tartaric acid. Used in medicine; sodium potassium tartrate (Rochelle salt) $NaKC_4H_4O_6 \cdot 4H_2O$, white solid, soluble. Used in medicine and as a source of tartrate.

Thiosulfate. Sodium thiosulfate, "hypo," $Na_2S_2O_3 \cdot 5H_2O$, white solid, soluble, formed by reaction of sodium sulfite and sulfur upon boiling, and then evaporating. Used (1) in photography as fixing agent to dissolve unchanged silver salt, (2) as a reducing agent and antichlor. See **Sodium Thiosulfate.**

Tungstate. Sodium tungstate $Na_2WO_4 \cdot 2H_2O$, white solid, soluble, by reaction of NaOH solution and tungsten trioxide upon boiling, and then evaporating. Used (1) in fireproofing fabrics, (2) as a source of tungsten for chemical reactions.

Uranate. Sodium uranate, uranium yellow Na_2UO_4, yellow solid, insoluble, formed by reaction of soluble uranyl salt solution and excess sodium carbonate solution. Used (1) in the manufacture of yellowish-green fluorescent glass, (2) in ceramic enamels, (3) as a source of uranium for chemical reactions.

Vanadate. Sodium vanadate, sodium orthovanadate Na_3VO_4, white solid, soluble, formed by fusion of vanadium pentoxide and sodium carbonate. Used (1) in inks, (2) in photography, (3) in dyeing of furs, (4) in inoculation of plant life.

The larger number of organic compounds of sodium are for great part derivatives of oxygen-containing compounds such as salts of organic acids (several of which are discussed above), alcoholic and phenolic compounds (carboxylates, alkoxides, phenoxides, etc.). However, in some cases, sodium derivatives of nitrogen-containing compounds, as sodium benzamide, $C_6H_5C(O)NHNa$, and sodium anilide, C_6H_5NHNa, contain sodium-nitrogen bonds, while even sodium-boron bonds exist in certain boron-containing compounds, as sodium triphenylborene $NaB(C_6H_5)_3$, and others; and in a number of compounds sodium is carbon-connected, as in methylsodium, CH_3Na, ethylsodium, C_2H_5Na, cyclopentadienylsodium, C_5H_5Na, and sodiumtriphenylmethane, $NaC(C_6H_5)_3$.

The organometallic compounds of sodium may be divided into two groups, differing in properties. One group, e.g., ethylsodium, consists of compounds that are colorless, insoluble in organic solvents, and that electrolyze readily in diethylzinc solution. Another group, e.g., benzylsodium, $C_6H_5CH_2Na$, are colored, and soluble in organic solvents.

Like all the alkali metals, sodium coordinates with salicylaldehyde. Its tetracovalent compounds, with those of potassium, are the more stable of the group, for the following reasons: (1) Increasing ionic size carries with it increasing electropositiveness and ease of ionization, which diminishes the tendency to coordinate. (2) The increasing distance of the nucleus from the coordinating electrons with increasing atomic volume makes it less likely that additional electrons will be held with ease. (3) On the other hand, there is an increase in the maximum coordination number with the elements of higher atomic number. These factors are in keeping with a maximum stability for the tetracovalent compounds occurring with sodium.

Sodium in Biological Systems. Sodium is essential to higher animals, which regulate the composition of their body fluids and to some marine organisms. The several important roles played by the sodium cation in biological systems, frequently in concert with the potassium cation, are described in entry on **Potassium and Sodium (In Biological Systems).**

See the list of references at the end of the entry on **Chemical Elements.**

SODIUM CARBONATE.   Leavening Agents.

SODIUM CHLORIDE.
NaCl, formula weight 58.44, white solid, cubic crystal structure, mp 800.6°C, bp 1,413°C, sp gr 2.165. Commonly called "salt," the mineral name for rock salt is *halite*. See also **Halite.** The compound is soluble in H_2O (35.7 g/100 g H_2O at 0°C; 39.8 g/100 g H_2O at 100°C), only slightly soluble in alcohol, and insoluble in HCl. Sodium chloride is produced in nearly all nations of the world, but some only have a sufficient supply for local needs. The leading salt-producing nations include the United States, China, the U.S.S.R., West Germany, France, the United Kingdom, India, Italy, Canada, and Mexico. In 28 states of the United States and in several provinces of Canada, salt occurs as bedded or domed deposits. Most of the rock salt produced in the United States comes from Michigan, New York, Texas, Ohio, Louisiana, and Kansas. Purity ranges from 97% NaCl for Kansas salt to 99% purity and higher for Louisiana salt. The main impurities are calcium sulfate (0.5–2%), dolomite, quartz, calcite, and traces of iron oxides. Natural rock salt is mined much as coal and usually marketed without purification, after crushing and screening. For most industrial and consumer requirements, the impurities are harmless. There is no evidence that bacteria exist in rock salt. Additionally, there is some solar salt production in the Great Salt Lake area of Utah and on the west coast. Salt deposits date back to past geologic ages and are believed to be the results of evaporated impounded sea water.

Purified salt for table and industrial processing requirements of a special nature is made by dissolving raw sodium chloride in H_2O and then evaporating the H_2O to form a final product. There are several types of evaporated salt, including *granulated salt* in which each crystal is a tiny cube, and *grainer* or *flake salt*, made up of irregularly shaped crystals, often thin and flaky and unusually soft. A process for producing evaporated salt is shown in the accompanying figure. Holes are drilled into the salt deposits, after which H_2O is pumped into the beds to create a brine which then is brought to the surface for refining. In this method, all insolubles are left in the bed. After some pretreatment to remove hardness and dissolved gases, the semi-pure brine is evaporated in multiple-effect vacuum pans. The salt crystallizes as perfect cubes of NaCl. In the system shown, each vacuum pan performs not only as an evaporator, but also as a boiler. The vapors from a preceding pan are used to heat the contents of the following pan. This system of heat economizing is possible because each succeeding pan in the series is under less pressure—hence the contents boil at a lower temperature. The lower pressure in succeeding pans results from condensation of the vapors as well as assistance from vacuum pumps. Crystal size is controlled by evaporation rate, the latter depending on the degree of vacuum, temperature, and agitation maintained.

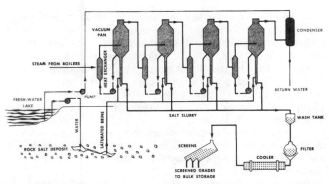

Multiple-effect vacuum pans used in production of sodium chloride from brine. The saturated brine is formed by pumping fresh water directly into the rock salt deposit, leaving insoluble materials in the deposit.

When grown to proper size, the crystals drop to the bottom of the pans and fall into the salt legs, from which they are drawn continuously in the form of a slurry. After washing, filtering, cooling, and screening, they are packaged. See also **Evaporation.**

Grainer salt is made by surface evaporation of brine in flat pans open to the atmosphere. Heat usually is furnished by steam pipes located a few inches below the tank bottom. Crystals form at the surface of the brine and are held there temporarily by surface tension. Thus, they grow laterally for awhile and form thin flakes. But, as they grow, they tend to sink and this process imparts a peculiar, hollow pyramid-like structure to them. Such crystals are called *hopper crystals*. Ultimately, the crystals sink to the bottom where they are scraped to one end of the pan. The crystals are fragile and during handling they break up, finally assuming a flake-like shape. Thus, the term *flake salt.*

In the *recrystallizer process* for making salt, advantage is taken of the fact that the solubility of NaCl increases with temperature whereas the solubility of the principal impurity, $CaSO_4$, decreases with temperature. In *solar* facilities, the raw brine is pumped into concentrating ponds where most of the H_2O is evaporated. Some of the impurities are precipitated out in this stage, after which the saturated brine is transferred to crystallization pounds where the salt crystallizes out at a high degree of purity. Since evaporation occurs at the surface of the ponds, hopper crystals are formed as in the grainer process, with flake salt being the final product.

Uses. Sodium chloride is a very high-tonnage material. In addition to its familiar use in the diets of man and animals, representing a small part of total production, large quantities are used by highway departments to control icy road conditions, in agriculture, and as a basic chemical raw material. The chemical industry consumes about two-thirds of the salt produced, the majority of it going to electrolytic plants. Some of the basic inorganic chemicals that require salt as a starting material include soda ash, calcium chloride, caustic soda, sodium sulfate, sodium bisulfate, HCl, sodium cyanide, sodium hypochlorite, and chlorine. See also **Chlorine; Sodium.**

Salt and the Diet

In food processing, the preservative and organoleptic qualities of salt are well established and it is fully appreciated why use of salt even to excess is attractive to food processors. Excessive usage is also habitual among people who "salt first and taste later." Dickinson in 1979 reported that just over 1 million tons

(900,000 metric tons) of salt are used in foods and in connection with eating in the United States in a given year. Nearly an additional 2 million tons (1.8 million metric tons) are used in the agricultural field, much of which is consumed by livestock. The total daily intake of the North American consumer as of 1980 is estimated to be in the range of 10–12 grams of salt, which reduces to a range of 3900–4700 milligrams of sodium. Highly salted snack foods, the consumption of which has increased markedly in many parts of the world during recent years, accounts for a significant consumption of salt. In addition, certain other food ingredients, such as monosodium glutamate and soy sauce, sometimes used in excess, also contribute to the average intake of sodium.

Sodium and chloride are not normally retained in the body even when there is a high intake. See also **Chloride.** Amounts consumed in excess of need are excreted, so that the level in the body is maintained within very narrow limits, as is also the chloride, regardless of intake. The primary route of excretion is via the urine, with substantial amounts also lost in sweat and feces. About 50% of the sodium in the human body is located in the extracellular body fluids; 10% inside the cells; and 40% in the bones. Chloride is found mainly in the gastric juice and other body fluids.

Essential though sodium is to the normal functioning of the human body, there has been considerable concern over the last few years about the amount of salt in the diet, this concern centering mainly on possible relationship between salt and hypertension (high blood pressure).

Hypertension afflicts more than 20% of the world population, with an estimated 24 million cases in the United States as of 1980. In 1976, Marx reported that in about 90% of these cases, the actual causes of hypertension cannot be pinpointed—this in face of the fact that research on the possible role of sodium in essential hypertension has been underway for 60 years or longer. Tests of unmedicated persons with essential hypertension indicate a lowering of blood pressure when sodium intake is restricted below one gram per day—and that the blood pressure rises again if additional sodium is taken. However, in other studies, some persons have retained a normal blood pressure level even when fed substantially increased amounts of salt (or other sodium-ion-furnishing substances). In 1976, Freis reported positive correlations between estimated average salt consumption of various ethnic populations and their incidence of hypertension. But such studies are complicated by many factors, including the inability to control or eliminate other possible causes of hypertension, such as obesity, genetic predisposition, general nutritional status, and potassium intake. It also has been generally proved extremely difficult to determine differences between individuals within these cultures.

Nevertheless, the concern remains on the part of a large number of professional people who feel that one day a definitive correlation will be made. And, with considerable awareness of the lay public in this regard, very definite pressures are being exerted on food processors to reduce salt usage and to more accurately label their merchandise in this regard.

Sodium Chloride and Energy

As pointed out by Wick (*Oceanus*, **22**, 4, 28, 1980), most of the energy in the oceans is bound in thermal and chemical forms. Although thermal energy is presently commanding the most attention, within the past few years another, rather unusual, form has received notice. Where rivers flow into the oceans a completely untapped source of energy exists—represented by a large osmotic pressure difference between fresh and salt water. If economical ways to tap these salinity gradients could be developed, large quantities of energy would be available.

SODIUM HYPOCHLORITE. Bleaching Agents.

SODIUM THIOSULFATE. $Na_2S_2O_3 \cdot 5H_2O$, formula weight 248.19, white crystalline solid, decomposes above 48°C, sp gr 1.685. Also known as "hypo" and sometimes misnamed "hyposulfite," sodium thiosulfate is very soluble in H_2O (301.8 parts in 100 parts H_2O at 60°C), soluble in ammonia solutions, and very slightly soluble in alcohol. When sodium thiosulfate is added to an acid, thiosulfuric acid $H_2S_2O_3$ may be formed, but only for an instant, immediately decomposing into sulfur and SO_2.

Sodium thiosulfate is used (1) to dissolve silver chloride, bromide, iodide in the photographic "fixing" bath, soluble sodium silver thiosulfate being formed plus sodium chloride, bromide, iodide, (2) in reaction with iodine in solution, sodium tetrathionate and sodium iodide being simultaneously formed, or with ferric salt solution, sodium tetrathionate and ferrous being simultaneously formed, (3) in reaction with chlorine as an "antichlor" forming sulfate and chloride. Sodium thiosulfate reacts with silver nitrate solutions yielding silver sulfide, brown precipitate, and with permanganate yielding manganous. Sodium amalgam changes sodium thiosulfate to sodium sulfide plus sodium sulfite.

Sodium thiosulfate is formed (1) by reaction of sodium sulfite solution and sulfur upon warming, (2) by reaction of sodium sulfite solid and sulfur upon heating, (3) by complex reaction of sulfur and sodium hydroxide solution upon warming—sulfur yields sodium sulfide plus sodium sulfite and the latter reacts with excess sulfur forming sodium thiosulfate, and the sodium sulfide present may be converted into sodium thiosulfate by passing in SO_2 until the solution changes from yellow to colorless.

There are numerous other thiosulfates, including potassium, magnesium, calcium, barium, mercury, lead, and silver. All are soluble in H_2O except Ba, Pb, and Ag thiosulfates.

Thiosulfates are commonly identified as follows:

1. Dilute acids precipitate sulfur from thiosulfates (difference from sulfides and sulfites).

2. Zinc sulfate and sodium hexacyanoferrate(II) give no color (difference from sulfites).

SOFTENING (Water). Water Conditioning; Water Treatment (Boiler).

SOIL CHEMISTRY. This field includes all aspects of the study of soil as a chemical system. The eight chemical elements in soils which generally surpass 1% by weight are oxygen, silicon, aluminum, iron, calcium, magnesium, potassium, and sodium; the eleven elements making up 0.2 to 1% include titanium, hydrogen, phosphorus, manganese, fluorine, sulfur, strontium, barium, carbon, chlorine, and chromium. The most abundant minerals present in less-weathered soils are quartz, feldspars, micas and colloidal layer silicates including vermiculite, chlorite, and montmorillonite. Calcareous soils contain calcite and dolomite. More-weathered soils contain larger amounts of more resistant minerals such as kaolinite, halloysite, allophane, hematite, goethite, gibbsite, anatase, pyrolusite, tourmaline, and zircon. The organic matter, or humus, content of soils varies from less than 1% to over 80%. Generally, upland soils range from 1 to 8% organic matter, while less well-drained soils are frequently higher.

Soils developed under coniferous forests often accumulate

acid organic matter at the surface; the resulting leaching through the soil of chelating organic acids bleaches (*podzolizes*) the mineral soil beneath. These soils are gray when plowed. Organic matter from hardwood trees and grasses which are high in bases, particularly calcium, accumulates in the soil and causes a dark color in the surface horizon. Poor drainage leads to the development of light-colored gray horizon within the soil column (*profile*), owing in part to the reduction of iron oxides to ferrous form. A bluish color is sometimes present, particularly when vivianite, $(Fe)_3(PO_4)_2 \cdot 8H_2O$, forms. Soluble soil salts, mainly chlorides and sulfates of sodium, calcium, and magnesium, when present in quantities over 0.1 to 0.7% cause a condition known as salinity or soil alkali. If much Na_2CO_3 is present, some organic matter is mobilized and together with FeS, colors the soil black, giving rise to the name black alkali.

The most reactive portion of the soil resides in colloidal organic matter, layer silicates, and hydrous oxides of iron, aluminum, and occasionally manganese and titanium. The colloids of soil have a negative electrostatic charge arising through carboxyls of organic compounds and through excess negative charge of oxygen in the silicate structure. The negative charge is neutralized by exchangeable cations, giving systems known as colloidal electrolytes. When these exchangeable ions, i.e., counterions, are hydrogen or aluminum, the colloids act as a moderately strong acid. Different colloids range in the strength of acidity as evidenced by the shapes of the titration curves, which are analogous to the shapes of those of weak and strong soluble acids. The colloids are hydrophilic and subject to flocculation in the presence of dilute salt solutions, owing to repression of the charge developed by dissociated cations. The flocculation is reversible.

Important chemical characteristics of the soil include the total exchange capacity for cations, expressed as total meq of cations per 100 gm of soil, and the base status, which is the percentage saturation of the negative charge with cations such as calcium, magnesium and sodium. The more productive soils are about 80% saturated with calcium and magnesium. Excessive hydrogen and aluminum saturation (much over 15%) is termed soil acidity. Excess sodium saturation (12% or more) leads to dispersiveness of the soil and poor productivity.

There are also positive charges associated with aluminum and iron colloids of soils. These charges give rise to phenomenon known as anion exchange capacity, which is mainly concerned with phosphorus chemistry; the usual soluble anions such as nitrate, chloride, and sulfate are little held. Synthetic organic soil conditioners are long-chain organic molecules with carboxyl charges along the chain which react with the positive charges of the soil particles. These colloidal molecules can bind the soil particles into aggregates. Natural humus of soil acts in a similar way until oxidized by soil organisms.

Analytical methods employed in soil chemistry include the standard quantitative methods for the analysis of gases, solutions, and solids, including colorimetric, titrimetric, gravimetric, and instrumental methods. The flame emission spectrophotometric method is widely employed for potassium, sodium, calcium, and magnesium; barium, copper and other elements are determined in cation exchange studies. Occasionally arc and spark spectrographic methods are employed.

The most commonly made chemical determination is that of soil pH measurement, as an indicator of soil acidity. The glass electrode has proved the most satisfactory method for soil pH measurement because the moistened soil rapidly equilibrates in contact with the glass surface, no reagents are added to the soil, and the soil CO_2 tension is not disturbed by bubbling through of gases. Colorimetric indicators are also employed.

Soils of pH 4.3 to 5 are highly acid, of pH of 5 to 6 are moderately acid, of pH 6.3 to 6.6 are very slightly acid, of pH 6.7 to 7.3 are considered neutral; soils of pH 8 to 9 are moderately alkaline, and of pH 9 to 11 are very alkaline. For acid soils, pH measurement serves as a guide to agricultural liming practices. For many crops, such as alfalfa, the soil is adjusted to pH 6.5 to 7 by the addition of ground limestone, the active ingredients of which are $CaCO_3$ in calcic limestone and $CaCO_3$ $MgCO_3$ in dolomitic limestone. A soil colloidal acid may be represented as HX. Then the liming reaction, by which the exchangeable calcium is increased, is

$$CaCO_3 + 2HX \rightarrow CaX_2 + CO_2 + H_2O$$

The proton donor, X, represents a variety of organic and inorganic donors, including the reaction $Al(OH_2)_6 \rightarrow Al(OH)(OH_2)_5 + H^+$. The reaction is hastened by fine grinding of the limestone, and liming materials are graded on the basis of fineness and $CaCO_3$ equivalence. Burned lime (CaO), marl ($CaCO_3$), and sugar refinery wastes [$Ca(OH)_2$] are also used in liming. For some crops, owing to disease susceptibility and preferences, soils are kept more acid, as low as pH 5.3. Calcium and magnesium, necessary to plant growth, are furnished to plants from exchangeable form.

Fine grains (finer than 20μ in diameter) of mica [$KAlSi_3Al_2O_{10}(OH)_2$], and potassium feldspar ($KAlSi_3O_8$) slowly undergo chemical weathering in soils with the release of potassium into exchangeable form. Other ions such as calcium, sodium, and iron are also released by mineral weathering. In subhumid and arid regions the release of potassium is fast enough for crop production but must be supplemented by the addition of potash fertilizer salts in more leached soils of humid regions.

Soil chemists have rapid chemical tests for measurement of the amounts of plant-available K, P, and N, as well as other elements in soils which are essential to plants. When the quantity of an element is too low for efficient crop production, it is added as fertilizer, such as KCl, $Ca(H_2PO_4)_2$, or ammonium or nitrate salts. Large chemical fertilizer industries are required for mining, refining, and preparation of chemical salts for soil application as fertilizers.

Extraction of soils for analysis of the readily available nutrients include replacement of exchangeable cations by salt solutions, dilute acids, and dilute alkalies such as $NaHCO_3$. Fluoride solutions are employed to repress iron, aluminum, and calcium activity during the extraction of phosphorus. Extraction of the soil solution is effected by displacement in a soil column, often through the application of pressure across a pressure membrane. The soil solution is analyzed by conductance and elemental analysis methods. Also, the total elemental analysis of soils is made by Na_2CO_3 fusion of the soil followed by classical geochemical analysis methods.

Organic compounds of great variety have accumulated in soils as residues from plant and animal life of the soil. The more unstable compounds of these residues are rapidly oxidized to CO_2 and H_2O by biochemical processes, while the more stable fractions accumulate. Conjugated ring compounds containing the elements C, H, O, N, P, S, and several other elements in small quantities accumulate in relatively stable organic and organomineral colloidal complexes. Lignin-like, phytin-like, and nucleoprotein-like compounds are included. Sorption of the organic matter on mineral colloid surfaces, particularly on layer silicates, such as montmorillonite, helps to stabilize the organic matter against biochemical oxidation. In tropical soils, high stability of soil organic matter is imparted by coatings of aluminum hydroxide and red ferric oxide. Organic and iron oxide colloids, when fairly abundant, stabilize the soil into porous aggregates through which ample air and water can circulate. Decomposition of soil organic matter, especially when hastened by tillage, gradually releases HNO_3, H_2SO_4, and H_3PO_4 in amounts which are highly significant in nutrition of crops. Much of the nitrogen, sulfur, and phosphorus required by crops is furnished in this way.

The oxidation potential of well-aerated soils is low (-0.5 V) and of reduced soils is high ($+0.30$ V). These relationships are sometimes expressed by soil scientists as reduction potentials or redox potentials, in which case the algebraic signs are the opposite. The oxidation potential is advantageously measured with a platinum-blackened electrode in the soil in place in the field. Moderately good aeration is a requirment of a productive soil. The oxidation status may also be tested in the field by rapid spot tests for ferric and ferrous iron in soils. Most of the dilute acid-soluble iron is in ferric form in well drained soils. Localized spots of decomposing organic matter are important in reducing small but important quantities of iron to ferrous form and manganese to divalent form so as to be available to plants. Moderately to highly alkaline soils sometimes have inadequate activity of the reduced forms of iron and manganese, particularly in the absence of sufficient organic matter. Small quantities of Cu, Zn, B, and Mo must be present in productive soils in forms which have enough activity to be available to growing plants.

—M. L. Jackson, University of Wisconsin, Madison, Wisconsin.

SOL AND SOLATION.   Colloid System.

SOLID STATE.   The physics and chemistry of the solid state embraces experimental investigations and theoretical interpretations of the physical behavior of matter in the solid phase.

Structure.   Most solids are crystalline because the energy of the ordered arrangement is less than for the disordered one. On the basis of the symmetry exhibited by the three-dimensional array of the atoms, crystals are categorized as belonging to one of seven crystal systems: triclinic, monoclinic, orthorhombic, tetragonal, hexagonal, cubic, and rhombohedral. Consideration of rotation-reflection axes leads to the further division of the seven systems into thirty-two crystal classes. Crystals may be classified according to the type of chemical binding present. Binding arises predominantly from electrostatic forces and quantum effects due to the motion of the atomic electrons. The five classes of binding with examples are: ionic (alkali halides), covalent (diamond), metallic (alkali metals), molecular (inert gases), and hydrogen-bonded (ice). Information about the arrangement of atoms in crystals is obtained by the diffraction of x-rays having wave lengths comparable with atomic spacings in solids. Extensive studies of amorphous solids and liquid-crystals accompany the development of uses for such substances.

Imperfections.   Actual crystals are not perfect, and it is known that imperfections play an essential role in crystal growth, diffusion, absorption, luminescence, and other physical processes. Among the imperfections are lattice vacancies and interstitial atoms; these have a marked effect on the optical and electrical properties of the crystal. Such defects are produced in solids irradiated by nucleons and electrons. In alkali halides, a deviation from stoichiometry produces coloring (color centers) and accompanying absorption phenomena. A large part of the optical and electrical behavior of semiconductors may be attributed to free electrons (or holes) liberated by foreign atoms in the lattice. Certain types of crystal irregularity that may be associated with missing or extra planes of atoms in part of the crystal are termed dislocations; these are particularly impor-

tant in determining mechanical properties and the mechanism of crystal growth. Impurity atoms, such as Cr in ruby, play an important role in solid state lasers and masers. These atoms are "pumped" to an energy level higher than the ground state and then are stimulated by electromagnetic radiation to emit the excess energy at the same frequency as the stimulating radiation and in phase, thus producing a coherent amplified beam.

Thermal Properties.   Theories of the specific heat of solids start from the assumption that the vibrational energy of a system of N atoms is equal to the energy of a system of $3N$ harmonic oscillators. Thus, the main problem in the theory of specific heat is the determination of the frequency spectrum of the oscillators. The Debye theory, based upon the vibrational modes of a continuous medium with a frequency range extending up to a cut-off frequency, predicts that the specific heat of the lattice can be approximated at very low temperatures by a contribution directly proportional to the cube of the absolute temperature T and as asymptotically approaching $3R$ (R = gas constant) at very high temperatures. The shape of the specific heat vs. T curve in the intervening range is determined by a parameter, the Debye temperature, which is characteristic of the lattice and which can be correlated with the lattice frequency spectrum. Experiment showed that in metals at low temperatures an additional specific heat directly proportional to T is to be added to the T^3 term predicted by Debye. This contribution is understood as arising from the conduction, or free, electrons in the solid, and thus this term is quite important in metals and detectable in some semiconductors. The thermal conductivity of a solid is made up of two contributions, one from the lattice and the other from the free electrons, with the latter term dominating in metals and the former in nonmetals. Actually the lattice contribution may be large, and thus the thermal conductivity of some semiconductors is comparable with that of metals.

Band Theory.   The behavior of electrons in metals varies from one metal to another, but one can treat metals on the basis of the free electron theory, according to which an appreciable fraction of the electrons in a metal specimen (of the order of one per atom) is able to move freely within the sample subject only to the potential barriers at the surfaces. More information about the behavior of all solids, including non-metals, is obtained by considering that an electron passing through a crystal undergoes a periodic variation in potential energy which is correlated with the periodicity of the crystal lattice. Whereas the simple free electron model permits all values of the electron energy, the introduction of a periodic potential yields forbidden energy ranges for which solutions representing an electron moving through the crystal do not exist.

Near the top or bottom of an allowed energy band, the energy is approximately a quadratic function of the wave number (2π times the reciprocal of the electron wave length). This dependence permits the determination of an effective mass which is used in describing the motion of the electron, or its associated wave packet, in applied electric or magnetic fields. In metals either allowed energy bands are only half-filled in the ground state or there is overlapping between filled and empty allowed bands; in either case the electrons can readily make transitions to empty allowed states and thus be accelerated by applied fields. In insulators the electrons completely fill the states in an allowed band and a sizable (forbidden) energy gap exists between the top of this filled band and the bottom of the next band of allowed energy states which are empty. Thus, a field can accelerate electrons only if a sizable amount of energy is supplied by thermal activation, optical excitation, or very strong applied electric field.

Semiconductors, which have electrical resistivities falling between the ranges ascribed to metals and to insulators, are characterized by (1) narrower forbidden energy gaps than insulators, making it relatively easy to stimulate *intrinsic* conduction thermally, optically, or electrically, and (2) the introduction of states into the forbidden band by the presence of appropriate impurity atoms (choice of impurity depends on the semiconductor, e.g., Group III or V elements in the Group IV semiconductors silicon and germanium). *Impurity* conduction occurs when electrons are excited from impurity levels just below the empty allowed band into that band, or when impurity levels just above the top of the filled band accept electrons from the filled band and leave in the filled band "holes," which act like free charge carriers of positive sign.

Effective mass values of electrons and holes have been experimentally determined for a number of solids by the technique of cyclotron resonance. This information, combined with other experimental results including magnetoresistance, optical absorption, and photoconductivity, yields a picture of the band structure as a function of the wave vector k (magnitude equal to the wave number and direction that of the electron momentum). Electron spin resonance gives values for the corrections due to spin-orbit interaction. Further data come from quantum phenomena (e.g., Shubnikov-de Haas effect) of the type in which magnetic field dependence of various properties such as magnetic susceptibility and electrical resistivity, measured at very low temperatures, shows an oscillatory character. All of these results are used to construct the surface of constant energy (Fermi surface) in k-space; this surface is characteristic of the solid and can be used to predict many of the physical properties of the material.

Electrical Properties. The electrical properties of a solid are primarily dependent upon the concentration of charge carriers (n) and the carrier mobility (μ), which is defined as the drift velocity acquired by a carrier in an electric field of unit intensity. An estimate of n is obtained by measuring the Hall coefficient, which is the ratio of the transverse electric field set up in the Y-direction divided by the product of the current density flowing in the X-direction and the magnetic field intensity applied in the Z-direction. In appropriate units the Hall co-efficient of a metal equals $1/(ne)$, where e is the electronic charge. In semiconductors the same equation applies, if carriers of only one sign are present, except that the right side is to be multiplied by a statistical factor of the order of unity. The sense of the Hall electric field indicates the sign of the charge carrier. The electrical conductivity is given by $ne\mu$, and conductivities due to positive and negative carriers are arithmetically additive. The mobility is experimentally determined by combining conductivity and Hall coefficinet measurments; theoretically the mobility is found by studying the collisions of charge carriers with lattice ions and impurities. In metals, in first approximation, n is temperature-independent and μ goes as $1/T$, so that the resistivity is directly proportional to T. As absolute zero is approached, the resistivity of a metal or alloy either approaches a constant value dependent on the disorder in the lattice, or the substance becomes a superconductor. In the latter case, the resistivity drops sharply, at a transition temperature, from the normal curve to a value immeasurably close to zero. In 1957 Bardeen, Cooper, and Schrieffer developed a fruitful theory according to which a quantum mechanical interaction of conduction electrons by pairs reduces their energy so that, in summation, there is a small energy gap between the normal and superconducting state that accounts for the phenomena characteristic of superconductivity. Niobium compounds in the superconducting state are used as the field windings of electromagnets to carry currents sufficient to generate fields of the order 100 kilogauss.

The resistivity of intrinsic semiconductors and alloys has the exponential temperature dependence given by proportion to exp $(E_g/2kT)$, where E_g is the width of the forbidden energy band and k is the Boltzmann constant. The resistivity of an impurity semiconductor is more complex in its temperature-dependence. Magnetoresistance refers to the increase observed in the resistance when a magnetic field is applied, usually transverse to the current flow. In a metal, in first approximation, the magnetoresistive ratio (resistance change divided by original resistance) is proportional to the square of the magnetic field strength with the proportionality factor containing μ^2. In semiconductors the behavior is more complicated, but the study of orientation effects has yielded essential information about the shapes of the energy bands.

Magnetic Properties. The magnetic characteristics of solids ultimately arise from the properties of orbital electronic motion and unpaired electron spins. A diamagnetic contribution is produced by an applied magnetic field in all atoms and ions, but this contribution may be more than balanced by a paramagnetic contribution. Diamagnetism may be thought of as an application of Lenz' Law to the orbital electronic motion. Diamagnetic susceptibilities are practically independent of temperature. Paramagnetism arises from the orientation of permanent magnetic dipoles with components parallel to the applied magnetic field. Since the orientation is influenced by thermal motion, paramagnetic susceptibilities are approximately inversely proportional to the absolute temperature. The permanent magnetic dipole moments arise from unpaired electron spins, especially in incompletely filled electron shells such as the $3d$ shell in elements 21 through 28 and the $4f$ shell in the rare earth elements. Ferromagnetism is characterized by a spontaneous magnetization such that the atomic magnetic moments throughout a small region (called a domain) are aligned to a high degree by a molecular field inherent in the material. This type of magnetization is a typical "cooperative" phenomenon, i.e., it arises from an interaction among atoms which can be attributed to exchange forces of a quantum mechanical nature. The application of an external magnetic field readily produces alignment of the domain moments with field and thus leads to the large magnetic induction characteristic of the ferromagnetic. If the sign of the exchange integral describing the interaction between atoms is negative instead of positive, ferromagnetism gives way to antiferromagnetism, a state in which neighboring spins are lined up antiparallel rather than parallel. The most readily observable result is that the susceptibility shows a maximum as a function of temperature. Ferrites (e.g., Fe_3O_4) demonstrate ferrimagnetism, which is attributed to antiparallel alignment of the two Fe^{3+} spins with the resultant molecular magnetic moment coming from the Fe^{2+} ion.

—V. A. Johnson, Purdue University, West Lafayette, Indiana.

SOLIDUS CURVE. A curve representing the equilibrium between the solid phase and the liquid phase in a condensed system of two components. The relationship is reduced to a two-dimensional curve by disregarding the influence of the vapor phase. The points on the solidus curve are obtained by plotting the temperature at which the last of the liquid phase solidifies, against the composition, usually in terms of the percentage composition of one of the two components.

SOLID WASTES. Wastes and Water Pollution.

SOLION. A small electrochemical oxidation-reduction cell consisting of a small cylinder containing a solution and divided into sections by platinum gauze, porous ceramics, or other materials. A type of solion for detecting sound waves consists of a potassium iodide-iodine solution in which the iodide ions are oxidized to triiodide ions at the anode, and the reverse process occurs at the cathode. The cell is constructed so that the sound waves cause agitation of the solution between the electrodes, and thus change the current. In addition to detection of sound, solions can be designed to detect changes in other conditions, such as temperature, pressure, and acceleration.

SOLUBILITY. A property of a substance by virtue of which it forms mixtures with other substances which are chemically and physically homogeneous throughout. The degree of solubility is the concentration of a solute in a saturated solution at any given temperature. The degree of solubility of most substances increases with a rise in temperature, but there are cases (notably the organic salts of calcium) where a substance is more soluble in cold than in hot solvents. See also **Solutions.**

SOLUBILITY PRODUCT. A numerical quantity dependent upon the temperature and the solvent, characteristic of electrolytes. It is the product of the concentrations of ions in a saturated solution and defines the degree of solubility of the substance. When the product of the ion concentrations exceeds the solubility product, precipitation commonly results. Strictly speaking, the product of the activities of the ions should be used to determine the solubility product, but in many cases the results obtained using concentrations, as suggested by Nernst, are correct.

SOLUBILIZATION. Defined loosely, solubilization is the enhancement of the solubility of one substance, the solubilizate, by another substance, the solubilizing agent or solubilizer. More strictly, it is a process occurring in the presence of a solvent, whereby one species, the solubilizing agent, diminishes the activity coefficient of another species, the solubilizate, and both species are soluble thereafter. J. W. McBain, who coined this term, used it to denote the dissolution of an otherwise insoluble material brought about by interaction with micelles, a type of colloid, present in the solvent. The definition given here, however, is more inclusive than his original concept, and could be extended logically to systems whose characteristics are remote or completely apart from colloidal behavior. Practice nevertheless limits the term to usage in which there is either a close or a marginal relationship to micelles, and the literature of solubilization refers chiefly to systems in which the solubilizers are micelle formers. For example, potassium laurate solubilizes hydrocarbons in water, and calcium xenylstearate solubilizes water in hydrocarbons because of the micelle-forming nature of the respective solubilizing agents. However, the striking similarity among interactions between various agents and both soluble and insoluble species makes it undesirably arbitrary to restrict the term solubilization rigidly to its original usage.

Because absolute insolubility does not exist in nature, insolubility must be considered a matter of degree. Consequently, if an *apparently* insoluble species, in unlimited excess, is in contact with a solvent it must have a finite concentration and activity in the solvent at equilibrium. A solubilizing agent added to the system may interact with this species by coordination, hydrogen bonding, dipole interaction, complex formation, or in some other manner. In any case, the interaction results in a decrease of the effective concentration, or activity, of this species. Accordingly, more of the solubilizate progressively dissolves until its activity returns to the initial equilibrium value in the pure solvent, whereupon the actvity coefficient is correspondingly less. If a species is freely soluble, or even infinitely miscible with a solvent, an interaction causes no *apparent* increase in the solubility of the species, but its activity, as evidenced by its osmotic behavior, nevertheless similarly decreases. The activity represents the tendency of the species to escape from the solution. Since solubility depends upon a balance between the opposing tendencies to enter and to leave the solution, the decreased activity is in effect equivalent to increased solubility. Solubilization is said to occur then, regardless of the independent solubility or insolubility in the pure solvent.

The salts of high-molecular weight organic acids are particularly important solubilizing agents. In nonpolar solvents such as hydrocarbons, they form colloidal aggregates known as association micelles. Most frequently such a micelle constitutes a limited number of salt monomers associated into a spheroidal cluster, with the polar ends of the salt monomers oriented toward the interior, and the nonpolar hydrocarbon ends at the periphery. Other polar species such as water, alcohol, acids, and dyes can be solubilized by these micelles in a variety of ways. In benzene solution, for example, zinc dinonylnaphthalene sulfonate can solubilize at least six moles of water for each equivalent weight of the salt present. The solution remains transparent, and no phase separation is observed. During the progressive addition of six moles of water per gram-equivalent of salt, the micelles expand to aggregations containing ten acid residues per unit, whereas the water-free micelles contain only seven. The water molecules are believed to be held in the polar core of the micelle where the environment is favorable to their retention.

Methanol, on the other hand, decreases the size of magnesium phenylstearate micelles. As methanol is solubilized by this salt in toluene solution, the micelle size decreases progressively from 23 salt monomers per aggregate to as little as 2 at a methanol concentration of 2% by weight. Each of these dimers is then associated with ten molecules of the alcohol. The partial pressure of methanol over the solution is demonstrably less than that over the salt-free methanol-toluene solution of equal methanol concentration. Rhodamine B dissolves very sparingly in pure benzene as the colorless and nonfluorescent base form. It is converted to the brilliantly fluorescent colored form by the addition of any of numerous micelle-forming solubilizers. No major changes in micelle size are believed to result from this solubilization, and it is postulated that a dye molecule replaces a monomer of the solubilizing agent in the matrix of the micelle.

In aqueous solutions, salts of high-molecular weight acids from micelles whose orientations are the reverse of those in nonpolar solvents. The hydrocarbon portions of the monomers, being insoluble in water, are oriented inward, whereas the ionic, or polar ends are oriented outward. Solubilization by these agents is complicated by dissociation of cations from surfaces of the aggregates and by the resulting surface charges developed. Both polar and nonpolar species such as hydrocarbons, dyes, alcohols, fats, organic acids, and a wide variety of soluble and insoluble species are solubilized in aqueous micellar solutions. Micelle enlargement is frequently said to follow from solubilization by these salts in aqueous solution, although the possibility of reduction of micelle size should not be excluded from consideration.

A nonpolar solubilizate such as hexane penetrates deeply into such a micelle, and is held in the nonpolar interior hydrocarbon environment, while a solubilizate such as an alcohol, which has both polar and nonpolar ends, usually penetrates less, with its polar end at or near the polar surface of the micelle. The vapor pressure of hexane in aqueous solution is diminished by the presence of sodium oleate in a manner analogous to that cited above for systems in nonpolar solvents. A 5% aqueous solution of potassium oleate dissolves more than twice the volume of propylene at a given pressure than does pure water.

Dimethylaminoazobenzene, a water-insoluble dye, is solubilized to the extent of 125 mg per liter by a 0.05 M aqueous solution of potassium myristate. Bile salts solubilize fatty acids, and this fact is considered important physiologically. Cetyl pyridinium chloride, a cationic salt, is also a solubilizing agent, and 100 ml of its $N/10$ solution solubilizes about 1 g of methyl ethyl-butyl either in aqueous solution.

Among other species that are good solubilizing agents are the nonionic compounds such as the polyethylene oxide-fatty acid condensates and the fatty esters of polyalcohols. A wide variety of nonionic solubilizing agents is possible, but most of those available are of variable composition. They can be effective in both aqueous and nonaqueous solutions.

The colloidal nature of some systems can disappear completely as solubilization proceeds. For instance, when methanol is solubilized by magnesium and sodium dinonylnaphthalene sulfonates, the aggregates decrease in size to a degree beyond which they can be considered micelles. In toluene solutions, micelles of these salts dissociate progressively on the addition of methanol increments until each of the particles in these solutions contains only one salt monomer when the methanol concentration reaches about 2% by weight. Probably the properties of some species which cause them to aggregate are those which make them good solubilizing agents, but it is evident that micelles are not a necessary condition for solubilization.

Accordingly, it is logical for solubilization to occur in systems which show no colloidal behavior, although frequently the effect in these cases is described by other proper terminology. Usually the term "solubilization" is applied in cases where the solubilizing agent is effective in small quantities, but arbitrary limitations of quantity might confuse the basic concept of solubilization. The terms cosolvency, hydrotropy, and "salting in" are used sometimes to describe effects which may be considered within the broad general scope of solubilization.

Applications of solubilization, although not always completely understood, range widely. A solubilizing agent can be used to bring an otherwise insoluble substance into solution where it is needed for a specific use, or it can be incorporated in a formulation to suppress the activity of an unwanted species which otherwise cannot be eliminated or prevented from occurring. In the pharmaceutical industry, drugs which are insoluble in pure water are solubilized by suitable agents to form homogeneous solutions. Dyes are solubilized for more efficient penetration and uniform coloring of fabrics. Soaps and detergents in aqueous solution are effective cleansing agents because they solubilize oily and greasy residues which may be flushed away from contaminated surfaces, although other effects may be equally important in the process. Removal of silver halides from photographic papers and films by aqueous fixing solutions may be considered solubilization by noncolloidal solubilizers. Certain oil-soluble salts dissolved in dry cleaning fluids can solubilize water. The water, which is solubilized in the micelles can in turn solubilize inorganic salts. The salts are then retained in the polar cores of the micelles where the water is held. This effect is referred to as secondary solubilization. In automotive fuels and lubricating oils, nonaqueous detergents are used to maintain engine cleanliness by solubilizing products of oxidation and combustion which tend to form sludges and gums, and to suppress the destructive effects of acids and other species generated in operation. Other solubilizing agents are used in these fluids to incorporate otherwise insoluble additives for oxidation and corrosion inhibition.

—Samuel Kaufman, Naval Research Laboratory, Washington, D.C.

SOLUTIONS. The equilibrium of a saturated solution represents a balance between the potentials and entropies of the molecules present in the two phases. These depend upon pressure, temperature, and the kind and strength of the attractions between the molecules. The attractions may be classified as interactions between ions, dipoles, metallic atoms, and the "electron clouds" of nonpolar molecules, differing among themselves in kind, in range, and in strength. The potential energy of the molecules of a nonpolar liquid is measured appropriately for the purpose of solubility relations by its energy of vaporization per cc, called its "cohesive energy density." The square root of this quantity will be used below as a "solubility parameter" δ.

We consider, first, the mutual solubility of two nonpolar liquids, whose molecules have practically equal sizes, and equal attractive and repulsive forces. When they are brought into contact, thermal agitation will cause mutual diffusion until the two species are uniformly distributed. The mixing process has produced maximum molecular disorder, and therefore entropy, which is given by the expression, for 1 mole of solution,

$$\Delta S^M = -R(x \ln x_1 + x_2 \ln x_2), \tag{1}$$

where R is the gas constant and x_1 and x_2 the respective mole fractions. The partial molal entropies of transfer of 1 mole from pure liquid to solution are

$$\bar{s}_1 - \bar{s}_1^0 = -R \ln x_1, \tag{2}$$

for component 1, and with subscript 2, for the other component.

The partial molal free energies of transfer are related to the *fugacities* in pure liquid, f^0 (vapor pressure corrected for deviation from the perfect gas law), and in solution, by the equations,

$$\bar{F}_1 - \bar{F}_1^0 = -R \ln (f_1/f_1^0). \tag{3}$$

and its counterpart.

Liquids such as are here postulated mix with no heat effect; therefore $F_1 - F_1^0 = -T(s_1 - s^0)$, etc.; therefore

$$f_1/f_1^0 = x_1 \quad \text{and} \quad f_2/f_2^0 = x_2 \tag{4}$$

which is Raoult's law, and defines the *ideal* solution.

If one of the components of an ideal solution, e.g., component 2, is a solid, its fugacity, f_2^s, is less than the fugacity of the pure, supercooled liquid, and limits the amount that can dissolve to $x_2 = f_2^s/f_2^0$. The ratio f_2^s/f_2^0, can be calculated from its melting point and heat of fusion.

Most solutions deviate from Raoult's law. The curved lines in Fig. 1 represent positive deviations, with $f_1/f_1^0 > x$. The ratio f_1/f_1^0 is called *activity*, and

$$f_1/f_1^0 = a, \quad \text{and} \quad a_1/x_1 = \gamma, \tag{5}$$

the *activity coefficient*.

Regular Solutions. The internal forces of a pair of liquids are seldom so nearly alike as to permit their mixture to obey Raoult's law very closely throughout the whole range of composition. In the absence of chemical interaction, the attraction between two different molecular species, provided their dipole moments are zero or small, is approximately the geometric mean of the attractions between the like molecules. Since a geometric mean is less than an arithmetic mean, the mixing is accompanied by expansion and absorption of heat. The partial molal heat of transfer per mole from pure liquid to solution is given with fair accuracy for many systems by the equation,

$$H_2 - H_2^0 = v_2 \phi_1^2 (\delta_2 - \delta_1)^2 \tag{6}$$

and its cognate. where $v \equiv$ molal volume, δ is a *solubility parameter*, the square root of the energy of vaporization per em³, and ϕ_2 is volume fraction.

Thermal agitation, except in the liquid-liquid critical region, suffices to give essentially maximum randomness of mixing, especially when one component is dilute, so that the entropy of mixing may be practically ideal, although the heat of mixing is not, and the partial molal free energy can be computed by combining the entropy and the heat terms, Eqs. (2), (5), and (6),

$$RT \ln a_2^s/x_2 = v_2\phi_1^2(\delta_1 - \delta_2)^2 \qquad (7)$$

This equation neglects the effects of expansion upon both the heat and the entropy, but the errors largely cancel when combined in Eq. (7).

A plot of a_2 vs. x_2 for symmetrical systems (i.e., $v_1 \approx v_2$) is shown in Fig. 1 for a series of values of the heat term. It

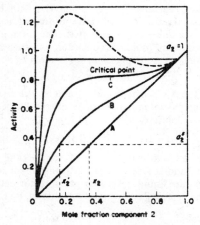

Fig. 1. Activity versus mole fraction for varying deviations from Raoult's law.

shows how the partial vapor pressure of a component of a binary solution deviates positively from Raoult's law more and more as the components become more unlike in their molecular attractive forces. Second, the place of T in the equation shows that the deviation is less the higher the temperature. Third, when the heat term becomes sufficiently large, there are three values of x_2 for the same value of a_2. This is like the three roots of the van der Waals equation, and corresponds to two liquid phases in equilibrium with each other. The criterion is that at the critical point the first and second partial differentials of a_2 and a_1 are all zero.

The presence of a dipole in one component adds a temperature-dependent component to its self-attraction and also induces a dipole in the other component. The effect can often by allowed for, for practical purposes, by an empirical adjustment of its solubility parameter.

If the dipole is hydrogen bonding, then this component is "associated," and it mixes less readily with a nonpolar second component.

If the components are, respectively, electron-donor and acceptor, or basic and acidic in the generalized sense of Gilbert Lewis, negative deviations from Raoult's law occur, with enhancement of solubility.

The effects of these various factors are well illustrated by solutions of iodine, I_2. In Fig. 2 are plotted the saturation values of log x_2 for iodine against log T. The slopes of the lines, when multiplied by R, give the entropy of transfer of iodine from solid to saturated solution. The solid lines are for violet solutions, from which chemical equilibria are absent. The posi-

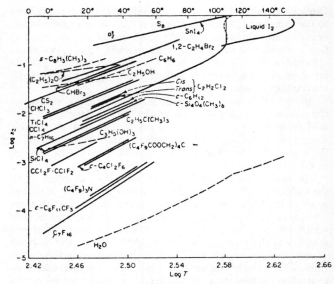

Fig. 2. Solubility of iodine.

tions of the lines are determined by the solubility parameters: how well is seen in the accompanying table, where δ-values are given for iodine in a spread of solvents calculated by means of Eq. (7) from the measured values of x_2. The broken lines indicate nonviolet solutions.

The factors that cause solutions of iodine to deviate from the behavior of regular solutions are illustrated in Fig. 3, in which values of the left hand member of Eq. (7) are plotted against those of the right for iodine solutions at 25°C; a_2^s is the activity of solid iodine; x_2 denotes measured solubility; v_2 is the extrapolated molal volume of liquid iodine, 59 cm³; ϕ_1 is the volume fraction of the solvent, ~1.0; $\delta_2 = 14.1$; δ_1 is the solubility parameter of the solvent. Illustrative values of x_2 and δ_1 are given in accompanying table.

δ-VALUES FOR I_2, 25°C

SOLVENT	MOLAL VOL. CC.	δ_1	$100x_2$
n-C_7F_{16}	227.0	5.7	.0185
$SiCl_4$	115.3	7.6	.499
Cyclo-C_6H_{12}	109.	8.2	.918
CCl_4	97.1	8.6	1.147
$TiCl_4$	110.5	9.0	2.15
CS_2	60.6	9.9	5.46
$CHBr_3$	87.8	10.5	6.16

The points on line A are all for regular solutions, conforming to Eq. (7) over large ranges of x_2. Line B starts with a point for iodine in cycyohexane, next a point for methylcyclohexane, followed by one for dimethylcyclohexane. The point below is for ethylcyclohexane. Line C is for normal alkanes, from $C_{16}H_{34}$ to C_5H_{12}; groups D and E are for branched alkanes. Displacements from line A increase with increasing ratios of —CH_3 to —CH_2. The reason for this is not clear.

Line F contains points for aromatics, from benzene at the top to mesitylene at the bottom. All complex with iodine, altering its color. Group G consists of CH_2Cl_2 and 1,1- and 1,2-

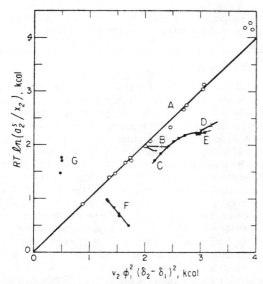

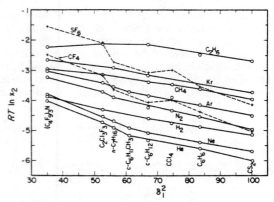

Fig. 3. Relation between energy of solution of iodine derived from measured solubility, x_2, and that calculated from solubility parameters.
Line A (beginning at lower left)
CS_2, $CHCl_3$, $TiCl_4$, cis-$C_{10}H_{18}$, trans-$C_{10}H_{18}$, CCl_4 c-C_6H_{12}, c-C_5H_{10}, $SiCl_4$, CCl_3CF_3, $CCl_2F \cdot CClF_2$, $C_4Cl_3F_7$, c-$C_4Cl_2F_6$, C_7F_{16}.
Line B (left to right)
c-C_6H_{12} (on line A), c-$C_6H_{11}C_2H_5$ (below), c-$C_6H_{11}CH_3$, c-$C_6H_{10}(CH_3)_2$.
Line C (left to right, normal paraffins)
$C_{16}H_{34}$, $C_{12}H_{26}$, C_8H_{18}, C_7H_{16}, C_6H_{14}, C_5H_{12}.
Line D (left to right)
2,3-$(CH_3)_2C_4H_8$, 2,2-$(CH_3)_2C_4H_8$.
Line E (left to right)
2,2,3-$(CH_3)_3C_4H_7$, 2,2,4-$(CH_3)_3C_5H_9$.
Line F (top to bottom)
C_6H_6, $C_6H_5CH_3$, p-$C_6H_4(CH_3)_2$, m-$C_6H_4(CH_3)_2$, 1,3,5-$C_6H_3(CH_3)_3$.
Group G (from top)
1,2-$C_2H_4Cl_2$, CH_2Cl_2, 1,1-$C_2H_4Cl_2$.

$C_2H_4Cl_2$, with strong dipoles, which enhance energy of vaporization without increasing solvent power for iodine.

Gases. Gas solubilities may be expressed as (1) volume of gas dissolved in unit volume of solvent, known as the *Ostwald coefficient*, designated by γ; (2) the volume of gas reduced to 0°C and 1 atmosphere dissolved in unit volume of solvent, known as the Bunsen coefficient, designated α; (3) the mole fraction, x; or (4) the moles per liter, c, dissolved at 1 atmosphere partial pressure. Henry's law, that the amount of gas dissolved is proportional to its partial pressure, holds rather well at moderate pressures in the basence of a chemical equilibrium. The fact that a substance is a gas at 1 atmosphere and ordinary temperatures indicates that its attractive forces are low and that consequently its solubility will be greater in solvents with low δ-values; also that solubility of different gases in the same solvent will be higher the higher the critical temperature of the gas.

The solubility of a number of gases at 1 atmosphere partial pressure and 25°C expressed as $RT \ln x_2$ is plotted in Fig. 4 against the squares of the solubility parameters of a number of solvents. A high amount of regularity is evident for all except the gases SF_6 and CF_4, whose molecules attract molecules of the solvents very selectively. Similar irregularity is evident in the case of the solvent $(C_4F_9)_3N$. In all other cases the positions of missing points could be predicted with confidence.

Fig. 4. Solubility of gases, log x_2 at 25°C and 1 atm versus square of solubility parameter of solvents.

Variations of solubility with temperature are illustrated in Fig. 4 for 10 gases in cyclohexane. The slopes of the lines times the gas constant R give values for the entropy of solution. In decending from C_2H_6 to He the entropy increases from -8.7 cal/deg mole to $+8.1$ partly from increases in entropy of dilution, $-R \ln x_2$, but also because the successive gases attract the surrounding solvent molecules less and less strongly, but since they have the same kinetic energy they finally almost blow bubbles permitting more freedom of motion to adjacent molecules of solvent.

The foregoing interesting phenomena are treated at length in "Regular and Related Solutions," by J. H. Hildebrand, J. M. Prausnitz, and R. L. Scott, Van Nostrand Reinhold, New York, 1970.

Solid Solutions. The formation of a solid solution requires not only attractive forces which are not too different, but also identical crystal structures. The latter condition is found most frequently among solids whose molecules are rotating, giving highly symmetrical crystals. See **Crystal.**

Metallic Solutions. In the absence of compounds, these follow the foregoing rules to a fair extent, but with added complications on account of the states of their electrons. The metals have a wide range of solubility parameters and exhibit many cases of incomplete miscibility in the liquid state.

Salt Solutions. The most obvious requirement necessary in a solvent for a salt is that it shall have a high dielectric constant, as is the case with water, liquid ammonia, hydrogen fluoride, and, in a smaller degree, methyl alcohol, in order to weaken the coulombic attraction of its ions for one another. It is possible to formulate the equilibrium between a solid salt and a solution of its ions by considering the changes in energy and entropy involved in vaporization of the solid to gaseous ions, and hydration of the ions. This would be relatively simple if the lattice energy of the solid and the hydration of the ions were solely electrostatic; but the process involves also van der Waals forces, polarization, covalent forces, hydrogen bonding, and entropy changes, which, in the case of water, are considerable, by reason of the ice structure persisting in water and the different structure of water of hydration. Consequently, such a breakdown of the problem, while it may serve to suggest comparisons, is better for explanation than for prediction.

The Periodic System offers the most useful guide by virtue of the trends it reveals; e.g., the decreasing solubility in water of the sulfates and the increasing solubility of the hydroxides of the elements of Group II in descending the group.

Liquid ammonia, because of its lower dielectric constant, is in general a much poorer solvent for salts than water; but this is offset to some extent toward salts of electron-acceptor, Lewis acid cations by its greater basic, electron-donor character.

Insight into the nature of electrolytes in water solutions is afforded by their effects in varying concentration upon the freezing point of water, Δt, at varying concentrations, m moles per 1000 grams. In Fig. 5 $\Delta t/m$ is plotted against m on a logarithmic scale.

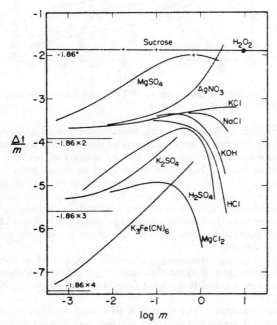

Fig. 5. Molal lowering of freezing points at different concentrations.

The molal lowering of nonelectrolytes is illustrated by sucrose and H_2O_2. These enter so easily into the hydrogen bonded structure of water that they give the theoretical lowering, 1.86° up to 0.1 M in the case of sucrose and to 10 M by H_2O_2.

Binary electrolytes, such as KCl, although completely ionized, even in the solid state, lower the freezing point less than $2 \times 1.86°$, even when as dilute as 10^{-3} M. This was at first attributed to incomplete ionization but is now explained by the long range of electrostatic forces. Note that Mg^{++} and SO_4^{--} are less independent than K^+ and Cl^-. $AgNO_3$, unlike KCl, etc., is a weak salt, and undissociated molecules increase rapidly with concentration. The ions nearer to an ion of one sign are those of opposite sign, therefore electric conductivity is less than the sum of ionic conductivities extrapolated to zero concentration.

This effect has been formalized in the concept of *ionic strength*, expressed as

$$I = \tfrac{1}{2}(m_1 z_1^2 + m_2 z_2^2 + m_3 z_3^2 + \cdots)$$

where z is the ionic charge. Applied to solutions of KCl, K_2SO_4 and $MgSO_4$ the values of I are respectively, 0.01, 0.03, and 0.04. Ionic strength is significant for dealing with the equilibrium and kinetic properties of an ion in mixtures of electrolytes.

Concentrated solutions are strongly affected by ionic hydration. Its strength depends upon ionic radius and charge, therefore it is in general stronger for cations than anions. K_2SO_4 and $MgSO_4$ both yield 3 ions, but the hydration is stronger for Mg^{++} than K^+, Na^+ than K^+, Na^+ than Ag^+. The line for K_2SO_4 ascends whereas the one for $MgSO_4$ plunges downward, (a) because the strong hydration of Mg^{++} diminishes the

coulombic attraction of SO_4^{--} and (b) because it ties up molecules of water, decreasing the amount of solvent.

—Joel H. Hildebrand, University of California, Berkeley, California.

SOLVENT. The term solvent generally denotes a liquid which dissolves another compound to form a homogeneous liquid mixture in one phase. More broadly, the term is used to mean that component of a liquid, gaseous, or solid mixture which is present in excess over all other components of the system. A *chemical solvent* is the term used for solvents in those instances where the process of solution is attended by a chemical reaction between the solvent and the solute. In contrast, a *physical solvent* is one that does not react with the solute. A *dissociating solvent* is one in which solutes that associate in many other solvents enter into solution as single molecules. For instance, various carboxylic acids associate and thus give abnormal elevations of the boiling point, abnormal depressions of the freezing point, etc., in many organic solvents; but in water, however, they do not associate. For this reason water is called a dissociating solvent for such solutes. A liquid that dissolves or extracts a substance from solution in another solvent without itself being very soluble in that other solvent is termed an *immiscible solvent*. A solvent whose constituent molecules do not possess permanent dipole moments and which do not form ionized solutions is termed a *nonpolar solvent*. *Polar solvents*, on the other hand, consist of polar molecules, that is, molecules that exert local electrical forces. In such solvents, acids, bases, and salts, that is, electrolytes, in general, dissociate into ions and form electrically conducting solutions. Water, ammonia, and sulfur dioxide are typical polar solvents. A *normal solvent* is one which does not undergo chemical association, namely, the formation of complexes between its molecules.

A *leveling solvent* is a solvent in which the acidity or basicity of a solute is limited (or leveled) by the acidity or basicity of the solvent itself. For example, the strongest acid which can exist in water is oxonium ion, H_3O^+. Consequently, even though HCl (for example) is intrinsically a much stronger acid than H_3O^+, its acidity in aqueous solution is "leveled" to that of H_3O^+ through the reaction $HCl + H_2O \rightleftharpoons H_3O^+ + Cl^-$. Likewise the very strong base KNH_2 is leveled in water to the basicity of OH^-.

$$KNH_2 + H_2O \rightleftharpoons K^+ + OH^- + NH_3$$

The solvents which are leveling to both acids and bases are self-ionized solvents, e.g., water, ammonia, alcohols, carboxylic acids, nitric acid, etc. Basic non-protonic solvents are leveling to acids, but not to bases (i.e., they are differentiating toward bases), e.g., pyridine, ethers, ketones, etc., since the strongest acid attainable is the protonated solvent molecule (e.g., $C_5H_5N + HCl \rightleftharpoons C_5H_5NH^+ + Cl^-$), whereas there is no corresponding basic species derived from the solvent. Though solvents leveling to bases but not to acids are in principle much more difficult to find, in practice, very strong acids like H_2SO_4 and $HClO_4$ are limiting to bases because the species HSO_4^- and ClO_4^-, which will be formed by almost any basic substance, are the strongest bases attainable in these solvents—$B^- + HClO_4 \rightleftharpoons HB + ClO_4^-$—whereas practically no other acid is capable of producing the cations $H_3SO_4^+$ and $H_2ClO_4^+$ in these solvents (i.e., they are differentiating toward acids).

Differentiating solvents are solvents in which neither the acidity of acids nor the basicity of bases is limited by the nature of the solvent. These solvents are not self-ionized. The aliphatic hydrocarbons and the halogenated hydrocarbons are such solvents.

In industry it is generally understood that solvents are simple or complex, pure or impure, compounds or mixtures of compounds (either natural or synthetic) which dissolve many water-insoluble products like fats, waxes, resins, etc., forming homogeneous solutions; that such organic solvents dissolve these water-insoluble products in various proportions depending on the solvent power of the solvent, the degree of solubility of the solute, and the temperature; and that the solute can be recovered with its original properties by the removal of the solvent from the solution. It is also understood in industry that there is a much more limited number of solvents which do not have the properties given above but which nevertheless are of considerable importance; they are the inorganic solvents like water, liquid ammonia, liquid metals, and the like.

Solvents have been classified on various arbitrary bases: (1) boiling point, (2) evaporation rate, (3) polarity, (4) industrial applications, (5) chemical composition, (6) proton donor and proton acceptor relationships, and (7) behavior toward a dye, Magdala Red. Thus on the basis of industrial application one can classify solvents as those for (1) acetylcellulose, (2) pyroxylin, (3) resins and rubber, (4) cellulose ether, (5) chlorinated rubber, (6) synthetic resins, and (7) solvents and blending agents for cellulose ester lacquers. Solvents classified according to chemical composition are noted below.

The term solvent action is understood to mean any process of making substances water-soluble; but in a broader interpretation the term is understood to be the phenomenon of making a substance soluble in a solvent. Solvent power, diluting power, solvency and similar expressions indicate the property of solvents to disperse the molecules of a solute or vehicle thereby causing a decrease in viscosity.

The most common solvent is water. Water dissolves a great many gases, liquids, and solids, and is much used for this purpose. Other liquids similarly dissolve many substances without reacting chemically with them. Important considerations in connection with the choice of solvent for a given case are (1) vapor pressure and boiling point, (2) solvent power under stated conditions of temperature, (3) ease and completeness of recoverability by evaporation and condensation, and completeness of seperation from dissolved material by evaporation, (4) heat of vaporization, (5) miscibility with water or other liquid, if present, (6) inertness to chemical reaction with the materials present, and with the apparatus, (7) inflammability and explosiveness, (8) odor and toxicity, (9) cost of solvent, loss in process, cost of recovering.

See also **Pollution (Air)**.

Colligative Properties of Solutions. When solute is added to a pure solvent, thus forming a solution, properties of the solvent are altered, including (1) osmotic pressure; (2) vapor pressure (lowered); (3) melting point (lowered); and (4) boiling point (elevated). These properties bar a relationship to the number of solute molecules in solution and not to the nature of the molecules. These phenomena are explained by enhanced tension in the solvent. Complete explanation of these changes is beyond the scope of this book, but reference is suggested to H. T. Hammel's article on "Colligative Properties of a Solution" (*Science*, **192**, 748–756, 1976).

SOLVOLYSIS. A generalized conception of the relation between a solvent and a solute (i.e., a relation between two components of a single-phase homogeneous system) whereby new compounds are produced. In most instances, the solvent molecule donates a proton to, or accepts a proton from a molecule of solute, or both, forming one or more different molecules. A particular case of special interest occurs when water is used as solvent, in which case the interaction between solute and solvent is called *hydrolysis*.

SORBITOL. Sweeteners.

SOYBEAN OIL. Vegetable Oils (Edible).

SOY PROTEIN. Protein.

SPECIFIC GRAVITY. For a given liquid, the specific gravity may be defined as the ratio of the density of the liquid to the density of water. Because the density of water varies, particularly with changes in temperature, the temperature of the water to which a specific gravity measurement is referred should be stated. In exacting, scientific observations, the reference may be to pure (double-distilled) water at 4°C (39.2°F). In engineering practice, the reference frequently is to pure water at 15.6°C (60°F). A value of unity is established for water. Thus, liquids with a specific gravity less than 1 are lighter than water; those with a specific gravity greater than 1 are heavier than water. From a practical standpoint, it usually is more meaningful to express the specific gravity of gases with reference to pure air rather than to pure water. Thus, for a given gas, the specific gravity may be defined as the ratio of the density of the gas to the density of air. Since the density of air varies markedly with both temperature and pressure, exacting observations should reflect both conditions. Common reference conditions are 0°C and 1 atmosphere pressure (760 torr; 760 millimeters Hg; 29.92 inches Hg).

Specific Gravity Scales. Arising essentially from a lack of communication between various scientific and industrial communities, a number of different specific gravity scales were formulated in earlier times and, because so much data and experience have been accumulated in terms of these scales, several methods of expressing specific gravity persist in common use. The most important of these scales are defined here.

API Scale—This scale was selected in 1921 by the American Petroleum Institute, the U.S. Bureau of Mines, and the National Bureau of Standards (Washington, D.C.) as the standard for petroleum products in the United States.

$$\text{Degrees hydrometer scale (at } 15.6°C; 60°F) = \frac{141.5}{\text{sp gr}} - 131.5$$

Balling Scale—This scale is used mainly in the brewing industry to estimate percent wort but also is used to indicate percent by weight of either dissolved solids or sugar liquors. Hydrometers are graduated in percent weight at 60°F or 17.5°C.

Barkometer Scale—This scale is used essentially in the tanning and tanning-extract industry. Water equals zero. Each scale degree equals a change of 0.001 in specific gravity. The following formula applies:

$$\text{Sp gr} = 1.000 \pm 0.001 \times (\text{degrees Barkometer})$$

Baumé Scale—This scale is used widely in connection with the measurement of acids and light and heavy liquids, such as syrups. The scale originally was proposed by Antoine Baumé, a French chemist, in 1768. The scale has been widely accepted because of the simplicity of the numbers which represent liquid specific gravity. Two scales are in use:

$$\text{For light liquids, } °Bé = \frac{140}{\text{sp gr}} - 130$$

$$\text{For heavy liquids, } °Bé = 145 - \frac{145}{\text{sp gr}}$$

The standard temperature for these formulas is 15.6°C (60°F).

To calibrate his instrument for heavy liquids, Baumé prepared a solution of 15 parts by weight of sodium chloride in water. On his hydrometer, Baumé marked zero at the point to which the float submerged in pure water; and he marked the scale 15 at the point to which the float submerged in the salt solution. He then divided the distance between the two marks into 15 equal spaces (or degrees as he termed them). In connection with liquids lighter than water, Baumé prepared a 10% sodium chloride solution. In this case, he marked the scale zero at the point to which the float submerged in the salt solution; and he marked the scale 10 at the point to which the float submerged in pure water. Thus, he created a scale which provided increasing numbers with decreases in density.

Users of the Baumé method found that the scale generally read 66 when the float was submerged in oil of vitriol. Thus, early manufacturers of hydrometers calibrated the instruments by this method. There were variations in the Baumé scale, however, because of lack of standardization in hydrometer calibration. Consequently, in 1904, the National Bureau of Standards made a careful survey and finally adopted the scales previously given for light and for heavy liquids.

Brix Scale—This scale is used almost exclusively by the sugar industry. Degrees on the scale represent percent pure sucrose by weight at 17.5°C (63.5°F).

Quevenne Scale—This scale is used for milk testing and essentially represents an abbreviation of specific gravity. For example, 20° Quevenne indicates a specific gravity of 1.020; 40° Quevenne, a specific gravity of 1.040, and so on. One lactometer unit approximates 0.29° Quevenne.

Richter, Sikes, and Tralles Scales—These are alcoholometer scales which indicate directly in percent ethyl alcohol by weight in water.

Twaddle Scale—This scale is the result of attempting to simplify the measurement of industrial liquids heavier than water. The range of specific gravity from 1.000 to 2.000 is divided into 200 equal parts. Thus, 1° Twaddle equals 0.005 sp gr.

An abridged compilation of specific gravity conversions is given in the accompanying table. The specific gravity (sp gr) of hundreds of materials are given throughout this volume.

SPECIFIC GRAVITY SCALE EQUIVALENTS

Specific Gravity 60°/60°F	°Baume	°API	Specific Gravity 60°/60°F	°Baume	°API
0.600	103.33	104.33	0.800	45.00	45.38
0.620	95.81	96.73	0.820	40.73	41.06
0.640	88.75	89.59	0.840	36.67	36.95
0.660	82.12	82.89	0.860	32.79	33.03
0.680	75.88	76.59	0.880	29.09	29.30
0.700	70.00	70.64	0.900	25.56	25.72
0.720	64.44	65.03	0.920	22.17	22.30
0.740	59.19	59.72	0.940	18.94	19.03
0.760	54.21	54.68	0.960	15.83	15.90
0.780	49.49	49.91	0.980	12.86	12.89
			1.000	10.00	10.00

Specific Gravity 60°/60°F	°Baume	°Twaddle	Specific Gravity 60°/60°F	°Baume	°Twaddle
1.020	2.84	4	1.500	48.33	100
1.040	5.58	8	1.520	49.61	104
1.060	8.21	12	1.540	50.84	108
1.080	10.74	16	1.560	52.05	112
1.100	13.18	20	1.580	53.23	116
1.120	15.54	24	1.600	54.38	120
1.140	17.81	28	1.620	55.49	124
1.160	20.00	32	1.640	56.59	128
1.180	22.12	36	1.660	57.65	132
1.200	24.17	40	1.680	58.69	136
1.220	26.14	44	1.700	59.71	140
1.240	28.06	48	1.720	60.70	144
1.260	29.92	52	1.740	61.67	148
1.280	31.72	56	1.760	62.61	152
1.300	33.46	60	1.780	63.54	156
1.320	35.15	64	1.800	64.44	160
1.340	36.79	68	1.820	65.33	164
1.360	38.38	72	1.840	66.20	168
1.380	39.93	76	1.860	67.04	172
1.400	41.43	80	1.880	67.87	176
1.420	42.89	84	1.900	68.68	180
1.440	44.31	88	1.920	64.98	184
1.460	45.68	92	1.940	70.26	188
1.480	47.03	96	1.960	71.02	192
			1.980	71.77	196
			2.000	72.50	200

NOTE: 60°F = 15.6°C

SPECIFIC HEAT. Sometimes called specific heat capacity. The quantity of heat required to raise the temperature of unit mass of a substance by one degree of temperature. The units commonly used for its expression are the unit mass of one gram, the unit quantity of heat in terms of the calorie.

Specific Heat at Constant Pressure. The amount of heat required to raise unit mass of a substance through one degree of temperature without change of pressure. Usually denoted by C_p, when the mole is the unit of mass, and c_p when the gram is the unit of mass.

Specific Heat of Constant Volume. The amount of heat required to raise unit mass of a substance through one degree of temperature without change of volume. Usually denoted by C_v, when the mole is the unit of mass, and c_v when the gram is the unit of mass.

SPERRYLITE. A mineral diarsenide of platinum, $PtAs_2$. Crystallizes in the isometric system. Hardness, 6–7; specific gravity, 10.58; color, white; opaque. Named after Francis L. Sperry, Sudbury, Ontario.

SPHALERITE BLENDE. Also known as zinc blende, this mineral is zinc sulfide, $(Zn, Fe)S$, practically always containing some iron, crystallizing in the isometric system frequently as tetrahedrons, sometimes as cubes or dodecahedrons, but usually massive with easy cleavage, which is dodecahedral. It is a brittle mineral with a conchoidal fracture; hardness, 2.5–4; specific gravity, 3.9–4.1; luster, adamantine to resinous, commonly the latter. It is usually some shade of yellow brown or brownish-black, less often red, green, whitish, or colorless; streak, yellowish or brownish, sometimes white; transparent to translucent. Certain varieties are phosphorescent or fluorescent. Sphalerite is the commonest of the zinc-bearing minerals, and is found associated with galena, chalcopyrite, tetrahedrite, barite, and fluorite, as a result of contact metamorphism, and as replacements and vein deposits.

There are very many European localities, including Saxony; Bohemia; Switzerland; Cornwall, in England; Spain; Sweden; Japan; and elsewhere. In the United States, sphalerite is found in Arkansas, Iowa, Wisconsin, Illinois, Colorado, New Jersey, Pennsylvania, Ohio, and especially in the area which includes parts of Kansas, Missouri, and Oklahoma. The word sphalerite is derived from the Greek, meaning treacherous, and its older name, blende, meaning blind or deceiving, refers to the fact that it was often mistaken for lead ore.

SPHENE. This mineral occurs as a yellow, green, gray, or brown calcium titanosilicate, corresponding to formula $CaTiSiO_5$, crystallizing in the monoclinic system. Fracture conchoidal to uneven; brittle; habit usually wedge-shaped and flattened crystals, also massive and lamellar; luster, resinous to adamantine; transparent to opaque; hardness, 5–5.5; specific gravity, 3.45–3.55.

Sphene is an accessory mineral of widespread occurrence in igneous rocks, and calcium-rich schists and gneisses of metamorphic origin, and very common in nepheline-syenites. In the United States sphene is found in Arkansas, California, New Jersey, New York; in Ontario and Quebec in Canada; and from Greenland, Brazil, Norway, France, Austria, Finland, U.S.S.R., Madagascar and New Zealand, as well as many other world localities.

SPINEL. The mineral spinel is one of a group of minerals which crystallize in the isometric system with an octahedral habit, and whose chemical compositions are analogous. These mineral are combinations of bivalent and trivalent oxides of magnesium, zinc, iron, manganese, aluminum, and chromium, the general formula being represented as $R''O \cdot R_2'''O_3$. The bivalent oxides may be MgO, ZnO, FeO, and MnO, and the trivalent oxides Al_2O_3, Fe_2O_3, Mn_2O_3, and Cr_2O_3. The more important members of the spinel group are spinel, $MgAl_2O_4$; gahnite, zinc spinel, $ZnAl_2O_4$, franklinite (Zn,Mn^{2+},Fe^{2+}) $(Fe^{3+},Mn^{3+})_2O_4$, and chromite, $Fe Cr_2O_4$. True spinel has long been found in the gem-bearing gravels of the Island of Ceylon and in limestones of Burma and Thailand.

Spinel usually occurs in isometric crystals, octahedrons, often twinned. It has an imperfect octahedral cleavage; conchoidal fracture; is brittle; hardness, 7.5–8; specific gravity, 3.58; luster, vitreous to dull; transparent to opaque; streak white; may be colorless, rarely through various shades of red, blue, green, yellow, brown, or black. These colors are doubtless due to small amounts of impurities. The clear red spinels are called spinel-rubies or balas-rubies and were often confused with genuine rubies in times past. Rubicelle is a yellow spinel. A violet-colored manganese-bearing spinel is called almandine spinel.

Spinel is found as a metamorphic mineral, also as a primary mineral in basic rocks, because in such magmas the absence of alkalies prevents the formation of feldspars, and any aluminum oxide present will form corundum or combine with magnesia to form spinel. This fact accounts for the finding of both ruby and spinel together. In addition to the localities mentioned above which yield beautiful specimens, spinel is found in Italy and Sweden and in the Malagasy Republic. Also in the United States in Orange County, New York, and in Sussex County, New Jersey, are many well-known spinel localities. Spinel is found also in Macon County, North Carolina, and in Canada in Quebec and Ontario.

The name spinel is derived from the Greek, meaning a spark, in reference to the fire-red color of the sort much used for gems. Balas ruby is derived from Balascia, the ancient name for Badakhshan, a region of central Asia situated in the upper valley of the Kokcha River, one of the principal tributaries of the Oxus.

SPIRO COMPOUND. Compound (Chemical).

SPODUMENE. The mineral spodumene is a lithium aluminum silicate corresponding to the formula $LiAlSi_2O_6$ and occurs in monoclinic prismatic crystals, occasionally of very large size. It also occurs massive. Spodumene has a perfect prismatic cleavage often very noticeable; uneven to splintery fracture; brittle; hardness, 6.5–7.5; specific gravity, 3–3.2; luster, vitreous to dull; color, grayish- to greenish-white, green, yellow and purple. Its streak is white; it is transparent to translucent. Spodumene is characteristically a mineral of the pegmatites, and it is found in Sweden, Ireland, the Malagasy Republic and Brazil. In the United States it is found especially in the pegmatites of Oxford County, Maine; in the towns of Goshen, Huntington and Chesterfield in western Massachusetts; at Branchville, Connecticut; in North Carolina; in South Dakota in huge crystals and in San Diego and Riverside Counties in California.

The name spodumene is derived from the Greek meaning ash-colored, particularly appropriate for the slightly weathered varieties. Hiddenite, the beautiful emerald-green or yellow-green spodumene that is used as a gem, was named for W. E. Hidden. Kunzite, named in honor of George F. Kunz, is a transparent lilac to rose-colored spodumene from the Malagasy Republic and California, and recently as magnificent, large gem crystals of both purple and yellow color from the Hindu-Kush Mountains, Nuristan Province in Afghanistan. Beautiful gem stones are cut from such crystals, but its easy cleavage discourages its use as a wearable gem. Spodumene alters rather readily to a mass of albite and muscovite. The commercial use of spodumene is chiefly as a source of lithium compounds.

See also **Lithium;** and **Lithium (For Thermonuclear Fusion Reactors.)**

SPRAY DRYING. A process used in the production of numerous chemical and food products. It is widely used in connection with the production of powdered milk and instant coffee preparations. The spray dryer is unique among dryers in that it dries a finely divided droplet by direct contact with the drying medium (usually air) in an extremely short retention time (3–30 seconds). This short contact time results in minimum heat degradation of the dried product, a feature that led to the popularity of the spray dryer in the food and dairy industries during its early development. In the case of coffee extract, water in the feed will range from 50 to 70%.

Atomization. Inasmuch as the spray dryer operates by drying a finely divided droplet, the feed to the dryer must be capable of being atomized sufficiently to ensure that the largest droplet produced will be dried within the retention time provided. There are different requirements on the degree of atomization needed to result in the desired product. These factors include minimizing the fines and/or coarse fractions, controlling particle dryness, and controlling bulk density. All commercial atomizers, whether of the centrifugal-wheel, pressure-nozzle, or other types, will produce a particle-size distribution that follows a probability curve. As the total energy input increases, the average particle size will decrease and the particle-size distribution will improve, i.e., the spread between the largest and smallest particles will be less.

Centrifugal-Wheel Atomizers. The wheel consists of a disk which is rotated at very high speed (1,700–50,000 revolutions per minute). See Fig. 1. Feed generally is introduced to the center, with centrifugal force dispersing the feed and throwing out a thin film to the periphery. As the film leaves the disk, it breaks up into a thread, which in turn forms droplets. The disk is located in the hot-air stream so that even though droplets are thrown toward the wall of the dryer, the hot air travels cocurrently and dries the particle sufficiently to prevent wall buildup upon contact. A spray dryer with a wheel atomizer

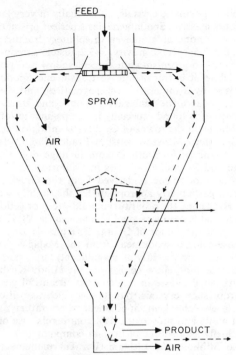

Fig. 1. Spray dryer with wheel atomizer. (1) Air outlet when drying chamber is used for initial separation.

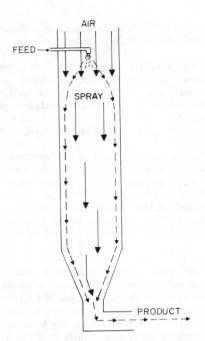

Fig. 2. Spray dryer featuring parallel flow.

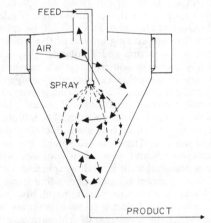

Fig. 3. Mixed-flow spray dryer.

must be relatively large in diameter and shorter than a dryer with pressure nozzles.

Pressure-Nozzle Atomizers. This system consists of an orifice placed after a fixed mechanism, called a core, swirl chamber, or whizzer, depending upon the manufacturer. A high-pressure pump moves the feed to the nozzle body at a pressure of from 250–8000 pounds per square inch (17–544 atmospheres). The feed slurry is pumped through high-pressure piping to the whizzer, where a spin is imparted to the fluid before it enters the nozzle orifices. This results in a hollow-cone spray which throws droplets either cocurrent or countercurrent to the air flow.

The flow pattern is such that a cocurrent spray dryer must be relatively long and small in diameter (Fig. 2), whereas a countercurrent dryer is shorter and larger in diameter. A third type, sometimes referred to as a mixed-flow dryer (Fig. 3), uses an air pattern similar to a cyclone collector, i.e., the spray is introduced at the upcoming air stream (countercurrent) and the particles transfer to the air sweeping the wall (cocurrent).

References

Charm, S. E.: "Fundamentals of Food Engineering," 3rd Edition, AVI, Westport, Connecticut (1978).

Flink, M. M.: "Energy Analysis in Dehydration Processes," *Food Technology*, **31**, 3, 77–84 (1977).

Harper, W. J., and C. W. Hall: "Dairy Technology and Engineering," AVI, Westport, Connecticut (1976).

STACHYOSE. Sweeteners.

STAINLESS STEEL. Iron Metals, Alloys, and Steels.

STALACTITE. A stalactite is a deposit of calcium carbonate which hangs icicle-like from the roof or wall of a limestone cavern, and is formed by the dipping of mineralized solutions. Corresponding columnar structures built upward from the floors of caves beneath the stalactites in a similar manner, are called stalagmites. *Stalactite* is derived from the Greek, meaning *to fall in drops; stalagmite*, from the Greek, meaning *that which drops.*

STANNITE (Mineral). This mineral is a sulfo-stannate of copper and iron, sometimes with some zinc, corresponding to the formula, Cu_2FeSnS_4. It is tetragonal; brittle with uneven fracture; hardness, 4; specific gravity. 4.3–4.5; metallic luster; color, gray to black, sometimes tarnished by chalcopyrite; treak, black; opaque. The mineral occurs associated with cassiterite, chalcopyrite, tetrahedrite, and pyrite, probably the result of deposition by hot alkaline solution. Stannite occurs in Bohemia; Cornwall, England; Tasmania; Bolivia; and in the United States in South Dakota. It derives its name from the Latin *stannum*, meaning *tin.*

STARCH. Chemically, starch is a homopolymer of α-D-glucopyranoside of two distinct types. The linear polysaccharide, amylose, has a degree of polymerization on the order of several hundred glucose residues connected by alpha-D-(1 → 4)-glucosidic linkages. The branched polymer, amylopectin, has a DP (degree of polymerization) on the order of several hundred thousand glucose residues. The segments between the branched points average about 25 glucose residues linked by alpha-D-(1 → 4)-glucosidic bonds, while the branched points are linked by alpha-D-(1 → 6)-bonds. See Fig. 1.

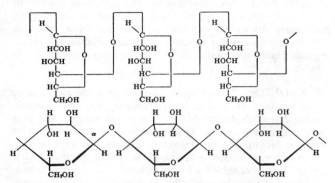

Fig. 1. A segment of the starch molecule.

Most cereal starches are made up of about 75% amylopectin and 25% amylose molecules. However, root starches are slightly higher in amylopectin, while waxy corn and waxy milo starch contain almost 100% amylopectin. At the other extreme, high amylose corn starch and wrinkled pea starches contain 60–80% amylose. The molecules of amylose and amylopectin are synthesized by enzymes inside the living cell in plastids known as amyyloplasts and are deposited as starch granules. These granules are microscopic in size, ranging from 3–8 micrometers in diameter for rice starch up to 100 micrometers for the larger potato starch granules. Corn starch usually falls in a range of 5–25 micrometers. An experienced observer usually can identify the genetic origin of a sample of starch by the size and shape of the granules. The granules are insoluble in cold water, but swell rapidly when heated to the gelatinization temperature range for the particular starch involved. As the granules swell, they lose their characteristic cross under polarized light and imbibe water rapidly until they are many times their original size. Upon continued heating or mechanical shear, the swollen granules begin to disintegrate and the viscosity, having reached a maximum, begins to decrease. However, there usually are some granules and some segments of granules that do not completely disperse in aqueous systems even under the most stringent conditions.

As the partially dissolved paste is cooled, the hydrated molecules and segments of granules begin to precipitate. In a dilute system (approximately 1%), the segments and molecules retrograde or precipitate. At higher concentrations, sufficient intermolecular and intersegment bonds form to fix the entire system into three-dimensional gel. The rigidity of this gel is affected by many factors, but the amylose content is perhaps the most significant. High amylose starches, when thoroughly cooked, form very rigid gels. Waxy corn or waxy milo starch paste form little, if any, gel structure when cooled.

While some wheat and potatoes are processed in the United States, over 90 % of all starch is produced from corn in what is called the *corn wet milling industry*. Close to one-quarter of a billion bushels of corn, representing about 5% of the total corn crop, is converted into wet-process products. The corn refining process is illustrated in Fig. 2. Shelled corn is delivered to the wet-milling plant in boxcars containing an average of

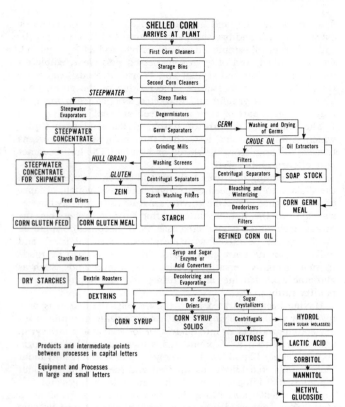

Fig. 2. The corn (maize) refining process. (*Corn Refiners Association, Inc.*)

2,000 bushels (50.8 metric tons) per car, and unloaded into a grated pit. The corn is elevated to temporary storage bins, and then to scale hoppers for weighing and sampling. The corn passes through mechanical cleaners designed to separate unwanted substances, such as pieces of cobs, sticks, and husks, as well as metal and stones. The cleaners agitate the kernels over a series of perforated metal sheets; the smaller foreign materials drop through the perforations, while a blast of air blows away chaff and dust, and electromagnets draw out nails and bits of metal. Coming out of the storage bins, the corn is given a second cleaning before going into very large "steep" tanks.

At this point, the use of water becomes an essential part of the corn refining process. The cleaned corn is typically moved into large wooden or metal tanks holding 2,000 to 6,000 bushels (50.8 to 152.4 metric tons), and soaked for 36 to 48 hours in circulating warm water 49°C (120°F) containing a small amount of sulfur dioxide to control fermentation and to facilitate softening. At the end of the steeping process, the steepwater contains much of the soluble protein, carbohydrates, and minerals of the corn kernel, and is drawn off as the first by-product of the process. Steepwater, unmodified and modified, is an essential nutrient for production of antibiotic drugs, vitamins, amino acids, and fermentation chemicals. It is also an effective growth supplement for animal feeds.

From the steeps, the softened kernels go through degerminating mills, which are designed not for fine grinding, but rather for tearing the soft kernels apart into coarse particles, freeing the rubbery oil-bearing germ without crushing it, and loosening the bran. The wet, macerated kernels then are sluiced into flotation tanks, called germ separators, or centrifugal hydrocyclones. The germs, lighter than the other components of the kernel, float to the surface, and are skimmed off. By oil expellers or extractors (heat and pressure) and by means of solvents, practically all of the oil is removed as another by-product to be settled, filtered, refined, and otherwise processed into clear, edible oil for salad dressing and frying, and "corn oil foods" or "soap stock" for soap manufacture. The residue of the germ, after oil-extraction, is ground and marketed as corn germ meal, or may become a part of corn gluten feed or meal.

The remaining mixture of starch, gluten, and bran (hull), which is finely ground, is washed through a series of screens to sieve the bran from the starch and gluten. The hull becomes part of corn gluten feed.

The remaining mixture of gluten and starch is pumped from the shakers to high speed centrifugal machines, which, because of the difference in specific gravity, separate the relatively heavier starch from the lighter gluten. After further processing, the protein-rich gluten is marketed as such, or becomes corn gluten meal, or may be mixed with steepwater, corn oil meal, and hulls to become corn gluten feed. Gluten may also be made to yield a highly versatile protein, *zein*; amino acids, such as glutamic acid, leucine, and tyrosine; and xanthophyll oil, for poultry rations.

Having been separated from the kernels, the starch is now ready for washing, drying, or further processing into numerous dry-starch products, or into dextrin, or for conversion into syrup and sugar. From a 56-pound (25 kilograms) bushel of corn, approximately 32 pounds (14.5 kilograms) of starch result, about 14.5 pounds (6.6 kilograms) of feed and feed products, about 2 pounds (0.9 kilogram) of oil, the remainder being water.

Starch Conversion. More than half of the total production of starch is converted into syrup dextrins or dextrose by acid hydrolysis and/or enzyme action or heat treatment.

Starch, mixed with water, and heated in the presence of weak hydrochloric acid, breaks down chemically by hydrolysis. If the hydrolysis or conversion of corn starch is interrupted before final conversion, a noncrystallizing corn syrup is obtained. Many varieties may be made by supplemental use of enzymes to meet specific functional requirements. The solids content is varied to suit the requirements of the users. Corn syrup is used in a wide variety of food products, including baby foods, breakfast foods, cheese spreads, chewing gum, chocolate products, confectionary, cordials, frostings and icings, peanut butter, sausage, and for numerous industrial products, including adhesives, dyes and inks, explosives, metal plating, plasticizers, polishes, textile finishes, and in leather tanning.

A process of converting starch to dextrose is shown in Fig. 3. The enzyme process shown overcomes flavor and color difficulties of the hydrochloric acid method. The enzyme is obtained by growing a mold (*Aspergillus phoenicis*, a member of the *Aspergillus niger* group). The mold yields the key glucoamylase as well as transglucosylase. The latter must be eliminated because it catalyzes the formation of undesirable glucosidic linkages. Through a special process, almost pure glucoamylase is obtained.

Purified starch slurry (30–40% solids), made from dent corn, is received in the converters from basic processing at the corn plant. A preliminary conversion using alpha-amylase enzyme or acid is carried out at 80–90°C (176–194°F), during which 15–25% of the starch is converted into dextrose. This thins the starch slurry, allowing easier addition of the glucoamylase enzyme. It also prevents formation of unhydrolyzable gelatinous material during the main conversion, and results in increased dextrose yields of from 3–4%. Thinning also reduces evaporation costs because starch concentrations of 30–40% can be handled compared with the 12–20% limit for the acid process. Before the main conversion, the starch-dextrose slurry is centrifuged to remove oil and protein by-products, which are processed for animal feed.

The slurry then goes to a 25,000-gallon (946-hectoliter) enzyme tank where, at pH 4.0–4.5 and 60°C, the major reaction with the glucoamylase takes place. It is a batch operation requiring about 72 hours. When conversion is complete, the batch (97–98.5% dextrose on a dry basis) is passed through a preliminary decolorizing filter of powdered carbon and then pumped on to the first of three evaporators. The remaining operations are evaporation and crystallization, followed by centrifuging, and rotary drying for dextrose crystals; and by a remelting and filtering process for handling of outsize crystals, the resulting liquid being returned to the third effect evaporator for reprocessing.

STAUROLITE. The mineral staurolite is a complex silicate of iron and aluminum corresponding to the formula $(Fe,Mg,Zn)_2Al_9Si_4O_{23}$ (OH) but somewhat varying and may carry magnesium or zinc. It is orthorhombic, prismatic, twins common, often producing cruciform crystals. It is a brittle mineral; fracture, sub-conchoidal; hardness, 7–7.5; specific gravity, 3.64–3.83; luster, subvitreous to resinous; color, dark brown, sometimes reddish to nearly black; grayish streak; tranlucent to opaque. Staurolite is a metamorphic mineral usually the result of regional rather than contact metamorphism, and is common in schists, phyllites and gneisses together with garnet, kyanite, and tourmaline.

Well-known European localities are in Switzerland and Brittany; and in the United States this mineral is common in the schists of New England, and those of the southern Alleghenies. Frequently the crystals are found loose in the soil after the disintegration of the country rock. The name staurolite is derived from the Greek meaning a cross, in reference to the twin crystals, the more nearly perfect crosses being somewhat in demand as curios.

STEAM DISTILLATION. Distillation.

STEAM REFORMING. Ammonia; Substitute Natural Gas (SNG).

STEARIC ACID AND STEARATES. Stearic acid $H \cdot C_{18}H_{35}O_2$ or $C_{17}H_{35} \cdot COOH$ or $CH_3(CH_2)_{16} \cdot COOH$ is a white solid, melting point 69° C, boiling point 383°C is insoluble in water, slightly soluble in alcohol, soluble in ether. Stearic acid may be obtained from glyceryl tristearate, present in many solid fats such as tallow, and in smaller percentage in semi-solid fats (lard) and liquid vegetable oils (cottonseed oil, corn oil), by hydrolysis. The crude stearic acid, after separation of the water solution of glycerol, is cooled to fractionally crystallize the stearic and palmitic acids, which are then separated by filtration (oleic acid in the liquid), and fractional distillation under diminished pressure. With sodium hydroxide, stearic acid forms sodium stearate, a soap. Most soaps are mixtures of sodium stearate, palmitate and oleate.

The following are representative esters of stearic acid: Methyl stearate $C_{17}H_{35}COOCH_3$, melting point 38°C, boiling point 215°C at 15 millimeters pressure; ethyl stearate $C_{17}H_{35}COOC_2H_5$, melting point 35°C, boiling point 200°C at 10 millimeters pressure; glyceryl tristearate [tristearin $C_3H_5(COOC_{17}H_{35})_3$], melting point 70°C approximately.

Stearic acid is used (1) in the preparation of metallic stearates, such as aluminum stearate for thickening lubricating oils, for waterproofing materials, and for varnish driers, (2) in the manufacture of "stearin" candles, and is added in small amounts to paraffin wax candles. As the glyceryl ester, stearic acid is

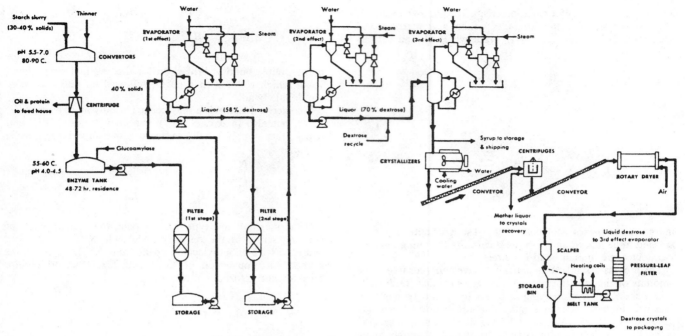

Fig. 3. Enzyme process for converting starch to dextrose.

one of the constituents of many vegetable and animal oils and fats.

STEELS AND STEELMAKING. Iron Metals, Alloys, and Steels.

STRECKER SYNTHESIS. Amino Acids.

STEPHANITE.
The mineral stephanite, silver antimony sulfide, Ag_5SbS_4, is found in short prismatic or tabular orthorhombic crystals. It is a brittle mineral; hardness, 2–2.5; specific gravity, 6.25; metallic luster; color, black; streak, black, opaque.

Stephanite occurs associated with other silver minerals and is believed to be primary in character. Localities are in Czechoslovakia, Saxony, the Harz Mountains, Sardinia; Cornwall, England; Chile; and Mexico. In the United States it is found in Nevada, where it is an important silver ore. It was named for the Archduke Stephan of Austria, mining director of that country at the time this mineral was first described.

STEREOCHEMISTRY.
Two molecules are said to be sterioisomers if they possess identical chemical formulas with the same atoms bonded one to another, but differ in the manner these atoms are arranged in space. Thus sec-butanol can exist in two forms, I and II, which cannot be superimposed on each other.

This particular example represents one class of stereoisomers known as enantiomers, which may be defined as two molecules that are mirror images but are nonetheless nonsuperimposable. Such molecules are said to possess opposite configuration. If these isomers are separated (resolved), the separate enantiomers have been found to rotate the plane of plane-polarized light. This phenomenon of optical activity has been known for well over a century. A 50–50 mixture of two enantiomers is optically inactive or racemic, since the rotation of light by one enantiomer is precisely compensated by the rotation of light in the opposite direction by the other enantiomer.

Physiological activity is closely related to configuration. Thus the left-rotating, or levo, form of adrenalin is over ten times more active in raising the blood pressure than is the right-rotating, or dextro form. Many organic chemicals essential to plants and animals are optically active. Enzymes, which catalyze chemical reactions in the body, are frequently programmed to accept only one enantiomer. All the essential amino acids, generally of the formula III, are of the levo type, although important

exceptions exist. Recently, several amino acids were found by NASA in a meteorite that presumably originated from the asteroid belt between Mars and Jupiter. The proof that the amino acids were extraterrestrial came from the fact that they were racemic. Any terrestrial contaminants from laboratory handling would have been optically active and levo.

Many examples of optically active molecules contain an asymmetric carbon atom, that is, one with four different groups attached, as in I-III. A wide variety of other atoms may also be asymmetric (IV-VI). An asymmetric center is by no means a necessary condition for enantiomerism. Well-known examples of nonsuperimposable mirror images without asymmetric atoms are allenes (VII), spiranes (VIII), biphenyls (IX), and various

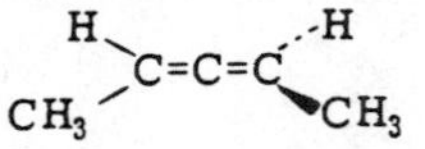

VII    VIII

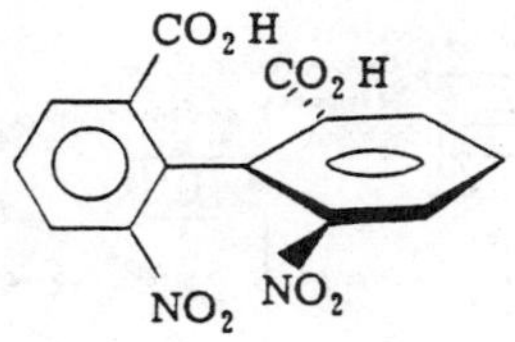

IX

inorganic complexes. Molecules that can support optical activity are said to be *chiral*, and to possess *chirality* (meaning *handedness*, since the human hand is chiral).

The process of converting optically active materials into equal amounts of the enantiomers is called *racemization*. Ordinarily this process requires breaking of bonds to form a symmetrical (*achiral*) intermediate, and reforming the bonds to generate the racemic material. For special cases such as phosphines (IV) and sulfoxides (VI), in which one "substituent," is a nonbonding electron pair, configurational inversion may occur without breaking any bonds. For such molecules the tetrahedral enantiomers may interconvert through a metastable planar intermediate.

Stereoisomers that are not enantiomers are called *diastereoisomers*. Three classes may be distinguished: configurational, geometrical, and conformational isomers. Configurational diastereomers include molecules with more than one chiral center. Thus 2,3-dichlorobutane can exist in three configurationally

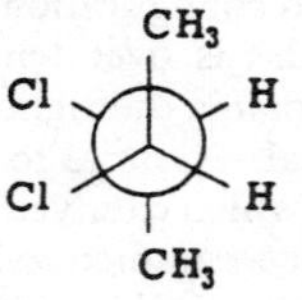

X 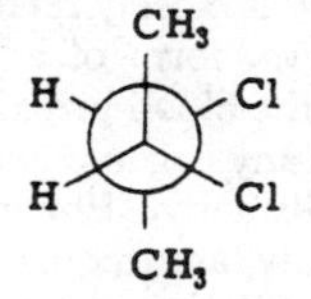   XI 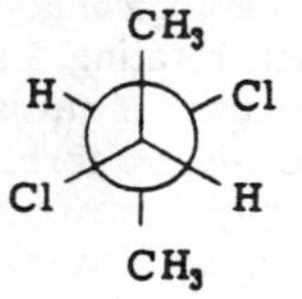   XII

different forms, X-XII. Although forms X and XI are enantiomers, XII is a stereoisomer that is not a mirror image of X or XI. It is therefore termed a diastereomer of X and XI. Even though there are two asymmetric centers in XII, it is superimposable on its mirror image. The molecule is therefore achiral. The term *meso* is applied to molecules that contain chiral centers but are achiral as a whole. The molecules of nature frequently have many asymmetric centers. If a molecule has n centers, there can be 2^n stereoisomers, although this number may be reduced if some of the diastereoisomers are *meso* or if certain ring constraints are present. Glucose is one of the aldohexose sugars, which contain four chiral centers (disregarding the phenomenon known as anomerism). The naturally occurring *dextro*-glucose is enantiomeric to *levo*-glucose, and diastereomeric to the other fourteen isomers.

Geometrical isomers differ in the arrangement of groups about certain bonds, rather than about a chiral center. 2-Butene may exist in *cis* (XIII) or *trans* (XIV) forms. In the former case the methyl groups lie on the same side of the double bond, and in the latter on opposite sides. Chirality is not important in geometrical isomerism. The isomers may be interconverted

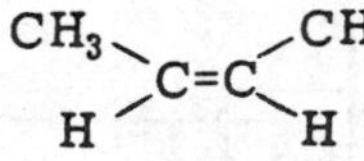

XIII 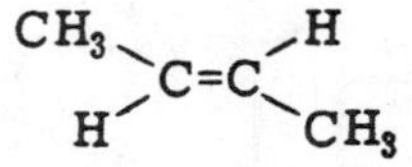   XIV

if the double bond is broken to leave a residual single bond about which rotation may occur, followed by reformation of the double bond.

Geometrical isomerism may occur not only in alkenes (XIII, XIV), but also in oximes (XV), azo compounds (XVI), and many other doubly bonded systems. More importantly, cyclic

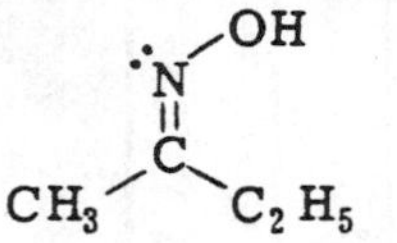

XV 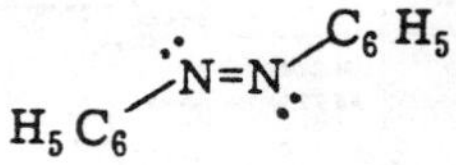   XVI

molecules exhibit this type of isomerism; the average plane of the ring serves as the reference. Molecule XVII is therefore named *cis*-1,2-dimethylcyclopropane, and XVIII is the *trans* isomer. Interconversion of XVII and XVIII would require breaking and reforming a ring bond.

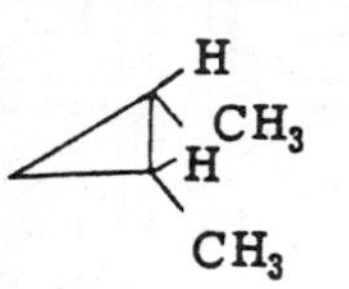

XVII 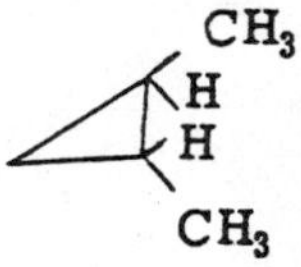   XVIII

Conformational isomers differ only in the arrangements of atoms obtainable by rotations about one or more single bonds. *meso*-2,3-Dichlorobutane can exist not only in the form XII, but also as the representation XIX. To take a simpler case,

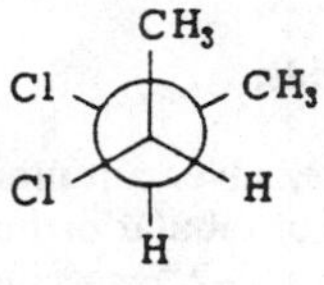

XIX 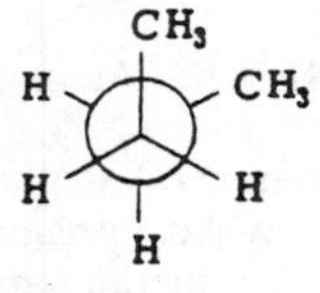   XX 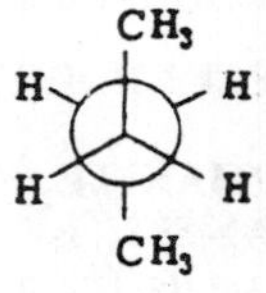   XXI

n-butane may exist in two conformational forms, known as the *gauche* (XX) and the *anti* (XXI). These isomers may interconvert by rotation about the C—C single bond, a process that requires an energy of only 3–5 kilocalories/mole. No bonds are broken, in contrast to the manner by which geometrical isomers are interconverted. In a more complicated but very common case, substituents on the six-membered cyclohexane ring may assume either the equatorial (XXII) or axial (XXIII) positions. These isomers may interconvert by *ring reversal*, Eq.

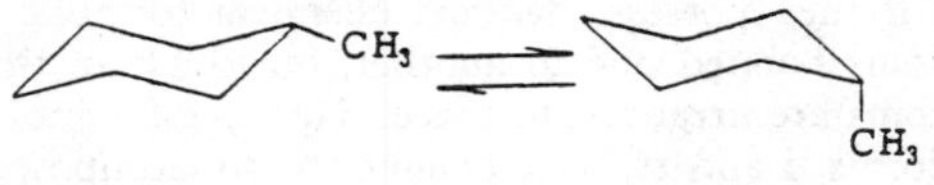

XXII   XXIII

(1), which consists of a sequence of single-bond rotations. β-D-(+)-Glucose has the specific conformation of XXIV, in which all the substituents are equatorial.

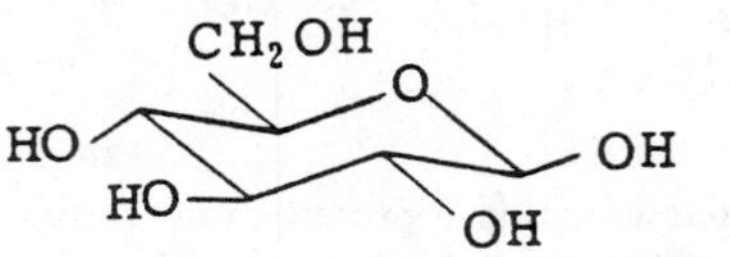

XXIV

Stereochemistry is one of the most important characteristics of a reaction mechanism. In the nucleophilic displacement of iodide ion on *sec*-butyl bromide, Eq. (2), the reaction is known to occur with inversion. If one disregards the identity of the

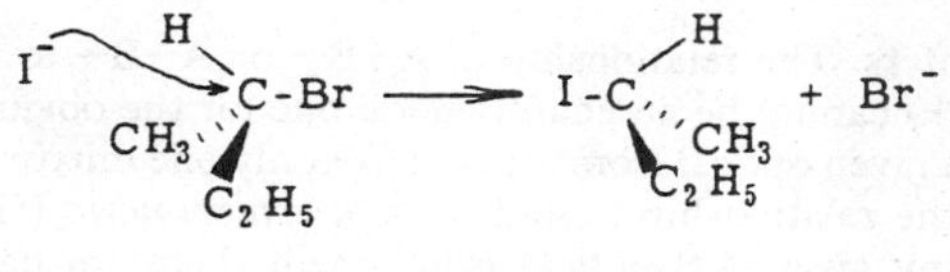

halogen, then the starting material and product have opposite configurations. Thus the iodide ion must attack the C—Br bond from the backside, thereby effecting an inversion of the chiral center.

Important mechanistic consequences may also derive from geometrical isomerism. *cis*-3,4-Dimethylcyclobutene may ring-open to form either *cis,trans*- or *trans,trans*-2,4-hexadiene. The methyl groups may rotate away from each other (disrotation, Eq. (3)) to form the *trans, trans* isomer. Alternatively, they

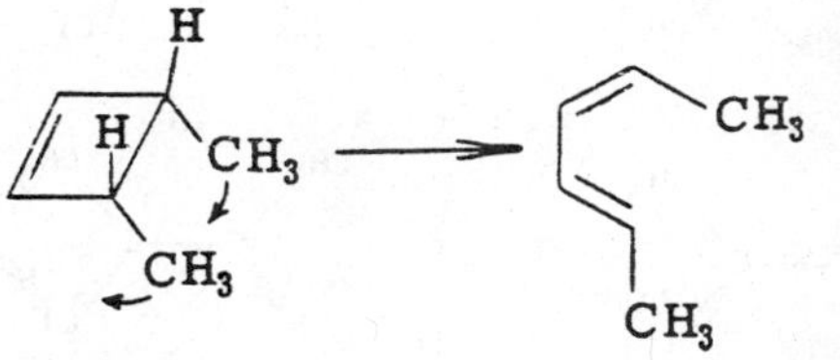

may rotate in the same direction (conrotation, Eq. (4)) to form the *cis,trans* isomer. An alternative disrotatory mode to form

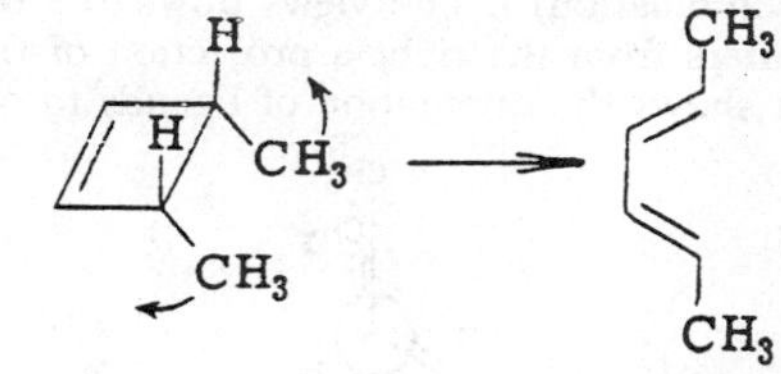

a *cis,cis* isomer by rotation of the methyl groups toward each other need not be considered because the methyl groups cannot pass by each other. Recent experimental and theoretical consideration of this problem has demonstrated that only the conrotatory mode is permitted for this ring opening. Thus *cis*-dimethyl-cyclobutene always forms only *cis,trans*-hexene.

—Joseph B. Lambert, Northwestern University, Evanston, Illinois.

STEREOISOMERISM. If one considers a molecular unit consisting of three unlike atoms A, B, and C which are connected (*bonded*) to each other to form a linear system, three arrangements are possible: A—B—C, A—C—B, and B—A—C. By employing the simple test of superimposability it is seen that these arrangements are nonidentical and are termed *structural isomers* (Gr. *isos*, equal). Structural isomers are thus chemical species that have the same molecular formula (the same number and types of atoms) but differ in the sequence in which the atoms are bonded. Of great significance is the fact that these isomers are separated by an energy barrier: in order to convert, for example, A—B—C into A—C—B an input of energy would be necessary to break the existing A—B and B—C bonds, followed by a rearrangement of the sequence of the atoms

to yield A—C—B. The rearrangement process is termed an *isomerization* and the magnitude of the energy required (the barrier) for the conversion has important consequences regarding the number of arrangements that may exist for structural isomers. This will be discussed shortly.

Turning to a four-atom arrangement consisting of A_2B_2, four structural isomers are possible if linearity of the system is again assumed: A—A—B—B; A—B—A—B; A—B—B—A; B—A—A—B. Each arrangement can be characterized in terms of interatomic distances (A to B, B to B, and A to A) and such distances would be distinctive for each isomer. It is not mandatory, however, that any one of these A_2B_2 combinations exists in a linear form and the consequences of nonlinearity will be viewed. If (for the ABBA case) planarity of the unit still exists but the internuclear angles A—B—B are set at, say, 120°, two arrangements are apparent:

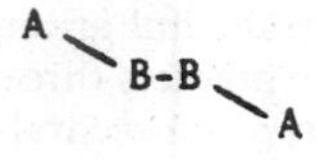

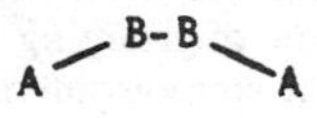

Arrangements **1** and **2** are clearly different forms of the same structural isomer and are called *stereoisomers* (Gr. *stereo*, solid or space). The relationship between **1** and **2** may be expressed in terms of the geometry of the molecule and hence these forms have also been called *geometrical isomers*. (Recalling the provisos set down for this system—planarity and angles—the difference between **1** and **2** is merely the location of one A with respect to the other in the unit.)

The statement made earlier about the energy requirement for the conversion of one structural isomer into another will now be examined. Conversion of **2** into **1** may be viewed in the simplest manner is involving a rotation of 180° about the B—B internuclear axis:

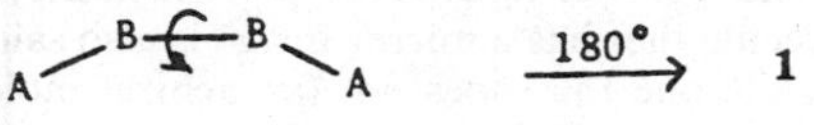

This operation involves no reshuffling of the atomic arrangement (or constitution) within the unit but does represent an isomerization. How readily such a conversion occurs then depends upon the size of the energy barrier to rotation. Looking at specific examples, the compound CH_3—N≡N—CH_3, exists in two different and stable stereochemical arrangements (**3**, *cis* and **4**, *trans*) corresponding to **1** and **2** above:

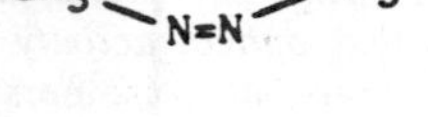

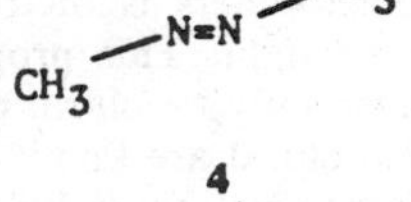

The C—N—N—C atoms are coplanar in each form and the interconversion of **3** and **4** requires a relatively high amount of energy. On the other hand, hydrogen peroxide (H_2O_2), which may be thought of as existing in similar *cis* and *trans* stereochemical arrangements (**5** and **6**), does *not* exhibit stereoisomer-

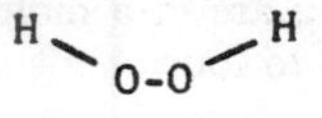

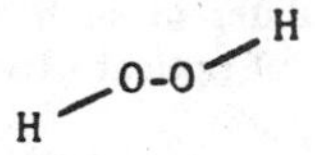

ism and only one isolable H_2O_2 is known. The difference in the two systems lies in the fact that rotation about the O—O single bond in H_2O_2 is relatively "free" (i.e., requires little energy) whereas the interconversion of **3** and **4** is a higher energy process that necessitates breaking a π bond before rotation may occur.

We now consider the case of four atoms (called *ligands*) that are attached to a center atom. If the general case consists of grouping A_{abcc}, square planar and tetrahedral structures, among others, may result:

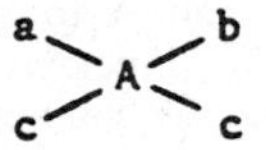

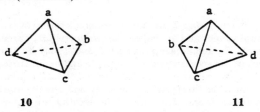

7 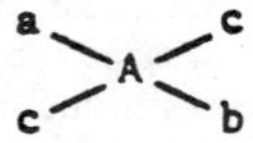   **8** 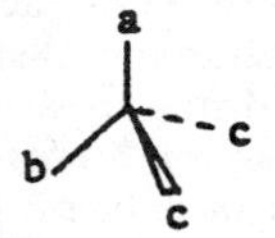   **9**

The square planar forms **7** and **8** are stereoisomers which are nonequivalent to the single nonplanar tetrahedral arrangement **9**. Now of grouping A_{abcd} is examined one predicts three square planar stereoisomers and the following *two* tetrahedral stereoisomers (**10** and **11**):

10   **11**

The relationship of **10** to **11** is that of the right hand to the left hand and these nonsuperimposable stereoisomers are mirror images or *enantiomers* (Gr. *enantio-*, opposite). Thus, *chiral* (Gr. *cheir*, hand) molecules are those which possess mirror images and arise when appropriate conditions of geometry and number and types of ligands are present in a system.

A molecule that has a mirror image is also said to be dissymmetric while one that does not (an achiral molecule) have an enantiomer is nondissymmetric. The classification of a given structure as dissymmetric or nondissymmetric is based upon the presence (or lack) of symmetry elements (axes, planes) in the structure.

It is important to note that in either **10** or **11** the magnitude of any internuclear angle (e.g., a—A—c), or any bond length (e.g., A—b), or the distance between any two ligands is exactly the same. This is not true in the case of the achiral square planar isomers that may be written for the A_{abcd} system.

The single most important physical property that differentiates enantiomers is their ability to rotate the plane of plane polarized light. This property is called *optical activity* and is displayed only by chiral molecules. Thus, stereoisomers which are also chiral are known as *optical isomers*. Chiral molecules that rotate polarized light in a clockwise fashion are termed *dextrorotatory* (*d*) while those that rotate the beam counterclockwise are *levorotatory* (*l*). Enantiomers have optical rotations of the same magnitude but of different signs (*d* or *l*).

The structures **10** and **11** denoted above contain a single chiral center A, the atom to which the ligands are attached. If two different such centers, A and B, are in a molecule the number of optical isomers is increased to four:

A±   A±
                           B±

one chiral center   two different chiral centers

where + and − refer to the handedness or *configuration* at the chiral center. In the AB system, each optical isomer will have a mirror image whose configurations are opposite at the

A+   A−   A+   A−
B+   B−   B−   B+

mirror images   mirror images

chiral centers. The relationship of A+B+ or A−B− to A+B− (or A−B+) cannot be an enantiomeric one for the obvious reason that a given optical isomer may have only one mirror image. Instead, the relationship is said to be *diastereomeric* (Gr. *dia*, apart). Any given AB optical isomer will therefore have one enantiomer and two diastereomers.

We again examine a specific case. In 3-chloro-2-butanol, CH_3CH—$CHCH_3$, and A±B± situation—exists whose four
         | |
         Cl   OH
isomers (**12–15**) are shown in three dimensions so that the mirror image relationship of **12** to **13** and **14** to **15** is readily apparent.

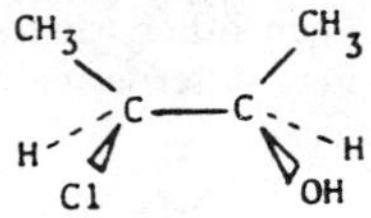

12 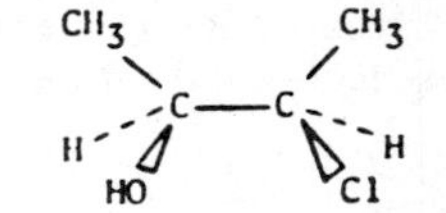   **13**

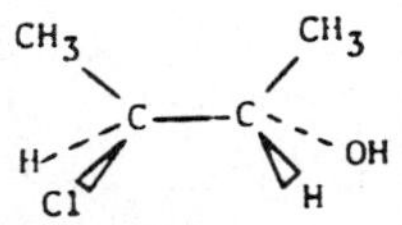

14 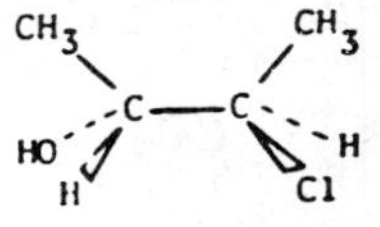   **15**

(The carbon atoms in the middle of the carbon chain are the chiral centers A and B.) Taking **12** (which is assumed to be the A+B+ combination) if one views down the bond axis of the chiral centers from the right a projection of the molecule results which shows the orientation of ligands to one another.

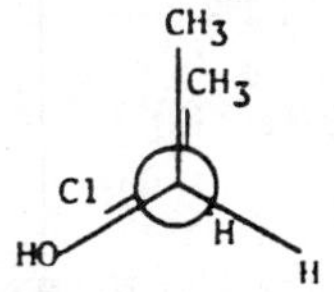

12

As noted earlier rotation may occur about single bonds in molecules and if a clockwise rotation of 180° is made about the bond axis of the chiral centers, a different form (*conformation*), **16**, results.

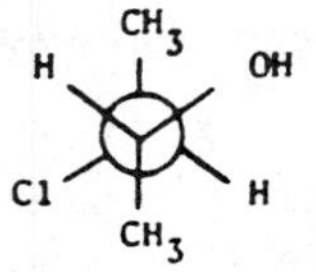

16

Conformations **16** and the *infinite number* of others that are obtained by rotation about the single bond in **12** are all nonidentical but the energy barrier separating them is small hence only one chiral compound having configurations ++ at the chiral centers may be isolated under normal conditions for **12**.

If two *identical* chiral centers are present in a molecular unit the number of stereoisomers is reduced to three: A+A+ and A−A− represent a pair of enantiomers but A+A− and A−A+ are identical arrangements. The +− form is said to be a *meso* or optically inactive diastereomer of the active forms A+A+ and A−A−.

The three tartaric acids may be used to illustrate the method employed in depicting three-dimensional molecules in two-dimension projections.

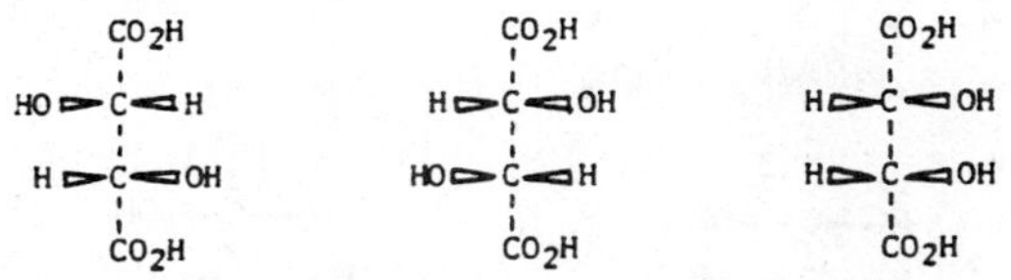

The top formulas show the three-dimensional relation of the groups along the main carbon chain (dashed groups lie below the plane of the paper, bold above) while the bottom formulas correspond to projections in two dimensions obtained by lifting the dashed substituents into the plane of reference and pushing the bold groups down into the plane. In this process a unique projection is obtained for each three-dimensional molecule. The symbols D and L under the formulas for the enantiomeric tartaric acids are notations used to relate the configuration at the bottom chiral center to that of a standard compound, glyceraldehyde.

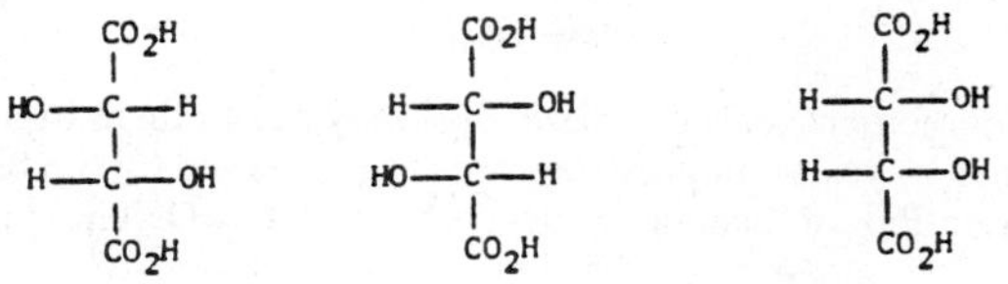

The presence of a chiral center is a sufficient, but not a necessary condition, for the existence of chirality in a molecule. For example, numerous biphenyl derivatives may exist in chiral pairs

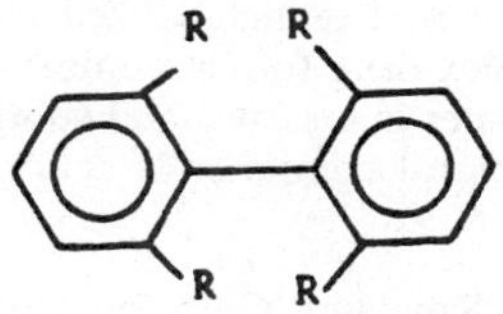

if the size of the R groups is large enough to restrict rotation about the single bond connecting the two rings. The restriction causes the rings to adopt a nonplanar orientation and raises the energy barrier to rotation about the connecting bond. No chiral center is present and the resulting enantiomers in the example following

are said to possess a *chiral axis* (coinciding with the connecting bond). Axial chirality is also found in the compounds known as allenes

e.g.

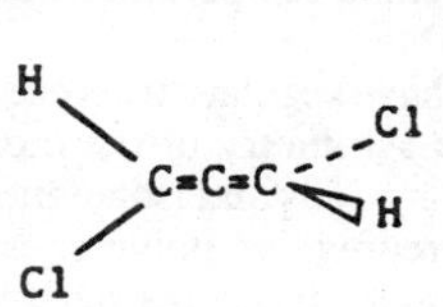

and in spiranes

e.g.

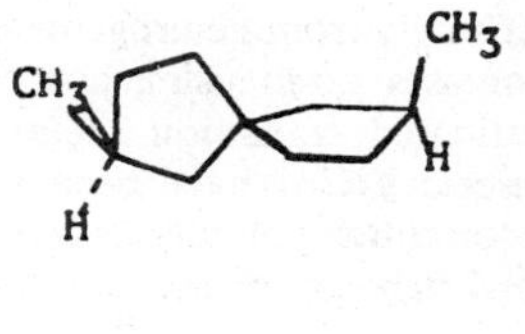

while chiral structures such as *trans*-cyclooctene

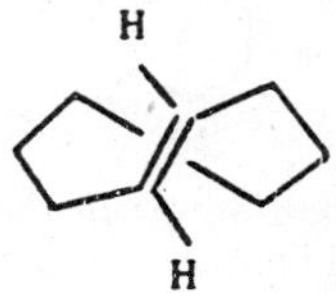

are said to possess a *chiral plane*. Each of the last three structures has a mirror image.

It is clear, then, that the number of chiral isomers that may exist for a given structural isomer is 2^n, where $n =$ the number of different chiral elements (centers, axes, or planes). When identical chiral elements are present, the 2^n formula does not hold.

The examples used above to illustrate centers of chirality included carbon atoms which were attached to four unlike ligands, resulting in localized tetrahedral geometry. Numerous other atoms may serve as chiral centers, however, and these include the Group IVA elements silicon, germanium, and tin; the Group VA elements nitrogen, phosphorus, antimony, and arsenic; and the Group VIA elements sulfur, selenium, and tellurium. Under conditions of bonding to three or four dissimilar ligands, chiral molecules containing these atoms as chiral centers may be isolated. Also, the geometric form about the chiral center need not be tetrahedral, for octahedral complexes of the transition metals or their ions (Co^{3+}, Cr^{3+}, etc.) may be chiral when substituted by the proper number and type of ligands.

Conformations in Six-Membered Rings. The nonplanar ring compound, cyclohexane, which contains six contiguous —CH_2— units, may exist in chair or boat forms (hydrogens are not shown):

These forms are conformations of the C_6H_{12} structure and are separated by energy barriers which restrict, but do not prohibit, interconversion of the conformations. Substitution of a group on the ring yields a single nonchrial structure that exists in two main conformations, one with the substituent R *axial* and the other *equatorial* to the main plane of the ring.

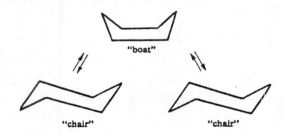

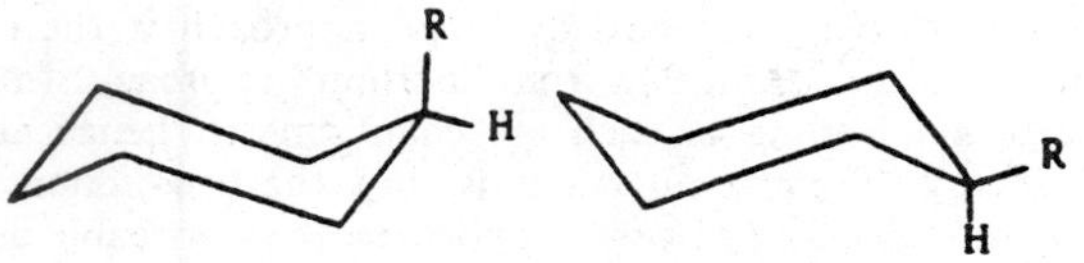

R *axial* (17)   R *equatorial* (18)

Conformation **18** is in a lower energy state and predominates in the equilibrium mixture. The introduction of a second substituent into the ring gives rise to three structural isomers (shown here in projection):

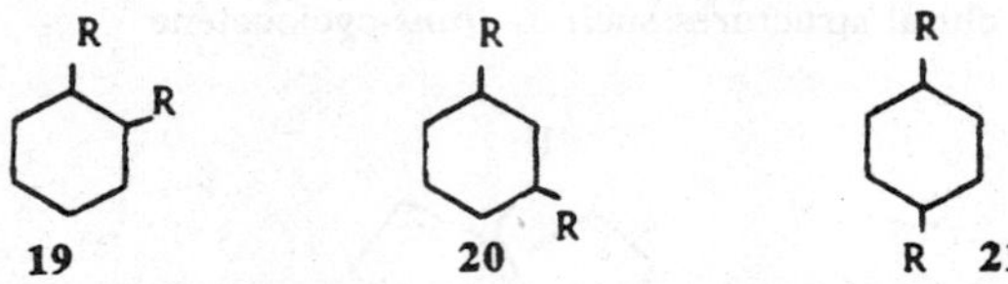

19   **20**   **21**

In **19** and **20**, both carbon atoms in the ring to which the R groups are bonded are identical chiral centers and hence one pair of enantiomers and one meso diastereomer exist for each structure. In **21** no chiral center is present but the isomers that are possible in this case (shown below in the preferred conformations)

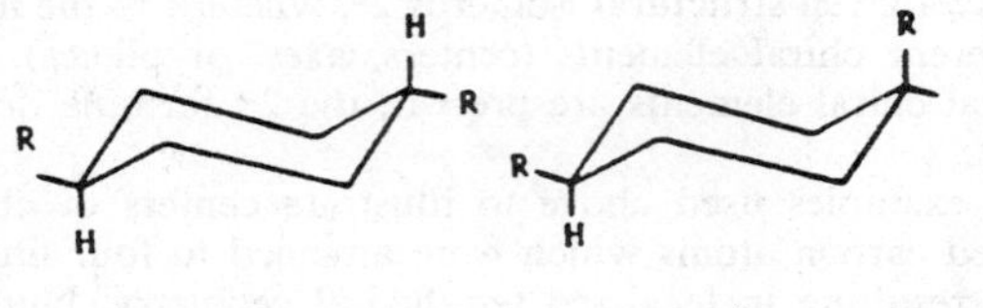

trans   *cis*

may be viewed as having a diastereomeric relationship.

Consequences of Molecular Chirality. A mixture containing an equal number of molecules of enantiomers is known as a *racemic modification*. The preparation and reactions of these modifications (as well as the individual enantiomers themselves) represent important aspects of the study of stereochemistry.

If a molecule $CH_3-\overset{\underset{\|}{O}}{C}-CH_2CH_3$ is converted into $CH_3-\overset{\underset{|}{OH}}{CH}-CH_2CH_3$ by some achiral reagent (e.g., hydrogen gas and a catalyst) the chiral center in the product molecule may be considered as being generated by approach of a hydrogen from the top *or* bottom "face" of the planar $C-C-O$ grouping in reactant molecule

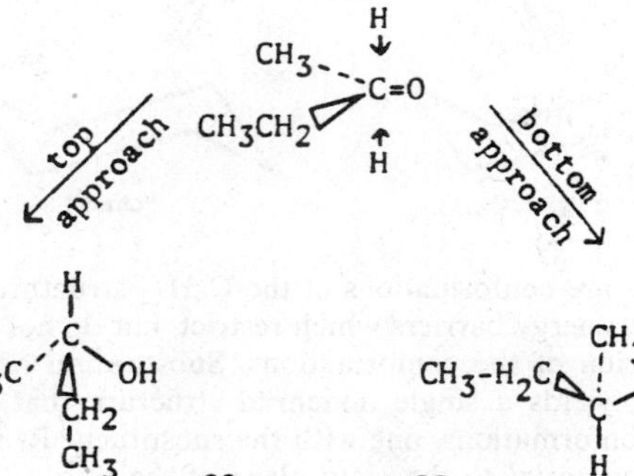

22   **23**

The molecule (**22**) produced by "top" approach is the enantiomer of that (**23**) resulting from "bottom" approach. In fact, the pathways leading to each are enantiomeric, hence are of equal energy. The overall result is thus the production of a racemic modification, since one approach is as probable as another.

Now if a similar reaction were conducted with a chiral substrate that has a preexisting chiral center (A+), the combinations of configurations at the centers in the product molecules would be A+B+ and A+B−. These are diastereomers, the pathways involving their information are diastereomeric (unequal energy!), and hence they are produced in unequal amounts. Such a case is illustrated as follows:

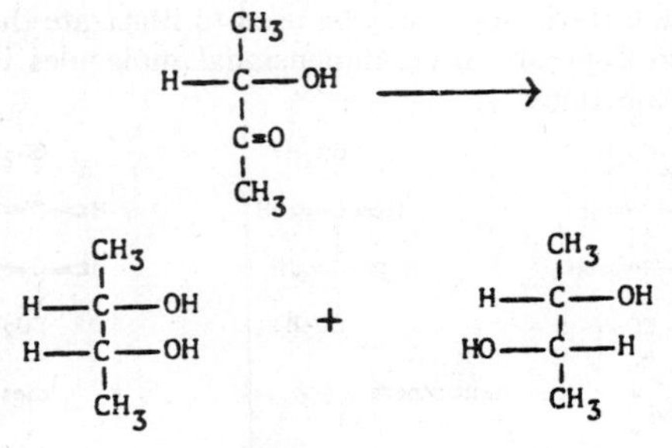

diastereomers

Diastereomeric reaction pathways may be obtained in numerous other ways. An interesting case is represented by the biological reduction of acetaldehyde-1-D, $CH_3-\overset{\underset{|}{D}}{C}=O$. The product is the chiral structure $CH_3-\overset{\underset{|}{D}}{CH}-OH$ and a racemic modification might be expected from the reduction. In fact, the transfer of a hydrogen during the enzymatic reduction (an enzyme is a large chiral molecule) to one face of the acetaldehyde is diastereomeric with the transfer to the opposite face, hence (very) unequal amounts of the enantiomers are formed. These interactions may be described as E+ .... A+ and E+ .... A−, where E+ represents the chirality of the enzyme and A+ or A− represent the incipient chirality in the reduced acetaldehyde molecules.

The last example reflects in a modest way the importance of the study of stereoisomerism. Biological conversions represent a glorious array of diastereomeric reactions and interactions: a given chiral amino acid is metabolized but its enantiomer is not; a certain complex drug (often a chiral molecule) alleviates pain but its enantiomer is inactive; and subtle changes in structure alter a given chiral compound's action completely in the human body.

—Alex T. Rowland, Gettysburg College, Gettysburg, Pennsylvania.

STEREOREGULAR POLYMERS. The properties of natural and synthetic high polymers and their applications in plastics, fibers, elastomers, adhesives and coatings are determined in large part by (a) their average molecular weights, (b) the forces between the long chain molecules and (c) geometrical considerations, especially the degree of regularity of repeating units. Molecular weight characteristics and forces have received attention for many years. In contrast, the importance of stereoregularity vs. irregularity of chemical substituents became fully appreciated only around 1950. In general, strong interchain forces and regularity promote normal crystallinity, along with high strength, high softening temperatures, hardness and insolubility in common solvents.

Some polymers have regular structure free of diastereoisomerism because of the symmetry of the monomers, such as vinylidene chloride, $CH_2=CCl_2$ and isobutene $CH_2=C(CH_3)_2$. However, the term stereoregular polymer is generally reserved for stereoregular polymer structures derived from unsymmetrical monomers which can be obtained by special ionic methods of polymerization (usually from heterogeneous systems). These polymerization processes often using complex catalysts such as Ziegler-Natta activated transition metal catalysts, e.g., from AlR_3 and polymeric $TiCl_3$, have been called stereoregulated, stereospecific or oriented polymerizations. In 1964 the writer suggested the word *stereopolymerization* to describe those special ionic polymerization systems for treating unsymmetrical ethylenic monomers to obtain either normally crystalline polymers with DDDDD regularity, permanently amorphous polymers with irregular DLDDL sequences or intermediate structures according to conditions chosen. Such control of polymer stereoisomerism was achieved first with vinyl alkyl ethers, but the first stereoregular polymers to become the basis of a major industry were the stereoregular or isotactic propylene polymers (heat-resistant molding plastics and fibers).

Crystallinity has been one of the principal effects by which stereoregularity or tacticity has been studied in polymers. However, as expected, not all stereoregular polymers are equally crystallizable. Differences in chemical reactivity, nuclear magnetic resonance and infrared have given useful information about stereoregularity. However, for comparing polymers from a given monomer x-ray diffraction and solubility data are most reliable for estimating tacticity.

Interest in controlling steric configurations and stereoisomerism in polymers by polymerization conditions developed only slowly. Staudinger and Schwalbach in 1931 suggested that invariable low crystallinity in polyvinyl acetate might be caused by diastereoisomerism, that is randomness in D and L positions of the acetate groups along the chain molecules[1]. Branching and deviations from head to tail addition were studied meanwhile as types of isomerism. However not until 1948 were examples of stereoregulated polymerizations of a vinyl-type monomer disclosed by Schildknecht and coworkers. In both early types of stereopolymerizations vinyl isobutyl ether diluted by liquid propane could be treated at low temperatures. Addition of gaseous boron fluoride gave very rapid polymerizations to rubberlike substantially amorphous high polymers. Careful addition of cold boron fluoride etherate, immiscible with liquid propane at −78°C or above produced a slow growth or proliferous polymerization to form normally crystalline polymers.

This suggested that it might be possible to prepare normally crystalline polymers from other unsymmetrical monomers of the type $CH_2=CHY$ in ionic heterogeneous systems.

The discovery of stereopolymerizations of 1-alkenes by use of Ziegler-Natta catalysts in heterogeneous systems in 1954, and subsequent studies of polymer structure, attracted worldwide attention to this field. Propylene and 1-butene, which are monoallylic compounds, had not been homopolymerized by conventional ionic or free radical conditions to give linear high polymers suitable as plastics or other synthetic materials. Short branches in polyethylenes had been shown to reduce crystallinity and hardness. Natta and coworkers demonstrated the ability of catalysts such as those from reaction of aluminum alkyls with titanium halides to form normally crystalline, surprisingly high softening polymers from propylene. A helix of three monomer units explained the regularity required for crystallization and the identity period observed from x-ray diffraction. Stereoregular isotactic polymers of crystal melting ranges shown in the accompanying table were prepared by slow heterogeneous ionic polymerizations using special catalysts at moderate pressures and temperatures.

Stereoregular propylene polymer plastics have outstanding utility, for example, heat resistance superior to that of polyethylenes in sterilizable hospital devices. By copolymerization and control of the degree of stereoregularity brittleness at low temperatures can be avoided. Stereoregular 1-butene and isobutylethylene polymers are also manufactured, but the isotactic polymers from styrene and from methyl methacrylate are too brittle for much use. Crystallizable polystyrenes also have been prepared by heterogeneous anionic polymerizations (Lewis basic catalysts) and crystalline methyl methacrylate polymers can be prepared using Grignard catalysts.

Soluble catalysts derived from organoaluminum compounds and vanadium halides promote formation of atactic elastomeric propylene polymers and copolymers such as ethylene-propylene-diene terpolymer rubbers (EPDM). Amorphous adhesive propylene homopolymers have some commercial use. Syndiotactic or DLDL propylene polymers have been reported but their structures and properties have not been completely established.

Isomeric isoprene polymers are formed biologically as natural rubber (cis-1,4) and balata or gutta-percha (largely trans-1,4). Modified Ziegler-Natta type catalysts, colloidal lithium or lithium alkyls (in absence of ethers) were found to give predominantly cis-1,4 polymer rubbers from isoprene. Cis-1,4-polybutadiene, so-called synthetic natural rubbers, have become important in tires and in graft copolymerization with styrene for high-impact plastics. Different conditions of polymerization give rigid trans-1,4-diene polymers resembling balata.

Although precise mechanisms of the stereopolymerizations are yet uncertain, several characteristics become evident. The reactions are predominantly heterogeneous, ionic reactions at low or moderate temperatures. The more stereoregular polymers grow as a separate phase upon the solid or immiscible liquid catalyst. However, some monomers such as vinyl isobutyl ether can form somewhat isotactic polymer fractions even from homogeneous solutions of Lewis acid and monomer. In contrast, the polymers from vinyl isopropyl ether, which apparently are stereoregular when obtained by slow growth polymerization using boron fluoride etherate catalysts, nevertheless do not crystallize readily. BF_3 can be used to form isotactic vinyl isobutyl ether polymers if it is applied in a separate phase of methylene chloride immiscible with liquid propane. Relatively polar solvents which favor separation of gegen ions from their growing macroions generally impair stereospecificity.

Although the Ziegler-Natta catalyst systems for stereopolymerization of 1-olefins were regarded by Natta, Mark, and others as examples of anionic polymerizations, the writer considers them as a special type of cationic polymerization. A consistent system relating monomer structure to response to catalyst types is only possible if propylene containing an electron repelling methyl group attached to the ethylene nucleus polymerizes with Lewis acid catalysts (cationic polymerization). Propylene as an allyl compound lacks sufficient electron withdrawal from the ethylene group to homopolymerize by free radical initiation (peroxide, azo catalysts or ultraviolet light) and it also lacks sufficient electron donation (as in isobutene) for homopolymerization by conventional cationic system. Ziegler-Natta catalysts have been observed to homopolymerize some other monoallyl compounds. An intensive study of the literature by the writer and Mabel D. Reiner showed no well-characterized homopolymers of high molecular weight obtained from monoallyl compounds by free radical or conventional ionic catalyst systems.

Transition metal catalysts for polymerization of 1-alkenes similar to those of Ziegler were developed in DuPont laboratories and have been called coordination catalysts.

Outside of vinyl addition polymerizations some crystallizable stereoregular polymers also have been prepared. An example is isotactic polymer from propylene oxide —CH CH₂O— made by

$$\underset{\underset{CH_3}{|}}{—CH} CH_2O—$$

CRYSTAL MELTING RANGES OF SOME STEREOREGULAR POLYMERS

Isotactic Stereopolymers	Melting Range	Stereopolymers	Approximate Maximum Melting Point
Propylene	165–176°C	Isotactic 1,2-butadiene	120°C
1-Butylene	120–136	Syndiotactic 1,2-butadiene	154
1-Amylene	60–70	Cis-1,4-butadiene	+1
Isopropylethylene	300–310	Trans-1,4-butadiene	148
Isobutylethylene	235–250	Cis-1,4-isoprene	22
Isoamylethylene	about 110	Trans-1,4-isoprene	65
4,4-Dimethyl-1-pentene	>380	Cis-1,4-(2,3-dimethyl butadiene)	190
Styrene	230–250	Trans-1,4-(2,3-dimethyl butadiene)	260
Isobutyl vinyl ether	100–130		
Methyl methacrylate	160		

using ferric chloride complex catalysts for proliferous type reactions.

After the demonstrations of preparation of stereoregular polymers having novel properties by means of special ionic methods, the possibilities of free radical methods were examined extensively. It must be concluded that in free radical systems the structures of homopolymers and copolymers can be little influenced by specific catalysts and other reaction conditions, but are determined largely by monomer structure. This is consistent with the relative uniformity of comonomer reactivity ratios in radical copolymerizations. However, it has been found possible to obtain somewhat more syndiotactic structure, DLDL, than normally obtained by radical reactions, at low temperatures and by selecting solvents. Examples are polyvinyl chlorides of higher than usual crystallinity from polymerizations at low temperature e.g., −50°C under ultraviolet light.

Although they do not crystallize, polyvinyl acetates prepared at low temperatures apparently are more syndiotactic since they yield more than usually crystalline polyvinyl alcohols by saponification. Monomers of high polarity such as vinyl trifluoracetate by radical polymerization can form relatively syndiotactic polymers from which more crystalline polyvinyl alcohols can be prepared by saponification.

—C. E. Schildknecht, Gettysburg College, Gettysburg, Pennsylvania.

STEROIDS. Organic compounds characterized from a structural standpoint by the cyclopentanophenanthrene nucleus as shown in Fig. 1. Biochemically, the steroids are closely related to the terpenes. Steroids occur widely in nature, both in animals

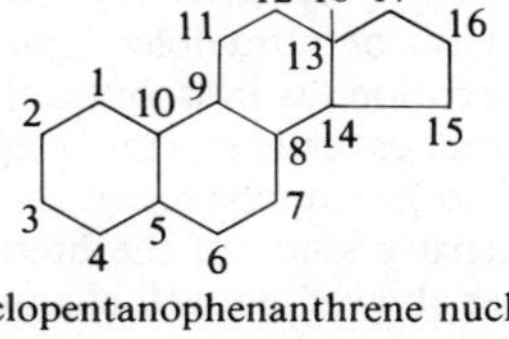

Cyclopentanophenanthrene nucleus.

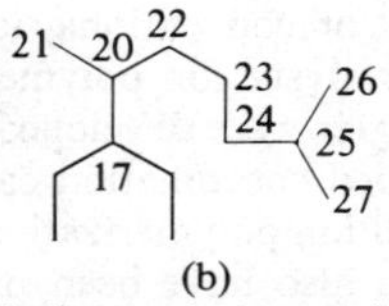

(b)
Side chain attached at position 17.

Fig. 1. Steroid molecule nucleus and side chain.

and plants. Many steroids are hormones, such as estrogens and cortisone, which are produced by the body's endocrine system and which are of great importance in the regulation of numerous body processes, such as growth and metabolism. Similarly, steroid hormones are important to several physiological processes within plants. Auxin, for example is a plant growth hormone that regulates longitudinal cell structure so as to permit bending of the stalk or stem in phototropic response. The most common animal steroid, a steroid alcohol (or *sterol*) is *cholesterol*, which is the precursor of bile acids, steroid hormones, and provitamin D_3. It should be stressed, however, that all steroids are not hormones; and all hormones are not steroids.

Over the last 35–40 years, *steroid therapy*, that is, the medical augmentation of steroid hormone insufficiencies in the body (as well as treating diseases and elements which result from an overabundance of certain steroid hormones in the body), is one of the major chapters in the history of medical progress. Similarly, the understanding of steroid chemistry has contributed markedly to plant biology, notably to plant breeding and the development of plant growth regulators. Commencing with the isolation of steroid hormones from natural sources, techniques were later developed to synthesize a number of hormones. Then, a further step involved the application of synthetic steroid hormones for which there are no known counterparts in nature. Some of the steroid hormones, including steroidal synthetics, of importance medically and to scientific investigations, are listed and described briefly in the accompanying table.

The history of steroid chemistry commenced with Mauthner, Windlaus, Wieland, Jacobs, Diels, and other organic chemists and biologists, who made early observations on the products of oxidation, aromatization, and other reactions of cholesterol, bile acids, and plant glycosides. The interrelationship between sterols and bile acids was recognized early during these investigations. Shortly after the steroid character of the female and male sex hormones had been established, x-ray crystallography demonstrated errors in early theories concerning molecular structure. Rosenheim and King, working with monomolecular layers and x-ray evidence gained by Bernal, formulated the concept of the cyclopentanophenanthrene nucleus.

Once the skeleton of the steroids was established, there remained the task of understanding the steric relationships of the molecules. With reference to Fig. 1, it will be noted that there are nine asymmetric carbon atoms in the steroid skeleton—C_5, C_{10}, C_9, C_8, C_{14}, and C_{13} in the ring system. There are also two asymetric carbons in the side chain attached at C_{17}. These are C_{20} and C_{25}. With reference to Figs. 1 and 2, the relative configuration of C_5 and C_{10}, of C_9 and C_8, and of C_{14} and C_{13}, determines whether the junctions between rings

REPRESENTATIVE STEROID HORMONES AND STEROIDAL SYNTHETICS[1]

ANTIINFLAMMATORY, ANTIALLERGIC, AND ANTIRHEUMATIC AGENTS (Adrenal Corticosteroids)

Betamethasone (9-fluoro-16β-methylprednisolone; 16β-methyl-11β,17α,21-trihydroxy-9α-fluoro-1,4-pregnadiene-3,20-dione). $C_{22}H_{29}O_5F$, mw = 329.5. Also, the *betamethasone acetate*, $C_{24}H_{31}O_6F$, mw = 434.5; and *betamethasone disodium phosphate*, $C_{22}H_{28}O_8FNa_2P$, mw = 516.4. Both of the latter compounds are used for treating carpal tunnel syndrome, the most common of the entrapment neuropathies. The median nerve is subjected to compression and possibly ischemia in the confined space between the carpal bones and the flexor retinaculum of the wrist.

Chloroprednisone acetate (6α-chloroprednisone acetate; 6α-chloro-$\Delta^{1,4}$-pregnidien-17β,21-diol-3,11,20-trione-21-acetate). Mw = 436.6. Multiple uses.

Corticosterone (11,21-dihydroxyprogesterone; Δ^4-pregnene-11β,21-diol-3,20-dione; 11β,21-dihydroxy-4-pregnene-3,20-dione). $C_{21}H_{30}O_4$, mw = 346.4. Multiple uses.

Cortisone (17-hydroxy-11-dehydrocorticosterone; 17α,21-dihydroxy-4-pregnene-3,11,20-trione; Δ^4-pregnene-17α,21-diol-3,11,20-trione; Kendall compound; Wintersteiner compound F). $C_{21}H_{28}O_5$, mw = 360.4. Multiple uses.

Desoxycorticosterone (deoxycorticosterone; 11-desoxycorticosterone; 21-hydroxyprogesterone; 4-pregnen-21-ol-3,20-dione; Kendall desoxy compound B; Reichstein substance Q). $C_{21}H_{30}O_3$, mw = 330.2. Also, the *desoxycorticosterone acetate* (DCA). $C_{23}H_{32}O_4$, mw = 372.4; and *desoxycorticosterone pivalate*, $C_{26}H_{38}O_4$, mw = 414.6. Multiple uses.

Dexamethasone (hexadecadrol; 9α-fluoro-16α-21-trihydroxy-16α-methyl-1,4-pregnadiene-3,20-dione). $C_{22}H_{29}FO_5$, mw = 392.4. Widely used in the treatment of benign intracranial hypertension, brain abscess, brain metastases, brain tumor, cerebral thrombosis, Cushing's syndrome, encephalitis, hypertensive encephalopathy, lumbar disk disease, meningococcal cerebral edema, shock, superior vena cava obstruction in cancer patients, ulcerative colitis.

Dichlorisone acetate (9α,11β-dichloro-1,4-pregnadiene-17α,21-diol-3,20-dione-21-acetate). $C_{23}H_{28}O_5Cl$, mw = 455.3.

Fluocinolone acetonide (6α,9α-difluoro-16α hydroxyprednisolone-16,17-acetonide). $C_{24}H_{30}O_6F_2$, mw = 452.50.

Fluorohydrocortisone (fludrocortisone; 9α-fluoro-11β,17α,21-trihydroxy-4-pregnene-3,20-dione). $C_{21}H_{29}O_5$, mw = 380.4. Used in treating Shy-Drager syndrome (parenchymatous degeneration of the central nervous system); also in treating orthostatic hypotension (a cause of temporary loss of consciousness when a person rises to an erect position). Also *fluorometholone* (9α-fluoro-11β,17α-dihydroxy-6α-methyl-1,4-pregnadiene-3,20-dione). $C_{22}H_{24}FO_4$, mw = 376.4; and *fluprednisolone* (6α-fluoroprednisolone), $C_{21}H_{27}FO_3$, mw = 378.4; and *flurandrenolone* (6-fluoro-16α-hydroxyhydrocortisone-16,17-acetonide), $C_{24}H_{33}O_6F$, mw = 436.5.

Hydrocortisone (cortisol; 11β,17α,21-trihydroxy-4-pregnene-3,20-dione). $C_{21}H_{30}O_5$, mw = 362.5. Used in treating adrenal insufficiency, notably in cancer patients, contact dermatitis, panhypopituitarism, psoriasis, shock, and urticaria. Also *hydrocortisone acetate* (cortisol acetate), $C_{23}H_{32}O_6$, mw = 404.5, used in treating rheumatoid arthritis; and *hydrocortisone sodium succinate* (11β,17α,21-trihydroxy-4-pregnene-3,20-dione-21-hydrogen succinate, sodium salt), $C_{25}H_{33}O_8Na$, mw = 484.5. The latter compound is used in treating ulcerative colitis.

Methylprednisolone (Δ^1-6α-methylhydrocortisone). $C_{22}H_{30}O_5$, mw = 374.5. Used in treating thrombocytopenia with intracranial hemorrhage, gram-negative bacteremia, posttransfusion purpura, and shock. Also *methylprednisolone sodium succinate*, $C_{26}H_{33}O_8Na$, mw = 496.5.

Paramethasone (6α-fluoro-16α-methylprednisolone). $C_{22}H_{30}O_5$, mw = 392.45. Also *paramethasone acetate*, $C_{24}H_{31}O_6F$, mw = 434.5.

Prednisolone (methacortandralone; 1,4-pregnadiene-3,20-dione-11β,17α,21-triol). $C_{21}H_{28}O_5$, mw = 360.4. Also *prednisolone phosphate sodium* (disodium prednisolone-21-phosphate), $C_{21}H_{27}Na_2O_8P$, mw = 484.4, used in treating ulcerative colitis. Also *prednosolone pivalate* (prednisolone trimethylacetate), $C_{26}H_{36}O_6$, mw = 444.6.

Prednisone (metacortandricin; 17α,21-dihydroxy-1,4-pregnadiene-3,11,20-trione). $C_{21}H_{26}O_5$, mw = 358.4. Used in the treatment of scores of ailments and diseases. To mention a few: acute erythroleukemia, acute gouty arthritis, acute pericarditis, aspiration pneumonitis, autoimmune hemolytic anemia, breast cancer, bronchial asthma, chronic hepatitis, dermatomyositis, desquamative interstitial pneumonia, Hodgkin's disease, hypercalcemia, immune neutropenia, lymphocytic leukemia, osteoporosis, hemoglobinuria, prostate cancer, psoriasis, radiation enteritis, rheumatoid arthritis, trichinosis, ulcerative colitis, usual interstitial pneumonia, viral anthropathies.

Triamcinolone (9α-fluoro-16α-hydrocyprednisolone). $C_{21}H_{27}FO_6$, mw = 394.4. Used in treating acute gouty arthritis and uremic pericarditis. Also *triamcinolone acetonide* (9α-fluoro-11β,21-dihydroxy-16α,17α-isopropylidenedioxy-1,4-pregnadiene-3,20-dione), $C_{24}H_{31}FO_6$, mw = 434.4. Used in treating acne vulgaris. Also *triamcinolone diacetate* (9α-fluoro-16α-hydroxyprednisolone-16,21-diacetate), $C_{25}H_{31}FO_8$, mw = 478.49.

ANDROGENS AND ANABOLIC AGENTS

Androsterone (3α-hydroxy-17-androstenone). $C_{19}H_{30}2$, mw = 290.4. Also *fluoxymesterone* (9α-fluoro-11β,17β-dihydroxy-17α-methyl-4-androsten-3-one), $C_{20}H_{29}FO_3$, mw = 336.4. Used in treating paroxysmal nocturnal hemoglobinuria. Also *aldosterone* (electrocortin; 18-formyl-11β,21-dihydroxy-4-pregnene-3,20-dione). $C_{21}H_{28}O_5$, mw = 360.4.

Hydroxydione sodium (21-hydroxypregnane-3,20-dione-21-sodium hemisuccinate). $C_{25}H_{35}O_6Na$, mw = 454.5.

Spironolactone (3-(30-oxo-7α-acetylthio-17βhydroxy-4-androsten-17α-yl)-propionic acid γ-lactone). $C_{24}H_{32}O_4S$, mw = 416.5. Used in treating congestive heart failure, hypertension, hypokalemia.

Methandrostenolone (17α-methyl-17β-hydroxy-1,4-androstadien-3-one). $C_{20}H_{28}O_2$, mw = 300.4.

Methylandrostenediol (MAD; methandriol; 17α-methyl-5-androsten-3β,17β-diol). $C_{20}H_{32}O_2$, mw = 304.4.

Methyl testosterone (17α-methyl-Δ^4-androsten-17-β-0 1-3-one). $C_{20}H_{30}O_2$, mw = 302.4.

Norethandrolone (17α-ethyl-19-nortestosterone). $C_{20}H_{30}O_2$, mw = 302.4. Also *oxandroline* (17β-hyroxy-17α-methyl-2-oxa-5α-androstane-3-one), $C_{19}H_{30}O_3$, mw = 306.4.

Oxymetholone (2-hydroxymethylene-17-α-methyldihydrotestosterone). $C_{21}H_{32}O_3$, mw = 332.4. Used in treating agnogenic myeloid metaplasia and hereditary angioedema. Also *prometholone* (2α-methyl-dihydro-testosterone propionate), mw = 360.5.

Testosterone (trans-testosterone; 17β-hydroxy-4-androsten-3-one). $C_{19}H_{28}O_2$, mw = 288.4. Used in treating acne vulgaris, impotence, polycystic ovary syndrome, male hypogonadism. Also *testosterone cypionate*, $C_{27}H_{40}O_3$, mw = 412.6; *testosterone enanthate*, $C_{26}H_{40}O_3$, mw = 400.6; *testosterone phenylacetate*, $C_{27}H_{34}O_3$, mw = 406.5; *testosterone propionate*, $C_{22}H_{32}O_3$, mw = 344.4.

REPRESENTATIVE STEROID HORMONES AND STEROIDAL SYNTHETICS[1] (*Continued*)

ESTROGENS

Equilenin (1,3,5–10,6,8-estrapentaen-3-ol-17-one). $C_{18}H_{18}O_2$, mw = 266.3. Also *equilin* (1,3,5,7-estratetraen-3-ol-17-one), $C_{18}H_{20}O_2$, mw = 268.3.

Estradiol (β-estradiol; dihydrofolliculin, dihydroxyestrin; 3,17-ephidhydroxyestratriene). $C_{18}H_{24}O_2$, mw = 272.3. Also *estradiol benzoate*, $C_{25}H_{28}O_3$, mw = 376.4; *estradiol cypionate*, $C_{26}H_{36}O_2$, mw = 396.6; *estradiol diprionate*, $C_{24}H_{32}O_4$, mw = 384.5.

Estriol (trihydroxyestrin; 1,3,5-estratriene-3,16α,17β-triol). $C_{18}H_{24}O_3$, mw = 288.3.

Estrone (folliculin; ketohydroxyestrin; 1,3,5-estratrien-3-ol-17-one). $C_{18}H_{22}O_2$, mw = 270.3. Also *estrone benzoate*, $C_{25}H_{26}O_3$, mw = 374.4.

Ethynyl estradiol (17-ethinyl estradiol; 17α-ethynyl-1,3,5-estratriene-3,17β-diol). $C_{20}H_{24}O_2$, mw = 296.4. Used to treat acne vulgaris, osteoporosis.

Mestranol (ethylestradiol-3-methylether; 3-methoxy-19-nor-17α-pregna-1,3,5-trien-20-yn-17-ol). $C_{21}H_{26}O_2$, mw = 310.4. Used in treating acne vulgaris.

PROGESTOGENS AND PROGESTINS

Acetoxypregnenolone (21-acetoxypregnenolone; 3-hydroxy-21-acetoxy-5-pregnen-20-one). $C_{23}H_{34}O_4$, mw = 374.5.

Anagestone acetate (6α-methyl-4-pregnen-17α-ol-20-one). $C_{24}H_{36}O_3$, mw = 372.6.

Chlormadinone acetate (6-chloro-Δ^{4,6}-pregnadiene-17α-ol-3,20-dione acetate). $C_{23}H_{29}ClO_4$, mw = 4.4.9.

Dimethisterone (17β-hydroxy-6α-methyl-17α-(prop-1-nyl)-androst-4-ene-3-one). $C_{23}H_{32}O_2 \cdot H_2O$, mw = 358.5. Also *ethisterone*, $C_{21}H_{28}O_2$, mw = 312.4.

Ethynodiol diacetate (19-nor-17α-pregn-4-en-20-yne-3β,17-diol diacetate). $C_{24}H_{32}O_4$, mw = 384.5. Used in treating acne vulgaris.

Flurogestone acetate (17α-acetoxy-9α-fluoro-11β-hydroxy-4-pregnene-3,20-dione). $C_{23}H_{31}O_5F$, mw = 406.5.

Hydroxymethylprogesterone (medroxyprogesterone; 17α-hydroxy-6α-methyl-4-pregnene-3,20-dione). $C_{22}H_{23}O_3$, mw = 344.5. Used for treating menopausal symptoms, secondary amenorrhea. Also, *hydroxymethylprogesterone acetate*, $C_{24}H_{34}O_3$, mw = 386.5. Used in treating hypogonadal females. Also *hydroxyprogesterone* (4-pregnen-17α-ol-3,20-dione), $C_{21}H_{30}O_3$, mw = 330.4. Also *hydroxyprogesterone caproate* (17α-hydroxy-4-pregnene-3,20-dione caproate), $C_{27}H_{40}O_4$, mw = 428.6.

Melengestrol acetate (MGA; 6-dehydro-17-hydroxy-6-methyl-16-methylene-progesterone acetate). $C_{25}H_{32}O_4$, mw = 396.51.

Norethindrone (norethisterone; 17α-ethynyl-17-hydroxy-19-nor-17α-4-en-20-yn-3-one). $C_{20}H_{26}O_2$, mw = 298.4. Used in treating acne vulgaris. Also *norethindrone acetate*, $C_{22}H_{28}O_3$, mw = 340.4. Also norethynodrel, $C_{20}H_{26}O_2$, mw = 298.4. Also *normethisterone*, $C_{19}H_{28}O_2$, mw = 288.4.

Pregnenolone (Δ^5-pregnen-3β-ol-20-one). $C_{21}H_{32}O_2$, mw = 308.4. Important in the synthesis of adrenal hormones.

Progesterone (progestin; progestone; Δ^4-pregnene-3,20-dione). $C_{21}H_{30}O_2$, mw = 314.4. Used in treating excessive uterine bleeding, hypogonadal females, menopausal symptoms, polycystic ovary syndrome, secondary amenorrhea.

DIURETIC, ANTIDURETIC AND LOCAL ANESTHETIC AGENTS

Aldosterone and spironoactone are described earlier in this list.

Hydroxydione sodium (21-hydroxypregnane-3,20-dione-21-sodium hemisuccinate). $C_{25}H_{35}O_6Na$, mw = 454.5

---

[1] This is an abridged list of steroid hormones. Some are much more important and widely used than others. Some are relatively recent to steroid therapy; others have been used more widely in the past than presently. See also **Hormones.**

a/b, b/c, and c/d, respectively, are *trans* or *cis*. According to an arbitrary convention, one designates the substituent groups as α or β depending upon whether they are situated below the plane of the molecule, when depicted in a certain way. Usually, in a structural diagram of a steroid molecule, a dotted line connection will be used between atoms to designate an alpha position; a regular solid line for a beta position.

Figure 2 illustrates one of the two most important configura-

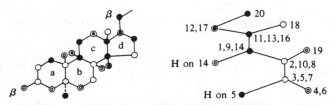

Fig. 2. Constellation of a saturated beta-sterol: ● = atoms in bottom plane; ● = atoms in second plane; ○ = atoms in third plane; ⊙ = atoms in top plane. Lateral view of molecule is shown at right.

tions of steroid skeletons, among 63 other possibilities, as they are found in nature. The side chain is usually attached in beta-position to C_{17}. The configurations on C_{20} and C_{24} have likewise

been determined and are known to produce steric isomerisms.

The sterols, from which the name of the entire group is derived, are monovalent alcohols with a secondary hydroxyl group on C_3 usually in beta-position. The best known representative is *cholesterol*. See Fig. 3. This compound forms esters with a

Fig. 3. Cholesterol.

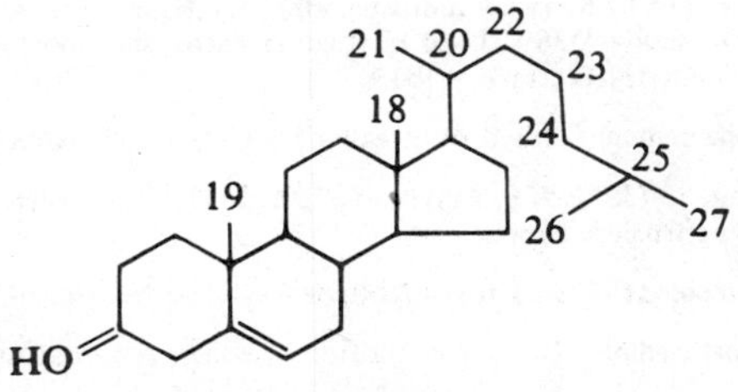

great variety of acids. Both the free and esterified sterols accompany the neutral fat and the phosphatides in most animal and plant fat. Upon alkaline hydrolysis, the other lipid constituents form fatty acid soaps; the fraction which remains insoluble in aqueous alkaline solution is called the *unsaponifiable* and consists primarily of sterols. Variations in the cholesterol content of blood in animals, particularly in humans, are of significance for the diagnosis of various diseases. See **Cholesterol.** Cholesterol is the principal sterol of all vertebrate animals. It is also found in some mollusks and in crustaceans, where it may be of alimentary origin.

Sterol Biogenesis. The finding that ingestion of radioactively labeled acetic acid leads to the synthesis of radioactive cholesterol was the first step in the elucidation of sterol biosynthesis. A growth factor for *Lactobacilli*, replaceable by acetic acid, was found to have the structure: $HOCH_2 \cdot CH_2 \cdot C(CH_3)(OH) \cdot CH_2 \cdot CO_2H$. This compound was termed *mevalonic acid.* Its close relationship to a trimer of acetic acid is evident. Six molecules of this C_6 acid polymerize, losing their carboxyl groups, to the linear isoprenol *squalene*, a hydrocarbon occurring in nature. Twelve of the carbon atoms (shown as circles in Fig. 4) originate from the carboxy groups of the original acetic acid,

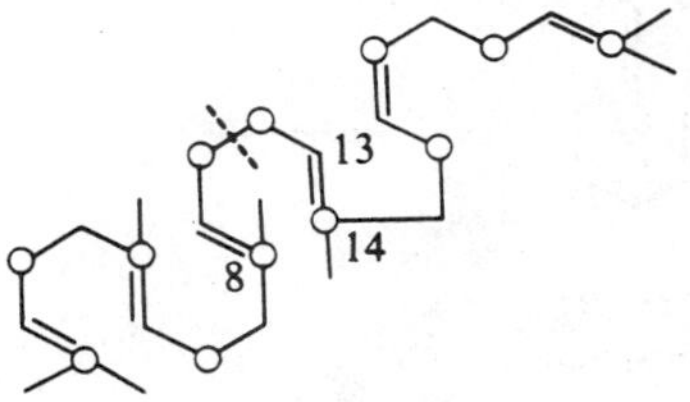

Fig. 4. Squalene.

the remaining 18 from the methyl groups. Squalene folds in the manner indicated in the figure and yields (with a two-step rearrangement of the methyl group from C_8 to C_{13}) *lanosterol*, a "protosterol" found in wool and fat. This protosterol loses three methyl groups in positions 4, 4, and 14 in the course of biosynthesis, yielding *zymosterol*, found in yeast, which is convertible to cholesterol. The gradual oxidative degradation of the side chain in cholesterol to bile acids and subsequently to the various steroid hormones in animals is well established and has been confirmed by ^{14}C tracer studies. Many of the enzymes operative during these hormone syntheses in the insertion of hydroxyl groups on individual carbon atoms have been separated and localized in various cell constituents. Major steps in sterol biogenesis are shown in Fig. 5.

Classification of Medically Important and Useful Steroid Hormones

In addition to the medical uses of steroid hormones for alleviating conditions brought about by insufficiencies or overabundances of any particular hormone of this class within the body and thus return the desirable hormone balance, there are numerous therapies which do not fall directly into these two categories. Rather, steroid hormones are used in connection with some ailments and diseases because of positive clinical results even though much remains to be learned concerning the details of their function.

Steroid hormones are difficult to classify because some of them serve large numbers of uses. The conventional approach places them into four categories.

Androgens and Anabolic Agents. The androgens are the male sex hormones. The androgenic hormones are synthesized in the body by the testis, the cortex of the adrenal gland, and, to a slight extent, by the ovary. The androgens have a number of sexually related functions. Androgens also serve as anabolic agents, i.e., nutrition of muscle and bone in both male and female persons. A number of androgens have been synthesized. Some of these reduce or eliminate the production of male secondary sex characteristics when administered to females (growth of facial hair, lowering of the voice, etc.)

Estrogens. Estrogen is a general term for female sex hormones. They are responsible for the development of the female secondary sex characteristics, such as the deposition of fat and the development of the breasts. Estrogens are produced by the ovary, and to a lesser degree, by the adrenal cortex and testis. Some synthetic *nonsteroid* compounds, such as diethylstilbestrol and hexestrol, have estrogenic activity.

Progesterone. Progesterone (Δ^4-pregnene-3,20-dione), $C_{21}H_{30}O_2$, is the female sex hormone secreted in the body by the corpus luteum, by the adrenal cortex, or by the placenta during pregnancy. It is important in the preparation of the uterus for pregnancy, and for the maintenance of pregnancy. Progesterone is believed to be the precursor of the adrenal steroid hormones.

Adrenal Corticosteroids. Among these hormones are compounds which have been found to be antiinflammatory, antiallergic, and antirheumatic agents and consequently are very important in steroid therapy. *Cortisone* was first applied in 1949 and became a major drug for treating rheumatoid arthritis, among other elements, soon thereafter. *Hydrocortisone* followed and a bit later several synthetic analogues (not found in the body) were developed. Well known among these is *prednisone*, found particularly effective in cases of diseases of collagen tissue, but also used in many other situations. Several examples are given in the accompanying table. See also **Hormones.** Steroid therapy throughout the years has had to cope with production of numerous side effects. Problems like this provide an incentive to continued vigorous research for new compounds.

In addition to the foregoing four major classifications (by application), there are also steroid hormones which are effective diuretic, antidiuretic, and local anesthetic agents. Among these are aldosterone, spironolactone, and hydroxydione sodium.

Bile Acids

The bile acids are monocarboxylic acids of the steroid group with 24 carbon atoms and 1–3 secondary hydroxyl groups. They occur in the bile of all vertebrates from the teleosts upward, mostly in peptidic conjugation with glycine or taurine. The bile acids are described in the entry on **Bile.**

STIBNITE. The mineral stibnite, antimony sulfide, Sb_2S_3, is found in radiated groups of acicular orthorhombic crystals or in other sorts of aggregates, as well as blades, also as columnar or granular masses. It shows a highly perfect pinacoidal cleavage; conchoidal fracture; hardness, 2; specific gravity, 4.63–4.66; luster, metallic and very brilliant on cleavage faces or freshly fractured surfaces. Its color is a steely gray; the streak very similar in color, may be covered with a black, sometimes iridescent tarnish.

Stibnite is the most common antimony mineral known and is the chief ore of that metal. It is a primary ore mineral and occurs with other antimony minerals and galena, sphalerite, and silver ores. It is found in Germany, Rumania, Czechoslovakia, Italy, Borneo, Peru, Japan, China, Mexico, and in the United States in California and Nevada.

The name *stibnite* is derived from the Latin word for antimony, *stibium*.

STILBITE. The mineral stilbite, $NaCa_2(Al_5Si_{13})O_{36} \cdot 14H_2O$, is a zeolite, the compound monoclinic crystals of which are usually grouped in approximately parallel positions, forming sheaflike aggregates, which have a soft pearly luster, whence the name stilbite from the Greek, meaning luster. The less commonly used term desmine is likewise from the Greek, meaning a bundle. Stilbite has one perfect cleavage; uneven fracture; is

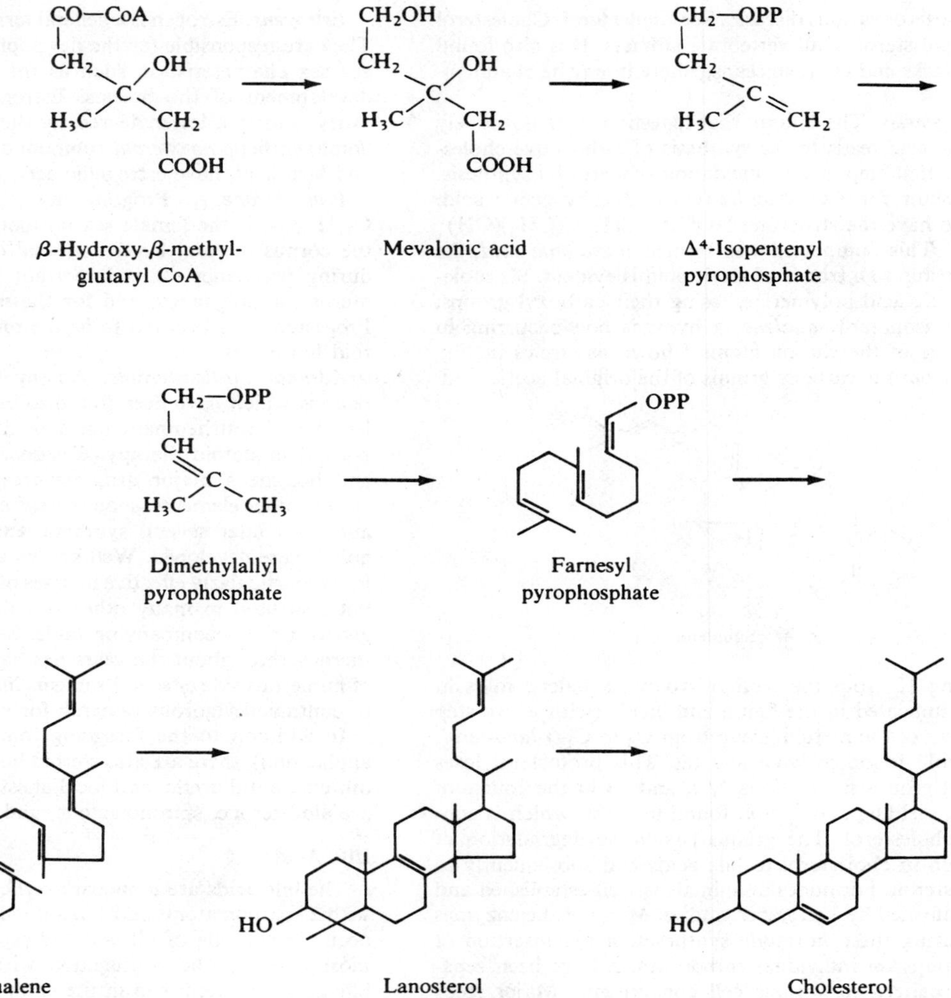

Fig. 5. Structures of key compounds involved in biogenesis of sterols.

brittle; hardness, 3.5–4; specific gravity, 2–2.2; luster, vitreous to pearly; color, usually white but may be brownish, yellowish, red or pink. Its streak is white, and it is transparent to translucent. Like the other zeolites stilbite occurs in cavities in basalts and traps, rarely in granites and gneisses. Of the many localities may be mentioned Trentino, Italy; the Harz Mountains; Valais, Switzerland; Arendal, Norway; the Ghats Mountains of India; and Mexico. The Triassic traps of New Jersey and Pennsylvania furnish specimens as do also rocks of the same age in Nova Scotia. This mineral sometimes is called *desmine*.

STOICHIOMETRY. The mathematics of chemical reactions and processes. It relates to all the quantitative aspects of chemical changes, both mass and energy. Stoichiometry is based on the absolute laws of conversion of mass and of energy and on the chemical law of combining weights. This basis makes stoichiometry as exact as any other branch of mathematics.

The law of conservation of mass dictates that, regardless of the nature of the changes undergone in a physical or chemical process, the total mass of all the materials in the system remains the same, even though the physical states and chemical compositions of the materials may change. Likewise, the law of conservation of energy is based upon the fact that the total energy in a reacting system remains constant even though the level or

form of the energy may change. In radioactive transformations, however, a slight correction must be applied to the law of the conservation of mass. Mass and energy have been found to be interconvertible, so that in general the total energy of the system remains constant even though there may be small mass changes.

The above concepts form the basis for weight and heat balance calculations. Such calculations are of great significance in engineering practice for the purpose of evaluating performance of existing operations or designing new manufacturing facilities and equipment.

The basic laws of conservation specifically state that matter or energy in a given system cannot be created or destroyed, and accordingly this requires that the following equality holds true:

$$\text{Input} = \text{output} + \text{accumulation}$$

For continuous, steady flow systems the change in in-process inventory is zero during any interval of time. In this case, therefore, the above expression reduces to the simplified form of input = output.

In making material weight balances the above relation may be applied to a single unit of the operation, or to the over-all operation with reference to the separate elements and/or the total mass entering and leaving the system. This method of analysis can best be exemplified by means of a synthetic problem: Let it be assumed that consideration is being given to a continuous, steady flow system, to which X pounds of material is fed per minute and from which Y pounds of useful product are analyzed to contain $a\%$ and $b\%$ by weight of a certain constituent, respectively. It is desired to determine the extent of unmeasured loss, Z pounds per minute, incurred from the system and the average concentration, c, of the said constituent in the waste stream.

First, it may be written that input = output. By dividing each side of this equality by an element of time, this relationship can then be transformed into the following expression:

$$\text{Rate of input} = \text{rate of output}$$

Then, a total weight balance can be written to express this statement of equality in terms of the quantities specified in the problem:

$$X = Y + Z \qquad (1)$$

A similar balance may also be written in terms of the constituent in question:

$$\frac{a}{100}X = \frac{b}{100}Y + \frac{c}{100}Z \qquad (2)$$

Finally, by algebraic solution of the two simultaneous equations it follows that

$$Z = X - Y \qquad (3)$$

and

$$c = \frac{aX - bY}{X - Y} \qquad (4)$$

The above example is only a simple illustration of a weight balance. Similarly, the reaction between elements and compounds may be symbolically expressed to portray the principle of conservation of matter. For example, if hydrogen is completely burned to water, the reaction between it and oxygen can be represented as follows:

$$(\text{Hydrogen}) + (\text{Oxygen}) \rightarrow (\text{Water})$$
$$H_2 + \tfrac{1}{2}O_2 \rightarrow H_2O$$
$$(2.02 \text{ Wgt. units}) + (16.00 \text{ Wgt. units}) \rightarrow (18.02 \text{ Wgt. units})$$

It would be found that these materials would always react in the same relative proportions to form water in an amount equal to the total weight of reactants. The relative weights indicated are equal to the molecular weights of the materials in question. Even if a reaction does not go to completion, the quantities which did react would be proportional to the combining weights expressed in the balanced chemical equation.

Since the element of time is usually involved as the basis of a stoichiometric calculation, proper quantitative deductions often depend on adequate knowledge of other laws or principles, such as those governing rates of reaction and those pertaining to chemical equilibria. When materials in the gaseous state are involved, the general gas laws are of great utility.

Another independent relation for a system is obtainable by applying the law of conservation of energy, which requires that energy input equals energy output. A valid equality of this type must include all forms of energy such as potential energy, kinetic energy, internal energy, flow work, electrical energy, etc. This type of equality results in the so-called "total energy balance." Another very useful but similar expression is the Bernoulli mechanical energy balance for steady mass flow of fluids. However, heat energy is very frequently the only primary effect in a process

so that, in such cases, the total energy balance can be simplified to the very advantageous expression of heat input equals heat output. This constitutes the basis for heat balances which, together with weight balances, are the most useful tools in any stoichiometric calculations.

Since chemical reactions involve combination of atoms or molecules to form new compounds or decomposition of compounds to form simpler ones, it is most convenient in stoichiometric calculations to employ molecular units rather than weight units. This particular kind of unit is called a "mole" and represents the quantity of substance numerically equal to its molecular weight. This weight quantity may be based on any system of weight units desired, and it is thus necessary to designate this basis by referring to pound moles, gram moles, etc.

A particular chemical reaction may be written to embody both laws of conservation of mass and energy as demonstrated below:

$$FeS + \tfrac{7}{4}O_2 \rightarrow \tfrac{1}{2}Fe_2O_3 + SO_2 + 268,000 \text{ Btu}$$

This equation states that 1 pound mole of ferrous sulfide reacts with 7/4 moles of oxygen to form $\tfrac{1}{2}$ mole of ferric oxide and 1 mole of sulfur dioxide, accompanied by a release of heat amounting to 268,000 Btu. However, to assign a specific meaning to the numerical value for this heat release, it is customary to specify a reference temperature and pressure for the reaction, these being 25°C and 1 atmosphere in the example cited. In making heat balance calculations, it is then convenient to choose these conditions as the datum level and then calculate the heat input and heat output quantities above or below the reference state.

—Walter C. Lapple, Alliance, Ohio.

STRIPPING COLUMN. Distillation.

STROMATOLITE.
A term that has been generally applied to variously shaped (often domal), laminated, calcareous sedimentary structure formed in a shallow-water environment under the influence of a mat or assemblage of sediment-binding blue-green algae that trap fine (silty) detritus and precipitate calcium carbonate and that commonly develop colonies or irregular accumulations of a constant shape, but with little or no microstructure. It has a variety of gross forms, from near-horizontal to markedly convex, columnar, and subspherical. Stromatolites were originally considered animal fossils, and although they are still regarded as "fossils" because they are the products of organic growth, they are not fossils of any specific organism, but rather they consist of associations of different genera and species of organisms that can no longer be recognized and named or that are without organic structures. An excellent treatise on stromatolites is "Stromatolites," M. R. Walter, Editor, Elsevier, New York (1976).

STRONG INTERACTION. Particles (Subatomic).

STRONTIANITE.
The mineral strontianite is strontium carbonate, $SrCO_3$, usually occurring in whitish-yellow or whitish-green masses of radiated acicular crystals, or in fibrous or granular form. When distinctly crystallized it is obviously orthorhombic, but such crystals are rare. It has a nearly perfect prismatic cleavage; uneven fracture; brittle; hardness, 3.5; specific gravity, 3.785; luster, vitreous; color, as above, also green, gray and colorless; streak, white; transparent to translucent. Strontianite occurs in veins chiefly in limestones, occasionally in the crystalline rocks, and usually associated with calcite and celestite. It

is found in the metalliferous veins in the Harz Mountains and Saxony. It is commercially important in Westphalia where it is mined for use in the beet sugar industry. In the United States, crystalline masses and geodes of strontianite are found in Schoharie County, New York, long a famous locality for this mineral.

STRONTIUM. Chemical element symbol Sr, at. no. 38, at. wt. 87.62, periodic table group 21 (alkaline earths), mp 770 $\pm 1°C$, bp 1384°C, density, 2.6 g/cm^3 (20°C). Below 215°C, elemental strontium has a face-centered cubic crystal structure; between 215 and 605°C, a hexagonal close-packed crystal structure; and above 605°C, a body-centered cubic crystal structure. Strontium is a silvery-white metal, soft as lead, malleable, ductile; oxidizes rapidly upon exposure to air; burns when heated in air emitting a brilliant light and forming oxide and nitride; reacts with water, yielding strontium hydroxide and hydrogen gas. Discovered by Hope and Klaproth in 1793, and isolated by Davy in 1808.

There are four stable isotopes, ^{84}Sr and ^{86}Sr through ^{88}Sr, and seven known radioactive isotopes, ^{82}Sr, ^{83}Sr, ^{85}Sr, and ^{89}Sr through ^{92}Sr, all with relatively short half-lives measurable in hours or days, except ^{90}Sr ($t_{1/2} = 26$ years). The latter isotope represents a hazard from nuclear blasting activities because of its long half-life, tendency to contaminate food products, such as milk, and retention in the body. See **Radioactivity.** In terms of abundance, strontium is 21st among the elements, occurring in the rocks of the earth's crust. In terms of content in seawater, the element ranks 11th, with an estimated 38,000 tons of strontium per cubic mile (8208 metric tons in a cubic kilometer) of seawater.

First ionization potential, 5.692 eV; second, 10.98 eV. Oxidation potentials: $Sr \rightarrow Sr^{2+} + 2e^-$, 2.89 V; $Sr + 2OH^- + 8H_2O \rightarrow Sr(OH)_2 \cdot 8H_2O + 2e^-$, 2.99 V. Other physical characteristics are given under **Chemical Elements.**

Strontium occurs chiefly as sulfate (celestite, $SrSO_4$) and carbonate (strontianite, $SrCO_3$) although widely distributed in small concentration. The commercially exploited deposits are mainly in England. The sulfate or carbonate is transformed into chloride, and the electrolysis of the fused chloride yields strontium metal.

As is to be expected from its high oxidation potential (2.89 V) strontium, like calcium and barium, reacts readily with all halogens, oxygen and sulfur toform halides, oxide and sulfide. See also **Celestite; Strontianite.** In all its compounds it is divalent. It reacts vigorously with H_2O to form the hydroxide, displacing hydrogen and it forms a hydride with hydrogen. Strontium hydroxide forms a peroxide on treatment with H_2O_2 in the cold. Strontium exhibits little tendency to form complexes; the ammines formed with NH_3 are unstable, the β-diketones and alcoholates are not well characterized, and the chelates formed with ethylenediamine and related compounds are the only representatives of the type. Common compounds of strontium are the following:

Strontium acetate $Sr(C_2H_3O_2)_2$, white crystals, soluble, formed by reaction of strontium carbonate or hydroxide and acetic acid.

Strontium carbide (acetylide) SrC_2, black solid, formed by reaction of strontium oxide and carbon at electric furnace temperature; the carbide reacts with water yielding acetylene gas and strontium hydroxide.

Strontium carbonate $SrCO_3$, white solid, insoluble ($K_{sp} = 9.4 \times 10^{-10}$), formed (1) by reaction of strontium salt solution and sodium carbonate or bicarbonate solution, (2) by reaction of strontium hydroxide solution and CO_2. Strontium carbonate decomposes at 1,200°C to form strontium oxide and CO_2, and

is dissolved by excess CO_2, forming strontium bicarbonate $Sr(HCO_3)_2$, solution.

Strontium chloride, $SrCl_2 \cdot 6H_2O$, white crystals, soluble, formed by reaction of strontium carbonate or hydroxide and HCl. Anhydrous strontium chloride, $SrCl_2$, absorbs dry NH_3 gas.

Strontium chromate, $SrCrO_4$, yellow precipitate ($K_{sp} = 3.75 \times 10^{-5}$) formed by reaction of strontium salt solution and potassium chromate solution.

Strontium cyanamide, $SrCN_2$, formed with the cyanide $Sr(CN)_2$, by heating strontium carbide at 1200°C with nitrogen.

Strontium hydride, SrH_2, white solid, formed by heating strontium metal or amalgam in hydrogen gas at 250°C. Compound is reactive with water, yielding strontium hydroxide and hydrogen gas.

Strontium nitrate, $Sr(NO_3)_2$, white crystals, soluble, formed by reaction of strontium carbonate or hydroxide and HNO_3.

Strontium oxide, SrO, white solid, mp about 2,400°C, reactive with H_2O to form strontium hydroxide ($K_{sp} = 3.2 \times 10^{-4}$); strontium peroxide, $SrO_2 \cdot 8H_2O$, white precipitate, by reaction of strontium salt solution and hydrogen or sodium peroxide, yields anhydrous strontium peroxide SrO_2, upon heating at 130°C in a current of dry air.

Strontium oxalate, SrC_2O_4, white precipitate ($K_{sp} = 5.6 \times 10^{-8}$) formed by reaction of strontium salt solution and ammonium oxalate solution.

Strontium sulfate, $SrSO_4$, white precipitate ($K_{sp} = 3.2 \times 10^{-7}$), formed by reaction of strontium salt solution and H_2SO_4 or sodium sulfate solution, insoluble in acids. On heating with carbon strontium sulfate yields strontium sulfide, SrS, while on boiling with sodium carbonate solution, $SrSO_4$ yields strontium carbonate.

Strontium sulfide, SrS, grayish-white solid (thermodynamic K_{sp} 500) reactive with water to form strontium hydrosulfide, $Sr(SH)_2$, solution. Strontium hydrosulfide is formed (1) by reaction of strontium sulfide and H_2O, (2) by saturation of strontium hydroxide solution with H_2S. Strontium polysulfides are formed by boiling strontium hydrosulfide with sulfur.

—Stephen E. Hluchan, Calcium Metal Products, Minerals, Pigments & Metals Division, Pfizer Inc., Wallingford, Connecticut.

STYRENE-BUTABIENE (SBR) RUBBER. Elastomers.

SUBATOMIC PARTICLES. Particles (Subatomic).

SUBLIMATION. The direct transition, under suitable conditions, between the vapor and the solid state of a substance. If solid iodine is placed in a tube and slightly warmed, it vaporizes and the vapor re-forms into crystals on the cooler parts of the tube. Many crystalline substances, both metallic and non-metallic, may be similarly sublimed in a vacuum; fairly large crystals of selenium have been thus prepared. The most familiar sublimates are frost and snow. As in the case of other changes of state, sublimation is accompanied by the absorption or evolution of heat, the quantity of which per unit mass is called the heat of sublimation of the substance. At pressures near the triple point the heat of sublimation is approximately equal to the sum of the heats of fusion and vaporization. In physical and chemical literature, it is customary to regard as sublimation only the transition from solid to vapor, not from vapor to solid; but metorologists do not make this distinction.

Sublimation plays a major role in the freeze-drying of foods. See also **Freeze-Drying.**

SUBSTITUTE NATURAL GAS (SNG). A rather general term for describing an artificially produced relatively-high Btu gas that compares favorably with natural gas as a fuel. SNG also may refer to synthetic natural gas; or, on some occasions, to synthesis gas. The latter generally connotes a specially constituted gas to be used as the raw material for a chemical process, such as an ammonia synthesis—thus ammonia synthesis gas, etc. The Btu value of natural gas typically lies within the range of 975 to 1,180 Btu per standard cubic foot (110–133 Calories/cubic meter). Thus, to substitute for and to compete with natural gas (where available), the artificially-produced gas must have a Btu content within this general range. Substitute or synthetic gases generally fall into two categories: (1) low-Btu-value gases with a Btu content of 400 to 600 Btu per cubic foot (50–67 Calories/cubic meter) or lower, sometimes suitable for combined-cycle power generation schemes or for subsequent enrichment to increase the Btu content; and (2) high-Btu-value gases (sometimes referred to as pipeline gases) which have a Btu content generally within the 950 to 1,050 Btu per cubic foot (107–118 Calories/cubic meter) range. Such gases, properly treated to remove traces of unwanted impurities and corrosives, can be introduced into transcontinental pipelines and handled essentially in the same manner as natural gas.

SNG is derived from coal, various petroleum fractions, and waste products. Gases produced from coal are described under **Coal.**

The most practical and economic source of raw material for producing SNG varies with the proximity to raw materials, the relative cost of raw materials, and by numerous other factors which affect the complex energy balance of a given nation and geographical location. Naphtha may make a logical choice of starting material in one area, whereas coal would be most logical in another area. Also, for some years to come—until SNG processes become better proved on a day-to-day operating basis—a somewhat more costly raw material, if available, may be the only practical answer. With proven processes, waste materials as a source of SNG is a very sensible approach, but in some areas the costs of collecting wastes (or the availability of sufficient waste products) may prohibit this approach. As proponents of various schemes and concepts have found upon undertaking detailed, practical development of concepts, a process will not necessarily be successful even though initial gross statistics "prove" the wisdom of the concept.

Catalytic Rich Gas (CRG) Process. This process was developed from the work of a team at the Gas Council's Midlands Research Station (MRS) (England) led by Dr. F. J. Dent. In the late 1950s, it became apparent that, due to the postwar increase in refining capacity in Europe, naphtha was becoming available as a potential feedstock for gas making and that its use would be more economical than coal carbonization, which was then the major source of fuel gas in the United Kingdom. The first semicommercial plant, producing 4 million standard cubic feet per day of rich gas, was commissioned in 1964. Within the next five years, nearly 40 units were installed in the United Kingdom for production of rich gas and town gas (470–500 Btu per standard cubic foot; 53–56 Calories/cubic meter). Plants were also installed in Japan, Italy, Brazil, and the United States.

The overall reactions which occur in the steam reforming of naphtha are:

(1) $4C_6H_{14} + 10H_2O \rightarrow 19CH_4 + 5CO_2$ Exothermic

(2) $CH_4 + H_2O \rightleftharpoons CO + 3H_2$ Endothermic

(3) $CO + H_2O \rightleftharpoons CO_2 + H_2$ Slightly Exothermic

At all practical temperatures, reaction (1) proceeds almost to completion; no significant quantities of higher hydrocarbons exist at the outlet of the CRG reactor.

Reactions (2) and (3) are reversible; the concentrations of the five components CH_4, H_2O, CO_2, CO and H_2 which result are governed by thermodynamic equilibrium. Raising the reaction temperature shifts the equilibrium for both reactions to the right. Thus at low temperatures the exothermic reaction (1) predominates, while at high temperatures the overall reaction is endothermic. At approximately 500–550°C the reaction is thermally neutral.

Naphthas boiling up to 185°C can be reformed at pressures up to 600 psig. Naphthas with final boiling point up to 240°C may be reformed at lower pressures. Higher olefin contents may be accepted provided that sufficient hydrogen is available in the recycle gas to saturate the feed in the desulfurization section. Higher aromatic contents may be accepted but the catalyst life will be reduced.

A typical rich gas leaving the CRG reactor has the following composition:

CO_2	23.0 mol.% (dry)
CO	0.7
H_2	12.8
CH_4	63.5
	100.0%
Calorific Value (Btu/standard bic foot)	675

Higher hydrocarbons are present in negligible quantities.

The calorific value of this gas is too high for direct use as town gas (470–500 Btu/standard cubic foot in the United Kingdom) and too low for SNG. However, by removing the CO_2 the calorific value is increased to about 870 Btu/standard cubic foot, which is useful for enriching lean gas (e.g. from an Imperial Chemical Industries (ICI) naphtha reformer) to town gas quality. Long has suggested that by enriching this gas with LPG (liquefied petroleum gas) a satisfactory SNG may be obtained.

Alternatively, the calorific value may be changed by bringing the components to a new equilibrium at a different temperature. In the Series "A" Process, part of the rich gas is further reformed at high temperature and remixed with the remaining rich gas. After water gas shift and partial CO_2 removal a 500 Btu/standard cubic foot product is obtained which is fully interchangeable with the town gas distributed in the United Kingdom.

If the subsequent stage is at a lower temperature, carbon oxides and hydrogen recombine to methane, increasing the calorific value. Following CO_2 removal, very little enrichment is required to achieve a product fully interchangeable with natural gas.

CRG catalyst is deactivated by low concentrations of sulfur and chlorine compounds. To achieve removal of sulfur to very low concentrations (less than 0.2 ppm), British Gas developed their own process which is always used in association with CRG catalyst. Organic sulfur compounds are hydrogenated to H_2S over nickelmolybdenum catalyst at about 380°C. The hydrogen is usually generated by reforming rich gas from the CRG reactor with added steam in a tubular reformer. Alternatively, gas from the reactor may be used directly, while in some plants it is normal to use CO_2-free town gas. The H_2S is then absorbed on zinc oxide or, in the case of many United Kingdom town gas plants, Luxmasse (hydrated ferric oxide).

Because of the large stream sizes involved, many of the SNG plants in the United States incorporate a bulk sulfur removal stage using a hydrofining process. The H_2S produced is commonly recovered as elemental sulfur by a Stretford plant, in which the hydrofiner off gas is washed with an aqueous alkaline solution which is regenerated by oxidation with air.

Chlorine compounds, if present in concentration higher than 1 ppm, not only deactivate the CRG catalyst but also interfere with the absorption of H_2S. Therefore they are removed by hydrogenation to HCl which is absorbed on a proprietary absorbent.

Reforming in the CRG process occurs adiabatically at 450–550°C at pressures up to about 600 psig (41 atmospheres). The reactor is a vertical cylindrical pressure vessel containing a bed of the special high-nickel catalyst which is supported on a grid or on inert ceramic balls. The gas flow is downwards through the bed and distributors are provided at inlet and outlet. A layer of ceramic balls on top of the bed prevents disturbance of the catalyst by the entering gas.

Normal practice is to install two reactors in parallel, of which one is working at any time. The catalyst charge in each vessel is designed for 3–6 months operation at full load. This system avoids unnecessary exposure of catalyst to high temperatures, minimizes the catalyst loss in the event of damage by maloperation and provides instant standby if such damage occurs.

The flow diagram for a rich gas plant producing gas with a calorific value of 710 Btu/standard cubic foot is shown in Fig. 1. The product is used to enrich lean gas from an ICI (Imperial Chemical Industries) naphtha reformer which has a calorific value of about 320 Btu/standard cubic foot to the town gas standard of 500 Btu/standard cubic foot (56 Calories/cubic meter). Typical gas analyses are given in Table 1.

In what is termed a Series "A" Process, part of the gas from the CRG reactor is reformed with additional steam and the resulting lean gas is reblended with the remaining rich gas. The mixed gas is then subjected to water gas shift and partial carbon dioxide removal, yielding a product with a calorific value of 470–500 Btu/standard cubic foot (53–56 Calories/cubic me-

TABLE 1. GAS ANALYSES IN RICH GAS PLANT

	RECYCLE GAS	GAS FROM CRGR	SCRUBBED RICH GAS
CO_2 (mol %)	0.9	20.3	13.5
CO	1.8	1.4	1.5
H_2	23.3	19.1	20.7
CH_4	74.0	59.2	64.3
	100.0	100.0	100.0
Calorific Value			
(Btu/standard cubic foot)	816	654	710
(Calories/cubic meter)	91.7	73.5	79.8

ter). Typical gas analyses are shown in Table 2. By varying the proportion of gas which flows to the tubular reformer and the degree of CO_2 removal, the characteristics of the product gas can be made interchangeable with any of the different standards employed by the United Kingdom Area Boards and Japanese and European gas companies. A variation of this process has been used in Italy. The calorific value of important Libyan LNG (1395 Btu/standard cubic foot; 157 Calories/cubic meter) was too high for direct use as pipeline gas. The LNG was, therefore, fractionated and the heavy ends (C_2H_6—C_6H_{14}) subjected to processing.

If the rich gas from the CRG reactor is passed over another bed of high-nickel catalyst at a lower temperature, the equilibrium of the five components is reestablished. Carbon oxides react with hydrogen to form methane and the calorific value of the gas is increased. It should be noted that this methanation step differs from that encountered in ammonia synthesis gas production; because of the high steam content the temperature rise is reduced and there is no possibility of temperature "run-

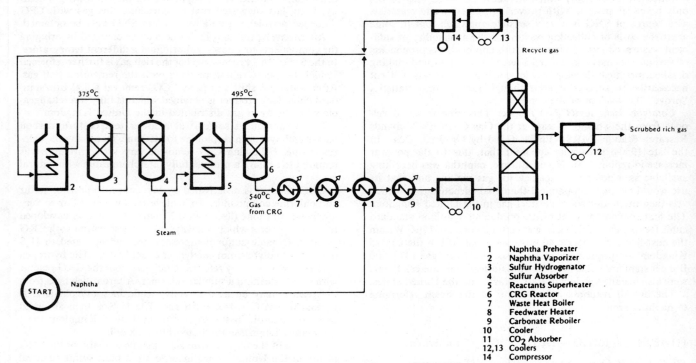

1	Naphtha Preheater
2	Naphtha Vaporizer
3	Sulfur Hydrogenator
4	Sulfur Absorber
5	Reactants Superheater
6	CRG Reactor
7	Waste Heat Boiler
8	Feedwater Heater
9	Carbonate Reboiler
10	Cooler
11	CO_2 Absorber
12,13	Coolers
14	Compressor

Fig. 1.   Typical flowsheet for rich gas plant. (*Woodall-Duckham Limited.*)

TABLE 2. GAS ANALYSES IN SERIES "A" PLANT

	RECYCLE GAS	RICH GAS	REFORMED GAS	MIXED GAS	CONVERTED GAS	PRODUCT GAS
CO_2 (mol %)	1.0	21.6	13.4	16.1	21.3	13.5
CO	3.4	0.9	13.7	9.6	2.7	3.0
H_2	60.9	15.3	59.0	44.9	48.4	53.2
CH_4	34.7	62.2	13.9	29.4	27.6	30.3
	100.0	100.0	100.0	100.0	100.0	100.0
Calorific Value						
(Btu/standard cubic foot)	550	670	370	466	438	480
(Calories/cubic meter)	61.8	75.3	41.6	52.4	49.2	53.9

away" as the exit temperature can never rise above the temperature corresponding to equilibrium at the inlet composition, i.e. the CRG exit temperature.

In order to minimize cold enrichment and to achieve a very low carbon monoxide content in the product, a second methanation stage is frequently employed. To achieve sufficient "driving force" to make the reaction proceed, the water vapor content is reduced by cooling the gas, rejecting condensate, and reheating to the required reaction temperature.

Table 3 shows the effect of the second methanation stage on product calorific value. While consumption of LPG is minimized, the capital cost is increased and the overall thermal efficiency is slightly reduced.

If part of the purified naphtha vapor from desulfurization is allowed to bypass the CRG reactor, it can be fully gasified by reaction with the hydrogen and steam in the rich gas. This reaction occurs at lower temperatures than the CRG reaction, and is known as hydrogasification. It has the advantage that the total steam requirement for the process is reduced, although with heavier feedstocks it may be necessary to add a little steam to the hydrogasifier in order to ensure that carbon is not formed by the Boudouard reaction. Since less makeup steam has to be generated from fired boilers, the overall efficiency is improved by 1–2%. The capital cost is slightly lower than that of the methanation route.

The calorific value of the product from hydrogasification is lower than that from single methanation, particularly with high carbon/hydrogen feedstocks because of the additional steam required. However, by adding a final methanator, the calorific value can be increased to that obtained from double methanation, again with increased capital cost and reduced efficiency. This process (Fig. 2) is used in the first operational SNG plant in the United States at Harrison, N.J. Typical gas analyses are given in Table 4.

Because of the lower temperature, the catalyst in the hydrogasifier, which is the same as that in the CRG reactor, is slowly deactivated by polymer formation. The activity may be recovered "in situ" by heating in hydrogen. Two hydrogasifiers are therefore provided in parallel so that regeneration can be carried out without interrupting production.

The CRG process is one of a range of processes developed by the British Gas Corporation for production of fuel gases. The range and application of these processes and their impact has been described by Hebden, illustrating the effect on capital cost of increasing the carbon/hydrogen ratio of the feedstock.

An alternative method of handling crude oil is the "energy refinery" in which crude is split into a number of fractions which can be treated by proven processes to yield two products, SNG and low sulfur fuel oil.

One such scheme is shown in Fig. 3. The advantages of using the CRG process as the final stage in the production of SNG are high efficiency and low capital cost, the predictable quality of the product gas, and the absence of by-products.

Methane-Rich Gas (MRG) Process. A process which produces methane gas from feedstock hydrocarbons, such as naphtha, liquefied petroleum gases (LPG), and refinery gas, developed by Japan Gasoline Co., Ltd., in collaboration with its affiliate, Nikki Chemical Co., Ltd., for application to town-gas facilities. The MRG process had its origin in the high-temperature hydrocarbon steam reforming technology. First efforts culminated in a successful installation in 1956. In the tendency to employ heavier hydrocarbons as feedstock, carbon formation in the low-temperature range posed a problem. The Japan Gasoline Co. took this up as a main research subject and continued the study of low-temperature-range reaction, concentrating especially on the difference of product gas properties and carbon formation according to reaction conditions and catalyst specifications. As a result, a new catalyst was developed which converts butane

TABLE 3. GAS ANALYSES—SNG PRODUCTION BY DOUBLE METHANATION

	RECYCLE GAS	1ST STAGE GAS	2ND STAGE GAS	3RD STAGE GAS	SCRUBBED GAS	PRODUCT GAS
CO_2 (mol %)	0.5	21.7	22.0	21.9	0.5	0.50
CO	1.5	0.8	0.1	0.1	0.1	0.04
H_2	86.5	12.8	3.7	0.4	0.5	0.53
CH_4	11.5	64.7	74.2	77.6	98.9	97.98
C_3H_8	—	—	—	—	—	0.95
	100.0	100.0	100.0	100.0	100.0	100
Calorific Value						
(Btu/standard cubic foot)	395	687	750	773	986	1000
(Calories/cubic meter)	44.3	77.2	84.3	86.9	110.8	112.4

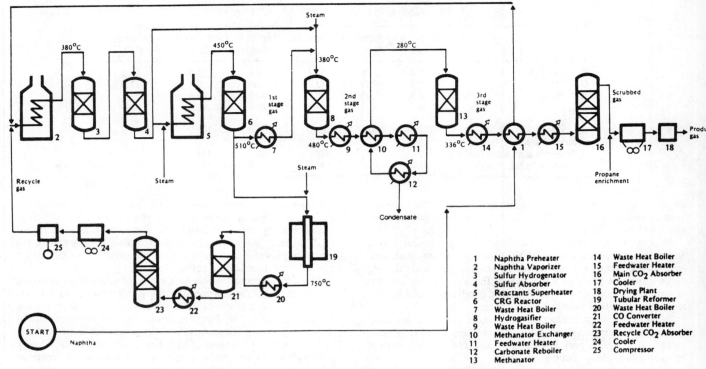

Fig. 2.   Typical flowsheet for SNG plant—hydrogasification route. (*Woodall-Duckham Limited.*)

TABLE 4.   GAS ANALYSES—SNG PRODUCTION BY HYDROGASIFICATION

	RECYCLE GAS	1ST STAGE GAS	2ND STAGE GAS	3RD STAGE GAS	SCRUBBED GAS	PRODUCT GAS
CO_2 (mol %)	0.5	21.8	21.7	21.9	0.5	~0.50
CO	1.6	0.7	0.6	0.1	0.1	~0.70
H_2	86.5	13.3	6.2	0.6	0.8	~0.79
CH_4	11.4	64.2	71.5	77.4	98.6	~97.56
C_3H_8	—	—	—	—	—	~1.08
	100.0	100.0	100.0	100.0	100.0	
Calorific Value						
(Btu/standard cubic foot)	395	683	733	772	984	1000
(Calories/cubic meter)	44.4	76.7	82.4	86.7	110.6	112.4

or naphtha to a gas consisting mainly of methane, hydrogen, and carbon dioxide, with a negligible amount of carbon monoxide. This was the first stage of development of the present MRG process.

In 1964, while developing practical applications for the town-gas industry, a pilot plant with a daily capacity of 15,000 cubic meters was built. Continuous test runs were conducted over a long term, in cooperation with Osaka Gas Co., Ltd., thus starting commercial production of equipment for the process. Based upon results of the aforementioned test runs, a commercial-size town-gas plant (200,000 cubic meters daily capacity) was constructed at the Hokkoh plant of Osaka Gas Co., Ltd. This was followed by two plants each of 500,000 cubic meters daily capacity in 1967 and 1969. Additional plants followed not only for town-gas uses, but also for petrochemical needs. Late in 1971, an MRG plant incorporating a wet methanation system went into operation at Keiyo Gas Co., Ltd., near Tokyo, with a capacity of 105,000 cubic meters per day. In late 1972, a complete MRG-based SNG plant, consisting of gasification,

methanation, and CO_2 removal sections was completed for the same firm, with a capacity of 200,000 cubic meters per day. In early 1974, Boston Gas Co. (U.S.) started up a 1,070,000 cubic meters per day SNG plant which employs a two-stage MRG gasification system.

The basic reactions of the MRG process consist of three stages: (1) hydrodesulfurization of sulfur compounds in the hydrocarbon feedstock; (2) low-temperature steam reforming (gasification) of desulfurized hydrocarbons; and (3) methanation reaction between hydrogen and carbon dioxide in methane gas available by gasification.

Sulfur compounds contained in hydrocarbon feedstock vary, depending on the types of crudes and their boiling points. Naphtha, for example, contains mainly mercaptans, disulfides, and thiophenes. Such sulfur compounds deteriorate the activity of the low-temperature steam-reforming MRG catalyst. They should be removed to some degree before the feedstock enters the system. Major reactions of the hydrodesulfurization step are:

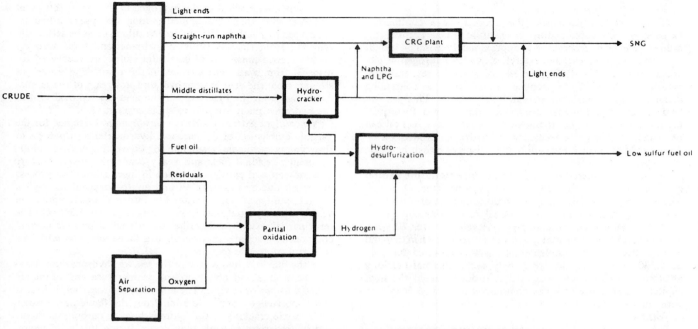

Fig. 3. SNG and low-sulfur fuel oil from crude. (*Woodall-Duckham Limited.*)

$$RSH + H_2 \longrightarrow RH + H_2S$$

$$R-S-R' + 2H_2 \longrightarrow RH + R'H + H_2S$$

$$R-S-S-R' + 3H_2 \longrightarrow RH + R'H + 2H_2S$$

$$\text{(thiophene ring)}-R + 3H_2 \longrightarrow RC_4H_9 + H_2S \quad \text{(alkyl group R not fixed to a specific carbon)}$$

Inasmuch as these are all exothermic reactions, low ambient temperatures are favorable from the standpoint of equilibrium theory, but in consideration of reaction rates, general processes are operated in a range of 350–400°C with the aid of a highly active catalyst (like Co-Mo or Ni-Mo), involving the side reactions:

$$CO_2 + 4H_2 \rightleftarrows CH_4 + 2H_2O$$

$$CO + 3H_2 \rightleftarrows CH_4 + H_2O$$

The foregoing reactions are highly exothermic and significantly raise reaction temperatures. The MRG process, however, does not involve such adverse side reactions with use of a special, selective hydrodesulfurizing catalyst (developed by Japan Gasoline Co. and Nikki Chemical). The MRG process uses part of product gas for hydrodesulfurization, and even if it contains only 20–25% hydrogen and as high as 20–23% carbon oxides, only the proper hydrodesulfurization reactions take place. The MRG process features a recycle use of product gas for hydrodesulfurization purposes without any special treatment.

To eliminate hydrogen sulfide formed in hydrodesulfurization reactions, two solutions are available: (1) fixation by H_2S contact with an adsorbent (zinc oxide) via the reaction: $ZnO + H_2S \rightarrow ZnS + H_2O$; and (2) physical removal by stripping. The hydrodesulfurization system is most economically practical with feedstocks containing less than 200 to 500 ppm sulfur. The removal of H_2S by stripping after hydrodesulfurization with an external hydrogen supply may be applied to naphtha stocks contaminated by trace metals as well as those high in sulfur.

Gasification by low-temperature steam-reforming reactions,

the heart of the MRG process, is carried out between liquid hydrocarbons and steam over catalyst to form methane, hydrogen, and carbon oxides. In order to increase the calorific value of product gas to the values similar to natural gas, methanation reactions are required. Hydrogen in product gas is reacted with CO_2 and CO to form methane, with only a small portion unconverted. Methanation reactions are:

$$CO + 3H_2 \rightleftarrows CH_4 + H_2O \quad 49.3 \text{ kcal/g-mole at } 25°C$$

$$CO_2 + 4H_2 \rightleftarrows CH_4 + 2H_2O \quad 9.8 \text{ kcal/g-mole at } 25°C$$

After methanation, the gas goes to a scrubber to remove CO_2 for further purification.

Because the MRG process is mainly based on steam reforming, the success of the process hinges on the reliable availability of steam. Steam should be controlled at a constant level somewhat above the projected requirements to assure continuous, effective reforming. Adequate steam must be on hand at all times to assure effective control of the steam/naphtha ratio. In summary, for the proper feedstocks and economic situations, the MRG process offers the following: (1) a wide variety of feedstocks can be used; (2) broad selection of calorific value of final product gas. Product gas is available in a range of calorific values from 5,500 to 9,400 kcal/cubic meter; (3) high-pressure operation. Many conventional town-gas plants which operate at near atmospheric pressure require an additional compressor to convey product gas through pipelines. The MRG process does not require any additional compressor, but does permit operation at high pressures—up to approximately 1,140 pounds per square inch gage (77.6 atmospheres), enabling high-pressure, long-distance transportation of product gas. (4) Noncomplex equipment is used. The use of a drum-type reactor makes the reactor design quite simple, resulting in a compact design of the overall system. In terms of product gas calorific value, the MRG reactor requires only $\frac{1}{6}$ to $\frac{1}{8}$ the area of a general coke oven and about $\frac{1}{3}$ that of a conventional high-temperature steam-reforming plant; (5) sulfur-resistant catalyst; and (6) high thermal efficiency. Operation at low temperatures and high pressures permits thermal efficiency as high as 92–96%, depending on desulfurized feedstock used.

Hydrocracking-Hydrogasification Process. In a continuous, two-step process developed by the Institute of Gas Technology, crude oil can be hydrocracked to approximately diesel oil weight and then made to react noncatalytically with hydrogen at elevated pressures (500–1,500 psig; 34–102 atmospheres) and temperatures 593–760°C) to produce methane-rich gas containing about 30% (volume) hydrogen and 10% ethane. This gas can be desulfurized and methanated to yield pipeline gas. By adjusting the conditions of hydrocracking, enough heavy fuel oil can be produced to provide feedstock for hydrogen production by partial oxidation, or a low-sulfur fuel oil product can be made if desired.

Interest has centered on plants to produce substitute natural gas from light distillate feedstocks, such as naphtha. However, when naphtha is in short supply, a process capable of converting the more plentiful crude and residual oils to pipeline gas is needed. A number of processes were developed over 20 years ago to provide supplemental gas in winter periods when demand is high. Because the supplemental gas was required for only 20 to 30 days during the year, only cyclic thermal cracking at atmospheric pressure was used in order to avoid the high capital cost of more complex continuous processes. However, with the need for base-load gas, a continuous process would be feasible.

The major difficulty associated with the production of pipeline gas from crude and residual oils is carbon deposition during the gasification step, which leads to reactor vessel plugging. One approach to this problem has been to conduct pressure gasification with hydrogen in a fluidized bed of coke, allowing carbon to deposit on the coke particles. To avoid accumulation of these deposits, a small amount of coke is continuously withdrawn from the bed. This technique, developed by the British Gas Council, gasifies the oil in a single reaction step. An approach to the problem of carbon deposition taken by the Institute of Gas Technology is to eliminate the carbon-forming materials in the heavy oil prior to hydrogasification by catalytic hydrocracking. In developing this concept, experiments were conducted on both the hydrocracking and the hydrogasification operations for a variety of feedstocks, ranging from kerosine to Bunker-C fuel oil. Distillate feeds required no hydrocracking.

A simplified flowsheet based on this concept (hydrocracking, separation, and hydrogasification) is shown in Fig. 4, where 250 billion Btu/day (63 billion Calories/day) of pipeline-quality gas is produced from 59,350 barrels/day of Taparito crude. The overall fuel efficiency of the process is 67%, allowing for all utility requirements, including oxygen production. The design of hydrogen plants based on partial oxidation of residual oils and the design of hydrocracking operations are well established arts. The only unusual component of the process, therefore, is the hydrogasification reactor itself. Offsites not shown on the flowsheet include an oxygen plant, sulfur-recovery facilities (Claus plant), power and steam generation equipment, and water treatment facilities.

At the hydrogasification conditions used, about 90% of the 360°C endpoint feed oil is gasified, yielding a raw gas containing about 52% methane and 10% ethane, with the remainder principally hydrogen. About 60% of the liquid products is benzene. Total liquid products are removed first by separation in a knockout drum and then by straw oil scrubbing. Benzene, the last traces of which are removed from the gas by activated carbon, is recovered and sold. Very heavy oil is used for plant fuel. Since the excess hydrogen in the gas is to be methanated with carbon dioxide, a small amount of carbon dioxide from the hydrogen plant is added to the gas prior to the removal of

hydrogen sulfide. Most of the hydrogen in the gas at this point will react with ethane during methanation. Because carbon dioxide is used for methanation, hydrogen sulfide must be selectively removed from the gas prior to methanation. Approximately 48,400 metric tons/year of elemental sulfur are recovered for disposal from waste gas streams in the plant. The waste gas streams from the acid-gas removal unit upstream of the methanator and from the first stage of the acid-gas removal unit in the hydrogen plant are sent to a Claus plant.

Noncatalytic Partial-Oxidation Gasification. Designed for the partial combustion or oxidation of hydrocarbons, a process of this type is particularly suitable for converting heavy, sulfur-containing residual fuels and heavy crude oils into a mixture of hydrogen and carbon monoxide in inert gases. The process is carried out by injecting oil and air (or oxygen) through a specially-designed burner assembly into a closed combustion vessel, where partial oxidation occurs at about 1,316°C. The term partial oxidation describes the net effect of a number of component reactions that occur in a flame supplied with less than stoichiometric oxygen.

In the fuel injection region of the reactor, hydrocarbons leaving the atomizer at about preheat temperature are intimately mixed with air or oxygen. The atomized hydrocarbon is heated and vaporized by back radiation from the flame and reactor walls. Some cracking of the hydrocarbons to carbon, methane, and hydrocarbon radicals may occur during this brief phase. When the fuel and air or oxygen reach the ignition temperature, part of the hydrocarbons reacts with oxygen in a highly exothermic reaction to produce carbon dioxide and water. Practically all available oxygen is consumed in this phase. The remaining hydrocarbons which have not been oxidized react with steam and the combustion products from reaction to form carbon monoxide and hydrogen. The carbon produced during gasification is recovered as a soot-in-water slurry.

Depending upon the desired heating value of the product gas, either oxygen or air may be used as oxidant. Nitrogen present in the air acts as a moderator for temperature control in the reactor and does not enter into the reactions. When either oxygen or air enriched with oxygen is used, a quantity of steam must be injected into the reactor for temperature moderation. Air oxidation alone requires no steam. The latter method produces a low heating-value fuel gas (approximately 120 Btu/standard cubic foot; 13.5 Calories/cubic meter) due to the presence of nitrogen. Oxygen feed produces a medium heating-value gas (approximately 300 Btu/standard cubic foot; 33.7 Calories/cubic meter).

The net products of the process are high-pressure steam, clean wastewater, and carbon-free fuel gas. While the high-pressure steam is saturated, pressures over 1,100 psig (74.8 atmospheres) have been commercially demonstrated and increases to substantially higher pressures in commercial practice are anticipated. Under any condition, using superheating, this steam is easily converted into an attractive feed for steam turbines. With appropriate design, oxygen-based oxidation units can be made almost entirely energy self-sufficient.

Gas from Solid Wastes. In one process, municipal refuse is charged at the top of a shaft furnace and is pyrolyzed as it passes downward through the furnace. Oxygen enters the furnace through tuyeres near the furnace bottom and passes upward through a 1,425–1,650°C combustion zone. The products of combustion then pass through a pyrolysis zone and exit at about 93°C. The offgas then passes through an electrostatic precipitator to remove flyash and oil formed during pyrolysis, both of which are recycled to the furnace combustion zone. The gas then passes through an acid absorber and a condenser. The

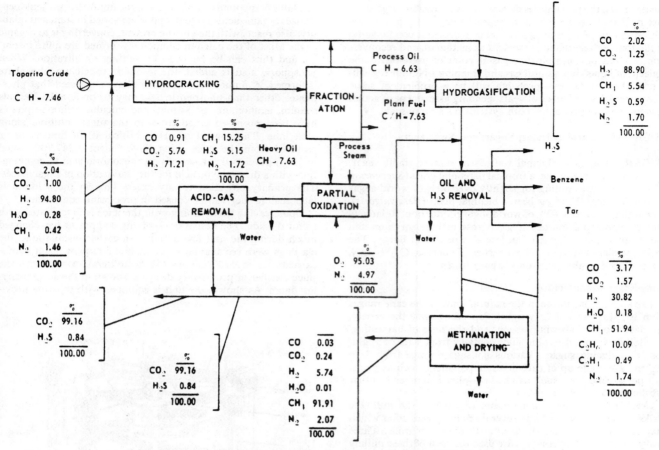

Fig. 4. Block flow diagram for pipeline gas production. (*Institute of Gas Technology.*)

clean fuel gas has a heating value of about 300 Btu/cubic foot (33.7 Calories/cubic meter) and a flame temperature equivalent to that of natural gas. As the solid waste passes downward through the furnace, it contacts the exiting pyrolysis products and traps a portion of the oil and flyash while itself losing moisture. After passing through the pyrolysis and combustion zones, the remaining solid waste is removed as a slag from the furnace bottom. The system has a net thermal efficiency of about 65% in converting solid waste to fuel gas. Process losses include energy losses in the conversion process and energy required for the operation of the onsite cryogenic gas separation unit for production of 95% oxygen needed by the system. The clean fuel gas is low in sulfur (about 15 ppm) and is essentially free of nitrogen oxides.

Another process involves the anaerobic digestion of a solid waste and water or sewage slurry at 60°C for 5 days to produce a methane-rich gas. Solid waste is prepared by shredding and air classification prior to being blended with water or sewage sludge to a 10 to 20% solids concentration. The slurry is heated and placed in a mixed digester for 5 days detention. The digestor gas is drawn off and separated into carbon dioxide and methane. The spent slurry from the digester is pumped through a heat exchanger to partially heat the incoming slurry prior to filtration. The filtrate is returned to the blender and the sludge is used as landfill. Heat addition to the refuse slurry is required to maintain the required digester temperature. This process is suited for use on sewage sludge, animal manures, and other high-moisture-content solid wastes. It is estimated that the pro-

cess reduces the volume of volatile solids by 75% while producing about 3,000 cubic feet (85 cubic meters) of methane per metric ton of incoming solid waste. The major residue is a sludge that requires landfilling or incineration. About 10% of the methane is consumed in heating the digester feed.

Methanol as Source of SNG. Methanol can be produced from a large range of feedstocks by a variety of processes. Natural gas, liquefied petroleum gas (LPG), naphthas, residual oils, asphalt, oil shale, and coal are in the forefront as feedstocks to produce methanol, with wood and waste products from farms and municipalities possible additional feedstock sources. In order to synthesize methanol, the main feedstocks are converted to a mixture of hydrogen and carbon oxides (synthesis gas) by steam reforming, partial oxidation, or gasification. The hydrogen and carbon oxides are then converted to methanol over a catalyst.

The concept of utilizing associated or natural gas for production of methanol which could be transported more economically than LNG from areas of surplus to areas of shortage was examined in the mid-1960s. At that time, the largest single-stream plant designed had a capacity of 900 metric tons of methanol per day. A fuel plant which might need to produce—say 22,500 metric tons per day of methanol was assessed on the economics basis of 25 times the small plant. It was quickly ascertained that methanol fuel delivered, for example, to the United States from the Middle East could not compete with local natural gas supplies which were than available at low cost within the United States. With shorter local supplies accompanied by much greater costs, the possibilities of economically feasible large-scale fuel methanol production now appear much more promising.

Conversion of natural gas (or other petroleum components) at the source to methanol, shipment as methanol, and reconversion of methanol into pipeline gas at point of use—versus the concept of liquefying natural gas and shipping LNG for regasification at point of use—probably will be a problem of some controversy for a number of years, pending assessment of actual system operating costs for both systems on a large-scale.

SUCROSE. Carbohydrates; Sugar; and Sweeteners

SUGAR. The two principal sources of sucrose (table sugar, saccharose) are sugarcane, a tropical perennial grass (*Saccharum officinarium*), accounting for slightly over 60% of world sugar production; and the sugar beet, a biennial plant (*Beta vulgaris*), accounting for nearly 40% of world sugar production. Relatively minor commercial sources of saccharose include sorghum and the sugar maple tree (sap). Sucrose also occurs in honey. The basic chemical and physical properties of sucrose, $C_{12}H_{22}O_{11}$, are described in the entry on **Carbohydrates.**

Cane Sugar Manufacture

The amount of sucrose in the natural juice of the cane ranges from 10% to nearly 17% (weight), depending upon the variety, the nature of the growing season, and the time of harvesting. In addition to sucrose, cane juice contains from 1 to 2.5% glucose or reducing sugars. Various nonsugars range from 1 to 3% and are made up of carbohydrate polymers, such as gums, and polysaccharides, such as pectins—plus a number of other substances in small quantities.

Crushing the Cane. Upon receipt of cane at the mill, the stalks are washed and cut into several smaller pieces, after which they are fed to a series (frequently three) roller mills. Three heavy, serrated roller crushers are used for each of these milling operations. Two of the rollers turn in opposite direction, while a third roller guides the flow of the stalks through the crushing operation. Where three sets of mills are used, the adjustment of the spacing between the crushing rollers will be wider for the first mill than the second mill, with the narrowest spacing for the third set of crushing rollers. The fibers are sprayed during crushing with a small amount of maceration water (from 5 to 20% of the weight of the cane). This facilitates extraction of sucrose. In some installations, the juice from the third mill is returned to the first and second mills as maceration water. The concentration of sucrose in juice from the first mill will usually be about 0.2% greater than that from the second mill; that of the third mill will be about 0.5% less than that of the first mill. However, the juice of the third mill will contain greater concentrations of gummy matter and some of the other impurities. The result of the total macerating action, which fully ruptures the plant cells, is a gray- to dark-green, cloudy juice that must be treated to effect a separation of impurities.

Liming and Clarifying. The ancient sugarmakers heated the raw juice and added ashes, causing a precipitation of many of the impurities, but the final product did not approach the purity of cane sugar marketed today which is one of the most highly purified compounds found in commerce. Over the past few centuries, lime has replaced the ashes and, during the past several decades, sulfur dioxide and phosphoric acid or phosphates also have been added to the total clarifying process. Sulfur dioxide bleach acts as an antimicrobial agent, and assists in the coagulation of such substances as albumin present in the juice. Further, the sulfur dioxide makes it possible to use more lime in the clarifying operation. Lime functions in several ways, forming

insoluble compounds with several of the impurities present, neutralizes organic acids present, and when added in sufficient quantity, also reacts with the glucose present, converting it to organic acids. Most of the calcium compounds formed are quite insoluble and thus can be removed by settling or filtration. When phosphoric acid is added, the insoluble tricalcium phosphate is formed. A number of researchers have suggested that phosphates other than phosphoric acid may be preferable. Various sodium, ammonium, potassium, and calcium orthophosphates have been proposed as additives to the sugar solution, along with lime. The control of pH is critical if all lime is to be removed from the juice. Jung (U.S. Patent 3,347,705 issued in 1967) developed the use of polyphosphoric acid in combination with a dicarboxylic acid for the clarification of sugar juices. The primary objection to any excess lime in the solution is later scaling that will be caused in processing equipment.

Upon leaving the crushing mill, the juice is first treated with sulfur dioxide. The design of clarifying equipment has changed much during the past few decades. In earlier installations, the clarifiers were rectangular or circular metal pans, each with a capacity up to 1200 gallons (45 hectoliters). A modern cane juice clarifier or proprietary design is shown in the accompanying figure. As shown, the unit is equipped with separate provi-

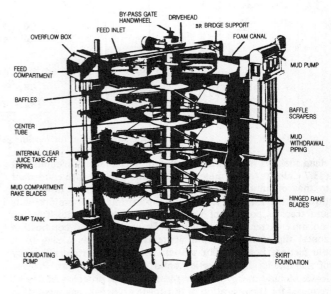

Modern cane juice clarifier. (*Dorr-Oliver Rapidorr 444™*)

sions in each compartment for feed, overflow takeoff and mud withdrawal which allows the unit to operate essentially as four totally independent clarifiers enclosed in a common housing. Juice is introduced as the top-center of each compartment through a hollow rotating center tube. This tube is fitted with a series of ports and scalpers that serve as feed introduction points. Located directly below each port, and attached to the center tube, are feed deflection baffles which insure uniform feeding and impede the natural tendency of the incoming juice to mix with the settled muds. Also attached to the center tube are the various sets of rake arms.

As the feed enters each compartment, it first strikes the deflection baffle, then flows outward at a decreasing velocity creating minimum turbulence. The various sets of rotating rake arms move the settled muds to the mud discharge boot located at the center of each tray. The mud is then withdrawn from each compartment separately. Overflow piping removes the clarified juice from each compartment independently at multiple points around the periphery of the clarifier, through a single overflow box where accurate flow distribution is easily maintained and controlled at one point. Standard capacities of the units range from 10,800 gallons (409 hectoliters) and a mud-thickening area of 312 square feet (29 square meters) to 140,800 gallons (5329 hectoliters) and a mud-thickening area of 4068 square feet (378 square meters).

Evaporation. After clarification and filtration, the juice goes to evaporators (vacuum pans), where upon concentration of the solution, small crystals grain out. Continuous evaporation produces a very thick mixture (*masscuite*), which is a mixture of sugar grains that are suspended in thick molasses. This mixture is centrifuged which throws off most of the molasses, leaving raw sugar, sometimes referred to as *centrifugal sugar*. At this point, the sugar is from 96 to 97% pure. The molasses may be reworked 2 or 3 times more to increase the yield of sugar. Although the remaining molasses may contain up to 50% sugar, the impurities present prevent any further formation of crystals. At this point the residue is called blackstrap and is further treated for use in animal feedstuffs. Molasses is described further a bit later.

Cane Sugar Refining. Raw cane sugar mills, as just described, produce the raw sugar. Refiners then further process the raw sugar into the more familiar white crystalline sugar. This 2-stage sugar production process for cane sugar stems from the economics of processing raw sugar in relatively small cane-producing regions and then refining the sugar on a much larger scale, usually thousands of miles closer to the markets. Traditionally, much of the raw sugar was imported from tropical, underdeveloped countries that lacked resources for constructing complete refineries. There are cases, however, where both processing and refining are performed at a single location or where the raw mill and refinery are adjacent and operated under one ownership.

The raw sugar as received at the refinery is mixed with sugar syrup for the purpose of dissolving the molasses residuals which still stick to the crystals. The heavy mixture resulting is sometimes called *magma*. This mix is centrifuged, after which the crystals are steam-treated and at this point are almost white. Again, the sugar crystals are dissolved in sugar syrup and, once again, are treated with lime and phosphoric acid in order to precipitate impurities present. From this operation, the effluent is filtered through bone char to yield a purified solution. Again, the solution is evaporated, crystallized, centrifuged, the final moisture content adjusted, after which the product is packaged.

Beet Sugar Production

The German chemist Marggraf discovered the presence of sugar in beets as early as 1747. Early laboratory methods to extract sugar from the beet proved overwhelmingly costly as contrasted with processing the traditional source, sugar cane. Little progress was made for over a half-century, when in 1802, another German chemist, Achard, found a way to extract sugar from the beet root on a relatively large scale. For a few years, a small manufacturing operation in Silesia prospered, mainly because of political factors that drove up the price of cane sugar. In 1812, Napoleon ordered an establishment of the beet sugar industry in France. Early attempts toward beet sugar extraction were made in the United States (Massachusetts) in 1838, followed by efforts over the subsequent 30 years in Illinois, Wisconsin, and California. The first real success in the United States

was achieved in the late 1870s by a factory in Alvarado, California.

The principal operations in sugar beet processing today include thorough washing of the beets, after which whirling knives slice the beets into thin strips, called *cossettes*. These are immersed in hot water where the sugar is removed from the beets by diffusion. The resulting solution is *raw juice*. This juice is purified in a process (carbonation), wherein lime and carbon dioxide are added to cause undesired impurities in the raw juice to precipitate out of the solution (as in the case of cane sugar previously described). This resulting, purified liquid is *thin juice*. Filtering and settling operations remove solid particles and impurities from the thin juice. This juice is concentrated by boiling off water to form *thick juice*. Further filtering ensures that all solid particles are eliminated. Sugar crystals are formed by boiling the thick juice under vacuum. The resulting mixture of crystals and liquid is known as *fillmass*. This mixture is spun and washed in high-speed centrifugals to separate the sugar crystals from the liquid. These crystals are now pure white sugar (sucrose). After further crystallization of the separated liquid, additional sugar and an important by-product (molasses) is obtained. The white sugar crystals are dried by tumbling in warm air in long rotating drums (granulators), after which the sugar is ready for market. The residue of the beets (*pulp*) is sold for livestock feed in either wet or dried form. Some molasses may be added to the pulp prior to drying.

References

Considine, D. M., and G. D. Considine, Editors: "Foods and Food Production Encyclopedia," Van Nostrand Reinhold, New York, 1982.
Junk, W. R., and H. M. Pancoast: "Handbook of Sugars for Processors, Chemists, and Technologists," Avi, Westport, Connecticut, 1973.
Shallenberger, R. S., and G. G. Birch: "Sugar Chemistry," Avi, Westport, Connecticut, 1975.

SULFITE PULP PROCESS. Pulp (Wood) Production and Processing.

SULFONAMIDE DRUGS. In 1935, Domagk, a German researcher, was the first to observe the clinical value of *prontosil*, a red compound derived from azo dyes. Para-aminobenzenesulfonamide was shown to be the effective portion of the prontosil molecule. This substance was given the name *sulfanilamide*. This was the first of a group of related drugs to receive wide clinical trial. It was found to be effective in the treatment of hemolytic streptococcal and staphylococcal infections. Within a short span of years, related drugs were synthesized and given clinical trials. These included *sulfapyridine*, *sulfathiazole*, *sulfaguanidine*, *sulfadiazine*, and *sulfamerazine*. These drugs acted by inhibiting the growth of bacteria rather than by killing organisms.

Even though numerous adverse side effects were observed over a period of time, the sulfonamides played an important role in medicine prior to the advent of the antibiotics. In recent years, the importance of the so-called *sulfa drugs* has diminished considerably, but for certain situations they are still considered important antimicrobials. Presently the sulfonamides are mainly used to treat uncomplicated urinary tract infections, including prostatitis, due to *E. coli*. They are also used to treat a number of noncardial infections. At one time the sulfa drugs were widely used in the treatment of meningococcal meningitis and bacillary dysentery. Unfortunately, the bacilli responsible for these diseases developed, over the years, a resistance to the drugs, severely reducing their efficacy.

Within the last few years, some new sulfa drugs have been introduced, including trimethoprim-sulfamethoxazole. This drug has broadened the scope in treatment of urinary tract infections derived from species in addition to *E. Coli*, namely, *Klebsiella*, *Enterobacter*, and *Proteus* species. This drug also is used for the treatment of acute otitis media in children, particularly those instances where strains of *H. influenzae* and *Str. pneumoniae* may be suspected. The drug is also used to treat systemic infections that may arise from chloramphenicol- and ampicillin-resistant *Salmonella*; as well as infections attributed to *Pneumocystis carinii*.

Also, the nature of sulfonamide compounds (relatively short duration of action, capability of entering into synergism with other drugs, poor absorbability, and topical effectiveness, not to mention relatively low cost) is taken advantage of in what are sometimes called *short-acting sulfonamides*. Short-acting sulfonamides include sulfisoxazole, sulfadiazine, and trisulfapyrimidines. An *intermediate-acting sulfonamide* in current use is sulfamethoxazole. This drug does tend to cause renal damage arising from sulfonamide crystalluria.

Sulfacetamide eyedrops continue to be used for treatment of superficial ocular infections. Sometimes silver-sulfadiazine cream is applied to burn surfaces to minimize or present bacterial growth, as well as preventing invasive infection.

The adverse effects of sulfonamides include hypersensitivity reactions, as manifested by rashes, photodermatitis (allergic reaction to light), so-called drug fever, nausea, and vomiting. These reactions occur with some frequency when sulfonamides are administered. Less frequently is crystalluria, previously mentioned, but with the risk lessened in the case of sulfisoxazole. Sulfa drugs also occasionally cause hemolytic anemia, agranulocytosis, hemolytic anemia, and kernicterus (in infants) when the drugs are given to nursing mothers. In rare instances, sulfa drugs may precipitate hepatitis, aplastic anemia, renal tubular necrosis, and certain blood disorders.

SULFONATION.

In its broadest sense, sulfonation includes all methods of converting organic compounds to sulfonic acids or sulfonates, containing the structural group C—SO$_2$—O or, in some cases, N—SO$_2$—O. The term is applied mainly to use of the common sulfonating agents, namely, concentrated sulfuric acid, oleum, and other reagents containing sulfur trioxide in labile form, and sulfur trioxide itself. The only other widely used method of making sulfonates is by the action of alkali metal sulfites on alkyl halides, known as the Strecker reaction.

Reaction of vegetable oils with concentrated sulfuric acid has also been called sulfonation, and the resulting products are known in industry as sulfonated oils, although they are predominantly sulfate esters characterized by the structural group C—O—SO$_3$. To chemists these are sulfated oils, and the formation of sulfuric acid esters from either alcohols or olefins is more properly termed *sulfation*.

Benzene reacts with concentrated sulfuric acid and with oleum, respectively, as follows:

$$C_6H_6 + H_2SO_4 \rightarrow C_6H_5SO_2OH + H_2O$$

$$C_6H_6 + SO_3 \rightarrow C_6H_5SO_2OH$$

The acid reaction product is benzenesulfonic acid, and its salts are called benzenesulfonates. Similar reactions of other aromatic compounds proceed readily over a wide range of conditions. The relatively great ease of sulfonation of aromatic compounds is a traditionally important property for distinguishing them from aliphatic compounds. Also side reactions are less common and yields are usually greater than in the sulfonation of aliphatics.

Reaction time, temperature, %SO$_3$ in the sulfonating reagent, and proportion of reagent to substrate are important variables in the sulfonation process. Agitation is also important, because of limited mutual solubility of the reacting materials. Special solvents and catalysts may be used, but are not often essential. Sulfonation of benzene beyond the monosulfonic stage yields mainly benzene-1, 3-disulfonic acid. The second stage of reaction requires a higher temperature or a stronger acid than the first. If the disulfonic acid is made with use of concentrated sulfuric acid, temperatures as high as 260°C are used, but it is more satisfactory to use a strong oleum at temperatures up to 100°C.

In sulfonation of more complex aromatic compounds, temperature may affect not only reaction rate but also the nature of the reaction product. For example, change of temperature in sulfonation of naphthalene can change the composition of the resulting monosulfonic acid from about 95% alpha isomer at room temperature to practically 100% beta isomer at 200°C.

Purification of sulfonated products, which is often difficult, is largely avoided by adjusting conditions of sulfonation so as to obtain a product suitable for practical use without purification or after mere drying. Inorganic sulfate, the chief impurity, can be held to a minimum by the judicious use of oleum. Adding water to the acid mix at the end of sulfonation often gives an upper layer containing nearly all of the sulfonic acid and a lower layer containing most of the excess of sulfuric acid. Conversion of the acid mix to calcium salts favors a more complete separation, since calcium sulfate is much less soluble in water than many calcium sulfonates.

At ordinary temperature, the gaseous paraffins are inert to concentrated sulfuric acid, but are slowly absorbed and sulfonated by oleum. Their susceptibility to sulfonation increases with temperature and with molecular weight, but direct sulfonation of aliphatic hydrocarbons is in general an unsatisfactory way of obtaining sulfonates. The preferred method is the Strecker reaction, as in the reaction:

$$CH_3I + Na_2SO_3 \rightarrow NaI + CH_3SO_3Na$$

A good yield of mixed sulfonic acids can be obtained by passing chlorine and sulfur dioxide through paraffin-base petroleum fractions, the main reaction being:

$$C_nH_{2n+2} + Cl_2 + SO_2 \rightarrow C_nH_{2n+1}SO_2Cl + HCl$$

The sulfonylchloride is easily hydrolyzed, e.g., during neutralization.

Olefins are more easily sulfonated than paraffins, but with complications. With concentrated sulfuric acid the main reaction is sulfation, as in formation of ethylsulfuric acid from ethylene. Both sulfation and sulfonation occur when oleum or sulfur trioxide act on olefins. For example, the vapor phase reaction of ethylene with sulfur trioxide gives mainly carbylsulfate:

$$C_2H_4 + 2SO_3 \rightarrow \begin{array}{l} CH_2SO_2O- \\ | \\ CH_2-O-SO_2 \end{array}$$

Complete hydrolysis of carbylsulfate yields isethionic acid (HOCH$_2$CH$_2$SO$_2$OH), a sulfonic acid of some importance in manufacture of wetting and cleansing agents.

Polymerization is a common side reaction in sulfonation of olefins, also oxidation of the hydrocarbon with formation of sulfur dioxide. Alicyclic hydrocarbons are somewhat more susceptible to sulfonation than their open chain analogues, with similar side reactions.

The action of sulfuric acid and related sulfonating agents on aliphatic compounds other than hydrocarbons is varying and complex. However, simple sulfation with negligible side reaction is possible in the case of most saturated primary alcohols, from which the alkyl sulfuric acids are easily formed and converted to neutral salts. The most important products made in this way are mixed sodium alkyl sulfates from higher alcohols, principally lauryl.

—A. S. Richardson, Proctor & Gamble Co., Cincinnati, Ohio.

SULFONES. Organic Chemistry; Sulfonamide Drugs.

SULFOXIDE. Organic Chemistry.

SULFUR. Chemical element symbol S, at. no. 16, at. wt. 32.064, periodic table group 6a, mp 112.8°C (rhombic), 119.0° (monoclinic), 120.0°C (amorphous), bp 444.7°C (all forms), sp gr, 2.07 (rhombic), 1.96 (monoclinic), 2.046 (amorphous). Atomic weight varies slightly because of naturally occurring isotopes 32, 33, 34, and 36, the total possible variation amounting to ±0.003.

The stable isotopes of sulfur are ^{32}S, ^{33}S, ^{34}S, and ^{36}S. There are three known radioactive isotopes, ^{31}S, ^{35}S, and ^{37}S, with ^{35}S having the longest half-life ($t_{1/2} = 87.1$ days). See **Radioactivity.** Electronic configuration is $1s^22s^22p^63s^23p^4$. Ionic radius, S^{2-}, 1.855 å, S^{6+}, 0.29 Å (*Pauling*). Covalent radius, 1.07 Å. In terms of abundance, sulfur ranks fourteenth among the elements occurring in the earth's crust, with an estimated 520 grams per metric ton. In seawater, the element ranks fifth, with an estimated 894 grams per metric ton.

First ionization potential 10.357 eV; second, 23.3 eV; third, 34.9 eV; fourth, 47.08 eV; fifth, 63.0 eV; sixth 87.67 eV. Oxidation potentials $H_2S(aq) \rightarrow S + 2H^+ + 2e^-$, -0.141 V; $H_2SO_3 + H_2O \rightarrow SO_4^{2-} + 4H^+ + 2e^-$, -0.20 V; $S + 3H_2O \rightarrow H_2SO_3 + 4H^+ + 4e^-$, -0.45 V; $SO_3^{2-} + 2OH^- \rightarrow SO_4^{2-} + H_2O + 2e^-$, 0.90 V; $S^{2-} \rightarrow S + 2e^-$, 0.508 V; $HS^- + OH^- \rightarrow S + H_2O + 2e^-$, 0.478 V. Other important physical properties of sulfur are given under **Chemical Elements.**

Sulfur has a large number of allotropes. The ordinary form, α-sulfur, is rhombic having a crystal unit cell composed of sixteen S_8 molecules. At 95.5°C it undergoes transition to β-sulfur, which is monoclinic and also has a molecular weight (in solution in carbon disulfide) corresponding to S_8. Four other monoclinic forms have been identified microscopically: γ-sulfur, prepared by heating α-sulfur to 150°C, cooling to 90°C, and inducing crystallization by friction, ρ-sulfur, S_6, prepared by extracting an acidulated sodium thiosulfate solution with toluene, as well as ν-sulfur, and δ-sulfur. There is also a tetrahedral form, θ-sulfur, crystallized from a carbon disulfide solution of rhombic sulfur treated with balsam. The first liquid form to appear is λ-sulfur, a pale yellow liquid, obtained on heating sulfur to 120°C. Above 160°C, this form changes to a viscous, dark brown liquid consisting mainly of μ-sulfur. A third liquid allotrope, π-sulfur is considered to exist in molten sulfur, in equilibrium with the other two forms, having its greatest concentration at about 180°C. Sulfur vapor has been shown to contain S_8, S_6, S_4, and S_2 molecules. Several other allotropes of sulfur have been produced, including two paramagnetic forms, purple and green in color, by low temperature processing.

Sulfur occurs as free sulfur in many volcanic districts, and may have been formed in part by sublimation, by decomposition of hydrogen sulfide, or metallic sulfides, or by organic agencies. It is often associated with limestones and gypsum. Sulfur is found in Spain, Iceland, Japan, Mexico, and Italy. It occurs especially in Sicily, which was the producer for the world until about the beginning of the twentieth century, when Herman Frasch, by inventing the superheated water method of mining sulfur, made available the great Louisiana and Texas deposits. This method of mining is at the same time a method of purifying sulfur, because in the process of heating, accompanying materials remain unmelted at the temperature at which sulfur melts and is drawn off. In the Louisiana and Texas deposits the sulfur is associated with gypsum, occurring in the caprock overlying the salt plugs that have pierced the strata underlying the Gulf coastal plain. In the United States, sulfur is also found in California, Colorado, Nevada, and Wyoming. Sulfur also occurs as (1) sulfides, e.g. cobaltite, iron disulfide, pyrite FeS_2, lead sulfide, galenite PbS, copper iron sulfide, copper pyrite $CuFeS_2$, zinc sulfide, zinc blende ZnS, mercury sulfide, cinnabar HgS; and (2) as sulfates, e.g., calcium sulfate, gypsum $CaSO_4 \cdot 2H_2O$, barium sulfate, barite $BaSO_4$. Several of these minerals are described under separate alphabetical entries.

Sulfur Production and Uses. The manufacture of sulfuric acid accounts for nearly 90% of all sulfur consumed. Of this, about 50% of the H_2SO_4 goes into fertilizer production, nearly 20% into chemical manufacture, 5% into pigments, about 3% each for iron and steel production and the manufacture of rayon and synthetic fibers, and about 2% for various petroleum processes. The balance of over 15% of sulfuric acid used represents scores of applications.

Sulfur Compounds.

Sulfur compounds are numerous, the element combining readily with oxygen, the halogens, hydrogen, nitrogen, as well as sulfur-organics.

Sulfur-Oxygen Compounds. Due to its $3s^23p^4$ electron configuration, sulfur, like oxygen, forms many divalent compounds with two covalent bonds and two lone electron pairs, but d-hydridization is quite common, to form compounds with oxidation states of 4+ and 6+.

A number of suboxides of sulfur have been reported, but in general their composition has not been clearly established. Polysulfur oxides of formula $S_{8-16}O_2$ are formed by reaction of hydrogen sulfide and sulfur dioxide. Also, when sulfur is burned with oxygen in very limited supply disulfur monoxide, S_2O is formed. This has the structure

A mixture of sulfur dioxide, SO_2, and sulfur vapor, at low pressure and with an electric discharge, forms sulfur monoxide, SO. Its presence is shown from its absorption spectrum, but upon separation it disproportionates at once to sulfur and SO_2. Sulfur sesquioxide, S_2O_3, is formed by reaction of powdered sulfur with anhydrous SO_3; S_2O_3 also disproportionates (at 20°C in nitrogen) to sulfur and SO_2. Sulfur dioxide, SO_2, is formed by the combustion in air or oxygen of sulfur and sulfur compounds generally, except those in which sulfur is in a higher state of oxidation. Sulfur dioxide has an O—S—O bond angle of 119.5°. The sigma bonds utilize essentially sulfur p orbitals, with dp hybridization for the pi bonds. Its oxidation to sulfur trioxide, SO_3, by atmospheric oxygen attains a significant rate only at higher temperatures, but can be materially increased by catalysts. Sulfur trioxide is also evolved from oleum on heating. It exists in the vapor state chiefly as the planar monomer, in which the oxygen atoms are spaced symmetrically (120° angles) about the sulfur atom, and it has S—O bond lengths of

1.43Å. Liquid SO_3 is partly trimerized, and exists in three physical forms.

Sulfur tetroxide is formed by reaction of pure oxygen and sulfur dioxide under the silent electric discharge. It is not obtained pure, but in a variable SO_3/SO_4 ratio, and as a polymerized white solid. Another peroxide, $(SO_2OOSO_2O)_x$, which is written as S_2O_7, is known.

Of the 16 oxyacids of sulfur that are recognized, only four have been isolated. The more important oxyacids of sulfur are: (1) Thiosulfurous acid, $H_2S_2O_2$, structure not established, existing only in compounds, an oxidizing agent for Fe^{2+}, H_2S and HI; (2) Sulfoxylic acid, H_2SO_2, existing only in salts and other compounds, e.g., $ZnSO_2$, SCl_2, $S(OR)_2$, structure probably

$$H \colon \overset{..}{\underset{..}{O}} \colon \overset{..}{\underset{..}{S}} \colon \overset{..}{\underset{..}{O}} \colon H$$

(3) Dithionous acid (or hydrosulfurous acid), $H_2S_2O_4$, existing only in compounds, widely used reducing agent, chiefly as the sodium salt, for organic substances, also reduces Sb^{3+}, Ag^+, Pb^{2+}, Cu^{2+} to the elements, structure

$$H \colon \overset{..}{\underset{..}{O}} \colon \overset{\overset{\displaystyle :\overset{..}{O}:}{}}{S} \quad \overset{\overset{\displaystyle :\overset{..}{O}:}{}}{S} \colon \overset{..}{\underset{..}{O}} \colon H$$

(4) Sulfurous acid, H_2SO_3, produced by hydration of SO_2, not isolated but existing in many salts, the sulfites and acid sulfites, and many organic compounds, including the dialkyl or diaryl sulfites and the alkyl or aryl sulfonic acid esters, which suggest two possible structures $(HO)_2SO$ and $H(HO)SO_2$, although the acid dissociation constants (first, 1.25×10^{-2}, and second, 5.6×10^{-8}) suggest the structure with only one unhydrogenated oxygen atom. Sulfurous acid and sulfites are fairly strong reducing agents, but the HSO_3^- ion may act as an oxidizing agent, as for formates and related compounds. Other compounds of SO_2 are the metabisulfites or pyrosulfites, containing the ion

$$\left[:\overset{..}{\underset{..}{O}} \colon \overset{\overset{\displaystyle :\overset{..}{O}:}{}}{S} \colon \overset{..}{\underset{..}{O}} \colon \overset{\overset{\displaystyle :\overset{..}{O}:}{}}{S} \colon \overset{..}{\underset{..}{O}} \colon \right]^-$$

which enters into equilibrium with water to form acid sulfite. (5) Thiosulfuric acid, $H_2S_2O_3$, existing only in compounds, the anion having the structure

$$\left[:\overset{..}{\underset{..}{O}} \colon \overset{\overset{\displaystyle \overset{..}{\underset{..}{S}}}{}}{\underset{\displaystyle :\overset{..}{O}:}{S}} \colon \right]^-$$

and widely used as a coordinating ion for forming complexes with metals; it also is an oxidizing agent, and is used in iodometric titrations. (6) Dithionic acid, $H_2S_2O_6$, existing only in compounds but stable in dilute solution at room temperature, and differing in its stability to hydrolysis and oxidation from the polythionates,

$$\left[:\overset{..}{\underset{..}{O}} \colon \overset{\overset{\displaystyle :\overset{..}{O}:}{}}{\underset{\displaystyle :\overset{..}{O}:}{S}} \quad \overset{\overset{\displaystyle :\overset{..}{O}:}{}}{\underset{\displaystyle :\overset{..}{O}:}{S}} \overset{..}{\underset{..}{O}} \colon \right]^-$$

(7) Polythionic acids, $H_2S_nO_6$, in which n has values of 3, 4, 5, 6 and others, some of which have been reported to have values indefinitely high (20–80), structure not established, though there is evidence that they consist of two sulfonic acid groups connected by a linear chain of sulfur atoms. An interesting property of the polythionates that are very rich in sulfur ($n > 20$) is their slight tendency to decompose to give free S. (8) Sulfuric acid, H_2SO_4, structure

$$H \colon \overset{..}{\underset{..}{O}} \colon \overset{\overset{\displaystyle :\overset{..}{O}:}{}}{\underset{\displaystyle :\overset{..}{O}:}{S}} \colon \overset{..}{\underset{..}{O}} \colon H$$

strong acid, formed by hydration of sulfur trioxide, completely dissociated (first ionization) in aqueous solutions up to 40%; above that concentration dissociation decreases and hydrate formation occurs. Both normal and acid sulfates are formed by metallic elements, though the products of their direct reaction with the acid vary with temperature. (9) Sulfuric acid dissolves SO_3, the product of a 1:1 ratio being pyrosulfuric or disulfuric acid, $H_2S_2O_7$, which forms the pyrosulfates, also obtainable by heating acid sulfates, structure $HO(O)(O)SOS(O)(O)OH$. Two series of alkali metal pyrosulfates are known: those formed from SO_3 and the metal sulfates and those formed from H_2SO_4 and the metal sulfates, which have the pyrosulfuric acid structure. (10) Peroxymonosulfuric acid is produced by addition of SO_3 to concentrated H_2O_2, its salts are fairly stable, and it has the structure $HOS(O)(O)OOH$. (11) Peroxydisulfuric acid is produced by reaction of concentrated H_2O_2 on H_2SO_4 or by electrolysis of acid sulfate solutions; its salts are fairly stable and it has the structure $HOS(O)(O)OOS(O)(O)OH$.

Hydrogen Sulfide. H_2S is a weak acid ($pK_{A_1} = 7.00$; $pk_{A_2} = 12.91$) stronger than water, but weaker than H_2Se, as expected from its position in the periodic system; its reducing strength exhibits the same relation. Its long use in analytical chemistry was due to the differential solubility of many sulfides with variation of the pH of an aqueous solution. Hydrogen persulfide, H_2S_2, structure HSSH, with an S—S bond distance of 2.05 Å, formed from an alkali metal polysulfide solution and HCl at low temperatures, is the first of a group of hydrogen polysulfides of the general formula, H_2S_x.

Sulfur Halides. Many are known. Those that have been identified and whose properties have been determined include the fluorine compounds, S_2F_2, SF_4, SF_6, S_2F_{10}; the chlorine compounds, S_2Cl_2, SCl_2, SCl_4; and the bromine compound, S_2Br_2. Sulfur chlorides of the general formula, S_nCl_2, are known up to $n \approx 20$. A similar series of cyanides, $S_n(CN)_2$, is known. Derivatives of SCl_4, e.g., SCl_3CN, have been prepared and the list of derivatives of SF_6 is rapidly growing, including $(SF_5)_2O$, SF_5Cl, SF_5Cl_3, $SF_4(CF_3)_2$, etc. All of them except the higher fluorides hydrolyze readily, are essentially covalent in character, and the simple compounds can be prepared directly from the elements, the activity of the halogen determining the product obtained.

Sulfur Oxyhalides. Four general compositions of oxyhalides or sulfur have been known for many years. In one of these, the thionyl halides, SOX_2, sulfur has a 4+ oxidation state. In the three others it has a 6+ oxidation state: thionyl tetrafluoride and the sulfuryl and pyrosulfuryl halides, SOF_4, SO_3X_2, and $S_2O_5X_2$, respectively. As in the case for the simple halides, no iodine compounds are known, but polyhalogen ones, such as SOFCl and SO_2FCl exist.

Isolable Oxysulfuranes. Sulfuranes, as described by Musher (1969), are compounds of sulfur(IV) in which four ligands are attached to sulfur and have in common with the rare gases, such as XeF_2, and electronic structure involving a formal expansion of the valence shell of the central atom from 8 to 10 electrons. Martin and Perozzi (1976) pointed out that the incorporation of oxygen ligands makes possible a wide range of new structural types that illustrate structure-reactivity relationships in a particularly illuminating way.

For many years, it was postulated that most types of sulfuranes were intermediates, not isolable compounds. However, the isolable halosulfuranes have been well established for many years. The first known of these, SCl_4, was prepared by Michaelis and Schifferdecker in 1873. In 1911, it was found that SF_4, while highly reactive, was thermally stable. However, the compound was not fully described until 1929. Development of SF_4

led to the creation of a family of stable fluorosulfuranes and their derivatives. It was found that the fluorines in these compounds can be replaced by aryl or perfluoroalkyl groups (Tyczkowski, 1953). Kimura and Bauer (1963) described the geometry of SF_4 as a distorted trigonal bipyramidal with two fluorines and a lone pair of electrons occupying equatorial positions, with the other two fluorines in apical positions. The postulated structures of SF_4 (a), and of a derivative (b) are shown below:

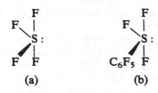

In the early 1970s, Sheppard, by reacting SF_4 with pentafluorophenyllithium, prepared an isolated sulfurane with four carbon-centered ligands, namely, *tetrakis*-(pentafluorophenyl)sulfurane, $(C_6F_5)_4S$. Martin and Perozzi (1976) prepared the first isolable diaryldialkyloxysulfurane. If protected against moisture, the researchers found the compound to be stable over an indefinite period at room temperature. Research in this interesting area continues, some of the details of which are well described by Martin-Perozzi (1976). Summarizing the situation, the researchers observe that the development of synthetic methods for oxysulfuranes has made a wide range of isolable compounds of ypervalent sulfur available for study. Structure-reactivity correlations are now becoming evident as a result of such study. The fact that oxygen is dicoordinate makes it possible to synthesize cyclic oxysulfuranes and to use the pronounced changes of reactivity which accompany cyclization to design new, potentially useful sulfurane reagents stable enough to allow isolation.

Sulfur-Nitrogen Compounds. Many of the sulfur-nitrogen compounds are sulfuric acid derivatives. Three of these compounds correspond to replacement of the hydrogen atoms of ammonia with one, two, and three —SO_3H radicals, the monosubstituted compound being aminesulfonic (sulfamic) acid, and being readily separated, the others known only in their salts, the aminedisulfonates (imidodisulfonates) and aminetrisulfonates (nitrilotrisulfonates). Other amines, such as hydroxylamine and hydrazine have similarly related compounds. (See **Hydrazine; Hydroxylamine.**) Diamino derivatives of the sulfoxy acids are also known, such as sulfamide $H_2NSO_2NH_2$. Imidosulfinamide, $HN(SONH_2)_2$ has been prepared by reaction of $SOCl_2$ and ammonia (also directly from SO_2 and ammonia), and a trimer of sulfimide $(O_2SNH)_3$, by ammoniation of SO_2Cl_2. It is cyclic in structure, composed of alternate $> NH$ and $> SO_2$ groups. Nitrosulfonates, containing the ion SO_3NO^- and dinitrososulfonates, containing $SO_3N_2O_2^{2-}$ are also known.

The most important sulfur-nitrogen compound is tetrasulfur tetranitride, S_4N_4, prepared in many ways, including the direct reaction of ammonia and sulfur. All data on its structure are in accord with a puckered eight-member ring, or a cage with N—S connections. The question as to whether there are also transannular N—N or S—S bonds has not been clearly settled. On hydrogenation it adds 4 H atoms, on fluorination it forms $S_4N_4F_4$, structure

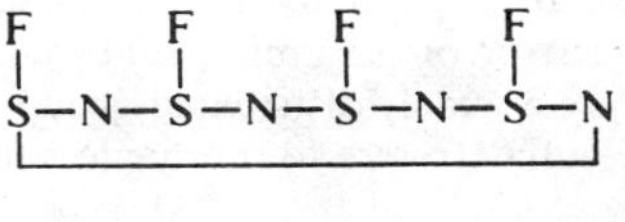

$$
\begin{array}{cccc}
\overset{\displaystyle F}{|} & \overset{\displaystyle F}{|} & \overset{\displaystyle F}{|} & \overset{\displaystyle F}{|} \\
S-N-S-N-S-N-S-N
\end{array}
$$

and SN_2F_2 the latter reacting with SNF to form SNF_3, structure F_2SNF. Other thiazyl compounds, prepared from S_4N_4 and the halogens or sulfur halides, include $(ClSN)_3$, S_4N_3Cl, S_4N_3Br, S_4N_3I. These last are salts, i.e., $[N_4S_3]X$, and salts of other anions can also be prepared. Other sulfur-nitrogen compounds known are SN_2, S_4N_2, S_5N_2, and S_2N_2, the last being formed by heating S_4N_4.

Thiocyanogen $(SCN)_2$ is formed by treatment of a metal thiocyanate with bromine in an organic solvent. It reacts with organic compounds in a manner completely analogous to the free halogens, lying between bromine and iodine in oxidizing power. The alkali metal and alkaline earth metal thiocyanates are prepared by fusing the cyanides with sulfur, and the other metal thiocyanates, as well as the organic ones, are usually prepared from the alkali metal thiocyanates. (See entries under Thio-.)

Many selenium analogs of thio compounds can be made, including $SeSO_3$, SO_3Se^{2-}, SSe^{2-}, etc.

In addition to carbon disulfide (odorless when pure), carbon subsulfide, $S=C=C=C=C=S$, an evil-smelling red oil and carbon monosulfide, $(CS)_x$, are known as well as COS, CSSe and CSTe. Because of its similarity to oxygen, and the reactivity of its acids, sulfur enters widely into organic compounds.

References

Kimura, K., and S. H. Bauer: *Jrnl. Chem. Phys.*, **39**, 3172 (1963).
Martin, J. C., and E. F. Perozzi: "Isolable Oxysulfuranes in Organic Chemistry," *Science*, 154–159 (1976).
Musher, J. I.: *Angew. Chem. Int. Ed. Engl.*, **8**, 54 (1969); and in *Sulfur Research Trends* (R. F. Gould, Editor), American Chemical Society, Washington, D.C. (1972).
Nickless, G.: "Inorganic Sulphur Chemistry," Elsevier, New York (1968).
Pryor, W. A.: "Mechanisms of Sulphur Reactions," McGraw-Hill, New York (1962).
Sheppard, W. A.: *Jrnl. Amer. Chem. Soc.*, **93**, 5597 (1971).
Suter, D. M.: "Organic Chemistry of Sulfur," Wiley, New York (1964).
Tyczkowski, E. A., and L. A. Bigelow: *Jrnl. Amer. Chem. Soc.*, **75**, 3523 (1953).

SULFUR (In Biological Systems). Sulfur in some form is required by all living organisms. It is utilized in various oxidation states, including sulfide, elemental sulfur, sulfite, sulfate, and thiosulfate by lower forms and in organic combinations by all. The more important sulfur-containing organic compounds include the amino acids (cysteine, cystine, and methionine, which are components of proteins); the vitamins thiamine and biotin; the cofactors lipoic acid anc coenzyme A; certain complex lipids of nerve tissues, the sulfatides; components of mucopolysaccharides, the sulfated polysaccharides; various low-molecular-weight compounds, such as glutathione and the hormones vasopressin and oxytocin; in many therapeutic agents, such as the sulfonamides and penicillins, as well as oral hypoglycemic agents, sometimes used in treatment of diabetes mellitus. Sulfhydryl groups of the cysteine residues in enzyme proteins and related compounds, such as hemoglobin, play a key role in many biocatalytic processes; sulfhydryl-disulfide interchange reactions involving the cysteine residues of proteins are critical events in the immune processes, in transport across cell membranes, and in blood clotting. The S—S bridges between these residues are important in the maintenance of the tertiary structure of most proteins.

The electronic structure of sulfur is such that a variety of oxidation states are readily obtainable. It can be said that a sulfur cycle exists in nature, as noted by

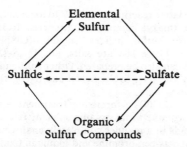

The oxidation and reduction of elemental sulfur and sulfide occur in different species of bacteria, e.g., the oxidation of sulfides via elemental sulfur to sulfate takes place in *Chromatia*, the alternative oxidation to sulfate in *Thiobacilli*. The reduction of sulfate to sulfide occurs in *Desulfovibrio*. The biosynthesis of organic sulfur compounds from sulfate takes place mainly in plants and bacteria, and the oxidation of these compounds to sulfate is characteristic of animal species and of heterotrophic bacteria.

The amino acids cysteine and cystine are interconverted by oxidation-reduction reactions, as shown by

$$
\begin{array}{ccc}
\text{S} \text{------} \text{S} & & \text{SH} \\
| \qquad\quad | & & | \\
\text{CH}_2 \quad \text{CH}_2 & & \text{CH}_2 \\
| \qquad\quad | & & | \\
\text{CNNH}_2 \; \text{CHNH}_2 + 2\text{H} \rightleftharpoons 2 & & \text{CHNH}_2 \\
| \qquad\quad | & & | \\
\text{COOH} \; \text{COOH} & & \text{COOH} \\
\text{(Cysteine)} & & \text{(Cystine)}
\end{array}
$$

Cystine was first isolated from a urinary calculus by Wollaston in 1805. It was shown to be a component of protein by Morner in 1899 and independently by Embden in 1900. Proof of its structure was given by Friedman in 1902. See also **Amino Acids; Coenzymes;** and **Vitamins.**

In the chain from soils to plants to humans, inorganic sulfur, or more accurately, the sulfate ion (SO_4^{2-}), is taken up by plants and converted within the plant to organic compounds (the sulfur amino acids). These amino acids combine with other amino acids to make up plant protein. When the plant is eaten by a human or by livestock animals, the protein is broken down and the amino acids are absorbed from the digestive tract and recombined in the proteins of the animal body. The most important feature of sulfur in the food chain is that plants use inorganic sulfur compounds to make sulfur amino acids, whereas animals and humans use the sulfur amino acids for their own processes and excrete inorganic sulfur compounds resulting from the metabolism of the sulfur amino acids.

Ruminants, such as cattle, sheep, and goats can use inorganic sulfur in their diets because the microorganisms in the rumen convert the inorganic sulfur into sulfur amino acids and these are then absorbed farther along in the digestive tract.

Soils very low in available sulfur are common in a number of regions of the world. In the United States, low-sulfur soils are frequently found in the Pacific Northwest and in some parts of the Great Lakes states. For many years, sulfur in the form of calcium sulfate was an accessory part of most commercial phosphate fertilizers, and this probably helped to prevent development of widespread sulfur deficiency in crops grown where these fertilizers were used. Volatile sulfur compounds from smoke, particularly before tight pollution controls, were an important source of sulfur for plants growing near industrial centers. In some cases, excessive sulfur in the air can cause injury to the plants. The trend toward high analysis fertilizers without sulfur and air pollution abatement diminishes some of the inad-

vertent sources of sulfur for plants and crops and creates a need for more deliberate use of sulfur-containing fertilizers.

The extent to which any plant will convert inorganic sulfur taken up from the soil into amino acids and incorporate these into protein is controlled by the genetics of the plant. Increasing the available sulfur in soils to levels in excess of those needed for optimum plant growth will not increase the concentration of sulfur amino acids in plant tissues. To meet the requirements for sulfur amino acids in human diets, the use of food plant species with the inherited ability to build proteins with high levels of sulfur amino acids is required in addition to that supplied by way of the soil.

Since animals tend to concentrate in their own proteins the sulfur amino acids contained in the plants they eat, such animal products (meat, eggs, and cheese) are valuable sources of the essential sulfur amino acids in human diets. In regions where the diet is composed almost entirely of foods of plant origin, deficiencies of sulfur amino acids may be critical in human nutrition. Frequently, persons in such areas (and also voluntary vegetarians) are also likely to suffer from a number of other dietary insufficiencies unless supplemental sources are used.

Diets of corn (maize) and soybean meal are usually fortified with sulfur amino acids for pigs and chickens. Sometimes fishmeal, a good source of sulfur amino acids, is added to the diets, or sulfur amino acids synthesized by organic chemical processes may be used.

Since ruminants can utilize a wide variety of sulfur compounds, any practice to increase the sulfur in plants may help to meet the requirements of these animals. Sheep appear to have a higher requirement for sulfur than most other animals, perhaps because wool contains a fairly high level of sulfur. Adding sulfur fertilizers to soils used to produce forage for sheep may improve growth and wool production, even though no increased yield of the forage crop per se may be noted.

Sulfate and Organic Sulfates. Inorganic sulfate ion (SO_4^{2-}) occurs widely in nature. Thus, it is not surprising that this ion can be used in a number of ways in biological systems. These uses can be divided primarily into two categories: (1) formation of sulfate esters and the reduction of sulfate to a form that will serve as a precursor of the amino acids cysteine and methionine; and (2) certain specialized bacteria use sulfate to oxidize carbon compounds and thus reduce sulfate to sulfide, while other specialized bacterial species derive energy from the oxidation of inorganic sulfur compounds to sulfate.

Among the variety of sulfate esters formed by living cells are the sulfate esters of phenolic and steroid compounds excreted by animals, sulfate polysaccharides, and simple esters, such as choline sulfate. The key intermediate in the formation of all of these compounds has been shown to be 3'-phosphoadenosine-5'-phosphosulfate (PAPS). This nucleotide also serves as an intermediate in sulfate reduction.

In organisms that utilize sulfate as a source of sulfur for synthesis of cysteine and methionine, the first step in the reduction process is the formation of PAPS. This is not surprising, since the direct reduction of sulfate ion itself is an extremely difficult chemical process. It is known that the reduction of esters and anhydrides occurs much more readily than the reduction of corresponding anions. Following activation, the sulfuryl groups of PAPS is reduced to sulfite ion (SO_3^{2-}) by reduced triphosphopyridine nucleotide (TPNH) and a complex enzyme system. Following the reduction of PAPS to sulfite, additional reduction steps readily produce hydrogen sulfide, which appears to be a direct precursor of the amino acid cysteine.

The sulfatases are a widely distributed group of enzymes that hydrolyze simple sulfate esters to inorganic sulfate.

Sulfur compounds, such as sulfur dioxide and sodium bisulfite, are used commercially to preserve the color of various food products, such as orange juice and dehydrated fruits and vegetables, e.g., apricots, carrots, peaches, pears, potatoes, and many others. Concentrated sulfur dioxide is used in wine-making to destroy certain bacteria. The color preservation of canned green beans and peas is enhanced by dipping the produce in a sulfite solution prior to canning.

References

Allaway, W. H.: "The Effect of Soils and Fertilizers on Human and Animal Nutrition," Cornell University Agricultural Experiment Station and U.S. Department of Agriculture, *Agriculture Information Bulletin 378*, Washington, D.C. (1975).

Baldridge, R. C.: "Sulfur Metabolism," in "The Encyclopedia of Biochemistry," (R. J. Williams and E. M. Lansford, Jr., Editors), Van Nostrand Reinhold, New York (1967).

Chichester, D. F., and F. W. Tanner, Jr.: "Antimicrobial Food Additives: Sulfur Dioxides and Sulfites," in "Handbook of Food Additives," (T. W. Furia, Editor), CRC Press, Boca Raton, Florida (1972).

Gray, P.: "The Encyclopedia of the Biological Sciences," Van Nostrand Reinhold, New York (1970).

Kirchgessner, M., Editor: "Trace Element Metabolism in Man and Animals," Institut für Ernährugsphysiologie, Technische Universität München, Freising-Weihenstephan, Germany (1978).

Luh, B. S., and J. G. Woodruff: "Commercial Vegetable Processing," AVI, Westport, Connecticut (1975).

Underwood, E. J.: "Trace Elements in Human and Animal Nutrition," 4th Edition, Academic, New York (1977).

SULFURIC ACID. Infrequently termed *oil of vitriol*, sulfuric acid, H_2SO_4, is a colorless, oily liquid, dense, highly reactive, and miscible with water in all proportions. Much heat is evolved when concentrated sulfuric acid is mixed with water and, as a safety precaution to prevent spluttering, the acid is poured into the water rather than vice versa. Sulfuric acid will dissolve most metals. The concentrated acid oxidizes, dehydrates, or sulfonates most organic compounds, sometimes causing charring. There are numerous commercial and industrial uses for H_2SO_4 and these include the manufacture of fertilizers, chemicals, inorganic pigments, petroleum refining, etching, as a catalyst in alkylation processes, in electroplating baths, for pickling and other operations in iron and steel production, in rayon and film manufacture, in the making of explosives, and in nonferrous metallurgy, to mention only some of its numerous uses. Because of its wide use industrially, some economists over the years have included sulfuric acid consumption among their economic indicators.

Most countries with significant industrial activity and particularly in chemicals production will have significant capacities for making sulfuric acid. In some countries, H_2SO_4 is the leading chemical in terms of tonnage production. Depending upon suppliers, H_2SO_4 is commercially available in a number of strengths, ranging from 77.7% H_2SO_4 (60° Baumé, sp. gr. 1.71) through 93.2% H_2SO_4 (66° Baumé), 98% H_2SO_4, 99% H_2SO_4, and 100% H_2SO_4 (sp. gr. 1.84).

Fundamentally, there are two kinds of sulfuric acid plants: (1) those that use the dry gas (sulfur burning) process; and (2) those that use the wet gas process. In the first type, the raw materials are elemental sulfur and water. In the second type, the sulfur dioxide feed may come from a variety of sources, including metallurgical smelters (copper, zinc, lead, etc.), pyrite roasters, waste acid decomposition furnaces, and hydrogen sulfide burners. In these plants, the SO_2 gas stream enters the acid plant containing a large amount of water vapor. The gas is usually hot (260–430°C) and dusty, and also may contain a number of impurities, such as fluorides, that could harm the catalyst in the contact section of the plant. These incoming gases thus require cooling and purification in a series of scrubbers and electrostatic precipitators, followed by drying prior to entering the contact section of the plant.

In either type of plant, sulfur dioxide is converted to sulfur trioxide in the contact portion of the plant. The reaction $SO_2 + \frac{1}{2}O_2 \rightarrow SO_3$ is effected by passing the SO_2 over a catalyst, usually vanadium pentoxide (V_2O_5). The catalyst in the converter vessel is usually in the form of small pellets and typically arranged in four layers. Provision is made for removal of the heat of reaction after each layer or stage. The catalyst may be used for a number of years with only a very moderate decrease in activity.

From this fundamental point, the sulfuric acid plant designer has a number of alternatives and options to consider. Two factors are of major import in sulfuric acid plant design today, namely, recovery and conservation of energy; and minimizing environmental impact. For example, in the relatively simple plants of a few years ago, the SO_2 needed to contact the catalyst but once and the absorption of the resulting SO_3 in water (a solution of sulfuric acid) could be handled in a single absorption tower. Recycling could be kept to a minimum. In the modern sulfuric acid plant, double contact (DC) of the gases with catalyst and double absorption (DA) of the gases is commonly practiced. Designs are available in numerous configurations, each offering various advantages in terms of energy conservation, pollution minimization, and initial and operating costs. A typical sulfur-burning DC/DA sulfuric acid plant is shown in accompanying diagram.

Of the approximately 40 million tons (36 million metric tons) of sulfuric acid manufactured in the United States per year, about 90% is used in the production of fertilizers and other inorganic chemicals. Much of the remaining 10% of H_2SO_4 is used by the petroleum, petrochemical, and organic chemical industries. Much of this latter acid is involved in recycling kinds of processes. As pollution regulations in various countries become more restrictive, spent acid may become a much more attractive raw material than has been the case in the past.

As pointed out by Sander and Daradimos (1978), a regeneration of sulfuric acid of high quality can only be attained by thermal decomposition back to sulfur dioxide at high temperatures, where all organic impurities are completely burned—followed by reprocessing the SO_2 gases by the contact process to concentrated acid or oleum.

Reactivity of Sulfuric Acid. Dilute sulfuric acid reacts: (1) with many hydroxides, e.g., sodium hydroxide, to yield two series of sulfates (the acid is dibasic), e.g., sodium sulfate or sodium hydrogen sulfate, depending upon the ratio of acid to base reacting, (2) with many ordinary oxides, e.g., magnesium oxide, to yield the corresponding sulfate, e.g., magnesium sulfate solution, (3) with some carbonates, e.g., zinc carbonate, to yield the corresponding sulfate, e.g., zinc sulfate solution plus carbon dioxide gas (calcium carbonate is soon coated by a layer of calcium sulfate, which prevents further reaction), (4) with some sulfides, e.g., ferrous sulfide, to yield the corresponding sulfate, e.g., ferrous sulfate plus hydrogen sulfide gas, (5) with many metals, e.g., zinc, if not too pure (but not copper), to yield the corresponding sulfate, e.g., zinc sulfate solution plus hydrogen gas, (6) with solutions of some salts to yield the corresponding sulfate, e.g., barium chloride, changed to barium sulfate precipitate, calcium citrate, malate, tartrate to calcium sulfate precipitate and the free organic acid in solution.

Higher strengths of sulfuric acid react similarly in kind to the cases of (1), (2), (3), (6) above, but not, in general, as in cases (4) and (5) above. Copper and concentrated sulfuric acid

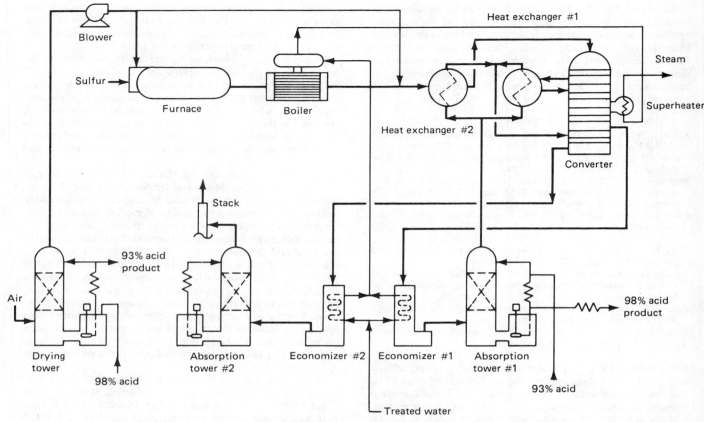

Representative sulfuric acid plant of the sulfur-burning, double-contact (DC)/double absorption (DA) type.

yield copper sulfate and sulfur dioxide gas. Iron reacts similarly, yielding ferric sulfate in the place of copper sulfate.

A number of other reactions of sulfuric acid are characteristic of its higher strengths. Concentrated sulfuric acid is thus (7) an oxidizing agent, and a further example is the oxidation of sulfur to sulfur dioxide (the reacting sulfuric acid is reduced to sulfur dioxide), (8) a sulfonating agent, e.g., naphthalene sulfonated to naphthalene sulfonic acids (mono-, alpha or beta, di- several), (9) an esterification agent, e.g., methyl alcohol esterified to dimethyl sulfate $(CH_3C)_2SO_2$, melting point $-32°C$, boiling point 189°C, or methyl hydrogen sulfate $CH_3O \cdot SO_2OH$, ethyl alcohol esterified to diethyl sulfate $(C_2H_5O)_2SO_2$, melting point $-26°C$, boiling point 208°C, or ethyl hydrogen sulfate $C_2H_5O \cdot SO_2OH$, (10) a dehydration agent, e.g., formic acid into carbon monoxide, sugar blackened with separation of carbon, (11) an addition agent, e.g., ethylene into ethyl hydrogen sulfate, (12) a non-volatile acid upon heating, e.g., with sodium chlorite or nitrate, hydrogen chloride or nitric acid, respectively, is volatilized and sodium sulfate or sodium hydrogen sulfate remains as a residue.

References

Bauer, R. A., and B. P. Vidon: "The Ugine Kuhlmann Pressure Process," *Chem. Eng. Progress*, **74**, 9, 68–69 (1978).

Browder, T. J.: "Improvements in Sulfuric Acid Processes," *Chem. Eng. Progress*, **73**, 3, 70–76 (1977).

Cameron, G. M., Nolan, P. D., and K. R. Shaw: "The CIL Process for Acid Manufacture," *Chem. Eng. Progress*, **74**, 9, 47–50 (1978).

Donovan, J. R., Palermo, J. S., and R. M. Smith: "Sulfuric Acid Converter Optimization," *Chem. Eng. Progress*, **74**, 9, 51–54 (1978).

Duros, D. R., and E. D. Kennedy: "Acid Mist Control," *Chem. Eng. Progress*, **74**, 9, 70–77 (1978).

Riedel, R. W., Knight, J. J., and R. E. Warner: "Alternatives in Sulfuric Acid Plant Design," *Chem. Eng. Progress*, **73**, 3, 55–60 (1977).

Sander, U.: "Waste Heat Recovery in Sulfuric Acid Plants," *Chem. Eng. Progress*, **73**, 3, 61–64 (1977).

Sander, U., and G. Daradimos: "Regenerating Spent Acid," *Chem. Eng. Progress*, **74**, 9, 57–67 (1978).

Smith, G. M.: "Energy Evaluation in Sulfuric Acid Plants," *Chem. Eng. Progress*, **73**, 3, 77–78 (1977).

Smith, G. M., and E. Mantius: "The Concentration of Sulfuric Acid," *Chem. Eng. Progress*, **74**, 9, 78–83 (1978).

Vanderland, D. F. R., and J. Rodda: "Corrosion in Sulfuric Acid Storage Tanks," *Chem. Eng. Progress*, **73**, 3, 65–69 (1977).

SULFUROUS ACID. H_2SO_3, formula weight 82.08, colorless liquid, prepared by dissolving SO_2 in H_2O. Reagent grade H_2SO_3 contains approximately 6% SO_2 in solution. As a bleaching agent, sulfurous acid is used for whitening wool, silk, feathers, sponge, straw, wood, and other natural products. In some areas, its use is permitted for bleaching and preserving dried fruits. The salts of sulfurous acid are sulfites.

Sulfurous acid is a strong reducing agent, being oxidized to H_2SO_4 (1) on standing in contact with air, (2) by chlorine, bromine, iodine, yielding HCl, HBr, or HI, respectively, (3) by HNO_3 or nitrous acid yielding nitric oxide, and (4) by permanganate. Sulfurous acid is itself reduced by zinc and dilute H_2SO_4 to H_2S. Sulfurous acid also may be formed by the reaction of a sulfite or bisulfite solution and an acid.

Sodium sulfite NA_2SO_3 and sodium hydrogen sulfite $NaHSO_3$ are formed by the reaction of sulfurous acid and NaOH or sodium carbonate in the proper proportions and concentrations. Sodium sulfite, when dry and upon heating, yields sodium sulfate and sodium sulfide. Sodium pyrosulfite (sodium metabisulfite)

$Na_2S_2O_5$ is a common sulfite. Crystalline sulfites are obtained by warming the corresponding bisulfite solutions. Calcium hydrogen sulfite $(Ca(HSO_3)_2$ is used in conjunction with excess sulfurous acid in converting wood to paper pulp. Sodium sulfite and silver nitrate solutions react to yield silver sulfite, a white precipitate, which upon boiling decomposes forming silver sulfide, a brown precipitate.

An esterification agent, sulfurous acid forms dimethyl sulfite $(CH_3O)_2SO$, bp $126°C$ and diethyl sulfite $(C_2H_5O)_2SO$, bp $161°C$. Sulfites give a white precipitate with barium chloride, soluble in HCl with evolution of SO_2. Sulfites decolorize iodine in acid solution.

SULFUROUS ACID DERIVATIVES. Herbicide.

SULFUR OXIDES (Pollution). Pollution (Air).

SULFUR (Vulcanization). Rubber (Natural).

SUNFLOWER OIL. Vegetable Oils (Edible).

SUPERACIDS. Acids and Bases.

SUPERCONDUCTORS.

The most spectacular property of a superconductor is the total disappearance of its electrical resistance when it is cooled below a critical temperature T_c. Very careful measurements show that the electrical resistance of a superconductor is at least a factor 10^{17} smaller than the resistance of copper at room temperature and may therefore for all practical purposes be taken to be zero. Some 25 elements and a vast number of alloys and compounds have so far been discovered to be superconducting; examples are In, Sn, V, Mo, Nb-Zr alloys and Nb_3Sn. Transition temperatures range from a few thousandths of a degree Kelvin (for certain Nb-Mo alloys) all the way to about 18 K for Nb_3Sn.

Another important property is the destruction of superconductivity by the application of a magnetic field equal to or greater than a critical field H_c. This H_c, for a given superconductor, is a function of the temperature given approximately by

$$H_c = H_0(1 - T/T_c^2) \qquad (1)$$

where H_0, the critical field at 0K, is in general different for different superconductors and has values from a few gauss to a couple of thousand gauss. For applied magnetic fields less than H_c, the flux is excluded from the bulk of the superconducting sample, penetrating only to a small depth λ into the surface. The value of λ (called the penetration depth) is in the range 10^{-5} to 10^{-6} centimeter. Thus the magnetization curve for a superconductor is

$$B \text{ (inside)} = 0 \text{ for } H < H_c$$

$$B \text{ (inside)} = B \text{ (outside) for } H > H_c$$

This magnetization behavior is reversible and cannot therefore be explained entirely on the basis of the zero resistance. The reversible magnetization behavior is called the Meissner effect.

The existence of the penetration depth λ suggests that a sample having at least one dimension less than λ should have unusual superconducting properties, and such is indeed the case. Thin superconducting films, of thickness d less than λ, have critical fields higher than the bulk critical field, approximately in the ratio of λ to d. This result follows qualitatively from the thermodynamics of the Meissner effect: the metal in the superconducting state has a lower free energy than in the normal state, and the transition to the normal state occurs when the energy needed to keep the flux out becomes equal to this free energy difference. But in the case of a thin film with $d < \lambda$, there is partial penetration of the flux into the film, and thus one must go to a higher applied field before the free energy difference is compensated by the magnetic energy.

It is clear that the existence of the critical field also implies the existence of a critical transport electrical current in a superconducting wire, i.e., that current I_c which produces the critical field H_c at the surface of the wire. For example, in a cylindrical wire of radius r, $I_c = \frac{1}{2}rH_c$. This result is called the Silsbee rule.

All of the above properties distinguish superconductors from "normal" metals. There is another very important distinction, which contains a clue to understanding some of the properties of superconductors. In a normal metal at 0 K, the electrons, which obey Fermi statistics, occupy all available states of energy below a certain maximum energy called the Fermi energy ζ. Raising the temperature of the metal causes electrons to be singly excited to states just above the Fermi energy. There is for all practical purposes a continuum of such excited energy states available above the Fermi energy. The situation is quite different in a superconductor; it turns out that in a superconductor, the lowest excited state for an electron is separated by an energy gap ϵ from the ground state. The existence of this gap in the excitation spectrum has been confirmed by a wide range of measurements: electronic heat capacity, thermal conductivity, ultrasonic attenuation, far infrared and microwave absorption, and tunneling. The energy gap is a monotonically decreasing function of temperature, having a value $\sim 3.5kT_c$ at 0 K (where k is the Boltzmann constant) and vanishing at T_c.

The superconducting state has a lower entropy than the normal state, and therefore one concludes that superconducting electrons are in a more ordered state. Without, for the present, inquiring more deeply into the nature of this ordering, one can state that a spatial change in this order produced say by a magnetic field will occur, not discontinuously, but over a finite distance ζ, which is called the *coherence length*. The coherence length represents the range of order in the superconducting state and is typically about 10^{-4} centimeter, though we shall see later that it can in some superconductors take much lower values and lead to some remarkable properties.

Measurements of the transition temperature on different isotopes of the same superconductor showed that T_c is proportional to $M^{-1/2}$, where M is the isotopic mass. This isotope effect suggests that the mechanism underlying superconductivity must involve the properties of the lattice, in addition to those of the electrons. Another indication of this is given by the behavior of allotropic modifications of the same element: white tin is superconducting, while grey tin is not, and the hexagonal and face-centered cubic phases of lanthanum have different transition temperatures. A third, and most striking, indication is that the current vs voltage characteristic of a superconducting tunneling junction shows a structure which is intimately related to the phonon spectrum of the superconductor.

The superconducting properties of alloys present a bewildering variety of phenomena. They show a great deal of magnetic hysteresis, with little indication of a perfect Meissner effect. The Silsbee rule is inapplicable, and the resistive transition occurs at fields generally very much higher than in pure superconductors. For example, a wire of Nb_3Sn can carry a current of 10^5 amperes/cubic centimeter in an applied field of 100 kilogauss, while a similar wire of lead would carry about 10^3 amperes/cubic centimeter in a field of only 100 gauss. When experiments are done using well-annealed (preferably single-crystal) alloys, it is found that the critical currents drop considerably, and the magnetic behavior becomes reversible but still quite unlike that of pure superconductors. The flux is excluded from the interior of the sample up to a well-defined field H_{c1}. When the applied field is raised further, flux begins to penetrate, even though the resistance remains zero, until a second critical field H_{c2} is reached, at which the flux penetration is complete, and normal resistance is abruptly restored.

The theory of superconductivity has developed along two lines, the phenomenological and the microscopic. The phenomenological treatment was initiated by F. London, who modified the Maxwell electromagnetic equations so as to allow for the Meissner effect. His theory explained the existence and order of magnitude of the penetration depth, and gave a qualitative account of some of the electrodynamic properties. The treatment was extended by V. L. Ginzburg and L. D. Landau, and by A. B. Pippard, who in particular emphasized the concept of the range of coherence. A. A. Abrikosov used these ideas to develop a model for alloy superconductors. He showed that if the electronic structure of the superconductor were such that the coherence length ξ becomes smaller than the penetration depth λ, one would get magnetic behavior similar to that observed in alloys, with two critical fields H_{c1} and H_{c2}. The problem of high critical currents in unannealed (or otherwise metallurgically imperfect) alloys and compounds is more complicated because it involves the interaction between the microscopic metallurgical structure and the superconducting properties. This is an area of great research activity because of the technological implication to be mentioned later.

The microscopic theory of superconductivity was initiated by H. Fröhlich, who first recognized the importance of the interactions of electrons with lattice vibrations and in fact predicted the isotope effect before its experimental observation. The detailed microscopic theory was developed by J. Bardeen, L. N. Cooper and J. R. Schrieffer in 1957, and represents one of the outstanding landmarks in the modern theory of solids. The BCS theory, as it is called, considers a system of electrons interacting with the phonons, which are the quantized vibrations of the lattice. There is a screened coulomb repulsion between pairs of electrons, but in addition there is also an attraction between them via the electron-phonon interaction. If the net effect of these two interactions is attractive, then the lowest energy state of the electron system has a strong correlation between pairs of electrons with equal and opposite momenta and opposite spin and having energies within the range $k\theta$ (where θ is the Debye temperature) about the Fermi energy. This correlation causes a lowering of the energy of each of these Cooper pairs (named after L. N. Cooper who first pointed out their existence on the basis of some general arguments) by an amount ϵ relative to the Fermi energy. The energy ϵ may be regarded as the binding energy of the pair, and is therefore the minimum energy which must be supplied in order to raise an electron to an excited state. We see thus that the experimentally observed energy gap follows from the theory. The magnitude ϵ_0 of the gap at 0 K is

$$\epsilon_0 \approx 4k\theta \exp\left(-\frac{1}{NV}\right)$$

where N is the density of electronic states at the Fermi energy and V is the net electron-electron interaction energy. The superconducting transition temperature T_c is given by

$$3.5kT_c \approx \epsilon_0$$

It has been shown that the BCS theory does lead to the phenomenological equations of London, Pippard and Ginzburg and Landau, and one may therefore state that the basic phenomena of superconductivity are now understood from a microscopic

point of view, i.e., in terms of the atomic and electronic structure of solids. It is true, however, that we cannot yet, *ab initio*, calculate V for a given metal and therefore predict whether it will be superconducting or not. The difficulty here is our ignorance of the exact wave functions to be used in describing the electrons and phonons in a specific metal, and their interactions. However, we believe that the problem is soluble in principle at least.

The range of coherence follows naturally from the BCS theory, and we see now why it becomes short in alloys. The electron mean free path is much shorter in an alloy than in a pure metal, and electron scattering tends to break up the correlated pairs, so that for very short mean free paths one would expect the coherence length to become comparable to the mean free path. Then the ratio $\kappa \approx \lambda/\xi$ (called the Ginzburg-Landau order parameter) becomes greater than unity, and the observed magnetic properties of alloy superconductors can be derived. The two kinds of superconductors, namely those with $\kappa < 1/\sqrt{2}$ and those with $\kappa > 1/\sqrt{2}$ (the inequalities follow from the detailed theory) are called respectively type I and type II superconductors.

In 1962, B. Josephson recognized the implications of the complex order parameter for the dynamics of the superconductor, and in particular when one considers a system consisting of two bulk conductors connected by a "weak link." This research led to the development of a series of weak link devices commonly called *Josephson junctions*. These devices hold much promise for achieving ultra high-speed computers where switching time is of the order of 10^{-11} second.

Good success also has been achieved in the use of certain type II superconductors, such as Nb-Zr and Nb-Ti alloys, and Nb_3Sn, in making electromagnets. In a conventional electromagnet employing normal conductors, the entire electric power applied to the magnet is consumed as Joule heating. For a magnet to produce 100 kilogauss in a reasonable volume, the power requirement can run into megawatts. In striking contrast, a superconducting magnet develops no Joule heat because its resistance is zero. Indeed, if such a magnet has a superconducting shunt placed across it after it is energized, the external power supply can be removed, and the current continues to flow indefinitely through the magnet and shunt, maintaining the field constant. Superconducting magnets have been constructed producing very strong fields in usable volumes. There is a natural upper limit to the critical field possible in such superconductors, given by the paramagnetic energy of the electrons (due to their spin moment) in the normal state becoming equal to the condensation energy of the Cooper pairs in the superconducting state. This leads to a limit of about 360 kilogauss for a superconductor with a T_c of 20 K.

Superconductor technology is also being applied aggressively in the design and construction of new electric generators which appear to have many advantages.

References

Cooper, L. N.: "Microscopic Quantum Interference in the Theory of Superconductivity," *Science*, **181**, 908–916 (1973).

Devreese, J. T., Evrard, R. P., and V. E. Van Doren, Editors: "Highly Conducting One-Dimensional Solids," Plenum, New York (1979).

Edmonds, J. S., and W. R. McCown: "Large Superconducting Generators for Electric Utility Applications," American Power Conf., Chicago, Illinois (April 1980).

Hulm, J. K., and B. T. Matthias: "High-Field, High-Current Superconductors," *Science*, **208**, 881–887 (1980).

Kircher, C. J., and M. Murakami: "Josephson Tunnel-Junction Electrode Materials," *Science*, **208**, 944–950 (1980).

McCown, W. R., and J. S. Edmonds: "300 MVA Superconducting

Generator," Intl. Conf. on Large High Voltage Electric Systems," Paris (August 27–September 4, 1980).

Savitskii, E., et al.: Superconducting Materials," Plenum, New York (1973).

Solymar, L.: "Superconductive Tunnelling and Applications," Wiley, New York (1972).

SUPERCOOLING.

The cooling of a liquid below its freezing point without the separation of the solid phase. This is a condition of metastable equilibrium, as is exhibited by solidification of the supercooled liquid upon addition of the solid phase, or the application of certain stresses, or simply upon prolonged standing.

SUPERFLUIDITY.

The term used to describe a property of condensed matter in which a resistance-less flow of current occurs. The mass-four isotope of helium in the liquid state, plus over 20 metallic elements, are known to exhibit this phenomenon. In the case of liquid helium, these currents are hydrodynamic. For the metallic elements, they consist of electron streams. The effect occurs only at very low temperatures in the vicinity of the absolute zero (−273.16°C or 0 K). In the case of helium, the maximum temperature at which the effect occurs is about 2.2 K. For metals, the highest temperature is in the vicinity of 20 K.

If one of the metals (commonly referred to as superconductors) is cast in the form of a ring and an external magnetic field is applied perpendicularly to its plane and then removed, a current will flow round the ring induced by Faraday induction. This current will produce a magnetic field, proportional to the current, and the size of the current may be observed by measuring this field. Were the ring (e.g., one made of lead) at a temperature above 7.2 K, this current and field would decay to zero in a fraction of a second. But with the metal at a temperature below 7.2 K before the external field is removed, this current shows no sign of decay even when observations extend over a period of a year. As a result of such measurements, it has been estimated that it would require 10^{99} years for the supercurrent to decay. Such persistent or "frictionless" currents in superconductors were observed in the early 1900s—hence they are not a recent discovery.

In the case of liquid helium, these currents are hydrodynamic, i.e., they consist of streams of neutral (uncharged) helium atoms flowing in rings. Since, unlike electrons, the helium atoms carry no charge, there is no resulting magnetic field. This makes such currents much more difficult to create and detect. Nevertheless, as a result of research carried out in England and the United States during the late 1950s and early 1960s, the existence of supercurrents in liquid helium has been established.

The concept of superfluidity is further developed under **Superconductors.**

SUPEROXIDES.

These compounds are characterized by the presence in their structure of the O_2^- ion. The O_2^- ion has an odd number of electrons (13) and, as a result, all superoxide compounds are paramagnetic. At room temperature all superoxides have a yellowish color. At low temperature many of them undergo reversible phase transitions which are accompanied by a color change to white. Superoxide compounds known to be stable at room temperature are:

Sodium superoxide	NaO_2
Potassium superoxide	KO_2
Rubidium superoxide	RbO_2
Cesium superoxide	CsO_2
Calcium superoxide	$Ca(O_2)_2$
Strontium superoxide	$Sr(O_2)_2$
Barium superoxide	$Ba(O_2)_2$
Tetramethylammonium superoxide	$(CH_3)_4NO_2$

The superoxides are generally prepared by one of three methods:

(1) Direct oxidation of the metal, metal oxide, or metal peroxide with pure oxygen or air. All alkali metal superoxides, with the exception of lithium have been prepared in this manner. The superoxides of potassium, rubidium, and cesium form quite readily upon direct oxidation of the molten metal in air or oxygen at atmospheric pressure. Attempts to prepare sodium superoxide under the same conditions result in the formation of sodium peroxide, Na_2O_2. As a result, it was generally felt, prior to 1949, that sodium superoxide was not stable enough to be synthesized. However, in 1949 this superoxide was prepared for the first time, in good yield and purity, by the direct oxidation of sodium peroxide at 490°C under an oxygen pressure of 298 atm. Sodium superoxide is now commercially available and is prepared by a high-temperature, high-pressure, direct oxidation of the peroxide. It is now known that the pale yellow color common in commercial grade sodium peroxide is due to the presence of 5 to 10% sodium superoxide.

(2) Oxidation of an alkali metal dissolved in liquid ammonia with oxygen. All the alkali metal superoxides have been prepared by this method. Although lithium superoxide (LiO_2) has not been isolated in a room temperature-stable form, it has been demonstrated that when lithium is oxidized in liquid ammonia at −78°C the superoxide does form and is stable at that temperature.

(3) Reaction of hydrogen peroxide with strong bases. Hydrogen peroxide can be caused to react with strong inorganic bases to form intermediate peroxide compounds which disproportionate to yield superoxides. The alkaline earth metal superoxides, and sodium, potassium, rubidium, cesium, and tetramethylammonium superoxide have been obtained via this process. Claims have also been made for the synthesis of lithium superoxide via this method; however, such claims have not been adequately substantiated.

Using the formation of potassium superoxide as an example, the reactions involved in this process are:

$$2KOH + 3H_2O_2 \rightarrow K_2O_2 \cdot 2H_2O_2 + 2H_2O$$

followed by

$$K_2O_2 \cdot 2H_2O_2 \rightarrow 2KO_2 + 2H_2O.$$

From the commercial point of view the most important of the superoxides is KO_2. This compound has been in large scale commercial production for many years. It is manufactured in very good yield and purity by air oxidation of the molten metal. This compound is utilized in self-contained breathing devices which are widely used in fire fighting operations and in mine rescue work. The function of the superoxide is to provide oxygen and to remove exhaled carbon dioxide. This unique capability of superoxides is explained by the following chemical reactions:

$$2KO_2(s) + HOH(v,l) \rightarrow 2KOH(s,soln) + 3/2O_2(g)$$

and

$$2KOH(s,soln) + CO_2(g) \rightarrow K_2CO_3(s,soln) + H_2O$$

where s = solid, v = vapor, l = liquid, g = gas, and soln = solution.

Up to 34% of the weight of potassium superoxide is available as breathing oxygen. The lower molecular weight NaO_2 is capable of supplying up to 43% of its weight as oxygen. Thus, sodium superoxide is a better oxygen storage compound. However, it has not been widely used due to its relatively high cost. The cost of KO_2 is much less. The use of superoxides for maintaining proper oxygen and carbon dioxide levels in the atmospheres of space vehicles, space stations, and submarines has been of some interest.

The handling and storage of superoxides requires care and caution. Chemically they are powerful oxidizing agents and strong bases and as a result, they react vigorously with acids and organic materials. All superoxides are extremely hygroscopic, thus their safe storage requires the use of tightly sealed, clean, dry containers.

The chemical bond between the superoxide ion, O_2^-, and the metal ion is ionic in nature. Melting points of potassium, rubidium and cesium superoxide have been determined, and in keeping with the ionic nature of the compounds, the melting temperatures are high, in the order of 400°C.

The most reliable technique for the analysis of superoxides is that developed by Seyb and Kleinberg. In this method the superoxide sample is treated with a mixture of glacial acetic acid and diethyl or dibutyl phthalate. The superoxide reacts with the acetic acid to yield oxygen, hydrogen peroxide, and potassium acetate. The amount of superoxide in the sample is related to the amount of oxygen evolved which is measured with a gas buret. The stoichiometry of the analytical reaction is:

$$2KO_2 + 2HC_2H_3O_2 \rightarrow 2KC_2H_3O_2 + H_2O_2 + O_2$$

It is important that a sufficiently dilute glacial acetic acid-diethyl phthalate mixture be used. Contact of undiluted glacial acetic acid with the superoxide will result in a violent and uncontrollable reaction.

As a result of the paramagnetic nature of superoxides, it is possible to determine their purity by means of paramagnetic susceptibility measurements. The use of this method is limited by its poor accuracy.

—A. W. Petrocelli, Westerley, Rhode Island.

SUPERSATURATED VAPOR. A vapor that remains dry, although its heat content is less than that of dry and saturated vapor at the same pressure. Supersaturation is an unstable condition, and is found in the steam emerging from the nozzles of a steam turbine. The abnormality of the phenomenon is similar to that of supercooling. Supersaturation of the steam probably results from the very rapid expansion of steam in the nozzle, permitting the traverse of a short distance before the condensation of moisture is completed. At a certain point, however, known as the Williams limit, the supersaturation vanishes, and the steam regains the wet state which would be normal in view of the pressure and the heat content. Supersaturation of vapor is impossible in the presence of numerous charged ions or dust particles.

SURFACE ACTIVE AGENTS. Detergents.

SURFACE CHEMISTRY. This topic deals with the behavior of matter, where such behavior is determined largely by forces acting at surfaces. Since only condensed phases, i.e., liquids and solids, have surfaces, studies in surface chemistry require that at least one condensed phase be present in the system under consideration. The condensed phase may be of any size ranging from colloidal dimensions to a mass as large as an ocean. Interactions between solids, immiscible liquids, liquids and solids, gases and liquids, gases and solids, and different gases on a surface fall within the province of surface chemistry.

Surface forces determine whether one material will wet and spread on a substrate, e.g., whether a liquid will wet a solid and spread into crevices and pores to displace air. This seemingly simple phenomenon is of cardinal importance in determining the strength of adhesive joints and of reinforced plastics; it establishes the printing and writing qualities of inks; lubricants will wet and spread over entire surfaces or be confined to limited working areas depending upon built-in wetting or nonwetting properties; ores are floated if the surrounding liquid is readily displaced by air bubbles; the dispersion of pigments in paints depends upon wetting of the individual particles by the liquid; the action of a foam breaker frequently depends upon its ability to spread on the foam; secondary oil recovery often involves displacement of oil from sand by water; wetting is also a factor in detergency; water and soil repellancy depend upon nonwetting.

Wetting or nonwetting often depends upon the adsorption of a solute at a surface or interface. The bulk liquid phase either advances or recedes, depending upon the nature of the solute and the condensed phase. However, there are many phenomena where adsorption is essential to the process but wetting is not a factor. For example, toxic gases and cigarette tars are removed by adsorption on suitable substrates; color bodies are removed from vegetable oils by adsorption on activated clays; heterogeneous catalysis requires the adsorption of reactants on the catalytic surface; dyeing of fabrics is an adsorption process; dispersions and emulsions are stabilized by the adsorption of suitable solutes and flocculated by the adsorption of other solutes; foaming depends upon adsorption; chromatography is a preferential adsorption process; the action of many corrosion inhibitors depends upon their adsorption on metal surfaces.

The spreading of an insoluble monolayer is a process analogous to adsorption with a number of specialized applications. Thus, cetyl alcohol is spread as a monolayer on reservoirs to retard the evaporation of water. Some antifoaming agents act by spreading as monolayers.

Because of the widespread applications of surface chemistry, practically all industries, knowingly or otherwise, make use of the principles of surface chemistry. Countless cosmetic and pharmaceutical products are emulsions—lotions, creams, ointments, suppositories, etc. Food emulsions include milk, margarine, salad dressings and sauces. Adhesive emulsions, emulsion paints, self-polishing waxes, waterless hand cleaners and emulsifiable insecticide concentrates are commonplace examples of emulsions, which fall within the province of surface chemistry. Other products which function in accordance with the principles of surface chemistry include detergents of every variety, fabric softeners, antistatic agents, mold releases, dispersants and flocculants.

Surface forces are merely an extension of the forces acting within the body of a material. A molecule in the center of a liquid drop is attracted equally from all sides, while at the surface the attractive forces acting between adjacent molecules results in a net attraction into the bulk phase in a direction normal to the surface. Because of unbalanced attraction at the surface, the tendency is for these molecules to be pulled from the surface into the interior, and for the surface to shrink to the smallest area that can enclose the liquid. The work required to expand a surface by 1 sq cm in opposition to these attractive forces is called the surface tension.

This concept applies equally well to solids. Molecules in a solid surface are also in an unbalanced attractive field and possess a surface tension or surface free energy. While the surface tension of a liquid is easily measured, this is much more difficult to do for a solid, since to increase the surface extraneous work must be done to deform the solid.

In the case of solid or liquid solutions it is frequently observed that one component of the solution is present at a greater concentration in the surface region than in the bulk of the solution. Thus, for an ethanol-water system, the surface region will contain an excess of ethanol. The concentration of water will be higher at the surface than in the bulk, if the solute is sulfuric acid. Molybdenum oxide dissolved in glass will concentrate at the surface of the glass. The concentrating of solute molecules at a surface is called adsorption.

If a clean solid is exposed to the atmosphere, molecules of one or more species present in the atmosphere will deposit on the surface. If the clean solid is immersed in a solution, molecules of one or another species present in the solution will be apt to concentrate at the solid-liquid interface. These phenomena are also referred to as *adsorption*.

All adsorption processes result from the attraction between like and unlike molecules. For the ethanol-water example given above, the attraction between water molecules is greater than between molecules of water and ethanol. As a consequence, there is a tendency for the ethanol molecules to be expelled from the bulk of the solution and to concentrate at the surface. This tendency increases with the hydrocarbon chain-length of the alcohol. Gas molecules adsorb on a solid surface because of the attraction between unlike molecules. The attraction between like and unlike molecules arises from a variety of intermolecular forces. London dispersion forces exist in all types of matter and always act as an attractive force between adjacent atoms and molecules, no matter how dissimilar they are. Many other attractive forces depend upon the specific chemical nature of the neighboring molecules. These include dipole interactions, the hydrogen bond and the metallic bond.

There is an additional explanation for the tendency of a solute such as ethanol to concentrate at the surface of a liquid, which originates with Langmuir. According to his "principle of independent surface action" each portion of a molecule behaves independently of other portions of the molecule in its attraction to other molecules or functional groups on a molecule. The attraction between the CH_3CH_2 portion of the ethanol molecule and water arises from relatively weak London dispersion forces, as compared with the additional attraction of strong hydrogen-bonding forces acting between the hydroxyl group of the alcohol and water. Hydrogen bonding is also responsible for the strong attraction between water molecules. As a consequence, not only is the alcohol concentrated at the surface, it is also oriented with the hydroxyl group toward the water and the hydrocarbon chain directed outward. Since the attraction between adjacent hydrocarbon molecules is less than that between adjacent water molecules, hydrocarbon liquids have lower surface tensions than water, and the surface tension of an aqueous alcohol solution is intermediate between that of liquid hydrocarbons and water.

As noted earlier, the phenomenon of adsorption is encountered in diverse applications. Medical applications are often the most complex and the least understood. For example, replacement hearts and kidney machines require plastics that can be kept in contact with human blood for long periods of time. However, foreign material in contact with blood results in clotting. The material first becomes coated with adsorbed protein. Some time later the clotting process begins, apparently due to activation of the Hageman factor, one of the proteins in blood, at the blood-material interface. The activation initiates a chain reaction that results in the conversion of fibrinogen to fibrin. It has been suggested that the Hageman factor is helical in form and that adsorption results in an unfolding of the protein helix with exposure of certain active sites which then initiate the clotting of blood. Other proteins are also adsorbed and their biological function may be altered, but little is known about this.

The ideal surface for contact with human blood is the surface of blood vessels, and the immediate surface contains heparinoid complexes. Heparin, a negatively charged polysaccharide, has been bonded to silicon rubber and other polymers. In one procedure, a quaternary ammonium compound is first adsorbed on the polymer substrate and heparin is in turn adsorbed on the positively charged surface. Chemical bonding of heparin has also been achieved. Such surfaces do not cause clotting of contacted blood.

As noted earlier, the phenomenon of adsorption is encountered in such diverse applications as the separation of components in chromatography, the removal of toxic gases by activated charcoal, heterogeneous catalytic reactions and the dyeing of fabrics. The surface area of solids is most commonly determined by the adsorption of nitrogen on the surface of the solid at −915°C. Nitrogen is assigned an area of 16.2 Å^2 per molecule. The method is due to Brunauer, Emmett and Teller and is referred to as the BET method for determining surface area.

The fundamental adsorption equation is due to Gibbs. In the case of a solution containing a single solute,

$$\Gamma = \frac{1}{RT}\left(\frac{\partial \gamma}{\partial \ln a}\right)_T$$

where Γ is the excess of solute at the surface as compared with the concentration of solute in the bulk liquid expressed in moles per sq cm, R is the gas constant, T is the absolute temperature γ is the surface tension of the solution, and a is the activity of the solute. Where the solution is sufficiently dilute, the concentration of the solute may be substituted for its activity.

The equation also applies to the adsorption of a gas on a solid. At low gas pressures, p, the equilibrium pressure of the gas can be substituted for a, the activity of the solute. The amount of gas adsorbed v/V is equivalent to the surface excess Γ, where v is equal to the volume of gas adsorbed per gram of solid and V is the molar volume of the gas. The total free energy change at constant pressure is $\Sigma \delta \gamma$, where Σ is the area per gram of solid.

When a drop of liquid is placed on the surface of a solid, it may spread to cover the entire surface, or it may remain as a stable drop on the solid. There is a solid-liquid interface between the two phases. In the case of liquids that do not spread on the solid, the bare surface of the solid adsorbs the vapor of the liquid until the fugacity of the adsorbed material is equal to that of the vapor and the liquid.

The equation relating contact angle to surface tension, generally ascribed to Young or Dupre, is

$$\gamma_{Se} = \gamma_{SL} + \gamma_L \cos \theta$$

where γ_{Se} is the surface tension of the solid covered with adsorbed vapor, γ_{SL} is the solid-liquid interfacial tension, γ_L is the surface tension of the liquid and θ is the contact angle.

As a general rule, organic liquids and aqueous solutions will spread on high-energy surfaces, such as the clean surfaces of metals and oxides. The rule has a number of exceptions. For one, certain organic liquids will deposit a low-energy film by adsorption on higher-energy surfaces over which the bulk liquid will not spread.

Zisman discovered that there is a critical surface tension characteristic of low-energy solids, such as plastics and waxes. Liquids that have a lower surface tension than the solid will spread on that solid, while liquids with a higher surface tension will not spread. Examples of critical surface tension values for plastic solids in dynes per cm are: "Teflon," 18; polyethylene, 31; polyethylene terephthalate, 43; and nylon, 42–46. As one indication of the way this information can be used in practical applications, one can consider the bonding of nylon to polyethylene. If nylon were applied as a melt to polyethylene, it would not wet the lower-energy polyethylene surface and adhesion would be poor. However, molten polyethylene would spread readily over solid nylon to provide a strong bond.

There are a large number of materials that exhibit a pronounced tendency to concentrate at surfaces and interfaces and thus alter the surface properties of matter. These materials are called surface-active agents or surfactants. Depending upon the manner in which they are used or the purpose they serve in specific applications, they may be referred to as detergents, emulsifying agents, foaming agents or foam stabilizers, antibacterial agents, fabric softeners, flotation reagents, antistatic agents, corrosion inhibitors, or by other names. There are two general ways of classifying surfactants. According to solubility, they are classified as water or oil soluble. The other classification is according to change type. Those that do not ionize are called nonionic surfactants. If they ionize and the surface-active ion is anionic, the material is an anionic surfactant. If the surface-active ion carries a positive charge, it is called a cationic surfactant.

Only molecules with certain specific types of configurations exhibit surface activity. In general, these molecules are composed of two segregated portions, one of which has low affinity for the solvent and tends to be rejected by the solvent. The other portion has sufficient affinity for the solvent to bring the entire molecule into solution. Water-soluble soaps are probably the oldest surfactants. The long hydrocarbon chain has a low affinity for water and is referred to as the hydrophobic or nonpolar portion of the molecule. The carboxylate group has a high affinity for water and is called the hydrophilic or polar portion.

—Lloyd Osipow, New York.

SURFACTANT. Detergents.

SURFACE TENSION.
Fluid surfaces exhibit certain features resembling the properties of a stretched elastic membrane; hence the term surface tension. Thus, one may lay a needle or a safety-razor blade upon the surface of water, and it will lie at rest in a shallow depression caused by its weight, much as if it were on a rubber air-cushion. A soap bubble, likewise, tends to contract, and actually creates a pressure inside, somewhat after the manner of a rubber balloon. The analogy is imperfect, however, since the tension in the rubber increases with the radius of the balloon, and the pressure inside, which would otherwise decrease, remains approximately constant; while the liquid "film tension" remains constant and the pressure in the bubble falls off as the bubble is blown.

Surface tension results from the tendency of a liquid surface to contract. It is given by the tension σ across a unit length of a line on the surface of the liquid. The surface tension of a liquid depends on the temperature; it diminishes as temperature increases and becomes 0 at the critical temperature. For water σ is 0.073 newtons/meter at 20°C, and for mercury, it is 0.47 newtons/meter at 18°C.

Surface tension is intimately connected with capillarity, that is, rise or depression of liquid inside a tube of small bore when the tube is dipped into the liquid. Another factor which is related to this phenomenon is the angle of contact. If a liquid is in contact with a solid and with air along a line, the angle θ between the solid-liquid interface and the liquid-air interface is called the angle of contact. See Fig. 1. If $\theta = 0$, the liquid is said to wet the tube thoroughly. If θ is less than 90°, the liquid rises in the capillary; and if more than 90°, the liquid does not wet the solid, but is depressed in the rube. For mercury on glass, the angle of contact is 140°, so that mercury is depressed when a glass capillary is dipped into mercury. The rise h of the liquid in the capillary is given by $h = 2\sigma \cos \theta / r\rho g$, where r is the radius of the tube, ρ the density of the liquid, and g is the acceleration due to gravity.

Surface tension can be explained on the basis of molecular theory. If the surface area of liquid is expanded, some of the molecules inside the liquid rise to the surface. Because a molecule inside a mass of liquid is under the forces of the surrounding molecules, while a molecule on the surface is only partly surrounded by other molecules, work is necessary to bring molecules from the inside to the surface. This indicates that force must be applied along the surface in order to increase the area

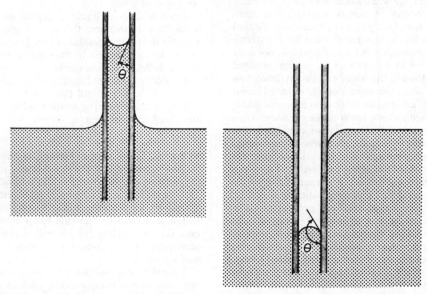

Fig. 1.   Interrelationship between surface tension and capillarity. (*Left*) angle $\theta < 90°$ (water); (*Right*) θ angle $< 90°$ (mercury).

of the surface. This force appears as tension on the surface and when expressed as tension per unit length of a line lying on the surface, it is called the surface tension of the liquid.

The molecular theory of surface tension was dealt with by Laplace (1749–1827). But, as a result of the clarification of the nature of intermolecular forces by quantum mechanics and of the more recent developments in the study of molecular distribution in liquids, the nature and value of surface tension have been better understood from a molecular viewpoint. Surface tension is closely associated with a sudden, but continuous change in the density from the value for bulk liquid to the value for the gaseous state in traversing the surface. See Fig. 2. As a result of this inhomogeneity, the stress across a strip parallel to the boundary—ρ_T per unit area—is different from that across a strip perpendicular to the boundary—ρ_T per unit area. This is in contrast with the case of homogeneous fluid in which the stress across any elementary plane has the same value regardless of the direction of the plane.

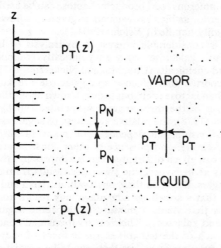

Fig. 2.   Stress relationships in surface tension

The stress ρ_T is a function of the coordinate z, the z-axis being taken normal to the surface and directed from liquid to vapor. The stress ρ_N is constant throughout the liquid and the vapor. Figure 2 shows the stress ρ_N and ρ_T. The stress $\rho_T(z)$ as a function of z is also shown in the left side of the figure.

Gibbs Formula for Surface Tension. The total differential of the surface tension γ in the variables temperature T, and chemical potentials μ_i is

$$d\gamma = -s^a dT - \sum_i \Gamma_i d\mu_i \tag{1}$$

where s^a is the entropy per unit area of the surface phase and Γ_i the adsorption of component i.

This formula is the analog for a surface phase, of the Gibbs-Duhem equation for a bulk phase.

Another basic formula also due to Gibbs which relates the surface tension to the thermodynamic functions of the surface phase is

$$\gamma = A^a - \sum_i \Gamma_i \mu_i \tag{2}$$

where A^a is the Helmholtz free energy (see **Free Energy (2)**) per unit area of the surface phase. This expression is the analog of the relation between the Gibbs free energy and the chemical potentials

$$G = \sum_i n_i \mu_i \tag{3}$$

valid for a bulk phase.

SUSPENSION. Colloid System.

SWEETENERS.   In terms of diet and nutrition, sweeteners fall into two major classifications—*nutritive* and *nonnutritive*.

Records reveal the strikingly large amounts of sucrose currently consumed *per capita* in various nations. Israel, 150 lb (68 kg); Bulgaria, 130 lb (59 kg); Australia, 119 lb (54 kg); New Zealand, 110 lb (49.9 kg); Costa Rica, 108 lb (49 kg); Cuba, 107 lb (48.5 kg); Switzerland, 106 lb (48.1 kg); United States, 102 lb (46.3 kg); Hungary, 99 lb (44.9 kg); Iceland, 98 lb (44.9 kg); Poland, 95 lb (43.1 kg); Sweden, 94 lb (43.1 kg); Austria, 92 lb (41.7 kg); Czechoslovakia, 92 lb (41.7 kg).

Nutritive Sweeteners

In addition to their sweetening power, nutritive sweeteners are effective preservatives in numerous foods, such as jams, jellies, syrups, candies, cured meats, among others. Sweeteners tie up water, essential for the growth of microorganisms, thus preventing or inhibiting spoilage. Nutritive sweeteners also serve as food for yeasts and other fermenting agents, so important to many properties. These properties are lost when nonnutritive sweeteners are used.

GLOSSARY OF NATURAL SUGARS

Following is an expanded alphabetical glossary of the principal nutritive sweeteners.

(*Dextrose*)—Synonymous with glucose (the preferred name). See *Glucose* (this list).

(*Dextrose Equivalent*)—A means for comparing one sugar with another. The total amount of reducing sugars expressed as dextrose (glucose) that is present in a given sugar syrup, calculated as a percentage of the total dry substance. One technique for determining D. E. in sugar syrups is the volumetric alkaline copper method.

(*Fructose*)—Also called *levulose* or *fruit sugar*, $C_6H_{12}O_6$. It is the sweetest of the common sugars, being from 1.1 to 2.0 times as sweet as sucrose. Fructose is generally found in fruits and honey. An apple is 4% sucrose, 6% fructose, and 1% glucose (by weight). A grape (*Vitis labrusca*) is about 2% sucrose, 8% fructose, 7% glucose, and 2% maltose (by weight) (Shallenberger, 1974). Commercially processed fructose is available as white crystals, soluble in water, alcohol, and ether, with a melting point between 103 and 105°C (217.4 and 221°F) (decomposition). Fructose can be derived by the hydrolysis of inulin; by the hydrolysis of beet sugar followed by lime separation; from cornstarch by enzymic or microbial action.

Dry crystalline fructose is reported to have a sweetness level of 180 on a scale in which sucrose is represented at 100 (Andres, 1977). In cool, weak solutions and at lower pH, sweetness value is reported to be 140–150. A neutral pH or higher temperatures, the sweetness level drops and at 50°C (122°F) sweetness equals that of a corresponding sucrose solution. A synergistic sweetness effect is reported between sucrose and fructose. A 40–60% fructose/sucrose mixture in a 100% water solution is sweeter than either component under comparable conditions (Unpublished report, University of Helsinki, 1972).

Andres (1977) reports that high-fructose corn (maize) syrup containing over 90% fructose has a sweetening power (varies with other ingredients in the formulation) of between 120 and

160 (sucrose = 100). Thus, in some formulations it may be possible to reduce the calorie content (resulting from sweetener used in the formulation) by 50%. The development area that has received the most attention with the high-fructose corn syrups has been in carbonated and still beverages. It is further reported that, in addition to permitting a lower calorie content, the fructose has the added potential of improving flavor in some foods, particularly those containing a fruit base. The caloric content per gram (solids) of the 90% fructose corn syrup is the same as that of other carbohydrates. Thus, calorie reduction is possible because of an increase in perceived sweetness.

In 1979, Cardello et al. researched the relative sweetness of fructose and sucrose in model solutions, lemon beverages, and white cake. Their findings, in part, included the observation that in distilled water and the less acidic beverages and citric acid solutions used in the experiment, magnitude estimates of sweetness of fructose were greater than sucrose by a factor of 1.6–1.9. However, they found that the greater relative sweetness of fructose declined with increasing sugar concentrations. In the more acidic beverages and citric acid solutions, no difference in perceived sweetness between fructose and sucrose was found at any sugar concentration. Similarly, no difference in the perceived sweetness of the two sugars was found when each was used as a sweetener in vanilla cake. The data suggest that the greatest sweetening advantage of fructose over sucrose occurs at the lower concentrations of the two sugars. Other researchers who have made relative sweetness tests of these two sugars include Dahlberg and Penczek (1941); Shallenberger (1963); Stone and Oliver (1969); Hyvönen et al. (1977); and Harris et al. (1978).

Combined effects of fructose and saccharin are discussed under *Saccharin* (this glossary).

(*Galactose*)—A monosaccharide commonly occurring in milk sugar or lactose. Formula, $C_6H_{12}O_6$.

(*Glucose*)—Also known as *grape sugar* or *dextrose*, this is the main compound into which other sugars and carbohydrates are converted in the human body and thus is the major sugar found in blood. Glucose is naturally present in many fruits and is the basic "repeating" unit of the starches found in many vegetables, such as potato. Purified glucose takes the form of colorless crystals or white granular powder, odorless, with a sweet taste. Soluble in water, slightly soluble in alcohol. Melting point is 146°C (294.8°F). Glucose finds many uses—confectionery, infant foods, brewing and winemaking, caramel coloring, baking, and canning. Glucose is derived from the hydrolysis of corn starch with acids or enzymes. Glucose is a component of invert sugar and glucose syrup. Glucose was first obtained (1974) from cellulose by enzyme hydrolysis.

Corn (maize) syrup is a sweetener derived from corn starch by a process that was first commercialized in the 1920s. Corn syrup is composed of glucose and a variety of sugars described as the *maltose series* of oligosaccharides. These syrups are not as sweet as sucrose, but are very often used in conjunction with sugar in confections and other food products. Additionally, in some instances, they have proven to be an effective replacement for sucrose in the brewing, soft drink, canning, and pet food industries.

(*Honey*)—A natural syrup which varies in composition and flavor, depending upon the plant source from which the nectar was collected by the honeybee, the amount of processing, and the duration of storage. The principal sugars contained in honey are fructose and glucose, the same components as in table sugar. There are minute amounts of vitamins and minerals in honey, but these are not usually considered in terms of calculating minimum requirements.

(*Invert Sugar*)—A mixture of 50% glucose and 50% fructose obtained by the hydrolysis of sucrose. Invert sugar absorbs water readily, and is usually only handled as a syrup. Because of its fructose content, invert sugar is levorotatory in solution, and sweeter than sucrose. Invert sugar is often incorporated in products where loss of water must be minimized. Commercially, invert sugar is obtained from the inversion of a 96% cane sugar solution. This sugar is used in various foods, in the brewing industry, confectionery field, and in tobacco curing.

(*Lactose*)—*Milk sugar* or *saccharum lactis*, $C_{12}H_{22}O_{11} \cdot H_2O$. Purified lactose is a white, hard, crystalline mass or white powder with a sweet taste, odorless. It is stable in air. Soluble in water; insoluble in ether and chloroform. Very slightly soluble in alcohol. The compound decomposes at 203.5°C (398.3°F). Lactose is derived from whey, by concentration and crystallization. Cow's milk contains about 5% lactose. Because of its relative lack of sweetening power, lactose is not considered a sweetener in the usual sense. It is used as a bulking agent in numerous food products. Lactose can be used effectively as a carrier for artificial sweeteners to give a free-flowing powder that is easily handled (Welch, 1965).

There is considerable interest in hydrolysis of lactose into glucose and galactose, both enzymatically (Mahoney et al., 1975), and chemically (Coughlin and Nickerson, 1975). In 1965, it was reported that glucose and galactose are known to be sweeter than lactose itself, but little is known about the sweetness of mixtures of the three sugars. As pointed out by Shah and Nickerson (1978), the relative sweetness of sugars is not a constant relationship, but depends upon many factors, including pH, temperature, and presence of other constituents. Mixtures of sugars can give a different sweetness impression than individual sugars alone; synergistic sweetness results from a combination of sugars. Based on these observations, Shah and Nickerson reasoned that a sweetness advantage may result from a partial hydrolysis that yields a mixture of the three sugars (lactose, glucose, and galactose). Their observations and conclusions, in part: In each of the test mixes, the amount of sucrose needed for equal sweetness was less than would be predicted from the relative sweetness of the individual sugars. At the concentration used in the mixes, glucose and galactose had nearly the same sweetness and slightly less than 60% that of sucrose. The data further indicated a synergistic enhancement of sweetness when the three sugars were mixed. The researchers further concluded that, in the overall, sugar mixtures equivalent to hydrolyzed lactose syrups have good sweetening properties and appear acceptable for use in foods. Complete hydrolysis appears unnecessary, for results were good at both 70% and 90% hydrolysis levels.

In another study, researcher Parrish et al. (1979) investigated the sweetness of *lactulose* relative to sucrose. Isomerization of lactose to lactulose was first accomplished by Montgomery and Hudson (1930). They recognized that lactulose is sweeter than lactose, but not as sweet as sucrose. They also found lactulose to be very soluble in water. Lee and Birch (1976) confirmed the sweetness observations of the earlier researchers. The partial findings of Parrish et al. (1979) confirm the greater sweetness and solubility of lactulose compared with lactose and suggest that lactulose could be a partial replacement for sucrose in certain food applications. Work is continuing at the laboratories (Eastern Regional Research Center, U.S. Department of Agriculture, Philadelphia, Pennsylvania).

(*Levulose*)—See *Fructose* (this glossary).

(*Maltose*)—Also known as *malt sugar*, maltose is a product of the fermentation of starches by enzymes or yeast. Barley malt, which is used as an adjunct in brewing, enhances the flavor and color of beer because of its maltose content. Maltose also is formed by yeast during breadmaking. Maltose is the most common reducing disaccharide, $C_{12}H_{22}O_{11} \cdot H_2O$, composed of two molecules of glucose. It is found in starch and glycogen. Purified maltose takes the form of colorless crystals, melting point, 102–103°C (215.6–217.4°F). Soluble in alcohol; insoluble in ether. Combustible. Maltose is used as a nutrient, sweetener, and culture medium.

(*Raffinose* and *Stachyose*)—These are sugars found in significant amounts in some foods, such as beans. These sugars are not digested in the stomach and upper intestine as are other disaccharides. They are fermented by bacteria in the lower digestive tract, producing gases and sometimes causing discomfort from flatulence. Raffinose is a trisaccharide composed of one molecule each of D(+)-galactose, D(+)-glucose, and D(−)-fructose, $C_{18}H_{32}O_{16} \cdot 5H_2O$. Raffinose is sometimes used in the preparation of other saccharides.

(*Sucrose*)—Table sugar, also known as *saccharose*, sucrose is a disaccharide, composed of two simple sugars, glucose and fructose, chemically bound together, $C_{12}H_{22}O_{11}$. Hard, white, dry crystals, lumps or powder. Sweet taste, odorless. Soluble in water; slightly soluble in alcohol. Solutions are neutral to litmus. Decomposes in range of 160–186°C (320–366.8°F). Combustible. Optical rotation = +33.6°. Derived from sugarcane or sugar beets and also obtainable from sorghum. Sucrose is the most abundant free sugar in the plant kingdom and has been used since antiquity (Mead and Chem, 1977). See **Sugar.**

(*Sugar Alcohols*)—These are polyols, chemically reduced carbohydrates. Important in this group are *sorbitol, mannitol, maltitol,* and *xylitol.*

Polyols are industrially important sugar substitutes and are used where their different sensory and functional properties make them most desirable (Dwivedi, 1978). Sorbitol, mannitol, and maltitol are considerably less sweet than sucrose, while xylitol has about the same sweetness. Xylitol is described separately a bit later in this glossary.

In addition to sweetness, some of the polyols have other useful properties. For example, although it contains the same number of calories per gram as other sweeteners, sorbitol is absorbed more slowly from the digestive tract than is sucrose (Wick et al., 1951). It is, therefore, useful in making foods intended for special diets. When consumed in large quantities, however (25–50 grams; 1–2 ounces) per day, sorbitol can have a laxative effect, apparently because of its comparatively slow intestinal absorption (Peters and Loch, 1958; Ellis and Krantz, 1943).

When sugar alcohols are ingested, the body converts them first to fructose, which does not require insulin to facilitate its entry into the cells. For this reason, ingesting these sweeteners (including fructose itself), does not cause the immediate increase in blood sugar level which occurs on eating glucose or sucrose. Within the body, however, the fructose is rapidly converted to other compounds, which *do* require insulin in their metabolism (IFT Expert Panel, 1978). One effect of this stepwise metabolism is to damp out the peaks in blood sugar levels which occur immediately after ingesting sucrose, but which are absent after ingesting fructose, even if the eventual total insulin requirements are the same (Brunzell, 1978; Talbot and Fisher, 1978).

Some health scientists are dubious about pursuing the apparent advantages of substituting fructose and the sugar alcohols for sucrose as a major sweetener, particularly for diabetics, until more research is done on their long-range nutritional and physiological aspects. Recent interest has also focused on these sweeteners because of their relatively low potential for causing dental carries. Studies have shown about a 30% reduction in dental caries in rats on sorbitol and mannitol diets, and virtually complete elimination of caries in rats when on xylitol diets (Grundberg et al., 1973; Bowen, 1978).

(*Xylitol*)—A 5-carbon sugar alcohol that occurs naturally in raspberries, strawberries, yellow plums, cauliflower, spinach, and several other plants (Washuttl et al., 1973). Although it is widely distributed, the low concentrations of xylitol within plants makes it uneconomic to extract the substance directly from plants. Consequently, xylitol must be commercially produced from xylan or xylose-rich precursors by way of chemical, enzymatic, or microbiological conversion. The most commonly used and highest yield xylan sources as of the early 1980s are birch tree chips. Other appropriate starting materials include beech and other hardwood chips, almond and pecan shells, cottonseed hulls, straw, cornstalks (maize), and corn cobs. As explained by Pintauro (1977), these xylan sources are routinely converted to xylitol by: hydrolysis of xylan to D-xylose; reduction of xylose to xylitol by pressure hydrogenation in the presence of a nickel catalyst; purification and crystallization of xylitol end-products. Numerous other approaches are currently in development.

Xylitol is of special interest inasmuch as its chemical backbone contains only 5 carbon atoms, compared with the 6-carbon backbone of other saccharides and the sugar alcohols. Xylitol is not utilized by most microorganisms and products made from it are usually unaffected by microbial attack. Xylitol is a normal intermediate product of carbohydrate metabolism in humans and animals, and it occurs naturally in plants as previously mentioned (Emodi, 1978). The use of xylitol in various food substances is being investigated by regulatory agencies in a number of countries. In addition to a number of companies in the United States that are interested in future large-scale production, similar interests are being actively pursued in Switzerland, Finland, and Germany (West), among others.

Physical and chemical properties of xylitol: Empirical formula, $C_5H_{12}O_5$. Structural formula:

$$HOCH_2 - \overset{\overset{\displaystyle H}{|}}{\underset{\underset{\displaystyle OH}{|}}{C}} - \overset{\overset{\displaystyle OH}{|}}{\underset{\underset{\displaystyle H}{|}}{C}} - \overset{\overset{\displaystyle H}{|}}{\underset{\underset{\displaystyle OH}{|}}{C}} CH_2OH$$

Molecular weight, 152.1. Crystalline, white, sweet, ordorless powder. Melting point, 93.4–94.7°C (200.1–202.5°F). Soluble in water; slightly soluble in ethanol and methanol. No optical activity.

Nonnutritive Sweeteners

Earlier in this entry, the circumstances behind a demand for sugar substitutes were briefly described. A number of nutritive sweeteners that are or can be used as substitutes for sucrose were just described. In terms of no-calorie or low-calorie sweeteners, after many decades of interest only two of these have been regarded as being reasonably successful, by processors and consumers alike. Saccharin and the cyclamates are far from perfect substitutes, but they proved reasonably workable for the processor and reasonably satisfactory for the consumer. Cyclamates were banned by some countries, including the United States, a few years ago because of carcinogenic indications. Similarly, as of the early 1980s, saccharin is under penetrating investigation in some countries. Much negative information concerning the compound has been disseminated and it is only due to pressure of a number of professional groups that outright banning has not occurred. But because of an indefinite future, research for new low-calorie sweeteners has been vigorous since the mid-1970s. A number of interesting substances have been revealed, but the drama of discovery has been essentially absent. Each substance presents its own problems—economic in terms of large-scale production, properties that are not quite right for the processor, other properties that may not please the consumer, and the ever-present fear that new substances may, at some future date, follow the pathways of the cyclamates and saccharin.

GLOSSARY OF NONNUTRITIVE SWEETENERS

Following is an expanded alphabetical glossary of the nonnutritive sweeteners that are under active investigation and/or development. Because banning has not been worldwide, and for historical comparative purposes, descriptions of the cyclamates and saccharin are included in the glossary. An extremely interesting approach to the sweetener problem (as well with other food additives) is described in separate entry **Polymeric Food Additives.**

(*Acetosulfam*)—A synthetic sweetener with a sweetness estimated at approximately 150 times that of sucrose. The German developers are continuing to test the product for possible toxicological properties.

(*Aspartame*)—A recent low-calorie sweetener. The relationship between sweetness of aspartame and sucrose is almost linear when plotted on a log-log scale. Aspartame is 182 times sweeter than a 2% sucrose solution, but only 43 times sweeter than a 30% solution (Cloninger and Baldwin, 1970). The clean, full sweetness of aspartame is similar to sucrose and complements other flavors. It has been reported as having no bitterness and that it blends well with food flavorings (Beck, 1974; McCormick, 1975; Pangborn and Larson, 1975). It has been suggested that aspartame may have a potentiating effect on certain fruitlike flavors (McCormick, 1975). In a study of this effect, researchers Balding and Korschgen (1979) reported that sensory scores indicated that both orange- and cherry-flavored beverages had a significantly higher intensity of fruit flavor than their counterparts when sweetened with sucrose. However, in strawberry-flavored beverages and in flavored gelatins, no such intensification of fruit flavor was found with aspartame.

The full name of aspartame is aspartylphenylalanine, a dipeptide that degrades to a simple amino acid. It has been reported as easily metabolized by humans. Synthesis of the compound in large-scale quantities has posed economic problems. The possible toxic effects of aspartame have been under study since the early 1970s. Earlier approval by the U.S. Food and Drug Administration for use of aspartame in dry food products (1974) was later withdrawn for further study.

(*D-6 Chlorotryptophan*)—A proprietary development claimed to have 1300 times the sweetness of sugar and without perceivable aftertaste or toxicity.

(*Cyclamate*)—Group name for synthetic, nonnutritive sweeteners derived from cyclohexylamine or cyclamic acid. The series includes sodium, potassium, and calcium cyclamates. Cyclamates occur as white crystals, or as white crystalline powders. They are odorless and in dilute solution are about 30 times as sweet as sucrose. The purity of commercially available compounds is approximately 98%. The titration curve for cyclamate is that of a very strong acid so that the sodium and calcium salts tend to be fairly neutral in character (Furia, 1972). The cyclamates were banned in the United States and in several other countries during the early 1970s. Prior and subsequent to the banning, some authorities questioned the procedures used for testing the products for possible carcinogenicity and other toxic properties. Findings included the possibility of causing testicular atropy. Atropy is attributed to cyclohexamine, which some persons may convert from ingestion of cyclamates. The investigations have not been completed to the satisfaction of numerous authorities. A concerned effort is still underway to have the cyclamates reconsidered for use in the United States.

(*Dihydrochalcone*)—Also known as *Neohesperidin*, this sweetener is derived from oranges and reported as being 1000 times as sweet as sucrose. Both this sweetener and a related product, *Naringin* (300 times as sweet as sucrose and derived from grapefruit) have been reported to have problems with off-tastes and blending in connection with several flavors (Krueger, 1979). The dihydrochalcones have undergone over 10 years of feeding studies by the Western Regional Research Laboratories of the U.S. Department of Agriculture. USDA has reported no indication of toxicity. Neohesperidin (neo-DHC) is a component of the peel of Seville Oranges, not produced in volume in this country. However, the compound can be produced by certain chemical processes on naringin dihydrochalcone prepared from grapefruit peel. Commercial interest in the production of these substances has been indicated. The compounds still are undergoing investigations for possible toxicity.

The National Institute of Dental Research (U.S.) has been seeking a nonnutritive sweetener for a number of years and has become interested in the dihydrochalcones. A number of synthetic analogs of the dihydrochalcones have been made (Dynapol, Palo Alto, California). It has been reported that two of them are better in many respects than saccharin or cyclamates, but somewhat inferior to sucrose in water solution equivalent to 8.25% sucrose. The slow onset and lingering aftertaste make them unsuitable for some food use, but the patterns they reveal are increasing an understanding of how sweeteners work at the molecular level. As of the late 1970s, one of the synthetic analogs holds promise for use in chewing gum, personal products, and as a table sweetener.

One explanation for the slow onset/lingering aftertaste effect of dihydrochalcone sweeteners apparently relates to a general property of phenols—they bind readily with proteins. Therefore, the compounds, with their phenolic groups, may bind rapidly and indiscriminately with protein in saliva. In that case, it would take time for the dihydrochalcone activity to use up available saliva protein before binding to taste receptor sites. This view is supported by the fact that the delayed onset effect does not happen when high concentrations of the compounds are tasted, furnishing adequate molecules for all available protein pairing.

Another significant finding is that three sites responsible for the sweetness have been located on dihydrochalcone molecules and analogs. They fit the model suggested by L. B. Kier, who proposed that surface configuration of a sweetener has three sites at specific angles to one another. Dynapol scientists also have observed that the dihydrochalcone molecules have two phenol rings, with part of the three activating sites distributed between them. Saccharin and cyclamates have only one similar site per molecule. The question is then raised—Is it possible that both of these compounds must pair up with a single receptor on the tongue to create the sweetness sensation?

As of the late 1970s, Dynapol scientists are working with a core concept, in which two or more of the sweetener nuclei are linked together through a common bridge. These sweeteners have less than 1% absorption in the intestines as determined by ^{14}C-labeled analogs in rats. See also entry on **Polymeric Food Additives.**

(*Glycyrrhizins*)—These are noncaloric sweeteners approximately 50 times as sweet as sugar and used as flavor enhancers under the GRAS (generally regarded as safe) classification in the United States. Glycyrhizins, which have a pronounced lico-rice taste, are used in tobacco, pharmaceutical and some confectionary products. They are available in powder or liquid form and with color, or as odorless, colorless products. These compounds are stable at high temperatures (132°C; 270°F) for a short time and thus can be used in bakery products. In some chocolate-based products, the sweetener has been used to replace up to 20% of the cocoa. The sweetener also has excellent foaming and emulsifying action in aqueous solutions. Typical products in which these sweeteners may have application include cake mixes, ice creams, candies, cookies, desserts, beverages, meat products, sauces, and seasonings, as well as some fruit and vegetable products. Generally available as malted and ammoniated glycyrrhizin.

(*Miracle Fruit*)—This product has been reported as used by African cultures for over 100 years. The fruit has the strange property of making sour foods taste sweet. The compound responsible for this characteristic has been determined to be *miraculin*. The product does not have approval for use in all countries.

(*Saccharin*)—A noncaloric sweetener that is about 300 times as sweet as sugar. The compound is manufactured on a large scale in a number of countries. The compound is *ortho-benzosulfimide*:

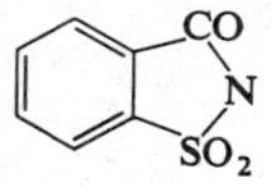

Saccharin was discovered in 1879 by I. Remsen and C. Fahlberg when they were researching the oxidation products of toluenesulfonamide. The most common forms of saccharin are sodium and calcium saccharin, although ammonium and other salts have been prepared and used to a very limited extent. The saccharins are white, crystalline powders, with melting points between 226 and 230°C (438.8 and 446°F). Soluble in amyl acetate, ethyl acetate, benzene, and alcohol; slightly soluble in water, chloroform, and ether. Saccharins are derived from a mixture of toluenesulfonic acids. They are converted into the sodium salts, then distilled with phosphorus trichloride and chlorine to obtain the ortho-toluenesulfonyl chloride, which by means of ammonia is converted into ortho-toluenesulfamide. This is oxidized with permanganate and treated with acid, and saccharin is crystallized out. In food formulations, saccharin is used mainly in the form of its sodium and calcium salts. Sodium bicarbonate may be added to provide improved water solubility.

Saccharin is no longer regarded as generally safe and is under penetrating analysis in some countries, including the United States and Canada.

Considerable research has gone forward in past years and, in fact, continues as regards the synergistic effects of saccharin when mixed with other sugars and sweeteners. In 1978, Hyvönen and associates studied the possible effects of fructose-saccharin and of xylitol-saccharin mixtures under conditions that normally correspond with the degree of sweetness desired in beverages, such as coffee and tea. In part, the researchers found that synergism between fructose and saccharin and between xylitol and saccharin were greatest when the sweeteners were almost equal in the mixture in relation to their sweetness at the prevailing temperatures. Mixtures of saccharin and fructose or xylitol without the aftertaste typical of saccharin were prepared. The sweetness of fructose-saccharin and xylitol-saccharin mixtures in coffee was enhanced compared with that in corresponding water solutions. The researchers concluded that coffee, tea, and juices can be prepared to conventional taste and sweetness standards,

using these mixtures, but with 40–70% less caloric content than when sucrose is used as the sweetener. The same researchers investigated the sweetening of soft drinks with mixtures of sugars and saccharin (Hyvönen et al. (1978)).

(*Serendipity Berry*)—Containing the protein *monoellin*, reputed to be some 2500 times as sweet as sucrose. The berry grows in West Africa. Numerous limitations of the product have been reported, including instability and early loss of sweetness properties, particularly when exposed to heat and low pH.

(*Stevioside*)—Derived from the roots of the herb *Stevia rebaudiana*, this compound has found limited use in Japan and a few other countries as a low-calorie sweetener having about 300 times the sweetening power of sucrose. The compound has not been investigated thoroughly by a number of countries with strong regulatory agencies and, therefore, is not on the immediate horizon for wide consideration as a sweetener.

(*Thalose*)—Not a sweetener per se, but rather it acts as an enhancer and can allow reduction of sugar usage by 10% in various products, such as bakery products, beverages, confections, ice creams, and preserves. As of the late 1970s, the ingredients used in preparing thalose are permitted under the GRAS category in the United States.

Formulating and Processing Problems

In using low-calorie sweeteners in various food products, the problems are not limited to flavor, but often much more importantly involve texture, acidity, storage stability, and preservability, among others. As pointed out by Salant (1972), acceptable nonnutritively sweetened products cannot be developed by the simple substitution of artificial sweeteners for sugars. Rather, the new product must be completely reformulated from the beginning. Three examples follow.

Jams, Jellies, and Preserves. Traditional products in this category contain 65% or more soluble solids. In low-calorie analogs, soluble solids range from 15 to 20%. Under these circumstances, commonly used pectins (high methoxyl content) do not suffice. Thus, special LM (low methoxyl) pectins must be used, along with additional gelling agents, such as locust bean gum, guar gum, and other gums and mucilagenous substances, some of which may require some masking. In the absence of sugar, a preservative such as ascorbic acid, sorbic acid, sorbate salts, propionate salts, or benzoate, usually is required to extent of about 0.1% (weight).

Soft Drinks. In addition to providing sweetness, sugar also functions to provide mouthfeel and to stabilize the carbon dioxide of soft drinks. To contribute to mouthfeel, the use of hydrocolloids and sorbitol has been attempted with limited success. Hydrocolloids also help to some degree with the problem of carbonation retention, but the principal solutions to this problem involve avoiding all factors which contribute to carbonation loss. Thus, the requirement for very well-filtered water to eliminate particulates as possible nucleation points; any substances that promote foaming must be avoided; any emulsifying agents used in connection with flavoring agents must be handled carefully to avoid foaming; carbonation should be carried out at low temperature (34°F; 1.1°C); and trace quantities of metals must be absent from the water.

Bakery Products. These foods are among the most difficult as regards the use of artificial sweeteners. A listing of the functions of sugar in baked goods beyond that of providing sweetness is indicative of these problems. Sugar contributes to texture in forming structures, in providing moist and tender crumb by counteracting the toughening characteristics of flour, milk, and egg solids. In the emulsification process required to retain gas during leavening, sugar is an effective accessory agent. Ingredients frequently used in bakery products to compensate for the absence of sugar include carboxymethylcellulose, mannitol, sorbitol, and dextrins, but generally, these have not been very satisfactory—either to processor or consumer. This remains a large area of challenge for the food processors and ingredient manufacturers.

Evaluating Synthetic Sweeteners. As pointed out by Crosby et al. (1978), evaluation of new sweeteners, unlike that of most functional food ingredients, is not possible using totally objective means. There are no general rules leading to structure/function relationships for all classes of sweeteners. The principal judgments must rely on human sensory panel tests. The training and administration of sensory panels for sweeteners is beyond the scope of this volume. Reference to the Swartz/Furia (1977) summary is suggested.

References

Andres, C.: "Alternate Sweeteners," *Food Processing*, **38**, 5, 50–52 (1977).

Crosby, G. A., DuBois, G. E., and R. E. Wingard, Jr.: "The Design of Synthetic Sweeteners," in "Drug Designm" Vol. VIII (E. J. Ariens, Editor), Academic, New York (1978).

Emodi, A.: "Xylitol: Its Properties and Food Applications," *Food Technology*, **32**, 28 (1978).

Hyvönen, L., et al.: "Fructose-Saccharin and Xylitol-Saccharin Synergism," *Journal of Food Science*, **43**, 1, 251254 (1978).

Hyvönen, L., Kovistoinen, P., and A. Ratilainen: "Sweetening of Soft Drinks with Mixtures of Sugars and Saccharin," *Journal of Food Science*, **43**, 5, 1580–1584 (1978).

IFT Expert Panel: "Sugars and Nutritive Sweeteners in Processed Foods," *Food Technology*, **33**, 5, 101–104 (1979).

Krueger, J.: "Alternative Sweeteners," *Processed Prepared Food*, **148**, 2, 95 (1979).

Parrish, F. W., et al.: "Sweetness of Lactulose Relative to Sucrose," *Journal of Food Science*, **44**, 3, 813–815 (1979).

Salant, A.: "Nonnutritive Sweeteners," in "Handbook of Food Additives," (T. E. Furia, Editor), CRC Press, Boca Raton, Florida (1972).

Shallenberger, R. S.: "Predicting Sweetness from Chemical Structure and Knowledge of the Chemoreception Mechanism of Sweetness," *Institute of Food Technologists Symposium*, Saint Louis, Missouri (1979).

Staff: "Sugar and Nutrition," *Processed Prepared Food*, **148**, 2, 96 (1979).

Staff: "Nonabsorbable Polymer-Leashed Additives," *Food Processing*, **40**, 1 (1979).

SYENITE. A coarse-grained, granular, therefore intrusive, igneous rock of the general composition of granite except that quartz is either absent or present in a relatively small amount. The feldspars are alkaline in character and the dark mineral is usually hornblende. Sodalime feldspars may be present in small quantities. The term syenite was originally applied to hornblende granite like that of Syene in Egypt from whence the name is derived. Syenite is not a common rock, some of the more important occurrences being, in the United States, in New England, Arkansas, Montana, and New York State (syenite gneisses), and elsewhere, in Switzerland, Germany, and Norway.

SYLVANITE. A mineral, a telluride of gold and silver approximating the formula $AgAuTe_4$. Sylvanite is monoclinic, occurring in bladed, columnar, and granular forms as well as arborescent and branching. It is a brittle mineral; hardness, 1.5–2; specific gravity, 8.16; luster, metallic; color and streak, steel gray to yellowish-gray. This mineral is found associated with gold and tellurides of gold and silver or with sulfides such as pyrite. It is found in Rumania, Australia, Colorado and California. It was named for Rumanian Transylvania where it was first found.

Krennerite is another telluride of gold and silver with a similar composition to sylvanite, but crystallizing in the orthorhombic system. Calaverite is a gold telluride with only a small silver content.

SYLVITE. A mineral potassium chloride, KCl, occurring in cubes, or as cubes modified by octahedra. Sylvite is therefore isometric. It has a perfect cubic cleavage; uneven fracture; is brittle, hardness, 2; specific gravity, 1.9; luster, vitreous; colorless when pure but may be white, bluish, yellowish, or reddish due to impurities. It is soluble in water. It is much rarer than halite and has been found as sublimates at Mt. Vesuvius and as bedded deposits at Stassfurt, Germany. Extensive deposits occur in sedimentary deposits in the Permian basin of southwestern New Mexico, near Carlsbad in the United States.

It is used as a source of potash salts. Potassium chloride was called by the early chemists *sal digestivus Sylvii*, whence the name of the mineral.

SYNDETS. Detergents.

SYNERESIS. Colloid System.

SYNTHESIS (Chemical). The process of building chemical compounds through a planned series of steps (reactions, separations, etc.). Synthesis usually is the method of choice (1) when the desired compound is not present in natural materials from which it can be isolated, (2) when the compound cannot be easily obtained from reacting readily available materials in a few simple steps; and (3) although a compound may be available within a natural complex, the economics of separation and purification are prohibitive, or often in the case of biochemicals, too little natural raw material is available to meet the demand.

Even more important, synthesis plays a key role in developing new, untried chemical structures which, on paper, appear to have properties that may be of great value, e.g., a new synthetic material, a new drug, or a new fuel. Chemicals by design from prior knowledge of related materials generally are created via the route of synthesis. Further, synthesis is fundamental to broadening the base of chemical knowledge. Sometimes unexpected results occur, i.e., compounds with unusual, unexpected, and often desirable practical chemical and/or physical properties.

Because of the hundreds of thousands of organic substances (already established, but many yet remaining to be "built"), organic synthesis predominates. Most of the synthetics (elastomers, fibers, and other polymers, coatings, films, adhesives, and numerous other products) that have appeared during the last 30–40 years resulted from research involving organic synthesis. Some of the early work in organic synthesis dealt with the creation of certain fatty acids and ketones. A few examples are given to provide an insight into the workings of synthesis.

In the following examples, only the main starting ingredients and products are shown. No attempt is made to indicate byproducts or the conditions of the reactions involved:

A. Target compound: Ethylpropylacetic acid,
$(C_2H_5)(C_3H_7)CH \cdot COOH$
1. Acetic anhydride $\longrightarrow$ ethyl acetate (+ alcohol)
2. Ethyl acetate $\longrightarrow$ ethyl acetoacetate (sodium + dilute acids)
3. Ethyl acetoacetate $\longrightarrow$ sodium derivative of ethyl acetoacetate (+ sodium ethoxide)
4. Sodium derivative of ethyl acetoacetate $\longrightarrow$ ethyl ethylpropyl acetoacetate (+ propyl iodide)

5. Ethyl ethylpropyl acetoacetate ⟶ ethylpropylacetic acid (concentrated alcohol and potash)

B. Target compound: Butyl acetone, $CH_3 \cdot CO \cdot CH_2 \cdot C_2H_4$

   1. 1–3 same as given in A

   4. Sodium derivative of ethyl acetoacetate ⟶ ethylbutylpropyl acetoacetate (+ butyl iodide)

   5. Ethylbutylpropyl acetoacetate ⟶ butyl acetone (+ dilute alcohol and potash)

C. Target compound: n-valeric acid, $CH_3 \cdot CH_2 \cdot CH_2CH_2COOH$

   1. Potassium chloroacetate ⟶ potassium cyanoacetate (+ potassium cyanide)

   2. Potassium cyanoacetate ⟶ ethyl malonate (+ alcohol and hydrogen chloride)

   3. Ethyl malonate ⟶ sodium derivative of ethyl malonate (+ sodium ethoxide)

   4. Sodium derivative of ethyl malonate ⟶ ethylpropyl malonate (+ propyl iodide)

The compounds on the right-hand side of intermediate reactions are often called *intermediates*. See also **Intermediate (Chemical).**

Some of the notable syntheses from the early history of the technique include:

Inorganic Syntheses

1746	Sulfuric acid (chamber process)
1800	Soda ash (Le Blanc process)
1861	Soda ash (Solvay process)
1890	Sulfuric acid (contact process)
1912	Ammonia (Haber-Bosch process)

Organic Syntheses

1828	Urea (Wohler)
1857	Mauveine (Perkin)
1869	Celluloid (Hyatt)
1877	Ethylbenzene (Friedel-Crafts)
1884	Rayon (Chardonnet)
1910	Phenolic resins (Baekeland)
1910	Neoarsphenamine (Ehrlich)
1920	Aldehydes, alcohols (Oxo synthesis)
1925	Insulin (Banting)
1927	Methanol
1930	Neoprene (Nieuwland)
1935	Nylon (Carothers)
1940	Styrene-butadiene rubber
1950	Polyisoprene

SYNTHESIS GAS. For a number of industrial organic syntheses that proceed in the gaseous phase, it is advantageous to prepare a chargestock to specification. When a mixture of gases is so prepared, the term *synthesis gas* is often used. Thus, there are several mixtures which qualify under this definition: (1) a mixture of H_2 and N_2 used for NH_3 synthesis; (2) a mixture of CO and H_2 for methyl alcohol synthesis; and (3) a mixture of CO, H_2, and olefins for the synthesis of oxo-alcohols. Ammonia synthesis gas is described briefly here.

The hydrogen required for NH_3 synthesis gas may be obtained in commercial quantities from coke oven water gas; from steam reforming of hydrocarbons; from the partial oxidation of hydrocarbon chargestocks; or from the electrolysis of H_2O. The nitrogen required may come from the introduction of air to the process, or where specifically required, pure nitrogen may be obtained from air separation plant. Since NH_3 synthesis occurs under high pressure, it is advantageous to generate the synthesis gas at high pressure and thus avoid additional high compression costs. For this and other economic situations, coke oven gas and hydrogen from electrolysis are eliminated. This leaves hydrocarbons as the logical choice.

In the steam-hydrocarbon reforming process, steam at temperatures up to 850°C and pressures up to 30 atmospheres reacts with the desulfurized hydrocarbon feed, in the presence of a nickel catalyst, to produce H_2, CO, CO_2, CH_4, and some undecomposed steam. In a second process stage, these product gases are further reformed. Air also is added at this stage to introduce nitrogen into the gas mixture. The exit gases from this stage are further purified to provide the desired 3 parts H_3 to 1 part N_2 which is the correct empirical ratio for NH_3 synthesis. See also **Ammonia.**

T

TACHYLYTE (or Tachylite). Pure tachylite is a natural, basic black glass which may form along the chilled contacts of dikes or sills. It also occurs as a rind on basic pillow lavas which have been suddenly chilled by plunging into water. Occasionally it forms entire flows from certain Hawaiian volcanoes.

TACONITE. Term for the ferruginous cherts associated with the iorn ores of Lake Superior district.

TALC. The mineral talc is a magnesium silicate corresponding to the formula $Mg_3Si_4O_{10}(OH)_2$ which occurs as foliated to fibrous masses, its monoclinic crystals being so rare as to be almost unknown. It has a perfect basal cleavage, the folia non-elastic although slightly flexible; it is sectile and very soft; hardness, 1; specific gravity, 2.5–2.8; luster, waxlike or pearly; color, white to gray or green; translucent to opaque. It has a distinctly greasy feel. Talc is a metamorphic mineral resulting from the alteration of silicates of magnesium like pyroxenes, amphiboles, olivine and similar minerals.

It is found chiefly in the metamorphic rocks, often those of a more basic type due to the alteration of the minerals above mentioned. Some localities are the Austrian Tyrol, the St. Gotthard district of Switzerland, Bavaria and Cornwall, England. In Canada talc is found in Brome County, Quebec and Hastings County, Ontario. In the United States well-known localities are to be found in Vermont, New Hampshire, Massachusetts, Rhode Island, New York, Pennsylvania, Maryland and North Carolina.

A coarse grayish-green talc rock has been called soapstone or steatite and was formerly much used for stoves, sinks, electrical switchboards, etc. Talc finds use as a cosmetic, for lubricants and as a filler in paper manufacturing. Most tailor's "chalk" consists of talc. The origin of the word talc is not definitely known.

TANTALITE. This black mineral (Fe, Mn)(Ta, Nb)$_2$O$_6$, is isomorphous with columbite and dimorphous with tapiolite. Tantalite occurs in pegmatites and is a principal ore of tantalum.

TANTALUM. Chemical element symbol Ta, at. no. 73, at. wt. 180.948, periodic table group 5b, mp 1996°C, bp 5427°C, density, 16.63 g/cm³ (solid at 20°C), 17.1 g/cm³ (single crystal). Elemental tantalum has a body-centered cubic crystal structure. Because of its high mp, it is considered a refractory metal. Tantalum is a slightly bluish metal, ductile, malleable, and when polished resembles platinum; burns upon heating in air; insoluble in HCl or HNO₃, but soluble in hydrofluoric acid, or in a mixture of hydrofluoric acid and HNO₃. The tough, impermeable oxide film formed on the metal when exposed to air makes tantalum the most resistant of all metals to atmospheric corrosion.

Tantalum was first identified by Ekeberg as a new element in yttrium minerals in 1802 and was first obtained in pure from by Berzelius in 1820 by heating potassium tantalofluoride with potassium. There is one naturally occurring stable isotope,

^{181}Ta. Isotope ^{180}Ta also occurs naturally (isotopic abundance of 0.012%), with a half-life of something greater than 10^7 years. At least nine other radioactive isotopes have been identified, ^{176}Ta through ^{179}Ta, and ^{182}Ta through ^{186}Ta. With exception of ^{179}Ta ($t_{1/2}$ = about 600 days), the remaining half-lives are expressed in minutes, hours, or days. Isotope ^{182}Ta has been used as a source of gamma rays. See **Radioactivity.** In terms of abundance, tantalum does not appear on the list of the first 36 elements that occur in the earth's crust and hence is relatively scarce. Also, tantalum does not appear on the list of the first 65 elements that are found in seawater.

Electron configuration is $1s^22s^22p^63s^23p^63d^{10}4s^24p^64d^{10}4f^{14}5s^25p^65d^36s^2$. First ionization potential, 7.7 eV. Oxidation potential: $2Ta + 5H_2O \rightarrow Ta_2O_5 + 10H^+ + 10e^-$, 0.71 V. Other physical properties of tantalum are given under **Chemical Elements.**

Tantalum is found in a number of oxide minerals, which almost invariably also contain niobium (columbium). The most important tantalum-bearing minerals are tantalite and columbite, which are variations of the same natural compound (Fe,Mn)(Ta, Nb)$_2$O$_6$. Much of the tantalum concentrates have been obtained as byproducts of tin mining; in recent years, tin slags, which are byproducts of smelting cassiterite ores, such as those found in the Republic of Congo, Nigeria, Portugal, Malaya, and Thailand, have been an important raw material source for tantalum.

The first successful industrial process used to extract tantalum and niobium from the tantalite-columbite containing minerals employed alkali fusion to decompose the ore, acid treatment to remove most of the impurities, and the historic Marignac fractional-crystallization method to separate the tantalum from the niobium and to purify the resulting K₂TaF₇. Most tantalum production now employs recovery of the tantalum and niobium values by dissolution of the ore or ore concentrate in hydrofluoric acid. Then the dissolved tantalum and niobium values are selectively stripped from the appropriately acidified aqueous solution and separated from each other in a liquid-liquid extraction process using methyl isobutyl ketone (MIBK) or other suitable organic solvent. The resulting purified tantalum-bearing solution is generally treated with potassium fluoride or hydroxide to recover the tantalum in the form of potassium tantalum fluoride, K₂TaF₇, or with ammonium hydroxide to precipitate tantalum hydroxide, which is subsequently calcined to obtain tantalum pentoxide, Ta₂O₅. Tantalum metal is generally obtained by sodium reduction of K₂TaF₇, although electrolysis of K₂TaF₇ and carbon reduction of Ta₂O₅ in an electric furnace have also been used. Tantalum metal can absorb large volumes of hydrogen during heating in a hydrogen-bearing atmosphere at an intermediate temperature range (450–700°C). The hydrogen is readily removed by heating in vacuum at higher temperatures.

Uses. Tantalum is used widely, although in small quantities, in the electronics industry for electrolytic capacitors, emitters, and getters. The corrosion resistance of tantalum has been compared with that of glass. Additionally, the metal has a high heat-transfer coefficient and is easy to fabricate. Consequently, tantalum finds use in equipment that must resist strong corrosive attack, as in the manufacture of hydrochloric acid, hydrogen peroxide, in chromium plating baths, in bromine heaters and stills, and in the preparation of corrosive fine chemicals, such as ethyl bromide. The metal also has been used in resistance heaters in very high-temperature furnaces and for some nuclear reactor parts.

Alloys. Tantalum is added to nickel and nickel-cobalt superalloys for gas-turbine and jet-engine parts. Several surgical applications for tantalum have developed because of the inertness of the metal to body fluids and the tolerance of the body for the metal. Tantalum may be placed in the skull or other body parts without rejection. Strips and screws made of tantalum are used for holding broken pieces of bone and tantalum wire mesh is used for surgical staples, braid for sutures, and reinforcements. Tantalum-base alloys are used for aerospace structures and space power systems, principally because of the high-temperature stability and strength of these alloys. They operate satisfactorily at temperatures in excess of 1,600°C. Small additions of zirconium to tantalum increases its tensile strength at normal temperatures and up to approximately 1,200°C. Also, when added in amounts of about 5%, hafnium, molybdenum, rhenium, tungsten, and vanadium also increase the strength of tantalum. The tensile strength of ternary alloys of tantalum (Ta with 30% Nb and 5% Zr or V) at room temperature is about 3X that of tantalum alone. A tantalum-tungsten alloy is used for fabricating springs for high temperature and high vacuum applications.

Chemistry and Compounds. As might be expected from its $5d^36s^2$ electron configuration, tantalum forms pentavalent compounds. In fact, they constitute the great majority of tantalum compounds, although the valences 2, 3, and 4 are known. However, the existence of the Ta^{5+} ion is very brief, since it readily coordinates with water, OH^-, and other anions or molecules. Tantalum is very resistant to chemical action, not being attacked by acids other than hydrofluoric acid, and by alkalis only upon fusion. Even fluorine and oxygen react only upon heating.

Tantalum pentoxide, formed by heating the metal with oxygen, reacts with hydrofluoric acid, alkali bisulfates or alkali hydroxides, forming tantalates with the latter. It reacts with a number of halogen compounds to give tantalum pentafluoride, pentachloride and pentabromide, TaF_5, $TaCl_5$, and $TaBr_5$. (Carbon tetrachloride is often used in this preparation of $TaCl_5$.) These compounds readily undergo hydrolysis, and may form oxyhalides, such as TaO_2F and $TaOBr_3$. They may be reduced, but with difficulty, $TaCl_5$ when heated with aluminum yielding the tetrachloride, $TaCl_4$. The trihalides, $TaCl_3$ and $TaBr_3$ have also been prepared. Tantalum(V) fluoride combines with other fluorides, notably the alkali metal fluorides, to yield complexes, such as K_2TaF_7 and Na_3TaF_8.

Other complexes of tantalum(V) are formed with oxygen-function compounds, such as o-dihydroxybenzene and acetylacetone.

In addition to Ta_2O_5, another oxide is known, TaO_2, which may be formed by active-metal reduction (as is the tetrachloride), except that the pentoxide is heated with magnesium rather than aluminum. It forms with alkali metals the metatantalates, $MTaO_3$, the orthotantalates, M_3TaO_4, and pyrotantalates, $M_4Ta_2O_7$, as well as such polytantalates as $M_8Ta_6O_{19}$, the latter requiring fusion with the alkali hydroxides.

The only known sulfide, which is produced by heating with carbon disulfide, is TaS_2, but at least two nitrides are known, TaN and Ta_3N_5, the latter being unstable.

The organometallic compounds of tantalum all involve oxygen bonding, with the exception of a dicyclopentadienyl compound, $(C_5H_5)_2TaBr_3$. The others are alkoxy compounds, such as $(C_2H_5O)_3TaCl_2$, $Ta(OC_2H_5)_5$, $Ta(OCH(C_2H_5)CH_3)_5$, etc., with the exception of bis(fluorosulfonyloxy)trichlorotantalane, $Cl_3Ta(OS(O_2)F)_2$.

See the references listed at the end of the entry on **Chemical Elements.**

—M. Schussler, Senior Scientist, Fansteel, North Chicago, Illinois.

TAR ACIDS AND BASES. Coal Tar and Derivatives.

TAR SANDS. Also called *bituminous sands* and *oil sands*, tar sands comprise a huge petroleumlike energy resource. The heavy, viscous petroleum substances which impregnate the tar sands are called *asphaltic oils*. Other names sometimes used for these oils are *maltha*, *brea*, and *chapapote*. The sands are composed of a mixture of 84–88% sand and mineral-rich clays, 4% water, and 8–12% bitumen. Bitumen is a dense, sticky, semisolid that is about 83% carbon. The substance does not flow at room temperature and is heavier than water. At higher temperatures, it flows freely and floats on water. Characteristics of tar sands important to mining, recovery, and processing include grain size, composition, sortability, porosity, permeability, and microscopic habitat.

Tar sands have been identified in a number of areas worldwide. The most important of these include the Athabasca deposit in northern Alberta, Canada; a deposit on the north bank of the Orinoco River in Venezuela; and a number of locations in the United States and Mexico, such as California and Utah.

Probably the greatest commercial efforts made to date to economically recover useful fuels from tar sands have been conducted by the Alberta Research Council in connection with the Athabasca deposits. The high-grade oil-bearing sands in Athabasca are mainly very fine to fine-grained (62.5–250 micrometers). However, coarser sands and local conglomerates do occur. Shale beds within the oil sands are composed of silt and clay materials and very infrequently contain an appreciable amount of oil. The mineralogy of tar sands, particularly of those found in the Athabasca fields of Canada, is extremely mature and stable, about 95% quartz grains, 2–3% mica flakes and clay minerals, and traces of other materials. The hardness of quartz and the accompanying abrasiveness of the sands causes severe wear on excavation, mining, and processing equipment. The clay minerals, predominantly kaolinte and illite present, tend to decrease the efficiency of extraction processes. Athabasca tar sands are reasonably well sorted, i.e., a large percentage of the grains are about the same size, contributing to ease of recovering and processing. The porosity of highgrade tar sands lies between 25 and 35%. This is considerably higher than most petroleum reservoir sandstones (5–20% on average). The sands contain no cementing agent to indurate the material and give it strength (the principal reason the material is called sand). This characteristic assists in handling, but experience has shown that precautions are required to prevent production of sand grains along with oil.

The permeability of many oil sands is quite high. Fluids can easily be transmitted through the sands. This is desirable because if the bitumen can be rendered less viscous, transmittal of fluids through the pore system to a production well can be facilitated. Saturation, a measure of the extent to which voids in a sediment are filled by various fluids, is relatively low in tar sands (about 36% by weight). The grains in tar sands generally are hydrophilic. The oil in the pores does not maintain direct contact with mineral grains, but rather each grain is surrounded by a thin film of water. This characteristic greatly facilitates the hot water extraction process.

In terms of processing and of final product quality and yield, tar sands are considered rich if they contain in excess of 10% bitumen by weight; moderate if bitumen is 6–10%; and lean if bitumen content is less than 6% by weight. Viscosity of bitumen as it occurs naturally determines relative ease or difficulty of recovery and affects subsequent processing options. The greatest cost is in the mining of the tar sands.

The hot-water extraction process has been the major approach used and dates back to the 1920s. The bituminous sand is passed through a conditioning drum where make-up water, steam, and caustic soda are added to adjust the solids content to about 70%, the temperature to 82–88°C, and the pH to 8.0–8.5. The resulting pulp is passed over a screen to remove oversize material and then into a primary separation cell where the oil floats to the top in the form of froth and the sand falls to the bottom and is removed, while part of the middlings are recycled and part drawn off to be treated by air flotation in scavenger cells. The bitumen froths from the separation cell and scavenger cells are combined and diluted with naphtha, and the fine mineral matter and water are removed in a two-stage centrifuge operation, after which the naphtha is recovered and recycled. The recovered bitumen is then subjected to an upgrading step to produce a marketable crude oil substitute.

Major deficiencies of bitumen extracted from the Alberta oil sands are low ratio of hydrogen to carbon and high sulfur content. Hydroprocessing and desulfurization are required to produce an acceptable synthetic crude oil. Four primary upgrading schemes have been evaluated, including delayed coking, fluid coking, residual hydrocracking, and solvent deasphalting. In-situ processing also has been considered for a number of years. In this approach, the objective is to soften the bitumen sufficiently so that it can be pumped to the surface. In an interwell scheme, a central injection well is created and surrounding this well are perhaps six adjacent wells some 60–150 meters from the central well. If there are no natural communications between the wells, artificial fracturing is required. In one version, steam injected into the central well moves outward toward the surrounding wells. Heated bitumen flows by gravity through the porous medium to collection points at the base of the surrounding collection walls. In another version, the formation is ignited in the vicinity of the injection well, with air supplied to support combustion and water to create a head of steam. Some bitumen is consumed in this procedure, but the majority of the bitumen is softened and flows in the manner previously described.

References

Abraham, H.: "Asphalts and Allied Substances," Vol. I, Van Nostrand Reinhold, New York, 1960.

Carrigy, M. S.: "Alberta Oil Sands," *Geoscience Canada*, **1**, 41–44 (1974).

Dickie, B., and M. Carrigy: "Fuel from Tar Sands," in "Energy Technology Handbook," (D. M. Considine, Editor), McGraw-Hill, New York, (1977).

Maugh, T. H.: "Tar Sands," *Science*, **199**, 756–760 (1978).

Maugh, T. H.: "U.S. Oil Sands," *Science*, **207**, 1191–1192 (1980).

Mossop, G. D.: "Geology of the Athabasca Oil Sands," *Science*, **207**, 145–152 (1980).

Redford, D. A., and A. G. Sinestock, Editors: "The Oil Sands of Canada–Venezuela," Spec. Vol. 17, Canadian Institute of Mining and Metallurgy, Montreal, Canada, 1978.

Stewart, G. A., and G. T. MacCallum: "Athabasca Oil Sands Guidebook," Canadian Society of Petroleum Geologists, Calgara, Alberta, Canada, 1978.

—Alberta Oil Sands Technology and Research Authority, Calgary, Alberta, Canada.

TARTARIC ACID. $(CHOHCO_2H)_2$, formula weight 150.09, white crystalline solid with four physical isomers, three of which are optically active: (1) dextro- and (2) levotartaric acid, both with same mp 168–170°C and sp gr 1.760, (3) racemic acid (dextrolevo), mp 205–206°C, sp gr 1.697, and (4) mesotartaric acid (inactive), mp 159–160°C, sp gr 1.737. Racemic acid crystallizes with one molecule of H_2O. All forms decompose before reaching the boiling point at atmospheric pressure. All forms are soluble in H_2O, slightly soluble in alcohol, and essentially insoluble in ether. Tartaric acid is a primary example of optical isomerism and one of the earliest compounds studied in this regard. Tartaric acid is a dibasic acid with two series of salts and esters.

Tartrates (like citrates) in solution change silver of ammonio-silver nitrate into metallic silver. Potassium hydrogen tartrate and calcium tartrate, on account of their solubility characteristics, are of importance in the separation and recovery of tartaric acid. The former salt is readily converted into the latter, and the resulting calcium tartrate plus dilute sulfuric acid yields tartaric acid plus calcium sulfate, and the latter may be separated by filtration. Tartaric acid may be obtained by evaporation of the filtrate. Ester: Diethyl tartrate $COOC_2H_5(CHOH)_2COOC_2H_5$, melting point 17°C, boiling point 280°C. Tartaric acid may be obtained (1) from some natural products, e.g., in the juice of grapes and acid fruits, often in conjunction with citric or malic acid; potassium hydrogen tartrate, "argol," in the residue of wine vats, (2) by synthesis.

Tartaric acid is used (1) in baking powders as potassium hydrogen tartrate ("cream of tartar") with sodium bicarbonate, (2) in medicine, e.g., potassium antimonyl tartrate ("Tartar emetic"), (3) in effervescent medicinal salts, (4) in blue printing as ferric tartrate, (5) in silvering mirrors—ammonio-silver nitrate yielding a smooth deposit of silver. Sodium potassium tartrate ("Rochelle salt," $NaKC_4H_4O_6 \cdot 4H_2O$) is used in medicine, and in the preparation of Fehling's solution, which is an alkaline cupric solution made by mixing copper sulfate solution, sodium potassium tartrate solution and sodium hydroxide solution, and is used as an oxidizing reagent in the case of many organic compounds, such as glucose and reducing sugars, and aldehydes, with which cuprous oxide, red to yellow precipitate, is formed.

See also **Isomerism**.

TAU PARTICLE. Discovered in 1975, the tau particle is a lepton with a mass of 1.8 GeV, almost twice that of the proton. Like other leptons, the tau particle is considered as pointlike. See **Particles (Subatomic).**

TCA CYCLE. Carbohydrates.

TECHNETIUM. Chemical element symbol Tc, at. no. 43, at. wt. 98.906, periodic table group 7b, mp 2171–2172°C, bp 4875–4877°C. The element does not occur in nature. The present location of technetium in the periodic table was vacant for many years, during which time several claims to having found the element were made, but never confirmed. One such claimant termed the element *measurium*. Technetium has been detected in certain stars and this discovery still remains to be resolved with current theories of stellar evolution and element synthesis.

The first isotope, ^{97}Tc, to be isolated was extracted by Perrier and Segre from molybdenum which had been bombarded with deuterons in the Berkeley cyclotron. The reaction was ^{96}Mo(d, n)^{97}Tc. The isotope with the longest half-life, ^{99}Tc ($t_{1/2}$ = 2.12 × 10^5 years) is found in relatively large amounts among the fission products of uranium. It is also produced by neutron irradiation of ^{98}Mo by the reaction ^{98}Mo(n, γ)^{99}Mo(β-decay)$^{99\,m}$Tc(isomeric transition)^{99}Tc. Significant quantities have been isolated and even larger quantities could be made available if applications for it were developed.

One proposal concerns a method for recovering technetium from nuclear fuel reprocessing waste solutions. ^{99}Tc has found some application in diagnostic medicine. Ingested soluble technetium compounds tend to concentrate in the liver and are valuable in labeling and in radiological examination of that organ. However, the more ideal nuclear properties of ^{99m}Tc have led to greater interest in medical diagnostics. By technetium labeling of suitable compounds (or blood serum components), diseases involving the circulatory system and organs other than the liver can be explored.

In all, to date, sixteen isotopes of technetium have been reported, with mass numbers 92–105, 107, and 108.

Superconductivity has been observed in technetium metal and in alloys based on technetium with additions of Pd, Os, Rh, Ru, Sn, V, Ti, Re, W, or C.

A study of the chemistry of technetium shows it to have, as expected, properties intermediate between those of its homologues manganese and rhenium, the resemblance to the latter being perhaps greater than to the former. Like rhenium, technetium apparently exists in (IV), (VI), and (VII) oxidation states. Pure technetium metal has been prepared by passing hydrogen gas at 1,000°C over the sulfide obtained by precipitation with H_2S from HCl solution. The metal has been shown to have the same crystal structure as rhenium and the adjacent elements osmium and ruthenium. Among its compounds are the ditechnetium heptasulfide, Te_2S_7, readily precipitated by H_2S from oxidized solutions, the corresponding oxide, Tc_2O_7 produced directly from the elements at higher temperatures, which reacts with NH_3 to form ammonium pertechnate, NH_4TcO_4, and the hexachloro complex ion, $TcCl_6^{2-}$, which like the corresponding rhenium ion, has a magnetic moment corresponding to three unpaired electron spins.

References

Cowan, G. A., and W. C. Haxton: "Solar Neutrino Production of Technetium-97 and Technetium-98," *Science*, **216**, 51–54 (1982).
Deutsch, E., et al.: "Heart Imaging with Cationic Complexes of Technetium," *Science*, **214**, 85–86 (1981).
Deutch, E. T.: "An Overview of Technetium Chemistry in Radiopharmaceuticals," American Pharmaceutical Assn. Annual Meeting, Washington, D.C. (April 1980).
Eckelman, W. C., and S. M. Levenson: "Pharmaceuticals Labeled with Technetium," *Int. J. Appl. Radiat. Isot.*, **28** (1–2), 67–82 (1977).
Horwitz, E. P., and W. H. Delphin: "Methods for Recovering Palladium and Technetium Values from Nuclear Fuel Reprocessing Waste Solutions," U.S. Patent (Application) (December 28, 1977).
Van Galenstraat, J.: "The Chemistry of Technetium and Rhenium," Elsevier, Amsterdam (1966).

—Robert Q. Barr, Director, Technical Information, Climax Molybdenum Company, Greenwich, Connecticut.

TEKTITE. A small (usually walnut-size), rounded, pitted, jet-black to olive-greenish or yellowish body of silicate glass of nonvolcanic origin, found usually in groups in several widely separated areas of the earth's surface and apparently bearing no relation to the associated geologic formations. Most tektites have uniformly high silica (68–82%) and very low water contents (average, 0.005%). Their composition is unlike that of obsidian and more like that of shale. They have various shapes, strongly suggesting modeling by aerodynamic forces and they average a few grams in weight. The largest found weighs 3.2 kilograms. Some authorities believe that tektites are of extraterrestrial origin, or alternatively the product of large hypervelocity meteorite impacts on terrestrial rocks. The term was proposed by Suess in 1900 who believed they were meteorites which at one time had undergone melting.

TELLURIUM. Chemical element symbol Te, at. no. 52, at. wt. 127.60, periodic table group 6a, mp 450–452°C, bp 1387–1393°C, density, 6.24 g/cm³ (crystalline form at 25°C), 6.00 (amorphous form at 25°C). Elemental tellurium has a hexagonal crystal structure with trigonal symmetry. Tellurium is a silvery-white brittle semimetal, stable in air, and in boiling water, insoluble in HCl, but dissolves in HNO_3, or aqua regia to form telluric acid. The element is dissolved by NaOH solution and combines with chlorine upon heating to form tellurium tetrachloride.

In observing a peculiar phase in gold ores of the Transylvania region, Franz Müller von Reichenstein first identified the element in 1782. There are several natural occurring isotopes ^{120}Te, ^{122}Te through ^{126}Te, ^{128}Te, and ^{130}Te. Nine radioactive isotopes have been identified ^{118}Te, ^{119}Te, ^{121}Te, ^{123}Te, ^{127}Te, ^{129}Te, and ^{131}Te through ^{133}Te. With exception of ^{123}Te which has a half-life something greater than 10^{13} years, all of the other radioactive isotopes have half lives measurable in terms of minutes, hours, or days.

In terms of abundance, tellurium does not appear on the list of the first 36 elements that occur in the earth's crust and hence is relatively scarce. Terrestrial abundance is estimated on the order of 0.002 ppm. Tellurium is found in seawater to the estimated extent of about 95 pounds per cubic mile (10.3 kilograms in one cubic kilometer) of seawater.

Electronic configuration is $1s^22s^22p^63s^23p^63d^{10}4s^24p^64d^{10}5s^25p^4$. First ionization potential, 9.01 eV; second, 18.6 eV; third, 30.5 eV; fourth, 37.7 eV; fifth, 59.95 eV. Oxidation potentials: $H_2Te(aq) \rightarrow Te + 2H^+ + 2e^-$, 0.69 V; $Te + 2H_2O \rightarrow TeO_2(s) + 4H^+ + 4e^-$, −0.529 V; $TeO_2(s) + 4H_2O \rightarrow H_6TeO_6(s) + 2H^+ + 2e^-$, −1.02 V; $Te^{2-} \rightarrow Te + 2e^-$, 0.92 V; $Te + 6OH^- \rightarrow TeO_3^{2-} + 3H_2O + 4e^-$, 0.02 V; $Te \rightarrow Te^{4+} + 4e^-$, −0.564 V.

Tellurium occurs chiefly as telluride in gold, silver, copper, lead, and nickel ores in Colorado, California, Ontario, Mexico, and Peru, and, infrequently, as free tellurium and tellurite (tellurium dioxide, TeO_2). The anode mud from copper and lead refineries or the flue dust from roasting telluride gold ores is treated by fusion with sodium nitrate and carbonate and the melt extracted with water. The resulting solution is acidified carefully with sulfuric acid, whereupon tellurium dioxide is precipitated; the dioxide is reduced to free tellurium by heating with carbon.

Uses. On the scale of most other commercial metals, the production of elemental tellurium is relatively limited—approximately 0.5 million pounds (2.27 million kilograms) annually. Commercial tellurium is marketed at a purity of about 99.7% although much purer forms are obtainable (up to 99.999%). The application of tellurium and tellurium compounds as catalysts is expanding. Small quantities are used in various electronic components, including solar cells, infrared detectors, emitters, and thermoelectric generators. Tellurium also is sometimes used as a dopant for semiconductor devices. The metal has been used in primer fuses for explosives. The main applications have been in metallurgy. Small additions of tellurium improve the machinability of low-carbon steels, stainless steels, and copper. The metal stabilizes the carbide in cast irons. Tellurium also helps to control pinhole porosity in steel castings. The very small addition (0.05%) of tellurium to lead improves a number of the physical properties of lead sheet, foil, and other shapes. To some extent, tellurium has been used as a curing agent and accelerator in rubber compounds.

Chemistry and Compounds. Tellurium occurs in the same periodic classification as sulfur, selenium, and polonium. Tellurium, unlike sulfur and selenium, has only two allotropic forms. Due to its $5s^25p^4$ electron configuration, tellurium, like sulfur and selenium, forms many divalent compounds with covalent bonds and two lone pairs, and d-hybridization is quite common, to form compounds with tellurium oxidation states of $+4$ and $+6$.

Tellurium dioxide, TeO_2, made directly from the element or by heating tellurous acid, H_2TeO_3, is a solid, subliming at 450°C, insoluble in H_2O, which dissolves in acids and alkalis, exemplifying the increasing metallic character with atomic weight of the main group 6 elements. Tellurium dioxide accepts a proton from strong acids to form the ion $TeOOH^+$. Dehydration of telluric acid at 400°C produces TeO_3, tellurium trioxide. It is not nearly as reactive as sulfur trioxide and selenium trioxide, but reacts with alkali hydroxides to form tellurates.

Tellurous acid, H_2TeO_3, can exist only in very dilute aqueous solutions (due to insolubility of TeO_2). It is a weak acid (ionization constants 2×10^{-3} and 2×10^{-8}). The salts of tellurous acid, the tellurites, may often be formed by reaction of TeO_2 with metal salts. Telluric acid, H_6TeO_6, is prepared by oxidation of tellurium with strong oxidizing agents, such as 30% hydrogen peroxide or boiling HNO_3 and catalyst. Various values of the ionization constants of telluric acid have been reported, on the order of 10^{-7} and 10^{-11}, but the best values would appear to be $pK_{A1} = 7.7$, $pK_{A2} = 11.0$, $pK_{A3} = 14.5$. Telluric acid is a quite strong oxidizing agent, forming halogens from hydrohalides in solution (except hydrogen fluoride). The alkali metal tellurates have the composition $M_2H_4TeO_6$, although metal tellurates with all H's replaced exist, such as Hg_3TeO_6 and Zn_3TeO_6.

Hydrogen telluride is a stronger acid than H_2S (ionization constants 2.27×10^{-3} and 10^{-11} (?) at 18°C) and is less readily obtained from tellurides than hydrogen sulfide from sulfides. Aluminum telluride, Al_2Te_3, requires heating with H_2O or dilute acids. In general the metal tellurides are prepared by direct combination of the elements. Those of the transition metals and the zinc, gallium and germanium families exhibit many instances of both well defined compounds and non-daltonide compositions, as well as substitutional solid solutions. Also six intermediate phases are found in the palladium-tellurium system, Pd_4Te, Pd_3Te, $Pd_{2.5}Te$, Pd_2Te, $PdTe$, and $PdTe_2$.

Tellurium hexafluoride, TeF_6, the only clearly defined tellurium hexahalide, is formed directly from the elements at 150°C, while at 0°C the product is mainly the decafluoride, Te_2F_{10}. TeF_6 is a relatively weak Lewis acid, forming complexes with pyridine and other nitrogen bases. Tellurium tetrahalides of all four halogens exist, the TeF_4 formed from $TeCl_2$ and fluorine, the $TeCl_4$ from $TeCl_2$ and chlorine, the $TeBr_4$ from bromotrifluoromethane, CF_3Br, and molten tellurium, and the TeI_4 directly from the elements. Tellurium dichloride, $TeCl_2$, is prepared by passing dichlorodifluoromethane, CF_2Cl_2, over molten tellurium. $TeCl_2$ is quite reactive, disproportionating to tellurium and $TeCl_4$, and useful in preparing other tellurium compounds. $TeBr_2$ is obtained by distillation of the mixture of tellurium and $TeBr_4$ obtained in the reaction between bromotrifluoromethane and tellurium. The tetrahalides of tellurium form many addition compounds with other halides.

Organotellurium compounds corresponding generally to those of sulfur and selenium are known. Although carbon ditelluride has not been prepared, COTe and CSTe have been.

Toxicity. Tellurium and compounds are toxic. Acceptable concentration limit for an 8-hour daily exposure to dust and fumes in air is 0.1 milligram of tellurium per cubic meter of air. Even exposure at this level may cause what is termed "garlic breath." Proper ventilation, appropriate hygienic practices, and good housekeeping should be observed in handling tellurium. Although elemental tellurium causes no apparent problems in contact handling, skin contact with soluble tellurium compounds must be avoided.

References

Bagnall, K. W.: "Chemistry of Selenium, Tellurium, and Polonium," Elsevier, Amsterdam (1966).

Carapella, S. C., Jr.: "Tellurium" in "Metals Handbook," Vol. 2, 9th Edition, 806–807, American Society for Metals, Metals Park, Ohio (1979).

Chizhikov, D. M., and V. P. Schastliviy: "Tellurium and Tellurides," (Translated from Russian by E. M. Elins), Collet's, Wellingborough, Northants, England (1970).

Cooper, W. C., Editor: "Tellurium," Van Nostrand Reinhold, New York (1971).

Sax, N. I.: "Dangerous Properties of Industrial Materials," 5th Edition, Van Nostrand Reinhold, New York, 1979.

—S. C. Carapella, Jr., ASARCO Incorporated, South Plainfield, New Jersey.

TEMPERATURE SCALES AND STANDARDS. Temperature is broadly defined as the degree of hotness or coldness of a body or an environment. In a narrower sense, temperature is the degree of hotness or coldness referenced to a specific scale. Temperature is an intensive quantity independent of the size of the system. Knowledge of the temperature of two bodies makes possible the prediction of the direction of heat flow when they are brought into contact. Heat will flow from the body at the higher temperature. This behavioral mode is the basis of the second law of thermodynamics.

Temperature Scales

There are two approaches that can be used in realizing the thermodynamic temperature scale. One is based on a heat engine operating according to the Carnot cycle. The other is based on the behavior of a perfect gas. The heat engine approach is impractical and is considered only in the development of theory. The perfect gas method is practical and has been used extensively for fundamental measurements ever since the concept was developed by Lord Kelvin (Sir William Thomson, 1824–1907) in 1848. Since there is no perfect gas to be found in nature, real gases are used and corrections applied for their departure from perfect gas behavior. The constant-volume gas thermometer is used, but because of the complexities and very taxing experimental difficulties involved, such thermometers are only to be found in large university or governmental standardizing laboratories, such as the National Bureau of Standards.

From the outset the value of a degree on the thermodynamic scale was established by assigning a span of 100° to the temperature difference between freezing and boiling water. This made the value of a degree agree with the Celsius (then called centigrade) degree. The Celsius scale was already in common use in science. However, Kelvin recognized the significance of using only one fixed point to establish the scale. It was not until 1954 that his suggestion was officially adopted.

Although the thermodynamic concept is only about 100 years old, thermometer makers had been building instruments for many years before that. In the early days there were as many arbitrary scales as there were makers. The confusion so created has persisted down to the present time. Material progress in resolving this state of confusion has been made through adoption of the International Temperature Scale which is now in the fourth generation. These various scales and their development are discussed in more detail in the sections that follow.

Fahrenheit Scale. The first empirical scale of temperature developed is named for its inventor, Daniel Gabriel Fahrenheit (1686–1736), who described it in a paper published in 1724. Fahrenheit used the ice point, designated 32°, and the body temperature of a healthy man, designated 96°, as fixed points in making scales for his thermometers. The fundamental interval, ice point to steam point, turned out to be 180°.

In the 1724 report of his invention Fahrenheit referred incorrectly to a third temperature as a fixed point which he designated 0°. He attained it with a mixture of ice, water, and sal ammoniac or sea salt. The temperature realized with such a system is actually a function of the relative proportions of the three ingredients and of the type of salt used.

The Fahrenheit scale is used extensively in English-speaking countries for meteorological, medical, and industrial purposes. Temperatures are denoted by the term *degrees Fahrenheit* and by the symbol °F.

Reamur Scale. This scale evolved from the invention prior to 1730 by René-Antoine Ferchalt de Réaumur (1683–1757). The initial concept was to use diluted wine as the thermometric liquid and to designate the ice point as 1000 and the boiling point of the liquid as 1080. In time this was changed so that the fundamental interval, ice point to steam point, while still 80°, was based on an ice-point designation of 0° and a steam-point designation of 80°. At present the scale is little used except in the brewing and liquor industries. Temperatures are denoted by the term *degrees Réaumur* and by the symbol °R.

Celsius (Centigrade) Scale. In 1742, Anders Celsius of Uppsala University (Sweden) reported on the use of thermometers in which the fundamental interval, ice point to steam point, was 100°. Celsius designated the ice point as 100° and the steam point as 0°. Subsequently Christin (1743) in Lyon, France, and Linnaeus (1745) in Uppsala independently interchanged the designations. For many years prior to 1948 it was known as the centigrade scale. In 1948 by international agreement it was renamed in honor of its inventor and to avoid an inconsistent connotation. The scale is used worldwide in scientific work and in medicine and for general purposes in non-English-speaking countries. Temperatures are denoted as *degrees Celsius* and by the symbol °C.

Thermodynamic Celsius Scale. This scale, called originally the *thermodynamic centigrade scale*, was developed from thermodynamic considerations, but was extremely difficult to reproduce in practice. The constant-volume gas thermometer with its extensive and elaborate ancillary equipment was used.

By international agreement the temperatures of melting ice and condensing water vapor, both under the pressure of one standard atmosphere and numbered 0° and 100°, respectively, were chosen to define the scale. Temperatures are denoted by *degrees Celsius thermodynamic* and by the symbol °C (therm).

Thermodynamic Kelvin Scale. The currently accepted theoretical scale has been named for Lord Kelvin who first enunciated the principle on which it is based. Thermodynamic temperature is denoted by the symbol T and the unit is the kelvin, symbol K. The kelvin is the fraction 1/273.16 of the thermodynamic temperature of the triple point of water. The triple point is realized when ice, water, and water vapor are in equilibrium. It is the sole defining fixed point of the thermodynamic Kelvin scale and has the assigned value 273.16 K.

Rankine Scale. This scale is the equivalent of the thermodynamic Kelvin scale, but expressed in terms of Fahrenheit degrees. Thus, the temperature of the triple point of water on the Rankine scale, corresponding to 273.16 K, is very nearly 491.69°Rankine.

The International Practical Temperature Scale

Over the years international agreements have been made to adopt scales which reproduced the thermodynamic Celsius and Kelvin scales as closely as possible within the limits of the state of the art. The scales have been given various names. The first was the *normal thermometric scale* (1887) based upon the constant volume hydrogen gas thermometer with a fundamental interval of 100° between the fixed points of melting ice, 0°C, and of condensing water vapor, 100°C. Both points were realized at one standard atmosphere pressure. This scale, commonly referred to as the *international hydrogen scale*, was reproduced for international comparisons by means of four liquid-in-glass thermometers. The range was limited to −35 to +100°C.

Improvements in materials and techniques made possible extension of the range with simultaneous increase in accuracy attainable. 1927 marked the adoption of the international temperature scale covering the range from the boiling point of oxygen to that of luminous incandescent bodies. Values were assigned to six fixed points, along with specifications for interpolation instruments, the platinum resistance thermometer and the platinum 10% rhodium-platinum thermocouple. Radiation constants for optical pyrometry were also adopted. This was a practical or working scale and was intended to reproduce as closely as possible the thermodynamic centigrade scale. If, in practice, it was desired to establish temperatures on the thermodynamic scale, measurements would be made on the international temperature scale and experimentally determined difference corrections could then be applied. These differences, however, were generally so small as to be of significance only in work of the very highest order of accuracy.

In 1948 a revision was made and the designation *degree Celsius* in place of *degree centigrade* was officially adopted. The scale changed slightly from the 1927 version, but at higher temperatures. Then in 1954 it was decided to define the scale by only one fixed point as had been recommended by Lord Kelvin a hundred years previously. The triple point of water was chosen, 273.16 K. A text revision was issued in 1960 covering these various changes, but it did not affect the values on the 1948 scale by as much as the experimental error of measurement. The name of the scale was changed to *international practical temperature scale of 1948*.

In 1968, the *international practical temperature scale* (IPTS 68) was adopted. Two changes of major significance were made at that time. The scale was extended below the oxygen point to effect standardization in the range 10–90 K. An updating over the range −183 to +1063°C based upon more refined gas thermometer measurements made since 1927 was accomplished.

The fixed points for the 1968 scale are given in Table 1, which is reproduced from the approved English translation of the official text. Although the maximum change in the 0–100°C portion of the scale is only 0.01° Celsius, at higher temperatures the changes are much more significant. For example, at the gold point the value is changed by 1.4° Celsius. The radiation constant was assigned a new value to bring the optical pyrometer portion of the range into closer accord with the thermodynamic scale. Table 2 shows the approximate differences between IPTS 68 and IPTS 48.

Temperature Standards

IPTS 68 as well as its predecessor international scales provided specifications for the interpolation instruments to be used at other than the fixed points in realizing the scale in practice. These instruments are the platinum resistance thermometer for use up to the antimony point, 630.74°C, and the platinum 10% rhodium-platinum thermocouple for use from 630.74 to 1064.43°C, the gold point. Above the gold point optical pyrometry is used and the Planck radiation constant is defined in the text of the scale. The official text has been published in *Metrologia*, volume 5, no. 2, April 1969.

TABLE 1. DEFINING FIXED POINTS OF THE IPTS-68 [a]

Equilibrium state	ASSIGNED VALUE OF INTERNATIONAL PRACTICAL TEMPERATURE	
	T_{68}(K)	t_{68}(°C)
Equilibrium between the solid, liquid, and vapor phases of equilibrium hydrogen (triple point of equilibrium hydrogen)	13.81	−259.34
Equilibrium between the liquid and vapor phases of equilibrium hydrogen at a pressure of 33 330.6 N/m² (25/76 standard atmosphere)	17.042	−256.108
Equilibrium between the liquid and vapor phases of equilibrium hydrogen (boiling point of equilibrium hydrogen)	20.28	−252.87
Equilibrium between the liquid and vapor phases of neon (boiling point of neon)	27.102	−246.048
Equilibrium between the solid, liquid, and vapor phases of oxygen (triple point of oxygen)	54.361	−218.789
Equilibrium between the liquid and vapor phases of oxygen (boiling point of oxygen)	90.188	−182.962
Equilibrium between the solid, liquid, and vapor phases of water (triple point of water) [c]	273.16	0.01
Equilibrium between the liquid and vapor phases of water (boiling point of water) [b,c]	373.15	100
Equilibrium between the solid and liquid phases of zinc (freezing point of zinc)	692.73	419.58
Equilibrium between the solid and liquid phases of silver (freezing point of silver)	1235.08	961.93
Equilibrium between the solid and liquid phases of gold (freezing point of gold)	1337.58	1064.43

[a] Except for the triple points and one equilibrium hydrogen point (17.042 K) the assigned values of temperature are for equilibrium states at pressure $p_0 = 1$ standard atmosphere (101 325 N/m²). In the realization of the fixed points small departures from the assigned temperatures will occur as a result of the differing immersion depths of thermometers or the failure to realize the required pressure exactly. If due allowance is made for these small temperature differences, they will not affect the accuracy of realization of the scale.

[b] The equilibrium state between the solid and liquid phases of tin (freezing point of tin) has the assigned value of $t_{68} = 231.9681$°C and may be used as an alternative to the boiling point of water.

[c] The water used should have the isotopic composition of ocean water.

TABLE 2. APPROXIMATE DIFFERENCES ($t_{68} - t_{48}$) IN KELVINS BETWEEN THE VALUES OF TEMPERATURE GIVEN BY THE IPTS OF 1968 AND THE IPTS OF 1948

t_{68}, °C	0	−10	−20	−30	−40	−50	−60	−70	−80	−90	−100
−100	0.022	0.013	0.003	−0.006	−0.013	−0.013	−0.005	0.007	0.012		
−0	0.000	0.006	0.012	0.018	0.024	0.029	0.032	0.034	0.033	0.029	0.022

t_{68}, °C	0	10	20	30	40	50	60	70	80	90	100
0	0.000	−0.004	−0.007	−0.009	−0.010	−0.010	−0.010	−0.008	−0.006	−0.003	0.000
100	0.000	0.004	0.007	0.012	0.016	0.020	0.025	0.029	0.034	0.038	0.043
200	0.043	0.047	0.051	0.054	0.058	0.061	0.064	0.067	0.069	0.071	0.073
300	0.073	0.074	0.075	0.076	0.077	0.077	0.077	0.077	0.077	0.076	0.076
400	0.076	0.075	0.075	0.075	0.074	0.074	0.074	0.075	0.076	0.077	0.079
500	0.079	0.082	0.085	0.089	0.094	0.100	0.108	0.116	0.126	0.137	0.150
600	0.150	0.165	0.182	0.200	0.23	0.25	0.28	0.31	0.34	0.36	0.39
700	0.39	0.42	0.45	0.47	0.50	0.53	0.56	0.58	0.61	0.64	0.67
800	0.67	0.70	0.72	0.75	0.78	0.81	0.84	0.87	0.89	0.92	0.95
900	0.95	0.98	1.01	1.04	1.07	1.10	1.12	1.15	1.18	1.21	1.24
1000	1.24	1.27	1.30	1.33	1.36	1.39	1.42	1.44			

t_{68}, °C	0	100	200	300	400	500	600	700	800	900	1000
1000		1.5	1.7	1.8	2.0	2.2	2.4	2.6	2.8	3.0	3.2
2000	3.2	3.5	3.7	4.0	4.2	4.5	4.8	5.0	5.3	5.6	5.9
3000	5.9	6.2	6.5	6.9	7.2	7.5	7.9	8.2	8.6	9.0	9.3

For routine measurements instruments of the same type may be used as working or secondary standards with their calibration corrections being determined by comparison with primary standard instruments. Other secondary standards such as liquid-in-glass thermometers similarly calibrated are commonly employed.

Platinum Resistance Thermometer. The text of IPTS 68 gives important details of construction of a standard platinum resistance thermometer. Such thermometers may be purchased from manufacturers who specialize in the production of instruments of this type. A thermometer to be used as a primary standard is calibrated by measuring its resistance at various fixed points. This can be done by the user or by a standardizing laboratory such as the National Bureau of Standards in the United States. Where traceability to establish accuracy of manufactured product is required, the national laboratory method is generally used.

Platinum–10% Rhodium-Platinum Thermocouple. The text of IPTS 68 gives important details of construction of a standard thermocouple. As with the resistance thermometer, calibration can be accomplished using fixed points or by having the service performed at a national standardizing laboratory.

Optical Pyrometer (Radiation Thermometer). National standardizing laboratories maintain their own standard optical pyrometers. Calibration of a user's instrument is best accomplished by comparison with such an instrument. When so calibrated, the pyrometer may be used for actual measurements or for the calibration of other working standards.

Secondary Standards and Calibration. Since a primary standard thermometer is an expensive instrument, the integrity of which is important, it is not commonly used for routine purposes. Instead, secondary standards are generally employed which in their turn are calibrated against a primary standard. These secondary standards may be of exactly the same design as the primary standards, but if damaged or destroyed, replaceability is not as significant. They may also be instruments of different design such as noble metal thermocouples of different composition, base metal thermocouples, or liquid-in-glass thermometers. In all cases they should be of a quality commensurate with their use. Calibrations should be conducted periodically against the primary standards to permit correction for aging effects and to evaluate the magnitude of alteration due to possible misuse.

References

Collier, R. D.: "Temperature Instrument Calibration Fundamentals," *ISA Temperature Symposium* (March 1982), Instrument Society of America, Research Triangle Park, North Carolina, 1982.

Staff: "Calibration and Test Services of The National Bureau of Standards," *NBS Spec. Publ. 250*, National Bureau of Standards, Washington, D.C., 1970.

Staff: "Calibration at Temperatures Other than Fixed Points," *ASTM E77–72*, American Society for Testing and Materials, Philadelphia, Pennsylvania, 1972.

Staff: "Evolution of the International Practical Temperature Scale of 1968," *ASTM STP 565*, American Society for Testing and Materials, Philadelphia, Pennsylvania, 1982.

TENORITE. A mineral oxide and ore of copper, CuO. Crystallizes in the monoclinic system. Hardness, 3.5; specific gravity, 6.45; color, gray to black with metallic luster. Named after M. Tenore (1780–1861), Naples.

TERBIUM. Chemical element symbol Tb, at. no. 65, at. wt. 158.92, eighth in lanthanide series in the priodic table, mp 1365°C, bp 3230°C, density, 8.230 g/cm³ (20°C). Elemental terbium has a close-packed hexagonal crystal structure at 25°C. Pure metallic terbium is a silvery-gray color, but oxidizes readily in moist air. It is the most reactive of the heavy rare-earth metals and must be handled and worked in a vacuum or inert atmosphere. When pure, the metal is malleable. There is one natural isotope of terbium, ^{159}Tb. The isotope is not radioactive and has a low acute-toxicity rating. Seventeen artificial isotopes have been identified. Little is known concerning the characteristics of terbium alloys and intermediate compounds.

Average content of the earth's crust is estimated at 0.9 ppm terbium, making this element the second least-abundant of the rare-earth elements. Even at this level, however, terbium is potentially more available than antimony, bismuth, cadmium, or mercury. The element was first identified by C. G. Mosander in 1843.

Electronic configuration is $1s^22s^22p^63s^23p^63d^{10}4s^24p^64d^{10}4f^85s^25p^65d^16s^2$. Ionic radius, Tb³⁺, 0.92 Å. Metallic radius, 1.783 Å. First ionization potential, 5.85 eV; second, 11.52 eV. Other physical properties of terbium are given under **Chemical Elements.**

Terbium occurs in apatite and xenotime and is derived from these minerals as a minor coproduct in the processing of yttrium. Processing involves organic ion-exchange or solvent extraction operations. Elemental terbium is produced by electrolytically reducing the oxide in a fused fluoride electrolyte, or by the calcium reduction of anhydrous TbF₃ in a sealed-bomb reaction. Both the oxides and the metal are available at 99.9% purity.

To date, the uses for terbium have been quite limited. Terbium-activated lanthanum oxysulfide Tb:La₂O₂S is a phosphor finding use as an image intensifier for x-ray screens. Terbium-activated indium borate Tb:InBO₃ phosphor emits an intense narrow green light (5,450–5,500Å) and has found use in information display systems where there are high ambient-light conditions. Future color television tubes may use terbium-activated yttrium silicate Tb:Y₂SiO₅ or yttrium phosphate Tb:YPO₄ green phosphors. They appear to be highly efficient post-deflection focused phosphors for elimination of the need for a shadow mask in a television tube. Although terbium oxide may be used as a stain for ceramics, the compound does not color glass. In soda-lime glass, a small quantity of terbium provides a strong green-blue fluorescence under ultraviolet radiation. The element currently is undergoing extensive studies to determine its future potential in photoconductive, semiconductor, and thermoelectric applications.

See the references listed at the ends of the entries on **Chemical Elements;** and **Rare-Earth Elements and Metals.**

NOTE: This 4th Edition entry was revised and updated by K. A. Gschneidner, Jr., Director, and B. Evans, Assistant Chemist, Rare-Earth Information Center, Energy and Mineral Resources Research Institute, Iowa State University, Ames, Iowa.

TEREPHTHALIC ACID. C₆H₄(COOH)₂, formula weight 166.13, crystalline solid sublimes upon heating, sp gr 1.510. The compound is almost insoluble in H₂O, only slightly soluble in warm alcohol, and insoluble in ether. Terephthalic acid (TPA) is a high-tonnage chemical, widely used in the production of synthetic materials, notably polyester fibers (poly-(ethylene terephthalate)).

There are several processes for making terephthalic acid on a large scale: (1) Benzoic acid, phthalic acid and other benzene-carboxylic acids in the form of alkali-metal salts, comprise the chargestock. In a first step, the alkali-metal salts (usually potassium) are converted to terephthalates when heated to a temperature exceeding 350°C. The dried potassium salts (of benzoic acid or o- or isophthalic acid) are heated in anhydrous form to approximately 420°C in an inert atmosphere (CO₂) and in the presence of a catalyst (usually cadmium benzoate, phthalate, oxide, or carbonate). The corresponding zinc compounds also have been used as catalysts. In a following step, the reaction products are dissolved in H₂O and the terephthalic acid precipitated out with dilute H₂SO₄. The yield of terephthalic acid ranges from 95 to 98%.

(2) Toluene, formaldehyde, HCl, calcium hydroxide, and HNO₃ comprise the chargestock. In step 1 of this process, the toluene is reacted with concentrated HCl at about 70°C along with paraformaldehyde. This accomplishes chloromethylation of approximately 98% of the toluene. In step 2, saponification of the chloromethyltoluene is effected with lime and H₂O under pressure and at about 125°C. The product is methylbenzyl alcohol. In step 3, the methylbenzyl alcohol is oxidized with HNO₃ (dilute) under a pressure of about 20 atmospheres and at a temperature of about 170°C. The main products are o-phthalic acid in HNO₃ solution and insoluble terephthalic acid.

(3) Paraxylene and air comprise the chargestock. These materials, along with a proprietary catalyst and solvent, are fed to

a liquidphase oxidation reactor, operated at moderate pressure and temperature. The reaction is:

$$C_6H_4(CH_3)_2 + 3O_2 \rightarrow C_6H_4(COOH)_2 + 2H_2O.$$

The design details of these processes are proprietary. There are several other processes which essentially are variations of the foregoing descriptions. See also **Intermediate (Chemical); Phthalic Acid;** and **Phthalic Anhydride.**

TERPENES AND TERPENOIDS. The class of organic compounds known as terpenes is characterized by the presence of the repeating carbon skeleton of isoprene:

$$\begin{array}{ccc} & CH_5 & \\ & | & \\ & C & \quad CH_2 \\ & \parallel & \diagup \\ CH_2 & & CH \end{array}$$

These compounds are widely distributed in nature. The name *terpene* used properly refers to the hydrocarbons which are exact multiples of the skeletal isoprene unit. However the name "terpene," sometimes used loosely, includes not only hydrocarbons but also other functional types of naturally occurring organic compounds which contain the reoccurring isoprene skeleton. In the strictest sense, the names *terpenoid* or *isoprenoid* should be used instead of the more loosely applied usage of terpene.

Terpenoids are divided into subclasses as follows:

Subclass Name	No. of Carbon Atoms	No. of Isoprene Skeletal Units
Hemiterpenoids	C_5	1
Monoterpenoids	C_{10}	2
Sesquiterpenoids	C_{15}	3
Diterpenoids	C_{20}	4
Sesterterpenoids	C_{25}	5
Triterpenoids	C_{30}	6
Tetraterpenoids	C_{40}	8
Polyterpenoids	$(C_5)^n$	n

Together they possess a wide variety of functional groups and structures. Nearly every common functional group is represented. Acyclic, monocyclic and polycyclic structures are observed. The greatest structural variation within a single subclass is to be found among sesquiterpenoids.

The combination of skeletal isoprene units in a regular fashion was exemplified early in the study of terpenoids by nearly all of these compounds. On the basis of this regularity the *regular isoprene rule* was formulated and was taken to mean that terpenoids would possess structures built from a regular "head-to-tail" arrangement of isoprene units. However, as structures of more terpenoids were elucidated, departures from the regular rule were observed. In time "irregular" structures were also accommodated through postulated rearrangement of a regular isoprenoid chain to the "irregular" isoprene skeleton. Rearrangements could occur subsequent to or concommitant with natural cyclization. Thus, the adaptation of the *regular isoprene rule* now finds expression in the *biogenetic isoprene rule*, a rule which is supported experimentally. Particularly significant examples of naturally occurring compounds conforming to the biogenetic isoprene rule are lanosterol, a triterpenoid alcohol associated with cholesterol in wool fat, and gibberelic acid, an important plant-growth regulating substance which is the product of a diterpenoid precursor.

The various terpenoid subclasses are not equally distributed in nature. Representatives of the low-molecular weight end of the terpenoid spectrum are seldom encountered as stable isolable natural products. Isoprene itself has not been detected in plants or animals, but the existence of two highly reactive hemiterpenoid substances in living cells is well established. These are the isomeric γ,γ-dimethylallyl pyrophosphate and isopentenyl pyrophosphate, the two being ubiquitous in living organisms and represent the fundamental isoprene building block in terpenoid biogenesis. One source of evidence for the existence of these hemiterpenoids is the presence of the γ,γ-dimethylallyl unit as a substituent of other classes of natural products, often phenols. The origin of these truly vital hemiterpenoids has been found to be mevalonic acid (3,5-dihydroxy-3-methylpentonoic acid), a six-carbon acid produced by the coenzyme A-assisted condensation of three moles of acetic acid. Decarboxylation and dehydration of mevalonic acid pyrophosphate are known to give the five-carbon unit of isopentenyl pyrophosphate. Isopentenyl pyrophosphate and γ,γ-dimethylallyl pyrophosphate are not only the links between the various subclasses of terpenoids but also biogenetically connect seemingly unrelated plant constituents such as steroids and some types of phenolics and alkaloids.

Monoterpenoids and to a lesser extent sesquiterpenoids are the chief components of the volatile oils readily obtained by the distillation of leaves, wood, and blossoms of a broad array of plants. Sesquiterpenoids are among the most universally distributed natural products. Iridolactone, a monoterpenoid, occurs as a defensive secretion of an ant species belonging to the genus *Iridomyrmex*. The iridoids, which is the general name given to the structural type exemplified by iridolactone, make up an important group of monoterpenoids. Another representative of this group is loganin which along with the amino acid tryptophan provides the carbon atoms of a group of indole alkaloids. Medically important quinine and reserpine are members of this group of alkaloids. The "resin acids" are a group of diterpenoid carboxylic acids which form the major nonvolatile part of natural resins often obtained from conifers. Examples are abietic, pimaric and isopimaric acids. Sesterterpenoids, the most recently discovered terpenoid subclass, are produced by insects and fungi. The tricyclic cereoplasteric acid has been isolated from the waxy coating secreted by the insect *Cereoplastes albolineatus*. The fungus responsible for the leaf spot disease in corn produces the acyclic geranylnerolidol and the tricyclic ophiobola-7,18-dien-3α-ol. Triterpenoids are found mainly in plants where they occur in resins and plant sap as free triterpenoids, esters or glycosides. A few are observed in animal sources, for example, the acyclic squalene and the tricyclic lanosterol. The connection between squalene and lanosterol is a vitally important one since these two triterpenoids, along with the hemiterpenoid isopentenyl pyrophosphate, are intermediates in the mevalonic acid based biogenesis of steroids. Tetraterpenoids are frequently referred to as *carotenoids*. These constitute a group of natural pigments containing long systems of conjugated carbon-carbon double bonds which are responsible for their color. β-Carotene is the principal pigment of the carrot but this pigment has been isolated from other plant sources also. See also **carotenoids.** Finally the presence of polyisoprenoids (natural rubber and gutta-percha) in nature shows that the isoprene unit, like the simple sugar and amino acids, has been used to form linear macromolecules.

The use of terpenoids, usually as mixtures prepared from plants, dates from antiquity. The several "essential oils" produced by distillation of plant parts contained the plant "essences." These oils have been employed in the preparation of perfumes, flavorings, and medicinals. Examples are: oils of clove (local anesthetic in toothache), lemon (flavoring), lavender (perfume), and juniper (diuretic). Usually essential oil production depends on a simple technology which often involves steam distillation of plant material. The perfume industry of Southern France uses somewhat more sophisticated procedures in the isolation of natural flower oils since these oils are heat sensitive. The separation of oils from citrus fruit residues in California and Florida is done by machine.

The oleoresinous exudate or "pitch" of many conifers, but mainly pines, is the raw material for the major products of the naval stores industry. The oleoresin is produced in the epithelial cells which surround the resin canals. When the tree is wounded the resin canals are cut. The pressure of the epithelial cells forces the oleoresin to the surface of the wound where it is collected. The oleoresin is separated into two fractions by steam distillation. The volatile fraction is called "gum turpentine" and contains chiefly a mixture of monoterpenes but a smaller amount of sesquiterpenes is present also. The nonvolatile "gum rosin" consists mainly of the diterpenoid resin acids and smaller amounts of esters, alcohols and steroids. "Wood turpentine," "wood rosin" and a fraction of intermediate volatility, "pine oil" are obtained together by gasoline extraction of the chipped wood of old pine stumps. "Pine oil" is largely a mixture of the monoterpenoids terpineol, borneol and fenchyl alcohol. "Sulfate turpentine" and its nonvolatile counterpart, "tall oil," are isolated as by-products of the kraft pulping process. "Tall oil" consists of nearly equal amounts of saponified fatty acid esters and resin acids.

Turpentine is used in syntheses by the chemical and pharmaceutical industries. It also is used as a paint thinner and as a component of polishes and cleaning compounds. Pine oil finds application as a penetrant, wetting agent and preservative, especially by the textile and paper industries, and as an inexpensive deodorant and disinfectant in specialty products. The resin acids are used in the production of ester gum, "Glyptal" resins and are indispensable in paper sizing.

A few individual terpenoids, as well as less expensive mixtures of these compounds, find practical applications. Some examples are: the diterpenoid Vitamin A, the sesquiterpenoid santonin (as an anthelmintic), and the pyrethrins, pyretholone esters of the monoterpenoid chrysanthemic acid (used as an insecticide). A number of sesquiterpenoid lactones of the germacranolide, guaianolide and elemanolide types have shown promise as tumor inhibitors.

The different terpenoid content of plants has served as a finger printing method helpful in botanical identification, especially in cases where differentiation by morphological characteristics has failed. The striking difference in the chemotaxonomy of Jeffery and ponderosa pines serves as an example. The turpentine from the former species consists almost entirely of the paraffinic hydrocarbon n-heptane. Turpentine from ponderosa pine consists largely of the monoterpenes β-pinene and Δ^3-carene.

Besides being of considerable biochemical and botanical interest and of importance in the industrial arts, the food, perfumary, and pharmaceutical trades, the terpenoids have been also a continuing challenge to the organic chemist. The earliest work on the volatile oils was very difficult since the oils were usually complex mixtures and as a consequence individual compounds were isolated as liquids of uncertain purity. Physical constants such as molecular refraction and melting points of solid derivatives, especially those from which the compound could be regenerated were of importance in structural investigations. Organic chemistry relied heavily on oxidative degradation techniques. Dehydrogenation of cyclic terpenoids to aromatic systems and the synthesis of these aromatics played an important role in structure determination. In recent times much of the previous difficulty of obtaining pure samples of terpenoids has been overcome through the use of various chromatographic techniques. Gas-liquid phase chromatography has been used to good advantage in separating both microgram quantities and much larger amounts of the more volatile monoterpenoids and sesquiterpenoids. Much of the type of information formerly obtained only be degradative procedures and dehydrogenations can now be obtained through mass spectrometry. Through nuclear magnetic resonance, infrared and ultraviolet spectrometry, structural information can be obtained in a small fraction of the time that was formerly needed to gain the same amount of information.

Acid-promoted cyclization of acylic terpenoids is common. Geraniol, or more readily its trans isomer nerol, can be cyclized with acids to p-menthane derivatives. More importantly, citral, 2,6-dimethyl-2-octen-8-al, when condensed with acetone gives pseudoionone. The latter when cyclized with acid gives a mixture of α- and β-isomers which in turn are used in the preparation of perfumes, as an intermediate in a number of industrial syntheses of Vitamin A, and also in a commercial synthesis of the plant hormone abscisic acid. Polyclic terpenoids are prone to rearrangement of the carbon skeleton. The acid catalyzed rearrangement of the monoterpene camphene to derivatives of isobornyl alcohol are well known and have been the subject of extensive theoretical studies. Diterpenoids and triterpenoids undergo "backbone" rearrangement through the migration of hydride and methyl groups. Frequently these migrations are stereospecific and are acid promoted.

Studies of terpenoid chemistry have also involved syntheses. Several of the complex sesquiterpenoid structures have been confirmed or, in some cases, correctly established through synthesis. Many elegant new general synthetic methods have been developed as a result of attempts to synthesize terpenoids. The chemistry of terpenoids present a continually growing area of chemical research, perhaps the equal of any in complexity, subtlety, and variety.

—Robert T. LaLonde, State University of New York, Syracuse, New York.

TESTOSTERONE. Hormones; Steroids.

TETRADYMITE. A mineral bismuth tellurium sulfide, corresponding to the formula Bi_2Te_2S. It is rhombohedral. Tetradymite occurs usually in gold quartz veins. It is found in Norway, Sweden, England, Bolivia, British Columbia; and in the United States in Virginia, North Carolina, Georgia, Montana, Colorado, and elsewhere. It derives its name from the Greek word meaning *fourfold*, in reference to the double twin crystals occasionally developed.

TETRAHEDRITE. A mineral of the composition, $(Cu,Fe)_{12}As_4S_{13}$, isomorphous with tennantite. The color ranges from steel-gray to iron-black. The mineral frequently contains cobalt, lead mercury, nickel, silver, or zinc in replacement of the copper. Tetrahedrite usually occurs in tetrahedral crystals associated with copper ores. The mineral is considered an important copper ore and sometimes is a valuable ore for silver. The mineral sometimes is referred to as *fahlore, gray copper ore*, and *stylotypite*.

THALLIUM. Chemical element symbol Tl, at. no. 81, at. wt. 204.39, periodic table group 3a, mp 303.3°C, bp 1460°C, density, 11.85 g/cm³ (20°C). Elemental thallium has a hexagonal close-packed crystal structure normally, but also exhibits a face-centered cubic crystal structure.

Thallium metal is bluish-gray upon fresh exposure, changing

to dark gray upon standing, this oxidation increased with temperature above 25°C; soft, and may be easily cut with a knife; malleable, but of low tenacity so that it must be extruded to form wire; HNO_3 is the best solvent; forms alloys with many metals, e.g., mercury, cadmium, zinc, silver, copper, magnesium.

The element was first identified by Sir William Crookes spectrographically in 1861. While seeking tellurium, Crookes observed the characteristic bright green lines in the emission spectrum of thallium. At just about the same time, A. Lamy identified the element. Thallium occurs naturally as ^{203}Tl and ^{205}Tl. Eleven radioactive isotopes have been identified, ^{198}Tl through ^{202}Tl, ^{204}Tl, and ^{206}Tl through ^{210}Tl. With the exception of ^{204}Tl, which has a half-life of 4.07 years, the other isotopes have relatively short half-lives expressed in minutes, hours, and days. See **Radioactivity**. Thallium is not considered an abundant element, estimates of occurrence in the earth's crust ranging from 0.3 to 3.0 ppm. In a list of 65 chemical elements found in seawater, thallium does not appear.

Electron configuration is $1s^2 2s^2 2p^6 3s^2 3p^6 3d^{10} 4s^2 4p^6 4d^{10} 4f^{14} 5d^2 5p^6 5d^{10} 6s^2 6p^1$. First ionization potential, 6.106 eV; second, 20.32 eV; third, 29.7 eV. Oxidation potentials: $Tl \rightarrow Tl^+ + e^-$, 0.336V; $Tl^+ \rightarrow Tl^{3+} + 2e^-$, -1.25 V; $Tl + OH^- \rightarrow TlOH + e^-$, 0.3445 V; $TlOH + 2OH^- \rightarrow Tl(OH)_3 + 2e^-$, 0.05 V. Other physical properties of thallium are given under **Chemical Elements**.

Thallium occurs in small amounts in pyrite, zinc blende, and hematite of certain localities, and in a few rare minerals in Sweden and Macedonia. For the recovery of thallium from flue dust of pyrite burners, the dust is boiled with H_2O, allowed to stand some time, filtered, and HCl added to the filtrate, whereupon crude thallous chloride is precipitated. This is purified by further treatment, and thallium metal obtained (1) by electrolysis of the sulfate solution or (2) by fusion of the chloride with sodium cyanide and carbonate.

Uses. Because thallium is recovered from smelting lead and zinc concentrates, it is available to fulfill any new uses up to several thousand pounds per year. To date, practical applications have been relatively limited. Thallium-activated sodium iodide crystals find use in photomultiplier tubes. It has been learned that thallium bromoiodide crystals transmit infrared radiation and that crystals of thallium oxysulfide detect infrared radiation. A combination of these crystals has been used in military communication systems. Because of their density, both thallous formate and thallous malonate have been used in the preparation of heavy-liquid sink-float solutions used in the gravity separation of minerals. Mixtures of thallium, arsenic, sulfur, and selenium form low-melting-point glasses for encapsulation of semiconductors has been under investigation. It has been found that the addition of small amounts of thallium to the counterelectrode alloy used in selenium rectifiers will improve the performance of the rectifiers. Claims have been made that the addition of a thallium salt to absorb traces of oxygen in tungsten-filament incandescent lamps will increase lamp life. Also, it has been shown that the addition of thallium to various glass formulations will improve optical properties and increase the refractive index.

For a number of years, thallium sulfate had been used in rodenticides. Some use of thallium has been made in connection with alloys for low-temperature applications, particularly for switches, seals, and thermometers. The ternary eutectic mercury-thallium-indium alloy has a freezing point of $-63.3°C$, while the binary eutectic mercury-thallium alloy has a freezing point of $-60°C$. These freezing points are considerably lower than that of mercury usually used for similar applications at higher temperatures. Mercury freezes at $-38.87°C$.

Toxicity. Thallium and thallium compounds are toxic and skin contact must be avoided. Impervious gloves and aprons should be worn and excellent ventilation and masks should be provided where dusts and fumes may be present.

Chemistry and Compounds. Thallium has two oxidation states: thallous, Tl^+ and thallic, Tl^{3+}. Because of low oxidation potential of thallium to form Tl^+, thallium is quite reactive, dissolving slowly in most dilute mineral acids to form thallium(I) solutions. The thallium(I) halides are insoluble in water, but thallium trihalides are soluble; the latter are formed by treatment of the thallium(I) halide in solution with the corresponding halogen. Thallium(III) iodide, however, does not exist, TlI_3 actually being $[Tl^+][I_3^-]$.

Thallium(III) compounds are readily reduced to the thallium(I) state (see difference in oxidation potentials above) and are thus fairly strong oxidizing agents. Thallium(I) compounds resemble those of the alkali metals in many respects, including a soluble, strong basic hydroxide (TlOH) ($K_B = 0.14$), a soluble carbonate (Tl_2CO_3), the formation of well crystallized salts, including those with complex anions, the formation of polysulfides (Tl_2S_5), and polyiodides (thus TlI_3 contains the monovalent metal ion, like rubidium iodide, RbI_3 and cesium iodide, CsI_3). Thallium(I) ion resembles silver in forming insoluble halides, sulfide and chromate. The thallium(I) ion forms only weak complexes (probably because of its larger size and low charge) but the thallium(III) ion forms strong ones. There are four complex chloro ions $[TlCl_4]^-$, $[TlCl_5]^{2-}$, $[TlCl_6]^{3-}$, and $[Tl_2Cl_9]^{3-}$, the last having the six-coordinated structure

$$\begin{bmatrix} Cl & & Cl & & Cl \\ Cl-Tl\cdots Cl-Tl\cdots Cl \\ Cl & & Cl & & Cl \end{bmatrix}^{3-}$$

The complex compounds include the chelates, such as the oxine chelate, and also such compounds as $Tl(TlCl_4)$, $Tl_3(TlBr_6)$, $TlCl_3 \cdot 3NH_3$, $14Rb_3TlBr_6 \cdot 16H_2O$ (here the presence of the ion $[Tl(H_2O)_8]^+$ has been shown), oxalates, such as $H[Tl(C_2O_4)_2]$ (dioxalatothallic acid), $K_2[Tl(C_2O_4)_2(NO_2)_2] \cdot H_2O$ and a number of complex hydrides, as well as the unstable binary hydride TlH_3.

The most readily prepared organometallic compounds are the dialkyl ones of the type R_2TlX, where X is an acid radical accompanying the ion $[R_2Tl]^+$. The trialkyl compounds of the type TlR_3 are immediately decomposed by H_2O, giving RH and R_2TlOH, in which the thallium atom is isoelectronic with the mercury atom in R_2Hg, and which is a strong base: $(C_2H_5)_2TlOH$, $K_B = 0.90$.

References

Carapella, S. C., Jr.: "Thallium," in "Metals Handbook," Vol. 2, 9th Edition, 807, American Society for Metals, Metals Park, Ohio (1979).
Klemert, R.: *Zeit. Erzbergbau*, **16**, 67–76 (1963).
Lee, A. G.: "The Chemistry of Thallium," Elsevier, Amsterdam (1971).

—S. C. Carapella, Jr., ASARCO Incorporated, South Plainfield, New Jersey.

THERMAL CRACKING. Cracking Process.

THERMAL INSULATION. Insulation (Thermal).

THERMODYNAMICS. The science of thermodynamics treats systems whose states are determined by thermal parameters, such as temperature, in addition to mechanical and electromagnetic parameters. By *system* we mean a geometric section of the universe whose boundaries may be fixed or varied, and which may contain matter or energy or both. The *state* of a system is a reproducible condition, defined by assigning fixed numerical values to the measurable attributes of the system.

These attributes may be wholly reproduced as soon as a fraction of them have been reproduced. In this case the fractional number of attributes determines the state, and is referred to as the *number of variables of state* or the *number of degrees of freedom* of the system.

The concept of *temperature* can be evolved as soon as a means is available for determining when a body is "hotter" or "colder." Such means might involve the measurement of a physical parameter such as the volume of a given mass of the body. When a "hotter" body, A, is placed in contact with a "colder" body, B, it is observed that A becomes "colder" and B "hotter." When no further changes occur, and the joint system involving the two bodies has come to equilibrium, the two bodies are said to have the same temperature. It is a fact of experience that two bodies which have been shown to be individually in equilibrium with a third, will be in equilibrium when placed in contact with each other, i.e., will have the same temperature. This statement is sometimes called the *zeroth law* of thermodynamics. The physical parameter used to specify the "hotness" of the third body might be adopted as a quantitative measure of temperature, in which case, that body becomes the thermometer.

From what has been said above it is apparent that temperature can only be measured at equilibrium. Therefore thermodynamics is a science of equilibrium, and a thermodynamic state is necessarily an equilibrium state. Furthermore, it is a macroscopic discipline, dealing only with the properties of matter and energy in bulk, and does not recognize atomic and molecular structure. Although severely limited in this respect, it has the advantage of being completely insensitive to any change in our ideas concerning molecular phenomena, so that its laws have broad and permanent generality. Its chief service is to provide mathematical relations between the measurable parameters of a system in equilibrium so that, for example, a variable like the pressure may be computed when the temperature is known, and *vice versa*.

It provides these relations with the aid of three postulates or *laws* (in addition to the zeroth law) which we shall now proceed to explain.

The *first law* of thermodynamics may be expressed in the following form,

$$dE = Dq - Dw \tag{1}$$

where dE represents a differential change in the *internal energy* of a system during some change defined by the passage of the system from one thermodynamic state to another. Dq and Dw are, respectively, the *heat* absorbed by the system and the *work* done by the system on its environment during the differential change. The small d preceding E in (1) implies that dE is an exact differential, or that E depends only on the state of the system, and is independent of the path of the change. The large D in Dq and Dw implies that in general this is not true for the *heat* and *work*.

To clearly understand what is meant by *heat* and *work* and to gain further insight into the meaning of (1), consider a system set up in such a manner that it can exchange energy only with its surroundings in the form of well defined mechanical or electrical work. Such an arrangement might involve surrounding the system by an *adiabatic* wall. Then it is observed that the system always undergoes the same change when a given amount of work is given to it or extracted from it. This prompts us to define a quantity, E, the internal energy, to be associated with each state of the system, whose change (and therefore whose value to within a constant) can be measured by the amount of work passing between system and environment when the change of state is performed under adiabatic conditions.

If, now, the same change of state (defined by the thermodynamic states between which the system passes) is conducted under nonadiabatic conditions, it is found that the work involved does not equal the previously measured change in E. Conservation of energy then demands that an exchange take place between the system and its environment which is not recognizable as work. This quantity of energy is defined as *heat*, and is nothing more than a discrepancy term designed to transform (1) into the equality which it is. Heat, therefore, has no measurable existence of its own. It is incorrect to say that a body contains a particular amount of heat, even though the quantity of heat passing in or out of a system is measurable because the work done by the system and its internal energy can both be measured.

The *second law* of thermodynamics may be expressed as follows,

$$T\,dS - Dq \geq 0 \tag{2}$$

Here T stands for the temperature and S denotes the *entropy* of the system. Dq has the same meaning as in (1). To fully understand the entropy as well as the inequality in (2) it is necessary to understand *reversible* and *irreversible* processes. A *reversible* process is one in which an infinitesimal change in driving force of the process completely reverses the process in all of its detailed aspects. For example, consider a gas enclosed in a cylinder by a weighted frictionless piston. Let the pressure exerted by the weighted piston be just insufficient (by an infinitesimal amount) to contain the gas in equilibrium. A slow expansion occurs which can be completely reversed by an infinitesimal change in driving force, i.e., by an infinitesimal increase in the weight of the piston.

Another example of a reversible process, more chemical in nature, is the slow discharge of an electrochemical cell, maintained slow by the action of an impressed electromotive force. A slight increase in this electromotive force not only reverses the flow of current, but also the chemical reaction of discharge, and causes a slow electrolysis to occur.

A distinguishing feature of a reversible process is its physical impossibility. It occurs infinitely slowly and may be aptly described as a sequence of equilibrium states. Nevertheless, calculations of phenomena as they would occur, were reversibility truly possible, are feasible (for example, it is possible to calculate the volume work performed by a gas as it passes through a sequence of equilibrium states or the electrical work performed by a cell as it discharges through a sequence of equilibrium states), and it is this aspect which lends importance to the concept of reversibility.

By contrast, an *irreversible* process is any real process, occurring at a finite rate, and with any degree of violence. In (2) the inequality refers to an irreversible, and the equality to a reversible change of state. As indicated by the use of d in front of S, the differential of S is exact, so that S is a function only of the thermodynamic state of the system. This is part of the postulate. Equation (2) also specifies how S is to be measured, for if the change of state is conducted reversibly,

$$dS = \frac{Dq}{T} \tag{3}$$

Measurement of the heat absorbed by a system in a change conducted reversibly then yields dS directly by division by T.

An immediate consequence of (2) is that in a system isolated from its surroundings, so that Dq is zero, any irreversible change is necessarily accompanied by an increase in entropy, while for a reversible change no variation in entropy occurs. Since irreversible changes are to be associated with real changes this is tantamount to the assertion that the entropy of an isolated system increases for real spontaneous changes so that it is maximized when the system achieves equilibrium, i.e., when no further real spontaneous change is possible. In line with this assertion, (2) shows that $dS = 0$, for a reversible change, i.e., one involving a system in equilibrium.

The quantity S seems somewhat more abstract than E. Logically, however, the two are on the same basis. E is generally more familiar than S since the latter is a thermal quantity, associated specifically with thermodynamics. A nonthermodynamic, molecular interpretation of S can be given. This asserts that S measures the logarithm of the relative probability of a given state. It is not surprising, then, that S increases in a spontaneous process.

In its early development thermodynamics was an engineering subject. A good deal of its inquiries were directed toward the efficiency of machines. The inequality, (2), tells immediately how to get the most work out of a system. Thus, by substituting the value of Dq, obtained through solution of (1), the following result is obtained,

$$DW \leq TdS - dE \qquad (4)$$

the inequality still referring to an irreversible process. Since the right member contains exact differentials only, its value is determined by the states between which the change occurs rather than by the path of the change. Therefore it is apparent that the left side (the work performed by the system) is maximized when the equality holds, i.e., when the system operates reversibly or when the change is conducted over such a path that it occurs reversibly.

A machine which converts heat into work generally operates in cycles, abstracting an amount of heat, Q_2, from a reservoir at a higher temperature, T_2, performing a certain amount of work, W, and returning an amount of heat, $-Q_1$, to a reservoir at the temperature, T_1. The efficiency, η, of the machine is defined as the fraction of the amount of heat absorbed at the higher temperature which is converted into work during one cycle. Thus

$$\eta = W/Q_2 \qquad (5)$$

After one cycle the machine returns to its original state, and, therefore the total changes in its internal energy, E, and entropy, S, are zero. From what has been said above, the maximum efficiency is achieved when the machine operates reversibly. But, then, since dS is zero and the total change of entropy is Q_2/T_2 plus Q_1/T_1, we have

$$\frac{Q_2}{T_2} + \frac{Q_1}{T_1} = 0 \qquad (6)$$

Furthermore, since dE is also zero, equation (1) demands that

$$Q_1 + Q_2 = W \qquad (7)$$

The simultaneous solution of (6) and (7) yields

$$\eta = \frac{W}{Q_2} = \frac{T_2 - T_1}{T_2} \qquad (8)$$

This shows that no machine operating in cycles can be 100% efficient, i.e., that it is impossible to convert heat entirely into work without effecting permanent changes. The latter statement is often taken as an alternative form of the second law.

The universal efficiency η, depending only, as it does, on the temperatures of the two reservoirs, can be used to define the thermodynamic scale of temperature. Thus, two reservoirs have the ratio of thermodynamic temperatures

$$\frac{K_2}{K_1} = \frac{1}{1 - \eta} \qquad (9)$$

where η is the efficiency of a machine (any machine) operating reversibly between them. If to (9) we add the requirement

$$K_2 - K_1 = 100 \qquad (10)$$

when the high-temperature reservoir is boiling water, and the low temperature is a mixture of ice and water, then K proves to be identical with T, the Kelvin temperature. The thermodynamic temperature has the advantage of being independent of the properties of any special substance.

The *third law* of thermodynamics asserts that *the entropy of a system at the absolute zero of temperature (T or $K = 0$) is zero, provided that the system is in its lowest energy state*. In reality this cannot be said to be strictly a law of thermodynamics, since it presumes an acquaintance with the detailed structure of the system, especially in regard to its spectrum of energy states. Despite these non-thermodynamic overtones, the third law is extremely useful in many applications. For example, if a system is observed to undergo a change at absolute zero involving a loss of entropy, one may conclude that originally it was not in its lowest energy state. Another application involves the calculation of the equilibrium constant of a chemical system from purely thermal measurements. In this brief exposition it is impossible to discuss the third law in greater detail.

Some of the most important applications of thermodynamics are achieved by substituting Dq from (1) into (2), i.e., by combination of the first and second laws. The result is

$$D\phi = dE + DW - T\,dS \leq 0 \qquad (11)$$

where the inequality still stands for an irreversible spontaneous process. $D\phi$ is used as a shorthand for the three terms on its right. $D\phi$ assumes a special form depending upon the work which the system is capable of doing. For example, if the system can only do volume work,

$$DW = p\,dV \qquad (12)$$

where p is the pressure and V the volume of the system. Then

$$D\phi = dE + p\,dV - T\,ds \leq 0 \qquad (13)$$

It is convenient to find some function ψ, dependent only the thermodynamic state of the system which imitates $D\phi$ when certain restricted changes of state are carried out. For example, let $\psi = E$. E initiates $D\phi$ for changes of state in which V and S are maintained constant, since then dV and dS are zero. Suppose the initial state of the system is one of equilibrium. Then no spontaneous change can occur and all possible changes are the reverse of spontaneous ones. Then

$$D\phi = (dE)_{S,V} > 0 \qquad (14)$$

for all possible changes, where the subscripts S and V indicate the maintained constancy of those variables. Equation (14) implies that if the initial state is one of equilibrium E is a minimum for all displacements along paths of constant entropy and volume.

The displacement involved is *not* one leading *out of equilibrium*, for then functions like S could not be given experimental meaning along its path. Rather, additional forces are added to the system against which the system can perform other than volume work, so that a new state of equilibrium is reached subject to these new forces. This will be elaborated below.

Another choice of ψ might be

$$\psi = F = E + pV - TS \qquad (15)$$

the so-called Gibbs' free energy. This imitates $D\phi$ along paths of constant temperature and pressure, and can be shown (in the same manner employed in the case of E) to be minimized along these paths in a state of equilibrium. This minimization may be expressed by

$$(dF)_{T,p} = 0 \qquad (16)$$

Here again a displacement from equilibrium is not implied. Rather an additional force is added leading to a new equilibrium. For example, suppose the system represents a solution involving a chemical reaction. If the reaction is set up in a chemical cell, no electromotive force will persist when chemical equilibrium has been achieved. The state of the system is determined by variables of state including temperature, pressure and composition. If, now, a new force is added in the form of an impressed electromotive force electrolysis occurs and a new state of equilibrium is achieved subject to the magnitude of the impressed electromotive force. The variables of state are greater by one (the E.M.F.) than in the original equilibrium state. It is to this kind of displacement that (16) refers.

It can be shown that the change in F, ΔF, can be measured by the electrical work performed in the displacement. Equation (16) can be used as a differential equation connecting the variables of state in the initial equilibrium state. So can (14), for that matter, by writing dF more explicitly in terms of the variations of variables of state.

In this manner it can be shown that when a solute is distributed between the two phases 1 and 2, at equilibrium, and F_1 and F_2 are the free energies of the two phases, then

$$\left(\frac{\partial F_1}{\partial n_1^1}\right)_{T,p,n^1} = \left(\frac{\partial F_2}{\partial n_1^2}\right)_{T,p,n^2} \qquad (17)$$

where n_1^1 and n_1^2 are the numbers of moles of the solute in phases 1 and 2, respectively, and n^1 and n^2 represent the mole numbers of other components. For many solutes it is observed that

$$\left(\frac{\partial F}{\partial n}\right)_{T,p} = K + RT \ln c \qquad (18)$$

where K and R are constants at any one temperature and pressure and c is the concentration of the solute. Substitution of (18) into (17) with appropriate attachment of subscripts yields

$$\frac{c_1^2}{c_1^1} = \exp\left[-\left(\frac{K_2 - K_1}{RT}\right)\right] = K \qquad (19)$$

and defines the partition coefficient k.

In closing, we note that since F is minimized at constant temperature and pressure in the equilibrium state, any reaction system which is in a state in which F is not minimized at the temperature and pressure in question drifts toward the state in which it is. A knowledge of F as a function of composition thus permits the prediction of the direction in which a particular reaction will proceed.

Many other applications are possible, for example, other thermodynamic potentials like F and E can be invented with consequent utility.

—Howard Reiss, University of California, Los Angeles, California.

THERMOELECTRIC COOLING.

Like conventional refrigeration systems, thermoelectric systems obey the laws of thermodynamics. Both in principle and result, thermoelectric cooling has much in common with conventional refrigeration systems; the main working parts are the freezer, condenser, and compressor. The freezer surface is where the liquid refrigerant boils, changes to vapor, and absorbs heat energy. The compressor circulates the refrigerant above ambient level. The condenser helps to discharge the absorbed heat into surrounding ambient. In thermoelectric refrigeration, the refrigerant in both liquid and vapor forms is replaced by two dissimilar conductors. The freezer surface becomes cold through absorption of energy by electrons as they pass from one semiconductor to another, instead of energy absorption by the refrigerant as it changes from liquid to vapor. The compressor is replaced by a direct-current power source which pumps the electrons from one semiconductor to another. A heat sink replaces the conventional condenser fins, discharging the accumulated heat energy from the system.

The components of a thermoelectric cooler are indicated by the cross section of a typical unit shown in Fig. 1. Thermoelectric

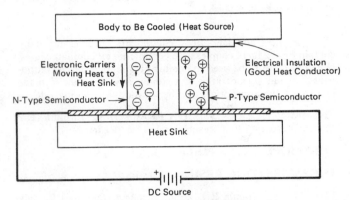

Fig. 1. Cross section of thermoelectric cooler.

coolers such as this are actually small heat pumps which operate on the physical principles well established over a century ago. Semiconductor materials with dissimilar characteristics are connected electrically in series and thermally in parallel, so that two junctions are created. The semiconductor materials are n- and p-type and are so named because either they have more electrons than necessary to complete a perfect molecular lattice structure (n-type), or not enough electrons to complete a lattice structure (p-type). The extra electrons in the n-type material and the holes left in the p-type material are called carriers and they are the agents that move the heat energy from the cold to the hot junction.

Heat absorbed at the cold junction is pumped to the hot junction at a rate proportional to carrier current passing through the circuit and the number of couples. Good thermoelectric semiconductor materials, such as bismuth telluride, greatly impede conventional heat conduction from hot to cold areas, yet provide an easy flow for the carriers. In addition, these materials have carriers with a capacity for carrying more heat. Only since the refinement of semiconductor materials in the early 1950s has thermoelectric refrigeration been considered practical for some applications.

In practical use, couples are combined in a module where they are connected in series electrically and in parallel thermally. See Fig. 2. Normally, a module is the smallest component available. The user can tailor quantity and size or capacity of the module to fit exact requirements without procuring more total capacity than is actually needed. Modules are available in a variety of sizes, shapes, operating currents, operating voltages, number of couples, and ranges of heat-pumping levels. The present trend is toward a larger number of couples operating at a low current.

Thermoelectric coolers find three basic applications: (1) use in electronic components, (2) in temperature control units, and

Bismuth Telluride
Elements with "N"
and "P" Type
Properties

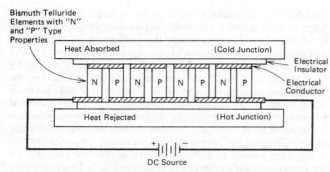

Fig. 2. Thermoelectric module assembly. Elements are electrically in series; thermally in parallel.

(3) in medical and laboratory equipment. Modules normally contain from 2 to 71 couples with ceramic-metal laminate plates. If modules are to be used in cooling chambers of large components, a total surface area of virtually any size can be made by placing the appropriate number of modules side by side.

THERMONUCLEAR FUSION REACTORS. Lithium (For Thermonuclear Fusion Reactors); Nuclear Reactor.

THERMOPLASTIC. Plastic

THERMOSETTING RESINS. Plastic.

THIAMINE (Vitamin B$_1$). Some earlier designations for this substance included aneurin, antineuritic factor, antiberiberi factor, and oryzamin. Thiamine is metabolically active as thiamine pyrophosphate (TPP), the formula of which is:

TPP functions as a coenzyme which participates in decarboxylation of α-keto acids. Dehydrogenation and decarboxylation must precede the formation of "active acetate" in the initial reaction of the TCA cycle (citric acid cycle):

$$CH_3-\overset{O}{\overset{\|}{C}}-COOH + NAD^+ + CoA$$

Pyruvic acid

$$\xrightarrow{\text{(FAD, TPP)}} CH_3-\overset{O}{\overset{\|}{C}}-CoA + CO_2 + NADH + H^+$$

Acetyl-CoA
"active acetate"

This reaction is a good example of the interrelationship of vitamin B coenzymes. Four vitamin coenzymes are necessary for this one reaction: (1) thiamine (in TPP) for decarboxylation; (2) nicotinic acid in nicotinamide adenine dinucleotide (NAD); (3) riboflavin in flavin adenine dinucleotide (FAD); and (4) pantothenic acid in coenzyme A (CoA) for activation of the acetate fragment.

TPP also mediates the oxidative decarboxylation of α-ketoglutaric acid, another intermediate of carboxydrate metabolism in the citric acid cycle. The nutritional requirement for thiamine increases as dietary carbohydrate increases because of a greater demand for TPP.

The structure of thiamine hydrochloride is:

In this form and as other salts, such as thiamine mononitrate, the vitamin is available as a dietary supplement.

Diseases and disorders resulting from a deficiency of thiamine include beriberi, opisthotonos (in birds), polyneuritis, hyperesthesia, bradycardia, and edema. Rather than a specific disease, beriberi may be described as a clinical state resulting from a thiamine deficiency. In body cells, thiamine pyrophosphate is required for removing carbon dioxide from various substances, including pyruvic acid. Actually, this is accomplished by a decarboxylase of which thiamine pyrophosphate is a part. Where thiamine is deficient, the process of oxidation necessary for converting food into energy is impeded, causing a variety of manifestations throughout the body. In so-called *dry* beriberi, pathologic alterations in neurons and nerve fibers occurs, leading in some instances to degeneration of peripheral nerves. This condition is termed *peripheral neuritis*, generally affecting the nerves in the arms and legs. There often is altered skin sensitivity to touch in the extremities and pain on pressure over large nerves. There is a gradual loss of muscle strength, which may lead to paralysis of a limb. In *wet* beriberi, there is a lessening of strength of the heart muscles. There is enlargement of the heart, dyspnoea, increased pulse rate, palpitation, and edema. Pathologically, degenerative changes are found in the nervous tissue, heart muscle, and gastrointestinal tract. In later stages, marked enlargement of the heart and liver may be noted. In one form of thiamine deficiency, Wernicke's syndrome may be noted. Here there is paralysis or weakness of the muscles that cause motion of the eyeball. Closely associated with thiamine deficiency are dietary problems of alcoholism. The psychotic disturbances of alcoholism, including delerium tremens, frequently respond to thiamine and other B complex vitamins. Injections of thiamine often produce dramatic improvements in persons suffering from beriberi. Beriberi sometimes occurs in infants who are breast-fed by mothers who suffer a thiamine deficiency. Beriberi remains of concern in the Orient where polished rice is a staple diet.

In cattle, a thiamine deficiency causes podioencephalomalcia (PEM), characterized by blindness, decreased feed intake, incoordination, failure of rumen to contract, spasms, and paralysis. In swine, a deficiency regards growth and sometimes causes cyanosis (insufficient oxygen in blood), enlarged heart, accompanied by fatty degeneration of heart muscles. Chicks suffer from paralysis of peripheral nerves, causing polyneuritis (head drawn back).

Relatively few natural foods are considered high in thiamine content. Those foods with the highest content include ham, rice bran, soybean flour, wheat germ, and yeast. Although there are exceptions, foods low in thiamine content include most fruits and fishes.

Commercial thiamine dietary supplements are prepared by synthesis: Pyrimidine + thiazole nuclei synthesized separately and then condensed; also built on pyrimidine with acetamidine. Precursors in the biosynthesis of thiamine include thiazole and pyrimidine pyrophosphate, with thiamine phosphate as an intermediate. In plants, production sites are found in grain and cereal germ.

Bioavailability of Thiamine. Factors which contribute to a lessening of thiamine bioavailability include: (1) cooking, inasmuch as the vitamin is heat labile and water soluble; (2) presence of certain enzymes in food, such as thiaminase for vitamin breakdown; (3) destruction by calcium carbonate, dibasic potassium phosphate, and manganous sulfate; (4) destruction by nitrites and sulfites; (5) diuresis and gastrointestinal diseases; (6) presence of live yeasts and alkalis. An increase in availability can result from: (1) presence of cellulose in diet which increases intestinal synthesis; (2) storage capacity in heart, liver, and kidney; (3) stimulation of bacterial synthesis in intestine (normally none).

Antagonists of thiamine include pyrithiamine, oxythiamine, and 2-*n*-butyl homologue. Synergists include vitamins B_2, B_6, B_{12}, and niacin, pantothenic acid, and somatotrophin (growth hormone).

Unusual features of thiamine as observed by some researchers include: (1) Exerts a hormonal function in plants, controlling root growth; (2) aids phosphorylation in liver, dephosphorylation in kidney; (3) easily poisoned by heavy metals, acetyl iodide; (4) plant and animal cocarboxylases are identical; (5) exerts a diuretic effect and is constipative; (6) can be allergenic on injection; (7) not available from intestinal bacteria; (8) blood contains most cocarboxylase in leukocytes.

Thiamine is soluble in water and easily destroyed by heat. These two properties account for appreciable losses of thiamine from processed and stored foods. An acid medium favors the retention of thiamine, whereas an alkaline medium is detrimental to retention.

Determination of Thiamine. Bioassay methods include yeast fermentation; polyneuritic rate of cure in rat; bacterial metabolism. Physicochemical methods include thiochrome fluorescence; polarography; chromatography; absorption in neutral and acid solutions.

References

Haresign, W., and D. Lewis: "Recent Advances in Animal Nutrition," Butterworth Group, Woburn, Massachusetts (1977).

Klutsky, J. F.: "Handbook of Vitamins and Hormones," Van Nostrand Reinhold, New York (1973).

Staff: "Food Chemicals Codex," National Academy of Sciences, Washington, D.C. (1981).

Staff: "Nutritional Requirements of Domestic Animals," separate publications on beef cattle, dairy cattle, poultry, swine, and sheep, periodically updated. National Academy of Sciences, Washington, D.C.

Watt, B. K., and A. L. Merrill: "Composition of Foods," *Agricultural Handbook 8*, U.S. Department of Agriculture, Washington, D.C. (1975).

Williams, R. R.: Thiamine and Beriberi," in "The Encyclopedia of Biochemistry," Van Nostrand Reinhold, New York (1967).

THIAZIDES.   Diuretics.

THIOALCOHOLS.   Mercaptans.

THIO- AND DITHIOCARBAMIC ACIDS.   Herbicide; Insecticide.

THIOCYANATES AND ISOTHIOCYANATES.   Herbicide; Insecticide.

THIOCYANIC ACID.   Aqueous solution of hydrogen thiocyanate, HSCN, formula weight 59.08, yellow solid below mp 5°C, unstable gas at room temperature. The acid is moderately stable only when dilute and cold. The salts of this acid are known as thiocyanates.

Thiocyanic acid is formed by reaction of barium thiocyanate solution and dilute sulfuric acid, and filtering off barium sulfate, or by the action of hydrogen sulfide on silver thiocyanate, filtering off silver sulfide.

Sodium, potassium, barium, or calcium thiocyanate may be made by reaction of sulfur and the corresponding cyanide by heating to fusion. Ammonium thiocyanate (plus ammonium sulfide) may be made by reaction of ammonia and carbon disulfide, a reaction which probably accounts for the presence of ammonium thiocyanate in the products of the destructive distillation of coal. This reaction corresponds to the formation of ammonium cyanate from ammonia and carbon dioxide.

Silver, lead, copper(I), and thallium(I) thiocyanates are insoluble and mercury(II), bismuth, and tin(II) thiocyanates slightly soluble. All of these are soluble in excess of soluble (e.g., ammonium) thiocyanate, forming complexes. Iron(III) thiocyanate gives a blood-red solution, used in detecting either Fe(III) or thiocyanate in solution, and is extracted from water by amyl alcohol. It is not formed in the presence of fluoride, phosphate and other strongly complexing ions.

When thiocyanic acid is treated with certain oxidizing agents, e.g., nitric acid, sulfuric acid and hydrocyanic acid are formed, but the action of lead tetraacetate on the acid, or of bromine in ether on lead(II) thiocyanate, gives thiocyanogen ("Rhodan") NCSSCN, a yellow, volatile oil, mp about −3°C, which polymerizes irreversibly at room temperature to insoluble, brick-red parathiocyanogen $(NCS)_x$. Thiocyanogen reacts with organic compounds like a free halogen. It liberates iodine from iodides. In water it is rapidly hydrolyzed to sulfuric and hydrocyanic acids. When thiocyanic acid is treated with reducing agents, e.g., aluminum and dilute hydrochloric acid, hydrogen sulfide plus carbon plus ammonium chloride are formed.

Esters.   Ethyl thiocyanate $C_2H_5 \cdot SCN$, colorless liquid, bp 142°C. Formed by reaction (1) of potassium thiocyanate and potassium ethyl sulfate, (2) of cyanogen chloride and ethanethiol. Oxidizable with fuming nitric acid to ethyl sulfonic acid $C_2H_5 \cdot SO_2OH$, and reducible with zinc and dilute sulfuric acid to ethane thiol C_2H_5SH. Ethyl isothiocyanate $C_2H_5 \cdot NCS$, colorless, odorous liquid, bp 132°C. Formed by reaction of ethyl amine and carbon disulfide (cf. the formation of ammonium thiocyanate from ammonia and carbon disulfide). Reducible to ethyl amine $C_2H_5NH_2$ plus methylene sulfide CH_2S. Allyl isothiocyanate ("mustard oil") $C_3H_5 \cdot NCS$ liquid, bp 151°C, odor of mustard, and causes blisters in contact with the skin.

THIOETHERS.   Hydrogen sulfide yields two classes of organic compounds: (1) hydrosulfides, and (2) sulfides. The sulfides are termed thioethers. A more general term, *thiols*, also is used. This term not only embraces thioethers, but also covers thioalcohols, sulfhydrates, and thiophenols.

Ethyl sulfide $(C_2H_5)_2S$, one of the better known thioethers, is an odorous, inflammable liquid, mp −102.1°C, bp 91.6°C, sp gr 0.837. The compound is insoluble in H_2O and soluble in alcohol and ether. It is prepared by distilling ethyl potassium sulfate with potassium sulfide. Chemically, ethyl sulfide behaves much like the ethers. For example, none of the hydrogen atoms can be displaced by metals and generally the compound is very inert. Additional thioethers can be prepared in a similar manner with the corresponding proper ingredients. Upon oxidation with HNO_3, thioethers are converted to sulfones. The latter are stable crystalline substances. An example is ethyl sulfone $(C_2H_5)_2SO_2$.

THIOKOL RUBBERS.   Elastomers.

THIOPHENE. $((CH:CH)_2)S$, formula weight 84.13, colorless liquid resembling benzene in odor, mp −30°C, bp 84°C, sp gr 1.070. Thiophene and its derivatives closely resemble benzene and its derivatives in physical and chemical properties. Thiophene is present in coal tar and is recovered in the benzene distillation fraction (up to about 0.5% of the benzene present). Its removal from benzene is accomplished by mixing with concentrated sulfuric acid, soluble thiophene sulfonic acid being formed. Thiophene gives a characteristic blue coloration with isatin in concentrated sulfuric acid.

Thiophene may be formed (1) by passing ethyl sulfide (diethyl sulfide) through a red-hot tube, (2) by reduction of sodium succinate and phosphorus trisulfide. Chlorine and bromine yield chloro- and bromo-substitution products, respectively, cold fuming nitric acid yields thiophene sulfonic acid. Thiophene aldehyde $C_4H_3S \cdot CHO$, liquid, bp 198°C, resembles benzaldehyde chemically rather than furfural. The corresponding primary alcohol and carboxylic acid are known. By comparison, where the sulfur atom of thiophene is occupied by oxygen, furane is the resulting compound. Where the sulfur atom of thiophene is occupied by a nitrogen group (NH), pyrrole is the resulting compound.

Benzothiophene $C_6H_4 \cdot (CH)_2S$ is a solid, mp 31°C, bp 221°C, with physical and chemical properties that resemble naphthalene. By comparison, where the sulfur atom of benzothiophene is occupied by oxygen, the resulting compound is benzofurane (coumarone). Where the sulfur atom of benzothiophene is occupied by a nitrogen group (NH), indole is the resulting compound.

THIOUREA. $(NH_2)_2CS$, formula weight 76.12, white crystalline solid, mp 180–182°C, decomposes before boiling at atmospheric pressure, sp gr 1.405. Thiourea is moderately soluble in H_2O, soluble in alcohol, and slightly soluble in ether. Sometimes referred to as thiocarbamide, sulfurea, and sulfocarbamide, thiourea may be considered chemically analogous to urea and is oxidized to urea by cold potassium permanganate solution. The compound is easily hydrolyzed to NH_3, CO_2, and H_2S. Upon long heating below the melting point, thiourea is transformed to ammonium thiocyanate. Thiourea is attractive for plastics manufacture because of the greater ease with which substitution can be made on the sulfur atom of thiourea than on the oxygen atom of urea.

Thiourea is formed by heating ammonium thiocyanate at 170°C. After about an hour, 25% conversion is achieved. With HCl, thiourea forms thiourea hydrochloride; with mercuric oxide, thiourea forms a salt; and with silver chloride, it forms a complex salt.

Symmetrical diphenyl thiourea (thiocarbanilide) $(C_6H_5NH)_2CS$ is a solid, mp 154°C. When heated with concentrated HCl, the compound yields aniline plus phenylisocyanate. Formed by the reaction of aniline and CS_2, symmetrical diethylthiourea $(C_2H_5NH)_2CS$ is a solid, mp 77°C.

In addition to its use in plastics manufacture, thiourea is used in some photographic processes and photocopying papers; in organic synthesis as an intermediate (drugs, dyes, cosmetics); in rubber accelerators; and as a mold inhibitor.

THIXOTROPIC SUBSTANCES. Rheology.

THORIANITE. The mineral of thorium oxide, ThO_2, is isomorphous with uraninite and occurs in black, nearly opaque cubic crystals in Ceylon and in the Malagasy Republic. Often containing rare-earth metals and uranium, the ore is strongly radioactive. Because of its radioactivity, it is valuable in helping to date the relative ages of rocks in which it occurs.

THORITE. The mineral thorite is a silicate of the rare element thorium and corresponds to the formula $ThSiO_4$. It is tetragonal and exhibits a prismatic cleavage. The original thorite was black in color with a specific gravity of 4.4–4.8. A variety orangite, so called from its orange-yellow color, has a specific gravity of 5.19–5.40. It has been found partly altered to thorite. Uranothorite contains uranium oxide. Thorite occurs in Norway in augite syenites. Thorite and orangite occur in Sweden, and orangite and uranothorite are found in the Malagasy Republic. Uranothorite is found in Ontario.

THORIUM. Chemical element symbol Th, at. no. 90, at. wt. 232.038, radioactive metal of the actinide series, mp 1740–1760°C, bp 4780–4800°C, density, 11.5–11.9 g/cm³ (17°C). Thorium metal is dark-gray, dissolves in HCl, is made passive in HNO_3, and is not affected by fusion with alkalis. The element combines with chlorine or sulfur at 450°C; with hydrogen or nitrogen at 650°C. All thorium-containing substances are radioactive. The element was discovered by J. J. Berzelius in 1829.

Electronic configuration is $1s^22s^22p^63s^23p^63d^{10}4s^24p^64d^{10}4f^{14}5s^25p^65d^{10}6s^26p^66d^27s^2$. Ionic radius, Th^{3+}, 1.08 Å, Th^{4+}, 0.95 Å (*Zachariasen*). Metallic radius, 1.7975 Å. First ionization potential, 5.7 eV; second, 16.2 eV; third, 29.4 eV. Oxidation potentials: $Th \rightarrow Th^{4+} + 4e^-$, 1.90 V; $Th + 4OH^- \rightarrow ThO_2 + 2H_2O + 4e^-$, 2.48 V.

The isotopes of thorium include mass numbers 223–234. Isotope ^{232}Th has a half-life of 1.39×10^{10} years. See **Radioactivity.** It emits an alpha particle and forms meso-thorium 1 (radium-228), which is also radioactive, having a half-life of 6.7 years, emitting a beta particle. Since ^{232}Th captures slow neutrons to form, by a series of nuclear reactions, ^{233}U, which is fissionable, thorium can be used as a fuel for nuclear reactors of the breeder type. Thorium occurs in earth minerals, an average content estimated at about 12 ppm. Findings of the *Apollo 11* space flight indicated that thorium concentrations in some lunar rocks are about the same as the concentrations in terrestrial basalts.

Thorium occurs in monazite sand in Brazil, India, North and South Carolina; this ore contains 3–9% thorium oxide, and is the chief source; thorium is also found in thorite containing about 60% oxide and in thorianite, about 80% oxide. When heated with concentrated H_2SO_4 the minerals form thorium sulfate, from which, by a series of reactions, thorium nitrate, the chief commercial compound, is obtained.

Thorium has the oxidation state of (IV) in all of its important compounds. Its oxide, ThO_2, and its hydroxide are entirely basic. The nature of the ions present in a number of solutions of the soluble compounds is not known with certainty. Complex ions involving sulfate are suggested by the increased solubility of the sulfate in solutions of the acid sulfates. Similarly, other complex ions are suggested by the solubility of the carbonate in excess alkali carbonate and of the oxalate in ammonium oxalate. Such ready complex ion formation is consistent with the high positive charge of the thorium(IV) ion.

Although the exact extent is not known accurately, hydrolysis of various salts is known to occur. Since the hydroxide is not precipitated, it is assumed that the hydrolysis product in some ion of the form $Th(OH)_2^{2+}$ or $ThOH^{3+}$. The solution chemistry of thorium is made more complex because of the hydrolytic phenomena observed and the polynuclear complex ions that are formed at low acidities and higher thorium concentrations.

Studies of the complex ions formed by Th^{4+} with various complexing anions have given much information. For example, the equilibria and ionic species involved in the chloride complexing of aqueous thorium have been studied through the method of measuring the distribution between H_2O and benzene containing thenoyltrifluoroacetone, with the conclusion that there is successive complexing involving the species $ThCl^{3+}$, $ThCl_2^{2+}$,

ThCl$_3^+$ and ThCl$_4$. Similarly all the intermediate chelate complex ions between thorium and acetylacetone exist in aqueous solution of proper acidity.

Thorium dioxide (face-centered cubic structure) is very insoluble in H$_2$O, but dissolves in acids to yield salts.

Thorium forms one series of halides, another one of oxyhalides, and also a series of double or complex halides. In general, stability of these compounds toward heat decreases as the atomic weight of the halogen increases. These compounds are often isostructural with the corresponding compounds of other actinide elements in the (IV) oxidation state.

Thorium metal reacts with hydrogen at moderately elevated temperatures to yield two hydrides: ThH$_2$, which has a pseudotetragonal body-centered unit containing two metal atoms, isomorphous or pseudoisomorphous with thorium carbide, zirconium hydride, and zirconium carbide. ThC$_2$, ZrH$_2$, and ZrC$_2$, and a hydride of approximate composition ThH$_{3.75}$ or ThH$_4$ possessing a unique cubic structure unrelated to that of the parent metal.

Thorium sulfide, ThS$_2$, is obtained by the action of H$_2$S or sulfur on thorium metal. The oxysulfide, ThOS, has been obtained in several ways, one of which is by the action of CS$_2$ on thorium dioxide at elevated temperatures. At 800°C and under pressure, sulfur combines with thorium to yield compounds with approximately the formulas ThS, Th$_2$S$_3$, and Th$_3$S$_7$. The first two have semimetallic properties and may be employed as ceramics for use with highly electropositive metals, whereas the last appears to be a polysulfide.

Anhydrous thorium sulfate, Th(SO$_4$)$_2$, is obtained by the action of concentrated H$_2$SO$_4$ on thoria (ThO$_2$). A solution of this salt deposits crystals of Th(SO$_4$)$_2$·9H$_2$O at about 15 C, Th(SO$_4$)$_2$·8H$_2$O near 24 C, and Th(SO$_4$)$_2$·4H$_2$O around 45 C. At 100 C other hydrates change to Th(SO$_4$)$_2$·2H$_2$O. In aqueous solution, the salt is considerably hydrolyzed to an oxysulfate—for instance, ThOSO$_4$·H$_2$O.

Thorium nitrate, Th(NO$_3$)$_4$·12H$_2$O, is obtained by dissolving thorium hydroxide in HNO$_3$.

Thorium orthophosphate, Th$_3$(PO$_4$)$_4$·4H$_2$O, is precipitated by adding a solution of sodium phosphate to an acidic solution of a thorium salt. Thorium pyrophosphate, ThP$_2$O$_7$·2H$_2$O, precipitates when an acidic solution of thorium nitrate is treated with one of tetrasodium pyrophosphate.

Thorium has been used as a fuel for nuclear reactors since it is a fertile material for the generation of fissionable uranium-233. Thorium oxide is used for gas mantles. The oxide also helps to control grain size in tungsten filaments and strengthens nickel alloys (TD nickel). Thorium is also used as an alloying addition in magnesium technology and as a deoxidant for molybdenum, iron, and other metals. Several applications for thorium are found in electronic technology.

References

Smith, J. F., et al.: "Thorium: Preparation and Properties," Iowa State University Press, Ames, Iowa (1975).

Smith, J. F.: "Thorium," in "Metals Handbook," 9th Edition, Vol. 2, American Society for Metals, Metals Park, Ohio (1979).

NOTE: See also the references listed at the end of the entry on **Chemical Elements.**

THORIUM SERIES. Radioactivity.

THREONINE. Amino Acids.

THULIUM. Chemical element symbol Tm, at. no. 69, at. wt. 168.934, twelfth in the lanthanide series in the periodic table,

mp 1545°C, bp 1950°C, density, 9.321 g/cm^3 (20°C). Elemental thulium has a close-packed hexagonal crystal structure at 25°C. Pure metallic thulium is gray, with no evidence of tarnishing up to a temperature of 200°C. Above 200°C, the element combines with oxygen, sulfur, nitrogen, carbon, and hydrogen and will form intermetallic compounds with metals heavier than iron. At higher temperatures, halogen gases react vigorously with the element to form trihalides. There is one natural isotope, ^{169}Tm. Seventeen artificial isotopes have been produced.

Average content of thulium in the earth's crust is estimated at 0.48 ppm, making this element the least abundant of the rare-earth elements. Even at this level, thulium is potentially more available than antimony, bismuth, cadmium, or mercury. The element was first identified by P. T. Cleve in 1879.

Electronic configuration is $1s^2 2s^2 2p^6 3s^2 3p^6 3d^{10} 4s^2 4p^6 4d^{10} 4f^{12} 5s^2 5p^6 5d^1 6s^2$. Ionic radius, Tm^{3+}, 0.87 Å. Metallic radius, 1.746 Å. First ionization potential, 6.18 eV; second, 12.05 eV. Other physical properties are given under **Rare-Earth Elements and Metals.**

Thulium occurs in apatite and xenotime and is derived from these minerals as a minor coproduct in the processing of yttrium. Processing involves organic ion-exchange, liquid-liquid, or solid-liquid techniques. Prior to the development of cation exchange resins capable of separating the chemically similar rare earths, thulium was practically unavailable in pure form. Thulium metal is made by the direct reduction of thulium oxide by lanthanum metal at high temperature in a vacuum.

Significant scientific and industrial applications for thulium and its compounds remain to be developed. In particular, the photoelectric, semiconductor, and thermoelectric properties of thulium and its compounds, mainly behavior in the near-infrared region of the spectrum, are being investigated. Thulium has been used in phosphors, ferrite bubble devices, and catalysts. Irradiated thulium (^{169}Tm) is used in some portable X-ray units.

See the references listed at the ends of the entries on **Chemical Elements;** and **Rare-Earth Elements and Metals.**

NOTE: This 4th Edition entry was revised and updated by K. A. Gschneidner, Jr., Director, and B. Evans, Assistant Chemist, Rare-Earth Information Center, Energy and Mineral Resources Research Institute, Iowa State University, Ames, Iowa.

THYROID HORMONES. Hormones; Iodine (In Biological Systems).

TIN. Chemical element symbol Sn, at. no. 50, at. wt. 118.69, periodic table group 4a, mp 231.97°C, bp 2270°C, density, 7.29 g/cm^3 (white tin at 15°C), 5.77 g/cm^3 (gray tin at 13°C), 6.97 g/cm^3 (liquid at mp). There are two allotropic forms of tin: (1) the more common soft, white beta tin has a body-centered tetragonal crystal form; (2) the brittle, gray alpha tin has a diamond-type cubic crystal form. The cubic form (α-tin), stable below 18°C, is an intrinsic semiconductor; at 161°C, white tin undergoes a transition to rhombic or γ-tin.

Tin is a silvery-white metal with a bluish tinge, softer than zinc and harder than lead; malleable, ductile at 100°C; can be powdered at 200°C, and upon exposure to temperatures below 18°C crumbles to a grayish powder due to the "tin pest," which is caused by transformation of white to gray tin (the reverse transformation may be brought about by heating gray tin to about 100°C); when a bar of tin is bent, a marked creaking sound is emitted due to the friction of the crystals; not oxidized on exposure to air at ordinary temperatures; burns to stannic oxide when heated to high temperatures in air or oxygen; soluble in HCl to form stannous chloride; converted by concentrated HNO$_3$ into insoluble beta-stannic acid; soluble in aqua regia to form stannic chloride; soluble in NaOH solution slowly to form sodium stannite and hydrogen gas; reacts with chlorine to form volatile stannic chloride. Discovery prehistoric.

Tin has the largest number of naturally occuring isotopes ^{112}Sn, ^{114}Sn through ^{120}Sn, ^{122}Sn, and ^{124}Sn. Five radioactive isotopes have been identified ^{111}Sn, ^{113}Sn, ^{121}Sn, ^{123}Sn, and ^{125}Sn. With exception of ^{121}Sn, which has a half-life of about 5 years, the half-lives of the other isotopes are comparatively short, expressed in minutes and days. Tin occurs in the earth's crust to the extent of about 40 grams/ton. It is estimated that a cubic mile of seawater contains about 15 tons of tin. First ionization potential 7.332 eV; second, 14.52 eV; third, 30.49 eV; fourth, 40.57 eV. Oxidation potential $Sn \rightarrow Sn^{2+} + 2e^-$, 0.406 V; $Sn^{2+} \rightarrow Sn^{4+} + 2e^-$, −0.14 V; $HSnO_2^- + 3OH^- + H_2O \rightarrow Sn(OH)_6^{2-} + 2e^-$, 0.96 V; $Sn + 3OH^- \rightarrow HSNO_{2-} + H_2O + 2e^-$, 0.79 V. Electronic configuration $1s^22s^22p^63s^2 3p^63d^{10}4s^24p^64d^{10}5s^25p^2$.

Other physical properties of tin are given under **Chemical Elements.**

Tin occurs as oxide (cassiterite, tin stone, stannic oxide, SnO_2), obtained commercially in Malaysia, Indonesia, Thailand, and Bolivia. The ore is concentrated and then roasted to oxide (83–88% stannic oxide). The product is treated in a blast furnace and crude tin recovered. Refining is conducted by electrolysis, or by fractional fusion.

Tin also occurs as complex sulfidic ores. The economic working of these ores is essentially confined to Bolivia. The ores include $SnS_2 \cdot Cu_2S \cdot FeS$ (stannite), SnS (herzenbergite), $SnS \cdot PbS$ (teallite), $2SnS_2 \cdot Sb_2S_3 \cdot 5PbS$ (franckeite), $Sn_6Pb_6 Sb_2S_{11}$ (cylindrite), $2SnS_2 \cdot 2PbS \cdot 2(FeZn)S \cdot Sb_2S_3$ (plumbostannite), and $4Ag_2S \cdot SnS_2$ (canfieldite).

Secondary tin is an important source of the metal. Tinplate scrap may be detinned electrolytically or chemically. The alkaline chemical process is the most widely used and involves a caustic solution which contains an oxidizing agent to remove both tin and the underlying iron-tin alloy from the steel. The solution formed then is either (1) crystallized to form sodium stannate, (2) electrolyzed to recover tin metal, or (3) acidified with CO_2, H_2SO_4, or acidic gases to precipitate hydrated tin oxide. There are over 85 secondary tin smelters in the United States, but only one primary smelter. The main primary tin smelters are located in the United Kingdom, Malaysia, and Thailand.

Uses. Not including Communist block nations, world consumption of tin is in excess of 187,000 metric tons annually. Principal free world uses are: (1) tinplate, 35%; (2) solder, 24%; (3) bronze, 9%; (4) other alloys, 8%; (5) tinning, 4%; (6) chemicals, 5%; (7) other uses, 15%. In the United States, tin consumption in 1979 was: (1) tinplate, 29%; (2) solder, 29%; (3) bronze and brass, 14%; (4) chemicals, 8%; (5) other alloys, 9%; (6) tinning, 4%; (7) other uses, 7%. In the United States, the major portion of tinplate is used in the making of cans. The advantages of tin for cans and food-processing equipment include its nontoxic nature, resistance to corrosive attack by acids and other aqueous solutions, and, when combined with other metals, strength.

Tin Plate. Tin coating may be applied to steel by (1) electroplating, usually as part of a highspeed, continuous process, or (2) by dipping cut sheets in a bath of molten tin. Electrolytic tin plate essentially is a sandwich in which the central core is strip steel. This core is thoroughly cleaned in a pickling solution prior to electroplating. The actual plating occurs as the strip moves through horizontal or vertical tanks containing electrolyte. The moving strip then is heated as it passes between high-frequency electric induction coils, whereupon the tin coating melts and flows to form a lustrous coat. The average thickness of tin on the end-product sheet is 0.00003 inch (0.0008 mm) on each side. A complex system of instrumentation is used to control process conditions and to inspect the moving sheet for any perforations in the plate. In hot-dip tinning, individual steel sheets are pickled and washed. A layer of hot palm oil is maintained on top of the molten tin bath to prevent oxidation of the molten tin by air and to prevent the molten tin from freezing too rapidly on the plate, thus providing a more even coating with a high luster.

Terne Plate. This is a sheet-steel product that is coated with an alloy of tin and lead. The coatings range from 50–50 mixtures of lead and tin to as low as 12% tin and 88% lead. Plate used for roofing normally is about 25% tin and 75% lead. In addition to roofing, terne plate is used in the manufacture of gasoline tanks for automotive vehicles, oil cans, and containers for solvents, resins, etc.

Alloys. Tin is widely used as both a major and minor ingredient of alloy metals. These applications are summarized in Tables

TABLE 1. REPRESENTATIVE SOFT SOLDERS

Composition, %				Temperature		Working Temperature Freezing
Tin	Lead	Antimony	Silver	Solidus	Liquidus	Range
80	20	0	0	183°C	203°C	20°C
70	30	0	0	183	192	9
60	40	0	0	183	189	6
50	50	0	0	183	216	33
49	50	1	0	186	210	24
40	60	0	0	183	234	51
39	60	1	0	186	230	44
30	70	0	0	183	252	69
29	70	1	0	186	252	66
20	80	0	0	183	273	90
20	78.75	0	1.25	180	270	90
19	80	1	0	185	273	88
10	88.50	0	1.50	178	290	112
10	90	0	0	183	297	114
5	95	0	0	270	311	41
2.5	97.5	0	0	301	319	18

Tin-bearing solders are considered soft solders as contrasted with the hard solders which contain substantial quantities of silver. However, small quantities of silver are added to some tin solders to increase strength of the resulting joint and to adjust working temperature range. Antimony also is added in some cases for the latter purpose. Wiping solders usually have a tin content ranging from 35 to 40%. Solders used in automotive-body work require a wide plastic (working temperature) range. Solders with a low tin content (below 25%) are generally used. For very low-temperature melting, bismuth and cadmium also may be added to tin-lead solders.

1, 2, and 3. Phosphor bronzes (Table 3) actually contain very little phosphorus, ranging from 0.03 to 0.50%, and hence the alloys are poorly designated. Tin bronzes is the better term. High-silicon bronzes contain about 2.8% tin; low-silicon bronzes about 2.0% tin. Gun metals are tin bronze casting alloys with a 5–10% zinc content. Some wrought copper-base alloys contain tin: (1) Inhibited Admiralty metal, 1% tin; (2) manganese bronze, 1% tin; (3) naval brass, 0.75% tin, (4) leaded naval brass, 0.75% tin. See also **Copper.**

Chemistry and Compounds. Tin forms two series of compounds: tin(II) or stannous; and tin(IV) or stannic compounds.

Tin(II) oxide, SnO, insoluble in water, is formed by precipitation of an SnO hydrate from an $SnCl_2$ solution with alkali and later treatment in water (near the bp and at constant pH).

TABLE 2.   REPRESENTATIVE BABBITT METALS

Ingredients	[a]SAE 10 %	SAE 11 %	SAE 12 %	SAE 13 %	SAE 14 %	SAE 15 %
Tin	90	86	88.25	4.5–5.5	9.25–10.75	0.9–1.25
Antimony	4–5	6–7.5	7–8.5	9.25–10.75	14–16	14.5–15.5
Lead	0.35	0.35	0.35	86	76	Remainder
Copper	4–5	5–6.5	2.25–3.75	0.50	0.50	0.6
Iron	0.08	0.08	0.08	—	—	—
Arsenic	0.10	0.10	0.10	0.60	0.60	0.8–1.10
Bismuth	0.08	0.08	0.08	—	—	—

[a] Society of Automotive Engineers.

It is amphiprotic, but only slightly acid, forming stannites slowly with strong alkalis. Sodium stannite is conveniently prepared from tin(II) chloride: $SnCl_2 + 3NaOH \rightarrow Na[Sn(OH)_3] + 2NACl$. Tin(IV) oxide, SnO_2, is much more acidic; it readily reacts with NaOH to form stannate ions, $Sn(OH)_6^{2-}$. In fact, no hydroxide of the formula $Sn(OH)_4$ ever has been obtained. The metal metastannates, e.g., $M^{II}SnO_3$, are generally made by fusion methods and have three-dimensional polymeric anions in which each tin atom is surrounded octahedrally by six oxygen atoms. There are, however, two forms of stannic acid, H_2SnO_3. The α-stannic acid is a white, gelatinous precipitate obtained by treating $SnCl_4$ with NH_4OH. The β-stannic acid (also called metastannic acid) is a white powder obtained by action of concentrated HNO_3 on tin; unlike the α-form, it is insoluble in concentrated acids and alkali metal hydroxides.

Tin forms dihalides and tetrahalides with all the common halogens. These compounds may be prepared by direct combination of the elements, the tetrahalides being favored. Like the halides of the lower main group 4 elements, all are essentially covalent. Their hydrolysis requires, therefore, an initial step consisting of the coordinative addition of two molecules of water, followed by the loss of one molecule of HX, the process being repeated until the end product, $H_2Sn(OH)_6$, is obtained. The most significant commercial tin halides are stannous chloride, stannic chloride, and stannous fluoride.

The increasingly electropositive character of main group 4 as tin is reached is evident from the fact that its hydrides are much less stable than those of silicon and germanium. Known are Sn_2H_6 and SnH_4, which is obtained by hydrolysis of magnesium stannide, Mg_2Sn, or by electrolysis of a phosphoric acid solution with a tin cathode.

Other inorganic compounds of tin include:

Nitrates. Stannous nitrate, $Sn(NO_3)_2$, white solid, by reaction of tin metal and dilute HNO_3, and crystallization, soluble in water with slight excess of HNO_3.

Sulfates. Stannous sulfate, a white powder soluble in water and sulfuric acid, is obtained commercially by action of H_2SO_4 on $SnCl_4$ or Sn. Stannic sulfate may be formed by the solution of stannic hydroxide in dilute H_2SO_4, or by action of oxidizing agents on stannous salts.

Sulfides. Stannous sulfide, SnS, dark brown precipitate, by reaction of stannous salt solution and H_2S, insoluble in sodium sulfide solution, but soluble in sodium polysulfide solution, forming sodium thiostannate; stannic sulfide, SnS_2, yellow precipitate, by reaction of stannic salt solution and H_2S, soluble in sodium sulfide solution, forming sodium thiostannate.

Organometallic Compounds. In common with the other elements of main group 4, tin forms many organometallic compounds; the range of possible combinations is virtually limitless. They include:

1. *Tetraorganotins*, R_4SN, prepared either by alkylation of tin halides with Grignard Reagents or alkyl lithium; by reaction of an organic halide with a tin-sodium alloy; by direct reaction of tin with an organic halide; or by reaction of stannic

TABLE 3.   REPRESENTATIVE TIN-BEARING BRONZES
Phosphor Bronzes

	1.25% Tin	4% Tin[c]	5% Tin	8% Tin	10% Tin
Copper[a]	98.75%	88.00%	95.00%	92.00%	90.00%
Melting point, °C	1,077	1,000	1,050	1,027	1,000
Tensile strength, 1,000 psi					
Hard sheet	65	58	81	93	100
Soft sheet	40	44	47	55	66
Rockwell hardness					
Hard sheet	75B	68B	87B	93B	97B
Soft sheet	60F	65F	73F	75F	55B
Electrical conductivity					
% IACS[b]	48	19	18	13	11
Thermal conductivity					
Btu(ft²)(ft)(°F) at 68°F	120	50	47	36	29
Major Uses	Electrical contact wire Messenger cable Flexible metal hose Pole-line hardware	Bearings Bushings Gears Pinions Shafts Screw-machine products Washers Valve parts	Bearings Bellows Bourdons Gears Rivets Springs Wire cloth Truss wire	Bearings Bellows Bourdons Fasteners Washers Springs Switch parts Chemical hardware	Heavy bars, plates Bridge and expansion plates Heavy springs

[a] Small amounts of zinc, lead, iron, antimony, and phosphorus also present.
[b] International Annealed Copper Standard.
[c] Free-cutting phosphor bronze.

chloride with alkyl aluminum compounds.

2. *Organotin halides*, $RSnX_3$, R_2SnX_2, and R_3SNX, prepared by disproportionation of the tetraorganotin with stannic halide or by direct alkylation of stannic halide.

3. *Organotin oxides*, R_2SnO or $(R_3SN)_2O$, prepared by treatment of the organotin halides with alkali.

4. *Stannoic acids*, $RSnOOH$.

The organotin halides and oxides are usually the intermediates used in the synthesis of other organotin derivatives, such as the organotin carboxylates, organotin sulfur-derivatives, organotin hydroxides, etc. The most significant commercial organotins include dibutyltin and dioctyltin carboxylates and sulfur derivatives, used as polyvinyl chloride (PVC) stabilizers and as catalysts in polymer systems; *bis*(tributyltin)oxide, triphenyltin fluoride, and tributyltin fluoride, used as antifoulants for marine paints, fungicides, bactericides, sanitizing agents, and wood preservatives; tricyclohexyltin hydroxide as an insecticide; triphenyltin hydroxide and triphenyltin acetate as agricultural fungicides; and dibutyltin dilaurate as a poultry anthelmintic.

References

Faulkner, C. J.: "The Properties of Tin," Tin Res. Inst. *Publ. 218* (1965).

Franklin, A. D., Olin, J. S., and T. A. Wertime, Editors: "The Search for Ancient Tin," Smithsonian Institution, Washington, D.C. (1978).

Hallas, L. E., Means, J. C., and J. J. Cooney: "Methylation of Tin by Estuarine Microorganisms," *Science*, 215, 1505 (1982).

Ingham, R. K., Rosenberg, S. D., and H. Gilman: "Organotin Compounds," *Chem. Rev.*, 60, 459–539 (1960).

Mantell, C. L.: "Tin: Its Mining, Production, Technology and Applications," 2nd Edition, Van Nostrand Reinhold, New York (1949).

Neuman, W. P.: "The Organic Chemistry of Tin," Wiley, New York (1970).

Staff: Various publications on tin and its compounds, including *Tin and Its Uses* (quarterly); *Tin International* (monthly); and statistical publications on tin (periodically), International Tin Research Institute, Middlesex, England.

Staff: "The Economics of Tin," 2nd Edition (1977); and "Statistical Supplement" (1980), Roskill Information Services Ltd., London.

Staff: "Annual Review of the World Tin Industry," Rayner-Harwill Ltd., London (1980).

Staff: "Tin Chemicals for Industry," International Tin Research Institute, Middlesex, England (1972).

Wright, P. A.: "Extractive Metallurgy of Tin," Elsevier, New York (1966).

—Marguerite K. Moran, M & T Chemicals Inc., Rahway, New Jersey.

TITANITE. A yellow or brown calcium silicotitanite, $CaTiSiO_5$, having a waxy luster, and often containing niobium (columbium), chromium, fluorine, and other elements. Titanite occurs in wedge-shaped monoclinic crystals, usually as an accessory mineral in granitis rocks and in calcium-rich metamorphic rocks. See also **Sphene.**

TITANIUM. Chemical element symbol Ti, at. no. 22, at. wt. 47.90, periodic table group 4b, mp 1650–1670°C, bp 3280°C, density, 4.507 g/cm³ (20°C). Elemental titanium has a hexagonal close-packed crystal structure below 885°C; above this temperature, it has a body-centered cubic crystal structure. Compact titanium is a white metal, when cold it is brittle and may be powdered, but at red heat may be forged and drawn into wire. Titanium exhibits some passivity in air to formation of coatings of oxide or nitride. At 610°C, titanium reacts with oxygen to form titanium dioxide; at 800°C, it reacts with nitrogen to form titanium nitride. Upon heating with chlorine, the metal forms titanium tetrachloride. Cold, dilute sulfuric acid readily dissolves the metal to form titanous sulfate. Hot, concentrated H_2SO_4 readily dissolves the metal to form titanic sulfate. The element was first identified by Gregor in 1789 and later named titanium by Klaproth (1795). A metal of 95% purity was not produced until 1887 when it was made by the reduction of titanium tetrachloride with sodium. The first commercial uses date back to 1860 when ferrotitanium was used as an alloying element in steel and a bit later as a deoxidizer in the production of steel. There are five natural isotopes, ^{46}Ti through ^{50}Ti, and three radioactive isotopes have been identified, ^{44}Ti, ^{45}Ti, and ^{51}Ti, the latter two having relatively short half-lives measured in minutes and hours. The isotope ^{44}Ti has a half-life of approximately 10^3 years, Titanium is relatively abundant, ranking 8th in the list of chemical elements occurring in the earth's crust. Titanium ranks 35th among the elements in terms of content in seawater, with an estimated 5 tons of titanium per cubic mile (1080 kilograms per cubic kilometer) of seawater.

Electronic configuration is $1s^2 2s^2 2p^6 3s^2 3p^6 3d^2 4s^2$. First ionization potential, 6.83 eV; second, 13.60 eV; third, 27.6 eV; fourth, 44.66 eV. Oxidation potential: $Ti + 2H_2O \rightarrow TiO_2 + 4H^+ + 4e^-$, 0.95 V; $Ti \rightarrow Ti^2 + 2e^-$, 1.75 V; $Ti^{2+} \rightarrow Ti^{3+} + e^-$, 0.37 V. Other physical properties of titanium are given under **Chemical Elements.**

Titanium occurs in practically all rocks and is an important constituent of many minerals. Only rutile TiO_2, however, is of commercial importance. The most important sources of this mineral are the sand dunes of Australia and Florida. Presently, Australia furnishes over 80% of the rutile requirements. Projects are underway to beneficiate the other major potential titanium source, e.g., ilmenite. The known reserves of ilmenite $FeTiO_3$ are estimate $50 \times$ greater than those of rutile. For mining the sand deposits for rutile, large floating dredge concentrators are used. Gravity concentration, followed by magnetic and electrostatic separation, yield a raw rutile of about 95% TiO_2 content. See also **Ilmenite; Rutile.**

Production of titanium metal first involves the preparation of $TiCl_4$, a colorless liquid. Rutile and coke are charged into a continuous chlorinator. Upon the addition of chlorine gas, $TiCl_4$ is yielded in an exothermic reaction. To separate the metal, the $TiCl_4$, in a separate process, is reacted with molten magnesium metal pigs at about 50°C. The products are magnesium chloride $MgCl_2$ and titanium metal sponge. The byproduct $MgCl_2$ is electrolyzed and the resulting magnesium and chlorine are recycled in the process. In another process, sodium metal is used instead of magnesium. And in still another process, the $TiCl_4$ may be electrolyzed.

Uses. The major uses for titanium are in various alloys, although unalloyed titanium finds some application. Titanium alloys are classified as alpha, alpha-beta, or beta, determined by the phases present in the alloy at room temperature. The alpha alloys usually result when the main elements present are the alpha stabilizers, e.g., oxygen, nitrogen, hydrogen, and carbon. Alpha-beta alloys and beta alloys contain increasing amounts of beta stabilizers, mainly vanadium, molybdenum, iron, chromium, manganese, tantalum, and niobium (columbium). The alpha-beta class of alloys normally has great room-temperature strength and may be heat treated. The annealed beta alloys show poor thermal stability over about 230°C, but do have good formability and weldability. The beta alloys may be age heat treated wherein some alpha phase is precipitated and this results in a very high room-temperature strength. The complexity of titanium alloys is brought about by the fact that the element is allotropic and undergoes a phase transformation at about 885°C, changing from one crystalline form to another as mentioned at the start of this entry.

The variations in strength and percent elongation for the three major types of alloys and for pure titanium are given in the accompanying table.

TITANIUM ALLOYS

ALLOY	TENSILE STRENGTH		YIELD STRENGTH		PERCENT ELONGATION
	psi	mPa	psi	mPa	
Pure titanium					
High purity (99.9%)					
Annealed	34,000	237	20,000	138	54
Commercial purity (99.0%)	79,000	545	63,000	435	27
Alpha alloy					
Ti-5Al-2.5 Sn					
Annealed	125,000	863	120,000	828	18
Alpha-beta alloy					
Ti-6Al-4V					
Annealed	135,000	932	120,000	828	11
Heat treated	170,000	1173	150,000	1035	7
Beta alloy					
Ti-3Al-13V-11Cr					
Heat treated	180,000	1242	170,000	1173	6

Classification of alloys by application:
Airframe Alloys
  Ti-75*A*, Ti-5Al-2.5Sn, Ti-6Al-6V-2Sn, Ti-6Al-4V, Ti-7Al-4Mo,
    Ti-4Al-3Mo-IV, Ti-8Mn, Ti-13V-11Cr-3Al
Engine Alloys
  Ti-8Al-1Mo-1V, Ti-5Al-2.5Sn, Ti-6Al-4V, Ti-6Al-2Sn-4Zr-2Mo,
    Ti-6Al-2Sn-4Zr-6Mo
Corrosion-resistant Alloys
  Ti-35*A*, Ti-50*A*, Ti-65*A*, Ti-0.2Pd, Ti-2Ni

Many diversified applications have been found for titanium and its alloys. Moreover, the number of these applications tends to increase steadily as greater production and improved processes reduce costs. At the present time, titanium is still an expensive material and is only used where its light weight, high strength, and corrosion resistance justify its cost. Aeronautical and missile design engineers find titanium and its alloys to be materials whose light weight and high strength, particularly at elevated temperatures (600°C+), give them many applications in aircraft and missile construction. A very high percentage of all titanium materials is used in these fields.

Titanium and its alloys are widely used in compressor blades, turbine disks and many other forged parts of the jet engine. Here they offer resistance to high temperature, as well as weight-saving. The latter quality is increasing their use in the structural airplane parts, ranging from engines and air frames to skin and fastenings. Titanium sheet finds application in shroud assemblies, cable shrouds and ammunition tracks. Titanium alloy sheet is formed into ribs for use as stiffeners, as well as fuselage frames and bulk heads. Other uses of titanium in aircraft include channel sections, flat rubbing strips, landing gear doors, hydraulic lines, baffles, tail cones, longerons, etc. Other uses of titanium alloys include bulk heads, ducts, fire walls, etc.

The light weight of titanium and its alloys, coupled with their corrosion resistance, has brought them into use in ships, especially naval ships. Here many investigations show the important advantages of the metal and its alloys as wet exhaust muffles for submarine diesel engines, and as meter disks, and heat exchanger tubes which offer improved service for widespread use in salt water. Military applications of titanium extend from cannon and guided missiles to light-weight armor-plate for tanks. These materials offer other weight savings in other parts of military vehicles, such as piston rods and transmissions, which may extend to the transportation industries generally.

Throughout the chemical industry, titanium is used extensively both in plant and in laboratory. Among important present-day applications are heat exchangers, autoclave heads, autoclave coils for cooling and heating, chemical processing racks, and valves and tanks where corrosion resistance is necessary.

Chemistry and Compounds. Due to its $3d^24s^2$ electron configuration, titanium forms tetravalent compounds readily, although the Ti^{4+} ion does not exist as such in aqueous solution, except at very low or high pH values, the common cation being hydrated TiO^{2+} (or more probably, $Ti(OH)_2^{2+}$). Many of the tetravalent compounds are largely covalent. There are also Ti (III) and a few Ti(II) compounds, the latter being very easily oxidized. Titanium dioxide, TiO_2, is well known both as a mineral, of which three structural forms exist, and as an industrial product obtained from ferrous titanate, $FeTiO_3$, ores or by oxidation of titanium(IV) chloride, $TiCl_4$. Moreover, the precipitate obtained by action of alkali metal hydroxides upon solutions of tetravalent titanium is a hydrated oxide. The latter is readily soluble in acids to form oxysalts, which are usually formulated in terms of the TiO^{2+} ion, without including its water of hydration, e.g., as $NaTiOPO_4$. The hydrated TiO^{2+} ion is not amphiprotic, in that it does not dissolve in alkali hydroxides; however, it does react on fusion with alkali carbonates to form such compounds as M_2TiO_3 and $M_2Ti_2O_5$, these compounds having been shown to be mixed oxides rather than titanates. The alkaline earth titanates have the face-centered perovskite structure, and barium titanate, widely used for its electrical properties, has been produced in other crystalline forms.

Lower oxides of titanium, Ti_2O_3 and TiO, have been produced by reduction of TiO_2.

All four of the common halogens form tetrahalides of titanium, $TiCl_4$ being a liquid at ordinary temperatures, while TiF_4, $TiBr_4$, and TiI_4 are solids. They are readily hydrolyzed, yielding as end products TiO_2 and the hydrogen halide, in the case of $TiCl_4$ an intermediate addition product of the type $H_2O \cdot TiCl_4$ is considered to be formed. This is in accordance with the behavior of $TiCl_4$ and $TiBr_4$ as Lewis acids to form such unstable adducts, not only with water, but with oxygen-function organic compounds. Likewise titanium chelates are formed with oxygen donor compounds such as acetylacetone.

The dihalides of titanium, formed by reduction of the tetrahalides, are vigorous reducing agents and unstable; $TiCl_2$ is inflammable in air. The trihalides, though more stable than the dihalides, are effective reducing agents. Ti(III) occurs in aqueous solutions as $Ti(H_2O)_6^{3+}$.

Normal oxyacid salts of titanium are unknown, but many basic salts, formulated as stated above, in terms of TiO^{2+}, though more or less hydrated, have been prepared.

Like the oxide, halides and sulfide, the nitride, boride, and carbide of titanium(IV) can be made by heating the elements together at high temperatures. The last three compounds are alloy-like in character, they can vary in composition without becoming unstable and they are extremely hard.

The halogen complexes are the most stable complex ions of titanium. The hexafluorotitanate ion, TiF_6^{2-} is very stable, as are the peroxo-complexes, containing $-Ti-O-O-$. The $TiCl_6^{2-}$ and $TiBr_6^{2-}$ complexes are less stable, except in concentrated solutions of the hydrogen halides. A number of compounds of the $TiCl_5^{2-}$ ion are known, especially of the higher alkali metals, e.g., $M_2TiCl_5 \cdot H_2O$.

References

Kuhlman, G. W., and T. B. Burganus: "Optimizing Thermomechanical Processing of Ti-10V-2Fe-3A1 Forgins," *Metal Progress*, **118**, 2, 30–35 (1980).

Lyman, W. S.: "Titanium," in "Metals Handbook," 9th Edition, American Society for Metals, Metals Park, Ohio (1979).

Weast, R. C., Editor: "Handbook of Chemistry and Physics," CRC Press, Boca Raton, Florida (Published annually).

TITANIUM DIOXIDE. TiO_2, formula weight 79.90, variously colored, depending upon source, but white when purified and sold in commerce. Decomposes at about 1,640°C before melting, density 4.26 g/cm³, insoluble in H_2O, soluble in H_2SO_4 or alkalis. Titanium dioxide is a very high-tonnage material and is the principal white pigment of commerce. The compound has an exceptionally high refractive index, great inertness, and a negligible color, all qualities that make it close to an ideal white pigment. Annual production approximates two million metric tons, of which nearly one-half of this amount is produced in the United States. Major uses of TiO_2 pigments are: (1) paint, 60%, (2) paper, 14%, (3) plastics and floor coverings, 12%, (4) printing inks, 3%, and (5) various applications including rubber, ceramics, roofing granules, and textiles, 11%.

Two major processes are used for producing raw titanium dioxide pigment: (1) the sulfate process, a batch process accounting for over half of current production, introduced by European makers in the early 1930s; and (2) the chloride process, a continuous process, introduced in the late 1950s and accounting for most of the new plant construction since the mid-1960s. The sulfate process can handle both rutile and anatase, but the chloride process is limited to rutile.

In the sulfate process, ilmenite (45–60% TiO_2) or a slag rich in titanium (70% TiO_2) obtained from electric smelting of ilmenite, is the feedstock. The raw materials first are digested: $FeTiO_3 + 2H_2SO_4 \rightarrow FeSO_4 + TiO \cdot SO_4 + H_2O$. In a second step, the concentrated liquor is nucleated, diluted with H_2O, and boiled until nearly all of the titanium has precipitated out in the form of flocculated titanium dioxide (anatase) hydrate: $TiO \cdot SO_4 + 2H_2O \rightarrow TiO_2 \cdot H_2O + H_2SO_4$. After filtering, the cake is leached under reducing conditions to remove residual iron. Conditioning agents are added, after which the hydrate is dried and calcined in a rotary kiln at approximately 900°C: $TiO_2 \cdot H_2O \rightarrow TiO_2 + H_2O$. The conditioning agents usually consist of a phosphate and a potassium salt, as well as zinc, antimony, and aluminum compounds. The purpose of these additions is to improve the final properties of the pigment, including color, photochemical stability, and dispersibility, as well as to catalyze the formation of rutile from the anatase hydrate.

In the chloride process, the feedstock must be high in titanium and low in iron. Mineral rutile (95% TiO_2) is best suited, but leucoxene (65% TiO_2) can be used. See also **Brookite.** An economical conversion of ilmenite for use as a chloride process feedstock has not been developed to date. The ore is mixed with coke and chlorinated at about 900°C in a fluidized bed. The principal product is titanium tetrachloride, but other impurities including iron also are chlorinated and thus must be removed by selective condensation and distillation. Up to this point, the process is similar to that of producing titanium metal as described under **Titanium.** By selective reduction prior to distillation, vanadium present is removed as $VOCl_3$. In the next step, the purified $TiCl_4$ reacts with oxygen at a temperature of about 1,000°C. The presence of $AlCl_3$ in this reaction promotes the formation of rutile instead of anatase. The two major steps are: (1) chlorination: $3TiO_2 + 4C + 6Cl_2 \rightarrow 3TiCl_4 + 2CO + 2CO_2$; and (2) oxidation: $TiCl_4 + O_2 \rightarrow TiO_2 + 2Cl_2$.

The chlorine is recycled. The raw titanium dioxide product generally is neutralized by washing in an aqueous solution of proper pH.

Many grades of titanium dioxide pigments are offered commercially. They range in crystal structure (anatase or rutile), particle shape and size, the type of hydrous oxide coating applied, and the type and quantity of additives applied. Generally, the commercial pigments contain 80–99% TiO_2, the remainder of the formulation comprised of alumina and silica hydrates. Nonpigmentary grades of titanium dioxide for the glass, welding-rod, electroceramic, and vitreous-enamel industries contain 99% TiO_2.

TOCOPHEROLS. Vitamin E.

TOLUENE. $C_6H_5CH_3$, formula weight 92.13, colorless, odorous liquid, mp −95°C, bp 110.8°C, sp gr 0.866. Toluene, a homolog of benzene (C_nH_{2n-6} series of aromatic hydrocarbons), essentially is insoluble in H_2O, but is fully miscible with alcohol, ether, chloroform, and many other organic liquids. Toluene dissolves iodine, sulfur, oils, fats, resins, and phosgene. When ignited, toluene burns with a smoky flame. Unlike benzene, toluene cannot be easily purified by crystallization.

Industrial-grade toluene distills between 108.6 and 112°C, is water-white and has a flash point of 2 to 5°C. The specific gravity ranges from 0.854 to 0.874. Toluene is a high-tonnage industrial chemical with United States production approximating 3 million tons/year. Petroleum sources account for about 95% of toluene production, the remainder coming from coal gas and coal tar. The dealkylation of toluene is a prime source of benzene, accounting for about one-half of toluene consumption. The production of diisocyanates from toluene is increasing. As a component of fuels, the use of toluene is lessening. Toluene takes part in several industrially-important syntheses. The hydrogenation of toluene yields methyl cyclohexane $C_6H_{11}CH_3$, a solvent for fats, oils, rubbers, and waxes. Trinitroltoluene (TNT) $C_6H_2(CH_3)(NO_2)_3$ is a major component of several explosives. When reacted with H_2SO_4, toluene yields *o*- and *p*-toluene sulfonic acids $CH_3C_6H_4SO_3H$. Saccharin is a derivative of the ortho acid; chloramine T (an antiseptic) is a derivative of the para acid. A widely used solvent for synthetic resins and rubber, monochlorotoluene $CH_3C_6H_4Cl$ is a derivative of toluene. Toluene also is used in the manufacture of benzoic acid, the latter an important ingredient for phenol production.

One modern toluene production process commences with mixed hydrocarbon stocks and can be used for making both toluene and benzene, separately or simultaneously. The process essentially is a combination of extraction and distillation. An aqueous dimethyl sulfoxide (DMSO) solution is passed countercurrently against the mixed hydrocarbon feed. A mixture of aromatic and paraffinic hydrocarbons serves as reflux.

In other industrially-important processes, toluene is a source of benzyl chloride $C_6H_5CH_2Cl$, benzal chloride $C_6H_5CHCl_2$, benzotrichloride $C_6H_5CCl_3$, benzyl alcohol $C_6H_5CH_2OH$, benzaldehyde, C_6H_5CHO, and sodium benzoate C_6H_5COONa.

TOPAZ. The mineral topaz is a silicate of aluminum and fluorine corresponding to the formula $Al_2SiO_4(F, OH)_2$. It is orthorhombic and its crystals are mostly prismatic terminated by pyramidal and other faces, the basal pinacoid being often present. Massive varieties are known. It has an easy and perfect basal cleavage hence for this reason gems or fine specimens should be handled with care to avoid developing cleavage flaws. The fracture is conchoidal to uneven; hardness, 8; specific gravity, 3,4–3.6; luster, vitreous; color, of typical topaz, wine or straw-yellow but may be colorless, white, gray, green, blue or reddish-yellow; transparent to translucent. When heated, yellow topaz often becomes a reddish pink.

Topaz is found associated with the more acid rocks of the granite and rhyolite type and may occur with fluorite and cassiterite. Topaz comes from many localities, a few of which are: the U.S.S.R. in the Urals and the Ilmen Mountains; Czechoslovakia, Saxony, Norway, Sweden, Japan, Brazil and Mexico. In the United States topaz has been found in Oxford County, Maine; Carroll County, New Hampshire; Fairfield County, Connecticut; El Paso and Chaffee Counties, Colorado; and in Texas, Utah and California.

The name topaz is derived from the Greek meaning to seek, which was the name of an island in the Red Sea that was difficult to find from which a yellow stone, now believed to be a yellowish-olivine, was obtained in ancient times. In the Middle Ages any yellow stone was called topaz, but now the name is properly applied only to the species here described.

TORBERNITE. An ore of uranium with the composition $Cu(UO_2)_2(PO_4)_2 \cdot 8-12H_2O$, green, radioactive, tetragonal, and isomorphous with autunite. Occurring in tabular crystals or in foliated form, the mineral is commonly a secondary mineral.

TOURMALINE. The mineral tourmaline is a complex silicate of aluminum and boron, but because of isomorphous replacements this mineral varies widely in chemical composition, iron, magnesium, and lithium entering into combination to a greater or less extent with the aluminum and boron. Its general formula is $(Na,Ca)(Mg,Fe^{2+},Fe^{3+},Al,Li)_3Al_6(BO_3)_3(Si_6O_{18})(OH,F)_4$. Tourmaline belongs to the hexagonal system, its crystals are usually prismatic, tending to be long and slender, often acicular. The crystals are ordinarily terminated with three faces of a rhombohedron and usually hemimorphic. The smaller crystals are frequently found in radial arrangement, and columnar masses are common. The prisms are usually three-, six-, or nine-sided heavy vertical striations producing a rounded effect.

Tourmaline is essentially without cleavage; fracture, conchoidal to uneven; brittle; hardness, 7–7.5; specific gravity, 3.03–3.25; luster, vitreous inclining to resinous; color, in common tourmaline black, bluish-black, brown, blue, green, red or pink, and in the transparent varieties colorless (rare), various shades of rose and pink, greens, blues and browns. The color arrangement in tourmaline is of considerable interest; bi-colored crystals are common and may be green at one end and pink at the other, or green on the outside, and pink within, which, in the case of transparent or translucent crystals, is very attractive.

The opaque black tourmaline is called schorl, a term which was applied to all tourmaline until 1703 when the word *tourmaline* was introduced, it being a corruption of the Ceylonese word, *turamali*. The origin of the word schorl is not known, but is perhaps Scandinavian, and is used to identify the iron-bearing black tourmalines; elbaites and liddicoatites tend to light shades of blue, red, green, and their bi-colored combinations; the brown colored tourmalines of varying shades of dark brown to yellow to nearly colorless are called dravites and uvites (with the exception of the black tourmalines found at Pierrepont, New York, which have been identified as uvites); the completely colorless variety, achroite, falls within the elbaite group. Small tourmalines are found in granites and some gneisses.

Due to the mineralizing action of magmatic vapors; tourmaline is found particularly well developed in pegmatites, and as a contact metamorphic mineral. A few of the important localities are: the Ural Mountains; Bohemia; Saxony; the Island of Elba; Norway; Devonshire and Cornwall, England; Greenland; the

Malagasy Republic. Magnificent elbaite crystals are obtained from Madagascar, Brazil, and Afghanistan; liddicoatite crystals from Madagascar. In the United States in Oxford and Androscoggin Counties, Maine; Grafton and Sullivan Counties, New Hampshire; Hampshire County, Massachusetts; Haddam and Fairfield Counties, Connecticut; St. Lawrence County, New York; Sussex County, New Jersey; Delaware County, Pennsylvania; and San Diego County, California.

TOXICOLOGY. The technology which deals with the study of poisons, their detection, and counteraction. Poison is a highly generalized term referring to living kinds of disease-producing agents, notably bacteria, the byproducts of which are called toxins; as well as to inorganic and organic chemical substances that produce ill effects in humans, ranging from some form of debilitation to almost instant death. Drugs, for example, when taken without guidance and/or in excess are poisons. Rather than to list the leading poisons encountered here, the toxicological properties of numerous materials are described as they appear alphabetically throughout this volume. An exhaustive reference on industrial poisons and contaminants is "Dangerous Properties of Industrial Materials," 5th Edition, N. I. Sax, Van Nostrand Reinhold, New York (1979).

TRACER (Radioactive). Radioactivity.

TRANSACTINIDE ELEMENTS. Chemical Elements.

TRANS-COMPOUND. Isomerism.

TRANSFERENCE NUMBER (Transport Number). Of a given ion in an electrolyte, the transference number is the fraction of total current carried by that ion.

TRANSURANIUM ELEMENTS. The chemical elements with an atomic number higher than 92 (uranium), commencing with 93 (neptunium) and through 103 (lawrencium) frequently are termed Transuranium elements. Any additional elements that may be officially identified will become a part of this series. See also **Actinide Series;** and **Chemical Elements.**

TRAVERTINE. Carbonated waters dissolve large amounts of calcium carbonate, especially under high temperature. Such waters reaching the earth's surface as hot springs often deposit the calcium carbonate in great quantities. This material is called *travertine* from the ancient name for Tivoli, Italy, where a very thick deposit occurs. Travertine may be compact, crystalline, fibrous, or, if rapidly deposited, spongy and porous. The less compact varieties are known as *tufa.* Travertine is being formed at the Mammouth Hot Springs, Yellowstone National Park and at many other localities. A banded travertine used as an ornamental stone is called *onyx marble* or *Mexican onyx.*

TREMOLITE. The mineral tremolite is a calcium-magnesium silicate corresponding to the formula $Ca_2Mg_5Si_8O_{22}(OH)_2$, belonging to the amphibole group. The replacement of magnesium by ferrous iron causes tremolite to approach actinolite in composition. Tremolite is monoclinic, developing bladed prismatic crystals, but it is frequently found in compact columnar, granular, or fibrous masses. The perfect prismatic cleavage at angles of 56° and 124° typical of this group is to be noted; hardness, 5–6; specific gravity, 2.9–3.1; luster, vitreous to silky; color, varies from white or whitish-gray through shades of green or greenish-yellow; transparent to opaque. Tremolite is formed as a result of contact metamorphism and occurs in marbles, dolomites, and schists. It may alter to talc. Tremolite is found in Switzerland, in the St. Gotthard region, being named for the Tremola Valley, and is common elsewhere in Europe. In the United States it occurs in Maine, Pennsylvania, and New York. In Canada tremolite has been found in Quebec and Ontario.

Hexagonite is a pinkish-purple variety of tremolite which contains a small amount of manganese. So called because it was at first believed to be hexagonal. It has been since shown to be monoclinic, and is found in St. Lawrence County, New York. Some nephrite and asbestos is tremolite.

TRIACETATE. Cellulose Ester Plastics (Organic); Fibers.

TRICARBOXYLIC ACID CYCLE. Carbohydrates.

TRIDYMITE. The mineral tridymite is, like quartz, silicon dioxide, SiO_2, but is a high-temperature variety, probably stable above 870°C. It has a conchoidal fracture; is brittle; hardness, 7; specific gravity, 2.28–2.33; vitreous luster; usually colorless and transparent. It is found chiefly in volcanic rocks of the more acidic types like rhyolite, tachyte, and andesite. It is not a particularly uncommon mineral, occurring in Germany, France, Italy, Japan, the Island of Martinique, Mexico, and in the United States in Wyoming and Washington. Tridymite is hexagonal but when heated to about 1,470°C passes into an isometric form, cristobalite, which was first noted in the andesitic lavas of the Cerro San Cristobal, Pachuca, Mexico, together with tridymite. Cristobalite has been found also in California and in Germany.

TRIPHYLITE. This mineral is a phosphate of lithium and ferrous iron, $LiFePO_4$. It crystallizes in the orthorhombic system but usually is characterized by large cleavable masses. The hardness is 4.5–5.0; specific gravity, 3.42–3.56, vitreous to resinous luster, translucent, and blue-gray color. The mineral occurs as a rare primary mineral in granitic pegmatites and, when available in large quantities, is a source of lithium. Worldwide occurrences include Bavaria, Finland, Sweden, and in the United States, New Hampshire, Maine, and South Dakota.

TRIPLE BOND (Carbon). Organic Chemistry.

TRITIUM. The radioactive isotope of hydrogen, with a mass number 3, is termed *tritium*. It is one form of heavy hydrogen; the other form being deuterium.

TUNGSTEN. Chemical element symbol W, at. no. 74, at. wt. 183.85, periodic table group 6b, mp 3410°C, bp 5660°C, density, 19.3 g/cm³. Two forms of metallic tungsten are known: α-tungsten, which has a body-centered cubic crystal structure, and β-tungsten, which has a face-centered cubic crystal structure. The metal exhibits the phenomenon of passivity so that it is quite resistant to chemical action even though it has a strong reducing action when a fresh surface is exposed, or in potentiometric titrations. Tungsten is a silvery-white-to-steel-gray, brittle, hard metal; not oxidized by air at ordinary temperature, but burns at high temperature, best dissolved by a mixture of hydrofluoric and HNO_3 acids. Tungsten has four naturally occurring isotopes, ^{180}W, and ^{182}W through ^{184}W. The isotope ^{180}W is radioactive with a half-life of approximately 3×10^{14} years. Six other radioactive isotopes have been identified, ^{176}W through ^{178}W, ^{185}W, ^{187}W, and ^{188}W. All have half-lives considerably less than 4 months. Tungsten does not occur in the free state and is a relatively scarce element, making up an estimated $7 \times 10^{-3}\%$ of the earth's crust. The tungsten content of seawater is estimated at about 950 pounds per cubic mile (103 kilograms per cubic kilometer) of seawater.

Although the tungsten mineral wolframite (iron manganese tungstate) was described as early as 1574, it was then mistaken as a mineral of tin. The term *tungsten* first appeared about 1758. K. W. Scheele identified tungstic oxide in 1781, after which the calcium mineral *scheelite* was named. The first metallic tungsten was produced by J. J. d'Elhuyar and F. d'Elhuyar in 1783 by the carbon reduction of the oxide. W. D. Cooldge obtained a patent in 1908 for making ductile tungsten wire for use in incandescent lamps. Diverse authorities accredit Scheele or the d'Elhuyar brothers for the discovery of the element. With exception of the United States, the element generally is referred to as *wolfram*.

Electronic configuration is $1s^22s^22p^63s^23p^63d^{10}4s^24p^64d^{10}$ $4f^{14}5s^25p^65d^46s^2$. Ionic radius, W^{+4}, 0.68 Å. Metallic radius, 1.3704_5 Å. Other physical properties of tungsten are given under **Chemical Elements.**

Usually tungsten minerals are found in pegmatites, sills, and batholiths. Minerals often accompanying tungsten minerals are cassiterite, quartz, feldspar, sulfides, arsenites, apatite, calcite, molybdenite, and bismuthinite. Several of these minerals are described under separate alphabetical entries. In order of decreasing magnitude, tungsten deposits occur in the People's Republic of China, the United States, Korea, Bolivia, Portugal, Burma, and Australia. Deposits also are found in at least ten other areas. In the United States, the most significant deposits are found in California, Nevada, South Carolina, Idaho, and Colorado. Tungsten concentration in the ores found in the United States run from 0.5 to 3% WO_3 (20 pounds of WO_3 contains about 15.9 pounds of W; 10 kilograms of WO_3 contains about 8 kilograms of W). High-purity tungsten metal is prepared by extracting tungsten from ore or by use of a strong alkali hydroxide solution at the boiling point. The alkali-metal carbonate or hydroxide thus obtained then is fused to form the water-soluble, alkali-metal tungstate. Where NH_4OH is used, the product is ammonium tungstate (NaOH yields sodium tungstate). The compound is reduced to metal powder. Conversion of the powder to massive metal is done by pressing, sintering, and mechanical working at high temperatures. In another process, the ore is fused with sodium carbonate and nitrate to yield sodium tungstate. Reduction of the oxide is accomplished by heating with carbon or hydrogen, whereupon tungsten metal is yielded.

Uses. Approximately one-half of the tungsten produced is in the form of sintered tungsten carbides. These compounds are used for cutting tools and wear-resistant parts. About 15% of production is consumed for making wire used in lamps and also for various shapes used in aerospace and defense products. Another 15% of production is used for high-temperature alloys and powder metallurgy. Approximately 10% goes into high-speed tools. The remaining production goes into a wide variety of applications.

Tungsten carbide WC is extremely hard (9.5 on the Mohs scale; diamond = 10) and has a melting point of 2,870°C. This combination of hardness and high-temperature stability makes it an excellent material for cutting tools. Additionally, the wear-resistant properties are excellent, accounting for the use of tungsten carbide for dies for hot and cold working of wire, rod and tubing, mining tools, snow-tire studs, and ball-point pens. For hard carbide tools and dies, tungsten carbide in the form of fine powder (1–10 micrometer particle size) is bonded with cobalt.

Special carbide tools also will often contain various percentages of titanium, tantalum, niobium (columbium), and hafnium carbides, along with the tungsten carbide. Chromium and vanadium carbides are also added to produce special, fine-grain size grades of cemented tungsten carbide-cobalt materials.

For hard-facing applications, fused tungsten carbide is used. Tungsten also forms the ditungsten carbide W_2C which has a melting point of 2,860°C. However, the term *tungsten carbide* usually refers to the mono compound. WC generally is made by combining tungsten metal powder with finely divided lampblack and the mixture then heated to about 1,500°C. A variety of tungsten powders are made which then are subjected to various powder metallurgy techniques to form numerous shapes with a wide range of characteristics. Tungsten carbide can also be manufactured by a so-called *menstrum* process which employs calcium carbide and aluminum metal to reduce scheelite via a thermite reaction, with the tungsten carbide recovered by acid washing.

Tungsten wire, including pure (unalloyed), doped (nonsag; potassium silicate and aluminum chloride or nitrate doped), and thoriated and zirconiated types are used extensively in applications, such as filaments for incandescent lamps, thermocouples, arc-lamp electrodes, electrochemical electrodes, and instrument springs. Tungsten disks are also used for electrical contacts, and tungsten is also used in glass-to-metal seals where the coefficient of thermal expansion of tungsten is close to that of hard borosilicate glass, and tungsten pads are used in connection with silicon semiconductors because of the high thermal conductivity of tungsten and good match of the coefficient of thermal expansion of tungsten with that of silicon.

Compositions of silver and tungsten and of copper and tungsten find application as electrical contacts where they are subject to severe arcing. As a shield or as containers for radioactive materials, heavy-metal alloys of tungsten alloyed with about 7% nickel and 4% copper are effective. The same alloys also find other uses where high density is required, as in gyroscope rotors, counterweights in aircraft, and self-winding watch parts. Alloys of cobalt, chromium, and tungsten also find use in cutting tools, dies, and wear-resistant parts. The function of tungsten in steel is that of forming stable carbides, strengthening ferrite, and refining the grain size for retaining high hardness at elevated temperatures—a requirement of highspeed steels.

Tungsten chemicals find limited use in inks, paints, enamels, dyes, and glass manufacture. Some tungsten compounds and their derivative phosphors find use in x-ray screens, television picture tubes, and luminescent light sources.

Chemistry and Compounds. In keeping with its $5d^46s^2$ electron configuration, tungsten forms many compounds in which its oxidation state is 6+, just like molybdenum. It forms divalent and tetravalent compounds to about the same extent as molybdenum, but its trivalent and pentavalent compounds are somewhat fewer. Its anion chemistry is closely akin to that of molybdenum.

Among the divalent compounds of tungsten, the diiodide, WI_2, dibromide, WBr_2, and the dichloride, WCl_2, are among the most clearly characterized; they all hydrolyze, although the iodide reacts only with warm water. Like molybdenum, tungsten(II) has a complex chloroion, $[W_6Cl_8]^{4+}$ which, however, is much more easily oxidized than its molybdenum analog.

Trivalent tungsten occurs rarely in simple compounds, other than certain high-temperature products such as one of the borides, WB, one of the phosphides, WP, and the complexes. Among the latter is the ion $[W_2Cl_9]^{3+}$ in which the two tungsten atoms participate in a Cl—W—Cl—W—Cl bridging structure, as they do in a W_2Cl_6 structure.

In addition to the tungsten(III) boride mentioned above, the element forms at least two other borides, W_2B and WB_2; it

forms a similar series of phosphides, W_2P, WP, and WP_2 as well as WO_2 (brown oxide), W_4O_{11} (blue oxide), and WO_3 (yellow oxide), and two sulfides WS_2 and WS_3. The tungsten(IV) oxide and sulfide are representative of the simple tetravalent compounds, which also include a tetrabromide, WBr_4, and tetraiodide, WI_4. Like the dihalides, these tetrahalides undergo hydrolysis quite readily.

Among the best known simple pentavalent tungsten compounds are the pentachloride, WCl_5, and the pentabromide, WBr_5. As is true of tungsten(IV), tungsten(V) forms complexes.

By far the greatest number of tungsten compounds are those in which the element is hexavalent. These include all common halides except the iodide, i.e., WF_6, WCl_6, and WBr_6, as well as a number of oxyhalides, WOF_4, $WOCl_4$, WO_2Cl_2, $WOBr_4$, and WO_2Br_2, the trioxide, trisulfide, diboride, and diphosphide already mentioned, various complexes and organometallic compounds, and the anions.

Tungsten trioxide dissolves in hot alkali metal hydroxide solutions to yield in more or less hydrated form, the tungstate ion, WO_4^{2-}. However, the ionic species that exist in solution are more complex than mere hydration of the WO_4^{2-} would indicate, and this is especially true of the compounds obtained from such solutions. There are, however, two simple forms of the orthotungstic acid: H_2WO_4, which is precipitated upon addition of HCl to a hot tungstate solution, and $H_2WO_4 \cdot H_2O$, which is similarly obtained from a cold solution. Neutralization of a tungstate solution under most conditions yields, upon crystallization, much more complex salts. The acidic groups condense, with elimination of water, to form complexes, that can be crystallized as salts, which can be regarded as derived from "poly" acids. When such salts have only one kind of metal atom (e.g., W) in their anions, they are called *isopoly acids*; when they have more than one kind, they are called *heteropoly acids*. The latter group comprises an entire field of tungsten chemistry (as well as that of molybdenum and other elements.); the tungstophosphates are important in analysis and other applications. Other examples are the heteropoly acid salts formed by tungsten with oxyanions of boron, silicon, germanium, tin, arsenic, titanium, zirconium, and hafnium. In particular, the 6-series and the 12-series, containing, respectively, 6 and 12 tungsten atoms per molecule, have been extensively investigated.

Other interesting compounds are the "tungsten blues," complex oxides of colloidal nature, obtained by reduction of tungstates in alkaline solution. At higher temperatures, reduction of tungstates (of main group 1 and 2 elements) by alkali metal, hydrogen, zinc, tungsten, or electrolysis, yields the semimetallic "tungsten bronzes," formulated as M_nWO_3, where M is the alkali metal and n is less than 1. They have cubical structures with W—O—W groups forming the sides, and the alkali or alkaline earth atom randomly located in the center of some of the cubes. The resulting extra electrons are considered to distribute over the entire structure, giving in metallic properties.

Tungsten forms many other complexes. Of particular interest are the octacyano complexes, containing eight cyanide, CN^- ions coordinated to a single tetravalent or pentavalent tungsten ion, $W(CN)_8^{4-}$ or $W(CN)_8^{3-}$, the latter being exceptionally stable, and forming octacyanotungstic(V) acid, $H_3[W(CN)_8] \cdot 6H_2O$, which is known in salts. Similar complexes are known for molybdenum, rhenium and osmium. The fluorocomplex of tungsten(VI) has the structure $[WF_8]^{2-}$ and forms salts with the higher alkali metals, potassium, rubidium and cesium.

Like molybdenum and chromium, tungsten forms a number of cyclopentadienyl compounds which are also carbonyls, e.g., $C_5H_5W(CO)_3H$, $C_5H_5W(CO)_3CH_3$, $C_5H_5W(CO)_3C_5H_5$, $C_5H_5(CO)WW(CO)C_5H_5$, and $C_5H_5(CO)_3WW(CO)_3C_5H_5$. Tungsten(VI) also forms several other organometallic compounds, e.g., $W(OC_6H_5)_6$ and $W(OC_6H_4CH_3)_6$, as well as a simple carbonyl, $W(CO)_6$.

References

Chelius, J., and M. Schussler: "1970 Metals Reference Issue," *Machine Design*, 42, 19, 87–88 (1970).

Smith, E. N.: "Macro Process for Direct Production of Tungsten Monocarbide," *Proceedings of Conference on Recent Advances in Hardmetal Production*, Loughborough, United Kingdom (September 1979).

Staff: "American Metal Market," Tungsten Section (February 16, 1970).

Staff: "American Metal Market," Tungsten Report (February 5, 1971).

Stevens, R. F., Jr.: "Mineral Facts and Problems," U.S. Department of Energy, Washington, D.C. (updated periodically).

Yih, S. W. H., and C. T. Wang: "Tungsten Sources, Metallurgy, Properties, and Applications," Plenum, New York (1979).

—M. Schussler, Senior Scientist, Fansteel, North Chicago, Illinois.

TURPENTINE.   Resins (Natural).

TURQUOISE.   The mineral turquoise is a hydrated phosphate of aluminum and copper. Its exact composition is doubtful, the formula may be expressed $CuAl_6(PO_4)_4(OH)_8 \cdot 5H_2O$; iron is often present. This mineral is found in minute triclinic crystals, but chiefly massive as seams and crusts. The fracture is conchoidal; hardness, 5–6; specific gravity, 2.6–2.8; luster, soft waxy; color, may be various shades of blue, bluish-green and green; essentially opaque. It takes a good polish and the sky-blue varieties have long been used as a gem material. Unfortunately many beautiful blue stones in time change their color to some greenish hue, usually not attractive, rendering them practically valueless.

For hundreds of years turquoise has been mined in Iran, where it is found with limonite filling crevices in a brecciated trachyteporphyry, and because it found its way into Europe through Turkey, it became known as turquoise, from the French word *turque*, Turkish. Other mines were worked by the Egyptians in ancient times on the Sinai Peninsula. Turquoise is also found in Siberia, Turkestan, Saxony and France; and in the United States in Arizona, California and New Mexico. A blue stone that has passed for turquoise is in reality odontolite, from the Greek meaning tooth, usually fossil teeth or bones colored with iron phosphate. Odontolite is softer than true turquoise,

has a somewhat higher specific gravity, 3.0–3.5, and may be distinguished by chemical tests, or by a microscopical examination which will reveal its organic structure.

TWINNING (Crystal).   A process in which a region in a crystal assumes an orientation which is symmetrically related to the basic orientation of the crystal. Usually, layers of atoms within this region are translated with respect to a basic plane (the twinning plane). Each atomic plane is displaced by a distance which is proportional to its distance from the twinning plane. Bands of metal (twin bands) thus assume a lattice structure which is the mirror image of the unchanged portion of the lattice. See also **Mineralogy.**

TWITCHELL PROCESS.   One of the very early methods for the hydrolysis of fats and oils on an industrial scale. The key to the process is *Twitchell's reagent*, which is prepared by interacting oleic acid, benzene, and H_2SO_4. The product sometimes is termed *sulfobenzenestearic acid*. The agent accelerates hydrolysis. Naphthalene may be used in place of benzene in the reagent with the assigned formula $C_{18}H_{35}O_2 \cdot C_{10}H_6SO_3H$. See also **Vegetable Oils (Edible).**

TYNDALL EFFECT.   A phenomenon first noticed by Faraday (1857). When a powerful beam of light is sent through a colloidal solution of high dispersity, the sol appears fluorescent and the light is polarized, the amount of polarization depending upon the size of the particles of the colloid. The polarization is complete if the particles are much smaller than the wavelength of the radiation.

TYROSINE.   Amino Acids.

TYUYAMUNITE.   An ore of uranium with the composition $Ca(UO_2)_2(VO_4)_2 \cdot 5$–$8H_2O$, which occurs in yellow incrustations as a secondary mineral. The mineral is orthorhombic. It occurs as a secondary mineral as incrustations on limestones, and as disseminated impregnations in sandstones. Found abundantly in the Western United States, at Grants, New Mexico, and in Wyoming, Utah, Colorado, Nevada, Arizona and Texas. Also at Tyuya Muyan in Turkestan, U.S.S.R.

U

ULEXITE. This mineral, a hydrated borate of sodium and calcium, $NaCaB_5O_9 \cdot 8H_2O$, is a product of crystallization in arid regions from shallow playas and lakes. Ulexite crystallizes in the triclinic system, but usually occurs in rounded masses of fine-fibered acicular crystals. The hardness is 2.5; specific gravity, 1.96; silky luster and white color. The mineral is found abundantly in Chile and Argentina and in Nevada and California in the United States. Ulexite is a source of boron.

ULTRAFILTRATION. This is a method for separating large dissolved molecules categorized as macromolecules or colloids from a host substance. In this size range, molecules usually are measured in micrometers or angstrom units. One angstrom $= 10^{-10}$ meter $= 10^{-4}$ micrometer. Ultrafiltration is applicable to sizes ranging from 10 to 1000 micrometers. Ultrafiltration membranes are anisotropic, i.e., they have a very thin skin, which is supported on a spongy sublayer of membrane material. The thin skin is the working part of the membrane, and separations take place at the skin surface. Flow of the solvent phases through the membrane skin is predominantly by a pore-flow mechanism. However, instead of a uniform pore structure, the membrane skin has a plurality of small, irregular passageways. See Fig. 1.

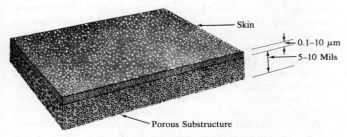

Fig. 1.   Anisotropic ultrafilter.

The separation of macromolecules takes place at the upstream side of the membrane. See Fig. 2. A dry cake is not obtained. Substantial velocity is maintained across the membrane surface—on the order of 2–10 feet per second (61–305 centimeters per second). Much of the research work reported prior to 1977 was based upon the use of cellulose acetate membranes, which are limited to operating temperatures, in most cases, below 50°C and at pH values between 3 and 8. In the late 1970s, a so-called second generation of membranes became available from a number of suppliers. The main limitations on the ultrafiltration system using the newer membranes is essentially whatever the system's piping and pumps can handle. In addition to improvements in temperature and pH tolerance, the new membranes have much higher permeation rates. For some feed solutions, initial rates greater than 100 gallons/square foot/day are not uncommon. (About 35 liters/square meter/day) This compares with the earlier cellulose acetate membranes of from 5 to 20

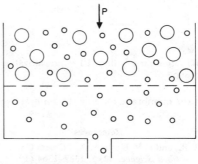

Fig. 2.   Ultrafiltration operating principle. Pressure forces solute of a preselected molecular weight through the permeable membrane while solutes not satisfying the criteria are held at the membrane surface.

gallons/square foot/day. (About 1.7 to 6.9 liters/square meter/day) Membrane life is quite long.

Considerable research activity continues in the area of using ultrafiltration for the modification and separation of protein substances. Several workers have reported using enzyme reactors to convert a cellulose substrate to glucose and the continuous removal of glucose through the ultrafiltration members. Some researchers have used an ultrafiltration cell to separate chymotrypsin hydrolyzed peptides from bovine whey protein. Removal of peptides from soybean protein after enzymatic hydrolysis in a ultrafiltration cell membrane reactor has been reported.

Ultrafiltration has been used for a number of years in cheese processing. In 1977, researchers reported on the concentration of pasteurized skim milk by ultrafiltration. Among other findings, the investigators noted that hot-pack cream cheese manufactured by ultrafiltration eliminated several time-consuming standard processing steps, since the cream cheese was standardized prior to fermentation. Ultrafiltered cheeses have a higher mineral content as the result of higher retention of calcium and phosphorus, associated with the micellar complex. The main difference in body and texture of cream cheese made by ultrafiltration and by conventional processes are in hardness and viscosity. Viscosity in ultrafiltered cream cheese is about 50% greater and hardness between 25 and 80% greater than those observed in commercial cream cheese samples.

Operation of the ultrafiltration process is shown in Fig. 3. A pressurized feed solution flows over the skin surface of a supported membrane. Under pressure, solvent and low-molecular-weight solutes pass through the membrane while larger macromolecules are retained in the system. The retained materials tend to collect on the surface of the membrane, forming a gel layer which limits the flux rate. To minimize the thickness of this gel layer, ultrafiltration systems are designed so that flow sweeps across the membrane surface. A pressurization feed pump feeds a recirculation loop, which includes a recirculation pump and the ultrafiltration modules. The recirculation provides

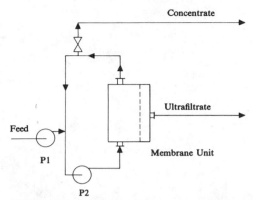

Fig. 3.   Ultrafiltration process flow. P1 = pressurization pump; P2 = recirculation pump.

flow through the membrane modules to maintain velocity across the membrane surfaces.

References

Covacevich, H. R., and F. V. Kosikowski: "Cream Cheese by Ultrafiltration," *Jrnl. of Food Science*, **42**, 5, 1362–1364 (1977).

Cunningham, S. D., et al.: "Cottonseed Protein Modification in an Ultrafiltration Cell," *Jrnl. of Food Science*, **43**, 5, 1477–1480 (1978).

Moretti, R. H., Perzuo, G., and Y. K. Park: "Enzymatic Hydrolysis of Soybean Proteins in a Continuous Ultrafiltration Reactor," *Paper 353*, Institute of Food Technologists 36 Annual Meeting, Anaheim, California (1976).

Setti, D.: "Developments in Membrane Separations," Institute of Food Technologists Symposium, Saint Louis, Missouri (1979).

ULTRAVIOLET LASER CHEMISTRY.   Photochemistry and Photolysis.

ULTRAVIOLET STABILIZERS.   When a polymer absorbs light, it may reradiate the absorbed energy at much longer wavelengths (heat), at slightly longer wavelengths (luminescence), or the energy may be transferred to another molecule. When none of these processes is operative, the absorbed light energy may cause bond breaking leading to degradation. By incorporating an ultraviolet absorbing compound into the plastic, it is possible to essentially eliminate all of the above processes, since the absorber even at concentrations as low as 0.5% can effectively compete for the incident ultraviolet radiation, thus protecting the plastic from degradation. A second approach to stabilization is to incorporate an additive which, though not an absorber in itself, can accept energy from the polymer substrate, and thus, leave the polymer intact. Since protection of plastics against light degradation can be achieved by these two mechanisms, the broader term *ultraviolet stabilizer* or *light stabilizer* is used to refer to such additives.

Ultraviolet absorbers continue to be the most widely used stabilizers. Such products must have long-term stability to ultraviolet light, be relatively nontoxic, heat stable, have little color, must not sensitize the substrate, and must be priced at levels which the plastics processor can tolerate. The principal classes of chemicals meeting these requirements at present are the 2-hydroxybenzophenones, and 2-(2'-hydroxyphenyl)benzotriazoles, substituted acrylates, and aryl esters. Typical compounds representative of these classes are 2-hydroxy-4-octoxybenzophenone, 2-(2'-hydroxy-5'-methylphenyl) benzotriazole, ethyl-2-cyano-3,3-diphenyl acrylate, dimethyl *p*-methoxybenzylidene malonate, and *p-tert*-octylphenyl salicylate.

The particular absorber to be used in a given application depends on several factors. One important criterion is whether the absorber will strongly absorb that portion of the ultraviolet spectrum responsible for degradation of the plastic under consideration. Compatibility, volatility, thermal stability, and interactions with other additives and fillers are other items that must be considered. When used in food wrappings, Food and Drug Administration approval must be obtained. While one or more of these considerations may rule out a given stabilizer or influence the choice of one class over another, the final selection must await the results of extensive accelerated and long-term tests.

At this point, it should be indicated that much effort has gone into the development of accelerated testing procedures. Many of the devices and techniques employed are based on knowledge gained in the evaluation of dyes, textiles, and rubber. For example, the carbon-arc Fade-O-Meter and the Xenon-arc Weather-O-Meter have been adapted from the dye field for use in plastics evaluation. Extensive use is also made of the fluorescent sunlamp, fluorescent blacklight, S-1 sunlamp, Hanovia lamp, and others. Such instruments are very useful for comparison of one stabilizer with others and for evaluating total stabilizing formulations in particular polymers. Nevertheless, no accelerated weathering device has yet been found which can accurately predict the outdoor weatherability of a broad range of polymers. Accelerated outdoor weathering is carried out in Phoenix, Arizona, where high levels of ultraviolet radiation occur and the temperature is high. To determine the lifetime under more humid conditions, tests are often conducted in the vicinity of Miami, Florida.

For extended outdoor applications, most polymers require some degree of light stabilization. There are wide variations in the inherent stability of different polymers ranging from less stable ones, such as polypropylene, to the highly light-stable poly(methyl methacrylate). Because of the dramatic growth of polyolefins, and particularly polypropylene, over the past several years, there has been an upsurge in requirements for ultraviolet absorbers. The hydroxybenzophenones, such as 2-hydroxy-4-octoxy benzophenone, have been widely used for stabilization of polypropylene. The benzotriazoles have also achieved commercial importance in this application. End uses of polypropylene requiring ultraviolet absorbers include upholstery fabrics, indoor-outdoor carpeting, lawn furniture, ropes, and various crates and boxes. Polyethylene is also stabilized with the hydroxy benzophenone absorbers. Applications include baskets, beverage cases, bags for fertilizer, and films for greenhouses.

Polystyrene light stabilization has been achieved with a variety of ultraviolet absorbers including the benzophenones, benzotriazoles, and salicylates. While yellowing of polystyrene occurs in many applications, it is particularly noticeable in diffusers used with fluorescent lights. This problem has been effectively solved by using ultraviolet light absorbers. In this instance, superior stabilization is achieved when the ultraviolet absorber is used in conjunction with specific antioxidants.

The hydroxybenzophenones, hydroxyphenylbenzotriazoles, and substituted acrylates are all used for stabilization of polyvinyl chloride. This polymer is growing at a substantial rate, and increasing uses are developing for light-stabilized grades. Among current uses, may be mentioned auto seat covers, floor tiles, light diffusers, vinyl-coated fabrics, siding, and exterior trim. Since the processing of polyvinyl chloride requires the use of a heat stabilizer, care must be exercised to avoid undesirable interactions between the heat and light stabilizers.

The stabilization of polyesters is generally achieved with the hydroxybenzophenone and hydroxyphenylbenzotriazole absorbers. The choice of absorber depends on the curing catalysts and promoters used. The stabilization of fire retardant grades of polyesters offers a greater problem than the standard grades because the halogenated monomer acids used are appreciably more sensitive to ultraviolet light than the unhalogenated acids (phthalic, isophthalic). Applications for light-stabilized polyesters include sheets for roofs and skylights and various surface coatings.

Cellulosic plastics are used in a number of outdoor applications with signs being one of the principal areas of use. This plastic can be stabilized reasonably well with the aryl esters of salicyclic acid. It is of interest to note that these esters undergo a photochemical rearrangement in the plastic to derivatives of hydroxy benzophenone. The hydroxy benzophenones may be added initially to effect stabilization.

As noted earlier, poly(methyl methacrylate) plastic has excellent resistance to ultraviolet radiation. Nevertheless, in long-term outdoor applications or in lighting fixtures, small amounts of ultraviolet absorbers are employed to retard the yellowing and degradation in physical properties which would otherwise occur.

The previous discussion illustrates how widely ultraviolet absorbers are used for stabilization of plastics against degradation by ultraviolet light. The second principal method for light stabilization is the use of energy transfer agents. Important stabilizers currently in use which function by this mechanism are nickel complexes of 2,2'-thiobis(4-tert-octylphenol). For example, the butyl amine adduct of this complex is widely used. The nickel salts of mono alkyl esters of 3,5-di-tert-butyl-4-hydroxy-benzyl phosphonic acid are also useful stabilizers. When color is not a consideration, nickel dialkyl dithiocarbamates, and nickel acetophenone oximes may be used. Thus far, the nickel stabilizers have been used primarily in polypropylene and to some extent in polyethylene. They are especially useful in polypropylene fibers since stabilization by energy transfer is less dependent on sample thickness than is stabilization by ultraviolet absorption. Some of the nickel stabilizers have the further advantage that they act as dye acceptors and thus aid printing and dyeing of fibers and other items made from polyolefins.

Thus far, the discussion on ultraviolet stabilizers has been concerned only with their use for stabilization of plastics. While this is the principal use, the stabilizers which function by absorption are also widely used to prevent ultraviolet light from damaging furniture, clothing, and other articles. For example, a thin plastic film (6–10 mil) containing a high concentration of absorber is useful for covering a store window exposed to the sun. Such a film absorbs the ultraviolet radiation and thus prevents damage to the articles behind the film. The clarity and lack of color of the film permits customers to see the articles readily. Surface coatings containing ultraviolet absorbers are used in the same way to protect items such as flooring and furniture. Ultraviolet absorbing coatings may also be used to protect plastics, but generally, it is more practical to incorporate the absorber in the plastic itself. The incorporation of ultraviolet absorbers into plastic sunglasses is important for protecting the eye from ultraviolet radiation damage. Suntan lotions contain compounds such as monoglyceryl p-aminobenzoate that permit the longer wave lengths (330–400 nm) in the ultraviolet which cause tanning to pass through to the skin. At the same time, the more highly energetic short wave lengths (290–330 nm) which cause burning are strongly absorbed. When no tanning is desired, creams containing hydroxy benzophenones may be used, since these products remove a high percentage of the 290–400 nm radiation.

Highly satisfactory formulations have been developed for light stabilization of a wide range of polymers. Studies are continuing not only toward empirical development of superior stabilizing formulations, but also toward understanding the mechanisms of the degradation and inhibition process involved. This dual approach can be expected to yield products which will meet the increasingly severe demands that will result as plastics find their way into new outdoor uses.

—W. B. Hardy, American Cyanamid Co., Bound Brook, New Jersey.

UPSILON PARTICLE. As of 1977, when the upsilon particle was discovered at the Fermi National Accelerator Laboratory, the particle was the heaviest to be identified. Discovery of upsilon prompted physicists to introduce a massive new quark, raising the number of quarks from four to five (but probably six). The upsilon has a mass three times greater than any subatomic entity previously detected. It was discovered in energetic collisions between protons and copper nuclei. With a mass at its lower energy state equivalent to 9.0 GeV and masses in excited states equivalent to 10 and 10.4 GeV, the upsilon particle has been interpreted by scientists as consisting of a massive new quark (fifth) bound to its antiquark. The experiment was later reinforced by research at the Deutsches Elektronen-Synchrotron (DESY), located near Hamburg. At Fermilab, the excited upsilon particle appeared as a resonance in the yield of muons generated in collisions between protons and nuclei. A discussion of the upsilon experiment is described by a principal scientist of the project, L. M. Lederman (*Sci. Amer.*, **239**, 4, 72–80, 1978). See also **Particles (Subatomic)**.

URALITE. A metamorphic mineral. It is well established that pyroxene rocks may be metamorphosed into hornblende rocks. If the hornblende thus produced is fibrous and retains the original form of the pyroxene, it is called *uralite*, and the process by which the change is brought about is called *uralitization*. It seems quite clear that uralitization is a chemical process which in many cases is accompanied by the generation of new minerals such as calcite, epidote, and magnetite. Uralite was first observed in rocks from the Ural Mountains, hence its name. See also **Hornblende.**

URANINITE. A mineral approximating the composition UO_2, but containing besides the higher oxide of uranium, UO_3, and oxides of lead, thorium, and rare earths. The uraninite usually occurs as cubic or cubo-octahedral crystals of specific gravity 7.5–10; when in masses of pitchy luster it is called pitchblende, specific gravity 6.5–9. All uraninites and pitchblende contain a minute amount of radium. It was in pitchblende obtained from the Joachimsthal in Czechoslovakia that Mme Curie discovered radium. Other localities for uraninite are in Saxony, Rumania, Norway, Cornwall, East Africa, and in the United States in the pegmatites of Connecticut Grafton Center, New Hampshire, North Carolina, and South Dakota, and in Gilpin County, Colorado. An important occurrence of pitchblende is at Great Bear Lake, Northwest Territories, Canada, where it has been found in large quantities associated with silver.

URANIUM. Chemical element symbol U, at. no. 92, at. wt. 238.03, periodic table group (actinides), mp 1131–1133°C, bp 3818°C, density, 19.05 g/cm³ (20°C). Uranium metal is found in three allotropic forms: (1) *alpha phase*, stable below 668°C, orthorhombic; (2) *beta phase*, existing between 668 and 774°C, tetragonal; and (3) *gamma phase*, above 774°C, body-centered cubic crystal structure. The gamma phase behaves most nearly like that of a true metal. The alpha phase has several nonmetallic features in its crystallography. The beta phase is brittle.

Prior to the production of artificially created elements, uranium was the highest in terms of atomic number and atomic weight. It was difficult to locate uranium in the periodic table of the elements, although chemically uranium resembles the elements of group 6b, namely, chromium, molybdenum, and tungsten. Subsequent to the production of the transuranium elements (atomic numbers 93 and higher), these elements, along with actinium (89), thorium (90), protactinium (91), and uranium (92) were placed in the actinide group of transition elements. They are similar in their mutual relations to the rare-earth group (lanthanum to lutetium). See **Periodic Table of the Elements.**

Earthly abundance of uranium is described a bit later. In terms of presence in seawater, no significant concentrations have been reported. In terms of cosmic abundance, uranium also is very scarce. The study by Harold C. Urey (1952), in which silicon was given a base figure of 10,000, the concentration of uranium was reported by the figure of 0.0002.

Uranium is a white metal, ductile, malleable, and capable of taking a high polish, but tarnishes rapidly on exposure to the atmosphere. Finely divided uranium burns upon exposure to air, and the compact metal burns when heated in air at 170°C. Uranium metal slowly decomposes water at ordinary temperatures and rapidly at 100°C; is soluble in HCl and in HNO_3; and is unattacked by alkalis. Chemically related to chromium, molybdenum, and tungsten, uranium, like thorium, is radioactive. In the radioactive decomposition, radium is formed. Discovered by Klaproth in 1789. See **Radioactivity.**

The element uranium found in nature consists of the three isotopes of mass numbers 238, 235, 234 with relative abundances 99.28, 0.71, and 0.006%, respectively.

The isotope ^{238}U is the parent of the natural uranium $4n + 2$ radioactive series, and the isotope ^{235}U is the parent of the natural actinium $4n + 3$ radioactive series.

The isotope ^{235}U has great importance because it undergoes the nuclear fission reaction with slow neutrons, and it has been separated in substantial amounts in nearly 100% isotopic composition.

Electronic configuration

$$1s^2 2s^2 2p^6 3s^2 3p^6 3d^{10} 4s^2 4p^6 4d^{10} 4f^{14} 5s^2 5p^6 5d^{10} 5f^3 6s^2 6p^6$$
$$6d^1 7s^2.$$

Ionic radii U^{4+} 0.89 Å; U^{3+} 1.04 Å (Zachariasen). Metallic radius 1.4318 Å (805°C). Oxidation potential $U + 2H_2O \rightarrow UO_2^{2+} + 4H^+ + 6e^-$, 0.82 V.

Uranium Reserves. Uranium has been known to be a distinct element since 1789, but apart from the small amount of its salts used in yellow pottery glazes, it remained more or less a laboratory curiosity until the 1920s, when the treatment of uranium ore for the recovery of its contained radium (for the treatment of cancer) began in Czechoslovakia and the Belgian Congo (now Zaire), followed by Canada in 1933. The separated uranium was mostly stockpiled or discarded.

After the development and successful explosion of the atomic bomb toward the end of World War II, an urgent search for workable uranium deposits was set in motion all over the world. The only high-grade deposits known to the western world were those in the countries just named as radium sources, but in view of the limited demand previously, serious exploration for uranium had never been undertaken. However, the offer of contracts by the U.S. Atomic Energy Commission, for fixed quantities at stated prices stimulated exploration for this hitherto largely ignored material.

TYPES OF NATURAL URANIUM RESOURCES

TYPE OF RESOURCE	ORE GRADE (Ppm Uranium)	PRINCIPAL KNOWN LOCATIONS
Vein deposits	10,000–30,000	Canada (near Great Bear Lake in the Northwest Territory) Western United States France Germany (East) U.S.S.R. Gabon Zaire Australia China
Vein deposits (pegmatites, unconformity deposits)	2,000–10,000	Canada (Saskatchewan) U.S.S.R. Australia
Fossil placers, sandstones	200–2,000	Canada (Ontario) Western United States Brazil Chile U.S.S.R. Japan Australia South Africa
Shales, phosphates	10–100	United States (Florida) Morocco Sweden U.S.S.R.
Pegmatites, other igneous and metamorphic deposits	1–10	Canada (Ontario) Greenland Brazil Spain U.S.S.R. India Angola Australia

Uranium is rather widely distributed throughout the world. See accompanying table. Deposits vary markedly in richness. Thus, reserves are usually evaluated in terms of the total cost per pound (kilogram) of U_3O_8 recovered. In terms of the future of nuclear power as presently constituted technologically, a number of authorities have expressed concern over the long-term availability of uranium by some countries, including the United States. At least one authority (Hays, 1979) has suggested that the United States, considering its present estimated uranium reserves, may not be able to support a light-water reactor program over double its present size through the year 2000. However, there are several factors which make the situation appear less bleak: (1) improvement of uranium efficiency in thermal reactors which consume more fissionable material than they breed, (2) design changes in new reactors can conserve uranium, (3) reprocessing of spent fuel and blanket materials, with the recovery of purified uranium and plutonium, (4) extending uranium supplies by using thorium in light-water reactors, and (5) ultimate development of fast breeder reactors which produce more fissionable material during operation than is originally furnished. More detail on the economics and politics of uranium can be found in several of the references listed.

Chemistry of Uranium. Uranium has four oxidation states (III), (IV), (V), and (VI), and the ions in aqueous solution are usually represented as U^{3+}, U^{4+}, UO_2^+, and UO_2^{2+}. The oxidation-reduction scheme, on the hydrogen scale (in which the potential for $\frac{1}{2}H_2 \rightarrow H^+$ is taken as zero) is indicated as follows:

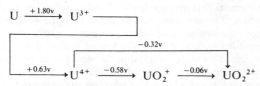

Oxidation-Reduction Potentials of Uranium Ions
(in 1-molar hydrochloric acid)

The UO_2^+ ion is unstable in solution and undergoes disproportionation to U^{4+} and UO_2^{2+}. A few solid compounds of this oxidation state are known, as for example, UF_5 and UCl_5.

The ion U^{3+} forms intense red solutions in H_2O and is oxidized by water at an appreciable rate. The rate of oxidation appears to increase with increasing ionic strength, although concentrated solutions are said to be stabilized by strong acids such as hydrochloric. Solutions of uranium(IV) are green, and uranium(VI) solutions, yellow.

The (IV) and (VI) are the important oxidation states and therefore the more important phases of the chemistry of uranium may be related to the two oxides UO_2 and UO_3, uranium dioxide and uranium trioxide. A series of salts such as the chloride and sulfate, UCl_4 and $U(SO_4)_2 \cdot 9H_2O$ is obtained from UO_2. The more common uranyl salts as $UO_2(NO_3)_2 \cdot 6H_2O$, UO_2Cl_2, and $UO_2SO \cdot nH_2O$ in which the UO_2^{2+} (uranyl ion) acts as a radical, are derived from UO_3. UO_3 is amphiprotic and forms a series of alkali and double alkali uranates and polyuranates of limited solubility, such as $Na_2U_2O_7$, $NaZn(UO_2)_3$ $(C_2H_3O_2)_9 \cdot 6H_2O$, $NaMg(UO)_2)_3(C_2H_3O_2)_9 \cdot 6H_2O$, etc.

The element uranium also exhibits a formal oxidation number of (II) in a few solid compounds, semimetallic in nature, such as UO and US. No simple uranium ions of oxidation state (II) are known in solution.

In addition to the three oxides, UO_2 (brown, cubic), U_3O_8 (greenish black, orthorhombic) and UO_3 (orange, hexagonal), which have been known for a long time, there are known to exist the monoxide, UO, and the pentoxide U_2O_5. There is also some evidence for the existence of U_4O_7 and U_6O_{17}. The phase relationships in the uranium-oxygen system are very complex because solid solutions are readily formed, so that it is possible to obtain uranium "oxides" with practically any composition intermediate between UO and UO_3, and with many crystal structures.

Uranyl peroxide, the formula of which is usually given as $UO_4 \cdot 2H_2O$, is formed by precipitation from solutions of uranyl nitrate by hydrogen peroxide. Alkali hydroxides, hydrogen peroxide, and sodium peroxide form soluble peroxyuranates, $Na_2UO_6 \cdot 4H_2O$ and $Na_4UO_8 \cdot 8H_2O$, when added to solutions of uranyl salts.

Two uranium carbides are known, the monocarbide, UC, and the dicarbide, UC_2. These can be prepared by direct reaction of carbon with molten uranium, or by reaction of carbon monoxide with metallic uranium at elevated temperatures. The sesquicarbide, U_2C_3, has been found to exist as a stable compound below about 1,800°C and can be produced by heating a mixture of UC and UC_2 between 1,250 and 1,800°C.

Uranium and nitrogen form an extensive series of compounds that can be prepared by direct action of nitrogen on the metal. Uranium mononitride, UN, is the lowest nitride of uranium. If the mononitride is treated with more nitrogen at atmospheric pressure, U_2N_3 is formed. With nitrogen under high pressure, UN_2 can be prepared, but it is difficult to obtain samples of UN_2 that are completely free of UN.

Uranium metal reacts with hydrogen at 250–300°C to form a well defined hydride, which resembles the rare-earth hydrides in many respects. The formula of this substance has been shown to be $UH_{3.00}$. The hydride undergoes decomposition with increasing temperature; the dissociation pressure of UH_3 is one atmosphere at 436°C.

Uranium tetrafluoride serves as a starting material for the preparation of the other fluorides. It is best prepared by hydrofluorination of uranium dioxide:

$$UO_2 + 4HF \xrightarrow{500°C} UF_4 + 2H_2O$$

Uranium trifluoride can be prepared by reduction of UF_4 with hydrogen at 1,000°C. Uranium hexafluoride, UF_6, white and orthorhombic is best obtained by direct fluorination of UF_4, green and monoclinic, although any uranium compound will yield UF_6 by reaction with fluorine at elevated temperatures:

$$UF_4 + F_2 \xrightarrow{350°C} UF_6$$

The hexafluoride can also be prepared by the interesting reaction:

$$2UF_4 + O_2 \xrightarrow{900°C} UF_6 + UO_2F_2$$

The intermediate fluorides $U_2F_9(UF_{4.5})$, $U_4F_{17}(UF_{4.25})$ and UF_5 are prepared by reaction of solid UF_4 and gaseous UF_6 under appropriate conditions of temperature and pressure.

Uranium hexafluoride is probably the most interesting of the uranium fluorides. Under ordinary conditions it is a dense, white solid with a vapor pressure of about 120 mm at room temperature. It can readily be sublimed or distilled, and it is by far the most volatile uranium compound known. Despite its high molecular weight, gaseous UF_6 is almost a perfect gas, and many of the properties of the vapor can be predicted from kinetic theory.

Uranium tetrachloride can be prepared by direct combination of chlorine with uranium metal or hydride; it can also be obtained by chlorination of uranium oxides with carbon tetrachloride, phosgene, sulfur chloride, or other powerful chlorinating agents. The trichloride is obtained by reaction of UCl_4 with hydrogen and the higher chlorides by reaction of UCl_4 and Cl_2. Uranium hexachloride, UCl_6 is a rather volatile, somewhat unstable substance. All of the uranium chlorides dissolve in or react readily with water to give solutions in which the oxidation state of the ion corresponds to that in the solid. All of the solid chlorides are sensitive to moisture and air.

The trichloride, tribromide and triiodide of uranium are obtained either by reaction of the elements or by treatment of UH_3 with the appropriate halogen acid. The thermal stability of the halides decreases as the atomic number of the halogen increases. No higher uranium bromides or iodides are known.

A series of oxyhalides of the type UO_2F_2, $UOCl_2$ UO_2Br_2, etc., are known. They are all water-soluble substances which become increasingly less stable in going from the oxyfluoride to the oxyiodide.

Uranyl ion forms complexes with many oxy anions. Both U(VI) and U(IV) compounds dissolve in alkali carbonate solutions with formation of carbonato complexes. Those of the larger alkali cations are only slightly soluble: $K_{sp} = 6 \times 10^{-5}$ for both $K_4[UO_2(CO_3)_3]$ and $(NH_4)_4[UO_2(CO_3)_3] \cdot 2H_2O$.

Aqueous solutions of uranium(III), uranium(IV), and uranium(VI) are readily obtained. Solutions of uranium(III) are blood-red in appearance; hydrogen is slowly evolved with the formation of uranium(IV) is a strong reducing agent and is easily oxidized to uranyl ion by oxygen, peroxide, and numerous other oxidizing agents. Uranyl solutions in turn may be reduced to uranium(IV) with sodium dithionite, zinc or cadmium amalgams, or by electrochemical or photochemical means.

Separation of Isotopes. Several methods are available for the separation of isotopes, including gaseous diffusion, centrifugation, electromagnetic methods, thermal diffusion, electrolytic methods, distillation, and chemical-exchange methods. The separation of ^{235}U from ^{238}U represented the first large-scale isotope-separation operation and, after considerable study, the principal plant utilized gaseous diffusion.

The gaseous diffusion method of isotope separation is based upon the difference in the rate of diffusion of gases which differ in density. Since the rate of diffusion of a gas is inversely proportional to the square root of its density, the lighter of two gases will diffuse more rapidly than the heavier and therefore the result of a partial diffusion process will be an enrichment of the partial product in the lighter component.

To separate isotopes by this process, they must be in the gaseous form. Therefore, the separation of isotopes of uranium required the conversion of the metallic uranium into a gaseous compound, for which purpose the hexafluoride, UF_6, was chosen. Since the atomic weight of fluorine is 19, the molecular weight of the hexafluoride of ^{235}U is $235 + (6 \times 19) = 349$, and the molecular weight of the hexafluoride of ^{238}U is $238 + (6 \times 19) = 352$. Since the rate of diffusion of a gas is inversely proportional to the square root of its density (mass per unit volume), the maximum separation factor for one diffusion process of the uranium isotopes is $\sqrt{352/349} = 1.0043$. Since only part of the gas can be allowed to diffuse, the actual separation factor is even less than this theoretical maximum.

From this small figure, it is apparent that many diffusion stages are necessary in the separation of ^{235}U from ^{238}U. The number originally calculated for the Oak Ridge plant was about 4,000. Other reasons are the small apertures demanded by diffusion processes (in this case less than .00001 centimeter in diameter), which reduce the rate of gas flow and demand a great barrier area for appreciable production.

The centrifugal method of isotope separation consists essentially of the passage of the mixture through a rapidly rotating force field, such as that of a rotating cylinder. If a current of mixed gases is passed into such a cylinder, moving parallel to the axis of rotation, the lighter gas will tend to concentrate near the axis, and the heavier gas, near the periphery. This is the principle of the cream separator; its successful application to separation of isotopes in the gaseous phase requires apparatus operating at very high speeds of rotation.

The electromagnetic method of isotope separation is based upon the principle of the mass spectrograph. As in that apparatus, a stream of charged particles is passed through a system of electric and magnetic fields. If the particles are ions of two or more isotopes of the same element, all bearing the same charge, the deflections produced by the fields will vary with the masses of the particles, and will thus provide a means for their separation. This method is especially effective for the separation of particles of a number of masses, and has been widely used for that purpose in research studies and in production-separation operations. The method is also used extensively in a number of research laboratories, particularly those of northern Europe, for the isotopic separation of individual radioactive nuclides that are to be used as sources in instruments, such

as beta- and gamma-ray spectrometers, in which measurements are made of the characteristics of ionizing radiations.

The thermal diffusion method of isotope separation has broad application to liquid-phase as well as gaseous-phase separations. The apparatus widely used for this purpose consists of a vertical tube provided with an electrically heated central wire. The gaseous or liquid mixture containing the isotopes to be separated is placed in the tube and heated by means of the wire. In such an apparatus, two effects act to separate the isotopes. Thermal diffusion tends to concentrate the heavier isotopes in the cooler outer portions of the system, while the portions near the hot wire are enriched in the lighter isotopes. At the same time, thermal convection causes the hotter fluid near the hot wire to rise, while the cooler fluid in the outer portions of the system tends to fall. The overall result of these two effects causes the heavier isotopes to collect at the bottom of the tube and the lighter at the top, whereby both fractions may be withdrawn.

The electrolytic method of isotope separation is of importance not only because of its present day uses, but also because of its historical interest. It was by this method that G. N. Lewis and his co-workers at the University of California obtained practically pure deuterium. Since deuterium oxide had been shown to be present in ordinary water, the conclusion was drawn that water (or rather the dilute aqueous solution) from electrolytic cells used for the production of hydrogen and oxygen by continuous electrolysis of water, should be richer in the heavier isotope (deuterium having a mass number of 2, as against 1 for protium). Starting with such residual water from an electrolytic cell, it was found that by repeated electrolysis a small residue consisting almost entirely of deuterium oxide (D_2O) was obtained. This process is still used for the separation of pure hydrogen isotopes, as well as for other purposes.

The distillation method of isotope separation has also been the basis of important research contributions. In the work on the hydrogen isotopes, it preceded the electrolytic separation methods discussed above. Following the suggestion by Birge and Menzel, of the possible presence of deuterium in ordinary hydrogen to the extent of 1 part in about 4,500, Urey, Brickwedde and Murphy, in 1931, began their search for this isotope. By evaporating about 4,000 milliliters of liquid hydrogen to a volume of about 1 milliliter, Brickwedde obtained a residue that gave conclusive spectroscopic evidence of the presence of deuterium. The distillation method of separation has been responsible for many other research contributions. Among them may be mentioned separation of the isotopes of oxygen, of mercury, zinc, potassium and chlorine.

The chemical exchange methods of isotope separation are of value, not only for that purpose, but also because they provide a direct means for the study of chemical reactions. A well-known example of an isotopic chemical exchange is the heavy water equilibrium:

$$H_2 + D_2O \leftrightarrows D_2 + H_2O$$

In this equation, the formula H_2 is used for the hydrogen isotope of mass number 1, which constitutes all but a small fraction of ordinary hydrogen; H_2O is the corresponding "light water"; D_2 is hydrogen of mass number 2 (deuterium) and D_2O is the corresponding heavy water. The double arrows indicate an equilibrium reaction, whereby under suitable conditions ordinary hydrogen reacts with heavy water to produce hydrogen of mass number 2 and light water. If one were to start either with the two reactants on the left of the arrows, or with the two on the right, the system at equilibrium would have all four present. However, in this system at equilibrium the reverse reaction predominates, so that the ratio of ^2H to ^1H (that is, the ratio of D_2O to H_2O) in the liquid phase is about three times as great as in the gas. Because of this differential reactivity, this method is useful in the separation of the two hydrogen isotopes.

Another equilibrium system is useful in the separation of ^{14}N and ^{15}N. It is represented by the equation:

$$^{15}NH_3 + {}^{14}NH_4NO_3 \leftrightarrows {}^{14}NH_3 + {}^{15}NH_4NO_3$$

This exchange reaction is conducted by the countercurrent flow of ammonia gas and ammonium nitrate solution (in water). The forward reaction is favored, resulting in the concentration of the ^{15}N in the ammonium nitrate in solution. The multi-state conduct of this reaction that is necessary for effective operation is accomplished by arranging later stages in which the enriched ammonium nitrate solution is divided into two parts. One part is treated with caustic soda to displace the enriched NH_3, which is then used in a second stage of the process with the other part of the NH_4NO_3 solution. Three or more stages may thus be used, until the desired concentration of ^{15}N has been effected.

Another method of isotope separation is by ion mobility, a process based on the difference in mobility of the ions in an electrolytic solution, under the influence of an electric field.

Laser Enrichment of Uranium. For nearly a decade, laser enrichment of uranium has been under development in the United States. It is estimated that one-quarter billion dollars have been invested in this research. It believed by many scientists that despite these high research costs and the high costs of constructing a laser enrichment plant of large scale, in the long run the laser processes will be capable of enriching uranium at costs substantially below those of present means. As explained by Lester (1980), nearly all laser-enrichment processes depend upon the *isotope shift* effect, the fact that in both molecular and simple atomic states, each isotope of an element absorbs light at its own characteristic set of frequencies. Thus, it is possible to excite only one of the isotopes in a mixture by exposing the mixture to light of a precision frequency. The excited species may then be made to enter into chemical reactions or to respond to physical stimuli while the unexcited species remain relatively inert. The fact that isotope separation could be brought about by photophysical or photochemical means was recognized some 60 or 70 years ago. However, the means did not show practical promise until invention of the laser about 20 years ago. Pioneering efforts were made by Exxon Nuclear Corporation and Avco Corporation in early 1971. These efforts concentrated on an atomic vapor laser isotope separation approach. U-235 atoms in a high-temperature uranium vapor stream are preferentially excited by laser photons. The atoms are further excited and ultimately ionized by the absorption of additional laser photons, and then deflected onto collector plates by external electric and magnetic fields. The unionized atoms, depleted in U-235, pass through these fields undeflected. A bit later, the Lawrence Livermore Laboratory conducted further research into this scheme. Reports from the aforementioned activities made in 1979 indicated that, although difficult engineering problems remain, large-scale production could probably commence by about 1990. The U.S. Government also has been funding development of a laserless advanced enrichment technique pioneered by TRW, Inc.

Principal advantages of laser and other advanced processing is the elimination of the numerous steps (diffusers, centrifuges, etc.) required in the techniques previously described in this entry. Advanced separation techniques also can be applied to the tailings from uranium ore refining.

References

Boyd, J., and L. T. Silver: "United States Uranium Position," *Pubn. PGC-NE-25*, American Society of Mechanical Engineers, Los Angeles, California (1977).

Boyd, W. R., Lowings, S. W., and E. M. Taft: "Programmable Logic Controllers in Solution Mining," *Instrumentation Technology*, **25**, 1, 47–52 (1978).

Carter, L. J.: "Uranium Mill Tailings," *Science*, **202**, 191–195 (1978).

Deffeyes, K. S., and I. D. MacGregor: "World Uranium Resources," *Sci. Amer.*, **242**, 1, 66–76 (1979).

Feiveson, H. A., von Hippel, F., and R. H. Williams: "Fission Power: An Evolutionary Strategy," *Science*, **203**, 330–337 (1979).

Gentry, R. V., et al.: "Radiohalos in Coalified Wood: New Evidence Relating to the Time of Uranium Introduction and Coalification," *Science*, **194**, 315–317 (1976).

Hafemeister, D. W.: "Nonproliferation and Alternative Nuclear Technologies," *Technology Review (MIT)*, **81**, 3, 58–62 (1979).

Hayes, E. T.: "Energy Resources Available to the United States, 1985 to 2000," *Science*, **203**, 233–239 (1979).

Krass, A. S.: "Laser Enrichment of Uranium: The Proliferation Connection," *Science*, **196**, 721–731 (1977).

Lester, R. K.: "Laser Enrichment of Uranium," *Technology Review (MIT)*, **82**, 8, 18–29 (1980).

Libby, L. M.: "The Uranium People," Charles Scribner's Sons, New York (1979).

Lieberman, M. A.: "U.S. Uranium Resources: An Analysis of Historic Data," *Science*, **192**, 431–436 (1976).

Lönnroth, M., Johansson, T. B., and P. Steen: "Sweden Beyond Oil: Nuclear Commitments and Solar Options," *Science*, **208**, 557–563 (1980).

Mattill, J. I.: "Laser Chemistry for New Energy," *Technology Review (MIT)*, **80**, 4, 57–58 (1978).

Metz, W. D.: "Laser Enrichment: Time Clarifies the Difficulty," *Science*, **191**, 1162–1163, 1193 (1976).

Nye, J. S., Jr.: "Balancing Nonproliferation and Energy Security," *Technology Review (MIT)*, **81**, 3, 48–57 (1979).

Rickard, C. L., and R. C. Dahlberg: "Nuclear Power: A Balanced Approach," *Science*, **202**, 581–584 (1978).

Robertson, J. A. L.: "The CANDU Reactor System: An Appropriate Technology," *Science*, **199**, 657–664 (1978).

Staff: "Projections of Energy Supply and Demand and Their Impacts," *Pubn. DOE/EIA-0036/2*, Vol. 2, Energy Information Administration, Department of Energy, Washington, D.C. (1978).

URBAN WASTES (As Energy Source). Biomass and Wastes as Energy Sources.

UREA. $H_2N \cdot CO \cdot NH_2$, formula weight 60.06, colorless crystalline solid, mp 132.7°C, sublimes unchanged under vacuum at its melting point, sp gr 1.335. Heating above the mp at atmospheric pressure causes decomposition, with the production of NH_3, isocyanic acid HNCO, cyanuric acid $(HNCO)_3$, biuret $NH_2CONHCONH_2$, and other products. Also known as carbamide, urea is very soluble in H_2O, soluble in alcohol, and slightly soluble in ether. The compound was discovered by Rouelle in 1773 as a constituent of urine. Historically, urea was the first organic compound to be synthesized from inorganic ingredients, accomplished by Wöhler in 1828. However, a century passed before the compound was manufactured on a large scale.

Because of the reactivity and versatility of its derivatives, urea is a very high-tonnage chemical. The compound and its derivatives are widely used in fertilizers, pharmaceuticls (e.g., barbiturates), and synthetic resins and plastics (urethanes). Although there are several chemical engineering approaches to the synthesis of urea, the principal reaction is that of combining NH_3 with CO_2 in a first step to form ammonium carbamate and, in a second step, dehydrating the ammonium carbamate to yield urea: (1) $2NH_3 + CO_2 \rightarrow NH_2COONH_4$, (2) $NH_2COONH_4 \rightarrow NH_2CONH_2 + H_2O$. The processing is complicated because of the severe corrosiveness of the reactants, usually requiring reaction vessels that are lined with lead, titanium, zirconium, silver, or stainless steel. The second step of the process requires a temperature of about 200°C to effect the dehydration of the ammonium carbamate. The processing pressure ranges from 160 to 250 atmospheres. Only about one-half of the ammonium carbamate is dehydrated in the first pass. Thus, the excess carbamate, after separation from the urea, must be recycled to the urea reactor or used for other products, such as the production of ammonium sulfate.

Some of the reactions of urea and derivatives include: (1) as a weak mono-acid base, urea forms stable salts, such as urea nitrate $CO(NH_2)_2 \cdot HNO_3$ and urea oxalate $2CO(NH_2)_2 \cdot H_2C_2O_4$; (2) urea reacts with malonic acid to form barbituric acid $CO(NHCO)_2CH_2$, the derivatives of which are barbiturates (sedative drugs); (3) with alcohols, urea reacts to form urethanes; (4) with formaldehyde, urea forms ureaforms which can be used as slow-release fertilizers and also as ingredients for adhesives and plastics; (5) with hydrogen peroxide, urea forms a useful crystalline oxidizing agent; (6) with straight-chain alkanes, urea forms crystalline complexes (clathrates) which are used in the petroleum industry for separating straight- and branched-chain hydrocarbons; (7) when heated rapidly to about 350°C in a fluidized bed at atmospheric pressure, urea decomposes to isocyanic acid and NH_3. The latter products, when passed over a catalyst at 400°C, yield melamine $(NCH_2)_3$ which is the triamide of cyanuric acid and widely used in plastics; (8) with acids or bases, urea hydrolyzes, yielding NH_3 and CO_2. Hydrolysis in aqueous solutions is accelerated by the presence of urease (an enzyme). This reaction frequently is used for the quantitative determination of urea; (9) upon heating aqueous solutions of urea, biuret is formed. When crystallizing urea from aqueous solutions, the presence of about 5% biuret alters the crystals from long needles to short rhombic prisms, the latter greatly enhancing the handling properties of the final product. A content of up to 1.5% biuret is satisfactory for most fertilizer applications, although for citrus fruits, coffee plants, and cherry trees, the biuret content must remain below 0.3%. As a feed supplement for ruminants, pure biuret has proved advantageous because of the slower release rate of NH_3 from biuret as compared with urea. See also **Fertilizer.**

Arginine-Urea Cycle (Ornithine Cycle). In adult animals, including humans, the characteristic tissue-specific levels of different enzymes are maintained by a dynamic balance between the independently controlled rates of biosynthesis and degradation of each enzyme. A dynamic rather than a static system most likely emerged because it enables organisms to adapt to widely different nutritional conditions and other environmental changes. Depending upon the physiological state of the animal at a given moment, amino acids derived from the hydrolysis of exogenous or endogenous protein may be predominantly utilized for *synthesis* of tissue-specific proteins or their carbon chains may be *metabolized* further to provide energy (ATP) or intermediates for synthesis of other cellular constituents. When the carbon chains of amino acids are utilized to provide energy, some provision must be made for disposal of the reduced nitrogen components. Animal tissues in general cannot tolerate accumulation of ammonia. Aquatic animals, which are surrounded by a convenient diluent, can simply excrete ammonia as rapidly as it is formed. In contrast, land-based animals have devised other solutions to this problem. They convert amino acid nitrogen and ammonia into nitrogen-rich, nontoxic compounds, such as *urea* and *uric acid*. These are then excreted at intervals. Synthesis of urea, the primary nitrogenous excretory product of mammals, is efficiently accomplished in the liver

by combining a portion of the already established pathway of arginine biosynthesis with the hydrolytic degradative enzyme, arginase.

Some of the enzymes required are widely distributed, but *ornithine carbamoyl-transferase* occurs only in the liver and thus the complete urea cycle occurs only in that organ.

In human liver, a given molecule of arginine has four possible metabolic fates: (1) It can be converted to argininosuccinic acid, (2) argininyl-sRNA, (3) ornithine plus urea, or (4) ornithine plus glycocyamine, the precursor of creatine. The flow along the pathway to (4) is regulated by feedback repression in which the steady-state level of the enzyme involved (arginine : glycine amidinotransferase) is regulated by the concentration of liver creatine. Thus, runaway synthesis of creatine is prevented. The flow along the pathway is regulated by supply and demand.

Experimentally, it has been observed that above a certain basal level, the quantity of urea excreted is proportional to the amount of ingested protein.

The enzyme urease was not discovered until 1926 (by Sumner). It was the first enzyme to be isolated as a crystalline protein. Sumner's accomplishment confirmed the then growing belief that enzymes, the biological catalysts, were indeed from the chemical standpoint protein molecules. Urease catalyzes the cleavage of urea to ammonia and carbon dioxide.

UREA AND THIOUREA DERIVATIVES.   Herbicide; Insecticide.

UREA-FORMALDEHYDE RESINS.   Amino Acids.

URINE.   The fluid secreted from the blood by the kidneys, stored in the bladder, and discharged by the urethra. In health, it is amber colored. About 1,250 milliliters of urine are excreted in 24 hours by normal humans, with specific gravities usually between 1.018 and 1.024 (extremes: 1.003–1.040). Flow ranges from 0.5–20 milliliters/minute with extremes of dehydration and hydration. Maximum osmolar concentration is 1,400, compared to plasma osmolarity of 300. In diabetes insipidus, characterized by inadequate antidiuretic hormone (ADH) production, volumes of 15–25 liters/day of dilute urine may be formed. In addition to the substances listed in the accompanying table, there are trace amounts of purine bases and methylated purines, glucuronates, the pigments urochrome and urobilin, hippuric acid, and amino acids. In pathological states, other substances may appear: proteins (nephrosis); bile pigments and salts (biliary obstruction); glucose, acetone, acetoacetic acid and beta-hydroxybutyric acid (diabetes mellitus). The U/P ratios of the substances in the table vary widely because of differential handling by the kidney. Quantitative knowledge of glomerular filtration, tubular reabsorption, and secretion of these requires an understanding of the concept of renal plasma clearance.

The rate at which a sustance (X) is excreted in the urine is the product of its urinary concentration, U_x (milligram/milliliter), and the volume of urine per minute, V. The rate of excretion $(U_x V)$ depends, among other factors, upon the concentration of X in the plasma, P_x (milligram/milliliter). It is therefore reasonable to relate $U_x V$ to P_x and this is called the clearance ratio: $(U_x \cdot V)/P_x$, or more generally, UV/P. This has the dimensions of volume and is in reality the smallest volume from which the kidneys can obtain the amount of X excreted per minute. The kidneys do not usually clear the plasma completely of X, but clear a larger volume incompletely. The clearance is therefore not a real, but a virtual volume. When substances are being cleared simultaneously, each has its own clearance rate, depending upon the amount absorbed from the glomerular filtrate or added by tubular secretion. The former will have the lower clearance, the latter the higher. Those cleared only by glomerular filtration will be intermediate, and their clearance will in effect measure the rate of glomerular filtration in milliliters/minute.

The best-known substance which can be infused into blood to provide a clearance equal to glomerular filtration rate is *inulin*, a polymer of fructose containing 32 hexose molecules (molecular weight, 5,200). Strong evidence indicates that it is neither reabsorbed nor secreted, is freely filterable, is not metabolized, and has no physiological influences. Its clearance in humans is 120–130 milliliters/minute. This is taken to be the glomerular filtration rate (*GFR*) or C_F (amount of plasma water filtered through glomeruli/minute). Besides inulin in the dog and other vertebrates, creatinine, thiosulfate, ferrocyanide, and mannitol also fulfill these requirements.

Knowing the glomerular filtration rate permits quantification of the amount of any substance freely filtered (C_F (milliliters/minute) $\times$ P_x (milligrams/milliliter)). Subtracting from this one minute's excretion, $U_x V$, would give the amount reabsorbed in milligrams/minute. A classical example is the glucose mechanism. At normal plasma concentrations, none or a trace appears in the urine. When plasma glucose is elevated to about 180–200 milligram percent (the "threshold"), the amount appearing in the urine begins to increase. As concentration is raised more, the nephrons become progressively saturated until the rate of reabsorption becomes constant and maximal. This indication of saturation of the transport system is referred to as the T_m ("tubular maximum—T_{mG}"). In humans, T_{mG} has the value of 340 milligrams/minute. Absorption occurs in the proximal convoluted tubules.

COMPOSITION OF 24-HOUR URINE IN THE NORMAL ADULT[a]

Substance	Amount (Grams)	U/P[b]
Urea	6.0–180.0 (nitrogen)	60.0
Creatinine	0.3–0.8 (nitrogen)	70.0
Ammonia	0.4–1.0 (nitrogen)	—
Uric acid	0.08–0.2 (nitrogen)	20.0
Sodium	2.0–4.0	0.8–1.5
Potassium	1.5–2.0	10.0–15.0
Calcium	0.1–0.3	—
Magnesium	0.1–0.2	—
Chloride	4.0–8.0	0.8–2.0
Bicarbonate	—	0.0–2.0
Phosphate	0.7–1.6 (phosphorus)	25.0
Inorganic sulfate	0.6–1.8 (sulfur)	50.0
Organic sulfate	0.06–0.2 (sulfur)	—

[a] Based upon data by White, Handler, Smith, and Stetten.
[b] U/P ratio = ratio of urinary to plasma concentration.

V

VACUUM. According to definition, a vacuum is a space entirely devoid of matter. The term is used in a relative sense in vacuum technology to denote gas pressures below the normal atmospheric pressure of 760 torr (1 torr = 1 mm Hg). The degree or quality of the vacuum attained is indicated by the total pressure of the residual gases in a vessel when it is pumped. The accompanying table shows generally accepted terminology for denoting various degrees of vacuum.

As pointed out by Fulcher, Rafelski, and Klein (*Sci. Amer.*, **241**, 6, 150–159, 1979), in the quantum field theories that describe the physics of elementary particles the vacuum becomes somewhat more complex than previously defined. Even in empty space, matter can appear spontaneously as a result of fluctuations of the vacuum. It may be pointed out, for example, that an electron and a positron, or antielectron, can be created out of the void. Particles created in this way have only a fleeting existence; they are annihilated almost as soon as they appear, and their presence can never be detected directly. They are called *virtual particles* in order to distinguish them from real particles. Thus, the traditional definition of vacuum (space with no real particles in it) holds. In their excellent paper, the aforementioned authors discuss how, near a superheavy atomic nucleus, empty space may become unstable, with the result that matter and antimatter can be created without any input of energy.

VACUUM DISTILLATION. Petroleum.

VALENCE. Chemical Elements; Molecule.

VANADINITE. The mineral vanadinite corresponds to the formula $Pb_5(VO_4)_3Cl$, being composed of lead chloride and lead vanadate in the proportion of 90.2% of the latter. It crystallizes in the hexagonal system, is usually prismatic, but the crystals are often skeletal or cavernous; it may be found in crusts. Its fracture is uneven; brittle; hardness, 2.75–3; specific gravity, 6.86; fresh fractures show a resinous luster; color, yellow, yel-

lowish-brown, reddish-brown, and red; streak, white to yellowish; translucent to opaque. Vanadinite, not a common mineral, occurs as an alteration product in lead deposits. It is found in the Urals, Austria, Spain, Scotland, Morocco, the Transvaal, Argentina, and Mexico. In the United States it occurs in Arizona, New Mexico, and South Dakota. It is interesting to note that this mineral was first described as a chromate upon its discovery in Mexico in 1801. It was not until the discovery of the element vanadium in 1830 that the true nature of this compound was known.

VANADIUM. Chemical element symbol V, at. no. 23, at. wt. 50.941, periodic table group 5b, mp 1880–2000°C, bp 3400°C, density, 6.10 g/cm³. Elemental vanadium has a body-centered cubic crystal structure. Vanadium is a silvery-white, very hard (7 on the Mohs scale), oxidizes upon exposure to air, burns upon ignition to form the pentoxide, V_2O_3, insoluble in HCl, slowly dissolves in hydrofluoric acid, HNO_3, or H_2SO_4 (hot, concentrated), or aqua regia. Insoluble in NaOH solution. The element was first reported by Andrés Manuel del Rio in 1801; later and separately reported by Nils Gabriel Sefström in 1830. There are two naturally occurring isotopes, ^{50}V (radioactive and with a half-life something greater than 10^{14} years and only present to the extent of 0.24% in natural substances) and ^{51}V (99.76 abundance percentage). Five other radioactive isotopes have been identified: ^{46}V through ^{49}V and ^{52}V. The isotope ^{49}V has a half-life of 330 days; ^{48}V has a half-life of 16.1 days; the others have half-lives of seconds or minutes. Vanadium ranks 22nd among chemical elements occurring in the earth's crust, an average composition of igneous rocks being 0.01% V. It is estimated that vanadium occurs in seawater to the extent of about 9.5 tons per cubic mile (2.1 metric tons per cubic kilometer).

Electronic configuration is $1s^22s^22p^63s^23p^63d^34s^2$. First ionization potential, 6.74 eV; second, 14.7 eV; third, 29.6 eV; fourth, 48.3 eV; fifth, 68.64 eV. Oxidation potentials: $V \rightarrow V^{2+} + 2e^-$, 1.5 V; $V^{2+} \rightarrow V^{3+} + e^-$, 0.255 V; $VO^{2+} + H_2O \rightarrow VO_2^+ +$

VARIOUS QUALITIES OF VACUUM AND PRESSURE RANGES

Quality of Vacuum	Pressure Range (torr)	Molecular Density, n (molecules/cubic centimeter)	Mean Free Path, λ (centimeters)
Coarse or rough vacuum	760–1	2.69×10^{19}–3.5×10^{16}	6.6×10^{-6}–5×10^{-3}
Medium vacuum	1–10^{-3}	3.5×10^{16}–3.5×10^{13}	5×10^{-3}–5
High vacuum	10^{-3}–10^{-7}	3.5×10^{13}–3.5×10^9	5–5×10^4
Very high vacuum	10^{-7}–10^{-9}	3.5×10^9–3.5×10^7	5×10^4–5×10^6
Ultrahigh vacuum	$< 10^{-9}$	$< 3.5 \times 10^7$	$> 5 \times 10^6$

NOTE: Molecular density is calculated from the equation $p = nkT$, where p is the pressure, n is the molecular density, i.e., number of molecules per cubic centimeter, k is the Boltzmann constant, and T is the absolute temperature. Mean free path is from the approximation equation for air: $\lambda = 5/p$ centimeters, where p is the pressure in millitorr.

$2H^+ + e^-$, -1.00 V. Other important physical characteristics of vanadium are given under **Chemical Elements.**

Vanadium occurs as patronite, containing vanadium pentasulfide, in Peru; as carnotite, potassium uranyl vanadate, in Colorado and Utah; as vanadinite, lead vanadate, in Arizona, New Mexico, the Republic of South Africa, and Zambia.

Ships burning Venezuelan or Mexican petroleum fuel oil recover vanadium oxide from the boiler and stack dust. In Italy, the refining of bauxite ore (for aluminum) yields vanadium, and in Germany some iron ores contain vanadium. Vanadium and radium ore is found in southwestern Colorado and southeastern Utah; in Arizona, a complex ore of gold, silver, and lead contains vanadium and molybdenum; and extensive deposits of phosphate rock in Idaho yield tonnage quantities of vanadium. The sulfide ore is roasted to remove sulfur, and the residue fused with sodium carbonate, forming sodium vanadate. This last is extracted with water and excess of sulfuric acid is added, causing precipitation of vanadium pentoxide, which is later reduced by carbon or aluminum at high temperatures.

Uses. Worldwide production of vanadium exceeds 30 million pounds (13.6 million kilograms) annually, up from 14 million pounds (6.4 million kilograms) in 1960. Most vanadium produced is consumed as ferrovanadium. This is made by the aluminum or silicon reduction of the oxide in the presence of iron in an electric-arc furnace. The product contains about 85% vanadium, 12% carbon, and 2% iron. Ductile vanadium metal also is produced in significant quantities, mainly by the calcium reduction of the oxide in a process developed by McKechnie and Seybolt. Pure vanadium oxide, calcium metal, and iodine are charged into a heavy-walled steel cylinder, excluding all moisture. After evacuation, heat is applied to initiate the reaction. Molten droplets of vanadium collect beneath the calcium oxide-calcium iodide slag and there form a single button or regulus. Ductile vanadium metal produced in this manner has an analysis of 99.7% vanadium, 0.10% oxygen, 0.04% nitrogen, 0.008% hydrogen, 0.04% iron, and 0.03% carbon. The metal is soft and ductile, can be hot- and cold-worked easily. Any heating must be done in vacuum or an inert atmosphere because the metal oxidizes readily. With exception of HNO_3, the metal withstands other acids and aerated saltwater better than most stainless steels. It has a comparatively low cross section for neutron capture and is of interest in the nuclear field. The density is 22% less than iron and 28% greater than titanium. The coefficients of thermal and electrical conductivity are higher than those of titanium.

Most vanadium is used by the steel industry. The addition of vanadium to steel causes the formation of vanadium carbide. The carbides are very hard and wear-resistant; they maintain a fine dispersion. The addition of very small quantities (0.02–0.08%) of vanadium enhances strength and toughness of the resulting steel. Many structural, plate, bar, and pipe steels contain vanadium in these amounts. For little additional cost, in comparison with plain carbon steels, there is a marked increase in performance, including higher strength in the as-rolled condition without heat treatment. Often, manganese and copper are added in small quantities along with the vanadium. Some sheet steels that are used for deep-drawing as in auto and home-appliance parts contain vanadium to suppress aging. For such steels, ferrovanadium is added to rimming steels, resulting in a good, nonaging, deep-drawing steel at a smaller cost than for aluminum-killed deep-drawing steel. Some large steel forgings contain vanadium to the extent of 0.5 to 0.15% with the object of improving the mechanical properties of the forgings. Vanadium is particularly effective in raising the strength and ductility of large steel castings and forgings when added in a small percent-

age. A large number of tool steels contain vanadium to the extent of 0.10 to 5.00%. In these steels, vanadium insures the retention of hardness and cutting ability at the high temperatures resulting from the rapid cutting of metals. The use of vanadium in cast iron controls the size and distribution of graphite flakes and thus improves strength and wear resistance. One of the most popular of the titanium-base alloys contains 4% vanadium and 6% aluminum. The addition ingredient for this titanium alloy, produced in large quantities, is a base 40:60 vanadium-alloy.

The use of certain vanadium compounds as catalysts has been increasing. Vanadium oxytrichloride is a catalyst in making ethylene-propylene rubber. Ammonium metavandate and vanadium pentoxide are used as oxidation catalysts, particularly in the production of polyamides, such as nylon, in the manufacture of H_2SO_4 by the contact process, in the production of phthalic and maleic anhydrides, and in numerous other oxidation reactions, such as alcohol to acetaldehyde, anthracene to anthraquinone, sugar to oxalic acid, and diphenylamine to carbazole. Vanadium compounds have been used for many years in the ceramics field for enamels and glazes. Colors are produced by various combinations of vanadium oxide and silica, zirconia, zinc, lead, tin, selenium, and cadmium. Vanadium intermediate compounds also are used in the making of aniline black used by the dye industry.

Chemistry and Compounds. The common oxidation states are: vanadous, V^{2+}, vanadic, V^{3+}, vanadyl, VO^{2+} or VO^{3+}; pervanadyl, VO_2^+; metavanadate, VO_3^-. There are also orthovanadates, VO_4^{3-}; pyrovandates, $V_2O_7^{4-}$; and complex polyvanadates. The latter group includes di, tri-, tetra-, and octavanadates. The hexavanadates are regarded as oxyvanadium(V) pentavanadates, containing the group $V_5O_{16}^{7-}$. Vanadium ions also form heteropolyacids with acids of molybdenum, tungsten, arsenic, phosphorus, silicon and tin. The peroxy vanadium ions include the diperoxorthovanadate ions $[VO_2(O_2)_2]^{3-}$ and the peroxovanadium(V) ions, $[V(O_2)]^{3+}$.

Among the simpler compounds of vanadium are:

Fluorides. Vanadium trifluoride, VF_3, green crystalline solid; vanadium tetrafluoride, VF_4, brownish-yellow crystalline solid; vanadium pentafluoride, VF_5, brownish-yellow crystalline solid, sublimes upon heating.

Chlorides: Vanadium dichloride, VCl_2, green crystalline solid, a strong reducing agent; vanadium trichloride, VCl_3, pink crystalline solid; vanadium tetrachloride, VCl_4, reddish-brown liquid, bp 148°C.

Bromide. Vanadium tribromide, VBr_3, green crystalline solid.

Iodides: Vanadium diiodide, VI_2, usually hydrated, green crystalline solid; vanadium triiodide, VI_3, brown crystalline solid.

Oxyhalides. These include VOF_2, VOF_3, $VOCl$, $VOCl_2$, $VOCl_3$, $VOBr$, and $VOBr_3$.

Hydroxides. Vanadium dihydroxide, $V(OH)_2$, brown precipitate by reaction of NaOH solution with hypovanadous acid (one of the most powerful of reducing agents) lavender solution; vanadous hydroxide, $V(OH)_3$, green precipitate by reaction of NaOH solution with vanadous salt, green solution.

Oxides. Vanadium monoxide VO, gray solid; vanadium trioxide V_2O_3, black solid; vanadium dioxide VO_2, dark blue solid; vanadium pentoxide V_2O_5, orange to red solid. The last is the most important oxide; formed by the ignition in air of vanadium sulfide, or other oxide, or vanadium; used as a catalyzer, e.g., the reaction SO_2 gas plus oxygen of air to form sulfur trioxide, and the oxidation of naphthalene by air to form phthalic anhydride.

Sulfides. Vanadium monosulfide VS; vanadium trisulfide V_2S_3, most stable; vanadium pentasulfide V_2S_5.

For further physical properties, consult "Vanadium" by Drangel and Martin in the "Metals Handbook," 9th Edition, American Society for Metals, Metals Park, Ohio, 1979.

VAN DER WAALS' EQUATION. Characteristic Equation.

VAN DER WAALS FORCES. In the most general sense, these are forces between atoms and/or molecules other than those leading to chemical bonding. While the distinction between these and chemical forces is not always sharp (e.g., there are loosely bound, so-called van der Waals molecules like O_4, HgA), the binding strength for the two types is ordinarily quite different: it is of the order of electron-volts in the case of chemical forces, millivolts in the case of van der Waals forces.

As a function of distance, van der Waals forces between *unexcited* structures (i.e., atoms or molecules in their normal states) usually show the behavior illustrated in the accompanying fig-

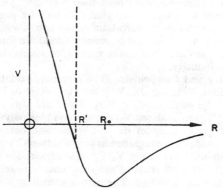

Typical intermolecular potential—potential energy versus distance of separation between molecules.

ure, where the potential energy (negative integral over force times displacement) is plotted against the distance of separation between the molecules. Beyond the point R_0 the energy curve rises, hence the forces are attractive. For smaller R the forces are repulsive.

The repulsive forces arise from the interpenetration of the electronic clouds surrounding the molecular nuclei. Because their calculation involves certain terms called exchange integrals, they are sometimes called *exchange forces*. Qualitatively, they are responsible for the mechanical rigidity, the "impenetrability" of the molecules and are sometimes approximated in computation by a straight vertical line at some characteristic distance R'. This distance is known as the "gas kinetic diameter" if the interaction takes place between similar molecules.

In the restricted sense, van der Waals forces are the forces which correspond to the attractive region beyond R_0 in the figure. If repulsive forces were absent, the constant b in van der Waals' equation would be zero; the constant a owes its existence to the attractive forces. Hence, it may be said by way of definition that van der Waals forces in the general sense are the interactions which cause deviations from the ideal gas law; in the specific sense they are the interactions giving rise to van der Waals' a. Henceforth, this article is devoted to the latter. Aside from the departures just mentioned, they are likewise responsible for surface tension, friction, changes of phase, cohesion of most solids, viscosity, the Joule-Thomson effect and many others.

The physical origin of these forces was once sought in gravitational interactions which, however, are now known to be very

much too small to produce such effects. The first modern conjecture was made by Debye (1920), who supposed that permanent multipoles, primarily quadrupoles, located in one of the structures, induced by electrostatic polarization small multipoles in the other. This *induction effect* does take place but is in general too small to account for the observations. Keesom, in 1921, presented a theory explaining van der Waals forces in terms of the alignment of permanent multipoles, which are shown to arrange themselves preferably in a manner leading to attraction. While this *alignment effect* is an important component of the van der Waals forces among polar molecules, it does not tell the whole story. In particular, neither induction nor alignment provides a way of explaining the van der Waals forces between symmetric atoms, such as the rare gases.

The principal missing factor in this account was provided by F. London in 1930. In developing the theory of what he called *dispersion forces*, he showed that quantum physics admits a mechanism crudely characterized as follows. The electrons in one atom (e.g., helium) revolve, as it were, in ignorance of those in the other so long as the atoms are very far apart. When they are brought together, they revolve more or less in phase in the two structures, the more so the closer they are together, and the result of this phase agreement is an attractive force. Dispersion forces are universally present in all interactions; they are not limited to molecules having permanent polarity.

The potential energy, V_D, arising from this effect has in general a much shorter range than the mutual energies of multipoles; for the simplest molecules (rare gases, O_2, N_2, etc.) they cease to be of importance at a distance of separation of 10^{-7} cm. They depend on distance in accordance with the formula

$$V_D = -\frac{A}{R^6} - \frac{B}{R^8} - \frac{C}{R^{10}} - \cdots$$

hence they disappear in a manner proportional to R^{-6} at large distances. In this expression, A/R^6 is said to represent the dipole-dipole, B/R^8 the dipole-quadrupole, and C/R^{10} the quadrupole-quadrupole effects. The latter were first calculated by the writer in 1931. Formally, the series has an indefinite number of terms, but its convergence is asymptotic, and it is generally inadvisable to retain terms beyond those written above.

Only for sufficiently symmetric molecules are A, B, and C independent of the orientation of the molecules in space. In general these parameters are functions of angles; hence the forces are not central forces. When more than two molecules interact, the dispersion effect is very nearly additive, i.e., the total energy is approximately a sum of pairwise components V_D. Additivity does not hold for the repulsive forces.

A general rule for the relative importance of van der Waals forces, for their dependence on angles or indeed distance, cannot be given. In the simple example of H_2O, at a distance of separation equal to the kinetic diameter, induction, dispersion, and alignment effects are all of comparable magnitude and function approximately in the ratio of $1:3:8$. In larger molecules the situation is very complex.

The forces between multipole molecules are not properly given by the figure, for they can be repulsive as well as attractive. But the average over-all orientations, weighted by the Boltzmann factor, is an attractive force. The dispersion effect is always attractive if the molecules are in their lowest energy states. When one or both of the molecules are excited, there are usually several modes of interaction, some attractive and others repulsive even at large distances of separation.

Information regarding the numerical values of van der Waals forces is mostly semiempirical, derived with the aid of theory from an analysis of chemical or physical data. Attempts to calculate the forces from first principles have had a measure of success only for the simplest systems, such as H—H, He—He and a few others. When judging the difficulties of such calculations, one must bear in mind that the energies sought are of the same order of magnitude as the *errors* in the best atomic energy calculations.

—Henry Margenau, Yale University, New Haven, Connecticut.

VAN SLYKE REACTION. Amino Acids.

VAN'T HOFF EQUATION.
A relationship representing the variation with temperature (at constant pressure) of the equilibrium constant of a gaseous reaction in terms of the change in heat content, i.e., of the heat of reaction (at constant pressure). It has the form:

$$\frac{d \ln K_p}{dT} = \frac{\Delta H}{RT^2}$$

in which K_p is the equilibrium constant at constant pressure, T is absolute temperature, R is the gas constant, and ΔH is the standard change in heat content, or, for ideal gases, the change in heat content.

VAN'T HOFF LAW.
A dissolved substance has the same osmotic pressure as the gas pressure it would exert in the form of an ideal gas occupying the same volume as that of the solution.

VAPORIZATION (Heat of).
The evaporation of a given mass of any liquid requires a definite quantity of heat, dependent upon the liquid and upon the temperature at which it evaporates. The quantity required per unit mass at a fixed temperature is called the heat of vaporization of the substance at that temperature. It may be measured by allowing the vapor to condense in a suitable calorimeter, the heat thus evolved, corrected for fall of temperature before and after condensation, being observed. (The heat evolved in condensing is equal to that absorbed when the liquid evaporates.) The result is often surprising. For example, the evaporation of water at the boiling point requires about 540 calories per gram, or more than five times the heat required to raise its temperature from freezing to boiling. The explanation is the large amount of energy necessary to separate the molecules against their cohesion, and the much smaller amount (about 7.4% of the whole) which is used in expanding the vapor against atmospheric pressure. At lower temperatures the value is still greater, because the cohesion is then more effective; with water, for each degree below the normal boiling point, about 0.6 calorie per gram must be added to the heat of vaporization. Trouton found that the heat of vaporization per mole for different liquids bears a nearly constant ratio to the absolute temperature of the boiling point.

A number of methods have been developed for measuring the heat of vaporization. In the Awberg and Griffith's method, the flow is continuous; the heat of evaporation of the liquid under investigation is transmitted to a stream of water flowing at constant rate and its increase in temperature is measured.

VAPOR PRESSURE.
The driving force behind the apparently universal tendency for liquids and solids to disperse into the gaseous phase. All known liquids and solids possess this fundamental property, although in some cases it is too minute to be measurable. A typical liquid will exert a vapor pressure which is constant and reproducible. This pressure is dependent only upon the temperature of the system, and increases with increasing temperatures.

The molecular theory explains the phenomenon of vapor pressure through molecular activity. The molecules of a liquid are in rapid motion, even though they are in contact with each other. This motion or activity increases with temperature. At the vapor-liquid interface, this motion results in diffusion of some molecules from the liquid into the vapor. The attraction between molecules is strong, and some of the molecules dispersed into the vapor return to the liquid. The net number of molecules escaping produces the vapor pressure. For all practical purposes, this vapor pressure can be assumed constant whether the system is at equilibrium or not, due to the extremely high rate of molecular diffusion at the interface of the two phases.

In solids, the attractive forces of the molecules are so dominant that each is more or less frozen in place. Some diffusion does occur, however, as evidenced by the evaporation of ice, the odor of moth balls, and the slow diffusion or alloy formation of some metals kept in intimate contact. This vapor pressure increases with temperature, but is also a function of the molecular arrangement of the solid. As some solids such as sulfur are heated and molecular rearrangements take place, forming another allotrope of the same element, the vapor pressure changes sharply as this rearrangement occurs. The diffusion of a solid into the vapor state is called sublimation. The energy required to accomplish this transfer is high, since it is the sum of the heat normally required to melt the substance as well as vaporize it.

Vapor pressure can only be exhibited when the molecular activity is at a low enough level to permit continuous contact of the molecules and thus formation of a liquid. The maximum temperature at which this is possible is a fundamental property, and is called the critical temperature. Above this temperature the material cannot be compressed to form a liquid, and only one phase results. This temperature is 374.0°C for water, and −240.0°C for hydrogen.

The fundamental relationship between temperature and vapor pressure can be derived from thermodynamic laws. With certain limiting assumptions, the Clausius-Clapeyron equation is most often applied:

$$\frac{dp}{dT} = \frac{qp}{RT^2}$$

where q is the heat of vaporization; in calories per gram mole; $\frac{dp}{dt}$ is the slope of vapor pressure versus temperature curve at the point in question in cm Hg per °C; p is the pressure in cm Hg; R is the gas constant in calories per °C per gram mole; and T is the temperature in °K. Because of these limiting assumptions, the integrated form of this equation is used in practice primarily as a guide to develop methods of correlating and plotting vapor pressure data.

An excellent and versatile empirical vapor pressure correlation was developed by F. D. Othmer, P. W. Maurer, C. J. Molinary, and R. C. Kowalski (*Ind. Eng. Chem.*, **49** (1957). The reference equation is:

$$\log P = \left(\frac{L}{L'}\right) \log P' + C$$

where L/L' is the ratio of molar latent heats of the substance in question to a reference substance. The ratio is the key to the accuracy of this method; it very neatly minimizes effects of deviations from ideality.

The vapor pressure of a solution containing a nonvolatile substance (e.g., salt in water) is lower than that of the pure liquid. This phenomenon can again be explained by interference of the liquid molecular activity by the dissolved substances. The relationship between this vapor pressure depression and the concentration of the dissolved substance is valid for most substances at low concentrations. It was found to be dependent on the relative numbers of molecules of the solute and the solvent, and allowed accurate determinations of molecular weights of unknown solutes. If the Clausius-Clapeyron equation given above is combined with the above concentration relationship, it can be shown that:

$$\Delta T = \frac{RT^2}{q} \cdot C$$

where ΔT is the elevation of the boiling point, and C is the mole ratio of solute to solvent. This defines the effect of any solute on the vapor pressure exhibited by any solvent of latent heat q.

In the same manner, the vapor pressure of one component of a solution of two liquids has a different relationship with temperature than if it were pure. For many liquid mixtures, such as most hydrocarbon mixtures, the vapor pressures of the components vary directly from that exhibited in the pure form as their molar concentration in the solution. This relationship is known as Raoult's law:

$$\text{Partial Pressure} = P_o x$$

where x is the molar concentration of the component in the liquid, the P_o is the vapor pressure of the pure component of the same temperature as the mixture. A mixture following this rule is called an "ideal" solution, and its total volume is the sum of its components' volumes.

When most mixtures are reduced by boiling, logically the more volatile component is removed faster and heavier material concentrates. There are important exceptions to this rule, and these are called azeotropic or constant boiling mixtures. In essence, due to association or some other reason for deviation from Raoult's law, there are certain mixtures where the boiling temperature is either above that of the high boiling component or below that of the more volatile component. As the composition changes, at some point the boiling temperature change has to reverse itself to approach the remaining major component. At this point the composition of the liquid and vapor are identical and the mixture behaves as a single component, boiling away with no change in temperature or composition, the CBM (constant boiling mixture). The composition of CBMs is frequently pressure sensitive.

If the gas phase above the liquid is also "ideal," the partial pressure of a component in this phase is equal to the total system pressure times the mole fraction of the component in the gas phase. This is called Dalton's law:

$$\text{Partial Pressure} = P_t y$$

Combining these two formulae, it can be seen that:

$$\frac{y}{x} = \frac{P_o}{P_t} = K$$

for any particular temperature. This relationship can largely define many very complex liquid mixtures if the pressures used in the correlation are corrected by experimental data for deviation from the ideal.

For work of any precision, it is essential that the correlative methods described above be only used to help fit experimental data, particularly if the system is at elevated pressure.

A different relationship results if two liquids are relatively immiscible in each other. Molecular interference is minimal, and the total pressure exerted is equal to the sum of those of the individual pure components. The fundamental property of vapor pressure is thus dependent on temperature and composition of the material considered. These known and reproducible relationships have great technical application.

Vapor pressure relations can be used to determine heats of solution, heats of sublimation, and heats of fusion. Problems dealing with the solution of gases of liquids and adsorption of gases by solids are best handled by vapor pressure concepts. In dealing with solutions of miscible liquids, the most simple and useful relationship involves plotting (for the most volatile component) the mole fraction y in the vapor against x, the mole fraction in the liquid. The ratio of y/x is called the phase equilibrium constant K, and is used for definition of bubble points and dewpoints of simple and complex hydrocarbon mixtures over temperature and pressure ranges to near the critical.

Some equations of state also define vapor pressure relationships. One of the most recognized is that developed by Benedict, Webb and Rubin due to its ability to predict P-V-T properties in the two-phase region plus describing behavior of the superheated vapor. (Ref., *J. Chem. Phys.*, **8**, 334 (1940); ibid., **10**, 747 (1942); *Chem. Eng. Progr.*, **47**, (1951).)

—Douglas L. Allen, Tuscola, Illinois.

VARIABLE (Process). The quantity or characteristic that is the object of measurement in an instrumentation or automatic control system. Other terms used include measurement variable, instrumentation variable, and process variable. The latter term is commonly used in the manufacturing industries. Numerous ways to classify variables have been proposed—by methods of measurement, by end-measurement objectives, and so on. One of the most convenient and meaningful classification is the physical and/or chemical nature of the variable, as follows:

Thermal Variables. These variables relate to the condition or character of a material dependent upon its thermal energy. Variables included are: *temperature, specific heat, thermal-energy variables* (enthalpy, entropy, etc.), and *calorific value.*

Radiation Variables. These variables relate to the emission, propagation, and absorption of energy through space or through a material in the form of waves; and by extension, corpuscular emission, propagation, and absorption. Variables included are: *nuclear radiation; electromagnetic radiation* (radiant heat, infrared, visible, and ultraviolet light; x- and cosmic rays; gamma radiation).

Force Variables, including: *total force, moment or torque,* and *force per unit area,* such as pressure, vacuum, and unit stress.

Rate Variables. These variables are concerned with the rate at which a body is moving toward or away from a fixed point. Time always is a component of a rate variable. Variables included are: *flow, speed, velocity,* and *acceleration.*

Quantity Variables. These variables relate to the total quantity of material that exists within specific boundaries. Variables include: *mass* and *weight.*

Physical Property Variables. These variables are concerned with the physical properties of materials with the exception of those properties which are related to chemical composition and direct mass and weight. Variables included are: *density* and *specific gravity, humidity, moisture content, viscosity, consistency,* and *structural characteristics,* such as hardness, ductility, and lattice structure.

Chemical-Composition Variables. These variables relate to the chemical properties and analysis of substances. A very abridged list of analysis variables would include: Identification and concentration of carbon dioxide, carbon monoxide, hydrogen, nitrogen, oxygen, water, hydrogen sulfide, nitrogen oxides, sulfur oxides, methane, ethylene, alcohol, and so on. Also included in this category is measurement of pH (hydrogen ion concentration) and redox measurements. See also **Analysis (Chemical)**.

Electrical Variables. Included here are those variables which are measured as the "product" of a process, as in the case of measuring the current and voltage of a generator, and also as part of an instrumentation system. Numerous transducers, of course, yield electrical signals which represent by inference some other variable quantity, such as a temperature or pressure. Variables in this class include: *electromotive force*, *electric current*, *resistance*, *conductance*, *inductance*, *capacitance*, and *impedance*.

Geometric Variables. These variables are related to position or dimension and relate to the fundamental standard of length. Variables include: *position*, *dimension*, *contour*, and *level* (as of a material in a tank or bin).

VEGETABLE OILS (Edible).

Vegetable oils are prepared from at least ten major oilseeds, plus a few other sources that may develop into high volume at some future date. There are both edible and nonedible vegetable oils. Technical and industrial vegetable oils, such as castorseed and tung oils, are used in nonfood applications. Some oils, such as linseed (flaxseed) and olive oil, find both food and nonfood applications, depending upon the manner in which they have been treated and refined. Oils such as soybean, sunflower, rapeseed, sesame, and safflower oils, are used predominantly in food processing and in the production of feedstuffs. Usually, in considering edible vegetable oils, the availability of animal fats, such as butterfat, lard, tallow, etc., and of fish and other marine oils, is noted because there is surprising versatility among all oils and fats, allowing significant substitutions in the same end-product. Perhaps this is best illustrated by the fact that coconut oil, corn (maize) oil, cottonseed oil, palm oil, groundnut (peanut) oil, safflower oil, and soybean oil all are or have been used in commercial margarines for the retail market. The processor of a given brand of margarine will select the oil or a combination of oils, considering price and availability, as well as desirable end-product characteristics (stick or brick; soft tub; diet or imitation, etc.), among other factors. The pressed cakes and meals remaining after extraction of oil from beans and seeds find wide use in animal feedstuffs.

The consumption of fats and oils by persons in the United States has increased steadily since the early 1960s. In 1963, 46.3 pounds (21 kilograms) were consumed per capita per year, rising to 54.4 pounds (24.7 kilograms) in the late-1970s. During this period, butter dropped from 13.9% of total fats and oils to 7.7%; lard dropped from 12.9% to 4%; margarine increased from 19.4% to 20.2%; shortenings of various kinds increased from 27.2% to 30.6%; and salad and cooking oils increased markedly from 26.6% to 37.5%.

On a worldwide basis, as of the early 1980s, the various sources of edible oils, in terms of volume of production, rank as follows: Soybeans (50.3%), cottonseed (16.2%), groundnuts (peanuts) (11.4%), sunflower seeds (7.9%), rapeseed (6.9%), copra (coconut) (2.9%), flaxseed (linseed) (1.8%), sesame seed (1.1%), palm kernels (0.9%), and safflower seed (0.6%).

Principal edible uses of these oils are found in cooking and salad oils; frying oils; margarine, mayonnaise, and salad dressings; bakery, cake mix, and pie shortenings; and whipped topping and other nondairy products, such as coffee creamers.

Definitions

Some of the common terms used in describing edible oils and fats are:

Saturated—The state in which all available valence bonds of an atom, especially carbon, are attached to other atoms. The straight-chain alkyls (paraffins) are typical saturated compounds. Several fatty acids found in oils and fats are listed in Table 1. Where no double or triple bonds are available, such compounds cannot be hydrogenated because there are no places for attaching additional hydrogen atoms.

Monounsaturated—The state where one double bond is present in the compound. Under proper conditions of hydrogenation, an additional hydrogen atom can be added, thus resulting in a saturated compound.

Polyunsaturated—The state where two or more double bonds are present in the compound. Again, under proper conditions of hydrogenation, one or more additional hydrogen atoms can be added, thus resulting in a compound that is less unsaturated or that is saturated, depending upon the number of double bonds originally available and the degree of hydrogenation effected.

Hydrogenation—Any reaction of hydrogen with an organic compound. In the case of fats and oils, this is the direct addition of hydrogen to double bonds of unsaturated molecules, resulting in a partially or a fully saturated product.

Hydrogenation can be accomplished with gaseous hydrogen under pressure in the presence of a catalyst (nickel, platinum, or palladium). The degree of saturation of an oil or its fatty acid substituents affects its fluidity (melting or softening point), density, refractive index, and general reactivity. Frequently, the degree of hydrogenation of an oil will be determined by measuring its refractive index. Vegetable and fish oils can be hardened or solidified by catalytic hydrogenation. Partial hydrogenation clarifies some oils and makes them odorless. Fatty oils, such as oleic acid, are converted into stearic acid by hydrogenation. Coconut oil, groundnut (peanut) oil, and cottonseed oils can be made to appear, taste, and smell like lard; or they can be made to resemble tallow. Sometimes, hydrogenated oils are referred to as synthetic shortenings. Generally, hydrogenated oils have higher melting points and lower iodine values than the natural, untreated oils.

For many years, hydrogenation was a batch operation, but in recent years, continuous hydrogenation processes have been displacing the batch methods, especially in large-scale facilities. Hydrogenation pressures vary as well as temperatures, but the latter are usually in the 93–135°C range. It is not uncommon to mix two or more oils prior to hydrogenation. Considerably more detail on the hydrogenation process is given in the entry on **Hydrogenation**.

Deodorizing—Crude oils tend to have undesirable odors and tastes. These are due to the presence of various volatile components. These are removed by passing steam through the heated oil under diminished pressure.

Winterizing—Because of widespread use of refrigeration, salad oils and other end-products prepared from vegetable oils must not become cloudy or solidify at relatively low storage temperatures. Thus, most salad oils differ from cooking oils, in that the latter, if stored under refrigeration, may slowly solidify and be difficult to pour from a container. To prevent cloudiness and solidification at low temperatures, the oils are "winterized." This involves subjecting the oils to low temperatures in stages, during which time crystals of high-melting fat are formed. A final low temperature of about 5.5°C is reached in the process, at which temperature the oil is allowed to stand for a considerable time, during which considerable crystallization occurs. Slow cooling and gentle agitation insure maximum removal of the high-melting fats. The crystalline material remaining after filtration is essentially stearine, with traces of wax, gums, and soaps. Improvements in the winterizing process during the past several years have reduced the time required

TABLE 1. CHARACTERISTICS OF VEGETABLE OIL FATTY ACIDS

Principal Fatty Acid	Formula Weight	Number of Double Bonds	Position of Bonds	Number of Carbon Atoms	Formula
SATURATED COMPOUNDS					
Caproic	116.09	0	—	6	$C_5H_{11}COOH$
Caprylic	144.21	0	—	8	$C_7H_{15}COOH$
Capric	172.26	0	—	10	$C_9H_{19}COOH$
Lauric	200.31	0	—	12	$C_{11}H_{23}COOH$
Myristic	228.36	0	—	14	$C_{13}H_{27}COOH$
Palmitic	256.42	0	—	16	$C_{15}H_{31}COOH$
Stearic	258.47	0	—	20	$C_{19}H_{39}COOH$
UNSATURATED COMPOUNDS					
Palmitolenic	254.42	1		16	$C_{15}H_{29}COOH$
Linolenic	278.42	3	cis-9, cis-12, cis-15	18	$C_{17}H_{29}COOH$
Linoleic	180.44	2	cis-9, cis-12	18	$C_{17}H_{31}COOH$
Oleic	282.45	1	cis-9	18	$C_{17}H_{33}COOH$

from 3–6 days down to about 5 hours. The more recent processes involve solvent extraction and centrifugation.

Iodine Value (or Number)—The percentage of iodine that will be absorbed by a chemically unsaturated substance, such as vegetable oil, in a given time under specified conditions. The iodine value is a measure of degree of unsaturation. Two methods are described in the "Food Chemicals Codex," the Hanus method and the Wijs method.

Saponification Number—The number of milligrams of potassium hydroxide required to hydrolyze one gram of sample of an ester (glyceride; fat) or mixture. This test is also described in the "Food Chemicals Codex."

Margarines

Regular margarines (by government regulations) contain at least 80% fat. The remaining content is ~16% water and small amounts (about 4%) of skim or nonfat dry milk and salt. There are several unsalted margarines available. Small amounts of emulsifying agents consisting of either lecithin and/or monoglycerides and diglycerides are contained in a number of commercially available margarines. The preservatives most commonly used include sodium benzoate, potassium sorbate, calcium disodium EDTA, isopropyl citrate, and citric acid. Most margarines in the United States are fortified with about 15,000 USP units of vitamin A and a few margarines also contain up to 2000 USP units of Vitamin D.

Margarines produced in the United States resemble butter in their proxmate composition. The great majority of margarines provide more polyunsaturated fatty acids than an equivalent weight of butter. Diet, imitation margarines were introduced about a decade ago and contain less than 80% fat. The fat content of light blends is about 60% and that of diet imitation margarines is about 40%. Lower fat content in these products is compensated for by higher water content.

A survey of some 40 margarines made in the United States in the late 1970s showed that stick margarines made from partially hydrogenated soybean and cottonseed oil were by far the largest category. Stick margarines made from corn (maize) oil and partially hydrogenated corn oil, and from partially hydrogenated soybean oil and liquid cottonseed oil ranked first and third, respectively. The survey showed that margarines with formulations that included liquid cottonseed oil or liquid and/or partially-hardened palm oil usually contained higher amounts of palmitic acid than did other margarines of the same type.

The stearic acid content was reasonably constant for most hard and soft margarines. Margarines made with coconut oil were extremely saturated and contained large amounts of lauric and myristic acids.

Properties of Major Edible Vegetable Oils

The major properties of nine of the principal edible vegetable oils are summarized in Table 2. For descriptions of the constituent acids, see **Arachidic Acid; Caproic Acid; Capric Acid; Lauric Acid; Linoleic Acid; Linolenic Acid; Myristic Acid; Oleic Acid; Palmitic Acid;** and **Stearic Acid and Stearates.**

Unconventional Sources of Oils and Proteins

Research efforts continue to find new sources of both food and nonfood oils. Some investigations have reported on the prospects of various gourds indigenous to western North America. Such cucurbits are vigorous and highly drought-resistant and produce large quantities of foliage and fruit containing seeds rich in protein and oil, with an extensive system of large, fleshy storage roots which contain starch. Varieties studied have included *Cucurbita foetidissima* and *C. digitata*, as well as varieties of hibiscus (okra). As pointed out by the researchers, the nonconventional protein sources as well as oils would be welcome in a time of protein deficits.

Investigators also have described the potential of the *jojoba plant*, a native of the Sonoran desert of Mexico, Arizona, and California. The desert shrub produces seeds about the size of groundnuts (peanuts) which contain a liquid wax, frequently called *jojoba oil*. The oil is similar to sperm whale oil in its suitability for a number of industrial applications. Sperm whale was placed on the endangered species list by a number of countries, including the United States, in 1971. Prospects of an edible oil from jojoba are undetermined, but the meal remaining after expressing jojoba seeds contains 30–35% protein which may have potential as a livestock feed. A new jojoba industry would mean a new, renewable resource that has not been exploited before and would offer economic relief to the people of the Sonoran desert region. Other researchers have recently shown a renewed interest in the lupin as a source of edible oils and proteins.

References

Aguilera, J. M., and A. Trier: "The Revival of the Lupin," *Food Technology*. **32**, 8, 70–76 (1978).

TABLE 2. CHARACTERISTICS AND PROPERTIES OF MAJOR VEGETABLE OILS

PROPERTY	COCONUT OIL	COTTONSEED OIL	GROUNDNUT (Peanut) OIL	PALM OIL	RAPESEED OIL
Color	White	Pale-yellow or yellowish-brown to dark ruby-red, or black-red.	Yellow to greenish-yellow	Yellow-brown	Brown (raw); yellow (refined)
State (at room temperature)	Semisolid (nondrying oil)	Liquid (semidrying oil)	Liquid (nondrying)	Solid (buttery)	Liquid (viscous)
Odor	Slight	Slight (when refined)	Nutlike	Agreeable	Characteristic
Melting point	77° – 90° F; 25° – 32° C	Slightly below 32° F; 0° C # (before winterizing process)	23° – 37° F; –5° – +3° C@	86° F; 30° C	32° F; 0° C$
Specific gravity	0.92	0.915 – 0.921	0.912 – 0.920	0.952	0.913 – 0.916
Saponification value	250 – 264	190 – 198	186 – 194	247.6	174
Iodine value	7 – 10	109 – 116	88 – 98	13.5	100.3

Principal constituents

COCONUT OIL — Glycerides of fatty acids of approximate composition:

Lauric acid	45 – 48%
Myristic acid	17 – 20%
Capric acid	6 – 8%
Palmitic acid	5 – 9%
Caprylic acid	5 – 7%
Oleic acid	4 – 8%
Stearic acid	2 – 5%
Arachidic acid	1%±
Palmitoleic acid	0.4%±
Unsaturated acids	8%
Saturated acids	92%
(Approximate)	

COTTONSEED OIL — Glycerides of fatty acids of approximate composition:

Linoleic acid	47 – 50%
Palmitic acid	26 – 27%
Oleic acid	18 – 19%
Stearic acid	2%
Palmitoleic acid	1%
Myristic acid	1%–
Capric acid	0.5%
Linolenic acid	0.4%±
Lauric acid	0.4%
Arachidic acid	0.2%±
Unsaturated acids	73%
Saturated acids	27%
(Approximate)	

GROUNDNUT (Peanut) OIL — Glycerides of fatty acids of approximate composition:

Oleic acid	40 – 52%
Linoleic acid	25 – 37%
Palmitic acid	8.5 – 10.5%
Stearic acid	2.5%±
Behenic acid	1.5%±
Lignoceric aci	3.0%
Stearic acid	1.5 – 2.5%
Arachidic acid	1.5 – 2.5%
Myristic	0.2%±
Unsaturated acids	82%
Saturated acids	18%
(Approximate)	

PALM OIL — Triglycerides of fatty acids of approximate composition:

Palmitic acid	37 – 47%
Oleic acid	31 – 44%
Lauric acid	4 – 8%
Stearic acid	2.5 – 5.5%
Myristic acid	0.8 – 1.5%
Linoleic acid	1 – 3%
Palmitoleic acid	0.2 – 0.4%
Arachidic acid	0.2 – 0.4%
Unsaturated acids	50%
Saturated acids	50%
(Approximate)	

There are some significant differences in composition of palm oil (from fleshy fruit pulp) and palm kernel oil.φ

RAPESEED OIL — High in unsaturated acids, especially oleic, linoleic, and especially oleic, linoleic, and erucic acids, the latter being 40% in older varieties, but more recent varieties have a lower erucic acid content.

| Soluble in | Alcohol, carbon disulfide, chloroform, ether. Immiscible in water. | Benzene, carbon disulfide, chloroform, ether. Slightly soluble in alcohol. | Carbon disulfide, chloroform, ether, petroleum ether. Insoluble in alkalies, but saponified by alkali hydroxides with formation of soaps. Slightly soluble in alcohol. Insoluble in water. | Alcohol, carbon disulfide. chloroform, ether. | Alcohol, carbon disulfide. chloroform, ether. |

TABLE 2.   CHARACTERISTICS AND PROPERTIES OF MAJOR VEGETABLE OILS (cont.)

PROPERTY	COCONUT OIL	COTTONSEED OIL	GROUNDNUT (Peanut) OIL	PALM OIL	RAPESEED OIL
Grades	Crude, refined, Ceyon, Manila	Crude, refined, prime summer yellow, bleachable, USP.	Crude, refined, edible, USP.	Crude, refined	
Derivation	Hydraulic press or expeller extraction from coconut meat, followed by alkali-refining, bleaching, and	From cotton seeds by hot-pressing or solvent extraction.	By pressing ground groundnut (peanut) meats or by extraction with hot or cold solvents. Purified by bleaching with fuller's earth or carbon. Hot-pressed oil may be allowed to stand to deposit stearin, prior to filtering.	Separation of fat from palm fruits by expression or centrifugation.	
Food uses	Margarine, hydrogenated shortenings, synthetic cocoa dietary supplements.	Margarine, shortening, salad oils and dressings, stabilizers.	Salad oils, mayonnaise, margarine. Sometimes a substitute for olive oil. Cooking oil.	Food shortenings, margarines, competes with soybean oil.	Salad dressings, margarine, substitute for soybean oil.
Special notations	See also entry on **Coconut.**	#Solidification range is 88° – 95°F (31° – 35°C) Flashpoint is 486°F (252°C) See also entry on **Cottonseed.**	@Flash point is 540°F (282°C) See also **Groundnut (Peanut).**	φApproximate composition of palm kernel oil is: 50% lauric acid, 15% myristic acid, 16% oleic acid, 7% palmitic acid with lesser amounts of capric and caprylic acids. See also **Palm Oil.**	$Flash point is 325°F (617°F) Autoignition temperature is 836°F (447°C). See also **Rape.**
Color	Straw (Nonyellowing)	Yellow	Pale Yellow	Pale Yellow	Pale Yellow
State (at room temperature)	Liquid	Liquid	Liquid (fixed drying oil)	Liquid (semidrying)	
Odor	Almost odorless	Almost odorless	Characteristic	Pleasant	
Melting point	0° – +13°F; –18° – –25°C	68° – 77°F; 20° – 25°C#	77° – 88°F; 22° – 31°C	3° – 1°F; –16° – –18°C	
Specific gravity	0.923 – 0.927	0.9187	0.924 – 0.929	0.924 – 0.926	
Saponification value	186 – 193	188 – 193	190 – 193	186 – 194	
Iodine value	140 – 152	103 – 114	137 – 143	130 – 135	

TABLE 2. CHARACTERISTICS AND PROPERTIES OF MAJOR VEGETABLE OILS (cont.)

PROPERTY	SAFFLOWER SEED OIL	SESAME SEED OIL	SOYBEAN OIL	SUNFLOWER OIL
Principal constituents	Basic fatty acid composition: Linoleic acid 78%± Oleic acid 13%± Stearic acid 3%± Palmitic acid 6%± Unsaturated acids 90% Saturated acids 10% (Approximate)	Basic fatty acid composition: Oleic acid 40%± Linoleic acid 44%± Other saturates 1%± Palmitic acid 9%± Stearic acid 4%± Other saturates 2%± Unsaturated acids 85% Saturated acids 15% (Approximate)	Basic fatty acid composition: Oleic acid 20 - 25% Linoleic acid 48 - 53% Linolenic acid 5 - 9% Stearic acid 3 - 5% Palmitic acid 9 - 12% Myristic acid 0.2%± Arachidic acid 0.2%± Palmitoleic acid 0.3% Lauric acid 0.1% Unsaturated acids 84% Saturated acids 16% (Approximate)	Basic fatty acid composition: Linoleic acid 60 - 75% Oleic acid 17 - 32% Palmitic acid 7 - 10% Stearic acid 4 - 8% Myristic acid 0.1% Arachidic acid 0.3% Linolenic acid 0.3%± Unsaturated acids 90% Saturated acids 10% (Approximate)
Soluble in	Benzene, carbon disulfide, chloroform, ether. Slightly soluble in alcohol.		Alcohol, carbon disulfide, chloroform, ether	Alcohol, carbon disulfide, chloroform, ether
Grades	Crude, refined	Edible should contain less than free fatty acids. Semi-refined, coast, USP.	Coast, refined (salad), crude, foots (for soapstock), clarified.	Crude, refined
Derivation	Hydraulic press or solvent extraction of seeds.	Pressing of seeds.	Oil for edible purposes is bleached with fuller's earth.	Expression from seeds.
Food uses	Dietetic foods, margarine, hydrogenated shortenings	Shortenings, salad oil, margarine	High-protein foods, margarine, salad dressings	Margarine, shortening
Special notations	Considered by some authorities as the most natural, nutritionally sound vegetable oil. See also **Safflower Oil.**	≠Solidifying point is 23° F (-5°C) See also **Sesame Seed Oil.**	See also **Soybean Processing.**	See also **Sunflower Seed and Oil.**

Anderson, B. A., Kinsella, J. E., and B. K. Watt: "Comprehensive Evaluation of Fatty Acids in Foods. II. Beef Products," *Jrnl of Amer. Dietet. Assn.*, **67**, 35 (1975).

Berry, J. W., et al.: "Cucurbit Root Starch: Isolation and Some Properties of Starches from *Cucurbita foetidissima*," *Jrnl. Agr. Food Che.*, **23**, 825 (1975).

Beuchet, L. R.: "Microbial Alterations of Grains, Legumes, and Oilseeds," *Food Technology*, **32**, 5, 193–197 (1978).

Brignoli, C. A., et al.: "Comprehensive Evaluation of Fatty Acids in Foods. V. Unhydrogenated Fats and Oils," *Jrnl. of Amer. Dietet. Assn.*, **68**, 224 (1976).

Deimel, R. W.: "The Promise of Jojoba," *Agr. Research*, **27**(3), 2 (1978).

Staff: "The Biological Effects of Polyunsaturated Fatty Acids," *Dairy Council Digest*, **46**, 31 (1975).

Staff: "Food Chemicals Codex," National Academy of Sciences, Washington, D.C. (Updated periodically).

VESUVIANITE. The mineral vesuvianite is a very complex silicate of calcium and aluminum with fluorine which may also contain varying amounts of boron, iron, lithium, magnesium, manganese, potassium, sodium and titanium. A suggested formula is $Ca_{10}Mg_2Al_4(SiO_4)_5(Si_2O_7)_2(OH)_4$. Its tetragonal crystals are usually short, somewhat stoutish, prisms, sometimes pyramids, but columnar to massive varieties are common. It is essentially without cleavage; fracture, uneven; brittle; hardness, 6–7; specific gravity, 3.3–3.5; luster, vitreous to greasy or resinous; color, commonly some shade of brown, green, or white, but may be reddish, bluish or yellowish; may be transparent, but is usually translucent.

This mineral was formerly called idocrase, having been named by Haüy from the Greek words meaning *form* and *mixture* because it resembled crystals of other species, scarcely a valid distinction. Werner gave it the name vesuvianite from Mt. Vesuvius where it was first found in blocks of limestone appearing as inclusions in the lava. Vesuvianite is not a constituent of the igneous rocks, but rather a contact metamorphic mineral resulting from the alteration of impure limestones and dolomites. It is usually associated with diopside, wollastonite, epidote, grossularite, and garnet. There are many localities worthy of mention among which are: The Urals, Czechoslovakia, Rumania, Trentino and Monzoni in Italy, as well as at Mt. Vesuvius and Mt. Somma, Switzerland, Mexico and Japan. In the United States vesuvianite is found in Androscoggin and York Counties in Maine; Orange and Warren Counties, New York; Sussex County, New Jersey; Garland County, Arkansas; and Riverside and Tulare Counties in California.

VINEGAR. Acetic Acid; Antimicrobial Agents (Foods).

VINYL ACETATE. Acetic Acid.

VINYL CHLORIDE. Chlorinated Organics.

VINYL ESTER RESINS. The vinyl ester resins are a relatively recent addition[1] to thermosetting-polymer chemistry. Superficially, they are similar to unsaturated polyester resins insofar as they contain ethylenic unsaturation and are cured through a free-radical mechanism, usually in the presence of a vinyl monomer, such as styrene. However, close examination of the chemistry and structure of the vinyl ester resins demonstrates several basic differences which lead to their unique characteristics.

Vinyl ester resins are manufactured through an addition reaction of an epoxy resin with an acrylic monomer, such as acrylic acid, methacrylic acid, or the half-ester product of an hydroxyalkyl acrylate and anhydride. In contrast, the polyester resins are condensation products of dibasic acids and polhydric alcohols. The relatively low-molecular-weight precise polymer structure of the vinyl ester resins is in contrast to the high-molecular-weight random structure of the polyesters.

Of particular importance in describing the difference between these two families of resins are the locations of the reactive unsaturation. In the polyester resin, these groups are located along the backbone of the polymer with terminal hydroxyl or carboxylic acid groups. In contrast, the vinyl ester resins contain no significant acidity but terminate in reactive vinyl ester groups. Because of the location of these reactive sites, the vinyl ester resins will homopolymerize as well as coreact with various vinyl monomers.

Resin Properties. Vinyl esters, because of their relatively low molecular weight and precise structure, can be characterized as low-viscosity, fast-wetting, consistent-reactivity products. Typical property profiles for some uncured vinyl ester resins are given in Table 1. Properties of some cured resins are given in Table 2. Typically, 6-month stability can be expected at 25°C (77°F) with decreased storage life under elevated temperatures. In general, anyone familiar with the proper storage and handling of unsaturated polyester resins and styrene monomer experiences no difficulty with these materials.

Cure Mechanism. The free-radical cure mechanism of the vinyl ester resins is well understood. In most respects, it is similar to that of the unsaturated polyester resins. To initiate the curing process, it is necessary to generate free radicals within the resin mass. Organic peroxides are the most common source of free radicals. These peroxides will decompose under the influence of elevated temperatures or chemical promoters, e.g., organometallics or tertiary amines, to form free radicals. Generation of free radicals also can be effected by ultraviolet or high-energy radiation applied directly to the resin system. The free radicals thus formed react to open the double bond of the vinyl

[1] The first literature reference was a patent issued in 1962 for a tooth-filling compound. Commercialization did not start until the late 1960s.

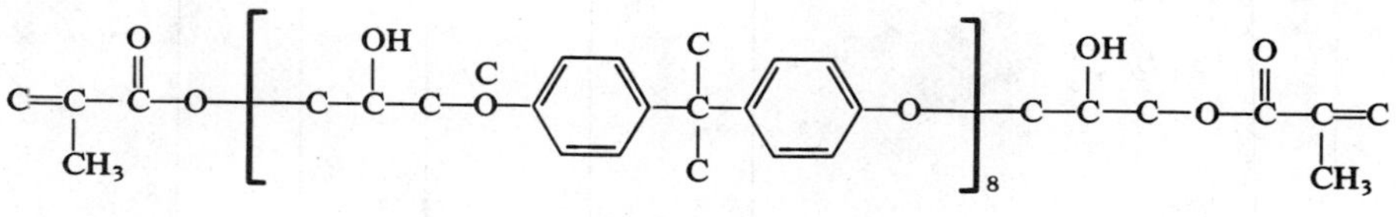

Typical Vinyl Ester Resin

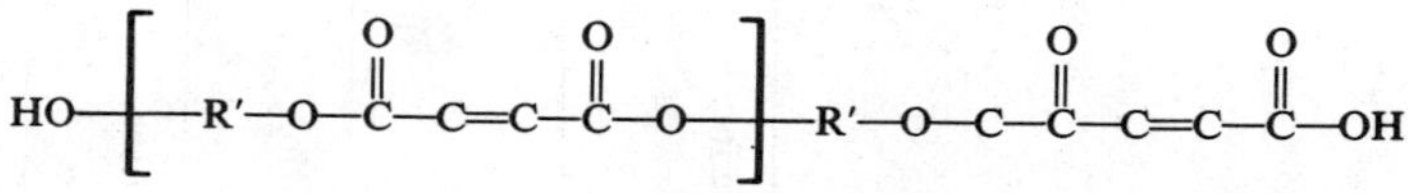

Typical Unsaturated Polyester Resin

TABLE 1. PROPERTIES OF TYPICAL UNCURED VINYL ESTER RESINS

PROPERTY	STANDARD RESIN	LOW-VISCOSITY RESIN
Monomer type	styrene	styrene
Level, %	45	45
Viscosity at 77°F (25°C), centipoises	550	200
Acid number	5	5
Specific gravity	1.04	1.04
SPI gel time (1% benzoyl peroxide), minutes		
at 82°C (180°F)	10	12
at 121°C (250°F)	1.4	1.5
Flash point (Tag open cup)		
°C	34	34
°F	93	93

SOURCE: The Dow Chemical Company.

group. Once opened, the resin vinyl group is highly reactive and rapidly combines with several more vinyl groups available from both the unreacted resin and the monomer. This exothermic reaction is rapidly carried to completion forming a 3-dimensional thermosetting network.

Applications. As might be expected from the wide variation in resin properties which can be built into the molecule, the vinyl ester resins find many applications. Chief among these are fiberglass-reinforced plastics where the inherent characteristics of the vinyl ester resin provide a cost and performance advantage over other materials. The largest application as of the early 1980s is in the manufacture of corrosion-resistant reinforced plastic structures. Due to the reduced number of ester groups within the resin structure, the corrosion-resistant vinyl ester resins are less prone to attack by hydrolysis than the bisphe-

TABLE 2. PROPERTIES OF TYPICAL CURED VINYL ESTER RESINS

CLEAR-CASTING PROPERTIES

Tensil strength, psi	12,000
megapascals	83
Tensile modulus, psi	500,000
megapascals	3,447
Ultimate elongation, %	5
Flexural strength, psi	18,000
megapascals	124
Flexural modulus, psi	450,000
megapascals	3,103
Yield compressive strength, psi	17,000
megapascals	117
Compressive modulus, psi	350,000
megapascals	2,413
Deflection at yield, %	7
Heat-distortion temperature, °C	101.7
°F	215.0
Barcol hardness	35

GLASS-REINFORCED LAMINATE PROPERTIES

Laminate thickness, inch	0.25
millimeters	6.3
Fiber-glass content, %	30
Tensile strength, psi	19,000
megapascals	131
Tensile modulus, psi	1,400,000
megapascals	9,653
Flexural strength, psi	22,000
megapascals	152
Flexural modulus, psi	1,000,000
megapascals	6,895

SOURCE: The Dow Chemical Company.

nol A-fumaric acid polyesters. In addition, the resilience of the vinyl ester resins (4–6% ultimate tensile elongation) results in a fabricated part which is less prone to damage during shipping, field erection, and service.

All fabrication methods commonly used in the manufacture of reinforced plastics can be used with the vinyl ester resins. In applications such as filament winding and bag molding, where fast wetting is important, significant increases in output can be realized. One technique which is particularly suited to vinyl ester resins is the manufacture of products using sheet molding compound (SMC). Here, the resin system (usually with a high filler loading) is combined in sheet form with the glass reinforcement and chemically thickened through the use of metal oxides, such as MgO. The SMC then is molded, usually in a matched-die molding operation, to give the desired final product, e.g., automotive parts, appliance housings, electrical structures, and panel configurations. Vinyl ester resins diluted with vinyl toluene monomer are used in the production of high-temperature electrical laminating systems.

References

The following United States patents, all assigned to The Dow Chemical Company, apply to various aspects of vinyl ester resins:

Bearden, C. R.: U.S. Patent 3,367,922, February 6, 1968.

Brewbaker, J. L., Nowak, R. M., and K. S. Dennis: U.S. Patent 3,836,000, September 17, 1974.

Jernigan, J. W.: U.S. Patent 3,466,259, September 9, 1969.

Jernigan, J. W.: U.S. Patent 3,548,030, December 15, 1970.

Najvar, D. J.: U.S. Patent 3,892,819, July 1, 1975.

Nowak, R. M., and T. O. Ginter: U.S. Patent 3,674,893, July 4, 1972.

Pennington, D. W., and J. H. Enos: U.S. Patent 3,887,515, June 3, 1975.

Swisher, D. H., and D. C. Garms: U.S. Patent 3,564,074, February 16, 1971.

—P. H. Cook, Dow Chemical U.S.A., an operating unit of The Dow Chemical Company, Freeport, Texas.

VINYL PAINTS. Paint and Finishes.

VINYLPYRIDINES. Pyridine and Derivatives.

VIRUS. Viruses are considered to be the smallest infectious agents capable of replicating themselves inside eukaryotic or prokaryotic cells. The majority of these extremely small infectious particles fall within a size range of about 0.02–0.25 micrometer and can only be visualized directly with the aid of an electron microscope.

In 1898, Loeffler and Frosch demonstrated that foot-and-mouth disease of cattle could be transferred by material passed through a filter capable of excluding bacteria. The new group of "organisms" subsequently became known as *filterable viruses*. Years of debate have centered around the question of whether viruses are living or nonliving and, although resolution of this is now considered to be simply a problem of semantics, several fundamental differences distinguish viruses from other organisms. Viruses, unlike true microorganisms, are not cells, do not replicate by binary fission, contain a genome consisting of only one type of nucleic acid (DNA or RNA, double or single stranded) and contain no organelles, such as mitochondria or ribosomes (except for the arenaviruses, which contain cellular ribosomes). Some virions contain special enzymes, such as transcriptase required for initiation of the viral growth cycle, not present in host cells.

Viruses are obligatory intracellular parasites and, as such, cannot replicate in cell-free media. Therefore, the study of viruses is normally carried out with cultured cells. These are classified as: (1) *primary cell cultures*; (2) *diploid cell strains*; and (3) *continuous cell lines*. Primary cell cultures are developed on tissue freshly removed from a plant or animal, contain several cell types capable of supporting replication of a wide range of viruses, but are limited to only a few cycles of cell division *in vitro*. Diploid cell strains contain cells of a single type which retain their original diploid chromosome number and are capable of up to about 100 divisions *in vitro*. Continuous cell lines consist of transformed or dedifferentiated cells of a single type which bear little resemblance to normal cells of that type. Continuous cell lines are capable of indefinite propagation *in vitro*.

Viral Structure. The mature virus particle, referred to as the *virion*, consists of a nucleic acid molecule(s) surrounded by a protein coat, the *capsid*. The capsid is composed of a number of *capsomeres* comprising one or more polypeptide chains and in some viruses surrounding a protein *core*. The capsid and enclosed nucleic acid together constitute the *nucleocapsid*. The viral capsid symmetry is characteristic of groups of virus and may be icosahedral (cubic) or helical. The icosahedron has 12 vertices and 20 faces, each an equilateral triangle. Nucleocapsids exhibiting helical symmetry consist of capsomeres and nucleic acid wound together into a spiral or helix. However, regardless of capsid symmetry, the actual virion may appear to be round, brick, or bullet-shaped. Some icosahedral and all helical viruses are enclosed in an outer *envelope* composed of lipoprotein which is derived directly from the virus-modified cellular membrane during release of the nucleocapsid from the infected cell by a process called *budding*. Enveloped viruses can usually be inactivated by ether, chloroform, or bile salts.

Nucleic acid extracted from purified virus using phenol or dodecyl sulfate is easily destroyed by the homologous nucleases which are present in normal sera or tissues. DNA is destroyed by the enzyme deoxyribonuclease; RNA by ribonucleases. This provides one means of identifying the type of nucleic acid. The intact virus is not affected by these enzymes.

Bacteriophage, a virus infecting bacterial cells, has a structure somewhat different from those previously described. A *head* contains the nucleic acid and the viral DNA passes through a *tail* during the infection process. In the *T*-even phages (Fig. 1), the tail consists of a tube surrounded by a *sheath* and is

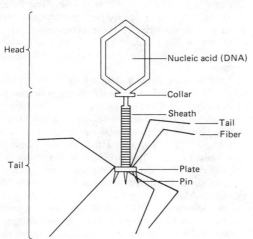

Head
Tail

— Nucleic acid (DNA)
— Collar
— Sheath
— Tail
— Fiber
— Plate
— Pin

Fig. 1.   *T*-even bacteriophage.

connected to a thin *collar* at the head end and a plate at the tip end. The sheath is capable of contraction and the plate possesses *pins* and *tail fibers* which are the organs of attachment of the bacteriophage to the wall of the host cell.

Viral Replication. In contrast to eukaryotic and prokaryotic cells which multiply by binary fission, viruses multiply by synthesis of their separate components, followed by assembly. Several stages are involved in viral replication:

(1) *Attachment or adsorption*. The virus becomes attached to the cell via specific receptors. Thus, cells lacking the receptors are resistant to attack.

(2) *Penetration*. With enveloped viruses, this step occurs when the virion's envelope fuses with the cellular membrane. Naked virions penetrate intact through the cellular plasma membrane and into the cell cytoplasm. Viruses may also enter the cell by cellular phagocytosis. Ordinarily, without a protein coat, viral nucleic acid is incapable of entering a cell, showing the importance of the coat in infectivity. The efficiency of infection with naked nucleic acid can be increased by the presence of basic polymers, such as DEAE-dextran, or by pretreating cells with hypertonic salt solution. Even under the most favorable conditions, however, the efficiency of infection is not more than 1% that of the corresponding intact virions.

(3) *Uncoating and eclipse*. Uncoating is detected by the lability of viral nucleic acids to nuclease after the artificial disruption of the cell. Eclipse is recognized by loss of infectivity of intracellular virions recovered from disrupted cells.

Once inside the cell, virulent viruses turn off cellular macromolecular synthesis and disaggregate cellular polyribosomes, thus favoring a shift to viral synthesis. These viruses cause the ultimate destruction of the infected cell. In contrast, moderate viruses may stimulate host DNA, mRNA and protein synthesis—this phenomenon may be of considerable importance in viral carcinogenesis.

In general, the DNA viruses multiply in the nucleus of the host cell. The viral DNA is transcribed in the nucleus and the resultant mRNA translated into proteins on cytoplasmic ribosomes. Depending upon the virus type, "early" or "late" proteins may be synthesized. These proteins may function as enzymes in replication of the viral DNA, as structural components of progeny virions, or as regulatory proteins. Replication of the viral DNA is semiconservative and, in general, depends upon viral proteins.

RNA viruses usually replicate in the cytoplasm and can be divided into five classes according to the nature of the RNA in the virion. *Class I* viruses contain a molecule of single-stranded RNA which acts as mRNA to be translated into viral proteins. The RNA is said to have *plus strand polarity*. The picornaviruses are an example. *Class II* viruses (e.g., paramyxoviruses) have a molecule of single-stranded RNA which cannot act as mRNA (minus strand polarity). A virion transcriptase synthesizes several complementary messenger molecules from which viral proteins are translated. *Class III* viruses (e.g., myxoviruses) contain single-stranded RNA of minus strand polarity, present in seven or more segments. A virion transcriptase transcribes each segment into a complementary messenger. *Class IV* viruses contain ten segments of double-stranded RNA which is transcribed into mRNA by a viral transcriptase. Representative of this group are the reoviruses. *Class V* viruses (e.g., leukoviruses) contain segmented single-stranded RNA of messenger polarity. Each RNA segment is transcribed into DNA by a *reverse transcriptase* present in the virion; mRNA is then transcribed from the DNA.

(4) *Assembly and release*. The assembly of the capsid and its association with nucleic acid is then followed by release of the virus from the cell. This may occur in different ways, depending upon the nature of the virus. Naked viruses may be released slowly and extruded without cell lysis, or released rapidly by disruption of the cell membrane. DNA viruses, which mature

in the nucleus, tend to accumulate within infected cells over a long period. Enveloped viruses generally acquire their envelope and leave the cell by *budding* through the nuclear or cytoplasmic membrane at a point where virus-specified proteins have been inserted. The budding process is compatible with cell survival.

Viral Classification

Several methods of viral classification are in use. Classification based upon epidemiological criteria, such as enteric or respiratory viruses, is useful, but, of more significance are schemes based upon the morphology of the virion (symmetry, envelope, etc.) and type of nucleic acid (DNA, RNA, number of strands, polarity, etc.).

The two groups of viruses, RNA and DNA, are further divided according to size, morphology and biological and chemical properties. Thus, the icosahedral RNA viruses which are ether stable are divided into the picornaviruses and the reoviruses. The name picornavirus comes from *pico* (meaning very small) and *rna* (indicating the type of nucleic acid). Included in the group are enteroviruses, such as polio, Coxsackie, foot-and-mouth, and echoviruses, among others, and also the rhinoviruses. The reoviruses (*respiratory enteric orphan virus*) cause inapparent infection in humans and other animals, and their relationship to spontaneous disease is uncertain. They are morphologically similar to the wound-tumor virus of clover, and a small cross-activity with this virus by means of complement fixation has been reported.

The arboviruses are those which multiply in both vertebrates and arthropods. The former serve as reservoirs and the latter primarily as vectors. Arbovirus is a somewhat arbitrary epidemiological classification which contains several heterogeneous groups. The togaviruses contain such entities as Eastern and Western equine encephalitis (EEE and WEE) and dengue and yellow fever viruses, which have mosquitoes as vectors. The arenaviruses comprise such agents as Lassa, Tacaribe, and lymphocytic choriomeningitis viruses. Although dangerous and difficult to study, the arboviruses appear to contain single-stranded RNA (positive polarity). They are ether sensitive and relatively unstable. The capsids are suggestive of icosahedral symmetry.

Myxoviruses, orthomyxoviruses, and paramyxoviruses are spherical or filamentous, enveloped single-stranded RNA viruses. The myxovirus group contains the influenza viruses which, in turn, have been separated into three distinct antigenic types designated A, B, and C. The genome of myxoviruses, unlike that of the paramyxoviruses, is segmented and it is this characteristic which is responsible for the devastating influenza pandemics which have occurred periodically. Influenza A viruses have undergone three major antigenic shifts since 1933 and each new variant is able to successfully infect populations of individuals, immune only to preexisting types. In the influenza pandemic of the winter of 1917–1918, over 20 million persons died worldwide with better than one-half million fatal cases in the United States. Over 50 million cases of influenza were reported in the United States in the 1968–1969 winter. These cases were attributed to a hitherto unknown variant, first isolated in Hong Kong (hence named "Hong Kong flu"). Some 20,000 and possibly as many as 80,000 deaths resulted from this influenza invasion and the side effects which it produced. In the 1972–1973 winter season, a much milder and minor variant, called the "London flu," caused well over 2000 deaths in the United States, particularly from complications, such as pneumonia. When combined with influenza, pneumonia is the fifth most serious public health problem in the United States. In terms of absenteeism, it is the number one problem.

Only Type A influenza virus has been found to be capable of producing *pandemics*. The influenza A virus is identified as a medium-size RNA virus, some 110 nanometers in diameter and delimited by a membrane of lipids and polysaccharides derived from the host cell and virus-specific protein. Five distinct protens have been identified, three of which are inside the virion. A schematic representation of the influenza virus emerging from a cell is given in Fig. 2.

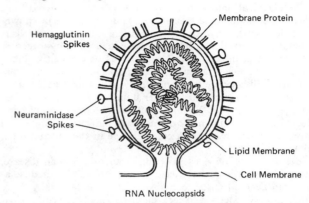

Fig. 2. Schematic representation of an influenza virus emerging from a cell. (*After Kilbourne.*)

It has been reported that the antigenic shifts are manifested in hemagglutinin and neuraminidase, two glycoproteins found on the surface of the influenza virion. It is suggested that the hemagglutinin binds the virus to the target cell and when the hemagglutinin function is inhibited (as by an antibody), the virus is no longer infective. It is believed that the neuraminidase cleaves a glycoside bond in the host membrane. This action frees the newly formed virus from the cell and, if inhibited, will not reduce the infectivity of the virus, but will deter the spread of virus particles to other cells.

The emergence of new influenza subtypes appears to be too abrupt to be explained fully by conventional concepts of mutation and the full story may rest in the very nature of the segmented viral genome. If a host cell is infected by two different subtypes of influenza virus at the same time, the genes from the subtype may undergo random reassortment in the cell, resulting not only in production of the two original subtypes, but of one or several subtypes as well. Each hybrid, of course, will have a different but full set of genes and recombination within the infected host can explain the large mutations which occur about once every decade. However, it is well established that only one influenza subtype can exist in humans at any given time. Also, that the emergence of a new subtype, such as the Hong Kong strain, is usually accompanied by the abrupt disappearance of the antecedent subtypes—thus allowing little, if any, opportunity for recombinations to occur within human cells. Of considerable interest, however, is the fact that several virus strains can exist simultaneously within animal hosts. In animals, the appearance of a new influenza virus strain is not necessarily accompanied by the disappearance of previously recognized strains. It has been established that there are at least two discrete subtypes of equine influenza, eight or more avian strains and two subtypes in swine. Thus, the postulation that recombination occurs in animals which share the general environs with humans. Some evidence of this may derive from the fact that most new subtypes appear to originate in Asia where animals and humans commonly inhabit the same building.

The paramyxovirus group includes the causative viruses of mumps, measles, parainfluenza, Newcastle disease, canine distemper, and several other diseases. These viruses are generally larger than the myxoviruses, are enveloped and pleomorphic, and contain one molecule of single-stranded RNA.

The rhabdoviruses, causative agents of rabies in humans and other animals, are also enveloped and contain a single strand of RNA. A peculiar bullet-shaped morphology disguises the helical nucleocapsid.

The chemical nature of many viruses, which either do not grow well or do not lend themselves to purification, is unknown. The Riley lactic dehydrogenase virus is a nonpathogenic virus which is recognized only by an increase in lactic dehydrogenase in the blood of infected mice. A lipovirus described by Chang causes marked degradation of infected cells and releases a lypogenic toxin dissociable from infectivity, which is capable of inducing fatty degeneration in other uninfected cells. A marked increase in the gamma globulin fraction of blood serum of mink infected with Aleutian mink disease is an indication of infection with a virus which causes a color change in the fur and often sickness and death.

Several groups of viruses, of importance in human disease, contain DNA. The adenoviruses, named for their original isolation from adenoid tissue, contain double-stranded DNA, have icosahedral symmetry and lack an envelope. This group of viruses, which multiply in the nucleus of infected cells, are usually associated with respiratory tract and eye infections, although it is now apparent that adenoviruses are not the etiological agents for the majority of acute viral respiratory infections. Although adenoviruses exhibit marked oncogenic (tumor causing) potential in animals, they are probably not oncogenic for humans.

The adenoviruses contain at least three protein moieties, and certain types are capable of inducing one or more new host antigens, such as tumor (T) antigens, the chemistry of which is at present unknown. The viral proteins can be separated by gel diffusion and correlated with results obtained by complement fixation. One moiety is the toxic protein which causes the host cell to degenerate. Another corresponds to the group antigen common to all 31 types of adenoviruses, and the third is the type-specific protein.

The papovaviruses (*pa*pilloma, *po*lyoma, *va*cuolating agent, SV40) are small nonenveloped icosahedral viruses which also replicate in the cell nucleus. The virion contains double-stranded DNA. Apart from causing several forms of warts, this group of viruses is of interest as models for understanding mechanisms of viral carcinogenesis.

The major herpes viruses (Greek: *herpein* = to creep) that infect humans are herpes simplex (Type I: fever blisters; Type II: genital lesions), varicella (chickenpox), zoster (shingles), cytomegalovirus, and Epstein-Barr viruses. A number of viruses which infect lower animals also belong to this group of enveloped icosahedral double-stranded DNA viruses. Most members of this group tend to produce latent infections with periodic recurrent disease. Two examples are fever blisters caused by herpes simplex I and shingles, the recurrent form of chickenpox. Cytomegalovirus causes a severe, often fatal, illness of newborns, usually affecting the salivary glands, brain, lungs, kidneys, and liver. Surpassing rubella virus, this is the most common viral cause of mental retardation. It has been estimated that cytomegalovirus (CMV) causes serious mental retardation of more than 3000 infants annually in the United States alone. In addition to mental retardation, the disease in infants may cause blindness and deafness. In about 90% of the infants infected with CMV, the disease can be detected only through urine examination. In about 10% of the cases, the disease is typified by enlargement of the spleen and liver, blood abnormalities and

hepatitis. Microcephaly (abnormally small head) is also sometimes an indication. CMV causes enlargement of the affected cells (cytomegaly). The disease is found throughout the world and it is believed that congenital infections result from a primary infection of the mother during pregnancy. CMV, like herpes viruses, probably persist in a latent stage for long spans of time. Immunosuppressed patients, such as those suffering from cancer or recipients of organ transplants, are also prone to infections with CMV.

The Epstein-Barr viruses play an etiological role in infectious mononucleosis, an acute infectious disease which affects lymphoid tissue throughout the body. A strong association of this virus with Burkitt's lymphoma and perhaps nasopharyngeal carcinoma also has been observed.

The pox viruses are the largest and most complex viruses of vertebrates and contain a large double-stranded DNA molecule. The virions are complex, brick-shaped particles, covered by several membrane layers of viral origin. Unlike other DNA viruses of mammals, pox viruses multiply in the cell cytoplasm. They can be divided into several groups on the basis of specific antigens, morphology and natural hosts. *Group I* consists of mammalian viruses, such as variola (smallpox), vaccinia, cowpox, ectromelia, and monkeypox. Of this group, variola or smallpox has caused the greatest human morbidity and mortality. However, because the virus has no animal reservoir, and is spread chiefly by human contact, the World Health Organization was able to announce in 1980 that, because of massive immunization campaigns, smallpox had been completely eradicated. Since that announcement, remaining stores of the virus have been destroyed to prevent laboratory accidents, such as the one in 1979 which took the life of a scientist. *Group II* comprises the tumor-producing viruses, the fibroma and myxoma viruses.

The hepatitic viruses appear to fall into two different groups of small icosahedral DNA viruses. *Type A* causes infectious hepatitis and is transmitted through the oral-intestinal route. *Type B* is transmitted by injection, usually of infected blood or its products.

Slow Viruses

During the last decade or two, there has been increasing speculation and some tentative evidence that so-called *slow viruses* may be operative and may be the underlying causes of a number of degenerative diseases, long poorly understood, such as multiple sclerosis and rheumatoid arthritis, among others. More recently, there have been increasing postulations of an association between viruses and diabetes. In fact, rather positive identification of slow viruses with some rare diseases has been established. Most investigators caution that the term "slow" should not necessarily be fully interpreted in terms of a virus *per se*, but equally if not completely with the manifestations of the virus. So-called slow virus infections are characterized by a long incubation period, followed by a protracted course of disease. The slowness may arise in some cases from the virus itself, but the slow pace also may be the result of weak but prolonged interactions between the virus and the host's immune system. It is also possible that these characterizations of slowness may not be attributable to viruses at all, but to some other unknown causative factors. Obviously as of this juncture, investigators are following a source of suspicion rather than a chain of hard evidence. Nevertheless, the case for the slow viruses is becoming increasingly convincing. The causative agents for at least four rare diseases, two in humans and two in animals, are sometimes referred to as "unconventional viruses."

One of these diseases in humans is *kuru*, encountered only

in the Fore people and their neighbors in New Guinea. The disease for many years was considered a genetic disease. However, it has been established that the disease can be transmitted to chimpanzees by injection of extracts from the brains of human kuru victims into the brains of chimpanzees. Kuru is a neurological disease with brain lesions located mainly in the gray matter. The cerebral cortex takes on a spongy appearance. The other human disease is Creutzfeld-Jakob disease, rare but of worldwide distribution. It involves the premature development of the mental deterioration sometimes seen in old age. It also has been established that it is caused by a transmissible agent that can infect chimpanzees and lower primates. One of the animal diseases referred to is *scapie*, known for over two centuries as a fatal disease among sheep. The other animal disease is *transmissible mink encephalopathy*, first discovered in Wisconsin in the late 1940s. A puzzling aspect of the unconventional slow viruses is the fact that they cannot be observed with an electron microscope. Another puzzling aspect is their apparent lack of antigenicity. To date, it has not been possible to demonstrate that any of these four "agents" will evoke production of antibodies. These unconventional slow viruses are not destroyed by ultraviolet radiation, and they are highly resistant to treatment with formalin or heat, but infectivity is destroyed by phenol or ether. Some investigators believe that these agents may incorporate a very small nucleus of the size range of the viroids (self-replicating infectious RNA molecules known to produce certain plant diseases).

Two slow infections of the human central nervous system—*progressive multifocal leukoencephalopathy* (PML) and *subacute sclerosing panencephalitis* (SSPE) are thought to be associated with conventional viruses. Although PML does not cause inflammation of the brain, it does produce demyelination, i.e., destruction of the layers of membranes surrounding nerve axons. Some investigators believe that the virus is a papovavirus (group of small viruses including human wart virus, simian virus 40, and the polyoma virus of mice). It is reasoned that in PML the virus destroys the cells needed for formation and maintenance of the myelin sheath. A conventional virus has been isolated from the brains of persons suffering from SSPE. An association between measles (in patients under two years of age) or immunization with a live measles virus vaccine and later development of SSPE has been shown. SSPE patients have unusually high titres of measles antibodies and affected brain cells have inclusions similar to those seen in measles infections.

Slow viruses are becoming increasingly suspect in the instances of much more common diseases, particularly the autoimmune diseases. An autoimmune disease may be defined as a disease wherein the immune system of the body does not direct its attack on an invading foreign substance, but instead at the body's own tissue. Many authorities consider rheumatoid arthritis and multiple sclerosis as autoimmune diseases. The precise causes of these diseases have remained obscure. Multiple sclerosis is a demyelinating disease and has variously been described as an autoimmune disease, a viral disease, or an autoimmune disease provoked by a virus. Epidemiological studies indicate that from 3 to 23 years may elapse between the time of exposure to the virus and the onset of symptoms. Further evidence points to involvement of a myxovirus. Measles virus is of this kind.

Possible Viral Connection to Diabetes. A Norwegian physician (J. Stang) in 1864 noted that diabetes developed in one of his patients within a short period after a mumps infection and was probably the first person to indicate a possible connection between viruses and diabetes. Over the years, numerous other connections have been attempted to relate diabetes with mumps, hepatitis, rubella, coxsackie, and influenza viruses, adenoviruses,

enteroviruses, and cytomegalovirus. One of the presumptions made is that viruses are understood to replicate in the pancreas. Commencing in the late 1950s, more substantive evidence has been given. Reports from Sweden in 1958 link juvenile diabetes with mumps infection. Reports from New York State in 1974 relate closely the cycles of incidence of mumps and those of juvenile diabetes. The study was based upon investigation of records for the period 1946–1971. Tentative conclusions indicate an average lag period of about 3.8 years between onset of diabetes and exposure to mumps and it is reasoned that this represents the time required for the virus to produce permanent damage to the pancreas. Other investigators have statistically linked diabetes to rubella (German measles). Some authorities suggest that the pancreas, along with other embryonic organs, may be damaged by the virus that causes congenital rubella. The cords of nearly 3,000 juvenile diabetics treated at King's College Hospital in London (1955–1968) have been studied and reveal a seasonal pattern to the onset of juvenile diabetes, striking a low incidence in June and a high incidence in October. Without presenting the details, conclusions are suggested that an association of viral infections with the juvenile form of diabetes is evident. However, the relationship, if any, has not been determined in the case of the maturity-onset form of diabetes.

Viral Diagnosis and Vaccination

Viral Diagnosis. Three major approaches to identification of viruses are commonly used: (1) *Microscopy*. Viruses may be observed directly by electron microscopy; viral antigens may be recognized in infected tissue by immunofluorescence, using virus-specific antisera; virus-induced pathology may be identified by light microscopy. (2) *Virus isolation*. Provisional viral identification may be based upon cytopathic effects produced in cell cultures infected with virus present in tissues or secretions of the patient. (3) *Serology*. Antibodies specific for a particular virus may be identified in a patient's serum. A very sensitive, accurate, and recently developed diagnostic approach is radioimmunoassay, which involves the use of an isotope labelled antibody or antigen.

Viral Vaccination. Vaccines, agents that elicit a specific antiviral immune response, have been very successful against smallpox, measles, rubella, poliomyelitis, and yellow fever, all of which are generalized diseases. Vaccines against diseases caused by respiratory tract viruses, where great antigenic diversity is found, have been less effective.

Vaccines may be prepared by rendering viruses harmless without affecting their immunogenicity. This can be done by either inactivating the virus, or by selecting avirulent mutants. The most successful vaccines are "living" avirulent viruses which possess the advantage of multiplying in the host and which usually require only a single dose to be effective. This leads to prolonged immunological stimulation similar to that which occurs in natural infection. Live vaccines, however, are subject to a number of problems, such as genetic instability and contamination by extraneous viruses. Inactivated viruses are usually produced by treatment with formaldehyde, which destroys infectivity. The major difficulty with inactivated viruses is the administration of sufficient viral antigen to induce a lasting immunity. In many cases, several injections must be given over a substantial period of time. The only inactivated viral vaccine in widespread use in humans is the influenza vaccine. The inactivated Salk polio vaccine has been largely replaced by the attenuated live-virus Sabin vaccine.

Interferon. Interferons (IFN) are proteins which evert virus nonspecific antiviral activities in cells through metabolic processes involving synthesis of both DNA and protein. The number of interferon-inducing substances has increased to include not only all of the major virus groups, but also bacterial and fungal products, nucleic acids, polymers, mitogens, and various low-molecular-weight substances. However, as interferons are induced by viruses and inhibit viral replication, viruses are usually considered to be natural inducers. The ability of viruses to induce interferon production depends upon the virus type. Some viruses, such as that responsible for Newcastle disease, are good inducers, while others, such as the adenoviruses, are regarded as poor inducers. Further, the type of cell used presents another factor in interferon production. In the whole animal, cells of the reticuloendothelial system are generally considered to be the major interferon producers. Recently, interferons have been classified into types on the basis of their antigenic specificities. Alpha and beta interferons (formerly called leukocyte and fibroblast, respectively) are acid stable and correspond to what have been called *Type I* IFNs (interferons). Gamma interferons (formerly called *Immune*) are acid labile and correspond to *Type II* IFNs.

Although interferon has been studied extensively for over a decade, the mechanism of its antiviral activity remains unclear. Considerable evidence exists to support the concept that interferon inhibits virus-specific protein synthesis, thus blocking viral replication in cells adjacent to the infected cell producing the interferon. There is no established reason to conclude, however, that interferon exerts antiviral action through a single mechanism. Interferon is probably one of the most important early determinants of recovery from a number of viral diseases.

Recent work has centered upon use of interferon as a therapeutic agent in humans and animals. In humans, local application of monkey interferon is effective in reducing the severity of vaccinia virus skin infections. Recent results with herpes keratitis and chronic hepatitis are promising. Interferon appears to be active against oncogenic viruses in the treatment of such cancers as osteogenic sarcoma, and at present it is only the limited availability of interferon that prevents more extensive testing.

References

Basu, R. N., Jezek, Z., and N. A. Ward: "The Eradication of Smallpox from India," World Health Organization, Publications Centre, Albany, New York (1980).
Butler, P. J. G., and A. Klug: "The Assembly of a Virus," *Sci. Amer.*, **239**, 5, 62–69 (1978).
Carter, W. A.: "Selective Inhibitors of Viral Functions," CRC Press, Boca Raton, Florida (1975).
Diener, T. O.: "Viroids and Viroid Diseases," Wiley, New York (1979).
Diener, T. O.: "Viroids: Structure and Function," *Science*, **205**, 859–866 (1979).
Gajdusek, D. C.: "Unconventional Viruses and the Origin and Disappearance of Kuru," *Science*, **197**, 943–960 (1977).
Gillespie, D., and R. C. Gallo: "RNA Processing and RNA Tumor Virus Origin and Evolution," *Science*, **188**, 802–811 (1975).
Henle, W., Henle, G., and E. T. Lennette: "The Epstein-Barr Virus," *Sci. Amer.*, **241**, 1, 48–59 (1979).
Marsh, R. F.: "The 1976 Nobel Prize for Physiology or Medicine," (relates to slow viruses), *Science*, **194**, 928–929 (1976).
Marshall, E.: "Mystery Disease of Naples is a Common Virus," *Science*, **203**, 980–981 (1979).
Marx, J. L.: "Persistent Infections: The Role of Viruses," *Science*, **196**, 151–152 (1977).
Melnick, J. L.: "The 1976 Nobel Prize for Physiology or Medicine," (relates to viral hepatitis) *Science*, **194**, 927–928 (1976).
Reddy, V. B., et al.: "The Genome of Simian Virus 40," *Science*, **200**, 494–502 (1978).
Staff: "Slow Viruses: I. Role in Persistent Disease," *Science*, **180**, 1351–1354 (1973).
Staff: "Slow Viruses: II. The Unconventional Agents," *Science*, **181**, 44–45 (1973).
Staff: "Interferon Nomenclature," *Amer. Soc. Microbiol. News.*, **46**, 9, 466–467 (1980).
Stewart, W. E., II: "The Interferom System," Springer-Verlag, New York (1979).

—Ann C. Vickery, Ph.D., College of Medicine, University of South Florida, Tampa, Florida.

VISCOSITY. Rheology.

VITAMIN. An organic compound that performs specific and necessary functions in humans, livestock, and other living organisms—even when present in very small concentrations, at the milligram or microgram/100 gram levels. The term *vitamine* was proposed by a Polish biochemist (Casimir Funk) in 1912 to designate substances required in trace amounts in the diet to prevent various nutritional-deficiency diseases.

Nearly all vitamins are associated in some way with the normal growth function as well as with the maintenance and efficiency of living things. In the case of some vitamins, various species are capable of synthesizing some of the vitamins from precursors that are present in the body. Synthesis is frequently by way of intestinal bacteria. In the case of vitamin D, substances in the skin combine with ultraviolet radiation from sunlight to yield the essential substance. Some vitamins, such as vitamin C, are specific, singular substances—in this case ascorbic acid. With other vitamins, there is a range of related compounds, as exemplified by the D, E, and K vitamins.

Because of inconsistencies in nomenclature, the B vitamins are not closely related as one might suspect. The B vitamins are different specific substances and hence the use of the letter B to designate them falsely indicates a degree of commonality which actually is not the case. Vitamin B_1 is thiamine, vitamin B_2 is riboflavin, vitamin B_6 is pyridoxine, vitamin B_{12} is cobalamin. Vitamins B_6 and B_{12}, for example differ markedly in function and structure. The alphabetical method of designation became complex and somewhat confusing as the various vitamins were recognized and studied over many years, during which period some substances were found to be identical with previously announced and described vitamins, and some substances were found not to be vitamins at all. Thus, over the years, the International Union of Pure and Applied Chemistry (I.U.P.A.C.) assigned new names to several of the vitamins.

The major vitamins are described in separate alphabetical entries. Titles used have been selected on the basis of the most frequently used designations as of the mid-1980s. In alphabetical order, these entries are:

Ascorbic Acid (Vitamin C)
Biotin
Choline and Cholinesterase
Folic Acid
Inositol
Niacin
Pantothenic Acid
Thiamine (Vitamin B_1)
Vitamin A
Vitamin B_2 (Riboflavin)
Vitamin B_6 (Pyridoxine)
Vitamin B_{12} (Cobalamin)
Vitamin D
Vitamin E (Tocopherols)
Vitamin K

The relationships between hormones and vitamins are described in the entry on **Hormones**.

Loss of Vitamins in Processing. During the last several years, much research has gone into determining the loss in effectiveness of vitamins as various foods are processed. It is a common

TABLE 1.   COMPARATIVE LOSSES OF VITAMINS FROM VEGETABLES
(Canning and Freeze Processing)

METHOD OF PRESERVATION	Value	LOSS OF VITAMINS AS COMPARED WITH VALUES OF FRESH-COOKED				
		Vitamin A	Thiamine (B_1)	Riboflavin (B_2)	Niacin	Ascorbic Acid (C)
Frozen, cooked (boiled), drained	mean	12%	20%	24%	24%	26%
	range	0–50%	0–61%	0–45%	0–56%	0–78%
Canned, drained solids	mean	10%	67%	42%	49%	51%
	range	0–32%	56–83%	14–50%	30–65%	28–67%

SOURCE: Fennema, *Food Technology*, **31**, 12, 3238 (1977).

tendency on the part of consumers to regard any fresh food as representing perfection in terms of nutritive including vitamin value and, conversely, to regard processed foods as nutritionally inferior. Under normal circumstances, these observations are true. Because fresh foods frequently are stored for several days at temperatures well above their freezing points, there are vitamin losses in unprocessed produce. Ascorbic acid content in vegetables, for example, can severely degrade during improper storage. The degradation of vitamin values depends a great deal upon the type of food substance, the particular vitamin, and the manner in which the raw food is processed. The consumer today also is protected by *vitamin fortified foods*, modified to compensate for vitamin effectiveness lost during processing, or, in some cases, to generally enrich the foods nutritionally. Losses of vitamins from fruits and vegetables during processing are tabulated in Tables 1 and 2.

References

NOTE: More detailed references pertaining to specific vitamins will be found at the ends of the previously mentioned entries.

Harris, R. H.: "Vitamins and Hormones," Academic, New York (1975).

Horwitz, W., et al.: "Vitamins and Other Nutrients," in "Official Methods of Analysis of the Association of Official Analytical Chemists," 12th Edition, page 816, AOAC, Washington, D.C. (1975).

Lund, D. B.: "Effect of Heat Processing on Nutrients," in "Nutritional Evaluation of Food Processing," 2nd Edition, AVI, Westport, Connecticut (1975).

Staff: "Nutritional Requirements of Domestic Animals," (Separate publications on Beef Cattle; Dairy Cattle; Poultry; Swine; and Sheep), National Academy of Sciences, Washington, D.C. (Periodically updated.)

Voigt, M. N., et al.: "Vitamin Assay by Microbial and Protozoan Organisms," *Jrnl. of Food Science*, **43**, 5, 1418–1421 (1978).

Watt, B. K., and A. L. Merrill: "Composition of Foods," *Agricultural Handbook 8*, U.S. Department of Agriculture, Washington, D.C. (1975).

VITAMIN A.   This substance also has been referred to as retinol, axerophthol, biosterol, vitamin A_1, antixerophthalmic vitamin, and anti-infective vitamin. The physiological forms of the vitamin include: retinol (vitamin A_1) and esters; 3-dehydroretinol (vitamin A_2) and esters; 3-dehydroretinal (retinine-2); retinoic acid; neovitamin A; neo-*b*-vitamin A_1. The vitamin is required by numerous animal species. All vertebrates and some invertebrates convert plant dietary carotenoids in gut to vitamin A_1, which is absorbed. Most animal species store appreciable amounts of the vitamin in their livers, have low concentrations in the blood, and undetectable quantities in most other tissues. A deficiency of the vitamin produces a variety of symptoms, the most uniform being eye lesions, nerve degeneration, bone abnormalities, membrane keratinization, reproductive failure, and congenital abnormalities. Toxic symptoms from large doses of vitamin A are readily produced in animals and humans. Overdosage may cause irritability, nerve lesions, fatigue, insomnia, pain in bones and joints, exophthalmia, and mucous cell formation in keratinized membranes.

The principal physiological functions of the vitamin include growth, production of visual purple, maintenance of skin and epithelial cells, resistance to infection, gluconeogensis, mucopolysaccharide synthesis, bone development, maintenance of myelin and membranes, maintenance of color and peripheral vision, maintenance of adrenal cortex and steroid hormone synthesis. Specific vitamin A deficiency diseases include xerophthalmia, nyctalopia, hemeralopia, keratomalacia, and hyperkeratosis.

In the rods of the retina, retinal is found combined with the protein *opsin*, the complex being called rhodopsin (visual purple). Although the entire series of reactions involved in dark vision has not been entirely worked out, the major steps in the cycle are quite clear. All-*trans*-retinol from the blood is oxidized by alcohol dehydrogenase (with NADP, nicotinamide adenine dinucleotide phosphate) to retinol which, in turn, is isomerized in the retina to 11-*cis*-retinal. This combines with opsin to form rhodopsin. On exposure to light, rhodopsin undergoes a sequence of changes with the eventual splitting off of retinal, which now has the all-*trans* configuration. This presumably can be reutilized in the retina by isomerization, or it can be reduced to retinol by alcohol dehydrogenase and returned to the circulation either as the free alcohol or as an ester.

The relatively recent observation that retinoic acid can replace retinol or retinal for normal growth of animals gave rise to further concepts in the biochemistry of vitamin A. Although retinoic acid cannot be demonstrated to be present normally in animal tissues, its formation by liver aldehyde dehydrogenase (NAD) and aldehyde oxidase has been accomplished so that the molecule must be considered in the general scheme of vita-

TABLE 2.   COMPARATIVE LOSSES OF VITAMINS FROM FRUITS

METHOD OF PRESERVATION	Value	LOSS OF VITAMINS AS COMPARED WITH VALUES OF FRESH PRODUCTS				
		Vitamin A	Thiamine (B_1)	Riboflavin (B_2)	Niacin	Ascorbic Acid (C)
Frozen (not thawed)	mean	37%	29%	17%	16%	18%
	range	0–78%	0–66%	0–67%	0–33%	0–50%
Canned, solids and liquids	mean	39%	47%	57%	42%	56%
	range	0–68%	22–67%	33–83%	25–60%	11–86%

SOURCE: Fennema, *Food Technology*, **31**, 12, 2328 (1977).

min A metabolism. When retinoic acid is given to animals as the only form of vitamin A, growth is normal, but the animals eventually become sterile and blind. This had led to the consideration that vitamin A may have at least three independent functions: (1) growth; (2) vision; and (3) reproduction.

The reversal of the oxidative pathway of vitamin A (retinol → retinal → retinoic acid) does not occur in the body. When retinoic acid is fed to animals, even in relatively large doses, there is no storage and, in fact, the molecule is rapidly metabolized and cannot be found several hours after administration. The metabolic products have not been fully identified. Several fractions from liver or intestine, isolated after administering retinoic acid marked with carbon-14, have been shown to have biological activity.

In 1912, Hopkins reported a factor in milk needed for the growth of rats. In 1913, Osborne and Mendel demonstrated milk factor is fat soluble; present in other fats also. McCollum and Davis, in 1913–1915, identified milk factor (fat-soluble A) in butter and egg yolk. In 1917, McCollum and Simmonds found xerophthalmia in rats due to lack of fat-soluble A. In 1920, Drummond renamed fat-soluble A *vitamin A*. In 1930, Moore determined carotene a precursor for vitamin A. See also **Carotenoids**. During 1930–1937, Karrer et al. isolated and synthesized vitamin A. In 1935, Wald reported visual purple in retina as a complex of protein and vitamin A.

Provitamin carotenoids are contained in numerous foods, but of varying concentrations. Those with a high content include carrot, dandelion greens, kohlrabi, liver (beef, calf, chicken, pig, sheep) liver oil (cod, halibut, salmon, shark, sperm whale), mint, palm oil, parsley, spinach, and turnip greens.

In higher plants, carotenoids are produced in green leaves. In animals, conversion of carotenoids to vitamin A occurs in the intestinal wall. Storage is in the liver; also the kidney in rat and cat. Target tissues are retina, skin, bone, liver, adrenals, germinal epithelium. Commercial vitamin A supplements are obtained chemically by extraction of fish liver; or synthetically from citral or beta-ionone.

Bioavailability of Vitamin A. Factors which may cause a decrease in the availability of vitamin A include: (1) liver damage; (2) impaired intestinal conversion of carotenes; (3) impaired absorption (low bile); (4) loss in food preparation (cooking and frying—heat oxidation); (5) presence of antagonists; (6) illness, causing increased destruction and excretion of the vitamin. Increases in availability may result from: (1) storage in body (liver); (2) factors which stimulate intestinal conversion of carotenes—tetraiodothyronine (thyroxine), insulin; (3) absorption aids—bile, fat; and (4) dietary protein which mobilizes vitamin A from storage in liver.

Antagonists of vitamin A include sodium benzoate, bromobenzene, citral, oxidized derivatives of vitamin A, excessive concentrations of thyroxine, estrogens, vitamin E (as regards membrane permeability). Synergists include vitamins B₂, B₁₂, and E, ascorbic acid, thyroxine, testosterone, melanocyte-stimulating hormone (MSH), and somatotrophin growth hormone.

Unusual features of vitamin A as observed by some investigators include: (1) Decreases serum cholesterol in large-quantity administration (chicks); (2) dietary protein required to mobilize liver reserves of vitamin A; (3) decreased quantities in tumors; (4) coenzyme Q_{10} accumulates in A-deficient rat liver; (5) ubichromenol-50 accumulation in A-deficient rat liver; (6) retinoic acid functions as vitamin A except for visual and reproductive functions; (7) anti-infection properties and anti-allergic properties; (8) decreases basal metabolism; (9) detoxification of poisons in the liver aided by vitamin A; and (10) vitamin A is involved in triose → glucose conversions.

References

Dennison, D. B., and J. R. Kirk: "Quantitative Analysis of Vitamin A in Cereal Products by High Speed Liquid Chromatography," *Jrnl. of Food Science*, **42**, 5, 1376–1379 (1977).

De Ritter, E.: "Newer Analytical Techniques—Vitamins," *Assoc. Food & Drug Off.*, **31**, 2, 94 (1967).

Haresign, W., and D. Lewis: "Recent Advances in Animal Nutrition," Butterworth Group, Woburn, Massachusetts (1977).

Kutsky, J. F.: "Handbook of Vitamins and Hormones," Van Nostrand Reinhold, New York (1973).

Preston, R. L.: "Typical Composition of Feeds for Cattle and Sheep—1978," Feedstuffs, A–2A (October 3, 1977); and 3A–10A (August 18, 1978).

Staff: "Food Chemicals Codex," National Academy of Sciences, Washington, D.C. (Revised periodically).

Staff: "Nutritional Requirements of Domestic Animals" (Separate publications on Beef Cattle; Dairy Cattle; Poultry; Swine; and Sheep), National Academy of Sciences, Washington, D.C. (Periodically updated).

Watt, B. K., and A. L. Merrill: "Composition of Foods," *Agricultural Handbook 8*, U.S. Department of Agriculture, Washington, D.C. (1975).

VITAMIN B₂ (Riboflavin). Some earlier designations for this substance included vitamin G, lactoflavin, hepatoflavin, ovoflavin, verdoflavin. The chemical name is 6,7-dimethyl-9-d-l' ribityl) isolloxazine. Riboflavin is a complex pigment with a green fluorescence. Riboflavin deficiency frequently accompanies pellagra and the typical lesions of both nicotinic acid and riboflavin deficiency are found in that disease. See also **Niacin**. Riboflavin, like nicotinic acid, forms an oxidation enzyme and, as such, acts as an oxygen carrier to the cell. The structure or riboflavin is:

$$CH_2-CHOH-CHOH-CHOH-CH_2OH$$

Disorders caused by a deficiency of riboflavin include anemia, cheilosis (a lip disorder); corneal vascularization, seborrheic dermatitis, and glossitis. Research leading to the current knowledge of riboflavin essentially commenced in 1917 when Emmet and McKim showed dietary growth factor for rats in rice polishings. In 1920, Emmet suggested the presence of several dietary growth factors in yeast concentrate, including the heat-stable component and B₁. The British Medical Research Council, in 1927, proposed that the designation B₂ be given to the heat-stable component. Warburg and Christian, in 1932, isolated yellow enzyme (containing riboflavin, FMN) from bottom yeast. In 1933, Kuhn isolated pure B₂ (riboflavin) from milk and recognized its growth-promoting activity. Several researchers (Kuhn, et al.; Karrer, et al.), in 1935, worked out the structure and synthesis of vitamin B₂, during which period it was named *riboflavin*. By 1954, Christie, *et al.* had determined the structure and synthesized riboflavin dinucleotide (FAD).

Riboflavin has been shown to be a constituent of 2 coenzymes: (1) Flavin mononucleotide (FMN); and (2) flavin adenine dinucleotide (FAD). The structures are shown on next page.

FMN was first identified as the coenzyme of an enzyme system that catalyzes the oxidation of the reduced nicotinamide coenzyme, NADPH (reduced NADP) to NADP (nicotinamide adenine dinucleotide phosphate). NADP is an essential coenzyme for glucose-6-phosphate dehydrogenase which catalyzes the oxidation of glucose-6-phosphate to 6-phosphogluconic acid. This reaction initiates the metabolism of glucose by a pathway other

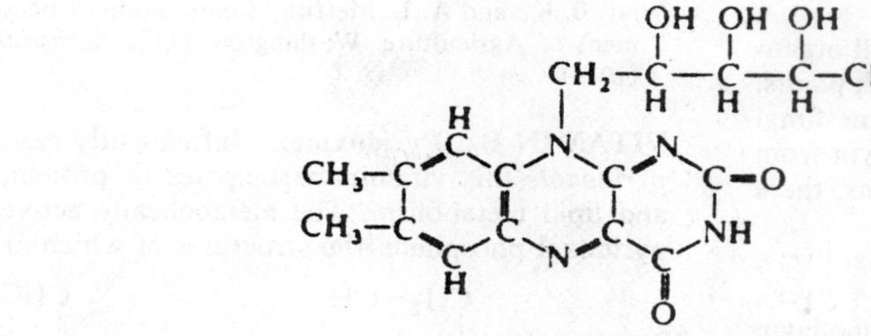

Flavin mononucleotide (FMN)

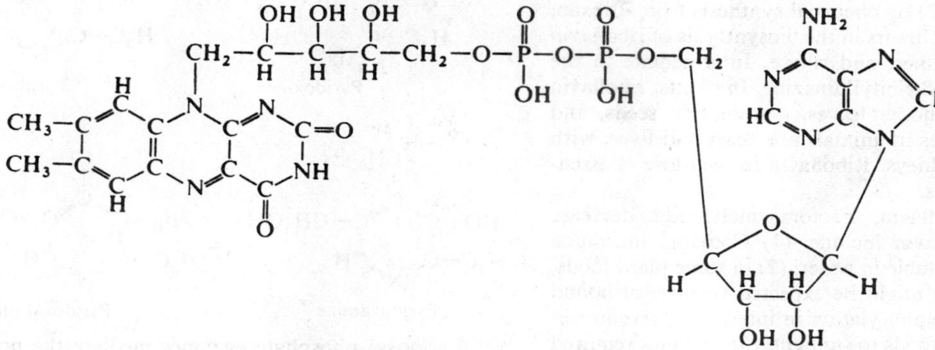

Flavin adenine dinucleotide (FAD)

than the TCA cycle (citric acid cycle). The alternative route is known as the phosphogluconate oxidative pathway, or the hexose monophosphate shunt. The first step is:

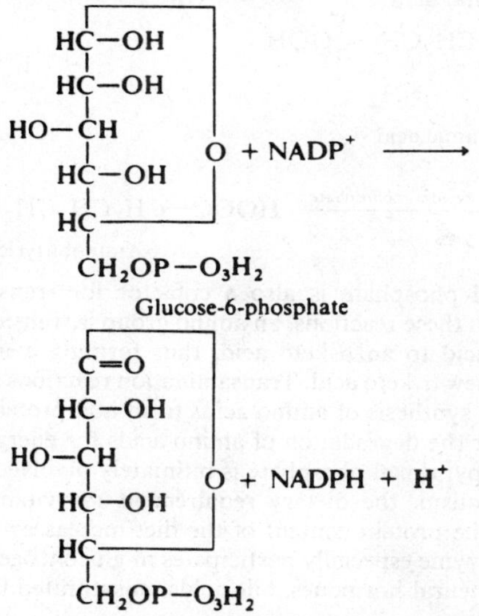

Glucose-6-phosphate

6-Phosphogluconolactone

In the biological oxidation-reduction system, reduced NAD (i.e., NADH) is reoxidized to NAD by the riboflavin-containing coenzyme FAD as shown by:

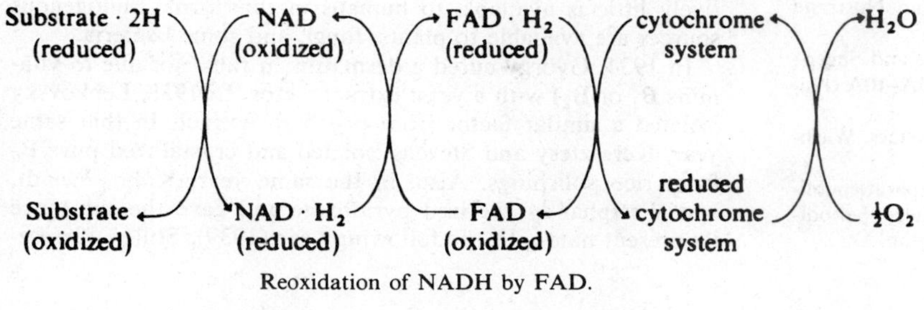

Most of the numerous other riboflavin-containing enzymes contain FAD. As shown by the foregoing diagram, FAD is an integral part of the biological oxidation-reduction system where it mediates the transfer of hydrogen ions from NADH to the oxidized cytochrome system. FAD can also accept hydrogen ions directly from a metabolite and transfer them to either NAD, a metal ion, a heme derivative, or molecular oxygen. The various mechanisms of action of FAD are probably due to differences in protein apoenzymes to which it is bound. The oxidized and reduced states of the flavin portion of FAD are:

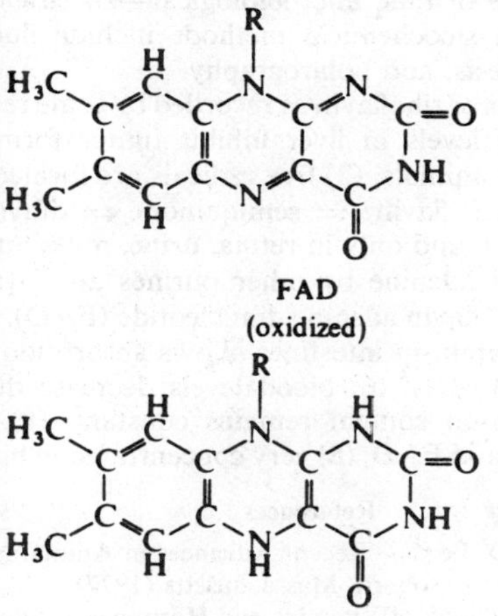

FAD
(oxidized)

FADH₂ (reduced)

Reoxidation of NADH by FAD.

See also entry on **Coenzymes.**

Distribution and Sources. Research indicates that all organisms require riboflavin. Endogenous sources exist in high plants, algae, some bacteria, and some fungi. All animals, some fungi and bacteria receive at least a partial supply of riboflavin from generation by intestinal bacteria. In the case of humans, there is a large dependence upon exogenous sources.

Foods high in riboflavin content include beef (kidneys, liver), calf (kidneys, liver), chicken (liver), pork (heart, kidneys, liver), sheep (kidneys, liver), and yeast (killed). Commercial riboflavin dietary supplements are prepared (1) by the fermentation process (bacteria or yeast); and (2) by chemical synthesis from alloxan, ribose, and o-xylene. Precursors in the biosynthesis of riboflavin include purines, pyrimidines, and ribose. Intermediate in the synthesis is 6,7-dimethyl-8-ribityllumazine. In plants, riboflavin production sites are found in leaves, germinating seeds, and root nodules. Storage sites in animals are heart and liver, with small amounts in the kidneys. Riboflavin in overdose is essentially nontoxic to humans.

Bioavailability of Riboflavin. Factors which tend to decrease the availability of riboflavin include: (1) Cooking, inasmuch as riboflavin is slightly soluble in water; (2) in some plant foods, availability is lower than might be expected because of bound forms; (3) decreased phosphorylation in intestines prevents absorption; (4) exposure of foods to sunlight; (5) enzymes required for breakdown are not present; (6) presence of gastrointestinal disease; and (7) diuresis. Riboflavin availability is increased by storage in heart, liver, and kidneys and by the presence of very actively producing intestinal bacteria.

Antagonists of riboflavin include isoriboflavin, lumiflavin, araboflavin, hydroxyethyl analogue, formyl methyl analogue, galactoflavin, and flavin-monosulfate. Synergists include vitamins A, B₁, B₆, and B₁₂, niacin, pantothenic acid, folic acid, biotin, tetraiodothyronine (thyroxine), insulin, and somatotrophin (growth hormone).

Determination of Riboflavin. Bioassay includes observance of the growth rate of rats; microbiological—*L. caseli,* and *L. mesenteroides.* Physicochemical methods include fluorimetry, paper electrophoresis, and polarography.

Unusual features of riboflavin as recorded by some researchers include: (1) High levels in liver inhibit tumor formation by azo compounds in animals; (2) free radicals are formed by light or dehydrogenation: flavine ⇌ semiquinone ⇌ dihydroflavin; (3) free vitamin is found only in retina, urine, milk, and semen; (4) substitution of adenine by other purines and pyrimidines destroys activity of flavin adenine dinucleotide (FAD); (5) phosphorylation of vitamin in intestines allows absorption as flavin mononucleotide (FMN); (6) blood levels decrease during life in humans; (7) brain content remains constant; (8) available in plants as FMN and FAD; (8) very concentrated in bull semen.

References

Haresign, W., and D. Lewis: "Recent Advances in Animal Nutrition," Butterworth Group, Woburn, Massachusetts (1977).

Kutsky, J. F.: "Handbook of Vitamins and Hormones," Van Nostrand Reinhold, New York (1973).

Plaut, G. W. E.: "Riboflavin," in *The Encyclopedia of Biochemistry* (R. J. Williams and E. M. Lansford, Jr., Editors), Van Nostrand Reinhold, New York (1967).

Preston, R. L.: "Typical Composition of Feeds for Cattle and Sheep–1977–1978," *Feedstuffs,* A-2-A (October 3, 1977); and 3A–10A (August 18, 1978).

Staff: "Food Chemicals Codex," National Academy of Sciences, Washington, D.C. (1972).

Staff: "Nutritional Requirements of Domestic Animals," separate publications on beef cattle, dairy cattle, poultry, swine, and sheep, periodically updated. National Academy of Sciences, Washington, D.C.

Watt, B. K., and A. L. Merrill: "Composition of Foods," U.S. Department of Agriculture, Washington, D.C., *Agricultural Handbook 8* (1975).

VITAMIN B₆ (Pyridoxine). Infrequently called *adermine* or *pyridoxol,* this vitamin participates in protein, carbohydrate, and lipid metabolism. The metabolically active form of B₆ is pyridoxal phosphate, the structures of which are:

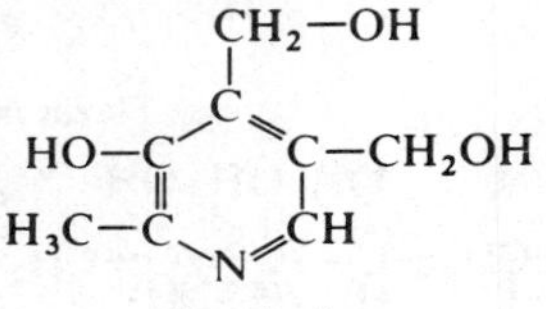

Pyridoxine
Pyridoxal 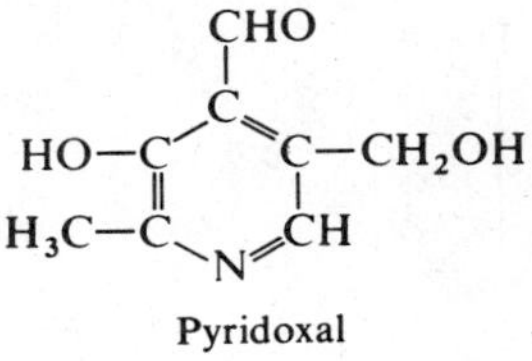
Pyridoxamine 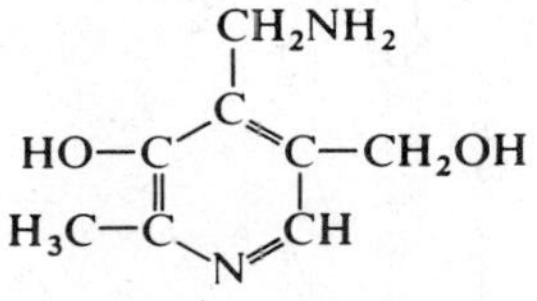
Pyridoxal phosphate 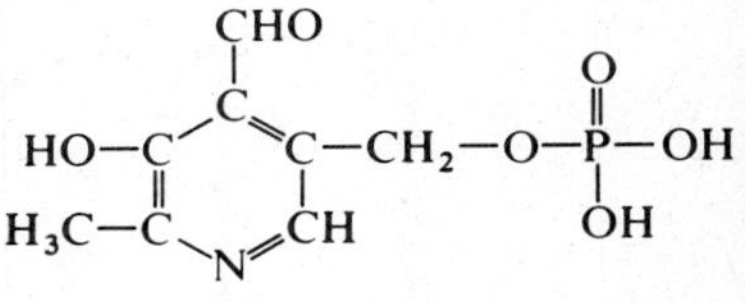

Pyridoxal phosphate enzymes mediate the nonoxidative decarboxylation of amino acids. This mechanism is of primary importance in bacteria, but it may be essential to proper function of the nervous system in humans by providing a pathway for the synthesis of a nerve impulse inhibitor, γ-amino-butyric acid from glutamic acid:

$$HOOC-CH_2CH_2CH-COOH$$
$$|$$
$$NH_2$$

Glutamic acid

$$\xrightarrow{\text{Pyridoxal phosphate}} HOOC-CH_2CH_2CH_2NH_2 + CO_2$$

γ-Aminobutyric acid

Pyridoxal phosphate is also a cofactor for transamination reactions. In these reactions, an amino group is transferred from an amino acid to an α-keto acid, thus forming a new amino acid and a new α-keto acid. Transamination reactions are important for the synthesis of amino acids from non protein metabolites and for the degradation of amino acids for energy production. Since pyridoxal phosphate is intimately involved in amino acid metabolism, the dietary requirement for vitamin B₆ increases as the protein content of the diet increases.

The coenzyme especially participates in gluconeogenesis, production of neural hormones, bile acids, unsaturated fatty acids, and porphyrins.

A deficiency of the vitamin can result in lymphopenia, convulsions, dermatitis, irritability, and nervous disorders in humans. A deficiency in monkey may cause arteriosclerosis; in rat, acrodynia. Research indicates that all animals require vitamin B₆. Bacteria in intestines generate some of this vitamin, but relatively little is available to humans in this form. Endogenous sources are available to plants, fungi, and some bacteria.

In 1934, György cured a dermatitis in rats (not due to vitamins B₁ or B₂) with a yeast extract factor. In 1938, Lepkovsky isolated a similar factor from rice bran extract. In that same year, Keresztesy and Stevens isolated and crystallized pure B₆ from rice polishings. Also, in the same year, Kohn, Wendt, and Westphal synthesized pyridoxine and gave the substance its present name. In the following year (1939), Stiller, Keresztesy, and Stevens established the structure of the vitamin. In 1945, Snell observed pyridoxal and pyridoxamine. The recognition of an establishment of B$_6$ requirements in humans was not achieved until 1953, by Snyderman et al.

In plants, the vitamin is present as pyridoxol-5-phosphate, pyridoxal-5-phosphate, or pyridoxamine-phosphate. In plants, production sites are found in fungi, cereal germ, and seeds.

Commercially the vitamin is available as a dietary supplement in the compound pyridoxine hydrochloride. The compound can be synthesized by condensing ethoxyacetylacetone with cyanoacetamide (method of Harris and Folkers); or from oxazoles.

Distribution and Sources. Most fruits and vegetables are low in pyridoxine content, although most nuts are quite high. Cereals and a number of other substances have low-to-medium content.

Bioavailability of Pyridoxine Factors which tend to decrease bioavailability of pyridoxine include: (1) administration of isoniazid; (2) loss in cooking (estimated at 30–45%), since the vitamin is water-soluble; (3) diuresis and gastrointestinal diseases; (4) irradiation. Availability can be increased by stimulating intestinal bacterial production (very small amount), and storage in liver. The target tissues of B$_6$ are nervous tissue, liver, lymph nodes, and muscle tissue. Storage is by muscle phosphorylase (in skeletal muscle, a small amount). It is estimated that 57% of the vitamin ingested per day is excreted. The vitamin exerts only limited toxicity for humans.

Precursors for biosynthesis of the vitamin include glycine, serine, or glycolaldehyde, although further research is required for further confirmation of these substances. Intermediates have not been identified. Antagonists of B$_6$ include 4-deoxypyridoxine, 4-methoxypyridoxine, toxopyrimidine, penicillamine, semicarbazide, and isoniazid. Synergists include ascorbic acid, biotin, epinephrine, folic acid, glucagon, niacin, norepinephrine, somatotropin (growth hormone), and vitamins B$_1$, B$_2$, and E.

Determination of Vitamin B$_6$. As pointed out by Gregory and Kirk (Department of Food Science and Human Nutrition, Michigan State University, East Lansing, Michigan), development of an adequate chemical procedure for the determination of biologically active forms of vitamin B$_6$ in foods has been a complex problem. Basic studies by Bonavita (1960), Toepfer et al. (1961), and Polansky et al. (1964) have demonstrated the feasibility of fluorometric measurement of pyridoxal (PAL), pyridoxamine (PAM), and pyrodixine (PIN) by conversion to PAL and reaction with potassium cyanide, forming the fluorphore, 4-pyridoxic acid lactone. Various fluorometric methods have been applied to vitamin B$_6$ compounds in biological materials (Fujiita et al., 1955; Contractor and Shane, 1968; Loo and Badger, 1969; Takanashi et al., 1970; Fiedlerova and Davidek, 1974; Chin, 1975). The results of Chin suggested that interfering compounds may be present in the PAL fraction after column chromatographic separation of the B$_6$ analogs by the procedure of Toepfer and Lehmann (1961). In the Gregory-Kirk (1977) study, methods for improving chromatographic separation and fluorometric determination of vitamin B$_6$ compounds in foods were investigated.

Traditionally, B$_6$ compounds also have been determined by bioassay, including rat and chicken growth assays.

References

Bonavita, V.: "The Reaction of Pyridoxal Phosphate with Cyanide and Its Analytical Use," *Arch. Biochem. Biophys.*, **88**, 366 (1960).

Chin, Y. P.: "Chromatographic Separation and Fluorometric Determination of Pyridoxa, Pyridoxamine, and Pyridoxine in Food System," M. S. thesis, Michigan State University, East Lansing, Michigan (1975).

Contractor, S. F., and B. Shane: "Estimation of Vitamin B$_6$ Compounds in Human Blood and Urine," *Clin. Chim. Acta*, **21**, 71 (1968).

Fiedlerova, V., and J. Davidek: "Fluorimetric Determination of Pyridoxal in Dried Milk," *Z. Lebensm. Unters-Forsch*, **155**, 277 (1974).

Kutsky, J. F.: "Handbook of Vitamins and Hormones," Van Nostrand Reinhold, New York (1973).

Staff: "Food Chemicals Codex," National Academy of Sciences, Washington, D.C. (Revised periodically).

Tryfiates, G. P.: "Vitamin B$_6$ Metabolism and Role in Growth," Food and Nutrition Press, Westport, Connecticut (1979).

Watt, B. K., and A. L. Merrill: "Composition of Foods," *Agricultural Handbook 8*, U.S. Department of Agriculture, Washington, D.C. (1975).

VITAMIN B$_{12}$ (Cobalamin). Sometimes also called cyanocobalamin, this vitamin is one of the more recent of the major B complex vitamins to be fully identified, with its structure not definitized (by Hodkin et al.) until 1955. The vitamin is required by most vertebrates, some protozoa, bacteria, and algae. Principal physiological functions include: (1) coenzyme in nucleic acid, protein, and lipid synthesis; (2) maintains growth; (3) participates in methylations; (4) maintains epithelial cells and nervous system (myelin sheath), (5) erythropoiesis (with folic acid); (6) leukopoesis. Deficiency diseases or disorders include retarded growth; pernicious anemia; megaloblastic anemia; macrocytic, hyperchromic anemia; glossitis; spinal cord degeneration; and sprue. The major physiological forms of B$_{12}$ available include hydroxocobalamin (vitamin B$_{12a}$) and aquocobalamin (vitamin B$_{12c}$).

In 1926, Minot and Murphy controlled pernicious anemia using liver. In 1944, Castle demonstrated instrinsic factor needed to control pernicious anemia with liver. Rickes et al., in 1948, isolated and crystallized factor in liver controlling pernicious anemia. In that same year, Smith and Parker crystallized and designated liver factor as vitamin B$_{12}$. West demonstrated, in 1948, clinical activity of vitamin B$_{12}$, and, in 1955, Hodgkin et al. determined the structure of the vitamin. This is shown below. Vitamin B$_{12}$ is the only vitamin with a metal ion. In this case, cobalt. Surrounding the cobalt is a macrocyclic corrin ring that is comprised of four nitrogen-containing, 5-membered rings joined through three methylene bridges. There is a similarity between this corrin ring and the dihydroporphyrin (chlorin) ring of chlorophyll.

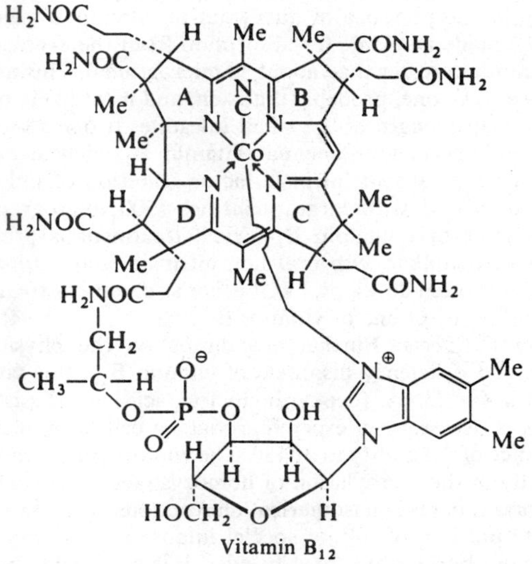

Vitamin B$_{12}$

Absorption of Vitamin B$_{12}$. This vitamin is not synthesized in animals, but rather it results from the bacterial or fungal fermentation in the rumen, after which it is absorbed and concentrated during metabolism. Among the known vitamins, this exclusive microbial synthesis is of great interest. One of the major results of vitamin B$_{12}$ deficiency is pernicious anemia. This disease, however, usually does not result from a dietary deficiency of the vitamin, but rather by an absence of a glycoprotein (*gastric intrinsic factor*) in the gastric juices that facilitates absorption of the vitamin in the intestine. Control of the disease hence is either by injection of B$_{12}$ or by oral administration of the intrinsic factor, with or without the vitamin injection.

There are two separate and distinct mechanisms for absorption of vitamin B$_{12}$. One mechanism is active, the other passive; both operate simultaneously. The active process is physiologically more important, since it is operative primarily in the presence of the small (1–2 micrograms) quantities of vitamin B$_{12}$ made available for absorption from the average meal. This special mechanism, perhaps uniquely necessary for vitamin B$_{12}$ because of its large size and polar properties, operates as follows: the normal gastric mucosa secretes a substance called the *intrinsic factor of Castle*, which combines with free vitamin B$_{12}$; the complex travels down the intestine to the ileum, where, in the presence of calcium and pH above 6, it attaches to "receptors" lining the wall of the ileal mucosa. Vitamin B$_{12}$ is then freed from intrinsic factor via a "releasing factor" mechanism of unknown nature, operating either at the surface of or within the ileal mucosal cell, and passes into the bloodstream. Thus, important requirements for normal absorption of vitamin B$_{12}$ from food are: (1) the vitamin must be freed from its peptide bonds in food; (2) the gastric mucosa must secrete an adequate quantity of intrinsic factor; (3) the ileal mucosa must be sufficiently normal both structurally and functionally so that vitamin B$_{12}$ may be absorbed across it.

Intrinsic factor is believed to be a glycoprotein or mucopolysaccharide with a molecular weight in the range of 50,000 and an end-group conformation like that of partly degraded blood group substance. The sole known role of intrinsic factor is to facilitate the transport of the large (molecular weight = 1,355) vitamin B$_{12}$ molecule across the wall of the ileal mucosa and into the blood-stream. Antibodies to intrinsic factor exists in the serum of approximately half of all patients with pernicious anemia.

The second mechanism for vitamin B$_{12}$ absorption is operative primarily in the presence of quantities of vitamin B$_{12}$ greater than those made available for absorption from the average diet (*i.e.*, quantities greater than about 30 micrograms). This mechanism is a passive one, probably diffusion, and most likely occurs along the entire length of the small intestine. It operates when patients with pernicious anemia (vitamin B$_{12}$ deficiency due to inadequate or absent intrinsic factor secretion of unknown cause) are treated with large quantities (500 micrograms or more daily) of oral vitamin B$_{12}$. Such treatment is probably better than treatment with oral hog intrinsic factor, to which refractoriness often develops, but it is not as certain as treatment with monthly injections of vitamin B$_{12}$.

Deficiency Effects. Further elucidating on the physiologic functions and deficiency disorders of vitamin B$_{12}$, this vitamin is required for DNA (deoxyribonucleic acid) synthesis and, therefore, is necessary in every reproducing cell in humans for maintenance of the ability to divide. The vitamin functions coenzymatically in the methylation of homocysteine to methionine. It is important in several isomerization reactions, and as a reducing agent, and is probably of special importance in anzymatic reduction of ribosides to deoxyribosides. It is involved in protein synthesis, partly via its role in the conversion of homocysteine to methionine; in fat and carbohydrate metabolism, partly via its role in the isomerization of succinate to methylmalonate

(which then may be decarboxylated to propionate), and in folate metabolism. Where these two vitamins interrelate, vitamin B$_{12}$ appears to serve as a coenzyme and folate as a substrate; such is true in the vitamin B$_{12}$-mediated transfer of a methyl group from N^5-methyltetrahydrofolic acid to homocysteine, which is thereby converted to methionine.

Vitamin B$_{12}$ is one of the most potent nutrients known; the minimal daily requirement for absorption by the normal adult is probably in the range of 0.1 microgram. This equals, for example, 1/500th of the minimal daily adult folate requirement, which is in the range of 50 micrograms.

As with all nutritional deficiencies, lack of vitamin B$_{12}$ may arise from inadequate ingestion, absorption, or utilization, and from increased requirement or increased excretion. Deficiency of vitamin B$_{12}$ produces megaloblastic (large germ cell) anemia, damage to the alimentary tract (glossitis being the most striking feature), and neurologic damage. The most classic neurologic sign of vitamin B$_{12}$ deficiency is decreased ability to perceive the vibration of a tuning fork pressed against the ankles. This finding is associated with damage to the posterior and lateral columns of the spinal cord, and also with damage to the peripheral nerves. This damage occurs because vitamin B$_{12}$ deficiency results in gradual deterioration of the myelin sheath, which is followed by deterioration of the axon. These processes occur slowly over months to years, and during this stage are reversible by treatment with vitamin B$_{12}$. However, when the nerve nucleus finally deteriorates, the neurologic damage becomes irreversible.

Particularly high in vitamin B$_{12}$ content are beef (brain), beef and lamb (kidney, liver), and pork (liver). Vitamin B$_{12}$ dietary supplements are often prepared commercially by the fermentation of *S. griseus*, *S. aureofaciens*, *Propionibacterium*, or as a byproduct of antibiotic production.

Certain species of bacteria and actinomycetes biosynthesize vitamin B$_{12}$. Precursors for this synthesis include glycine—corrin nucleus; δ-aminolevulinic acid—corrin nucleus; and methionine—corrin nucleus. Intermediates during the synthesis include porphobilinogen, α-D-ribosides of benzimidazole; 5,6-dimethylbenzimidazole; and α-ribazole. Antagonists of vitamin B$_{12}$ include methylamide, ethylamide, anilide, lactone derivatives, pteridine, nicotinamide. Synergists include ascorbic acid, biotin, folic acid, pantothenic acid, thiamine, and vitamins A and E.

Bioavailability of Vitamin B$_{12}$. Factors which tend to decrease the availability of this vitamin include: (1) cooking losses, since the vitamin is heat labile; (2) cobalt deficiency in ruminants; (3) intestinal malabsorption or parasites; (4) lack of intrinsic factor; (5) intestinal disease; (6) aging; (7) vegetarian diet; (8) excretion in feces; (9) gastrectomy. Factors which help to increase availability include: (1) administration of sorbitol; (2) synthesis by intestinal bacteria (not normally); (3) reduced temperature; and (4) presence of food in the stomach.

Although vitamin B$_{12}$ is essentially considered nontoxic, polycythemia has been reported from excessive dosages. From 30 to 60% of the vitamin is stored in the liver; the remainder is found in the kidneys, lungs, and spleen. Target tissues are the central nervous system, kidneys, myocardium, muscle, skin, and bone.

Unusual features of vitamin B$_{12}$ observed by some investigators include: (1) Cyanide group is an artifact of preparation; (2) the only vitamin synthesized in appreciable amounts only by microorganisms (possibly in tumors); (3) only vitamin with a metal ion; (4) works with glutathione; (5) glutathione content decreased on B$_{12}$ deficiency; (6) mitosis retarded in B$_{12}$ deficiency; (7) requires intrinsic factor (enzyme) for oral activity; (8) increases tumor size (Rous sarcoma); (9) diamagnetic properties; (10) no acidic or basic groups revealed on titration (no pK_a).

Additional Sources of B_{12}. Fermented soybean and fish products have been found to contain B_{12} (Lee et al., 1958). Nutritionally significant amounts of B_{12} also were found in the Indonesian fermented products, *ontjom* and *tempeh* (Liem et al., 1977). The microbial production of vitamin B_{12} in *kimchi*, Korean fermented vegetables, including cabbage, have been reported (Lee et al., 1958; Kim et al., 1960). The strain producing the vitamin during the fermentation was identified as *Bacillus megaterium*. As reported by Ro, Woodburn, and Sandine (1979) (Foods and Nutrition Department and Department of Microbiology, Oregon State University, Corvallis, Oregon), inoculation of fermented foods with strains known to produce vitamin B_{12} has been evaluated as a vitamin enrichment method. Soybean paste inoculated with *Bacillus megaterium* and fermented was found to contain increased vitamin levels (Choe et al., 1963; Ke et al., 1963). *Propionibacterium* species widely used in the industrial production of vitamin B_{12} (Wuest and Perlman, 1968), have been recommended for vitamin fortification of some dairy products. Karlin (1961) fortified *kefir* with vitamin B_{12} by the addition of *Propionibacterium* to the kefir grains. Kruglova (1963) prepared vitamin-enriched curds from pasteurized cow's milk by fermentation with equal parts of cultures of lactic acid and propionic acid bacteria (2.5% each). The curds had approximately 10 times more vitamin B_{12} than when produced in the usual way with only lactobacilli. In 1979, Ro, Woodburn, and Sandine undertook to increase the vitamin B_{12} content in the production of kimchi. Changes in the ascorbic acid content during the kimchi fermentation were also observed. The details of their findings are reported in the reference cited.

Determination of Vitamin B_{12}. Microbial (using *L. leichmanii*, *O. malhamensis*, *E. gracilis*, etc.) bioassay methods are used, as are checking the effects of curative doses on experimental animals (chick, rat, etc.). Physicochemical methods used include spectrophotometry, polarography, and isotope dilution.

References

Choe, C. E., Lee, S. K., and Y. S. Chung: "Studies on the Vitamin B_{12} Contents of Fermented Soybean Inoculated with *Bacillus megasterium*," *The Army Research and Testing Laboratory*, **2**, 22, Seoul, Korea (1963).

Haresign, W., and D. Lewis: "Recent Advances in Animal Nutrition," Butterworth Group, Woburn, Massachusetts (1977).

Karlin, R.: "The Fortification of Kefir with Vitamin B_{12} by the Addition of *Propionibacterium shermanii*," *Soc. Biol. Paris*, **155**, 1309 (1961) (*Dairy Sci. Abstract 24*, No. 1475).

Ke, S. Y., Chung, Y. S., and K. H. Lee: "Studies on Fermented Food Product by *Bascillus megaterium* group 1, Nutrition Evaluation of Fermented Soybean by *Bacillus megaterium*," *Korean Journal Microbiol.*, **1**, 26 (1963).

Krautmann, B. A., and C. R. Zimmerman: "A New Look at the Vitamins in Layer Feeds," *Feedstuffs*, 22–23 (August 21, 1978).

Kruglova, L. A.: "Enriching Dairy Products with Vitamin B_{12} by the Use of Propionic Acid Bacteria," *Izv. Timirjazev. Sel'skohoz. Akad.*, **4**, 208 (1963). (*Nutr. Abstract Rev. 33*, No. 2470).

Kutsky, J. F.: "Handbook of Vitamins and Hormones," Van Nostrand Reinhold, New York (1973).

Lee, T. Y., et al.: "Biochemical Studies on Korean Fermented Foods. 9. Variation of Vitamin B_{12} during the Kimchi Fermentation Period," *Report of National Chemistry Laboratories, Korea*, **7**, 18 (1958).

Preston, R. L.: "Typical Composition of Feeds for Cattle and Sheep—1977–1978," *Feedstuffs*, A–2A (October 3, 1977); and 3A–10A (August 18, 1978).

Ro, S. L., Woodburn, M., and W. E. Sandine: "Vitamin B_{12} and Ascorbic Acid in Kimchi Inoculated with *Propionibacterium freudenreichii* ss. *shermanii*," *Journal of Food Science*, **44**, 3, 873–877 (1979).

Staff: "Food Chemicals Codex," National Academy of Sciences, Washington, D.C. (1972).

Staff: "Nutritional Requirements of Domestic Animals" (Separate publications on beef cattle; dairy cattle; poultry; swine; and sheep) National Academy of Sciences, Washington, D.C. (Periodically updated).

Watt, B. K., and A. L. Merrill: "Composition of Foods," *Agriculture Handbook 8*, U.S. Department of Agriculture, Washington, D.C. (1975).

Wuest, H. M., and D. Perlman: "Industrial Preparation and Production," in "The Vitamins: Chemistry, Physiology, Pathology, Assay," 2nd Edition, Vol. II (W. H. Sebrell, Jr., and R. S. Harris, Editors), Academic, New York (1968).

VITAMIN D. Although the term *vitamin D*, is convenient to use in discussions of nutrition, this singular term is unsatisfactory when used in a strict biochemical context, because there are different substances capable of performing vitamin D nutritional functions, namely, the promotion of growth, including bone growth, and prevention of rickets in young animals. With reference to generalized terms used over the years in the development and refining of knowledge of related substances, such terms as *antirachitic vitamin*, rachitamin, rachiasterol, cholecalciferol, activated 7-dehydrocholestrol, etc. have been used. As pointed out later, some of these terms remain quite appropriate.

As a brief introductory summary, vitamin D substances perform the following fundamental physiological functions: (1) Promote normal growth (via bone growth); (2) enhance calcium and phosphorus absorption from the intestine; (3) serve to prevent rickets; (4) increase tubular phosphorus reabsorption; (5) increase citrate blood levels; (6) maintain and activate alkaline phosphatase in bone; (7) maintain serum calcium and phosphorus levels. A deficiency of D substances may be manifested in the form of rickets, osteomalacia, and hypoparathyroidism. Vitamin D substances are required by vertebrates, who synthesize these substances in the skin when under ultraviolet radiation. Animals requiring exogenous sources include infant vertebrates and deficient adult vertebrates.

The most important or at least the best known members of the family of D vitamins are vitamin D_2 (calciferol), which has the structure indicated in abbreviated form below and can be produced by ultraviolet irradiation of ergosterol, and vitamin D_3, which may be produced by the irradiation of 7-dehydrocholesterol.

Vitamin D_2

Vitamin D_3

Nomenclature. Subscript numerals have a different connotation in connection with vitamin D substances than is true, for example, with B vitamins. Vitamins B_1, B_2, B_6, B_{12}, etc., represent individual substances which have little or no chemical resemblance to each other and perform different metabolic functions. The various vitamin D's, however, have very similar structures, differing only in the side chains, and perform the same functions.

Biochemical Requirements. There are several unique features exhibited by the D vitamins. First, they are not required nutritionally at all if the organism has access to ultraviolet light (which is present in sunlight). Some animals, kept away from ultraviolet light, require so little D vitamins that the need cannot be demonstrated using ordinary diets. Rats, for example, exhibit a need for D vitamins when the calcium/phosphorus ratio in the diet is about 5:1 but not when it is the more usual 1:1. Chickens, on the other hand, exhibit a need even when the calcium/phosphorous ratio is "normal" (1.5:1).

Different species of animals respond distinctively to the different members of the vitamin D family. The most striking example of this is the fact that vitamin D_2 (calciferol) has practically no vitamin D activity for chickens. Rats respond about equally to D_2 and D_3. Human beings respond both to D_2 and D_3. Information as to how various animals react to the other less known forms of vitamin D is largely lacking and for practical reasons is not sought after.

Members of the vitamin D family are extremely difficult to isolate and identify in pure form from any source. Fish liver oils are rich sources, and vitamins D_2 and D_3 have been isolated from them. Most ordinary foods are such poor sources in terms of amounts present, that the presence of D vitamins in them has not been demonstrated. Sterols which can be converted into some form of vitamin D by ultraviolet light are, however, widespread, and it may be inferred that D vitamins are often present even when their presence has never been demonstrated.

The requirements of animals for D vitamins in terms of actual weight are extremely small. It is estimated that human beings need about 400 international units of vitamin D per day. Since an international unit of vitamin D corresponds to 0.025 microgram of crystalline vitamin D, this means that the daily human requirement is about 0.01 milligram. Foods can contain as little as 0.02 parts per million of vitamin D and yet furnish an ample supply on the basis of the foregoing estimate.

Excessive dosages of D vitamins have caused excessive calcification and damage (hypervitaminosis). The full story of vitamin D dosage remains obscure. It has been observed, for example, that some "susceptible" children do not respond to the usual doses, but require 5,000–10,000 units per day to keep them free from rickets. There are other children that are afflicted with "vitamin D-resistant rickets" who do not respond even to these high doses, but may do so when doses of the order of 500,000–1,000,000 units are administered. Although unclear, it would seem that in some individuals the vitamin D has difficulty in getting through to where it is needed.

For many years it has been recognized that all cells need calcium to function because their growth and development is related with changes in their intracellular calcium content. Reasoning further, it has been postulated that calcium may serve as a cellular regulatory agent. Growing interest has been shown by investigators, in a steroid that is derived from vitamin D and that regulates the amount of calcium in the animal's blood. This substance has been referred to as a hormone. It is 1,25-dihydroxyvitamin D_3 and is metabolized from vitamin D. In response to a skeletal need for calcium, the hormone is secreted by the kidney and transported to the intestine and bones. Many authorities believe that parathyroid hormone is involved in signaling the kidney to release 1,25-$(OH)_2D_3$. Hypoparathyroid patients lack parathyroid hormone and fail to make 1,25-$(OH)_2D_3$. The result is an abnormally low concentration of calcium in the blood, producing severe bone disease. DeLuca and associated have used 1,25-$(OH)_2D_3$ along with calcium to correct deficits in serum calcium concentrations of a limited number of patients. Corticosteroid therapy of long duration is known to produce bone disease. Corticosteroids are frequently administered to persons with rheumatoid arthritis, systemic lupus erythematosis, and asthma, in addition to persons who have received transplants. Some investigators have found that large doses of vitamin D tend to overcome the adverse effects of the corticosteroids. Findings to date essentially are the results of clinical applications rather than based upon a more detailed knowledge of the molecular mechanisms that operate in the metabolism of D vitamins.

Chronology of Vitamin D Substances. In 1918, Mellanby produced experimental rickets in dogs. In 1919, Huldschinsky ameliorated rachitic symptoms in children with ultraviolet radiation. Hess, in 1922, showed that liver oils contain the same antirachitic factor as sunlight. In that same year, McCollum increased calcium deposition in rachitic rats with cod liver oil factor. In 1924, Steenbook and Hess demonstrated irradiated foods have antirachitic properties. It was in 1925 that McCollum named antirachitic factor as vitamin D. In 1931, Angus isolated crystalline vitamin D (calciferol). In 1936, Windaus isolated vitamin D_3 (activated 7-dehydrocholesterol).

Rickets. Vitamin D deficiency (also calcium deficiency) produces a condition known in children as *rickets* and in adults as *osteomalacia*. The bones and teeth of children with rickets are poorly formed and soft. A child with rickets frequently has malformed limbs, especially bowlegs. Blood clotting may be impaired, and in extreme cases, there may be disturbances of the nervous system. An improvement in the level of calcium in the diet, along with vitamin D or parathyroid extract when required, brings about a hardening of the bones, but leaves them misshapen if deformity has already occurred. Adults, particularly pregnant or nursing women, also require vitamin D because calcium and phosphorus are continually dissolving from bones; and vitamin D is necessary for their utilization. Rickets is not to be confused with the entirely unrelated Rickettsial group of diseases (Rocky Mountain fever, etc.) that are of virus origin.

Sources. Animal sources predominate in terms of high vitamin D content. Particularly high are liver oils from bonito, cod, halibut, herring, lingcod, sablefish, sea bass, soupfin shark, swordfish, and tuna. Milk is commonly fortified with vitamin D substances in many countries.

Bioavailability. Factors which tend to cause a decrease in available vitamin D substances include: (1) liver damage; (2) presence of antagonists; (3) presence of phytin in gut; (4) low bile salts in gut; (5) high pH in gut; (6) destruction of intestinal flora; and (7) excretion in feces. Factors that enhance availability include: (1) Storage in liver and skin; (2) absorption aids, such as bile salts; (3) decrease in pH of lower intestine; and (4) irradiation by ultraviolet. Antagonists of vitamin D include toxisterol, phytin, phlorizin, cortisone, cortisol, thyrocalcitonin, and parathormone. Synergists include niacin, parathormone (concentration dependent), and somatotrophin (growth hormone).

Dosages exceeding 4000 I.U./day may cause varying degrees of toxicity in humans. Symptoms include anorexia, nausea, thirst, and diarrhea. There also may be polyuria, muscular weakness, and joint pains. Serum calcium increases and calcification of soft tissues (arteries, muscle) may commence. Arterial lesions and kidney injury have been noted in rats.

In the biosynthesis of vitamin D substances, precursors include cholesterol (skin + ultraviolet radiation) in animals; ergosterol (algae, yeast + ultraviolet radiation). Intermediates in the biosynthesis include pre-ergocalciferol, tachysterol, and 7-dehydrocholesterol. Provitamins in very small quantities are generated in the leaves, seeds and shoots of plants. In animals, the production site is the skin. Target tissues in animals are bone, intestine, kidney, and liver. Storage sites in animals are liver and skin.

Commercial vitamin D dietary supplements are prepared by

the irradiation of ergosterol, 7-dehydrocholesterol; or by extraction of fish liver oils.

Unusual features of vitamin-D substances noted by some investigators include: (1) Vitamin has hormonal qualities due to internal synthesis; (2) vitamin D_2 has little activity for chickens—various species differ in response to the vitamin; (3) vitamin D substances may play a role in aging calcification phenomena, especially in skin; (4) the vitamin can mimic rickets with a high-calcium-low-phosphorus diet; (5) the vitamin can mimic osteomalacia under the same conditions; (6) the vitamin is absorbed through skin; (7) the vitamin activates transport of heavy metals by intestinal cells; (8) the vitamin has an exceptionally long half-life (days to weeks); (9) furred and feathered animals obtain some vitamin D as the result of grooming and licking; (10) fishes are believed to obtain vitamin D from marine invertebrates; (11) the vitamin has been found useful in the treatment of lead poisoning.

Determination of Vitamin D. Bioassay techniques involve testing rats on antirachitic qualities. An important physicochemical method involves reaction with antimony trichloride.

See also entries on **Calcium; Hormone; Lipids;** and **Phosphorus.**

References

Haresign, W., and D. Lewis: "Recent Advances in Animal Nutrition," Butterworth Group, Woburn, Massachusetts (1977).
Kutsky, J. R.: "Handbook of Vitamins and Hormones," Van Nostrand Reinhold, New York (1973).
Pollin, D., and R. K. Ringer: "25-Hydroxy-D_3 and Graded Levels of Phosphorus: Effect on Egg Production and Shell Quality," *Feedstuffs*, 40–41 (October 24, 1978).
Preston, R. L.: "Typical Composition of Feeds for Cattle and Sheep—1977–1978," *Feedstuffs*, A–2A (October 3, 1977); and 3A–10A (August 18, 1978).
Staff: Food Chemicals Codex," National Academy of Sciences, Washington, D.C. (1972).
Staff: "Nutritional Requirements of Domestic Animals," in separate publications on beef cattle, dairy cattle, poultry, swine, and sheep, periodically updated. National Academy of Sciences, Washington, D.C.
Staff: "Vitamin D: Investigations of Steroid Hormone," *Science*, page 713 (February 21, 1975).
Watt, B. K., and A. L. Merrill: "Composition of Foods," *Agriculture Handbook 8*, U.S. Department of Agriculture, Washington, D.C. (1975).
Williams, R. J.: "Vitamin D" in "The Encyclopedia of Biochemistry," (R. J. Williams and E. M. Lansford, Jr., Editors), Van Nostrand Reinhold, New York (1967).

VITAMIN E. Sometimes referred to as the *antisterility vitamin*, factor X (an earlier designation), chemically vitamin E is alpha-tocopherol, the structure of which is:

Alpha-tocopherol

Active analogues and related compounds include: DL-α-Tocopherol; L-α-tocopherol; esters (succinate, acetate, phosphate), and β, ζ_1, ζ_2-tocopherols. The principal physiological forms are D-α-tocopherol, tocopheronolactone, and their phosphate esters.

The physiological functions of vitamin E substances include: (1) biological antioxidant; (2) normal growth maintenance; (3) protects unsaturated fatty acids and membrane structures; (4) aids intestinal absorption of unsaturated fatty acids; (5) maintains normal muscle metabolism; (6) maintains integrity of vascular system and central nervous system; (7) detoxifying agent; and (8) maintains kidney tubules, lungs, genital structures, liver, and red blood cell membranes.

In livestock and laboratory animals, a deficiency of vitamin E substances may cause degeneration of reproductive tissues, muscular dystrophy, encephalomalacia, and liver necrosis. Considerable research is required to fully determine supplementation of livestock diets unless typical symptoms of a deficiency appear. Symptoms have appeared where there are selenium deficiencies in the soil and where there are excessive levels of nitrates in the soil. *White muscle* is the term used to describe a condition of muscular dystrophy in cattle.

In 1922, Evans and Bishop reported *dietary factor X* needed for normal rat reproduction. In that same year, Matill found dietary factor X in yeast and lettuce. Evans et al., in 1923, found factor X in alfalfa, butterfat, meat, oats, and wheat. The designation *factor X* was changed to *vitamin E* by Sure in 1924. In 1936, Evans et al. demonstrated that vitamin E belongs to the tocopherol family of compounds. During that year, these researchers isolated several active tocopherols and found α-tocopherol to be the most active of the number. Fernholz, in 1938, determined the structure of vitamin E. It was first synthesized by Karrer during that same year. During the interim between 1938 and 1956, several tocopherols were identified and studied. It was in 1956 that Green observed the eighth in the family of tocopherols.

The tocopherols were identified as naturally occurring oily substances and the first three were characterized as alpha, beta, and gamma forms, the biological activity of which decreased in that order.

Vitamin E substances are necessary for the normal growth of animals. Without vitamin E, the animals develop infertility, abnormalities of the central nervous system, and myopathies involving both skeletal and cardiac muscle. The antioxidant activity of the tocopherols is in reverse order to that of their vitamin activity. Muscular tissue taken from a deficient animal has an increased rate of oxygen utilization. The tocopherols are so widely distributed in natural foods that a spontaneous deficiency is infrequent unless diseases of the gastrointestinal or biliary system hinder absorption. Symptoms indicating a vitamin E deficiency include: red blood cell hemolysis, creatinuria, xanthomatosis and cirrhosis of gall bladder, steatorrhea (in young), cystic fibrosis of pancreases (in young), poorly developed muscles. Rats, dogs, monkeys, and chickens display muscular dystrophy; myocardial degeneration is observed in dogs and rabbits; resorption of fetus, degeneration of germ epithelium, disturbance of estrus cycle are observed in rats; hepatic necrosis is shown in rats. Encephalomalacia and vascular degeneration are manifested in chickens.

Role of Vitamin E in Humans. The fundamental needs for vitamin E in humans have long been established. There are factors associated with this vitamin, however, that have created controversy and disagreement among highly qualified professional people. Although nearly every vitamin, at one time or other, has been used unwisely (in retrospect) in the treatment of human diseases, perhaps no other vitamin substance has aroused more discussion among clinicians than vitamin E. Because deficient animals develop a form of myopathy, it was natural to test the therapeutic efficacy of vitamin E in various forms of progressive muscular dystrophy and in diseases of the reproductive system. Enthusiastic claims have been made and refuted by investigators. From the standpoint of solid evidence, as of the early 1980s, the principal advantage of administering vitamin E lies exclusively in those instances where a vitamin

E malabsorption syndrome exists. Associated with this fundamental situation are hemolytic anemia of premature infants; diseases caused by poor fat and oil absorption; and intermittent claudication (limping). A 1979 Institute of Food Technologists Food Safety and Nutrition Panel reported no incidence of vitamin E deficiency. Three underlying reasons were cited for this: (1) ample storage in adipose tissue; (2) slow elimination from the body; and (3) prevalence in foods. Significant amounts are present in vegetable oils and margarine (70% of the average daily intake), cereal products, fish, meat, eggs, dairy products, and leafy green vegetables.

Cure-all claims for the vitamin appear to stem from the vitamin's antioxidant properties and subsequent ability to neutralize harmful free radical products of oxidation. This has led to vitamin E administration for diseases of the circulatory, reproductive, and nervous systems, increased athletic and sexual endurance, and protection against aging and air pollution effects. Although some claims have been verified by animal studies, evidence is not conclusive for humans. Elderly individuals have resorted to vitamin E in hopes of slowing the aging process. The idea is not unfounded, for in the laboratory, the nutrient neutralizes radicals normally contributing to aging pigment formation. Neutralization within humans, however, remains unproven.

Sources. Oily substances are, by far, the best natural sources of vitamin E. These include corn (maize) oil, cottonseed oil, margarine, safflower oil, soybean oil, and wheat germ oil. Production sites for vitamin E biosynthesis occur in nuts, seeds, cereal germ, green leaves, and legumes. Biosynthesis also occurs in some microorganisms. Precursors for biosynthesis include mevalonic acid and phenylalanine (probably these compounds with side chains). Considerable more research is required to pinpoint the exact precursors. Tocotrienol occurs as an intermediate in the biosynthesis.

Commercial production of vitamin E tocopherols is by way of molecular distillation from vegetable oils.

Antagonists of the tocopherols include alpha-tocopherol, quinone, oxidants, cod liver oil, and thyroxine. Synergists include ascorbic acid, estradiol, somatotrophin (growth hormone), testosterone, and vitamins A, B_6, B_{12}, and K.

Bioavailability of Vitamin E. Factors which tend to reduce availability of the vitamin include: (1) presence of antagonists; (2) mineral oil ingestion; (3) presence of vitamin E oxidation products; (4) occurrence with other less-active analogues; (5) excessive excretion in feces; (6) impaired fat absorption; (7) chemical binding in foods; (8) cooking losses (the vitamin is heat and oxygen labile); (9) losses in frozen storage, steatorrhea, and variability of natural sources. Factors which may increase absorption include: (1) storage of vitamin in adipose and muscle tissues; (2) esterification, which increases stability; (3) use of unprocessed fresh food sources; and (4) absorption aids, such as bile salts.

Storage sites for the tocopherols in the body include muscle and adipose tissues and the liver. Target tissues include the adrenals, pituitary, kidney, genital organs, muscles, liver, lungs, and bone marrow.

Unusual features of vitamin E substances as observed by various investigators include: (1) The vitamin may be involved in aging mechanisms by protecting unsaturated fatty acids and membrane against free radicals; (2) only D-isomers occur naturally; (3) vitamin E is replaceable by selenium salts in therapy of rat and pig liver necrosis, and chick exudative diathesis; (4) vitamin E is replaceable by coenzyme Q (see also **Coenzymes**) and antioxidants for certain symptoms of vitamin E deficiency, but not for all, e.g., red blood cell hemolysis, resorption gestation

not affected; (5) species differences in response to vitamin E treatment of similar symptoms, e.g., muscular dystrophy—positive in rabbits, negative in humans; (6) other tocopherols are only slightly active as compared with vitamin E; (7) vitamin content is decreased in tumors.

Alpha-Tocopherol and Nitrosamine Formation. Because of the growing concern, commenced in the late 1970s, as regards the formation of N-nitrosamines, such as dimethylnitrosamine and N-nitrosopyrrolidine, upon cooking of certain meat products cured with sodium nitrite, a number of investigators began studies to find materials which may inhibit nitrosamine formation. Reporting in late 1978, W. J. Mergens and a team of investigators (Hoffmann-LaRoche Inc., Nutley, New Jersey) observed that N-nitrosopyrrolidine has been found in fried bacon, but not in raw bacon (Fazio et al., 1973; Fiddler et al., 1974), apparently because of the influence of heat in accelerating the reaction of nitrite with the amine group of proline or its decarboxylated product, pyrrolidine, formed in frying (Archer et al., 1976; Hwang and Rosen, 1976). The effect of ascorbic acid in inhibiting nitrosamine formation has been demonstrated by various workers both *in vitro* and *in vivo* (Mirvish et al., 1972, 1973; Kamm et al., 1973, 1975; Greenblatt, 1973; Ivankovic et al., 1973).

The promising contribution of adding tocopherol to bacon, along with sodium ascorbate, to inhibit nitrosamine formation undertaken by Mergens and associates is reported in detail in the Mergens et al. reference (1978).

Determination of Vitamine E. Bioassay methods include measurements of quantity required to prevent fetal resorption; and for red blood cell hemolysis (in rat). Measurements also are made of liver storage in the chick. Physicochemical methods used include colorimetric two-dimensional paper chromatography.

References

Archer, M. C., et al.: "Nitrosamine Formation in the Presence of Carbonyl Compounds," *IARC Scientific Publication 14*, International Agency for Research on Cancer, Lyon, France (1976).

Cort, W. M., Mergens, W., and A. Greene: "Stability of Alpha- and Gamma-Tocopherol: Fe^{3+} and Cu^{2+} Interactions," *Journal of Food Science*, **43**, 3, 797–802 (1978).

Fazio, T., et al.: "Nitrosopyrrolidine in Cooked Bacon," *Journal of Assoc. Offic. Anal. Chem.*, **56**, 919 (1973).

Fiddler, W., et al.: "Some Current Observations on the Occurrence and Formation of N-Nitrosamines," *Proceedings, 18th Meeting of Meat Research Workers*, Guelph, Ontario, Canada (1972).

Greenblatt, M.: "Ascorbic Acid Blocking of Aminopyrine Nitrosation in NZO/BI Mice," *Journal of Nat. Cancer Inst.*, **50**, 1055 (1973).

Hwang, L. S., and J. D. Rosen: "Nitrosopyrrolidine Formation in Fried Bacon," *Journal of Agric. Food Chem.*, **24**, 1152 (1976).

Ivankovic, S., et al.: "Verhütung von Nitrosamidbedingtem Hydrocephalus durch Ascorbinsaure nach praenataler Gabe von Aethylharnstoff und Nitrite an Ratten," *Z. Krebsforsch*, **79**, 145 (1973).

Kamm, J. J., et al.: "Protective Effect of Ascorbic Acid on Hepatoxicity Caused by Sodium Nitrite plus Aminopyrine," *Proceedings, National Academy of Sciences*, **70**, 747 (1973).

Kamm, J. J., et al.: "Inhibition of Amine-Nitrite Hepatotoxicity by Alpha-Tocopherol," *Toxicol, and App. Pharmacol*, **41**, 575 (1977).

Kendall, J. D.: "Supplemental Vitamin E; Selenium Effects Studied," *Feedstuffs*, 13–21 (June 20, 1977).

Kutsky, R. J.: "Handbook of Vitamins and Hormones," Van Nostrand Reinhold, New York (1973).

Mergens, W. J., et al.: "Stability of Tocopherol in Bacon," *Food Technology*, **32**, 11, 40–44, 52 (1978).

Pensabene, J. W., et al.: "Effect of Alpha-Tocopherol Formulations on the Inhibition of Nitrosopyrrolidine Formation in Model Systems," *Journal of Food Science*, **43**, 3, 801–802 (1978).

Staff: "IFT Food Safety and Nutrition Panel Report on Vitamin E," Institute of Food Technology, Chicago, Illinois (1979).

Watt, B. K., and A. L. Merrill: "Composition of Foods," *Agriculture Handbook 8*, U.S. Department of Agriculture, Washington, D.C. (1975).

VITAMIN K. Sometimes referred to as the *antihemorrhagic vitamin*, and, earlier in its development, the *prothrombin factor* or *Koagulations-vitamin*, vitamin K is a substituted derivative of naphthoquinone and occurs in several forms. The designation, *phylloquinone*, or K_1, refers to 2-methyl-3-phytyl-1,4-naphthoquinone; the designations *farnoquinone, prenylmenaquinone*, or K_2, refer to 2-difarnesyl-3-methyl-1,4-naphthoquinone. *Menadione*, sometimes called *oil-soluble vitamin K_3*, is 2-methyl-1,4-naphthoquinone. The structure of phylloquinone is:

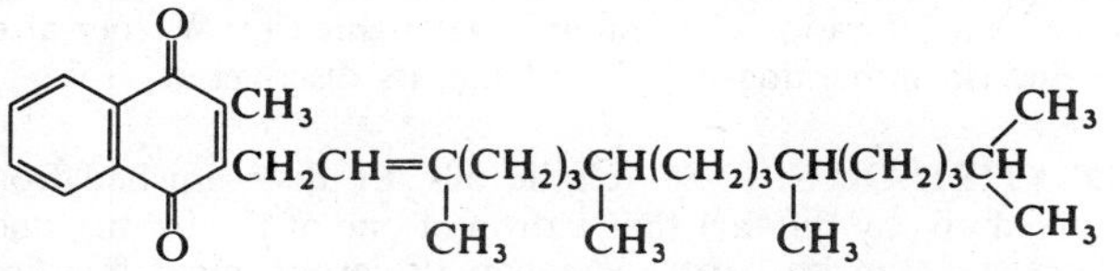

Generally, when vitamin K substances are absent or deficient in the diet of animals, including humans, a hemorrhagic disorder will appear. Young fowls that are allowed to continue on a deficient diet for extended periods will ultimately die of internal hemorrhage, or from extensive bleeding from small external wounds. Fowls experience difficulty in absorbing vitamin K from the intestine, whereas humans, rats, and dogs absorb it readily and normally obtain their requirement from intestinal bacteria without need of dietary supplementation. If, however, bacterial synthesis is inhibited by the use of sulfa drugs or certain antibiotics, the disease will develop unless the diet is supplemented with some form of vitamin K. When there is a decrease in the amount of bile salts in the intestine, as in obstructive jaundice, vitamin K is absorbed in such small amounts that the disease will also ensue. The use of vitamin K also is suggested to control and prevent the disease in premature babies. Vitamin K_1 is also able to reverse the hemorrhagic condition resulting from the administration of dicumarol to animals.

It has been reported that vitamin K_1 and several of the vitamin K_2 homologues are capable of restoring electron transport in solvent-extracted or irradiated bacterial and mitochondrial preparations. Other reports suggest that vitamin K is concerned with the phosphorylation reactions accompanying oxidative phosphorylation. The capacity of these compounds to exist in several forms, i.e., quinone, quinol, chromanol, etc., appears to strengthen the proposal that links them to oxidative phosphorylation. Information has suggested that vitamin K acts to induce prothrombin synthesis. Since prothrombin has been shown to be synthesized only by liver parenchymal cells, in the dog, it would appear that the proposed role for vitamin K is not specific for only prothrombin synthesis, but applicable to other proteins.

In 1929, Dam reported chicks on a synthetic diet develop hemorrhagic conditions. In 1935, Dam named vitamin K as the missing factor in synthetic diets. In that same year, Almquist and Stokstad demonstrated the presence of vitamin K in fish meal and alfalfa. In 1939, Dam and Karrer isolated vitamin K from alfalfa; and, in that same year, Doisy isolated K_1 from alfalfa, K_2 from fish meal, and demonstrated differences between the two substances. Also, in 1939, MacCorquodale, Cheney, and Fieser determined the structure of vitamin K_1. In that same year, Almquist and Klose synthesized vitamin K_1 for the first time. In 1941, Link et al. discovered dicumarol, an anticoagulant and antagonist of vitamin K.

In addition to compounds previously mentioned, active analogues and related compounds include menadiol diphosphate, menadione bisulfite, phthicol, synkayvite, menadiol (vitamin K_4), and compounds designated as vitamins K_5, K_6, and K_7.

Many species require vitamin K. The vitamin is frequently administered to poultry via feedstuffs. Intestinal bacteria, normally functioning, supply the vitamin to the human body.

In the therapy of deep venous thrombosis, heparin is commonly administered. This drug takes effect immediately to prevent further thrombus formation. However, heparin is regarded as a hazardous drug and possibly may be the leading cause of drug-related deaths in hospitalized patients who are relatively well. Usually administered intravenously, preferably by pump-driven infusion at a constant rate rather than by intermittent injections, sometimes may cause major bleeding, which is particularly hazardous if intracranial. The action of heparin can be terminated almost immediately by intravenous injection of protamine sulfate, but where there may be less urgency, vitamin K_1 may be used. The vitamin preparation may be administered intravenously, intramuscularly, or subcutaneously.

Vitamin K is also an antagonist of warfarin, which is sometimes used in rodenticides. Pets that have been exposed to warfarin-containing poisons may be saved from death by internal hemmorhaging through the immediate administration of vitamin K.

Vitamin K is sometimes used in the treatment of viral hepatitis.

It has been found that vitamin K analogues possess an ability to insert themselves into the oxygen-binding cleft of hemoglobin. This may result in hemolysis (dissolution of red blood corpuscles with liberation of their hemoglobin).

See also **Anticoagulants.**

Sources. Some fruits, vegetables, and nuts, as well as meat products, contain good sources of K vitamins. Intestinal bacteria, *M. phlei*, synthesize it. Particularly good sources of vitamin K are beef kidney and liver, cabbage, cauliflower, pork, soybean, and spinach.

Commercial production of vitamin K is by column chromatography of fish meal extracts. In biosynthesis, precursors include polyacetic acid (ring); acetate (side chain). Intermediates include dehydroquinic acid (ring); farnesol (side chain).

Bioavailability of Vitamin K. Factors which decrease availability of the vitamin include: (1) biliary obstruction; (2) liver damage—cirrhosis, toxins; (3) poor food preparation (vitamin is strong-acid, alkali, light, and reduction labile); (4) impaired lipid absorption in gut; (5) presence of antagonists; (6) ingestion of mineral oil; (7) sterilization of gut with antibiotics and sulfa drugs; and (8) excessive excretion in feces. Availability may be increased by way of storage in the liver and absorption aids, such as bile salts.

Antagonists of vitamin K substances include dicoumarol, sulfonamides, antibiotics, α-tocopherol quinone, dihydroxystearic acid glycide, salicylates, iodine, warfarin. Synergists include ascorbic acid, somatotrophin (growth hormone), and vitamins A and E.

General symptoms of a vitamin K deficiency include hypoprothrombinemia, increased bleeding and hemorrhage, increased clotting time, and neonatal hemorrhage. Internal hemorrhage is a symptom in chicks. Usually the vitamin is nontoxic, but, in humans, very excessive dosages can cause thrombosis, vomiting, and porphyrinuria. Target tissues are liver and vascular system. Small quantities are stored in liver.

Determination of Vitamin K. A vitamin K deficient chick assay may be made; or physicochemical techniques, including polarographic methods, spectrophotometry of pure solutions, and prothrombin time determinations, may be used.

References

Dallam, R. D.: "Vitamin K Group," in "The Encyclopedia of Biochemistry," (R. J. Williams and E. M. Lansford, Jr., Editors), Van Nostrand Reinhold, New York (1967).

Haresign, W., and D. Lewis: "Recent Advances in Animal Nutrition," Butterworth Group, Woburn, Massachusetts (1977).

Kutsky, R. J.: "Handbook of Vitamins and Hormones," Van Nostrand Reinhold, New York (1973).

Madsen, F. C., and C. B. Atwater: "Vitamin K Active Compounds Used in Feedstuffs," *Feedstuffs*, 24–25 (November 7, 1977).

Staff: "Nutritional Requirements of Domestic Animals" (Separate Publications on beef cattle; dairy cattle; poultry; swine; and sheep) National Academy of Sciences, Washington, D.C. (Periodically updated).

VITAMIN K (Blood Coagulation). Anticoagulants.

VITREOUS STATE. When certain liquids are cooled fairly rapidly, crystals do not form at a definite temperature, but the viscosity of the liquid increases steadily until a glassy substance is obtained. A glass may be thought of as a disordered amorphous solid, or as a supercooled liquid which only devitrifies into the crystalline state after extremely long standing. Glasses are optically isotropic, which explains their value in optical instruments. The property of forming a glass is possessed particularly by the oxides of silicon, boron, germanium, arsenic, phosphorus, etc., and by many organic compounds, especially those containing several hydroxyl groups per molecule. See accompanying figure.

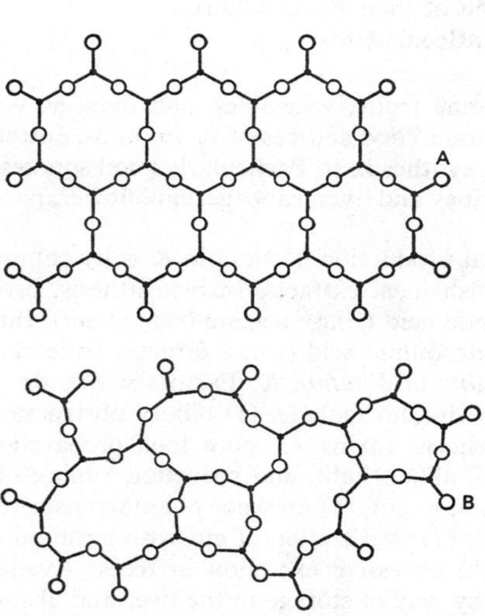

Two-dimensional diagram showing (A) an oxide of composition X_0O_3 in the crystalline form; and (B) the same oxide in the vitreous state.

VIVIANITE. The mineral vivianite is a hydrous iron phosphate, $Fe_3(PO_4)_2 \cdot 8H_2O$, its monoclinic crystals are usually prismatic or bladelike but may be in massive forms. Vivianite has one perfect cleavage; hardness 1.5–2; specific gravity 2.58–2.68; luster, pearly on cleavage faces, otherwise vitreous; colorless, when freshly exposed, but becoming blue or brownish with the alteration of the ferrous to ferric iron; transparent to translucent. Vivianite is an associate of pyrrhotite, pyrite and copper and tin ores. It is found also in clay beds forming the so-called blue iron earth which is common and of wide distribution in peat bogs. Vivianite is found in Rumania; Bavaria; Cornwall in England; and elsewhere in Europe; Australia; Bolivia; and Greenland. In the United States it occurs in New Jersey, Delaware, and Colorado. This mineral was named by Werner after the English mineralogist J. G. Vivian, its discoverer.

VOLATILE OILS. The volatile oils are distinguished from the fixed oils by the fact that a drop of one of the former does not leave a spot on paper. Members of certain plant families, such as the *Labiatae*, contain a larger percentage of such oils than do other families. But volatile oils are in no sense restricted to any small group, nor are they found only in certain tissues. Sometimes, certain parts may be principally used for the oils, as the seeds of the *Umbelliferae*.

Various methods are used in extracting the oils from the plant tissue. Many are distilled with water or steam, the oil being carried over with the distillate. In others, as for example oil of bitter almonds, the oil develops in the tissues only after fermentation. It is then obtained by distillation. Another method, and one especially used for more delicate and valuable oils, is called *enfleurage*. In this method the flowers containing the oil are spread as a thin layer over a layer of lard or olive oil. The latter absorbs the delicate oil in the flowers, after which distillation may separate the volatile oil from the other.

VOLATILE OILS (Flavoring). Flavorings.

VOLATILITY PRODUCT. The product of the concentrations of two or more ions or molecules that react to produce a volatile substance. The volatility product is analogous to the solubility product, except that, when it is exceeded, the substance escapes from the system by volatilization rather than precipitation. As with the solubility product, if any of the reacting ions or molecules have a numerical coefficient greater than one, then the concentration term of that ion or molecule is raised to the corresponding power.

VOLUME (Standard). The volume occupied by one gram molecular weight of a gas at 0°C and a pressure of 1 standard atmosphere.

W

WAFER (Silicon). Microstructure Fabrication; Semiconductor.

WALDEN INVERSION. Rearrangement (Organic Chemistry).

WARFARIN. Anticoagulants.

WASTES AND WATER POLLUTION. Although there are exceptions, such as the application of control chemicals in agriculture, most modern pollution problems are the result of disposing gaseous, liquid, and solid wastes into the environment. For study, the field falls into three classes—air, water, and soil pollution. In terms of history, water pollution and abatement technology extends backward for many decades. Air pollution and abatement technology is a more recent concern, but still dates back several decades. In recent years, one of the most active areas of concern has been soil contamination as the result of improperly designed and managed operations, such as land fills. Chemistry has played a very active role in devising equipments and reactions to neutralize the effects of waste disposal on the environment. As with several other fields of chemistry, the technology has been closely enmeshed with chemical analytical instrumentation with remarkable progress made toward identifying pollutants in very small concentrations (parts per million and parts per billion). For example, chromatography has played a leading role in making pollution surveys and in the establishment of effective abatement programs. See **Analysis (Chemical).**

Water Pollution

The pollution of the waters of the earth commenced shortly after the waters were originally formed. The present dissolved solids content of the oceans, for example, represents natural water pollution that has taken place ever since land masses above sea level appeared. Pollution of waters by people (*anthropogenic*) extends back many centuries as well, resulting from the convenience offered by rivers, lakes, and ultimately the oceans to carry away and "hide" substances that people desired to dispose. It may be surprising to note, but some rivers were much more highly polluted a couple of centuries ago than they are today even though the populations close to those rivers has increased many times during the interim. The cleanup of the Thames in the vicinity of London is an example. Nevertheless, water pollution, particularly when coupled with water shortages in numerous areas of the world, continues to be a major problem. Extensive industrialization, particularly when concentrated geographically, and agricultural practices have been major sources of water pollution in recent years. Some forms of water pollution are more dramatic than important in the total scale of pollution. Acid rain is one of the more recent discoveries in terms of water pollution, the latter usually considered to arise from the dumping of liquid and solid wastes into streams. With acid rain, the pollution is caused by gaseous effluents from plants that may be a thousand or more miles away from the polluted lakes.

A fact sometimes overlooked in this recent period of environmental emphasis is that very extensive steps have been made over the years to control water pollution. Water and waste treatment and water purification represent, in the main, well established and refined technologies. Possibly the most recent (appearing essentially a few decades ago on a large scale) technology that has and will contribute in the future to lessening water pollution is that of *water reuse*—the recycling and in-plant treatment of water for manufacturing and processing purposes.

Treatment Methods. The method of treating waste waters is determined by the nature of the pollutants present. Major categories include: (1) Suspended solids; (2) oils and greases; (3) organic matter; (4) dissolved metals; and (5) toxic chemicals in liquid, solid, and gaseous phases. Industrial activities which create these kinds of pollutants are summarized in Table 1.

Suspended Solids. Large suspended solids and trash are removed by screening devices, such as a bar screen. Further removal of gross suspended solids can be effected on a revolving drum covered with a wire or cloth screen. Sedimentation is an operation which removes suspended solid particles from a liquid stream by gravitational settling. There are two major types of sedimentation: (1) Thickening; and (2) clarification. In thickening, the primary purpose is to maximize the concentration of solids removed from the waste; the main objective of clarification is to minimize the solids concentration of the treated effluent. Both clarifiers and thickeners are commonly used for primary treatment in waste treatment facilities. These operations permit easy removal of about 65% of the suspended solids and from 35 to 40% of the BOD_5, the latter defined as the 5-day biochemical oxygen demand. In primary treatment, coagulating chemicals are not usually added. Depending upon the waste characteristics; greater removal could be experienced if such chemicals were added. Design rates for primary clarifiers range between 500 and 1,400 gallons/day/square foot (176 and 492 liters/day/square meter) of clarification area.

Thickening units in waste treatment plants are normally sized for average flows on the basis of overflow rate/unit area and detention time. Operating units of this type range from 5 feet to as large as 500 feet (1.5–164 meters) in diameter, with flow rates ranging from a few gallons to many thousands of gallons/minute. Detention time may vary from a few minutes to several days.

Flocculation can effectively agglomerate very fine particles and thus increase their settling rate for gravity removal. Flocculation is the process of bringing together fine particles so that they agglomerate. In most cases, fine suspended solids are stable and will not agglomerate without chemical treatment to destabilize them so that they will stick together when flocculated.

The coagulant most often used is alum, $Al_2(SO_4)_3 \cdot 18H_2O$. When alum is added to a suspension, the aluminum ion reacts with the alkalinity in the water to form polymeric aluminum hydrolysis species, more commonly referred to as an *alum floc*. For water treatment with low suspended solids (less than 50 milligrams/liter), an alum dose of 10 to 50 milligrams/liter is

TABLE 1.   SOURCES, CATEGORIES, AND TREATMENT OF INDUSTRIAL WATER POLLUTANTS

INDUSTRY	CATEGORY OF POLLUTANT	CHARACTERISTICS	TYPE OF TREATMENT
Metal finishing, plating, rayon processing, steel mills, tanneries	Dissolved metals	Generally cations of Al, Cr, Cu, Fe, or Zn in low-pH solution	Precipitate with lime, followed by sedimentation
Tanneries, plating, metal finishing		Chromates	Reduce with ferrous iron sulfate or sulfur dioxide; then precipitate with lime, followed by sedimentation
Plating, foundry	Toxic materials	Cyanide (generally with metal complexes)	Oxidize with chlorine or hypochlorite; then treat with lime for precipitating metals
Mining, phosphate, steel mills, power plants (fly ash), beet-sugar processing (beet washing), pulp (hydraulic debarking), foundry		Dense, rapid settling	Plain sedimentation
Pulp and paper, textile, petroleum and petrochemical, food plants, steel mills, mining, chemical plants	Suspended solids	Colloidal	Chemical coagulation followed by sedimentation or flotation
Petroleum and petrochemical, laundry, meat packing, machining (cutting oil), aircraft or railroad-car washing, dairies, food plants		Oily material or light-weight solids	Flotation (with chemical treatment if necessary)
Beet- and cane-sugar plants, dairies, meat packing, pulp and paper, canning, chemical plants, brewing, petroleum and petrochemical, tanneries	Organic matter	Vary with industry; some are easily oxidized biologically; others require special techniques	Trickling filter; activated sludge: conventional, high-rate, contact stabilization, aerobic digestion
Chemical and Petrochemical		Very strong organic wastes	Anaerobic treatment followed by aerobic treatment

commonly used. In order to increase the settling rate of the slow-settling alum floc, a polymer or a weighting agent is added. For clarification of waste water and phosphorus removal, alum doses of 50 to 300 milligrams/liter can be used. In general for clarification using alum, an overflow rate of 0.5 to 1 gallon/(minute)(square foot) [20–40 liters (minute)/(square meter)] is used.

Iron salts, such as ferric sulfate, ferrous sulfate, and ferric chloride can be used in water and waste-water clarification. The reactions are similar to those with aluminum salts. Iron salts have a wider pH range of application. Normally, alum is most effective in a pH range of 5.5 to 7, ferric iron in a pH range of 5 to 9, and ferrous iron above a pH of 9.

Lime generally is not considered an effective coagulant, but it is used in some water clarification processes. Lime, $Ca(OH)_2$, does not produce a floc like that with aluminum and iron salts. When added to water, it reacts with bicarbonate alkalinity and phosphorus compounds and adjusts the pH to cause precipitation of calcium carbonate and/or calcium hydroxyl apatite and magnesium hydroxide. Lime is used extensively to soften hard waters. If the pH of the water is raised to 10.5 or higher, floccu-

lant magnesium hydroxide is formed. This assists in the clarification process. If the pH is not raised to this value, a coagulant such as alum or iron salt will be required to assure clarification. In the treatment of a waste stream for clarification and phosphorus removal, a lime dosage of 100 to 500 milligrams/liter is usually used.

Polyelectrolytes are natural or synthetic long-chained (sometimes branched) organic macromolecules with a multiplicity of ionizable functional groups. When they are placed in water, the type of charge and degree of ionization of these functional groups will determine the charge character of the polyelectrolyte. Cationic polyelectrolytes have an excess of positive site (attracting anions) over negative sites (attracting cations) or only positive sites. Anionic polyelectrolytes have an excess of negative sites over positive sites or only negative sites. Nonionic polyelectrolytes have an equal distribution of positive and negative sites, or no ionized sites. Many types of polyelectrolytes have been used as coagulants, but the interaction of polyelectrolytes and suspended solids which achieves coagulation is such that normally there is a relatively narrow dosage of polyelectrolytes which will effectively coagulate a given type and concentration of particles. However, polyelectrolytes are extensively used as flocculating aids, promoting growth and strength of floc particles. Dosages of polyelectrolytes of 0.25 to 5.0 milligrams/liter have been used to aid flocculation of coagulated suspended solids in concentrations of 20 to 200 milligrams/liter.

Activated silica sols have been used both as coagulants and flocculation aids, but mainly the latter. The nature of their interaction with suspended solids is somewhat analogous to that of polyelectrolytes. Activated silica sols have been used with and without alum to achieve clarification in lime water-softening plants. Approximately 0.5 to 5.0 milligrams/liter may be required.

Various coagulation reactions are summarized in Table 2.

TABLE 2. COMMON COAGULATION REACTIONS USED IN WATER CLARIFICATION

$$Na_2Al_2O_4 + Ca(HCO_3)_2 + 2H_2O \rightarrow$$
$$2Al(OH)_3 + CaCO_3 + Na_2CO_3$$

$$Al_2(SO_4)_3 \cdot K_2SO_4 + 3Ca(HCO_3)_2 \rightarrow$$
$$2Al(OH)_3 + K_2SO_4 + 3CaSO_4 + 6CO_2$$

$$Al_2(SO_4)_3 + 6NaOH \rightarrow 2Al(OH)_3 + 3Na_2SO_4$$

$$Al_2(SO_4)_3 \cdot (NH_4)_2SO_4 + 3Ca(HCO_3)_2 \rightarrow$$
$$2Al(OH)_3 + (NH_4)_2SO_4 + 3CaSO_4 + 6CO_2$$

$$Al_2(SO_4)_3 + 3Na_2CO_3 + 3H_2O \rightarrow$$
$$2Al(OH)_3 + 3Na_2SO_4 + 3CO_2$$

$$Al_2(SO_4)_3 + 3Ca(HCO_3)_2 \rightarrow$$
$$2Al(OH)_3 + 3CaSO_4 + 6CO_2$$

$$Fe_2(SO_4)_3 + 3Ca(HCO_3)_2 \rightarrow$$
$$2Fe(OH)_3 + 3CaSO_4 + 6CO_2$$

$$FeSO_4 + Ca(OH)_2 \rightarrow Fe(OH)_2 + CaSO_4$$

$$4Fe(OH)_2 + O_2 + 2H_2O \rightarrow 4Fe(OH)_3$$

Where the solids to be removed have a specific gravity close to that of water, flotation can be an effective process. Most wastes which contain oils or greases are effectively clarified by flotation because of the specific gravity of the oils, regardless of whether they are separated as an oil phase or after being sorbed by a coagulant.

Biological Treatment. Either trickling filters or an activated-sludge process can be used for the biological treatment of waste waters. In either case, microorganisms are caused to grow under favorable conditions of oxygenation and nutrients if required with a balance maintained between the ingestion and digestion of the food on which the microorganisms live. The food is the dissolved or suspended organic matter in the waste being treated. The food value of the waste is usually expressed in terms of the 5-day biochemical oxygen demand, BOD_5. In the treatment processes, the microorganisms absorb this material and secrete enzymes to digest and utilize the absorbed material, thereby purifying the wastes. The microorganisms consist of bacteria, protozoa, and other microscopic forms of life which occur in sewage, river mud, and soil. Given the proper amount of food, free oxygen, and a favorable environment, the organisms will grow and reproduce, forming the slime on trickling filters or flocculant growths. Development of a healthy biological growth requires that the organisms be fed a balanced diet. Bacterial growth can be maintained with any waste containing BOD_5, providing certain essential elements are present. Some industrial

waste waters are deficient in nitrogen and phosphorus. Many high-carbohydrate wastes, as from grain products and paper-mill processes, as well as oil-refinery wastes, require the addition of nitrogen and phosphorus to assure the treatment. Packing plant wastes, in contrast, contain a high concentration of protein material and nitrogen, well in excess of the base requirement. Nutrient requirements are frequently expressed in terms of the BOD-nitrogen-phosphorus ratio. Wastes having a ratio of $100:5:1$ will usually ensure adequate nutrition.

Trickling filtration is a process in which surfaces, such as rocks, are coated with slime growth of bacteria and other microorganisms which absorb and oxidize dissolved and organic matter. The operations involved in a trickling filter included: (1) an active biological growth forms on the stone or contacting surface; (2) dissolved and colloidal organic matter is absorbed by this growth; (3) the absorbed substances are digested by the microorganisms and oxidized or assimilated to promote further growth; and (4) as the biological growth accumulates, it gets too heavy to adhere to the surface, sloughs off, goes out in the flow, and settles in a final clarifier. Oxygen is supplied by spraying the waste above the surface.

The activated sludge process is comprised of two phases: (1) adsorption and absorption of the polluting material; and (2) oxidation and digestion of the absorbed substances. The actual removal of pollutants from the liquid and their absorption by the activated sludge occurs within a few minutes of contact between sewage and healthy activated sludge. The oxidation of the absorbed material takes much longer. In conventional processes, the two operations take place in the same aeration tank. There are numerous variations of the activated sludge process.

There are some impurities in waste waters which cannot be removed by conventional treatments of the type just described. Some of the other processes are listed in Table 3.

Air Pollution

This topic is discussed in considerable detail under **Pollution (Air)**.

Soil Pollution

The disposal of conventional solid wastes is described in considerable detail in the entry on **Biomass and Wastes as Energy Sources**.

Radioactive Wastes. The wastes associated with nuclear reactors fall into two categories: (1) *commercial wastes*—the result of operating nuclear-powered electric generating facilities; and (2) *military wastes*—the result of reactor operations associated with weapons manufacture. Because the fuel in plutonium production reactors, as required by weapons, is irradiated less than the fuel in commercial power reactors, the military wastes contain fewer fission products and thus are not as active radiologically or thermally. They are nevertheless hazardous and require careful disposal.

Nuclear power plants use fuel rods with a life span of about three years. Each year, roughly one-third of spent fuel rods are removed and stored in cooling basins, either at the reactor site or elsewhere. Typical modern nuclear power plants discharge about 30 tons of the spent fuel per reactor per year. Comparatively little of the radioactive wastes, as is currently reliably known worldwide, has been processed for return to the fuel cycle. Actually, fuel reprocessing causes a net increase in the volume of radioactive wastes, but, as in the case of military wastes, they are less hazardous in the longterm. Nevertheless, the wastes from reprocessing also must be disposed with great care.

TABLE 3.   VARIOUS IMPURITIES AND PROCESSES USED TO REMOVE THEM

IMPURITIES	PROCESS
Iron, manganese, and hydrogen sulfide	Oxidation (aeration) and precipitation; Cl_2 and alkali may be added
Carbon dioxide, methane	Degasification
Oxygen, nitrogen, carbon dioxide	Vacuum degasifier
Calcium and magnesium	Sodium-cycle cation exchange
Bicarbonate, carbonate, and hydroxyl alkalinity	Dealkalization by ion exchange, followed by aeration and pH adjustment
All ionized salts, acids, or bases	Deionization via ion exchange resins
75–95% ionized salts from higher solids waters	Reverse osmosis
Ionized salts from brackish waters down to approximately 300 ppm	Electrodialysis
Calcium, magnesium, iron, manganese, turbidity, organic matter	Cold lime and hot lime-soda, softening-precipitation, settling and filtration
Turbidity and suspended matter, oil, color, colloidal silica	Coagulation, using sludge contact clarifiers followed by filtration
Turbidity and suspended matter, organic matter, iron, manganese	Filtration: sand, multimedia, diatomaceous earth, and carbon filters
Organic matter, bacteria, tastes and odors, hydrogen sulfide	Breakpoint or superchlorination.

Spent fuel from a reactor contains unused uranium as well as plutonium-239 which has been created by bombardment of neutrons during the fission process. Mixed with these useful materials are other highly radioactive and hazardous fission products, such as cesium-137 and strontium-90. Since reprocessed fuels contain plutonium, well suited for making nuclear weapons, concern has been expressed over the possible capture of some of this material by agents or terrorists operating on behalf of unfriendly governments that do not have a nuclear weapons capability.

Radioactive Wastes Inventory. The amount of all radioactive wastes produced since the dawn of the atomic age in 1942, particularly as produced by those countries in the communist block of nations or with centrally controlled governments, are difficult to estimate with any degree of reliability. It is even more difficult to ascertain the manner in which these wastes have been disposed or possibly held for final disposition.

The situation in the United States is reasonably well known. To date, the accumulation of hazardous radioactive wastes from commercial reactors is comparatively modest, constituting some 2800–3000 metric tons. But assessing a continued gradual expansion of nuclear power plants, this may increase to 18,000 to 20,000 tons by the late 1980s and perhaps increase 80,000–100,000 tons by the year 2000. Wastes are produced, of course, in direct proportion to the total operating hours of nuclear reactors. If there were 500 large reactors in operation by the year 2000, a very optimistic assumption as of the early 1980s, the projection is that a total of 125,000 metric tons would have been produced from commercial operations by the year 2000.

As of the late 1970s, it was reported that some 75 million gallons of military wastes had been produced and were in underground tank storage at the Hanford, Savannah River, and Idaho Falls installations monitored by the Energy Research and Development Administration. Further, it has been estimated that by the year 2000, if military wastes are solidified, the wastes would occupy a space of some 11 million cubic feet (311,520 cubic meters).

Converting, the highest figure previously given for commercial wastes (after solidification) by the year 2000 to volume, this amounts to some 8,824,000 cubic feet (249,896 cubic meters). If one adds the two sources of wastes, this gives a total volume figure of just short of 20 million cubic feet (about 560,000 cubic meters). For comparison, it is interesting to note that this represents just a little over 21% of the volume of the Great Pyramid of Cheops, built some 4500–6500 years ago.

Categories of Wastes by Content. In addition to the two source categories previously mentioned, radioactive wastes are classified* in accordance with their content:

- *High-level wastes* contain 99.9% of the nonvolatile fission products, 0.5% of the uranium and plutonium, and all the actinides formed by transmutation of the uranium and plutonium in the reactors. Among the actinides are neptunium and americium. High-level wastes are either the aqueous wastes resulting from reprocessing; or the spent-fuel rods to be disposed of in the absence of reprocessing.

- *Cladding wastes* are comprised of solid fragments of zircalloy and stainless steel cladding (tube in which the fuel is placed) and other structural elements of the fuel assemblies remaining after the final cores have been dissolved.

- *Low-level transuranic wastes* are solid or solidified materials which contain plutonium or other long-lived alpha-particle emitters in known or suspected concentrations higher than 10 nanoCuries per gram and external radiation levels after packaging sufficiently low to allow direct handling.

- *Intermediate level transuranic wastes* are solids or solidified materials that contain long-lived alpha-particle emitters at concentrations greater than 10 nanoCuries per gram and which have, after packaging, typical surface dose rates between 10 and 1000 mrems/hour due to fission product contamination.

- *Nontransuranic low-level wastes* are diverse materials which are contaminated with low levels of beta- and gamma-emitting isotopes, but which contain less than 10 nanoCuries of long-lived alpha activity per gram.

On-Site Storage. Operators of nuclear reactors commonly gage the space for on-site spent-fuel storage in two ways: (1) A pool with "normal discharge capacity" can accommodate the volume of spent fuel that may be discharged from a reactor

* As proposed by Shoup. See reference list.

during an annual refueling—or about one-third of the total core. (2) A storage pool with "full core reserve" can store all of the fuel in an operating nuclear power plant. Most utilities find this desirable because this increases flexibility in dealing with maintenance problems. With the outlook for fuel reprocessing somewhat bleak over the past few years, utility operators have had to expand their pool storage capacity by going to high-density storage racks. These allow fuel rods to be closely packed without overheating. Some transfers of spent fuel among power plant sites also have been reported, but reliable information has been difficult to obtain.

Away-from-Reactor Storage (AFR). Although interim measures for on-site storage may not cause severe problems until the mid-1980s or a bit later, depending upon local situations, before the end of the present decade, several facilities might have to close. The social reaction to this, of course, could be mixed. It would appear that, as of the mid-1980s, there are not many policy options: (1) If the situation essentially goes unattended or kept in planning for over a few years, nuclear power facilities will become inoperative one by one. Further, a stalemate would be less than encouraging to capital investment in completing or commencing new generating facilities. (2) If it is agreed among government officials and a majority of the scientific community that planning for permanent disposition (including a policy decision on spent fuel reprocessing) should be continued for 10, 20, or even 30 more years in an effort to reduce the risk of permanent storage to an extremely low level, but also assuming that it is not desirable to allow operating plants to go down for the lack of spent fuel storage capacity, then an interim storage plan should be agreed upon in the very near future and construction of such facilities immediately commenced. (3) Reflecting the message of an A.I.Ch.E. policy statement, commence with reprocessing and permanent storage facilities posthaste.

Option (2) was suggested in the mid-1970s (Retrievable Surface Storage Facility) at a most inopportune time in terms of sociological reaction. It was a costly proposal, but all solutions will be costly. It was viewed by some persons as a way to gain more time in making critical decisions for the long term. It was viewed by others as a makeshift arrangement that simply would delay facing up to the problem for a few more decades. In fact, it was one of several scientifically motivated approaches.

Current Planning Activity. Even though some 6000 studies may have been completed, these studies continue—and this can be partly explained by the array of options available. Reference is made to Fig. 1.

The Carter Administration, during the early part of 1980, adopted a "Consultation and Concurrence" policy. This includes many of the recommendations of the Interagency Review Group on nuclear waste management, charted by the President in 1977. The new policy stipulates that state governments are to have a "continuing role in decision-making with regard to the federal government's actions in nuclear waste disposal." As pointed out by Lee (1980), the idea of giving state governments a role commensurate with federal executive agencies is so old that it had to be rediscovered. Although beyond the scope of this book, Lee develops the differences between consultation and concurrence in connection with the waste problem and points out that the weaknesses of the concurrence approach and suggests an alternative institutional framework for locating a waste repository. In essence, a siting jury that provides representation for state and local interests, while still maintaining a high level of technical review, would be more effective than to place decision making at state levels where a full and accurate

understanding of the technology involved may not be present. It is suggested that this proposal could be tested in the siting of away-from-reactor storage facilities for spent nuclear fuel.

Trends. As of the mid-1980s, there is a trend away from spent fuel reprocessing, although advocates of this procedure are strongly represented and have developed convincing arguments. There is a strong trend toward permanent deposition in geologic structures, with salt domes high on the list of preference of several authorities. The trend is toward solidification and thus volume reduction of both liquid and semi-solid (sludge) wastes, then placing these solid wastes in canisters which, in turn, would be buried in subterranean cavities. Such deposits would essentially be nonretrievable in terms of current technology, although some future generation might develop methods to retrieve them if the energy were required. There appears to be a strong trend toward encapsulation of containers within glass or ceramics. This would require calcination and vitrification.

The concept of a "sequence of barriers" has developed strongly during the past few years. The first barrier is the form in which radioactive materials are embedded—vitrification, calcination, etc. The requirements for the first barrier are that it not be corrosive and possess excellent thermal stability and mechanical integrity. Wastes generate much heat during their initial decade of confinement. This affects decisions as regards the wasteform and the second barrier, the frequently mentioned canister which encapsulates the wasteform. The principal function served by the canister is protection of the material during the collection and transportation (to geologic site) phases. The canisters also should provide excellent protection of their contents for a minimum of 50 years, just in case it is desired to retrieve the wastes at some future date. Canisters must resist corrosive chemicals, they must withstand extremely high radiation fluxes caused by fission-product decay and the heat generated by the decaying wastes. It is interesting to note that an unprotected stainless steel canister will not resist structural deterioration arising from salt brines for that long a period. Provision for cooling canisters, either by air or water, must be provided. The canisters should be designed to permit maximum heat transfer and, currently, the cylinder and annulus configurations are preferred. A third barrier would be the geologic site itself, obviously impervious to water penetration and in a seismically stable location. To accomplish all of the foregoing needs (and more), consideration is being given to phasing the waste storage procedure, possibly storing the canisters for water cooling during the first few years, after which air cooling would suffice.

Commenting briefly on other proposed methodologies, as shown in Fig. 1, (a) in *solution-mined cavities*, it is proposed that chemical solutions would be used to mine cavities in appropriate media, such as rock salt; (b) in the *drilled-hole matrix*. A series of large-diameter holes would be drilled into the geologic media to depths up to 2 kilometers to form a grid of holes. The solid wastes would be packed into these holes, then sealed. (c) In the *rock-melting concept*, liquid wastes (no solidification) are poured into a subterranean cavity which would be created by an underground explosion. (d) In the *hydrofracture concept*, liquid radioactive wastes are converted into a type of grout (cement or cement-like materials used). This grout is pumped under high pressure into shale as deep as 1 kilometer. The pressure of the operation causes the underlying shale to fracture and the wastes fill up the cracks so formed. This procedure has been used for years in the petroleum field. See also **Petroleum.** (e) In the *polar ice* concept, the wastes would melt through the ice (although this approach would require consider-

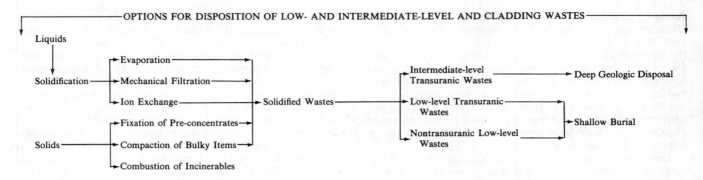

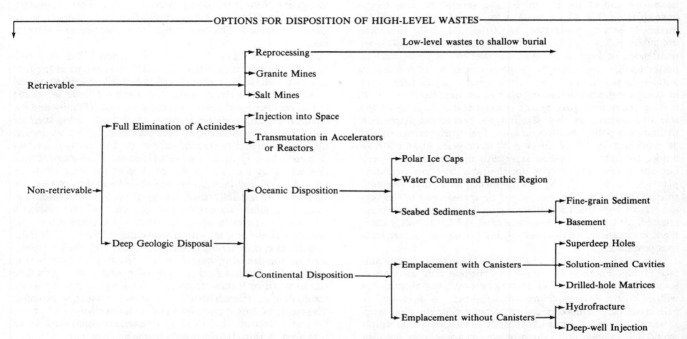

Panorama of options for consideration in the disposition of radioactive wastes. (Top portion of chart: options for disposition of low- and intermediate-level and cladding wastes; Bottom portion: options for disposition of high-level wastes.

able new technology); or the wastes would be placed on the surface of the ice or anchored within the ice. Advantages include long distances from populations, excellent for thermal cooling. Disadvantages include extensive transport and poor retrievability. This method is not high on the list of choices mainly because of too many unknown factors that will require considerable research and experimentation. (f) *Oceanic disposition*, in addition to the polar ice cap concept, are subduction zones and other deep sea trenches and rapid sedimentation areas. A recent "Seabed Disposal Program Annual Report" states, "Placing high-level wastes on the seafloor, i.e., in the water column, effectively puts the waste contained directly into the biosphere. Since it is difficult to conceive of a practical man-made wasteform/container system that would survive without releasing radionuclides for hundreds of thousands of years in a marine environment, one must assume the radioactive material would eventually enter the ecosystem." The sub-seabed sediments in the central North Pacific are loosely packed, fine-grained, deepsea "red clays" and have not been fully dismissed from continuing investigation.

Much greater detail on most of these proposed methods will be found in the references listed. Some of these references have excellent bibliographies which will lead the interested reader to fine detail.

References

Water Pollution

Afghan, B. K., and D. Mackay: "Hydrocarbons and Halogenated Hydrocarbons in the Aquatic Environment," Plenum, New York (1979).

Chapra, S. C., and A. Robertson: "Great Lakes Eutrophication: The Effect of Point Source Control of Total Phosphorus," *Science,* **196,** 1448–1449 (1977).

Cornett, R. J., and F. H. Rigler: "Hypolimnetic Oxygen Deficits: Their Prediction and Interpretation," *Science,* **205,** 580–581 (1979).

Eckenfelder, W. W., Jr.: "Principles of Water Quality Management," C.B.I. Publishers, Boston, Massachusetts (1980).

Hanrahan, D.: "Hazardous Wastes: Current Problems and Near-term Solutions," *Technology Review* (*MIT*), **82,** 2, 20–31 (1979).

Kominek, E. G.: "Waste Treatment," in "Chemical and Process Technology Encyclopedia," (D. M. Considine, Editor), McGraw-Hill, New York (1974).

Lehman, A. (editor): "Lakes: Chemistry, Geology, Physics," Springer-Verlag, New York (1978).

Maugh, T. H., II: "Restoring Damaged Lakes," *Science*, **203**, 425–427 (1979).

Robbins, L. A.: "Liquid-Liquid Extraction: A Pretreatment Process for Wastewater," *Chem. Eng. Progress*, **76**, 10, 58–61 (1980).

Rzoska, J. (editor): "The Nile: Biology of an Ancient River," Junk, The Hague (1976).

Schindler, D. W.: "Evolution of Phosphorus Limitations in Lakes," *Science*, **195**, 260–262 (1977).

White, G. F.: "Environment," *Science*, **209**, 183–190 (1980).

Radioactive Wastes

AAAS Symposium: "The Geologist in the Nuclear Fuel Cycle," Denver, Colorado, American Association for the Advancement of Science, Washington, D.C. (1977).

Angino, E. F.: "High-Level and Long-Lived Radioactive Waste Disposal," *Science*, **198**, 885–890 (1977).

Bondietti, E. A., and C. W. Francis: "Geologic Migration Potentials of Technetium-99 and Neptunium-237," *Science*, **230**, 1337–1340 (1979).

Bonniaud, R., and C. Sombret: "Continuous Vitrification of Radioactive Wastes," *Chem. Eng. Progress*, **72**, 3, 47–48 (1976).

Carter, L. J.: "Radioactive Wastes," *Science*, **195**, 661–666 (1977).

Chapman, C. C., Blair, H. T., and W. F. Bonner: "Waste Vitrification at Battelle Northwest," *Chem. Eng. Progress*, **72**, 3, 58–60 (1976).

Cohen, B. L.: "The Disposal of Radioactive Wastes from Fission Reactors," *Sci. Amer.*, **236**, 6, 2–25 (1977).

Eister, W. K.: "Materials Considerations in Radioactive Waste Storage," *Nuclear Technology*, **32** (1977).

Hanrahan, D.: "Hazardous Wastes," *Technology Review (MIT)*, **82**, 2, 20–31 (1979).

Jakimo, A., and I. C. Bupp: "Nuclear Waste Disposal," *Technology Review (MIT)*, **80**, 5, 64–72 (1978).

Johnson, C. D.: "Modern Radwaste Systems: An Overview," *Chem. Eng. Progress*, **72**, 3, 43–46 (1976).

Kearney, M. S., and R. D. Walton, Jr.: "Long-Term Management of High Level Wastes," *Chem. Eng. Progress*, **72**, 3, 61–62 (1976).

Kerr, R. A.: "Nuclear Waste Disposal: Alternatives to Solidification in Glass Proposed," *Science*, **204**, 289–291 (1979).

Kerr, R. A.: "Geologic Disposal of Nuclear Wastes," *Science*, **204**, 603–606 (1979).

Lee, K. N.: "A Federalist Strategy for Nuclear Waste Management," *Science*, **208**, 679–684 (1980).

Lee, K. N., Klemka, D. L., and M. E. Marts: "Electric Power and the Future of the Pacific Northwest," Chaps. 6 and 7, Univ. of Washington Press, Seattle (1980).

Legler, B. M., and G. R. Bray: "Concentration and Storage of High Level Wastes," *Chem. Eng. Progress*, **72**, 3, 52–53 (1976).

Resen, L.: "Nuclear Waste Disposal: The Problem in Perspective," *Chem. Eng. Progress*, **75**, 3, 57 (1979).

Shoup, R. L.: "International Waste Management Symposium," *Nuclear Safety*, **18**, 4, (1977).

Staff: "High-level Nuclear Wastes in the Seabed?" *Oceanus*, **20**, 1 (1977).

Staff: "Report to the President," *TID-29442*, Interagency Review Group on Nuclear Waste Management, National Technical Information Service, Springfield, Virginia (1979).

Staff: "Consultation and Concurrence," *ONWI-87*, Office of Nuclear Waste Isolation, Department of Energy, Columbus, Ohio (1980).

Talbert, D. M., Editor: "Seabed Disposal Program Annual Report," Sandia Laboratories, New Mexico (Annual summaries).

van Geel, J. N. C., Eschrich, H., and E. J. Detilleux: "Conditioning High Level Radioactive Wastes," *Chem. Eng. Progress*, **72**, 3, 49–51 (1976).

Welty, R. K.: "Solar Evaporation of Fluoride Wastes," *Chem. Eng. Progress*, **72**, 3, 54–57 (1976).

WASTES (As Energy Sources). Biomass and Wastes as Energy Sources.

WASTE (Nuclear) Management. Nuclear Reactor.

WATER. A colorless (blue in thick layers) liquid, H_2O, odorless, tasteless, melting point 0°C (one of the standard temperature points), boiling point 100°C at 760 millimeters of mercury pressure (another standard temperature point).

The boiling point of water increases with increasing pressure: 100.366°C at 770 mm; 120.6°C at 1,520 mm; 180.5°C at 7,600 mm. The boiling point decreases with decreasing pressure: 99.360°C at 750.0 mm; 99.255°C at 740.0 mm; 98.877°C at 730.0 mm; 81.7°C at 380 mm; 46.1°C at 76.0 mm.

At 0°C, the density of water is 0.99987 gram per milliliter. at 8°C, 0.99988; at 15°C, 0.99913; at 16°C, 0.99897; at 17.5°C, 0.99871; at 20°C, 0.99823; at 25°C, 0.99707; at 40°C, 0.99224; at 50°C, 0.99807; at 75°C, 0.97489; at 100°C, 0.95838; at 120°C, 0.9434.

The critical temperature of water is 374°C; critical pressure, 217.7 atmospheres; critical density, 0.4 gram per cubic centimeter.

The viscosity at 0°C is 0.01792 poise (dyne-second per square centimeter), specific viscosity, 1.000. At 20°C, the viscosity is 0.01005 poise, specific viscosity, 0.561. At 50°C, the viscosity is 0.00549 poise, specific viscosity, 0.307. At 75°C, the viscosity is 0.00380 poise, specific viscosity, 0.212. At 100°C, the viscosity is 0.00284 poise, specific viscosity, 0.158.

The surface tension of water against air at 0°C is 75.6 dynes per centimeter. At 10°C, the surface tension is 74.22; at 20°C, 72.75; at 30°C, 71.18; at 60°C, 66.18; and at 100°C, 58.9.

The specific heat of water is 1.000000 at 15°C (standard of specific heat). At 0°C, the specific heat is 1.00874; at 25°C, 0.99765; at 35°C, 0.99743 (minimum); at 50°C, 0.99829; at 65°C, 1.00001; at 80°C, 1.00239; at 100°C, 1.00645; at 120°C, 1.016; at 180°C, 1.04.

The electrical conductivity of water at 18°C is 0.04×10^{-6} reciprocal ohms (measurements of Kohlraush and Heydweiller, 1902); of pure water in equilibrium with air, 0.8×10^{-6}; of ordinary distilled water, about 5×10^{-6}.

The dielectric constant of water (specific inductive capacity) is 81.07 at 18°C.

Pure water, when free of dissolved gases, may be heated above 100°C (even up to 180°C) without boiling, but upon further heating, boiling with explosive violence may occur. Steam at 100°C occupies a volume 1,700 times greater than water at 100°C. Pure water, when not agitated, may be cooled somewhat below 0°C without freezing, but upon further cooling it congeals with an increase of volume (density of ice, 0.917) exerting great force, when confined, but if in intimate contact with water at atmospheric pressure, the freezing temperature is 0°C. The vapor pressure of ice and of water is 4.579 millimeters at 1 atmosphere pressure and 0°C. The triple point (ice, water, and water vapor) is +0.007°C in vacuum. When water is compressed to about 20,000 atmospheres and then cooled, other varieties of ice, all denser than water, are formed. Ice II is 12% denser; Ice III is 3% denser. At least six varieties of ice are known.

In ice I, ice II, and ice III, each oxygen atom is surrounded tetrahedrally by four other oxygen atoms, the difference between the three forms being largely in some distortion of the linkages, since the O—O distance varies little from the 2.76 value of ice I. Each oxygen atom has 2 hydrogen atoms quite close (about 1 Å) to it. At lower temperatures, and presumably higher pressures (forms V, VI, and VII, the water molecules with four hydrogen bonds are more in evidence.

Liquid water exhibits the same tendency toward increased bonding at lower temperatures. While individual water molecules have a non-linear structure, there is association between

H_2O molecules by hydrogen bonding, the degree of association being greater at lower temperatures. Based upon the statistical mechanical treatment of Frank and Wen, liquid water may be regarded as a mixture of hydrogen bonded clusters and unbonded molecules. Other research characterizes this model in terms of 5 species: unbonded molecules, tetrahydrogen bonded molecules in the interior of cluster; and surface molecules connected to the cluster by 1, 2, or 3 hydrogen bonds.

The chemical properties of water change with temperatures at high temperatures. The reaction, $2H_2O \rightleftharpoons 2H_2 + O_2$, shows an appreciable shift to the right, reaching 0.8% at 2,000°C, and increasing rapidly above that temperature. At ordinary temperatures, the equilibrium, $2H_2O \rightleftharpoons H_3O^+ + OH^-$, is important because it enables water to act either as a proton donor or acceptor. With stronger acids, water can act as a proton acceptor:

$$HCl + H_2O \rightleftharpoons H_3O^+ + Cl^-$$

$$HBr + H_2O \rightleftharpoons H_3O^+ + Br^-$$

$$HSO_4^- + H_2O \rightleftharpoons H_3O^+ + SO_4^{2-}$$

With stronger bases, water can act as a proton donor:

$$H_2O + CO_3^{2-} \longrightarrow HCO_3^- + OH^-$$

$$H_2O + NH_3 \longrightarrow NH_4^+ + OH^-$$

Although the ions shown are written as CO_3^{2-}, HCO_3^-, Cl^-, etc., they of course are more or less solvated by the water, i.e., they have water molecules attached to them by ion-dipole bonds, since water is a polar compound. One of the most strongly marked properties of water is its behavior as an electrolytic solvent, which is due to its high dielectric constant. The energy of separation of two ions is an inverse function of the dielectric constant of the solvent. Some of the parameters of water are shown in Fig. 1.

Greatly oversimplified, the water molecule may appear as

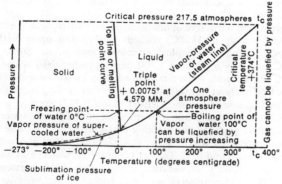

Fig. 1. Pressure-temperature diagram for water.

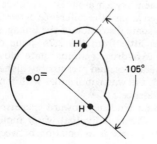

Fig. 2. Schematic representation of arrangement of electrical charges in water molecule.

shown in Fig. 2 which indicates the equilibrium position of the oxygen atom and the hydrogen atoms, i.e., the equilibrium position of the positive and negative charges of the molecule. Because of this orientation, the water molecule has a strong tendency to be oriented in an electrical field. The dipole moment depends upon the magnitude of the charge separation within the molecule, and in the water molecule, the separation is large. Thus, water may be described as having an exceptionally large dipole moment and consequently a large dielectric constant. On the basis of ascribing a dielectric constant of 1 for a vacuum, the dielectric constant of water is 80; i.e., in water, 2 electrical charges will attract or repel each other with only 1/80th as much strength as would be the case in a vacuum. This accounts, at least in part, for the remarkable ability of water to dissolve substances, particularly materials whose molecules are held together primarily by ionic bonding. The bonding arrangements within the water molecule also account for the exceptional cohesive power exhibited in water's high surface tension and the outstanding ability of water to adhere strongly to a variety of materials (the property of wetting). Bonding also accounts for the manner in which water crystals, e.g., snowflakes, are formed and for the maximum density of water (4°C), below which water assumes less dense forms, causing ice to float. Bonding is responsible for the exceptional heat capacity and exceptionally high latent heats of fusion and evaporation of water.

The molecular behavior of various molecular types in electrolytes is shown in Fig. 3. This behavior is of particular significance to the role of water in biological systems. See also **Molecule.**

Heavy water, also known as deuterium oxide, D_2O, is water in which the hydrogen of the water molecule consists entirely of the heavy-hydrogen isotope having a mass number of 2. The density of heavy water is 1.1076 at 20°C. Heavy water has been used as a moderator in nuclear reactors as well as a coolant.

Uses. Water is such a common substance that its importance and versatility are usually taken for granted. Included among the major ways in which water is important would be: (1) As a raw material for incorporation into final products without chemical change; (2) as a raw material for undergoing chemical change; (3) as a transport and conveyance medium with water acting as a solvent or carrier of solutions and suspensions in and out of reactions and physical-change operations—at an industrial as well as biochemical level; (4) as a heating and cooling medium over the wide temperature range from below normal freezing temperature (brine solutions, for example) to those of superheated steam; (5) as an energy-storage medium; (6) as a gathering medium for waste products; (7) as a cleaning medium; (8) as a shield against heat and nuclear radiation (heavy water); (9) as a convenient standard in terms of temperature, density, viscosity, and other units; and (10) with exception of a few situations where the presence of water is hazardous, as a fire-fighting medium.

Water Metabolism in Vertebrates. Those vertebrates that now inhabit land, seas, and brackish and fresh waters have survived because they have developed homeostatic mechanisms that enable them to cope with considerable variation in the content and availability of water, sodium, potassium, and chloride in their external environment. These mechanisms prevent life-threatening changes in their internal environment by (1) assuring that the cells are bathed by fluid with the same osmotic concentrations as themselves; and (2) by preventing major qualitative changes in the intra- and extracellular content of these ions or water. Regardless of species, one is impressed not by the differences, but by the similarities in the ionic composition of their intra- and extracellular fluids. The water content of the fat-free tissues of all vertebrates ranges between 70 and 80%.

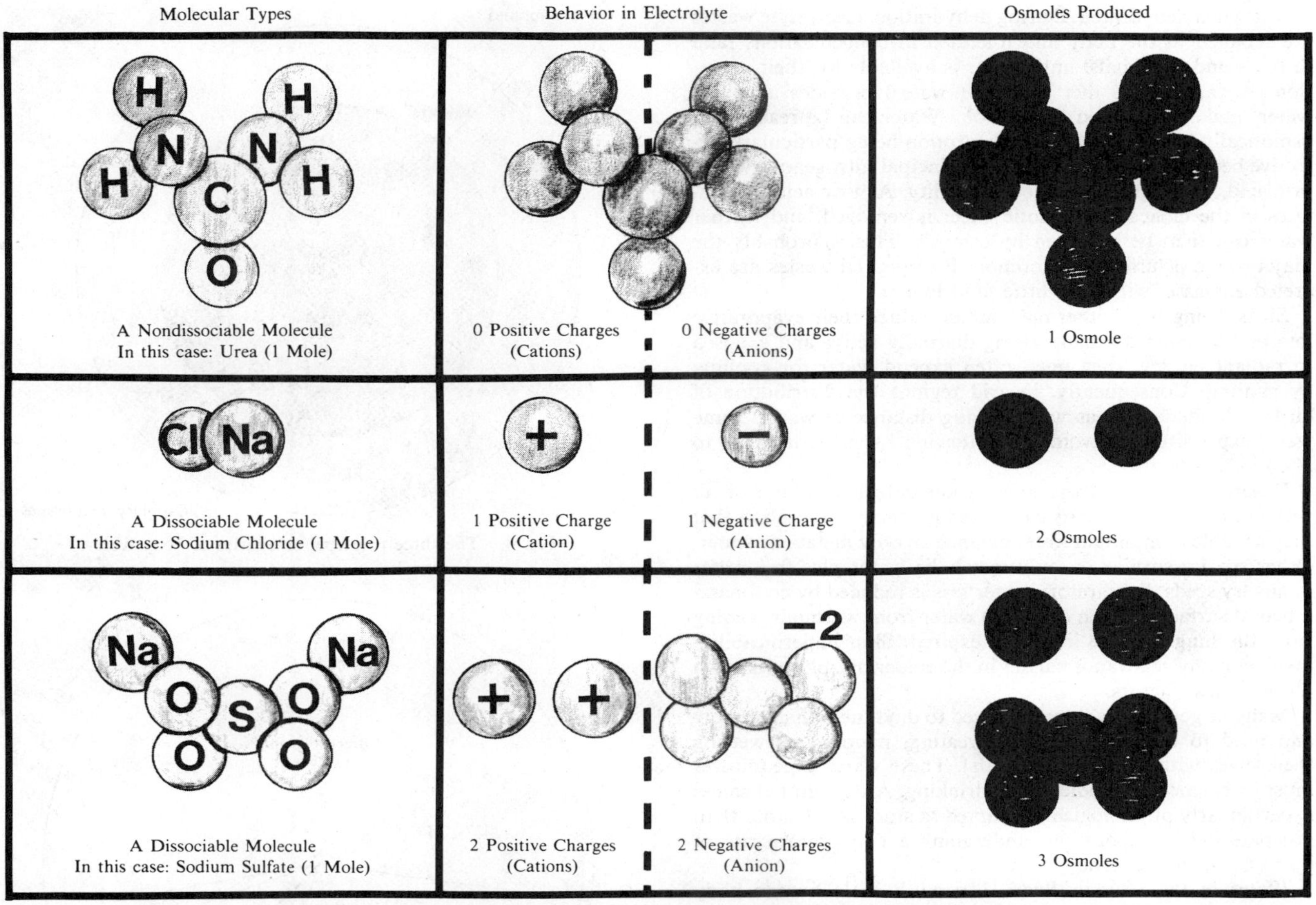

Molecular Types	Behavior in Electrolyte		Osmoles Produced
A Nondissociable Molecule In this case: Urea (1 Mole)	0 Positive Charges (Cations)	0 Negative Charges (Anions)	1 Osmole
A Dissociable Molecule In this case: Sodium Chloride (1 Mole)	$+$ 1 Positive Charge (Cation)	$-$ 1 Negative Charge (Anion)	2 Osmoles
A Dissociable Molecule In this case: Sodium Sulfate (1 Mole)	$+$ $+$ 2 Positive Charges (Cations)	-2 2 Negative Charges (Anion)	3 Osmoles

Fig. 3.   Behavior of various molecular types in electrolytes. (*After Maffly*.)

Water diffuses freely along its concentration gradient (osmosis) throughout all body tissues. Therefore, any deviation of the osmotic pressure of intra- or extracellular fluids, by either withdrawal or addition of water, causes an immediate movement of water from the more dilute to the more concentrated solution until osmotic equilibrium is reestablished.

Water is lost from the body of mammals by evaporation across the skin and in the expired air, urine, and feces. The more arid the environment, the more a mammal must be able to reduce water loss and tolerate longer periods of water dehydration and hypertonicity of its body fluids.

According to Chew (1970), vertebrates fall into several groups in terms of how they maintain their water balance. *Fishes and amphibians* in fresh water are very hypertonic to their medium and must counteract a continual dilution of their body fluids. Water influx is reduced by the relative impermeability of the skin, and is balanced by diuresis. Electrolytes lost in this urine are replaced in food eaten and by absorption through gill surfaces (fishes) and skin (amphibians). *Marine elasmobranches* are unique in maintaining themselves slightly hypertonic to sea water by retention of urea (2 ± %) in their body fluids, making their osmotic pressure more than twice that of marine teleosts. *Marine teleosts* and terrestrial tetrapods face the continual problem of counteracting desiccation due to osmotic loss to a hypertonic medium or to evaporation. Mammals are the most effective of vertebrates in conserving urine water, by concentrating the

urine, which is achieved by reabsorption of water in the kidney tubules.

Terrestrial tetrapods adjust by avoidance of evaporative stress, reduction of evaporative and urinary water losses, and temporary toleration of hyperthermia or hypernatremia. Antidiuretic hormone (ADH) from the neurohypophysis is very important in enhancing uptake of water through the skin (amphibians), reduction in glomerular filtration (amphibians, reptiles, birds), and increase in tubular reabsorption of water (mammals).

Water balance processes are best developed in species inhabiting deserts, where little drinking water is available and climatic conditions accentuate evaporation.

Certain toads and frogs survive in deserts, needing open water only for breeding, largely by remaining dormant during dry periods. Evaporation is greatly retarded in a cool, damp burrow, and urine volume is reduced by 98–99% (filtration antidiuresis), but urine remains hypotonic. Urinary water may be recycled through the body by reabsorption from the bladder. Dormant animals tolerate a loss of 50–60% of their body water. They emerge during rains, and in their dehydrated state quickly reabsorb water through the skin.

Terrestrial reptiles also avoid considerable evaporation by being quiescent in burrows much of the time. Also, their skin is more impermeable than that of amphibians, although water is still lost in expired air. Hydrated lizards have low urine filtration rate (urine always hypotonic), amd may become almost anuric when dehydrated. During dehydration, electrolyte wastes are retained in the body and tolerated in concentrations fatal to birds and mammals, until water is available for their excretion. A carnivorous diet (70 ± % water) provides adequate water intake while food is available. Water can be reabsorbed osmotically from the cloaca, reabsorption being particularly effective because of the nature of the principal nitrogenous waste, uric acid, which has a very low solubility. As uric acid precipitates in the cloaca, its osmotic effect is removed, and further water can then be absorbed by osmosis. This is/probably the major value of uric acid excretion. Precipitated wastes are excreted enmasse, with very little fluid loss.

Birds, being homeothermic, cannot reduce their evaporative loss by becoming dormant. Being diurnally active and exposed to radiant energy, they must often expend water for cooling, by panting. Consequently, in arid regions the distribution of birds is limited to areas within flying distance of water. Some water expenditure is avoided by allowing hyperthermia (up to 3°C) in the daytime.

Desert rodents lead the most water-independent life of all vertebrates. Kangaroo rats can so reduce their evaporation that they are able to maintain water balance on only metabolic water. Other species survive on only metabolic water plus free water in air-dry seeds. Respiratory water loss is reduced by cool nasal mucosal surfaces, which condense water from warm air coming from the lungs, before it can be expired. Skin impermeability involves a physical vapor barrier in the epidermis, plus unknown physiological factors.

Many larger mammals are exposed to daytime radiant energy and need to dissipate heat by sweating, panting, or wetting themselves with saliva (marsupials). These water expenditures must be balanced periodically by drinking. A dehydrated camel is particularly physiologically adapted to store heat (rather than dissipate it by evaporation, undergoing a temperature rise of up to 6° in the daytime.

Water in the Human Body. The adult male human body contains about 60% (weight) of water and the adult female body about 50%. The large amount of body water is compartmentalized, each compartment being bounded by membranes. It has been estimated (Edelman and Leibman, 1959) that 55% of this water is contained within cells bounded by cell membranes. The remaining, extracellular water or fluids (ECF) is made up of a relatively small volume of plasma (7.5% of total body water in the vascular tree), and the remaining 37.5% in nonplasma and located outside the vascular tree. The latter includes interstitial water (20%), another 15% in bone and dense connective tissue, and 2.5% in secretions. These numbers are shown graphically in Figures 4 and 5.

Two driving forces control the movement of water in the body, namely, hydrostatic pressure and osmotic pressure. Because transmembrane pressures are so low, it is not believed that hydrostatic pressure plays a role in the movement of water across cell membranes. On the other hand, hydrostatic pressure resulting from heart action creates a gradient of about 20 millimeters of mercury pressure across the capillary walls. The principles of osmosis are described in the entry on **Osmotic Pressure**. Normally, when describing osmotic pressure, reference is made to salt solutions of differing concentrations separated by a membrane. Concentrations are expressed in terms of solute in solvent (water). When thinking in terms of body water, one usually considers the addition of solute to the water as a dilution of the pure water rather than as an increase in solute content of the water. For example, pure water contains 55.5 moles per kilogram (about 55,500 mmoles/liter). Body fluids, that is, intracellular fluid (ICF) and extracellular fluid (ECF), contain ap-

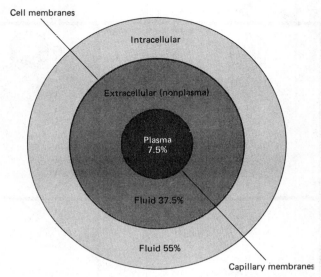

Fig. 4. The three principal categories of body water.

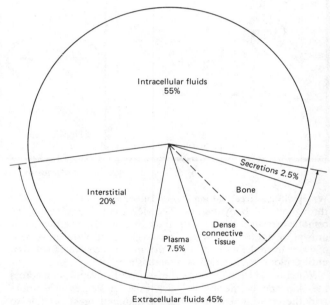

Fig. 5. Pie chart showing approximate volumetric proportions of body fluids.

proximately 99.5% water molecules and 0.5% solute molecules. This equals about 55,200 mmoles of water per liter; and 300 mmoles of solute per liter. *Osmolality* is a measure of the concentration of osmoles present in the water. If the osmolality is low, the concentration of water is high; if the osmolality is high, the concentration of water is low. Movement of water across the cell membranes occurs because of differences in concentration, the movement always being from a phase of high water concentration to one of lower concentration. Movement is "downhill," so to speak, and the motivating force is called *osmotic pressure*. This movement can be counteracted by the application of an opposing force, notably hydrostatic pressure.

It is interesting to note that the main cation present in ICF is the potassium ion, whereas the principal cation in ECF is the sodium ion. The role of potassium and sodium ions in the biological system is described in the entry on **Potassium and Sodium (In Biological Systems)**.

Water Intoxification. Either an increased intake of water or a decreased output of water can cause an excess of body water. Because healthy kidneys have outstanding ability to increase water excretion, a condition of water intoxification usually occurs because of an inability to excrete water (*hyponatremia*). Frequently impairment may not be the result of disease or damage to the kidneys per se, but rather due to faulty processing of stimuli by the kidneys. Excessive renal reabsorption of water can result from the action of antidiuretic hormone (ADH) or by the excessive reabsorption of sodium in the proximal tubules. Under such conditions, the excretion of water and sodium will be low. Retention of salt and water causes an expansion of the ECF, usually resulting in edema and effusions.

Water Deficiency. This condition occurs when water output exceeds intake. Water is continually lost by way of the lungs, skin, and kidneys and thus a deficiency of body water will occur if a critical minimal supply is not maintained. Decreased intake when water is available is uncommon. Very rarely, a brain malfunction may interfere with one's sense of thirst. Increased output of water can result from many causes. For example, a person with diabetes insipidus who lacks ADH (antidiuretic hormone) or a person whose kidneys do not respond normally to ADH, as in instances of nephrogenic diabetes insipidus, will increase water output. Other diseases which may cause excess excretion of water include osmotic diuresis, hypercalcemia, hypokalemia, chronic pyelonephritis, and sickle cell anemia, among others. Excessive water losses are also associated in some cases with advanced age and in some burn cases. Two clinical features are good measures of dehydration—weight loss of the patient and an elevation of the serum sodium concentration. In situations of dehydration, the body initiates mechanisms which manipulate the transfer of water from one compartment to the next, retaining water in those cells and organs where it is most needed.

In cases of severe dehydration, rehydration must be brought about carefully and in steps. The usual practice is to make an initial estimate of water or electrolytes that require replacement and then administering one-half the amount of the deficit, after which another series of measurements is made, followed by replacement of one-half of the estimate—until a satisfactory ultimate balance has been attained. If rehydration is not gradual, some organs, such as the brain, may take up water beyond normal requirements and this can result in cerebral edema. Also, in the case of acute dehydration, permanent brain damage can occur as the result of a shrinking brain size tearing away vessels causing cerebral hemorrhages.

Water and Macromolecules. Over eons of time, processes have appeared for making the great variety of macromolecules required by living organisms. Most of these developments have occurred in a water medium, or in one having a high water vapor content. So it is not surprising that the majority of the macromolecules involved in the life processes are hydrophilic, in different degrees. One may find minor exceptions in the cases of certain fats and lipids. It should be noted that in living organisms, the hydrophilic macromolecules of one kind often join with those of other kinds to produce the useful structures required in the life process, i.e., membranes.

Decades ago, biochemists recognized that it was difficult to remove all water from a large number of macromolecular materials. The term *bound water* was coined to explain the great affinity which many of these materials showed for water, particularly the proteins. Biochemists were convinced that biological behavior, at least in part, resulted from the amount of bound water contained in the macromolecular structure. As an example, water held in plant structures so that it did not freeze in below-freezing temperatures was considered to be found. At the time these ideas found favor, the method of lyophilization or quick freeze-drying had not been perfected. Lyophilization removes the bulk of the water held in biological substances without destroying their structures or their activity.

In more recent years, lyophilized proteins have been further dried to a constant weight in a high vacuum and then studied as they adsorbed water vapor. The heats of adsorption of the first water vapor molecules were considerably higher than the values obtained as the adsorption approached that of the saturated vapor. These results indicated that the first molecules to be adsorbed were on the most water-loving or active sites. Such adsorbed molecules on the higher-energy sites would be desorbed last in a high vacuum. Without going into considerable detail, the results of this line of research led to the conclusion that the earlier concept of bound water was unfounded.

Many biological systems depend in part on their degree of hydration. Most biological membranes are hydrophilic, but should the membrane be a multi-layered one, the hydration of the different layers may differ markedly. A great many of the membranes used by living organisms are known to be very selective in what passes through them. For many years the theory of the role of some membranes in pumping water into the cells they surround even against high osmotic gradients due to salt concentration has maintained among most biologists. The fascinating field of membrane biophysics has shed much light concerning the hydration of biological macromolecules.

As pointed out by Kolata (1976), a primary difference between living and dead cells is that living cells selectively retain certain ions, such as potassium, and exclude others, such as sodium. The water in dead cells reflects the conditions of those solutions surround them. The conventional explanation is that this difference is due to ion "pumps" in membranes, pumps which are purported to use cell energy to transport some ions into and other ions out of the cell. There is another school of thought, however, which denies that such pumps exist. This school claims that ions are excluded from cells on the basis of their low solubilities in cellular water, except when specific charged sites with which the ions can associate are available. It is maintained that cell water has a different structure than either liquid water or ice and that it is this special structure that affects the solubility of various ions in it. This school also suggests that there is evidence that ion pumps are thermodynamically impossible, requiring more energy than is available to the cell. The school also claims that nuclear magnetic resonance (NMR) studies show that cell water is more structured than liquid water and less structured than ice. This, it is believed, would affect the solubilities of ions in the cell and could account for selective ion exclusions. A number of investigators, including advocates of pumps, have agreed that cell water may have some ordered structure that makes it different from liquid water. Most investigators do not question that cell water is likely to be structured, but do ask to what extent it is structured and what the physiological importance of this structure may be. Further details on some of investigations along these lines that took place during the 1970s and earlier are given in the Kolata (1976) reference. Considerably more research, as of the 1980s, is required to resolve the differences of these two schools of thought.

Degassed Water. It is interesting to note that during the past several years, Soviet scientists have raised some intriguing points concerning water, not all of which have held up under intensive scrutiny. Some readers may recall the discussions of the early 1970s concerning the "discovery" of polywater or so-called anomalous water by some Soviet scientists. Observations of this water did not meet with accepted criteria for physical properties. For a while, it was considered by some scientists to be of a polymeric nature, but as the result of subsequent numerous exchanges of views and a careful scrutiny of the water it was found to be ordinary water which had a large concentration of dissolved minerals.

In 1978, other Soviet scientists proposed that meltwater (from freshly melted snows) carries certain biological properties not known in ordinary water. As reported by Maugh (1978), the Soviet scientists theorize that meltwater retains some of the order that is characteristic of frozen water and that this increased order alters vital reaction rates within cells. This tends to tie in with the concept of structured water discussed briefly later.

In the course of investigation at the Institute of Fruit-Growing and Vine-Growing (Kazakhstan, U.S.S.R.), investigators tested the relative growing rates, qualities, and yields of various plants subjected to meltwater, tap water, and boiled water. Although the plants responded in a superior way to the meltwater, as compared with tap water, it was found (more or less accidentally) that the plants responded even better to quickly-cooled boiled water. Thus, the experimenters concluded that the superior action on plants derived from the fact that some of the waters tested (meltwater and quickly-cooled boiled water) contained less dissolved gases. In other words, the claim is made that any degassed water is superior when administered to plants. Igor Zelepukhin suggests that the conductivity of degassed water is decreased considerably and there are comparative increases in density, viscosity, surface tension, energy of intermolecular interaction, and internal pressure, factors that may enhance the value of water. Zelepukhin thus observed further that degassed water bears a closer resemblance to the fluid in cells than does ordinary tap water. Experiments in utilizing degassed water for cattle and livestock are proceeding. Other Soviet investigators have observed that concrete prepared with degassed water is 8–10% stronger than when ordinary water is used. It should be stressed that, as of the early 1980s, these observations have not as yet been accepted by the general scientific community. Considerably more research is required toward the development of hard, convincing data.

Water Problems. Not associated with the characteristics of water, but rather with its exploitation, there are the large problems of water supply and water pollution. Some of the very desirable characteristics of water for some applications pose difficult problems in its use for others. Included are: (1) Interference of water with the operation of electrical communications and equipment; (2) freezing in pipelines, including water vapor in pneumatic transmission lines; (3) as a medium of disease transmission and corrosion; (4) the relatively easy manner in which water is polluted (because it is such an excellent solvent) and the difficulty and energy required to remove dissolved impurities. See also **Chemical Elements; Desalination; Fluorine; Ocean Water;** and **Wastes and Water Pollution.**

References

Ambroggi, R. P.: "Water," *Sci. Amer.*, **243**, 3, 100–117 (1980).

Chew, R. M.: "Water Metabolism in Mammals," in "Physiological Mammalogy," (W. V. Mayer and R. G. Van Gelder, Editors,") Academic, New York (1963).

Culp, G. L., and R. L. Culp: "New Concepts in Water Purification," Van Nostrand Reinhold, New York (1974).

Dodge, P. R., Crawford, J. D., and J. H. Probst: "Studies in Experimental Water Intoxification," *Arch. Neurol.*, **3**, 513 (1960).

Draganic, I. G., and D. Zorica: "The Radiation Chemistry of Water," Academic, New York (1971).

Edelman, I. S., and J. Leibman: "Anatomy of Body Water and Electrolytes," *Am. J. Med.*, **27**, 256 (1959).

Edelman, I. S., Leibman, J., and M. P. O'Meara: "Interrelations between Serum Sodium Concentration, Serum Osmolarity and Total Exchangeable Sodium, Total Exchangeable Potassium and Total Body Water," *J. Clin. Invest.*, **37**, 1236 (1958).

Gehm, H. W., and J. I. Bregman: "Water Resources and Pollution Control," Van Nostrand Reinhold, New York (1976).

Hays, R. M., and S. D. Levine: "Pathophysiology of Water Metabolism," in "The Kidney," Vol. I (B. M. Brenner and F. C. Rector, Jr., Editors), Saunders, Philadelphia (1976).

Horne, R. A., Editor. "Water and Aqueous Solutions: Structure, Thermodynamics, and Transport Processes," Wiley, New York (1972).

Kolata, G. B.: "Water Structure and Ion Binding: A Role in Cell Physiology?" *Science*, **192**, 1220–1222 (1976).

Langgård, H., and W. O. Smith: "Self-Induced Water Intoxication without Predisposing Illness," *N. Engl. J. Med.*, **266**, 378 (1962).

Maffly, R. H., and A. Leaf: "The Potential of Water in Mammalian Tissues," *J. Gen. Physiol.*, **42**, 1257 (1959).

Maffly, R. H.: "The Body Fluids: Volume, Composition, and Physical Chemistry," in "The Kidney," Vol. I (B. M. Brenner and F. C. Rector, Jr., Editors), Saunders, Philadelphia (1976).

Maugh, T. H., II: "A Wonder Water from Kazakhstan," *Science*, **202**, 414 (1978).

Miller, M., and A. M. Moses: "Drug-Induced States of Impaired Water Excretion," *Kidney Int.*, **10**, 96 (1976).

Stillinger, F. H.: "Water Revisited," *Science*, **209**, 451–457 (1980).

Wen, F. C., McLaughlin, T., and J. L. Katz: "Photoinduced Nucleation of Water Vapor," *Science*, **200**, 769–771 (1978).

Wiener, A.: "The Role of Water in Development," McGraw-Hill, New York (1972).

WATER CONDITIONING. This term is used broadly to cover all processes used in treating water to remove or reduce undesirable impurities to specified tolerances. Treatment of boiler water is a specialized technology and is described separately under **Water Treatment (Boiler).**

Water impurities and the treatments required may be grouped under three headings: (1) suspended matter, color and organic matter, (2) dissolved mineral matter, and (3) dissolved gases; (1) may include sediment, turbidity, color, microorganisms, tastes, odors and other organic matter. (2) consists chiefly of the bicarbonates, sulfates and chlorides of calcium, magnesium and sodium. Small amounts of silica and alumina are commonly present. Other constituents which may be present are iron, manganese, fluorides, nitrates, potassium, and some mine water and surface waters (contaminated with mine drainage or trade wastes) may be acid, usually with sulfuric acid. Dissolved gases which may be present are oxygen, nitrogen, carbon dioxide, hydrogen sulfide and methane.

Sedimentation. With waters containing large amounts of coarse, easily settled, suspended matter, sedimentation (plain sedimentation) is often of value in reducing the load on the filters and effecting economies in amounts of chemicals used for coagulation. Sedimentation may be carried out in sedimentation tanks or basins or in reservoirs. Detention periods vary over a wide range—from a few hours up to one or more months.

Coagulation. Coagulation is employed to form, by cataphoresis and entanglement, larger aggregates with the turbidity, color, microorganisms, and other organic matter present in the water. These larger particles, known as the "floc," may then be removed by filtration through a sand or Anthrafilt filter or by settling and filtration. The coagulant employed is either an aluminum or iron salt, usually the surface. Aluminum sulfate is the most widely used coagulant. Others are ferric sulfate, ferrous sulfate (must be oxidized by air or chlorine) and sodium aluminate. Most favorable pH values for aluminum coagulants usually range from 5.5 to 6.8 and for iron from 3.5 to 5.5 and above

9.0 but there are exceptions. Coagulation aids are ground clay (not too finely pulverized) and activated silica.

Filtration. Filtration is effected by flowing the coagulated or coagulated and settled water downward through a bed of fine filter sand or Anthrafilt is either a pressure type or gravity type filter. Flow rates in industrial practice range up to 3 gpm per sq. ft. (122 liters/min/sq. meter) of filter bed area while in municipal practice maximum flow rate is usually 2 gpm per sq. ft. (81.5 liters/min/sq. meter).

Chlorination. Chlorination is the most widely used disinfecting or sterilizing process. Where daily water requirements are not large, it is common practice to use a hypochlorite, but for large plants liquefied chlorine gas is used. Chlorination may be practiced before filtration (prechlorination), after filtration (post-chlorination), or both before and after.

Taste and Odor Removal. Except for sulfur waters, most tastes and odors are organic in nature. Activated carbon is widely used for their removal. In powdered form, it may be added to the water being treated in coagulation and settling equipment. In such installations, aeration is frequently used as preliminary treatment. In granular form, it is used in filters (activated carbon filters or purifiers). As substances producing tastes and odors are usually extremely small in amount, activated carbon filters are frequently operated for 6 months to one or more years before replacement of bed is necessary.

Hardness Removal (Water Softening). *Sodium Cation Exchanger (Zeolite) process.* This is the most widely used water-softening process in industrial, commercial, institutional and household applications. Hard water is softened by flowing it, usually downward, through a bed (2 feet to over 8 feet in thickness) of a granular or bead type sodium cation exchanger in either a pressure-type (most widely used) or gravity-type water softener. As water comes in contact with sodium cation exchanger, hardness (calcium and magnesium ions) is taken up and held by the exchanger which gives up to the water an equivalent amount of sodium ions. At end of softening run (4 to over 24 hours in industrial practice and 1 to over 2 weeks in household use), softener is cut out of service, regenerated and returned to service ($\frac{1}{2}$ to $1\frac{1}{2}$ hours). Regeneration is effected in 3 steps: (1) backwashing to cleanse and hydraulically regrade the bed (2) salting with specified amount of common salt (sodium chloride) solution, usually 10 to 15% in strength, which removes calcium and magnesium from the exchanger and restores sodium to it and (3) rinsing to remove calcium and magnesium chlorides and excess salt.

Hydrogen Cation Exchanger Process. Calcium, magnesium, sodium and other cations are removed by flowing water (usually downward) through a bed (2 feet to over 8 feet in thickness) in an acid-proof pressure-type (most widely used) or gravity-type shell. As water comes in contact with hydrogen cation exchanger, calcium, magnesium, sodium and other cations are taken up by the exchanger which gives up to the water an equivalent amount of hydrogen ions. At the end of operating run (4 to over 24 hours), unit is cut out of service, regenerated and returned to service ($1\frac{1}{4}$ to 2 hours). Regeneration is effected in 3 steps (1) backwashing to cleanse and hydraulically regrade the bed (2) acid treatment with sulfuric or hydrochloric acid which removes metallic cations from the bed and restores hydrogen to it and (3) rinsing to remove salts (sulfates or chlorides) and excess acid. The carbon dioxide formed from the bicarbonates may be reduced to below 5 to 10 ppm by aeration and the sulfuric and hydrochloric acids formed from chlorides and sulfates may be (1) neutralized with an alkali (usually caustic soda) (2) neutralized by sodium bicarbonate content of a sodium cation exchanger softened water (in which case aeration follows

neutralization) or (3) removed by an anion exchanger.

Cold Lime (or Lime Soda) Process. Chemicals used may be (1) lime plus a coagulant or (2) lime plus soda ash plus a coagulant. Dosages vary according to composition of raw water and result desired such as (a) calcium alkalinity reduction, (b) calcium and magnesium alkalinity reduction, (c) reduction of total hardness without excess chemicals and (d) excess chemical treatment. Precipitates produced are calcium carbonate and magnesium hydroxide. Rated residuals without excess chemicals are 35 ppm for calcium and 33 ppm for magnesium, both expressed as calcium carbonate. Operating results will range between these and theoretical solubilities. With excess chemicals, total hardness may be lowered to 16 ppm. The process is best carried out in the sludge blanket-type of equipment in which the treated water is filtered upward through a suspended blanket of previously formed sludge. Detention periods range from one to two hours. Usually, treated water is filtered before going to service but where small amounts of turbidity are unobjectionable, filters may be omitted.

Hot Lime Soda Process. In this process, treatment with lime and soda ash is carried out at temperatures around the boiling point in closed, steel pressure tanks. Heating is usually accomplished with exhaust steam and pressures most widely used range from 5 to 10 psig, but higher pressures up to but seldom above 20 psig are also used. At these temperatures, the reactions proceed swiftly and precipitates formed are larger than those in cold lime soda process so no coagulant is needed. Detention period is one hour and deaeration effected in primary heater is sufficient to lower dissolved oxygen content to 0.3 milliliter per liter which is sufficient for low-pressure boilers. For high-pressure boilers, either an integral or separate deaerator is used and this will bring the dissolved oxygen down to less than 0.005 ml/1. With 20 to 30 ppm excess soda ash, the hardness will be reduced to 25 ppm. Softening to practically zero hardness may be effected by either (1) two-stage hot lime soda phosphate treatment in which effluent from hot lime soda softener is treated with sodium phosphate or (2) the filtered effluent is passed through a sodium cation exchanger. Anthrafilt filters are usually employed with hot process softeners.

Demineralization (Deionization). Metallic cations are removed by a hydrogen cation exchanger. Anions are removed by an anion exchanger. Depending on the hookup used, the carbon dioxide formed from the bicarbonates may be removed mechanically by an aerator, degasifier or vacuum deaerator, or chemically by a strongly basic anion exchanger. Strongly basic anion exchangers will remove both strongly ionized acids, such as sulfuric and hydrochloric, and weakly ionized acids, such as silicic and carbonic. Weakly basic anion exchangers will remove only strongly ionized acids.

Iron and Manganese Removal. In clear, deep ground waters, iron and/or manganese may occur as soluble, colorless, divalent bicarbonates. These may be removed (1) by oxidation plus settling (if necessary) plus filtration (2) by cation exchange with sodium or hydrogen cation exchangers or (3) filtration through an oxidizing (manganese zeolite) filter. In (1) addition of an alkali or lime may be needed to build up the pH value so as to speed up the oxidation. Iron and/or manganese in organic (chelated) form may usually be removed by coagulation, settling and filtration. In acid waters, these metals may be removed by neutralization (plus increase of pH), aeration, settling and filtration.

Fluoride Removal. Fluorides may be reduced to below 1 ppm by filtration through a bed of a specially prepared, granular bone char (bone black). Regeneration is effected with caustic soda solution followed by treatment with dilute phosphoric acid.

Dissolved Gases. *Oxygen and Nitrogen* may be removed (1) hot in a deaerating heater (deaerator) or (2) cold in a vacuum deaerator. *Carbon dioxide* may be removed in (1) an aerator, (2) a deaerating heater or (3) a vacuum deaerator or it may be neutralized with lime or an alkali or by filtration through a bed of granular calcite. *Hydrogen sulfide* may be removed by (1) aeration followed by chlorination, (2) treatment with flue gas plus aeration followed by chlorination or (3) filtration through an oxidizing manganese zeolite filter (household use). If sulfur content and pH values are high, (1) may effect but little removal but, in some cases, with fairly long detention periods, sulfur bacteria may effect notable reductions.

—Eskel Nordell, Edison, New Jersey.

WATER (Desalination). Desalination.

WATER (Electrolysis). Hydrogen (Fuel).

WATER GAS. Coal; Substitute Natural Gas (SNG).

WATER GAS SHIFT REACTION. Ammonia.

WATER TREATMENT (Boiler). One of the most critical and exacting requirements for pretreating water prior to industrial use is found in connection with the operation of modern boilers. Many of the basic principles of water treatment are encountered for this application.[1] The advantages of modern boilers can be realized to the fullest only if proper attention is given to water treatment. No boiler can operate efficiently or dependably if its heat-transfer surfaces are allowed to foul with scale or if corrosion is permitted to occur.

Water treatment must include conditioning of the:

1. Raw-water supply.
2. Condensate returns from process steam or turbines.
3. Boiler water.

Proper conditioning will result in:

1. Freedom from deposits on internal surfaces.
2. Absence of corrosion of internal surfaces.
3. Prevention of carry-over of boiler water solids into the steam, caused by foaming and/or high total dissolved solids.

Some definitions of the water terminology in various parts of the boiler cycle are desirable. Steam that is condensed and returned to the boiler system is termed condensate. Steam lost due to process requirements, blowdown or leakage out of the system, has to be replaced; the replacement water added to the system is termed makeup water. The condensate together with the makeup water comprise the feedwater to the boiler. In some plants only a small percentage of condensate is returned; in others, almost all the steam generated is recovered as condensate. Feedwater enters the boiler and is evaporated into steam, leaving behind solids to concentrate in the boiler water. If the concentration of solids in the boiler water exceeds certain limits, the quality of steam can be impaired by carry-over. Also, boiler-water solids may settle out on the boiler surfaces as sludge. The concentration of solids in the boiler water can be controlled

---

[1] Abstracted, with permission, from "Steam/Its Generation and Use," 39th edition, Babcock & Wilcox, New York (1978).

by removing a portion of the water either intermittently or continuously. This bleeding of a portion of the boiler water from the drum is termed blowdown.

In Universal-Pressure boilers, there are no drums to concentrate the boiler-water salts and impurities, and blowdown is not utilized. Purification takes place by continuously passing all or part of the condensate through demineralizers in a process called condensate polishing.

The treatment of raw water, condensate, feedwater and boiler water, and the subjects of carry-over and steam purity are considered in detail in the sections which follow.

Raw-Water Treatment

Water never exists in the pure form. All natural waters contain varying amounts of dissolved and suspended matter. The type and amount of matter in water varies with the source, such as lake, river, well or rain, and also with the section of the country.

As rain, water brings into solution the atmospheric gases of oxygen, nitrogen and carbon dioxide. As it percolates through the soil, it dissolves and picks up many minerals harmful to boiler operation. Surface waters frequently contain organic matter that must be removed before the water is satisfactory for use in a boiler.

Suspended solids are those that do not dissolve in water and can be removed or separated by filtration. Examples of suspended solids are mud, silt, clay and some metallic oxides.

Dissolved solids are those which are in solution and cannot be removed by filtration. The major dissolved materials in water are silica, iron, calcium, magnesium and sodium. Metallic constituents occur in various combinations with bicarbonate, carbonate, sulfate and chloride radicals. In solution these materials divide into their component parts called ions, which carry an electrical charge. The metal ions carry a positive charge and are referred to as cations. The bicarbonate, carbonate, sulfate and chloride ions are negatively charged and are referred to as anions.

Scaling occurs when calcium or magnesium compounds in the water (*water hardness*) precipitate and adhere to boiler internal surfaces. These hardness compounds become less soluble as temperature increases, causing them to separate from solution. Scaling causes damage to heat-transfer surfaces by decreasing the heat-exchange capability. The result is overheating of tubes, followed by failure and equipment damage.

Porous deposits will allow concentration of boiler-water solids. This concentration of boiler-water solids, particularly if strong alkalies are present, will result in severe corrosion of the tube surfaces.

Since water impurities cause boiler problems, careful consideration must be given to the quality of water in the boiler. External treatment of water is required when the amount of one or more of the feedwater impurities is too high to be tolerated by the boiler system.

The selection of equipment for raw-water preparation should only be made after a careful analysis of the raw-water composition, quantity of makeup required, boiler type and operating pressure. Generally, the first step in the water processing involves coagulation and filtration of the suspended material. Natural settling in quiescent water will remove relatively coarse suspended solids. The required settling time depends on specific gravity, shape and size of particles and currents within the settling basin. This process can be speeded up by coagulation. Coagulation is the process by which finely divided materials are combined by the use of chemicals to produce large particles capable of rapid settling. Typical coagulant chemicals are alum and iron sulfate. The preliminary treatment involves chlorination of the water for the destruction of the organic matter. Several manufacturers offer equipment to operate on a completely automated basis, as illustrated in Fig. 1.

Following coagulation, settling and chlorination, the water should be passed through filters. Filtration removes the finely divided suspended particles not removed in the coagulation and settling tanks. Special equipment such as activated-charcoal filters may be necessary to remove the final traces of organic and excess chlorine. After the removal of the suspended material in the raw water, the hardness of scale-forming materials are still present in solution. Further treatment is required to remove these materials. This treatment consists of precipitating the hardness constituents and/or exchanging the hardness for non-hardness constituents in a process called ion exchange. Brief descriptions of each of these processes follow. It is recommended that assistance by obtained from a water consultant in order to select the best process and equipment for a specific installation.

Sodium-Cycle Softening. This process, called sodium zeolite softening, utilizes resin materials that have the property of exchanging the hardness constituents, calcium and magnesium, for sodium. The process continues until the sodium ions become depleted or, conversely, the resin capacity to absorb the calcium and magnesium no longer exists. When this occurs, the resin is said to be exhausted, and is regenerated by passing a solution of salt through it.

Water, after passing through the zeolite process, contains as much bicarbonate, sulfate and chloride as the raw water, only the calcium and magnesium having been exchanged for the sodium ions. There is no reduction in the overall amount of dissolved solids and neither is there a reduction of alkalinity content. When it is necessary to reduce the amount of total dissolved solids, zeolite must be coupled with other methods such as hot lime zeolite softening. The reduction of alkalinity is discussed under *Hot Lime Zeolite–Split Stream Softening.*

Hot-Lime Zeolite Softening. In this process hydrated lime is employed to react with the bicarbonate alkalinity of the raw water. The precipitate is calcium carbonate and is filtered from the solution. To reduce silica, the natural magnesium of the raw supply can be precipitated as magnesium hydroxide, which acts as a natural absorbent for silica. These reactions are carried out in a vat or tank that is located just head of the zeolite softener tank. The effluent from this tank is filtered and then introduced into the zeolite softener. There is always some residual hardness leakage from the hot-process softener to be removed in the final zeolite process. The hot lime process operates at about 220°F (104°C). At this temperature the potential for the exchange of sodium for hardness ions is greater than at ambient temperature, and the result is a lower hardness effluent than is achieved at ambient temperatures. This system is shown schematically in Fig. 2.

Hot Lime Zeolite–Split Stream Softening. Many raw waters softened by the first two processes would contain more sodium bicarbonate than is acceptable for boiler feedwater purposes. Sodium bicarbonate will decompose in the boiler water to give caustic soda. Caustic soda in high concentrations is corrosive and promotes foaming. The American Boiler Manufacturers Association has adopted the standard that the alkalinity content should not exceed 20% of the total solids of the boiler water. Split stream softening provides a means for reducing the alkalinity content.

This requires a second zeolite tank that has a zeolite resin in the hydrogen form in addition to the usual tank with the resin in the sodium form. The two tanks are operated in parallel. In one tank, calcium and magnesium ions are replaced by hydrogen ions. The effluent from this tank with the resin in hydrogen form is on the acid side and has a lower total-solids content. The total flow can be proportioned between the two tanks to produce an effluent with any desired alkalinity as well as excellent hardness removal. When the hydrogen resin is exhausted, it is regenerated with acid.

Demineralization and Evaporation. At drum pressures over 1000 psi (68 atm), demineralization or evaporation of the makeup water is generally desirable. A water that closely approaches theoretical chemical purity can be obtained by either of these processes.

Evaporation as a source of purified water does not involve ion exchange. It is actually a distillation process, consisting of evaporation, leaving most of the solids behind, and recondensation of the purified water. While evaporated water is quite satisfactory, economics generally favors demineralization.

Demineralization, like the zeolite process, involves ion exchange. The metal ions are replaced with hydrogen ions by means of the process and equipment described for the hydrogen-zeolite system (see *Hot Lime Zeolite—Split Stream Softening,* previously described). In addition, the salt anions (bicarbonate, carbonate, sulfate and chloride) are replaced by hydroxide ions

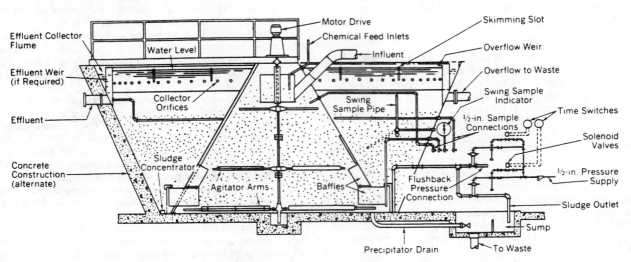

Fig. 1. Sludge contact softener. (*Betz "Handbook of Industrial Water Conditioning."*)

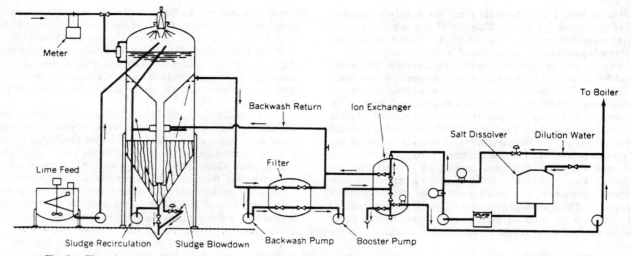

Fig. 2. Flow sheet of typical hot lime zeolite softening process. (*Betz "Handbook of Industrial Water Conditioning."*)

by means of a specially prepared resin saturated with hydroxide ions.

The two types of resins can be located in separate tanks. In this system, the two tanks are operated in series in a cation-anion sequence. The anion resin is regenerated after exhaustion with a solution of sodium hydroxide. The cation resin is regenerated with an acid, either hydrochloric or sulfuric. Some leakage of cations always occurs in a cation exchanger, resulting in leakage of alkalinity from the anion exchanger.

In another arrangement, known as the mixed-bed demineralizer, the two types of resins are mixed together in a single tank. In the mixed-bed demineralizer, cation and anion exchanges take place virtually simultaneously, resulting in a single irreversible reaction that goes to completion. Regeneration is possible in a mixed bed because the two resins can be hydraulically separated into distinct beds. The cation resin is approximately twice as dense as the anion resin. Resins can be regenerated in place or sluiced to external tanks for this purpose.

The raw water for drum-type boilers operating above 2000 psi (136 atm) drum pressure and for once-through units should be prepared by passing water through a mixed-bed demineralizer as a final step before adding to the cycle.

The effluent from demineralization is approximately neutral. With nearly all salts removed, the problem of chemical control of the boiler water is minimized.

Treatment of Condensate

In most cases, condensate does not require treatment prior to reuse. Makeup water is added directly to the condensate to form boiler feedwater. In some cases, however, especially where steam is used in industrial processes, the steam condensate is contaminated by corrosion products or by the in-leakage of cooling water or substances used in the process. Hence steps must be taken to reduce corrosion or to remove the undesirable substances before the condensate is recycled to the boiler as feedwater.

The presence of acidic gases in steam makes the condensate acidic with consequent corrosion of metal surfaces. In such cases, the corrosion rate can be reduced by feeding to the boiler water chemicals that produce alkaline gases in the steam. The addition of neutralizing and filming amines to boiler water or to condensate to minimize corrosion by condensate and feedwater is discussed later under *Control of* pH.

Many types of contaminants can be introduced to condensate by various industrial processes. They include liquids, such as oil and hydrocarbons, as well as all sorts of dissolved and suspended materials. Each installation must be studied for potential sources of contamination. The recommendations of a water consultant should be obtained to assist in determining corrective treatment.

Fig. 3 shows a condensate purification system used in a paper-

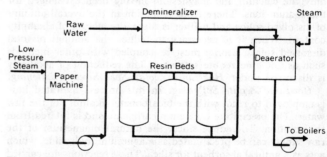

Fig. 3. Condensate purification system. (*Cochrane Div., Crane Company.*)

mill boiler cycle. The resin beds not only remove dissolved impurities by ion exchange but also serve as filters to remove suspended solids. It is necessary to backwash and regenerate these resin beds periodically. Several types of condensate purification systems are available from various vendors. Some of these are capable of operation at temperatures as high as 300°F (149°C).

One such example of a condensate purification system is the use of ion exchange equipment as shown in Fig. 3 for a typical paper-mill boiler cycle. The resin beds not only remove dissolved impurities by ion exchange, but also serve as filters to remove suspended solids, such as products of corrosion. It is necessary to backwash and regenerate such resin beds periodically. These resin beds can be purchased for in-place regeneration or for regeneration in external tanks. External regeneration facilitates more efficient removal of suspended metal oxides from resin beds.

Where significant quantities of magnetic oxide or other magnetic species are present in the condensate-feedwater system, the application of an electro-magnetic filter (EMF) to effectively remove these suspended solids has been proved to be quite successful in both operation and performance. The EMF, Fig. 4,

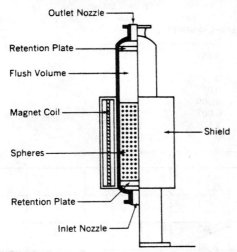

Fig. 4. Sectional view of electromagnetic filter (EMF).

consists of a pressure vessel, coil, spheres, and a power control unit. The pressure vessel is constructed of a non-magnetic material and contains a featherbed of magnetizable spheres of approximately $\frac{1}{4}$-in. (6.5-mm) diam. The pressure vessel is surrounded by a magnetic coil which is supplied with direct current from the power control unit. Flow is upward through the filter both during operation and when flushing the filter, thereby minimizing and simplifying the piping and valving system. Some other advantages that the electro-magnetic filter offers are low pressure drop through the filter, minimum quantity of flush water, entire backwash process only takes several minutes and no chemicals are required in its use.

Laboratory analysis of flush water indicates that some non-magnetic iron oxide may also be retained by the EMF. The non-magnetic iron removal is believed to be due to the presence of magnetic/non-magnetic composite particles, which are magnetically attracted by the filter.

Condensate-Polishing Systems. Demineralizer systems, installed for the purpose of purifying condensate, are known as condensate-polishing systems. A condensate-polishing system is a requisite to maintain the purity required for satisfactory operation of once-through boilers (see *Water Treatment for Universal-Pressure Boilers*). High-pressure drum-type boilers (over 2000 psi; 136 atm) can and do operate satisfactorily without condensate polishing. However, many utilities recognize the benefits of condensate polishing in high-pressure plants, including:

1. Improved turbine capability and efficiency.
2. Shorter unit start-up time.
3. Protection from the effects of condenser leakage.
4. Longer intervals between acid cleanings.

Two types of condensate-polishing systems are available, both capable of removing suspended material, such as corrosion products, as well as ionized solids.

Deep-bed demineralizers operate at flow rates of 40–60 gpm per sq ft (1624–2441 liters/minute/square meter) of bed cross section. This type requires external regeneration facilities. The deep-bed system has a higher initial cost with possible lower operating costs, especially during initial unit start-up. Its greater

capacity for removing ionized solids permits continued operation with small amounts of condenser leakage. The deep-bed system is usually operated with the cation resin in the hydrogen form, but the ammonium form can also be used. The cost of regenerating the resin in the ammonium form is greater than for the hydrogen form, but the time period between regenerations is much longer.

The cartridge-tubular type, such as the Powdex system, uses smaller amounts of disposable resins, eliminating the need for regeneration. The Powdex system uses cation resin in the ammonium form. Because of the many considerations involved, an evaluation of alternate types should be made before a system is selected for any given installation.

Use of the EMF with ion exchange equipment in a condensate purification system provides the ultimate in condensate polishing. Location of the EMF upstream of the resin beds offers the advantages of longer operating periods for the beds, thereby reducing the frequency of bed regeneration, and, subsequently, lesser costs for regeneration chemicals.

Treatment of Feedwater

The following discussion outlines what is required to produce and maintain the quality of feedwater recommended in Table 1.

The pre-boiler equipment, consisting of feedwater heaters, feed pumps and feed lines, is constructed of a variety of materials, including copper, copper alloys, carbon steel, and phosphor bronzes. To reduce corrosion, the makeup and condensate must be at the proper pH level and free of gases such as carbon dioxide and oxygen. The optimum pH level is that which introduces the least amount of iron and copper corrosion products into the boiler cycle. This optimum pH level should be established for each installation. It generally ranges between 8.0 and 9.5

Control of pH. The control of corrosion in the condensate system is generally accomplished by adding one of the following chemicals:

1. Neutralizing amines–ammonia, morpholine, cyclohexylamine and hydrazine.
2. Filming amines–octadecylamine acetate.

Neutralizing amines are volatile alkalizers that distill with the steam and neutralize acids that form in the condensate. Hydrazine, which is also an excellent oxygen scavenger, is included with the volatile alkalizers. It decomposes in the boiler, forming ammonia, hydrogen and nitrogen. The ammonia provides pH control in the condensate. The use of hydrazine as an oxygen scavenger is discussed later under *Chemical Scavenging of Oxygen by Hydrazine*.

Selection of the proper alkalizer should be considered for each plant to minimize the pickup of iron and copper in the condensate and feedwater. Optimum conditions should be established by tests during the early operation of the unit. Factors to be considered in selecting the alkalizer are steam temperature, makeup requirements, and carbon-dioxide concentration in the condensate.

Condensate corrosion rates are increased by the partial pressure of carbon dioxide in the steam. Carbon dioxide originates in the breakdown of carbonates in the boiler water. If steam has a high carbon-dioxide content, filming amines should be considered as a corrective. Filming amines reduce corrosion rates by forming a protective coating on the surfaces contacted by the steam and condensate. Since this is a surface phenome-

TABLE 1.   RECOMMENDED LIMITS OF SOLIDS IN BOILER FEEDWATER

DRUM PRESSURE	BELOW 600 psi (41 atm)	600 TO 1,000 psi (41–68 atm)	1,000 TO 2,000 psi (68–136 atm)	OVER 2,000 psi (136 atm)
Total solids, ppm			0.15	0.05
Total hardness (as ppm CaCO$_3$)	0	0	0	0
Iron, ppm	0.1	0.05	0.01	0.01
Copper, ppm	0.05	0.03	0.005	0.002
Oxygen, ppm	0.007	0.007	0.007	0.007
pH	8.0–9.5	8.0–9.5	8.5–9.5	8.5–9.5
Organic	0	0	0	0

non, the amount of metal surface to be protected is more important economically than the concentration of gas in the steam. Control of filming amine feed rate is critical. The protective film will not form if the feed is insufficient while excessive feed of this waxlike substance can plug the flow passages of the equipment.

In units with a high percentage of makeup feed, it may not be necessary to add chemicals specifically for pH control. When the makeup is treated by the lime-soda process or a lime-soda-zeolite system, the effluent is normally within the recommended pH range or slightly higher. If pH exceeds the recommended limits, the high alkalinity may lead to foaming and carry-over. For corrective action, see *Hot Lime Zeolite—Split Stream Softening*, previously discussed.

Control of Oxygen. The presence of gases, particularly oxygen, leads to corrosion of the boiler and cycle equipment. This type of attack will occur in an operating boiler as well as an improperly stored idle boiler. The consequent effect of dissolved oxygen in feedwater is pitting of the internal surfaces. This is most prevalent in the economizer, the steam drum and the supply tubes. The pitting may be general or selective. In either case, if allowed to proceed unchecked, it will adversely affect the reliability of the unit and shorten its service life.

The most logical approach to the prevention of corrosion by gases is to expel them from the system at the first opportunity. The usual method is by means of a deaerating heater. This equipment must be kept in prime operating order over the complete load range. If the deaerator operates under vacuum at low loads, the entrance of air must be prevented. Oxygen concentrations at the deaerator outlet should be consistently less than 0.007 ppm. As a further assurance against the destructive effect of dissolved oxygen, a residual quantity of an oxygen-scavenging compound should be maintained in the system.

Chemical scavenging of oxygen by sodium sulfite. Most operators use sodium sulfite for the chemical scavenging of oxygen. Fig. 5 shows the recommended sulfite concentration as a function of boiler pressure. The amount of sulfite that can be safely carried decreases as pressure increases. At the high temperatures associated with the higher pressures, sulfite decomposes into acidic gases that can cause increased corrosion. Consequently, sulfite should not be used at pressures greater than 1800 psi. On boilers having spray attemperation, the sodium sulfite should be added after the attemperator take-off point. About 8 ppm is required to remove 1 ppm of oxygen.

Chemical scavenging of oxygen by hydrazine. Hydrazine is an alternate scavenger, offering two principal advantages:

1. The decomposition and dissolved-oxygen reaction products of hydrazine are volatile. Consequently, they do not

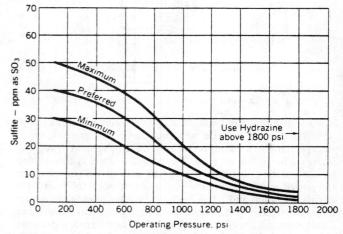

Fig. 5.   Usually recommended sulfite residual in boiler water.

increase the dissolved-solids content of the boiler water, nor do they cause corrosion where steam is condensed.
2. Experience has shown that condensate pH will usually stabilize in the range of 8.5–9.5 if a 0.06-ppm hydrazine residual is maintained at the boiler inlet. This eliminates the need for pH treatment of the condensate-feedwater.

It is apparent that a residual of 0.06 ppm of hydrazine will provide only a limited protection against oxygen entering the boiler. Thus it is not practicable to utilize this scavenger as a substitute for an airtight system.

Changes in Feedwater Treatment. Any changes in feedwater treatment or boiler water conditions can have troublesome results. Changes should therefore be made gradually and with close observation. For instance, if sulfite treatment is to be replaced by hydrazine, initial dosage should be small and changes in the iron and copper concentration in the feedwater should be carefully monitored. If iron and copper concentrations in the feedwater and boiler water increase significantly, load should be reduced and blowdown increased. It may require days or weeks for conditions to stabilize, so results must be observed and evaluated over a significant period.

Treatment of Boiler Water for Natural-Circulation Units

Direct treatment of boiler water, usually referred to as internal treatment, is used (1) to prevent scale formations caused by hardness constituents, and (2) to provide pH control to prevent corrosion. Treatment that is incorrect or inadequate in either respect can lead to tube failures and result in costly unscheduled outages. The permissible limits on contaminants entering the boiler and also on treatment chemicals that can be added to the boiler decrease with rising boiler pressures.

Fig. 6 shows the relationship between dissolved solids in boiler

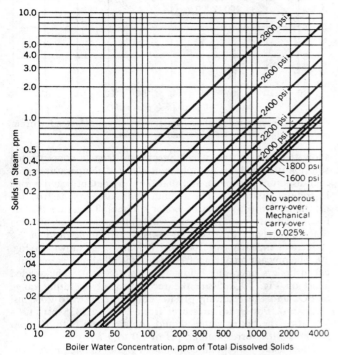

Fig. 6. Solids in steam versus dissolved solids in boiler water.

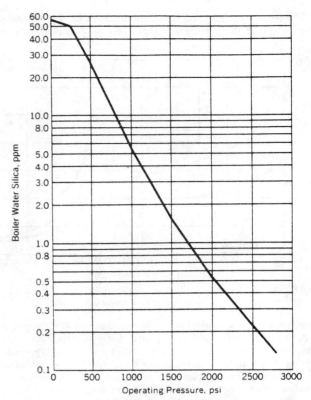

Fig. 7. Recommended maximum silica concentration in boiler water at pH 9.5 (drum-type boilers).

water and solids in steam at various drum operating pressures. This correlation agrees reasonably well with both laboratory and field data. If a boiler-water total-solids concentration of 15 ppm is assumed, Fig. 6 indicates that 15 ppb solids would be expected in the steam at 2400 psi (163 atm) drum pressure, while at 2800 psi (190 atm) drum pressure about 75 ppb would be expected in the steam. Fig. 7 indicates the great reduction in silica concentration in boiler water that must occur as pressures increase if silica in the steam is to be limited to 20 ppb. Experience has demonstrated that a concentration of 20 ppb will pass through the superheater and turbine without deposition. This curve is valid for a boiler water pH of 9.5. At the higher pressures, boiler water additives must be reduced to low levels in order to avoid deposits on turbine parts.

There are four methods of internal treatment in common use on natural-circulation drum-type boilers:

1. Phosphate-hydroxide (conventional-treatment).
2. Coordinated phosphate.
3. Chelant.
4. Volatile.

The method of treatment is generally dictated by the pressure range of the unit.

Methods 1 and 2 are intended to control the boiler water pH and to precipitate the calcium and magnesium compounds as a flocculent sludge, so that they can be removed in the boiler blowdown rather than being deposited on heat-transfer surfaces.

Method 1 maintains an excess of hydroxide alkalinity. The effects of alkalinity are discussed later under *Steam Purity*. Method 3 involves the addition of a complex metal-chelant compound such as ethylenediamine-tetraacetic acid (Na_4EDTA) or nitrilotriacetic acid (NTA). In Method 4, as the name implies, no solid chemicals are added to the boiler or pre-boiler cycle. The pH of the boiler water and condensate cycle is controlled by adding a volatile amine.

In Methods 1 through 3, either sulfite (up to 1800 psi; 122 atm) or hydrazine can be used as the oxygen scavenger. Above 1800 psi (122 atm), or with the volatile treatment (Method 4), hydrazine is used.

Phosphate-Hydroxide (Conventional-Treatment) Method. This is the most prevalent method of treatment for industrial boilers operating below 1000 psi (68 atm). It involves the addition of phosphate and caustic to the boiler water. Caustic is added in sufficient quantity to maintain a pH of 10.5 to 11.2. A boiler treated with caustic and phosphate is less sensitive to upsets than with other methods of feedwater control.

The primary purpose of phosphate addition is to precipitate the hardness constituents. The calcium reacts with phosphate under the proper pH conditions to precipitate calcium phosphate as calcium hydroxyapatite, $Ca_{10}(PO_4)_6(OH)_2$. This is a flocculent precipitate that tends to be less adherent to boiler surfaces than simple tricalcium phosphate, which is precipitated below 10.2 pH. Caustic reacts with magnesium to form magnesium hydroxide or brucite, $Mg(OH)_2$. This precipitate is formed in preference to magnesium phosphate at a pH above 10.5 as it is less adherent.

The recommended phosphate concentration for a given boiler operating pressure is shown in Fig. 8. At the higher pressures,

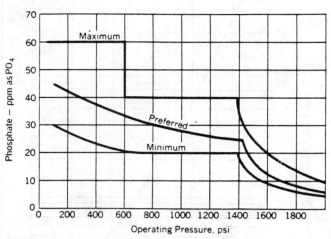

Fig. 8.   Recommended phosphate concentration in boiler water at various boiler operating pressures (phosphate-hydroxide treatment).

comparatively low phosphate residuals must be maintained in order to avoid appreciable phosphate "hideout." Hideout is the term used to identify the phenomenon of the temporary disappearance of phosphate in the boiler water upon increase in load and its reappearance upon load reduction. The recommended alkalinity as a function of pressure is given in Fig. 9.

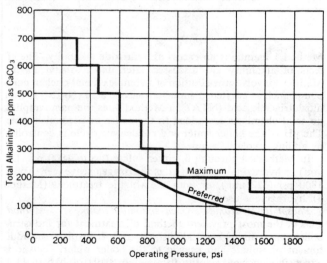

Fig. 9.   Recommended alkalinity of boiler water at various boiler operating pressures (phosphate-hydroxide treatment).

Phosphate hideout does not appear to be as important below 1500 psi (102 atm) and even at this pressure, phosphate concentrations of 12 to 25 ppm as PO_4 can be carried without appreciable hideout. Either sulfite or hydrazine may be used to scavenge oxygen.

Coordinated Phosphate Method. In this method of treatment, no free caustic is maintained in the boiler water. Fig. 10 shows the phosphate concentration versus the resulting pH when trisodium phosphate is dissolved in water. Recent laboratory tests

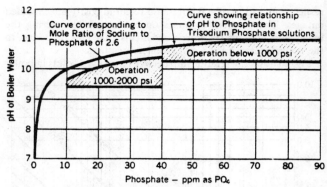

Fig. 10.   Recommended phosphate content of boiler water for drum boilers, using coordinated phosphate treatment.

show that the crystals which precipitate from a concentrated solution of trisodium phosphate at elevated temperatures contain disodium phosphate and that the supernatant liquid is rich in sodium hydroxide. The sodium hydroxide can destroy the magnetite protective film on boiler surfaces. To assure that no free caustic is present, a boiler-water phosphate concentration that corresponds to a sodium-to-phosphate mole-ratio of 2.6 is recommended above 1000 psi (68 atm), as shown in Fig. 10. The precaution against free hydroxide alkalinity is less critical in boilers operating below 1000 psi (68 atm). The shaded areas on Fig. 10 indicate the recommended operating range of PO_4 and the resulting pH for boiler pressures to 2000 psi (136 atm). For drum pressures from 2000 to 2835 psi (136–192 atm) the boiler water should contain from 3 to 10 ppm Na_3PO_4 with corresponding pH of 9.0 to 9.7.

When using the regular commercial grades of chemicals, caution should be used in calculating the weights needed to provide the proper mole ratios. Commercial phosphates commonly are in the form of $Na_3PO_4 \cdot 12H_2O$ and $Na_2HPO_4 \cdot 7H_2O$. A mixture of 65% $Na_3PO_4 \cdot 12H_2O$ and 35% $Na_2HPO_4 \cdot 7H_2O$ corresponds to a mole ratio of Na to PO_4 of 2.6. If the pH is too low, it may be corrected by increasing the ratio of trisodium to disodium phosphate. If the pH is too high, the ratio should be decreased.

Use of Chelants. This method of water treatment has become popular in recent years with industrial boiler operators. These organic agents react with the residual divalent metal ions, calcium, magnesium and iron, in the feedwater to form soluble complexes. The resultant soluble complexes are removed through continuous blowdown. This method of treatment has been used in boilers operating as high as 1500 psi (102 atm) although present B&W recommendations limit its use to units below 1000 psi (68 atm).

Certain precautions are necessary in using this treatment. The chelating agents do not chelate ferric iron or copper. The presence of chelating agents and oxygen together, in the boiler or pre-boiler cycle, must be avoided. During operation, deaeration must be good at all times, and measures must be taken to protect the boiler from oxygen at all times during off-line periods.

Experience indicates that it is difficult to control chelant feed based on chelant residual in the boiler water. Excess chelant will attack clean boiler surfaces. B&W therefore recommends that chelant feed be based on known quantities of hardness and iron present in the feedwater, with the objective of maintaining a residual approximating 1 ppm of chelant in the boiler water. To protect the boiler from upsets resulting from heat-exchanger leakage or makeup plant overrun, a phosphate residual of 15 to 30 ppm should be maintained in the boiler water. The boiler internal surfaces should be inspected whenever opportunity permits. If sludge deposits accumulate, the chelant feed should be increased by 1 to 2 ppm. If the boiler is found to be exceptionally clean and shiny surfaces are in evidence, the chelant feed should be decreased. A light gray dust on the internal boiler surfaces appears to characterize the ideal condition.

For handling chelants, chemical feed piping must be made of stainless steel or some other corrosion-resistant material.

Volatile Treatment. This method of treatment may be used for units operating above 2000 psi (136 atm) drum pressure. In this method, no solid chemicals are added to either the boiler or pre-boiler cycle. By eliminating solid treatment, the volatile carry-over of solids is eliminated and consequently turbine deposits are avoided. Cycle pH is controlled at 9.0 to 9.5 with a volatile amine such as ammonia. Hydrazine is added as an oxygen scavenger in quantity sufficient to provide a concentration of 20 to 30 ppb at the economizer inlet.

With volatile treatment, the feedwater must not contain hardness of condenser-leak constituents. Since no phosphate is present to remove hardness, any contamination assumes major importance. Prompt detection and remedial action is required. Failure to take such action endangers the future availability of the unit. A condensate-polishing system in the cycle is the best insurance against condenser leakage and hardness constituents.

Steam Purity. The trend toward higher pressures and temperatures in steam power plant practice imposes a severe demand on steam-purification equipment for elimination of troublesome solids in the steam. Carry-over may result from ineffective mechanical separation and from the vaporization of boiler-water salts. Total carry-over is the sum of the mechanical and vaporous carry-over of all impurities.

Mechanical carry-over is the entrainment of small droplets of boiler water in the separated steam. Since entrained boiler-water droplets contain solids in the same concentration and proportions as the boiler water, the amount of impurities in steam contributed by mechanical carry-over is the sum of all impurities in the boiler water multiplied by the moisture content of the steam. Foaming of the boiler water results in gross mechanical carry-over. The common causes of foaming are excessive boiler-water solids, excessive alkalinity or the presence of certain forms of organic matter, such as oil.

Maintaining dissolved solids at the level required to prevent foaming requires continuous or periodic blowdown of the boiler. Table 2 gives the recommended total solids concentration for

TABLE 2. LIMITS FOR TOTAL SOLIDS CONTENT IN BOILER WATER (Drum Boilers)

DRUM PRESSURE		TOTAL SOLIDS (ppm)
(psi)	(atm)	
0–300	0–20.4	3500
301–450	20.5–30.6	3000
451–600	30.7–40.8	2500
601–750	40.9–51.0	2000
751–900	51.1–61.2	1500
901–1000	61.3–68.0	1250
1001–1500	68.1–102.0	1000
1501–2000	102.1–136.1	750
over 2000	over 136.1	15

the prevention of excessive carryover at various operating pressures. Most operators find it convenient and advisable to run well below these limits. Exceeding them may endanger the superheater, the turbine, or the process application.

High boiler-water alkalinity tends to increase carryover, particularly in the presence of an appreciable quantity of suspended matter. This effect may be corrected by various methods, dependent on the cause of the high alkalinity. For example, if trisodium phosphate is being added to the boiler water, a less alkaline phosphate, such as disodium or monosodium phosphate will help in reducing alkalinity.

The presence of oil in boiler water is intolerable, as it causes foaming and carry-over. Steps should be taken to prevent its entry into the feedwater system or leakage through pressure seals and joints. Organic antifoaming agents are a recent development with some successful application. However, their use should not be considered a cure-all.

Spray water for use in a spray attemperator should be of the highest quality. Solids entrained in the spray water enter the steam and can cause troublesome deposits on superheater tubes and turbine blades.

Carry-over of volatile silica is generally a problem only at pressures of 1000 psi (68 atm) or above, although it can be encountered at pressures as low as 600 psi (41 atm). For the protection of the turbine, it is important that silica carryover be prevented in this pressure range by adherence to the silica limits of Fig. 7.

The prevention of vaporous carry-over is much more difficult than the correction of mechanical carry-over. The only effective method is to reduce the solids concentration in the boiler water.

Controls for Water Conditioning

The safe and efficient operation of boilers at pressures over 1000 psi (68 atm) requires continuous monitoring of the water conditioning system. Early detection of any contamination entering the system is essential, so that immediate corrective action can be taken before the boiler and its related equipment are damaged.

Electrical conductance, the reciprocal of resistance, affords a rapid means of checking for contamination in a water sample. Electrical conductance of a water sample is the measure of its ability to conduct an electric current. It can be related to the ionizable dissolved solids in the water. A single instrument will measure and record important conductivities of the water from as many as twenty different locations in the system. The electrical conductivity signal can be used to actuate alarm systems or to operate equipment in the water system. The micromho $(1 \times 10^{-6}$ mho) is normally the unit of measurement. For most salts in low concentrations, 2 micromhos is equal to 1 ppm concentration when corrected to 77°F (25°C).

Ammonia or amines used for pH control affect the conductivity. To obtain an accurate indication of solids, a cation ion exchanger removes the volatile alkalizers and converts the salts to their corresponding acids. Seven micromhos are equivalent to 1 ppm concentration for most salts.

For boilers with operating pressures over 1000 psi (68 atm) cation conductivity of the condensate should normally run between 0.2 and 0.5 micromhos. A reading above this limit indicates the presence of condenser leakage or contamination from some other source. The source of the contamination should be investigated and remedied at the first opportunity. However, when a cation conductivity limit of 1.0 is reached, the internal water treatment and blowdown must be changed appropriately.

Dissolved oxygen should be monitored at the condensate pump discharge and the deaerator outlet. Sulfite or hydrazine can be used for oxygen scavenging. Over 1800 psi (122 atm) drum pressure, sulfite should not be used; only hydrazine is recommended. Sulfite or hydrazine can be added to the condensate on a manual or automated basis.

Feedwater pH is monitored at the condensate pump discharge and the economizer inlet. Chemical-injection pumps are usually adjusted manually to maintain the proper pH for the conventional and coordinated phosphate water-treatment systems. Where volatile water treatment is used, pH can be controlled automatically by using conductivity to transmit signals to the ammonia injection pumps. It is generally preferable to use conductivity rather than pH to transmit signals to the ammonia pumps. Conductivity equipment has been found to be more reliable for this purpose and the linear, rather than logarithmic relationship to concentration, enables better control. Ammonia should be added at the hotwell effluent, or, if condensate polishing is used, at the effluent of the demineralizing system.

Hydrogen should be monitored at the economizer inlet and the superheater outlet. A hydrogen analyzer-recorder can actuate an alarm when the hydrogen concentration of the feedwater or steam deviates from the safe value, which is specified for the plant. Deviation from the normal hydrogen concentration can indicate that corrosion is taking place within the water-steam system.

Automated equipment is commercially available for the continuous on-stream analysis of the critical constituents of the boiler water, such as hardness, phosphate, iron, copper and silica. Most laboratory analytical procedures that depend on the development of a color, and then measuring the intensity of that color to indicate the concentration of the constituent in the water sample, can be put on an automatic basis.

Water Treatment for Universal-Pressure Boilers

Satisfactory operation of the once-through boiler and associated turbine requires that the total solids in the feedwater be

less than 0.05 ppm. Table 3 lists recommended maximum limits for feedwater contaminants and typical values obtained during operation.

TABLE 3.   RECOMMENDED LIMITS OF SOLIDS IN FEEDWATER FOR UNIVERSAL-PRESSURE BOILERS

	MAXIMUM LIMIT	TYPICAL CONCENTRATIONS
Total solids	0.050 ppm	0.020 ppm
Silica as SiO_2	0.020 ppm	0.002 ppm
Iron as Fe	0.010 ppm	0.003 ppm
Copper as Cu	0.002 ppm	0.001 ppm
Oxygen as O_2	0.007 ppm	0.002 ppm
Hardness	0.0 ppm	0.0 ppm
Carbon dioxide	0.0 ppm	not measured
Organic	0.0 ppm	0.002 ppm
pH	9.2–9.5	9.45

Recommended limits should be low because all solids in the feedwater will either deposit in the boiler or be carried over with the steam to the turbine. Consequently, water-treatment chemicals must be volatile. All cycles should have condensate-polishing systems to meet the limits shown in Table 3. A schematic diagram is shown in Fig. 11. Laboratory tests as well as field studies show that high-flow-rate condensate-polishing systems [25 to 50 gal per min per sq ft (1015–2030 liters/minute/square meter) of cross-sectional bed area] perform as filters of suspended material and ionized particles. Ammonia is added to control the pH in the system. Fig. 12 indicates the amount of ammonia required, in terms of ppm or solution conductivity, to give a certain pH in the system. Hydrazine is added to the cycle for oxygen scavenging.

Most of the iron entering the boiler originates in the condensate-feedwater cycle downstream of the polishing demineralizers or in the shell side of feedwater heaters where drips bypass

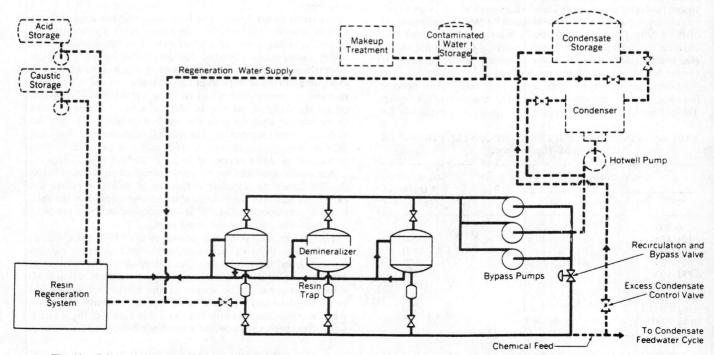

Fig. 11.   Schematic diagram of condensate-polishing system with high-quality makeup treatment (four-bed ion exchange or equivalent).

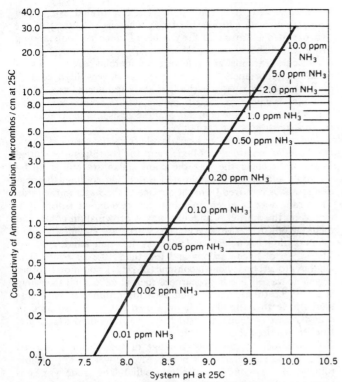

Fig. 12. Theoretical relationship between conductivity and pH for ammonia solutions.

the polishing demineralizers. Studies on a number of installations with carbon-steel feedwater heaters have shown that iron pickup can be minimized by operating with feedwater pH in the range of 9.3 to 9.5. The best pH for minimizing iron pickup should be determined for each cycle during the first several months of operation.

Ammonia is injected downstream of the condensate polishers and controlled from a sample taken far enough downstream of the injection point to assure good mixing. Hydrazine is generally fed at the exit of the condensate-polishing system and/or at the boiler feed-pump suction. Automatic controls are available to regulate the positive displacement pumps that meter ammonia and hydrazine introduction. The signal to the pump-controller for ammonia usually comes from a specific conductivity-recording instrument that is compensated for temperature changes of the cycle water. Hydrazine feed is frequently automatic, utilizing an analyzer and controller. Hydrazine residuals of 10–20 ppb are normally maintained at the boiler inlet.

Prior to plant start-up, either initially or after long outages, water must be circulated through the condensate-polishing system to reduce the dissolved material and suspended particles. The cation conductivity of the cycle water must be reduced to less than 1.0 micromho before a fire is lighted in the unit. Temperatures are not allowed to exceed 550°F (288°C) at the convection pass outlet until the iron levels are less than 100 ppb at the economizer inlet. Cation-conductivity and suspended-iron requirements are generally met after 4 to 5 hours of circulation with cycles having a bypass arrangement.

Many units have instrumentation that will trip the unit in case of excessive feedwater contamination. Trip-limit recommendations are based on the measurement of cation conductivity at the boiler inlet. An actual unit trip is usually preceded by alarms at the hotwell discharge to warn the operator of feedwa-ter contamination and possible load reduction. In setting feedwater trip limits, protection of both the boiler and the turbine must be considered. A common arrangement consists of two cation conductivity alarm devices, both required to read high to initiate the trip sequence. A conductivity of 2 micromhos for five minutes or 5 micromhos for two minutes, results in a unit trip. Properly installed and maintained, these trip devices are highly reliable.

WAVELLITE. The mineral wavellite is a hydrous phosphate of aluminum, formula $Al_3(PO_4)_2(OH)_3 \cdot 5H_2O$. It is orthorhombic but crystals are of rare occurrence as it is ordinarily found in crusts or radial aggregates, sometimes fibrous. Its hardness is 3.25–4; soecific gravity, 2.36; may be of various colors, gray, blue, green, yellow, black, or colorless. It has a vitreous luster, and is translucent. This mineral is of secondary origin, probably formed by waters bearing phosphoric acid which have acted on aluminum minerals. Wavellite is found in Saxony, Bavaria, Devonshire, from where it was originally described; and in the United States in Chester and Cumberland Counties, Pennsylvania; and Montgomery and Garland Counties, Arkansas. It was named after its discoverer, Dr. Wavel.

WAXES. The English term *wax* is derived from the Anglo-Saxon *weax*, which was the name applied to the natural material gleaned from the honeycomb of the bee. In modern times the term *wax* has taken on a broader significance, as it is generally applied to all waxlike solids, natural or synthetic, and to liquids when they are composed of monohydric alcohol esters. Unlike the ordinary oils of animal and vegetable origin and the animal tallows, the waxes, with a few exceptions, are free from glycerides, which are common constituents of oils and fats. Bayberry wax is a vegetable tallow which happens to have all the physical characteristics of a wax, and has always been classed as such.

Animal and Vegetable Waxes. The most important insect wax from an economic viewpoint is *beeswax*, secreted by the hive-bee. Wax scales are secreted by eight wax glands on the underside of the abdomen of the worker bee. These wax wafers are used by the bee in building its honeycomb. From 1½ to 3 pounds of wax can be obtained from the combs when they are scraped. The crude wax must be rendered and refined before it can be sold as "yellow beeswax." When this is bleached, it is known as "white beeswax."

The chemical components of beeswax are alkyl esters of monocarboxylic acids (71–72%), cholesteryl esters (0.6–0.8%), coloring matter (0.3%), lactone (0.6%), free alcohols (1–1½%), free wax acids (13.5–14.5%), hydrocarbons (10.5–11.5%), moisture and mineral impurities (0.9–2%). Myricyl palmitate ($C_{46}H_{92}O_2$) is the principal constituent of the simple alkyl esters (49–53%); the simple esters include alkyl esters of unsaturated fatty acids. The complex esters include hydroxylated esters the chief component of which is believed to be ceryl hydroxypalmitate, $C_{42}H_{84}O_3$. The principal free wax acid component is cerotic acid ($C_{26}H_{52}O_2$). The principal hydrocarbon is hentriacontane ($C_{31}H_{64}$).

The uses of beeswax are many, including church candles, electrotypers and pattern makers wax, cosmetic creams, adhesive tape, munition shells, modelling of flowers, shoe paste constituent, etc. The United States consumes about 8 million pounds of beeswax annually, more than half of which it imports from foreign countries.

Although there are many other kinds of insect waxes, only two are of economic importance namely, shellac wax and Chinese insect wax. Shellac wax is derived from the lac insect, a parasite that feeds on the sap of the lac tree indigenous to India. The commercial wax is not ordinarily the native Indian lac wax, but is a by-product recovered from the dewaxing of shellac spar varnishes. Lac wax melts at 72–80°C, whereas commercial shellac wax melts at 80–84.5°C. Its high melting point and dielectric properties favor its use in the electrical industry for insulation. Chinese insect wax is the product of the scale insect.

The land animal waxes are either solid or liquid. *Woolwax*, derived from the wool of the sheep, is of great economic value. It is better known as anhydrous lanolin, and is of a stiff, soft, solid consistency. The only representative of liquid animal wax is "mutton bird oil" obtainable from the stomach of the mutton bird.

The unsaponifiables of woolwax, known as "woolwax alcohols," are in considerable demand by cosmetic and pharmaceutical industries. Woolwax has a great affinity for water, of which it will absorb 25 to 30%. Refined woolwax is kneaded with water to produce a water-white, colorless ointment, known as hydrous lanolin or "lanolin USP." Anhydrous lanolin is widely used in cosmetic creams, since it is readily absorbed by the skin. It is also used in leather dressings and shoe pastes, as a superfatting agent for toilet soap, as a protective coating for metals, etc. United States consumption of wool wax is about 1.5 million lb/year.

The marine animal waxes are both solid and liquid. The solid marine animal waxes are represented by a wax of considerable economic importance, namely *spermaceti*, derived from a concrete obtained from the head of the sperm whale. The liquid waxes of marine animals are represented by sperm oil obtained from the blubber and cavities in the head of the sperm whale. Spermaceti is the wax used in the candle which defines our unit of candle power; it is used chiefly as a base for ointments, cerates, etc. Sperm oil contains a considerable amount of esters made up of unsaturated alcohols and acids, both of which are susceptible to hydrogenation. Hydrogenated sperm oil is the equivalent of spermaceti wax and harder than the commercial pressed spermaceti. Both yield cetyl alcohol as the unsaponifiable. There is a fairly large demand for cetyl alcohol in the manufacture of lipstick, shampoo, and other cosmetics. Sperm oil itself is an excellent lubricant for lubricating spindles of cotton and woolen mills, or wherever there is need for a very light, limpid, nongumming lubricant.

The waxes obtained from plants occur in the leaves, stems, barks, fruit, flowers, and roots. The leaves of palm trees furnish wax of great economic importance. Particularly is this true of the product furnished by harvesting the leaves of the carnauba palm. The wax is removed from the leaves by sundrying, trenching, threshing and beating; the powdered wax is melted in a clay or iron pot over a fire, strained, cast into blocks, and broken into chunks for shipment from Brazil. *Carnauba wax* dissolves well in hot turpentine and/or naphtha, from which solvents it gels on cooling; it has a good solvent retention power. Its hardness, luster, and favorable behavior with solvents make it a highly valued ingredient in shoe pastes, floor polishes, carbon paper, etc. A small amount of carnauba, such as 2.5%, when added to paraffin will raise the melting point of the latter enormously (e.g., from 130 to 170°F; 54 to 77°C), making it a very useful ingredient in the production of inexpensive high-melting blended waxes. United States consumption is provided by imports from Brazil which amount to over 11 million lb/year.

The chemical composition of carnauba wax comprises 84–85% of alkyl esters of higher fatty acids. Of these esters only 8–9% (wax basis) are simple esters of normal acids. The other esters are acid esters 8–9%, diesters 19–21%, and esters of hydroxylated acids 50–53% (was basis) of which about one-third

are unsaturated. It is the hydroxylated saturated esters that give carnauba its extreme hardness, whereas the esters of the hydroxylated unsaturated fatty acids produce the outstanding luster to polishes.

Ouricury, carandá, and raffia are commercial palm leaf waxes of lesser importance. Ouricury wax has a very high content of esters of hydroxylated carboxylic acids and is used as a substitute for carnauba in carbon papers, etc. Carandá and raffia waxes have a very low contents of these acids and make unsatisfactory substitutes for carnauba.

The most important wax obtainable from the stems of plants is *candelilla*, obtained in Mexico and the southwestern United States. To recover the wax the plant stalks are pulled up by the roots and boiled in acidulated water. On cooling, the congealed wax is removed from the surface of the water in the tank. The crude wax is given an additional refinement before it si placed on the market. Candelilla wax is brownish in color, and melts at 66–78°C. Most vegetable waxes are essentially alkyl esters of aliphatic acids; candelilla, on the other hand, contains 51 to 59% of hycrocarbons and less than 30% of esters. The chief hydrocarbon is hentriacontane ($C_{31}H_{64}$), common to other vegetable waxes. The hydrocarbons melt at 68°C, and the esters at 88–90°C. Candelilla is often used in conjunction with carnauba in leather dressings, floor waxes, etc. It is also used in sound records, electrical insulators, candle compositions, etc. Imports average about 1600 tons per year.

Because of the enormous tonnage of sugar cane processed in Cuba and elsewhere, it is possible to recover an appreciable tonnage of *sugarcane wax* as a by-product. The crude wax contains about one-third each of wax, resin, and oil, and hence needs considerable refinement by selective solvents before it can become of value for industrial use. The refined sugarcane wax is dull yellow in color, melts at 79–81°C and is hard and brittle. It has a durometer hardness of 85–96. It is chemically composed of 78–82% of wax esters, 14% free wax acids, 6–7% free alcohols, and 3–5% hydrocarbons. A proportion of the esters are sterols—sitosterol and stigmasterol—combined with palmitic acid, which are responsible for the good emulsification properties of the wax itself in the preparation of polishes and the like. The proportions of sterols in the refined wax is far less than in the crude wax.

Of waxes obtained from fruits, *japanwax* is the only one of great economic importance, particularly to the Asiatic countries. The wax occurs as a greenish coating on the kernels of the fruit of a small sumac-like tree. Japanwax is actually a vegetable tallow, since it is comprised of 90–91% of glycerides. Peculiarly the glycerides include 3–6.5% (wax basis) of alkyl esters of dicarboxylic acids as well as monocarboxylic acids. The chief dicarboxylic acid is known as japanic acid $[(CH_2)_{19}(COOH)_2]$ which is present with lower as well as higher homologs. The dicarboxylic acids have 19 to 23 carbon atoms, whereas the monocarboxylic acids of the simple glycerides present have 16 to 20 carbons.

The textile industries in the past have been large users of japanwax since it is a source of emulsifying softening agents. Other industries using japanwax include those engaged in the manufacture of rubber, soap, polishes, pomades, leather dressings, cordage, etc. Japanwax is a relatively soft but firm wax, which melts at 48.5–54.5°C. About 3000 tons are normally produced per year in China, and twice that amount in Japan.

Other fruit waxes include *bayberry wax*, used in making Christmas candles since the days of the Pilgrams. The wax of rice bran is coming into commercial use, but waxes of the cranberry, apple, grapefruit, etc. are only of academic interest.

Waxes from grasses include bamboo leaf wax, esparto wax,

and hemp fiber wax. Esparto wax is a hard, tough wax with a melting point of 73–78°C, and is the most important grass wax. Most of the esparto wax produced is consumed in the British Isles. It is chiefly useful as a substitute for carnauba. Waxes obtained from roots of various species of plants are minute in quantity and of no economic importance.

Mineral Waxes. The fossil waxes are associated with fossil remains which have not been bituminized, that is, converted to hydrocarbons by geological change. A fossil wax, chemically speaking, is composed largely of saponifiables, such as wax acids and esters. Fossil waxes of nearly pure ester composition are occasionally found in fragments of prehistoric plant life, still in a state of preservation as to the original wax constituents. Not far removed from fossil wax of the pure ester composition is *montan wax*, a natural mineral wax which is essentially an ester wax that has undergone partial bituminization. Montan wax is commercially extracted from the nonasphaltic insoluble pyrobitumen with which it is associated, by means of selective solvents such as alsohol and benzene, or by means of benzene alone. Crude montan wax is black and contains about 30% of resins, which is reduced to 10% or less upon refinement. The chemical components of montan wax (deresinified) are alkyl esters of fatty acids (40%), alkyl esters of hydroxy fatty acids (18%), free wax acids (18%), free monohydric alcohols (3%), resins (<12%), and ketones (<10%). There are a number of industrial uses of montan wax: electrical insulation, leather finishes, polishes, carbon papers, shoe pastes, brewer's pitch, etc. The crude wax is also used as a basic material in the manufacture of many synthetic waxes where its montanic acid content is utilized in making derivatives.

The principal source of the montan wax consumed in the United States is imports in annual amounts of some 3.4 million lb from Germany and Czechoslovakia, where it is derived from lignites and brown coals. Also, one California company extracts it from lignite.

Peat wax has somewhat the same composition as montan wax. It has only been produced on a limited scale in Ireland, where it is processed from the native peat. It has asphaltic constituents that tend to make it incompletely miscible with paraffin waxes.

The earth waxes are naturally occurring mineral waxes consisting of hydrocarbons with some oxygenated resinous bodies which can be eliminated by ordinary refining procedures. The earth wax of great economic importance is *ozocerite*, which originally was called ceresin wax. Important sources are the Carpathian mountains in Europe and to a far lesser extent Utah. Chemically speaking, ozocerite has hydrocarbons of a type different from those found in paraffin wax, giving it unique physical properties. The melting point of pure Galician ozocerite is 73°C. It is less soluble in organic solvents than paraffin. When added to paraffin wax in amounts of 15% or thereabouts, it will reduce the paraffin crystals to micro size and improve the tensile strength. Crude imported ozocerite has a dielectric constant of 2.37–2.43, the refined 2.03, and domestic (Utahwax) 2.63. Ozocerite is used in the electrical industry, paste polishes, cosmetics, wax flowers, crayons, etc. Of all the waxes it has the greatest affinity for oil.

Petroleum Waxes. Petroleum is the largest single source of hydrocarbon waxes. The largest single use of petroleum waxes is in paper coatings which require about 53% of the total. The second largest use is in candles. The third greatest use in electrical equipment. In contrast, in the early 1950's the largest single use for petroleum waxes was in the manufacture of paper containers for dairy products, and the second was in waxed wrappers for bread. Both outlets are now dominated by plastics.

Crude petroleums differ greatly in both the nature of their hydrocarbons (paraffinic, aromatic, naphthenic, etc.), as well as in their available content of wax. Wax distillates are obtainable with the batch-type, continuous-type, and pipe-still processes, but not from the cracking process. There are, broadly speaking, three principal types of wax encountered in crude oil, namely *paraffin wax*, *slop wax*, and *petrolatum*. The ordinary procedure in producing slack wax is to pump the paraffin distillate at a temperature of 80–100°F (26.7–38°C) to the paraffin sheds (wax plant), where it is allowed to repose in tanks to promote settling at a temperature between 0 and 32°F (−17.8–0°C). It is then pumped through a bank of cooling units (wax chillers) to hydraulic presses, which squeeze out the wax from the chilled distillate. The product is a soft solid known as *slack wax*. Slack wax finds uses in the industries, but most of it is sweated, pressed, and further refined to produce the various grades of fully refined paraffin wax of commerce. Some of the slack may be "pudged" to the extent that it still contains several percent of oil; it is then called *scale wax*. Most of the scale waxes produced have a melting point (drop) of 126–130°F (52–54°C) (ASTM) and are used in waterproofing thread in the fabrication of cotton duck and canvas, waxing kraft papers, builders' papers, cement bag stock, roofers' felt, car liners, and match splints. Scale wax of very low oil content and higher melting point is used in the manufacture of crayons.

Paraffin wax contains 14 hydrocarbons ranging from $C_{18}H_{38}$ to $C_{32}H_{66}$, solidifying between 27.0 and 68.9°C (80.5 and 156.0°F; 26.9–68.8°C). *Petrolatum wax*, which is a microcrystalline wax, has hydrocarbons ranging from $C_{34}H_{70}$ to $C_{43}H_{88}$, inclusive. Its solidifying range is 71.0–83.8°C (159.7–182.7°F). *Slop wax* (by-product from the heavy distillate in the coking process) has 13 hydrocarbons ranging from C_{26} to C_{43}, solidifying between 55.7 and 83.3°C (132.2 and 182.0°F). *Rod wax* (collected from the sucker rods in the field) has 8 hydrocarbons ranging from C_{35} to C_{41}, solidifying at 73.9 to 82.5°C (165.0–180.5°F).

Fully refined paraffin wax as regularly offered in the market is graded according to its melting point. There are also special refined grades offered by some refining companies, such as the so-called hard block fully refined paraffins with melting points of 138–140°F (58.8–60°C), and 143–145°F (61.6–62.7°C). The tensile strength of a paraffin wax is greatly influenced by the oil content. Ordinarily a well-refined paraffin of 130°F (54°C) melting point will have a tensile strength of about 250 psi (1.7 MPa). The addition of 1 or 2% of an oil-absorbent wax of microcrystalline structure will increase the tensile strength to 350 psi (2.4 MPa).

Microcrystalline Waxes. Microcrystalline petroleum waxes are characterized not only by microcrystalline structure but by very high average molecular weight, manifested by a much higher viscosity than that of paraffin wax. The chlorophyll present in plants is considered to be a microcrystalline wax.

Microcrystalline waxes are obtained as by-products from (a) the dewaxing of "lube oil raffinates," (b) the deoiling of petrolatum produced from deasphaltic residual oil, or (c) the deasphalting and deoiling of settlings of tanks holding crude oil in the oil field. These types of microcrystalline waxes are sometimes referred to as "motor-oil wax," "residual oil microcrystalline wax," and "tank-bottom microcrystalline wax," respectively. They have also been referred to as "micro wax," "petrolatum wax," and "petroleum ceresin," respectively.

The *micro waxes* are graded with 145–150°F (63–66°C) and 160–165°F (71–74°C) ASTM melting points, and are refined by selective solvent extraction from the crude wax, a "mobile slurry." A yield of 25–27% of refined wax of the lower melting point is claimed from S.A.E. 20 motor-oil distillate. These waxes are paraffin-like. *Petrolatum waxes* are of a 145–175°F (63–80°C) melting point range. The *petroleum ceresins* which are refined from deposits taken from tanks near the wells, called lease tanks, or in the refinery storage, have melting points which range between 165 and 195°F (74–91°C). In the solvent dewaxing processes the solvents for effectively separating the microcrystalline waxes vary with the refinery methods and the character of the feed stock.

A microcrystalline wax derived from petroleum may be defined as a solid hydrocarbon mixture, of average molecular weight range of 490 to 800, considerably higher than that of paraffin wax, which is 350 to 420. The viscosity (SUS at 210°F; 99°C) of a microcrystalline wax is within the range of 45 to 120 seconds. The lower limit corresponds to 5.75 and the upper limit to 25.1 centistokes at 210°F (99°C). The penetration value (ASTM) is of wide variation, namely 3 to 33, although sticky oily laminating waxes are encountered with as high a penetration as 60. Microcrystalline waxes have an occluded oil content which is not easily set free as it is in paraffin waxes. Therefore, a microcrystalline wax which has an oil content of 1 to 4% is virtually a dry wax. A microcrystalline wax which shows a penetration 20 to 30, which is desirable for many needs, will have an oil content of 5.5 to 10.5%.

When a microcrystalline wax is added to melted paraffin it acts like a solute with paraffin as the solvent; the melting point of the blend is greatly elevated, and the crystallization of the paraffin is depressed. The behavior is that of a two-phase system until about 15% of the microcrystalline wax has been added. With the addition of 15 percent of the petrolatum wax (M.P. 188°F; 86.6°C) the melting point of the paraffin is elevated from 130 to 160°F (54 to 71°C). For many industrial uses microcrystalline wax is admixed with paraffin wax.

The uses of microcrystalline waxes include adhesives, barrel lining, beater size for paper stocks, beer can lining, carbon papers, cheese coatings, cosmetic creams, drinking cups, electrical insulation, floor wax, fruit coating, glass fabric impregnation, heat sealing compounds, laminants for paper, ordnance packing, paper milk bottles, shoe and leather treatments, vegetable coatings, wax emulsions, wax figures and toys, and other miscellaneous purposes.

Synthetic Waxes. These include the following types:

(1) Long-chain polymers of ethylene with OH or other stop-length groupings at end of chain. An example is polyethylene wax of about 2000 molecular weight.

(2) Long-chain polymers of ethylene oxide combined with a dihydric alcohol, namely polyoxyethylene glycol, ("Carbowax").

(3) Chlorinated naphthalenes, ("Halowaxes").

(4) Waxy polyol ether-esters, as for example, polyoxyethylene sorbitol.

(5) Synthetic hydrocarbon waxes prepared by the water-gas synthesis in which carbon monoxide (CO) is reduced by hydrogen (H_2) under pressure, at a predescribed temperature, by means of a catalytic agent. (Fischer-Tropsch waxes "F-T 200" and "F-T 300").

(6) Wax-like ketones, straight-chain and cyclic: (a) Symmetrical ketones produced by the catalytic treatment of the higher fatty acids. (b) Unsymmetrical ketones produced by the Friedel-Crafts' condensation of fatty acids and the like with cyclic hydrocarbons. Examples of the straight-chain ketones are laurone, palmitone, and stearone, and of the cyclic ketones are phenoxyphenyl heptadecyl ketone.

(7) Amide derivatives of fatty acids. The length of the chain may be increased by heating the fatty acid, e.g., stearic acid, with an amino alcohol.

(8) Imide (*N*) condensation products that are wax-like are those of the condensation reaction of one mole of phthalic anhydride with one mole of a primary aliphatic amine to produce a phthalimide. Phthalimide waxes are used in polishes and carbon paper.

(9) Polyoxyethylene fatty acid esters are produced by the reaction of polyethylene glycols with fatty acids. The commercial products are waxy solids which include "Carbowax 4000 (Mono) Stearate." Some of the products act as plasticizers and lubricants for plastics. The polyethylene glycols are soluble in water.

(10) Miscellaneous synthetic waxes (unclassified).

In addition to the above waxes there is a group of synthetic wax-like emulsifiable materials extensively employed in the industries. They are the polyhydric alcohol fatty acid esters, such as ethylene glycol monostearate, glyceryl monostearate, glycerol distearate, and a number of others.

—Albin H. Warth, Cape May, New Jersey.

WHEY PROTEIN. Protein.

WIGNER NUCLIDES. A special case of mirror nuclides. Pairs of odd mass number, isobars for which the atomic number and the neutron number differ by one, and in which the numbers of protons and neutrons are so related that each member of the pair would be transformed into the other by exchanging all neutrons for protons and vice versa.

WILLEMITE. The mineral willemite is a zinc silicate, Zn_2SiO_4, occurring in hexagonal prisms, as masses or scattered grain. It is a brittle mineral with conchoidal fracture; hardness, 5.5; specific gravity, 3.9–4.2; subvitreous luster; usually some shade of yellow, yellowish-green, green, or reddish-brown, but may be colorless, white, or blue to nearly black; transparent to opaque. Much willemite is strongly fluorescent in yellow or yellowish-green hues. Willemite occurs associated with other zinc materials in Belgium, Algeria, Zaire, South West Africa, and Greenland. In the United States, except for three occurrences, one in Colorado, one in New Mexico, and one in Utah, Sussex County, New Jersey, is the only locality in the United States for willemite and is the only one in which that mineral is found in quantity. Here it is found associated with zincite and franklinite, forming an important ore of zinc. It was named by the French mineralogist, Michel Lévy, in honor of King William the First of the Netherlands.

WILLGERODT REACTION. This reaction, discovered in 1887, is conducted by heating a ketone, for example $ArCOCH_3$, with an aqueous solution of yellow ammonium sulfide (sulfur dissolved in ammonium sulfide), and results in formation of an amide derivative of an arylacetic acid and in some reduction of the ketone. The dark reaction mixture usually is refluxed with

$$ArCOCH_3 + (NH_4)Sx \longrightarrow ArCH_2CONH_2 + ArCH_2CH_3$$

alkali to effect hydrolysis of the amide, and the arylacetic acid is recovered from the alkaline solution. Although the yields are not high, the process sometimes offers the most satisfactory route to an arylacetic acid, as in the preparation of 1-acenaphthylacetic acid from 1-acetoacenaphthene, a starting material made in 45% yield by acylation of the hydrocarbon with acetic acid and liquid hydrogen fluoride. The product is obtained in better yield and is more easily purified than that from an alternate process consisting in hypochlorite oxidation, conversion to the acid chloride, and Arndt-Eistert reaction.

$$C_{12}H_9COCH_3 \xrightarrow[78\%]{\substack{1.\ KOCl \\ 2.\ SOCl_2}}$$

1-Acetoacenaphthene

$$C_{12}H_9COCl \xrightarrow{CH_2N_2}$$

$$[C_{12}H_9COCHN_2] \xrightarrow[64\%,\ from\ acid\ chloride]{Ag_2O,\ Na_2S_2O_3}$$

Diazo ketone

$$C_{12}H_9CH_2COOH$$

1-Acenaphthylacetic acid

A modification of the Willgerodt reaction that simplifies the procedure by obviating the necessity of a sealed tube or autoclave consists in refluxing the ketone with a high-boiling amine and sulfur (Schwenk, 1942). Morpholine, so named because of a relationship to an early erroneous partial formula suggested for morphine, is suitable and is made technically by dehydration of diethanolamine. The reaction is conducted in the absence of water, and the reaction product is not the amide but the thioamide; this, however, undergoes hydrolysis in the same manner to the arylacetic acid.

—L. F. and Mary Fieser, Harvard University, Cambridge, Massachusetts.

WITHERITE. The mineral witherite is barium carbonate, $BaCO_3$, crystallizing in the orthorhombic system. It is interesting to note that at 811°C it changes to the hexagonal system, and at 982°C it appears to become isometric. It has a rather imperfect prismatic cleavage; uneven fracture; hardness 3–3.7; specific gravity, 4.29; luster, vitreous to resinous; color, white to yellowish or grayish; streak, white; transparent to translucent. Witherite is found in veins, and often is associated with galena, as at Alston Moor, Cumberland, England. Associated with barite at Freiberg, Saxony, and at Lexington, Kentucky. Named in honor of Dr. William Withering, an English botanist.

WITTIG REACTION. This reaction provides an excellent method for the conversion of a carbonyl compound to an olefin:

$$(C_6H_5)_3 \xrightarrow{CH_3Br}$$

Triphenylphosphine

$$(C_6H_5)_3\overset{+}{P}CH_3(Br^-) \xrightarrow[-C_6H_6-Li\ Br]{C_6H_5Li}$$

Methyltriphenyl-phosphonium bromide

$$(C_6H_5)_3P{=}CH_2 \longleftrightarrow (C_6H_5)_3\overset{+}{P}{-}\overset{-}{C}H_2$$

Wittig Reagent

$$(C_6H_5)_2C{=}O$$
$$(C_6H_5)_3P \cdots CH_2 \longrightarrow \underset{O}{(C_6H_5)_3P} + \underset{C(C_6H_5)_3}{CH_2}$$
$$O \cdots C(C_6H_5)_2$$

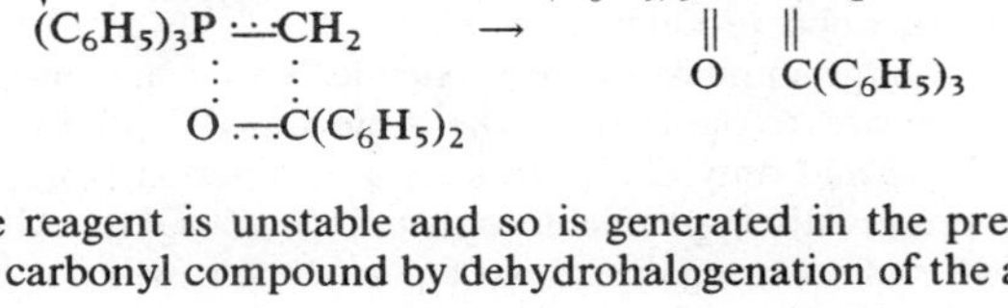

The reagent is unstable and so is generated in the presence of the carbonyl compound by dehydrohalogenation of the alkyl-triphenylphosphonium bromide with phenyllithium in dry ether in a nitrogen atmosphere. There are various modifications, such as the phosphonate, in which diethylbenzylphosphonate, cinnamaldehyde, and sodium methoxide yield 1,4-diphenylbutadiene.

Dr. Georg Wittig, University of Heidelberg, was awarded the 1979 Nobel Prize for chemistry for his work in organic synthesis. Dr. Herbert C. Brown of Purdue University also participated in the joint award, but for separate work in organic synthesis.

More details on the early development of the Wittig reaction, dating back to the 1940s, is given in *Science*, **207**, 42–44 (1980).

WOLFF-KISHNER REACTION. This method of reduction was discovered independently in Germany (Wolff, 1912) and in Russia (Kishner, 1911). A ketone (or aldehyde) is converted into the hydrazone, and this derivative is heated in a sealed tube or an autoclave with sodium ethoxide in absolute ethanol.

$$\ce{>C=O} \xrightarrow{H_2NNH_2} \ce{>C=NNH_2} \xrightarrow[200°]{NaOC_2H_5}$$

$$\ce{>CH_2 + N_2}$$

After preliminary technical improvements, Huang Minlon (1946) introduced a modified procedure by which the reduction is conducted on a large scale at atmospheric pressure with efficiency and economy. The ketone is refluxed in a high-boiling water-miscible solvent (usually di- or triethylene glycol) with the aqueous hydrazine and sodium hydroxide to form the hydrazone; water is then allowed to distil from the mixture till the temperature rises to a point favorable for decomposition of the hydrazone (200°); and the mixture is refluxed for three or four hours to complete the reduction.

—L. F. and Mary Fieser, Harvard University, Cambridge, Massachusetts.

WOOD. A vascular tissue which occurs in all higher plants. The most important commercial sources of wood are the gymnosperms, or softwood trees and the dicotyledonous angiosperms, or hardwood trees. Botanically, wood serves the plant as supporting and conducting tissue, and it also contains certain cells which serve in the storage of food. The trunks and branches of trees and shrubs are composed of wood, except for the very narrow cylinder of pith in the center and the bark which covers the outside. Botanists refer to wood by its Greek name, *xylem*.

A new cylinder of wood is laid down each year around the previously formed wood in the tree. This new growth originates in the cambium, a very narrow growing layer, which elaborates both the wood and the bark, and which separates these two tissues from each other. Each year's growth of wood forms a new concentric ring in the woody-stem, as viewed in cross-section. These are termed the annual rings. Each of these has an inner part (toward the pith) which is laid down in the early growing season and is termed the spring wood and an outer layer (towards the bark) which is laid down later and is known as summer wood. The cell walls of the latter are often thicker, forming a denser structure than the spring wood.

Most of the cells of wood are long, narrow hollow fibers and tubular-shaped cells arranged with their long axes parallel to the axis of the tree trunk. Certain food storage cells lie in radial bands, termed wood rays, which are perpendicular to the tree axis. The walls of this complex system of plant cells form the basic framework and material of all wood substance. All wood substance is composed of two basic chemical materials, *lignin*, and a polysaccharidic system, which is termed *holocellulose*. The latter embraces *cellulose* and the *hemicelluloses*, a mixture of pentosans, hexosans and polyuronides, and in some instances small amounts of pectic materials. Wood cell wall tissue also always retains small amounts of mineral matter (ash).

The outer portion of the cell wall, known as the primary wall, is heavily lignified. The intercellular substance, termed the middle lamella, is mainly lignin. The lignin of the middle lamella and primary walls thus serves as a matrix in which the cells are imbedded. Dissolution and removal of the lignin results in separation of the wood fibers. This is the underlying principle in the manufacture of chemical pulps from wood for paper or other cellulose products. **Papermaking and Finishing; and Pulp (Wood) Production and Processing.**

About one-fourth or more of the lignin is in the middle lamella-primary wall complex. The remainder is within the holocellulose system of the cell walls.

Besides the cell wall tissue, which is the basic material of all wood substance, wood contains a variety of materials, many of which may be extracted by selected solvents. These extraneous components lie mainly within the cavities (lumen of the cells and on the surfaces of the cell walls. These "extraparietal substances" include a wide range of chemically different materials, such as essential oils, aliphatic hydrocarbons, fixed oils, resin acids, resinols, tannins, phytosterols, alkaloids, dyes, proteins, water-soluble carbohydrates, cyclitols, and salts of organic acids. The amount and composition of these extraneous substances vary greatly. The occurrence of certain of these substances is often very specific as to genera or species. Generally, however, the total amount of the extraneous components is only a few percent of the total weight of the wood.

The chemical composition of the extractive-free wood, i.e., the cell wall substance varies less than do the extractives. However, it is by no means constant, there being major differences between hardwoods and softwoods, and often between different genera and even between species. There is even some variability in chemical composition within the same log. For example, there is more lignin in the thinner-walled spring wood than in the summer wood, and more lignin in wood ray cells than in the tracheids or fibers. The heart-wood tissue often contains greater deposits of extraneous (extractive) components than the sapwood. There are major differences between most softwoods (conifers) and most temperate zone hardwoods (broad-leaf trees). Usually the softwoods have greater amounts of extraneous components extractable by organic solvents. Generally the lignin content of softwoods is higher than that of hardwoods, *viz.*, in the order of 25–30% compared to 17–24%. Also there is a major difference in the pentosan content between these two groups of woods, the hardwoods containing usually about 17–22% and the softwoods about 8–14%. There are exceptions to these generalizations, however.

The components of the cell wall substance of wood are exceedingly difficult to separate. Separations are rarely complete and generally bring about drastic chemical changes, especially in the lignin and molecular size degradation of the polysaccharides. To a considerable extent the components are apparently interpenetrating polymer systems. The long-chain linear polysaccharides tend to be parallel to the fiber axis and to form areas of varying degrees of crystallinity, as shown by x-rays and other physical and chemical properties. The lignin appears to be an amorphous tridimensional polymer.

Wood forms one of the world's most important chemical raw materials. It is the primary source of cellulose for the pulp and paper and cellulose industries. These industries are well up in the group of 10 major industries of the United States. For paper, rayon, films, lacquers, explosives and plastics, which comprise the greatest chemical uses of wood, it is the cellulose component (plus certain amounts of hemicellulose) of wood that is of value. The lignin forms a major industrial waste as a by-product of the paper and cellulose industries. Its major use is in its heat value in the recovery of alkaline pulping chemicals. A variety of minor uses for lignin have been developed, such as for the manufacture of vanillin, adhesives, plastics, oil-well drilling compounds and fillers for rubber.

Wood wastes from the lumber and woodworking industries form a great potential source of sugars and alcohol by acid hydrolysis of its polysaccharides followed by fermentation. Wood hydrolysis processes, however, are not yet economically competitive with other sources of sugars and alcohol in this country and many other areas of the world.

Wood is also an industrial source of charcoal, tannin, rosin, turpentine, and various other essential oils and pharmaceutical products.

—Edwin C. Jahn, State College of Forestry, Syracuse, New York.

WOOD (As Energy Source). Biomass and Wastes as Energy Sources.

WOOD PRESERVATIVE. A material applied to wood to prevent its destruction by fungi, wood-boring insects, marine borers and fire. A common characteristic of these materials is toxicity to those organisms that attack wood, or in the case of fire retardants the ability to control combustion in terms defined by the Underwriters Laboratory. In addition, a satisfactory wood preservative must also (a) be capable of penetrating wood, (b) remain in the wood for extended periods without losing its effectiveness due to chemical breakdown, (c) be harmless to humans and animals, (d) be noncorrosive and, (e) be available in quantity at a reasonable cost. For certain uses, the preservative may be required to be colorless, odorless, nonswelling and paintable.

The principal wood preservatives in use today are classified as (a) preservative oils, (b) toxic chemicals in organic solvents, and (c) water-soluble salts.

The most important of the preservative oils is coal-tar creosote and its solutions in the form of creosote-coal tar and creosote petroleum. Coal-tar creosote is defined by the American Wood Preservers Association as—"A distillate of coal tar produced by high-temperature carbonization of bituminous coal; it consists principally of liquid and solid aromatic hydrocarbons and contains appreciable quantities of tar acids and tar bases; it is heavier than water and has a continuous boiling range of at least 125°C, beginning at about 200°C." This material is one of the oldest wood preservatives and is regarded by many as the best substance known for protection against all forms of wood-destroying organisms. For normal service conditions, minimum retentions of 8–10 lb/ft³ (128–161 kg/m³) of wood penetrated are sufficient. For extreme conditions, retentions as high as 35 lb/ft³ (561 kg/m³) are desirable. Currently, coal-tar creosote and creosote solutions are used as preservatives for about 60% of all wood products treated.

Of the large number of toxic chemicals that are oil-soluble, only three are recognized by the American Wood Preservers Association and only one of these is a commercially important wood preservative, pentachlorophenol (C_6Cl_5OH). Although "penta" is effective against fungi and insects, it will not protect against marine borers, and hence cannot be used as a wood preservative for salt water installations.

The two most common solvents for carrying penta into the wood are a relatively high-boiling No. 2 fuel oil and a very low-boiling liquified petroleum gas, butane. When fuel oil is used it remains in the wood and although the end product is brighter and cleaner than creosote-treated wood, it has a somewhat oily character and cannot readily be painted. When LP gas is used as the solvent, the butane is recovered and the penta is deposited in the wood as a dry crystalline material, hence the wood retains its color and is readily paintable.

Water-borne preservatives are divided into two categories. One group which includes acid copper chromate, chromated zinc chloride, copperized chromated zinc arsenate and fluor-chrome-arsenate-phenol is used where the wood is not subjected to excessive leaching. The second group, ammoniacal copper arsenite and three types of chromated copper arsenate which react to become practically water insoluble, are used at about 0.6 lb/ft³ (9.6 kg/m³) when wood is placed in ground contact under severe service conditions.

Water-borne preservatives penetrate wood easily and the solution presents no problem in flammability or health hazards. Disadvantages of water soluble preservatives include the swelling and shrinking of the treated material, reduction in bending strength and stiffness as a result of failure to redry following treatment, and less protection against weathering and mechanical wear than provided by either preservative oils or oil soluble chemicals in which the solvent remains in the wood.

In addition to the general preservative categories discussed there is a fourth group known as "proprietary preservatives" which are composed of various combinations of toxic materials and solvents. These are sold under trade names and in some cases are protected by patents.

While wood preservatives can be applied by simple means such as brushing, spraying and cold soaking, well over 90% of all commercial wood treatment is by one of the "pressure processes." Although the details of the individual processes vary, the type of equipment and general procedure used are similar. Treatment takes place in closed cylinders 6–9 feet (1.8–2.7 m) in diameter and up to 180 feet (54 m) in length. The wood to be treated is placed in the retort, submerged in the preservative and subjected to pressures in the order of 200 psi (13.6 atm). The preservative is often at an elevated temperature and in some cases the wood is given a preliminary pressure or vacuum period prior to admitting the preservative into the retort.

The most satisfactory nonpressure process is known as "thermal" treatment. The wood to be treated is immersed in a preservative at an elevated temperature. This causes the air in the wood cells to expand so that when the wood is transferred into a preservative bath of lower temperature, the air contracts forming a partial vacuum and atmospheric pressure forces the liquid into the wood.

Although the fire retardant treatment of wood is essentially the injection of water soluble salts into the wood by pressure treating methods, it deserves separate mention if for no other reason than the higher salt retention (4lbs/cu. ft. vs. 0.6 lbs/cu. ft.) (64 kg/m³ vs 9.6 kg/m³) required. Most fire retardant treatments are proprietary and are specified by the Underwriters Laboratory on the basis of flame spread ratings. The latter's list of building materials will give details of the ratings including those of a recently introduced leach-resistant fire retardant suitable for outdoor service. Common fire-retardant chemicals include diammonium phosphate, ammonium sulfate, sodium tetraborate and boric acid which are used in various combinations.

—William T. Nearn, Weyehaeuser Company, Seattle, Washington.

WOOD PULP.   Pulp (Wood) Production and Processing.

WOOL.   The natural, highly crimped fiber from sheep, wool is one of the oldest fibers from the standpoint of use in textiles. Minute scales on the surface of the fibers allow them to interlock and are responsible for the ability of the fiber to *felt*, a phenomenon responsible for felt cloth and mill-finished worsteds. Crimpiness in wool is due to the open formation of the scales. Fine merino wool has 24 crimps per inch (~10 per centimeter). Luster of the fiber depends upon the size and smoothness of the scales. The basic wool protein, *keratin*, comprises molecular chains that are linked with sulfur. When sulfur is fed to sheep in areas deficient of the element, the quality of the wool improves. Wool fibers that fall below 3 inches (7.5 centimeters) in length are known as *clothing wool*; fibers 3–7 inches (7.5–17.8 centimeters) long are referred to as *combing wools*. The wool-fiber diameter ranges from 0.0025 to 0.005 inch (0.06–0.13 millimeter). See also **Fibers.**

WULFENITE.   The mineral wulfenite is lead molybdate corresponding to the formula $PbMoO_4$, analyses showing that a part of the lead may be replaced by calcium. Wulfenite crystallizes in the tetragonal system usually in thin tabular forms, but is also found massive. It is a brittle mineral; hardness, 2.75–3; specific gravity, 6.5–7; luster, adamantine to resinous; color, yellowish to green or red, may be whitish or grayish; transparent to translucent. Wulfenite is a secondary mineral found in association with other lead minerals such as galena, and pyromorphite. It is believed to have been formed, at least in part, by the action of waters containing molybdenum salts on cerussite, anglesite, and pyromorphite.

Especially important localities are in Yugoslavia, Czechoslovakia, Morocco, Zaire, New South Wales and Mexico. In the United States it has been found in Phoenixville, Pennsylvania, and in the Organ Mountains, New Mexico; Yuma County, Arizona; Box Elder and Salt Lake Counties, Utah; and in Clark and Eureka Counties, Nevada. Wulfenite was named in honor of F. X. von Wülfen, an Austrian mineralogist of the eighteenth century.

WURTZITE.   A mineral zinc sulfide, (Zn, Fe)S, similar to sphalerite. Crystallizes in the hexagonal system. Hardness, 3.5–4; specific gravity, 3.98; color, brownish-black with resinous luster. Named after Adolphe Würtz, France.

WURTZ REACTION.   A method of synthesizing hydrocarbons discovered by Wurtz (1855) consists in treatment of an alkyl halide with metallic sodium, which has a strong affinity for bound halogen and acts on methyl iodide in such a way as to strip iodine from the molecule and produce sodium iodide. The reaction involves two molecules of methyl iodide and two atoms of sodium:

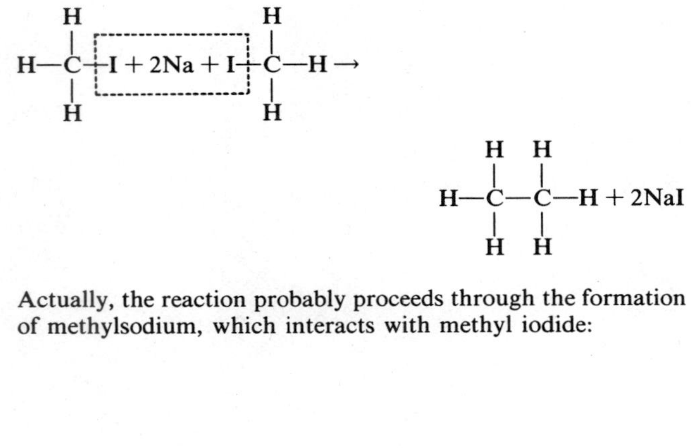

Actually, the reaction probably proceeds through the formation of methylsodium, which interacts with methyl iodide:

$$CH_3I \xrightarrow{Na} CH_3Na \xrightarrow{CH_3I} CH_3CH_3$$

The Wurtz reaction can be applied generally to synthesis of hydrocarbons by the joining together of hydrocarbon residues of two molecules of an alkyl halide (usually the bromide or iodide). With halides of high molecular weight the yields are often good, and the reaction has been serviceable in the synthesis of higher hydrocarbons starting with alcohols found in nature, for example:

$$2C_{20}H_{41}Br \xrightarrow[31\% \text{ yield}]{Na} C_{40}H_{82}$$

Dihydrophytyl bromide   Perhydrolycopene

$$2n\text{-}C_{16}H_{33}I \xrightarrow[70-80\% \text{ yield}]{Mg \text{ (ether)}} C_{32}H_{66}$$

Cetyl iodide   n-Dotriacontane

Cetyl iodide (from the alcohol of spermaceti wax) has been converted into the C_{32}-hydrocarbon both by the action of sodium amalgam in alcohol-ether and, as shown in the equation, with use of magnesium in place of sodium.

A general expression for the Wurtz synthesis is:

$$2RX + 2Na \rightarrow R \cdot R + 2NaX$$

Alkyl halide   Alkane

It might appear that the synthesis could be varied by use of two different alkyl halides, with the linking together of the two hydrocarbon fragments, for example:

$$CH_3CH_2CH_2CH_2I + ICH_2CH_2CH_3 \xrightarrow{2Na}$$

n-Butyl iodide   n-Propyl iodide

n-Heptane

$$CH_3CH_2CH_2CH_2CH_2CH_2CH_3$$

The reaction mixture, however, contains many millions of molecules of each halide, and there is nearly as much opportunity for interaction of like as of unlike molecules. Some butyl iodide molecules will react with molecules of propyl iodide and yield heptane as pictured, but some will combine with other molecules of the same kind and produce octane. The total result can be represented as follows:

$$CH_3CH_2CH_2CH_2CH_2I +$$

$$CH_3CH_2CH_2I \xrightarrow{2Na} \begin{cases} n\text{-}C_6H_{14}, \text{ b.p. } 69° \\ n\text{-}C_7H_{16}, \text{ b.p. } 98° \\ n\text{-}C_8H_{18}, \text{ b.p. } 126° \end{cases}$$

The reaction affords a mixture of which the unsymmetrical product n-heptane can be expected to constitute no more than one-half, and since the three hydrocarbon components are similar and do not differ greatly in boiling pont, isolation of even a small amount of n-heptane in a moderately homogeneous condition would obviously be difficult. It is therefore impracticable to utilize an unsymmetrical Wurtz reaction in synthesis, for the inevitable result is:

$$RX + R'X \rightarrow RR' + RR + R'R'$$

—L. F. and Mary Fieser, Harvard University, Cambridge, Massachusetts.

X

XANTHAN GUM. A very high-molecular-weight polysaccharide produced by pure culture fermentation of glucose by *Xanthamonas campestris*. The substance is readily soluble in hot or cold water, imparting a high viscosity at low concentrations. The solutions are pseudoplastic, with viscosity decreasing rapidly as shear rate increases. Heat, acid, and salt have little effect on the stability of its solutions. The substance is compatible with most other hydrocolloids, including starch. Xanthan gum undergoes a unique gel reaction with locust bean gum to produce a synergistic increase in viscosity. The gum is used as a thickening, suspending, emulsifying, and stabilizing agent in foods and has a number of important nonfood uses as well.

In 1974, the Northern Regional Research Center (Peoria, Illinois) of the U.S. Department of Agriculture and the Kelco Company were joint recipients of the Institute of Food Technologists award for the development and commercialization of xanthan gum. As early as 1956, researchers discovered unusual water-thickening abilities of a substance produced by the bacterium *Xanthamonas campestris*. They chemically identified the substance and named it *Polysaccharide B-1459* after the culture number. A food additive regulation for its use (when it was renamed xanthan gum) was issued in 1959. The process was brought into commercial production in 1964.

Glucose from starch is fermented by the aforementioned bacteria to produce xanthan gum, which is recovered by precipitation with isopropyl alcohol, then washed, dried, and milled. Among several applications, xanthan gum is used in pourable salad dressings for its emulsifying properties; in frozen foods for its freeze-thaw stabilizing effect; in juice drinks for its suspending properties; and in creamed cottage cheese for its stabilizing properties, pseudoplasticity, and mouthfeel. The gum makes a stable cream dressing that clings to the curd, but shears easily and thus does not have a gummy texture. The gum is also used as a stabilizer for frozen desserts and as a suspending agent in liquid feed supplements for cattle feeding and in milk replacers for calves.

It also has been found that xanthan gum makes the proteins in nonwheat breads more extensible and thus produces better breads.

See the list of reference in the entry on **Colloidal Systems.** Other gums and mucilages are described in the entry on **Gums and Mucilages.**

XANTHOPHYLLS. Carotenoids.

XENON. Chemical element symbol Xe, at. no. 54, at. wt. 131.30, periodic table group 0 (inert or noble gases), mp -107.1 $\pm 2.5°C$, bp $-107.1°C$, density 3.5 g/cm^3 (liquid at $-109°C$). Specific gravity compared with air is 4.561. Solid xenon has a face-centered cubic crystal structure. At standard conditions, xenon is a colorless, odorless gas and does not form stable compounds under normal conditions with any other element. Due to its low valence forces, xenon does not form diatomic molecules, except in discharge tubes. It does form compounds under highly favorable conditions, such as excitation in discharge tubes, or pressure in the presence of a powerful dipole. Xenon forms a hydrate much more readily than argon, at a pressure slightly above 1 atm at 0°C. The element also forms addition compounds with a number of organic substances, such as $Xe \cdot 2C_6H_5OH$ with phenol, which has a dissociation pressure of 1 atm at 4°C.

In 1962, the compound xenon platinum hexafluoride was synthesized by Bartlett. Later in the same year, Classen confirmed the synthesis and prepared the first binary compound of an inert gas, xenon tetrafluoride, a stable crystalline compound, mp about 90°C. The compound was prepared by heating a 5 : 1 mixture of fluorine and xenon to 400°C, then cooling it rapidly to room temperature. Since the original research, additional xylene-fluorine compounds have been reported, including XeF_2, XeF_4, XeF_6, XeF_8, $SeSiF_6$, XeO_2F_2, and Na_4XeO_6, as well as the hydrate. By heating xenon and fluorine at 250°C above 10 atm (up to 170 atm), Weinstock, Weaver, and Krop obtained XeF_5 and XeF_6 in an equilibrium mixture. D. F. Smith, S. M. Williamson, and C. W. Koch have reported the preparation of XeO_3 from the hexafluoride tetrafluoride. XeO_3 is a white, crystalline, explosive compound.

Xenon also forms compounds, possibly clathrates, with certain substances in nonstoichiometric proportions. Crystalline compounds with benzene or hydroquinone, formed under 40 atm pressure, contain about 26% xenon by weight. Alkaline hydrolysis of XeF_6 produces salts of octavalent xenon. No persistent divalent or tetravalent compounds are found in aqueous solution, but the former is intermediate in hydrolysis of the fluorides, and the latter in reactions of XeO_3 with $XeF_2 \cdot H_2O$, and various organic compounds.

Xenon occurs in the atmosphere to the extent of approximately 0.00087%, making it the least abundant of the rare or noble gases in the atmosphere. In terms of abundance, xenon does not appear on lists of elements in the earth's crust because it does not exist in stable compounds under normal conditions. However, xenon because of its limited solubility in H_2O, is found in seawater to the extent of approximately 950 pounds per cubic mile. Commercial xenon is derived from air by liquefaction and fractional distillation. There are nine natural isotopes ^{124}Xe, ^{126}Xe, ^{128}Xe, through ^{132}Xe, ^{134}Xe, and ^{136}Xe, and seven radioactive isotopes ^{123}Xe, ^{125}Xe, ^{127}Xe, ^{133}Xe, ^{135}Xe, ^{137}Xe, and ^{138}Xe, all with relatively short half-lives, the longest ^{127}Xe with a half-life of about 36 days. See also **Radioactivity.** First ionization potential, 12.127 eV; second, 21.1 eV; third, 32.0 eV. Van der Waals radius 2.20 Å Electronic configuration $1s^22s^22p^63s^23p^63d^{10}4s^24p^64d^{10}5s^25p^6$.

Uses. Xenon finds principal application in special electronic devices and lamps. Xenon, in a vacuum tube, produces a beautiful blue glow when excited by an electrical discharge. Xenon lamps have been developed which provide a constant light (described as sunlight-plus-north-sky light) even when there are significant voltage changes. Thus, the lamps do not require voltage regulators. For a given wattage, xenon lamps have been found to deliver a greater light output than incandescent lamps. For example, an 800-watt xenon lamp will produce 2000 lumens as compared with a 1000-watt incandescent lamp that produces only about 200 lumens. Xenon also has found application in certain lasers. It has been learned that xenon produces mild anesthesia, but it cannot be used for surgery because the quantity required would cause asphyxiation. Xenon is used in bubble chambers, probes, and other applications where its high molecular weight is of advantage in the atomic energy field. Potentially, xenon is of interest as a gas for ion engines.

Ramsey and Travers first found the element when investigating the properties of liquid air in 1898.

See the list of references at the end of the entry on **Chemical Elements.**

XI PARTICLE. A hyperon with a rest mass energy of about 1318.4 MeV, an isospin quantum number $\frac{1}{2}$, an angular momentum spin quantum number $\frac{1}{2}$, and a strangeness quantum number 2. Symbol, Ξ

X-RAY ANALYSIS. X-rays occupy that portion of the electromagnetic spectrum between 0.01 and 100 angstroms (Å). Their range of approximate quantum energy is from 2×10^{-6} to 2×10^{-10} erg, or from 10^6 to 100 eV. Important x-ray analytical methods are based upon: (1) fluorescence; (2) emission; (3) absorption; and (4) diffraction. These methods are used qualitatively and quantitatively to determine the element content of complex mixtures and to determine exactly the atomic arrangement and spacings of crystalline materials.

Source of X-rays. X-rays are emitted by atoms which are bombarded with energetic electrons. This results from two separate effects: (1) deceleration of high-speed electrons as they pass through matter and (2) ionization of individual atoms which abruptly stop the electrons. The first effect results in a continuous-type spectrum; the second effect results in characteristic line spectra.

Continuous Spectrum. The bulk of x-radiation arising from electron bombardment is the continuous spectrum. If an individual electron is abruptly decelerated, but not necessarily stopped, in passing through or near the electric field of a target atom, the electron will lose some energy ΔE, which appears as an x-ray photon of frequency $\vartheta = \Delta E / h$, where h is Planck's constant. An electron may experience several such decelerations before it is finally stopped, emitting x-ray photons of widely different energy and wavelength. A few electrons will be stopped in a single process, losing their entire energy and emitting an x-ray photon having the exact energy of the incident electron.

X-Ray Spectral Lines. These result when the incident electrons knock orbital electrons out of an atom. If an ejected electron is from one of the inner orbits of the atom (Fig. 1), an electron from an outer shell will fall to the inner orbit to fill the vacancy. The decrease in potential energy of this electron in approaching the nucleus results in the emission of an x-ray photon having an energy exactly equal to that lost by the electron. The wavelength λ for such photons is related to ΔE by $\lambda = ch / \Delta E$, where c is the velocity of electromagnetic energy and h is Planck's constant. Because the energy of orbital electrons is quantized, the x-ray photons can have only certain definite wavelengths which are characteristic of the atom. This situation is somewhat analogous to the more familiar ultraviolet and visible-emission spectra of materials, the difference being that the optical spectra are the result of electron transitions between energy levels of just the outermost electrons of the atoms.

X-rays resulting from an electron transition filling an electron vacancy in the innermost shell of an atom are known as K x-rays or K lines; those from the L shell are known as L lines, and so on.

Generation of X-rays. An important component in an x-ray analytical device is an x-radiation generator. A high-vacuum Coolidge type tube, wherein electrons are emitted from a heated tungsten filament and accelerated by a high voltage to an anode (target) is a common source of x-rays. See Fig. 2.

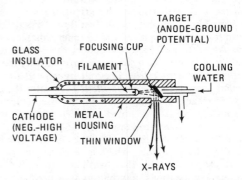

Fig. 2.   High-voltage, high-vacuum x-ray tube.

A wide variety of tubes is available. All high-power (high-current) commercial tubes employ a water-cooled anode. Tubes of this type have been built with ratings up to 10 kilowatts.

Detection of X-rays. Detectors include (1) Geiger-Mueller tube, (2) ionization chambers, (3) scintillation counters, (4) proportional counter, (5) electron-multiplier tubes, and (6) nondispersive detectors using cooled lithium-drifted Si detectors. See Fig. 3.

X-ray Crystallography. X-rays penetrating below the surface of crystalline materials are scattered by the individual parallel layers of atoms; each atomic layer acts as a new, although weak, source of x-rays. To be reinforced in a given direction at an angle θ (Fig. 4), the spacing d between crystal planes must be rigorously related to the wavelength of the radiation. At a given angle, x-rays of one definite wavelength will be constructively reinforced. These variables are related by Bragg's law.

$$n\lambda = 2d \sin \theta$$

where n is an integer. Note that θ is measured relative to the crystal face rather than to the perpendicular.

In addition to fulfilling this wavelength requirement, the energy will be diffracted only at an angle equal to the angle of the incident x-rays, independent of wavelength; otherwise, destructive interference is possible, and the "reflected" energy is negligible.

X-Ray Analyzer, Electron-Microprobe. Advantages of this

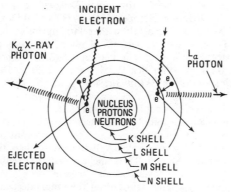

Fig. 1.   Origin of x-ray spectra due to electron bombardment.

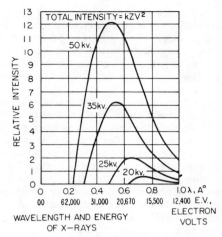

Fig. 3.   Continuous x-ray spectrum of tungsten ($Z = 74$) at various tube voltages.

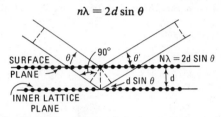

Fig. 4.   Reflection of x-rays from internal crystal plane.

analytical instrument include: (1) analysis can be confined to very small (microsamples) amounts of materials; (2) the particular material to be analyzed need not be physically separated from its surrounding materials, as is often required with many analytical methods; and (3) through the development of associated instrumentation, diagnostic techniques, and information displays, the method can be quite fast. Limits of detection in solid solution are from approximately 0.005 to 0.5%, depending upon the elements and sample matrixes involved. See Fig. 5.

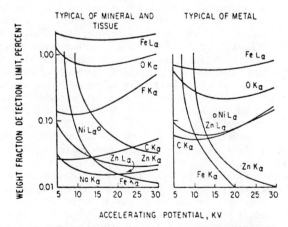

Fig. 5.   Typical weight-fraction detecting limits of electron-microprobe x-ray analyzer.

Concentrations as low as 10^{-16} gram may be measured.

Mainly used for metallurgical studies, nonmetallics also may be analyzed when samples are properly prepared. Biological applications include tooth and bone samples, cytochemical problems and staining techniques, physiochemical problems, and studies in pathology. Relative weight-fraction-detection limits

for most elements in biological specimens are in the general range of 0.01 to 0.10%. Electronics industry applications include studies of diffusion phenomena, electrical-contact surfaces, interfaces on transistors, and microcircuitry analysis.

As shown by Fig. 6, electrons from an electron gun are di-

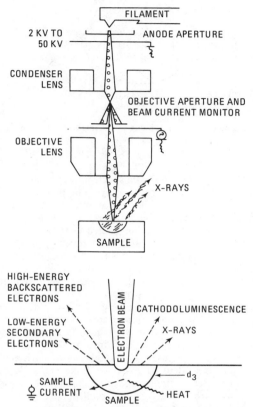

Fig. 6.   Electron gun and probe-forming lens system of an integrated electron probe. (*Bausch & Lomb.*)

rected to the sample through an electron optical system. Once the electron beam strikes the sample, a number of signal sources are activated, including (1) high-energy backscattered electrons, (2) low-energy secondary electrons, (3) cathodoluminescence, and (4) x-rays. Some heat also is generated within the sample. Volume d_3 of the specimen is that *volume from which x-rays are emitted*.

The x-rays produced may be detected nondispersively by a proportional counter whose output may be separated as a function of energy by a pulse-height analysis system into the various wavelength components. Better detection sensitivities, however, can be obtained through the use of a fully focusing diffracting-crystal spectrometer in conjunction with a proportional detector and the necessary pulse-height analyzer. As shown by Fig. 7, the necessary condition for fully focusing optics is to have the x-ray source, the crystal, and the detector slit all placed on a common circle. This geometry requires that the diffracting-crystal planes be bent to the diameter of the Rowland circle.

Spherical aberration at the detector slit is minimized by further grinding the crystal surface to fit the radius of the Rowland circle. With the resultant Johansson optics, the crystal radius is fixed, and the 2θ range of the spectrometer is scanned by moving the crystal radially away from the source and, at the same time, rotating it into the detector to achieve a true focus throughout the spectrometer range. X-rays particularly of a

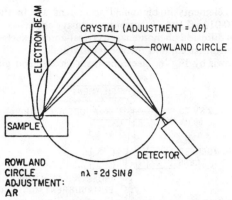

Fig. 7.   Geometry of a fully focusing diffracting-crystal spectrometer.

wavelength greater than 2 Å, and electrons are highly absorbed in an air atmosphere. Thus, the spectrometer must be enclosed in a vacuum of the order of 10^{-5} torr. The present wavelength range of interest extends from approximately 1 to 100Å. Diffracting crystals to cover this range must provide broad wavelength coverage, high diffraction efficiencies for high peak-count intensities, good resolution, and good resulting peak-to-background ratios. Crystals that meet these objectives include lithium fluoride, ammonium dihydrogen phosphate (ADP), ethylenediamine-D-tartrate. (EDT), quartz, and sodium chloride.

Detectors. Of the three commonly used x-ray detectors—(1) Geiger counter, (2) scintillation counter, and (3) proportional counter—the latter is used most frequently for electron-probe microanalysis. In the wavelengths from 1 to 10 Å, sealed proportional counters may be used. For longer-wavelength analysis—in the range from 10 to 93 Å—the thinnest possible detector window is required to limit spectral attenuation. Nitrocellulose windows have proved successful. Nondispersive detection systems using cooled Li-drifted Si are also applicable.

An *optical microscope* is required in the system to provide the analyst with a means of reference to identify various sample areas for analysis. Sample stages may hold single or multiple samples and are provided with means for moving the sample in x, y, and z planes without breaking the system vacuum. After the point of interest is located on the specimen, the data may be read out in a number of ways: (1) quantitative and seimquantitative information may be obtained by processing the x-ray detector signal through a rate meter to a strip-chart or X-Y recorder; (2) scaler systems also provide direct readout of quantitative data integration; and (3) for operational convenience, a data translator and typewriter or teletype printout system may be connected directly to record digital-counter information as hard copy.

X-ray Fluorescence Analysis. One of several types of spectrochemical techniques now used for laboratory analysis. The method is nondestructive. The characteristic x-ray spectrum of each element bears a simple direct relationship to the atomic number. The relation of the wavelength λ to the atomic number Z is

$$\frac{1}{\lambda} \, aZ^2 \quad \text{(Moseley's law)}$$

Since the x-ray spectral lines come from the inner electrons of the atoms, the lines are not related to the chemical properties of the elements or to the compounds in which they may reside. Because the characteristics of the x-ray spectra are associated

with energies released through transitions of electrons within the inner shells of the atom, the spectra are simple. Most practical x-ray fluorescence analysis involves the detection of radiation release through electron transitions from outer shells to the K shell (K spectra), outer shells to the L shell (L spectra) and, in very few cases, from outer shells to the M shell (M spectra).

The simplest form of energy source available for commercial instrumentation is that obtained from an x-ray tube. For samples containing predominantly low-atomic-number elements, as in cement raw mix, the most efficient excitation is accomplished by using an x-ray tube target material of relatively low atomic number, such as chromium. Elements having higher atomic numbers are most effectively excited by high-atomic-number targets, such as tungsten or platinum. An optimum target material is rhodium for the analysis of a broad range of elements. The x-ray tube irradiates the sample which, in turn, emits characteristic fluorescent radiation of its atoms.

Once x-ray fluorescence is produced from the sample by means of an x-ray tube, appropriate components of the instrument (Fig. 8) separate this radiation into its characteristic wave-

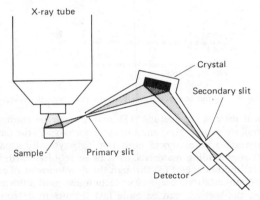

Fig. 8.   Optical path for one monochromator used in modern analyzer.

lengths, detect the energy emitted from each excited atom, and produce a signal that is representative of the number of atoms (concentration) of the elements in the sample. Typical excitation conditions (x-ray tube) are 50 kilovolts, 35 milliamperes. Bragg's law of x-ray diffraction is satisfied by the condition $N\lambda = 2d \sin \theta$, where λ is the wavelength and θ is the angle of incidence and diffraction of x-rays from a crystal whose lattice spacing is defined by d and N, the order of harmonic of the diffraction. For almost all fluorescence analysis, $N = 1$. The usable x-ray spectrum normally extends from 0.1 to about 20 Å. A helium or vacuum path is required for x-ray analysis of elements with atomic numbers lower than $Z \cong 24$ (wavelengths longer than 2.3 Å).

Instrumentation can provide for simultaneous elemental analysis using fixed, preselected x-ray detection channels and scanners or a goniometer to provide one or more channels that can be tuned to a wide wavelength coverage. Up to 30 monochromator positions are possible. Typical crystal materials covering the practical wavelength range of 1 to 20 Å are lithium fluoride, silicon oxide, sodium chloride, EDT, and ADP. Optimum analytical data for the elements of interest are obtained by using fully focusing Johansson curved and ground crystals.

An optical diagram of a Johansson curved-crystal spectrometer is given in Fig. 9. Each spectrometer of an x-ray quantometer may be equipped with optimum crystal-detector combinations for specific determinations in a wide variety of matrixes, includ-

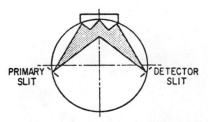

Fig. 9. Optical diagram of Johannson curved-crystal spectrometer.

ing steel, aluminum, copper-base materials, ores, cement, and slags—in both liquid and solid states.

The diffracted x-radiation is detected by Geiger, proportional, or semiproportional detectors. See Fig. 10. The detector of each

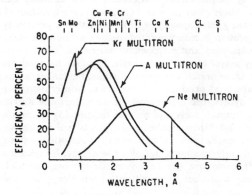

Fig. 10. Relative efficiency of x-radiation detectors.

monochromator generates pulses which are a measure of the intensity of radiation of each wavelength. The pulses are filtered through a discriminator in order to avoid undesired interferences. Pulses shrinking due to an increase of frequency of pulses is automatically compensated. Collected pulses are transferred to a computer for processing and output. See Fig. 11.

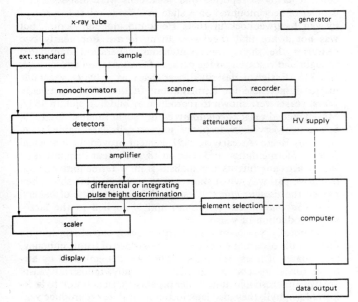

Fig. 11. System configuration of x-ray fluorescence. (*Bausch & Lomb.*)

The transition from laboratory to automated instrument to achieve high-speed continuous analysis of dry or wet materials

primarily involves the sample-handling and presentation hardware.

Limits of detectability for the desired elemental analyses vary depending upon the matrix, elements, methods of sample preparation, and quality of instrumentation applied. Generally, these are on the order of 1 to 100 parts per million. The limit of detectability, however, is only one criterion in evaluating methods of analysis. The time of analysis is important, particularly in production and process control laboratories. In multi-element spectrometers, it is possible to perform as many as 30 simultaneous elemental determinations in from 20 to 120 seconds, depending upon the material being analyzed.

—W. G. Shequen, Bausch & Lomb, Sunland, California.

XYLENE. $C_6H_4(CH_3)_2$, formula weight 106.16. There are three xylenes, ortho-, meta-, and para-xylene. Sometimes referred to as dimethylbenzenes, the xylenes have the following key physical properties: All of these compounds are insoluble

o-Xylene	mp	−25°C	bp	144°C	sp gr 0.881
m-Xylene		−47.4°C		139°C	0.867
p-Xylene		13.2°C		138.5°C	0.861

in H_2O, soluble in alcohol, and *o*-xylene and *m*-xylene are miscible in all proportions with ether; *p*-xylene is very soluble in ether.

The xylenes are very high-tonnage industrial chemicals and are raw materials or intermediate materials for numerous synthetic fibers, resins, and plastics. A large amount of *p*-xylene goes into polyester fiber production, while substantial quantities of *o*-xylene are consumed by the manufacture of phthalic anhydride. The prime source of xylenes are petroleum refinery reformate streams in conjunction with benzene and toluene extraction. The xylenes occur mixed in these streams.

When naphtha or naphthenic gasoline fractions are catalytically reformed, they usually yield a C_8 aromatics stream that is comprised of mixed xylenes and ethylbenzene. It is possible to separate the ethylbenzene and *o*-xylene by fractionation. It is uneconomic to separate the *m*- and *p*-xylenes in this manner because of the closeness of their boiling points. To accomplish the separation, a Werner-type complex for selective absorption of *p*-xylene from the feed mixture may be used. Or, because of the widely different freezing points of the two xylene isomers, a process of fractional crystallization may be used. To boost the *p*-xylene yield, the filtrate from the crystallization step can be catalytically isomerized.

XYLITOL. Sweeteners.

Y

YAG AND YIG. Synthetic yttrium iron and yttrium aluminum garnets, respectively. These materials were developed in the mid-1960s. They are pressed and sintered polycrystalline ceramics and are made by a solid-state reaction of Y_2O_3 with iron oxide or aluminum oxide. Garnets operate in microwave bandpass (filters) circulators and isolators in telephone, radar, and space-communication networks. The original electronic use led to the development of single-crystal yttrium aluminum garnets which approach the brilliance and hardness of diamond. Yttrium oxide is the base for neodymium-doped laser crystals.

See also **Neodymium; Rare-Earth Elements and Metals;** and **Yttrium.**

YEASTS AND MOLDS. These are very important plant organisms that make both positive and negative contributions to mammalian life processes. Their plus and minus values are particularly noted in connection with the production and storage of food products.

Taxonomy Plants that lack true roots, stems, or leaves, and that are without highly organized conducting systems are called simple plants and they make up the phylum Thallophyta. The two subdivisions of Thallophyta are *algae* and *fungi*. Algae have chlorophyll; fungi do not. Fungi utilize carbohydrates that are synthesized by green plants. Fungi are classified as *parasites* or *saprophytes*, the latter obtaining food from nonliving organic material. The science of fungi is *mycology*.

The fungi include: *Slime molds* (Myxomycophyta); *algal fungi* (Phycomycetes); *sac fungi* (Ascomycetes); and *club fungi* (Basidiomycetes); among others. The yeasts and most of the molds with which food products are associated in some way are of the Ascomycetes variety.

The varied interface between the Ascomycetes and foods ranges from their very negative, parasitic habit on certain crops (exemplified by Chestnut blight fungus; or ergot on grains); their cause of certain foodborne diseases; to their positive use in connection with the production (notably by fermentation) of major food products, such as bread and wine.

The term *yeast* is used to describe a relatively small number of Ascomycetes fungi (a few hundred—Lodder et al. (1970) classified 349 fermentative species, of which relatively few are used industrially), as compared with the vast number of other fungi (several thousand) that have been identified.

The term *mold* does not have a clearcut mycological definition. While the yeasts have a preponderantly positive value in food production and utilization, the molds make some positive contributions (some cheeses, etc.), but they participate in many more negative ways (leaf molds on plants; blue-green molds on fruits; machinery mold; causes of some foodborne diseases; etc.). Some authorities identify all or most molds as *saprophytes*, as previously defined. This identification, however, is not fully satisfactory in terms of the present loose ways in which the word *mold* is used in the professional literature. Molds do play an important role in the *biodegradation* of unwanted substances.

Yeasts

The importance of the economically useful yeasts can be attributed to two main factors: (1) *Fermentation*—the transformation of simple sugars and other organic chemicals to other, more desirable chemicals; and (2) *respiratory (oxidative) metabolism*—the great capacity of some yeasts for a protein synthesis during growth in richly aerated media containing a wide variety of carbonaceous and nitrogenous nutrients. Thus, yeasts serve in many ways: (1) As living cells, they are biocatalysts in the production of bread, wine, beer, distilled beverages, among other important food products. (2) As dried, nonfermentative whole cells or hydrolyzed cell matter, yeasts contribute nutrition and flavor to human diets and animal rations. (3) As producers of vitamins and other biochemicals, yeasts are a rich source of enzymes, coenzymes, nucleic acids, nucleotides, sterols, and metabolic intermediates. See also **Coenzyme; Enzymes;** and **Enzyme Preparations.** (4) As a versatile biochemical tool, yeasts aid research studies in nutrition, enzymology, and molecular biology.

Background. It is estimated that the arts of making wine, leavened bread, and beer were practiced more than 4000 years ago. The phenomena for producing these foods were attributed to "yeast" at an early age. In many languages, the word for yeast describes the visible effects of fermentation, as observed in the expansion of bread dough and the accumulation of froth or barm on the surface of fermenting juices and mashes. Historically, it has been reported that yeast cells were first seen in a droplet of beer mounted on a slide in a crude microscope used by van Leeuwenhoek in 1680. He found globular bodies, but was not aware that they were living forms. For nearly two centuries the theory of spontaneous generation dominated thought and research on the causes of fermentation and disease. In 1818, Erxleban described beer yeast as a living vegetable matter responsible for fermentation. In the following twenty years, yeasts were shown to reproduce by budding and, in 1837, Meyen named yeast *Saccharomyces*, or "sugar fungus." By 1839, Schwann observed "endospores" in yeast cells, later named *ascospores* by Reese. As early as 1857, Pasteur proved the biological nature of fermentation and later, in 1876, Pasteur demonstrated that yeast can shift its metabolism from a fermentative to an oxidative pathway when subjected to aeration. This shift, then named the *Pasteur effect*, is especially characteristic of bakers' yeast (*Saccharomyces cerevisiae*) and is applied in the large-scale production of yeasts.

Botanically, yeasts form a heterogeneous group of saprophytic forms of life occuring naturally on the surface of fruits, in honey, in exudates of trees, and in soil. They are disseminated by airborne dusts, insects, and animals. Typically, industrial yeasts are oval, microscopic, unicellular organisms. In addition to lacking chlorophyll, they also lack locomotion. They reproduce vegetatively by budding and sexually by spore formation (ascospores) within the mother cell or ascus. These properties place them in the family *Endomycetaceae* of the class Ascomycetes, as previously mentioned. Among the most important industrial yeasts are:

Saccharomyces cerevisiae (alcoholic beverages and bread)
S. cerevisiae var. *ellipsoideus*; *S. bayanus*; and *S. beticus* (wines)
S. uvarium (formerly called *S. carlsbergensis*) (beer, ale, etc.)
Kluyveromyces fragilis (formerly called *S. fragilis*) (whey disposal)

Food and feed yeast production employ several species in the family *Cryptococcaceae*: *Candida utilis*; *C. tropicalis*; and *C. japonica*, which are cultivated on plant wastes (wood sugars, molasses, stillage); and *C. lipolytica*, which converts hydrocarbons to yeast protein.

Properties of Yeasts. The cell structure of *S. cerevisiae*, as observed in optical microscope, reveals a rigid cell wall, a colorless, granular cytoplasm, and one or more vacuoles. Dimensions of a typical bakers' yeast are about 4–6 by 7–10 micrometers. Electron microscopy of ultrathin sections of a yeast cell show the microstructures, including: birth and bud scars on the cell wall, plasmalemma (cytoplasmic membrane), nucleus, mitochondria, vacuoles, fat globules, cytoplasmic matrix and volution or polyphosphate bodies. The cell walls of bakers' yeast contain 30–35% glucan (yeast cellulose), 30% mannan (yeast gum), which is bound to protein (about 7%), 1–2% chitin, 8–13% lipid material, plus inorganic components, largely phosphates.

Gross chemical composition of compressed bakers' yeast is approximately 70% moisture. The dry matter is made up of 55% protein (N × 6.25), 6% ash, 1.5% fat, and the remainder mostly polysaccharides, including about 15% glycogen and 8% trehalose.

Food yeast, molasses-grown, is dried to about 5% moisture and has the same chemical composition as bakers' yeast. In terms of micrograms per gram of yeast, the vitamin content is: 165 thiamine; 100 riboflavin; 590 niacin; 20 pyridoxine; 13 folacin; 100 pantothenic acid; 0.6 biotin; 160 para-aminobenzoic acid; 2710 choline; and 3000 inositol. Yeast crude protein 80% amino acids; 12% nucleic acids; and 8% ammonia. The latter components lower the true protein content to 40% of the dry cell weight.

Yeast protein is easily digested (87%) and provides amino acids essential to human nutrition. Most commercial yeasts show the following pattern of amino acids, as percent of protein: 8.2% lysine; 5.5% valine; 7.9% leucine; 2.5% methionine; 4.5% phenylalanine; 1.2% tryptophan; 1.6% cystine; 4% histidine, 5% tyrosine; and 5% arginine. The usual therapeutic dose of dried yeast is 40 grams/day, which supplies significant daily needs of thiamine, riboflavin, niacin, pyridoxine, and general protein.

The ash content of food yeasts ranges from 6 to 8% (dry basis), consisting mainly of calcium, phosphorus, and potassium. Contained in quantities of less than 1% are magnesium, sulfur, phosphorus, and sodium. At the microgram level are included iron, copper, lead, manganese, and iodine.

Triglycerides, lecithin, and ergosterol are the main constituents of yeast lipid (fat). Oleic and palmitic acids predominate in yeast fat. These resemble the composition of common vegetable fats. Ergosterol, the precursor of calciferol (vitamin D_2) varies from 1 to 3% of yeast dry matter.

Metabolic Activity. This is generally associated with the familiar alcoholic fermentation in which theoretically 100 parts of glucose are converted to 51.1 parts of ethyl alcohol (ethanol),

48.9 parts of carbon dioxide (CO_2), and heat. In addition, however, the anaerobic reaction also yields minor byproducts in small amounts—mainly glycerol, succinic acid, higher alcohols (fusel oil), 2,3-butanediol, and traces of acetaldehyde, acetic acid, and lactic acid. Fusel oil is a mixture of alcohols, including *n*-propyl, *n*-butyl, isobutyl, amyl, and isoamyl alcohols.

Respiratory activity of oxidative dissimilation is characteristic of many species of yeasts. During their aerobic growth, sugar is oxidized to carbon dioxide and water, with release of large amounts of energy (about 680 kcal when complete oxidation occurs). Aerobiosis produces a variety of byproducts, some in unusually high concentration, such as acetic acid, succinic acid, zymonic acid, polyhydric alcohols (glycerol, erythritol, etc), extracellular lipids, carotenoid pigments in shades of red and yellow, black pigment (melanin), and capsular polysaccharides (phosphomannan).

Production. Well over 85,000 tons (76,500 metric tons) of yeast dry matter are produced in the United States alone each year. About 75% of this is in the form of bakers' yeast, the remaining 25% represents about equal amounts of food yeast and feed yeast. This production issues from four types of manufacture: (1) Bakers' yeast is grown batchwise in aerated molasses solutions. (2) *Candida utilis* is obtained from wood pulp mill spent liquid. (3) *K. fragilis* is grown batchwise in cottage cheese whey. (4) Dried yeast is recovered as spent beer yeast. Worldwide production of all types of food and feed yeast is estimated at more than 450,000 dry tons (405,000 metric tons) per year.

The process for growing bakers' yeast is a model system for the propagation of microorganisms. The process commences with a laboratory culture of a pure strain of *S. cerevisiae*. Seed yeast is developed in successively larger volumes of nutrient solutions, beginning with a Pasteur flask and ending in a fermenting tank containing as much as 40,000 gallons (1514 hectoliters) of sterilized and diluted molasses maintained at 30°C. During the highly aerated growth period, minerals are added, pH is adjusted to 4.5, and diluted molasses is continuously fed in proportion to the increase in cell mass. Under ideal conditions, yeast cells may double in number every 2.5 hours, converting more than half (56.7%) of the sugar supplied to cell components. Biosynthesis of cell matter requires an equal amount of oxygen. To produce 100 weight units of yeast dry matter with 50% protein content requires about 400 weight units of molasses, 25 weight units of aqua ammonia, 15 weight units of ammonium sulfate, and 7 weight units of monobasic ammonium phosphate. For each 100 pounds of dry yeast, 75,000 cubic feet of air are required; for 100 kilograms of dry yeast, 4683 cubic meters of air are needed.

Fermentation Processes. The biochemistry of alcoholic fermentation involves a series of internal enzyme-mediated oxidation-reduction reactions in which glucose is degraded via the Embden-Meyerhof-Parnas pathway. See **Carbohydrates; Glycolysis.**

Some typical reactions performed by yeasts are listed in the accompanying table.

Post-Processing Spoilage of Food by Yeasts. Microbiological spoilage of food is a competitive process occurring among yeasts, bacteria, and molds. Yeasts normally play a small role in spoilage because they constitute only a small portion of the initial population, because they grow slowly in comparison with most bacteria, and because their growth may be limited by metabolic substances produced by bacteria. Evidence does not show food poisoning being caused by the presence of spoilage yeasts in foods. The byproducts of metabolism are not considered toxic and, while there are a few yeasts that may be considered patho-

REACTIONS IN WHICH YEASTS PARTICIPATE[1]

Type of Reaction	Number of Known Reactions	Examples of Reactions
Reduction	156	Diacetyl to acetoin (in beer); cinnamic aldehyde to cinnamic alcohol.
Decarboxylation	21	Malic acid to lactic acid (in wines); amino acids to amines (histamine and tyramine accumulate in soft cheeses because of surface growth of *Torulopsis candida* and *Debaryomyces kloeckera*).
Deamination	17	Examples of deamination and decarboxylation include conversion of amino acids to fusel oil (leucine to isoamyl alcohol, isoleucine to amyl alcohol, and phenylalanine to phenyl ethanol). Fusel oil formation is a normal function of all yeast fermentations (in alcoholic beverages, levels range from trace to 2200 parts per million). Deamination: Glutamic acid to gamma-OH-butyric acid (*S. cerevisiae*).
Oxidation	14	Acids, alcohols, sugars, hydrocarbons. Also stepwise: alcohol to aldehyde or acid (sake yeast).
Esterification	10	Ethyl acetate (*Hansenula anomala*).
Condensation	9	Acetaldehyde to acetoin; acetaldehyde to pyruvic acid to alpha-acetolactic acid.
Hydrolysis	5	Starch hydrolysis: By *Endomycopsis fibuligera*; artichoke starch (inulin) by *Kluyveromyces fragilis*.
Amination	1	—

Examples of Lesser-known Properties and Characteristics of Yeasts

Type of Reaction	Yeasts Involved	Examples
Lipolysis	*Candida lipolytica; C. rugosa; Torulopsis sphaerica*	Mainly in butter, margarine, and cheese.
Proteolysis	*C. lipolytica; T. sphaerica*	Especially on soft cheese surfaces.
Pectinolysis	*Saccharomyces kluyveri; Kluyveromyces fragilis; Hansenula anomala*	Softening of olives and cherries in brines, followed by formation of gas pockets in fruit. Strains of wine yeasts contain polygalacturonases which, during fermentation of grape juice, participate in the solubilization of pectin.
Acid formation	Species of Brettanomyces, Hansenula, Pichia, Saccharomyces	As a contaminant in wines, Brettanomyces spp. forms a higher concentration of volatile acids (also isobutyric and isovaleric acids) than *S. cerevisiae*. Pichia species and other yeasts are responsible for acetic acid production in brines of domestic green olives; *not* lactobacilli, as assumed for years (Vaughn *et al.*, 1976).
Pigmentation	*Rhodotorula glutinis.* Sporobolomyces spp.	Carotenoids (pink, red); *Rhodotorula glutinis* causes pink sauerkraut and discolors the surface of high-moisture cheeses.
Esterification	Species of Hansenula, Kluyveromyces, Brettanomyces	Ethyl acetate and ethyl lactate in cottage cheese and shredded Mozzarella cheese.
Turbidity formation	In wine: *Saccharomyces bailii, S. chevalieri*, Brettanomyces spp. In beer: *S. diastaticus, S. bayanus*	In soft drinks; in wine: *S. bailli, S. chevalieri;* in beer: *S. diastaticus, S. bayannus*.

[1]Based, in part, upon Wallen *et al.* (1959); Peppler (1977). See references listed.

genic, they are not known to be responsible for foodborne infections or intoxications (Walker, 1977; Peppler, 1977). The metabolism of yeasts can result in the development of unnatural flavors and odors and changes in pH because of the utilization of organic acids important in fermented foods. Many yeasts are capable of utilizing lactic, acetic, and citric acids, which are essential for production of flavor and for preservation of some foods. Decreased concentration of these acids causes an increased pH and produces conditions that favor growth of spoilage bacteria. Normally, several factors interact to establish conditions that favor growth and subsequent spoilage of foods by yeasts. Examples of food spoilage by yeasts are given in the lower half of the accompanying table.

Although yeasts are abundant in nature, notably on leaves and in the soil, they do not compete with bacteria and molds as sources of major problems. But in assessing the potential for spoilage by yeasts, one must consider those factors which are favorable or unfavorable to the growth and multiplication of yeast populations. Availability of oxygen is important. No yeast is known that can grow under strictly anaerobic conditions; thus, all require oxygen to be present in some proportion. Temperature is also an important factor. Yeasts are not heat resistant in the sense that bacterial endospores are. Most yeasts cannot withstand temperatures above 65–70°C. The majority of yeast species have an optimum temperature for growth between 20 and 30°C. Even though yeasts are very resistant to low temperatures and can survive frozen storage, they normally are not a major problem in spoilage of frozen or refrigerated foods. In refrigerated foods, psychrotrophic bacteria rapidly outnumber the yeasts, the latter constituting a minor part of the initial population. However, where antibiotics or ionizing radiation have been used to reduce or inhibit bacterial growth, then yeast population may predominate. If the definition of a *psychrophilic* organism is one that has an optimum temperature below 20°C, then there are a number of strains of psychrophilic yeasts.

The survival and growth of all organisms is dependent on the presence of some water in the environment. Yeasts generally require more moisture for growth than molds, but less than bacteria. Systems containing high concentrations of sugar or salt have low a_w values or high osmotic pressure (Walker, 1977). Organisms growing in such systems are usually referred to as being *osmophilic*. A number of investigators have not been able to find any yeast that is clearly osmophilic. Windisch (1969) suggested the term *osmotolerant* to describe some yeasts. It is important to note that yeasts which normally will not grow in a high-sugar environment may appear if the substance has been exposed to the air sufficiently to absorb water and dilute spots or edges of the material, thus favoring yeast growths. Species such as *Saccharomyces rouxii*, *S. rouxii* var. *polymorphous*, and *S. mellis* (Lodder, 1952) are capable of spoiling high-sugar foods, such as honey, maple sugar, sugar cane syrups, molasses, fruit syrups, candy, crystallized fruits, jams, jellies, and dried fruits. It is not uncommon for dried figs and prunes to be covered with a white coating of yeast during storage. The foods have a typical fermented odor and may contain gas pockets. This coating normally is a mixture of sugar and yeasts, with the principal yeast being *Schizosaccharomyces octosporus*.

Molds

The most critical factor in mold growth is the availability of sufficient moisture. Molds are widely distributed throughout nature. Chemicals for the destruction of molds, including those in the soil, are mentioned in the entry on **Antimicrobial Agents (Foods)**. In addition to food spoilage, various molds can be destructive of various materials, notably of substances prepared from animal skins, such as leather, book covers, shoes, and materials prepared from vegetable substances, such as paper and wood products. In warm, humid areas, mold (mildew) grows on wooden structures, including painted surfaces, unless these have been chemically treated. In the case of foods, molds cause innumerable problems during the full cycle of food production.

Storage temperature of food is less important than the presence of moisture, since fungi can grow and produce toxins over a wider span of temperatures than can any other microorganisms. Most species are able to grow at an a_w of 0.8–0.88; while xerophilic types can grow at an a_w of 0.65–0.75. A relative humidity of 70–90% establishes suitable moisture equilibrium for initiation of mold growth and toxin formation in numerous food products. See **Activity Coefficient.**

Unfortunately, to the would-be consumer of food, fungal contamination is not always immediately indicated and thus may not be an apparent reason either for the rejection of food by humans or the rejection of feedstuffs by livestock. Concentrations of mold organisms are not always easy to see or to smell. Further, the relative stability of most mycotoxins to heat precludes the use of cooking as a detoxifying procedure. Among the most aggressive of the molds are species of *Candida*, *Aspergillus*, *Rhizopus*, and *Mucor*. Several of the fungi are not pathogenic to healthy humans, but may be virulent pathogens in debilitated persons, or those treated with broad-spectrum antibacterial drugs or immunosuppressive measures (Lechevalier and Pramer, 1971). An example of this type of organism is *Cryptococcus neoformans* (Davis, 1973). Many of the fungi produce disease through infection rather than through mycotoxin production, but this usually requires predisposing factors.

Milner and Geddes pointed out in 1946 that the aflatoxigenic organism *Aspergillus flavus* is a common soil fungus throughout the world, and aflatoxin-contaminated food has been noted in many countries for a number of years (Wogan, 1966). Crops such as groundnuts (peanuts), in intimate contact with soils, are likely to contact the necessary mold inoculum for aflatoxin production, resulting in toxin formation upon storage in air.

Several investigators have shown that toxic strains of *Fusarium*, *Cladosporium*, *Penicillium*, and *Mucor* can sporulate and grow at temperatures well below 0°C. It also has been shown that toxin formation can be associated with overwintering of grain in the field. In some varieties of barley, such as Siri and Mala, the incidence of *Aspergillus* and *Penicillium* species is related solely to temperature and moisture storage conditions and is not correlated to the percentage of unripe grains. Low humidities tend to predispose hosts to invasion by molds by causing them to lose turgor. Mold is the cause of serious problems in connection with rice. The predominant molds present in wild rice during fermentation curing are *Mucor*, aflatoxigenic and nontoxigenic *Aspergillus*, *Penicillium*, and *Rhizopus* species. Most of the B_1 aflatoxin is in the hulls, and the level of the toxin can be reduced by parching, which serves to reduce, but not fully destroy molds.

Contamination of soybeans with *Aspergillus flavus* is found in approximately 50% of commercial samples. Fortunately, the incidence of aflatoxin from this route is quite low. Moisture at the time of maturity, development of the seed in a closed pod, and binding of zinc by phytic acid are suggested as reasons for resistance of soybeans to aflatoxin production.

The most active sugar contaminants are molds, particularly *Aspergillus*. Penicillium growth on cheddar cheese results in a pH gradient, increasing to a pH of 8 at the surface. Mycotoxins diffuse through foods and are not removed when the surface molds are removed. Therefore, molds of unknown variety or origin should not be ingested. On the other hand, many of the common molds found as contaminants of Western foods are used in Asian fermented foods and beverages, principally as sources of flavor.

Degradation of Aflatoxin. Numerous researchers have investigated a variety of chemical and physical means for degrading aflatoxin. These include the use of irradiation, heat, acids and bases, oxidizing agents, bisulfite, and biological agents. The practical application of ultraviolet radiation to date has not proved successful. Aflatoxin is quite heat-stable. Numerous investigators have concluded that although heat can degrade aflatoxins, it is not an effective and economically feasible means for inactivating these toxins when once present in foods or feeds.

Among oxidizers used have been sodium hypochlorite, potassium permanganate, and hydrogen peroxide. Bisulfite is commonly used in the wet milling of corn (maize) and in processing wines, fruit juices, jams, and dried fruits.

Machinery Mold. This mold generally refers to the buildup of the organism *Geotrichum candidum* Link on food-contact factory equipment in processing plants. The term *dairy mold* is used in the processing of milk to identify this mold.

Since passage of the original Federal Food and Drugs Act of 1906 (United States), the presence of machinery mold in any food processing plant is a violation of regulations. Antimicrobial agents have made it possible to keep machinery clear of this mold. The mold can be particularly bothersome in tomato and pineapple processing plants. See **Antimicrobial Agents.**

References

Anderson, A. W.: "The Significance of Yeasts and Molds in Foods," *Food Technology*, **31**, 2, 47–51 (1977).

Batra, L. R., and P. D. Millner: "Some Asian Fermented Foods and Beverages Associated with Fungi," *Mycologia*, **66**, 942 (1974).

Chipley, J. R., and M. L. Cremer: "Microbiological Problems in the Food Service Industry," *Food Technology*, **34**, 10, 59–68 (1980).

Davis, B. D., et al.: "Microbiology," Harper and Row, Hagerstown (1973).

Draughon, F. A.: "Effect of Plant-Derived Extenders on Microbiological Stability of Foods," *Food Technology*, **34**, 10, 69–74 (1980).

Eisenberg, W. V., and S. M. Cichowica: "Machinery Mold—Indicator Organism in Food," *Food Technology*, **31**, 2, 52–56 (1977).

Frank, J. F., et al.: "Control of Aflatoxin Production during Fermentation of Wild Rice," *Jrnl. Milk Food Technol.*, **38**, 73 (1975).

Hultin, H. O., and M. Milner, Editors: "Postharvest Biology and Biotechnology," Food & Nutrition Press, Westport, Connecticut (1979).

Ito, K. A., and G. R. Bee: Microbiological Hazards Associated with New Packaging Techniques," *Food Technology*, **34**, 10, 78–80 (1980).

Lindroth, S., Niskanen, A., and O. Pensala: "Patulin Production During Storage of Blackcurrant, Blueberry and Strawberry Jams Inoculated with *Penicillium expansum* Mold," *Jrnl. of Food Science*, **43**, 5, 1427–1429 (1978).

Lodder, J.: "The Yeasts, A Taxonomic Study," 2nd Edition, North-Holland Publishing Company, Amsterdam (1970).

Marth, E. H., and M. P. Doyle: "Molds: Degradation of Aflatoxin," *Food Technology*, **33**, 1, 81–87 (1979).

Milner, M., and W. F. Geddes: "Grain Storage Studies. III. The Relation between Moisture Content, Mold Growth, and Respiration of Soybeans," *Cereal Chem.*, **23**, 225 (1946).

Peppler, H. J.: "Yeast Properties Adversely Affecting Food Fermentations," *Food Technology*, **31**, 2, 62–65 (1977).

Lucas, G. B.: "The War Against Blue Mold," *Science*, **210**, 147–153 (1980).

Phaff, H. J., Miller, M. W., and E. M. Mrak: "The Life of Yeasts," 2nd Edition, Harvard Univ. Press, Cambridge, Massachusetts (1978).

Robach, M. C.: "Use of Preservatives to Control Microorganisms in Food," *Food Technology*, **34**, 10, 81–85 (1980).

Roeder, G. S., et al.: "The Origins of Gene Instability in Yeast," *Science*, **209**, 1375–1380 (1980).

Rowley, D. B., and A. Brynjolfsson: "Potential Uses of Irradiation in the Processing of Food," *Food Technology*, **34**, 10, 69–74 (1980).

Sauer, F.: "Control of Yeasts and Molds with Preservatives," *Food Technology*, **31**, 2, 66–67 (1977).

Shetty, K. J., and J. E. Kinsella: "Preparation of Yeast Protein Isolate with Low Nucleic Acid by Succinylation," *Jrnl. of Food Science*, **44**, 3, 633–638 (1979).

Walker, H. W.: "Spoilage of Food by Yeasts," *Food Technology*, **31**, 2, 57–61 (1977).

Windisch, S.: "Studies on Osmotolerant Yeasts," in "Yeasts" (A. Kocva-Kratochiflova, Editor), Vydavatel'stvo Slovenskej Akadmine Vied, Bratislava (1969).

Wogan, G. N.: "Chemical Nature and Biological Effects of the Aflatoxins," *Bacterio. Rev.*, **30**, 460 (1966).

YEAST INHIBITORS. Antimicrobial Agents (Foods).

YTTERBIUM. Chemical element symbol Yb, at. no. 70, at. wt. 173.04, thirteenth in the lanthanide series in the periodic table, mp 819°C, bp 1196°C, density, 6.966 g/cm³ (20°C). Elemental ytterbium has a face-centered cubic crystal structure at 25°C. Pure metallic ytterbium is silvery-gray and is stable in moist or dry air up to 200°C, after which oxidation occurs. The metal is readily dissolved by dilute and concentrated mineral acids. The metal dissolves in liquid NH_3 to yield a dark-blue color. There are seven natural isotopes, ^{168}Yb, ^{170}Yb through ^{174}Yb, and ^{176}Yb. Ten artificially produced isotopes have been identified. Ytterbium is one of the least abundant of the rare-earth group elements and is 53rd among all elements occurring in the earth's crust. The element was first identified by J. D. G. Marignac in 1878.

Electronic configuration is $1s^22s^22p^63s^23p^63d^{10}4s^24p^6$-$4d^{10}4f^{13}5s^25p^65d^16s^2$. Ionic radius, Yb^{2+}, 0.93 Å; Yb^{3+}, 0.86 Å. Metallic radius, 1.940 Å. First ionization potential, 6.25 eV; second, 12.17 eV. Other physical properties of ytterbium are given under **Rare-Earth Elements and Metals.**

The principal sources of ytterbium are euxenite, gadolinite, monazite, and xenotime, the latter being the most important. Ytterbium is separated from a mixture of yttrium and the heavy Lanthanides by using the sodium amalgam reduction technique. Ytterbium metal is obtained by heating a mixture of lanthanum metal and ytterbium oxide under high vacuum. The ytterbium sublimes and is collected on condenser plates whereas the lanthanum is oxidized to the sesquioxide.

To date, the major uses of ytterbium have been in applied and fundamental research. The element and its compounds have been used in magnetic "bubble" domain devices (ytterbium orthoferrite), in phosphors to convert infrared to visible light, in lasers, and radioisotope ^{169}Yb has found application in portable industrial and medical radiographic units.

NOTE: This 4th Edition entry was revised and updated by K. A. Gscheidner, Jr., Director, and B. Evans, Assistant Chemist, Rare-Earth Information Center, Energy and Mineral Resources Research Institute, Iowa State University, Ames, Iowa.

YTTRIUM. Chemical element symbol Y, at. no. 39, at. wt. 88.905, periodic table group 3b, mp 1522°C, bp 3338°C, density, 4.469 g/cm³ (20°C). Most of the properties of yttrium are similar to those of the heavy rare-earth elements, falling between gadolinium and erbium. Elemental yttrium has a close-packed hexagonal crystal structure at 25°C. The pure metallic yttrium is silvery-gray, retaining a luster in air up to about 400°C, above which it oxidizes to Y_2O_3. The metal is dissolved by most mineral acids, but is relatively inert in a 1:1 mixture of concentrated HNO_3 and 48% hydrofluoric acid. Yttrium is capable of working by normal tools; during the period 1958–1962, approximately 20,000 pounds (9072 kilograms) of the metal were fabricated into rod, sheet, and tubing for nuclear reactor development programs. The metal is immiscible with liquid or solid uranium metal and alloys and has a low thermal-neutron-absorption cross section and a low acute-toxicity rating. Since the early 1960s, however, yttrium has been replaced by other more available and more cost effective metal alloys. The natural isotope of yttrium is ^{89}Y. Fourteen artificial isotopes have been identified. See **Radioactivity.**

In terms of abundance, yttrium is present on the average of 33 ppm in the earth's crust and potentially is as plentiful as cobalt. The element was first identified by Fredrich Wohler in 1828.

Electronic configuration is $1s^22s^22p^63s^23p^63d^{10}4s^24p^64d^1 5s^2$. Ionic radius, Y^{3+}, 0.893 Å. Metallic radius, 1.801 Å. First ionization potential, 6.38 eV; second, 12.24 eV. Other physical properties of ytrrium are given under **Chemical Elements;** and **Rare-Earth Elements and Metals.**

Residues from uranium mining operations in Canada have been a major source of yttrium. Xenotime (YPO_4) found in Malaysia is another source. Some apatite deposits are unusually rich in yttrium and it also is found in gadolinite, euxenite, and samarskite.

In recovering yttrium, mixed rare-earth minerals or wastes are dissolved in HNO_3 or H_2SO_4. Liquid-liquid organic ion-exchange solvent extraction cells then separate a pure yttrium fraction, usually precipitated as an oxalate and then calcined to the oxide. Current capacity exists for the production of over 5,000 pounds (2268 kilograms) of pure yttrium compounds annually. The metal is obtained by metallothermic reduction, using calcium mixed with YCl_3 or YF_3 in a sealed retort at a temperature in excess of 1,550°C. Yttrium metal and alloys of yttrium and cobalt, or yttrium and magnesium have been deposited out of a molten electrolyte BaF_2-LiF-YF_3.

Aside from the earlier extensive use of yttrium in nuclear reactor hardware, the main use developed in 1965 with the discovery of a new red phosphor for use in color television picture tubes. Y_2O_3 is the host matrix which collects energy from the cathode ray tube electron beam. Transfer of this energy to europium oxide Eu_2O_3 in vacuum causes europium to emit a chromatically true visible red light. The europium-activated yttrium oxide red phosphor $Eu:Y_2O_3$ is now widely used in this application. Y_2O_3 also is used in synthetic yttrium-iron and yttrium-aluminum garnets which find use as microwave bandpass filters, circulators, and isolators in electronic and communications circuitry. Yttrium oxide also is the base for neodymium-doped laser crystals. The compound is used in nickel-based superalloys made by powder metallurgy techniques. Y_2O_3 is effective as a metallurgical dispersion hardening agent. Yttrium metal also is specified in several cobalt-base superalloys in which it improves hot corrosion (sulfidation) resistance at high temperatures. When used in iron-chromium-aluminum alloys, yttrium improves workability and adds resistance to sag when the alloys are used as electrical-heating elements. The metal also is applied as cladding to rotating turbine engine parts to obtain superior oxidation resistance. Yttrium also has been used in permanent magnets, YCo_5, and shows great promise. These magnets are second only to $PrCo_5$ as the most powerful permanent magnet materials developed to date, far exceeding alnico and other more conventional materials.

Chemistry and Compounds. Yttrium hydroxide, $Y(OH)_3$ is precipitated by NH_3 from solutions of yttrium salts. It differs in properties from the lanthanide hydroxides, both structurally on x-ray diffraction, and chemically, as in its ready absorption of atmospheric carbon dioxide. Yttrium oxide, Y_2O_3, is obtained by heating the hydroxide or oxyacid salts; it forms mixed oxides when heated with other oxides, such as Fe_2O_3 and TiO_2. Yttrium also forms a peroxide, Y_4O_9, obtained in hydrated form by treatment of yttrium solutions with hydrogen peroxide.

All four halides are known. The fluoride, YF_3, is readily formed by action of a fluoride on a solution of the nitrate. There is an oxyfluoride, YOF, formed by high-temperature, low-pressure heating of the mixed oxide and fluoride. Complex fluorides, containing $[YF_6]^{3-}$ exist; the cryolite minerals are of this composition. The group, $-YF_4$ is found in double salts formed by YF_3 and the alkali fluorides. The chloride, YCl_3, is formed as a hydrate by action of HCl solution upon the hydroxide. It may be dehydrated by slow heating; rapid heating gives the oxychloride, YOCl. The bromide, YBr_3, is formed as a hydrate by action of HBr upon the hydroxide. Its dehydration requires heating under vacuum. The iodide, YI_3 is best prepared from the anhydrous chloride, by action of HI and I_2.

Both normal and mixed carbonates are known. The former is precipitated, as $Y_2(CO_3)_3 \cdot 3H_2O$ from Y^{3+} solutions by alkali metal carbonates, which in excess dissolve the precipitate to form a soluble hydrated double carbonate. The oxycarbonate is also a double molecule, $3Y_2(CO_3)_3 \cdot 2Y(OH)_3$, formed by action of CO_2, upon the hydroxide.

The nitrate, $Y(NO_3)_3$ exists as a number of hydrates. The hexahydrate formed by action of HNO_3 upon the hydroxide, is dehydrated at 100°C to give the trihydrate and the anhydrous salt. However, other hydrates are known, as well as double nitrates (especially with the lanthanide elements) and oxynitrates, of the general formula, $xY_2O_3 \cdot yN_2O_5 \cdot zH_2O$.

Yttrium hydroxide forms a hydrated sulfate with H_2SO_4, which is dehydrated on heating. It forms double sulfates with alkali and ammonium sulfates.

The carbide, formed from the oxide and carbon in the electric furnace, appears to have the composition, YC_2. It yields acetylene and other hydrocarbons upon hydrolysis.

NOTE: This 4th Edition entry was revised and updated by K. A. Gschneider, Jr., Director, and B. Evans, Assistant Chemist, Rare-Earth Information Center, Energy and Mineral Resources Research Institute, Iowa State University, Ames, Iowa.

Z

ZEOLITE GROUP. To the zeolite group of minerals belong a number of hydrous silicates of aluminum which also ordinarily contain sodium or calcium, but rarely they may carry barium, strontium, magnesium, and potassium. These minerals are not related crystallographically as they occur in the isometric, orthorhombic, hexagonal and monoclinic systems, but they are all characterized by the presence of water, up to 10 or 20%, which is easily released with the application of heat. They are all rather soft minerals, hardness, 3.5–5.5; of low specific gravity, 2.0–2.5, and they will decompose readily upon treatment with acid, most of them yielding a gelatinous mass.

The easy fusion, together with the rapid expulsion of water, is responsible for the name of this interesting group; it is derived from the Greek words to boil and a stone, hence zeolite, "a boiling stone." The zeolites are secondary minerals, usually found filling fissures and cavities in the more basic igneous rocks as basalt, and gabbro, but occasionally in the more acidic types as granite or in gneisses. The following members of the zeolite group are described under **Analcime; Chabazite; Harmotome; Heulandite; Natrolite; Phillipsite; Scolecite;** and **Stilbite.**

ZEOLITE SOFTENING (Boiler Water). Water Treatment (Boiler).

ZEOLYTIC CATALYST. Petroleum.

ZINC. Chemical element symbol Zn, at. no. 30, at. wt. 65.38, periodic table group 2b, mp 419.57°C, bp 907°C, density, 7.1 g/cm^3. Elemental zinc has a close-packed hexagonal crystal structure. There are five stable isotopes, ^{64}Zn, ^{66}Zn through ^{68}Zn, and ^{70}Zn. Six radioactive isotopes have been identified, ^{62}Zn, ^{63}Zn, ^{65}Zn, ^{69}Zn, ^{71}Zn, and ^{72}Zn. With exception of ^{65}Zn which has a half-life of 245 days, the half-lives of the other isotopes are measured in minutes and hours.

Zinc is a bluish-white metal, malleable and ductile at 150°C, but at 180°C it changes rapidly so that at 205°C, it may be easily powdered; remains lustrous in dry air, but is slightly tarnished in moist air or in water; burns upon heating to vaporization with a bluish flame, forming zinc oxide; soluble in acids—slowly when pure, but rapidly on contact with copper or platinum; soluble in alalis. Discovery prehistoric.

Zinc ranks 27th in order of abundance of the chemical elements in the earth's crust, an estimated 0.004% content of igneous rocks on the average. It is estimated that a cubic mile of seawater contains about 48 tons of zinc (10.4 metric tons in a cubic kilometer).

Electronic configuration is $1s^22s^22p^63s^23p^63d^{10}4s^2$. First ionization potential, 9.391 eV; second, 17.89 eV. Oxidation potential: $Zn \rightarrow Zn^{2+} + 2\bar{e}$, 0.762 V; $Zn + 4OH^- \rightarrow ZnO_2^{2-} + 2H_2O + 2\bar{e}$, 1.216 V. Ionic radius, Zn^{2+}, 0.75 Å. Metallic radius, 1.33245 Å. Other physical properties of zinc are given under **Chemical Elements.**

Zinc and lead usually occur together in nature as sulfides. Earlier separation processes involved the fine grinding of the combined sulfides and then treating the particles with chemical reagents to cause one sulfide to be preferentially wetted and thus the two sulfides separated by the froth flotation process. In a first stage, the lead sulfide is floated while the zinc sulfide sinks to the bottom of the tank. In the second stage, the process is reversed and the zinc sulfide is floated. Gangue and other nonmetals collect at the bottom of the tank. The separated sulfides are dewatered to a 6–8% moisture content and are referred to as the *zinc concentrate* and the *lead concentrate*.

A major zinc ore is ZnS (sphalerite) which frequently occurs with the major lead ore PbS (galena). The lead-zinc ores usually contain recoverable quantities of copper, silver, antimony, and bismuth as well. Major deposits of this type are worked in Australia, the United States, Canada, Mexico, Peru, Yugoslavia, and the Soviet Union. Two other important zinc ores are $ZnCo_3$ (smithsonite) and iron-zinc-manganese oxide (franklinite). Several of these minerals are described under separate alphabetical entries.

Extractive Metallurgy of Zinc[1]

As of the early 1980s, zinc production throughout the world is based almost exclusively on two processes:

(1) *Electrolytic process*, in which oxidized zinc concentrates are leached in sulfuric acid and then electrolyzed to plate SHG (special high-grade zinc metal) on the cathode and to regenerate the acid on the anode.

(2) *ISP* (*Imperial smelting process*), a combined lead/zinc process in which oxidized concentrates are reduced with coke in a shaft furnace and the zinc vapor collected in a lead splash condenser.

The remaining, previously conventional retort and electrothermic plants are under both environmental and economic pressures. Many of those plants remaining have opted to rely on larger percentages of secondary feeds to stay competitive.

The ISP evolved to fill a very special niche in nonferrous metallurgy because of its capability of treating lead-zinc concentrates which may also contain appreciable amounts of copper. An estimated 10–15% of the free world's zinc production is now supplied by eleven ISP plants presently operating. The concentrate is normally oxidized in a sintering machine to produce a feed for the blast furnace where the zinc oxide is reduced with coke. Some effort has been underway to develop a hot briquetting operation to produce a suitable feed without sintering. Other efforts to improve the economic competitiveness of the process include air preheat and the use of an oxygen-enriched blast to reduce coke consumption. The application of the process expanded rapidly during the 1960–1968 period. However, no ISP plant has been built in the Western countries since 1972. A market shift to a preference for SHG zinc and the rapidly increasing environmental pressures have contributed to stifling further adoption of the process.

---

[1] Information for this topic furnished by C. O. Bounds, St. Joe Minerals Corporation, Monaca, Pennsylvania

Although the electrolytic zinc process can trace its industrial history back over 60 years, only during the last 5 to 10 years has it become the industry standard, commanding presently over 75% of the free world's zinc capacity. While the electrolytic zinc process has been varied to meet the demands of the particular feed, the flow sheet always contains the basic steps of roasting, leaching, solution purification, and electrolysis. The zinc concentrate is oxidized with air to produce acid-soluble zinc oxide (calcine), and sulfur dioxide-containing off gas suitable for conversion to acid, as well as byproduct steam. The Vieille-Montagne/Lurgi fluidized bed roaster is the industry standard with ever-larger capacity roasters being installed.

The roaster product is leached with spent electrolyte (sulfuric acid) under near-neutral conditions to dissolve most of the zinc, copper, and cadmium, but little of the iron. The leach residue solids are releached in hot, strong acid to dissolve more zinc since it attacks the otherwise insoluble zinc ferrites. The iron which is also dissolved in this second leach is then precipitated as jarosite, goethite, or hematite. The development of these iron precipitation techniques permitted the use of the hot, strong acid leach and an increase in zinc extraction from about 87% to greater than 95%. Simultaneously, the hot acid leach frequently generates a leach residue rich enough in lead and silver to provide significant byproduct value, as well as increased recovery of cadmium and copper.

The neutral solution is purified to remove impurities more noble than zinc, e.g., cadmium, copper, cobalt, nickel, arsenic, antimony, and germanium. The purification is accomplished by cementation in two or more steps with the addition of zinc dust. Generally, at least one cementation step is conducted at high temperature with arsenic, antimony, or copper-arsenic added. Cadmium is usually recovered in the metallic state and copper, nickel, and cobalt are recovered as sludges if present in sufficient quantities.

Zinc is extracted from the purified solution in cells using lead/silver alloy anodes and aluminum cathodes at a current density of 38–60 amperes/square foot (400–650 amperes/square meter). The product is normally SHG zinc, particularly if strontium carbonate is added and/or lead/silver anodes of greater than 0.596 silver content are used. After deposition of 24 to 72 hours, the cathodes are removed from the cells, the zinc stripped by automatic machines in modern plants, and melted and cast for market. The move to automated handling of large cathodes was a major factor in lowering the overall labor requirement in producing zinc.

The following technical developments have led to the emergence of the electrolytic process:

1. Adoption of high-capacity, low-labor fluidized bed roasters
2. Adoption of continuous schemes for leaching and solution purification, allowing more automation and lower operating costs
3. Improved raw material utilization via hot acid leaching and iron precipitation as previously described
4. Construction of higher-capacity plants with larger equipment
5. Adoption of mechanized cathode handling/stripping with dramatically lower labor demands
6. Improved byproduct recovery via the hot acid leach.

With exception of leach residue (jarosite, goethite, etc.) disposal, the process is environmentally sound.

The electrolytic process is inherently a somewhat energy inefficient (being based on electrical rather than directly on fossil energy) and capital-intensive operation. A return to pyrometallurgical smelting is conceivable if the environmental concerns are addressed and a noncoke (or at least low-coke) process is developed. New hydrometallurgical developments include the near-commercial Sherritt-Gordon pressure leach to eliminate roasting and the generation of sulfur dioxide-rich gases, and laboratory experiments with chloride-based leaching/electrolysis.

Production Tonnage. Worldwide production of slab zinc (1979) was slightly over 7 million short tons (6.3 million metric tons). The leading producers were the U.S.S.R. (16.5% of the total), followed by Japan (12.3%); Canada (9.1%); the United States (7.6%); Germany (West) (5.5%); Australia (4.8%); Belgium (4.0%); France (3.9%); Poland (3.5%); Italy (3.2%); Spain (2.7%); Mexico (2.5%); and China (2.5%), these countries accounting for over 75% of the total production. Other countries, in descending order, with over 100,000 short tons (90,000 metric tons) are Finland, the Netherlands, Yugoslavia, Bulgaria, and North Korea.

Uses of Zinc

Some concept of the major uses for zinc can be gleaned from Table 1. These data indicate the consumption patterns in the

TABLE 1.   CONSUMPTION OF SLAB ZINC IN THE UNITED STATES (1979)

APPLICATIONS OF ZINC		PERCENT OF TOTAL TONNAGE CONSUMED
Galvanizing		40.8
sheet and strip	26.5	
wire and wire rope	2.3	
tube and pipe	4.1	
fittings (tube and pipe)	0.6	
tanks and containers	0.3	
structural shapes	2.2	
fasteners	0.5	
pole-line hardware	0.4	
fencing, wire cloth and netting	1.5	
other	2.4	
In Brass and Bronze		13.8
sheet, strip, and plate	6.6	
rod and wire	5.0	
tube	0.6	
casting and billets	0.2	
copper-base ingots	0.6	
other copper-base products	0.8	
In Zinc-base Alloys		28.0
die casting alloy	27.4	
dies and rod alloy	0.1	
slush and sand casting alloy	0.5	
Other Uses		17.4
rolled zinc	2.2	
zinc oxide	3.5	
other applications	11.7	
Total		100.0

SOURCE: U.S. Bureau of Mines

United States as of 1979. Slab zinc is available in three grades, as specified by the American Society for Testing and Materials (Table 2).

Zinc Coating. Constituting the largest single use of the metal, zinc coating is accomplished mainly by dipping the product in molten zinc or by electroplating (electrogalvanizing). Hop dip galvanizing employs chiefly the less pure grades of zinc. The life of galvanized material is proportional to the thickness

TABLE 2. GRADES OF SLAB ZINC (ASTM B6)

	COMPOSITION (Percent)			
GRADE OF ZINC	Lead (maximum)	Iron (maximum)	Cadmium (maximum)	Zinc (minimum by difference)
Special High Grade	0.003	0.003	0.003	99.990
High Grade	0.03	0.02	0.02	99.90
Prime Western	1.4	0.05	0.20	98.0

NOTES: When specified for use in manufacture of rolled zinc or brass, aluminum is held to 0.005% maximum.
Tin in Special High Grade zinc is held to 0.001% maximum.
Aluminum in Prime Western zinc is held to 0.05% maximum.
SOURCE: American Society for Testing and Materials.

of the coating, and recent developments in both electrogalvanizing and hop dip galvanizing have been toward application of heavy coatings that will withstand deformation without peeling.

Because of its relatively high electropotential (position in the emf series of metals), zinc can provide electrolytic protection against corrosion of several common metals, notably products made of iron and steel. Advantage of this characteristic also is taken in the use of zinc as the anode material for a number of types of batteries, power packs, and fuel cells. In providing a protective coating over ferrous metals, the attack of corrosive materials on the zinc produces a relatively inert reaction-product film which deters destruction of the underlying zinc and base metal. When the coating is broken, as may result from mechanical means such as scratching, abrading, etc., the zinc, having a higher electropotential than the ferrous metals, slowly is expended in furnishing the required protective current. Thus, serious corrosion is delayed for a long period, providing long useful life to the zinc-coated products except in the most adverse cases, or where a poor selection of construction materials was initially made.

The most widely used form of zinc coating is effected by hot-dip galvanizing in which steel sheets, coils, structurals, hardware, wire, and other forms are dipped in a bath of molten zinc. The zinc readily adheres to a previously cleaned (pickled and thoroughly washed) iron or steel surface. The thickness of the coating is controlled by manipulating process temperatures, time the underlying metals are in contact with the molten bath, and mechanical means used. Products that are commonly hot-dip galvanized include roofing, siding, transmission towers, highway guardrails, light poles, culverts, and fencing. Other forms of zinc coating include flame-spraying or metallizing, flake galvanizing, sherardizing or cementation, plasma arc spraying, vacuum metallizing, and also painting with materials that contain zinc pigments.

Electrogalvanizing makes it possible to apply a zinc coating to such products as steel strip, wire, conduits, hardware, etc., in a high-speed fashion through electroplating. One-side zinc coating of steel sheet for automotive applications by electrodeposition is a notable innovation in recent years. One advantage of electroplating is that the products to be coated are not subject to thermal conditions which may alter dimensions and shape. Electrodeposits range in thickness from 0.00015 inch (0.004 millimeter) to 0.001 inch (0.25 millimeter). Sherardizing is accomplished by heating the ferrous materials to be coated at a temperature of about 350°C in contact with zinc dust in a closed vessel. The resulting coating consists of iron-zinc alloys. It has been found that sherardized coatings match the corrosion resistance of electrogalvanized or hot-dip coatings of equivalent thickness.

Die Castings. The major use of zinc as a structural material is in alloys for pressure die casting. Development of the modern zinc die casting alloys was directly related to use of Special High Grade zinc, with the addition of particular alloying constituents held within close limits and control of impurities. For die castings it is essential to use this extra pure grade of zinc to ensure extremely low iron, lead, cadmium, and tin contents. It is also necessary to limit these same impurities in the metals added to make the desired zinc alloy composition. Only by such control can zinc die castings be produced that are stable in dimensions and properties.

The impurities lead, cadmium, and tin, if present in castings in amounts greater than the established maximums (0.005% lead; 0.004% cadmium; 0.003% tin) cause subsurface network corrosion. These limits are close to critical values. Iron is held to 0.10% maximum to prevent excessive skimming losses and machining problems.

Zinc alloys are low in cost of metal per casting, are easy to die cast, are cast at low temperatures, have strength greater than all other die casting metals except the copper alloys, lend themselves to casting within close dimensional limits, permit the thinnest sections yet produced, and are machined at minimum cost. Their resistance to surface corrosion is adequate in a wide range of applications. Prolonged contact with moisture results in formation of white corrosion products, but surface treatments can be applied that largely prevent formation of such products.

Limiting service conditions for standard zinc-base die casting alloys are as follows: At temperatures slightly above 95°C (200°F) their tensile strength is reduced 30% and their hardness 40%. At subzero temperatures some embrittlement occurs, but impact strength is still in the same range as that of aluminum and magnesium die casting alloys at normal service temperatures. At room temperature, impact strength of die castings is much higher than that of aluminum or magnesium die castings or iron sand castings.

All die castings have at least a light flash at the die parting, and those requiring movable cores will have some flash around the cores. Flash is also formed around ejector pins at the points at which they make contact with the casting.

Although it often is cheaper to cut threads than to cast them, for many pieces cast threads usually are more economical. Male threads usually are made with a parting parallel to the axis; this leaves a flash at the parting. The flash can be removed with a shaving tool in some instances, but in others a chasing operation is necessary.

Zinc die castings are invariably cast within quite close dimensional limits, but some machining is commonly required in addition to removal of flash, even though it may consist only of such simple operations as punching, drilling reaming, or tapping of holes. Zinc die castings can be soldered or welded, but ordinarily neither is used except for special applications or repair.

Many of the finishes applied to other types of metal products can also be applied to zinc die castings, although some differences in formulation as well as occasional differences in method of application may be desirable. The types of finishes applicable to zinc die castings include mechanical finishes—buffed, polished, brushed, and tumbled; electrodeposited finishes—copper, nickel, chromium, brass, silver, and black nickel; chemical finishes—chromate, phosphate, molybdate and black nickel; organic finishes—enamel, lacquer, paint and varnish; and plastic finishes. Electrodeposited coatings of virtually any metal capable of electrodeposition can be applied to zinc die castings.

The automotive industry uses by far the largest number of zinc alloy die castings. Some of the most important mechanical parts made include carburetors, bodies for fuel pumps, windshield-wiper parts, speedometer frames, grilles, horns, heaters, and parts for hydraulic brakes. The electrical industry probably uses a larger diversity of die castings than the automotive industry. Such parts are used in washing machines, oil burners, stokers, motor housings, vacuum cleaners, electric clocks, and kitchen equipment and utensils. Zinc die castings are used in business machines—typewriters, recording machines, picture projectors, vending machines, accounting machines, cash registers, cameras, slicing machines, garbage disposers, gasoline pumps, hoists, and drink mixers. Building hardware, padlocks, toys and novelties also consume a substantial percentage of the total production of zinc-base die castings.

Much more detail on zinc metal products and applications will be found in the Horvick (1979) reference listed. Zinc alloys with copper are described under **Copper.**

Chemistry and Compounds. In virtually all of its compounds zinc exhibits the +2 oxidation state, although compounds of zinc(I) have been reported in the gaseous phase. Zinc is readily oxidized in the presence of hydroxide ions, e.g., by H_2O, this behavior being attributed to the stability of the $Zn(OH)_4^{2-}$ ion. Like other transition group 2 elements zinc has a marked tendency to form covalent structures, e.g., ZnO and ZnS.

Zinc oxide, formed by oxidation of the metal, dissolves in acids to yield Zn^{2+} ions, and in alkalies to form $Zr(OH)_4^{2-}$ ions. It reacts slowly with moist CO_2 to form the oxycarbonate, $5ZnO\cdot2Co_2\cdot4H_2O$. Addition of an alkali metal hydroxide to a solution of a zinc salt does not precipitate the hydroxide, producing instead hydroxyzincate salt or a precipitate of flocculant zinc oxide. However, the hydroxide can be obtained from a sodium zincate solution on dilution and standing. Zinc hydroxide is amphiprotic, yielding Zn^{2+} ions with acids and $Zn(OH)_4^{2-}$ ions with (excess of) alkali hydroxides. Zinc peroxide is produced by treating a zinc chloride solution with sodium peroxide at a pH of 9.5. It is unstable, decomposing slowly upon standing. Zinc oxide is used extensively in rubber, paints, and chemicals. Expanding uses include exterior latex paints, particularly alkyd-modified latex, the photocopy paper field, and as a substitute for mercury in mildew and fungus prevention formulations.

Unlike the other zinc halides, zinc fluoride, ZnF_2, is only slightly soluble in cold water. The anhydrous halides are prepared by direct union of the elements. In solution, zinc chloride, bromide, and iodide exhibit anomalous conductance properties attributed to undissociated molecules and complex ions. On heating these solutions, halogen acids, HX, are evolved, leaving oxyhalides in the fused residue. The zinc halides readily form double salts with halides elements of main groups 1, 2, and sometimes 3 and 4. See also **Bromine.**

Zinc oxycarbonate is formed by the reaction of suspensions of the oxide or hydroxide with CO_2; to produce the normal carbonate, a very rapid stream of carbon dioxide and, usually, a somewhat higher pH is required.

Zinc also forms both nitrates and oxynitrates (in various hydrated forms) but the oxynitrate is less stable. It is formed by heating the hexanitrate, or treating the nitrate solution with NH_3.

Zinc forms a wide variety of other salts, many by reaction with the acids, though some can only be obtained by fusing the oxides together. The salts include arsenates (ortho, pyro, and meta), the borate, bromate, chlorate, chlorite, various chromates, cyanide, iodate, various periodates, permanganate, phosphates (ortho, pyro, meta, various double phosphates), the selenate, selenites, various silicates, fluosilicate, sulfate, sulfite, and thiocyanate.

Zinc sulfide, selenide, and telluride are more pronouncedly covalent than the oxide. They can be prepared from the elements, or in the case of first two, by the action of H_2S or hydrogen selenide upon zinc solutions. Zinc nitride, Zn_3N_2, prepared from zinc dust and NH_3, hydrolyzes readily to NH_3 and the oxide. Two zinc phosphides are known; Zn_3P_2, formed by heating the elements, yields phosphine with acids; ZnP_2 and ZnHP have also been prepared.

One of the features of the chemistry of zinc is the fact that it is among the elements having a large number of complex compounds, mostly with coordination numbers of four and tetrahedral, but some with coordination numbers of 6, such as those of ethylenediamine which contain the ion $[Zn(en)_3]^{2+}$. The large number of halogen double salts have already been cited. The marked donor ability of oxygen toward zinc is evident from the number of basic salts, the existence of the zincates (containing ZnO_2^{2-} and also $Zn(OH)_4^{2-}$, the latter in strong alkali), and the formation of such chelate complexes as acetylacetonates and dioxalato complexes. Sulfur is a better donor than oxygen, so that addition compounds of the type $(R_2S)_2ZnX_2$ are formed from dialkyl sulfides and zinc halides, while thiourea forms chelate complexes containing $[Zn(th)^2]^{2+}$. The ready reactions with ammonia, as with amines, give large numbers of complexes; those with ammonia include diammines, triammines and tetrammines, containing $[Zn(NH_3)_2]^+$, $[Zn(NH_3)_3]^{2+}$ and $[ZN(NH_3)_4]^{2+}$ respectively.

Prominent among the carbon donor complexes are the cyanides, principally compounds of $[Zn(CN)_4]^{2-}$ and $[Zn(NH_3)]^-$ is also known. Other carbon donor complexes are the triethyl and tetraethyl complexes, which are readily electrolyzed.

References

Andre, J.: "Zinc," San Francisco Mining Convention, San Francisco, California (September 16–20, 1972).

Barry, G. S.: *CIM Bulletin*, pages 156–159 (March 1975).

Boersma, J., and J. G. Noltes: "Organic Coordination Chemistry," International Lead-Zinc Research Organization, Inc., New York (1968).

Horvick, E. W.: "Selection and Application of Zinc and Zinc Alloys, "Properties of Zinc and Zinc Alloys," in "Metals Handbook," 9th Edition, Vol. 2, American Society for Metals, Metals Park, Ohio (1979).

Gordon, A. R., and R. W. Pickering: *Metallurgical Transactions B*, Vol. 6B, pages 43–53 (March 1975).

Mathewson: "Zinc, The Metal, Its Alloys and Compounds," *ACS Monograph 142*, American Chemical Society, Washington, D.C. (1959).

Meisel, G. M.: *Journal of Metals*, pages 25–32 (August 1974).

Staff: "Metalurgiya (Sofia)," translated by Sunlone, Inc., Arlington Heights, Illinois (1978).

Staff: "A Mine to Market Outline: Zinc," Zinc Institute Inc., New York (revised periodically).

Udrycki, A., and D. Krupkowa: *Chemical Age of India*, **26**, 8, 597–614 (August 1975).

—E. W. Horvick, Zinc Institute Inc., New York

ZINC (In Biological Systems). Zinc was one of the first of the trace elements to be essential for both plants and animals, and yet problems of zinc nutrition are still of pressing importance. Evidence of zinc deficiency in crops is being recognized in new areas and the use of zinc in fertilizers has increased steadily in recent years. A dry, cracked condition of the skin of pigs (*parakeratosis*) has been a zinc deficiency problem to pork producers. Diseases and syndromes that have been attributed to zinc deficiency in the human diet include loss of appetite, loss of sense of taste, and delayed healing of burns and other wounds. It is interesting to note that application of zinc-containing ointments to promote healing is an old practice in human medicine. Laboratory animals deficient of zinc may be subject to serious reproductive problems, including infertility of males, failure of conception or implantation of the embryo, difficult births, and deformed offspring. The extent to which zinc deficiency is a primary cause of reproductive problems in farm animals and humans is not thoroughly understood and research continues.

Patients of both sexes with sickle cell anemia have been reported to be zinc deficient. Administration of zinc to males with this disease has been shown to improve the hypogonadism and the short stature associated with this deficiency. Administration of small dosages of zinc sulfate is part of the therapy in treating certain types of sickle cell anemia.

Zinc deficiency in crops is frequently observed where fields have been graded to smooth them so that irrigation water can be applied more uniformly. Where the topsoil is cut away from small areas of these fields, such crops as corn (maize) and beans may be stunted and many leaves will be white instead of the usual green. If zinc fertilizers, supplying as little as 10 pounds of zinc per acre (11.2 kilograms per hectare) are applied, bumper crops may be grown on these soils. Citrus trees are commonly fertilized with zinc.

When zinc fertilizers are used on soils deficient in zinc, crop production may be increased even though the zinc concentration in the plant tissues and especially in the seed show no increase. With higher levels of zinc fertilization, the zinc concentration in plants may increase. Some evidence shows that the value of food and feed crops as sources of dietary zinc can be improved by using zinc fertilizers at rates exceeding those required for optimal plant growth. However, very high rates of zinc fertilization can depress crop yields.

The zinc contained in plants is not fully utilized by animals. Diets high in calcium and phosphorus have been associated with poor digestibility of dietary zinc. Diets with large amounts of soy protein are particularly likely to require extra zinc fortification for livestock. Meat is an important source of zinc for human diets. Where supplementation of zinc is indicated, zinc sulfate, zinc oxide, and zinc carbonate are commonly used.

Zinc in Metabolism. Zinc was first recognized as a trace element—then referred to as a *growth factor* for *Aspergillus niger* by Raulin (1869). Evidence for a specific biochemical role of zinc was first obtained by Keilin and Mann in 1940, when the metal was shown to be a stoichiometric component of bovine carbonic anhydrases. The findings of many other zinc-containing enzymes during the interim have indicated a diverse biological role for zinc.

The high affinity of zinc for nitrogeneous and sulfur-containing ligands seems chiefly responsible for the occurrence of zinc in a wide variety of biological compounds, such as proteins, amino acids, nucleic acids, and porphyrins. Operationally, the enzymes which are affected by zinc can be considered in two groups: (1) zinc metalloenzymes, and (2) zinc metal-enzyme complexes. Zinc metalloenzymes incorporate zinc so firmly in the protein matrix that they can generally be considered as an entity. Under reasonably mild conditions, the metal and protein moiety are isolated together and exhibit an integral stoichiometric relationship. On the same basis, a strict correlation is preserved between metal and enzyme activity, allowing the inferential identification of a specific biological function of zinc *in vivo*. Zinc metal-enzyme complexes, in contrast, comprise enzymes which are activated *in vitro* by the addition of zinc ions. The loose association and the relative lack of metal ion specificity render it difficult in many cases to assign specific biological significance to zinc *in vivo*.

In zinc metalloenzymes, zinc is a selective stoichiometric constituent and is essential for catalytic activity. It is frequently present in numerical correspondence with the number of active enzymatic sites, coenzyme binding sites, or enzyme subunits. Removal of zinc results in loss of activity. Inhibition by metal complexing agents is a characteristic feature of zinc metalloenzymes. However, no direct relationship holds between the inhibitory effectiveness of these agents and their affinity for ionic zinc. Although zinc is the only constituent of zinc metalloenzymes *in vitro*, it can be replaced by other metals *in vitro*, such as cobalt, nickel, iron, manganese, cadmium, mercury, and lead, as in the case of carboxypeptidases.

Zinc is a ubiquitous component of animal and plant tissue. In vertebrates, most organs, including pancreas, contain 20–30 micrograms of zinc per gram of wet tissue. Liver, voluntary muscle, and bone hold about double this amount. Zinc contents ranging from 100 to 1000 micrograms/gram weight have been measured in islet tissues of certain teleost fishes. Correlation between zinc content and insulin storage suggest a parallelism, but evidence is wanting for zinc-insulin complexes *in vivo*. The highest zinc content determined among mammals is found in the *tapetium lucidum cellulosum* of adult fox seals—up to 150,000 micrograms/gram. Human blood contains 7–8 micrograms zinc/milliliter. About 12% of this is present in serum, 3% in leukocytes, and 85% in erythrocytes. In these compartments, zinc occurs as part of zinc proteins and zinc metalloenzymes. In erythrocytes, zinc is correlated to carbonic anhydrase activity.

Underwood (1977) reported that high levels of dietary zinc interfere with the normal absorption and metabolism of several minerals. Earlier, Stewart and Magee (1964) reported that experience with laboratory animals indicated that 7500 parts per million (ppm) zinc in a purified diet lowered the concentrations of calcium and phosphorus in bone. Other investigators showed that animals fed the same level of zinc were anemic and had deformed, fragile erythrocytes. The livers contained reduced amounts of ferritin and lower concentrations of iron in the ferritin. Whanger and Weswig (1971) showed that in rats fed 2000 and 4000 ppm zinc, liver copper concentration was decreased. The most sensitive responses to zinc have been obtained with animals receiving minimally required or suboptimal levels of copper (Hill and Matrone, 1962; Campbell and Mills, 1974; and Murthy et al., 1974). Based upon these earlier findings, Hamilton, et al. (1979) investigated possible zinc interference with copper, iron, and manganese in young Japanese quail (*Coturnix coturnix japonica*). The objective of the investigators was to identify the minimal level of excess dietary zinc that would produce physiological and metabolic deviations from normal and to define some of the most sensitive zinc-mineral interactions. Because other workers had found a sensitive zinc-copper antagonism, the Hamilton team studied the effects of supplemental zinc at copper levels of marginal deficiency. Data from the study showed that adequacy of copper intake is important when supplemental zinc is consumed either as a dietary supplement or in foods fortified with zinc. It was reported that results of the findings may be important for the general human population, whose dietary intake of many minerals, including copper, does not usually exceed the requirement or may be marginally deficient (Milne et al., 1978; Harland et al., 1978; Klevay, 1978). See also **Copper (In Biological Systems).**

References

Allaway, W. H.: "The Effect of Soils and Fertilizers on Human and Animal Nutrition," Cornell University Agricultural Experiment Station, *Agricultural Information Bulletin 378*, U.S. Department of Agriculture, Washington, D.C. (1975).

Campbell, J. K., and C. F. Mills: "Effects of Dietary Cadmium and Zinc on Rats Maintained on Diets Low in Copper," *Proceedings Nutr. Society*, **33**, 1, 15A (Abstract) (1974).

Gray, P.: "The Encyclopedia of the Biological Sciences," pages 571, 942, Van Nostrand Reinhold, New York (1970).

Hamilton, R. P., et al.: "Zinc Interference with Copper, Iron and Manganese in Young Japanese Quail," *Journal of Food Science*, **44**, 3, 738–741 (1979).

Harland, B., Prosky, L., and J. Vanderveen: "Nutritional Adequacy of Current Levels of Zn, et al., in the American Food Supply for Adults, Infants, and Toddlers," in "Trace Element Metabolism in Man and Animals" (M. Kirchgessner, Editor), page 311, Institut für Ernährungsphysiologie, Technische Universität München, Freising-Weihenstephan, Germany (1978).

Hesse, G. W.: "Chronic Zinc Deficiency Alters Neutronal Function of Hippocampal Mossy Fibers," *Science*, **205**, 1005–1007 (1979).

Klevay, L. M. "Dietary Copper and the Copper Requirements in Man," in "Trace Element Metabolism in Man and Animals" (M. Kirchgessner, Editor), page 307, Institut für Ernährungsphysiologie, Technische Universität München, Freising-Weihenstephan, Germany (1978).

Milne, D. B., et al.: "Dietary Intakes of Copper, Zinc, and Manganese by Military Personnel," *Fed. Proc.*, **37**, 894 (Abstract) (1978).

Murthy, L., et al.: "Interrelationships of Zinc and Copper Nurtriture in the Rat," *Jrnl. of Nutrition*, **104**, 1458 (1974).

Stewart, A. K., and A. C. Magee: "Effect of Zinc Toxicity on Calcium, Phosphrous and Magnesium Metabolism of Young Rats," *Jrnl. of Nutrition*, **82**, 287 (1964).

Underwood, E. J.: "Trace Elements in Human and Animal Nutrition," 4th Edition, Academic, New York (1977).

Whanger, P. D., and P. H. Weswig: "Effect of Supplementary Zinc on the Intracellular Distribution of Hepatic Copper in Rats," *Jrnl. of Nutrition*, **101**, 1093 (1971).

ZINCITE. This mineral, (Zn,Mn)O, is an ore of zinc and occurs in considerable quantities at Franklin Furnace, New Jersey, where it is associated with willemite and franklinite. Its hexagonal crystals are rare, as it usually occurs massive, foliated, or in coarse to fine grains. When the crystals are observable, it reveals a perfect cleavage parallel to the base of the prism. The fracture is conchoidal. The mineral has a hardness of 4; specific gravity, 5.684; luster, subadamantine to vitreous; orange-yellow streak; color, red to orange-yellow; translucent to opaque. Zincite also has been found in Poland, Tuscany, Spain, Saxony, and Tasmania.

ZINC OXIDE PIGMENT. Paint and Finishes.

ZINC PHOSPHATE COATINGS. Conversion Coatings.

ZIRCON. This mineral is zirconium silicate, $ZrSiO_4$, and is the chief ore of zirconium. Zircon occurs in square tetragonal prisms, although sometimes it assumes pyramidal or irregular forms. The mineral may be found in some beach and river placer deposits, but generally it is associated as an accessory mineral in siliceous igneous rocks, crystalline limestones, schists, and gneisses—and in sedimentary rocks derived from the foregoing. Zircon is without good cleavage; is brittle, with a conchoidal fracture; hardness, 7.5; specific gravity, 3.7–4.7; luster, adamantine, brilliant; color, green, yellow-green, golden-yellow, red, red-brown, brown, and blue. The name zircon derives from an Arab word *zarqun*, meaning *vermilion*, or perhaps from the old Persian *zargun*, meaning *golden-colored*.

Zircon occurs in the Ural Mountains; Trentino, Monte Somma, and Vesuvius, Italy; Arendal, Norway; Ceylon, India; Thailand; at the Kimberley mines, Republic of South Africa; the Malagasy Republic; and in Canada in Renfrew County, Ontario, and Grenville, Quebec. In the United States, zircon is found at Litchfield, Maine; Chesterfield, Massachusetts; in Essex, Orange, and St. Lawrence Counties, New York; Henderson County, North Carolina; the Pikes Peak district of Colorado; and Llano County, Texas.

Gem quality crystals from Ceylon (Sri Lanka) have been known for many years. They range from colorless to brownish orange, yellow, dark red, to light reddish-violet. Heat treated zircons provide a beautiful stone of light blue color. Colorless stones are used as a diamond substitute.

ZIRCONIUM. Chemical element symbol Zr, at. no. 40, at. wt. 91.22, periodic table group 4b, mp 1853°C, by 4376°C, density, 6.47 g/cm^3 (single crystal). Metallic zirconium is allotropic. Up to about 863°C, the alpha phase (hexagonal close-packed) is stable; above this temperature, the metal assumes the beta phase (body-centered cubic). The most common impurity, oxygen, tends to stabilize the alpha phase. Zirconium metal exhibits passivity in air due to the formation of adherent coatings of oxide and nitride. Even without the coating, it is resistant to the action of weak acids and acid salts, but dissolves in HCl (warm) or sulfuric acid slowly, and more rapidly if F^- is present, forming compounds of ZrO^{2+} ions, or fluorozirconates in the last case.

Crystalline zirconium of high purity is a white, soft, ductile, and malleable metal, but that of 99% purity, when obtained at high temperature, is hard and brittle. Pure zirconium has a combination of properties which make it valuable as a structural material for nuclear reactors. In addition to low neutron capture, zirconium has good strength at high temperatures, corrosion resistance to high-velocity coolants, avoidance of formation of high-activity isotopes, and resistance to mechanical damage from neutron radiation. Amorphous zirconium is a bluish-black powder. At about 500°C, zirconium burns in air; heated in hydrogen, it forms the hydride; heated in nitrogen, a nitride; and heated in chlorine, the tetrachloride. On the laboratory scale, zirconium metal may be produced by the reduction of the chloride, oxide, or potassium zirconium fluoride with sodium metal.

Zirconium was first identified by Klaproth in 1789. The first crude powder was made by Berzelius in 1824 by reducing potassium fluorozirconate with potassium. A sample with a purity of 98% was not produced until the 1950s. There are five natural isotopes, ^{90}Zr through ^{92}Zr, ^{94}Zr, and ^{96}Zr. Six radioactive isotopes have been identified, ^{87}Zr through ^{89}Zr, ^{93}Zr, ^{95}Zr, and ^{97}Zr. With the exception of ^{93}Zr, with a half-life of 1.1 × 10^6 years, the other isotopes have half-lives expressed in minutes, hours, or days. Zirconium is ranked 19th in abundance of the chemical elements occurring in the earth's crust, with an estimated average content of zirconium in igneous rocks of 0.026%.

First ionization potential 6.95 eV; second, 13.97 eV; third, 24.00 eV; fourth, 33.83 eV. Oxidation potentials Zr + $2H_2O \rightarrow ZrO_2 + 4H^+ + 4e^-$, 1.43 V; Zr + $4OH^- \rightarrow ZrO(OH)_2 + H_2O + 4e^-$, 2.32 V. Electronic configuration

$1s^2 2s^2 2p^6 3s^2 3p^6 3d^{10} 4s^2 4p^6 4d^2 5s^2$. Ionic radius Zr^{+4} 0.80 Å. Metallic radius 1.5895 Å.

The most important ore for production of zirconium metal is zircon $ZrSiO_4$ which occurs in several regions in the form of a beach sand, often mixed with silica, ilmenite, and rutile. A floating-dredge technique is used in the mining operation. Early phases of beneficiation often take place on the dredge. Nearly all of the silica is separated by means of spiral concentrators, with the ilmenite and rutile removed by magnetic and electrostatic separators. The purest concentrates are used for metal production; others for refractories. Direct chlorination of the ore is the most modern method of extraction. In a simple reaction, water-soluble zirconium tetrachloride is yielded. Liquid-liquid extraction in several states is required for the removal of hafnium. With this process, zirconium containing less than 50 ppm hafnium can be produced. Ammonium thiocyanate is used to complex the zirconium while hafnium is extracted by a methyl ethyl ketone solvent. In a similar system, HNO_3 is used in the aqueous phase and tributyl phosphate as the solvent. After separation, the two metals are precipitated as their sulfates or hydroxides. Calcination yields a pure ZrO_2. For production of pure metal, the pure ZrO_2 is chlorinated to $ZrCl_4$. Sometimes this is sublimed for additional purification. The zirconium tetrachloride in the gaseous phase is reacted with molten magnesium, forming zirconium metal and magnesium chloride. There are several minor variations of these processes. For example, sodium may replace magnesium.

Uses. The most important application of zirconium is in the formulation of the base metal in the alloy of 98% zirconium, 1.5% tin, 0.35% iron-chromium-nickel, and 0.15% oxygen. This alloy is widely used in water-cooled nuclear reactors because of its excellent corrosion resistance up to about 350° in H_2O, and its low neutron cross section. Currently, about 90% of the zirconium produced is used for this application. The excellent corrosion resistance of zirconium to both strong acids and alkalis, particularly its resistance to strong caustic solutions at all concentrations and temperatures, is attracting increasing attention for application in chemical processing equipment.

Chemistry and Compounds. Due to its $4d^2 5s^2$ electron configuration, zirconium forms tetravalent compounds readily, although the Zr^{4+} ion does not exist as such in aqueous solution, except at very low pH values, the common cation being hydrated ZrO^{2+} $[Zr(OH)_2^{2+}]$. Many of the tetravalent compounds are partly covalent. There are also less stable Zr(III) compounds. The remarkably close similarity in chemical properties to those of hafnium is due to the identical outer electron configuration and the almost identical ionic radius, the relatively low radius value for Hf^{4+} being due to the lanthanide contraction.

Zirconium oxide, ZrO_2 is widely known, both as a mineral, baddeleyite, and as an industrial product obtained from zircon, $ZiSO_4$. Moreover, the precipitate obtained by action of alkali hydroxides upon solutions of tetravalent zirconium is a hydrated oxide. The latter is readily soluble in acids to form oxysalts, which are usually formulated in terms of the ZrO^{2+} ion, without including its water of hydration, e.g., as $\cdot ZrO(H_2PO_4)_2$. The hydrated ZrO^{2+} ion is not amphiprotic; it does not dissolve in alkali hydroxides. While it does react with alkali carbonate fusions, the compounds formed have been shown to be mixed oxides rather than zirconates.

All four of the halides of zirconium and the common halogens are known and are solids at ordinary temperatures. They are readily hydrolyzed, to form hydrated oxyhalides such as $ZrOCl_2 \cdot 8H_2O$ or $ZrOBr_2 \cdot 8H_2O$. The tetraiodide yields both $ZrOI_2 \cdot 8H_2O$ and $ZrI(OH)_3 \cdot 3H_2O$, and this last composition probably represents best the structure of all these compounds.

Like titanium tetrahalides, $ZrCl_4$ and $ZrBr_4$ act as Lewis Acids to form adducts, though of lesser stability, with H_2O and oxygen-function compounds, such as alcohols, ethers, and carboxy compounds generally. These include chelates formed with such compounds as 1,2-dihydroxy-benzene and acetylacetone, in which oxygen atoms act as electron donors. Zirconium also forms very stable complexes with $POCl_3$.

The trihalides of zirconium, like the dihalides of titanium, are extremely strong reducing agents, reacting even with H_2O.

Zirconium nitride, Zr_3N_4, is made by ammoniating the tetrachloride to yield $Zr(NH_3)_4Cl_4$, which yields the nitride on heating. The nitride, like the boride and carbide, are alloy-like in character, with high fusing points, extreme hardness, and subject to considerable variation in composition. Thus Zr_3N_4 may vary in composition to ZrN without material change in its properties.

Unlike titanium, zirconium forms a few normal tetravalent salts, such as a tetranitrate and a tetrasulfate, as well as its more common basic salts. However, the normal salts readily undergo hydrolysis to form the basic salts.

Like titanium, zirconium forms halogen complexes, the most stable of which is the hexafluoride, ZrF_6^{2-}, as well as $ZrCl_6^{2-}$, and $ZrBr_6^{2-}$, which are less stable, except in concentrated solutions of the hydrogen halides. Zirconium also forms stable peroxy complexes, containing $>Zr-O-O-$. Unlike titanium, zirconium forms a heptafluoride ion, ZrF_7^{3-}, which is quite stable.

References

Buckley, D. H., and R. L. Johnson: "Relation of Lattice Parameters to Friction Characteristics of Beryllium, Hafnium, Zirconium, and Other Hexagonal Metals in Vacuum," *NASA TND-2670*, National Aeronautics and Space Administration, Washington, D.C. (1965).

Douglass, D. L.: "The Metallurgy of Zirconium," *Atomic Energy Review Supplement*, Nuclear Regulatory Agency, Washington, D.C. (1971).

Lustman, B., and F. Kerze: "The Metallurgy of Zirconium," McGraw-Hill, New York (1955).

Taylor, A., and B. J. Kagle: "Crystallographic Data on Metal and Alloy Structures," Dover, New York (1963).

Webster, R. T.: "Zirconium," in "Metals Handbook," 9th Edition, Vol. 2, American Society for Metals, Metals Park, Ohio (1979).

ZOISITE. This mineral is a hydrous aluminum silicate corresponding to formula $Ca_2Al_3(Si_3O_{12})(OH)$, crystallizing in the orthorhombic system. Clinozoisite (monoclinic) is its isomorphous counterpart. Zoisite occurs as prismatic crystals, usually deepened striated vertically, and as compact or columnar masses. Perfect prismatic cleavage; brittle; uneven to conchoidal fracture; hardness, 6.5–7; specific gravity, 3.355; luster, vitreous to pearly; transparent to translucent; color, grayish-white, green, pink (the manganese-rich variety, thulite), and blue to purple (tanzanite).

Zoisite occurs in crystalline schists which are products of regional metamorphism of basic igneous rocks rich in plagioclase, the calcium-rich feldspar, also in argillaceous calcareous sandstones, thulite from quartz veins, pegmatites are metamorphosed impure limestones and dolomites.

Tanzanite is the blue to purple variety of zoisite and represents a recent discovery of this heretofore unknown variety from Tanzania. It occurs here as excellent transparent crystals from which fine gems have been cut.

ZSIGMONDY REAGENT. A reagent for colloids which is a red colloidal solution of metallic gold obtained by reducing auric chloride by formaldehyde in the presence of an alkali. When mixed with sodium chloride, this reagent becomes blue because of an agglomeration of the particles of gold, but this color change is prevented by the presence of an adequate amount of certain other colloids. They can be classified according to the amount required to prevent the color change. See **Colloid Systems.**

ZWITTERION. An ion carrying charges of opposite sign, which thus constitutes an electrically neutral molecule with a dipole moment, looking like a positive ion at one end and a negative ion at the other. Most aliphatic amino acids form such dipolar ions, hence react with both strong acids and strong bases.

ZYMOLYTIC REACTION. A chemical reaction catalyzed by an enzyme, especially a reaction involving bond rupture or splitting, usually a hydrolysis.

INDEX